S0-BKX-028

COMPLEX SYSTEMS

COMPLEX SYSTEMS

5th International Workshop on Complex Systems

Sendai, Japan *25 – 28 September 2007*

EDITORS

Michio Tokuyama
Tohoku University
Sendai, Japan

Irwin Oppenheim
Massachusetts Institute of Technology
Cambridge, Massachusetts, U.S.A.

Hideya Nishiyama
Tohoku University
Sendai, Japan

All papers have been peer reviewed.

SPONSORING ORGANIZATION

21st Century COE Program, International COE of Flow Dynamics

Melville, New York, 2008
AIP CONFERENCE PROCEEDINGS ■ VOLUME 982

Editors

Michio Tokuyama
WPI Advanced Institute for Materials Research
and Institute of Fluid Science
Tohoku University
2-1-1 Katahira, Aobaku
Sendai 980-8577
JAPAN
E-mail: tokuyama@fmail.ifs.tohoku.ac.jp

Irwin Oppenheim
Department of Chemistry
Massachusetts Institute of Technology
Cambridge, MA 02139
U.S.A.
E-mail:irwin@mit.edu

Hideya Nishiyama
Institute of Fluid Science
Tohoku University
2-1-1 Katahira, Aobaku
Sendai 980-8577
JAPAN
E-mail: nishiyama@ifs.tohoku.ac.jp

L.C. Catalog Card No. 2008920655

ISBN 978-0-7354-0501-1
ISSN 0094-243X

Printed in the United States of America

CONTENTS

I. SUPERCOOLED LIQUIDS AND GLASS TRANSITION

II. COMPLEX FLUIDS

III. POLYMER DYNAMICS

IV. NANO-MEGA SCALE FLOW DYNAMICS IN COMPLEX SYSTEMS

V. CROSS DISCIPLINARY PHYSICS

PREFACE

This volume contains a collection of 150 papers presented at the 5th International Workshop on Complex Systems. It was held at Sendai International Center, Sendai, Japan, from 25 to 28 September 2007. This workshop was financially supported by the 21st century COE program on Flow Dynamics.

The main topics of this meeting were as follows:

I. Supercooled liquids and Glass Transition
II. Complex Fluids
III. Polymer Dynamics
IV. Nano-mega Scale Flow Dynamics in Complex Systems
V. Cross Disciplinary Physics

There were 87 participants from 22 foreign countries and 92 from Japan. There were 25 invited talks, 60 oral presentations, and 73 poster presentations.

This is one of a series of International Workshop on Complex Systems. The main purpose of this workshop is to provide an opportunity for an international group of experimentalists, theoreticians, and computational scientists, who are mainly working on complex systems, to gather together and discuss their recent advances, and thus to find a new perspective towards further developing statistical physics. Another important aspect of the meeting is the presence of Japanese graduate students. The meeting was very lively and enjoyable. Especially, the discussions among experimentalists, theoreticians, and computational scientists were interesting and effective. We believe that these proceedings reflect at least partially the atmosphere of this meeting. The workshop was very successful in broadening and deepening the relationships between scientists of different countries. We hope that the success of the meeting in training young scientists and establishing contacts among leaders in the scientific field will enable us to hold this workshop every year.

On behalf of the organizing committee, we would like to thank all the participants for their efforts to make their presentations of the highest quality. We also thank the reviewers of the submitted papers for their contributions to the quality of this volume. Our thanks are also due to all members of the 21st Century COE Program "International COE of Flow Dynamics" Office for carrying the financial burden and providing the conference rooms and facilities.

Last, but not least, we wish to thank Mr. Taku Kudo, Mr. Masayoshi Takahashi, and Ms. Saiko Yumura for their businesslike effort in the preparation of this volume, and all the others who took such good care of the many chores that had to be done for the smooth operation of the Workshop.

Michio Tokuyama
Professor, WPI Advanced Institute for Materials Research
Professor, Institute of Fluid Science
Tohoku University

Irwin Oppenheim
Emeritus, Department of Chemistry, Massachusetts Institute of Technology

Hideya Nishiyama
Professor, Institute of Fluid Science, Tohoku University

Sendai, September 2007

Organization of Workshop

Chairperson:

Michio Tokuyama	(WPI Advanced Institute for Materials Research and Institute of Fluid Science, Tohoku University, Japan)

Organizing Committee:

Irwin Oppenheim	(Co-Chairperson, Department of Chemistry, M.I.T., USA)
Masayuki Imai	(Department of Physics, Ochanomizu University, Japan)
Yasuhiro Shiwa	(Graduate School of Sci. and Tech., Kyoto Inst. of Tech., Japan)
Akira Suzuki	(Graduate School of Science, Tokyo University of Science, Japan)
Michio Tokuyama	(WPI Advanced Institute for Materials Research and Institute of Fluid Science, Tohoku University, Japan)
Tadashi Toyoda	(Department of Physics, Tokai University, Japan)

Local Committee:

Mitsuhiro Akimoto	(Faculty of Science, Tokyo University of Science Yamaguchi, Japan)
Yoshihisa Enomoto	(Dept. of Mechanical Engineering, Nagoya Inst. of Tech., Japan)
Yoon Hwae Hwang	(Dept. of Nanomaterials, Pusan National University, South Korea)
Tadashi Muranaka	(Science div. General Education, Aichi Inst. of Tech., Japan)
Yayoi Terada	(Institute of Fluid Science, Tohoku University, Japan)

Secretariat:

Taku Kudo	(Institute of Fluid Science, Tohoku University, Japan)
Masayoshi Takahashi	(Institute of Fluid Science, Tohoku University, Japan)
Saiko Yumura	(Institute of Fluid Science, Tohoku University, Japan)

Invited Speakers

Supercooled Liquids and Glass Transition

S.-H. Chen	Massachusetts Institute of Technology
M. Descamps	University Lille1
A. Inoue	Tohoku University
K. L. Ngai	Naval Research Laboratory
M. Oguni	Tokyo Institute of Technology
G. Szamel	Colorado State University
H. Tanaka	University of Tokyo
G. Tarjus	Université Pierre et Marie Curie
M. Tokuyama	Tohoku University

Complex Fluids

G. S. Grest	Sandia National Laboratories
P. Harrowell	Univeristy of Sydney
H. Löwen	Heinrich-Heine-Universität Düsseldorf
I. Oppenheim	Massachusetts Institute of Technology
H. E. Stanley	Boston University
P. Tartaglia	Università di Roma “La Sapienza”
D. Weitz	Harvard University

Polymer Dynamics

R. G. Larson	University of Michigan
D. Richter	IFF Forschungszentrum Juelich

Nano-Mega Scale Flow Dynamics in Complex Systems

P. Sheng	Hong Kong University of Science and Technology
J. J. Lowke	Commonwealth Scientific and Industrial Research Organisation
J. Jeništa	Academy of Sciences of the Czech Republic
O. P. Solonenko	Institute of Theoretical and Applied Mechanics
P. Proulx	Universite de Sherbrooke
A. Ivlev	Max Planck Institute for Extraterrestrial Physics
H.-P. Li	Tsinghua University

The 5th International Workshop on Complex Systems
September 25-28, 2007 Sendai, Japan

PART I

SUPERCOOLED LIQUIDS AND GLASS TRANSITION

Universal Behavior near the Glass Transitions in Fragile Glass-Forming Systems

Michio Tokuyama[1,2]

[1]WPI Advanced Institute for Materials Research, Tohoku University, Sendai 980-8577, Japan
[2]Institute of Fluid Science, Tohoku University, Sendai 980-8577, Japan

Abstract. The slow dynamics of a single particle in different fragile systems near the glass transition is analyzed from a unified point of view based on the mean-field theory (MFT) recently proposed by the present author. It is shown that the experimental data and the simulation results for the mean-square displacement are collapsed into a master curve given by MFT if a reduced long-time self-diffusion coefficient (or a universal parameter) in different systems has the same value. It is also shown that the long-time self-diffusion coefficient for any systems is well described by a non-singular function predicted by MFT. The solutions of the mode-coupling equations are also analyzed by MFT and are shown to disagree with the simulation results from a new viewpoint.

Keywords: Fragile systems, Glass transition, Long-time self-diffusion coefficient, Supercooled liquid, Universality
PACS: 64.70.Pf, 64.70.Dv, 61.20.Gy, 83.10.Mj

INTRODUCTION

Understanding of the glass transition is one of the pioneering problems encountered in a wide variety of fields, such as science and technology, which deal with complex systems [1, 2, 3]. Although it is about 80 years since the glass transition has been observed thermodynamically, our understanding is still incomplete. With recent progress in science and technology, however, the relaxation processes of viscous liquids near the glass transition are extensively studied by experiments and computer simulations [4, 5].

In this paper, we focus only on the glass transition of fragile glass-forming systems, including colloidal suspensions. We study the relaxation processes of those systems and analyze the experimental data and the simulation results for the mean-square displacement from a unified point of view based on the mean-field theory (MFT) recently proposed by the present author [6, 7, 8]. Thus, we show that the properties of the relaxation processes in those systems are remarkably universal, especially in the following points. The first point is that any states of the systems are uniquely determined by a reduced long-time self-diffusion coefficient (or a universal parameter u). The second is that any data for the mean-square displacements in different systems can be described by a single master curve given by MFT at a given value of u. The last is that the data of the long-time self-diffusion coefficients in different systems obey the non-singular functions predicted by MFT, although they are partially described by a singular function predicted by MFT. These might suggest the possibility of existence of a rather simple mechanism near the glass transition, which is able to describe universal features in many different fragile systems.

The mode-coupling equations for the mean-square displacement were recently solved numerically by Flenner-Szamel [9] and also by Voigtmann et al [10] by using the static structure factor obtained by their simulations. Those solutions are also analyzed by MFT from the same standpoint as that used for analyses of the experimental data and the simulation results. Thus, we show that the mean-square displacements given by the mode-coupling equations are inconsistent with the simulation results around the β-relaxation stage at a given value of u and also that the resultant long-time self-diffusion coefficients do not agree with those of the simulations.

We begin in Sec. II by reviewing the mean-field theory, which is used in the present paper. In Sec. III we show the universal behavior among different systems from a unified point of view based on the MFT. We conclude in Sec. IV with a summary.

TWO TYPES OF MODEL SYSTEMS

We consider two types of systems. One is a molecular system (M) and another is a suspension of colloids (S). Both systems contain N particles of interest with mass m_i and radius a_i of ith particle in the total volume V at temperature T. Let $\boldsymbol{X}_i(t)$ denote the position vector of ith particle at time t. In (M), it obeys the Newton equation

$$m_i\frac{d^2}{dt^2}\boldsymbol{X}_i(t) = \sum_{j(\neq i)} \boldsymbol{F}(\boldsymbol{X}_{ij}), \tag{1}$$

where $\boldsymbol{F}(\boldsymbol{X}_{ij})$ is a force acting on the ith particle from other particles, and $\boldsymbol{X}_{ij} = \boldsymbol{X}_i - \boldsymbol{X}_j$. Equation (1) holds

CP982, *Complex Systems, 5th International Workshop on Complex Systems*
edited by M. Tokuyama, I. Oppenheim, and H. Nishiyama

on the time scale longer than $t_0(=a/v_0)$, where $v_0(=(dk_BT/m)^{1/2})$ is an average particle velocity and a the average radius, d being a spatial dimensionality. On the other hand, in (S), the particles are immersed in an equilibrium liquid with viscosity η and temperature T and undergo a Brownian motion on the time scale of order t_B, where $t_B(=m/\zeta_0)$ denotes a Brownian relaxation time, $\zeta_0(=6\pi\eta a)$ being a friction constant. Hence the position vector $\boldsymbol{X}_i(t)$ is described by the Langevin equation [11, 12]

$$\begin{aligned} m_i\frac{d^2}{dt^2}\boldsymbol{X}_i(t) &= -\sum_{j=1}^{N}\boldsymbol{\zeta}(\boldsymbol{X}_{ij}(t))\cdot\frac{d}{dt}\boldsymbol{X}_j(t) \\ &+ \sum_{j(\neq i)}\boldsymbol{F}(\boldsymbol{X}_{ij})+\boldsymbol{R}_i(t). \end{aligned} \tag{2}$$

Here the random force $\boldsymbol{R}_i(t)$ obeys a Gaussian, Markov process with zero mean and satisfies the fluctuation-dissipation relation

$$<\boldsymbol{R}_i(t)\boldsymbol{R}_j(t')>=2k_BT\boldsymbol{\zeta}(X_{ij}(t))\delta(t-t'), \tag{3}$$

where the brackets $<\cdots>$ denote an equilibrium ensemble average. The first term of Eq. (2) indicates the friction force and the coefficient $\boldsymbol{\zeta}_{ij}(t)$ represents the renormalized friction tensor given by

$$\boldsymbol{\zeta}(\boldsymbol{X}_{ij})=\zeta_0\left[(\boldsymbol{1}+\boldsymbol{g}(t))^{-1}\right]_{ij}. \tag{4}$$

Here the tensor $\boldsymbol{g}_{ij}(t)(i\neq j)$ indicates the hydrodynamic interactions between colloids i and j, and $\boldsymbol{g}_{ii}=0$. Up to order $(a_i/|\boldsymbol{X}_{ij}|)^4$, it can be written in terms of the Oseen tensor $\boldsymbol{g}^O_{ij}$ and the dipole tensor $\boldsymbol{g}^D_{ij}$ as

$$\boldsymbol{g}_{ij} = \boldsymbol{g}^O_{ij}+\boldsymbol{g}^D_{ij}, \tag{5}$$

$$\boldsymbol{g}^O_{ij} = \frac{3}{4}\frac{a_i}{|\boldsymbol{X}_{ij}|}\left(\boldsymbol{1}+\frac{\boldsymbol{X}_{ij}}{|\boldsymbol{X}_{ij}|}\frac{\boldsymbol{X}_{ij}}{|\boldsymbol{X}_{ij}|}\right), \tag{6}$$

$$\boldsymbol{g}^D_{ij} = \frac{1}{4}\frac{a_i(a_i^2+a_j^2)}{|\boldsymbol{X}_{ij}|^3}\left(\boldsymbol{1}-3\frac{\boldsymbol{X}_{ij}}{|\boldsymbol{X}_{ij}|}\frac{\boldsymbol{X}_{ij}}{|\boldsymbol{X}_{ij}|}\right). \tag{7}$$

In this paper, we are only interested in the single-particle dynamics near the glass transition, whose space-time scales are much larger than those of microscopic processes. Then, the useful physical quantities to describe the relaxation of a single particle near the glass transition are given by the self-intermediate scattering function

$$F_S(q,t)=<\exp[i\boldsymbol{q}\cdot\{\boldsymbol{X}_i(t)-\boldsymbol{X}_i(0)\}]>, \tag{8}$$

and the mean-square displacement

$$M_2(t)=<|\boldsymbol{X}_i(t)-\boldsymbol{X}_i(0)|^2>, \tag{9}$$

both of which are related through the relation

$$F_S(q,t)\simeq\exp\left[-\frac{q^2}{2d}M_2(t)+\frac{q^4}{2}\left(\frac{M_2(t)}{2d}\right)^2\alpha_2(t)\right], \tag{10}$$

where $\alpha_2(t)$ is the non-Gaussian parameter [13]. In the following, we only discuss the time evolution of $M_2(t)$.

MEAN-FIELD THEORY

In this section we briefly summarize and discuss the mean-field theory of the glass transition (MFT) recently proposed by the present author [6, 7, 8].

The mean-field theory consists of two essential points.

(i) The mean-field equations for the mean-square displacement

(ii) The singular and non-singular long-time self-diffusion coefficients

Mean-Field Equations

The mean-square displacement $M_2(t)$ is described by a nonlinear equation

$$\frac{d}{dt}M_2(t)=2dD_S^L(p)+2d[s(t)-D_S^L(p)]e^{-M_2(t)/\ell(p)^2}, \tag{11}$$

where $D_S^L(p)$ is a long-time self-diffusion coefficient and p is a control parameter, such as a volume fraction ϕ, an inverse temperature $1/T$, and a square of an external magnetic field $\boldsymbol{H}$. Here the mean-free path $\ell(p)$ is a length in which a particle can move freely without any interactions between particles. Although it is originally related to the static structure factor $S(q)$ [6], it is determined by a fitting with data here. The short-time behavior is only different between type (S) and type (M) since the system is governed by a short-time diffusion process for (S) and by a ballistic motion for (M). In fact, for short times, the function $s(t)$ is given by

$$s(t)=\begin{cases} D_S^S(p) & \text{for (S)}, \\ (v_0^2/d)t & \text{for (M)}, \end{cases} \tag{12}$$

where $D_S^S(p)$ denotes the short-time self-diffusion coefficient. Equation (11) can be solve to give a formal solution

$$\begin{aligned} M_2(t) &= 2dD_S^Lt \\ &+ \ell^2\ln\left[e^{-\frac{2dt}{\tau_\beta}}+\kappa\left\{1-\left(1+c\frac{2dt}{\tau_\beta}\right)e^{-\frac{2dt}{\tau_\beta}}\right\}\right] \end{aligned} \tag{13}$$

with

$$\kappa = \begin{cases} \frac{\tau_\beta}{\tau_f} & \text{for (S)}, \\ \frac{1}{2d^2}\left(\frac{\tau_\beta}{\tau_f}\right)^2 & \text{for (M)}, \end{cases} \tag{14}$$

where $c = 0$ for (S) and $c = 1$ for (M). Here $\tau_\beta (= \ell^2/D_S^L)$ denotes a time for a particle to diffuse over a distance of order ℓ with the diffusion coefficient D_S^L and is identical to the so-called β-relaxation time. The mean-free time τ_f is given by

$$\tau_f = \begin{cases} \ell^2/D_S^S & \text{for (S)}, \\ \ell/v_0 & \text{for (M)}. \end{cases} \tag{15}$$

Within τ_f, therefore, each particle can move freely without any interactions between particles. As shown in the previous paper Ref. [8], the mean-free path ℓ is uniquely determined by $D_S^L/d_0 (= \tau_f/\tau_\beta)$, where d_0 has a dimension of a diffusion constant given by

$$d_0 = \begin{cases} D_S^S(p) & \text{for (S)}, \\ av_0 & \text{for (M)}. \end{cases} \tag{16}$$

Hence the solution (13) suggests that the dynamics is described by only one parameter D_S^L/d_0 if the length and the time are scaled by a and $t_0 (= a^2/d_0)$, respectively. Here the parameter D_S^L/d_0 denotes the long-time correlation effects due to the many-body interactions between particles.

The solution (13) suggests three different time scales; τ_f, τ_β, and a long diffusion time $\tau_L (= a^2/D_S^L)$, where $\tau_f \ll \tau_\beta \ll \tau_L$. In fact, one can find the following asymptotic forms:

$$M_2(t) \simeq \begin{cases} \ell^2 \ln\left[1 + (2d)^{1-c}\left(\frac{t}{\tau_f}\right)^{1+c}\right] & \text{for } t \ll \tau_f, \\ 2dD_S^L t & \text{for } \tau_\beta \ll t. \end{cases} \tag{17}$$

Hence we have two stages; an early stage [E] for $t \ll \tau_f$, where

$$M_2(t) \simeq \begin{cases} 2dD_S^S t & \text{for (S)}, \\ (v_0 t)^2 & \text{for (M)}, \end{cases} \tag{18}$$

and a late stage [L] for $\tau_\beta \ll t$, where

$$M_2(t) \simeq 2dD_S^L t, \quad \text{for (S) and (M)}. \tag{19}$$

As shown in the previous paper [7], for $p \geq p_s$ there exists a new time stage, the so-called β-relaxation stage [β] for $\tau_f \ll t \ll \tau_L$, where p_s depends on the systems. In fact, one can find one more time scale, the caging time τ_γ, where $\tau_\gamma \ll \tau_\beta$. On the time scale of order τ_γ, each particle behaves as if it is trapped in a cage which is mostly formed by neighboring particles. This is the so-called cage effect. The β stage is separated into two stages; a fast β stage [β_f] for $\tau_f \ll t \ll \tau_\beta$ and a slow β stage [β_s] for $\tau_\gamma \ll t \ll \tau_L$. On a time scale of order τ_β, the particles can escape their cages and on a time scale of order τ_L, they finally obey a long-time diffusion process. By expanding $M_2(t)$ in powers of $\ln(t/\tau_\gamma)$ or $\ln(t/\tau_\beta)$ on each stage, one can then find the following asymptotic forms:

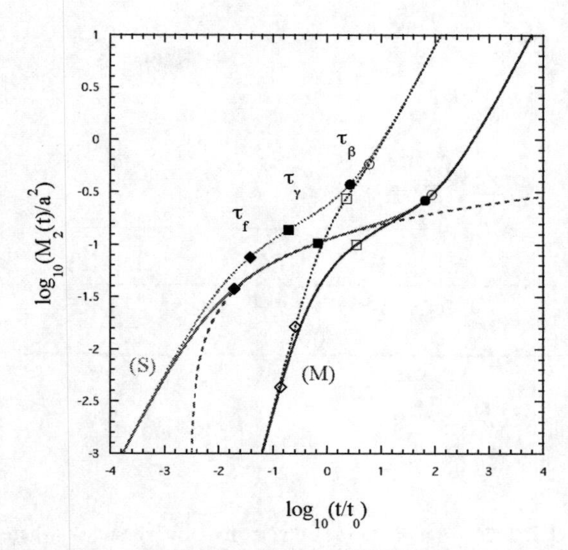

FIGURE 1. (Color online) A log-log plot of the mean-square displacement $M_2(t)$ versus time at various diffusion coefficients; $u = 1.890$ (dotted lines), where $\ell/a = 0.1948$ (S) and 0.2582 (M), and u=3.565 (solid lines), where $\ell/a = 0.1386$ (S) and 0.1388 (M). The filled symbols indicate the times τ_f (◇), τ_β (○), and τ_γ (□) for (S), while the open symbols for (M). The dashed line indicates the logarithmic growth for (S) given by Eq. (20) at u=3.565.

$$M_2(t) \simeq \begin{cases} A_\gamma + \ell^2(1+c)\ln\left(\frac{t}{\tau_\gamma}\right) + B_\gamma\left(\frac{t}{\tau_\gamma}\right)^{b_\gamma} & \text{for } [\beta_f], \\ A_\beta + B_\beta\left(\frac{t}{\tau_\beta}\right)^{b_\beta} & \text{for } [\beta_s], \end{cases} \tag{20}$$

where $A_\alpha = \ln[1 + (2d)^{1-c}(\tau_\alpha/\tau_f)^{1+c}]$. Here B_α is a positive constant and b_α a time exponent to be determined. We mention here that in stage [β_f] the logarithmic growth dominates the dynamics for all systems since $B_\gamma \ll 1$, while in stage [β_s] the power-law growth dominates the dynamics. We should also note here that since $b_\beta \geq 1$, the power-law behavior in stage [β_s] is a super-diffusion type and is different from that of von Schweidler type. In Fig. 1, a log-log plot of $M_2(t)$ is shown at the same value of D_S^L/d_0 for (S) and (M). For comparison, the logarithmic growth given by Eq. (20) is also shown at stage [β_f], where we set $B_\gamma = 0$. As D_S^L/d_0 decreases, the time separation among the characteristic times, τ_f, τ_β, and τ_γ, becomes large.

The single-particle dynamics is determined by only one parameter D_S^L/d_0. Hence it is convenient to introduce

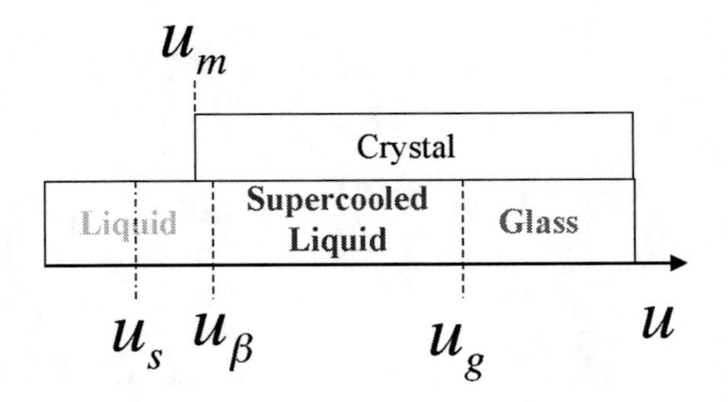

FIGURE 2. (Color online) Phase diagram versus u, where $u_\beta \simeq 2.530$, $u_g \simeq 5.038$, $u_s \simeq 0.536$ for (S) and 1.0 for (M), and $u_m \simeq 1.43$ for (S) and 2.15 for (M).

a new parameter u by

$$u = \log_{10}(d_0/D_S^L). \tag{21}$$

As is discussed later, u could be a universal parameter since all the dynamics in any fragile systems coincide with each other at a fixed value of u. As p is increased, the supercooled state and the glassy state appear at p_β (or u_β) and p_g (or u_g), respectively, where $p_g > p_\beta > p_s$ ($u_g > u_\beta > u_s$). Analyses of various data suggest that $u_\beta \simeq 2.530$, $u_g \simeq 5.038$, and $u_s \simeq 0.536$ for (S) and 1.0 for (M) [8]. In Fig. 2, the phase diagram is shown versus u. For comparison, the melting point u_m above which the crystal phase appears is also shown, where $u_m \simeq 1.43$ for (S) and 2.15 for (M) [14].

Long-Time Self-Diffusion Coefficient

We next discuss the long-time self-diffusion coefficient $D_S^L(p)$. As discussed in Ref. [7], it is well described by a singular function of p as

$$\frac{D_S^L}{\Delta_0} = \frac{d_0}{\Delta_0}\frac{1-Cp}{1+\varepsilon\frac{d_0}{\Delta_0}\left(\frac{p}{p_c}\right)\left(1-\frac{p}{p_c}\right)^{-2}}, \tag{22}$$

where $D_0(= k_BT/6\pi\eta a)$ is a free diffusion constant, p_c a singular point, ε a positive coefficient to be determined, and

$$\Delta_0 = \begin{cases} D_0 & \text{for (S)}, \\ av_0 & \text{for (M)}. \end{cases} \tag{23}$$

The singular part of Eq. (22) results from the long-time correlation effects due to the many-body interactions between particles. Here the factor C denotes the long-time coupling effect between different interactions. The original form of Eq. (22) was derived analytically for the suspension of monodisperse hard spheres by Tokuyama and Oppenheim (TO) [11, 12], where the singular part was obtained by the correlation effect due to the long-range hydrodynamic interactions and C was calculated from the coupling between the direct interactions and the short-range hydrodynamic interactions, leading to $C = 9/32$, $\varepsilon = 1$, and $p_c = (4/3)^3/[7\ln 3 - 8\ln 2 + 2] \simeq 0.57184\cdots$. We note here that the parameter ε describes the softness of interactions, where $\varepsilon = 1.0$ for hard spheres and $\varepsilon > 1$ for others.

Since the experiments and the simulations show a non-singular behavior for D_S^L, we next transform Eq. (22) into a non-singular function. The simplest way to do this is to introduce a transformation of p into a new parameter p' by [15]

$$p' = p + 10^{-\alpha}\frac{p}{p_c(p_c-p)}, \tag{24}$$

where α is a positive constant to be determined. In fact, solving Eq. (24) for p, one obtains

$$p = \frac{1}{2}[p' + p_c' - \sqrt{(p'+p_c')^2 - 4p_cp'}], \tag{25}$$

where $p_c' = p_c + 10^{-\alpha}/p_c$. Inserting this solution into Eq. (22), one can then find a non-singular function of p'. The exponent α is determined by the main mechanism and does not depend on the details of the systems, while p_c depends on the details. In Fig. 3, a log plot of D_S^L is shown versus volume fraction. As an example, data for two different hard-sphere systems with 6% size polydispersity are also shown. One is the experimental data for a suspension of hard-sphere colloids (HSC) by van Megen et al [16, 17] and another the simulation results for a hard-sphere fluid (HSF) by Tokuyama et al [18], where the control parameter p is given by the volume fraction ϕ. The main interactions in (HSC) are direct interactions between particles and hydrodynamic interactions between particles, while those in (HSF) are direct interactions only. The values of fitting parameters ε and α are listed in Table 1, where $\varepsilon = 1.0$ for hard spheres. For comparison, the simulation results for HSF with 0 %

TABLE 1. The fitting parameters ϕ_c, ε, α, and C for different systems.

Systems	ϕ_c	ε	α	C
TO Theory	0.57184	1.0	4.86	9/32
HSC Experiment (6%)	0.5560	1.0	4.86	9/32
HSF Simulation (6%)	0.5843	1.0	4.86	0
HSF Simulation (0%)	0.5828	1.0	4.86	0
HSF Simulation (15%)	0.5923	1.0	4.86	0

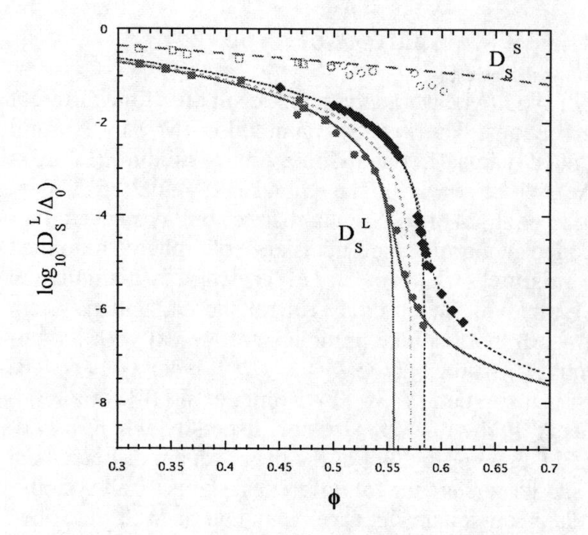

FIGURE 3. (Color online) Non-singular behavior of the long-time self-diffusion coefficient $D_S^L(\phi)$ versus volume fraction. The dashed line indicates the theoretical results from Refs. [11]. The solid lines indicate the mean-field fitting results for the experimental data [16] of colloidal suspension with 6% size polydispersity, where $\phi_c \simeq 0.556$, $\varepsilon = 1.0$, and $\alpha \simeq 4.86$, and the dotted lines for the simulation results [18] of hard-sphere fluid with 6% size polydispersity, where $\phi_c \simeq 0.5843$, $\varepsilon = 1.0$, and $\alpha \simeq 4.86$. The long-dashed line indicates the theoretical short-time self-diffusion coefficient [11]. The filled circles indicate the experimental data from Ref. [16] and the filled squares from Ref. [17]. The filled diamonds indicate the simulation results from Ref. [18]. The open circles indicate the experimental data from Ref. [16] and the open square from Ref. [17].

and 15% size polydispersities [14] are also listed in Table 1. It turns out that the polydispersity affects only the singular point and the exponent α is determined by the direct interactions only. As the polydispersity increases, ϕ_c increases, while ε and α are constant. By comparing the experimental data [16] with the simulation results [18], we can thus predict that the long-time hydrodynamic interactions play a role in reducing ϕ_c from 0.5843 to 0.556, where the short-time hydrodynamic interactions lead to $D_S^S(\phi)$.

UNIVERSAL BEHAVIOR NEAR THE GLASS TRANSITION

In this section, we analyze the mean-square displacements obtained in eight different systems from a unified point of view based on MFT and explore universal behavior near the glass transition.

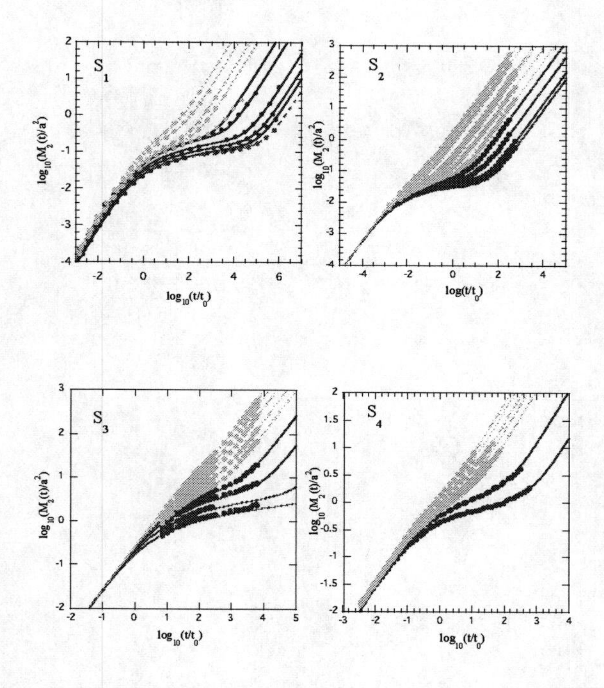

FIGURE 4. (Color online) A log-log plot of $M_2(t)$ versus time for different systems. The filled circles indicate the experimental data from Ref. [16] for (S_1), Ref. [21] for (S_3), Ref. [22] for (S_4), and the simulation results from Ref. [19] for (S_2). The solid lines indicate the mean-field results at a supercooled state ($u_\beta \leq u < u_g$), the dashed lines at a glass state ($u_g \leq u$), and the dotted lines at a liquid state ($u < u_\beta$).

Analyses of Type (S)

We analyze the results for three experimental data and one simulation result. The first is an experiment (S_1) by van Megen et al [16] for colloidal suspensions of neutral hard spheres with 6% polydispersity (neutral colloids), where the control parameter p is given by the volume fraction ϕ. The second is a Brownian dynamics simulation (S_2) by Flenner et al [9, 19] for a colloidal binary mixture (LJ colloids) with the Lennard-Jones potentials $U_{\alpha\beta}(r) = 4\varepsilon_{\alpha\beta}[(\sigma_{\alpha\beta}/r)^{12} - (\sigma_{\alpha\beta}/r)^6]$, where $p = 1/T$ and indices α, β run on the particle types A and B. Here the parameters $\sigma_{\alpha\beta}$ and $\varepsilon_{\alpha\beta}$ are chosen as in Ref. [20]. The third is an experiment (S_3) by Hwang et al [21] for the suspension of magnetic colloidal chains interacting through magnetic dipole moments, where p is given by the square of the dimensionless external magnetic field, Γ. The last is an experiment (S_4) by König et al [22] for binary two-dimensional 50% mixtures of super paramagnetic colloidal particles interacting through magnetic

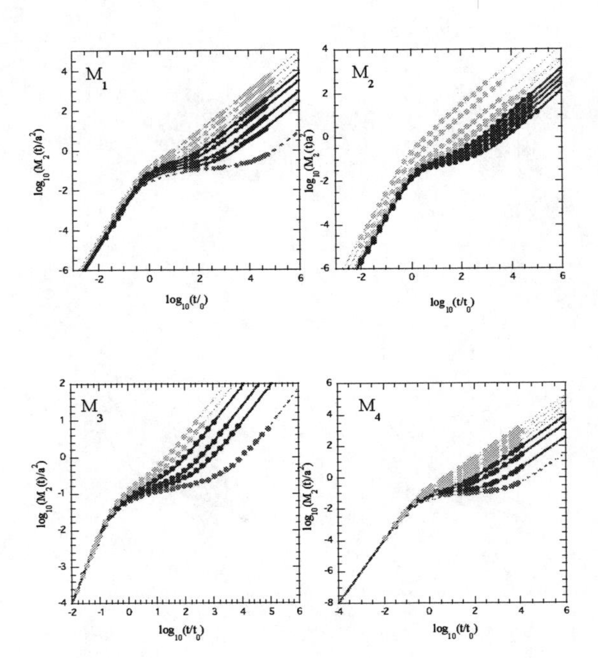

FIGURE 5. (Color online) A log-log plot of $M_2(t)$ versus time for different systems. The filled circles indicate the simulation results from Ref. [19] for (M_1), Ref. [24] for (M_2), Ref. [10] for (M_3), and Ref. [14] for (M_4). The details are the same as in Fig. 4.

dipole moments, where $p = \Gamma$. In Fig. 4, a log-log plot of the mean-square displacement is shown versus time for different systems. For long times, one can first obtain the long-time self-diffusion coefficient D_S^L for different systems by using the relation given by Eq. (19). By adjusting the parameter ℓ appropriately, one can then fit the mean-field equation given by Eq. (13) with the mean-square displacements for the experiments and the simulations. In all systems, the mean-field theory describes the data well. However, we should note here that the simulation results slightly deviate from the mean-field equation in fast β-stage $[\beta_f]$ at a strong supercooled state and a glass state. This is because the systems do not reach an equilibrium state yet within computational times and are trapped into a nonequilibrium metastable state. Hence it would take a infinite time for the system to reach an equilibrium state.

Analyses of Type (M)

In type (M), we analyze the results for four different simulations. The first is a simulation (M_1) by Narumi et al [23] for a Lennard-Jones binary mixture (LJ mixture), where $p = 1/T$. The second is a simulation (M_2) by Gallo et al [24] for a Lennard-Jones binary mixture embedded in an off-lattice matrix of soft spheres (confined LJ mixture), where $p = 1/T$. The third is a simulation (M_3) by Voigtmann et al [10] for the molecular system of nearly hard-sphere particles (soft cores) with the potential $U_{ij}(r) = k_B T(r/(a_i + a_j))^{-36}$, where $p = \phi$. The last is a simulation (M_4) by Kohira et al [14] for a hard-sphere fluid with 15% size polydispersity, where $p = \phi$. In Fig. 5, a log-log plot of the mean-square displacement is shown versus time for different systems. In all systems, their mean-square displacements can be well described by Eq. (13) by adjusting ℓ appropriately. Similarly to (S), slight deviations from the mean-field equation are found in stage $[\beta_f]$ at a strong supercooled state and a glass state since the systems do not reach an equilibrium state yet within computational times.

Non-Singular Behavior of D_S^L

In Fig. 6, the fitting values of the long-time self-diffusion coefficient D_S^L in (S_2) are shown versus (a) $1/T$ and (b) T_c/T. The fitting results are obtained from two different simulations on the Lennard-Jones binary colloids. One is a simulation by Flenner-Szamel where $N = 10^3$. Another is a simulation by Kimura-Tokuyama where $N = 10^4$. Both fitting results are well described by the mean-field non-singular function

$$\frac{D_S^L(T)}{D_0} = \frac{1}{1 + \varepsilon \frac{D_S^S(\phi)}{D_0} \frac{T}{T_c} \left(1 - \frac{T}{T_c}\right)^{-2}} \tag{26}$$

with the transformation given by Eq. (25), where $D_S^S(\phi)/D_0 \simeq 0.24$. The fitting parameters are listed in Table 2. Here the short-time self-diffusion coefficient

TABLE 2. The fitting parameters T_c, ε, α, and D_S^S/D_0 for (S_2).

Systems	T_c	ε	α	D_S^S/D_0
Flenner-Szamel	0.4698	46.77	2.05	0.24
MCT solution	0.9515	46.77	–	0.24
Kimura-Tokuyama	0.4698	46.77	2.05	0.24
Narumi-Tokuyama	0.4698	46.77	2.05	–

$D_S^L(\phi)$ results from the short-time hydrodynamic interactions between particles and depends only on the volume fraction ϕ. The comparison of the fitting value 0.24 with TO theory [11] suggests that the effective diameter is $0.9334\sigma_{AA}$. This short-time behavior is clearly seen by

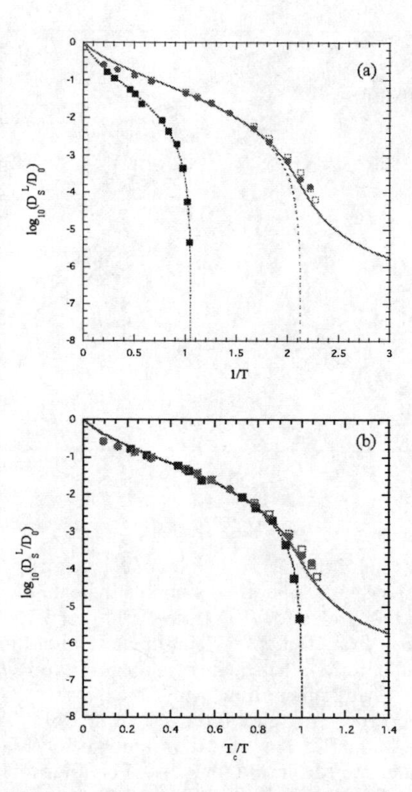

FIGURE 6. (Color online) A log plot of D_S^L versus (a) $1/T$ and (b) T_c/T. The filled circles indicate the fitting results from Ref. [19] and the open squares from Ref. [9]. The filled squares indicate the fitting results for the solutions of the mode-coupling equation from Ref. [9]. The solid lines indicate the mean-field non-singular function given by Eq. (26) with Eq. (25). The dotted and the dashed lines indicate the singular function given by Eq. (26).

comparing the Lennard-Jones binary colloids (S_2) with the Lennard-Jones binary mixtures (M_1). In Fig. 7, the fitting results in both systems are shown. The results in (M_1) are well described by the mean-field non-singular function

$$\frac{D_S^L(T)}{av_0} = \frac{1}{1+\varepsilon\frac{T}{T_c}\left(1-\frac{T}{T_c}\right)^{-2}}. \tag{27}$$

All the common fitting parameters are the same as those in (S_2). Hence only difference between two systems is due to the short-time self-diffusion coefficient D_S^L.

Flenner and Szamel [9] have also solved the mode-coupling equation for the mean-square displacement numerically by using the static structure factor obtained by their simulations. The fitting results for those solutions are also shown in Fig. 6. We should note here that the mode-coupling results are well described by the mean-field singular function given by Eq. (26), although they

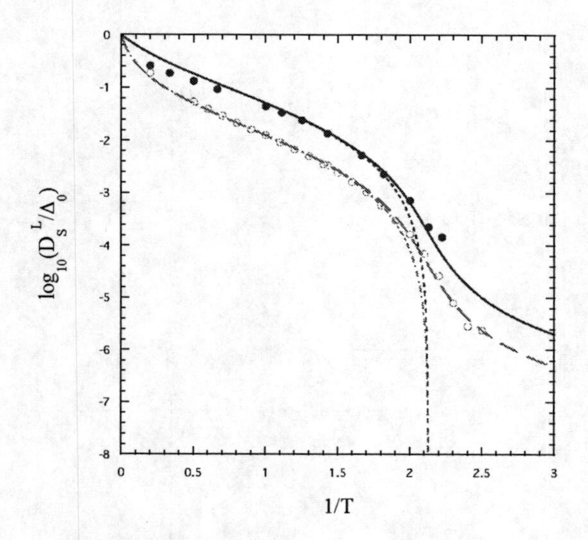

FIGURE 7. (Color online) A log plot of D_S^L versus $1/T$ for different systems, (S_2) and (M_1). The open circles indicate the fitting results from Ref. [23]. The long-dashed line indicates the mean-field non-singular function given by Eq. (27) with Eq. (25). The dot-dashed lines indicate the singular function given by Eq. (27). The details are the same as in Fig. 6.

do not agree with the simulation results. Since the simulation results and the mode-coupling results are described by Eq. (26), those results versus T_c/T are collapsed into a single curve shown in Fig. 6 (b). Thus, it turns out that the simulation results obey a non-singular function, while the mode-coupling results obey a singular function.

In Fig. 8, the fitting values of the long-time self-diffusion coefficient D_S^L in (M_3) are shown versus (a) ϕ and (b) ϕ/ϕ_c. The fitting results are well described by the mean-field singular function

$$\frac{D_S^L(T)}{av_0} = \frac{1}{1+\varepsilon\frac{\phi}{\phi_c}\left(1-\frac{\phi}{\phi_c}\right)^{-2}}. \tag{28}$$

The fitting parameters are listed in Table 3. Voigtmann et al [10] have also solved the mode-coupling equation numerically by using the static structure factor obtained by their simulations. The fitting results for those solutions are also shown in Fig. 8. The mode-coupling results are well described by the mean-field singular function given by Eq. (28), although they do not agree with the simula-

TABLE 3. The fitting parameters ϕ_c, ε, and α for (M_3).

Systems	ϕ_c	ε	α
Voigtmann et al	0.592	1.1	4.86
MCT solution	0.516	1.1	–

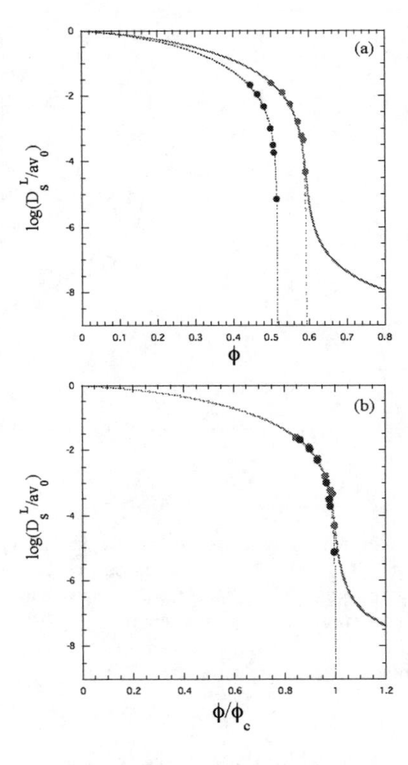

FIGURE 8. (Color online) A log plot of D_S^L versus (a) ϕ and (b) ϕ/ϕ_c. The filled circles indicate the fitting results from Ref. [10] and the filled squares indicate the fitting results for the solutions of the mode-coupling equation from Ref. [10]. The solid lines indicate the mean-field non-singular function given by Eq. (28) with Eq. (25). The dotted and the dashed lines indicate the singular function given by Eq. (28).

tion results. Since the simulation results and the mode-coupling results are described by Eq. (28), a plot of those results versus ϕ/ϕ_c is also collapsed on a single curve shown in Fig. 8 (b). In order to show whether the simulation results obey a non-singular function or not, one must perform the simulations for higher volume fractions.

In Fig. 9, the fitting results in seven different systems are shown versus p/p_c. The fitting parameters are listed in Table 4. It turns out that the parameter ε becomes large as the softness increases, while the exponent α depends only on the main mechanism. The fitting results for the MCT solutions are also plotted in Fig. 9. Finally, we note that although all the results are well described by the non-singular function given by Eq. (22) with Eq. (25), MCT results obey the singular function given by Eq. (22).

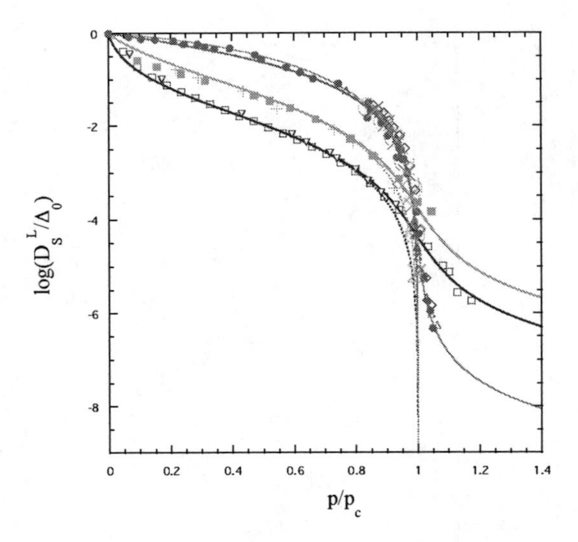

FIGURE 9. (Color online) A log plot of D_S^L versus p/p_c for seven different systems. The filled symbols indicate the fitting results in (S); the circles for (S_1) from Refs. [16, 17] and the squares for (S_2) from Ref. [19]. The open symbols indicate the fitting results in (M); the squares for (M_1) from Ref. [23], the triangles for hard-sphere fluid with 6% size polydispersity from [18], the down triangles for (M_2) from Ref. [24], the circles for (M_3) from Ref. [10], and the diamonds for (M_4) from [14]. $+$ indicates the MCT solutions for LJ colloids and $\times$ for soft cores. The solid line indicates the mean-field non-singular function and the dotted the mean-field singular function.

TABLE 4. The fitting parameters ε, α, and p_c.

Systems	ε	α	p_c
Neutral colloids	1.0	4.86	0.5560
Hard spheres (6%)	1.0	4.86	0.5843
Hard spheres (15%)	1.0	4.86	0.5923
Soft cores (10%)	1.1	4.86	0.5920
LJ colloids	46.77	2.05	2.1285
LJ mixture	46.77	2.05	2.1285
Cofined LJ mixture	46.77	2.05	2.9070

Mean-Free Path ℓ

In Fig. 10, we now plot the fitting values of the mean-free path ℓ/ℓ_0 versus the universal parameter u given by Eq. (21) for nine different systems, including the hard-sphere fluid with 6% size polydispersity [18]. Here ℓ_0 indicates the smallest characteristic length, such as a radius and a mean distance between particles [8]. As u increases, the mean-free path ℓ decreases in each system. Within error, it seems to obey a single curve in each system, leading to a unique function of u, $\ell/\ell_0 = f(u)$. Here the function $f(u)$ in each system coincides with each other in a supercooled state and in a glass state, while it is different from the other in a liquid state.

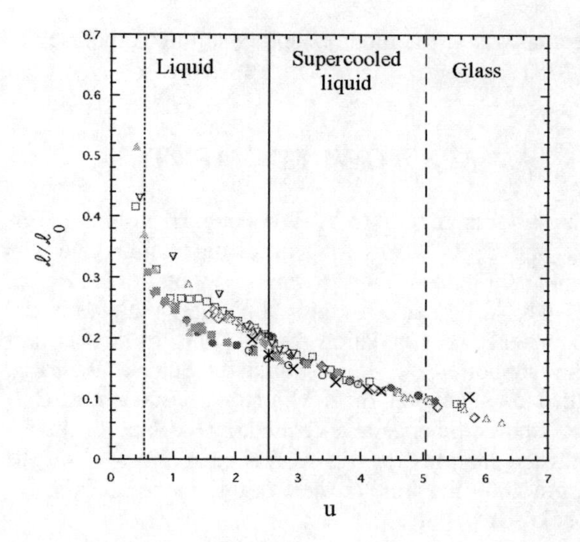

FIGURE 10. (Color online) A plot of the mean-free path ℓ/ℓ_0 versus u for different systems, where ℓ_0 denotes the smallest characteristic length. The filled symbols stand for (S) and the open symbols for (M). The filled triangles indicate the fitting results for (S_3) and the filled down triangles for (S_4). The details are the same as in Fig. 9.

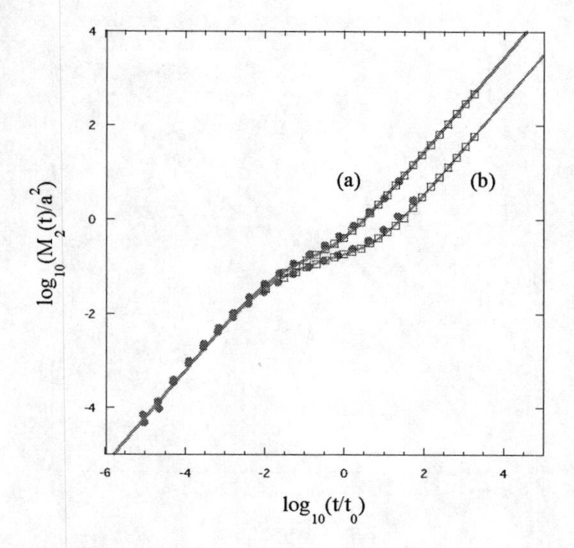

FIGURE 11. (Color online) A log-log plot of the mean-square displacement $M_2(t)$ versus t for different systems (S_1) and (S_2). The open circles indicate the results for (S_1) from Ref. [16] and the open squares for (S_2) from Ref. [19] at (a) $u \simeq 1.347$, where (○) $\phi = 0.502$ and (□) $T = 1.0$, and (b) $u \simeq 2.273$, where (○) $\phi = 0.538$ and (□) $T = 0.6$. The solid lines indicate the mean-field results given by Eq. (13).

This would be explained by the following reason. Since the non-local, collective motion dominates the system in both supercooled and glass states, the details of each system do not affect the mean-free path ℓ. On the other hand, in a liquid state the local motion dominates the system. Hence ℓ strongly depends on the details. In fact, ℓ in (S) is always shorter than that in (M) because the particle undergoes diffusive motion in (S) for a short time, while it does ballistic motion (M).

Master Curve For $M_2(t)$

As shown in the previous section, the mean-free path ℓ is uniquely determined by the parameter u. Hence any data for the mean-square displacement $M_2(t)$ in different systems are expected to be collapsed into a master curve if the value of u is the same. As examples, we first compare the mean-square displacement $M_2(t)$ in (S_1) where the volume fraction ϕ is the control parameter to that in (S_2) where the temperature T is the control parameter. We have two cases, (a) $u \simeq 1.347$, where $\phi = 0.502$ and $T = 1.0$, and (b) $u \simeq 2.273$, where $\phi = 0.538$ and $T = 0.6$. In Fig. 11, a log-log plot of $M_2(t)$ is shown versus time. Thus, the single-particle dynamics in the experiment for the suspension of neutral colloids is shown to coincide with that in the simulation for the suspension of Lennard-Jones colloids at a given value of u. Next, we compare the mean-square displacement $M_2(t)$ in (M_3) where ϕ is the control parameter to that in (M_4) where ϕ is the control parameter. We have two cases, (a) $u \simeq 2.215$, where $\phi = 0.55$ for (M_4) and $\phi = 0.50$ for (M_3), and (b) $u \simeq 2.886$, where $\phi = 0.57$ for (M_4) and $\phi = 0.55$ for (M_3). In Fig. 12, a log-log plot of $M_2(t)$ is shown versus time. Thus, the single-particle dynamics in the simulation for the hard-sphere fluid with 15% size polydispersity is shown to coincide with that in the simulation for the soft-core system at a given value of u. In general, one can conclude that any pair of the systems in each type can be compared at a given value of u and can be described by a master curve given by the mean-field equation.

SUMMARY

In this paper, we have analyzed the data for several different systems near their glass transitions from a unified point of view based on the mean-field theory. Thus, we have first shown that there exist the universal features among widely different glass-forming systems near their glass transitions. The key parameter is the reduced long-time self-diffusion coefficient D_S^L/d_0 or the universal parameter $u(= \log_{10}(d_0/D_S^L))$. When the value of u is the same for any different systems, the properties of the processes are the same as each other. In fact, the mean-free path ℓ is uniquely determined by u in each type (S) or

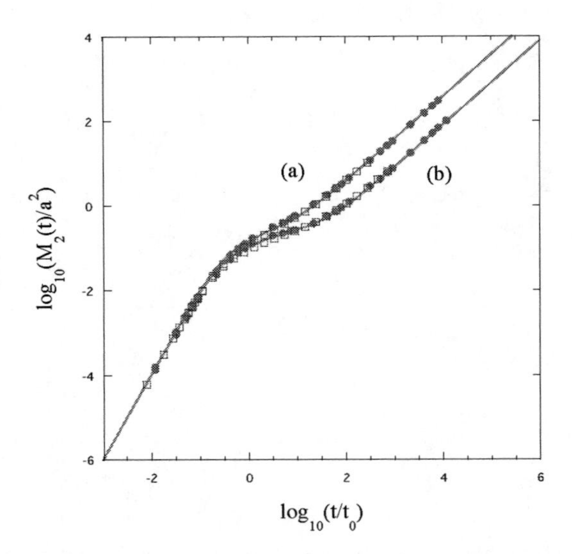

FIGURE 12. (Color online) A log-log plot of the mean-square displacement $M_2(t)$ versus t for different systems (M_3) and (M_4). The open circles indicate the results for (M_4) from Ref. [14] and the open squares for (M_3) from Ref. [10] at (a) $u \simeq 2.215$, where (○) $\phi = 0.55$ and (□) $\phi = 0.50$, and (b) $u \simeq 2.886$, where (○) $\phi = 0.57$ and (□) $\phi = 0.55$. The details are the same as in Fig. 12.

(M) (see Fig. 10). Hence the data for the mean-square displacements between two different systems have been shown to be collapsed into a master curve given by the mean-field solution given by Eq. (13). Thus, one can predict that the state of the system is determined only by u, where the supercooled point u_β and the glass transition point u_g are independent of the systems. This suggests that any systems can be converted into the other systems at a given value of u [15].

Next, we have shown that the data for the long-time self-diffusion coefficient in different fragile systems are well described by the non-singular function given by Eq. (22) with Eq. (25). We have also analyzed the numerical solutions of the mode-coupling equations calculated by Flenner-Szamel [9] and also by Voigtmann et al [10] by MFT. Analyses showed that both results for the long-time self-diffusion coefficient disagree with the simulation results since their singular points are quite different from those of the simulation results, although both the simulation and the mode-coupling results are partially described by the singular function given by Eq. (22).

Finally, we mention that the discrepancy between the simulation results and the mode-coupling solutions is also seen in the mean-free path ℓ (see Fig. 10). This means that even if u is the same, the mode-coupling solutions do not agree with a master curve around the β-relaxation stage, which can describe the simulation results well. This will be discuss elsewhere together with the analyses of the mode-coupling solutions themselves by MFT.

ACKNOWLEDGMENTS

The author is grateful to Y.-H. Hwang, H. König, Y. Terada, and E. R. Weeks for fruitful discussions. He also thanks E. Flenner, Y.-H. Hwang, H. König, G. Szamel, and Th. Voigtmann for providing him with their experimental and simulation results. This work was partially supported by Grants-in-aid for Science Research with No. 18540363 from Ministry of Education, Culture, Sports, Science and Technology of Japan. Numerical computations for this work were performed on the Origin 2000 machine at the Institute of liquid Science, Tohoku University.

REFERENCES

1. C. A. Angell, K. L. Ngai, G. B. McKenna, P. F. McMillan, and S. W. Martin, *J. Appl. Phys.* **88**, 3113-3157 (2000).
2. P. G. Debenedetti and F. H. Stillinger, *Nature* **410**, 259-267 (2001).
3. K. Binder and W. Kob, in *Glassy Materials and Disordered Solids* (World Scientific, Singapore, 2005).
4. Proceedings of the 4th Int. Discussion Meeting on Relaxation in fragile Systems, edited by K. L. Ngai, J. Non-Cryst. Solids **307-310** (2002).
5. Proceedings of the 3rd Int. Symposium on Slow Dynamics in fragile Systems, edited by M. Tokuyama and I. Oppenheim (AIP, New York, 2004) , CP**708**.
6. M. Tokuyama, *Phys. Rev. E* **62**, R5915-R5918 (2000).
7. M. Tokuyama, *Physica A* **364**, 23-62 (2006).
8. M. Tokuyama, *Physica A* **378**, 157166 (2007).
9. E. Flenner and G. Szamel, *Phys. Rev. E* **72**, 031508-1-031508-15 (2005).
10. Th. Voigtmann, A. M. Puertas, and M. Fuchs, *Phys. Rev. E* **70**, 061506-1-061506-19 (2004).
11. M. Tokuyama and I. Oppenheim, *Phys. Rev. E* **50**, R16-R19 (1994).
12. M. Tokuyama and I. Oppenheim, *Physica A* **216**, 85-119 (1995).
13. A. Rahman, K. S. Singwi, and A. Sjölander, *Phys. Rev.* **126**, 986-996 (1962).
14. E. Kohira, Y. Terada, and M. Tokuyama, *Rep. Inst. Fluid Science* **19**, 91-94 (2007).
15. M. Tokuyama, T. Narumi, and E. Kohira, *Physica A* **385**, 439-455 (2007).
16. W. van Megen, T. C. Mortensen, S. R. Williams, and J. Müller, *Phys. Rev. E* **58**, 6073-6085 (1998).
17. W. van Megen and S. M. Underwood, *J. Chem. Phys.* **91**, 552-559 (1989).
18. M. Tokuyama and Y. Terada, *J. Phys. Chem. B* **109**, 21357-21363 (2005).
19. Y. Kimura and M. Tokuyama, *Rep. Inst. Fluid Science* **19**, 85-90 (2007).
20. W. Kob and H. C. Andersen, *Phys. Rev. Lett.* **73**, 1376-1379 (1994).

21. Y.-H. Hwang and X.-L. Wu, *Phys. Rev. Lett.* **74**, 2284-2287 (1995).
22. H. König, R. Hund, K. Zahn, and G. Maret, *Eur. Phys. J. E* **18**, 287-293 (2005).
23. T. Narumi and M. Tokuyama, *Rep. Inst. Fluid Science* **19**, 73-78 (2007).
24. P. Gallo, R. Pellarin, and M. Rovere, *Phys. Rev. E* **67**, 041202-1-041202-7 (2003).

Universal Secondary Relaxation of Water in Aqueous Mixtures, in Nano-Confinement, and in Hydrated Proteins

K. L. Ngai*, S. Capaccioli§, and N. Shinyashiki+

*Naval Research Laboratory, Washington, DC 20375-5320, USA
§INFM-CNR, SOFT, Piazzale A. Moro 2, I-00185 Roma, & Dipartimento di Fisica, Università di Pisa, Largo B. Pontecorvo 3, I-56127, Pisa, Italy
+Department of Physics, Tokai University, Hiratsuka, Kanagawa 259-1292, Japan

Abstract. From a large volume of experimental data of relaxation of water in various aqueous mixtures, in different forms of nano-confinement, and in two hydrated proteins, we show the presence of a secondary relaxation in all these systems that have similar characteristics. This ubiquitous secondary relaxation originates from water in the systems and is the analogue of the universal Johari-Goldstein secondary relaxation of glass-forming substances in general. Like all Johari-Goldstein secondary relaxation, this one from water bears an intimate and important relation to the primary structural relaxation, and hence it plays a fundamental role in glass transition or function of these water containing systems.

Keywords: Aqueous mixtures, Dynamics, Glass Transition, Hydrated proteins, Relaxation, Water

INTRODUCTION

Water is the most abundant substance in our planet Earth, and it is present in significant proportion in all living organisms. Despite the water molecule is simple compared with most other glass-forming substances, its thermodynamic and kinetic properties are rather odd and have been difficult to clarify experimentally. Consequently, controversies abound in the studies of water and its mixtures with other substances [1]. Help in gaining insight into the true nature of the dynamics of water in aqueous mixtures came recently from the experimental data of binary mixtures of van der Waals liquids that yield unambiguous information on the dynamics of each component [2-6]. For each component, there is a particular secondary relaxation (named after Johari-Goldstein (JG)) [7] with properties that are intimately related to that of the primary relaxation of the same component or the other component [8-11]. For example, the secondary relaxation time $\tau_{JG,p}$ as well as its dielectric relaxation strength $\Delta\varepsilon_{JG,p}$ of component p changes their temperature dependences when crossing the glass transition temperature, T_{gp}, of the same component. The primary relaxation time of component p usually have the Vogel-Fulcher T-dependence, and T_{gp} is the temperature at which $\tau_{\alpha,p}$ attains a long time such as 10^3 s. At temperatures below T_{gp}, $\tau_{JG,p}$ has the usual Arrhenius T-dependence characteristic of secondary relaxation. But, this Arrhenius dependence does not continue to temperatures above T_{gp}, where $\tau_{JG,p}$ assumes a different and significantly stronger temperature dependence. The value of $\tau_{JG,p}$ at T=T_{gp} anti-correlates with T_{gp}, decreasing with increasing T_{gp}. Another example is provided by the invariance of the ratio $\tau_{\alpha,p}/\tau_{JG,p}$ to various combinations of temperature and pressure P while maintaining $\tau_{\alpha,p}$ constant [12,13].

In the limit of vanishing concentration of the other component, mixtures become neat glassformers and the relaxation time τ_{JG} of JG relaxation of neat glassformers is strongly correlated with the primary relaxation time τ_α in the same ways as in mixtures plus more. The separation between the JG secondary relaxation time τ_{JG} and the α-relaxation time τ_α measured by [log(τ_α)-log(τ_{JG})] for any fixed log(τ_α) is strongly correlated with n when many glass-formers were considered [8-10]. Here n is the fractional exponent appearing in the Kohlrausch function describing the non-exponential time dependence of the α-relaxation of neat glassformers,

CP982, *Complex Systems, 5th International Workshop on Complex Systems*
edited by M. Tokuyama, I. Oppenheim, and H. Nishiyama

$$\phi(t) = \exp[-(t/\tau_\alpha)^{1-n}],\ 0<n<1 \quad (1)$$

In fact, $[\log(\tau_\alpha)-\log(\tau_{JG})]$ is approximately given by the expression, $n[11.7 + \log(\tau_\alpha)]$ for any T and P. This relation is derivable from the coupling model (CM) [8-10,14,15], because the primitive relaxation time τ_0 of the CM is a good indicator of the most probably JG relaxation time τ_{JG}. One immediate consequence of this relation between τ_α and τ_{JG} is that the well known crossover of T-dependence of τ_α from VF above T_g to Arrhenius below T_g is transferred from τ_α to τ_{JG} (or vice versa) in a qualitative manner. This can be seen directly by rewriting the relation as

$$\log\tau_{JG}(T) \approx [(1-n)\log\tau_\alpha(T)-11.7n]. \quad (2)$$

From this relation, it is clear that the temperature dependence of $\tau_{JG}(T)$ mimics that of $\tau_\alpha(T)$, albeit weaker due to the factor $(1-n)$ on the right-hand-side of Eq.(2). A comprehensive review of these extraordinary properties indicating the fundamental role played by JG relaxation in glass transition can be found in Ref.[8], where it was shown that the properties can be explained by the coupling model (CM). The experimental facts all indicate that a fully successful theory of glass transition must take into consideration the time/frequency dependence of the primary structural relaxation and the JG secondary relaxation, such as the proposed approach based on the CM [16].

Advance in understanding the dynamics of water was made recently by the finding [17] of the universal presence of the JG relaxation of the water component in many different aqueous mixtures [18], and having the same properties as the JG relaxation of a component in binary mixtures of van der Waals liquids. The knowledge of the component dynamics gained from the study of the van der Waals mixtures can be transferred to elucidate and interpret the experimental data on the dynamics of water in aqueous mixtures, in nano-confinement, and in hydrated proteins. This is the objective of this paper. The exercise illustrates the benefit of 'cross-field fertilization', *i.e.*, the application of the clear understanding achieved in the study of one class of glass-forming materials to another class.

RELATION BETWEEN α- AND JG β-RELAXATIONS IN MIXTURES OF VAN DER WAALS LIQUIDS

Experiments which probe selectively the relaxation dynamics of a component in a binary homogeneous mixture are most informative. This ideal situation was achieved in broadband dielectric relaxation measurements by choosing the probed component to have much larger dipole moment than the other component. Oligomers of styrene and polystyrene mix well with many glass-formers, and are ideal for being used as the unseen component because the dipole moment of the styrene repeat unit is very small. Measurements were made on the dynamics of two polar rigid small molecules, *tert*-butylpyridine (TBP) (T_g=164 K) and quinaldine (QN) (T_g=180 K) in mixtures with tristyrene (T_g=234 K), with concentration ranging from 5 wt.% to 100% of TBP and QN [4,5,12]. The α- and JG β-relaxations of TBP or QN are found in the dielectric spectra. For mixtures with concentrations below 50% of 2-picoline, 60% of TBP, and 40% of QN, the JG relaxation appears as a well resolved peak in the dielectric loss spectrum. The α- (JG-) loss peak frequency $\nu_{\alpha,p}$ ($\nu_{JG,p}$), and hence also the corresponding $\tau_{\alpha,p}$ ($\tau_{JG,p}$) is determined unequivocally without the need of using any assumed procedure to fit the spectrum. The directly obtained $\tau_{JG,p}$ of TBP and QN in 16% and 5% mixtures with tristyrene all have the characteristic change of T-dependence when crossing the glass transition temperature T_{gp} of the minority component defined by $\tau_{\alpha,p}(T_{gp})=10^3$ s as shown in Fig.1. Below T_{gp}, the temperature dependence of $\tau_{JG,p}$ is Arrhenius, but it changes to a much stronger temperature dependence at higher temperatures after crossing T_{gp}. Furthermore, the dielectric relaxation strength of the JG relaxation, $\Delta\varepsilon_{JG}(T)$, exhibits a stronger increase with temperature above T_{gp} than below (not shown here), a feature found also for JG relaxation in neat glass-formers. At 5 wt.% of QN, effect of concentration fluctuation in the mixture in broadening the frequency dispersion of the α-relaxation of QN is negligible. The fit of the spectrum by the one-sided Fourier transform of the Kohlrausch function (eq.1) gave $(1-n_p)$ and $\tau_{\alpha,p}(T)$. From these two experimental parameters, $\tau_{JG,p}(T)$ of QN was determined by Eq.2. There is good agreement between the calculated $\tau_{JG,p}(T)$ and the experimental $\tau_{JG,p}(T)$ for all temperatures above T_{gp} (see Fig.1). The general properties of the JG relaxation, demonstrated by examples taken from the two mixtures in Fig.1, and also elsewhere for picoline in tristyrene [16], follow as immediate consequences of Eq.2 and hence are explained by the coupling model. Details can be found in several published works [3-6,8-10,15,17].

In the following sections, we show that the same properties are also found in the relaxation of aqueous mixtures, water in nano-confinement, and hydrated proteins, all of which belong to a different class of systems than the mixtures of van der Waals liquids discussed above. Due to the extraordinary nature of water, the dynamics of pure water as well as aqueous mixtures are hard to interpret, leading to controversy. Naturally, the transparent results of van der Waals liquids (Fig.1) together with the satisfactory explanation given for them should help in guiding us to correctly interpret the experimental data of aqueous mixtures. This will be carried out in the following sections, and the success serves to demonstrate the payoff in applying well understood property in a class of glass-formers to another class where the situation is still unclear.

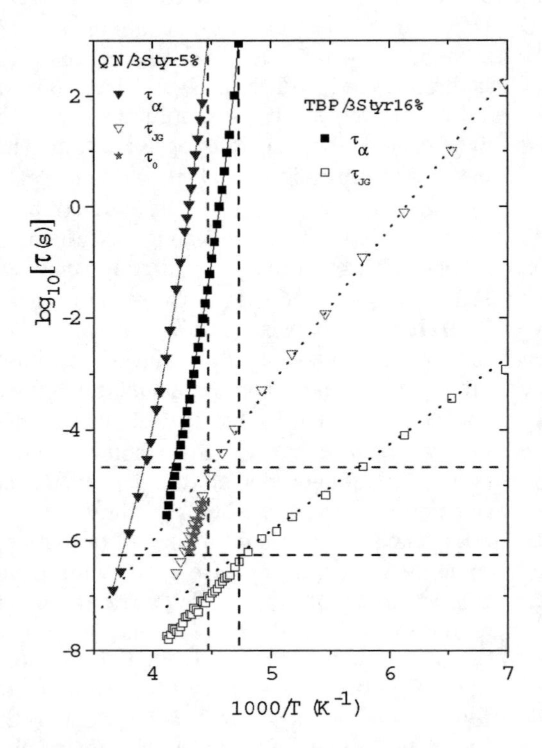

FIGURE 1. (Color online) Relaxation map for mixtures of the polar molecules (16 wt.%) TBP (black squares) and (5 wt.%) QN (blue triangles) in tristyrene. Full and open symbols represent the α- and JG β-relaxation times respectively. Continuous lines are VF equation fits. Dotted lines are Arrhenius fits. Dashed vertical and horizontal dashed lines show the crossover at T_g of the temperature dependence of $\tau_{JG,p}(T)$. The red stars indicate the $\tau_{JG,p}(T)$ of QN calculated by Eq.(2) using known value of (1-n) from the fit to the dielectric spectra for several temperatures.

RELATION BETWEEN α- and JG β-RELAXATIONS IN AQUEOUS MIXTURES

Small in size and strong affinity for hydrogen bonding, water does not behave in exactly the same way as other glass-formers in mixtures. In aqueous mixtures the other component is usually hydrophilic and susceptible to hydrogen bonding with water. Examples include 1-propanol, ethylene glycol (EG), ethylene glycol oligomers (EGO), tri-propylene glycol, glycerol, xylitol, sorbitol, poly(ethylene glycol) (PEG) [17,18], poly(vinylpyrrolidone) (PVP) [19] and poly(vinyl methylether) (PVME) [20]. Broadband dielectric relaxation spectroscopy were used together with differential scanning calorimetry (DSC) to characterize the dynamics of these aqueous mixtures. Representative results from some aqueous mixtures are shown in Fig.2.

In all aqueous solutions, a faster β-relaxation identified as the JG relaxation of water in the mixture is observed by broadband dielectric relaxation. The JG relaxation time, $\tau_{JG}(T)$, of the water component in these mixtures is shown as a function of 1000/T in Fig.2. Blue open circles are for 35 wt.% water/6EG, red open diamonds for 35 wt.% water/PEG400, brown open inverted triangles for 35 wt.% water/PEG600, magenta filled squares for 50 wt.% water/PVME, green open triangles for 50 wt.% water/PVP. The results of τ_α from dielectric measurements are shown for mixtures of 35 wt.% water with the ethylene glycol oligomer, 6EG (blue closed circles), and 35wt.% water with PEG400 (red closed diamonds). The slower relaxation has the VF T-dependence of α-process which is due to the cooperative motions of the hydrogen-bonded water and solute molecules that give rise to the glass transition of the mixture at T_{gm} defined by $\tau_\alpha(T_{gm})=10^3$ s. In the other mixtures, 35 wt.% water/PEG600, 50 wt.% water/PVME, and 50wt.% water/PVP, the presence of conductivity contribution at low frequencies made the determination of the dielectric τ_α impossible. However, the presence of the α-relaxation in all mixtures was assured by thermal analysis, and T_{gm} is taken here as the glass transition temperature obtained by DSC.

The temperature dependence of τ_{JG} is Arrhenius at lower temperatures but it crosses over to stronger temperature dependence above T_{gm}. In Fig.2, the location of 1000/T_{gm} is indicated by a vertical line for each mixture. It can be seen by inspection of the figures that the change of the temperature dependence of τ_{JG} occurs near T_{gm} in all cases. This general crossover property of the T-dependence of τ_{JG} of water in aqueous mixtures is exactly analogous to that of TBP, QN, and picoline in mixtures with tri-styrene shown in Fig.1, and in neat glass-formers including xylitol and sorbitol [8,9]. Hence the same interpretation holds, and we can conclude that the observed faster process is the JG secondary relaxation of water in the mixtures above and below T_{gm}. This conclusion also invalidates the suggestion by others that the observed faster process is a merged α- and β-process at temperatures above T_{gm}.

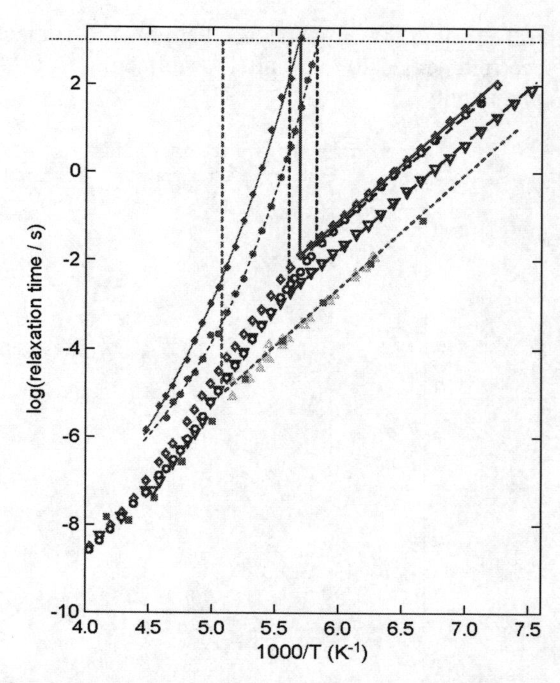

FIGUERE 2. (Color online) Relaxation map, $\log_{10}(\tau)$ vs 1000/T, of the JG β-relaxation of the water component and some α-relaxation in several aqueous mixtures [17,18]. The location of $1000/T_{gm}$ for each mixture is indicated by the vertical line. For 35 wt.% water mixture with 6EG and PEG400, data of τ_α are shown by (blue) filled circles and (red) filled diamonds respectively, and T_{gm} is defined as the temperature at which $\tau_\alpha(T_{gm})=10^3$ s as indicated. For 35 wt.% water mixture with PEG600, and 50% water mixtures with PVME, the dielectric data of τ_α are not available and in this cases T_{gm} is taken from calorimetric data. For 50 wt.% water mixture with PVP, T_{gm} from calorimetric data is close to that of 50% water mixtures with PVME. It is not shown to avoid crowding. The JG relaxation time, $\tau_{JG}(T)$, of the water component are indicated by (blue) open circles for 35 wt.% water/6EG, open (red) diamonds for 35 wt.% water/PEG400, open (brown) inverted triangles for 35 wt.% water/PEG600, filled (magenta) square for 50 wt.% water/PVME, open (green) triangles for 50 wt.% water/PVP. Note that in all mixtures $\tau_{JG}(T)$ has Arrhenius temperature below T_{gm}, but a stronger temperature dependence above T_{gm}. The crossing of temperature dependence at T_{gm} is clearly indicated by the vertical line drawn at T_{gm} terminating the line fitting the Arrhenius dependence for each mixture. The same color is used for the two lines for ease in identifying the pair.

Whenever data are available, the dielectric strength, $\Delta\varepsilon_{JG}$, of the JG-process of water in aqueous mixtures increases monotonically with temperature with an elbow shape that indicates a change of slope at T_g (for experimental data, see Ref.17). This property is typical of secondary JG relaxation of a component in non-aqueous mixtures shown before in Fig.1, and also in neat glass-formers [8,21]. Hence, it can be considered as another indication that the process is the secondary JG relaxation of the water component in the aqueous mixtures.

In non-aqueous binary mixtures, both components have their own primary α-relaxation and glass transition temperature, which have been detected by dielectric, mechanical, and NMR spectroscopies as two separate α-loss peaks in some cases, or two different T_g's by calorimetry. In aqueous solutions discussed above, there is only one α-loss peak in the dielectric spectra, and a single T_{gm} found by calorimetry. The difference between the component α-dynamics of aqueous solutions and non-aqueous solutions is due to hydrogen bonding of the water with the other hydrophilic components. Consequently, neither the water nor the other component can be separately identified in the cooperative motion, and there is only one combined structural α-relaxation.

At temperatures below T_{gm} of the mixture, τ_{JG} of all mixtures have Arrhenius temperature dependences (see Fig.2) with activation energies around 50 kJ/mol, which is not much larger than the energy of breaking two hydrogen bonds before translation or rotation of the molecule can occur (57 kJ/mol according to Ref.22). This property also indicates that the process in all these aqueous mixtures is the JG relaxation of the water component.

Although the JG-processes in various mixtures are of the same nature, their τ_{JG}s are not expected to have the same value because the concentration of water as well the chemical structure of the solute is different. Despite these differences, τ_{JG} does not vary by much (Fig.2). On the other hand, $\tau_\alpha(T)$ of these mixtures can be immensely different because of their different glass transition temperatures T_{gm}.

THE JG β-RELAXATION OF WATER IN NANO-CONFINEMENT

Molecular Sieves

The dynamics of water confined in molecular sieves had been studied by means of dielectric spectroscopy [23] and by adiabatic calorimetry [24]. In a cylindrical pore with diameter of 10 Å, two major relaxation processes were observed (see Fig.2b in Ref.23, and Fig.3 of the present paper). According to the authors of Ref.23, the much more intense slower process (open squares) is contributed by most of the water molecules in 10 Å pores, and is likely due to strong interactions with the hydrophilic inner surfaces of molecular sieves. Its relaxation time, τ_{sMS}, increases with decreasing temperature to attain 100 s at about 170 K. The relaxation time of the weak faster process, τ_{fMS} (closed squares), has Arrhenius temperature dependence at low temperatures below about 175 K. The magnitude of τ_{fMS} is nearly the same as the relaxation time $\tau_{6Å}$ (closed circles) of the 6 Å ultrathin water layer (two molecular layers) confined in fully hydrated Na-vermiculite clay measured by dielectric spectroscopy [25], and the JG relaxation time, $\tau_{JG}(T)$, of the water component in the aqueous mixtures as exemplified in Fig.3 by the data of 50% water mixtures with PVME (closed diamonds). Like τ_{JG} of this aqueous mixture, the Arrhenius temperature dependence of τ_{fMS} does not persist at higher temperatures, but change to stronger temperature dependence at temperatures above 180 K. This crossover of temperature dependence of τ_{fMS} is the same as what we have seen in τ_{JG} of the JG-process in various non-aqueous and aqueous mixtures (see Figs.1 and 2), and is a general property of JG β-relaxation. The striking similarity of this feature of τ_{fMS} and that shown by τ_{JG} leads us to interpret the observed faster process in a molecular sieve as the JG β-relaxation associated with slower process, which is the α-process of water strongly interacting with the hydrophilic inner surfaces of the molecular sieve. Such a relation between the two processes is further supported by applying the coupling model (CM) relation between τ_{JG} and τ_{α} given by Eq.(2) to the present case of τ_{fMS} and τ_{sMS} for water confined in 10 Å pores. Using the CM relation adapted for the present case,

$$\tau_{sMS}(T) = [t_c^{-n}\tau_{fMS}(T)]^{1/(1-n)}, \qquad (3)$$

we calculate τ_{sMS} from the data of τ_{fMS} (closed triangles) by assuming $t_c \approx 2$ ps as other molecular glass-formers and $(1\text{-}n)\equiv\beta_K=0.69$. With this choice of the Kohlrausch exponent, $\beta_K=0.69$, the calculated τ_{sMS} (closed triangles) are in approximate agreement with the experimental values (open squares) as shown in Fig.3. The value of $\beta_K=0.69$ chosen is not far from the value of 0.65 reported by Swenson et al. [26] as the Kohlrasuch exponent obtained by means of quasi-elastic neutron scattering after averaging the result from various scattering vectors.

More recent dielectric data [27] of τ_{fMS} of water confined in molecular sieves MCM-41 C10 with pore diameter 2.14 nm at hydration levels with H = 12wt%, 22 wt%, 55 wt% are also shown in Fig.3. The change of T-dependence of τ_{fMS} at the temperature indicated by the vertical solid line is made clear by the two lines with different slopes. The single open inverted triangle is the value of τ_{sMS} deduced from the dielectric spectrum at 220 K after subtracting off the conductivity.

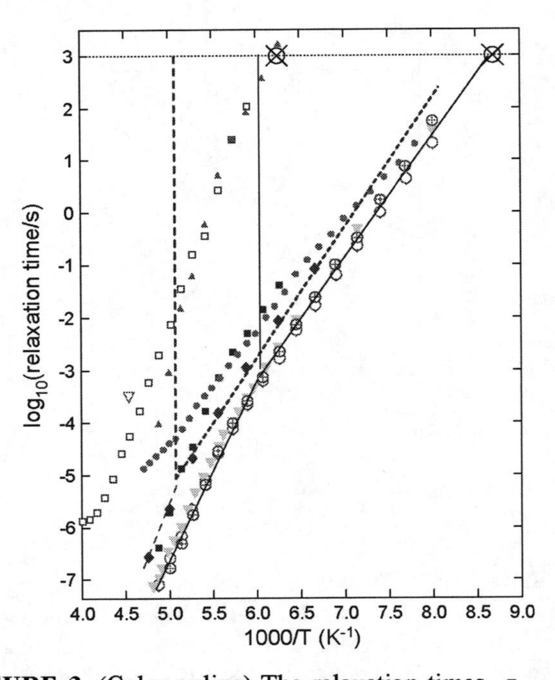

FIGURE 3. (Color online) The relaxation times, τ_{sMS} and τ_{fMS}, of a slower (open squares) and faster (closed squares) relaxation of water confined in molecular sieves of cylindrical pore with diameter of 10 Å [data from Ref.23]. The Arrhenius temperature dependence of τ_{fMS} changes to stronger temperature dependence at temperatures above approximately T_g=180 K where τ_{sMS} will reach about 10^3 s. The values of τ_{sMS} calculated by the coupling model eq. (3) are shown by closed triangles (see text). Shown also are more recent dielectric data [Ref.27] of τ_{fMS} of water confined in molecular sieves MCM-41 C10 with pore diameter 2.14 nm at hydration levels with H = 12wt% (open circles), 22 wt% (closed inverted triangles), 55 wt% (open circles with + inside). The change of T-dependence of τ_{fMS} is made clear by the two lines with different slopes. The single open inverted triangle is the value of τ_{sMS} deduced from the dielectric spectrum at 220 K after subtracting off the conductivity. For comparison, we plot the dielectric relaxation time (closed circles) of 6 Å thick layer of water in fully hydrated Na-vermiculite clay [data from Ref.25]. The change of T-dependence of τ_{fMS} of water confined in molecular sieves at T_g is like that found for τ_{JG} of non-aqueous mixtures and in various aqueous mixtures at T_{gm}, illustrated here by τ_{JG} of the 50% water mixtures with PVME (closed diamonds, data from Ref.20), and the two broken lines to indicate the change at T_{gm} determined by DSC (location indicated by the vertical

broken line). The two larger circles (⊗) are the relaxation times of τ_{sMS} and τ_{fMS} of water confined within 1.1, 1.6 and 1.8 nm nano-pores of silica MCM-41 obtained by adiabatic calorimetry having the value of 10^3 s at T=115 and 165 K respectively as found by adiabatic calorimetry in [Ref.24]. The triangles are τ_{sMS} calculated from τ_{fMS} by the coupling model described in Ref.[17].

Adiabatic calorimetry was also used to study the enthalpy relaxation of water confined within MCM-41 nano-pores [24]. Again a faster and a slower relaxation of water confined within 1.2, 1.6 and 1.8 nm nano-pores of silica MCM-41 have been found. Their relaxation times, τ_{sMS} and τ_{fMS}, have the value of 10^3 s at T=115 and 165 K respectively, and are shown in Fig.3 by the two larger circles (⊗). The location of τ_{sMS} from adiabatic calorimetry is about the same as that from dielectric relaxation [27]. Also, the location of τ_{fMS} from adiabatic calorimetry is nearly the same as that obtained by extrapolating the Arrhenius dependence of τ_{fMS} determined by dielectric relaxation [27] to 10^3 s. The agreement shows that τ_{fMS} from both methods belong to the same process. Interestingly, this process of water in molecular sieves has the about same activation enthalpy as τ_{JG} of the 50% water mixtures with PVME or PVP (not shown) and the other aqueous mixtures shown in Fig.2, and $\tau_{6Å}$ of the 6 Å water layer confined in fully hydrated Na-vermiculite clay. Moreover, the τ_{fMS} from adiabatic calorimetry located at T=115 K is nearly the same as value obtained by extrapolation of the Arrhenius dependence of τ_{JG} of the 50% water mixtures with PVME or with PVP to lower temperatures. These coincidences indicate that τ_{fMS} from adiabatic calorimetry is an intrinsic feature of water, independent of the mixtures and the manner in which it is confined. These seem to rule out the alternative interpretation by Oguni and coworkers [24] that τ_{fMS} at 115 K is caused by the interfacial water molecules and that τ_{sMS} at =165 K by the internal ones surrounded only by other water molecules.

Other Forms Of Nano-Confinement

Two relaxation processes of pure water confined in aged silica hydrogels were also observed dielectrically [28]. Their relaxation times are shown in Fig.4 as a function of reciprocal temperature. The relaxation time, τ_{fC}, of the faster relaxation has Arrhenius temperature dependence in the lower temperature range with an activation enthalpy of about 50 kJ/mol. The resemblance of these properties of the faster relaxation to that of some of the JG-processes of water in aqueous mixtures suggests that it is the JG β-relaxation of water confined in hydrogels. The slower relaxation has VFTH temperature dependence for its relaxation time, τ_{sC}, like that of typical α-process. It was suggested [28] that this relaxation originates from water molecules within the pores that do not interact strongly with the matrix and behave collectively. Another possibility is interaction of water with silica and silica being modified by hydrolysis. Whichever is the ultimate interpretation of the slower relaxation, it is an α-relaxation, and the faster relaxation with Arrhenius temperature dependence may be its JG β-relaxation. Their relaxation times, τ_{sC} and τ_{fC}, may be related by the CM equation (3) just like the relation between τ_{sMS} and τ_{fMS} of water confined in a molecular sieve shown in the previous Figure 3. In fact, when we compare τ_{sC} with τ_{sMS}, and τ_{fC} with τ_{fMS} in Fig.4, the values of each pair are similar, indicating the origins of the fast process as well as the slow process are similar in the two cases.

We consider here also the relaxation of water confined in a hydrogel of PHEMA studied by Pathmanathan and Johari (PJ) [22]. Within the experimental frequency range from 10 to 10^5 Hz, again there are two relaxation processes. The faster relaxation of the 38.6 wt% water has Arrhenius temperature dependence for it relaxation time, τ_{fPJ}, with activation energy of 60.8 kJ/mol, as shown in Fig.4 by the Arrhenius fit to the data (dashed line) but without the data themselves to avoid crowding. It can be seen from this figure that the relaxation times τ_{fPJ} are near τ_{fC} of the faster process found in water confined in aged silica hydrogels as well as τ_{fMS} of water confined in a molecular sieve. This coincidence suggests that τ_{fPJ} is the JG relaxation time of water in the PHEMA hydrogel. The slower process of water in PHEMA hydrogels can only be observed at higher temperatures dielectrically. At lower temperatures, the d.c. conductivity dominates and preempts the observation of the slow process. Its relaxation time, τ_{sPJ}, determined from the dielectric loss spectra at several temperatures are shown as open triangles in Fig.4. Again, at the same temperature, τ_{sPJ} is about the same as τ_{sC} or τ_{sMS}.

Calorimetric measurements of H_2O and D_2O in hydrogels of PHEMA by Hofer et al. [29] show water has a very weak endothermic step at 132 ± 4 K for H_2O and another endothermic step at 162 ± 2 K (H_2O) and 165 ± 2 K (D_2O), all for a heating rate of 30 K/min. The lower temperature, 132 ± 4 K, is nearly the same as T_g of hyperquenched glassy water [30,31]. The activation energy of the α-relaxation of hyperquenched glassy water is about 55 kJ/mol, which is also not far from 60.8 kJ/mol for τ_{fPJ}. Moreover, from the calorimetric data, PJ deduced that τ_{fPJ}=53 s at

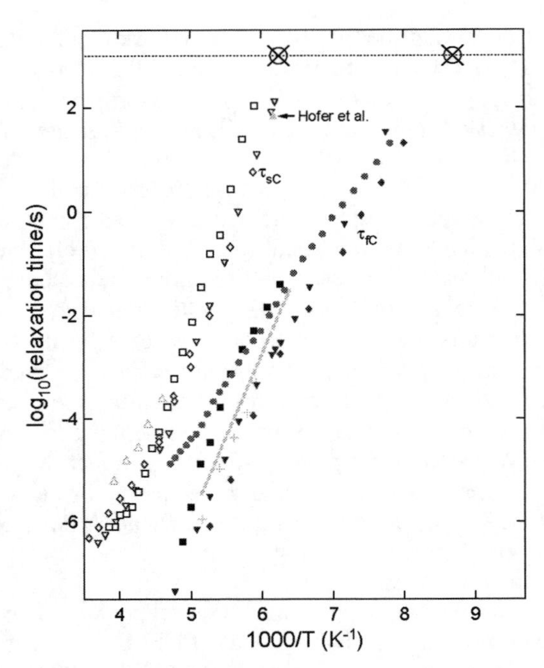

FIGURE 4. (Color online) The relaxation times, τ_{sC} (open diamonds, and open inverted triangles for dry samples) and τ_{fC} (closed diamonds, and closed inverted triangles for dry samples), of a slower and faster relaxation of water confined in aged silica hydrogels [data from Ref.28]. The relaxation time, τ_{fC}, of the faster relaxation has Arrhenius temperature dependence in the lower temperature range with an activation enthalpy of about 50 kJ/mol. Shown for comparison are the τ_{sMS} (open squares) and τ_{fMS} (closed squares), of the slower and faster relaxations of water confined in molecular sieves of cylindrical pore with diameter of 10 Å [23]; τ_{JG} (closed circles) of 6 Å thick layer of water in fully hydrated Na-vermiculite clay [25]; τ_{α} or τ_{sPJ} (open triangles) and τ_{JG} or τ_{fPJ} (dashed line) of mixtures of 38.6 wt.% of water with PHEMA [dielectric data from Ref.22]; τ_{α} (closed triangle) of 34% and 42% of water with PHEMA [DSC data from Ref.29]; τ_{JG} or τ_{fPJ} (plus signs) of mixtures of 50 wt.% of water with PHEMA [data from Ref.22]. The two larger circles (⊗) with relaxation times of two relaxations both equal to 10^3 s detected by adiabatic calorimetry at T=115 and 160 K for water in the pores of silica gel with average diameter of 1.1 nm [32].

135 K. The larger endothermic step at 162 K at the heating rate of 30 K/min indicates the relaxation time of the slower process, τ_{sPJ}, is also 53 s at 162 K. This result, shown by a lone closed triangle located at 1000/T=6.17 in Fig.4 and indicated there that it comes from Hofer et al., seems to be the continuation of τ_{sPJ} obtained by dielectric measurements at higher temperatures, as could be suggested by drawing a line to connect them. Can be seen by inspection of Fig.4, τ_{sPJ}, τ_{sMS}, and τ_{sC} all have comparable values over the same temperature region where they increase towards long time-scale of vitrification. The activation enthalpy of τ_{sPJ} from calorimetric relaxation near 162 K is about 120 kJ/mol, twice the activation energy of τ_{fPJ}. From the contrasting properties of τ_{sPJ} and τ_{fPJ} discussed above, we suggest in PHEMA hydrogels that the slower process is the relaxation of water interacting with PHEMA, like water interacting with the inner surface of molecular sieves [23] or with the silica in silica hydrogels [28]. The faster process could be considered as the JG β-relaxation of water in PHEMA hydrogels. Again τ_{sPJ} and τ_{fPJ} are related by an analogue of the CM equation (3), and there is a change of the T-dependence of τ_{fPJ} at the temperature where τ_{sPJ} reaches the long time of 10^3 s like found for other mixtures (see Figs.1-3).

New study of water confined in the pores of silica gel with average diameter of 1.1 nm. by adiabatic calorimetry reported again the presence of a faster and a slower process [32]. We do not discuss the data of water confined in 6, 12 and 52 nm pores of silica gel in Ref.[32] because of reported large fraction of ice formed in the central portion of these pores, which could complicated the observed results. The relaxation times of the two processes both equal to 10^3 s were detected by adiabatic calorimetry at T=115 and 160 K. They are shown by the two larger circles (⊗) on the top in Fig.4. The relaxation time of the slower process is coincident with the values of τ_{sC}, and τ_{sMS} when they are extrapolated to lower temperatures. The relaxation time of the faster process agrees approximately with τ_{fPJ}, τ_{fC} and $\tau_{6Å}$ (as well as τ_{JG} of the 50% water mixtures with PVME) when their Arrhenius dependences are extrapolated down to 115 K. Again, the Arrhenius activation enthalpies of all these processes are about the same. Thus, the two processes of water confined in 1.1 nm pores of silica gel observed by adiabatic calorimetry have to be interpreted in the same way. In particular, the faster one is an intrinsic process of water, independent of the form of confinement or mixtures. These facts are at odds with the interpretation [32] that the fast process is caused by the interfacial water molecules bonded to the silanol groups of the wall of the pores.

HYDRATION WATER AND PROTEIN DYNAMICS

Water is important for biological systems to function. Dehydrated proteins cannot function, but a layer of water surrounding them fully activates the protein functionality [33]. Neutron scattering experiments [34] have shown in hydrated proteins that the rate of change of the mean-square-displacement with temperature increases abruptly past a temperature approximately equal to 220 K. This change of dynamics of proteins is believed to be triggered by their strong coupling with the hydration water through the hydrogen bonding. Thus, for protein functions, it is important to understand how water affects conformation dynamics of the protein, and the dynamics of water molecules in the hydration layer surrounding proteins as well.

In this section, we use the knowledge we have acquired on the dynamics of aqueous mixtures in the previous section to understand the dynamics of hydration water and the role it has for protein dynamics. In particular we address the recently published dielectric relaxation data of a sample containing an equal weight fraction of myoglobin, water and glycerol (i.e. 50 wt% water in the solvent, and a total solvent content h = 2) published by Swenson and coworkers [35]. Several relaxation processes were found in the dielectric spectra. The relaxation times of four such processes are shown in Fig.5 together with some of the aqueous mixtures, 35 wt.% water/PEG400, 50 wt.% water/PVME, and 50 wt.% water/PVP, shown before in Fig.2. One of the two slower processes, with T-dependence like that of Vogel-Fulcher, are likely originating from some cooperative protein conformational fluctuations, giving rise to glass-transition of the hydrated protein. By inspection of Fig.5, at approximate 200 K, the relaxation times of the slow relaxations are of the order of 10^3 s. Thus, T_g of the hydrated myoglobin is near or slightly higher than 200 K.

At temperatures below some T_x which is near $T_g \approx$ 200 K, the relaxation time of the fastest relaxation process in the hydrated myoglobin, τ_{fMY}, (black closed circles) is nearly the same as τ_{JG} of water in the aqueous mixtures of water with the polymers, PVME and PVP (see Fig.5). Moreover, as a function of temperature, τ_{fMY} exhibits a crossover from stronger temperature dependence at high temperatures to approximately Arrhenius dependence at lower temperatures, again like the behavior of τ_{JG} of water in the aqueous mixtures. The temperature at which the crossover takes place is at $T_x \approx T_g \approx 200$ K. Therefore, by analogy, the fastest relaxation in the mixture of myoglobin with 50 wt% water/glycerol solvent is naturally identified as the JG relaxation of the hydration water. The second fastest process has been attributed to motions of polar side groups of the protein [35].

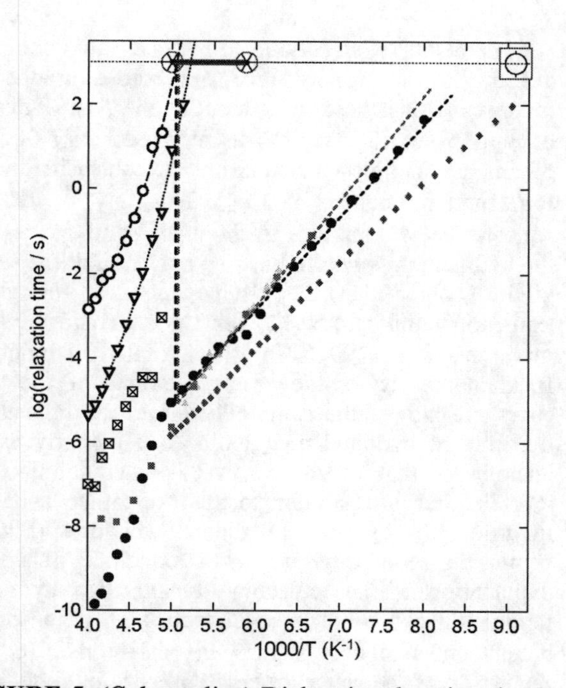

FIGURE 5. (Color online) Dielectric relaxation times of four processes observed by Swenson et al. in a sample containing an equal weight fraction of myoglobin, water and glycerol [35]. The relaxation times of the JG relaxation process of water (black closed circles), motions of polar side groups of the protein (black squares with crosses inside), and conformational protein fluctuations (black open circles and inverted triangle) [35]. The light black curves are VF fits. The broken black vertical line indicates the temperature T_g of the hydrated myoglobin where one of the VF fits reaches 10^3 s. T_g is nearly coincident with T_x, the temperature above which the JG relaxation time assumes a stronger T-dependence than the Arrhenius dependence below it. The JG relaxation time, $\tau_{JG}(T)$, of the water component in 50 wt.% water/PVME are indicated by filled (magenta) square, in 50 wt.% water/PVP by closed (green) triangles, and they are exactly the same data as shown in Fig.2. The change of T-dependence of τ_{JG} of 50 wt.% water/PVME occurs at T_{gm} (located by the vertical magenta line) which is accidentally almost the same as T_x. Closed diamonds are $\tau_{JG}(T)$ for mixture of 20 wt.% water with propanol. The large symbols on top all having relaxation times of 10^3 s are from adiabatic calorimetric measurements of a 20% (w/w) aqueous solution of bovine serum albumin [Ref.43]. The two larger circles (⊗) connected by the thick line circle indicate the range from 170 to 200 K of a slower relaxation found in the quenched sample. The open circle and open square are the relaxation occurring at T=110 K found for the quenched and annealed samples.

The result, $T_x \approx T_g$, found in hydrated myoglobin is exactly the same as we have seen before in aqueous mixtures. This is best brought out by the T-dependence of τ_{JG} of 50 wt.% water/PVME and the location of T_{gm} of this aqueous polymer mixture, presented before in Fig.2 and reproduced in Fig.5. Not only τ_{JG} of the aqueous PVME mixture is about the same as τ_{fMY} of the hydrated myoglobin, but also they exhibit the same kind of change of the T-dependence at the glass transition temperature, T_{gm} for the aqueous polymer mixture and $T_g \approx 200$ K for the hydrated myoglobin. Incidentally, the crossover temperatures of the two cases are almost the same. Therefore, the relaxation dynamic of hydrated myoglobin have a nearly exact analogy to that of the 50 wt.% water solution of PVME, and are similar to that of other aqueous mixtures (Fig.2), and non-aqueous mixtures (Fig.1). From the similarity we may conclude that the explanation of the molecular dynamics in hydrated myoglobin can be taken over from that of the other better understood systems. In short, the JG β-relaxation of the water in the 50/50 wt% of water and glycerol solvent in hydrated myoglobin is the precursor and governor of both the local protein process and the cooperative conformational fluctuations in the protein responsible for the glass transition-like behavior of the protein.

Our interpretation of the fastest relaxation process of the hydrated myoglobin in Fig.5 with relaxation time, τ_{fMY}, differs from the authors of Ref.35. They interpret the observed hydration water solvent process is a merged α- and β-process at temperatures above $T_x \approx T_g \approx 200$ K. This interpretation is incorrect because the hydration water process is exactly the analogue of the JG β-relaxation process in aqueous and nonaqueous mixtures (Figs.1 and 2), and evidences from the other systems have shown that the process is entirely one and the same JG relaxation above and below the crossover temperature. Our interpretation also implies that the cooperative α-relaxation of the 50/50 wt% of water/glycerol solvent of myoglobin may also have been observed in the dielectric spectra of Ref.35. Its relaxation times could be located in Fig.5 here at or near the data points represented by the squares (with crosses inside) and interpreted in Ref.35 as due to the motion of polar side groups of myoglobin. For the size and temperature dependence of the cooperative α-relaxation times of some water/glycerol mixtures, see Fig.5 in Ref.17.

A protein process with Arrhenius temperature dependence was observed below $T_g \approx T_x \approx 200$ K (not shown here in Fig.5, see Fig.4 of Ref.35), and was attributed to local non-cooperative protein motions [35], or the β-relaxation of the hydrated protein. It has the same activation energy as the JG relaxation of the hydration water in the same temperature range where it is Arrhenius. This supports the suggestion by Fenimore *et al.* [36] that the local protein motions are related to the β-relaxation in the solvent.

Calorimetric measurements have been widely used in studying the molecular motions in hydrated proteins. Examples from publications of earlier years are cited as Refs.[37-42]. Here we discuss the results of a recent study [43]. The relaxation processes of another hydrated protein, a 20% (w/w) aqueous solution of bovine serum albumin (BSA), were obtained by adiabatic calorimetry at temperatures ranging from 80 to 300 K [43]. Measurements were made on sample after quenching it from 300 down to 80 K and on another sample after it has been annealed. The relaxation times of processes found are shown in Fig.5. The two larger circles (⊗) connected by a thick line indicate the range from 170 to 200 K of a slower relaxation found in the quenched sample with relaxation time τ_{sBSA} equal to 10^3 s. The circle and the open square represent the fast relaxation with relaxation time τ_{fBSA} equal to 10^3 s when T=110 K found for the quenched and annealed samples respectively. We can see from Fig.5 that τ_{fBSA} is nearly the same as τ_{fMY} of hydrated myoglobin or τ_{JG} of aqueous mixtures when their Arrhenius T-dependences are extrapolated down to 110 K. The agreement between τ_{fBSA} and τ_{fMY} and with τ_{JG} of aqueous mixtures, and τ_{fPJ}, τ_{fC} and $\tau_{6Å}$ of water in nano-confinement is an indication of the ubiquitous nature and universal property of the JG relaxation of water in all the systems discussed in this paper. The importance roles it plays in glass transition and functions are expected.

CONCLUSION

From the plethora of experimental data of water in various aqueous mixtures, in different forms of nano-confinement, and in hydrated proteins, we have found a ubiquitous fast relaxation originating from water that has the same characteristics. The properties of this fast relaxation are similar in many respects to those of the well studied Johari-Goldstein (JG) secondary relaxation of neat glassformers and non-aqueous mixtures of two glass-formers. These evidences lead us to conclude that the fast relaxations observed in the various water containing systems are the JG secondary relaxation originating from the water component. Previous studies of glass-formers unrelated to water and aqueous mixtures have shown the fundamental roles played by the JG β-relaxation in glass transition. Its fundamentals roles are exemplified by several general quantitative relations of its relaxation time and relaxation strength with the structural α-relaxation [8-

10]. Without considering the JG β-relaxation and how the structural α-relaxation is spawned from it, a theory of glass transition cannot be complete and adequate. This is because the experimental facts showing the relations that exist between the JG β-relaxation and the structural α-relaxation are left unexplained. As far as we know, the coupling model is the only model that satisfies this criterion for a viable theory of glass transition [8-10,16], and provides explanation of the dynamics of non-aqueous [3-6,12,13] and aqueous mixtures presented here [see also 17]. This progress should be helpful in achieving deeper understanding of relaxation and glass transition in hydrated proteins. The results given in this paper are encouraging. More effort along the present line of approach, leading to further development in this important research area, is a worthwhile undertaking in the future.

ACKNOWLEDGMENTS

The work at NRL was supported by the Office of Naval Research, and at the Università di Pisa by MIUR (PRIN 2005: "Aging, fluctuation and response in out-of-equilibrium glassy systems").

REFERENCES

1. P.G. Debenedetti, *J. Phys.: Condens. Matter* **15**, R1669-R1726 (2003).
2. T. Blochowicz, E. A. Rössler, *Phys. Rev. Lett.* **92,** 225701-1-225701-4 (2004).
3. S. Capaccioli, K.L. Ngai, *J.Phys.Chem.B* **109,** 9727-9735 (2005).
4. S. Capaccioli, K. Kessairi, D. Prevosto, M. Lucchesi, K.L. Ngai, *J.Non-Cryst.Solids*, **352**, 4643-4649 (2006)
5. S. Capaccioli, K. Kessairi, D. Prevosto, M. Lucchesi, P. Rolla. *J.Phys.: Condens. Matter* **19**, 205133-1-205133-10 (2007).
6. K.L. Ngai, S. Capaccioli, *J.Phys.Chem.B* **108,** 11118-11125 (2004).
7. G.P. Johari and M. Goldstein, *J. Chem. Phys.* **53**, 2372-2381 (1970). G.P. Johari, *Annals New York Acad. Sci.* **279**, 117-130 (1976). G. P. Johari, *J. Non-Cryst. Solids,* **307-310**, 317-325 (2002).
8. K.L. Ngai, R. Casalini, S. Capaccioli, M. Paluch, C.M. Roland, *Adv.Chem.Phys. Part B,* Volume **133**, Chapter 10, 497-593 (2006).
9. K.L. Ngai and M. Paluch, *J. Chem. Phys.*, **120,** 857-873 (2004).
10. K.L. Ngai, *J.Chem.Phys.* **109**, 6982-6994 (1998).
11. R. Böhmer et al. *Phys.Rev.Letters,* **97**, 135701-1-135701-4 (2006).
12. S. Capaccioli, K. Kessairi, D. Prevosto, M. Lucchesi, P. Rolla, submitted to *Phys.Rev.Lett.*, and in this Volume.
13. M. Mierzwa, S. Pawlus, M. Paluch, K.L. Ngai, *J.Chem. Phys.*in press (2008).
14. K.L. Ngai and K.Y. Tsang, *Phys.Rev.E* **60**, 4511-4517 (1999).
15. K.L. Ngai, S. Capaccioli, *J. Phys.: Condens. Matter* **19,** 205114-1-205114-25 (2007).
16. K.L. Ngai, *J.Non-Cryst.Solids* **351,** 2635-2645 (2005).
17. S. Capaccioli, K.L. Ngai and N. Shinyashiki, *J.Phys.Chem.B* **111**, 8197-8209 (2007).
18. For data, see N. Shinyashiki, *et al.*, *J.Phys.: Condens. Matter*, **19**, 205113-1-205113-12 (2007).
19. S.K. Jain, G.P. Johari, *J.Phys.Chem.*, **92**, 5851-5857 (1988).
20. S. Cerveny, J. Colmenero, A. Alegria, *Macromolecules*, **38**, 7056-7063 (2005).`
21. G.P. Johari, G. Power, J.K. Vij, *J.Chem.Phys.*, **116**, 5908-5909 (2002); **117**, 1714-1719 (2002).
22. K. Pathmanathan, G. P. Johari, *J. Polym. Sci. Polym. Phys. B*, **28,** 675-681 (1990). *J. Chem. Soc. Faraday Trans.* **90,** 1143-1149 (1994).
23. H. Jansson, J. Swenson, *Eur.Phys.J.E* **12**, S51-54 (2003).
24. M. Oguni, Y. Kanke, and S. Namba, Abstract submitted to this Workshop and article in this Volume.
25. R. Bergman, J. Swenson, *Nature*, **403**, 283 (2000). For a critique of this paper, see G.P. Johari, *Chem.Phys.* **258**, 277 (2000).
26. J. Swenson, H. Jansson, W.S. Howells, S. Longeville, *J.Chem.Phys.* **122**, 084505-1-084505-2 (2005).
27. J. Hedström, J. Swenson, R. Bergman, H. Jansson, and S. Kittaka, *Eur. Phys. J. Special Topics* **141**, 53-59 (2007).
28. M. Cammarata, M. Levantino, A. Cupane, A. Longo, A. Martorana, F. Bruni, *Eur. Phys. J. E* **12**, S63-69 (2003).
29. K. Hofer, E. Mayer, G.P. Johari, *J. Phys. Chem.* **94**, 2689-2695 (1990); **95**, 7100-7112 (1991).
30. G.P. Johari, A. Hallbrucker, E. Mayer, E. *Nature*, **330**, 552-559 (1987); *J.Chem.Phys.* **92**, 6742-6749 (1990).
31. G.P. Johari, G. Astl, E. Mayer, *J. Chem. Phys.* **92**, 809-817 (1990).
32. M. Oguni, S. Maruyama, K. Wakabayashi, A. Nagoe, *Chem.Asian J.* **2**, 514-520 (2007).
33. R. B. Gregory, editor "Protein-solvent interactions," Marcel Dekker, New York (1995).
34. W. Doster, et al., *Nature* **337**, 754-756 (1989).
35. J. Swenson, H. Jansson, J. Hedsröm, R. Bergman, *J. Phys.: Condens. Matter* **19,** 205109-1-205109-9 (2007).
36. P. W. Fenimore, Hans Frauenfelder, B. H. McMahon, R. D. Young, *Proc. Natl Acad. Sci.* **101,** 14408-14413 (2004).
37. G. Sartor, E. Mayer, and G. P. Johari. *Biophys. J.* **66,** 249-258 (1994).
38. J. L. Green, J. Fan, and C. A. Angell, *J. Phys. Chem.* **98,** 13780-13790 (1994).
39. G. Sartor, et al.. *Biophys. J.* **69**, 2679-2694 (1995).
40. C. Inoue, and M. Ishikawa, *J. Food Sci.* **65**, 1187-1193 (2000).
41. A. Cupane, M. Leone, and V. Militello *Biophys.Chem.* **104**, 335-344 (2003).
42. Yuji Miyazaki, Takasuke Matsuo, and Hiroshi Suga, *J. Phys. Chem. B*, **104**, 8044-8052 (2000).
43. K. Kawai, T. Suzuki, and M. Oguni, *Biophys.J.* **90,** 3732-3738 (2006).

Kinetics and Control of Liquid-Liquid Transition

Hajime Tanaka*, Rei Kurita*, and Ken-ichiro Murata*

*Institute of Industrial Science, University of Tokyo, Meguro-ku, Tokyo 153-8505, Japan

Abstract. Recently it was revealed that even a single-component liquid can have more than two liquid states. The transition between these liquid states is called "liquid-liquid transition". This phenomenon has attracted a considerable attention because of its counter-intuitive character and the fundamental importance for our understanding of the liquid state of matter. The connection between the liquid-liquid transition and polyamorphism is also an interesting issue. In many cases, liquid-liquid transitions exist in a region which is difficult to access experimentally. Because of this experimental difficulty, the physical nature and kinetics of the transition remains elusive. However, a recent finding of liquid-liquid transition in molecular liquids opens up a possibility to study the kinetics in detail. Here we report the first detailed comparison between experiments and a phenomenological theory for the liquid-liquid transition of a molecular liquid, triphenyl phosphite. Both nucleation-growth-type and spinodal-decomposition-type liquid-liquid transformation are remarkably well reproduced by a two-order-parameter model of liquid that regards the liquid-liquid transition as the cooperative formation of locally favored structures. This may shed new light on the nature and the dynamics of the liquid-liquid transition. We also show evidence that this second order parameter controls the fragility of the liquid. We also discuss a possibility of controlling liquid-liquid transition by spatial confinement. Remaining open questions on the nature of the transition are also discussed.

Keywords: Confinement, Glass Transition, Liquid-liquid transition

INTRODUCTION

Recently, a mass of experimental, numerical, and theoretical evidence of liquid-liquid transition (LLT) has been accumulated [1, 2, 3, 4, 5, 6, 7, 8, 9, 10, 11, 12, 13], namely, the existence of more than two distinct liquid states for a single-component substance. For example, several atomic liquids were suggested as candidates of liquids having LLT [1]. For molecular liquids, Mishima et al. found an amorphous-amorphous transition in water [2]. Computer simulations also suggest the existence of LLT(s) in water [2, 3]. On the basis of these findings the connection of amorphous-amorphous transition and LLT in water was suggested and actively studied [2, 3], although the role of mechanical stress involved in amorphous-amorphous transition may make the connection a bit obscure [14]. In many cases, LLT is accompanied by a large change in the physical properties of liquid even though the component is exactly the same. This means that the problem of LLT is linked to the fundamental question of what physical factors control the properties of a liquid. Unfortunately, however, the transition is located at high pressure and high temperature (e.g., for atomic liquids) or hidden by crystallization (e.g., for water) in the above examples. This makes detailed experimental studies difficult.

Since the finding of so-called glacial phase in triphenyl phosphite (TPP) by Kivelson and his coworkers, there have been intensive researches on the nature of the glacial phase [15, 16, 17, 18, 19, 20, 21, 22, 23, 24, 25, 26]. Some of them favored the LLT scenario [15, 17, 22, 24, 25, 26], whereas the others did not [16, 18, 19, 20, 21, 23]. Kivelson and his coworkers reported that a peak appears at the wavenumber $q \sim 1.1$ Å^{-1} and this peak position is similar to that of a Bragg peak of the c-axis of the crystal [16]. Hedoux et al. proposed on the basis of X-ray scattering measurements that liquid I transforms into a mixture of nano-crystallites and the untransformed liquid (liquid I) [18] and does not into a new liquid (liquid II). We found several pieces of experimental evidence supporting the LLT scenario in molecular liquids, TPP [27, 28] and n-butanol [29], although there still remains controversy on the nature of the transition. For example, we found two types of the transformation from liquid I to liquid II [27, 28, 29]: For TPP, (i) the nucleation-growth (NG)-type LLT observed when we quench TPP and anneal it between 223 K and 215.5 K and (ii) the spinodal-decomposition (SD)-type LLT which occurs in the unstable region below the spinodal temperature T_{SD} = 215.5 K. We also found that the correlation length ξ diverges as $\xi = \xi_0[(T_{SD} - T)/T_{SD}]^{-\nu}$ where ξ_0 = 60 nm and ν = 0.5. This is an indication of critical phenomena associated with LLT and thus we concluded that there is a second order parameter controlling LLT [27, 30, 31, 32]. We also confirmed that this transformation from liquid I to liquid II accompanies a transition from a fragile to a strong liquid [34]. Thus, LLT does affect fundamental physical properties of the liquid. Furthermore, the finding of LLT in n-butanol [29], whose molecular shape and chemical structure are very different from those of TPP, suggests not only a possible abundance of LLT in many molecular liquids, but also

CP982, *Complex Systems, 5th International Workshop on Complex Systems*
edited by M. Tokuyama, I. Oppenheim, and H. Nishiyama

the generality of the nature and kinetics of LLT. This means that a coarse-grained phase field description [30] may be useful as anticipated from its success in critical dynamics [35].

Encouraged by this, here we make a detailed comparison of numerical simulation results based on a phase field model of LLT [33], which we called "two-order-parameter model" [30], and experimental results of TPP. A link between liquid-liquid transformation and the nature of glass transition is also discussed. We also discuss the effects of spatial confinement on the liquid-liquid transition of TPP [36].

Two-order-parameter model of liquid-liquid transition

First we describe the spirit of our two-order-parameter (TOP) model of liquid [30]. Density ρ that is usually believed to be the only order parameter describing the state of liquid. Density order parameter indeed controls the gas-liquid transition. We introduce an additional order parameter, which we call bond order parameter S, to describe LLT. We proposed that S is the fraction (or, number density) of locally favored structures (LFS). If a molecule has anisotropic interactions such as covalent and hydrogen bonding, there should exist long-lived, locally favored structures in a liquid. For water and Si, LFS have the tetrahedral order [37]. For spherical particles interacting with the Lennard-Jones potential, on the other hand, it is widely recognized that an icosahedral arrangement is locally favored [4]. Although the structure of LFS is material specific, the existence of LFS in a liquid itself may be universal. For simplicity, we take a two-state model, which is composed of normal-liquid and locally favored structures, as the minimal model [30, 38, 39]. Then the free energy associated with fluctuations of ρ and S around their average values $\bar{\rho}$ and $\bar{S}$, which we denote $\delta\rho$ and δS respectively, is approximately given by [30]

$$\begin{aligned} f(\delta\rho, S) = \int dr[&\frac{\tau}{2}\delta\rho^2 + \frac{\kappa}{2}\delta S^2 + \frac{b_4}{4}\delta S^4 \\ &-C_\rho \delta\rho(\bar{S} + \delta S) - C_S(\bar{\rho} + \delta\rho)\delta S \\ &+\frac{K_\rho}{2}|\nabla\delta\rho|^2 + \frac{K_S}{2}|\nabla\delta S|^2]. \end{aligned}$$

Here $\tau = \beta(\bar{\rho}^2 K_T)^{-2}$, $\bar{\rho}$ is the average density and K_T is the isothermal compressibility. $\kappa = b_2(T - T_s^*)$, T_s^* is a spinodal temperature of bond ordering without the coupling to ρ, and b_2 and b_4 are positive constants. C_ρ and C_S are constants, which express the strength of the coupling between the two order parameters.

Reflecting the conserved and non-conserved nature of $\delta\rho$ and δS respectively, the time evolution of $\delta\rho$ and δS obeys the following equations:

$$\frac{\partial\delta\rho}{\partial t} = L_\rho \nabla^2\left[-K_\rho\nabla^2\delta\rho + \frac{h(\delta\rho,\delta S)}{\delta\rho}\right] + \zeta_\rho \quad (1)$$

$$\frac{\partial\delta S}{\partial t} = -L_S\left[-K_S\nabla^2\delta S + \frac{h(\delta\rho,\delta S)}{\delta S}\right] + \zeta_S \quad (2)$$

where L_ρ and L_S are kinetic coefficients and ζ_ρ and ζ_S are thermal noises. In our model, LLT is driven by cooperative ordering of δS and $\delta\rho$ is just a slave variable of δS.

Experimental

The sample used was TPP, which was purchased from Acros Organics and used after extracting only a crystallizable part. We carefully avoided moisture to prevent chemical decomposition. We observed the transformation process from liquid I to liquid II with phase contrast microscopy. We prepared a sample cell, which was made of two parallel thin glass plates. The spacing between the two cover glasses was controlled to be between 1 μm and 20 μm by using monodisperse glass beads as spacers. To study effects of confinement, we also prepared a wedge-shaped sample cell, whose wedge angle θ satisfies $\tan\theta = 0.0075$, for optical microscopy observation. The temperature was controlled within ± 0.1 K by a computer-controlled hot stage (Linkam LK-600PH) with a cooling unit (Linkam L-600A). The intensity distribution function and the structure factor $F(q)$ were calculated from an optical microscopy image, using the digital image analysis [40].

Broadband dielectric measurements were performed for a frequency range from 10 mHz to 1 MHz with SI1260 Impedance/Gain-Phase Analyzer, which was controlled by a computer. Data accumulation was carried out by using solatron analytical SMaRT system. We measured the dielectric response at 40 frequency points in the frequency range from 10 mHz to 1 MHz Under this condition, it took 5 min to get a dielectric spectrum with a good signal-to-noise ratio. Thus, the time resolution of our measurement system is 5 min, which is fast enough to follow the process of LLT since LLT proceeds very slowly over several hours. Finally, we analyzed the data obtained with the Agilent VEE program.

Simulation Method

We solve Eqs. (1) and (2) by an explicit Euler method after scaling the length and time by the correlation length of density fluctuations ξ_ρ and the characteristic time of the lifetime of the fluctuations $L_\rho\tau/\xi_\rho^2$, respectively. The system size is 256×256 (2D). The initial equilibrium state (at $t = 0$) is carefully prepared by a long enough run with thermal noises. The important parameters are $\Xi = \xi_S/\xi_\rho$, $\alpha = L_S\xi_\rho^2/L_\rho$, $C = (C_\rho + C_S)/\tau$, $\Theta = \kappa/\tau$ and $E = C_S\bar{\rho}/\tau$ where ξ_S is the correlation length of S fluctuations without coupling to ρ. C and Θ corresponds to the strength of the coupling and the temperature, respectively. We set the parameters as $\Xi = 2$, $\alpha = 0.1$, $C = 0.1$ and $E = 0.03$; and Θ is the only variable parameter. The spinodal temperature Θ_{SD} is $\Theta = 0.182$. We added the Gaussian noise terms ζ_ρ and ζ_S satisfying the fluctuation-dissipation theorem [see Eqs. (1) and (2)].

Comparison of experimental and simulation results

Here we consider only the SD-type LLT, which occurs when a liquid is quenched into an unstable region below T_{SD} [27, 29, 28]. First we summarize our experimental findings. The pattern evolution during SD-type LLT is shown in Fig. 1(a). We followed the temporal change of the structure factor $F(q)$, which is calculated as the power spectrum of the Fourier transform of a phase-contrast microscopy image. We found that $F(q)$ has a peak, which apparently resembles the behavior observed in usual spinodal decomposition of a binary mixture. In the early stage, the peak wavenumber q_p is constant with time and the peak intensity $F(q_p)$ grows exponentially with time (see Fig. 2). This is reminiscent of the Cahn's linear regime. In the intermediate stage, q_p decreases as $q_p \sim t^{-0.5}$. This domain coarsening is characteristic of the SD-type ordering of a non-conserved order parameter [35]. Finally, $F(q_p)$ decreases since a liquid becomes homogeneous liquid II.

Our simulation results are fully consistent with these experimental results. At $\Theta = 0.1$, the SD-type ordering proceeds. Figure 1(b) shows the evolution of density fluctuations. In the beginning, the amplitude of fluctuations grows with time and thus the contrast increases [Fig. 1(b)]. Then, the domain size increases and the contrast increases until $t = 220$. Then a liquid becomes more homogeneous, which leads to the decrease in the contrast after $t = 220$. Finally, a liquid becomes homogeneous liquid II. This captures all the essential features of pattern evolution observed with phase-contrast microscopy for TPP and n-butanol.

We confirmed that $F(q)$ calculated from the spatial density distribution, has a distinct peak, consistent with the experimental results. This comes from the conserved nature of $\delta\rho$, which is coupled to a non-conserved order parameter S governing LLT. Figure 2 shows the temporal change of the peak wavenumber q_p of $F(q)$ and the peak intensity $F(q_p)$. In the early stage, q_p is constant and the peak intensity $F(q_p)$ increases exponentially with

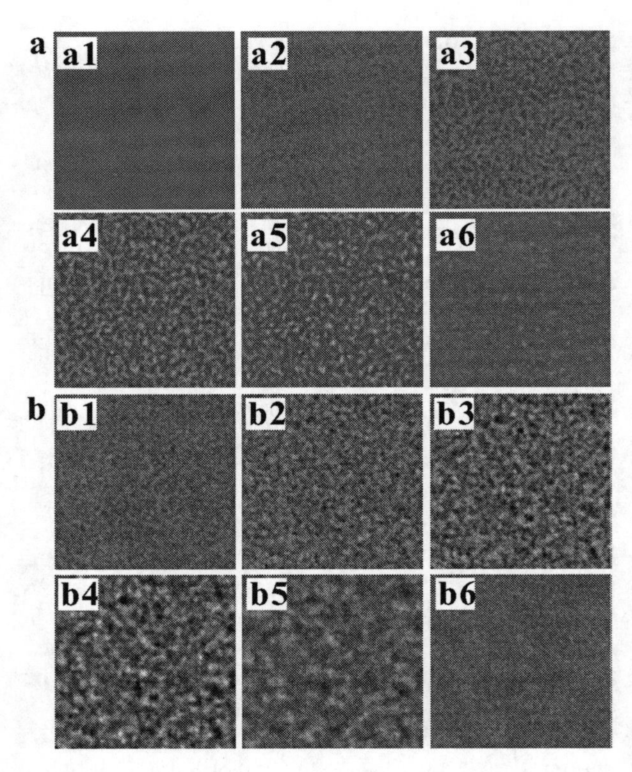

FIGURE 1. Temporal evolution of density fluctuations during SD-type LLT. (a) simulation results at $\Theta = 0.1$. t =0 (a1), 80 (a2), 120 (a3), 180 (a4), 240 (a5) and 350 (a6). (b) experimental results for TPP at 214 K. (b1) 0 min, (b2) 120 min, (b3) 150 min, (b4) 180 min, (b5) 210 min and (b6) 270 min. The size of each image is 120 μm × 120 μm.

time, which is characteristic of the linear stage of SD known as the Cahn's linear regime [35]. The correlation length ξ $(= 1/\sqrt{2}q_p)$ in the early stage is estimated as $\xi = 1.414$. $F(q)$ prior to the initiation of LLT is well described by an Ornstein-Zernike (OZ) structure factor, $F(q) = F(0)/(1 + q^2\xi^2)$. The fitting of this function to $F(q)$ yields $\xi_S = 1.471$ Note that without coupling of S to ρ, ξ_S should be 2. In our simulation, density fluctuations reflect fluctuations of S. In the intermediate stage, the domain size R increases as $R \sim t^{0.5}$, again consistent with the experimental results [29, 28]. This indicates the SD-type ordering of a nonconserved order parameter [35]. For the ordering of a nonconserved order parameter, the interface motion is described by the so-called Allen-Cahn equation: $v = dR/dt = -L\kappa$, where R is the domain size, v is the interface velocity, L is the kinetic coefficient, and κ is the mean curvature of the interface ($\sim 1/R$). This relation yields the domain coarsening law of $R \sim \sqrt{Lt}$. Finally, $F(q_p)$ decreases since a liquid becomes homogeneous liquid II (homogeneous S).

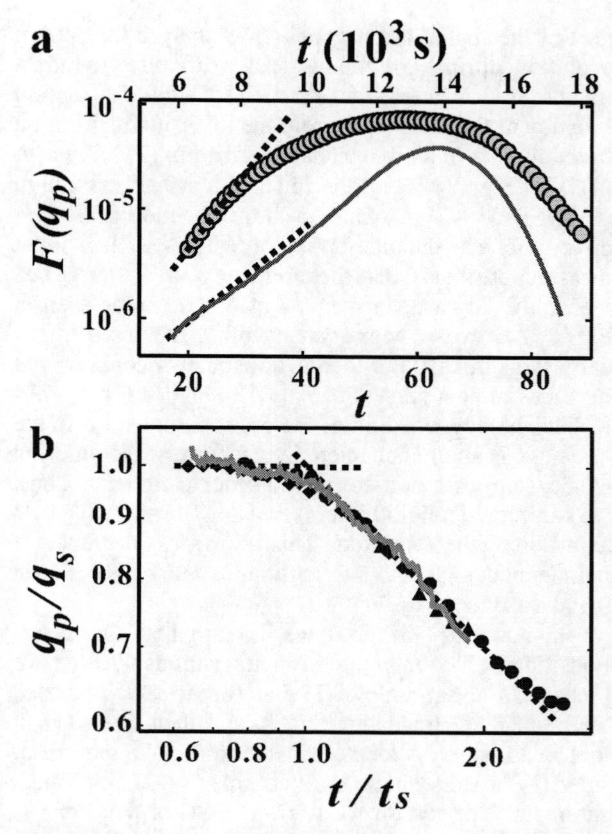

FIGURE 2. Kinetics of SD-type LLT. (a) Temporal change of the peak intensity $F(q_p)$ (semi log plot) for simulation at $\Theta = 0.1$ (open triangle) and experiment in TPP at 214 K (open circle). (b) Temporal change of the normalized peak wavenumber q_p/q_s (q_s: the peak wavenumber of fluctuations) as a function of the normalized time t/t_s (t_s: the onset time of domain coarsening) (log-log plot). The gray line is the simulation result, whereas the filled circles, triangles and diamonds are the experimental results of TPP at 212 K, 213 K and 214 K, respectively.

Control of the fragility of a liquid by the bond order parameter

Here we study the glassy state of a liquid undergoing LLT. We know that (i) the glass-transition temperature T_g of liquid I is 205 K, while that of liquid II is 212 K and (ii) liquid I is much more fragile than liquid II. Fragility is a parameter characterizing the glass-transition behavior and it is also closely related to the topological structure of the energy landscape [5]. Thus we can say that the question of what physical factor controls the fragility of liquid is one of the most fundamental questions that remain unsolved in glass science and may be a clue to understanding the origin of the glass transition itself. Empirically it is known that a liquid with a stronger network-forming ability is stronger, or less fragile [4, 5]. This gives us an impression that the fragility is a material-specific quantity. Contrary to this, the above observation in TPP suggests a possibility that the order parameter controlling LLT, S, also controls the fragility of liquid. TPP may provide us with a model system of the fragile-to-strong transition because it is experimentally accessible and it exhibits a clear LLT.

Furthermore, the continuous LLT found in TPP has the following remarkable feature. Considering that the bond order parameter is non-conserved, its spatial inhomogeneity decays during a rather early stage of the transformation and the system quickly becomes homogeneous. Then the average value of the order parameter S still gradually increases with time toward the final value S_{II}. This provides us with a unique opportunity to control the physical properties of the homogeneous liquid such that the fragility varies continuously from those of liquid I to those of liquid II.

Our strategy to prepare a liquid with different fragility is as follows. When we quench TPP below the spinodal temperature $T_{SD} = 215.5$ K, the liquid I continuously transforms into liquid II [27, 28]. This transformation from liquid I to liquid II is completed after more than 10 h. Provided that the order parameter governing LLT is the fraction of locally favored structures (LFSs), S, liquid I has a low fraction of LFSs (S_I), while liquid II has a high fraction of LFSs (S_{II}). During the continuous transformation, we can obtain a liquid state with any S between S_I and S_{II}. Since liquid I is fragile and liquid II is strong, we expect to be able to change the fragility of the liquid in a wide range by using LLT.

The process of LLT is followed by differential scanning calorimetry (DSC) measurements. We previously confirmed that the continuous LLT below T_{SD} does not accompany any crystallization and the liquid and glassy states produced by the continuous LLT are not contaminated by small crystallites [27]. Thus, the heat flow signal $dH(t)/dt$ purely reflects the rate of heat released upon the transformation from liquid I to liquid II. The total heat released during the annealing time t_a, $H(t_a)$, is then given by the time integration of the heat flow signal:

$$H(t_a) = \int_0^{t_a} \left(\frac{dH(t)}{dt}\right) dt. \tag{3}$$

In our view, $H(t)$ reflects the transformation of normal liquid structures with random configurations to LFS, namely, $H(t) \propto S(t) - S(0)$. Thus we define the scaled bond order parameter $\tilde{S}$ as $\tilde{S}(t) = [S(t) - S(0)]/[S(\infty) - S(0)] \equiv H(t)/H(\infty)$. Note that $S(0) = S_I$ and $S(\infty) = S_{II}$: $\tilde{S}$ is zero for liquid I and 1 for liquid II. The temporal evolution of $\tilde{S}$ is plotted against the annealing time t_a for $T_a = 213$ K in Fig. 3. The behavior is well explained by the theoretical prediction for the temporal evolution of the non-conserved order parameter for a spinodal case

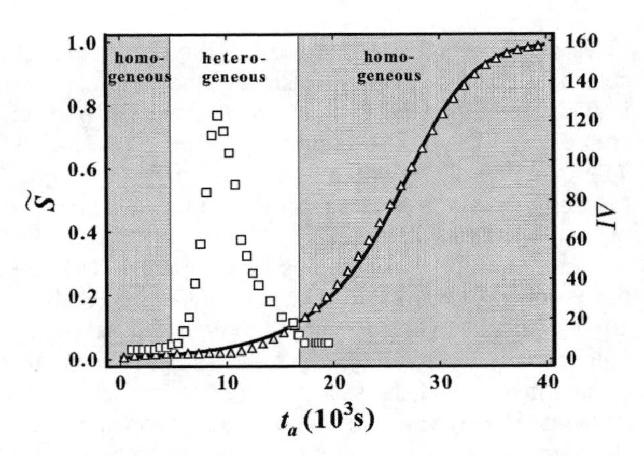

FIGURE 3. Temporal change of $\tilde{S}$ (open triangles) and the full width of the half maximum of the intensity distribution of an image, ΔI, (open squares) for $T_a = 213$ K. $\tilde{S}$ is calculated from the heat flow measured by DSC [see Eq. (3)]. The solid line is the fitted theoretical curve for the $\tilde{S}$ data [see Eq. (4)]. The fitting result is excellent. On the other hand, ΔI was estimated as follows. We observed the pattern evolution during the continuous LLT with phase-contrast microscopy and obtained the density distribution function $P(\rho)$ from the intensity distribution function $P(I)$ of the images (see Ref. [28] for the details). Note that the density follows the change of the fraction of LFS (S) through the coupling between ρ and S [27, 28]. Thus $P(I)$ reflects the distribution of S. Thus ΔI represents the degree of the spatial heterogeneity of S. At the early stage of LLT, ΔI drastically increases with time, which means that S grows heterogeneously in space. But ΔI decreases after 9000 s and thus a system becomes more homogeneous. When the annealing time exceeds 18000 s, the system becomes almost completely homogeneous. This can be confirmed form the fact that ΔI becomes almost zero. Note that ΔI for $t_a > 18000$ s is almost the same as that at $t_a = 0$ s, where the sample is in a homogeneous liquid I state.

[35, 30]:

$$\tilde{S}(t) \sim \frac{S(\infty)[1+((\frac{S(\infty)}{S(0)})^2-1)e^{-2\gamma_0 t}]^{-1/2}-S(0)}{S(\infty)-S(0)}, \quad (4)$$

where γ_0 is the growth rate of S. From the fitting, γ_0 and $S(\infty)/S(0)$ are determined to be 1.8×10^{-4} s^{-1} and 1.9×10^2, respectively. The good fit of Eq. (4) to the data strongly supports the conclusion derived from our previous work [27, 28] that the order parameter governing LLT is non-conserved. This result tells us that a liquid indeed changes from liquid I to liquid II continuously. Thus, we should be able to prepare a glassy state of TPP with various $\tilde{S}$ between 0 and 1, by quenching the liquid annealed for various periods far below T_g.

However, we should note that the spinodal-decomposition-type liquid-liquid transformation leads to a heterogeneous spatial distribution of S in the early stage [28]. $\tilde{S}$, which is estimated from the heat flux, is the average value of $\tilde{S}$ over the sample. To evaluate the degree of the spatial heterogeneity, we analyze the pattern evolution during continuous LLT with phase-contrast microscopy. We can evaluate $P(\rho)$, the distribution function of the density, ρ, from the intensity distribution function $P(I)$ of an image, as described in [28] (see also [41]). In Fig. 3 we plot the full width at half maximum (FWHM) ΔI of $P(I)$. Since ρ is the function of S, $P(I)$ directly reflects the distribution of S. Thus, ΔI is a direct measure of the spatial heterogeneity of S: larger ΔI means the larger heterogeneity of S. As can be seen in Fig. 3, ΔI, or the heterogeneity of S, increases in the early stage of LLT, but then ΔI drastically decreases and becomes almost zero after $t_a = 18\times10^3$ s for T_a=213 K. This homogenization of the order parameter S in the late stage is another characteristic feature of the ordering of a system of a non-conserved order parameter. Thus, we can regard a liquid after $t_a = 18\times10^3$ s (or $\tilde{S}\geq 0.1$) as a homogeneous liquid. This allows us to prepare a homogeneous glassy state with any value of $\tilde{S}$ between 0.1 and 1.0 (see Fig. 3).

We now explain how we investigated the glass-transition behavior of liquids with various $\tilde{S}$. First we prepared a glassy state of TPP as follows: We annealed TPP at 213 K for a duration t_a and then quenched it to 150 K, which is located far below T_g. We then measured T_g of this sample by a complex (AC) DSC measurement. The reason we use AC DSC is that crystallization occurs as soon as a sample passes through T_g and becomes a liquid state. If we use conventional DSC with a slow scanning rate (< 5 K/min), the crystallization heat signal overlaps the glass transition heat signal. The heat flux obtained by AC calorimetric measurements can be divided into the two parts: the reversible dH'/dt and the non-reversible part $dH"/dt$. The reversible and the non-reversible part correspond to the heat capacity and the relaxation part, respectively. The average part $dH/dt = dH'/dt + dH"/dt$ is equivalent to the heat flux of a DC calorimetric measurement. The heat flux of the glass transition contributes to the reversible part, whereas that of the crystallization to the non-reversible part. Thus, AC DSC allows us to distinguish the heat signal of glass transition from that of crystallization.

A sample, which was prepared as a glassy state, was heated from 150 K to 310 K by using AC DSC with an alternating heating rate, whose average was 2 K/min and whose amplitude and period were, respectively, 0.16 K and 30 s. We measured the reversible part of the heat flow upon heating, which is shown in Fig. 4. The reversible part, which is proportional to the heat capacity, exhibits a steplike change across T_g. The temperature width of this steplike change ΔT_g is given by $T_g^H - T_g^L$, where T_g^H and T_g^L are the high and low edge temperature of the glass transition, respectively. In our DSC measurements, T_g^H is the temperature where the structural relaxation time

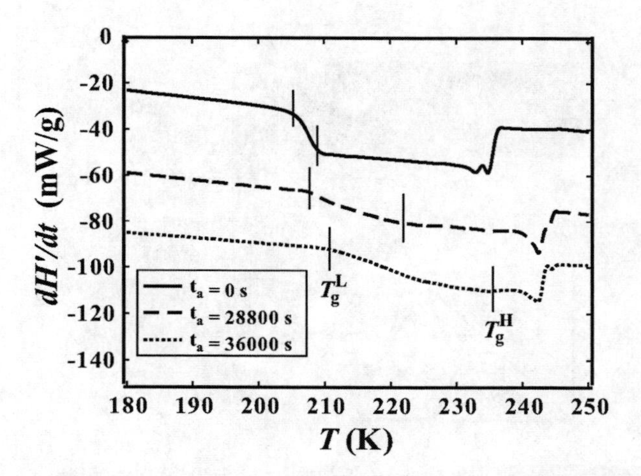

FIGURE 4. The reversible part of the heat flow (dH'/dt) during the heating process of TPP annealed at 213K for $t_a =$ 0 s, 28800 s, and 36000 s. The measurements were made by AC calorimetry. A sample was heated with an alternating rate, whose average was 2 K/min and whose period and amplitude were 30 s and 0.16 K, respectively. A different value of t_a means a different value of $\tilde{S}$: For $t_a = 0$ s, 28800 s, and 36000 s, $\tilde{S}$ is 0, 0.66, and 0.95, respectively. We can clearly see that the steplike change near T_g becomes broader with an increase in t_a: $\Delta T_g = T_g^H - T_g^L$ increases with t_a.

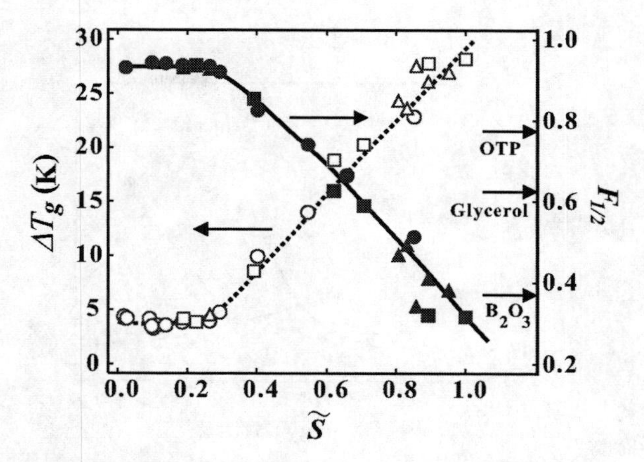

FIGURE 5. Relation of $\tilde{S}$ to ΔT_g (open symbols) and $F(1/2)$ (filled symbols). We estimated ΔT_g and $F(1/2)$ for T_a = 213 K (circles), 214 K (squares), and 215 K (triangles), all of which are located below T_{SD}. The curves are the guide for eye. The results for a heterogeneous state are not included in the plot. ΔT_g increases with an increase in $\tilde{S}$. The relation between ΔT_g and $\tilde{S}$ are independent of T_a between 213 and 215 K. $F_{1/2}$ decreases with an increase in $\tilde{S}$. This means that the fragility of TPP continuously increases from one of the most fragile liquids to a very strong liquid (stronger than $B_2O_3, F_{1/2} \sim 0.38$) with an increase in $\tilde{S}$.

$\tau = 1$ ms and T_g^L is the temperature where $\tau = 30$ s. ΔT_g is known to be negatively correlated with the fragility [42]: The smaller ΔT_g is, the more fragile a liquid is. This is because the temperature sensitivity of the structural relaxation time τ is larger for a more fragile liquid. In other words, ΔT_g is determined by the steepness of the temperature dependence of τ. We can clearly see the increase of ΔT_g with an increase in the annealing time t_a, or, S.

We also made similar measurements for other samples, which were prepared by annealing TPP at $T_a = 214$ K and 215 K, both of which are below T_{SD}, and quenching to 150 K. Note that $H(\infty)$ is constant to within a few % for T_a =213-215 K: $H(\infty) = 21.72$, 21.37, 21.09 J/g for T_a =213, 214, and 215 K, respectively. Figure 5 shows the relation between $\tilde{S}$ and ΔT_g for all the T_a's. We find that ΔT_g monotonically increases with an increase in $\tilde{S}$ on the same master curve for all the annealing temperatures.

Richert and Angell [42] proposed that the quantity $F_{1/2} = 2(T_g/T_{1/2}) - 1$ is a good measure of the fragility of liquid: The larger $F_{1/2}$ is, the stronger the liquid is. Here $T_{1/2}$ is defined as the temperature at which $\tau = 10^{-6}$ s. Note also that $\tau = 10^2$ s at T_g. Furthermore, Ito et al. [43] established the following empirical relation between ΔT_g and $F_{1/2}$: $\Delta T_g/T_g = 0.151(1-F_{1/2})(1+F_{1/2})$. We substitute the values of $\Delta T_g/T_g$ into this relation to estimate $F_{1/2}$. Figure 5 shows the relation between $F_{1/2}$ and $\tilde{S}$. On the left axis we indicate three materials whose values of $F_{1/2}$ are known. Surprisingly, a liquid of $\tilde{S} > 0.8$ is stronger than B_2O_3 ($F_{1/2} = 0.38$), whereas liquid I is extremely fragile and more fragile than a typical fragile molecular liquid, orthoterphenyl (OTP). Figure 5 clearly indicates that the liquid becomes stronger with an increase in $\tilde{S}$.

Our experimental results indicate that (i) LLT is controlled by the bond order parameter S, which is nonconserved, and (ii) a liquid with larger S is stronger. Now we consider these findings on the basis of our two-order-parameter (TOP) model of liquid since this model may explain both LLT and glass-transition phenomena in the same framework [44, 45, 46, 47, 48, 49]. The two order parameters refer to S and the density ρ. Our TOP model relies on the fact that a liquid generally has a tendency to form short-range bond order and thus the structure of a liquid becomes locally more ordered with decreasing temperature; i.e., a liquid is in a state of long-ranged disorder, but it can locally possess short-range bond order. This short-range bond ordering is due to specific interactions between liquid atoms (or molecules) that have the symmetry-selective nature. We already demonstrated (see also [27, 28]) that the TOP model can explain the kinetics of LLT quite well (see Fig. 3 and the related discussion). We can say that LLT may be a result of cooperative bond ordering [30].

There are few remaining crucial questions on the

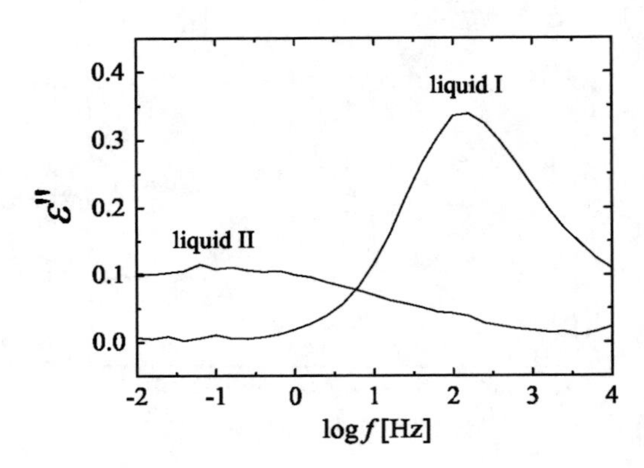

FIGURE 6. Frequency dependence of the imaginary part of the complex permittivity, ε'', of liquid I and liquid II.

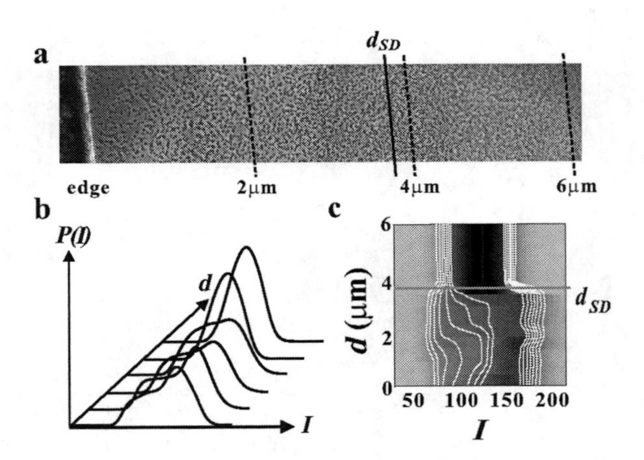

FIGURE 7. Process of liquid-liquid transformation under spatial confinements. (a) Phase-contrast microscopy image. A top view of a wedge-shaped sample cell filled with TPP. $T_a = 214$ K and $t = 150$ min. The black region on the left side is the edge of the wedge-shaped cell. The sample thickness is indicated with dashed lines below the image. (b) The d-dependence of the intensity (I) distribution function $P(I)$ at $T_a = 214$ K and $t = 150$ min. $P(I)$ at large d has only one peak, suggesting that the transformation pattern is SD-type. (c) Contour plot of the d-dependence of $P(I)$. For larger I, the color is more black. The white lines are the contours of $P(I)$. We can see $P(I)$ suddenly changes its shape at d_{SD} (see also the solid line in (a)): For $d > d_{SD}$ SD-type LLT is observed, whereas for $d < d_{SD}$ NG-type one is observed. .

fragility. Dielectric measurements tell us that the dielectric spectra for liquid II is much broader than for liquid I, as shown in Fig. 6 (see also Ref. [22]). This is contrary to the empirical relation that a more fragile liquid has a broad distribution of the structural relaxation time. This may be related to hierarchic ordering in the liquid and the unusually long bare correlation length. Providing answers to these questions may be quite important for the understanding of the nature of LLT.

Effects of geometrical confinement on LLT

It is widely known that phase transition can be seriously affected by a spatial confinement [50] when the correlation length of the relevant order parameter, ξ, becomes comparable to the characteristic length scale of confinement, d. This finite-size effect on phase transitions is important for its own sake, but also for applications to nanotechnology. Here we report the systematic experimental study on geometrical confinement effects on LLT [36]. We found a significant decrease of both the spinodal and the binodal temperature of LLT with a decrease in d. The correlation length ξ is found to obey the following scaling relation: $\xi = \xi_0(d)[(T_{SD}(d) - T)/T_{SD}(d)]^{-\nu}$, where ξ_0 is the bare correlation length and $\nu = 0.5$ is the mean-field critical exponent for ξ. Practically, this divergent behavior is quite useful for controlling LLT by spatial confinement. Because of the unusually large value of ξ_0, ξ can very easily reach a length scale of microns near T_{SD}. This means that a spatial confinement of the order of microns should be enough to see a significant finite-size effect on LLT in TPP. Furthermore, we confirmed that there is little surface wetting effects on LLT by using a wedge-shaped cell. Studying finite-size effects is quite useful for elucidating the nature of fluctuations of the order parameter, or, revealing the nature of the order parameter. In particular, it is not clear why the bare correlation length ξ_0 of this system is so long compared to the molecular size (~ 1 nm) and what it represents for. This is one of the most mysterious problems, which may be intimately related to the nature of LLT. Finite-size effects on LLT might provide us with key information on these difficult, but important questions.

First we show a pattern, which was observed in a wedge-shaped cell at $T = 214$ K and $t = 150$ min during LLT, in Fig. 7(a). Here t is the time after temperature quench. The line on the left side is the edge of the wedge-shaped cell, where $d = 0$ μm. We confirmed that for a thick enough sample (bulk) the transformation pattern is SD-type at this temperature [27, 28]. Due to the spatial confinement effects, however, the transformation behavior clearly depends on d (see Fig. 7(a)). The transformation pattern is SD-type at large d, however, droplets of liquid II nucleate and grow with time for d, which is smaller than the critical thickness d_{SD}. We can see the increase of the droplet number density with an increase in d. To analyze this change in a more quantitative manner, we calculated the intensity distribution function $P(I,d)$ from an image [41] as a function of d. Here $P(I,d)$ represents an average of $P(I,h)$ over a region $d - \delta d < h < d + \delta d$, where h is the spacing be-

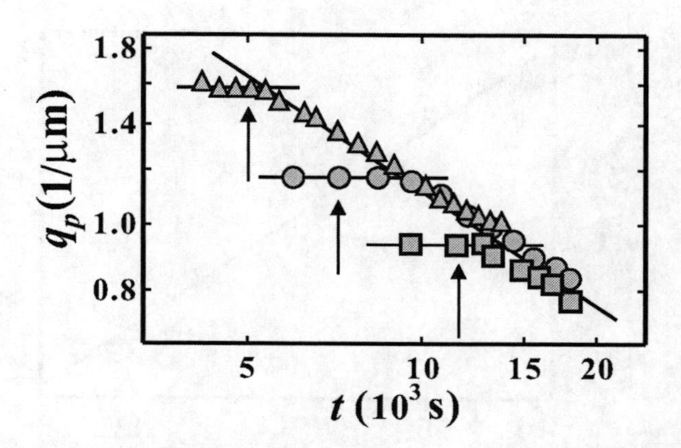

FIGURE 8. Kinetics of SD-type LLT under spatial confinements. Temporal change of q_p at 212 K for d = 10 μm (triangles), 2.5 μm (circles), and 2 μm (squares). q_p in the linear stage does depend on d, whereas it collapses on the same power law in the late stage.

tween the upper and lower glass and we set δd to be 0.15. Figure 7(b) shows the d-dependence of the shape of $P(I,d)$. We can see two types of shapes of $P(I,d)$: For small d, $P(I)$ has a peak with a distinct shoulder indicative of the appearance of a small additional peak. The existence of the two peaks in $P(I,d)$ indicates that liquid I and liquid II coexist during the transformation. This is a characteristic feature of NG-type LLT [35]. For large d, on the other hand, $P(I,d)$ has only a broad peak, suggesting that the density distribution during the transformation is continuous. This is a characteristic feature of SD-type LLT [35]. Figure 7(c) indicates the intensity map of $P(I,d)$. For small d, $P(I,d)$ has a peak around I = 150 and has a gradual slope around I = 100, which corresponds to the shoulder. We can see a distinct change in the shape of $P(I,d)$ at $d = d_{SD} = 3.8$ μm. This means that $T_{SD}(d) = 214$ K at $d = 3.8$ μm. This sharp change of the behavior at d = 3.8 μm is consistent with the mean-field nature of LLT, namely, a shape boundary between NG and SD behavior [28].

Next we studied the kinetics of SD-type LLT under a planar confinement at various T_a's to obtain the d-dependence of the critical behavior of the correlation length ξ. We calculated $F(q)$ from each image, using the digital image analysis [40] and extracted the peak wavenumber q_p. Figures 8 shows the temporal change of q_p for three different d's. q_p is constant with time and the peak intensity $F(q_p)$ increases exponentially (not shown) in the initial stage of SD-type LLT: Cahn's linear regime. Then, q_p decreases as $q_p \sim t_a^{-0.5}$. This time exponent is consistent with the prediction of the Allen-Cahn relation, $dR/dt \sim \sqrt{Lt}$ (L being the kinetics coefficient), which describes the domain interface motion in a system of a non-conserved order parameter [35]. We can see that the temporal change of q_p for various d's almost collapses on the same power law in the late stage, which indicates that L is independent of d.

On the other hand, q_p in the linear regime does depend on d. This means that the correlation length ξ is a function of d since we have the relation of $\xi = 1/\sqrt{2}q_p$ in the linear regime [35]. Figure 9(a) shows $\xi(T)$ for various values of d. We confirmed that $\xi(T,d)$ is well fitted by

$$\xi(T,d) = \xi_0(d)[(T_{SD}(d) - T)/T_{SD}(d)]^{-\nu}. \quad (5)$$

We obtained that ν is 0.50 ± 0.02 for all the cases, which means that LLT obeys a mean field theory under the confinement as in bulk. This is consistent with the fact that the mean field theory is not affected by the spatial dimensionality. Then, we fixed $\nu = 0.5$ to reduce the number of the adjustable parameters in the fitting of the above function to the data and to obtain the d-dependence of $\xi_0(d)$ accurately. We independently determined $T_{SD}(d)$ from optical microscopic measurements (see, e.g., Fig. 7). Figure 9(b) shows the d-dependence of ξ_0, which is obtained from the fitting. We can clearly see that $\xi_0(d)$ decreases with a decrease in d. We also found the following empirical relation:

$$\xi(\infty) - \xi(d) = \xi(\infty)\exp(-d/d_c), \quad (6)$$

where d_c is the characteristic decay length and is determined as 2.79 μm by the fitting. This decrease in the mesoscopic length ξ_0 may reflect the suppression of hierarchical ordering in a liquid induced by specific interactions. If this is the case, it may share some common physics with the suppression of the hydrogen-bonded network in confined water under a stronger confinement [51]. This may have a significant implication on the question of what is the origin of the long bare correlation length. We speculate some hierarchical structural ordering takes place in this liquid and it plays a crucial role in LLT. For an extremely strong confinement, ξ_0 might becomes short enough to break the mean-field nature of the transition.

Figure 10 plots $\Delta T_{SD} \equiv T_{SD}(\infty) - T_{SD}(d)$, where $T_{SD}(\infty) = 215.5$ K [27], as a function of $d/\xi_0(d)$. Usually, this shift of the critical temperature is plotted against d/a to check the prediction of Fisher scaling [50]. Note that in ordinary systems the bare correlation length $\xi_0 = a$ is associated with the size of the elementary length scale of a system such as the size of constituent molecules or atoms. Thus, it is not expected that ξ_0, or a, depends on d. However, in our case, the bare correlation length, which is an intermediate length scale much larger than the molecular size a, does depend upon d. Using d/ξ_0 as a key parameter instead of d/a, we recover $\Delta T_{SD} \propto [d/\xi_0(d)]^{-\lambda}$ with $\lambda = 2$, which is consistent with the Fisher's scaling $\lambda = 1/\nu$ (note $\nu = 0.5$ in our case). We expect that this generalized scaling relation

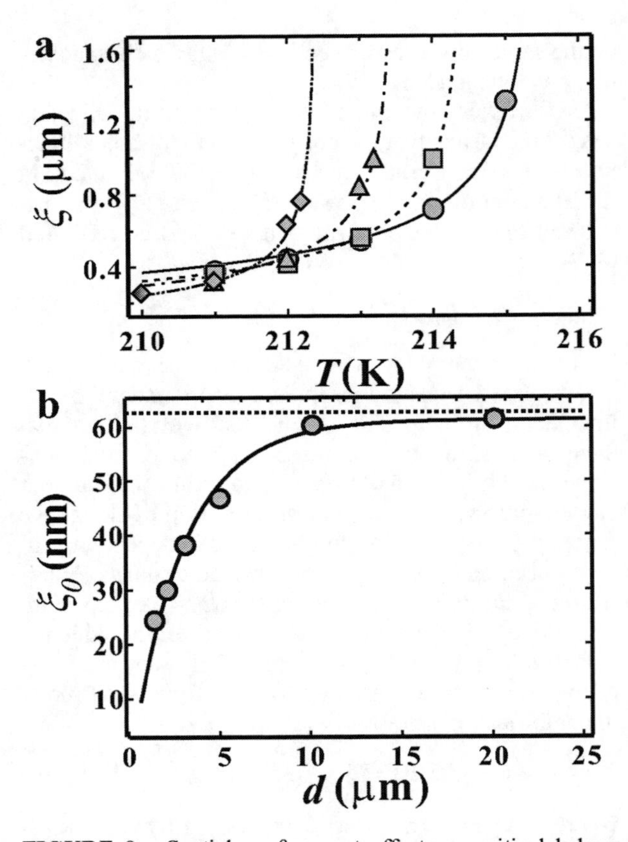

FIGURE 9. Spatial confinement effects on critical behavior. (a) d-dependence of the critical divergence of ξ; circle: $d = 10$ μm, square: $d = 5$ μm, triangle: $d = 3$ μm, and diamond: $d = 2$ μm. The T-dependence of ξ can be well fitted by $\xi_0(d)[(T_{SD}(d) - T)/T_{SD}(d)]^{-\nu}$. In the fitting, ξ is independently determined by optical microscopy observation (see Fig. 1), ν is fixed to 0.5 (see text), and thus ξ_0 is the only adjustable parameter. The position of the vertical lines represents $T_{SD}(d)$ determined by optical microscopy observation. (b) d-dependence of ξ_0. We can see that ξ_0 decreases with decreasing d. The dashed line is a bulk value for ξ_0. The line represents the fitting of the following function: $\xi(\infty) - \xi(d) = \xi(\infty)\exp(-d/d_c)$. Here the characteristic decay length d_c is determined as 2.79 μm by the fitting.

$\Delta T_{SD} \propto [d/\xi_0(d)]^{-1/\nu}$ may hold universally. This tells us that the strength of the spatial confinement and the degree of the shift of the critical point should be controlled by the ratio of the characteristic length of a system, d, to the bare correlation length, ξ_0, even if ξ_0 also depends on d.

SUMMARY and open questions

In sum, our TOP model remarkably well reproduces all the basic features of LLT observed experimentally. However, there remain fundamental questions, which include (i) what does S represent microscopically in real sys-

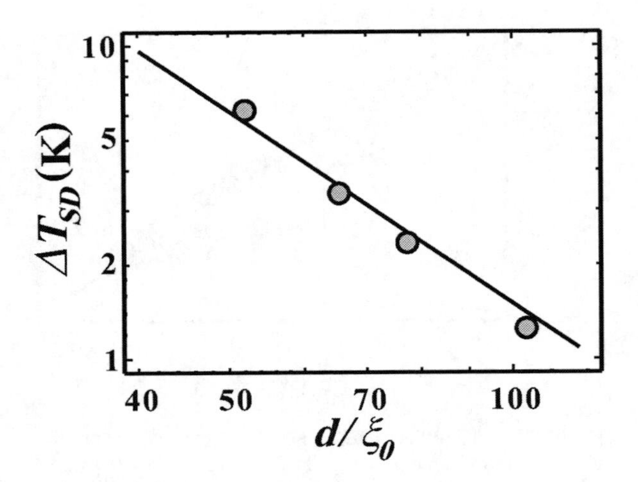

FIGURE 10. Shift of the spinodal and binodal temperature by spatial confinement. d-dependence of ΔT_{SD}. The solid line represents the prediction of the Fisher scaling: $\Delta T_{SD} \propto (d/\xi_0(d))^{-\lambda}$, where $\lambda = 1/\nu = 2$.

tems? and (ii) what is the microscopic origin of the cooperativity of the bond ordering? To answer these questions, microscopic-level information on LFS is highly desirable.

We also succeeded in continuously controlling the fragility of a liquid in the range of the fragility parameter m between $m = 154$ and $m = 34$ by using continuous LLT. Liquid I is more fragile than OTP, while liquid II is stronger than B_2O_3. Our study clearly indicates that the fragility is not a material-specific quantity; namely, it is not related to the type of the interparticle potential in a direct manner. We find that the fragility is controlled by the same order parameter controlling LLT, which we believe is the fraction of LFS. The observed behavior is qualitatively consistent with the prediction of the TOP model of a liquid-glass transition [45, 46, 47, 48, 49]. Our experimental findings shed new light on a fundamental question of what physical factor controls the fragility of liquid and contribute to our deeper understanding of liquid-glass transition.

We also revealed the spatial confinement effects on LLT. The strength of confinement is found to be characterized by d/ξ_0. We can easily shift T_{SD} as well as T_{BN} more than several K just by the micron-order confinement of a system. We expect nanoscale confinements may have significant effects on LLT and the mesoscopic structure of a liquid, whose length scale can be characterized by ξ_0. Our finding may have significant implications not only on the physical understanding of LLT, but also on nanoscience. For example, we may be able to control LLT by the degree of spatial confinement. Since liquid I and II have very different physical properties such as density, refractive index, and viscosity, this may open up a new possibility to control the physical properties of a liquid by confinement-induced LLT.

Finally, we mention some important open questions. First we should mention that there still remains controversy on the nature of the transition. Our study supports the LLT scenario. However, there are some mysterious characteristics in liquid II. They include (i) the broad dielectric spectra despite of the strong nature of liquid II (see Ref. [22] and Fig. 6), (ii) the unusually long bare correlation length for a molecular liquid [28], and (iii) the existence of a small-angle scattering peak [20, 21]. Interestingly, the liquid II of n-butanol also have features similar to points (i) and (ii). All these points might arise from a common origin. The liquid II might be an exotic state of liquid matter. Further studies are highly desirable for clarifying the nature of liquid II.

REFERENCES

1. M. C. Wilding, M. Wilson and P. F. McMillan, *Chem. Soc. Rev.* **35**, 964-986 (2006).
2. O. Mishima and H. E. Stanley, *Nature* **396**, 329-335 (1998).
3. P. G. Debenedetti, *J. Phys.: Condens. Matter* **15**, R1669-R1726 (2003).
4. P. G. Debenedetti, *Metastable Liquids* (Princeton University Press, Princeton, 1997)
5. C. A. Angell, *Science* **267**, 1924-1935 (1995).
6. P. H. Poole, T. Grande, C. A. Angell, and P. F. McMillan, *Science* **275**, 322-324 (1997).
7. Y. Katayama, T. Mizutani, W. Utsumi, O. Shimomura, M. Yamakata and K. Funakoshi, *Nature* **403**, 170-173 (2000).
8. V. V. Brazhkin, and A. G. Lyapin, *J. Phys.: Condens. Matter* **15**, 6059-6084 (2003).
9. T. Morishita, *Phys. Rev. Lett.* **87**, 105701-1-105701-4 (2001).
10. P. F. McMillan, *J. Mater. Chem.* **14**, 1506-1512 (2004).
11. S. K. Deb, M. Wilding, M. Somayazulu, and P. F. McMillan, *Nature* **414**, 528-530 (2001).
12. S. Sastry and C. A. Angell, *Nat. Mater.* **2**, 739-743 (2003).
13. S. Aasland and P. F. McMillan, *Nature* **369**, 633-636 (1994).
14. H. Tanaka, *Europhys. Lett.* **50**, 340-346 (2000).
15. I. Cohen, A. Ha, X. Zhao, M. Lee, T. Fischer, J. Strouse, and D. Kivelson, *J. Phys. Chem.* **100**, 8518-8526 (1996).
16. B. G. Demirjian, G. Dosseh, A. Chauty, M. L. Ferrer, D. Morineau, C. Lawrence, K. Takeda, and D. Kivelson, *J. Phys. Chem. B* **105**, 2107-2116 (2001).
17. M. Mizukami, K. Kobashi, M. Hanaya, and M. Oguni, *J. Phys. Chem. B* **103**, 4078-4088 (1999).
18. A. Hedoux, O. Hernandez, J. Lefebvre, Y. Guinet, and M. Descamps, *Phys. Rev. B* **60**, 9390-9395 (1999).
19. A. Hedoux, Y. Guinet, P. Derollez, O. Hernandez, L. Ronan, and M. Descamps, *Phys. Chem. Chem. Phys.* **6**, 3192-3199 (2004).
20. C. Alba-Simionesco and G. Tarjus, *Europhys. Lett.* **52**, 297-303 (2000).
21. B. E. Schwickert, S. R. Kline, H. Zimmermann, K. M. Lantzky, and J. L. Yarger, *Phys. Rev. B* **64**, 045410-1-045410-6 (2001).
22. J. Senker and E. Rössler, *Chemical Geology* **174**, 143-156 (2001).
23. D. Kivelson and G. Tarjus, *J. Non-Crystal. Solids* **307-310**, 630-636 (2002).
24. J. Senker, J. Sehnert, and Correll, *J. Am. Chem. Soc.* **127** 337-349 (2005).
25. Q. Mei, P. Ghalsaki, C. Benmore, and J. L. Yarger, *J. Phys. Chem. B* **108**, 20076-20082 (2004).
26. Q. Mei, J. E. Siewenie, C. Benmore, P. Ghalsasi, and J. L. Yarger, *J. Phys. Chem. B* **110**, 9747-9750 (2003).
27. H. Tanaka, R. Kurita, and H. Mataki, *Phys. Rev. Lett.* **92**, 025701-1-025071-4 (2004)
28. R. Kurita and H. Tanaka, *Science* **306**, 845-848 (2004).
29. R. Kurita and H. Tanaka, *J. Phys.: Condens. Matter* **17**, L293-L302 (2005).
30. H. Tanaka, *J. Phys.: Condens. Matter* **11**, L159 (1999); *Phys. Rev. E* **62**, 6968-6976 (2000).
31. R. Kurita, Y. Shinohara, Y. Amemiya and H. Tanaka, *J. Phys.: Condens. Matter* **19**, 152101-152108 (2007).
32. R. Kurita and H. Tanaka, *Phys. Rev. B* **73**, 104202-1-14202-5 (2006).
33. R. Kurita and H. Tanaka, *J. Chem. Phys.* **126**, 204505-1-204505-8 (2007).
34. R. Kurita and H. Tanaka, *Phys. Rev. Lett.* **95**, 065701-1-065701-4 (2005).
35. A. Onuki, *Phase Transition Dynamics* (Cambridge University Press, Cambridge, 2002).
36. R. Kurita and H. Tanaka, *Phys. Rev. Lett.* **98**, 235701-1-235071-4 (2007).
37. H. Tanaka, *Phys. Rev. B* **66**, 064202-1-064202-8 (2002).
38. S. Strässler and C. Kittel, Phys. Rev. A **139**, 758-760 (1965).
39. E. Rappoport, *J. Chem. Phys.* **46**, 2891-2895 (1967).
40. H. Tanaka, T. Hayashi and T. Nishi, *J. Appl. Phys.* **59**, 3627-3643 (1986).
41. H. Tanaka and T. Nishi, *Phys. Rev. Lett.* **59**, 692-695 (1987)
42. R. Richert and C. A. Angell, *J. Chem. Phys.* **108**, 9016-9026 (1998).
43. K. Ito, C. T. Moynihan, and C. A. Angell, *Nature* **398**, 492-495 (1999).
44. H. Tanaka, *AIP Conference Proceedings* **708**, 3rd International Symposium on Slow Dynamics in Complex Systems (edited by M. Tokuyama and I. Oppenheim), pp. 541-546 (2004).
45. H. Tanaka, *J. Phys.: Condens. Matter* **10** L207-L214 (1998).
46. H. Tanaka, *J. Chem. Phys.* **111**, 3163-3174 (1999).
47. H. Tanaka, *J. Chem. Phys.* **111**, 3175-3182 (1999).
48. H. Tanaka, *J. Phys.: Condens. Matter* **15**, L491-L498 (2003).
49. H. Tanaka, *J. Non-Cryst. Solids* **351**, 3371-3384, 3385-3395, 3396-3413 (2005).
50. M. E. Fisher and M. N. Barder, *Phys. Rev. Lett.* **28**, 1516-1519 (1972).
51. U. Raviv, P. Laurat and J. Klein, *Nature* **413**, 51-54 (2001).

Thermal Properties of the Water Confined within Nanopores of Silica MCM-41

M. Oguni[1], Y. Kanke[1], and S. Namba[2]

[1] *Department of Chemistry, Graduate School of Science and Engineering, Tokyo Institute of Technology, 2-12-1 O-okayama, Meguro-ku, Tokyo 152-8551, Japan*
[2] *School of Science and Engineering, Waseda University, Totsuka-machi, Shinjuku-ku, Tokyo 169-8050, Japan*

Abstract. Adiabatic calorimetry was carried out of the water confined within the pores of silica MCM-41 with diameters of 1.2, 1.6, and 1.8 nm. Glass transitions were found at T_g = 115 and 165 K in the 1.2 nm case, 117, 165, and 205 K in the 1.6 nm, and 118 and 210 K in the 1.8 nm, respectively. The transition at T_g = 115-118 K was interpreted as caused by the freezing of the rearrangement of the water molecules located on the pore wall and interacting with silanol groups, and those at T_g = 165 and 205-210 K of the water molecules located in the center of the pores. It was noticed that the T_g increased discretely with increasing the pore diameter from 115 to 165 to 210 K. This indicates that the T_g and therefore the activation energy for the rearrangement are strongly connected with the development of the hydrogen-bond network and furthermore the number of the hydrogen bonds formed by each water molecule. It was suggested that the bulk water undergoes the glass transition at T_g = 210 K rather than at 136 and 165 K debated hitherto and shows a change from fragile to strong behaviors in the relaxation times with cooling down to 210 K.

Keywords: Calorimetry, Glass transition, Heat capacity, MCM-41, Water
PACS: 61.43.Fs, 64.70.Pf, 65.20.fw, 65.60.+a, 65.80.+n

INTRODUCTION

Water plays essential roles in many natural phenomena on the Earth. While water molecule is simple in the structure, the aggregate of the molecules shows rather abnormal behaviors. It is well known that liquid water exhibits a maximum density at 277 K and many thermodynamic properties such as heat capacity and thermal expansivity increase in their anomalousness with decreasing the temperature below 0 °C [1]. However, water has a tendency to crystallize easily; bulk water crystallizes usually at around −10 °C. Therefore, it is hard to probe directly the thermodynamic and kinetic properties at low temperatures, and those properties have been examined by using a variety of contrived sample-preparation and observation methods. As one method, water droplets were formed in an emulsified sample: Heat capacities, measured by a differential scanning calorimetry (DSC), of the droplets showed an anomaly apparently following a critical-point equation with the critical temperature T_c = 228 K on cooling [2]. Even with such a contrivance, the water crystallized below 235 K. Whether any kind of phase transition takes place at around 228 K or not is therefore unclear yet. In view that gaseous and liquid states are changed continuously by taking a roundabout of their critical point, vapor-deposited amorphous solid water (ASW) was obtained on a substrate at 80-100 K. Splaying water droplets on the cold substrate was also tried successfully (hyper-quenched glassy water, HGW). Heat capacities of the ASW and HGW samples were measured by DSC [3-5], and of the ASW by an adiabatic calorimetry [6]. The results have been interpreted mainly as indicating that the glass transition takes place at around 135 K [3,4,6]. Dielectric data of ASW and HGW were also analyzed to indicate the T_g = 136 K [7]. On the other hand, a possibility of T_g = 160-165 K was discussed by Angell and his group based on DSC [5] of ASW and heat release behavior of HGW [8]. Recently McClure *et al.* indicated from thermal desorption experiments of layered nm-scale films of labeled ASW that the glass transition occurs above 160 K [9]. Since those samples crystallize above 150 K, no decision concerning which of the two interpretations is appropriate for the T_g of supercooled water has been derived from the works.

Liquid state can be stabilized, as compared with the crystalline one, by reducing the size of the system [10]. We observed the crystallization of the water even when it was confined within silica-gel pores above 2 nm in the average diameter, but succeeded in keeping water from the crystallization within pores with 1.1 nm in the average [11,12]. Two glass transitions were found to occur at 115-130 K and 160 K; the former was interpreted as due to the freezing of the rearrangement of the water molecules located on the

CP982, *Complex Systems, 5th International Workshop on Complex Systems*
edited by M. Tokuyama, I. Oppenheim, and H. Nishiyama

pore wall, namely interacting with silanol groups, and the latter of the water molecules located in the center of the pores, namely surrounded only by other water molecules. Therefore, the latter T_g was indicated as connected to the property of bulk water; the value is close to the 165 K claimed by Angell *et at.* [8]. However, arrayed along the diameter of 1.1 nm are only three water molecules. Although this suggests that the water molecules show their inherent property as far as surrounded by other water molecules, the number of three seems too small. In this respect, it is of interest to look into the detailed behavior of the water confined within pores with diameters between 1.1 and 2 nm. Considering that the water crystallizes above 2 nm, the situation resembling the bulk water is potentially realized within pores of the size immediately before the crystallization occurs inevitably. Here we studied the thermal behaviors of the water confined within MCM-41 pores by using an adiabatic calorimeter [13]; MCM-41 is known to have the pores with reasonably constant diameter [14].

CALORIMETRIC DETERMINATION OF THE PRESENCE OF A GLASS TRANSITION

The presence of a calorimetric glass transition is ordinarily identified through finding a heat-capacity jump in DSC. However, it is rather difficult to identify it when the heat-capacity jump is small or occurs in a wide temperature range. Adiabatic calorimetry is then a powerful tool in the identification of the glass transition even in such situations since enthalpy relaxation is directly observed. Ordinarily, the calorimetry is carried out by heating the sample in the intermittent way under adiabatic conditions [13,15]: The sample temperature is followed for *ca.* 10 min to determine the initial temperature T_i, increased at a rate of *ca.* 0.1 $\mathrm{Kmin^{-1}}$ with the supply of electrical energy ΔE, and followed again to determine the final one T_f. The gross heat capacity is evaluated by an equation $C = \Delta E/(T_f - T_i)$. When the sample absorbs or releases heat to reach the equilibrium state, spontaneous temperature drift $\mathrm{d}T/\mathrm{d}t$ is observed during the thermometry periods. The enthalpy relaxation rate is then evaluated by an equation $(-\mathrm{d}H/\mathrm{d}t) = C(\mathrm{d}T/\mathrm{d}t)$.

Figure 1 shows diagrams illustrating how the enthalpy relaxation is observed in the glass transition region. As shown in Fig. 1(a), the equilibrium configurational enthalpy of the sample decreases with decreasing the temperature. The characteristic, relaxation time τ for the configuration of molecules to reach its equilibrium state elongates with the temperature decrease. Then the enthalpy relaxation rates display characteristic dependence on temperature

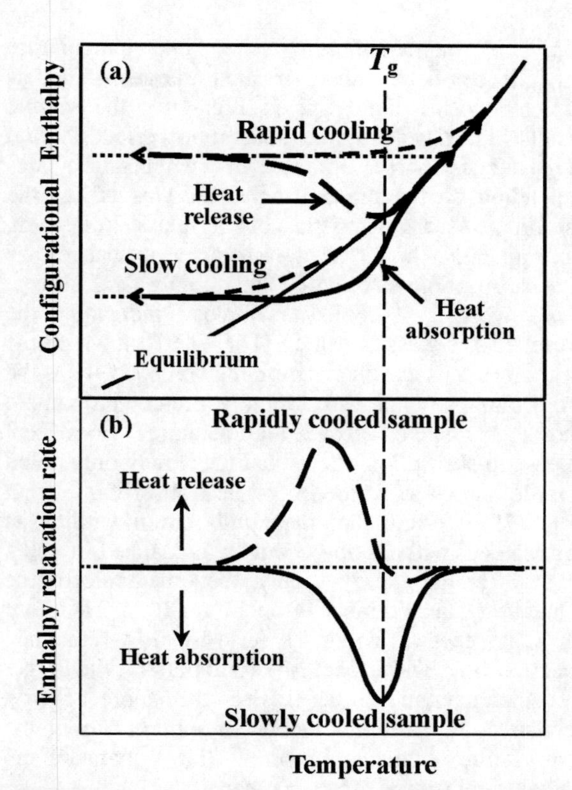

FIGURE 1. Diagrams representing the enthalpy relaxation occurring in the glass transition region, and the calorimetric method to determine the glass transition temperature T_g.

in association with the pre-cooling rates of the sample. When the sample is pre-cooled rapidly *e.g.* at 10 $\mathrm{Kmin^{-1}}$, the configurational structure is frozen at the one corresponding to a high temperature and the associated enthalpy is rather high. As the temperature is increased at about 0.1 $\mathrm{Kmin^{-1}}$ in the calorimetry as indicated with a broken line, the structure and therefore enthalpy relax as a heat-release process towards the equilibrium state, cross the equilibrium line at around T_g, depart downwards the line in the opposite direction, and soon return the equilibrium state. Then, the temperature dependence of the enthalpy relaxation rates is such as depicted with a broken line in Fig. 1(b); the sample shows a rather large heat release effect. On the other hand, when pre-cooled slowly *e.g.* at 10 $\mathrm{mKmin^{-1}}$, the sample is frozen with a configurational structure and the associated enthalpy corresponding to a relatively low temperature. As the temperature is increased in the calorimetry as indicated with a thick solid line in Fig. 1(a), the structure and enthalpy remain in the frozen state till a rather high temperature with long relaxation times, often start to relax as a heat absorption process after crossing the equilibrium line, give the maximum heat-absorption effect at around T_g, and gradually approach the equilibrium state. The temperature dependence of the enthalpy relaxation rates is such as depicted with a solid line in Fig. 1(b); the sample results in a rather large heat-absorption effect. A kind of hysteresis loop is obtained in the relaxation rates depending on the pre-cooling rates. This is just the calorimetric characteristic of a glass transition. Consequently, the observation of a set of heat-release and -absorption effects for the rapidly and slowly, respectively, pre-cooled samples indicates the presence of a glass transition [16]. The T_g at which the τ becomes 1 ks has been determined empirically as the temperature at which the rapidly pre-cooled sample shows a change (against temperature) from heat release to absorption effects and the slowly pre-cooled sample shows a maximum of heat-absorption effect [13, 17]. In view that the equilibration within the sample cell, without any anomaly, is reached in 100 s or so after the energy supply, the time scale of the relaxation detected ranges from 10^6 to 10^2 s. Therefore the calorimetry works as a very low-frequency spectroscopy. This aspect is very important; the longer the characteristic time scale, the more clearly separated the different modes of motion are in the temperature ranges where the characteristic enthalpy-relaxation effects appear.

EXPERIMENT

The silica MCM-41 was prepared in the procedure described previously [14]. Alkyl tri-methyl ammonium salts were used as the surfactants where the carbon numbers n of the alkyl groups were 10, 12, and 14. The pore size was analyzed by a nitrogen-gas absorption method, and the average diameters of the pores were 1.2, 1.6, and 1.8 nm for the surfactants with n = 10 and 12 and of a mixture of 12 and 14, respectively.

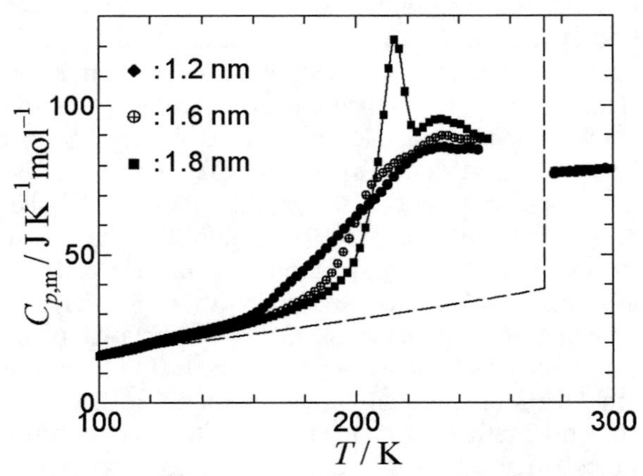

FIGURE 2. Molar heat capacities of the water confined within nano-pores of silica MCM-41.

Distilled water was used as a sample. The water was degassed through a repeated freezing and thawing process, and introduced into the pores of the silica which had been dried for 20 h at 200 °C in vacuum beforehand.

Heat capacities were measured by an intermittently heating method using an adiabatic calorimeter reported previously [13]. The sample prepared above was loaded into a calorimeter cell and the cell was sealed with an indium wire under an atmosphere of helium gas.

The amount of the water loaded was estimated, after the heat capacity measurements, by drying the used sample at 80 °C and weighing the extracted water. The amount of the bulk water included was evaluated from the enthalpy required for the fusion, since the fusion of the bulk ice was clearly separated from the anomalies due to the water confined within the pores.

RESULTS AND DISCUSSION

Figure 2 shows the results of molar heat capacities of the water confined within the pores of 1.2, 1.6 and 1.8 nm in diameter. No crystallization was observed in the cases of 1.2 and 1.6 nm pores. About 5 % of water crystallized within the 1.8 nm pores as estimated from the enthalpy of fusion. No data was plotted in the range between 250 and 275 K, because the bulk ice present showed its fusion effect and it was difficult to derive the heat capacities of water within the pores in the range.

Figure 3 shows the rates of spontaneous enthalpy release and absorption observed in the thermometry periods. The behavior in the 1.2 nm case is quite similar to that obtained in the case of silica gel with 1.1 nm pores [12]: According to the interpretation, the glass transition at T_g=115 K is recognized as caused by the interfacial water molecules and that at T_g=165 K by the internal ones surrounded only by other water molecules. A strange behavior was observed in the case of 1.6 nm pores: Another glass transition appeared around 205 K, and the enthalpy release/absorption effects at 165 K were reduced appreciably. The features became more remarkable in the case of 1.8 nm pores: The glass transition wholly jumped from 165 K to 210 K. In association with the jump in the T_g, the rise or jump in C_p moved from 160 K to 200-210 K as seen in Fig. 1: When compared among the curves for three different pore diameters, the C_p rise around 160 K is appreciable in the 1.2 nm pore case, becomes very small in the 1.6 nm, and does much smaller in the 1.8 nm. Accordingly, the heat capacity values around 180 K decrease with increasing the diameter. On the other hand, around 200 K, a jump appears in the 1.6 nm pore case and

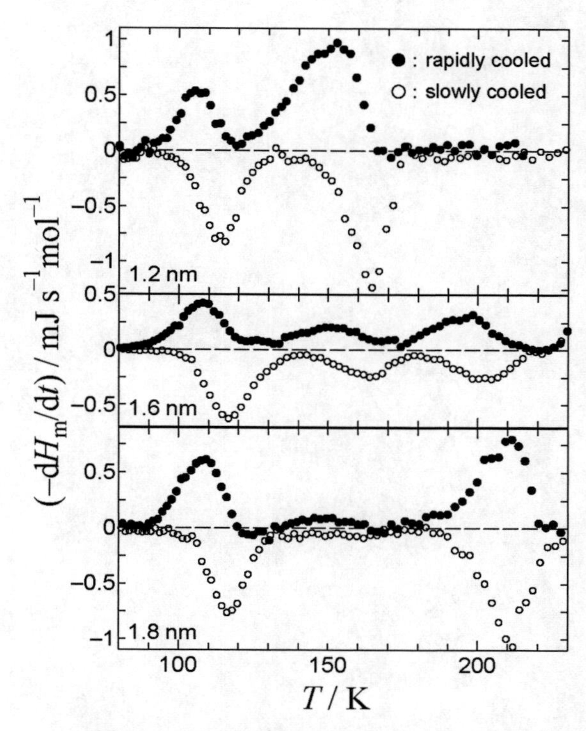

FIGURE 3. Spontaneous enthalpy release and absorption rates of the water confined within nanopores of silica MCM-41.

becomes sharper and larger in the 1.8 nm as seen from the heat capacities below 200 K and above 220 K. It is claimed that, with increasing the pore size, the T_g of the water located in the pore center shows a jump from 165 K to 210 K.

The increase in T_g, namely in activation energy, probably reflects the development to more complete hydrogen-bond network. The development is considered as consequently causes the crystallization of the water confined within (2 nm)-in-diameter pores. Here, it is noticed that the T_g increased discretely with increasing the pore diameters; the property quite differs from that found in simple van der Waals liquids. The relaxation time in the classical process is expressed by an Arrhenius equation [18,19]; $\tau = \tau_0 \exp(\Delta\varepsilon_a/k_BT)$, where it is reasonably assumed [18,19] that $\tau_0 = 10^{-14}$ s as indicated with a short dashed line in Fig. 4. Substituting T_g = 115, 165, and 210 K yields the $\Delta\varepsilon_a$ = 37, 54, and 68 kJmol^{-1}, respectively. Provided that the hydrogen bond energy is around 17 kJmol^{-1} [20], the $\Delta\varepsilon_a$ corresponds to the energy required for breaking the 2, 3, and 4, respectively, bonds. This suggests that the $\Delta\varepsilon_a$ and therefore the T_g are strongly connected with the number of the hydrogen bonds formed by each water molecule. Then, it is hard to consider even for bulk water the T_g higher than 210 K. This consideration leads to suggestions

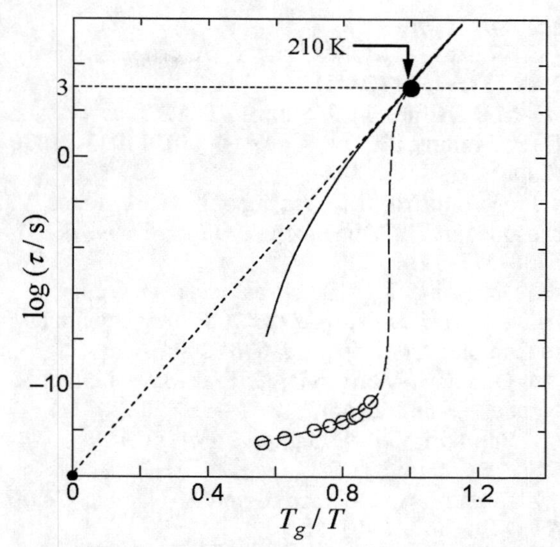

FIGURE 4. Expected temperature dependence of relaxation times of bulk water: solid circle, predicted from the present work; open circles, taken from literature [21]; solid line, temperature dependence expected for the relaxation times of the water confined within the 1.8 nm pores; dashed line, temperature dependence expected for the relaxation times of bulk water.

first that the glass transition of bulk water occurs at 210 K rather than at 136 and 165 K debated hitherto, and second that the relaxation times probably exhibit a change from fragile to strong behaviors with cooling down to 210 K as indicated with a long dashed line in Fig. 4.

ACKNOWLEDGMENTS

This work was financially supported partly by a Grant-in-Aid for Scientific Research (Grant 18350003) from the Ministry of Education, Culture, Sports, Science, and Technology, Japan.

REFERENCES

1. F. Franks, *A Comprehensive Treatise*, Plenum Press, New York, **1972**.
2. C. A. Angell, M. Oguni, and W. J. Sichina, *J. Phys. Chem.* **86**, 998-1002 (1982).
3. A. Hallbrucker, E. Mayer, and G. P. Johari, *J. Chem. Phys.* **93**, 4986-4990 (1989).
4. G. P. Johari, A. Hallbrucker, and E. Mayer, *Nature* **330**, 552-553 (1987).
5. D. R. MacFarlane and C. A. Angell, J. Phys. Chem. **88**, 759-762 (1984).
6. M. Sugisaki, H. Suga, and S. Seki, *Bull. Chem. Soc. Jpn.* **41**, 2591-2599 (1968).
7. G. P. Johari, *J. Chem. Phys.* **122**, 144508/1-10

(2005).
8. V. Velikov, S. Borick, and C. A. Angell, *Science* **294**, 2335-2338 (2001).
9. S. M. McClure, D. J. Safarik, T. M. Truskett, and C. B. Mullins, *J. Phys. Chem.* **B 110**, 11033-11036 (2006).
10. T. Takamuku, M. Yamagami, H. Wakita, Y. Masuda, and T. Yamaguchi, *J. Phys. Chem.* **B 101**, 5730-5739 (1997).
11. S. Maruyama, K. Wakabayashi, and M. Oguni: *AIP Conference Proceedings* **708**(Slow Dynamics in Complex Systems), 675-676 (2004).
12. M. Oguni, S. Maruyama, K. Wakabayashi, and A. Nagoe, *Chem. – Asian J.* **2**, 514-520 (2007).
13. H. Fujimori and M. Oguni, *J. Phys. Chem. Solids* **54**, 271-280 (1993).
14. S. K. Jana, A. Mochizuki, and S. Namba, *Catal. Surveys Asia* **8**, 1-13 (2004).
15. E. F. Westrum, Jr., G. T. Furukawa and J. P. McCullough, *Experimental Thermodynamics* vol **1**, 133 (ed. by McCullough J P and Scott D W, London: Butterworths) (1968).
16. H. Suga and S. Seki, *Faraday Discuss. Chem. Soc.* **69**, 221-240 (1980).
17. M. Oguni, T. Matsuo, H. Suga, and S. Seki, *Bull. Chem. Soc. Jpn.* **50**, 825-833 (1977).
18. H. Fujimori and M. Oguni, *Solid State Commun.* **94**, 157-162 (1995).
19. M. Oguni, *J. Non-Cryst. Solids* **210**, 171-177 (1997).
20. G. C. Maitland, M. Rigby, E. B. Smith, and W. A. Wakeham, *Intermolecular Forces*, Clarendon Press, Oxford, 1987.
21. C. A. Angell, *J. Phys. Chem.* **97**, 6339-6341 (1993).

Dynamic Crossover Phenomenon in Confined Supercooled Water and Its Relation to the Existence of a Liquid-Liquid Critical Point in Water

Sow-Hsin Chen[1], F. Mallamace[1,2], L. Liu[1], D. Z. Liu[1], X. Q. Chu[1], Y. Zhang[1], C. Kim[1], A. Faraone[3], C.-Y. Mou[4], E. Fratini[5], P. Baglioni[5], A. I. Kolesnikov[6], and V. Garcia-Sakai[3]

[1]*Department of Nuclear Science and Engineering, Massachusetts Institute of Technology, Cambridge, MA 02139*
[2]*Dipartmento di Fisica and CNISM, Universita' di Messina, Vill, S. Agata CP 55, 98166 Messina, Italy*
[3]*NIST Center for Neutron Research, Gaithersburg, MD 20899-8562 and Department of Material Science and Engineering, University of Maryland, College Park, MD 20742*
[4]*Department of Chemistry, National Taiwan University, Taipei, Taiwan*
[5]*Department of Chemistry and CSGI, University of Florence, 50019 Italy*
[6]*Intense Pulsed Neutron Source Division, Argonne National Laboratory, Argonne, Illinois 60439*

Abstract. We have observed a Fragile-to-Strong Dynamic Crossover (FSC) phenomenon of the α-relaxation time and self-diffusion constant in confined supercooled water. The α-relaxation time is measured by Quasielastic Neutron Scattering (QENS) experiments and the self-diffusion constant by Nuclear Magnetic Resonance (NMR) experiments. Water is confined in 1-d geometry in cylindrical pores of nanoscale silica materials, MCM-41-S and in Double-Wall Carbon Nanotubes (DWNT). The crossover phenomenon can also be observed from appearance of a Boson peak in Incoherent Inelastic Neutron Scattering experiments. We observe a pronounced violation of the Stokes-Einstein Relation at and below the crossover temperature at ambient pressure. Upon applying pressure to the confined water, the crossover temperature is shown to track closely the Widom line emanating from the existence of a liquid-liquid critical point in an unattainable deeply supercooled state of bulk water. Relation of the dynamic crossover phenomenon to the existence of a density minimum in supercooled confined water is discussed. Finally, we discuss a role of the FSC of the hydration water in a biopolymer that controls the biofunctionality of the biopolymer.

Keywords: Density minimum of water, Dynamic crossover phenomenon in hydration water, Dynamic crossover phenomenon of confined water, Liquid-liquid critical point of water, Quasielastic neutron scattering, the Widom line
PACS: 61.20.Lc, 61.12.Ex, 61.20.Ja, 64.70.Pf

INTRODUCTION

The study of supercooled and glassy water is motivated by the well known observation of anomalous behavior in thermodynamic as well as transport properties in bulk liquid water, that at ambient temperature and pressure, although quantitatively small, becomes increasingly significant at supercooled temperatures [1, 2]. It has been found that the extrapolated thermodynamic response functions and characteristic relaxation times appear to diverge, according to power laws, at a singular temperature $T_s = 228$ K [3]. Although this anomaly has sparked an enormous interest in the scientific community, a coherent explanation of the apparent singularity in supercooled water has not yet emerged. The basic reason for this is the fact that T_s is buried below the homogeneous nucleation temperature of water, $T_H = 235$ K [4], in an inaccessible temperature range for bulk supercooled water. This hampers a direct experimental investigation of the thermodynamics and the dynamics in the critical region in order to confirm, or to rule out one of the proposed scenarios, for example, the liquid-liquid phase transition and the associated second (liquid-liquid) critical point in water [2, 5].

While many methods can be used to measure the macroscopic properties of water inside biological, geological and engineering systems, experimental techniques capable of determining the structure and dynamics of water molecules under nanometer-scale

CP982, *Complex Systems, 5th International Workshop on Complex Systems*
edited by M. Tokuyama, I. Oppenheim, and H. Nishiyama

confinement are scarce. Neutron scattering is a method of choice because of the extraordinarily large neutron incoherent scattering cross section of hydrogen atoms, rendering high sensitivity to hydrogen motion unmatched by optical and x-ray spectroscopy [6]. Furthermore, judicious H-D substitution or application of high magnetic fields and neutron polarization analysis can enhance significantly the contrast between targeted hydrogen groups against the host medium for structural determination. The spatio-temporal range that neutron scattering method probes encompasses the 0.1 ~ 100 Å and 10^{-4} ~ 20 *ns* realm that matches well the length and time scale of short-to-long range order structure, molecular diffusion and atomic vibrations in water. Additionally, the measured neutron spectra, expressed as the time Fourier transform of the Intermediate Scattering Function (ISF), can be quantitatively compared with those calculated by computer Molecular Dynamics (MD) simulations or theoretical modeling.

Besides being relevant for many industrial and biological applications, water confined in nanoporous matrices and forming the hydration layer on the surface of biopolymers allow us to enter into the inaccessible temperature range for supercooled bulk water. Therefore, both the structure and dynamics of water in confined geometries have been studied using MD simulations and different experimental techniques [7]. In particular, previous neutron scattering experiments [8, 9] clearly showed that the ISF of water in vycor glass exhibits the α-relaxation at long time at a lower equivalent temperature, much the same as supercooled bulk water, as shown in an MD simulation of SPC/E model [10].

Search for the predicted [5] first-order liquid-liquid transition line and its end point, the second low-temperature critical point [1, 2] in water, has been hampered by intervention of the homogeneous nucleation process. However, by confining water in nanopores of mesoporous silica MCM-41-S with cylindrical pores of 14 Å diameter, we have been able to study the dynamical behavior of water in a temperature range down to 160 K, without crystallization. Using high-resolution QENS method and Relaxing-Cage Model (RCM) [11] for the data analyses, we determine the temperature and pressure dependencies of the average translational relaxation time $<\tau_T>$ for the confined supercooled water [12, 13, and 14].

In this article, we first summarize the neutron scattering results on water confined in 1-d cylindrical pores of MCM-41-S as functions of temperature and pressure. In particular, we identify a well-defined dynamic crossover temperature for each pressure. As a result, we are able to trace a curve in the *P-T* plane [13] which when combined with the result of an MD simulation [15] is interpreted to be the Widom line emanating from the possible liquid-liquid critical point C_{LL} long sought by researchers [1, 2]. We then give a new experimental result on a Small-Angle Neutron Scattering (SANS) measurement of the absolute density of D_2O in pores of MCM-41-S. In this experiment, we discover for the first time the existence of density minimum at 210 K besides the well-known density maximum at 284 K (for D_2O). We then show the temperature derivative of the density exhibits a pronounced peak at 235 K which gives the strong evidence of the Widom temperature at ambient pressure for D_2O. We also discuss the recent experimental result obtained for H_2O confined in hydrophobic Double-Wall Carbon Nanotube (DWNT) which shows the depression of the dynamic crossover temperature T_L by 35K as compared to that of the H_2O confined in hydrophilic MCM-41-S.

As an independent confirmation of the dynamic crossover temperatures observed by QENS technique, we also quote the results of an NMR self-diffusion constant measurement done in collaboration with the group of Professor Francesco Mallamace, Physics Department of the University of Messina, Italy. These results when plotted in the form of $\log(1/D)$ *vs.* $1/T$ give the identical crossover temperature T_L within experimental errors [16, 17].

In the second part of the article, we discuss the consequence of the dynamic crossover observed in the 2-d confined hydration water on the conformational flexibility of the biopolymers which sets the low temperature limit of biological activities. This temperature is sometimes called the glass transition temperature of biopolymers.

INCOHERENT NEUTRON SCATTERING SPECTROSCOPY

QENS and Inelastic Neutron Scattering (INS) techniques offer many advantages for the study of single particle dynamics of confined water. The main reason is that the total scattering cross section of hydrogen is much larger than that of atoms in for example silica or carbon, composed of oxygen and silicon or carbon in the confined substrates. Furthermore, the neutron scattering of hydrogen atoms is mostly incoherent so that QENS and INS spectra reflect, essentially, the self-dynamics of the hydrogen atoms in water. Combining this dominant cross section of hydrogen atoms with the use of spectrometers having different energy resolutions, we can study the molecular dynamics of water in a wide range of time-scale, encompassing picoseconds to tens of nanoseconds. In addition, investigating different Q values (Q being the magnitude of the wave vector transfer in the scattering) in the range from $0.2\,\text{Å}^{-1} \leq Q \leq 2.0\,\text{Å}^{-1}$, the spatial characteristics of water dynamics can be investigated at the sub-nanometer level.

It can be shown generally [18] that the double differential scattering cross section is proportional to the self-dynamic structure factor of hydrogen atoms $S_H(Q,E)$ through the following relation:

$$\frac{d^2\sigma_H}{d\Omega dE} = N\frac{\sigma_H}{4\pi\hbar}\frac{k_f}{k_i}S_H(Q,E) \qquad (1)$$

where $E = E_i - E_f = \hbar\omega$ is the energy transferred by the neutron to the sample; $\hbar\vec{Q} = \hbar\vec{k}_i - \hbar\vec{k}_f$, the momentum transferred in the scattering process; and N, the number of scattering centers in the scattering volume. The self-dynamic structure factor, $S_H(Q,E)$ embodies the elastic, quasielastic and inelastic scattering contributions. It can be expressed as a Fourier transform of the self-ISF of a typical hydrogen atom according to:

$$S_H(Q,E) = \frac{1}{2\pi\hbar}\int_{-\infty}^{\infty} dt e^{-i\omega t} F_H(Q,t) \quad (2)$$

$F_H(Q,t)$ is the density-density time correlation function of the tagged hydrogen atom being measured by the neutron scattering. It is, thus, the primary quantity of theoretical interest related to the experiment. It can be calculated by a model, such as the RCM, and by an MD simulation based on a phenomenological potential model of water.

Elastic Scattering ($E = 0$)

For analysis of the elastic incoherent scattering intensity from hydrogen atoms when they are bound in space, it can be shown that

$$S_H(Q,0) = B\exp\left(-Q^2 < u_H{}^2 >\right) \qquad (3)$$

where $< u_H{}^2 >$ is the projection of the mean-square displacement of the hydrogen atoms in the direction of Q vector, and B, a constant. Therefore, $< u_H{}^2 >$ can be determined experimentally by measuring the peak height of $S_H(Q,0)$ as a function of Q.

Inelastic Scattering ($E \neq 0$)

From the inelastic scattering intensity dominated by incoherent scattering from hydrogen atoms, the Q-dependant vibrational Density-Of-States (Q-DOS) of hydrogen atoms can be obtained by

$$G_H(Q,E) = \frac{2M_H}{\hbar^2}\frac{E}{n(E)+1}\left\langle \frac{\exp\left(Q^2\left\langle u_H{}^2\right\rangle\right)}{Q^2}S(Q,E)\right\rangle,$$

where M_H is the mass of hydrogen atom and $n(E)$ is the Bose-Einstein distribution function, and <...> represents the average over all observed Q values.

The true hydrogen DOS is obtained in the $Q \to 0$ limit of the $G_H(Q,E)$. In practice, $Q \to 0$ limit means $Q < 1\,\text{Å}^{-1}$ in the case of water:

$$G_H(E) = \lim_{Q=0} G_H(Q,E). \quad (4)$$

Quasielastic Scattering ($E \approx 0$)

In principle, the single-particle dynamics of bulk or confined water should include both the translational and the rotational motions of a rigid water molecule. Given the fact that in the process of QENS data analysis, we only focus our attention to ISF with $Q \leq 1.1\,\text{Å}^{-1}$, we can safely neglect the contribution of the rotational motion to the total dynamics [19], which means $F_H(Q,t) \approx F_T(Q,t)$, where $F_T(Q,t)$ is the translational part of the ISF.

Relaxing-Cage Model (RCM) of the Single-Particle Dynamics of Water

During the past several years, we have developed the RCM for the description of the translational and the rotational dynamics of water at supercooled temperatures. This model has been tested with MD simulations of SPC/E water, and has been found to be accurate. It has been used to analyze many QENS data from supercooled bulk water as well as interfacial water [20-24].

On lowering the temperature below the freezing point, around a given water molecule, there is a tendency to form a hydrogen-bonded, tetrahedrally coordinated first and second neighbor shells (cage). At short times, less than 0.05 *ps*, the center of mass of a water molecule performs vibrations inside the cage. At long times, longer than 1.0 *ps*, the cage eventually relaxes and the trapped particle can migrate through the rearrangement of a large number of particles surrounding it. Therefore, there is a strong coupling between the single particle motion and the density fluctuations of the fluid. The mathematical expression of this physical picture is the so-called RCM.

The RCM assumes that the short-time translational dynamics of the tagged (or the trapped) water molecule can be treated approximately as the motion of the center of mass in an isotropic harmonic potential well provided by the mean field generated by its neighbors. We can, then, write the short time part of the translational ISF in the Gaussian approximation, connecting it to the velocity auto-correlation function, $\langle \vec{v}_{CM}(t)\cdot\vec{v}_{CM}(0)\rangle$, in the following way:

$$F_T^s(Q,t) = \exp\left(-\frac{Q^2}{2}\langle r_{CM}^2(t)\rangle\right) = \exp\left(-Q^2\left[\int_0^t (t-\tau)\langle \vec{v}_{CM}(0)\cdot\vec{v}_{CM}(\tau)\rangle d\tau\right]\right) \quad (5)$$

Since the translational density of states, $Z_T(\omega)$, is the time Fourier transform of the normalized center of mass velocity auto-correlation function, one can express the mean squared deviation, $\langle r_{CM}^2(t)\rangle$ as follows:

$$\langle r_{CM}^2(t)\rangle = \frac{2}{3}\langle v_{CM}^2\rangle\int_0^\infty d\omega\frac{Z_T(\omega)}{\omega^2}(1-\cos\omega t) \quad (6)$$

where $\langle v_{CM}^2\rangle = \langle v_x^2\rangle + \langle v_y^2\rangle + \langle v_z^2\rangle = 3v_0^2 = 3\frac{k_B T}{M}$ is the average center of mass square velocity, and M is the mass of water molecule.

Experiments and MD results show that the translational harmonic motion of a water molecule in the cage gives rise to two peaks in $Z_T(\omega)$ at about 10 and 30~meV, respectively [25]. Thus, the following Gaussian functional form is used to represent approximately the translational part of the density of states:

$$Z_T(\omega) = \frac{(1-C)\omega^2}{\omega_1^2\sqrt{2\pi\omega_1^2}}\exp[-\frac{\omega^2}{2\omega_1^2}] + \frac{C\omega^2}{\omega_2^2\sqrt{2\pi\omega_2^2}}\exp[-\frac{\omega^2}{2\omega_2^2}] \quad (7)$$

Moreover, the fit of MD results using Eq. (7) gives $C = 0.44$, $\omega_1 = 10.8$ THz, and $\omega_2 = 42.0$ THz. Using Eqs. (6-7), we finally get an explicit expression for $F_T^s(Q,t)$:

$$F_T^s(Q,t) = \exp\left\{-Q^2 v_0^2\left[\frac{(1-C)}{\omega_1^2}\left(1-\exp(-\frac{\omega_1^2 t^2}{2})\right) + \frac{C}{\omega_2^2}\left(1-\exp(-\frac{\omega_2^2 t^2}{2})\right)\right]\right\} \quad (8)$$

Eq. (8) is the short-time behavior of the translational ISF. It starts from unity at $t = 0$ and decays rapidly to a flat plateau determined by an incoherent Debye-Waller factor $A(Q)$, given by:

$$A(Q) = \exp\left\{-Q^2 v_0^2\left[\frac{(1-C)}{\omega_1^2} + \frac{C}{\omega_2^2}\right]\right\} = \exp[-Q^2 a^2/3] \quad (9)$$

where a is the root mean square vibrational amplitude of the water molecules in the cage, in which the particle is constrained during its short-time movements. According to MD simulations, $a \approx 0.5$ Å is fairly temperature independent [10].

On the other hand, the cage relaxation at long-time can be described by the standard α-relaxation model, according to the Mode-Coupling Theory (MCT), with a stretched exponential having a structural relaxation time τ_T and a stretch exponent β. Therefore, the translational ISF, valid for the entire time range, can be written as a product of the short time part and a long time part:

$$F_T(Q,t) = F_T^s(Q,t)\exp\left[-\left(\frac{t}{\tau_T}\right)^\beta\right] \quad (10)$$

The fit of the MD generated $F_T(Q,t)$ using Eq. (10) shows that τ_T is Q-dependent, obeying the power-law:

$$\tau_T = \tau_0(aQ)^{-\gamma} \quad (11)$$

where γ is ≤ 2, with a slight dependency on Q, and $\beta < 1$ is slightly Q dependent as well. In the $Q \to 0$ limit, one should approach the diffusion limit, where $\gamma \to 2$ and $\beta \to 1$. Thus the translational ISF can be written as: $F_T(Q,t) = \exp[-DQ^2 t]$, D being the self-diffusion coefficient. In QENS experiments, this low Q limit is not usually reached, and both β and γ can be considered Q-independent in the limited Q range of $0 \leq Q \leq 1$ [22 and 23].

We define a Q-independent average translational relaxation time

$$\langle \tau_T\rangle = (\tau_0/\beta)\Gamma(1/\beta) \quad (12)$$

which is a convenient quantity to be extracted from the experimental data by the fitting process of RCM. This quantity can be identified to be proportional to the α-relaxation time which dominates the long-time decay of the ISF in low temperature water. Combining Eqs. (1), (8), and (10), we can calculate the theoretical values of $S_H(Q,\omega)$ and compare it directly with its experimental spectral data.

In actual QENS experiment, we have to take into account the signal coming from the bound hydrogen atoms in silanol ($Si(OH)_4$) on the pore surfaces of the silica sample. Denoting the fraction of the elastic scattering coming from the bound hydrogen atom by p we can analyze the experimental data according to the following model:

$$S(Q,\omega) = pR(Q_0,\omega) + (1-p)FT\{F_H(Q,t)R(Q_0,t)\} \quad (13)$$

where $F_H(Q,t) \approx F_T(Q,t)$ is the ISF of hydrogen atoms which defines the quasielastic scattering, $R(Q_0,t)$ is the experimental resolution function, and the symbol FT denotes the Fourier transform from time t to frequency ω.

MESOPOROUS SILICA SAMPLES

Micellar templated mesoporous silica matrices MCM-41-S have 1-d cylindrical pores arranged in 2-d hexagonal arrays. Similar to synthesizing MCM-48-S [26], to make the mesoporous materials together with short chain surfactant (C_{12}TMAB), we employed small quaternary ammonium ions, TEAOH, to separately develop a zeolitic nanocluster as the silica precursor. In this way, we will get MCM-41-S with smaller pore sizes and stronger silica walls than traditional ways [27, 28].

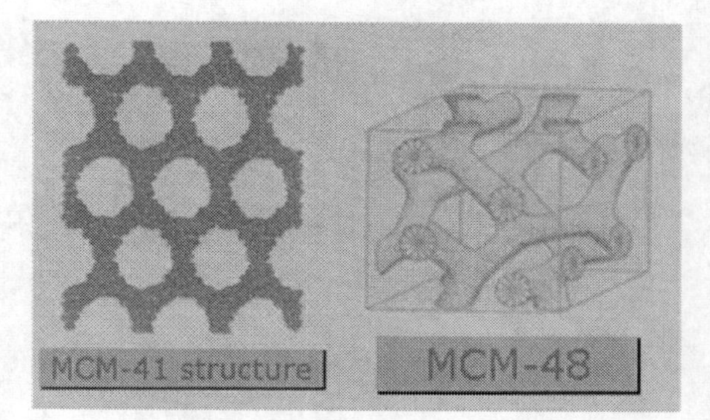

FIGURE 1. Microscopic structures of MCM-41-Sand MCM-48-S

Characterization of Silica Samples

The synthesized samples were characterized using X-Ray powder Diffraction (XRD), nitrogen absorption–desorption, and Differential Scanning Calorimetry (DSC).

XRD were acquired with a Scintag X1 diffractometer, using CuKα (λ=0.154 nm). The XRD patterns of the samples show that the MCM-41-S had hexagonal (P6mm) symmetry, as shown in Figure 1. All the samples exhibited high hydrothermal stability.

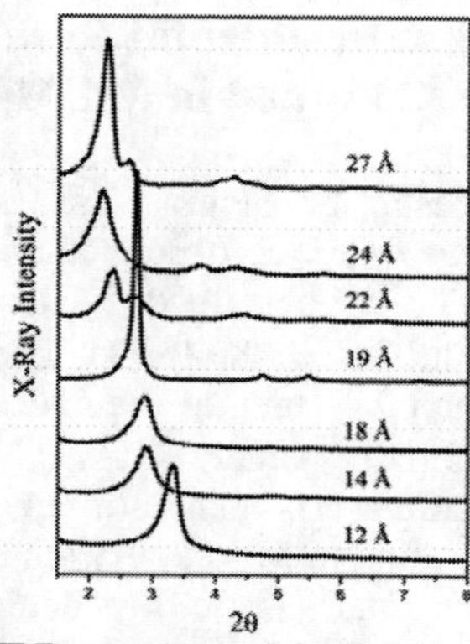

FIGURE 2. XRD patterns of MCM-41-S samples with different pore sizes taken at room temperature and ambient pressure. The sharp XRD peaks show that the samples are well-ordered. The (1 0 0) peak (the highest peak) is related to the d-spacing according to the rule: $2d\sin(\theta) = \lambda$, where $\lambda = 1.54\,\text{Å}^{-1}$ and θ is in degree. The number above each curve indicates the average pore diameter of that sample.

The melting/freezing behavior of water in the samples (fully hydrated) was checked by DSC measurements (Figure 3). According to the Gibbs–Thomson equation, the melting point of a small crystal is proportional to the crystal size, which, in this case, is equal to the pore size of the material. Thus, one expects that the liquid state of water would persist to very low temperature if the pore size can be decreased further. In Figure 3, the DSC curve shows the melting points which are specified by the temperatures at the positions of the inverted peaks. For samples with pore size 19Å, we do see a small peak near 0°C which is due to the water outside the nanochannel (unconfined). For samples having pore sizes 18Å, we do not see any abrupt melting transition near 0°C, indicating that there is no water residing outside the channel. However, we seem to see a gradual change of enthalpy from −100°C to −50°C. This could be due to some second-order transition or glass transition.

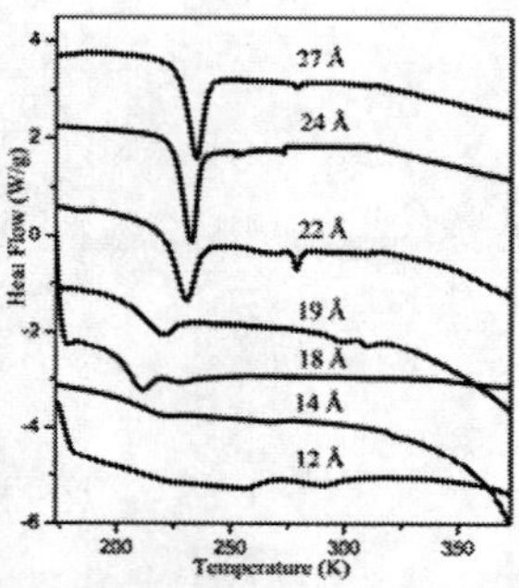

FIGURE 3. DSC curves of water inside MCM-41-S samples having different pore sizes (indicated by the numbers above the curves). The sharp negative-going peaks signal the freezing temperatures. It is noted that the samples with pore size ≤ 18Å do not show an obvious freezing peak.

QENS Experimental Results of Water Confined in MCM-41-S

The quasielastic broadening has been analyzed according to the RCM as described in the previous section. $F_H(Q,T)$ is described in terms of seven parameters (from Eqs. 1, 8, 10, 11, and 13): C, ω_1, ω_2, τ_0, γ, β and p. Three of them are related to the short-time dynamics, namely C, ω_1, and ω_2. The short-time dynamics is not strongly temperature dependent, according to MD simulation results. On the other hand, the quasielastic broadening is mostly determined by the long-time dynamics. Therefore, the values of C, ω_1, and ω_2 are fixed according to the MD simulation results [14].

The remaining four parameters, namely p, τ_0, γ, and β, can then be determined from the analysis of a group of low-Q QENS spectra. We report the results of our analysis with the values of p fixed to their plateau values. We show in Figure 4, as an example, two sets (temperature series) of QENS area-normalized spectra taken at HFBS and DCS spectrometers. As can be seen, RCM analysis agrees well with experimental spectra in all investigated cases. It is remarkable that using four parameters we were able to reproduce the data from 9 (DCS), and 7 (HFBS) constant angle spectra. On the other hand, it is clearly shown in the figure that the quasielastic broadening is strongly temperature dependent. The width of the peak is progressively sharpened as temperature is lowered. Thus for T ≤ 240K, the 0.8 μeV resolution of HFBS is necessary to obtain useful data.

The fitting of the data allowed us to extract parameters describing the translational dynamics of water and calculate the average translational relaxation time, $\langle \tau_T \rangle$, which shows the FSC phenomenon and that is the central result of QENS experiments.

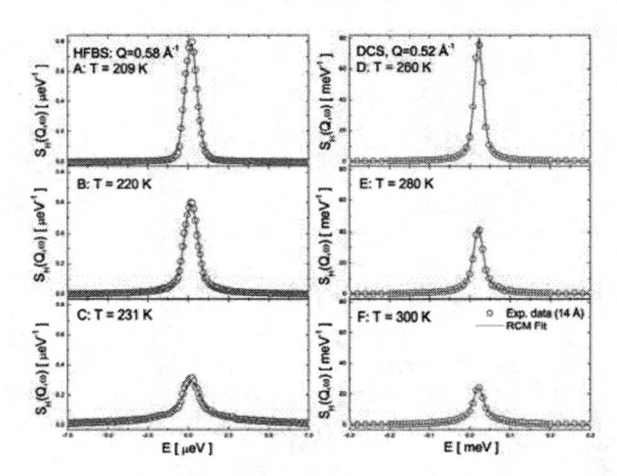

FIGURE 4. (Color online) These typical QENS spectra of hydrated MCM-41-S-14 sample show that the RCM analysis agrees well with experimental data. The left-hand panels show the spectra taken from HFBS with resolution of 0.8 μeV. The right-hand panels are the data taken from DCS with resolution of 20 μeV. The solid circles are the experimental data; the continuous lines, the RCM fit.

Pressure dependency of FSC phenomena in confined water

We have shown previously that at ambient pressure, for fully hydrated MCM-41-S of pore size ≤ 18Å, $\langle \tau_T \rangle$ exhibits a FSC with the same crossover temperature T_L = 225K within error bars [24]. This shows that FSC is independent of pore size within this range from 18Å down to 10Å.

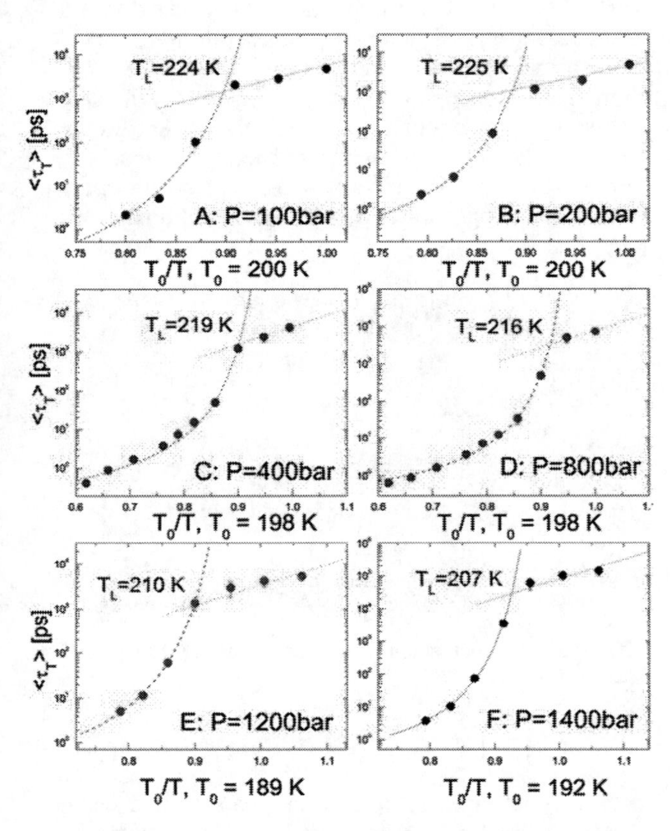

FIGURE 5. (Color online) Summary of the data from pressure range where a well-defined FSC is observed. Temperature dependence of $\langle \tau_T \rangle$ plotted in $\log(\langle \tau_T \rangle)$ *vs.* T_0/T scale. Data from 100, 200, 400, 800, 1200, 1400 bars are shown in panels A, B, C, D, E, and F, respectively.

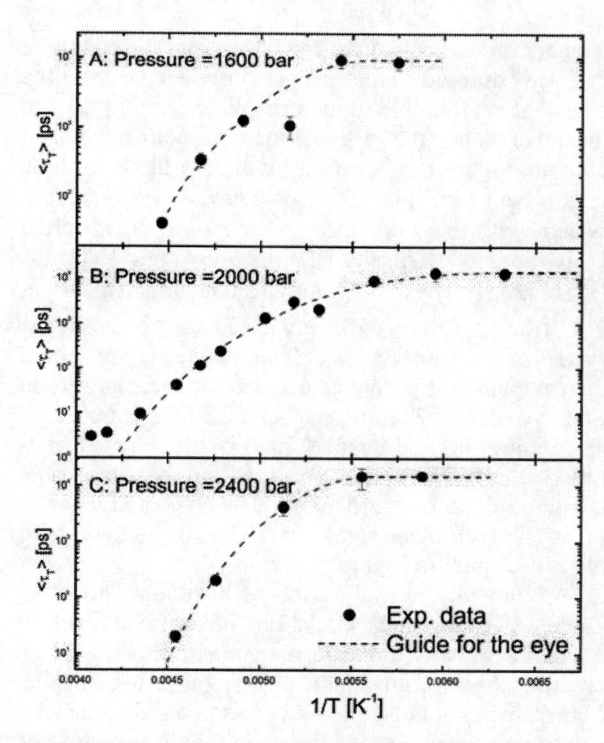

FIGURE 6. (Color online) Summary of the data from pressure range where no well-defined FSC is observed. Temperature dependence of $\langle \tau_T \rangle$ plotted in $\log(\langle \tau_T \rangle)$ *vs.* $1/T$ scale. Data from 1600, 2000, and 2400 bars are shown in panels A, B, and C, respectively.

In Figures 5 and 6, we exhibit the temperature variation of $\langle \tau_T \rangle$ for water molecules as a function of pressure. It is seen that the panels in Figure 5 all show clearly a FSC from a Vogel-Fulcher-Tammann (VFT) law to an Arrhenius law. This crossover, is the signature of a FSC predicted by Ito et al [38], and now extends into finite pressures. The crossover temperature, T_L, as the crossing point of the VFT law and the Arrhenius law, is calculated by $1/T_L = 1/T_0 - (Dk_B)/E_A$. However, in Figure 6, the cusp-like transition becomes rounded off and there is no clear-cut way of defining the FSC temperature. Note that while we have done more measurements at high temperatures at 2000 bar pressure, shown in Figure 6B, there is still a hint of fragile behavior at high enough temperature.

The Liquid-Liquid Coexistence Line and the Associated Widom Line

According to the liquid-liquid phase transition hypothesis [5], to explain the anomalies of thermodynamic and transport properties of supercooled water, one postulates the existence of a first order phase transition line between two phases of liquid water: a Low-Density Liquid (LDL) and a High-Density Liquid (HDL). This line is called Liquid-Liquid (L-L) coexistence line and terminates at an L-L critical point T_C. It is to be noted that if the state point is on the L-L coexistence line, one has a two-phase liquid consisting of a mixture of HDL and LDL, just as the gas and liquid phases are coexisting on the liquid-gas coexistence line (refer to the $p-\rho$ diagram of superheated water, Fig. 2.2 of Ref. [29]). The L-L coexistence line extends into the one-phase region after terminating at the critical point [30]. This extended line, not being real, is the so-called Widom line or the critical isochore. The Widom line is defined as a straight line in the Pressure-Temperature (*P-T*) plane, starting from the critical point C$^{|}$(P_C, T_C) and extending into the one-phase region, with the same slope as that of the L-L coexistence line at (P_C, T_C). Even though this line is an imaginary line, experiments of superheated water show that many thermodynamic quantities and transport coefficients, such as the isothermal compressibility, thermal-expansion coefficient, isobaric specific heat capacity, isochoric specific heat capacity, speed of sound, thermal conductivity, shear viscosity, and thermal diffusivity [29], show a peak when crossing the Widom line at a constant pressure.

Summarizing all the experimental results of fully hydrated MCM-41-S-14 under pressure, we show in a *P-T* plane, in Figure 7 [13], the observed pressure dependence of T_L and its estimated continuation, denoted by a dash line, in the pressure region where no clear-cut FSC is observed. One should note that the T_L line has a negative slope, parallel to TMD line, indicating a lower density liquid on the lower temperature side. This T_L line also approximately tracks the T_H line, and terminates in the upper end when intersecting the T_H line at 1600 bar and 200K, at which point the character of the dynamic transition changes. We shall discuss the significance of this point next.

Since T_L determined experimentally is a dynamic crossover temperature, it is natural to question whether the system is in a liquid state on both sides of the T_L, and if so, what would the nature of the high-temperature and low-temperature liquids be? Sastry and Angell have recently shown by an MD simulation that at a temperature $T \approx 1060$K (at zero pressure), below the freezing point 1685K, the supercooled liquid silicon undergoes a first-order liquid-liquid phase transition, from a fragile, dense liquid to a strong, low-density liquid with nearly tetrahedral local coordination [31]. Prompted by this finding, we may like to relate, in some way, our observed T_L line to the L-L transition line, predicted by MD simulations of water [5] and speculating on the possible location of the low-temperature critical point.

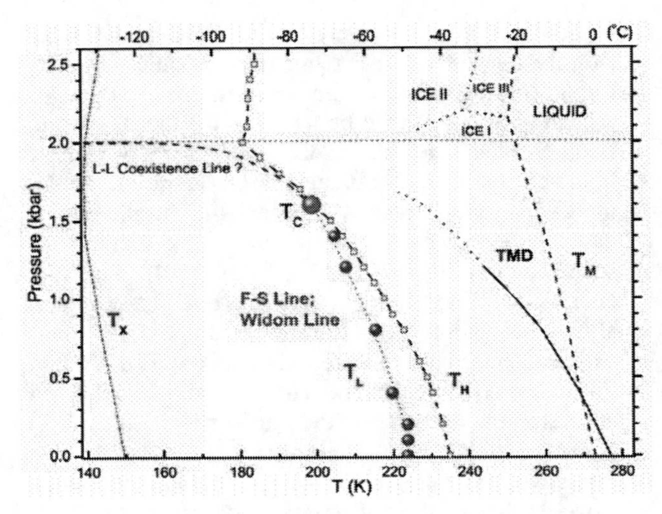

FIGURE 7. (Color online) The pressure dependence of the measured FSC temperature, T_L, plotted in the P-T plane (solid circles). Also shown are the homogeneous nucleation temperature line, denoted as T_H, crystallization temperatures of amorphous solid water, denoted as T_X, and the temperature of maximum density line, denoted as TMD, taken from known phase diagram of bulk water.

According to our INS experiments [24], water remains in disordered liquid state both above and below the FSC at ambient pressure. Furthermore, our analysis of the FSC for the case of ambient pressure indicates that the activation energy barrier for initiating the local structural relaxation is E_A = 4.89 Kcal/mol = 20.47 KJ/mol for the low-temperature strong liquid. Yet, previous INS experiments of stretch vibrational band of water [35] indicate that the effective activation energy of breaking a hydrogen bond at 258K (high-temperature fragile liquid) is 3.2 Kcal/mol = 13.40 KJ/mol. Therefore, it is reasonable to conclude that the high-temperature liquid corresponds to the high-density liquid (HDL) where the locally tetrahedrally coordinated hydrogen bond network is not fully developed, while the low-temperature liquid corresponds to the low-density liquid (LDL) where the more open, locally ice-like hydrogen bond network is fully developed [36].

It is appropriate now to address the possible location of the second critical point [5]. Above the critical temperature T_C and below the critical pressure P_C, we expect to find a one-phase liquid with a density ρ, which is constrained to satisfy an equation of state: $\rho = f(p,T)$. If an experiment is done by varying temperature T at a constant pressure $p < p_C$, ρ will change from a high-density value (corresponding to HDL) at sufficiently high temperature to a low-density value (corresponding to LDL) at sufficiently low temperature. Since the fragile behavior is associated with HDL and the strong behavior with LDL, we should expect to see a clear FSC as we lower the temperature at this constant p. Therefore, the cusp-like FSC we observed should then occur when we cross the so-called Widom line in the one-phase region [29]. On the other hand, if the experiment is performed in a pressure range $p > p_C$, corresponding to the two-phase region and crossing the L-L coexistence line, the system will be consisting of mixture of different proportions of HDL and LDL as one varies T. In this latter case, $\langle \tau_T \rangle$ *vs.* $1/T$ plot will not show a clear-cut FSC (the transition will be washed out) because the system is in a mixed state. The above picture would then explain the dynamical behavior we showed in Figs. 6 and 7. In Figs. 6 and 7, a clear FSC is observed up to 1400 bar and beyond 1600 bar the crossover is rounded off. From this observation, the reasonable location of the L-L critical point is estimated to be at $P_C = 1500 \pm 100$ bar and $T_C = 200 \pm 10$K, shown by a big round point in Figure 7.

Additionally, in a recent MD simulation using TIP5P, ST2, and Jagla Model Potential of Xu et al [15], a small peak was found in the specific heat C_P when crossing the Widom line at a constant p. Meanwhile, Maruyama et al conducted an experiment on adiabatic calorimetry of water confined within nano-pores of silica gel [37]. It was found that water within 30Å pores was well prevented from crystallization, and also showed a small C_P peak at 227K at ambient pressure. This experimental result further supports that the FSC we observed at 225K at ambient pressure is caused by the crossing of the Widom line in the one-phase region above the critical point [15].

Observation of Density Minimum in D_2O

Very recently, using SANS method, we discovered that there exists a density minimum in MCM-41-S confined water (D_2O) at 210K besides the well-known density maximum at 284K [38]. Figure 8 is taken from a more recent paper [39] which shows the temperature dependence of D_2O density from 300K down to 140K. The important result comes out of the plot of $(\partial\rho/\partial T)_P$ *vs.* T shown in Fig. 8 (B). This quantity, which is proportional to the thermal expansivity, exhibits a pronounced peak at T_L=235K which can be identified as the Widom temperature of D_2O at ambient pressure. This fact further strengthens the argument given in last section from the FSC phenomenon in favor of the existence of the Widom line in supercooled confined water.

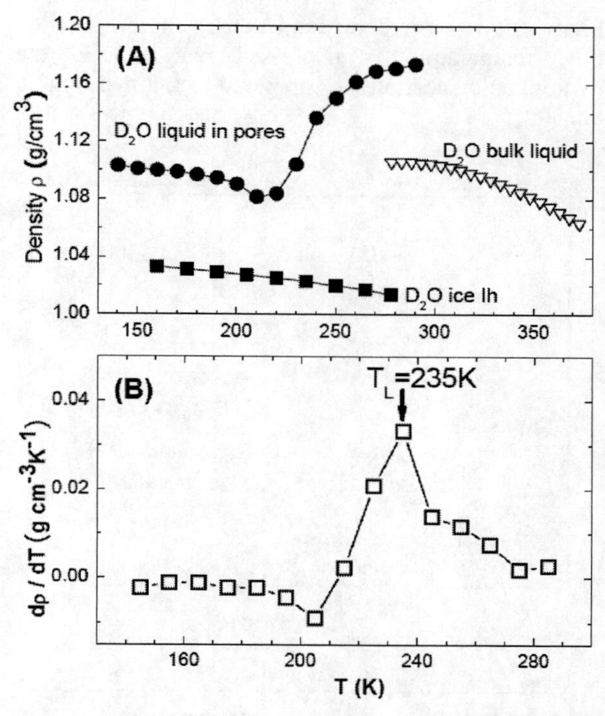

FIGURE 8. (A): Average D_2O density inside the 19Å pore MCM-41-S measured by SANS method as a function of temperature. The open triangular symbols are the density data for bulk D_2O taken from CRC handbook. It shows that the density of confined D_2O is 8% higher than bulk D_2O at room temperature. (B): The curve of $(\partial\rho/\partial T)_P$ *vs.* T shows a peak at T_L=235 K.

The Violation of Stokes-Einstein Relation

Another dynamical parameter can be used to show the FSC phenomena in confined water, namely, the self-diffusion coefficient D [15]. D can be measured by NMR method for the confined water [16]. Top panel of Figure 9 shows a log($1/D$) *vs.* 1/T plot which nicely exhibit the FSC phenomenon similar to the QENS results shown in bottom panel of Figure 9.

Since we experimentally measure independent D and α-relaxation time $\langle\tau_T\rangle$ which is proportional to the shear viscosity η, we should be able to check the validity of the well-known Stokes-Einstein Relation (SER) which is $D = k_B T / 6\pi\eta R$. If SER is valid, then $D\langle\tau_T\rangle/T$ should be constant in temperature. Figure 10 gives such a plot.

It is interesting to note that the SER starts to break down at the crossover temperature T_L=225K. The more striking plot can be made by using the scaling plot proposed by Jung, Garrahan, and Chandler [41]. This plot shows that the scaling exponent ξ abruptly changes from a value 0.74 to 2/3 at the FSC temperature agreeing with the prediction.

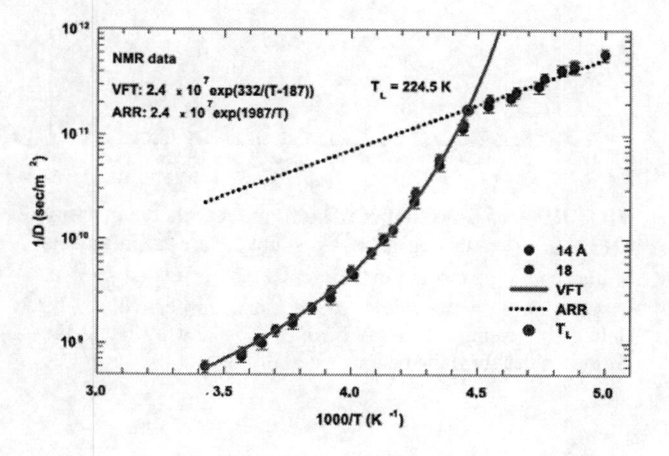

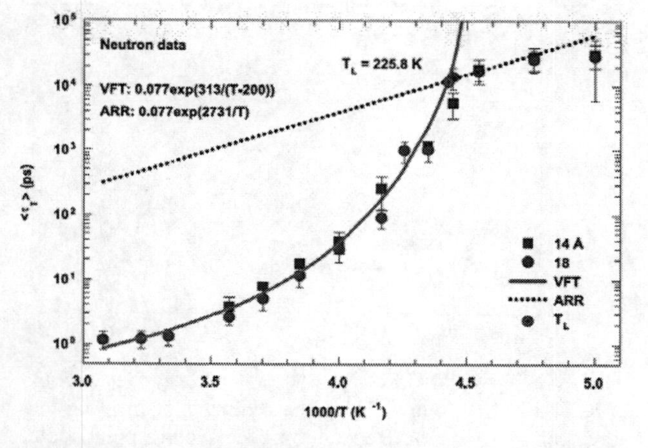

FIGURE 9. (Color online) Upper panel shows, for the fully hydrated MCM-41-S samples with diameter 14Å and 18Å, the inverse of the self-diffusion coefficient of water measured by NMR as a function of $1/T$ in a log-linear scale. The solid line denotes the fit of the data to a VFT relation. The short dotted line denotes the fit to an Arrhenius law with the same prefactor $1/D_0$. Lower panel reports, as a function of $1/T$, the average translational relaxation time $\langle\tau_T\rangle$ obtained from QENS spectra in the same experimental conditions of the NMR experiment. The dashed line denotes the VFT law fit, and the dotted line the Arrhenius law with the same prefactor. In both panels, fitting parameters are shown.

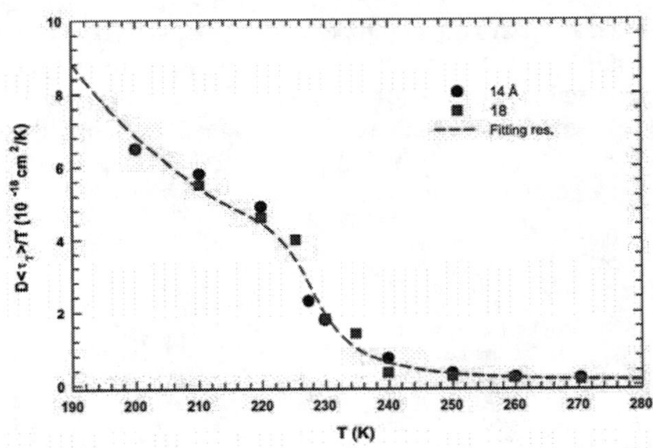

FIGURE 10. (Color online) The figure reports the quantity $D\langle\tau_T\rangle/T$ as a function of T. Dots and squares represent its values coming from the experimental data of D and $\langle\tau_T\rangle$ in samples with diameter 14Å and 18Å, respectively. The dotted line represents same quantity obtained by using the fitting values from the data reported in Fig. 9.

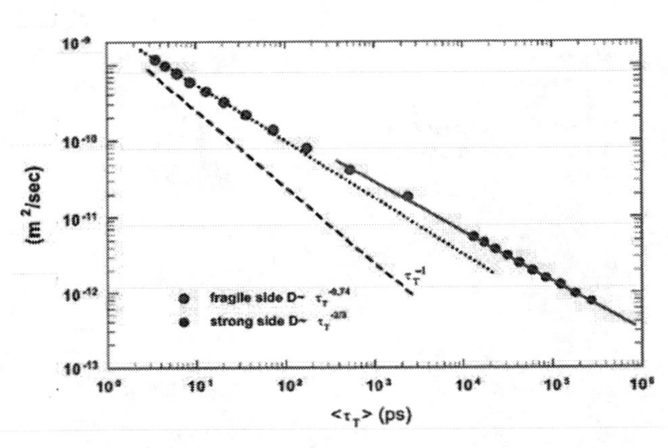

FIGURE 11. (Color online) The scaling plot in a log-log scale of D *vs.* $\langle\tau_T\rangle$. Red dots are data corresponding to temperatures above T_L, i.e. when water is in the fragile glass phase, whereas blue dots correspond to the strong Arrhenius region. Two different scaling behaviors exist above and below the temperature of the Fragile-to-Strong Transition (FST). In fragile regime, the scaling exponent is $\xi \sim 0.74$ (dotted line) and ~2/3 in the strong side (solid line). Dashed line represents the situation in which the SER holds, $D \sim \tau^{-1}$.

Difference in the Crossover Temperature Between Hydrophilic and Hydrophobic Confinement Substrates

One can use the model systems: MCM-41-S (hydrophilic substrate) and Double-Wall Carbon Nanotubes (hydrophobic substrate) to check the dependency of the FSC temperature on the degree of hydrophilicity of the surface of the confining matrices.

Figure 12 is an experimental demonstration that the FSC temperature is depressed by 35K in the hydrophobic substrate as compared to the hydrophilic one [42].

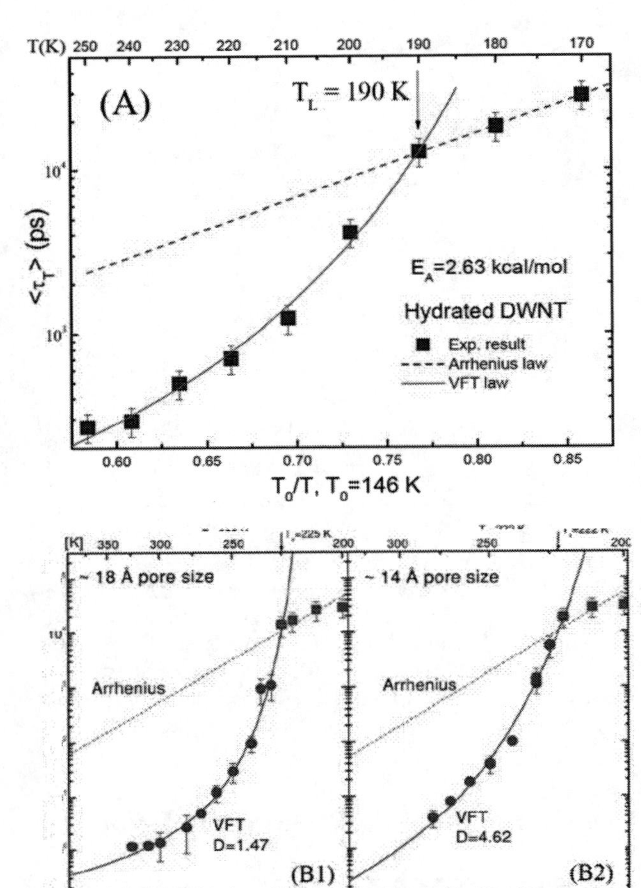

FIGURE 12. (Color online) Panel (A) shows $\langle\tau_T\rangle$ from fitting of the quasielastic spectra of water confined in DWNT (inner diameter 16 Å) by RCM plotted in a log scale against $1/T$. It shows a well-defined cusp-like dynamic crossover behavior occurring at T_L=190K. The solid line represents fitted curves using the VFT law, while the dashed line is the fitting according to the Arrhenius law. For comparison, panels (B1) and (B2) show the results of our previous experiments on supercooled water confined in porous silica material MCM-41-S with two different pore sizes, which show that the crossover temperature T_L is insensitive to confinement pore sizes. From the results shown in the upper and lower panels, we estimate that the water confined in a hydrophobic substrate (DWNT) has a lower dynamic crossover temperature by $\Delta T_L \approx 35$K, as compare to that in a hydrophilic substrate (MCM-41-S).

Appearance of Boson Peak Below the Crossover Temperature

The INS technique also allows us to observe the FSC phenomena through the appearance of a Boson peak. It is seen that this peak starts to appear at a temperature of 220K, which is near the crossover temperature 225K for the confined water [43].

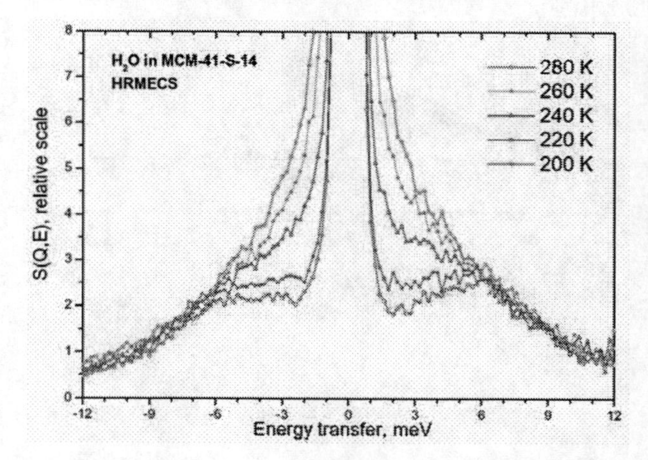

FIGURE 13. (Color online) Self-dynamic structure factor $S(Q,E)$ is obtained at HRMECS at IPNS. The plot shows $S(Q,E)$ *vs.* E summed over a range of Q values. It is seen that the Boson peak at E=6meV is visible at 220K and 200K. Above 220K, the peak gradually merges into the quasielastic peak.

HYDRATION WATER IN BIOPOLYMERS

Water molecules in a protein solution may be classified into three categories: (i) the bound internal water; (ii) the surface water, i.e., the water molecules that interact with the protein surface strongly; and (iii) the bulk water. The bound internal water molecules, which occupy internal cavities and deep clefts, are extensively involved in the protein-solvent H-bonding and play a structural role in the folded protein itself. The surface water, which is usually called the hydration water, is the first layer of water that interacts with the solvent-exposed protein atoms of different chemical character, feels the topology and roughness of the protein surface, and exhibits the slow dynamics. Finally, water that is not in direct contact with the protein surface but continuously exchanges with the surface water has properties approaching that of bulk water. The hydration water is believed to have an important role in controlling the biofunctionality of the protein. In this article, we shall present some neutron scattering results of hydrated protein powder. In this case, the hydration water represents the water in category (i) and (ii) mentioned above.

Functions of many globular proteins generally show sharp slowing-down around the temperatures between 200 and 240 K [44, 45]. Experimental [46-48] and computational [49, 50, 58] results show a sharp increase of $< x^2 >$ of hydrogen atoms in proteins at about T_C = 220 K, which suggests that the dynamic transition (sometime called the glass transition) may be occurring in the proteins at this temperature. There is strong evidence that this dynamic transition of protein is solvent-induced, since the hydration water of a protein also shows a kind of dynamic transition around similar temperature [51-53].

In our previous research [54], it was demonstrated that this dynamic transition of hydration water on lysozyme protein was, in fact, a FSC at 220 K. We also showed that the hydration water on B-DNA exhibits a similar FSC at T_L = 222 K [55]. More recently, we have studied a case of hydration water in RNA, which shows the FSC at a similar temperature, T_L = 220 K [56]. These results suggest the universality of T_C in biopolymers in spite of the difference in chemical backbone structure of proteins, DNA, and RNA.

FSC Phenomena in Hydration Water of Biopolymers

QENS studies have been made on hydration water of lysozyme, DNA, and RNA, respectively.

Figure 14 shows the log$\langle \tau_T \rangle$ *vs.* 1/*T* plot of the average translation relaxation time $\langle \tau_T \rangle$ of hydration water in each case. At high temperatures, above $T_L \approx 220$ K, $\langle \tau_T \rangle$ obeys the VFT law, namely, $\langle \tau_T \rangle = \tau_0 \exp[DT_0/(T-T_0)]$, where D is a dimensionless parameter providing the measure of fragility and T_0 is the ideal glass transition temperature. Below T_L, the temperature dependence of $\langle \tau_T \rangle$ switches to an Arrhenius behavior which is written as $\langle \tau_T \rangle = \tau_0 \exp(E_A/RT)$, where E_A is the activation energy for the relaxation process and R is the gas constant. This dynamic crossover from a super-Arrhenius (the VFT law) to the Arrhenius behaviors is cusp-like and thus it sharply defines the crossover temperature T_L, much more sharply than that indicated by the Mean Squared atomic Displacement (MSD) $\langle x^2_{H_2O} \rangle$, shown in the figure 15.

Recently, Kumar et al. has shown by MD simulations that protein glass transition occurs at the temperature of dynamic crossover in the diffusivity of hydration water and also coincides with the maxima of the isobaric specific heat C_P [58].

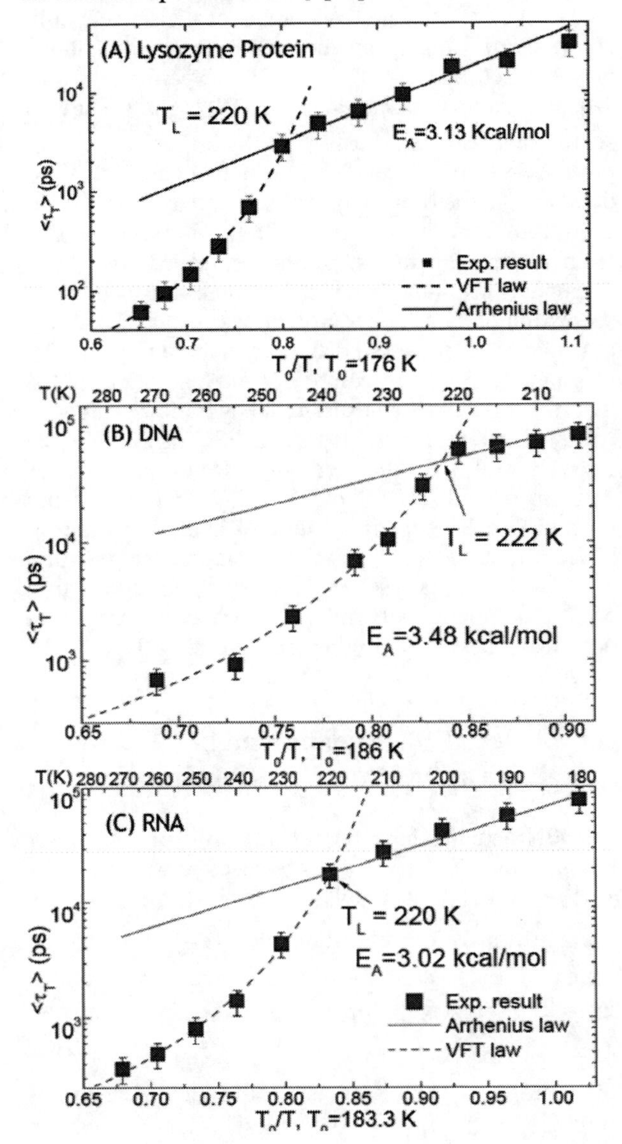

FIGURE 14. The extracted $<\tau_T>$ from fitting of QENS spectra by RCM plotted in the log scale *vs.* $1/T$. Panels (A), (B), and (C) show clearly well defined cusp-like dynamic crossover behavior in each case. The dashed line represents fitted curves using the VFT law, while the solid line is the fitting according to the Arrhenius law. FSC temperatures are respectively, (A) 220 K, hydration water in lysozyme; (B) 222 K, hydration water in DNA; and (C) 220 K, hydration water in RNA [54-56]. All three temperatures are essentially the same within the experimental error of 5 K suggesting universality of T_L in hydration water of all biopolymers.

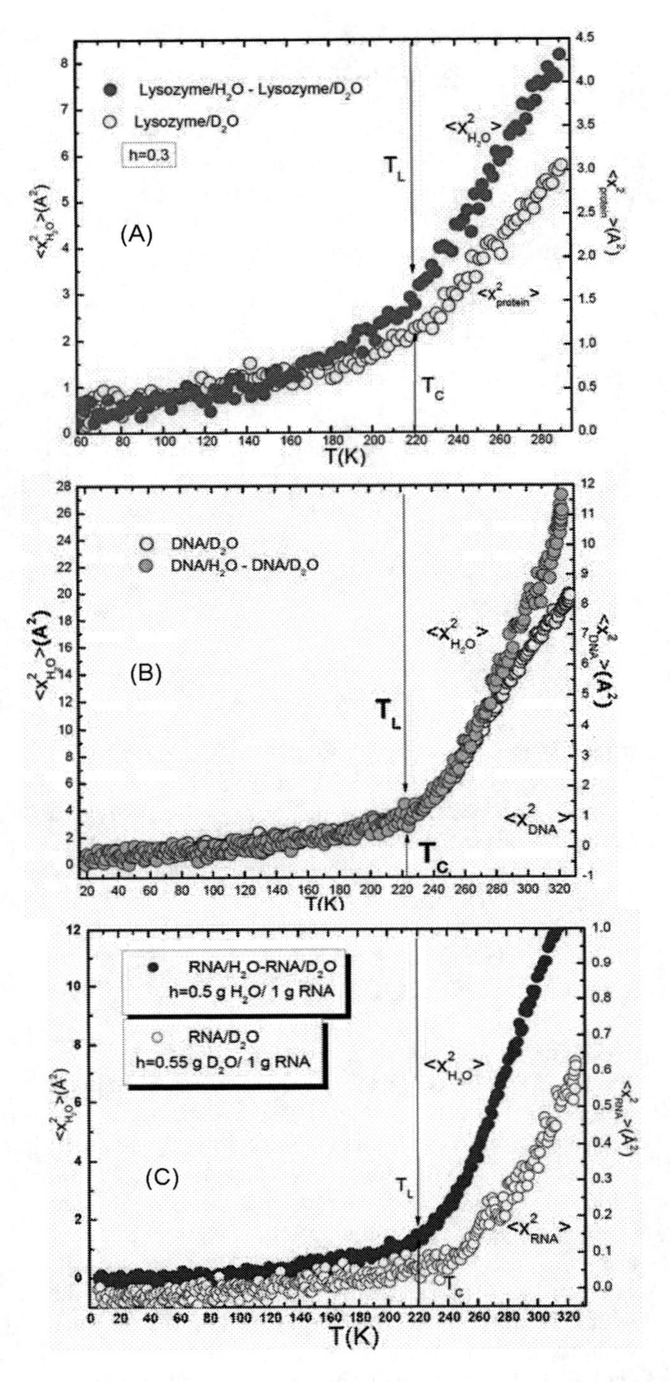

FIGURE 15. (Color online) Each panel shows the temperature dependence of the MSD of hydrogen atoms in both the biopolymer and its hydration water, respectively. (A): MSD of hydrated lysozyme; (B): MSD of the hydrated DNA; (C): MSD of the hydrated RNA. The arrow signs indicate the approximate positions of the crossover temperature in both the biopolymer (T_C) and its hydration water (T_L). Note that the scale on the left hand side is for MSD of the hydration water and that on the right hand side is for the biopolymer.

MSD of Hydrogen Atoms in Hydration Water and in Biopolymers

Figure 15 shows the MSD of hydrogen atoms in the hydration water and in three biopolymers, respectively, lysozyme, DNA, and RNA.

It can be seen that all the MSD's show a linear behavior at lower temperatures, but above certain temperature, the slope abruptly increases. This change in slope, thus, defines a crossover temperature T_L.

It should be noted that the crossover temperatures in each biopolymer and its hydration water are approximately the same, implying that the dynamic crossover in a biopolymer is triggered by the dynamic crossover in its hydration water. Since the crossover temperatures in all the hydration water are the same, this leads to a conclusion that the so-called glass transition temperature in all biopolymers is universal.

Possibility that Hydration Water Drives the Dynamic Transition in Biopolymer

We have shown before that above T_L the structure of hydration water is predominately in its high-density form (HDL), which is more fluid. But below T_L, it transforms to predominately the low-density form (LDL), which is less fluid. The concepts of the existence of HDL and LDL in deeply supercooled water have been extensively discussed in the recent literature [2, 57]. Thus this abrupt change in the mobility of hydration water apparently induces the change in the energy landscape of biopolymer, which causes the dynamic transition (or the glass transition) in the biopolymer.

The coincidence of the transition temperature in proteins [54], DNA [55], tRNA [57], and RNA [56] reinforce the plausibility that the dynamic transitions are not the intrinsic properties of the biomolecules themselves but are imposed by the hydration water on their surfaces.

SUMMARY

In this article, we describe the application of incoherent neutron scattering and NMR spectroscopy methods to investigate the FSC phenomenon in 1-d confined supercooled water. We discuss the cases of water confined in nanometer-scale porous silica materials MCM-41-S and double-wall carbon nanotubes.

We discuss the plausibility that the FSC phenomenon in supercooled water arises from crossing of the Widom line emanating from the existence of a liquid-liquid critical point at $P_C \approx 1500$ bars and $T_C \approx 200$ K. The first evidence comes from an MD simulation [15], and the second one from the prominent peak shown by a $(\partial\rho/\partial T)_P$ *vs.* T plot in our recent experiment (Fig. 8 (B)) [39]. Consequences of the FSC phenomenon include the breakdown of the Stokes-Einstein Relation at the crossover temperature T_L and the appearance of Boson peak at and below T_L. Finally, we show that the FSC temperature depends sensitively on the hydrophilicity of the confining surfaces.

We also briefly discuss the FSC phenomenon in hydration water (a 2-d confinement) of three biopolymers. We suggest that the FSC of hydration water in a biopolymer triggers the dynamic transition in the biopolymer itself.

ACKNOWLEDGMENTS

The research at MIT is supported by DOE Grants DEFG02-90ER45429 and 2113-MIT-DOE-591. Work at ANL was performed under the auspices of the US DOE-BES, Contract No. DE-AC02-06CH11357. This work utilized facilities supported in part by the National Science Foundation under Agreement No. DMR-0454672. The identification of any commercial product or trade name does not imply endorsement of recommendation by the National Institute of Standards and Technology (NIST). We benefited from affiliation with EU-Marie-Curie Research and Training Network on Arrested Matter.

REFERENCES

1. P. G. Debenedetti and H. E. Stanley, *Phys. Today* **56**, 40-46 (2003).
2. P. G. Debenedetti, *J. Phys.: Condes. Matter* **15**, R1669-R1726 (2003).
3. R. J. Speedy and C. A. Angell, *J. Chem. Phys* **65**, 851-858 (1976).
4. D. H. Rasmussen and A. P. MacKenzie, "Interactions in the water-polyvinylpyrrolidone system at low temperatures" in *Water Structure at the Water-Polymer Interface*, edited H.H. Jellinek, Plenum Press, NY, 1972, 126
5. P. H. Poole, F. Sciortino, U. Essmann, and H. E. Stanley, *Nature (London)* **360**, 324-328 (1992).
6. K. Skosld and D. L. Price, *Methods of Experimental Physics Vol. A-C*, London: Academic Press, 1986.
7. Proceeding of the 2nd International Workshop on Dynamics in Confiement, organized by B. Frick, M. Koza, R. Zorn, ILL, France, *European Phys. J. E* **12**, no. 1 (2003).
8. M. -C. Bellissent-Funnel, S. Longeville, J. -M. Zanotti, and S. -H. Chen, *Phys. Rev. Lett.* **85**, 3644-3647 (2000).
9. J. -M. Zanotti, M. -C. Bellissent-Funnel, and S. -H. Chen, *Phys. Rev. E.* **59**, 3084-3093 (1999).
10. P. Gallo, F. Sciortino, P. Tartaglia, and S. -H. Chen, *Phys. Rev. Lett* **76**, 2730-2733 (1996).

11. S. -H. Chen, C. Liao, F. Sciortino, P. Gallo, and P. Tartaglia, *Phys. Rev. E* **59**, 6708-6714 (1999).
12. A. Faraone, L. Liu, C. -Y. Mou, C. -W. Yen, and S. -H. Chen, *J. Chem. Phys.* **121**, 10843-10846 (2004).
13. L. Liu, S. -H. Chen, A. Faraone, C. -W. Yen, and C. -Y. Mou, *Phys. Rev. Lett.* **95**, 117802-117802 (2005).
14. L. Liu, S. -H. Chen, A. Faraone, C.-W. Yen, C.-Y. Mou, A. I. Kolesnikov, E. Mamontov, and J. Leao, *J. Phys.: Cond. Matter* **18**, S2261-S2284 (2006).
15. L. Xu, P. Kumar, S.V. Buldyrev, S.H. Chen, P.H. Poole, F. Sciortino, and H.E. Stanley, *PNAS USA* **102**, 16558-16562 (2005).
16. F. Mallamace, M. Broccio, C. Corsaro, A. Faraone, U. Wanderlingh, L. Liu, C. -Y. Mou, and S. -H. Chen, *J. Chem. Phys* **124**, 161102-161105 (2006).
17. F. Mallamace, S. -H. Chen, M. Broccio, C. Corsaro, V. Crupi, D. Majolino, V. Venuti, P. Baglioni, E. Fratini, and C. Vannucci, and H. E. Stanley, *J. Chem. Phys.* **127**, 045104-045109 (2007).
18. *Spectroscopy in Biology and Chemistry: Neutron, X-ray, Laser*, edited by S. -H. Chen and S. Yip, London: Academic Press, 1974.
19. S. -H. Chen, "Quasi-Elastic and Inelastic Neutron Scattering and Molecular Dynamics of Water at Supercooled Temperature" in *Hydrogen Bonded Liquids*, edited J. C. Dore and J. Teixeira, Kluwer Academic Publishers, 1991, pp. 289-332
20. E. Fratini, S. -H. Chen, P. Baglioni, and M. -C. Bellissent-Funel, *Phys. Rev. E* **64**, 020201-020204 (R) (2001).
21. E. Fratini, S. -H. Chen, P. Baglioni, and M. -C. Bellissent-Funel, *J. Phys.Chem* **106**, 158-166 (2002).
22. A. Faraone, S. -H. Chen, E. Fratini, P. Baglioni, L. Liu, and C. Brown, *Phys. Rev. E* **65**, 040501-040503 (2002).
23. A. Faraone, L. Liu, C. -Y. Mou, P. -C. Shih, J. R. D. Copley, and S. -H. Chen, *J. Chem. Phys* **119**, 3963-3971 (2003).
24. L. Liu, A. Faraone, C. -Y. Mou, C. -W. Yen, and S. -H. Chen, *J. Phys: Cond. Matter* **16**, S5403-S5436 (2004).
25. M. -C. Bellissent-Funel, S. -H. Chen, and J. M. Zanotti, *Phys. Rev. E* **51**, 4558-4569 (1995).
26. P. C. Shih, H. P. Lin, and C. -Y. Mou, *Stud. Surf. Sci. Catal.* **146**, 557-560 (2003).
27. Y. Lin, W. Zhang, and T. J. Pinnavaia, *J. Am. Chem. B* **122**, 8791-8792 (2000).
28. R. Ryoo, S. H. Joo, and J. M. Kim *J. Phys. Chem. B* **103**, 7435-7440 (1999).
29. M. A. Anisimov, J. V. Sengers, and J. M. H. Levelt Sengers, "Near-critical behavior of aqueous systems" in *Aquesous Systems at Elevated Temperatures and Pressures: Physical Chemistry in water, Steam and Hydrothermal Solutions*, edited D.A. Palmer, R. Fernandez-Prini and A.H. Harvey, Elsevier Ltd., 2004.
30. Private communication with L. Xu, S. Buldyrev, and H. E. Stanley. The authors wish to acknowledge conversations with these researchers on Mar. 27, 2005 when they drew our attention to this interpretation of the T_L line.
31. S. Sastry, C. A. Angell, *Nature Materials* **2**, 739-743 (2003).
32. H. Kanno, R. J. Speedy, and C. A. Angell, *Science* **189**, 880-881 (1975).
33. H. E. Stanley, "Mysteries of water" in *The Nato Science Series A* **305**, edited M. -C. Bellissent-Funel (1999).
34. C. A. Angell, S. Borick, and M. Grabow, *J. Non-Crys. Sol.* **463**, 205-207 (1996).
35. M.A. Ricci and S.-H. Chen, *Phys. Rev. A* **34**, 1714-1719 (1986).
36. A. K. Soper et al., *Phys. Rev. Lett.* **84**, 2881-2884 (2000).
37. W. Doster, S. Cusack, and W. Petry, *Nature (London)* **337**, 754-756 (1989); W. Doster, S. Cusack, and W. Petry, *Phys. Rev. Lett.* **65**, 1080-1083 (1990).
38. D. Liu, Y. Zhang, C. -C. Chen, C. -Y. Mou, P. H. Poole, and S. -H. Chen, *PNAS USA* **104**, 9570-9574 (2007).
39. D. Liu, Y. Zhang, Y. Liu, J. Wu, C. -C. Chen, C. -Y. Mou, and S. -H. Chen, "Absolute Density Measurement of Confined Water by Small Angle Neutron Scattering Method" (to be published).
40. S. -H. Chen, F. Mallamace, C. -Y. Mou, M. Broccio, C. Corsaro, A. Faraone, and L. Liu, *PNAS USA* **103**, 12974-12978 (2006).
41. Y.-J. Jung, J. P. Garrahan, and D. Chandler, *Phys. Rev. E* **69**, 061205-061211 (2004).
42. X.-Q. Chu, A. I. Kolesnikov, A. P. Moravsky, V. Garcia-Sakai, and S. -H. Chen, *Phys. Rev. E* **76**, 021505-021510 (2007).
43. S. -H. Chen, L. Liu, A. I. Kolesnikov (to be published).
44. B. F. Rasmussen, A. M. Stock, D. Ringe, and G. A. Petsko, *Nature* **357**, 423-424 (1992).
45. R. M. Daniel, J. C. Smith, M. Ferrand, S. Hery, R. Dunn, and J. L. Finney, *Biophys. J.* **75**, 2504-2507 (1998).
46. W. Doster, S. Cusack, and W. Petry, *Nature* **337**, 754-756 (1989).
47. M. Ferrand, A. J. Dianoux, W. Petry, and G. Zaccai, *PNAS USA* **90**, 9668-9672 (1993).
48. A. M. Tsai, D. A. Neumann, and L. N. Bell, *Biophys. J.* **79**, 2728-2732 (2000).
49. M. Tarek and D. J. Tobias, *Phys. Rev. Lett.* **88**, 138101-138104 (2002).
50. A. L. Tournier, J. Xu, and J. C. Smith, *Biophys. J.* **85**, 1871-1875 (2003).
51. A. Paciaroni, A. R. Bizzarri, and S. Cannistraro, *Phys. Rev. E* **60**, R2476-R2479 (1999).
52. G. Caliskan, A. Kisliuk, and A. P. Sokolov, *J. Non-Crys. Sol.* **307-310**, 868-873 (2002).
53. P. W. Fenimore, H. Frauenfelder, B. H. McMahon, and F. G. Parak, *PNAS USA* **99**, 16047-16051 (2002).
54. S. -H. Chen, L. Liu, E. Fratini, P. Baglioni, A. Faraone, and E. Mamontov, *PNAS USA* **103**, 9012-9016 (2006).
55. S. -H. Chen, L. Liu, X. Chu, Y. Zhang, E. Fratini, P. Baglioni, A. Faraone, and E. Mamontocv, *J. Chem. Phys.* **125**, 171103-171106 (2006).
56. X. Chu, E. Fratini, P. Baglioni, A. Faraone and S. -H. Chen, "Observation of a Dynamic Crossover in RNA Hydration Water which Triggers the Dynamic Transition in the Biopolymer" (to appear in *Phys. Rev. E*)
57. C. A. Angell, *Ann. Rev. Phys. Chem.* **55**, 559-583 (2004).
58. P. Kumar, Z. Yan, L. Xu, M. G. Mazza, S. V. Buldyrev, S.-H. Chen, S. Sastry, and H. E. Stanley, *Phys. Rev. Lett.* **97**, 177802-177806 (2006).

The Glass Transition of Driven Molecular Materials

M. Descamps+, J.F. Willart, and A. Aumelas

Laboratoire de Dynamique et Structure des Matériaux Moléculaires. UMR. CNRS 8024. and Therapeutic Material Group, ERT 1066. University LILLE1. Bat P5 - 59655 Villeneuve d'Ascq CEDEX, France
+ Marc.descamps@univ-lille1.fr

Abstract. There are many cases of practical interest where materials are maintained in nonequilibrium conditions by some external dynamical forcing: typical examples of these *driven materials* are provided by irradiation, grinding, extrusion…Contrary to usual phase transitions which are properly addressed by thermal equilibrium states, equilibrium and irreversible thermodynamics, no such general framework is available for driven systems. The purpose of this paper is to show some examples of phase transformations in driven molecular materials. These materials are considered because they are extremely sensitive to external disturbances and are generally very good glass formers. This allows investigating more easily a broad range of the parameters which possibly influence the nature of the end product. We will examine mainly the effect of grinding. Contrary to other materials, metals or minerals, systematic investigations of transformations induced by grinding of molecular materials have not yet been done despite the practical and fundamental interests of such investigations in pharmaceutical and agro-chemical science. We will address several modes of interconversions between crystalline and glassy states of the same compound. We will further discuss specific processing effects on the physical state of the glass itself. It will be shown from these investigations that rationalization and possibilities of prediction are emerging. The use of effective temperature concepts to describe the end product of milling will be discussed. These findings may be of general concern for driven materials of any chemical nature.

Keywords: Driven materials, Glass transition, Milling, Molecular crystals, Solid-Solid transition
PACS: 61.43Fs, 64.70 P, 64.70 K, 64,70 kt

INTRODUCTION

Industrial processing may lead to a variety of physical state transformations of materials. Typical examples are provided by molecular compounds – either pharmaceutically active ingredients or excipients – exposed to pharmaceutical processing (e.g. milling, drying extrusion…). These operations may induce intentionally or unintentionally the slipping of crystalline substances into the amorphous glassy state. Disordered solids and amorphous materials are of interest in pharmaceutical and food formulations because such states may have favorable biopharmaceutical properties i.e. enhanced solubility and dissolution capabilities. The drawback is their intrinsic physical and chemical instability. Also surface amorphization may act to increase water uptake and promote stickiness and collapse of the material. Sometimes processing acts in totally opposite direction promoting recrystallization of the amorphous state or inducing polymorphic conversion between crystalline phases [1,2]. In addition to the usual thermodynamic parameters, temperature and pressure (as at thermal equilibrium), materials processed under such conditions are exposed to some external dynamical forcing which maintain them in non equilibrium conditions. The end product of so called *driven materials* [3] is in a steady state of a dynamical system rather than a thermal equilibrium state. Contrary to metallic compounds driven molecular materials have not yet been the subject of intense investigations despite their practical interest and physical originality.

In this presentation we explore several aspects of dynamically transformed molecular materials via milling.

The first point concerns the possibility to predict the nature of the end product as a function of the dynamic forcing parameters. The antagonistic effect of the temperature and forcing intensity will be clarified on several examples. The important question of the possible existence of transient intermediate states developing during the forcing operations will be addressed and discussed. The nanostructural path leading to amorphization or polymorphic transformation of crystals will be described.

CP982, *Complex Systems, 5th International Workshop on Complex Systems*
edited by M. Tokuyama, I. Oppenheim, and H. Nishiyama

The second point concerns the glassy state itself. A careful analysis of the glass transformation domain of compounds amorphized due to a dynamical forcing shows specific dynamic heterogeneities which allow proposing a mechanism of the amorphization process. Comparison with the glassy states obtained by cooling of the melt shows differences in their relaxational behavior [2,4].

Another interesting possibility to manipulate the solid state with forcing techniques is offered by applying forcing to the glassy state itself. Since in that case the noise level is not merely fixed by the temperature, but also by the frequency and amplitude of the forcing, it is possible to reach new regions of the energy landscape of the glass. Such regions of the configurational phase space would be otherwise inaccessible, or accessible through very specific quenching and annealing conditions. It will be shown that this offers possibility either to accelerate aging or rejuvenate the glass. Sometimes it allows reaching megabassins of the energy landscape which prepare nucleation of new polymorphic crystalline varieties [2,5].

The possibility to use *effective temperature* concepts to describe non-equilibrium systems is a question which is actively debated (see for example references reported in [26] and [30]). We will attempt to use several of theses approaches to rationalize our experimental findings. This allows suggesting that a combination of these concepts would be necessary to fully describe driven glasses.

EXPERIMENTAL

Ball-milling experiments were performed in a high energy planetary mill (Pulverisette 7 – Fritsch) at room temperature or at 0°C by placing the whole device in a temperature controlled chamber. Samples were sealed in Zirconium vials of 45 cm^3 volume with seven balls (Ø 15 mm) of the same materials. The vials are fixed onto a rotation disc and rotate with the same rotation speed Ω on the opposite direction to the disc. The important feature is that the intensity of the mechanical treatment is an increasing function of Ω. The apparatus was specially modified to grind hygroscopic samples in a controlled atmosphere. Prior to the milling procedure, the vials were flushed with dry nitrogen gas in order to prevent hydration of the powder during experiments. To avoid any overheating of the samples, grinding was performed alternating milling periods (typically 10 min) with pause periods (typically 5 min). By doing this, the external vial temperature was kept below 50°C. The analysis of the ground samples were performed by X-ray diffraction and DSC.

AMORPHIZATION

Pharmaceutical literature describes numerous examples of organic compounds like piroxicam [4], budesonide [5], sucrose [6,7], lactose [8,9], trehalose [10]...which become partially or totally amorphous when they are submitted to mechanical treatments. We are interested here in anhydrous trehalose, a disaccharide well known for its efficient biopreservation properties of proteins. This crystalline compound (β form) is submitted to mechanical treatment in a planetary mill at room temperature during several hours alternating milling periods of 20 min with pause periods of 10 min in order to limit the warm-up of the powder in the jars. The figure 1 shows the X-ray diffraction patterns (XRPD) and the DSC scans recorded upon heating (5 $°C.min^{-1}$) of trehalose before and after milling. The detailed results are available in [10]. The XRPD patterns of trehalose recorded after 26 h of effective milling is composed of a diffusion halo which reveals that the compound is X-ray amorphous. The presence of a glass transition at 120°C followed by a recrystallization at higher temperature observed on the DSC scan of the milled powder confirms this amorphization. We are thus facing a crystal to glass transformation in a solid state. We can notice that the glass transition temperature of the milled compound is very close to that of the glass obtained by a conventional melting/quenching process. Thereafter, more details about the calorimetric signature of the glass transition which appears after milling will be given. They will shed some light on the amorphization mechanism. Anhydrous lactose, another disaccharide very used as excipient, also amorphizes in the same conditions [8]. It has been demonstrated that the obtained amorphous compound is only in the initial molecular α variety. There is no trace of mutarotation – to the β variety – as it is expected at high temperature [11]. This is an indirect evidence that amorphization by milling isn't due to a simple local warm-up due to the formation of the so-called "hot points" as it is sometimes suggested. As for trehalose, the experiments were performed at a temperature well below the glass transition temperature of the compound ($Tg_{lactose} \approx 110$ °C). This point seems to be a relevant factor to determine the final state resulting from milling.

TRANSFORMATION BETWEEN CRYSTALLINE FORMS

Pharmaceutical literature also reports a lot of observations about crystal-crystal transformations between different polymorphic forms resulting of milling. It is for example the case of: chloramphenicol [12], cimetidine [13], sulfamerazine [14], indomethacin [15], phenylbutazone [16], sorbitol [17], mannitol [17]…for which a conversion from the stable phase to a metastable phase is observed.

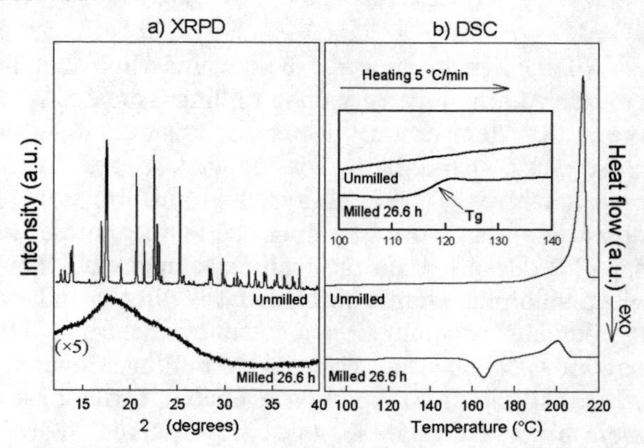

FIGURE 1. X-ray diffraction patterns of trehalose recorded at room temperature before milling (a) and after 26 hours of milling treatment with Ω = 400 rpm (b). DSC heating curves (5°C/min) of crystalline trehalose before milling (c) and after 20 hours of milling treatment with Ω = 400 rpm (d). The curve in the inset is a close up view of the glass transition temperature domain of curve d.

The inverse situation is described for caffeine, compound for which a quick conversion from the metastable phase I to the stable form II is reported [18]. We demonstrated that in the case of polymorphic transformations, final crystalline state is independent of the initial physical state. Mannitol presents a complex polymorphism with three polymorphic forms whose increasing stability order at room temperature is δ, α, β. The milling at room temperature of the most stable form β and the less stable form δ both induce – for the same milling conditions (temperature and intensity) – a transformation towards the intermediate stability form α (Figure 2) [19]. So, a transformation towards an intermediate stability state stabilized by milling is observed. After stopping milling, it is noticed that the metastable phase α quickly transforms towards the more stable phase β.

A similar crystal-crystal transformation was found when grinding another polyol: the sorbitol [17]. Mannitol and sorbitol can be obtained in the amorphous state by the melting/quenching process. It must be noted that sorbitol – as mannitol – has its glass transition temperature Tg below the milling temperature ($Tg_{sorbitol}$ ≈ 0°C [20], $Tg_{mannitol}$ ≈ 12°C [21]).

DUALITY OF TRANSFORMATIONS

The previous results suggest that a low milling temperature or more exactly a milling temperature lower than the glass transition temperature favors the amorphization of the compounds. Recent milling experiments performed at different temperatures with fananserine clearly show this point. Fananserine is a pharmaceutical compound which presents a complex

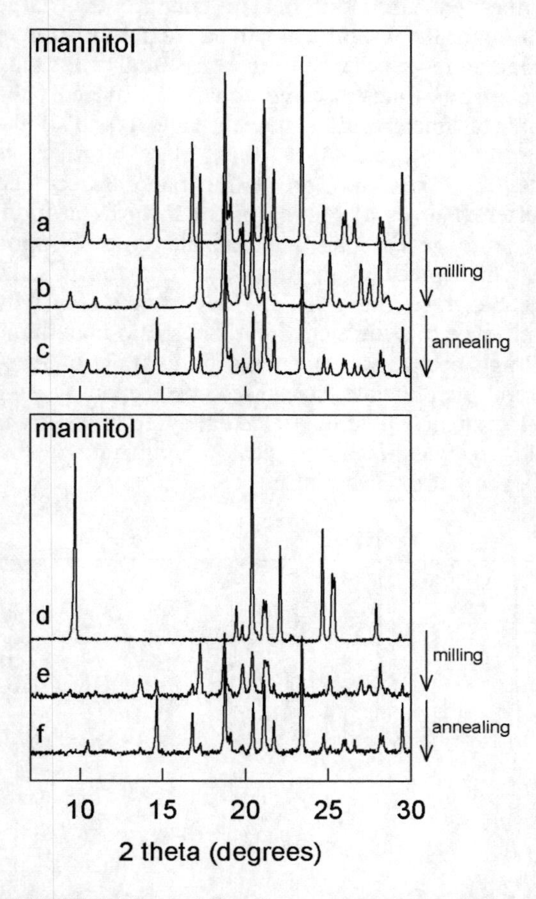

FIGURE 2. X-ray diffraction patterns of mannitol recorded at room temperature:
(a), (b) and (c) correspond respectively to mannitol β before milling, after milling (3 h), and after a 8 days annealing at RT.
(d),(e) and (f) correspond respectively to mannitol δ before milling, after milling (3 h), and after a 1 days annealing at RT.
Upon milling, the forms β and δ both undergo a polymorphic transformation toward the form α. Upon annealing the form α then reverses toward the most stable form β.

polymorphism with at least four crystalline forms. The milling at room temperature of the most stable forms III and IV induces for both a polymorphic transformation towards the least stable form I. Decreasing the milling temperature of 25 °C is sufficient for the milling to induce an amorphization of the fananserine rather than a polymorphic transformation. This result is illustrated on the figure 3. We can see on the XRPD pattern the diffusion halo which characterizes amorphous compounds and on the DSC scan the Cp jump which proves the vitrification. The glass transition temperature of fananserine is Tg=19 °C (heating at 5 °C.min^{-1}). For this compound, the modification of the milling temperature was thus performed on either side of Tg. This proves a large sensibility of the response at a dynamic perturbation in a temperature zone where the structural relaxation characteristic times evolve quickly. Indeed, the dynamic characterization of metastable liquid of the fananserine showed that this glass former is fragile [22]. The X-ray diffraction patterns recorded for different times of milling of the form IV at room temperature reveals an intermediate amorphization stage which precedes the transformation towards the metastable crystalline phase I. This is clearly shown by the existence of a transient amorphous halo underlying the Bragg peaks. Moreover, the DSC scans of these different intermediate states recorded upon heating reveal a small Cp jump immediately followed by a small recrystallization peak confirming the vitrification of the material.

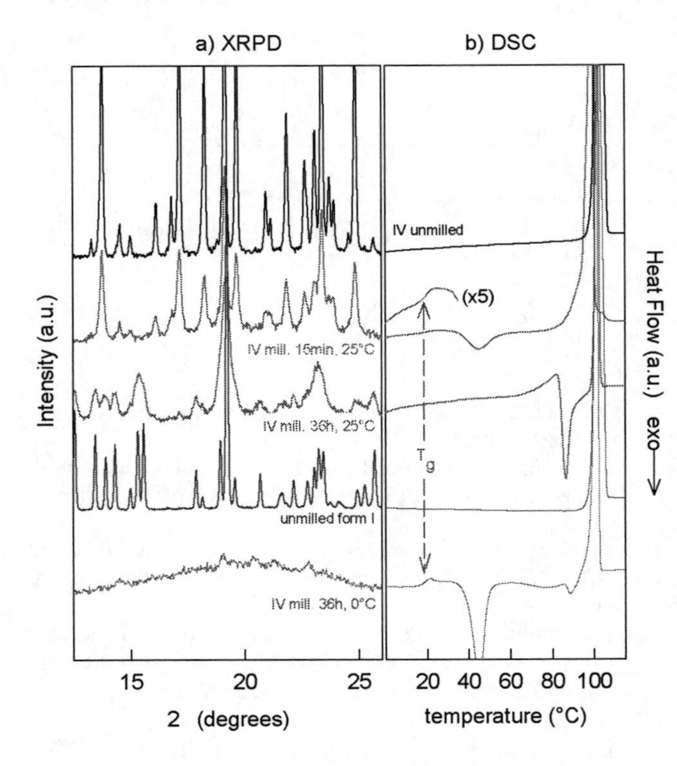

FIGURE 3. a) X-ray diffraction patterns of fananserine form IV recorded at room temperature after different milling times at different milling temperatures b) corresponding DSC heating curves (5°C/min). A close up view of the glass transition temperature underlines a transient amorphization.

RELEVANCE OF THE DRIVEN MATERIAL CONCEPT

All the results reported below underline that the amorphization propensity upon milling increases when the milling temperature decreases. Besides, there are clear clues –particularly for the lactose and for all cases where a crystal-crystal transformation is observed – that, during milling, the temperature of the samples doesn't reach the melting temperature. Thus, when amorphization is observed, it is directly induced by defaults accumulation created by shearing. This new physical state is stabilized by milling. Generally, when milling is stopped, irreversible evolutions to more ordered crystalline states are observed. It looks as if, during milling, the temperature couldn't counterbalance the mechanically generated disorder. The increasing amorphization efficiency of milling for decreasing milling temperatures is well taken into account by the "driven material concept" developed by G. Martin and P. Bellon [3]. This model was initially proposed to describe transformations under irradiation [3]. According to this model, the compound can be viewed as exploring its configuration space under the effect of two distinct mechanisms which act in parallel: the usual thermally activated molecular motions which operate in the absence of the external forcing and, on the other hand, ballistic jumps which are generated by the external forcing. The originality of this approach is to take into account the overall physical state itself. The idea is that the competition between, on the one hand, the disorder induced by milling and, on the other hand, the diffusion effects – which tend to restore the system to lower energy configurational states – is modelized for the configuration of the total system and not only locally. It is noticed that diffusion can also be accelerated by the perturbation.

Closely following the Martin's approach, this concept can be simply illustrated using a kinetic Ising model. For the simple case of a relaxation of the parameter order S and for a second order transition occurring at Tc with no milling, the equation of relaxation in presence of milling can be written (in the range of temperature where, under milling, S is infinitely small):

$$\frac{dS}{dt} = D'\left(-S + \left(\frac{T_c}{T}\right)S\right) - D_B S \qquad (1)$$

- the first term of right member is that of the equation of motion in the molecular field approximation (after linearisation of the hyperbolic tangent). (See for example reference [23]). In this expression the heath bath jump frequency term D’ is enhanced because external forcing modifies the noise level and may also generates extra defects.
- the second term describes the effect of disordering with a "ballistic" jump frequency D_B, (rate of ballistic jump) independent of the temperature but which depends on the milling intensity.

In stationary regime (dS/dt = 0), eq (1) possesses non-zero solutions only for temperatures below a critical temperature under milling:

$$T_c' = \frac{T_c}{\left(1+\frac{D_B}{D'}\right)} \qquad (2)$$

This is obviously equivalent to a rescaling of the physical interactions:

$$J_{eff} = \frac{J}{\left(1+\frac{D_B}{D'}\right)} \qquad (3)$$

That can be extended to situations which need to express the exchange integral in the reciprocal space [30]

Otherwise stated the physical state of the system which is ground at the real temperature T is that of the system in absence of milling at an effective temperature:

$$T_{eff} = T\left(1+\frac{D_B}{D'}\right) \qquad (4)$$

The rate of thermal jump D’ decreases when the temperature decreases. Fig.4. represents typical variations of T_{eff} as a function of T. At high temperature thermal jumps are efficient enough to restore an equilibrium so that $T \cong T_{eff}$. It means that milling has no practical influence on the physical state. However at low temperature thermal restoration becomes less efficient and T_{eff} eventually increases. It may thus overcome the equilibrium phase transition temperature.

Even if the kinetics of melting are less simple to modelize, such an approach can explain amorphization. It occurs at low enough milling temperature or (and) rather high milling intensity when T_{eff} overcomes the value of the melting temperature. It also reproduces consistently what has been experimentally observed to make amorphization easier i.e. a decrease of the milling temperature.

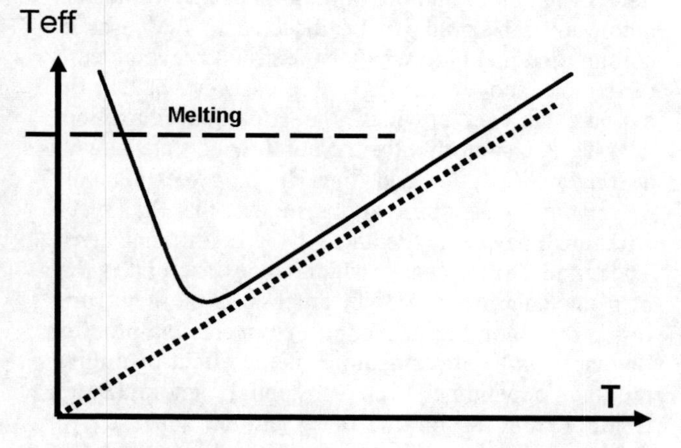

FIGURE 4. SEE DESCRIPTION IN THE TEXT

THE CASE OF MILLING INDUCED CRYSTAL TO CRYSTAL CONVERSIONS

More difficult is to explain the observed crystal-crystal polymorphic transitions using the above-mentionned approach since polymorphs form monotropic sets. In these cases no equilibrium transitions between the different polymorphs are expected. It is also to be noticed that there is no group to subgroup symmetry relation between these phases. The transitions are totally reconstructive. It is thus not possible to explain the conversion induced by milling by the argument that Teff would rich a temperature higher than that of the equilibrium transition. Furthermore the conversion is sometimes observed from a metastable to a more stable state; that would certainly need to admit some decrease of Teff under milling which is not predicted by eq (4) It is to be remembered that such milling induced crystal conversions are correlated to the fact that milling is performed at temperatures higher than the glass transition of the compound, or very close to it. For the milling experiments which could be performed at temperatures very close to Tg – as for fananserine and indomethacin – we have observed some transient stage of amorphization preceding the polymorphic crystalline transformation. The formation of a non crystalline solid in the initial stage of formation of phase I was also observed during the ball milling of sulfathiazole form III [1,4]. The mechanisms of transformation during grinding of other drugs [24], cephalexin, chloramphenicol palmitate and indomethacin were found to proceed by going through amorphous states of the material followed by transformation to another polymorph. Such a transient amorphization could not be detected in the cases of sorbitol and mannitol which were ground several teens of degrees above their Tg's. We can suspect that the extreme slowing down of the structural relaxations near Tg is responsible for the amorphous state to have a longer life time and thus being experimentally detectable. These observations suggest that the crystal structure change is mediated by the amorphization. The simplest mechanism which comes to mind is that of a nucleation and growth process of the new form inside the amorphized fraction. An interesting proof of this is given by investigations of the effect of milling on the cimetidine [13]. Substantial amorphization resulted from the milling of cimetidine which shows four crystalline modifications. Form A was found to transform in 84 hours of grinding into form D after a transient amorphization if 0.1% of phase D is admixed during the grinding process. The kinetics of phase transformation could be significantly accelerated by increasing the amount of added phase D. A 1:1 mixture of modifications A and D transformed completely to D within 6h. This supports the idea that nucleation out of the amorphous content was essential for the path of phase transformation [25].

The transformation would thus involve a two step process:

- the first is an amorphization according to the driven material concept described above
- the second is the milling assisted crystallization out of the amorphized part of the compound. An effective temperature argument can not justify such a transformation in the same way than the initial amorphization. We may imagine that milling effectively pushes the compound to an higher effective temperature Teff. But that time we are forced to admit that the system does not transform because it is displaced with regard to an equilibrium transition. However it is reasonable to assume that milling effectively pushes the system up to temperatures where nucleation becomes more favourable. This temperature is situated between Tg and the equilibrium melting temperature. This second part of the process does not come out directly from a milling-modified relaxation equation of an order parameter like in (1). The driven activation concept of Barrat et al. [26], to be evoked below, may be that time relevant. It is worth to be noticed that such a modification of the activation energy is still present in the Martin's like expression (1) through a modified D' frequency jump term.

HETEROGENEITIES IN VITRIFIED TREHALOSE OBTAINED BY MILLING

The milling induced crystal-crystal conversions would thus need the formation of heterogeneities in the sample. Recent investigations of the effect of grinding on trehalose bring unexpected indications of such heterogeneities. This in turn provides information about the way mechanical activation leads to the vitrification of the compound. Figure 5. provides evidence that the amorphous glassy state obtained by milling the crystal state is heterogeneous. It shows the modulated DSC signals upon reheating: the heat flow, reversing heat flow and non reversing heat flow.

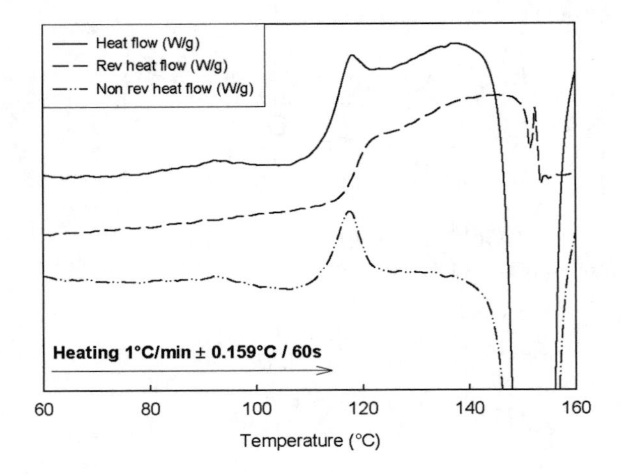

FIGURE 5. Heat flow (—), reversing heat flow (- - -) and non reversing heat flow (¯ .. ¯ .. ¯) of anhydrous trehalose amorphized by milling recorded at 1°C/min ± 0.159°C / 60s.

The average heat flow signal shows complex evolutions in the zone of Tg: the Cp jump which characterizes the glass transition is at about 120°C. It is followed by some endothermic drift before a crystallization exotherm is detected at 150 °C. These complex temperature evolutions are resolved in the reversing and non-reversing components. The reversing heat flow enables us to see the glass transition more clearly since dehydration and crystallization phenomena do not give any contribution to this signal (except for heat capacity changes). It is clearly seen on this signal a first Cp jump at 120°C followed by a second jump at 134°C which reveals a very clear second glass transition. This jump corresponds to the average heat flow drift. It can be checked that it has no non-reversing origin. This glass transition appears at a temperature higher than that of the glass transition of a quench cooled trehalose glass. It reveals the dynamic heterogeneity of the glass resulting from milling. These observations furthermore suggest that amorphization by milling occurs through a two step mechanism.

EVOLUTION OF THE GLASSY STATE UNDER MILLING: DRIVEN ACTIVATION

Milling a sample which has been first prepared in the glassy state by quench cooling offers an original non equilibrium situation to search new evidence of an effective temperature for the driven material. The pharmaceutical ingredient indomethacine is a very good glass former whose recrystallization properties are rather well documented [24].

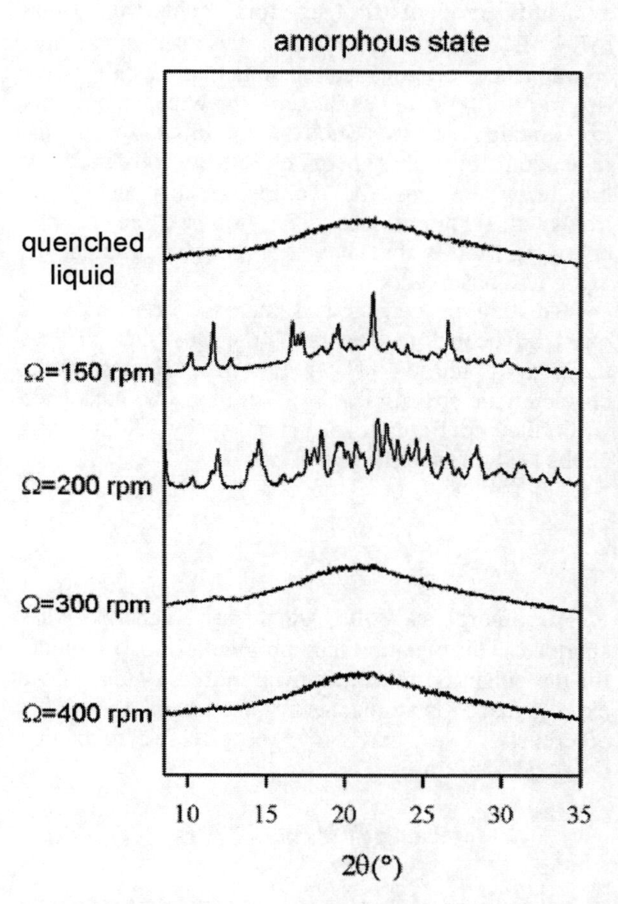

FIGURE 6. X-ray diffraction patterns of glassy indomethacine milled at room temperature with different milling intensities.

Systematic investigations of the effects of milling the glassy state of indomethacine have been reported in [2]. They consisted in performing milling of the quench-cooled glass, at room temperature (about 30°C below Tg), with various grinding intensities (by varying Ω). Steady states were reached upon milling which were found to depend on the rotation discspeed Ω as follows (figure 6):

- For low Ω value (150 rpm), the steady state is that of a sample partially crystallized in the stable γ form,
- For medium Ω value (200 rpm), the steady state is that of a sample entirely crystallized in the metastable α form,
- For high Ω values (300 – 400 rpm), ball milling maintains the sample in a fully amorphous state.

It is interesting to notice that for the two lower milling energies we observe a crystallization of the amorphous state under milling. There is further a correlation between the nature of the recrystallized phases, the intensity of grinding, and the temperatures where these phases have their maximum of recrystallization rate in the absence of milling. The stable phase γ whose recrystallization is enhanced at lower milling intensity has also its maximum rate of recrystallization at lower temperature. Slightly higher milling intensity is correlated to the appearance of phase α which also recrystalizes at higher temperature in the non milled glass. Milling the amorphous state at high intensity has no apparent effect on the nature of the physical state which remains amorphous. This is coherent with other observations resulting from milling crystalline indomethacine at nitrogen temperature [15] or high intensity which also induce amorphization. These last features agree with the driven material conception of a material pushed to an effective temperature higher than melting, and this independently of the initial physical state of the compound.

As for the case of the crystal-crystal transformations described above, it is impossible to directly interpret the milled enhanced crystallizations using the same effective temperature concept by which the system would be displaced with regard to an equilibrium phase transition temperature. The correlations existing between the milling intensities and recrystallization temperatures strongly suggest the possibility to consider an effective temperature under milling which is related to the structural relaxation of the amorphous state itself. The latter is enhanced by milling at least if the milling intensity is not too high – and promote nucleation, even when the real temperature is lower than Tg. Crystallization of a given phase thus occurs when the effective temperature reaches the temperature domain of most rapid recrystallization. If we adopt an Eyring activation process description this is equivalent to a rescaling of the activation energy ΔE in an Arrhenius contribution.

$$\tau(T) \propto \exp\left(\frac{-\Delta E}{k_B T}\right) \quad (5)$$

It is to be noticed that such an enhancement of the dynamics under external driving was also implicitly considered in the D' term of the Martin equation (1).

Activated dynamics in a glassy system undergoing shear deformation was recently studied numerically by Ilg and Barrat [26]. They have shown that the introduction of an effective temperature in an Arrhenius representation of the slow relaxation processes is close to the effective temperature that can be determined from the fluctuation-dissipation relation. In this latter definition the effective temperature is defined as the ratio between the response function and the correlation function [31]. It differs from the heat bath temperature for out-of-equilibrium systems. The idea is that the mechanical activation provides additional noise to the normal thermal one.

If high intensity milling of the glass does not induce a phase change, the nature of the milled amorphous however appears to be modified after long enough milling. Upon heating the later recrystalizes toward the metastable α phase. Crystallization toward the γ phase is only observed upon reheating the briefly milled amorphous. The situation is similar to that observed for the recrystallization of amorphous indomethacine initially obtained by cooling of the liquid with different cooling rates [27]. Higher cooling rate eventually promotes a recrystallization towards α, as does a sample milled during a long time. Long milling is thus able to push the amorphous to a state similar to one of higher fictive temperature.

DISCUSSION. CONCLUSION.

We have shown that mechanical milling can induce various changes of the physical state of molecular materials. These changes may even appear antagonistic. We can summarize the overall range of behavior which may be observed as follows:

- amorphization (Only for grinding performed below Tg)
- A crystal to crystal transformation (for grinding performed at higher temperature). If grinding is performed near Tg a transient amorphous state can be observed.
- Grinding the glass can either induce a recrystallization (for low grinding intensity) or drive the amorphous to states of high fictive temperature.

These different behaviors can be made reasonably rational by appealing to an effective increase of the temperature under milling. We have however shown that operationally it is possible to refer to two different types of effective temperatures in order to describe the physical state produced by grinding. These two effective temperatures are respectively related to a rescaling of the interaction energy J which drives the phase transition ("Martin effective temperature") and a rescaling of the activation energy ΔE ("Barrat effective temperature").

The consequences of the rescaling of these two energies or equivalently of the temperatures are apparently different. They correspond either to a change of the phase transition temperature (J→J') or to a modification of the structural relaxation times (ΔE→ΔE'). This latter change by itself may have consequences on a nucleation rate and may thus give rise in turn to the appearance of a new phase but only for kinetic reasons. Such a modification of the relaxation times has been effectively observed by measuring in situ the volume relaxation of an irradiated selenium glass at a temperature slightly below Tg [32]. In this latter case an acceleration of the aging was observed.

Creation of defects and increase of disorder is often advocated to "explain" the physical changes induced by milling [1]. In the case of glasses the characteristic structural relaxation time was correlated with the configurational entropy by Adam and Gibbs [28] through the expression:

$$\tau = \tau_0 \exp\left(\frac{C}{TS_c}\right) \quad (6)$$

An amorphous solid which has higher S_C has shorter τ. This equation may provide a heuristic model for the enhanced relaxation in a milled a glass. If we express that S_C is an increasing function of the number of defects, n, we may write the effective relaxation time under milling:

$$\tau_{eff} = \tau(T)\exp(-C'\Delta n) \quad (7)$$

Where Δn is the excess of defects created by milling and $C' = C\,(\partial S_C/\partial n) / T.S_C^2$ [29]. Such an approach is enough to account for transformations resulting from milling at not too high intensity and not too low temperature: the shortening of the structural relaxation time is able either to accelerate aging or to promote nucleation of crystalline phases. The situation is opposite in the case of high milling intensity below Tg. We have mentioned that it results a glass with a high fictive temperature. In the case of irradiation at temperatures much lower than Tg, it has also been observed a rejuvenation of the glass i.e. an effect contrary to that resulting from irradiation near Tg which hyperstabilizes the glass [29, 32].

All these observations suggest that some competition occurs between:

- a thermally activated effect – even enhanced by the driving mechanism – which is predominant at high temperature and low milling intensity

and - a ballistic term which induces additional disorder in the glass without enough thermal compensation at low temperature and high milling intensity.

There is no hope that a single order parameter may describe a glass and it is certainly difficult to adapt directly the driven material concept in the form of a simple equation like (1). However a combination of the Martin [3] and Barrat [26] concepts could certainly help providing a unified description of driven glasses. The need for such a theoretical development in this area where industrial formulation problems catches up with non-equilibrium physics is acute.

ACKNOWLEDGMENTS

We gratefully acknowledge discussions with G. Martin and P. Bellon. and the support of Sanofi-Aventis for the PhD of A. Aumelas.

REFERENCES

1. T. P. Shakhtshneider and V.V. Boldyrev, in *Reactivity of molecular solids*, edited by E. Boldyreva and V. Boldyrev (E. Boldyreva, V. Bold UK, 1999), pp. 271-312.yrev, John Wiley & Sons, Chichester,
2. S. Desprez and M. Descamps, *Journal of Non-Crystalline Solids* **352**, 4480-4485 (2006).
3. G. Martin and P. Bellon, *Solid State Physics, New York* **50**, 189-331 (1997).
   G. Martin, Phys.Rev.B **30**, 1424 (1984).
4. T. P. Shakhtshneider, *Solid State Ionics* **101-103**, 851-856 (1997).
5. E. Dudognon, J. F. Willart, V. Caron, F. Capet, T. Larsson, and M. Descamps, *Solid State Communications* **138** (2), 68-71 (2006).
6. I. Tsukushi, O. Yamamuro, and T. Matsuo, *Solid State Communications* **94** (12), 1013-1018 (1995).
7. J. Font, J. Muntasell, and E. Cesari, *Materials Research Bulletin* **32** (12), 1691-1696 (1997).
8. J. F. Willart, V. Caron, R. Lefort, F. Danede, D. Prevost, and M. Descamps, *Solid State Communications* **132**, 693-696 (2004).
9. M. Otsuka, H. Ohtani, N. Kaneniwa, and S. Higuchi, *Journal-of-pharmacy-and-pharmacology* **43** (3), 148-153 (1991).
10. J. F. Willart, A. De Gusseme, S. Hemon, G. Odou, F. Danede, and M. Descamps, *Solid State Communications* **119** (8-9), 501-505 (2001).
11. Ronan Lefort, Vincent Caron, Jean-Francois Willart, and Marc Descamps, *Solid State Communications* **140** (7-8), 329-334 (2006).
12. M. Otsuka and N. Kaneniwa, *Journal of pharmaceutical sciences* **75** (5), 506-511 (1986).
13. A. Bauer-Brandl, *International Journal of Pharmaceutics* **140** (2), 195-206 (1996).
14. G. G. Zhang, C. Gu, M. T. Zell, R. T. Burkhardt, E. J. Munson, and D. J. Grant, *Journal of pharmaceutical sciences* **91** (4), 1089-1100 (2002).
15. M. Otsuka, T. Matsumoto, and N. Kaneniwa, *Chemical and pharmaceutical bulletin* **34** (4), 1784-1793 (1986).
16. T. Matsumoto, J. Ichikawa, N. Kaneniwa, and M. Otsuka, *Chemical and Pharmaceutical Bulletin* **36** (3), 1074-1085 (1988).
17. J. F. Willart, J. Lefebvre, F. Danede, S. Comini, P. Looten, and M. Descamps, *Solid State Communications* **135**, 519-524 (2005).
18. Jukka Pirttimaki, Ensio Laine, Jarkko Ketolainen, and Petteri Paronen, *International Journal of Pharmaceutics* **95**, 93-99 (1993).
19. M. Descamps, J.F. Willart, E. Dudognon, and V. Caron, *Journal of Pharmaceutical Sciences* **96** (5), 1398-1407 (2007).
20. C.A. Angell, R.C. Stell, and Z. Sichina, *J. Phys. Chem.* **86**, 1540 (1982).
21. L. Yu, D. S. Mishra, and D. R. Rigsbee, *Journal of pharmaceutical sciences* **87** (6), 774-777 (1998).
22. A. De Gusseme, L. Carpentier, J. F. Willart, and M. Descamps, *Journal of Physical Chemistry B* **107** (39), 10879-10886 (2003).
23. E. Stanley, *"Introduction to phase transition and critical phenomena", Clarendon Press*, (1971).
24. M. Otsuka, K. Otsuka, and N. Kaneniwa, *Drug Development and Industrial Pharmacy* **20** (9), 1649-1660 (1994).
25. H.G. Brittain, *Journal of Pharmaceutical Sciences* **91** (7), 1573-1580 (2002).
26. P. Ilg and J. L. Barrat, *Epl* **79** (2), NIL_54-NIL_58 (2007).
27. V. Andronis and G. Zografi, *Journal-of-non-crystalline-solids* **271** (3), 236-248 (2000).
28. G. Adam and J.H. Gibbs, *Journal of Chemical Physics* **43**, 139 (1965).
29. Roberto Calemzuk, University of Grenoble, 1983.
30. R.A. Enrique and P. Bellon Phys.Rev.B **70**, 224106 (2004).
31. L. Cugliandolo, J. Kurchan, and L. Peliti, Phys.Rev. **E55,** 3898 (1997). J. Kurchan, Nature **433**, 222 (2005).
32. R. Calemczuk and E. Bonjour, J. Physique – Lettres **42**, L-501-L-502, (1981). R. Calemczuk and E.Bonjour, journal of Non-Cristalline-Solids, **43**, 427-432, (1981).

Diagrammatic Approach to the Dynamics of Interacting Brownian Particles: Mode-Coupling Theory, Generalized Mode-Coupling Theory, and All That

Grzegorz Szamel

Department of Chemistry, Colorado State University, Fort Collins, CO 80523, USA.

Abstract. Recently, we proposed a diagrammatic formulation of a theory for the time dependence of density fluctuations in equilibrium systems of interacting Brownian particles [J. Chem. Phys. **127**, 084515 (2007)]. We derived a series expansion for the time-dependent density correlation function in which successive terms are represented by diagrams with two, three, and four-leg vertices. We analyzed the structure of the diagrammatic series and investigated a diagrammatic interpretation of reducible and irreducible memory functions. We showed that the one-loop self-consistent approximation for the latter function coincides with the mode-coupling approximation for Brownian systems that was derived previously using a projection operator approach. Here we review these results and then we investigate a diagrammatic interpretation of a generalized mode-coupling theory that we introduced some time ago [Phys. Rev. Lett. **90**, 228301 (2003)].

Keywords: Brownian Dynamics, Diagrammatic Expansion, Mode-Coupling Theory
PACS: 82.70.Dd, 64.70.pv, 64.70.qd

INTRODUCTION

There has been a lot of interest in recent years in the dynamics of interacting Brownian particles [1, 2]. It has been stimulated by ingenious experimental studies which provided detailed information about the microscopic dynamics of concentrated colloidal systems [3]. In particular, the colloidal glass transition has emerged as a favorite, model glass transition to be studied and mode-coupling theory [4, 5] has become accepted as the theory for the dynamics of concentrated suspensions and their glass transition [7].

The acceptance of the mode-coupling theory overlooks its well known problems. The most important, fundamental problem is the *ad-hoc* nature of the approximation that lies at the heart of the mode-coupling theory: the factorization approximation for a pair-density correlation function. Once this approximation is used, there is no obvious way to extended and/or improve the theory. Incidentally, this feature of the mode-coupling theory is most troubling for Brownian systems since for such systems local density is the only conserved quantity and thus couplings to other modes cannot be invoked. The difficulty with improving upon the factorization approximation partially originates from the fact that the original derivations of the mode-coupling theory for both Newtonian [4] and Brownian [5] systems used the projection operator technique [6]. This technique is very well suited to derive the so-called memory function representation that is the starting point of the mode-coupling theory. However, the projection operator technique cannot be straightforwardly applied to investigate dynamic processes which are neglected within the factorization approximation.

In addition to the above described fundamental problem, the mode-coupling theory has also more practical problems. It is often stated [7] that the predictions of the theory differ from experimental and simulational results by about 10%. For example, it is known that the theory systematically overestimates the so-called dynamic feedback effect and as a result it underestimates by approximately 10% the glass transition volume fraction for a colloidal hard-sphere system [7]. More recently, it has been shown that the theory makes an error of approximately 17% in the dynamical length scale for a polydisperse quasi-hard-sphere system [8]. However, for some other quantities, the difference between the predictions of the mode-coupling theory and experimental and/or simulational results is much bigger than the oft quoted 10-20%. For example, the theory greatly underestimates decoupling of temperature dependence of the self-diffusion coefficient and the structural relaxation time [9], it underestimates the so-called non-Gaussian parameter by about one order of magnitude [9], it qualitatively underestimates deviations of the self part of the van Hove function from a Gaussian distribution [9, 10], and it fails to describe the observed non-trivial wave vector dependence of the relaxation time of the self-intermediate scattering function [9]. It has to be admitted that the latter three examples are concerned with observables that seem to be, in one way or another, related to the existence of the so-called dynamic heterogeneities [11, 12]. The mode-coupling theory relies upon a factorization approximation and therefore does not include any dynamic

CP982, *Complex Systems, 5th International Workshop on Complex Systems*
edited by M. Tokuyama, I. Oppenheim, and H. Nishiyama

fluctuations. Thus, it is not able to properly describe dynamic heterogeneities. On the other hand, all of the above discussed quantities can be obtained from standard two-point functions. Therefore, the failure of the mode-coupling theory to reproduce these experimental findings at least qualitatively suggests that the oft quoted 10-20% accuracy with which the theory describes the time-dependence of collective and self intermediate scattering functions is overly optimistic.

It is quite surprising that in spite of these fundamental and practical problems with the mode-coupling theory, until recently there has been relatively little work on its generalization and/or extension[1]. It has been only during the last 4 years that several different approaches have appeared that try to go beyond the standard mode-coupling theory.

Our approach [14] to address some of the problems of the mode-coupling theory, and its later extensions [15, 16], has been christened generalized mode-coupling theory [16]. The essence of this approach is to delay the factorization approximation of the standard mode-coupling theory. Our starting point was the same as that of the standard mode-coupling theory: the memory function representation of the density correlation function. The memory function can be expressed in terms of a time-dependent pair-density correlation function. We used a projection operator method and derived an exact, formal equation of motion for this function. Next, we applied a factorization approximation to its evolution operator. In this way we obtained a closed set of equations for the density correlation function and the memory function. This approach predicts an ergodicity breaking transition similar to that predicted by the standard mode-coupling theory, but at a higher density.

Our approach was subsequently extended by Wu and Cao [15]. They derived a hierarchy of mode-coupling-like equations and argued that this hierarchy leads to a rapidly convergent predictions for the ergodicity breaking transition density.

Finally, Mayer, Miyazaki and Reichman [16] used the spirit of our approach to avoid the factorization approximation altogether. They considered a hierarchy of so-called schematic models that was inspired by our and Wu and Cao's results. The simplicity of this hierarchy allowed Mayer, Miyazaki and Reichman to show that successive mode-coupling-like closures push the ergodicity breaking transition towards bigger values of the coupling constant. Furthermore, they showed that in the limit of complete hierarchy there is no ergodicity breaking for any finite value of the coupling constant.

We should also mention here two other alternative attempts to go beyond the standard mode-coupling theory. Both of these approaches incorporate activated hopping events in a very explicit way. The first one has been developed by Schweizer and Saltzman [17]. It is based on a non-linear Langevin equation for the single-particle motion derived by Schweizer [18]. This equation describes the competition between a systematic force, derived from an effective, non-equilibrium free energy and a fluctuating force. For large enough densities the systematic force describes localization of the particle and, in the absence of the fluctuating force, it leads to a mode-coupling-like ergodicity breaking transition. The fluctuating force facilitates barrier hopping events, restores ergodicity and replaces the sharp transition by a smooth crossover. The second approach is due to Bagchi, Wolynes and collaborators [19]. Loosely speaking, it incorporates activated events by adding a hopping vertex to the usual mode-coupling equations.

In this contribution we are concerned with the physical content of the generalized mode-coupling approach. The standard mode-coupling theory is known to be equivalent to a self-consistent single-loop approximation [7]. Our goal is to investigate whether the generalized mode-coupling approximation of Ref. [14] is diagrammatically proper (*i.e.* whether it corresponds to a re-summation of a class of diagrams) and whether it adds additional diagrams or removes some of the diagrams included in the standard mode-coupling theory. To address this question we use a recently developed diagrammatic approach [20] to the dynamics of interacting Brownian particles. Briefly, in Ref. [20] we introduced a set of basis functions that consisted of orthogonalized many-particle densities. Next, we derived a hierarchy of equations of motion for time-dependent correlations of orthogonalized many-particle density correlation functions. We simplified the exact hierarchy by keeping only the lowest order cluster expansion terms. An iterative solution of the simplified hierarchy can be represented by diagrams with two, three and four-leg vertices. We analyzed the structure of the diagrammatic series for the time-dependent density correlation function and obtained diagrammatic expressions for reducible and irreducible memory functions. We showed that the one-loop self-consistent approximation for the latter function coincides with the mode-coupling approximation for Brownian systems that was derived previously using a projection operator approach.

The starting point of this contribution is the diagrammatic expansion derived in Ref. [20]. The series expansion can be expressed in terms of un-labeled time-ordered and un-labeled time-unordered diagrams. Either form can be used as a starting point of the memory function analysis. Thus we obtain two different expressions

[1] The so-called extended mode-coupling theory [13] that was proposed shortly after the formulation of the standard version of the theory relies upon coupling to current modes that are absent in Brownian systems. Thus, the extended mode-coupling theory is not applicable for Brownian systems.

for the irreducible memory function. The expression in terms of the time-unordered diagrams leads to the standard mode-coupling theory in a very natural way. A subset of contributions from the series expansion in terms of time-ordered diagrams leads to the generalized mode-coupling theory of Ref. [14].

DIAGRAMMATIC SERIES FOR THE DENSITY CORRELATION FUNCTION

We consider a system of N interacting Brownian particles in volume V. The average density is $n = N/V$. The brackets $\langle ... \rangle$ indicate a canonical ensemble average at temperature T. As shown in Ref. [20], after some preliminary calculations it is convenient we take the thermodynamic limit, $N \to \infty, V \to \infty, N/V = n = const$. This limit is implicit in all formulas in this contribution.

We define the time dependent density correlation function as

$$\langle n(\mathbf{k}_1;t) n^*(\mathbf{k}_2) \rangle , \tag{1}$$

with $n(\mathbf{k}_1;t)$ being the Fourier transform of the microscopic density fluctuation at time t,

$$n(\mathbf{k}_1;t) = \sum_{j=1}^{N} e^{-i\mathbf{k}_1 \cdot \mathbf{r}_j(t)} - \left\langle \sum_{j=1}^{N} e^{-i\mathbf{k}_1 \cdot \mathbf{r}_j(t)} \right\rangle , \tag{2}$$

and $n(\mathbf{k}_2) \equiv n(\mathbf{k}_2;t=0)$. To write down the diagrammatic series it is convenient to express the density correlation function in terms of the so-called response function $G(k;t)$,

$$\theta(t) \langle n_1(\mathbf{k}_1;t) n_1^*(\mathbf{k}_2) \rangle = n G(k;t) S(k) (2\pi)^3 \delta(\mathbf{k}_1 - \mathbf{k}_2). \tag{3}$$

Note that due to the translational invariance, the correlation function $\langle n_1(\mathbf{k}_1;t) n_1^*(\mathbf{k}_2) \rangle$ is diagonal in the wave vector space. The response function is related to the usual collective intermediate scattering function $F(k;t)$,

$$F(k;t) = G(k;t) S(k). \tag{4}$$

The diagrams that occur in a series expansion for the response function consist of the following elements:

- response function $G(k;t)$:
- bare response function $G_0(k;t)$:
- "left" vertex $\mathcal{V}_{12}$:
- "right" vertex $\mathcal{V}_{21}$:
- four-leg vertex $\mathcal{V}_{22}$:
- $(2\pi)^3\delta$ vertex:

The bare response function $G_0(k;t)$ is defined as

$$G_0(k;t) = \theta(t) \exp(-D_0 k^2 t / S(k)), \tag{5}$$

and the explicit expressions for the three- and four-leg vertices are:

$$\mathcal{V}_{12}(\mathbf{k}_1;\mathbf{k}_2,\mathbf{k}_3) = D_0 (2\pi)^3 \delta(\mathbf{k}_1 - \mathbf{k}_2 - \mathbf{k}_3) \mathbf{k}_1 \cdot (c(k_2)\mathbf{k}_2 + c(k_3)\mathbf{k}_3) \tag{6}$$

$$\mathcal{V}_{21}(\mathbf{k}_1,\mathbf{k}_2;\mathbf{k}_3) = n D_0 (2\pi)^3 \delta(\mathbf{k}_1 + \mathbf{k}_2 - \mathbf{k}_3) S(k_1) S(k_2) \times (c(k_1)\mathbf{k}_1 + c(k_2)\mathbf{k}_2) \cdot \mathbf{k}_3 S(k_3)^{-1} \tag{7}$$

$$\mathcal{V}_{22}(\mathbf{k}_1,\mathbf{k}_2;\mathbf{k}_3,\mathbf{k}_4) = n D_0 (2\pi)^3 S(k_1) S(k_2) \delta(\mathbf{k}_1 + \mathbf{k}_2 - \mathbf{k}_3 - \mathbf{k}_4) \times (c(k_1)\mathbf{k}_1 + c(k_2)\mathbf{k}_2) \cdot (c(k_3)\mathbf{k}_3 + c(k_4)\mathbf{k}_4). \tag{8}$$

Finally, the $(2\pi)^3\delta$ vertex is equal to $(2\pi)^3\delta(\mathbf{k}_1 - \mathbf{k}_2)$.

We refer to the leftmost bare response function as the left root, and to the other bare response functions as bonds. The left root is labeled by a wave vector and the bonds are unlabeled. We consider two diagrams to be topologically equivalent if there is a way to assign labels to unlabeled bonds so that the resulting labeled diagrams are topologically equivalent[2]. To evaluate an unlabeled diagram one assigns wave vectors to unlabeled bonds, integrates over all wave vectors (with a $(2\pi)^{-3}$ factor for each integration) except the wave vector corresponding to the left root, integrates over all intermediate times, and divides the result by a symmetry number of the diagram (*i.e.* the number of topologically identical labeled diagrams that can be obtained from a given unlabeled diagram by permutation of the bond labels).

There are two equivalent ways to write down a diagrammatic series for the response function. First, we can express the response function in terms of time-ordered diagrams. These diagrams have the following property: at each intermediate time, in addition to a three- or four-leg vertex there are also $(2\pi)^3\delta$ vertices on each bare response function present (see the fourth diagram in Fig. 1). The response function is given by the following series:

$G(k;t) =$ (9)
sum of all topologically different time-ordered diagrams with a left root labeled k, a right root, G_0 bonds, $\mathcal{V}_{12}$, $\mathcal{V}_{21}$, $\mathcal{V}_{22}$ and $(2\pi)^3\delta$ vertices,

[2] Two labeled diagrams are topologically equivalent if each labeled bond in one diagram connects vertices of the same type as the corresponding labeled bond in the other diagram [21].

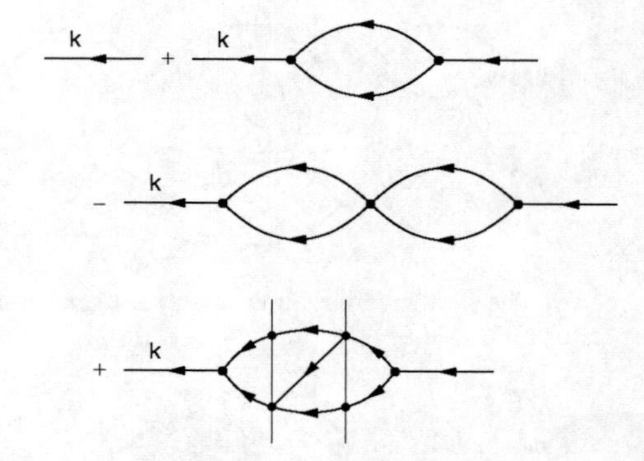

FIGURE 1. First few diagrams in the series (9) for the response function $G(k;t)$. The thin vertical dashed lines in the fourth diagram emphasize that the three-leg vertices and the $(2\pi)^3\delta$ vertices correspond to the same intermediate times.

in which diagrams with odd and even numbers of $\mathcal{V}_{22}$ vertices contribute with overall negative and positive sign, respectively.

The first few time-ordered diagrams contributing to the series (9) are shown in Fig. 1.

The response function can also be expressed in terms of time-unordered diagrams [20]. These diagrams are obtained by integrating over δ functions associated with $(2\pi)^3\delta$ vertices and summing up diagrams with different orderings of intermediate times corresponding to commuting vertices. The response function is then given by the following series:

$$G(k;t) = \tag{10}$$

sum of all topologically different time-unordered diagrams with a left root labeled k, a right root, G_0 bonds, $\mathcal{V}_{12}$, $\mathcal{V}_{21}$ and $\mathcal{V}_{22}$ vertices, in which diagrams with odd and even numbers of $\mathcal{V}_{22}$ vertices contribute with overall negative and positive sign, respectively.

The first few time-unordered diagrams contributing to the series (10) are shown in Fig. 2.

MEMORY FUNCTIONS: REDUCIBLE AND IRREDUCIBLE

The Dyson equation has the usual form

$$\begin{aligned} G(k;t) &= G_0(k;t) + \int dt_1 dt_2 \int \frac{d\mathbf{k}_1}{(2\pi)^3} G_0(k;t-t_1) \\ &\times \Sigma(\mathbf{k},\mathbf{k}_1;t_1-t_2) G(k_1;t_2), \end{aligned} \tag{11}$$

where Σ is the self energy. Diagrammatic representation of the Dyson equation is showed in Fig. 3. Due to the

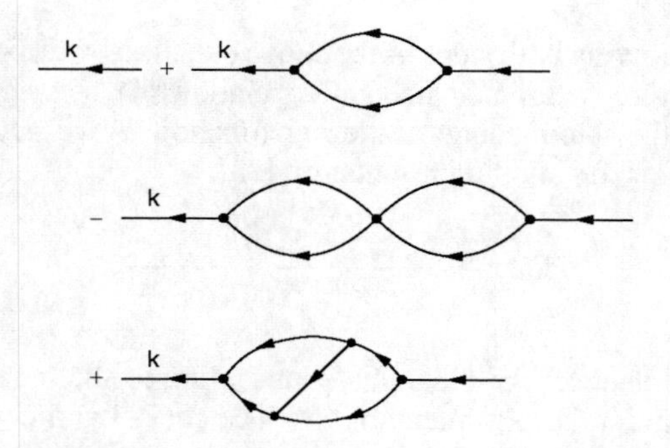

FIGURE 2. First few diagrams in the series (10) for the response function $G(k;t)$ [20].

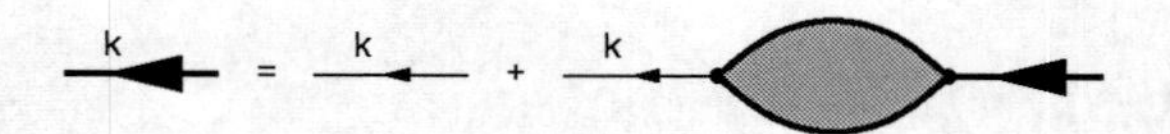

FIGURE 3. Diagrammatic representation of the Dyson equation, Eq. (11) [20].

translational invariance the self-energy is diagonal in wave vector,

$$\Sigma(\mathbf{k},\mathbf{k}_1;t) \propto (2\pi)^3\delta(\mathbf{k}-\mathbf{k}_1). \tag{12}$$

It follows from the Dyson equation that the self-energy Σ is a sum of diagrams that do not separate into disconnected components upon removal of a single bond. We note that the diagrams contributing to the self-energy start with $\mathcal{V}_{21}$ vertex on the right and end with $\mathcal{V}_{12}$ vertex on the left. To define the memory function for a Brownian system we factor out parts of these vertices. First, we define memory matrix $\mathbf{M}$ by factoring out $\mathbf{k}$ from the left vertex and $(D_0/S(k_1))\mathbf{k}_1$ from the right vertex,

$$\Sigma(\mathbf{k},\mathbf{k}_1;t) = D_0\mathbf{k}\cdot\mathbf{M}(\mathbf{k},\mathbf{k}_1;t)\cdot\mathbf{k}_1 S(k_1)^{-1}. \tag{13}$$

Due to the translational and rotational invariance memory matrix $\mathbf{M}$ is diagonal in the wave vector and longitudinal. Thus we can define memory function M through the following relation

$$\mathbf{M}(\mathbf{k},\mathbf{k}_1;t) = M(k;t)\hat{\mathbf{k}}\hat{\mathbf{k}}(2\pi)^3\delta(\mathbf{k}-\mathbf{k}_1). \tag{14}$$

Using Eq. (13) and (14) we can obtain the following equation from the Laplace transform of the Dyson equation,

$$G(k;z) = G_0(k;z) + G_0(k;z)\frac{D_0k^2}{S(k)}M(k;z)G(k;z). \tag{15}$$

Eq. (15) can be solved with respect to (w.r.t.) response function $G(k;z)$. Using the definition of bare response function G_0 we obtain

$$G(k;z) = \frac{1}{z + \frac{D_0k^2}{S(k)}(1-M(k;z))}. \tag{16}$$

Multiplying both sides of the above equation by the static structure factor and using the relation (4) between G and the intermediate scattering function F we get the memory function representation [22],

$$F(k;z) = \frac{S(k)}{z + \frac{D_0 k^2}{S(k)}\left(1 - M(k;z)\right)}. \quad (17)$$

To analyze the diagrams contributing to the memory function it is convenient to introduce cut-out vertices:

$$\begin{aligned} \mathbf{V}^{\mathrm{c}}_{12}(\mathbf{k}_1;\mathbf{k}_2,\mathbf{k}_3) &= D_0(2\pi)^3\delta(\mathbf{k}_1-\mathbf{k}_2-\mathbf{k}_3) \\ &\times \left(c(k_2)\mathbf{k}_2 + c(k_3)\mathbf{k}_3\right) \quad (18) \end{aligned}$$

$$\begin{aligned} \mathbf{V}^{\mathrm{c}}_{21}(\mathbf{k}_1,\mathbf{k}_2;\mathbf{k}_3) &= n(2\pi)^3\delta(\mathbf{k}_1+\mathbf{k}_2-\mathbf{k}_3)S(k_1)S(k_2) \\ &\times \left(c(k_1)\mathbf{k}_1 + c(k_2)\mathbf{k}_2\right). \quad (19) \end{aligned}$$

These vertices are obtained by factoring out $\mathbf{k}_1$ from vertex $\mathscr{V}_{12}$ and $(D_0/S(k_3))\mathbf{k}_3$ from vertex $\mathscr{V}_{21}$.

It should be noted that

$$\begin{aligned} &\mathscr{V}_{22}(\mathbf{k}_1,\mathbf{k}_2;\mathbf{k}_3,\mathbf{k}_4) \\ &= \int \frac{d\mathbf{k}'}{(2\pi)^3}\mathbf{V}^{\mathrm{c}}_{21}(\mathbf{k}_1,\mathbf{k}_2;\mathbf{k}')\cdot\mathbf{V}^{\mathrm{c}}_{12}(\mathbf{k}';\mathbf{k}_3,\mathbf{k}_4). (20) \end{aligned}$$

The diagrammatic rules for functions $\mathbf{V}^{\mathrm{c}}_{12}$ and $\mathbf{V}^{\mathrm{c}}_{21}$ are as follows:

- "left" cut-out vertex $\mathbf{V}^{\mathrm{c}}_{12}$: $<$
- "right" cut-out vertex $\mathbf{V}^{\mathrm{c}}_{21}$: $>$

and we refer to wave vector $\mathbf{k}_1$ in $\mathbf{V}^{\mathrm{c}}_{12}(\mathbf{k}_1;\mathbf{k}_2,\mathbf{k}_3)$ and $\mathbf{k}_3$ in $\mathbf{V}^{\mathrm{c}}_{21}(\mathbf{k}_1,\mathbf{k}_2;\mathbf{k}_3)$ as roots of these vertices.

It follows from the definition of the memory matrix $\mathbf{M}$ that

$\mathbf{M}(\mathbf{k},\mathbf{k}_1;t) =$ (21)
sum of all topologically different time-ordered diagrams which do not separate into disconnected components upon removal of a single bond, with vertex $\mathbf{V}^{\mathrm{c}}_{12}$ with root $\mathbf{k}$ on the left and vertex $\mathbf{V}^{\mathrm{c}}_{21}$ with root $\mathbf{k}_1$ on the right, G_0 bonds, $\mathscr{V}_{12}$, $\mathscr{V}_{21}$, $\mathscr{V}_{22}$ and $(2\pi)^3\delta$ vertices, in which diagrams with odd and even numbers of $\mathscr{V}_{22}$ vertices contribute with overall negative and positive sign, respectively.

The first few diagrams in the series for $\mathbf{M}$ are showed in Fig. 4.

It is obvious that a similar diagrammatic expression for the memory matrix $\mathbf{M}$ can be also obtained in terms of time-unordered diagrams (see Sec. VI of Ref. [20] for details).

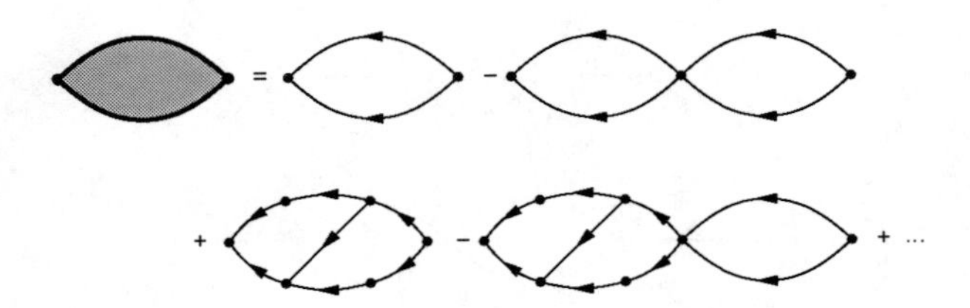

FIGURE 4. The first few diagrams in series expansion for memory matrix $\mathbf{M}$.

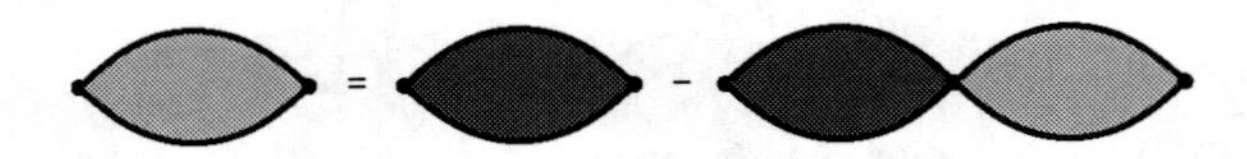

FIGURE 5. Memory matrix $\mathbf{M}$ can be represented as a sum of $\mathbf{M}^{\mathrm{irr}}$ and all other diagrams [20]. The latter diagrams can be re-summed and it is easy to see that as a result we get the second diagram at the right-hand side.

The series expansion for $\mathbf{M}$ consists of diagrams that are one-propagator irreducible (*i.e.* diagrams that do not separate into disconnected components upon removal of a single bond) but not all of these diagrams are completely one-particle irreducible. Some of the diagrams contributing to $\mathbf{M}$ separate into disconnected components upon removal of a $\mathscr{V}_{22}$ vertex (and bonds attached to this vertex). The examples of such diagrams are the second and the fourth diagrams on the right-hand side of the equality sign in Fig. 4.

We define the irreducible memory matrix $\mathbf{M}^{\mathrm{irr}}$ as a sum of only those diagrams in the series for $\mathbf{M}$ that do not separate into disconnected components upon removal of a single $\mathscr{V}_{22}$ vertex. Diagrammatically, we can represent memory matrix $\mathbf{M}$ as a sum of $\mathbf{M}^{\mathrm{irr}}$ and all other diagrams. The latter diagrams can be re-summed as showed in Fig. 5. Using Eq. (20), we can introduce an additional integration over a wave vector and then we see that the diagrammatic equation showed in Fig. 5 corresponds to the following equation,

$$\begin{aligned} &\mathbf{M}(\mathbf{k},\mathbf{k}_1;t) = \mathbf{M}^{\mathrm{irr}}(\mathbf{k},\mathbf{k}_1;t) \quad (22) \\ &- \int dt_1 \int \frac{d\mathbf{k}_2}{(2\pi)^3}\mathbf{M}^{\mathrm{irr}}(\mathbf{k},\mathbf{k}_2;t-t_1)\cdot\mathbf{M}(\mathbf{k}_2,\mathbf{k}_1;t_2) \end{aligned}$$

Again, we use translational and rotational invariance to introduce the irreducible memory function M^{irr},

$$\mathbf{M}^{\mathrm{irr}}(\mathbf{k},\mathbf{k}_1;t) = M^{\mathrm{irr}}(k;t)\hat{\mathbf{k}}\hat{\mathbf{k}}(2\pi)^3\delta(\mathbf{k}-\mathbf{k}_1), \quad (23)$$

Taking Laplace transform of Eq. (22) and then using Eq. (23) we obtain

$$M(k;z) = M^{\mathrm{irr}}(k;z) - M^{\mathrm{irr}}(k;z)M(k;z). \quad (24)$$

This equation can be solved w.r.t. memory function M. Substituting the solution into Eq. (17) we obtain a representation of the intermediate scattering function in terms

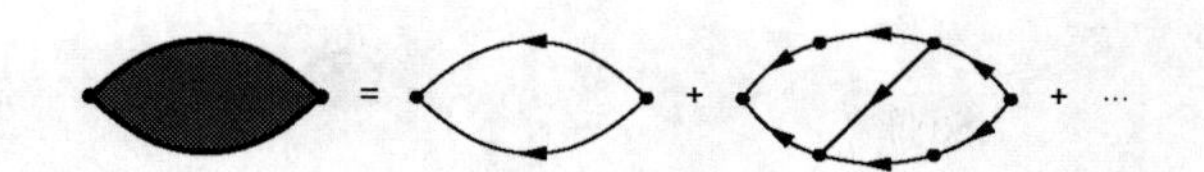

FIGURE 6. The first few diagrams in series expansion for the irreducible memory matrix $\mathbf{M}^{\text{irr}}$.

of the irreducible memory function,

$$F(k;z) = S(k)G(k;z) = \frac{S(k)}{z + \frac{D_0 k^2}{S(k)\left(1+M^{\text{irr}}(k;z)\right)}}. \tag{25}$$

Eq. (25) was first derived by Cichocki and Hess [23] using a projection operator approach.

Diagrammatically,

$$\mathbf{M}^{\text{irr}}(\mathbf{k},\mathbf{k}_1;t) = \tag{26}$$

sum of all topologically different time-ordered diagrams which do not separate into disconnected components upon removal of a single bond or a single $\mathscr{V}_{22}$ vertex, with vertex $\mathbf{V}^{\text{c}}_{12}$ with root $\mathbf{k}$ on the left and vertex $\mathbf{V}^{\text{c}}_{21}$ with root $\mathbf{k}_1$ on the right, G_0 bonds, $\mathscr{V}_{12}$, $\mathscr{V}_{21}$, $\mathscr{V}_{22}$ and $(2\pi)^3\delta$ vertices, in which diagrams with odd and even numbers of $\mathscr{V}_{22}$ vertices contribute with overall negative and positive sign, respectively.

The first few diagrams in the series for $\mathbf{M}^{\text{irr}}$ are shown in Fig. 6.

It is obvious that the series expansion in terms of time-unordered diagrams can be analyzed in the same way. The result of such analysis is a series expansion for the irreducible memory matrix $\mathbf{M}^{\text{irr}}$ in terms of time-unordered diagrams (see Sec. VI of Ref. [20]).

STANDARD MODE-COUPLING APPROXIMATION

To obtain the standard mode-coupling expression for the memory function it is convenient to start from a series expression for $\mathbf{M}^{\text{irr}}$ in terms of time-unordered diagrams. The simplest re-summation of this series includes diagrams that separate into two disconnected components upon removal of the $\mathbf{V}^{\text{c}}_{12}$ and $\mathbf{V}^{\text{c}}_{21}$ vertices. It is easy to see that in such diagrams each of these components is a part of the series for the response function G. Summing all such diagrams we get a one-loop diagram (*i.e.* the first diagram shown on the right-hand side in Fig. 6) but with G_0 bonds replaced by G bonds, see Fig. 7. Thus, we get one-loop self-consistent approximation for the memory matrix,

$$\mathbf{M}^{\text{irr}}(\mathbf{k},\mathbf{k}_1;t) \approx \mathbf{M}^{\text{irr}}_{\text{one-loop}}(\mathbf{k},\mathbf{k}_1;t) = \tag{27}$$

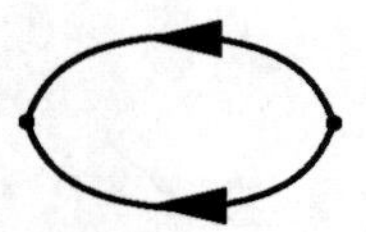

FIGURE 7. Re-summation of diagrams that separate into two disconnected components upon removal of the $\mathbf{V}^{\text{c}}_{12}$ and $\mathbf{V}^{\text{c}}_{21}$ vertices leads to a one-loop diagram with G bonds [20].

$$\int \frac{d\mathbf{k}_2 d\mathbf{k}_3}{2(2\pi)^6} \mathbf{V}^{\text{c}}_{12}(\mathbf{k};\mathbf{k}_2,\mathbf{k}_3) G(k_2;t) G(k_3;t) \mathbf{V}^{\text{c}}_{21}(\mathbf{k}_2,\mathbf{k}_3;\mathbf{k}_1).$$

The factor 2 in the denominator is the symmetry number of the single-loop diagram.

Using explicit expressions (18-19) for the cut-out vertices we easily show that (27) leads to the following expression for the irreducible memory function

$$M^{\text{irr}}(k;t) \approx M^{\text{irr}}_{\text{one-loop}}(k;t) = \tag{28}$$

$$\frac{nD_0}{2}\int \frac{d\mathbf{k}_1}{(2\pi)^3}\left[\hat{\mathbf{k}}\cdot\left(c(k_1)\mathbf{k}_1 + c(|\mathbf{k}-\mathbf{k}_1|)(\mathbf{k}-\mathbf{k}_1)\right)\right]^2$$

$$\times S(k_1) S(|\mathbf{k}-\mathbf{k}_1|) G(k_1;t) G(|\mathbf{k}-\mathbf{k}_1|;t) \equiv M^{\text{irr}}_{MCT}(k;t)$$

As indicated above, the one-loop self-consistent approximation coincides with the standard mode-coupling approximation, *i.e.* both approximations result in exactly the same expression for the irreducible memory function.

GENERALIZED MODE-COUPLING APPROXIMATION

To derive the generalized mode-coupling expression for the memory function which was first obtained in Ref. [14], we start from a series expression for $\mathbf{M}^{\text{irr}}$ in terms of time-unordered diagrams. In Fig. 6 we showed two diagrams that contribute to $\mathbf{M}^{\text{irr}}$; these two diagrams and a few other ones are shown in Fig. 8.

To facilitate the analysis of the diagrams contributing to $\mathbf{M}^{\text{irr}}$ we introduce two definitions. First, we define a two-particle reducible region. This region is a time interval (time slice) that consists of two independent bare response functions or a time instant that consist of a four-leg vertex and a $(2\pi)^3\delta$ vertex. Second, we define a two-particle irreducible region to be a time interval (time slice) that is not two-particle reducible [3].

The first diagram in Fig. 8 consists of one two-particle reducible region. The second diagram consists of two two-particle reducible regions separated by a two-particle irreducible region, *etc.* For all diagrams in Fig.

[3] These definitions are inspired by the definition of a higher order memory function which was introduced in Ref. [14]; the two-particle irreducible region constitutes a contribution to a self-energy for a pair-density correlation function

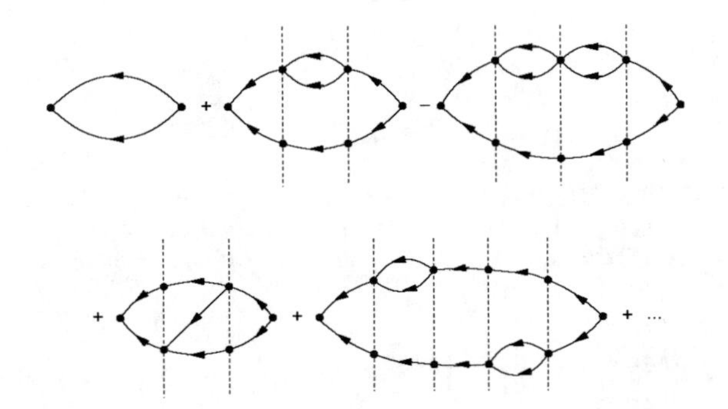

FIGURE 8. The diagrams contributing to irreducible memory matrix $\mathbf{M}^{\text{irr}}$. The thin vertical dashed lines denote the limits of two-particle irreducible regions. Note that the two-particle reducible region that separates two successive two-particle irreducible regions can be a time instant (*e.g.* in the third diagram) or a time interval (*e.g.* in the fifth diagram).

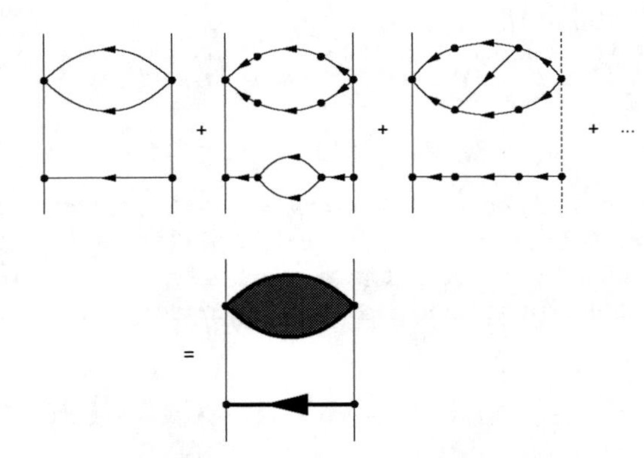

FIGURE 9. First few diagrams that are included in a re-summation which is the first part of the two-step procedure leading to the generalized mode-coupling approximation of Ref. [14].

8, thin vertical dashed lines denote limits of two-particle irreducible regions.

In the language of the present diagrammatic approach the generalized mode-coupling approximation of Ref. [14] consists of the following two steps: first, for a given set of the two-particle reducible regions we sum a class of diagrams that have different two-particle irreducible regions. Then, we sum over all possible arrangements of the two-particle reducible regions. The final result is an approximate self-consistent expression for the memory function in terms of the response function and the memory function.

Specifically, we sum contributions to the two-particle irreducible regions that correspond to diagrams with the structure shown in Fig. 9. These diagrams consist of two disconnected parts. The first one starts with the $\mathbf{V}^{c}_{21}$ vertex, ends with the $\mathbf{V}^{c}_{12}$ vertex, and is one-particle irreducible (*i.e.* it does not separate into disconnected components upon removal of a single bond or a single $\mathcal{V}_{22}$ vertex). The second part starts with a bare response function, ends with a bare response function, and possibly contains vertices and other bare response functions in between.

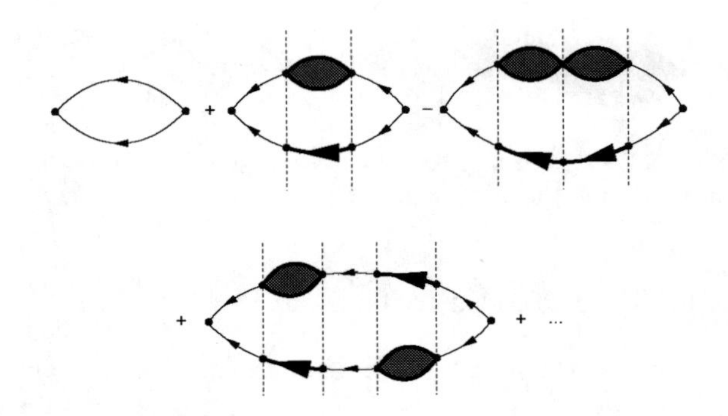

FIGURE 10. First few diagrams that are included in a re-summation which is the second part of the two-step procedure leading to the generalized mode-coupling approximation of Ref. [14].

After the re-summation we get a two-particle irreducible region that consists of two disconnected components. The first component is the irreducible memory matrix and the second one is the full response function.

It can be easily shown that the re-summation over different two-particle reducible regions (see Fig. 10 for the first few contributions to this re-summation) leads to the following equation

$$M^{\text{irr}}(k;z) \approx \frac{nD_0}{2}\int \frac{d\mathbf{k}_1 d\mathbf{k}_2}{(2\pi)^3}\delta(\mathbf{k}-\mathbf{k}_1-\mathbf{k}_2) \qquad (29)$$

$$\times \frac{\left[\hat{\mathbf{k}}\cdot(c(k_1)\mathbf{k}_1+c(k_2)\mathbf{k}_2)\right]^2 S(k_1)S(k_2)}{z+\left[\frac{D_0k_1^2/S(k_1)}{1+LT\left(M^{\text{irr}}(k_1;t)G(k_2;t)\right)}+(1\leftrightarrow 2)\right]} \equiv M^{\text{irr}}_{GMCT}(k;z).$$

Here $LT\left(M^{\text{irr}}(k_1;t)G(k_2;t)\right)$ denotes the Laplace transform of the product of the irreducible memory function and the response function.

Eqs. (25) and (29) form a closed system of equations for the response function and the irreducible memory function. These equations predict an ergodicity breaking transition that is similar to the ergodicity breaking transition predicted by the standard mode-coupling theory (*i.e* to the transition predicted by Eqs. (25) and (28)). The transition predicted for the hard-sphere interaction potential was briefly discussed in Ref. [14]. It was shown that a non-trivial result for the nonergodicity parameter $f(k) = \lim_{t\to\infty} G(k;t)$ appears at $\phi_{GMCT} = .549$ (here ϕ is the volume fraction; it is related to the number density n via the relation $\phi = n\pi\sigma^3/6$ where σ is the hard sphere diameter). We note that for the hard-sphere interaction potential the standard mode-coupling theory predicts the transition at $\phi_{MCT} = .525$ whereas the accepted experimental value is $\phi_{exp} \approx .58$. At the transition, according to both the generalized and the standard mode-coupling approaches, $f(k)$ has a jump. Also, $f(k)$ predicted by Eqs. (25) and (29) at the transition (*i.e.* at $\phi_{GMCT} = .549$) is similar to that of predicted by the standard mode-coupling theory at its transition, $\phi_{MCT} = .525$.

DISCUSSION

The main result presented here is the diagrammatic derivation of the generalized mode-coupling theory presented in Ref. [14]. The very existence of this derivation shows that this theory is diagrammatically proper, *i.e.* it corresponds to a re-summation of a class of diagrams in the expansion for the irreducible memory function. In addition, it is clear that the generalized mode-coupling theory contains fewer diagrams than the standard mode-coupling theory. The latter theory corresponds to a re-summation of *all* the diagrams that separate into two disconnected components upon removal of the leftmost and rightmost vertices. In contrast, the generalized mode-coupling theory corresponds to a re-summation of a sub-class of these diagrams: the diagrams included in the generalized mode-coupling theory, loosely speaking, have the structure shown in Fig. 10. The important point is that while these diagrams separate upon removal of the leftmost and rightmost vertices, they do not factorize. This follows from the fact that these diagrams are time-ordered, *i.e.* the three and four-leg vertices connecting response functions in one part of any of these diagrams occur at the same time instant as $(2\pi)^3\delta$ vertices connecting response functions in the other part.

It can be noted that, at least as far as the location of the ergodicity breaking transition is concerned, the generalized mode-coupling theory is superior to the standard theory. This occurs in spite of the fact that the former theory re-sums fewer diagrams. This fact is not unprecedented: for short-range repulsive interaction potentials the Percus-Yevick integral equation for the equilibrium pair-correlation function is more accurate than the HNC equation in spite of including fewer diagrams [24].

ACKNOWLEDGMENTS

I gratefully acknowledge the support of NSF Grant No. CHE 0517709.

REFERENCES

1. See, *e.g.*, K. Kroy, M.E. Cates, and W.C.K. Poon, *Phys. Rev. Lett.* **92**, 148302-1–148302-4 (2004); J.M. Brader, Th. Voigtmann, M.E. Cates, and M. Fuchs, *Phys. Rev. Lett.* **98**, 058301-1–058301-4 (2007).
2. For a general introduction see J.K.G. Dhont, *An Introduction to Dynamics of Colloids* (Elsevier, New York, 1996).
3. See, *e.g.*, R. Besseling, E. R. Weeks, A. B. Schofield, and W. C. Poon, *Phys. Rev. Lett.* **99**, 028301-1–028301-4 (2007); C. R. Nugent, K. V. Edmond, H. N. Patel, and E. R. Weeks, *Phys. Rev. Lett.* **99**, 025702-1–025702-4 (2007); for a recent review see L. Cipelletti and L. Ramos, *J. Phys. Cond. Matter* **17** R253–R285 (2005).
4. W. Götze, in *Liquids, Freezing and Glass Transition*, J.P. Hansen, D. Levesque, and J. Zinn-Justin, eds. (North-Holland, Amsterdam, 1991).
5. For a derivation for a colloidal system see G. Szamel and H. Löwen, *Phys. Rev. A* **44**, 8215–8219 (1991).
6. See, *e.g.*, R. Zwanzig, *Nonequilibrium Statistical Mechanics*, (Oxford, New York, 2002).
7. M. E. Cates, *Ann. Henri Poincare* **4**, S647-S661 (2003).
8. Th. Voigtmann, A. M. Puertas, and M. Fuchs, *Phys. Rev. E* 70, 061506-1–061506-19 (2004).
9. E. Flenner and G. Szamel, *Phys. Rev. E* **72**, 031508-1–031508-15 (2005).
10. P. Chaudhuri, L. Berthier, and W. Kob, *Phys. Rev. Lett.* **99**, 060604-1–060604-4 (2007).
11. M. Ediger, *Annu. Rev. Phys. Chem.* **51**, 99–128 (2000).
12. For a recent review of so-called single-particle dynamic fluctuations see K.S. Schweizer, *Current Opin. Colloid Interface Sci.*, in press.
13. W. Götze and L. Sjögren, *Z. Phys. B* **65**, 415–427 (1987); S.P. Das and G.F. Mazenko, *Phys. Rev. A* **34**, 2265–282 (1986); see also, R. Schmitz, J. W. Dufty, and P. De, *Phys. Rev. Lett.* **71**, 2066–2069 (1993). The extended mode-coupling approach for Newtonian systems has recently been critically analyzed by M.E. Cates and S. Ramaswamy [*Phys. Rev. Lett.* **96**, 135701-1–135701-4 (2006)].
14. G. Szamel, *Phys. Rev. Lett.* **90**, 228301-1–228301-4 (2003).
15. J. Wu and J. Cao, *Phys. Rev. Lett.* **95**, 078301-1–078301-4 (2005).
16. P. Mayer, K. Miyazaki, and D.R. Reichman, *Phys. Rev. Lett.* **97**, 095702-1–095702-4 (2006).
17. K.S. Schweizer and E.J. Saltzman, *J. Chem. Phys.* **119**, 1181–1196 (2003).
18. K.S. Schweizer, *J. Chem. Phys.* **123**, 244501-1–244501-13 (2005).
19. S.M. Bhattacharya, B. Bagchi, and P.G. Wolynes, *Phys. Rev. E* **72**, 031509-1–031509-12 (2005); arXiv: cond-mat/0702435.
20. G. Szamel, *J. Chem. Phys.* **127**, 084515-1–084515-13 (2007).
21. H.C. Andersen, in *Statistical Mechanics*, Part *A: Equilibrium Techniques*, B.J. Berne, ed. (Plenum, New York, 1977); J.-P. Hansen and I.R. McDonald, *Theory of Simple Liquids*, (Academic, London, 1986).
22. W. Hess and R. Klein, *Adv. Phys.* **32**, 173–283 (1983).
23. B. Cichocki and W. Hess, *Physica A* **141**, 475–488 (1987).
24. J.P. Hansen and J.R. McDonald, *Theory of Simple Liquids*, 2nd ed. (Academic, London, 1986).

Peculiar Behavior of the Secondary Dielectric Relaxation in Propylene Glycol Oligomers near the Glass Transition

K. Grzybowska, A. Grzybowski, and M. Paluch

Institute of Physics, Silesian University, Uniwersytecka 4, 40-007 Katowice, Poland

Abstract. We present results of broadband dielectric measurements of dipolypropylene glycol at ambient pressure in the wide temperature range from 123 to 263 K. The well resolved secondary γ-relaxation exhibits the anomalous behavior near the glass transition which is reflected in the presence of the minimum of the temperature dependence of the γ-relaxation time. To describe the peculiar behavior of γ process in dipolypropylene glycol we propose to use, besides the minimal model of Dyre and Olsen, the nonmonotonic relaxation kinetics model formulated by Ryabov *et al.*

Keywords: Dielectric spectroscopy, Glass transition, Minimal model, Nonmonotonic relaxation kinetics model, Secondary relaxation.
PACS: 64.70.Pf, 61.25.Em, 61.43.Fs, 77.22.Gm

INTRODUCTION

Recent studies have shown that some liquids exhibit a curious and very intriguing behavior of their molecular dynamics when they are reaching the glassy state. During vitrification the reorientations of entire molecules become very slow and the structural relaxation times tend exponentially to infinity, while molecular motions reflected in the secondary relaxation unexpectedly become faster up to some local minimal value of its relaxation time. These phenomena are quite surprisingly since the secondary process were considered for a long time to reflect local motion of the molecule, and then considered to be quite insensitive to the glass transition phenomenon. Recently the existence of such a minimum in the temperature dependence of secondary relaxation times $\tau_\gamma(T)$ near the glass transition has been experimentally confirmed for several polypropylene glycols (PPG) at ambient and high pressure [1,2,3,4]. The minimum in the dependence $\tau_\gamma(T)$ is clearly formed and surprising deep for dipropylene glycol (DPG) and becomes shallower with increasing molecular weight of PPG [4].

The minimal model (MM) [1] based on an asymmetric double-well potential with a barrier U and an asymmetry Δ has been successfully applied to describe the temperature dependences of the secondary γ-relaxation times (Eq. 1),

$$\tau_\gamma = \tau_0 \exp\left(\frac{2U + \Delta}{2kT}\right)\operatorname{sech}\left(\frac{\Delta}{2kT}\right) \quad (1)$$

where k denotes the Boltzmann constant, τ_0 is the preexponential factor. The potential parameters U and Δ are temperature dependent in the liquid state,

$$U = U_0 + akT \ , \ \Delta = \Delta_0 - bkT \quad (1a)$$

and fixed in glass,

$$U = U_0 + akT_f \ , \ \Delta = \Delta_0 - bkT_f \quad (1b)$$

where U_0, Δ_0, a, b, T_f are fitting parameters. The parameter T_f is a temperature below which the liquid structure is frozen in.

In this work we propose to adopt an alternative method to analyze the peculiar behavior of γ process, earlier worked out by Ryabov *et al.* to describe the minimum in the temperature dependence of relaxation time for confined systems [5, 6].

EXPERIMENTAL

Herein we consider dipropylene glycol (DPG) which was purchased from Aldrich. Isobaric measurements of the complex dielectric permittivity $\varepsilon^*(\omega) = \varepsilon'(\omega) - i\varepsilon''(\omega)$ were made using the Novo-Control Alpha dielectric spectrometer at ambient pressure in the broad band from 10^{-2} Hz to 10^6 Hz. The sample temperatures in the range from 123.15 to 263.15 K were controlled by a Quatro System using a nitrogen gas cryostat. The temperature stability was better than 0.1 K.

CP982, *Complex Systems, 5th International Workshop on Complex Systems*
edited by M. Tokuyama, I. Oppenheim, and H. Nishiyama

RESULTS AND DISCUSSION

Examples of the isothermal dielectric loss, ε'', are presented in Fig.1 for DPG over the frequency range of $10^{-2}< f <10^{6}$ Hz. At ambient pressure the dielectric loss spectra exhibit dc-conductivity, a main α-relaxation process, an excess wing (EW) and a well resolved secondary γ relaxation. The EW becomes more prominent at elevated pressure as well as with physical aging and it is regarded as an unresolved secondary β process [7,8].

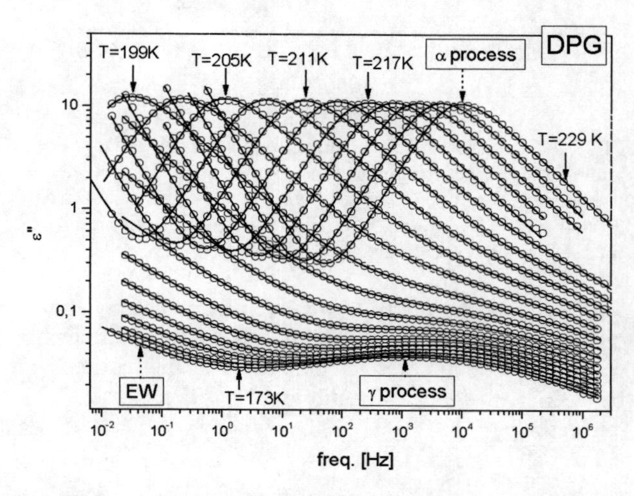

FIGURE 1. Selected dielectric loss spectra at ambient pressure and at different temperatures for DPG. The solid lines indicate the fits of the dielectric spectra to Eq. 2.

In order to obtain the temperature dependencies of relaxation times the best fits of entire spectra were found. After subtraction of the dc conductivity the complex permittivity $\varepsilon^*(\omega)$ data of DPG are well described (see Fig. 1) by superposition of Havriliak-Negami (HN) and Cole-Cole (CC) functions:

$$\varepsilon^*(\omega) = \varepsilon'(\omega) - i\varepsilon''(\omega)$$
$$= \varepsilon_\infty + \frac{\Delta\varepsilon_\alpha}{\left[1+(i\omega\tau_\alpha)^{\lambda_\alpha}\right]^{\delta_\alpha}} + \frac{\Delta\varepsilon_\gamma}{1+(i\omega\tau_\gamma)^{\lambda_\gamma}} \quad (2)$$

for α and γ processes, respectively. Here ε_∞ is the high frequency limit permittivity, $\Delta\varepsilon_\alpha$ and $\Delta\varepsilon_\gamma$ are the relaxation strength of α and γ processes, respectively. The parameter τ_α denotes the HN relaxation time, whereas τ_γ is the CC relaxation time. The exponents λ_α, δ_α, and λ_γ are shape parameters. An example of spectra approximation is shown in Fig. 2 for dielectric loss at 199 K.

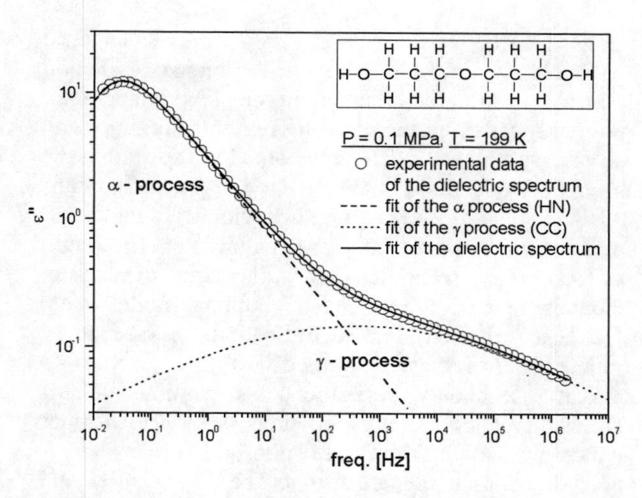

FIGURE 2. Fitting procedure for dielectric spectra of DPG at ambient pressure and T=199K. The solid line indicates the fit of entire spectrum obtained by the sum of HN (dashed line) and CC (dotted line) functions (Eq. 2) which represent α and γ relaxations, respectively.

The γ relaxation of DPG exhibits an intriguing behavior in the vicinity of glass transition temperature T_g. As temperature is reduced a molecular dynamics becomes unexpectedly faster until the loss peak frequency reaches its maximum value, and then it becomes slower in accordance with the Arrhenius law (see Figs. 3 and 4). Such a behavior is reflected by the presence of a minimum in the temperature dependence of γ-relaxation times as can be seen in Fig. 4.

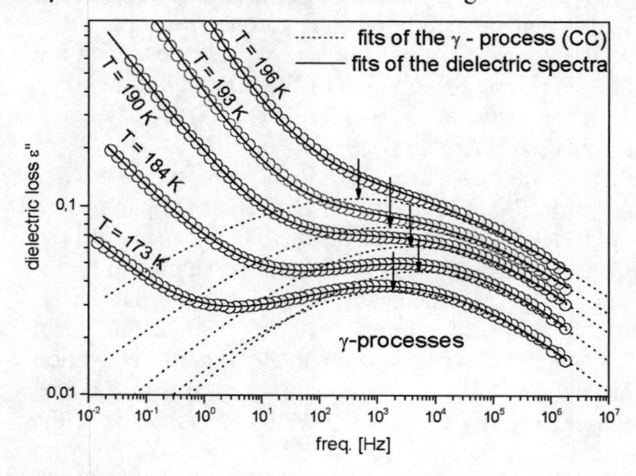

FIGURE 3. Selected dielectric loss spectra of the γ relaxation near the glass transition at ambient pressure and at different temperatures for DPG. The solid lines depict the fits of the dielectric spectra. The dotted lines represent the fits of the γ process. The arrows indicate loss γ-peak maxima obtained by fitting the dielectric spectra, showing the intriguing behavior of γ relaxation.

Consequently, in the vicinity of T_g the dependence $\tau_\gamma(T)$ clearly deviates from the Arrhenius prediction which was usually assumed in the low temperature range as well as in the coupling region of primary and secondary processes. This phenomenon shows that the secondary relaxation is sensitive to the glass transition.

To explain this interesting behavior one should find a molecular origin of the γ process in DPG. The origin of secondary relaxations is still under discussion. However, using the extended coupling model (CM) [9] a criterion for identification of secondary relaxation has been suggested [10]. It enables to classify secondary relaxations as genuine Johari-Goldstein (JG) or non-JG processes. The former concerning both rigid and nonrigid molecules is considered as a precursor of primary relaxation and originates from the motion of entire molecules or essentially all molecular parts. However, the non-JG relaxation reflects rotations of rather some atoms groups in molecules. According to the criterion we have recently established that the γ process in DPG is the non-JG relaxation, whereas the EW can be treated as the hidden JG relaxation [8]. We have suggested that the γ relaxation originates from some reorientations of terminal molecular units containing OH groups and is strongly dependent on hydrogen-bonded network [4] which is created by intra- and intermolecular H-bonds between the OH groups as well as the ether and OH groups. Therefore, the sensitivity of the γ process in DPG to the glass transition seems to be related to the strong effect of the thermodynamic conditions on hydrogen bonding.

Previously [4] we have successfully described the dependence $\tau_\gamma(T)$ in terms of the minimal model (Eq. 1) with parameters U_0=-67.31kJ/mol, Δ_0=24.58kJ/mol, a=69.38, b=12.0, T_f=186K, τ_0=6.5·10^{-17}s. The fit of experimental data for DPG to the minimal model is shown in Fig. 4 as the solid line. The MM parameters yield the linearly decreasing dependence of the potential barrier U with decreasing temperature in liquid (Eq. 1a), whereas in the glassy state the barrier reaches its fixed minimal value U =39.7kJ/mol (Eq. 1b) which is very close to the Arrhenius activation energy, 39.9kJ/mol, obtained from fitting in the low temperature range. The temperature dependence of the potential barrier U reflects the observed slow down in the γ relaxation with increasing temperature near T_g.

Herein we propose to apply an alternative model to describe the peculiar behavior of γ process in DPG, earlier formulated by Ryabov *et al.* [5,6] for confined systems. The nonmonotonic relaxation kinetics model (NRKM) based on the older approach of Macedo and Litovitz [11] considers the relaxation phenomenon as a result of two simultaneous events:

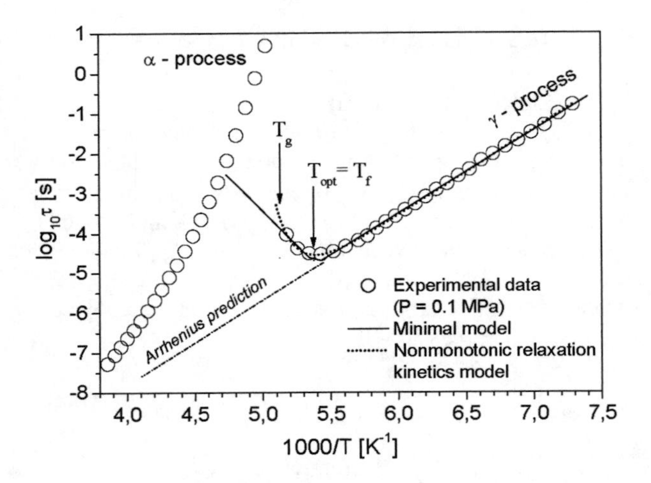

FIGURE 4. Dielectric relaxation map for DPG. The solid and dotted lines indicate the approximations of the temperature dependence of τ_γ obtained by using the minimal model (Eq. 1) and the nonmonotonic relaxation kinetics model (Eq. 7), respectively.

(i) A relaxing entity has enough thermal activation energy to reorient. The probability of attaining sufficient energy by the entity to overcome the potential barrier E_a between two local equilibrium positions can be expressed by

$$p_a \cong \exp\left(-\frac{E_a}{kT}\right). \tag{3}$$

(ii) There is sufficient free volume in the vicinity of the moving molecule to occur the reorientation. The probability of finding a defect of the volume υ_d necessary for the molecular reorientation can be described by

$$p_d \cong \exp\left(-\frac{\upsilon_d}{\upsilon_f}\right) \tag{4}$$

where υ_d is close to the molecule volume, whereas υ_f is the average free volume per defect. Assuming that the number of defects varies with temperature in accordance with the Boltzmann law,

$$N = N_0 \exp\left(-\frac{E_d}{kT}\right), \tag{5}$$

where N_0 is the maximum possible number of defects and E_d is the energy of defect formation, and estimating the volume system as V=$(\upsilon_d$+$\upsilon_f)N$≈$\upsilon_f N$ due to $\upsilon_d \ll \upsilon_f$, one can derive the temperature dependence of the probability of defect formation,

$$p_d \cong \exp\left(-C\exp\left(-\frac{E_d}{kT}\right)\right) \quad (6)$$

where the constant $C=\upsilon_d N_0/V$.

Then, the relaxation time of the system, $\tau_\gamma \sim 1/p$, where $p=p_a p_d$ is the probability of the relaxation act, can be expressed by using Eqs. 3 and 6 as

$$\tau_\gamma = \tau_0 \exp\left[\frac{E_a}{kT} + C\exp\left(-\frac{E_d}{kT}\right)\right]. \quad (7)$$

We have described the temperature dependence of the γ-relaxation times for DPG by using the NRKM (Eq. 7) with four fitting parameters E_a=40.93kJ/mol, E_d=56.10kJ/mol, $C=4.5\cdot10^{15}$, $\tau_0=4.2\cdot10^{-17}$s. It is shown in Fig. 4 that the dotted line obtained from the fitting procedure approximates quite well the experimental points. As has been mentioned we suggested that the γ relaxation is the non-JG process and originates from some reorientations of terminal molecular units containing hydroxyl groups. This idea and the mechanism proposed in the NRKM lead us to the further conclusion. In the case of the γ process in DPG the reorientations of OH groups require the thermal activation energy E_a according to the Arrhenius law. However, hydroxyl groups are mainly responsible in DPG for forming strongly hydrogen bonded network. Therefore, to enable the reorientations of the terminal units of DPG molecules some defects have to occur in the network near the molecules, whereas the energy E_d is necessary for such a defect formation.

Both the considered models (Eqs. 1 and 7) are some generalizations about the Arrhenius law, $\tau_\gamma = \tau_0 \exp(E/kT)$, which well describes the dependences of secondary relaxation times in low temperatures, thus some parameters of the models are analogous and their values should be nearly the same for a given material. Comparing the values of parameters obtained from fitting the temperature dependence of the γ relaxation times for DPG by means of the considered models one can notice that the barrier of double-well potential in the MM for the glassy state (U =39.7kJ/mol) and the activation energy in the NRKM (E_a=40.93kJ/mol) are very close to each other and to the Arrhenius activation energy (E=39.9kJ/mol). Similarly, the values of preexponential factors τ_0 are approximately the same. Moreover, the optimal temperature $T_{opt} = E_d / k\ln(CE_d/E_a)$ established by using the NRKM [6], at which the dependence $\tau_\gamma(T)$ reaches its local minimum, is the same for DPG as the fictive temperature T_f=186K found within the framework of the MM. Consequently, one can state that the approximations are consistent to each other. Both the low temperature range and the minimum in the dependence $\tau_\gamma(T)$ can be satisfactory described by using each of the models. However, the considered models can provide us with slightly different kinds of information. The MM refers to changes in the structure of the relaxing system, whereas the NRKM considers the events contributing to the relaxation act. Therefore, we think that further tests and analyses in terms of the two models will cause our better understanding of the mechanisms governing the secondary processes and their peculiar behaviors near the glass transition, especially that we have obtained promising preliminary results also by using the NRKM for several other polypropylene glycols.

ACKNOWLEDGEMENTS

This research was supported by the State Committee for Scientific Research KBN, grants no. 1 P03B 075 28 and N202 148 32/4241, Poland.

REFERENCES

1. J. C. Dyre and N. B. Olsen, *Phys. Rev. Lett.* **91**, 155703-155706 (2003).
2. S. Pawlus, S. Hensel-Bielowka, K. Grzybowska, J. Zioło, and M. Paluch, *Phys. Rev. B* **71**, 174107-174110 (2005).
3. K. Grzybowska, A. Grzybowski, M. Paluch, and S. Cappacioli, *J. Chem. Phys.* **125**, 044904-044911 (2006).
4. K. Grzybowska, A. Grzybowski, J. Ziolo, S. J. Rzoska, and M. Paluch, *J. Phys.: Condens. Matter* **19**, 376105-376115 (2007).
5. Ya. Ryabov, A. Gutina, V. Arkhipov, and Yu. Feldman, *J. Phys. Chem. B* **105**, 1845-1850 (2001).
6. Ya. Ryabov, A. Puzenko, Y. Feldman, *Phys. Rev. B* **69**, 014204-014213 (2004).
7. R. Casalini and C. M. Roland, *Phys. Rev. B* **69**, 094202-094208 (2004).
8. K. Grzybowska, S. Pawlus, M. Mierzwa, M. Paluch, and K. L. Ngai, *J. Chem. Phys.* **125**, 144507-144514 (2006).
9. K. L. Ngai, *J. Phys.: Condens. Matter* **15**, S1107-S1125 (2003).
10. K. L. Ngai and M. Paluch, *J. Chem. Phys.* **120**, 857-873 (2004).
11. P. B. Macedo and T. A.Litovitz, *J. Chem. Phys.* **42**, 245-256 (1965).

Statistical Mechanics of Time Independent Non-Dissipative Nonequilibrium States

Stephen R. Williams and Denis J. Evans

Research School of Chemistry, Australian National University, Canberra, ACT 0200, Australia

Abstract. Amorphous solids are typically nonergodic and thus a more general formulation of statistical mechanics, with a clear link to thermodynamics, is required. We present a rigorous development of the nonergodic statistical mechanics and the resulting thermodynamics for a canonical ensemble, where the $6N$ dimensional phase space contains a set of distinct nonoverlapping domains. An ensemble member which is initially in one domain is assumed to remain there for a time long enough that the distribution within the domain is Boltzmann weighted. The number of ensemble members in each domain is arbitrary. The lack of an a priori specification of the number of members in each domain is a key differences between the work presented here and existing energy landscape treatments of the glass transition. Another important difference is that the derivation starts with the phase space distribution function rather than an equilibrium expression for the free energy. The utility of this newly derived statistical mechanics is demonstrated by deriving an expression for the heat capacity of the ensemble. Computer simulations on a model glass former are used to provide a demonstration of the validity of this result which is different to the predictions of standard equilibrium statistical mechanics.

Keywords: glass, statistical mechanics, nonergodicity
PACS: 64.70.Pf, 05.70.-a, 05.40.-a

INTRODUCTION

A history dependent amorphous solid is typically in a state which is evolving so slowly that it may be treated as stationary, on any time scale of interest, and hence it appears nondissipative. For glass, in contrast to very viscous liquids [1], the oldest undisturbed samples dating back over some two thousand years, show no evidence of flow due to gravity [2, 3]. A feature of such a material is its nonergodic character.

To understand this nonergodic character it is necessary to consider an ensemble. Imagine that we prepare a large number of systems by an identical protocol. This could involve a large number of initially ergodic samples, at a relatively high temperature, in an identical thermodynamic state, say the same temperature T, number of particles N and volume V. At some stage we subject each of these systems to an identical, macroscopically controlled, protocol such as a temperature quench. Given that we wait a sufficient time after this quench, we will have a large number of identically prepared nondissipative amorphous solids. However from the microscopic point of view the various ensemble members may be significantly different due to nonergodicity. If we increase the number of particles, N, in each ensemble member the relative macroscopic differences will become smaller, presumably as $\sim 1/\sqrt{N}$, and if the ensemble members are large enough they will all seem macroscopically identical.

The available $6N$ dimensional phase space may be decomposed into a number of distinct domains. The coordinates of a single ensemble member may be followed with time to obtain a trajectory. For glass a given trajectory remains confined to a given domain. Within each of these domains the dynamics is ergodic, it is the inability of trajectories to change domains which is responsible for the nonergodicity. The number of ensemble members in each of these domains will depend on the macroscopic protocol by which the ensemble was prepared, i.e. the ensemble has a history dependence. So from the microscopic point of view, the state of an ensemble member will depend on the domain D_α that the trajectory occupies, in addition to the other state variables N, V and T.

The energy landscape [4] treatment of Stillinger and Weber [5, 6, 7] does not allow for population levels in the various domains to vary from the equilibrium Boltzmann weightings. Thus this treatment is not suitable for applying to a system which has a history dependence. For this reason the approach has been extended by the addition of a single fictive parameter [8]. Sciortino [8] has pointed out that this extension is not able to accommodate the observation that systems, with the same macroscopic parameters but different preparative histories, may have different macroscopic properties. Further he identifies the problem of recovering a thermodynamic description for a system, where the domains have arbitrary population levels. Recently we have addressed this problem, focusing of the isothermal isobaric ensemble [9]. Here we provide a more concise account in the canonical ensemble, and focus on specifying the theory. For a more comprehensive analysis and justification of the various assumptions see ref. [9].

CP982, *Complex Systems, 5th International Workshop on Complex Systems*
edited by M. Tokuyama, I. Oppenheim, and H. Nishiyama

THEORY

Equations of Motion and The Phase Space Distribution Function

We will use the Nosé Hoover [10] equations of motion. These equations preserve the canonical distribution. In contrast to the standard Hamiltonian equations of motion, a single trajectory is able to access the range of energy levels present in the canonical ensemble. It is known that these equations of motion do not lead to any artifacts, in the systems linear response to an external field, and that the various correlation functions are effected by the thermostat at most by $\mathscr{O}(1/N)$ [11]. The equations of motion are

$$\begin{aligned} \dot{\mathbf{q}}_i &= \frac{\mathbf{p}_i}{m} \\ \dot{\mathbf{p}}_i &= \mathbf{F}_i - \zeta \mathbf{p}_i \\ \dot{\zeta} &= \left(\frac{\sum_{i=1}^{N} \mathbf{p}_i \cdot \mathbf{p}_i / m_i}{3Nk_B T} - 1 + \frac{1}{N} \right) \frac{1}{\tau_\zeta^2}, \end{aligned} \quad (1)$$

where $\mathbf{q}_i$ & $\mathbf{p}_i$ are the position and momentum vectors for the ith particle, k_B is Boltzmann's constant, m_i is the particle mass, $\mathbf{F}_i$ is the force on the ith particle due to the other particles and T is the temperature. The temperature may be expressed in terms of the average kinetic energy $3(N-1)k_B T = \sum_{i=1}^{N} \langle \mathbf{p}_i \cdot \mathbf{p}_i \rangle / m_i$.

We can use the Liouville equation to obtain the following particular time independent solution for the distribution function,

$$f(\mathbf{\Gamma}, \zeta) \sim \exp(-\beta H_E(\mathbf{\Gamma}, \zeta)) \quad (2)$$

$$H_E(\mathbf{\Gamma}, \zeta) = H_0(\mathbf{\Gamma}) + \frac{3}{2} N k_B T \zeta^2 \tau_\zeta^2 \quad (3)$$

$$H_0(\mathbf{\Gamma}) = \sum_{i=1}^{N} \mathbf{p}_i \cdot \mathbf{p}_i + \Phi(\mathbf{q}) \quad (4)$$

where $\mathbf{\Gamma} = (\mathbf{q}, \mathbf{p})$ is the $6N$ dimensional phase space vector representing all of the particle postions and momenta and $\Phi(\mathbf{q})$ is the total interparticle potential energy. If the system is ergodic we obtain the expression for the distribution function,

$$f(\mathbf{\Gamma}, \zeta) = \tau_\zeta \sqrt{\frac{3}{2} \frac{N}{\pi}} \exp\left(-\frac{3}{2} N \zeta^2 \tau_\zeta^2 \right) f_0(\mathbf{\Gamma}), \quad (5)$$

where the $f_0(\mathbf{\Gamma})$ is the standard canonical distribution function,

$$f_0(\mathbf{\Gamma}) = \frac{\exp(-\beta H_0(\mathbf{\Gamma}))}{\int_D d\mathbf{\Gamma} \exp(-\beta H_0(\mathbf{\Gamma}))}, \quad (6)$$

where the integral is limited by the domain D and,

$$H_0(\mathbf{\Gamma}) = \sum_{i=1}^{N} \mathbf{p}_i \cdot \mathbf{p}_i + \Phi(\mathbf{q}), \quad (7)$$

is the Hamiltonian in the absence of the thermostat. Note that in Eq. 5 the standard canonical distribution function is statistically independent from the rest of the distribution function which depends only on ζ.

Our assertion is that the distribution function for a history dependent, nonergodic, solid features multiple domains. We have provided strong evidence, in the form of computer simulation results, along with convincing theoretical arguments for this assertion elsewhere [9]. It is only the positional coordinates for which the domains differ and the distribution function is,

$$\begin{aligned} f(\mathbf{\Gamma}, \zeta) &= \tau_\zeta \sqrt{\frac{3}{2} \frac{N}{\pi}} \exp\left(-\frac{3}{2} N \zeta^2 \tau_\zeta^2 \right) f(\mathbf{\Gamma}) \\ f(\mathbf{\Gamma}) &= \sum_{i=1}^{N_D} w_\alpha s(\mathbf{\Gamma}, D_\alpha) f_\alpha(\mathbf{\Gamma}) \\ \sum_{\alpha=1}^{N_D} w_\alpha &= 1 \end{aligned} \quad (8)$$

where N_D is the total number of domains and D_α is the αth domain which limits the positional coordinate vector $\mathbf{q}$ to a contiguous region of phase space and each component of the momentum to $\pm\infty$. The domains do not overlap so each allowed value of $\mathbf{\Gamma}$ is uniquely associated with one of them. This allows us to introduce the switch $s(\mathbf{\Gamma}, D_\alpha)$ which is set to unity if $\mathbf{\Gamma}$ is inside the domain D_α and zero otherwise. The weight w_α is determined by the portion of ensemble members which reside in the domain D_α. The distribution function for the αth domain is given by,

$$f_\alpha(\mathbf{\Gamma}) = \frac{\exp(-\beta H_0(\mathbf{\Gamma}))}{\int_{D_\alpha} d\mathbf{\Gamma} \exp(-\beta H_0(\mathbf{\Gamma}))}. \quad (9)$$

An ensemble average may now be computed using Eqs. 8 & 9. The average of some phase variable $B(\mathbf{\Gamma})$ is,

$$\begin{aligned} \langle B(\mathbf{\Gamma}) \rangle &= \int_D d\mathbf{\Gamma} B(\mathbf{\Gamma}) f(\mathbf{\Gamma}), \\ &= \sum_{\alpha=1}^{N_D} w_\alpha \int_{D_\alpha} d\mathbf{\Gamma} f_\alpha(\mathbf{\Gamma}) B(\mathbf{\Gamma}), \\ &= \sum_{\alpha=1}^{N_D} w_\alpha \langle B(\mathbf{\Gamma}) \rangle_\alpha \end{aligned} \quad (10)$$

where D is a contiguous domain formed from the sum of all the domains D_α.

Thermodynamic Connection

As it turns out the multi-domain distribution function, Eqs. 8 & 9, defines the temperature through the parameter $\beta = 1/(k_B T)$. To obtain a connection with thermodynamics it remains to obtain an expression for the entropy. This necessitates the introduction of robust domains. Consider the nth derivative of the function

$$A_\alpha = -k_B T \ln \int_{D_\alpha} d\mathbf{\Gamma} \exp(-\beta H_0(\mathbf{\Gamma})) \tag{11}$$

taken with respect to a state variable. If any change in the domain D_α makes no contribution to the derivative then the domain is robust to order n. If we wish to calculate a variable that is related to the thermodynamic potential by a first derivative, i.e. the internal energy, then the domains must be robust to at least first order. If we wish to calculate a second order quantity, such as the heat capacity, then the domains must be robust to at least second order. An investigation of domain robustness can be found in reference [9]. There an independent test of the robustness condition was carried out using the transient fluctuation theorem [12]. This showed, at temperatures below the glass transition, that the domains were robust to finite changes in the temperature or pressure.

The standard expression for the entropy is

$$S = -k_B \langle \ln(f(\mathbf{\Gamma})) \rangle . \tag{12}$$

Combining Eqs. 8, 10 & 12 we obtain,

$$\begin{aligned} S &= -k_B \left[\sum_{\alpha=1}^{N_D} w_\alpha \int_{D_\alpha} d\mathbf{\Gamma} f_\alpha(\mathbf{\Gamma}) \ln f_\alpha(\mathbf{\Gamma}) \right. \\ &\quad \left. + \sum_{\alpha=1}^{N_D} w_\alpha \ln w_\alpha \right] . \end{aligned} \tag{13}$$

We note that this expression does not take account of indistinguishable particles. If we wished to do this we would need to subtract the sum, $k_B \sum_{i=1}^{N_{sp}} N_i!$, from the entropy, S, given by Eq. 13, where N_{sp} is the number of chemical species and N_i is the number of particles of the ith species. As we are not presently interested in the mixing effects of multi-component systems we will not worry about this here.

The next step is to obtain an expression for the thermodynamic potential which, in the case of the canonical ensemble, is the Helmholtz free energy. The Helmholtz free energy is defined as,

$$A \equiv U - TS, \tag{14}$$

where $U = \langle H_0(\mathbf{\Gamma}) \rangle$ is the internal energy. We now define the interdomain entropy as,

$$S_D = -k_B \sum_{\alpha=1}^{N_D} w_\alpha \ln(w_\alpha), \tag{15}$$

and combining Eqs. 9, 10, 13 & 15 we obtain

$$\begin{aligned} S &= S_D + \sum_{\alpha=1}^{N_D} w_\alpha \langle H_0(\mathbf{\Gamma}) \rangle_\alpha / T \\ &\quad + k_B \sum_{\alpha=1}^{N_D} w_\alpha \ln \int_{D_\alpha} d\mathbf{\Gamma} \exp(-\beta H_0(\mathbf{\Gamma})). \end{aligned} \tag{16}$$

Combining Eqs 14 & 16 we arrive at the microscopic expression for the free energy

$$\begin{aligned} A &= -k_B T \sum_{\alpha=1}^{N_D} w_\alpha \ln \int_{D_\alpha} d\mathbf{\Gamma} \exp(-\beta H_0(\mathbf{\Gamma})) - T S_D \\ &= \sum_{\alpha=1}^{N_D} w_\alpha A_\alpha - T S_D . \end{aligned} \tag{17}$$

As it turns out the free energy given by Eq. 17 is a minimum when the weights are consistent with the Boltzmann distribution, i.e. when

$$w_\alpha = \frac{\int_{D_\alpha} d\mathbf{\Gamma} \exp(-\beta H_0(\mathbf{\Gamma}))}{\sum_{\alpha=1}^{N_D} \int_{D_\alpha} d\mathbf{\Gamma} \exp(-\beta H_0(\mathbf{\Gamma}))} . \tag{18}$$

If the weights are given by Eq. 18 then Eq. 17 becomes the same as the standard equilibrium expression for the free energy in an ergodic system. It can be mathematically proved that the free energy is a minimum here, when Eq. 18 is obeyed, by the same method shown in detail for the isothermal isobaric ensemble in reference [9]. In general a history dependent ensemble is defined by the weights, w_α, in addition to the temperature T and the volume V.

If we reversably change the temperature T or the volume V, while holding the domain weights fixed, the entropy Eq. 16 obeys the second law equality, $\Delta Q = T \Delta S$ where ΔQ is the change in heat. Thus if we subject the ensemble to a reversable change in the temperature, from Eq. 14 we obtain,

$$\left. \frac{dA}{dT} \right|_{V, N_i, w_\alpha} = -S. \tag{19}$$

We can easily verify Eq 19 by taking the derivative of Eq 17 and comparing the result with Eq. 16. If we subject the ensemble to a reversable change in the volume from Eq. 14 we obtain,

$$\left. \frac{dA}{dV} \right|_{T, N_i, w_\alpha} = -\langle P \rangle . \tag{20}$$

To verify this we take the derivative of Eq. 17 to obtain

$$\begin{aligned} \left. \frac{dA}{dV} \right|_{T, N_i, w_\alpha} &= \sum_{\alpha=1}^{N_D} w_\alpha \left. \frac{dA_\alpha}{dV} \right|_{T, N_i} \\ &= -\sum_{\alpha=1}^{N_D} w_\alpha \langle P \rangle_\alpha = -\langle P \rangle . \end{aligned} \tag{21}$$

Application to the Heat Capacity

The internal energy U can be obtained from a knowledge of the free energy, Eq. 14, by taking the derivative,

$$U = -k_B T^2 \frac{d(\beta A)}{dT}\bigg|_{V,N_i,w_\alpha} . \tag{22}$$

An expression for the nonergodic heat capacity can now be obtained by combining Eq. 17 & 22 thus,

$$C_V = \frac{dU}{dT}\bigg|_{V,N_i,w_\alpha} = \frac{1}{k_B T^2} \sum_{\alpha=1}^{N_D} w_\alpha \left(\langle H_0^2 \rangle_\alpha - \langle H_0 \rangle_\alpha^2 \right) . \tag{23}$$

In the case of an ergodic system there will be only one domain, $w_1 = 1$, and Eq. 23 will become the same as the standard well known equilibrium formula for the heat capacity, i.e.

$$C_V = \frac{1}{k_B T^2} \left(\langle H_0^2 \rangle - \langle H_0 \rangle^2 \right) . \tag{24}$$

For a nonergodic ensemble the theory presented here is unambiguous about how the heat capacity must be computed, i.e. by Eq. 23 rather than by 24.

SIMULATION, RESULTS AND DISCUSSION

As a demonstration and test of the theory we will compute the heat capacity from computer simulations on a model glass former using both Eqs. 23 & 24 as well as by central difference.

Simulation Details

We employ a binary mixture of nonadditive WCA particles to form a model glass former for which crystallisation is strongly frustrated. This is simply a WCA version of the Kob-Andersen mixture [13] which we have employed previously [9, 14]. The pairwise additive potential is

$$u_{ij}(r_{ij}) = 4\varepsilon_{\alpha\beta} \left[\left(\frac{\sigma_{\alpha\beta}}{r_{ij}} \right)^{12} - \left(\frac{\sigma_{\alpha\beta}}{r_{ij}} \right)^{6} + \frac{1}{4} \right] , \tag{25}$$

for $r_{ij} < \sqrt[6]{2}\sigma_{\alpha\beta}$ and $u_{ij} = 0$ otherwise, where the species identities of particles i and j, either A or B, are denoted by the subscripts α and β. The energy parameters are set $\varepsilon_{BB} = 0.5\,\varepsilon_{AA}$, $\varepsilon_{AB} = 1.5\,\varepsilon_{AA}$ and the particle interaction distances $\sigma_{BB} = 0.88\,\sigma_{AA}$, $\sigma_{BB} = 0.88\,\sigma_{AA}$. The energy unit is ε_{AA}, the length unit is σ_{AA} and the time unit is $\sqrt{m\sigma_{AA}^2/\varepsilon_{AA}}$ with both species having the same mass m. The resulting temperature unit is ε_{AA}/k_B. The composition is set to $X = N_B/N_A = 0.2$, the number of particles is $N = N_A + N_B = 108$ and the density is set to $N/V = 1.255$ which coincides with the average density of the constant pressure simulations in reference [9] at the nominal glass transition temperature (using data on the diffusion). Those constant pressure, $P = 14$, simulations found the nominal glass transition temperature to be $T = 0.435$. Assuming the equivalence of ensembles we use this as the glass transition temperature here.The equations of motion were integrated using a fourth order Runge-Kutta method [15]. The time step used was $dt = 0.002$.

An ensemble, formed from 100 simulations, was computed. The time to equilibrate after each temperature change was $t_{eq} = 2,400$ and the time used for averaging at each temperature was $t_p = 12,000$. Each of the ensemble members was initially equilibrated at a temperature of $T = 5$, it was then quenched to a temperature of $T = 2.5$ by instantly changing the parameter T in Eq. 1 and equilibrated. Fluctuation data was then recorded at a series of temperatures[1].

Results and Discussion

The heat capacity was calculated three different ways: by the standard *ergodic average* Eq. 24, by the *nonergodic average* Eq. 23 and by using the *central difference* formula for the change in internal energy[2]. In the case of the nonergodic average, each simulation was interpreted as being confined to a single domain and the time average for that simulation was considered to be equivalent to the domain's ensemble average. Thus by obtaining a time averaged estimate of, $\langle H_0^2 \rangle_\alpha - \langle H_0 \rangle_\alpha^2$, for each simulation and then averaging these together we are able to obtain a result representative of Eq. 23. The results are shown in Fig. 1, where for the higher temperatures, $T \geqslant 0.7$, it can be seen that all three methods, for calculating the heat capacity, give consistent results. In this ergodic region there is only one domain and thus Eq. 23 and Eq. 24 are equivalent. As the temperature is lowered below $T = 0.7$ results indicative of a loss of ergodicity may be observed. At the temperature of approximately $T = 0.6$ all three methods diverge. If the simulations were run for considerably longer times the temperature where this divergence is first observed would be lowered [16]. The divergence is due to a given trajec-

---

[1] T = 2.5, 2, 1.75, 1.5, 1.25, 1, 0.9, 0.8, 0.7, 0.6, 0.55, 0.5, 0.45, 0.4375, 0.425, 0.4, 0.375, 0.35, 0.3

[2] $C_V(T) = (\langle H_0(T+\Delta T)\rangle - \langle H_0(T-\Delta T)\rangle)/(2\Delta T)$

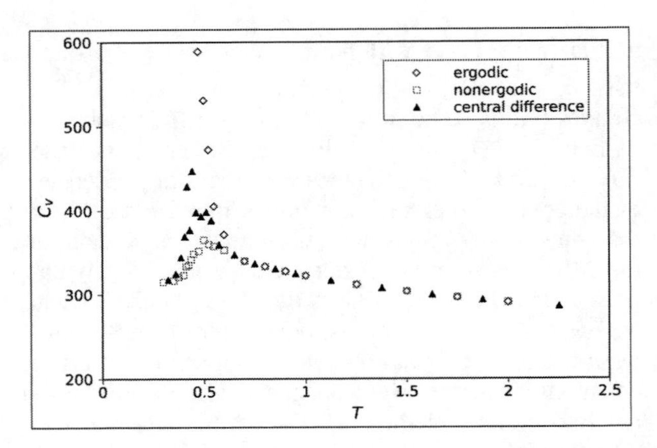

FIGURE 1. (Color online) The heat capacity, C_V, calculated using Eq. 24 (ergodic), Eq. 23 (nonergodic), and the central difference formula.

tory only being able to explore a limited amount of phase space, constituting a domain, over the time scale of the simulation. The amount of time necessary to explore an amount of phase space, representative of the entire available phase space, grows dramatically as the temperature is decreased. While the data sets from the central difference calculations and from the nonergodic average show a peak in the heat capacity, the naive use of the ergodic average does not.

In the vicinity of the peak the central difference and nonergodic average give different results. This is due to the domains not being robust in this region [9]. As the temperatrue is reduced further these two methods for calculating the heat capacity once again converge. In contrast the ergodic average continues to increase as the temperature is reduced, even at the lowest temperatures, in a manor that is not related to the actual heat capacity of the material. At very low temperatures we see the heat capacity coincides with the expectation from the Dulong-Petit law [17], $C_V = 3Nk_B$, which gives the heat capacity for a system where the degrees of freedom may be treated as independent. The expectation that at low temperatures, the potential energy surface of the glass may be approximated as a set of independent harmonic oscillators, upon a suitable orthogonal transformation [4] is consistent with this.

CONCLUSIONS

A rigorous development of equilibrium statistical mechanics and thermodynamics has been presented for nonergodic nondissipative systems in the canonical ensemble. The development requires the introduction of robust phase space domains. The conditions necessary for robustness have been discussed in more detail and subject to independent tests involving transient fluctuation relations elsewhere [9]. Here we have shown that this approach results in a correct calculation of the heat capacity at low temperatures for molecular dynamics simulations on a model glass former. While we have focused on amorphous solids here, the approach is applicable and necessary for many other solid materials such as polycrystalline states. Indeed one could argue that, in the case of solid materials, it is the norm rather than the exception that this formulation of equilibrium statistical mechanics is required.

ACKNOWLEDGMENTS

We thank the Australian Partnership for Advanced Computing (APAC) for computational facilities and the Australian Research Council (ARC) for Funding. We thank Henk van Beijeren for helpful discusions.

REFERENCES

1. R. Edgeworth, B. J. Dalton, and T. Parnell, *Eur. J. Phys.* **5**, 198–200 (1984).
2. E. D. Zanotto, *Am. J. Phys.* **66**, 392–395 (1998).
3. E. D. Zanotto, and P. K. Gupta, *Am. J. Phys.* **67**, 260–262 (1999).
4. D. J. Wales, *Energy Landscapes*, Cambridge University Press, Cambridge, 2003.
5. F. H. Stillinger, and T. A. Weber, *Phys. Rev. A* **25**, 978–989 (1982).
6. F. H. Stillinger, and T. A. Weber, *Science* **225**, 983–989 (1984).
7. P. G. Debenedetti, and F. H. Stillinger, *Nature* **410**, 259–267 (2001).
8. F. Sciortino, *J. Stat. Mech.* p. P05015 (2005).
9. S. R. Williams, and D. J. Evans, *arXiv:cond-mat/0706.0753* (2007). *J. Chem. Phys.* in press DOI: 10.1063/1.2780161
10. W. G. Hoover, *Phys. Rev. A* **31**, 1695–1697 (1985).
11. D. J. Evans, and G. P. Morriss, *Statistical Mechanics of Nonequilibrium Liquids.*, Academic, London, 1990.
12. D. J. Evans, and D. J. Searles, *Adv. Phys.* **51**, 1529–1585 (2002).
13. W. Kob, and H. C. Andersen, *Phys. Rev. E* **51**, 4626–4641 (1995).
14. S. R. Williams, and D. J. Evans, *Phys. Rev. Lett.* **96**, 015701 (2006).
15. J. C. Butcher, and G. Wanner, *Appl. Numer. Math.* **22**, 113 (1996).
16. C. C. Yu, and H. M. Carruzzo, *Phys. Rev. E* **69**, 051201 (2004).
17. L. D. Landau, and E. M. Lifshitz, *Statistical Physics*, Pergamon Press, 1968.

Effect of Nanoscopic Confinement on the Microscopic Dynamics of Glass-Forming Liquids and Polymers Studied by Inelastic Neutron Scattering

R. Zorn*, M. Mayorova*, D. Richter*, A. Schönhals†, L. Hartmann**, F. Kremer** and B. Frick‡

*IFF, Forschungszentrum Jülich, D-52425 Jülich, Germany
†Federal Institute for Materials Research and Testing, D-12205 Berlin, Germany
**University Leipzig, D-04103 Leipzig, Germany
‡Institut Laue-Langevin, F-38042 Grenoble, France

Abstract. In this article we present inelastic neutron scattering (INS) experiments on different systems of confined glass-formers. The aim of these experiments is to study the influence of spatial restriction on the microscopic dynamics related to the glass transition. Such results could be helpful for the detection of a currently speculated cooperativity length of the glass transition. The glass-forming component is either a molecular liquid or a polymer. The confining matrices are 'hard' (silica glass, silicon) or 'soft' (microemulsion droplets). For some experiments the confining structure could be spatially oriented. Except for the soft confinement the naïvely expected acceleration effect could only be found at low temperatures where INS experiments are difficult because of the long relaxation times. A clear effect of confinement could be observed for the glass-typical low energy vibrations (boson peak). This effect seems to be completely different for soft and hard confinement. Surprisingly, the experiments on oriented nanopores did not show any signs of an anisotropy of the dynamics.

Keywords: Alpha relaxation, Boson peak, Confinement, Cooperativity length, Glass transition
PACS: 61.12.-q, 61.20.Lc, 64.70.Pf, 63.50.+x

INTRODUCTION

The influence of spatial confinement on the dynamics of liquids is of wide interest, both for application and for fundamental understanding. In the quest for a possible cooperativity length which might underly the physics of glass formation it seems attractive to study glass-forming systems in confinement. This subject of interest has motivated many experiments on low molecular and polymeric glass-forming systems spatially confined e.g. within pores or films [1].

Inelastic neutron scattering (INS) plays an important rôle in the observation of the dynamics in confined liquids. Firstly (and this is an argument for glass-forming liquids in general), the dynamics is detected on a true microscopic length scale (Å to nm). This means that the molecular mechanism of a motion can be directly observed and not via a macroscopic quantity, e.g. a susceptibility. Also important thermodynamic quantities as the vibrational density of states (VDOS) can be measured directly by INS.

A technical advantage of neutron scattering (NS) is that it is isotope sensitive. This is commonly employed by exploiting the high incoherent NS cross section of hydrogen nuclei. By replacing hydrogen by deuterium in the confining matrix unwanted scattering can be reduced. It has also to be noted that incoherent NS is additive if multiple scattering is avoided. Therefore, the separation of matrix and confined-liquid scattering does not need any effective medium calculations but is a simple subtraction.

On the other hand it has to be mentioned that INS is a rather intricate technique. The measurements usually are more time-consuming than other spectroscopic methods and require large-scale neutron sources. This is aggravated here because it is necessary to perform background measurements in the matrix material or fully deuterated samples to subtract its scattering.

EXPERIMENTAL RESULTS

α Relaxation

The first experiments using inelastic neutron scattering (INS) to study the effects of confinement of the microscopic dynamics of a liquid were done on the system Gelsil/salol [2, 3]. Salol (phenylsalicylate) is a good glass-former ($T_g = 219$ K) which was already used before in confinement studies by dielectric spectroscopy [4] because of its high electric dipole moment [5]. Gelsil is a controlled porous silica glass which was commercially available with pore diameters 2.5, 5.0, 7.5, and 20 nm. In

CP982, *Complex Systems, 5th International Workshop on Complex Systems*
edited by M. Tokuyama, I. Oppenheim, and H. Nishiyama

all experiments shown here the inner surface of the Gelsil glass was covered by a deuterated silane in order to reduce interaction with salol via hydrogen bonds.

Fig. 1 shows the incoherent intermediate scattering function $S(Q,t)$ of salol in bulk and confined to 2.5 nm pores. $S(Q,t)$ can be calculated by Fourier transform from the INS data $S(Q,\omega)$ and reflects the decay of correlation by *self* diffusion. In the Gaussian approximation it would be

$$S(Q,t) = \exp\left(-\frac{\langle r^2(t)\rangle Q^2}{6}\right) \qquad (1)$$

where $\langle r^2(t)\rangle$ is the mean-square displacement (MSD) of the scatterers (mostly protons) within the time t and $\vec{Q} = \vec{k}_\mathrm{f} - \vec{k}_\mathrm{i}$ the difference of the incident and final wave vector of the scattered neutrons.

What can be seen immediately comparing the relaxation for both cases is that the relaxation in confinement is more stretched or broader. It is more difficult to see whether the relaxation is faster or slower. Indeed, at $T = 260$ K the relaxation seems to start at shorter times and stretch out to longer times for the confined liquid. This was confirmed by fitting a model based on Kohlrausch functions with a distribution of relaxation times:

$$S(Q,t) = f_Q \cdot \int \exp\left(-(t/\tau)^\beta\right) g(\tau)\mathrm{d}\tau. \qquad (2)$$

(For the bulk only a single Kohlrausch function was fitted.) Because it would be impossible to derive the complete distribution $g(\tau)$ from the data, model assumptions were used [3]. Fig. 2 shows the resulting distributions $g(\tau)$ compared to the single Kohlrausch time for the bulk. It can be seen that for low temperatures the major part of the distribution lies on the short-time side of the bulk value while for high temperatures all times are shifted towards larger values.

In later studies it turned out that it was also possible to fill the pores of Gelsil by highly flexible polymers, polypropyleneglycol (PPG), polydimethylsiloxane (PDMS) [6], and polymethylphenylsiloxane (PMPS) [7]. Fig. 3 shows the results of elastic scans for PMPS (bulk $T_g = 212$ K) in three different-sized confinements. For an elastic scan an INS spectrometer is set to detect neutrons scattered with an energy transfer less than the instrument's resolution. This quantity is scanned varying e.g. the temperature. By this kind of measurement a quick overview of the dynamics on a large temperature range can be obtained. >From the data an approximate value of the MSD $\langle r^2(t_\mathrm{res})\rangle$ can be evaluated on the timescale t_res corresponding to the resolution of the instrument, here for a backscattering spectrometer about 1.3 ns.

It can be seen that in this system for all temperatures where a difference is detectable the relaxation seems to

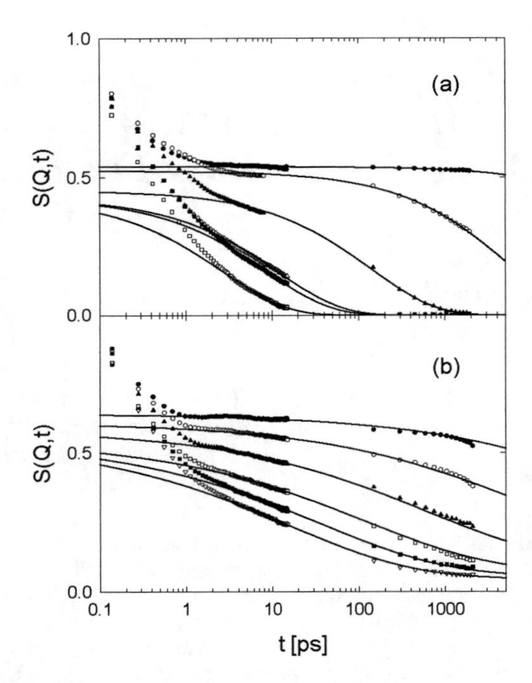

FIGURE 1. (a) Intermediate scattering function $S(Q,t)$ of bulk salol determined by Fourier transform. Temperatures: 245, 260, 280, 302, 319, and 339 K (top to bottom). (b) The same representation of $S(Q,t)$ for salol confined in 2.5 nm pores. Temperatures: 245, 260, 270, 280, 300, 320, and 339 K (top to bottom). The curves are fits with a single Kohlrausch law for (a) and a distribution as described in the text for (b). Adapted from [3].

be the slower the smaller the confinement is[1]. Also a step at lower temperatures in the MSD is visible. This step belongs to the rotation of the methyl side-group of PMPS which gets mobile at temperatures far below T_g. Considering the MSD, this side-group motion is completely unaffected by the confinement.

On the same material (PMPS/Gelsil) INS spectra were taken and fitted similarly to those in Fig. 1. An effective characteristic time was extracted in order to allow the comparison with other spectroscopic methods [8]. Fig. 4 shows a relaxation map of the INS data together with data from dielectric and thermal spectroscopy. It can be seen that all three methods fit well together. For confinement of 5 nm and smaller the usual Vogel-Fulcher temperature dependence (curved in this plot) turns into an Arrhenius (straight lines). This is what is expected from theories involving a 'cooperativity length' and what is

[1] The same kind of experiments was done on PDMS. There, the MSD was even more reduced for all confinement sizes suggesting that the dynamics in the pores is restricted to about 1 Å. We consider this result as a peculiarity of PDMS possibly related to its tendency to crystallise [9]. For this reason we will omit these experiments from the discussion.

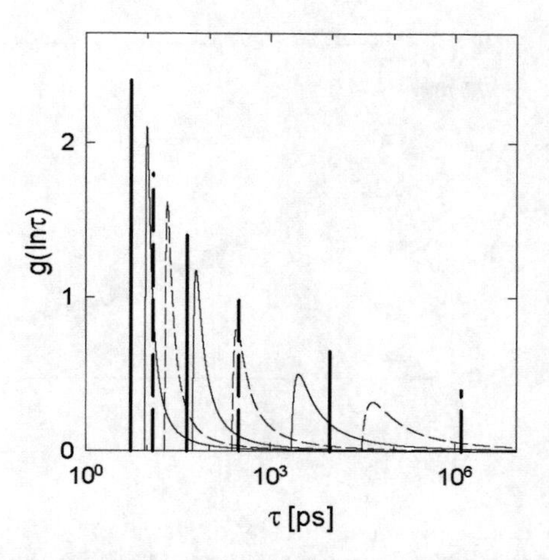

FIGURE 2. Relaxation times for the fit of the bulk salol $S(Q,t)$ data (thick lines) and relaxation time distributions for the corresponding fits of confined salol (thin curves). Temperatures: 339, 320, 300, 280, 260, and 245 K (left to right). For clarity every second line/curve is dashed.

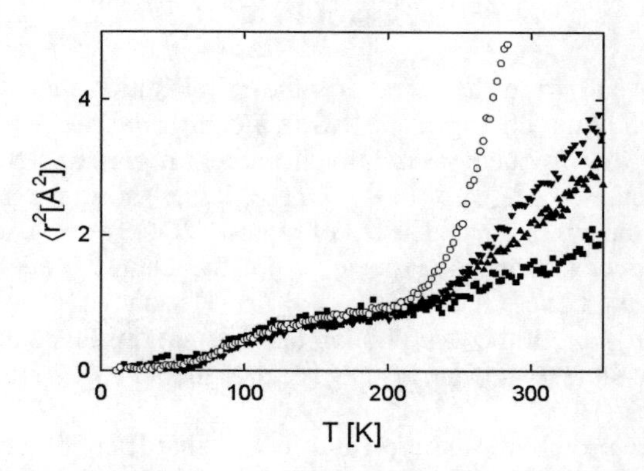

FIGURE 3. Mean square displacement from elastic scans ($t \approx 1.3$ ns) of PMPS confined in Gelsil glass and in bulk. Confinement sizes: 2.5 nm (■), 5.0 nm (▲), 7.5 nm (▼), bulk (○). With kind permission of Springer Science and Business Media adapted from [7].

often found in the experiment. Here, this leads to a crossover of time scales at 230 K/1 kHz. This explains the apparent contradiction that methods as INS which work at high temperatures and short time scales report a slowing-down while methods for longer relaxation times report an acceleration of the dynamics on the same system.

The experiments shown so far were all done on *hard* confinement, i.e. using matrices which are in a solid state. Nevertheless, comparisons using polymer films have shown that the confinement effect on vitrification is more expressed for free surfaces than for solid interfaces [10]. It is clearly impossible to construct free-surface nanofilms or -particles of molecular liquids because they would either collapse by their surface tension or evaporate. As close as one can get to a free surface

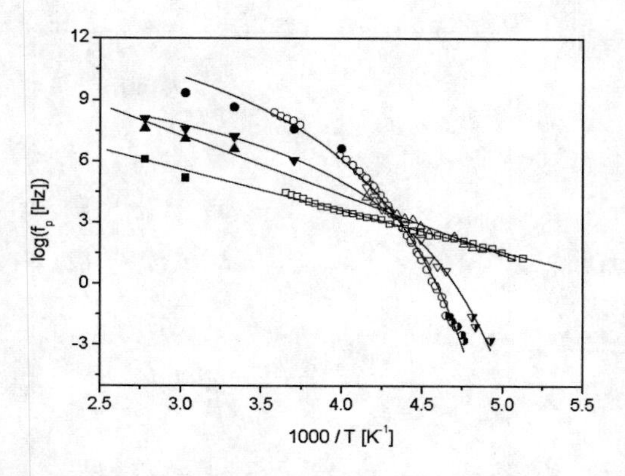

FIGURE 4. Map of the α relaxation in confined and bulk PMPS. The filled symbols are from inelastic neutron scattering, the hollow symbols from dielectric spectroscopy, and the half-filled from thermal spectroscopy. Confinement sizes: 2.5 nm ■, 5.0 nm (▲, 7.5 nm (▼), bulk (●). With kind permission of Springer Science and Business Media adapted from [8].

is the use of *soft* confinement. In soft confinement the glass-forming liquid is surrounded by a matrix which is less viscous than the glass-former itself. In practice this can be realised by a droplet microemulsion where the droplet cores are a glass-forming liquid, here propyleneglycol (PG, bulk $T_g = 167$ K), and the surrounding liquid (here: decalin) has a lower T_g, 135 K [11]. Another advantage of these systems is that the matrix does not exert a significant pressure on the core. This avoids interpretation problems whether confinement effects exist or the change of T_g is just a consequence of changed density.

For neutron scattering it is possible to extract the desired signal from the liquid cores by subtracting spectra of two samples where in one case the core (PG) is protonated and in the other it is deuterated. (The remaining components of the microemulsion are deuterated in both cases.) Fig. 5 shows the MSD from elastic scans done in this way [12]. For comparison the figure shows in addition MSDs for bulk PG and for PG in a hard confinement constituted by nanoporous silicon [13]. For the discussion it has to be noted that in the case of soft confinement the diameter is about 1.7 nm while for the hard confinement it is 4.7 nm.

An unavoidable complication in these soft confinement systems is that the droplets move freely in the matrix liquid. Therefore, the dynamics observed in the laboratory frame of reference is a superposition of the microscopic dynamics of the glass-forming liquid within the droplet and the Brownian motion of the droplet itself. It is reasonable to assume that both dynamics are statistically independent and their MSDs simply add up. We calculated the Brownian diffusion contribution from the viscosity of decalin with one single free parameter for all temperatures. Fig. 5 shows the variants of the mini-

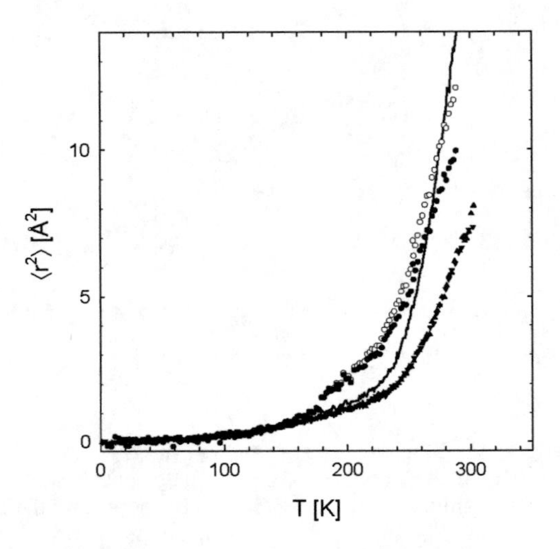

FIGURE 5. Mean-square displacements of propyleneglycol in bulk (line), soft confinement (○●), and hard confinement (▲▼). For the soft confinement two variants of the data evaluation are included: without correction of droplet diffusion (○) and with the maximal allowed correction (●). Note that there are two runs included for the hard confinement: The upwards pointing triangles (▲) correspond to $\vec{Q}$ perpendicular to the pores, the downwards pointing (▼) to the parallel orientation[2].

mal and maximal value of this parameter which induces an uncertainty of up to 2 Å^2 at the highest temperatures shown.

It can be seen that (in contrast to the data from hard confinement displayed in the same plot) the dynamics clearly leads to a larger MSD over a large temperature range. Only for temperatures > 270 K this difference seems to revert.

Concerning the data from hard confinement it is worth to mention the special structure of the sample. The matrix is made from a silicon wafer into which nanoscopic holes are etched by an electrochemical procedure [14]. An interesting feature is that these holes are preferentially growing along the 100 direction. The orientation of the pores could be confirmed by a preliminary small-angle neutron scattering experiment although it is not yet quantified. Nevertheless, for any preferred orientation one would intuitively expect an anisotropy of the dynamics. Surprisingly, the MSDs as well as INS spectra $S(\vec{Q},\omega)$ (not shown here) do not exhibit any difference for a change of the orientation of $\vec{Q}$ with respect to the sample axis.

[2] In order to detect a possible orientation dependence only one detector angle (90°) was used for PG in porous silicon while for the bulk and the microemulsion the whole Q range was used. This is a different definition than for the other curves and may cause slightly different absolute values of $\langle r^2\rangle$.

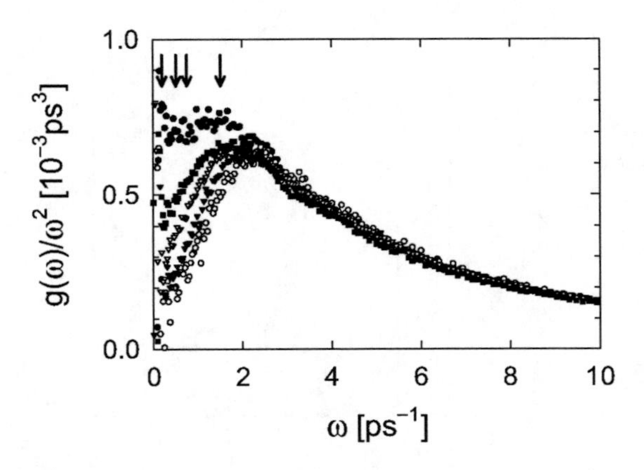

FIGURE 6. Vibrational density of states at $T = 100$ K for the samples with 2.5 nm (○), 5.0 nm (▼), 7.5 nm (▽), and 20 nm (■) pore size and the bulk salol sample (●). The arrows indicate the frequencies where the cut-off is expected from simple sound wave considerations. From [3].

Boson Peak

Apart from the effects on the α relaxation the other ubiquitous finding from INS is a confinement effect on the low frequency vibrational modes in glasses. In the frequency range $\nu = 0.3\ldots2$ THz glasses show an excess of the vibrational density of states (VDOS) $g(\omega)$ with respect to what is expected from the Debye model of sound waves, namely $g(\omega) \propto \omega^2$. Because $S(Q,\omega) \propto g(\omega)/\omega^2$ in the one-phonon approximation, this excess is visible as a peak in INS spectra, the so-called boson peak (BP).

Fig. 6 shows the typical effect of (hard) confinement on the BP [2, 3]: While for bulk salol the peak is not very pronounced, in confinement there seems to be a cut-off towards lowest frequencies. The consequence is that the BP is more clearly visible for the confined material. This effect could also be confirmed by nuclear inelastic absorption [15]. It is surprising that—as in the case here—this effect often extends to very large confinement sizes. Although a description of the BP as modified sound waves would lead qualitatively to such a cut-off, the locations of that cut-off would be different for sizes > 2.5 nm (see arrows in Fig. 6).

Another surprise came with the study of the VDOS in *soft* confinement [12] (Fig. 7). There, the BP is completely washed out and the VDOS at lowest frequency is *enhanced* with respect to the bulk. So the effect is more or less opposite to that in hard confinement. For comparison Fig. 7 also shows the results of the same liquid (PG) in hard confinement (nanoporous silicon). In the hard confinement the effect is as expected with the exception that the VDOS seems to be by an overall factor

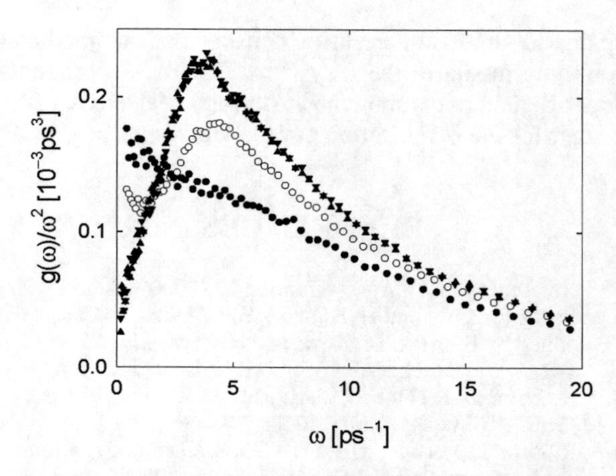

FIGURE 7. Vibrational density of states at $T = 100$ K of bulk propylenеglycol (∘), PG in soft confinement (•), and in nanoporous silicon (▲▼). The upwards pointing triangles (▲) correspond to $\vec{Q}$ perpendicular to the pores, the downwards pointing (▼) to the parallel orientation.

higher than in the bulk[3]. From Fig. 7 one can also see that the BP effect does not depend on orientation with respect to the pores too.

DISCUSSION

Soft vs. Hard Confinement

For the α relaxation it can be seen that there is a general tendency that for high temperatures it is slower in confinement while for low temperatures it is faster. This can be explained by a superposition of a 'genuine' confinement effect causing an acceleration and an effect due to immobilisation at the surface. Because the former effect is expected to become stronger at lower temperature when the cooperativity length reaches the magnitude given by the confinement dimensions, it may dominate at low temperatures.

The exact temperature where the scale of confinement and surface effects is tipped depends strongly on the method used and the system studied. Inspection of the relaxation time distribution from inelastic neutron scattering (INS) in the system Gelsil/salol (Fig. 2) leads to $\approx T_g + 50$ K [4]. For the polymeric system Gelsil/PMPS (Fig. 4) the relaxation map from different methods gives a cross-over at a lower temperature, $\approx T_g + 10$ K. On the other hand for the same system the mobility defined by the mean-square displacement (MSD) in the nanosecond range (Fig. 3) indicates slowing-down at all temperatures. This is evident because at the cross-over point the characteristic time is in the millisecond range. For the soft confinement (Fig. 5) the MSDs indicate a higher mobility in confinement up to very high temperatures $\approx T_g + 100$ K, while for the same liquid (propyleneglycol) in hard confinement the result is similar to the hard-confined polymeric system. Qualitatively, this agrees with the results by dielectric spectroscopy and solvation dynamics [11] which found acceleration in soft confinement (for $T < T_g + 20$ K) and deceleration in hard matrices. Nevertheless, the dielectric data there suggest a cross-over to slowing down in the soft confinement at a lower temperature $\approx T_g + 50$ K.

Altogether, the results show that there is a clearer tendency to acceleration of the α relaxation in soft confinement than it is in hard. This is well understandable because the soft interface is not expected to hinder the molecular motion as much as a solid surface e.g. in a nanoporous silica glass. It is questionable whether the simple argument that a capping of the cooperativity length ξ inevitably leads to an acceleration holds for a solid rough interface. In the presence of a hard interface one rather expects that a layer of thickness ξ becomes immobile as soon as one molecule of the cooperatively rearranging unit (CRU) is blocked. This would lead to an Arrhenius behaviour but at lower rates (as in Fig. 4) and not higher as in the usual reasoning.

Another clear difference between soft and hard confinement can be found in the vibrational density of states (VDOS). The usual cut-off of (boson peak) modes at lowest frequencies does not take place in the soft confinement. This is on the first glance surprising because the measurements were done at 100 K, clearly below T_g for *both* components. So also the matrix is 'hard' during the measurement. Nevertheless, the elastic moduli of a vitreous organic material are an order of magnitude lower than those of a silica glass. Therefore, such a difference is expected following the spirit of theories explaining the boson peak from fluctuations of the elastic constants [16].

An uncertainty of the present results (as well as the earlier study [11] on the same system) is the extremely small droplet size (1.7 nm diameter). Experiments with microemulsions with larger droplets comparable to pores in hard-confinement systems are being planned; but the technical problem is that systems which fulfill the T_g conditions and are *stable* microemulsions are difficult to construct.

---

[3] This is not a contradiction to the mode sum rule because frequencies > 10 THz are not observable on the INS instruments used here. So there could be a compensation by a reduction in the unobserved part of $g(\omega)$. On the other hand the VDOS data sets are from INS spectrometers with strongly different resolution so that a systematic normalisation error cannot be excluded.

[4] A closer inspection of the dielectric spectroscopy data on the same system [5] also shows such a cross-over but at $T_g + 37$ K.

Orientation Effect

Neither concerning the α relaxation nor the VDOS (boson peak, BP) we could find any anisotropy of the dynamics in a matrix with oriented pores. In this respect it has to be noted that we do not expect any *direct* influence of the confinement by its size limiting the maximal diffusional displacement. Because the confinement sizes were always more than an order of magnitude larger than the length defined by $2\pi/Q$ or $\sqrt{\langle r^2 \rangle}$ this trivial effect cannot be seen by our INS experiments. Nevertheless, it would be possible that the confinement leads to an *indirect* effect, e.g. by a surface induced structuring of the liquid which allows easier motion in one direction than another.

The absence of such an anisotropy in the α relaxation supports the idea of CRUs: The effect of confinement on the surface particles may be anisotropic—a facilitation of movement in a certain direction for soft confinement or a blocking perpendicular to a hard surface. But because all molecules within a CRU have different directions in a cooperative movement the total effect is either an isotropic acceleration or an isotropic deceleration. Because we cannot 'label' surface molecules in INS this would be an interesting subject to study by molecular dynamics simulation.

For the BP the missing anisotropy of the VDOS in confinement confirms the result from size scaling (Fig. 6) that a simple sound wave picture is not sufficient. For an anisotropic resonating cavity one would expect that modes perpendicular to the long axis are suppressed up to higher frequencies. Such a difference cannot be seen in Fig. 7. Therefore, models which assume a region close to the interface which is depleted of 'BP modes' may be closer to the reality [3].

SUMMARY

Comparing the results it becomes clear that the crucial question of confined dynamics—faster or slower?—cannot be answered in a universal way. The answer depends on the system studied, the temperature, and the spectroscopic method used. In general, at low temperatures dynamics is accelerated while it is decelerated at high temperatures. This may be the result of a genuine confinement effect modified by a surface-induced slowing-down. In this respect, the crucial experiment is done using 'soft' confinement in microemulsion droplets. As expected, the surface effect is less important there shifting the faster-slower cross-over to much higher temperature. We also observe an opposite influence of the soft confinement on the low frequency vibrational spectrum (boson peak) compared to the hard. This may be due to more similar elastic constants of confined and confining media in these systems. Finally, experiments with oriented pores show no anisotropy of the dynamics, neither for the α relaxation nor the vibrations.

REFERENCES

1. References in *J. Phys. IV France* **10** (2000) edited by B. Frick, R. Zorn, and H. Büttner; *Eur. Phys. J. E* **12** (2003) edited by B. Frick, M. Koza, and R. Zorn; *Eur. Phys. J. ST* **141** (2007) edited by B. Frick, M. Koza, and R. Zorn.
2. R. Zorn, D. Richter, L. Hartmann, F. Kremer, and B. Frick, *J. Phys. IV France* **10**, Pr 7-83–86 (2000).
3. R. Zorn, L. Hartmann, B. Frick, D. Richter, and F. Kremer, *J. Non-Cryst. Solids* **307–310**, 547–554 (2002).
4. F. Kremer and A. Schönhals (editors), *Broadband Dielectric Spectroscopy*, Springer, Berlin, 2002, ISBN 978-3-540-43407-8.
5. M. Arndt, R. Stannarius, H. Groothues, E. Hempel, and F. Kremer, *Phys. Rev. Lett.* **79**, 2077–2080 (1997).
6. A. Schönhals, H. Goering, Ch. Schick, B. Frick, and R. Zorn, *Eur. Phys. J. E* **12**, 173–178 (2003).
7. A. Schönhals, H. Goering, C. Schick, B. Frick, and R. Zorn, *Coll. Polym. Sci.* **282**, 882–891 (2004).
8. A. Schönhals, H. Goering, Ch. Schick, B. Frick, M. Mayorova, and R. Zorn, *Eur. Phys. J. ST* **141**, 255–259 (2007).
9. A. Schönhals, H. Goering, C. Schick, B. Frick, and R. Zorn, *J. Non-Cryst. Solids* **351**, 2668–2677 (2005).
10. J. A. Forrest, K. Dalnoki-Verres, and J. R. Dutcher, *Phys. Rev. E* **56**, 5705–5716 (1997).
11. Li-Min Wang, Fang He, and R. Richert, *Phys. Rev. Lett.* **92**, 95701-1–4 (2004).
12. R. Zorn, M. Mayorova, B. Frick, and D. Richter, submitted to *Soft Matter* (2007).
13. M. Mayorova, R. Zorn, and B. Frick, ILL experimental report 6-05-677 (2007).
14. B. Coasne, A. Grosman, C. Ortega, and M. Simon, *Phys. Rev. Lett.* **88**, 256102-1–4 (2002); D. Wallacher, N. Künzner, D. Kovalev, N. Knorr, and K. Knorr, *Phys. Rev. Lett.* **92**, 195704-1–4 (2004).
15. T. Asthalter, M. Bauer, U. van Bürck, I. Sergueev, H. Franz, and I. Chumakov, *Hyperfine Interact.* **144/145**, 77–83 (2002).
16. W. Schirmacher, *Europhys. Lett.* **73**, 892–898 (2006).

Interdependence of Primary and Secondary Relaxations in Glass-Forming Systems Undergoing Compression and Cooling

S. Capaccioli*, D. Prevosto[§], K. Kessairi[§], M. Lucchesi[§], and P. A. Rolla[§]

*INFM-CNR, SOFT, I-00185 Roma & Physics Dept., University of Pisa, Largo Pontecorvo 3, 56127 Pisa Italy
[§]INFM-CNR, Polylab & Physics Dept., University of Pisa, Largo Pontecorvo 3 , 56127 Pisa Italy

Abstract. Dynamics of the primary and the Johari-Goldstein (JG) secondary relaxations for polar molecules as a component in mixtures and also neat glass-forming systems have been investigated by means of broadband dielectric measurements at ambient pressure and elevated pressures P and various temperatures T. The ratio of their relaxation times was found invariant to different combinations of P and T keeping the primary relaxation time constant. In addition, isochronal spectra, obtained with different T-P combinations, superimposed. Several other properties of the JG relaxation were also found to be related to or correlated with the primary relaxation. Our results indicate the fundamental role played by the JG relaxation in glass transition.

INTRODUCTION

Dynamics of glass-forming systems are characterized by motions taking place on a wide range of length and time scales. The cooperative motions at the origin of the structural α-relaxation involve an increasing number of molecules and slow down dramatically when the glass transition is approached by decreasing temperature T or increasing pressure P (i.e., density). Non-cooperative local dynamic processes involving single or few molecules, or even parts of them, like secondary relaxations, are faster and less dependent on T and P than the α-relaxation. In recent years, there is an increasing interest on the possible fundamental role played by the secondary relaxation in glass transition, which is usually considered to be solely responsible by the much slower primary structural relaxation. Such possibility was suggested by relations and correlations found between various properties of secondary β- and primary α-relaxations [1,2,3,4,5,6]. Some secondary relaxations do not have such relation to the primary relaxation, and hence they are not relevant for glass transition. For those that have [4], they are now referred to as the Johari-Goldstein (JG) secondary or β-relaxation to honor their important discovery of secondary relaxation even in totally rigid molecules that have no internal degree of freedom [7]. JG β-relaxation can be considered an universal feature of glassy systems, as shown by its ubiquitous presence in various classes of glass-forming materials including simple molecular liquids, molten salts, plastic crystals and metallic glasses [8, 9]. In general, only few theories or models assume a correlation between α- and JG β- relaxation to account for glass-forming dynamics [10, 11]. In particular, the Coupling Model (CM) [11] provides a quantitative relation between the α-process dispersion and the time scale of α- and β-relaxation. In fact, according to CM, the JG β-process time scale can be identified with that of the primitive relaxation, a local motion that is the precursor of the α-relaxation [8,10]: $\tau_{JG} \approx \tau_0$. Since CM gives also a quantitative relation, due to many-molecules dynamics, between the primitive τ_0 and the α-relaxation time τ_α, then the following relation is predicted:

$$\tau_{JG} \approx \tau_0 = \tau_\alpha^{(1-n)} t_c^{\ n} \qquad (1)$$

where t_c=2 ps and n is the coupling parameter, which is related to the stretching of KWW function reproducing the structural peak, n=1-β_{KWW}:

$$\phi(t) = \exp\left[-(t/\tau_\alpha)^{1-n}\right], \quad 0<n<1 \qquad (2)$$

n is a measure of the extent of the non-exponentiality of the relaxation and hence of the intermolecular coupling. According CM, a close relation is expected

CP982, *Complex Systems, 5th International Workshop on Complex Systems*
edited by M. Tokuyama, I. Oppenheim, and H. Nishiyama

between time scales of slow (τ_α) and fast (τ_{JG}) relaxation and the dispersion of structural dynamics (β_{KWW}). Higher is the intermolecular coupling (n) larger is the time scale separation of the JG peak from the structural one, i.e. τ_α/τ_{JG}. Moreover, this relation is expected to hold also at different T and P. In fact, it has been proven for several glass-former that, once τ_α(T,P) is chosen constant for different combinations of T and P, then the stretching parameter β_{KWW} (and so n) is constant [12]. Therefore, once τ_α(T,P) is fixed for different combinations of T and P, then the coupling parameter n is constant, and consequently from eq.(1) τ_{JG} should be constant too. It is noteworthy that the strong correlation of eq.(1) between τ_α and τ_{JG} can explain the anomalous change of dynamics at T_g recently reported in the T-P behavior of JG relaxation. For instance, the well known crossover of T-dependence of τ_α from Vogel-Fulcher above T_g to Arrhenius below T_g is transferred from τ_α to τ_{JG} (or vice versa) in a qualitative manner: in fact the T-P dependence of τ_{JG} mimics that of τ_α, although weaker due to the exponent (1-n). This issue has been extensively discussed in some recent papers [13, 14]. A comprehensive review of the properties indicating the fundamental role played by JG relaxation in glass transition can be found in Ref.[15], where a rationale for the experimental findings was also provided in the framework of CM.

The aim of the present paper is to show new experimental results on different glass-formers (low molecular glass-forming neat systems and binary mixtures) over wide range of T and P, providing clear evidences that a strong correlation between slow (α-) and fast (JG β-) dynamics exists. In fact, the ratio of their relaxation times was found invariant to different combinations of P and T keeping the primary relaxation time constant. In addition, isochronal spectra, obtained with different T-P combinations, superimposed.

EXPERIMENTAL

Oligomers of styrene mix well with many glass-formers, and are ideal for being used as the non active component in dielectric measurements because the dipole moment of the styrene repeat unit is very small. We investigated the dynamics of mixtures containing the polar rigid small molecule quinaldine (QN) (T_g=180 K), and of the neat system poly-phenyl-glycidyl-ether (PPGE) (T_g=262 K). All these compounds were purchased from Aldrich. Quinaldine was mixed, at concentration ranging from 5 to 100 wt.%, with tristyrene (T_g =234 K) (obtained from PSS) and oligostyrene Mw=800g/mol (PS800) (T_g= K, obtained from Scientific Polymer Product). In this paper we will present only the low concentration cases (5-10% wt.) in order to have negligible fluctuation concentration effects, that, for higher concentrations, give an additional broadening to the α-relaxation peak, preventing a correct calculation of n [16,17,19].
Dielectric measurements, both at atmospheric and at high pressure were carried out using a Novocontrol Alpha-Analyzer (ν=10^{-5}-10^{7}Hz). For atmospheric pressure measurements, a parallel plate capacitor separated by a quartz spacer (geometric capacitance ~ 90 pF) and filled by the sample was placed in the nitrogen flow Quatro cryostat. For high pressure measurements, a sample-holder multi-layer capacitor (geometric capacitance ~ 30pF) was isolated from the pressurizing fluid (silicon oil) by a Teflon membrane. The dielectric cell was then placed in a Cu-Be alloy high pressure chamber, provided by UNIPRESS, connected to a manually operated pump with a pressure intensifier able to reach 700 MPa, and controlled in the interval 200–360 K within 0.1 K by means of a thermally conditioned liquid flow. Further experimental details can be found in Refs [16, 17].

RESULTS AND DISCUSSION

In our experiments on binary mixtures of rigid polar molecules in apolar matrices we probed selectively the relaxation dynamics of the component having the larger dipole moment. Even al low concentration, only the polar molecules are observed in the dielectric spectra because the dipole moment of the host molecules is so small that their motions make no contribution. Similar systems have been previously studied at ambient pressure and interesting results were obtained on the α- and β- dynamics of the polar component [16-19]. In our experiment, both temperature T and applied pressure P were varied over wide ranges. Elevated pressure slows down the α-relaxation and increases its relaxation time τ_α, but the increase can be compensated by raising temperature. Naturally, widely different combinations of P and T can be found to have the same α-relaxation time τ_α, although there are significant variations in the density.
As an example, a selection of loss spectra of 5%wt. QN in tristyrene obtained along two different thermodynamic paths is shown in Fig.1. The α- and β-relaxation peaks are well resolved in the dielectric spectra for any T-P conditions. For some of them, the T-P condition is able to yield the same frequency of the loss maximum (and so τ_α) for structural process. Not only τ_α is the same, but also the shape (or the dispersion) of the α-peak is invariant in that condition, a fact already reported in ref. [12] for several glass-

forming systems. Unexpectedly, also the JG β-relaxation (that shifts to lower frequency on increasing pressure) has the same frequency location of the maximum and almost the same shape, although its dielectric strength is slightly different. Summarizing, pressure and temperature seem to have a similar effect on α- and β- dynamics, although the T-P dependences for both processes are different. Therefore, these results could imply a strong correlation between the two processes and lend strong support to the interdependency of the JG β-relaxation and the α-relaxation, suggested by a recent NMR study [5].

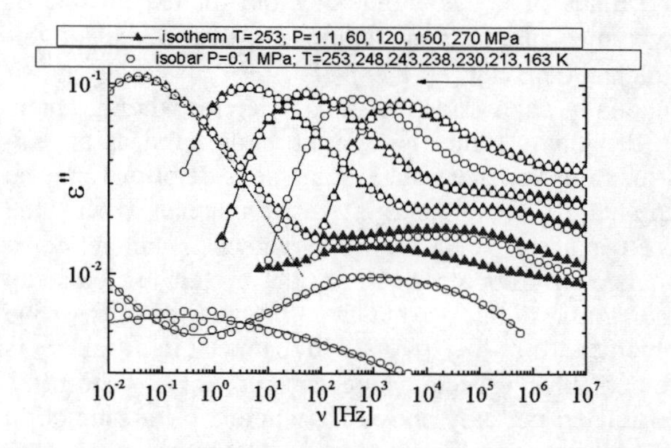

FIGURE 1: (Color onilne) Selected loss spectra for mixtures of the polar molecule (5 wt.%) QN in tristyrene. Open circles and full triangles represent the isobaric (P=0.1 MPa) and the isothermal (253 K) measurement respectively. The arrow indicates the direction of increasing pressure or decreasing temperature. Continuous lines are HN equation fits. Dotted line is a KWW fit.

The results in Fig.1 are not peculiar of that system. We found similar evidences in all the mixtures we studied. The fact that, simultaneously with τ_α, the most probably JG β-relaxation times τ_{JG} are the same for all these combinations of P and T is valid also when the applied pressure is very high and the density variation leads to an increase of T_g of almost 40%. A remarkable example is shown in Fig.2, where loss spectra for seven T-P combinations taking a constant τ_α=0.67 s are collected. The dispersion of α-peak does not change with P, and also the location of the maximum of JG β-relaxation is constant. That can be easily rationalized by using CM equation (eq.1): in fact, n is not changing and from that the quantitative prediction for the frequency of the primitive process is a constant value, in good agreement with the experimental results. Excepting for the JG dielectric strength (that is slightly different for different P), the loss spectra are superimposing quite well, even if the temperature at which they are obtained are very different. This fact cannot be obtained if α- and β- processes would be independent processes, originated by unrelated motions. It has to be pointed out that also the distribution of time (i.e. the shape) of JG β-relaxation does not change appreciably: it is an indication of the strong correlation between the dispersion of the two processes too, as recently shown by Richert for several supercooled liquids [21].

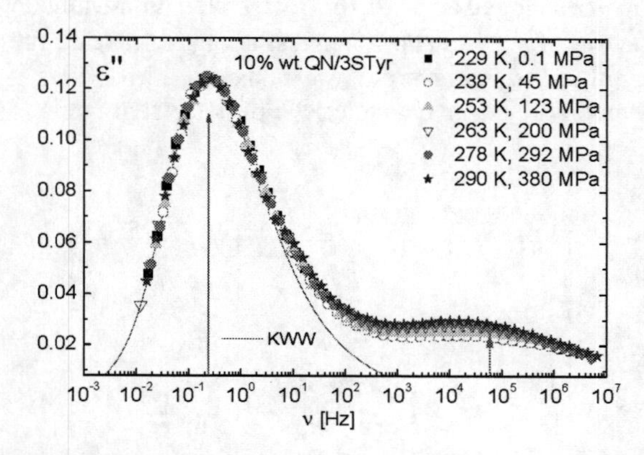

FIGURE 2: (Color onilne) Selected loss spectra for mixtures of the polar molecule (10 wt.%) QN in tristyrene, obtained at different combinations of T and P, taking a constant τ_α=0.67 s. Continuous line is a KWW fit, β_{KWW}=0.5. Left and right vertical arrows indicate the frequency of the maximum of α-peak and that predicted for the CM primitive process, respectively.

Until now, all the evidences have been obtained only by comparison of spectra, i.e. avoiding any model dependent analysis. Anyway, important indications can come also from obtaining the time scales of α- and JG β-relaxation processes from fitting dielectric spectra. τ_α and τ_{JG} were determined by using a superposition of an Havriliak-Negami (HN) function (for the α-relaxation) and a Cole-Cole (CC) function (for the JG β-relaxation), and by calculating the frequency f_m of the loss peak maximum ($\tau=1/(2\pi f_m)$) of each process. Different approaches, based on the Williams ansazt procedure, are also possible [18], but preliminary analysis on our data gave similar results to those of the first procedure.

The τ_α and τ_{JG} of 10% QN in tristyrene obtained from the spectra measured isobarically at three different pressures, 0.1, 181 and 380 MPa, are plotted against 1000/T in Fig.3a. The data of τ_α and τ_{JG} determined by fitting spectra in isothermal condition at four different temperatures (238, 253, 263, 278 K) are plotted as a function of P in Fig.3b. For each isobar (isotherm) in

Fig.3a (Fig.3b), we located the temperature (pressure) at which τ_α is fixed at a certain value and we determined the corresponding value of τ_{JG} (results in Figs.3a and 3b are related to the data of Fig.2). The values of τ_{JG} are falling around a constant value within the errors ($\log_{10}\tau_{JG}$=-5.45 by using a least square optimization procedure). Thus, the ratio τ_α/τ_{JG} is invariant to changes in the combinations of P and T that keep τ_α constant. This results can be applied for any chosen value of τ_α or τ_{JG}. This remarkable properties, together with the fact that n is an unique function of τ_α, i.e. n(τ_α), suggest a way to rescale the overall α-β dynamics plotting relaxation times versus a temperature scale characteristic of the system.

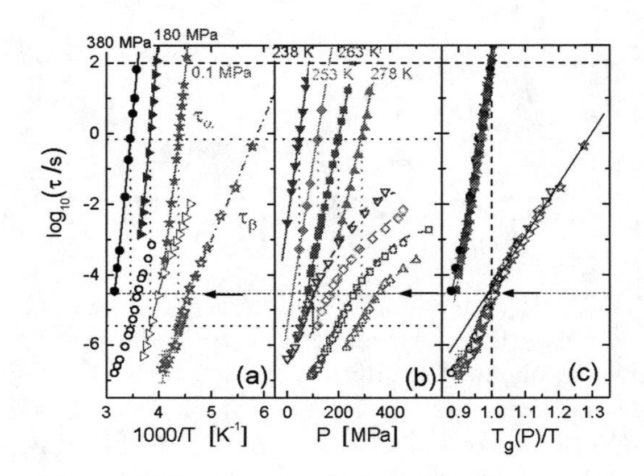

FIGURE 3: (Color onilne) Relaxation map for mixtures of 10 wt.% QN in tristyrene. Full and open symbols represent the α- and JG β-relaxation times respectively. (a) Isobaric measurements: continuous lines are VF equation fits, dashed lines are Arrhenius fits. (b) Isothermal measurements: continuous lines are VF-like equation fits, dashed lines are guides for eyes. Dotted vertical and horizontal lines show the constant ratio of τ_α/τ_{JG} for the data of Fig.2. (c) Master curve obtained by plotting data of plot (a) and (b) versus the normalized quantity T_g(P)/T. Dashed horizontal and vertical lines show the T_g locus, continuous line is an Arrhenius fit. In (a-b-c) the horizontal arrows show the crossover at T_g of the T-P dependence of τ_{JG}.

For instance, one can normalize the temperature scale by using the value of the glass transition temperature attained at the value P at which the measurement is performed (for both isothermal and isobaric measurements). This plot is shown in Fig3.c. The α-relaxation times are superimposed, as it is expected, since the steepness index (or fragility according to Angell definition) does not change appreciably with pressure (it actually decreases by less than 10%). As also n does not change appreciably close to T_g, from eq.(1) a master curve is expected also for τ_{JG}, as shown by the results in Fig.3c. This master plot occurs for all the other QN mixtures investigated and also for the neat epoxy system PPGE (not shown here for lack of space). From the plot in Fig.3c it is also clear that the change of T-P behavior reported on crossing T_g or P_g line for several glass-formers (marked by horizontal arrows also in Fig.2a-b) is reflecting the freezing of the structural process. For the isobaric measurements, the temperature dependence of τ_{JG} is Arrhenius $\tau_{JG}=\tau_\infty\exp(E_{JG}/RT)$ at lower temperatures but it crosses over to a stronger temperature dependence above T_g. Applying eq.(1), we can conclude that the change of dynamics of τ_{JG} is mimicking the related change of dynamics of structural α-process that has a Vogel-Fulcher behavior $\tau_\alpha=\tau_\infty\exp[B/(T-T_0)]$ above T_g and it gradually attains an Arrhenius behavior when it enters in the glassy state, that can be considered as an iso-structural frozen state. Similar considerations can be applied to the isothermal measurements where the change of slope of $\log_{10}(\tau_{JG})$ versus P can reflect a change in compressibility of the system on crossing the glassy state. Anyway, the master curve we obtained for the overall dynamics in Fig.3.c is remarkable, especially because τ_α or τ_{JG} resulted correlated not only above T_g (where a clear prediction comes from eq.1) but also in glassy state. In fact, below T_g, a quantitative and rigorous check of eq.(1) is not possible, as τ_α can only be inferred or extrapolated from data obtained in the liquid state.

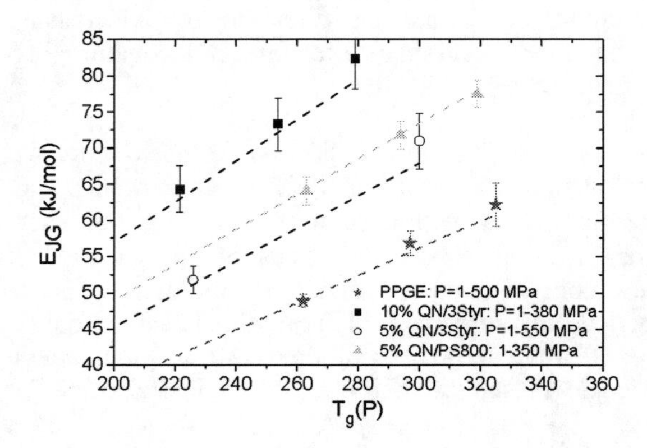

FIGURE 4: (Color onilne) Activation energy of the JG process in the glassy state for different isobaric measurements plotted versus the related T_g(P): QN in tristyrene (5% and 10% wt. are indicated by open circles and solid squares, respectively), 5% wt. QN in PS800 (solid triangles), PPGE (solid stars). Dashed lines are the predictions according to eq.(3).

A possible check can come from relation previously derived for glass-forming systems at ambient pressure.

For instance, an interesting relation was noticed: the ratio between the activation energy E_{JG} of the JG β-relaxation time τ_{JG} in the glassy state and the glass transition temperature T_g (a quantity related to α-process) is slightly constant for several systems [22]. This finding was extended to several other systems and rationalized by using CM prediction [20]. The ratio was found to depend on:

$$E_{JG}/RT_g = 2.303(2 - 13.7n - \log_{10}\tau_\infty) \quad (3)$$

where τ_∞ is the pre-factor of the Arrhenius fit to JG times. This relation can be extended to measurements done at high pressure: if $\log_{10}(\tau_\infty)$ and $n(T_g)$ do not change appreciably with pressure, a constant value is expected also when the pressure is varied and isobaric measurements are done. This is evident from the data of 10% wt. QN in tristyrene of Fig.3c, where different sets of results in the glassy state are quite superposed. The same constant ratio with pressure has been found also for the other systems investigated (see Fig.4), i.e., 5% wt. QN in tristyrene and PS800, PPGE, having different values of $n(T_g)$ and τ_∞.

Summarizing, the master curve of Fig3.c show a remarkable relation between structural and JG relaxation both above and below T_g. The interdependency of the two processes is so demonstrated. Eq.(1) provides a rationale for the quantitative relation above T_g. Concerning the master curve that holds also below T_g, this is possible since the activation energy E_{JG} of JG relaxation scales with P proportional to $T_g(P)$. This fact cannot be only due to a difference in density, that reflects in an *uncorrelated* way on structural and secondary relaxation. For PPGE, for instance, while T_g varies from 262 to 326 K in the range 0.1-500 MPa, ΔE_{JG} increases by 25%: the ratio $\Delta E_{JG}/(RT_g)$~22.5 is almost constant over the whole pressure range, while density is increased almost by 8%. So, the activation energy of this intermolecular secondary process *in the glassy state* is scaling with the glass transition temperature, that is characteristic of the primary relaxation of the *supercooled liquid.*

CONCLUSION

Dynamics of polar rigid molecules dissolved in apolar solvents of a neat epoxy system were studied by means of broadband dielectric spectroscopy. The α- and the true JG β-processes were well resolved in the spectra. Temperature T and applied pressure P were varied over wide ranges in the present experiment. As expected, for each investigated system, the JG β-relaxation was strongly affected by pressure both in the supercooled liquid and glassy state. Moreover, analyzing the T and P behavior of α- and β- processes, a clear correlation was found between the structural, the JG relaxation time and the dispersion of the structural relaxation (i.e. its Kohlrausch parameter). In particular we found, over a broad *T-P* range, that, for a fixed value of the α-relaxation time, the dispersion (i.e. the stretching parameter) of α-relaxation was constant, independent of thermodynamic (*T-P*) conditions: in other words, the shape of the structural relaxation function depended only on its time scale and not on *T* and *P* separately. This result agrees with what recently found for many glass-formers [12], but in our case, unexpectedly, also the ratio of α- and β- relaxation times was found invariant to different combinations of *P* and *T* keeping the primary relaxation time constant. In fact, spectra obtained at very different T-P conditions but with the same frequency of loss maxima almost superposed in both α- and β- time scale range. In addition, several other properties of the JG relaxation were found to be related to or correlated with the primary relaxation. For instance, the β-relaxation time showed, in correspondence with the glass transition, a deviation from its behavior in the glassy state for both isobaric and isothermal variation. The overall scenario for α- and β- relaxation can be represented in an unique way, simply by plotting all the data versus a reduced parameter (i.e. $T_g(P)/T$). Such a master curve has many important implications, for instance that the activation energy of JG relaxation is directly proportional to $T_g(P)$. The results indicate the fundamental role played by the JG relaxation as a precursor of the structural relaxation in glass transition, a role often overlooked by common theories. On the other hand, all these evidences can be rationalized in the framework of the Coupling Model (CM) [10].

ACKNOWLEDGEMENTS

The work at the Università di Pisa was supported by by MIUR (PRIN 2005: "Aging, fluctuation and response in out-of-equilibrium glassy systems").

REFERENCES

1. K. L. Ngai, *J. Chem. Phys.* **109**, 6982-6994 (1998).
2. G. P. Johari, G. Power, and J. K. Vij, *J. Chem. Phys.* **116**, 5908-5909 (2002). ibid. **117**, 1714-1722 (2002).
3. R. Brand, P. Lunkenheimer, and A. Loidl, *J.Chem.Phys.* **116**, 10386-10401 (2002).
4. K. L. Ngai, M. Paluch, *J. Chem. Phys.* **120**, 857-873 (2004).
5. R.Böhmer, G.Diezemann, B.Geil, G. Hinze, A. Nowaczyk, M. Winterlich, *Phys.Rev.Lett.* **97**, 135701-4 (2006).

6. R. Casalini and C.M. Roland, *Phys.Rev.Lett.* **91**,015702-4 (2003)
7. G. P. Johari and M. Goldstein, *J. Chem. Phys.* **53**, 2372-2388 (1970)
8. K.L. Ngai, *Physica A* **261**, 36-50 (1998).
9. K.L. Ngai, *J.Non-Cryst.Solids* **351,** 2635-2642 (2005) and references therein.
10. K.L. Ngai, *J. Phys.: Condens. Matter* **15**, S1107-S1125 (2003).
11. J.Y. Cavaille, J. Perez, and G. P. Johari, *Phys. Rev. B* **39**, 2411-2422 (1989).
12. K.L. Ngai, R. Casalini, S. Capaccioli, M. Paluch, C.M. Roland, *J. Phys. Chem. B* **109**, 17356-17360 (2005).
13. K.L. Ngai, S. Capaccioli, N. Shinyashiki, in this volume.
14. S. Capaccioli, K.L. Ngai, N. Shinyashiki, *J.Phys.Chem. B* **111**, 8197-8209 (2007).
15. K.L. Ngai, R. Casalini, S. Capaccioli, M. Paluch, C.M. Roland, *Adv.Chem.Phys. Part B,* Volume 133, chapter **10**, 497-593 (2006).
16. S. Capaccioli, K. Kessairi, D. Prevosto, M. Lucchesi, K.L. Ngai, *J.Non-Cryst.Solids*, **352**, 4643-4648 (2006)
17. S. Capaccioli, K. Kessairi, D. Prevosto, M. Lucchesi, P. Rolla. *J.Phys.: Condens. Matter*, **19**, 205133-10 (2007).
18. T. Blochowicz, E. A. Rössler, *Phys. Rev. Lett.* **92,** 225701-4 (2004).
19. S. Capaccioli, K.L. Ngai, *J.Phys.Chem.B* **109,** 9727-9735 (2005).
20. K.L. Ngai, S. Capaccioli, *Phys. Rev. E.* **69,** 031501-4 (2004).
21. LM Wang and R. Richert, *Phys. Rev. B.* at press (2007).
22. A. Kudlik, C. Tschirwitz, S. Benkhof, T. Blochowicz and E. Rössler, *Europhys.Lett.* **40**, 649-654 (1997).

Molecular Motions in Amorphous Pharmaceuticals

Ana R. Brás, Madalena Dionísio, and Natália T. Correia*

Requimte, Departamento de Química, FCT, Universidade Nova de Lisboa
2829-516 Caparica, Portugal

Abstract. Dielectric Relaxation Spectroscopy (10^{-1}Hz to 10^{6}Hz) was used in order to get relevant information regarding the different modes of motion, from supercooled liquid down to the glassy state, present in the glass-forming Ibuprofen. This pharmaceutical drug easily avoids crystallization on cooling from the melt, forming a glass below the dielectric T_g(100s)=226K, and undergoes cold crystallization upon further heating. In the supercooled state, attained from above, a dominating relaxation process associated with the glass transition was detected. The relaxation time temperature dependence of this α process presents the usual VFT type curvature found in non-pharmaceutical glass formers, which steepness gave a fragility index of m= 93 allowing to classify ibuprofen as a fragile glass-former. In the glassy state, two secondary relaxation processes were found, β and γ, in decreasing temperature location. Analysis of the spectra using the Coupling Model confirms that the β process is the genuine Johari-Goldstein relaxation probably playing a role in cold crystallization of ibuprofen.

Keywords: Dielectric relaxation, Ibuprofen, Molecular mobility, Pharmaceutical amorphous

INTRODUCTION

It is well recognised that a disordered amorphous material dissolves faster and has a greater solubility than the corresponding ordered crystalline solid [1,2]. Consequently, the amorphous form of a drug often shows an improved bioavailability. However, the amorphous state is a non-equilibrium state and, therefore, it is unstable. If the molecular motions that originate this instability, are not retarded over a meaningful pharmaceutical timescale, a significant variation in some of the key physicochemical properties of the drug may occur [1,3,4]. In this context, the knowledge of the timescales of molecular motions in amorphous systems, *i.e.*, the knowledge of the relaxation map that characterizes the molecular dynamics in a given material, is needed for profiting from the advantages of the amorphous state and is an important requirement for a safe storage and use of amorphous pharmaceutical solids [5,6].

Dielectric Relaxation Spectroscopy (DRS) is a suitable tool to monitor molecular mobility covering an extraordinarily extended dynamic range from 10^{-6} to 10^{12} Hz, enabling us to gain a wealth of information on the dynamics of bound dipoles and mobile charge carriers [7].

Recently, the DRS technique has been introduced to pharmaceutical sciences [8-14] where it was recognized that the same concepts used to describe the general behaviour of non-pharmaceutical glass formers apply to pharmaceutical counterparts. Due to its sensibility, this technique is able to analyse amorphous phases as well as to detect almost all physico-chemical transformations, such as amorphous-crystalline, crystalline-crystalline and hydrate-anhydrous transformations occurring in pharmaceutical materials (small organic molecules, polymers, drug delivery composites…).

In the present work we report a dielectric study of a low molecular weight pharmaceutical glass former: (±)-Ibuprofen; it is a poorly water soluble drug which belongs to the category of 2-arylpropanoic acid (see structure in Fig. 1) showing non-steroidal analgesic, antipyretic and anti-inflammatory properties [15]. Very recently, while this work was been performed, Johari *et al.* published a DRS study of several pharmaceutical materials including ibuprofen [16]. Nevertheless, since it covers a more extended frequency range, the present work can represent a valuable contribution on the characterization of ibuprofen regarding molecular dynamics, where an additional relaxation process is revealed, as well as conductivity relaxation of charged carriers.

* Author to whom correspondence should be addressed; e-mail: n.correia @dq.fct.unl.pt

CP982, *Complex Systems, 5th International Workshop on Complex Systems*
edited by M. Tokuyama, I. Oppenheim, and H. Nishiyama

EXPERIMENTAL

Ibuprofen was purchased from Sigma (catalogue number I4883 (CAS 15687-27-1), 98% GC assay) with a molecular weight of 206.29 and a melting point 350-351K, being provided as a racemic mixture of (+)-ibuprofen and (-)-ibuprofen. It was used without further purification. For simplicity, hereinafter the studied (±)-ibuprofen (Fig. 1) will be referred to as ibuprofen. Its calorimetric glass transition temperature, T_g(onset)=228K, was determined by means of differential scanning calorimetry using a heating rate of 5K/min, after cooling the melt to a temperature at least 50K below T_g at 5K/min without crystallization [17].

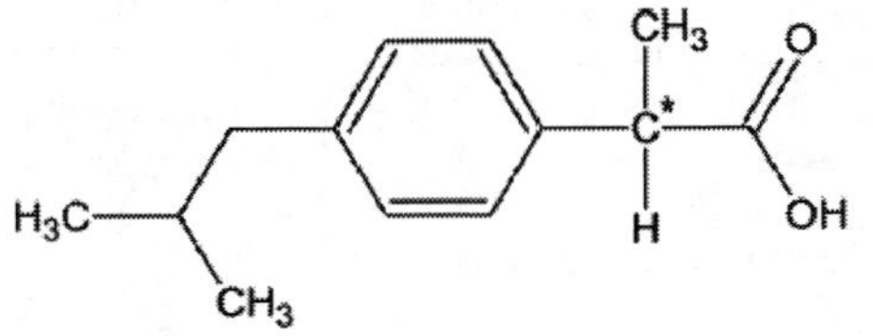

FIGURE 1. Chemical structure of Ibuprofen; C* is a chiral atom.

The complex permittivity was measured using a broadband impedance analyzer, Alpha-N analyzer from Novocontrol GmbH, covering a frequency range from 10^{-1}Hz to 1MHz. The sample with two silica spacers of 50μm thick was placed between two gold plated electrodes (diameter 20mm) of a liquid parallel plate capacitor (BDS 1308). The temperature control was performed within ±0.5K, with a Quatro Cryosystem. Novocontrol GmbH supplied all the used modules.

Isothermal dielectric spectra were recorded on cooling from melt down to the vitrified state (between 353K and 143K), allowing characterizing ibuprofen without being influenced by partial crystallization as observed by Johari et *al.* [16]. Measurements were also performed upon increasing temperature coming from 143K up to 353K. More details of the experimental methods and of the sample preparation are described in [17].

The real and imaginary parts of the complex dielectric constant ($\varepsilon^*(\omega)=\varepsilon'(\omega)-i\varepsilon''(\omega)$) were fitted by a sum of the well-known Havriliak-Negami (HN) function [18]

$$\varepsilon^*(\omega)=\varepsilon_\infty+\sum_j \frac{\Delta\varepsilon_j}{\left[1+\left(i\omega\tau_{HN_j}\right)^{\alpha_{HNj}}\right]^{\beta_{HNj}}} \qquad (1)$$

where j is the index over which the relaxation processes are summed, ε_∞ is the high frequency limit of the real part $\varepsilon'(\omega)$, ω, is the angular frequency $\omega=2\pi f$ (f, the frequency of the outer electric field), τ_{HN} the relaxation time that is related to the reciprocal of maximal loss frequency, $\Delta\varepsilon$ is the dielectric relaxation strength; α_{HN} and β_{HN} are fractional shape parameters ($0<\alpha_{HN}\leq1$ and $0<\alpha_{HN}.\beta_{HN}\leq1$) describing the symmetric and asymmetric broadening of the dielectric spectrum. The specific case $\alpha_{HN}=\beta_{HN}=1$ gives the Debye relaxation law; $\beta_{HN}=1$, $\alpha_{HN}\neq1$ corresponds to the so-called Cole-Cole equation, whereas the case $\alpha_{HN}=1$, $\beta_{HN}\neq1$ corresponds to the Cole-Davidson formula.

At temperatures higher than 258K data is influenced by low frequency conductivity contribution, thus an additional term ($\sigma_{DC}/\omega\varepsilon_o$) was added to the dielectric loss $\varepsilon''(\omega)$, where σ_{DC} is related to the pure d.c. conductivity of the sample and ε_0 is the dielectric permittivity of vacuum. This conductivity can be due *i*) to intrinsic ionic impurities that, in the ultraviscous melt, migrate over extended trajectories between the electrodes [19], and/or *ii*) to proton diffusion along hydrogen bonds when a hydrogen-bonded network forms in the melt [16].

Characteristic relaxation times, that are model independent, were obtained by HN fits to the loss profiles and calculating the peak related relaxation time $\tau_{max}=1/\omega_{max}=1/2\pi f_{max}$, according to the following equation [20]:

$$\tau_{max}=\tau_{HN}\times\left[\sin\left(\frac{\alpha_{HN}\pi}{2+2\beta_{HN}}\right)\right]^{-1/\alpha_{HN}}\left[\sin\left(\frac{\alpha_{HN}\beta_{HN}\pi}{2+2\beta_{HN}}\right)\right]^{1/\alpha_{HN}} \qquad (2)$$

based on the HN relaxation time τ_{HN} and shape parameters α_{HN} and β_{HN}.

RESULTS AND DISCUSSION

The dielectric relaxation spectrum of ibuprofen presents five distinct features [17]: the d.c. conductivity, a primary (α) relaxation peak, a Debye-type (D) relaxation and two secondary processes (β and γ). In the following, we will focus on the main α-relaxation and the two secondary processes.

Figures 2 and 3 show representative dielectric loss spectra (ε'' plotted as a function of the frequency *f*) of ibuprofen taken isothermally at some selected temperatures in the range from 143K to 261K. In the lowest temperature range (163K-143K), *i.e.*, in the glassy state, a broad and low intense secondary process is observed-designated as γ (Fig. 2(a)). At temperatures between 191K and 228K, another secondary relaxation process of even lower intensity is detected, named β (Fig. 3). In the supercooled liquid state this process becomes partially submerged under the main relaxation process (α) associated with cooperative motions that are in the origin of the glass transition (Fig. 2(b)). This α-process partially hides in the low frequency flank, a less intense process named D (discussed in [17]). In the supercooled regime, the shape of the α-loss peak is basically invariant allowing it to be fitted with constant α_{HN} and β_{HN} shape parameters, respectively 0.81 and 0.56.

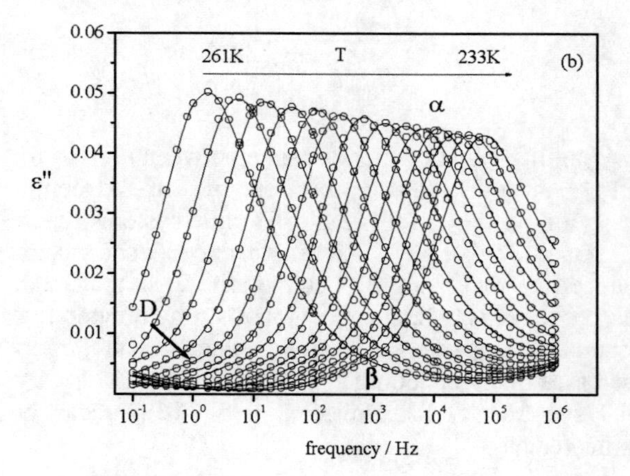

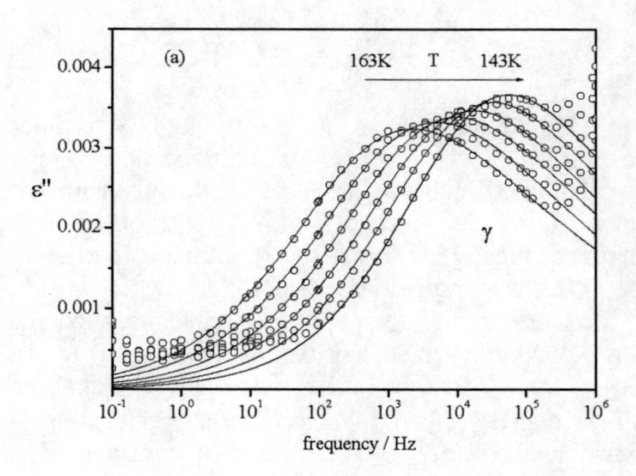

FIGURE 2. Dielectric loss spectra of ibuprofen at different temperatures: (a) from 163K to 143K (in steps of 4K) and (b) between 261K and 233K (in steps of 2K). Solid lines are the overall fitting curves (Eq. (1)).

The non-Debye nature of the relaxation response in non-ideal supercooled liquids can also be described, in time domain, by the well-known Kohlrausch-Williams-Watts (KWW) function given by [21,22]:

$$\phi(t) = \exp\left[-\left(\frac{t}{\tau_{KWW}}\right)^{\beta_{KWW}}\right] \quad (3)$$

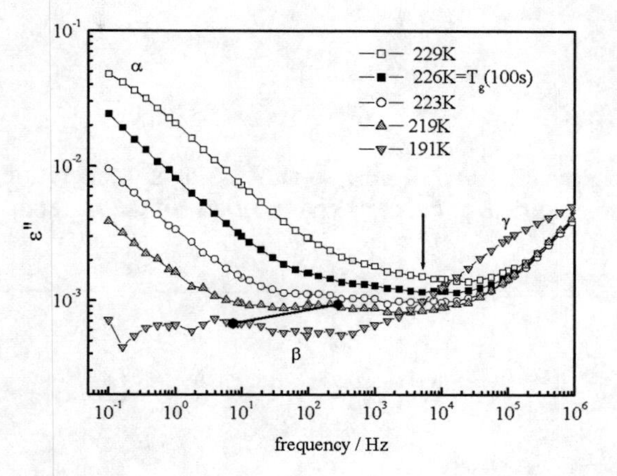

FIGURE 3. Dielectric loss spectra (in log-log scale) at the indicated temperatures evidencing the low amplitude β-process; lines are only guide for the eyes. Filled dark circles point out the location of the calculated frequency of the Johari-Goldestein (JG) process, $f_{JG}=1/(2\pi\tau_{JG})$, at 191K and 219K in reasonable agreement with the maxima of the detected β-process; τ_{JG} was estimated from Arrhenius equation with $E_a = 24RT_g$ and $\tau_\infty=10^{-14}$s. The vertical arrow indicates the location of f_{JG} at $T=T_g$ calculated using Eq. (8) (for details see text).

where t is time, τ_{KWW} is a characteristic relaxation time and β_{KWW} is the called "stretching parameter" ($0<\beta_{KWW}\leq 1$) that quantifies the non-exponentiality of the time decay ($\beta_{KWW}=1$ for the Debye response). The disadvantage of this equation is its transformation into the frequency domain through a Fourier transform that cannot be solved analytically. From the α_{HN} and β_{HN} shape parameters of the α-loss peaks it is possible to estimate the β_{KWW} exponent through the empirical correlation proposed by Alegría and co-workers [23,24]:

$$\beta_{KWW} = (\alpha_{HN}\beta_{HN})^{\frac{1}{1.23}} \quad (4)$$

From Eq. (4), using $\alpha_{HN}=0.81$ and $\beta_{HN}=0.56$, β_{KWW} was estimated as 0.52 in excellent agreement with the value reported in ref. [16] (β_{KWW} = 0.54), that was directly obtained from the fit of the loss peak with the stretch exponential function.

The temperature dependence of the relaxation times, *i.e*, the relaxation map, corresponding to the α and the two secondary processes detected in both glassy and supercooled liquid states is shown in Fig. 4.

The non-linear temperature dependence of τ_α in the vicinity of the glass transition can be described by means of Vogel-Fulcher-Tamman (VFT) equation [25-27]:

$$\tau(T) = \tau_\infty \exp\left(\frac{B}{T - T_0}\right) \qquad (5)$$

where τ_∞, B and T_0, the so-called Vogel temperature, are empirical parameters characteristic of the material.

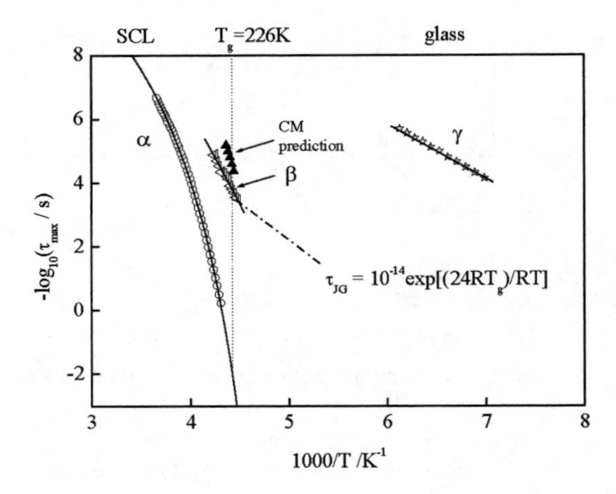

FIGURE 4. Arrhenius plot of dielectric relaxation times $\tau_{max}=(2\pi f_{max})^{-1}$, obtained from data collected on cooling, from the supercooled liquid (SCL) to the glassy state of ibuprofen. There are two secondary relaxations (γ and β) in addition to the primary α-relaxation. Lines are the Arrhenian and VFT fits. Filled up triangles indicate the $\tau_{JG}=\tau_{J0}$ estimated, at five temperatures just above T_g, from the coupling model (CM) relation (Eq. (8)), with n=0.48 and the corresponding $\tau_\alpha(T)$ given by the VFT fit. There is a good correspondence between the calculated τ_{JG} and the experimental τ_β. Dashed dot line is the predict JG relaxation time in the glassy state. Vertical doted line points out dielectric estimated $T_g(\tau=100s)$.

Commonly, a temperature dependence of the relaxation times according VFT-equation is regarded as a sign of molecular cooperativity often associated with the glass transition dynamics (main α relaxation). From the VFT parameters (B=612 K, $\tau_\infty=1.3\times10^{-14}$s, T_0=187K) it is possible to estimate the glass transition temperature, T_g, using the criterion $\tau(T_g)$=100s [28,29], as T_g=226K in very good agreement with the calorimetric glass transition temperature, T_g(onset)=228K.

The curvature of the $\log\tau_\alpha$ *vs.* 1/T plot, *i.e.*, the degree of deviation from Arrhenius-type temperature dependence provides a useful classification of glass formers [28,30]. Materials are called "fragile" if their τ_α(T) dependence markedly deviates from the Arrhenius behavior and "strong" if τ_α(T) is close to the latter. In other words, fragility indicates how fast the relaxation changes as T_g is approached from above. Thus, strong systems are those which show a strong resistance against structural changes upon cooling in their supercooled regime.

A quantitative measure of the fragility can be obtained from the steepness index *m* according the following equation (6):

$$m = \left.\frac{d\log_{10}\tau(T)}{d(T_g/T)}\right|_{T=T_g} \qquad (6)$$

Fragility values typically range between *m* = 16 for strong systems (those that shown an Arrhenius behavior), and *m* ~ 200 [28] for fragile systems where a marked deviation to Arrhenian dependence occurs induced by high cooperativity molecular rearrangements, which reflects in a strong temperature dependent apparent activation energy that largely increases on approaching T_g from above [31].

Using the VFT expression in Eq. (6), *m* can be estimated as:

$$m = \frac{BT_g}{\ln 10(T_g - T_0)^2} \qquad (7)$$

that, for ibuprofen, taking the B and T_0 values determined for the α relaxation as studied on cooling gives a fragility index of *m*=93 analogous with the value found for sorbitol (m=93) and other non-pharmaceutical glass formers [28], allowing to classify ibuprofen as a fragile glass former.

Concerning the temperature dependences of the two secondary (γ and β) relaxation processes (Fig. 4), they were fit by the Arrhenius equation: $\tau(T)=\tau_\infty\exp(E_a/k_BT)$ (E_a - activation energy, τ_∞ - pre-exponential factor, k_B - Boltzman's constant, T – absolute temperature); the activation energies were estimated as 35kJ.mol^{-1} and 110kJ.mol^{-1} and the pre-exponential factors as 1.4×10^{-17}s and 6.5×10^{-30}s, for the γ and β processes, respectively.

In order to test if any of these two secondary relaxation processes is the "universal" [32] Johari-Goldstein (JG) process [33,34], entailing the motion of a molecule as a whole, the coupling model (CM) [35] was applied to calculated the relaxation time for the JG process, τ_{JG}, at T_g. CM predicts a strong correlation between τ_{JG} and the primitive relaxation time τ_0 of the coupling model, that is related to the relaxation time of the main α-process, at the same temperature, according Eq. (8):

$$\tau_{JG} \approx \tau_0 = (t_C)^n (\tau_\alpha)^{1-n} \quad (8)$$

where the coupling parameter n=1-β_{KWW} and t_C has the approximate value of $2x10^{-12}$s for molecular liquids [32]. Using β_{KWW}=0.52, the estimated value of $\tau_{JG}(T_g)$ is $3x10^{-5}$s in excellent agreement with τ_β as depicted in Fig. 4 (filled up triangle) that also contains τ_{JG} values estimated for higher temperatures still exhibiting a good agreement with τ_β. The frequency determined from $\tau_{JG}(T_g)$ is indicated by an arrow in Fig. 3 close to the location where the low amplitude β-relaxation observed in the glassy state partially merges under the incoming α-process.

Furthermore, the maximum location of the glassy β-process was estimated from the Arrhenius equation using the correlation between the activation energy for a JG process and the glass transition temperature proposed by Kudlik *et al*, $E_{a,JG}=24RT_g$ [36] and replacing the pre-exponential factor τ_∞ by $1x10^{-14}$s. The obtained values are indicated in Fig. 3 between 191K and 219K as full dark circles revealing reasonable agreement with the experimental maxima of the glassy β-process. The CM criterion is not observed for the γ-relaxation since it is not a JG-relaxation; it has intramolecular origin.

The existence of a JG relaxation can play a fundamental role in the cold crystallization observed on heating ibuprofen from the glassy to the supercooled liquid state. In Fig. 5 are presented the isochronal plots of ε'', taken at 10^5Hz from isothermal measurements upon heating between 143K and 353K; gray symbols (down triangles) are relative to the first cooling run and full dark symbols (up triangles) concern to the subsequent heating run.

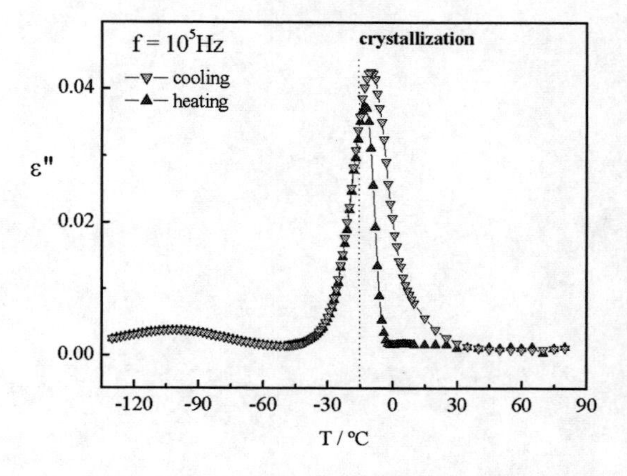

FIGURE 5. Isochronal plots of imaginary part of complex permittivity, taken at 10^5Hz from isothermal measurements. The plots show the mobility in both glassy and supercooled liquid states, further evidencing the occurrence of cold crystallization upon heating. This figure nicely illustrates the sensitivity of the dielectric spectroscopy to monitor amorphous-crystalline transformation.

Until 258K (indicated by a vertical dashed line) no apparent changes are observed between the two runs. For temperatures higher than 258K, a decrease in both real (not shown) and imaginary parts of the complex permittivity is observed with the consequent decline of the dielectric strength of the α-process. This is due to the reduction of the amorphous phase whose dipoles become immobilized in crystals and thus, dielectrically inactive. Given that the sample prevented crystallization in the previous cooling run, this is an evidence of cold crystallization since the temperature of the system was increased from the glassy state, where the sample was in a completely amorphous state, to the temperature of crystallization.

It is generally recognized that the crystallization process is composed of nucleation and crystal growth, being expected that the maximum rate of the crystal nucleation is located at lower temperatures relative to that of the crystal growth. In the present case, given that crystallization is only observed on heating and at temperatures above the glass transition, it seems reasonable to conclude that nucleation occurs somewhere between the glassy state and the onset of the α-relaxation. The study of the general effects of crystallization on the molecular relaxations of ibuprofen and the role of the sub-glass JG process as a pre-condition to crystallization [37] will be reported soon after.

ACKNOWLEDGMENTS

A. R. Brás acknowledges Fundação para a Ciência e Tecnologia (FCT, Portugal) for a PhD grant SFRH/BD/23829/2005.

REFERENCES

1. B. C. Hancock and G. Zografi, *J. Pharm. Sci.* **86**, 1–12 (1997).
2. D. Q. M. Craig, P. G. Royal, V. L. Kett, and M. L. Hopton, *Int. J. Pharm.* **179**, 179–207 (1999).
3. V. Andronis and G. Zografi, *Pharm. Res.* **15**, 835-842 (1998).
4. E. Shalaev, M. Shalaeva, and G. Zografi, *J. Pharm. Sci.* **91**, 584-593 (2001).
5. S. L. Shambling, X. Tang, L. Chang, B. C. Hancock, and M. J. Pikal, *J. Phys. Chem. B* **103**, 4113-4121 (1999).
6. N. T. Correia, J. J. Moura Ramos, M. Descamps, and G. Collins, *Pharm. Res.* **18**, 1767-1774 (2001).
7. F. Kremer and A. Schönhals, *Broadband Dielectric Spectroscopy*, Berlin: Springer, 2002.

8. D. Q. M. Craig, *Dielectric Analysis of Pharmaceutical Systems,* London: Taylor & Francis, 1995.
9. G. P. Johari, S. Kim, and R. M. Shanker, *J. Pharm. Sci.* **94**, 2207-2223 (2005).
10. M. Descamps, N. T. Correia, P. Derollez, F. Danede, and F. Capet, *J. Phys. Chem. B.* **109**:16092-16098 (2005).
11. L. Carpentier, R. Decressain, A. De Gusseme, C. Neves and M. Descamps, *Pharm. Res.* **23**, 798-805 (2006).
12. L. Carpentier, R. Decressain, S. Desprez, and M. Descamps, *J. Phys. Chem. B* **110,** 457-464 (2006).
13. J. Alie, J. Menegotto P. Cardon, H. Duplaa, A. Caron, C. Lacabanne, and M. Bauer, *J. Pharm. Sci.* **93**, 218-233 (2004).
14. K. Kamiński, M. Paluch, J. Ziolo, and K. L. Ngai, *J. Phys.:Condens. Matter* **18**, 5607-5615 (2006).
15. R. Bhushan and J. Martens, *Biomed. Chromatogr.* **12,** 309-316 (1998).
16. G. P. Johari, S. Kim, and R. M. Shanker, *J. Pharm. Sci.* **96**, 1159–1175 (2007).
17. A. R. Brás, A. M. M. Antunes, M. M. Cardoso, J. P. Noronha, M. Dionísio, and N. T. Correia, submitted.
18. S. Havriliak and S. Negami, *Polymer* **8**, 161 (1967); S. Havriliak and S. Negami, *J. Polym. Sci. C* **16**, 99 (1966).
19. A. K. Jonscher, *J. Phys. D: Appl. Phys.* **32**, R57-R70 (1999).
20. A. Schönhals and F. Kremer, "Analysis of dielectric spectra", in *Broadband Dielectric Spectroscopy*, edited by F. Kremer and A. Schönhals, Berlin: Springer Verlag, 2002, chap. 3.
21. R. Kohlrausch, *Pogg. Ann. Phys (III)* **12**, 393-399 (1847).
22. G. Williams, and D. C. Watts, *Trans. Faraday. Soc.* **66**, 80-85 (1966).
23. F. Alvarez, A. Alegría, and J. Colmenero, *Phys. Rev. B* **44**, 7306-7312 (1991).
24. A. Alegría, E. Guerrica-Echevarría, L. Goitiandía, I. Telleria, J. Colmenero, *Macromolecules* **28**, 1516-1527 (1995).
25. H. Vogel, *Phys. Zeit.* **22**, 645-646 (1921).
26. G. S. Fulcher, *J. Am. Ceram. Soc.* **8,** 339-355 (1925).
27. G. Tammann and G. Hesse, *Z. Anorg. Allg. Chem.* **156**, 245-257 (1926).
28. R. Böhmer, K. L. Ngai, C. A. Angell, and D. J. Plazek, *J. Chem. Phys.* **99**, 4201-4209 (1993).
29. C. T. Moynihan, P. B. Macebo, C. J. Montrose, *et a.l.*, *Ann. N. Y. Acad. Sci.* **279,** 15-35 (1976).
30. C. A. Angell, *J. Non-Cryst. Solids* **131-133**, 13-31 (1991). C. A. Angell, *J. Res. Natl. Inst. Stand. Technol.* **102**, 171-185 (1997).
31. P. G. Debenedetti and F. H. Stillinger, *Nature* **410**, 259-267 (2001).
32. K. L. Ngai, E. Kamińska, M. Sekuła, and M. Paluch, *J. Chem. Phys.* **123**, 204507-1 – 204507-7 (2005).
33. G. P. Johari and M. Goldstein, *J. Chem. Phys.* **53**, 2372-2388 (1970).
34. G. P. Johari and M. Goldstein, *J. Chem. Phys.* **55**, 4245-4252 (1971).
35. K. L. Ngai, *J. Phys.: Condensed Matter* **15**, S1107-S1125 (2003).
36. A. Kudlik, C. Tschirwitz, T. Blochowicz, S. Benkhof, and E. Rossler, *J. Non-Cryst. Solids* **235-237,** 406-411 (1998).
37. M. Hatase, M. Hanaya, and M. Oguni, *J. Non-Cryst. Solids* **333**, 129-136 (2004).

Direct Observation of Low-Energy Clusters in a Colloidal Gel

C. Patrick Royall*, Stephen R. Williams†, Takehiro Ohtsuka**, and Hajime Tanaka**

*School of Chemistry, University of Bristol, Bristol, BS8 1TS, UK.
†Research School of Chemistry, The Australian National University, Canberra, ACT 0200, Australia.
**Institute of Industrial Science, University of Tokyo, 4-6-1 Komaba, Meguro-ku, Tokyo 153-8505, Japan.

Abstract. We investigate local structure in a colloidal gel formed by Brownian dynamics simulation. With a novel topological algorithm, we show that the network of this gel is in fact largely comprised of the same structures known to be energy minima for isolated clusters. Despite some hints from the radial distribution function, we find no evidence even of local crystallisation, even though the equilibrium state is gas-FCC crystal coexistence.

Keywords: Colloids, Gels, Dynamical Arrest
PACS: 82.70.Dd

INTRODUCTION

Colloidal gels are well known as everyday systems which exhibit dynamical arrest [1]. Of particular interest are colloid-polymer mixtures, where the interactions between the colloids can be precisely tuned in model experimental systems [1]. Recent advances in experimental techniques have opened the possibility of analysing colloidal gels at the single-particle level [2, 3, 4], which has provided new insights into gel structure. While much work, both experimental [2, 5] and simulation [6, 7] has focussed on the dynamics and longer-ranged structure [2, 3], relatively little attention has been paid to the role of local structure in gel formation.

This in sharp contrast to glasses, where although the main focus has been upon possible dynamical mechanisms underlying arrest, the role of local structure can be traced back at least as far as F.C. Frank [8], who, in the 1950s, proposed that icosahedra, which are likely to occur in simple atomic systems such as Lennard-Jones liquids since they are a lower potential energy state than face-centred cubic crystals, and may inhibit crystallisation because of their five-fold symmetry.

In the case of colloid-polymer mixtures, gelation is identified with arrested phase separation [9, 10], at least in the absence of long-ranged electrostatic repulsions [3, 10]. Thus we may draw a parallel between gels and glasses, in the sense that the network of the gel is locally dense. Consider the phase diagram outlined in Fig. 1. The underlying equilibrium phase is gas-FCC crystal coexistence. While the gel state exhibits arrested phase separation, the network itself should be locally rather dense. Indeed, in these dense regions, we may expect cystallites to form, or perhaps structures akin to the icosahedra in the case of glasses [8].

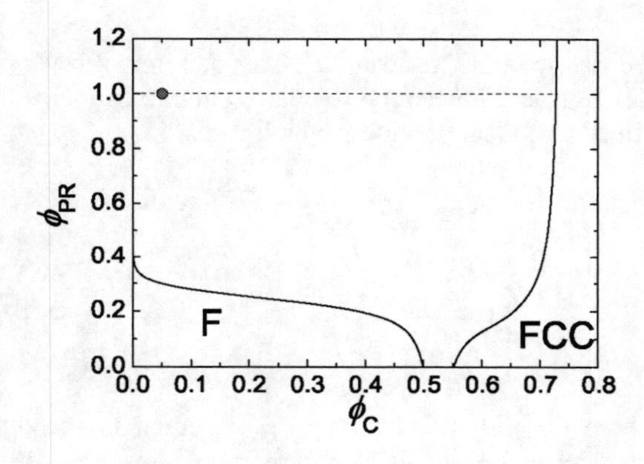

FIGURE 1. Color online. Equilibrium phase diagram calculated from free volume theory [11]. ϕ_C is the colloid volume fraction and $\phi_P = 4\pi R_G^3/3$ is the polymer volume fraction where R_G is the polymer radiius of gyration. The circle denotes the simulation state point, located deep in the phase-separating region. The tie line connects a very dilute gas and crystal of around 0.73 colloid volume fraction. Here the polymer-colloid size ratio $q = 0.18$. F and FCC refer to fluid and FCC cyrtsal phase respectively.

While icosahedra may be expected in glasses, exactly what might we expect in a colloidal gel? It is important to note that colloid polymer mixtures typically interact via short ranged attractions [12]. Now for these short ranged interactions, icosahedra are not favoured over FCC. In fact, if we consider icosahedra to be isolated *clusters* interacting via the Lennard-Jones potential, then in fact we should look for clusters which interact under a potential appropriate to colloidal gels.

The seminal theory of colloid-polymer mixtures is that

CP982, *Complex Systems, 5th International Workshop on Complex Systems*
edited by M. Tokuyama, I. Oppenheim, and H. Nishiyama

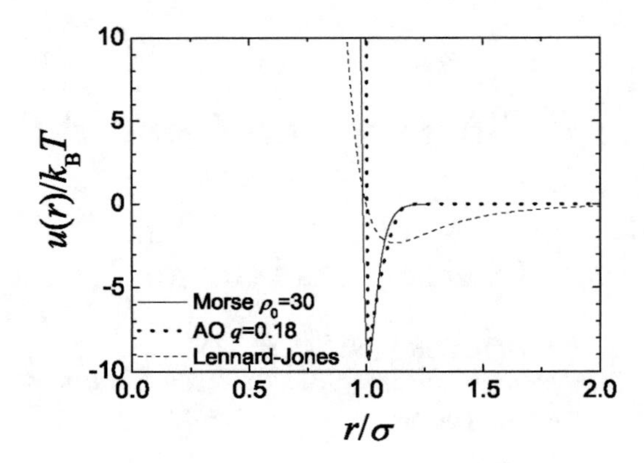

FIGURE 2. Comparison of interactions. There is little difference between the Morse ($\rho_0 = 30$) and AO $q = 0.18$ interactions. We therefore assume that structures which are potential minima for the Morse interaction are also minima for the AO potential. Also shown is the longer-ranged Lennard-Jones interaction.

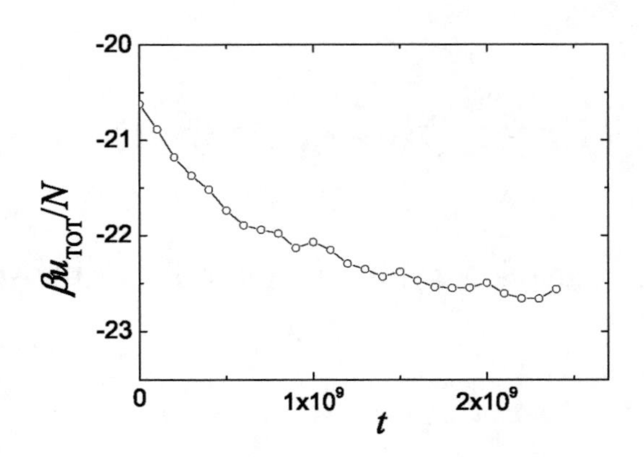

FIGURE 3. The potential energy per particle during the simulation. t is in units of attempted MC moves.

of Asakura and Oosawa [12]. This AO model leads to a pair interaction between two hard colloidal spheres in a solution of ideal polymers which reads

$$\beta u_{AO}(r) = \begin{cases} \infty & \text{for} r < \sigma \\ \frac{\pi(2R_G)^3 z_{PR}}{6}\frac{(1+q)^3}{q^3} & \\ \times\{1 - \frac{3r}{2(1+q)\sigma} & \\ + \frac{r^3}{2(1+q)^3\sigma^3}\} & \text{for} r \geq \sigma < \sigma + (2R_G) \\ 0 & \text{for} r \geq \sigma + (2R_G) \end{cases} \tag{1}$$

where the polymer fugacity z_{PR} is equal to the number density ρ_{PR} of ideal polymers in a reservoir at the same chemical potential as the colloid-polymer mixture. Thus within the AO model the effective temperature is inversely proportional to the polymer reservoir concentration. The polymer-colloid size ratio $q = 2R_G/\sigma$ where R_G is taken as the polymer radius of gyration, and σ is the colloid diameter.

Now for $q > 0.3$, the AO interaction is sufficiently long-ranged that gas-liquid phase separation is thermodynamically stable. Shorter interaction lengths, ie smaller q, lead to gels [1]. We are thus motivated to use a shorter ranged interaction, $q = 0.18$, noting that we have recently shown that Eq. (1) gives a very good agreement with an experimental system [13]. The AO interaction is plotted in Fig. 2. Also plotted are the Morse [14] and longer-ranged (6-12) Lennard-Jones interactions. The Morse interaction reads

$$\beta u_M(r) = \varepsilon e^{\rho_0(1-r/\sigma)}(e^{\rho_0(1-r/\sigma)} - 2) \tag{2}$$

where ε is the well depth and ρ_0 is a range parameter. In Fig. 2, the Morse potential is plotted with $\rho_0 = 30$. The difference between the AO potential for $q = 0.18$ and the Morse potential for $\rho_0 = 30$ are slight. Since the local energy minima of isolated clusters interacting via the Morse potential are documented [15], we take these results and assume that these same structures are also energy minima for clusters interacting via the AO potential.

We use Brownian dynamics [16] to simulate a colloidal gel, acting under the realistic [13] Asakura-Oosawa potential Eq. (1). We introduce a methodology to analyse the structure of the gel network in terms of its constituent Morse clusters, the topological cluster classification (TCC) [17]. Noting that the ground state of this system is gas-FCC crystal coexistence (Fig. 1) [11], we also investigate the possibility of local crystal-like structures.

METHODOLOGY

Simulation

Brownian dynamics is well-established as a means to tackle systems such as colloidal gels [18]. It is further well known that in the limit of short step length, Monte-Carlo (MC) simulation [19] reduces to Brownian dynamics. Recently, we found that in fact the step length need not be very small, provided it is smaller than all the lengthscales in the system under consideration, and, crucially, that the 'clock' ticks only when a MC move is accepted [16]. One attraction of this method is that discontinuous potentials, such as Eq. (1) may be treated very easily.

Here, we use a Monte-Carlo scheme, with a steplength of 0.01σ to simulate a colloidal gel. At the level of this work, we are simply interested in generating a non-equilibrium structure that resembles a colloidal gel, noting that in any case, the influence of hydrodynamic interactions has truly profound consequences for the structure of colloidal gels [20]. A full description of hydrodynamic interactions lies well outwith the scope of this work.

We simulated the system for 2×10^9 attempted MC moves. The average energy per particle is shown in Fig 3. This corresponds to several minutes for a comparable experimental system [13]. A colloidal gel is a non-equilibrium state, and so the energy per particle continues to decrease throughout the run. However, for the final 30% of the run, the thermal energy fluctuations are comparable to the energy decrease. Furthermore, the difference between the radial distribution functions $g(r)$ plotted in Fig. 4. for $t = 2.0 \times 10^9$ and $t = 2.5 \times 10^9$ MC steps are very small indeed, suggesting that the system is not evolving significantly at these times. In the following, we analyse data from the end of the simulation run $t \approx 2.5 \times 10^9$. We used $N = 1000$, and the state point simulated is shown in Fig. 1, $\phi_C = 0.05$, $\phi_P = 1.0$, which is deeply quenched to inhibit phase separation. Here the polymer volume fraction is defined as $\phi_P = 4\pi R_G^3/3$.

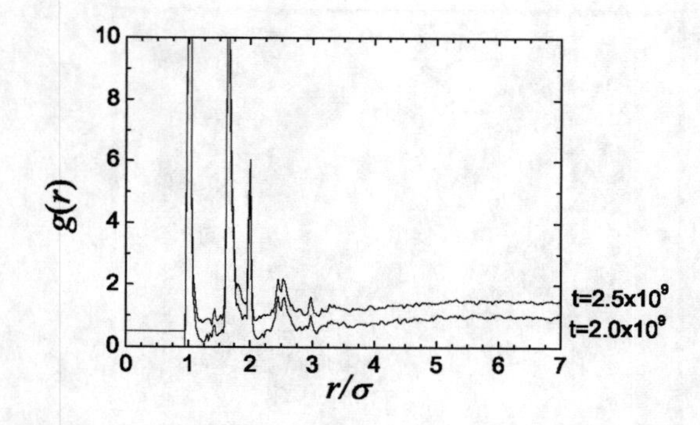

FIGURE 4. Radial distribution functions $g(r)$ for $t = 2.0 \times 10^9$ and $t = 2.5 \times 10^9$ MC steps. Note the strong peak-splitting, particularly in the case of the second peak. The $g(r)$ for these two times are almost indistiguishable. $t = 2.5 \times 10^9$ MC steps data are shifted upwards by 0.5 for clarity.

The topological cluster classification

The structure of the ground state clusters for the Morse potential has been calculated, both as a function of m the cluster size, and ρ_0 the interaction range parameter. Following the notation of Doye *et. al.* [15], these 'Morse clusters' are termed mA for the cluster which is the ground state for the longest ranged potential, and mB, and so on, at increasingly shorter interaction ranges, where m is the number of particles in the cluster. In the case of $\rho_0 = 30$, a rather short-ranged interaction, the ground state clusters are thus 5A, 6A, 7A, 8B, 9B, 10B, 11F, 12E, and 13B. 13A is the icosahedron. Those clusters we have identified in this gel are pictured in Fig. 6. More information on the cluster geometries and ground state energies may be found in [15].

We have developed a novel algorithm to identify structures which are topologically equivalent to all the Morse clusters for $m < 13$, ie we also include 8A, 9A, etc. We now provide a brief description of our method, which we term the topological cluster classification (TCC). More details are provided in [17]. In the TCC we start with the bond network between all the particles. The range of the interaction sets a natural bondlength, 0.18σ [18]. All the shortest path three, four and five membered rings are identified [21]. These rings are then classified in terms of those which have an additional particle bonded to all the particles in the ring and those which have two or no such additional particles. We call these the basic clusters, into which many of the larger clusters can be decomposed. A given particle may be a member of more than one basic cluster, i.e. basic clusters may overlap. We use this strategy to identify all the topologically distinct 'Morse clusters' with thirteen or less particles which occur in a bulk system. In addition we identify the FCC and HCP thirteen particle clusters in terms of a central particle and its twelve nearest neighbours. If a particle was found to be part of more than one cluster size, it was labelled as part of the larger cluster size, and the association with the smaller cluster size was ignored. We do not consider 11D or 12C, in the notation of [15], as these are not distinct in the topological cluster classification.

RESULTS

As mentioned above, radial distribution functions are shown in Fig. 4. The structure is typical for a colloidal gel with a strong peak at contact. The $g(r)$ rise to unity by around $r \sim 6\sigma$ which provides a measure of the network lengthscale. However we focus more on the local structure, noting that the second and third peaks show a strong degree of splitting. In the case of denser systems, this peak splitting has been interpreted as the onset of crystallisation [22]. However, as we shall see below, although we search explicitly for local crystalline order, none whatsoever is found. Thus peak splitting must have some other interpretation in this case.

A snapshot of the system is shown in Fig. 5. Different shades correspond to different cluster sizes m. The gel network is seen to be almost largely comprised of these clusters. As might be inferred from Fig 1, there are rather few particles which are not bound to the network. Furthermore, a mere 8.7% of particles are not part of

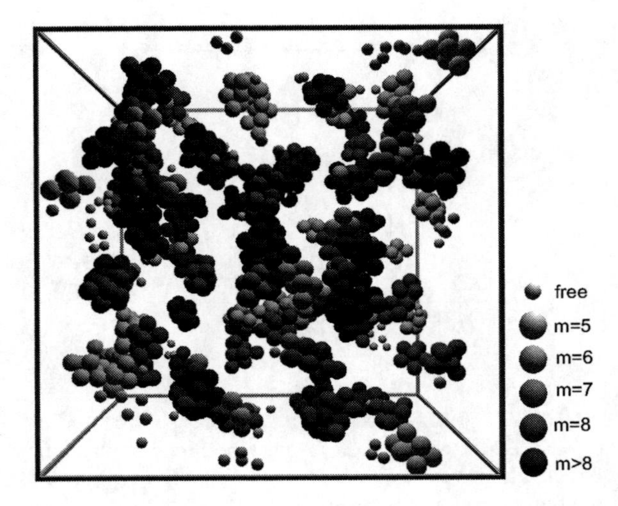

FIGURE 5. Snapshot of the simulation. Different shades denote clusters of different m, as shown on the right. All particles are shown actual size, except for 'free' particles which are not part of any cluster, which are shown 60% actual size.

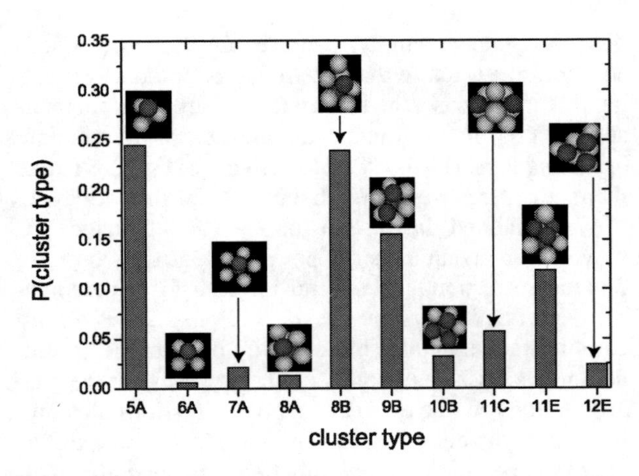

FIGURE 6. Color online. Distribution of cluster types. The population of each cluster type is shown as a fraction of the whole. Recall that one particle may be a member of more than one cluster type, we only consider the larger cluster type in this case.

any cluster. Thus it is reasonable to conclude that the gel network may be largely decomposed into isolated clusters.

A more quantitative picture emerges if we consider the cluster type distribution in Fig. 6. The most striking observation is that no FCC- nor HCP-like local environments are found, although our topological cluster classification includes these structures [17]. Even locally, therefore, this gel is very far from the gas-FCC coexistence that forms its equilibrium state. We also see that the system appears to be dominated by 5A, 8B, 9B and 11E. With the exception of 11E, these are ground state clusters for the short-ranged Morse potential. We find no 10A, 11A, 11B, 12A, 12B, nor 12D. These are global potential energy minimum structures for longer-ranged Morse potentials, so their absence is perhaps not too surprising. Likewise $m = 13$ is not found, so we find no $m = 13$ icosahedra (13A), which might be expected in the case of longer ranged interactions such as Lennard-Jones [8, 15].

CONCLUSIONS

We have shown that a colloidal gel may be deconstructed in terms of small clusters. This suggests that locally, for $m < 13$, groups of particles can lower their potential energy quite effectively. Considering larger lengthscales, the system becomes dynamically arrested, unable to reach gas-FCC crystal equlibrium. Our methodology may also be applied to experimental colloidal gels, and perhaps even to colloidal glasses, which may now be resolved at the single-particle level.

Furthermore we have shown that even though the radial distribution function may exhibit features which have been interpreted as signs of crystallisation , in the form of peak splitting [22], in fact other structures, namely low-energy clusters can also give rise to the same behaviour. Thus measures such as the radial distribution function which average over many particles may be insufficient to provide a complete picture of local structure, even in amorphous systems such as colloidal gels.

ACKNOWLEDGMENTS

This work was partly supported by a Grant-in-Aid for JSPS Fellowships for Overseas Reseachers and also a Grant-in-Aid from the Ministry of Education, Culture, Sports, Science, and Technology, Japan. CPR acknowledges The Royal Society for financial support.

REFERENCES

1. W. C. K. Poon, *J. Phys. Condens. Matter.* **14**, R859–R880 (2002).
2. A. Dinsmore, and D. Weitz, *J. Phys. Condens. Matter* **14**, 7581–7597 (2002).
3. A. I. Campbell, V. J. Anderson, J. S. van Duijneveldt, and P. Bartlett, *Phys. Rev. Lett.* **94**, 208301 (2005).
4. Y. Gao, and M. Kilfoil, *Phys. Rev. Lett.* **99**, 078301 (2007).
5. A. Dinsmore, V. Prasad, I. Wong, and D. A. Weitz, *Phys. Rev. Lett.* **96**, 185502 (2006).
6. A. Puertas, M. Fuchs, and M. Cates, *Phys. Rev. Lett.* **88**, 098301 (2002).
7. A. Puertas, M. Fuchs, and M. Cates, *J. Chem. Phys.* **121**, 2813–2822 (2004).

8. F. C. Frank, *Proc. R. Soc. Lond. A.* **215**, 43–46 (1952).
9. M. Miller, and D. Frenkel, *Phys. Rev. Lett.* **90**, 135702 (2003).
10. E. Zaccarelli, *J. Phys. Cond. Matter* **19**, 323101 (2007), URL `http://arxiv.org/abs/cond-mat/0705.3418v1`.
11. H. N. W. Lekkerkerker, W. C. K. Poon, P. N. Pusey, A. Stroobants, and P. B. Warren, *Europhys. Lett.* **20**, 559–564 (1992).
12. S. Asakura, and F. Oosawa, *J. Chem. Phys.* **22**, 1255–1256 (1954).
13. C. P. Royall, A. Louis, and H. Tanaka, *J. Chem. Phys.* **127**, 044507 (2007).
14. P. Morse, *Phys. Rev.* **34**, 57–64 (1929).
15. J. P. K. Doye, D. J. Wales, and R. S. Berry, *J. Chem. Phys.* **103**, 4234–4249 (1995).
16. C. Royall, J. Dzubiella, S. Schmidt, and A. van Blaaderen, *Phys. Rev. Lett.* **98**, 188304 (2007).
17. S. R. Williams, *(http://arxiv.org/abs/0705.0203)* **arXiv:0705.0203v1 [cond-mat.soft]** (2007).
18. K. G. Soga, J. R. Melrose, and R. C. Ball, *J. Chem. Phys.* **108**, 6026–6032 (1998).
19. D. Frenkel, and B. Smit, *Understanding Molecular Simulation: from Algorithms to Applications*, New York: Academic, 2001.
20. H. Tanaka, and T. Araki, *Europhys. Lett.* **79**, 58003 (2007).
21. D. S. Franzblau, *Phys. Rev. B* **44**, 4925–4930 (1991).
22. T. M. Truskett, S. Torquato, S. Sastry, P. G. Debenedetti, and F. H. Stillinger, *Phys. Rev. E* **58**, 3083–3088 (1998).

Experimental Determination of Structural Relaxation in Trehalose-Water Solutions by Inelastic Ultraviolet Scattering

S. Di Fonzo[1], C. Masciovecchio[1], F. Bencivenga[1], A. Gessini[1], D. Fioretto[2], L. Comez [2], A. Morresi[3], M. E. Gallina[3], O. De Giacomo[4], and A. Cesàro[4]

[1]*Sincrotrone Trieste, Strada Statale 14 km 163.5, Area Science Park, I-34012 Trieste, Italy*
[2]*Dipartimento di Fisica, Università di Perugia, Via Pascoli, I-06100 Perugia, Italy*
[3]*Dipartimento di Chimica, Università di Perugia, Via Elce di Sotto, 8, I-06100 Perugia, Italy*
[4]*Dipartimento di Biochimica Biofisica e Chimica delle Macromolecole, Università di Trieste, Via L.Giorgieri 1, I-34127 Trieste, Italy*

Abstract. We report Brillouin ultraviolet scattering measurements on trehalose-water solutions in a wide range of concentrations (φ = 0-0.74). A complete set of data as a function of temperature (-10 °C $\leq$ T $\leq$ 100 °C) has been obtained for each concentration. The T- φ evolution of the system has been analyzed in terms of energy position and linewidth of inelastic peaks. These results have been used to derive the structural relaxation time of the system, τ, which was found in the tens of ps timescale. Its T-dependence can be described with an Arrhenius activation law, and, most importantly, a significant slowing down of the relaxation dynamics has been observed as trehalose concentration was increased. At low- φ , the activation energy of the relaxation has been found to be consistent with literature data for pure water, and comparable with intermolecular hydrogen bond (HB) energy. This evidence strongly supports the hypothesis that the main microscopic mechanism responsible for the relaxation process in trehalose solutions lies in the continuous rearrangement of the HB network. Finally, the results are discussed in terms of the evolution of the system upon increasing trehalose concentration, in order to provide a complete description of the viscoelastic stiffening in real biological conditions.

Keywords: Inelastic ultraviolet scattering, Structural relaxation, Trehalose solutions
PACS: 62.60.+v , 67.57.Jj , 87.64.Cc

INTRODUCTION

The bioprotective properties of sugar trehalose (α-D-glucopyranosyl-α-D-glucopyranoside), a naturally-occurring glass-forming and non-reducing disaccharide with chemical formula $C_{12}H_{22}O_{11}$, have been known for almost two decades and are still a matter of extensive interest[1-8] both from an academic point of view, and for the potential applicative fall-out. Trehalose has been identified at fairly high concentrations in several natural organisms (i.e. seeds, plants, fungi, bacteria, insects hemolynph and several invertebrate animals). If harsh environmental conditions occur (for istance drought, freezing and ionizing radiations), these species start to produce trehalose at fairly high concentrations and, under a mechanism which is still controversial, are capable of surviving for a long time in a state in which their metabolic activity becomes quiescent or even undetectable (anhydrobiosis).

Trehalose bioprotective skills have been attributed to several different mechanisms and factors. Four major hypotheses have been proposed to explain the stabilizing effect of trehalose on biostructures. The *water-replacement* hypothesis[1,2,3] suggests that, upon drying, sugars can substitute water molecules by forming a water-like hydrogen bond (HB) network, thus preserving the native structure of proteins in the absence of water.. The *vitrification* hypothesis[4] suggests that sugars found in anhydrobiotic systems protect biostructures through the formation of glasses, thereby reducing structural fluctuations and preventing protein denaturation or mechanical stresses. Furthermore, the higher glass transition temperature (T_g = 121 °C in the pure anhydrous sugar) compared to similar disaccharides (e.g. sucrose: T_g = 68.5 °C) and other protectants (xylitol, sorbitol glucose), could explain its greater efficiency. The *destructuring-effect on the water HB* hypothesis[5] suggests that trehalose, compared to other disaccharides, facilitates a more extended hydration and binds water molecules more strongly, thus preventing ice formation and the

CP982, *Complex Systems, 5th International Workshop on Complex Systems*
edited by M. Tokuyama, I. Oppenheim, and H. Nishiyama

subsequent irreversible damage of biosystems. Finally the *reversible-dehydration* hypothesis[6,7,8] refers to the formation of trehalose dihydrate nano-crystals and their subsequent reversible transformation into an anhydrous state during water removal, enhancing the biosystem stability.

In the past few years, a consensus has emerged in attributing trehalose's bioprotectant action to a synergic combination of the above factors, which cumulatively contribute to: i) a drastic slowing-down of molecular motions ensuring both structural conservation as well as chemical integrity; ii) a preservation of native conformations and morphologies of cellular biostructures during the dehydration-rehydration cycles.

Even though the exact mechanism responsible for bioprotective effectiveness is still unknown, it is clear that a complex array of interactions at structural, physiological and molecular levels may play an important role. Therefore, a paramount, multidisciplinary investigation is still required in order to characterize structural and molecular interactions on binary trehalose-water solutions and on ternary trehalose-protein-water solutions.

In the present work, propagating collective excitations in binary trehalose-water solutions have been determined by Brillouin Inelastic Ultra-Violet Scattering (IUVS) at λ=244 nm incident photon wavelength. Measurements have been carried out in a wide range of trehalose concentrations ($0<\varphi<74$ weight %) and temperatures (-10 °C<T<+100 °C). The T-φ evolution of the system has been analyzed in terms of energy positions and linewidths of inelastic excitations in the dynamic structure factor, $S(Q,\omega)$. These results have been further correlated with the structural relaxation time τ, of the system, and its evolution with increasing concentration and decreasing temperature is discussed in details.

In general, if the relaxation timescale of the system is much shorter than the period of the acoustic wave, $\tau << T_w$, the internal fluid rearrangements are so fast that the system can reach its local equilibrium configuration within the period of density wave propagation, which therefore propagates adiabatically: i.e. over successive local equilibrium states. This is the so-called hydrodynamic or *viscous regime* that characterizes the low-frequency acoustic propagation in simple fluids. If density waves with $T_w<<\tau$ are probed, internal fluid rearrangements are indeed too slow and, consequently, the system cannot equilibrate within a propagation period. In this case, the energy carried by the density wave cannot be efficiently dissipated and, therefore, it propagates elastically: i.e. with substantially reduced energy losses. This limiting behavior is commonly referred to as the *elastic regime*. In the intermediate, *viscoelastic, regime,* T_w is comparable with τ and, therefore, relaxation and density wave propagation mechanisms are strongly coupled[9]. The occurrence of a relaxation phenomenon can be experimentally observed either by varying the period of the probed density (sound) wave, or by varying the relaxation time itself. The former task can be accomplished, e.g., by changing the momentum transfer, $Q\propto 1/T_w$, while the latter goal is usually achieved by varying the sample temperature. This paper will show that trehalose-water solutions in real biological conditions as a function of both T and φ can be studied, at best, with IUVS since, thanks to the unique Q-range probed by this technique, $T_w\approx\tau$ and, therefore, sensitivity to the relaxation processes is maximized. adding important new knowledge about these complex biophysical systems.

EXPERIMENTAL DESCRIPTION

Sample Preparation

The solutions were prepared by weight, dissolving Trehalose dihydrate Sigma-T9449[10], with purity greater than 99%, in double distilled and deionized water and by stirring and heating until the corresponding dihydrate solubility point was reached. Table 1 summarizes the different φ-values probed in our experiments and the corresponding number of water molecules per trehalose, n*.

TABLE 1. Measured samples: weight fraction, φ, number of water molecules per trehalose, n*, explored temperature ranges, glass transition temperature, T_g, average density, ρ_{av}, and momentum transfer values, Q_{av}, are reported.

ϕ	n^*	T_m (°C)	T_M (°C)	T_g (°C)	ρ_{av} (Kg/³)	Q_{av}
0	∞	1.5	30	-135	997.8	0.0712
0.15	107	0.0	89.7	-126.4	1046.5	0.0722
0.32	41	-4.2	89.8	-114.7	1125.0	0.0738
0.38	31	15.4	95.1	-108.6	1148.9	0.0744
0.42	26	-2.5	95.1	-103.4	1176.7	0.0750
0.47	21	-11.15	69.9	-96.9	1214.0	0.0757
0.59	13	-3.15	86.9	-79.6	1268.3	0.0769
0.74	7	-12.75	94.9	-45.6	1357.4	0.0788

The explored conditions include φ-T values outside the thermodynamic stability limits of the homogeneous liquid phase. Figure 1 reports the phase diagram of water-trehalose solutions.

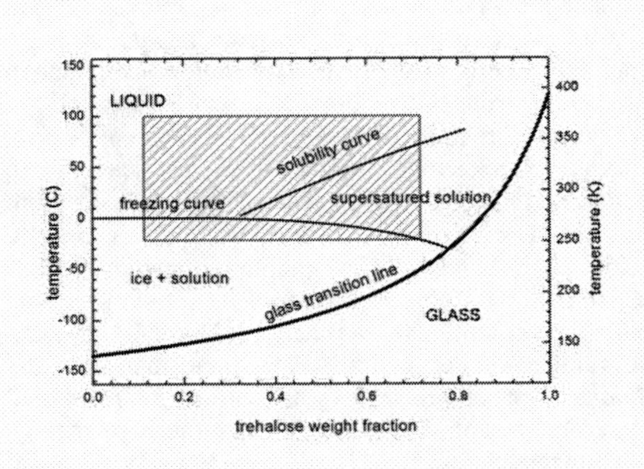

FIGURE 1. Phase diagram of water-trehalose solutions. Freezing and solubility curves are from Ref. 11 and 12. Glass transition line is calculated according to the Gordon–Taylor equation[13] with the data of ref 14. The shaded region indicates the φ-T ranges investigated in the present work.

The shaded region indicates the investigated φ-T ranges. The solubility curve of dihydrate trehalose, freezing and glass transition lines are indicated as well.

IUVS Measurements

IUVS measurements were carried out at the beamline 10.2 of the Elettra Synchrotron radiation Laboratory in Trieste[15]. The UV source was a frequency-doubled, CW output Argon Ion laser delivering a single-mode beam at λ=244 nm (typical power ≈ 5 mW). The solutions were contained in a 2 mL cell placed in the beamline experimental chamber. This cell had two optically polished sapphire windows (3 mm thickness, 20 mm diameter) sealed by Teflon O-rings. The beam optical path-length in the cell was 10 mm. The sample spot size in the focal plane was 300x200 μm^2 (FWHM). The signal, scattered in the nearly backward direction (θ= 172°), was energy-analyzed by a normal incidence Czerny-Turner analyzer[16] and finally detected by a Peltier cooled CCD detector with 13.5x13.5 μm^2 pixel size and 20% quantum efficiency. The momentum transfer was calculated using the equation: $Q = \frac{4\pi}{\lambda} n \sin(\frac{\theta}{2})$, where the refractive index n was directly measured with an accuracy of the order of $5e^{-4}$ using a new method based on a modification of the fixed angle of incidence method for a hollow prism[17]. The overall experimental energy and momentum resolutions were typically around 1.4 GHz ($\Delta E/E \approx 10^{-6}$) and 0.004 nm^{-1} ($\Delta Q/Q \approx$ 0.05), respectively. Spectra were collected by varying the sample temperature in the ranges indicated in table 1. A liquid nitrogen cryostat and a sample heater were used to scan the temperature, which was controlled with a K-type thermocouple. Temperature stability was greater than 0.1 °C during each spectrum acquisition (≈ 30 minutes). A moderate vacuum (10^{-6} mbar) was applied in the sample chamber during cooling in order to avoid air condensation.

RESULTS AND DISCUSSION

ϕ-T Dependencies of Raw Data

In Figure 2 (left panels), a set of representative IUVS spectra for samples corresponding to ϕ=0.74, ϕ=0.42 and ϕ=0.15 are indicated. The occurrence of a relaxation process that matches the reciprocal frequency of the longitudinal modes can be directly observed.

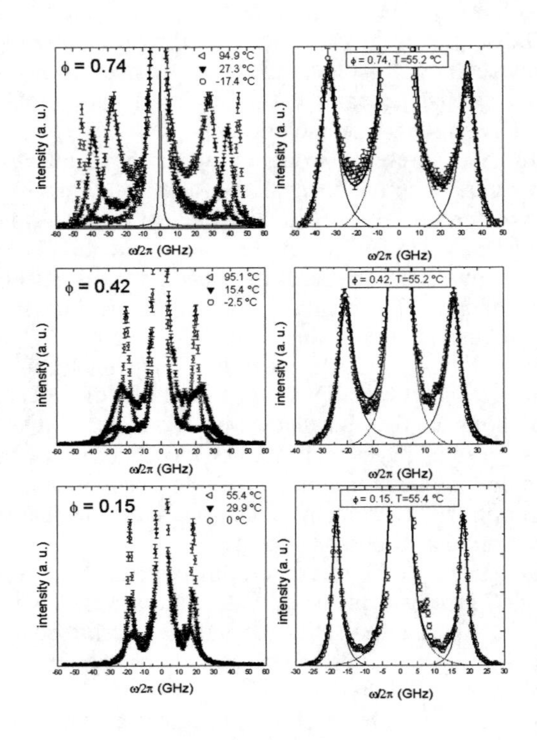

FIGURE 2. Left panels (top to bottom): representative IUVS spectra at φ=0.74, φ=0.42 and φ=0.15. The solid line shown with φ=0.74 is the instrumental resolution function. Right panels: three examples of experimental spectra, at T=55.2 °C, compared with the respective best fit results (heavy solid line). The elastic (dotted line) and inelastic (solid line) contributions are indicated as well.

In particular, a shift of inelastic peak positions toward higher energies is found for ϕ=0.42 and 0.74, while almost constant positions are observed at ϕ=0.15. Moreover, the energy positions of inelastic peaks at constant temperature notably increase with trehalose concentration. A scrutiny of the T-dependence of inelastic peak widths show, as the temperature decreases, a sharpening for φ=0.74 and a broadening for φ=0.42 and φ=0.15. In the frame of the viscoelastic theory, the former case can be interpreted as a transition, undergone by inelastic excitations, from the maximum of the structural relaxation to the elastic limit. On the other hand, the latter two cases can be interpreted as the transition of the longitudinal mode away from the viscous (adiabatic) limit to the maximum of the structural relaxation.

Figure 3 reports representatives IUVS spectra at weight concentrations φ=0.74, 0.59, 0.42 and 0.15 and at the biological temperature of T=26.1 °C. At this temperature, the addition of trehalose shifts the maximum of the inelastic peak toward higher energies and shrinks its width, thus indicating that the internal degrees of freedom become too slow to efficiently dissipate the energy of the acoustic wave, as is expected for a solid-like, i.e. frozen, medium.

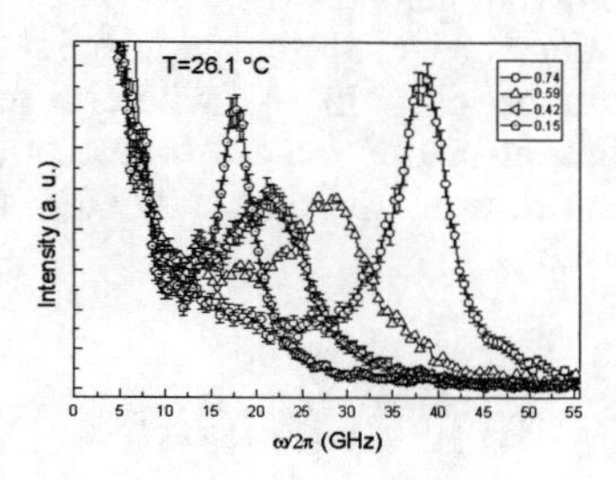

FIGURE 3. Representative IUVS spectra at weight concentrations φ=0.74, 0.59, 0.42 and 0.15 at T=26.1 °C.

Spectral Analysis and Fitting Procedure

According to the standard theory of Brillouin scattering in liquids and glasses[18], the scattered intensity is directly proportional to the dynamic structure factor, S(Q,ω). In order to extract quantitative information on the T-φ dependencies of the lineshape parameters, S(Q,ω) has been modeled by the sum of a δ-function and a DHO function[19], accounting for elastic and inelastic contributions, respectively:

$$S(Q,\omega) = A\,\delta(\omega) + B\frac{\omega_b^{\,2}(Q)\,\Gamma_b(Q)}{(\omega^2-\omega_b^{\,2}(Q))^2+\omega^2\Gamma_b^{\,2}(Q)} \qquad (1)$$

where ω_b and Γ_b correspond to the position and line width (FWHM) of inelastic excitations. The intensity A of the central spectral line is mostly due to spurious scattering from the sample environment. The model reported in Eq.(1) has been properly convoluted with the instrumental resolution function and scaled by an arbitrary intensity factor. A flat background contribution, mostly due to the electronic noise of the detection system, was previously subtracted from the raw data. The fit procedure consists of a χ^2 minimization based on a standard nonlinear least squares Levenberg-Marquardt algorithm[20]. The quality of our results can be appreciated in Figure 2 (right panels), where the measured spectra (open circles) are compared with the best fit lineshape (heavy solid line).

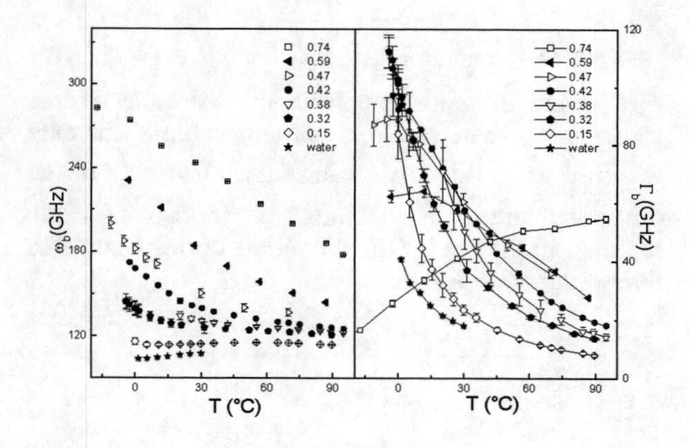

FIGURE 4. T-dependence of ω_b and Γ_b for some selected φ-values. Solid lines in the right panel are guides for the eye. Moreover, this dispersion shifts towards higher temperatures with increasing φ.

The T-dependencies of ω_b and Γ_b are reported in Figure 4 for some selected φ-values. Both of these hypersonic parameters show clear temperature dispersion, highlighting the presence of an active relaxation process in the experimental window.

In the spectral region around Brillouin lines, the parameters ω_b and Γ_b are related to the real (M') and imaginary (M'') parts of the longitudinal elastic modulus, respectively[21]:

$$M'(\omega_b) = \frac{\rho\omega_b^{\,2}}{Q^2} \quad \text{and} \quad M''(\omega_b) = \frac{\rho\omega_b\Gamma_b}{Q^2} \qquad (2)$$

Both of these parameters are sensitive to the presence of an active relaxation whose timescale is of the order of ω_b^{-1}. According to the memory function framework[9, 22], as a first approximation the quantities $M'(\omega_b)$ and $M''(\omega_b)/\omega_b$ can be written as:

$$M'(\omega_b) = (c_s Q)^2 + \frac{\Delta\xi^2}{1+\xi^2} \quad \text{and} \quad \frac{M''(\omega_b)}{\omega_b} = \frac{\Delta\tau}{1+\xi^2} \qquad (3)$$

where c_s is the adiabatic sound velocity and ξ is the dimensionless quantity $\omega_b\tau$. In order to derive Eqs. (3)

the following assumption were made: (i) there is only an active relaxation process in the considered frequency window, (ii) this relaxation is described by a Debye relaxation function [9,22], i.e. an exponential time decay and (iii) the specific heat ratio, γ, is equal to 1, thus leading to negligible contributions of thermal fluctuations. The first two assumptions are consistent with those used in previous experiments[23-26], while the latter is justified by thermodynamic data for pure water[27]. Equating Eqs. (2) and (3), the following expression for the relaxation time can be derived:

$$\tau = \frac{1}{\Gamma_b}\left(1 - \frac{c_s^2 Q^2}{\omega_b^2}\right) \tag{4}$$

Eq. (4) is meaningful only if $\omega_b \neq c_s Q$: i.e. for $\xi \neq 0$, where the departure from the adiabatic regime takes place. Therefore, it is not applicable to the data for pure water, since we found that $\omega_b \approx c_s Q$ in the probed T-range. The results of this analysis are reported in Figure 5, where the data corresponding to pure water are from Ref. 24.

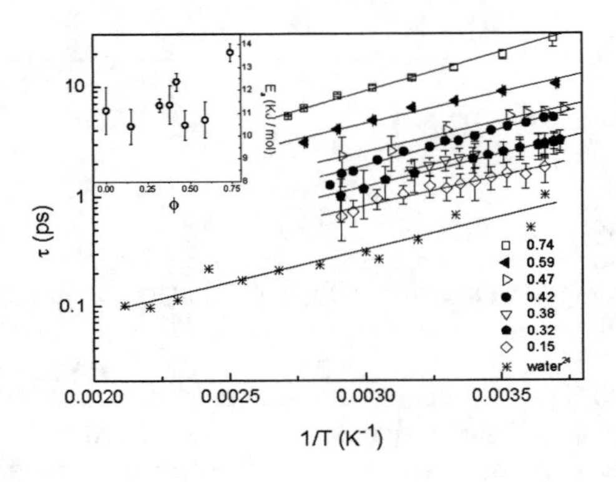

FIGURE 5. Structural relaxation time as a function of the inverse temperature. Data for pure water are taken from Ref. 24.

The c_s-values for trehalose solutions were taken from literature data, after a proper φ- interpolation[28]. This Figure shows that by adding trehalose a progressive slowing-down of τ occurs at a given temperature. Furthermore, in the thermodynamic range spanned in the present experiment, the data can be described by an exponential (Arrhenius) T-dependence: $\tau(T) = \tau_0 e^{\frac{E_a}{k_B T}}$. The obtained values for E_a ,close to *12 kJ/mol* , are not appreciably dependent on the concentrations of the solutions, and are in agreement with experimental determinations for pure water.[24] This result suggests that the whole dynamic process remains substantially unperturbed in quite a large range of sugar concentration and is still governed by molecular H-bonds[29].

A similar conclusion has been reached from recent Neutron scattering experiments and MD simulations of glucose in water at concentrations of up to 50% (5 m)[30]. The authors claim that the first hydration shell, dominated by HB, is not significantly perturbed even at high concentrations and that the long-range structure of water is not affected by the solute. They further address their attention to the real meaning of water "structure" in this context. While this study provides clear support to the "pseudo-ideal" behavior of glucose-water systems (up to 5 m), it also calls for a better rationalization of the properties of the sugar "family", including trehalose, at higher concentrations. As far as the results of the present study are concerned, an increase of E_a of about 2 kJ/mol was found at the highest concentration, although this point needs further experimental verification. Experiments in this direction are under way, as well as other investigations focused on covering the higher concentrations, up to pure undercooled trehalose.

In particular, provided that a robust theoretical formalism can be used in the whole compositional range, the present results demonstrate that IUVS can give consistent information on the dynamics of water-trehalose solutions.

CONCLUSIONS

The results of this study concur with the general trend of a strengthening of aqueous systems upon increasing sugar concentration, in this case trehalose. At molecular level, the activation threshold of the relaxation can be straightforwardly related to the continuous rearrangement of hydrogen bonds in the solution. Whether this development of mechanical properties is sufficient for the full explication of the bioprotective functions of trehalose up to more concentrate systems is still a matter of discussion.

Apparently, a certain amount of experimental evidence at low moisture has now been accumulated for a parallel formation of nano-crystals of dihydrate crystals during dehydration of trehalose and other sugars[6-8,31]. Therefore, as the potentially plasticizing water is captured in the dihydrate crystalline cage, the remaining dry amorphous trehalose enters the glassy state and the whole matrix becomes much firmer.

The relevance of the above data, within the hypotheses made in the present work, relies on the observed changes in the dynamics of the trehalose-water solutions upon increasing trehalose concentration. Although a continuous variation is observed, due to its modest magnitude it can hardly be responsible for the expected straightening in natural conditions. However, at ambient temperature trehalose-water solutions enter the glassy phase upon the removal of water molecules because of both evaporation and/or crystallization in confined nano-crystals. It is straightforward to conclude that both processes produce the desired results, but the extent of the resulting firmness and the rate of evolution could be very different.

Of course, although our data cover a wide concentration range, experiments still need to be extended up to anhydrous trehalose, as well as in a wider wavelength range. In the future our laboratory plans to enrich this type of information and to investigate the subtle differences in homologous disaccharides.

ACKNOWLEDGMENTS

We gratefully acknowledge the Structural Biology Lab group (Elettra) for the hospitality and access to instrumentation; in particular Ivet Krastanova and Theodora Zlateva, for their patient support during sample preparation. We would also like to thank Riccardo Comin for his valuable contribution during the experiments. ODG is gratefull to the University of Trieste for the financial support of the Ph.D. Fellowship.

REFERENCES

1. J. F. Carpenter and J. H Crowe,. *Biochemistry* ***28***, 3916-3922 (1989).
2. J. H. Crowe and L. M. Crowe, *Nature Biotechnology* ***18***, 145-146 (2000).
3. M. A. Villareal, S. B. Diaz, E. A. Disalvo, G. G. Montich *Langmuir* ***20***, 7844- 7851 (2004).
4. S.-P. Ding, J. Fan, J.L. Green, Q. Lu, E. Sanchez, C. A Angell, *J. Thermal Anal.* **47**, 1391-1405 (1996).
5. C. Branca, S. Magazù, G. Maisano,P. Migliardo, *J. Chem. Phys.* **111**, 281-287 (1999).
6. A. Cesàro, *Nature Materials* **5**, 593-594 (2006).
7. F. Sussich, R. Urbani, F. Princivalle, A Cesàro, *J. Amer. Chem. Soc* **31**, 7893-7899 (1998).
8. F. Sussich C. Skopec, J. Brady, A. Cesàro, *Carbohyd. Res.*, **334**, 165-176 (2001).
9. J. P. Boon and S. Yip, *Molecular Hydrodynanics*, New-York, Dover, 1991 .
10. D(+)- Trehalose Dihydrate T 9449-100G from Sigma (USA) , Batch # 114K7064.
11. P. M. Mehl, *J. Therm. Anal.*, **49**, 817-822 (1997).
12. D. P. Miller, J. J. de Pablo, H. Corti, *Pharm. Res.*, **14**, 578-590 (1997).
13. M. Gordon, J. S. Taylor, *J. Appl. Chem.*, **2**, 493-500, (1952); in the calculation of T_g we assumed a value of k = 5.2 and 394.2 K and 138 K for the glass transition temperatures of pure trehalose and water, respectively. The values for the maximal freeze concentration have been taken from Ref.14.
14. T. Chen, A. Fowler, M. Toner, *Cryobiology*, **40**, 277-282 (2000).
15. C. Masciovecchio, D. Cocco, A. Gessini, *Proceedings of 8th International Conference of Synchrotron Radiation Instrumentation, San Francisco, California,* AIP Conference Proceedings, **705,** 1190-1196 (2004).
16. M. Czerny, A. F. Turner, *Z. Physik*, **61**, 792-797 (1930).
17. A. Zaidi, Y. Makdisi, K. S. Bhatia, I. Abutahun, *Rev. Scient. Instr.*, **60**, 803-805 (1989).
18. B. J. Berne, R. Pecora, *Dynamic Light Scattering*, New York: J. Wiley & Sons Ed., 1976.
19. B. Fak, B. Dorner, *Institute Laue Langevin (Grenoble France) Tech. Rep. No. 92FA008S* (1992).
20. J. J. More, *Numerical Analysis, Lecture Notes in Mathematics*, Vol. 630, Berlin: G. A. Watson Ed., Springer-Verlag, 105-161, 1977.
21. D. Fioretto, L. Comez, G. Socino, L. Verdini, S. Corezzi, P. A. Corezzi, P. A. Rolla, *Phys. Rev. E*, **59**, 1899-1907 (1999).
22. U. Balucani, M. Zoppi, *Dynamics of the Liquid State*, Oxford: Clarendon Press, 1994.
23. A. Cunsolo, M. Nardone, *J. Chem. Phys.*, **105**, 3911-3917, (1996).
24. G. Monaco, A. Cunsolo, G. Ruocco, F. Sette, *Phys. Rev. E*, **60**, 5505-5521 (1999).
25. F. Bencivenga, A. Cunsolo, M. Krisch, G. Monaco, G. Ruocco, F. Sette, *Phys. Rev. E*, **75**, 051202 (2007).
26. F. Bencivenga, A. Cunsolo, M. Krisch, G. Monaco, L. Orsingher, G. Ruocco, F. Sette, A. Vispa, *Phys. Rev. Lett.*, **98**, 085501 (2007).
27. *Release on the IAPWS Formulation 1995 for the Thermodynamic Properties of Ordinary Water Substance for General and Scientific Use*, Fredericia, Denmark, September 1996.
28. S. Magazù, P. Migliardo, A. M. Musolino, M. T. Sciortino, *J. Phys. Chem B*, **101**, 2348-2351 (1997).
29. D. Eisenberg, W. Kauzmann, *The Structure and Properties of Water*, London: Oxford Press, 1979, Chapt 4.6.
30. P. E. Mason, G. W. Neilson, J. E. Enderby, M.-L. Saboungi, J. W. Brady, J. *Phys. Chem. B*, **109**, 13104-13111 (2005).
31. F. Sussich, Ph. D. Thesis, University of Trieste, 2004.

Direct Crystal to Glass Transformations of Trehalose Induced by Milling, Dehydration and Annealing

J. F. Willart, M. Descamps, and V. Caron

Laboratoire de Dynamique et Structure des Matériaux Moléculaires, UMR CNRS 8024, ERT 1066
University of Lille 1, Bât. P5, 59655 Villeneuve d'Ascq, France

Abstract. In this paper, we present a short review of the different solid state vitrification routes of trehalose recently identified by our group. Three routes are investigated: the mechanical milling of the stable crystalline anhydrous form, the rapid dehydration of the dihydrate form, and the sub Tg annealing of the polymorphic form α. The investigations have been performed by powder x-ray diffraction and differential scanning calorimetry.

Keywords: Solid state amorphization, Glass transition, Molecular materials
PACS: 61.43.-j, 61.43.Fs, 65.60.+a, 6150K

INTRODUCTION

Amorphous solids are generally obtained by the rapid quench of their liquid phase. However, there exist other amorphization routes which do not require the melting of the compound like the mechanical milling [1,2] or the dehydration of hydrate forms [3,4]. Moreover, when milling or dehydration is carried out below the glass transition temperature (Tg) of the corresponding liquid states they can provide very unusual examples of direct crystal to glass transformations [5-7]. These vitrification routes are particularly effective for molecular materials because of their rather weak elastic constants and their numerous crystalline hydrate forms [8]. Moreover, they appear to be a useful alternative to the usual thermal quench of the liquid since organic compounds often undergo chemical degradations upon heating.

However, the solid state vitrification processes raise many fundamental questions. We can wonder in particular if the glasses obtained by these routes have the same nature and the same characters (dynamics, local structure, energy landscape…) than those obtained by the usual thermal quench of the liquid state. Another point is also to identify the universal mechanisms which govern these solid state vitrification processes. The study of crystalline samples which can be amorphized and vitrified by different routes independent of one another should greatly help the general understanding of the solid state vitrification processes. We thus present in this paper a short review of the solid state vitrification possibilities of trehalose ($C_{12}H_{22}O_{11}$). This disacharide has been found to reach a glassy amorphous state by at least 3 independent routes in addition to the usual quench of the melt. These routes are: the milling of its crystalline anhydrous form β [9], the dehydration of its dihydrate form T_{2H_2O} [10] and the sub Tg annealing of its crystalline polymorphic form α [11].

EXPERIMENTAL

α–α anhydrous trehalose (form β) and trehalose dihydrate (T_{2H_2O}) were purchased from Fluka. Both were more than 99% pure and were used without further purification.

The ball milling was performed in a high energy planetary mill (Pulverisette 7 – Fritsch) at room temperature (RT) and under a dry nitrogen atmosphere. We used ZrO_2 milling jars of 45 cm^3 with seven balls (Ø=15 mm) of the same material. 1 g of material was placed in the planetary mill corresponding to a ball / sample weight ratio of 75:1. The rotation speed of the solar disk was set to 400 rpm which corresponds to an average acceleration of the milling balls of 5 g.

The powder X-ray diffraction (XRD) experiments were performed at RT with an Inel CPS 120 diffractometer ($\lambda_{CuK\alpha}$ = 1.540 Å) equipped with a 120° curved position sensitive detector coupled to a

CP982, *Complex Systems, 5th International Workshop on Complex Systems*
edited by M. Tokuyama, I. Oppenheim, and H. Nishiyama

4096 channel analyser. The samples were placed into Lindemann glass capillaries (∅ = 7 mm).

The differential scanning calorimetry (DSC) experiments were performed with the 2920 microcalorimeter of TA Instruments and the DSC7 of perking Elmer. During all the measurements the sample was placed in an open cell (container with no cover) and was flushed with highly pure nitrogen gas. Temperature and enthalpy readings were calibrated using pure indium at the same scan rates used in the experiments.

VITRIFICATION BY MILLING OF THE CRYSTALLINE FORM β

We have investigated the transformations of trehalose β upon high energy mechanical milling. Figure 1 shows the x-ray diffraction pattern of the stable crystalline form β before and after a 30 hour milling process. After milling the Bragg peaks characterizing the crystalline state have clearly disappeared and the x-ray diffraction pattern is found to be fully similar to that of the quenched liquid suggesting an amorphization of the milled sample.

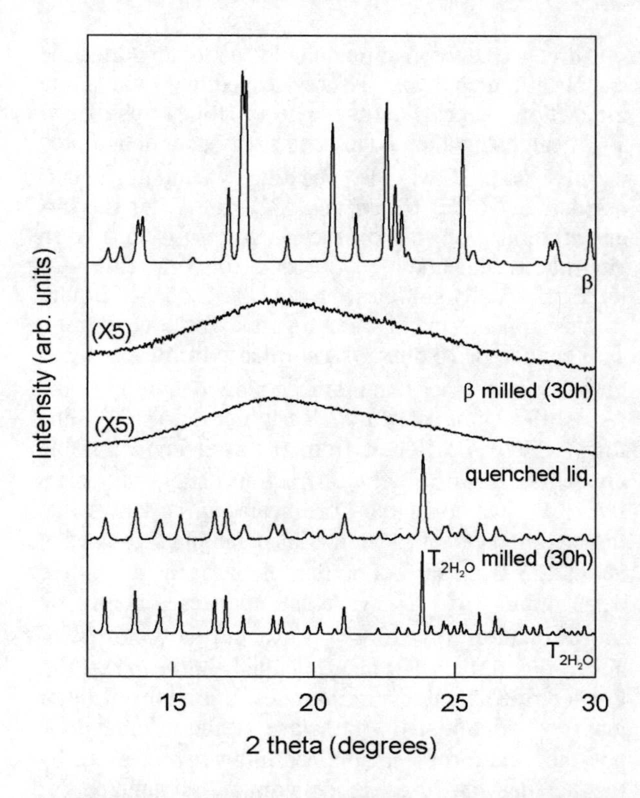

FIGURE 1. X-ray diffraction patterns recorded at RT. From top to bottom: non milled form *β*, form β after a 30 hour milling process, quenched liquid trehalose, T_{2H_2O} after a 30 hour milling process, non milled T_{2H_2O}.

Figure 2 shows the DSC heating scan (run 3) of the milled sample. It exhibits clearly a Cp jump characteristic of a glass transition which occurs exactly at the same temperature than that of the quenched liquid, followed at higher temperature by an exothermic recrystallization. This recrystallization is an indirect signature of the amorphization which has occurred during the milling process, and the Cp jump indicates that the amorphous material has a glassy character. Such a thermal behavior reveals that the milling has given rise to a real glassy state, and not to a mere nanocrystalline material resulting from a strong size reduction of the crystallites.

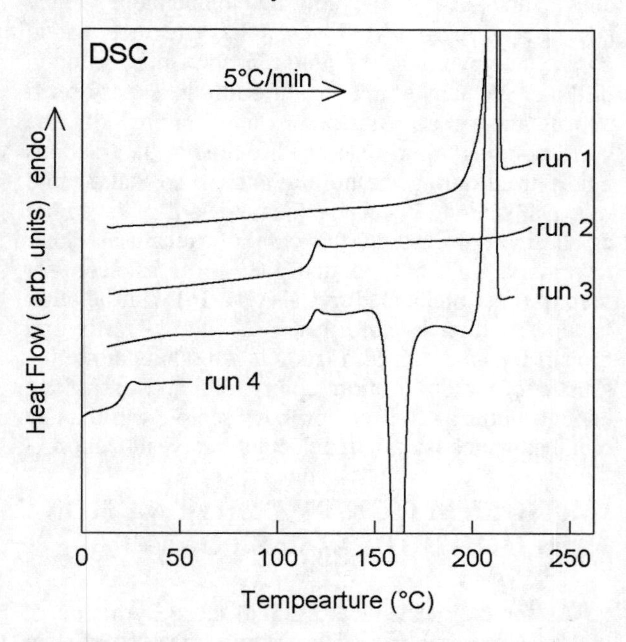

FIGURE 2. DSC scans recorded upon heating (5°C/min). run 1: non milled form β, run 2: quenched liquid trehalose, run 3: Form β after a 30 hour milling process, run 4: T_{2H_2O} melted and quenched in an hermetic pan.

A most interesting point is that the dihydrate form of trehalose (T_{2H_2O}) has a totally different behavior upon milling. Figure 1 shows the x-ray diffraction pattern of crystalline T_{2H_2O} before and after a 30 hour milling process. Contrary to the case of the anhydrous crystal, the Bragg peaks broaden but do not disappear. This indicates that the milling only reduces the size of the crystallites but does not induce the solid state vitrification of the sample.

The conditions required to amorphize a crystal upon milling are not yet clearly established. However, it appears empirically that amorphization mainly occurs when the milling is performed below the glass transition temperature (Tg) of the corresponding liquid state. In the case of trehalose, the glass transition of the anhydrous compound is located at Tg (β) = 120°C ie far above the milling temperature (RT). On the other hand, because of the plasticizing effect of its structural water molecules, the glass transition of the dihydrate form is expected to occur at a much lower temperature, which is confirmed by the DSC run 4 of figure 2. This run has been recorded upon heating, and after melting and quench of a T_{2H_2O} sample in an hermetic pan. It clearly reveals a Cp jump at Tg (T_{2H_2O}) = 20°C corresponding to the glass transition of an amorphous trehalose / water mixture having the stoechiometric composition of trehalose dihydrate. The forms β and T_{2H_2O} have thus their glass transition temperature above and below the milling temperature which suggests strongly that Tg is a key thermodynamic parameter governing the amorphization process upon milling. We may think that if milling is performed above the glass transition temperature of the corresponding liquid state, the life time of the fractions amorphized during the milling is short so that a rapid recrystallization occurs either toward the initial crystalline form as in the case of trehalose, either towards a different polymorphic form as seen for several other molecular crystals [12-16]. On the other hand, if milling is performed well below Tg, the life time of the amorphized fraction is long in regard to the kinetic of amorphization upon milling. For a long enough milling a fully amorphous sample can thus be obtained which is durably reluctant to crystallization.

VITRIFICATION BY DEHYDRATION OF THE DIHYDRATE FORM T_{2H_2O}

As for the mechanical milling, the dehydration of hydrated molecular crystals often induces solid state transformations [3,17,18]. Despite many recent investigations, the understanding and the control of these transformations remain empirical. For dehydration, the perturbation is no longer extrinsic to the sample as in the case of milling. On the contrary, the perturbation is intrinsic to the sample since the destabilization of the crystalline lattice is due to the removal of the water molecules and to the breaking of the hydrogen bonds that they often develop with the other molecules. As a result, a structural reorganization of the sample occurs, either toward a crystalline anhydrous from (stable or metastable), either towards an amorphous state. Since trehalose has a dihydrate form, we have investigated in details its transformations upon dehydration in order to check if this process could provide another solid state vitrification route in addition to the milling route.

Our results have clearly revealed a duality between an amorphization process and a polymorphic transformation in this compound during the stage of water removal. This duality appears to be very

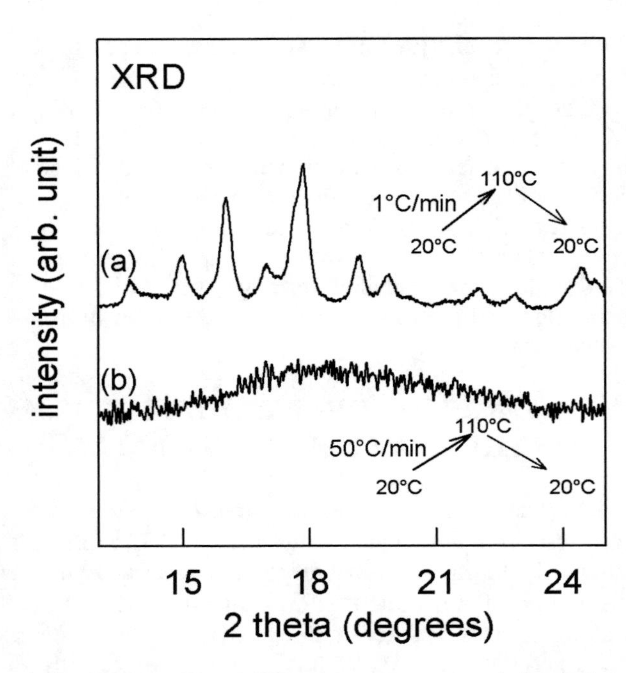

FIGURE 3. X-ray diffraction patterns recorded at RT for two anhydrous trehalose samples obtained by heating previously T_{2H_2O} up to 110°C at 1°C/min (a) and at 50°C/min (b).

sensitive to the thermal treatment used to dehydrate the sample. Figures 3 and 4 show respectively the X-ray diffraction patterns and the heating DSC scans of two T_{2H_2O} samples which have been previously dehydrated rapidly and slowly by heating them at 110°C respectively at 50°C/min and 1°C/min. After the fast dehydration the x-ray diffraction pattern is that of an amorphous state and the DSC scan (run 3) reveals a Cp jump at Tg fully similar to that of the quenched liquid. It thus appears clearly that upon a fast dehydration T_{2H_2O} undergoes a direct transformation from crystal to glass. On the other hand, after a slow dehydration the x-ray diffraction pattern is clearly that of a crystalline form which is different from the well known stable crystalline form β (x-ray diffraction pattern shown in figure 1). Moreover the DSC scan (run1) shows that the melting peak of the dehydrated sample is located at about 125°C, ie far below that of the form β. A slow dehydration of T_{2H_2O} thus induces clearly a transformation toward a metastable polymorph of anhydrous trehalose which is called the form α [10]. For intermediate dehydration rates, a mixture made of glassy amorphous trehalose and crystalline polymorphic form α is obtained in proportion fixed by the heating rate used to dehydrate the sample. For instance, the DSC run 2 of figure 4 recorded after a dehydration performed at 10°C/min shows clearly the thermodynamic signatures of both the glass transition of amorphous trehalose and the melting of form α. As

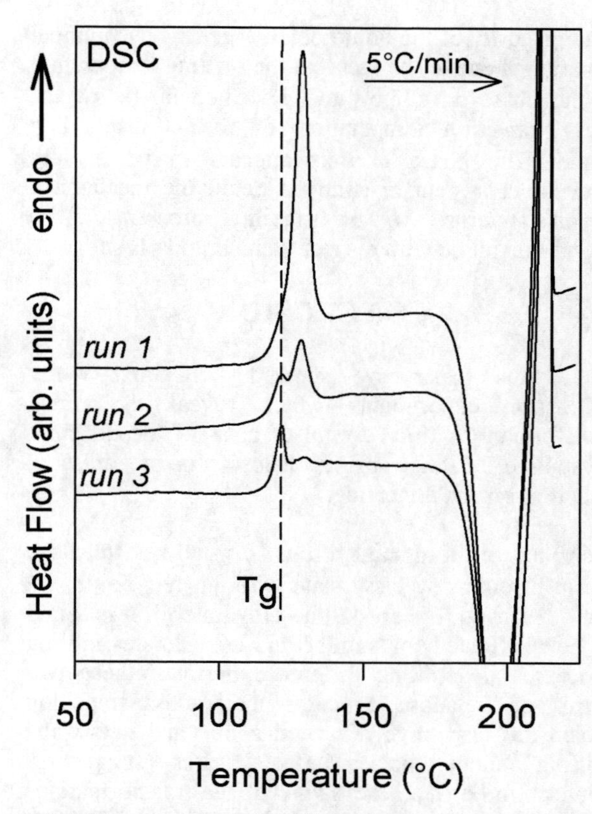

FIGURE 4. Heating DSC scans (5°C/min) of three anhydrous trehalose samples obtained previously by heating T_{2H2O} up to 110°C at 1°C/min (run 1), at 10°C/min (run 2) and at 50°C/min (run 3)

a result, the nature of the transformation of T_{2H_2O} induced by dehydration appears to be governed by the rate of water removal. This suggests that the structural evolutions of T_{2H_2O} upon dehydration result from an athermal driving mechanism similar to that of mechanical milling or irradiation [19].

VITRIFICATION BY ISOTHERMAL ANNEALING OF THE POLYMORPHIC FORM α

The melting of the polymorphic form α shown in the previous section occurs just above the glass transition temperature. Such a close to Tg melting is very unusual and suggests that the polymorphic form α is in fact unstable with respect to the amorphous state, and thus transforms toward this amorphous state as soon as the molecular mobility is high enough, ie: just above Tg. To reveal this instability of the phase α we have followed the structural and thermodynamic evolutions of this phase during long isothermal annealing below the glass transition temperature, and thus below its apparent melting temperature.

Figure 5b shows the DSC heating scans of the samples after increasing annealing times at 105°C, 110°C and 117°C. We can see that for increasing annealing time at 110°C, the melting peak of the form α progressively drops while the Cp jump characteristic of the glass transition develops. This behaviour clearly reveals a slow isothermal vitrification of the material. After 325 hours of annealing the melting peak has totally disappeared indicating that the vitrification process is completed. In this view, the kinetic of vitrification is directly related to the molecular mobility so that it is very sensitive to the annealing temperature. It is, for instance, much slower at 105°C where after 188h of annealing the glass transition only start to be detectable through a shouldering in the left hand wing of the melting peak (figure 5). On the other hand, the kinetic of amorphization is much more rapid at 117°C since the overall sample is vitrified in less than 50 hours.

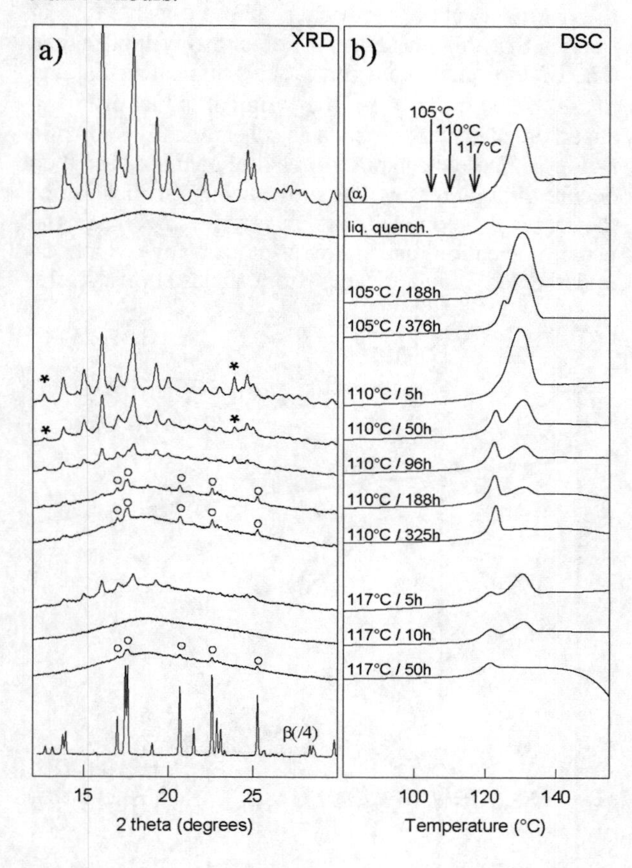

FIGURE 5. X-ray diffraction patterns (a) and heating (5°C/min) DSC scans (b) of the phase α of trehalose recorded after different annealing times at 105°C, 110°C and 117°C. The annealing times are reported on the left hand side of the thermograms. Data for some pure forms of trehalose (α, β, and quenched liquid) are also reported for comparison. Stars and open circles mark respectively the main Bragg peaks of T_{2H_2O} and form β.

Figure 5a shows the evolutions of the x-ray diffraction pattern of phase α during the annealing at 110°C and 117°C. The progressive disappearance of the Bragg peaks confirms the isothermal amorphization process. It must also be noted that some x-ray diffraction patterns show additional Bragg peaks revealing traces of T_{2H_2O} and form β. The traces of T_{2H_2O} mainly appear for short annealing times when a large amount of phase α is still remaining. Since this phase is very hygroscopic [20-22], some reversion toward the T_{2H_2O} form cannot be fully avoided during the management of the x-ray diffraction experiments which follow the annealing stages. On the other hand, the traces of form β mainly appear for the longest annealing times. They reveal a very slow nucleation of the most stable crystalline form in the amorphized fraction of the sample during the sub-Tg annealing.

As a conclusion, the spontaneous vitrification of the form α upon isothermal annealing below Tg implies that this phase has a free enthalpy higher than that of the amorphous state. This indicates that the phase α is in fact in the original situation of a superheated crystalline state as schematically shown in figure 6. The metastability breaking of this phase then occurs through a slow but spontaneous vitrification of the material below Tg. It is exactly the inverse of the usual nucleation and growth process by which an undercooled liquid escapes from metastability. Here,

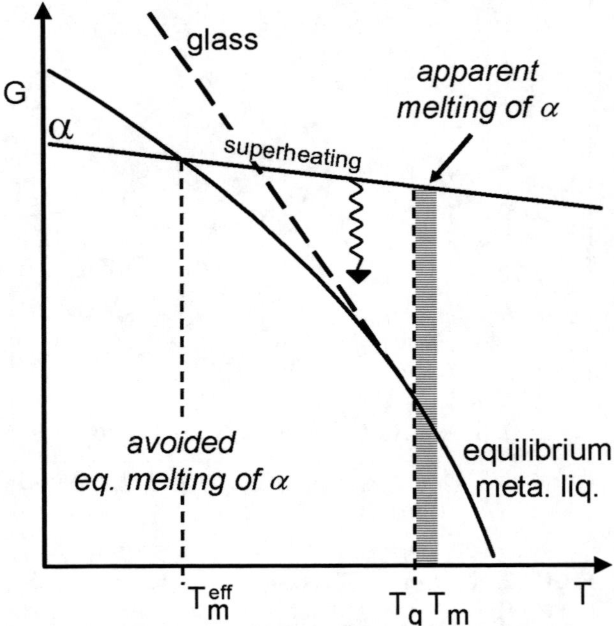

FIGURE 6. Schematic free enthalpy diagram illustrating the superheating condition of the polymorphic form α between its equilibrium melting temperature and the glass transition point.

we have the nucleation and growth of liquid droplets, and this mechanism is made very slow by the very high viscosity of the liquid below Tg. This exceptional situation is due to the fact that the melting temperature of the phase α is likely to be located far below the glass transition temperature of the corresponding liquid. The phase α thus appears to be a very convenient system to study in detail the mechanism which governs the metastability breaking of a crystalline phase with respect to the liquid phase.

CONCLUSION

In this paper, we showed DSC and X-ray diffraction experiments which reveal the direct transformations from crystal to glass of the different crystalline forms of trehalose upon milling, dehydration and annealing.

Upon milling, anhydrous crystalline trehalose (form β) enters a glassy state within a few hours. On the other hand, the crystalline dihydrate form is found to be unaffected for similar and even longer milling processes. A striking difference between these two forms of trehalose concerns their glass transition which are respectively located above and below the milling temperature. These features suggest a connection between the nature of the transformations induced by milling and the relative position of Tg with respect to the milling temperature.

Upon rapid dehydration, T_{2H_2O} is found to transform into an anhydrous compound which is x-ray amorphous and shows a Cp jump at Tg characteristic of a glassy state. This indicates that the sample has been vitrified upon dehydration. On the other hand, a slow dehydration is found to induce a polymorphic transformation toward a metastable polymorphic form called "α". The nature of the transformation upon dehydration thus appears to depend on the dehydration rate itself.

Upon annealing below Tg, the crystalline form α obtained by slow dehydration of T_{2H_2O} vitrifies slowly but spontaneously, indicating that the phase α is in a very unusual superheating situation. This behavior suggests that the effective melting temperature (T_m^{eff}) of the phase α is likely to be located far below the glass transition temperature (T_g) of this compound. The high viscosity of the liquid trehalose between T_m^{eff} and Tg is thus invoked to explain the long life time of the phase α in this temperature range.

Our results show that glassy trehalose can thus be obtained by three independent solid state vitrification routes in addition to the usual thermal quench of the liquid state. These possibilities make trehalose a very promising system to understand further the physics of the solid state vitrification.

ACKNOWLEDGMENTS

The authors thank F. Danede and F. Capet for their useful assistance in the management of the x-ray diffraction experiments.

REFERENCES

1. G. Martin and P. Bellon, *Solid State Physics, New York* **50**, 189-331 (1997).
2. H.G. Brittain, *Journal of Pharmaceutical Sciences* **91** (7), 1573-1580 (2002).
3. A. K. Galwey, *Thermochimica Acta* **355** (1-2), 181-238 (2000).
4. N. Onodera, H. Suga, and S. Seki, *Bulletin of the Chemical Society of Japon* **41**, 2222-2222 (1968).
5. I. Tsukushi, O. Yamamuro, and T. Matsuo, *Solid State Communications* **94** (12), 1013-1018 (1995).
6. J. Font, J. Muntasell, and E. Cesari, *Materials Research Bulletin* **32** (12), 1691-1696 (1997).
7. J. F. Willart, V. Caron, R. Lefort, F. Danede, D. Prevost, and M. Descamps, *Solid State Communications* **132**, 693-696 (2004).
8. M. Nagahama, H. Suga, and O. Andersson, *Thermochimica acta* **363** (1-2), 165-174 (2000).
9. J. F. Willart, A. De Gusseme, S. Hemon, G. Odou, F. Danede, and M. Descamps, *Solid State Communications* **119** (8-9), 501-505 (2001).
10. J. F. Willart, A. De Gusseme, S. Hemon, M. Descamps, F. Leveiller, and R. Rameau, *Journal of Physical Chemistry B* **106** (13), 3365-3370 (2002).
11. J.F. Willart, A. Hédoux, Y. Guinet, F. Danède, L. Paccou, F. Capet, and M. Descamps, *Journal of Physical Chemistry B* **110** (23), 11040-11043 (2006).
12. M. Otsuka, T. Matsumoto, and N. Kaneniwa, *Chemical and pharmaceutical bulletin* **34** (4), 1784-1793 (1986).
13. T. Matsumoto, J. Ichikawa, N. Kaneniwa, and M. Otsuka, *Chemical and Pharmaceutical Bulletin* **36** (3), 1074-1085 (1988).
14. A. Bauer-Brandl, *International Journal of Pharmaceutics* **140** (2), 195-206 (1996).
15. T.P. Shakhtshneider and V.V. Boldyrev, *Drug Development and Industrial Pharmacy* **19** (16), 2055-2067 (1993).
16. J. F. Willart, J. Lefebvre, F. Danede, S. Comini, P. Looten, and M. Descamps, *Solid State Communications* **135**, 519-524 (2005).
17. S. Petit and G. Coquerel, *Chemistry of materials* **8** (9), 2247-2258 (1996).
18. A. Saleki-Gerhardt, J.G. Stowell, S.R. Byrn, and G. Zografi, *Journal of Pharmaceutical Sciences* **84** (3), 318-323 (1996).
19. J. F. Willart, F. Danede, A. De Gusseme, M. Descamps, and C. Neves, *Journal of Physical Chemistry B* **107** (40), 11158-11162 (2003).
20. F. Sussich, C. Skopec, J. Brady, and A. Cesaro, *Carbohydrate Research* **334** (3), 165-176 (2001).
21. M. Rani, R. Govindarajan, R. Surana, and R. Suryanarayanan, *Pharmaceutical Research* **23** (10), 2356-2367 (2006).
22. T. Furuki, A. Kishi, and M. Sakurai, *Carbohydrate research* **340** (3), 429-438 (2005).

Extraordinarily Stable Organic Glasses Prepared by Vapor Deposition: Dependence of Stability and Dynamics upon Deposition Temperature

Stephen F. Swallen and M. D. Ediger

University of Wisconsin-Madison, Department of Chemistry, Madison, WI 53706

Abstract. We have recently prepared the most stable glasses ever produced in the laboratory. Using vapor deposition in a temperature range somewhat below T_g, thin films of tris-naphthylbenzene can be prepared which show very slow kinetic relaxation upon isothermal heating. An optimal deposition temperature of $T_g - 50$ K is found, giving films with structural relaxation times exceeding the dielectric relaxation τ_α by a factor of 300 at T_g. The slow response to superheating permits a direct study of the transformation of a stable glass to a supercooled liquid. This melting occurs heterogeneously, with localized, large regions undergoing the transformation on timescales ranging over two orders of magnitude.

Keywords: Diffusion, Glass transition, Spatially heterogeneous dynamics, Stable glass
PACS: 61.12.Ha, 61.43.-j, 64.70.Pf, 66.30.Hs

INTRODUCTION

Glassy materials play a critical role in many technologies, including optics, plastics, and even pharmaceuticals. Given this importance, there has been significant research into the nature of glasses, to elucidate how such materials can be prepared and how they behave over time. Despite this, the nature of the glass transition is still poorly understood. Due in part to the inherent sluggishness of the molecular motions, the fundamental causes of kinetic arrest which give rise to a glass have remained an open issue.

We have recently reported the preparation of the most stable glasses ever produced in the laboratory. [1] We have used physical vapor deposition to make thin films of low molecular weight organic molecules which show remarkable thermodynamic and kinetic stability. They have low fictive temperatures, high relaxation onset temperatures, and very long relaxation times. As prepared, these materials are very nearly in equilibrium with the (metastable) supercooled liquid at temperatures well below the bulk glass transition temperature, T_g. Traditional methods of producing such glasses by slowly cooling from the high temperature liquid [2] would require many thousands of years. In contrast, the vapor deposition process can be completed in minutes to hours.

In addition to possible technological applications of such materials, these stable glasses are uniquely suited to answer fundamental questions regarding transitions between the glass and liquid states. Their high kinetic stability causes the "melting" of a glass to a supercooled liquid just above T_g to be slowed enough to allow direct measurement of the underlying processes on a molecular length scale; these time and length scales have previously been inaccessible by experimental methods. New questions regarding the glassy state can be posed. What are the molecular processes by which a stable glass relaxes to equilibrium upon heating? Does this "melting" occur isotropically throughout the material by spinodal decomposition, or locally through nucleation and growth? What are the length and time scales associated with these processes? How does this structural relaxation compare to less stable glasses, which typically exhibit Fickian diffusive motion?

For the stable glasses that we have recently prepared, the timescales of these processes are

CP982, *Complex Systems, 5th International Workshop on Complex Systems*
edited by M. Tokuyama, I. Oppenheim, and H. Nishiyama

experimentally accessible by a range of measurement techniques. This work will highlight recent discoveries made regarding the kinetic stability of these materials. The most critical parameter controlling the glass stability is substrate temperature, $T_{substrate}$, during the vapor deposition process. By carefully choosing this value in a range somewhat below T_g, the kinetic stability can be optimized. We investigate the strong dependence upon this parameter, then use these systems to shed light on the glass to liquid "melting" process that occurs during isothermal annealing.

We have used neutron reflectivity to measure the concentration profiles of isotopically labeled multilayer thin films of tris-naphthylbenzene (TNB). Neutrons show a significant contrast between protons and deuterons, giving a highly sensitive means of following structural relaxation during annealing. Figure 1 shows neutron reflectivity data for a TNB film prepared at $T_{substrate}$ = 278 K (= T_g – 65 K). The Bragg peaks at wavevectors above q = 0.17 $Å^{-1}$ indicate initially very sharp 2 - 3 nm interfacial broadening. The wavevector $q = 2\pi/\lambda$, where λ is the wavelength of the structural organization investigated at the given wavevector. The top inset shows a typical sample structure composed of 5 protio/deuterio-TNB bilayers, with each layer a uniform 30 nm thick. A schematic of TNB is also shown. A series of samples, prepared at a range of $T_{substrate}$ values, were isothermally annealed near T_g in the neutron beam. These *in situ* measurements provided direct observations of isothermal structural evolution.

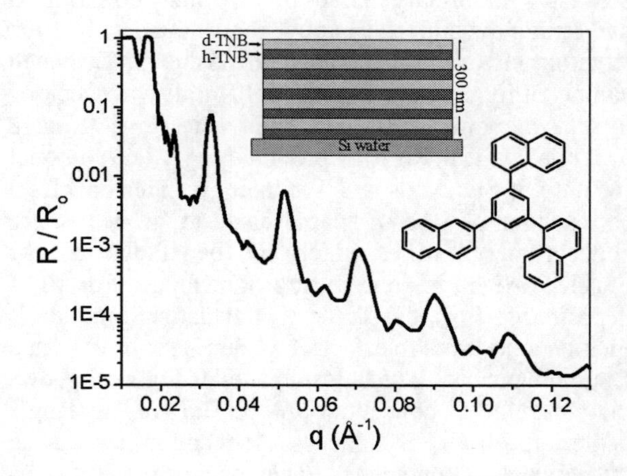

FIGURE 1. Neutron reflectivity data for a sample vapor deposited at $T_{substrate}$ = 277 K. The critical edge of the silicon wafer appears at q = 0.013 A^{-1}. Top inset shows multilayer sample structure. Right inset shows schematic of tris-naphthylbenzene (TNB) molecular structure.

RESULTS AND DISCUSSION

Deposition Temperature Dependence of Kinetic Stability

The influence of substrate temperature on kinetic stability is shown in Figure 2. The time-dependent structure factor, S(q;t), is plotted for samples with $T_{substrate}$ between T_g – 3 K and T_g – 150 K, while being annealed at 345 K (= T_g + 3 K). This is calculated from the amplitude of the third harmonic Bragg peak (q = 0.03 $Å^{-1}$) after subtraction of background and normalization. It is found that films deposited within a few degrees of the bulk T_g behave as ordinary glasses, with prompt, single exponential relaxation functions. The structural relaxation for these samples is well described by Fickian diffusion, with the concentration profile evolving with Gaussian broadening of the interface widths. At lower temperatures the relaxation kinetics slow dramatically. A maximum in the structure factor decay time is seen at T_g – 50 K, with relaxation time more than an order of magnitude longer than for glasses deposited near T_g. Below this optimal temperature, kinetic stability is lessened, eventually falling below even that of the ordinary high temperature glass. The inset summarizes the kinetic data, giving the 1/e decay time of the structure factor, $\tau_{1/e}$, as a function of $T_{substrate}$. These changes in relaxation times are also manifested by greatly increasing times before the onset of structural relaxation. The ability to superheat these materials is a hallmark of kinetic stability; the molecules are packed in such low temperature configurations that they cannot undergo thermally induced local rearrangements for long periods. In contrast to supercooled liquids and ordinary glasses which rapidly relax, these stable materials transform to the liquid sufficiently slowly to allow direct measurement of the glass to liquid transition *in situ*.

Changes to the functional form of the structure factor are also observed to increase at lower temperatures, with a maximum at $T_{substrate}$ = T_g – 50 K. S(q;t) becomes progressively more super-exponential, even after the onset of relaxation. Decay of the initially sharp concentration profile proceeds by increasingly non-Fickian mechanisms, with the structure factor decaying almost linearly in time for the most stable materials. Structural relaxation becomes dominated by fundamentally different molecular mechanisms, which may have a wide spatial and temporal distribution. Furthermore, the S(q;t) exhibits plateaus, indicating long periods during which structural evolution temporarily ceases. These arise from kinetic mechanisms that are either non-continuous in time, or that vary spatially with separate time scales. Because dynamics measured by neutron reflectivity are a spatial average over the whole sample, it is not an ideal technique for measurement of specific local events. However, we will show that, when coupled with additional data below, these measurements provide evidence for spatially heterogeneous dynamics during the evolution of a stable glass to a supercooled liquid.

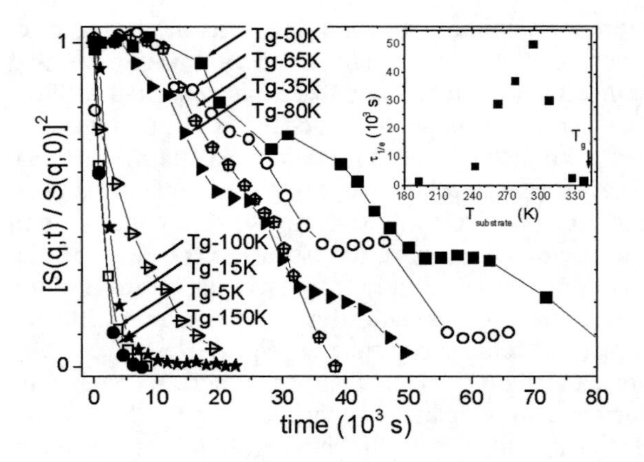

FIGURE 2. Time-dependent structure factor for films deposited at given substrate temperatures. All samples were annealed at 345 K. S(q;t) determined from third harmonic Bragg peak intensity at q = 0.03 Å^{-1}. The inset shows the decay time $\tau_{1/e}$ as a function of substrate temperature measured from each structure factor curve.

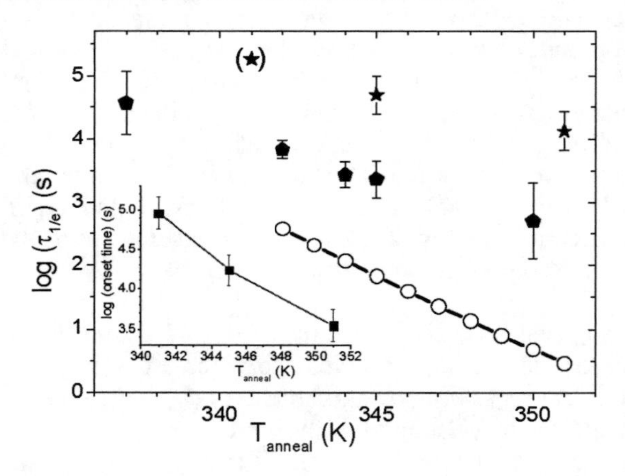

FIGURE 3. Neutron reflectivity 3rd harmonic 1/e decay time of the time-dependent structure factor for stable glasses ($T_{substrate}$ = T_g -50 K, stars), and ordinary glasses ($T_{substrate}$ = T_g – 5 K, pentagons), at annealing temperatures T_{anneal}. Lowest temperature stable glass value is estimated from extrapolation of data to long times. Open circles are VTF fit to dielectric relaxation data [3]. Inset is time before onset of structural relaxation for stable glasses. Values are determined for Bragg peak at q = 0.03 Å^{-1}.

The relaxation kinetics of both stable and ordinary glasses were studied as a function of annealing temperature. Figure 3 plots $\tau_{1/e}$ for films prepared at T_g – 50 K (stars) and at T_g – 5 K (circles) at a range of annealing temperatures around T_g. The remarkable kinetic stabilization of the most stable samples can be seen in comparison to τ_α, the structural relaxation time determined from dielectric relaxation measurements [3] (solid line is VFT fit to data). At T_g, these relaxation times exceed τ_α by a *factor of 300*. The concentration profiles persist for orders of magnitude longer than molecular relaxation processes in ordinary glasses. The inset shows the onset time before any decay of S(q;t) is seen at each annealing temperature: *i.e.*, before any measurable relaxation occurs on molecular length scales. Thus these films can be superheated for long periods. This quiescent period is more than $100\tau_\alpha$ at T_g, and $1000\tau_\alpha$ at T_g + 9 K.

Structural Evolution of Stable Glasses is Spatially Heterogeneous

The ability to superheat these stable materials makes them particularly well suited for the study of the glass-liquid transformation. By slowing down the melting process to time scales at which structural information is readily accessible, the underlying mechanisms can be elucidated. Comparisons of structural relaxation rates at many length scales for ordinary and stable glasses, directly measured by neutron reflectivity, provide the key to these questions.

As we have shown earlier, [4] structural relaxation processes in ordinary glasses of TNB are well modeled by Fickian dynamics. Figure 4 shows the decay of the structure factor for a film vapor-deposited at T_g, which can be fit by a single exponential for all wavevectors. Fits to the raw reflectivity data were done using a program called Reflfit, provided by the National Institute of Standards and Technology. This enables a conversion of the q-space data to a real-space concentration profile, which can be studied as the profile evolves in time during annealing. For films deposited at $T_{substrate} = T_g - 5$ K, translational motion is Fickian, and layer interfaces broaden over time with a Gaussian profile. The bottom panel of Figure 4 shows the calculated concentration profile of a single deuterio/protio-TNB bilayer, at several times during the annealing process. The decay of S(q;t) is wavevector-dependent, with $\tau_{1/e}$ varying as q^{-2}. Thus, the extracted diffusion coefficient in these systems, $D = 1/2q^2\tau_{1/e}$, is distance invariant.

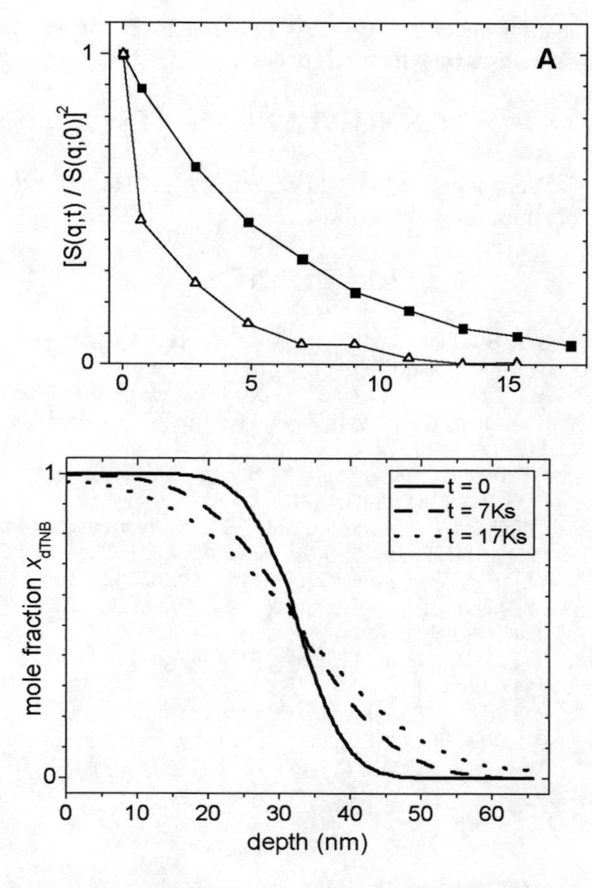

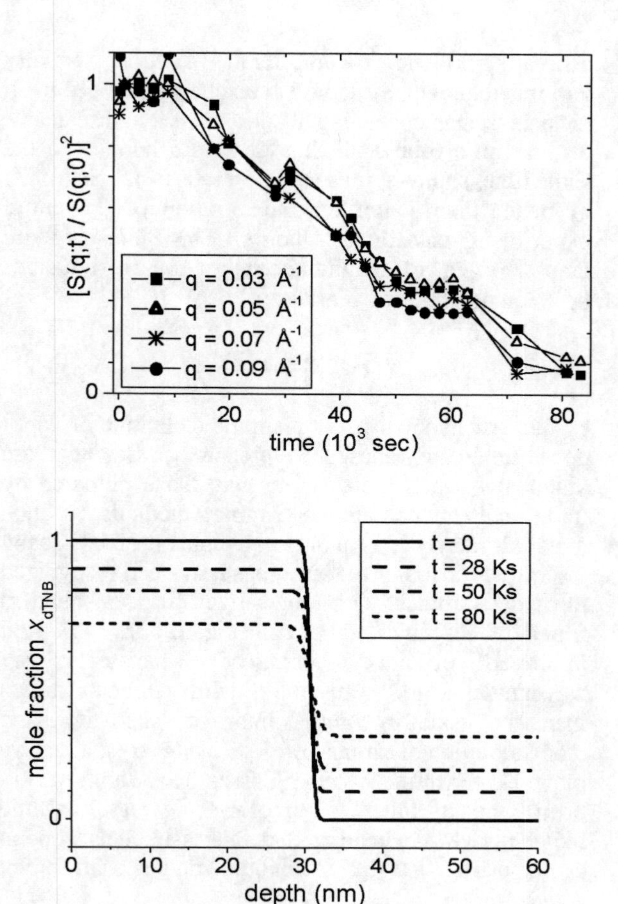

FIGURE 4. (top) Time-dependent structure factors for odd harmonics of an ordinary glass sample prepared at $T_{substrate}$ = 339 K and annealed at 342 K. Squares are from the third harmonic Bragg peak at q = 0.03 $Å^{-1}$, open triangles are 5th harmonic at q = 0.05 $Å^{-1}$. (bottom) Concentration profiles calculated at annealing times of t = 0, 7000 s, and 17,000 s. Profile is shown for a single deuterio/protio-TNB bilayer within a multilayer sample.

FIGURE 5. (top) Time-dependent structure factors for odd harmonics of a stable sample prepared at $T_{substrate}$ = 294 K and annealed at 345 K. (bottom) Concentration profiles calculated at annealing times t = 0, 28,000 s, 50,000 s, and 80,000 s. Profile is shown for a single deuterio/protio-TNB bilayer within a multilayer sample.

The relaxation kinetics for TNB samples vapor deposited at $T_{substrate} = T_g - 50$ K are fundamentally different. Figure 5 shows that for these stable glasses, S(q;t) decays uniformly at all q values. The loss of structural organization occurs simultaneously at all distances up to at least 20 nm. This indicates that large regions undergo rapid molecular reorganization, destroying the initial concentration gradient, while other regions remain static. This has the effect of reducing the neutron reflectivity contrast between protio- and deuterio-TNB layers, while retaining the sharp interfaces between isotopically labeled layers. The bottom panel of Figure 5 shows the resulting concentration profiles during isothermal evolution at 345 K.

These results provide insight into the evolution of a stable glass to a supercooled liquid above T_g. Locally large regions, exceeding 20 nm in size, rapidly melt to the lower density and lower viscosity liquid. The surrounding matrix remains temporarily fixed as a rigid glass, with additional regions melting at progressively longer times. The ordinary glass evolves in a manner consistent with translational motion in high temperature liquids. In contrast, the stable glass has regions of liquid melt which grow with time. The transition to a liquid does not happen uniformly throughout the film, with a gradual, isotropic relaxation to the liquid. Rather, the glass-liquid transition is localized, with regions enlarging over time. This suggests a mechanism of nucleation and growth. Unfortunately this data does not provide details on nucleation density or the growth rates of individual regions, but does provide a means of estimating kinetic relaxation rates in these spatially varying regions. Structural relaxation must occur on distance scales exceeding all measured lengthscales (> 20 nm) on a time scale shorter than 5 hours. At the same time, some regions retain an interfacial sharpness of better than 3 nm for the duration of the entire experiment, exceeding 24 hours. This indicates more than 2 orders of magnitude separation in dynamics between neighboring regions.

SUMMARY

We have shown that the simple technique of vapor deposition can be used to prepare glasses that are kinetically much more stable than those prepared by traditional quenching and aging methods. Such materials may be technologically useful across many disciplines. Similar behavior has now been recognized in optoelectronics, with longer-lived devices found in films of organic light emitting diodes [5] and improved efficiency of lead-germanate planar waveguides. [6] In the pharmaceutical arena, increased stability may improve shelf-life and bioavailability of drugs by decreasing crystallization rates. [7] Extremely stable glasses have very recently been seen in other low molecular weight organic molecules [8], indicating that this is likely a general characteristic of the technique, rather than being material specific. Work is ongoing to elucidate the mechanism by which this occurs.

ACKNOWLEDGMENTS

We acknowledge NSF Chemistry (CHE-605136) for supporting this work.

REFERENCES

1. S. F. Swallen, K.L. Kearns, M. K. Mapes et al., *Science* **315**, 353-356 (2007).
2. A.J. Kovacs, *Fortschr Hochpolym-Forsch.* **3**, 394 (1963).
3. R. Richert, K. Duvvuri, and L.-T. Duong, *J. Chem. Phys.* **118**, 1828-1836 (2003).
4. S.F. Swallen, M.K. Mapes, Y.S. Kim et al., *J. Chem. Phys.* **124** (18), 184501 (2006).
5. J. Salbeck, F. Weissortel, and J. Bauer, *Macromol. Symp.* **125**, 121-132 (1998); S. A. VanSlyke, C. H. Chen, and C. W. Tang, *Appl. Phys. Lett.* **69** (15), 2160-2162 (1996).
6. M. Dussauze, A. Giannoudakos, L. Velli et al., **127**, 034704 (2007).
7. B. C. Hancock and M. Parks, *Pharm. Res.* **17** (4), 397-404 (2000).
8. K. Ishii, H. Nakayama, T. Okamura et al., *J. Phys. Chem. B* **107** (3), 876-881 (2003); K. Ishii and H. Nakayama, *J. Non-Cryst. Sol.* **353** (13-15), 1279-1282 (2007).

Structural Relaxation of Vapor-Deposited Amorphous Molecular Systems: Dependence on the Deposition Temperature

Kikujiro Ishii and Hideyuki Nakayama

Department of Chemistry, Gakushuin University, 1-5-1 Mejiro, Toshimaku, Tokyo 171-8588 Japan

Abstract. The effect of the deposition temperature on the structural relaxation of vapor-deposited amorphous molecular systems was discussed on the basis of previously-reported observations and recent experimental results on the molar volume of glassy ethylbenzene. The latter examples confirmed the universality of the observations by Swallen et al. that anomalously stable glass states were obtained by the vapor deposition of their compounds at temperatures close to T_g.

Keywords: Amorphous, Crystallization, Glass, Glass transition, Structural relaxation, Vapor deposition
PACS: 61.43.Er, 61.43.Fs, 82.56.Na, 64.70.Pf, 68.03.Fg.

INTRODUCTION

Amorphous molecular systems have attracted our attention for their possibility to show dynamic behaviors different from those of classical glasses made of inorganic materials. The key issues are the anisotropy of the molecular shape and the molecular deformation seen for compounds with flexible molecular structures [1]. Owing to the diversity of molecular structures, a variety of phenomena, from crystallization to glass transition, are seen for amorphous molecular systems when they are annealed by temperature elevation. In the title of this paper, we use the term *amorphous* rather than *glassy*, since we will include the systems in the following discussion which directly crystallize from amorphous states without undergoing glass transition.

From the viewpoint of structural chemistry, we aimed to clarify the structure and motional characteristics of the molecules in the systems. We thus chose simple organic compounds as the target. Most of such organic compounds, however, easily crystallize in the cooling process, if we used the usual liquid-quenching method to prepare their amorphous states. We thus adopted the method of the vapor deposition onto cold substrates. The structures of the obtained samples were studied in situ by Raman scattering, light interference, and X-ray diffraction during the annealing process by the temperature elevation with a constant rate.

Hikawa and coworkers [2] have pointed out that vapor-deposited glassy samples have enthalpies higher than the glasses prepared by the liquid-quenching method. Our previous studies on the molar volume of vapor-deposited glasses of mono-alkyl benzenes [3] gave the results which support the high enthalpy of the samples just after the deposition. We also systematically studied the relaxation behavior of binary molecular systems as a function of composition [4,5], and drew a model diagram in which the relaxation paths leading to the glass transition and direct crystallization respectively bifurcate at a high-enthalpy state point of the amorphous state.

Recently, Swallen and coworkers [6] reported that 1,3-bis-(1-naphthyl)-5-(2-naphthyl)benzene (T_g = 347 K) and indomethacin (T_g = 315 K) gave glasses with anomalous kinetic stabilities when the vapor was deposited onto a substrate controlled near T_g–50 K. These results imply that there are some mechanism of molecular condensation in which stable local intermolecular arrangements are formed but the macroscopic structure is disordered. To examine the universality of their results, we are studying the relaxation behaviors of amorphous molecular systems vapor-deposited at temperatures close to T_g. In this paper, we summarize our previous experimental results related to the effect of the deposition temperature on the relaxation of amorphous molecular systems. We also report our recent studies on the molar volume of glassy ethylbenzene, and discuss the importance of dynamics of the molecule at the deposition on the sample surface.

CP982, *Complex Systems, 5th International Workshop on Complex Systems*
edited by M. Tokuyama, I. Oppenheim, and H. Nishiyama

EXPERIMENTAL

The vapor deposition onto cold substrates and the in situ optical or X-ray measurements were carried out using the vacuum chambers reported in our previous papers [3,7,8] except the case of anthracene that was studied earlier [9]. The samples were deposited in most of the cases with a rate of about 0.3 μm/min to the thickness of about 10 μm. The substrate was a gold-plated copper block in the optical measurements, and was a silicon crystal with the (1 0 0) face in the X-ray measurements.

The Raman spectra were measured using the 514.5 nm light from an Ar^+ laser with an incident power of 40 mW. A single monochromator equipped with a CCD detector was use for excitation in combination with a cut filter to diminish the elastically scattered laser light. Light interference in the vapor-deposited film samples was monitored using the above laser light and a silicon photocell. The incident angle of the laser light was 60 degree from the normal line of the substrate.

RESULTS AND DISCUSSION

Substrate Temperature and the Effective Deposition Temperature

Before discussing the effect of the deposition temperature on the structure and relaxation behavior of the resultant amorphous materials, we first recall our previous observations on the gauche mole fraction x_g of 1,2-dichloroethane (DCE) during the vapor deposition onto the substrates of different temperatures [10]. DCE has two kinds of conformation isomers, gauche and trans, which are easily distinguished by their Raman spectra. It has been known that the relative concentrations of these isomers are determined mainly by the dielectric constant of the media surrounding the molecules [11]. In the previous studies on the vapor-deposited DCE, we found that x_g was larger as the substrate temperature was higher, and also found that x_g of the samples deposited below 80 K increased when the samples were annealed by raising the temperature (see Fig. 3 which will be discussed later).

The above phenomena seen for x_g of DCE provided us a tool to detect the change of the effective temperature at the top surface of the sample during the vapor deposition. We monitored the apparent x_g of the sample by Raman measurement during the vapor deposition with a rate of 0.3 μm/min. It was found that x_g was almost constant or slightly increasing during the deposition of the room-temperature vapor to the sample thickness up to 10 μm. The results indicate that the difference between the effective temperature at the top surface of the sample and that of the metal substrate was less than a few degree Kelvin at most even for the substrate temperature at 30 K. Therefore, we will assume in the following discussion that the temperature of the top surface of the sample was the same as the temperature of the substrate. This assumption may be valid for the vapor deposition with a rate smaller than the above value.

Metastable Intermolecular Conformation appeared in Amorphous Anthracene

In our early studies on the vapor-deposited amorphous anthracene, we examined the difference of Raman spectra of the samples prepared at different deposition temperatures [9]. We reproduce in Fig. 1 the features of the small-wavenumber Raman spectra. Four spectra for which the deposition temperatures are indicated in the figure are those of amorphous anthracene soon after the vapor deposition. The spectrum of the molten liquid at 500 K and that of crystalline powder at 9 K are also shown in the same figure. The six Raman bands seen in the spectrum of the crystalline powder are assigned to the three kinds of librational lattice vibrations each of which are split into a pair of bands owing to the interaction of two molecules in a unit cell. On the other hand, there is no definite structure in the spectrum of the liquid state. This indicates the lack of the Fourier component of a specific vibrational frequency, and reflects the disordered structure of the liquid.

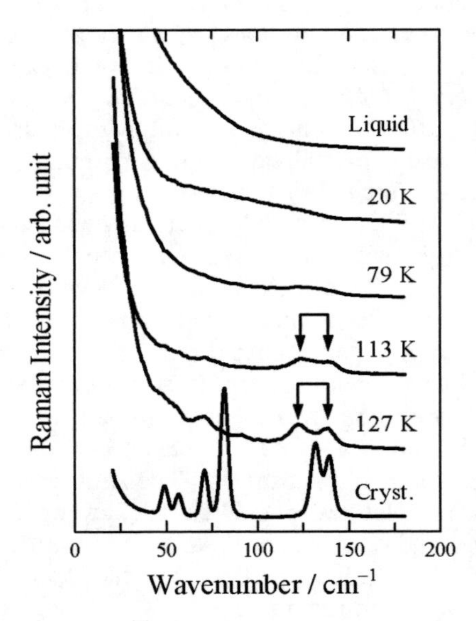

FIGURE 1. Small-wavenumber Raman spectra of amorphous anthracene vapor-deposited at different temperatures. Spectrum of crystalline powder at 9 K and that of molten liquid at 500 K are also shown for comparison.

Among the four spectra of the amorphous samples, the spectrum of the sample deposited at 20 K shows a feature similar to that of the molten liquid, indicating the disordered structure of the sample. However, the spectra for the samples deposited at 79, 113, and 127 K show weak hump(s) in the region around 130 cm^{-1}. Apparently, these humps correspond to the bands seen for the crystalline sample in the same wavenumber region. Thus the humps seen for the amorphous samples are considered to arise from the molecular motions similar to those in the crystal (the librations around the longest two-fold axis of the anthracene molecule). However, the splitting of the humps of the amorphous samples deposited at 113 and 127 K is obviously larger than that of the bands of the crystal. This indicates that the intermolecular interaction related to the above librational motion was stronger in these amorphous samples than in the crystal.

In our studies of amorphous anthracene, we also observed the fluorescence of the vapor-deposited samples [9]. The sample deposited at 115 K showed a featureless broad fluorescence spectrum peaked around 470 nm. This was attributed to the so-called excimer that is a molecular pair on which the excitation energy is delocalized. Interestingly, there appeared vibrational structures in the fluorescence spectrum when we raised the sample temperature to about 220 K. This phenomenon was attributed to the crystallization of the sample, since the samples deposited at similar temperatures showed changes of Raman spectra being accompanied with the appearance of definite lattice-vibrational bands.

The above characteristics of Raman and fluorescence spectra observed for vapor-deposited amorphous anthracene indicate the formation of a metastable local molecular conformation in amorphous samples when the deposition temperature is rather high but low enough to prevent the crystallization of the sample. Interestingly, the crystallization temperature seemed to be higher for the samples in which the more stable local conformation appeared. These observations coincide with those made by Swallen et al. for the glasses deposited at temperatures close to T_g [6], and are considered to be related to the relaxation in the intermolecular conformation localized around a molecule that has just landed on the top surface of the sample.

Differences in Optical Transparency and Induction Time for Crystallization of Samples Deposited at Different Temperatures

We have previously studied the crystallization behaviors of amorphous samples of mono-halogenated benzenes vapor-deposited at different temperatures [7,12]. Figure 2 shows the interference fringes recorded during the deposition of chlorobenzene vapor onto the substrates at different temperatures. At 78 K, the deposited sample showed a constant interference fringe indicating the formation of a transparent film with a homogeneous thickness more than 10 μm. At the lower substrate temperatures, however, the fringes showed a decay in its amplitude indicating the loss of transparency as the film thickness was increased. This implies the appearance of optical inhomogeneity in the sample when the deposition temperature was low.

From the X-ray studies with raising the sample temperature stepwise, the amorphous samples of fluorobenzene (FB), chlorobenzene (CB), and bromobenzene (BB) were found to crystallize in a temperature region characteristic to each compound almost irrespective to the deposition temperature. However, by the detailed studies with the temperature-jump method, we found that the CB and BB samples deposited at higher temperatures (78 K for CB and 80 K for BB) showed some induction time before the crystallization started, while the samples deposited at 25 K started the crystallization immediately when we raised the temperature to the crystallization region.

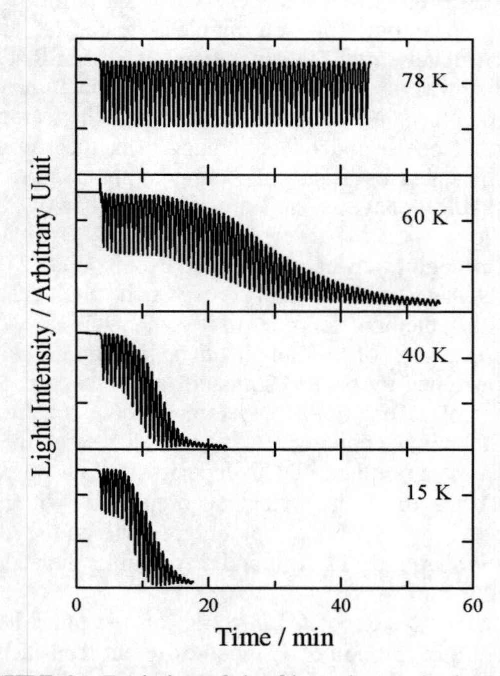

FIGURE 2. Evolution of the fringe due to the laser light interference in the film samples during the vapor deposition of chlorobenzene onto metal substrate at different temperatures.

It is a general tendency, irrespective of the kind of the compound, that samples with a good transparency are obtained by the deposition at rather high temperatures, but the decay of the interference fringe is serious at lower temperatures. We consider that the transparent samples are formed at higher temperatures because the molecules deposited on the sample surface have some forbearance time to seek the potential energy minimum, while they are frozen rapidly at lower temperatures leaving a small space in their neighborhood. These inferences are in harmony with the facts that the transparent samples deposited at high temperatures have induction times to crystallize while the opaque samples deposited at low temperatures start immediately the crystallization when the sample temperature is raised to the crystallization region.

Deposition-Temperature Dependences of Gauche Mole Fraction and Relaxation Behavior of Amorphous 1,2-Dichloroethane

We have mentioned in the first topic of the results and discussion that we have studied the gauche mole fraction x_g of amorphous DCE [10]. Three representative series of the data are plotted in Fig. 3. They are the evolution of x_g of the samples deposited at three different temperatures. The data point of the lowest temperature for each sample indicates the initial x_g immediately after the deposition. It is seen that the initial x_g was larger as the deposition temperature was higher. It is also noticed that x_g of the samples deposited at 30 and 69 K increased as the samples were annealed by raising the temperature. These are remarkable behaviors, and are interpreted with (1) a very low energy barrier for the mutual structural transformation between gauche and trans, and (2) the slightly smaller Gibbs energy of gauche than that of trans in the condensed phase of DCE. Very interestingly, x_g of the samples deposited at 30 and 69 K approached the value estimated for the supercooled liquid of DCE just before they underwent crystallization showing a sudden decrease in x_g. Therefore, amorphous DCE deposited below 80 K is considered to relax being accompanied with the increase in x_g. Such a sample might undergo the glass transition around 112 K, if it did not suffer the sudden crystallization.

The sample deposited at 82 K, on the other hand, showed an evolution of x_g much different from that of other samples deposited below 80 K. First, its x_g was almost constant, and started gradually a decrease, and finally fell to zero, the value of the crystal. This behavior may be explained as follows. The sample deposited at 82 K had a fairly large density at the initial stage after the deposition. This suppressed the transformation from trans to gauche. When the temperature passed the region of 90 K, the existing trans molecules started to form the crystalline seeds or embryos, and crystallization took place around 105 K. Such critical relaxation and crystallization behaviors of amorphous DCE in the deposition-temperature range around 80 K will be discussed elsewhere [13].

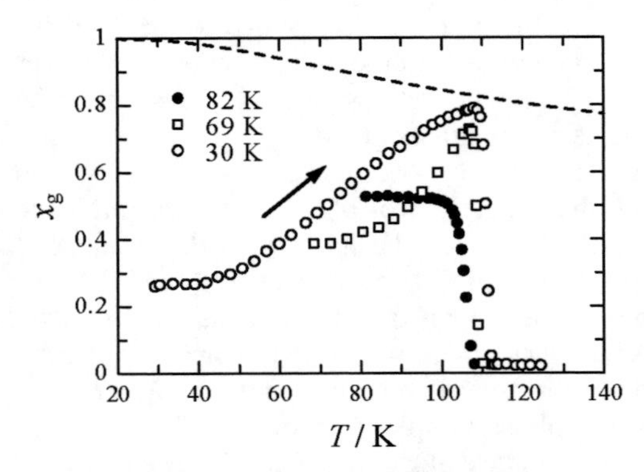

FIGURE 3. Evolution of the gauche mole fraction x_g of 1,2-dichloroethane samples deposited on a substrate at different temperatures during the annealing by the temperature elevation. Dashed curve indicates the extrapolation of x_g estimated for the supercooled liquid.

Excess Volume of Glassy Ethylbenzene and its Change due to Structural Relaxation

To examine the universality of the observations made by Swallen and coworkers on their glassy samples [6], we studied the relaxation behavior of glassy ethylbenzene vapor-deposited at different temperatures [14]. We monitored the change of the interference fringe of the film samples deposited on the metal substrate during the annealing by the temperature elevation, and estimated the molar volume V_m of the sample by the method previously reported [3].

The results are summarized in Fig. 4. For the sample deposited at 78 K, the feature almost reproduces the result of the previous work [3]. In the initial annealing process, the sample showed a thermal expansion keeping the framework of the solid structure. Then, V_m started to decrease around 108 K owing to the structural relaxation just below the glass transition, and approached the value of the liquid (dashed line). The difference between the data point and the dashed line at each temperature indicates the excess volume of the glass as compared with the liquid.

The initial V_m of the sample deposited at 88 K showed a smaller value than that of the samples deposited at 78 K. Such a difference is in harmony with the early results of thermal measurements on butyronitrile glasses [2].

The most remarkable results in Fig. 4 are those for the sample deposited at 105 K. Surprisingly, the initial V_m after the deposition was smaller than that of liquid, and the sample showed only a slight increase of V_m as the temperature was raised for annealing. By raising the temperature further above the T_g region around 116 K, V_m finally showed a sudden increase around 119 K, and converged into the line of the liquid. This behavior is very similar to the enthalpy changes observed for the glasses studied by Swallen et al. [6], and indicates the formation of an ethylbenzene glass with an anomalous kinetic stability by the deposition at 105 K (about 0.9 times T_g). Obviously, the sudden increase of V_m around 119 K is not due to the relaxation accompanied with the decrease of enthalpy, but is attributed to the relaxation accompanied with the increase of entropy.

The above tendency that a sample deposited at a higher temperature has a smaller volume is in harmony with the early results by the thermal measurements [2], but it is an unexpected anomalous finding that the volume can be smaller than that of liquid when the deposition temperature is close to T_g. It should be also noted that the initial thermal expansion of the sample deposited at 105 K is obviously much smaller than those of the samples deposited at lower temperatures. This seems to indicate that the sample deposited at 105 K had a dense and solid structure, while those deposited at lower temperatures had less dense and rather flexible structures. If we turn to the problem of the mechanism of the glass formation, the above observations are considered to be related to the dynamics of the local relaxation around a molecule just after its landing on the top surface of the sample. This issue is worth further studies.

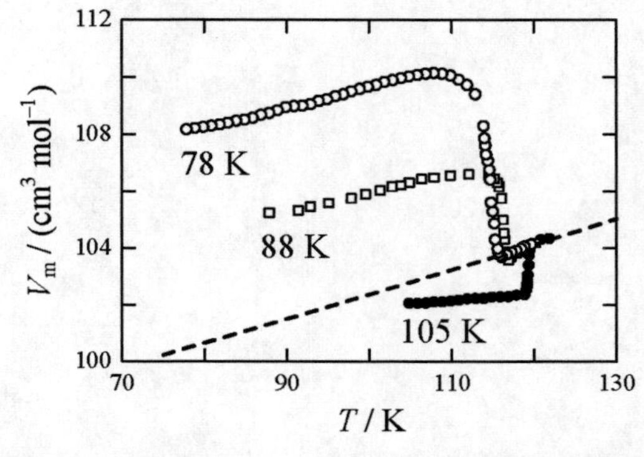

FIGURE 4. Evolution of molar volume of ethylbenzene samples vapor-deposited at different temperatures indicated in the figure due to the annealing by temperature elevation. Dashed line is the extrapolation of the molar volume estimated for the supercooled liquid.

CONCLUSION

We have seen so far the deposition-temperature dependence of the structure and/or relaxation behavior of vapor-deposited amorphous molecular systems including the results of our recent studies on glassy ethylbenzene. The remarkable feature common to all the examples is that more stable amorphous states are obtained by the vapor deposition onto the substrate at higher temperatures provided that the temperature is low enough to form the amorphous state. This leads sometimes to the formation of dense and transparent samples that need some induction time to begin the crystallization. If the molecule of a compound is asymmetry and has a large degree of the freedom of deformation, it is sometimes possible to deposit it at temperatures close to T_g without suffering the crystallization. Such amorphous systems can be more dense than supercooled liquid at the same temperature. They do not undergo the glass transition expected for the usual glass state of the compound, but exhibit a novel relaxation phenomena accompanied with an increase in entropy and turn into the liquid. The mechanism of the appearance of such stable amorphous molecular systems is considered to be a critical issue of the dynamics of the molecules during the deposition of the vapor.

ACKNOWLEDGMENTS

The authors appreciate the contributions made by the former students in our research group who have joined the work on the amorphous molecular systems reported in the previous papers. The authors also appreciate the contributions made by Mr. Shin Hirabayashi and Mr. Ryo Moriyama in the studies of the deposition-temperature dependence of the molar volume of glassy ethylbenzene.

REFERENCES

1. K. Ishii, M. Kawahara, Y. Yagasaki, Y. Hibino, and H. Nakayama, *J. Phys. D* **26**, B193-B197 (1993).
2. H. Hikawa, M. Oguni, and H. Suga, *J. Non-Cryst. Solids*, **101**, 90-100 (1988).
3. K. Ishii, H. Nakayama, T. Okamura, M. Yamamoto, and T. Hosokawa, *J. Phys. Chem. B*, **107**, 876-881 (2003).
4. M. Murai, H. Nakayama, and K. Ishii, *J. Ther. Anal. Calor.* **69**, 953-959 (2002).
5. K. Ishii, M. Takei, M. Yamamoto, and H. Nakayama, *Chem. Phys. Lett.*, **398**, 377-383 (2004).
6. S. F. Swallen, K. L. Kearns, M. K. Mapes, Y. S. Kim, R. J. McMaphon, M. D. Ediger, T. Wu, L. Yu, and S. Satija, *Science*, **315**, 353-356 (2007).
7. K. Ishii, M. Yoshida, K. Suzuki, H. Sakurai, T. Shimayama, and H. Nakayama, *Bull. Chem. Soc. Jpn.* **74**, 435-440 (2001).
8. K. Ishii, H. Nakayama, T. Yoshida, H. Usui, and K. Koyama, *Bull. Chem. Soc. Jpn.* **69**, 2831-2838 (1996).

9. K. Ishii, H. Nakayama, Y. Yagasaki, K. Ando, and M. Kawahara, *Chem. Phys. Lett.* **222**, 117-122 (1994).
10. K. Ishii, Y. Kobayashi, K. Sakai, and H. Nakayama, *J. Phys. Chem. B*, **110** , 24827-24833 (2006).
11. K. B. Wiberg, T. A. Keith, M. J. Frisch, and M. Murcko, *J. Phys. Chem.*, **99**, 9072-9079 (1995).
12. H. Nakayama, S. Ohta, I. Onozuka, Y. Nakahara, and K. Ishii, *Bull. Chem. Soc. Jpn.* **77**, 1117-1124 (2004).
13. N. Yasuda, K. Inoue, H. Nakayama, and K. Ishii, poster presentation G_TE-P8, this workshop.
14. K. Ishii, H. Nakayama, S. Hirabayashi, and R. Moriyama, to be published.

Dielectric Spectroscopy on Propylene Glycol/Poly(Vinyl Pyrrolidone) Solutions: Polymer and Solvent Dynamics in Hydrogen-Bonding Systems

Anna Spanoudaki*, Naoki Shinyashiki†, Apostolos Kyritsis*, and Polycarpos Pissis*

*Department of Applied Mathematics and Physics, National Technical University of Athens,
Heroon Polytechneiou 9, 15780 Athens, Greece
†Department of Physics, Tokai University, Hiratsuka, Kanagawa 259-1292, Japan

Abstract. The dielectric behavior of solvent-rich solutions of poly(vinyl pyrrolidone) in propylene glycol has been studied through broadband dielectric spectroscopy. Measurements covered a broad temperature range both in the liquid and the glassy state of the samples and concentrations from 0% to 60% wt. of poly(vinyl pyrrolidone). Two distinct glass transitions are observed dielectrically, evidencing decoupling of the segmental motion of PVP and the rotation of the propylene glycol molecules. The decoupling appears not to be full, since the dielectric strength data hint to a participation of propylene glycol molecules in the relaxation connected with the glass transition attributed to poly(vinyl pyrrolidone). Interactions between the two components are expressed through the concentration dependency of the relaxation times, as well as through the marked broadening of the propylene glycol relaxation.

Keywords: Dielectric spectroscopy, Glass transition, Polymer solutions
PACS: 64.70.Pf, 61.25.Hq

Polymer-solvent interactions, the dynamics of solvent molecules and the segmental motion of polymer chains are open subjects in the physics of polymer solutions. Since the glass transition is a cooperative phenomenon it is commonly assumed that in miscible two-component systems, which exhibit no phase-seperation, the molecules of both species show a joint relaxation. This is commonly considered a criterion for miscibility. However, it has been shown through dielectric as well as calorimetric studies [1, 2, 3, 4] that this is not the case when the components have very distinct relaxation times. Instead, although they may be highly miscible, and the system looks macroscopically homogeneous, the two components relax independently and two distinct glass transitions are observed [5].

The influence of the presence of one component on the dynamics of the other is still expressed through plasticization/antiplasticization of the slower/faster component and further through the significant broadening of the relaxations in the dielectric spectrum or of the glass transition step in the heat capacity in calorimetry measurements [3]. It is attributed to microscopic heterogeneity, resulting in concentration fluctuations in the local environments of the relaxing units. It is of topological rather than dynamical origin and is caused by the chain connectivity of polymers which does not allow a uniform distribution of solvent and solute molecules, even in perfectly miscible systems [6] . For this reason it is present in polymer blends [7] , but not in small molecule mixtures [6] .

Polymers/small molecule solutions often exhibit two glass transitions and are suitable model systems to study the dynamics under the conditions of miscibility and dynamical heterogeneity. Especially interesting is the case of polymers in polar solvents, since the choice of solvent allows to control the strength of the polymer-solvent interactions. Further, the dynamics of hydrogen-bonding liquid/polymer solutions provide fundamental information that can be used to understand the dynamics and function of biopolymers. Dielectric spectroscopy should allow to follow the temperature dependent development of the dynamics. There is, however, in such experiments an inherent difficulty to extract accurate data or even observe the polymer relaxation. The reason lies at the strong relaxation process of the solvent and, more importantly, on the high dc conductivity of such solutions, that mask the usually weaker response of the polymer.

Poly(vinyl pyrrolidone) (PVP) is an amorphous and nontoxic synthetic polymer, which has a randomly coiled and highly flexible chain behavior in polar solvents. Due to the formation of hydrogen bonds, PVP is highly miscible with water and alcohols. The dielectric behavior of PVP in solution has been studied extensively in water, mono- and di-hydroxyl alcohol and ethylene glycol oligomer solutions [4, 8, 9, 10, 11, 12, 13, 14, 15, 16, 17, 18, 19, 20, 21]. Furthermore, it has also been studied in dimethyl sulfoxide [22], tetramethylurea [22, 23], N-methyl-2-pyrrolidone [18], and pyrrolidone [24], with a concentration variation, by dielectric spectroscopy.

In the present work we attempt to clarify the dynamics

CP982, *Complex Systems, 5th International Workshop on Complex Systems*
edited by M. Tokuyama, I. Oppenheim, and H. Nishiyama

of both polymer and solvent in hydrogen-bonding polymer solutions. With this aim, we performed broadband dielectric spectroscopy on PVP solutions in propylene glycol over a broad frequency and temperature spectrum, covering both the liquid and the glassy state. The emphasis is given on solvent-rich samples. We study the relaxation spectra and the concentration dependence of the segmental motion of PVP and the rotation of the propylene glycol molecules in terms of the effects of the polymer to the dynamics of the solvent and vice versa.

EXPERIMENTAL

Poly(vinyl pyrrolidone) (PVP) with an average molecular weight M_w=40000 and propylene glycol monomer (PG) were purchased from Sigma. Solutions of PVP in PG were prepared, with PVP concentrations that varied from 0 to 60% wt., and were known with an accuracy better than 0.05%. Prior to mixing PVP was kept in dry atmosphere, in a desiccator with P_2O_5.

Dielectric measurements were performed in the frequency range of 0.02 Hz to 3 MHz and at temperatures between $-150°$ and 60°C, extended to 200°C for pure PVP. The viscous polymer solution and silica spacers with a diameter of 50 μm were sandwiched between brass electrodes. Pure poly(vinyl pyrrolidone) was dried in vacuum for more than 24 h and pressed under 10 tons to form a pellet, which was placed between the brass electrodes. The measurements were performed using an Alpha dielectric analyzer and the temperature was controlled using a Quatro cryosystem (Novocontrol).

The spectra were analysed (in the cases that this was feasible) fitting to the imaginary part, ε'', of the dielectric function with a sum of Havriliak-Negami functions [25], which are commonly used for modelling of dielectric spectra. The dc conductivity was taken into consideration using a term $i\sigma_{dc}/(\varepsilon_0 2\pi f)$. Alternatively, the ε'-derivative representation [26] has been employed,

$$\varepsilon''(f) \approx \varepsilon''_{der}(f) = -\frac{\pi}{2}\frac{\vartheta\varepsilon'}{\vartheta\ln(2\pi f)} \quad , \qquad (1)$$

which is free from the dc conductivity (but not conductivity related polarization) response and allows the clearer observance of peaks masked by the conductivity in the ε'' representation.

RESULTS AND DISCUSSION

Pure Poly(Vinyl Pyrrolidone)

Figure 1 shows the imaginary part, ε'', of the dielectric function of dried PVP at selected temperatures. At low frequencies and high temperatures the dc conductivity dominates, observed as a constant slope of -1 in the log-log scale of the plot. At temperatures lower than 20°C a rather symmetrically shaped secondary relaxation, named hereafter γ_{pvp}, appears, moving to lower frequencies as temperature is reduced. At intermediate temperatures, two more relaxation processes are present in the spectrum (labelled here α_{pvp}, β_{pvp} in order of increasing frequency). Due to their mutual overlap and also due to the strong contribution of the dc conductivity, these relaxations are not easily discerned. Analysis of the spectra through fitting or direct reading of the peak frequencies in the derivative representation, can give reliable information on the relaxation frequencies. In all cases, the results for the peak maxima were in very good agreement. The relaxation plot is presented in Fig. 6.

The very high apparent activation energy of α_{pvp}, together with the good agreement of the extrapolation of its trace with the calorimetric glass transition values from the literature [27] and our own differential calorimetry measurements (T_g=163°C, see relevant data point in Fig. 6) support the attribution of α_{pvp} to the segmental relaxation of PVP, related to the glass transition. Concerning β_{pvp}, Karabanova *et al.* [27] attribute it to a second glass transition at lower temperatures, undergone by molecules that find themselves in pseudohexagonally packed areas of the polymer [28]. Based on the Arrhenius-like behavior [29] and the lack of a second glass transition step in the calorimetry data, we propose that β_{pvp} is rather due to the rotational motion of the side chain, which carries a dipole moment. The size of the side chain and the considerable free volume required for the rotation makes it improbable that this movement is connected with the very fast γ_{pvp} peak. Jain and Johari [30] attribute γ_{pvp} to the orientational relaxation of water molecules still remaining in the sample after the drying procedure.

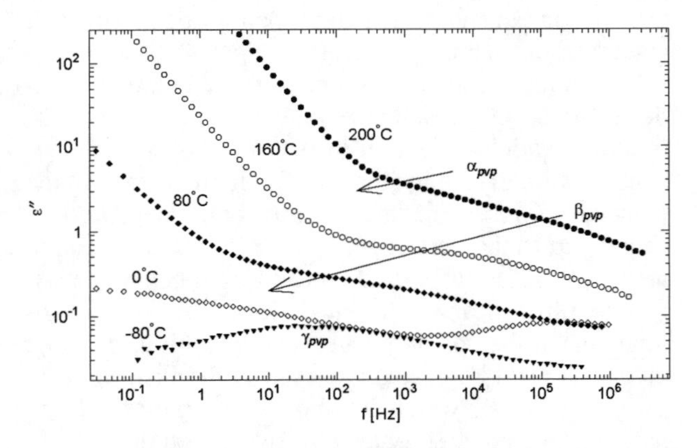

FIGURE 1. Imaginary part, ε'', of the dielectric permittivity as a function of frequency for pure PVP at selected temperatures

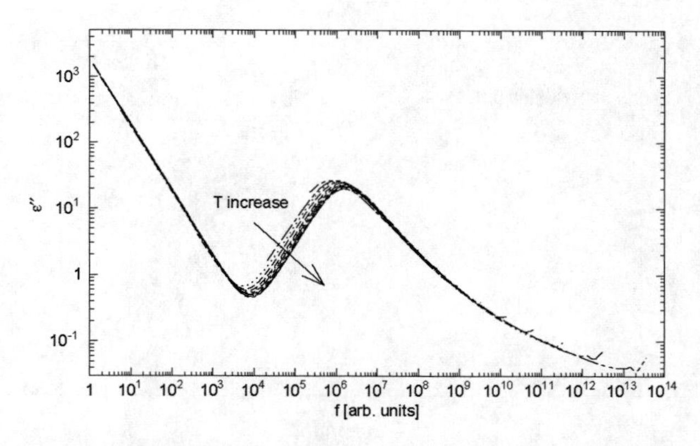

FIGURE 2. Master curve of ε'' for pure PG at the temperatures, at which the α_{pg} peak maximum is within our frequency window ($-100°$C to $-40°$C)

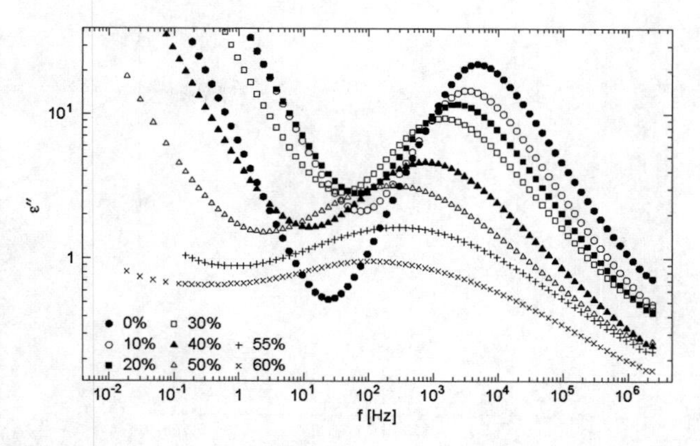

FIGURE 3. Imaginary part ε'' of the dielectric permittivity for pure PG, and the PG/PVP solutions at $-70°$C

Pure Propylene Glycol

Apart from the dc conductivity, pure propylene glycol shows in the frequency range of the measurements only the main relaxation, α_{pg}, related to its glass transition. It is observed as a marked peak visible between $-100°$C and $-40°$C and moving to lower frequencies with decreasing temperature. The high frequency flank of this peak, changes at about three decades higher than the frequency of the peak maximum to a lower slope. This is the so-called "excess wing" already observed for PG and other low molecular weight glass-formers [31, 32, 33]. In pure PG, as in other small molecule liquids, the main relaxation has an almost temperature independent peak shape. The frequency of the peak and the dc conductivity show the same temperature dependence over many orders of magnitude. The above are shown clearly in Fig. 2, a plot of the imaginary part of the spectra of PG at temperatures between $-100°$C and $-40°$C, scaled in frequency to same dc conductivity, that is, translated parallel to the horizontal axis, until the dc-conductivity slopes coincide. With decreasing temperature the dielectric strength of the relaxation increases slightly, as expected on the basis of its cooperative nature. According to Fig. 2, it is more specifically the low frequency modes of α_{pg} that gain in dielectric strength. This assymetrical amplification causes also the slight deviation of the peak from the master curve. Fitting the Havrilliak-Negami function to the data we could derive the exact dependence of the dielectric strength on temperature.

PG/PVP Solutions

Figure 3 shows a comparative plot of the dielectric response of the different samples at $-70°$C. At this temperature, the α-relaxation of PG is well inside our frequency window preceded at low frequencies by the slope related to the dc conductivity. Pure PVP is more than 200°C underneath the glass transition. Its local γ relaxation is present in the spectrum but its dielectric response is masked in the spectra of the solutions by the much stronger one of PG, which has at this temperature a similar relaxation frequency (see Fig. 6). Starting from the pure PG and with increasing PVP concentration α_{pg} moves to lower frequencies, indicating a slowing down of the PG molecular motion as a result of the interaction with the much slower or even arrested PVP (Fig. 4). The relaxation peak also becomes broader. The broadening hints to the existence of inhomegeneities in the local environment of each relaxing unit. Chain connectivity is relevant also in the case of polymer solute in small molecule solvent as the present and other studies show [3]. Although, as explained also in the introduction, polymer-solvent interactions are not necessary to observe the broadening, we cannot exclude them as a factor that intensifies it. Similar studies with other solvents would give information on the role played by each factor.

In order to study the relaxational behavior of PVP in solutions, it is preferable to use the derivative representation of the dielectric data, since it is free from the dc conductivity. Figure 5 shows representatively the spectrum of the PG solution with 55% wt. PVP content at selected temperatures in the derivative representation. Similar data have been obtained for the other samples. The steep rise at low frequencies is due to the polarization effects connected with free carriers. The slowest observed relaxation, labelled α_{pvp}, is related to the polymer segmental relaxation. It is followed by a weaker peak, present also only in the spectra of the PVP containing samples. This relaxation could be attributed to the local β_{pvp} motion plasticised by PG. Nevertheless, plasticization to such an extend is not characteristic of local relaxations. More data is necessary in order to identify it with

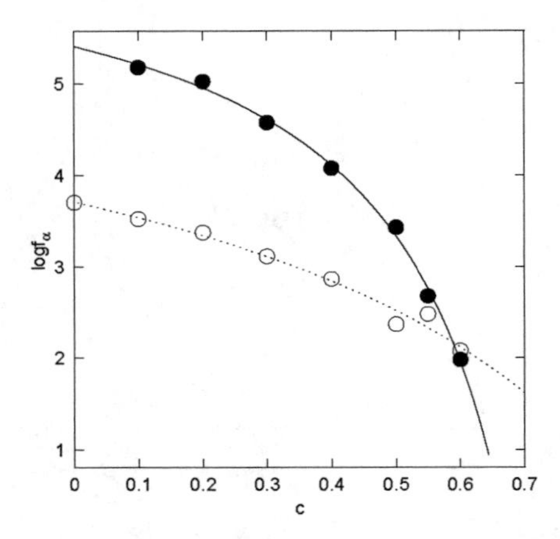

FIGURE 4. Relaxation frequency of α_{pg} (at $-70°$C, open symbols) and α_{pvp} (at 20°C, full symbols) as a function of the PVP concentration. The curves are fits to the Mashimo equation [34].

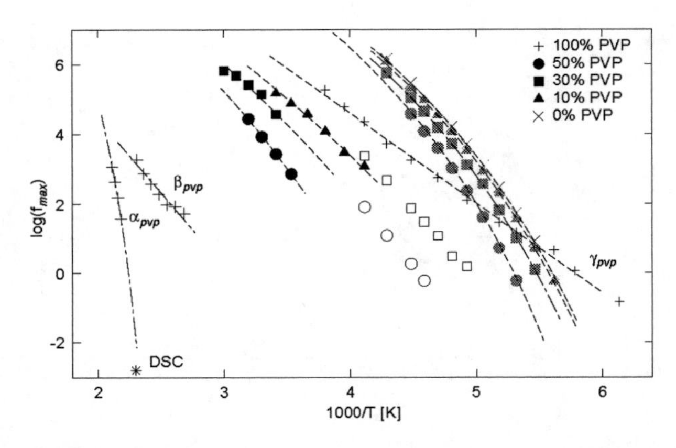

FIGURE 6. The curves are fits to the Vogel-Fulcher-Tammann [29], the straight lines to the Arrhenius equation. The crosses correspond to pure PVP as noted. For the other samples, the grey symbols correspond to α_{pg}, the full symbols to α_{pvp} and the open symbols to the third relaxation. For clarity reasons, the data for only selected samples have been depicted.

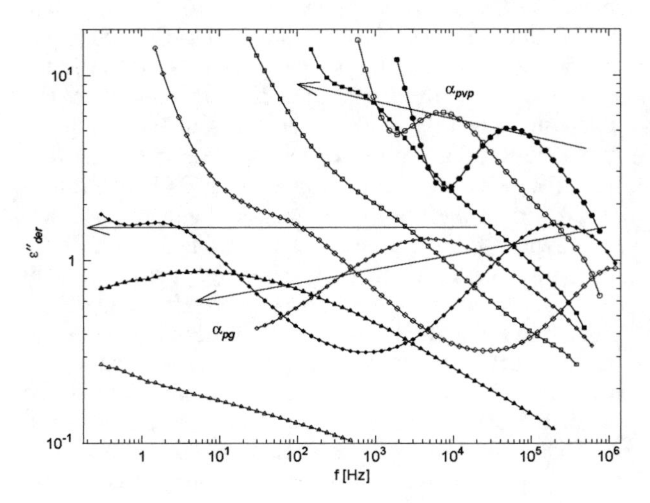

FIGURE 5. Derivative representation of the dielectric spectrum for the PG solution with 55% wt. PVP content at temperatures from $-100°$C to 60°C with a step of 20°C

certainty.

The characteristic frequencies of the different relaxations can be obtained either by fitting a sum of Havrilliak-Negami terms to the data, or in a more direct and model-independent way by reading the maxima of the peaks. The first option is possible both in the ε'' and in the ε''-derivative representation, while the second only in the latter one, since in the former the PVP-peak maxima are masked by the dc conductivity. Since the results are fully compatible, we present in Fig. 4 and 6 the peak maxima as read from the derivative plots.

According to the relaxation map, although PG and PVP are highly miscible and coexist in a soluted state, they appear to show two distinct main relaxations with characteristic times that can differ by more than four decades at the same temperature. In other words, they undergo separate glass transitions. Nevertheless, the relaxational behavior does not remain unaffected. The influence of the presence of PVP on the main relaxation α_{pg} of the solute has already been discussed. Even stronger is the acceleration of α_{pvp} by the presence of PG. The dependence of the relaxation frequencies f_α at constant temperature on the PVP concentration are shown in Fig. 4. The strong dependence of $f_{\alpha,pvp}$ on the presence of the solvent is in accordance with the formula proposed by Mashimo [34], despite the fact that Mashimo addressed the case of non-polar solvents, indicating that the reduction of the relaxation rate does not have to be attributed to polymer-splvent intaractions, but may be explained in terms of the distribution of the molecules in the solution. [36].

We have extrapolated the traces of the two main relaxations in the Arrhenius plot using a VFT equation, in order to estimate the glass transition temperatures $T_{g,pg}$, $T_{g,pvp}$ connected with the PG and PVP relaxations respectively. The results are presented in Fig. 7. In the same plot, the curve corresponding to the Fox equation [35] for well miscible systems is also shown, for comparison. At least at low concentrations, the estimated glass transition data connected with PVP show a very good agreement with the concentration dependence predicted by the Fox equation and confirmed experimentally for the common glass transition of the components in many miscible systems with unique glass transition, despite the existence here of the second glass transition attributed to PG. The temperature of the latter is much less affected by the PVP presence.

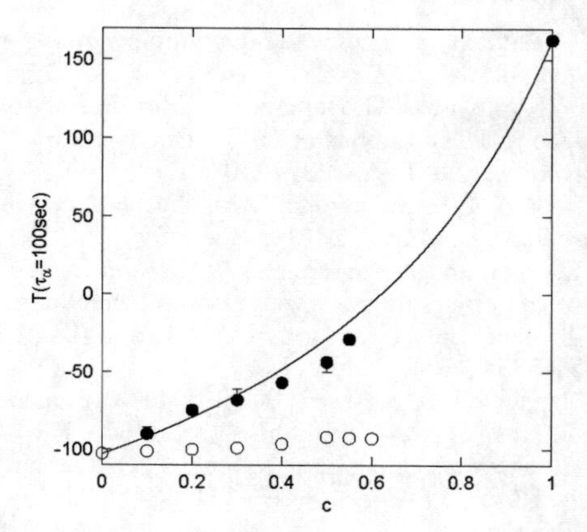

FIGURE 7. Glass transition temperatures as a function of PVP concentration. Full symbols: $T_{g,pvp}$, open symbols: $T_{g,pg}$. The data were estimated by extrapolating the best fit of the Vogel-Fulcher-Tammann equation to the data. The pre-exponential factor was set constant to 10^{13}, while the error bars show the uncertainty if its real value varies from 10^{12} to 10^{14}. Where no error bars are visible, the uncertainty is smaller than the symbol size.

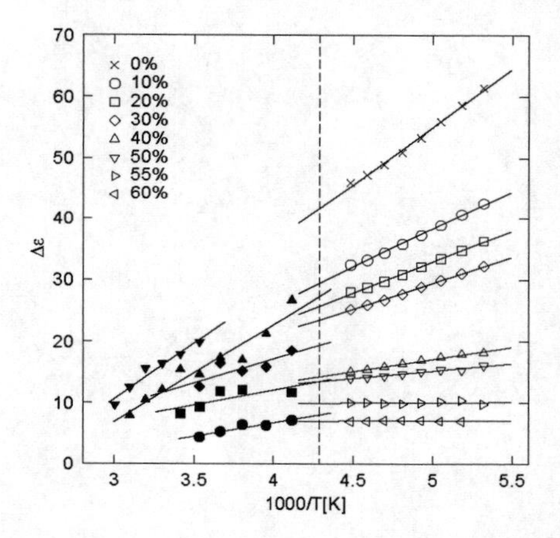

FIGURE 8. Dielectric strength of α_{pg} (open symbols) and α_{pvp} (full symbols) as a function of the inverse temperature. The full lines are linear fits to the data. The dashed line denotes −40°C.

Figure 8 shows $\Delta\varepsilon$ as a function of the inverse temperature for the different samples. For both relaxations it increases with decreasing temperature as expected due to their cooperative nature. The increase of the dielectric strength, $\Delta\varepsilon_{pg}$, of the segmental relaxation of the solute with decreasing temperature becomes less pronounced as PVP content is increased, until for the samples with 55% and 60% wt. PVP it becomes almost temperature independent (Fig. 8).

In order to compare the PVP concentration dependence of the dielectric strength of the two main relaxations, we estimated the dielectric strength at −40°C.

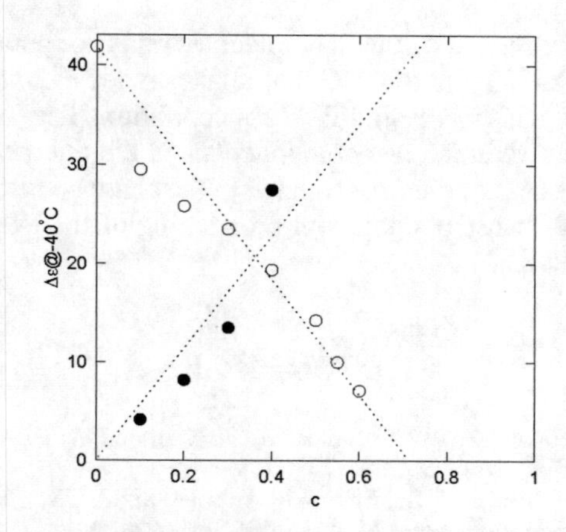

FIGURE 9. Estimated dielectric strength, $\Delta\varepsilon$, of α_{pg} (open symbols) and α_{pvp} (full symbols) at −40°C as a function of the PVP concentration. The dashed lines are guides for the eye.

This was done by extrapolating the linear fits to the data, as shown in Fig. 8. The results are presented in Fig. 9. At any constant temperature $\Delta\varepsilon_{pg}$ is reduced and $\Delta\varepsilon_{pvp}$ increases with increasing PVP content, even if these values are normalised to the concentrations. (Changes proportional to the concentration would follow the diagonals of the plot.) The hydrogen bonds between PG and PVP have the result that PVP during its segmental relaxation drags along also the attached to it PG molecules. Thus their dipole moments contribute to the effective $\Delta\varepsilon_{pvp}$. One can therefore conclude, that a part of the solvent molecules moves cooperatively with the polymer, in the same way that this happens in the case of miscible systems with a single glass transition temperature. The rest relax almost independendly, undergoing a separate glass transition, the dynamics of which are not strongly affected by the interactions with the polymer. This explains also the strong decrease of $\Delta\varepsilon_{pg}$ with PVP concentration.

CONCLUSIONS

The dielectric spectra of poly(vinyl pyrrolidone) (PVP), propylene glycol (PG) and solvent-rich solutions of the two substances were studied in a broad temperature range covering both the liquid and the glassy state. The data show a decoupling of the PVP and PG relaxations, resulting in the existence of two distinct glass transitions. The relaxation rate of PVP is strongly concentration dependent, the polymer being plasticised through the solvent molecules. In very dilute solutions, PVP has a glass transition temperature very close to the one of the solvent. PG is antiplasticised and subsequently its T_g rises with increasing PVP concentration, but to a much lesser extend. The analysis of the relaxation strengths reveals that the decoupling of the PG and PVP relaxations is not full, since a part of the PG molecules relax cooperatively with PVP. The part of PG that relaxes independently decreases with increasing PVP concentration. The interactions between the two components and the microheterogeneity of the solutions (related to the chain connectivity of PVP) cause a significant broadening of the PG main relaxation.

REFERENCES

1. G. Floudas, W. Steffen, E. W. Fischer, and W. Brown, *J. Chem. Phys.*, **99**, 695-703 (1993).
2. G. Floudas, J. S. Higgins, F. Kremer, and E. W. Fischer, *Macromolecules*, **25**, 4955-4961 (1992).
3. N. Taniguchi, O. Urakawa, and K. Adachi, *Macromolecules*, **37**, 7832-7838 (2004).
4. N. Shinyashiki, R. J. Sengwa, S. Tsubotani, H. Nakamura, S. Sudo, and S. Yagihara, *J. Phys. Chem. A*, **110**, 4953-4957 (2006).
5. F. Alvarez, A. Alegria, and J. Colmenero, *Macromolecules*, **30**, 597-604 (1997).
6. T. P. Lodge, and T. C. B. McLeish, *Macromolecules*, **33**, 5278-5284 (2000).
7. D. A. Savin, A. M. Larson, and T. P. Lodge, *J. Polym. Sci., Part B: Polym. Phys. Ed.*, **42**, 1155-1163 (2004).
8. N. Shinyashiki, N. Asaka, S. Mashimo, and S. Yagihara, *J. Chem. Phys.*, **93**, 760-764 (1990).
9. N. Miura, N. Shinyashiki, and S. Mashimo, *J. Chem. Phys.*, **97**, 8722-8726 (1992).
10. N. Shinyashiki, Y. Matsumura, N. Miura, S. Yagihara, and S. Mashimo, *J. Phys. Chem.*, **98**, 13612-13615 (1994).
11. N. Shinyashiki, S. Yagihara, I. Arita, and S. Mashimo, *J. Phys. Chem. B*, **102**, 3249-3251 (1998).
12. N. Shinyashiki, and S. Yagihara, *J. Phys. Chem. B*, **103**, 4481-4484 (1999).
13. U. Kaatze, *Adv. Mol. Relax. Processes*, **7**, 71-85 (1975).
14. U. Kaatze, O. Gottman, R. Podbielski, R. Pottel, and U. Terveer, *J. Phys. Chem.*, **82**, 112-120 (1978).
15. F. Wang, R. Pottel, and U. Kaatze, *J. Phys. Chem. B*, **101**, 922-929 (1997).
16. B. Y. Zaslavsky, L. M. Miheeva, M. N. Rodnikova, G. V. Spivak, V. S. Harkin, and A. U. Mahmudov, *J. Chem. Soc., Faraday Trans. 1*, **85**, 2857-2865 (1989).
17. E. Dachwitz, *Z. Naturforsch. A*, **45**, 126-134 (1990).
18. S. S. N. Murthy, *J. Phys. Chem. B*, **104**, 6955-6962 (2000).
19. M. Tyagi, and S. S. N. Murthy, *Carbohydr. Res.*, **341**, 650-662 (2006).
20. N. Asaka, N. Shinyashiki, T. Umehara, and S. Mashimo, *J. Chem. Phys.*, **93**, 8273-8275 (1990).
21. M. Stockhausen, and M. Z. Abd-El-Rehim., *Z. Naturforsch. A*, **48**, 1229-1230 (1994).
22. S. Sudo, S. Tobinai, N. Shinyashiki, and S. Yagihara, in *CP832, Flow Dynamics: The Second International Conference on Flow Dynamics*, edited by M. Tokuyama, and S. Maruyama, 2006, p. 149-152.
23. J. B. Hasted, *Water, a Comprehensive Treatise*, Plenum, New York, 1972.
24. U. Kaatze, *J. Chem. Eng. Data*, **34**, 371-374 (1989).
25. S. Havriliak, and S. Negami, *J. Polym. Sci. C*, **14**, 99 E17 (1966).
26. M. Wübbenhorst, and J. van Turnhout, *J. Non-Cryst. Solids*, **305**, 40-49 (2002).
27. L. V. Karabanova, G. Boiteux, O. Gain, G. Seytre, L. M. Sergeeva, E. D. Lutsyk, and P. A. Bondarenko, *J. Appl. Polym. Sci.*, **90**, 1191-1201 (2003).
28. Y. Nishio, T. Haratani, and T. Takahashi, *J. Polym. Sci. Part B: Polym. Phys.*, **28**, 355-376 (1990).
29. F. Kremer, and A. Schönhals, *Broadband Dielectric Spectroscopy*, Springer Verlag, Berlin Heidelberg, 2003.
30. S. K. Jain, and G. . P. Johari, *J. Chem. Phys.*, **92**, 5851-5854 (1988).
31. S. Capaccioli, K. Kessairi, D. Prevosto, M. Lucchesi, and K. L. Ngai, *J. Non-Cryst. Solids*, **352**, 4643-4648 (2006).
32. T. Blochowicz, C. Gainaru, P. Medick, C. Tschirwitz, and E. A. Rössler, *J. Chem. Phys.*, **124**, 134503-1-134503-11 (2006).
33. K. L. Ngai, P. Lunkenheimer, C. Leon, U. Schneider, R. Brand, A. T. Loidl, *J. Chem. Phys.*, **115**, 1405-1413 (2001).
34. S. Mashimo, *J. Chem. Phys.*, **67**, 2651-2658 (1977).
35. T. J. Fox, *Bull. Am. Phys. Soc.*, **1**, 123-125 (1956).
36. N. Shinyashiki, D. Imoto, and S. Yagihara, *J. Chem. Phys. B*, **111**, 2181-2187 (2007).

Effects of Disaccharide Sugars on Dynamics of Water Molecules: Dynamic Light Scattering and Dielectric Loss Spectroscopy Studies

Jeong-Ah Seo, Hyun-Joung Kwon, Hyung Kook Kim, and Yoon-Hwae Hwang*

Department of Nanomaterials & BK21 Nano Fusion Technology Division,
Pusan National University, Miryang 627-706, South Korea

Abstract. We studied the effects of disaccharide sugars (trehalose, sucrose, and maltose) on the dynamics of water molecules in sugar-water mixtures. We measured the acoustic phonons in sugar-water mixtures with different sugar contents by using a Sandercock Tandem 6-pass Febry-Petor interferometer and found that the Brillouin peak positions shifted to higher frequencies as the sugar concentration increased. We also measured the dielectric loss of hydrogen bonds in water molecules in sugar-water mixtures by using a Network analyzer with different sugar contents. The loss peak position in the dielectric loss spectra moved to lower frequencies as the sugar contents increased. The trehalose-water mixture showed the largest Brillouin peak shift and relaxation time change with increasing sugar content among three disaccharides indicating that the effect of trehalose on the dynamics of water molecules is the strongest. This unique property of trehalose sugar might be the origin of the superior bio-protection ability of trehalose.

Keywords: Dielectric loss spectra, Disaccharide, Dynamic light scattering, Dynamic reducer, Trehalose
PACS: 64.70. Fd, 64.70.Pf, 77.22.Gm

INTRODUCTION

Sugar is an important component of the biological system and the system of sugar and sugar containing materials are a matter of common interest to many researchers. The properties of sugar glasses, especially disaccharides glasses, are important because disaccharide glasses are known to be very effective materials for cryopreservation and anydrobiosis[1].

Some organisms can survive in very cold or dry condition because they contain disaccharide sugars such as sucrose or trehalose and can be reversibly glassy[2]. There are many studies about the protection ability of disaccharide sugars during drying and ensuing storage [3,4,5]. They reported several possible processes which maybe responsible for these effects by which this protection occurs, such as water replacement process[6,7,8], vitrification[9,10], and dynamic reducer[11].

In this paper, we studied the effects of sugars on water molecules as a dynamic reducer. To this end, we studied the effects of disaccharide sugars on the dynamics of water molecules in sugar-water mixtures by using dynamic light scattering and dynamic loss spectroscopy techniques. We found that as the weight percent of sugar increased, the speed of acoustic phonon in sugar-water mixture increased and the relaxation time of hydrogen bonds in water molecules increased. Among the three disaccharides trehalose showed the strongest effect on the sound velocity and the hydrogen bond in water molecules. The unique property of trehlose sugar might be the origin of its superior bio-protection ability.

EXPERIMENTS

The disaccharide sugars used in this study were sucrose, maltose, and trehalose. Trehalose disaccharide was sponsored by the Cargill Co. in U.S.A. and the other two disaccharides were purchased from Sigma Chemical Co. Three disaccharide sugars were used without further purification. The glass phase of all sugar-water mixtures was prepared by using a microwave method [12]. The weight percent of water in sugar-water mixtures was measured by using a moisture analyzer (Satorius MA 100, Gemnany) with 0.1 wt.% precision. All measurements were performed at room temperature with different sugar contents ranging from 0 wt.% to 50 wt.%. We used the sugar concentration

CP982, *Complex Systems, 5th International Workshop on Complex Systems*
edited by M. Tokuyama, I. Oppenheim, and H. Nishiyama

up to 50 wt.% due to the solubility limit [13] because the water activity is supersaturated above the solubility limit.

The back scattering geometry was used in a Brillouin spectroscopy measurement [14]. The incident beam was a vertically polarized 488 nm blue light of Ar ion laser (Coherent I90-C) with 500 mW power. The VV polarized scattered light was measured by using six-pass Sandercock tandem Fabry-Perot interferometer (JAS scientific instruments). A cylindrical shape glass cell was used. The size of the sample cell was 12 mm in diameter and 20 mm in height and the cell was sealed with a silicon plug.

The frequency dependent dielectric loss spectra with different water contents were carried out by using a vector network analyzer ((HP 8720D, Hewlett-Packard U.S.A.) with the help of a coaxial cable (HP 85070B, Dielectric Probe Kit) at frequencies ranging from 0.05 GHz to 20 GHz. The sample cell used in this measurement was a cylindrical glass cell. The volume of the sample cell was 20 ml. All measurements were repeated more than 3 times.

Results and Discussion

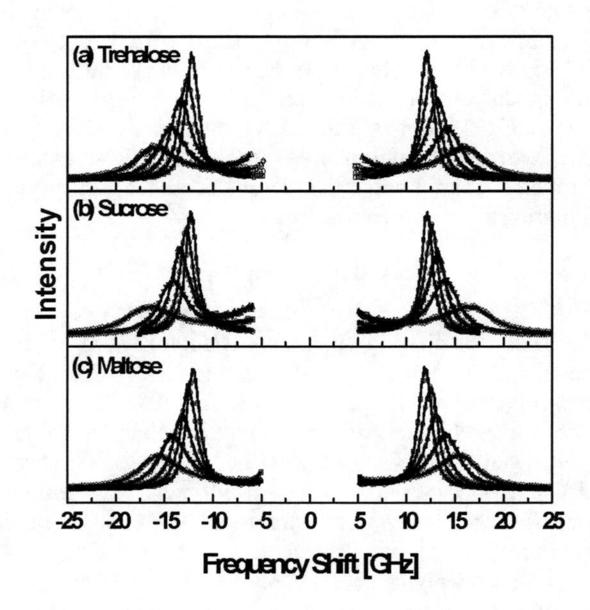

FIGURE 1. The VV component of Brillouin spectra of sugar-water mixtures ((a) trehalose, (b) sucrose, and (c) maltose) with different water contents at room temperature.

Figures 1(a)-(c) shows the Brillouin doublets of three different disaccharide-water mixtures at room temperature. In the measurements, the central peak was cut by the shutter to avoid a strong elastic laser line. In these figures, symbols and solid lines represent the measured data and fits to simple Lorentzian function, respectively. We can observe in Figs. 1(a)-(c) that the quality of fit is excellent.

In a Brillouin scattering experiment, the Brillouin peak is naturally convoluted by Gaussian-shaped laser lines (instrumental factor). Therefore, it is necessary to deconvolute the measured data or to modify the fitting function with the laser line. In our data analysis process, we convoluted the Lorentzian function with Gaussian shaped laser lines.

As we can see in Figs. 1(a)-(c), the Brillouin peak shift increased as the concentration of sugar increased. It implies that the sound velocity in the sugar-water mixture increased as the concentration of sugar increased.

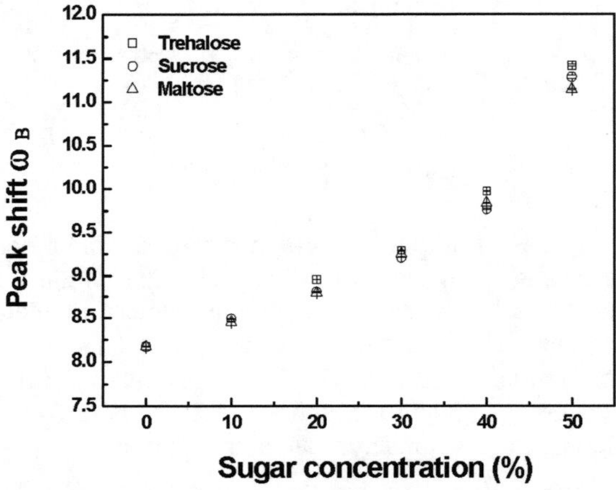

FIGURE 2. Brillouin peak shifts with different sugar contents in sugar-water mixtures. Open squares, open circles, and open triangles represent trehalose-, sucrose-, and maltose-water mixtures, respectively. The trehalose-water mixture showed the largest Brillouin peak shift.

Figure 2 shows the Brillouin peak shifts with different sugar contents in sugar-water mixtures. In this figure, different symbols represent three disaccharide sugars i.e.; open squares, open circles, and open triangles represent trehalose-, sucrose-, and maltose-water mixtures, respectively. As we can observe in Fig. 2, the amount of the Brillouin peak shift of trehalose-water mixture was the largest. Between sucrose- and maltose-water mixtures, the difference in Brillouin peak shift is not big enough. We can also observe that the difference in the Brillouin peak shift between three different sugar-water mixtures becomes evident as the sugar contents in the mixture increased.

We also measured the dielectric loss spectra of thehydrogen bond in water molecules in sugar-water mixtures by using a vector network analyzer at frequencies ranging from 0.05 GHz to 20 GHz at room temperature. Figures 3(a) – (c) show the dielectric loss spectra of trehalose-, sucrose-, and maltose-water mixtures with different water contents. The symbols and solid lines in these figures represent the experimental data and fits to the Havriliak-Negami(HN) function [15] defined by

$$\varepsilon_D(\omega) = \varepsilon_\infty + \frac{\varepsilon_0 - \varepsilon_\infty}{\left[1+(i\omega\tau)^a\right]^b} \quad (1)$$

Here, a and b are the exponents, τ is the relaxation time, and ω is the frequency. As we can see in Figs. 3(a)-(c), the quality of fit was excellent.

It is well known that in the microwave region, the dielectric loss spectrum of water can be described by a Debye-type relaxation function. In the case of distilled water, the resonant frequency at room temperature is known to be 18.4 GHz [16]. In our data, we observed that the resonant frequency of pure water (0 wt.% sugar) was 19.2 GHz at 25 °C which shows good agreement with the reference value. As the sugar concentration increased, the fitting function started to deviate from a simple Debye function and we were able to describe the dielectric loss spectrum by using HN function as shown in Eq. (1). Other types of fitting functions were also possible but we could obtain the best fitting results with the HN function. In the case of pure water, we observed that the exponents a and b were very close to unity.

As we can observe in Figs. 3(a)-(c) the dielectric loss peak moved to higher frequencies as the water contents increased or sugar concentration decreased. In other words, the relaxation time increased as the sugar contents increased since the loss peak position is inversely proportional to the relaxation time.

Figure 4 shows the relaxation time of hydrogen bond of water molecules in sugar-water mixtures. As we can see in Fig. 4, the relaxation time increased as the sugar content increased. We also observed in Fig. 2 that the sound velocity of the sugar water mixture increased as the sugar contents increased. These observed phenomena were consistent with each other because sound velocity is known to be proportional to the viscosity of the sample. We can also observe in Fig. 4 that the relaxation time change in the trehalose-water mixture was the largest. The relaxation times of sucrose- and maltose-water mixtures are very similar. We can again observe in Fig. 4 that the difference in relaxation times between the three different sugar-water mixtures becomes evident as the sugar contents in the mixture increased

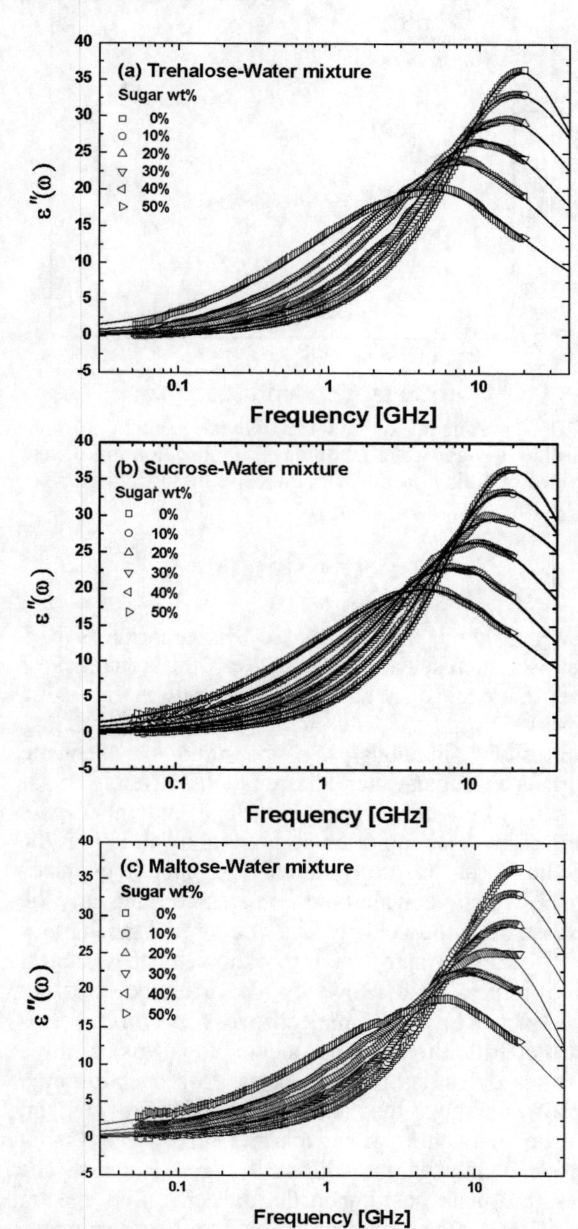

FIGURE 3. The dielectric loss spectra of disaccharide-water mixtures ((a) trehalose, (b) sucrose, and (c) maltose) with different water contents at room temperature. The frequency range used in this measurement was from 0.05 GHz to 20 GHz which can cover the resonant frequency of hydrogen bond in water molecules.

Based on our findings, we believe that the effect of trehalose disaccharide on the dynamics of water molecules is the strongest among three disaccharides. Therefore, we speculate that this unique characteristic of trehalose sugar might be the origin of the superior bio-protection ability of trehalose.

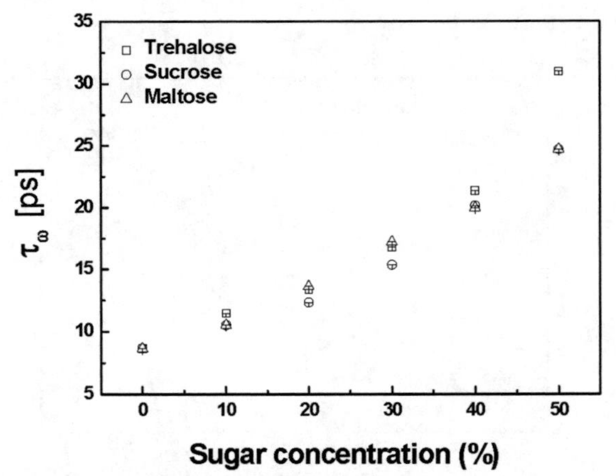

FIGURE 4. The relaxation times of hydrogen bond of water molecule in sugar-water mixtures. The relaxation times were obtained by fitting the dielectric loss spectra to HN function.

CONCLUSION

We studied the effects of disaccharide sugars (maltose, sucrose, and trehalose) on the dynamics of water molecules in sugar-water mixtures by using dynamic light scattering and dynamic loss spectroscopy methods. We measured the acoustic phonons in sugar-water mixture with different sugar contents by using Sandercock Tandem 6-pass interferometer at room temperature and found that the Brillouin peak positions shifted to higher frequencies as the sugar concentration increased due to the increase of sound velocity. Among three disaccharides, the trehalose-water mixture showed the largest Brillouin peak shift. However, the difference was not as distinct as that in the dielectric loss spectra. It seems that the Brillouin scattering technique is less sensitive to water dynamics than dielectric loss spectroscopy. We also measured the dielectric loss of hydrogen bond in water molecules in sugar-water mixtures by using Network analyzer with different sugar contents. The relaxation peak position in the dielectric loss spectra moved to lower frequencies as the sugar contents increased. Similar to the case of the Brillouin experiment, the trehalose-water mixture showed the largest relaxation time change among three disaccharides. Therefore, we concluded based on our findings that the effect of trehalose on the dynamics of water molecules is stronger than other two disaccharides. The longer relaxation time in the case of trehalose sugar may be the origin of its superior bio-protection ability.

ACKNOWLEDGMENTS

This work was supported by a Korean Research Foundation grants (KRF-2006-005-J02802 and KRF-2006-005-J02803).

REFERENCES

* Corresponding author: yhwang@pusan.ac.kr

1. S. Magazu, G. Maisano, F. Migliardo and C. Mondelli, *Biopys J.* **86**, 3241-3249 (2004).
2. A. Patist, H. Xoerb, *Colloids and Surfaces B: Biointerfaces* **40**, 107-113 (2005).
3. J. H. Crowe, L. M. crowe, and D. Champman, *Science*, **223**, 701 (1984).
4. K. Koster, M. S. Web, G. Bryant, and D. V. Lynch, *Biochemical Biophysics Acta*, **1193**, 143 (1994).
5. M. F. Mazzobre, and M. P. Buera Chirife, *J. of Biotechnology Progress*, **13**, 195 (1997).
6. P. Bordat, A. Lerbret, J-P. Demaret, F. Affouard, and M. Descamps, *Europhysics Letter*, **65**, 41 (2004).
7. R. Giangiacomo, *Food Chemistry*, **96**, 371 (2006).
8. A. Lerbret, P. Bordat, F. Affouard, Y. Guiner, A. Hedoux, L. Paccou, D. Prevost, and M. Descamps, *Carbohydrate Research*, **340**, 881 (2005).
9. J. H. Crowe, J. F. Carpenter, and L. M. Crowe, *Annual Review Physiology*, **60**, 73 (1998).
10. J. H. Crowe, S. B. Leslie, and L. M. Crowe, *Cryobiology*, **31(4)**, 355 (1994).
11. Y. Choi, K. W. Cho, K. Jeong, and S. Jung, *Carbohydrate Research*, **341**, 1020 (2006).
12. Jeong-Ah Seo, Jioung Oh, Dong Jin Kim, Hyung Kook Kim, and Yoon-Hwae Hwang, *J. Non-Crstal. Solids*, **333**, 111-114 (2004).
13. M. V. Galmarini, J. Chirife, M. C. Zamorea, A. Perez, *LWT-Food Science and Technology*, Accepted 10 April, (2007).
14. B. J. Berne, and R. Pecora, *Dynamic Lighe Scattering : With Applications to Chemistry, Biology, and Physics*, Dover Publication Inc., New York, 2000, pp. 233-248.
15 . S. Havriliak Jr., S. Negami, *Polymer*, **8**, 161 (1967).
16. R. Buchner, J. Barthel, J. Stauer, Chem. *Phys. Lett.*, **306**, 57-63 (1999).

Heterogeneous Structure and Ionic Transport Properties of Silver Chalcogenide Glasses

Junichi Kawamura, Naoaki Kuwata, and Takanari Tanji

Institute of Multidisciplinary Research for Advanced Materials,
Tohoku University
Katahira 2-1-1, Aobaku, Sendai, 980-8577, JAPAN

Abstract. Silver chalcogenide glasses is a new example of the importance of concentration fluctuation in supercooled liquid state to ionic conductivity, which remains as a nano to micro phase separation in glassy state. The frozen heterogeneous structures of Ag-$GeSe_3$ glasses have been investigated precisely by using FE-SEM and EPMA analysis. The variation of the local structure is studied by NMR and Raman scattering. From these structure analyses, the Ag-Ge-Se and Ag-Ge-S system have a bistable structure in supercooled liquid state, which is edge sharing $GeSe_4$ based network with Se-Se chains and corner sharing $GeSe_4$ network combined with Ag without Se-Se bonding. During the glass forming process, the homogeneous liquid separates into these two different structures to form micro heterogeneous structure in the glassy state. The silver ionic transport is only possible in the latter composition region, which is separated with each other below the composition x=0.3 and is connected with each other to form percolation path above x=0.3. The previously reported conductivity jump at x=0.3 is the consequence of the percolation transition and is expressed by generalized effective medium approximation (GEMA).

Keywords: Chalcogenide glass, Generalized effective medium approximation, Heterogeneous structure, Ionic conductivity, Micro phase separation, Percolation, Superionic conductor glass
PACS: 66.10.-Ed, 66.30.Dn, 66.30.hh, 66.30.Pa

INTRODUCTION

Ionic transport in supercooled liquid and glass is not only an interest of fundamental physics but also a big topic in application to ionics devices such as sensors, batteries, atom switches etc [1]. The ionic conductivity can be seen as a good prove of density, bond energy and concentration fluctuations in supercooled liquid and glasses. The density fluctuation couples with the ionic transport through the free-volume formation in supercooled liquid state, which is frozen in at the glass transition temperature and affect the ionic conductivity in the glassy state [2,3]. Energy fluctuation relating to chemical bond rearrangement in liquid state is another important factor to affect the ionic transport [4,5]. Another factor which has been discussed for two decades is a concentration fluctuation in the liquid state, which might be frozen in glass to form an inhomogeneous structure or a micro phase separation. Such an inhomogeneous structure is observed by small angle X-ray or neutron scattering in many fast ion conductor glasses [6,7]. Based on this picture, some models based on percolation theory have been proposed [8,9]. However the percolation model has not been accepted widely so far, since the predicted percolation threshold is not seen in conventional fast ion conductor glasses [10].

We have demonstrated the role of the inhomogeneous structure on the ionic transport in glass based on the results of organic-inorganic hybrid superionic conductor glasses, where a clear percolation threshold is observed in ionic conductivity as well as a first sharp diffraction peak (FSDP) of ca. 1nm is seen in small-angle X-ray scattering (SAXS) [11-13].

In the quest for the glasses exhibiting a percolation transition, we have encountered some silver chalcogenide glasses [14,15], where a peculiar transition from semiconductor to superionic conductor is observed at a concentration, although no crystallization is seen in X-ray diffractions.

For example, the ionic conductivity of $Ag_x(GeSe_3)_{1-x}$ glasses depends strongly on the concentration of the amount of silver. The sudden increase in conductivity is observed at silver composition x=0.3, below which the glass is a hole conducting semiconductor of 10^{-13} S/cm and jumps to a fast silver ionic conduction of 10^{-4} S/cm above x=0.3 [14].

Recent studies on the thermodynamic and morphology analysis by high resolution field emission

CP982, *Complex Systems, 5th International Workshop on Complex Systems*
edited by M. Tokuyama, I. Oppenheim, and H. Nishiyama

scanning electron microscope (FE-SEM) technique have revealed a phase separation of this material, which affects the ionic and transport in the chalcogenide glasses [15-17].

In this paper is reported our recent results on the structure and ionic transport properties of silver chalcogenide glasses, which reveals an ultimate effect of the concentration fluctuation in supercooled liquid state to the fast ion transport in the glass.

EXPERIMENTS

The glasses for experiments were prepared from 99.999 % pure Ag, Ge, Se, which were melt in a vacuum silica tube at 950℃ for 24 h followed by quenching with water. The samples were characterized by X-ray diffraction (XRD) to confirm the glass forming region, differential scanning calorimetry (DSC) for detecting their glass transition temperatures. The conductivity had been reported previously and is confirmed. The morphology of the glasses were investigated by a field-emitted scanning electron microscope (FE-SEM) and an electron prove micro analyzer (EPMA). The average compositions were analyzed by fluorescent X-ray analysis and the local composition is evaluated by EPMA. The local structure of the glasses were investigated by ^{77}Se magic-angle spinning nuclear magnetic resonance (MAS-NMR), ^{109}Ag NMR and Raman scatterings.

RESULTS AND DISCUSSION

DSC analysis indicates the glass transition temperatures Tg of these glasses are in the range from 200 ℃ to 245 ℃, and the crystallization peaks are between 300 ℃ to 430 ℃. XRD patterns of the quenched samples are shown in figure 1, which is in good agreement with the previous report [14], although a crystal phase of Ag_8GeSe_6 appears above x=0.727. A clear first-sharp diffraction peak (FSDP) is seen at 2θ=15° for the glassy samples below x=0.4, which disappears above x=0.571. The FSDP of these glasses are attributed to the quasi-layer type intermediate ordering of edge sharing $GeSe_4$ networks. The disappearance of the FSDP at high concentration of silver is the indication of the network rearrangement to corner sharing $GeSe_4$ by the interaction to silver ions, which is also supported by ^{77}Se NMR and Raman scattering as shown later.

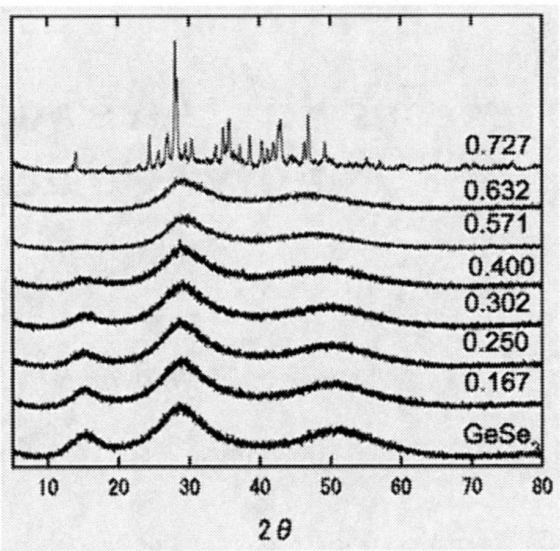

FIGURE 1. XRD patterns of $Ag_x(GeSe_3)_{1-x}$ glasses; x values are shown in the figure.

In spite of the small change in XRD patterns, the conductivity of the $Ag_x(GeSe_3)_{1-x}$ glasses jumps from 10^{-12} S/cm to 10^{-5} S/cm at x~0.3 as shown in figure 2 [14].

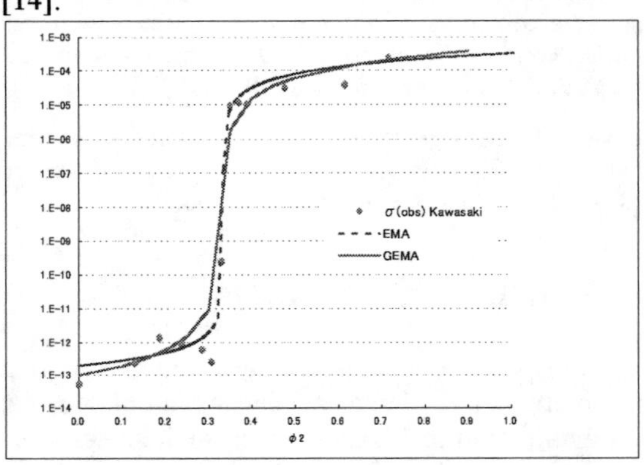

FIGURE 2. The composition dependence of the electrical conductivity of $Ag_x(GeSe_3)_{1-x}$glasses plotted as a function of volume fraction of assumed high conductivity precipitate of x=0.75. Solid and broken curves are calculations by GEMA (eq.2) and EMA (eq. 1) model: see text.

Morphology of the samples observed by FE-SEM is shown in Fig. 3 for x=0.3 and x=0.4, where the conductivity jumps to fast ionic conductor region. Many bright droplets of about a few μm are observed in dark matrix at x=0.3 and reversely many dark droplets are distributed in bright matrix at x=0.4. From detailed EPMA analysis of these morphology, it was found that the dark region contains no silver and is close to the pure $GeSe_3$ glass. On the other hand, the bright region is rich in silver whose composition is close to x=0.571. At lower concentration of silver, the droplet size is decreased to 10 nm or less.

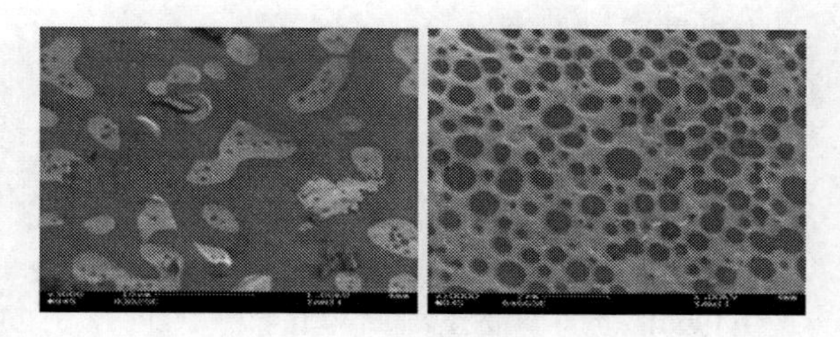

FIGURE 3. FE-SEM Images of $Ag_{0.3}(GeSe_3)_{0.7}$ (left) and $Ag_{0.4}(GeSe_3)_{0.6}$ (right) glasses.

Thus, the origin of the conductivity jump can be attributed to a percolation transition of a nano-scale phase separation of the glasses. This is probably due to the immiscibility gap in supercooled liquid state, through which the sample passed in the glass formation process.

The macroscopic conductivity of heterogeneous medium can be modeled by an effective medium approximation (EMA), in which the each constituent domain with conductivity σ_i is considered to be embedded in the effective medium with conductivity of σ_m, which can be evaluated from the following EMA equation [18];

$$\left\langle \frac{\sigma_i - \sigma_m}{\sigma_i + 2\sigma_m} \right\rangle = \sum_i \eta_i \frac{\sigma_i - \sigma_m}{\sigma_i + 2\sigma_m} = 0, \qquad (1)$$

where, η_i is the volume fraction of the component i. The fitted values are shown by dashed curve in figure 2, where the parameters are chosen to be $\sigma_1 = 5\text{x}10^{-14}$ Scm^{-1} and $\sigma_2 = 2\text{x}10^{-4}$ Scm^{-1} . The composition of the high conductive phase x_0 is estimated to be 0.75.

A more precise form of the extension of EMA to be consistent with percolation theory has been proposed by McLachlan et al, as a generalized effective medium approximation (GEMA) [19,20], which is expressed as,

$$\phi_1 \frac{\sigma_1^{1/s} - \sigma_m^{1/s}}{\sigma_1^{1/s} + A\sigma_m^{1/s}} + \phi_2 \frac{\sigma_2^{1/t} - \sigma_m^{1/t}}{\sigma_2^{1/t} + A\sigma_m^{1/t}} = 0, \qquad (2)$$

$$A = (1 - \phi_c)/\phi_c$$

where, *s* and *t* are the parameters to mach with critical exponents of percolation transition from high conductivity and low conductivity sides. *A* is also a parameter to fit the observed percolation threshold of ϕ_c. In the original EMA of eq. (1) corresponds to $s=t=1$ and $A=2$, which results in $\phi_c=1/3$. The fitting to eq. (2) is also shown in figure 2 by solid curve, where x_1 and x_2 are the same as above and $A=2$, but $\sigma_1 = 10^{-13}$ Scm^{-1}, $\sigma_2 = 3\text{x}10^{-4}$ Scm^{-1} and $s=t=2.0$ are used. The both results are fairly in good agreement with the experiments although the estimated composition (x_0=0.75) is larger than that of determined EPMA analysis (x_0=0.571). It is concluded that the percolation concept is applicable to this clearly phase separated glasses.

The existence of heterogeneous structure of the silver chalcogenide glass is relating to the dynamical concentration fluctuation in supercooled liquid state, where the two locally stable structures are competing with each other to have two minima in free energy diagram. It results in a spinodal or a binodal decomposition during the glass formation.

The bistability or micro phase separation tendency is originated from local bond fluctuations in the liquid state, which can be detected by atomic scale fluctuation analysis of the glasses. NMR analysis indicates the modification of the local structure by the addition of silver to $GeSe_3$ glass. The ^{109}Ag NMR of the glass indicate only one peak at around 950 ppm which decreases to 920 ppm at x=0.571 as shown in figure 4. This may seem contradictive to the phase separation of the glass. But, it is explained that only the bright region in figure 3 contains silver ions and almost no silver is solved in the dark region as was found in EPMA analysis.

Raman scattering and ^{77}Se NMR indicate that the glass containing large amount of silver has different local bonds between the Ge and Se in comparison with low concentration glasses. ^{77}Se NMR shows two components corresponding to Ge-Se and Se-Se bonds as shown in fig. 5, whose fractions are evaluated from the deconvolution of the spectra and shown in figure 6; the error of the fraction is about 5%. The component 1 of Ge-Se bond increases above x=0.4, and component 2 of Se-Se bond decreases to almost zero above x=0.4. Raman scattering also suggests the disappearance of Se-Se bonds and edge sharing $GeSe_4$ units. The disappearance of FSDP in figure 1 and the composition dependence of the glass transition temperature Tg are also consistent with the rearrangement of the glass network at around x=0.4 from edge sharing to corner shearing networks.

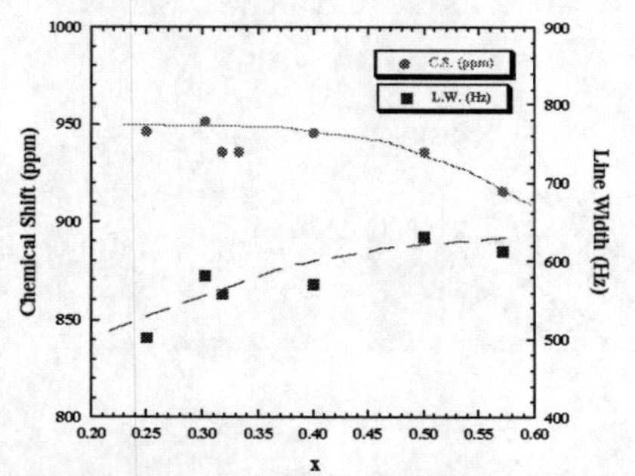

FIGURE 4. ^{109}Ag NMR chemical shift and line width of $Ag_x(GeSe_3)_{1-x}$ glasses; curves are guides for the eye.

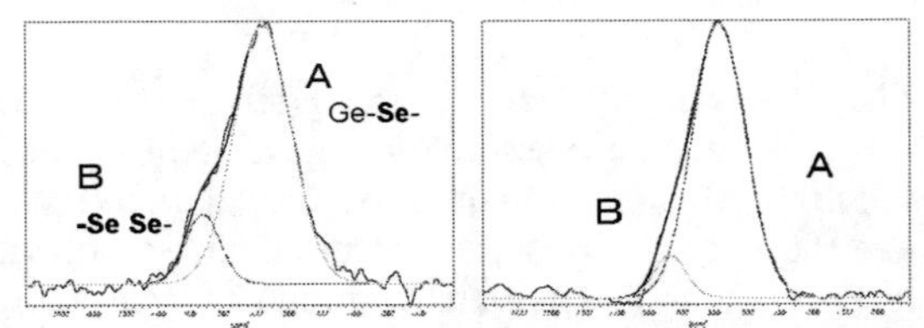

FIGURE 5. ^{77}Se NMR spectrum of $GeSe_3$ (left) and $Ag_{0.4}(GeSe_3)_{0.6}$ (right) glasses.

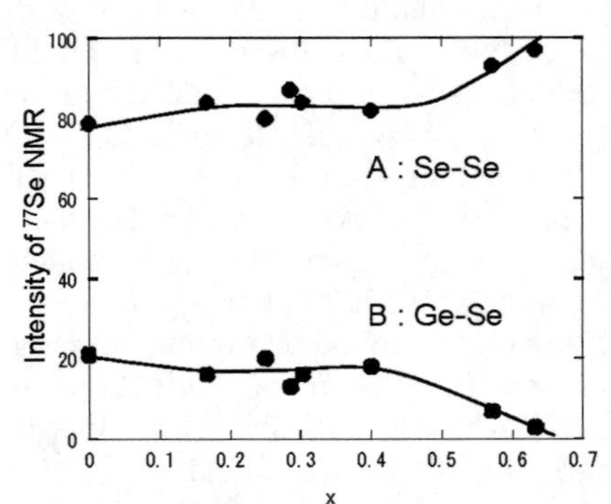

FIGURE 6. ^{77}Se NMR intensity ratios of A (Se-Se bonds) and B (Ge-Se bonds) of $Ag_x(GeSe_3)_{1-x}$ glasses as a function of the composition x; curves are guides for the eye.

From these structural data, we can conclude that the Ag-GeSe$_3$ system has bistable local structures, which is (i) an edge sharing based $GeSe_2$ like structure with excess Se-Se chains and (ii) an corner sharing based Ag_xGeSe_3 networks without Se-Se chains. At high temperatures above 950 °C where the sample was melt, the liquid is almost homogeneous, although it has a large concentration fluctuation. In the cooling process of glass formation, the concentration fluctuation grows up to nano to micron scale and is frozen into a structural heterogeneity in the glass below the Tg. This frozen heterogeneity is in large scale in the composition around $Ag_{0.3}(GeSe_3)_{0.7}$ to give a observable phase separation and discontinuous conductivity jump at x=0.3. This explanation is schematically shown in figure 7.

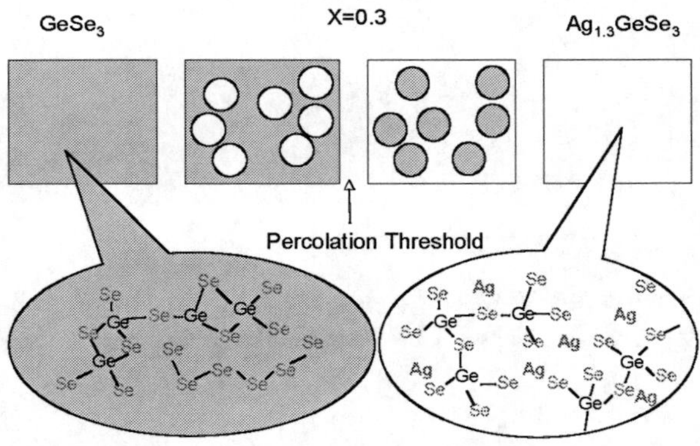

FIGURE 7. A schematic structure model of Ag-GeSe$_3$ glasses;

REFERENCES

1. *Solid State Ionics for Batteries*, Vol. , edited by T. Minami (Springer-Verlag, Tokyo, 2005).
2. J. Kawamura and M. Shimoji, *Mater.Chem.Phys.* **23,** 99 (1989).
3. J. Swenson and L. Borjesson, *Phys.Rev.Lett.* **77,** 3569 -3572(1996).
4. J. L. Souquet, M. Levy, and M. Duclot, *Solid State Ionics* **70,** 337-345 (1994).
5. M. Aniya, *Recent Res.Devel.Phys.Chem.Solids* **1,** 99-120 (2002).
6. J. Swenson, R. L. McGreevy, L. Borjesson, and J. D. Wicks, *Solid State Ionics* **105,** 55-65 (1998).
7. S. Adams and J. Swenson, *Phys. Rev. B* **6305,** 11 (2001).
8. M. D. Ingram, M. A. Mackenzie, W. Muller, and M. Torge, *Solid State Ionics* **40/41,** 671-675 (1990).
9. A. Bunde, *Solid State Ionics,* **75,** 147-155 (1995).
10. S.W. Martin, *Solid State Ionics* **51,** 19-26 (1992).
11. R. Asayama, N. Kuwata, and J. Kawamura, *Flow Dynamics: The second international Conference on Flow Dynamics, Sendai, Japan, 16-18, Nov. 2005 (AIP Press).* , 176-179 (2006).
12. J. Kawamura, N. Kuwata, and Y. Nakamura, *Solid State Ionics* **113-115,** 703-709 (1998).
13. J. Kawamura, R. Asayama, N. Kuwata, and O. Kamishima, in *Physics of Solid State Ionics*, edited by H. T. T.Sakuma (Transworld Research Network, Kerala, India, 2005), pp. 193-246.
14. M. Kawasaki, J. Kawamura, Y. Nakamura, and M. Aniya, *Solid State Ionics* **123,** 259-269 (1999).
15. A. Pradel, N. Kuwata, and M. Ribes, *J.Phys.Cond.Matt.* **15,** S1561-1571 (2003).
16. Y. Wang, M. Mitkova, D. G. Georgiev, S. Mamedov, and P. Boolchand, *J.Phys.Cond.Matt.* **15,** S1573-S1584 (2003).
17. A. Piarristeguy, J. M. C. Garrido, M. A. Ure, M. Fontana, and B. Arcondo, *Journal of Non-Crystalline Solids* , 353, 3314-3317 (2007).
18. S. Kirkpatrick, *Rev.Mod.Phys.* **45,** 574-588 (1973).
19. D. S. McLachlan, M. Blaszkiewicz, and R. E. Newnham, *J. Am. Ceram. Soc.* **73,** 2187-2203 (1990).
20. K. Nozaki and T. Itami, *J. Phys.: Condens. Matter* **18,** 2191-2198 (2006).

Can the Frequency Dependent *Isobaric* Specific Heat be Measured by Thermal Effusion Methods?

T. Christensen, N. B. Olsen, and J. C. Dyre

DNRF centre "Glass and Time", IMFUFA, Department of Sciences, Roskilde University, Postbox 260, DK-4000 Roskilde, Denmark

Abstract. It has recently been shown that plane-plate heat effusion methods devised for wide-frequency specific-heat spectroscopy do not give the isobaric specific heat but rather the so-called longitudinal specific heat. Here it is shown that heat effusion in a spherical symmetric geometry also involves the longitudinal specific heat

Keywords: Glass Transition, Longitudinal Specific Heat, Specific Heat Spectroscopy, Thermal Effusion
PACS: 64.70.Pf

INTRODUCTION

The frequency-dependent specific heat is one of the most fundamental thermoviscoelastic response functions characterizing relaxation of liquid structure in highly viscous liquids. The measurement of this quantity as an alternative to enthalpy relaxation studies was conceived more than two decades ago [1, 2].

When the specific heat is frequency dependent one faces experimentally the problem of separating out the trivial frequency dependence from the slow propagation of heat. This can be solved in two ways. One can go to the thermally thin limit [2, 3] where the sample is small compared to the thermal diffusion length l_D. l_D is inversely proportional to the square root of the frequency and typically 0.1mm even at 1Hz. So this condition can be difficult to fulfill over a wide frequency range unless the sample is very small [3].

The other - effusion - approach is to choose the sample size much larger than the thermal wavelength. This is the thermally thick limit and it is easier to realize over a wide frequency range [4]. Both methods have to take due account of the stresses coming from the material supporting the sample associated with different thermal expansion coefficients. The effusion method suffers additionally from thermal stresses produced within the liquid itself. It is only the latter problem we consider in this paper.

In the effusion methods one typically produces a harmonically varying heat current $\mathrm{Re}\{P_\omega e^{i\omega t}\}$ at a surface in contact with the liquid. The corresponding temperature response $\mathrm{Re}\{T_\omega e^{i\omega t}\}$ on the very same surface is measured. Since the response is linear in the stimulus it is convenient to introduce the complex thermal impedance

$$Z = \frac{T_\omega}{P_\omega} \tag{1}$$

The thermal impedance of a sample of volume V and volume specific heat c is

$$Z = \frac{1}{i\omega cV} \tag{2}$$

in the thermally thin limit.

If on the other hand planar thermal waves effuses from a plate of area A into a liquid the thermal impedance is

$$Z = \frac{1}{A\sqrt{i\omega c\lambda}} \tag{3}$$

in the thermally thick limit [1]. Here λ is the thermal conductivity.

It has allways tacitly been assumed that measurements done at ambient pressure are isobaric and that the c of formula (3) is c_p. However it was formerly stated [5] and recently shown [6] that the ordinary heat diffusion with a complex diffusion constant does not describe the experimental situation adequately. For unidirectional heat effusion it was shown that the effective specific heat measured is the so-called longitudinal specific heat $c_l(\omega)$ which is between the isochoric, $c_V(\omega)$ and isobaric, $c_p(\omega)$ specific heats. Denoting the adiabatic and isothermal bulk moduli by $K_s(\omega)$ and $K_T(\omega)$ respectively and the shear modulus by $G(\omega)$ one can write the adiabatic and isothermal longitudinal moduli as $M_s(\omega) = K_s(\omega) + 4/3G(\omega)$ and $M_T(\omega) = K_T(\omega) + 4/3G(\omega)$. Now $c_l(\omega)$ is related to $c_V(\omega)$ as [6]

$$c_l(\omega) = \frac{M_s(\omega)}{M_T(\omega)} c_V(\omega), \tag{4}$$

whereas $c_p(\omega)$ is related to $c_V(\omega)$ as

$$c_p(\omega) = \frac{K_s(\omega)}{K_T(\omega)} c_V(\omega). \tag{5}$$

CP982, *Complex Systems, 5th International Workshop on Complex Systems*
edited by M. Tokuyama, I. Oppenheim, and H. Nishiyama

In an easily flowing liquid $G(\omega)$ is negligible since $1/\omega$ is large compared to the Maxwell relaxation time, τ_M, and there is no difference between c_l and c_p. However in a highly viscous liquid near the dynamic glass transition the shear modulus becomes comparable to the bulk moduli and the diffence between $c_l(\omega)$ and $c_p(\omega)$ becomes significant.

We show below that the same is true for heat effusion in a spherical symmetric geometry. Here one also obtains $c_l(\omega)$. Furthermore in spherical geometry one can also get the heat conductivity λ and thus get $c_l(\omega)$ absolutely. The planar unidirectional method in fact gives only the effusivity, $\sqrt{\lambda c_l(\omega)}$; that is $c_l(\omega)$ is determined only to within a proportionality constant.

THERMAL AND MECHANICAL COUPLING

The General Equations

One cannot treat the diffusion of heat independently of the associated creation of strains or stresses. Let the temperature field, $T(\mathbf{r},t)$ be described in terms of the small deviation $\delta T(\mathbf{r},t) = T(\mathbf{r},t) - T_0$ from a reference temperature T_0 and denote the displacement field by $\mathbf{u} = \mathbf{u}(\mathbf{r},t)$. Dealing with relaxation is most conveniently done in the frequency domain. Thus time dependence of the fields is given by the factor e^{st}, $s = i\omega$. Considering only cases where inertia can be neglected the equations that couple temperature and displacement are [6]

$$M_T\nabla(\nabla\cdot\mathbf{u}) - \beta_V\nabla\delta T - G\nabla\times(\nabla\times\mathbf{u}) = 0 \quad (6)$$

$$c_V s\delta T + \beta_V T_0 s\nabla\cdot\mathbf{u} - \lambda\nabla^2\delta T = 0. \quad (7)$$

Here the isochoric pressure coefficient $\beta_V(\omega)$ is defined in the constitutive equation for the trace of the stress tensor σ

$$\frac{1}{3}\mathrm{tr}(\sigma) = K_T\nabla\cdot\mathbf{u} - \beta_V\delta T. \quad (8)$$

The Isobaric Case

If the trace of the stress tensor is constant in time then the term $\beta_V T_0 s\nabla\cdot\mathbf{u}$ in equation (7) becomes $T_0\beta_V^2/K_T\delta T$. Since $T_0\beta_V^2/K_T = c_p - c_V$ equation (7) now becomes the ordinary heat diffusion equation

$$s\delta T = D_p\nabla^2\delta T, \quad (9)$$

decoupled from the displacement field and with a diffusion constant involving the isobaric specific heat

$$D_p = \frac{\lambda}{c_p} \quad (10)$$

It is usually assumed that thermal experiments on liquids with a completely or partially free surface will be at isobaric conditions. However this is only true if the shear modulus G can be neglected compared to bulkmodulus K_T. This condition fails near the glass transition and the full coupled problem of equations (6) and (7) has to be considered.

Radial Heat Effusion from a Spherical Surface into an Infinite Media

We would like to show here that the inherent problem of measuring c_p is not only confined to one-dimensional heat flow in the geometry considered in [6] where the associated displacement field is forced to be longitudinal.

The longitudinal specific heat also emerges in the thermal impedance against effusion out from a sphere. In the spherically symmetric case $\nabla\times\mathbf{u}$ vanishes. If we denote differentiation with respect to r by a prime (6) and (7) becomes

$$M_T(r^{-2}(r^2u)')' - \beta_V\delta T' = 0 \quad (11)$$

$$c_v s\delta T + T_0\beta_V s r^{-2}(r^2u)' - \lambda r^{-2}(r^2\delta T')' = 0. \quad (12)$$

Define now the longitudinal specific heat,

$$c_l \equiv c_V + T_0\frac{\beta_V^2}{M_T}, \quad (13)$$

the heat diffusion constant,

$$D = \frac{\lambda}{c_l}, \quad (14)$$

and the wave vector

$$k = \sqrt{\frac{s}{D}}. \quad (15)$$

We thermally perturb the system by a harmonically varying heat current density $j_q e^{st}$ with $s = i\omega$ at the surface of radius r_1. If we impose the boundary conditions of vanishing fields at infinity and a hard core, $u(r_1) = 0$ then the coupled solution is

$$\delta T(r) = \frac{k}{sc_l}\frac{(kr_1)^2}{(1+kr_1)kr}e^{-k(r-r_1)}j_q \quad (16)$$

$$u(r) = \frac{\beta_V}{M_T c_l s}(\frac{r_1}{r})^2(1 - \frac{1+kr}{1+kr_1}e^{-k(r-r_1)})j_q. \quad (17)$$

The total thermal impedance thus becomes

$$Z \equiv \frac{\delta T(r_1)}{4\pi r_1{}^2 j_q} = \frac{1}{4\pi\lambda r_1}\frac{1}{1+kr_1} \quad (18)$$

or

$$Z = \frac{1}{4\pi\lambda r_1} \frac{1}{1+\sqrt{i\omega c_l(\omega)/\lambda r_1^2}} \quad (19)$$

It should be noted that in solving the same problem on the basis of the ordinary heat diffusion equation (9) one arrives at (19) but with c_p instead of c_l. It is thus seen that in doing specific heat spectroscopy by effusion in a spherical geometry one obtains again the longitudinal specific heat and not the isobaric specific heat.

One can also consider the case of a soft core, $\sigma_{rr}(r_1) = 0$. Although the displacement field is altered compared to the case of a hard core, the expression for the thermal impedance is still found to be given by (19).

The DC-limit gives the heat conductivity,

$$Z \rightarrow \frac{1}{4\pi\lambda r_1} \quad \text{for} \quad \omega \rightarrow 0 \quad (20)$$

The high-frequency limit is in concordance with the one-dimensional result,

$$Z \rightarrow \frac{1}{4\pi r_1^2 \sqrt{i\omega c_l(\omega)\lambda}} \quad \text{for} \quad \omega \rightarrow \infty \quad (21)$$

since short thermal waves cannot "see" the curvature of the sphere. It is seen that effusion in spherical geometry in fact gives information on two properties, the heat conductivity and the heat capacity, whereas the unidirectional effusion only gives the effusivity. This is because a characteristic length scale, the radius of the heat-producing spherical surface, is involved. Effusivity in spherical geometry thus makes it possible to derive the heat capacity absolutely. However the practical usable frequency range will be more limited for a given sensitivity since the contribution from c_l in (19) will vanish at low frequencies. At high frequency the possibility of modelling the contribution to the thermal impedance from the heat-producing device itself will also put a limit.

In real plane-plate effusion experiments the finite width of the plate gives rise to boundary effects when the heat diffusion length becomes comparable to the plate width. The deviation from the simple formula (3) is dependent on the ratio between these to quantities. Since a length now appears in the problem this deviation again gives the possibility of determining λ separately. This has been addressed perturbatively [7] on the basis of the ordinary heat diffusion equation (9), but not with the more exact coupled thermomechanical equations (6) and (7). Thus in fact it seems that of the two simple idealized models - the planar and the spherical - of heat effusion including the thermomechanical coupling the spherical may be the one that mostly resembles its practical realisation.

CONCLUSION

These examples - the unidirectional and the spherical geometry - seem to show that it is inherently difficult to get the *isobaric* specific heat directly from effusivity measurements. However another well-defined quantity, the *longitudinal* specific heat can be found.

REFERENCES

1. N. O. Birge and S. R. Nagel, *Phys. Rev. Lett.* **54**, 2674-2677 (1985).
2. T. Christensen, *J. Physique Colloq.* **46**, C8-635-C8-637 (1985).
3. H. Huth, A. A. Minakov, A. Serghei, F. Kremer and C. Schick, *Eur. Phys. J. Special Topics* **141**, 153-160 (2007).
4. N. O. Birge, *Phys. Rev. B* **34**, 1631-1642 (1986).
5. T. Christensen and N. B. Olsen, *Prog. Theor. Phys. Suppl.* **126**, 273-276 (1997).
6. T. Christensen, N. B. Olsen and J. C. Dyre, *Phys. Rev. E* **75**, 041502-1-041502-11 (2007).
7. N. O. Birge, P. K. Dixon and N. Menon *Thermochim. Acta* **304**, 51-65 (1997).

Density of States Simulations of Various Glass Formers

Jayeeta Ghosh and Roland Faller

Department of Chemical Engineering and Materials Science, University of California–Davis, One Shields Ave, Davis, CA, 95616.

Abstract. The behavior of glassy systems in bulk and especially in confined geometries has received considerable attention over the last decades because of the technological importance and inherent complexity of the systems near or below the transition temperature. Confined glasses have been studied using different theoretical and experimental techniques which helped shape our understanding; but still huge gaps remain.

In this work we are using the Wang–Landau Monte Carlo approach to study different model glasses. General Monte Carlo fails to sample all relevant regions of phase space; the application of this method gives us the opportunity to directly estimate the density of states and consequently any other thermodynamic properties. We can calculate properties in different ensembles using the same simulation runs. This random walk algorithm is designed to visit all energy states with equal probability to produce a flat histogram. We can estimate the density of states on the fly whenever any energy state is visited. We perform multiple simulations in overlapping energy regions and finally join them after proper scaling to obtain the overall density of states; the global density of states of the glass former is then known to within a constant.

We apply this technique to a model binary Lennard Jones glass which is a well tested model, as well as for the first time to a realistic glass forming system, the small organic glass former Ortho–terphenyl (OTP). For OTP we start from a united atom model and derive systematically a coarse–grained representation by replacing each phenyl ring with one interaction site. We apply the Iterative Boltzmann Inversion for this purpose. This method relies on the structure of the atomistic model, mainly the radial distribution function (RDF). One needs to Boltzmann invert the atomistic RDF to obtain an initial guess for the non–bonded potential. Then using this potential for the preliminary coarse grained run gives a first set of RDFs to compare with the atomistic target RDFs. We then iterate this procedure until the structures coincide. After calibration of the mesoscale system against the atomistic model we estimate the density of states for this small organic glass former.

We find that the properties of the bulk Lennard Jones model show very good agreement with literature values. The atomistic and coarse grained representations of Ortho–terphenyl in the bulk are in good agreement with experiments. Freestanding films show a lower glass transition than the bulk value for both models.

Keywords: Binary Lennard-Jones glass, Density of states, Glass, molecular dynamics, Monte Carlo, Ortho-terphenyl
PACS: 61.43.Fs,02.70.Uu,02.70.Ns

INTRODUCTION

The liquid to amorphous solid, i.e. glass, transformation is an unresolved problem in condensed matter research. Many theories address different aspects of the glassy and super cooled states of matter [1, 2, 3, 4, 5, 6]. Experimentally, this transition is concomitant with a strong increase in all relevant time scales posing severe challenges for experiments. Computer modeling can add valuable insight. A number of theoretical models which have been proposed to explain the transition can be tested by simulations. Several theoretical models base on the assumption that the transition is triggered by an underlying ideal thermodynamic glass transition which is believed to occur at the Kauzmann temperature T_K where only one minimum energy basin of attraction is accessible to the system [7]. This transition would then be overshadowed by a kinetic transition. The mode–coupling theory on the other hand explains the glass transition exclusively from a kinetic standpoint [8]. Potential energy landscape and frustration–limited domain theories shed additional light into this matter [9, 10, 11]. But a general consensus is still missing.

A variety of simulation approaches have been used for glassy systems. Often very low temperature, i.e. below the transition, simulations are avoided and replaced by extrapolations [11, 12]. This cannot always guarantee to infer the correct system behavior in the low energy and temperature range. The time scales below the mode coupling temperature T_{MCT} or the actual transition temperature T_{g} are too long to be sampled by standard simulation techniques. Here we are using Density of States Monte Carlo [13, 14] to study glass transition in different models of glass forming materials.

DENSITY OF STATES MONTE CARLO

The Density of States Monte Carlo (DOS) technique is very successful keeping in mind that it is only a few years old [13, 14, 15, 16, 17, 18, 19, 20, 21]. The purpose of the DOS algorithm is to sample all energy space with a probability $P(E)$ which is independent of energy, in order to cover all energy ranges with equal accuracy.

CP982, *Complex Systems, 5th International Workshop on Complex Systems*
edited by M. Tokuyama, I. Oppenheim, and H. Nishiyama

Note that we focus here on energy exclusively but generalizations to other extensive variables are straightforward [17]. In a Monte Carlo simulation the probability $P(E)$ is approximated by a histogram of energies $H(E)$. Fundamentally, the probability of sampling a given energy value in a Monte Carlo simulation is a product of a weighting factor $w(E)$ and the density of states (or multiplicity) of the energy $\Omega(E)$.

$$P(E) = \alpha w(E)\Omega(E) \tag{1}$$

where the proportionality factor α is determined by normalization. The weighting factors are in most cases prescribed by the statistical mechanical ensemble which we want to sample. If the simulation, however, realizes a flat histogram the weight $w(E)$ has to be proportional to the inverse of the density of states. Statistical mechanics tells us that by obtaining the density of states our problem is solved from a thermodynamic point of view as all other properties can be derived from $\Omega(E)$. Moreover, if we have a flat histogram we cover the energy landscape homogeneously and do not get stuck in deep minima.

Technically, the DOS algorithm starts with a guess of the density of states $\Omega_0(E)$ (which is often set to be uniform $\Omega_0(E) = c$) and performs a random walk in energy space using a Metropolis–like acceptance criterion

$$p = \min\left\{1, \frac{W^R_{new}\Omega'(E_{old})}{W^R_{old}\Omega'(E_{new})}\right\}. \tag{2}$$

W^R is the Rosenbluth weight of the corresponding Monte Carlo move which has to be taken into account for biased techniques. After every move the current estimate of the density of states for the energy the simulation is visiting $\Omega(E_{visit})$ is updated by multiplication with a factor $f > 1$. The ensuing violation of detailed balance decreases as the updated factor decreases during the course of the simulation. f decreases whenever the collected energy histogram is flat to a prescribed tolerance. Eventually the convergence factor approaches $f = 1$ and detailed balance is restored [13]. The Wang–Landau algorithm yields the density of states $\Omega(E)$ without using a particular ensemble during the simulation. This provides the unique opportunity to analyze the results from the very same simulation in different ensembles without biasing the accuracy towards any one ensemble as there is no "best" or "natural" ensemble to analyze the data. From the density of states we can get the entropy as

$$S(E) = k_B \ln\Omega(E) \tag{3}$$

and this serves as direct connection to thermodynamics. One caveat is in order here, the algorithm yields the density of states only to within an arbitrary constant as we only enforce the histogram to be flat at an arbitrary value. We only know the entropy exactly if we know the degeneracy of the ground state (or any other state). Otherwise we obtain an entropy which is shifted by a constant

$$S'(E) = S(E) + S_0. \tag{4}$$

In the microcanonical (NVE) ensemble we use the definition of temperature

$$T^{-1}(E) = \partial_E S(E)|_{N,V}. \tag{5}$$

to connect to thermodynamics as there is no temperature in the DOS simulation itself. The additive constant in the entropy is irrelevant here and for any other thermodynamic property, except the thermodynamic potentials. Inverting equation 5 yields $E(T)$ and $c_V = T\partial_T S|_V$ or $c_V = \partial_T E(T)|_V$ leads to the (isochoric) heat capacity in the microcanonical ensemble. The derivatives as well as any other analysis are done numerically after a completed DOS simulation. Only then the density of states is known with the desired accuracy. This means a running average over the simulation *does not* yield the correct value for any observable as we are *not* following a thermodynamic ensemble.

Most applications of the Density of States algorithm use the canonical ensemble to calculate properties. In the canonical (NVT) ensemble the average of any property can be calculated from

$$\langle A(T)\rangle = \frac{\sum A(E)\Omega(E)\exp(-\beta E)}{\sum\Omega(E)\exp(-\beta E)}, \beta = (k_B T)^{-1} \tag{6}$$

if the property A was recorded during the simulation as a function of energy. Energy and heat capacity can be calculated without additional measurements. They read

$$\langle E(T)\rangle = \frac{\sum\Omega(E)E\exp(-\beta E)}{\sum\Omega(E)\exp(-\beta E)} \tag{7}$$

$$\langle c_V(T)\rangle = \frac{\sum(E - \langle E(T)\rangle^2)\Omega(E)\exp(-\beta E)}{\sum\Omega(E)\exp(-\beta E)}. \tag{8}$$

Again we can not just use a simulation average. Different ensembles are only required to be equivalent in the thermodynamic limit of infinitely large systems. This limit does not apply in a computer simulation with a finite system size. So we expect differences between different ensembles.

MODELS AND TECHNICAL DETAILS

Binary Lennard–Jones Glass

Lennard–Jones mixture models are widely used in understanding behaviors of glasses by simulation [22, 23, 24, 25, 20, 26]. The model we are using here consists of a binary mixture of Lennard–Jones particles, 80% A and

20% B [23]. 250 particles are used in our calculations. The interaction parameters between particles of species A and B are $\varepsilon_{AA} = 1.0$ and $\sigma_{AA} = 1.0$, $\varepsilon_{BB} = 0.5$ and $\sigma_{BB} = 0.88$, and $\varepsilon_{AB} = 1.5$ and $\sigma_{AB} = 0.8$. The density is $1.204\sigma_{AA}^{-3}$. All quantities are in reduced units: length in units of σ_{AA}, temperature in units of ε_{AA}/K_B, and time in units of $(\sigma_{AA}^2 m/\varepsilon_{AA})^{1/2}$. We use particle identity swap [25] and simple translation moves for this system. For this model in bulk it was difficult to get flat histograms over the wide energy range from -1850 to -1400 which covers the relevant temperatures. First we performed simulations in 18 small overlapping energy windows ($-1850 \rightarrow -1800$, $-1825 \rightarrow -1775$ and so on). After obtaining the DOS in the individual windows we joined them by shifting the DOS to overlap in the common energy ranges which resulted in a singled DOS and started a DOS run in the whole range. We did not try to push for $\log f < 10^{-5}$ for the whole energy range but after we obtained the DOS for the whole range we continued to update it at $\log f = 10^{-5}$ for more than 2 million steps. Once the density of states covered almost the full energy range we obtained the final DOS from which the subsequent analysis started.

We used a similar approach for our simulations of the freestanding films. We model freestanding films by dropping the periodicity in the z direction while there are periodic boundary conditions in the xy plane. We kept the volume fixed. So essentially we have hard non interacting vacuum wall in the Z-direction. We performed simulations in 8 small overlapping energy windows ($-1650 \rightarrow -1550$, $-1600 \rightarrow -1500$ and so on) and joined the segmental density of states to get the overall density of states. We then simulated the whole energy range ($-1650 \rightarrow -1200$) at once. Once we get the overall density of states in this large window we carry out subsequent analysis to compare the data with bulk.

Ortho–Terphenyl

Ortho–terphenyl is a small organic glass former and a good test system to study the glass transition [27, 28]. The molecule consists of three benzene rings with the two outer rings connected in ortho positions of the central one. We first perform atomistic molecular dynamics simulations using GROMACS v3.2 [29]. Although simple in structure devising a well–suited computer model for OTP is non–trivial as we need a detailed model to represent the benzene rings in their proper orientation correctly and of course the model is required to be simple as glass transition studies are notorious for their computational demand. Our atomistic model bases on the model by Kudchadkar et al. [30] and is described in detail in [28]. Our OTP system consists of 800 molecules in a cubic box under periodic boundary conditions with a 2 fs time step in the NPT ensemble. The system was studied at 11 different temperatures from below the glass transition temperature to well above. Details of the simulation conditions as well as the results are again described in [28].

As such detailed simulations are very time consuming we develop a systematically coarse–grained model for OTP with fewer interaction sites. We use the Iterative Boltzmann Method (IBM) for this purpose. OTP is the first small molecule studied using this technique [31]. IBM is a successful approach for the structural coarse–graining of polymers [32, 33]. It is designed to optimize a mesoscale model to reproduce the structure of an atomistic simulation. The method bases on radial distribution functions (RDFs) obtained from an atomistic simulation. In the limit of an infinitely diluted solution of a homogeneous system one could use the potential of mean force (PMF) gained by Boltzmann inverting the pair distribution function to get an interaction potential between superatoms. However, in concentrated solutions, melts, or glasses the structure is defined by an interplay of the PMF with local packing effects. Thus, a direct application of the potential of mean force is not correct. Still the PMF can be used by the Iterative Boltzmann Method. For details of the technique the reader is referred to recent reviews on this subject [34, 35, 36, 37].

In the meso–scale model of Ortho–terphenyl each benzene ring is replaced by a single interaction center. Thus, OTP is modeled as a trimer of type 1–2–1. This means that the two outer rings are required to have the same interaction. This leads to three interdependent radial distribution functions to be optimized [33]. We cannot assume any mixing rules a priori. For a full description of the optimization see [27, 31]. The mapping satisfies the basic requirements [36] that the distances between bonded super–atoms follow well–defined single peak distributions. The single peak distribution is modeled by a single Gaussian curve, which defines a harmonic bond potential. Our meso–scale model does not use any other intra–molecular potentials. To keep the correct angle between the three rings, we use a fictitious bond potential between the outer rings. Cross dependencies between different potentials are neglected. For the non–bonded potential we use the Iterative Boltzmann Inversion described above.

Iteration was deemed converged when all coarse–grained RDFs were within 5% of the atomistic target, i.e. the maximum difference was not larger than 5% of the value. The optimization was performed independently at temperatures in the glassy (230 K) and liquid region (300 K). We use the DLPOLY [38] molecular dynamics simulation package for the meso–scale simulations, i.e. we use molecular dynamics for the necessary simulations during the optimization. In every iteration we ensured that the simulation is equilibrated by monitoring the dependence of the radial distribution function on simulation time. We took the final configuration from the atomistic simulation at various temperatures to initialize the mesoscale simulation with 800 OTPs. The meso–scale simulations have been performed under NVT conditions with density equal to the atomistic density at that condition.

After calibration of the mesoscale model against the atomistic model we use this model to calculate the density of states for this system. Unlike the Lennard–Jones model this system has intra–molecular energy and additionally a tabulated potential for the non–bonded energy which we incorporated into the our code for the density of states algorithm. We include two types of particle movement. In one of them we translate the whole molecule with center of mass motion and in other we rotate the molecule along either the X, Y, or Z axis in which the center of mass remains fixed. We do not change the bond lengths or angles for our three atom OTP model. We verified our code against the coarse grained molecular dynamics results. We also recalculated the binary Lennard–Jones systems using a tabulated Lennard–Jones potential. A detailed discussion of the model and the algorithm will be published elsewhere [39].

RESULTS

We recently studied the Lennard–Jones system using Density of states Monte Carlo in two independent studies [20, 26]. We here compare the relative density of states of the bulk system with free standing films for the binary Lennard-Jones glass in Fig 1. The DOS values have been shifted arbitrarily in the ordinates to have a similar range of density of states as we are anyway getting density of states only to within an arbitrary constant. It is clear that the relevant energies for the free standing film are about 200 energy units higher than for the bulk as the surface particles have less interactions. Otherwise the shape of the curves is very similar.

We determined the energy versus temperature curve from the density of states data using the canonical ensemble. We show in figure 2 the comparison of the energy against temperature for bulk and film using the binary Lennard Jones glass. The energy curves are very similar with slight kinks hinting towards glass transitions at different temperatures. It appears that the transition temperature in the film is lower compared to the bulk.

We find a peak in the heat capacity for the bulk model [26] and a kink in the energy versus temperature. As we find only weak structural changes and no structural transition we can conclude that we find a glass transition in both systems. This glass transition is actually very close and in agreement with earlier data of mode

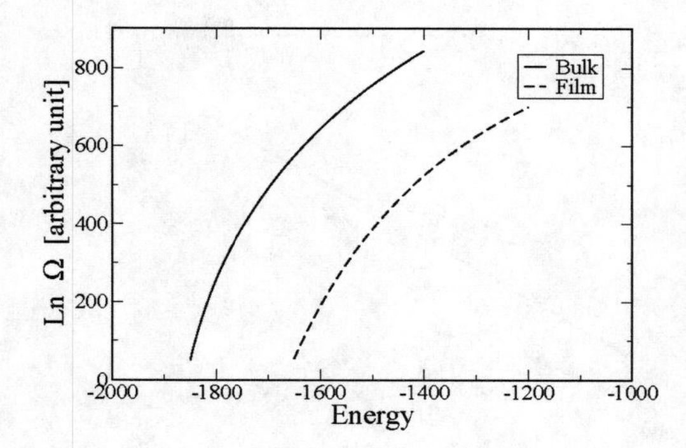

FIGURE 1. Comparison of the logarithm of the density of states between bulk and freestanding film. The DOS values have been shifted arbitrarily in the ordinates to cover a similar range.

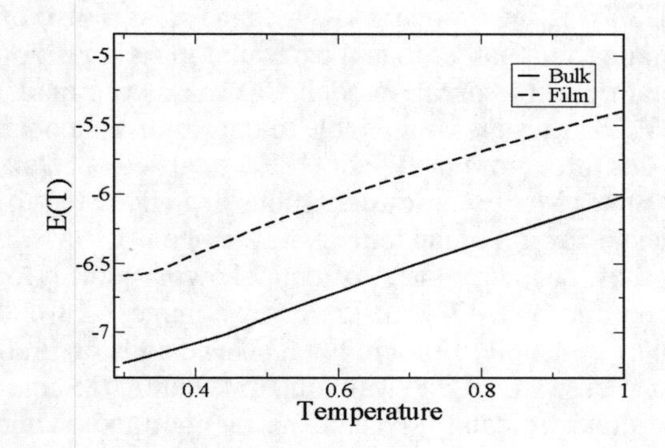

FIGURE 2. Canonical ensemble internal energies for bulk and film using the Lennard–Jones model

coupling temperature data [40] for bulk system. We see an increasing trend of density undulations near the surface with higher energies. But at lower energy this effect is negligible which negates the possibility of crystal formation. We could not compare the film data as no literature values exist. We studied properties in both canonical and microcanonical ensembles, and they show qualitatively similar results but the quantitative data are slightly different which is expected due to finite size effects.

Our atomistic model for ortho–terphenyl yields the glass transition temperature as defined by the density versus temperature curve to be around 260 K [28] compared to the experimental value of 243 K [41, 42]. The glass transition temperature for the freestanding film of OTP is also reduced compared to the bulk. For the mesoscale model we could not use one universal optimized potential for wide range of temperatures [31]. The potential obtained in the liquid range produces a crystal structure

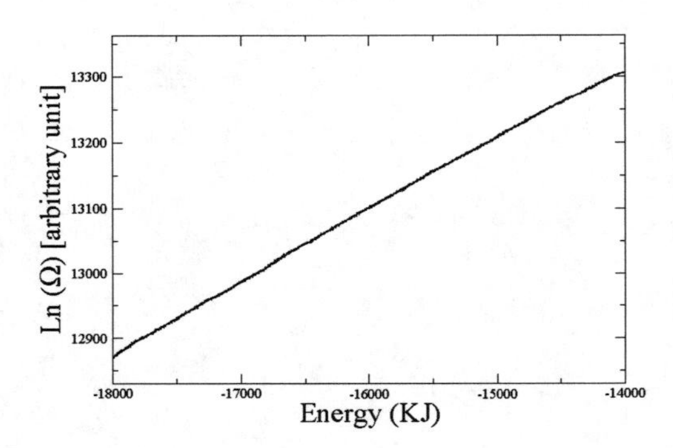

FIGURE 3. Relative density of states of ortho-terphenyl in the energy range $-18,000$ kJ/mol to $-14,000$ kJ/mol

in the low temperature range. So we needed a separate optimized potential for the glassy range. Consequently we can only define a range into which the glass transition temperature falls rather than a particular glass transition temperature in mesoscale model. Our mesoscale model for OTP gives results comparable to the atomistic model. Using this three atom model for OTP we carried out Density of States Monte Carlo simulations in a wide range of energies to cover a broad temperature range.

We show the early stage of data showing density of states for the bulk OTP in the energy range $-18,000$ kJ/mol to $-14,000$ kJ/mol in Fig 3. This data is preliminary in a sense we need to wait long time before the equilibration and we can only calculate the thermodynamic properties after we have completed the simulation.

CONCLUSIONS

We show that Density of States Monte Carlo technique is a powerful tool to model the glass transition in various models. We are able to compare bulk and film data for small model glasses. We also show that we can combine the density of states technique with systematic coarse–graining to elucidate the thermodynamics of realistic glass formers. In the future it would be interesting to increase the system size in the mesoscale model of OTP and calculate the density of states to see effects of finite size scaling on glasses. This would have important implications for the determination of the nature of the transition. With a larger system size and higher simulation time we will be able to find if there is any domain decomposition or in other words if we see dynamic heterogeneity both in bulk and in films.

ACKNOWLEDGMENTS

This work was supported by the U.S. Department of Energy, Office of Advanced Scientific Computing Research through an Early Career Principal Investigator Award (grant DE–FG02–03ER25568) and a faculty research grant from the University of California, Davis. This research also used resources of the National Energy Research Scientific Computing Center, which is supported by the Office of Science of the U.S. Department of Energy under Contract No. DE–AC03–76SF00098

REFERENCES

1. M. L. Williams, R. F. Landel, and J. D. Ferry, *J. Am. Chem. Soc* **77**, 3701 – 3707 (1955).
2. D. Turnbull, and M. H. Cohen, *J. Chem. Phys* **34**, 120 – 125 (1961).
3. J. H. Gibbs, *J. Chem. Phys.* **25**, 185 – 186 (1956).
4. G. Adam, and J. H. Gibbs, *J. Chem. Phys* **43**, 139–146 (1965).
5. F. H. Stillinger, *Science* **267**, 1935–1939 (1995).
6. M. Goldstein, *J. Chem. Phys.* **51**, 3728 – 3739 (1969).
7. W. Kauzmann, *Chem. Rev.* **43**, 219–256 (1948).
8. W. Götze, and L. Sjögren, *Rep Prog Phys* **55**, 241–376 (1992).
9. D. Kivelson, and G. Tarjus, *Philosophical Magazine B* **77**, 245–256 (1998).
10. D. Kivelson, and G. Tarjus, *J. Chem. Phys.* **109**, 5481–5486 (1998).
11. G. Tarjus, D. Kivelson, and P. Viot, *J. Phys. Condens. Matter* **12**, 6497–6508 (2000).
12. B. Coluzzi, G. Parisi, and P. Verocchio, *J Chem Phys* **112**, 2933–2944 (2000).
13. F. Wang, and D. P. Landau, *Phys. Rev. Lett.* **86**, 2050–2053 (2001).
14. F. Wang, and D. P. Landau, *Phys Rev E* **64**, 056101–1–056101–16 (2001).
15. T. S. Jain, and J. J. de Pablo, *J Chem Phys* **116**, 7238–7243 (2002).
16. N. Rathore, and J. J. de Pablo, *J Chem Phys* **116**, 7225–7230 (2002).
17. Q. Yan, R. Faller, and J. J. de Pablo, *J Chem Phys* **116**, 8745–8749 (2002).
18. Q. Yan, and J. J. de Pablo, *Phys Rev Lett* **90**, 035701–1–035701–4 (2003).
19. M. Troyer, S. Wessel, and F. Alet, *Phys Rev Lett* **90**, 120201–1–120201–4 (2003).
20. R. Faller, and J. J. de Pablo, *J Chem Phys* **119**, 4405–4408 (2003).
21. M. S. Shell, P. G. Debenedetti, and A. Z. Panagiotopoulos, *J Chem Phys* **119**, 9406–9411 (2003).
22. T. A. Weber, and F. H. Stillinger, *Phys. Rev. B* **31**, 1954–1963 (1985).
23. W. Kob, and H. C. Andersen, *Phys Rev Lett* **73**, 1376–1379 (1994).
24. F. Sciortino, W. Kob, and P. Tartaglia, *Phys. Rev. Lett.* **83**, 3214–3217 (1999).
25. T. S. Grigera, and G. Parisi, *Phys Rev E* **63**, 045102(R)–1–045102(R)–4 (2001).

26. J. Ghosh, B. Y. Wong, Q. Sun, F. R. Pon, and R. Faller, *Molecular Simulation* **32**, 175–184 (2006).
27. J. Ghosh, and R. Faller, "Modelling of the Glass Transition of Ortho-terphenyl in Bulk and Thin Films," in *Mechanics of Nanoscale Materials and Devices*, edited by A. Misra, J. Sullivan, H. Huang, K. Lu, and S. Asif, MRS, Warrendale, PA, 2006, vol. 924E of *Mater. Res. Soc. Symp. Proc.*, pp. 0924–Z03–21.
28. J. Ghosh, and R. Faller, *J Chem Phys* **125**, 044506–1–044506–11 (2006).
29. E. Lindahl, B. Hess, and D. van der Spoel, *J. Mol. Model.* **7**, 306–317 (2001).
30. S. R. Kuchadkar, and J. M. Wiest, *J Chem Phys* **103**, 8566–8576 (1995).
31. J. Ghosh, and R. Faller, *Molecular Simulation* **33**, 759–767 (2007).
32. D. Reith, H. Meyer, and F. Müller-Plathe, *Comput. Phys. Commun.* **148**, 299–313 (2002).
33. Q. Sun, and R. Faller, *J Chem Theor Comp* **2**, 607–615 (2006).
34. F. Müller-Plathe, *Soft Materials* **1**, 1–31 (2003).
35. R. Faller, *Polymer* **45**, 3869–3876 (2004).
36. R. Faller, *Coarse–Grain Modelling of Polymers*, Wiley-VCH, 2007, vol. 23 of *Rev. Comp. Chem.*, chap. 4, pp. 233–262.
37. Q. Sun, J. Ghosh, and R. Faller, State point dependence and transferability of potentials in systematic structural coarse graining (2007), submitted to Greg Voth: Coarse-Graining of Condensed Phase and Biomolecular Systems, Taylor and Francis.
38. W. Smith, and T. Forester, *J. Molec. Graphics* **14**, 136–141 (1996).
39. J. Ghosh, and R. Faller, A comparative study of the density of states of binary lennard jones glasses in bulk and freestanding films (2007), in preparation.
40. R. Yamamoto, and W. Kob, *Phys Rev E* **61**, 5473–5476 (2000).
41. R. J. Greet, and D. Turnbull, *J. Chem. Phys* **46**, 1243 – 1251 (1967).
42. M. Naoki, and S. Koeda, *J. Phys. Chem* **93**, 948–955 (1989).

Event-Driven Simulation of the Dynamics of Hard Ellipsoids

Cristiano De Michele*, Rolf Schilling†, and Francesco Sciortino*

*Dipartimento di Fisica and INFM-CRS Soft, Università di Roma La Sapienza, P.le A. Moro 2, 00185 Roma, Italy
†Johannes-Gutenberg-Universitat Mainz, D-55099 Mainz, Germany

Abstract. We introduce a novel algorithm to perform event-driven simulations of hard rigid bodies of arbitrary shape, that relies on the evaluation of the geometric distance. In the case of a monodisperse system of uniaxial hard ellipsoids, we perform molecular dynamics simulations varying the aspect-ratio X_0 and the packing fraction ϕ. We evaluate the translational D_{trans} and the rotational D_{rot} diffusion coefficient and the associated isodiffusivity lines in the $\phi - X_0$ plane. We observe a decoupling of the translational and rotational dynamics which generates an almost perpendicular crossing of the D_{trans} and D_{rot} isodiffusivity lines. While the self intermediate scattering function exhibits stretched relaxation, i.e. glassy dynamics, only for large ϕ and $X_0 \approx 1$, the second order orientational correlator $C_2(t)$ shows stretching only for large and small X_0 values. We discuss these findings in the context of a possible pre-nematic order driven glass transition.

Keywords: Computer simulation, Glass transition, Hard ellipsoids, Mode coupling theory, Nematic order
PACS: 64.70.Pf,61.20.Ja,61.25.Em,61.20.Lc

INTRODUCTION

Particles interacting with only excluded volume interaction may exhibit a rich phase diagram, despite the absence of any attraction. Simple non-spherical hard-core particles can form either crystalline or liquid crystalline ordered phases [1], as first shown analytically by Onsager [2] for rod-like particles. Although detailed phase diagrams of several hard-body (HB) shapes can be found in literature [3, 4, 5, 6] less detailed information are available about dynamics properties of hard-core bodies and their kinetically arrested states.

The slowing down of the dynamics of the hard-sphere system on increasing the packing fraction ϕ is well described by mode coupling theory (MCT) [7], but on going from spheres to non-spherical particles, non-trivial phenomena arise, due to the interplay between translational and rotational degrees of freedom. The slowing down of the dynamics can indeed appear either in both translational and rotational properties or in just one of the two. Hard ellipsoids (HE) of revolution [1, 8] are one of the most prominent systems composed by hard body anisotropic particles.

The equilibrium phase diagram, evaluated numerically two decades ago [9], shows an isotropic fluid phase (I) and several ordered phases (plastic solid, solid, nematic N). The coexistence lines show a swallow-like dependence with a minimum at the spherical limit $X_0 = 1$ and a maximum at $X_0 \approx 0.5$ and $X_0 \approx 2$ (cf. Figure 1). Application to HE [10] of the molecular MCT (MMCT) [11, 12] predicts also a swallow-like glass transition line. In addition, the theory suggests that for $X_0 \lesssim 0.5$ and $X_0 \gtrsim 2$, the glass transition is driven by a precursor of nematic order, resulting in an orientational glass where the translational density fluctuations are quasi-ergodic, except for very small wave vectors q. Within MCT, dynamic slowing down associated to a glass transition is driven by the amplitude of the static correlations. Since the approach of the nematic transition line is accompanied by an increase of the nematic order correlation function at $q = 0$, the non-linear feedback mechanism of MCT results in a glass transition, already before macroscopic nematic order occurs [10]. In the arrested state, rotational motions become hindered.

We perform event-driven (ED) molecular dynamics simulations, using a new algorithm [13, 14], that differently from previous ones [15, 16], relies on the evaluation of the distance between objects of arbitrary shape and that will be illustrated shortly.

The outline of the manuscript is as follows. In the next section we illustrate shortly the new algorithm, that we proposed for simulating objects of arbitrary shape. Then in Sec. we illustrate the model used and we give all the details of the simulations we performed. In Sec. we show the results concerning the dynamics of the system investigated. The final section contains our conclusions.

ALGORITHM DETAILS

An Event-driven Algorithm for Hard Rigid Bodies

In an ED simulation the system is propagated until the next event occurs, where an event can be a collision between particles, a cell crossing (if linked lists are used),

CP982, *Complex Systems, 5th International Workshop on Complex Systems*
edited by M. Tokuyama, I. Oppenheim, and H. Nishiyama

etc. All these events must be ordered with respect to time in a calendar and insertion, deletion and retrieving of events must be performed as efficiently as possible.

One elegant approach has been introduced twenty years ago by Rapaport [17], who proposed to arrange all events into an ordered binary tree (that is the calendar of events), so that insertion, deletion and retrieving of events can be done with an efficiency $O(\log N)$, $O(1)$ and $O(\log N)$ respectively, where N is the number of events in the calendar. We adopt this solution to handle the events in our simulation; all the details of this method can be found in [17]. Our ED algorithm can be schematized, as follows:

1. Initialize the events calendar (predict collisions, cells crossings, etc.).
2. Retrieve next event $\mathscr{E}$ and set the simulation time to the time of this event.
3. If the final time has been reached, terminate.
4. If $\mathscr{E}$ is a collision between particles A and B then:
   (a) change angular and center-of-mass velocities of A and B (see [18]).
   (b) remove from the calendar all the events (collisions, cell-crossings) in which A and B are involved.
   (c) predict and schedule all possible collisions for A and B.
   (d) predict and schedule the two cell crossings events for A and B.
5. If $\mathscr{E}$ is a cell crossing of a certain rigid body A:
   (a) update linked lists accordingly.
   (b) remove from calendar all events (collisions, cell-crossings) in which A is involved.
   (c) predict and schedule all possible collisions for A using the updated linked lists.
6. go to step 2.

All the details about linked lists can be found again in Ref. [17], where an ED algorithm for hard spheres is illustrated. We note also that in the case of HE, according to [18], if the elongation of HE is big (i.e. one axes is much greater or much smaller than the others), the linked list method becomes inefficient at moderate and high densities. In fact, given one ellipsoid A, using the linked lists, collision times of A with all ellipsoids in the same cell of A and with all the 26 adjacent cells have to be predicted. In the case of rotationally symmetric ellipsoids, the number of time-of-collision predictions grows as X_0^2, if $X_0 > 1$ and as X_0, if $X_0 < 1$ [18]. To overcome this problem, we developed also a new nearest neighbours list (NNL) method. At the begin of the simulation we build an oriented bounding parallelepiped (OBP) around each ellipsoid and we only predict collisions between ellipsoids having overlapping OBP. In other words given an ellipsoid A, all the ellipsoids having overlapping OBP are the NNL of A. In addition the time-of-collisioni t_i of each ellipsoid with its corresponding OBP is evaluated and the NNL [1] of each HE is rebuilt at the time $t = \min_i\{t_i\}$. The most time-consuming step in the case of hard rigid bodies is the prediction of collisions between HE, hence in the following a short description of the algorithm, that we developed, is given.

Distance between two Rigid Bodies

Our algorithm for predicting the collision time of two rigid bodies relies on the evaluation of the distance between them. The surfaces of two rigid bodies, A and B, are implicitly defined by the following equations:

$$f(\mathbf{x}) = 0 \tag{1a}$$

$$g(\mathbf{x}) = 0 \tag{1b}$$

In passing we note that in the case of uniaxial hard ellipsoids, we have $f(\mathbf{x},t) = {}^t(\mathbf{x}-\mathbf{r}_A(t))X_A(t)(\mathbf{x}-\mathbf{r}_A(t)) - 1$ and a similar expression holds for $g(\mathbf{x})$, where $\mathbf{r}_A$ is the position of the center-of-mass of A and $X_A(t)$ is a matrix, that depends on the orientation of the A (see [18]). It is assumed that, if a point $\mathbf{x}$ is inside the rigid body A (B), then $f(\mathbf{x}) < 0$ ($g(\mathbf{x}) < 0$), while if it is outside $f(\mathbf{x}) > 0$ ($g(\mathbf{x}) > 0$). Then the distance d between these two objects can be defined as follows:

$$d = \min_{\substack{f(\mathbf{x}_A)=0 \\ g(\mathbf{x}_B)=0}} \|\mathbf{x}_A - \mathbf{x}_B\| \tag{2}$$

Equivalently, the distance between these two objects can be defined as the solution of the following set of equations:

$$f_{\mathbf{x}_A} = -\alpha^2 g_{\mathbf{x}_B} \tag{3a}$$

$$f(\mathbf{x}_A) = 0 \tag{3b}$$

$$g(\mathbf{x}_B) = 0 \tag{3c}$$

$$\mathbf{x}_A + \beta f_{\mathbf{x}_A} = \mathbf{x}_B \tag{3d}$$

where $\mathbf{x}_A = (x_A, y_A, z_A)$, $\mathbf{x}_B = (x_B, y_B, z_B)$, $f_{\mathbf{x}_A} = \frac{\partial f}{\partial \mathbf{x}_A}$ and $g_{\mathbf{x}_B} = \frac{\partial f}{\partial \mathbf{x}_B}$. Eqs. (3b) and (3c) ensure that $\mathbf{x}_A$ and $\mathbf{x}_B$ are points on A and B, eq.(3a) provides that the normals to the surfaces are anti-parallel, and eq.(3d) ensures that the displacement of $\mathbf{x}_A$ from $\mathbf{x}_B$ is collinear to the normals of the surfaces. Equations (3) define extremal points of d; therefore, for two general HB these equations can have multiple solutions, while only the smallest one is the actual distance. To solve such equations iteratively is

[1] This is a particular case of the collision between rigid bodies.

therefore necessary to start from a good initial guess of $(\mathbf{x}_A, \mathbf{x}_B, \alpha, \beta)$ to avoid finding spurious solutions.

In addition note that if two HB overlap slightly (i.e. the overlap volume is small) there is a solution with $\beta < 0$ that is a measure of the inter-penetration of the two rigid bodies; we will refer to such a solution as the "negative distance" solution.

Newton-Raphson Method for the Distance

The set of equations (3) can be conveniently solved by a Newton-Raphson (NR) method, as long as first and second derivatives of $f(\mathbf{x}_A)$ and $g(\mathbf{x}_B)$ are well defined. This method, provided that a good initial guess has been supplied, very quickly reaches the solution because of its quadratic convergence [19]. If we define:

$$\mathbf{F}(\mathbf{y}) = \begin{pmatrix} f_{\mathbf{x}_A} + \alpha^2 g_{\mathbf{x}_B} \\ f(\mathbf{x}_A) \\ g(\mathbf{x}_B) \\ \mathbf{x}_A + \beta f_{\mathbf{x}_A} - \mathbf{x}_B \end{pmatrix} \tag{4}$$

Eqs. (3) become:

$$\mathbf{F}(\mathbf{y}) = 0 \tag{5}$$

where $\mathbf{y} = (\mathbf{x}_A, \mathbf{x}_B, \alpha, \beta)$.

Given an initial point $\mathbf{y}_0$ we build a sequence of points converging to the solution as follows:

$$\mathbf{y}_{i+1} = \mathbf{y}_i + \mathbf{J}^{-1}\mathbf{F}(\mathbf{y}_i) \tag{6}$$

where $\mathbf{J}$ is the Jacobian of $\mathbf{F}$,i.e. $\mathbf{J} = \frac{\partial \mathbf{F}}{\partial \mathbf{y}}$.

The matrix inversion, required to evaluate $\mathbf{J}^{-1}$, can be done making use of a standard LU decomposition [19]. This decomposition is of order $N^3/3$, where is N the number of equations (8 in the present case). Finally we note that the set of 8 equations (3) can be also reduced to 5 equations, eliminating $\mathbf{x}_A$ or $\mathbf{x}_B$, using Eq. (3d) .

Prediction of the Time-of-collision

The collision (or contact) time of two rigid bodies A and B is, the smallest time t_c, such that $d(t_c) = 0$, where $d(t)$ is the distance as a function of time between A and B. For finding t_c we perform the following steps:

1. Bracket the contact time using *centroids* (this technique will not be discussed here, see [16, 18] for the details)[2]
2. Overestimating the rate of variation of the distance with respect to time ($\dot{d}(t)$), refine the bracketing of the collision time obtained in 1.
3. Find the collision time to the best accuracy using a Newton-Raphson on a suitable set of equations for the contact point and the contact time.[3]

Bracketing of the Contact Time Overestimating $\dot{d}(t)$.

It can be proved that an overestimate of the rate of variation of the distance is the following:

$$\dot{d}(t) \leq \|\mathbf{v}_A - \mathbf{v}_B\| + \|\mathbf{w}_A\| L_A + \|\mathbf{w}_B\| L_B \tag{7}$$

where the dot indicates the derivation with respect to time, $\mathbf{r}_A$ and $\mathbf{r}_B$ are the centers of mass of the two rigid bodies, $\mathbf{v}_A$ and $\mathbf{v}_B$ are the velocities of the centers of mass, and the lengths L_A, L_B are such that

$$L_A \geq \max_{f(\mathbf{r}')\leq 0} \{\|\mathbf{r}' - \mathbf{r}_A\|\} \tag{8a}$$

$$L_B \geq \max_{g(\mathbf{r}')\leq 0} \{\|\mathbf{r}' - \mathbf{r}_B\|\} \tag{8b}$$

Using this overestimate of $\dot{d}(t)$, that will be called $\dot{d}_{max}$ from now on, an efficient strategy, to refine the bracketing of the contact time, is the following:

1. If t_1 and t_2 bracket the solution, set $t = t_1$.
2. Evaluate the distance $d(t)$ at time t.
3. Choose a time increment Δt as follows:
$$\Delta t = \begin{cases} \frac{d(t)}{\dot{d}_{max}}, & \text{if} d(t) > \varepsilon_d; \\ \frac{\varepsilon_d}{\dot{d}_{max}}, & \text{otherwise.} \end{cases} \tag{9}$$
where $\varepsilon_d \ll \min\{L_A, L_B\}$
4. Evaluate the distance at time $t + \Delta t$.
5. If $d(t+\Delta t) < 0$ and $d(t) > 0$, then $t_1 = t$ and $t_2 = t + \Delta t$, find the collision time/point via NR (see Sec.) and terminate (collision will occur after t_1).
6. if both $0 < d(t+\Delta t) < \varepsilon_d$ and $0 < d(t) < \varepsilon_d$, there could be a "grazing" collision between t and $t+\Delta t$ [18] (distance is first decreasing and then increasing). To look for a possible collision, evaluate the distance $d(t+\Delta t/2)$ and perform a quadratic interpolation of the 3 points ($(t, d(t))$, $(t+\Delta t/2, d(t+\Delta t/2)$, $(t+\Delta t, d(t+\Delta t)$). If the resulting parabola has zeros, set t_1 to the smallest zero (first collision will occur near the smallest zero), and find the collision time/point via NR and terminate (see Sec.).

---

[2] A centroid of a given ellipsoid A with center $\mathbf{r}_A$ is the smallest sphere centered at $\mathbf{r}_A$ that that encloses A

[3] Alternatively you can use any one-dimensional root-finder for the equation $d(t) = 0$.

7. Increment time by $t \to t + \Delta t$.
8. if $t > t_2$ terminate (no collision has been found).
9. Go to step 2.

Finally we note that if the quadratic interpolation fails, the collision will be missed, anyway if ε_d is enough small all "grazing" collisions will be properly handled, i.e. all collisions will be correctly predicted.

Set of Equations to Find the Contact Time and the Contact Point

The contact time, to the best possible accuracy, can be found solving the following equations:

$$f_{\mathbf{x}}(\mathbf{x},t) = -\alpha^2 g_{\mathbf{x}}(\mathbf{x},t) \tag{10a}$$

$$f(\mathbf{x},t) = 0 \tag{10b}$$

$$g(\mathbf{x},t) = 0 \tag{10c}$$

where now f and g depend also on time because the two objects move, and the independent variables are the contact time and the contact point. Again, a good way to solve such a system is using the NR method, very similarly to what we did for evaluating the geometric distance. NR for Eqs. (10) is again very unstable unless a very good initial guess is provided, but the bracketing evaluated using $\dot{d}_{max}$ is sufficiently accurate, provided that ε_d is enough small (typically $\varepsilon_d \lessapprox 1E-4$ is a good choice to give a good initial guess for this NR and to avoid "grazing" collisions).

Finally we want to stress that this algorithm will be exploited fully, when hard bodies of arbitrary shape will be simulated, and that with minor changes it can also be used to simulate hard bodies decorated with attractive spots, as it has been done in the past for the specific case of hard-spheres [20, 21]. Work along these directions is on the way, and in particular a system of super-ellipsoids has been already successfully simulated using the present algorithm.

METHODS

We perform an extended study of the dynamics of monodisperse HE in a wide window of ϕ and X_0 values, extending the range of X_0 previously studied [15]. We specifically focus on establishing the trends leading to dynamic slowing down in both translations and rotations, by evaluating the loci of constant translational and rotational diffusion. These lines, in the limit of vanishing diffusivities, approach the glass-transition lines. We also study translational and rotational correlation functions, to search for the onset of slowing down and stretching in the decay of the correlation. We simulate a system of $N = 512$ ellipsoids at various volumes $V = L^3$ in a cubic box of edge L with periodic boundary conditions. We chose the geometric mean of the axis $l = \sqrt[3]{ab^2}$ as unit of distance, the mass m of the particle as unit of mass ($m = 1$) and $k_B T = 1$ (where k_B is the Boltzmann constant and T is the temperature) and hence the corresponding unit of time is $\sqrt{ml^2/k_B T}$. The inertia tensor is chosen as $I_x = I_y = 2mr^2/5$, where $r = \min\{a,b\}$. The value of the I_z component is irrelevant [22], since the angular velocity along the symmetry (z-) axis of the HE is conserved. We simulate a grid of more than 500 state points at different X_0 and ϕ.To create the starting configuration at a desired ϕ, we generate a random distribution of ellipsoids at very low ϕ and then we progressively decrease L up to the desired ϕ. We then equilibrate the configuration by propagating the trajectory for times such that both angular and translational correlation functions have decayed to zero. Finally, we perform a production run at least 30 times longer than the time needed to equilibrate. For the points close to the I-N transition we check the nematic order by evaluating the largest eigenvalue S of the order tensor $\mathbf{Q}$ [23], whose components are:

$$Q_{\alpha\beta} = \frac{3}{2}\frac{1}{N}\sum_i \langle (\mathbf{u}_i)_\alpha (\mathbf{u}_i)_\beta \rangle - \frac{1}{3}\delta_{\alpha,\beta} \tag{11}$$

where $\alpha\beta \in \{x,y,z\}$, and the unit vector $(\mathbf{u}_i(t))_\alpha$ is the component α of the orientation (i.e. the symmetry axis) of ellipsoid i at time t. The largest eigenvalue S is non-zero if the system is nematic and 0 if it is isotropic. In the following, we choose the value $S = 0.3$ as criteria to separate isotropic from nematic states.

RESULTS AND DISCUSSION

Isodiffusivity Lines

From the grid of simulated state points we build a corresponding grid of translational (D_{trans}) and diffusional (D_{rot}) coefficients, defined as:

$$D_{trans} = \lim_{t\to+\infty} \frac{1}{N}\sum_i \frac{\langle \|\mathbf{x}_i(t) - \mathbf{x}_i(0)\|^2 \rangle}{6t} \tag{12}$$

$$D_{rot} = \lim_{t\to+\infty} \frac{1}{N}\sum_i \frac{\langle \|\Delta\Phi_i\|^2 \rangle}{4t} \tag{13}$$

where $\Delta\Phi_i = \int_0^t \omega_i dt$, $\mathbf{x}_i$ is the position of the center of mass and ω_i is the angular velocity of ellipsoid i. By proper interpolation, we evaluate the isodiffusivity lines, shown in Fig. 1. Results show a striking decoupling of the translational and rotational dynamics. While the translational isodiffusivity lines mimic the swallow-like shape of the coexistence between the isotropic liquid and the crystalline phases (as well as the MMCT prediction for the glass transition [10]), rotational isodiffusivity lines reproduce qualitatively the shape of the I-N coexistence. As a consequence of the the swallow-like shape, at large fixed ϕ, D_{trans} increases by increasing the particle's anisotropy, reaching its maximum at $X_0 \approx 0.5$ and $X_0 \approx 2$. Further increase of the anisotropy results in a decrease of D_{trans}. For all X_0, an increase of ϕ at constant X_0 leads to a significant suppression of D_{trans}, demonstrating that D_{trans} is controlled by packing. The iso-rotational lines are instead mostly controlled by X_0, showing a progressive slowing down of the rotational dynamics independently from the translational behavior. This suggests that on moving along a path of constant D_{trans}, it is possible to progressively decrease the rotational dynamics, up to the point where rotational diffusion arrest and all rotational motions become hindered. Unfortunately, in the case of monodisperse HE, a nematic transition intervenes well before this point is reached. It is thus stimulating to think about the possibility of designing a system of hard particles in which the nematic transition is inhibited by a proper choice of the disorder in the particle's shape/elongations. We note that the slowing down of the rotational dynamics is consistent with MMCT predictions of a nematic glass for large X_0 HE [10], in which orientational degrees of freedom start to freeze approaching the isotropic-nematic transition line, while translational degrees of freedom mostly remain ergodic.

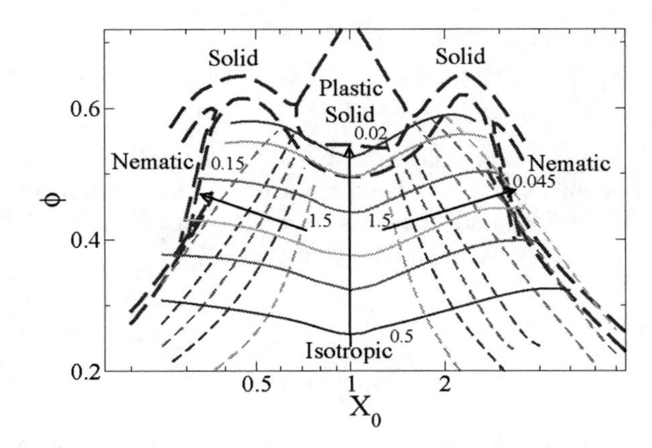

FIGURE 1. [Color online] Isodiffusivity lines. Solid lines are isodiffusivity lines from translational diffusion coefficients D_{trans} and dashed lines are isodiffusivities lines from rotational diffusion coefficients D_{rot}. Arrows indicate decreasing diffusivities. Left and right arrows refer to rotational diffusion coefficients. Diffusivities along left arrow are: 1.5, 0.75, 0.45, 0.3, 0.15. Diffusivities along right arrow are: 1.5, 0.75, 0.45, 0.3, 0.15, 0.075, 0.045. Central arrow refers to translational diffusion coefficients, whose values are: 0.5, 0.3, 0.2, 0.1, 0.04, 0.02. Thick Long-dashed curves are coexistence curves of all first order phase transitions in the phase diagram of HE evaluated by Frenkel and Mulder (FM) [9] Solid lines are coexistence curves for the I-N transition of oblate and prolate ellipsoids, obtained analytically by Tijpto-Margo and Evans [6] (TME).

Orientational Correlation Function

To support the possibility that the slowing down of the dynamics on approaching the nematic phase originates from a close-by glass transition, we evaluate the self part of the intermediate scattering function $F_{self}(q,t) = \frac{1}{N}\langle \sum_j e^{i\mathbf{q}\cdot(\mathbf{x}_j(t)-\mathbf{x}_j(0))}\rangle$ and the second order orientational correlation function $C_2(t)$ defined as [15] $C_2(t) = \langle P_2(\cos\theta(t))\rangle$, where $P_2(x) = (3x^2-1)/2$ and $\theta(t)$ is the angle between the symmetry axis at time t and at time 0. The $C_2(t)$ rotational isochrones are found to be very similar to rotational isodiffusivity lines.

These two correlation functions never show a clear two-step relaxation decay in the entire studied region, even where the isotropic phase is metastable, since the system can not be significantly over-compressed. As for the well known hard-sphere case, the amount of over-compressing achievable in a monodisperse system is rather limited. This notwithstanding, a comparison of the rotational and translational correlation functions reveals that the onset of dynamic slowing down and glassy dynamics can be detected by the appearance of stretching.

We note that F_{self} shows an exponential behaviour close to the I-N transition ($X_0 = 3.2,\ 0.3448$) on the prolate and oblate side, in agreement with the fact that translational isodiffusivities lines do not exhibit any peculiar behaviour close to the I-N line [14]. Only when $X_0 \approx 1$, F_{self} develops a small stretching, consistent with the minimum of the swallow-like curve observed in the fluid-crystal line [24, 25], in the jamming locus as well as in the predicted behavior of the glass line for HE [10] and for small elongation dumbbells [26, 27]. Opposite behavior is seen for the case of the orientational correlators. C_2 shows stretching at large anisotropy, i.e. at small and large X_0 values, but decays within the microscopic time for almost spherical particles. In this quasi-spherical limit, the decay is well represented by the decay of a free rotator [28, 14]. Previous studies of the rotational dynamics of HE [15] did not report stretching in C_2, probably due to the smaller values of X_0 previously investigated and to the present increased statistic which allows us to follow the full decay of the correlation functions.

In summary C_2 becomes stretched approaching the I-N transition while F_{self} remains exponential on approaching the transition. To quantify the amount of stretching in C_2, we fit it to the function $A\exp[-(t/\tau_{C_2})^{\beta_{C_2}}]$ (stretched exponential) for several state points and we show in Fig.

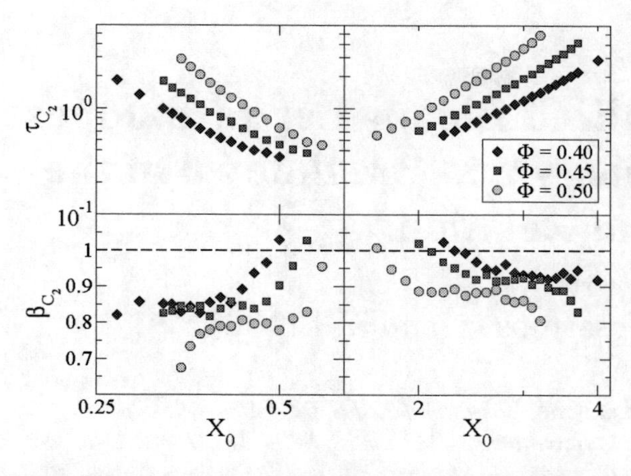

FIGURE 2. [Color online] β_{C_2} and τ_{C_2} are obtained from fits of C_2 to a stretched exponential for $\phi = 0.40, 0.45$ and 0.50. Top: τ_{C_2} as a function of X_0. Bottom: β_{C_2} as a function of X_0. The time window used for the fits is chosen in such a way to exclude the microscopic short times ballistic relaxation. For $0.588 < X_0 < 1.7$ the orientational relaxation is exponential.

2 the X_0 dependence of τ_{C_2} and β_{C_2} for three different values of ϕ. In all cases, slowing down of the characteristic time and stretching increases progressively on approaching the I-N transition.

CONCLUSIONS

In summary we have applied a novel algorithm for simulating hard bodies to investigate the dynamics properties of a system of monodisperse HE and we have shown that clear precursors of dynamic slowing down and stretching can be observed in the region of the phase diagram where a (meta)stable isotropic phase can be studied. Despite the monodisperse character of the present system prevents the possibility of observing a clear glassy dynamics, our data suggest that a slowing down in the orientational degrees of freedom — driven by the elongation of the particles — is in action. The main effect of this shape-dependent slowing down is a decoupling of the translational and rotational dynamics which generates an almost perpendicular crossing of the D_{trans} and D_{rot} isodiffusivity lines. This behavior is in accordance with MMCT predictions, suggesting two glass transition mechanisms, related respectively to cage effect (active for $0.5 \lessapprox X_0 \lessapprox 2$) and to pre-nematic order ($X_0 \lessapprox 0.5$, $X_0 \gtrapprox 2$) [10].

ACKNOWLEDGMENTS

We acknowledge support from MIUR-PRIN.

REFERENCES

1. M. P. Allen, "Liquid Crystal Systems," in *Computational Soft Matter: From Synthetic Polymers to Proteins*, edited by N. Attig, K. Binder, H. Grubmüller, and K. Kremer, John von Neumann Institute for Computing, 2004, vol. 23, pp. 289–320.
2. L. Onsager, *Ann. N. Y. Acad. Sci.* **51**, 627–659 (1949).
3. J. D. Parsons, *Phys. Rev. A* **19**, 1225–1230 (1979).
4. S.-D. Lee, *J. Chem. Phys.* **89**, 7036–7037 (1989).
5. A. Samborski, G. T. Evans, C. P. Mason, and M. P. Allen, *Mol. Phys.* **81**, 263–276 (1994).
6. B. Tijpto-Margo, and G. T. Evans, *J. Chem. Phys.* **93**, 4254–4265 (1990).
7. W. Götze, "Aspects of Structural Glass Transition," in *Liquids, Freezing and the Glass Transition*, edited by J. P. Hansen, D. Levesque, and J. Zinn-Justin, North-Holland, 1991, pp. 287–499.
8. G. S. Singh, and B. Kumar, *Ann. Phys.* **294**, 24–47 (2001).
9. D. Frenkel, and B. M. Mulder, *Molec. Phys.* **55**, 1171–1192 (1991).
10. M. Letz, R. Schilling, and A. Latz, *Phys. Rev. E* **62**, 5173–5178 (2000).
11. T. Franosch, M. Fuchs, W. Götze, M. R. Mayr, and A. P. Singh, *Phys. Rev. E* **56**, 5659–5674 (1997).
12. R. Schilling, and T. Scheidsteger, *Phys. Rev. E* **56**, 2932–2949 (1997).
13. C. De Michele, A. Scala, R. Schilling, and F. Sciortino, *J. Chem. Phys.* **124**, 104509–1–104509–7 (2006).
14. C. De Michele, R. Schilling, and F. Sciortino, *Phys. Rev. Lett.* **98**, 265702–1–265702–4 (2007).
15. M. P. Allen, and D. Frenkel, *Phys. Rev. Lett.* **58**, 1748–1750 (1987).
16. A. Donev, F. H. Stillinger, and S. Torquato, *J. Comp. Phys* **202**, 737–764 (2005).
17. D. C. Rapaport, *The Art of Molecular Dynamics Simulation*, Cambridge University Press, 2004.
18. A. Donev, F. H. Stillinger, and S. Torquato, *J. Comp. Phys* **202**, 765–793 (2005).
19. W. H. Press, B. P. Flannery, S. A. Teukolsky, and W. T. Vetterling, *Numerical Recipes in Fortran*, Cambridge University Press, 1999.
20. C. De Michele, S. Gabrielli, P. Tartaglia, and F. Sciortino, *J. Phys. Chem. B* **110**, 8064–8079 (2006).
21. C. De Michele, P. Tartaglia, and F. Sciortino, *J. Chem. Phys.* **125**, 204710–1–204710–8 (2006).
22. M. P. Allen, D. Frenkel, and J. Talbot, *Molecular Dynamics Simulations Using Hard Particles*, North-Holland, 1989.
23. S. C. McGrother, D. C. Williamson, and G. Jackson, *J. Chem. Phys.* **104**, 6755–6771 (1996).
24. P. N. Pusey, and W. van Megen, *Phys. Rev. Lett.* **59**, 2083–2086 (1987).
25. W. G. Hoover, and F. H. Ree, *J. Chem. Phys* **49**, 3609–3617 (1968).
26. S.-H. Chong, and W. Götze, *Phys. Rev. E* **65**, 041503–1–041503–17 (2002).
27. S.-H. Chong, A. J. Moreno, F. Sciortino, and W. Kob, *Phys. Rev. Lett.* **94**, 215701–1–215701–4 (2005).
28. C. Renner, H. Löwen, and J. L. Barrat, *Phys. Rev. E* **52**, 5091–5099 (1995).

Molecular Dynamics of Generalized Binary Lennard-Jones Systems: Effects of Anharmonicity and Breakdown of the Stokes-Einstein Relation

J. Habasaki*, F. Affouard**, M. Descamps**, and K. L. Ngai***

*Tokyo Institute of Technology, 4259 Nagatsuta-cho, Yokohama 226-8502, JAPAN
**Laboratoire de Dynamique et Structure des Materiaux Moleculaires, UMR CNRS 8024, Universite Lille 1, 59655, Villeneuve d'Ascq Cedex FRANCE
*** Naval Research Laboratory, Washington DC 20375-5320, USA

Abstract. We have performed molecular dynamics simulations (MD) of three glass-forming binary Lennard-Jones (LJ) mixtures in order to clarify the mechanism responsible for the breakdown of the Stokes-Einstein relation and the role of the anharmonicity of the interaction potential in determining the structural and dynamical properties of glasses. The changes in the potential parameter of one of the species of the mixture are found to affect the behaviours of the other species. Similar correlations between fractal dimension of random walks and diffusivity has been found for all systems and species. This means that the change in geometrical correlations among successive motion is a dominant determining factor of the dynamics. At high temperatures, the mutual interception of paths in a certain time scale occurs already for the more harmonic system, while this kind of events is rare in the anharmonic systems. These differences are responsible to the different diffusivity and coupling of the systems studied. The mechanisms are similar to that giving rise to the mixed alkali effect of ion transport in glasses, where the interception of the jump paths plays the dominant role in the slowing down of the dynamics.

Keywords: Binary Lennard-Jones, Glass transition, Similarity to the mixed alkali effect
PACS: 61.43.Hv, 64.70.Pf, 66.20.+d

INTRODUCTION

Despite a great interest in the past several decades as evidenced by the developments of many models and theories, the glass transition problem remains elusive. A complete and satisfactory understanding of the glassy state and the dynamics leading to glass formation has not emerged.

Diffusion coefficient D or shear viscosity η are key parameters to follow the evolution of the transport properties of glass-formers from high temperatures (where the structural relaxation time $\tau\approx$ pico-nanosecond, $\eta \approx 10^{-2}$ Poise) to low temperatures close to T_g (where $\tau\approx 10^2$ s, $\eta \approx 10^{13}$ Poise). Several experimental studies have determined the diffusion coefficients and the shear viscosity of various glass-formers, and these works have found deviations from the Stokes-Einstein relation at temperatures below about 1.2 times T_g. In this temperature range other characteristic dynamical properties also emerge: stretched exponential time dependence of the relaxation, $\exp[-(t/\tau)^{\beta}]$, non-Arrhenius behaviour of the relaxation times. The breakdown of the Stokes-Einstein relation and the Debye-Stokes-Einstein relation (*i.e.*, the decoupling of the translational from the rotational dynamics) are still not completely understood although explanations based on spatial dynamical heterogeneities and different coupling parameters have been proposed.

We have performed molecular dynamics simulations (MD) of glass-forming binary Lennard-Jones (LJ) mixtures in order to clarify the mechanisms of the slowing down of the dynamics, breakdown of the Stokes-Einstein relation and the role of the anharmonicity of the interaction potential on different structural and dynamical properties of glasses. In this paper, the dynamics of the particles near the glass transition temperature are determined as function of the waiting time distribution (temporal term) and effective jump distance (spatial term) resulting from the correlation among jumps. Contribution of these two terms in the dynamics has been separated using the fractal dimension analysis of trajectories.

CP982, *Complex Systems, 5th International Workshop on Complex Systems*
edited by M. Tokuyama, I. Oppenheim, and H. Nishiyama

METHODS

Molecular Dynamics Simulations (MD) has been performed for binary Lennard-Jones (LJ) particles systems composed of 1500 particles (1200 species A and 300 species B) with three different interaction potentials, characterized by different values of q and p of type A in the following expression.

$$V(r) = \frac{E_0}{(q-p)}\left(p\left(\frac{r_0}{r}\right)^q - q\left(\frac{r_0}{r}\right)^p\right) \quad (1)$$

In the model I, q=12 and p=11. In the model III, q=8 and p=5. The parameters for q=12 and p=6 for model II corresponds to the well known Kob-Anderson model [1]. The anharmonicity of the potential for A-A interaction is increasing in the order of I (12-11), II (12-6) and III (8-5). Details for computational methods and models are given in previous publications [2, 3]. In these works, the onset temperature of the non-exponential time dependence and non-Arrhenius relaxation, T_A, and the critical temperature of the mode-coupling theory, T_c, have been determined. In reduced units, T_A is 1.7, 1.0 and 0.5, and T_c is 0.695, 0.435 and 0.262 for systems I, II and III respectively. The 'fragility' obtained from several methods is increasing in order of I, II and III [2].

TABLE 1. Parameters of the Lennard-Jones potentials in the Kob-Andersen model [1]. $(\sigma = \frac{r_0}{2^{1/6}})$.

Interaction	A-A	B-B	A-B
E_0	1.0	0.5	1.5
σ	1.0	0.88	0.8

Fractal dimension of random walks, d_w, which is a measure of the complexity of the motion, is determined as follows. When the length of trajectories obtained by simulation after a run of time t_{run} is measured by using a divider of length L, N is the number of times required to cover all the trajectories. The dimension, d_w, is defined by

$$N = AL^{-d_w} \quad , \quad (2)$$

where, A is a constant.

Viscosity of the system was determined from the fluctuations of the stress tensors for the same runs used to calculate other properties by equilibrium molecular dynamics simulations. The results for system II are consistent with the previous results [4] obtained by non-equilibrium molecular dynamics simulations.

RESULTS AND DISCUSSION

Difference in the Diffusivity

Changes of the diffusion coefficients in these three models are shown in Figure 1 as a function of the inverse of the temperature. Not only the diffusivity of type A particles, D_A, but also the diffusivity of the type B particles, D_B, also changes with the modification of A-A interactions. The D_B increases in the order of I, II and III. The difference of the diffusivity of A and B is also decreasing in this order.

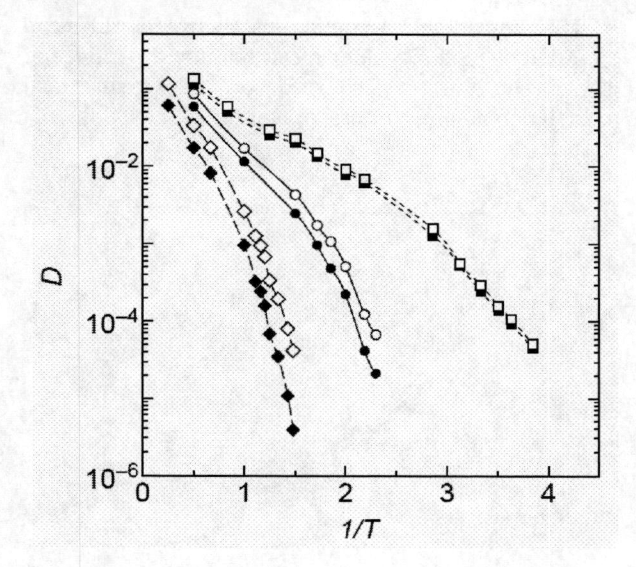

FIGURE1. Temperature dependence of the diffusion coefficients in the three models. Diamonds:model I, Circles: model II and Squares: model III. Filled symbols are for A particles and the corresponding open symbols are for B particles. Curves are for a guide of eyes.

Temporal and Spatial Aspects of the Dynamics

We have determined the fractal dimension of the random walks for trajectories of type A and B particles over wide temperature regions. The dimensions d_{w1} and d_{w2} are obtained for shorter (L<1) and longer ($1 \leq L<3$) length scale regions, respectively. The value is expected to be 2 for a free random walk and becomes larger when the trajectory becomes complicated due to localized motions. When d_w equals to 1, the particle is moving on a straight line. The value can be larger than 3 due to folding of trajectories and this means strong localization.

The value d_{w1} of type A is always larger than B at all temperatures for all models. So that, type A is more localized than type B is.

We have plotted the diffusion coefficients against d_{w1} values in Figure 2. With decreasing temperature, the d_{w1} value increases and there are remarkable correlations between diffusion coefficients D and the values of d_{w1}. Furthermore, this correlation holds regardless the difference of models and type of particles. That is, almost data points for both A and B for all models at all temperatures are found to collapse onto a master curve. Similar trend is also observed for the plot of diffusion coefficient D against d_{w2} as shown Figure 3.

The remarkable correlations between diffusion coefficients and d_w values mean that the diffusivity of these particles is governed by the geometrical correlation among successive motions.

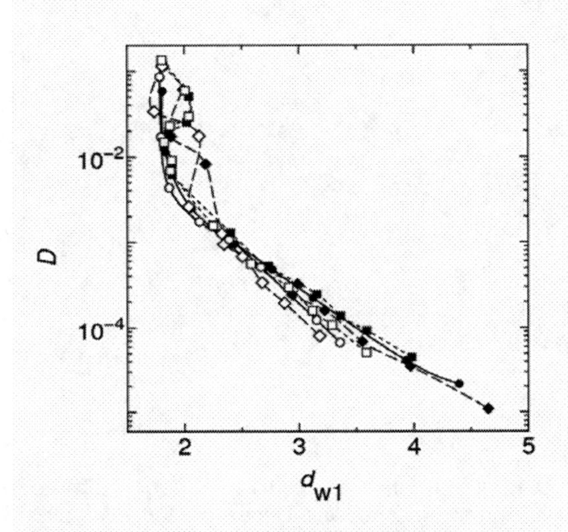

FIGURE 2. Diffusion coefficients are plotted against fractal dimension of the random walk for a short length scale region, dw 1. Symbols are the same as in Fig. 1.

That is, the slowing down of the diffusivity is directly related to the direction of motions. Thus the main causes of changes in diffusivity in type B are also geometrical and d_w is a good measure of the changes in the diffusivity. Since the power law exponent of the MSD, θ corresponds to $2/d_w$, this description is equivalent to the characterization by the changes in the exponent of the power law, if the temporal contribution of the dynamics can be neglected. Then, this can also be compared to the changes in the exponents in the stretched exponential relaxation, although the larger temporal effect is expected in the relaxation function compared with MSD. There are small systematic deviations between behaviors of type A and type B particles in Figure 3, which suggests contribution of temporal term for acceleration of type B particles in each system. Temporal acceleration of the dynamics concerned with forward correlated motion and accompanied with cooperative jumps has been observed for the dynamics of ions in the lithium silicate glass [5].

Evidence of Interception of Paths Observed in the Distinct Part of the van Hove functions

The large contribution of the geometrical effect can be also confirmed by the distinct part of the van Hove functions. The function was examined to learn more of the changes in the possible connection between the paths for each i and j (A-A, B-B and A-B) pairs.

$$G_d^{\alpha,\beta}(\vec{r},t)=\frac{1}{N_\alpha}\sum_{i=1}^{N_\beta}\sum_{j=1}^{N_\beta}\left\langle \vec{r}-\vec{r}_i^{\,\alpha}(0)+\vec{r}_j^{\,\beta}(t)\right\rangle \quad , \quad (3)$$

where in the summations the self-term $i=j$ is to be left out if $\alpha=\beta$. N_α and N_β are the number of particles of species α and β, respectively.

When the motion consists of clear jumps, a new peak, appears at around $r=0$, where the type i particles are located at $t=0$.

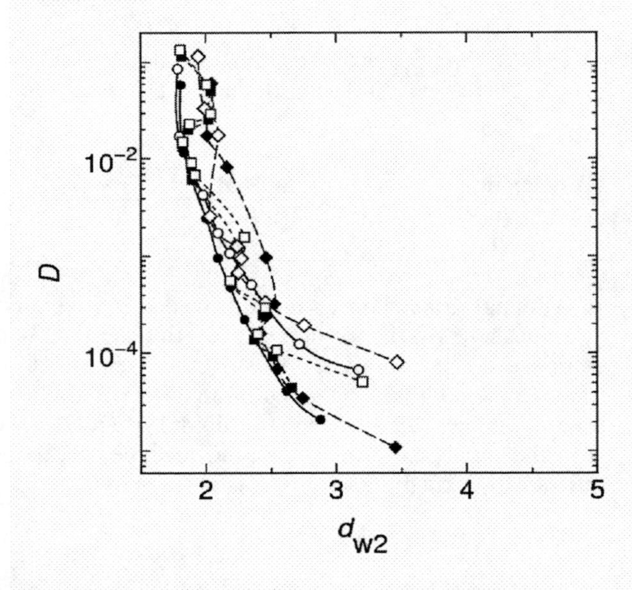

FIGURE 3. Diffusion coefficients are plotted against fractal dimension of the random walk for a long length scale region, d_{w2}. Marks are the same as in Fig. 1.

The function reveals that the jump motion of particles begins at higher temperature in the order of I, II and III. With decreasing temperatures, jump motion becomes clearer.

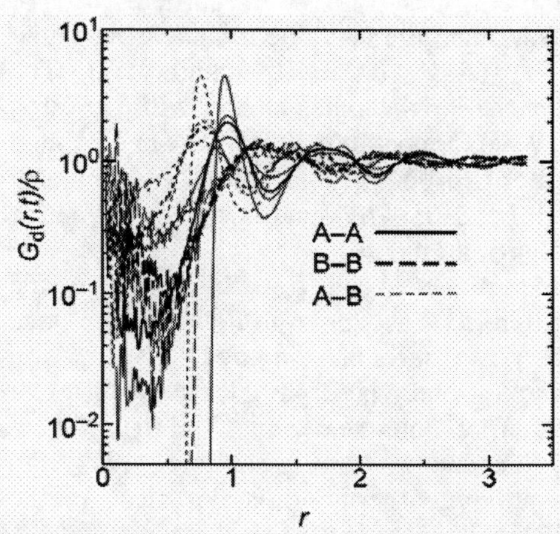

FIGURE 4. Distinct part of the van-Hove function for system I at T=1. t=0, 46.4, 92.9, 185.7 for upper to lower in the original peaks for each pairs. The contribution at around r=0 is due to jumps occurred during these times. Thus the jump paths are intercepting each other in this time scale and this resulted in an increase of d_w values. In this figure, $r = \|\vec{r}\|$.

Stronger localization of type A in the model I is a natural result of the harmonic type interaction potential. Due to the localization, A-A jump and A-B jump is already rare at high temperature region in model I as found in the distinct part of the van Hove functions. The long life time of the structures concerning with A-A and A-B pairs means that the possible paths of type B are cut by type A particles in a certain time scale and this resulted in a suppression of the motion of B. Contribution of this suppression of motion of B to the decrease of diffusivity is large, because of the long length scale. That is, the behavior of type B for long length scale is affected by the immobile particles. Therefore, the diffusivity of B is the lowest in model I The suppression is not complete due to the motion concerning with the changes in A-B pairs and possible acceleration of B-B motions by the cooperativity among B type ions in their own paths, and therefore, the motion of type B in the model I is most decoupled from the motion of type A. These features are consistent with the observation in d_{w1} and d_{w2}.

The function for every pair becomes rapidly flat in model III at this temperature region. Therefore, motion is not a clear jump in model III and there is no clear caging in the time scales comparable to that shown in Figure 4. In this case, concept of the path is not meaningful. The type A particle can visit the place previously occupied by type B and *vice versa.* With lowering temperature, jump character becomes clearer and mutual interception of paths occurs. In Fig. 1, the data are plotted against the $1/T$. Actual temperature dependence of the mobility for the model III expressed as the activation energy is not necessarily the largest even in the low temperature region. The steep slope has been observed when the $1/D$ is plotted against T_{ref}/T [2], where T_{ref} is the reference temperature near T_g. The steeper slope in low temperature region is concomitant with the slower decrease of D at high temperature region, since the range of the diffusivity is almost fixed in the plot.

These changes in dynamics may be described by those concerning with the waiting time distribution as proposed in several models [6,7], since the localization in a certain region can be regarded as the waiting time of the jumps. However, in our explanation, origin of the slowing down of B following to the localization of type A ions is essentially geometrical and the effect of localization propagates to the other particles through the cooperative character of the motions. Of course, mean jump rate is a function of temperature and this can contribute to the changes in the dynamics with temperatures; however, the difference of motion of type A and B or the difference among models can be well explained by the changes in the geometrical term, that is d_w.

This mechanism of the slowing down due to the interception of path is not necessarily limited to the mixture such as the binary systems. Even for a single component system, it can work when the dynamics are dynamically heterogeneous.

It is worth to note that these findings in LJ systems are similar to that observed in the MD simulation of silicate glasses including mixed alkali effect (MAE) [8], where the suppression of dynamics of the host alkali ions by foreign alkali ions are explained by the interception of the jump paths, whereby the long range (cooperative) jumps are suppressed considerably. For the slowing down of the dynamics in the mixed alkali system (MAE), we have noted that the effect should be common to other systems with and without framework structures. Similarity observed in LJ systems, this clearly means that the existence of the chain structures is not essential for the slowing down, although it can play a role in keeping the structure of paths for long times. Small number of immobile ion affects the mobility of majority ions where the longer scale motion is suppressed. In MAE, the effect is larger when the size ratio of alkali ions is larger.

When we compared models I, II and III, anharmonicity of the interaction of A-A increases in this order. The additional works to distinguish the role of anharmonicity and the size of particles are in progress since the repulsive parts of the potential functions are also changed by the modification of potential parameters.

Breakdown of Stokes-Einstein Laws

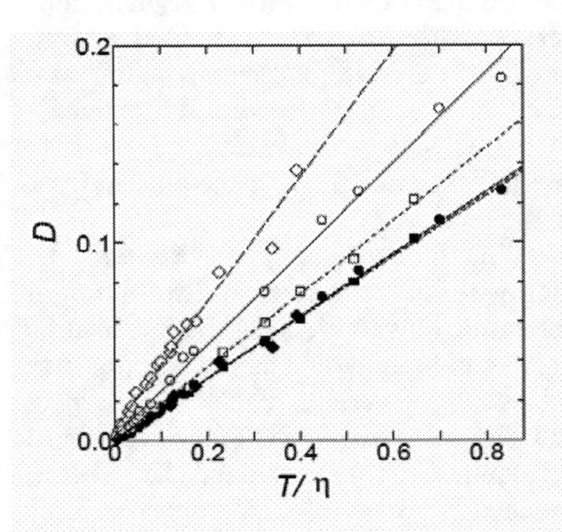

FIGURE 5. Diffusion coefficients are plotted against T/ η. Marks are the same as in Fig. 1.

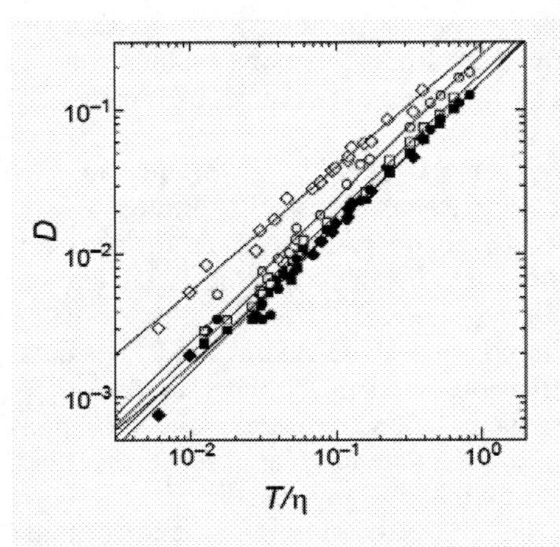

FIGURE 6. Diffusion coefficients are plotted against T/ η in a double logarithmic plot. Symbols are the same as in Fig. 1 .

It is known that the deviation of the Stokes-Einstein law occurs near the glass transition temperature in several experiments or in simulations [9-11]. We have checked the validity of the Stokes Einstein law. In the stick boundary conditions,

$$D = \frac{k_B T}{3\pi\eta d} \quad . \tag{4}$$

In Figure 5, diffusion coefficients of both type of particles in three systems are plotted against T/ η, A nearly linear line is observed in the figure and therefore the law apparently seems to hold for each species. From the law, we have calculated the hydrodynamic diameter *d*. Lines for A particles of three models are almost coincident with each other, regardless the different mobility. The values for d_A of I, II and III obtained from the slopes are (0.69, 0.68 and 0.68), while the slopes for type B particles do not coincident each other and d_B value is increasing in the order of I, II and III (0.33, 0.46 and 0.57). The deviation from the law becomes clearer with lowering temperature. When temperature is lower than T_A The d values in model II drops [4] and similar trend is also found for both I and III. The trend is also observed for the ratio d_A /d_B for all models.

When the data in Figure 5 are presented in the double logarithm plots as shown in Figure 6, we have found that the power laws are the good representation of the data. Type A and type B particles show different slopes and therefore, both the d values and the ratio are temperature dependent and saturate at high temperature, although the trend is not clear in the linear scale due to small values of *D*.

That is, the following fractional relation holds for each species.

$$D = A'(\frac{k_B T}{\eta})^{\gamma} \quad , \tag{5}$$

where A' is a constant.

The slopes in the Figure 6 are nearly 1 for type A particles, which is more localized than type B. Slopes for the type B particles depend on the models. The smallest slope is found in model I, where the motion is most decoupled from the motion of type A particles. That is, diffusion dynamics for faster type B particles deviate from that expected from the viscosity according to the law, when the motion is decoupled. In Figure 6, behaviors of A type particles in three models can be represented by the almost the same line in the double logarithmic plot and hence this linear power law seems to be reasonable to represent the general feature of the dynamics.

In several experiments or in simulations of glass forming liquids, the fractional-Stokes-Einstein law in a following form is known to hold,

$$D \propto \eta^{-\xi}, \tag{6}$$

This relation also approximately holds for each species. Further study for the lower temperature regimes in these models and other systems will be necessary to judge the validity of equation (5) or (6).

CONCLUSION

Generalized binary Lennard-Jones models have been examined by MD simulations. Modification of the potential parameters for A-A interactions resulted in the modification of the mobility of type B particles. Diffusion coefficients of both types of particles of three models are found to collapse onto a master curve, when they are plotted against fractal dimension of random walks. That is, the coefficients are governed by the geometrical correlations among successive motions. The changes due to the interception of the paths for a certain time scale are also observed in the distinct part of the van Hove functions.

Stokes-Einstein law approximately holds at high temperature regions. Deviation occurs at lower temperatures, where the fractional power law is more suitable description of the dynamics.

In model I, immobile particles are introduced by the more harmonic potentials of A-A interactions. In this case, interception of the jump paths for a certain time scales by the immobile particles causes the slowing down of type B particles. However, the motion is not completely suppressed and therefore decoupled from the diffusivity of A type particles. Diffusivity of B can be larger than that expected from the viscosity of the whole system. When the decoupling is large, the situation is similar to ions in a high conductivity glass embedded in a framework with much longer relaxation time.

In model III, the potential for A-A is more anharmonic and the motion of type A and B is more coupled with each other. The deviation from the Stokes-Einstein is the smallest in this case. The values d_w for B is similar to d_w of A and therefore D_A and D_B are also similar. The effect of the interception of the path is the minimum in this case at a higher temperature region.
Situation for the model II is between I and III.

ACKNOWLEDGMENTS

This research was partly supported by the Ministry of Education, Science, Sports and Culture, Japan, Grant-in-Aid for Scientific Research (C), 19540396, 2007-2009.
The work performed at the Naval Research Laboratory was supported by the Office of Naval Research.
Some of the authors (F. A. and M. D.) wish to acknowledge the use of the facilities of the IDRIS (Orsay, France) and the CRI (Villeneuve d'Ascq, France) where calculations were carried out. This work was supported by the INTERREG III (FEDER) program (Nord-Pas de Calais/Kent).

REFERENCES

1. W. Kob and H. C. Andersen, *Phys. Rev. E***51**, 4626-4641 (1995).
2. P. Bordat, F. Affouard, M. Descamps and K. L. Ngai, *Phys. Rev. Lett.*, **93**, 105502(1-4) (2004).
3. P. Bordat, F. Affouard, M. Descamps and K. L. Ngai, *J. Non-Cryst. Solids,* **352**, 4630-4634. (2006).
4. P. Bordat, F. Affouard, M. Descamps and F. Mueller-Plathe, *J. Phys. Cond. Matter,* **15**, 5397-5407 (2003).
5. J. Habasaki and Y. Hiwatari, *Phys. Rev. E***59**, 6962-6966 (1999).
6. T. Odagaki and Y. Hiwatari, *Phys. Rev. A***41**, 929-937 (1990); A. Yoshimori and T. Odagaki, *J. Phys. Soc. Jpn,* **74**, 1206-1213 (2005).
7. F. W. Schmidlin, *Phys. Rev. B***16**, 2362-2385 (1977).
8. J. Habasaki, I. Okada and Y. Hiwatari, *J. Non. Cryst. Solids,* **183,** 12-21(1995); J. Habasaki and K. L. Ngai, *Phys. Chem. Chem. Phys.*, **9**, 4673-4689 (2007).
9. E. Rössler, *Phys. Rev. Lett.*, 65, 1595-1598 (1990).
10. L. Andreozzi, A. Di Schino and D. Leoporini, *Europhys. Lett.* **38,** 669-674(1997).
11. K. L. Ngai, *J. Phys. Chem. B,* **110,** 26211-26214 (2006).

Diffusion, Structural Relaxation and Rheological Properties of a Simple Glass Forming Model: A Molecular Dynamics Study

Fathollah Varnik

Max-Planck Institut für Eisenforschung, Max-Planck Straße 1, 40237 Düsseldorf, Germany
e-mail: f.varnik@mpie.de

Abstract. Via large scale molecular dynamics simulations, we study diffusion in melts undergoing strong shear flow and its relation to the rheological response in a well established glass forming model system, namely the 80:20 binary Lennard-Jones system first introduced by Kob and Andersen [W. Kob and H.C. Andersen, PRL **73**, 1376 (1994)]. In previous works [F. Varnik JCP **125**, 164514 (2006) and F. Varnik and O. Henrich PRB **73**, 174209 (2006)], the interplay between the dynamics of structural relaxation on the length scale of the average interparticle distance and the stress response of the model was studied. Here we focus on the large scale dynamics under homogeneous shear by evaluating the time dependence of the mean square displacements for temperatures ranging from the supercooled state to far below the mode coupling critical temperature of the model. Particularly long simulations are performed allowing an accurate determination of the diffusion constant. For low temperatures and at not too high shear rates, the mean square displacements exhibit the well known two step relaxation behavior with a long time diffusive motion along the spatial directions perpendicular to the flow. In the flow direction, on the other hand, a third regime follows the diffusive motion, where Taylor dispersion with the typical t^3 time dependence clearly dominates the long time behavior of the particle displacements. At the lowest studied temperatures, the cross over from the diffusive regime to the regime where the contribution of Taylor dispersion becomes significant, occurs at length scales of the order of a particle diameter but is shifted towards progressively larger displacements as temperature increases.

Keywords: Diffusion, Driven glassy systems, Non equilibrium molecular dynamics, Structurall relaxation
PACS: 64.70.Pf,05.70.Ln,83.60.Df,83.60.Fg

INTRODUCTION

Suspensions of spherical colloidal particles are ubiquitous in nature, every day life and industrial applications and thus represent an important class of soft materials. They are soft in the sense that often a small perturbation is sufficient in order to significantly disturb their structure and dynamics [1].

While a metallic solid can resist a load of millions of Pascal, it is often sufficient to shake a colloidal solid in order to make it flow [2]. This forced change of the physical state of the material is called 'shear melting'. In addition to this shear melting, there appears another interesting phenomena, that of 'shear thinning', a decrease of the shear viscosity upon an increase of the applied shear rate [3, 4].

Interestingly, an exactly opposite behavior, namely an *increase* of viscosity with increasing shear rate is also observed when studying the rheology of soft materials [5, 6]. This phenomena is generally related to hydrodynamic effects. In the present work, however, we are studying the opposite limit where Brownian motion yields the dominant contribution to the stress (see also below). Therefore, we will not consider shear thickening here.

Remarkably, shear melting and shear thinning phenomena are observed not only in colloidal systems [2, 7, 8] but also in the case of a variety of other soft materials such as emulsions [9, 10], polymers [4, 11, 12], foams [13, 14]. The practical importance of shear melting and shear thinning phenomena can hardly be overemphasized, since, without these effects, it would be nearly impossible to process many of the soft materials used e.g. in the food industry, pharmaceutical products and paints.

In the shear melted regime, experiments show evidence for shear thinning due to the presence of freely slipping two dimensional crystalline layers [15, 16]. Simulations also show a shear thinning regime below a "critical" shear rate, $\dot{\gamma}_c$, followed by a transition to a string-like order for $\dot{\gamma} > \dot{\gamma}_c$ [17].

On the other hand, studies of disordered suspensions of hard spheres show that shear thinning and shear melting phenomena may also occur in the absence of a crystalline structure [4, 18, 19**?**]. Similar observations have also been made in light scattering echo studies of (disordered) dense emulsions [10]. Brownian dynamics simulations show that shear thinning in concentrated colloidal suspensions is related to the fact that, in the limit of low shear rates, the main contribution to the shear stress originates from the Brownian motion of colloidal particles and that this contribution decreases with $\dot{\gamma}$ [20].

CP982, *Complex Systems, 5th International Workshop on Complex Systems*
edited by M. Tokuyama, I. Oppenheim, and H. Nishiyama

The above mentioned disordered soft materials are characterized by a dramatic increase of the structural relaxation times upon an increase of density or a decrease of temperature [21, 22] while at the same time their static structure hardly changes. This anomalous increase of the structural relaxation times upon keeping the fluid-like (amorphous) structure is called the glass transition [23, 24].

Bearing this in mind, it then appears natural to foot a theoretical description of the rheological properties of the so called 'soft glassy materials' [1] in general and colloidal suspensions in particular on an approach capable of adequate description of the glass transition in the absence of external forces. The challenge would then be to include the action of a dissipative external force into the theory.

This is the route chosen almost simultaneously and independently by Fuchs and Cates [25, 26] on the one hand and Miyazaki and Reichman [27, 28] on the other hand. Both groups extended the well known mode coupling theory of the glass transition (MCT) [23] in order to include the effects of a temporally constant and homogeneous shear. Following a similar philosophy, Berthier, Barrat and Kurchan extended the well studied spin glass model to take into account the influence of a dissipative external force [29].

In its simplest version, the mode coupling theory starts from the idea that, in a supercooled liquid, the nearest neighbors of a particle form a 'cage' around it which gradually solidifies upon an increase of density or a decrease of temperature until it fully freezes. This defines the ideal glass transition of MCT and the corresponding 'critical point' (critical density, ρ_c, or critical temperature, T_c) [23].

In the non-equilibrium version of the MCT, the effect of shear is taken into account via advection (flow-mediated transport) of density fluctuations towards larger wave vectors, i.e. smaller length scales. Colloidal particles thus need to explore smaller regions in order for correlation function of density fluctuations to decay. As a result, density-density correlations decay faster compared to the non-driven case [30].

Despite its similarity, the MCT-based approach of Fuchs and Cates is different from that of Miyazaki and Reichman. While the latter group focused on temporal fluctuations around the steady state, Fuchs and Cates tried to compute all properties of the stationay state via transient fluctuations [25, 26].

The main goal of the present work is to study some basic issues such as homogeneity of shear, the effect of Taylor dispersion as well as possible anisotropy in the particle dynamics. In view of recently developed experimental techniques allowing a detailed study of the dynamics of individual particles [31], a computer simulation study of these issues has obtained renewed interest, since it provides a complementary approach to the topic.

After an introduction of the model and the simulation method in the next section, we focus on the combined effects of shear and temperature on the dynamics of particle motion in section . In particular, the effect of Taylor dispersion on particle mobility will be discussed. It will be shown that Taylor dispersion dominates the system dynamics parallel to the flow direction at times considerably longer than the time necessary for the decay of nearest neighbor cage. A summary compiles our results.

A GLASS FORMING MODEL

The present molecular dynamics simulations are performed using a generic glass forming system, the so called Kob-Andersen model [32]. It consists of an 80:20 binary mixture of Lennard-Jones particles (whose types we call A and B) at a total density of $\rho = \rho_A + \rho_B = 1.2$ and in a cubic box of length $L = 10$ (see Fig. 1 for a snapshot of the simulation box).

A and B particles interact via $U_{LJ}(r) = 4\epsilon_{\alpha\beta}[(d_{\alpha\beta}/r)^{12} - (d_{\alpha\beta}/r)^6]$, with $\alpha, \beta = A, B$, $\epsilon_{AB} = 1.5\epsilon_{AA}$, $\epsilon_{BB} = 0.5\epsilon_{AA}$, $d_{AB} = 0.8 d_{AA}$, $d_{BB} = 0.88 d_{AA}$ and $m_B = m_A$. The potential was truncated at twice the minimum position of the LJ potential, $r_{c,\alpha\beta} = 2.245 d_{\alpha\beta}$. The parameters ϵ_{AA}, d_{AA} and m_A define the units of energy, length and mass. All other quantities reported in this paper are expressed as a combination of these units. The unit of time, for example, is given by $\tau_{LJ} = d_{AA}\sqrt{m_A/\epsilon_{AA}}$ and that of stress by ϵ_{AA}/d_{AA}^3.

Equations of motion are integrated using a discrete time step of $dt = 0.005$. The system density is kept con-

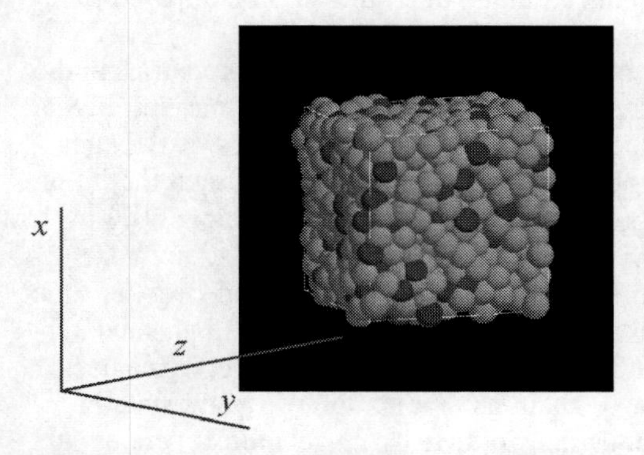

FIGURE 1. A snapshot of the simulated binary lennard-Jones system. The system consists of a total of 1200 particles composed of 20% small (red) and 80% large (blue) ones. Interaction parameters are tuned such that small particles prefer to be close to the large ones (see the text for more details).

stant at the value of 1.2 for all simulations whose results are reported here. This density is high enough so that no voids occur at low temperatures and low enough so that system dynamics remains sensitive to a variation of temperature (see references [33, 34] for effects of high density/pressure on the liquid-glass transition).

The present model was found suitable for an analysis of many aspects of the mode coupling theory of the glass transition [35, 36]. In particular, at a total density of $\rho = 1.2$, equilibrium studies of the model showed that the growth of the structural relaxation times at low temperatures could be approximately described by a power law as predicted by the ideal MCT, $\tau_{\text{relax}} \propto (T - T_c)^{-\gamma_{\text{MCT}}}$. Here, $T_c = 0.435$ is the mode coupling critical temperature of the model and γ_{MCT} is the critical exponent. For the present binary Lennard-Jones system, numerical solution of ideal MCT equations yields a value of $\gamma_{\text{MCT}} \approx 2.5$ [37]. A similar value is also obtained for a binary mixture of soft spheres [38].

Simulation results are averaged over 10 independent runs. For this purpose, ten independent samples are equilibrated at a temperature of $T = 0.45$ (above T_c) and serve as starting configurations for all simulated temperatures and shear rates. The temperature is controlled via Nosé-Hoover thermostat [39, 40]. It is set from $T = 0.45$ to the desired value at the beginning of shear, whereby only the y-component of particle velocities is coupled to the heat bath (x being the streaming and z the shear gradient directions).

The temperature quench is done only in one step, i.e. without a continuous variation from T_{start} to T_{end}. However, as the numerical value of T is changed, it takes a time of the order of the velocity autocorrelation time for the new temperature to be established. During this period of time the Maxwell distribution of velocities undergoes changes in order to adapt itself to a distribution determined by the new temperature. This time is of order unity (in reduced units) and quite short compared to all other relevant timescales in the problem.

Since we are interested in the system properties in the steady state, we must make sure that transient effects do not affect our results. Previous studies of the stress-strain relation of the same model showed that the initial transient behavior is limited to strains below 50% [41]. Indeed, by shifting the time origin in measurements of various correlation functions, we verified that the time translation invariance was well satisfied in sheared systems for strains larger than 50%. Therefore, we neglected strains $\gamma < 100\%$ in a computation of steady state properties such as density-density correlation functions and mean square displacements.

A series of simulations were performed using the above mentioned initial configurations equilibrated (in the absence of shear) at a temperature of $T = 0.45$. The maximum strain reached during these simulations was 7.8 (780%). In the steady state, correlation functions were averaged both over independent runs and over time origins distributed equidistantly along each run.

We also performed a large number of additional simulations with initial configurations equilibrated at a higher temperature of $T = 0.5$. This allowed us to generate a large number of 'low-cost' equilibrated samples, since the simulation time necessary for equilibration was considerably reduced in this way. Comparisons of various quantities obtained within these simulations agreed perfectly with previous ones thus emphasizing again the fact that steady state is reached during all our simulations. The length of all this new set of simulation runs was equal to $11.8/\dot{\gamma}$.

Data is saved on disk using a combined linear and logarithmic time sampling thus allowing accurate calculation of mean square displacements both at short, intermediate and long times. By doing so, we were able to compute the diffusion coefficient with a high precision and determine its dependence on temperature and rate of shear.

RESULTS

Recent studies of the present model in the glassy state showed that the system may exhibit shear-localization if the shear rate is imposed by using a conventional Couette cell with moving atomistic walls [42]. A shear banding is, however, undesired in the context of present analysis since we are interested in the effects of a homogeneous shear. On the other hand, simulations of the present model using the so-called Lees-Edwards boundary conditions along with the SLLOD equations of motion first proposed by Evans and coworkers [43] show that a linear velocity profile forms across the system thus leading to a spatially constant velocity gradient [44]. We, therefore, also used this approach for our simulations. Within this simulation method, we do indeed observe a spatially constant velocity gradient in all studied cases (see e.g. Fig. 2 in Ref. [45]).

In the glassy phase, large particle displacements can be regarded as a manifest of shear melting of an amorphous solid. While in the quiescent state a particle is practically eternally trapped in its local environment (the 'cage" formed by its nearest neighbors) it can explore far larger distances as the system is exposed to an external drive such as a shear motion. Thus, the particles in a driven glass are not localized but behave as in a fluid with the possibility to explore the whole available space.

In order to underline this and some other important features, we show in the left panel of Fig. 2 typical particle trajectories in the vorticity plane at three different shear rates of $\dot{\gamma} = 10^{-5}$, 10^{-4} and 10^{-3} for two temperatures of $T = 0.45$ (supercooled state) and $T = 0.2$

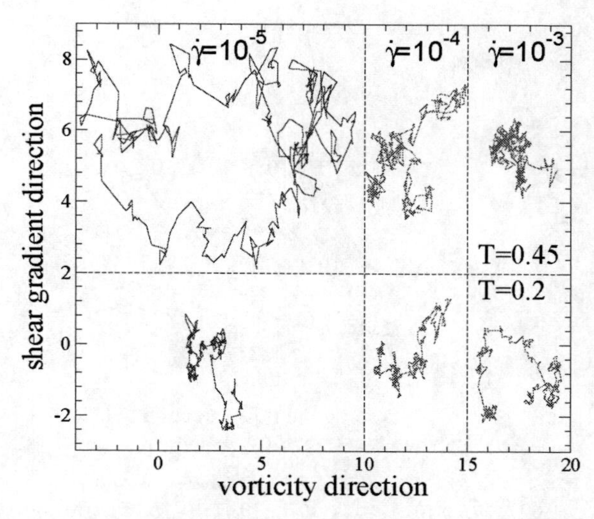

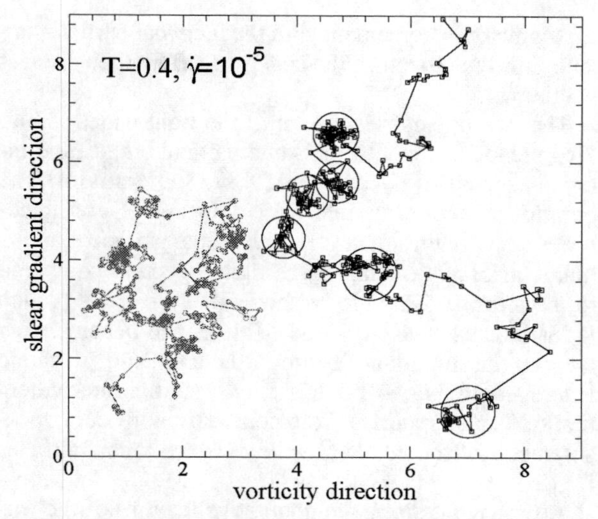

FIGURE 2. Left: Typical particle trajectories in a plane perpendicular to the flow (vorticity plane) during a total strain of 11.8 (1180%). Upper three trajectories correspond to a temperature of $T = 0.45$ (supercooled state) and the lower ones to $T = 0.2$ (glassy phase). The shear rate varies from one column to another as indicated in the panel. Irrespective of shear rate, a particle explores a region of equal extend in the glassy phase suggesting that the inverse shear rate is the only relevant time scale. In sharp contrast to this behavior, particle displacements in the supercooled state strongly increase upon decreasing shear rate for a constant amount of total strain. This is a manifest of the enhanced role of thermal fluctuations at lower $\dot{\gamma}$. Right: A closer look to the trajectory of a particle (here a particle of type A) in the vorticity plane. The particle motion consists of long periods of localized motion followed by short periods of relatively fast displacement. Large open circles serve to draw the attention to this type of dynamics. Note that the size of circles compares well to the particle diameter (here roughly equal to unity). If we associate the large scale motion with escaping from the nearest neighbor cage, this would imply that cage relaxation occurs at length scales close to the particle diameter. Similar results are reported from three dimensional imaging experiments (via confocal microscopy) on sheared colloidal particles [31]. The left trajectory corresponds to the same particle at a different time interval (shifted to the left by an arbitrary amount for better visibility). It serves to emphasize that a particle exhibiting fast diffusive motion and thus contributing significantly to the system's relaxation dynamics at large length scales, can develop a considerably slower motion at a later time interval.

(glass[1]). In all the cases shown, the observation time interval corresponds to $11.8/\dot{\gamma}$ (i.e. 1180% strain). As seen from this panel, there is a significant change in the way how shear affects particle mobility in the supercooled state as compared to the glass.

In the glassy phase ($T = 0.2$), the selected particle explores spatial domains of roughly equal size for all shear rates investigated. This is in contrast to the supercooled state ($T = 0.45$), where the spatial extension of the visited region increases upon decreasing the shear rate.

A way to rationalize this observation is the following. In the glassy phase, thermal fluctuations are quite weak and the resulting particle mobility is practically negligible. Particle dynamics in the glassy phase is, therefore, fully determined by the imposed shear, $1/\dot{\gamma}$ being the unique relevant time scale in the process. In the supercooled state, on the other hand, the particle motion results from the combined effects of thermal fluctuations and shear.

While at high shear rates the imposed flow dominates the system dynamics, thermal fluctuations gain importance as $\dot{\gamma}$ decreases. The increasing additional mobility seen at lower shear rates within the same amount of total strain nicely visualizes these enhanced thermal effects upon decreasing $\dot{\gamma}$. This observation is also in line with our previous studies of the decay of incoherent scattering function, $\Phi_q(t)$ [45]. In this reference it was shown that, at a temperature of $T = 0.45$, effects of shear on $\Phi_q(t)$ (shear thinning) progressively weaken as shear rate is reduced, becoming fully negligible for $\dot{\gamma} \leq 10^{-5}$, thus underlining further the growing role of thermal fluctuations as compared to shear induced dynamics.

The right panel of Fig. 2 illustrates a particle trajectory in a slightly more detail for a temperature of $T = 0.4$. It emphasizes the intermittent nature of the particle motion. The particle rattles in a relatively small region for relatively long time (note that each two successive points are a strain of 2% apart; The time is, therefore, proportional to the number of points) followed by rather large jumps between individual 'traps' (highlighted as circles). The typical size of a trap compares well with the parti-

[1] The best fit result for the mode coupling critical temperature of the model under homogeneous shear is T_c=0.4 [46]. Note that this value lies slightly below $T_c = 0.435$ obtained from the studies of the equilibrium dynamics of the model in the supercooled state [35]. A discussion of this issue can be found in [46].

cle diameter in agreement with the idea that particles are temporarily arrested in the cage formed by their nearest neighbors.

The two trajectories shown in the right panel of Fig. 2 correspond in fact to the same particle but at time intervals separated by a time of $8/\dot{\gamma}$ (800% strain). The duration of each time interval is the same. Both trajectories show intermittency. However, the particle exhibits much faster motion during the right trajectory compared to the left one. This nicely shows that a particle, which is fast during a given period of time, can become slow later on, an important feature in the context of dynamic heterogeneity [47, 48]. Interestingly, similar intermittent motion is also reported from confocal microscopy measurements of sheared hard-sphere like colloidal particles [31].

After having discussed qualitative features of the particle motion in some details. We now turn our attention to the effect of temperature and shear rate on mean square displacements (MSD) along all relevant spatial directions. For a given direction, say x, the MSD is computed as $\delta x^2(t) = \langle (x(t+t_0) - x(t_0))^2 \rangle$, where $\langle \cdots \rangle$ stands for statistical average (here average over independent runs and independent starting points, t_0, in the steady state). For this purpose we first show in the upper panel of Fig. 3 mean square displacements along the flow direction versus time for a temperature of $T = 0.1$ (glassy phase). The shear rate covers 4 decades from $\dot{\gamma} = 10^{-1}$ to $\dot{\gamma} = 10^{-5}$ (note also that the data cover 8 decades in time.).

At very short times, all mean square displacements scale as t^2 regardless of shear rate. This is the so called ballistic regime where particles do not feel the presence of their neighbors. They thus fly with a constant velocity[2]. At not too high shear rates, an intermediate regime occurs characterized by a plateau in MSD.

The presence of a plateau in mean square displacements is a clear signature of temporal arrest of particles in the cage formed by their neighbors. As the plateau regime is left, the MSD gradually develops a linear dependence on time indicative of long time diffusive motion. For still longer times, a new regime occurs, where MSD grows as t^3, i.e. even faster than the ballistic motion t^2.

This t^3 dependence (known as Taylor dispersion) is a clear result of the flow effects combined with transverse diffusion along the shear gradient direction. As a particle moves a distance dz along the shear gradient direction, it

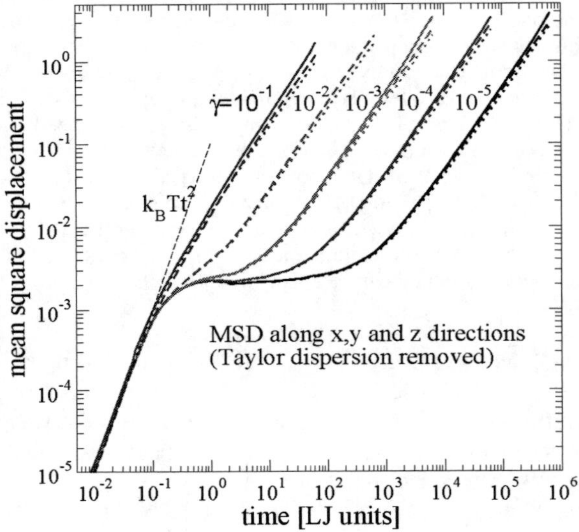

FIGURE 3. Effects of a homogeneous shear on particle mobility. Mean square displacements (MSD) are shown for a temperature of $T = 0.1$ (glassy phase) for some typical shear rates simulated. The upper panel shows the MSD along the flow direction. The same data are also shown in the lower panel after the subtraction of the contribution of Taylor dispersion. In addition to this, results obtained along two other spatial directions are also included. A systematic anisotropy is observed in the particle mobility. This effect is, however, small compared to the effect of shear rate.

obtains an extra velocity of $dz(t)\dot{\gamma}$ from the surrounding fluid. Its motion along the flow direction is thus modified via this flow induced contribution. A simple derivation (see e.g. [48]) shows that $\langle [x(t) - x(0)]^2 \rangle = 2Dt[1 + (t\dot{\gamma})^2/3]$. Here D is the diffusion coefficient along the neutral (vorticity) direction *in the presence of the flow*.

[2] Note that ballistic regime is a special feature of atomistic model systems as is the case in the present studies. In the presence of stochastic forces such as the fluctuating forces exerted by solvent molecules on a colloidal particle, this ballistic regime is replaced by the so called short time diffusive motion with a diffusion coefficient determined by the solvent viscosity (and independent of the concentration of the suspension).

The latter point is important. If the system shear thins under the flow (as is the case for the present model), it will have a higher probability for a motion along the velocity gradient direction thus enhancing the effect of the flow on the particle motion along the flow direction. In order to check this idea, we have first determined the self diffusion coefficient along the y (vorticity) direction for all studied shear rates. Indeed, we find that the above formula for MSD along the x (flow) direction yields an excellent description of our data if the shear rate dependent diffusion coefficient is used.

One can correct for the effect of Taylor dispersion by removing from particle displacements all the contributions to the motion along the flow arising from the above discussed flow gradient induced acceleration (deceleration), i.e. via introducing $\tilde{x}(t) = x(t) - \int_0^t dt'(v(z(t')) - v(z_0)) = \dot{\gamma}\int_0^t dt'(z(t') - z(0))$ [31]. The lower panel of Fig. 3 depicts results on the means square displacements after applying this correction procedure to the flow direction. As seen from this panel, the entire t^3 dependence disappears via this procedure and the MSD along the flow direction takes a form very similar to the MSD along the other two spatial directions.

Slight but systematic deviations are, however, observed in the lower panel of Fig. 3 when comparing the mean square displacements along x, y and z directions. Simulations performed with a smaller integration step dt, allow us to exclude any anisotropic effect related to the choice of dt.

On the other hand, by comparing the results at lower shear rates and high temperatures, we observe that these deviations disappear as soon as system dynamics becomes rather independent of the imposed shear. The above observation of a slightly anisotropic behavior seems, therefore, to be related to a corresponding shear induced perturbation of the system structure.

Next we address in this report the relation between diffusion and rheological response. This is an interesting issue, since experiments [49, 50, 47] as well as computer simulations [51, 52, 53] report a violation of the so called Stokes-Einstein (SE) relation (which states that diffusion coefficient is inversely proportional to shear viscosity) upon approaching the glass transition.

For this purpose, we have computed the diffusion coefficient for all simulated temperatures and shear rates via linear regression fits to the mean square displacements while neglecting the transients. The shear viscosity η, on the other hand, is obtained from the flow curves [46] via $\eta = \sigma/\dot{\gamma}$, where σ is the shear stress. The thus obtained shear viscosity $\eta(\dot{\gamma})$ is depicted in Fig. 4 versus diffusion coefficient $D(\dot{\gamma})$ (the shear rate $\dot{\gamma}$ is the parameter of the plot).

Although shear viscosity as well as diffusion coefficient vary by roughly 4 decades, both quantities remain approximately inversely proportional to one another. Nevertheless, slight deviations from the perfect line become visible at temperatures around $T_c = 0.4$. Interestingly, these deviations are practically negligible both at high and low temperatures.

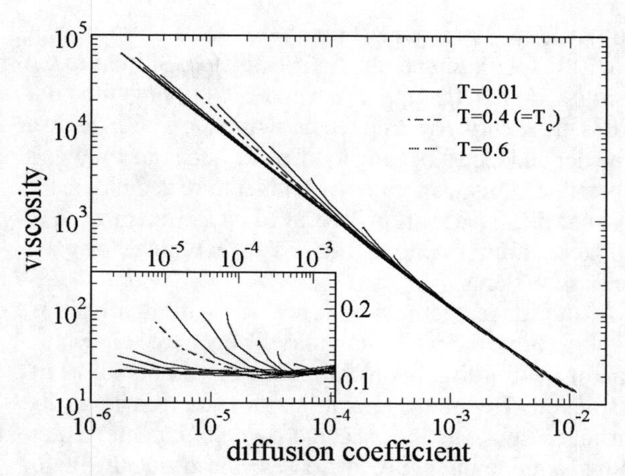

FIGURE 4. Shear viscosity, η, as determined from the knowledge of the shear stress ($\eta = \sigma/\dot{\gamma}$) versus self diffusion coefficient, D. The parameter of the plot is the shear rate. High (low) viscosities correspond to low (high) shear rates. Inset: ηD versus D. A straight horizontal line corresponds the prediction of the Stokes-Einstein relation. Deviations from the SE-relation seem to occur only at 'intermediate' temperatures (close to $T_c = 0.4$) and low shear rates only (note that the data for high and low temperatures overlap for all shear rates, making a clear distinction of the curves rather difficult). Note, however, that ηD varies (at most) by a factor of 2.

In order to examine this observation further, we plot in the inset of Fig. 4 the product of both quantities. A perfectly validity of the Stokes-Einstein relation would lead to a horizontal line. Obviously, the product of the shear viscosity and diffusion coefficient slightly deviates from a straight line at low shear rates for some temperatures around T_c (note, however, that ηD varies (at most) by a factor of 2).

Interestingly, practically no deviations from the Stokes-Einstein occur in the case of the lowest simulated temperatures. Furthermore, deviations from the SE relation observed at temperatures close to T_c roughly correspond to (low) shear rates where the system dynamics is less affected by shear. This suggests that shear tends to restore the Stokes-Einstein relation. It is worth noting that a similar behavior is also reported from molecular dynamic simulation results of Yamamoto and Onuki [48] who studied the dynamics of bond-breakage in a Lennard-Jones system slightly different from the present model.

Finally, we discuss the effect of temperature on mean square displacements. Figure 5 shows mean square displacements measured along the flow (upper panel) and the vorticity (lower panel) directions for a shear rate of $\dot{\gamma} = 10^{-5}$ for all temperatures investigated. Similar to the results shown in the upper panel of Fig. 3, three time domains are identified along the flow direction with distinct time dependencies. Again, a diffusive behavior follows a short time ballistic motion. As temperature decreases, the onset of diffusive motion is delayed by the occurrence of a plateau at intermediate times. This is typical for glass forming systems.

At the longest simulated times, the particle displacements exhibit the well-known t^3 dependence, characteristic of Taylor dispersion. At the lowest temperatures investigated, Taylor dispersion starts to dominate as the particle displacements exceed a particle diameter. Interestingly, as temperature increases, the onset of Taylor dispersion dominated motion is delayed and occurs at still larger mean square displacements.

In contrast to the flow direction, particle displacements along the y (vorticity) direction, depicted in the lower panel of Fig. 5, show the ballistic and diffusive regimes only thus allowing the use of a far larger time interval for a determination of the diffusion coefficient. Indeed, diffusion coefficients used in Fig. 4 are obtained using the MSD along the y direction.

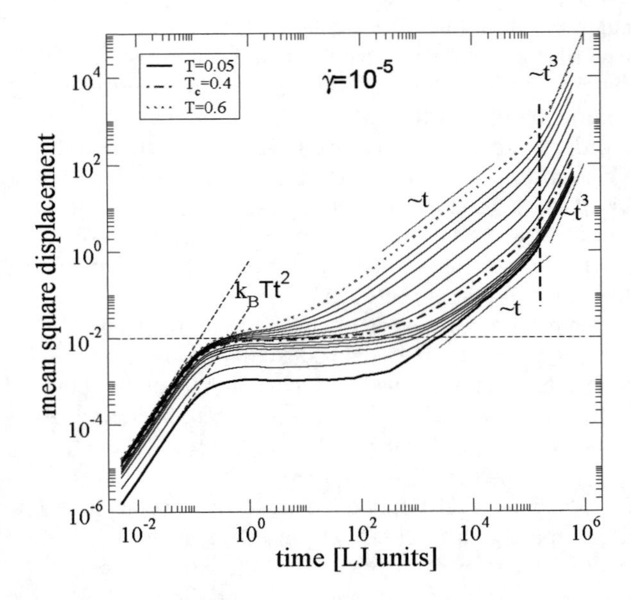

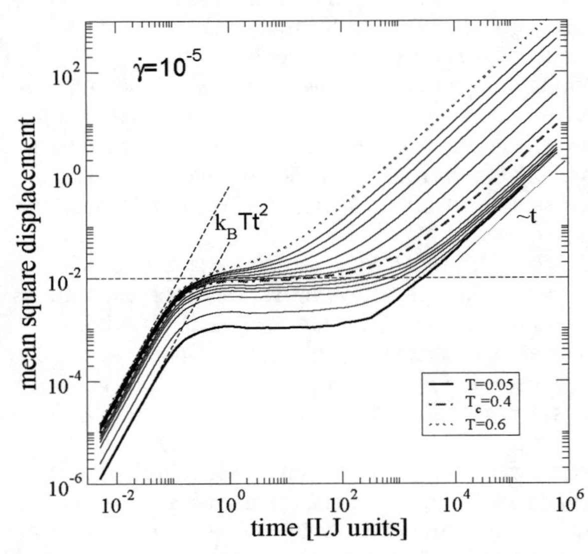

FIGURE 5. Mean square displacements at a (low) shear rate of $\dot{\gamma} = 10^{-5}$ (LJ units) for a wide range of temperature from the supercooled state down to the glassy phase. The shear rate is $\dot{\gamma} = 10^{-5}$. Similar to the upper panel of Fig. 3, three regimes of ballistic motion ($\sim t^2$), diffusion ($\sim t$) and Taylor dispersion ($\sim t^3$) are identified. At low temperatures, the ballistic and diffusive regimes are separated by a plateau whose width grows with decreasing temperature, a characteristic feature of glass forming systems. A thick dashed line roughly delimits the cross over from diffusion to Taylor dispersion. As temperature increases, diffusive regime extends to larger displacements. Lower panel: The same data but now measured along the vorticity direction. The temperatures shown are $T =$ 0.05, 0.1, 0.2, 0.26, 0.3, 0.34, 0.4, 0.41, 0.42, 0.43, 0.44, 0.45, 0.5, 0.55, 0.6

CONCLUSION

Results of large scale molecular dynamics simulations on a homogeneously driven glass forming model, a well established 80:20 binary Lennard-Jones system first introduced by Kob and Andersen [32] are reported. While previous works [46, 45] focused on flow curves and their relation to the dynamic of structural relaxation on the length scale of nearest neighbor distance, we investigate here the dynamics of diffusion with a particular emphasize on Taylor dispersion and possible anisotropy of diffusion. Furthermore, the validity of the Stokes-Einstein relation is also examined by direct measurements of the shear viscosity (via shear stress) and the long time diffusion coefficient (via mean square displacements).

In the glassy state and at not too high shear rates, the mean square displacements exhibit the well known two step relaxation behavior with a diffusive motion at long times along the spatial directions perpendicular to the flow.

In the flow direction, on the other hand, a third regime follows the diffusive motion, where Taylor dispersion with the typical t^3 time dependence clearly dominates the long time behavior of the particle displacements. At low temperatures (glassy phase), the cross over from the diffusive regime to the regime where the contribution of Taylor dispersion becomes significant, occurs at length scales of the order of a particle diameter but is shifted towards progressively larger lengths as temperature increases.

This underlines the subdominant role of Taylor dispersion at times characteristic for the structural relaxation on the length scale of nearest neighbor distance.

As to the possible flow anisotropy, we address this issue via a comparison of the means square displacements along all the three spatial directions x, y and z. Prior to the comparison, we remove the contribution of Taylor dispersion from the MSD along the flow direction. Small but systematic deviations between particle displacements along different spatial directions are seen, suggesting small changes in the system ¡s structure.

Our results on the diffusion coefficient and its relation to the shear viscosity suggest a violation of the Stokes-Einstein relation. This seems to occur at low but not too low temperatures, where the inherent system dynamics is slow but not fully frozen and for sufficiently low shear rates, so that thermal fluctuations still yield the dominant contribution to the particle dynamics. On the other hand, the Stokes-Einstein relation is observed to hold in all cases where the system dynamics is dominated by the flow. In agreement with previous simulation studies [48], this suggests that a homogeneous shear tends to restore the Stokes-Einstein relation.

ACKNOWLEDGMENTS

I am indepted to M. Fuchs and O. Henrich for careful reading of this manuscript and their constructive comments. During this work, F.V. was supported by the Max-Planck Grant Multi-Scale Materials Modeling of Condensed Matter (MMM).

REFERENCES

1. R. G. Larson, *The structure and Rheology of Complex Fluids*, Oxford University Press, New York, 1999.
2. G. Petekidis, D. Vlassopoulos, and P. N. Pusey, *Faraday Discuss.* **123**, 287–302 (2003).
3. C. E. Chaffey, *Colloid & Polymer Sci.* **255**, 691–698 (1977).
4. H. M. Laun, R. Bung, S. Hess, W. Loose, . Hess, K. Hahn, E. Hädicke, R. Hingmann, F. Schmidt, and P. Lindner, *J. Rheol.* **36**, 743–787 (1992).
5. W. J. Frith, P. d'Haene, R. Buscall, and J. Mewis, *J. Rheol.* **40**, 531–548 (1996).
6. C. B. Holmes, M. E. Cates, M. Fuchs, and P. Sollich, *J. Rheol.* **49**, 237–269 (2005).
7. G. Petekidis, D. Vlassopoulos, and P. N. Pusey, *J.Phycs: Condens. Matter* **16**, S3955–S3963 (2004).
8. M. Fuchs, and M. Ballauff, *J. Chem. Phys.* **122**, 094707–1–094707–6 (2005).
9. T. G. Mason, J. Bibette, and D. A. Weitz, *J. Colloid Interf. Sci.* **179**, 439–448 (1996).
10. P. Hébraud, F. Lequeux, J. Munch, and D. Pine, *Phys. Rev. Lett.* **78**, 4657–4660 (1997).
11. M. Kröger, and S. Hess, *Phys. Rev. Lett.* **85**, 1128–1131 (2000).
12. S. Barsky, and M. O. Robbins, *Phys. Rev. E* **65**, 021808–1–021808–7 (2002).
13. G. Debrégeas, H. Tabuteau, and J.-M. di Meglio, *Phys. Rev. Lett.* **87**, 178305–1–178305–4 (2001).
14. F. D. Cruz, F. Chevoir, D. Bonn, and P. Coussot, *Phys. Rev. E* **66**, 051305–1–051305–7 (2002).
15. B. J. Ackerson, and N. A. Clark, *Phys. Rev. Lett.* **46**, 123–126 (1981).
16. B. J. Ackerson, *J. Rheol* **34**, 553–590 (1990).
17. W. Xue, and G. S. Grest, *Phys. Rev. Lett.* **64**, 419–422 (1990).
18. G. Petekidis, P. N. Pusey, A. Moussaid, S. Egelhaaf, and W. C. K. Poon, *Physica A* **306**, 334–342 (2002).
19. A. M. G. Petekidis, and P. N. Pusey, *Phys. Rev. E* **66**, 051402–1–051402–13 (2002).
20. T. Phung, J. Brady, and G. Bossis, *J. Fluid Mech.* **313**, 181–207 (1996).
21. W. van Megen, and I. Swook, *Adv. Coll. Interf. Sci.* **21**, 119–194 (1984).
22. W. van Megen, and S. Underwood, *Phys. Rev. E* **47**, 248–261 (1993).
23. W. Götze, "Freezing and the Glass Transition," in *Liquids, Freezing, and the Glass Transition*, edited by J. P. Hansen, D. Levesque, and J. Zinn-Justin, North-Holland, Amsterdam, 1991, p. 287–503.
24. W. Götze, and L. Sjögren, *Rep. Prog. Phys.* **55**, 241–341 (1992).
25. M. Fuchs, and M. E. Cates, *Phys. Rev. Lett.* **89**, 248304–1–248304–4 (2002).
26. M. Fuchs, and M. E. Cates, *Faraday Discuss.* **123**, 267–286 (2003).
27. K. Miyazaki, and D. Reichman, *Phys. Rev. E* **66**, 050501(R)–1–050501(R)–4 (2002).
28. K. Miyazaki, D. Reichman, and R. Yamamoto, *Phys. Rev. E* **70**, 011501–1–011501–14 (2004).
29. L. Berthier, J.-L. Barrat, and J. Kurchan, *Phys. Rev. E* **61**, 5464–5472 (2000).
30. O. Henrich, F. Varnik, and M. Fuchs, *J.Phys.: Condens. Matter* **17**, S3625–S3630 (2005).
31. R. Besseling, E. R. Weeks, A. B. Schofield, and W. C. K. Poon, *Phys. Rev. Lett.* **99**, 028301–1–028301–4 (2007).
32. W. Kob, and H. Andersen, *Phys. Rev. Lett.* **73**, 1376–1379 (1994).
33. F. Starr, M.-C. Bellissent-Funel, and H. Stanley, *Phys. Rev. E* **60**, 1084–1087 (1999).
34. S. Sastry, *Phys. Rev. Lett.* **85**, 590–593 (2000).
35. W. Kob, and H. C. Andersen, *Phys. Rev. E* **51**, 4626–4641 (1995).
36. W. Kob, and H. C. Andersen, *Phys. Rev. E* **52**, 4134–4153 (1995).
37. M. Nauroth, and W. Kob, *Phys. Rev. E* **55**, 657–667 (1997).
38. J.-L. Barrat, and A. Latz, *J. Phys. Condens. Matter* **2**, 4289–4295 (1990).
39. S. Nosé, *J. Chem. Phys.* **81**, 511–519 (1984).
40. W. G. Hoover, *Phys. Rev. A* **31**, 1695–1697 (1985).
41. F. Varnik, L. Bocquet, and J.-L. Barrat, *J. Chem. Phys.* **120**, 2788–2801 (2004).

42. F. Varnik, L. Bocquet, J.-L. Barrat, and L. Berthier, *Phys. Rev. Lett.* **90**, 095702–1–095702–4 (2003).
43. D. J. Evans, and G. P. Morriss, *Statistical Mechanics of Non Equilibrium Liquids*, Academic Press, London, 1990.
44. L. Berthier, and J.-L. Barrat, *J. Chem. Phys.* **116**, 6228–6242 (2002).
45. F. Varnik, *J. Chem. Phys.* **125**, 164514 –1–164514–10 (2006).
46. F. Varnik, and O. Henrich, *Phy. Rev. B* **73**, 174209–1–174209–5 (2006).
47. F. H. Stillinger, and J. A. Hodgdon, *Phys. Rev. E* **50**, 2064–2068 (1994).
48. R. Yamamoto, and A. Onuki, *PhysRev. E* **58**, 3515–3529 (1998).
49. F. Fujara, B. Geil, H. Silescu, and G. Fleischer, *Z. Phys. B* **88**, 195–204 (1992).
50. M. D. Ediger, C. A. Angell, and S. R. Nagel, *J. Phys. Chem.* **100**, 13200–13212 (1996).
51. J.-L. Barrat, J.-N. Roux, and J.-P. Hansen, *Chem. Phys.* **149**, 197–208 (1990).
52. W. Kob, C. Donati, S. J. Plimpton, P. H. Poole, and S. C. Glotzer, *Phys. Rev. Lett.* **79**, 2827–2830 (1997).
53. J. Horbach, and W. Kob, *Phys. Rev. B* **60**, 3169–3181 (1999).

Incorporating Activated Hopping Processes into the Mode-Coupling Theory for Glassy Dynamics

Song-Ho Chong

Institute for Molecular Science, Okazaki 444-8585, Japan

Abstract. A new attempt for extending the idealized mode-coupling theory (MCT) for glass transition is presented which incorporates activated hopping processes via the dynamical theory originally developed to describe diffusion-jump processes in crystals. The approach adapted here to glass-forming liquids treats hopping as arising from vibrational fluctuations in quasi-arrested state where particles are trapped inside their cages, and the hopping rate is formulated in terms of the Debye-Waller factor characterizing the structure of the quasi-arrested state. The resulting expression for the hopping rate takes an activated form, in which the barrier height is "self-generated" in the sense that it is present only in those supercooled states where the dynamics exhibits a well defined plateau. It is discussed how such a hopping rate can be incorporated into MCT so that the sharp nonergodic transition predicted by the idealized version of the theory is replaced by a rapid but smooth crossover. Preliminary result for the hard-sphere system is also presented.

Keywords: Glass transition, Hopping process, Mode-coupling theory
PACS: 64.70.Pf, 61.20.Lc

Understanding the microscopic origin of the evolution of slow dynamics in glass-forming liquids is one of the most challenging problems in condensed matter physics. During the past decades the research in this field has been strongly influenced by the idealized mode-coupling theory (MCT) for glass transition [1]. Indeed, extensive tests of the theoretical predictions against experimental and computer-simulation results suggest that MCT deals properly with some essential features of supercooled liquids [2, 3]. On the other hand, a well-recognized limitation of the idealized MCT is the predicted divergence of the α-relaxation time at a critical temperature or density – also referred to as the idealized glass transition or the nonergodic transition – which is not observed in experiments and computer simulations. In reality there are slow activated processes called hopping which restore ergodicity. Extended version of MCT [4] aims at incorporating these processes, but its applicability has been restricted to schematic models. This is largely because of the presence of subtraction term in the expression for the hopping kernel which might violate the positiveness – a fundamental property – of correlation spectrum.

There have been relatively few other attempts to go beyond the idealized MCT [5, 6, 7], and incorporating the activated hopping process within a microscopic approach has been a major unsolved problem. In this paper, new such an attempt is presented which is motivated by ideas from the dynamical theory developed for describing diffusion-jump processes in crystals [8]. The approach adapted here to glass-forming liquids treats hopping as arising from vibrational fluctuations in quasi-arrested state where particles are trapped inside their cages, and the hopping rate is formulated in terms of the Debye-Waller factor charactering the structure of the quasi-arrested state. The resulting expression for the hopping rate takes an activated form, in which the barrier height is "self-generated" in the sense that it is present only in those supercooled states where the dynamics exhibits a well defined plateau. It will be discussed how such a hopping rate can be incorporated into MCT.

We start from surveying basic features of the idealized theory [1]. A system of N atoms of mass m distributed with density ρ shall be considered. Structural changes as a function of time t are characterized by density correlators $\phi_q(t) = \langle \rho_{\vec{q}}^* e^{i\mathscr{L}t} \rho_{\vec{q}} \rangle / NS_q$. Here $\rho_{\vec{q}} = \sum_i \exp(i\vec{q}\cdot\vec{r}_i)$ with $\vec{r}_i$ referring to ith particle's position denotes density fluctuations for wave vector $\vec{q}$; $\mathscr{L}$ the Liouville operator; $\langle \cdot \rangle$ the canonical averaging for temperature T; $S_q = \langle \rho_{\vec{q}}^* \rho_{\vec{q}} \rangle / N$ the static structure factor; $q = |\vec{q}|$. Within the Zwanzig-Mori formalism one can derive based on a projection operator $\mathscr{P}$ onto $\rho_{\vec{q}}$ [9]

$$\phi_q(z) = -1 / [z + K_q(z)]. \tag{1}$$

Here and in the following, we use the convention $f(z) = i \int_0^\infty dt\, e^{izt} f(t)$ with $\mathrm{Im}\, z > 0$ for the Laplace transform. The function $K_q(t)$ is defined as

$$K_q(t) = q^2 \langle j_{\vec{q}}^{\mathrm{L}*} e^{i\mathscr{Q}\mathscr{L}\mathscr{Q}t} j_{\vec{q}}^{\mathrm{L}} \rangle / NS_q, \tag{2}$$

in terms of longitudinal current density fluctuations $j_{\vec{q}}^{\mathrm{L}} = \mathscr{L}\rho_{\vec{q}}/q$ evolving under the generator $\mathscr{Q}\mathscr{L}\mathscr{Q}$, where $\mathscr{Q} \equiv 1 - \mathscr{P}$. The Zwanzig-Mori equation for $K_q(t)$ is obtained via a projection operator onto $j_{\vec{q}}^{\mathrm{L}}$:

$$K_q(z) = -\Omega_q^2 / [z + \Omega_q^2 m_q(z)]. \tag{3}$$

CP982, *Complex Systems, 5th International Workshop on Complex Systems*
edited by M. Tokuyama, I. Oppenheim, and H. Nishiyama

Here $\Omega_q^2 = q^2(k_BT/m)/S_q$ with k_B denoting Boltzmann's constant. The memory kernel $m_q(t)$ describes correlations of fluctuating forces. Under the mode-coupling approximation, the fluctuating forces are approximated by their projections onto the subspace spanned by pair-density modes $\rho_{\vec{k}}\rho_{\vec{p}}$. The factorization approximation for dynamics of the pair-density modes yields the idealized-MCT expression for the kernel to be denoted as $m_q^{\rm id}(t)$:

$$m_q^{\rm id}(t) = \int d\vec{k}\, V(\vec{q};\vec{k},\vec{p})\, \phi_k(t)\, \phi_p(t). \tag{4}$$

Here $\vec{p} = \vec{q} - \vec{k}$ and the vertex V is determined by S_q [1]. The idealized-MCT equations, consisting of Eqs. (1), (3) and (4), exhibit a bifurcation for $\phi_q(t \to \infty) = f_q$ [1]. If a control parameter, say n, is smaller than a critical value n_c, the correlator relaxes towards $f_q = 0$ as expected for an ergodic liquid. However, for $n \geq n_c$, density fluctuations arrest in a disordered solid, quantified by a Debye-Waller factor $f_q > 0$.

It is the factorization approximation that leads to the sharp nonergodic transition at n_c. Therefore, in extending the idealized theory, this approximation has to be avoided, and one has to consider corrections: $m_q(t) = m_q^{\rm id}(t) + \Delta m_q(t)$. The extended MCT of Götze and Sjögren [4] can be considered as a theory for such corrections, and yields $\Delta m_q(z) = m_q^{\rm id}(z)\delta_q(z)m_q(z)$ with the hopping kernel $\delta_q(z)$, i.e., $m_q(z)$ is given by

$$m_q(z) = m_q^{\rm id}(z) \,/\, [1 - \delta_q(z)\, m_q^{\rm id}(z)]. \tag{5}$$

Our new attempt for extending the idealized MCT is also based on this expression for $m_q(z)$, and is motivated by the following observation: substituting Eq. (5) into Eq. (3) yields for small frequencies [2]

$$K_q(z) = \delta_q(z) - \Omega_q^2 \,/\, [z + \Omega_q^2 m_q^{\rm id}(z)]. \tag{6}$$

Dropping $\delta_q(z)$, this equation reduces to the one of the idealized MCT: approaching n_c from the liquid side, $m_q^{\rm id}(z)$ becomes larger, and the current correlator $K_q(z)$ vanishes at $n = n_c$ leading to the nonergodic transition. In the presence of $\delta_q(z)$, on the other hand, the second term in Eq. (6) becomes not important when $m_q^{\rm id}(z)$ becomes large, and there holds $K_q(z) \approx \delta_q(z)$. This term takes over and hinders the currents from vanishing, thereby preventing the density fluctuations from being arrested completely. The long-time dynamics of $\phi_q(t)$ in this case is thus determined by $\delta_q(z)$ for small z. This observation raises a possibility of constructing an approximate theory for $\delta_q(z)$ described below with which a new extended MCT can be formulated.

To this end, we first derive a rate for a hopping process in which an atom at $\vec{r}_i$ jumps to a nearby site $\vec{r}_i'$ separated by a distance $|\vec{r}_i' - \vec{r}_i| \approx d$ of particle's diameter d. This will be done based on the dynamical theory which has been originally developed to describe diffusion-jump processes in crystals [8]. The approach adapted here to glass-forming liquids treats hopping as arising from vibrational fluctuations (phonons) in quasi-arrested state where particles are trapped inside their cages. Let us suppose that the quasi-arrested state – the ideal glass state in the sense of the idealized MCT [1] – can be described as a frozen, irregular lattice, as is usually assumed in structural models for glasses such as random packings [10] (see also Ref. [11]). Each particle then has a well defined equilibrium position $\vec{R}_i$ within the lifetime of the quasi-arrested state, and we introduce the displacement from the equilibrium position via $\vec{r}_i = \vec{R}_i + \vec{u}(\vec{R}_i)$. The essential feature of the hopping process is that a jumping atom passes through the barrier presented by those neighbors that bar a direct passage to the new site. The criterion that determines whether or not a given fluctuation is sufficient to cause a jump is therefore concerned with the relative displacements of the atom and the saddlepoint. We thus employ as a "reaction coordinate" [8]

$$x(t) = \left[\vec{u}(\vec{R}_i + \vec{s}, t) - \vec{u}(\vec{R}_i, t)\right] \cdot \hat{\vec{s}}, \tag{7}$$

and assume that a hopping occurs when $x(t)$ exceeds a critical value x^*, which measures the size of fluctuation needed to effect a jump. Here $\vec{s}$ denotes the equilibrium saddlepoint position with respect to $\vec{R}_i$, and the scalar product selects only those fluctuations directed towards $\hat{\vec{s}} = \vec{s}/s$. Each phonon displaces a hopping atom towards the saddlepoint. Since the phonon phases are random, the displacements may occasionally coincide in such a way that a hopping process occurs. The hopping rate $w_{\rm hop}$ is calculated from such a probability per unit time using the Kac formula [12]. A tractable expression is then obtained with the isotropic Debye approximation, $w_{\rm hop} = (1/2\pi)(3/5)^{1/2}\omega_{\rm D}\exp[-3mv^2\delta^2/2k_BT]$, in terms of the sound velocity v [8]. Here $\omega_{\rm D} = k_{\rm D}v$ with the Debye wave number $k_{\rm D} = (6\pi^2\rho)^{1/3}$, and $\delta \equiv x^*/s$. Notice that the sound velocity here refers to the one in the quasi-arrested state, which is renormalized by the Debye-Waller factors [13, 14]. To emphasize this, the sound velocity shall be expressed in terms of the (longitudinal) elastic modulus: $v = \sqrt{M_{\rm L}/(\rho m)}$. The modulus $M_{\rm L} = M_{\rm L}^0 + \delta M_{\rm L}$ consists of the equilibrium value $M_{\rm L}^0 = \rho(k_BT)/S_0$ and an additional contribution for the quasi-arrested state, for which MCT yields $\delta M_{\rm L} = \rho(k_BT)\int dk\, V_{\rm L}(k) f_k^2$ with $V_{\rm L}$ determined by S_q [15]. The hopping rate then reads

$$w_{\rm hop} = \frac{1}{2\pi}\left(\frac{3}{5}\right)^{\frac{1}{2}} \omega_{\rm D} \exp\left[-\frac{3M_{\rm L}}{2\rho k_BT}\delta^2\right], \tag{8}$$

with $\omega_{\rm D} = k_{\rm D}(M_{\rm L}/\rho m)^{1/2}$. The pre-exponential factor represents a mean attack frequency while the exponential term, which takes an activated form, gives the probability that the system is found at the critical displacement x^*. In addition, the barrier height for the hopping is "self-generated" in the sense that it is determined by the plateau heights f_q of the coherent density correlators, and is present only in those supercooled states where the dynamics exhibits a well defined plateau.

The parameter $\delta = x^*/s$ remains undetermined within the treatment in Ref. [8]. Here we estimate this parameter based on the following argument. We first notice that the long-time limit of the mean-squared displacement $\delta r^2(t)$ at the critical point $n = n_c$ of the idealized MCT – referred to as the critical localization length – provides the largest length scale (the Lindemann length) for the dynamics of a particle trapped inside the cage [16]. We therefore set $x^* = \sqrt{\delta r^2(t \to \infty)}$ evaluated at $n = n_c$ within the idealized MCT. Concerning s, we notice that, from the symmetry of the hopping process under consideration, it is approximately given by half of the hopping distance to a nearby site. Let us introduce the weighted average a of the nearest-neighbor distances via $a = \int_0^{r_{\min}} dr\, r\, [N(r)/N_c]$. Here $r_{\min}$ denotes the first minimum of the radial distribution function $g(r)$ defining the first shell; $N(r)dr$ with $N(r) = 4\pi r^2 \rho g(r)$ gives the mean number of particles at a distance between r and $r + dr$; $N_c = \int_0^{r_{\min}} dr N(r)$ is the coordination number of the first shell. This quantity is an analogue of the lattice spacing, and we set $s = a/2$. The resulting value $\delta^2 = 0.102$ for the hard-sphere system at the critical point (see below) is very close to $\delta^2 \approx 0.1$ which is known to provide a remarkably satisfactory account of migration properties observed in a wide variety of crystals [8].

We next relate the hopping rate w_{hop} to $\delta_q(z)$. Our discussion becomes simpler if the tagged particle density correlator $\phi_q^s(t)$ – the self part in $\phi_q(t)$ – is considered, so this case shall be studied first. (Quantities referring to the tagged particle shall be marked with the superscript s.) In the absence of the hopping kernel, the idealized kernel $m_q^{s\,\text{id}}(t)$ exhibits the plateau for long times whose height is given by $C_q^s = f_q^s/(1 - f_q^s)$ [1], i.e., there holds $m_q^{s\,\text{id}}(z) = -C_q^s/z$ for small z. Substituting this into Eq. (6) yields $K_q^s(z) = \delta_q^s(z) + z/C_q^s$ for small z. The α process of $\phi_q^s(t)$ in the presence of the hopping kernel is thus determined by $\phi_q^s(z) = -f_q^s/[z + f_q^s \delta_q^s(z)]$ (*cf.* Eq. (1)). To express $\delta_q^s(z)$ in terms of w_{hop}, let us consider a situation where the hopping from $\vec{r}$ to $\vec{r}\,'$ characterized by a rate $w_{\text{hop}}^{(\vec{r} \to \vec{r}\,')}$ dominates the long-time dynamics. Then, the van Hove correlation function $\phi_{\vec{r}}^s(t)$ – the inverse Fourier transform of $\phi_q^s(t)$ having the physical meaning of the probability of finding the tagged atom at $\vec{r}$ and t [9] – obeys the simple rate equation

$$\dot{\phi}_{\vec{r}}^s(t) = \sum_{\vec{l}} \left[w_{\text{hop}}^{(\vec{r}+\vec{l} \to \vec{r})} \phi_{\vec{r}+\vec{l}}^s(t) - w_{\text{hop}}^{(\vec{r} \to \vec{r}+\vec{l})} \phi_{\vec{r}}^s(t) \right]. \quad (9)$$

Since only hoppings with $|\vec{l}| \approx a$ are relevant and there is no site dependence in the rate we formulated, there holds $\dot{\phi}_{\vec{r}}^s(t) = w_{\text{hop}} \sum_{|\vec{l}| \approx a} [\phi_{\vec{r}+\vec{l}}^s(t) - \phi_{\vec{r}}^s(t)]$. Fourier transforming this yields $\dot{\phi}_q^s(t) = -w_{\text{hop}} \sum_{|\vec{l}| \approx a} [1 - e^{-i\vec{q}\cdot\vec{l}}] \phi_q^s(t)$. Assuming that $\vec{l}$ are oriented at random, the summation $\sum_{|\vec{l}| \approx a}$ is given by the orientational average multiplied by the number of sites satisfying $|\vec{l}| \approx a$, which is approximated by the coordination number N_c. This leads to $\dot{\phi}_q^s(t) = -w_{\text{hop}} N_c [1 - \sin(qa)/(qa)] \phi_q^s(t)$, implying that

$$\delta_q^s(z) = i\, w_{\text{hop}} N_c [1 - \sin(qa)/(qa)] / f_q^s. \quad (10)$$

Concerning the collective hopping kernel, one understands from Eqs. (2) and (6) that it can be expressed as $\delta_q(z) = \delta_q^s(z)/S_q + \delta_q^{\text{dist}}(z)$ where $\delta_q^{\text{dist}}(z)$ refers to the distinct part. In the present work, the distinct part in $\delta_q(z)$ shall be neglected, i.e.,

$$\delta_q(z) = \delta_q^s(z)/S_q. \quad (11)$$

Equations (1), (3), (5) with (4), (10), and (11) constitute our new extended MCT.

Let us present preliminary result for a system of hard spheres of diameter d. All equilibrium quantities are specified by the packing fraction $\varphi = \pi \rho d^3/6$, and S_q is evaluated within the Percus-Yevick theory [9]. This model has been studied in detail based on the idealized MCT [17]. The units of length and time shall be chosen so that the diameter d and the thermal velocity are unity. Figure 1 shows $\phi_q(t)$ for φ close to the critical value $\varphi_c \approx 0.516$ of the idealized MCT [17]. It is seen that the curves labeled A and B are not affected by the hopping dynamics. The solid curve C from the extended theory exhibits the same decay up to $\log_{10} t \approx 3$ as the corresponding dashed curve from the idealized MCT, but the slope of the solid curve around $\log_{10} t \approx 3$ is slightly larger than for the dashed curve and the α relaxation time is smaller. The effects from hoppings are drastic for $\varphi > \varphi_c$ where the idealized theory predicts the arrested dynamics at the plateau (dashed curves labeled C' and B'), whereas the corresponding solid curves from the extended theory relax to zero for long times.

ACKNOWLEDGMENTS

This work was supported by Grant-in-Aids for scientific research from the Ministry of Education, Culture, Sports, Science and Technology of Japan (No. 17740282).

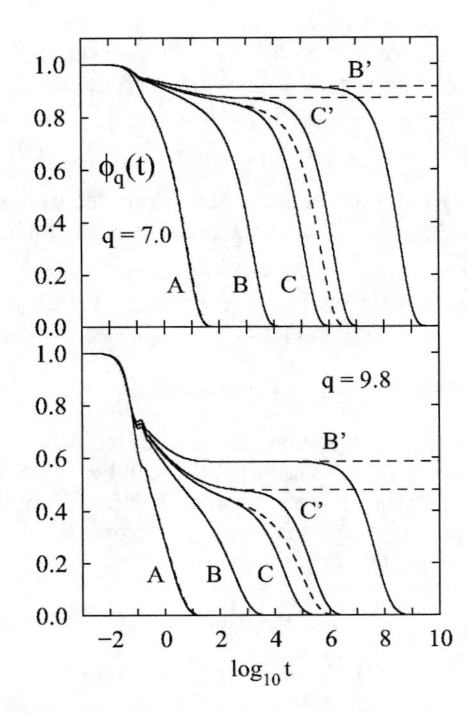

FIGURE 1. Coherent density correlators $\phi_q(t)$ as a function of $\log_{10}t$ for packing fractions $\varphi = \varphi_c(1+\varepsilon)$ close to the critical point $\varphi_c \approx 0.516$ of the idealized MCT: the labels A, B, C refer to negative $\varepsilon = -10^{-n}$ with $n = 1,2,3$ respectively, whereas the primed labels B', C' to positive $\varepsilon = 10^{-n}$ with $n = 2, 3$. The wave numbers are $q = 7.0$ (upper panel) and $q = 9.8$ (lower panel) which respectively correspond to the first peak and first minimum positions of S_q. Solid curves refer to the results from the present extended theory, whereas dashed curves to the ones from the idealized MCT.

REFERENCES

1. W. Götze, "Aspects of structural glass transitions," in *Liquids, Freezing and Glass Transition*, edited by J.-P. Hansen, D. Levesque, and J. Zinn-Justin, North-Holland, Amsterdam, 1991, pp. 287-503.
2. W. Götze, and L. Sjögren, *Rep. Prog. Phys.* **55**, 241-376 (1992).
3. W. Götze, *J. Phys.: Condensed Matter* **11**, A1–A45 (1999).
4. W. Götze, and L. Sjögren, *Z. Phys. B* **65**, 415-427 (1987).
5. K. Kawasaki, *Physica A* **208**, 35-64 (1994).
6. K. S. Schweizer, and E. J. Saltzman, *J. Chem. Phys.* **119**, 1181-1196 (2003).
7. S. M. Bhattacharyya, B. Bagchi, and P. G. Wolynes, *Phys. Rev. E* **72**, 031509-1-031509-12 (2005).
8. C. P. Flynn, *Phy. Rev.* **171**, 682-698 (1968).
9. J.-P. Hansen, and I. R. McDonald, *Theory of Simple Liquids*, Academic Press, London, 1986, 2nd edn.
10. C. H. Bennett, *J. Appl. Phys.* **43**, 2727-2734 (1972).
11. J.-L. Barrat, W. Götze, and A. Latz, *J. Phys.: Condens. Matter* **1**, 7163-7170 (1989).
12. M. Kac, *Am. J. Math.* **65**, 609-615 (1943).
13. W. Götze, and M. R. Mayr, *Phys. Rev. E* **61**, 587-606 (2000).
14. S.-H. Chong, *Phys. Rev. E* **74**, 031205-1-031205-23 (2006).
15. W. Götze, and M. Sperl, *J. Phys.: Condens. Matter* **15**, S869-S879 (2003).
16. M. Fuchs, W. Götze, and M. R. Mayr, *Phys. Rev. E* **58**, 3384-3399 (1998).
17. T. Franosch, M. Fuchs, W. Götze, M. R. Mayr, and A. P. Singh, *Phys. Rev. E* **55**, 7153-7176 (1997).

Nonlinear Susceptibility and Dynamical Length Scale of Glassy Systems

K. Miyazaki*,†, G. Biroli**, J-P. Bouchaud‡, and D. R. Reichman†

*The Research Institute of Kochi University of Technology, Tosa Yamada, Kochi 782-8502, Japan
†Department of Chemistry, Columbia University, 3000 Broadway, New York, NY 10027, USA
**Service de Physique Théorique, Orme des Merisiers CEA Saclay, 91191 Gif sur Yvette Cedex, France
‡Service de Physique de l'État Condensé, Orme des Merisiers CEA Saclay, 91191 Gif sur Yvette Cedex, France

Abstract. It is known that spatially heterogeneous dynamical structures are the cause of the drastic slowing down of atomic motion near the glass transition. Whereas tremendous progress has been made towards characterizing such structures by numerical and experimental studies, no microscopic theory has been developed for the multipoint correlation functions that characterize such behavior. We have extended the standard mode-coupling theory to the calculation of such quantities and succeeded in describing the growing dynamical structures.

Keywords: Dynamical heterogeneities, Glass transition, Mode-coupling theory
PACS: 64.70.Pf,05.70.Ln

INTRODUCTION

The glass transition is distinct from other critical phenomena in that the relaxation time and viscosity grow sizably with no sign of a macroscopic length scale in static or even simple dynamical correlation functions. Recently, extensive numerical simulations and experiments have shown that the viscous slowing-down of the glassy systems is associated with dynamical heterogeneities and a growing dynamical length scale[1, 2]. Such a length scale can be detected and quantified via multipoint correlation functions, or nonlinear susceptibilities, such as the four-point correlation function $\chi_4(t)$, rather than a conventional two-point correlation functions, such as the intermediate correlation function $F(k,t)$ which smears out spatial heterogeneities of trajectories when averaged over the space. Multipoint correlation functions and an associated length scale have been recently quantified by numerical simulations[3, 4, 5, 6] and by experiments[7, 8].

Arguably the only first principles theory to describe the slow dynamics of supercooled liquids near the glass transition is the mode-coupling theory (MCT). MCT successfully describes behaviors of two point correlation functions and transport coefficients quantitatively at a certain temperature regime above the real glass transition point. It has been argued, however, that the freezing transition which MCT predicts is due to the local caging phenomena and therefore MCT is not able to describe collectively growing dynamical length scales. But this is rather surprising since a diverging relaxation time is expected to be associated with the growing number of particles participating collective motions. In 2000, Franz and Parisi have analyzed the four-point correlation function for the p-spin mean field spin glass model and showed that it diverges near the dynamic critical temperature in a similar manner as simulated $\chi_4(t)$ for supercooled liquids[9]. Since the dynamical equation for the two-point correlation function for this model is mathematically equivalent with MCT for supercooled liquids, it is natural to expect that MCT should also be capable to describe the multipoint correlation functions and therefore the growing dynamical length scale. Besides, the fact that the exponents of the algebraic time dependence of simulated $\chi_4(t)$ are close to the MCT β exponents[10] hinted that MCT can capture this growing length scale. Such an idea has been put forward by Biroli and Bouchaud within field theoretic formulation of MCT[11]. They have demonstrated that MCT indeed can describe four-point correlation function and the growth of dynamical length scale near the freezing temperature. However, their argument is based on field theory and is not directly related to the standard MCT for liquids originally developed using the Mori-Zwanzig projection operator method[12], as suggested by recent works[13, 14, 15].

The aim of this work is to develop a first-principles theory to describe multipoint correlation function based on MCT using the standard projection operator method and to make a quantitative prediction for the growing cooperative length scale[16].

CP982, *Complex Systems, 5th International Workshop on Complex Systems*
edited by M. Tokuyama, I. Oppenheim, and H. Nishiyama

INHOMOGENEOUS MCT

In this work, we focus on the 3-point density correlation function defined by

$$\chi_{\mathbf{k}}(q,t) = \frac{1}{N}\langle \delta\rho_{\mathbf{q}}(t)\delta\rho_{\mathbf{k}}(t)\delta\rho_{-\mathbf{k}-\mathbf{q}}(0)\rangle, \qquad (1)$$

where $\delta\rho_{\mathbf{q}}$ is the density fluctuation and N is the total number of particles in the system. The 3-point density correlation function is the lowest order correlation function which can detect the dynamical length scale. The equation of motion for this function can be derived using the linear response theory. The original linear response theory relates the average quantities (the first moment) in the presence of the weak external perturbation to the 2-point correlation function (the second moment) in thermal equilibrium. This can be extended to the higher order. The 3-point correlation function in equilibrium can be expressed in terms of the linear response of 2-point correlation function in the presence of the external field. We consider a liquid subject to an external field $h_{\mathbf{q}}$. The intermediate scattering function of the system $F(\mathbf{k}_1,\mathbf{k}_2,t) = N^{-1}\langle\delta\rho_{\mathbf{k}_1}(t)\delta\rho_{\mathbf{k}_2}(0)\rangle_h$, where $\langle\cdots\rangle_h$ denotes the ensemble average in the presence of the external field, is now a function of two wavevectors due to violation of translational invariance in the presence of the external field. $\chi_{\mathbf{k}}(q,t)$ can be given by the response function for $F(\mathbf{k}_1,\mathbf{k}_2,t)$ as

$$\chi_{\mathbf{k}}(q,t) \approx \left.\frac{\partial F(\mathbf{k},\mathbf{k}+\mathbf{q},t)}{\partial h_{\mathbf{q}}}\right|_{h_{\mathbf{q}}=0}. \qquad (2)$$

In this expression, we disregard an extra term which involves a complicated memory term but will not affect the critical behavior of the correlation function of our interest. Eq.(2) reduces the problem to deriving the equation of motion for $F(\mathbf{k}_1,\mathbf{k}_2,t)$. One can derive MCT equation for $F(\mathbf{k}_1,\mathbf{k}_2,t)$ by following the standard procedure based on the Mori-Zwanzig projection operator formalism[17]. Projecting out all microscopic variables onto two gross variables of total momentum $\mathbf{J}_{\mathbf{k}} = \sum_{i=1}^{N}\mathbf{p}_i\exp[i\mathbf{k}\cdot\mathbf{r}_i]$ and density field $\delta\rho_{\mathbf{k}} = \sum_{i=1}^{N}\exp[i\mathbf{k}\cdot\mathbf{r}_i] - \langle\rho\rangle_h$, where $\mathbf{r}_i$ and $\mathbf{p}_i$ are the position and momentum of the i-th particle, and using the decoupling approximation for the memory kernel of the generalized Langevin equation derived for the correlation function of density and momentum, we arrive at

$$\begin{aligned}&\frac{\partial^2}{\partial t^2}F(\mathbf{k}_1,\mathbf{k}_2,t) + \int d\mathbf{k}'_1\,\Omega^2(\mathbf{k}_1,\mathbf{k}'_1)F(\mathbf{k}'_1,\mathbf{k}_2,t)\\ &+\int_0^t dt'\int d\mathbf{k}'_1\,M(\mathbf{k}_1,\mathbf{k}'_1,t-t')\frac{\partial F(\mathbf{k}'_1,\mathbf{k}_2,t')}{\partial t'} = 0,\end{aligned} \qquad (3)$$

In this expression, $\Omega^2(\mathbf{k}_1,\mathbf{k}_2) \equiv k_B T/m\mathbf{k}_1\cdot\mathbf{k}_2\langle\rho_{\mathbf{k}_1-\mathbf{k}_2}\rangle S^{-1}(\mathbf{k}_1,\mathbf{k}_2)$ is the Einstein frequency where m is the mass of particle and $S(\mathbf{k}_1,\mathbf{k}_2)$ is the static structure factor. $M(\mathbf{k}_1,\mathbf{k}_2,t)$ is the memory kernel whose explicit expression is given by

$$\begin{aligned}M(\mathbf{k}_1,\mathbf{k}_2,t) = \frac{1}{2}\int d\mathbf{q}_1\int d\mathbf{q}_2\int d\mathbf{p}_1\int d\mathbf{p}_2\int d\mathbf{k}'_1\\ \mathscr{V}_{\mathbf{k}_1,\mathbf{q}_1,\mathbf{q}_2}F(\mathbf{q}_1,\mathbf{p}_1,t)F(\mathbf{q}_2,\mathbf{p}_2,t)\mathscr{V}^{\dagger}_{\mathbf{k}'_1,\mathbf{p}_1,\mathbf{p}_2}S^{-1}(\mathbf{k}'_1,\mathbf{k}_2),\end{aligned} \qquad (4)$$

where $\mathscr{V}_{\mathbf{k},\mathbf{q}_1,\mathbf{q}_2}$ is the vertex which is a function of the static structure factor. The structure of eq.(3) and (4) is the same as the standard MCT except that the translational invariance is not assumed and we refer to this equation as inhomogeneous MCT (IMCT). Eq.(3) is a generalization of MCT to spatially inhomogeneous situations which can be applied to wide variety of situations such as confined systems and systems under the influence of gravity.

Taking the derivative of eq.(3) with respect to the external field $h_{\mathbf{q}}$ and then setting $h_{\mathbf{q}} = 0$, one derives the MCT equation for $\chi_{\mathbf{k}}(q,t)$;

$$\begin{aligned}&\frac{\partial^2\chi_{\mathbf{k}}(q,t)}{\partial t^2} + \frac{k_B T k^2}{mS(k)}\chi_{\mathbf{k}}(q,t)\\ &+\int_0^t dt'\,M_0(k,t-t')\frac{\partial\chi_{\mathbf{k}}(q,t')}{\partial t'} + \int_0^t dt'\frac{k_B T\rho k}{m|\mathbf{k}+\mathbf{q}|}\\ &\times\int\frac{d\mathbf{k}'}{(2\pi)^3}v_{\mathbf{k}}(\mathbf{k}',\mathbf{k}-\mathbf{k}')v_{\mathbf{k}+\mathbf{q}}(\mathbf{k}-\mathbf{k}',\mathbf{q}+\mathbf{k}')\\ &\times\chi_{\mathbf{k}'}(q,t-t')F(|\mathbf{k}-\mathbf{k}'|,t-t')\frac{\partial F(|\mathbf{k}+\mathbf{q}|,t')}{\partial t'}\\ &= \mathscr{S}_{\mathbf{k}}(q,t)\end{aligned} \qquad (5)$$

where $F(k,t)$ is the intermediate scattering function in the absence of the external field and the solution of the standard MCT. $M_0(k,t)$ and $v_{\mathbf{k}}(\mathbf{q},\mathbf{p}) = \hat{\mathbf{k}}\cdot\mathbf{q}c(q) + \hat{\mathbf{k}}\cdot\mathbf{p}c(p)$ ($c(q)$ is a direct correlation function) are the memory kernel and vertex function of the standard MCT, respectively. The source term $\mathscr{S}_{\mathbf{k}}(q,t)$ is a function of $F(k,t)$ and and $S(k)$ does not affect the critical properties of $\chi_{\mathbf{k}}(q,t)$ and the dynamical length scale.

CRITICAL PROPERTIES OF $\chi_{\mathbf{k}}(q,t)$

Extensive numerical analysis is required to integrate eq.(5) to obtain the dynamic behavior of $\chi_{\mathbf{k}}(q,t)$. However, it is possible to extract out the critical behavior of $\chi_{\mathbf{k}}(q,t)$ analytically if the temperature is close to MCT freezing transition point T_c. This can be done by noticing that eq.(5) has the form of the linear matrix equation for $\chi_{\mathbf{k}}(q,t)$ and the part which dictates the critical properties has the similar form as the stability matrix which determines the scaling behavior of $F(k,t)$ at small $\varepsilon \equiv |T/T_c - 1|$ for the standard MCT. The results are listed below. The detail of calculation is given elsewhere[16, 18].

First let us consider the limit of $\mathbf{q}=0$. At the β regime, we have

$$\chi_{\mathbf{k}}(q=0,t)=\frac{S(k)h(k)}{\sqrt{\varepsilon}}g_\beta(0,t/\tau_\beta) \tag{6}$$

and at the α regime,

$$\chi_{\mathbf{k}}(q=0,t)=\frac{1}{\varepsilon}g_{\alpha,k}(t/\tau_\alpha), \tag{7}$$

where $\tau_\beta=\varepsilon^{-1/2a}$ and $\tau_\alpha=\varepsilon^{-1/2a-1/2b}$ are the β and α relaxation times, respectively. a and b are standard MCT exponents and $h(k)$ is the critical amplitude. $g_\beta(0,x)$ behaves as x^a ($x\ll 1$) and x^b ($x\gg 1$), whereas $g_{\alpha,k}(x)\propto S(k)h(k)x^b$ ($x\ll 1$).

For $q\neq 0$ case, similar but more involved analysis yields that for the β regime

$$\chi_{\mathbf{k}}(q,t)=\frac{1}{\sqrt{\varepsilon}+\Gamma q^2}S(k)h(k)g_\beta\left(\frac{\Gamma q^2}{\varepsilon},\frac{t}{\tau_\beta}\right) \tag{8}$$

and for the α regime,

$$\chi_{\mathbf{k}}(q,t)=\frac{\Theta(\Gamma q^2/\varepsilon)}{\sqrt{\varepsilon}(\sqrt{\varepsilon}+\Gamma q^2)}g_{\alpha,k}\left(\frac{t}{\tau_\alpha}\right), \tag{9}$$

where Γ is a material-dependent constant, $\Theta(x)$ is a regular function which satisfies $\Theta(x=0)\neq 0$ and $\Theta(x)\sim 1/x$ for $x\gg 1$, and $g_{\alpha,k}(x)=S(k)h(k)x^b$ ($x\ll 1$) and $g_{\alpha,k}(x)\sim 0$ ($x\gg 1$). These results imply that, at the α regime, $\chi_{\mathbf{k}}(q,\tau_\alpha)\sim 1/q^4$ and the correlation length scales as $\xi\sim\sqrt{\Gamma}\varepsilon^{-1/4}$. In other words, the critical exponent is $\nu=1/4$.

In order to confirm these analytical scaling behavior, let us solve eq.(5) numerically. The full solution of $\chi_{\mathbf{k}}(q,t)$ of eq.(5) can be obtained by substituting the solution of the standard MCT for $F(k,t)$ into eq.(5) and by the direct numerical integration. But this is a rather cumbersome task and detailed analysis is given elsewhere[18]. Instead, here we shall focus on a simplified version of eq.(5) where all $\mathbf{k}$-dependence of eq.(5) is neglected. This is in the same spirit as the Leutheusser equation which neglects $\mathbf{k}$-dependence of the standard MCT for $F(k,t)$[19]. The Leutheusser equation is simpler than the standard MCT but still preserves all qualitative mathematical features of the full $\mathbf{k}$-dependent solution. For this purpose, we replace $F(k,t)$ by $F(t)$ (the solution of the Leutheusser equation), $M_0(k,t)$ by $4\lambda F^2(t)$, where λ is a coupling constant, $\chi_{\mathbf{k}}(q,t)$ by $\chi(q,t)$, and finally the memory kernel appeared in the fourth term of the left hand side of eq.(5) by $8\lambda(1-q^2)\chi(q,t)F(t)$. Only the lowest order q-dependence in the memory term has been retained since we are interested in the long

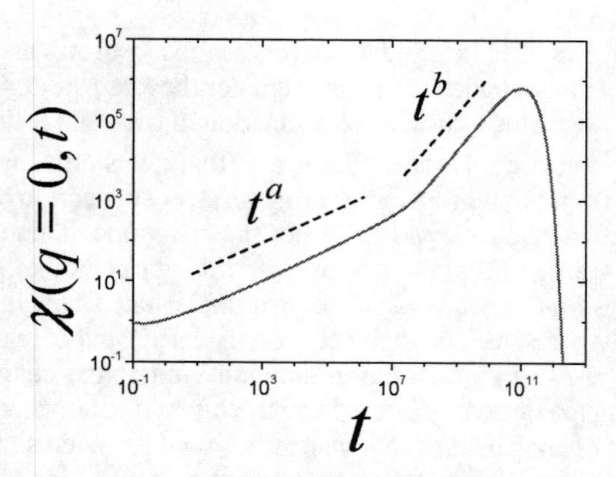

FIGURE 1. Numerical solution of the schematic version of eq.(5) for $q=0$ and for $\varepsilon=10^{-6}$. The algebraic growth with the early and late β exponents are indicated by dotted lines.

range correlation. The full numerical solution of this simplified schematic equation is shown in Fig.1 for $q=0$. $\varepsilon\equiv 1-\lambda/\lambda_c$ is fixed at 10^{-6} ($\lambda_c=1$ in our case). The figure shows that $\chi(0,t)$ grows as t^a at the early β regime ($t<\tau_\beta$) and t^b at the late β regime ($t>\tau_\beta$) and reaches maximum at $t=\tau_\alpha$ followed by rapid decay. Furthermore, $\chi(0,\tau_\alpha)$ grows as $1/\varepsilon$. These results agree with analytic results of eqs.(6) and (7). This is also consistent with the behavior of $\chi_4(t)$ for binary mixtures obtained by molecular dynamic simulations[10].

For $q\neq 0$, $\chi(q,t)$ shows far richer behavior. Fig.2 is the plot of the q-dependence of $\chi(q,t=\tau_\alpha)$ at the α regime for several ε's. The wavevector is scaled by $\varepsilon^{-1/4}$. This clearly shows that $\chi(q,\tau_\alpha)$ does not follow

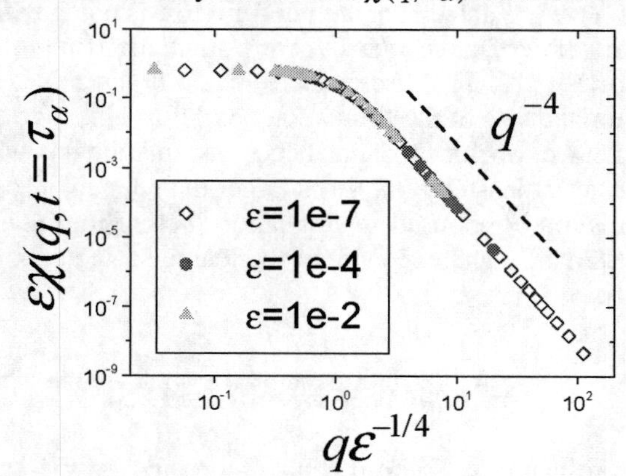

FIGURE 2. q-dependence of $\chi(q,t=\tau_\alpha)$ for several ε's in the α regime.

the Orstein-Zernike law but instead decays as q^{-4} and the dynamic length is scaled as $\xi\sim\varepsilon^{-1/4}$. In other words, relevant exponents are $\eta=-2$ and $\nu=1/4$, rather than $\eta=0$ and $\nu=1/2$ which mean-field theory predicts for standard critical phenomena. Further numerical analysis shows that, in the early β regime, $\chi(q,t)$ is given by the Orstein-Zernike form $\sim 1/(1+q^2\xi^2)$, while the height of grows as t^a (see Fig.1). In this regime, ξ grows as $t^{a/2}$ and is independent of ε. As we enter the late β regime at $t = \tau_\beta$, ξ stops growing as a function of time and follows $\varepsilon^{-1/4}$ scaling. Besides, the form of $\chi(q,t)$ starts changing from Orstein-Zernike form to a steeper function of q, eventually develops q^{-4} tail at the α regime. These results suggest that the size of the correlated region of particles first grows while keeping a fractal-like shape in the early β regime and then between the late β and α regime ($\tau_\beta < t < \tau_\alpha$), this region gradually fatten and changes its shape into a more compact structure while keeping the size of the cluster. This interesting scenario seems to be consistent with a recent observation of molecular simulations reported in Refs.[20, 21].

CONCLUSIONS

In summary, we have extended the standard MCT to the inhomogeneous situation which allows us to calculate higher oder correlation functions and the dynamic correlation length scale. Our results show that as the MCT freezing point T_c (λ_c for Leutheusser schematic equation) approaches the growing length scale ξ scales as $|T/T_c - 1|^{-1/4}$ which is very distinct from the standard critical phenomena (at the mean field level). The shape of $\chi(q,t)$ as a function of q is Orstein-Zernike form at the early β stage where the *size* of fractal-shaped correlated regions grows. Between the late β and α regime, it metamorphoses to a function with q^{-4} tail where the correlated region becomes more compact. The exponents which we predicted here are not fully consistent with the observation of molecular dynamic simulations reported so far[10, 22]. The discrepancies could be attributed to the limited size of the simulation box. >From a crude estimate, we conclude that one needs a simulation box with a side of at least several times larger than typical size of simulation boxes used in simulations before in order to access the critical exponent of the length scale quantitatively.

ACKNOWLEDGMENTS

This research has been partially supported by EU contract HPRN-CT-2202-00307 (DYGLAGEMEM) (GB) and by grants NSF CHE-0505939 and NSF DMR-0403997 (KM, DRR).

REFERENCES

1. M. D. Ediger, *Annu. Rev. Phys. Chem.* **51**, 99–128 (2000).
2. M. M. Hurley, and P. Harrowell, *Phys. Rev.* E **52**, 1694–1698 (1995).
3. R. Yamamoto, and A. Onuki, *Phys. Rev.* E **58**, 3515–3529 (1998).
4. G. Parisi, *J. Phys. Chem.* B **103**, 4128–4131 (1999).
5. C. Bennemann, C. Donati, J. Baschnagle, and S. C. Glotzer, *Nature* **399**, 246–249 (1999).
6. S. C. Glotzer, *J. Non-Cryst. Solids* **274**, 342–355 (2000).
7. L. Berthier, G. Biroli, J.-P. Bouchaud, L. Cipelletti, D. El Masri, D. L'Hôte, F. Ladieu, and M. Pierno, *Science* **310**, 1797–1800 (2005).
8. O. Dauchot, G. Marty, and G. Biroli, *Phys. Rev. Lett.* **95**, 265701-1-265701-4 (2005).
9. S. Franz, and G. Parisi, *J. Phys.: Condens. Matter* **12**, 6335–6342 (2000).
10. C. Toninelli, M. Wyart, L. Berthier, G. Biroli, and J.-P. Bouchaud, *Phys. Rev.* E **71**, 041505-1-041505-20 (2005).
11. G. Biroli, and J.-P. Bouchaud, *Europhys. Lett.* **67**, 21–27 (2004).
12. W. Götze, "Aspects of Structural Glass Transition," in *Les Houches, Session LI, "Liquids, Freezing and Glass Transition"*, edited by J. P. Hansen, D. Levesque, and J. Zinn-Justin, Elsevier Science Publishers B. V., (1989), pp. 765–942.
13. K. Miyazaki, and D. R. Reichman, *J. Phys.* A **38**, L343–L355 (2005).
14. A. Andreanov, G. Biroli, and A. Lefèvre, *J. Stat. Mech.* , P07008-1-P07008-54 (2006).
15. B. Kim, and K. Kawasaki, *J. Phys.* A **40**, F33–F42 (2007).
16. G. Biroli, J.-P. Bouchaud, K. Miyazaki, and D. R. Reichman, *Phys. Rev. Lett.* **97**, 195701-1-195701-4 (2006).
17. R. Zwanzig, *"Nonequilibrium Statistical Mechanics"*, Oxford University Press, Oxford, (2001).
18. G. Biroli, J.-P. Bouchaud, K. Miyazaki, and D. R. Reichman, (unpublished).
19. E. Leutheusser, *Phys. Rev.* A **29**, 2765–2733 (1984).
20. G. A. Appignanesi, J. A. R. Fris, R. A. Montani, and K. W., *Phys. Rev. Lett.* **96**, 057801-1-057801-4 (2006).
21. C. Donati, J. F. Douglas, W. Kob, S. J. Plimpton, P. H. Poole, and S. C. Glotzer, *Phys. Rev. Lett.* **80**, 2338–2341 (1998).
22. N. Lačević, F. W. Starr, T. B. Schrøder, and S. C. Glotzer, *J. Chem. Phys.* **119**, 7372–7387 (2003).

Calorimetric Study of Kinetic Glass Transition in Metallic Glasses

Y. Hiki* and H. Takahashi**

*Faculty of Science, Tokyo Institute of Technology, 39-3-303 Motoyoyogi, Shibuya-ku, Tokyo 151-0062, Japan
** Institute of Applied Beam Science, Graduate School of Science and Engineering, Ibaraki University, 4-12-1 Nakanarusawa, Hitachi 316-8511, Japan

Abstract. Differential scanning calorimetry (DSC) experiments were carried out for a bulk metallic glass (BMG), $Zr_{41.2}Ti_{13.8}Cu_{12.5}Ni_{10.0}Be_{22.5}$, below and above the glass transition temperature T_g. The T_g values were determined from the DSC curves. A wide range of heating rate, q=dT/dt= 0.1-100 K/min, was adopted for the experiment, and the q dependence of the apparent T_g was investigated. As q was decreased, the value of T_g decreased rapidly, then more slowly, and seemed to approach a constant value at low q. The experimental result of this kinetic glass transition phenomenon was analyzed on the basis of the relaxation process occurring in the transition temperature range.

Keywords: Calorimetry, Glass transition, Heating-rate dependence, Metallic glass
PACS: 61.43.Fs, 64.70.Pf, 65.60.+a, 81.05.Kf

INTRODUCTION

In glassy materials, a variety of relaxation phenomena are observed in various physical properties. The glass transition is also commonly considered as a relaxation phenomenon. An evidence of this consideration is that the experimentally-observed glass transition temperature T_g depends on the heating or cooling rate adopted in the experiment. The glass transition is thus "kinetic" in its nature, and the study of the phenomenon is interesting.

Various kinds of mechanical methods can be used to investigate the relaxation near the glass transition. For example, the measurements of shear viscosity, internal friction, sound velocity and attenuation are adopted. Our studies concerned for metallic glasses are cited here as examples [1-5]. In the present study, a thermometric method was attempted to be used for the investigation. As one of the most convenient methods, the differential scanning calorimetry (DSC) measurement was adopted. Namely, the temperature of the specimen is increased from a low temperature up to a temperature above the expected glass transition temperature, the DSC curve is observed, and the apparent glass transition temperature T_g is determined. The heating rate q in the experiment is widely changed, and by analyzing the T_g-vs-q data the kinetic glass transition is investigated.

There are a variety of glasses or amorphous materials. In the present study, a kind of bulk metallic glasses (BMG) is chosen as the test material. BMG alloys have high glass-forming ability (GFA). Namely, such alloys can be vitrified even when they are cooled slowly. There are a variety of BMG alloys composed of various kinds of elements. Interest in these materials is increasing because of their scientific and technological importance [6-9]. The study of the kinetic glass transition in such a kind of glasses seems to be quite important. We choose a Zr-base alloy as the specimen material.

EXPERIMENTAL

The specimen material used in the present study is the alloy $Zr_{41.2}Ti_{13.8}Cu_{12.5}Ni_{10.0}Be_{22.5}$, which is one of BMGs with the highest GFA. The supplied bulk specimen is cut into small pieces for the DSC experiments. No specific heat treatment is made for the specimens. New specimen is always used in every DSC experiment. A preliminary measurement is carried out for determining the glass transition temperature T_g and crystallization temperature T_x with the standard heating rate of 10 K/min. The obtained values are T_g=621.8 K and T_x=712.9 K. The temperature difference T_x–T_g is quite large. This is a characteristic of BMGs with high GFA [8-9]. In such a material, the DSC curve near T_g can be observed without the disturbance of the crystallization effect. This is an advantage of using such glasses for the DSC study.

CP982, *Complex Systems, 5th International Workshop on Complex Systems*
edited by M. Tokuyama, I. Oppenheim, and H. Nishiyama

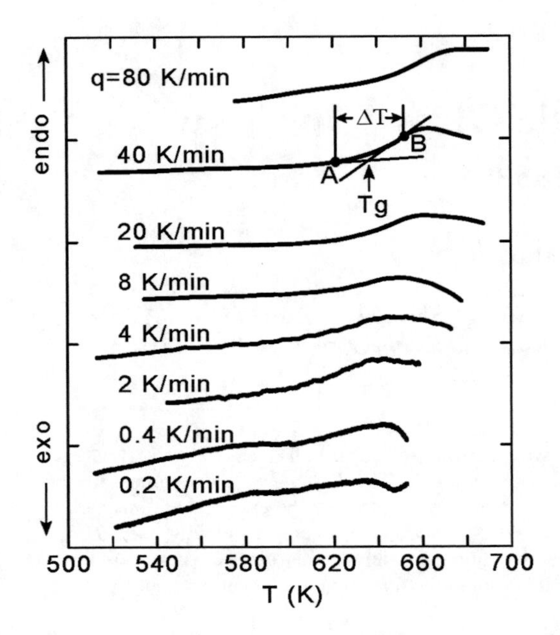

FIGURE 1. Recorded DSC curves with various heating rate q (K/min) for $Zr_{41.2}Ti_{13.8}Cu_{12.5}Ni_{10.0}Be_{22.5}$.

Calorimetric measurements are carried out using a commercial instrument, DSC8230 (Rigaku Co., Ltd, Tokyo). The specification of the apparatus is as follows: measurable temperature range is -150 °C – 725 °C; the DSC range is ±100 μW – ±100 mW/FS; the maximum heating rate is 150 °C/min; the maximum specimen volume is 100 μl. The DSC data are digitally recorded and displayed on a chart. In the present measurements, about 10 mg of specimen enclosed in an Al pan is used. A constant Ar gas flow is supplied through the specimen chamber. The DSC measurements are performed in a temperature range below and above the expected glass transition temperature. Here, a wide range of heating rate, q=dT/dt= 0.1-100 K/min, is adopted. The long time scale (small q) measurements are especially important and are performed very carefully. The recorded DSC curves for various heating rate q are illustrated in Fig. 1.

As a typical example, note the curve q=40 K/min. As the temperature is increased, the curve is initially enhanced almost linearly, starts to increase more rapidly at the point A, and the rapid increase ends at the point B. The conventional "thermometric" T_g can be obtained by using the following method. The background part at low temperature and the rapid rise part at higher temperature are extrapolated, and the value of T_g is determined as the intercept of the two extrapolations. The glass transition is considered to be essentially occurring in the temperature range between A and B with the temperature width ΔT. The DSC curves for other cases with different q are sometimes not so typical. However, the overall behaviors of the curves are almost the same. Using all of the DSC curves we determine the values of T_g. Note here that the temperature width ΔT seems to be not so much different in different q cases. Namely, the glass transition occurs in the same temperature range.

RESULTS AND ANALYSIS

The determined glass transition temperature T_g (K) is plotted against the heating rate q (K/min) in Fig. 2. The value of T_g decreases with decreasing q, and the change is nonlinear. Namely, as q is decreased, T_g decreases rapidly, then more slowly, and seems to approach a constant value at low q. The data are analyzed in the following.

The characteristics of mechanical relaxation in fragile glasses are here described. The following is a generally accepted argument [10]. In the range of temperature T above T_g, the viscoelastic property of the material is liquid-like (hydrodynamic regime) where the Vogel-Tammann-Fulcher relation holds for the relaxation time:

$$\tau = A\exp[B/(T-T_0)], \tag{1}$$

where A, B, and T_0 are constants. At lower temperatures the viscoelasticity of thermal-activation type showing the Arrhenius relation is preferable (hopping regime):

$$\tau = \tau_0\exp(E/RT), \tag{2}$$

and here τ_0 is the pre-exponential factor, R is the gas constant, and E is the activation energy. A crossover of the two regimes occurs at a temperature near and somewhat above T_g. These characteristics have been well certified through measuring the temperature dependence of shear viscosity for various kinds of glasses [11-14].

At present, it is presumed that we are observing the relaxation in the hydrodynamic regime in the present T_g-vs-q experiment. This is the idea proposed by Brüning and Samwer (BS) [15]. Then, by using the VTF relation, the following formula is obtained [15]:

$$T_g = T_{go} + A/\ln(B/q). \tag{3}$$

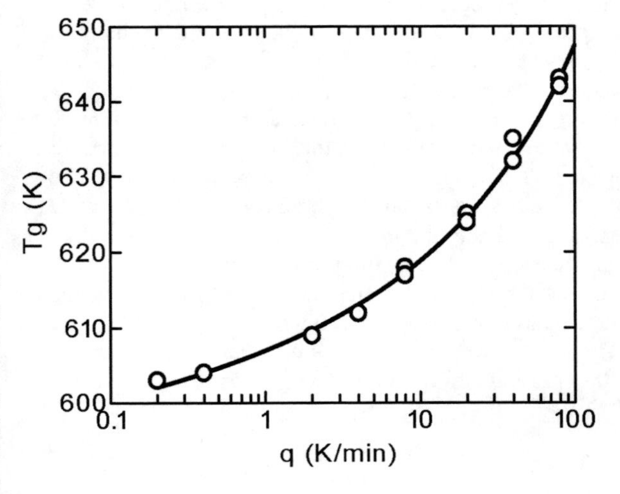

FIGURE 2. Observed glass transition temperature T_g (K) versus heating rate q (K/min) with a parameter-fitted curve.

TABLE 1. Parameter values for metallic glasses.

No.	Composition	T_{go} (K)	A (K)	logB(Ks^{-1})
1	$Pd_{40}Ni_{40}P_{19}Si_{1}$	518	538	3.6
2	$Pd_{77.5}Si_{16.5}Ag_{6}$	576	770	4.6
3	$La_{55}Al_{25}Ni_{20}$	338	2000	6.9
4	$Zr_{41.2}Ti_{13.8}$ $Cu_{12.5}Ni_{10.0}Be_{22.5}$	578	228	1.7

We tried to use Eq. (3) for analyzing our experimental result. The data shown in Fig. 2 are fitted to this expression with regarding T_{go}, A, and B as fitting parameters. The fitted curve is shown in the figure, and the fitting seems to be quite reasonable. The obtained values of the parameter values are given in Table 1 in the line 4, together with those for other alloys obtained by BS (lines 1-3) These parameter values are much different in different alloys. It is noted that for an alloy with large A value, the log$B$ value is also large. Meanwhile, no correlation is seen between these two parameter values and the T_{go} value. Meanings of these results can possibly be considered. However, only the facts are cited here. We are now continuing the kinetic glass transition experiment using other glassy materials. The discussion concerned will be delayed until a plenty of experimental data are accumulated in the near future.

DISCUSSION

Some of the studies by other authors concerning the kinetic glass transition problem are cited and considered. Brüning and Samwer [15] performed the heating-rate experiments for several metallic, inorganic, and polymer glasses, and presented a plenty of experimental data. They analyzed their data by their formulation based on the hydrodynamic regime. Gutzow and Schmelzer [16] introduced a formulation for the cooling-rate dependence of kinetic glass transition temperature based on the hopping-type relaxation. Johnson and his collaborators [17-19] carried out the DSC study for several kinds of metallic glasses. They showed that the glass transition behaviors depended on the heating rate, but detailed analysis of the data was not given. Several experimental studies on the kinetic glass transition were also made by other authors [20-24]. The materials used were mainly inorganic glasses, and metallic glass was not included. These authors analyzed their DSC data by adopting the hopping-like relaxation models. Finally, a theoretical study on the kinetic glass transition is cited. Ishii [25] constructed a formalism based on the hopping model using a nonequilibrium relaxation mode approach. He adopted our DSC data for checking his formalism.

After reviewing these studies, it seems that many investigators analyzed experimental data based on the relaxation of hopping type. We also adopted the same idea for analyzing the present experimental data [26]. However, in the present article, we adopted the VTF formalism for the analysis. This is because the data obtained by BS are to be compared with ours, and they have adopted that formalism. Further consideration and re-analysis of the data shall be made in the future.

REFERENCES

1.Y. Hiki, T. Yagi, A. Aida, and S. Takeuchi, *J. Alloys Comp.* **355**, 42-46 (2003).

2 Y. Hiki, *Mat. Sci. Eng.* A **370**, 253-259 (2004).

3.Y. Hiki, T. Yagi, T. Aida, and S. Takeuchi, *Mat. Sci. Eng.*A **370**, 302-306 (2004).

4 Y. Hiki, T. Aida, and S. Takeuchi, *Proc. 3rd Int. Sym. on Slow Dynamics in Complex Systems*, American Institute of Physics, NY, 2004, pp. 661-662.

5. Y. Hiki, M. Tanahashi, and S. Takeuchi, *Mat. Sci. Eng.* A **442**, 287-291 (2006).

6.W. L. Johnson, *Mater. Sci. Forum* **225/227**, 35-50 (1996).

7.W. L. Johnson, *MRS Bull.* October, 42-56 (1999).

8. A. Inoue, *Mater. Trans. JIM* **36**, 866-875 (1995).

9. A. Inoue, *Acta mater.* **48**, 279-306 (2000).

10. C. A. Angell, *J. Phys. Chem. Solids* **49**, 863-871 (1988).

11. H. Kobayashi, Y. Hiki, and H. Takahashi, *J. Appl. Phys.* **80**, 122-130 (1996).

12. H. Takahashi, Y. Hiki, and H. Kobayashi, *J. Appl. Phys.* **84**, 213-218 (1998).

13. H. Kobayashi, H. Takahashi, and Y. Hiki, *J. Non-Cryst. Solids* **290**, 32-40 (2001).

14. Y. Hiki, and Y. Kogure, *Recent Res. Devel. Non-Crystalline Solids* **3,** 199-231 (2003).

15. R. Brüning, and K. Samwer, *Phys. Rev.* B **46**, 11318-11322 (1992).

16 I. Gutzow, and J. Schmelzer, *The Vitreous State*, Springer, Berlin, 1995, pp. 83-86.

17. R. Busch, Y. J. Kim, and W. L. Johnson, *J. Appl. Phys.* **77**, 4039-4043 (1995).

18. R. Busch, W. Lu, and W. L. Johnson, *J. Appl. Phys.* **83**, 4134-4141 (1998).

19. S. C. Glade, R. Busch, D. S. Lee, W. L. Johnson, R. K. Wunderlich, and H. J. Fecht, *J. Appl. Phys.* **87**, 7242-7248 (2000).

20. C. T. Moynihan, A. J. Easteal, J. Wilder, and J. Tucker, *J. Phys. Chem.* **78**, 2673-2677 (1974).

21. C. Bergman, I. Avramov, C. Y. Zahra, and J.-C. Mathieu, *J. Non-Cryst. Solids* **70**, 367-377 (1985).

22. I. Avramov, and I. Gutzow, *J. Non-Cryst. Solids* **104**, 148-150 (1988).

23. A. Hallbrucker, and G. P. Johari, *Phys. Chem. Glasses* **30**, 211-214 (1989).

24. W. Pascheto, M. G. Parthun, A. Hallbrucker, and G. P. Johari, *J. Non-Cryst. Solids* **171**, 182-190 (1994).

25. T. Ishii, *Proc. 3rd Int. Sym. on Slow Dynamics in Complex Systems,* American Institute of Physics, NY, 2004, pp. 643-646.

26. Y. Hiki, and H. Takahashi, *Fourteenth Symposium on Thermophysical Properties*, Boulder, CO, 2000, unpublished.

Low-Energy Collective Excitations in Glassy and Supercooled Liquid Silica Probed by Inelastic X-ray Scattering

S. Hosokawa*, T. Ozaki*, S. Tsutsui†, and A. Q. R. Baron†,**

*Center for Materials Research using Third-Generation Synchrotron Radiation Facilities, Hiroshima Institute of Technology, Hiroshima 731-5193, Japan
†SPring-8/JASRI, Hyogo 679-5198, Japan
**SPring-8/RIKEN, Hyogo 679-5148, Japan

Abstract. The dynamic structure factor $S(Q,\omega)$ of glassy and suppercooled liquid SiO_2 was measured by a high energy-resolution inelastic X-ray scattering (IXS) spectrometer installed at BL35XU of the SPring-8, in a wide temperature range up to 1300 °C, a wide momentum transfer Q range from 1.3 to 37.2 nm^{-1}, and an energy transfer ω range of $\pm$40 meV. The acoustic-like propagating excitations can be seen even at the room temperature unlike the earliest IXS experiment, and become distinct with increasing temperature as in a previous IXS result, which continues up to the supercooled liquid phase. The temperature dependence of the features of the low-energy excitations is discussed within the context of buckling motions of the distorted ring structure.

Keywords: Boson peak, Collective dynamics, Disordered system, Glass transition, Inelastic excitation
PACS: 63.50.+x, 61.10.Eq

INTRODUCTION

Boson peak, appeared as an excess peak at about 5 meV in the vibrational density of states, is commonly observed in glasses, and its origin has widely been discussed for more than five decades. Features and the origin of Boson peak were often discussed with the relation of a microscopic and/or intermediate range structure, and sometimes explained together with the origin of the first sharp diffraction peak (FSDP) which is characteristic for the glasses. Besides localized or propagating vibrational modes, the low energy excitations of glasses can be involved also with relaxational and/or deformational motions of restricted structures. This fact has made the exploration of the low energy modes (including the Boson peak) of glasses difficult and complicated.

The origin of the Boson peak still remains open to question, and the proposed ideas seem to be still controversial. An inelastic X-ray scattering (IXS) measurement [1] gave an insight that the propagating modes must contribute to the Boson peak. On the contrary, an inelastic neutron scattering (INS) study [2] showed that the Boson peak is almost independent of momentum transfer Q, and they speculated that this low-energy excitation originates from a two-level system based on buckling motions of the oxygen atoms. Based on a recent IXS experimental result [3], highly localized transverse acoustic excitation mode is again thought as the origin of the Boson peak.

However, the IXS was measured at high temperatures of 750 – 1000 °C in the small Q range up to 16 nm^{-1}, while the INS was performed at room temperature beyond several nm^{-1}. Therefore, the spectra between the IXS and INS are hardly compared, although the difference of the scattering crosssections of Si and O is helpful for the understandings of the low-energy excitations.

In this study, we measured IXS at six temperatures from 20 to 1300 °C including the supercooled liquid region in a wide Q range of 1.3 – 37.2 nm^{-1} and an energy transfer ω range of $\pm$40 meV for obtaining a overall feature of the low-energy collective excitations appeared in glassy and supercooled SiO_2.

EXPERIMENTAL PROCEDURE

The IXS experiments were performed at the beamline BL35XU in the SPring-8 using a high energy-resolution IXS spectrometer [4]. A monochromatized beam of 3.5 $\times$ 10^9 photons/s was obtained from a cryogenically cooled Si(111) double crystal followed by a Si (11 11 11) monochromator operating in extremely backscattering geometry (89.975°, 21.75 keV). The same backscattering geometry of twelve two-dimensionally curved Si analyzers was used for the energy analysis of the scattered X-ray photons. The energy resolution was determined by the scattering from a Plexiglas sample and values of 1.6-1.9 meV (FWHM) were found for the detecting systems depending on the analyzer crystals. The Q resolution was set to be about $\pm$ 0.30 nm^{-1}.

The sample thickness was about 0.7 mm, being

CP982, *Complex Systems, 5th International Workshop on Complex Systems*
edited by M. Tokuyama, I. Oppenheim, and H. Nishiyama

slightly thinner than an 1/e absorber. The purity of the sample was 99.9999 %. It was placed in a vessel equipped with single-crystal Si windows. The high temperature up to 1300 °C was achieved using an Mo resistance heater, and monitored with two W-5%Re/W-26%Re thermocouples. The IXS experiments were carried out at about twenty Q values between 1.3 and 37.2 nm^{-1} covering an energy transfer range of about ± 40 meV, at temperatures of 20, 400, 700, 1000, 1200 °C in the glassy state and 1300 °C in the supercooled liquid phase. Experiments without sample were separately performed, and used for the background corrections.

RESULTS

Figure 1 shows the logarithmic plots of selected $S(Q,\omega)$ spectra of SiO_2 at the selected temperatures of (a) 20 °C, (b) 1000 °C (glassy phase), and (c) 1300 °C (supercooled liquid phase). The Q values are indicated at the right-hand side of each spectrum. The typical resolution functions are given by the dashed curves.

At room temperature of 20 °C in Fig. 1(a), $S(Q,\omega)$ in the very low Q region shows highly damped shoulders of the acoustic-like longitudinal propagating excitation modes. This was not observed in the earliest IXS experiment by Foret et al. [5], who used an IXS spectrometer with a worse energy resolution of about 3.0 meV, which may be the reason of the difference from the present experimental result.

With increasing Q, the acoustic-like excitations damp heavily, and almost the resolution function-like spectra continue up to about $Q = 15$ nm^{-1}, where the FSDP of SiO_2 locates in the static structure factor $S(Q)$. Beyond there, the excess signals at the tails of central line become visible, i.e., a broad low-energy excitation appears at about $5-7$ meV with a long tail, which seems to be the so-called Boson peak. It should be noted that in the previous INS experiment by Nakamura et al. [2], clear low-energy excitation signals can be detected at the whole measured Q values from 5 to 60 nm^{-1}, while the present IXS shows small low-energy excitations in the limited Q range. The difference may be due to the difference of the scattering crosssections of the elements, i.e., the X-ray scattering crosssection of O atoms in the IXS is much smaller than neutron scattering crosssection in the INS compared to the corresponding Si values.

At $T = 1000$ °C in the high-temperature glassy state, the magnitudes of the acoustic-like and the low-energy excitations drastically increase as shown in Fig. 1(b). In the low Q region below 4 nm^{-1}, distinct propagating excitation modes are seen as large shoulders at both the side of the central line. With increasing Q, the excitation modes become broad but are still observable up to 15 nm^{-1} in the present experiment. The recent IXS experiment by Ruzicka et al. was carried out up to $Q = 16.5$ nm^{-1} in a wide ω range up to 115 meV at the same temperature, and the propagating excitation modes are visible in the whole Q range measured. The present data are in good agreement with theirs within the overlapped $Q-\omega$ region.

At higher temperature of 1300 °C in the supercooled liquid region, the general feature of the $S(Q,\omega)$ spectra depicted in Fig. 1(c) is similar to those of the high temperature glass at 1000 °C given in Fig. 1(b). However, the magnitudes of the acoustic-like and the low-energy excitations increase further, and the acoustic-like excitation modes become clear peaks.

Figure 2 shows temperature dependence of $S(Q,\omega)$ spectra of glassy and supercooled liquid SiO_2 at $Q =$ 28.3 nm^{-1} to observe the evolution of the low-energy excitation in detail. The temperatures are 20, 400, 700, 1000 °C for the glassy phase, 1200 °C near the glass-transition region, and 1300 °C for the supercooled liquid state, given at the upper-right position of each spectrum. With increasing temperature, the excess signals appeared at the tails of the central peaks grow up. With the further increase of the temperature, the magnitude of the excess signals seems to be saturated at the glass-transition and supercooled liquid temperature ranges, but a decrease was not observed in the liquid phase. In the high Q region, in general, similar evolutions of the low-energy excitations are seen.

DISCUSSION

Firstly we shall discuss the acoustic-like propagating excitations in the glassy and supercooled liquid SiO_2. The energies and widths of the acoustic-like excitation modes were obtained by fitting the data using a damped harmonic oscillator (DHO) model [6] for the inelastic excitations and a delta function for the quasielastic line. The excitation energies obtained for the $S(Q,\omega)$ data of the room temperature glass are given as bars in Fig. 1(a). From the obtained excitation energy data, one can derive the velocity of sound of 6000 ± 300 m/s, which almost coincides the hydrodynamic value obtained from a Brillouin scattering measurement [7].

For the high-temperature glass data, fits were performed using a single DHO for the inelastic excitations at low Q values below 3 nm^{-1}. At Q values from 4 to about 10 nm^{-1}, however, a double DHO model was applied to obtain the parameters of the longitudinal and transverse propagating excitation modes as Ruzicka et al. performed [3], and the obtained excitation energies are given as bars in Fig. 1(b), which are in good agreement with the previous IXS work [3] within the overlapped $Q-\omega$ ranges. The velocities of the longitudinal and transverse sounds were calculated from the data in the low Q region

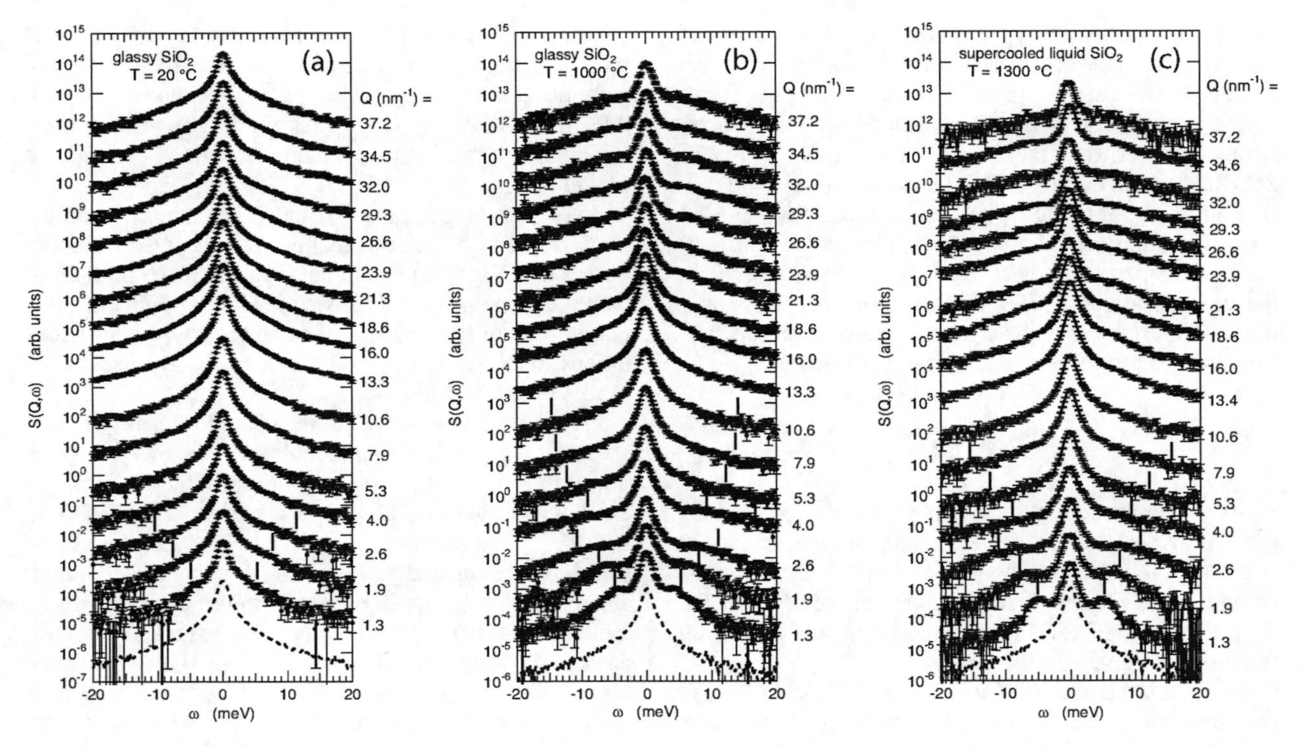

FIGURE 1. Logarithmic plots of selected $S(Q,\omega)$ spectra of SiO_2 at the selected temperatures of (a) 20 °C, (b) 1000 °C (glassy phase), and (c) 1300 °C (supercooled liquid phase). The Q values are indicated at the right-hand side of each spectrum. The typical resolution functions are given by the dashed curves. The bars below 2.6 nm^{-1} represent the excitation energies of the acoustic-like longitudinal modes obtained by a single DHO model, and those beyond 4.0 nm^{-1} the longitudinal and transverse modes estimated by a double DHO model. The longitudinal modes beyond 5.3 nm^{-1} locate outside of the figure.

to be 6500 ± 300 and 3700 ± 400 m/s, respectively, and in relatively good agreement with the hydrodynamic values obtained from Brillouin scattering [7].

The same procedure was applied to the $S(Q,\omega)$ data of the supercooled liquid SiO_2 at $T = 1300$ °C. The velocities of the longitudinal and transverse sounds were calculated from the data to be 6500 ± 300 and 3700 ± 400 m/s, respectively, and again in consistent with the hydrodynamic values interpolated from the Brillouin scattering measurement [7] for glassy and liquid SiO_2. The details of the acoustic-like collective excitations including their damping will be discussed elsewhere.

Next, we shall discuss the low-energy excitation modes appeared mainly in the high Q range. At room temperature, the present IXS and the previous INS [2] studies were carried out at the same $Q-\omega$ ranges. In both the experiments, excess signals of the low-energy excitations can be visible, but their intensities are different from each other and the observable Q region as well. In order to solve this contradiction between two methods, the origin of low-energy excitations should be discussed. Based on the INS results, Nakamura et al. [2] argued that a large averaged mean-square displacement of SiO_2 glass originates from a double-well potential model, where it is assumed that atoms or a group of atoms have two almost equivalent equilibrium positions in glass. An intuitive picture of the double-well potential model based on buckling motion was proposed by Nakayama [9], where buckling motions of disordered ring structure in glassy SiO_2 may produce a double-well potential with a shallow barrier height between two potential wells. Since the buckling motions of the rings mainly happen at the O site, the magnitude of the inelastic signals measured by IXS would be suppressed by comparing the INS results because the X-ray scattering crosssection of the O atoms in the IXS is much smaller than that in the INS. Thus, it can be concluded that the model of a two-level system based on buckling motions of the O atoms does not contradict the present IXS result.

As seen in Fig. 2, the magnitudes of the low-energy excitation signals drastically increase with increasing temperature. In order to obtain the vibrational density of state, however, the temperature effect should be taken into account. Figure 3 shows temperature dependence of the low-energy collective excitations in glassy and supercooled liquid SiO_2 obtained by subtracting the delta-function like quasielastic peak, and then taking the Bose factor into account. At room temperature, the low-energy

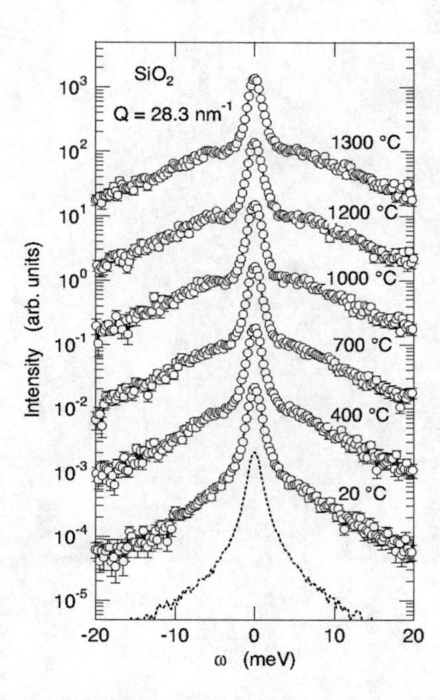

FIGURE 2. Temperature dependence of $S(Q,\omega)$ spectra of glassy and supercooled liquid SiO_2 at $Q = 28.3\ \mathrm{nm}^{-1}$. The temperatures are given at the upper-right position of each spectrum.

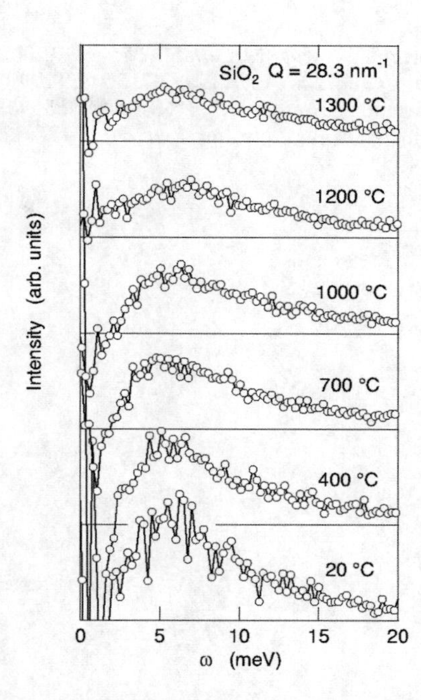

FIGURE 3. Temperature dependence of the low-energy excitations in glassy and supercooled liquid SiO_2 obtained by subtracting the delta-function like quasielastic peak, and then taking the Bose factor into account.

excitation spectrum has a peak at about 5.5 meV with a long tail towards the higher ω side up to about 20 meV. This feature is in good agreement with the INS data given in Fig. 6 of Ref. [2].

With increasing temperature, the intensity of the low-energy excitation modes gradually decreases, and the peak position seems to gradually shift towards the higher ω side up to the glass transition temperature of about 1200 °C, which is consistent with vibrational density of state results obtained from a Raman scattering [8] and an INS [10]. Across the glass transition temperature, however, the intensity remains unchanged but the peak position slightly shifts towards the lower ω side, which may be due to the softening of the buckling mode caused by crossing the glass transition temperature.

CONCLUSION

The $S(Q,\omega)$ of glassy and suppercooled liquid SiO_2 was measured by a high energy-resolution IXS spectrometer installed at BL35XU of the SPring-8, in a wide temperature range up to 1300 °C. The acoustic-like propagating excitations can be seen even at the room temperature unlike the earliest IXS experiment, and become distinct with increasing temperature as in a previous IXS result, which continues up to the supercooled liquid phase. The temperature dependence of the features of the low-energy excitations is discussed within the context of buckling motions of the distorted ring structure.

ACKNOWLEDGMENTS

The authors greatly appreciate Professor K. Matsuishi of the University of Tsukuba for helpful discussion. This work was financially supported by Nippon Sheet Glass Foundation for Materials Science and Engineering. The SiO_2 sample was provided by Dr. A. Takada and Dr. T. Minematsu of Asahi Glass Co. Ltd. The IXS experiments were performed at the beamline BL35XU of the SPring-8 with the approval of the Japan Synchrotron Radiation Research Institute (Proposal No. 2004B0597 and 2005B0124).

REFERENCES

1. P. Benassi et al., *Phys. Rev. Lett.* **77**, 3835-3838 (1996).
2. M. Nakamura et al., *J. Non-Cryst. Solids* **293-295**, 377-382 (2001).
3. B. Ruzicka et al., *Phys. Rev.* B **69**, 100201-1-4 (2004).
4. A. Q. R. Baron et al., *J. Phys. Chem. Solids* **61**, 461-465 (2000).
5. M. Foret et al., *Phys. Rev. Lett.* **77**, 3831-3834 (1996).
6. B. Fåk and B. Dorner, *Institut Laue-Langevin Reports*, 92FA008S (1992).

7. A. Polian et al., *Europhys. Lett.* **57**, 375-381 (2002).
8. A. Fontana et al., *Europhys. Lett.* **47**, 56-62 (1999).
9. T. Nakayama, *J. Phys. Soc. Jpn.* **68**, 3540-3555 (1999).
10. U. Buchenau et al., *Phys. Rev. Lett.* **60**, 1318-1321 (1988).

Glass Transition Behaviors of Ethylene Glycol – Water Solutions Confined within Nano-Pores of Silica Gel

Atsushi Nagoe and Masaharu Oguni

Department of Chemistry, Graduate School of Science and Engineering, Tokyo Institute of Technology, 2-12-1 O-okayama, Meguro-ku, Tokyo 152-8551, Japan

Abstract. Enthalpy relaxation properties of the ethylene glycol (EG) aqueous solutions confined within silica-gel void spaces of 1.1 nm in the average void thickness and 6, 12 and 52 nm in their average diameters were examined by an adiabatic calorimetry to understand the glass transition behavior of the solutions and the rearrangement processes of the molecules. The glass transition temperature T_g of EG was found to decrease with adding the water molecules which are mobile under the condition lacking in the full hydrogen-bond network. Meanwhile, the T_g in the water-rich region showed a rise towards pure water; after a phase separation in a 25 mol% (x = 0.25) EG solution, the T_g was 160 K which was higher than that derived by extrapolating the composition dependence to pure water. The T_g = 160 K is the same as observed in the pure water confined within 1.1 nm voids; this indicates the validity of the interpretation that the glass transition at 160 K of the confined water originates from the internal, but not interfacial, water molecules. A risk is also indicated of extrapolating the composition dependence of the T_g to pure water from the concentrated water solution range.

Keywords: Ethylene glycol, Glass transition, Hydrogen bond, Water
PACS: 61.25.Em, 61.46.Fg, 64.70.Pf, 65.20.+w, 65.60.+a

INTRODUCTION

Where water displays its glass transition is our great concern. The glass transition temperature T_g of the amorphous water obtained by rapid quenching has been claimed to be 136 K by DSC measurements [1]. But this value was questioned [2] and some research groups reassigned it to about 165 K [3]. Since the amorphous water prepared as such crystallizes quickly above 150 K, the issue of where the real T_g of water is has not been solved yet. Confining water within nm-size space enables to avoid its crystallization [4]. Swenson et al. speculated that the glass transition observed at around 136 K in the bread and so on with water is attributed to the β-like relaxation of the water and further that the α glass transition would be present at around 165 K [4]. This speculation is also obscure without decisive experimental results.

Recently, we have succeeded in observing two glass transitions at T_g = 115 and 160 K in the water confined within silica-gel void spaces of 1.1 nm average void thickness by an adiabatic calorimetry

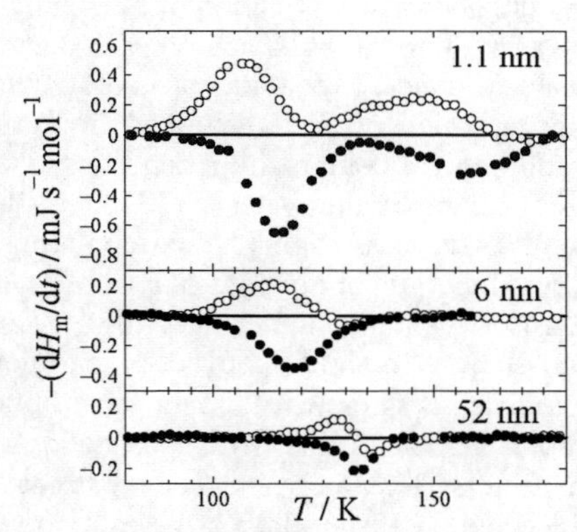

FIGURE 1. Temperature dependence of the rates of spontaneous heat release or absorption observed in the measurements [5]: 1.1 nm, 6 nm, and 52 nm are the average void thickness or pore diameters. ○, cooled rapidly at around 5 Kmin^{-1} before the measurements, ●, cooled slowly at 10 mKmin^{-1}. The systematic heat-release or -absorption effects depending on the cooling rates are characteristic of a glass transition. The glass transition temperature T_g was determined according to an empirical relation [6] that the slowly cooled sample shows the maximum heat-absorption rate at the T_g where the relaxation time becomes 1 ks.

CP982, *Complex Systems, 5th International Workshop on Complex Systems*
edited by M. Tokuyama, I. Oppenheim, and H. Nishiyama

(Fig. 1) [5]. When confined within larger pores of average diameters than 2 nm, most of the water was judged to crystallize except the interfacial water molecules which are located on the pore wall of silica gel. The glass transition related to this interfacial water was observed in 119~132 K even in the cases where the water molecules located in the pore center crystallized. Therefore, the glass transition at 160 K in the water confined within silica-gel void spaces of 1.1 nm was interpreted as due to the internal water molecules that were surrounded only by water molecules. It was suggested that the glass transition of bulk supercooled water takes place at 160 K as well. But the glass transition at 160 K was observed only within silica-gel void spaces of 1.1 nm where only one molecule can be accommodated except interfacial molecules and which is quite different in the size from the bulk water. Moreover, one might imagine a possibility that the observed glass transitions at 160 K and 115 K are α and β relaxation processes respectively of the confined water. And the view that the relaxation times of the interfacial and internal molecules in the confined water are different from each other is not so common at present. In such situations, knowledge about the relaxation behavior of the water confined within larger space is indispensable to confirm the above interpretation of ours [5].

The glassy state of bulk sample can be realized rather easily in concentrated aqueous solutions. Many research groups have, in fact, estimated the T_g of water by extrapolating its composition dependence to pure water; they suggested the water's T_g is around 136 K [2]. Recently, Cerveny et al. [7], extrapolating the T_g data of propylene glycol aqueous solutions, suggested T_g = 162-165 K which is higher than 136 K. However, this extrapolation method involves a problem; the composition range where the T_g data were obtained is considerably high and the hydrogen-bond network structure peculiar to water is expected to be absent. Angell expected an abrupt drop in T_g value potentially to occur by mixing a small quantity of second component into pure water [8] in analogy to the behavior of SiO_2 which makes a tetrahedral network structure as well. In order to examine this possibility, it is required to obtain the T_g data in the very low second-component concentration close to pure water.

In this study, we tried to obtain information about the glass transition behavior of dilute aqueous solutions closer to bulk pure water. Ethylene glycol (EG) - water solutions were used and confined within silica-gel pores of 1.1, 12, and 52 nm in diameters. Enthalpy relaxation properties were examined by an adiabatic calorimetry. It is aimed to confirm our interpretation on the glass transition behavior of the water confined within void spaces of 1.1 nm [5] and to observe the behaviors of the aqueous solutions at very low EG concentrations in which water usually crystallize easily.

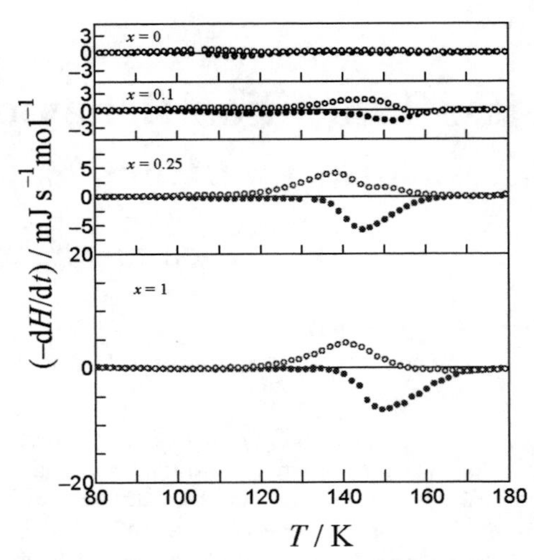

FIGURE 2. Temperature dependence of the rates of spontaneous heat release or absorption in the EG aqueous solutions confined within voids of 1.1-nm average thickness with different EG mole fractions x.

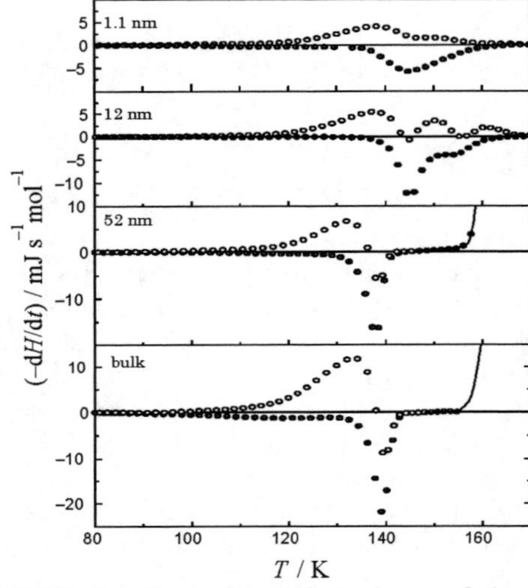

FIGURE 3. Temperature dependence of the rates of spontaneous heat release or absorption in the x = 0.25 EG aqueous solutions confined within the pores of different average diameters.

RESULTS AND DISCUSSIONS

Figure 2 shows temperature dependence of the rates of spontaneous heat-release or -absorption of the EG aqueous solutions, within pores of 1.1-nm average void thickness, with different EG mole fractions x. Open and filled circles represent the results of the samples pre-cooled rapidly at about 10 $Kmin^{-1}$ and slowly at about 20 $mKmin^{-1}$, respectively, from 200 K to the lowest temperature of the measurements. The

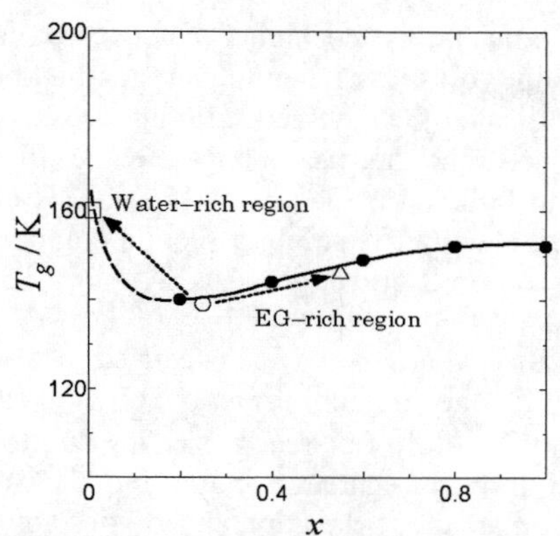

FIGURE 4. Composition dependence of the glass transition temperatures of EG aqueous solutions: ●, taken from ref 10; ○, result of the $x = 0.25$ EG aqueous solution before the phase separation; □, △, result after the phase separation. The present T_g (○) is a little lower than the previous DSC values [10] because of the difference in the characteristic times of the calorimetric methods.

heat-release and -absorption phenomena definitely depend on the pre-cooling rates and show the characteristics to be found in glass transitions. The T_g of the pure EG ($x = 1$) confined within the void spaces is 148 K, almost the same as that of bulk EG [9], and no difference is found between the relaxation properties of the internal and interfacial molecules without double heat-release or -absorption effects. This observation is consistent with the result of no pore-size dependence reported in the dielectric relaxation times by Huwe et al. [11]. Small heat-release or –absorption effects due to a glass transition appeared around 115 K in the x = 0, 0.1, and 0.25 samples. This glass transition is recognized as due to water molecules because no effect was detected in pure EG. The heat-release or -absorption phenomena around 140 K in the $x = 0.25$ sample are considered as composed of the superposition of those of EG and water as understood from the results shown in Fig. 3.

Open and filled circles in Fig. 3 represent the rates of spontaneous heat release or absorption of the x = 0.25 EG aqueous solutions pre-cooled rapidly at about 10 Kmin^{-1} and slowly at about 20 mKmin^{-1}, respectively, from 160 K to the lowest temperature of the measurements. Average diameters of the pores are written at the upper left in each portion of the figure. Glass transitions around 115 K are found in all the cases and understood as due to one part of water molecules. Both in the 1.1 and 12 nm samples, definitely double heat-release effects are found with their peaks at around 138 and 152 K. These are considered as ascribed to EG molecules and the other part of water molecules respectively, as stated in detail below, as a result of a phase separation which proceeded even in the cooling procedure on account of the pore wall present. The EG molecules both in bulk and 52-nm samples are interpreted to undergo a glass transition around 139 K; this indicates that the T_g of EG molecules decreases with mixing water molecules as shown as a composition dependence of T_g in Fig. 4.

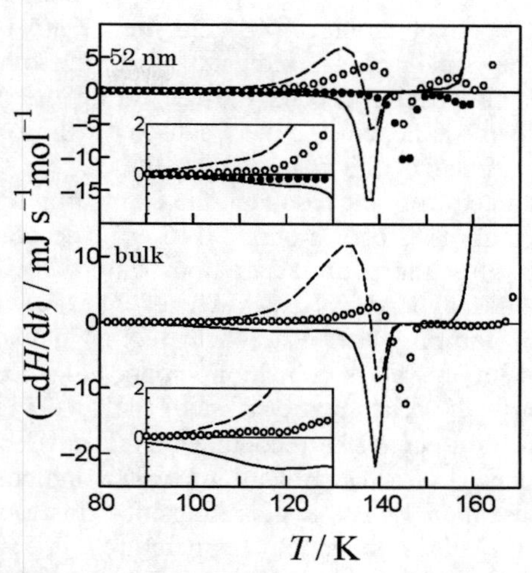

FIGURE 5. Effect of ageing at 160 K on the temperature dependence of the rates of spontaneous heat-release or absorption in the $x = 0.25$ EG aqueous solutions confined within 52-nm pores and in bulk. Solid and dashed lines represent the temperature dependence of the rates before the ageing treatment (see Fig. 3). Insets show the dependence around 115 K on an enlarged scale.

Large heat-release effect appeared around 160 K both in 52-nm and bulk samples; the corresponding effect in the 12-nm sample was detected likewise as a small one. After ageing at 160 K for 12 h, the glass transition behaviors of the 52-nm and bulk samples changed drastically. Fig. 5 shows the changes observed. The glass transition at 139 K of the respective solutions disappeared and two new glass transitions appeared instead in the 52-nm sample; one appeared at T_g = 146 K which is close to the T_g of pure EG, and the other was found at T_g = 160 K. The T_g = 160 K is higher than T_g = 148 K in pure EG and equal to the T_g observed in pure water within voids of 1.1-nm average thickness (see Fig. 1). These changes indicate that the heat-release effect around 160 K was due to a phase separation leading to the formation of two regions; EG-rich and water-rich regions. The ageing treatment therefore enhanced the phase separation. It is reasonable to consider from the result that the T_g = 146 K glass transition originates from the EG-rich region and the T_g = 160 K glass transition from the water-rich region. In the bulk sample, only the T_g = 146 K glass transition was observed. The absence of the T_g = 160 K glass transition is interpreted potentially as the water-rich region grew to cause the crystallization to ice. These understandings are completely consistent with the previous interpretation on the results of the pure water confined within the voids of 1.1-nm average thickness [5]. The heat-release effect around 160 K in the 12-nm case was rather small. This indicates that the phase separation process proceeded mostly during the pre-cooling. Almost no change was caused to the other relaxation effects by the treatment.

It is noticed from the results in the 52 nm pores that the relaxation effect around 115 K decreased drastically after the phase separation while the effect around 160 K appeared. This evidences that the water molecules forming the water-rich region have no relation with the glass transition around 115 K; in other words, the relaxation process of the T_g = 115 K glass transition cannot be recognized as, if any, the β one intrinsic to the aggregate of water molecules which cause the T_g = 160 K glass transition. Instead, in general, the glass transition temperatures of water molecules should be understood as determined by the local environment surrounding them. The glass transition at 160 K would be due to the water molecules developing more hydrogen-bonds and the transition at 115 K to the water molecules having the ethylene groups of EG molecules as their neighbors.

In the 1.1 and 12 nm samples, the phase separation occurs during the pre-cooling, as stated above. This is understood as water molecules tend to aggregate in the pore center to form hydrogen bonds among them, and consequently as EG molecules are pushed out on the pore wall where Si-O-Si bonds with a hydrophobic attribute are found. The hydrogen-bonding process of water molecules thus tend to proceed even during the cooling in the both pores of small diameters.

Both the two regions formed through the phase separation in the 52 nm pores have higher T_g values, 146 K and 160 K, than T_g =138 K of well-mixed solution. This reflects the composition dependence of T_g shown in Fig. 4. The T_g of pure EG decreases gradually with increasing water content; the presence of the water molecules, as its neighbors, which form insufficient hydrogen-bonds and are more mobile than EG is understood to make the rearrangement of the EG molecules easier. The EG-rich region is interpreted to have approached pure EG in the composition through the phase separation. In view that the T_g of the water-rich region increased very much, the T_g should change drastically in the range close to pure water as shown in Fig. 4. This is understood as pure water develops a peculiar hydrogen-bond network in the range. It is evidenced that the extrapolation of the T_g values in the concentrated aqueous solution range to pure water composition is quite risky.

CONCLUSIONS

We observed the glass transition taking place in the water-rich region which was formed through a phase separation in the rather large 52 nm pores. The T_g = 160 K is the same as the higher T_g observed in pure water within voids of 1.1-nm average thickness. It strongly indicates that the relaxation process of T_g = 160 K is due to the internal water molecules, the glass transition in bulk water occurs at 160 K as well, and the T_g drops steeply by adding a second component in the dilute composition range. The T_g = 160 K is the same temperature as expected by Velikov et al [3]. Glass transition at around 115 K is caused by the water molecules in the neighborhood of EG's ethylene groups, and therefore not recognized as due to the β process in the aggregate of water molecules. It is concluded that the relaxation properties of water molecules depend only on the local environment surrounding them, namely a degree of the development of hydrogen-bond network. The steep change in the T_g in the dilute aqueous solutions is reasonably connected with the accelerating development/collapse of the network.

ACKNOWLEDGMENTS

We would like to thank Fuji Silysia Chemical Ltd., Japan for providing us kindly with the silica gel. This work was financially supported partly by a Grant-in-Aid for Scientific Research (Grant 18350003) from the Ministry of Education, Culture, Sports, Science, and Technology, Japan.

REFERENCES

1. A. Hallbrucker, E. Mayer, and G. P. Johari, *Phil. Mag.* B **60**, 179-187 (1989).
2. D. R. MacFarlane, and C. A. Angell, *J. Phys. Chem.* **88**, 759-762 (1984).
3. V. Velikov, S. Borick, and C. A. Angell, *Science* **294**, 2335-2338 (2001).
4. S. Cerveny, G. A. Schwartz, R. Bergman, and J. Swenson, *Phys. Rev. Lett.* **93**, 245702-1-245702-4 (2004).
5. M. Oguni, S. Maruyama, K. Wakabayashi, and A. Nagoe, *Chem. Asian J.* **2**, 514-520 (2007).
6. H. Fujimori and M. Oguni, *J. Phys. Chem. Solids* **54**, 271-280 (1993).
7. S. Cerveny, G. A. Schwartz, A. Alegria, R. Bergman, and J. Swenson, *J. Chem. Phys.* **124**, 194501-1-194501-9 (2006).
8. C. A. Angell, *Chem. Rev.* **102,** 2629-2650 (2002).
9. K. Takeda, O. Yamamuro, I. Tsukushi, T. Matsuo, and H. Suga, *J. Mol. Str.* **479**, 227-235 (1999).
10. C. A. Angell, J. M. Sare, and E. J. Sare, *j. Chem. Phyis.* **82**, 2623-2629 (1978).
11. A. Huwe, F. Kremer, J. Kärger, P. Behrens, W. Schwieger, G. Ihlein, and Ö. Weiß, and F. Schüth , *J. Mol. Liq.* **86**, 173-182 (2000).

Ordering and Glass-Transition Behaviors of the Water Confined between the Silicate Layers in Na(H)-RUB-18

Keisuke Watanabe and Masaharu Oguni

Department of Chemistry, Graduate School of Science and Engineering, Tokyo Institute of Technology 2-12-1 Ookayama, Meguro-ku, Tokyo, 152-8551, Japan

Abstract. Adiabatic calorimetry was carried out for Na(H)-RUB-18 crystals with and without the water confined between the silicate layers. The hydrated sample was prepared by allowing the Na cations of Na-RUB-18 to be ion-exchanged in a 0.2 moldm^{-3} HCl aqueous solution. A heat-capacity peak was found at 194 K and two glass transitions at T_g = 96 K and 130 K in the hydrate, while no anomaly was detected in the anhydride. The heat capacity peak and the glass transition around 100 K were reported in the hydrated Na-RUB-18 crystal as well. On the other hand, the glass transition at 130 K was newly found. The difference in the T_g values indicates that the hydrogen-bond networks formed are different between the water molecules coordinated to Na-ions and confined in H-RUB-18.

Keywords: Calorimetry, Glass transition, Nano channel, Phase transition, Water
PACS: 61.43.Gt, 64.70.Md, 64.70.Pf, 65.40.-b, 65.80.+m

INTRODUCTION

A water molecule has a simple molecular structure, however the behavior of the water as a molecular aggregate changes dramatically depending on the environment where the water molecules are located. Bulk water easily crystallizes below 273 K, but the water confined within pores in diameter below 2 nm forms a glassy state [1]. This change has been interpreted to reflect that the development of a hydrogen-bond network of water molecules depends on the size of the aggregate, but the detailed structural and dynamic characters are unclear. Confined water is, meanwhile, present ubiquitously on the earth in various kinds of nano spaces such as in layered clay silicates, zeolites, biological membranes, and so on. Detailed understanding on the static and dynamic properties of the confined water is not only intriguing in itself but also important from the viewpoints of applied chemistry, biology, and geology.

Na-RUB-18, $Na_8[Si_{32}O_{64}(OH)_8]32.0H_2O$, is one of crystalline silicates with a layered structure. The crystal structure including the positions of water molecules was determined at room temperature by powder X-ray diffractometry and Rietvelt refinement [2]: Sodium cations and water molecules are confined in the two-dimensional nano-spaces between the silicate layers. The thickness of the void space is of the size in which two or three water molecules are arrayed. However, the progress of the orientational ordering of the water molecules and the dynamic properties at low temperatures are unknown yet, and only few studies have been done about H-RUB-18 crystal in which the above sodium ions are replaced by protons. The replacement is expected to change the hydrogen-bond network formed among the water molecules. In the present study, it is aimed to investigate the static and dynamic behaviors of the confined water at low temperatures by using an adiabatic calorimetry and to compare the results with those in the Na-RUB-18 crystal [3].

EXPERIMENTAL

The hydrated Na-RUB-18 crystal was synthesized in the procedure reported by Kosuge et al. [4]. Na(H)-RUB-18 was obtained by ion-exchanging the Na cations to protons; the exchange was executed by stirring the Na-RUB-18 polycrystals in a 0.2 moldm^{-3} HCl aqueous solution. The exchange was expected to have occurred partially. Heat capacity measurements were carried out for the Na(H)-RUB-18 polycrystals with extra bulk water by using an adiabatic calorimeter [5]. The amount of the extra bulk water was estimated from the enthalpy of the fusion observed at 273 K in the calorimetry.

The anhydride Na(H)-RUB-18 was prepared by removing the water molecules of the hydrated sample

CP982, *Complex Systems, 5th International Workshop on Complex Systems*
edited by M. Tokuyama, I. Oppenheim, and H. Nishiyama

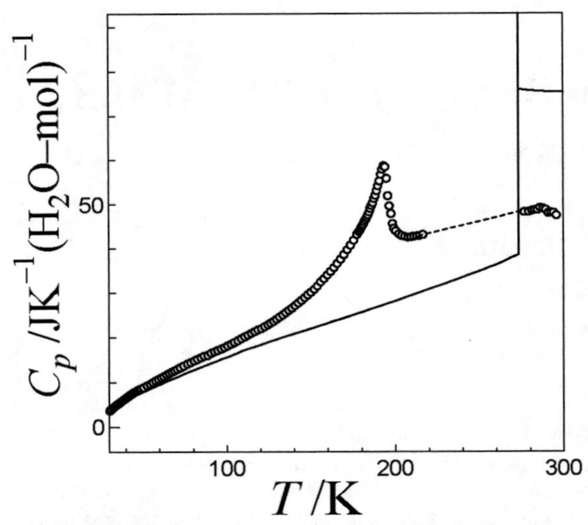

FIGURE 1. Temperature dependence of molar heat capacities of the water confined between silicate layers of Na(H)-RUB-18

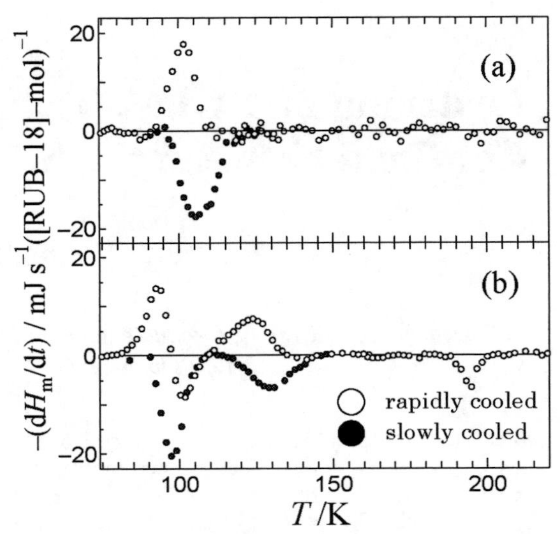

FIGURE 2. Temperature dependence of spontaneous enthalpy-release and -absorption rates observed in (a) Na-RUB-18 and (b) Na(H)-RUB-18: open circles, sample pre-cooled rapidly at ca. 2 Kmin-1; filled circles, sample pre-cooled slowly at ca. 20 mKmin-1.

with drying it in vacuum at 297 K for 3 days after the measurements of the hydrate had been finished. The amount of the water included within the silicate layers was evaluated by subtracting that of the extra bulk water from the total amount of water removed by the drying.

RESULTS AND DISCUSSION

Figure 1 shows the molar heat capacities of the water confined within the voids in Na(H)-RUB-18. The heat capacities were derived by subtracting the contributions of the bulk water and the anhydride sample from the values of the hydrated Na(H)-RUB-18. Solid lines represent the heat capacity curves of bulk pure ice and water. A heat capacity peak was observed at 194 K. Because no anomaly was found in the anhydride, it was interpreted as attributed to the structural ordering of the water molecules confined within the voids. The long low-temperature tail of the heat-capacity peak and the temperature 194 K of the peak are quite similar to those observed in the hydrated Na-RUB-18. This indicates that the same ordering of the water molecules proceeds in both the crystals. In view that the position of each water molecule was determined as fixed by Vortmann et al. [2], it is concluded that only an orientational degree of freedom remains as mobile in the water molecules. The anomaly is thus assigned as a phase transition originating from the orientational ordering of the water molecules as reported previously [3].

Figure 2 shows the rates of spontaneous heat-release or -absorption effects observed while the hydrated Na-RUB-18 (a) and Na(H)-RUB-18 (b) samples were heated intermittently for the heat capacity measurements. Two sets of the heat-release and -absorption effects depending on the pre-cooling rates were found around 100 K and 130 K in the latter sample (see Fig. 2(b)). Each set of the effects is characteristic of a glass-transition phenomenon. The transition temperatures where the relaxation time becomes 1 ks were determined to be T_g = 96 K and 130 K, respectively, as the temperature at which the endothermic relaxation rate for the slowly cooled sample showed its maximum. This empirical correspondence has been observed to hold to a good approximation [5, 6]. The former glass-transition is assigned as attributed to the freezing of the rearrangement of the water molecules coordinated to Na cations, because the glass transition was observed around 100 K also in the Na-RUB-18 sample (see Fig. 2(a) [3]). This indicates that the ion exchange, replacing Na cations with protons, was incomplete and most of the Na cations were left between silicate layers. The latter glass-transition at 130 K is assigned as the freezing of the water molecules in the absence of Na cations. While all the water molecules exist as interfacial ones in the Na-RUB-18 crystal with making contact with the silicate wall, part of the water molecules exist potentially as surrounded only by other water molecules in the H-RUB-18 crystal. The transitions at T_g = 96 K and 130 K are therefore considered as originating respectively from the interfacial water molecules coordinated to Na-ions and from the ones surrounded by other water molecules in the region with protons. It is implied that the water molecules in H-RUB-18 potentially form a two-dimensionally extended hydrogen-bond network within the layered nano-space. The activation energy of the rearrangement process giving $T_g = 130$ K can be estimated to be $\Delta\varepsilon_a = 42$ kJmol^{-1} according to a relation $\Delta\varepsilon_a = 39RT_g$ [7]. This activation energy corresponds roughly to the energy to break two or three hydrogen bonds with a normal OH···O bond length of 0.27 – 0.28 nm.

The phase transition at 194 K in the Na(H)-RUB-18 crystal showed a heat absorption effect. In view that no corresponding effect was detected in the Na-RUB-18 crystal, it would be related with the aspect that the replacement of the Na-ions by protons was only partial so that the orientational ordering of the water molecules could not proceed completely homogeneously over the crystal. Basically, the orientation of each water molecule below 194 K should be the same between in the Na(H)-RUB-18 and Na-RUB-18 crystals, but the presence of the protons replacing Na-cation sites would prohibit the homogeneous progress of the phase transition over the whole crystal.

REFERENCES

1. M. Oguni, S. Maruyama, K. Wakabayashi, and A. Nagoe, *Chem. Asian. J.*, **4** (4), 514-520 (2007)
2. S. Vortmann, J. Rius, S. Siegmann, and H. Gies, *J. Phys Chem. B*, **101,** 1292-1297 (1997).
3. K. Watanabe and M. Oguni, *Bull. Chem. Soc. Jpn.* **80** (9), 1758-1763 (2007).
4. K. Kosuge and A. Tsushima, *J. Chem. Soc., Chem. Commun.* 2427-2428 (1995).
5. H. Fujimori and M. Oguni, *J. Phys. Chem. Solids*, **54**, 271-280 (1993).
6. M. Oguni, T. Matsuo, H. Suga, and S. Seki, *Bull. Chem. Soc. Jpn.*, **50** (4), 825-833 (1977).
7. M. Oguni, *J. Non-Cryst. Solids*, **210**, 171-177 (1997).

Novel Effects of Confinement and Interfaces on the Glass Transition Temperature and Physical Aging in Polymer Films and Nanocomposites

John M. Torkelson[1,2,*], Rodney D. Priestley[1], Perla Rittigstein[1], Manish K. Mundra[2], and Connie B. Roth[1]

[1]*Dept. of Chemical and Biological Engineering and* [2]*Dept. of Materials Science and Engineering, Northwestern University, Evanston, Illinois 60208 USA*

Abstract. Recently, it has become evident that the magnitude of the glass transition (T_g)-confinement effect depends strongly on the polymer repeat unit and that the magnitude of the physical aging rate can be dramatically reduced relative to neat polymer when attractive polymer-nanofiller interactions are present in well-dispersed nanocomposites. However, in neither case has a quantitative, fundamental understanding been developed. By studying polymers with different chain backbone stiffness, e.g., polystyrene (PS) vs. polycarbonate (PC) vs. polysulfone (PSF) and that lack attractive interactions with the substrate interface, we show that the T_g-confinement effect is the weakest in the polymer with the least stiff backbone (PS) and strongest in the polymer with the most stiff backbone (PSF). These results are consisten with the notion that, other things being equal, a larger requirement by the polymer for the cooperativity of the segmental mobility that is associated with the glass transition will result in a greater reduction of T_g near the free surface of the film and thus the average T_g across the film. A quantitative understanding of the causes behind the effects of confinement on the glass transition and physical aging of nanocomposites with dispersed nanofiller is prevented by the complex nanoparticle distribution. Here, we have developed model silica nanocomposites with a single, quantifiable interlayer spacing equal to the film thickness separating two silica slides. Comparisons show that the model nanocomposites yield results consistent with the complex real nanocomposites. This provides the possibility to conduct studies that will allow for an understanding of how the separation distance between nanofiller interfaces impacts the glass transition and physical aging.

Keywords: Glass transition, Nanocomposites, Physical aging, Thin polymer films
PACS: 64.70.kj, 64.70km, 64.70.pj

INTRODUCTION

The study of the effects of confinement on the properties of amorphous polymers began more than a dozen years ago.[1,2] Since that time, it has been observed that the glass transition temperature, T_g, can undergo major changes relative to the bulk T_g ($T_{g,\text{bulk}}$) in polymer films that are confined to a nanoscale thickness. In freely standing films, T_g decreases with decreasing nanoscale thickness.[3,4] However, in substrate-supported films, T_g may decrease or increase with decreasing nanoscale thickness.[1,2,5-10] In those cases in which T_g decreases with confinement, it is understood that the effect is associated with how the presence of the free surface locally reduces the requirement for cooperative segmental mobility not simply at the free surface but some tens of nanometers away from the free surface.[7] In those cases in which T_g increases with confinement, it is understood that the effect is associated with how attractive polymer-substrate interfacial interactions, e.g., hydrogen bonds, increase the requirement for cooperativity in the segmental mobility associated with T_g.[5,6,9] These attractive interfacial interactions not only perturb the T_g response within the interior of the nanoscopically confined film but also must do so in a way that overcomes the tendency of the free surface to lead to a reduction of the effective average T_g of the confined film.

In contrast to the many studies devoted to the effect of film thickness on T_g, far less research has focused on other confinement effects in amorphous polymers. For example, few studies have been devoted to characterizing and understanding how confinement via addition of well-dispersed nanofiller or nanoparticles to polymers affects T_g.[11-15] The results

CP982, *Complex Systems, 5th International Workshop on Complex Systems*
edited by M. Tokuyama, I. Oppenheim, and H. Nishiyama

associated with these nanocomposites studies are largely consistent with the results on confined films, i.e., free surfaces tend to reduce T_g while attractive interactions at the nanofiller or nanoparticle interface tend to increase T_g. Likewise, there has been little study of the effect of confinement in either thin-film geometries or nanocomposites with low nanoparticle loadings on physical aging[14-20] or the change in properties associated with the relaxation of non-equilibrium, glassy polymers toward equilibrium.

Here we provide results on the effects of confinement in both thin-film geometries and nanocomposites on T_g and physical aging in a variety of amorphous polymers. In combination with previously reported results, these results support several key conclusions associated with effects of confinement on glass-forming polymers. First, other things being equal and in the absence of attractive interfacial interactions, an increase in the requirement for cooperativity in the segmental motion associated with T_g caused by structural features in the polymer repeat unit leads to a stronger decrease in polymer T_g with decreasing film thickness. Second, polymer-nanocomposites with well-dispersed nanoparticles yield inherently complex systems for fundamental study because of the broad distribution of distances between nanoparticle interfaces. However, in the case of polymer-silica nanocomposites, "model nanocomposites" may be produced in which the interlayer spacing can be tuned allowing the simplified study of the effect of confinement on both the T_g and physical aging responses of nanocomposites. Third, nanocomposites with less than 1 vol% of well-dispersed nanofiller that forms attractive interfacial interactions with the polymer can result in glassy materials with strong suppression of physical aging.[14,15] This has important implications for a previously unrecognized technological application of nanocomposites.

EXPERIMENTAL

Sample preparation and measurements for assessing T_gs and physical aging by fluorescence are described in refs. 7-9, 14-17 and 21-23. Poly(methyl methacrylate) (PMMA), poly(2-vinyl pyridine) (P2VP) and silica used in the real and model nanocomposites studies are those described in refs. 14 and 15. The bisphenol-A polycarbonate (PC) (nominal M_n = 17,300 g/mol, M_w/M_n = 1.65, and $T_{g,\mathrm{bulk}}$ = 415 K by DSC) and bisphenol-A polysulfone (PSF) (nominal M_n = 20,400 g/mol, M_w/M_n = 3.30, and $T_{g,\mathrm{bulk}}$ = 463 K by DSC) were purchased from Scientific Polymer Products and used as received.

RESULTS AND DISCUSSION

Figure 1 shows the effect of confinement in supported films on T_g - $T_{g,\mathrm{bulk}}$ for three polymers that lack significant attractive interactions with the silica substrate: polystyrene (PS), PC, and PSF. The backbone chain stiffness, and thereby the requirement for cooperativity in the segmental mobility associated with T_g in these three polymers, increases in the order PS < PC < PSF. This is also the order in which the T_g-confinement effect increases. In particular, in 32-nm-thick films on silica, T_g - $T_{g,\mathrm{bulk}}$ = -8 K in PS, -12 K in PC, and -24 K in PSF. Previous studies comparing the roles of side-chain steric effects (PS, poly(4-methyl styrene) and poly(*t*-butyl styrene) systems[8]) and addition of low molecular weight diluents to polymer films[22,23] have indicated that the cooperativity requirements for segmental mobility associated with T_g are strongly correlated with the magnitude of the T_g-confinement effect in supported polymer films. The results in Figure 1 make evident that the strong correlation between the cooperativity requirements in the segmental mobility associated with T_g and the

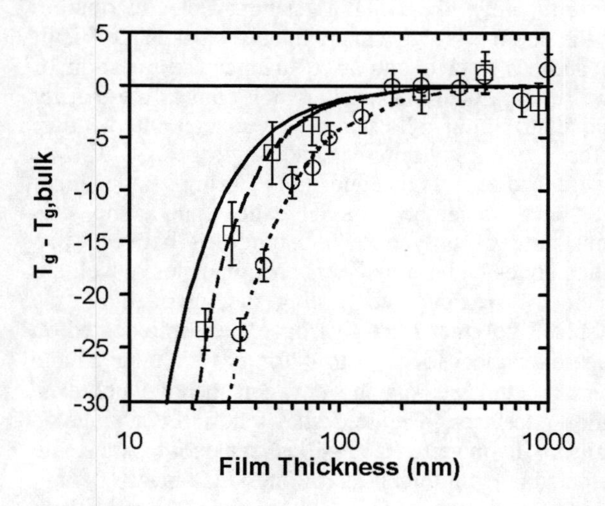

Figure 1. T_g - $T_{g,\mathrm{bulk}}$ as a function of thickness for PS (bold curve is best fit to data in ref. 8) PC (square) and PSF (circles) films as measured via fluorescence of pyrene dopant or dye label, 7-(dimethylamino)-1-methylquinolinium tetrafluoroborate dopant, and intrinsic fluorescence, respectively. The dashed curve represents a fit to PC data using the empirical relation originally proposed by Keddie et al.,[1] yielding the parameter values A = 5.0 nm and δ = 1.84. The dotted curve represents a fit to PSF data using the same empirical relation, yielding the parameter values A = 5.4 nm and δ = 1.60.

observed magnitude of the T_g-confinement effect is sustained when the variable being changed is the relative stiffness of the polymer chain backbone.

The effect of confinement on T_g can also be observed in various polymer nanocomposites as long as the nanofiller is well dispersed. When polymer nanocomposites possess wetted polymer-nanoparticle interfaces, the strength of the attractive interactions at the interfaces is key in defining the magnitude of the increase in T_g observed with increasing nanofiller content and thereby decreasing average interparticle spacing. For example, when 0.4 vol% of 10-15 nm diameter silica nanospheres (with hydroxyl groups on the surfaces) are well-dispersed in PS, the T_g is unchanged from $T_{g,bulk}$.[14,15] However, when the same amount of silica nanospheres is dispersed in poly(methyl methacrylate) (PMMA) and poly(2-vinyl pyridine), the T_g values are increased relative to $T_{g,bulk}$ by ~ 5 K in PMMA and ~ 10 K in P2VP. These differences can be explained by the lack of attractive interfacial interactions in PS-silica nanocomposites and stronger interfacial interactions being present in nanocomposites made with P2VP than with PMMA. Unfortunately, the complex nanoparticle distribution in the polymer-silica nanocomposites makes it difficult to address in a quantitative manner the question of how interparticle distances, which control the density and distribution of attractive interfacial interactions, affect T_g values within the nanocomposites.
In order to address this question in a simplified manner, we have developed model silica nanocomposites consisting of polymer films confined between two silica slides. These model nanocomposites, which are made by layering and healing the interface of two identical polymer films that have been spin coated on separate silica slides, allow for easy tuning of the known interlayer spacing between the two polymer-silica interfaces. Figure 2 shows how the average T_g values in model P2VP-silica nanocomposites are affected by film thickness (or intersilica-layer spacing). These results are compared to the T_g-confinement effects observed in P2VP films supported on silica slides and possessing a polymer-air interface (free surface). It is noteworthy that the replacement of a free surface by a second silica interface can yield dramatic changes in the length scale over which confinement affects average T_g. For example, T_g - $T_{g,bulk}$ = 5 K in both ~ 100-nm-thick P2VP films supported on silica and 500-nm-thick P2VP model nanocomposites. Also, T_g - $T_{g,bulk}$ = 21 K in both 30- to 40-nm-thick P2VP films supported on silica and 200-nm-thick P2VP model nanocomposites. Related effects have also been observed in PMMA-silica model nanocomposites.[15] Because the attractive interfacial interactions are weaker in silica nanocomposites made with PMMA than with P2VP,

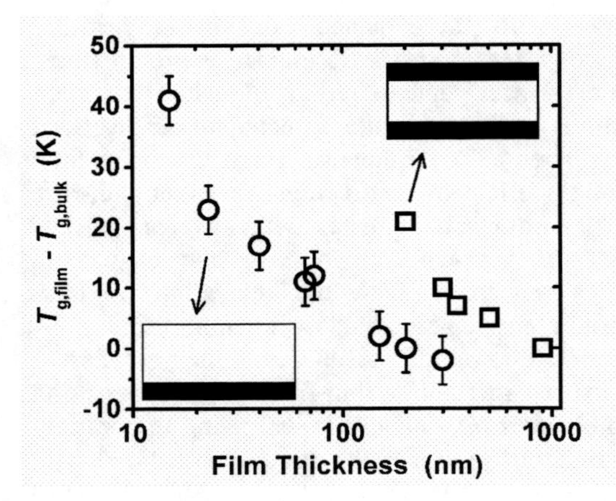

Figure 2. T_g - $T_{g,bulk}$ of ultrathin films supported on silica and 'model' nanocomposites. Thickness dependence of the T_g deviation of P2VP supported films (circles) and pyrene-doped P2VP 'model' nanocomposites (squares). With both the P2VP supported films and the P2VP 'model' nanocomposites, the errors in the data (error bars or 'smaller than symbol size' in the absence of error bars) are due to the fitting required to obtain T_g.

the impact of nanoparticles or silica interlayer spacing on T_g is significantly smaller in PMMA systems.

Attractive interfacial interactions can also have significant impact on the rate of physical aging observed in polymer nanocomposites relative to neat, bulk polymer. Using multilayer-fluorescence methods, we have previously shown that there is a very strong suppression, by more than an order of magnitude, of physical aging deep in the glassy state (T_{aging} = 305 K) in the 25-nm-thick region of a bulk PMMA film that is in contact with a silica slide interface.[17] Similar strong suppressions of physical aging are observed in both real and model P2VP-silica nanocomposites relative to neat PMMA. Using fluorescence to measure initial (over 8 hr after quenching from above T_g) physical aging rates at 303 K,[16,17] we find that the physical aging rate in the well-dispersed 0.4 vol% P2VP-silica nanocomposites is 20-25% of that in the neat, bulk P2VP. With a model P2VP-silica nanocomposite of the same T_g as that of the 0.4 vol% P2VP-silica nanocomposite, we find that the physical aging rate in the model nanocomposite is 4-8% of that in the neat, bulk P2VP. Strong suppression of physical aging has also been observed with PMMA-silica nanocomposites. Thus, both the real and model nanocomposites demonstrate strong suppressions of physical aging relative to neat, bulk polymer. This suggests that nanocomposites may have an important, novel technological application related to the production of glassy-state polymeric materials with properties that are much less subject to physical aging than neat polymers.

In order to obtain longer-term aging measurements and complementary information regarding aging and T_g dynamics in confined glassy polymers and nanocomposites, we are extending our methods beyond fluorescence to include ellipsometry, differential scanning calorimetry, and dielectric spectroscopy. We are also expanding the variety of nanocomposites and model systems beyond silica nanospheres and substrates to include carbon-based nanofillers, e.g., polymer-decorated single-wall carbon nanotubes. Preliminary data using these other methods and nanofillers are consistent with those obtained by fluorescence, i.e., relative to neat, bulk polymers, aging rates can be reduced substantially in polymer nanocomposites containing very small loadings of well-dispersed nanofiller.

REFERENCES

1. J. L. Keddie, R. A. L. Jones and R. A. Cory, *Europhys. Lett.* **27**, 59-64 (1994).
2. J. L. Keddie, R. A. L. Jones and R. A. Cory, *Faraday Discuss.* **98**, 219-230 (1994).
3. J. A. Forrest, K. Dalnoki-Veress and J. R. Dutcher, *Phys. Rev. E* **56**, 5705-5716 (1997).
4. C. B. Roth and J. R. Dutcher, *Eur. Phys. J. E* **12**, S103-S107 (2003).
5. J. H. van Zanten, W. E. Wallace and W. L Wu, *Phys. Rev. E* **53**, R2053-R2056 (1996).
6. Y. Grohens, L. Hamon, G. Reiter, A. Soldera and Y. Holl, *Eur. Phys. J. E* **8**, 217-224 (2002).
7. C. J. Ellison and J. M. Torkelson, *Nature Mater.* **2**, 695-700 (2003).
8. C. J. Ellison, M. K. Mundra, and J. M. Torkelson, *Macromolecules* **38**, 1767-1778 (2005).
9. M. K. Mundra, C. J. Ellison, R. E. Behling, and J. M. Torkelson, *Polymer* **47**, 7747-7759 (2006).
10. M. Alcoutlabi and G. B. McKenna, *J. Phys.: Condens. Matter* **17**, R461-R524 (2005).
11. B. J. Ash, L. S. Schadler and R. W. Siegel, *Mater. Lett.* **55**, 83-87 (2002).
12. J. Berriot, H. Montes, F. Lequeux, D. Long and P. Sotta, *Macromolecules* **35**, 9756-9762 (2002).
13. F. D. Blum, E.N. Young, G. Smith and O. C. Sitton, *Langmuir* **22**, 4741-4744 (2006).
14. P. Rittigstein and J. M. Torkelson, *J. Polym. Sci. Part B: Polym. Phys.* **44**, 2935 (2006).
15. P. Rittigstein, R. D. Priestley, L. J. Broadbelt and J. M. Torkelson, *Nature Mater.* **6**, 278-282 (2007).
16. R. D. Priestley, L. J. Broadbelt and J. M. Torkelson, *Macromolecules* **38**, 654-657 (2005).
17. R. D. Priestley, C. J. Ellison, L. J. Broadbelt and J. M. Torkelson, *Science* **309**, 456-459 (2005).
18. Y. Huang and D. R. Paul, *Macromolecules* **39**, 1554-1559 (2006).
19. S. Kawana and R. A. L. Jones, *Eur. Phys. J. E* **10**, 223-230 (2003).
20. H. B. Lu and S. Nutt, *Macromolecules* **36**, 4010-4016 (2002).
21. C. B. Roth, K. L. McNerny, W. F. Jager, and J. M. Torkelson, *Macromolecules* **40**, 2568-2574 (2007).
22. C. J. Ellison, R. L. Ruszkowski, N. J. Fredin and J. M. Torkelson, *Phys. Rev. Lett.* **92**, 095702-1-095702-4 (2004).
23. M. K. Mundra, C. J. Ellison, P. Rittigstein, and J. M. Torkelson, *Eur. Phys. J. Special Topics* **141**, 143-151 (2007).

Dynamic Nature of the Liquid-Liquid Transition of Triphenyl Phosphite Studied by Simultaneous Measurements of Dielectric and Morphological Evolution

Ken-ichiro Murata, Rei Kurita, and Hajime Tanaka

Institute of Industrial Science, University of Tokyo, Meguro-ku, Tokyo 153-8505, Japan

Abstract. We performed broadband dielectric measurements for the process of liquid-liquid transformation in triphenyl phosphite (TPP). According to our dielectric measurements, the static dielectric constant monotonically decreases and the distribution of the relaxation time becomes broader during the liquid-liquid transformation from liquid I to II. The direct comparison with morphological evolution provides key information on the dynamical and structural evolution during LLT.

Keywords: Dielectric spectroscopy, Glass transition, Liquid liquid transition
PACS: 64.70.Ja, 64.70.P-, 64.60.My, 77.22.Gm

INTRODUCTION

It is usually believed that any single component substance has only one liquid state. Contrary to this common belief, it has recently been revealed that even a single component liquid can have more than two liquid states [1]. The transition between these liquid states is called "liquid-liquid transition (LLT)". Recently we discovered convincing pieces of evidence for the existence of LLTs in typical molecular liquid, triphenyl phosphte (TPP) and n-butanol [2], which take place at ambient pressure. This easy accessibility to LLT allows us to investigate kinetics of LLT in detail by using various experimental techniques. We found two types of the transformation pattern; one is nucleation-growth (NG) type, the other spinodal-decomposition (SD) type (Fig.1). We proposed a two order parameter (TOP) model [3] to describe the kinetics of LLT. TOP model suggest that the number density of locally favored structures (LFS) is the key order parameter of LLT. Recently numerical simulations based on a coarse grained phase field approach [4] show a good agreement of the prediction with our experimental results. TOP model successfully explains the kinetics of LLT at least on a macroscopic level.

However, the microscopic mechanism of LLT, including the details of locally favored structures, is still missing. In order to investigate microscopic changes during LLT, we performed real-time dielectric measurements of the process of LLT for TPP. This method allows us to follow the time evolution of dielectric relaxation during LLT. A comparison between dielectric evolution and the previously measured pattern and heat evolution provides detailed information on the kinetics of LLT.

EXPERIMENTAL

The sample used is triphenyl phosphite TPP purchased from Acros Organics and used without further purification. Since water decomposes TPP, we avoid a contact of TPP with moisture using dried nitrogen gas. The temperature was controlled within 0.1 K by a computer-controlled hot stage (Linkam LK-600PH) with a cooling unit (Linkam L-600A). We performed real-time measurements of the dielectric relaxation during the liquid-liquid transition of TPP. Figure 2 shows the experimental setup to realize the real-time measurement of dielectric spectroscopy during LLT. This electrode formed a parallel plate capacitor: the sample thickness was fixed to 10 μ m and the cell capacitance was 90 F. We also observed the pattern evolution of LLT with phase-contrast microscopy. Broadband dielectric measurements were performed for a frequency range from 10 mHz to 1 MH to MHz with SI1260 Impedance/Gain-Phase Analyzer, which was controlled by a computer. Data accumulation was carried out by using solatron analytical SMaRT sys-

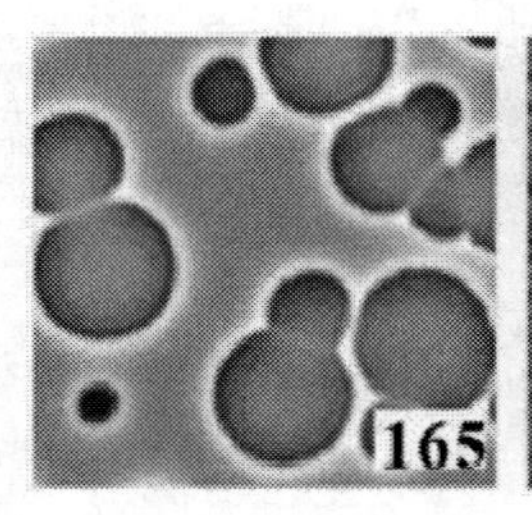

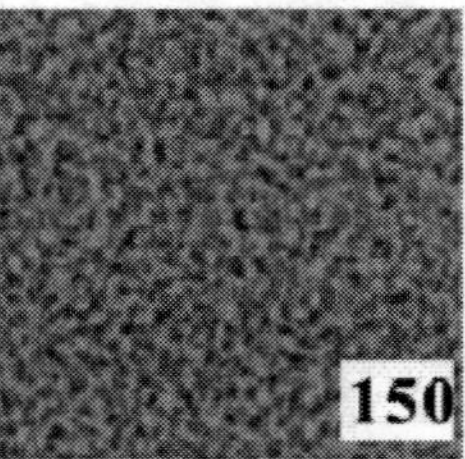

FIGURE 1. Two types of transition kinetics in LLT; (left) nucleation and growth type transformation at 219K (165min). (right) spinodal decomposition type transformation at 214K (150min). The size of both images is 120-120 μm^2.

CP982, *Complex Systems, 5th International Workshop on Complex Systems*
edited by M. Tokuyama, I. Oppenheim, and H. Nishiyama

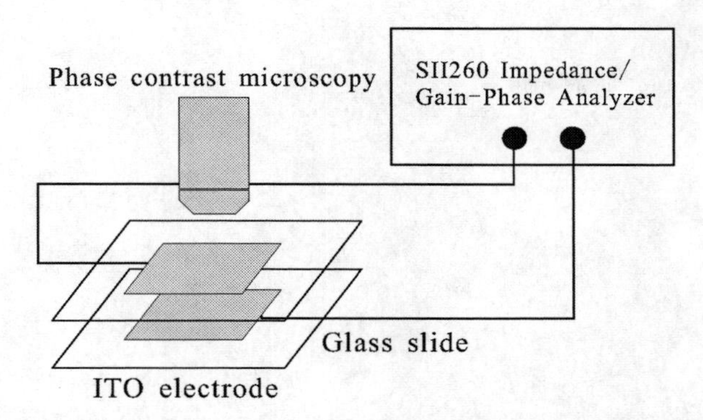

FIGURE 2. Block diagram of the measurement system.

tem. We measured the dielectric response at 40 frequency points in the frequency range from 10 mHz to 1 MHz Under this condition, it took 5 min to get a dielectric spectrum with a good signal-to-noise ratio. Thus, the time resolution of our measurement system is 5 min, which is fast enough to follow the process of LLT since LLT proceeds very slowly over several hours. Finally, we analyzed the data obtained with the Agilent VEE program.

RESULT AND DISCUSSION

Here we show results of dielectric measurements at 214K. According to our previous study [5], the spinodal temperature of LLT is located at 215.5 K. Below this temperature, spinodal decomposition type LLT takes place. We prepared liquid II of TPP while annealing a sample at 214 K. The transformation from liquid I to liquid II is accomplished after about 400 min at 214K. Figure 3 shows the frequency dependence of the complex permittivity of both liquid I and II at 214 K. It is found that the static dielectric constant monotonically decreases and the distribution of the relaxation time becomes broader during the liquid-liquid transformation from liquid I to II. Our result agrees well with previous dielectric study performed by Dvinskiih *et al* [6]. We apply the Havriliak-Negami function (see Eq. (1)) for describing the spectra shape of liquid I, whereas the Cole-Cole function (see Eq. (2)) for describing the spectra shape of liquid II:

$$\varepsilon^*(\omega)_{\mathrm{liqI}} = \varepsilon_\infty + \frac{\Delta\varepsilon_1}{(1+(j\omega\tau_1)^{\gamma_1})^{\alpha_1}} \quad (1)$$

$$\varepsilon^*(\omega)_{\mathrm{liqII}} = \varepsilon_\infty + \frac{\Delta\varepsilon_2}{1+(j\omega\tau_2)^{\beta_2}} \quad (2)$$

where ε_∞ is a high frequency limit of the dielectric constant, $\Delta\varepsilon$ is the relaxation strength and τ is the relaxation time. α and γ are the symmetric and asymmetric parameter for the Havriliak-Negami function. β is the symmetric parameter for Cole-Cole function. Before the analysis, the contribution of the dc conductivity to the spectra was removed. The adjustable parameters of liquid I and II were determined by fitting as follows; for the liquid I, ε_∞ is 3.408, $\Delta\varepsilon_1$ is 1.32 and $\log\tau_1$ is -2.61. The shape parameter α_1 and γ_1 are 0.5 and 0.812. For the liquid II, ε_∞ is 3.44, $\Delta\varepsilon_2$ is 0.801 and $\log\tau_2$ is 0.764. The shape parameter β_2 is 0.36.

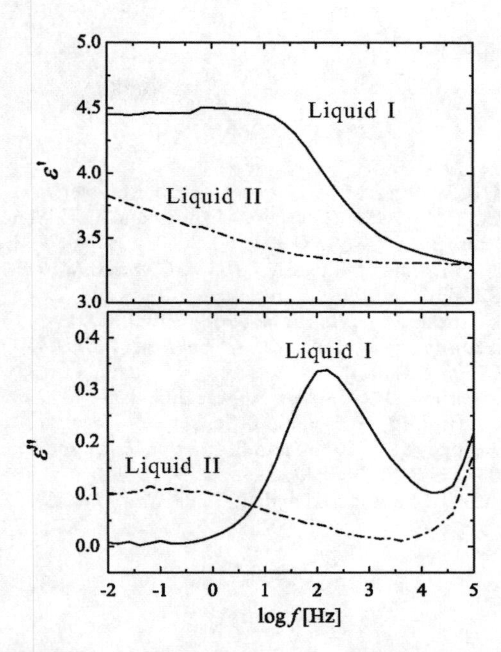

FIGURE 3. Frequency dependence of complex permittivity of liquid I and liquid II. Liquid I can be fitted by Havriliak-Nagami function, whereas liquid II by Cole-Cole function.

It is noteworthy that liquid II shows a very wide distribution of the relaxation time in comparison to a usual molecular liquid. Previous dielectric study [6] also indicated such broadness of the relaxation time. Recently, Kurita *et al.* [5] reported unusually long bare correlation length of 60 nm, which are estimated from the temperature dependence of density fluctuation under SD-type transformation. In addition, Alba-Simionesco *et al.* [7] also suggested the existence of a mesoscale structure in liquid II by using neutron scattering experiments. The existence of such mesoscopic structures might be related to the unusual broadness of the relaxation time. Results of more detailed measurements will be presented elsewhere in the near future.

ACKNOWLEDGMENTS

This work was partially supported by a grant-in-aid from the Ministry of Education, Culture, Sports, Science and

Technology, Japan.

REFERENCES

1. M. C. Wilding, M. Wilson and P. F. McMillan, *Chem. Soc. Rev.* **35**, 964-986 (2006); O. Mishima and H. E. Stanley, *Nature* **396**, 329-335 (1998).
2. R. Kurita and H. Tanaka, *J. Phys.: Condens Matter* **17**, L293-L302 (2005).
3. H. Tanaka, *Phys. Rev. E* **62** 6968-6976, (2000).
4. R. Kurita and H. Tanaka, *J. Chem. Phys* **126**, 204505-1-204505-8 (2007).
5. R. Kurita and H. Tanaka, *Science* **306**, 845-848 (2004).
6. S. Dvinskiih, G. Benini, J. Senker, M. Vogel, J. Wiedersich, A. Kudlik and E. Rössler, *J. Phys. Chem. B* **103**, 1727-1737 (1999).
7. Ch. Alba-Simionesco and G. Tajus, *Europhys. Lett* **52**, 297-303 (2000).

Crystallization and Structural Relaxation of Vapor-Deposited Amorphous 1,2-Dichloroethane: Dependence on the Deposition Conditions

Naohiro Yasuda, Katsunobu Inoue, Hideyuki Nakayama, and Kikujiro Ishii

Department of Chemistry, Gakushuin University, 1-5-1 Mejiro, Toshimaku, Tokyo 171-8588 Japan

Abstract. Structural relaxation and crystallization of amorphous 1,2-dichloroethane prepared by vapor deposition were studied by X-ray diffraction and light interference in the film samples. We estimated the characteristic temperature T_r, that represents the start of the structural relaxation, and the crystallization temperature T_c from the change of diffraction patterns during the temperature elevation with a constant rate. T_r increased while T_c decreased with increasing the deposition temperature T_d. These T_d dependences of T_r and T_c were almost linear and then crossed each other around T_d = 85 K. It was found that amorphous samples were hardly formed in this temperature region but seemed obtainable by increasing the deposition rate.

Keywords: Amorphous, Crystallization, Glass transition, Structural relaxation, Vapor deposition, X-ray diffraction
PACS: 61.43.Er, 82.56.Na, 64.70.Pf, 87.15.nt, 68.03.Fg, 61.05.cp

INTRODUCTION

Amorphous organic materials are expected to be useful in fabricating the optical or electronic devices [1]. However, their structures and characteristics against relaxation or crystallization have not yet been understood well. We have been studying such properties of vapor-deposited amorphous samples of simple organic compounds by vibrational spectroscopy [2-6], X-ray diffraction [4-7], and light interference [6,8,9]. Recently, we found curious relaxation phenomena for amorphous 1,2-dichloroethane (DCE) by Raman spectroscopy [10]. DCE takes trans and gauche conformation isomers. Mole fractions of these isomers are sensitive to the sample state. We used this property of DCE as a probe of the molecular environment in amorphous systems.

We reproduce in Fig. 1 the evolution of gauche mole fraction x_g of amorphous samples deposited at different temperatures [10]. The initial x_g immediately after the deposition was lager as the deposition temperature was higher. x_g of the samples deposited at 30 and 69 K increased as the temperature was elevated approaching the value estimated for the supercooled liquid. The abrupt decreases in x_g around 110 K are due to the crystallization. On the other hand, the sample deposited at 82 K crystallized without the

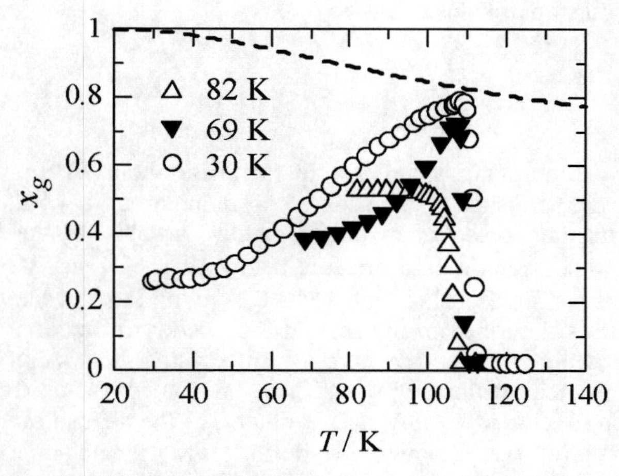

FIGURE 1. Evolution of the mole fraction of gauche isomer in the amorphous samples of DCE deposited at different substrate temperatures during the annealing by temperature elevation. Dashed line represents the data estimated for supercooled liquid.

increase of x_g. These results are summarized as follows. The initial structure and successive relaxation process of amorphous DCE depend on the deposition temperature. The local intermolecular conformations appearing in the relaxation process are inferred to be different from that of the stable crystal. Therefore, we

CP982, *Complex Systems, 5th International Workshop on Complex Systems*
edited by M. Tokuyama, I. Oppenheim, and H. Nishiyama

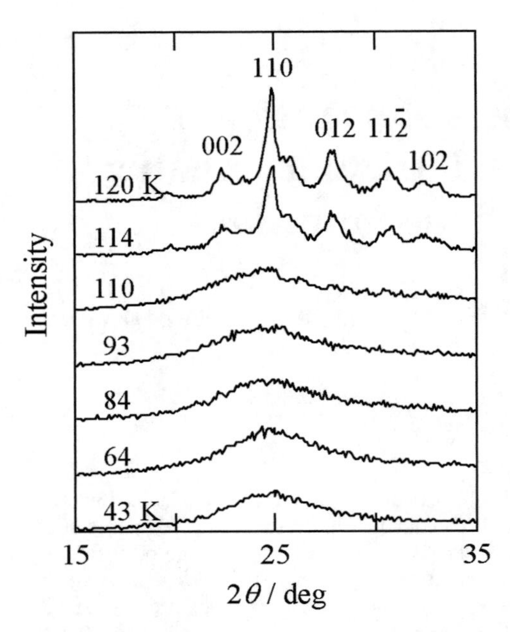

FIGURE 2. Evolution of X-ray diffraction pattern of the sample deposited at 43 K during the temperature elevation. Miller indices were shown for principal peaks.

studied the relaxation process by X-ray diffraction and light interference to obtain more detailed information about amorphous DCE.

EXPERIMENTAL

Amorphous samples with thickness about 10 μm were prepared by vapor deposition onto a cold substrate made of a silicon single crystal with the (100) face at a base pressure below 10^{-5} Pa. Usually, we set the deposition rate to be 0.3 μm/min except the cases described in the section 3-2. X-ray diffraction and light interference were measured using the system reported previously [4]. A Cu X-ray source was used at 40 kV and 20 mA. The evolution of the diffraction pattern was monitored during the temperature elevation with a constant rate 0.28 K/min. Interference of laser light in the sample was also monitored to see the change in the optical properties of the sample [6,8].

RESULTS AND DISCUSSION

Deposition-Temperature Dependence of Relaxation Behavior

Figure 2 shows the evolution of X-ray diffraction pattern of the sample deposited at 43 K during the temperature elevation. The initial samples immediately

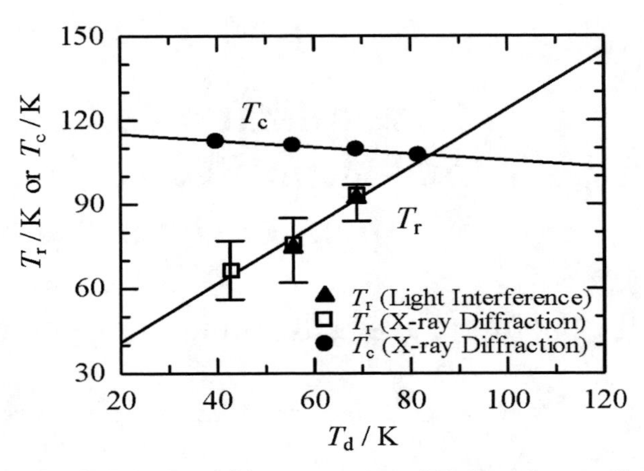

FIGURE 3. Deposition-temperature (T_d) dependences of the crystallization temperature T_c and the characteristic temperature T_r that represents the start of the structural relaxation.

after the deposition showed a broad diffraction hump characteristic to the amorphous structure. The broad hump increased its width and decreased its peak position angle with increasing temperature. When the temperature reached around 112 K, sharp Bragg peaks appeared abruptly. The finally observed diffraction pattern corresponded to that of the reported crystal structure at 110 K [11]. From the appearance of the Bragg peaks, we estimated the crystallization temperature T_c. This temperature almost agreed with the Raman results [10].

We found that the increasing rate of the width of the broad hump changed around 65 K during the annealing process. We found also that the interference of the laser light showed a turnover on the fringe pattern around this temperature. We previously discussed the relation between the movement of the state point on the fringe and the structural change of the sample [8,9]. According to the previous discussion, it is considered that the present sample showed first a thermal expansion almost keeping the initial amorphous structure generated on vapor deposition, and then started the structural relaxation at the above turning point, being accompanied with the sample shrinkage. Thus, we estimated the characteristic temperature T_r, which is considered to mark the start of the structural relaxation, from the observed changes in the diffraction pattern and light-interference fringe.

We carried out the same measurements for the samples deposited at several temperatures from 43 to 100 K. Samples deposited at temperatures below 82 K showed similar evolution of the diffraction pattern to that of Fig. 2. However, the samples deposited at temperatures above 85 K showed Bragg peaks already during the deposition.

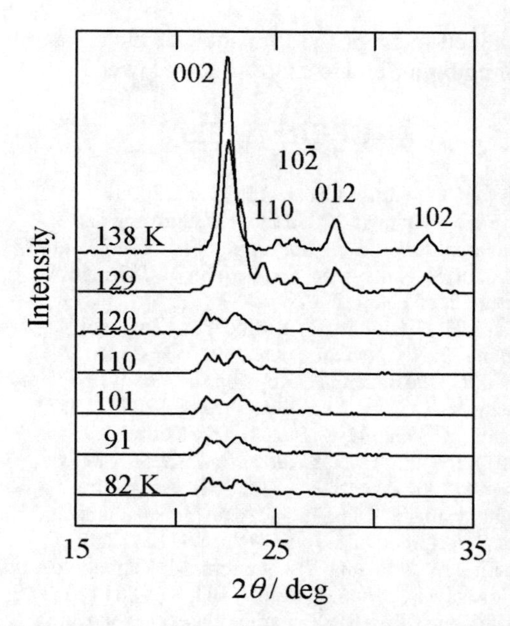

FIGURE 4. Evolution of X-ray diffraction pattern of the sample the deposited at 82 K during the temperature elevation. Deposition rate was 0.1 μm/min.

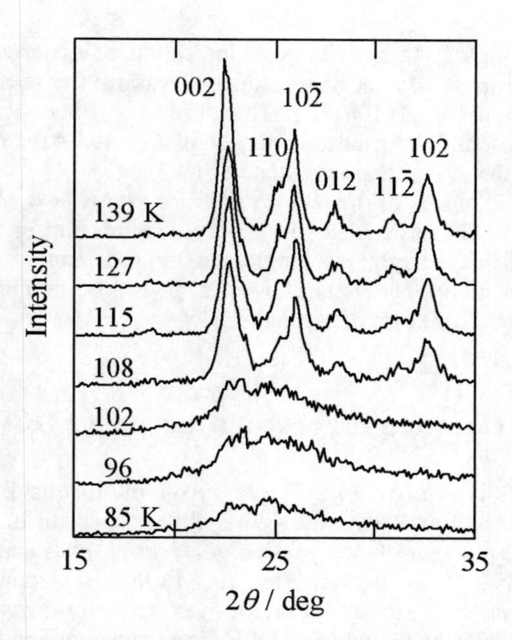

FIGURE 5. Evolution of X-ray diffraction pattern of the sample deposited at 85 K during the temperature elevation. Deposition rate was 1.2 μm/min.

Figure 3 summarizes T_r and T_c observed for the amorphous samples deposited at different T_d. For T_r, two types of the data which were estimated by the change in the diffraction pattern and the state-point motion on the light-interference fringe respectively are indicated. Obviously, T_r is high when T_d is high. T_c shows, on the other hand, a slight tendency to decrease as T_d is higher. The T_d dependences of T_r and T_c are almost represented by linear lines that cross each other around T_d = 85 K. This implies that samples deposited in this or higher temperature regions may not undergo structural relaxation before the crystallization. We have already observed a curious behavior of the change in the gauche mole fraction for a sample deposited at 82 K (see Fig. 1) [10]. Thus, we studied further the annealing behavior of samples deposited at temperatures in the crossing region of T_r and T_c as described below.

Deposition-Rate Dependence of Relaxation Behavior

The results summarized in Fig. 3 suggested that the preparation of amorphous sample at temperatures in the crossing region of T_r and T_c may be sensitive to the deposition condition. Thus, we examined the structure and relaxation process of the samples prepared with different deposition rates in the crossing region.

As described above, the evolution of X-ray diffraction pattern of the sample deposited at 82 K with a deposition rate 0.3 μm/min was almost similar to that of Fig. 1. We then examined a slower deposition rate 0.1 μm/min at 82 K. The results are shown in Fig. 4. For this sample, weak Bragg peaks appeared already during the deposition. These peaks did not correspond to those of the known crystal structure. When the temperature reached around 129 K, the diffraction pattern suddenly changed being accompanied with a large intensity increase. The new peaks almost coincided with those of the reported crystal structure, although the complete coincidence needed the further temperature elevation.

Apparently, the diffraction pattern at 138 K is different in the relative intensities of the Bragg peaks from those at 120 K in Fig. 2. This difference is attributed to the difference in the distribution of the microcrystal orientation in the sample. Similar phenomena were seen in Fig. 5 described below. On the other hand, the above unknown diffraction pattern seen below 129 K is considered to indicate the formation of a metastable crystal phase at the sample surface by the slow deposition at 82 K.

Next, we tried to prepare an amorphous sample at 85 K with a faster deposition rate. Figure 5 shows the evolution of the X-ray diffraction pattern of the sample deposited with a deposition rate 1.2 μm/min. The initial sample immediately after the deposition showed the broad hump characteristic to the amorphous state, although it showed a weak indication of underlying structures. By the temperature elevation, the sample crystallized at 108 K. The observed Bragg peaks corresponded to those of the known crystal structure, and the T_c was that expected in Fig. 3.

The results of the above two trials imply, at least in the DCE case, that the crystallization during the deposition might be suppressed by accelerating the deposition. This issue, however, should be examined using other kinds of compounds.

Relaxation Process of Amorphous DCE

It is seen in Fig. 1 that x_g of amorphous DCE deposited at 30 and 69 K was almost constant in the initial stage of temperature elevation, and started gradually an increase by the further temperature elevation. Similar behavior was seen for samples deposited at 43 and 56 K [10]. The temperature region in which x_g started the increase with each sample almost corresponds to T_r in Fig. 3. Taking this fact into consideration, the relaxation of amorphous DCE by the temperature elevation is considered to take place as follows. First, the sample shows a thermal expansion almost keeping the initial amorphous structure, and then starts the structural relaxation being accompanied with the sample shrinkage and the conformation change from trans to gauche. However, when the temperature is raised up to the region around 110 K, the samples suffer sudden crystallization owing probably the steep decrease in the macroscopic viscosity.

Obviously, the sample deposited at a higher T_d maintains its initial amorphous structure up to a higher T_r. Namely, it is considered that more stable local structure is made by the deposition at a higher temperature. This tendency is in harmony with the results of other compounds [12,13].

The curious behavior of x_g in Fig. 1 observed for the sample deposited at 82 K may be explained also with a context described above. Namely, the sample was fairly dense at the initial stage on deposition, and the conformation change from trans to gauche was prohibited in the bulk of the sample. Thus the sample did not show the increase of x_g until it underwent finally the crystallization.

Finally, the weak T_d dependence of T_c is a critical issue. We have previously pointed out that the samples deposited at higher temperatures tend to form more dense structures. This seems to suppress the increase of x_g and to cause higher T_r. Then, the mole fraction of trans isomer is slightly larger in the samples deposited at higher temperatures when the sample temperature is raised to the crystallization region. This is considered to favor the formation of the crystalline seeds or embryos, and to cause slightly lower T_c.

REFERENCES

1. Shirota, Y. *J. Mater. Chem.*, **15,** 75-93 (2005).
2. K. Ishii, M. Nukaga, Y. Hibino, S. Hagiwara, and H. Nakayama, *Bull. Chem. Soc. Jpn.*, **68,** 1323-1330 (1995).
3. M. Murai, H. Nakayama, and K. Ishii, *J. Ther. Anal. Calor.*, **69,** 953-959 (2002).
4. K. Ishii, H. Nakayama, T. Yoshida, H. Usui, and K. Koyama, *Bull. Chem. Soc. Jpn.*, **69**, 2831-2838 (1996).
5. K. Ishii, H. Nakayama, K. Koyama, Y. Yokoyama, and Y. Ohashi, *Bull. Chem. Soc. Jpn.*, **70**, 2085-2091 (1997).
6. K. Ishii, M. Yoshida, K. Suzuki, H. Sakurai, T. Shimayama, and H. Nakayama, *Bull. Chem. Soc. Jpn.*, **74**, 435-440 (2001).
7. H. Nakayama, S. Ohta, I. Onozuka, Y. Nakahara, and K. Ishii, *Bull. Chem. Soc. Jpn.*, **77**, 1117-1124 (2004).
8. K. Ishii, H. Nakayama, T. Okamura, M. Yamamoto, and T. Hosokawa, *J. Phys. Chem. B*, **107**, 876-881 (2003).
9. K. Ishii, M. Takei, M. Yamamoto, and H. Nakayama, *Chem. Phys. Lett.*, **398**, 377-383 (2004).
10. K. Ishii, Y. Kobayashi, K. Sakai, and H. Nakayama, *J.Phys. Chem. B*, **110**, 24827-24833 (2006).
11. R. Boese, D. Bläser, and T. Haumann, Zeitschrift für Kristallographie, **198**, 311-312 (1992).
12. K. Ishii, and H. Nakayama, oral presentation, this workshop.
13. K. Ishii, H. Nakayama, S. Hirabayashi, and R. Moriyama, to be published.

Colloidal Glass Formation in Polymer Nanocomposites

Benjamin J. Anderson and Charles F. Zukoski†

Dept. of Chemical & Biomolecular Engineering, University of Illinois at Urbana-Champaign
Urbana, IL 61801 USA

Abstract. The microstructure of 44 nm diameter silica particles in a polyethylene oxide (PEO) melt are assessed through the experimental measurement of interparticle structure factors $S(q,\phi)$. Measurements are made using side bounce ultra small angle x-ray scattering (SBUSAXS) of intermediate to highly filled PEO melts. Results show an increase in the effective particle size by a shift in the first peak of $S(q,\phi)$ to lower q when compared to theoretical $S(q,\phi)$ based solely on particle volume exclusion. The shift is attributed to an increase in the effective particle size due to adsorbed polymer chains giving a larger excluded volume per particle. Suspension flow properties are observed to obey theories based on volume exclusion when assuming the adsorbed polymer adds $3.1R_g$ to the particle diameter. As the particle concentration is increased, a new state appears that displays glassy characteristics.

Keywords: Glass, Nanocomposite, Filled Polymers, Scattering
PACS: 61.10.Eq, 62.10.+s, 66.20.+d

INTRODUCTION

Filled polymers are an important class of materials in which particulate matter is dispersed in a polymer matrix to improve strength and durability over the neat polymer. Filler dispersion is of considerable importance in the manufacture of rubber, nanocomposites, biomaterials, insulating and conductive polymers, and coatings. The quality of the dispersion has an effect on the final material properties. Processing of polymer resin materials is performed in the melt. Control over filler dispersion is therefore most important in the melt, and final material properties will depend on the ability to engineer the desired dispersion state.

The development of systematic studies aimed at understanding the microscopic physio-chemical effects in filled polymers that manifest themselves in macroscopic material properties is problematic. Relaxation times are long for high molecular weight polymers with the result that commercial filled polymers are unlikely to reach equilibrium such that the state of dispersion is dominated by mixing conditions. To achieve equilibrium and provide a basis to compare experimental observations with models, we study silica nanoparticle dispersions in low MW polymer. We employ dispersions of 44 nm monodisperse silica nanoparticles in a polyethylene oxide (PEO) melt. Through measurement of the particle second virial coefficient accessible through scattering experiments, we find the dispersions are thermodynamically stable.[1] The stability is believed to be due to the absorption of polymer segments to the particle surface that provide a net interparticle repulsion.[2]

In this study, we are interested in how the polymer matrix influences suspension microstructure at intermediate to high loading and the effect of the filler on the mechanical properties of the melt. Below in the experimental section, we describe sample preparation methods and our experimental procedure of measuring filler microstructure and filled melt viscosity and elasticity. In the results and discussion section, we report and discuss our findings that these mixtures undergo a glass transition at volume fractions similar to that expected for hard particles of size slightly larger than core silica size.

† Author to whom correspondence should be addressed. Email address: czukoski@uiuc.edu

CP982, *Complex Systems, 5th International Workshop on Complex Systems*
edited by M. Tokuyama, I. Oppenheim, and H. Nishiyama

EXPERIMENTAL

Sample preparation has been previously described in Ref. 1. In this study, 44 nm Stöber silica particles were dispersed in PEO of 1000 MW purchased from Sigma-Aldrich. PEO1000 is below the entanglement MW of PEO ($M_{e,PEO} \approx 2000$). Silica and PEO are contrast matched which reduces the Van Der Waals forces between the particles leaving particles to only interact via volume exclusion and polymer mediated interactions.[3]

SBUSAXS measurements are made at a sample temperature of 75°C.[4] Background intensity was accounted for by measuring the scattered intensity of neat PEO. The scattering intensity of the PEO is subtracted off of the sample scattering leaving only scattering due to the silica filler. This is based on the fact that the scattering from the silica dominates leading to our assumption that cross scattering terms between the polymer and filler can be neglected and the dispersion can be viewed as an effective one component system.[5]

The scattering intensity of a single component dispersion is given by

$$I(q) = \phi V \Delta\rho_e^2 P(q) S(q,\phi). \quad (1)$$

ϕ is the filler volume fraction, V is the volume of a single particle, and $\Delta\rho_e$ is the electron scattering length density of silica over that of the PEO dispersing phase. The variable q is the scattering vector, $q = (4\pi/\lambda)\sin(\theta/2)$ where λ is the wavelength of incident x-rays and θ is the scattering angle. $P(q)$ is the form factor accounting for intraparticle scattering interference and $S(q,\phi)$ is the structure factor accounting for interparticle scattering interference. In the dilute particle limit, the structure factor goes to unity and the scattering equation reduces to $I(q) = \phi V \Delta\rho_e^2 P(q)$.

The form factor for spherical particles is given by

$$P(q) = \left(3\frac{\sin(qD/2) - (qD/2)\cos(qD/2)}{(qD/2)^3}\right)^2. \quad (2)$$

We account for modest polydispersity in particle size by calculating $P(q)$ for a size distribution by employing a Gaussian diameter distribution to calculate an average form factor for a population of particles with a mean diameter $\bar{\sigma}$ and standard deviation ν . Experimental scattering of dilute suspensions are fit to the scattering equation utilizing Eq. (2) integrated over a Gaussian distribution for the average form factor to determine a scattering size and standard deviation in the polymer melt. The fitting procedure has three adjustable parameters: particle diameter, standard deviation, and electron contrast density.

As the volume fraction of filler goes up, the structure factor measures the spatial distribution of filler particles in inverse space. Experimental structure factors for concentrated suspensions (cs) are calculated by dividing the intensity of a concentrated suspension by the intensity of a dilute suspension (ds). In the dilute limit, the structure factor goes to unity since particle positions are uncorrelated and scattering is dominated by intraparticle scattering leaving $S(q,\phi)$ for the concentrated suspension,

$$S(q,\phi) = \frac{I_{cs}(q,\phi_{cs})}{I_{ds}(q,\phi_{ds})}\frac{\phi_{ds}}{\phi_{cs}}, \quad (3)$$

typically $\phi_{ds} \le 0.05$.

Rheological experiments have been performed using a constant stress C-VOR rheometer with cup and bob and cone and plate geometries. The cup and bob was used for measuring the shear viscosity of liquid like samples. The bob diameter is 14 mm and the gap is 0.7 mm. A 20 mm diameter, 4° cone and plate was used to measure the elastic modulus of highly concentrated samples. Measurements were made at a sample temperature of 75°C.

RESULTS & DISCUSSION

Figure 1A shows filler structure factors for moderately low to high filler concentration. The filler structure displays liquid like character at all concentrations up to a maximum loading of 50%. We see a sharp initial growth in structure indicated by the magnitude of the peak $S(q^*)$ where q^* is the location of the peak and mild growth thereafter.

From extrapolation to $S(0,\phi)$, the change in the osmotic pressure of the filler which is called the compressibility is accessible through Eq. (4) and plotted in Figure 1B.

$$\lim_{q\to 0}\frac{1}{S(q,\phi)} = \frac{V}{k_B T}\frac{\partial \Pi}{\partial \phi} \quad (4)$$

Π is the osmotic pressure, k_B is Boltzmann's constant, and T refers to temperature. As expected, the filler are more incompressible upon addition of filler. When compared to a model hard particle suspension in a low MW solvent at the same particle concentration, we notice that the filler are more incompressible than if the filler interacted only via volume exclusion interactions. The data indicate that the particle osmotic pressure increases faster with volume fraction than expected for hard particles. At higher filler concentration, the osmotic pressure is a

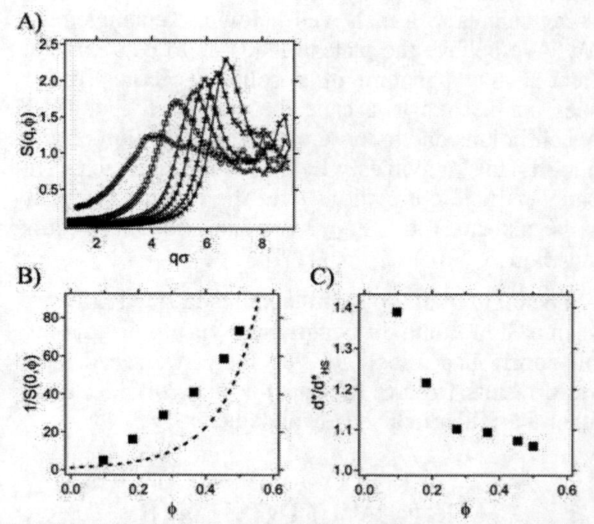

FIGURE 1. A) Experimental filler structure factors of filled PEO1000 versus dimensionless scattering vector for volume fractions of 0.096 [o], 0.183 [□], 0.273 [Δ], 0.365 [#], 0.454 [⧗], 0.500 [*]. B) Filler compressibility [■] versus volume fraction compared to the compressibility of a hard particle suspension [dashed line]. C) The average separation between two filler particles relative to a hard particle suspension ($d^*/d^*_{HS} = q^*_{HS}/q^*$) [■] versus filler volume fraction.

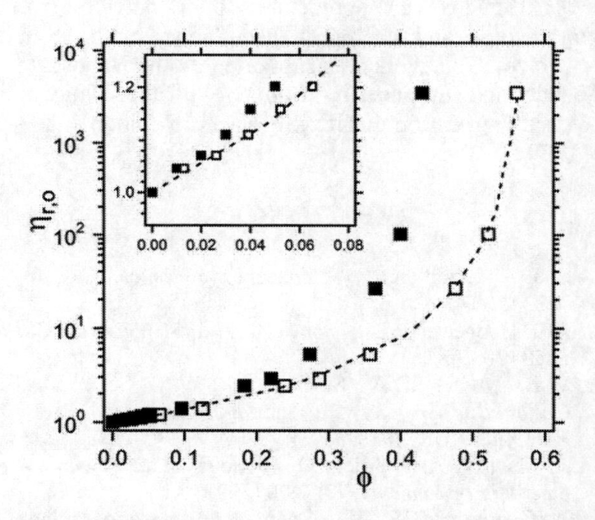

FIGURE 2. The zero shear rate relative viscosity of PEO1000 versus filler volume fraction [■] and effective filler volume fraction [□] assuming the polymer adds 3.1Rg to the filler diameter. The dashed line is a smooth curve of experimental hard sphere data of Cheng et. al. [6].

weaker function of volume fraction than expected for hard spheres. These results demonstrate that the mixture of polymer and particles cannot be thought of as hard particles suspended in a continuum even though $2R_g/\sigma = 0.06$. The effects of internal configuration of the polymer and its tendency to associate with the particle play a role in determining microstructure.

In Figure 1C, the average interparticle separation, $d^* = 2\pi/q^*(\phi)$, relative to spacing expected for hard spheres at the same volume fraction is seen to decrease at intermediate filler concentration and then diminish much more slowly at high concentrations. The initial rapid reduction of the interparticle separation occurs over a volume fraction range where the osmotic pressure grows rapidly. At low volume fractions density fluctuations giving rise to $S(0,\phi)$ are clearly suppressed by the presence of polymer suggesting that the polymer gives rise to an effective repulsion between the particles. However, the effective one component particle interactions are soft such that at elevated volume fraction, nearest neighbor cages approach separations expected for particles with effective sizes of the core silica particles while the osmotic pressure approaches that expected for hard particles of the same size.

The trends in these data point to the possible presence of a long range repulsion between two particles giving the appearance of particles with a larger effective size. We can account for the larger size of the filler by assuming that the PEO adsorbs to the silica surface forming a shell of bound polymer around a particle such that polymer chains lose configurational entropy when forced closer together giving rise to an effective repulsion which causes significant filler structure at moderately low volume fractions. The weaker growth in the correlation of filler structure and in the osmotic pressure can be attributed to the softness of the bound shell which increases the distribution of the average distance between two particles at a particular filler concentration.

Measurement of the zero shear rate relative viscosity $\eta_{r,o}$ of the filled PEO is seen to rise with the addition of particles as Einstein predicts [Fig. 2]. When comparing the viscosity to that expected for hard spheres[6], we find the elevation in the viscosity of filled PEO to be greater than would be expected for a suspension of particles interacting only through volume exclusion. This supports the conclusion that the filler volume is enhanced by a bound polymer shell which increases the hydrodynamic size of the particles. Yet, whereas there appeared to be little similarity between the microstructure of the filler in PEO to that of hard particles, the zero shear rate relative viscosity collapses onto the model hard particle data when applying $\phi_{eff} = \left[\left(3.1R_g + \sigma\right)/\sigma\right]^3 \phi$. Whereas filler

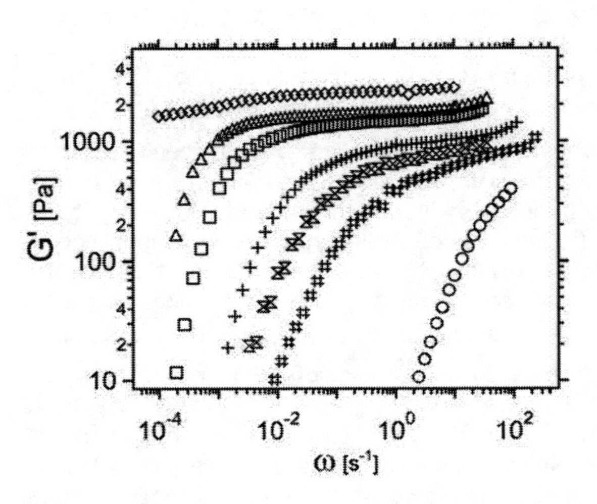

FIGURE 3. The elastic modulus of filled PEO1000 versus frequency for filler volume fractions of 0.400 [o], 0.435 [#], 0.440 [x], 0.444 [+], 0.451 [□], 0.454 [Δ], 0.460 [◇].

thermodynamic interactions based on the osmotic compressibility show little similarity to hard spheres, filler hydrodynamic interactions are well modeled by hard spheres when increasing the excluded volume of the particles due to the adsorption of polymer by $3.1R_g$ where R_g is the polymer radius of gyration. The scaling of a bound layer with R_g has been substantiated for confined polymer melts between two strongly adsorbing plates.[7]

At high filler concentrations, we see a divergence in the viscosity at a volume fraction lower than expected for a low MW continuous phase due to the larger effective particle size. The divergence in the viscosity coincides with the buildup of a plateau in the elastic modulus [Fig. 3]. This is a common feature of nanocomposites upon the addition of filler particles. The emergence of a plateau can either be a polymer or a particle effect. High MW polymers exhibit a plateau modulus due to entanglement topological constraints. In this regime, sheared polymer chains realize that they are entangled. Over time chains escape entanglements via reptation and begin to flow yielding terminal behavior.

Dense particle suspensions can exhibit a plateau modulus as the concentration approaches the glass transition. In dense suspensions, particles become localized in a cage of nearest neighbor particles arresting flow. Eventual escape from the cage yields terminal flow. The glass transition of hard particles occurs at a volume fraction $\phi_g \approx 0.58$. Mode coupling theories has been shown to capture the flow behavior of hard particle dense suspensions. PEO1000 is not entangled; therefore, we do not expect entanglement dynamics to explain the relaxation behavior. It is also unlikely that the particles would be introducing entanglements to a melt well below the entanglement MW. We believe the plateau modulus to be a particle effect and a signature of a colloidal glass. Longer times are needed to escape the cage at higher filler concentrations due to more particle localization. Fig. 1 supports this hypothesis by showing an increase in spatial particle correlations over the volume fractions of the plateau. From Fig. 3, we can estimate a glass transition $\phi_g \approx 0.46$.

The buildup of a plateau modulus in filled polymer systems is a common occurrence, yet its origins are still poorly understood. In this study, we see that its origin results from the formation of a colloidal glass rather than filler induced entanglements.

ACKNOWLEDGMENTS

SBUSAXS data was collected at the X-ray Operations and Research beamline 32ID-B at the Advanced Photon Source (APS), Argonne National Laboratory. The APS is supported by the U.S. Department of Energy, Office of Science, Office of Basic Energy Sciences, under Contract No. DE-AC02-06CH11357. We appreciate our collaborative relationship and helpful discussions with Ken Schweizer. This work was supported by the Nanoscale Science and Engineering Initiative of the National Science Foundation under NSF Award Number DMR-0117792.

REFERENCES

1. B. J. Anderson and C. F. Zukoski, *Macromolecules*, 40, 5133 (2007).
2. J. B. Hooper and K. S. Schweizer, *Macromolecules*, 38, 8858 (2005).
3. W.B. Russel, D.A. Saville, and W.R. Schowalter, *Colloidal Dispersions*, Cambridge University Press, Cambridge, UK, 1989.
4. J. Ilavsky, A. J. Allen, G. G. Long et al., *Review of Scientific Instruments*, 73, 1660 (2002).
5. A. George and W. W. Wilson, *Acta Crystallographica Section D-Biological Crystallography*, 50, 361 (1994); A. M. Kulkarni, A. P. Chatterjee, K. S. Schweizer et al., *Physical Review Letters*, 83, 4554 (1999); D. Rosenbaum, P. C. Zamora, and C. F. Zukoski, *Physical Review Letters*, 76, 150 (1996); D. F. Rosenbaum and C. F. Zukoski, Journal of Crystal Growth, 169, 752 (1996).
6. Zhengdong Cheng, Jixiang Zhu, Paul M. Chaikin et al., *Physical Review E*, 65, 041405 (2002).
7. Steve Granick and Hsuan-Wei Hu, *Langmuir*, 10, 3857 (1994).

Fragility Variation of Lithium Borate Glasses Studied by Temperature-Modulated DSC

Yu Matsuda[a, b], Yasuteru Fukawa[a], Mitsuru Kawashima[a], and Seiji Kojima[a]

[a] Graduate School of Pure and Applied Sciences, Univ. of Tsukuba, Tsukuba, Ibaraki, 305-8571, Japan
[b] JSPS Research Fellow

Abstract. The fragility of lithium borate glass system has been investigated by Temperature-Modulated Differential Scanning Calorimetry (TMDSC). The frequency and temperature dependences of dynamic specific heat have been observed in the vicinity of a glass transition temperature T_g. It is shown that the value of the fragility index m can be determined from the temperature dependence of the α-relaxation times observed by TMDSC, when the raw phase angle is properly corrected. The composition dependence of the fragility has been also discussed.

Keywords: Dynamic heat capacity, Fragility, Glass transition, Lithium borate glass, Relaxation time, TMDSC
PACS: 64.70.pf, 65.60.+a, 67.40.Fd

INTRODUCTION

Dynamics of supercooled liquid and glass transition has been one of the central issues in condensed-matter physics. In a glass forming liquid, a characteristic relaxation time of α-relaxation related to the cooperative rearrangement process in liquid drastically increases near a glass transition temperature T_g. To understand the origin of the dramatic slowing down of the relaxation dynamics near T_g is an important challenge [1].

In order to classify the variation of relaxation process in glass-forming materials, the concept of "fragility" was introduced by Angell [2]. The classification is based on an T_g-scaled Arrhenius plot (Angell plot) of the temperature dependence of α-relaxation time $\tau(T)$ or viscosity. "Strong" liquids obeys the Arrhenius law, whereas "fragile" liquids display pronounced deviations from the Arrhenius law as temperatures approach T_g. To evaluate degrees of fragility in a more quantitative way, the fragility index m is used defined by

$$m = \left| \frac{d \log_{10} \tau}{d(T_g / T)} \right|_{T=T_g} . \quad (1)$$

In other word, a value of m is determined as the slope of the $\tau(T)$ curve at $T = T_g$ in the Angell plot.

Alkali oxide borate family is interesting glass-forming system, since its physical properties vary markedly with the alkali oxide content. Especially, the fragility of lithium borate glasses (LiB) shows the strongest composition dependence among alkali borate family. Pure B_2O_3 glass belongs to a "strong" glass former, while the fragility of LiB markedly increases with Li_2O content [3, 4]. Chryssikos *et al.* studied the composition dependence of the fragility of LiB [4]. They determined the values of the fragility index by using the literature data of the viscosity measurement. However, the range of composition studied was particularly limited up to about 25 mol% Li_2O. The reason of this is that since the glass-forming ability of LiB is poorer as the Li_2O content increases, it is difficult to perform a viscosity measurement. In addition, the high ionic conductivity of Li ion also obstructs the conventional dielectric spectroscopy.

Recent development of a Temperature-Modulated DSC (TMDSC) technique can perform the specific heat spectroscopy, and thus to study slow dynamics near T_g. The TMDSC can be the alternative technique to viscosity and dielectric measurements near T_g. Carpentier and Descamps *et al.* demonstrated the ability of the TMDSC as a spectroscopy in the study for the organic and saccharic glass-forming materials [5, 6]. Within our knowledge, there is no paper to study slow dynamics of inorganic glasses, whose T_g is much higher than organic or saccharic compound, by means of TMDSC in the light of spectroscopy. The purpose of the present paper is to investigate slow dynamics of LiB near T_g and evaluate a fragility index by the measurements of a complex specific heat $C_p^*(\omega)$ as a function of frequency. In addition, the

CP982, *Complex Systems, 5th International Workshop on Complex Systems*
edited by M. Tokuyama, I. Oppenheim, and H. Nishiyama

values of the fragility index applied or not applied to a phase correction are compared.

EXPRIMENTAL

Sample Preparation

Chemical formula of lithium borate glasses (LiB) is denoted by $xLi_2O \bullet (100-x)B_2O_3$, where x is a molar concentration of Li_2O to B_2O_3. The glass samples were prepared by the "Solution method" with high homogeneity in order to investigate the inherent nature of binary system. Actual composition of the samples was analyzed by using a potentiometric titration. The detail description for the sample preparation has been written in [7, 8, 9].

Temperature-Modulated DSC Technique

In 1993, a Temperature-Modulated DSC (TMDSC) was developed by Reading *et al* [10]. In TMDSC, a small sinusoidal temperature modulation is superimposed on a linear heating rate (underlying heating rate) used in a conventional DSC.

A modulated heating rate β is described as

$$\beta = \beta_o + A_\beta \cos(\omega \cdot t), \quad (2)$$

where β_o is underlying heating rate, A_β amplitude of the temperature modulation in the heating rate and ω angular frequency of the modulation. Since the modulated heating rate contains linear and periodic parts, a resultant heat flow can be separated to the response to the linear part and the response to the periodic part. The latter can be written as

$$HF_{periodic} = A_{HF} \cos(\omega \cdot t - \varphi), \quad (3)$$

where $HF_{periodic}$ is the heat flow response to the periodic heating rate, A_{HF} the amplitude and φ the phase angle between the periodic heating rate and the heat flow response. Figure 1 shows the temperature dependence of the A_β and A_{HF}, from which an absolute values of a complex specific heat $|C_p^*(\omega)|$ can be obtained by the ratio of A_β and A_{HF} by

$$\left|C_p^*(\omega)\right| = A_{HF} / (m \cdot A_\beta), \quad (4)$$

where m is the mass of a sample. By using the phase lag φ, real and imaginary parts can be obtained by

$$C_p' = \left|C_p^*\right| \cos\varphi \quad C_p'' = \left|C_p^*\right| \sin\varphi \quad (5)$$

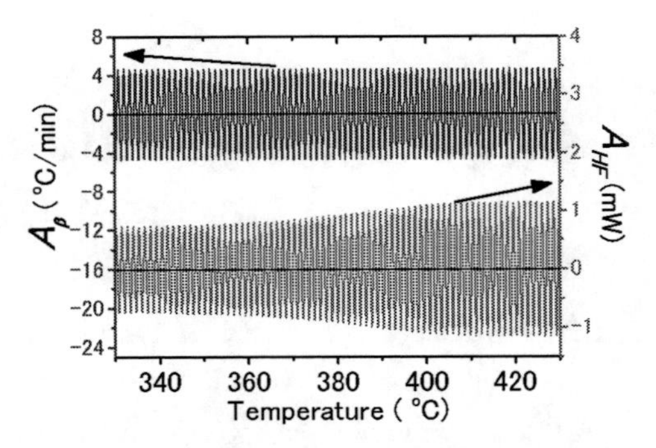

FIGURE 1. (Color online) The modulation in the heating rate (A_β) and the heat flow response to the modulation (A_{HF}). X = 10, β_0 = 1.0 °C/min and P = 80 sec (DSC2920).

Experimental Conditions

The TMDSCs (DSC 2920, Q 200, TA Instruments) were used for all of experiments. The enthalpy and temperature are calibrated by a heat of fusion of In. The value of $|C_p^*(\omega)|$ was calibrated by using a sapphire. A sample was heated through T_g with the temperature profile; underlying heating rate (β_0) is 1.0 °C/min and modulation period (P) 30 – 100 sec (DSC 2920); β_0 is 0.5 °C/min and P 20 – 200 sec (Q200).

Although the phase angles before and after glass transition should be zero, the experimental results show non-zero phase angle. The deviation comes from the heat transfer effects. Therefore, the raw phase angle was corrected by the method suggested by Weyer *et al.* [11]. The detailed description of the method, raw and corrected phase angles at the modulation period 100 sec was presented in [8, 9].

In shorter period experiments, an additional correction was required because a baseline of raw phase angle was not flat, while a baseline of the phase at the longer period was flat. Firstly, a baseline of phase angle was checked by performing the experiments using empty pan or sapphire disk. After the substraction of the baseline from the raw data, the Weyer's method was, then, applied.

RESULTS AND DISCUSSION

Figure 2 a) shows the $|C_p^*(\omega)|$ and raw phase angle in which the phase angle baseline determined by the sapphire disk run has been already subtracted. Modulation period is 40 sec. Dotted line in a) shows the contribution of a heat transfer to the raw phase angle, which is proportional to the measured heat capacity and scaled to fit to the raw phase angle [11].

The C_p' and C_p'' are shown in Fig. 2 b) as a function of temperature. Both of them have been

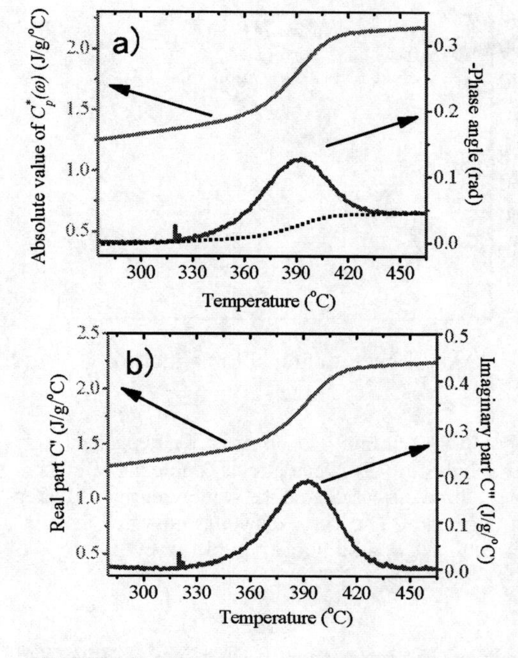

FIGURE 2. (Color online) a) Temperature dependences of $|C_p^*(\omega)|$ and raw phase angle. The dotted line is the corrected factor calculated by Weyer's method. b) Temperature dependences of real and imaginary parts. X = 10, β_0 = 1.0 °C/min and P = 40 sec (DSC2920).

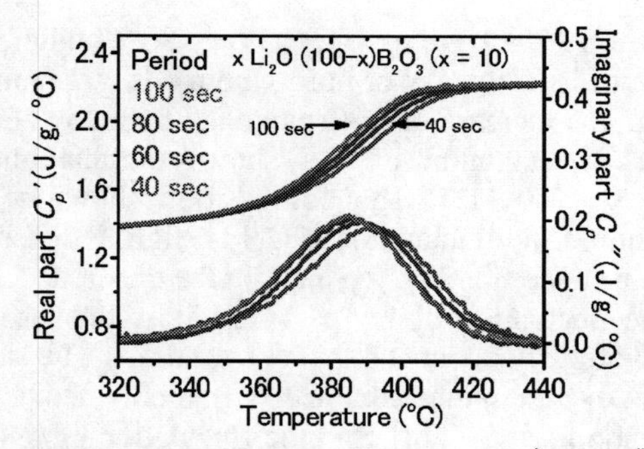

FIGURE 3. Frequency dependences of $C_p{}'$ and $C_p{}''$ and raw phase angle. X = 10, β_0 = 1.0 °C/min and P = 40 - 100 sec (DSC2920). (Color online).

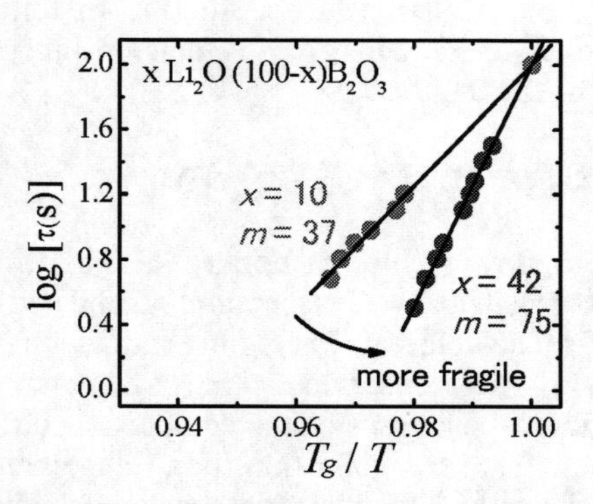

FIGURE 4. (Color online) T_g-scaled Arrhenius plot (Angell plot), where T_g is defined as the temperature when τ is 100 sec. The T_g is determined by the extrapolation from $\tau(T)$ vs. $1/T$ plot.

calculated by eqs (5) by using $|C_p^*(\omega)|$ and corrected phase angle.

The change of the frequency of the modulation enables us to investigate a frequency dependence of dynamic specific heat. Figure 3 shows the frequency dependences of the $C_p{}'$ and $C_p{}''$. The peak temperature of $C_p{}''$ shifts higher as the modulation period becomes shorter. The characteristic relaxation time τ of the system is the most probable at the peak temperature with the constant modulation period. Thus, the temperature dependence of $\tau(T)$ can be given by

$$\tau = \frac{1}{\omega} = \frac{P}{2\pi} . \quad (6)$$

The obtained values of $\tau(T)$ of x = 10 and 42 are plotted against T_g-scaled reciprocal temperature, as shown in Figure 4. The T_g is defined as the temperature when τ becomes 100 sec. The value of T_g is calculated by the extrapolation from the present experimental data.

The fragility index m defined by eq. (1) is determined by a slop of $\tau(T)$ in Fig. 4 under the assumption of the linear evolution. This assumption is valid in the present study because the relaxation mode in the vicinity of T_g ($0.96 < T_g/T < 1$) is investigated by the TMDSC experiments. As shown in Fig.4, the crossover from "strong" to "fragile" with the increase of Li_2O content is clearly observed.

The value of m of x = 10 determined in this study is compared to those in the reference studied by a viscosity measurement [4] (Fig. 5). The value applied to the phase correction described in EXPERIMENTAL is in good agreement with the reference value. It is also shown that without proper phase correction, the value studied by TMDSC is not comparable; one always underestimates the fragility.

The composition dependence of the fragility of LiB is shown in Fig. 5. The values increase with the increase of Li_2O content and may show the broad peak between x = 42 and x = 58. Although exact peak composition is not determined in the present study, it is firstly revealed that the fragility trend of LiB has the maximum. To understand this trend, the fragility model proposed by Vilgis is useful, in which the degree of fragility connects the fluctuation of coordination number (CN) in a system [12]. For instance, network glasses with a fixed covalent bonding such as vitreous silica belong to "strong" glass former, whereas the CN of o-terphenyl (OTP) varies from 11 – 16, thus OTP belongs to "fragile". In the case of LiB, the CN of boron atom changes from 3 to 4 with the increase of Li_2O content, proven by NMR [13]. The fluctuation of CN shows the maximum around x = 40 [13]. The composition showing the maximum in the fluctuation of CN is slightly different from that in the fragility. To explain the difference, not only the fluctuation of CN of boron atom but also of oxygen atom must be taken into account. The non-bridging oxygen atoms are created gradually above the composition x = 30 [14], and the fluctuation of CN of oxygen atom may show the maximum between x = 42 and x = 58. The quantitative calculation of the fluctuation of CN of LiB system is in progress. The fragility trend of LiB clarified for the first time by using the TMDSC will give us new insight to the origin(s) of the fragility.

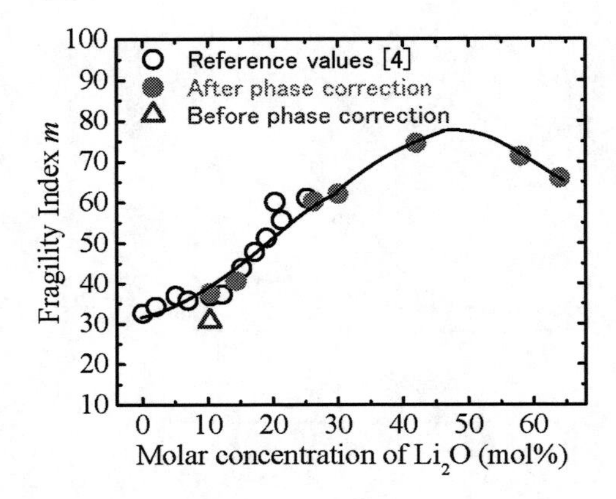

FIGURE 5. (Color online) The composition dependence of the fragility index of LiB. Open circles denote the reference values [4]. Closed circles denote the values determined after the phase correction. The triangle denotes the value without phase correction. The solid line is a guide for eyes.

CONCLUSION

The fragility of lithium borate glasses (LiB) has been investigated by Temperature-Modulated DSC (TMDSC). The value of the fragility index m has been determined by observing the frequency and temperature dependences of the complex specific heat in the vicinity of the glass transition temperature, T_g. The 2 step method for the phase angle correction for the short period experiment has been described. It is shown that the value of m with proper phase angle correction is good agreement with the reference value. The range of the composition studied has been extended, and the composition dependence of the fragility shows the broad maximum above x = 40 mol% Li_2O. The TMDSC technique as a spectroscopy is a powerful tool to investigate slow dynamics in the vicinity of T_g.

ACKNOWLEDGMENTS

The authors gratefully acknowledged helpful discussion with Prof. Masao Kodama, Department of Applied Chemistry - Sojo Univ., for the sample preparation and characterization. (Y M) is also thankful for the JSPS Research Fellowship for Young Scientists (19•574).

REFERENCES

1. C. A. Angell, K. L. Ngai, G. B. McKenna, P. F. McMillan and S. W. Martin, *J. Appl. Phys.*, **88**, 3113-3157 (2000).
2. C. A. Angell, *J. Non-Cryst. Solids*, **131-133**, 13-31 (1991).
3. S. Kojima, V. N. Novikov and M. Kodama, *J. Chem. Phys.*, **113**, 6344-6350 (2000).
4. G. D. Chryssikos, J. A. Duffy, J. M. Hutchinson, M. D. Ingram, E. I. Kamitsos and A. J. Pappin, *J. Non-Cryst. Solid*, **172-174**, 378-383 (1994).
5. L. Carpentier, O. Bustin and M. Descamps, *J. Phys. D: Appl. Phys.*, **35**, 402-408 (2002).
6. A. De Gusseme, L. Carpentier, J. F. Willart and M. Descamps, *J. Phys. Chem*, **B 107**, 10879-10886 (2003).
7. M. Kodama, T. Matsushita and S. Kojima, *Jpn. J. Appl. Phys.*, **34** , 2570-2574 (1995).
8. Y. Matsuda, C. Matsui, Y. Ike, M. Kodama and S. Kojima, *J. Therm. Anal. Cal.*, **85**, 725-730 (2006).
9. Y. Matsuda, Y. Fukawa, C. Matsui, Y. Ike, M. Kodama and S. Kojima, *Fluid Phase Eq.*, **256**, 127-131 (2007).
10. M. Reading, A. Luget and R. Wilson, *Thermochim. Acta*, **238**, 295-307 (1994).
11. S. Weyer, A. Hensel and C. Schick, *Thermochim. Acta*, **305**, 267-275 (1997).
12. T. A. Vilgis, *Phys. Rev.*, **B 47**, 2882-2885 (1993).
13. G. E. Jellison, S. A. Feller and P. J. Bray, *Phys. Chem. Glasses*, **19**, 52-53 (1978).
14. E. I. Kamitsos, A. P. Patsis, M. A. Karakassides and G. D. Chryssikos, *J. Non-Cryst. Solids*, **126**, 52-67 (1990).

Separation of Dynamics in the Free Energy Landscape

Toru Ekimoto, Takashi Odagaki, and Akira Yoshimori

Department of Physics, Kyushu University, Fukuoka 812-8581, Japan

Abstract. The dynamics of a representative point in a model free energy landscape (FEL) is analyzed by the Langevin equation with the FEL as the driving potential. From the detailed analysis of the generalized susceptibility, fast, slow and Johari-Goldstein (JG) processes are shown to be well described by the FEL. Namely, the fast process is determined by the stochastic motion confined in a basin of the FEL and the relaxation time is related to the curvature of the FEL at the bottom of the basin. The jump motion among basins gives rise to the slow relaxation whose relaxation time is determined by the distribution of the barriers in the FEL and the JG process is produced by weak modulation of the FEL.

Keywords: Glass transition, Free energy landscape
PACS: 64.70.Pf

INTRODUCTION

It is well known that various characteristic behaviors in thermodynamic and dynamic properties are observed in the vitrification process and there have been many efforts to explain these characteristic behaviors. The free energy landscape theory[1, 2] is one of the most promising approaches for unifying understanding of the glass transition. This is a clear contrast to the mode coupling approach[3] which focuses only on the dynamical anomalies or to the potential energy landscape picture[4] which provides a theoretical frame work to evaluate thermodynamic properties in the complex random systems.

In the FEL picture, a thermodynamic potential defined in a configurational space of all constituents plays an essential role. We expect a flat landscape at high temperatures and a rugged one at low temperatures as recently shown by a first principle calculation based on the density functional theory[5]. A schematic behavior of the FEL is depicted in Fig. 1.

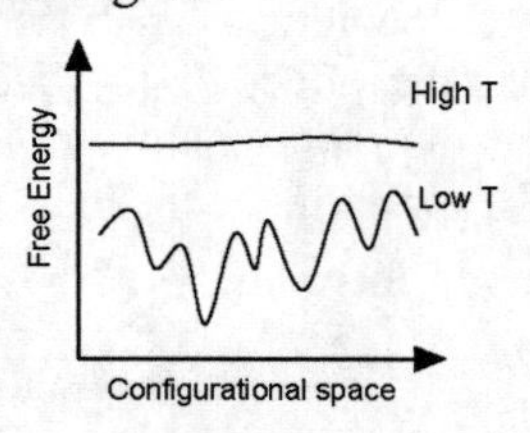

FIGURE 1. A schematic picture of the FEL at high and low temperatures.

Assuming a stochastic motion in the rugged landscape, Tao et al.[6] showed that a rapid change of the specific heat occurs when the representative point begins to be trapped in a basin and that the glass transition can be understood as a transition from annealed to quenched dynamics of the representative point. The theoretical predictions are consistent with experimental observations[7].

As for the dynamics, single particle dynamics has been studied within the trapping diffusion model which is a projection of the dynamics in the FEL onto the coordinate of a single particle [8]. It is known, however, that there exit several relaxation processes in supercooled liquids such as slow (α), Johari-Goldstein (JG: slow β) and fast (fast β) processes[9], and it is a challenging problem to derive these processes on the basis of the FEL picture.

In this paper, we investigate the dynamics of the representative point on the FEL on the basis of the Langevin equation[2] and demonstrate that the three relaxation processes can well be understood by the FEL picture. In next section, the model we investigate is explained. Results for α and fast β processes are presented in Sec. 3. In Sec. 4, the JG process is investigated and conclusion is given in Sec.5

MODEL

We focus on a tagged particle in the FEL and represent the FEL along the path of the particle so that the FEL can be plotted on one dimensional space. In order to obtain clear relation between relaxation processes and the characteristics of the FEL, we employ a simple model system where the one dimensional FEL is given by

$$\Phi(x) = \varepsilon \sin(x/\ell) + \varepsilon A \sin(Bx/\ell), \qquad (1)$$

where ℓ is the period of the FEL and A and B are positive constants. We assume $A << 1$ and $B >> 1$ so that the second term on the right hand side of Eq. (1) gives a small modulation on the smooth landscape determined by the first term. Figure 2 shows the model FEL for $A = 1/5, B = 10$.

CP982, *Complex Systems, 5th International Workshop on Complex Systems*
edited by M. Tokuyama, I. Oppenheim, and H. Nishiyama

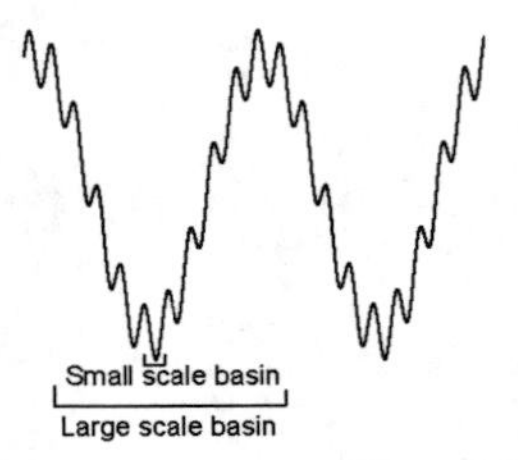

FIGURE 2. The model FEL (1) for $A = 1/5$ and $B = 10$.

As an ansatz, we employ the Langevin equation for the dynamics of the representative point on the FEL [2]

$$\zeta \frac{dx}{dt} = -\frac{d\Phi}{dx} + R(t), \tag{2}$$

where the random force $R(t)$ is assumed to be the Gaussian white noise

$$< R(t_1)R(t_2) >= 2\zeta k_B T \delta(t_1 - t_2). \tag{3}$$

We solve Eq. (2) numerically and investigate the characteristics of the dynamics through the imaginary part of the self-part of the genelarized susceptibility $\chi_s''(q,\omega) = \omega S_s(q,\omega)$, where the dynamical structure factor $S_s(q,\omega)$ is given by

$$S_s(q,\omega) = \frac{1}{2\pi}\int_{-\infty}^{\infty} e^{i\omega t} < e^{-iq[x(t)-x(0)]} > dt \tag{4}$$

$$\simeq \lim_{t_{obs}\to\infty} \frac{1}{2\pi}\frac{1}{t_{obs}}\left|\int_0^{t_{obs}} e^{iqx(t)-i\omega t} dt\right|^2. \tag{5}$$

Here t_{obs} is the time of the numerical integration of Eq. (2).

Although we incorporate the temperature effect only in the random force in this paper, we expect to see the essential features of the separation of slow relaxations. We will discuss effects of the temperature dependence of the FEL in Sec. 5.

α AND FAST β PROCESSES

In this section, we analyze the dynamics on a smooth model FEL by setting $A = 0$ in Eq. (1). We solved Eqs. (1), (2) and (3) by the Euler method and obtained the generalized susceptibility. Figures 3 (a) and (b) show the frequency dependence of $\chi_s''(q,\omega)$ for four different temperatures, where $q\ell = 0.5$.

At high temperature $T/T_0 = 100$ ($T_0 = 2\varepsilon/k_B$), $\chi_s''(q,\omega)$ shows a single peak, where the representative point does not feel the structure of the FEL and moves freely in the space. As the temperature is decreased, the peak becomes wider and at the same time the position of the peak is shifted to lower frequency side. (See $T/T_0 = 0.2$ in Fig. 3 (a)). At $T/T_0 = 0.1$, $\chi_s''(q,\omega)$ shows two peaks: One at the lower frequency side can be regarded as α peak corresponding to the α relaxation and the other at the higher frequency side is related to the fast β process. When the temperature is reduced further ($T/T_0 = 0.01$), the α moves out from the frequency window and only the fast β process is observed.

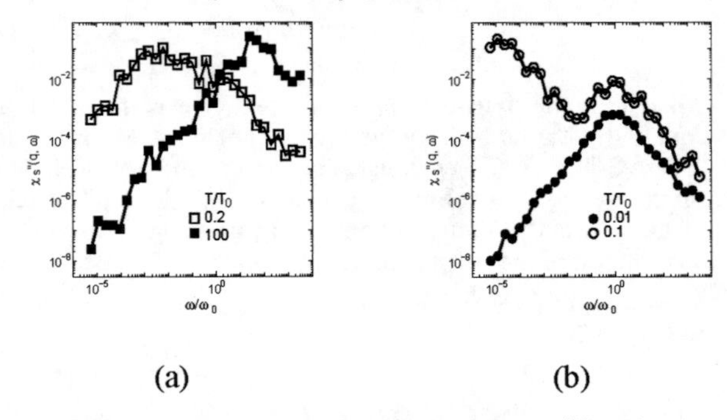

FIGURE 3. The frequency dependence of $\chi_s''(q,\omega)$ for (a) $T/T_0 = 100$ and 0.2 and (b) $T/T_0 = 0.1$ and 0.01. $q\ell = 0.5$, $\omega_0 = \varepsilon/\zeta\ell^2$ and $T_0 = 2\varepsilon/k_B$.

We determined the relaxation time of slow and fast processes from the peak position, whose temperature dependence is shown in Fig. 4.

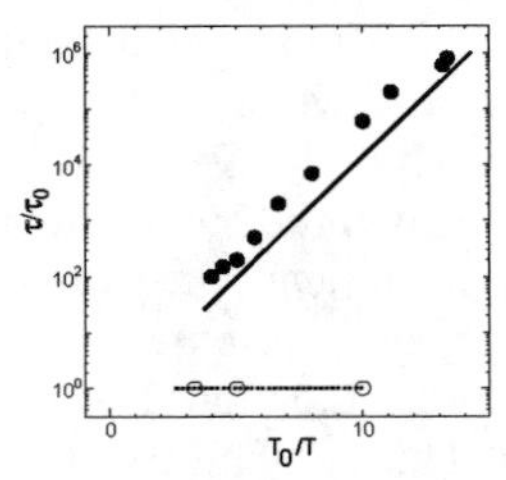

FIGURE 4. The temperature dependence of the relaxation time of the slow (solid circles: τ_α) and fast β (open circles: τ_β) processes, where $q\ell = 0.5$. Solid and dotted curves show the theoretical estimations Eq. (9) and Eq. (8), respectively.

We now make theoretical analysis of the model. First we consider the effect of the deep basins when $k_BT/\varepsilon << 1$ and the representative point is trapped in one basin. We can approximate the local minimum of the FEL by a harmonic form $\Phi(x) \sim \varepsilon\left(-1 + \frac{x^2}{2\ell^2}\right)$. Then, Eq. (2) reduces to

$$\zeta \frac{dx}{dt} = -\frac{\varepsilon}{\ell^2}x + R(t). \tag{6}$$

This equation can be readily solved and we find the self part of dynamical structure factor:

$$S_s(q,\omega) = \frac{1}{2\pi}\frac{k_BT\ell^2q^2}{\varepsilon}\frac{2\left(\frac{\varepsilon}{\zeta\ell^2}\right)}{\omega^2 + \left(\frac{\varepsilon}{\zeta\ell^2}\right)^2}. \tag{7}$$

Thus the relaxation time is given by

$$\tau_\beta = \frac{\zeta \ell^2}{\varepsilon}, \tag{8}$$

which is shown in Fig. 4 by the dotted line. Equation (8) indicates that τ_β is determined by the curvature of the basin and it has the same temperature dependence with the curvature[2].

The slow relaxation is determined by the distribution of the height of the FEL and we can estimate the temperature dependence of τ_α by a simple argument[10, 11]

$$\tau_\alpha = \frac{\left(\int_0^\ell e^{-u/k_BT} dx\right)\left(\int_0^\ell e^{u/k_BT} dx\right)}{D\ell^2 q^2} \equiv \frac{1}{D' q^2}, \tag{9}$$

where $D \equiv k_BT/\zeta$, $u \equiv \varepsilon \sin(x/\ell)$. The numerator of this expression shows influences of the distribution of height of the FEL and consequently the effective diffusion constant D' is modified. The solid line in Fig. 4 shows the relation (9) which agrees quite well with the numerical results. In the present model, the height of the FEL is assumed to be constant and the solid line follows the simple Arrehenius law. If the spatial distribution and temperature dependence of the barriers are incorporated appropriately, then the main relaxation time will follow the Vogel-Fulcher law.

JOHARI-GOLDSTEIN PROCESS

Recent experiments[12] and the molecular dynamics simulation[13] have suggested that the JG process may be originated from rotational motion of molecules. If such dynamics exists, then the FEL will have some modulation due to the orientation of molecules. This effect can be implemented in the model FEL (1) by setting a finite value to parameter A. Setting $A = 1/5$ and $B = 10$ in Eq. (1), we calculated the generalized susceptibility for five different temperatures, which are shown in Figs. 5 (a) and (b).

At high temperature (open squares:$T/T_0 = 100$) in Fig. 5 (a), the susceptibility shows one peak at $\omega/\omega_0 \sim 10^2$. As the temperature is reduced (open circles:$T/T_0 = 0.5$), the peak moves to the lower frequency region. As the temperature is reduced further (solid squares:$T/T_0 = 0.1$), the peak is split into two peaks; one at the lower frequency side is the α process and the other at the higher frequency side is the β process. The α peak at the lower frequency side cannot be seen in this frequency window at $T/T_0 = 0.04$ (open diamonds in Fig. 5 (b)), and the β peak splits again into two peaks, forming the fast β peak at the higher frequency side and the slow (JG) β peak at the lower frequency side. These two peaks are well separated at much lower temperature $T/T_0 = 0.02$ (solid circles in Fig. 5 (b)).

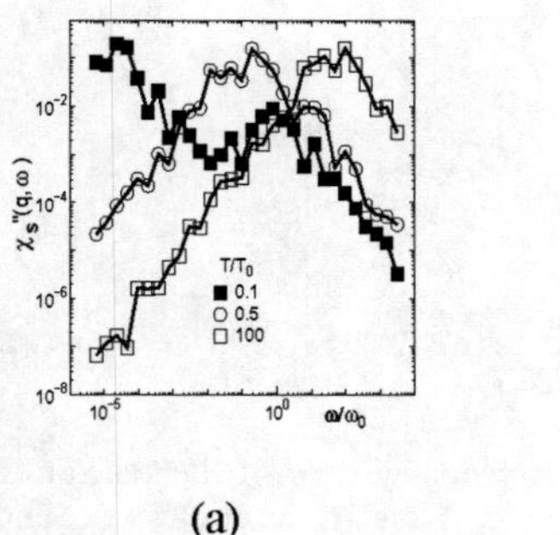

(a)

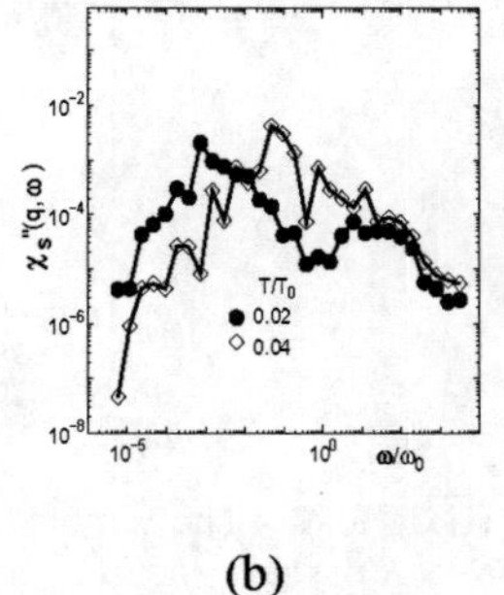

(b)

FIGURE 5. The frequency dependence of $\chi_s''(q,\omega)$ for (a) $T/T_0 = 100$, 0.5 and 0.1 and (b) $T/T_0 = 0.04$ and 0.02. $\omega_0 = \varepsilon/\zeta\ell^2$ and $T_0 = 2\varepsilon/k_B$.

We defined the relaxation time for each process from the peak position of the susceptibility. Figure 6 shows the temperature dependence of the relaxation time of the three relaxation processes. Using the same argument in Sec. 3, we conclude that the curvature of the FEL around a small scale basin determines the temperature dependence of the fast process which is the oscillatory motion within the basin. The relaxation time is given by

$$\tau_\beta = \frac{\zeta \ell^2}{\varepsilon A B^2} \tag{10}$$

as before, which coincides well with the numerical results (open circles) as shown by the dotted curve in Fig. 6. The height of the large scale basins of the FEL determines the temperature dependence of the slow process as before and the relaxation time τ_α is again given by Eq. (9). The solid line in Fig. 6 agrees quite well with numerical results (solid circles).

In order to see the effect of the modulation of the FEL, we analyze the Fokker-Planck equation derived from the Langevin equation (2) when the representative point is trapped in one of the large scale basins. Assuming the large scale basin to be a harmonic form, we can derive the relaxation time as[11]

$$\tau_{JG} = \frac{\left(\int_0^{\ell'} e^{-u'/k_BT} dx'\right)\left(\int_0^{\ell'} e^{u'/k_BT} dx'\right)}{D\ell'^2 \beta\varepsilon}, \tag{11}$$

where $u' = \varepsilon A sin(Bx/\ell)$ and $\ell' = \ell/B$. The dashed line in Fig.6 shows relation (11) which agrees well with numerical results (open squares).

CONCLUSION

We have presented a phenomenological explanation of dynamical characteristics of the glass transition on the basis of the FEL picture. The confined stochastic motion within a basin gives rise to the fast β process and

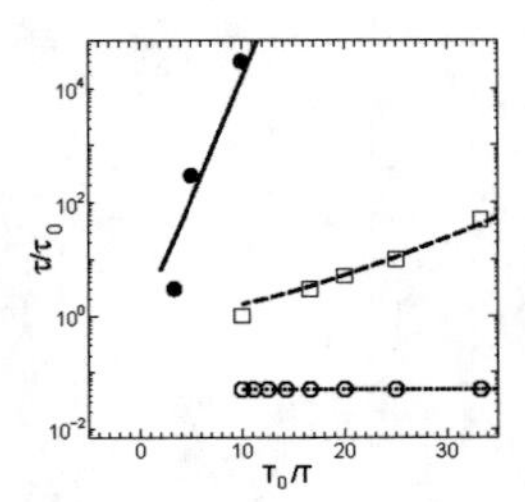

FIGURE 6. The temperature dependence of the relaxation time for the slow (solid circles: τ_α), fast (open circles: τ_β) and JG (open squares; τ_{JG}) processes. $q\ell = 0.5$, and $T_0 = 2\varepsilon/k_B$. The solid, dotted and dashed lines represent approximations Eq. (9), Eq. (10) and Eq. (11), respectively.

its relaxation time is determined by the curvature of the bottom of the basin. The jump motion among basins determines the slow relaxation and short scale modulation of the FEL is related to the JG process. Once the FEL is constructed as a function of temperature, we can obtain various processes by analyzing the Langevin equation or equivalently corresponding Fokker-Planck equation. In particular, the FEL depends on the temperature, and as the temperature is reduced, the FEL acquires deep basins and high barriers. Therefore, the average waiting time for the jump motion among the basins gets longer as the temperature is reduced, and when it reaches the observation time, then the system vitrifies as predicted by Odagaki[14].

It is, therefore, important to emphasize that the FEL picture can explain the thermodynamic and dynamic characteristics of glass transition in a unified description and the complete understanding of the glass transition will be achieved in the FEL approach.

ACKNOWLEDGMENTS

This work was supported in part by the Grant-in-Aid for scientific Research from the Ministry of Education, Culture, Sports, Science and Technology.

REFERENCES

1. T. Odagaki, T. Yoshidome, A. Koyama, and A. Yoshimori: J. Non-Crys. Solids **352**, 4843-4846 (2006).
2. T. Odagaki and T. Ekimoto: J. Non-Crys. Solids **353**, 3928-3931 (2007)
3. W. Götze: in J.P. Hansen, D. Levesque, J. Zinn-Justin (Eds.), Liquids Freezing and the Glass Transition, North Holland, 1989, p.287; W. Götze, J. Phys. Condens. Matter 11 (1999) A1-A45.
4. F. Sciortino: J. Stat. Mech. P05015 (2005).
5. T. Yoshidome, A. Yoshimori, and T. Odagaki: J. Phys. Soc. Jpn. **75**, 054005-1-054005-4 (2006); T. Yoshidome, T. Odagaki and A. Yoshimori, in preparation.
6. T. Tao, A. Yoshimori, and T. Odagaki: Phys. Rev. E **64**, 046112-1-046112-5 (2001); T. Tao, A. Yoshimori, and T. Odagaki: Phys. Rev. E **66**, 041103-1-041103-5 (2002); T. Odagaki, T. Tao, and A. Yoshimori: J. Non-Crys. Solids. **307-310**, 407-411 (2002); T. Odagaki, T. Yoshidome, T. Tao, and A. Yoshimori: J. Chem. Phys. **117**, 10151-10155 (2002); T. Tao, T. Odagaki, and A. Yoshimori: J. Chem. Phys. **122**, 044505-1-044505-5 (2005).
7. O. Yamamuro, I. Tsukushi, A. Lindqvist, S. Takahara, M. Ishikawa, and T. Matsumoto: J. Phys. Chem. B **102**, 1605-1609 (1998).
8. T. Odagaki and Y. Hiwatari: Phys. Rev. A **41**, 929-937 (1990); A. Yoshimori and T. Odagaki: J. Phys. soc. Jpn. **74**, 1206-1213 (2005).
9. M. D. Ediger, C. A. Angell, and S. R. Nagel: J. Phys. Chem. **100**, 13200-13212 (1996).
10. K. Hayashi and S. Sasa: Phys. Rev. E **69**, 066119-1-066119-6 (2004).
11. T. Ekimoto, A. Yoshimori, and T. Odagaki, in preparation.
12. L. Wu: Phys. Rev. B **43**, 9906-9915 (1990).
13. M. Higuchi, J. Matsui, and T. Odagaki: J. Phys. Soc. Jpn. **72**, 178-184 (2003).
14. T. Odagaki, Phys. Rev. Lett. **75**, 3701-3704 (1995).

The Defect Diffusion Model, Glass Transition and the Properties of Glass-Forming Liquids

J. T. Bendler,* J. J. Fontanella,* M. F. Shlesinger,† and M. C. Wintersgill*

*Physics Department, U.S. Naval Academy, Annapolis, MD 21402-5026, USA
†Physical Sciences Division, Office of Naval Research, ONR-30, 875 N. Randolph St., Arlington VA 22203

Abstract. Both the history and current state of the defect diffusion model (DDM) are described. The description includes how the DDM accounts for the glass transition via rigidity percolation. It is shown how the DDM can be used to represent experimental data including viscosity, dielectric relaxation and ionic conductivity. This includes the temperature and pressure variation and the relative contributions of volume and temperature. In addition, a simplified expression for the ratio of the isochoric to isobaric activation energy-enthalpy is presented.

Keywords: Defect Diffusion Model, Dielectric Relaxation, Ionic Conductivity, Viscosity, High Pressure
PACS: 64.70.Pf, 66.20.+d, 77.22.Gm, 66.10.Ed

INTRODUCTION

The first defect diffusion model (DDM) was introduced in 1960 by Glarum.[1] That model was one-dimensional and the nearest defect diffused to a waiting dipole by a steady hopping motion. It was assumed that (1) instantaneous dipole reorientation occurs as soon as a site is encountered by a defect, (2) diffusion is one-dimensional, (3) only the diffusion of the defect nearest to the dipole at $t = 0$ is capable of producing relaxation. Bordewijk then allowed any defect executing normal diffusion to trigger the relaxation.[2] He considered one- and three-dimensional defect diffusion and derived the following Kohlrausch-Williams-Watts (KWW) relaxation function for two special cases of the exponent α'

$$\phi(t) = A\exp(-(t/\tau)^{\alpha'}) \tag{1}$$

where $\alpha' = 0.5$ in one dimension and $\alpha' = 1.0$ in three-dimensions. The value $\alpha' = 1.0$ corresponds to Debye relaxation. At that point in time, the non-Debye-like character of relaxations associated with the glass-transition seemed to rule out the DDM. However, a decade later, it was shown how the DDM can lead to the general KWW relaxation function.[3-6] In those papers, the concept that the glass transition occurs when rigidity percolates was considered. Later it was shown how the DDM leads to the following equation for the temperature dependence of the dielectric relaxation time[7]

$$\tau_{\mathrm{DDM}} = A_\tau \exp\left(\frac{D}{(T - T_C)^{1.5\gamma}}\right). \tag{2}$$

This is frequently a 3/2 power law since often $\gamma=1$.[11] Equation (2) is similar to the original Vogel-Tammann-Fulcher (VTF) equation[8]

$$\tau_V = A_V \exp\left(\frac{D_V}{(T - T_V)}\right). \tag{3}$$

The 3/2 power law entails the use of a critical temperature, T_C, which is the temperature at which mobile single defects (MSDs) would disappear if the glass transition did not intervene. MSDs disappear by clustering into immobile clustered single defects (ICSDs). In the simplest model, two MSDs cluster to form a dimer. More recently, the DDM was extended to include the effect of high pressure[9] and a physical interpretation of fragility was provided.[10] The model yields in certain cases the widely used first power VTF law of eq. (3)[11] and provides a simple understanding of the linear relationship between the pressure derivative and logarithmic temperature derivative of transport properties.[11] The DDM has also been applied to interpret the variation of free volume with temperature inferred from positron annihilation lifetime

CP982, *Complex Systems, 5th International Workshop on Complex Systems*
edited by M. Tokuyama, I. Oppenheim, and H. Nishiyama
2008 American Institute of Physics 978-0-7354-0501-1/08/$23.00

spectroscopy (PALS) data[12] and was used to rationalize widely different pre-exponentials and exponents in the Vogel-type laws that materials exhibit.[13] Reference (13) provides qualitative explanations for the liquid-liquid transition, T_B, the differing values of the critical temperature, T_C, that materials can exhibit for different physical measurements, the origin of secondary relaxations such as the β process, the excess wing and the boson peak.[13] The "merging" of the α and β relaxations is shown to be a natural consequence of the defect picture.[13] Another result is that the pressure dependence of the melting temperature and glass transition temperatures for a given material should be closely related since similar defect populations are responsible for both phenomena.[14,15] Despite the success of the theory, some confusion exists concerning the nature of the DDM. For example, in the Introduction to their papers, Roland et al.[16] and Floudas et al.[17] group the DDM with traditional or conventional free volume theories (CFVT), similar to those described by Ferry.[18] One characteristic of CFVT is that relaxation rates are constant if volume is held constant. Therefore, according to CFVT, if x is a kinetic quantity such as dielectric relaxation time, τ, conductivity, σ, or reciprocal viscosity, R, $(\partial x/\partial T)_V = 0$. In fact, the DDM accounts for the experimental result that, in general, $(\partial x/\partial T)_V \neq 0$.[19] One way to quantify the relative contributions of volume and temperature to a kinetic parameter is via the ratio of the isochoric activation energy, E_V or E_V*,

$$E_V = E_V{}^* = \left(\frac{\partial \ln x}{\partial (1/T)}\right)_V \tag{4}$$

to the isobaric activation enthalpy, E_P or H*.

$$E_P = H^* = \left(\frac{\partial \ln x}{\partial (1/T)}\right)_P . \tag{5}$$

For CFVT, then, $E_V{}^*/H^* = 0$ which is not supported by experiment. The authors have recently shown how the DDM can quantitatively account for the experimental values of $E_V{}^*/H^*$.[20] In the present note, new calculations of $E_V{}^*/H^*$ are presented along with the implications concerning the glass transition.

WHAT ARE THE DEFECTS?

The mobile defects of the DDM were originally conceived to be mobile packets of free volume.[1] As pointed out by Bendler and Shlesinger, specific candidates include vacancies, high-energy conformers in polymers, dangling bonds and 5-coordinated oxygen in water.[21] A more general definition of a defect is that it comprises any intermolecular region where the unoccupied volume is larger or smaller than the average intermolecular unoccupied volume regions in the material. Thus, defects may be associated with fluctuations in atomic positions in a material. MSDs are often regions where the free volume is larger than average and ICSDs are then regions where the free volume is smaller than the average.[12] The important implication, then, is that mobile regions near MSDs are often regions where the local density is smaller than average. By contrast, ICSDs usually represent regions of higher than average local density. The higher-than-average local density regions might sometimes be referred to as rigid. Dynamic heterogeneity, then, is a natural feature of the DDM.

All of the properties of a material's glass transition temperature, T_g, are determined by the nature of the defects and how they interact. As an example of an application of the DDM to glass-forming materials, in Fig. 1, T_g is plotted vs. monomer volume for a number of polymers and glass-forming liquids.[17]

It is clear that T_g generally increases with an increase in monomer volume. The DDM interpretation of Fig. 1 is that materials formed from smaller atomic volumes have lower cohesive energies and should be more defective or have more weakly interacting defects and thus T_C and hence T_g will be lower. Of course, as expected, the correlation is not perfect. For example, it is unlikely that there is a perfect correlation between T_g and T_C for different

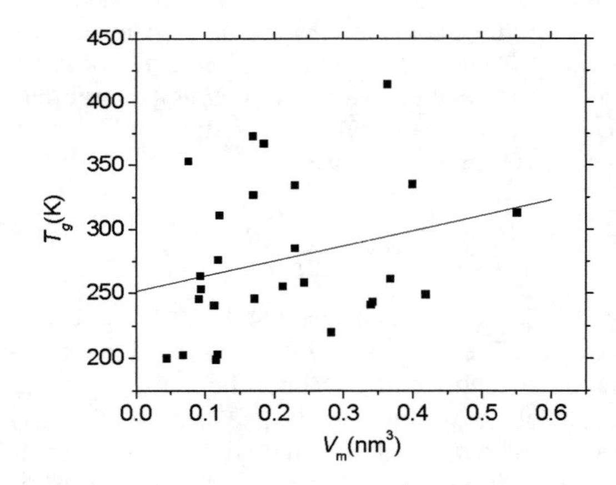

FIGURE 1. Glass transition temperature vs. monomer volume for polymers and glass-forming liquids given in the paper by Floudas et al.[17] Also shown is the best-fit straight line.

materials. If temperature is decreased, T_C is the temperature where MSDs would disappear. On the other hand, T_g is the temperature at which rigidity percolates. Since T_g is higher than T_C, T_g always intervenes before T_C is reached. In that sense there is a relationship between T_g and T_C but it is not expected to be simple. There is more involved in defect clustering (and hence the value of T_C) than just molecular volume. As pointed out elsewhere,[10,20] the critical temperature will depend as well on the number of nearest neighbors and the attractive interaction (free) energy between molecules and MSDs. In a given material, where both T_g and T_C vary with pressure, a better correlation is expected.

APPLICATIONS OF THE DDM

In the DDM, R, τ, and σ can be written[20]

$$x_{DDM,i} = F_i \exp\left(\frac{BT_C^{0.5\eta}}{(T-T_C)^{0.5\eta}\left(V(T,P)/V_o\right)}\right) \tag{6}$$

where T is the absolute temperature, P is the pressure and V is the volume of the material. Each dynamical variable has a different pre-exponential, F_i. In particular,

$$F_\tau = A_\tau \tag{7}$$

$$F_\sigma = \frac{A_\sigma}{T\left(V(T,P)/V_o\right)^{1/3}} \tag{8}$$

and

$$F_R = \frac{A_R}{T}\left(V(T,P)/V_o\right)^{2/3}. \tag{9}$$

As an application of the model, viscosity data for dibutyl phthalate (DBP) will now be described. *PVT* data were obtained both from the analysis by Cook et al.[22] and the data of Bridgman.[23] Equations (6) and (9) were then used to fit the data using techniques described elsewhere.[19] The resultant values of the fitting parameters for η=2 are $\log_{10}(A_R)$=3.89, B=6.14, $T_C(0)$=150.3K, $(\partial T_C/\partial P)_{P=0}$=119.7K/GPa and $(\partial^2 T_C/\partial P^2)_{P=0}$=-268.9K/GPa^2. The best-fit curves are shown in Fig. 2. The agreement between the model calculation and the experimental values is good.

The DDM prediction for the ratio E_V^*/H^* is[20]

$$\left(\frac{E_V\,*}{H\,*}\right)_{DDM} \approx \frac{1-\dfrac{\alpha_P}{\kappa_T}\dfrac{T}{T_C}\dfrac{\partial T_C}{\partial P}}{1+\dfrac{\alpha_P(T-T_C)}{0.5\eta}\dfrac{V_o}{V}}. \tag{10}$$

Equation (10) has been used previously to calculate values of E_V^*/H^* for R in glycerol, σ in poly(propylene glycol) containing $LiCF_3SO_3$, and τ in poly(vinyl acetate) and compare them with experiment.[20] In the present note, results are reported for R in DBP, τ_α (dielectric) in Parel® elastomer (mostly high molecular weight polypropylene oxide) (PPO) and σ in highly crosslinked poly(dimethyl siloxane-block-ethylene oxide) (PDMS-EO) containing $NaCF_3COO$. The results are shown in Fig. 3. Good agreement between theory and experiment is found. The disparity between theory and experiment for PDMS-EO is due, at least in part, to approximate *PVT* data. (*PVT* data for PDMS were used since none appear to be available for crosslinked PDMS-EO.)

The relative contributions of volume/density and temperature have been discussed by several groups.[16,17,24-26] The DDM can be used to further understand the relative contributions of volume and temperature. The dominant factor in eq. (10) is the pressure (and hence volume) dependence of T_C. The larger the value of $\partial T_C/\partial P$, the smaller is the ratio E_V^*/H^* and therefore the smaller is the contribution of temperature. What controls $\partial T_C/\partial P$ is the tendency of defects to cluster and this is in turn governed by the volume decrease upon clustering.[20] The data show that dynamics in PDMS-EO are predominantly controlled

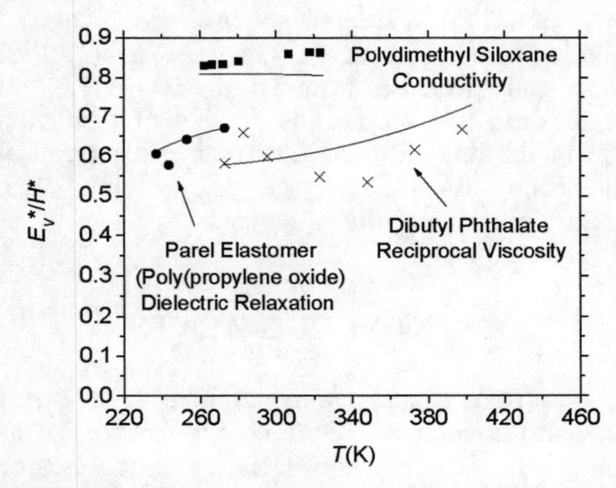

FIGURE 2. E_V^*/H^* vs. temperature for a different physical property for each of three materials.

by temperature while dynamics in the other materials are controlled by temperature to a lesser extent. Since PDMS-EO is quite rigid and incompressible due to a high degree of cross linking, its volume changes more slowly with pressure changes. The defects in PDMS-EO are also likely to be nearly volume-conserving, and thus will experience less volume change upon clustering than do defects in PPG or DBP. H-bonding in glycerol also results in a more dense, rigid fluid, leading to a large value of E_V^*/H^*[20] compared with the "softer" DBP fluid where only van der Waals forces and dipoles are important.

Simplification of eq. (10) is achieved by evaluation at $T=T_g$. This leaves a factor of T_g-T_C in the denominator and we let $T_g-T_C\approx50$K. Next, first power behavior is assumed, $\eta=2$, and the small temperature and pressure variation in V is ignored by setting $V=V_o$. With these approximations, the denominator may be set equal to one since the thermal expansion coefficient is about 10^{-3} K^{-1}. Finally, it is assumed that T_g scales with T_C so that $\partial \ln T_C/\partial P = \partial \ln T_g/\partial P$.[9] This gives

$$\left(\frac{E_V*}{H*}\right)_{DDM} \approx 1-\frac{\alpha_P}{\kappa_T}\frac{\partial T_g}{\partial P}. \qquad (11)$$

Equation (11) is identical to eq. (16) in ref. (26) and makes it easy to understand trends. For example, it has been pointed out that a very small value of $E_V*/H*$, 0.25, is observed for poly(2,6 dimethyl-1,4-phenylene oxide) (PPO®).[24] As expected from eq. (11), PPO® has a very large value of ($\partial T_g/\partial P$), 840 K/GPa. This is to be compared with glycerol where ($\partial T_g/\partial P$) is small, on the order of 35 K/GPa.[14] Correspondingly, for the relatively incompressible glycerol the value of $E_V*/H*$ is very large, 0.8, in agreement with eq. (11). Finally, the data in Table 1 of ref. 24 are of interest. For blends of PPO with polystyrene (PS), $E_V*/H*$ gradually increases from 0.25 to 0.64 as ($\partial T_g/\partial P$) systematically decreases from 840 to 360 K/GPa.[24] It is clear, then, that eq. (11) is capable of predicting trends in the ratio E_V*/H. The underlying physical phenomenon is the change of volume that accompanies the clustering of defects.

ACKNOWLEDGMENTS

This work was supported in part by the U. S. Office of Naval Research. JTB gratefully acknowledges support by the Department of Defense-Army Research Office (Grant No. DAAD19-01-1-0482). The authors would like to thank Frank P. Pursel for helpful discussions.

REFERENCES

1. S. H. Glarum, *J. Chem. Phys.*, **33**, 639-643 (1960).
2. P. Bordewijk, *Chem. Phys. Lett.*, **32**, 592-596 (1975).
3. M. F. Shlesinger and E. W. Montroll, *Proc. Nat. Acad. Sci.*, **81**, 1280-1283 (1984).
4. J. Klafter, M. F. Shlesinger, *Proc. Natl. Acad. Sci. USA*, **83**, 848-851 (1986).
5. J. T. Bendler and M. F. Shlesinger, *Macromolecules*, **18**, 591-592 (1985).
6. J. T. Bendler and M. F. Shlesinger, *J. Mol. Liq.*, **36**, 37-46 (1987).
7. J. T. Bendler and M. F. Shlesinger, *J. Stat. Phys.*, **53**, 531-541 (1988).
8. H. Vogel, *Physik Z.*, **22**, 645 (1921); V. G. Tammann and W. Hesse, *Z. Anorg. Allg. Chem,.* **156**, 245-257 (1926); G. S. Fulcher, *J. Amer. Ceram. Soc.*, **8**, 339 (1925).
9. J. T. Bendler, J. J. Fontanella and M. F. Shlesinger, *Phys. Rev. Letters*, **87**, 195503-1-195503-4 (2001).
10. J. T. Bendler, J. J. Fontanella and M. F. Shlesinger, *J. Chem. Phys.*, **118**, 6713-6716 (2003).
11. J. T. Bendler, J. J. Fontanella, M. F. Shlesinger, and M. C. Wintersgill, *Electrochim. Acta*, **49**, 5249-5252 (2004).
12. J. T. Bendler, J. J. Fontanella and M. F. Shlesinger, J. Bartoš, O. Šauša, J. Krištiak *Phys. Rev. E*, **71**, 031508-1-031508-10 (2005).
13. J. T. Bendler, J. J. Fontanella and M. F. Shlesinger, *J. Non-Crys. Solids*, **352**, 4835-4842 (2006).
14. A. Reiser, G. Kasper, *Europhysics Letters*, **76**, 1137-1143 (2006).
15. K. R. Booher, J. J. Fontanella, S. Passerini, C. A. Edmondson, *Solid St. Ionics*, **177**, 2687-2690 (2006).
16. C. M. Roland, K. J. McGrath, R. Casalini, *J. Non-Cryst. Solids*, **352**, 4910-4914 (2006).
17. G. Floudas, K. Mpoukouvalas, P. Papadopoulos, *J. Chem. Phys.*, **124**, 074905-1-074905-5 (2006).
18. J. D. Ferry, *Viscoelastic Properties of Polymers*, Wiley, New York, 1961, pp. 218-235.
19. J. T. Bendler, J. J. Fontanella and M. F. Shlesinger, and M. C. Wintersgill, *Electrochim. Acta*, **48**, 2267-2272 (2003).
20. J. T. Bendler, J. J. Fontanella and M. F. Shlesinger, in preparation.
21. J. T. Bendler and M. F. Shlesinger, *J. Phys. Chem.*, **96**, 3970-3973 (1992).
22. R. L. Cook, H. E. King, Jr., C. A. Herbst, and D. R. Herschbach, *J. Chem. Phys.*, **100**, 5178-5189 (1994).
23. P. W. Bridgman, *Proc. Am. Acad. Arts. Sci.*, **67**, 1-27 (1932).
24. C. M. Roland, R. Casalini, *Macromolecules*, **38**, 8729-8733 (2005).
25. M. L. Ferrer, C. Lawrence, B. G. Demirjian, D. Kivelson, C. Alba-Simionesco, G. Tarjus, *J. Chem. Phys.*, **109**, 8010-8015 (1998).
26. M. Naoki, H. Endou, K. Matsumoto, *J. Phys. Chem.*, **91**, 4169-4174 (1987).

Network Motif of Water

Masakazu Matsumoto, Akinori Baba[*], and Iwao Ohmine

Research Center for Materials Science and Chemistry Department, Nagoya University, Furo-cho, Chikusa, Nagoya, 464-8602 JAPAN, and

[*]Department of Earth and Planetary Sciences, Faculty of Science, Kobe University, Rokkodai, Nada, Kobe, 657-8501 JAPAN

Abstract. The network motif of water, vitrite, is introduced to elucidate the intermediate-range order in supercooled liquid water. Unstrained vitrites aggregate in supercooled liquid water to form very stable domain. Hydrogen bond rearrangements mostly occur outside the domain, so that the dynamical heterogeneity also stands out. Pre-peak in the structure factor of low density amorphous ice is reproduced by inter-vitrite structure factor. The vitrite can therefore be regarded as a plausible building block of the intermediate-range order and heterogeneity in supercooled liquid water and low-density amorphous ice.

Keywords: Amorphous water; Low-density water; Network motif; Water structure.
PACS: 61.43.Bn, 64.70.Ja, 64.60.aq, 61.20.Gy

INTRODUCTION

The past ten years of research in supercooled liquid have produced major theoretical and experimental advancements. Anomalies of liquid water such as expansion below 4 degree C, for example, are now considered to arise from the properties of metastable forms of water. [1] The expansion accelerates below the melting point, and it is hypothecated to become the low-density amorphous ice if crystal nucleation does not intercept. This expanding "low-density liquid water" has structural similarities with the low-density amorphous ice (LDA). [2]

Recent study revealed that any liquid that expands as it cools must have two distinct phases. [3] Mishima reported that there are actually two different amorphous ice phases, low- and high-density amorphous ices (LDA and HDA), and observed the first-order phase transition between them. [4] Two amorphous ices even share the interface when they coexist. [5] That is, the two "phases" have distinct structural difference, though both are random phases. In this case, LDA is more ordered phase than HDA.

Then, what kind of order emerges in LDA? What is the origin of the expansion in supercooled liquid water? Does the heterogeneity in liquid water really come from hypothetical liquid-liquid coexistence? To answer these questions, first of all, we must understand the microscopic structure of LDA and its difference from HDA. Structure of LDA is characterized by the tetrahedral local order and the intermediate-range order in the hydrogen bond network (HBN). Such a structure, also known as continuous random network (CRN), has been studied for a long time as the model of tetrahedral semiconductors, and its short-range order was assessed in terms of Voronoi polyhedra, coordination number, rings, etc. The existence of intermediate-range order was also assessed in terms of low configurational entropy, a distinct pre-peak in the structure factor, etc. [6,7]: nevertheless, its structure is still not identified. [8]

We characterize the network topology of supercooled liquid water by introducing the network motif (NM) called "vitrite". The vitrite is not a random rubble of the network but a non-overlapping tile of the mosaic, i.e. vitrites are found to aggregate each other to fill large part of LDA. [9] We also try to explain the intermediate-range order observed by diffraction experiments in terms of spatial correlation between vitrites.

METHOD

Determination of Network Motif

Water is a network-forming substance. HBN of liquid water is fully percolated 3-dimensionally. We here use terms "edge" and "vertex" when only the network topology is considered, while "bond" and

CP982, *Complex Systems, 5th International Workshop on Complex Systems*
edited by M. Tokuyama, I. Oppenheim, and H. Nishiyama

"node" for the real structure embedded in 3-dimensional space. In terms of complex network, the order (i.e. number of edges at a vertex) of HBN is about 4 on average, so that the network is not scale-free. Bond length of HBN is almost fixed, so HBN topology is not a small world network. Nevertheless, HBN has own characteristics in both geometry and topology.

There are more than ten different crystal ice phases, extraordinarily large variety for small molecule like water. Surprisingly, water molecules have four hydrogen bonds (HB) and Pauling's ice rule is satisfied in all the ice phases even under super-high pressure. [10] Thus the ice phases are identified not by local coordination number but by the difference in the HBN topology. As for the network geometry, water molecule prefers tetrahedral local order (TLO). Constraints in both geometry and topology yield the cage-like structure that is common motif of HBN of water. For example, HBN of the hexagonal ice is built of cage-like 12-mer. (Fig. 1(a)) Cubic ice Ic is built of network motifs with 10 water molecules. (Fig. 1(b)) Clathrate hydrates are also built of polyhedral cages with 20-30 water molecules. (Fig. 1(c)) Such cage-like structures are also found in the liquid water, especially at supercooled state. [11] Cage-like structure contains a void, and that the increase of such structures is considered to be the origin of expansion when liquid water is cooled. [12]

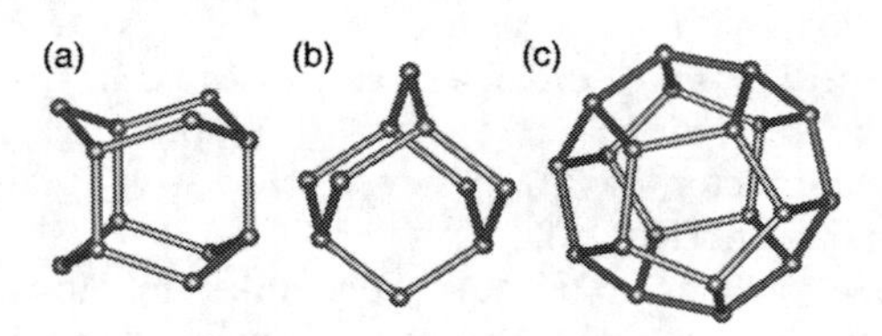

FIGURE 1. Cage-like structure of (a) hexagonal ice, (b) cubic ice, and (c) clathrate hydrate are illustrated. A ball and stick correspond to a water molecule and a hydrogen bond, respectively.

Any kind of given subgraph in HBN of water can be searched by means of some graph-matching algorithms. Thus we can enumerate the population of ice motifs in liquid water. When a set of unit structures are given, one can even build up the possible crystal structures combinatorially, that has been done for the family of zeolites. Some of them are isomorphic with ice network. [13]

Such a strategy is, however, not always helpful to understand the total topology of the random network of supercooled liquid water. While ice motif is not popular in liquid water, there are too ample kinds of motifs possible for matching templates. Although the precedent works also demonstrated the locally preferred structures (e.g. vitron, amorphon, etc.), it seems unreasonable to choose the limited set of local structures as the representative of glassy material. Instead, some systematic method to tessellate the network and to enumerate NM is wanted.

We introduce "vitrites", a family of network motifs satisfying the common topological conditions. A vitrite is defined as a NM built of surface rings encapsulating a void. It must satisfy the following conditions; (1) each edge is shared by two rings, (2) each vertex is shared by 2 or 3 rings, and (3) the graph obeys the Euler's formula, F-E+V=2, where F, E, and V are number of rings, edges, and vertices, respectively. First condition certifies the closed surface. Second condition comes from the characteristics of the network with TLO. It is purely a topological definition but still guarantee the hollow geometry by the third condition. See Ref. 9 for more detail on the definition of the vitrites. Major vitrites in water at low temperature are illustrated in Fig. 2.

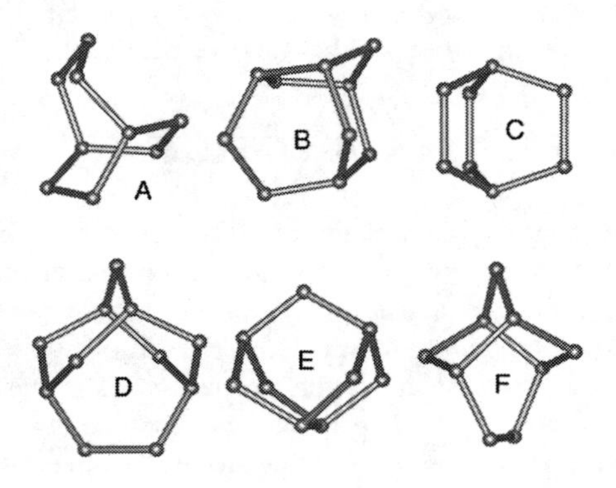

FIGURE 2. Six typical vitrites in liquid water are illustrated. A ball and a stick correspond to a water molecule and a hydrogen bond, respectively. All vitrites found in HBN are available online at the vitrite database: http://vitrite.chem.nagoya-u.ac.jp.

Molecular Dynamics Simulation

Molecular dynamics simulations are performed to obtain the equilibrium water configurations in the temperature range between 200K and 360K under zero and 3,000 atm. Simulation details are described in the previous paper. [9] In the present work, Nada's 6-site water model is used. [14] All the results are reconfirmed by the calculations with TIP4P/2005 water model. [15] The topological properties of the HBN are analyzed in terms of the inherent structures to eliminate the ambiguity over the hydrogen bond criteria.

RESULTS

Distribution of Network Motifs

Topological differences among normal liquid water, water at low temperature and water under high pressure are elucidated by their NM statistics.

In supercooled liquid water, each node (water molecules) prefers to be in regular-tetrahedral local coordination. This preference restricts the variety of intermediate-range network topology. Water at low temperature thus has almost defect-free network consisting mainly of 5- to 7-membered rings, and is filled with stable vitrites with small distortion. (Fig. 3) Hydrogen bonds belonging to unstrained vitrites actually have much longer lifetime than the average.[9] Although NM of ice is also stable, it is not quite popular in supercooled liquid water because of its high symmetry number.

At higher temperature and under high pressure of 3,000 atm, number of the vitrites decreases; they cover only a half of the HB network, and the other half consists of entangled topologies where vitrite-like compact NM is absent.[9]

FIGURE 3. (Color Online) All the vitrites found in the hydrogen bond network in a snapshot at 200K is illustrated, where water molecules and hydrogen bonds are not drawn and the vitrites are depicted by translucent hulls. Vitrites of the same color are identical topologically. A small gap is put between vitrites in order to show the internal structure. The network is totally tessellated into vitrites.

Vitrite Aggregation

In supercooled liquid water, vitrites consist of 5- to 7-membered rings predominate because their surface rings can reduce strains (angular distortion) when they are in the appropriate conformations (boat, chair, and so on). Such vitrites aggregate by sharing the surface rings of the same conformation to form very stable network with TLO.[9] Inside the aggregate, a casual network rearrangement at a bond affects the topology of all the vitrites sharing the bond and might result in increase of strain. Hence, network rearrangement would be as difficult as that inside crystal ice. Actually, most rearrangements take place at the surface of the aggregate but not in the aggregate. (Fig. 4) Thus the vitrite aggregation emerges heterogeneity of both HBN structure and rearrangement in supercooled water, i.e., "ice-like" domain is topologically identified by the aggregation of vitrites. Note that hydrogen bonds in vitrite aggregates also have longer lifetime.[9]

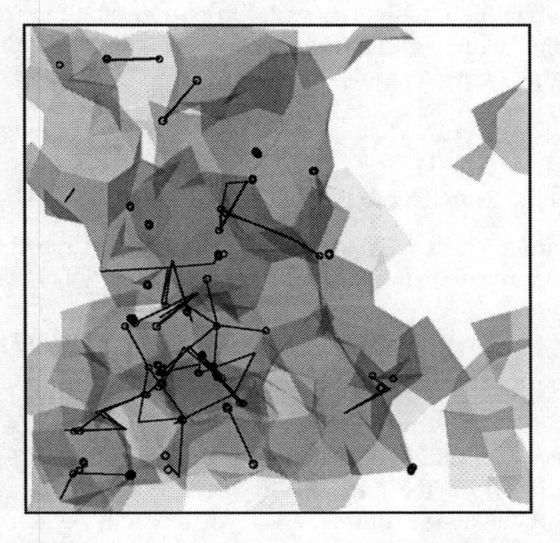

FIGURE 4. Positions of the surface rings of the vitrite aggregates, hydrogen bond rearrangements, and network "defects" during 10 ps, are drawn with translucent flakes, lines, and black dots, respectively. Temperature is 230K and pressure is 0 atm. Water molecules are omitted. All 10 snapshots during 10 ps are overlaid in the picture, so the dark area of flakes corresponds the high probability of finding a surface ring. Top-right region, where no flakes, lines, nor dots are drawn, indicates that the space is filled by the vitrite aggregates without "defects" and no HB rearrangements happen there. Note that the heterogeneity in this picture lasts far longer than 1 ns.

First Sharp Diffraction Peak

The concept of vitrites is also useful to interpret the "First Sharp Diffraction Peak (FSDP)" observed in the x-ray structure factor (SF) of LDA ice. [16] FSDP is found to exist in all network-forming ionic liquids, like $ZnCl_2$ and SiO_2. [17] This peak is not attributed to any real-space partial pair distribution functions. [18] Instead, it is considered to correspond to the correlation length among cavities, which exist in an amorphous network. Barker and co-workers applied the Voronoi analysis on LDL and HDL water and defined the position of voids at vertices of Voronoi polyhedra. They have found that the peak of the void-void SF corresponds quite closely to FSDP. [18] It must be noted, however, that there is no one-to-one correspondence between a void and a physical "cavity" in the network structure.

The origin of the void-void correlation, i.e. of FSDP, is well explained in terms of the spatial correlation among the vitrites. A vitrite has an "empty volume", a cavity, in it. By assigning the virtual atoms at the centers of vitrites and their weighting factors proportional to the vitrite's "volumes" (i.e. effective number of molecules constructing the vitrite. See Ref. 9.) on the corresponding virtual atoms, then we can calculate the inter-vitrite pair correlation function (that is, the pair correlation function of these volume-weighted virtual atoms) as

$$g_{ff}(r) = \frac{V}{4\pi r^2 N_f^{\ 2}} \left\langle \sum_i \sum_{j \neq i} w_i w_j \cdot \delta(r - r_{ij}) \right\rangle \quad (1)$$

where w_i is the "volume" of ith vitrite and $N_f = \sum_i w_i$. The inter-vitrite SF is defined by the Fourier transform of this inter-vitrite pair correlation function $g_{ff}(r)$, such as:

$$S_{ff}(k) = x_f + x_f^{\ 2} \rho \hat{g}_{ff}(k) \quad (2)$$

where x_f is coverage by vitrites, i.e. $x_f = N_f / N$.

The inter-vitrite SF thus obtained, shown in Figure 5, has a distinguished single peak at around 1.6Å^{-1}. [18] FSDP of the inter-vitrite SF becomes very small for higher temperature. FSDP of oxygen-oxygen SF of water yields the same behavior. When the volume-weighted virtual atoms are placed at the center of the vitrites, FSDP, which is apparent in oxygen-oxygen SF, vanishes in the total (i.e. oxygen + virtual atoms) SF (Figure 5), [19] showing that this peak indeed arises from the distribution of vitrites.

ACKNOWLEDGMENTS

This work is partially supported by the Grant-in-Aid for Scientific Research on "Meso-timescale dynamics of crystallization" (Grant No. 14,077,210), "Chemistry on Homogeneous Nucleation" (Grant No. 15,685,002), and "Water Dynamics" (Grant No. 1,400,100) from the Ministry of Education, Science and Culture. Calculations were carried out partly by using the supercomputers at Nagoya University Information Technology Center and at Research Center for Computational Science of Okazaki National Institute.

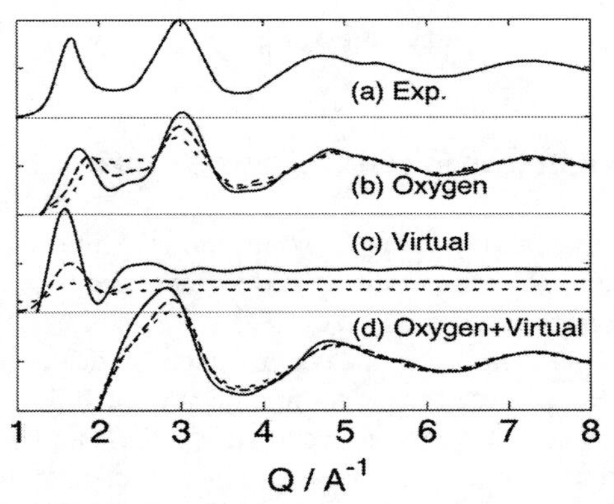

FIGURE 5. SFs from experiments and those obtained from the simulations. (a) X-ray SF of LDA ice of D_2O at 77K taken from Ref. 16. (b) SF between Oxygen atoms. (c) SF between virtual atoms. (d) SF of both Oxygen and virtual atoms. Solid, dashed, and dotted lines correspond to 230K, 260K, and 290K, respectively, and pressure is zero. Instantaneous structures (instead of inherent structures used in other graphs) are used in order to compare with experimental data.

REFERENCES

1. O. Mishima and H. E. Stanley, *Nature* **396**, 329-335 (1998).
2. M.-C. Bellissent-Funel, J. Teixeira, and L. Bosio, *J. Chem. Phys.* **87**, 2231-2235 (1987).
3. F. Sciortino, E. La Nave, and P. Tartaglia, *Phys. Rev. Lett.* **91**, 155701 (2003).
4. O.Mishima, *J. Chem. Phys.* **100**, 5910-5912 (1994).
5. O. Mishima and Y. Suzuki, *Nature* **419**, 599-603 (2002).
6. E. Whalley, D. D. Klug, and Y. P. Handa, *Nature* **342**, 782-783 (1989).
7. D. R. Barker, M. Wilson, and P. A. Madden, *Phys. Rev. E* **62**, 1427-1430 (2000).
8. J. R. Errington et al, *Phys. Rev. Lett.* **89**, 215503 (2002).
9. M. Matsumoto, A. Baba, and I. Ohmine, *J. Chem. Phys.*, **127**, 134504 (2007).
10. D. Eisenberg and W. Kauzmann, *The structure and properties of water*, Oxford Univ. Press, London, 1969, pp.79-92.
11. F.H.Stillinger, *Science* 209, 451-457 (1980).
12. P. G. Debenedetti, *J. Phys.: Condens. Matter* **15**, R1669-R1726 (2003).
13. O. D. Friedrichs, A. W. M. Dress, *Nature* (London) **400**, 644-647 (1999).
14. H. Nada and J.-P J.M. van der Eerden, *J. Chem. Phys.* **103**, 7401-7413 (2003).
15. J. L. F. Abascal and C. Vega, *J. Chem. Phys.* **123**, 234505 (2005).
16. A. Bizid, L. Bosio, A. Defrain, and M. Oumezzine, *J. Chem. Phys.* **87**, 2225-2230 (1987).
17. P. S. Salmon, *Proc. R. Soc. London, Ser. A* **437**, 591-606 (1992); **445**, 351-365 (1994).
18. D. R. Barker, M. Wilson, and P. A. Madden, *Phys. Rev. E* **62**, 1427-1430 (2000).
19. M. Popescu, *Journal of Optoelectronics and Advanced Materials* **5**, 1059-1068 (2003).

A FDR-Preserving Field Theory for Interacting Brownian Particles: One-Loop Theory and MCT

Bongsoo Kim* and Kyozi Kawasaki†

*Institute for Molecular Science, Okazaki 444-8585, Japan
†Electronics Research Laboratory, Fukuoka Institute of Technology, Fukuoka 811-0295, Japan

Abstract. We develop a field theoretical treatment of a model of interacting Brownian particles. We pay particlular attention to the requirement of the time reversal (TR) invariance and the flucutation-dissipation relationship (FDR). Previous field theoretical formulations [1, 2] were found to be inconsistent with this requirement [3, 4]. The method used in the present formulation [5] is a modified version of the auxilliary field method due originally to Andreanov, Biroli and Lefevre (ABL) [4]. We recover the correct diffusion law when the interaction is dropped as well as the standard mode coupling equation in the one-loop order calculation for interacting Brownian particle systems.

Keywords: FDR, Interacting brownian particles, Loop expansion, MCT, Time-reversal symmetry
PACS: 64.70Pf

The only existing successful first-principle theory of structural glass transition [6], the mode coupling theory (MCT) [7, 8, 9, 10], is beset with absence of controllable approximation characterized by smallness parameter. Some years ago one of us attempted to remedy the situation by introducing and working out a dynamical fluid model with a Kac-type long range interaction [11] of appropriate form among elements of the reference fluid, which is anticipated to exhibit glassy behavior [12]. As is well-known, this model has a smallness parameter which is the inverse force range of the Kac potential measured in units of inverse microscopic length scale of the reference fluid. However, the difficulty with this work is the inadequacy of the expansion scheme which violated the detailed balance originating from the time-reversal (TR) invariance of the model equation.

Recently a great deal of attention [13] is being paid to go beyond the standard MCT. This stimulated developments of satisfactory perturbative calculational methods. Particularly noteworthy is the work of ABL [4] where complications associated with the nonlinear TR transformation of the variable set (namely, the density field $[\rho]$ and its conjugate $[\hat{\rho}]$) are avoided by introducing auxiliary variable set $[\theta]$, $[\hat{\theta}]$ which linearizes the TR transformation. The theory was applied to both interacting Brownian particle system and to continuum nonlinear fluid model [1]. However, although ingenious is the whole approach, consequences of the theory worked out so far have yielded some unsatisfactory features as follows.

- The equation for the nonergodicity parameter gives non-trivial result even for non-interacting Brownian particle systems.
- The memory integrals entering the equation for the density-density correlation function are ill-behaved.

Recently for interacting Brownian particle system we have proposed [5] a new set of auxiliary fields still denoted as $[\theta]$, $[\hat{\theta}]$ which are defined slightly differently from ABL. However, consequences are drastically different so that the two unsatisfactory features mentioned above now disappear. Here we present a brief account of our proposal. Our starting point is the dynamical density field model of interacting Brownian particle system, which is expressed as an action integral containing the density field $[\rho]$ and its conjugate field $[\hat{\rho}]$. This action is shown to be invariant under a certain nonlinear TR transformation. This transformation can be converted into a linear one by introducing a conjugate pair of auxiliary fields $[\theta]$ and $[\hat{\theta}]$. The resulting action integral is divided into the Gaussian and non-Gaussian parts, each of which is separately invariant. However, certain terms coming from the both parts cancel when summed over. Consideration of this fact is essential to recover a simple diffusion law for the nonequilibrium averaged density in non-interacting case. We develop a renormalized perturbation theory (RPT) for interacting cases, and recover within one-loop order the standard MCT equation for the density-density correlation function.

We start with the following Langevin equation for the density field $\rho(\mathbf{r},t)$ of interacting Brownian particles

$$\partial_t \rho(\mathbf{r},t) = \nabla \cdot \left(\rho(\mathbf{r},t)\nabla \frac{\delta F[\rho]}{\delta \rho(\mathbf{r},t)}\right) + \eta(\mathbf{r},t) \quad (1)$$

where the Gaussian thermal noise $\eta(\mathbf{r},t)$ has zero mean and variance of the form

$$\langle \eta(\mathbf{r},t)\eta(\mathbf{r}',t')\rangle = 2T\nabla\rho(\mathbf{r},t)\cdot\nabla'\delta(\mathbf{r}-\mathbf{r}')\delta(t-t') \quad (2)$$

where the Boltzmann constant k_B is set to unity, and T is the temperature of the system. Note that the noise

CP982, *Complex Systems, 5th International Workshop on Complex Systems*
edited by M. Tokuyama, I. Oppenheim, and H. Nishiyama

is multiplicative: the noise correlation depends on the density variable. In (1), $F[\rho]$ is the free energy density functional which takes the following form:

$$F[\rho] = T\int d\mathbf{r}\rho(\mathbf{r})\Big(\ln\frac{\rho(\mathbf{r})}{\rho_0}-1\Big) + \frac{1}{2}\int d\mathbf{r}\int d\mathbf{r}'\,\delta\rho(\mathbf{r})U(\mathbf{r}-\mathbf{r}')\,\delta\rho(\mathbf{r}') \quad (3)$$

where $\delta\rho(\mathbf{r},t) \equiv \rho(\mathbf{r},t)-\rho_0$ is the density fluctuation around the equilibrium density ρ_0. In (3) the first (second) term is the ideal-gas (interaction) part of the free energy, $F_{id}[\rho]$ ($F_{int}[\rho]$). Dean [14] has derived the above nonlinear Langevin equation for the microscopic density of system of interacting Brownian particles with pair potential $U(\mathbf{r})$. Kawasaki [15] has also obtained the same form of Langevin equation for the coarse-grained density with $U(\mathbf{r})$ replaced by $-Tc(\mathbf{r})$, $c(\mathbf{r})$ being the direct correlation function, by adiabatically eliminating the momentum field in the fluctuating hydrodynamic equations [1] of simple liquids. For this case, the free energy density functional (3) takes the Ramakrishnan-Yussouff (RY) form [16].

The corresponding action density $s[\rho,\hat{\rho}]$ is given by

$$s[\rho,\hat{\rho}] = i\hat{\rho}\Big[\partial_t\rho-\nabla\cdot\Big(\rho\nabla\frac{\delta F}{\delta\rho}\Big)\Big]-T\rho(\nabla\hat{\rho})^2 \quad (4)$$

where the last term comes from the average over multiplicative thermal noise η. The above action governs the stochastic dynamics of the density variable. In the derivation of (4) employing Ito calculus makes the Jacobian of transformation constant. The dynamic action of this form appearing in the above equation with the RY free energy functional was first written down in [2].

It is a crucial observation of ABL that the action (4) becomes invariant under the TR transformation

$$\begin{aligned}\rho(\mathbf{r},-t) &= \rho(\mathbf{r},t)\\ \hat{\rho}(\mathbf{r},-t) &= -\hat{\rho}(\mathbf{r},t)+\frac{i}{T}\frac{\delta F}{\delta\rho(\mathbf{r},t)}\end{aligned} \quad (5)$$

The standard FDR between the response function $R(\mathbf{r},t)$ and the density correlation function $C(\mathbf{r},t)$ is then readily derived from the action density (4) using the TR transformation (5)

$$R(\mathbf{r},t) = -\Theta(t)\frac{1}{T}\partial_t C(\mathbf{r},t) \quad (6)$$

where $\Theta(t)$ is the Heaviside step function. In (6) the response function $R(\mathbf{r},t;\mathbf{r}'t')$ is given by [3]

$$R(\mathbf{r},t;\mathbf{r}'t') = i\Big\langle\rho(\mathbf{r},t)\nabla'\cdot\Big(\rho(\mathbf{r}',t')\nabla'\hat{\rho}(\mathbf{r}',t')\Big)\Big\rangle \quad (7)$$

The unconventional form of the response function reflects the multiplicative nature of the original Langevin equation (1) and (2).

The transformation (5) is intrinsically nonlinear due to $F_{id}[\rho]$ in (3). As elucidated by ABL [4], this intrinsic nonlinearity of the TR transformation is the underlying reason why the FDR is not preserved order by order in RPT developed for the action (4). The inconsistency of the FDR with the perturbation expansion would then be resolved if the transformation is properly linearized.

With the form of the free energy given in (3), the TR transformation (5) takes the following explicit form

$$\begin{aligned}\rho(\mathbf{r},-t) &= \rho(\mathbf{r},t)\\ \hat{\rho}(\mathbf{r},-t) &= -\hat{\rho}(\mathbf{r},t)+i\hat{K}*\delta\rho(\mathbf{r},t)+if(\delta\rho),\\ \hat{K}*\delta\rho(\mathbf{r},t) &\equiv \int d\mathbf{r}'K(\mathbf{r}-\mathbf{r}')\delta\rho(\mathbf{r}',t)\end{aligned} \quad (8)$$

where $f(\delta\rho) = -\sum_{n=2}^{\infty}(-\delta\rho/\rho_0)^n/n$ is the contribution of the non-Gaussian (higher than the quadratic) part of $F_{id}[\rho]$, and the kernel $K(\mathbf{r})$ is defined as $K(\mathbf{r}) \equiv (\delta(\mathbf{r})/\rho_0+U(\mathbf{r})/T)$.

The entire ignorance of $f(\delta\rho)$ in (8) is tantamount to approximating $F_{id}[\rho]$ to the Gaussian form

$$F_{id}^G[\rho] \simeq \frac{T}{2\rho_0}\int d\mathbf{r}(\delta\rho(\mathbf{r},t))^2 \quad (9)$$

In this case, the transformation (8) becomes linear and consequently the FDR would be preserved by a RPT order by order. However, the contribution of (9) to the body force generates the following two terms in the dynamic equation

$$\nabla\cdot\Big(\rho\nabla\frac{\delta F_{id}^G[\rho]}{\delta\rho}\Big) = T\nabla^2\rho+\frac{T}{\rho_0}\nabla\cdot(\delta\rho\nabla\rho) \quad (10)$$

While the first term is the anticipated diffusion term, the second nonlinear term gives rise to a spurious contribution, incorrectly yielding a nontrivial result even in the absence of particle interaction [3, 4, 17]. One thus should take into account the full logarithmic form of $F_{id}[\rho]$ in (3).

A natural way to make the transformation (8) linear is to introduce a new field $\theta(\mathbf{r},t)$ defined as

$$\theta(\mathbf{r},t) \equiv f(\delta\rho(\mathbf{r},t)) = \frac{1}{T}\frac{\delta F_{id}}{\delta\rho}-\frac{\delta\rho}{\rho_0} \quad (11)$$

Note that the definition (11) differs from that of ABL in that whereas in the work of ABL the auxiliary field involves the *full* free energy

$$\theta_{ABL}(\mathbf{r},t) \equiv \frac{1}{T}\frac{\delta F}{\delta\rho(\mathbf{r},t)} = \hat{K}*\delta\rho(\mathbf{r},t)+f(\delta\rho(\mathbf{r},t)), \quad (12)$$

the eq. (11) limits the variable $\theta(\mathbf{r},t)$ to the nonlinear part of the transformation. The constraint (11) leads to the ideal-gas contribution to the body force

$$\nabla\cdot\left(\rho\nabla\frac{\delta F_{id}}{\delta\rho}\right)=T\nabla\cdot\left(\rho\nabla\left(\frac{\delta\rho}{\rho_0}+\theta\right)\right)=T\nabla^2\rho$$
$$+\ \frac{T}{\rho_0}\nabla\cdot(\delta\rho\nabla\rho)+\rho_0T\nabla^2\theta+T\nabla\cdot(\delta\rho\nabla\theta) \quad (13)$$

On the other hand, since due to cancellation of the two nonlinear effects the entire ideal-gas contribution to the dynamics should be of pure diffusion

$$\nabla\cdot\left(\rho\nabla\delta F_{id}/\delta\rho\right)=T\nabla\cdot\left(\rho\nabla\ln(\rho/\rho_0)\right)=T\nabla^2\rho,$$

the sum of the last three terms in (13) should vanish:

$$(T/\rho_0)\nabla\cdot(\delta\rho\nabla\rho)+\rho_0T\nabla^2\theta+T\nabla\cdot(\delta\rho\nabla\theta)=0 \quad (14)$$

With the constraint (11), the action density (4) takes the new form which is decomposed into the Gaussian part $s_g[\rho,\hat{\rho},\theta,\hat{\theta}]$ and the non-Gaussian part $s_{ng}[\rho,\hat{\rho},\theta,\hat{\theta}]$ as

$$\begin{aligned}
s[\rho,\hat{\rho},\theta,\hat{\theta}] &\equiv s_g[\rho,\hat{\rho},\theta,\hat{\theta}]+s_{ng}[\rho,\hat{\rho},\theta,\hat{\theta}]\\
s_g[\rho,\hat{\rho},\theta,\hat{\theta}] &\equiv i\hat{\rho}\left[\partial_t\rho-T\nabla^2\rho-\underline{\rho_0T\nabla^2\theta}\right.\\
&- \left.\rho_0\nabla^2\hat{U}*\delta\rho\right]-T\rho_0(\nabla\hat{\rho})^2+i\hat{\theta}\theta\\
s_{ng}[\rho,\hat{\rho},\theta,\hat{\theta}] &\equiv i\hat{\rho}\left[-\nabla\cdot\left(\delta\rho\nabla\hat{U}*\delta\rho\right)\right.\\
&- \left.\underline{\frac{T}{\rho_0}\nabla\cdot(\delta\rho\nabla\rho)}-\underline{T\nabla\cdot(\delta\rho\nabla\theta)}\right]\\
&- T\delta\rho(\nabla\hat{\rho})^2-i\hat{\theta}f(\delta\rho) \qquad (15)
\end{aligned}$$

The corresponding linear transformation under which the new action density (15) becomes invariant is given by

$$\begin{aligned}
\rho(\mathbf{r},-t) &= \rho(\mathbf{r},t)\\
\hat{\rho}(\mathbf{r},-t) &= -\hat{\rho}(\mathbf{r},t)+i\hat{K}*\delta\rho(\mathbf{r},t)+i\theta(\mathbf{r},t)\\
\theta(\mathbf{r},-t) &= \theta(\mathbf{r},t)\\
\hat{\theta}(\mathbf{r},-t) &= \hat{\theta}(\mathbf{r},t)+i\partial_t\rho(\mathbf{r},t) \qquad (16)
\end{aligned}$$

It is easy to show that the modulus of the associated transformation matrix O is unity ($\det O=-1$). The Gaussian and nonGaussian actions are *separately* invariant under the transformation (16). Though with the constraint (11) the three underlined terms in (15) vanish when summed together, each of them should be kept in explicit calculation of RPT since its presence is crucial for the separate invariance of the Gaussuan and non-gaussian actions, which enables one to construct the FDR-preserving RPT from these actions. Nevertheless it is shown in the one-loop order that the ultimate effect of these three underlined terms is their cancellation.

The response function corresponding to the new form of the action (15) takes the form

$$R(\mathbf{r},t;\mathbf{r}',t')=\frac{i}{T}\left\langle\delta\rho(\mathbf{r},t)\,\hat{\theta}(\mathbf{r}',t')\right\rangle \qquad (17)$$

Using (17) and taking correlation of the last member of (16) with $i\delta\rho(\mathbf{r},t)/T$ one can recover the FDR (6).

Developing a standard RPT for the action (15) we obtain the following nonperturbative closed dynamic eq. for the density correlation function $C(\mathbf{k},t)$ with the one-loop expressions for the self-energies

$$\begin{aligned}
\dot{C}(\mathbf{k},t) &= -\frac{\rho_0Tk^2}{S(\mathbf{k})}C(\mathbf{k},t)+\frac{1}{S(\mathbf{k})}\Sigma_{\hat{\rho}\hat{\rho}}\otimes C(\mathbf{k},t)\\
&- \Sigma_{\hat{\rho}\hat{\theta}}\otimes\dot{C}(\mathbf{k},t),\\
\Sigma_{\hat{\rho}\hat{\rho}}(\mathbf{k},t) &= \frac{T^2}{2}\int_{\mathbf{q}}\left[V^2(\mathbf{k},\mathbf{q})-\frac{k^2}{\rho_0}V(\mathbf{k},\mathbf{q})\right]\\
&\times\ C(\mathbf{q},t)C(\mathbf{k}-\mathbf{q},t),\\
\Sigma_{\hat{\rho}\hat{\theta}}(\mathbf{k},t) &= \frac{T}{2\rho_0^2}\int_{\mathbf{q}}V(\mathbf{k},\mathbf{q})C(\mathbf{q},t)C(\mathbf{k}-\mathbf{q},t),\\
V(\mathbf{k},\mathbf{q}) &\equiv \mathbf{k}\cdot\mathbf{q}c(\mathbf{q})+\mathbf{k}\cdot(\mathbf{k}-\mathbf{q})c(\mathbf{k}-\mathbf{q}) \qquad (18)
\end{aligned}$$

where $A\otimes B(\mathbf{k},t)\equiv\int_0^t ds A(\mathbf{k},t-s)B(\mathbf{k},s)$ and $\int_{\mathbf{q}}\equiv\int d\mathbf{q}/(2\pi)^3$. Note that (18) reduces to the diffusion eq. $\dot{C}(\mathbf{k},t)=-Tk^2C(\mathbf{k},t)$ in the absence of interaction ($c(\mathbf{k})=0$). The equation (18) is not quite the same as the standard MCT equation: while the convolution integral in MCT equation contains only the time derivative of the density correlation function, the first part of the convolution integral in (18) involves the density correlation function itself, instead of its time derivative. This contrasting feature is reminicient of that between the reducible and irreducible memory functions [18] appearing in the dissipative stochastic systems. It has been shown for a general class of dissipative stochastic systems obeying the detailed balance condition [19] that the memory function in the exact eq. for the correlation function obtained from the projection operator method can be further reduced to the so-called irreducible memory function. In particular, the derived exact dynamic eq. for the correlation function of the slow variable $A(t)$ is given by

$$\begin{aligned}
\dot{C}_A(t) &= -|E_A|C_A(t)+M_A\otimes C_A(t),\\
C_A^L(z) &= C_A(0)\left[z+|E_A|-M_A^L(z)\right]^{-1} \qquad (19)
\end{aligned}$$

where $C_A^L(z)$ is the Laplace transform of $C_A(t)$, and $|E_A|^{-1}$ is a characteristic short relaxation time scale in the system. The memory function $M_A(t)$ turns out to be reducible to another memory function, the irreducible memory function $M_A^{ir}(t)$:

$$\begin{aligned}
M_A(t) &= M_A^{ir}(t)-|E_A|^{-1}M_A\otimes M_A^{ir}(t),\\
M_A^L(z) &= M_A^{L,ir}(z)\left[1+|E_A|^{-1}M_A^{L,ir}(z)\right]^{-1} \qquad (20)
\end{aligned}$$

Note that $M_A^{L,ir}(z=0)$ can grow indefinitely when the global relaxation time grows indefinitely in contrast to $M_A^L(z=0)$. The eqs. (19) and (20) lead to the dynamic eq. for $C_A(t)$

$$\begin{aligned}\dot{C}_A(t) &= -|E_A|C_A(t) - |E_A|^{-1}M_A^{ir}\otimes\dot{C}_A(t),\\ C_A^L(z) &= C_A(0)\left[z+\frac{|E_A|}{1+|E_A|^{-1}M_A^{L,ir}(z)}\right]^{-1} \quad (21)\end{aligned}$$

Using this formal correspondence between the reducible memory function and (18), we can rewrite (18) into the corresponding form of (21) where the convolution integral involves only the time derivative of the density correlation function:

$$\begin{aligned}\dot{C}(\mathbf{k},t) &= -\frac{\rho_0 T k^2}{S(\mathbf{k})}C(\mathbf{k},t) - \mathscr{M}\otimes\dot{C}(\mathbf{k},t),\\ \mathscr{M}(\mathbf{k},t) &\equiv \Big(\Sigma_{MC} + \mathscr{M}\otimes\big(\Sigma_{MC} - \Sigma_{\hat{\rho}\hat{\theta}}\big)\Big)(\mathbf{k},t),\\ \Sigma_{MC}(\mathbf{k},t) &\equiv \frac{1}{\rho_0 T k^2}\Sigma_{\hat{\rho}\hat{\rho}}(\mathbf{k},t) + \Sigma_{\hat{\rho}\hat{\theta}}(\mathbf{k},t)\\ &= \frac{T}{2\rho_0}\int_{\mathbf{q}} V^2(\hat{\mathbf{k}},\mathbf{q})C(\mathbf{q},t)C(\mathbf{k}-\mathbf{q},t) \quad (22)\end{aligned}$$

where $\hat{\mathbf{k}} \equiv \mathbf{k}/k$. The kernel $\mathscr{M}(\mathbf{k},t)$ corresponds to the irreducible memory function: $\mathscr{M}(\mathbf{k},t) = M_A^{ir}(t)/|E_A|$ with $|E_A| = \rho_0 T k^2/S(\mathbf{k})$. The kernel $\Sigma_{MC}(\mathbf{k},t)$ is the same as the one appearing in the MCT.

Now in (22) one should retain only the first term $\mathscr{M}(\mathbf{k},t) = \Sigma_{MC}(\mathbf{k},t)$ since it is the only one-loop (two-particle irreducible) diagram (when iterated, the convolution term generates higher-loops). With $\mathscr{M}(\mathbf{k},t) = \Sigma_{MC}(\mathbf{k},t)$, the first eq. of (22) leads to the standard MCT equation:

$$\dot{C}(\mathbf{k},t) = -\frac{\rho_0 T k^2}{S(\mathbf{k})}C(\mathbf{k},t) - \Sigma_{MC}\otimes\dot{C}(\mathbf{k},t) \quad (23)$$

Therefore we obtain the standard MCT equation in the one loop order.

In summary, we analysed a field theoretical model for interacting Brownian particle systems with a view to develop a RPT consistent with the FDR. This is made particularly transparent by introduction of a conjugate pair of auxiliary field variables thereby linearizing the TR transformation. For non-interacting particle systems we recover a simple diffusion law as one expects, which is not the case in some recent works [3, 4]. For interacting particle cases, we recover the standard MCT in the one-loop order of the RPT.

Having worked out the one loop calculation, the next natural step is to undertake higher loop calculations. This is anticipated to be an audacious task: one needs to pay meticulous attention to all possible terms (or diagrams) up to desired higher order. Still this is worthwhile since no systematic calculation of corrections to the standard MCT is yet available which is crucial to theoretically assess successes and inadequacies of the standard MCT, and to push the theory beyond the current MCT. In order to systematize higher order calculations, one needs to identify a proper smallness parameter (or parameters) of the expansion. We also may undertake extensions of our approach to treat multibody correlations and to genuinely non-equilibrium problems like aging and colloids under external shear flow.

ACKNOWLEDGMENTS

KK was supported by Grant-in-Aid for Scientific Research (C) Grant 17540366 by Japan Society for the Promotion of Science (JSPS). BK thanks Prof. Fumio Hirata, Prof. Song-Ho Chong, and other group members for warm hospitality during his sabbatical stay at the Institute for Molecular Science. BK acknowledges a financial support by the Institute for Molecular Science.

REFERENCES

1. S. P. Das and G. F. Mazenko, *Phys. Rev. A* **34**, 2265-2282 (1986).
2. K. Kawasaki and S. Miyazima, *Z. Phys. B: Condens. Matter* **103**, 423-431 (1997).
3. K. Miyazaki and D. R. Reichman, *J. Phys. A* **38** L343-L355 (2005).
4. A. Andreanov, G. Biroli, and A. Lefèvre, *J. Stat. Mech.* P07008-1-P07008-53 (2006).
5. B. Kim and K. Kawasaki, *J. Phys. A* **40**, F33-F42 (2007).
6. H. C. Andersen, *PNAS* **102**, 6686-6691 (2005).
7. W. Götze *J. Phys. Condens. Matter* **11**, A1-A45 (1999).
8. S. P. Das, *Rev. Mod. Phys.* **76**, 785-851 (2004).
9. D. R. Reichman and P. Charbonneau, *J. Stat. Mech.* P05013-1-P05013-23 (2005).
10. K. Miyazaki, *Glass transition and Mode Coupling Theory; Recent progress and Fundamental Questions, Bussei Kenkyu* **88**, 621-720 (2007) (in Japanese).
11. K. Kawasaki, *J. Stat. Phys.* **110**, 1249-1304 (2003).
12. K. Loh, K. Kawasaki, A. Bishop, T. Lookman, A. Saxena, and Z. Nussinov, *Phys. Rev. E* **69**, 010501-1-010501-4 (2004).
13. L. Berthier, G. Biroli, J. -P. Bouchaud, W. Kob, K. Miyazaki, and D. R. Reichman, *J. Chem. Phys.* **126**, 184503-1-184503-21 (2007); **126**, 184504-1-184504-21 (2007), and references therein.
14. D. Dean, *J. Phys. A* **29**, L613-L617 (1996).
15. K. Kawasaki, *Physica A* **208**, 35-64 (1994).
16. T. V. Ramakrishnan and M. Youssouff, *Phys. Rev. B* **19**, 2775-2794 (1979).
17. R. Schmitz, J. W. Dufty, and P. De, *Phys. Rev. Lett.* **71**, 2066-2069 (1993).
18. B. Cichocki and W. Hess, *Physica A* **141**, 475-488 (1987).
19. K. Kawasaki, *Physica A* **215**, 61-74 (1995); *J. Stat. Phys.* **87**, 981-988 (1997).

Relationship between the Diffusive Coefficient and the Specific Heat for Lennard-Jones Binary Mixture

Takayuki Narumi* and Michio Tokuyama†,**

*Graduate School of Tohoku University, Sendai, 980-8579, Japan
†WPI Advanced Institute for material Research, Tohoku University, Sendai 980-8577, Japan
**Institute of Fluid Science, Tohoku University, Sendai, 980-8577, Japan

Abstract. We perform the extensive molecular dynamics simulations on the Kob-Andersen type Lennard-Jones binary mixtures. We study not only dynamical behavior near the glass transition but also the static one, including the specific heat. We then analyze the simulation results from a unified point of view suggested by Tokuyama. The view suggests the definition of the glass transition by using the long-time self-diffusion coefficient. Thus, we show that the temperature at the peak of the specific heat coincides with the glass transition temperature calculated from the long-time self-diffusion coefficient. We also discuss the relationship between the specific heat and the long-time self-diffusion coefficient from a new standpoint.

Keywords: Glass transition, Kob-Andersen model, Long-time self-diffusion coefficient, Specific heat
PACS: 64.70.Q-, 64.70.P-, 65.40.Ba, 64.70.kj, 64.70.qd, 05.20.Jj

INTRODUCTION

Glass and supercooled liquid have been studied since antiquity, and we have much knowledge of them [1, 2, 3]. Near the glass transition temperature, glass-forming liquids show slow dynamical behavior: Their viscosity of glass forming liquids increases continuously but rapidly. Moreover, their relaxation functions (e.g. intermediate scattering functions) exhibit some complicated relaxations. However, the mechanisms of the slow dynamics have not been elucidated theoretically yet.

The specific heat also changes drastically at the vicinity of the glass transition point [4, 5, 6]. While the specific heat diverges to infinity in the second-order transition phenomena, it does not in the glass transition. Thus, it is considered that the glass transition is dynamical crossover. In order to understand the mechanisms of the glass transition and slow dynamics, we need to link statics and dynamics like the Adam-Gibbs approach [7].

The viscosity decreasing corresponds to the increasing of the diffusion coefficient. Tokuyama suggests analytic forms of the long-time self-diffusion coefficient (LSD) for fragile glass forming systems (singular [8] and non-singular type [9]). The equations contain a parameter p_c which means a singular point of the diffusion coefficient. However, it is unclear whether a singular point of the diffusion coefficient exists or not. We do not know yet what proper meaning the p_c has.

We also suggest the definition of the glass transition by using the LSD [9]. The purpose of this proceedings is to investigate the validity for the definition via the specific heat at constant volume. We perform the molecular dynamics simulations and measure the specific heat at constant volume (statics) and the LSD (dynamics) for the Kob-Andersen Lennard-Jones binary mixture model [10]. For the Kob-Andersen model, the glass transition temperature is obtained as $T_g = 0.4376$ from the LSD analysis [9]. Note that this temperature corresponds to the glass transition LSD ($\hat{D}_g = 9.16 \times 10^{-6}$), which is based on the experiment performed by van Magen et. al. [11].

MEASUREMENTS

The long-time self-diffusion coefficient (LSD) D_s^L is represented by

$$D_s^L = \lim_{t\to\infty} \frac{M_2(t)}{2d}, \tag{1}$$

where $M_2(t)$ denotes the mean-square displacement and d the spatial dimensionality. The LSD depends on an appropriate control parameter p (e.g. volume fraction, inverse temperature, and so on). Tokuyama has theoretically derived an equation of the LSD for hard-sphere suspensions as [12]

$$D_s^L(\phi) = D_s^S(\phi)\frac{1-9\phi/32}{1+(D_s^S(\phi)/D_0)(\phi/\phi_c)(1-\phi/\phi_c)^{-2}}, \tag{2}$$

where ϕ denotes the volume fraction, ϕ_c ($\simeq 0.5718$) the critical volume fraction, D_s^S the short-time self-diffusion coefficient, and D_0 the diffusion coefficient of a single particle. The term $9\phi/32$ appears due to the coupling between the hydrodynamic interactions among particles and the direct interactions [12]. Tokuyama considers that the LSD has the universal form which is similar to Eq.

CP982, *Complex Systems, 5th International Workshop on Complex Systems*
edited by M. Tokuyama, I. Oppenheim, and H. Nishiyama

(2), and then extends Eq. (2) to molecular systems as [8]

$$\hat{D}(p) = \frac{D_s^L(p)}{av_0} = \frac{1}{1+\varepsilon(p/p_c)(1-p/p_c)^{-2}}, \quad (3)$$

where a denotes the radius of the molecules, and v_0 the unit velocity. We note that ε and p_c should be treated as fitting parameters in Eq. (3), whereas D_s^S, D_0, and ϕ_c are not fitting parameters in Eq. (2). It is unclear whether the singular point p_c of the LSD exists or not. Results from simulation studies show that the singular point does not exist [13, 14].

Tokuyama also suggests a non-singular type equation of the LSD [9]. To fit the non-singular equation to simulation or experimental data, we use a new fitting parameter α in addition to ε and p_c. Because the singular point p_c is needed even for non-singular fitting, we consider that it has physical meanings even if the singular point does not exist. Using the non-singular fitting line, we can calculate the glass transition point which corresponds to $\hat{D}_g = 9.16 \times 10^{-6}$.

The specific heat c_V per particle at constant volume is defined by

$$c_V := N^{-1}(\partial E/\partial T)_V, \quad (4)$$

where N denotes the number of particle, E the internal energy, T the temperature, and V the volume. When the internal energy is represented by $E = K + \Phi$ where K denotes the kinetic energy and Φ the potential energy, the specific heat can be represented by

$$c_V = c_V^K + c_V^\Phi = dk_B/2 + c_V^\Phi, \quad (5)$$

where c_V^K the specific heat due to the kinetic energy, c_V^Φ that due to the potential energy, and k_B the Boltzmann constant. Because of the equipartition law of energy, one can show that $c_V^K = dk_B/2$. Moreover, $c_V^\Phi = dk_B/2$ in the harmonic oscillator approximation, and thus one can obtain $c_V = dk_B$ in solid states, which is known as the Dulong-Petit law[1]. The behavior of the specific heat is determined by the potential energy part since the specific heat due to the kinetic energy is constant.

MODEL

We consider binary-mixture molecular systems which consist of two kinds of particles (particle A and particle B) where the numbers of A particles and B particles are given by N_A and N_B, respectively. The difference between A and B particle is particle size, but we set them as the same mass m. The particles are confined in a certain three-dimensional space with the volume V constant, and we neglect the interaction with the boundary of the domain. We investigate those systems in which the control parameter is given by the inverse temperature; $p = \beta(= 1/k_BT)$.

The motion of particles is described by the classical Newtonian equation

$$m\frac{d^2}{dt^2}\vec{X}_i(t) = -\nabla_i \sum_{j\neq i} u_{\eta\xi}(X_{ij}), \quad (6)$$

where ∇_i denotes the derivative with respect to $\vec{X}_i$, $X_{ij} = |\vec{X}_i - \vec{X}_j|$, and the pair interaction $u_{\eta\xi}(r)$ is described by the Lennard-Jones binary mixture interaction

$$u_{\eta\xi}(r) = 4\varepsilon_{\eta\xi}\left[\left(\frac{\sigma_{\eta\xi}}{r}\right)^{12} - \left(\frac{\sigma_{\eta\xi}}{r}\right)^{6}\right], \quad (7)$$

where $\sigma_{\eta\xi}$ denotes one of the Lennard-Jones potential parameters which corresponds to the size of the particles, $\varepsilon_{\eta\xi}$ another parameter which has energy dimension and corresponds to the depth of the potentials, and η and ξ a kind of particle; $\{\eta, \xi\} \in \{A, B\}$. We adopt the Kob-Andersen model [10] in which the parameters of Eq. (7) are given by

$$\frac{\sigma_{AB}}{\sigma_{AA}} = 0.8, \quad \frac{\sigma_{BB}}{\sigma_{AA}} = 0.88, \quad \frac{\varepsilon_{AB}}{\varepsilon_{AA}} = 1.5, \quad \frac{\varepsilon_{BB}}{\varepsilon_{AA}} = 0.5, \quad (8)$$

and the number density is set as $1.2\sigma_{AA}{}^{-3}$. This model simulates a metallic alloy $Ni_{80}P_{20}$ [16].

To perform molecular dynamics simulations, we employ the Lennard-Jones units. Length is scaled with σ_{AA}, energy with ε_{AA}, temperature with ε_{AA}/k_B, and time with $\tau = \sigma_{AA}\sqrt{m/48\varepsilon_{AA}}$. The unit velocity v_0 is represented by $v_0 = \sigma_{AA}/4\tau = \sqrt{3\varepsilon_{AA}/m}$. Moreover, the numbers of particles are given by $N_A = 8780$ and $N_B = 2196$. The particles are in a cubic cell with the length $L = 20.89\sigma_{AA}$. The equation of motion is integrated by the velocity Verlet method with a time step 0.01τ. The periodic boundary conditions are employed. The cutoff distance of the interactions for each combination of particles is set as $2.5\sigma_{AA}$. An initial configuration is a random configuration. In order to observe equilibrium data, we firstly wait for time which is ten times longer than the relaxation time for each temperature (preparing calculations), and then we measure physical variables (main calculations). Temperature is adjusted by the velocity scaling method in the preparing calculations.

RESULT AND DISCUSSION

Fig. 1 shows the results of the mean-square displacement of the only A particle. In the range of the inverse temperature less than $\beta = 2.2$ (the temperature more than

[1] In the lower temperature region, the Dulong-Petit law does not hold and the specific heat is a function of the square of the temperature (Debye model).

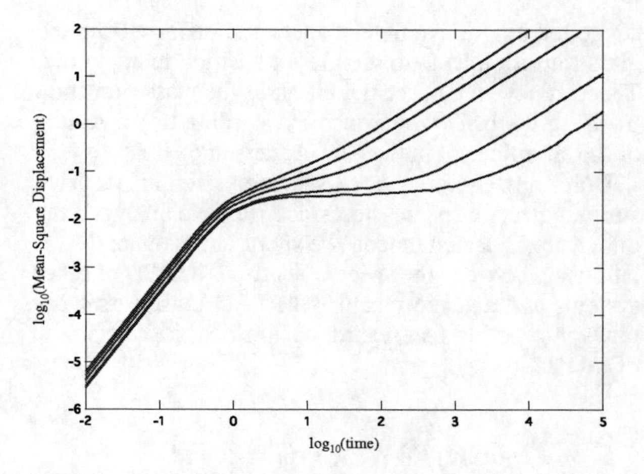

FIGURE 1. (Color online) Log-log plot of the mean-square displacement of the A particle for $T = 1.0, 0.83, 0.67, 0.50$, and 0.45 (from top to bottom).

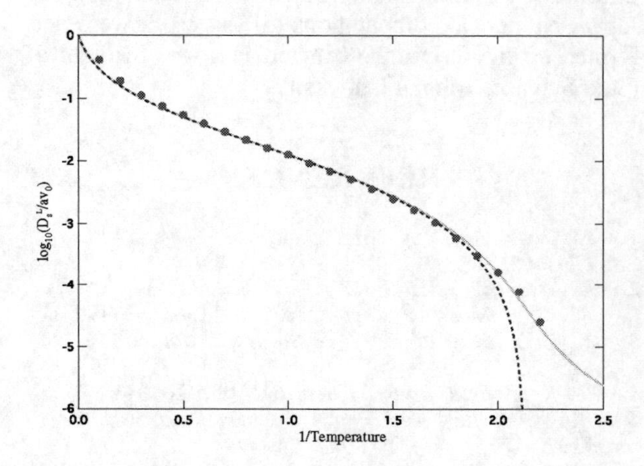

FIGURE 2. (Color online) Plot of the common logarithm of the LSD versus the inverse temperature. The red circles indicate the simulation results, the dashed blue line the singular type fitting (Eq. (3)), and the solid green line the non-singular type fitting [9]. The parameters are $\varepsilon = 46.75$, $\beta_c = 2.13$, and $\alpha = 2.05$.

$T = 0.455$), those results are steady-state within error, and then we regard them as equilibrium state. The slope of the log-log plot is unity in large time scale that means the dynamics is dominated by the diffusive motion. On the other hand, in $\beta \geq 2.3$ ($T \leq 0.435$) we have not obtained steady-state results yet. It is unclear whether our calculation time is lacking or the systems show aging.

We can measure the long-time self-diffusion coefficient (LSD) from the long-time limit of the mean-square displacement (Eq. (1)). The results of the LSD are shown in Fig. 2. We note that the LSD is scaled with av_0 in Fig. 2. In higher temperature (lower inverse temperature) region, the both singular and non-singular fitting are in good agreement with the simulation results. However, the singular line diverges from the simulation results, while the non-singular line is in good agreement with them. Considering that the simulation results are equilibrium, it seems that the LSD does not have singular effect but non-singular behavior. Thus, the singular point β_c looks superficial. Nevertheless, the non-singular fitting, which contains the superficial singular point β_c as a fitting parameter, is in good agreement with the results. Moreover, the relationship the LSD and the control parameter scaled with the singular point p_c shows universality [8, 15]. Therefore, we consider that β_c (in general, p_c) is still important.

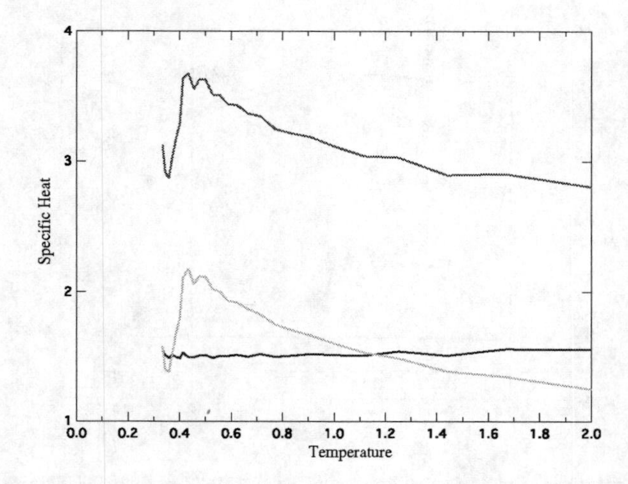

FIGURE 3. (Color online) Plot of the specific heat per particle at constant volume versus the temperature. The red line indicates the specific heat given from Eq. (4), the blue one that due to the kinetic energy, and the green one that due to the potential energy.

As above mentioned, we suggest the definition of the fragile glass transition by using the LSD [9]. According to the classification, the system in which the LSD $\hat{D}$ is less than $\hat{D}_g = 9.16 \times 10^{-6}$ is glassy state. The glass transition LSD $\hat{D}_g$ corresponds to $T_g = 0.4376$ ($\beta_g = 2.2851$) in the Kob-Andersen model. This transition temperature is calculated by the non-singular LSD fitting line. It is considered that the results of the mean-square displacement in $\beta > \beta_g$ are out of equilibrium.

Fig. 3 shows the results of the specific heat per particle at constant volume. Those results are scaled with the Boltzmann constant. We measure them from Eq. (4). The energy is average over $10^4\tau$. While the mean-square displacement is not steady-state in $T < 0.434$, the internal, kinetic, and potential energy are equilibrium.

The dimensionless specific heat due to the kinetic energy is constant value 1.5, and it is reasonable. On the other hand, they due to the internal and potential energy increase as the temperature decreases, but rapidly

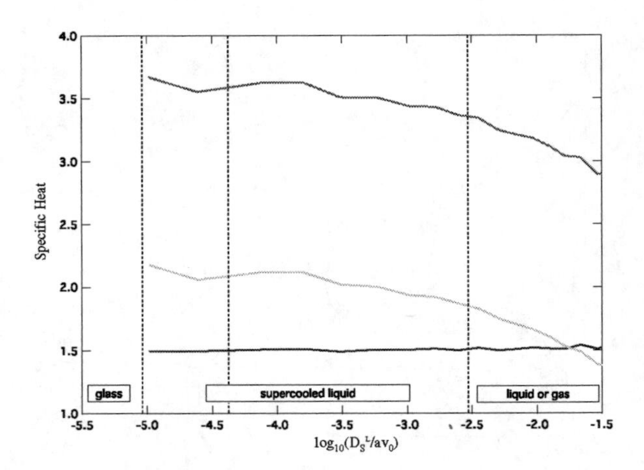

FIGURE 4. (Color online) Plot of the specific heat per particle at constant volume versus the common logarithm of the LSD. The details are the same as in Fig. 3. The vertical dotted lines denote the supercooled point (right, $\hat{D}_\beta = 2.95 \times 10^{-3}$), the (superficial) singular point (center, $\hat{D}_c = 4.22 \times 10^{-5}$), and the glass transition point (left, $\hat{D}_g = 9.16 \times 10^{-6}$) [9].

decrease below $T = 0.43$. It indicates that dynamics links to statics because the peak of the specific heat (statics) corresponds to the glass transition point $T_g = 0.4376$ obtained from the LSD (dynamics).

The growth of the specific heat due to the potential energy represents the importance of the spatial structuring at the vicinity of the glass transition point. It is consistent in view of the dynamical heterogeneity [17].

We suggest that the LSD has the universal form described by Eq. (3). It means that we treat the LSD as an universal parameter. Fig. 4 shows the relationship between the specific heat and the LSD. The specific heat does not change drastically at the supercooled LSD $\hat{D}_\beta$ and the singular LSD $\hat{D}_c$. As shown in Ref. [18], we should investigate and compare to another systems.

SUMMARY

We have investigated the specific heat at constant volume and the long-time self-diffusion coefficient (LSD) of the Kob-Andersen type Lennard-Jones binary mixture fluids by using molecular dynamics simulations.

Our statement is that the LSD has an universal form for fragile glass-forming systems [8] and we treat the LSD as an universal parameter [18]. When we regard it as an universal parameter, the glass transition LSD is $\hat{D}_g = 9.16 \times 10^{-6}$ [9]. The definition is based on dynamical analysis and the glass transition LSD $\hat{D}_g$ is equivalent over fragile glass-forming liquids. It strongly indicates that the glass transition is based on dynamical behavior. Moreover, using the non-singular type LSD suggested by Tokuyama [9], one can transform from the glass transition LSD to the glass transition temperature. Thus, we reveal that the definition of the glass transition by using the LSD is reasonable according to the results of the specific heat in the Kob-Andersen model.

Unfortunately, our above statements might not have strong powers of persuasion since we have analyzed the only Kob-Andersen model. We should investigate the relationship between the specific heat and the LSD of other systems and analyze whether the LSD is universal parameter or not. In that regard, what is needed is accuracy of calculation.

ACKNOWLEDGMENTS

This work was partially supported by Grants-in-Aid for Science Research with No. 18540363 from Ministry of Education, Culture, Sports, Science and Technology of Japan. Numerical computations for this work were performed on the Origin 2000 machine at the Institute of Fluid Science, Tohoku University.

REFERENCES

1. M. D. Ediger, C. A. Angell, and S. R. Nagel, *J. Phys. Chem.* **51**, 13200–13212 (1996).
2. C. A. Angell, K. L. Ngai, G. B. McKenna, P. F. McMillan, and S. W. Martin, *J. Appl. Phys.* **88**, 3113–3157 (1998).
3. P. G. Debenedetti and F. H. Stillinger, *Nature* **410**, 259–267 (2000).
4. B. Wunderlich, *J. Phys. Chem.* **64**, 1052–1056 (1960).
5. S. S. Chang and A. B. Bestul, *J. Chem. Phys.* **56**, 503–516 (1972).
6. A. V. Granato, *J. Non-Cryst. Solids*, **307-310**, 376–386 (2002).
7. G. Adam and J. H. Gibbs, *J. Chem. Phys.* **43**, 139–146 (1965).
8. M. Tokuyama, *Physica A* **364**, 23–62 (2006).
9. M. Tokuyama, T. Narumi, and E. Kohira, *Physica A* **385**, 439–455 (2007).
10. W. Kob and H. C. Andersen, *Phys. Rev. E* **51**, 4626–4641 (1995).
11. W. van Magen, T. C. Mortensen, S. R. Williams, and J. Müller, *Phys. Rev. E* **58**, 6073–6085 (1998).
12. M. Tokuyama and I. Oppenheim, *Physica A* **216**, 85–119 (1995).
13. E. Flenner and G. Szamel, *Phys. Rev. E* **72**, 031508 (2005).
14. M. Tokuyama and Y. Terada, *Physica A* **375**, 18–36 (2007).
15. T. Narumi and M. Tokuyama, *Rep. Inst. Fluid Science* **19**, 73–78 (2007).
16. T. A. Weber and F. H. Stillinger, *Phys. Rev. B* **31**, 1954–1963 (1985).
17. M. D. Ediger, *Annu. Rev. Phys. Chem.* **51**, 99–128 (2000).
18. M. Tokuyama, *Physica A* **378**, 157–166, (2007).

Critical Fluctuations of Time-Dependent Magnetization in Ordering Processes near the Disorder-Induced Critical Point

Hiroki Ohta and Shin-ichi Sasa

Department of Pure and Applied Sciences, University of Tokyo, 3-8-1 Komaba, Tokyo 153-8902, Japan

Abstract. We study a scale-free behavior in a random field Ising model from the viewpoint of dynamical properties. By focusing on fluctuations of dynamical events, we find that the intensity of the fluctuations diverges at a critical point. Moreover, we calculate a new dynamical exponent that characterizes the divergence.

Keywords: Critical, Dynamical, Random field Ising model
PACS: 75.60.Ej, 05.70.Jk, 64.70.Pf

INTRODUCTION

Scale-free behaviors such as earthquakes, acoustic emissions in a deformed complex material and Barkhausen noise exhibit distinctive power-law behaviors of quantities of space and time [1, 2, 3, 4]. All these systems possess frustration and also they are driven by a slowly varying external field. As a simple model of such systems, a random field Ising model (RFIM) under a slowly varying magnetic field has been investigated in order to elucidate the essential mechanism of the scale-free phenomena. By performing numerical experiments of the RFIM under a slowly varying magnetic field, a scale-free behavior was found at a certain parameter value, which is called *disorder-induced critical phenomena*.

It has been conjectured since the report in Ref. [5] that fluctuations of magnetization in equilibrium states does not exhibit a power-law divergence. A conjecture similar to that was presented in Ref. [6]. This raises a naive question whether the disorder-induced critical phenomena are characterized in terms of fluctuations of magnetization.

In this study, we present a positive answer to this question. Our key idea is to notice the recent extensive studies for critical fluctuations of dynamical events near an ergodicity breaking transition in glassy systems [7, 8, 9, 10]. In particular, we consider ordering processes of the magnetization from a special initial condition. We have numerically found that fluctuations of the time-dependent magnetization exhibit a critically divergent behavior near the disorder-induced critical point.

MODEL

We study dynamical properties of a RFIM with zero-temperature. Let us consider a three-dimensional cubic lattice $\Lambda \equiv \{i = (x,y,z) | 1 \le x,y,z \le L\}$. A spin variable $\sigma_i \in \{-1, 1\}$ is defined at each site $i \in \Lambda$. The Hamiltonian is described by

$$H = -J \sum_{\langle i,j \rangle} \sigma_i \sigma_j - \sum_{i \in \Lambda} (h_i + h) \sigma_i, \tag{1}$$

where $\langle i,j \rangle$ represents a nearest-neighbor pair of sites, h is a constant external field and the random field h_i obeys a Gaussian distribution

$$P(h_i) = \frac{1}{\sqrt{2\pi R^2}} \exp(-\frac{h_i^2}{2R^2}). \tag{2}$$

The time evolution of the spin variables is described by the Monte Carlo method. The time unit is L^3 Monte carlo steps per site.

ORDERING BEHAVIORS

From now, we numerically study ordering processes of magnetization

$$\hat{m}(t) \equiv \frac{1}{N} \sum_{i \in \Lambda} \sigma_i(t) \tag{3}$$

from the initial condition $\hat{m}(0) = -1$ in which all the spins are downward. As preliminary calculations, we measure

$$m(t) \equiv \langle \hat{m}(t) \rangle \tag{4}$$

for several values of (R, h), where $\langle A \rangle$ represents the average of a physical quantity A with respect to the stochastic time evolution and quenched disorder.

CP982, *Complex Systems, 5th International Workshop on Complex Systems*
edited by M. Tokuyama, I. Oppenheim, and H. Nishiyama

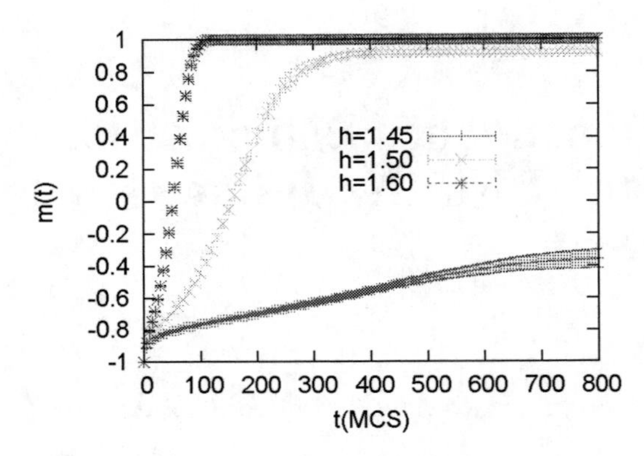

FIGURE 1. Ordering process of magnetization, $m(t)$. $L=40$ and $R=2.16$

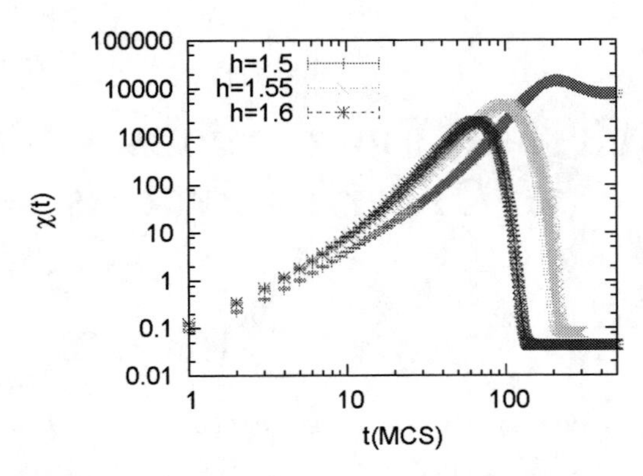

FIGURE 2. $\chi(t)$ for three values of h. $R=2.16$.

Here, we briefly review the disorder-induced critical phenomena in the RFIM. Let us consider the quasi-static operation of the external field from $h=-\infty$ to $h=\infty$ with a given R. In this operation, $h_c(R)$ is defined as a characteristic value of h at which a spin flipping in a region over the whole system first occurs. Then, R_c is defined as the maximum value below which $h_c(R)$ exists. Their actual values were numerically found as $(R_c, h_c(R_c)) \simeq (2.16, 1.44)$ [5], which is called disorder-induced critical point. It was reported that power-law behaviors of the size distribution of avalanches were observed in the quasi-static operation of the magnetic field when $R=R_c$. Such power-law behaviors are called the disorder-induced critical phenomena.

Typical samples of $m(t)$ for the case $R=2.16$ are displayed in Fig. 1. It is observed that the magnetization $m(t)$ quickly approaches the equilibrium value $m_{\rm eq}$ from -1 when h is sufficiently large and that the ordering process becomes slower for the system with smaller h. Then, by decreasing h further, we find that $m(t)$ does not reach the state with the equilibrium value $m_{\rm eq}$ when $h \le h_c$.

DIVERGENT FLUCTUATIONS OF DYNAMICAL EVENTS

Now, we focus on the behaviors near the disorder-induced critical point (R_c, h_c). That is, by fixing R as R_c, we investigate the system with several values of h near h_c. Following our motivation, we are interested in fluctuations of $\hat{m}(t)$. It should be noted that our study is concerned with *fluctuations of relaxation events*. To our knowledge, there have been no such arguments on statistical properties near the disorder-induced critical point.

Since a simple quantity characterizing the fluctuations of $\hat{m}(t)$ is given by

$$\chi(t) \equiv N(\langle \hat{m}(t)^2 \rangle - \langle \hat{m}(t) \rangle^2), \tag{5}$$

we first demonstrate the graphs of $\chi(t)$ for a few values of h in Fig. 2. It is observed that $\chi(t)$ has a peak at a time τ and that both $\chi_{\rm m} = \chi(\tau)$ and τ increase when h approaches $h_c (\simeq 1.44)$.

Thus, for the systems with $L=10$, 20, and 40, finite-size scaling is performed with $\theta \simeq 0.7$ and $\alpha \simeq 1.7$ (Fig. 3). we conjecture a scaling form

$$\tau(h,L) = L^{\alpha} F_{\tau}\left(\left| \frac{h-h_c}{h_c} \right| L^{\frac{1}{\theta}} \right), \tag{6}$$

by using the scaling function F_τ. Considering the asymptotic law $F_\tau(z) \simeq z^{-\zeta}$ (with $\zeta \simeq 1.3$) in the regime $z \gg 1$, we expect the following critical behavior in the large size limit:

$$\tau \simeq (h-h_c)^{-\zeta}. \tag{7}$$

In a manner similar to that in the analysis of $\tau(h,L)$, we assume a form of the finite-size scaling as follows:

$$\chi_m(h,L) = L^{\beta} F_{\chi}\left(\left| \frac{h-h_c}{h_c} \right| L^{\frac{1}{\theta}} \right). \tag{8}$$

Indeed, from Fig. 3, we determine β and the scaling function F_χ, where $\beta \simeq 3.0$ and we find the asymptotic relation $F_\chi(z) \simeq z^{-\gamma}$ (with $\gamma \simeq 2.1$) in the regime $z \gg 1$. We thus obtain

$$\chi_{\rm m} \simeq (h-h_c)^{-\gamma}. \tag{9}$$

Moreover, with using the exponent θ in (6) and (8), we can naturally conjecture the following asymptotic behavior of some dynamical length scale ξ:

$$\xi \simeq (h-h_c)^{-\theta}. \tag{10}$$

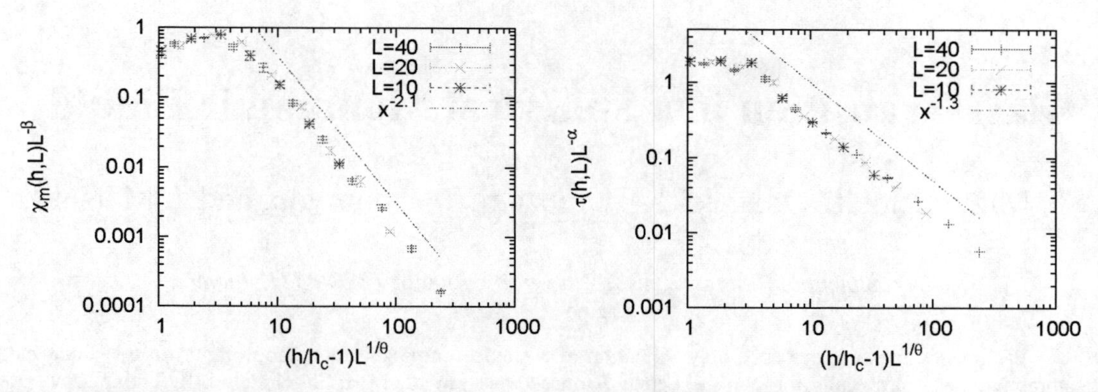

FIGURE 3. Finite-size scaling for $\chi_m(h,L)$ and $\tau(h,L)$. The three graphs for systems with different sizes are collapsed into a single curve. $h_c = 1.44$ and $R = 2.16$.

The exponent characterizing the length scale appearing in a finite-size scaling in terms of avalanches was obtained in Ref. [11, 6], where the values of the exponent are close to that of θ in our study. The relation among these results will be studied in future. We also note that the growing manner of the dynamical length scale can be directly extracted from the correlation function of the spin flips at a certain time near τ. (See Ref. [12] for details.)

ACKNOWLEDGMENTS

The authors thank K. Hukushima for useful comments on this work. This work was supported by a grant from the Ministry of Education, Science, Sports and Culture of Japan, No. 19540394.

REFERENCES

1. J. P. Sethna, K. A. Dahmen and C. R. Myers, *Nature* **410**, 242–250 (2001).
2. M. C. Miguel, A. Vespignani, S. Zapperi, J. Weiss and J. R. Grasso, *Nature* **410**, 667–671 (2001).
3. G. Durin and S. Zapperi, *Phys. Rev. Lett.* **84**, 4705–4708 (2000).
4. F. Pazmandi, G. Zarand, and G. T. Zimanyi, *Phys. Rev. Lett.* **83**, 1034–1037 (1999).
5. O. Perkovic, K. A. Dahmen and J. P. Sethna, *Phys. Rev. B* **59**, 6106–6119 (1999).
6. F. J. Perez-Reche and E. Vives, *Phys. Rev. B* **70**, 214422-1–214422-14 (2004).
7. R. Yamamoto and A. Onuki, *Phys. Rev. E* **58**, 3515–3529 (1998).
8. L. Berthier, G. Biroli, J. P. Bouchaud, L. Cipelletti, D. E. Masri, D. L. Hote, F. Ladieu and M. Pierno, *Science* **310**, 1797–1800 (2005).
9. O. Dauchot, G. Marty and G. Biroli, *Phys. Rev. Lett.*, **95**, 265701-1–265701-4 (2005).
10. D. Chandler, J. P. Garrahan, R. L. Jack, L. Maibaum, and A. C. Pan, *Phys. Rev. E* **74**, 051501-1–051501-9 (2006).
11. J. H. Carpenter and K. A. Dahmen, *Phys. Rev. B* **67**, 020412(R)-1–020412(R)-4 (2003).
12. H. Ohta and S. Sasa, *cond-mat/0706.3629*.

Glass Transition in a Spherical Monatomic Liquid

T. Mizuguchi, T. Odagaki, M. Umezaki, T. Koumyou, and J. Matsui

Department of Physics, Kyushu University, Fukuoka 812-8581, Japan

Abstract. We present the first evidence of vitrification of a single-component monatomic liquid in two dimensions, where atoms interact isotropically through the Lennard-Jones-Gauss potential [M. Engel and H.-R. Trebin, Phys. Rev. Lett. **98**, 225505 (2007)]. With the use of a molecular dynamics simulation, we quench the system from a melt to a state well below the melting temperature. At lower temperatures, the state of this system does not have any long range order nor show any structural relaxation. We obtain various physical quantities such as the specific heat, the intermediate scattering function and the mean square displacement during the cooling process and demonstrate that the structural relaxation of the monatomic liquid shows characteristic behaviors of the common glass transition.

Keywords: Glass transition, Lennard-Jones-Gauss potential, Monatomic liquid
PACS: 64.70.P-,05.20.Jj,05.10.Gg

INTRODUCTION

Since the first discovery of the glass transition in 1923 [1], numerous theoretical and experimental efforts have been made to understand the transition. However, the complete understanding of the glass transition is still yet to be achieved. The difficulties in understanding the glass transition arise from the fact that almost all glass forming materials consist of many constituents. Even in computer simulations, two components are needed to produce quasi stable glassy state. There have been considerable but unsuccessful attempts to produce a single component glass. If a single component glass could be formed even in a computer, understanding of the glass transition will advance significantly since theoretical analysis of a single component system is much easier and can be scrutinized carefully through comparison with (computer) experiments.

Recently, Engel and Trebin [2] reported that quasi-crystals and complex crystals can be realized as thermodynamically stable states for a single-component monatomic system where atoms interact through the Lennard-Jones-Gauss (LJG) potential.

In this paper, we report for the first time that the glassy state of a single-component monatomic system can be obtained by rapid cooling of the LJG liquid. In next section, we explain the model potential and the method of molecular dynamics (MD) simulation. In Sec. 3, the static structure and thermodynamic properties are investigated. A brief report for dynamical properties is given in Sec. 4 and Sec. 5 is devoted to discussion.

MODEL POTENTIAL AND METHOD OF MD SIMULATION

The Lennard-Jones-Gauss potential consists of the Lennard-Jones potential and a pocket represented by the Gaussian function [2], where r is the distance between two atoms.

$$V(r)=\varepsilon_0\left\{\left(\frac{r_0}{r}\right)^{12}-2\left(\frac{r_0}{r}\right)^{6}-\varepsilon\exp\left[-\frac{(r-r_G)^2}{2\sigma^2}\right]\right\} \quad (1)$$

Figure 1 shows the LJG potential. Note that this kind of effective potential is known to exist in many metallic alloys [3].

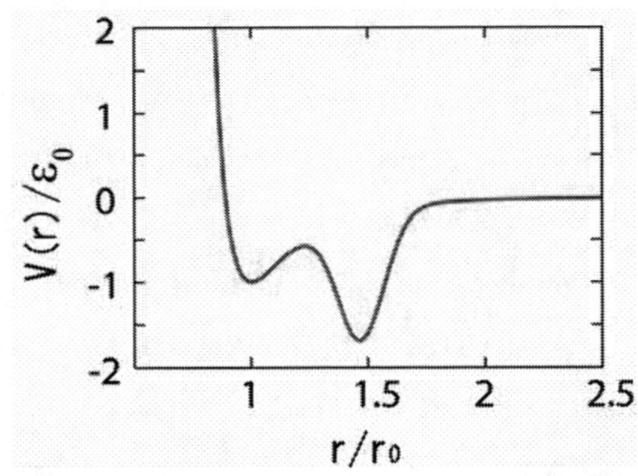

FIGURE 1. (Color online) The LJG potential is plotted as a function of the distance between two atoms.

CP982, *Complex Systems, 5th International Workshop on Complex Systems*
edited by M. Tokuyama, I. Oppenheim, and H. Nishiyama

In the following discussion, the unit of length, energy and time are r_0, ε_0 and $t_0 \equiv (m r_0^2 / \varepsilon_0)^{1/2}$, respectively.

We investigate the vitrification process for 2,500 atoms on two dimensional space with periodic boundary conditions, where the LJG parameters are set to $r_G = 1.47$, $\varepsilon = 1.5$, $\sigma^2 = 0.02$. This set of parameters produces the pentagon-triangle phase in equilibrium [2]. The size of the simulation cell is chosen so that the density becomes $\rho = 1$.

We employ the Verlet algorithm for the MD simulation with time step $0.01 t_0$. We prepare a liquid at $T^* \equiv k_B T / \varepsilon_0 = 2.2$ as the initial state and then cool the system down to a target temperature at quenching rate $0.01T^*$/step with the velocity scaling. After equilibrating some time, we measure various physical quantities.

STATIC STRUCTURE AND SPECIFIC HEAT

Figure 2 shows the atomic configuration in the real space and the static structure factor at $T^* = 0.1$, where the system is annealed at this temperature for 10^7 time steps with the velocity scaling.

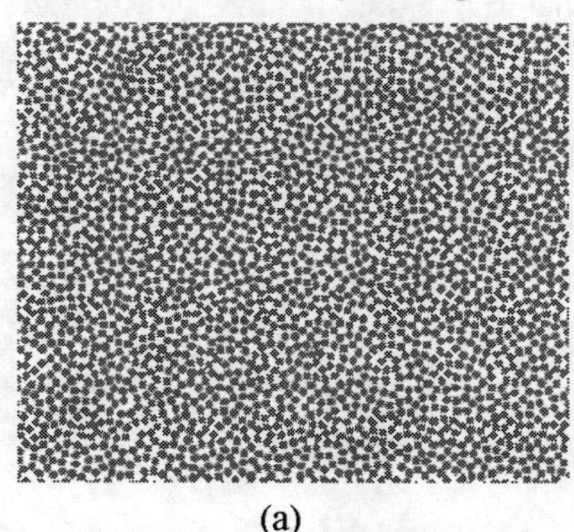

(a)

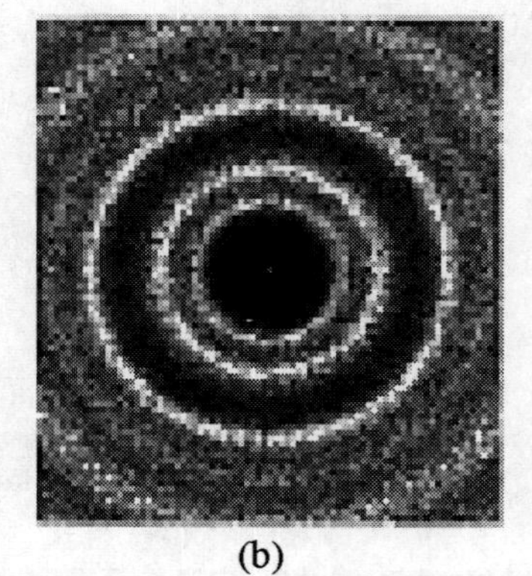

(b)

FIGURE 2. (Color online) (a) the atomic configuration and (b) the static structure factor at $T^* = 0.1$.

Apparently, no long range order is observed and the state does not show any structural relaxation at least for the period we observed. Thus, the state we obtained can be a glassy state.

In the cooling process, we measured the total energy of the system as quenched. Circles in Fig. 3 show the temperature dependence of the total energy of the system as quenched. Each of circles corresponds to an individual quenching from the liquid state.

Calculating the temperature derivative of the energy numerically, we obtained the specific heat as shown in Fig. 4. The abrupt change in the specific heat at $T^* = 0.35$ resembles the specific heat of glass forming materials [4]. It should be emphasized here that the amorphous state produced by quenching can relax to more stable state when the temperature is high. In fact, the state below $T^* \approx 0.2$ seems to be stable glass, but the amorphous state above $T^* \approx 0.2$ tends to transform into a mixture of quasi-periodic regions. We confirm this transformation from the change of the atomic configuration and the static structure factor. The timescale of the transformation depends on temperatures. At higher temperature (T*~0.4), it's the order of 10^4 time steps, at lower temperatures (T*~0.3), it's the order of 10^6 time steps.

We examined the total energy in the heating process from the glassy state at $T^* = 0.2$. The system is heated after equilibrating 5×10^5 time steps with the velocity scaling from the glassy state as quenched to $T^* = 0.2$.

The squares in Fig. 3 show the energy of the system obtained by heating as a function of temperature. In the heating process, the system transforms its structure to a more stable mixture of quasi-periodic regions. The melting was observed at $T^* \approx 0.44$.

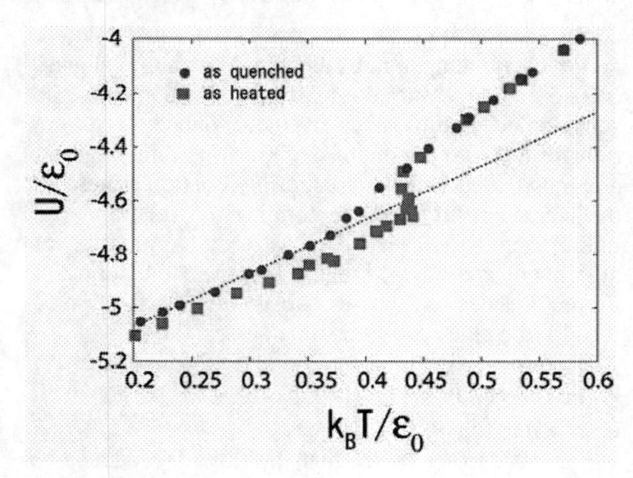

FIGURE 3. (Color online) Temperature dependence of the total energy.
Circles are the energy of the system as quenched and squares are the energy measured in the heating process. The dashed line is fitted to the low-temperature points as quenched.

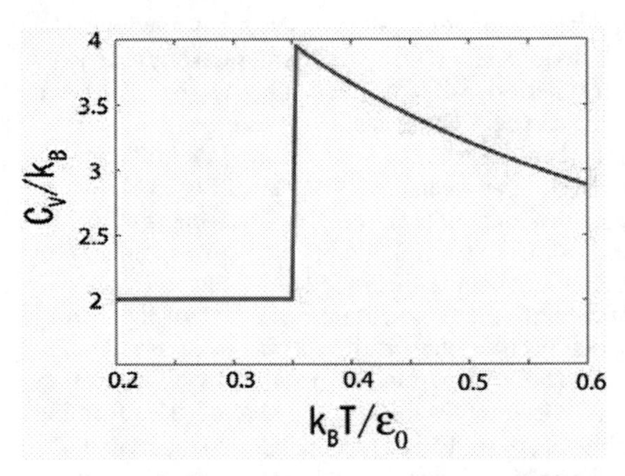

FIGURE 4. (Color online) Temperature dependence of the specific heat as quenched.

It will be important to see how the structure is transformed from the amorphous state as quenched to the mixture of quasi-periodic regions. Research in this direction is underway.

DYNAMICAL PROPERTIES

In order to see the dynamical characteristics of the vitrification process, we first calculated the self part of the intermediate scattering function $F_S(q,t)$

$$F_S(\mathbf{q},t) = \frac{1}{N}\left\langle \sum_i e^{i\mathbf{q}[\mathbf{r}_i(t)-\mathbf{r}_i(0)]} \right\rangle \qquad (2)$$

of the amorphous state as quenched. Figure 5 shows the time dependence of $F_S(q,t)$ for the liquid state $T^* = 0.495$ and three supercooled states $T^* = 0.401, 0.361, 0.353$ where $\mathbf{q}r_0 = (8.7, 0)$, corresponding to the second peak of the static structure factor. These temperatures are above the nominal glass transition temperature. As already stated above, the supercooled liquid states at these temperatures are transformed into the mixture of quasi-periodic regions, and thus the intermediate scattering function represents the dynamics in this structural transformation. The initial state in Figure 5 is already a mixture of amorphous and ordered regions. In the time period of Figure 5, those two phases coexist and the fraction of the ordered phase grows as time passes.

We also obtained the mean square displacement for four temperatures $T^* = 0.495, 0.401, 0.361, 0.353$ which is shown in Fig. 6.

It is interesting to see that the time dependence of the intermediate scattering function and the mean square displacement show two step relaxation in the super cooled region at lower temperatures. This indicates that the structural transformation takes place with the same dynamics of atoms as in glass forming systems.

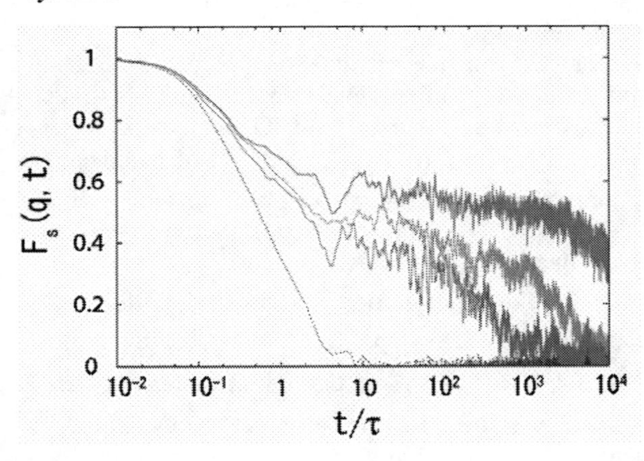

FIGURE 5. (Color online) The self part of the intermediate scattering function for four different temperatures. Four curves correspond to $T^* = 0.353, 0.361, 0.401, 0.495$ from top to bottom.

DISCUSSION

We presented strong evidences of vitrification of a monatomic liquid in two dimensions which consists of single-component atoms interacting isotropically via the LJG potential. The system quenched to temperatures far below the melting temperature keeps the amorphous structure for fairly long time. At T*=0.1, the amorphous structure keeps over 10^7 time steps. If this time scale is longer than the observation time, then we can consider the system as a glass.

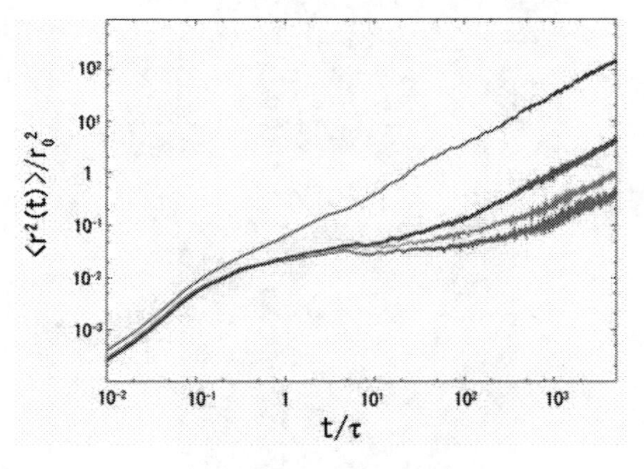

FIGURE 6 (Color online) dependence of mean square displacement. four curves correspond to $T^* = 0.495, 0.401, 0.361, 0.353$ from top to bottom.

We measured several thermodynamic and dynamic quantities for supercooled liquid and amorphous state as quenched and showed that these quantities exhibit characteristic behaviors of the glass transition even though the system may be transformed into the mixture of quasi-periodic regions for higher temperatures we studied. Further detailed investigation of this process will reveal the origin of slow relaxations in glass forming materials.

Our preliminary calculation shows that the LJG liquid in three dimensions can also be vitrified by rapid quenching [5] which will be reported in the near future.

ACKNOWLEDGMENTS

This work was supported in part by the Grant-in-Aid for scientific Research from the Ministry of Education, Culture, Sports, Science and Technology.

REFERENCES

1. G. E. Gibson and W. F. Giauque, *J. Am. Chem. Soc.* 45, 93-104 (1923)
2. M. Engel and H.-R. Trebin, *Phys. Rev. Lett.*, **98**, 225505-1—225505-4 (2007)
3. Al-Lehyani, M. Widom, Y. Wang, N. Moghadam, G. M. Stocks and J. A. Moriaty, *Phys. Rev. B* 64, 075109-1—075109-7 (2001).
4. O. Yamamuro, I. Tsukushi, A. Lindqxist, S. Takahara, M. Ishikawa, and T. Matsuo, *J. Phys. Chem. B* **102**, 1605-1609 (1998).
5. Vo Van Hoang, T. Odagaki and M.Engel (in preparation)

Link between Vitrification and Crystallization in Two Dimensional Polydisperse Colloidal Liquid

T. Kawasaki, T. Araki, and H. Tanaka

Institute of Industrial Science, University of Tokyo, 4-6-1 Komaba, Meguro-ku, Tokyo 153-8505, Japan

Abstract. Glasses are formed if crystallization is avoided upon cooling or increasing density. However, the physical factors controlling the ease of vitrification and nature of the glass transition remain elusive. We use numerical simulations of polydisperse hard disks to tackle both of these longstanding questions. Here we systematically control the polydispersity in two-dimensional colloidal simulations, i.e., the strength of frustration effects on crystallization. We demonstrate that medium-range crystalline order grows in size and lifetime with an increase in the colloid volume fraction or a decrease in polydispersity (or, frustration). We find a direct relation between medium-range crystalline ordering and the slow dynamics that characterizes the glass transition. This suggests an intriguing scenario that the strength of frustration controls both the ease of vitrification and nature of the glass transition at least in this system. Vitrification may be a process of hidden crystalline ordering under frustration. This not only provides a physical basis for glass formation, but also an answer to a longstanding question on the structure of amorphous materials: "order in disorder" may be an intrinsic feature of a glassy state of material. Thus our scenario makes a natural connection between structure and dynamics in glass-forming materials, although its relevance should be carefully checked for 3D glass-forming liquids.

Keywords: Crystallization, Dynamic heterogenity, Frustration, Glass transition, Simulation
PACS: 64.70.Pf,64.60.My,61.20.Ja,81.05.Kf

INTRODUCTION

Despite a long history of research, the origin of glass transition and the slow dynamics associated with it remain elusive. One of the most drastic changes occurring near the glass transition is the slowing down of dynamics (or, the increase in the viscosity) more than 10 orders of magnitude while accompanying little change in the static structure [1, 2]. Recently it was revealed both experimentally and numerically [3, 4, 5, 6, 7] that dynamic heterogeneity develops upon cooling toward the glass transition temperature T_g. There are various ideas to explain this emergence of dynamic heterogeneity. We recently proposed that it may be the development of medium-range crystalline order that makes dynamics heterogeneous and leads to the slowing down of the dynamics approaching T_g [8]. This idea may be supported by an experimental finding by Eckstein et al. [9]. Here we check the relevance of this proposal by using a standard glass former, namely, a polydisperse colloidal liquid as a model system [10, 11, 12].

We introduced the polydispersity as a source of frustration effects on crystallization in colloidal dispersion. In this model, the glass-forming ability can continuously be controlled by changing the degree of polydispersity. The higher the polydispersity, the higher the nucleation barrier. Experimentally polydisperse colloidal dispersions have often be used as a model glass forming liquid and the importance of polydispersity in glass-forming ability has recently been emphasized. Here we study the the roles of frustration effects on crystallization in vitrification by means of standard Brownian dynamics simulations of polydisperse colloidal systems.

METHOD

In our simulations, the motion of a diffusing colloidal particle is described by the following Langevin equation,

$$m_0 \frac{d\vec{v}_i}{dt} = \vec{F}_i^I - \zeta \vec{v}_i + \vec{F}_i^B, \tag{1}$$

where $\vec{v}_i$ is the velocity of i-th particle. m_0 and ζ are, respectively, mass and friction coefficient, both of which are set to be constant in this study for simplicity. $\vec{F}_i^I$ is the inter-particle force acting on i-th particle, which is calculated as $\vec{F}_i^I = -\sum_{k \neq i} \frac{\partial}{\partial \vec{r}_i} U_{ik}(|\vec{r}_{ki}|)$. Here $\vec{r}_i$ represents the position of i-th particle and $\vec{r}_{ki} = \vec{r}_k - \vec{r}_i$. We employ the Weeks-Chandler-Andersen repulsive potential [14]:

$$U_{ik}(r) = \begin{cases} 4\varepsilon \left\{ \left(\frac{\sigma_{ik}}{r}\right)^{12} - \left(\frac{\sigma_{ik}}{r}\right)^{6} + \frac{1}{4} \right\} & (\text{for } r < 2^{\frac{1}{6}} \sigma_{ik}) \\ 0 \ (\text{otherwise}), \end{cases} \tag{2}$$

where $\sigma_{ik} = (\sigma_i + \sigma_k)/2$ and σ_i represents the size of particle i. With this potential we can construct hard sphere or disk-like simulations. We introduce the Gaussian distribution of particle size. Its standard deviation Δ is regarded as polydispersity; $\Delta = \sqrt{(\langle \sigma^2 \rangle - \langle \sigma \rangle^2)} / \langle \sigma \rangle$,

CP982, *Complex Systems, 5th International Workshop on Complex Systems*
edited by M. Tokuyama, I. Oppenheim, and H. Nishiyama

where $\langle x \rangle$ means the average of variable x_i among all the particles. The thermal noise $\vec{F}_i^B$ satisfies the following fluctuation-dissipation theorem:

$$\langle \vec{F}_i^B(t) \cdot \vec{F}_k^B(t') \rangle = 4\zeta k_B T \delta_{ik} \delta(t-t'), \quad (3)$$

where k_B is Boltzmann's constant and T is the temperature. Hereafter we scale the length and time by $\langle \sigma \rangle$ and $t_0 = 0.075 m_0/\zeta$, respectively. We simulate $N = 1024$ (occasionally, $N = 4096$) particles in a 2D fixed square cell (L^2) under a periodic boundary condition. The volume fraction ϕ is defined as $\phi = \frac{1}{L^2} \sum_i^N \pi(\sigma_i/2)^2$. We solve Eq. (1) with an explicit Euler scheme with a time increment $\Delta t = 0.01 t_0$. In a liquid-glass transition of colloidal dispersions, the temperature is fixed and the control parameter for the transition is ϕ and Δ [10, 11, 12]. (Those colloidal particles are repulsive hard spheres or disks. Thus the temperature is not so relevant for the transition. In such repulsive systems, increasing ϕ plays the same role as decreasing temperature in the molecular liquids.) Therefore for all simulations shown below, the temperature is fixed at $k_B T/\varepsilon = 0.025$ and we change the degree of ϕ and Δ for getting various states (liquid, crystal, and glass).

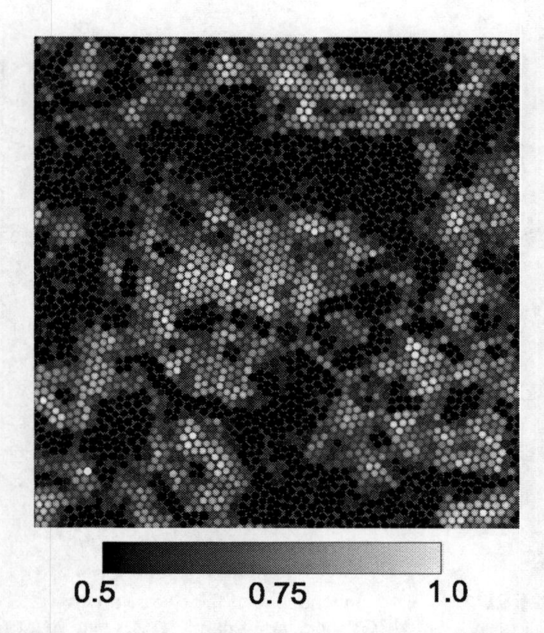

FIGURE 1. A snapshot of a polydisperse colloidal system of $\phi = 0.631$ for $\Delta = 9\%$. The brightness of particles represents the degree of the sixfold bond-orientational order Ψ_6 (see the color bar).

RESULTS

First we consider the state behavior of this system as a function of the degree of polydispersity Δ . A system of smaller polydispersity ($\Delta \leq 8\%$) exhibits (quasi-)long-range ordering upon increasing ϕ. The physical nature of this 2D ordering has been intensively discussed [15]; however, it is still a matter of debate. Here we do not dwell into this problem since it is out of the scope of this paper. For larger $\Delta (\geq 9\%)$, we observe vitrification instead of such (quasi-)long-range ordering. Thus hereinafter, we focus on the systems on $\Delta (\geq 9\%)$.

We can see for $\Delta = 9\%$ that medium-range crystalline order (MRCO) grows in size with an increase in the colloid volume fraction (see Fig. 1). MRCO (locally ordered structure) is characterized by the following sixfold bond-orientational order parameter (see, e.g., [13, 15])

$$\bar{\Psi}_6^i = \frac{1}{\tau_\alpha} \int_{t'}^{t'+\tau_\alpha} dt \left| \frac{1}{n_i} \sum_{m=1}^{n_i} e^{j6\theta_m^i} \right|, \quad (4)$$

where n_i is the number of nearest neighbors of particle i, $j = \sqrt{-1}$, and θ_m^i is the angle between $(\vec{r}_m - \vec{r}_i)$ and the x-axis, where particle m is a neighbor of particle i. Note that $\Psi_6^i = 1$ means the perfect hexagonal arrangement of six nearest-neighbor particles around particle i and $\Psi_6^i = 0$ means a random arrangement.

We also calculated the number of particles belonging to i-th domain of high MRCO ($\Psi_6^i \geq 0.75$) as N_i. Then the average particle number belonging to high MRCO domains is given by $N_c = \sum_{i=1}^n N_i^2 / \sum_{i=1}^n N_i$, where n is the number of high MRCO domains. Then we determined the characteristic size of MRCO as $\xi = \sqrt{N_c}$. We confirmed that the results are not sensitive to the choice of the threshold value of Ψ_6^i The size of MRCO ξ is increasing with an increase in ϕ.

We also estimate the ξ_4 from the four-point density correlation function, following [6]. It has been well established that ξ_4 characterizes the size of dynamic heterogeneity of translational particle motion. Here we describe the four-point density correlation function analysis. First a time-dependent order parameter that measures the number of overlapping particles in two configurations separated by a time interval t is defined as

$$Q(t) = \sum_{i=1}^{N} \sum_{k=1}^{N} w(|r_i(0) - r_k(t)|), \quad (5)$$

where $w(r) = 1$ for $r \leq b$ whereas $w(r) = 0$ for $r > b$ and we set $b = 0.4$. The structural relaxation can then be characterized by the variance of $Q(t)$ as

$$\chi_4(t) = \frac{L^2}{N^2} \left[\langle Q(t)^2 \rangle - \langle Q(t) \rangle^2 \right]. \quad (6)$$

This χ_4 is related to the four-point density correlation function $g_4(t)$ (see below on its definition) as

$$\chi_4(t) = \int dr\, 2\pi r g_4(r,t). \quad (7)$$

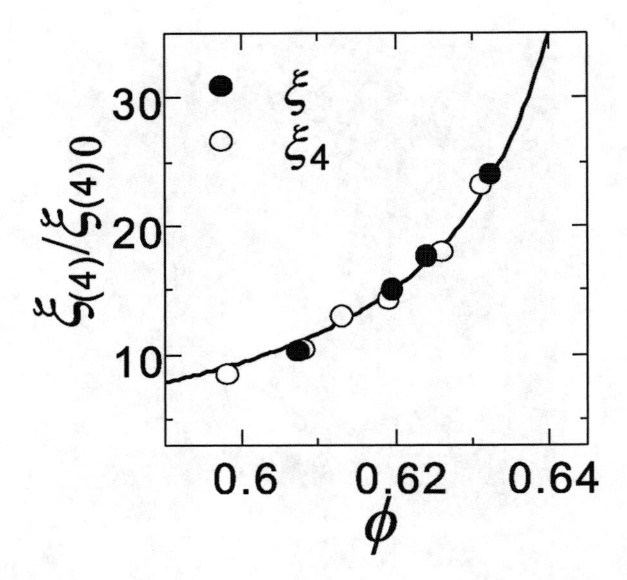

FIGURE 2. The ϕ-dependence of the characteristic size ξ of clusters of high MRCO and the dynamic coherence length ξ_4 calculated from the four-point density correlation function for $\Delta = 9\%$. For all polydispersities, ξ/ξ_0 and ξ_4/ξ_{40} both grow as ϕ increases. Furthermore, we can see that $\xi/\xi_0 \simeq \xi_4/\xi_{40}$. The solid lines are the fitted curves of $\xi = \xi_0 \left[\left(\frac{1}{\phi} - \frac{1}{\phi_0}\right)/\frac{1}{\phi_0}\right]^{-1}$. Here ϕ_0 is independently determined by the Vogel-Fulcher fitting to τ_α[18].

Since $\chi_4(0)$ and $\chi_4(\infty)$ should be 0, $\chi_4(t)$ is expected to have a peak at a time τ_H, where the dynamic heterogeneity becomes most pronounced.

Then we calculated the correlation length ξ_4 as follows. We define the following pair correlation function of overlapping particles:

$$\begin{aligned} g_4^{ol}(r,t) &= \frac{1}{N\rho}\langle \sum_{iklm} \delta(r - r_k(0) + r_i(0)) \\ &\times w(|r_i(0) - r_k(t)|) w(|r_l(0) - r_m(t)|)\rangle \\ &= g_4(r,t) + \left\langle \frac{Q(t)}{N} \right\rangle^2. \end{aligned}$$

To extract ξ_4, we fitted the following Ornstein-Zernike function to the Fourier transformation of the self part of g_4^{ol} ($i = k$; $l = m$), $S_4^{ol}(q,t)$, at τ_H:

$$S_4^{ol}(q,\tau_H) = \frac{S_0}{1 + (\xi_4 q)^2}. \quad (8)$$

ξ_4 obtained from this S_4^{ol}.

With an increase in ϕ, both ξ and ξ_4 increase . The ϕ-dependence of ξ (ξ_4) is well fitted by the following power law[16, 17]

$$\xi_{(4)} = \xi_{(4)0}\left[\left(\frac{1}{\phi} - \frac{1}{\phi_0}\right)/\frac{1}{\phi_0}\right]^{-2/d} \quad (d = 2). \quad (9)$$

Furthermore, we found the relation $\xi/\xi_0 \cong \xi_4/\xi_{40}$, indicating that the dynamic heterogeneity of translational motion, which is usually characterized by ξ_4, is indeed caused by medium-range crystalline ordering.

SUMMARY

Our results together with our previous study [8] suggest an intriguing scenario that the strength of frustration controls both the ease of vitrification and nature of the glass transition. Vitrification may be a process of hidden crystalline ordering under frustration. This not only provides a physical basis for glass formation, but also an answer to a longstanding question on the structure of amorphous materials: "order in disorder" is an intrinsic feature of a glassy state of material. Thus our scenario makes a natural connection between structure and dynamics in glass-forming materials.

REFERENCES

1. C.A. Angell, *Science* **267**, 1924-1935, (1995).
2. P.G. Debenedetti and F.H. Stillinger, *Nature* **410**, 259-267 (2001).
3. D. Perera and P. Harrowell, *Phys. Rev. E* **59**, 5721-5743 (1999).
4. M.M. Hurley and P. Harrowell, *J. Chem. Phys.* **105**, 10521-10526 (1996).
5. R. Yamamoto and A. Onuki, *J. Phys. Soc. Jpn.* **66**, 2545-2548 (1997).
6. F.W. Lačević N, Starr, T.B. Schroder and Glotzer SC, *J. Chem. Phys.* **119**, 7372-7387 (2003).
7. A. Widmer-Cooper and P. Harrowell, *Phys. Rev. Lett.* **96**, 185701-1-185701-4 (2006).
8. H. Shintani, H. Tanaka, *Nature Phys.* **2**, 200-210 (2006).
9. E. Eckstein, J. Qian, R. Hentschke, T. Thurn-Albrecht, W. Steffen and E. W. Fischer, *J. Chem. Phys.* **113**, 4751-4762 (2000).
10. S.R. Williams, I.K. Snook and W. van Megen, *Phys. Rev. E* **64** 021506-1-021506-7 (2001).
11. E.R. Weeks, J.C. Crocker, A.C. Levitt, A. Schofield, and D.A. Weitz, *Science* **287**, 627-631 (2000).
12. S. Auer and D. Frenkel, *Nature* **413**, 711-713 (2001).
13. T. Hamanaka and A. Onuki, *Phys. Rev. E* **74**, 011506-1-011506-7 (2006).
14. J.D. Weeks, D. Chandler and H. C. Andersen, *J. Chem. Phys.* **54**, 5237-5247 (1971).
15. K. Binder, S. Sengupta and P. Nielaba, *J. Phys: Condens. Matter* **14**, 2323-2333 (2002).
16. T.R. Kirkpatrick, D. Thirumalai, P.G. Wolynes, *Phys. Rev. A* **40** 1045-1054 (1989).
17. H. Tanaka, *J. Chem. Phys.* **111**, 3163-3174 (1999).
18. T. Kawasaki, T. Araki, and H. Tanaka, *Phys. Rev. Lett.* (in press).

Relationship between the Slow Dynamics and the Structure of Supercooled Liquid

Hiroshi Shintani and Hajime Tanaka

Institute of Industrial Science, University of Tokyo, 4-6-1 Komaba, Meguro-ku, Tokyo 153-8505, Japan

Abstract. Here we present a new molecular dynamics simulation model with the rotating degree of freedom, where we can systematically control the stability and number density of locally favored structures (short-range ordering), i.e. the frustration effects against crystallization (long-range ordering). This model can cover from crystallization to vitrification just by changing the degree of frustration and enable us to investigate the roles of the frustration in vitrification. We found medium-range crystalline ordering in a supercooled state of this model liquid, which plays important roles in the slow dynamics.

Keywords: Glass transition, Molecular dynamics simulations
PACS: 64.70.P-,61.20.Gy,83.10.Mj

INTRODUCTION

The structural relaxation time of a supercooled liquid drastically increases and finally a liquid become a non-ergodic glassy state[1, 2]. Despite intensive efforts over more than several decades, the mechanism of this phenomena remains elusive. The relationship between the growing length scale of dynamic heterogeneity[3, 4] in a supercooled liquid and the slowing down of the structural relaxation has been actively discussed because it may provide a clue to our understanding of the liquid-glass transition. However, the physical factor that controls dynamic heterogeneity has not yet been clarified.

Contrary to popular models of the liquid-glass transition, we consider that vitrification is intrinsically linked to crystallization and proposed a "two order parameter model"[5]. According to this model, vitrification and the resulting slow dynamics are due to frustration between long-range density ordering toward crystallization and short-range bond ordering toward the formation of locally favored structures.

Based on this idea, we introduced a new type of interaction potential between particles, which can directly control the tendency of short-range bond ordering, namely, the strength of frustration against crystallization, and investigated the dynamics of liquids by means of molecular dynamics simulations. Here we describe the details of our simulation model, which is designed to investigate the roles of crystallization in the glass transition.

MODEL

First we explain a new interaction potential between spherical particles, which can directly and systematically control the strength of frustration against crystallization. The potential is expressed by a superposition of two functions as

$$U(r_{ij},\Omega_{ij}) \;=\; \bar{U}(r_{ij})+\Delta U(r_{ij},\Omega_{ij}),$$

where r_{ij} is the distance between two particles i and j and Ω_{ij} expresses the orientation between the particles. The first term of the potential is an isotropic potential which tends to maximize the density of a liquid and encourage crystallization. The second term is an anisotropic potential which leads to the formation of locally favored structures. In this paper, we adopted the Lennard-Jones potential as the isotropic potential $\bar{U}_{ij}$

$$\bar{U}_{ij}(r_{ij}) \;=\; 4\varepsilon\left[\left(\frac{\sigma}{r_{ij}}\right)^{12}-\left(\frac{\sigma}{r_{ij}}\right)^{6}\right],$$

where ε is a well depth of the potential and σ is a diameter of the particle. The anisotropic part of the potential is given by

$$\begin{aligned}\Delta U(r_{ij},\Omega_{ij}) &= -4\varepsilon\Delta\left(\frac{\sigma}{r_{ij}}\right)^{6}\left[h\left(\frac{\theta_i-\theta_0}{\theta_c}\right)\right.\\ &\qquad \left.+h\left(\frac{\theta_j-\theta_0}{\theta_c}\right)-\frac{64}{35\pi}\theta_c\right]\\ h(x) &= 1-3x^2+3x^4-x^6 \quad (-1<x<1);\\ h(x) &= 0 \quad (x\le -1, \quad x\ge 1)\end{aligned}$$

Here θ_i is an angle between the relative vector $\mathbf{r}_{ji}=\mathbf{r}_j-\mathbf{r}_i$ and the unit vector $\mathbf{u}_i$ which represents the orientation of the axis of particle i. The θ_j is an angle between the relative vector $\mathbf{r}_{ij}=\mathbf{r}_i-\mathbf{r}_j$ and the unit vector $\mathbf{u}_j$ (see Fig. 1). The function $h(\frac{\theta-\theta_0}{\theta_c})$ ($\theta_0=126°$ and $\theta_c=53.1°$) has maximum for $\theta=126°$. Thus, this term stabilizes the locally favored structure of five-fold symmetry in Fig. 1.

CP982, *Complex Systems, 5th International Workshop on Complex Systems*
edited by M. Tokuyama, I. Oppenheim, and H. Nishiyama

The parameter Δ controls not only the stability of the locally favored structure, but also the strength of frustration against crystallization because the symmetry of locally favored structure is not consistent with that of the crystalline structure.

SIMULATION METHODS

Next we describe the details of our simulation. The number of particles is $N = 1024$. We used periodic boundary conditions. We cut off the potential and smoothen its derivative at 3.0σ. All results are shown in the following reduced units. We used m (mass of particle), σ, and ε as the basic unit of mass, length, and energy, respectively. Thus, the mass m, the inertia moment I, the temperature T, the pressure P, the distance r, and the time t are scaled as: $m \equiv m/m = 1, I \equiv I/m\sigma^2, T \equiv k_BT/\varepsilon$ (k_B: the Boltzmann's constant), $P \equiv P\sigma^2/\varepsilon, r \equiv r/\sigma$, and $t \equiv t/\tau$ ($\tau = \sqrt{m\sigma^2/\varepsilon}$).

To control the temperature (T) and pressure (P) of a system, we employed the Nose-Poincaré-Andersen (NPA) thermostat [6] and extended the NPA method to include the rotational degree of freedom. Our modified Nose-Poincaré-Andersen Hamiltonian(H_{NPA}) is given by

$$
\begin{aligned}
H_{NPA}(t) &= (H_{NA}(t) - H_{NA}(0))s \\
H_{NA} &= \sum \frac{p^2}{2mVs^2} + \sum \frac{p_\phi^2}{2Is^2} + U(\mathbf{q}V^{1/2}, \{\phi_i\}) \\
&\quad + gk_BT\ln s + \frac{\Pi_V^2}{2M_V} + \frac{\Pi_s^2}{2M_s} + PV
\end{aligned}
$$

where ϕ_i is an angle between x axis and u_i, g is the degree of freedom of this system, s is a variable to control temperature, V is a volume of the system, Π_V is a momentum conjugate to V, Π_s is a momentum conjugate to s, M_V is a mass of V, and M_s is a mass of s. Here we used $I = 0.04$, M_V=0.0001, and M_s=1.0. We confirmed the equipartition of the energy into the translational and the rotational degrees of freedom. We also proved that the above NPA method extended to a system with the rotational degree of freedoms constitutes the NPT ensemble.

RESULTS AND DISCUSSION

We investigate the phase behavior of this system as a function of Δ by studying the temperature (T) dependence of the volume V of the system at $P = 0.5$ for a fixed cooling rate $Q \equiv dT/dt = 0.0001$ (see the Fig. 2). A system with the Lennard-Jones potential $\bar{U}(r)$ which corresponds to $\Delta = 0.0$ melts at $T = 0.46$ at $P = 0.5$. The stronger the frustration is, the lower the melting temperature is. But for weak frustration (i.e., for small Δ (≤ 0.5)), a system crystallizes upon cooling while accompanying a distinct jump in the volume. For strong frustration (i.e., for large Δ (≥ 0.6)), on the other hand, there is no jump in the volume and there is a jump in dV/dT (V is the volume) and dU/dT (U is the potential energy) around T_g. We investigate the temperature dependence of the radial distribution function and confirm that there is no long-range correlation even at a low temperature. This indicates that a system is vitrified even for the same cooling rate Q under strong frustration. Energetic frustration against crystallization prevents crystal nucleation.

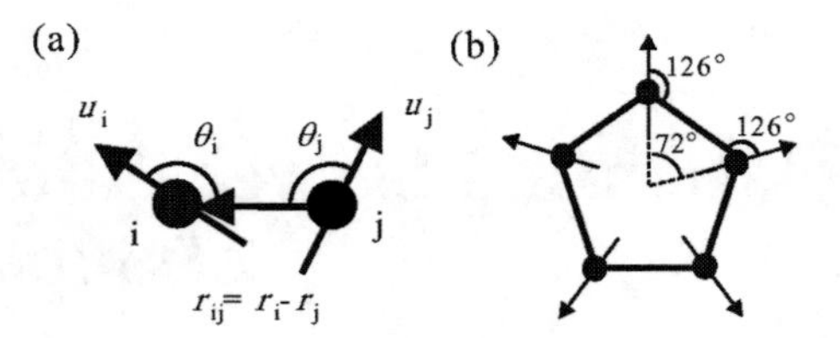

FIGURE 1. **(a)**: A definition of variables of our potential. A configuration where θ_i and θ_j are equal to $\theta_0 = 126°$ is energetically favored. **(b)**: A locally favored structure preferred by our potential. All the bonds of this structure form the most stable angle. Note that if particle i and particle j rotate freely with each other, the average of $\Delta U(r,\Omega)$ becomes 0.

We also studied dynamical properties of this system for $\Delta \geq 0.6$ by calculating intermediate scattering function and rotational autocorrelation function. We found that the correlation functions can be fit with stretched exponential and the temperature dependence of relaxation time obeys the Vogel-Fulcher law. Thus, we may conclude that this system well reproduces all essential features of the liquid-glass transition, which were known from previous studies.

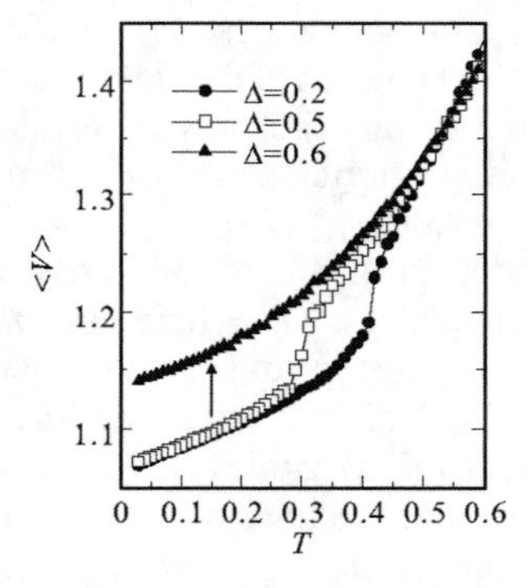

FIGURE 2. The T-dependence of the volume of the system for various Δ upon cooling. The arrows indicates the location of T_g. For $\Delta \leq 0.5$, we see a step in the volume, reflecting the crystallization into a crystal. For $\Delta = 0.6$, on the other hand, there is no step and only the slope of dV/dT changes at T_g.

Using this model, we investigate the dynamics of liquids and find that there exist medium-range crystalline order (MRCO) in supercooled liquids the life time of which is finite, but longer than the structural relaxation time of a liquid (see Figure 3). The characteristic size of MRCO is proportional to the dynamic coherence length obtained by the analysis using the four-point time correlation function. This indicates that MRCO may be a cause of dynamic heterogeneity [7].

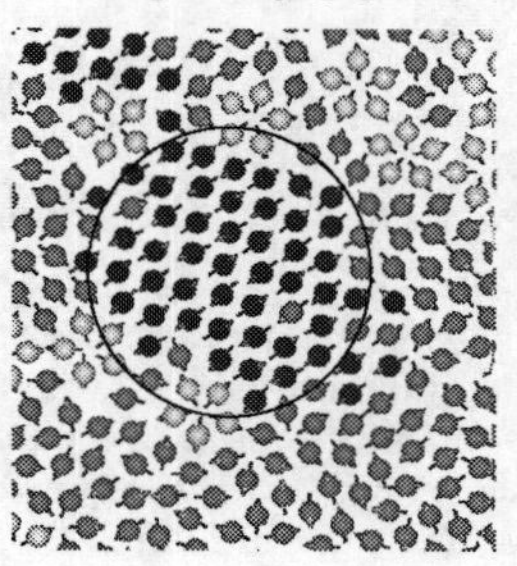

FIGURE 3. The snapshot of a supercooled liquid ($T = 0.17$). The arrow in a particle is the directional vector of the particle.The area in the open circle possesses medium-range antiferromagnetic crystalline order (MRCO).

We also investigate the relationship between fragility of liquids and the growth of MRCO by changing frustration against crystallization.(see Figure 4). It is clear that the stronger the frustration is, the stronger the liquid is. This result is consistent with the prediction of two order parameter model and indicates that there is an intrinsic correlation between the slow dynamics of supercooled liquids and frustration against crystallization.

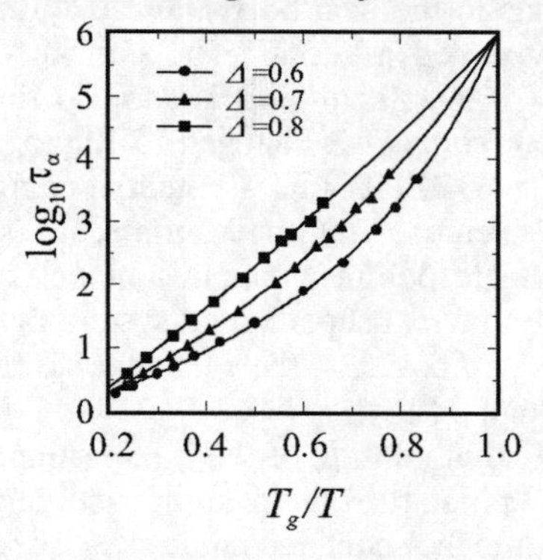

FIGURE 4. Angell plot for several liquids with different degrees of frustration. Δ represents the strength of frustration against crystallization.

SUMMARY

We developed a novel model of liquid that can cover from crystallization to vitrification just by changing the degree of frustration. Our results indicate that there is an intrinsic correlation between the slow dynamics of supercooled liquids and frustration against crystallization. Our model enables us to investigate the roles of the frustration in vitrification. Our study suggests that the growth of MRCO is related to the slowing down of the dynamics of a supercooled liquid and there may be an intrinsic link between the glass transition and crystallization.

ACKNOWLEDGMENTS

This work was partially supported by a grant-in-aid from the Ministry of Education, Culture, Sports, Science and Technology, Japan.

REFERENCES

1. M. D. Ediger, C. A. Angell, and S. R. Nagel, *J. Phys. Chem.* **100**, 13200-13212 (1996).
2. P. G. Debenedetti, *Metastable Liquids*, Princeton: Princeton University Press, 1997.
3. H. Sillescu, *J. Non-Cryst. Solids*, **243**, 81-108 (1999).
4. R. Richert, *J. Phys.Condens. Matter*, **14**, R703-R738 (2002)
5. H. Tanaka, *J. Chem. Phys.* **111**, 3163-3174 (1999).
6. J. B. Sturgeon and B. B. Laird, *J. Chem. Phys.* **112**, 3474-3482 (2000).
7. H. Shintani and H. Tanaka, *Nature Physics*, **2**, 200-206 (2006).

A Molecular Dynamics Simulation of a Supercooled System with 3,200 Model Polymers

T. Muranaka

Aichi Institute of Technology, 1247 Yachigusa, Yagusa-cho, Toyota, 470-0392, Japan

Abstract. A model system is studied via Molecular Dynamics (MD) simulation. The system consists of 3,200 molecules placed in the rigid cubic cell with the periodic boundary condition. A molecule has 100 united atoms which are assumed the CH_2. The molecule has the bonds between the united atoms nearby, the bends to the next bond, and the torsional potential. The motions of the united atoms in the model system have at least three time stages before the time region of the system relaxation. The first stage is marked by the ballistic motion. The second stage is marked by the motion in some cage. The third stage is marked by the collective motion with the surrounding united atoms. The fourth stage is marked by the elementary process which causes the system relaxation for long time.

Keywords: Computer simulation, Glass, Molecular dynamics
PACS: 78.55.Qr

INTRODUCTION

The β relaxation in the glass forming materials has been known for long time [1]. But the principle for this phenomenon has not been expressed clearly yet. On the other side, the α relaxation in the glass forming materials has been thought by common consent. That is because of an elementary process which causes the system relaxation for long time [2, 3]. I have focused on the secondary (β) relaxation from the long time side and an atomistic point of view. It has been found that the correlated motion of atoms plays an important role for the spatial heterogeneity in a two-dimensional model glass formed by the soft-sphere (disk) mixtures [4]. But the correlated motion of atoms in three-dimensional model glass was not found yet. I think that the correlation of atoms becomes large for the two-dimension and the system size is very important to find the correlated motion of atoms. Then I have studied the large system with 320,000 united atoms in three-dimension and the model system formed by the molecules with some internal structure to correlate the united atoms each other. I have assumed that the correlated motion of atoms is the origin of the β relaxation in the glass forming materials, because the time region of this motion existed in second time region from the long time.

SYSTEM

A MD simulation is performed for a model supercooled liquid. The system consists of 3,200 model molecules, and it is in a rigid cube with the periodic boundary condition. The linear dimension of the simulation cell is 20nm and the mass of each united atoms is 14 g/mol. This corresponds to a density of 0.9 g/cm^3 for all temperature. The volume is fixed during all the simulations. A MD step is 1.0 fs to vibrate the united atoms normally.

A model molecule has 100 united atoms which are assumed the CH_2, and the united atoms are connected with each other to form a normal molecule. The bonds have the bends to the next bonds, and the torsional potentials are adopted as usual. The Lennard-Jones potential ($\varepsilon_{LJ} = 0.1984$ kcal/mol) is adopted for the inter molecules, but somewhat changed to improve calculation speed. The DREIDING [5] potential parameters are used in this research. This model molecule is far away from the realistic polymer, but this model system becomes closely the real supercooled system than the system which is composed of the spherical atoms with the soft-core potential [2, 4].

The initial configuration of the liquid sample is made up as below. In first, the united atoms randomly are scattered all over the face centered cubic sites. In second, the united atoms are connected slowly during the temperature controlled MD simulation [6]. In last, the bends and the torsional potentials are adopted during the MD simulation. The initial configuration of each temperature is obtained by annealing that liquid sample for some time and changing the temperature immediately. After annealing MD simulation for 1,000,000 MD steps, the NVE ensemble for each temperature is completed.

The system is big, but I doubt it is sufficient to find the correlated motions of atoms. For my environment, 3,200 model molecules are maximum number to study long time relaxation.

CP982, *Complex Systems, 5th International Workshop on Complex Systems*
edited by M. Tokuyama, I. Oppenheim, and H. Nishiyama

RESULTS

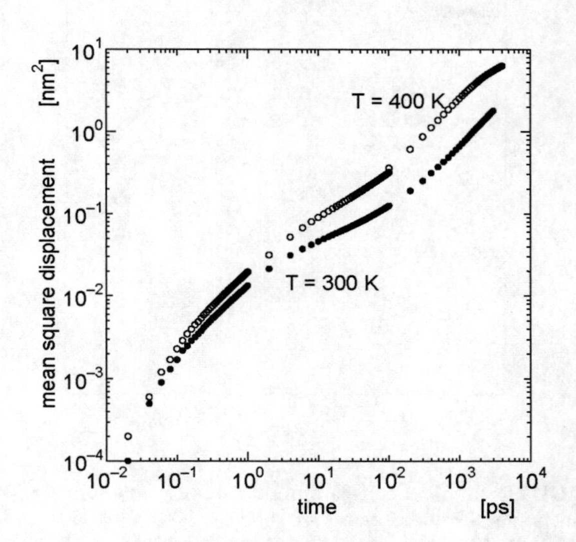

FIGURE 1. A log-log plot of the mean-square displacements of the united atoms at 300K (•) and 400K (○).

The figure 1 shows a log-log plot of the Mean-Square Displacement (MSD) vs. time at the temperature 400K and 300K. MSD is calculated as below.

$$\mathrm{MSD}(t) = < |\mathbf{r}_i(t) - \mathbf{r}_i(0)|^2 >, \quad (1)$$

where $\mathbf{r}_i(t)$ means the position of the i-th united atom at time t, $<$ and $>$ means the average for the initial times and all united atoms. In the worst statistic, the curve is obtained by averaging over 10 initial times, and 320,000 united atoms. There are four steps relaxations for this time region.

Shortest Time Region

In the first time region (before 0.1 ps), the united atoms move directly to obey the law of inertia.

The figure 2 shows a Velocity Auto-correlation Function (VAF) vs. time at the temperature 400K, 350K, and 300K. VAF is calculated as below.

$$\mathrm{VAF}(t) = \frac{< \mathbf{v}_i(t) \cdot \mathbf{v}_i(0) >}{< \mathbf{v}_i(0) \cdot \mathbf{v}_i(0) >}, \quad (2)$$

where $\mathbf{v}_i(t)$ means the velocity of the i-th united atom at time t.

In the shortest time region before 0.1 ps, the all united atoms move in similar way, even if the temperature is different. The time is about 0.04 ps, which the united atoms move to the opposite direction first. The united atoms are moving in the complex way, but the origin of this motion is simple. The bonds between the united atoms nearby are influenced for this time region.

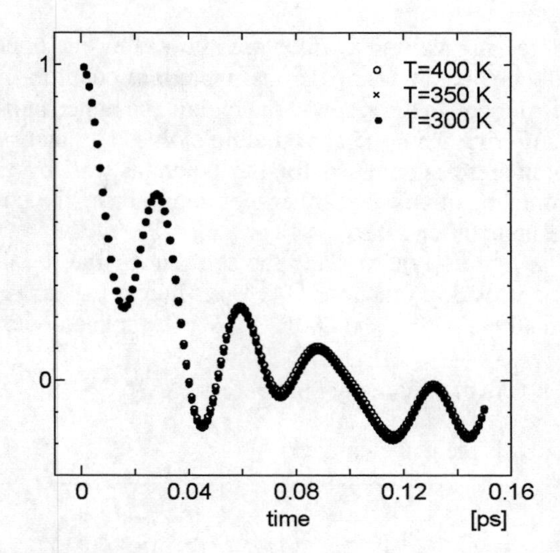

FIGURE 2. The velocity auto-correlation function of the united atoms are shown at 300K (•), 350K (×), and 400K (○).

Second Time Region

In the second time region (after 0.1 ps upto 1 ps), the united atoms move in some small cage of the surrounding atoms.

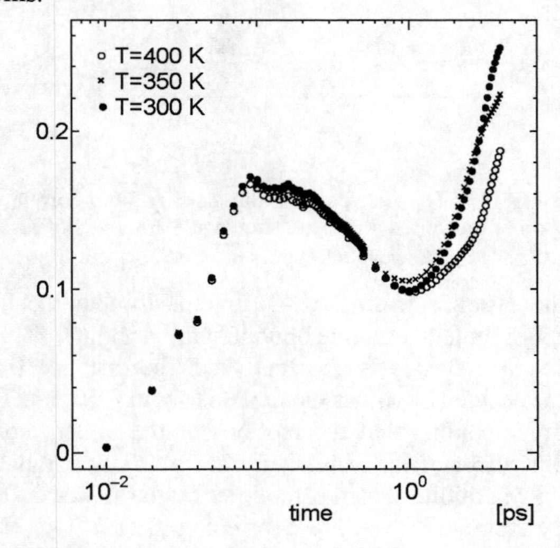

FIGURE 3. The Non-Gaussian Parameter of the united atoms are shown at 300K (•), 350K (×), and 400K (○).

The figure 3 shows a Non-Gaussian Parameter (NGP) vs. time at the temperature 400K, 350K, and 300K. NGP is calculated below.

$$\mathrm{NGP}(t) = \frac{3}{5} \frac{< |\mathbf{r}_i(t) - \mathbf{r}_i(0)|^4 >}{< |\mathbf{r}_i(t) - \mathbf{r}_i(0)|^2 >^2} \quad (3)$$

>From this figure, it is found that the united atoms don't move like gaussian before 0.1 ps. But NGP decreases between 0.1 ps and 1.0 ps. The united atoms are

scattered in small area which are allowed by the bonds nearby and so on. Many thermal vibrations occur in this time window. In the second time region, the united atoms are suffering from the surrounding atoms. The motions of atoms are suppressed for the potential wall by the surrounding atoms. The other motion produces the non-gaussianity after 1.0 ps.

The figure 4 and 5 show the self-part of the density auto-correlation function (DAF) vs. time at the temperature 400K, 350K, and 300K. DAF is calculated below.

$$\mathrm{DAF}(\mathbf{q},t) = < \exp[i\mathbf{q}\cdot\{\mathbf{r}_i(t)-\mathbf{r}_i(0)\}] >, \quad (4)$$

where $\mathbf{q}$ is the wave number.

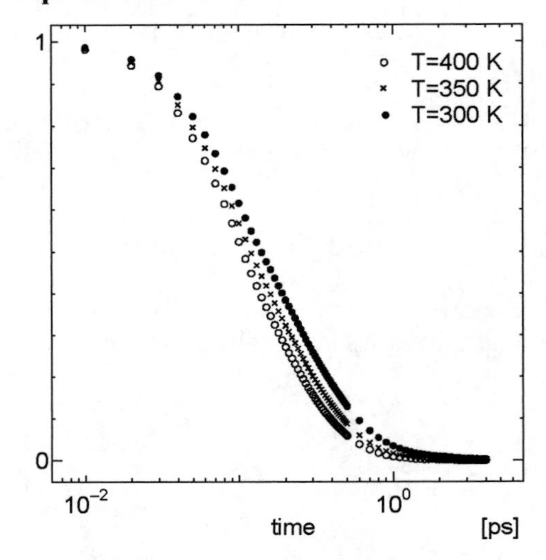

FIGURE 4. The self-part of the density auto-correlation function of the united atoms are shown at 300K (•), 350K (×), and 400K (∘). The wave number is $q = 2\,\pi/0.15$ [nm^{-1}].

The distance from the i-th united atom to the first neighbor united atom (the bond length) is about 0.15 nm. If the crystal state is accomplished, the distance from one molecule to another molecule is about 0.50 nm. The figure 4 exhibits that the motions of the united atoms within 0.15 nm have finished before 1.0 ps. And the figure 5 exhibits that the motions of the united atoms in the range 0.60 nm are continuing after 1.0 ps. The relaxation of the united atoms in this rage begins about 1.0 ps.

Third Time Region

In the third time region (after 1 ps upto 100 ps), the united atoms move with the surrounding atoms. This motion twists the original configuration, even though at low temperature. The united atoms move from the original positions. To reveal the motion of this, I have

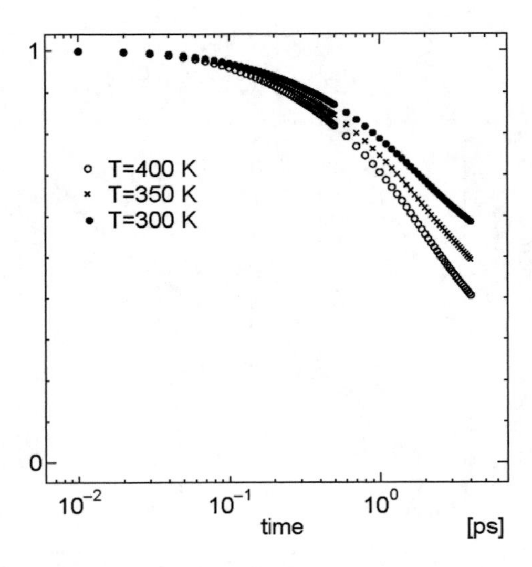

FIGURE 5. The self-part of the density auto-correlation function of the united atoms are shown at 300K (•), 350K (×), and 400K (∘). The wave number is $q = 2\,\pi/0.60$ [nm^{-1}].

introduced the correlated motion coefficient [4]

$$\mathrm{CM_i}(r_c,t) = \frac{1}{N_i}\sum_{r_{ij}\le r_c}^{N_i} \frac{\Delta\mathbf{r}_i(t)\cdot\Delta\mathbf{r}_j(t)}{<\Delta\mathbf{r}_i(t)\cdot\Delta\mathbf{r}_i(t)>}, \quad (5)$$

where $\Delta\mathbf{r}_i(t)$ is the displacement vector of the i-th united atom for an elapsed time t, a denominator $<\Delta\mathbf{r}_i(t)\cdot\Delta\mathbf{r}_i(t)>$ is the mean square displacement for an elapsed time t, and N_i is the number of the united atoms within the distance r_c from the tagged (i-th) united atom at $t=0$.

The figure 6 shows the correlated motion coefficient of the united atoms vs. time at 300K, 350K, and 400K. Three distant ranges are exhibited within $r_{ij} < 0.2$ nm, 0.2 nm $< r_{ij} <$ 0.4 nm, and 0.4 nm $< r_{ij} <$ 0.6 nm, where r_{ij} means the distance from the tagged (i-th) united atom at $t=0$ to the j-th united atoms at $t=0$.

This figure shows that the correlated motion within 0.2 nm after 0.1 ps suddenly increases, and the correlated motion with another molecule gradually increases after 0.5 ps. The value 0.2 is high for the correlated motion coefficient, because the value is calculated by all united atoms which contain the inactive united atoms.

According to the correlated motion, the united atoms move from the original position. But the configurations of atoms aren't changed for this motion. The surroundings atoms move as same as the tagged united atom. This motion doesn't break the original configuration, and diffuse the united atoms.

The last time region is after 100 ps. In this long time region, the united atoms break the original configurations by the surrounding atoms. The united atoms diffuse from the original positions. The configuration of atoms

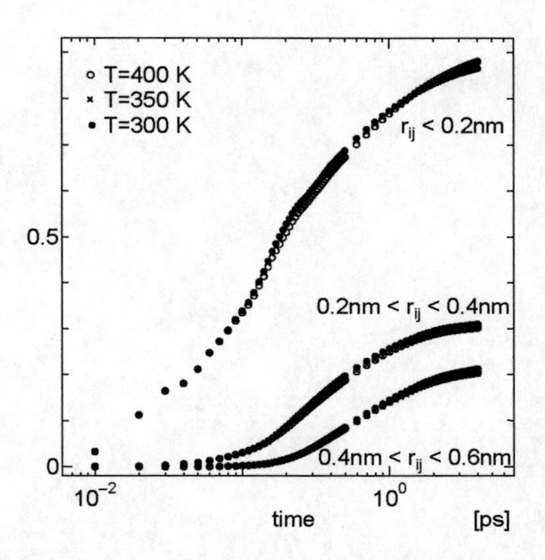

FIGURE 6. The correlated motion coefficient of the united atoms are shown at 300K (•), 350K (×), and 400K (∘). Three distant ranges are exhibited within $r_{ij} < 0.2$ nm, 0.2 nm $< r_{ij} <$ 0.4 nm, and 0.4 nm $< r_{ij} <$ 0.6 nm, where r_{ij} means the distance from the tagged (i-th) united atom at $t = 0$ to the j-th united atoms at $t = 0$

never returns the original configuration before this motion. Because this breaking configuration motion is very rare event, the diffusion is hidden in shorter time region at the low temperature.

CONCLUSION

The figure 7 shows that the schematic images of the motions of a united atom, where an arrow means the motion of the united atoms, and a circle means the area allowed by the surrounding atoms.

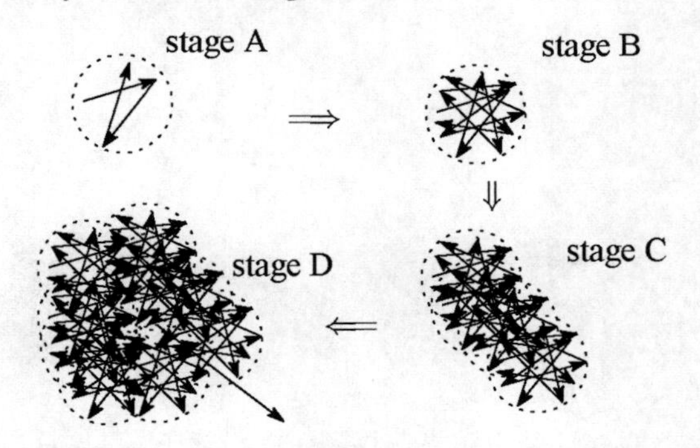

FIGURE 7. The schematic images of motions of the united atoms are shown. An arrow means the motion of the united atoms. A circle means the area allowed by the surrounding atoms. There are 4 stages corresponding to 4 time regions.

The stage A corresponds to the shortest time region before 0.1 ps. The united atoms move directly to obey the law of inertia. One united atom exists in a part of small cage of the surrounding atoms.

The stage B corresponds to the time region after 0.1 ps and before 1.0 ps. The united atoms move about in some small cages of the surrounding atoms, so the Gaussianity is restored in this time region.

The stage C corresponds to the time region after 1 ps and before 100 ps. The united atoms move with the surrounding atoms. The collective motion of atoms influences the relaxation of the system in this time region. The spatial heterogeneity becomes large.

The stage D corresponds to the time region after 100 ps. I think that the collective motions of united atoms occur toward many directions. And it may happen that the united atoms break the original configurations by the surrounding atoms. I can't show the data of this long time.

It seems that the story of the motions of atoms is common to the system which is inactive and dense. There is no evidence that the system is made up by the molecules. I suppose that the relaxation of the configuration for the molecules exists beyond this simulation time. The phenomenon of the decrease of the potential energy is observed in the MD simulation beyond 20 nanoseconds at the temperature 600K.

The correlated motion coefficient became over 0.5 within the distance of 2.0 σ_1 in the previous two-dimensional system [4]. In the present three-dimensional system, it becomes below 0.3 over the distance of 0.4 nm. Then this type of motion of atoms may not play an important role in this system. I should get the cool sample which doesn't decrease the potential energy.

REFERENCES

1. Y. Ishida, K. Yamafuji, and K. Shimada, *Kolloid-Z. u. Z. Polymere* **200**, 49 (1965).
2. H. Miyagawa, Y. Hiwatari, B. Bernu, and J. P. Hansen, *J. Chem. Phys.* **88**, 3879–3886 (1988).
3. E. Bartsch, M. Antonietti, W. Schupp, and H. Sillescu, *J. Chem. Phys.* **97**, 3950–3963 (1992).
4. T. Muranaka, and Y. Hiwatari, *Phys. Rev. E* **51**, R2735–R2738 (1995).
5. S. L. Mayo, B. D. Olafson, and W. A. Goddard III, *J. Phys. Chem.* **94**, 8897–8909 (1990).
6. S. Nosé, *J. Phys. Chem.* **81**, 511–519 (1984).

PART II

COMPLEX FLUIDS

Liquid Polyamorphism: Some Unsolved Puzzles of Water in Bulk, Nanoconfined, and Biological Environments

H. E. Stanley*, Pradeep Kumar*, Giancarlo Franzese†, Limei Xu*, Zhenyu Yan*, Marco G. Mazza*, S.-H. Chen**, F. Mallamace‡, and S. V. Buldyrev§*

*Center for Polymer Studies and Dept. of Physics, Boston University, Boston, MA 02215 USA
†Departament de Fisica Fonamental, Univ. de Barcelona, Diagonal 647, Barcelona 08028, SPAIN
**Nuclear Science and Engineering Dept., Mass. Inst. of Technology, Cambridge, MA 02139 USA
‡Dipartimento di Fisica, Univ. Messina, Vill. S. Agata, C.P. 55, I-98166 Messina, ITALY
§Department of Physics, Yeshiva University, 500 West 185th Street, New York, NY 10033 USA

Abstract. We investigate the relation between changes in dynamic and thermodynamic anomalies arising from the presence of the liquid-liquid critical point in (i) Two models of water, TIP5P and ST2, which display a first order liquid-liquid phase transition at low temperatures; (ii) the Jagla model, a spherically symmetric two-scale potential known to possess a liquid-liquid critical point, in which the competition between two liquid structures is generated by repulsive and attractive ramp interactions; and (iii) A Hamiltonian model of water where the idea of two length/energy scales is built in; this model also displays a first order liquid-liquid phase transition at low temperatures besides the first order liquid-gas phase transition at high temperatures. We find a correlation between the dynamic fragility crossover and the locus of specific heat maxima $C_P^{\max}$ ("Widom line") emanating from the critical point. Our findings are consistent with a possible relation between the previously hypothesized liquid-liquid phase transition and the transition in the dynamics recently observed in neutron scattering experiments on confined water. More generally, we argue that this connection between $C_P^{\max}$ and the dynamic crossover is not limited to the case of water, a hydrogen bonded network liquid, but is a more general feature of crossing the Widom line, an extension of the first-order coexistence line in the supercritical region. We present evidence from experiments and computer simulations supporting the hypothesis that water displays polyamorphism, i.e., water separates into two distinct liquid phases. This concept of a new liquid-liquid phase transition is finding application to other liquids as well as water, such as silicon and silica. We also discuss related puzzles, such as the mysterious behavior of confined water and the "skin" of hydration water near a biomolecule. Specifically, using molecular dynamics simulations, we also investigate the relation between the dynamic transitions of biomolecules (lysozyme and DNA) and the dynamic and thermodynamic properties of hydration water. We find that the dynamic transition of the macromolecules, sometimes called a "protein glass transition", occurs at the temperature of dynamic crossover in the diffusivity of hydration water, and also coincides with the maxima of the isobaric specific heat C_P and the temperature derivative of the orientational order parameter. We relate these findings to the hypothesis of a liquid-liquid critical point in water. Our simulations are consistent with the possibility that the protein glass transition results from a change in the behavior of hydration water, specifically from crossing the Widom line.

Keywords: High-density liquid, Liquid-liquid critical point, Liquid-liquid phase transition, Low-density liquid
PACS: 61.20.Ja, 61.20.Gy

Background

One "mysterious" property of liquid water was recognized 300 years ago [1]: although most liquids contract as temperature decreases, liquid bulk water begins to expand when its temperature drops below 4°C. Indeed, a simple kitchen experiment demonstrates that the bottom layer of a glass of unstirred iced water remains at 4°C while colder layers of 0°C water "float" on top (cf., Fig. 1 of Ref. [2]). The mysterious properties of liquid bulk water become more pronounced in the supercooled region below 0°C [3, 4, 5]. For example, if the coefficient of thermal expansion α_P, isothermal compressibility K_T, and constant-pressure specific heat C_P are extrapolated below the lowest temperatures measurable they would become infinite at a temperature of $T_s \approx 228$ K [3, 6].

Water is a liquid, but glassy water—also called amorphous ice—can exist when the temperature drops below the glass transition temperature T_g. Although it is a solid, its structure exhibits a disordered molecular liquid-like arrangement. *Low-density* amorphous ice (LDA) has been known for 60 years [7], and a second kind of amorphous ice, *high-density* amorphous ice (HDA) was discovered in 1984 [8, 9, 10]. HDA has a structure similar to that of high-pressure liquid water, suggesting that HDA may be a glassy form of high-pressure water [11, 12], just as LDA may be a glassy form of low-pressure water. Water has at least two different amorphous solid forms, a phenomenon called *polyamorphism* [13, 14, 15, 16, 17, 18, 19], and recently additional forms of glassy water have been the focus of active experimental and computational investigation [20, 21, 22, 23, 24, 25, 26, 27, 28].

CP982, *Complex Systems, 5th International Workshop on Complex Systems*
edited by M. Tokuyama, I. Oppenheim, and H. Nishiyama

Current Hypotheses

Many classic "explanations" for the mysterious behavior of liquid bulk water have been developed [29, 30, 31, 32, 33, 34], including a simple two-state model dating back to Röntgen [35] and a clathrate model dating back to Pauling [36]. Three hypotheses are under current discussion:

(i) The *stability limit hypothesis* [37], which assumes that the spinodal temperature line $T_s(P)$ between two liquids with different densities in the pressure-temperature ($P-T$) phase diagram connects at negative P to the locus of the liquid-to-gas spinodal for superheated bulk water. Liquid water cannot exist when cooled or stretched beyond the line $T_s(P)$.

(ii) The *singularity-free hypothesis* [38], considers the possibility that the observed polyamorphic changes in water resemble a genuine transition, but is not. For example, if water is a locally-structured transient gel comprised of molecules held together by hydrogen bonds whose number increases as temperature decreases [39, 40, 41], then the local "patches" or bonded sub-domains [42, 43] lead to enhanced fluctuations of specific volume and entropy and negative cross-correlations of volume and entropy whose anomalies closely match those observed experimentally.

(iii) The *liquid-liquid (LL) phase transition hypothesis* [44] arose from MD studies on the structure and equation of state of supercooled bulk water and has received some support [45, 46, 47, 48, 49, 50, 51, 52, 53]. Below the hypothesized *second* critical point the liquid phase separates into two distinct liquid phases: a low-density liquid (LDL) phase at low pressures and a high-density liquid (HDL) at high pressure (Fig. 1). Bulk water near the known critical point at 647 K is a fluctuating mixture of molecules whose local structures resemble the liquid and gas phases. Similarly, bulk water near the hypothesized LL critical point is a fluctuating mixture of molecules whose local structures resemble the two phases, LDL and HDL. These enhanced fluctuations influence the properties of liquid bulk water, thereby leading to anomalous behavior.

Selected Experimental Results

Many precise experiments have been performed to test the various hypotheses discussed in the previous section, but there is as yet no widespread agreement on which physical picture—if any—is correct. The connection between liquid water and the two amorphous ices predicted by the LL phase transition hypothesis is difficult to prove experimentally because supercooled water freezes spontaneously below the homogeneous nucleation temperature T_H, and amorphous ice crystallizes above the crystallization temperature T_X [54, 55, 56]. Crystallization makes experimentation on the supercooled liquid state between T_H and T_X almost impossible. However, comparing experimental data on amorphous ice at low temperatures with that of liquid water at higher temperatures allows an indirect discussion of the relationship between the liquid and amorphous states. It is found from neutron diffraction studies [12] and simulations that the structure of liquid water changes toward the LDA structure when the liquid is cooled at low pressures and changes toward the HDA structure when cooled at high pressures, which is consistent with the LL phase transition hypothesis [12]. The amorphous states (LDA and HDA) are presently considered to be "smoothly" connected thermodynamically to the liquid state if the entropies of the amorphous states are small [57, 58], and experimental results suggest that their entropies are indeed small [59].

In principle, it is possible to investigate experimentally the liquid state in the region between T_H and T_X during the extremely short time interval before the liquid freezes to crystalline ice [17, 56, 60]. Because high-temperature liquid bulk water becomes LDA without crystallization when it is cooled rapidly at one bar [16, 61], LDA appears directly related to liquid water. A possible connection between liquid bulk water at high pressure and HDA can be seen when ice crystals are melted using pressure [17]. Other experimental results [56] on the high-pressure ices [34, 62] that might demonstrate a LL first-order transition in the region between T_H and T_X have been obtained.

Selected Results from Simulations

Water is challenging to simulate because it is a molecular liquid and there is presently no precise yet tractable intermolecular potential that is universally agreed on. Nevertheless there are some distinct advantages of simulations over experiments. Experiments cannot probe the "No-Man's land" that arises in bulk water from homogeneous nucleation phenomena, but simulations have the advantage that they can probe the structure and dynamics well below T_H since nucleation does not occur on the time scale of computer simulations. Of the three hypotheses above, the LL phase transition hypothesis is best supported by simulations, some using the ST2 potential which *exaggerates* the real properties of bulk water, and others using the SPC/E and TIP4P potentials which *underestimate* them [44, 63, 64, 65, 66, 67]. Recently, simulations have begun to appear using the more reliable TIP5P potential [68, 69, 70]. The precise location of the LL critical point is difficult to obtain since the continuation of the first order line is a locus of maximum compressibility [63, 64, 66].

Further, computer simulations may be used to probe the *local* structure of water. At low temperatures, many water molecules appear to possess one of two principal local structures, one resembling LDA and the other HDA [63, 44, 65, 71]. Experimental data can also be interpreted in terms of two distinct local structures [72, 73, 74].

Understanding "Static Heterogeneities"

The systems in which water is confined are diverse—including the rapidly-developing field of artificial "nanofluidic" systems (man-made devices of order of nanometer or less that convey fluids). Among the special reasons for our interest in confined water is that phenomena occurring at a given set of conditions in bulk water occur under perturbed conditions for confined water [75, 76, 77, 78, 79, 80, 81, 82, 83, 84, 85, 86, 87, 88]. For example, the coordinates of the hypothesized LL critical point lie in the experimentally-inaccessible No-Man's land of the bulk water phase digram, but appear to lie in an accessible region of the phase diagrams of both two-dimensionally and one-dimensionally confined water [89, 90]. Simulations have been carried out to understand the effect of purely geometrical confinement [91, 92, 93, 94, 95, 96] and of the interaction with hydrophilic [97, 98, 99, 100, 101] or hydrophobic [102, 103, 104, 105] surfaces. It is interesting also to study the effects that confinement may have on the phase transition properties of supercooled water [96], in order to clarify the possible presence of a LL phase transition in the water. Recent work on the phase behavior of confined water suggests a sensitive dependence on the interaction with the surfaces [104], as a LL phase transition appears to be consistent with simulations of water confined between two parallel flat hydrophobic walls [94]. Works are in progress to extend this work to hydrophilic pores, such as those in Vycor glasses or biological situations, and to hydrophobic hydrogels, systems of current experimental interest [106, 107, 108, 109, 110, 111, 112, 113, 114, 115, 94, 116, 117, 118, 119, 120, 121].

Potentials with Two Characteristic Length Scales: Physical Arguments

A critical point appears if the pair potential between two particles of the system exhibits a minimum, and Fig. 1a sketches the potential of such an idealized system.

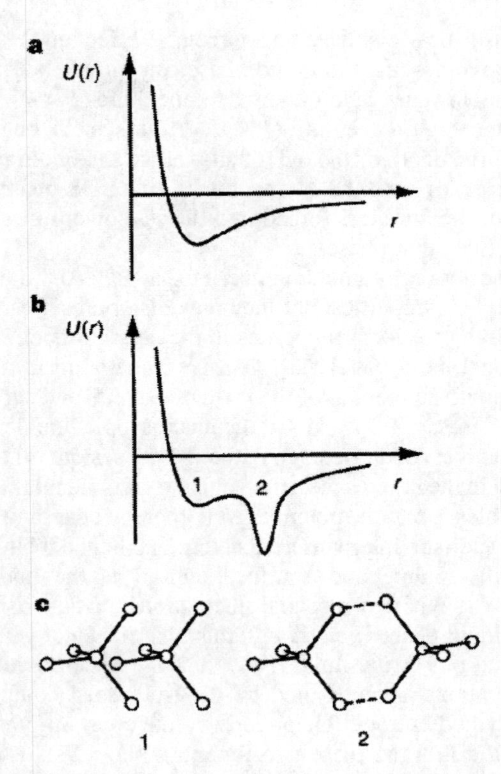

FIGURE 1. (a) Idealized system characterized by a pair interaction potential with a single attractive well. At low enough T ($T < T_C$) and high enough P ($P > P_C$), the system condenses into the "liquid" well shown. (b) Idealized system characterized by a pair interaction potential whose attractive well has two sub-wells, the outer of which is deeper and narrower. For low enough T ($T < T_{C'}$) and low enough P ($P < P_{C'}$), the one-phase liquid can "condense" into the narrow outer "LDL" sub-well, thereby giving rise to a LDL phase, and leaving behind the high-density liquid phase occupying predominantly the inner subwell. (c) Two idealized interaction clusters of water molecules in configurations that may correspond to the two sub-wells of (b).

At high temperature, the system's kinetic energy is so large that the potential well does not have an effect, and the system is in a single "fluid" (or gas) phase. At low enough temperature ($T < T_C$) and large enough pressure ($P > P_C$) the fluid is sufficiently influenced by the minimum in the pair potential that it can condense into the low specific volume liquid phase. At lower pressure ($P < P_C$), the system explores the full range of distances—the large specific volume gas phase.

If the potential well has the form shown in Fig. 1b—the attractive potential well of Fig. 1a has now bifurcated into a deeper outer sub-well and a more shallow inner sub-well. Such a two-minimum ("two length scale") potential can give rise to the occurrence at low temperatures of a LL critical point at ($T_{C'}$,$P_{C'}$) [122]. At high temperature, the system's kinetic energy is so large that the two sub-wells have no appreciable effect on the thermodynamics and the liquid phase can sample both sub-wells. However, at low enough temperature ($T < T_{C'}$) and not too high a pressure ($P < P_{C'}$) the system must respect the depth of the outer sub-well so the liquid phase "condenses" into the outer sub-well (the LDL phase). At higher pressure it is forced into the shallower inner sub-well (the HDL phase).

The above arguments concern the average or "thermodynamic" properties, but they may also be useful in anticipating the local properties in the neighborhood of individual molecules [123]. Consider, again, an idealized fluid with a potential of the form of Fig. 1a. and suppose that T is, say, $1.2\ T_C$ so that the macroscopic liquid phase has not yet condensed out. Although the system is not entirely in the liquid state, small clusters of molecules begin to coalesce into the potential well, thereby changing their characteristic interparticle spacing (and hence their local specific volume) and their local entropy, so the fluid system will experience spatial fluctuations characteristic of the liquid phase even though this phase has not yet condensed out of the fluid at $T = 1.2\ T_C$. Specific volume fluctuations are measured by the isothermal compressibility and entropy fluctuations by the constant-pressure specific heat, so these two functions should start to increase from the values they would have if there were no potential well at all. As T decreases toward T_C, the magnitude of the fluctuations (and hence of the compressibility and the specific heat) increases monotonically and in fact diverges to infinity as $T \to T_C$. The cross-fluctuations of specific volume and entropy are proportional to the coefficient of thermal expansion, and this (positive) function should increase without limit as $T \to T_C$.

Consider an idealized fluid with a potential of the form of Fig. 1b, and suppose that T is now *below* T_C but is 20 percent *above* $T_{C'}$, so that the LDL phase has not yet condensed out. The liquid can nonetheless begin to sample the two sub-wells and clusters of molecules will begin to coalesce in each well, with the result that the liquid will experience spatial fluctuations characteristic of the LDL and HDL phase even though the liquid has not yet phase separated. The specific volume fluctuations and entropy fluctuations will increase, and so the isothermal compressibility K_T and constant-pressure specific heat C_P begin to diverge. Moreover, if the outer well is narrow, then when a cluster of neighboring particles samples the outer well it has a larger specific volume and a smaller entropy, so the anticorrelated cross-fluctuations of specific volume (the isothermal expansion coefficient α_P) is now *negative*, and approaching $-\infty$ as T decreases toward $T_{C'}$.

Now if by chance the value of $T_{C'}$ is lower than the value of T_H, then the phase separation discussed above would occur only at temperatures so low that the liquid would have frozen! In this case, the "hint" of the LL critical point C' is the presence of these local fluctuations whose magnitude would grow as T decreases, but which would never actually diverge if the point C' is never actually reached. Functions would be observed experimentally to increase as if they would diverge to ∞ or $-\infty$ but at a temperature below the range of experimental accessibility.

Now consider not the above simplified potential, but rather the complex (and unknown) potential between nonlinear water molecules. The tetrahedrality of water dictates that the outermost well corresponds to the ordered configuration with lower entropy. Thus although we do not know the actual form of the intermolecular potential in bulk water, it is not implausible that the same considerations apply as those discussed for the simplified potential of Fig. 1b. Indeed, extensive studies of such pair potentials have been carried out recently and the existence of the LL critical point has been demonstrated in such models [124, 47, 48, 125, 50, 51, 52, 53, 143, 144, 145, 146, 147].

To make more concrete how plausible it is to obtain a bifurcated potential well of the form of Fig. 1b, consider that one can crudely approximate water as a collection of 5-molecule groups called Walrafen pentamers (Fig. 1c) [73]. The interaction strengths of two adjacent Walrafen pentamers depends on their relative orientations. The first and the second energy minima of Fig. 1b correspond to the two configurations of adjacent Walrafen pentamers with different mutual orientations (Fig. 1c).

The two local configurations—#1 and #2 in Fig. 1c—are (i) a high-energy, low specific volume, high-entropy, non-bonded #1-state, or (ii) a low-energy, high specific volume, low-entropy, bonded #2-state. The difference in local structure is also the difference in the local structure between a high-pressure crystalline ice (such as ice VI or ice VII) and a low-pressure crystalline ice (such as ice I_h) [34] (Fig. 1c).

The region of the P-T plane along the line continuing from the LDL-HDL coexistence line extrapolated to higher temperatures above the second critical point is the locus of points where the LDL on the low-pressure side and the HDL on the high-pressure side are continuously transforming—it is called the Widom line, defined to be the locus of points where the correlation length is maximum. Near this line, two different kinds of local structures, having either LDL or HDL properties, "coexist" [71, 126, 127]. The entropy fluctuations are largest near the Widom line, so C_P increases to a maximum, displaying a λ-like appearance [128]. The increase in C_P [58] resembles the signature of a glass transition as suggested by mode-coupling theory [129, 130, 131]. Careful measurements and simulations of static and dynamic correlation functions [126, 132, 133, 134, 135] may be useful in determining the exact nature of the apparent singular behavior near 220K.

Potentials with Two Characteristic Length Scales: Tractable Models

The above discussion is consistent with the possible existence of two well-defined classes of liquids: simple and water-like. The former interact via spherically-symmetric non-softened potentials, do not exhibit thermodynamic nor dynamic anomalies. We can calculate translational and orientational order parameters (t and q), and project equilibrium state points onto the (t,q) plane thereby generating what is termed the Errington-Debenedetti (ED) order map [136, 43]. In water-like liquids, interactions are orientation-dependent; these liquids exhibit dynamic and thermodynamic anomalies, and their ED "order map" is in general two-dimensional but becomes linear (or quasi-linear) when the liquid exhibits structural, dynamic or thermodynamic anomalies.

Hemmer and Stell [137] showed that in fluids interacting via pairwise-additive, spherically-symmetric potentials consisting of a hard core plus an attractive tail, softening of the repulsive core can produce additional phase transitions. This pioneering study elicited a considerable body of work on so-called core-softened potentials which can generate water-like density and diffusion anomalies [137, 138, 139, 140, 141, 142, 143, 144, 145, 146, 147, 151, 152, 153, 154, 155, 156]. This important finding implies that strong orientational interactions, such as those that exist in water and silica, are not a necessary condition for a liquid to have thermodynamic and dynamic anomalies.

A softened-core potential has been used [122] to explain the isostructural solid-solid critical point present in materials such as Cs and Ce, for which the shape of the effective pair potential obtained from scattering experiments is "core-softened" [4, 122, 140, 157, 158]. Analytical work in 1D suggests a LL phase transition, and the existence at $T=0$ of low and high density phases. Recent work using large-scale MD simulations reports anomalous behavior in 2D as well [140, 142]. In 3D we showed that a squared potential with a repulsive shoulder and an attractive well displays a phase diagram with a LL critical point and no density anomaly [143, 144, 145, 146]. The continuous version of the the same shouldered attractive potential shows not only the LL critical point, but also the density anomaly [147, 156]. We use the soft-core potential to investigate the relationship between configurational entropy S_{conf} and diffusion coefficient D. Recent work using the SPC/E potential [159] suggests that the temperature-density dependence of S_{conf} may correlate with D, and that the maximum of S_{conf} tracks the density maxima line.

Two questions arise naturally from this emerging taxonomy of liquid behavior. First, is structural order in core-softened fluids hard-sphere or water-like? Second, is it possible to seamlessly connect the range of liquid behavior from hard spheres to water-like by a simple and common potential, simply by changing a physical parameter?

In recent work, Yan et al. [160, 161, 162] used a simple spherically-symmetric "hard-core plus ramp" potential to address the first question. They found that this core-softened potential with two characteristic length scales not only gives rise to water-like diffusive and density anomalies, but also to an ED water-like order map, implying that orientational interactions are not necessary in order for a liquid to have structural anomalies. We are investigating the evolution of dynamic, thermodynamic and structural anomalies, using the ratio λ of hard core and soft core length scales as a control parameter. We hope to show that the family of tunable spherically-symmetric potentials so generated evolves continuously between hard sphere and water-like behavior; if successful, this will be the first demonstration that essential aspects of the wide range of liquid behavior encompassed by hard spheres and tetrahedrally-coordinated network-formers can be systematically traversed by varying a single control parameter. We will study the equation of state, diffusion coefficient, and structural order parameters t and q. Our calculations seem to reveal a negative thermal expansion coefficient (static anomaly) and an increase of the diffusion coefficient upon isothermal compression (dynamic anomaly) for $0 \leq \lambda < 6/7$. As in bulk water, the regions where these anomalies occur are nested domes in the (T,ρ) or (T,P) planes, with the "thermodynamic anomaly dome" contained within the "dynamic anomaly dome." The ED order map evolves from water-like to hard-sphere-like upon varying between 4/7 and 6/7. Thus, we traverse the range of liquid behavior encompassed by hard spheres ($\lambda = 1$) and water-like ($\lambda \sim 4/7$) by simply varying the ratio of hard to soft-core diameters.

Further work is needed to establish whether a ratio of competing length scales close to 0.6 is generally associated with water-like anomalies in other core-softened potentials, e.g., achieving two characteristic length scales by using a linear combination of Gaussian [163] potentials of different widths.

In sum, motivated by the need to better understand the phenomenon of liquid polyamorphism [164, 165, 166], we are carrying out a systematic study of the effects of λ and the ratio of characteristic energies on the existence of a LL transition, the positive or negative slope of the line of first-order LL transitions in the (P,T) plane, and the relationship, if any [143, 144], between the LL transition and density anomalies. We will perform calculations in parallel for both confined and bulk water.

Understanding "Dynamic Heterogeneities"

Recent Experiments on Confined Water

Simulations and experiments both are consistent with the possibility that the LL critical point, if it exists at all, lies in the experimentally inaccessible No-Man's land. If this statement is valid, then at least two reactions are possible:

(i) If something is not experimentally accessible, then it does not deserve discussion.

(ii) If something is not experimentally accessible, but its influence *is* experimentally accessible, then discussion is warranted.

Option (ii) has guided most research thus far, since the manifestations of a critical point extend far away from the actual coordinates of that point. Indeed, accepting option (i) means there is nothing more to discuss. However if we confine water, the homogeneous nucleation temperature decreases and it becomes possible to enter the No-Man's land and, hence, search for the LL critical point. In fact, recent experiments at MIT and Messina by the Chen and Mallamace groups demonstrate that for nanopores of typically 1.5 nm diameter, the No-Man's land actually ceases to exist—one can supercool the liquid state all the way down to the glass temperature. Hence studying confined water offers the opportunity of directly testing, for the first time, the LL phase transition hypothesis.

In fact, using two independent techniques, neutron scattering and NMR, the MIT and Messina groups found a sharp kink in the dynamic properties (a "dynamic crossover") at the same temperature $T_L \approx 225$K [90, 167]. Our calculations on *bulk* models [168] are not inconsistent with one tentative interpretation of this dynamic crossover as resulting from the system passing from the high-temperature high-pressure "HDL" side of the Widom line (where the liquid might display fragile behavior) to the low-temperature low-pressure "LDL" side of the Widom line (where the liquid might display strong behavior). By definition, the Widom line—defined to be the line in the pressure-temperature plane where the correlation length has its maximum—arises only if there is a critical point. Hence interpreting the MIT-Messina experiments in terms of a Widom line is of potential relevance to testing experimentally, for *confined* water, the liquid-liquid critical point hypothesis.

The interpretation of the dynamic crossover could have implications for nanofluidics and perhaps even for natural confined water systems, e.g., some proteins appear to undergo a change in their flexibility at approximately the same temperature T_L that the MIT-Messina experiments identify for the dynamic crossover in pure confined water.

Possible Significance of the Widom Line

The conjectured interpretation of the MIT-Messina experiments relies on the concept of the Widom line, a concept not widely appreciated even though it has been known by experimentalists dating back to the 1958 Ph.D. thesis of J. M. H. Levelt (now Levelt-Sengers) [171]. Since a Widom line arises only from a critical point, if the MIT-Messina experiments can be rationalized by a Widom line then they are consistent with the existence of a LL critical point in confined water.

By definition, in a first order phase transition, thermodynamic functions discontinuously change as we cool the system along a path crossing the equilibrium coexistence line [Fig. 2(a), path β]. However in a *real* experiment, this discontinuous change may not occur at the coexistence line since a substance can remain in a supercooled metastable phase until a limit of stability (a spinodal) is reached [4] [Fig. 2(b), path β].

If the system is cooled isobarically along a path above the critical pressure P_c [Fig. 2(b), path α], the state functions continuously change from the values characteristic of a high temperature phase (gas) to those characteristic of a low temperature phase (liquid). The thermodynamic response functions which are the derivatives of the state functions with respect to temperature [e.g., C_P] have maxima at temperatures denoted $T_{\max}(P)$. Remarkably these maxima are still prominent far above the critical pressure [171, 172], and the values of the response functions at $T_{\max}(P)$ (e.g., $C_P^{\max}$) diverge as the critical point is approached. The lines of the maxima for different response functions asymptotically approach one another as the critical point is approached, since all response functions become expressible in terms of the correlation length. This asymptotic line is sometimes called the Widom line, and is often regarded as an extension of the coexistence line into the "one-phase regime."

Suppose now that the system is cooled at constant pressure P_0. (i) If $P_0 > P_C$ ("path α"), experimentally-measured quantities will change dramatically but continuously in the vicinity of the Widom line (with huge fluctuations as measured by, e.g., C_P). (ii) If $P_0 < P_C$ ("path β"), experimentally-measured quantities will change discontinuously if the coexistence line is actually seen. However the coexistence line can be difficult to detect in a pure system due to metastability, and changes will occur only when the spinodal is approached where the gas phase is no longer stable.

In the case of water—the most important solvent for biological function [173, 174]—a significant change in dynamical properties has been suggested to take place in deeply supercooled states [175, 176, 177, 178]. Unlike other network forming materials [179], water behaves as a fragile liquid in the experimentally accessible window [176, 180, 181]. Based on analogies with other

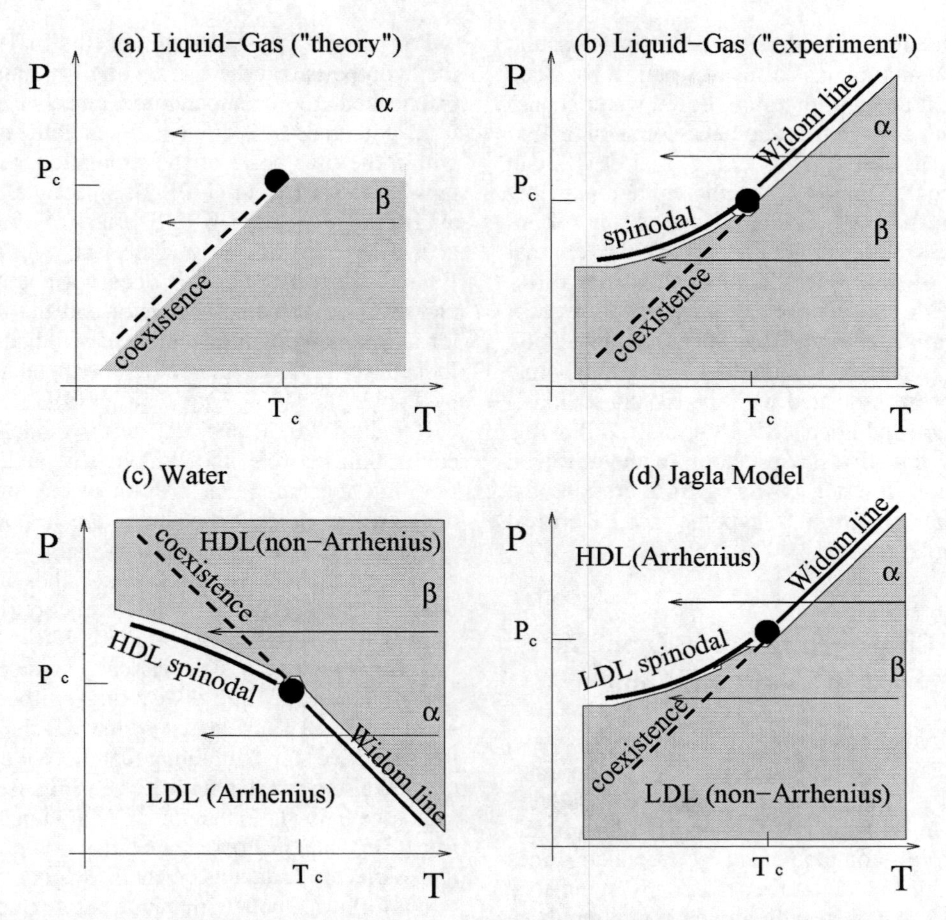

FIGURE 2. (a) Schematic phase diagram for the critical region associated with a liquid-gas critical point. Two features display mathematical singularities: the critical point and the liquid-gas coexistence. (b) Same, with the addition of the gas-liquid spinodal and the Widom line. Along the Widom line, thermodynamic response functions have extrema in their T dependence. (c) A hypothetical phase diagram for water of possible relevance to recent confined water neutron scattering experiments [169, 167, 170, 90]. (d) A sketch of the $P-T$ phase diagram for the two-scale Jagla model.

network forming liquids and with the thermodynamic properties of the amorphous forms of water, it has been suggested that, at ambient pressure, liquid water should show a crossover between fragile behavior at high T to strong behavior at low T [182, 177, 138, 139, 183] in the deep supercooled region of the phase diagram below the homogeneous nucleation line. This region may contain the hypothesized LL critical point [44], the terminal point of a line of first order LL phase transitions. Recently, dynamic crossovers in confined water were studied experimentally [96, 184, 169, 90] since nucleation can be avoided in confined geometries. Also, a dynamic crossover has been associated with the LL phase transition in both silicon and silica [185, 186]. In the following, we offer a very tentative interpretation of the observed fragility transition in water as arising from crossing the Widom line emanating from the hypothesized LL critical point [186] [Fig. 2, path α].

Methods Employed to Study Dynamic Crossovers in Confined Water

Using MD simulations [187], we study three models, each of which has a LL critical point. Two of the models, (the TIP5P [68] and the ST2 [188]) treat water as a multiple site rigid body, interacting via electrostatic site-site interactions complemented by a Lennard-Jones potential. The third model is the spherically symmetric "two-scale" Jagla potential with attractive and repulsive ramps which has been recently studied in the context of LL phase transitions and liquid anomalies [157, 151]. For all three models, we evaluate the loci of maxima of the relevant response functions, K_T and C_P, which coincide close to the critical point and give rise to the Widom line. We carefully explore the hypothesis that, for all three potentials, a dynamic crossover occurs when the Widom line is crossed.

For TIP5P we find a LL critical point [69, 70], from which the Widom line develops [Fig.4(a)]. The coexistence curve is negatively sloped, so the Clapeyron equation implies that the high-temperature phase is a high-density liquid (HDL) and the low-temperature phase is a low-density liquid (LDL). The diffusion coefficient D is evaluated from the long time limit of the mean square displacement along isobars [Fig.4(b)]. We find that isobars crossing the Widom line (path α) show a clear crossover (i) from a *non-Arrhenius behavior at high* T [which can be well fitted by a power law function $D \sim (T - T_{\rm MCT})^{\gamma}$], consistent with the mode coupling theory predictions [189]), (ii) to an *Arrhenius behavior at low* T [which can be described by $D \sim \exp(-E_a/T)$]. The crossover between these two functional forms takes place when crossing the Widom line.

For paths β, crystallization occurs in TIP5P [69], so the hypothesis that there is no fragility transition cannot be checked at low temperature. Hence we consider a related potential, ST2, for which crystallization is absent within the time scale of the simulation. Simulation details are described in [190]. This potential also displays a LL critical point [44, 190] [Fig.4(c)]. Along paths α a fragility transition may take place, while along paths β the T dependence of D does not show any sign of crossover to Arrhenius behavior and the fragile behavior is retained down to the lowest studied temperature. Indeed, for paths β, the entire T dependence can be fit by a power law $(T - T_{\rm MCT})^{\gamma}$ [Fig.4(d)].

If indeed TIP5P and ST2 water models support the connection between the Widom line and the dynamic fragility transition, it is natural to ask which features of the water molecular potential are responsible for the properties of water, especially because water's unusual properties are shared by several other liquids whose inter-molecular potential has two energy (length) scales such as silicon and silica [185, 191, 186]. Hence we also investigate the two-scale spherically symmetric Jagla potential [157, 138, 139], displaying—without the need to supercool—a LL coexistence line which, unlike water, has a positive slope, implying that the Widom line is now crossed along α paths with $P > P_C$ [Fig.4(e)]. We verify a crossover in the behavior of $D(T)$ when the Widom line ($C_P^{\rm max}$ line) is crossed, such that at high temperature, D exhibits an Arrhenius behavior, while at low temperature it followa a non-Arrhenius behavior, consistent with a power law [Fig.4(f)]. Along a β path ($P < P_C$), $D(T)$ appears to follow the Arrhenius behavior over the entire studied temperature range. Thus we test that the dynamic crossover coincides with the location of the $C_P^{\rm max}$ line, extending the conclusion of the TIP5P and ST2 potentials to a general two-scale spherically symmetric potential.

Hamiltonian Model of Water

In Ref. [148], we investigated the generality of the dynamic crossover in a Hamiltonian model of water which displays a liquid-liquid phase transition at low temperatures. We consider a cell model that reproduces the fluid phase diagram of water and other tetrahedral network forming liquids [50, 51, 52, 53]. For sake of clarity, we focus on water to explain the motivation of the model. The model is based on the experimental observations that on decreasing P at constant T, or on decreasing T at constant P, (i) water displays an increasing local tetrahedrality [149], (ii) the volume per molecule increases at sufficiently low P or T, and (iii) the O-O-O angular correlation increases [74], consistent with simulations [150].

The system is divided into cells $i \in [1, \ldots, N]$ on a regular square lattice, each containing a molecule with volume $v \equiv V/N$, where $V \geq Nv_0$ is the total volume of the system, and v_0 is the hard-core volume of one molecule. The cell volume v is a continuous variable that gives the mean distance $r \equiv v^{1/d}$ between molecules in d dimensions. The van der Waals attraction between the molecules is represented by a truncated Lennard-Jones potential with characteristic energy $\varepsilon > 0$

$$U(r) \equiv \begin{cases} \infty & \text{for} \quad r \leq R_0 \\ \varepsilon \left[\left(\frac{R_0}{r}\right)^{12} - \left(\frac{R_0}{r}\right)^{6} \right] & \text{for} \quad r > R_0 \,, \end{cases} \tag{1}$$

where $R_0 \equiv v_0^{1/d}$ is the hard-core distance [148].

Each molecule i has four bond indices $\sigma_{ij} \in [1, \ldots, q]$, corresponding to the nearest-neighbor cells j. When two nearest-neighbor molecules have the facing σ_{ij} and σ_{ji} in the same relative orientation, they decrease the energy by a constant J, with $0 < J < \varepsilon$, and form a bond, e.g. a (non-bifurcated) hydrogen bond for water, or a ionic bond for SiO_2. The choice $J < \varepsilon$ guarantees that bonds are formed only in the liquid phase. Bonding and intramolecular (IM) interactions are accounted for by the two Hamiltonian terms

$$\mathscr{H}_{\mathrm{B}} \equiv -J \sum_{\langle i,j \rangle} \delta_{\sigma_{ij}\sigma_{ji}}, \tag{2}$$

where the sum is over n.n. cells, $0 < J < \varepsilon$ is the bond energy, $\delta_{a,b} = 1$ if $a = b$ and $\delta_{a,b} = 0$ otherwise, and

$$\mathscr{H}_{\mathrm{IM}} \equiv -J_\sigma \sum_i \sum_{(k,\ell)_i} \delta_{\sigma_{ik}\sigma_{i\ell}}, \tag{3}$$

where $\sum_{(k,\ell)_i}$ denotes the sum over the IM bond indices (k,l) of the molecule i and $J_\sigma > 0$ is the IM interaction energy with $J_\sigma < J$, which models the angular correlation between the bonds on the same molecule. The total energy of the system is the sum of the van der Waals interaction and Eqs. (2) and (3).

We find that different response functions such as C_P, α_p (see Fig. 3) show maxima and these maxima increase and seem to diverge as the critical pressure is approached, consistent with the picture of Widom line that we discussed for other water models in the sections above. Moreover we find that the temperature derivative of the number of hydrogen bonds dN_{HB}/dT displays a maximum in the same region where the other thermodynamic response functions have maxima; suggesting that the fluctuations in the number of hydrogen bonds is maximum at the Widom line temperature T_W.

To futher test if this model system also displays a dynamic crossover as found in the other models of water, we study the total spin relaxation time of the system as a function of T for different pressures. We find that for $J_\sigma/\varepsilon = 0.05$ (*liquid-liquid critical point scenario*) the crossover occurs at $T_W(P)$ for $P < P_{C'}$ [Fig. 5(a)]. For completeness we study the system also in the case of *singularity free scenario*, corresponding to $J_\sigma = 0$. For $J_\sigma = 0$ the crossover is at $T(C_P^{\max})$, the temperature of $C_P^{\max}$ [Fig. 5(b)].

We next calculate the Arrhenius activation energy $E_{\mathrm{A}}(P)$ from the low-T slope of $\log\tau$ vs. $1/T$ [Fig. 6(a)]. We extrapolate the temperature $T_{\mathrm{A}}(P)$ at which τ reaches a fixed macroscopic time $\tau_{\mathrm{A}} \geq \tau_{\mathrm{C}}$. We choose $\tau_{\mathrm{A}} = 10^{14}$ MC steps > 100 sec [131] [Fig. 6(b)]. We find that $E_{\mathrm{A}}(P)$ and $T_{\mathrm{A}}(P)$ decrease upon increasing P in both scenarios, providing no distinction between the two interpretations. Instead, we find a dramatic difference in the P dependence of the quantity $E_{\mathrm{A}}/(k_B T_{\mathrm{A}})$ in the two scenarios, increasing for the LL critical point scenario and approximately constant for the singularity free scenario [Fig. 6(c)].

Glass Transition in Biomolecules

Next we explore the hypothesis [211] that the observed glass transition in biomolecules [198, 212, 213, 214, 215, 216, 217, 218, 219, 220, 221, 222, 223, 224] is related to the liquid-liquid phase transition using molecular dynamics (MD) simulations. Specifically, Kumar et al. [211] studied the dynamic and thermodynamic behavior of lysozyme and DNA in hydration TIP5P water, by means of the software package GROMACS [225] for (i) an orthorhombic form of hen egg-white lysozyme [226] and (ii) a Dickerson dodecamer DNA [227] at constant pressure $P = 1$ atm, several constant temperatures T, and constant number of water molecules N (NPT ensemble).

The simulation results for the mean square fluctuations $\langle x^2 \rangle$ of both protein and DNA are shown in Figure 7(a). Kumar et al. calculated the mean square fluctuations $\langle x^2 \rangle$ of the biomolecules from the equilibrated configurations, first for each atom over 1 ns, and then averaged over the total number of atoms in the biomolecule. They find that $\langle x^2 \rangle$ changes its functional form below $T_{\mathrm{p}} \approx 245$ K, for *both* lysozyme [Fig. 7(a)] and DNA [Fig. 7(b)].

Kumar et al. next calculated C_P by numerical differentiation of the total enthalpy of the system (protein and water) by fitting the simulation data for enthalpy with a fifth order polynomial, and then taking the derivative with respect to T. Figures 8(a) and 8(b) display maxima of $C_P(T)$ at $T_{\mathrm{W}} \approx 250 \pm 10$ K for both biomolecules.

Further, to describe the quantitative changes in structure of hydration water, Kumar et al. calculated the local tetrahedral order parameter Q [136, 160, 161, 162] for hydration water surrounding lysozyme and DNA. Figures 8(c) and 8(d) show that the rate of increase of Q has a maximum at 245 ± 10 K for lysozyme and DNA hydration water respectively; the same temperatures of the crossover in the behavior of mean square fluctuations.

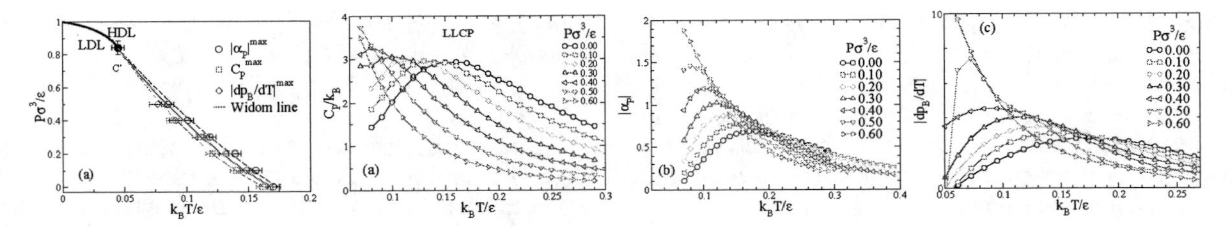

FIGURE 3. (a) Phase diagram below T_{MD} line shows that $|dp_B/dT|^{max}$ (◇) coincides with the Widom line $T_W(P)$ (solid line) within error bars: C' is the HDL-LDL critical point, end of first-order phase transition line (thick line) [148]; symbols are maxima for $N = 3600$ of $|\alpha_P|^{max}$ (○), C_P^{max} (□), and $|dp_B/dT|^{max}$ (◇); upper and lower dashed line are quadratic fits of $|\alpha_P|^{max}$ and C_P^{max}, respectively, consistent with C'; $|\alpha_P|^{max}$ and C_P^{max} are consistent within error bars. Maxima are estimated from panels (b), (c) and (d), where each quantity is shown as functions of T for different $P < P_{C'}$. In (e) $|dp_B/dT|^{max}$ is the numerical derivative of p_B from simulations in (d).

Upon cooling, the diffusivity of hydration water exhibits a dynamic crossover from non-Arrhenius to Arrhenius behavior at the crossover temperature $T_\times \approx 245 \pm 10$ K [Figure 8(e)]. The coincidence of $T_\times$ with T_p within the error bars indicates that the behavior of the protein is strongly coupled with the behavior of the surrounding solvent, in agreement with recent experiments [198]. Note that $T_\times$ is much higher than the glass transition temperature, estimated for TIP5P as $T_g = 215$K. Thus this crossover is not likely to be related to the glass transition in water.

The fact that $T_p \approx T_\times \approx T_W$ is evidence of the correlation between the changes in protein fluctuations [Figure 7(a)] and the hydration water thermodynamics [Figure 8(a)]. Thus these results are consistent with the possibility that the protein glass transition is related to the Widom line (and hence to the hypothesized liquid-liquid critical point). Crossing the Widom line corresponds to a continuous but rapid transition of the properties of water from those resembling the properties of a local HDL structure for $T > T_W(P)$ to those resembling the properties of a local LDL structure for $T < T_W(P)$. A consequence is the expectation that the fluctuations of the protein residues in predominantly LDL-like water (more ordered and more rigid) just below the Widom line should be smaller than the fluctuations in predominantly HDL-like water (less ordered and less rigid) just above the Widom line.

The quantitative agreement of the results for both DNA and lysozyme [Figures 3 and 4] suggests that it is indeed the changes in the properties of hydration water that are responsible for the changes in dynamics of the protein and DNA biomolecules. Our results are in qualitative agreement with recent experiments on hydrated protein and DNA [228] which found the crossover in side-chain fluctuations at $T_p \approx 225$ K.

Outlook

It is possible that other phenomena that appear to occur on crossing the Widom line are in fact not coincidences, but are related to the changes in local structure that occur when the system changes from the "HDL-like" side to the "LDL-like" side. In this work we concentrated on reviewing the evidence for changes in dynamic transport properties, such as diffusion constant and relaxation time. Additional examples include: (1) a breakdown of the Stokes-Einstein relation for $T < T_W(P)$ [192, 193**?** , 195, 196, 197], (2) systematic changes in the static structure factor $S(q)$ and the corresponding pair correlation function $g(r)$ revealing that for $T < T_W(P)$ the system more resembles the structure of LDL than HDL, (3) appearance for $T < T_W(P)$ of a shoulder in the dynamic structure factor $S(q,\omega)$ at a frequency $\omega \approx 60$ cm$^{-1} \approx 2$ THz [198, 199], (4) rapid increase in hydrogen bonding degree for $T < T_W(P)$ [200, 201], (5) a *minimum* in the density at low temperature [202], and (6) a scaled equation of state near the critical point [203]. It is important to know how general a given phenomenon is, such as crossing the Widom line which by definition is present whenever there is a critical point. Using data on other liquids which have local tetrahedral symmetry, such as silicon and silica, which appear to also display a liquid-liquid critical point and hence must possess a Widom line emanating from this point into the one-phase region. For example, we learned of interesting new work on silicon, which also interprets structural changes as arising from crossing the Widom line of silicon [204]. It might be interesting to test the effect of the Widom line on simple model systems that display a liquid-liquid critical point, such as two-scale symmetric potentials of the sort recently studied by Xu and her collaborators [205] or by Franzese [147] and Barros de Oliveira and coworkers [156].

Very recently, Mallamace and his collaborators succeeded in locating the Widom line by finding a clearcut maximum in the coefficient of thermal expansion, at

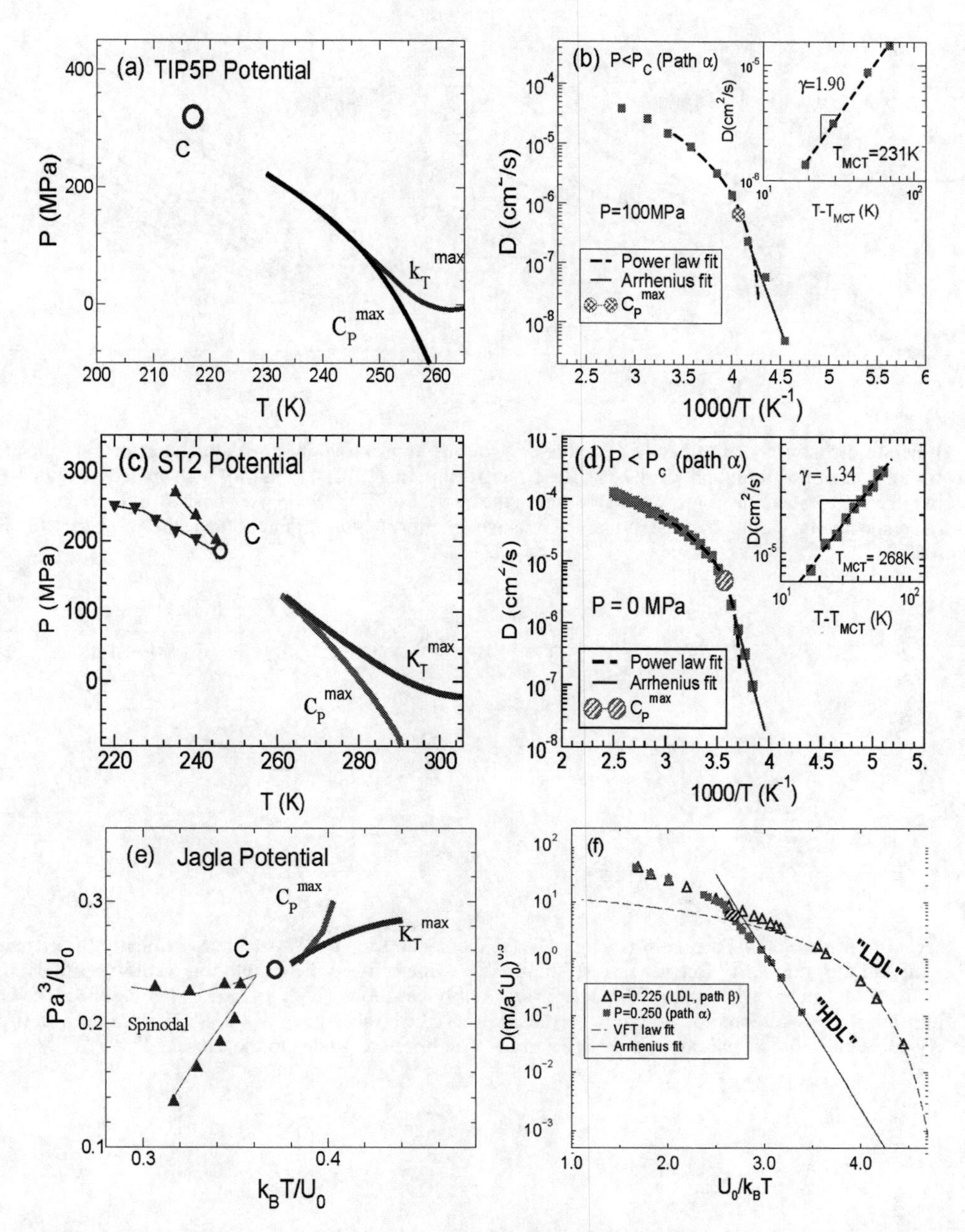

FIGURE 4. (a) Relevant part of the phase diagram for the TIP5P potential, showing the liquid-liquid critical point C at $P_c = 320$ MPa and $T_c = 217$ K , the line of maximum of isobaric specific heat $C_P^{\max}$ and the line of maximum of isothermal compressibility $K_T^{\max}$. (b) D as a function of T for $P = 100$ MPa (path α). At high temperatures, D behaves like that of a non-Arrhenius liquid and can be fit by $D \sim (T - T_{MCT})^{\gamma}$ (also shown in the inset) where $T_{MCT} = 220$ K and $\gamma = 1.942$, while at low temperatures the dynamic behavior changes to that of a liquid where D is Arrhenius. (c) The same for the ST2 potential. The liquid-liquid critical point C is located at $P_c = 246$ MPa and $T_c = 146$ K. (d) D as a function of T for $P = 100$ MPa (path α). At high temperatures, D behaves like that of a non-Arrhenius liquid and can be fit by $D \sim (T - T_{MCT})^{\gamma}$ (also shown in the inset) where $T_{MCT} = 239$ K and $\gamma = 1.57$, while at low temperatures the dynamic behavior changes to that of a liquid where D is Arrhenius. (e) Phase diagram for the Jagla potential in the vicinity of the liquid-liquid phase transition. Shown are the liquid-liquid critical point C located at $P_c = 0.24$ and $T_c = 0.37$, the line of isobaric specific heat maximum $C_P^{\max}$, the line of isothermal compressibility $K_T^{\max}$, and spinodal lines. (f) D as a function of T for $P = 0.250$ (squares, path α) and $P = 0.225$ (triangles, path β). Along path α, one can see a sharp crossover from the high temperature Arrhenius behavior $D \approx \exp(-1.53/T)$ with lower activation energy to a low temperature Arrhenius behavior $D \approx \exp(-6.3/T)$ with high activation energy, which is a characteristic of the HDL. Along path β, there is no sharp changes near the critical point, because the liquid remains in the LDL phase. However, near the glass transition, LDL exhibits a non-Arrhenius behavior characterized by the VFT fit at very low temperature.

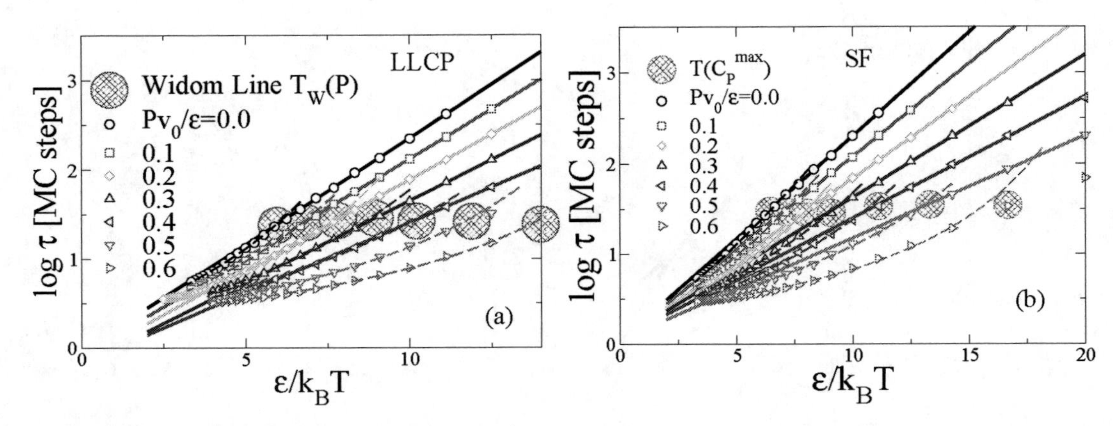

FIGURE 5. Dynamic crossover—large hatched circles of a radius approximately equal to the error bar—in the orientational relaxation time τ for a range of different pressures. (a) The LL critical point (LLCP) scenario, with crossover temperature at $T_W(P)$. (b) The singularity free (SF) scenario, with crossover temperature at $T(C_P^{\max})$. Solid and dashed lines represent Arrhenius and non-Arrhenius fits, respectively. Notice that the dynamic crossover occurs at approximately the same value of τ for all seven values of pressure studied.

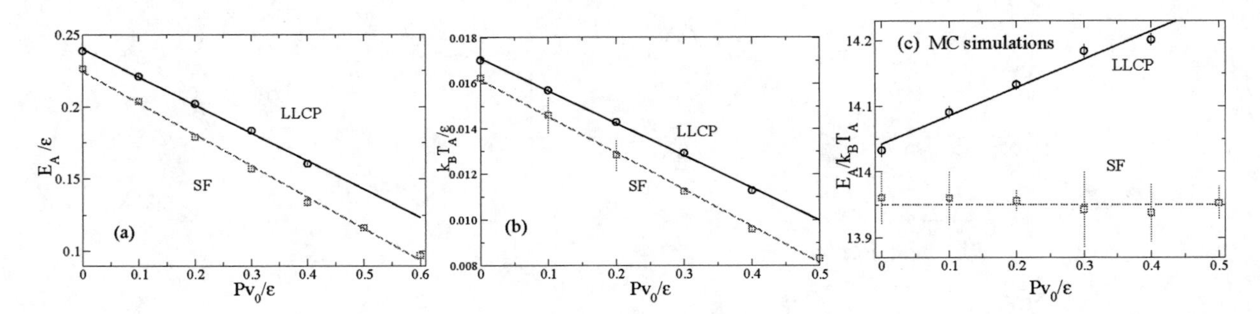

FIGURE 6. Effect of pressure on the activation energy E_{A}. (a) Demonstration that E_{A} decreases linearly for increasing P for both the LL critical point and the singularity free scenarios. The lines are linear fits to the simulation results (symbols). (b) T_{A}, defined such that $\tau(T_{\mathrm{A}}) = 10^{14}$ MC steps > 100 sec [131], decreases linearly with P for both scenarios. (c) P dependence of the quantity $E_{\mathrm{A}}/(k_B T_{\mathrm{A}})$ is different in the two scenarios. In the LL critical point (LLCP) scenario, $E_{\mathrm{A}}/(k_B T_{\mathrm{A}})$ increases with increasing P, and it is approximately constant in the singularity free (SF) scenario. The lines are guides to the eyes.

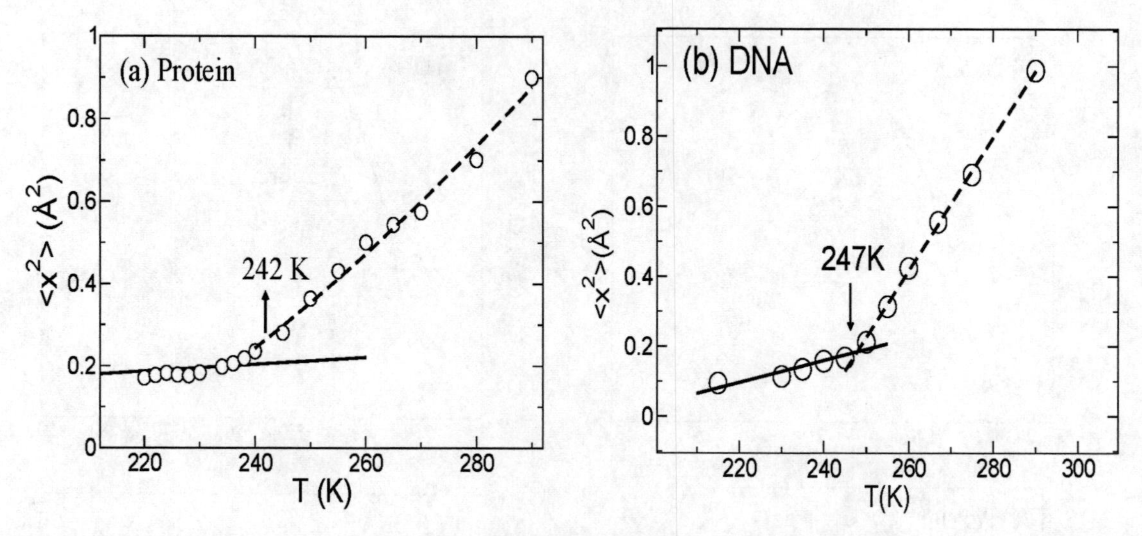

FIGURE 7. Mean square fluctuation of (a) lysozyme, and (b) DNA showing that there is a transition around $T_p \approx 242 \pm 10$ K for lysozyme and around $T_p \approx 247 \pm 10$ K for DNA. For very low T one would expect a linear increase of $\langle x^2 \rangle$ with T, as a consequence of harmonic approximation for the motion of residues. At high T, the motion becomes non-harmonic and we fit the data by a polynomial. We determine the dynamic crossover temperature T_p from the crossing of the linear fit for low T and the polynomial fit for high T. We determine the error bars by changing the number of data points in the two fitting ranges.

$T_W \approx 225$K [206, 207, 208], which remarkably is the same temperature as the specific heat maximum [209]. Also, private discussions with Jacob Klein reveal a possible reason for why confined water does not freeze at -38C, the bulk homogeneous nucleation temperature: Klein and co-workers [210] noted that confined water behaves differently than typical liquids in that it water does not experience the huge increase in viscosity characteristic of other strongly confined liquids. They interpret this experimental finding as arising from the fact that strong confinement hampers the formation of a hydrogen bonded network, and we know from classic work of Linus Pauling that without the extensive hydrogen bonded network, water's freezing temperature will be depressed by more than 100 degrees. Thus confinement reduces the extent of the hydrogen bonded network and hence lowers the freezing temperature, but leaves the key tetrahedral local geometry of the water molecule itself unchanged.

Acknowledgments

This work has been supported by the NSF and DOE.

REFERENCES

1. R. Waller, trans., *Essayes of Natural Experiments* [original in Italian by the Secretary of the Academie del Cimento]. Facsimile of 1684 English translation (Johnson Reprint Corporation, New York, 1964).
2. H. E. Stanley, "Unsolved Mysteries of Water in its Liquid and Glass Phases" [edited transcript of Turnbull Prize lecture], *Materials Research Bulletin* **24**, No. 5, 22–30 (May 1999).
3. C. A. Angell, M. Oguni, and W. J. Sichina, "Heat Capacity of Water at Extremes of Supercooling and Superheating," *J. Phys. Chem.* **86**, 998–1002 (1982).
4. P. G. Debenedetti and H. E. Stanley, "The Physics of Supercooled and Glassy Water," Physics Today **56**[issue 6], 40–46 (2003).
5. O. Mishima and H. E. Stanley, "The relationship between liquid, supercooled and glassy water" [invited review article], Nature **396**, 329-335 (1998).
6. R. J. Speedy and C. A. Angell, "Isothermal Compressibility of Supercooled Water and Evidence for a Thermodynamic Singularity," *J. Chem. Phys.* **65**, 851–858 (1976)
7. E. F. Burton and W. F. Oliver, "The Crystal Structure of Ice at Low Temperatures," *Proc. Roy. Soc. London Ser. A* **153**, 166–172 (1936).
8. O. Mishima, L. D. Calvert, and E. Whalley, "'Melting' Ice I at 77 K and 10 kbar: A New Method of Making Amorphous Solids," *Nature* **310**, 393–395 (1984).
9. H.-G. Heide, "Observations on Ice Layers," *Ultramicroscopy* **14**, 271–278 (1984).
10. O. Mishima, L. D. Calvert, and E. Whalley, "An Apparently First-Order Transition between Two Amorphous Phases of Ice Induced by Pressure," *Nature* **314**, 76–78 (1985).
11. M. C. Bellissent-Funel, L. Bosio, A. Hallbrucker, E. Mayer, and R. Sridi-Dorbez, "X-Ray and Neutron Scattering Studies of the Structure of Hyperquenched Glassy Water," *J. Chem. Phys.* **97**, 1282–1286 (1992).
12. M. C. Bellissent-Funel and L. Bosio, "A Neutron Scattering Study of Liquid D_2O," *J. Chem. Phys.* **102**, 3727–3735 (1995).

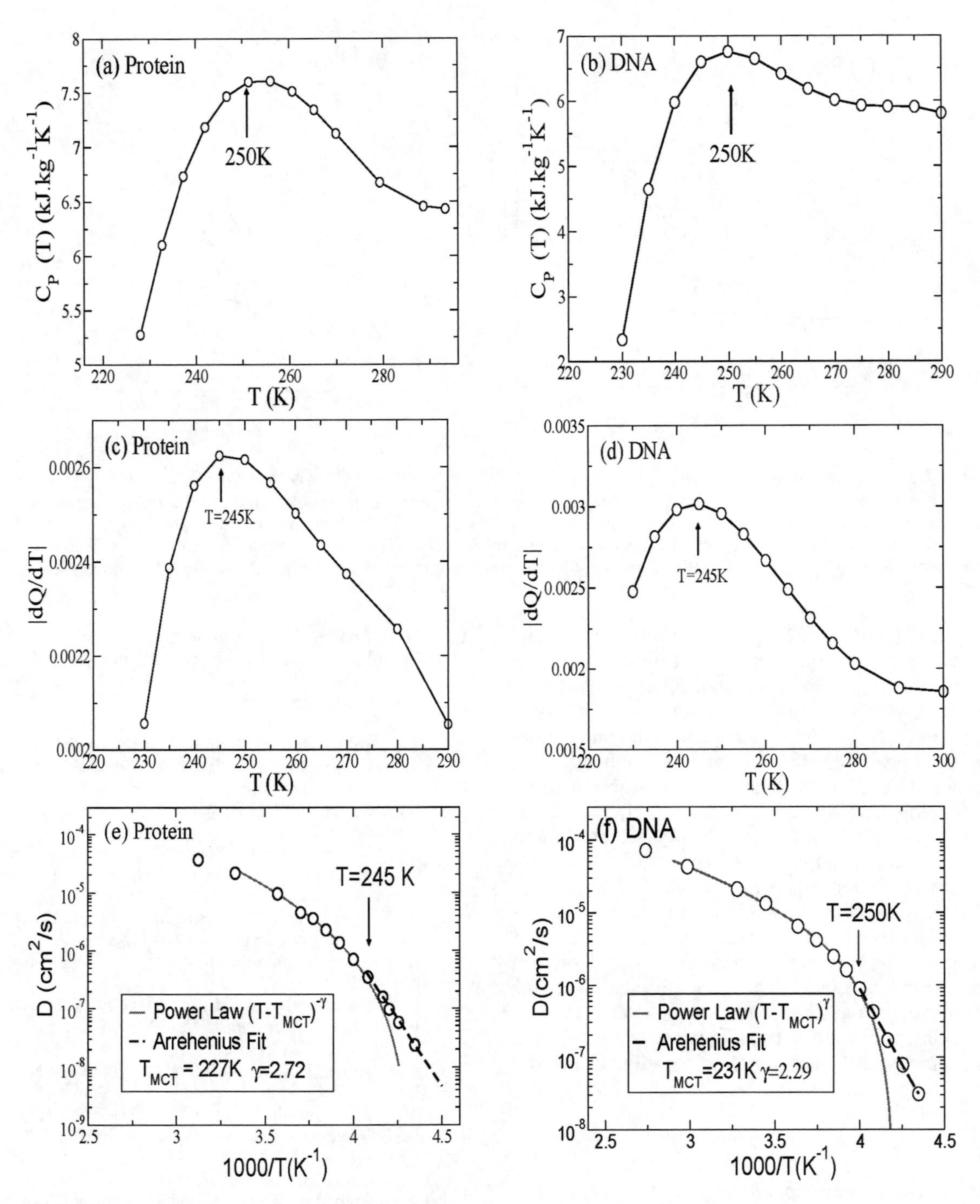

FIGURE 8. The specific heat of the combined system (a) lysozyme and water, and (b) DNA and water, display maxima at 250 ± 10 K and 250 ± 12 K respectively, which are coincident within the error bars with the temperature T_p where the crossover in the behavior of $\langle x^2 \rangle$ is observed in Figure 7. Derivative with respec t to temperature of the local tetrahedral order parameter Q for (c) lysozyme and (d) DNA hydration water. A maximum in $|dQ/dT|$ at Widom line temperature suggests that the rate of change of local tetrahedrality of hydration water has a maximum at the Widom line. Diffusion constant of hydration water surrounding (e) lysozyme, and (f) DNA shows a dynamic transition from a power law behavior to an Arrhenius behavior at $T_\times \approx 245 \pm 10$ K for lysozyme and $T_\times \approx 250 \pm 10$ K for DNA, around the same temperatures where the behavior of $\langle x^2 \rangle$ has a crossover, and C_P and $|dQ/dT|$ have maxima.

13. K. H. Smith, E. Shero, A. Chizmeshya, and G. H. Wolf, "The Equation of State of Polyamorphic Germania Glass: A Two-Domain Description of the Viscoelastic Response," *J. Chem. Phys.* **102**, 6851–6857 (1995).
14. P. H. Poole, T. Grande, F. Sciortino, H. E. Stanley, and C. A. Angell, "Amorphous Polymorphism," *J. Comp. Mat. Sci.* **4**, 373–382 (1995).
15. J. L. Finney, "Water? What's So Special About It? Phil. Trans. R. Soc. Lond. B: Biol. Sci. **359**, 1145–1163 (2004).
16. E. Mayer, "Hyperquenched Glassy Bulk Water: A Comparison with Other Amorphous Forms of Water, and with Vitreous but Freezable Water in a Hydrogel and on Hydrated Methemoglobin," in *Hydrogen Bond Networks*, edited by M.-C. Bellissent-Funel and J. C. Dore (Kluwer Academic Publishers, Dordrecht, 1994), pp. 355–372.
17. O. Mishima, "Relationship between Melting and Amorphization of Ice," *Nature* **384**, 546–549 (1996).
18. G. P. Johari, A. Hallbrucker, and E. Mayer, "Two Calorimetrically Distinct States of Liquid Water below 150 Kelvin," *Science* **273**, 90–92 (1996).
19. U. Essmann and A. Geiger, "Molecular-Dynamics Simulation of Vapor-Deposited Amorphous Ice," *J. Chem. Phys.* **103**, 4678–4692 (1995).
20. T. Loerting, C. Salzmann, I. Kohl, E. Mayer, and A. Hallbrucker, "A Second Structural State of High-Density Amorphous Ice at 77 K and 1 bar," *Phys. Chem. Chem. Phys.* **3**, 5355 (2001).
21. J. L. Finney, D. T. Bowron, A. K. Soper, T. Loerting, E. Mayer, and A. Hallbrucker, "Structure of a New Dense Amorphous Ice," *Phys. Rev. Lett.* **89**, 503–506 (2002).
22. J. S. Tse, D. D. Klug, M. Guthrie, C. A. Tulk, C. J. Benmore, and J. Urquidi, "Investigation of the Intermediate- and High-Density Forms of Amorphous Ice by Molecular Calculations Dynamics and Diffraction Experiments," Phys. Rev. B **71**, 214107 (2005).
23. I. Brovchenko, A. Geiger, and A. Oleinikova, "Multiple Liquid-Liquid Transitions in Supercooled Water," *J. Chem. Phys.* **118**, 9473–9476 (2003).
24. I. Brovchenko, A. Geiger, and A. Oleinikova, "Liquid-Liquid Phase Transitions in Supercooled Water Studied by Computer Simulations of Various Water Models," J. Chem. Phys. **123**, 044515 (2005).
25. P. Jedlovszky and R. Vallauri, "Liquid-Vapor and Liquid-Liquid Phase Equilibria of the Brodholt-Sampoli-Vallauri Polarizable Water Model," J. Chem. Phys. **122**, 081101 (2005).
26. J. A. White, "Multiple Critical Points for Square-Well Potential with Repulsive Shoulder," *Physica A* **346**, 347–357 (2004).
27. J. K. Christie, M. Guthrie, C. A. Tulk, C. J. Benmore, D. D. Klug, S. N. Taraskin, and S. R. Eliot, "Modeling the atomic structure of very high-density amorphous ice," Phys. Rev. B **72**, 012201 (2005).
28. J. L. Finney, A. Hallbrucker, I. Kohl, A. K. Soper, and D. T. Bowron, "Structures of High and Low Density Amorphous Ice by Neutron Diffraction," Phys. Rev. Lett. **88**, 225503 (2002).
29. D. Eisenberg and W. Kauzmann, *The Structure and Properties of Water* (Oxford University Press, New York, 1969).
30. J. D. Bernal and R. H. Fowler, "A Theory of Water and Ionic Solution, with Particular Reference to Hydrogen and Hydroxyl Ions," *J. Chem. Phys.* **1**, 515–548 (1933).
31. J. A. Pople, "Molecular Association in Liquids. II. A Theory of the Structure of Water," *Proc. Roy. Soc. Lond. Ser. A* **205**, 163–178 (1951).
32. H. S. Frank and W.-Y. Wen, "Structural Aspects of Ion-Solvent Interaction in Aqueous Solutions: A Suggested Picture of Water Structure," *Disc. Faraday Soc.* **24**, 133–140 (1957).
33. G. Némethy and H. A. Scheraga, "Structure of Water and Hydrophobic Bonding in Proteins: I. A Model for the Thermodynamic Properties of Liquid Water," *J. Chem. Phys.* **36**, 3382–3400 (1962).
34. B. Kamb, "Ice Polymorphism and the Structure of Water," in *Structural Chemistry and Molecular Biology*, edited by A. Rich and N. Davidson (Freeman, San Francisco, 1968), pp. 507–542.
35. W. C. Röntgen, "Ueber die constitution des flüssigen wassers," *Ann. d. Phys. u. Chem.* **45**, 91–97 (1892).
36. L. Pauling, "The Structure of Water," in *Hydrogen Bonding*, edited by D. Hadzi (Pergamon Press, New York, 1959), pp. 1–5.
37. R. J. Speedy, "Stability-Limit Conjecture: An Interpretation of the Properties of Water," *J. Phys. Chem.* **86**, 982–991 (1982).
38. S. Sastry, P. Debenedetti, F. Sciortino, and H. E. Stanley, "Singularity-Free Interpretation of the Thermodynamics of Supercooled Water," Phys. Rev. E **53**, 6144–6154 (1996). Citations: 125
39. H. E. Stanley, J. Teixeira, A.Geiger, and R.L.Blumberg, "Interpretation of the unusual behavior of H_2O and D_2O at low temperature: Are Concepts of Percolation Relevant to the 'Puzzle of Liquid Water'?" Invited Talk, INTERN'L CONFERENCE ON STATISTICAL MECHANICS, published in Physica A **106**, 260–277 (1981).
40. H. E. Stanley, "A Polychromatic Correlated-Site Percolation Problem with Possible Relevance to the Unusual behavior of Supercooled H_2O and D_2O," J. Phys. A **12**, L329-L337 (1979).
41. H. E. Stanley and J. Teixeira, "Interpretation of The Unusual Behavior of H_2O and D_2O at Low Temperatures: Tests of a Percolation Model," J. Chem. Phys. **73**, 3404–3422 (1980).
42. A. Geiger and H. E. Stanley, "Low-density patches in the hydrogen-bonded network of liquid water: Evidence from molecular dynamics computer simulations" *Phys. Rev. Lett.* **49**, 1749–1752 (1982).
43. J. R. Errington, P. G. Debenedetti, and S. Torquato, "Cooperative Origin of Low-Density Domains in Liquid Water," Phys. Rev. Lett. **89**, 215503 (2002).
44. P. H. Poole, F. Sciortino, U. Essmann, H. E. Stanley, "Phase Behavior of Metastable Water," *Nature* **360**, 324–328 (1992).
45. E. G. Ponyatovskii, V. V. Sinitsyn and T. A. Pozdnyakova, "Second Critical Point and Low-Temperature Anomalies in the Physical Properties of Water," *JETP Lett.* **60**, 360–364 (1994).
46. C. T. Moynihan, "Two Species/Nonideal Solution Model for Amorphous/Amorphous Phase Transition." *Mat. Res. Soc. Symp. Proc.* **455**, 411–425 (1997).
47. P. H. Poole, F. Sciortino, T.Grande, H. E. Stanley and C. A. Angell, "Effect of Hydrogen Bonds on the

Thermodynamic Behavior of Liquid Water," Phys. Rev. Lett. **73**, 1632–1635 (1994).
48. S. S. Borick, P. G. Debenedetti, and S. Sastry, "A Lattice Model of Network-Forming Fluids with Orientation-Dependent Bonding: Equilibrium, Stability, and Implications from the Phase Behavior of Supercooled Water," *J. Phys. Chem.* **99**, 3781–3793 (1995).
49. C. F. Tejero and M. Baus, "Liquid Polymorphism of Simple Fluids within a van der Waals Theory," *Phys. Rev. E* **57**, 4821–4823 (1998).
50. G. Franzese and H. E. Stanley, "A theory for discriminating the mechanism responsible for the water density anomaly", *Physica A* **314**, 508 (2002).
51. G. Franzese and H. E. Stanley, "Liquid-liquid critical point in a Hamiltonian model for water: analytic solution", *J. Phys.: Cond. Mat.* **14**, 2193 (2002).
52. G. Franzese, M. I. Marqués, and H. E. Stanley, "Intramolecular coupling as a mechanism for a liquid-liquid phase transition", *Phys. Rev. E.* **67**, 011103 (2003).
53. G. Franzese and H. E. Stanley, "The Widom line of supercooled water", *J. Phys.: Cond. Mat.* **19**, 205126 (2007).
54. H. Kanno, R. Speedy, and C. A. Angell, "Supercooling of Water to -92^oC under Pressure," *Science* **189**, 880–881 (1975).
55. O. Mishima, "Reversible first-order transition between two H_2O amorphs at -0.2 GPa and 135 K," *J. Chem. Phys.* **100**, 5910–5912 (1994).
56. O. Mishima and H. E. Stanley "Decompression-Induced Melting of Ice IV and the Liquid-Liquid Transition in Water," *Nature* **392**, 164–168 (1998).
57. E. Whalley, D. D. Klug, and Y. P. Handa, "Entropy of Amorphous Ice," *Nature* **342**, 782–783 (1989).
58. G. P. Johari, G. Fleissner, A. Hallbrucker, and E. Mayer, "Thermodynamic Continuity between Glassy and Normal Water," *J. Phys. Chem.* **98**, 4719–4725 (1994).
59. R. J. Speedy, P. G. Debenedetti, R. S. Smith, C. Huang, and B. D. Kay, "The Evaporation Rate, Free Energy, and Entropy of Amorphous Water at 150K," *J. Chem. Phys.* **105**, 240–244 (1996).
60. L. S. Bartell and J. Huang, "Supercooling of Water below the Anomalous Range near 226 K," *J. Phys. Chem.* **98**, 7455–7457 (1994).
61. P. Brüggeller and E. Mayer, Complete Vitrification in Pure Liquid Water and Dilute Aqueous Solutions," *Nature* **288**, 569–571 (1980).
62. P. W. Bridgman, The Pressure-Volume-Temperature Relations of the Liquid, and the Phase Diagram of Heavy Water," *J. Chem. Phys.* **3**, 597–605 (1935).
63. P. H. Poole, U. Essmann, F. Sciortino, and H. E. Stanley, "Phase Diagram for Amorphous Solid Water," *Phys. Rev. E* **48**, 4605–4610 (1993).
64. H. Tanaka, "Phase Behaviors of Supercooled Water: Reconciling a Critical Point of Amorphous Ices with Spinodal Instability," *J. Chem. Phys.* **105**, 5099–5111 (1996).
65. S. Harrington, R. Zhang, P. H. Poole, F. Sciortino, and H. E. Stanley, "Liquid-Liquid Phase Transition: Evidence from Simulations," *Phys. Rev. Lett.* **78**, 2409–2412 (1997).
66. F. Sciortino, P. H. Poole, U. Essmann, and H. E. Stanley, "Line of Compressibility Maxima in the Phase Diagram of Supercooled Water," *Phys. Rev. E* **55**, 727–737 (1997).
67. S. Harrington, P. H. Poole, F. Sciortino, and H. E. Stanley, "Equation of State of Supercooled SPC/E Water," J. Chem. Phys. **107**, 7443–7450 (1997).
68. W. L. Jorgensen, J. Chandrasekhar, J. Madura, R. W. Impey, and M. Klein, "Comparison of Simple Potential Functions for Simulating Liquid Water," *J. Chem. Phys.* **79**, 926 (1983).
69. M. Yamada, S. Mossa, H. E. Stanley, F. Sciortino, "Interplay Between Time-Temperature-Transformation and the Liquid-Liquid Phase Transition in Water," Phys. Rev. Lett. **88**, 195701 (2002).
70. D. Paschek, "How the Liquid-Liquid Transition Affects Hydrophobic Hydration in Deeply Supercooled Water," Phys. Rev. Lett. **94**, 217802 (2005).
71. E. Shiratani and M. Sasai, "Molecular Scale Precursor of the Liquid-Liquid Phase Transition of Water," *J. Chem. Phys.* **108**, 3264–3276 (1998).
72. M.-C. Bellissent-Funel, "Is There a Liquid-Liquid Phase Transition in Supercooled Water?" *Europhys. Lett.* **42**, 161–166 (1998).
73. H. E. Stanley, S. V. Buldyrev, M. Canpolat, O. Mishima, M. R. Sadr-Lahijany, A. Scala, and F. W. Starr, "The Puzzling Behavior of Water at Very Low Temperature" [opening paper in Proc. International Meeting on Metastable Fluids, Bunsengesellschaft] Physical Chemistry and Chemical Physics (PCCP) **2**, 1551–1558 (2000).
74. A. K. Soper and M. A. Ricci, "Structures of High-Density and Low-Density Water", *Phys. Rev. Lett.* **84**, 2881 (2000) and references cited therein.
75. A. C. Mitus, A. Z. Patashinskii, and B. I. Shumilo, "The Liquid-Liquid Phase Transition," *Phys. Lett.* **113A**, 41–44 (1985).
76. A. C. Mitus and A. Z. Patashinskii, "The Liquid-Liquid Phase Transition. 1. Statistical-Mechanics Description," *Acta Physica Polonica A* **74**, 779–796 (1988).
77. R. Zangi and A. E. Mark, "Bilayer Ice and Alternate Liquid Phases of Confined Water," J. Chem. Phys. **119**, 1694–1700 (2003).
78. R. Zangi, "Water Confined to a Slab Geometry: A Review of Recent Computer Simulation Studies," J. Phys.-Cond. Mat. **16**, S5371–S5388 (2004).
79. P. M. Wiggins, "Role of Water in Some Biological Processes," *Microbiological Reviews* **54**, 432–449 (1990).
80. P. M. Wiggins, "High and Low-Density Water in Gels," *Prog. Polym. Sci.* **20**, 1121–1163 (1995).
81. M.-C. Bellissent-Funel, J.-M. Zanotti, and S. H. Chen, "Slow Dynamics of Water Molecules on the Surface of a Globular Protein," *Faraday Discuss.* **103**, 281–294 (1996).
82. V. Crupi, S. Magazu, D. Majolino, P. Migliardo, V. Venuti, and M.-C. Bellissent-Funel, "Confinement Influence in Liquid Water Studied by Raman and Neutron Scattering," *J. Phys. Cond. Matter* **12**, 3625–3630 (2000);
83. M.-C. Bellissent-Funel, "Hydration in Protein Dynamics and Function," *J. Mol. Liq.* **84**, 39–52 (2000).
84. M.-C. Bellissent-Funel, S. H. Chen, and J. M. Zanotti, "Single-Particle Dynamics of Water-Molecules in Confined Space," *Phys. Rev. E* **51**, 4558–4569 (1995).
85. M.-C. Bellissent-Funel, J. Teixeira, K. F. Bradley, and S.

H. Chen, "Dynamics of Hydration Water in Protein," *J. de Physique I* **2**, 995–1001 (1992).
86. S. H. Chen, P. Gallo, and M.-C. Bellissent-Funel, "Slow Dynamics of Interfacial Water," *Canadian J. Phys.* **73**, 703–709 (1995).
87. Z. Dohnálek, G. A. Kimmel, R. L. Ciolli, K. P. Stevenson, R. S. Smith, and B. D. Kay, "The Effect of the Underlying Substrate on the Crystallization Kinetics of Dense Amorphous Solid Water Films," *J. Chem. Phys.* **112**, 5932–5941 (2000).
88. T. M. Truskett, P. G. Debenedetti, and S. Torquato, "Thermodynamic Implications of Confinement for a Waterlike Fluid," J. Chem. Phys. **114**, 2401–2418 (2001).
89. J.-M. Zanotti, M.-C. Bellissent-Funel, and S.-H. Chen, "Experimental Evidence of a Liquid-Liquid Transition in Interfacial Water," Europhys. Lett. **71**, 91–97 (2005).
90. L. Liu, S.-H. Chen, A. Faraone, C.-W. Yen, and C. Y. Mou, "Pressure Dependence of Fragile-to-Strong Transition and a Possible Second Critical Point in Supercooled Confined Water," Phys. Rev. Lett. **95**, 117802 (2005).
91. M. E. Green and J. Lu, "Monte-Carlo Simulation of the Effects of Charges on Water and Ions in a Tapered Pore," J. Coll. Int. Sci. **171**, 117–126 (1995).
92. K. Koga, X. C. Zeng, and H. Tanaka, "Freezing of Confined Water: A Bilayer Ice Phase in Hydrophobic Nanopores," *Phys. Rev. Lett.* **79**, 5262–5265 (1997).
93. J. Slovak, K. Koga, H. Tanaka, and X. C. Zeng, "Confined Water in Hydrophobic Nanopores: Dynamics of Freezing into Bilayer Ice," *Phys. Rev. E* **60**, 5833–5840 (1999).
94. K. Koga, X. C. Zeng, and H. Tanaka, "Effects of Confinement on the Phase Behavior of Supercooled Water," *Chem. Phys. Lett.* **285**, 278–283 (1998).
95. K. Koga, H. Tanaka, and X. C. Zeng, "First-Order Transition in Confined Water between High-Density Liquid and Low-Density Amorphous Phases," to appear in *Nature*.
96. R. Bergman and J. Swenson, "Dynamics of Supercooled Water in Confined Geometry," *Nature* **403**, 283–286 (2000).
97. J. Teixeira, J. M. Zanotti, M.-C. Bellissent-Funel, and S. H. Chen, "Water in Confined Geometries," *Physica B* **234**, 370–374 (1997).
98. P. Gallo, "Single Particle Slow Dynamics of Confined Water," *Phys. Chem. Phys.* **2**, 1607–1611 (2000).
99. P. Gallo, M. Rovere, M. A. Ricci, C. Hartnig, and E. Spohr, "Non-Exponential Kinetic Behavior of Confined Water," *Europhys. Lett.* **49**, 183–188 (2000).
100. P. Gallo, M. Rovere, M. A. Ricci, C. Hartnig, and E. Spohr, "Evidence of Glassy Behavior of Water Molecules in Confined States," *Philos. Mag. B* **79**, 1923–1930 (1999).
101. M. Rovere, M. A. Ricci, D. Vellati, and F. Bruni, "A Molecular Dynamics Simulation of Water Confined in a Cylindrical SiO_2 Pore," *J. Chem. Phys.* **108**, 9859–9867 (1998).
102. M.-C. Bellissent-Funel, R. Sridi-Dorbez, and L. Bosio, "X-Ray and Neutron Scattering Studies of the Structure of Water at a Hydrophobic Surface," *J. Chem. Phys.* **104**, 10023–10029 (1996).
103. J. Forsman, B. Jonsson, and C. E. Woodward, "Computer Simulations of Water between Hydrophobic Surfaces: The Hydrophobic Force," *J. Phys. Chem-US* **100**, 15005–15010 (1996).
104. M. Meyer and H. E. Stanley, "Liquid-Liquid Phase Transition in Confined Water: A Monte Carlo Study," *J. Phys. Chem. B* **103**, 9728–9730 (1999).
105. P. A. Netz and T. Dorfmuller, "Computer Simulation Studies on the Polymer-Induced Modification of Water Properties in Polyacrylamide Hydrogels," *J. Phys. Chem. B* **102**, 4875–4886 (1998).
106. J.-M. Zanotti, M.-C. Bellissent-Funel, and S. H. Chen, "Relaxational Dynamics of Supercooled Water in Porous Glass," *Phys. Rev. E* **59**, 3084–3093 (1999).
107. M.-C. Bellissent-Funel and J. Teixeira, "Structural and Dynamic Properties of Bulk and Confined Water," in *Freeze-Drying/Lyophilization of Pharmaceutical and Biological Products*, edited by L. Rey and J. C. May (Marcel Dekker, New York, 1999), Chapter 3, pp. 53–77.
108. H. Tanaka and I. Ohmine, "Large Local Energy Fluctuations in Water," *J. Chem. Phys.* **87**, 6128–6139 (1987).
109. I. Ohmine, H. Tanaka, and P.. G. Wolynes, "Large Local Energy Fluctuations in Water. II. Cooperative Motions and Fluctuations," *J. Chem. Phys.* **89**, 5852–5860 (1988).
110. H. Tanaka and I. Ohmine, "Potential Energy Surfaces for Water Dynamics: Reaction Coordinates, Transition States, and Normal Mode Analyses," *J. Chem. Phys.* **91**, 6318–6327 (1989).
111. I. Ohmine and H. Tanaka, "Potential Energy Surfaces for Water Dynamics. II. Vibrational Mode Excitations, Mixing, and Relaxations," *J. Chem. Phys.* **93**, 8138–8147 (1990).
112. I. Ohmine and H. Tanaka, "Dynamics of Liquid Water: Fluctuations and Collective Motions," in *Molecular Dynamics Simulations*, edited by F. Yonezawa (Springer Verlag, Berlin, 1991), pp. 130–138.
113. I. Okabe, H. Tanaka, and K. Nakanishi, "Structure and Phase Transitions of Amorphous Ices," *Phys. Rev. E* **53**, 2638–2647 (1996).
114. H. Tanaka, "Phase Diagram for Supercooled Water and Liquid-Liquid Transition," in *ACS Symposium Series on Experimental and Theoretical Approaches to Supercooled Liquids*, edited by J. Fourkas (ACS, 1997), Chap. 18, pp. 233–245.
115. T. Kabeya, Y. Tamai, and H. Tanaka, "Structure and Potential Surface of Liquid Methanol in Low Temperature: A Comparison of Hydrogen Bond Network in Methanol with Water," *J. Phys. Chem. B* **102**, 899–905 (1998).
116. Y. Tamai and H. Tanaka, "Effects of Chain on Structure and Dynamics of Supercooled Water in Hydrogel," *Phys. Rev. E* **59**, 5647–5654 (1999).
117. H. Tanaka, R. Yamamoto, K. Koga, and X. C. Zeng, "Can Thin Disk-Like Clusters Be More Stable Than Compact Droplet-Like Clusters?" *Chem. Phys. Lett.* **304**, 378–384 (1999).
118. J. Slovak, K. Koga, H. Tanaka, and X. C. Zeng, "Confined Water in Hydrophobic Nanopores: Dynamics of Freezing into Bilayer Ice," *Phys. Rev. E* **60**, 5833–5840 (1999).
119. G. T. Gao and X. C. Zeng, and H. Tanaka, "The Melting Temperature of Proton-Disordered Hexagonal Ice: A Computer Simulation of TIP4P Model of Water," *J.*

Chem. Phys. **112**, 8534–8538 (2000).
120. M.-C. Bellissent-Funel, "Status of Experiments Probing the Dynamics of Water in Confinement," Eur. Phys. J. E **12**, 83–92 (2003).
121. M.-C. Bellissent-Funel, "Hydrophilic-Hydrophobic Interplay: From Model Systems to Living Systems," C. R. Geoscience **337**, 173–179 (2005).
122. P. C. Hemmer and G. Stell, "Fluids with Several Phase Transitions," *Phys. Rev. Lett.* **24**, 1284–1287 (1970).
123. M. Canpolat, F. W. Starr, M. R. Sadr-Lahijany, A. Scala, O. Mishima, S. Havlin and H. E. Stanley, "Local Structural Heterogeneities in Liquid Water under Pressure," Chem. Phys. Lett. **294**, 9–12 (1998).
124. S. Sastry, F. Sciortino, and H. E. Stanley, "Limits of Stability of the Liquid Phase in a Lattice Model with Water-Like Properties," *J. Chem. Phys.* **98**, 9863–9872 (1993).
125. C. J. Roberts, A. Z. Panagiotopulos, and P. G. Debenedetti, "Liquid-Liquid Immiscibility in Pure Fluids: Polyamorphism in Simulations of a Network-Forming Fluid," *Phys. Rev. Lett.* **77**, 4386–4389 (1996).
126. E. Shiratani and M. Sasai, "Growth and Collapse of Structural Patterns in the Hydrogen Bond Network in Liquid Water," *J. Chem. Phys.* **104**, 7671–7680 (1996).
127. H. Tanaka, "Fluctuation of Local Order and Connectivity of Water Molecules in Two Phases of Supercooled Water," *Phys. Rev. Lett.* **80**, 113–116 (1998).
128. C. A. Angell, J. Shuppert, and J. C. Tucker, "Anomalous Properties of Supercooled Water: Heat Capacity, Expansivity, and Proton Magnetic Resonance Chemical Shift from 0 to -38 C," *J. Phys. Chem.* **77**, 3092–3099 (1973).
129. F. Sciortino, L. Fabbian, S.-H. Chen, and P. Tartaglia, "Supercooled Water and the Kinetic Glass Transition: 2. Collective Dynamics," *Phys. Rev. E* **56**, 5397–5404 (1997).
130. F. Sciortino, P. Gallo, P. Tartaglia, and S.-H. Chen, "Supercooled Water and the Glass Transition," *Phys. Rev. E* **54**, 6331–6343 (1996).
131. P. Kumar, G. Franzese, S. V. Buldyrev, H. E. Stanley, "Molecular Dynamics Study of Orientational Cooperativity in Water", *Phys. Rev. E* **73**, 041505 (2006).
132. Y. Xie, K. F. Ludwig, G. Morales, D. E. Hare, and C. M. Sorensen, "Noncritical Behavior of Density Fluctuations in Supercooled Water," *Phys. Rev. Lett.* **71**, 2051–2053 (1993).
133. F. Sciortino, P. Poole, H.E.Stanley and S. Havlin, "Lifetime of the Hydrogen Bond Network and Gel-Like Anomalies in Supercooled Water," *Phys. Rev. Lett.* **64**, 1686–1689 (1990).
134. A. Luzar and D. Chandler, "Hydrogen-Bond Kinetics in Liquid Water," *Nature* **379**, 55–57 (1996); A. Luzar and D. Chandler, "Effect of Environment on Hydrogen Bond Dynamics in Liquid Water," *Phys. Rev. Lett.* **76**, 928–931 (1996).
135. F. W. Starr, J. K. Nielsen, and H. E. Stanley. "Fast and Slow Dynamics of Hydrogen Bonds in Liquid Water," Phys. Rev. Lett. **82**, 2294–2297 (1999); F. W. Starr, J. K. Nielsen, and H. E. Stanley, "Hydrogen Bond Dynamics for the Extended Simple Point Charge Model of Water," Phys. Rev. E **62**, 579–587 (2000).
136. J. R. Errington and P. G. Debenedetti, "Relationship between Structural Order and the Anomalies of Liquid Water," Nature **409**, 318–321 (2001).
137. J. M. Kincaid, G. Stell and C. K. Hall, "Isostructural Phase Transitions Due to Core Collapse: I. A One-Dimensional Model," *J. Chem. Phys.* **65**, 2161 (1976).
138. E. A. Jagla, "A Model for the Fragile-to-Strong Transition in Water," *J. Phys. Cond. Mat.* **11**, 10251–10258 (1999).
139. E. A. Jagla, "Low-Temperature Behavior of Core-Softened Models: Water and Silica Behavior," Phys. Rev. E **63**, 061509 (2001).
140. M. R. Sadr-Lahijany, A. Scala, S. V. Buldyrev and H. E. Stanley, "Liquid State Anomalies for the Stell-Hemmer Core-Softened Potential," *Phys. Rev. Lett.* **81**, 4895–4898 (1998).
141. A. Scala, M. R. Sadr-Lahijany, N. Giovambattista, S. V. Buldyrev, and H. E. Stanley, "Water-Like Anomalies for Core-Softened Models of Fluids: Two Dimensional Systems," *Phys. Rev. E* **63** 041202 (2001).
142. A. Scala, M. Reza Sadr-Lahijany, N. Giovambattista, S. V. Buldyrev, and H. E. Stanley, "Applications of the Stell-Hemmer Potential to Understanding Second Critical Points in Real Systems," [Festschrift for G. S. Stell] J. Stat. Phys. **100**, 97–106 (2000).
143. G. Franzese, G. Malescio, A. Skibinsky, S. V. Buldyrev, and H. E. Stanley, "Generic Mechanism for Generating a Liquid-Liquid Phase Transition," *Nature* **409**, 692–695 (2001).
144. G. Franzese, G. Malescio, A. Skibinsky, S. V. Buldyrev, and H. E. Stanley, "Metastable Liquid-Liquid Phase Transition in a Single-Component System with only one Crystal Phase and No Density Anomaly," *Phys. Rev. E* **66**, 051206 (2002).
145. A. Skibinsky, S.V. Buldyrev, G. Franzese, G. Malescio, and H. E. Stanley, "Liquid-liquid phase transitions for soft-core attractive potentials", *Phys. Rev. E* **69**, 061206 (2004).
146. G. Malescio, G. Franzese, A. Skibinsky, S. V. Buldyrev, and H. E. Stanley, "Liquid-liquid phase transition for an attractive isotropic potential with wide repulsive range", *Phys. Rev. E* **71**, 061504 (2005).
147. G. Franzese, "Differences between discontinuous and continuous soft-core attractive potentials: the appearance of density anomaly", *J. Mol. Liq.* **136**, 267 (2007).
148. P. Kumar, G. Franzese, and H. E. Stanley, "Predictions of Dynamic Behavior Under Pressure for Two Scenarios to Explain Water Anomalies" (submitted), cond-mat/0702108.
149. G. D'Arrigo et al., J. Chem. Phys. **75**, 4264 (1981); C. A. Angell and V. Rodgers *ibid.* **80**, 6245 (1984).
150. E. Schwegler et al., Phys. Rev. Lett. **84**, 2429 (2000); P. Raiteri et al., *ibid.* **93**, 087801 (2004) and references cited therein.
151. P. Kumar, S. V. Buldyrev, F. Sciortino, E. Zaccarelli, and H. E. Stanley, "Static and Dynamic Anomalies in a Repulsive Spherical Ramp Liquid: Theory and Simulation," Phys. Rev. E **72**, 021501 (2005).
152. P. G. Debenedetti, V. S. Raghavan, and S. S. Borick, "Spinodal Curve of some Supercooled Liquids," *J. Phys. Chem.* **95**, 4540–4551 (1991).
153. V. B. Henriques and M. C. Barbosa, "Liquid Polymorphism and Density Anomaly in a Lattice

Gas Model," Phys. Rev. E **71**, 031504 (2005).

154. B. Guillot and Y. Guissani, "Polyamorphism in Low Temperature Water: A Simulation Study," J. Chem. Phys. **119**, 11740–11752 (2003).

155. V. B. Henriques, N. Guisoni, M. A. Barbosa, M. Thielo, and M. C. Barbosa, "Liquid Polyamorphism and Double Criticality in a Lattice Gas Model," Molec. Phys. **103**, 3001–3007 (2005).

156. A. Barros de Oliveira, G. Franzese, P. A. Netz, and M. C. Barbosa, "Water-like hierarchy of anomalies in a continuous spherical shouldered potential", *J. Chem. Phys.* (in press), arXiv:0706.2838.

157. E. A. Jagla, "Core-Softened Potentials and the Anomalous Properties of Water," *J. Chem. Phys.* **111**, 8980–8986 (1999).

158. T. H. Hall, L. Merril, and J. D. Barnett, "High Pressure Polymorphism in Cesium," *Science* **146**, 1297–1299 (1964).

159. A. Scala, F. W. Starr, E. La Nave, F. Sciortino and H. E. Stanley, "Configurational Entropy and Diffusivity of Supercooled Water," Nature **406**, 166–169 (2000).

160. Z. Yan, S. V. Buldyrev, N. Giovambattista, and H. E. Stanley "Structural Order for One-Scale and Two-Scale Potentials," Phys. Rev. Lett. **95**, 130604 (2005).

161. Z. Yan, S. V. Buldyrev, N. Giovambattista, P. G. Debenedetti, and H. E. Stanley, "Family of Tunable Spherically-Symmetric Potentials that Span the Range from Hard Spheres to Water-like Behavior," Phys. Rev. E **73**, 051204 (2006).

162. Z. Yan, S. V. Buldyrev, P. Kumar, N. Giovambattista, P. G. Debenedetti, and H. E. Stanley, "Structure of the First- and Second-Neighbor Shells of Simulated Water: Quantitative Relation to Translational and Orientational Order," Phys. Rev. E **76**, 051201 (2007).

163. N. Giovambattista, P. J. Rossky, and P. G. Debenedetti, "Effect of Pressure on the Phase Behavior and Structure of Water Confined between Nanoscale Hydrophobic and Hydrophilic Plates," Phys. Rev. E **73**, 041604 (2006).

164. C. A. Angell, "Amorphous Water," *Ann. Rev. Phys. Chem.* **55**, 559–583 (2004).

165. P. H. Poole, T. Grande, C. A. Angell, and P. F. McMillan, "Polymorphic Phase Transitions in Liquids and Glasses," Science **275**, 322–323 (1997).

166. J. L. Yarger, and G. H. Wolf, "Chemistry: Polymorphism in Liquids," Science **306**, 820–821 (2004).

167. F. Mallamace, M. Broccio, C. Corsaro, A. Faraone, U. Wanderlingh, L. Liu, C.-Y. Mou, and S.-H. Chen, "The Fragile-to-Strong Dynamic Crossover Transition in Confined Water: NMR Results," J. Chem. Phys. **124**, 161102 (2006).

168. L. Xu, P. Kumar, S. V. Buldyrev, S.-H. Chen, P. H. Poole, F. Sciortino, and H. E. Stanley, "Relation between the Widom Line and the Dynamic Crossover in Systems with a Liquid-Liquid Critical Point," Proc. Natl. Acad. Sci. **102**, 16558–16562 (2005).

169. A. Faraone, L. Liu, C.-Y. Mou, C.-W. Yen, and S.-H. Chen, "Fragile-to-Strong Liquid Transition in Deeply Supercooled Confined Water," J. Chem. Phys. **121**, 10843–10846 (2004).

170. L. Liu, "Study of Slow Dynamics in Supercooled Water by Molecular Dynamics and Quasi-Elastic Neutron Scattering," Ph.D. thesis, M.I.T., September 2005.

171. J. M. H. Levelt, *Measurements of the Compressibility of Argon in the Gaseous and Liquid Phase*, Ph.D. Thesis (University of Amsterdam, Van Gorkum and Co., 1958).

172. M. A. Anisimov, J. V. Sengers, and J. M. H. Levelt-Sengers, "Near-Critical Behavior of Aqueous Systems," in *Aqueous System at Elevated Temperatures and Pressures: Physical Chemistry in Water, Stream and Hydrothermal Solutions*, edited by D. A. Palmer, R. Fernandez-Prini, and A. H. Harvey (Elsevier, Amsterdam, 2004).

173. M.-C. Bellissent-Funel, ed., *Hydration Processes in Biology: Theoretical and Experimental Approaches* [Proc. NATO Advanced Study Institutes, Vol. 305] (IOS Press, Amsterdam, 1999).

174. G. W. Robinson, S.-B. Zhu, S. Singh, and M. W. Evans, *Water in Biology, Chemistry, and Physics: Experimental Overviews and Computational Methodologies* (World Scientific, Singapore, 1996).

175. C. A. Angell, R. D. Bressel, M. Hemmatti, E. J. Sare, and J. C. Tucker, "Water and Its Anomalies in Perspective: Tetrahedral Liquids With and Without Liquid-Liquid Phase Transitions," Phys. Chem. Chem. Phys. **2**, 1559–1566 (2000).

176. P. G. Debenedetti, "Supercooled and Glassy Water," J. Phys.: Condens. Matter **15**, R1669–R1726 (2003).

177. C. A. Angell, "Water-II is a Strong Liquid," *J. Phys. Chem.* **97**, 6339–6341 (1993).

178. F. W. Starr, C. A. Angell, and H. E. Stanley, "Prediction of Entropy and Dynamic Properties of Water below the Homogeneous Nucleation Temperature," Physica A **323**, 51–66 (2003).

179. J. Horbach and W. Kob, "Static and Dynamic Properties of a Viscous Silica Melt," *Phys. Rev. B* **60**, 3169–3181 (1999).

180. E. W. Lang and H. D. Lüdemann, "Anomalies of Liquid Water," *Angew Chem. Intl. Ed. Engl.* **21**, 315–329 (1982).

181. F. X. Prielmeier, E. W. Lang, R. J. Speedy, H. D. Lüdemann, "Diffusion in Supercooled Water to 300-MPA," *Phys. Rev. Lett.* **59**, 1128–1131 (1987) .

182. K. Ito, C. T. Moynihan, and C. A. Angell, "Thermodynamic Determination of Fragility in Liquids and a Fragile-to-Strong Liquid Transition in Water," *Nature* **398**, 492–495 (1999).

183. H. Tanaka, "A New Scenario of the Apparent Fragile-to-Strong Transition in Tetrahedral Liquids: Water as an Example," *J. Phys.: Condens. Matter* **15**, L703–L711 (2003).

184. J. Swenson, H. Jansson, W. S. Howells, and S. Longeville, "Dynamics of Water in a Molecular Sieve by Quasielastic Neutron Scattering," J. Chem. Phys. **122**, 084505 (2005).

185. S. Sastry and C. A. Angell, "Liquid-Liquid Phase Transition in Supercooled Silicon," *Nature Materials* **2**, 739–743 (2003).

186. I. Saika-Voivod, P. H. Poole, and F. Sciortino, "Fragile-to-Strong Transition and Polyamorphism in the Energy Landscape of Liquid Silica," *Nature* **412**, 514–517 (2001).

187. D. C. Rapaport, in *The Art of Molecular Dynamics Simulation* (Cambridge University Press, Cambridge, 1995).

188. F. H. Stillinger and A. Rahman, "Molecular Dynamics Study of Temperature Effects on Water Structure and

Kinetics," *J. Chem. Phys.* **57**, 1281–1292 (1972).
189. W. Götze and L. Sjögren, "Relaxation Processes in Supercooled Liquids," *Rep. Prog. Phys.* **55**, 241–376 (1992).
190. P. H. Poole, I. Saika-Voivod, and F. Sciortino, "Density Minimum and Liquid-Liquid Phase Transition," J. Phys.: Condens. Matter **17**, L431–L437 (2005).
191. N. Jakse, L. Hennet, D. L. Price, S. Krishnan, T. Key, E. Artacho, B. Glorieux, A. Pasturel, and M.-L. Saboungi, "Structural Changes on Supercooling Liquid Silicon," Appl. Phys. Lett. **83**, 4734–4736 (2003).
192. L. Xu, F. Mallamace, F. W. Starr, Z. Yan, S. V. Buldyrev, and H. E. Stanley, "Interpretation for the Breakdown of Stokes-Einstein Relation in Water" (preprint).
193. P. Kumar, S. V. Buldyrev, S. L. Becker, P. H. Poole, F. W. Starr, and H. E. Stanley, "Relation between the Widom line and the Breakdown of the Stokes–Einstein Relation in Supercooled Water" Proc. Natl. Acad. Sci. **104**, 9575–9579 (2007).
194. P. Kumar, S. V. Buldyrev, and H. E. Stanley "Space-Time Correlations in the Orientational Order Parameter and the Orientational Entropy of Water" (submitted).
195. S.-H. Chen, F. Mallamace, C.-Y. Mou, M. Broccio, C. Corsaro, and A. Faraone, "The Violation of the Stokes-Einstein Relation in Supercooled Water," *Proc. Nat. Acad. Sciences USA* **103**, 12974–12978 (2006)
196. M. G. Mazza, N. Giovambattista, F. W. Starr, and H. E. Stanley, "Relation between Rotational and Translational Dynamic Heterogeneities in Water," Phys. Rev. Lett. **96**, 057803 (2006).
197. M. G. Mazza, N. Giovambattista, H. E. Stanley, and F. W. Starr, "Connection of Translational and Rotational Dynamical Heterogeneities with the Breakdown of the Stokes-Einstein and Stokes-Einstein-Debye Relations in Water," Phys. Rev. E **76**, 031202 (2007).
198. S.-H. Chen, L. Liu, E. Fratini, P. Baglioni, A. Faraone, and E. Mamontov, "The Observation of Fragile-to-strong Dynamic Crossover in Protein Hydration Water," *Proc. Natl. Acad. Sci. USA* **103**, 9012 (2006).
199. F. Mallamace, S.-H. Chen, M. Broccio, C. Corsaro, V. Crupi, D. Majolino, V. Venuti, P. Baglioni, E. Fratini, C. Vannucci, and H. E. Stanley, "The Role of the Solvent in the Dynamical Transitions of Proteins: The Case of the Lysozyme-Water System," J. Chem. Phys. **127**, 045104 (2007); F. Mallamace, C. Branca, M. Broccio, C. Corsaro, N. Gonzalez-Segredo, H. E. Stanley, and S.-H. Chen, "Transport Properties of Supercooled Confined Water," Euro. Phys. J. **xx**, xx (2008).
200. P. Kumar, F. W. Starr, S. V. Buldyrev, and H. E. Stanley, "Effect of Water-Wall Iteraction Potential on the properties of Nanoconfined Water," Phys. Rev. E **75**, 011202 (2007).
201. P. Kumar, S. V. Buldyrev, and H. E. Stanley, "Dynamic Crossover and Liquid-Liquid Critical Point in the TIP5P Model of Water," in *Soft Matter under Extreme Pressures: Fundamentals and Emerging Technologies*, edited by S. J. Rzoska and V. Mazur [Proc. NATO ARW, Odessa, Oct. 2005] (Springer, Berlin, 2006).
202. D. Liu, Y. Zhang, C.-C. Chen, C.-Y. Mou, P. H. Poole, and S.-H. Chen, "Observation of the Density Minimum in Deeply Supercooled Confined Water," Proc. Natl. Acad. Sci. **104**, 9570–9574 (2007).
203. D. A. Fuentevilla and M. A. Anisimov, "Scaled Equation of State for Supercooled Wataer near the Liquid-Liquid Critical Point," Phys. Rev. Lett. **97**, 195702 (2006).
204. T. Morishita, "How Does Tetrahedral Structure Grow in Liquid Silicon on Supercooling?" *Phys. Rev. Lett.* **97** (2006) 165502.
205. L. Xu, S. V. Buldyrev, C. A. Angell, and H. E. Stanley, "Thermodynamics and Dynamics of the Two-Scale Spherically Symmetric Jagla Ramp Model of Anomalous Liquids," *Phys. Rev. E* **74** 031108 (2006).
206. F. Mallamace, C. Branca, M. Broccio, C. Corsaro, C.-Y. Mou, and S.-H. Chen, "The Anomalous Behavior of the Density of Water in the Range 30 K $< T <$ 373 K," C. Y. Mou, and S.-H. Chen, Proc. Natl. Acad. Sci. USA **104**, 18387–18391 (2007).
207. F. Mallamace, C. Corsaro, M. Broccio, C. Branca, N. González-Segredo, J. Spooren, S.-H. Chen, and H. E. Stanley, "NMR Evidence of a Sharp Change in a Measure of Local Order in Deeply Supercooled Water" (in press).
208. S.-H. Chen, F. Mallamace, L. Liu, D. Z. Liu, X. Q. Chu, Y. Zhang, C. Kim. A. Faraone, C.-Y. Mou, E. Fratini, P. Baglioni, A. I. Kolesnikov, and V. Garcia-Sakai, "Dynamic Crossover Phenomenon in Confined Supercooled water and its relation to the existence of a liquid-liquid critical point in water," AIP Conf Proc. (in press).
209. S. Maruyama, K. Wakabayashi, and M. Oguni, *AIP Conf. Proc.* **708**, 675–676 (2004).
210. U. Raviv, P. Laurat, and J. Klein, Fluidity of water confined to subnanometre films, Nature 413 (2001) 51–54.
211. P. Kumar, Z. Yan, L. Xu, M. G. Mazza, S. V. Buldyrev, S.-H. Chen. S. Sastry, and H. E. Stanley, "Glass Transition in Biomolecules and the Liquid-Liquid Critical Point of Water," Phys. Rev. Lett. **97**, 177802 (2006).
212. J. M. Zanotti, M.-C. Bellissent-Funel, and J. Parrello, *Biophys. J.* **76**, 2390 (1999).
213. D. Ringe, G. A. Petsko, *Biophys. Chem.* **105**, 667 (2003).
214. J. Wang, P. Cieplak and P. A. Kollman, *J. Comp. Chem.* **21**, 1049 (2000); E. J. Sorin and V. S. Pande, *Biophys. J.* **88**, 2472 (2005).
215. B. F. Rasmussen, M. Ringe, and G. A. Petsko, *Nature* **357**, 423 (1992).
216. D. Vitkup, D. Ringe, G. A. Petsko, and M. Karplus, *Nat. Struct. Biol.* **7**, 34 (2000).
217. M. Yamada, S. Mossa, H. E. Stanley, F. Sciortino, Phys. Rev. Lett. **88**, 195701 (2002).
218. A. P. Sokolov, H. Grimm, A. Kisliuk and A. J. Dianoux, *J. Chem. Phys.* **110**, 7053 (1999).
219. W. Doster, S. Cusack, and W. Petry, *Nature* **338**, 754 (1989).
220. J. Norberg and L. Nilsson, *Proc. Natl. Acad. Sci. USA* **93**, 10173 (1996).
221. M. Tarek and D. J. Tobias, *Phys. Rev. Lett.* **88**, 138101 (2002); *Biophys. J.* **79**, 3244 (2000).
222. H. Hartmann, F. Parak, W. Steigemann, G. A. Petsko, D. R. Ponzi, H. Frauenfelder, *Proc. Natl. Acad. Sci. USA* **79**, 4067 (1982).
223. A. L. Tournier, J. Xu, and J. C. Smith, *Biophys. J.* **85**, 1871 (2003).
224. A. L. Lee and A. J. Wand, *Nature* **411** 501 (2001).

225. E. Lindahl, B. Hess, and D. van der Spoel, *J. Molecular Modeling* 7, 306 (2001).
226. P. J. Artymiuk, C. C. F. Blake, D. W. Rice, K. S. Wilson, *Acta Crystallogr.* **B 38**, 778 (1982).
227. H. R. Drew, R. M. Wing, T. Takano, C. Broka, S. Tanaka, K. Itakura and R. E. Dickerson, *Proc. Natl. Acad. Sci. USA* **78**, 2179 (1981).
228. S.-H. Chen, L. Liu, X. Chu, Y. Zhang, E. Frattini, P. Baglioni, A. Faraone, and E. Mamontov, *J. Chem. Phys.* **125**, 171103 (2006).

Theory of Multi-Time Correlations: Application to Pump-Probe Experiments of Reorientation in Anisotropic Cavities

I. Oppenheim* and T. Keyes†

*Department of Chemistry, MIT, Cambridge, MA 02139
†Department of Chemistry, Boston University, Boston, MA 02215

Abstract. A simple generalized Langevin theory is presented for multi-variable, multi-time correlations and applied to the four-dipole, two-time functions arising in pump-probe spectroscopy. More specifically, the case of dipole relaxation in an anisotropic coordinate frame which itself reorients very slowly with respect to the laboratory, representing a small molecule confined to the interior of a protein, is considered.

Keywords: Anisotropic rotation, Multi-time correlations, Nonlinear response, Nonlinear spectroscopy
PACS: 05.20.-y,61.20.Lc,68.49.Uv,78.47.+p

INTRODUCTION

There is a very large literature on equilibrium time correlation functions of two variables. Because only time differences are relevant in equilibrium, these are functions of one time argument. Very little is known about multi- (more than two) variable, multi- (more than one) time correlations, which provide more detailed probes of dynamics, notably of non-Gaussian effects and dynamical heterogeneity, and are required for interpretation of the current generation of non-linear pump-probe spectroscopic experiments.

Here we present a simple theory of multi-time correlations. For a first application, the correlations of four Cartesian components of a unit vector along the axis of a molecule reorienting in an anisotropic environment are considered.

The theoretical interpretation of non-linear spectroscopic experiments is based on non-linear response theory, which has been studied in classical systems by Oppenheim [1] and using quantum mechanics by Mukamel [2]. In both of these theories there are subtle contributions in higher order [1] which must be taken into account. This is not the main thrust of this article, however.

Spectra are expressed as averaged nested commutators of dipole or polarizability operators at different times, which can be expanded as linear combinations of time-ordered, computationally intractable quantal correlations. A classical approximation may be obtained by replacing the commutators with nested, multi-time Poisson brackets, which are also extremely challenging computationally. Thus, progress requires expressing the response functions in terms of multiple-time, primarily classical correlation functions. This has been done by Mukamel et al [2] for the quantal correlations, invoking correlations of the fluctuating transition frequency. Space, Keyes et al [3] relate the response function to a quantum-corrected, multi-time correlation of the classical dipole or polarizability.

We will focus our attention on how to obtain simple, accurate and straightforward expressions for the relevant multi-time correlation functions using mode coupling theory. These correlation functions have been studied by Mukamel et al [2, 4], Sillescu [5], Okumura [6], Reichman [7], Keyes et al [8], Jonas [9], and Ma and Stratt [10], among others. None of these authors take into account the subtleties that can arise with correlation functions involving products of fluctuating forces and their long-time behavior and therefore their results are, at best, approximate.

The only authors who have paid attention to these subtleties are von Zon and Schofield [11]. Our results are similar to theirs but are more straightforward and easier to use. Much of our notation will follow that in Ref [11] and we will make use of the properties of mode coupling theories described therein, and in previous papers by Machta and Oppenheim [12], Schofield, Lim and Oppenheim [13], and Schofield and Oppenheim [14]. The mode coupling techniques were introduced by Kawasaki [15] and utlized by Goetze et al [16] among others.

The techniques are useful when the system has slow and fast variables and we are interested in the time dependencies of the slower variables.

CP982, *Complex Systems, 5th International Workshop on Complex Systems*
edited by M. Tokuyama, I. Oppenheim, and H. Nishiyama

MULTI-TIME CORRELATION FUNCTIONS

We denote the set of linear slow variables by $\underset{\sim}{A}$. Usually the $\underset{\sim}{A}$ are collective variables, although the specific example considered is slow rotation of a single molecule. Individual linear modes, A_a, include the hydrodynamics conserved variables and any other slow variables, for all relevant wavevectors. Remembering that all the products of slow variables are slow, we introduce the multi-linear modes $\underset{\sim}{Q}_\ell$ where ℓ denotes the order of the mode and orthognalize the various orders so that $\langle \underset{\sim}{Q}_j \underset{\sim}{Q}_\ell \rangle = \Delta_{\ell j}$, where $\langle\rangle$ denotes an equilibrium ensemble average. With ℓ specified the mode is still a vector because it contains the different combinations of linear variables, including multiple 'intermediate' wavevectors. An individual mode is denoted $Q_{\alpha\ell}$. The summation convention will be employed in matrix and vector multiplication. In the general formalism we leave it implicit that, in an average of products, the rightmost variable should be complex conjugated.

The equation of motion is

$$\dot{\underset{\sim}{Q}}_\ell(t) = \underset{\sim}{M}_{\ell\ell'} \star \underset{\sim}{Q}_{\ell'}(t) + \underset{\sim}{K}_\ell(t), \tag{1}$$

where the fluctuating force, $\underset{\sim}{K}_\ell(t)$, is defined by

$$\underset{\sim}{K}_\ell(t) = e^{(1-P)Lt}(1-P)L\underset{\sim}{Q}_\ell, \tag{2}$$

where L is the Liouville operator and P is a projection operator onto the set of multi-linear slow variables such that, for an arbitrary dynamical variable B,

$$PB = \langle B\underset{\sim}{Q}_\ell\rangle \star \langle \underset{\approx}{QQ}\rangle^{-1}_{\ell\ell'} \star \underset{\sim}{Q}_{\ell'}. \tag{3}$$

Thus, $\underset{\sim}{K}_\ell(t)$ is orthogonal to all the slow variables in the sense that

$$\langle \underset{\sim}{K}_\ell(t)\underset{\sim}{Q}_{\ell'}\rangle = 0 \tag{4}$$

$$\langle \underset{\sim}{K}_\ell(t)\underset{\sim}{Q}_{\ell'}(\tau)\rangle = 0, \quad t-\tau > \tau_m, \tag{5}$$

and τ_m is a molecular (fast) relaxation time. Eq 5 can be generalized to

$$\langle \underset{\sim}{K}_\ell(t)B[\{\tau\}]\rangle = 0 \tag{6}$$

where B is any product of slow and/or fast variables at a variety of times $\{\tau\}$ as long as the latest time τ_{max} obeys the inequality, $t-\tau_{max} > \tau_m$. The quantity $\underset{\sim}{M}$ is the hydrodynamic matrix and is given, to a very good approximation, by

$$\underset{\sim}{M}_{\ell\ell'} = [\langle \dot{\underset{\sim}{Q}}_\ell \underset{\sim}{Q}_{\ell''}\rangle - \int_0^\infty d\tau \langle \underset{\sim}{K}_\ell(\tau)\underset{\sim}{K}_{\ell''}\rangle] \star \langle \underset{\approx}{QQ}\rangle^{-1}_{\ell''\ell'}. \tag{7}$$

The formal solution of Eq 1 is

$$\underset{\sim}{Q}_\ell(t) = e^{Mt}_{\ell\ell'} \star \underset{\sim}{Q}_{\ell'} + \int_0^t d\sigma e^{M(t-\sigma)}_{\ell\ell'} \star \underset{\sim}{K}_{\ell'}(\sigma) \tag{8}$$

or more generally

$$\underset{\sim}{Q}_\ell(t) = e^{M(t-\tau)}_{\ell\ell'} \star \underset{\sim}{Q}_{\ell'}(\tau) + \int_\tau^t d\sigma e^{M(t-\sigma)}_{\ell\ell'} \star \underset{\sim}{K}_{\ell'}(\sigma) \tag{9}$$

for $\tau \le t$.

For $t' > t$, we have

$$\langle \underset{\sim}{K}_\ell(t)\underset{\sim}{Q}_{\ell'}(t')\rangle = \int_0^{t'} d\sigma e^{M(t'-\sigma)}_{\ell'\ell''} \star \langle \underset{\sim}{K}_\ell(t)\underset{\sim}{K}_{\ell''}(\sigma)\rangle. \tag{10}$$

The correlation function $\langle \underset{\sim}{K}_\ell(t)\underset{\sim}{K}_{\ell'}(\sigma)\rangle$ can be adequately represented by

$$\langle \underset{\sim}{K}_\ell(t)\underset{\sim}{K}_{\ell'}(\sigma)\rangle = \Gamma_{\ell\ell'}\delta(t-\sigma) \tag{11}$$

where Γ is a time independent coefficient. Thus, Eq 10 becomes

$$\langle \underset{\sim}{K}_\ell(t)\underset{\sim}{Q}_{\ell'}(t')\rangle = e^{M(t'-t)}_{\ell\ell''}\Gamma_{\ell\ell''}, \tag{12}$$

which has a long time decay.

Furthermore, as has been shown by Schramm and Oppenheim [17], products of fluctuating forces must be considered with great care since they may exhibit long-time behavior. In particular, $K(t)K(t+\tau)$ is not a fast variable for $\tau < \tau_m$.

To obtain some insight into this behavior consider the correlation function $\langle A_a(t)A_b(t+\tau)\rangle$, which can be written

$$\begin{aligned}\langle A_a(t)A_b(t+\tau)\rangle &= \langle A_a A_b(\tau)\rangle = e^{M\tau}_{bd}\langle A_a A_d\rangle = \\ \langle A_a(t)e^{M\tau}_{b\ell} \star \underset{\sim}{Q}_\ell(t)\rangle &+ \langle A_a(t)\int_t^{t+\tau} d\sigma e^{M(t+\tau-\sigma)}_{b\ell} \star \underset{\sim}{K}_\ell(\sigma)\rangle \\ = e^{M\tau}_{bd}\langle A_a A_d\rangle &+ \langle A_a(t)\int_t^{t+\tau} d\sigma e^{M(t+\tau-\sigma)}_{b\ell} \star \underset{\sim}{K}_\ell(\sigma)\rangle,\end{aligned} \tag{13}$$

where we invoked the orthogonality of the linear to the non-linear modes. Clearly, the second term on the RHS of the third line of Eq 13 must be zero. This follows from the calculation

$$A_a(t) = e^{Mt}_{a\ell'}Q_{\ell'} + \int_0^t d\sigma e^{M(t-\sigma)}_{a\ell'}K_{\ell'}(\sigma) \tag{14}$$

which when inserted into the term yields

$$\int_0^t d\phi e^{M(t-\phi)}_{a\ell'} \int_t^{t+\tau} d\sigma e^{M(t+\tau-\sigma)}_{b\ell}\langle K_{\ell'}(\phi)K_\ell(\sigma)\rangle, \tag{15}$$

which since both ϕ and σ must be close to t becomes

$$e^{M\tau}_{b\ell}\int_0^t d\phi \int_t^{t+\tau} d\sigma \langle K_{\ell'}(\phi)K_\ell(\sigma)\rangle. \tag{16}$$

Using the approximation

$$\langle K_{\ell'}(\phi)K_{\ell}(\sigma)\rangle \sim \Gamma_{\ell'\ell} \lim_{\varepsilon\to 0} \varepsilon^{-1/2} e^{(\phi-\sigma)^2/\varepsilon}, \quad (17)$$

Eq 16 is proportional to $\varepsilon^{1/2}$ and becomes negligible. We shall use this fact in all of our succeeding calculations.

Note that if we carried out the calculation differently, i.e. we integrated from 0 to $t+\tau$, we obtain

$$\langle A_a(t)A_b(t+\tau)\rangle = e^{Mt}_{a\ell} e^{M(t+\tau)}_{b\ell'} \langle Q_\ell Q_{\ell'}\rangle$$
$$+\int_0^t d\phi \int_0^{t+\tau} d\sigma e^{M(t-\phi)}_{a\ell} e^{M(t+\tau-\sigma)}_{b\ell'} \langle K_\ell(\phi)K_{\ell'}(\sigma)\rangle \quad (18)$$

and the term involving products of Ks cannot be neglected.

At last we turn our attention to the two time correlation function $C(t_1,t_2)$ where

$$C_{abc}(t_1,t_2) = \langle A_a A_b(t_1) A_c(t_1+t_2)\rangle =$$
$$e^{Mt_2}_{c\alpha^\ell} \star \langle A_a[A_b Q_{\alpha^\ell}](t_1)\rangle. \quad (19)$$

We now use the fact that products of slow variables must be slow and must be expressable in terms of modes. Thus

$$A_b Q_{\alpha^\ell} = \langle A_b Q_{\alpha^\ell} Q_{\gamma^{\ell'}}\rangle \star \langle QQ\rangle^{-1}_{\gamma^{\ell'}\gamma^{\ell'}} \star Q_{\gamma^{\ell'}}. \quad (20)$$

Because of the orthogonality of the Q, the only non vanishing terms are $\ell' = \ell+1, \ell, \ell-1$. The maximal terms are obtained when the correlation functions factorize. Therefore, Eq 20 becomes

$$A_b Q_{\alpha^\ell} = Q_{b\alpha^\ell} + \langle A_b A_{\alpha_1} A_{\gamma_1}\rangle \star \langle QQ\rangle^{-1}_{\gamma_1\gamma_1} \star Q_{\gamma_1\alpha^{\ell-1}}$$
$$+\langle A_b Q_{\alpha_1}\rangle \star Q_{\alpha^{\ell-1}}. \quad (21)$$

Substitution of Eq 21 into Eq 19 yields

$$C_{abc}(t_1,t_2) = e^{Mt_2}_{c\alpha^\ell}[e^{Mt_1}_{b\alpha^\ell,a'} +$$
$$e^{Mt_1}_{\gamma_1\alpha^{\ell-1},a'} \langle A_b A_{\alpha_1} A_{\gamma_1}\rangle \star \langle QQ\rangle^{-1}_{\gamma_1\gamma_1}$$
$$+e^{Mt_1}_{\alpha^{\ell-1},a'} \langle A_b A_{\alpha_1}\rangle]\langle A_a A_{a'}\rangle. \quad (22)$$

In the above, an $\ell' th$ order mode denoted by γ^ℓ is factorized into a linear mode A_{γ_1} and $Q_{\gamma^{\ell-1}}$. Correspondingly, subscript $\gamma_1\alpha^{\ell-1}$ denotes an $\ell' th$ order mode formed by the product of A_{γ_1} and the $(\ell-1)'th$ order mode obtained by stripping off A_{α_1} from Q_{α^ℓ}. Note that $\underset{\sim}{Q_1} = \underset{\sim}{A}$.

We now search for the leading terms in Eq 22 using the facts that $e^{Mt}_{\ell\ell'}$ is of order 1 for $|\ell'| \le |\ell|$ and of order $N^{|\ell|-|\ell'|}$ for $|\ell'| > |\ell|$, where $||$ gives the mode order. Also $Q_0 = 1$ and $e^{Mt_2}_{c1}$ or $e^{Mt_1}_{1a'}$ are zero. Thus $e^{Mt_2}_{c\alpha^\ell} = 0$ for $\ell = 0$, $\sim O(1)$ for $\ell = 1$, $\sim O(1/N)$ for $\ell = 2$, etc; $e^{Mt_1}_{b\alpha^\ell,a'} \sim 1$ for $\ell = 0,1,2$ etc; $e^{Mt_1}_{\gamma_1\alpha^{\ell-1},a'} \sim 1$ for ℓ = 1, 2, etc; $e^{Mt_1}_{\alpha^{\ell-1},a'}$ is zero for $\ell = 0$ or 1, and ~ 1 for $\ell = 2$. The leading terms are of $O(N)$ and are

$$C_{abc}(t_1,t_2) = e^{Mt_2}_{c,\alpha} e^{Mt_1}_{b\alpha,a'} \star \langle A_a A_{a'}\rangle +$$
$$e^{Mt_2}_{c,\alpha} e^{Mt_1}_{\gamma,a'} \star \langle QQ\rangle^{-1}_{\gamma\gamma'} \star \langle A_b A_\alpha A_{\gamma'}\rangle \star \langle A_a A_{a'}\rangle +$$
$$e^{Mt_2}_{c,\alpha_1\alpha_2} \star \langle A_b A_{\alpha_1}\rangle \star e^{Mt_1}_{\alpha_2,a'} \star \langle A_a A_{a'}\rangle. \quad (23)$$

In order to proceed, we need to simplify the propagators, e^{Mt}, and will use techniques described in detail in Ref [14]. We will also need to make the wave vector dependences of the various modes explicit.

It is easier to use the Laplace transforms of the propagators to simplify the results. We consider

$$G_{\ell\ell'}(z) \equiv \mathscr{L} e^{Mt}_{\ell\ell'} = (z\underline{I} - \underline{\underline{M}})^{-1}_{\ell\ell'} \quad (24)$$

where $\mathscr{L}$ denotes the Laplace transform and the matrix $\underline{\underline{M}}$ has elements in the space of modes of all orders. The resummation proceeds by writing

$$M_{\ell\ell'} = M^D_{\ell\ell'} + M^O_{\ell\ell'} \quad (25)$$

where $\underline{\underline{M}}^D$ is zero unless $|\ell| = |\ell'|$ and all the wave vectors in ℓ are equal to those in ℓ'; $\underline{\underline{M}}^O$ is off diagonal.

Resummation for $\ell = \ell' = 1$ yields

$$G_{a_k,b_k}(z) = (z\underline{I} - \tilde{\underline{\underline{M}}})^{-1}_{a_k,b_k} =$$
$$(z\underline{I} - \underline{\underline{M}}^D - \underline{\underline{\Theta}}(z))^{-1}_{a_k,b_k} \quad (26)$$

where the matrix $\tilde{M}$ has elements in 1-1 space only, as does $\underline{\underline{\Theta}}$. The explicit form of $\underline{\underline{\Theta}}$ is given in the paper by Schofield, Lim and Oppenheim [13].

The diagonal elements in 2-2 space have the form

$$e^{Mt}_{a_{k-q},a'_q:b_{k-q},b'_q} = e^{Mt}_{a_{k-q},b_{k-q}} e^{Mt}_{a'_q,b'_q} \quad (27)$$

Therefore

$$G_{a_{k-q},a'_q:b_{k-q},b'_q}(z) = \mathscr{L}[\mathscr{L}^{-1}(z\underline{I} - \underline{\underline{M}})^{-1}_{a_{k-q},b_{k-q}}$$
$$\mathscr{L}^{-1}(z\underline{I} - \underline{\underline{M}})^{-1}_{a'_q,b'_q}] \quad (28)$$

and is easily evaluated.

Again by approximate resummation we find

$$G_{a_{k-q},a'_q:b_k}(z) = G_{a_{k-q},a'_q:c_{k-q},c'_q}(z)$$
$$M^0_{c_{k-q},c'_q:b'_k} G_{b'_k,b_k}(z) \quad (29)$$

and

$$G_{a_k:b_{k-q},b'_q}(z) = G_{a_k,a'_k}(z)$$
$$M^0_{a'_k:c_{k-q},c'_q} G_{c_{k-q},c'_q:b'_{k-q},b_q}(z) \quad (30)$$

All of these propagators can be evaluated in a straightforward manner.

The procedure we have outlined can be easily extended to higher order multi-time correlation functions. We shall briefly go through the steps for the three time, four point function

$$C_{abcd}(t_1,t_2,t_3) = \langle A_a A_b(t_1) A_c(t_1+t_2) A_d(t_1+t_2+t_3)\rangle \tag{31}$$

The actual form of the third order response function that arises from non-linear response theory has subtractions of the factorization of Eq 31. These subtractions are

$$\langle A_a A_b(t_1)\rangle\langle A_c A_d(t_3)\rangle + \langle A_a A_c(t_1+t_2)\rangle \times \langle A_b A_d(t_2+t_3)\rangle + \langle A_a A_d(t_1+t_2+t_3)\rangle\langle A_b A_c(t_2)\rangle \tag{32}$$

These terms are of order N^2 for special choices of the wave vectors. When they are subtracted from Eq 31 we obtain a result of order N.

The first step in the procedure is

$$C_{abcd}(t_1,t_2,t_3) = e^{Mt_3}_{d\alpha^\ell} \star \langle A_a A_b(t_1)[A_c(t_2) Q_{\alpha^\ell}(t_2)]\rangle. \tag{33}$$

We use the analog of Eq 21 to obtain

$$C(t_1,t_2,t_3) = [e^{Mt_3}_{d,\alpha^\ell} e^{Mt_2}_{c\alpha^\ell,\beta^{\ell'}} + e^{Mt_3}_{d,\alpha^\ell} e^{Mt_2}_{\theta_1\alpha^{\ell-1},\beta^{\ell'}} \langle QQ\rangle^{-1}_{\theta_1\theta_1}\langle A_c A_{\alpha_1} A_{\theta_1}\rangle + e^{Mt_3}_{d,\alpha^\ell}\langle A_c A_{\alpha_1}\rangle e^{Mt_2}_{\alpha^{\ell-1},\beta^{\ell'}}] \langle A_a[A_b Q_{\beta^{\ell'}}](t_1)\rangle \tag{34}$$

One more use of Eq 21 yields for the first term on the RHS of Eq 34

$$(1) = e^{Mt_3}_{d,\alpha^\ell} e^{Mt_2}_{c\alpha^\ell,\beta^{\ell'}} e^{Mt_1}_{b\beta^{\ell'},a'}\langle A_a A_{a'}\rangle + e^{Mt_3}_{d,\alpha^\ell} e^{Mt_2}_{c\alpha^\ell,\beta^{\ell'}} e^{Mt_1}_{\theta\beta^{\ell'-1},a'}\langle QQ\rangle^{-1}_{\theta\theta}\star\langle A_b A_{\beta_1} A_\theta\rangle\star\langle A_a A_{a'}\rangle + e^{Mt_3}_{d,\alpha^\ell} e^{Mt_2}_{c\alpha^\ell,\beta^{\ell'}} e^{Mt_1}_{\beta^{\ell'-1},a'}\langle A_b A_{\beta_1}\rangle\langle A_a A_{a'}\rangle. \tag{35}$$

The leading terms are

$$e^{Mt_3}_{d,\alpha} e^{Mt_2}_{c\alpha,\beta} e^{Mt_1}_{b\beta,a'}\langle A_a A_{a'}\rangle + e^{Mt_3}_{d,\alpha} e^{Mt_2}_{c\alpha,\beta} e^{Mt_1}_{\theta,a'}\langle QQ\rangle^{-1}_{\theta\theta}\star\langle A_b A_b A_\theta\rangle\star\langle A_a A_{a'}\rangle + e^{Mt_3}_{d,\alpha} e^{Mt_2}_{c\alpha,\beta_1\beta_2} e^{Mt_1}_{\beta_2,a'}\langle A_b A_{\beta_1}\rangle\langle A_a A_{d'}\rangle. \tag{36}$$

The first two terms are of order N. The last term is of order N^2, but is cancelled by one of the subtractions in Eq 32.

The second term on the RHS of Eq 34 is similar. The only interesting behavior occurs in the third term, that is:

$$e^{Mt_3}_{d,\alpha^\ell}\langle A_c A_{\alpha_1}\rangle e^{Mt_2}_{\alpha^{\ell-1},\beta^{\ell'}}\langle A_b A_{\beta_1}\rangle e^{Mt_1}_{\beta^{\ell'-1},a'}\langle A_a A_{a'}\rangle. \tag{37}$$

The leading term is for $\ell = 2, \ell' = 2$ and becomes

$$e^{Mt_3}_{d,\alpha_1\alpha_2}\langle A_c A_{\alpha_1}\rangle e^{Mt_2}_{\alpha_2,\beta_1\beta_2}\langle A_b A_{\beta_1}\rangle e^{Mt_1}_{\beta_2,a'}\langle A_a A_{a'}\rangle. \tag{38}$$

which is of order N.

Although the algebra becomes more complicated as we look at higher multi-time correlations, the procedure is straightforward.

APPLICATION TO FOUR-DIPOLE ORIENTATIONAL CORRELATION

As a first example we consider the correlation of four Cartesian components of a unit vector along the axis of a symmetric top molecule, representing the orientational contribution to the dipole, which dominates certain IR pump-probe experiments. The vector components may be expressed with the first-rank spherical harmonics, Y_{1m}, and for the case of rotational diffusion, assumed in the following, the basis functions are all the spherical harmonics, Y_{lm}.

The example represents a considerable simplification of, and illustrates the flexibility of, the general theory. With single particle variables, there are no wavevector or $O(N)$ considerations, so many of the steps discussed previously do not enter. Nevertheless, if reorientation is the slow process, our fundamental idea - that the variable at t_{n+1} is propagated from t_n and combined with the variable at t_n to obtain new members of the basis set at t_n - is easily applied. Products of spherical harmonics are themselves linear combinations of spherical harmonics, so the complete set, $\{Y_{lm}\}$, also includes all the product variables, making reorientation a particularly convenient target for our approach. The rules for coupling of angular momenta yield the restrictions on ℓ' in Eq 20, in a completely different context.

In an isotropic system the propagator does not couple the different Y_{lm}, and the well-known results for their autocorrelations are easily obtained. Even so, multi-time correlations are nontrivial. A more interesting situation is found in the presence of anisotropy, where the different basis functions do couple. Thus, we will first show how to calculate a correlation of four spherical harmonics for a molecule executing rotational diffusion in an anisotropic environment.

The experiment we have in mind [18] is that of a small molecule in an anisotropic cavity in a protein, e.g., CO in the heme pocket in myoglobin. To demonstrate our method we will obtain the relevant correlations in the anisotropic cavity-fixed coordinate frame, denoted with 'a', at $t_1 = 0$, for pump and probe pulses polarized along z and α, respectively,

$$C^a_{zz\alpha\alpha}(t_2,t_3) = \langle z(0)z(0)\alpha(t_2)\alpha(t_2+t_3)\rangle_a \tag{39}$$

and the anisotropy,

$$\Delta C^a(t_2,t_3) = C^a_{zzzz}(t_2,t_3) - C^a_{zzxx}(t_2,t_3), \quad (40)$$

where z and α denote Cartesian components of the unit vector along the molecular axis.

The experiment is carried out in the laboratory coordinate frame, which is isotropic on average, as the orientation of the anisotropic cavity is random; it is assumed that reorientation of the protein is very slow. Lab-frame correlations are related to correlations in the anisotropic protein-frame by rotational transformations. In the last step these transformations and the complete calculation will be outlined, saving the details for future work.

1. Rotational diffusion and single and multi-time correlations in an anisotropic system

The random forces, $\partial Y_{lm}/\partial t$, are given by functions of the molecular orientation, $\Omega = (\theta,\phi)$, times the angular velocities, $\dot{\theta}$ and $\dot{\phi}$. In rotational diffusion, the angular velocities are fast variables, and time integrals of the random force correlation are given by same-time averages of the functions of Ω times time-integrals of the angular velocity correlations, obeying

$$\int dt \langle \dot{\theta}(t)\dot{\theta}(0)\rangle = D_R \quad (41)$$

$$\int dt \langle \dot{\phi}(t)\dot{\phi}(0)\rangle = 3/2\ D_R \quad (42)$$

where D_R is the rotational diffusion coefficient, and the integral of the cross-correlation vanishes.

We will incorporate anisotropy into the same-time average of an arbitrary orientational function, $f(\Omega)$, in the anisotropic frame, as follows,

$$\langle f(\Omega)\rangle_a = \langle f(\Omega)\rangle + a_1\langle f(\Omega)Y_{10}(\Omega)\rangle + a_2\langle f(\Omega)Y_{20}(\Omega)\rangle, \quad (43)$$

where $\langle\rangle$ with no subscript denotes an isotropic average, and the coefficients, a_1 and a_2, control the dipolar and quadrupolar character of the anisotropy, respectively. Then, while the spherical harmonics all have zero isotropic average,

$$\langle Y_{1,0}\rangle_a = a_1/4\pi;\ \langle Y_{2,0}\rangle_a = a_2/4\pi, \quad (44)$$

and the new basis functions are $\delta Y_{lm} = Y_{lm} - \langle Y_{l,m}\rangle_a$. The coefficient of the projection of an arbitrary orientational variable B onto δY_{lm} is

$$P(B,lm) = \langle B\ \delta Y_{l'm'}\rangle_a \langle \delta Y\,\delta Y\rangle_a^{-1}(l'm',lm), \quad (45)$$

where $\langle \delta Y\,\delta Y\rangle$ is the matrix of same-time averages and the summation convention is employed.

With the above results the transport matrix may be constructed,

$$\underline{\underline{M}} = -\int_0^\infty dt \langle \dot{\underset{\sim}{Y}}(t)\dot{\underset{\sim}{Y}}(0)\rangle_a \star \langle \delta\underset{\sim}{Y}\,\delta\underset{\sim}{Y}\rangle_a^{-1}, \quad (46)$$

and the propagator matrix obeys $\underline{\underline{G}}(t) = exp(\underline{\underline{M}}t)$. The correlation of four δY denoted a-d in order of increasing time, separated by intervals t_1, t_2 t_2, in a shorthand notation representing l and m with a single index, is

$$C^a_{a-d}(t_1,t_2,t_3) = G_{d,i}(t_3)P(\delta Y_i\delta Y_c,j) \times G_{j,k}(t_2)P(\delta Y_k\delta Y_b,n)G_{n,p}(t_1)\langle \delta Y_p\delta Y_a\rangle_a, \quad (47)$$

and the ordinary time correlation obeys

$$C^a_{ab}(t) = G_{b,i}(t)\langle \delta Y_i\delta Y_a\rangle_a. \quad (48)$$

Now consider some example calculations. The anisotropy under consideration couples different l, but not m. Practically, the basis set must be truncated. To examine the behavior of the z-component of the molecular axis in the anisotropic frame, let the basis set be $(\delta Y_{10},\delta Y_{20},\delta Y_{30})$. Even in only three dimensions, the M matrix is a complicated function of a_1 and a_2. To quadratic order in the parameters, its elements are

$$-M^{(2)}_{11}/D_R = 2 - 0.7570a_2 + 0.0955a_1^2 + 0.06821a_2^2 \quad (49)$$

$$-M^{(2)}_{12}/D_R = 0.2523a_1 + 0.1819a_1a_2 \quad (50)$$

$$-M^{(2)}_{13}/D_R = 0.4955a_2 - 0.06251a_1^2 + 0.1042a_2^2 \quad (51)$$

$$-M^{(2)}_{21} = -0.7570a_1 + 0.6821a_1a_2 \quad (52)$$

$$-M^{(2)}_{22}/D_R = 6 - 0.5407a_2 + 0.06821a_1^2 + 0.5749a_2^2) \quad (53)$$

$$-M^{(2)}_{23}/D_R = 0.4955a_1 + 0.2381a_1a_2 \quad (54)$$

$$-M^{(2)}_{31}/D_R = -1.9821a_2 + 0.2501a_1^2 + 0.6251a_2^2 \quad (55)$$

$$-M^{(2)}_{32}/D_R = -0.9911a_1 + 0.8038a_1a_2 \quad (56)$$

$$-M^{(2)}_{33}/D_R = 12 - 0.5046a_2 + 0.2455a_1^2 + 0.5760a_2^2 \quad (57)$$

The anisotropy-independent diagonal terms 2, 6 and 12 represent the well known $l(l+1)D_R$ decay rate of isotropic Debye diffusion. To leading order in the parameters, quadrupolar anisotropy slows reorientation via the negative a_2 terms along the diagonal, while dipolar anisotropy dominates the coupling of the spherical harmonics. With $a_1 = 0$, δY_{20} decouples from the other two variables. The decay of the first-rank autocorrelation, Fig 1, slows markedly with increasing a_2. The smaller, opposite effect with increasing a_1 is a consequence of coupling in the faster second-rank dynamics, as seen through the propagator element, G_{12}, in Fig 2. While coupling requires nonzero a_1, it is strongly increased in amplitude, and slowed, by finite a_2. Anisotropy has a smaller effect on the second-rank autocorrelations, Fig 3, with a_1 alone causing a slowing and a_2 a slight speedup.

The multi-linear $C^a_{zz\alpha\alpha}(t_2,t_3)$ in the anisotropic frame are obtained from Eq 47 with $t_1 = 0$, and recognizing that the product of basis functions at $t_1 = 0$ is itself a basis function. Although this does not provide a complete

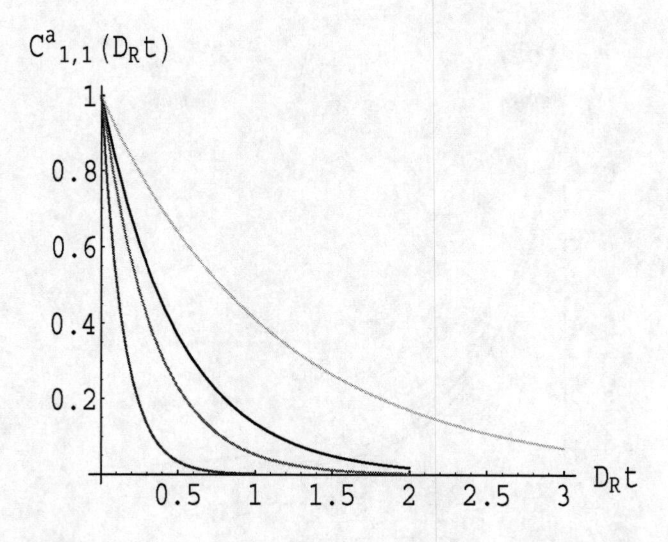

FIGURE 1. (Color online) Normalized first rank autocorrelation, $C^a_{1,1}(D_R t)$, vs $D_R t$, for $m = 0$ and, top to bottom, (a_1, a_2) pairs (0,2) (green), (0,0) (black), (2,0) (red), (2,2) (blue).

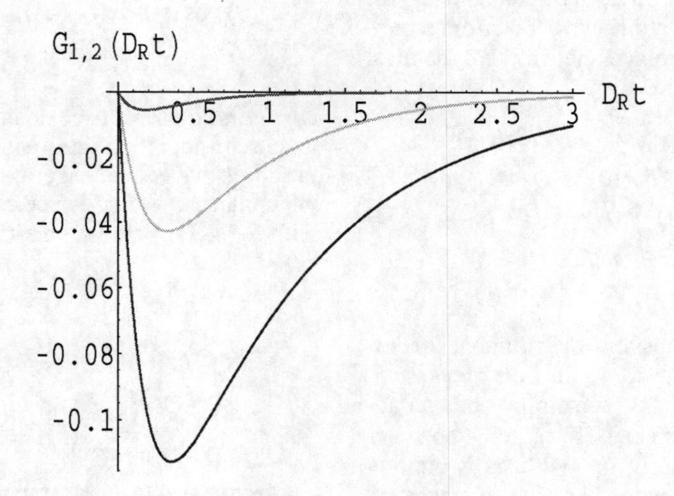

FIGURE 2. (Color online) First-second rank cross propagator, $G_{12}(D_R t)$, vs $D_R t$, for $m = 0$ and, top to bottom, (a_1, a_2) pairs (2,0) (red), (1,1) (green), (2,2) (blue).

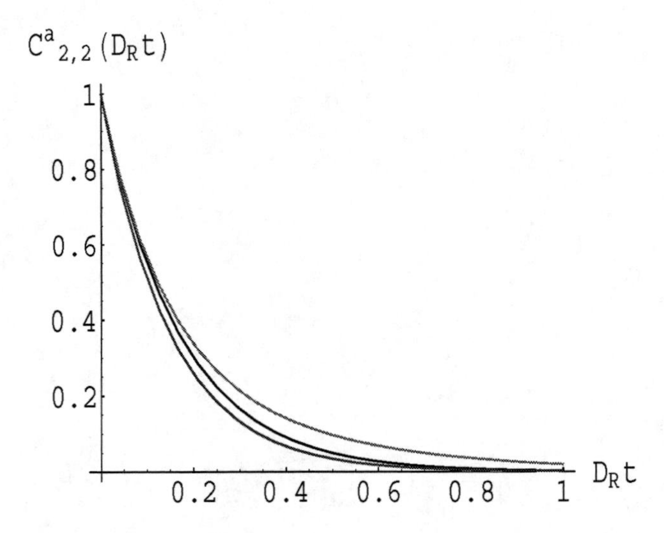

FIGURE 3. (Color online) Normalized second rank autocorrelation, $C^a_{2,2}(D_R t)$, vs tD_R, for $m = 0$ and, top to bottom, (a_1, a_2) pairs (2,0) (red), (0,0) (black), and superimposed, (0,2) (green), (2,2) (blue).

solution of the lab-frame problem, it demonstrates our method. Note that the correlations have lower-order contributions due to the nonzero averages of the dynamical variables,

$$C^a_{zzzz}(t_2,t_3) = \langle z^2\rangle_a \langle z\rangle_a^2 + \langle z^2\rangle_a \langle \delta z(0)\delta z(t_3)\rangle_a$$
$$+\langle z\rangle_a(\langle \delta z^2(0)\delta z(t_2)\rangle_a + \langle \delta z^2(0)\delta z(t_2+t_3)\rangle_a)$$
$$+\langle \delta(z^2(0))\ \delta z(t_2)\ \delta z(t_2+t_3)\rangle_a \quad (58)$$
$$C^a_{zzxx}(t_2,t_3) = \langle z^2\rangle_a \langle x(0)x(t_3)\rangle_a$$
$$+\langle \delta(z^2(0))\ x(t_2)\ x(t_2+t_3)\rangle_a, \quad (59)$$

where we have used time-translational invariance for correlations between t_2 and t_2+t_3. In an isotropic system $\langle z^2\rangle$ already has a value of 1/3; anisotropy adds an additional contribution and also causes $\langle z\rangle_a$ to be nonzero. Furthermore, the correlation of δz^2 with δz is entirely due to anisotropy. The fluctuating parts and averages are expressed as:

$$z^2 = \frac{4}{3}\sqrt{\frac{\pi}{5}}\ \delta Y_{20} + \frac{1}{3}(1+\frac{a_2}{\sqrt{5\pi}}) \quad (60)$$

$$z = 2\sqrt{\frac{\pi}{3}}\ \delta Y_{10} + \frac{a_1}{2\sqrt{3\pi}} \quad (61)$$

$$x = \sqrt{\frac{2\pi}{3}}(Y_{1-1} - Y_{1,1}), \quad (62)$$

where for $m \neq 0$, $Y = \delta Y$.

We immediately obtain

$$\langle \delta(z^2(0))\ \delta z(t_2)\ \delta z(t_2+t_3)\rangle_a = \frac{16\pi}{9}\sqrt{\frac{\pi}{5}}G_{10,i}(t_3)$$
$$P(\delta Y_i \delta Y_{10}, \delta Y_j) G_{j,k}(t_2) \langle \delta Y_k \delta Y_{20}\rangle_a \quad (63)$$

$$\langle \delta(z^2(0))\ \delta x(t_2)\ \delta x(t_2+t_3)\rangle_a = -\frac{16\pi}{9}\sqrt{\frac{\pi}{5}}G_{11,i}(t_3)$$
$$P(\delta Y_i \delta Y_{1-1}, \delta Y_j) G_{j,k}(t_2) \langle \delta Y_k \delta Y_{20}\rangle_a, \quad (64)$$

where in Eq 64 we have noted that, since the anisotropy under consideration does not couple different m, Eq 62 produces two equivalent cross terms in $\delta x(t_2)\ \delta x(t_2+t_3)$ which have m=0 and hence can form a nonzero average with δY_{20}. The relevant one-time correlations are

$$\langle \delta z(0)\delta z(t)\rangle_a = \frac{4\pi}{3}G_{10,i}(t)\langle \delta Y_i \delta Y_{10}\rangle_a \quad (65)$$

$$\langle \delta x(0)\delta x(t)\rangle_a = -\frac{4\pi}{3}G_{11,i}(t)\langle \delta Y_i \delta Y_{1-1}\rangle_a \quad (66)$$

$$\langle \delta z^2(0)\delta z(t)\rangle_a = \frac{8\pi}{3\sqrt{15}}G_{10,i}(t)\langle \delta Y_i \delta Y_{20}\rangle_a. \quad (67)$$

Observe that in an isotropic system the different δY do not couple and a known result is immediately recovered,

$$\langle \delta(z^2(0))\ \delta\alpha(t_2)\ \delta\alpha(t_2+t_3)\rangle =$$
$$G_{1,1}(t_3)G_{2,2}(t_2)\langle \delta(\alpha)^2 \delta(z^2)\rangle. \quad (68)$$

We may calculate C_{zzzz} with the basis set used previously, $(\delta Y_{10}, \delta Y_{20}, \delta Y_{30})$, since only m=0 contributes. The true multi-linear part, $\langle \delta(z^2(0))\ \delta z(t_2)\ \delta z(t_2+t_3)\rangle_a$ or equivalently $\langle \delta Y_{20}\ \delta Y_{10}(t_2)\ \delta Y_{10}(t_2+t_3)\rangle_a$, since we normalize to unity at $t_2 = t_3 = 0$, is shown in the next four figures. Figs 4 and 5, for (a_1, a_2) pairs (0,0) and (2,0), primarily illustrate the effect of coupling. The isotropic case, Fig 4, simply displays the well known product behavior of Eq 68, with a faster second-rank t_2 decay rate of $6D_r$ and a slower first rank t_3 decay rate of $2D_r$. With

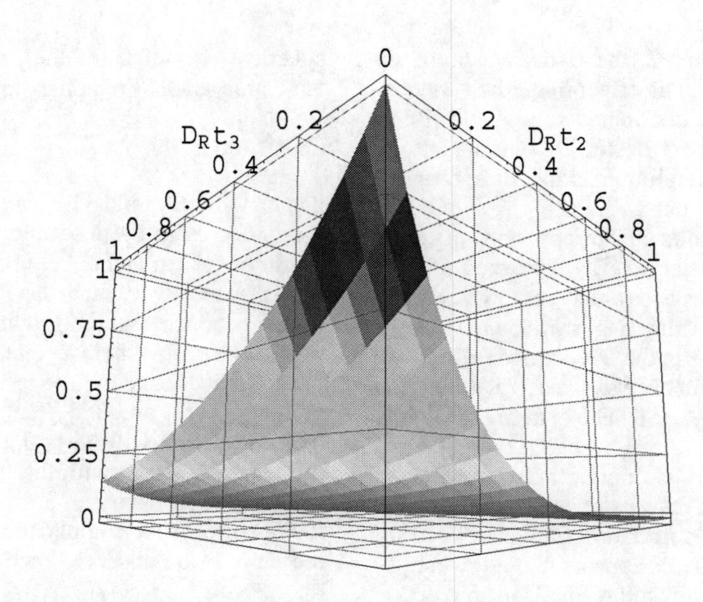

FIGURE 4. (Color online) Two-time *zzzz* correlation, $\langle \delta Y_{20}\ \delta Y_{10}(t_2)\ \delta Y_{10}(t_2+t_3)\rangle_a$, for isotropic case, a_1=0, $a_2 = 0$, vs $D_R t_2$ and $D_R t_3$.

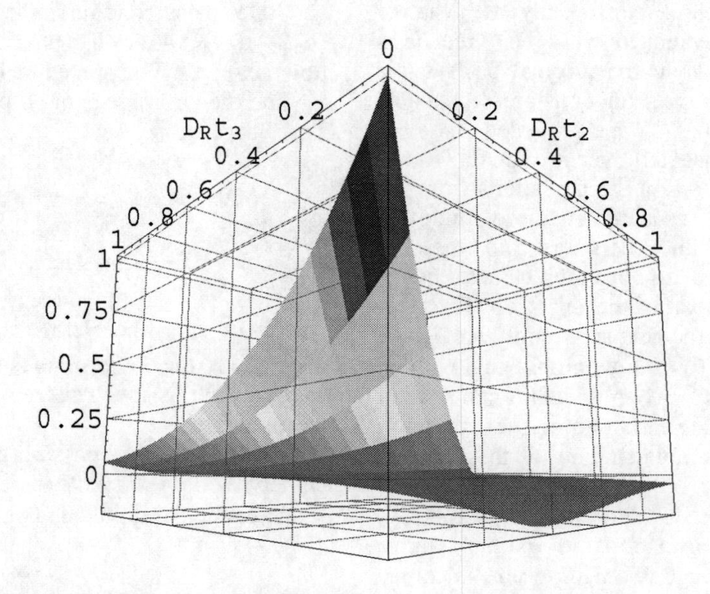

FIGURE 5. (Color online) Two-time *zzzz* correlation, $\langle \delta Y_{20}\ \delta Y_{10}(t_2)\ \delta Y_{10}(t_2+t_3)\rangle_a$, for a_1=2, $a_2 = 0$, vs $D_R t_2$ and $D_R t_3$.

dipolar anisotropy only, a_2=2, the t_3 decay speeds up slightly but there is a dramatic effect upon the t_2 dependence, with a negative excursion and subsequent rise to zero giving an overall slower decay. Dipolar anisotropy acts primarily through coupling, mixing in a negative slower, first-rank contribution.

The effect of quadrupolar anisotropy only, (0,2), is seen in Fig 6. Here coupling to δY_{20} is turned off but, compared to the isotropic system, the t_3 decay is greatly slowed, due to changes in the diagonal elements of the M matrix. Note the change in the $D_R t_3$-range from 0-1 in the isotropic case to 0-2, but even so the correlation is not fully decayed at $D_r t_3$=2, $D_r t_2$=0. The t_2 decay is slightly speeded up.

With both dipolar and quadrupolar anisotropy, (2.2), Fig 7, the very slow t_3 decay associated with nonzero a_2 remains. The negative correlation found for (2,0) for intermediate t_2 and short t_3 disappears, but a low amplitude slow decay in t_2, compared to the isotropic decay, persists. This may be viewed as the (2,0) t_2 decay counteracted by the speedup found for (0,2).

Because the different m do not couple, in calculating the multi-linear part of C^a_{zzxx} consistently with that of C^a_{zzzz}, Eq 64, $G(t_3)$ is evaluated with the three basis functions (Y_{11}, Y_{21}, Y_{31}) while the original basis set is used for $G(t_2)$. Recall, $m = 1$ functions are propagated over t_3, then multiplied by Y_{1-1} and projected onto $m = 0$, and the result is propagated over t_2. For the values of (a_1, a_2) under consideration the deviations from the isotropic case are much weaker then those found for $zzzz$, and mainly act to dilute their signature when the anisotropy is calculated. Thus for completeness, Fig 8, we show the multi-linear part of the anisotropy for a_1=1, a_2=4 only; it is shown in the next subsection that $a_2 \sim 4$ is indicated for unbound CO in myoglobin, and with no information we choose $a_1 = 1$ to maintain coupling of the modes. Compared to the earlier figures, the relatively large a_2 causes a considerable slowing of the decay in both time arguments.

2. Lab frame correlations

Experiments probe correlations in the lab frame, which is isotropic on average. We will present a detailed calculation of the anisotropy in our next paper, and here just outline some key points. First note that

$$\langle(\hat{\mu}\cdot\hat{\mu})\ (\hat{\mu}(t_2)\cdot\hat{\mu}(t_2+t_3))\rangle = 3C_{zz}(t_3) = 3C_{zzzz}(t_2,t_3)+6C_{zzxx}(t_2,t_3), \quad (69)$$

where we have used the equivalence of the lab-frame axes. With this result the anisotropy may be rewritten in a simpler form,

$$\Delta C(t_2,t_3) = \frac{3}{2}C_{zzzz}(t_2,t_3) - \frac{1}{2}C_{zz}(t_3). \quad (70)$$

The lab-frame variables at $t = 0, t_2$, and t_2+t_3 are expressed as a sum of products of three corresponding spherical harmonics of the probe molecule axis in the lab-frame, each of which is further expressed as

$$Y_{lm}(\Omega^L_{mol}(t)) = D^l_{mm'}(\Omega^L_p(t))Y_{lm'}(\Omega^P_{mol}(t)), \quad (71)$$

where $\Omega^L_{mol}, \Omega^L_p$ and Ω^P_{mol} represent the orientation of the molecule in the lab frame, the protein cavity-fixed frame in the lab frame, and the molecule in the protein frame, respectively, and the $D^l_{mm'}$ are the Wigner rotation functions. Assuming that the very slow reorientation of the protein is independent of motion of the probe molecule in the cavity, the resulting average of products of 3 $D^l_{mm'}(\Omega^L_p)$ and 3 $Y_{lm'}(\Omega^P_{mol})$ is factorized. The former correlation will be evaluated with isotropic Debye diffusion, the latter with the methods just described for the anisotropic cavity.

For spectroscopic applications the protein dynamics is too slow to be observed, and $D^l_{mm'}(\Omega^L_p(t))$ may be replaced by $D^l_{mm'}(\Omega^L_p(t=0))$ for all times, a considerable simplification. Helbing et al found that the anisotropy for unbound CO in the heme pocket of myoglobin, at $t_3 = 0$, decayed to a finite value, not to zero, for large t_2. Under these conditions the factors $\langle\mu_z(0)\mu_z(0)\rangle$ and $\langle\mu_z(t_2)\mu_z(t_2)\rangle$ are well separated in time and the correlation factorizes. The procedure is to first average for fixed Ω^L_p and then average over all protein orientations. Thus, Eq 70 becomes

$$\Delta C(t_2 \gg t_{mol}, t_3=0) = \frac{1}{4\pi}\int d\Omega^L_p \left(\frac{3}{2}\langle\mu_z^2\rangle_{\Omega^L_p}\langle\mu_z^2\rangle_{\Omega^L_p} - \frac{1}{2}\langle\mu_z^2\rangle_{\Omega^L_p}\right) \quad (72)$$

where $\langle\mu_z^2\rangle_{\Omega^L_p}$ is the average at fixed protein orientation, and t_{mol} is the longest relaxation time in the protein frame, which is much shorter than the time for protein reorientation.

In this cases it is possible to avoid the Wigner machinery. Letting $\hat{\alpha}$ denote the protein-fixed axes, the lab-frame molecular dipole unit vector obeys

$$\mu_z = (\hat{\mu}\cdot\hat{\alpha})(\hat{\alpha}\cdot\hat{Z}) \quad (73)$$

where $\hat{Z}$ denotes the lab z-axis. Since the three directions in the protein frame remain independent, with x and y equivalent,

$$\langle\mu_z^2\rangle_{\Omega^L_p} = \langle\mu_z^2\rangle_p(\hat{z}\cdot\hat{Z})^2 + (\langle\mu_x^2\rangle_p)((\hat{x}\cdot\hat{Z})^2+(\hat{y}\cdot\hat{Z})^2) \quad (74)$$

where lower-case unit vectors are protein frame axes and subscript p denotes an average in the protein-fixed frame. With elementary properties of vectors Eq 74 is rewritten,

$$\langle\mu_z^2\rangle_{\Omega^L_p} = \langle\mu_x^2\rangle_p + (\langle\mu_z^2\rangle_p - \langle\mu_x^2\rangle_p)(\hat{z}\cdot\hat{Z})^2, \quad (75)$$

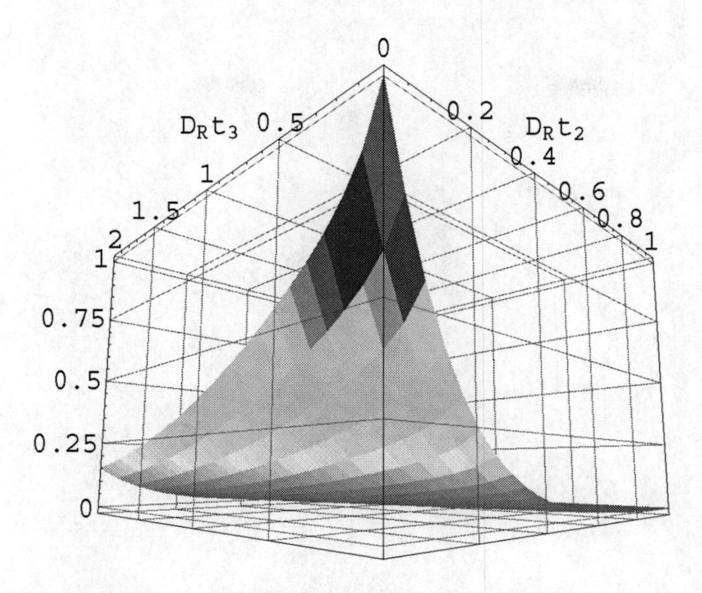

FIGURE 6. (Color online) Two-time *zzzz* correlation, $\langle \delta Y_{20}\ \delta Y_{10}(t_2)\ \delta Y_{10}(t_2+t_3)\rangle_a$, for a_1=0, $a_2=2$, vs $D_R t_2$ and $D_R t_3$.

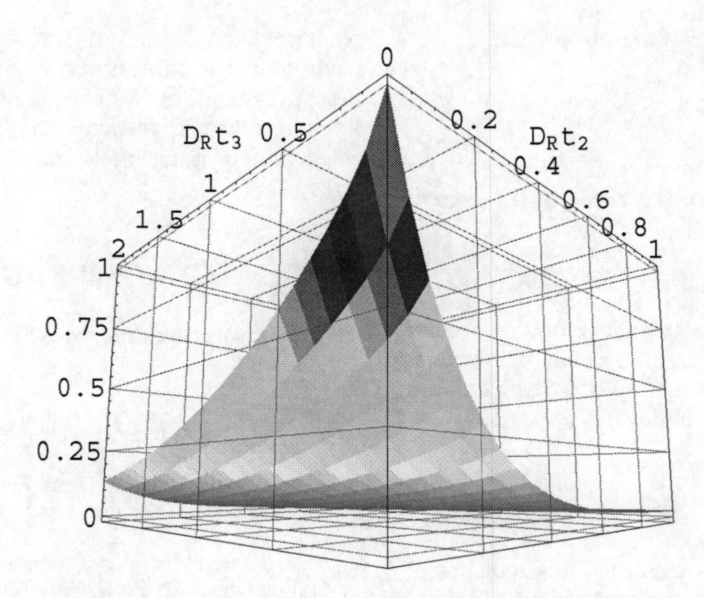

FIGURE 7. (Color online) Two-time *zzzz* correlation, $\langle \delta Y_{20}\ \delta Y_{10}(t_2)\ \delta Y_{10}(t_2+t_3)\rangle_a$, for a_1=2, $a_2=2$, vs $D_R t_2$ and $D_R t_3$.

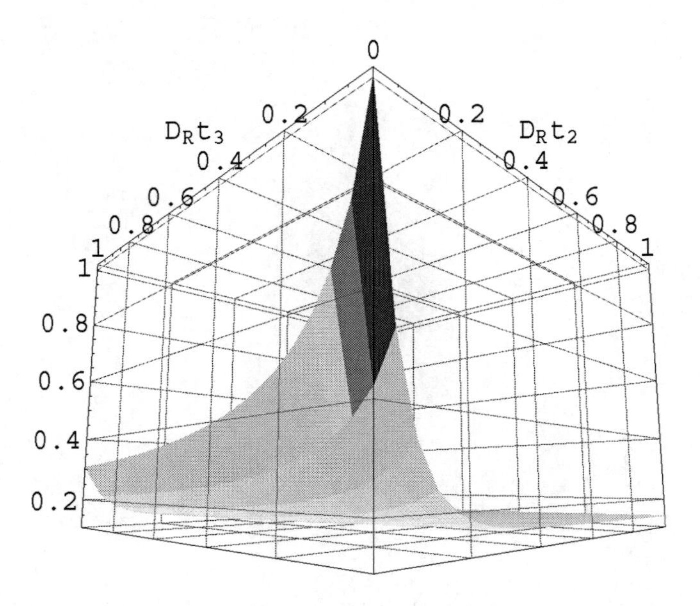

FIGURE 8. (Color online) Multi-linear contribution to the molecular-frame anisotropy for a_1=1, $a_2 = 4$, vs $D_R t_2$ and $D_R t_3$.

and it follows from our model of anisotropy that

$$\langle \mu_z^2 \rangle_p = \frac{1}{3}(1+\frac{a_2}{\sqrt{5\pi}});\ \langle \mu_x^2 \rangle_p = \frac{1}{3}(1-\frac{a_2}{2\sqrt{5\pi}}). \quad (76)$$

Inserting Eqs 75 and 76 into Eq 72, one need only average over the directions of the z-axis of the protein frame. The result is

$$\Delta C(t_2 \gg t_{mol}, t_3 = 0)/\Delta C(0,0) = a_2^2/(20\pi). \quad (77)$$

Thus we obtain the decay of the anisotropy to a finite value, and connect that value to the model parameters. Experimental [18] results for unbound CO in myoglobin are in the range of 0.25-0.35, indicating $a_2 \sim 4$.

DISCUSSION

Multi-time correlations are increasingly important in statistical mechanics, but are poorly understood. Here we have presented a general theory based on projection operators; this approach requires a timescale separation and identification of the slow variables. This can be done rigorously for the collective, hydrodynamic multi-linear variables, and the derivation proceeds accordingly. However it is applicable to any problem with identifiable slow variables, as is demonstrated for single-molecule reorientation.

In future work we will complete the calculation of lab-frame orientational correlations and compare with experiment. The examples used three basis functions, and the effect of including more basis functions will be considered. Note that while our focus is multi-time correlations, we have given a theory of reorientation in an anisotropic environment, and this is also virgin territory, even for ordinary correlations. We hope to explain the recent simulation results of Chowdhary and Ladanyi [19] on the reorientation of water molecules at water-hydrocarbon interfaces.

ACKNOWLEDGMENTS

This work was supported by NSF Grant CHE-0352026.

REFERENCES

1. I. Oppenheim, Prog. of Theor. Phys. (Supplement, Ed. by K. Kawasaki, Y. Kuramoto and H. Okamoto), 369-379 (1990).
2. S. Mukamel, *Principles of non-linear Optical Spectroscopy*, Oxford University Press, New York (1995).
3. R. De Vane, C. Ridley, B. Space and T. Keyes, J. Chem. Phys. **119**, 6073-6083 (2003); R. De Vane, C. Ridley, B. Space and T. Keyes, J. Chem. Phys. **121**, 3688-3699 (2004); R. DeVane, C. Ridley, B. Space and T. Keyes, Phys. Rev. E **70**, 050101-1-050101-9 (2004); R. DeVane, C. Ridley, B. Space and T. Keyes, J. Chem. Phys. **123**, 194507-1-194507-11 (2005); R. DeVane, C. Kasprzyk, B. Space and T. Keyes, J. Phys. Chem. B **110**, 3773-3786 (2006).
4. V. Chernyak and S. Mukamel, J. Chem. Phys. **108**, 5812-5822 (1998).
5. H. Sillescu, J. Chem. Phys. **104**, 4877-4885 (1996).
6. K. Okumura and Y. Tanimura, J. Chem. Phys. **107**, 2267-2279 (1997).

7. R. A. Denny and D. R. Reichman, Phys. Rev. E **63**, 65101-1-65101-12 (2001).
8. T. Keyes and J. T. Fourkas, J. Chem. Phys. **112**, 287-297 (2000); J. Kim and T. Keyes, Phys. Rev. E **65**, 061102-1-061102-12 (2002); J. Kim and T. Keyes, J. Chem. Phys. **122**, 244502-1-244502-13 (2005).
9. D. Jonas, Ann. Rev. Phys. Chem. **54**, 425-435 (2003).
10. A. Ma and R. M. Stratt, J. Chem. Phys. **116**, 4972-4980 (2002).
11. R. van Zon and J. Schofield, Phys. Rev. E **65**, 011106-1-011106-13 (2001).
12. J. Machta and I. Oppenheim, Physica A **112**, 361-371 (1982).
13. J. Schofield, R. Lim and I. Oppenheim, Physica A **181**, 89-97 (1992).
14. J. Schofield and I. Oppenheim, Physica A **187**, 210-222 (1992).
15. K. Kawasaki, Ann. Phys. (N.Y.) **61**, 1-36 (1970).
16. W. Goetze and L. Sjogren, Rep. Prog. Phys. **55**, 241-255 (1992).
17. P. Schramm and I. Oppenheim, Physica A **137**, 81-96 (1986).
18. J. Helbing, K. Nienhaus, G. Nienhaus and P. Hamm, J. Chem. Phys. **122**, 124505-1-124505-10 (2005).
19. J. Chowdhary and B. M. Ladanyi, preprint (2007).

Driven Colloidal Mixtures and Colloidal Liquid Crystals

H. Löwen, H. H. Wensink, and M. Rex

Institut für Theoretische Physik II: Weiche Materie, Heinrich-Heine-Universität Düsseldorf, Universitätsstraße 1, D-40225 Düsseldorf, Germany

Abstract. Colloids driven by an external field are ideal model systems for non-equilibrium problems. In particular, a binary driven system driven by a constant external force exhibits a non-equilibrium phase transition towards lane formation. In oppositely charged colloidal mixtures there are several intermediate states with lateral crystalline order. Driving tracer particles through quiescent liquid crystals generates rhythmic clustering at high drives. Driven binary mixtures of liquid crystals and spheres exhibit again lane formation of pure phases containing particles driven alike with a local biaxiality at the driven interface between pure spheres and pure rods. The results presented are based on Brownian dynamics computer simulation and theory.

Keywords: Colloids, Driven systems, Lane Formation, Liquid crystals, Non-equilibrium
PACS: 82.70.Dd, 05.40.-a, 61.20.Ja

INTRODUCTION

Colloidal suspensions are excellent model systems for equilibrium and non-equilibrium situations. In particular, the recent development of well-characterized particles and their control by external fields like optical tweezers have it made possible to expose colloids to controlled and tailored non-equilibrium conditions [1]. One possible set-up consists of dragging a tracer particle through a quiescent suspension which will destroy the colloidal order along its trace. The driving speed can be imposed by the external force field acting on the tagged bead particle. Another possibility is to consider a binary mixture of oppositely driven particles. This can e.g. be realized by using oppositely charged particles in an electric field [2] or by a combination of gravity and magnetic response [3].

In this paper we review recent progress in this field and concentrate mainly on results obtained by Brownian dynamics computer simulations. First we discuss a binary mixtures with identical interactions, the different species just interact differently with the external field imposed. The formation of lanes is discussed. Then we turn to oppositely charged particle where the like-species interact repulsively while the cross-interaction between different species is attractive. We subsequently consider colloidal liquid crystals. Dragging a spherical tracer particle through a nematic suspension perpendicular to the director will give rise to rhythmic clustering of rods in front of the moving particle. Finally we expose a binary mixture of rods and spheres to a constant driving force which acts differently on the two species and thus we obtain again lanes of rods and spheres driven alike. At the interface there is a further alignment which can be traced back to an equilibrium effect under confinement.

DRIVEN BINARY MIXTURES OF SPHERICAL PARTICLES WITH REPULSIVE CROSS-INTERACTION

In this case, the symmetric Yukawa-interaction has been studied in great detail both in two and three spatial dimensions [5]. In this case, without driving field, all particles interact exactly via the same interparticle potential. For increasing driving force, the system goes from a steady state with anisotropic correlations towards a state where macroscopic *lanes* are formed. The transition possesses a significant hysteresis, hence it can be classified as a first-order non-equilibrium phase transition. For very long times the system tends to exhibit macroscopic phase separation where two blocks of particles are sliding against each other, each block containing particles driven alike.

There is a reentrance effect for increasing density at fixed driving force: at small densities, there is no lane formation, at intermediate densities lanes are formed and at very high densities jammed configurations dominate which destroy again the lanes [6]. A dynamical density functional theory [7] describes the trends quantitatively well as compared to the Brownian dynamics computer simulations data [6]. These behavior is reminiscent to pedestrian dynamics [8].

The scenario described above is stable for asymmetric Yukawa interactions [9]. It occurs also for low temperatures where the starting state is a mixed binary crystal [10]. Finally we remark that it is irrelevant whether both species are driven or only one [5, 11]. If the two driving forces acting on the two particle species are tilted relative to each other, there is again lane formation but with tilted lanes [10]. A very strong repulsive cross-

CP982, *Complex Systems, 5th International Workshop on Complex Systems*
edited by M. Tokuyama, I. Oppenheim, and H. Nishiyama

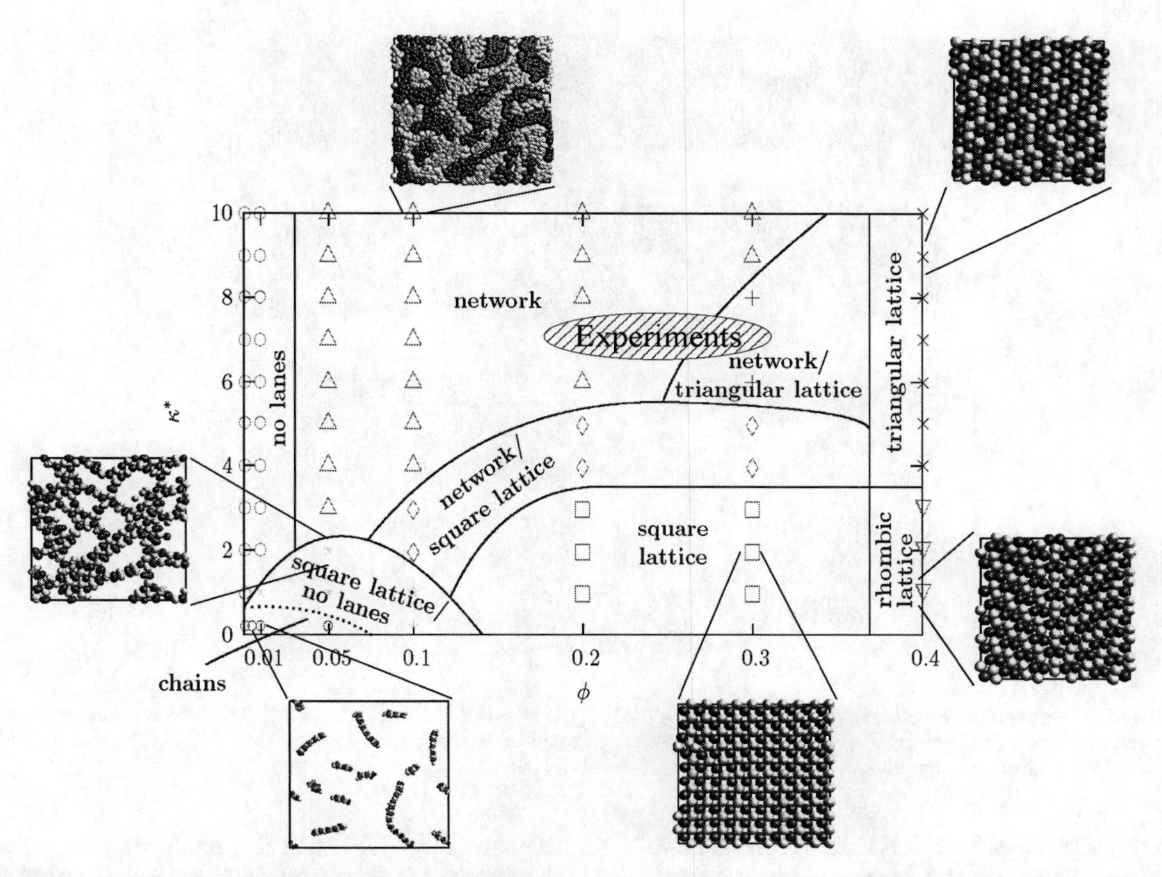

FIGURE 1. Non-equilibrium steady-state phase diagram for a constant driving force. A simulation snapshot of the projection of the particle coordinates perpendicular to the driving field is displayed for each different state. We find a variety of different phases involving lane chains at small volume fraction and low screening and lanes with two-dimensional crystalline order perpendicular to the field at high volume fraction. The lateral crystalline order can be a square, triangular or rhombic lattice. The hatched region indicates the parameters for which experimental results are published by Leunissen *et al.* [2]. For more details we refer to Ref. [4].

interaction leads to macro-phase separation. The time evolution of phase separation from a initially mixed state has been studied under an external drive in Ref. [12]. The Rayleigh-Taylor-like instability at a fluid-fluid interface, on the other hand, has been considered in Ref. [13].

DRIVEN BINARY MIXTURES OF SPHERICAL PARTICLES WITH ATTRACTIVE CROSS-INTERACTION

Recent experiments on oppositely charged particles in an electric field [2] have motivated studies of a Yukawa model with attractive cross-interaction. Using Brownian dynamics computer simulations in three dimensions, a wealth of different lateral structures were found resulting from a competition between Coulomb interaction and friction and sliding lanes [4]. The steady states involve lateral network-like structures and square, triangular and rhombic lattices of oppositely driven lanes. In two spatial dimensions, on the other hand, there is just an alternating stripe formation [14].

The steady state phase diagram is summarized in Figure 1 [4]. Here a fixed external force is used and the results are shown for different screening parameters κ^* and volume fractions ϕ. Here, the interaction is a superposition of a (softened) core of diameter σ and a Yukawa interaction $\pm U_0 \exp(-\kappa^* r/\sigma)/r$. The parameters are: $U_0/\sigma k_B T = 50$, $F_0\sigma/k_B T = 236$, where $k_B T$ denotes the thermal energy and F_0 is the external driving forces induced by the electric field.

The network structure was also found in real space experiments of driven oppositely charged colloids [2]. This is confirmed by simulation by using estimated model parameters suitable to describe the experimental situation, see the star symbol in Figure 1. Still a comparison in the full parameter space has to be performed. Furthermore a theory, e.g. in the spirit of dynamical density functional theory [6, 7] is still missing in this case.

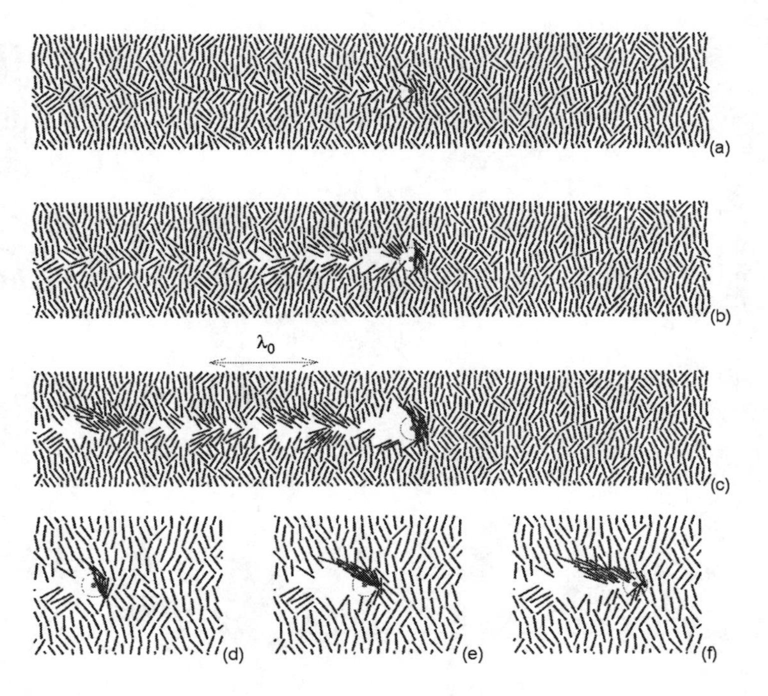

FIGURE 2. Snapshots of the nematic structure at (a) $v^* = 100$, (b) $v^* = 1000$ and (c) $v^* = 4000$. The driven sphere is indicated by the encircled dot. The sequence (d)-(f) depicts the growth and rotation of a single rod cluster at $v^* = 4000$. The arrow indicates a cluster-cluster correlation length $\lambda_0 = 5L_0$. Figure taken from Ref. [15].

DRIVING A SPHERE THROUGH A NEMATIC COLLOIDAL CRYSTAL

The Brownian dynamics can readily be generalized to rod-like particles [16, 17]. Then two translational diffusion constants, parallel and perpendicular to the orientation of the particle have to be considered as well as on orientational diffusion constant. For large rod aspect ratios, the suspensions can form a nematic state where the director is oriented.

Recently an interesting *rhythmic clustering* behavior was found in a driven system [15] with Yukawa resp. Yukawa segment interactions. In particular, a sphere was dragged by constant speed through a two-dimensional nematic crystal. The driving direction was perpendicular to the nematic director. At high driving strengths, the driven sphere generates clusters of rods which rhythmically grow and dissolve with a characteristic frequency. This corresponds to a characteristic length along the trace of the driven sphere. The simulation data are in accordance with a dynamical scaling theory. A typical simulation snapshot is shown in Figure 2.

The parameters are $U_0/\sigma k_B T = 0.065$ for the segment-segment interaction, $\sigma/L_0 = 16.7$ is the ratio of the rod length L_0 and diameter σ and $\kappa^*\sigma = 0.36$. We simulated $N = 900$ rods each composed of 13 segments. The interaction between the tracer sphere and the rods is of the form $1/r^{12}$ and the amplitude is tuned such that the sphere cannot penetrate the rods. Furthermore, $v^* = vL_0/D_0^T \gg 1$ is the speed of the dragged tracer sphere expressed in units of the typical translational diffusion speed via D_0^T, the average translation diffusion coefficient of the rod. The dimensionless density of the systems is $NL_0^2/A = 3.0$ (A is the area of the rectangular simulation box) which is high enough to ensure a 2D bulk nematic configuration with a director pointing along the vertical axis of the box.

Clearly a cluster of rods is formed in front of the driven particles which dissolves and leaves a clustered structure in the "wake" of the driven sphere. Still an experimental verification of this effect is missing but it is in principle possible to prepare nematic colloidal liquid crystals and drag tracer particles through them.

DRIVEN BINARY MIXTURES OF REPULSIVE SPHERES AND RODS

Finally one may consider a binary mixtures of spheres and rods and drive the rods by a constant force acting on their center of mass. We have performed Brownian dynamics computer simulations in two spatial dimensions by using a repulsive Yukawa interaction, similar to that in Ref. [15]. In detail, the segment-segment interactions and

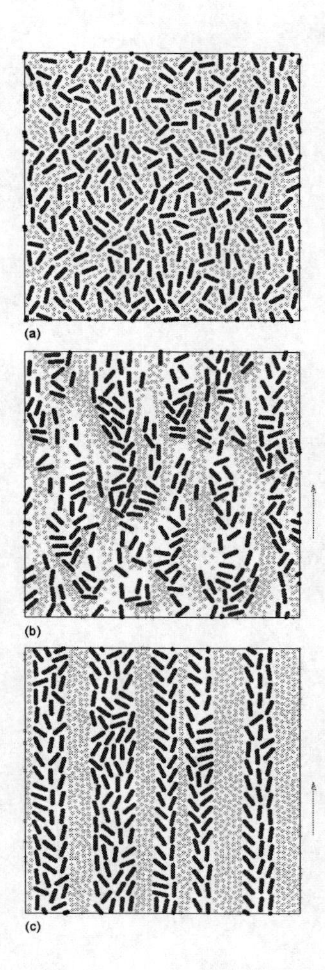

FIGURE 3. Snapshots of the driven rod-sphere mixture at constant driving force. (a) Equilibrium state at $t = 0$. (b) Intermediate driven state at $t = 1.4$. (c) Steady-state laning structure at $t = 45$. The direction of the applied force is indicated by the arrow. The time t is measured in units of the typical Brownian translational relaxation time of a single rod.

the rod aspect-ratio are the same as described in the previous paragraph. Here, the cross interaction between the spheres and the rod segment is identical to the segment-segment interaction of the rods. Furthermore, we have $N_{rods}/N_{spheres} = 0.25$ with a total of 1250 particles using 10 Yukawa segments for each rod. The driving force acting on both species is fixed at $F_0 L_0 / k_B T = 100$. Note that the rod density is smaller than in the previous case such as to guarantee a thermodynamically stable reference mixture with an isotropic rod configuration.

In this system, sufficiently strong drives lead again to lane formation similar to that discussed in the 2nd paragraph for spheres. However, there is an additional aligning effect in the lanes of the driven rods. The lanes slide against each other and at this sliding interface a homeotropic ordering of the rods is observed. This means that the rods are oriented perpendicular to (or make a typical angle with) this interface which can clearly been seen from the simulation snapshots shown in Figure 3. This effect can be explained from a static concept and stems from a subtle interplay between and external potential imposed by the quasi-static soft walls composed of the spheres and the orientation-dependent pair potential of the rods. Moreover, upon averaging over the rod orientations we find that the local orientational structure within the lanes is in fact biaxial. This effect is also observed in strongly confined rods in equilibrium, see e.g. [18]. In conclusion, the drive here produces a *self-organized confinement* and confinement-induced effects, such as tilting and biaxiality, of the rods take place.

CONCLUSIONS

Colloidal suspensions which are driven by an external field exhibit interesting non-equilibrium phase transitions such as laning, rhythmic clustering and self-organized confinement. In principle the results of Brownian dynamics computer simulations can be checked by real-space experiments but they serve also as benchmark data to test theories such as the dynamical density functional theory.

There is still a need to understand the influence of hydrodynamic interactions mediated by the solvent. Fortunately the leading long-ranged contribution is screened for electrophoresis [19], therefore one might hope that they are not relevant for charged colloids in an electric field while it seems that they are more important for sedimentation [20]. Certainly more work is needed in this direction following more sophisticated simulation schemes, e.g. as proposed by Yamamoto and coworkers [21].

Finally it would be very interesting to study the lane formation in a confined geometry between parallel plates. It was found that jammed configurations are significantly stabilized by confinement relative to lane formation [22]. More simulations to study systematically the effect of confinement are needed. It would also be interesting to perform an instability analysis based on dynamical density functional theory for the confined case following the strategy put forward in Ref. [7].

ACKNOWLEDGMENTS

We thank A. van Blaaderen, M. Leunissen, D. Cleaver and R. Yamamoto for helpful discussions. This work was financially supported by the DFG under SFB-TR6 (project section D1).

REFERENCES

1. H. Löwen, *J. Phys.: Condens. Matter* **13**, R415-423 (2001).
2. M. E. Leunissen, C. G. Christova, A. P. Hynninen, C. P. Royall, A. I. Campbell, A. Imhof, R. van Roij, and A. van Blaaderen, *Nature* **437**, 235-240 (2005).
3. M. Köppl, P. Henseler, A. Erbe, P. Nielaba, and P. Leiderer, *Phys. Rev. Lett.* **97**, 208302-1-4 (2006).
4. M. Rex, and H. Löwen, *Phys. Rev. E* **75**, 051402-1-10 (2007).
5. J. Dzubiella, G. P. Hoffmann, and H. Löwen, *Phys. Rev. E* **65**, 021402-1-8 (2002).
6. J. Chakrabarti, J. Dzubiella, and H. Löwen, *Phys. Rev. E* **70**, 012401-1-4 (2004).
7. J. Chakrabarti, J. Dzubiella, and H. Löwen, *Europhys. Lett.* **61**, 415-421 (2003).
8. R. Jiang, D. Helbing, P. K. Shukla, and Q. S. Wu, *Physica A* **368**, 567-574 (2006).
9. H. Löwen, and J. Dzubiella, *Faraday Dicussions* **123**, 99-105 (2003).
10. J. Dzubiella, and H. Löwen, *J. Phys.: Condens. Matter* **14**, 9383-9395 (2002).
11. C. Reichhardt, and C. J. Olson-Reichhardt, *Phys. Rev. E* **74**, 011403-1-7 (2006).
12. H. Löwen, C. N. Likos, R. Blaak, S. Auer, V. Froltsov, J. Dzubiella, A. Wysocki, and H. M. Harreis, "Colloidal suspensions driven by external fields," in *AIP Conference Proceedings*, edited by M. Tokuyama, and I. Oppenheim, 2004, vol. 708, pp. 3-7.
13. A. Wysocki, and H. Löwen, *J. Phys.: Condens. Matter* **16**, 7209-7224 (2004).
14. R. R. Netz, *Europhys. Lett.* **63**, 616-622 (2003).
15. H. H. Wensink, and H. Löwen, *Phys. Rev. Lett.* **97**, 038303-1-4 (2006).
16. H. Löwen, *Phys. Rev. E* **50**, 1232-1242 (1994).
17. T. Kirchhoff, H. Löwen, and R. Klein, *Phys. Rev. E* **53**, 5011-5022 (1996).
18. A. Chrzanowska, P. I. C. Teixeira, H. Ehrentraut, and D. J. Cleaver, *J. Phys.: Condens. Matter* **13**, 4715-4726 (2001).
19. D. Long, and A. Ajdari, *Eur. Phys. J. E* **4**, 29-32 (2001).
20. C. P. Royall, J. Dzubiella, M. Schmidt, and A. van Blaaderen, *Phys. Rev. Lett.* **98**, 188304-1-4 (2007).
21. K. Kim, Y. Nakayama, and R. Yamamoto, *Phys. Rev. Lett.* **96**, 208302-1-4 (2006).
22. D. Helbing, I. J. Farkas, and T. Viscek, *Phys. Rev. Lett.* **84**, 1240-1243 (2000).

The Structure and Thermodynamic Stability of Reverse Micelles in Dry AOT/Alkane Mixtures

Adam Wootton, Francois Picavez, and Peter Harrowell

School of Chemistry, University of Sydney, Sydney, New South Wales 2006, Australia

Abstract. Monte Carlo simulation studies of reverse micelles of an anionic surfactant, sodium AOT, in a non-polar solvent provide strong evidence that, in the absence of water, these clusters are charge ordered polyhedral shells. The stabilizing energy of these clusters is so large that the entropy of mixing is, in comparison, inconsequential and we predict that, if all waters of hydration could be removed (something not yet accomplished for the sodium salt) then AOT would be insoluble in nonpolar solvents.

INTRODUCTION

The self assembly of amphiphiles in non-aqueous media represents an important and outstanding puzzle. In the absence of water, we cannot invoke the hydrophobic effects that govern aqueous self assembly into micelles. In water, these small aggregates represent a balance between two opposing fragile effects (i.e of the order of k_BT) : the entropy of mixing, on one hand, and the hydrophobic interaction between the surfactant's nonpolar 'tails' and the water, on the other. No such general explanation currently exists for reverse micelles (micelles with the non-polar groups on the exterior) that are observed in organic solvents. The best studied case is that of the anionic surfactant sodium bis (2-ethyl-hexyl) sulfosuccinate (commonly known as sodium Aerosol OT or sodium AOT). Many surfactants are not oil-soluble, so compounds like sodium AOT represent a significant but select subset of amphiphiles, typically characterised by branched alkane chains. There has been considerable interest in using reverse micelles or, more specifically, the microemulsions obtained by swelling these reverse micelles with water, as microscopic reaction vessels. While there has been a number of theoretical studies of the reverse AOT micelles (summarised below), most have focussed on the properties of the confined water rather than the stability of the self assembled structure itself. The goal of this paper is to explore the energetics and thermodynamics of micelle formation in dry sodium AOT/alkane mixtures. In doing so we shall draw on our recent work in modeling the stability of ionic clusters with significant size asymmetry [1].

The experimental situation has been described in the comprehensive study by Ekwall, Mandell and Fontell [2]. Of particular significance we note that pure sodium AOT does not form a crystalline solid, rather it appears as an inverse hexagonal structure. Note that whereas the crystal structure has no connection with the normal micelle structure, it is, as we shall demonstrate, closely associated with that of the reverse micelle. In a sense, we shall argue, that reverse micelles are best regarded as representing a partial disruption of the crystalline structure. The forces that dominate crystallisation, hence, dominate reverse micelle formation. We are not aware of this point being made previously.

Salaniwal et al [3,4] have performed simulations on inverse microemulsions of sodium AOT and water (w $\geq$ 10) in supercritical CO_2. Senapati et al [5] also used CO_2 as a solvent but with phosphate fluoro-surfactants. Self assembly of roughly spherical micelles of the experimentally observed size was reported. The orientation and location of the confined waters were observed to be strongly influenced by the electric field of the ions.

Comprehensive MD simulations have also been performed to study the stability of micelles with respect to aggregation [6] and the optimum solvation shells of AOT molecules [7]. Both of these studies reported good agreement with experimental observations. Other studies have used MD studies on surfactants with Ca^{2+} as the cations and CO_3^{2-} as the ionic head group [8,9]. These highly charged ions

CP982, *Complex Systems, 5th International Workshop on Complex Systems*
edited by M. Tokuyama, I. Oppenheim, and H. Nishiyama

were tightly bound and were reported essentially as solid cores as the ions were tightly bound.

Early simulations of reverse micelles used a simplified tail model in which the tails where represented as a soft repulsive sphere [10]. The tail has been represented by beads and springs [11-13] and flexible chains of soft sphere [14]. All studies emphasise that the key contribution of the tails is to prevent the head groups moving into the core of the micelle. It must be noted that these simulations were performed with w >> 1 and completely neglecting the tail would have permitted the complete hydration of the head group.

A number of models of reverse micelles have been reported in which the chain is discarded to be replaced by some constraint on the head group. In refs. [15-17] the head groups are constrained to lie on a spherical shell while the counterions and water are constrained to the interior. These simulations, for $1 \leq w \leq 10$, confirmed the distinction of bound and unbound waters. The artifacts arising from the restriction of the waters to the inside of the sphere was acknowledged in this work. Another variant pins the anions to the surface of the cluster and allows the cluster shape to vary while optimizing the positions of the counter ions. [18].

An early study by Kotlarchyk et al [19] attempted to fit the results of small angle neutron scattering (SANS) to a model. These workers noted that the centre of the micelle was going to be densely filled. They proposed that the head groups would form an icosahedron (the counter ions were ignored) with the water occupying the internal space.

THERMODYNAMICS OF CLUSTERING IN AN IDEAL SOLUTION

We shall develop the theory of reverse micelle stability along similar lines to those followed in the theory of normal micelles in aqueous solution as presented by Ben-Shaul and Gelbart [20]

We shall consider reverse micelles containing N ionic surfactant molecules (with the associated counter ion bound). Assuming that dilute solution theory holds for the surfactant we can write the chemical potential of the N-cluster to be

$$\mu_N = \mu_N^{o,\rho} + k_B T \ln \rho_N \qquad (1)$$

where $\mu_N^{o,\rho} = -k_B T \ln(q_N / V)$ with q_N the partition function of the N-cluster and $\rho_N = n_N/V$, the number density of N-clusters. It is useful to refer concentrations back to the total amount of surfactant. To this end we define the mole fraction of surfactants to be found in the n_N N-cluster as

$$X_N = N n_N / (N_{tot} + Q) \qquad (2)$$

where N_{tot} and Q are the total number of surfactant and organic solvent molecules respectively. Note that

$$\sum_N X_N = X \qquad (3)$$

where $X = N_{tot}/(N_{tot}+Q)$ is the total mole fraction of surfactant present and represents the important experimental control parameter.

We now need to re-express the chemical potential in terms of X_N. We first note that

$$X_N = \frac{N\rho_N}{\rho_{tot}} \qquad (4)$$

where $\rho_{tot} = N_{tot}/V$. Substituting, Eq. xx into Eq. xx gives

$$\mu_N = \mu_N^o + k_B T \ln(X_N / N) \qquad (5)$$

where $\mu_N^o = \mu_N^{o,\rho} + k_B T \ln \rho_{tot}$. While strictly μ_N^o retains a dependence on the surfactant concentration through ρ_{tot}, for small concentrations $\rho_{tot} \approx \rho_{solvent}$ and, on this basis, we shall neglect this X dependence in the following analysis. At equilibrium we have

$$\mu_N = N\mu_1 \qquad (6)$$

so that

$$\mu_N^o + k_B T \ln(X_N / N) = N(\mu_1^o + k_B T \ln X_1) \qquad (7)$$

We can rearrange this relation to give the following expression for the cluster fraction X_N,

$$X_N / N = X_1^N \exp[\beta N(\mu_1^o - \mu_N^o / N)] \qquad (8)$$

Following Israelachvili *et al* [21], we shall define the mean cluster size as

$$\overline{N} = \frac{\sum_{N>1} N^2 x_N}{\sum_{N>1} x_N} \qquad (9)$$

This average cluster size can also be written as

$$\overline{N} = \frac{\partial \ln(S - x_1)}{\partial \ln x_1} \qquad (10)$$

The variance σ_N^2 can be written as

$$\sigma_N^2 = \frac{\partial \overline{N}}{\partial \ln x_1} = \overline{N}\frac{\partial \overline{N}}{\partial \ln(S - x_1)} \qquad (11)$$

THE CHEMICAL POTENTIAL OF THE SURFACTANT CLUSTER (WHERE WE TREAT DRY REVERSE MICELLES AS A PROBLEM OF IONIC CLUSTERING IN THE VACUUM).

To calculate the chemical potential difference $\Delta_N = \mu_N^o - N\mu_1^o$ between the cluster of N surfactants and N isolated surfactants we shall introduce three simplifications. While each sounds more monstrous than the last, we beg the reader's patience as we shall seek to justify these assumptions *a postori* on the basis of the energy scales involved. The three simplifications are: i) we shall replace the solvent by a vacuum, ii) neglect any contribution from the tails, and iii) model the sulfonate head groups as a spherical ion.

The neglect of the solvent can be relaxed with the use of the generalised Born model of Still and coworkers [22] but we do not expect any significant change to the major conclusions of this study. In this calculation we shall treat the tails of the surfactant as simply part of the nonpolar solvent. Our reasoning is that there would appear to be little change to the environment and degree of constraint of the chain from that of the isolated surfactant to that when the surfactant is incorporated on the surface of a small cluster with high curvature. (Neglect of the chains would become a problem as the curvature changed sign.) As described below, the surfactant tails are not required to constrain the AOT anions to the surface of the cluster, this being automatically accomplished (in the absence of water) by the size asymmetry of the ions. Probably the most serious of our simplifications is the neglect of the shape and charge distribution in the anionic head group.

From Eqs. 1 and 5 we have

$$\mu_N^o = -k_B T \ln(q_N / V) + k_B T \ln \rho_{tot} \qquad (12)$$

Following ref. [23], we can divide the aggregate partition function q_N into the momenta and configurational contributions so that

$$q_N = \frac{8\pi^2 V \exp(-E_N / k_B T)}{\Lambda^{3N}} \qquad (13)$$

where the $8\pi^2$ represents the rotational configuration space of the aggregate (or surfactant) and $\Lambda = \left(\frac{h^2}{2\pi m k_B T}\right)^{1/2}$ is the thermal de Broglie wavelength with m being the mass of the individual ion pair. We have assumed that the ions in the aggregate bind so strongly that, at the temperatures of interest, we can neglect vibrations in the configurational contribution and assume that the aggregate structure is simply the minimum energy structure, with that minimum energy being E_N. We acknowledge a logical inconsistency here in that, strictly, we should not treat the vibrations classically in the case of such tight binding. We model the sulfonate head group as a soft spherical ion. The following interaction potential was used:

$$V_{ij} = \frac{\alpha}{r_{ij}^{12}} - \frac{\beta}{r_{ij}^{6}} + \frac{q_i q_j}{r_{ij}} \qquad (14)$$

where the charges on the sulfonate and sodium ions are q = -1 and +1, respectively, and $\alpha_{Na\text{-}Na}$ = 0.021025 $kJmol^{-1}nm^{12}$, $\alpha_{AOT\text{-}AOT}$ = 2.6341 $kJmol^{-1}nm^{12}$, $\beta_{Na\text{-}Na}$ = 0.7206 x 10^{-4} $kJmol^{-1}nm^{6}$ and $\beta_{AOT\text{-}AOT}$ = 0.002617 $kJmol^{-1}nm^{6}$. The cross terms are obtained using $\alpha_{ij} = \sqrt{\alpha_{ii}\alpha_{jj}}$.

The ratio of the radius of the sulfonate ion over that of the sodium cation is 1.5 (taken as $(\alpha_{AOT\text{-}AOT}/\alpha_{Na\text{-}Na})^{1/12}$). We have recently completed studies of clusters of ions with large size asymmetries [1]. Our main observation was that large size asymmetry (and cluster sizes with N < 120) results in energy minima corresponding to polyhedral shells rather than recognizable fragments of the bulk crystal phase. The relevance of this result in the context of reverse micelles is that the polyhedral shell automatically stabilizes the tail-bearing anions at the surface of the cluster without the need for the explicit inclusion of the tail or some equivalent constraint. We also note that the counter ions, rather than making up a dissociated double layer, form an integral part of the ordered cluster.

The size asymmetry of the sodium and sulfonate ions is similar to that of the AgI system that we have studied previously. The gallery of minimum energy clusters for the sodium is shown in Fig.1. Briefly, these minima represent the lowest energy structure obtained after applying a conjugate gradient minimisation to the potential energy to the complete set of possible clusters of size N consisting of trivalent vertices. This complete set was generated using an algorithm *CaGe* [24]. For clusters of N > 48 we have assumed that the lowest energy structure is a rod.

ESTIMATING THE CMC AND AVERAGE CLUSTER SIZE

We shall begin by considering an ionic cluster of N ion pairs (each pair being a head group and counterion). It is worth pausing here to emphasise the essential conclusion from the previous consideration of aggregation energetics. In the absence of water, reverse AOT micelles are ionic clusters. In stark contrast to the case of normal micelles, the aliphatic chains serve no specific role in stabilising aggregation in nonpolar solvents. As we shall discuss in Section 5, the main role of the branched tails is to *destabilize* the crystal state and so enhance surfactant solubility in nonaqueous phases.

The average potential energy of the lowest energy ion clusters at T = 300K is well described by a fitting function provided by Eq.15. This form has been discussed previously [21] as arising from the contributions of the rod ends (b) and the 'barrel' (Na) and can be written as

$$\frac{E_N}{N} = \frac{b}{N} - a \tag{15}$$

where a = 5.8 eV and b = 2.8 eV. Note 5.8eV corresponds, at 300K, to 200kT, a contribution well in excess of the contributions neglected such as the solvent polarizability and the chain entropy.

Based on the estimates described above, we shall approximate the chemical potential μ_N^o for the N-cluster as

$$\mu_N^o \approx \frac{E_N}{N} \tag{16}$$

Israelachvili et al [21] have considered the equilibrium distribution of micelles for the case when the chemical potential for the N-cluster can be written as $\mu_\infty^o + b/N$ (i.e similar to our Eq.15). In this case one can write

$$\frac{X_N}{N} = e^{-\alpha} Y^N \tag{17}$$

where $\alpha = b/kT$ and

$$Y = X_1 \exp[(\mu_1^o - \mu_\infty^o)/kT] \tag{18}$$

The total amphiphile concentration S can be written as

$$S = X_1 + e^{-\alpha} \sum_{N=2}^{\infty} N Y^N \tag{19}$$

The sum in Eq. 19 can be performed analytically to give

$$S = X_1 + \frac{2Y^2 e^{-\alpha}}{1-Y}\left(1 + \frac{Y}{2(1-Y)}\right) \tag{20}$$

Using Eqs.10 and 19 we have the following expression for the average cluster size

$$\overline{N} = 2 + \frac{Y}{1-Y}\left(1 + \frac{1}{2(1-Y)+Y}\right) \tag{21}$$

Given a value of S, the total concentration of amphiphile, and explicit expressions for the cluster chemical potentials, Eq.20 provides, in principle, as explicit solution for X_1, the concentration of free surfactant. In the case where the energetics favours large clusters, we can obtain an approximate analytical expression for X_1. From Eq.17 we can see that for X_N to be significant at large N, $Y \approx 1$. In this case we can approximate S and $\overline{N}$ by

$$S \approx X_1 + \frac{e^{-\alpha}}{(1-Y)^2} \tag{22}$$

and

$$\overline{N} \approx \frac{2}{1-Y} \tag{23}$$

From Eq.22 we have

$$X_1 \approx (1-(Se^{\alpha})^{-1/2})\exp\left[\left(\mu_{\infty}^{o} - \mu_1^{o}\right)/kT\right] \tag{24}$$

For all S > 10^3 $e^{-\alpha}$, this gives us $X_1 \approx \exp\left[\left(\mu_{\infty}^{o} - \mu_1^{o}\right)/kT\right]$. This result is consistent with our assumption that Y ≈ 1. More importantly, it predicts that the surfactant is insoluble at any accessible temperature. Consideration of the average cluster size underscores this conclusion. We can use Eq.22 to eliminate Y in Eq.23 to give

$$\overline{N} \approx 2\sqrt{Se^{\alpha}} \tag{25}$$

Taking solutions of 10^{-4} % and 1% by weight of sodium AOT in decanol, Eq. 25 gives an estimate of the number of surfactants per cluster of 10^{21} and 10^{22}, respectively. Such large rods would certainly aggregate and fall out of solution. Clearly the large ionic stabilization energy precludes solubility of such ionic surfactants in a nonpolar medium.

CONCLUSIONS

The AOT clusters are ordered along the tubes and insoluble in organic solvents. Should it be possible to produce a hexagonal solid consisting of a single domain of aligned tubes then we predict that x-ray scattering would observe the periodicity of the ion ordering. The tubes which pack to form the hexagonal solid phase are stabilised by the asymmetry in ion size. This purely ionic effect acts quite independently of any additional stabilization of the tubes arising from excluded volume effects of the chains.

To our knowledge sodium AOT has not been dried to less than 0.7 water molecules per surfactant. We conclude that this trace water is essential for AOT solubility and must act to break up the extended tubes. Preliminary Monte Carlo simulations of the ionic clusters studied here in the presence of a small amount of water strongly supports this proposition. We examine the role of waters of hydration n stabilizing reverse AOT micelles in a paper currently in preparation.

The role of chains in reverse micelles is quite different from their role in aqueous micelles. In reverse micelles they function only to disrupt the solid phase. This means that, whereas in normal micelles the degree of branching of the alkane chains is of relatively minor significance with respect to micelle stability, such branching is all important in micelles in organic solvents.

ACKNOWLEDGMENTS

We would like to thank Greg Warr for valuable and timely guidance. AW acknowledges the support of the Gritton and Joan Clark scholarships.

REFERENCES

1. A.Wootton and P.Harrowell, *J.Chem.Phys* **121**, 7440-7442 (2004); *J.Chem.Phys.B* **108**, 8412-8418 (2004).
2. P. Ekwall, L. Mandell and K. Fontell, *J. Coll. Int. Science* **33**, 215-235 (1970).
3. S.Salaniwal, S.T.Cui, H.D.Cochran and P.T.Cummins, *Langmuir*, **17**, 1784-1792 (2001).
4. S.Saliniwal, S.K.Kumar and A.Z.Panagiotopoulos, *Langmuir* **19**, 5164-5168 (2003).
5. S.Senapati and M.Berkowitz, *J.Chem.Phys.* **118**, 1937-1944 (2003).
6. M. Alaimo and T. Kumosinski, *Langmuir* **13**, 2007-2018 (1997).
7. B.Derecskei, A. Derecskei-Kovacs and Z. Schelly, *Langmuir* **15**, 1981-1992 (1999).
8. J.Griffiths and D.Heyes, *Langmuir* **12**, 2418-2424 (1996).
9. D.Tobias and M.Klein, *J.Phys.Chem.* **100**, 6637-6648 (1996).
10. D.Brown and J.Clarke, *J.Phys.Chem.* **92**, 2881-2888 (1988).
11. B.Smit, P.Hilbers, K.Esselink, L.Rupert, N. Van Os and A.Schlijper, *Nature* **348**, 624-625 (1990).
12. B.Smit, P.Hilbers, K.Esselink, L.Rupert, N. Van Os and A.Schlijper, *J.Phys.Chem.* **95**, 6361-6368 (1991).
13. B.Smit, K.Esselink, P.Hilbers, P.Van Os, l.Rupert and I.Szleifer, *Langmuir* **9**, 9-11 (1993).
14. S.Karaboni and R. O'Connell, *Langmuir* **6**, 905-911 (1990).
15. J.Faeder and B.Ladanyi, *J.Phys.Chem.B* **104**, 1033-1046 (2000).
16. J.Faeder, M.Albert and B.Ladanyi, *Langmuir* **19**, 2514-2520 (2003).
17. J.Faeder and B.Ladanyi, *J.Phys.Chem.B* **105**, 11148-11158 (2001).
18. A.Bulavchenko, A.Bastishchev, E.Batishcheva and V.Torgov, *J.Phys.Chem B* **106**, 6381-6389 (2002).
19. M.Kotlarchyk, J.Huang and S.Chen, *J.Phys.Chem.* **89**, 4382-4386 (1985).
20. A. Ben-Shaul and W. M. Gelbart, in *Micelles, Membranes, Microemulsions and Monolayers,* ed. A. Ben-Shaul, W. Gelbart and D. Roux, Berlin Spinger, 1994, p.1-28.

21. J. N. Israelachvili, D. J. Mitchell and B. W. Ninham, *J. Chem. Soc. Faraday Trans. 2* **72**, 1525-1530 (1976).
22. W.C.Still, A.Tempczyk, R.Hawley and T.Hendrickson, *J.Am.Chem.Soc.* **112**, 6127-6131 (1990).
23. R.Nagarajan and E.Ruckenstein, *Langmuir* **7**, 2934-2969 (1991).
24. The CaGe software was developed in the Department of Mathematics at the University of Bielefeld. At the time of writing; the software is available from http://www.mathematik.uni-bielefeld.de/ CaGe/Archive/.

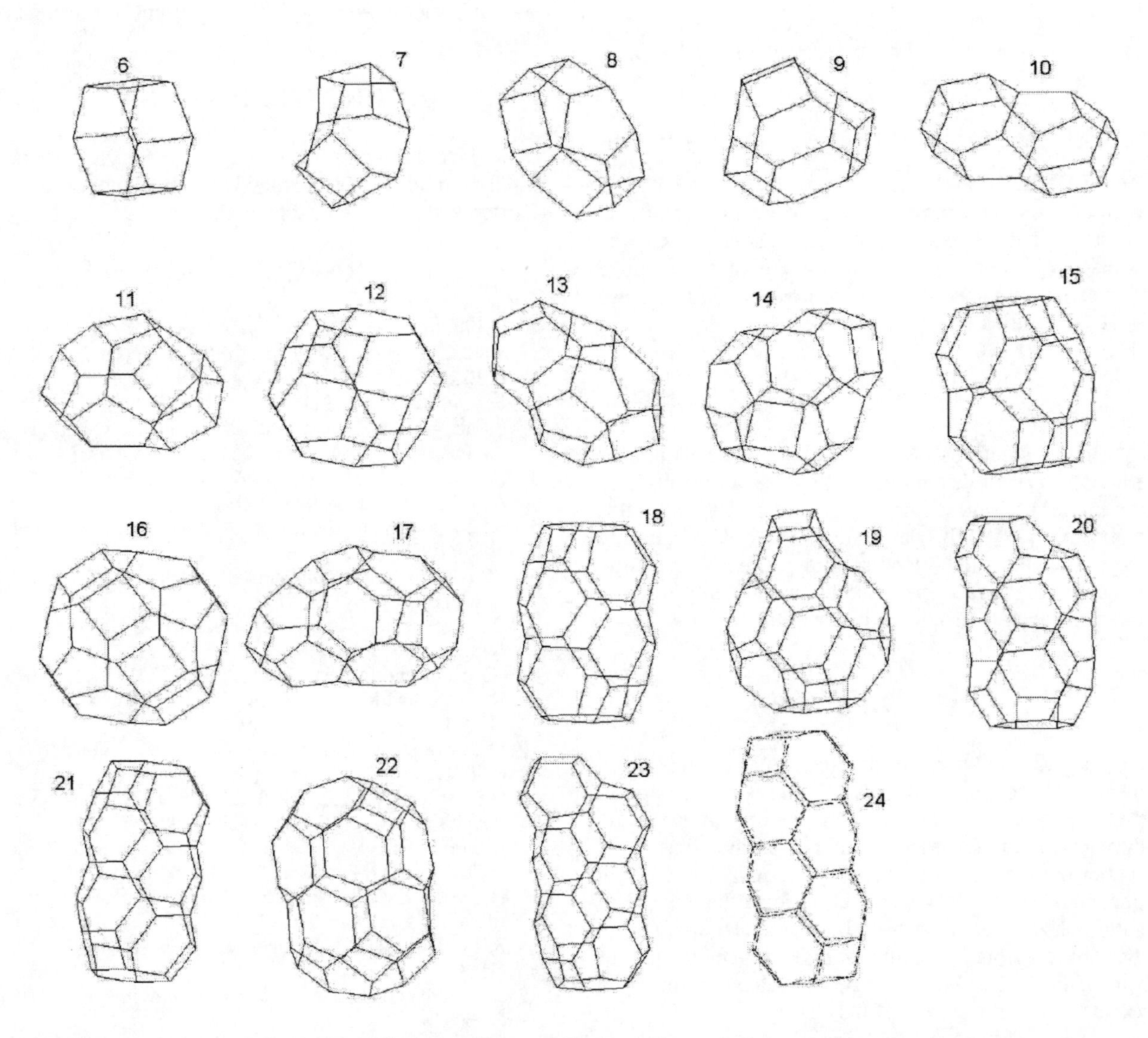

FIGURE 1. Structures of the lowest energy clusters for the sodium-sulfonate system as described in the text. N refers to the number of surfactants or, equivalently, the number of anion-cation pairs.

Models of Gel-Forming Colloids

P. Tartaglia

Dipartimento di Fisica and SMC-INFM-CNR: Center for Statistical Mechanics and Complexity, P.le Aldo Moro 5, I-00185 Roma, Italy

Abstract. We study the phenomenon of structral arrest in colloidal suspensions, with emphasis on the equilibrium and non-equilibrium routes to gel formation. We describe a number of model systems, treated with molecular dynamics simulations, which give rise to equilibrium gels in a wide region of volume fractions of the dispersed phase. These models have in common the anisotropy of the mutual interactions and show a set of characteristic features, i.e. the drastic reduction of the unstable two-phase region ('empty liquids'), the Arrhenius form decay for the diffusivity, the typical behavior of strong glasses, and the approach to an ideal gel state due to bonding among the particles. We also show that anisotropic interactions allow self-assembly in colloids to be treated using an equilibrium liquid-state approach.

Keywords: Colloids, Gels, Glass transition
PACS: 82.70.Dd, 61.20.Ja, 82.70.Gg

INTRODUCTION

Structural arrest refers in general to various phenomena which have in common a marked slowing down of the dynamics. They include aggregation and cluster formation, percolation and glass transition. In the last few years the study of the glass transition of colloidal systems in the supercooled region revealed a series of new phenomena which were interpreted as the possibility of including in a single framework glass and gel transition in liquids [1].

The study of the glass transition in supercooled liquids was initiated in hard-sphere systems, where the cage effect is the relevant physical mechanism for the structural arrest. Experimental, theoretical and simulation approaches gave a reasonable picture of the phenomenon. In particular the mode-coupling theory (MCT) of supercooled liquids has proven to be able to catch the essential features of the glass transition and to give the most accurate predictions for the behavior of the relevant physical quantities [2]. In the case of colloids an interaction potential with a hard core and an attractive tail brings in the usual phase separation with a critical point, but also a new mechanism of glass formation. The first attempt to study attractive colloids at high volume fractions was essentially theoretical, in fact MCT was able to predict a new mechanism of structural arrest. Besides the repulsive glass transition due to excluded volume effects, a line of attractive liquid-glass appears, the driving mechanism of which is the bonding between the particles [3, 4, 5]. The resulting glass phase was also hypothesized to give rise to a colloidal gel [4]. Since the attractive glass line extends to a wide range of volume fractions of the dispersed phase, the colloid could also give rise to a low-density gel, similar to the gelling due to cluster formation and aggregation at very low densities. In this paper we will summarize a number of recent investigations aimed at describing the mechanisms of gel formation in colloidal systems.

THE BOUNDARIES OF THE LIQUID STATES

Relevant properties of a fluid are essentially related to the hard-core repulsion between the particles when they approach each other. When adding an attractive component to the repulsion a pocket of liquid states appears in the phase diagram between the gas and the crystal phases, together with a region of phase coexistence of gas and liquid, with the associated critical point. The usual limits of existence of a liquid in thermodynamic equilibrium are given by the two-phase region at lower densities and the crystal at higher densities. If we take into account metastable states, then a liquid exists in the region limited by the spinodal curve on one side and by the glass transition line of a supercooled liquid on the other side. This is what happens in a molecular fluid, where the repulsive and attractive parts of the inter-particle potential have similar range. The situation is different when considering colloidal suspensions, where the attractive range can be rather small compared to the particle diameter. In this case the liquid phase disappears, i.e. the coexistence region between the low-density phase (colloid-poor phase) and the high-density one (colloid-rich phase) tends to become metastable. The relation between the spinodal and the glass lines was studied, in the case of a Lennard-Jones systems, by Sastry and coworkers [6, 7] using the potential landscape approach in order to derive the ideal glass transition and integral equations to get the

CP982, *Complex Systems, 5th International Workshop on Complex Systems*
edited by M. Tokuyama, I. Oppenheim, and H. Nishiyama

spinodal. A point of intersection of the two lines was thus unambiguously determined.

In order to control if the same situation is present in colloidal systems we have studied the behavior of a model binary system with hard-core repulsion followed by a short-ranged square-well attractive potential $(i,j = 1,2)$

$$V_{ij}(r) = \begin{cases} \infty & r < \sigma_{ij} \\ -u_0 & \sigma_{ij} < r < \sigma_{ij} + \Delta_{ij} \\ 0 & r > \sigma_{ij} + \Delta_{ij} \end{cases} \quad (1)$$

where $\sigma_{ij} = (\sigma_i + \sigma_j)/2$ and Δ_{ij} is the square-well width. We use an equal number of particles ($N_1 = N_2 = 1000$) and a slight mismatch of the diameters of the two components ($\sigma_1 = 1.2$ and $\sigma_2 = 1.2$) to avoid crystallization. In order to approach the so-called Baxter limit [8] of vanishing Δ_{ij} and u_0 diverging, we have used very small values of the range Δ_{ij}, i.e. $\varepsilon = \Delta_{ij}/(\sigma_{ij} + \Delta_{ij}) = 0.03$. We assume σ_{11} as length unit, u_0 as energy unit and the Boltzmann constant $k_B = 1$. The Baxter limit allows an analytic solution of the Ornstein-Zernike equation using the Percus-Yevick closure [8] and was also studied in a numerical simulation of Miller and Frenkel [9]. The result of these approaches [10] is that, below values of $\varepsilon \approx 0.10$, it is possible with good accuracy to scale the coexistence curve of the system on the master curve of Baxter, when using the appropriate scaled variables. The temperature is scaled using the relative second virial coefficient $B_2^* = B_2/B_{HS}$, where $B_2 = (B_2^{11} + 2\,B_2^{12} + B_2^{22})/4$ with

$$\frac{B_2^{ij}}{4} = \frac{\pi}{6}\sigma_{ij}^3\{1 - (e^{\beta u_0} - 1)[(1-\varepsilon)^{-3} - 1]\} \quad (2)$$

and $B_{HS} = \pi[\sigma_{11}^3 + 2\,\sigma_{12}^3 + \sigma_{22}^3]/6$.

The ideal glass transition line has been studied with a simulation of the dynamics of the colloidal system and by evaluating the diffusion coefficient D for long times. We used the loci of constant D, the iso-diffusivity lines, extrapolating which to the limit $D \to 0$ we get the arrest line. Mathematically the extrapolation, according to MCT, follows a power law in temperature or volume fraction of the dispersed phase. The generic result is that in the case of colloids the glass line intersects the binodal line below the critical point, in a way analogous to molecular fluids. A direct evaluation of the binodal line, using the distribution of the particles in sub-volumes of the the total volume, although not very accurate, confirms the previous results. Fig. 1 shows the relative positions of the binodal and the glass-liquid transition lines. The interesting consequence of the situation described above is that it is possible, with a deep quench in the two-phase region, to generate a glassy phase as opposed to the dense liquid one gets when quenching to a temperature above the intersection between binodal and glassy lines.

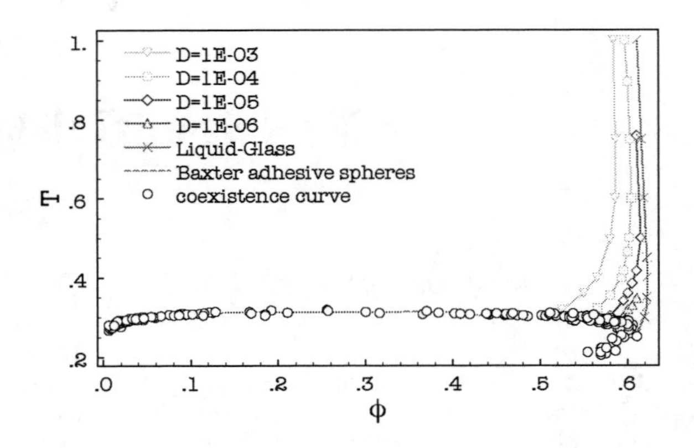

FIGURE 1. (Color online) Phase diagram of a short-range square-well binary system. Points on the coexistence curve are shown, together with the iso-diffusivity lines for various values of D. The extrapolated liquid-glass line intersects the binodal near $\phi \approx 0.6$ and $T \approx 0.26$. The dashed line refers to the simulation of the Baxter adhesive spheres model [9].

This has been interpreted as a non-equilibrium route to the structural arrest characteristic of a glass phase, which was identified with a gel phase of the colloidal suspension by many authors [1].

The formation of an equilibrium gel in colloids is not possible, unless one is able to extend the glass line to lower values of the volume fraction. Since it seems to be impossible to avoid the crossing of the binodal, the only possibility remains to reduce the size and extension of the two-phase region. The efforts made in this direction using experiments, theory and simulations are the topics of the following Sections.

LIMITED VALENCE MODELS

A first attempt to reduce the effect of the attractive interactions between two colloidal particles, and therefore to decrease the size of the two-phase region, is to consider limited valence models [11, 12], inspired by the work of Speedy and Debenedetti [13, 14]. The interaction is considered to be hard-sphere followed by a square well attraction if the number n of particles surrounding a given one is less than a predetermined number n_{max}, $n < n_{max}$, but only hard-core when $n > n_{max}$. This gives rise to an effective many-body interaction which limits the number of neighbors of a particle when n_{max} is assumed to be less than 12, the maximum number of neighbors allowed by geometrical packing. As a result, the driving force for phase separation is significantly reduced, and open structures are favored. We find that, for $n_{max} < 6$, the system can access regions usually dominated by phase separation and experience a dynamical slowing down by several orders of magnitude, thus entering the gel regime.

Remarkably, the system undergoes the fluid-gel transition in equilibrium, so that the process is fully reversible.

We performed event-driven molecular dynamics simulations of particles interacting via a square-well potential of the type described above, with the additional constraint that particles can form a maximum number of bonds. We report in Fig. 2 the evolution of the spinodal line for different values of n_{max} in the (ϕ, T) plane. We estimate the spinodal line by bracketing it with the last stable state point and the first phase separating state point along each isochore. The unstable area in the phase diagram shrinks on decreasing n_{max} , showing that the additional constraint opens up a significant portion of phase space, where the system can be studied in equilibrium one-phase conditions. In fact it is possible to study the dynamics of the model at very low T, where the lifetime of the interparticle bond increases, stabilizing for longer and longer time intervals the percolating network. When the bond lifetime becomes of the same order as the observation time, the system will behave as a disordered solid. It is worth stressing that in the present model there is no thermodynamic transition associated with the onset of a gel phase.

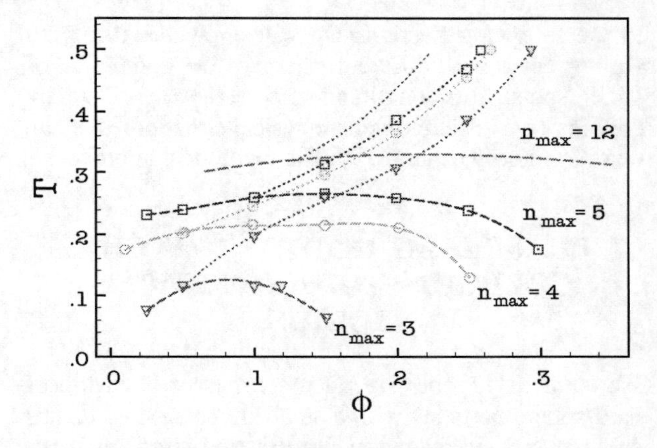

FIGURE 2. (Color online) Phase diagram of the limited valence model for various values of n_{max}, showing phase coexistence (dashed lines) and corresponding percolation (dotted lines) loci. Lines are only for guidance.

To quantify these statements, we studied the dynamics for all temperatures in the region where the system is in a single phase. We found that there are two distinct arrest lines for the system, a glass line at high packing fraction, and a gel line at low ϕ and T. In the (ϕ, T) plane the former is rather vertical and controlled by the volume fraction, while the latter is rather horizontal and controlled by temperature. Dynamics on approaching the glass line along isotherms exhibit a power-law dependence on ϕ, while dynamics along isochores follow an activated Arrhenius dependence. The gel has clearly distinct properties from those of both a repulsive and an attractive glasses, and a gel to glass crossover occurs in a fairly narrow range in ϕ along low-T isotherms. Fig. 3 shows the logarithm of the diffusivity D as a function of $1/T$, for $n_{max} = 4$. Arrhenius behavior of D is observed at all ϕ at low T. The activation energy is around 0.55. Since at the lowest studied temperatures the structure of the system is already essentially T independent, there is no reason to expect a change in the functional law describing the $T \to 0$ dynamics. In this respect, the true arrest of the dynamics is located along the $T = 0$ line, limited at low ϕ by the spinodal and at high ϕ by crossing of the repulsive glass transition line. This peculiar behavior is possible only in the presence of limited valence, since when such a constraint is not present, phase separation preempts the possibility of accessing the $T \to 0$ Arrhenius window.

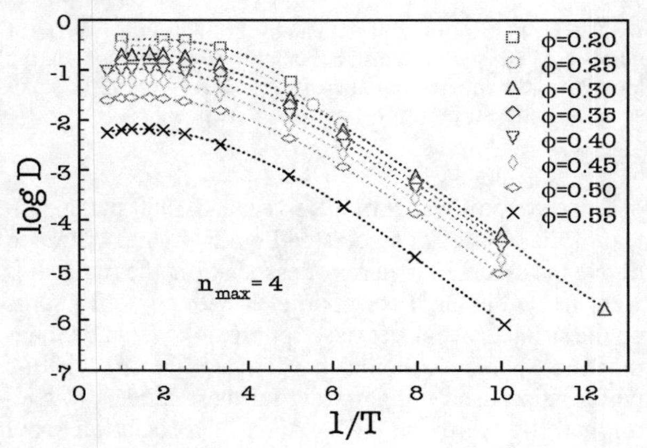

FIGURE 3. (Color online) Arrhenius plot for the diffusivity for the $n_{max} = 4$ model along lines at constant volume fraction, showing activated behavior.

PATCHY COLLOIDS AND WERTHEIM THERMODYNAMIC PERTURBATION THEORY

The limited valence model introduced earlier has a precise experimental realization in the so-called patchy colloids. In fact the possibilities offered by the physicochemical manipulation of colloidal particles, due to the precise control of the interparticle potential, has made it possible to design colloidal molecules, particles decorated on their surface by a predefined number of attractive sticky spots, i.e., particles with specifically designed shapes and interaction sites [15]. A model version we used for simulation purposes, but close to the real colloid molecules, is represented in Fig. 4. To characterize patchy colloids it is necessary to be able to determine their phase diagram, and in particular to predict the regions in which clustering, phase separation, or even gelation are expected. The theory of the physical properties

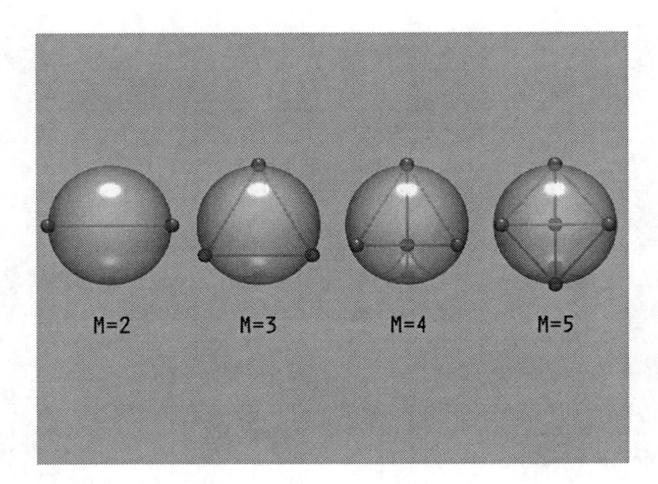

FIGURE 4. (Color online) Patchy colloids with different number M of sticky spots. Schematic representation of the location of the square-well interaction sites, centers of the small spheres on the surface of the hard-core particle.

of these systems was formulated around 1980 in the context of the physics of associated liquids. In the attempt to derive the essential features of association, the molecules were treated as hard-core particles with attractive spots on the surface, a realistic description of the recently manufactured patchy colloidal particles. A thermodynamic perturbation theory appropriate for these models was introduced by Wertheim [16] to describe association under the hypothesis of absence of closed ring configurations, and that a sticky site on a particle cannot bind simultaneously to two (or more) sites on another particle. Such a condition can be naturally implemented in colloids, using an appropriate choice of the range of attraction, due to the relative large size of the particle compared to the range of the sticky interaction. Wertheim thermodynamic perturbation theory (WTPT) provides an important starting point to study the phase diagram of this new class of colloids, and in particular the role of the patches number.

The WTPT describes the free energy of associating liquids made by molecules with fixed valence, i.e. a small number of patchy interacting sites. The main assumption in the theory is that molecules form open clusters without closed bond loops, a condition that can be realized with patchy particles when the average functionality is small. The free-energy due to bonding, to be added to the hard-sphere contribution, is given by

$$\frac{\beta A_{bond}}{N} = M\,ln(1-p_b) - \frac{M}{2}p_b. \tag{3}$$

The theory predicts that the bond probability p_b can be calculated from the chemical equilibrium between two non-bonded sites forming a bonded pair. For the present model, such relation reads

$$\frac{p_b}{(1-p_b)^2} = M\rho\Delta \tag{4}$$

where

$$\Delta = 4\pi \int dr_{12}\, r_{12}^2\, g_{HS}(r_{12})\, \langle f_{ij}(12)\rangle_{\omega_1,\omega_2}. \tag{5}$$

The Mayer $f_{ij}(12)$-function between two arbitrary sticky sites i and j — respectively on particle 1 and 2, separated by $\mathbf{r}_{12}^{ij}$ and interacting through the square-well potential V_{SW} — is

$$f_{ij}(12) = exp\left[-\frac{V_{SW}(\mathbf{r}_{12}^{ij})}{k_B T}\right] - 1. \tag{6}$$

$\langle f_{ij}(12)\rangle_{\omega_1,\omega_2}$ represents an angle average over all orientations of the two particles at fixed relative distance r_{12}, and $g_{HS}(12)$ is the reference hard-sphere radial distribution function in the linear approximation

$$g_{HS}(r_{12}) = \frac{1-0.5\phi}{(1-\phi)^3} - \frac{9\phi(1+\phi)}{2(1-\phi)^3}(r_{12}-1). \tag{7}$$

In the following Sections we will apply the theory to various cases with different values of the number M of sticky spots, in order to shows how the use of patchy colloids gives rise to diverse physical situations that can be satisfactorily described by the theoretical approach.

PHASE DIAGRAMS FOR PATCHY COLLOIDS. TOWARDS EMPTY LIQUIDS.

We studied [17] the general properties of a system of hard-sphere particles with a small number M of identical short-ranged, square-well attraction sites per particle, distributed on the surface with the same geometry as the recently produced patchy colloidal particles [15]. We identify the number of possible bonds per particle as the key parameter controlling the location of the critical point, as opposed to the fraction of surface covered by attractive patches. We summarize here results of extensive numerical simulations of this model in the grand-canonical ensemble to evaluate the location of the critical point of the system in the (ϕ, T) plane as a function of M. We complement the simulation results with the evaluation of the region of thermodynamic instability according to the WTPT reported above. Both theory and simulation confirm that, on decreasing the number of sticky sites, the critical point moves toward smaller ϕ and T values. We note that while adding to hard spheres a spherically symmetric attraction creates a liquid-gas critical point which shifts toward larger ϕ, on decreasing the range of interaction the opposite trend is present when the number of interacting sites is decreased. Simulations and theory also provide evidence that for binary mixtures of particles with two and three sticky spots (where $\langle M \rangle$, the average M per particle can be varied continuously down to two by changing the relative concentration of the two species) the critical point shifts continuously toward vanishing. This makes it possible to realize equilibrium liquid states with arbitrary small ϕ (empty liquids), a case which cannot be realized using spherical potentials.

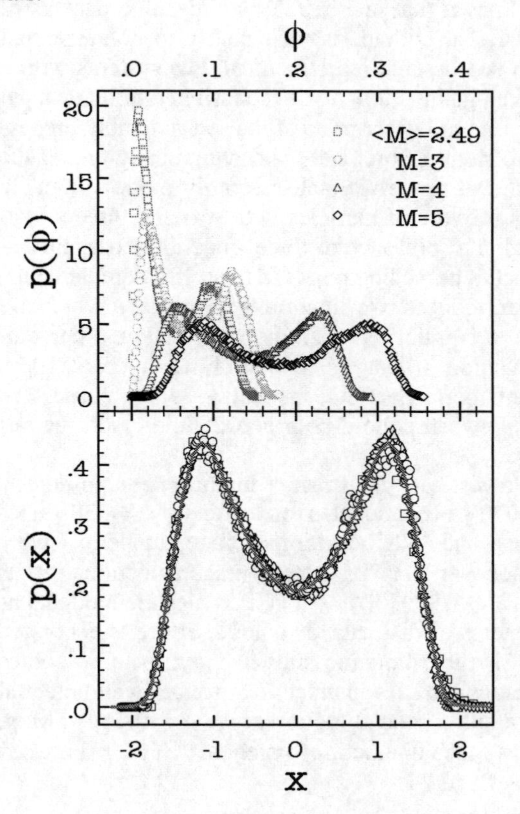

FIGURE 5. (Color online) Order parameter distributions. In the upper panel volume fraction fluctuations distribution $P(\phi)$ at the critical point for various M values. The lower panel shows the order parameter x fluctuations distribution $P(x)$ for all studied cases, compared with the expected distribution (full line) for systems at the critical point of the Ising universality class [18].

To locate the critical point, we performed grand-canonical Monte Carlo simulations (MC) and histogram reweighting [18] for $M = 5$, 4 and 3 and for binary mixtures of particles with $M = 3$ (fraction α) and $M = 2$ (fraction $1-\alpha$) at five different compositions, down to $< M > \equiv 3\alpha + 2(1-\alpha) = 2.05$. We implement MC steps composed by 500 random attempts to rotate and translate a random particle and one attempt to insert or delete a particle. On decreasing $< M >$, numerical simulations become particularly time-consuming, since the probability of breaking a bond $\sim e^{1/T_c}$ becomes progressively small. To improve statistics, we average over 15-20 independent MC realizations. After choosing the box size, the T and the chemical potential μ of the particle(s), the grand-canonical simulation evolves the system toward the corresponding equilibrium density. If T and μ correspond to the critical point values, the number of particles N and the potential energy E of the simulated system show ample fluctuations between two different values. The linear combination $x \sim N + sE$ (where s is named field mixing parameter) plays the role of order parameter of the transition. At the critical point, its fluctuations are found to follow a known universal distribution, i.e. (apart from a scaling factor) the same that characterizes the fluctuation of the magnetization in the Ising model [18]. Fig. 5 (upper panel) shows the resulting volume fraction fluctuations distribution $P(\phi)$ at the estimated critical temperature T_c and critical chemical potentials for several M values. The distributions, whose average is the critical packing fraction ϕ_c, shift to the left on decreasing M and become more asymmetric, signalling the progressive increasing role of the mixing field. In the lower panel, the calculated fluctuations of x, $P(x)$, are compared with the expected fluctuations for systems in the Ising universality class. They provide evidence that the critical point has been properly located and the transition belongs to the Ising class. Data show a clear monotonic trend toward decreasing T_c and ϕ_c on decreasing M.

Fig. 6 shows a quantitative comparison of the numerical and theoretical estimates for the critical parameters T_c and ϕ_c performed using WTPT. Theory predicts quite accurately T_c but underestimates ϕ_c, nevertheless clearly confirms the M dependence of the two quantities. The overall agreement between Wertheim theory and simulations reinforces our confidence in the theoretical predictions and supports the possibility that, on further decreasing $< M >$, a critical point at vanishing ϕ can be generated.

WTPT allows us also to evaluate the locus of points where $(\partial P/\partial V)_T = 0$, which provides, at the mean field level, the spinodal locus. Fig. 7 shows the predicted spinodal lines in the (ϕ, T) plane for several M values. On decreasing M also the liquid spinodal boundary moves to lower values, suggesting that the region of stability of the liquid phase is progressively enhanced. It will be desirable to investigate the structural and dynamical properties of such empty liquids by experimental and numerical work on patchy colloidal particles.

We can therefore conclude that for particles interacting with attractive spherical potentials, phase separation always destabilizes the formation of a homogeneous arrested system at low temperature. Instead with patchy particles and small $< M >$, disordered states in which particles are interconnected in a persistent gel network

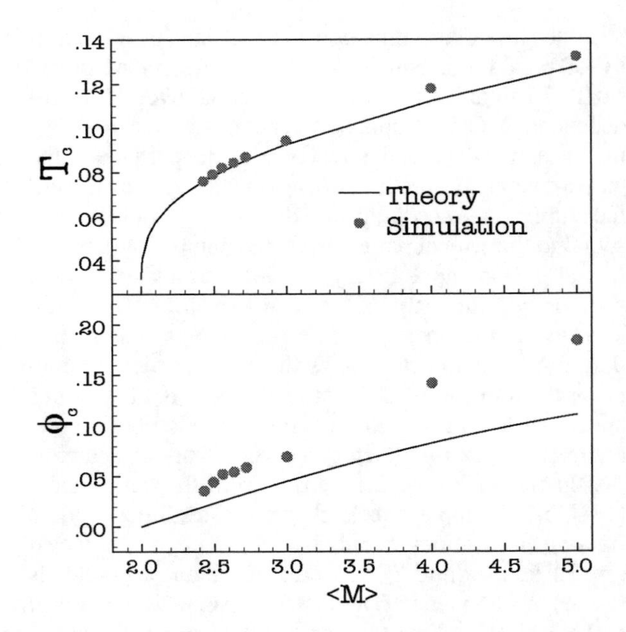

FIGURE 6. (Color online) Comparison between theoretical and numerical results for patchy particles with different number of sticky spots. Upper and lower panels shows respectively the location of the critical values T_c and ϕ_c.

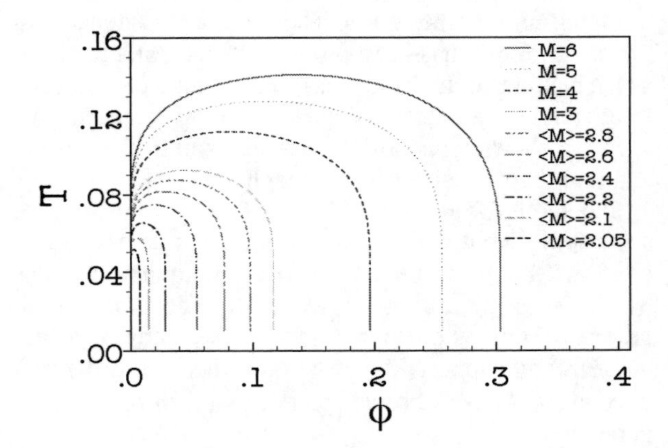

FIGURE 7. (Color online) Spinodal curves calculated according to WTPT for the patchy particles models for several values of $< M >$.

can be reached at low T, without encountering phase separation. Indeed, at such low T, the bond-lifetime will become comparable to the experimental observation time. Under these conditions, a dynamic arrest phenomenon at small ϕ will take place. It will be thus possible to approach dynamic arrest continuously from equilibrium and to generate a state of matter as close as possible to an ideal gel.

MIXTURES OF PATCHY COLLOIDS. WERTHEIM AND FLORY-STOCKMAYER.

Patchy colloids made of particles with two sticky spots self-assemble only in chains [19], while systems with $M > 2$ form ramified clusters and show a phase separation at low volume fractions. It is also interesting to study the process of formation of clusters or, in other words, the mechanism of self-assembly of the colloidal particles. It is well known that one of the most peculiar aspects of the physics of colloids is their ability to organize spatially in self-assembling structures. For systems with a small average functionality it is possible to provide a parameter free full description of the self-assembly process [20]. We study theoretically and numerically one of the simplest, but not trivial self-assembly process, namely a binary mixture of particles with two and three attractive sites. The presence of three -functional particles — which act as branching points in the self-assembled clusters — introduces two important phenomena which are missing in equilibrium chain polymerization: a percolation transition, where a spanning cluster appears, and a region of thermodynamic instability, where phase separation between colloid-poor and colloid-reach regions appears.

We investigate a binary mixture composed by $N_2 = 5670$ bi-functional particles and $N_3 = 330$ three-functional ones. The resulting average number of sticky spots per particle, i.e. the average functionality, is $\langle M \rangle = (2N_2 + 3N_3)/(N_2 + N_3)$. Particles are modeled as hard-spheres of diameter σ, whose surface is decorated by two or three bonding sites at fixed locations. Sites on different particles interact via a square-well potential V_{SW} of depth u_0 and attraction range $\delta = 0.119\sigma$. More precisely, the interaction potential $V(\mathbf{1},\mathbf{2})$ between particles **1** and **2** is

$$V(\mathbf{1},\mathbf{2}) = V_{HS}(\mathbf{r}_{12}) + \sum_{i=1,n_1} \sum_{j=1,n_2} V_{SW}(\mathbf{r}_{12}^{ij}) \quad (8)$$

where V_{HS} is the hard-sphere potential and $\mathbf{r}_{12}$ and $\mathbf{r}_{12}^{ij}$ are respectively the vectors joining the particle-particle centers and the site-site (on different particles) locations; n_i indicates the number of sites of particle i. The well width δ is chosen to ensure that each site, due to steric effect, is engaged at most in one interaction. Distances are measured in units of σ. Temperature is measured in units of the potential depth (i.e. Boltzmann constant $k_B = 1$). In the studied model, bonding is properly defined: two particles are bonded when their pair interaction energy is $-u_0$. This means that the potential energy of the system is proportional to the number of bonds. The lowest energy state of the system (the ground state energy) coincides with configurations in which all bonds are formed, i.e

$E_{gs} = -u_0\ (2N_2 + 3N_3)/2$. As a result, the bond probability can be calculated as the ratio of the potential energy E and E_{gs}, $p_b = E/E_{gs}$. Pairs of bonded particles are assumed to belong to the same cluster. We have performed standard Monte Carlo MC Metropolis simulations at several T and ρ. An MC step is defined as an attempted move per particle. A move is defined as a displacement of a randomly selected particle in each direction of a random quantity distributed uniformly between $\pm 0.05\ \sigma$ and a rotation around a random axis of a random angle uniformly distributed between ± 0.1 radiant.

We describe in what follows the comparison between the simulation and the WTPT.

Bond probability. The WTPT theory describes the free energy of associating liquids made by molecules with fixed valence, i.e. a small number of patchy interacting sites, and forming open cluster without closed bond loops. The comparison between the theoretical predictions and the numerical data for the T and ρ dependence of p_b is shown in Fig. 8. Data show clearly that the theory is able to predict precisely p_b (or equivalently the system potential energy) in a wide T and ρ range. At low T, $p_b \to 1$, and the system approaches a fully bonded disordered ground state configuration.

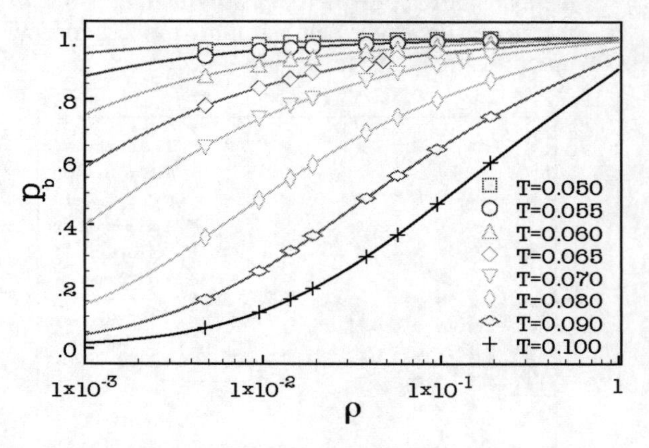

FIGURE 8. (Color online) Temperature and volume fraction dependence of the bond probability p_b. Points are simulation results based on Monte Carlo simulations. Lines are parameter-free predictions based on the WTPT. Note that at low T the system reaches a fully bonded configuration.

Combining Wertheim and Flory-Stockmayer theories. To derive information on the structure of the system and the connectivity of the aggregates we combine the Wertheim and the Flory-Stockmayer (FS) theories [21, 22], both relying on the absence of closed bonding loops. The Wertheim prediction for p_b can thus be consistently used in connection with the FS approach to predict the T and ρ dependence of the cluster size distributions.

The number of clusters (per unit volume) containing l bi-functional particles and n three-funtional ones can be written [21, 22] as

$$\begin{aligned} \rho_{nl} &= \rho_3 \frac{(1-p_b)^2}{p_3 p_b} [p_3 p_b (1-p_b)]^n [p_2 p_b]^l w_{nl} \\ w_{nl} &= 3 \frac{(l+3n-n)!}{l!n!(n+2)!} \end{aligned} \quad (9)$$

where $p_3 \equiv 3N_3/(2N_2+3N_3)$ and $p_2 = 1 - p_3$ are the probabilities that a randomly chosen site belongs to a three-functional or to a two-functional particle, p_b is given by Eq. (4), and w_{nl} is a combinatorial contribution [21, 22]. Distributions are normalized in such a way that $\sum_{ln,l+n>0} (l+n)\rho_{nl} = \rho_2 + \rho_3$. As shown in Fig. 9, on decreasing T, the ρ_{nl} distribution becomes wider and wider and develops a power-law tail with exponent -2.5, characteristic of loopless percolation [23]. On further decreasing T, the distribution of finite size clusters progressively shrinks, since most of the particles attach themselves to the infinite cluster. Data show that Eq. 9, with no fitting parameters, predicts extremely well the numerical distributions at all state points, both above and below percolation.

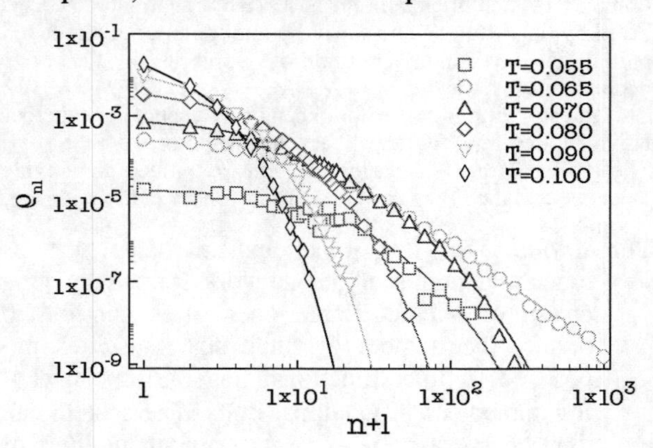

FIGURE 9. (Color online) Cluster size distribution for $\rho\sigma^3 = 0.04$ for some of the investigated T. Points are simulation data and lines are the corresponding theoretical curves. The number of finite clusters (per unit volume) containing l bi-functional units plus n three-functional ones.

In the framework of the FS approach it is also possible to evaluate the number of clusters $\rho_c \equiv \sum_{nl} n_{nl}$ as a function of p_b, irrespectively of the cluster size. Below percolation, in the absence of bonding loops, the relation between ρ_c and p_b is linear, since each added bond decreases the number of clusters by one. Above percolation the relation crosses to a non-linear behavior, so that the number of clusters becomes one when $p_b = 1$. As shown in Fig. 10, the simulation data conform perfectly to the theoretical expectation both below and above percolation.

This suggests that, when the average functionality is small, bonding loops in finite size clusters can always be neglected. This agreement, which covers the entire range of p_b values, implies that closed loops of bonds are statistically less favored than the corresponding open structure.

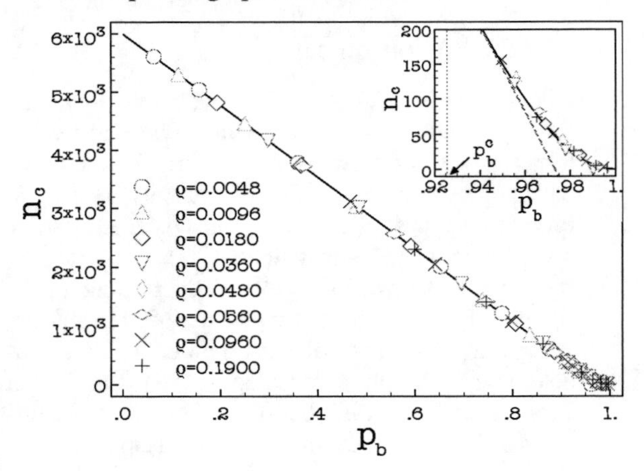

FIGURE 10. (Color online) The number of finite-size clusters n_c (irrespective of their size) as a function of the bond probability. Symbols are simulation result and solid lines FS predictions. Below percolation, in the absence of bonding loops, n_c is given by the difference between the total number of simulated particles N and the number of bonds N_b, since each added bond decreases n_c by one, i.e. $n_c = N - N_b = N - p_b(2N_2 + 3N_3/2)$. This linear relation (dashed line) can be extended above percolation (dotted line), but never beyond the point where $n_c < 1$. The inset enlarges the region of large p_b values, to provide evidence that the FS approach is valid over the entire p_b range.

Percolation. The three-functional particles act as branching points of the network formed by long chains of two-functional ones. It is possible to predict the number of finite size clusters composed of n three-functional units, irrespective of the number of bifunctional units. The system can thus be considered as a one-component fluid of three-functional particles forming clusters, in which the bonding distance between the three-functional particles is given by the length of the chains formed by the bi-functional units. Following again FS, it is possible to predict the p_b value at which the systems develops a percolating structure: when $p_b \geq p_b^c \equiv 1/(1+p_3) = 0.9256$, an infinite cluster is present in the system. The percolation line is thus the locus of points in the phase diagram such that $p_b(T,\rho) = p_b^c$, with $p_b(T,\rho)$ given in Eq.(4).

Liquid-gas phase separation. The WTPT predicts a phase separation into two phases of different density and connectivity at small ρ for any non-vanishing amount of branching point [17]. The theoretical spinodal curve, the line separating the stable (or metastable) state points from the unstable ones, is defined as the locus of points such that the volume V derivative of the pressure P vanishes, i.e. $(\partial P/\partial V)_T = 0$. It is located below the percolation line and the two lines merge asymptotically for $T \to 0$ and $\rho \to 0$ as shown in Fig. 11. The configurations for the two investigated state points located inside the spinodal are characterized by a bimodal distribution of the density fluctuations and a very large value of the small-angle structure factor, indicating a phase-separated structure.

Maximum of the specific heat C_V. As in chain polymerization [19], the theory predicts a line of constant volume specific heat C_V maxima, calculated from the inflection point in the p_b as a function of T curves (see Fig. 8), which also agrees very well with the simulation results. The line of C_V extrema is also shown in Fig. 11. The presence of a maximum in C_V is a characteristic of bond-driven assembly, and the locus of maxima in the phase diagram is one of the precursors of the self-assembly process for low functionality particles.

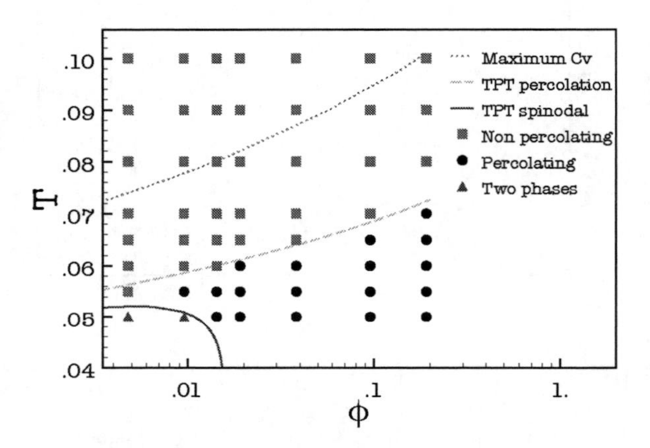

FIGURE 11. (Color online) Phase diagram of the 2-3 mixture. Lines are theoretical predictions, the continuous line is the spinodal curve obtained from the WTPT equation of state; the dotted line the locus of maximum specific heat C_V; the dashed line the percolation locus. Points are simulation results, white and black squares are the non-percolating and the percolating equilibrium states. Triangles are non-equilibrium states in the two-phase region.

CONCLUSIONS

In conclusion, we summarize the relevant new facts that have emerged from the study of systems where a reduction of the attractive interactions leads to network-forming phases. This result is obtained by means of surface patches which furnish anisotropic coupling between the colloidal particles.

Phase separation. Particles strongly interacting with simple spherical potentials always phase-separate in colloid-rich and colloid-poor phases. The phase boundary intersects the liquid-glass transition line.

Empty liquids. It is possible to significantly reduce the extent of the unstable coexistence region, the colloid-rich region is characterized by a very small volume fraction, and can be called an empty liquid.

Limited valence. Particles strongly interacting with limited valence, such as patchy particles with highly directional interactions, form equilibrium open structures – network forming liquids, glasses or gels.

Self-assembly. With an appropriate choice of the directional interactions, self-assembly can be treated as an equilibrium liquid-state problem.

Arrested states at low volume fraction. Directional interactions and limited valence are essential ingredients for increasing the range of the liquid state, and in particular to generate arrested states at low volume fraction.

Strong liquids. The low temperature liquid state, when moving along isochores, behaves like a strong liquid in the sense of Angell's classification.

Ideal equilibrium gel. When the temperature tends to vanish, the strong liquid turns into an ideal equilibrium gel.

Gelation and glass formation. Gels and strong liquids share common features. The arrest transition can be simultaneously interpreted in terms of equilibrium gelation by the colloid community and of strong-glass formation by the supercooled liquid community.

ACKNOWLEDGMENTS

This work has been performed in the Rome group mainly in collaboration with F. Sciortino, and many people have contributed to it in various periods of time, namely E. Bianchi, G. Foffi, J. Largo, C. De Michele, E. La Nave, F. Romano and E. Zaccarelli. Collaboration with S. V. Buldyrev, J. F. Douglas, A. J. Moreno, I. Saika-Voivod is also gratefully acknowledged. We acknowledge support from MIUR-Prin and MCRTN-CT-2003-504712.

REFERENCES

1. F. Sciortino, and P. Tartaglia, *Advances in Physics* **54**, 471–524 (2005).
2. W. Götze, in *Liquids, Freezing and the Glass Transition*, edited by J. P. Hansen, D. Levesque, and J. Zinn-Justin, Les Houches Summer Schools of Theoretical Physics Session LI 1989, North-Holland, Amsterdam, 1991, pp. 287–503.
3. L. Fabbian, W. Götze, F. Sciortino, P. Tartaglia, and F. Thiery, *Phys. Rev. E* **59**, R1347– (1999) and **60**, 2430– (1999).
4. J. Bergenholz, and M. Fuchs, *Phys. Rev. E* **59**, 5706– (1999).
5. K. Dawson, G. Foffi, M. Fuchs, W. Götze, F. Sciortino, M. Sperl, P. Tartaglia, Th. Voigtmann, and E. Zaccarelli *Phys. Rev. E* **63**, 011401 (2000).
6. S. Sastry, *Phys. Rev. Lett.* **85**, 590– (2000).
7. S. S. Ashwin, G. I. Menon, and S. Sastry, *Europhys. Lett.* **75**, 922–928 (2006).
8. R. J. Baxter, *J. Chem. Phys.* **49**, 2770– (1968).
9. M. A. Miller, and D. Frenkel, *Phys. Rev. Lett.* **90**, 135702-1–135702-4 (2003).
10. G. Foffi, C. De Michele, F. Sciortino, and P. Tartaglia, *Phys. Rev. Lett.* **94**, 078301-1–078301-4 (2005).
11. E. Zaccarelli, S. V. Buldyrev, E. La Nave, A. J. Moreno, I. Saika-Voivod, F. Sciortino, and P. Tartaglia, *Phys. Rev. Lett.* **94**, 218301-1–218301-4 (2005).
12. E. Zaccarelli, I. Saika-Voivod, S. V. Buldyrev, A. J. Moreno, P. Tartaglia, and F. Sciortino, *J. Chem. Phys.* **124**, 124908-1–124908-14 (2006).
13. R. J. Speedy, and P. G. Debenedetti, *Mol. Phys.* **81**, 237– (1994).
14. R. J. Speedy, and P. G. Debenedetti, *Mol. Phys.* **88**, 1293– (1996).
15. V. N. Manoharan, M. T. Elsesser, and D. J. Pine, *Science* **301**, 483– (2003).
16. M. Wertheim, *J. Stat. Phys.* **35**, 19 (1984).
17. E. Bianchi, J. Largo, P. Tartaglia, E. Zaccarelli, and F. Sciortino, *Phys. Rev. Lett.* **97**, 168301-1–168301-4 (2006).
18. N. B. Wilding, *J. Phys. Condens. Matter* **9**, 585– (1997).
19. F. Sciortino, E. Bianchi, J. F. Douglas, and P. Tartaglia, *J. Chem. Phys.* **126**, 194903-1–194903-10 (2007).
20. E. Bianchi, P. Tartaglia, E. La Nave, and F. Sciortino, "A Fully Solvable Equilibrium Self-Assembly Process: Fine Tuning the Clusters Size and the Connectivity in Patchy Particle Systems" *J. Phys. Chem. B* **in press**, (2007).
21. P. J. Flory, *Principles of polymer chemistry*, Cornell University Press, Ithaca and London, 1953.
22. W. H. Stockmayer, *J. Chem. Phys.* **11**, 45–55 (1943).
23. M. Rubinstein, and R. H. Colby, *Polymer Physics*, Oxford University Press Inc., New York, 2003.

Molecular Simulations of Nanoparticles in an Explicit Solvent

Gary S. Grest*, Pieter J. in 't Veld*, and Jeremy B. Lechman*

*Sandia National Laboratories, Albuquerque, NM 87185 USA

Abstract. Results of large scale equilibrium and non-equilibrium molecular dynamics (NEMD) simulations are presented for nanoparticles in an explicit solvent. The nanoparticles are modeled as a uniform distribution of Lennard-Jones particles, while the solvent is represented by standard Lennard-Jones particles. Unlike hard sphere models, the nanoparticles and solvent do not phase separate for disparate sizes of nanoparticles and solvent, which allows us to study the static and dynamic properties of nanoparticle suspensions with an explicit solvent. Here we present results for nanoparticles of size 5 to 20 times that of the solvent for a range of concentrations from 7 to 40% volume fraction. The nanoparticles are found to segregate to the liquid/vapor interface or repel the interface depending on the relative strength of the nanoparticle/nanoparticle and nanoparticle/solvent interactions. Results from NEMD simulations suggest that the shear rheology of the suspension depends only on the nanoparticle concentration not the size of the nanoparticle, even for nanoparticles as small as 5 times that of the solvent.

Keywords: Nanoparticle, Rheology, Simulations, Suspensions
PACS: 82.70.Dd 83.80.Hj 82.70.-y

INTRODUCTION

Nanoparticles have great potential for being used to tailor material function and performance in differentiating technologies because of their profound effect on thermophysical, mechanical and optical properties. The most feasible way to disperse particles in a bulk material or control their packing at a substrate is through fluidization in a carrier that can be processed by any number of methods, including spin, drip and spray coating, fiber drawing, casting followed by solidification through solvent evaporation. However, processing particles as concentrated, fluidized suspensions into useful products remains an art largely because the effect of particle shape and volume fraction on fluidic properties and suspension stability remains unexplored in the regime where particle-particle interaction mechanics are prevalent.

Most previous simulations of colloids have modeled the solvent implicitly due to limitations of computational resources. The most common approach has been to treat the particles by Brownian dynamics coupled to a heat bath [1, 2, 3]. However, in so doing, one neglects long range hydrodynamics. One approach to include hydrodynamic interactions with an implicit solvent is Stokesian Dynamics [4] or related methods [5, 6]. Perhaps the most elegant way of treating the hydrodynamics of spherical particles, it is also a computationally expensive technique; although advances have been made recently to achieve greater efficiency and flexibility in terms of boundary conditions [7, 8]. It also is limited to spherical or possibly mildly aspherical particles. Alternatively one can treat the background solvent in a coarse-grained particle fashion. One successful approach for particle-fluid suspensions is the lattice Boltzmann method [9]. Another is to treat the background solvent by dissipative particles (DPD) [10, 11, 12, 13] or stochastic rotation dynamics (SRD) particles [14, 15, 16, 17]. Each of these particle based methods introduces an effective coarse-graining length scale that is smaller than the colloids but much larger than the natural length scales of a microscopic solvent. Finite-Element approaches have also been proposed [18] which allow for greater generality in modeling suspension flows, but are typically limited in use due to the computational cost. All of these approaches rely on the fact that the colloidal particle is much larger than the solvent and therefore there is a clear separation of time and length scales. For nanoparticles in the 2-20 nm size, treating the background as a continuum may not always be adequate. For example, it does not account for local packing of the solvent around the nanoparticles, which can increase the effective radii of the nanoparticles and strongly affect the properties of the suspension. Thus it is important to be able model nanoparticle suspensions in an explicit solvent to address even such simple questions as to how large should the nanoparticles be to treat the solvent as a continuum or how strongly do changes in the relative interactions between nanoparticles and between nanoparticles and the solvent effect rheology.

Before beginning simulations of colloids, particularly in the presence of a solvent, one must ask how best to describe the particles and what interactions to include. The most prevalent model of colloidal particles is to treat them as hard spheres. This is a computationally efficient approach but is not suitable for modeling colloids in an

CP982, *Complex Systems, 5th International Workshop on Complex Systems*
edited by M. Tokuyama, I. Oppenheim, and H. Nishiyama

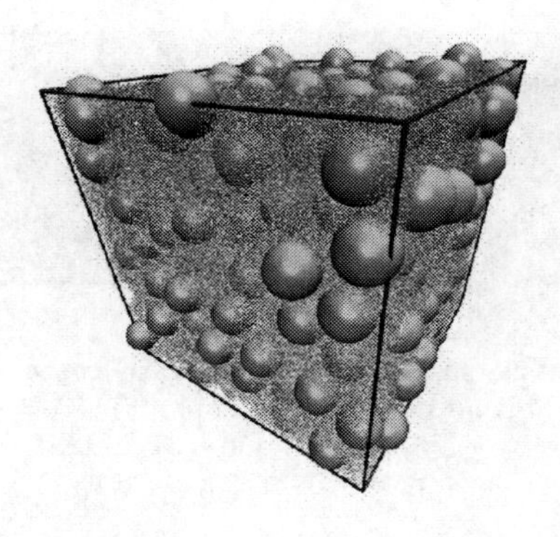

FIGURE 1. Sample simulation cell with 333 nanoparticles of radii $a = 10\sigma$ at volume fraction $\phi_v = 39\%$ in an explicit solvent. Solvent is shown as points.

explicit solvent since hard spheres strongly phase separate even for relatively small differences in size [19]. For this reason most simulations of hard spheres treat the solvent implicitly, usually by Brownian dynamics. To solvate the nanoparticles in an explicit solvent, it is critical to include an attractive component of the interaction between the nanoparticle and the solvent. This can be done in a variety of ways. One approach, which is particularly useful for aspherical nanoparticles, is to treat each nanoparticle as a collection of atoms [20, 21, 22, 23, 24]. Interactions between particles then follow directly from pair interactions between its constituent atoms. This has the advantage that it is straightforward to construct nanoparticles of varying shape [24], but can quickly become computationally intractable as even small nanoparticles can contain hundreds to thousands of atoms.

A computationally efficient approach is to treat the nanoparticles as large particles interacting with an effective potential which depends only on the distance between their centers. The simplest such effective potential which has been used to model the interaction between nanoparticles is by a Lennard-Jones (LJ) interaction shifted to the surface of the nanoparticle [25, 26, 27]. However this potential does not capture the true interaction between nanoparticles as the range for which the interactions are important relative to thermal fluctuations becomes increasingly small with increasing particle size. A more realistic approach is to treat each nanoparticle as being made of a uniform distribution of atoms, similar to treating them as a collection of atoms, except that, since the atoms are uniformly distributed, the effective potential can usually be determined analytically [28, 29]. For most interaction potentials, such as the LJ potential, the integrated potential can easily be determined analytically as well as the interaction between the solvent and nanoparticle. As shown in this paper, this integrated form of the potential gives a physically motivated approach to model nanoparticle suspensions. As a demonstration we present results for the interfaces properties of nanoparticles at the liquid/vapor interface and effect of nanoparticles on the rheological properties of suspensions for nanoparticles of varying size and concentration.

In the next section we present details of the model as well as computational details. We briefly review enhancements to our parallel molecular dynamics simulation code LAMMPS [30] which have made these simulations possible. In Sec. 3 we present results for the bulk and interfacial properties of nanoparticle suspensions. These results show that nanoparticles can locally phase segregate to the liquid/vapor interface depending on interactions and concentrations. In Sec. 4, we present results of our non-equilibrium molecular dynamics (NEMD) simulations for the viscosity as a function of nanoparticle size and concentration. In Sec. 5 we briefly summarize our results and future work.

MODEL AND METHODOLOGY

The nanoparticles are assumed to be comprised of uniformly distributed atoms interacting with the Lennard-Jones (LJ) potential

$$U_{LJ}(r_{12}) = 4\varepsilon_{\rm nn}\left[\left(\frac{\sigma_{\rm n}}{r_{12}}\right)^{12} - \left(\frac{\sigma_{\rm n}}{r_{12}}\right)^{6}\right], \qquad (1)$$

where r_{12} is the distance between two LJ atoms [28]. Here $\varepsilon_{\rm nn}$ is the interaction energy between LJ atoms in two nanoparticles and $\sigma_{\rm n}$ is the diameter of the LJ atoms which make up the nanoparticles. For spherical nanoparticles, the total interaction between nanoparticles can then be calculated analytically by integrating over all the interacting LJ atoms within the two particles [28]. The total interaction between nanoparticles

$$U_{\rm nn}(r_{12}) = U^{\rm A}_{\rm nn}(r_{12}) + U^{\rm R}_{\rm nn}(r_{12}), \qquad (2)$$

where r_{12} is the distance between the centers of the nanoparticles and $U^{\rm A}_{\rm nn}(r_{12})$ is the standard attractive interaction between colloidal particles, first derived by Hamaker [31]. For particles of equal radii a,

$$U^{\rm A}_{\rm nn}(r_{12}) = -\frac{A_{\rm nn}}{6}\left[\frac{2a^2}{r_{12}^2 - 4a^2} + \frac{2a^2}{r_{12}^2} + \ln\left(\frac{r_{12}^2 - 4a^2}{r_{12}^2}\right)\right]. \qquad (3)$$

The Hamaker constant $A_{\rm nn} = 4\pi^2\varepsilon_{\rm nn}\rho_1\rho_2\sigma_{\rm n}^6$, where ρ_1 and ρ_2 are the number density of LJ atoms within each sphere. For small separation between particles $h = (r_{12} - 2a) \ll a$, $U^{\rm A}_{\rm nn}(r_{12}) = -\frac{A_{\rm nn}}{12}\frac{a}{h}$ [32]. The repulsive compo-

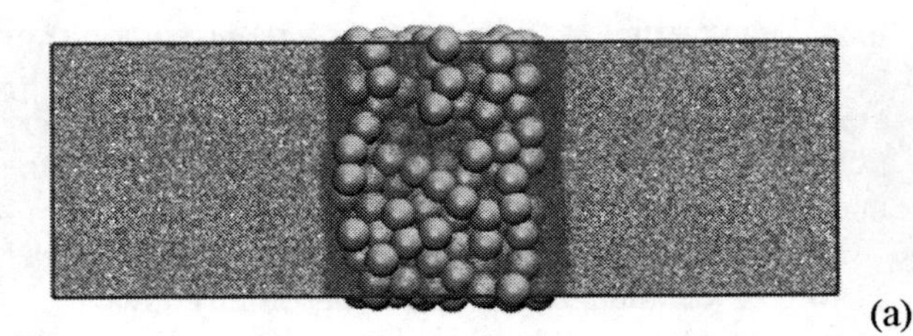

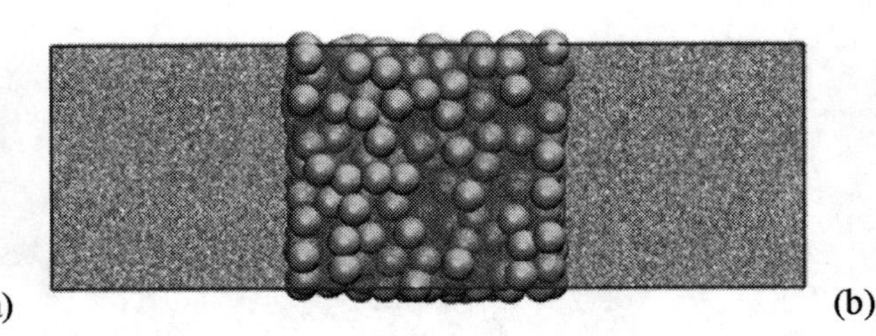

FIGURE 2. Sample simulation cell of liquid/vapor coexistence of a nanoparticle suspension with 500 nanoparticles of radii $a = 5\sigma$ in an explicit solvent for $\phi_v \approx 25\%$ in the center of the liquid phase. Solvent is shown as points. (a) $A_{nn} = 4\pi^2\varepsilon$ and $A_{ns} = 24\pi\varepsilon$ (b) $A_{nn} = \varepsilon$ and $A_{ns} = 12\varepsilon$. The dimensions of the simulation cell are $L_x = L_y = 111.2\sigma$ and $L_z = 333.6\sigma$.

nent of the interaction $U^{R}_{nn}(r_{12})$ is [28]

$$U^{R}_{nn}(r_{12}) = \frac{A_{nn}}{37800}\frac{\sigma_n^6}{r_{12}}\Bigg[\frac{r_{12}^2 - 14r_{12}a + 54a^2}{(r_{12}-2a)^7} + \frac{r_{12}^2 + 14r_{12}a + 54a^2}{(r_{12}+2a)^7} - 2\frac{(r_{12}^2 - 30a^2)}{r_{12}^7}\Bigg], \tag{4}$$

For $h \ll a$, $U^{R}_{nn}(r_{12}) = -\frac{A_{nn}}{2520}\frac{a\sigma^6}{h^7}$. Equations 3 and 4 reduce to the standard LJ potential, Eq. 1, in the limit $a \to 0$ and $\frac{4}{3}\pi a^3 \rho_i = 1$.

The interaction $U_{ns}(r_{12})$ between the LJ solvent atoms and the nanoparticle be determined by integrating the interaction between a LJ solvent atom and the LJ atoms within the particle,

$$U_{ns}(r_{12}) = \frac{2\,a^3\,\sigma_{ns}^3\,A_{ns}}{9\,(a^2 - r_{12}^2)^3}\Bigg[1 - \frac{(5\,a^6 + 45\,a^4\,r_{12}^2 + 63\,a^2\,r_{12}^4 + 15\,r_{12}^6)\,\sigma_{ns}^6}{15\,(a - r_{12})^6\,(a + r_{12})^6}\Bigg], \tag{5}$$

where $A_{ns} = 24\pi\varepsilon_{ns}\rho_1\sigma_{ns}^3$. Here ε_{ns} is the interaction between a solvent atom and an atom in the nanoparticle and $\sigma_{ns} = (\sigma_n + \sigma_s)/2$, where σ_s is the size of a LJ solvent atom. The interaction between solvent atoms is the same Lennard-Jones interaction given in Eq. (1) with $\varepsilon_{ss} = \varepsilon$ and $\sigma_s = \sigma$. Note that, unlike most interaction potentials, $U_{nn}(r_{12})$ and $U_{ns}(r_{12})$ depend on two lengths: the size of the atoms making up the nanoparticle σ_n and the radii of the nanoparticle a.

All of the molecular dynamics simulations were carried out using the LAMMPS simulation package [30] which has been modified to model nanoparticle suspensions [33]. Recent improvements to the algorithm, including multi-region neighbor lists and enhanced communications, have made it possible to model systems of disparate particle sizes. For example the speed up over previous versions of the code for nanoparticles of radii $a = 10\sigma$ in an explicit solvent of LJ atoms is 200-400 depending on nanoparticle concentration. This has allowed us, for the first time, to model hundreds of nanoparticles in an explicit solvent as illustrated in Figs. 1 and 2.

For equilibrium simulations Newton's equations of motion were integrated using a velocity-Verlet algorithm coupled weakly to a heat bath [34, 35]. The integration time step $\delta t = 0.005\tau$, where $\tau = \sigma(m/\varepsilon)^{1/2}$ and m is the mass of the solvent atoms. The damping constant $\Gamma = 0.01\tau^{-1}$. The interactions between LJ atoms were cutoff at $r_c = 3.0\sigma$, between nanoparticles at $5.0a$ and between LJ solvent atoms and nanoparticles at $a + 4.0\sigma$. The pair correlations $g(r)$ were close to 1 for $r > r_c$. We studied nanoparticle suspensions with nanoparticles of radii $a = 2.5$, 5.0 and 10.0σ and mass $m_p = \frac{4}{3}\pi a^3 \rho_1$ with $\rho_1\sigma^3 = 1.0$. All simulations were carried out at temperature $T = \varepsilon/k_B$ and pressure $P = 0.1\varepsilon/\sigma^3$. The number of nanoparticles varied from $N_n = 125$ to 2100 and LJ solvent atoms from a few hundred thousand to a million depending on the nanoparticle concentration. The volume fraction of nanoparticles $\phi_n = N_n\frac{4}{3}\pi a^3/V$ varied from 7 to 40%, where V is the volume of simulation cell.

The initial states were made by first equilibrating pure colloidal systems of various sizes and concentrations. Randomly placed colloidal particles with $a = 0$, in a system with the desired volume fraction, were dynamically grown to their final size by using a hard core repulsive Lennard-Jones pair potential. Once the particles reached their final size, this system was run for 10^5 time steps using a purely repulsive Lennard-Jones potential to assure that the nanoparticles are well dispersed. Without the solvent, the nanoparticles form a dense solid for most of the values of A_{nn} studied here for $T = \varepsilon/k_B$. While the phase diagram for pure nanoparticle system has yet to be determined, the liquid/gas critical point $T_c \approx 0.13A_{nn}/k_B$ for $a = 5\sigma$ [36] for a interaction cutoff $r_c = 5a$. The resulting system was then combined with a pre-equilibrated system – with the same dimensions – of Lennard-Jones solvent particles by deleting overlapping solvent particles. Finally, the resulting suspensions were equilibrated for 10^6 time steps in an NPT ensemble until the desired pressure was reached, after which the volume was fixed. Figure 1 shows a typical equilibrated configuration for $a = 10\sigma$ for $\phi_v = 39\%$.

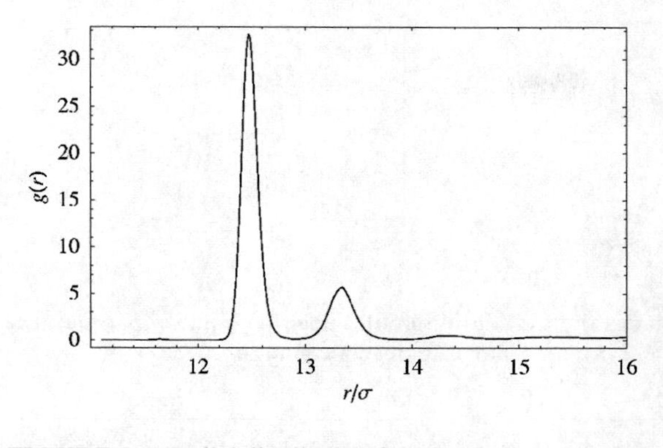

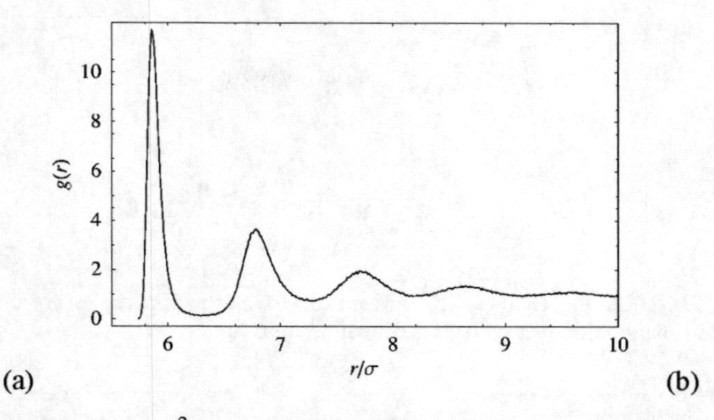

FIGURE 3. Radial distribution function for $a = 5\sigma$, $\phi_v = 25\%$ for $A_{\mathrm{nn}} = 4\pi^2\varepsilon$ and $A_{\mathrm{ns}} = 24\pi\varepsilon$ for(a) nanoparticle-nanoparticle and (b) nanoparticle-solvent.

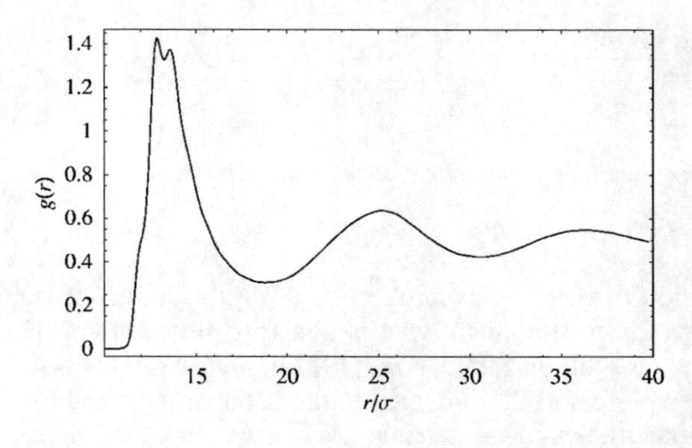

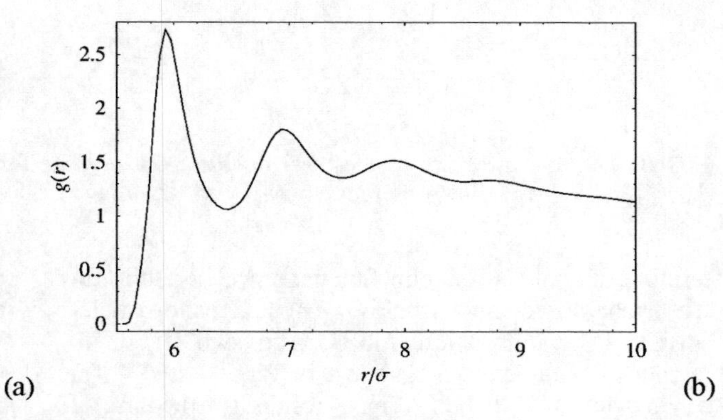

FIGURE 4. Radial distribution function for $a = 5\sigma$, $\phi_v = 25\%$ for $A_{\mathrm{nn}} = \varepsilon$ and $A_{\mathrm{ns}} = 12\varepsilon$ for (a) nanoparticle-nanoparticle and (b) nanoparticle-solvent.

Liquid/vapor initial states were constructed by a similar method as described above with regards to the nanoparticles. After equilibration of the ideal nanoparticle suspension, the length of the simulation cell in the z-direction was enlarged by a factor of 3.0. An additional 10% of Lennard-Jones solvent atoms were added to each interface to allow for the fact that the Lennard-Jones solvent has a high vapor density. Figure 2 shows an example of an equilibrated liquid-vapor configuration for nanoparticles of radii $a = 5\sigma$.

Shear simulations were performed using non-equilibrium molecular dynamics (NEMD), which is implemented by deforming the simulation box shape in a Parrinello-Rahman fashion [37, 38]. Specifically, nanoparticle suspensions were sheared under constant volume in an NVT ensemble, adjusted with the SLLOD equations of motion [39, 40, 41]. Nanoparticle suspension of different sizes and concentrations were studied for shear rates ranging from 10^{-2} to $10^{-6}\ \tau^{-1}$.

Depending on the values of the Hamaker constant, the nanoparticles can either be dispersed in the solvent or aggregate. For nanoparticles made of the same Lennard-Jones atoms as the solvent ($\varepsilon_{\mathrm{nn}} = \varepsilon_{\mathrm{ns}} = \varepsilon$ and $\sigma_{\mathrm{n}} = \sigma$) at density $\rho_1\sigma^3 = 1.0$, $A_{\mathrm{nn}} = 4\pi^2\varepsilon$ and $A_{\mathrm{ns}} = 24\pi\varepsilon$. As seen from the pair correlation functions in the next section, this corresponds to strong interaction between the solvent and the nanoparticles. By separately varying Hamaker constants A_{nn} and A_{ns} it is possible to control the relative strength of the nanoparticle/nanoparticle and nanoparticle/solvent interaction and to study the interface of the various interactions. In most nanoparticle suspensions, the nanoparticles are coated with short surfactants to avoid flocculation, which in our case is modeled by reducing A_{nn}. To reduce the number of parameters we set $\sigma_{\mathrm{n}} = \sigma_{\mathrm{s}} = \sigma$ throughout.

EQUILIBRIUM RESULTS

Bulk

After equilibration at constant pressure $P = 0.1\varepsilon/\sigma^3$, we carried out simulations at constant volume to de-

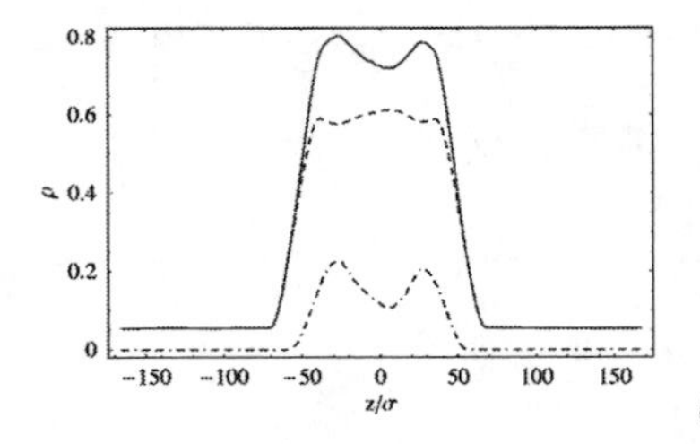

(a)

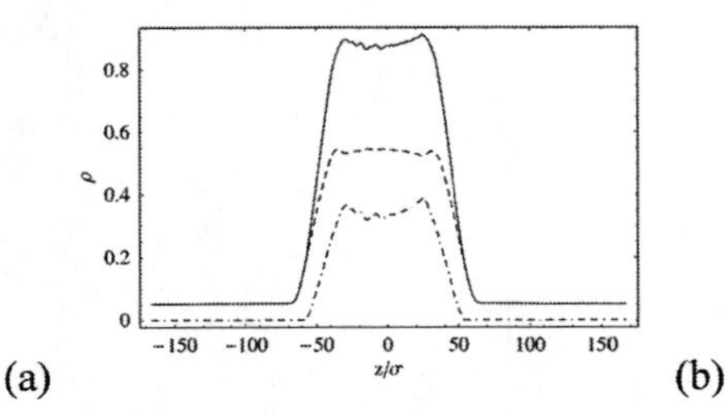

(b)

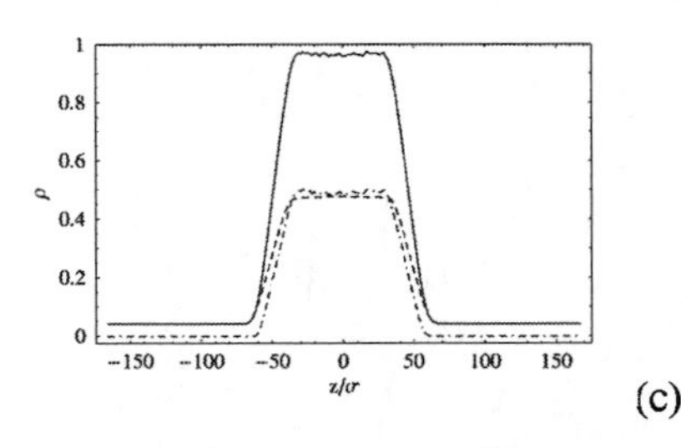

(c)

FIGURE 5. Total (solid), solvent (dash), and nanoparticle (dot-dash-dash) mass density profiles near the liquid/vapor interface for nanoparticle suspensions of radii $a = 5\sigma$ and $A_{\rm nn} = 4\pi^2\varepsilon$ and $A_{\rm ns} = 24\pi\varepsilon$ for three different concentrations.

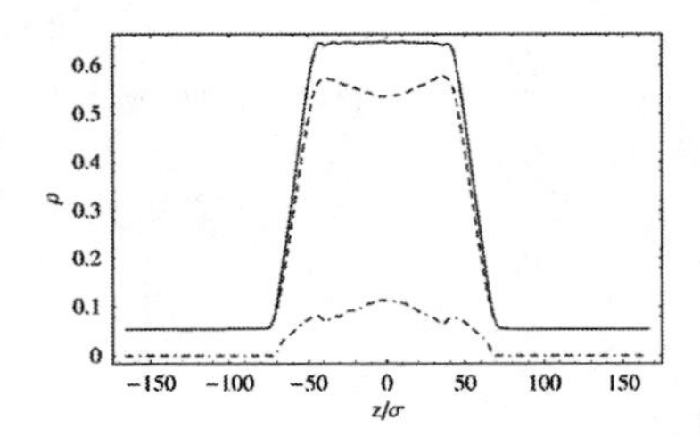

(a)

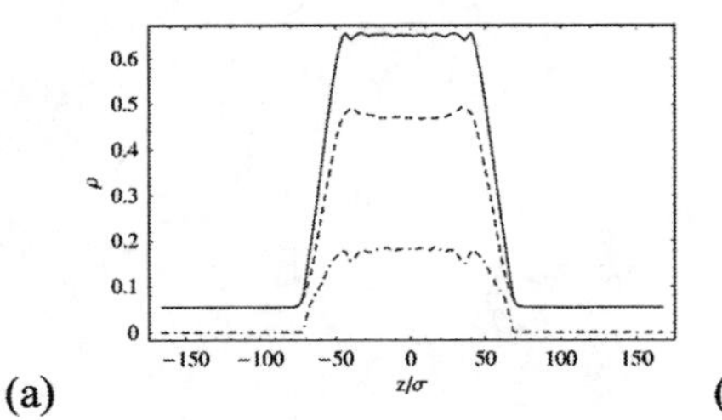

(b)

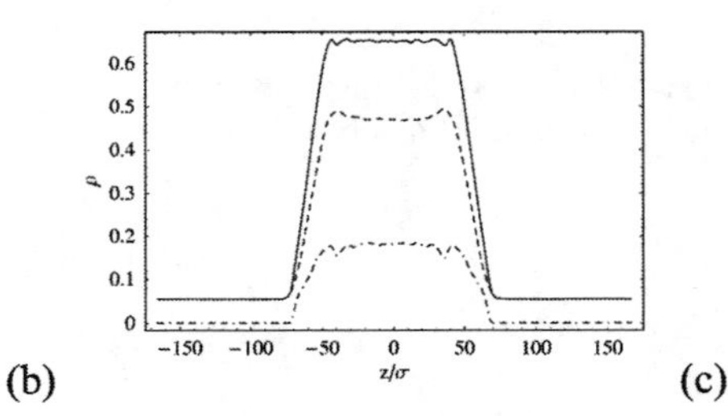

(c)

FIGURE 6. Mass density profiles near liquid/vaport interface for nanoparticle suspensions of radii $a = 5\sigma$ for $A_{\rm nn} = \varepsilon$ and $A_{\rm ns} = 12\varepsilon$ for three different concentrations. Labels are same as in Fig. 5

termine the pair correlation functions $g(r)$. Results for the nanoparticle-nanoparticle $g_{\rm nn}(r)$ and nanoparticle-solvent $g_{\rm ns}(r)$ correlations functions for two sets of interactions parameters are shown in Figs. 3 and 4 for $a = 5\sigma$ for $\phi_v = 25\%$. In Fig. 3 results for the case of a LJ nanoparticle of density $\rho_1\sigma^3 = 1$ in a LJ solvent, $A_{\rm nn} = 4\pi^2\varepsilon$ and $A_{\rm ns} = 24\pi\varepsilon$ are shown. Note that there is almost zero probably that two nanoparticles are closer than $\sim 12.2\sigma$ even though $U(r) = 0$ at $r = 10.46\sigma$, consistent with the fact that there is at least one layer of solvent particles surrounding each nanoparticle. This can also been seen from the strong peak in $g_{ns}(r)$ at $r = 5.5\sigma$, where the minimum of attractive well for $U_{ns}(r)$ is at $r = 5.86\sigma$. This strong attraction is also seen for the all the systems studied ($\phi_v = 7\%$ to 39%). From visual observation of the configurations, the low concentration systems studied, $\phi_v = 7\%$ and 12%, form a colloid gel with a highly non-uniform distribution of nanoparticles. For the two higher concentration studied, $\phi_v = 25\%$ and 39%, the nanoparticles are uniformly distributed as seen in Fig. 2a.

As the interaction between nanoparticles in reduced this strong correlation between nanoparticles disappears. As shown in Fig. 4 for $A_{\rm nn} = \varepsilon$, $A_{\rm ns} = 12.0\varepsilon$, the pair correlation between nanoparticles is close to what one expects for soft, repulsive particles. In this case the nanoparticles are very weakly interacting, essentially hard sphere-like, as the temperature T is much greater than the liquid/vapor critical temperature T_c for pure nanoparticles. Reducing the nanoparticle-solvent interaction further, for example to $A_{\rm ns} = 6.0\varepsilon$, led to phase separation, similar to what is observed for hard sphere mixtures of disparate size [19]. In this case the solvent prefers itself compared to the nanoparticles and the nanoparticles phase separate. While we have not determined the complete phase boundary, phase separation also occurs when the nanoparticle-nanoparticles interact too strongly compared to the nanoparticle-solvent and solvent-solvent interactions.

Liquid-Vapor Interfaces

The interfacial density profiles for $A_{\rm nn} = 4\pi^2\varepsilon$ and $A_{\rm ns} = 24\pi\varepsilon$ is shown in Fig. 5 and for $A_{\rm nn} = \varepsilon$ and $A_{\rm ns} = 12\varepsilon$ is shown in Fig. 6. The density profiles in Fig. 5b correspond to the snapshot shown in Fig. 2a, while the density profiles in Fig. 6b correspond to the snapshot shown in Fig. 2b. For both systems there are no nanoparticles in the vapor phase. The vapor phase density corresponds to that of the pure solvent system. For the strongly interacting case shown in Fig. 5, at low nanoparticle concentrations, the nanoparticles avoid the interface, where as at higher concentrations the distributions for both nanoparticles and solvent are uniform. The fact that the nanoparticles avoid the liquid/vapor interface in this case is not surprising since at the temperature of these simulations, $T = 1.0\varepsilon/k_B$, the pure nanoparticle system without any solvent is solid and has a much higher surface tension than that of the Lennard-Jones

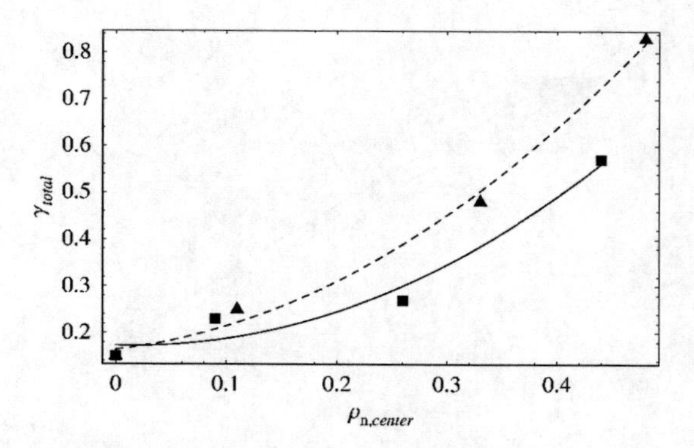

FIGURE 7. Surface tension (in units of ε/σ^2) as a function of nanoparticle density in the liquid phase at Hamaker constants $A_{nn} = 12.9\varepsilon$, $A_{ns} = 43.1\varepsilon$ (squares) and $A_{nn} = 4\pi^2\varepsilon$, $A_{ns} = 24\pi\varepsilon$ (triangles).

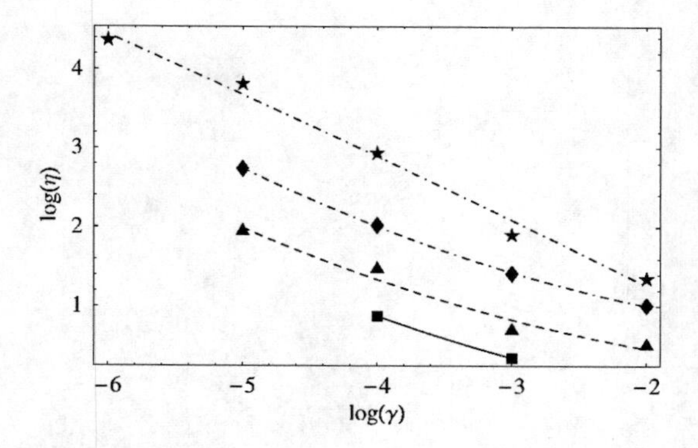

FIGURE 8. Viscosity as a function of shear rate at radii $a = 5\sigma$ for volume fractions of 7% (squares), 12% (triangles), 25% (diamonds), and 39% (stars).

solvent. The liquid/vapor surface tension for this system along with a second strongly interacting case ($A_{nn} = 12.9\varepsilon$ and $A_{ns} = 43.1\varepsilon$) is shown in Fig. 7 as a function of the nanoparticle concentration in the center of the liquid phase. For comparison the surface tension for the pure LJ solvent $\gamma = 0.15 \pm 0.01\varepsilon/\sigma^2$. Figure 7 shows a sublinear increase in surface tension with increasing nanoparticle concentration as one expects when one mixes two systems with disparate surface tensions. The lower surface tension liquid, in this case the Lennard-Jones solvent, goes to the interface, particularly for low nanoparticle concentrations. Also, as is apparent from the same figure, a systemically larger surface tension is observed for stronger nanoparticle and solvent interactions.

Results for the weakly interacting nanoparticle system ($A_{nn} = \varepsilon$ and $A_{ns} = 12.0\varepsilon$) are shown in Fig. 6. In this case, except for the lowest concentration the nanoparticle and solvent concentrations are uniform. The liquid/vapor tension is nearly independent of concentration and only slightly larger than that of the pure solvent for these three concentrations.

SHEAR RHEOLOGY

The effect of particle size on the shear rheology was evaluated for three sized nanoparticles, $a = 2.5$, 5.0, and 10.0σ for nanoparticle concentrations ranging from $\phi_v =$ 6% to 40%. Shear rates ranged from 10^{-2} to $10^{-6}\tau^{-1}$. For $T = \varepsilon/k_B$ and $P = 0.1\varepsilon/\sigma^3$, the shear viscosity of the LJ solvent is $\eta = 1.1 \pm 0.1 m/\tau\sigma$. Here we present results for the case $A_{nn} = 4\pi^2\varepsilon$ and $A_{ns} = 24\pi\varepsilon$, which due to the strong interaction between the solvent and the nanoparticles is expected to have display the largest increase in viscosity. The effect of varying A_{nn} and A_{ns} on the suspension rheology will be discussed elsewhere [42].

Results for $a = 5\sigma$ for four concentrations are shown in Fig. 8. Clearly most of the results are in the shear thinning regime and have not reached the Newtonian, shear rate independent regime. Even for the lowest concentration studied there is a significant increase compared to the background solvent. Comparison of different size nanoparticles are shown in Fig. 9 for $\phi_v = 25\%$ and 39%. Note that at least in the shear thinning regime – which we can explore with the present simulations – the viscosity is only weakly dependent on the size of the nanoparticles even for nanoparticles as small as $a = 2.5\sigma$. This result is also seen in a plot of shear viscosity versus concentration ϕ_v for fixed shear as shown in Fig. 10 for $\gamma = 10^{-3}$ and $10^{-4}\tau^{-1}$.

These results that the shear viscosity is at most weakly dependent on the size of the nanoparticles are consistent with the results of Knauert *et al.* [24] who found that at 5% volume fraction there was very little difference between the shear viscosity for size shaped nanoparticles – an icosahedra, a rod and a sheet. In this simulation, the nanoparticles interacted more weakly with the solvent than for the case studied here, however in both the dependence on nanoparticle size was surprisingly weak. From these results it appears at concentrations where jamming and interactions of the nanoparticles do not play a significant role, the size and shape of the nanoparticle is not very critical. Further simulations with composite nanoparticles are presently underway to test this result [42].

SUMMARY

Here we have presented the first large scale simulations of nanoparticle suspensions for hundreds of nanoparticles in an explicit solvent. The nanoparticles are treated

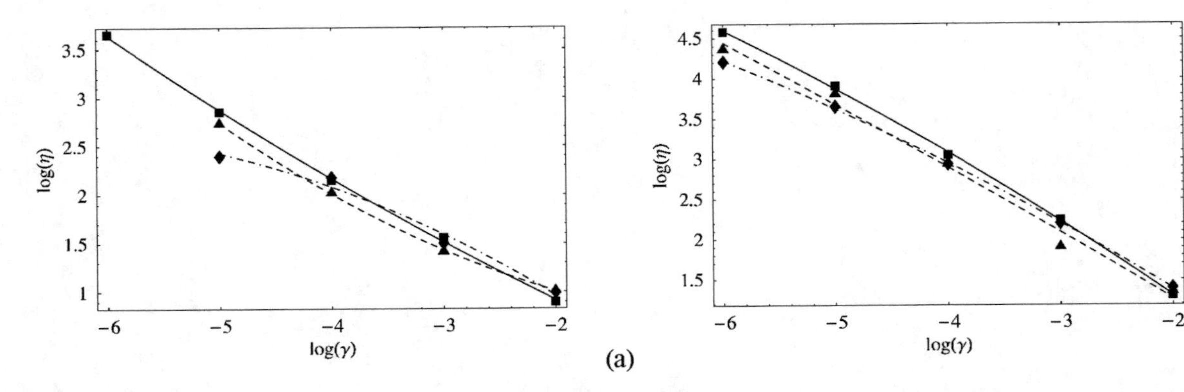

FIGURE 9. Viscosity as a function of shear rate for volume fractions ϕ_v of (a) 25% for radii $a = 2.5\sigma$ (squares), 5.0σ (triangles), and 10σ (diamonds) and (b) 39% for $a = 5.0\sigma$ and 10.0σ and 41% for $a = 2.5\sigma$.

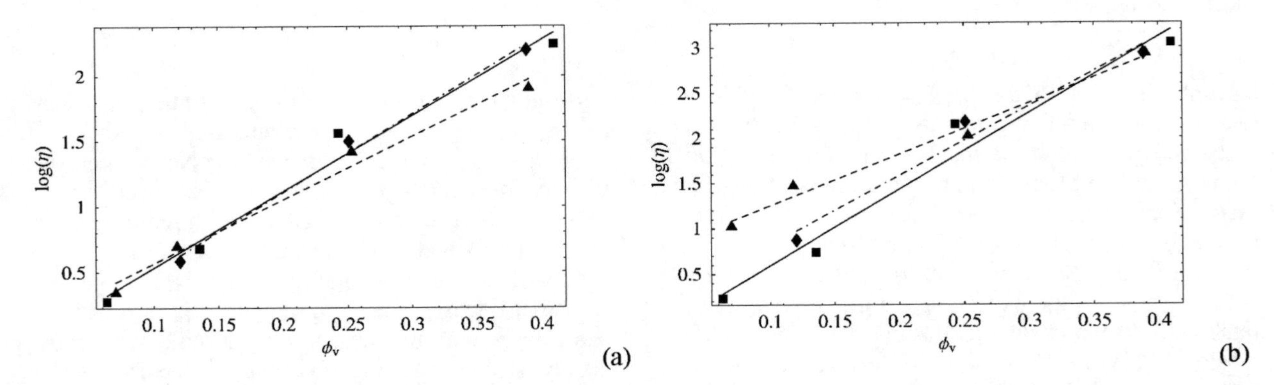

FIGURE 10. Viscosity as a function of volume fraction for radii $a = 2.5\sigma$ (squares), 5.0σ (triangles), and 10σ (diamonds) for (a) $\gamma = 10^{-3}$ and (b) $10^{-4}\tau^{-1}$.

as a uniform distribution of Lennard-Jones atoms for which all the relevant interactions can be determined analytically. Using a newly modified version of the molecular dynamics simulation package LAMMPS [33], we studied nanoparticles from size 5 to 20 times that of the solvent for a range of concentrations. For nanoparticles made of the Lennard-Jones atoms of densities comparable to that of a Lennard-Jones solid, the nanoparticles and solvent interaction was so strong that each nanoparticle was coated with a layer of solvent atoms. For low concentrations, this system formed a colloidal gel. For weakly interacting nanoparticles, a minimum nanoparticle-solvent interaction strength was needed to keep the nanoparticles in solution. Without this attractive interaction between the nanoparticles and the solvent, the nanoparticles phase separated from the solvent, similar to what occurs for hard sphere mixtures of disparate size. Phase separation also occurred when the nanoparticle-nanoparticle interaction dominated. Results from NEMD simulations showed the somewhat surprising result that the size of the nanoparticle had minimum effect on the shear viscosity of the solution, even for nanoparticles as small as 5 times that of the solvent.

One interesting future direction of this work is to model polymer nanocomposites. Following previous studies of polymer nanocomposites [20, 21, 22, 23, 24, 29], the polymer can be modeled as a coarse-grained Lennard-Jones bead-spring chain [34, 35]. Previous studies of aspherical particles have built the nanoparticle out of Lennard-Jones atoms. By combining nanoparticles of different sizes using the model outlined here, one can build aspherical nanoparticles with a smaller number of atoms. Initial studies of two dumbbell integrated nanoparticle in an explicit solvent are presently underway.

ACKNOWLEDGMENTS

Sandia is a multiprogram laboratory operated by Sandia Corporation, a Lockheed Martin Company, for the United States Department of Energy's National Nuclear Security Administration under Contract DE-AC04-94AL85000.

REFERENCES

1. D. L. Ermak and J. A. McCammon, *J. Chem. Phys.* **69**, 1352-1360 (1978).
2. P. Strating, *Phys. Rev. E* **59**, 2175-2187 (1999).
3. D. R. Foss and J. F. Brady, *J. Fluid Mech.* **407**, 167-200 (2000).
4. J. F. Brady and G. Bossis, *Annu. Rev. Fluid Mech.* **20**, 111-157 (1988).
5. R. C. Ball and J. R. Melrose Jr., *Adv. Colloid Interface Sci.* **59**, 19-30 (1995).
6. A. A. Catherall, J. R. Melrose Jr., and R. C. Ball, *J. Rheology* **44**, 1-25 (2000).
7. A. Sierou and J. F. Brady, *J. Fluid Mech.* **448**, 115-146 (2001).
8. J. P. Hernandez-Ortiz, J. J. de Pablo, and M. D. Graham, *Phys. Rev. Lett.* **98**, 140602 (2007).
9. A. J. C. Ladd and R. Verberg, *J. Stat. Phys.* **104**, 1191-1251 (2001).
10. P. J. Hoogerbrugge and J. M. V. A. Koelman, *Europhys. Lett.* **199**, 155-160 (1992).
11. P. Espanol, *Phys. Rev. E* **57**, 2930-2948 (1998).
12. W. Dzwinel, K. Boryczko, and D. A. Yuen, *J. Colloid Interface Sci.* **258**, 163-173 (2003).
13. V. Pryamitsyn and V. Ganesan, *J. Chem. Phys.* **122**, 104906 (2005).
14. A. Malevanets and R. Kapral, *J. Chem. Phys.* **110**, 8605-8613 (1999).
15. A. Malevanets and R. Kapral, *J. Chem. Phys.* **112**, 7260-7269 (2000).
16. M. Hecht, J. Harting, T. Ihle, and H. J. Herrmann, *Phys. Rev. E* **72**, 011408 (2005).
17. J. T. Padding and A. A. Louis, *Phys. Rev. E* **74**, 031402 (2006).
18. R. Glowinski, T. W. Pan, D. D. Joseph, and J. Periaux, *J. Comput. Phys.* **169**, 363-426 (2001).
19. M. Dijkstra, R. van Roij, and R. Evans, *Phys. Rev. E* **59**, 5744-5771 (1999).
20. F. W. Starr, T. B. Schoder, and S. C. Glotzer, *Macromolecules* **35**, 4481-4492 (2002).
21. F. W. Starr, J. F. Douglas, and S. C. Glotzer, *J. Chem. Phys.* **119**, 1777-1788 (2003).
22. G. D. Smith, D. Bedrov, L. Li, and O. Byutner, *J. Chem. Phys.* **117**, 9478-9489 (2002).
23. S. Sinsawalt, K. L. Anderson, R. A. Vaia, and B. L. Farmer, *J. Polym. Sci. B Polym. Phys.* **41**, 3272-3282 (2003).
24. S. T. Knauert, J. F. Douglas, and F. W. Starr, *J. Polm. Sci Part B Polym. Phys.* **45**, 1882-1897 (2007).
25. C. Powell, N. Fenwick, F. Bresme, and N. Quirke, *Colloids Surf. A* **206**, 241-251 (2002).
26. D. Gersappe, *Phys. Rev. Lett.* **89**, 058301 (2002).
27. S. R. Challa and F. Van Swol, *Phys. Rev. E* **73**, 016306 (2006).
28. R. Everaers and M. R. Ejtehadi, *Phys. Rev. E* **67**, 041710 (2003).
29. T. Desai, P. Keblinski, and S. K. Kumar, *J. Chem. Phys.* **122**, 134910 (2005).
30. S. Plimpton, *J. Comp. Phys.* **117**, 1-19 (1995).
31. H. C. Hamaker, *Physica* **4**, 1058 (1937).
32. J. N. Israelachvili, *Intermolecular and Surface Forces*, Academic Press, 1991, 2nd edn.
33. P. J. in 't Veld, S. Plimpton, and G. S. Grest, submitted to *Comp. Phys. Comm.* 2007.
34. G. S. Grest and K. Kremer, *Phys. Rev. A* **33**, 3628-3631 (1986).
35. K. Kremer and G. S. Grest, *J. Chem. Phys.* **92**, 5057-5086 (1990).
36. M. A. Horsch, P. J. in 't Veld, J. B. Lechman, and G. S. Grest, submitted to *J. Chem. Phys.* 2007.
37. M. Parrinello and A. Rahman, *J. Appl. Phys.* **52**, 7182-7190 (1981).
38. M. Parrinello and A. Rahman, *J. Chem. Phys.* **76**, 2662-2666(1982).
39. D. J. Evans and G. P. Morriss, *Comp. Phys. Rep.* **1**, 297-343 (1984).
40. J. W. Rudisill and P. T. Cummings, *Rheology Acta* **30**, 33-43 (1991).
41. M. E. Tuckerman, C. J. Mundy, S. Balasubramanian, and M. L. Klein, *J. Chem. Phys.* **106**, 5615-5621 (1997).
42. P. J. in 't Veld, M. K. Petersen, and G. S. Grest, in preparation.

From Molecular Solutions to Fragile Gels: Dynamics of Rigid Polymers in Solutions

Yunfei Jiang[1], Uwe H. F. Bunz[2], and Dvora Perahia[1]

[1]Chemistry Department, Clemson University, Clemson SC 2934-0973 USA
[2]School of Chemistry and Biochemistry, Georgia Institute of Technology, Atlanta GA 30313 USA

Abstract. The dynamic characteristics of highly rigid polymers in their different association modes, as discerned from neutron spin echo and NMR studies are discussed. Highly rigid polymers assemble to form supramolecular hierarchal structures that exhibit well defined electro-optical characteristics. The dynamic processes of the molecules within these agglomerates are a key in controlling the effective conjugation length and thus their optical properties. Among their association modes are micellar structures, liquid crystalline phases and gels. Using small angle neutron scattering together with neutron spin echo, and NMR, the different association modes of dialkyl poly (*para* phenyleneethynylene (PPE) in cyclohexane and their dynamics were investigated. In molecular solutions, the chains assume a worm-like conformation. With decreasing temperatures the chains associate to form aggregates that jam into each other and form fragile gels. The dynamics of the PPE molecules in the molecular solutions is characterized predominantly by center of mass diffusion, coupled to undulations of the backbone and a motion of side chains, whereas in the aggregated phase the center of mass is constrained while the undulations are maintained. As gel is formed, the motion of the aggregates is constrained, however the undulations of the chains within the aggregates is retained. The solvent molecules, though are sensitive to the presence of the polymer chains, retains a large degree of freedom.

Keywords: Fragile matter, Gels, NSE, Polymers
PACS: 82.35.Lr, 82.56.Lz, 83.80.Kn, 83.80.Rs

INTRODUCTION

Gels formed by highly conjugated polymers often exhibit well-defined electro-optical response. The association mode of the polymers and the dynamics of the macromolecules on multiple length scales fine-tune their electronic levels. [1,2] In general, polymers with conjugated backbones are intrinsic organic semiconductors. They are candidates for active components in organic based electro-optical devices including organic light-emitting diodes (OLED)s, photovoltaic cells, transistors, detectors /sensors and liquid crystal based devices.[3-12] some are solid-state devices and others are liquid and liquid crystalline materials encapsulated in different cells.

Multiple factors determine their electro-optical response among them is the conjugation length, or the length of the segment over which the electrons can be described in terms of bands. The conjugation length is determined predominantly by the chemical structure and the conformation of the macromolecules. In addition to the basic chemical structure, the dynamics of the system and the association mode of the polymers affect their conformation.

We have recently shown that dialkyl PPEs (Figure 1) form optically active gels when immersed in toluene at concentrations bellow their liquid crystalline critical threshold. Small angle neutron scattering SANS) revealed that at high temperatures in toluene, the polymer chains are extended with a persistence length that is often extended across the entire molecular dimensions. As the temperature decreases, association into flat aggregates takes place. [13-15]

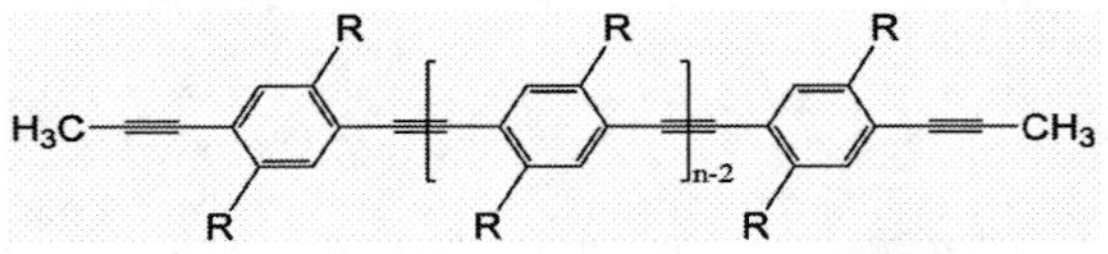

FIGURE 1. The chemical structure of poly (*para* phenyleneethynylenes (PPEs). R represents different groups.

The aggregates increase in size until they jam into each other and form a fragile gel. Figure 2 introduces a schematic description of the process. This gel is essentially a fragile phase where a small shear breaks the network. The conformation of the polymeric backbone differs in the molecular solutions, the micellar ones, and the gels. It is evident by the colorless nature of the molecular solutions and the yellow color in the solutions that consists of clusters

CP982, *Complex Systems, 5th International Workshop on Complex Systems*
edited by M. Tokuyama, I. Oppenheim, and H. Nishiyama

that correspond to different separation of the electronic levels.

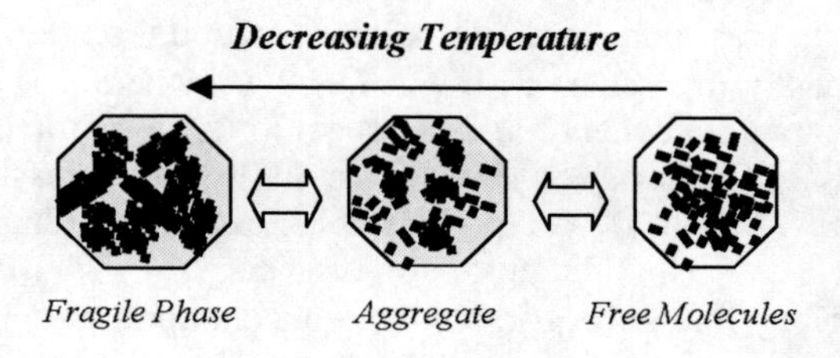

FIGURE 2. A schematic representation of the different assembling modes of PPE in toluene.

There are varieties of open questions regarding the driving force to association as well as the factors that control the stability of the clusters. There is very limited understanding however, of the dynamic processes that take place within the different phases. While many polymers form gels in presence of solvents, the fragile nature of this gel is more in line with a colloidal complex fluid rather than a polymeric solution.

A wide range of soft materials including colloids, emulsions and foams undergo constrain of their motion at low temperatures or high volume fractions. [16-19] The constrained gel-like phase, or 'jammed' phase, is characterized by a solid-like rheological response while the local dynamics is often rather fast in comparison with that of molecules in a solid. In contrast to solids, in this type of 'fragile matter', changing the direction of applied stress will release the constraint. In colloidal systems (rigid particles with no electrostatic interaction), the balance between the thermal energy ~kT of a system at a given temperature T (where k is the Boltzmann constant) and the attractive interaction between particles determines the state of the system, i.e. fluid or constrained.

The same phenomena have been observed in solutions of polymeric micelles and star polymers, where cooling or heating leads to formation of gel-like phases. [13-15,19,20]. Similar to colloidal solutions, the energy balance controls the state of the system. However, changing temperature in a complex system, which consists of aggregates, does not just change the inter-particle interactions and kT, but also affects the association mode. Relatively small changes in temperature may be accompanied by significant changes in the microstructure of the system. Thus the formation of a constrained phase is a combination of multiple factors. The assembling of particles, in particular, polymeric chains into aggregates and the dynamics of the entire aggregate define the overall state of the complex fluid.

Neutron Spin Echo (NSE) and nuclear magnetic resonance (NMR) measurements of PPE in toluene, [18, 20] revealed that the solvent has been highly confined to the di-alkyl PPE molecule. The similarity in the chemical structure of the solvent and the backbone of the polymer drive further association of the solvent via $\pi-\pi$ stacking. As aggregation takes place, significant amount of solvent is trapped within the aggregates affecting the motion of the PPE molecule on the length scale of the side chain and the rigid segment. The gellation process is accompanied by a constrained of motion on the length scale of the aggregates. The dynamic processes measured in toluene are those of a strongly coupled polymer-solvent complex fluid.

The present study focuses on the structure and dynamic processes of PPE in cyclohexane, a solvent that readily dissolves the polymer, however its interaction with the chains is limited to the Van der Waals energy order of magnitude. The dynamic characteristics of the polymer and the solvent span multiple length and time scales. Nuclear magnetic resonance has been used to study the dynamics of the solvent and neutron spin echo provided insight into the dynamics of the PPE molecule.

EXPERIMENTAL

Materials

The poly(*para* phenyleneethynylene) (PPE) with the nonyl side chain was synthesized by alkyne metathesis. Details are given elsewhere. [21] The number of repeating unit of dinonyl PPE has been determined by gel permeation chromatography (GPC) as 140 and M_w/M_n=3.1. Solutions of dinonyl PPE in deuterated toluene (Cambridge Isotope Lab) was prepared for the measurements of small angle neutron scattering (SANS) and neutron spin echo (NSE).

Neutron Scattering

NSE and SANS experiments were carried out at NCNR of NIST. SANS experiments were carried out on NG3 with $\lambda = 6$ Å. Two-Dimension data were collected and then integrated into one-dimensional patterns. At a single temperature, data were collected at two q ranges, where the detector was placed at 1.5 and 13.1 m from the samples. The momentum transfer vector q is defined as: $q=(4\pi/\lambda)\sin\Theta$ is the wave vector in the z direction, λ is the wave length and Θ is the incidence angle. The data measured at two separate detector distances to cover as large as possible q range, were combined. Samples were encapsulated in a 2mm thick, 1 cm diameter, quartz cells. The temperature was controlled using a water bath to ± 0.5°C. All the data are normalized to transmission of the sample and to the scattering of the solvent at the measured temperature.

NSE experiments were performed on NG5 spectrometer using an incoming neutron wavelength of λ = 6Å and 8Å (FWHM of 10%). NSE measures directly the intermediate scattering function S(q,t), which is displayed in normalized form S(q,t)/(S(q,0). The wave vector settings were 0.045, 0.085, 0.15 $Å^{-1}$ and the Fourier time ranges were 0.045ns < t < 17ns for 0.15 $Å^{-1}$ and 0.1 ns < t < 40ns for 0.085 and 0.045 $Å^{-1}$, respectively. The temperature was controlled using a water bath to ±0.5°C. The solvent was measured independently in an identical temperature range to that of the solutions. The scattering function as a function of time $S(q,t) = S(q,0) \exp(-\Gamma t)$, is analyzed in terms of a correlation time Γ. Depending on the length scale, this correlation time corresponds to a different process within the complex fluid.

NMR Measurements

NMR T_1 measurements were carried out on a Jeol 500MHz for protons. Nitrogen was bubbled through the samples and the samples were sealed, to avoid any effects of oxygen on the relaxation times. The temperature was controlled to ± 0.20°C. NMR signals of the solvent, the terminal CH_3 on the side chains are clearly resolved, were those of the CH_2s are overlapping. NMR signal of the PPE backbone was broad and weak as expected form a polymer backbone in confined geometry. Here in the data for the solvent are presented.

RESULTS AND DISCUSSION

The present study follows the dynamics of PPE in its various association modes in cyclohexane, a good solvent whose interaction with the polymer are limited to Van der Waals ones. Similar to toluene solutions, the PPEs molecules associate into aggregates that form a fragile gel. The conformation of the molecules however differs form that in toluene.

Structural Studies

SANS profiles of a 2.06wt.% solution of dinonyl PPE in cyclohexane as a function of temperature, are shown in Figure 3. The line shape of the pattern at 45°C differs significantly from those measured at lower temperatures. The pattern at 45°C corresponds to molecular solutions. At lower temperatures, the patterns correspond to assemblies of chains. No significant differences are observed in the scattering functions as gelation takes place. This is similar to our previous studies in toluene.[20] An upturn is observed at low q, corresponding to a network formation.

Focusing first on the pattern at 45°C, a slope α= 1.4 is measured. In dilute polymer solution, scattering intensity $I(q)$ scales with q as $I(q) \sim q^{-\alpha}$ where α is an effective exponent, and is specific to a particular polymer chain. [18, 22-24] For a flexible Gaussian chain, $I(q) \sim q^{-2}$, for a chain with excluded volume, $I(q) \sim q^{-1.66}$, and for rigid rod polymers, $I(q) \sim q^{-1}$. For PPE in toluene, α~1 for similar concentrations, i.e, the backbone in solution is rigid, whereas for cyclohexane, an enhanced flexibility is observed.

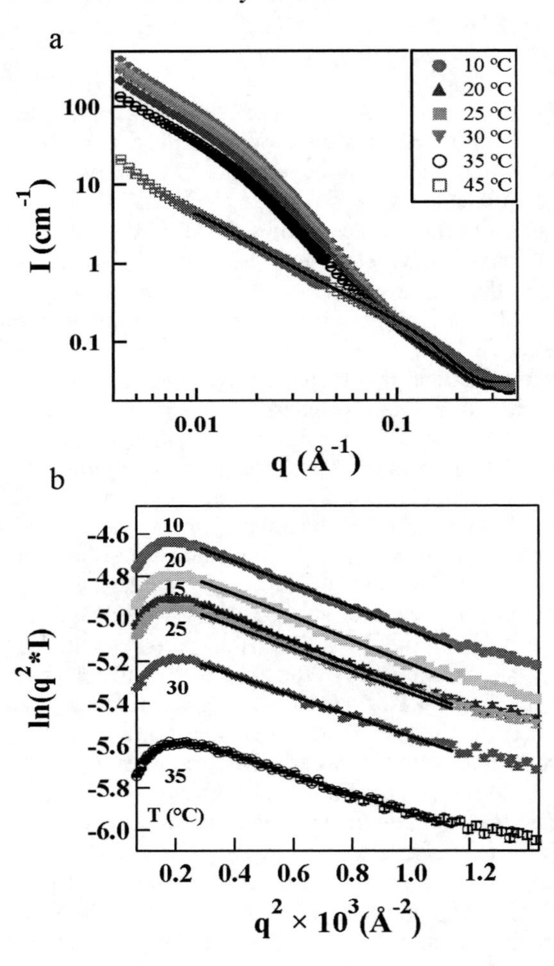

FIGURE 3. The SANS profiles of a 2.06wt.% in cyclohexane as a function of temperature. a) data normalized to the scattering of the solvent (b)A Kratkey plot of the Small angle region, where the lines correspond to a Guenier fitting.

The value of α demonstrates that PPE molecules in cyclohexane are semi-flexible and they assume a worm-like conformation. For worm-like objects with the length of the chains significantly larger than their cross-sectional radius, the total scattering function is described by [25, 26]:

$$I_{WC}(q,L,b,R) = c\Delta\rho^2 M S_{WC}(q,L,b) P_{CS}(q,R)$$

where c is the concentration, $\Delta\rho^2$ is the scattering length density contrast, M is the average molecular weight, and $S_{WC}(q,L,b)$ is the scattering function of single semi-flexible chain . P_{CS} is the form factor of a rigid rod and is given by:

$$P_{CS}(q,R) = [\frac{2J_1(qR)}{qR}]^2$$

where $J_1(x)$ denotes a first order Bessel function. The model for worm like object is presented as a line through the pattern at 45°C.

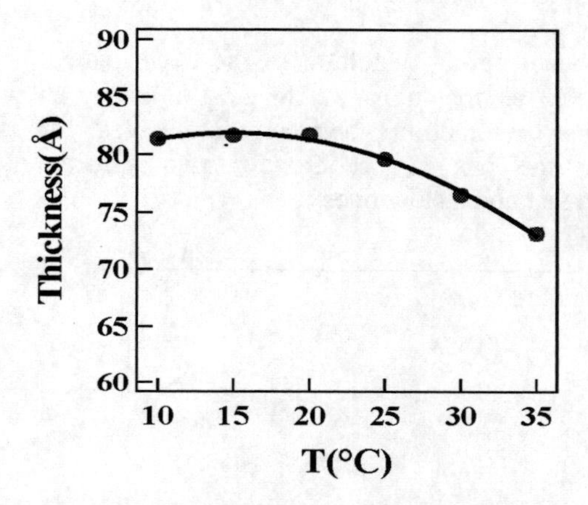

FIGURE 4. The thickness of PPE aggregates in cyclohexane.

The model, marked as a solid line in Figure 3a, provides a cross section of 12.0Å , a Kuhn length of 119 Å and an overall length of 1200 Å. The 12 Å radius is consistent with the dimension of a PPE backbone with stretched out side chains. The Kuhn length of 119 Å corresponds to 17 repeating units of PE. 1200 Å is longer than a fully extended PPE molecule of 90 repeating units, which is ca. 630 Å. This dimension is consistent with an instantaneous aggregation to form longer structures. This is further supported by the upturn in the intensity observed at low q.

The patterns at lower temperatures are characteristics of flat aggregates with the thickness T much thinner then their X-Y dimension. The scattering intensity is proportional to the thickness of the aggregates and is given by:

$$I(q) = N_P A \frac{2\pi(\Delta\rho)^2 T^2}{q^2} \exp(-q^2 R_T^2); qR_T < 1$$

where $R_T = \frac{T}{\sqrt{12}}$, and A is the area of the particle cross section.[25-26] The thickness has been extracted from the plot of $\ln(q^2 * I)$ versus q^2, where the slope and the intercept are correlated to the thickness and volume fraction of the scattering particle since

$$\ln(q^2 * I(q)) = \ln(2\pi(\Delta\rho)^2 N_P A T^2) - q^2 R_T^2 .$$

The thickness of these aggregates is ~80Å. It decreases slightly with increasing temperature, until the aggregates break up. The aggregates appear to be more tightly packed with less solvent molecules incorporated. While the onset of association is at ~35°C, gelation takes place at 20°C, where the thickness becomes constant. Similar association pattern is observed over a concentration range of 0.8wt% to 6wt%.

Dynamics Study

The dynamics of solvent as well as the PPE molecules were followed as a function of temperature in the molecular solution, the micellar phase and the fragile gel.

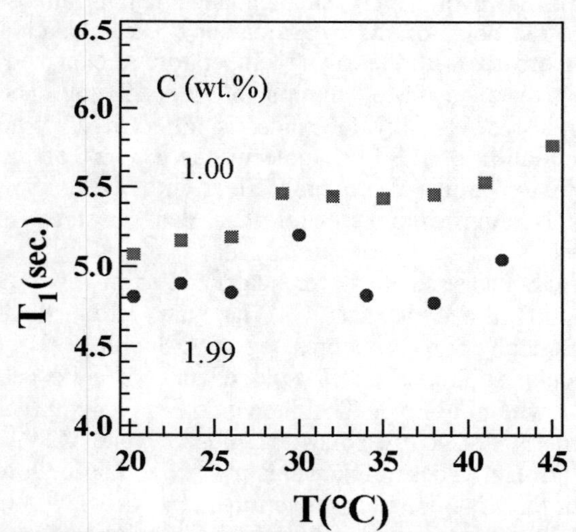

FIGURE5. The spin lattice relaxation time of cyclohexane, measured at a temperature range similar to that of the structural data.

The time scale that would capture the dynamics of the solvent is within the range of NMR. The spin-spin and spin-lattice relaxation times T_1 and T_2 reflects the dynamic processes of the solvent that would

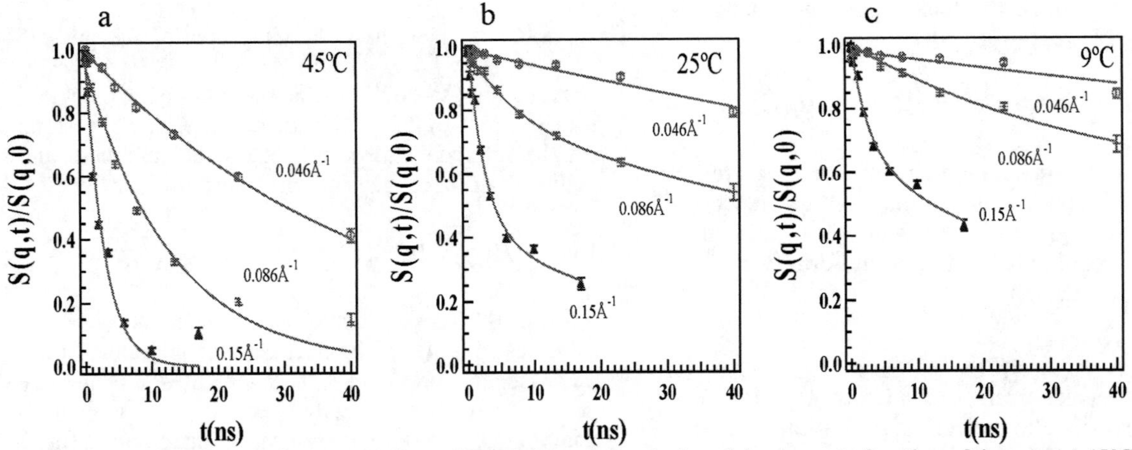

FIGURE 6. NSE patterns of a 1 wt. % dinonyl PPE(140) solution in d-cyclohexane as a function of time at: (a) 45°C, (b) 25°C, (c) 9°C. The symbols represent the experimental data and the lines correspond to different models.

contribute to magnetic relaxation including rotational and translational diffusion. For non viscous Newtonian liquids $T_1 \sim T_2$. [27] T_1 of the solvent as a function of temperature for two concentrations is presented in Figure 5. No significant changes are observed as association takes place. A small decrease in relaxation times is observed at the gellation point, which is attributed to an overall constrained of the aggregates. This is however a very small change that is almost within the noise of the measurement. The relaxation time increases with increasing temperature as expected, indicative of a faster motion of the solvent. The dynamics of the cyclohexane is affected by the concentration of the PPE molecule as a result of the VdW interactions. The overall effect on the relaxation times is significantly smaller then that observed in toluene.

NSE measurements were carries out on a 1wt% PPE in d-cyclohexane. The data for three representative temperatures, corresponding to a molecular solution, micellar phase, and a fragile gel, are shown in Figure 6. The three q ranges 0.046, 0.086 and 0.15 Å^{-1}, correspond to 140, 87 and 42 Å, respectively. These dimensions are above and bellow that of the Khun length, as determined by SANS.

With decreasing temperature, a significant constrain of motion is observed at all q ranges. Molecular solutions are characterized by a fast decay on shorter length scales, whereas on the length scale of the rigid segment of the backbone, a slower decay is observed. As association takes place the dynamics is constrained on all length scale. As a gel is formed, hardly any relaxation takes place on the large length scale, however the chains retain a significant amount of motion on shorter length scale.

Analyzing the data with a single exponential $S(q,t)/S(q,0) = A \exp[-D_{eff}(q)\, q^2\, t]$ at 25°C and 45°C yields reasonable good matching to the experimental data. However, as constrains are imposed either by an instantaneous interaction between molecules or aggregates, one has to account for two separate processes, a fast and a slow ones.

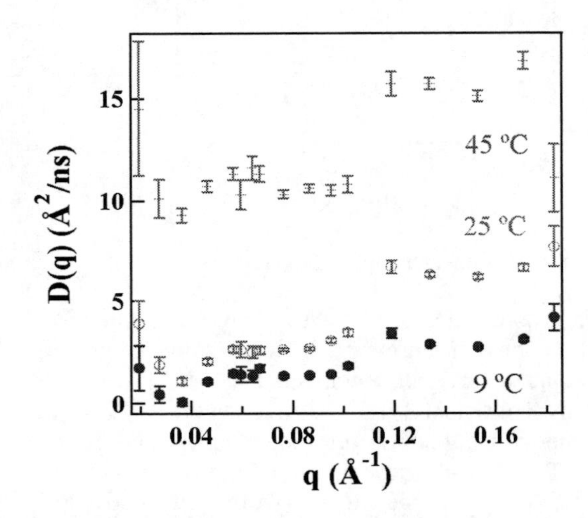

FIGURE 7. Effective diffusion coefficients as derived from the different models describe in the text.

Then, $S(q,t)/S(q,0) = A_1 \exp[-D_{eff}(q)\, q^2\, t] + A_2 \exp[-D'_{eff}(q)\, q^2\, t]$ to include both fast and slow dynamics. The effective diffusion extracted from a single exponential diffusion for the 45°C data and the slow component for the micellar and gels as calculated from the double exponents are described in Figure 7. The fast component of the 25 and 9°C lies with in the range of the motion of 45°C and is omitted for clarity.

As a function of q in all phases there are two regions bellow and above q=0.12, where above this value, the effective diffusion is higher. This region may depict the motion of the side chains that is less constrained then the backbone of the polymer in all phases.

Both the molecular solutions and the micellar phases are Newtonian liquids whereas the transition into the gel phase is accompanied by a change in viscosity. On the time scale of the NMR measurements hardly any restriction of motion of the solvent has been observed, though there are macroscopic changes in the system as the gellation takes place.

The smallest q range reflects the center of mass diffusion in the molecular solutions. In the complex fluids that consist of aggregates it reflects the center of mass diffusion of the clusters. Zooming in, in long range an undulation is observed in all phases, where it is further constrained as aggregates are formed and jamming takes place. We assign this dynamic mode to undulation of the polymer chain in and out of the plane or in other words buckling. The buckling is faster in molecular solutions and slows down as the chain is constrained. Since all aggregates contain solvent, this dynamic mode is not fully arrest, but only slows down as the temperature decreases. The highest q range encapsulates the dynamics of the side chains. The translational diffusion is affected as aggregation and gellation takes place, however the rotational motion is retained. This is consistent with our NMR measurements of the relaxation time of the CH_3 group of the side chains (not shown inhere).

In order to obtain further insight into the cooperative motion of the PPE backbone its side chains and the solvent the data were analyzed using a Zim model. [28-33] The coherent intermediate scattering function is scaled by the Zimm time $(q^3t)^{2/3}$ and plotted against for the PPE molecular solution , micellar solution and gel in Figure 8. This model have yield good agreement in molecular solutions where the molecules are free to diffuse as well as in the gel, where the overall motion of the network in constrained. In the micellar solution though, the Zimm model could not describe the dynamics. In this intermediate range where the micelles co-exist with molecular solutions, a more complex description would be needed to capture the coupled motion in the system.

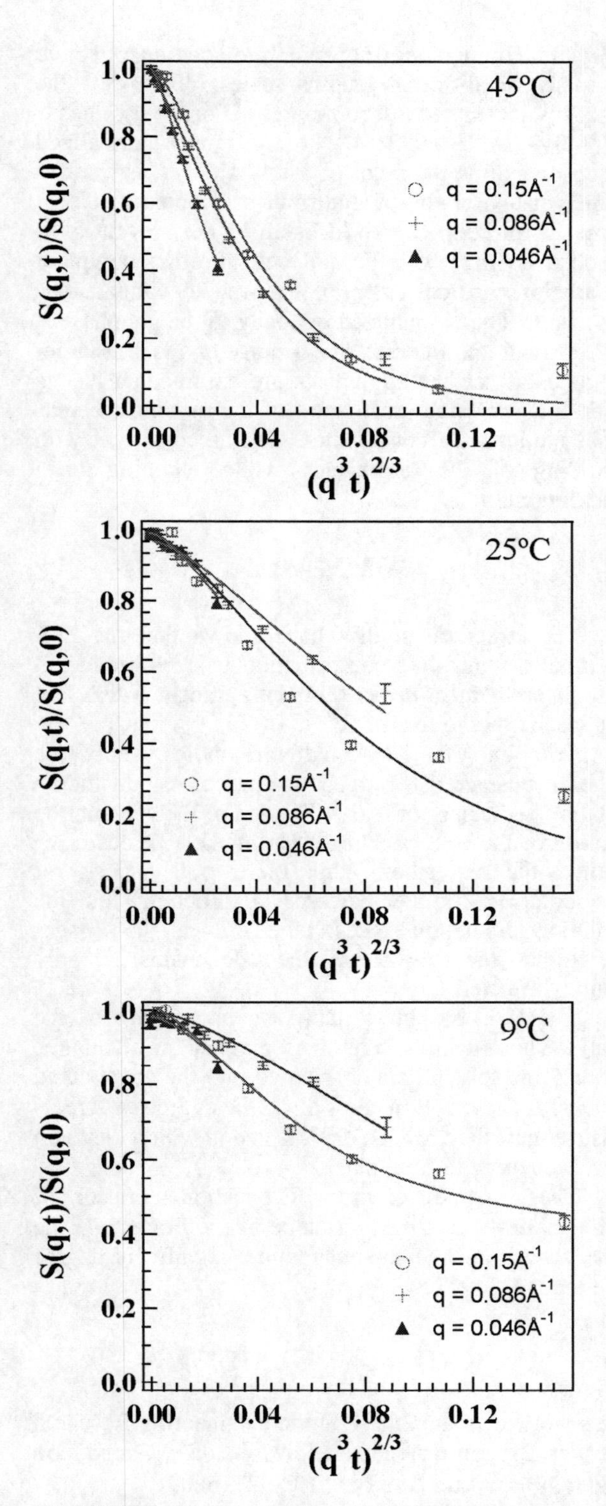

FIGURE 8. Zimm model at three temperatures, in the molecular solution at 45°C, in the micellar phase at 25°C and in the gel phase at 9°C.

The Zimm model however provides an effective viscosity on different lengths scales. For 45°C the viscosity increases with zooming in. For the q values of 0.043 $Å^{-1}$: 0.083 $Å^{-1}$: 0.15$Å^{-1}$ the normalized viscosity exhibit the ratios of 1:1.47:1.62. This result is some what surprising. We attribute the increase in local viscosity in cooperative diffusion of the solvent as it associated with the PPE molecule. Further computer simulations are currently on the way to validate this assertion. The Normalized viscosity in the gel phase is 3.7 in small and intermediate q range and is 3 at larger q values. Since the aggregates are confined, it results with enhance large scale arrest of the motion, whereas local motions are not affected. This is consistent with jamming of the aggregates while retaining local buckling motion.

SUMMARY

The structural studies have shown that the PPE assumes a worm-like conformation in cyclohexane. It associates to form dense thin flat aggregates that jam into each other to form a gel.

Combining the NMR and NSE studies we observe several characteristic dynamic processes, including rotation and center of mass diffusion of the solvent, the motion of the PPE backbone that of the side chains as well as the aggregates. While the overall viscosity of the complex fluids increases significantly as the gellation transition takes place, it is no longer Newtonian the solvent and the side chains are only slightly affected.

The PPE backbone undulates in and out of the plane. This buckling motion is retained in all phases since some solvent is incorporated into the constrained modes. As association takes place the aggregates freely diffuse until they reach a critical size in which they jam into each other to form a gel.

The relaxation of the PPE backbone reflects the changes in the aggregation mode of the PPE while the side chains and the solvent retain a significant degree of motional freedom.

ACKNOWLEDGEMENTS

This work was supported by the Engineering Research Centers Program of the National Science Foundation under NSF Award Number EEC-9731680.

This work utilized facilities supported in part by the National Science Foundation under Agreement No. DMR-0454672. We acknowledge the support of the National Institute of Standards and Technology, U.S. Department of Commerce, in providing the neutron research facilities used in this work.

We would like to express our gratitude to Drs Steve Klein, Bouallem Hamuoda and Antonio Faraone from NCNR for their assistance.

REFERENCES

1. Y. Jiang, C. Szymanski, U. H. F. Bunz, J. McNeill, D. Perahia; *Submitted.*
2. C. J. Szymanski, Y. Jiang, U. H. F. Bunz, D. Perahia, J. McNeill, *Submitted.*
3. J. H. Burroughes, D. D. C. Bradley, A. R. Brown, R. N. Marks, K. Mackay, R. H. Friend, P. L. Burns, and A. B. Holmes, *Nature* **347** (6293), 539-541 (1990).
4. J. L. Bahr, E. T. Mickelson, M. J. Bronikowski, R. E. Smalley, and J. M. Tour, *Chemical Communications* (2), 193-194 (2001).
5. G. Yu, J. Gao, J. C. Hummelen, F. Wudl, and A. J. Heeger, *Science* **270** (5243), 1789-1791 (1995).
6. J. K. Mwaura, M. R. Pinto, D. Witker, N. Ananthakrishnan, K. S. Schanze, and J. R. Reynolds, *Langmuir* **21** (22), 10119-10126 (2005).
7. J. H. Burroughes, C. A. Jones, and R. H. Friend, *Nature* **335** (6186), 137-141 (1988).
8. R. Zeineldin, M. E. Piyasena, T. S. Bergstedt, L. A. Sklar, D. Whitten, and G. P. Lopez, *Cytometry Part* A **69A** (5), 335-341 (2006).
9. J. H. Moon, W. McDaniel, and L. F. Hancock, *Journal of Colloid and Interface Science* **300** (1), 117-122 (2006).
10. J. H. Wosnick, C. M. Mello, and T. M. Swager, *Journal of the American Chemical Society* **127** (10), 3400-3405 (2005).
11. W. Y. Huang, S. Matsuoka, T. K. Kwei, and Y. Okamoto, *Macromolecules* **34**, 7166-7171 (2001).
12. D. Steiger, P. Smith, and C. Weder, Macromolecular Rapid Communications **18** (8), 643-649 (1997).
13. D. Perahia, R. Traiphol, and U. H. F. Bunz, *J. Chem. Phys.* **117** (4), 1827-1832 (2002).
14. D. Perahia, X. S. Jiao, and R. Traiphol, *Journal of Polymer Science Part B-Polymer Physics* **42** (17), 3165-3178 (2004).
15. D. Perahia; R. Traiphol, Rakchart; J. N.Wilson; U. H.-F. N. Rosov, *PMSE Preprints*, **93** 228-229 (2005).
16. A. J. Liu and S. R. Nagel, *Nature* **396**, 21-22 (1998).

17. S. O'Hern, S. A. Langer, A. J. Liu, and S. R. Nagel, *Physical Review Letters* **86**, 111-114 (2001).

18. M. E. Cates, J. P. Wittmer, J. P. Bouchaud, and P. Claudin, *Physical Review Letters* **81** (9), 1841-1844 (1998).

19. M. Kapnistos, D. Vlasoopoulos, G. Fytas, K. Mortensen, G. Fleischer, and J. Roovers, *Physical Review Letters* **85**, 4072- (2000).

20. Yunfei Jiang, Uwe H. F. Bunz, and Nicholas Rosov and Dvora Perahia *Submitted*

21. Kloppenburg, L.; Jones, D.; Claridge, J. B.; zur Loye, H. C.; Bunz, U. H. F. *Macromolecules* **1999**, *32*, 4460-4463.

22. O. Glatter and O. Kratky, *Small Angle X-ray Scattering*. (Academic, New York, 1982).

23. Guinier and O. Kratky, *Small-Angle Scattering of X-rays*. (Wiley, New York, 1955).

24. J. S. Higgins and H. C. Benoit, *Polymers and Neutron Scattering*. (Clarendon Press) Oxford, 1994).

25. S. R. Kline, *Journal of Applied Crystallography* ***39***, 895-900 (2006).

26. http://www.ncnr.nist.gov/programs/sans/data/red_anal.html.

27. D. E. Woessner - Molecular Physics, 1977 Taylor & Francis *Molecular Physics* **34**, 4, 899-920 (1977)

28. D. Richter *Journal of Applied Crystallography,* 40(S1), s28-s33 (2007).

29. D. Richter, *Journal of the Physical Society of Japan* , 75(11), 111004/1-111004/12 (2006).

30. D. Richter, *Diffusion in Condensed Matter*, 513-553 (2005).

31. M. Monkenbusch, J. Allgaier, D. Richter, J. Stellbrink, L. J. Fetters, and A. Greiner, *Macromolecules* **39** (26), 9473-9479 (2006).

32. T. Kanaya, N. Takahashi, K. Nishida, H. Seto, M. Nagao, and Y. Takeba, *Physica B-Condensed Matter* **385**, 676-681 (2006).

33. T. Kanaya, N. Takahashi, K. Nishida, H. Seto, M. Nagao, and T. Takeda, *Physical Review E* **71** 011801-1 011801-7 (2005).

How a Colloidal Paste Flows
– Scaling Behaviors in Dispersions of Aggregated Particles under Mechanical Stress –

R. Botet*, B. Cabane†, M. Clifton**, M. Meireles**, and R. Seto*

*Laboratoire de Physique des Solides, Université Paris-Sud, CNRS-UMR8502, Orsay F-91405, France
†Laboratoire PMMH, ESPCI, 10 Rue Vauquelin, Paris Cedex 05 F-75231, France
**Laboratoire de Génie Chimique, Université Paul Sabatier, CNRS-UMR5503, Toulouse F-31062, France

Abstract. We have developed a novel computational scheme that allows direct numerical simulation of the mechanical behavior of sticky granular matter under stress. We present here the general method, with particular emphasis on the particle features at the nanometric scale. It is demonstrated that, although sticky granular material is quite complex and is a good example of a challenging computational problem (it is a dynamical problem, with irreversibility, self-organization and dissipation), its main features may be reproduced on the basis of rather simple numerical model, and a small number of physical parameters. This allows precise analysis of the possible deformation processes in soft materials submitted to mechanical stress. This results in direct relationship between the macroscopic rheology of these pastes and local interactions between the particles.

Keywords: Granular matter, Numerical simulations, Pastes
PACS: 05.70-a, 05.20Gg, 05.70Fh, 05.50+q

INTRODUCTION

The case of sticky granular matter is far from new. It is defined as solid particles dispersed in a fluid phase, with hard-core and short-range attractive interaction between the particles. It includes such important topics as slurries [1], cements [2], ink [3], paints [4], all kind of pastes [5], sandcastles [6], or blood [7]. Last, they form the bulk of industrial and city effluents. Wet granular materials [8] form a generic class of sticky granular matter. The flow of these materials, when they are submitted to mechanical stress, is the most important issue common to all these examples.

An important point to notice, is that most of the particle interactions in the dense colloidal suspensions are noncentral forces, as they originate in surface interactions. This is worth recalling here that systems with central and noncentral forces may behave quite differently [10]. Moreover, these interactions are non-permanent, as they can be destroyed by stress. Material deformations are then irreversible and dissipative, what corresponds to a class of actual challenging computational problems.

Deformation of low volume fraction of dispersed particles is governed by particular scaling laws. This can be exemplified through an ideal gas of particles compressed under controlled energy. In this case, the Poisson's adiabatic law applies, namely :

$$P \propto \phi^{\gamma} , \tag{1}$$

relating pressure P to volume fraction ϕ. The value of the adiabatic exponent γ is in the range $1.3 \sim 1.7$, depending on the way the gas particles reallocate energy.

A fractal aggregate in a fluid medium is an example of a non-ideal dispersion. The correlation length is then infinite, hence the system is at a critical point, and ϕ is the order parameter. Moreover, the thermodynamical field conjugated to the volume fraction is the external pressure. The general "magnetic" relation describing how the system loses criticality when a small field is applied, writes here as the scaling law :

$$P \propto \phi^{\delta} , \tag{2}$$

with δ a critical exponent. This relation, deduced from the general theory of the critical phenomena, is formally similar to (1).

At last, for the highly concentrated solid particle dispersion in a continuous phase, the particles form complex disordered patterns and the forces propagate along particular paths - possibly fractal -, of connected particles. Therefore, response to external forces depend on the internal structure of the network made with the solid parts. Rheology of such strongly-flocculated dispersions is complex, and little is known about possible general laws relating stress and volume fraction [9].

The purpose of the present work is to discuss numerical models of disordered systems of hard particles interacting through noncentral, non-permanent, forces, and to understand how such systems deform and flow when they are submitted to external stresses.

CP982, *Complex Systems, 5th International Workshop on Complex Systems*
edited by M. Tokuyama, I. Oppenheim, and H. Nishiyama

NUMERICAL MODEL

We will discuss the model as a system of hard particles embedded in the ordinary 3-dimensional continuous space [11]. It was also studied in the 2-dimensional space, as it may represent experimental situations as well (colloidal dispersions between glass plates [12]). Variants were considered to know which features of the model are relevant or not, or to adapt the numerical model to various experimental situations. They are briefly mentionned below.

As a matter of fact, the basic model is issued from ideas of the discrete element method [13], used for granular matter modeling. Namely, each particle is regarded as an individual hard element and actual microscopic forces result from pair interaction. Hence, the solid phase is essentially granular matter, except that the interaction forces are *not* due to friction.

The Dispersion

Basics.–

The incoming solid matter dispersed in the liquid consists in monodispersed hard spheres of radius a, and confined into a finite box, with periodic boundary conditions along two perpendicular directions, while fixed boundary conditions are used in the third direction.

Variants.–

Other convex particle shapes can be considered (e.g. platelets), with additional technical complications for the treatment of the possible overlaps. A simple way is to combine several spheres through unbreakable bonds in order to make individual particles of about the desired shape. In addition, the particles can be made rigid or deformable according to the stiffness (infinite or finite) of the permanent bonds.

Another point is polydisperse material. Tries with Gaussian distributions show that this is irrelevant for the constitutive equations. But more special distributions (e.g. Pareto distribution) have not been studied so far.

Of course, periodic boundary conditions can be replaced by fixed boundary conditions if needed (e.g. simulation of a colloidal paste between two plates).

The Bonds

Basics.–

Short-range attractive interaction between the particles results from Van der Waals interactions, local chemical bonds (e.g. polycations), and screened electrostatic forces. Formation of such attractive bonds between the particles is modelled by creation of massless harmonic springs – whose stiffness is characteristic of the potential curvature of the interaction potential –, whenever the distance between two particle surfaces is less than the equilibrium value, l_o. This defines the energy unit E_o as the energy needed to compress the spring completely, as well as the pressure unit $P_o \equiv E_o/\pi a^3(l_o/2a)$. For E_o of order $1k_BT$ at room temperature, and a in the range $3 \sim 5$ nm, one obtains P_o of order 1 bar.

During rearrangement of the system, a spring may break whenever its length exceeds the threshold, l_{max}. Such breakable springs is the numerical materialization of the chemical bonds. The ratio between the energy of a spring at the rupture threshold and the energy to compress it completely (namely : $E_d/E_o = (l_{max}/l_o - 1)^2$, which is called the reduced disruptive energy) is a fundamental non-dimensional parameter of the model.

Variants.–

Distribution of stiffness can be considered straightforwardly. This can be particularly important in the case where two or more different chemical counterions are used for flocculation.

Non-harmonic springs can be considered as well. But the relevant feature is the bottom of the attractive potential energy, therefore anharmonicity is not expected to play significant role.

Bonds Location

Basics.–

In the real systems, the number of bonds between two particles is limited, either because the bonding energy is finite or because of excluded volume effect on the chemical bonds. We take this constraint into account by considering that springs can attach only at definite locations, called *pins*, on the surface of each particle. Only one bond can be attached to a given pin at the same time. The pins are defined randomly (with the uniform distribution) for each particle, once and for all at the beginning of each simulation. Consequently, each particle is entirely represented by the list of its pins

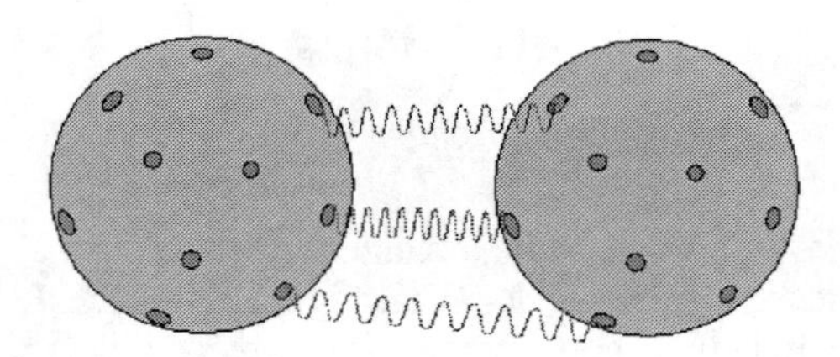

FIGURE 1. Schematic view of two spheres connected by a few springs. Pins are marked as small chips on the surface of each sphere. Two bonds cannot attach to the same pin, and a given bond connects two different spheres. Typically $200-500$ pins per particle surface are used.

(Fig. 1), and its local frame which translates and rotates with respect to the overall reference frame of the box.

Variants.–

In some conditions, non-uniform spatial distribution of the pins can be considered. This could be the case for more complex geometries, for example with platelets for which one can clearly control the mechanism of aggregation through the distribution of the pins (e.g. pins localized on the largest faces of the platelets will result in stack arrangement, while 2-dimensional structures will appear if they are localized on the edges of the platelets).

Movement Equations

Basics.–

When pressure is applied to the particle network, particles are submitted to the pressure forces, which, for spheres, are central in nature (contrary to the forces due to stretching of the bonds, which are essentially noncentral). In the frame of the box, the full equations of motion for the spherical particle i (radius a, mass m), submitted to force $\vec{F_i}$ and moment $\vec{M_i}$, are:

$$\begin{aligned} m\frac{d\vec{v}_i}{dt} &= \vec{F_i} + \lambda\left(\vec{v}_f - \vec{v}_i\right) , \qquad (3) \\ \frac{5}{2}ma^2\frac{d\vec{\omega}_i}{dt} &= \vec{M_i} + \frac{4}{3}a^2\lambda\left(\vec{\omega}_f - \vec{\omega}_i\right) , \end{aligned}$$

The coefficient λ is the proportionality constant between the drag force on a particle and its relative translational velocity with respect to the fluid in the Stokes regime. Therefore, $\vec{v}_i$ and $\vec{\omega}_i$ denote the translational and angular velocity of the particle, and $\vec{v}_f, \vec{\omega}_f$ the corresponding macroscopic velocities of the fluid at the location of particle i. The proper derivation of the fluid velocities $\vec{v}_f, \vec{\omega}_f$ is indeed intricate. It was achieved through Boltzmann-on-lattice simulations. Nevertheless, the present work focuses on the quasi-static regime where the relative velocities are negligible, then derivation of the fluid velocities is not needed. More precisely, the quasi-static regime is characterized by characteristic compression time much larger than the relaxation time of the overall structure. Within such approximation, the system evolves through molecular dynamics (e.g. Verlet algorithm [14]), according to (3) where the drag forces are neglected. Pressure is then applied by small incremential steps, and mechanical relaxation of the structure is achieved before applying the next pressure step.

Variants.–

An alternative for the dynamics of such a system in the quasi-static regime, consists in replacing the classical equations (3) by a Monte Carlo procedure. In this approach, a particle of the system is chosen randomly. This particle is moved randomly (random translation + random rotation) if the change of energy is consistent with the Metropolis condition [14]. If the shift is effective, relaxation is performed, *i.e.* bonds are destroyed or created according to the rules governing the bonds. All this sequence is repeated until statistical equilibrium is reached.

Since the deformation process is governed by energies, one has to define forces through gradients. For example, the pressure force P is defined through the equation $\Delta E = -P\Delta H$, where ΔE is the difference of the system energy for a decrease ΔH of the system height. Note that the total system energy is the sum of the energies of all the bonds and of the energies previously released in the system when bonds are broken.

Initial Conditions

Basics.–

Before applying the pressure, one builds the system by adding N particles in the box. We use standard reaction-limited cluster-cluster aggregation (RCCA) model [15] to generate randomly aggregates. This model is known to correctly describe experimental flocculation of colloidal particles – such as silica [16], polystyrene [17], or metallic [18] colloids – in the conditions where the aggregation rate is limited by the time it takes by the clusters to form a bond. The model generates an ensemble of disordered fractal aggregates, of fractal dimension $D_f = 2.1$.

Once an aggregate is generated, it is inserted randomly at top of the box. The aggregate is then gently settled onto bottom of the box, or onto existing particles, without deformation of its structure nor overlap.

Variants.–

Any alternative pre-aggregation process can be used as well. One important issue is to know if the proper sizes of the initial aggregates may play a role in the compaction process. Indeed, the size of the initial fractal clusters defines a correlation length in the system, what can be quite important for the subsequent collective deplacement of the particles. This question is yet unsolved.

Below, we discuss an example (flow around fixed obstacle) where pre-aggregation is not considered, all the particles being placed randomly at the beginning, with a definite volume fraction.

HOMOGENEOUS COMPRESSION

Numerical Model Results

A sketch of the visual aspect of the particle system (bonds are not represented) is shown in Fig.2 for three different pressures. these are projections onto a plane, so the system appears more dense than it actually is. One can note that, except for small statistical fluctuations, the systems appear to be spatially homogeneous. This result – which can be made more quantitative by a study of the average volume fraction through slabs [11] –, is well known in compression experiments [19].

Double-logarithmic plot for the external pressure versus the volume fraction of the particles is shown in Fig.3. Several sets of data are represented, all of them obtained from numerical simulations of systems with $\simeq 500$ particles, 200 pins per particle. Four sets of values of E_d are shown, namely $E_d/E_o = 0.04, 4, 9$, and in the range $360 \sim 10000$. The latter case corresponds to rupture lengths comparable to the box size.

Two power-law behaviors

$$P/P_o \propto \phi^{\delta}\ , \qquad (4)$$

are clear on the Fig.3. One with a slope $\delta \simeq 4.4$ will be called elastic as it corresponds to small amount of bond breaks. This is *not* exactly elastic behavior because bonds are created through compression advancement, but the created bonds are essentially permanent at this stage. The other exponent is $\delta \simeq 1.7$, and is recovered in the mid-range of values of the volume fraction, namely $0.07 < \phi < 0.5$, for any value of the disruptive energy E_d larger than E_o. This corresponds to plastic behavior as the rate of creation-destruction process for the bonds, is maximum [11].

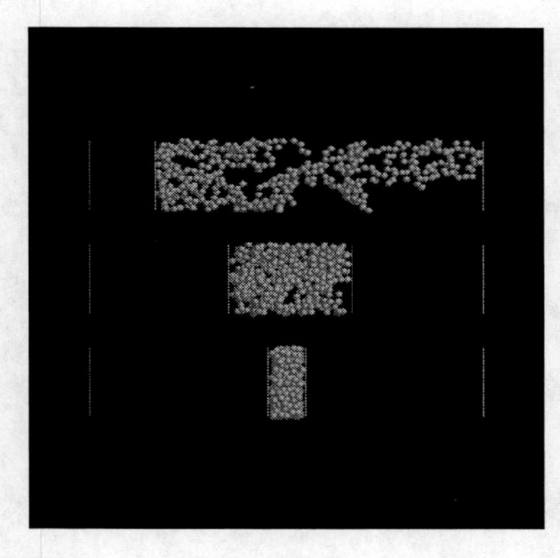

FIGURE 2. Three pictures (projection) of the same system during compaction under uniform pressure. The value of E_d/E_o is here equal to 4, and the total number of particles is 500. Volume fraction are respectively 0.06, 0.30 and 0.63 from top to bottom. The dashed lines visualize the initial and actual planes where external pressure is applied. Periodic boundary conditions are used on all other sides.

An important point to notice here is that the equations (4) are constitutive equations dependent only on the material [11]. Unlike the very beginning of the compression curve, (4) do not depend, in the statistical sense, on the initial conditions, on the pressure increment (if small enough) or on the periodic or fixed boundary conditions (if the system is large enough). This point is really nontrivial and has been discussed in details in [11] through intensive numerical simulations. Even the proper values of some parameters, such as the number of pins per particle (provided it is large enough), or the polydispersity of the particle radius (if narrow), are not relevant. Such universality allows direct comparison between results of the numerical simulations and experimental data on compression of colloidal pastes. This was done successfully in recent works [19, 20, 21].

Theoretical Arguments

The Elastic Behavior.–

When bond breaking is unlikely to occur (at the very beginning of the compression process or because of the large value of the disruptive energy), the system behaves as a disordered elastic network. The response of the system is due to the presence of resistant columns whose structure is a consequence of the initial fractal morphology of the individual aggregates. Given a homogeneous arrangement of such disordered fractal aggregates, the

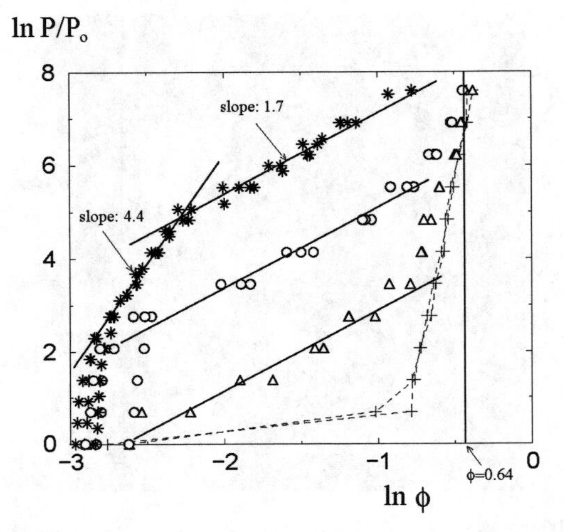

FIGURE 3. Double-logarithmic plot of the reduced pressure P/P_o vs the volume fraction ϕ for various values of the disruptive energy, E_d : $E_d/E_o = 4$ (triangles), $E_d/E_o = 9$ (circles) and various values of E_d/E_o above 360 (stars). A dashed line is used for the smallest values of E_d (namely $E_d/E_o = 0.04$), which exhibits fragile behavior as discontinuous jumps in ϕ from 0.01 to 0.5 at $P/P_o \simeq 1.3$. Full lines are the power-law behaviors (4) with exponents 4.4 and 1.7 for the elastic and plastic behaviors respectively.

overall elastic modulus K, of the system, writes [22] :

$$K \propto \phi^{(3+x)/(3-D_f)} , \tag{5}$$

with the fractal dimension of the backbone of the individual clusters $x \simeq 1.4$. This leads to $\delta \simeq 4.8$ in (4). Alternative derivations were proposed [23], leading to values of exponent δ in between 4 and 5. Precise analysis of the theoretical arguments, with comparison with the numerical simulations, remains to be done.

The Plastic Behavior.–

Rupture of the resistant columns occur if the applied pressure is larger than the threshold $P^\star$ for which the density of elastic energy stored in a column equals the average energy needed to break up all the bonds linking two neighboring particles of the column [24]. The latter energy is independent on the volume fraction : $P^\star \propto \sqrt{K}$ with a proportionality constant independent on ϕ. On the other hand, the effective stiffness of a disordered column is [25] : $K = nka/N_r R_\perp^2$, with $N_r \sim H^{d_{min}}$ the number of particles involved in the minimal chain throughout a percolating system of size H (and $d_{min} \simeq 1.4$ [26]). The distance $R_\perp$ follows simple power-law $R_\perp \propto H$ for the isotropic chains [27]. This results in the power-law dependence of the critical pressure $P^\star$ with the volume fraction :

$$P^\star \propto \phi^{1+d_{min}/2} . \tag{6}$$

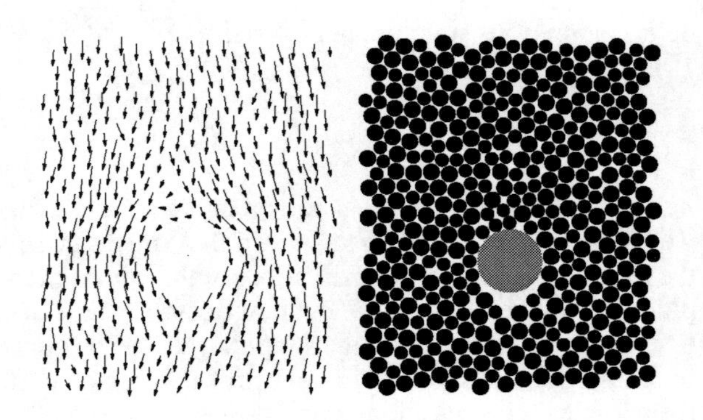

FIGURE 4. Central part of a 2-dimensional colloidal dispersion flowing around an intruder in a box with periodic boundary conditions. The flow goes regularly from top to bottom and the intruder is fixed. Two sizes of particles are used in order to avoid cristallization. On the left, location of the particles (drawn as black circles) at a given time of the stationary state. The bonds between particles are not drawn. The still intruder is the large grey circle. On the right, instantaneous velocity of the particles for the same system as on the left.

Interpretation of (6) in terms of the current pressure P, can be done by the following argument [28]. When P reaches the threshold $P^\star(\phi)$, the resistant column of height H breaks into two or more fragments, and the volume fraction ϕ increases by elastic deformation of the next resistant column. As this process goes on, the system passes through a series of discrete states $(P^\star(\phi), \phi)$. If the disordered system is large enough, the states $(P^\star(\phi), \phi)$ are close to each other, and :

$$P \propto \phi^{1+d_{min}/2} , \tag{7}$$

is then expected, with the exponent $1 + d_{min}/2 \approx 1.7$.

FLOW AROUND OBSTACLE

We present here another experimental situation related to the same problematics. Let us consider a colloidal dispersion confined in between two plates. The paste flows because of pressure gradient, and a fixed intruder (here, a disk) is placed in the center of the system. This geometry was recently studied for the granular matter [29]. The particles are put randomly at the beginning of the experiments (real or numerical).

Contrary to the pure granular matter, short-range attractive interactions exist now between the sticky particles. This induces strong correlations in the fluid, as seen on Fig.4. In particular, the proper size of the void created downstream after the intruder depends essentially on the properties of the non-permanent microscopic springs. The average flow pattern is another example of a macroscopic field depending on the microscopic properties of the material.

CONCLUSION

We presented in this paper a numerical model of sticky granular systems, based on simple physical features, working with a limited number of parameters. The model allows realistic approach of the static and dynamic behaviors of these complex collective systems – generically called pastes –, which exhibit plastic behavior and non-linear response to mechanical stress. This numerical tool allows to study in details the close relationship between the microscopic interactions and macroscopic deformation or flow of a paste.

During compression of a colloidal dispersion, two stages are clear and related to particular scaling laws. In the first one, very few interparticle bonds are broken. Displacement of colloidal aggregates within the structure lead to the collapse of the largest voids, while the smallest voids and the local structure remain unchanged. In the second stage, the compression causes the rupture of bonds everywhere in the system and the collapse of voids of any size. As a result, the less-dense regions of the aggregates are compressed, and they form a homogeneous dispersion. Meanwhile, the denser cores of the aggregates are pushed through this soft material, collect more particles, and turn into dense space-filling lumps.

During flow of a paste, the interparticle bonds generate strong correlation effects in the Non-newtonian complex fluid, resulting in self-organized structures such as large voids around obstacles.

ACKNOWLEDGMENTS

This work was supported by GDR 2980 (CNRS) "Structuration, consolidation et drainage de colloïdes : de l'ingénierie des surfaces • celles des procédés (PROSURF)". The authors thank P. Aimar, P. Bacchin, C. Bourgerette, B. Lartige, P. Levitz, L. Michot and E. Kolb for stimulating discussions and valuable comments.

REFERENCES

1. R. G. Gillies and C. A. Shook *Canad. J. Chem. Eng.*, 78, 709-716 (2000).
2. Z. Saada, J. Canou, L. Dormieux, J. C. Dupla, and S. Maghous *Int. J. Numer. Anal. Methods Geomech.*, **29**, 691-711 (2005).
3. H. Kimura, Y. Nakayama, A. Tsuchida and T. Okubo, *Colloids and Surface B: Biointerfaces*, **5**6, 236-240 (2007).
4. K. Holmberg *Handbook of Applied Surface and Colloid Chemistry*, vol.1, John Wiley and sons Inc., Publisher, New York, 2001.
5. P. Coussot *Rheometry of Pastes, Suspensions, and Granular Materials: Applications in Industry and Environment*, John Wiley and sons Inc., Publisher, Hoboken, 2005.
6. T. C. Halsey and A. J. Levine *Phys. Rev. Lett.* **8**0, 3141-3144 (1998).
7. S. L. Diamond, *Biophys. J.* **8**0, 1031-1032 (2001).
8. S. Herminghaus *Advances in Physics* **5**4, 221-244 (2005).
9. R. Buscall, P. D. A. Mills, R. F. Stewart, D. Sutton, L. R. White and G. E. Yates *J. Non-Newtonian Fluid Mech.* **2**4 183-202 (1987).
10. S. Feng, P. N. Sen, B. I.Halperin and C. J. Lobb, *Phys. Rev. B* **3**0 5386-5389 (1984).
11. R. Botet and B. Cabane, *Phys. Rev. E* 70, 031403-1-031403-11 (2004).
12. C. Allain and L. Limat, *Phys. Rev. Lett.* 74, 2981-2984 (1995).
13. P. A. Cundall and O. D. L. Strack, *Geotechnique* **2**9 47-65 (1979).
14. D. C. Rapaport, *The art of molecular dynamics simulation*, 2nd edition, Cambridge University Press, Publisher, 2004.
15. R. Jullien and R. Botet, *Aggregates and Fractal Aggregates*, World Scientific, Publisher, Singapore, 1987.
16. Z. Zhou and B. Chu, *J. Colloid Interface Sci.* **1**46, 541-555 (1991).
17. V. Oles, *J. Colloid Interface Sci.* **1**54, 351-358 (1992).
18. D. A. Weitz, J. S. Huang, M. Y. Lin and J. Sung, *Phys. Rev. Lett.* **5**4, 1416-1419 (1985).
19. J.-B. Madeline, M. Meireles, C. Bourgerette, R. Botet, R. Schweins and B. Cabane, *Langmuir* **2**3, 1645-1658 (2007).
20. J.-B. Madeline, M. Meireles, J. Persello, C. Martin, R. Botet, R. Schweins and B. Cabane, *Pure Appl. Chem.* **7**7 1369-1394 (2005).
21. C. Parneix, *Agrégats colloïdaux destinés au renforcement des élastomères*, PhD thesis, Université de Franche-Comté, Besançon, France (2006).
22. W. D. Brown and R. C. Ball, *J. Phys. A* **1**8, L517-L521 (1985).
23. R. Buscall, P. D. A. Mills, J. W. Goodwin, D. W. J. Lawson, *Chem. Soc. Faraday Trans* 1 **8**4, 4249–4260 (1988);
Wei-Heng Shih, Wan Y. Shih, Seong-Il Kim, Jun Liu and Ilhan A. Aksay, *Phys. Rev. A* **4**2, 4772-4776 (1990);
M. Chen, W. B. Russell, *J. Colloid Interface Sci.* **1**41, 564-577 (1991);
Hua Wu and M. Morbidelli, *Langmuir* **1**7, 1030-1036 (2001).
24. S. O. Gregg *The Surface Chemistry of Solids*, Chapman and Hall, Publisher, London, 1965.
25. Y. Kantor and I. Webman, *Phys. Rev. Lett.* **5**2, 1891-1894 (1984).
26. J. Vannimenus, in *Physics of Finely Divided Matter*, edited by N. Boccara and M. Daoud, Springer-verlag, Publisher, Proc. in Physics 5, Berlin, 1985, p. 317.
27. A. A. Potanin, *J. Colloid Interface Sci.* **1**57, 399-410 (1993) ;
A. A. Potanin and W. B. Russel, *Phys. Rev. E* **5**3, 3702-3709 (1996).
28. R. Buscall and L. R. White, *J. Chem. Soc. Faraday Trans.* I **8**3, 873-891 (1987).
29. E. Kolb, J. Cviklinski, J. Lanuza, P. Claudin, and E. Clément, *Phys. Rev. E* **6**9, 0313006-1-0313006-5 (2004).

Nanoparticle Retardation in Semidilute Polymer Solutions

Remco Tuinier*, Jan K. G. Dhont*, Takashi Taniguchi†, and Tai-Hsi Fan**

*Forschungszentrum Jülich, Institut für Festkörperforschung, Soft Matter, 52425 Jülich, Germany
†Polymer Science and Engineering, Yamagata University, 4-3-16 Jonan, Yonezawa 992-8510, Japan
**Department of Mechanical Engineering, University of Connecticut, Storrs, CT 06269-3139, USA

Abstract. The flow of a solution containing semidilute nonadsorbing polymers near a flat wall and past a sphere is considered. We focus on polymer concentrations beyond overlap. Near a flat wall we numerically compute the slip thickness for chains with excluded volume. For the flow of a polymer solution past a sphere we compute Stokes flow with nonuniform viscosity numerically and compare it with recent analytical results [Fan, Dhont and Tuinier, *Phys. Rev. E* **75**, 011803 (2007)] based on a step function approximation. We show the regime where the analytical results are valid and to what extent nanoparticle motion is retarded.

Keywords: Colloid, Depletion, Diffusion, Fluid dynamics, Polymer, Slip
PACS: 66.20.+d, 82.70.Dd, 61.25.Hq

INTRODUCTION

In many situations, flowing solutions with dissolved or dispersed polymers, micelles or colloids are in contact with surfaces. As examples we mention the flow through pipes in industry, the measurements of the viscosity of a colloidal dispersion or polymer solution using capillary rheometry [1], pore flow [2] or using polymer solutions to sweep out oil out of natural porous media. Here we show how the presence of polymer chains influences transport phenomena.

In polymer solutions the polymer chains either adsorb to a surface or are depleted from it, depending on the segment surface affinity [3]. We focus on the case of non-adsorbing polymers in solution [4]. Unless polymers are significantly attracted to the surface, they do not adsorb. Due to a loss of configurational entropy the segment concentration close to the surface is then smaller as in the bulk [5], see Fig. 1. Near a surface a depletion layer then exists with a thickness close to the radius of gyration in dilute solutions, and to the correlation length in semidilute polymer solutions [6].

Overlap of depletion layers results in attractive forces [4, 7] between particles in colloidal [8] and biological systems such as actin networks [9, 10] or virus dispersions [11]. Depletion forces induce phase transitions in colloid-polymer mixtures [12, 13, 14], which is well understood [15, 16, 17, 18, 19].

When a solution with nonadsorbing polymers flows near a surface the presence of the depletion layer affects the flow pattern [20, 21]. Here we consider two cases: flow near a single wall and flow around a translating sphere. The first situation is relevant for estimating the slip effects, the second one for estimating the transport properties such as diffusion [22, 23, 24] or sedimentation

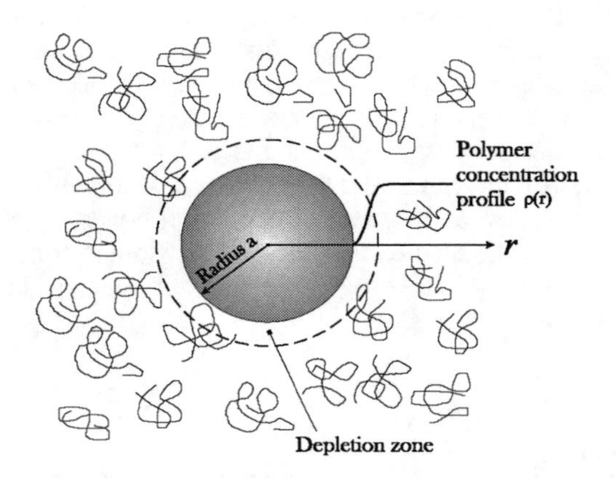

FIGURE 1. Sketch of a sphere surrounded by polymer chains in solution. The dashes indicate the depletion layer.

[25] of nanoparticles (e.g. colloids or proteins) through a solution containing macromolecules. Further, in microrheology [26, 27, 28, 29] it is essential to understand the flow induced by a colloid through a complex medium filled with polymer chains.

Since the 1980s many studies have been performed on dilute nanospheres translating through a polymer solution, see e.g. Refs [30, 31, 32, 33, 34]. The friction coefficient is often interpreted in terms of the Stokes approximation in a pure solvent $6\pi\eta a$, with sphere radius a. The obtained viscosity η lies, however, in between the values of a pure solvent η_0 and the viscosity of the polymer solution, η_p. This means the particle is effectively slipping through the polymer solution since it experiences a viscosity that is smaller than the bulk viscosity. Nanoparticles are thus only partially retarded in their motion by polymers in solution. In this paper we will quantify this phenomenological slip behavior.

CP982, *Complex Systems, 5th International Workshop on Complex Systems*
edited by M. Tokuyama, I. Oppenheim, and H. Nishiyama

SLIP EFFECT CAUSED BY A DEPLETION LAYER NEAR A FLAT WALL

When a solution containing nonadsorbing polymers is sheared next to a flat wall at constant shear stress σ we have: $\sigma = \dot{\gamma}(x)\eta(x)$, with shear rate $\dot{\gamma}$ and (dynamic) viscosity η. Due to the presence of the depletion layer we are dealing with local shear rates and viscosities that depends on the position from the wall surface x. The velocity profile $v(x)$, which can be found from $\mathrm{d}v(x) = \dot{\gamma}(x)\mathrm{d}x$, then follows as:

$$v(x) = \int_0^x \mathrm{d}x' \frac{\sigma}{\eta(x')}. \quad (1)$$

Once the viscosity profile is known, one can derive [20] the (apparent) slip velocity v_s and slip or Navier length b of which we here only consider its absolute value. Previously, we have shown that it is a fair approximation, at least for shear rates that are small compared to the inverse of the longest polymer relaxation time scale, to assume the viscosity profile follows the relative segment density profile $\rho(x)$ [20]. The local density profile near a wall can be calculated using statistical thermodynamics based on the Edwards equation [35] with proper boundary conditions at the nonadsorbing surface. It reads [36, 37]:

$$\rho(x) = \tanh^2\left(\frac{x}{\delta}\right), \quad (2)$$

with the depletion thickness δ characterizing the length scale over which polymer chains are depleted from the wall. In Eq. (2), ρ=0 at the wall ($x = 0$) and reaches ρ=1 in the bulk ($x = \infty$).

Here we wish to describe slip effects at concentrations beyond overlap. For the viscosity of the polymer solution in the bulk η_p we take the often used semi-empirical Martin equation: $\eta_\mathrm{p}/\eta_0 = 1 + [\eta]c_\mathrm{b}\exp(k_H[\eta]c_\mathrm{b})$, which describes the polymer concentration dependence up to the concentrated regime. Here c_b is the bulk concentration, k_H the Huggins constant for which 0.5 is a common value for many polymer solutions, $[\eta]$ is the intrinsic viscosity, which is about $1.2/c^*$, with c^* the polymer overlap concentration. Now we replace c_b in the Martin equation with $c_\mathrm{b}\rho(x)$ to arrive at the viscosity profile $\eta(x)$:

$$\frac{\eta(x)}{\eta_0} = 1 + [\eta]c_\mathrm{b}\rho(x)\mathrm{e}^{k_H[\eta]c_\mathrm{b}\rho(x)}. \quad (3)$$

At the wall it follows $\eta = \eta_0$ and in the bulk Eq. (3) becomes the Martin equation. We also need to account for the concentration dependence of the depletion thickness in Eq. (2) for which we use a recent accurate analytical expression [6]:

$$\frac{\delta}{\delta_0} = \left[1 + 2.91([\eta]c_\mathrm{b})^{1.54}\right]^{-\frac{1}{2}}, \quad (4)$$

valid for polymer solutions where the segments have excluded volume (ev) interactions. In this expression δ_0 is the depletion thickness in the dilute limit that equals $1.071\, R_\mathrm{g}$ for ev chains [38]. It was shown Eq. (4) accurately describes simulations and experiments [6].

The resulting velocity profile of Eq. (1) can now be obtained numerically for interacting polymer solutions by inserting Eq. (3) with Eqs. (2) and (4). For two polymer concentrations we plot indicative velocity profiles $v(x)$ as a function of the distance from the wall in Fig. 2.

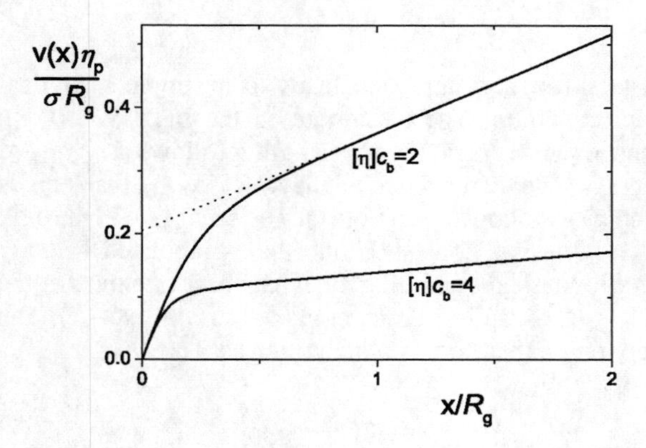

FIGURE 2. Normalized fluid velocity profiles near a flat wall of a polymer solution with excluded volume chains under simple shear flow for two polymer concentrations as indicated (solid curves). The dashed line is the linearly extrapolated fluid velocity, indicating the slip velocity v_s at the wall surface (x=0).

>From the velocity profile the slip velocity v_s and Navier length b follow by linear extrapolation of the velocity in the bulk to the wall surface [20]. The extrapolated ordinate-intercept equals v_s, while the abscissa-intercept defines the Navier length b. The result for the slip or Navier length is plotted as the solid curve in Fig. 3. We compare these results for excluded volume chains with those for ideal or ghost chains which do not self-interact. For ghost chains the absolute Navier length can be obtained analytically [20]:

$$b = \delta_0\sqrt{[\eta]c_\mathrm{b}}\tan^{-1}(\sqrt{[\eta]c_\mathrm{b}}). \quad (5)$$

Beyond the overlap concentration ($[\eta]c_\mathrm{b} \approx 1.2$) we observe the slip thickness exceeds the polymer's radius of gyration. For large polymer concentrations ($[\eta]c_\mathrm{b} > 3$) the Navier length increases strongly for polymer solutions where the segments have excluded volume interactions. For the ghost chains the increase is more gradual. For polymer chains with excluded volume (ev), b attains values that are an order of magnitude larger than the polymer's R_g for cases with c_b values about five times c^*. For large polymer concentrations the depletion thickness shrinks according to Eq. (4). Therefore one might expect the Navier length would decrease as well. A more dominant effect is however the strong increase of the bulk viscosity with polymer concentration beyond overlap. The Navier length difference between ghost and excluded volume chains is only small near and below the overlap concentration.

An analytical expression for b for ev chains can be obtained by assuming the depletion layer can be simplified using a step function:

$$\eta(x) = \begin{cases} \eta_0 & \text{for} \quad 0 \leq x \leq \delta \\ \eta_\mathrm{p} & \text{for} \quad x > \delta \end{cases} \tag{6}$$

A step-function approach leads to a simple analytical approximation. The shear rate in the bulk $\dot{\gamma} = \sigma/\eta_\mathrm{p}$ must equal v_s/b, so the slip length b follows as $v_s\eta_\mathrm{p}/\sigma$. At $x = \delta$ one may define the velocity v_δ that equals $\delta\sigma/\eta_0$ as follows from Eq. (1). The shear rate in the bulk $\dot{\gamma} = \sigma/\eta_\mathrm{p}$ is a constant along the extrapolated velocity profile (dashed line in Fig. 2). That shear rate must equal $(v_\delta - v_s)/\delta$, so it follows $\eta_\mathrm{p} v_s/\sigma = \delta(\eta_\mathrm{p}/\eta_0 - 1)$. The slip length therefore can be written as:

$$b = \delta\left(\frac{\eta_p}{\eta_0} - 1\right), \tag{7}$$

Upon inserting δ from Eq. (4) and the Martin equation for η_p we obtain the result plotted as the dotted curve. It is clear the step-function approach overestimates the slip effect. This deviation can be corrected by accounting for a continuous polymer segment density profile.

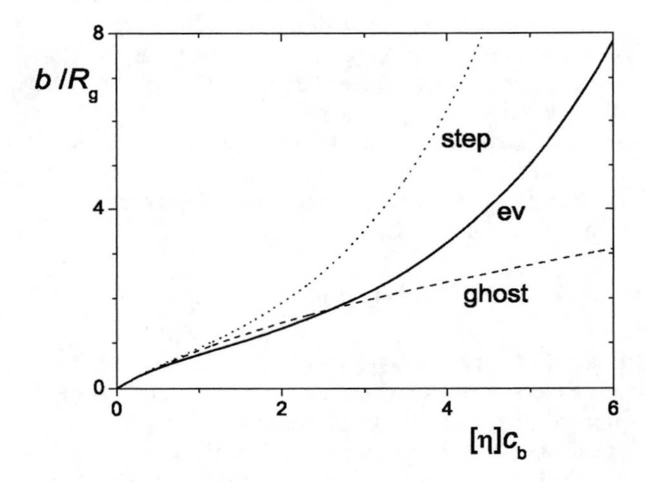

FIGURE 3. Slip thickness as a function of the polymer concentration in the bulk for a polymer solution under simple shear flow. Results are given for ghost chains (dashed curve) [20], chains with excluded volume (solid curve) and using the step function approach (Eq. (7); dotted curve).

TRANSLATIONAL MOTION OF A SPHERE MEDIATED BY A DEPLETION LAYER

As a sphere translates through a polymer solution it moves though a fluid with nonuniform viscosity. The question then arises: what is the effective viscosity such a sphere experiences? This effective viscosity directly identifies how strongly nanopshere motion is retarded. The long-time diffusion coefficient of a sphere can be measured using several experimental techniques of which dynamic light scattering (DLS) is a common one [22, 30, 31, 34]. Often the obtained diffusion coefficient is interpreted in terms of the Stokes-Einstein relation $D = kT/f^\mathrm{t}$, where the translational friction coefficient $f^\mathrm{t} = 6\pi\eta^\mathrm{eff}a$, with sphere radius a and effective viscosity η^eff. In a pure solvent $\eta^\mathrm{eff} = \eta_0$. In a polymer solution with bulk viscosity η_p one naively expects $\eta^\mathrm{eff} = \eta_p$. In practice experimental results indicate $\eta_0 < \eta^\mathrm{eff} < \eta_p$.

Recently, we presented analytical results for f^t [21, 39]. We included the depletion layer by a simplified two-layer approximation to represent the viscosity profile and incorporated this into Stokes' stream function theory. We found that [21, 39]:

$$f^\mathrm{t} = 6\pi\eta_s a g^\mathrm{t}, \tag{8}$$

with the following analytical expression for the correction function g^t:

$$g^\mathrm{t} = \frac{1}{\Pi}\left[2(2+3\lambda)(1+\tilde{\delta})^6 - 4(1-\lambda)(1+\tilde{\delta})\right], \tag{9}$$

where $\lambda = \eta_0/\eta_\mathrm{p}$, $\tilde{\delta} = \delta/a$ and with

$$\Pi = 2(2+3\lambda)(1+\tilde{\delta})^6 - 3(3+2\lambda)(1-\lambda)(1+\tilde{\delta})^5 \\ +10(1-\lambda)(1+\tilde{\delta})^3 - 9(1-\lambda)(1+\tilde{\delta}) + 4(1-\lambda)^2.$$

We expect this result is accurate for a depletion layer thickness that is small with respect to the sphere size but will deviate for large values of δ.

Therefore we investigated whether we could make progress and find solutions for the translational friction of a sphere through a polymer solution while accounting for the viscosity *profile*. In order to do so we used Eq. (3) but with the density profile near a wall $\rho(x)$ replaced by the polymer segment density profile around a sphere $\rho(r)$ [37] (see Fig.1 where the density profile is indicated):

$$\rho(r) = \left[1 - \frac{a}{r} + \frac{a}{r}\tanh\left(\frac{r-a}{\delta}\right)\right]^2, \tag{10}$$

with r the position from the sphere center. Next, the Stokes equation with nonuniform viscosity profile,

$$\nabla p = \eta\nabla^2\mathbf{v} + \nabla\eta \cdot \left[\nabla\mathbf{v} + (\nabla\mathbf{v})^T\right], \tag{11}$$

was solved. In this equation η is the local viscosity that depends on local concentration ρ, p is the pressure field, $\mathbf{v}$ is the velocity, and $\nabla\mathbf{v}+(\nabla\mathbf{v})^T$ is the strain rate tensor. Commonly, the axisymmetric Stokes flow problem induced by the translational motion of the particle is analyzed by using the Stokes stream function [40, 41], which we extended in order to account for a local viscosity profile. After replacing the velocity components by the Stokes stream function φ and applying the trial solution $\varphi=\sin^2(\theta)f(r)$, the following 4th-order differential equation for the radial function $f(r)$ is obtained:

$$0=f^{(4)}+\frac{2\eta'}{\eta}f'''-\left(\frac{4}{r^2}+\frac{2\eta'}{r\eta}-\frac{\eta''}{\eta}\right)f'' \quad (12)$$
$$+\left(\frac{8}{r^3}-\frac{2\eta'}{r^2\eta}-\frac{2\eta''}{r\eta}\right)f'-\left(\frac{8}{r^4}-\frac{8\eta'}{r^3\eta}-\frac{2\eta''}{r^2\eta}\right)f,$$

with vanishing far-field and no-slip boundary conditions. It turns out we can express the correction function g^{t} as:

$$g^{\mathrm{t}}=\frac{1}{9}\left[4+4f''(r=a)-2f'''(r=a)\right]. \quad (13)$$

Equation (12) was solved numerically and the derivatives at the sphere surface yield g^{t} using (13).

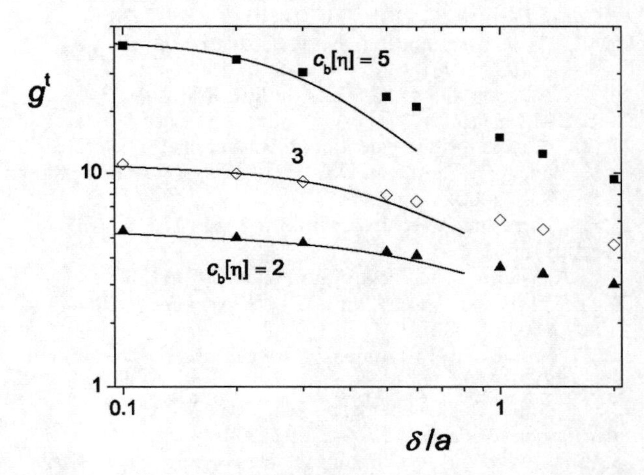

FIGURE 4. Correction function to the translational friction coefficient g^{t} as a function of the depletion thickness δ, which is normalized by the sphere radius a for three polymer concentrations as indicated. Data: numerical results. Solid curves follow the analytical two-layer results of Eq. (9).

Some numerical results obtained are plotted in Fig. 4 as data points. The solid curves follow the step-function two-layer model of Eq. (9). For $\delta/a<0.3$, the two-layer approximation is very accurate. For larger values of the depletion thickness significant deviations occur, especially if δ gets close to a.

The slip length was also analyzed using the two-layer model [39] and we found in the limit of thin depletion layers:

$$b=\delta\left(\frac{\eta_{\mathrm{p}}}{\eta_0}-1\right)(1-3\frac{\delta}{a}+...), \quad (14)$$

Apart from curvature correction terms, of which the first-order is $-3\delta/a$, this result is identical to Eq. (7).

Experimentally, one often uses DLS and uses the obtained translational diffusion coefficient in a dilute colloidal sphere suspension to compute an *effective* diameter d^{eff} through the Stokes-Einstein (SE) equation:

$$D=\frac{kT}{3\pi\eta_{\mathrm{p}}d^{\mathrm{eff}}}. \quad (15)$$

>From the above it becomes clear that using the Stokes-Einstein relation directly, with $d^{\mathrm{eff}}=d$, is incorrect. However, we can now predict the obtained effective diameter through:

$$d^{\mathrm{eff}}=dg^{\mathrm{t}}\frac{\eta_0}{\eta_{\mathrm{p}}}. \quad (16)$$

In Fig. 5 we show how d^{eff} depends on polymer concentration for two polymer to sphere size ratios, $R_g/a=0.1$ (solid curve) and 0.5 (dashed curve). It is clear that d^{eff} starts to deviate significantly from d for polymer concentrations beyond overlap, the more so for relatively larger polymer chains. The depletion layer results in slip so that the sphere experiences an effective viscosity that is smaller than the bulk viscosity. The *apparent* diameter therefore drops when slip effects affect sphere diffusion. For very thick depletion layers the slip effects are expected to be very significant. Under such conditions we need to solve g^{t} numerically as indicated.

At the moment we are trying to find a simple prescription for the numerical data for larger depletion thickness. We will also report on the correct function for rotational motion with variable viscosity profile [42].

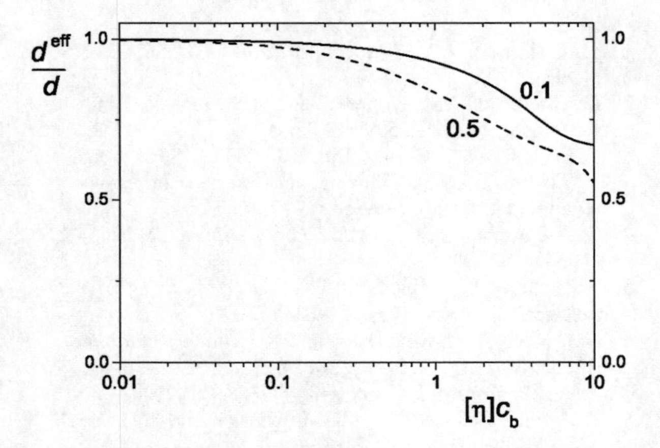

FIGURE 5. Effective diameter as a function of polymer concentration for various size ratios R_g/a as indicated.

CONCLUDING REMARKS

We have shown how depletion layers in solutions with semidilute nonadsorbing polymer chains induce slip effects near a planar wall and a spherical surface. At a flat wall the velocity profile near a flat wall was computed by inserting a viscosity profile, based on the polymer depletion segment density profile, into the equations of motion. We accounted for excluded volume effects by using the Martin equation for the concentration-dependence of the viscosity and a correct concentration dependence of the depletion thickness. The obtained slip length is more than a magnitude larger than the coil size for just a few times the overlap concentration.

As a sphere translates through a semidilute polymer solution containing nonadsorbing chains the depletion layers affect the friction coefficient. It appears nanoparticles are less retarded by polymers in their motion as can be expected on the basis of the bulk viscosity of the polymer solution. We compare recent analytical expressions for a simple two-layer model with numerical results where the entire viscosity profile is accounted for. It is shown the theory is only accurate for relatively thin depletion layers. The resulting slip thickness is, for big spheres, shown to be in full agreement with the result near a flat wall. It is demonstrated how the *effective* diameter can be calculated as a function of relative size of the chains and polymer concentration.

T.-H. Fan is grateful for the financial support of this work from the University of Connecticut Research Foundation. We thank P.N. Pusey, N.J. Wagner, M. Fuchs, M.H.G. Krüger and G. Nägele for useful discussions.

REFERENCES

1. H.A. Barnes, *J. Non-Newton. Fluid Mech.*, **56**, 221-251 (1995).
2. C. Cheikh, G.J.M. Koper, and T.G.M. van de Ven. *Langmuir*, **22**, 5991-5993 (2006).
3. G.J. Fleer, M.A. Cohen Stuart, J.M.H.M. Scheutjens, T. Cosgrove, and B. Vincent, *Polymers at Interfaces*, Chapman and Hall, London, 1993.
4. S. Asakura and F. Oosawa. *J. Chem. Phys.*, **22**, 1255-1256 (1954).
5. J.F. Joanny, L. Leibler, and P.G. De Gennes. *J. Polymer Sci.: Polym. Phys.*, **17**, 1073-1084 (1979).
6. G.J. Fleer, A.M. Skvortsov, and R. Tuinier. *Macromol. Theory Sim.*, **16**, 531-540 (2007).
7. A. Vrij. *Pure Appl. Chem.*, **48**, 471-483 (1976).
8. R. Verma, J.C. Crocker, T.C. Lubensky, and A.G. Yodh. *Macromolecules*, **33**, 177-186 (2000).
9. M. Hosek and J.X. Tang. *Phys. Rev. E*, **69**, 051907-1-051907-9 (2004).
10. R. Tharmann, M.M.A.E. Claessens, and A.R. Bausch. *Biophys. J.* **90**, 2622-2627 (2006).
11. Z. Dogic, K.R. Purdy, E. Grelet, M. Adams, and S. Fraden. *Phys. Rev. E*, **69**, 051702-1-051702-9 (2004).
12. S. M. Ilett, A. Orrock, W. C. K. Poon, and P. N. Pusey. *Phys. Rev. E*, **51**, 1344-1351 (1995).
13. A. Moussaïd, W. C. K. Poon, P. N. Pusey, and M. F. Soliva. *Phys. Rev. Lett.*, **82**, 225-228 (1999).
14. W.C.K. Poon. *J. Phys: Condens. Matter*, **14**, R859-R880 (2002).
15. A.P. Gast, C.K. Hall, and W.B. Russel. *J. Colloid Interface Sci.*, **96**, 251-267 (1983).
16. H.N.W. Lekkerkerker, W.C.K. Poon, P.N. Pusey, A. Stroobants, and P.B. Warren. *Europhys. Lett.*, **20**, 559-564 (1992).
17. E.J. Meijer and D. Frenkel. *J. Chem. Phys.*, **100**, 6873-6887 (1994).
18. D.G.A.L. Aarts, R. Tuinier, and H.N.W. Lekkerkerker. *J. Phys: Condens. Matter*, **14**, 7551-7561 (2002).
19. P.G. Bolhuis, A.A. Louis, and J-P. Hansen. *Phys. Rev. Lett.*, **89**, 128302-1-128302-4 (2002).
20. R. Tuinier and T. Taniguchi. *J. Phys: Condens. Matter*, **17**, L9-L14 (2005).
21. R. Tuinier, J.K.G. Dhont, and T.-H. Fan. *Europhys. Lett.*, **75**, 929-935 (2006).
22. G.S. Ullmann, K. Ullmann, R.M. Lindner, and G.D.J. Phillies. *J. Phys. Chem.*, **89**, 692-700 (1985).
23. K. Kang, J. Gapinski, M.P. Lettinga, J. Buitenhuis, G. Meier, M. Ratajczyk, J.K.G. Dhont, and A. Patkowski. *J. Chem. Phys.*, **122**, 044905-1-044905-13 (2005).
24. K. Kang, A. Wilk, A. Patkowski and J.K.G. Dhont. *J. Chem. Phys.*, **126**, 214501-1-214501-17 (2007).
25. X. Ye, P. Tong, and L.J. Fetters. *Macromolecules*, **31**, 5785-5793 (1998).
26. T.G. Mason and D.A. Weitz. *Phys. Rev. Lett.*, **74**, 1250-1253 (1995).
27. J.C. Crocker, M.T. Valentine, E.R. Weeks, T. Gisler, P.D. Kaplan, A.G. Yodh, and D.A. Weitz. *Phys. Rev. Lett.*, **85**, 888-891 (2000).
28. L. Starrs and P. Bartlett. *Faraday Disc.*, **123**, 323-334 (2003).
29. T.A. Waigh. *Rep. Progr. Phys.*, **68**, 685-742 (2005).
30. T.H. Lin and G.D.J. Phillies. *J. Phys. Chem.*, **86**, 4073-4077 (1982).
31. W. Brown and R. Rymdén. *Macromolecules*, **21**, 840-846 (1988).
32. D. Gold, C. Onyenemezu, and W.G. Miller. *Macromolecules*, **29**, 5700-5709 (1996).
33. S.P. Radko and A. Chrambach. *Biopolymers*, **4**, 183-189 (1997).
34. G.H. Koenderink, S. Sacanna, D.G.A.L. Aarts, and A.P. Philipse. *Phys. Rev. E*, **69**, 021804-1-021804-12 (2004).
35. S.F. Edwards and K.F. Freed. *J. Phys. A*, **2**, 145-150 (1969).
36. P.G. De Gennes. *Scaling Concepts in Polymer Physics*, Cornell University Press, Ithaca, 1979.
37. G.J. Fleer, A.M. Skvortsov, and R. Tuinier. *Macromolecules*, **36**, 7857-7872 (2003).
38. A. Hanke, E. Eisenriegler, and S. Dietrich. *Phys. Rev. E*, **59**, 6853-6878 (1999).
39. T.-H. Fan, J.K.G. Dhont, and R. Tuinier. *Phys. Rev. E*, **75**, 011803-1-011803-10 (2007).
40. G.G. Stokes. *Trans. Camb. Phil. Soc.*, **9**, 8-106 (1851).
41. J. Happel and H. Brenner. *Low Reynolds Number Hydrodynamics*, Prentice-Hall, New York, 1965.
42. T.-H. Fan, B. Xie, and R. Tuinier, accepted for publication in *Phys. Rev. E* (2008).

Geometrical Classification of Spaghetti-Like Nanoclusters

Acep Purqon, Ayumu Sugiyama, Hidemi Nagao, Masako Takasu, and Kiyoshi Nishikawa

Division of Mathematical and Physical Science, Graduate School of Natural Science and Technology, Kanazawa University, Kakuma, Kanazawa 920-1192, JAPAN

Abstract. Spaghetti-like nanoclusters show irregular shapes. We investigate their shapes by using the concept of symmetry and isotropy. The Symmetry-S evaluates the degree of symmetry of a cluster implying aggregate orderness, while, the Isotropy-I evaluates the degree of parallelism of a cluster. To investigate cluster dynamics in detail, we perform molecular dynamics simulation for POPC and POPE lipids for 300 K and 340 K. >From the simulations, the clusters are not easy to configure $S \approx 0$; which implies that the cluster shapes are neither sphere nor rod shapes; simply disorder shapes. However, at some times, the clusters show similar shapes with definite shapes implying some regions or classifications. For the reasons, we classify the irregular shapes in spaghetti-like nanoclusters by using geometrical classification as physical meaning of the concepts of symmetry and isotropy. We find, at least, four cluster modes: sphere-like, rod-like, cone-like, and monolayer-like. We also use geometrical classification as diagnostics of stability or anomalous behaviour and discuss the dependence of cluster shapes to temperature, number of lipids and odd-even number of lipids.

Keywords: Isotropy, Spaghetti-like nanoclusters, Symmetry
PACS: 87.15.He, 87.15.Ya

INTRODUCTION

Cluster shapes may reflect the underlying physical process driving a phenomenon. In other circumstances, the shapes may act as diagnostics of anomalous behavior [1]. One of the interesting cluster shapes is the clusters which are formed by self-assembly such as surfactants, lipids, or amphiphiles aggregates [2, 3]. Their molecular interactions can form various shapes such as micelles [4, 5], vesicles [6, 7] and membranes [8] as shown in Figure 1.a. In micellar size, they also can form some definite structures such as spherical, cylindrical, rod-like [9], disk-like, worm-like [10], and so on.

There are some reports about the cluster shapes from either experiment or simulation. In experiment by using Atomic Force Microscopy (AFM), Zou *et al.* [11] have observed three stable states in nanostructure of amphiphiles. The stable states are globule-like (sphere-like), stripe-like (rod-like) and spaghetti-like micelles (irregular shapes resembling spaghetti). The irregular shapes of spaghetti-like nanoclusters also occur in small micelles less than 32 surfactants. From numerical simulations, Marrink *et al.* [12] have discussed dynamics of micelles in relation to scaling of cluster properties for six different cluster sizes: single surfactants, small clusters(2-4 surfactants), intermediate clusters(5-8), large clusters(9-15), small micelles(16-31), and micelles(32 or more). In our previous work [13], we investigated structures and dynamics of small micelle less than 32 surfactants showing spaghetti-like shapes. An open question is what shape are the spaghetti-like micelles and how can these shapes be classified.

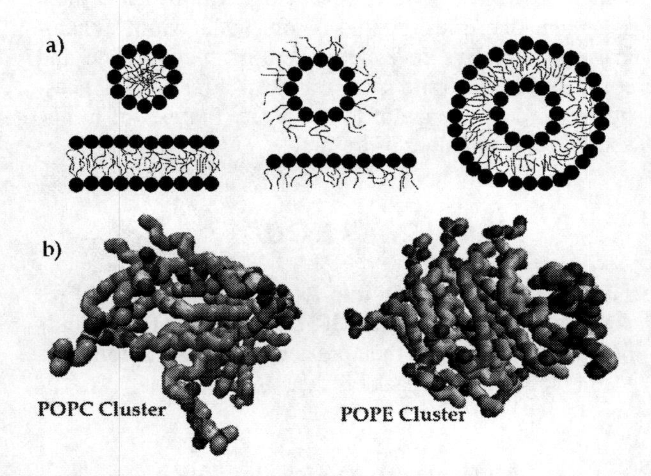

FIGURE 1. a) Surfactant aggregates can form various shapes such as micelle, inverse micelle, monolayer, vesicle, and membrane bilayer. b) A cluster of POPC and POPE.

To classify the irregular shapes, we continue our previous work and explore the parameters of Symmetry-S and Isotropy-I to evaluate symmetry and isotropy for some cluster systems. The Symmetry-S measures the degree of symmetry or aggregate orderness, while, the Isotropy-I measures the degree of parallelism of each lipid to one another in a cluster. From the results, they imply some types of dynamics and classes of structure in small surfactant aggregates. Interestingly, we do not find $S \approx 0$, which implies that the micelles are neither spherical nor

CP982, *Complex Systems, 5th International Workshop on Complex Systems*
edited by M. Tokuyama, I. Oppenheim, and H. Nishiyama

rod-like in shape. Instead the micelles are disordered and resemble spaghetti-like structures. This geometry may imply the stability of flexible structure.

To study the geometry of the spaghetti-like nanoclusters, we simulate POPC (1-palmitoyl-2-oleoyl- phosphatidycholine) and POPE (1-palmitoyl- 2-oleoyl- phosphatidylethanolamine) for 300 K and 340 K by using molecular dynamis (MD) simulations with AMBER force field [14] as shown in Figure 1.b. There are several methods to simulate micelles such as coarse grain simulations [15], Brownian dynamics simulations [16], MD simulations [17], and Monte Carlo simulations [18]. However, since we are interested in detailed information of the tail dynamics, we use MD simulations.

In this study, we determine the geometrical classification in spaghetti-like nanoclusters as part of our studies on structure and dynamics of spaghetti-like nanocluster systems. First, we investigate and classify some geometrical classification based on symmetry and isotropy. Second, we discuss stable structures of each system concerning the geometrical classification versus their stability. Third, we discuss static and dynamic classification. From statical approach, we compare with definite shapes such as sphere-like, rod-like and so on, while, from dynamical approach, we investigate certain interval time and compare with definite shapes as well. Finally, we investigate the dependence of temperature, number of lipids, and odd-even number of lipids.

METHOD

This section, we divide into two subsections. First, we present the details of our MD simulations of spaghetti-like nanoclusters. We then present results for Symmetry-S and Isotropy-I for each cluster.

Molecular Dynamics Simulation

We used POPC simulation results from our previous work [13] and we continued for POPE calculation by the same methods. We carried out the MD simulations of 16 POPC lipids and 16 POPE lipids for 300 K and 340 K at a constant pressure condition (NPT) with periodic boundary using Amber force field 03. The chemical structure of POPC and POPE lipid are shown in Figure 2.a. We used 11,326 and 10,100 TIP3P water molecules, respectively. We adopted the AM1-bcc charge and used 8 Å cutoff radii for non-bond interactions. After performing 1 ns (1 MD step=2 fs with SHAKE algorithm) for optimization and equilibration, we performed 7 ns of MD simulation for the analysis. After equilibration, the box size fluctuation is less than 0.2 % throughout the simulation.

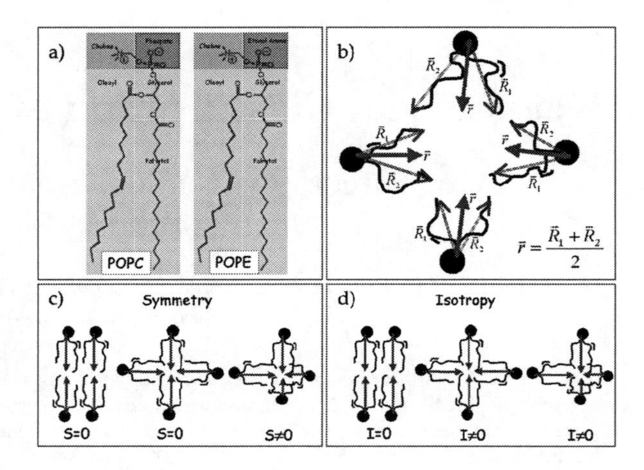

FIGURE 2. a) Chemical structure of POPC and POPE lipid b) Phosphate atom as center point and carbon at each edge of tail as other points. c) Symmetry is defined as the summation of all $\vec{r}$. The more similar the length of vector $\vec{r}$ is, the lower the S-value is. d) Isotropy is defined as the summation of each vector product $\vec{r}$ with every other vector. The lower the I-value is, the more parallel the structure is.

Classification of Spaghetti-like Nanoclusters

For geometrical analysis, we simply defined vector $\vec{R}_1(t)$ representing a vector of the unsaturated tail at time t, and vector $\vec{R}_2(t)$ representing a vector of the saturated tail at time t. We defined $\vec{r}(t)$ representing the average between vector $\vec{R}_1(t)$ and $\vec{R}_2(t)$ as shown in Figure 2.b.

We defined Symmetry-S to represent whether the structures are symmetric or asymmetric (irregular) in shape. The time-dependent symmetry $S(t)$ is expressed as

$$S(t) = \frac{1}{N} |\sum_{i=1}^{N} \vec{r}_i(t)|, \quad (1)$$

where $\vec{r}_i(t)$ indicates the mean vector of the two tail vectors $\vec{R}_{1i}$ and $\vec{R}_{2i}$ of the i-th POPC or POPE lipid.

The S-value measures the degree of symmetry or regularity. For example, $S = 0$ is geometrically high symmetry or ordered aggregates as shown in Figure 2.c. The more similar the length of vector $\vec{r}$ is, the lower the S-value is. Low-S represents a high symmetry dimension such as nearly spherical shape, rod, cylinder and so on. In contrast, high-S represents an asymmetric structure.

We defined Isotropy-I to represent the degree of parallelism of each lipid to one another in the cluster. The time-dependent isotropy $I(t)$ is expressed as

$$I(t) = \frac{1}{N} \sqrt{\sum_{\substack{j,k=1,\\ j<k}}^{N} |\vec{r}_j(t) \times \vec{r}_k(t)|}. \quad (2)$$

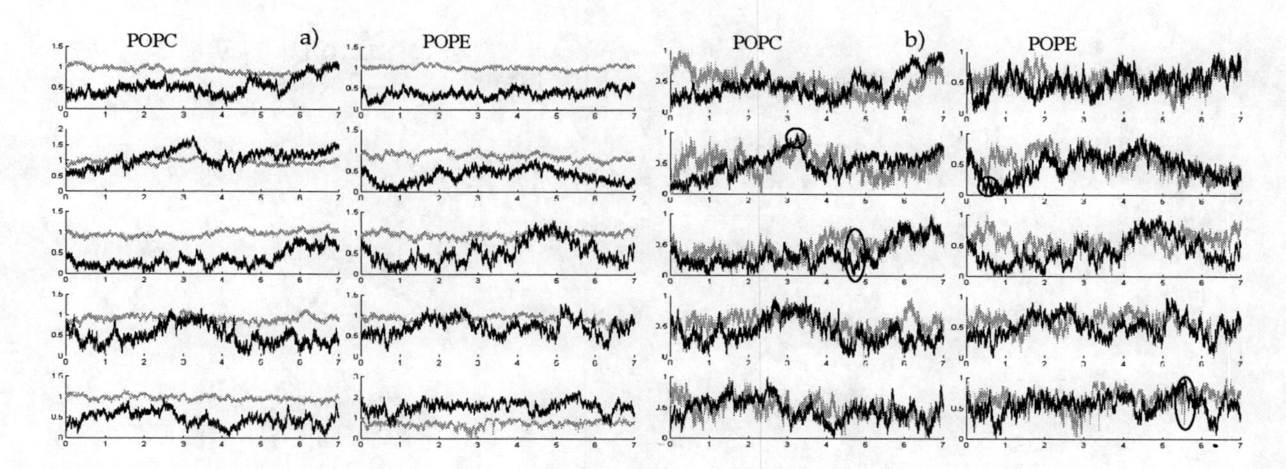

FIGURE 3. a) Graphics of symmetry-isotropy dynamics. Black is the symmetry of $S(t)$ and gray is the isotropy of $I(t)$. The graphics consist of POPC1 to POPC5 and POPE1 to POPE5, respectively. b) Graphics of normalized scale of symmetry-isotropy dynamics. Circled marks indicate cluster modes. For examples, high-S and high-I at 3.2 ns of POPC2, low-S and high-I at 4.5 ns of POPC3, low-S and low-I at 0.7 ns of POPE7, and high-S and low-I at 5.5 ns of POPE10. POPE3 almost show opposite relation between symmetry and isotropy.

The lower the value of I is, the more parallel the structure is. For example, $I = 0$ represents a parallel structure or same direction as shown in Figure 2.d. For example, a rod structure has low-S and low-I, while a sphere has low-S and high-I and irregular shapes have high-S and high-I. In simple cases, we assume sphere structure has properties such as low-S and high-I, whereas rod structure, ellipsoid, cylindrical structure have properties such as low-S and low-I. In short, $S(t)$ concerns shape fluctuation and $I(t)$ concerns the parallelism of each lipid to one another.

TABLE 1. Simulation results of spaghetti-like nanoclusters. There are 10 cluster samples for various conditions.

Name	N	T
POPC1	6	300
POPC2	4	300
POPC3	6	300
POPC4	6	340
POPC5	10	340
POPE1	9	300
POPE2	7	300
POPE3	7	340
POPE4	7	340
POPE5	2	340

RESULTS AND DISCUSSION

>From the simulation results, we obtain some clusters as shown in Table 1; namely POPC1 to POPC5 and POPE1 to POPE5. While, from Eq.1 of $S(t)$ and from Eq.2 of $I(t)$, we obtain $S-I$ dynamics for each cluster as shown in Figure 3.a.

We start by analyzing the dynamics. It is important to note that the S-value and I-value have different scale. For easy comparison, we normalized the scale of each $S-I$ dynamics as shown in Figure 3.b. From the normalized scale, we find some interesting cluster modes as shown by the circled marks. For example, we find high-S and high-I at 3.2 ns of POPC2, low-S and high-I at 4.5 ns of POPC3, low-S and low-I at 0.7 ns of POPE2, and high-S and low-I at 5.5 ns of POPE5. Interestingly, symmetry and isotropy is not always in opposite relation such as if high-S then low-I or if low-S then high-I. It implies that they have certain configurations or classifications.

To investigate the distribution of their dynamics, we calculated the $S-I$ distribution and normalized distribution as shown in Figure 4. For detail distribution information, we arranged the results in Table 2. >From Table 2, we can easily see that POPC3 is the most symmetric and POPE5 is the most asymmetric. In addition, POPE5 is the most isotropic and POPE1 is the most unisotropic.

They also show interesting temperature dependence. For examples, among 6 lipids of POPC, POPC4 at 340 K has wider distribution interval than POPC1 and POPC2 at 300 K. In addition, among 7 lipids of POPE, POPE2 at 300 K has a narrow distribution than POPE3 and POPE4, which demonstrates that the temperature affects S-value. In other words, the temperature affects the fluctuation of the symmetry.

In the same way, for the I-value, the cluster show an interesting temperature dependence. For example, among 6 lipids of POPC, POPC4 at 340 K has wider distribution than POPC1 and POPC2 at 300 K. In addition, among 7

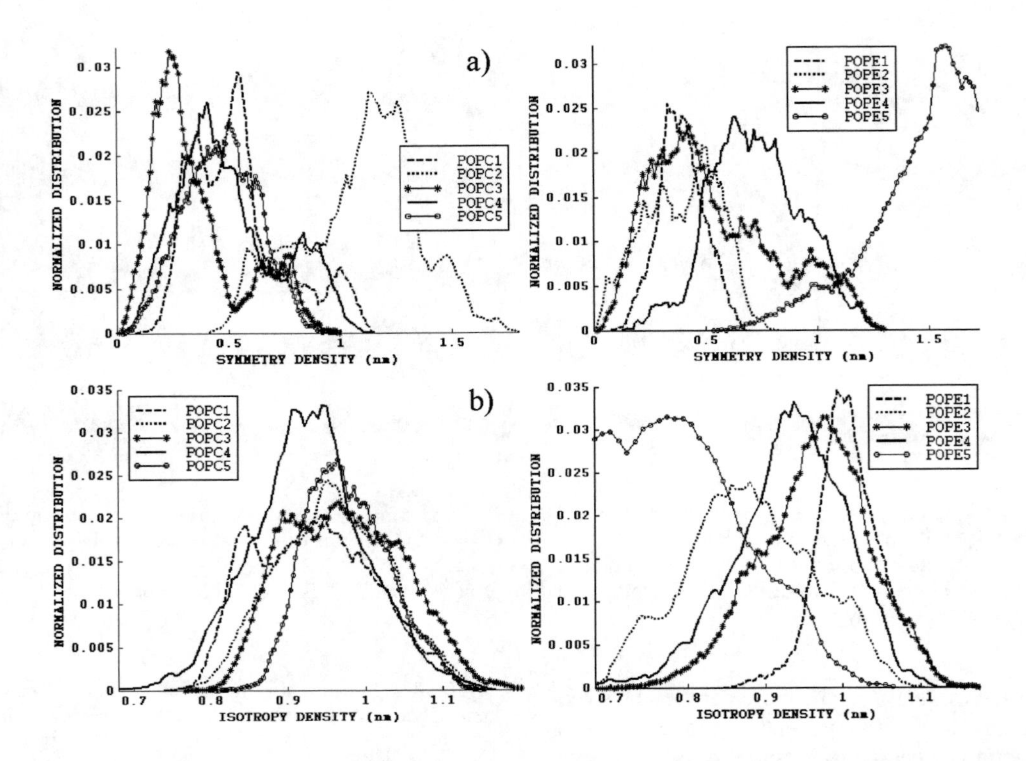

FIGURE 4. Normalized Symmetry-Isotropy distributions. a) Symmetry density for POPC1 to POPC5 and POPE1 to POPE5. b) Isotropy density for POPC1 to POPC5 and POPE1 to POPE5.

TABLE 2. The detail information of the symmetry and isotropy distribution for all samples of spaghetti-like nanoclusters.

Symmetry			
Name	*S*-value	Peak	Range
POPC1	0.546	0.030	1.071
POPC2	1.145	0.027	1.391
POPC3	0.239	0.032	0.986
POPC4	0.411	0.026	1.137
POPC5	0.507	0.024	0.950
POPE1	0.332	0.026	0.613
POPE2	0.487	0.021	0.818
POPE3	0.426	0.023	1.284
POPE4	0.634	0.024	1.176
POPE5	1.575	0.032	1.809
Isotropy			
Name	*I*-value	Peak	Range
POPC1	0.956	0.020	0.411
POPC2	0.952	0.025	0.415
POPC3	0.965	0.022	0.449
POPC4	0.945	0.033	0.555
POPC5	0.964	0.027	0.356
POPE1	0.994	0.035	0.314
POPE2	0.881	0.024	0.482
POPE3	0.975	0.032	0.498
POPE4	0.937	0.033	0.532
POPE5	0.773	0.032	1.029

lipids of POPE, POPE2 at 300 K has more narrow distribution than POPE3 and POPE4, where implies the temperature also affects *I*-value as it did for *S*-value.

Interestingly, isotropy is not depend on size or number of lipids. The behavior can be explained by the saturation effect behavior as follow:

Using a simple two-dimensional model as an approximation, we assume identical vectors $\vec{r}_i$ are symmetrically configured with equal θ in two dimensional space for simplicity. In this approximation,

$$I = \frac{r}{N}\sqrt{\sum_{i=1}^{N}(N-i)|sin\ i\theta|}, \qquad (3)$$

where I is the isotropy parameter, r is the length of lipid, N is the number of lipids, and θ is the angle between lipids.

We can then calculate the *I*-value for some number of lipids as shown in Table 3. From the Table 3, it is clear that after certain number of lipids, *I*-value has nearly same value. It looks like that isotropy is not dependent on the number of lipids.

In simple concept, *S*-value and *I*-value show opposite relation as shown in Figure 5.a for POPE3. It clearly can be shown, when *S*-value is low then *I*-value is high and when *S*-value is high then *I*-value is low. Nevertheless, in other circumstances, the relation is not so simple as we have shown in Figure 3.b.

TABLE 3. The dependence of Isotropy-*I* to number of lipids in small surfactant aggregates using Equation 3. After a certain value for the number of lipids, *I*-value saturates and does not change. This is one of explanation for Figure 1.a that the *I*-value appears to be stable in certain values, whereas, it actually fluctuates. This is the reason for the normalized scales.

Number of lipids (N)	I	I/N
2	0	0
4	1	0.25
6	3.2	0.53
7	3.91	0.56
9	5.05	0.56
10	5.55	0.56

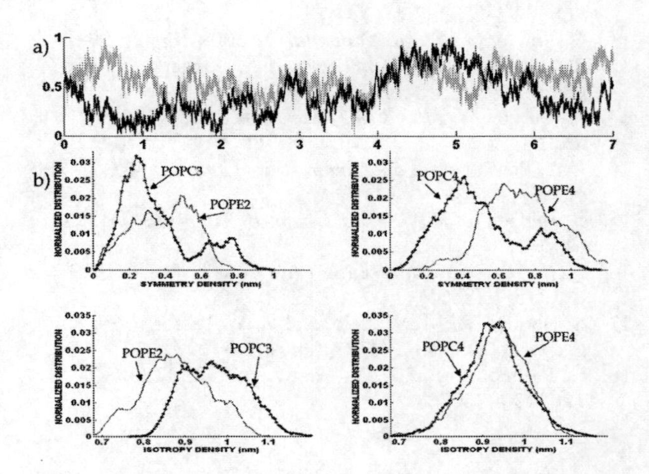

FIGURE 5. a) A sample from POPE3 dynamics that show opposite relation between symmetry and isotropy. When symmetry is low then isotropy is high and when symmetry is high then isotropy is low. However, this relationship does not hold in all cases. b) The density comparison and the dependence for odd-even number of lipids. Symmetry and isotropy density between POPC3 (6 lipids at 300 K) and POPE2 (7 lipids at 300 K). Symmetry and isotropy density between POPC4 (6 lipids at 340 K) and POPE4 (7 lipids at 340 K).

Another interesting aspect of the symmetry in shapes is the dependence of odd-even number of lipids and temperature with respect to *S*-value and *I*-value. We investigate some conditions as shown in Figure 5.b. For example, we compare POPC3 (6 lipids) and POPE2 (7 lipids). POPC3 is more symmetric than POPE2, whereas, POPE2 is more isotropic than POPC3. They indicate that even number of lipids have more symmetric than odd

TABLE 4. The classification for four cluster mode and typical nanoclusters.

S–I	Low	High
Low	Rod-like (POPE3)	Sphere-like (POPC3)
High	Cone-like (POPE5)	Monolayer-like (POPC2)

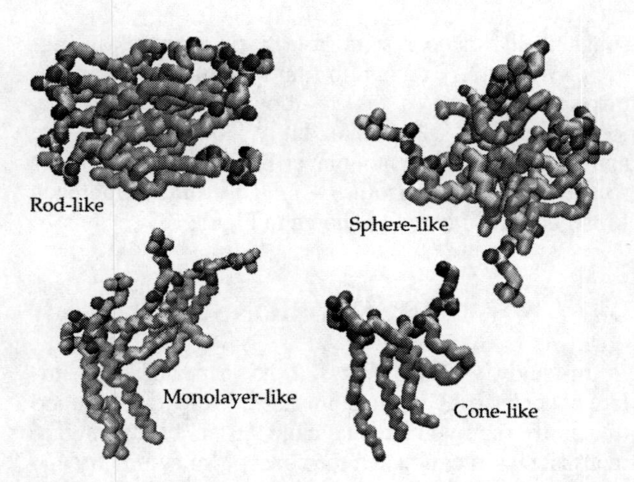

FIGURE 6. Classification for four cluster modes from some points in their dynamics from static approach. The comparison with definite shapes are as follow: low-*S*-low-*I* is rod-like, low-*S*-high-*I* is sphere-like, high-*S*-low-*I* is cone-like and high-*S*-high-*I* is monolayer-like.

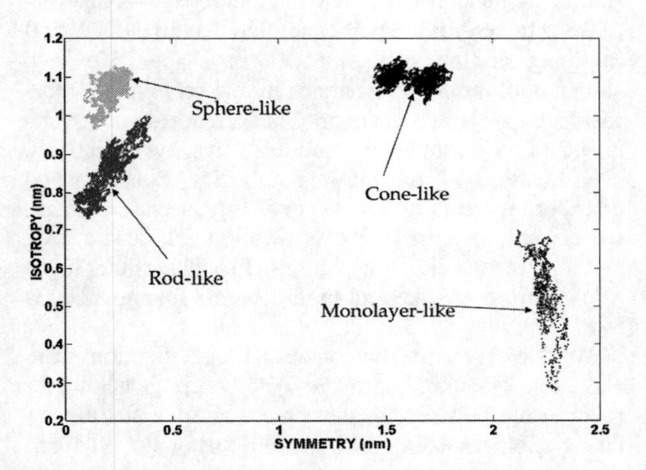

FIGURE 7. The symmetry versus the isotropy and the classification for four cluster modes and some typical structures in some time interval from dynamical approach in real scale as follow: low-*S*-low-*I* for rod-like, low-*S*-high-*I* for sphere-like, high-*S*-low-*I* for monolayer-like and high-*S*-high-*I* for cone-like.

number of lipids, and odd number of lipids have more isotropic than even number of lipids.

In addition, we compare them with the dependence to the temperature. For example, for 340 K, POPC4 (6 lipids) and POPE4 (7 lipids), POPC4 is more symmetric than POPE4, whereas isotropy is not so different or nearly same. They indicate even number of lipids is more symmetric than odd number of lipids, while, as noted in previous section in some conditions, isotropy looks like that it is not depend on number of lipids.

Finally, we show some cluster modes representing

spaghetti-like nanoclusters from static approach in Figure 6. We classify them into four type of structure combination as shown in Table 4. Low-S-low-I is rod-like, low-S-high-I is sphere-like, high-S-low-I is cone-like and high-S-high-I is monolayer-like. Finally, we show some classifications from $S-I$ of dynamical approach for some time interval as shown in Figure 7.

CONCLUSIONS

In this study, we have analyzed the shapes of spaghetti-like nanoclusters. We have presented an analysis method to classify the spaghetti-like nanoclusters based on geometrical classification and the concepts of symmetry and isotropy. From the results, both of parameters have confirmed that symmetry concerns with shape fluctuation and isotropy concerns with the parallelism of each lipid to one another.

>From the S-value dynamics, we have found that all results are generally not easy to configure $S \approx 0$. Nevertheless, in certain interval time, they have nearly $S \approx 0$ implying similar with some definite shapes. We have shown both the $S-I$ dynamics in real scale and in normalized scale for comparison. They are not always in opposite the $S-I$ relation implying that they have fuzzy classifications. >From analysis of the $S-I$ dynamics and distribution based on simple geometrical classification, we classify at least four classification. These are low-S-low-I for rod-like, low-S-high-I for sphere-like, high-S-low-I for cone-like and high-S-high-I for monolayer-like.

We have discussed geometrical classification from static and dynamic approaches. Both approaches imply relationship between cluster shapes and the stability of clusters. We have also discussed the cluster shape dependence to temperature, number of lipids and odd-event number of lipids. They indicate even number of lipids have more symmetric than odd number of lipids. Although our starting point is from small surfactant aggregates, the method can be used for large aggregates as well. In the future work, it is interesting to classify other region for other fuzzy region of cluster modes.

ACKNOWLEDGMENTS

H.N. is grateful for financial support from the Ministry of Education, Science and Culture of Japan (grant 19029014).

REFERENCES

1. M.E. Fisher, *Physica D* **38**, 112-118 (1989).
2. H. Heerklotz *et al.*, *J. Phys. Chem. B* **101**, 639-645 (1997).
3. K. John and M. Bar, *Phys. Rev. Lett.* **95**, 198101-1-198101-4 (2005).
4. N. Yoshii, K. Iwahashi, and S. Okazaki, *J. Chem. Phys.* **124**, 184901-1-184901-6 (2006).
5. N. Yoshii and S. Okazaki, *Chem. Phys. Lett.* **425**, 58-61 (2006).
6. S. Rasi, F. Mavelli, and P.L. Luisi, *J. Phys. Chem. B* **107**, 14068-14076 (2003).
7. W. Wang *et al.*, *Macromolecules* **37**, 9114-9122 (2004).
8. J. Prost and R. Bruinsma, *Europhys. Lett.* **33**, 321-326 (1996).
9. A. Kroeger *et al.*, *Macromolecules* **40**, 105-115 (2007).
10. M.R. Lopez-Gonzalez *et al.*, *Phys. Rev. Lett.* **93**, 268302-1-268302-4 (2004).
11. B. Zou *et al.*, *Chem. Commun.* **9**, 1008-1009 (2002).
12. S.J. Marrink, D.P. Tieleman, and A. E. Mark, *J. Phys. Chem. B* **104**, 12165-12173 (2000).
13. A. Purqon *et al.*, *Chem. Phys. Lett.* **443**, 356-363 (2007).
14. D.A. Pearlman *et al.*, *Comp. Phys. Commun.* **91**, 1-41 (1995).
15. S. Nielsen *et al.*, *J. Phys. Chem. B* **107**, 13911-13917 (2003).
16. H. Noguchi and M. Takasu, *Phys. Rev. E* **64**, 041913-1-041913-7 (2001).
17. S. Bogusz, R. M. Venable, and R. W. Pastor, *J. Phys. Chem. B* **105**, 8312-8321 (2001) .
18. C. Pierleoni *et al.*, *Phys. Rev. Lett.* **96**, 128302-1-128302-4 (2006) .

Numerical Simulation of Flow Patterns of Spherical Particles in a Semi-Cylindrical Bin-Hopper

Chuen-Shii Chou and Ang-Fen Lee

Powder Technology R&D Laboratory, Department of Mechanical Engineering, National Pingtung University of Science & Technology, Pingtung 912, Taiwan

Abstract. This study numerically investigated the flow patterns of spherical particles in a semi-cylindrical bin-hopper with (or without) a semi-conical obstacle using the 3-D model based on discrete element method. By referring to the method of Johanson [1], the placement of a semi-conical obstacle in the semi-cylindrical bin-hopper was determined. The numerical predictions of flow patterns were visualized by the computer software Auto CAD. This research results not only show the quasi-stagnant zone adjacent to the wall can be shrunken, but also present the method to visualize the flow patterns of any portion of particles in the semi-cylindrical bin-hopper at a certain time step.

Keywords: Bin-hopper, Discrete element method, Semi-conical obstacle, Semi-cylindrical

INTRODUCTION

Granular materials (or bulk solids) in fuel, chemical, pharmaceutical and food industries are typically processed through hoppers, bins, silos or bunkers. In the 1950s, Jenike differentiated mass flow and core flow, which is alternatively called funnel flow. In a mass flow silo, all material is flowing; however, a funnel flow silo contains regions in which material does not move.

One primary requirement for hopper design is consistent flow during discharge. However, most existing hoppers are not designed consistent flow during discharge and are later used to store different materials. Therefore, rigid objects called flow corrective inserts are among a number of means used inside a hopper to alleviate flow problems [1].

In addition to serving as a flow corrective element in a silo, an obstacle can be utilized as a splitter in an induced-roll separator, which is frequently applied for dry granulated material. Moreover, a blending silo with an obstacle has been used to blend or homogenize substantial amounts of bulk solids and minimizes fluctuations in product quality. Therefore, granular flows in a hopper (or bin) with an obstacle and an eccentric (or a centric) discharge have attracted the interest of researchers.

Particular theoretical, numerical and experimental approaches have been adopted to examine flow patterns, wall stresses or insert load for a silo with an insert [1-8]. The references cited are all considered the granular flow in a symmetrical bin, bin-hopper, or wedge hopper with a symmetrical isosceles triangle insert or hexahedron insert. Relatively fewer studies have investigated the granular flows in a cylindrical bin-hopper with a conical obstacle, and have presented a method to visualize the flow patterns of any portion of particles in the bin-hopper. The purpose of this research is to integrate the tools (such as DEM and Auto CAD) to visualize the flow patterns of any portion of particles in the bin-hopper at a certain time step.

MATHEMATICAL MODEL

DEM is a numerical model capable of describing the flow behavior of particulate material. Cundall and Strack [9] proposed the DEM model and carried out its first application to particulate systems. The principle of DEM is to track, in a time stepping simulation, the trajectory and rotation of each particle in the system to determine its position and orientation, and then to calculate the interactions between the particles themselves and also between the particles and their environment [10].

The calculations performed in DEM alternate between the application of Newton's second law to the particles and a force-displacement law at the contacts of the particle (Fig. 1). The DEM numerical simulation

CP982, *Complex Systems, 5th International Workshop on Complex Systems*
edited by M. Tokuyama, I. Oppenheim, and H. Nishiyama

consists of four major steps: (1) contact detection, (2) contact processing (i.e. applying interaction laws to calculate forces and moments at all particle-particle and particle-wall contacts), (3) particle motion (i.e. applying Newton's second law to determine particles' accelerations, velocities, and positions), and (4) updating [9,11].

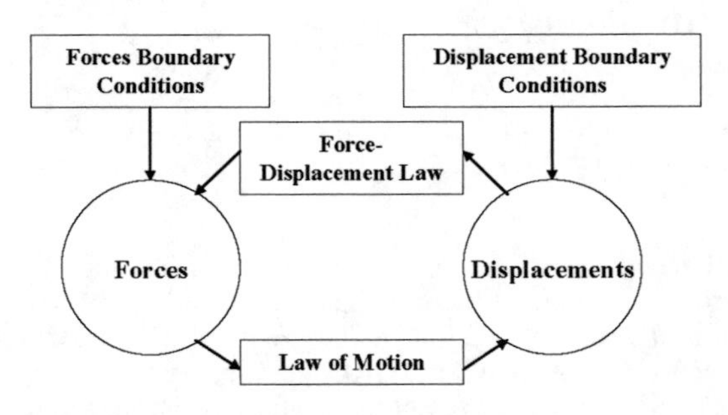

FIGURE 1. The calculation procedures of DEM

Contact Detection

At each time step in the simulation, the DEM model must first execute the procedure to detect which particles and boundaries (or walls) are in contact. Besides of the contact between particles, in this 3-D DEM model for dry granular flows of spheres, the particle-wall contact consists of four probable situations: (1) the contact between a particle and a point-boundary, (2) the contact between a particle and a line-boundary, (3) the contact between a particle and a plane-boundary, and (4) the contact between a particle and multi-plane-boundary (Fig. 2). Additionally, the neighbour list bookkeeping scheme [10,11] is used to record those particle pairs, which are checked for possible contact. The principles of the particle-particle contact detection and the particle-wall contact detections were explicated [11].

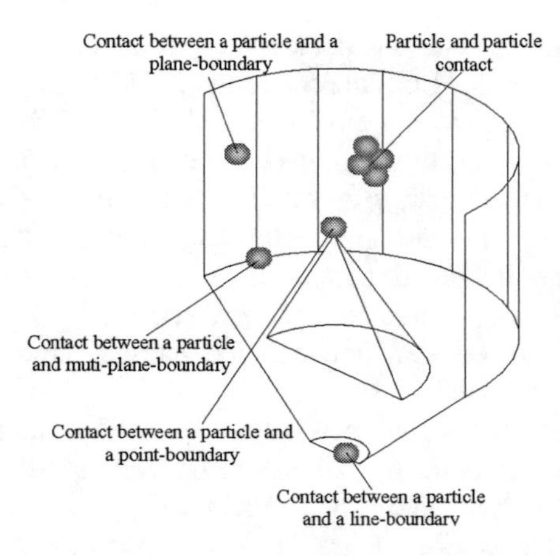

FIGURE 2. (Color Online) The schematic of possible contacts.

Contact Processing

The contact force model (force-displacement law) implemented in the present model is same as that used by Cundall and Strack [9]. This model consists of elastic contributions (linear spring), which represent the repulsive force between any two particles and are proportional to the relative displacement between particles and the stiffness of spring, and viscous damping contributions (dashpot), which dissipate a proportion of the relative kinetic energy and are proportional to the rate of relative displacement and the damping coefficient (Fig. 3(a)-(b)). In Fig. 3(a)-(b), k_{nb}, c_{nb}, k_{sb}, c_{sb}, and μ_b represent normal stiffness, normal damping coefficient, tangential stiffness, tangential damping coefficient, and friction coefficient between particles, respectively. For the Nth time step, the force in normal direction is modeled as viscous-elastic, and the force in the tangential direction is modeled as viscous-elastic below the friction limit (i.e., the Mohr-Coulomb law).

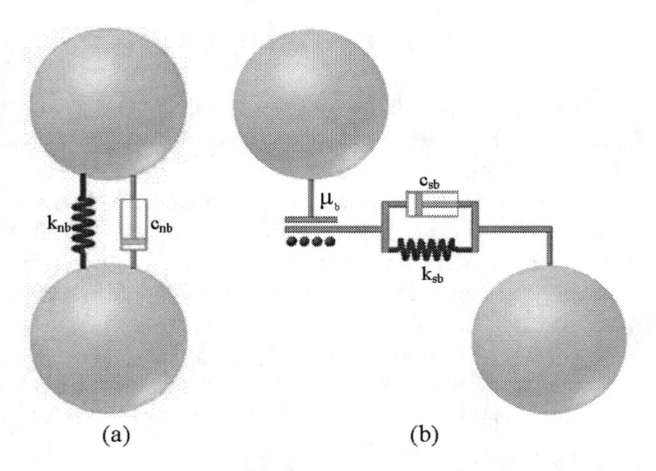

FIGURE 3a-b. (Color Online) The contact force model between particles. (a) Normal direction, (b) tangential direction.

Moreover, the interaction between the particle and its boundary (e.g., a bin-hopper wall) is modeled in a similar manner to a particle-particle contact. For particle-wall contact as shown in Fig. 4(a)-(b), k_{nw}, c_{nw}, k_{sw}, c_{sw}, and μ_w represent normal stiffness, normal damping coefficient, tangential stiffness, tangential damping coefficient, and friction coefficient between particle and wall, respectively.

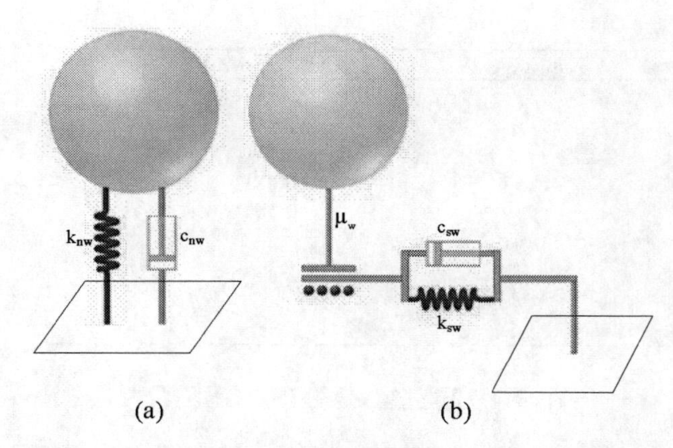

FIGURE 4a-b. (Color Online) The contact force model between the particle and the wall. (a) Normal direction, (b) tangential direction.

Governing Equations for the Particles

Once all of the resultant forces and moments acting on each spherical particle were obtained, the new velocities and positions of each spherical particle could be calculated by numerically integrating Newton's second law. Equations of the translational and rotational motion are given, respectively, by

$$m\frac{d^2r_i}{dt^2} = F_i^C + mg_i \tag{1}$$

$$\frac{d\omega_i}{dt} = \frac{T_i}{I} \tag{2}$$

In Eq. (1), i, m, r_i, F_i^c and g_i represent the ith direction in the coordinate, the particle mass, the ith component of the particle position vector, the ith component of the total contact force and the ith component of the gravity, respectively. In Eq.(2), ω_i, T_i and I denote the ith component of the angular velocity, the ith component of the net torque due to the tangential contact force and the moment of inertia of the particle.

Making use of the half-step "Leap-Frog" finite difference method, and assuming the acceleration of a sphere over a time step Δt is constant, the integration of Eq. (1) leads to an expression for translational velocity

$$(\dot{r}_i)_{t+\frac{\Delta t}{2}} = (\dot{r}_i)_{t-\frac{\Delta t}{2}} + \left[\frac{(F^c)_i}{m} + g_i\right]_t \Delta t \tag{3}$$

where "." is the derivative with respect to time. This new value for velocity is used to update the position of the sphere using further numerical integration

$$(r_i)_{t+\Delta t} = (r_i)_t + \left[(\dot{r}_i)_{t+\frac{\Delta t}{2}}\right]\Delta t \tag{4}$$

By the same token, the sphere angular velocity at time $t+(\Delta t/2)$ and rotation at time $t+\Delta t$ can be determined.

NUMERICAL SIMULATION

Procedures

The steps of the numerical simulation are as follows: (i) generate the model for numerical simulation using computer software Auto CAD; (ii) determine the coordinates of each spherical particle and the boundaries using Auto LISP; (iii) determine the contact force using the model described above; (iv) calculate the updated velocities and positions of each spherical particle by using Eqs. (3) and (4); (v) repeat the steps (iii)-(iv) until the number of iteration exceeds the value set before executing computer code. Finally, the flow status of spherical particles in the bin-hopper at a certain time step can be visualized using Auto CAD in the follow-up processing phase. Figure 5 presents the flow chart of the numerical simulation. Table 1 presents the input for numerical simulation.

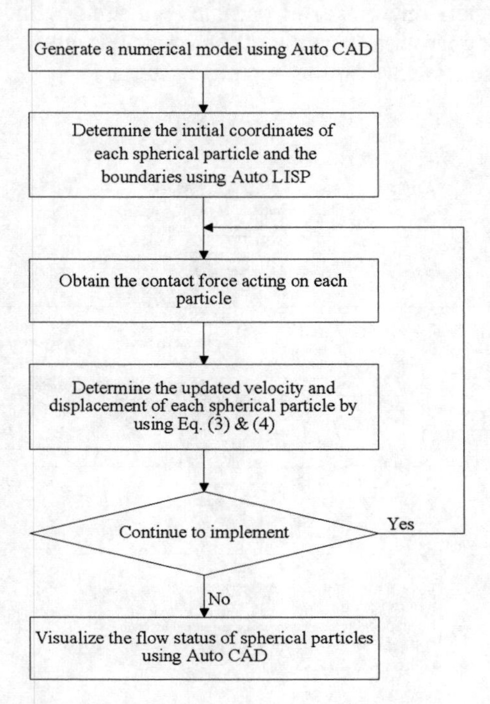

FIGURE 5. Flow chart of the numerical simulation

Table 1. Input for computer simulation

Parameter	Particle-particle		Particle-wall
Stiffness K_n [N/m]	100		1000
Stiffness K_s [N/m]	100		1000
Damping coefficient C_n [Ns/m]	0.5		0.01
Damping coefficient C_s [Ns/m]	0.0001		0.00001
Coefficient of friction μ	0.5		0.14
Time step (sec)		20×10^{-4}	
Mass of granular (kg)		0.000123	
Density (kg/m^3)		1086.6	
Gravity (m/s^2)		9.81	

Assumptions

The domain defined in the numerical simulation was a three-dimensional semi-cylindrical bin-hopper with a hopper half angle of 40°, a bin diameter of 30 mm, a bin height of 450 mm, a hopper height of 144 mm, and an exit diameter of 30 mm. The placement of conical insert with an insert half angle 25° was determined using the method of Johanson [1].

FOLLOW-UP PROCESSING

Four output modes were developed to visualize the flow patterns of particles at a certain time step in this study. (1) All the particles in a semi-cylindrical bin hopper with (or without) an insert; (2) half of the particles in a semi-cylindrical bin hopper with (or without) an insert; (3) a vertical layer of particles in a semi-cylindrical bin hopper with (or without) an insert; (4) a horizontal layer of particles n a semi-cylindrical bin hopper with (or without) an insert. Figure 6 shows the procedures of follow-up processing.

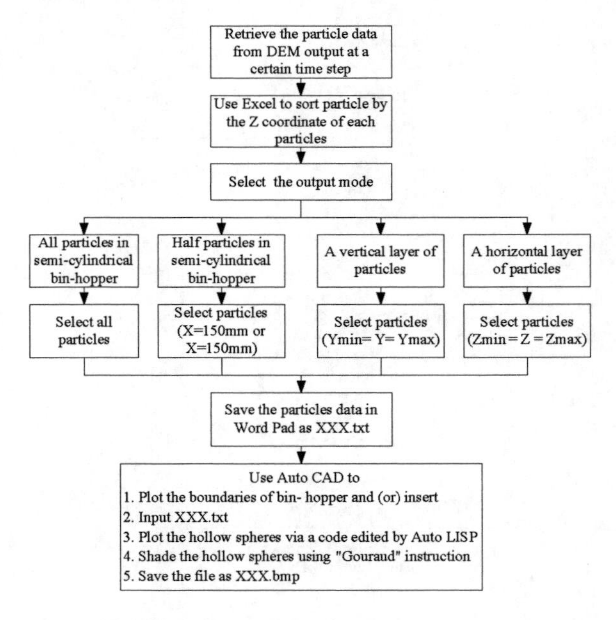

FIGURE 6. Flow chart of the follow-up processing

RESULTS AND DISCUSSION

Figure 7 shows the axonometric views of the flow history of all particles in a semi-cylindrical bin-hopper without ((a)-(c)), and with ((d)-(f)) a semi-conical insert. Figure 8 shows the axonometric views of the flow history of half of particles (X coordinate of particle center ≤ 150 mm) in a semi-cylindrical bin-hopper without ((a)-(c)), and with ((d)-(f)) a semi-conical insert. Figure 9 shows the axonometric views of the flow history of a vertical layer of particles (50 mm ≤ Y coordinate of particle center ≤ 70 mm) in a semi-cylindrical bin-hopper without ((a)-(c)), and with ((d)-(f)) a semi-conical insert. Additionally, Fig. 10 shows the axonometric views of the flow history of a horizontal layer of particles (194 mm ≤ Z coordinate of particle center ≤ 224 mm) in a semi-cylindrical bin-hopper without ((a)-(c)), and with ((d)-(f)) a semi-conical insert.

Quasi-stagnant zones ascended along the wall in the semi-cylindrical bin-hopper without an insert (Fig. 7(c)). The flow history of purple granules (Fig. 7 (f)) reveals that the semi-conical insert partitions granular flows into two equal flowing streams in the semi-cylindrical bin-hopper when the material begins to be withdrawn. Rather than quasi-stagnant zones, the narrow wall shear layers arose along the wall due to the placement of the semi-conical insert placed in the semi-cylindrical bin-hopper.

The particles adjacent to the wall separate from the other particles during discharge (Fig. 8(c)). Comparing Fig. 10(b) with Fig. 10(e), the semi-conical obstacle in the semi-cylindrical bin-hopper can disrupt the contact force network, and help the spherical particles stack loosely.

CONCLUSION

The method presented in this study may help understand and visualize the complex granular flow in the bin-hopper, or other facilities of bulk solid handling.

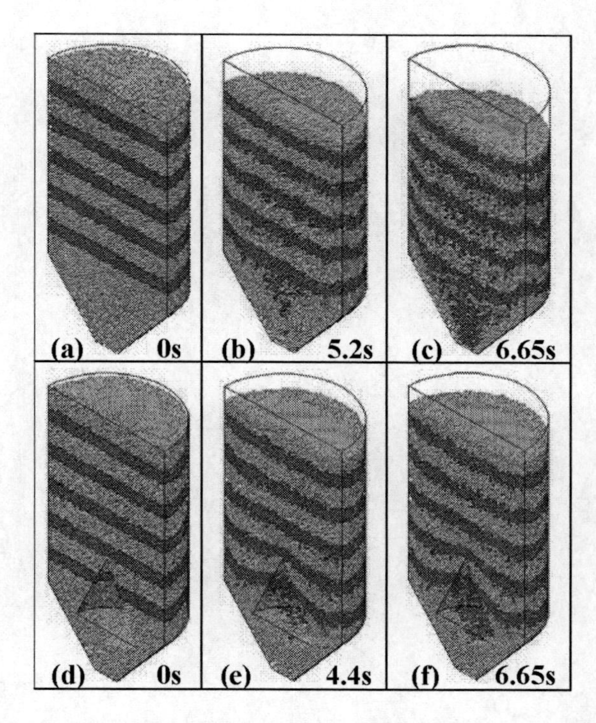

FIGURE 7. (Color Online) The axonometric view of flow history of all particles in a semi-cylindrical bin-hopper. (a)-(c) without an insert, (d)-(f) with an insert.

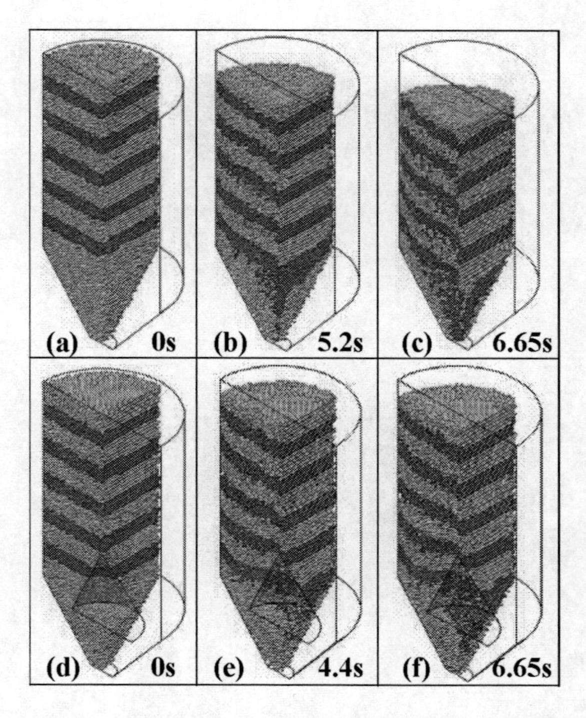

FIGURE 8. (Color Online) The axonometric view of flow history of half of particles in a semi-cylindrical bin-hopper. (a)-(c) without an insert, (d)-(f) with an insert.

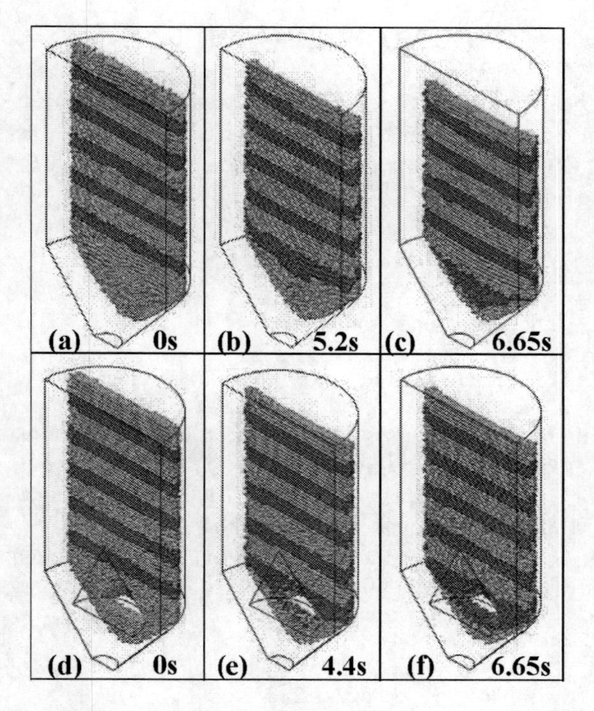

FIGURE 9. (Color Online) The axonometric view of flow history of a vertical layer of particles in a semi-cylindrical bin-hopper. (a)-(c) without an insert, (d)-(f) with an insert.

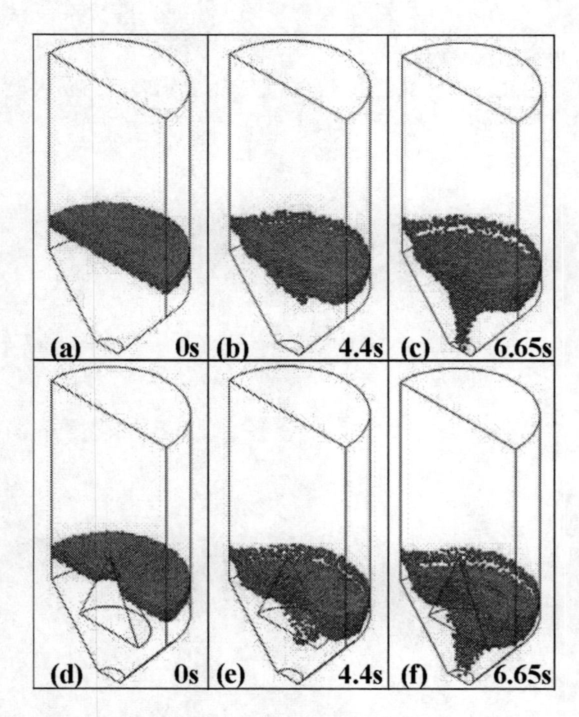

FIGURE 10. (Color Online) The axonometric view of flow history of a horizontal layer of particles in a semi-cylindrical bin-hopper. (a)-(c) without an insert, (d)-(f) with an insert.

ACKNOWLEDGEMENTS

The authors gratefully acknowledge the financial support from the National Science Council of the R.O.C. for this work through projects NSC 95-2218-E-002-035.

REFERENCES

1. J. R. Johanson, *Powder Technology*, 1, pp. 328-333 (1967/68).
2. H. Tsunakawa and R. Aoki, *Powder Technology*, 11, pp. 237-243 (1975).
3. H.G. Polderman, A.M. Scoot, and J. Boom, *Int. Chem. Eng. Symp. Ser.*, 91, pp. 227-240 (1985).
4. U. Tüzün, and R. M. Nedderman, *Chemical Engineering Science*, 40, pp. 325-336 (1985).
5. R. Moriyama, and T. Jotaki, *Powder Handling Process*, 1, pp. 353-355 (1989).
6. J. Strusch, and J. Schwedes, *Bulk Solids Handling*, 14, pp. 505-512 (1994).
7. C. S. Chou, and T. L. Yang, *Advanced Powder Technology*, 15, pp. 567-582 (2004).
8. C. S. Chou, C. Y. Tseng, and T. L. Yang, *Journal of Chinese Institute of Engineer*, 29, pp. 1-12 (2006).
9. P.A. Cundall and O.D.L. Starck, *Geotechnique*, 29, pp. 47-65 (1979).
10. B.N. Asmar, P.A. Langston, A.J. Matchett, and J.K. Walter, *Computers and Chemical Engineering*, 26, pp. 785-802 (2002).
11. C. S. Chou, A. F. Lee, and C.H. Yeh, *Particle & Particle System Characterization*, 24, 3, pp. 210-222 (2007).

Power-Law Fluctuations at the Order-Disorder Transition in Colloidal Suspensions under Shear Flow

Masamichi J. Miyama and Shin-ichi Sasa

Department of Pure and Applied Science, University of Tokyo, Komaba, Tokyo, 153-8902, Japan

Abstract. We investigate order-disorder transitions in non-equilibrium systems. Specifically, we consider colloidal suspensions under shear flow by performing Brownian dynamics simulations. We characterize the order-disorder transition in this system in terms of a statistical property of the time-dependent maximum value of structure factor. We report that it exhibits power-law fluctuations, which appears only in the ordered phase.

Keywords: Brownian dynamics simulations, Colloidal suspensions, Order-disorder transitions under shear flow
PACS: 05.40.-a, 05.70.Fh, 83.80.Hj

INTRODUCTION

It is known that colloidal suspensions exhibit an order-disorder transition under equilibrium conditions. In particular, crystalline structure of the particles, which is so-called "colloidal crystal", is observed in the ordered phase. According to statistical mechanics, the ordered (crystal) phase is defined as the translational symmetry breaking of the configuration of the particles. For example, one can detect this symmetry breaking with the examination whether the structure factor, which is determined by the configuration of the particles, has Dirac's delta function peaks or not. Since the structure factor corresponds to the intensity of the Bragg reflection in laboratory experiments, one can detect the order-disorder transition with measuring the Bragg peaks.

Unlike in the case under equilibrium conditions, there is no systematic understanding of the order-disorder transition of non-equilibrium systems from the viewpoint of statistical mechanics, because non-equilibrium statistical mechanics has not been established as yet. Recalling the history of equilibrium statistical mechanics, one finds that the order-disorder transition has played a prominent role in its establishment. Thus, from the viewpoint of exploring non-equilibrium statistical mechanics, it is worthwhile to study a typical example, such as the order-disorder transition in colloidal suspensions under shear flow.

With regard to crystalline structures in colloidal suspensions under shear flow, several studies have already been reported [3, 4]. Ackerson and Clark focused on the shear-induced disordering phenomena with using light scattering [2]. This is so-called 'shear-melting' and it is investigated and explained by Harrowell et al. with several simulations [5, 6]. These results suggest the existence of the ordered phase even under shear flow. However, the configuration of the particles under stationary shear flow is continuously changing with time. Thus, in this situation, by considering the ensembles generated by the time evolution, we expect that the translational symmetry in the shear gradient direction recovers. With regard to the rest two dimensions of freedom, there are no positional order in two dimensional systems according to Mermin-Wagner's theorem. Here, recalling that 'crystal' is defined as the translational symmetry breaking, we can conclude that there is no crystal under stationary shear flow. Thus, when we consider the order-disorder transition under shear flow, we cannot rely on the concept of 'crystal' in equilibrium systems. In other words, in order to understand the phenomena observed in colloidal suspensions under shear flow, we recall the difference between crystal in equilibrium situations and the ordered phase under shear flow again, before we investigate the shear-disordering and other non-equilibrium phenomena under shear flow. We then find that it is not enough to focus on the 'static' feature, such as the snapshot of the configuration of particles, long-time average of the intensity of static structure factor and so on. Now, the objective of this study is to find a useful and thorough characterization of the ordered state for systems under shear flow, especially, with focusing on the dynamical feature. We attempt this characterization by performing Brownian dynamics simulations of the colloidal suspensions under shear flow.

MODEL

We consider a system consisting of N colloidal particles suspended in a fluid where the stationary planar shear flow is realized. The effect of shear flow is represented as the external force acting on each particle depending on its position, and now we do not go into the specifics how the stationary planar shear flow is introduced as

CP982, *Complex Systems, 5th International Workshop on Complex Systems*
edited by M. Tokuyama, I. Oppenheim, and H. Nishiyama

well as in real experiments. The system is confined to a cubic cell with a length L. The x-axis and z-axis are chosen to be the directions of the shear velocity and velocity gradient, respectively. We impose Lees-Edwards periodic boundary conditions [7, 8] to avoid peculiarities near the boundaries of the cell.

We assume that Langevin dynamics can describe the motion of the colloidal particles. Concretely, the force exerted from the fluid is represented by the Stokes force and Gaussian noise. In other words, we neglect the so-called hydrodynamic effects. Then, the particle positions $r_i(t) = (x_i(t), y_i(t), z_i(t))$, where $1 \leq i \leq N$, obey the Langevin equations

$$\eta \frac{\mathrm{d}r_i}{\mathrm{d}t} = -\sum_{j \neq i} \nabla U(|r_i - r_j|) + \eta \dot{\gamma} z_i(t) e_x + \xi_i(t), \quad (1)$$

where η is a friction coefficient; $\dot{\gamma}$, the shear rate; and e_x, the unit vector that is parallel to the x-axis. The variable $\xi_i(t) = (\xi_i^x(t), \xi_i^y(t), \xi_i^z(t))$ represents the Gaussian noise that satisfies

$$\left\langle \xi_i^\alpha(t) \xi_j^\beta(t') \right\rangle = 2\eta k_\mathrm{B} T \delta_{ij} \delta^{\alpha\beta} \delta(t - t'). \quad (2)$$

Here, k_B is the Boltzmann constant and T is the temperature. The superscripts α and β represent the Cartesian components. Each pair of particles interacts via a screened Yukawa potential

$$U(r) = \begin{cases} U_0 \sigma \dfrac{\exp(-\kappa(r-\sigma))}{r} - U_{r_\mathrm{c}}, & \text{if } r \leq r_\mathrm{c}, \\ 0, & \text{otherwise.} \end{cases} \quad (3)$$

with

$$U_{r_\mathrm{c}} = U_0 \sigma \frac{\exp(-\kappa(r_\mathrm{c} - \sigma))}{r_\mathrm{c}}, \quad (4)$$

where r is the distance between the particles; r_c, the cutoff length that simplifies the calculation in numerical simulations; and κ, a Debye screening parameter. U_{r_c} is introduced for avoiding the anomaly of energy caused by the cutoff length r_c of the interaction potential.

In this study, all the quantities are converted to dimensionless forms by setting $\sigma = U_0 = \eta / k_\mathrm{B} T = k_\mathrm{B} = 1$. We estimate the correspondence between these parameters and those of real experimental systems to be $\sigma \sim 10^2$ nm, $U_0 \sim 10 k_\mathrm{B} T$ and $k_\mathrm{B} T / \eta = 10^{-11} \mathrm{m}^2 \mathrm{s}^{-1}$. In our simulation, we assume that $\kappa\sigma = 5.8$, $L/\sigma = 10$, $r_\mathrm{c} = 2.5$ and $N\sigma^3/L^3 = 1$. In other words, the Debye screening length κ^{-1} corresponds to 17 nm. The typical values of the parameters used in our simulations are $T \sim 0.1$ and $\dot{\gamma} = 0.001$, and this situation corresponds to experimental systems wherein the temperature is 300 K and the shear rate is 1 s^{-1}. The systems that have the values are available by laboratory experiments.

In our simulations, we discretized (1) with the time step $\Delta t = 0.0025$. Note that in the arguments below, $\langle \cdots \rangle$ represents the statistical average in the steady states. In the calculation we performed, we estimated $\langle A \rangle$ to be $\int_{t_0}^{t_0+\tau} dt A(t)/\tau$, where t_0 and τ were chosen as that larger than 10^3, because we had confirmed that the relaxation time is approximately 10^2.

RESULTS

In order to characterize the order-disorder transition under shear flow, we focus on the dynamical feature relating on the structure of the particles. Thus, as the order parameter, we investigate the time-dependent value $s_\mathrm{m}(t)$ of the first maximum of the structure factor determined by a configuration of the particles $\{r_i(t)\}$ with time t. Concretely, in our simulation, $s_\mathrm{m}(t)$ is defined as follows. First, we measure the radial distribution function $g(r; \{r_i\})$ for a given configuration $\{r_i\}$ by the relation

$$\hat{g}(r; \{r_i\}) = \frac{1}{4\pi\rho r^2} \frac{\mathrm{d}n(r)}{\mathrm{d}r}. \quad (5)$$

Here, $\mathrm{d}n(r)$ is the average number of particles at distances between r and $r + \mathrm{d}r$ from any particle, and its average is taken over all the particles. Using $\hat{g}(r; \{r_i\})$, we express

$$s_\mathrm{m}(t) = \max_k S_\mathrm{L}(k), \quad (6)$$

with

$$S_\mathrm{L}(k) = 1 + 4\pi\rho \int_0^{L/2} \mathrm{d}r\, (\hat{g}(r; \{r_i\}) - 1) \frac{\sin(kr)}{kr} \frac{\sin(2\pi r/L)}{2\pi r/L}. \quad (7)$$

The term $\sin(2\pi r/L)/(2\pi r/L)$ appended in the integrand is the window function to reduce the termination effects resulting from the finite upper limit [9, 10].

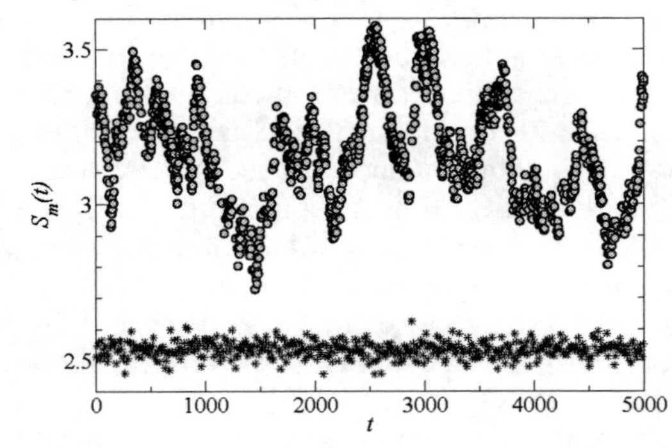

FIGURE 1. $s_\mathrm{m}(t)$ for $T = 0.14$ (open circles) and $T = 0.18$ (stars).

In figure 1, typical data of $s_\mathrm{m}(t)$ in the cases of two different temperatures are displayed. We can see apparently in this figure that the long-time average of $s_\mathrm{m}(t)$ ($\langle s_\mathrm{m}(t) \rangle \sim 3.2$) in the low temperature case ($T = 0.14$) is higher than that ($\langle s_\mathrm{m}(t) \rangle \sim 2.55$) in the high temperature case ($T = 0.18$). With reference to equilibrium cases, the value 3.2 in the low temperature case is higher than the value 2.85 that is regarded as the criterion of the crystallization called Hansen-Verlet's rule [11]. Thus, we can conclude that the system in $T = 0.14$ at $\dot{\gamma} = 0.001$ is in the highly ordered state than the that in $T = 0.18$. Moreover, we can observe that $s_{\rm m}(t)$ exhibits a significantly larger fluctuation in the low temperature case ($T = 0.14$) as compared to that in the high temperature case ($T = 0.18$). In order to characterize the difference between the two cases quantitatively, we consider the spectra of the fluctuations, which are defined by

$$\tilde{S}_m(\omega) = \left\langle \left| \int_{-\infty}^{\infty} \mathrm{d}t\, s_{\rm m}(t) \exp(-i\omega t) \right|^2 \right\rangle. \quad (8)$$

Note that $\langle \cdots \rangle$ represents the statistical average in steady states.

In figure 2, $\tilde{S}_m(\omega)$ for three different temperatures with the fixed shear rate $\dot{\gamma} = 0.001$ is showed. The shapes of the power spectra shown in figure 2 indicate a distinct transition at a certain temperature $T_{\rm c}$. Indeed, for the temperatures $T = 0.14$ and 0.16, the power-law behaviour $\tilde{S}_m(\omega) \simeq \omega^{-2}$ is observed in the frequency regime $\omega \geq 2\pi\dot{\gamma}$. Moreover, focusing on the behaviour in the low frequency regime $\omega \leq 2\pi\dot{\gamma}$, we observe the $1/f$-type fluctuation. Meanwhile, the spectrum becomes flat at the temperature $T = 0.18$.

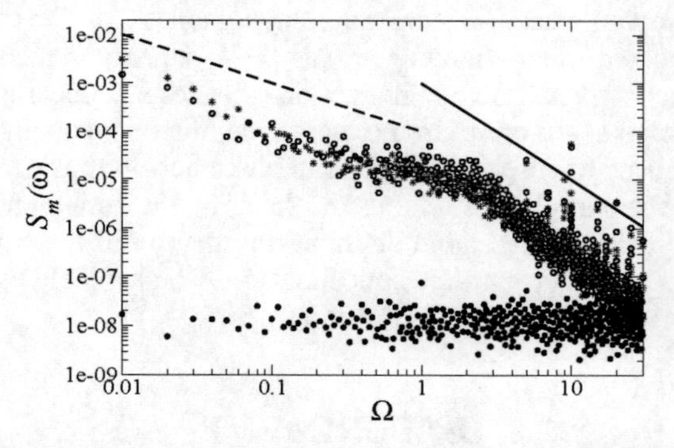

FIGURE 2. Spectra $\tilde{S}_{\rm m}(\omega)$ as a function of $\Omega = \omega/2\pi\dot{\gamma}$. $T = 0.14$ (open circles), 0.16 (stars) and 0.18 (closed circles). The solid line represents the ω^{-2} slope, and the dashed line denotes the ω^{-1} slope.

CONCLUSIONS

We have characterized the order-disorder transition of colloidal suspensions under shear flow by employing dynamical features of the structure factor because there exists no 'crystal' under the shear flow. In order to understand the ordered phase, we have defined a new order parameter $\tilde{S}_{\rm m}(\omega)$, and we have found power-law fluctuations which are observed only in the ordered phase. In the disordered (high temperature) case, such power-law fluctuations disappear, and only white-noise type fluctuations can be observed. In this manner, we conclude that we can detect the order-disorder transition by measuring $\tilde{S}_{\rm m}(\omega)$.

ACKNOWLEDGMENTS

We thank K. Kaneko and K. Hukushima for their helpful comments. This work was supported by a grant (No. 19540394) from the Ministry of Education, Science, Sports and Culture of Japan.

REFERENCES

1. M.J. Miyama and S. Sasa, arXiv:0706.4386 (2007).
2. B.J. Ackerson and N.A. Clark, *Phys. Rev. Lett.*, **46**, 123-126 (1981).
3. A. Tsuchida, E. Takyo, K. Taguchi and T. Okubo, *Colloid Polym. Sci.* **282**, 1105-1110 (2004).
4. P. Holmqvist, M.P. Lettinga, J. Buitenhuis and J.K.G. Dhont, *Langmuir*, **21**, 10976-10982 (2005).
5. S. Butler and P. Harrowel, *J. Chem. Phys.* **103**, 4653-4671 (1995).
6. S. Butler and P. Harrowel, *Phys. Rev. E* **52**, 6424-6430 (1995).
7. A.W. Lees and S.F. Edwards, *J. Phys. C* **5**, 1921-1928 (1972).
8. D.J. Evans and G.P. Morriss, *Statistical Mechanics of Nonequilibrium Liquids*, Academic Press, New York, 1990, pp. 133-146.
9. E.A. Lorch, *J. Phys. C* **2**, 229-237 (1969).
10. G. Guitiérrez and J. Rogan, *Phys. Rev. E*, **69**, 031201 (2004).
11. J.-P. Hansen and L. Verlet, *Phys. Rev.*, **184**, 151-161 (1969).

Colloidal Phase Behavior in a Binary Mixture

H. Guo*, J. Rasink*, P. Schall*, T. Narayanan†, and G. H. Wegdam*

*Van der Waals-Zeeman Institute, University of Amsterdam, Valckenierstraat 65, Amsterdam, The Netherlands
†European Synchrotron Radiation Facility, F-38043, Grenoble, France

Abstract. We present the observation of fluid-solid phase transition in a close density matched system of charge stabilized polystyrene spheres suspended in a binary mixture. The microstructure and dynamics of crystallization are studied by Small Angle X-ray Scattering (*SAXS*) and real space imaging in the confocal microscope. The reversible phase transitions are induced by using temperature as control parameter. The structure factor from SAXS can characterize the solid phase as fcc crystal; the radial distribution function of particles from confocal microscope can determine the particle pair potential in the dilute system in the qualitative manner. The colloids in a binary liquid mixture offer the opportunity to manipulate the particle interactions to understand the nature of weakly attractive systems.

Keywords: Colloids, Crystallization, Pair potential
PACS: : 82.70.Dd

INTRODUCTION

The structure and dynamics of colloidal systems interacting via an attractive potential is a topic interest. Such colloidal systems are known to assume phases that can be classified by comparison with the phases displayed by atoms in various states of matter. An attractive interaction with controllable magnitude and range can be induced in several different ways, most notably by adding a non-adsorbing polymer to a hard-sphere repulsive colloidal system[1]. Charge stabilized colloidal particles suspended in certain binary mixtures provide an alternative route for tuning the interaction in a controllable and reversible manner [2, 3]. This reversible aggregation of colloids near the phase separation temperature T_{cx}, was first observed by Beysens and Esteve more than twenty years ago[4]. They suggested that the aggregation is induced by the preferential adsorption of one of the liquid components onto the colloid surface near T_{cx}. The true nature of the interaction leading to aggregation is still under the discussion. Clearly, the adsorption layer makes the effective interaction between the colloids strongly temperature dependent. It leads to a solvent mediated attractive interaction against the repulsive screened coulomb interaction of the charge stabilized colloids. A rich variety of phase diagram behavior is expected which remains still largely unveiled. The attractive aspect of the solvent mediated interaction contributes not just the control of the colloidal interaction but also the use of temperature as a continuous control parameter.

In this paper we report Small Angle X-ray Scattering and Confocal microscopy measurements on the charge stabilized polystyrene spheres suspended in quasibinary system, 3-methylpyridine, water and heavy water. We have chosen this system because the density of the polystyrene particles can be matched with the density of solvent as much as possible in the relevant temperature range by tuning the ratio of heavy water and water. The "aggregation" can now be studied without gravity pulling and distorting the aggregated state. In the q space, the evolution of microstructure is illustrated by the change of structure factor. The face centered cubic crystal has been observed when the temperature approaches T_{cx}. The evolution of microstructure reveals the systematic development of the attractive interaction prior to the appearance of crystal structure. In the real space, the image analysis are used to observe the colloidal phase behavior as we alter the particle interactions by changing the temperature. We measure the radial distribution function of the particles at the low particle concentration, which enable us to calculate the potential of mean force[5].

EXPERIMENT

Our system consists of charge stabilized polystyrene spheres suspended in a mixture of 3-methylpyridine(3MP), water and heavy water with weight fractions of C_{3MP}=0.25, C_{H_2O}=0.5625, C_{D_2O}=0.1875. For this liquid mixture without the colloids the phase separation takes place at 65°C [3]. The water to heavy water ratio is chosen such that the density of particles and liquid is closely matched in the temperature region where the aggregation process takes place. For the SAXS measurement, the particles have a diameter of 105 nm, and an effective surface charge density - 0.04 $\mu C/cm^2$, measured by electrophoresis. The 3-methylpyridine and heavy water were purchased from Aldrich. The water is deionized Millipore water(MilliQ).

CP982, *Complex Systems, 5th International Workshop on Complex Systems*
edited by M. Tokuyama, I. Oppenheim, and H. Nishiyama

The samples are made with volume fractions of the polystyrene particles: ϕ = 0.0025, 0.005, 0.010 and 0.015.The sample without the particle has considered as reference. The samples are put into flat borosilicate capillaries with an optical path length of 0.5 mm, which are flame sealed to avoid solvent evaporation. The glass surfaces are thoroughly cleaned with chromic acid and rinsed with Millipore water to guarantee a clean OH surface and avoid contaminations that might act as preferential nucleation sites.the capillaries are mounted inside a thermostatted oven, which ensures a temperature stability better than 2 mK. For the confocal microscopy measurement, we use a suspension of fluorescent polystyrene latex spheres (*PLS*) of diameter $\sigma = 2.0\ \mu m$ and fluorescent silica spheres of diameter $\sigma = 1.6\mu m$ in the same composition of quasi-binary solvent mixture of 3-methylpyridine (3MP), water, and heavy water. Our suspensions typically have a volume fraction ϕ =0.001 and 0.0025. The SAXS measurements were performed at the High Brilliance beam line(ID02) at the European Synchrotron Radiation Facility in Grenoble. Typical SAXS profiles after background substraction before and after the transition are shown in Fig1. The images are taken from a Zeiss Axiovert 200 confocal microscope. The objective of the microscope, either an EC Plan-Neofluar 100x / 1.30 Oil or a Plan-Apochromat 63x / 1.40 Oil, is aimed at a stage where a sample cell containing a colloidal suspension is glued to a Peltier element. The Peltier allows temperature control to within 0.1 K. The stage can be moved freely in the plane above the objective, and the objective itself can be moved vertically, so that any point in the sample cell can be observed.

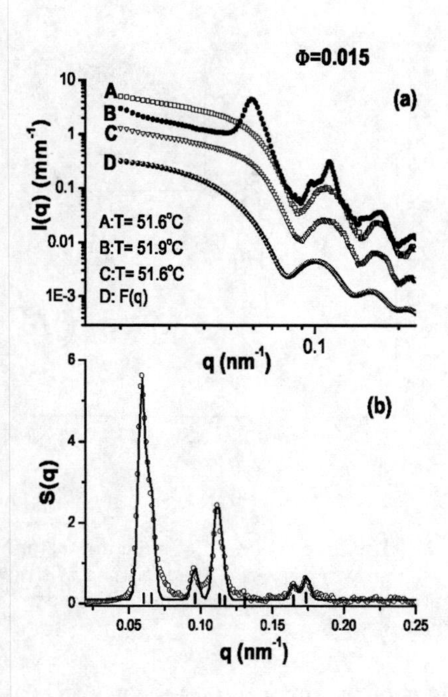

FIGURE 1. (a) Temperature evolution of SAXS intensity profile in the vicinity of T_a. The different scattering curves have been displaced for the sake of clarity. (b)The structure factor corresponding curve C together with a model curve for fcc crystals

RESULTS

Typical intensity profile $I(q)$ before and after the phase transition is shown in Figure1. The spectrum A taken at T=51.6 °C shows only a weak modulation of the form factor (*curveD*), typical for a dilute gas phase. The measured form factor is determined by the spectrum of volume fraction of particle 0.0005. Within 0.3 degree we observe the appearance of several orders of pronounced peaks, which indicate a phase transition from gas to solid (B). We obtain the structure factor of the solid phase by dividing the scattering function B by the measured form factor. The result is shown in fig1(b). We observe peaks characteristic for a crystalline solid. Firstly, the height of the first maximum of S(q) is above the static freezing criterion (~ 2.85) [6]. The peaks are located at q-values in the ratios $\sqrt{3} : \sqrt{4} : \sqrt{8} : \sqrt{11} : \sqrt{12}$. These peak positions are characteristic of the face centered cubic (fcc) lattice. From the absolute values of q we calculate the lattice spacing to be 184 nm. Fitting the peaks with Gaussian functions with a full width at half maximum of Δq =0.005nm^{-1}. This corresponds to a size of the crystalline domains (l_c) of $l_c \approx 2\pi/\Delta q$ = 1250 nm. The crystalline grains are preferentially oriented with their [111] planes parallel to the glass wall. In the 2-D images of the scattered intensity in the q_x - q_y plane rings consisting of a mixture of speckles and Laue spots with hexagonal symmetry. The crystallization is reversible process, the crystal can be melt when we decrease the temperature. Spectrum C taken when we cool the sample back to T=51.6°C,the same shape as before crystallization. The reversible nature of this transition suggests that it occurs at the secondary minimum of the interaction potential as opposed to the primary minimum in conventional colloidal aggregation.

For a quantitative evaluation of the crystal growth process, we keep the temperature constant as soon as the crystal peaks appear, and record diffraction patterns every 40 seconds. In fig2 we show four structure factors during the growth process. The peaks grow in height, while the width and the position remain same. In the inset we display $S(q_m)$ as a function of time. The amplitude increases continuously until it saturates. The characteristic time (τ, in which the amplitude reaches its satura-

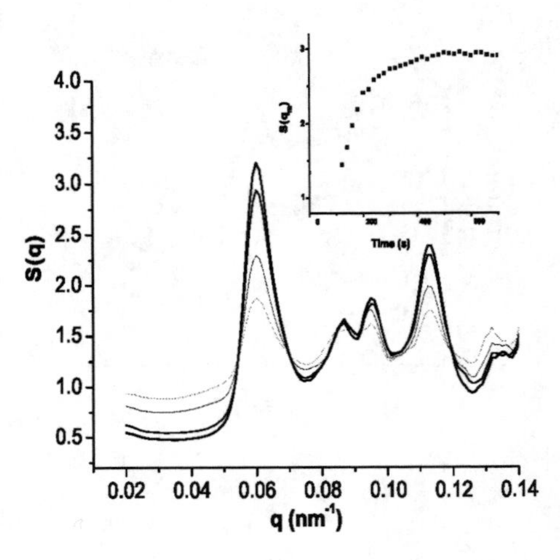

FIGURE 2. The time evolution of structure factor $S(q)$ during the crystal growth process. The amplitude of structure factor $S(q_m)$ progressively increases as function time in the inset.

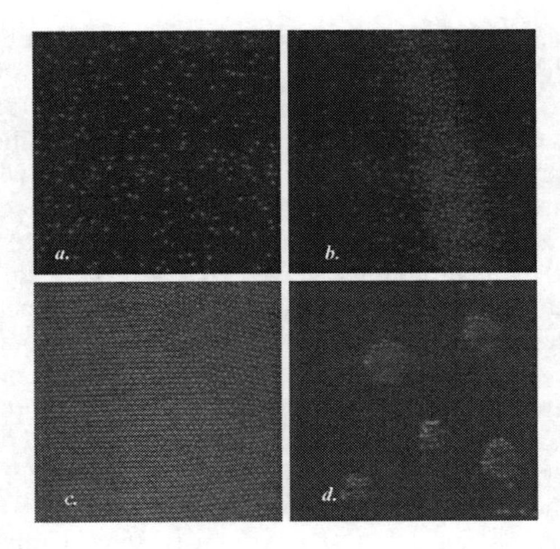

FIGURE 3. **a.** At 20°C, a homogeneous suspension, **b.** At 48°C,the system undergoes some changes.This cluster of particles was moving rapidly through the plane. **c.**At 54°C, we observe the formation of what appear to be two-dimensional crystal structures. **d.** After phase separation, droplets are formed. Taken at 64.5°C.

tion value is of the order of 600 seconds. The diffusion constant measured by dynamic light scattering is D = $4.6\mu m^2/s$. The time τ then corresponds to a length scale for several millimeters. Assuming that the diffusion limited growth process, the numbers seem to be reasonable.

Figure3 shows the confocal microscope images of colloidal phase behavior in a binary mixture as function of temperature. We choose silica particle image as representative sample to present the real space study. As we increase the temperature from 20°C to 47°C, we observe no significant change in the behavior of our particles in the horizontal plane. However, while we increase the temperature, the particles under the microscope go out of focus, we need to lower the focus position to get a sharp image. Apparently the distance of the plane to the capillary wall becomes smaller as temperature increases, this is an indication of 3MP adsorption onto the particles and the capillary wall, screening the Coulomb repulsion, and reducing the separation at which it balances with the gravitational force.
At 48°C, the colloidal system separates into relatively dense and dilute phases. The transition is sudden, with groups of particles contracting to clusters that move around in the plane before dispersing again. After several minutes, the particles settle into a new equilibrium.
Around 54°C, in the dense areas, we observe the formation of what appear to be two-dimensional crystalline structures. As we increase temperature further, we observe more layers being added to the structures, making them three-dimensional. This indicates that the attraction between the particles becomes significant compared to the forces confining the particles to one plane. As we increase temperature beyond 64.5°C, the phase separation temperature of the binary liquid mixture. We observe $(Fig3d)$the formation of spherical aggregates of colloids or droplets.

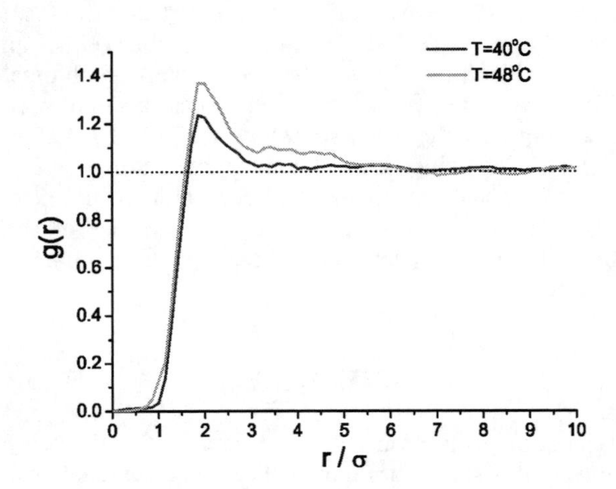

FIGURE 4. Measured $g(r)$ for particles in binary solvent, at different temperatures corresponding to the images above.

We analyze our data in the IDL environment.The pair distribution function $g(r)$ from this series of measurements are shown in fig4. We observe a phase separation at 48°C, this change is reflected in our $g(r)$ measurement. This change in $g(r)$ enabled us to qualitatively observe a phase transition at the same temperature where we observed the phase transition occurring from the image. By tuning temperature, we are able to play with interaction between the particles to induce the phase transition. $g(r)$ can be quantitative measurement to calculate the potential.

We have observed the formation of solid colloidal phase in equilibrium with a dilute gas phases as temperature control parameter. This transition is fully reversible. The interparticle interaction in the close vicinity of the transition can be shown by the $g(r)$. The solid phase can be characterized with face center cubic crystal. The dynamics of crystallization is determined as the diffusion limited mechanism. The temperature in this model plays the key role to tune the interaction. On the one hand, it is used to modify the adsorption process to change the screening length. On the other hand,the correlation length of binary mixture also change as function of temperature to tune the attractive potential induced by the solvent mediated effect. It is a good candidate system to study weakly attractive interaction.

ACKNOWLEDGMENTS

The research has been supported by the Foundation for Fundamental Research on matter (FOM) ,which is financially supported by Netherlands Organization for Scientific Research (NWO).We thanks ESRF for the beamtime and technical support.

REFERENCES

1. P.N.Pusey ,in Liquids,Freezing and Glass transition,edited by J.-P.Hansen, D.Levesque,and J.Zinn-Justin (North-Holland, Amsterdam, 1991), P.763
2. D.Beysens, D.Esteve, Phys Rev.Lett,**54** 2123-2126 (1985).
3. T.Narayanan, Phys. Reports , **249** 135-218 (1994).
4. D.Beysens , T.Narayanan, Journal of Statistical Physics. **95** 997-1008 (1999).
5. Y.L.Han, D.G.Grier, Phys.Rev.lett,**91** 038302-038305 (2003).
6. H.Lowen, Phys.Rep. **237**, 249-324 (1994).

Dielectric Relaxation Phenomena in Hydrogen-Bonded Liquids

Yoshiki Yomogida and Ryusuke Nozaki

Department of Physics, Faculty of Science, Hokkaido University
Sapporo 060-0810, Japan

Abstract. The complex permittivities of some monohydric alcohols and dihydric alcohols have been measured in the frequency range from 500 MHz to 20 GHz at temperatures between 273 K and 313 K. Analysis of dipole-dipole correlation factors indicates that cluster structures formed through intermolecular hydrogen bonds exist in these liquids and the structures of diols are slightly different from those of mono-alcohols. The steric hindrance in molecules has a significant influence on the main dielectric relaxation process, while the number of OH groups is not an important factor. The relaxation time of the main process is not related to the macroscopic viscosity.

Keywords: Dielectric relaxation, Dynamical structure
PACS: 61.25.Em, 77.22.Gm

INTRODUCTION

Hydrogen-bonded liquids have been extensively studied because of their interesting behavior and important roles in many fields. Monohydric alcohols (mono-alcohols) and dihydric alcohols (diols) are useful in the study of hydrogen-bonded liquids because they can create convenient conditions by changing the number of OH groups, length of the carboxyl chain and their structures. Dielectric spectroscopy, which measures the response of the complex permittivity to time-dependent electric fields, is one of the most suitable methods for investigating the molecular dynamics of liquids and has been used on mono-alcohols and diols and their mixtures [1-9]. It is expected that the hydrogen bonds form cluster structures and the molecular dynamics involves in collective motions within large cluster in these liquids. Our aim is to obtain a reasonable description of dielectric relaxation based on the collective reorientation of these molecules.

In this study, firstly, the complex permittivities of some n-alcohols (1-propanol, 1-butanol, and 1-pentanol) and diols (1,3-propanediol, 1,4-butanediol, and 1,5-pentanediol) with OH groups at the chain ends, were measured to obtain information on the influence of the number of OH groups on the dielectric spectra. Then, dielectric measurements were performed on 2-butanol, 2-methyl-1-propanol 1,2-butanediol, and 1,3butanediol, which have three carbon atoms, to study the effect of the steric hindrance in the molecules. Finally, their dielectric properties are discussed.

EXPERIMENTAL

The chemicals obtained from Kishida and Kanto Chemical were used as received. The mono-alcohols used were 1-propanol, 1-butanol, 2-butanol, 2-methy-1-propanol, and 1-pentanol and the diols used were 1,3-propanediol, 1,2-butanediol, 1,3-butanediol, 1,4-butanediol, and 1,5-pentanediol.

In order to cover the frequency ranges between 1 MHz and 20 GHz, two different experimental setups were used. From 1 MHz to 500 MHz, the complex permittivity measurements were performed by employing a network analyzer (HP4195A). The coaxial sample cell was composed of an outer conductor with an inner diameter of 3.5 mm and a center conductor with an outer diameter of 2 mm. The length of the outer conductor was 10 mm. The values of the complex permittivity were obtained from reflection measurements with the sample cell located at the end of a coaxial line. From 500 MHz to 20 GHz, time domain reflectometry [10,11] was employed. A flat-end capacitor cell using a 2 mm semirigid coaxial waveguide was used in this measurement. The temperature of the sample cells in both experimental setups was controlled between 273 K and 313 K with an accuracy of 0.1 K by using a water jacket.

CP982, *Complex Systems, 5th International Workshop on Complex Systems*
edited by M. Tokuyama, I. Oppenheim, and H. Nishiyama

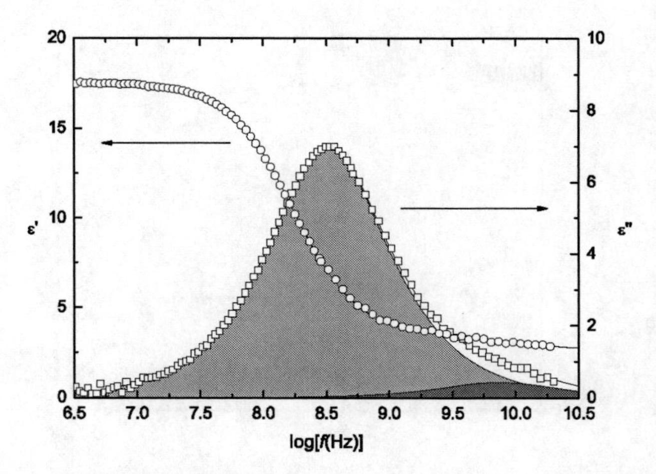

FIGURE 1. Complex permittivity of 1-butanol at 298K

We also measured the viscosity using Canon-Fenske viscometers at 298K.

RESULTS

Figures 1 and 2 show the complex permittivities of 1-butanol and 1,4-butanediol at 298 K. The superposition of the Debye function [12] for the main relaxation process and the Cole-Cole function [13] for the high-frequency process could describe the complex permittivities of all materials measured in this study. In this case, the complex permittivity is given by

$$\varepsilon^*(\omega) = \frac{\Delta\varepsilon_1}{1+i\omega\tau_1} + \frac{\Delta\varepsilon_2}{1+(i\omega\tau_2)^{\delta_2}} + \varepsilon_\infty . \quad (1)$$

The parameter $\Delta\varepsilon$ is the relaxation strength, τ is the relaxation time, and δ is the shape parameter that represents the systematic broadening of the relaxation time. ε_∞ is the high-frequency limiting permittivity and ω is the angular frequency. The subscripts 1 and 2 represent the main and additional high-frequency relaxation processes, respectively. However, further investigations are required to estimate the high-frequency process accurately and this topic is not discussed in this paper.

Other options were also examined to describe the dielectric spectra, which are not given in detail for the sake of simplicity. The Davidson-Cole function does not adequately fit the data of all the samples.

DISCUSSION

Figures 3 and 4 show the Arrhenius diagrams of the main relaxation process. According to Eyring's

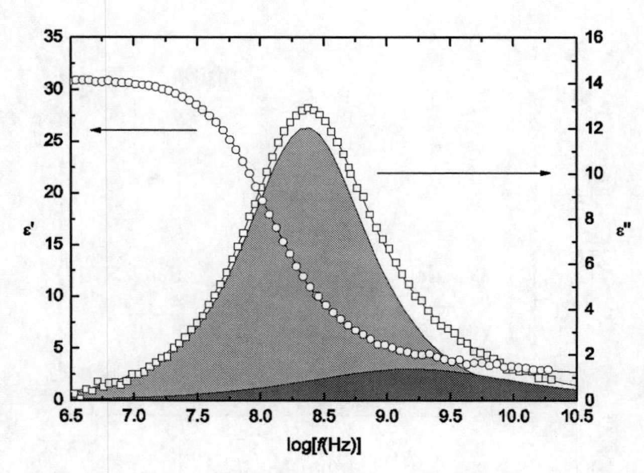

FIGURE 2. Complex permittivity of 1,4-butanediol at 298K

theory of rate process, the relaxation frequency f_m is given by

$$\log f_m = \log f_0 + \frac{\Delta E}{2.303RT} . \quad (2)$$

Here, R is the gas constant, T is the absolute temperature, ΔE is the activation energy associated with the dipole orientation, and f_0 represents the attempt frequency at $T \to \infty$. The activation energy of the main process is determined by Eq. (4) and the values are given in Table 1. The ΔE of the diols is slightly larger than that of the corresponding mono-alcohols. For example, ΔE of 1-butanol is 31.39 kJ/mol and that of 1,4-butanediol is 36.61 kJ/mol. On the other hand, when the data of the samples that have three carbon atoms are compared, it is found that the steric hindrance in molecules is a very important factor for the dipole orientation in these liquids, and the molecules that have a straight shape have low activation energy. For example, ΔE of 2-butanol is 40.37 kJ/mol and that of 2-methyl-1-propanol is 35.55 kJ/mol.

TABLE 1. Static permittivity ε_s of all samples at 298 K and Activation energy of main dielectric process.

Sample	ε_s	ΔE [kJ/mol]
1-propanol	20.56	29.91
1-butanol	17.52	31.39
2-butanol	16.53	40.37
2-methyl-1-propanol	17.29	35.55
1-pentanol	15.02	33.61
1,3-propanediol	34.66	32.63
1,2-butanediol	22.25	44.49
1,3-butanediol	28.56	43.27
1,4-butanediol	31.14	36.61
1,5-pentanediol	26.47	38.98

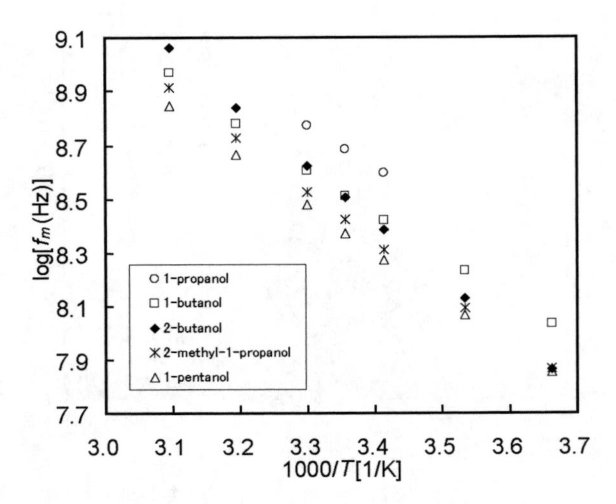

FIGURE 3. Arrhenius diagram for mono-alcohols

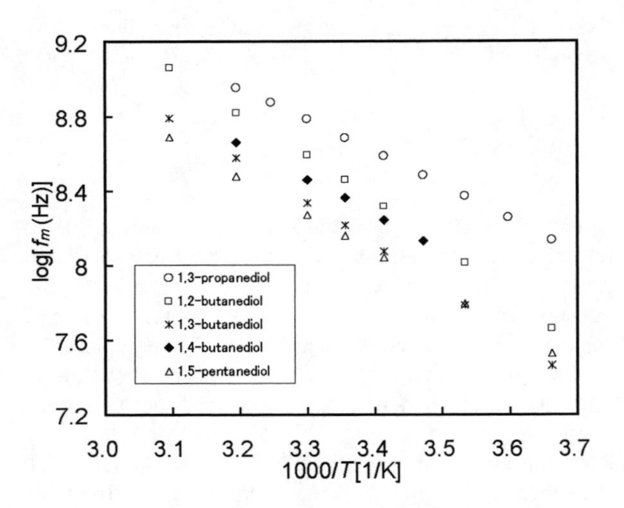

FIGURE 4. Arrhenius diagram for diols

The Fröhlich theory [14] expresses the static permittivity ε_s as follows:

$$\varepsilon_s - \varepsilon_0 = \frac{3\varepsilon_s}{2\varepsilon_s + \varepsilon_\infty}\left(\frac{\varepsilon_\infty + 2}{3}\right)^2 \frac{N_A c g \mu^2}{3\varepsilon_0 k_B T}. \quad (3)$$

Here, ε_∞ is the high frequency permittivity, N_A is Avogadro's number, k_B is the Boltzmann constant, c is the concentration of the dipole molecules, μ is the electric dipole moment in the gaseous state, g is the dipole-dipole correlation factor, and ε_0 is the vacuum permittivity. The dipole-dipole correlation factor can be calculated using Eq. (3). However, there is no available electric dipole moment of diols in the gaseous state. In this case, we use the value 2.52 D for 1,3-propanediol, 2.50 D for 1,4-butanediol, and 2.45 D [15] for 1,5-pentanediol in the nonpolar solvent. It

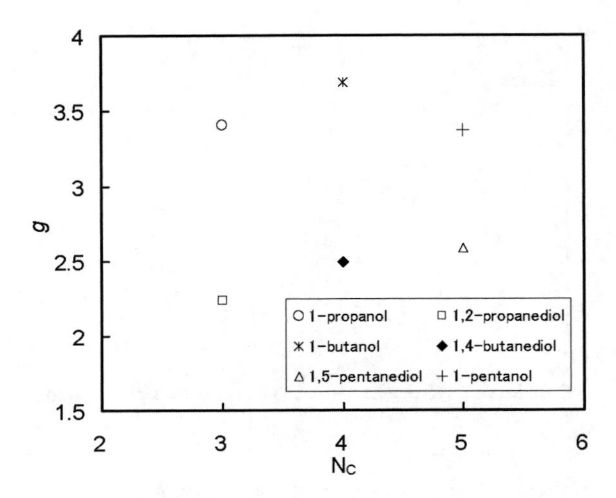

FIGURE 5. Dipole-dipole correlation factor g plotted against number of carbon atoms N_C of mono-alcohol and diol molecules.

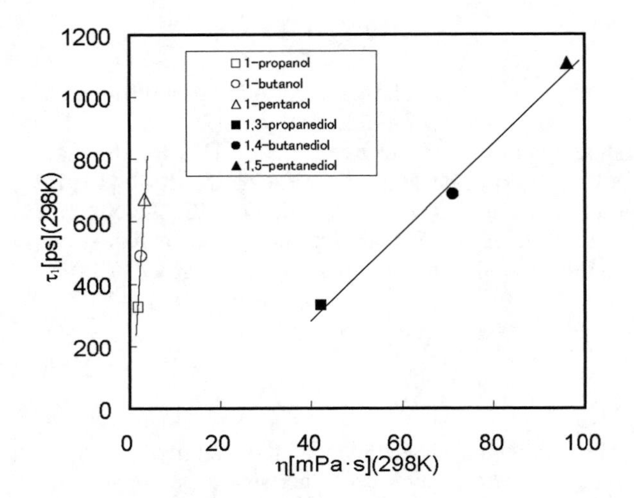

FIGURE 6. Relaxation time τ_1 plotted as a function of viscosity measured at 298K

is reported that the dipole moment of ethylene glycol in the gaseous state is almost the same as that in the nonpolar solvent [15]. In the calculation of g, ε_∞ is set equal to n^2, where n is the refractive index. Although the values of g thus obtained are not very accurate, we can obtain the relative values and discuss their tendency.

Figure 5 shows the dependence of the g factor on the number of carbon atoms of mono-alcohol and diol molecules. It is clear that there is a local structure and parallel alignment of the dipole moments of these molecules. Monohydric alcohol is reported to have a chain-like cluster connected with hydrogen bonds [16]. In the case of diols, the hydrogen-bonded cluster might have some branches because a molecule may have two hydrogen bonds. These branches must disturb the parallel alignment of the dipole moments and reduce the dipole-dipole correlation factor. This view is consistent with the facts indicated in Fig. 5.

In Fig. 6, the relaxation times of the main dielectric relaxation process are plotted against the macroscopic viscosity η at 298 K. The viscosities of the diols are much higher than those of the mono-alcohols because the simultaneous breaking of more than one hydrogen bond is required for the translational motion of the diol molecule. On the other hand, the values of the dielectric relaxation time τ_1 at 298 K and the activation energy ΔE of the diols are comparable to those of the mono-alcohols. This result indicates that the OH groups of diols orientate independently.

CONCLUSION

In the present study, complex permittivity measurements of some monohydric and dihydric alcohols have been performed and the data obtained have been analyzed. While mono-alcohols have chain-like clusters and parallel alignment of dipole moments, diol molecules, which have two OH groups, may form network structures through intermolecular hydrogen bonds. In this structure, each OH group in a diol molecule orientates independently and the steric hindrance in a molecule significantly affects the orientation of an OH group.

As mentioned previously, the deviations from the Debye-type relaxation process toward the high frequencies are observed in the complex permittivities of all the samples. Although these features are treated as an additional Cole-Cole-type relaxation process in this study, future investigation on high-frequency regions are necessary to estimate the deviations accurately. However, it is noteworthy that this systematic research provides a considerable amount of information on the molecular dynamics of mono-alcohols and diols.

ACKNOWLEDGMENTS

This study was supported by the 21st Century Center of Excellence (COE) program “Topological Science and Technology” of Hokkaido University from the Japan Society for the Promotion of Science (JSPS) and the Sasakawa Scientific Research Grant from The Japan Science Society.

REFERENCES

1. M. W. Sagal, *J. Chem. Phys.* **36**, 2437-2442 (1962).
2. E. Ikeda, *J. Phys. Chem.* **75**, 1240-1246 (1971).
3. E. Noreland, B. Gestblom, and J. Sjöblom, *J. Solut. Chem.* **18**, 303-312 (1989).
4. J. Barthel, K. Bachhuber, R. Buchner, and H. Hetzenauer, *Chem. Phys. Lett.* **165**, 369-373 (1990).
5. R. Buchner and J. Barthel, *J. Mol. Liq.* **52**, 131-144 (1992).
6. F. Wang, R. Pottel, and U. Kaatze, *J. Phys. Chem. B* **101**, 922-929 (1997).
7. P. Petong, R. Pottel, and U. Kaatze, *J. Phys. Chem. A* **103**, 6114-6121 (1999).
8. S. Schwerdtfeger, F. Köhler, R. Pottel, and U. Kaatze, *J. Chem. Phys.* **115**, 4186-4194 (2001).
9. S. Sudo, N. Shinyashiki, Y. Kitsuki, and S. Yagihara, *J. Phys. Chem. A* **106**, 458-464 (2002).
10. R. Nozaki and T. K. Bose, *IEEE Trans. Instrum. Meas.* **39**, 945-951 (1990).
11. R. Nozaki, *Solid State Phys. (Tokyo)* **28**, 505-513 (1993).
12. P. Debye, Polar molecules, Dover Publications Inc.: New York, 1929.
13. K. S. Cole and R. H. Cole, *J. Chem. Phys.* **9**, 341-351 (1941).
14. H. Fröhlich, Theory of Dielectrics, Claredon: Oxford, 1958.
15. A. L. McClellan, Tables of Experimental Dipole Moments, San Francisco: W. H. FREEMAN AND COMPANY, 1963.
16. A. M. Planner, *Acta. Cryst.* **A33**, 433-437 (1977).

Dynamic Sound Scattering Investigation of the Dynamics of Sheared Particulate Suspensions

Anatoliy Strybulevych*, Tomohisa Norisuye†, Matthew Hasselfield*, and J. H. Page*

*Department of Physics and Astronomy, University of Manitoba, Winnipeg R3T 2N2, Canada
†Department of Macromolecular Science and Engineering, Graduate School of Science & Technology, Kyoto Institute of Technology, Matsugasaki, Sakyo-ku, Kyoto 606-8585, JAPAN

Abstract. We show how Dynamic Sound Scattering (the acoustical analog of Dynamic Light Scattering) can be used to measure the average velocity profile and the particle velocity variance of neutrally buoyant suspensions in the Couette geometry. The motion of the particles was determined from the temporal autocorrelation function of the scattered field fluctuations in a single speckle spot. In contrast to theories that assume isotropic particle velocity fluctuations, it is shown that "suspension temperature" (defined as the average kinetic energy contained in the particle velocity fluctuations) is a tensor, whose components have different magnitudes. For particle volume fractions up to 0.4, components of the fluctuations parallel to the flow and in the vertical direction are much larger than in the radial direction. For higher concentrations the fluctuation anisotropy decreases.

Keywords: Dynamic sound scattering, Particle dynamics, Suspensions
PACS: 43.20.Fn, 43.35.Yb, 47.57.E-, 47.80.Cb, 82.70.-y

INTRODUCTION

The flow properties of suspensions of solid particles in a liquid have been intensively investigated over the last several decades. The flow of concentrated suspensions often produces particle migration, when an initially uniform distribution of the particles becomes non-uniform. In granular-flow based models (e.g. see [1]), migration of the particles is attributed to gradients in so-called "suspension temperature", defined as the average kinetic energy of particle velocity fluctuations due to interparticle collisions. To explore and understand the fluid dynamics of such suspension under shear, accurate measurement of flow velocities is essential. Such measurements need to be non-intrusive, because any probe inside the suspension may perturb the flow or even change the structure.

Laser Doppler velocimetry (LDV) is the most popular non-intrusive technique [2] to measure velocity profiles. Velocity is obtained from the Doppler shift of laser light scattered by seeding particles moving with the fluid. Particle imaging velocimetry (PIV) may also be used for this purpose. However, concentrated suspensions may be not transparent enough to use LDV and PIV. Nuclear magnetic resonance (NMR) imaging technique [3] can be used for opaque media, but this technique is quite expensive and time consuming.

On the other hand, ultrasound may propagate through optically opaque media. When an ultrasonic wave travels through inhomogeneous media, it gets scattered when the acoustical impedances (product of density and sound velocity) of the particles and surrounding fluid are different. An Ultrasonic Velocity Profiler (UVP) has been developed and successfully used for in-line flow rate and velocity profile measurements [4]. The technique is based on time-domain cross-correlation of ultrasonic speckle signals backscattered by the seeded particles.

Another approach, Dynamic Sound Scattering (DSS) (the acoustical analog of Dynamic Light Scattering) utilizes the temporal autocorrelation function of the scattered field fluctuations in a single speckle spot to measure motion of the particles [5]. The field correlation function is defined as

$$g_1(\tau) = \frac{\int \psi^*(T)\psi(T+\tau)dT}{\int |\psi(T)|^2 dT}, \qquad (1)$$

where $\psi(T)$ is the scattered ultrasonic field measured at time T; its decay results from the change in phase of the scattered ultrasonic field due the particle motion. As discussed by Cowan et al. [5], the field correlation function can be written

$$g_1(\tau) = \langle \exp[-i\Delta\phi_p(\tau)] \rangle = \langle \exp[-i\vec{q}\cdot\Delta\vec{r}_p(\tau)] \rangle, \qquad (2)$$

where $\Delta\phi_p(\tau)$ is the phase change of the scattered waves due to the pth particle's motion, $\vec{q} = \vec{k}' - \vec{k} = = 2k\sin(\theta/2)$ is the scattering wave vector, θ is the scattering angle, $\Delta\vec{r}_p(\tau)$ is the change in position of the pth particle during the time interval τ, and $\langle ... \rangle$ denotes time average. Since $\Delta\phi_p(\tau) = \langle \Delta\phi_p(\tau) \rangle + \delta\phi_p(\tau)$, after using a cumulant expansion, the real part of the correlation func-

CP982, *Complex Systems, 5th International Workshop on Complex Systems*
edited by M. Tokuyama, I. Oppenheim, and H. Nishiyama

tion for ballistic particle motion, where $\Delta\vec{r}_p(\tau) = \vec{V}_p\tau$, is

$$g_1(\tau) \simeq \cos[2k\sin(\theta/2)\langle V_{\vec{q}}\rangle\tau] \times \exp[-2k^2\sin^2(\theta/2)\langle\delta V_{\vec{q}}^2\rangle\tau^2]. \quad (3)$$

Thus, the time dependence of the correlation function is determined by the motion of the particles. Since the scattering angle can be controlled by the scattering geometry, and the magnitude of the wave vector k can be measured independently, both the average particle velocity along $\vec{q}$, $\langle V_{\vec{q}}\rangle$, and the variance (or "suspension temperature"), $\langle\delta V^2\rangle = \langle\delta V_{\vec{q}}^2\rangle - \langle V_{\vec{q}}\rangle^2$, can be independently determined by measuring the correlation function.

In this paper, we describe experiments to measure particle dynamics of neutrally buoyant suspensions of uniform glass beads in the Couette geometry using DSS.

MATERIALS AND METHODS

The samples consisted of uniform glass beads (radius 63 $\pm$11 mm, density 2.220 kg/m^3) immersed in a density-matched liquid consisting of a mixture LST (a low viscosity aqueous solution of lithium heteropolytungstates) and water. The measurements were carried out for volume fractions ϕ of 0.2, 0.3, 0.4, 0.45, and 0.5 at a temperature of 27°C. The dynamics were studied under steady shear flow at shear rate,$\dot{\gamma}$, from 0.26 to 8.8 s^{-1} and across the entire gap between the stator and rotor, thereby enabling a complete investigation of the dynamics to be carried out. The experiments were performed using a Couette geometry where the inner cylinder is the rotor, the outer cylinder is the stator, and the sample volume has an inner and outer radius of 54 and 58 mm respectively, corresponding to a gap width, l, of 4 mm. The rotor went straight to the bottom of the cell with a bottom bearing, thereby eliminating any low-shear-rate fluid reservoir at the bottom.

To measure the correlation function in these experiments, the following procedures were followed. The sample cell and ultrasonic transducer were immersed in a water tank, since water is a convenient low-attenuation medium in which the placement of the transducers relative to the cell can be easily controlled. This also allowed precise control of the sample temperature, by regulating the temperature of the water bath. Pulsed experiments were used, in which a regular train of short ultrasonic pulses was sent to the sample, with a repetition time that was generally of order 0.5 - 5 ms depending on the speed of the dynamics. The ultrasonic scattering measurements were performed in reflection mode using a focusing transducer with central frequency of 2.25 MHz, whose orientation relative to the cell was varied. Experiments were carried out for the following orientations (see Fig. 1), so that all three orthogonal components of the particle motions could be measured: radial (beam axis perpendicular to the cell wall), "tangential-radial" (beam axis rotated in the horizontal plane perpendicular to the cylindrical axis (z) with incidence angles $\phi_T = 3, 5$ and 7.5°), and "neutral-radial" (beam axis at $\theta_T = 10$ and 15° to the r-axis in the rz or vertical plane). To obtain more complete information during one run, experiments were performed using the gated digitization feature of a GageScope oscilloscope card, which allows the entire scattered ultrasonic waveform as a function of propagation time inside the cell to be captured. The advantage of this approach is more efficient data acquisition, allowing measurements of the particle motions at all depths inside the cell to be measured simultaneously. The data were then saved to the computer hard drive and the process repeated up to 20 times, depending on the statistics desired. The correlation function was calculated numerically using a FFT approach from the digitized data.

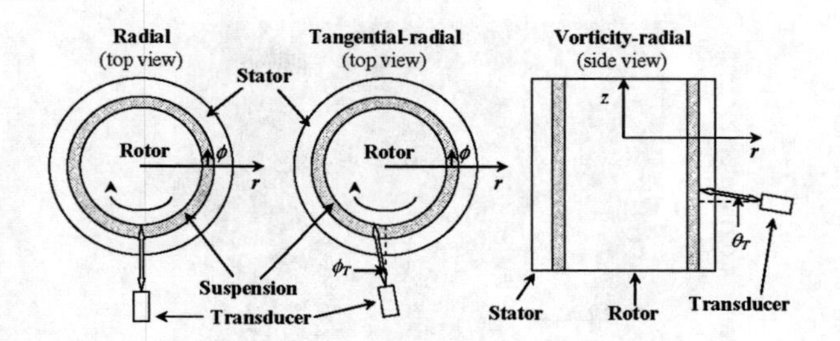

FIGURE 1. Orientations of the transducer for measurements of the three orthogonal components of the particle motions.

RESULTS AND DISCUSSION

An example of the scattered waveforms for a 30 % suspension of glass beads in LST/water, recorded using the GageScope for many repetitions of the input pulse, is shown in Fig. 2. The transducer was oriented in the "tangential-radial" direction and the mean shear rate was 4.145 s^{-1}. The figure shows an image plot of the wave field, with the y-axis representing distance from the stator determined from the pulse propagation time. In the image plot, white indicates wave crests and black represents the wave troughs. Near the stator, the wave crests are almost tangential to the fluctuation time axis, indicating that the phase of the scattered waves barely changes, corresponding to very slow motion of the beads, while deeper into the cell, the wave crests tilt markedly downwards, indicating increasing rapid motion. Note that the white and black bands are not continuous in the image plot, a signature of fluctuations in the particle velocities. This behavior is also captured in Fig. 3, which shows the scattered wave field versus fluctuation time T at three

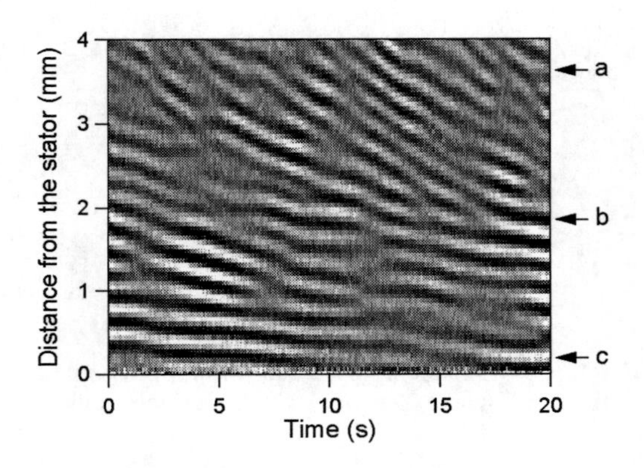

FIGURE 2. Image plot of the scattered wave field as a function of distance inside the cell and fluctuation time T.

depths inside the cell, labeled a, b, and c, and indicated by arrows in Fig. 2. It is evident that the field fluctuations are more rapid near the rotor than the stator, and are neither perfectly regular in T nor completely random. The corresponding field autocorrelation functions at these depths are shown in Fig. 4, where the data have been averaged over 10 sequences similar to those plotted in Fig. 3 to improve the statistics. As the propagation time of the pulse and the depth inside the cell increases, the autocorrelation function decays more rapidly, confirming that the motion of the beads is also more rapid. The shape of the correlation function, a Gaussian decay modulated by a sinusoidally varying term, indicates that the bead motion is one of particles moving predominantly in one direction, with substantial fluctuations about this average velocity. This shape is exactly what is predicted for this type of motion (see Eq. (4)). By fitting Eq. (4) to the measured autocorrelation functions, both the average velocity $\langle V_{\vec{q}} \rangle$ and the velocity variance $\langle \delta V^2 \rangle$ can be accurately determined along the direction given by the scattering wave vector $\vec{q}$.

Figure 5 compares the field autocorrelation functions for three orientations of the transducer: radial, "neutral-radial" with $\theta_T = 15°$, and "tangential-radial" with $\phi_T = 5°$. The correlation functions were measured in the middle of the gap, for the 30% glass bead suspension sheared at $\dot{\gamma} = 4.145\,\mathrm{s}^{-1}$. It is clear from the figure that only in the "tangential-radial" direction is there a negative dip in the correlation function, indicating only in the tangential direction parallel to the flow is the average velocity of the beads non-zero. The solid curves in the figure are fits of Eq. (4) to the data, with $\langle V_{\vec{q}} \rangle = 0$ for both the radial and neutral directions. Excellent fits are obtained, giving reliable measurements of the average particle velocity in the "tangential-radial" direction and the velocity variance along all three measurement directions.

To extract the components of the velocity and variance along the radial, tangential and neutral directions, it is necessary to account for the refraction of ultrasound at the cylindrical cell walls and for the propagation of not only the longitudinal waves in the wall but also shear waves generated by mode conversion at the interface. Since a focusing transducer is used, the input beam contains a range of angles, and so it is also necessary to average over these angles. For a given direction we measure

$$2k\langle \vec{V} \rangle_{meas} = P_r\langle V_r \rangle + P_\phi \langle V_\phi \rangle + P_z\langle V_z \rangle \quad \text{and} \qquad (4)$$

$$4k^2\langle(\delta V^2\rangle_{meas} = Q_r\langle \delta V_r^2 \rangle + Q_\phi \langle \delta V_\phi^2 \rangle + Q_z\langle \delta V_z^2 \rangle \qquad (5)$$

where r, ϕ and z refer to the cylindrical polar coordinates of the cylindrical cell, $\langle \vec{V} \rangle_{meas}$ and $\langle(\delta V^2\rangle_{meas}$ are the fitted values from the correlation functions, assuming backscattering. The parameters $[P_r, P_\phi, P_z]$ and $[Q_r, Q_\phi, Q_z]$ were then calculated for a transducer of specified size, position and orientation, for a cylindrical cell of given acoustic properties and dimensions, and for a sample of known acoustic properties (which depend on particle concentration).

Figure 6 shows the variation $\langle V \rangle$ and $\langle \delta V^2 \rangle$ with distance inside the cell for the 30% suspension obtained from different transducer orientations. The figure shows that the average velocity across a Couette gap at all three orientations is linear, and good agreement between the predictions and experiments are found. Despite the quite wide gap, the shear profile is linear in the cell for the shear rates used in experiments, and there is no evidence of slippage of the suspension near the stator or rotor, since the average particle velocity varied from zero at the stator to a value equal to the wall velocity at the rotor.

The dependence of "suspension temperature" components on volume fraction for four different shear rates is presented in Fig. 7. In comparing the components in

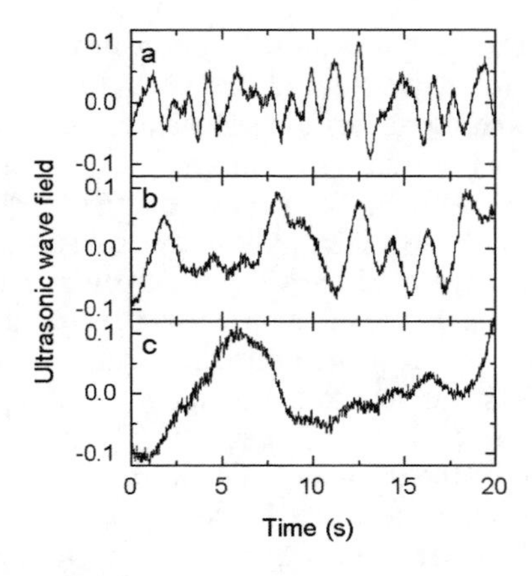

FIGURE 3. Field fluctuations as a function of time at the depths a, b, and c indicated by the arrows in Fig. 2.

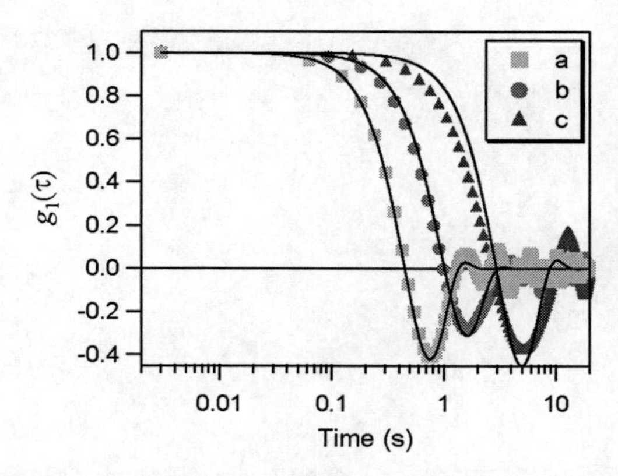

FIGURE 4. Corresponding field autocorrelation functions for the data shown in Fig. 3.

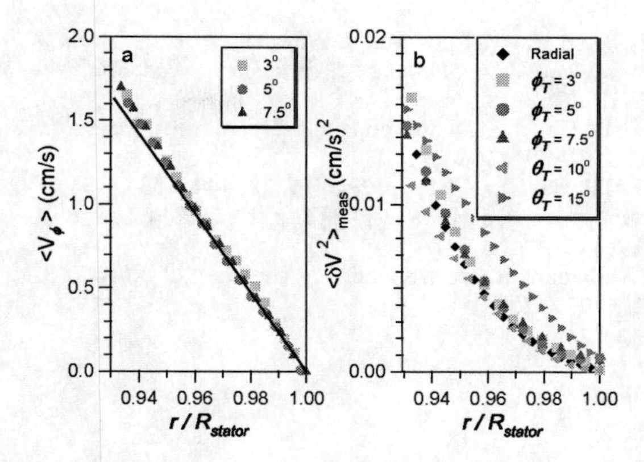

FIGURE 6. Comparison of (a) average velocity and (b) variances profiles for different transducer orientations with theoretical predictions (solid line) for a 30% dispersion.

these four graphs, it is clear that the tangential component at particle volume fractions up to 0.4 is the largest, followed by the vertical and then the radial components. For higher concentrations the fluctuation anisotropy decreases. It is also apparent that each of the three components varies differently with particle concentration. The radial components is almost independent of particle concentration; it appears to increase slightly and then drops significantly for highest concentration. The tangential component monotonically decreases and the vertical component has a maximum at volume fraction of 0.3. Similar behavior was observed by Shapley et al. [6].

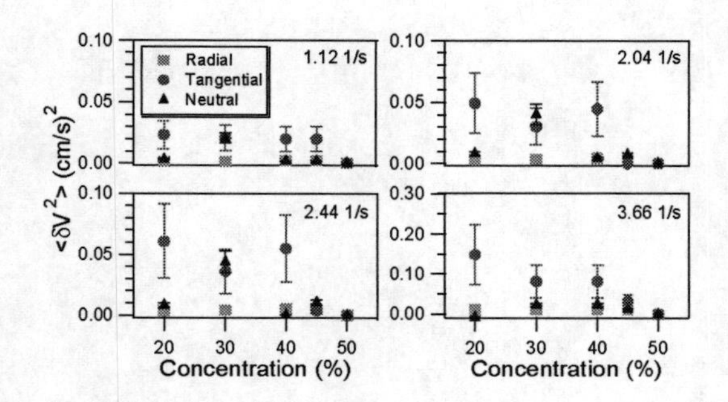

FIGURE 7. Components of the variance as function of particle volume fraction for different shear rate.

CONCLUSIONS

In this paper, we demonstrate that Dynamic Sound Scattering is a powerful new technique for investigating the dynamics of sheared particulate suspensions. It was found that the average particle velocity increases linearly with distance from the stator, indicating that the average motion conforms to simple shear flow. The fluctuations in the particle velocities are quite large, indicating that the particles are not confined to streamlines but continue to fluctuate substantially during steady flow. For particle volume fractions up to 0.4, components of the fluctuations parallel to the flow and in the vertical direction are much larger than in the radial direction. For higher concentrations the fluctuation anisotropy decreases.

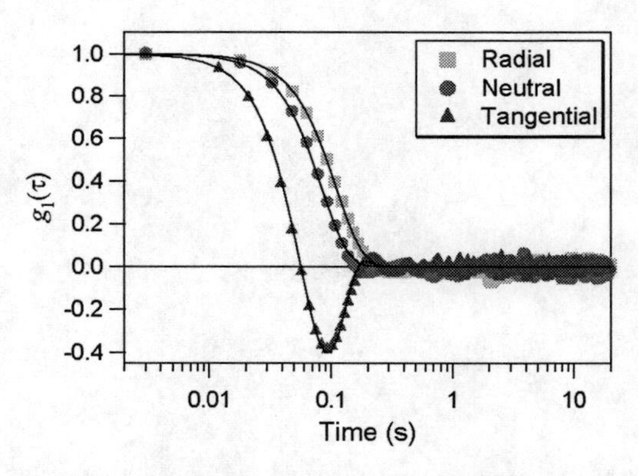

FIGURE 5. Temporal field autocorrelation functions of the scattered fields for the three transducer orientations indicated in the legend.

ACKNOWLEDGMENTS

Support of this research by the Canadian Space Agency is gratefully acknowledged.

REFERENCES

1. P. R. Nott, and J. F. Brady, *J. Fluid Mech.* **275**, 157–199 (1994).

2. F. Durst, A. Melling, and J. Whitelaw, *Principles and Practice of Laser-Doppler Anemometry*, Academic, New York, 1981.
3. A. W. Chow, S. W. Sinton, and J. H. Iwamiya, *Phys. Fluids* **6**, 2561–201 (1994).
4. Y. Takeda, *Exp. Therm. Fluid Sci.* **10**, 444–453 (1995).
5. M. Cowan, J. Page, and D. Weitz, *Phys. Rev. Lett.* **85**, 453–456 (2000).
6. N. Shapley, R. Brown, and R. Armstrong, *J. Rheol.* **48**, 255–279 (2004).

Influence of Hydrodynamics on Relaxation Time of Dynamics in Colloidal Crystal

T. Ueno*, K. Yamazaki*, Y. N. Ohshima†, and I. Nishio*

*Department of Physics and Mathematics, Aoyama Gakuin University, Fuchinobe 5-10-1, Sagamihara, Kanagawa, 229-8558 Japan

†Incubation Center, PENTAX Corporation, Shirako 1-9-30, Wako, Saitama, 351-0101 Japan

Abstract. We analyzed particle motions in colloidal crystals by digital video microscopy, and obtained the dispersion relations of amplitude and relaxation time for normal relaxation modes of particle dynamics. Because the observed crystals were soft, the soft mode which had larger amplitude and longer relaxation time appeared in Γ-M. Such soft mode, however, could not be reproduced at all by a simple overdamped Langevin theory in which the hydrodynamic interaction and long-range interaction were not incorporated. Therefore we attempted to introduce the influence of flow into the equation of motion by using the Oseen tensor. The results of the modified model were in quantitative agreement with the experimental results. We also found that it was essential to consider the negative force constant between 3rd-nearest neighbor particles for explaining the soft mode.

Keywords: Colloidal crystal, Hydrodynamic interaction
PACS: 61.66.-f, 63.20.Dj, 82.70.Dd

INTRODUCTION

Colloidal crystals are known as a visible model of crystals. A fundamental difference between colloidal crystals and atomic crystals, however, is the existence of viscous fluid. Hydrodynamic interaction in colloidal crystal has been investigated by the use of light scattering measurements for a number of years [1, 2, 3]. In recent years, a single particle motion has been analyzed in real space by digital video microscopy [4, 5, 6]. To explain the dynamics of colloidal crystal, we have developed the overdamped bead-spring lattice (OBS) model [7, 8].

In the OBS model, colloidal crystal is regarded as a bead-spring lattice immersed in viscous media. Because the system is immersed in viscous liquid, the motion of the particles is overdamped and can hardly oscillate. So each normal vibration mode is transformed into each normal relaxation mode. As the result, the motion of the particles is described as superposition of the normal relaxation modes with eigen-amplitude and eigen-relaxation time.

Using this model, we have succeeded in explaining mean-square displacement of a particle (MSD, which is identical to autocorrelation function) in colloidal crystals. However, dispersion relations of the amplitude and the relaxation time could not be reproduced by using the same viscous drag coefficient and spring constant that could fit the observed MSD [9]. This inconsistency was caused by neglecting two considerable effects in the simple OBS approximation. One is the long-range and direct interaction from farther particles. And the other is the influence of hydrodynamic couplings between particles.

Here, we present the necessity of such interactions to explain the soft mode of particle motions in soft colloidal crystal. The basic equation of motion on lattice dynamics is almost the same expression that was reported by Hurd *et al.* [1]. However, it is quite different in following three points; (i) long-range effect from farther particles than third-nearest neighbor is incorporated, (ii) continuum approximation should not be applied to our video microscopy with high magnification, and (iii) it is not necessary to be simultaneous with the Navier-Stokes equation for our observation which was recorded at slow videoflame rate of 30 fps. The outcomes were in quantitative agreement with the experimental results, and the dispersion curves of amplitude and relaxation time could be reproduced by completely the same fitting parameters for MSD.

THEORETICAL BACKGROUND

The motion of particles in colloidal crystal is determined by following four forces; (i) elastic force from particles at neighbor lattice sites, (ii) viscous drag force from solvent, (iii) hydrodynamic force of flow which is produced by surrounding particles, and (iv) random force from solvent molecules. Because the system is immersed in viscous media, the motion of the particles is overdamped, so the inertia term can be neglected. Therefore, the equation of motion can be expressed by the following

CP982, *Complex Systems, 5th International Workshop on Complex Systems*
edited by M. Tokuyama, I. Oppenheim, and H. Nishiyama

overdamped Langevin equation:

$$0 = \sum_{\mathrm{p}}^{\text{neighbors}} C_{\mathrm{p}}\left[\boldsymbol{u}_{\boldsymbol{r}_{\mathrm{j}}+\boldsymbol{n}_{\mathrm{p}}}(t) - \boldsymbol{u}_{\boldsymbol{r}_{\mathrm{j}}}(t)\right] - \gamma\frac{d}{dt}\boldsymbol{u}_{\boldsymbol{r}_{\mathrm{j}}}(t) + \gamma \sum_{\mathrm{p}}^{\text{neighbors}} \Omega(\boldsymbol{r}_{\mathrm{a}},\boldsymbol{n}_{\mathrm{p}})\frac{d}{dt}\boldsymbol{u}_{\boldsymbol{r}_{\mathrm{j}}+\boldsymbol{n}_{\mathrm{p}}}(t) + \boldsymbol{f}_{\mathrm{j}}(t). \quad (1)$$

Here $\boldsymbol{r}_{\mathrm{j}}$ represents the equilibrium position of jth particle, $\boldsymbol{u}_{\boldsymbol{r}_{\mathrm{j}}}(t)$ is the displacement of a particle from $\boldsymbol{r}_{\mathrm{j}}$ at time t, and $\boldsymbol{n}_{\mathrm{p}}$ stands for a vector directed from $\boldsymbol{r}_{\mathrm{j}}$ to an another equilibrium position of neighbor particle p. C_{p} is the spring constant at the average distance $\boldsymbol{n}_{\mathrm{p}} - \boldsymbol{r}_{\mathrm{j}}$, γ is the viscous drag coefficient of a particle, and $\Omega(\boldsymbol{r}_{\mathrm{a}},\boldsymbol{n}_{\mathrm{p}})$ represents the preaveraged Oseen tensor for the particles with radius $\boldsymbol{r}_{\mathrm{a}}$ at the equilibrium distance $\boldsymbol{n}_{\mathrm{p}} - \boldsymbol{r}_{\mathrm{j}}$ [10, 11]. $\sum_{\mathrm{p}}^{\text{neighbors}}$ stands for the summation by neighbor particles, and $\boldsymbol{f}_{\mathrm{j}}(t)$ is the fluctuating force exerted on jth particle.

Under the preaveraging approximation to the Oseen tensor, and performing a discrete Fourier transform to the displacement, autocorrelation function $c(\tau)$ can be described as the superposition of the normal relaxation modes with eigen-amplitude $\langle|\boldsymbol{U}_{\boldsymbol{q}_{\boldsymbol{k}}}(t)|^2\rangle_t$ and eigen-relaxation time $T_{\boldsymbol{q}_{\boldsymbol{k}}}$.

$$\begin{aligned} c(\tau) &= \frac{1}{N}\sum_{\boldsymbol{k}}\langle \boldsymbol{U}_{\boldsymbol{q}_{\boldsymbol{k}}}(t+\tau)\cdot \boldsymbol{U}^*_{\boldsymbol{q}_{\boldsymbol{k}}}(t)\rangle_t \\ &= \frac{1}{N}\sum_{\boldsymbol{k}}\langle|\boldsymbol{U}_{\boldsymbol{q}_{\boldsymbol{k}}}(t)|^2\rangle_t \exp\left(-\frac{\tau}{T_{\boldsymbol{q}_{\boldsymbol{k}}}}\right), \quad (2) \\ \boldsymbol{U}_{\boldsymbol{q}_{\boldsymbol{k}}}(t) &= \frac{1}{\sqrt{N}}\sum_{\boldsymbol{r}_{\mathrm{j}}}\boldsymbol{u}_{\boldsymbol{r}_{\mathrm{j}}}(t)\exp(-i\boldsymbol{q}_{\boldsymbol{k}}\cdot\boldsymbol{r}_{\mathrm{j}}). \end{aligned}$$

Where $\boldsymbol{q}_{\boldsymbol{k}}$ is a wavevector of kth normal relaxation mode, N is the number of particles and thus the total number of the independent modes, and τ is the delay time.

Applying the fluctuation-dissipation theorem and the principle of equipartition of energy, amplitude and relaxation time are obtained as follows:

$$\langle|\boldsymbol{U}_{\boldsymbol{q}_{\boldsymbol{k}}}(t)|^2\rangle_t = \frac{dk_{\mathrm{B}}T}{D(\boldsymbol{q}_{\boldsymbol{k}})}, \quad (3)$$

$$T_{\boldsymbol{q}_{\boldsymbol{k}}} = \frac{\gamma D'(\boldsymbol{q}_{\boldsymbol{k}})}{nD(\boldsymbol{q}_{\boldsymbol{k}})}. \quad (4)$$

Here d is the dimension of system, k_{B} is the Boltzmann constant, and T is the temperature of system. $D(\boldsymbol{q}_{\boldsymbol{k}})$ is widely known as the dynamical matrix in classical theory of the harmonic crystal [12],

$$D(\boldsymbol{q}_{\boldsymbol{k}}) = \sum_{\mathrm{p}}^{\text{neighbors}} C_{\mathrm{p}}\left[1-\cos(\boldsymbol{q}_{\boldsymbol{k}}\cdot\boldsymbol{n}_{\mathrm{p}})\right]. \quad (5)$$

$D'(\boldsymbol{q}_{\boldsymbol{k}})$ is also the form factor similar to $D(\boldsymbol{q}_{\boldsymbol{k}})$, corresponds to the dissipation matrix called by Hurd *et al.* [1], and contains the preaverage Oseen tensor $H_{\mathrm{p}} = n\Omega(\boldsymbol{n}_{\mathrm{p}})$;

$$D'(\boldsymbol{q}_{\boldsymbol{k}}) = \sum_{\mathrm{p}}^{\text{neighbors}} \left[1-H_{\mathrm{p}}\cos(\boldsymbol{q}_{\boldsymbol{k}}\cdot\boldsymbol{n}_{\mathrm{p}})\right]. \quad (6)$$

n represents the number of neighbor particles. These equations indicate that the actual diffusion coefficient $D_H(\boldsymbol{q}_{\boldsymbol{k}})$ for the particle in colloidal crystal can be deduced from the amplitude and the relaxation time as

$$\frac{1}{D_H(\boldsymbol{q}_{\boldsymbol{k}})} = \frac{T_{\boldsymbol{q}_{\boldsymbol{k}}}}{\langle|\boldsymbol{U}_{\boldsymbol{q}_{\boldsymbol{k}}}(t)|^2\rangle_t} = \frac{\gamma D'(\boldsymbol{q}_{\boldsymbol{k}})}{ndk_{\mathrm{B}}T} = \frac{D'(\boldsymbol{q}_{\boldsymbol{k}})}{nD_0}, \quad (7)$$

and D_0 stands for the free diffusion coefficient in steady water. Because $D'(\boldsymbol{q}_{\boldsymbol{k}})$ contains H_{p}, the relaxation time and the diffusion coefficient are influenced by hydrodynamic couplings, and depend on $\boldsymbol{q}_{\boldsymbol{k}}$ and the configuration of lattice, too.

Using above amplitude and relaxation time, MSD is obtained as below:

$$c_{\mathrm{MSD}}(\tau) \equiv \langle|\boldsymbol{u}_{\boldsymbol{r}_{\mathrm{j}}}(t+\tau) - \boldsymbol{u}_{\boldsymbol{r}_{\mathrm{j}}}(t)|^2\rangle_{t,\mathrm{j}}$$

$$= \frac{2dk_{\mathrm{B}}T}{N}\sum_{\boldsymbol{k}}\frac{1}{D(\boldsymbol{q}_{\boldsymbol{k}})}\left[1-\exp\left(-\frac{nD(\boldsymbol{q}_{\boldsymbol{k}})}{\gamma D'(\boldsymbol{q}_{\boldsymbol{k}})}\tau\right)\right]. \quad (8)$$

MATERIALS AND METHODS

Experimental procedure was scarcely changed in the previous paper [9]. The observed colloidal suspension consisted of polystyrene latex (particle size $2r = 2.06$ [μm] in diameter) dispersed in deionized water. Polystyrene latex, distilled water, and ion exchange resin were sealed in a glass cell, and colloidal crystals were formed as the result of deionization [13]. The cell was made of a glass tube cut its bottom off and coverslip glued to the bottom. Lattice type of the observed colloidal crystals was face-centered cubic (FCC), and its (111) plane faced to the bottom surface of the cell. The bottom layer of the crystal nearest to the coverslip could be observed through the coverslip by an inverted optical microscope at room temperature. The sample was polycrystal and the fraction was observed.

These images were recorded by DVD at 30 frames per second. The recorded images were replayed at 30 frames per second and were digitized using personal computer (PC) with a video-capture board. The digitized images were stored in the PC's hard disk. The particle recognition, the tracing of the particles, and MSD were calculated by the PC.

To observe cooperative relaxation mode, we recorded the displacement of 168 particles in colloidal crystal, performed a discrete Fourier transform to these observed data, and calculated the correlation functions of the amplitude of each Fourier component. Because of long-time tail of the velocity autocorrelation function, however, these correlation functions could not be fitted well by single exponential functions [14, 15]. Therefore each relaxation time was obtained from the initial slope which could be regarded as single exponential decay (see figure 1). Each amplitude could be simply obtained from the value of the correlation function at $\tau = 0$.

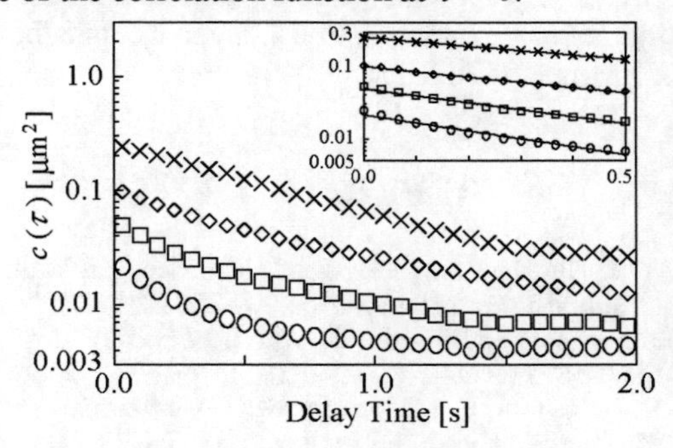

FIGURE 1. Autocorrelation functions of the observed displacements. The number of all modes is 168. The plots are the parts of them. The inset displays the shorter time region which could be fitted well by single exponential function.

RESULTS AND DISCUSSION

The observed motion of colloidal particles was two dimensional, however, the amplitude shown in figure 2 is converted to the value per one dimension. And also the fitted line was calculated by eq. (3) with $d = 1$. The fitting parameters are shown in table 1. C_p is tensor in eqs. (1) and (5), here the parameters are scalars. The reason is that the two eigenvalues of the two eigenvectors parallel to FCC (111) plane are equivarent when the tensor is diagonalized. The eigenvalue is equal to diagonal component minus non-diagonal one.

1st, 3rd, 4th, 7th, and 9th[a]-nearest neighbor particles are located on the observed plane, and 1st, 4th, and 9th[a]-nearest neighbor particles form a line. The influence of 4th-nearest neighbor particle was as large as the effect of 1st-nearest one. But, in contrast, C_{3rd} and C_{7th} are negative. These negative force constants mean that the effect of particles located on such position makes the crystal unstable. In this experiment, the soft crystal is expanded to its limit, and therefore the displacement and the amplitude become larger. On the other hand, it was not necessary to consider the influence of 2nd, 5th, 6th, 8th, and 9th[b]-nearest neighbor particles that do not exist on FCC (111) plane.

The anomalous soft mode appears along the direction Γ-M. These are too large amplitude to be fitted simultaneously with the other curves by nonnegative force constant. This reveals that there are some directions in which the observed crystal is fragile.

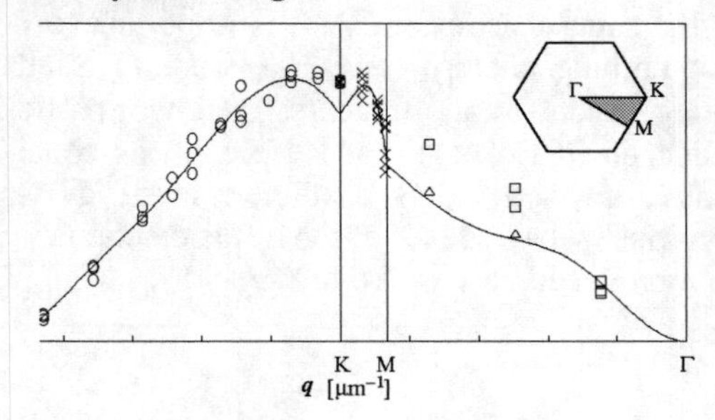

FIGURE 2. The symbols indicate dispersion curve of the observed amplitude along high-symmetry lines for FCC (111). The symmetry points are labeled according to the convention [16]. The triangles are the anomalous soft mode. The line is the fitting curve by eq. (3).

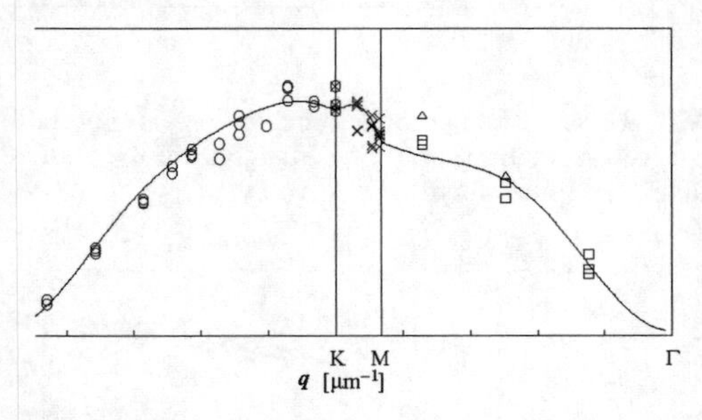

FIGURE 3. q-dependence of the observed diffusion coefficient. The ratio of the observed relaxation time / the observed amplitude for each mode are plotted. The line is the fitting curve calculated by eq. (7).

TABLE 1. Parameters used for fitting curves.

p	C_p [fN/μm] *	H_p [Ns/m] †
1st	8.80	7.20
3rd	$-0.50C_{1st}$	0.00
4th	$1.00C_{1st}$	$0.50H_{1st}$
7th	$-0.03C_{1st}$	0.00
9th[a]	$0.13C_{1st}$	0.00

* $C_{2nd} = C_{5th} = C_{6th} = C_{8th} = C_{9th^b} = 0.00$
† $H_{2nd} = H_{5th} = H_{6th} = H_{8th} = H_{9th^b} = 0.00$

Figure 3 shows q-dependence of the diffusion coefficient calculated from the ratio of the observed relaxation time divided by the observed amplitude of each mode. Diffusion coefficient for a particle is constant in steady water, but it is found to depend on q as the result of hydrodynamic interaction in actual colloidal crystal. These results indicate that diffusion coefficient is large for small q and small for large q. When the particles move to the same direction, i.e. for small q mode, flow helps the cooperative motion, so diffusion coefficient become lager. For large q, where neighbor particles move to the opposite direction, flow disturbs the movement, so diffusion coefficient is reduced for such mode.

The fitted line was calculated by eq. (7), and the fitting parameters are shown in table 1. H_p is scalar because of the same reason for C_p. It is important to point out that nothing but H_{1st} and H_{4th} were required for the fitting. And 1st and 4th-nearest neighbor particles form a line on FCC (111) plane. Here γ was equal to 1.95 times the viscous drag coefficient in steady water and constant. $\boldsymbol{q}$-dependence of the diffusion coefficient could be reproduced very well by $D'(\boldsymbol{q_k})$.

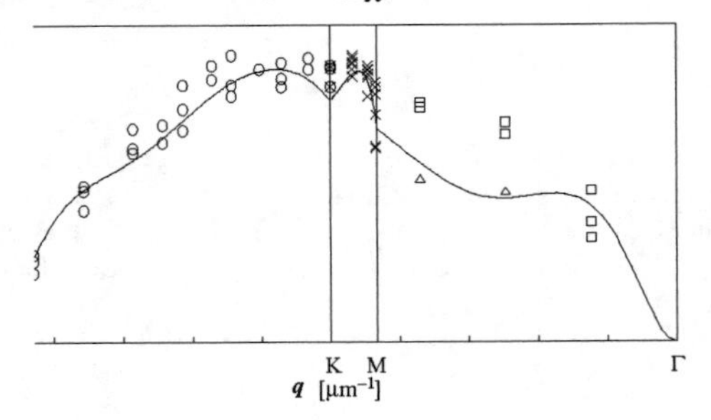

FIGURE 4. The symbols indicate dispersion relation of the observed relaxation time. The triangles are the anomalous soft mode. The line is the fitting curve by eq. (4).

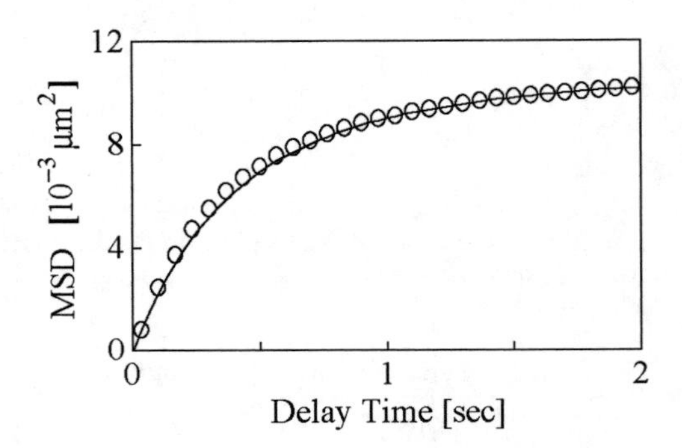

FIGURE 5. The circles are the observed MSD. The line is the fitting curve by eq. (8).

The dispersion relation of the relaxation time and the MSD were reproduced naturally by exactly the same fitting parameters (see figure 4 and 5). In figure 4, the anomalous soft mode appears distinctly in Γ-M, too. It is essential for explaining such mode of relaxation time to consider the influence of hydrodynamic interaction and the long-range effect with negative force constant.

In this paper, we do not present the case of the other samples, however, the ratios of C_p/C_{1st} and H_p/H_{1st} for $p \geq 2$ were universal to the dynamics pararell to FCC (111) plane. Therefore the adjustable parameters are just γ, C_{1st}, and H_{1st}. And also such soft mode is always observed in every sample we have studied. Here hydrodynamic wall effect is not considered, however, the results of numerical calculation based on the modified OBS model are in quantitative agreement with the experimental results. And to treat hydrodynamic interaction simply in the observed plane is sufficient to comprehend the dynamics of colloidal crystal analyzed by our video microscopy. We have already shown that hydrodynamic wall effect can be neglected concerning the motion parallel to the coverslip, too [8].

ACKNOWLEDGMENTS

We thank A. C. Yamada for helpful lecture. Financial supports from Aoyama Gakuin University and the Research Institute of AGU.

REFERENCES

1. A. J. Hurd, N. A. Clark, R. C. Mockler, and W. J. O'Sullivan, *Phys. Rev. A* **26**, 2869-2881 (1982).
2. J. Derksen, and W. van de Water, *Phys. Rev. A* **45**, 5660-5673 (1992).
3. B. V. R. Tata, P. S. Mohanty, M.C. Valsakumar, and J. Yamanaka, *Phys. Rev. Lett.* **93**, 268303-1-4 (2004).
4. J. Bongers and H. Versmold, *J. Chem. Phys.* **104**, 1519-1523 (1996).
5. J. A. Weiss, A. E. Larsen, and D. G. Grier, *J. Chem. Phys.* **109**, 8659-8666 (1998).
6. P. Keim, G. Maret, U. Herz, and H. H. von Grunberg, *Phys. Rev. Lett.* **92**, 215504-1-4 (2004).
7. Y. N. Ohshima and I. Nishio, *J. Chem. Phys.* **114**, 8649-8658 (2001).
8. Y. N. Ohshima, K. E. Hatakeyam, M. Satake, Y. Homma, R. Washidzu, and I. Nishio, *J. Chem. Phys.* **115**, 10945-10954 (2001).
9. T. Ueno, K. Yamazamki, and I. Nishio, "Hydrodynamic Effects of Particle Motions in Colloidal Crystal," in *FLOW DYNAMICS*, edited by M. Tokuyama and S. Maruyama, AIP Conference Proceedings **832**, American Institute of Physics, New York, 2006, pp.287-291.
10. B. H. Zimm, *J. Chem. Phys.* **24**, 269-281 (1956).
11. J. G. Kirkwood and J. Riseman, *J. Chem. Phys.* **16**, 565-573 (1948).
12. N. W. Ashcroft and N. D. Mermin, "Crystal Theory of the Harmonic Crystal," in *Solid State Physics*, Saunders College, Philadelphia, 1976, Chap. 22, pp.422-450.
13. A. Kose, M. Ozaki, Y. Kobayashi, K. Takano, and S. Hachisu, *J. Colloid Interface Sci.* **44**, 330-338 (1973).
14. B. J. Alder and T. E. Wainwright, *Phys. Rev. A* **1**, 18-21 (1970).
15. K. Ohbayashi, T. Kohno, and H. Utiyama, *Phys. Rev. A* **27**, 2632-2641 (1983).
16. L. B. Bouckaert, R. Smoluchowski, and E. Wigner, *Phys. Rev* **50**, 58-68 (1936).

DSC Study of Light and Heavy Water Confined within Mesoporous Silica Pores

Yoshihito Nishioka and Hiroki Fujimori

Graduate School of Integrated Basic Sciences, Nihon University, 3-25-40 Sakurajosui, Setagaya-ku, Tokyo, Japan

Abstract. Differential scanning calorimetry measurements of light water (H_2O) and heavy water (D_2O), confined within mesoporous silica of pore diameter 1.4–7.1 nm, were performed. The fusion temperature of water within the pores decreased with decreasing pore diameter. Crystallization of water within the pores was not observed in samples with pore diameters of 2.2 and 1.4 nm.

Keywords: DSC, Heavy water, Light water, Mesoporous silica
PACS: 65.20.JK

INTRODUCTION

Water is one of the most common substances on the earth, and about 60% of the human body is made of water. It has many interesting properties in physics and chemistry, e.g., water becomes amorphous ice by hyperquenching [1]. There are two types of amorphous ice—high-density amorphous (HDA) and low-density amorphous(LDA) [2]. Using computer simulation, Mishima *et al.* discovered two similar types of liquid water—high-density liquid (HDL) and low-density liquid (LDL)—and determined that a liquid–liquid transition occurs between the two types [3, 4]. However, HDL and LDL have not been experimentally observed, because supercooled-liquid water and amorphous ice crystallize at below 236 and above 136 K, respectively, at atmospheric pressure. Observing HDL and LDL requires the preparation of water that does not crystallize at low temperature. The crystallization of the liquid confined within nanopores is restrained, because the long-range interaction between molecules disappears within the pores, and the surface tension of the liquid increases due to the pore walls. Oguni *et al.* found that the fusion temperature of water decreased within silica gels and that water confined within the silica gels divided into two parts—an internal and an interfacial region [5, 6]. Silica gel pores are spherical; therefore, there is a large variation in pore diameter. On the other hand, mesoporous silicas, such as MCM, FSM, and MSU-H, have highly ordered cylindrical channels and pore diameter can be easily controlled—in the order of nanometers. Differential scanning calorimetry (DSC) measurements were performed for light water (H_2O) and heavy water (D_2O) confined within mesoporous silica pores.

EXPERIMENT

The following mesoporous silicas were used: FSM16 (pore diameter: 3.1 nm), FSM12 (2.2 nm), and FSM8 (1.4 nm) supplied from Fuji Silysia Chemical Co. Ltd, and MSU-H (7.1 nm) purchased from the Sigma-Aldrich Co. Ltd. Table 1 presents the characteristics of FSM and MSU-H. The mesoporous silicas were dried at 393 K under atmospheric pressure for three days or more and then at 393 K under vacuum for three hours or more. Samples were prepared by loading distilled H_2O or D_2O into the pores of the mesoporous silica in a vacuum line by distillation. The rates of filling of water per unit pore volume are summarized in Table 1. DSC measurements were performed with a SEIKO Instruments DSC120 at a scanning rate of 7 K min^{-1} in the temperature range of 110 to 300 K. Sealed aluminum pans were used in all experiments. An empty aluminum pan was used as the reference.

CP982, *Complex Systems, 5th International Workshop on Complex Systems*
edited by M. Tokuyama, I. Oppenheim, and H. Nishiyama

TABLE 1. Characterization of mesoporous silica and experimental results of water confined within the pores

silica	Pore diameter (nm)	Pore volume (mL g^{-1})	H_2O Rate of filling	H_2O T_{fus} (K)	H_2O $\Delta_{fus}H$ (kJ mol^{-1})	H_2O c	D_2O Rate of filling	D_2O T_{fus} (K)	D_2O $\Delta_{fus}H$ (kJ mol^{-1})	D_2O c
MSU-H	7.1	0.91	0.46	257	3.3	0.54	0.50	260	4.8	0.55
FSM16	3.1	0.90	0.67	229	1.5	0.25	0.50	232	1.2	0.19
FSM12	2.2	0.78	0.68	—	—	—	0.51	—	—	—
FSM8	1.4	0.34	0.70	—	—	—	0.87	—	—	—

RESULTS AND DISCUSSION

Figures 1 and 2 show the DSC scanning curves of H_2O and D_2O, respectively, confined within mesoporous silica pores. In all the samples, the heat anomalies of ice fusion (peak A) were observed at around 273 and 276 K, for H_2O and D_2O, respectively. They were due to the presence of ice on the surface of the mesoporous silica.

For H_2O and D_2O confined within MSU-H and FSM16, other heat anomalies of ice fusion were observed (peak B). They were due to the presence of the ice within the pores. The temperatures and enthalpies of fusion related to peak B are summarized in Table 1. The water within the pores is classified into interfacial and internal water. Because it is known that the interfacial water does not undergo crystallization into ice, the peak B corresponds to the fusion of the internal water. A crystallization ratio, c, of the water confined within the pores is calculated by $(\Delta_{fus}H)^{obs}/(\Delta_{fus}H)^{calc}$, where $(\Delta_{fus}H)^{obs}$ is the enthalpy of fusion obtained by DSC, and $(\Delta_{fus}H)^{calc}$ is the enthalpy of fusion of bulk water at the fusion temperature, T_{fus}, obtained from the peak B. $(\Delta_{fus}H)^{calc}$ is calculated by

$$(\Delta_{fus}H)^{calc} = \Delta_{fus}H(T_0) - \int_{T_{fus}}^{T_0} \Delta C_p(T)\mathrm{d}T,$$

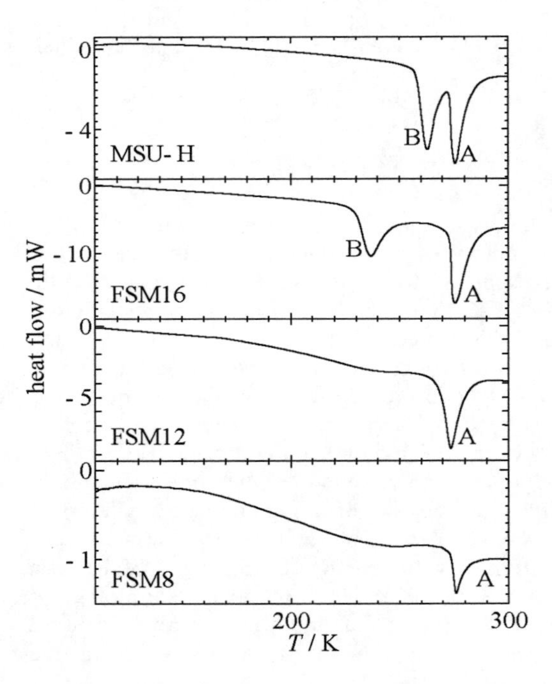

FIGURE 1. DSC curves of H_2O confined within mesoporous silica pores.

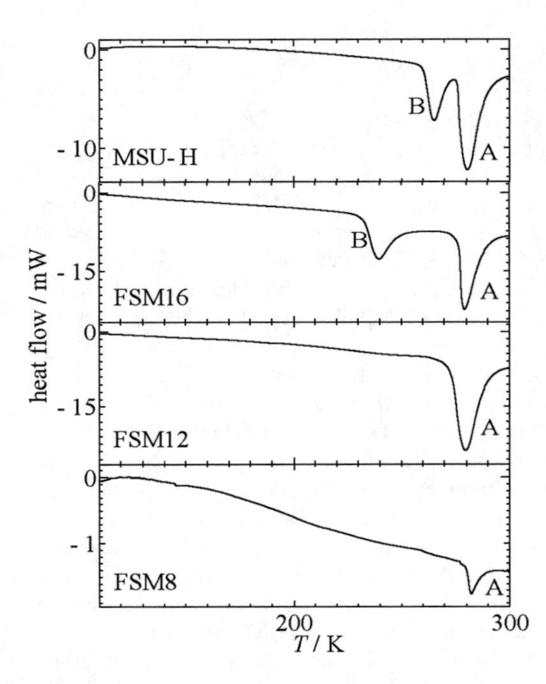

FIGURE 2. DSC curves of D_2O confined within mesoporous silica pores.

where the $\Delta_{fus}H\ (T_0)$ is the enthalpy of fusion of H_2O at 273 K or of D_2O at 276 K, and $\Delta C_p(T)$ is the difference in the heat capacities of water and ice. $\Delta C_p(T)$ is calculated by $a(T_0-T)^4$, where $a = 2.82 \times 10^{-8}$ $JK^{-5}mol^{-1}$ for H_2O, and $a = 3.14 \times 10^{-8}$ $JK^{-5}mol^{-1}$ for D_2O [7]. Table 1 summarizes the crystallization ratios. The decrease in the crystallization ratio with decreasing pore diameter indicates a reduction in the amount of internal water. The ratios of the volume of the internal water to the water confined within the FSM pores (pore diameters: 1.9–4.1 nm) were found to be 0.46–0.50, calculated from the results of NMR experiments [8]. The crystallization ratio for H_2O within the FSM16 pores obtained in this study was smaller than that found in literature 8. This may be either due to the insufficient filling rate of water, or because the enthalpy of the internal water is different from that of the bulk water.

For the water confined within FSM12 and FSM8, heat anomalies of ice fusion within the pores were not clearly observed; however, small heat anomalies were observed over a wide temperature range. The similar anomalies were observed by adiabatic calorimetry for water confined within 1.1 nm pores of the silica gels [5]. This indicates that almost all of the water within the pores remains in the liquid state above 110 K. This suggests two possibilities—the internal water remains in the liquid state or there is no internal water in the FSM12 and FSM8 pores. However, because the ratios of internal water calculated by Mori et al. are 0.46–0.50 [8], it is verified that the internal water exists as a liquid, at least in the FSM12 pores.

If water exists as a supercooled liquid at low temperature, and its crystallization to ice is not observed, it must undergo glass transition. In fact, the glass transition of water confined within silica gel pores was detected by adiabatic calorimetry [5, 6]. However, in this study, the glass transition was not observed in all samples because the heat capacity jump originated the glass transition was very small [5, 6]. We need a high-precision measurement to observe the glass transition.

CONCLUDING REMARKS

The thermal properties of H_2O and D_2O confined within the mesoporous silica were disclosed by DSC measurements. Figure 3 shows the depression of fusion temperature plotted against the inverse pore-diameter. Since the interfacial water does not undergo crystallization into ice, the thickness of the interfacial water must be subtracted from the pore diameter. However, as the exact thickness has not been obtained, we plot the depression of fusion temperature against the inverse pore-diameter of mesoporous silica. The depression of fusion temperature increased with decreasing the pore diameter. The similar dependency was observed by x-ray diffraction measurements of light water confined within MCM-41 [9]. However the depression of fusion observed in this study was higher than in literature 9 at the same pore-diameter. This may be caused by the difference of the chemical structures on the surface of mesoporous silica. The small difference of the depression of fusion temperature existed between H_2O and D_2O, too. Two possibilities can be considered as this reason. One is due to the difference of the size of the internal water. The thickness of the interfacial water will be changed the interaction between water molecules and the surface of mesoporous silica. D_2O may tend to be influenced of surface compared with H_2O. The other is due to the difference of the molecular interactions in the internal water. The difference of the interaction produces 3.81 K in the difference of the fusion temperature in bulk water. The interaction may depend also on a size of the internal water. In order to discuss the detail of the thermal properties of H_2O and D_2O confined within pores, we need to estimate the size of the internal water using by other experiments.

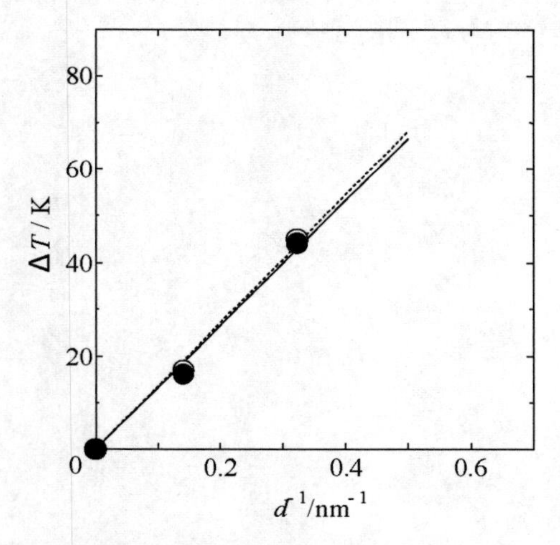

FIGURE 3. Depression of fusion temperature plotted against the inverse pore-diameter. Solid and open circles represent the depressions of fusion temperature of H_2O and D_2O, respectively. Solid and dotted lines represent the fitted straight-lines for H_2O and D_2O, respectively.

ACKNOWLEDGMENTS

We would like to thank Fuji Silysia Chemicals Ltd. for supplying us FSM. This work was financially supported partly by a grant from the College of Humanities and Sciences, Nihon University and by a Grant-in-Aid for Scientific Research from the Ministry of Education, Culture, Science and Technology, Japan.

REFERENCES

1. H. Kanno, R. Speedy, C. A. Angell, *Science*, **189,** 880-881 (1975).
2. O. Mishima, L. D. Calvert, E. Whalley, *Nature*, **310**, 393-395 (1984).
3. O. Mishima and H. E. Stanley, *Nature*, **396**, 329-335 (1998).
4. O. Mishima, *Phys. Rev. Lett.*, **85**, 334-336 (2000).
5. M. Oguni, S. Maruyama, K. Wakabayashi, and A. Nagoe, *Chem. Asian J.*, **2,** 514-520 (2007).
6. S. Maruyama, K. Wakabayashi, and M. Oguni, "Thermal Properties of Supercooled Water Confined within Silica Gel Pores," in *Slow Dynamics in Complex Systems*, edited by M. Tokuyama and I. Oppenheim, AIP Conference Proceedings 708, American Institute of Physics, New York, 2004, pp.675-676.
7. D. H. Rasmussen and A. P. Mackenzie, *J. Chem. Phys.*, **59**, 5003-5013 (1973).
8. T. Mori, T. Ueda, K. Miyakubo, and T. Eguchi (private communication).
9. K. Morishige and K. Kawano, *J. Chem. Phys.*, **110,** 4867-4872 (1999).

Micro-Brillouin Scattering Study of Acoustic Properties of Protein Crystals

Eiji Hashimoto, Yuichiro Aoki, Yuichi Seshimo,
Keita Sasanuma, Yuji Ike, and Seiji Kojima

Graduate School of Pure and Applied Sciences, University of Tsukuba, Tsukuba Ibaraki, 305-8573, Japan

Abstract. Polymorphism and dehydration process are studied in lysozyme crystals. Three kinds of crystals with tetragonal, orthorhombic, monoclinic systems are successfully grown by two liquids method. The dehydration process of a tetragonal crystal is investigated by the micro-Brillouin scattering techniques. This process is discussed by the Avrami–Erofe'ev equation.

Keywords: Dehydration, Lysozyme, Polymorphism, Sound velocity
PACS: 62.20.-X, 81.10.Dn, 82.30.Rs, 87.14.Ee

INTRODUCTION

Hydration mechanism plays a central role in sustaining the three dimensional structure and biological activity of protein molecules both in solution and in crystal. The effects of water molecules in protein crystals have been focused on hydration and intramolecular and intermolecular interactions in the crystals. The effects of water on protein crystals have been extensively studied , however, most of the studies on elastic properties of protein crystals have been carried out in solutions or wet conditions. Therefore, we report the dehydration process of a protein crystal by micro-Brillouin spectroscopy.

Brillouin spectroscopy has been a reliable method of noncontact probe of elastic properties. Elastic properties provide us microscopic interaction on the proteins and have been studied by several authors [1, 2]. The Brillouin scattering measurement enables us to determine the sound velocity, elastic constant and damping of sound waves. It is known to be a powerful tool to explore the elastic property of high frequency gigahertz range and of extremely wide temperature range. Furthermore, it only requires a small volume of sample to be studied. Therefore, it is quite suitable for the substance of limited size. In the present study, we investigate protein crystal growth and polymorphism, in addition the temperature dependence of acoustic properties in protein crystals by micro-Brillouin scattering [3].

EXPERIMENTAL

The experimental setup of a micro-Brillouin scattering instrument is shown as FIGURE 1. The Brillouin scattering spectra were measured at the backward scattering geometry. The Sandercock-type 3 + 3 pass tandem Fabry-Perot Interferometer (FPI) was combined with microscope and operated to acquire spectra of scattered light. A free spectral range of 30 GHz was applied and the Brillouin spectra were measured in ± 25 GHz frequency range. The sample temperature was controlled within $\pm$ 0.1 °C and increased from room temperature with a cryogenic cell (LINKAM HTMS600).

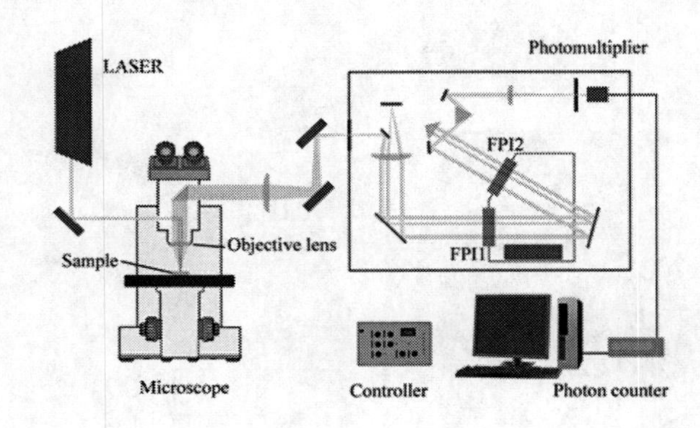

FIGURE 1. Schematic diagram of the micro-Brillouin scattering instrument [3].

CP982, *Complex Systems, 5th International Workshop on Complex Systems*
edited by M. Tokuyama, I. Oppenheim, and H. Nishiyama

CRYSTAL GROWTH AND POLYMORPHISM

Tetragonal hen egg white lysozyme (HEWL) crystals were grown by two liquids method that employing an insoluble and dense liquid [4]. Some crystals were grown on the interface of two liquids of lysozyme solution and dense liquid such as Fluorinate. The solutions were consisted of 25 mg/ml lysozyme and 5 % (w/v) NaCl in 50 mM acetate buffer solution (pH = 4.5) at 25 ℃. All crystals were grown in few days and show the crystal habit of {110} and {101} planes. The polarizing micrograph of as-grown crystal is shown in FIGURE 2. (a). The laser beam was incident perpendicular to the {110} habit plane, which was the favored growth crystallographic planes of an as-grown crystal.

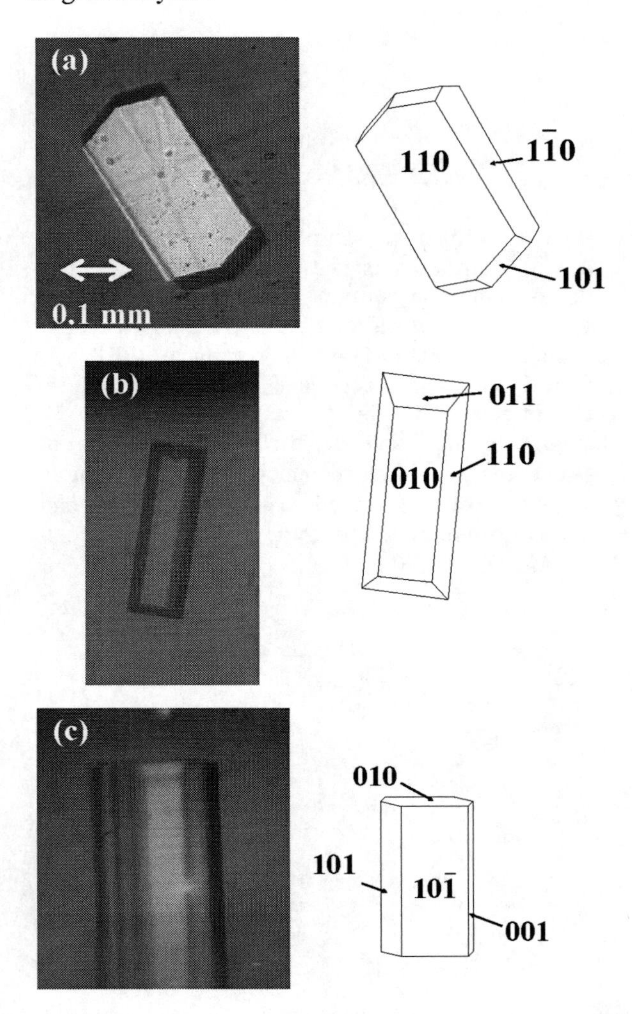

FIGURE 2. (Color online) **(a)** A tetragonal HEWL crystal; **(b)** An orthorhombic HEWL crystal; **(c)** A monoclinic HEWL crystal.

We also have succeeded making orthorhombic and monoclinic HEWL crystals by two liquids method. The solutions for orthorhombic crystals were grown from 55 mg/ml lysozyme and 3 % (w/v) NaCl in 50 mM acetate buffer solution (pH = 4.5) at 38 °C, and for monoclinic crystals were grown from 45 mg/ml lysozyme and 3 % (w/v) NaCl in 50 mM acetate buffer solution (pH = 4.5) at 25 °C. The polarizing micrographs of as-grown crystals are shown in FIGURE 2. (b) and (c).

Preliminary measurements of three kinds of crystals show that the values of longitudinal sound velocity are similar. This fact suggests that the strength of intermolecular interaction is very close among three different crystals systems.

In the present study, we report the dehydration process of a tetragonal HEWL crystal.

DEHYDRATION PROCESS OF A TETRAGONAL HEWL CRYSTAL

The observed Brillouin spectrum of a HEWL crystal showed one broad longitudinal acoustic (LA) mode as shown in FIGURE 3. Brillouin shift increases and full width at half maximum (FWHM) gradually becomes narrow as time progresses. Brillouin shift and FWHM were determined from the spectra. The longitudinal sound velocity, V_l, was calculated from the Brillouin shift ν as shown by equation

$$\nu = qV_1 / 2\pi , \tag{1}$$

and the scattering wave vector q is written by

$$q = 2\pi n \sin(\theta / 2) / \lambda , \tag{2}$$

where n, λ and θ are the values of refractive index of the sample, the wavelength of the incident beam and scattering angle, respectively. The value of refractive index of a HEWL crystal [5] was used. The LA mode absorption coefficient α is related to FWHM of Brillouin component by the equation, $\alpha = \pi\Gamma/V_l$ where Γ is FWHM. The time dependence of sound velocity with a constant temperature is plotted as shown in FIGURE 4. About 40 °C, The figure indicates that the sound velocity drastically increases after 20 minutes, and becomes close to the horizontal asymptote. In contrast, sound velocity at 30 °C moderately increases and reaches the constant value which is slightly lower than that of 40 °C.

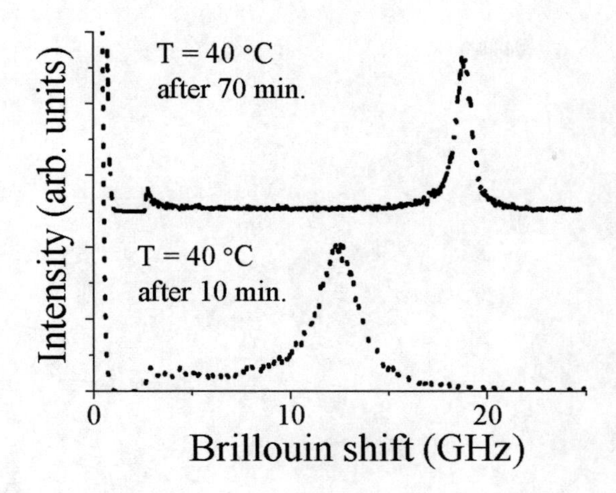

FIGURE 3. Observed Brillouin spectra of a HEWL crystal which include the scattering from the longitudinal acoustic phonon.

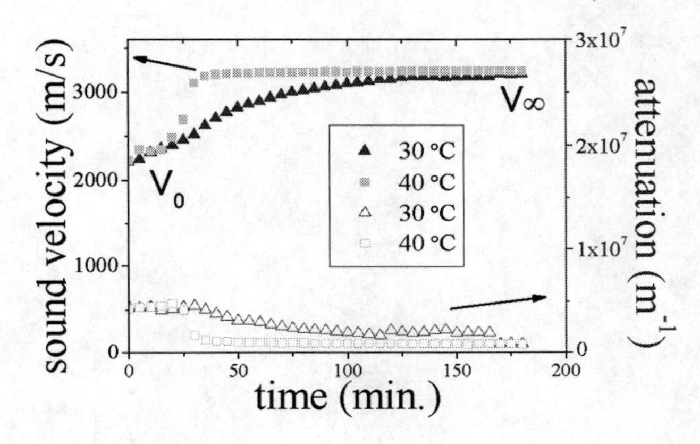

FIGURE 4. (Color online) Time dependence of V_l of a HEWL crystal along the <110> direction.

It is clear that the behavior of time dependence is different between 30 °C and 40 °C. FIGURE 5. (a) plots r as a function of time during the dehydration at 30 °C and 40 °C. The dehydration kinetics can be explained by using various kinetics equations, which have found the best application in solid-state reactions. Several kinetics equations or models have been attempted, including growth controlled reactions (the Avrami–Erofe'ev equation), phase boundary controlled reactions, diffusion controlled reactions, power-law equations, and equations based on the order of reactions [6, 7]. From the data fitting, the best fit was found with the two-dimensional phase boundary-controlled reaction, as shown in FIGURE 5. (b), known as the contracting area equation

$$1-(1-r)^{1/2} = kt \ , \tag{3}$$

where k is the dehydration rate constant and t is time, and r is assumed to the next equation.

$$r = \frac{V - V_0}{V_\infty - V_0} \ , \tag{4}$$

where V is sound velocity. The above equation states a mechanism for the solid-state reactions that are controlled by the advancement of a two dimensional phase boundary from the outside of a crystal. The studies of dehydration process on orthorhombic and monoclinic systems are in progress.

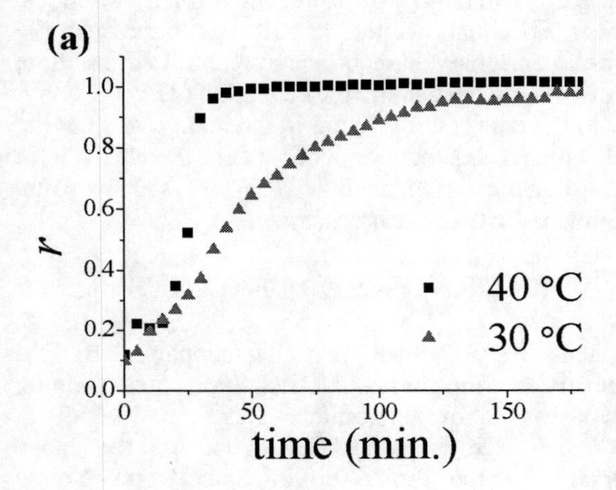

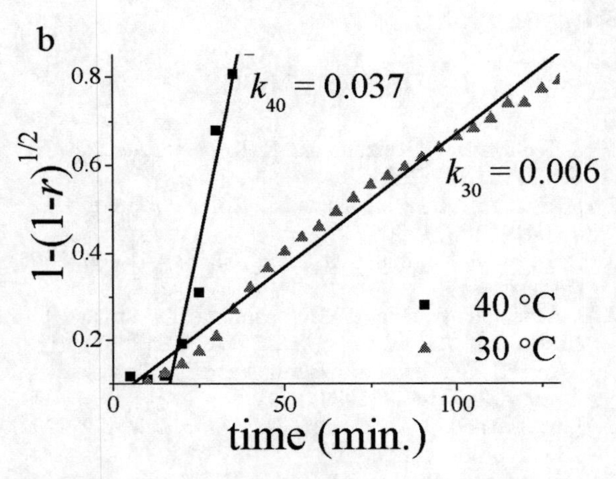

FIGURE 5. (Color online) **(a)** Normalized fraction of anhydrous lysozyme crystals, r, as a function of time at 30 °C and 40 °C; **(b)** dehydration of lysozyme crystals evaluated according to the contracting area equation of the solid-state reaction. The scatter symbols are the experimental data and the solid lines are the linear fits.

CONCLUSION

Polymorphism and dehydration process are studied in lysozyme crystals. Three kinds of crystals, tetragonal, orthorhombic, monoclinic systems are successfully grown by two liquids method. In addition, by using the micro-Brillouin scattering method, we have observed the time dependence of sound velocity and absorption in tetragonal HEWL crystals. About 40 °C, the sound velocity drastically increases, while sound velocity at 30 °C moderately increases and reaches the constant value which is slightly lower than that of 40 °C.

It is possible that the dehydration kinetics is investigated by using various kinetics equations. We attempt reliable kinetics equations or physical models, including growth controlled reactions (the Avrami-Erofe'ev equation), diffusion controlled reactions, power-law equations. the best fit was found with the two-dimensional phase boundary-controlled reaction, known as the contracting area equation (3) .

In the connection, we are in the middle of studying the time dependence of sound velocity in orthorhombic and monoclinic HEWL crystals by using the micro-Brillouin scattering method.

ACKNOWLEDGEMENTS

This research was partially supported by the Ministry of Education, Science, Sports and Culture, Grant-in-Aid for Exploratory Research, 19656005, 2007 and the Research Foundation of the Japan Society for the Promotion of Science for Young Scientists, 18·3811.

REFERENCES

1. M. Tachibana, H. Koizumi and K. Kojima: *Phys. Rev.* E **69** 051921-1-5 (2004).
2. H. Koizumi, M. Tachibana and K. Kojima: *Phys. Rev.* E **73** 041910-1-7 (2006).
3. Y. Ike and S. Kojima: *J. Aco. Soc. Jpn.* **61** 461-466 (2005). (in Japanese)
4. H. Adachi, T. Watanabe, M. Yoshimura, Y. Mori and T. Sasaki: Jpn. *J. Appl. Phys.* **41** L 726-L 728 (2002).
5. B. Cervelle, F. Cesbron, J. Berthou and P. Jolles: *Acta Cryst.* A**30** 645-648 (1974).
6. H. -B. Liu and X. -C. Zhang: *Chem. Phys. Lett.* **429** 229-233 (2006).
7. J. H. Sharp, G. W. Brindley and B. N. Narahari Achar: *J. Am. Ceram. Soc.* **49** 379-382 (1966).

Dynamical Properties on the Thermal Denaturation of Lysozyme-Trehalose Solution

Keita Sasanuma, Yuichi Seshimo, Eiji Hashimoto, Yuji Ike, and Seiji Kojima

Graduate School of Pure and Applied Sciences, University of Tsukuba Tsukuba, Ibaraki 305-8573, Japan

Abstract. We studied the dynamics of lysozyme solution and the bioprotective effect of trehalose. Thermodynamics and elastic properties related to the thermal denaturation were investigated by Modulated-temperature DSC (MDSC) and Brillouin scattering. By MDSC measurements, it is found that the thermal stability of lysozyme depends on the trehalose concentration, and trehalose suppresses the denaturation induced by pH change. The sound velocity of lysozyme-trehalose solution is studied as a function of trehalose concentration. With increasing trehalose concentration, the number density of tetrahedral structure of water molecules decreases. Furthermore, we reveal the interaction between lysozyme and trehalose increases, especially around room temperature. We suggerst that trehalose molecules tend to associate lysozyme by preferential hydration, and the trehalose-induced bioprotection phenomenon may result from the mechanical suppression of lysozyme unfolding.

Keywords: Brillouin scattering, Elastic property, Modulated-temperature DSC, Protein, Solution, Thermodynamics
PACS: 87.14.Ee, 82.60.Lf, 62.60.+v, 78.35.+c

INTRODUCTION

The biological activity of proteins depends on native structure in three dimensions, but most proteins may denature under extreme conditions (temperature, high pressure, desiccation). The conformational stability is influenced by not only intramolecular interactions of protein but also protein-solvent interactions such as hydrogen bonding, hydrophobic, and electrostatic interactions. There are many works which investigate the stability of biomaterial with the addition of denaturant and salt etc. A variety of sugars are in particular known to enhance the stability of biomaterials. Trehalose, a nonreducing disaccharide composed of two α,α-linked D-glucopyranose units, appears to be one of the most effective protectants. Both in vivo and in vitro, trehalose protects proteins and membranes from damage due to dehydration, heat, or cold. However, despite the significant amount of experimental data on trehalose, no clear picture of the mechanism responsible for its stabilizing properties has emerged yet.

Lysozyme, an enzyme with a molecular weight of 14,400 Da and a pI of 10.7, was used as a model protein. Lysozyme is one of the most famous globular proteins and a large number of works have demonstrated various properties. In the present study, we investigated the thermodynamics and elastic properties of lysozyme aqueous solutions at the various concentrations of trehalose and tried to make clear the bioprotection mechanism induced by the influence of trehalose on the thermal denaturation of protein.

EXPERIMENTAL

Hen egg white lysozyme (HEWL) was purchased from Sigma and Seikagaku Kogyo, Co., Ltd. D-(+)-Trehalose was purchased from Fluka. Acetic acid and sodium acetate were from Wako Pure Chemical Industries, Ltd.

Thermal analysis was used by Modulated-temperature Differential Scanning Calorimetry (MDSC) technique. The MDSC experiments were performed using the MDSC 2920 microcalorimeter of TA instruments. Measurement conditions were underlying heating rate, 1 °C/min, modulation amplitude, ±0.5 °C and modulation period, 60 s.

We measured GHz ultrasonic sound velocity by Brillouin scattering. The light source was a green YAG laser (λ=532 nm, 50 mW). The Brillouin scattering spectra were measured using the Sandercock-type 3 + 3 pass tandem Fabry-Perot Interferometer (FPI) at the backward scattering geometry. A free spectral range of 30 GHz was applied, and the Brillouin spectra were measured between ±25

CP982, *Complex Systems, 5th International Workshop on Complex Systems*
edited by M. Tokuyama, I. Oppenheim, and H. Nishiyama

GHz frequency range. The sample temperature was controlled from room temperature with a cryostat cell (LINKAM HTMS600). Temperature stability was within ±0.1 °C in the whole experiment. Refractive index at 532 nm was measured by the minimum deviation-angle method in order to determine the sound velocity.

RESULTS AND DISCUSSION

MDSC Measurements

Figure 1 shows the reversing C_p of the lysozyme solution at pH 4.5 in 0.1 M acetate buffer. The application of a modulated temperature program in MDSC decomposes the total C_p signal into reversing and non-reversing signals [1]. The endotherms were seen in reversing signals, while the exothermic events were in non-reversing signals. Although the analytical approach of thermal denaturation of protein by MDSC is developing, the total C_p signal was approximately equal to a combination of endotherms by unfolding of protein and overlapping exothermic events by irreversible reaction, such as aggregation [2]. We obtained the denaturated temperature T_m from the endothermic peak of reversing C_p.

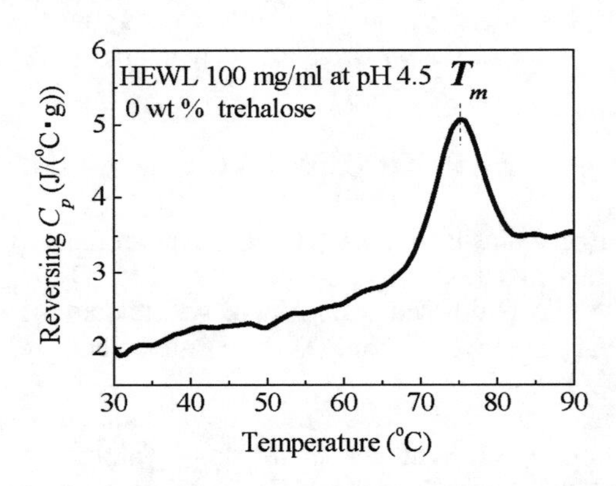

FIGURE 1. Reversing C_p of a lysozyme 100 mg/ml solution at pH 4.5 in 0.1 M acetate buffer.

Trehalose concentration and pH dependences of denaturated temperature T_m were shown in Fig. 2. T_m of lysozyme solution without trehalose decreases with the increasing pH of buffer in the studied pH range. Changes in pH can cause changes in the electrostatic environment of the protein leading to charge effects and protonation changes during thermal unfolding. Decrease in T_m as observed by us at high pH may be attributed to changes in the electrostatic environment, resulting in dimerization of lysozyme [3]. At low buffer concentrations, in contrast, charge-shielding effects result in reduced electrostatic interactions. However, this graph indicates trehalose addition suppressed the decrease of T_m for pH change and that T_m increased with trehalose concentration. Additionally, the reversibility of lysozyme unfolding is seen in the wide pH range. Their non-specific stabilizing effect is usually attributed to the mechanism of preferential exclusion [4]. Preferential interaction, preferential hydration and preferential exclusion are terms used to explain the effects of additives on protein stability.

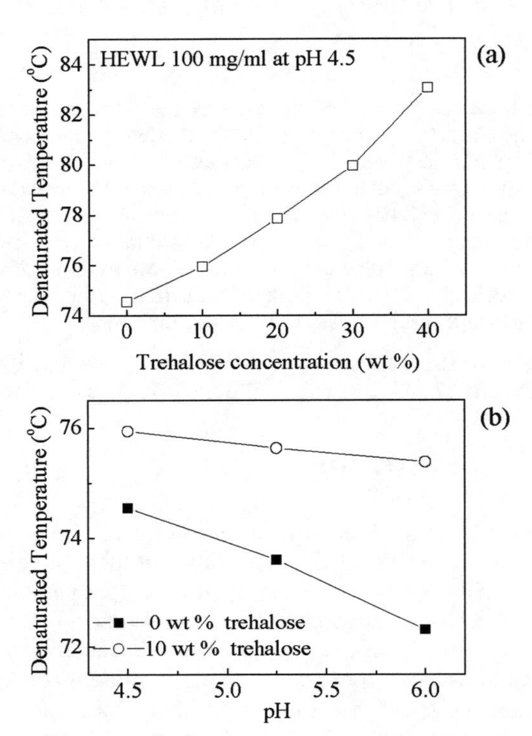

FIGURE 2. Denaturated Temperature T_m of lysozyme 100 mg/ml solutions as functions of pH and trehalose concentration.
(a) Open squares indicate the trehalose concentration dependence at pH 4.5.
(b) The pH dependence of the lysozyme solutions with trehalose (full squares) and without trehalose (open circles).

We studied the nature of the transformation at denaturated temperature T_m by MDSC technique. The effect of time on the transformation was investigated by keeping at T_m. In the new set of experiments, the sample was heated at 1 °C/min from room temperature to T_m of each sample, while still being subjected to the above given temperature modulation, and kept at that temperature. The isothermal behaviors of lysozyme and lysozyme-trehalose solutions are plotted in Fig. 3. The measured reversing C_p values were found to decrease with time. These slow changes are the overall kinetics of the irreversible denaturation process, not reversible unfolding process [5]. In Fig. 3, Trehalose addition is induced fast decrease in our studied measurement condition. To describe the decrease in reversing C_p more clearly, the shape of the curves in Fig. 3 obeys the case in which the relaxation function $\psi(t)$ is a exponential function,

$$C_p^*(t) = C_p^*(\infty) + \left[C_p^*(0) - C_p^*(\infty)\right] \Psi(t), \quad (1)$$

$$\Psi(t) = \exp(-\frac{t}{\tau}), \quad (2)$$

where $C_p*(t)$, $C_p*(\infty)$, and $C_p*(0)$ in eq. (1) are the values of reversing C_p at time t, at an infinite time, and at an initial time respectively, and τ in eq. (2) is the relaxation time. The $C_p*(0)$ value was the reversing C_p at T_m. Obtained data are in good agreement with the above equations. Fig. 4 shows the relaxation time τ vs. the trehalose concentration and the denaturated temperature of each sample. These plots indicate the relaxation time τ decreases with trehalose concentration, while is also dependent on the measured temperature equal to T_m. The plots of the logarithmic relaxation time against $1/T_m$ are straight line. To understand the dependence, we were analyzed by the Arrhenius equation given by,

$$\tau = \tau_0 \exp(-\frac{E_a}{k_B T_m}), \quad (3)$$

where k_B is Boltzmann constant, τ_0 is the pre-exponential factor, and E_a. is the activation energy. The obtained plots fitted by the eq. (3). The results are τ_0 is 1.0×10^{-9} s, and E_a is 88 kJ/mol. Evidently, the overall kinetics of the irreversible denaturation process may be affected by temperature rather than trehalose addition, therefore the Arrhenius equation holds.

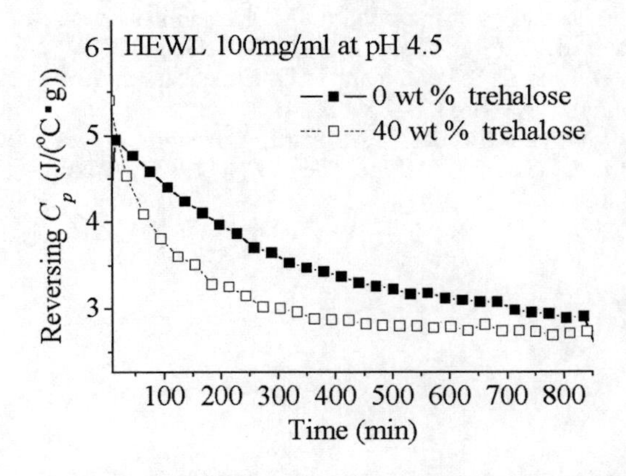

FIGURE 3. The plots of reversing C_p against time of lysozyme 100 mg/ml solution at pH 4.5.

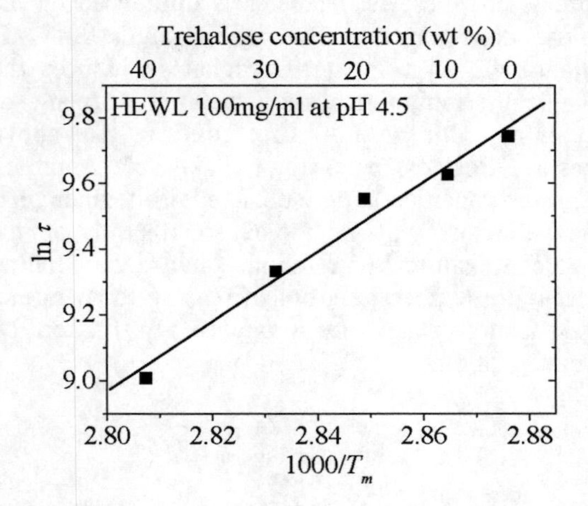

FIGURE 4. These plots are the relaxation time τ as functions of trehalose concentration and denaturated temperature.

Brillouin Scattering Investigations

Temperature dependence of the sound velocity of lysozyme solutions in buffer at pH 4.5 and in the different trehalose concentration was shown in Fig. 5. No anomalous behavior was observed on the thermal denaturation. The plots obtained were progressively changed, leading to the shifts of the velocity maximum temperature toward lower values, and were similar in behavior of trehalose-water binary system [6]. This phenomenon has already been observed in many electrolytic solutions, and it can be related to the building up of more compact trehalose-imposed structure. In other words, the number of tetrahedral structure of water gradually decreases. To investigate the relative contributions of lysozyme, we calculated the normalized sound velocity V_{LTS}/V_{TS} which was the sound velocity of lysozyme-trehalose solutions V_{LTS} divided by that of trehalose solutions V_{TS}. These temperature dependences were shown in Fig. 6. The normalized sound velocity linearly decreases with increasing temperature, and the slope becomes sharp with increasing trehalose concentration. These results indicate the interaction between lysozyme and solvent becomes stronger with the increase of trehalose concentration at room temperature range. However the difference against the concentration becomes smaller at higher temperature.

Three major hypotheses (water–trehalose hydrogen bond replacement, coating by a trapped water layer, and mechanical inhibition of the conformational fluctuations) have been proposed to explain the stabilizing effect of trehalose on proteins [7]. However currently available experimental techniques have often led to contradictory conclusions. Additionally the importance of these mechanisms is influenced by the trehalose concentration. The recent studies of MD simulation [7, 8] suggest that trehalose induces the mechanical inhibition of the conformational fluctuations. Our results also confirm the above suggestion. Besides, increasing trehalose concentration, the elastic properties largely change with temperature. These behaviors may be the reason that, lysozyme associates trehalose by preferential hydrogen bonding around room temperature, but increasing temperature, the associated states are progressively broken by thermal agitation.

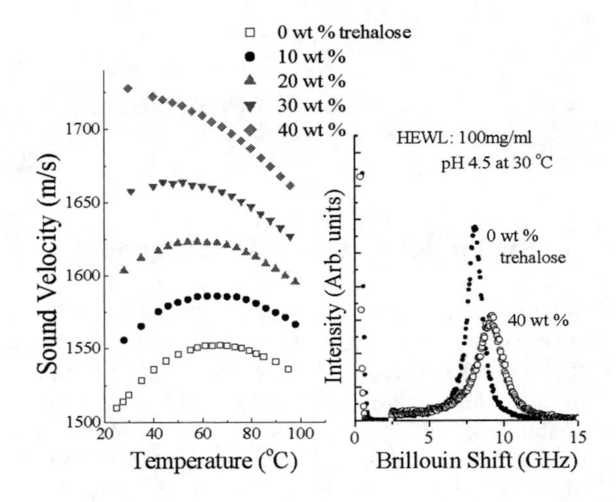

FIGURE 5. (Color online) Temperature dependence of sound velocity of the lysozyme 100 mg/ml solutions at pH 4.5 and the obtained Brillouin scattering spectra is the right.

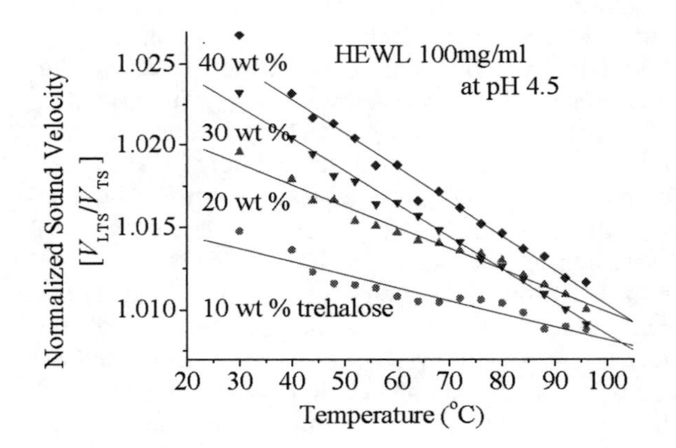

FIGURE 6. (Color online) Temperature dependences of the normalized sound velocity of the lysozyme 100 mg/ml solutions at pH 4.5.

CONCLUSION

We have studied the thermodynamics and elastic properties of lysozyme-trehalose aqueous solutions above room temperature by Modulated-temperature DSC and Brillouin Scattering measurements. The denaturated temperatures of lysozyme solutions become higher with increasing trehalose concentration. The denaturation caused by pH change is suppressed by the influence of trehalose. The time dependences of isothermal denaturation at T_m are more affected by temperature rather than trehalose concentration, and the relaxation time obeys the Arrhenius like behavior vs. temperature.

The sound velocity behavior of lysozyme-trehalose solution is discussed by the fact that the number density of tetrahedral structure of bulk water decreases gradually. The interaction between lysozyme and trehalose molecules becomes stronger, especially around room temperature. Therefore, our results suggest that lysozyme and trehalose form the associated clusters. The thermostabilization induced by trehalose is attributed to the mechanical suppression of the conformational fluctuations of lysozyme tertiary structure.

REFERENCES

1. A. A. Lacey, D. M. Price and M. Reading, "Theory and practice of Modulated Temperature Differential Scanning Calorimetry" in *Modulated Temperature Differential Scanning Calorimetry*, edited by M. Reading and D. J. Hourston, Springer, Dordrecht, 2006, pp. 1-82.
1. B. Aniket, Y. Paulos and B. Ajay, *Int. J. Pharm.*, **309**, 146 -156 (2006).
3. S. Branchu, R.T. Forbes, P. York, H. Nyqvist, *Pharm. Res.*, **16**, 702–708 (1999).
4. W. Wang, *Int. J. Pharm.*, **185**, 129-188 (1999).
5. G. Salvetti, E. Tombari, L. Mikheeva and G. P. Johari, *J. Phys. Chem. B*, **106**, 6081-6087 (2002).
6. S. Magazu, P. Migliardo, A. M. Musolino and M. T. Sciortino, *J. Phys. Chem. B*, **101**, 2348-2351 (1997).
7. R. D. Lins, C. S. Pereira and P. H. Hünenberger, *Struct. Func. Bioinf.*, **55**, 177-186 (2004).
8. A. Lerbret, P. Bordat, F. Affouard, A. Hédoux, Y. Guinet and M. Descamps, *J. Phys. Chem. B*, **111**, 9410-9420 (2007).

Brillouin and Raman Scattering Study of Ethylene Glycol Aqueous Solutions

Y. Seshimo, Y. Ike, and S. Kojima

Graduate School of Pure and Applied Sciences, University of Tsukuba Tsukuba, Ibaraki 305-8573, Japan

Abstract. We studied the cluster structure of ethylene glycol aqueous solutions by Brillouin and Raman scattering. We measured the ultrasonic sound velocity of the sample by Brillouin scattering. From the concentration dependence of the sound velocity, we studied the cluster structure in the solution. We showed that the number of H_2O molecule neighboring a EG molecule becomes a little higher with increasing temperature and the intermolecular interaction between EG and H_2O molecules weakened with increasing temperature. In Raman scattering study, We studied the hydrogen bond in the solution using the OD stretching band. We revealed that the strength of the hydrogen bond is independent of the EG concentration.

Keywords: Cluster structure, Ethylene glycol aqueous solutions, Glass transition, Hydrogen bond
PACS: 43.35.Bf, 43.30.Es, 78.35.+c

INTRODUCTION

To understand the properties of mixture of lower alcohol (ethanol, ethylene glycol, glycerol, etc) and water is very important in biophysics and chemical physics.

Ethylene glycol $(HO)CH_2CH_2(OH)$, EG and glycerol are well known cryoprotectants because of their high glass forming tendency. Cryoprotectant is a material that enables cells to cryopreserve. Usually, the water in the cell crystallizes below a melting point, and ice crystals cause destruction of the cell. This is a problem on freeze preservation [1].

Vitrification might provide a general solution to the problem of organ preservation because this procedure inhibits both the direct and indirect damaging effects of freezing. EG aqueous solutions have been used as cryoprotectant for sperm of domestic animals [2].

However, the fundamental physical properties on glass transition are not fully understood. It is very important to investigate nonequilibrium physical phenomena.

Glass transition phenomenon is related to cluster structure formed by hydrogen bond. We studied the elastic properties and cluster structure of EG aqueous solutions by Brillouin scattering and the hydrogen bond in the solution by Raman scattering.

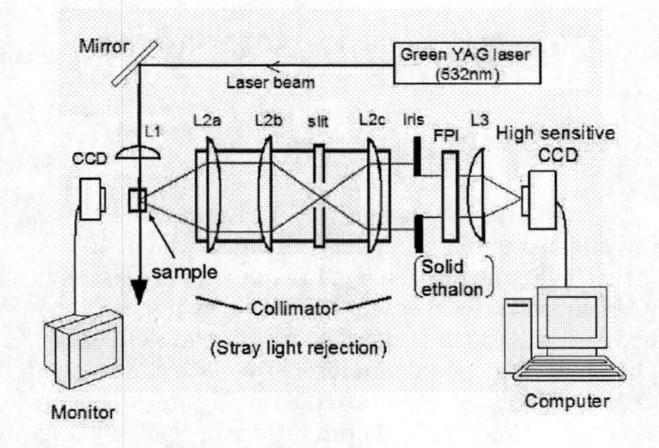

FIGURE 1. (Color online) Instrumental diagram of Brillouin scattering.

EXPERIMENTAL

Ethylene glycol aqueous solutions (0-100 mol%) were prepared. Pure water and ethylene glycol were purchased from Wako Pure Chemical Industries, Ltd. Deuterated water was purchased from Isotec Inc. The purity of EG and D_2O were 99.5 % and 99.9 %, respectively.

In Raman scattering study, We used R-3000, Raman Systems, Inc. The acquisition time of every sample was 120 second. We investigated the hydrogen

CP982, *Complex Systems, 5th International Workshop on Complex Systems*
edited by M. Tokuyama, I. Oppenheim, and H. Nishiyama

bond in solutions at room temperature by Raman scattering. Because of the difficulty in investigating OH stretching vibration, we used 5 mol% deuterated water $(H_2O)_{0.95}$-$(D_2O)_{0.05}$ in the Raman scattering measurements [4]. By measuring OD stretching vibration, we can obtain the essential contribution of hydrogen bond in the peak.

Brillouin scattering of EG aqueous solutions was studied using an angular dispersive Fabry-Perot interferometer (ADFPI) and a tandem multi pass FPI[3]. We show the instrumental diagram of ADFPI Brillouin scattering in FIGURE 1. Quick measurements become possible with the combined system of the ADFPI and a highly sensitive CCD detector. The light source for Brillouin scattering was a green YAG laser (λ=532 nm, 50mW).

We measured the temperature dependence of GHz ultrasonic sound velocity of solutions by Brillouin scattering from -150 °C to 100 °C. We used an angular dispersive Fabry-Perot interferometer (ADFPI) from 25 °C to 100 °C, a tandem multi pass FPI from -150 °C to 20 °C. The refractive index of the sample was measured at 532 nm by the minimum deviation-angle method.

RESULTS AND DISCUSSION

Raman Scattering

In Raman scattering measurements, We investigated the OD stretching vibration which reflects the strength of hydrogen bond in the solution. Usually, it is difficult to measure essential hydrogen bond because of Fermi resonance and coupling effects of OH stretching vibration. Horning *et. al.,* studied the OD stretching vibration of HDO of 5 mol% D_2O in H_2O.[4] In this system, Fermi resonance and coupling effects are fully inhibited.

We measured the concentration dependence of Raman spectra of EG aqueous solutions at 25 °C as shown in FIGURE 2. In the spectra, the peaks around 2530 cm^{-1} show the OD stretching vibration bands of HDO in the solution. We show the concentration dependence of OD band in FIGURE 3. The spectra in FIGURE 3 was corrected by subtracting EG contribution from the raw data to discuss the essential peak shift of OD stretching vibration. The increase of the peak position indicates the destruction of hydrogen bonding network, while the decrease indicates the formation of hydrogen bond [5]. We revealed that the peak position of EG aqueous solutions doesn't change. This means that the strength of hydrogen bond in solution is independent of the EG concentration.

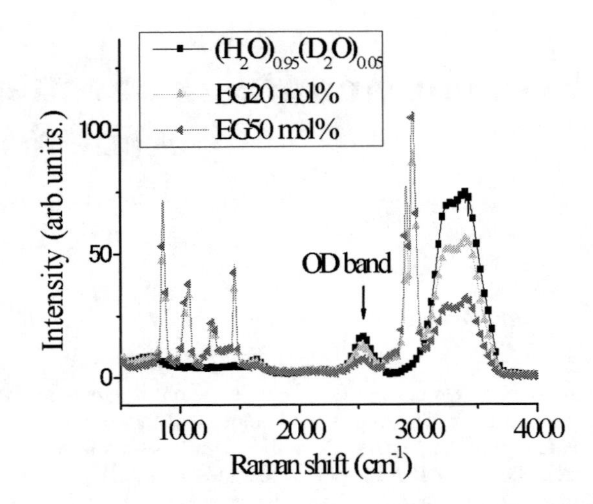

FIGURE 2. (Color online) Raman spectra of the EG aqueous solutions.

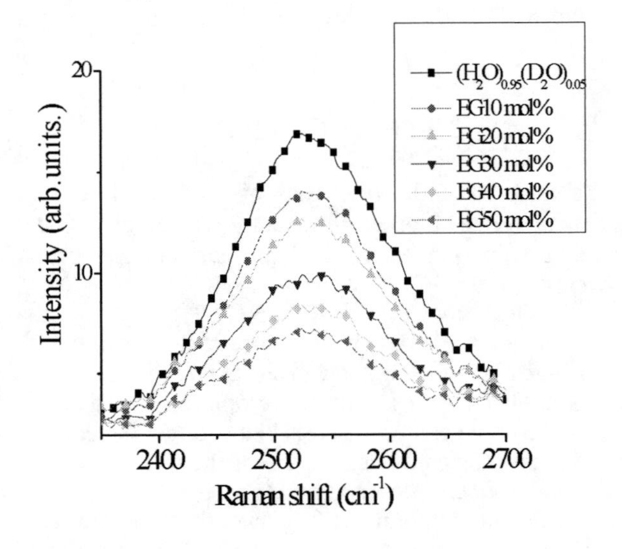

FIGURE 3. (Color Online) The concentration dependence of OD stretching vibration of EG aqueous solutions.

Brillouin Scattering

We measured the temperature dependences of the sound velocity of the EG aqueous solutions by Brillouin scattering. The Brillouin spectra of EG aqueous solutions at room temperature are shown in FIGURE 4.

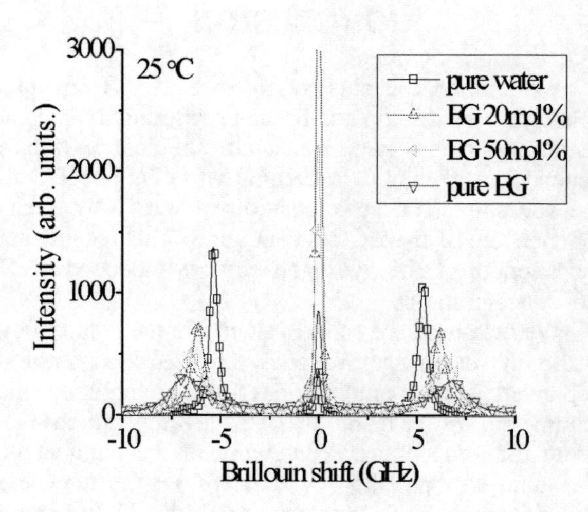

FIGURE 4. (Color Online) Brillouin spectra of EG aqueous solutions.

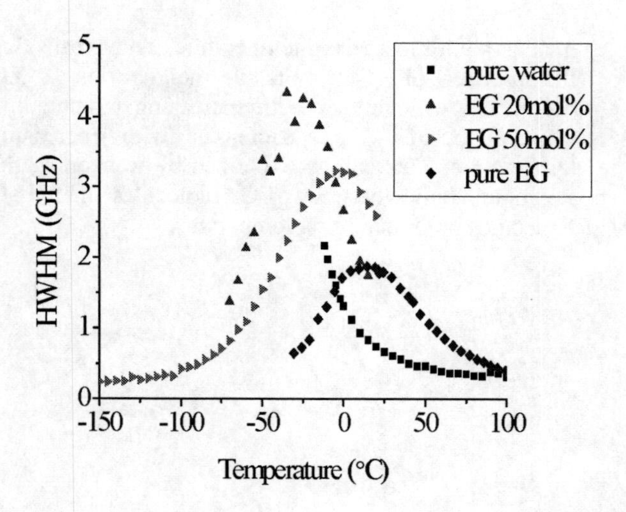

FIGURE 6. (Color Online) Temperature dependence of HWHM of EG aqueous solutions.

We can obtain the information of sound velocity in the solution from the Brillouin shift, and the half width at half maximum (HWHM) related to the attenuation constant. We show the temperature dependences of sound velocity and the attenuation constant of EG aqueous solutions in FIGUREs 5 and 6. Pure water and EG crystallized at about -12 °C and -30 °C, respectively. From the temperature dependence of sound velocity of EG aqueous solutions, we discussed the cluster structure, containing EG and H_2O molecules.

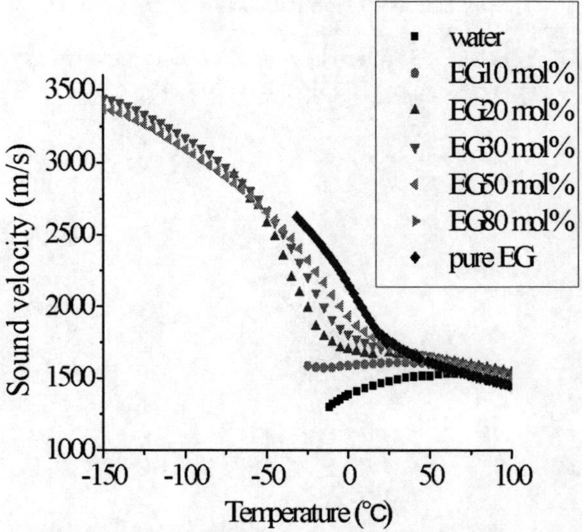

FIGURE 5. (Color Online) Temperature dependence of ultrasonic sound velocity of EG aqueous solutions.

The change of the sound velocity reflects the intermolecular interaction and cluster structure. If the sound velocity shows linear change with increasing concentration, it suggest no interaction between EG molecules and H_2O molecules. However, the sound velocity shows local maximum at a concentration. This can be explained by the cluster structure consists of EG and H_2O molecules. We analyzed the results of the sound velocity, by the following equation.

$$V(x_{EG})=x_{EG}V_{EG}+(1-x_{EG})V_{water}+x_{EG}(1-x_{EG})^{\alpha}V_{cluster} \quad (1)$$

This equation takes the effect of cluster structure into account. In the equation, x_{EG}, V_{EG}, V_{water}, α, $V_{cluster}$ are mol fraction of EG, sound velocity of EG, sound velocity of water, number of water neighbering a EG molecule, sound velocity of the cluster structure consist of EG and H_2O, respectively. The first term reflects the contribution of associated EG molecules, the second one reflects the contribution of associated H_2O molecules, and the third one reflects the contribution of cluster structure consist of EG and H_2O.

From the results of Raman scattering, we showed that strength of the hydrogen bond is independent of the EG concentration. In the fitting procedure, we assume that α and $V_{cluster}$ are independent on the concentration. The result of fit is shown in FIGURE 7.

We determined the parameters α and $V_{cluster}$ at 25 °C. We revealed the value of α= 2.51 and $V_{cluster}$= 1178 m/s. We consider that the EG-H_2O cluster consists of two EG molcules and five H_2O molecules.

The temperature dependence of α and $V_{cluster}$ is shown in FIGURE 8. We revealed that the value of α

increases with increasing temperature. This means that the number of H_2O molecule neighboring a EG becomes a little higher with increasing temperature. The value of $V_{cluster}$ decreases with increasing temperature. This can be explained by weakening the hydrogen bond between EG molecules and H_2O molecules with increasing temperature.

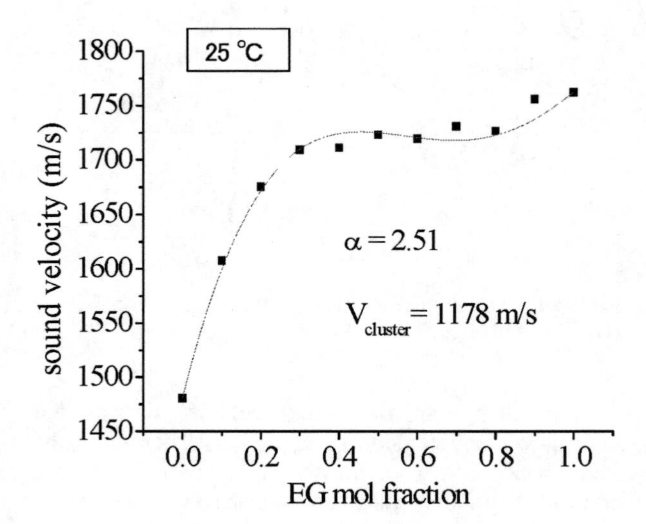

FIGURE 7. (Color Online) Result of fit of the sound velocity at 25 °C.

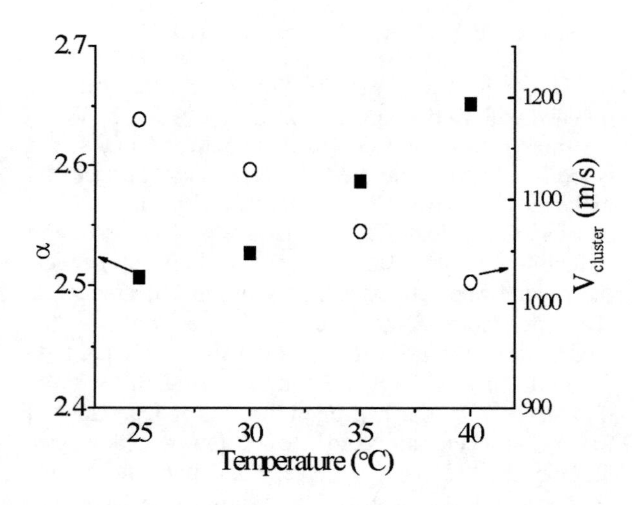

FIGURE 8. (Color Online) Temperature dependence of α and $V_{cluster}$.

CONCLUSION

We studied the cluster structure of EG aqueous solutions by Brillouin and Raman scattering. In Raman scattering study, we measured the concentration dependence of the OD stretching vibration of HDO in the solutions. The peak position of the OD band is independent of the EG concentration. This means that the strength of the hydrogen bond is independent of EG concentration.

We measured the sound velocity in the solution by Brillouin scattering. We analyzed the concentration dependence of sound velocity, and determined the parameters α= 2.51 and $V_{cluster}$= 1178 m/s in eq. (1). From the temperature dependence of the parameters, we shows that the number of H_2O molecule neighboring a EG becomes a little higher with increasing temperature and the combination between EG molecules and H_2O molecules weaken with increasing temperature.

In this study, we measured that the sound velocity related to the cluster structure. Brillouin scattering is a n important tool to investigate the dynamical properties of aqueous solutions.

REFERENCES

1. C. Gao, G. Y. Zhou, Y. Xu and T. C. Hua, Thermochimica acta, **435**, 38-43 (2005).
2. A. Baudot and V. Odagescu, Cryobiology, **48**, 283-294 (2004).
3. Y. Ike, S. Tsukada and S. Kojima, Rev. Sci. Inst. **78**, 076104-076106 (2007).
4. T. T. Wall and D. F. Horning, J. Chem. Phys. **43**, 2079-2087 (1965).
5. K. Yonehama, S. Amornthep and H. Kanno, Cryobiology and Cryotechnology, **51**, 151-154 (2005).

Terahertz Time Domain Spectroscopy of Glass-Forming Materials

S. Kojima, Y. Ike, Y. Seshimo, K. Fukushima*, and R. Fukasawa*

Graduate School of Pure and Applied Sciences, University of Tsukuba, Tsukuba, Ibaraki 305-8573, Japan
**Tochigi Nikon Corporation, 770 Midori, Otawara 324-8625, Japan*
e-mail: kojima@bk.tsukuba.ac.jp

Abstract. The terahertz dynamics of glass-forming liquids has been studied by terahertz time domain spectroscopy. The real and imaginary parts of dielectric constants have been determined accurately by the transmission measurements using the dual-thickness geometry. The complex dielectric spectra of aqueous solutions of ethylene glycol clearly show the existence of relaxation processes at room temperature. As the concentration of ethylene glycol increases, the relaxation time of the main relaxation process becomes longer. It is attributed to the destruction of the hydrogen-bonded tetrahedral structure of water and the formation of clusters consists of water and ethylene glycole molecules.

Keywords: Terahertz, Spectroscopy, Dielectric constant, Glass, Relaxation
PACS: 64.70.Pf, 77.22.Gm, 78.20.Ci

INTRODUCTION

The recent development of the generation of coherent THz radiation by a femto-second pulse laser enables the very promising Terahertz Time-Domain Spectroscopy (THz-TDS). It has the better signal to noise ratio at frequency below 3 THz in comparison with the conventional Fourier transform infrared (FTIR) spectroscopy. This time-gated coherent nature makes possible to measure its phase delay accurately and to determine the real and the imaginary parts of complex dielectric constants in the THz region independently. Since the terahertz dynamics of condensed matters has been not yet fully understood, THz-TDS has been applied to study THz dynamics of low frequency optical phonon, phonon-polariton, exciton-poparition and other low energy excitations such as a boson peak of glassy materials including biomaterials such as protein.

Cooperative motions in liquid-glass transitions are the current topics of non-equilibrium statistical physics. It has been extensively studied with much interest on the dynamics of the glass-forming materials theoretically and experimentally. However, the present knowledge especially on THz dynamics such as fast relaxation process and boson peak is far from the satisfaction [1-3]. Currently aqueous solutions of ethylene glycol attract attention as the promising cryo-protectant and bioprotectant materials for cryo-preservation of cells or tissues because of their strong glass-forming tendency at low temperatures. However, the mechanism of their glass-forming tendency is still unclear. In the present study the fast relaxation processes of aqueous solutions of ethylene glycol are studied by THz-TDS.

EXPERIMENTAL

The THz time domain spectrometer used is shown in FIGURE 1. The femtosecond pump pulses with the wavelength of 780 nm and the power of 1.0 W were generated by a Ti sapphire laser (Maitai) with a repetition rate of 80 MHz. The reduced pump pulses of about 20 mW are focused by an objective lens of high magnification onto the biased gap of 5 μm on the photoconductive antenna made by a low-temperature grown GaAs. The emitted THz radiation is collimated and focused by a paraboloidal mirror into a sample cell. The transmitted THz signals from a sample are collimated again, and focused by another paraboloidal mirror onto a detector of a GaAs photoconductive antenna of the same type as the emitter. The received signals are gated using a part of incident femtosecond pulses. An ammeter and a lock-in amplifier are used to measure the photocurrent induced by the transmitted

CP982, *Complex Systems, 5th International Workshop on Complex Systems*
edited by M. Tokuyama, I. Oppenheim, and H. Nishiyama

radiation field through a sample. The time-dependent waveform of transmitted signals is accurately reproduced by changing the delay time of a reference beam. The real and imaginary parts of complex dielectric constants in the frequency domain are determined from the time dependent amplitude and the phase of a transmission spectrum using the Fourier transformation in the frequency range between 0.1 THz and 1.5 THz. All samples were measured at room temperature.

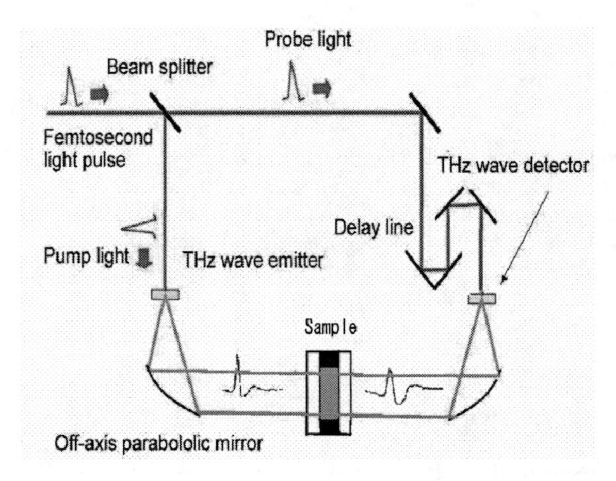

FIGURE 1 Block diagram of terahertz time domain spectrometer.

One of unsolved problems of THz-TDS was the uncertainty of complex dielectric constants of frequency domain determined by Fourier transformation of time domain amplitude and phase signals because of multiple splutions. In the present study, it can be solved by the dual-thickness geometry method, and it becomes possible to obtain a unique solution [4]. When the multiple reflection does not occur, the following equation holds for the transmission ration $t(\omega)$ between two samples with different thickness d_1 and d_2 (FIGURE 2),

$$t(\omega) = \frac{t_{sam-2}}{t_{sam-1}}$$

$$= \exp\{-\frac{k\omega}{c}(d_2 - d_1) + i\frac{(n-1)\omega}{c}(d_2 - d_1)\} \qquad (1)$$

where c is the light velocity, and the real part $n(\omega)$ and imaginary part $k(\omega)$ of refractive index of a sample to be studied are related to real part $\varepsilon'(\omega)$ and imaginary part $\varepsilon''(\omega)$of complex dielectric constant.

$$\varepsilon'(\omega) + i\varepsilon''(\omega) = (n(\omega) - ik(\omega))^2 \ . \qquad (2)$$

The second advantage of this geometry is that the equation (1) is independent of the parameters of optical windows of a sample cell if the thickness of the windows of two optical cells with different spacing is equal. The real part $n(\omega)$ and imaginary part $k(\omega)$ of refractive index are derived uniquely from the following equations (3) and (4),

$$\theta(\omega) = \theta_1(\omega) - \theta_2(\omega) \ , \qquad (3)$$

$$n(\omega) = \frac{c}{\omega(d_2 - d_1)}\theta(\omega) + 1, \qquad (4)$$

$$k(\omega) = -\frac{c}{\omega(d_2 - d_1)}\ln t(\omega) \ . \qquad (5)$$

where $t_i(\omega)$ and $\theta_i(\omega)$ are Fourier transforms of time dependent transmission and phase delay for sample i (i=1,2), respectively.

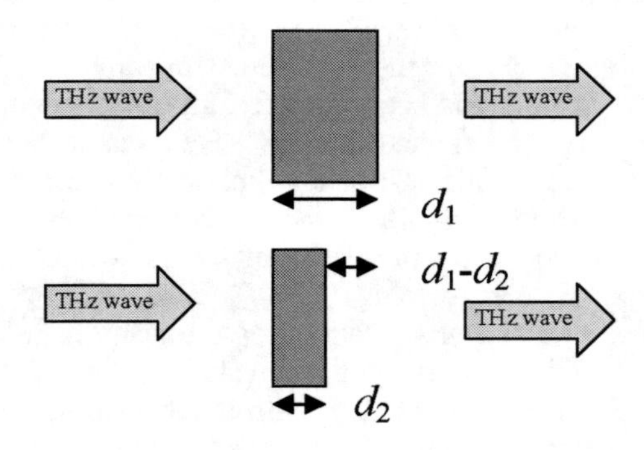

FIGURE 2 The dual-thickness geometry of THz-TDS. Two samples with two different spacing d_1 and d_2 are measured in the same condition at the same optical geometry. It is possible to obtain a unique solution of real and imaginary parts of dielectric constants.

DETERMINATION OF COMPLEX DIELECTRIC CONSTANT

Since the present THz-TDS apparatus has been newly designed and constructed, the calibration is necessary. Therefore, we studied water at first as a standard sample to confirm the accuracy of the present THz spectrometer. The purified water was measured at the dual-thickness geometry with d_1=100 μm and d_2=50 μm. The observed spectra of real and imaginary parts of complex dielectric constant are shown in FIGURES 3 and 4, respectively. The recent THz study by Yu et al. fitted the dielectric spectra in the THz range by the double Debye model [4]. In this model it holds that

$$\varepsilon(\omega) = \varepsilon_\infty + \frac{\varepsilon_S - \varepsilon_2}{1 + i\omega\tau_1} + \frac{\varepsilon_2 - \varepsilon_\infty}{1 + i\omega\tau_2}, \quad (6)$$

where ε_∞ is the high frequency limit of dielectric constant and ε_S-ε_2 and ε_2-ε_∞ are the relaxation strength for slow and fast relaxation processes, respectively. τ_1 and τ_2 are the relaxation time for slow and fast relaxation processes, respectively. The slow process relates the cooperative motion of the tetrahedral structure of water molecules (a water "cage"), while the fast process is attributed to the fast relaxation motion of a water molecule inside a tetrahedral cage.

In FIGURES 3 and 4 the solid line shows the calculated curve by eq. (6) using the parameters reported in ref. 4. The results of the present measurement of complex dielectric constant of water are in good agreement with those reported by Yu et al within experimental uncertainty.

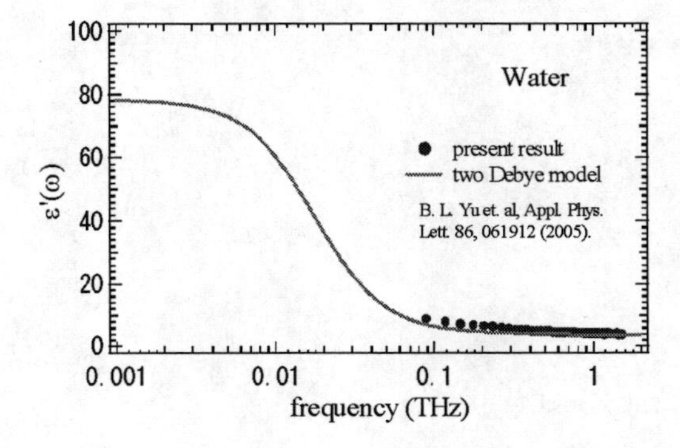

FIGURE 3 Real part of dielectric constant of water. The solid line is the calculated result by eq.(6) using the parameters reported in ref. 4.

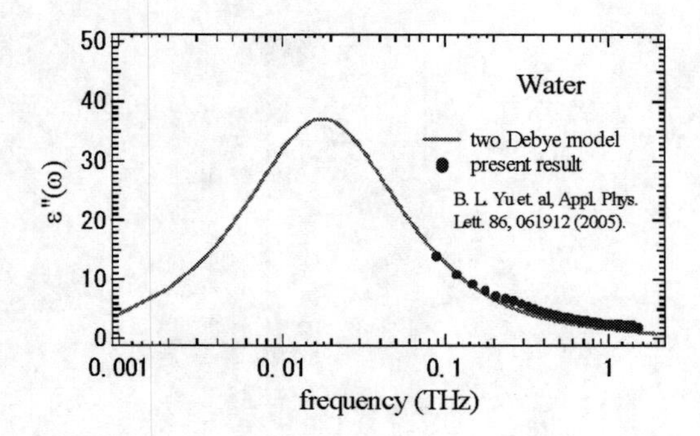

FIGURE 4 Imaginary part of dielectric constant of water. The solid line is the calculated result by eq.(6) using the parameters reported in ref. 4.

Then we studied the glass-forming ethylene glycol aqueous solutions. The real and imaginary parts of complex dielectric spectra were measured as a function of ethylene-glycol mole fraction. The complex dielectric spectra of aqueous solutions o clearly show the existence of slow and fast relaxation processes at room temperature as shown in FIGURES 5 and 6. The concentration dependence of these relaxation processes is related to the associated clusters related to the cooperative motion which controls the glass-forming tendency. As the concentration of ethylene glycol increases, the relaxation time of the slow relaxation process becomes longer by the formation of clusters consists of water and ethylene glycole molecules, and enhances the glass-forming tendency.

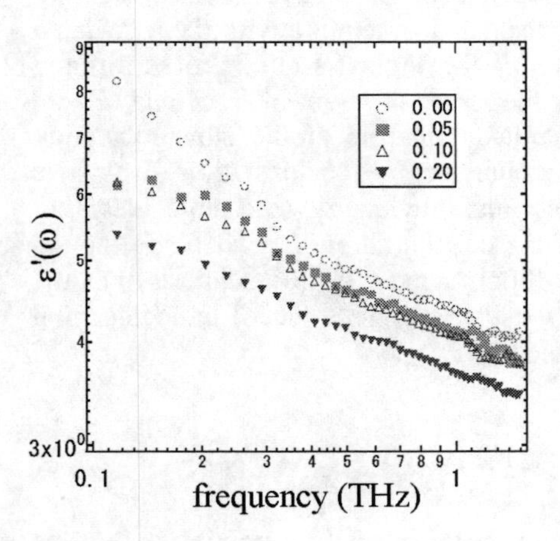

FIGURE 5 The real part of the dielectric constant of ethylene glycole aqueous solution measured by the transmission at the dual thickness geometry. The numbers denote the mol fraction of ethylene glycol.

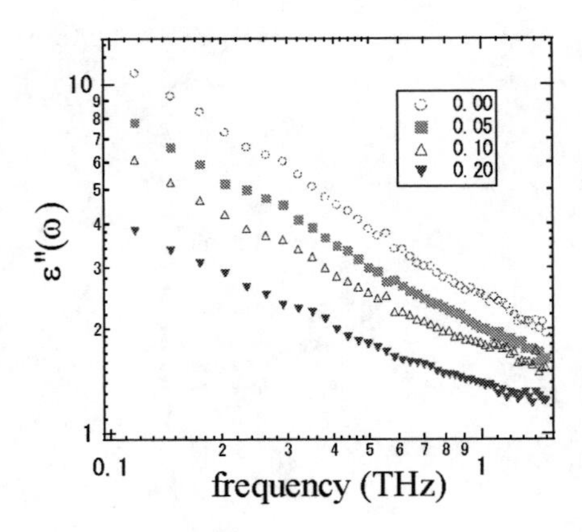

FIGURE 6 The imaginary parts of the dielectric constant of ethylene glycole aqueous solutions determined by the measurement of the transmission through a sample at the dual thickness geometry. The numbers denote the mol fraction of ethylene glycol in solutions.

CONCLUSION

Terahertz time domain spectroscopy has been applied to study fast relaxation processes of glass-forming materials. To determine uniquely the real and imaginary parts of complex dielectric constants in the frequency domain, the dual-thickness geometry has been used for all experiments. The complex dielectric spectra of aqueous solutions of ethylene glycol clearly show the existence of relaxation processes related to cooperative motion at room temperature. The concentration dependence of these relaxation processes is studied and discussed in relation with the associated clusters which play the dominant role in glass-forming tendency. As the concentration of ethylene glycol increases, the relaxation time of the slow relaxation process becomes longer, by the formation of clusters consists of water and ethylene glycole molecules. Our latest study of the concentration dependence of sound velocity by Brillouin scattering [6] suggests that the cluster consists of two ethylene glycol molecules and five water molecules.

ACKNOWLEDGMENTS

This research was partially supported by the Ministry of Education, Science, Sports and Culture in Japan, Grant-in-Aid for Scientific Research (A), 2004-2006, No.15654057.

REFERENCES

1. S. Kojima H. Kitahara, S. Nishizawa, and M. Wada Takeda, *J. Mol. Struc.*, **651,** 285-288 (2003).
2. Y. Ike, Y. Seshimo and S. Kojima, *Cryobiology and Cryotechnology*, **52,** 105-108 (2006).
3. S. Kojima, H. Kitahara, S. Nishizawa, Y. Yang and M. Wada Takeda, *J. Mol. Struc.*, **744,** 243-246 (2005).
4. S. P. Maickan et al., J. Opt. B: *Quantum Semiclass. Opt.*, **6,** S786-S795 (2004).
5. B. L. Yu, Y. Yang, F. Zeng, X. Xin and R. R. Alfano, *Appl. Phys. Letters.*, **86,** 061912-1-3 (2005).
6. Y. Seshimo, Y. Ike and S. Kojima, *AIP Proc.* (to be published)

Novel Finite Difference Lattice Boltzmann Model for Gas-Liquid Flow

Long Wu, Michihisa Tsutahara, and Shinsuke Tajiri

Graduate School of Science and Technology, Kobe University
Rokko-dai, Nada-ku, Kobe, 657-8501, Japan

Abstract. A novel model for the finite difference lattice Boltzmann method is proposed to simulate gas-liquid two-phase flow. The effect of the large density difference is incorporated by applying acceleration modification which is deduced from macroscopic dynamics in this model. Compressibility of the fluid is adjustable. Surface tension effects are included by introducing a body force term based on the Continuum Surface Force (CSF) method. The recoloring step is replaced by a anti-diffusion scheme. Here we present results for liquid column collapse and droplet splashing on a thin liquid film. Results are in good agreement with experiment.

Keywords: Droplet splashing, Lattice Boltzmann method, Liquid column collapse, Multiphase flow
PACS: 47.11.-j, 47.61.Jd

INTRODUCTION

In recent years, the lattice Boltzmann method (LBM) has been developed as an alternative and powerful tool for multiphase flow. Several LB models for the simulation of multiphase flow have been developed. In this study, the immiscible LBGK model (ILBGK) proposed by Gunstensen *et al.*[1] is improved to simulate the gas-liquid flow for large density ratio. In order to achieve stable computation for large density ratios, a finite difference lattice Boltzamann method (FDLBM) was employed in this study.

NUMERICAL METHOD

FDLBM is derived from the continuous Boltzmann equation with the collision term in the Bhatnagar-Gross-Krook (BGK) approximation for multi-component system

$$\frac{\partial f_i^k}{\partial t} + \mathbf{e}_i \cdot \nabla f_i^k + \mathbf{a} \cdot \nabla_{\mathbf{e}_i} f_i^k = -\frac{1}{\tau^k}\left(f_i^k - f_i^{k,eq}\right) \qquad (1)$$

where $f^{k,eq}$ is the equilibrium distribution function for the kth fluid. $\mathbf{e}_i$ is the lattice velocity in the ith direction. For the 9-speed model, we have the discretized the force term [2] as follows

$$\mathbf{a} \cdot \nabla_{\mathbf{e}} f^k = -3\omega_i \rho^k \left[\frac{1}{c^2}(\mathbf{e}_i - \mathbf{u}) + 3\frac{(\mathbf{e}_i \cdot \mathbf{u})}{c^4}\mathbf{e}_i\right] \cdot \mathbf{a} \qquad (2)$$

In Eq.(1), τ_k is the relaxation time for each kind of fluid. The viscosity of the fluid is

$$\nu = \tau/3 \qquad (3)$$

where τ is the relaxation time for the color blind fluid and is defined as $\tau = \sum \rho^k \tau^k / \rho$. These collision and relaxation rules lead to the macroscopic mass and momentum equations as

$$\partial_t \rho + \mathbf{u}\nabla \cdot \rho = 0 \qquad (4)$$

$$\partial_t \mathbf{u} + \mathbf{u} \cdot \nabla \mathbf{u} = -\frac{1}{\rho}\nabla p + \nu \nabla^2 \mathbf{u} \ . \qquad (5)$$

The densities ρ^k for different phases, the total density ρ, the momentum $\rho\mathbf{u}$ and the pressure p are obtained from the following equations:

$$\rho^k = \sum_{i=0}^{N} f_i^k = \sum_{i=0}^{N} f_i^{k,eq} \qquad (6)$$

$$\rho = \sum_k \rho^k \qquad (7)$$

$$\rho\mathbf{u} = \sum_{i=0}^{N} \mathbf{e}_i f_i = \sum_{i=0}^{N} \mathbf{e}_i f_i^{eq} \qquad (8)$$

$$p = \frac{1}{3}\rho \qquad (9)$$

CP982, *Complex Systems, 5th International Workshop on Complex Systems*
edited by M. Tokuyama, I. Oppenheim, and H. Nishiyama

The time integration is performed by the second order Runge-Kutta method. The third-order upwind scheme is employed for space discretization in FDLBM calculation. We modify Gunstensen's model [1] by replacing the perturbation step with a direct force term and change the recoloring step by an anti-diffusion scheme.

In order to incorporate the interfacial tension effect, we force a local pressure gradient across the interface by an additional force term. A popular model for the interfacial tension force is the Continuum Surface Force (CSF) model, which was developed by Brackbill *et al.* [3]. The interfacial tension force F is defined as

$$\mathbf{F} = \sigma\kappa \frac{\nabla C}{|\nabla C|} \tag{10}$$

where σ is the interfacial tension coefficient. $C(\mathbf{x},t)$ is the color field and κ is the interface curvature.

The other improvement is the modification of the recoloring step by implementing Latva-Kokko anti-diffusion scheme [4]. The following redistributions for the red and blue particles (two phases) are performed after the collision step.

$$f_i^r = \frac{\rho_r}{\rho_r + \rho_b} f_i + \beta \frac{\rho_r \rho_b}{(\rho_r + \rho_b)^2} f_i^{eq(0)} \cos\varphi \tag{11}$$

$$f_i^b = \frac{\rho_b}{\rho_r + \rho_b} f_i - \beta \frac{\rho_r \rho_b}{(\rho_r + \rho_b)^2} f_i^{eq(0)} \cos\varphi \tag{12}$$

where f_i and $f_i^{eq(0)}$ are the color-blind distribution functions and zero-velocity equilibrium distribution functions going to ith direction respectively. φ, defined in Eq.(13), is the angle between the color gradient and the particle velocities $\mathbf{e}_i$. β is the parameter relating to the tendency of the two fluids to separate.

$$\cos\varphi\big|_i = \frac{\mathbf{G}\cdot\mathbf{e}_i}{|\mathbf{G}||\mathbf{e}_i|} \tag{13}$$

The usefulness of our improved model for easily and accurately simulating multiphase flow has been tested previously [5]. The spurious velocities are significantly decreased and the side effect of a recoloring step is removed.

Until now, we presented an improved model for the simulation of multiphase flow with same molecular mass $m_b = m_r$, which are usually set as 1 for simplicity ($\rho = \rho m$). However, the mass density ratio of liquid-gas systems is usually larger than 100. For the purpose of recovering correct Navier-Stokes equations for gas-liquid flow, the correctional force must be incorporated.

When Eq.(1) is solved by conventional model ($m_b = m_r$), the macroscopic equations can be derived from FDLBM through the Chapman-Enskog expansion procedure as

$$\frac{\partial \mathbf{u}}{\partial t} + \mathbf{u}\cdot\nabla\mathbf{u} = -\frac{1}{\rho}\nabla p + \nu\nabla^2\mathbf{u} \tag{14}$$

Assuming $\mathbf{a} = \partial\mathbf{u}/\partial t$, then, the acceleration obtained by conventional LBGK can be written as

$$\mathbf{a} = -\mathbf{u}\cdot\nabla\mathbf{u} - \frac{1}{\rho}\nabla p + \nu\nabla^2\mathbf{u}. \tag{15}$$

However, for the real flow field, molecular mass m may be very different, so the correct N-S equation should be written as

$$\mathbf{a}' = -\mathbf{u}\cdot\nabla\mathbf{u} - \frac{1}{\rho m}\nabla p + \nu\nabla^2\mathbf{u} \tag{16}$$

where $m = \sum \rho^k m_k / \rho$ is molecular mass for color blind fluid. The acceleration modification is given as

$$\Delta\mathbf{a} = \mathbf{a}' - \mathbf{a} = -\frac{1}{\rho m}\nabla p - \left(-\frac{1}{\rho}\nabla p\right). \tag{17}$$

Then, this modification can be introduced to Eq.(1) as a force term to realize the computation on two phase flow with large density ratio.

After introducing our modification, the sound speed of the system is $c_s = \sqrt{\Delta p/\Delta(\rho m)}$. If the molecular mass is much larger then 1, the sound speed is very small, which can lead to a large compressibility error for the lattice Boltzmann method. In addition, the sound speed of liquid phase should be much larger than that of the gas phase and to maintain incompressibility. To achieve this, a strong interaction between molecules should be introduced for the liquid phase.

As is well known, for an non-ideal gas or liquid, a small change of density leads to large pressure fluctuations. However, the mechanism of this relation is still not clear except those near critical point. Since our aim is to establish an incompressible model, it is not necessary to use the exact equation of state for non-ideal gas or liquid. In the present work, a numerical technique is employed to include the effect of large pressure fluctuation of liquid phase by incorporating the linear correction to the equation of state

$$p' = c_s^2 \rho_{ref} + \gamma c_s^2 (\rho - \rho_{ref}) \tag{18}$$

where, γ is used for controlling the intensity of compressibility and adjusting the sound speed (here the sound speed is changed to $c_s = \sqrt{\Delta p'/\Delta(\rho m)} = \sqrt{\gamma/m}\sqrt{\Delta p/\Delta\rho}$). ρ_{ref} is the reference density of the fluid and set to be 1. With the new definition of pressure, Eq.(17) can be written as

$$\Delta\mathbf{a} = \mathbf{a}' - \mathbf{a} = -\frac{1}{\rho m}\nabla p' - \left(-\frac{1}{\rho}\nabla p\right) \tag{19}$$

The efficiency of the current method on maintaining the incompressibility was checked by simulations on Poiseuille flow and unsteady Womersley flow in our previous work [6]. The results show that the proposed model can really decrease the compressibility error by using a proper parameter γ ($\gamma = 3m$ is used in this study).

RESULTS AND DISCUSSIONS

Liquid Column Collapse

The collapse of a liquid column was simulated schematically as shown in Fig.1. The non-dimensional gravitational acceleration, g, is set to -0.0001. The viscosity is set to $\nu_l = 0.0001$ for liquid phase (water) and $\nu_g = 0.0075$ for gas phase (air). The mass densities for liquid and gas are $m_l = 1000$ and $m_g = 1$ respectively. All the solid boundaries are assumed to be slippery walls.

Figure 2 shows the velocity vector and profiles of collapsing liquid column in a dimensionless time ($T = nt\sqrt{|g|/a}$) for $n^2 = H/a = 2$ at T=0.89443, (ii) T=1.7890, (iii) T=2.6834, and (iv) T=4.4721. Figures 3 and 4 show the comparisons of the instantaneous leading-edge position X and height of liquid column Z between our simulations and experiment for $n^2 = H/a = 1,2,4$. The experimental data were provided by Martin and Moyce [7]. It can be seen from Figs. 3 and 4 that the calculated X and Z at various values of n^2 are in good agreement with the experiment data. Using the present new model, the liquid column collapse problem was simulated successfully.

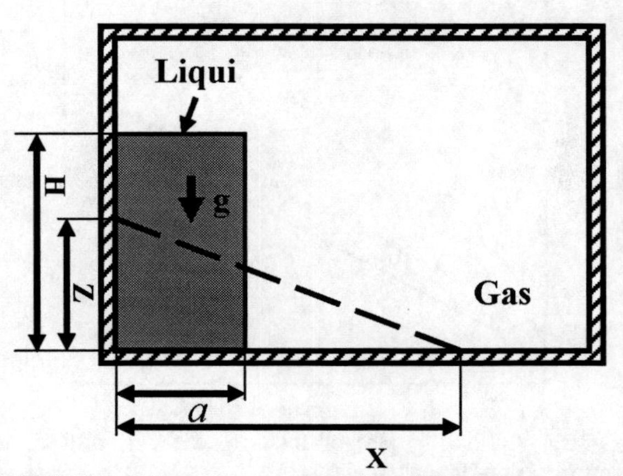

FIGURE 1. Schematic representation of the computation for the liquid column collapse.

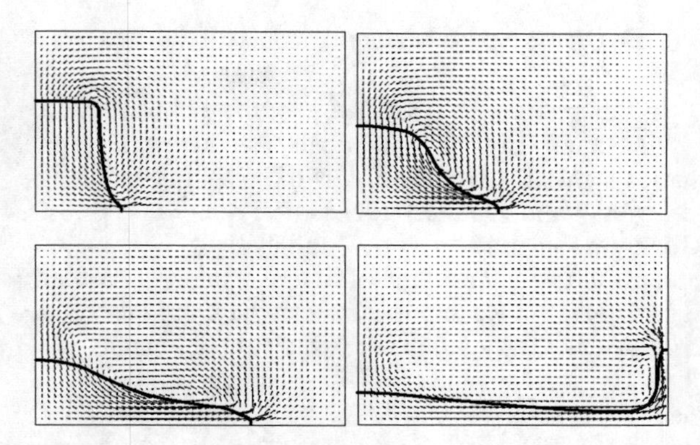

FIGURE 2. The time series of profiles of the collapsing liquid.

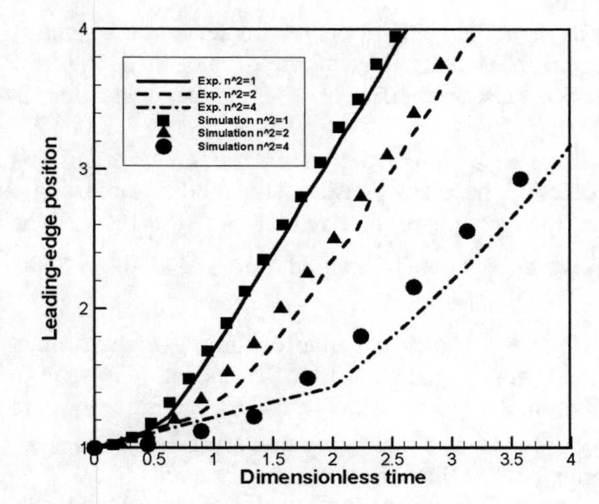

FIGURE 3. Comparison of the instantaneous leading-edge position X between the simulated results and experiment for water system.

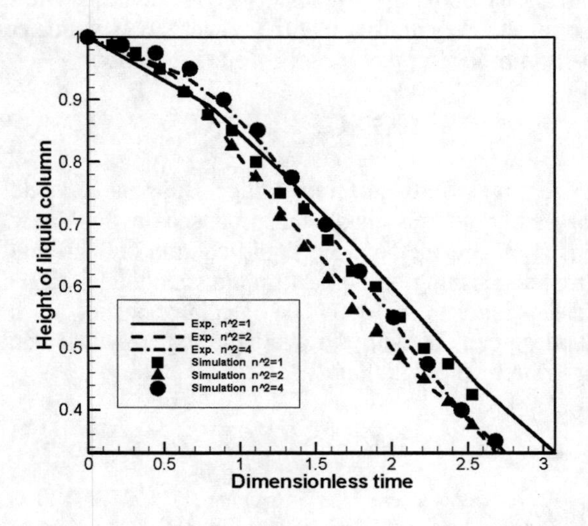

FIGURE 4. Comparisons of the instantaneous height of liquid column between the simulated results and experiment for water stystem.

Droplet Splashing on a Liquid Film

In this section, the early stage of droplet impact on a thin liquid film is numerically studied. Previous studies show that the power law of splashing radius is generally valid at short times after the impact [8], which can be used to verify our method. The major dimensionless parameters of interest are the Weber number and the Reynolds number which are based on the properties of liquid phase, which are defined as

$$We_l = \frac{2\rho_l m_l U^2 R}{\sigma} \tag{20}$$

$$Re_l = \frac{2\rho_l m_l UR}{\mu_l} = \frac{2UR}{\nu_l} \tag{21}$$

Here, U is the velocity of the droplet at the instant of impact, R is the radius of the droplet, and σ is the surface tension coefficient. The characteristic time is $2R/U$.

Figure 5 shows the schematic diagram of this problem. The mass density ratio of the liquid phase and the gas phase is fixed at $m_l/m_g = 1000$. The Weber number and Reynolds are set to $We_l = 8000$ and $Re_l = 500$.

Figure 6 shows the time evolution of the droplet and the thin liquid film after the instant of impact, at which the non-dimensional time $Ut/2R$ is set to zero when the droplet impinging just starts. A liquid sheet coming out of the neck is clearly observed.

Figure 7 shows the log-log plot of the spread factor r/2R against the non-dimensional time $Ut/2R$. The straight line corresponds to the power law $r = \sqrt{2RUt}$ and the symbols are the results of simulations. From the results, we can find that the power law is validated at short time after the impact by our simulations.

CONCLUSIONS

A novel finite different lattice Boltzmann model for gas-liquid flow has been proposed in this study. Based on this method, the liquid column collapse and droplet splashing on liquid film are simulated with the density ratio up to 1000. The calculated results are in good agreement with the available experimental data or theoretical predictions.

REFERENCES

1. A. K. Gunstensen, D. H. Rothman, S. Zaleski and G. Zanetti, *Phys. Rev. A* **43**, 4320-4327 (1991).
2. L. S. Luo, *Phys. Rev. E*, **62**, 4982-4996 (2000).
3. J. U. Brackbill, D. B. Kothe and C. Zemach, *J. Comput. Phys.*, **100**, 335-354 (1992).
4. M. Latva-Kokko and D. H. Rothman, *Phys. Rev. E*, **71**, 056702 1-8 (2005).
5. L. Wu, M. Tsutahara and S. Tajiri, "Numerical Simulations of Droplet Formation in a Cross-junction Microchannel by the Lattice Boltzmann Method" *Int. J. Numer. Methods Fluid.*(in press).
6. L. Wu, M. Tsutahara and S. Tajiri, *J. Fluid Sci. Technol.*, **2**, 35-44 (2007).
7. J. C. Martin and W. J. Moyce, *Philos. Trans. Roy. Soc. Lond. A*, **244**, 312-324 (1952).
8. C. Josserand and S. Zaleski, *Phys. Fluids*, **15**, 1650-1657 (2003).

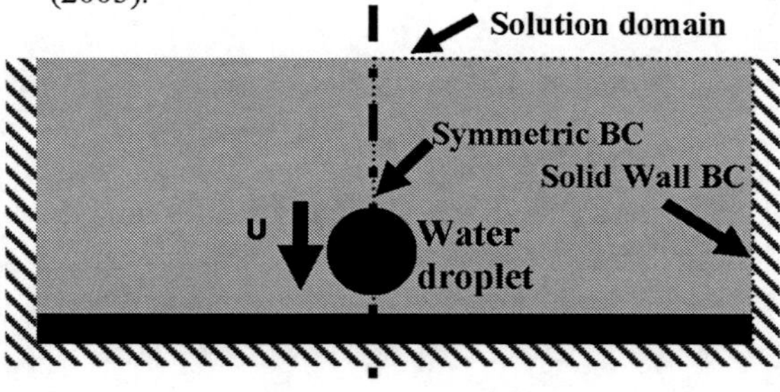

FIGURE 5. Schematic representation of a droplet splashing on a liquid film.

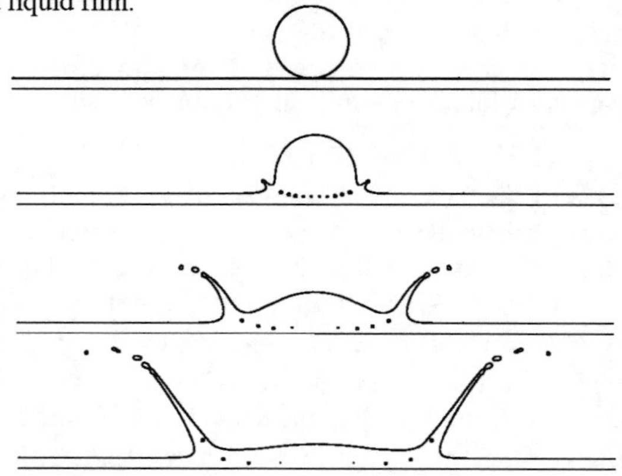

FIGURE 6. The time evolution of droplet splashing.

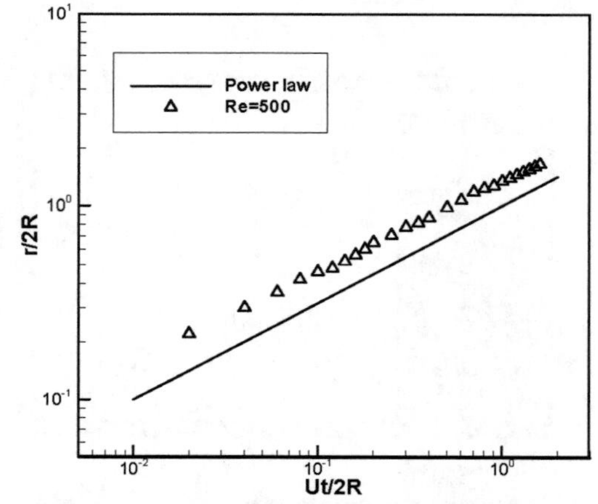

FIGURE 7. Log-log plot of the spread factor r/2R as a function of non-dimensional time.

Similarities in the Dynamics of Suspensions of Monodisperse Colloidal Chains with Different Lengths Confined in the Thin Films

Yayoi Terada* and Michio Tokuyama†,*

*Institute of Fluid Science, Tohoku University, Sendai 980-8577, Japan
†WPI Advanced Institute for Materials Research, Tohoku University, Sendai, 980-8577, Japan

Abstract. We perform the extensive Brownian dynamics simulations on the suspensions of the monodisperse magnetic colloidal chains confined in the thin films at several different area fractions. It is shown that the long-time self-diffusion coefficients of the colloidal chains with different chain lengths converge on the single master curve, even though the area fraction of the chains is different. We also show the phase diagram of the suspensions. The value of the melting point depends on not only the number of colloidal particles N_z within one chain but also the area fraction σ for the suspensions of the colloidal chains.

Keywords: Dilute suspensions, Long-time self-diffusion coefficient, Magnetic colloidal chains
PACS: 47.57.-s, 47.57.E-, 64.60.-i, 66.10.Cb, 83.80.Hj

INTRODUCTION

When the dilute suspensions of the magnetic colloidal particles are confined in the thin films and the external magnetic field is applied perpendicular to them, the colloidal chains parallel to the field are formed. The diffusive motions of the magnetic colloidal chains have been investigated in the bulk system [1, 2] and confined in the thin film [3, 4]. The monolayer colloidal suspensions have also been investigated [5, 6]. In addition to that, the anisotropic diffusion of the orientated liquid-crystals like chains has been also investigated [7]. As the external magnetic field is increased, the diffusion of those colloidal chains slows down, leading to a decrease of the long-time self-diffusion coefficient of the chains. Recently we found that the long-time self-diffusion coefficient of the chains obeys a singular function of a control parameter proportional to the square of the applied field strength at the constant area fraction $\sigma = 0.03$. The singular point is also shown to be inversely proportional to the square of the number of magnetic colloidal particles within one chain.

In this paper, we perform the extensive Brownian dynamics simulations on the suspensions of the monodisperse magnetic colloidal chains at several different area fractions. We discuss the diffusive motion of the chains projected on the films. By analyzing the simulation results from a unified standpoint based on the mean-field theory proposed recently by Tokuyama [8, 9], we have shown that there exist similarities not only in the dynamics of the chains but also in the spatial distribution of the chains with different chain lengths at the different area fractions. We also discuss the phase diagram of those suspensions for the different chain lengths at the different area fractions.

SIMULATION MODEL

We consider the colloidal chains dispersed in an equilibrium solvent with a viscosity η at temperature T. Each colloidal chain consists of N_z identical magnetic colloidal particles with radius a and the magnetic susceptibility χ. The suspension is confined in the thin film with thickness $L_z(= 2aN_z)$.The dipole moment of colloidal particle i in the αth chain under the external magnetic field $\boldsymbol{H}(= H\boldsymbol{e}_z)$ is given by $\boldsymbol{m}_{\alpha_i}(\simeq \frac{4}{3}\pi a^3 \mu_0 \chi H)$, where $\boldsymbol{e}_z$ denotes the unit vector whose direction is parallel to the magnetic field and μ_0 the absolute permeability of vacuum. The time evolution of the position of the αth chain which consists of N_z colloidal particles projected on the film, $\boldsymbol{X}_\alpha$, is described by the stochastic diffusion equation on the time scale t_c,

$$\frac{d}{dt}\boldsymbol{X}_\alpha(t) = \frac{1}{\gamma N_z}\sum_{\beta\neq\alpha}\sum_{i=1}^{N_z}\sum_{j=1}^{N_z}\boldsymbol{F}^d(\boldsymbol{r}_{\alpha_i\beta_j}) + \boldsymbol{V}_\alpha(t), \quad (1)$$

where $t_c(= a^2/D_c)$ denotes the structural-relaxation time for a chain to diffuse over a distance a, $D_c(= D_0/N_z)$ the diffusion constant of the single colloidal chain formed by N_z colloidal particles, $D_0(= k_BT/\gamma)$ the diffusion constant of the single colloidal particle, $\gamma(= 6\pi\eta a)$ the friction constant of the particle, and $\boldsymbol{r}_{\alpha_i\beta_j} = \boldsymbol{r}_{\alpha_i} - \boldsymbol{r}_{\beta_j}$ [10]. Here $\boldsymbol{r}_{\alpha_i}$ is the position of particle i in the αth chain. The magnetic force of particle i in the αth chain

CP982, *Complex Systems, 5th International Workshop on Complex Systems*
edited by M. Tokuyama, I. Oppenheim, and H. Nishiyama

from particle j in the βth chain, $\boldsymbol{F}^d(\boldsymbol{r}_{\alpha_i\beta_j})$, is given by $\boldsymbol{F}^d = -(\partial/\partial \boldsymbol{r}_{\alpha_i})U$, where magnetic dipole potential U is described by

$$\begin{aligned} & U(\boldsymbol{r}_{\alpha_i\beta_j}) \\ & = \frac{1}{4\pi\mu_0}\left\{-\frac{\boldsymbol{m}_{\alpha_i}\cdot\boldsymbol{m}_{\beta_j}}{|\boldsymbol{r}_{\alpha_i\beta_j}|^3}+\frac{3(\boldsymbol{m}_{\alpha_i}\cdot\boldsymbol{r}_{\alpha_i\beta_j})(\boldsymbol{m}_{\beta_j}\cdot\boldsymbol{r}_{\alpha_i\beta_j})}{|\boldsymbol{r}_{\alpha_i\beta_j}|^5}\right\}, \\ & = Ja^3k_BT\left\{-\frac{1}{|\boldsymbol{r}_{\alpha_i\beta_j}|^3}+\frac{3(\boldsymbol{e}_z\cdot\boldsymbol{r}_{\alpha_i\beta_j})(\boldsymbol{e}_z\cdot\boldsymbol{r}_{\alpha_i\beta_j})}{|\boldsymbol{r}_{\alpha_i\beta_j}|^5}\right\}. \quad (2) \end{aligned}$$

Here $J(=\frac{4}{9}\pi a^3\mu_0 H^2\chi^2/k_BT)$ denotes the dimensionless magnetic field energy. The z positions of all colloidal particles are fixed because the film thickness is equal to the length of the chains and the strength of the magnetic field is enough strong to keep the rodlike chains parallel to the field. Hence the magnetic force term $\boldsymbol{F}^d(\boldsymbol{r}_{\alpha_i\beta_j})$ in xy plane is finally obtained by

$$\begin{aligned} & \boldsymbol{F}^d(\boldsymbol{r}_{\alpha_i\beta_j}) \\ & = Ja^3k_BT\frac{1}{|\boldsymbol{r}_{\alpha_i\beta_j}|^4}\left(\boldsymbol{e}_{\alpha_i\beta_j}-5\left(\boldsymbol{e}_z\cdot\boldsymbol{e}_{\alpha_i\beta_j}\right)^2\boldsymbol{e}_{\alpha_i\beta_j}\right), (3) \end{aligned}$$

where $\boldsymbol{e}_{\alpha_i\beta_j}=\frac{\boldsymbol{r}_{\alpha_i\beta_j}}{|\boldsymbol{r}_{\alpha_i\beta_j}|}$. The random stochastic velocity $\boldsymbol{V}_\alpha(t)$ obeys a Gaussian, Markov random process and satisfies the following relations $<\boldsymbol{V}_\alpha(t)>=0$ and

$$<\boldsymbol{V}_\alpha(t)\boldsymbol{V}_\beta(t')> \;=\; 2D_c\delta_{\alpha,\beta}\delta(t-t')\boldsymbol{E}, \quad (4)$$

where the brackets denote the ensemble average and $\boldsymbol{E}$ unit matrix.

It is convenient to introduce the dimensionless control parameter Γ which is defined by the average dipole energy between two particles over the average distance $\ell_0(=a/\sqrt{\sigma})$ on the film in [8, 6]

$$\Gamma = \frac{U(\ell_0)}{k_BT} = \frac{4}{9}\frac{\pi a^3\mu_0 H^2\chi^2\sigma^{3/2}}{k_BT} = J\sigma^{3/2}. \quad (5)$$

SIMULATION RESULTS

Our simulation cell consists of $N_{xy}=2{,}500$ and $10{,}000$ chains in a rectangular cell with the periodic boundary conditions. The simulation is done for $N_z=1$, 2, 3, 4, 5, 10, 15, and 20 at $\sigma=0.03$ ($\phi=0.02$), $\sigma=0.01$ ($\phi=0.0067$), and $\sigma=0.003$ ($\phi=0.002$), where ϕ denotes the volume fraction of the colloidal particles. The total number of the colloidal particles N is from $2{,}500$ for (N_z, N_{xy}) $=$ (1, 2500) to $150{,}000$ for (N_z, N_{xy}) $=$ (15, 10000). The chains on xy plane are set on the random configuration initially. After waiting the enough long time to equilibrate the system, we observe the physical quantities. In the present simulation, the cutoff distance of the force calculation is set as $21\ell_0$ and the time interval of the difference equation of Eq. (1) $0.001t_D$, where ℓ_0 is the average distance between the nearest chains on xy plane.

Figure 1 shows the phase diagram at the different area fractions (a) $\sigma=0.03$ and (b) 0.003. At the short chain region, the melting point of the chain $\Gamma_m(N_z)$ decreases rapidly as N_z is increased on the both area fractions. $\Gamma_m(N_z)$ decreases gradually for $N_z>5$. Even though the area fraction of the system in Fig. 1(a) is 10 times bigger than that in Fig. 1(b), both phase diagrams are similar. The only difference between them is that the melting point of the chains ($N_z \geqq 2$) for $\sigma=0.03$ is slightly bigger than that for $\sigma=0.003$.

The Γ dependence of the long-time self-diffusion coefficient of the chains with different lengths D_S^L is shown at $\sigma=0.03$, 0.01 and 0.003 in Fig. 2. On the monolayer colloidal suspensions ($N_z=1$), Γ dependences of D_S^L with different area fractions converge on the single master curve. On the other hand, Γ dependence of D_S^L for the chains ($N_z\geq 2$) depends on not only the chain lengths N_z but also the area fractions σ. The intensity of the many-body dipole interaction among the chains, which determines the value of D_S^L, is affected by the chain lengths and the average distance between the chains. The dimensionless parameter Γ is however given by only the dipole energy between two particles in the same plane. It is necessary to consider the dipole energy between the chains. Therefore we propose the dimensionless control parameter Γ_{N_z} of the chains by considering the repulsive force between the αth chain and the βth chain whose distance is the average distance ℓ_0 on the film,

$$\Gamma_{N_z} \;=\; \Gamma\frac{\sum_{i=1}^{N_z}\sum_{j=1}^{N_z}F^d(\boldsymbol{r}_{\alpha_i\beta_j})}{F_1^d(\ell_0)}, \quad (6)$$

where $F_1(\ell_0)$ denotes the force between the colloidal particles whose distance is ℓ_0 in the same plane. Data converge on a single master curve, if D_S^L/D_c is plotted versus Γ_{N_z} in Fig. 3. The similarity in the long-time self-diffusion coefficient is found on not only the monolayer suspensions but also the suspensions of the colloidal chains. Here D_S^L is proposed by Tokuyama in [8, 9],

$$D_S^L(\Gamma_{N_z}) \;=\; \frac{D_c(N_z)}{1+\varepsilon\left(\frac{\Gamma_{N_z}}{\Gamma_c}\right)^A\left(1-\frac{\Gamma_{N_z}}{\Gamma_c}\right)^{-B}}. \quad (7)$$

Γ_c is the singular point where D_S^L becomes zero theoretically. It is here given by $\Gamma_c=99.7$ from our previous simulation results. The parameters A, B, and ε are also given by

$$A \;=\; 1/2,\;\; B \;=\; 1,\;\; \varepsilon \;=\; 4.53.$$

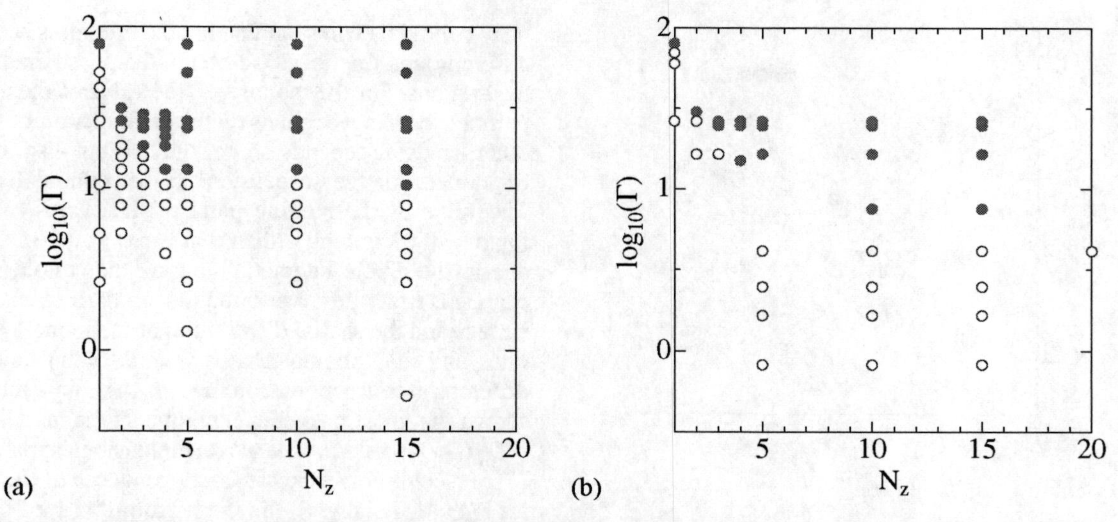

FIGURE 1. Phase diagram of the monodisperse colloidal chains (a) at $\sigma = 0.03$ and (b) $\sigma = 0.003$. Open circles denotes the liquid region and closed circles the crystal region.

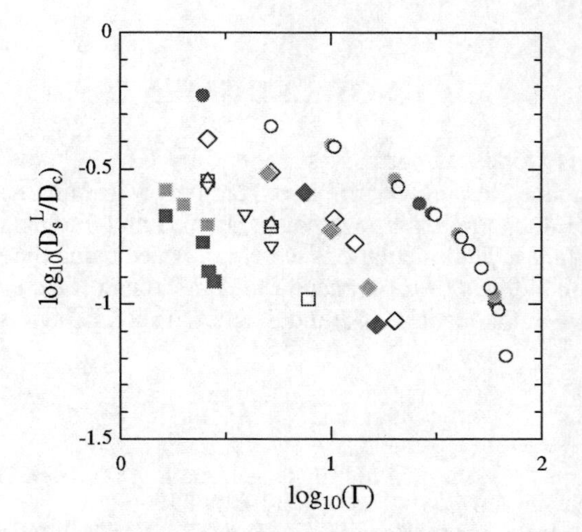

FIGURE 2. (Color online) Γ dependence of the long-time self-diffusion coefficient of the chains with different lengths for $\sigma = 0.03$, 0.01, and 0.003. Circles denote the simulation results for $N_z = 1$, diamonds $N_z = 2$, squares $N_z = 5$, triangles $N_z = 10$, and inverted triangles $N_z = 15$. Open symbols denote the simulation results for $\sigma = 0.03$, gray filled symbols $\sigma = 0.01$, and dark red filled symbols $\sigma = 0.003$.

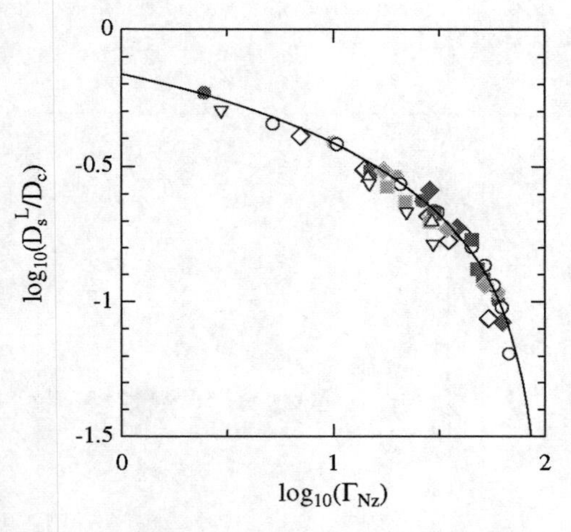

FIGURE 3. (Color online) Γ_{N_z} dependence of the long-time self-diffusion coefficient of the chains with different lengths for $\sigma = 0.03, 0.01$, and 0.003. The symbols are the same as Fig.2 The solid line is given by the mean-field theory of Eq. (7) [8, 9].

Our simulation results are in good agreement with the theoretical prediction in Fig.3.

Finally we show the similarities in the dynamics and spatial distribution of chains, if the values of the relative long-time self-diffusion coefficient to the diffusion constant of the single chain D_S^L/D_c are the same. The examples are (a) $(\Gamma, N_z, \sigma) = (57.15,\ 1,\ 0.03)$ and (b) $(15, 2, 0.01)$ with $D_S^L/D_c = 0.115$. The mean-square displacements on the film $M_2(t)(=<|X_\alpha(t) - X_\alpha(t=0)|^2>)$ are plotted versus time in Fig. 4. $M_2(t)$ with different lengths and area fractions converge on a single master curve, if M_2 and t are scaled by ℓ_0^2 and $t_c^l(=\ell_0^2/D_c)$. In Fig. 4, we also plot the theoretical results proposed by Tokuyama in [8, 9]. $M_2(t)$ for the suspensions is given theoretically by

$$M_2(t) \quad = \quad \ell^2 \ln\left[1 + \left(\frac{D_S^S}{D_S^L}\right)\{e^{4D_S^L t/\ell^2} - 1\}\right], \quad (8)$$

where ℓ indicates the mean-free path and D_S^S the short-time self-diffusion coefficient. Here we could replace D_S^S by D_c. Our simulation results are in good agreement with the mean-field theory. The radial distribution functions

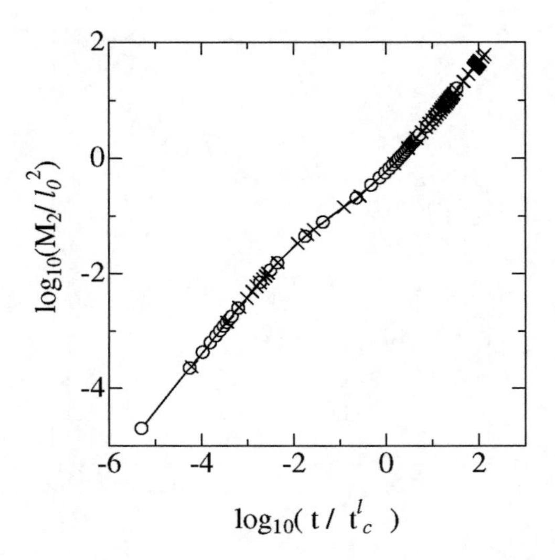

FIGURE 4. Time evolution of mean-square displacement of the chains. The symbols are the simulation results for $D_S^L/D_c = 0.115$ at $(\Gamma, N_z, \sigma) = (57.15,\ 1,\ 0.03)$ (open circles) and (15, 2, 0.01) (crosses). Solid line indicates the theoretical result of Eq. (8) [8, 9].

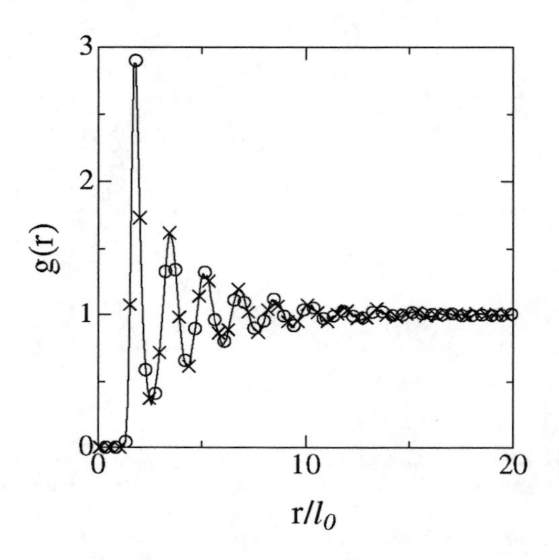

FIGURE 5. Radial distribution functions of the chains. The symbols are the same as in Fig. 4. The solid line also indicates the simulation result at $(\Gamma, N_z, \sigma) = (15,2,0.01)$ for a guide to eyes.

of them are also the same in Fig. 5. We note that the characteristic length is the average distance between the chains ℓ_0 substitute for the radius of the chains a here.

SUMMARY

The extensive Brownian dynamics simulations are performed on the suspensions of the monodisperse magnetic colloidal chains confined in the thin films at several different area fractions. We obtain the phase diagram of the systems. For the monolayer, the value of the melting point Γ_m does not depends on the area fraction σ. On the other hands, it depends on not only N_z but also the area fraction σ for the suspensions of the colloidal chains. The value of the melting point of the chains ($N_z \geqq 2$) for $\sigma = 0.03$ is bigger than that for $\sigma = 0.003$. This is due to the dipole interaction among the colloids in the chains. Despite this, we found the similarities in the dynamics and the spatial distribution of the monolayer colloids and the colloidal chains with different lengths at different area fractions confined in the thin films. It is shown that most data converge on a single master curve if D_S^L/D_c is plotted versus novel dimensionless parameter Γ_{N_z}. It seems however that Γ_{N_z} dependence of the chains for $N_z = 15$ is slightly different from that for $N_z \leq 10$. This is attributed to the dynamical many-body interactions among the particles in the long chains. It needs the further consideration.

ACKNOWLEDGMENTS

This work was partially supported by Grants-in-aid for Science Research with No.(C)18540363 from Ministry of Education, Culture, Sports, Science and Technology of Japan. The calculations were performed using the SGI Altix3700Bx2 in Advanced Fluid Information Research Center, the Institute of Fluid Science, Tohoku University.

REFERENCES

1. S. H. Patrick Ilg, Martin Kröger, and A. Y. Zubarev, *Phys. Rev. E* **67**, 061401-1–061401-8 (2003).
2. P. Ilg, and M. Kröger, *Phys. Rev. E* **72**, 031504-1–031504-7 (2005).
3. Y.-H. Hwang, S. Ahn, and S. Park, "The Self and Relative Dynamics of Magnetic Chains in a Quasi-Two Dimensitonal Aqueous Solution," in *The 5th International Workshop on Similarity in Diversity*, 1999, pp. 121 – 129.
4. Y. Hwang, and X. l. Wu, *Phys. Rev. Lett.* **74**, 2284–2287 (1995).
5. K. Zahn, R. Lenke, and G. Maret, *Phys. Rev. Lett.* **82**, 2721–2724 (1999).
6. K. Zahn, and G. Maret, *Phys. Rev. Lett.* **85**, 3656–3659 (2000).
7. D. Baalss, and S. Hess, *Z. Naturforschung . A* **43**, 662–670 (1988).
8. M. Tokuyama, *Physica A* **364**, 23–62 (2006).
9. M. Tokuyama, *Physica A* **378**, 157–166 (2007).
10. Y. Terada, and M. Tokuyama, "Slow Dynamics of Magnetic Colloidal Chains Confined in Thin Film under a Magnetic Field," in *Special Issue:Complex Systems*, edited by M. Tokuyama, 2007, vol. 19 of *Rep. . Inst. Fluid Sci., Tohoku Univ.*, pp. 61–65.

Size Effects on Short-Time Self-Diffusion in Dilute Highly-Charged Colloidal Suspensions

T. Furubayashi*, M. Tokuyama†,**, and Y. Terada**

*Graduate School of Tohoku University, Sendai, 980-8579, Japan
†WPI Advanced Institute for Materials Research, Tohoku University, Sendai, 980-8577, Japan
**Institute of Fluid Science, Tohoku University, Sendai, 985-8577, Japan

Abstract. We treat not only macro ions as charged colloids but also small ions as counter ions dispersed in a solvent by Brownian-dynamics simulation. The size and charge of particles are changed, while the other parameters are fixed. We compare the simulation results with theoretical results by Tokuyama [Physca A 352 (2005) 252-264]. The results are in good agreements with the theory. Thus, it is shown that the short-time self-diffusion coefficient is decreased as the radius is increased.

Keywords: Charged colloid, Counter ions, Diffusion coefficient, Mean-square displacement, Short-time self-diffusion
PACS: 47.57.-s, 47.57.eb, 47.57.J-

INTRODUCTION

Usually, a colloid is neutral, but ionized in a solvent. There are many charged colloidal suspensions in nature. The protein dispersed in solvent is one of examples. There are not only charged colloids but counter ions occurred by detaching. It is considered that some phenomena like the protein foldings are affected strongly by counter ions. In such systems, particles are affected by Coulomb interactions, random forces from solvent, direct interactions, and hydrodynamic interactions. Especially, the Coulomb interactions among ions are dominant. They affect the diffusion coefficient of charged colloids.

The screen Coulomb repulsion is suggested by Derjaguin-Landau-Verwey-Overbeek (DLVO) theory [1]. However, there are some observations that cannot be explained by DLVO theory. Recently, Tata *et. al.* observed the colloids cluster formation in dilute highly charged colloidal suspensions [2] and the gas-solid coexistence [3]. If the repulsive Coulomb interactions between charged colloids were dominant in the system, they would not occur. It is considered that there are some attractive interactions between charged colloids by the effects of the counter ions. Thus, it is important that we take into consideration correctly the Coulomb interaction between charged colloids and counter ions.

In this paper, we perform the Brownian-dynamics simulation on the three-dimensional dilute highly-charged colloidal system including two species. The system is enough dilute to ignore the hydrodynamic interactions. It is found the effects of the charge and the size of macro ion on the short-time diffusion. We also compare the present results with Tokuyama's theory which describes the theory of the short-time self-diffusion in the dilute suspensions of highly-charged colloids [4].

SIMULATION

The system is considered that there exist two types of particles dispersed in the solvent at room temperature. One is the macro ion with negative charge $-Ze$ and radius a, and the other the small ion with positive charge qe and radius $a_c(< a)$, where e is the unit charge. The system satisfies the charge neutrality, so that the number of each particles, N_m and N_s, and the valences of ions have the relation as $ZN_m = qN_s$. Usually, there is a relation as $Z \gg q$, so that we should set as $N_m \ll N_s$. Although we want to observe the behavior of macro ions, we have to treat small ions more than macro ions and treat the time scale of small ions that is much smaller than that of macro ions. Thus, these calculations need high efficiency computers like super computers. It is assumed that the system is the bulk space of a large system, and therefore we employ the periodical boundary condition. Then we have to take into consideration the effects from particles in the image cells because the Coulomb interaction is the long range interaction, so that the Coulomb forces between particles are calculated by Ewald's method [5].

The ions of type α obey the Langevin equations

$$m\frac{d^2}{dt^2}\boldsymbol{X}_i^{\alpha}(t) = -\gamma_{\alpha}\frac{d}{dt}\boldsymbol{X}_i^{\alpha}(t) + \boldsymbol{F}_i^{\alpha}(t) + \boldsymbol{R}_i^{\alpha}(t), \quad (1)$$

where $\boldsymbol{X}_i^{\alpha}$ is the position vector of ith ion α, $\gamma_{\alpha}(= 6\pi\eta a_{\alpha})$ the friction constant, η a viscosity, and $\boldsymbol{R}$ the random force. Here, α indicates a type of ion, m for macro ion and s for small ion. $\boldsymbol{F}_i^{\alpha}$ is the sum of the

CP982, *Complex Systems, 5th International Workshop on Complex Systems*
edited by M. Tokuyama, I. Oppenheim, and H. Nishiyama

Coulomb forces and the repulsive forces from all particles,

$$F_i^\alpha(t) = \sum_\beta \sum_{j \neq i}^{N_\beta} F_{ij}^{\alpha\beta}(t) + \sum_\beta \sum_{j \neq i}^{N_\beta} F_{ij}^{\prime\alpha\beta}(t), \tag{2}$$

where $F_{ij}^{\alpha\beta}(t)(= -\nabla_i U^{\alpha\beta}(|X_i^\alpha - X_j^\beta|))$ is the Coulomb force between ith ion α and jth ion β, $U_{\alpha\beta}(X)$ the Coulomb potential between ion α and ion β, and $F_{ij}^{\prime\alpha\beta}(t)$ the repulsive force of the short distance between ith ion α and jth ion β. $R_i^\alpha(t)$ obeys the Gaussian Markov process and satisfies the fluctuation-dissipation relation

$$< R_i^\alpha(t) R_j^\beta(t') > = 2\gamma_\alpha k_B T \delta(t-t') \delta_{ij} \delta_{\alpha\beta} \mathbf{1}, \tag{3}$$

where k_B is the Boltzmann constant, and T the temperature.

On the time scale of order $t_D^s = (a_s^2/D_0^s)$, where D_0^s is the single small ion diffusion constant, the inertial terms in Eq.(1) can be negligible. Then, Eq.(1) reduces to

$$\frac{d}{dt} X_i^\alpha = \frac{1}{\gamma_\alpha} F_i^\alpha(t) + f_i^\alpha(t), \tag{4}$$

$$< f_i^\alpha(t) f_j^\beta(t') > = 2D_0^\alpha \delta(t-t') \delta_{ij} \delta_{\alpha\beta} \mathbf{1}. \tag{5}$$

The initial positions of particles are random. We show three cases, case (a) : the charge of a macro ion $Z = 360$ and the radius of a macro ion $a = 55.4$nm, case (b) : $Z = 360$ and $a = 166.2$nm, and case (c) : $Z = 120$ and $a = 55.4$nm to investigate the size effects and the charge effects on the diffusion process in the systems. The other parameters, the numbers of each particle N_m, N_s, the valence of small ions q, the volume fraction of macro ion ϕ, and the ratio of the radius a/a_c are fixed as in Table 1. The ratio of a in case (a) and in case(b) is $a_a/a_b = 3$ as same ratio of Z in case (a) and in case(c) is $Z_a/Z_c = 3$, and therefore we can compare the effects of the size with of the charge.

TABLE 1. Conditions of simulations, $q = 3$, $\phi = 0.00125$, $N_m = 50$, and $a/a_s = 6$

case	Z	a (nm)	D_S^S/D_0^m	D_0^m (m^2/s) 10^{-12}	σ (m^{-2}) 10^{-3}
(a)	360	55.4	0.0457	4.43	0.166
(b)	360	166.2	0.238	1.48	1.49
(c)	120	55.4	0.713	4.43	0.498

RESULTS AND DISCUSSION

The mean-square displacement $M_2(t)$ of the macro ions is given by

$$M_2(t) = \frac{1}{N_m} \sum_{i=1}^{N_m} \langle |X_i(t) - X_i(0)|^2 \rangle. \tag{6}$$

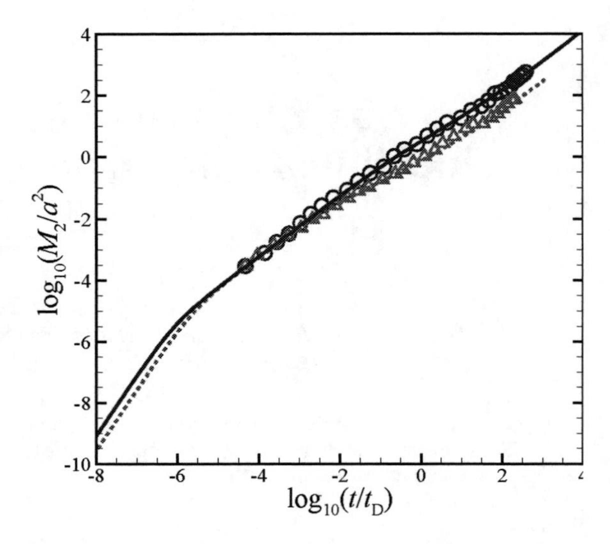

FIGURE 1. (Color online) A log-log plot of mean-square displacement of macro ions versus time. Triangles denotes the simulated results and doted line the theoretical result for case(a), and open circles the simulated results and solid line the theoretical result for case(b).

It is convenient to introduce a Laplace transform $M_2[s]$ of $M_2(t)$ by $M_2[s] = \int_0^\infty M_2(t) e^{-st} dt$. By taking into account of the many-body effects due to the small ions, Tokuyama has derived its explicit form as [4]

$$M_2[s] = \frac{2dD_0^m}{s^2[st_B + 1 + \Lambda[s]]}, \tag{7}$$

where

$$\Lambda[s] = \frac{\Delta}{1 + \sqrt{st_\lambda}}, \tag{8}$$

$$\Delta = \frac{D_0^m}{D_0^s} \frac{Z}{q} \left(\frac{\phi \Gamma^3}{3} \right)^{1/2}, \tag{9}$$

d is a dimension number, $D_0^m (= k_B T/6\pi\eta a)$ the single macro ion diffusion constant, $t_B (= m/\gamma_m)$ a Brownian relaxation time for a macro ion, $t_\lambda (= \lambda_D^2/D_0^s)$ a time for small ions to diffuse over a distance of order Debye length λ_D, $\Gamma = Zql_B/a$ the coupling parameter between macro ions and small ions, $l_B (= e^2/\varepsilon k_B T)$ the Bjerrum length, and ε a dielectric constant. The short-time self-diffusion coefficient D_S^S is given by

$$D_S^S = \lim_{st_\lambda \to 0} \frac{s^2 M_2[s]}{2d} = \frac{D_0^m}{1 + \Delta}, \tag{10}$$

From Eq.(6), we have the asymptotic forms

$$M_2(t) \simeq \begin{cases} (dk_B T/m) t^2 & \text{for } t \ll t_B \\ 2dD_0^m t & \text{for } t_B \leq t \ll t_\lambda \\ 2dD_S^S t & \text{for } t_\lambda \ll t \leq t_\gamma \end{cases}, \tag{11}$$

where $t_f(= 9t_B/2)$ is a relaxation time of the momentum contained in the fluid volume of size a, and $t_\gamma(= a^2/D_S^S)$ a time for dressed macro ions to diffuse over a distance of order a. Here, a dressed macro ion means a macro ion affected by small ions localized near around it. The first term of Eq.(11) results from a Brownian motion, the second from a free diffusion, and the last from a dressed diffusion. t_B is the characteristic time of the kinetic stage, but we see the suspension-hydrodynamic stage in the simulation. Then the characteristic times in the simulation are a structural relaxation time $t_D(= a^2/D_0^m)$ for a macro ion to diffuse over a distance a and a structural relaxation time $t_D^c(= a_s^2/D_0^s)$ for a small ion to diffuse over a distance a_s. Therefore we compare the simulation with the theory after $t > t_D^c$.

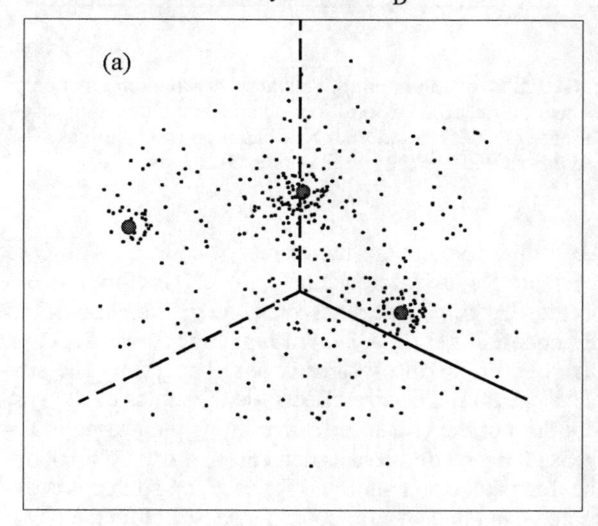

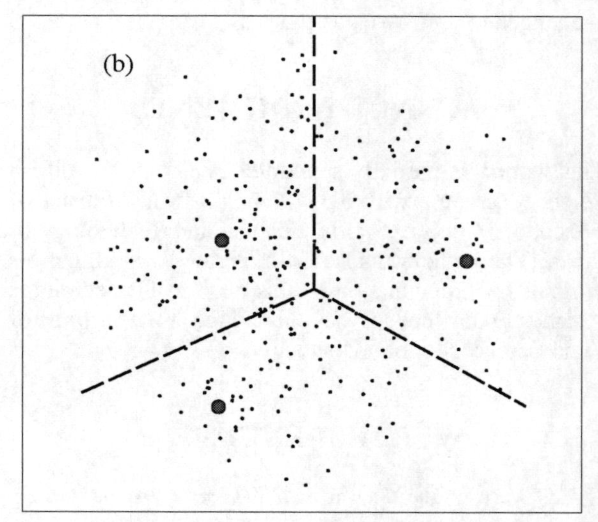

FIGURE 2. (Color online) The snap shots around a macro ion in case (a) and case (b). Large spheres are macro ions and small spheres are small ions.

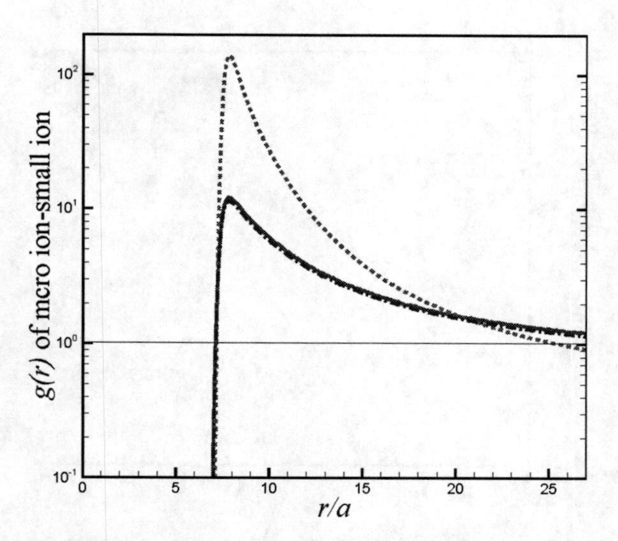

FIGURE 3. (Color online) The radial distribution function of macro ion-small ion. Dashed line is the result for case(a), solid line the result for case(b), and dash doted line the result for case (c).

A couple of results of simulation and Tokuyama's theory [4] are shown in Fig. 1 that shows time dependence of the mean-square displacement of macro ion for case (a) and case (b). Within our simulation time, we find that the simulated and the theoretical results agree well. Here, $D_S^S = 0.0457D_0^m$ in case (a), and $D_S^S = 0.238D_0^m$ in case (b). We can find what is the difference between case (a) and case (b) in Fig. 2 and Fig. 3. Fig. 2 shows the snap shots around some macro ions. It is observed that many small ions are localized around a macro ion in case (a) and small ions are distributed uniformly in case (b). Fig. 3 shows the radial distribution functions $g_{ms}(r)$ of macro ion-small ion obtained by our simulations. The horizontal axis is scaled by the radius of macro ion a of each condition. We find that the peak value of $g_{ms}(r)$ in case (a) is much larger than in case (b) at $r/a = 1.2 \sim 1.4$. It means that there are many small ions near a macro ion like Fig. 2. These are because the surface charge $\sigma(= Ze/4\pi a^2)$ is increased as the radius is decreased, and the higher surface charge makes the interaction strong. Many trapped small ions make a macro ion slow. Thus, D_S^S of macro ion is decreased as the size of macro ion is increased.

Next, we see the charge effects by comparing the results for case (a) with that for case (c). There are two simulated and two theoretical results in Fig. 4 that shows time dependence of $M_2(t)$ of macro ion for case (a) and case (c). In the early stage, there are no difference between results of theory. It is considered that the macro ions are not affected by the charge in this scale. After $t > t_D \cdot 10^{-4}$, the simulated results agree well with theoretical results at both conditions. The D_S^S obtained by Eq.(10) for case (c) is $0.713D_0^m$, much larger than for

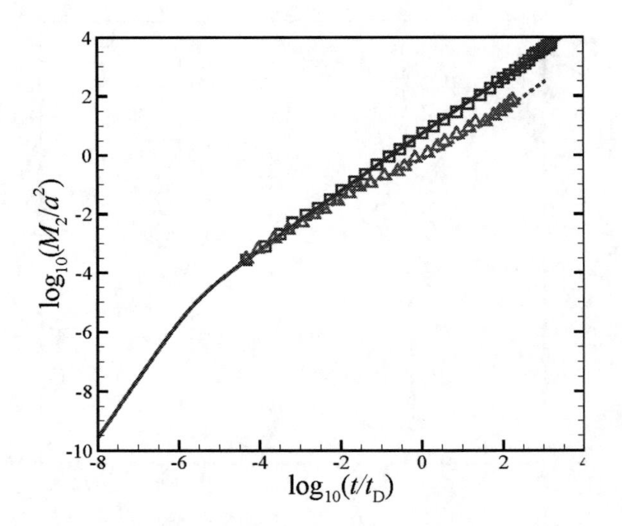

FIGURE 4. (Color online) A log-log plot of mean-square displacement of macro ions versus time. Open triangles denote the simulated results and solid line the theoretical result for case (a), and open squares the simulated results and dashed line the theoretical result for case (c).

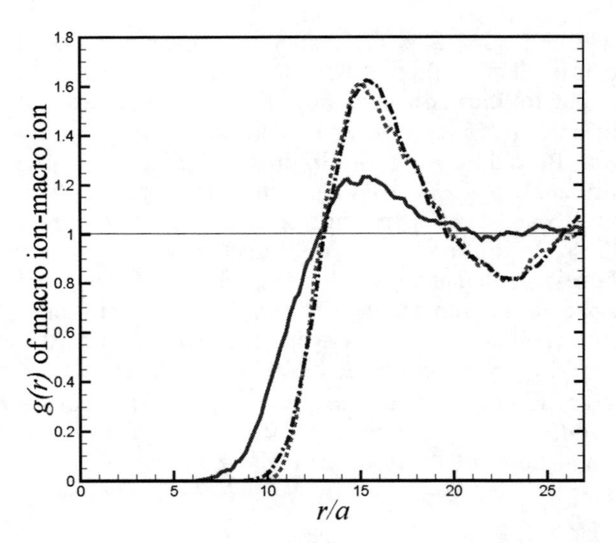

FIGURE 5. (Color online) The radial distribution function of macro ion-macro ion. Solid line denotes the simulation result for case (a), dash doted line the simulation result for case (b), and dash doted line the simulation result for case (c).

case (a). The reason why they are so different is the distribution of small ions. We find that the value of $g(r)_{\mathrm{ms}}$ at $r/a = 1.2 \sim 1.4$ in case (c) is very small in Fig. 3 same as case (b). It means that the small ions are not affected strongly by the charge of macro ions, and are distributed uniformly. Then, the macro ions are not dressed and can move freely. Thus, the short-time self-diffusion coefficient $D_{\mathrm{S}}^{\mathrm{S}}$ of macro ion is increased as the charge of macro ion is decreased. Fig. 5 shows the radial distribution function $g_{\mathrm{mm}}(r)$ of macro ion-macro ion. In all cases, we find some peaks in $g_{\mathrm{mm}}(r)$ at similar positions. These are characteristics of a liquid phase. There is no difference of the value of the first peak of $g_{\mathrm{mm}}(r)$ between case (a) and case (b), and the positions where the peaks exist. It means that the structure of the system on case (a) is different from on case (b).

CONCLUSION

In this paper, we have compared the simulated results with the theoretical results, and results in some conditions. From Eq.(9), (10), Fig. 1, and Fig. 4, it can be said that the valence of macro ion has larger effect than the size in short-time self-diffusion. However, as the charges are same, the size is important. From Fig. 2 and Fig. 3, it is observed that the size of macro ions strongly affects the distribution of small ions. As the size is small, the surface charge is strong and many small ions are localized around a macro ion. Thus, the macro ions are dressed and the short-time self-diffusion coefficient $D_{\mathrm{S}}^{\mathrm{S}}$ is decreased. According to compare the result for case (b) with case (c), there are no difference on $g_{\mathrm{ms}}(r)$. Therefore it is expected that $g_{\mathrm{ms}}(r)$ depends on $(Z/a)^n$. We should find the equation of $g_{\mathrm{ms}}(r, Z, a)$. From Fig. 5, we find that the structure of the system is not changed by the size of macro ions. The size affects the dynamics of the system but not the structure. However, as the charge is decreased, the second peak is decreased. It is very small but the first peak is not vanished for case (c), so that it is expected that there are attractive gas phase before the phase is changed to repulsive gas (normal gas) phase.

ACKNOWLEDGMENTS

This work was partially supported by Grants-in-Aid for Science Research with No. 18540363 from Ministry of Education, Culture, Sports, Science and Technology of Japan. The authors thank T. Shimura. We modified his programs. Computer simulations were performed on the Altix 3000 machine and SX-8 machine at the Institute of Fluid Science, Tohoku University.

REFERENCES

1. E. J. Verwey, and T. Overweek, *"Theory of the Stability of Lyophobic Colloids"*, Elsevier, Amsterdam, 1948.
2. B. V. R. Tata, and S. S. Jena, *Solid State Communication*, **139**, 562–580 (2006).
3. B. V. R. Tata *et al.*, *Phys. Rev. Lett.*, **78**, 2660–2663 (1997).
4. M. Tokuyama, *Physica A*, **352**, 252–264 (2005).
5. P. Ewald, *Ann. Phys.* , **64**, 253–287(1921) .

Layering Phenomena Driven by Rotating Magnetic Field in Ferrofluid

Y. Yamada and Y. Enomoto

Department of Environmental Technology, Nagoya Institute of Technology, Nagoya 466-8555, Japan

Abstract. Magnetic fluid is a compound material of colloidal liquid, consisting of a base suspension liquid, nano sized magnetic particles such as Co-ferrite or Ba-ferrite which have their own magnetic moment, and surfactant.

In order to calculate the microstructure formation process of magnetic fluid, we use the Langevin-type microscopic equation of motion and perform the molecular dynamics simulation. In the study, we investigate physical force originated by the magnetic and surrounding fluid flow and torque.

We consider the rotating field effect, and obtain layering structures in various ranges of the rotating frequency and packing density. These structures are found in ER and MR fluid, however, not yet found in MF.

Keywords: Computer simulation, Magnetic fluids, Microstructure, Nanoparticles, Rotating magnetic field
PACS: 82.70.-y, 83.80.Gv, 89.75.Kd

INTRODUCTION

Ferrofluid is a colloidal suspension of magnetic nano particles of *Fe* or magnetite dispersed into water or organic liquid, etc. by using the surfactant, so that it has an interesting property.

The method of supplying liquid fuel to a spaceship in the no-gravitation space was developed in the early 1960's by NASA as a part of space science. The purpose of "transportation by magnetic force using *Fe* nanoparticles mixing in the fuel," is the basis for the invention of ferrofluid.

A drastic change of viscosity appears by applying the magnetic field. Because of its ferromagnetism, it becomes easy to transport this magnetic fluid. These properties are applied to the space suit, the computer seal, the audio speaker, the semiconductor, electronics, and telecommunication, etc.

It is known that electro-rheological (ER) and the magneto-rheological (MR) fluid create the seat-like constructions when subjected to the rotation electro or magnetic field by the current research [1],[2].

As for ER and MR, the moment of each particle synchronizes with the external field because of the paramagnetic characteristic. On the other hand, considering a case for ferromagnetic colloid, the moment of the particle synchronize with the torque of the suspension, which causes a phase shift between the particle and the external magnetic field. This paper addresses the issue of whether the magnetic fluid (MF) causes a sheet-like formation under the rotating field by using the computer simulation.

THE MODEL

The model of the magnetic fluid for the simulation of our study is explained as follows. We assume that spherical particles are dispersed in a suspension in a cubic space.

We use the Stokes resistance term for fluid suspension. As for an external condition, we considered an external magnetic field which is rotating in the x-y plane. The model magnetic fluid system studied here is supposed to consist of N interacting and uniformly sized particles with a point magnetic dipole moment, suspended in an incompressible Newtonian fluid of viscosity η. The ferro magnetic particles are coated with surfactant, which prevents the particles from sticking to one another. The center of mass of the i-th particle is $\vec{r}_i (1 \le i \le N)$ and the magnitude of the magnetic dipole moment, m_0. The unit vector of the dipole moment, n_i, is fixed with a particle.

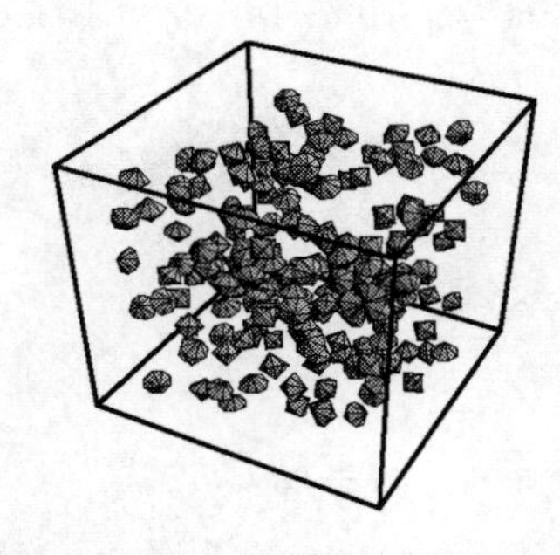

FIGURE 1. Initial conditions used in this simulation $N = 200$, $L_x = L_y = L_z = 20$ ($\phi = 0.1$)

CP982, *Complex Systems, 5th International Workshop on Complex Systems*
edited by M. Tokuyama, I. Oppenheim, and H. Nishiyama

We assume a cubic system; $L_x = L_y = L_z = 16 \sim 50$(volume fraction$\phi$: $0.2 \sim 0.01$) with a periodic condition, and the unit size a as the radius of the particle.

As for an initial state, we start with a randomly aligned particles(Fig.1). We assume a rotating magnetic field in the x-y plane; $\vec{H}(t) = H_0(\hat{\mathbf{i}} \sin \omega t + \hat{\mathbf{j}} \cos \omega t)$, where H_0 is the magnitude of the magnetic field.

Furthermore, we assume a simple ideal situation: (1) the coating effect is modeled as a repulsive force between particles; (2) the surrounding fluid effect is introduced only as a quasistatic Stokes drag force and other hydrodynamics effects are completely neglected; (3) inertia effects for both translational and rotational motions are neglected.

These simplified situations can be easily realized for most real magnetic fluids, in the case of a low Reynolds number at low particle volume fractions [2]-[3].

Under the above situations, the equations of motion of $\mathbf{r}_i = x_i\hat{\mathbf{x}} + y_i\hat{\mathbf{y}} + z_i\hat{\mathbf{z}}$ (translational motion) and $\mathbf{n}_i$ (rotational motion) for interacting spherical magnetic particles can be described as at the zero temperature [6]-[8].

$$\frac{d\vec{r}_i}{dt} = \frac{1}{\xi_t}[\sum_{j(\neq i)}^{N} (\vec{F}_{ij}^1 + \vec{F}_{ij}^2)], \tag{1}$$

$$\frac{d\vec{n}_i}{dt} = \frac{1}{\xi_r}(\sum_{j(\neq i)}^{N} \vec{T}_{ij}^1 + \vec{T}_{ij}^2) \times \vec{n}_i, \tag{2}$$

where the $\vec{F}_{ij}^1$ is the force on the i-th particle due to the dipole-dipole interaction between particles i and j. $\vec{F}_{ij}^2$ is the short range repulsive force due to surfactant. $\vec{T}_{ij}^1$ is the dipole induced torque and $\vec{T}_{ij}^2$ is the torque induced by the external magnetic field, with the translational drag force coefficient $\xi_t \equiv 6\pi a\eta$, and the rotational drag force coefficient $\xi_r \equiv 8\pi a^3\eta$. These forces are given by [4],[6]

$$\begin{aligned}\vec{F}_{ij}^1 &= F_0\frac{a^4}{r_{ij}^4}[-(\vec{n}_i \cdot \vec{n}_j)\vec{e}_{ij} + 5(\vec{n}_i \cdot \vec{e}_{ij})(\vec{n}_j \cdot \vec{e}_{ij})\vec{e}_{ij} \\ &\quad -(\vec{n}_j \cdot \vec{e}_{ij})\vec{n}_i - (\vec{n}_i \cdot \vec{e}_{ij})\vec{n}_j],\end{aligned} \tag{3}$$

$$\vec{F}_{ij}^2 = AF_0\frac{a^{13}}{r_{ij}^{13}}\vec{e}_{ij}, \tag{4}$$

$$\vec{T}_{ij}^1 = \frac{aF_0}{3}\frac{a^3}{r_{ij}^3}[\vec{n}_i \times \vec{n}_j - 3(\vec{n}_j \cdot \vec{e}_{ij})\vec{n}_i \times \vec{e}_{ij}], \tag{5}$$

$$\vec{T}_{ij}^2 = m_0 H\vec{n}_i \times \hat{z} \tag{6}$$

with the dipolar force strength $F_0 = 3m_0^2/(4\pi\mu_0 a^4), r_{ij} = |\vec{r}_i - \vec{r}_j|$, and $\vec{e}_{ij} = (\vec{r}_i - \vec{r}_j)/r_{ij}$. Here, for simplicity, we assume $\vec{F}_{ij}^2$ is the Lennard-Jones (LJ) -type repulsive force. In Eq. (4) the coefficient A is determined such that $\vec{F}_{ij}^1 + \vec{F}_{ij}^2 = \vec{0}$ for two particles i and j under the condition of parallel head to tail alignment (i.e., $\vec{n}_i = \vec{n}_j = \vec{e}_{ij}$ and $r_{ij} = 2a$), and for the spherical particles case, $A = 2^{-16}$. The random force $\vec{R}_i$ in Eq. (1) and the random torque $\vec{N}_i \equiv \vec{n}_i \times \vec{R}_i^r$ in Eq. (2) are due to the thermal noise which obey the fluctuation- dissipation theorem given by [2].

Finally, we introduce scale units of some physical quantities: the unit of length as a, force as F_0, time as $t_0 \equiv a\xi_t/F_0$, and magnetic field as $H_0 \equiv 4aF_0/(3m_0)$. These units are estimated as $F_0 \sim 10^{-9}$N, $t_0 \sim 10^{-6}$s, and $H_0 \sim 10^6$ A/m for the case of $a = 5 \times 10^{-9}$m, $m_0 = 10^{-23}$T m^3, $\eta = 0.1$Pa s, and $T = 0$K.

The resulting equations include the important dimensionless parameters: H/H_0. Using these dimensionless parameters, it is convenient to study the parametric properties of the model system.

As the numerical method, we use the simple Euler method. Time step, Δt is chosen small enough to avoid overlapping between particles.

SIMULATION RESULTS

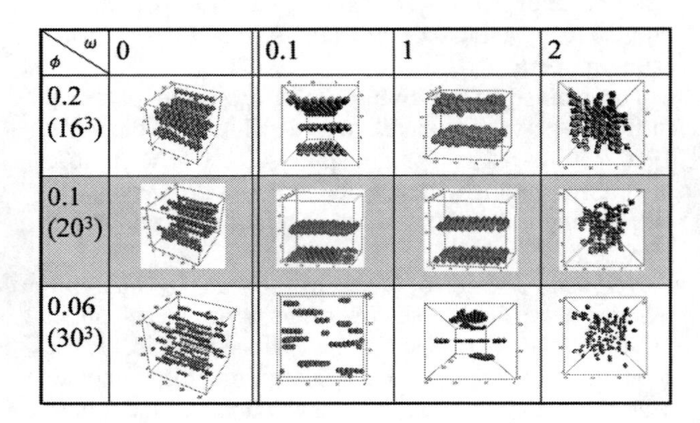

FIGURE 2. Aggregation pattern of $N = 200$, $L_x = L_y = L_z = 16, 20, 30$ and $\omega = 0, 0.1, 1$ and 2.

As Fig. 2 shows, we obtain the sheet structure as well as ER and MR as for MF under the certain range of frequency of the magnetic field ω and particles volume fraction ϕ. The chain structure is formed when $\omega = 0$ and the sheet structure can be observed with the range of $0 < \omega \leq 1.0$ and $0.01 < \phi < 0.1$. At smaller volume fraction($\phi = 0.06$), the small fractions of the sheet (two-dimensional cluster on the x-y plane) are observed. On the other hand, at the larger volume fraction ($\phi = 0.2$), the thick layer structure appears.

Here, we explore the sheet formation process at the modest volume fraction ($\phi = 0.1$) by comparing the different values of rotating frequency ($\omega = 1.0$ and $\omega = 1.1$). Figure 3 shows the formation process of sheet structure at $\phi = 0.1$ and $\omega = 1.0$. On the other hand, Figure 4 shows the formation process of $\phi = 0.1$ and $\omega = 1.1$. The result shows that at the larger value of $\omega (> 1.1)$, no sheet structure can be seen. By comparing Fig. 3 and Fig. 4, we estimate that the threshold value for formation of sheet structure is located in the region of $1.0 < \omega \leq 1.1$.

First, we present the reason for the sheet structure. In the present system, each particle rotates in the *x-y* plane due to the external rotating magnetic field. At the same time, neighboring particles tend to make chains due to the head-tail alignment effect. When a particle orients perpendicularly to chain clusters due to the rotating magnetic field, the particle attracts particles in other chains which are also rotated, and the particles achieve a lower energy state due to the dipole-dipole interaction between particles. This repetition of the competition of interparticle interaction and rotating magnetic field effect, finally lead to form a two dimensional plane.

On the other hand, Halsey et al. researched of the "Rotary" ER effect, the relation between electrical torques and disk rotation, reveals that particles repel in the direction of perpendicular to the plane of the field, and attract in the plane of the field [1]. They consider a monodispersed suspension of spherical particles of radius a in an insulating fluid and assume that the particles have a simple dielectric responses, so that their dipole moment $\vec{p}$ obeys

$$\vec{p} = \beta a^3 \vec{E},$$

where β is polarizability and $\vec{E}$ is the instantaneous electric field.

They suppose that the instantaneous interaction energy u between two dipoles $\vec{p}^{(1)}$ and $\vec{p}^{(2)}$ separated by a distance r is

$$u(\vec{r}) = -\vec{p}^{(1)} \frac{3r_i r_j - r^2 \delta_{ij}}{r^5} \vec{p}^{(2)}$$

and suppose that $\vec{p}^{(1,2)}(t) = \beta a^3 \vec{E}_0(t)$; thus the dipole moment is proportional to the applied field $\vec{E}_0(t)$– contributions of other dipoles to the local field are neglected.

And apply $\vec{E}_0(t)$ is the rotating fields,

$$\vec{E}_0(t) = E_0(\hat{x}\cos\omega t + \hat{y}\sin\omega t)$$

. Introducing this form into the equation of $u(\vec{r})$, and average over time, the instantaneous interaction energy u between dipoles

$$u(\vec{r}) = -\frac{\beta^2 a^6}{2r^3}(1 - 3\cos^2\theta)$$

where θ is the angle between the normal of the plane of the rotating electric field and $\vec{r}$. Since this equation is the opposite of the resulting averaged interaction for a uniaxial field $\vec{E}_0(t) = \vec{E}_0 f(t)$:

$$u(\vec{r}) = -\frac{\beta^2 a^6 E_0^2 < f^2(t) >}{r^3}(3\cos^2\theta - 1),$$

where θ is now the angle between $\vec{r}$ and the field direction. Thus the particles repel in the direction perpendicular to the plane of the field, and attract in the plane of the field.

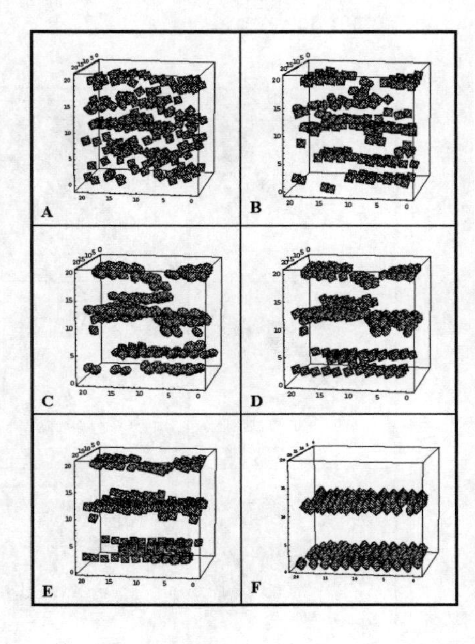

Figure 3. Time evolution of the microscopic structures of $\phi = 0.1$, $\omega = 1$. $A: t = 101\frac{t}{t_0} \times \Delta t$, $B: t = 301\frac{t}{t_0} \times \Delta t$, $C: t = 501\frac{t}{t_0} \times \Delta t$, $D: t = 701\frac{t}{t_0} \times \Delta t$, $E: t = 901\frac{t}{t_0} \times \Delta t$ and $F: t = 5,001\frac{t}{t_0} \times \Delta t$

This seems to be helpful to support our model, however, we would like to note that in the result of our simulation, a particle rotates individually in the seat.

On the other hand, the experiment result shows that the whole disk of particles aggregation rotates after the particles cohere[1]. This is because we do not focus the motion of the gravitation center but of the each particle's motion. As for the reason why the simulation result shows the different phenomena from the result of the experiment, we consider that our program focuses on the formation process so that we rather treat all the particles equally, to see the each particle's motion [8].

In our model, each particle rotates individually induced by the external rotating field, and forms chain structures by the interacting force and gradually separate into the upper and lower regions to aggregate in sheets.

Thus, we can conclude that the point dipoles in three-dimensional space under the rotating magnetic field tends to locate in the two-dimensional plane, at a certain range of the rotating frequency. This can be explained by the expansion of the head-tail alignment.

Secondly, as shown in Fig.4, in the range $1.1 \leq \omega$, we cannot observe the sheet-like structures. Thus the threshold value of ω (ω_c) should be $1.0 < \omega_c \leq 1.1$.

When the rotating frequency ω exceeds $\omega_c \leq \omega$, chain structures cannot be formed.

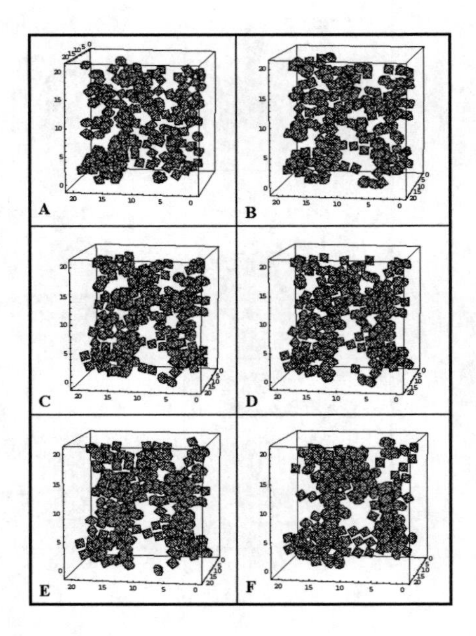

FIGURE 4. Time evolution of the microscopic structures of $\phi = 0.1$, $\omega = 1.1$. $A : t = 101\frac{t}{t_0} \times \Delta t$, $B : t = 301\frac{t}{t_0} \times \Delta t$, $C : t = 501\frac{t}{t_0} \times \Delta t$, $D : t = 701\frac{t}{t_0} \times \Delta t$, $E : t = 901\frac{t}{t_0} \times \Delta t$ and $F : t = 5{,}001\frac{t}{t_0} \times \Delta t$

CONCLUSIONS

This paper has discussed the microstructure of a magnetic fluid under a rotating magnetic field. The structure is independent of the magnitude of magnetic force. We have observed the sheet structure in a range of the rotating frequency $0.01 < \omega < 1.0$.

We propose that the reason for these phenomena is because the particles achieve a lower energy state by taking a head-tail alignment, due to the dipole-dipole interaction between particles. Thus first, the particles tends to form the chain structure along the magnetic direction. In the present system, each particle rotates in the x-y plane due to the effect of the external rotating magnetic field. When a particle orients perpendicularly to the chain cluster, the particle attracts a particle in another chain, which is also perpendicular to the chain, and form a two dimensional plane finally(as shown in Fig.5).

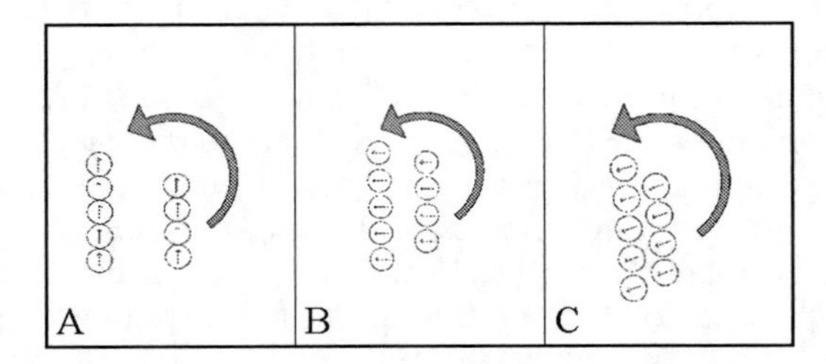

FIGURE 5. Process of the sheet formation from chain clusters. Two chains (A) effected by rotating field (B), align in plane finally(C).

The distance between two planes will be studied by changing the particle density and rotating frequency. When the frequency is large ($\omega > 1.1$), the sheet structures are no longer observed. Thus we can extract conclusions that the threshold value of frequency ω_c exists between $1.0 < \omega_c \leq 1.1$.

REFERENCES

1. T.C.Halsey et al. *Int. J. of Mod. Phys.* **10**, 3019-3027 (1996).
2. V.V.Murashov and G.N.Patey *J. of Chem.Phys.* **112**, 9828-9833 (2000).
3. G.N. Coverdale, R.W. Chantrell, A. Satoh, R. Vitech, *J. Appl. Phys.* **81**, 3818-3820 (1997).
4. Rosensweig R.E., *Ferrohydrodynamics* New York: Cambridge University Press, 1987.
5. M.I. Shliomis, K.I. Morozov, *Phys. Fluids* **6**, 2855-2861 (1994).
6. S. Odenbach, *Magnetoviscous Effects in Ferrofluids*, Springer, Berlin, 2002.
7. J.P. McTague, *J. Chem. Phys.* **51**,133-136 (1969).
8. Y. Enomoto, K. Oba, *Physica A* **309**, 15-25 (2002).

A Theory of Electrical Conductivity of Pseudo-Binary Equivalent Molten Salt

Shigeki Matsunaga[1], Takahiro Koishi[2], and Shigeru Tamaki[3]

[1]*Nagaoka National College of Technology, Nagaoka 940-8532, Japan*
[2]*Department of Applied Physics, Faculty of Engineering, University of Fukui, Japan*
[3]*Department of Physics, Faculty of Science, Niigata University, Japan*

Abstract. Many years ago, Sundheim proposed the "universal golden rule" by experiments, i.e. the ratio of the partial ionic conductivities in molten binary salt is equal to the inverse mass ratio of each ions, $\sigma^+/\sigma^- = m^-/m^+$. In the previous works, we have proved this relation by the theory using Langevin equation, and by molecular dynamics simulations (MD). In this study, the pseudo binary molten salt NaCl-KCl system is investigated in the same theoretical framework as previous works as the serial work in molten salts. The MD results are also reported in connection with the theoretical analysis.

Keywords: Langevin equation, Molecular dynamics, NaCl-KCl, Partial conductivity,Velocity correlation function,
PACS: 66.10.-x, 66.10.Ed

INTRODUCTION

In previous papers, it has been confirmed that the partial dc conductivities of cation and anion in the equivalent molten salt are always equal to the inverse mass ratio of the constituents and the effective friction constants acting on the cation and anion are the same using a generalized Langevin equation.[1,2] Furthermore, it has been suggested that a combination of generalized Langevin equation and damped oscillator equation is useful to analyze a short-time behaviors of ions.
As a natural extension, these fruitful methods and results prompt us to investigate the conductivities in a more complex system such as a binary equivalent molten salt exhibited by $[AX]_{1-c}[BY]_c$. In this paper, we will present an extended application onto the binary molten salts of NaCl-KCl system. In the next section, we will summarize some important results obtained in the former articles.[1,2,3]

A BRIEF SURVEY ON PREVIOUS RESULTS

An equivalent molten salt composed of the density $n^+ = n^- = n_0$ (= N/V_0) and of the charge $z^+ = -z^- = z$ was considered, where N being the total number of cation or anion in the volume V_0. The current density at the time t can be written as,

$$\mathbf{j}(t) = \mathbf{j}^+(t)[= \Sigma_{i=1}^{n+} z^+ e\mathbf{v}_i^+(t)] + \mathbf{j}^-(t)[= \Sigma_{k=1}^{n-} z^- e\mathbf{v}_k^-(t)] \quad (1)$$

According to the linear response theory,[2, 3] the dc conductivity in such a molten salt is given by the following formula,

$$\sigma = \sigma^+(=\sigma^{++}+\sigma^{+-}) + \sigma^-(=\sigma^{-+}+\sigma^{--}) \quad (2)$$

Conductivity coefficients σ^{++}, σ^{+-} and σ^{--} are expressed as follows,

$$\sigma^{\alpha\beta} = (1/3k_BT)\int_0^\infty \langle \mathbf{j}^\alpha(t)\cdot \mathbf{j}^\beta(0)\rangle dt \quad (3)$$

where ($\alpha,\beta = +, -$). All current correlation functions in these integrations are replaced by the form of velocity correlation function $\langle \mathbf{v}_i^\alpha(t)\cdot \mathbf{v}_j^\beta(0)\rangle$ ($\alpha, \beta = +, -$) in relation to (1). In the previous paper,[2] we have obtained that the Taylor expansion for velocity correlation function is expressed as follows,[4,5]

$$\langle \mathbf{v}_i^\alpha(t)\cdot \mathbf{v}_j^\beta(0)\rangle = \langle \mathbf{v}_i^\alpha(0)\cdot \mathbf{v}_j^\beta(0)\rangle\{1-(t^2/2)(\alpha^0/3\mu) + (\text{higher order over } t^4)\} \quad (4)$$

$$1/\mu = 1/m^+ + 1/m^- \quad (5)$$

where m^+ and m^- are masses of cation and anion.

CP982, *Complex Systems, 5th International Workshop on Complex Systems*
edited by M. Tokuyama, I. Oppenheim, and H. Nishiyama

The zero-th order term in the Taylor expansion of velocity correlation function is expressed as follows,

$$\langle \mathbf{v}_i^+(0)\cdot \mathbf{v}_j^+(0)\rangle=(3k_BT/m^+)(1+m^+/m^-)\ (i=j \text{ and } i\neq j) \quad (6)$$
$$\langle \mathbf{v}_k^-(0)\cdot \mathbf{v}_l^-(0)\rangle=(3k_BT/m^-)(1+m^-/m^+)\ (k=l \text{ and } k\neq l) \quad (7)$$

and

$$\langle \mathbf{v}_i^+(0)\cdot \mathbf{v}_k^-(0)\rangle = -3k_BT/(m^- + m^+) \quad (i\neq k) \quad (8)$$

On the other hand, a generalized Langevin equation for cation or anion under an infinitesimal external field $\mathbf{E}$ is written as follows,

$$m^\pm d\mathbf{v}_i^\pm(t)/dt = -m^\pm \int_0^t \gamma^\pm(t-t')\mathbf{v}_i^\pm(t')dt'+\mathbf{R}_i^\pm(t)+z^\pm e\mathbf{E} \quad (9)$$

where $\gamma^\pm(t)$ is the memory function in relation to the friction force acting on the cation or anion, and $\mathbf{R}_i^\pm(t)$ is the stochastic or random fluctuating force acting on the ion i. Under the condition that the total momentum of the system is conserved zero, we have also obtained that the long wavelength limit of Laplace transformation of the memory functions for cation and anion are the same[1,2], *i.e.*, using (5),

$$\gamma^{\sim+}(0)=\gamma^{\sim-}(0)\equiv \gamma^\sim(0) = (\alpha^0/3\mu)^{1/2} \quad (10)$$

and

$$\alpha^0 = n_0 \int_0^\infty\{ \partial^2 \phi^{+-}(r)/\partial r^2 + (2/r)\partial \phi^{+-}(r)\}g^{+-}(r)\cdot 4\pi r^2dr \quad (11)$$

where $\phi^{+-}(r)$ and $g^{+-}(r)$ are the pair potential and pair distribution function between cation and anion. From these, the partial dc conductivities σ^+ and σ^- are expressed as follows,

$$\sigma^\pm = (n_0z^2e^2/m^\pm)/\gamma^\sim(0) \quad (12)$$

It is apparent that the quantity $\gamma^\sim(0)$ plays a role of effective friction constant of ions.

CURRENT CORRELATION FUNCTIONS

We are now in a position to extend the above theory onto the binary molten salts of $[NaCl]_{1-c}[KCl]_c$. The partial and total current densities are written as follows,

$$\mathbf{j}^+ = \mathbf{j}_{Na}^+ + \mathbf{j}_K^+,\quad \mathbf{j}^- = \mathbf{j}_{Cl}^-,\quad \mathbf{j} = \mathbf{j}_{Na}^+ + \mathbf{j}_K^+ + \mathbf{j}_{Cl}^- \quad (13)$$

Therefore the partial dc conductivity σ^+ in its Kubo formulae is written as follows,

$$\sigma^+ = (1/3k_BT)\int_0^\infty\langle \mathbf{j}^+(t)\cdot \mathbf{j}(0)\rangle dt \quad (14)$$

where

$$\langle \mathbf{j}^+(t)\cdot \mathbf{j}(0)\rangle = \langle \mathbf{j}_{Na,i}^+(t)\cdot (\mathbf{j}_{Na,j}^+(0)+\mathbf{j}_{K,j'}^+(0)+\mathbf{j}_{Cl,l}^-(0))\rangle + \langle \mathbf{j}_{K,i'}^+(t)\cdot (\mathbf{j}_{Na,j}^+(0)+\mathbf{j}_{K,j'}^+(0)+\mathbf{j}_{Cl,l}^-(0))\rangle$$

$$= \{\langle \mathbf{j}_{Na,i}^+(t)\cdot \mathbf{j}_{Na,j}^+(0)\rangle + \langle \mathbf{j}_{Na,i}^+(t)\cdot \mathbf{j}_{K,j'}^+(0)\rangle + \langle \mathbf{j}_{Na,i}^+(t)\cdot \mathbf{j}_{Cl,l}^-(0)\rangle\} + \{\langle \mathbf{j}_{K,i'}^+(t)\cdot \mathbf{j}_{Na,j}^+(0)\rangle + \langle \mathbf{j}_{K,i'}^+(t)\cdot \mathbf{j}_{K,j'}^+(0)\rangle + \langle \mathbf{j}_{K,i'}^+(t)\cdot \mathbf{j}_{Cl,l}^-(0)\rangle\} \quad (15)$$

In a similar way,

$$\sigma^- = (1/3k_BT)\int_0^\infty\langle \mathbf{j}^-(t)\cdot \mathbf{j}(0)\rangle dt \quad (16)$$

In the following sections, however, we will firstly obtain the velocity correlation functions corresponding to current correlation functions and thereafter the latter are obtained. The velocity correlation functions are already expanded as a series function of time t as shown in equations (4).

Zero-th Order Velocity Correlation Functions in Relation to the Partial Conductivities

First of all, we will obtain the coefficients of the zero-th order of velocity correlation functions. To do so, we have to consider the momentum conservation in this system, in relation to the charge neutrality.

$$\Sigma_{i=1}^{(1-c)N}\mathbf{p}_{Na,i}^+(0) + \Sigma_{i'=1}^{cN}\mathbf{p}_{K,i'}^+(0) = -\Sigma_{k=1}^{N}\mathbf{p}_{Cl,k}^-(0) \quad (17)$$

Multiplying $\mathbf{v}_{Cl,l}^-(0)$ on this equation and taking ensemble average, we have

$$(1-c)m_{Na}\langle \mathbf{v}_{Na,i}^+(0)\cdot \mathbf{v}_{Cl,l}^-(0)\rangle + cm_K\langle \mathbf{v}_{K,i'}^+(0)\cdot \mathbf{v}_{Cl,l}^-(0)\rangle = -m_{Cl}\langle \mathbf{v}_{Cl,k}^-(0)\cdot \mathbf{v}_{Cl,l}^-(0)\rangle \quad (18)$$

The distribution of Cl^- ions around a Na^+ ion may, more or less, differ from that around a K^+ ion, because the attractive part of the inter-ionic potential between Na^+-Cl^- pair seems stronger than that between K^+-Cl^- pair. Taking this fact in mind, we can define the following relation,

$$f(c)(m_{Na}/m_K)\langle \mathbf{v}_{Na,i}^+(0)\cdot \mathbf{v}_{Cl,k}^-(0)\rangle=\langle \mathbf{v}_{K,i'}^+(0)\cdot \mathbf{v}_{Cl,k}^-(0)\rangle \quad (19)$$

where f(c) is a parametric function. Inserting (19) into (18) we have

$$(1/m_{Cl})m_{Na}\{(1-c) + cf(c)\}\langle \mathbf{v}_{Na,i}^{+}(0)\cdot \mathbf{v}_{Cl,k}^{-}(0)\rangle = -\langle \mathbf{v}_{Cl,k}^{-}(0)\cdot \mathbf{v}_{Cl,l}^{-}(0)\rangle \quad (20)$$

In order to obtain $\langle \mathbf{v}_{Na,i}^{+}(0)\cdot \mathbf{v}_{Cl,k}^{-}(0)\rangle$ we have to apply a simple statistical mechanics described in the next section.

Ensemble Averages of Velocity Correlation Functions

According to the statistical mechanics, the following relation is well-known.

$$\langle \mathbf{p}_{Cl,k}^{-}\cdot \partial H/\partial \mathbf{p}_{Cl,k}^{-}\rangle = 3k_BT \quad (21)$$

where H is the Hamiltonian of the system and composed of the kinetic energies and potential ones. We assume that the potential energy terms are approximated as the pair-wise interaction ones. Considering that the fraction of Cl^- ions to the total ions is equal to 1/2 and those of Na^+ ions and K^+ ions are equal to (1/2)(1−c) and (1/2)c, respectively, we have

$$\langle \mathbf{p}_{Cl,k}^{-}\cdot \partial H/\partial \mathbf{p}_{Cl,k}^{-}\rangle = (1/2)\{m_{Cl}\langle\langle \mathbf{v}_{Cl,k}^{-}\cdot \mathbf{v}_{Cl,l}^{-}\rangle\rangle-(1-c)m_{Na}\langle\langle \mathbf{v}_{Na,i}^{+}\cdot \mathbf{v}_{Cl,k}^{-}\rangle\rangle- c(m_{Cl}m_{Na}/m_K)\langle\langle \mathbf{v}_{K,i'}^{+}\cdot \mathbf{v}_{Cl,k}^{-}\rangle\rangle\} \quad (22)$$

where $\langle\langle\ \rangle\rangle$ means the ensemble average of velocity correlation functions per two particles and all these double brackets are equal to the twice of the ensemble average per one particle, that is, $\langle\langle\ \rangle\rangle = 2\langle\ \rangle$. Then we have,

$$\langle \mathbf{v}_{Na}^{+}(0)\cdot \mathbf{v}_{Cl}^{-}(0)\rangle = -3k_BT/[m_{Na}\{(1-c)+ cf(c)\} +(1-c)m_{Cl} +cf(c)(m_{Na}m_{Cl}/m_K)] \quad (23)$$

Here and in the following, we omit the suffixes of ion's numbering, for simplicity. In a similar way, we have

$$\langle \mathbf{v}_{K}^{+}(0)\cdot \mathbf{v}_{Cl}^{-}(0)\rangle=-3k_BT(m_{Na}/m_K)f(c)/[m_{Na}\{(1-c)+ cf(c)\}+(1-c)m_{Cl} +cf(c)(m_{Na}m_{Cl}/m_K)] \quad (24)$$

Equations (23) and (24) tell us that the zero-th order velocity correlation function between cation and anion is related to all constituents of the system. Using equations (23) and (24), we have the following zero-th order current correlation,

$$\langle \mathbf{j}_{Na}^{+}(0)\cdot \mathbf{j}_{Cl}^{-}(0)\rangle=(1/2n_0)(6k_BT)(n_0^2e^2)(1-c)(m_{Cl}/m_{Cl})/D \quad (25)$$

$$\langle \mathbf{j}_{K}^{+}(0)\cdot \mathbf{j}_{Cl}^{-}(0)\rangle =(1/2n_0)(6k_BT)(n_0^2e^2)cf(c)\{(m_{Cl}m_{Na}/m_K)/m_{Cl}\}\}/D \quad (26)$$

$$\langle \mathbf{j}_{Cl,}^{-}(0)\cdot \mathbf{j}_{Cl l}^{-}(0)\rangle=(1/2n_0)(6k_BT)(n_0^2e^2)(m_{Na}/m_{Cl})/D \quad (27)$$

where

$$D = [m_{Na}\{(1-c)+ cf(c)\}+(1-c)m_{Cl} +cf(c)(m_{Na}m_{Cl}/m_K)]$$

Therefore, the summation of zero-th order current correlation functions incorporating with the partial conductivity for anion is described as follows,

$$\langle \mathbf{j}_{Cl}^{-}(0)\cdot \mathbf{j}_{Cl}^{-}(0)\rangle + \langle \mathbf{j}_{Na}^{+}(0)\cdot \mathbf{j}_{Cl}^{-}(0)\rangle + \langle \mathbf{j}_{K}^{+}(0)\cdot \mathbf{j}_{Cl}^{-}(0)\rangle = (3k_BT)(n_0e^2/m_{Cl}) \quad (28)$$

This relation is related to a kind of sum rule indicating the conduction ability of Cl^- ion described by (n_0e^2/m_{Cl}). On the analogy of equation (28), we can naturally infer the following relations,

$$\{\langle \mathbf{j}_{Na}^{+}(0)\cdot \mathbf{j}_{Na}^{+}(0)\rangle+\langle \mathbf{j}_{Na}^{+}(0)\cdot \mathbf{j}_{K}^{+}(0)\rangle+\langle \mathbf{j}_{Na}^{+}(0)\cdot \mathbf{j}_{Cl}^{-}(0)\rangle\} = (3k_BT)\{n_0e^2/m_{Na}\}(1- c) \quad (29)$$

$$\{\langle \mathbf{j}_{K}^{+}(0)\cdot \mathbf{j}_{Na}^{+}(0)\rangle+\langle \mathbf{j}_{K}^{+}(0)\cdot \mathbf{j}_{K}^{+}(0)\rangle+\langle \mathbf{j}_{K}^{+}(0)\cdot \mathbf{j}_{Cl}^{-}(0)\rangle\} = (3k_BT)\{n_0e^2/m_K\}c \quad (30)$$

At the present stage, we have no prescription to verify $\langle \mathbf{v}_{Na}^{+}(0)\cdot \mathbf{v}_{Na}^{+}(0)\rangle$, $\langle \mathbf{v}_{Na}^{+}(0)\cdot \mathbf{v}_{K}^{+}(0)\rangle$ and $\langle \mathbf{v}_{K}^{+}(0)\cdot \mathbf{v}_{K}^{+}(0)\rangle$, without using relations of (29) and (30). However, in next section, we will derive these quantities by using molecular dynamics (MD) simulation.

Result of Molecular Dynamics Simulation

In this section, we obtain the velocity correlations by MD simulation. The details of the MD procedure are same as that of the previous works [6,7,8,9]. The used pair potentials are Huggins-Mayer type,

$$\phi_{ij}(r) = z_iz_je^2r^{-1} + b_{ij}\exp(-r/\rho) + c_{ij}r^{-6} + dij\, r^{-8} \quad (31)$$

The symbols are used as the ordinal meaning. The parameters are taken from the literature[10]. The total number of atoms is 512 (128Na + 128K + 256Cl) in a cubic cell under the periodic boundary condition.

In the numerical calculation, the velocity correlation functions for particle i and j, *i.e.* $\langle \mathbf{v}_i^{\alpha}(t)\cdot \mathbf{v}_j^{\beta}(0)\rangle$ (α, β= +, −), are obtained by the following procedure; (a) after the sufficient thermal equalization, the coordinates of the atoms are sampled for n_s MD time steps, *i.e* $n_s\Delta t$; (b) the inner product are made between the sample coordinates and the another n_s sets of coordinates of atoms obtained after one MD time step, Δt, next the average is calculated by dividing by the number of samples n_s and the number of atoms of i and j, then we have the numerical value of the velocity correlation function at time $(n_s + 1)\Delta t$; (c) the previous procedure (b) is repeated until the convergence of the velocity correlation functions is sufficient. If we set the time $n_s\Delta t$ as the initial time t=0, then we have the numerical value of the velocity correlation function as a function of t. In this calculation, we adopt the value ns = 3000, and $\Delta t = 2.0\times10^{-15}$ s.

The obtained velocity correlation functions in molten phase at 1500K are shown in Figure 1, which are normalized as $\langle \mathbf{v}_{Na}^{+}(0)\cdot \mathbf{v}_{Na}^{+}(0)\rangle = 1$. As seen in Figure 1 at t=0, Eq.(18) and related equations based on the momentum conservation law are confirmed , *i.e.* the following relation as valid, as,

$$(1-c)m_{Na}\langle \mathbf{v}_{Na}^{+}(0)\cdot \mathbf{v}_i^{\alpha}(0)\rangle + cm_K\langle \mathbf{v}_{K,}^{+}(0)\cdot \mathbf{v}_i^{\alpha}(0)\rangle = -m_{Cl}\langle \mathbf{v}_{Cl,k}^{-}(0)\cdot \mathbf{v}_i^{\alpha}(0)\rangle \qquad (32)$$

where (i = Na, K, Cl) and (α, β = +, −). We have also confirmed the exchange relations, $\langle \mathbf{v}_i^{\alpha}(0)\cdot \mathbf{v}_j^{\beta}(0)\rangle = \langle \mathbf{v}_j^{\beta}(0)\cdot \mathbf{v}_i^{\alpha}(0)\rangle$. Following the definition of current correlation functions.(25)-(27), the concentration factor c and (1-c) are included in the conductivity coefficients, then we obtain the relations,

$$m_{Na}\,\sigma_{Na\,\xi} + m_K\,\sigma_{K\,\xi} = m_{Cl}\,\sigma_{Cl\,\xi} \qquad (33)$$

where (ξ = Na, K, Cl). Applying Eq.(2) in the present ternary case, *i.g.*,

$$\sigma_{Na} = \sigma_{Na\,Na} + \sigma_{Na\,K} + \sigma_{Na\,Cl} \qquad (34)$$

we obtain the relation between partial conductivities.

$$m_{Na}\,\sigma_{Na} + m_K\,\sigma_K = m_{Cl}\,\sigma_{Cl} \qquad (35)$$

We have already confirmed corresponding relations in ternary systems of molten noble metal halides by non-equilibrium MD[6,7]. However, in order to treat this relation more strictly, the discussion of second order correlation functions is required as mentioned in the previous sections. We would discuss this point in the near future.

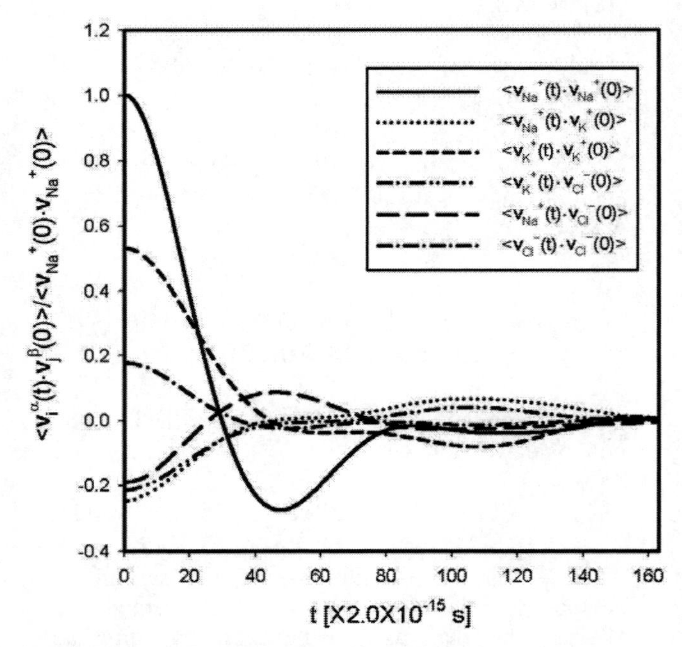

FIGURE 1. Velocity correlation functions of molten $(NaCl)_{0.5}(KCl)_{0.5}$ at 1500K obtained by MD.

ACKNOWLEDGMENTS

One of the authors (SM) expresses his thanks to the financial support from the Nippon Sheet Glass Foundation for Material Science and Engineering. This study was supported by the Grant-in-Aid for Science Research from the Ministry of Education, Science and Culture.

REFERENCES

1. T. Koishi and S. Tamaki, *J. Phys. Soc. Japn.* **68**, 964-971 (1999).
2. T. Koishi, S. Kawase and S. Tamaki, *J. Chem. Phys.* **116**, 3018-3026 (2002).
3. T. Koishi and S. Tamaki, *J. Chem. Phys.*, **121**, 333 -340 (2004).
4. D. C. Douglass, *J. Chem. Phys.* **35**, 81-90 (1961).
5. D. A. McQuarrie, *Statistical Mechanics,* Indiana University Press, Bloomington, 1976.
6. S. Matsunaga and P. A. Madden , *J. Phys.: Condens. Matter,* **16**, 181-194 (2004).
7. S. Matsunaga, *Solid State Ionics*, **176,** 1929-1940 (2005).
8. S. Matsunaga, T. Koishi and S. Tamaki, *Matarials Science and Engineering A*, **449-451**, 693-698 (2007).
9. S. Matsunaga, T. Koishi and S. Tamaki, *Molecular Simulation,* **33**, 613-621 (2007).
10. B. Larsen and T. Forland, *Molecular Physics*, **26**, 1521-1532 (1973).

Ordering Dynamics of Lamellar Patterns in Cooperative Systems with a Small-World Topology

R. Imayama and Y. Shiwa

Statistical Mechanics Laboratory, Kyoto Institute of Technology, Matsugasaki, Sakyo-ku, Kyoto 606-8585, Japan

Abstract. We have studied numerically the ordering dynamics of lamellar patterns in two-dimensional small-world media. It is found that increase of the density of random shortcuts causes the coarsening of lamellar domains to become slower and eventually get stuck in a frozen configuration. We discuss the underlying mechanism responsible for this frozen state.

Keywords: Coarsening, Lamellar patterns, Small-world networks
PACS: 64.60.Cn, 47.54.+r, 05.65.+b, 89.75.Fb

INTRODUCTION

We study ordering dynamics of lamellar patterns in two-dimensional systems on square lattices which have a small-world topology [1]. The small-world (SW) topology has a few random long-range connections (shortcuts) linking spatially distant units of the systems, and is neither completely random nor completely local but lies between these two extremes. For instance, neurobiological systems consisting of a large number of neurons possess such SW properties, and the observation of intercellular spiral waves in the brain is noteworthy in particular [2].

In this paper, we investigate the Swift-Hohenberg (SH) equation [3] for the spontaneous lamellar pattern formation with increasing density of random long-range connections. Although the SH equation was originally proposed as a simplified model of convective roll patterns, it now constitutes a popular paradigm of pattern formation exhibiting many features common to natural patterns [4]. In appropriate units, our model reads

$$\frac{\partial \psi_i}{\partial t} = \epsilon \psi_i - g\psi_i^3 - (\nabla^2 + k_0^2)^2 \psi_i. \qquad (1)$$

Here ψ_i defined at each site i represents the order parameter field, and the positive constants g and k_0 are phenomenological parameters. The ϵ is the bifurcation parameter, with the transition to lamellar structures of spatial period of order k_0^{-1} appearing for $\epsilon > 0$ in the absence of shortcuts. The Laplacian operator in the present case consists of two terms:

$$\nabla^2 = (\nabla^2)^{nn} + (\nabla^2)^{sc},$$

where $(\nabla^2)^{nn}$ is the Laplacian on the original square lattice, while $(\nabla^2)^{sc}$ is the one on the random part of the SW network. Both of these Laplacians are fully characterized by the adjacency matrix A_{ij}, taking the value $A_{ij} = 1$ if the vertices i and j are connected by an edge, whereas $A_{ij} = 0$ otherwise. Thus, for any quantity X_i on the site i of the network

$$\nabla^2 X_i = \sum_j (A_{ij} - k_i \delta_{ij}) X_j, \qquad (2)$$

where k_i is the degree of the vertex i; $k_i = \sum_j A_{ij}$.

METHODOLOGY

The SW network is constructed as follows [5]. We start with a two-dimensional square lattice. We then add shortcuts between pairs of vertices chosen uniformly at random but we do not remove any bonds from the regular lattice. More than one bond between any two vertices as well as any bond connecting a vertex to itself are prohibited. In particular, we add with probability p one shortcut for each bond on the original lattice, so that an average coordination number z is given by $z = 4(1+p)$. As p is increased, the SW network interpolates between the regular lattice ($p = 0$) with only geographical neighbors in contact, and a random graph ($p = 1$) where short and long range connections are equally likely.

We have performed simulations based on Eq. (1) for different choices of the shortcut parameter p. The two-dimensional space is divided into 256×256 square cells with periodic boundary conditions. The cell size is set to be unity. In order to allow easier exploration of the long-time regime of pattern formation, we utilized the cell-dynamical-system method [6]. Initial conditions for ψ were the spatially random distribution between ± 0.01. For the parameter values used in the simulation, the aspect ratio Γ as defined by the system length versus $2\pi/k_0$ was always $\Gamma \approx 40$. Hence the finite-size effects are assumed to be negligible in our simulations.

CP982, *Complex Systems, 5th International Workshop on Complex Systems*
edited by M. Tokuyama, I. Oppenheim, and H. Nishiyama

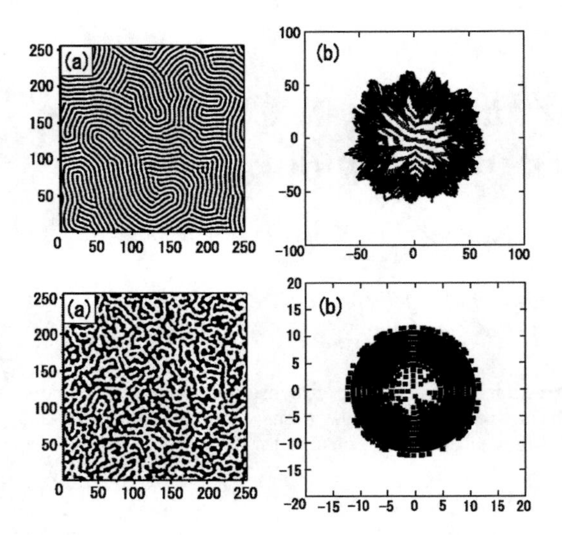

FIGURE 1. Typical patterns at late times ($t = 398107$) obtained by numerical simulations of Eq. (1) on the 256×256 SW lattice. Top panels: the lamellar pattern for $p = 0.001$; bottom panels: the inhomogeneous pattern for $p = 0.2$. The left column (a) shows patterns as seen in Euclidean space while the right column (b) in geodesic space. The white region represents positive values of the order parameter ψ and the black one negative ψ.

GENERAL RESULTS

There were obviously three different types of patterns occurring in different regions in the $\epsilon - p$ plane. We found the lamellar pattern and the inhomogeneous pattern as given in Fig. 1 as well as the trivial homogeneous pattern. These three regions of existence are in accordance with the mean-field theoretical analysis [7] in which random links are replaced with their expected number, i.e., each site is linked to all others with a link of strength $4p/N$, where N is the total number of vertices in the graph (network).

What cannot be qualitatively accounted for by the mean field theory is that the inhomogeneous state attained at the late stage of time evolution, as given in Fig. 1(bottom), is frozen, i.e., the coarsening stops at larger times and patterns get pinned. A similar observation was made recently [8] from the Glauber dynamics of the Ising model on SW networks. It was suggested that the strong connectivity of SW produces frozen metastable states that are disordered at large scales.

Let us now concentrate on the so-called small world regime, that corresponds to the case $p \lesssim 0.1$. Here we observed lamellar patterns. We computed the circularly-averaged structure factor $S(k,t)$ defined by

$$S(k,t) = \left\langle \tilde{\psi}(\boldsymbol{k},t)\tilde{\psi}^*(\boldsymbol{k},t) \right\rangle,$$

where $\tilde{\psi}(\boldsymbol{k},t)$ is the Fourier transform of the order parameter, and the orientation of the wave vector $\boldsymbol{k}$ is averaged over. To remove any effect due to the finiteness of the ratio of the thickness of domain walls to the domain size, we calculated $S(k,t)$ after the data were hardened using the transformation $\psi \to \text{sgn}\psi$. We fitted $S(k,t)$ to a squared Lorentzian form

$$S(k,t) = a^2/[(k^2-b)^2+c^2]^2,$$

and extracted the full width at half maximum, $\delta k(t)$, and the peak height, $S_p(t)$, of the structure factor at time t. The characteristic length scale (domain size) may then be defined by $1/\delta k(t)$. Let us remark in passing that we have also performed a fit to a Lorentzian form. In the SW regime, however, the Lorentzian fit shows a systematic deviation in the peak and tail regions.

The coarsening processes can be monitored by studying the time evolution of the structure factor $S(k,t)$. The $S(k,t)$ is sharply peaked around the wave number $k = k_0$, and narrowing of the profile and increase of the peak intensity are observed to occur as time evolves. This behavior represents the growth of the characteristic length scale over which the lamellas are ordered. Figure 2 (a) shows the time variation of the width δk of $S(k,t)$.

The peak intensity S_p of $S(k,t)$ was also found to exhibit the similar behavior. At late times these quantities obey the power law:

$$S_p \sim t^{\alpha}, \quad 1/\delta k \sim t^{\beta} \tag{3}$$

with the growth exponents α and β. These exponents change when one varies the disorder parameter p. It is demonstrated in Fig. 2 (b) that $\alpha \simeq \beta$, the exponents decreasing with p. In particular, a quick decrease in the exponents occurs in the small-world regime. There is a little increase in the growth exponents in the region $p \lesssim 0.001$. However, we attribute the apparent increase to a statistical error. Accordingly, in the small-p region we suppose that the exponent stays almost at the same value as in the case of the regular lattice.

Finally when p becomes larger, with $0.1 < p < 0.3$, we have observed the peak position of the asymptotic structure factor is shifted towards smaller value of $k \neq 0$. Therefore, the evolution of the system still corresponds to the lamella phase. However, in this case the growth becomes very slow, and with further increase of p, coarsening stops and the system gets stuck in a frozen configuration. The frozen configurations are elucidated below.

SUMMARY AND DISCUSSION

Our study has revealed the following distinctive features: (i) In the SW regime ($p \lesssim 0.1$), the coarsening of domains becomes slower with the addition of shortcuts, making

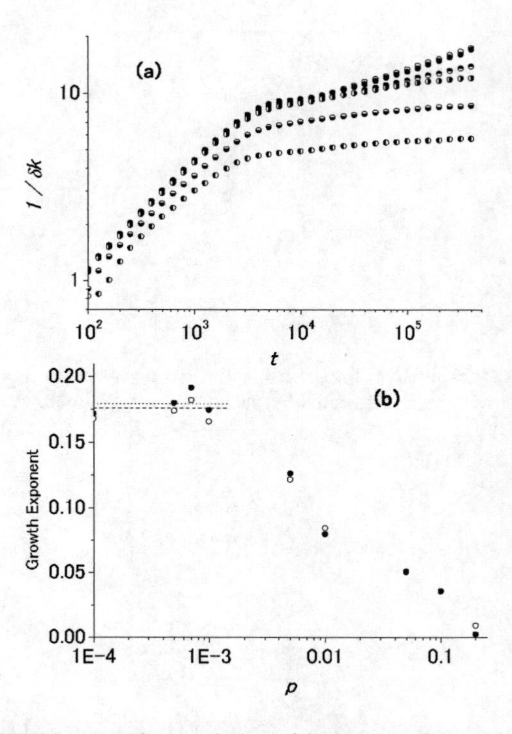

FIGURE 2. (a) Time evolution of the inverse of the width δk of the structure factor for lamellar formation. Different symbols are data for $p = 0, 0.001, 0.005, 0.01, 0.05$ and 0.1 from top to bottom. (b) The p dependence of the growth exponents, α and β, of the peak intensity S_p (open circles) and the inverse of the width δk (filled circles) of the structure factor for lamellar formation. The dashed and dotted lines indicate respectively the values of α and β for $p = 0$.

the lamellar ordering harder to achieve. Recall here the topological properties of SW networks. The addition of shortcuts shortens the average distance between nodes in the network, yet it results in a destruction of the cohesiveness (clustering). One can then deduce that in order to provide the earlier development of the pattern, the larger clustering coefficient is required.
(ii) In contrast, for larger $p \gtrsim 0.7$ the situation is completely reversed. In this case the system equilibrates over a rather short time, the ψ approaching its equilibrium value exponentially fast. Now the important ingredient to enhance the coarsening toward the homogeneous state is naturally the smaller average network distance.
(iii) The regime $0.1 < p < 0.7$. Here the system gets stuck in a frozen disordered state with coexisting domains of opposite (i.e., positive and negative) ψ's, and a fully ordered state is not reached.

In order to identify the underlying physical mechanism for freezing, we first draw the observed lamellar state in geodesic space. The required transformation to the geodesic space from Euclidean space can be accomplished by the replacement of $\boldsymbol{r}_{ij}$, a vector distance from a vertex i to a vertex j:

$$\boldsymbol{r}_{ij} \rightarrow \ell_{ij}\boldsymbol{r}_{ij}/|\boldsymbol{r}_{ij}|, \tag{4}$$

where ℓ_{ij} is the shortest path length from the vertex i to j. Figure 1 (upper b) shows the morphology at $p = 0.001$ as seen in the geodesic space. We found that black regions in this figure enclosing the central lamellar portion are formed by those vertices (to be called "influenced nodes") having long-range connections to the vertex (say vertex C) at the origin whose respective shortest path length to the vertex C is shortened by the addition of shortcuts. Consequently, the buildup of the lamellar state in these regions is prevented by the influenced sites. Figure 3 depicts the field ψ at the *uninfluenced* nodes for several different values of the shortcut density in Euclidean space. One sees that the lamellar region shrinks with increasing p, and it is clearly demonstrated that the freezing is tied to a new length scale, ξ_s, which characterizes the typical length over which the shortest path length is unaffected by the presence of shortcuts. Specifically, freezing emerges if the shortcut density is sufficiently high and

$$\xi_s \lesssim \lambda_p, \tag{5}$$

with λ_p being the inter-lamella distance.

The above finding is in accord with a proposed scenario for slow (glassy) dynamics [9] that the competition of interactions on different length scales causes the emergence of an exponentially large number of metastable states, which then results in a self-generated randomness and thus glassiness.

In conclusion, we have studied phase ordering dynamics of spatially periodic patterns. We showed that spatial disorder in the form of the SW topologies (connections) gives rise to arrested or frozen configurations. Whether the essential physics presented in this paper can be extrapolated to other spatially extended systems with different kinds of pattern formations merits further investigation, and we will address this problem in a separate paper [10].

REFERENCES

1. D. J. Watts and S. H. Strogatz, Nature (London) **393**, 440-442 (1998).
2. M. E. Harris-White, S. A. Zanotti, S. A. Frautschy and A. C. Charles, J. Neurophysiol. **79**, 1045-1052 (1998); T. I. Netoff, R. Clewley, S. Arno, T. Keck and J. A. White, J. Neurosci. **24**, 8075-8083 (2004).
3. J. Swift and P. C. Hohenberg, Phys. Rev. A **15**, 319-328 (1977); P. C. Hohenberg and J. B. Swift, Phys. Rev. A **46**, 4773-4785 (1992).
4. M. C. Cross and P. C. Hohenberg, Rev. Mod. Phys. **65**, 851-1112 (1993).
5. M. E. J. Newman and D. J. Watts, Phys. Rev. E **60**, 7332-7342 (1999).

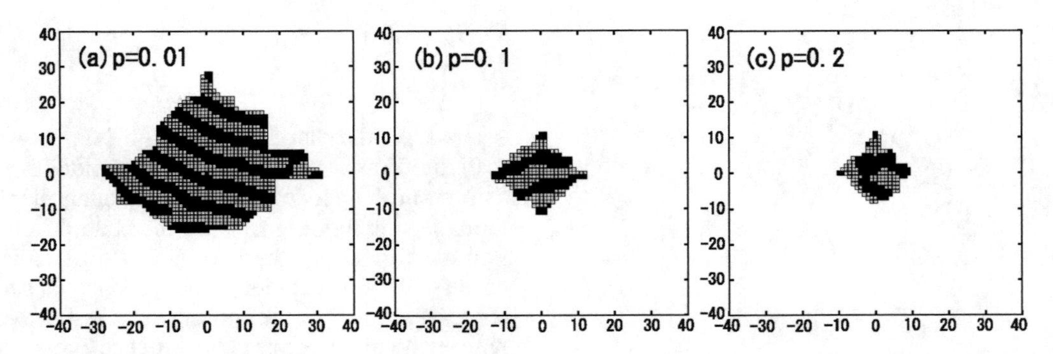

FIGURE 3. Lamellar-state landscapes in Euclidean space at late times ($t = 398107$) for (a) $p = 0.01$; (b) $p = 0.1$; (c) $p = 0.2$. Each site (black or gray) is connected to the node at the origin with its shortest path unaffected by the addition of SW connections. The gray region represents positive values of ψ and the black one negative ψ.

6. Y. Oono and S. Puri, Phys. Rev. Lett. **58**, 836-839 (1987); for more recent references, see A. Shinozaki and Y. Oono, Phys. Rev. E **48**, 2622-2654 (1993).
7. R. Imayama and Y. Shiwa, unpublished.
8. D. Boyer and O. Miramontes, Phys. Rev. E **67**, 035102(R)–1-4 (2003).
9. T. R. Kirkpatrick and D. Thirumalai, Phys. Rev. Lett, **58**, 2091-2094 (1987); J. Schmalian and P. G. Wolynes, *ibid.* **85**, 836-839 (2000).
10. R. Imayama and Y. Shiwa, to be published in Eur. Phys. J B.

Volume-Energy Correlations in the Slow Degrees of Freedom of Computer-Simulated Phospholipid Membranes

Ulf R. Pedersen*, Günther H. Peters†, Thomas B. Schrøder*, and Jeppe C. Dyre*

*DNRF Centre "Glass and Time," IMFUFA, Department of Sciences, Roskilde University, Postbox 260, DK-4000 Roskilde, Denmark
†Center for Membrane Biophysics (MEMPHYS), Department of Chemistry, Technical University of Denmark, DK-2800 Lyngby, Denmark

Abstract. Molecular dynamics simulations of phospholipid membranes reveals striking correlations between equilibrium fluctuations of volume and energy on the nanosecond time-scale. Volume-energy correlations have previously been observed experimentally at the phase transition between the L_α phase and the L_β phase, but not in the fluid L_α phase. The correlations are investigated in four membranes, with correlation coefficients ranging between 0.81 and 0.89. An experimentally single parameter test is proposed.

Keywords: Molecular dynamics simulation, Phospholipid membrane, Single parameter description
PACS: 05.40.-a, 64.60.Cn, 68.05.Cf

Biological membranes are an essential part of living cells. They not only act as passive barrier between outside and inside, but play an active role in various biological mechanisms. The major constituent of biological membranes are phospholipids. Pure phospholipid membranes often serve as a models for the more complex biological membranes. Close to physiological temperatures membranes undergo a transition from the fluid L_α phase (often referred to as the biologically relevant phase) to an ordered gel phase L_β. In the melting regime, at the transition temperature T_m, the response functions such as heat-capacity, volume expansion coefficient and area expansion coefficient increase dramatically. Also, the characteristic time for the collective degrees of freedom increases and becomes longer than milliseconds. Heimburg et. al. found that the response functions are connected so a single function describes the temperature dependence of all of them. Experiments indeed shows that heat capacity and volume expansion coefficient of DMPC can be superimposed at T_m [1] (see also [2, 3]).

The fluctuation-dissipation (FD) theorem connects response functions to equilibrium fluctuations. The isobaric heat capacity c_p can be calculated from enthalpy fluctuations as follows: $c_p = \langle(\Delta H)^2\rangle/(Vk_BT^2)$, where $\langle\ldots\rangle$ is an average in the constant temperature and pressure ensemble and Δ is deviation from the average value. Similarly, volume fluctuations are connected to the isothermal volume compressibility by the expression $\kappa_T = \langle(\Delta V)^2\rangle/(Vk_BT)$. If the response functions were described by a single parameter, fluctuations are also described by a single parameter [1, 4] and the microstates were connected via the relation $\Delta H_i = \gamma^{vol}\Delta V_i$. At constant pressure this relation applies if and only if $\Delta E_i = \gamma^{vol}\Delta V_i$ (where E is energy), which is the relation investigated below. This situation is referred to as a that of a single-parameter description [4]. A single-parameter description applies to a good approximation for several models of van der Waals bonded liquids as well as for experimental super-critical argon [5].

Molecular dynamics simulation is not a good method for investigate the "single parameter"-ness of membranes at T_m, since the relaxation time for collective motions exceeds typical simulation times. However, we show below that a single parameter description is not only feasible at the phase transition, but also in the fluid L_α phase approaching T_m. We discusss how this can be tested in experiments.

It is not trivial that a single parameter is sufficient. Simulations of water and methanol show no "single parameter"-ness, since competing interactions destroy the correlation. Contributions to volume and energy fluctuations from hard core repulsion compete with directional hydrogen bonds [5]. Membranes are complex anisotropic systems and one cannot give an a priori reason for why volume and energy should be correlated.

A membrane may be perturbated via three thermodynamic energy bonds. A change of the enthalpy dH can be written as a contribution from a thermal bond, a mechanical volume bond and mechanical lateral (or area) energy bond, $dH = dE + pdV + \Pi dA$ where E is internal energy, p is pressure, V is volume, Π is surface tension of the membrane and A is the membrane area. The natural ensemble to consider is the constant T, p and Π ensemble since water will act as a reservoir. If a single parameter controls the microstates, for all states i one would have $\Delta E_i = \gamma^{vol}\Delta V_i = \gamma^{area}\Delta A_i$ where the γ's are constants.

CP982, *Complex Systems, 5th International Workshop on Complex Systems*
edited by M. Tokuyama, I. Oppenheim, and H. Nishiyama

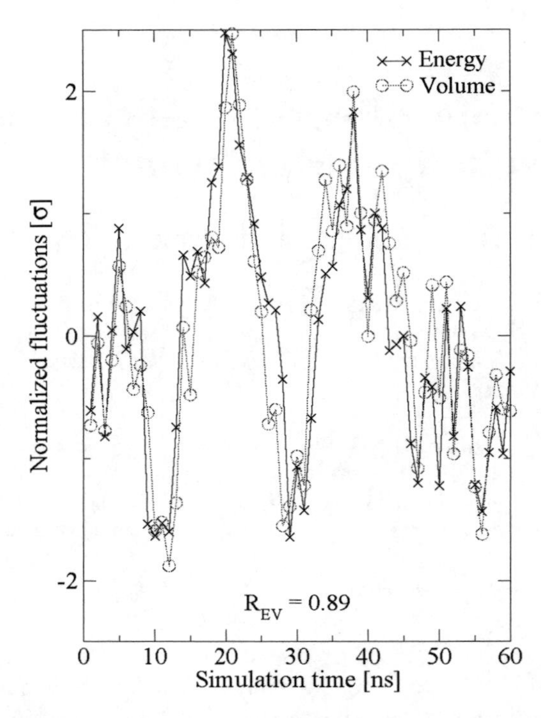

FIGURE 1. (Color online) Normalized fluctuations of energy (×) and volume (○) for a DMPC membrane at 310 K. Each data point represents a 1 ns average. Energy and volume are correlated with a correlation coefficient of $R = 0.89$.

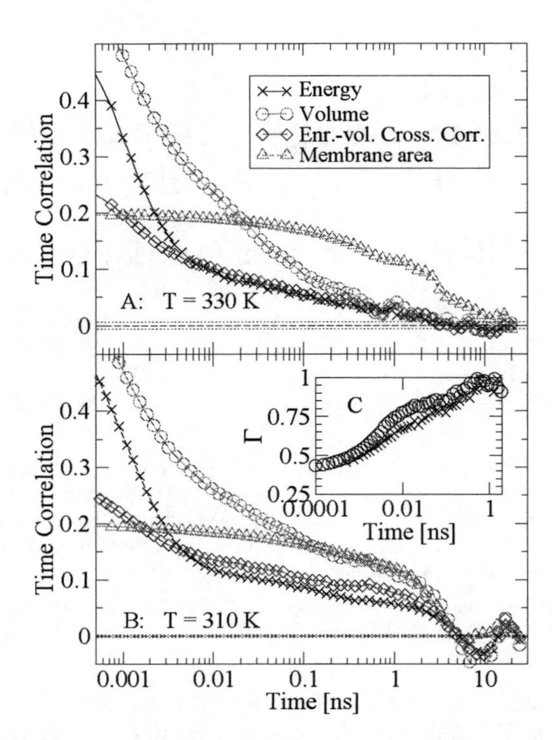

FIGURE 2. (Color online) Time correlation functions of potential energy C_{EE} (×), total volume C_{VV} (○), membrane area C_{AA} (△) and cross correlation between energy and volume C_{EV} (◇). Time correlation for membrane areas are scaled by a factor 0.2. Panel A shows data for a DMPC membrane at 330 K and panel B for a DMPC membrane at 310 K. The inset (C) displays $\Gamma = C_{EV}/\sqrt{C_{EE}C_{VV}}$ at 310 K (○) and 330 K (×). Γ approaches unity at $t \simeq 1$ ns meaning that volume and energy become strongly correlated on this timescale. Color online.

In general, the microstates may of course be controlled by several parameters. An interesting question is how many parameters are sufficient to describe the membrane thermodynamics to a good approximation. This question is addressed below by investigating molecular-dynamics simulations of different phospholipid membranes.

An overview of the simulated systems is found in table 1. The simulated systems include different head groups (both charged and zwitterions) and temperatures. All simulations was carried out in the constant pressure, temperature ensemble. The membranes are fully hydrated and in the fluid L_α phase. The simulations were performed using the program NAMD [6] and a modified version of CHARMM27 all hydrogen parameter set [7, 8]. More simulation details have previously been published [8, 9].

Striking correlations between time-averaged equilibrium fluctuations of volume and energy on the nanosecond time-scale of a DMPC membrane at 310 K are shown on Figure 1. If $\overline{E(t)}$ is the energy averaged over 1 nanosecond and $\overline{V(t)}$ is the volume averaged over 1 nanosecond then

$$\Delta\overline{E(t)} \simeq \gamma^{vol}\Delta\overline{V(t)} \tag{1}$$

where $\gamma^{vol} = \sigma_E/\sigma_V$ is a constant, σ is the standard deviation and Δ is difference from the thermodynamical average value. Equivalent results are found for the remaining membranes. Table 1 shows correlation coefficients between volume and energy, $R = \langle\Delta\overline{E}\Delta\overline{V}\rangle/\sqrt{\langle(\Delta\overline{E})^2\rangle\langle(\Delta\overline{V})^2\rangle}$ ranging between 0.81 and 0.89.

The correlation depends on the time scales considered. This can be investigated by evaluating $\Gamma(t) = C_{EV}(t)/\sqrt{C_{EE}(t)C_{VV}(t)}$ where $C_{AB}(t) = \langle\Delta A(\tau)\Delta B(\tau + t)\rangle/\sqrt{\langle(\Delta A)^2\rangle\langle(\Delta B)^2\rangle}$ is a time correlation function. $\Gamma(t) = 0$ implies that energy at time τ is uncorrelated with volume at time $t+\tau$ whereas $\Gamma(t)$ close to unity implies strong correlation. $\Gamma(t)$ is plotted in the inset of figure 2. At short time (picoseconds) Γ is around 0.5 but approaches unity at $t \simeq 1$ ns.

The "single parameter"-ness between volume and energy is connected to the experimental finding of Heimburg et. al. [1] since the slow collective degrees of freedom (at $t > 1$ ns) are the same ones which give rise to the dramatic changes of the response functions at T_m. There are some indications of this. Figure 2 shows time correlation functions of energy, volume and membrane area of

TABLE 1. Data from simulations of fully hydrated phospholipid membranes of 1,2-Dimyristoyl-sn-Glycero-3-Phosphocholine (DMPC), 1,2-Dimyristoyl-sn-Glycero-3-Phospho-L-Serine with sodium as counter ion (DMPS-Na) and hydrated DMPS (DMPSH). The columns lists temperature, correlation coefficient between volume and energy, average lateral area per lipid, simulation time in equilibrium (used in data analysis) and total simulation time.

	T **[K]**	R	A_{lip} **[Å^2]**	t **[ns]**	t_{tot} **[ns]**
DMPC	310	0.885	53.1	60	114
DMPC	330	0.806	59.0	50	87
DMPS-Na	340	0.835	45.0	22	80
DMPSH	340	0.826	45.0	40	77

DMPC membranes at 330 K and 310 K. The time constant and the magnitude of the slow fluctuations increase when the the temperature is decreased. γ^{vol} in equation 1 is 9.3×10^{-4} cm^3/J. This is the same order of magnitude as $\gamma^{vol} = 7.721 \times 10^{-4}$ cm^3/J calculated from experimental data of $C_p(T)$ and $\kappa_T^{vol}(T)$ at T_m [1].

Both the volume and energy time correlation functions show a two step relaxation at 310 K for DMPC (Figure 2B). As temperature is lowered toward T_m the separation is expected to be more significant. It makes sense to divide the dynamics into two separated processes, a fast and a slow collective process. Our simulations suggest that slow degrees of freedom can be described by a single parameter.

To see the "single parameter"-ness of the L_α phase, the fast degrees of freedom have to be filtered out. Experiments must deal with macroscopic samples where fluctuations are small and difficult to measure (the relative fluctuations scales as $1/\sqrt{N}$ where N is the number of molecules). It is therefore difficult to do the same analysis as we have done here. It is easier to measure response functions.

Fast fluctuations can be filtered out by measuring frequency dependent response functions. The slow collective degrees of freedom will show as a separate peak. The frequency dependent Prigogine-Defay ratio $\Lambda_{Tp}(\omega)$ has been suggested as a test quantity for "single parameter"-ness [4]. If $c_p''(\omega)$, $\kappa_T''(\omega)$ and $\alpha_p''(\omega)$ are the imaginary parts of the isobaric specific heat capacity per volume, isothermal compressibility and isobaric expansion coefficient respectively then

$$\Lambda_{Tp}(\omega) = \frac{c_p''(\omega)\kappa_T''(\omega)}{T_0[\alpha_p''(\omega)]^2}. \quad (2)$$

In general $\Lambda_{Tp} \geq 1$ and only have the value of unity if a single parameter describes the fluctuations. $1/\sqrt{\Lambda_{Tp}}$ can be related to a correlation coefficient.

In summary, we found strong volume-energy correlations of the slow collective degrees of freedom in molecular dynamics simulations of different phospholipid membranes in the L_α phase. An experimental test was suggested.

ACKNOWLEDGMENTS

The authors would like to thank Thomas Heimburg, Søren Toxværd and Nick P. Bailey for fruitful discussions and comments.

This work was supported by the Danish National Research Foundation Centre for Viscous Liquid Dynamics "Glass and Time". GHP acknowledges financial support from the Danish National Research Foundation via a grant to the MEMPHYS-Center for Biomembrane Physics. Simulations were performed at the Danish Center for Scientific Computing at the University of Southern Denmark.

REFERENCES

1. T. Heimburg, *Biochim. Biophys. Acta* **1415**, 147–162 (1998).
2. T. Heimburg, and A. D. Jackson, *Proc. Natl. Acad. Sci.* **102**, 9790–9795 (2005).
3. H. Ebel, P. Grabitz, and T. Heimburg, *J. Phys. Chem. B* **105**, 7353–7360 (2001).
4. N. L. Ellegaard, T. Christensen, P. V. Christiansen, N. B. Olsen, U. R. Pedersen, T. B. Schrøder, and J. C. Dyre, *J. Chem. Phys.* **126**, 074502–1–074502–8 (2007).
5. U. R. Pedersen, T. Christensen, T. B. Schrøder, and J. C. Dyre, *arXiv.org cond-mat/0611514* (2006).
6. J. C. Phillips, R. Braun, W. Wang, J. Gumbart, E. Tajkhorshid, E. Villa, C. Chipot, R. D. Skeel, L. Kale, and K. Schulten, *J. Comput. Chem.* **26**, 1781–1802 (2005).
7. N. Foloppe, and A. D. MacKerell, *J. Comput. Chem.* **21**, 86–104 (2000).
8. U. R. Pedersen, C. Leidy, P. Westh, and G. H. Peters, *Biochim. Biophys. Acta* **1758**, 573–582 (2006).
9. U. R. Pedersen, G. H. Peters, and P. Westh, *Biophysical Chemistry* **125**, 104–111 (2007).

Local Structural Effects on Intermolecular Vibrations in Liquid Water: The Instantaneous-Normal-Mode Analysis

K. H. Tsai* and Ten-Ming Wu†

*Dept. of Electrophysics, National Chiao-Tung University, Hsin-Chu 300, Taiwan
†Institute of Physics, National Chiao-Tung University, Hsin-Chu 300, Taiwan

Abstract. Currently, the designations for the low-frequency vibrational spectrum of liquid watrer are still diversified. In this paper, the water molecules simulated by the SPC/E model are classified into subensembles, characterized by their local structures, which are specified in two different ways: the geometry of Voronoi polyhedron or the H-bond configuration. Using the instantaneous normal mode (INM) analysis for these subensembles, we investigate the effects of local structure on the low-frequency INM spectrum of liquid water. From the contributions of these subensembles to the translational INM spectrum, our results provide insights into the geometric effects of local structure and the H-bond configuration on intermolecular vibrations in liquid water.

Keywords: H-bond, Instantaneous normal modes, Raman spectrum, Voronoi polyhedron
PACS: 43.35.Ei, 78.60.Mq

INTRODUCTION

Many anomalous properties of liquid water are associated with the H-bonds between molecules [1]. Generally, the H-bonds in liquid water form a network, which can be considered as a mixture of local structures with different H-bonds. In static structure, the H-bond network can be described by various models, from the two-state network model to the percolation model of H-bond. In dynamics at a short time scale ($10^{-15} \sim 10^{-12}$ s), thermal motions of molecules cause the H-bonds of a molecule breaking and reforming during time evolution. Thus, the short-time dynamics of liquid water is essentially dominated by the local geometry of the H-bond network.

Below 400 cm^{-1}, the Raman spectrum of liquid water shows two broad bands around 60 and 180 cm^{-1} [2, 3]. However, the designations for the two bands are still in dispute, especially for the band around 60 cm^{-1}. Walrafen *etal.* attributed the 60 cm^{-1} band to the bending mode of molecular triplets, which are a central molecule and its two H-bonded neighbors, and the 180 cm^{-1} band to the stretching mode between two neighboring molecules [4]. The bending and stretching modes are actually related to the molecular motions perpendicular and parallel to a H-bond, respectively. In other interpretation, the two bands are resulted from the restricted translations of molecules, hindered by the neighbors around each molecule [5].

Recently, the 60 cm^{-1} is designated by a different interpretation: Arising from the frustrated translation caused by the cage effect, the band is not necessarily associated with the H-bonds in liquid water, but a general result to all kinds of dense liquids [6]. Indicated by a comment for the interpretation, with intensities crucially determined by the H-bonds, the 60 and 180 cm^{-1} bands are, respectively, contributed by the transverse and longitudinal dynamics of nearest neighboring oxygen pairs, but the transverse dynamics is not necessarily described as the bending motions of molecular triplets [7].

Developed in the past decade, the theory of intantaneous normal modes (INMs) provides a microscopic description for the collective dynamics of a liquid [8]. In terms of projection operators defined by the INM eigenvectors, the molecular contributions to the collective dynamics of liquids can be extracted [9]. Using projection operators defined for the H-bonded clusters, Vallauri and co-workers [10] show that for liquid water the INMs with frequencies around 70 cm^{-1} are dominated by the bending motions of H-bonded molecular triplets, whereas the INMs with frequencies from 185 to 240 cm^{-1} are by the stretching motions of the triplets. In this paper, we use the INM formalism, but in a different approach, to investigate the effects of local structures on the low-frequency intermolecular vibrations in liquid water.

METHOD: THE INM FORMALISM

We have simulated 256 SPC/E water molecules in a cubic box with periodic boundary conditions at density $\rho = 1.0gcm^{-3}$ and temperature $T = 300K$ [11]. The Lennard-Jones (LJ) potential between two oxygens are truncated at the half of the box length and shifted upward to make the first derivative of the potential continuous at the cutoff. The Coulomb interactions are treated by the Ewald summation. In our simulations, the leap-frog

CP982, *Complex Systems, 5th International Workshop on Complex Systems*
edited by M. Tokuyama, I. Oppenheim, and H. Nishiyama

algorithm is used with a time step of 1 fs. After 300,000 time steps from an initial lattice configuration, we start to collect data for 1000 configurations at every 1000 steps.

In order to examine the local structurs of molecules, we classify all molecules in the simulated liquid water into different subensembles, according to (a) the geometry of Voronoi polyhedron (VP) constructed for oxygens [12] and (b) the H-bond configuration, which is defined as the numbers of the donated and accepted H-bonds attching to a molecule [13]. We use the energy definition for a H-bond: Two water molecules are connected by a H-bond as the energy between them is less than $-12KJmol^{-1}$ [11]. With this definition, the distributions of the H-bonds in our simulated system are not much different from those obtained with other definition [10].

For molecules in subensemble L, we define the translational projection operator $P^L_{\alpha T}$ of INM α to be

$$P^L_{\alpha T} = \sum_{j=1}^{N} \sum_{\mu=1}^{3} U^2_{\alpha,j\mu} \Theta_j(L), \tag{1}$$

where N is the total number of molecules in the system. $\Theta_j(L) = 1$ if molecule j belongs to the subensemble L and 0 otherwise [11]. $U_{\alpha,j\mu}$ is the eigenvector component of INM α for molecule j and Cartesian coordinate μ of its center of mass. The averaged contribution to the normalized translational INM density of states (DOS) $D_T(\omega)$ by a molecule in subensemble L is defined as

$$D^L_T(\omega) = \frac{1}{3N_L} \left\langle \sum_{\alpha=1}^{6N} \delta(\omega - \omega_\alpha) P^L_{\alpha T} \right\rangle, \tag{2}$$

where N_L is the average number of molecules in subensemble L [14]. By summing all classified subensembles, we have the sum rule $D_T(\omega) = \sum_L \chi_L D^L_T(\omega)$, where $\chi_L = N_L/N$ is the average number fraction of subensemble L.

THE GEOMETRIC EFFECTS

The Voronoi polyhedron of a particle in a liquid is a generalization of the Wigner-Seitz unit cell in a crystal [12, 15, 16]. The geometric shape of a Voronoi polyhedron can be characterized by a dimensionless parameter: the asphericity $\eta = A^3/36\pi V^2$, where A and V are the total surface area and volume of the polyhedron, respectively. For a sphere, $\eta = 1$. As the geometry of the polyhedron becomes more aspherical, η increases. We classify the molecules of each configuration into four subensembles, named as asphericity group (AG) I to IV with increasing η value [11]. Shown in Fig.1A are the oxygen-oxygen radial distribution functions $g_{OO}(r)$ of the four AGs and in the inset are the length distributions of

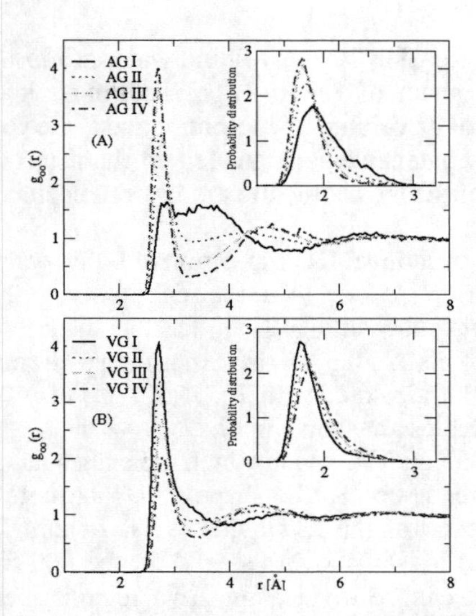

FIGURE 1. (Color online) $g_{OO}(r)$ distributions for molecules in AG (A) and VG (B) subensembles. The solid, dotted, dashed, and dot-dashed lines are for group I, II, III and IV, respectively. The insets show the normalized distributions of the H-bond length for molecules in corresponding subensemble.

the H-bonds attaching to molecules in each subensemble. For AG IV, the sharp first peak in $g_{OO}(r)$ and almost four nearest neighbors of each molecule, obtained by integrating $g_{OO}(r)$ up to the first minimum, indicate that the tetrahedral arrangement is the most possible local structure around a molecule. For AG I, the plateau in the first shell of $g_{OO}(r)$ and a longer H-bond length in average manifest that the molecules in AG I have more random local structures and are connected to their neighbors with weaker H-bonds.

Similarly, according to the volumes of Voronoi polyhedra, we divide the molecules into four different subensembles, named as volume group (VG) I to IV, with increasing the value of V [14]. Shown in the inset of Fig. 1B, the average length of the H-bonds attaching to molecules increases with V. In Fig.1B, a sharp first maximum in $g_{OO}(r)$ of VG I suggest that the molecules have a compact structure in the first shell. The peak value of $g_{OO}(r)$ diminishes fast with increasing the value of V. Beyond the second shell, the four $g_{OO}(r)$ become almost indistinguishable. This manifests that the local structure of a water molecule is more sensitive with the asphericity of its Voronoi polyhedron than the volume.

To examine the correlations between the subensembles of the AGs and the VGs, we have calculated the conditional probability χ_{ML} to find a molecule in subensemble AG M for the molecules which belong to subensmble VG L. Indicated by the data in Table I, the order of magnitude of χ_{ML} is roughly independent of the VG which the molecules belong to, so the AGs and the VGs are weakly correlated. This is consistent with the conclusion given in Ref. [15] that for liquid water at room temperature the shape of a Voronoi polyhedron is almost independent of its volume. Therefore, η and V are considered to be two independent parameters to characterize the geometry of the local structure around a molecule in liquid water.

The translational INM spectrum of liquid water covers a range from 650 cm^{-1} in the real branch to 250 cm^{-1} in the imaginary branch [11], and the shape of the real branch looks simliar as the experimental Raman spectrum [17]. Calcaulted with Eq.(2), $D_T^L(\omega)$ of AG and VG subensembles are shown in Fig.2.

For AG I and VG IV, in which molecules have weaker and longer H-bonds, the shape of D_T^L is more like the INM spectra of the LJ liquids [8?]. According to the variations of $D_T^L(\omega)$ with η and V, the real-frequency spectrum can be divided into two regions: below 150 cm^{-1}, where $D_T^L(\omega)$ increases by either decreasing η or increasing V, and above 150 cm^{-1}, where the variation of $D_T^L(\omega)$ with η or V is reversed. The low- and high-frequency regions generally correspond to the bending and stretching modes of molecular triplets in liquid water, respectively [10].

We pay attention to the effects of local structure on the peak around 65cm^{-1} in $D_T^L(\omega)$. The position of the peak is insensitive to both η and V of a Voronoi polyhedron. However, the peak value is much affected by V than η. That is, the larger the Voronoi volume of a molecule, the more significant the contribution of the molecule to the 65 cm^{-1} peak. A possible explanation for the local-volume effect on the 65 cm^{-1} peak is that a larger local volume occupied by the central molecule of a triplet would favor the bending motion of the triplet; on the contrary, a smaller local volume would restrict, or even "freeze", the bending motion.

TABLE 1. Averaged number fraction $\chi_L = N_L/N$ of each VG and the conditional probability $\chi_{ML} = N_{ML}/N_L$, where both L and M can be I,II,III and IV. N_L is the averaged number of molecules in VG L and N_{ML} is that of the molecules which belong to both VG L and AG M. $\tilde{V} = V/V_{ave}$ is the volume V scaled by the averaged molecular volume of liquid water. The ranges of the four AGs are given in [11].

	VG I	VG II	VGIII	VGIV
$\tilde{V}$	~ 0.84	$0.84 \sim 1.0$	$1.0 \sim 1.24$	$1.24 \sim$
χ_L	0.039	0.510	0.419	0.032
χ_{ML}				
AG I	0.010	0.011	0.021	0.043
AG II	0.765	0.685	0.652	0.602
AG III	0.223	0.296	0.306	0.287
AG IV	0.002	0.008	0.021	0.067

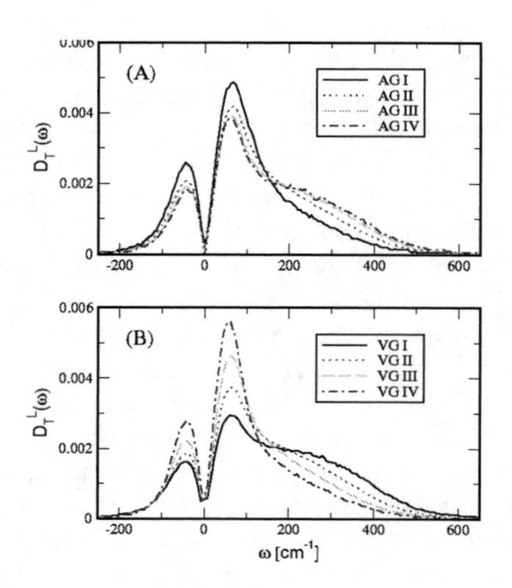

FIGURE 2. (Color online) Geometric effects of local structure on the translational INM spectrum: $D_T^L(\omega)$ calculated for subensembles of AGs (A) and VGs (B).

By either increasing η or decreasing V, a hump appears in $D_T^L(\omega)$ above 150 cm^{-1}. The occurrence of the hump can be interpreted by the stretch of a H-hond. For a molecule with a smaller V or a larger η, the H-bonds attaching to the molecule is short in bond length and strong in strength, so the stretching motion of the H-bonds are active and contribute more to $D_T^L(\omega)$.

THE H-BOND EFFECTS

In the SPC/E model [18], the interactions between two water molecules are composed of two parts: the pair-wise additive site-site Coulomb potentials of the point charges at the H and O coordinates, with $q_H = 0.4238e$ and $q_O = -0.8476e$, where e is the proton charge, and a LJ potential between the two O atoms. To investigate whether the low-frequency translational INMs are related to the H-bonds or not, we calculate $D_T(\omega)$ for a reduced system, with the same configurations generated by our simulations but all charges at the H and O sites reduced by a common factor γ. As $\gamma = 1$, the reduced system is exactly the SPC/E model; as $\gamma = 0$, the interactions in the system are reduced to the LJ potentials between O atoms. In Fig.3, the peak value at 65cm^{-1} drops dramatically as the γ is reduced from 1 to 0.5, but does not change much as γ is further reduced from 0.5 to 0. On the other hand, the fraction of water molecules having four H-bonds is over 40% at $\gamma = 1$, only about 5% at $\gamma = 0.75$, and almost zero at $\gamma = 0.5$. This result gives a strong evidence that the 65 cm^{-1} peak in $D_T(\omega)$ is partially associated with the H-bonds.

Shown in Fig. 4 is the dependence of $D_T^L(\omega)$ on n, the number of the H-bonds attaching to a water molecule.

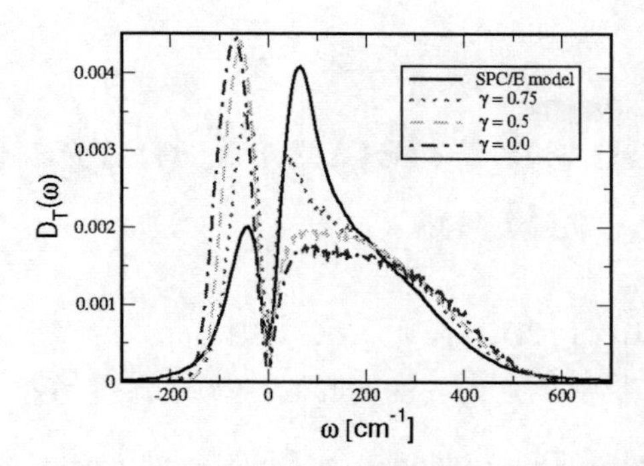

FIGURE 3. (Color online) $D_T(\omega)$ calculated for the SPC/E model (solid line) and the systems with the charges of the O and H atoms reduced by a common factor γ. The dotted, dashed and dot-dashed lines are for $\gamma = 0.75, 0.5$ and 0, respectively.

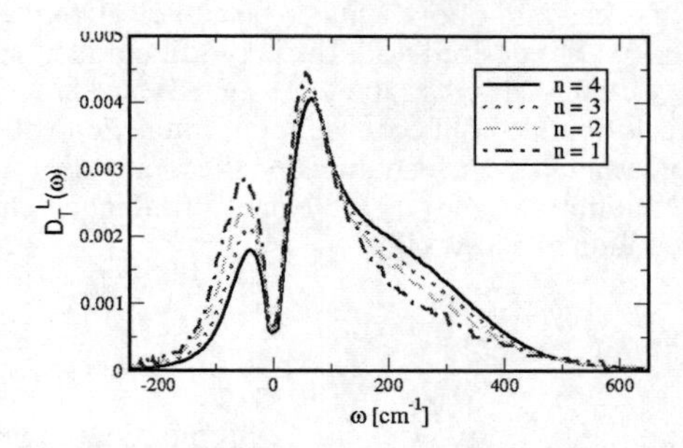

FIGURE 4. (Color online) $D_T^L(\omega)$ calculated for subensembles of molecules having n H-bonds.

Clearly, there are two main effects on $D_T^L(\omega)$ as n is varied from 1 to 4: First, the position of the low-frequency peak increases from $55cm^{-1}$ to $65cm^{-1}$ but the value of the peak has a small decrease. The shift of the peak position implies that the force constant of the bending mode increases with the number of the H-bonds, and the amount of the shift is roughly in the same order as the change of the 60 cm^{-1} band observed in the Raman spectrum with increasing temperature [2]. We suggest that the variations of the 60 cm^{-1} band in the INM and Raman spectra are consistent due to the same physical origin, which is associated with the local network of the H-bonds in liquid water. Secondly,as n increases from 1 to 4, a hump in $D_T^L(\omega)$ between 150 cm^{-1} and 450 cm^{-1} grows. The strong dependence of the hump on n is consistent with the picture attributing the INMs at the corresponding frequencies to the stretch of a H-bond.

CONCLUSIONS

The INM formalism with the projection approach is a useful tool to investigate the effects of local structure on intermolecular vibrations in liquid water. We evidence that for liquid water the INMs below $300cm^{-1}$ are partially associated with H-bonds. Our results are consistent with the interpretation that the INMs near $60cm^{-1}$ and those around $200cm^{-1}$ are dominated by the bending modes of molecular triplets and stretching modes of oxygen pairs, respectively. In the translational INM spectrum, the position of the peak $60cm^{-1}$ is influenced by the number of the H-bonds attaching to a molecule and the intensity of the peak by the geometry of local structures, especially the local volume. On the other hand, a hump around $200cm^{-1}$ grows as the strength and the number of the H-bonds attaching to a water molecule increases.

ACKNOWLEDGMENTS

TM Wu acknowledge supports from the National Science Council of Taiwan, under grant No. NSC 95-2112-M-009-027-MY2.

REFERENCES

1. F. Franks, *Water: a Comprehensive Treatise*, Vol. 1-7 (Plenum Press, New York, 1972-1982).
2. G. E. Walrafen, M. R. Fisher, M. S. Hokmabadi, and W. H. Yang, *J. Chem. Phys.* **85** 6970-6982 (1986).
3. I. Ohmine and H. Tanaka, *Chem. Rev.* **93**, 2545-2566 (1993).
4. G. E. Walrafen, Y. C. Chu, and G. J. Piermarini, *J. Phys. Chem.* **100** 10363-10372 (1996).
5. K. Mizoguchi, Y. Hori, and Y. Tominaga, *J. Chem, Phys.* **97** 1961-1968 (1992).
6. J. A. Padro and J. Marti, *J. Chem. Phys.* **118** 452-453 (2003).
7. A. De Santis, A. Erocli, and D. Rocca, *J. Chem. Phys.* **120** 1657-1658 (2004).
8. T. M. Wu and R. F. Loring, *J. Chem. Phys.* **97**, 8568-8575 (1992).
9. M. Buchner, B. Ladanyi, and R. Stratt, *J. Chem. Phys.* **97** 8522-8535 (1992).
10. G. Garberoglio, R. Vallauri, and G. Sutmann, *J. Chem. Phys.* **117** 3278-3288 (2002).
11. S. L. Chang, T. M. Wu, and C. Y. Mou, *J. Chem. Phys.* **121** 3605-3612 (2004).
12. Y. L. Yeh and C. Y. Mou, *J. Phys. Chem.* B **103** 3699-3705 (1999).
13. S. Myneni, *et al*, *J. Phys: Conden. Matt.* **14** L213-L219 (2002).
14. K. H. Tsai and T. M. Wu, *Chem. Phys. Lett.* **417** 389-394 (2006).
15. G. Ruocco, M. Sampoli, and R. Vallauri, *J. Chem. Phys.* **96** 6167-6176 (1992).
16. R. Ruocco, M. Sampoli, A. Torcini, and R. Vallauri, *J. Chem. Phys.* **99** 8095-8104 (1993).
17. D. M. Carey and G. M. Korenowski, *J. Chem. Phys.* 108 2669-2675 (1998).
18. H. J. C. Berendsen, J. R. Grigera, and T. P. Straatsma, *J. Phys. Chem.* **91**, 6269-6271 (1987).

Effects of Multivalent Salt Addition on Effective Charge of Dilute Colloidal Solutions

Tzu-Yu Wang[a], Yu-Jane Sheng[b], and Heng-Kwong Tsao[*]

[a]Department of Chemical and Materials Engineering, National Central University, Jhongli, Taiwan 320,Republic of China

[b]Department of Chemical Engineering, National Taiwan University, Taipei 106, Taiwan, Republic of China

[*]Department of Chemical and Materials Engineering, National Central University, Jhongli, Taiwan 320,Republic of China

Abstract. The effective charge Z^* is often invoked to account for the accumulation of counterions near the colloid with intrinsic charge Z. Although the ion concentrations c_i are not uniform in the solution due to the presence of the charged particle, their chemical potentials are uniform everywhere. Thus,on the basis of ion chemical potential, effective ion concentrations c_i*, which can be experimentally measured by potentiometry, are defined with the pure salt solution as the reference state. The effective charge associated with the charged particle can then be determined by the global electroneutrality condition. In terms of the charge ratio $\alpha = Z^* /Z$, the effects of added salt concentration, counterion valency, and particle charge are examined. The effective charge declines with increasing salt concentration and the multivalent salt is much more efficient in reducing the effective charge of the colloidal solution. Moreover, the extent of effective charge reduction is decreased with increasing intrinsic charge for a given concentration of added salt.

Keywords: Colloid, Effective charge, Multivalent salt

INTRODUCTION

A colloidal dispersion, consisting of many charged particles and small ions, is a very complicated system. To describe the equilibrium and dynamic properties of colloidal solutions, the concept of effective charge is commonly adopted in the literature.[1–4] The essential idea associated with charge renormalization is that counterions accumulate in the vicinity of the surface of the colloid carrying intrinsic charge Z because of strong electrostatic coupling. That is, the electrostatic attraction is large compared to the thermal energy k_BT. Consequently, the decorated object (charged particle plus counterions)may be regarded as a single entity which possesses an effective charge, Z^*. The effective charge (in absolute value) can be much less than the intrinsic charge, $Z^* \ll Z$. The determination of Z^* is dependent on which physical property is considered. It is frequently regarded as an adjustable parameter in a fit of experimental data with approximated models. For instance, the effective charge can be inferred from voltammetry, electrophoresis, or small-angle neutron scattering.[5–8] The concept of effective charge basically characterizes a battle fought between energy and entropy in minimizing the free energy of a solution of mobile charges in the neighborhood of charged particles. For a charged sphere of size a, the electric potential energy of a counterion on the surface is finite $\psi_s \sim -Z/a$. The entropy associated with the counterion is proportional to $k_BT \ln V$ with the available volume V, At infinite dilution($V \rightarrow \infty$), the isolated, charged sphere is unable to bind a counterion at finite temperature and the effective charge is the intrinsic charge, $Z^*=Z$. However, in all practical colloidal systems, the counterion entropy is finite, $\sim k_BT \ln c$, owing to finite counterion concentration c. As a result, it is expected that charge renormalization takes place eventually when the electrostatic energy gain overcomes the counterion entropy loss.

CELL MODEL AND SIMULATION

It is very common to adopt a spherical Wigner-Seitz (WS) cell to investigate the physical properties associated with colloidal solutions.[1–4] A short description of Monte Carlo (MC) simulation is given below. The simulation details can be seen elsewhere.[9] The system simulated in the present work comprises a charged sphere of radius a fixed at the center of a spherical cell with radius R and a collection of hard spheres with valency $\pm z_i$=1, 2, or 3. The diameter of the counterion d is assumed 0.4 nm and is taken as the unit for the spatial length, At 298 K, the dimensionless energy parameter in Coulombic interaction $(l_B/d)z_iz_j/|\mathbf{r}_i - \mathbf{r}_j|$ is thereby $E^* = l_B/d = 1.785$. The simulations were conducted under conditions of constant volume V, temperature T, and total number of ions N.

RESULTS & DISCUSSION

Our simulation results indicates that when monovalent salt is added, the loss of counterion entropy is responsible for the decrease of the effective charge. Moreover, the counterion fluctuation-correlation effect may provide additional

CP982, *Complex Systems, 5th International Workshop on Complex Systems*
edited by M. Tokuyama, I. Oppenheim, and H. Nishiyama

attraction between counterions and colloids. Figure 1 also shows that for a given amount of salt addition, the effect of salt addition on Z^* is more significant for a colloid with smaller Z than for one with larger Z. Since a colloid with larger intrinsic charge has higher free counterion concentration, one expects that more amount of salt is required to provide added counterions, which can neutralize the particle charge. Thus the extent of the change of the charge ratio, $\Delta\alpha$, can be assumed as a function of the ratio of added salt concentration to the intrinsic counterion concentration, c_s/c_c (or c_s/Z). As shown in the inset of Fig. 1, the data points for three different values of Z fall into a single curve, confirming this relationship. Note that $\Delta\alpha=\alpha(c_s=0)-\alpha(c_s)$ with $\alpha=Z^*/Z$ and the concentration range of added salt in Fig. 1 corresponds to $c_s \leq c_c$. The fact that the change in α grows with c_s/Z instead of c_s indicates that at high enough salt concentration, one may have $\alpha(Z_1) < \alpha(Z_2)$ for $Z_1 < Z_2$ even though $\alpha_0(Z_1) > \alpha_0(Z_2)$, where $\alpha_0=\alpha(c_s=0)$. Moreover, the approximately linear relation between $\Delta\alpha$ and c_s/c_c at higher values of c_s/c_c reveals that the effective charge may approach zero essentially at high enough salt concentration.

Addition of multivalent salt can reduce the effective charge much more efficiently than addition of monovalent salt. As shown in Fig. 2, the effective charge is plotted against the salt concentration for various multivalent salts with Z=100 ($c_c \approx$ 10 mM). The effective charge drops veryquickly upon addition of 3:1 salt. For example, the effective charge becomes $Z^*/Z \approx 0.02$ for 3 mM 3:1 salt, but is still $Z^*/Z \approx 0.14$ for 10 mM 1:1 salt. According to the PB theory, the effect of salt addition is usually understood in terms of the ionic strength, $I=(c_+z_+^2+c_-z_-^2)/2$. If the effectiveness of reducing Z^* is proportional to the ionic strength, then the equivalent concentrations are 6:4:3:1 for 3:1, 2:2, 2:1, and 1:1 salts, respectively, for the same salt concentration. That is,

$$\alpha(z_c)=\alpha_0-\Delta\alpha(z_c=1)\left[\frac{I(z_c)}{I(z_c=1)}\right]$$

for a specified c_s. This simple model would predict that $Z^*/Z \approx 0$ for 3 mM 3:1 salt based on the result of monovalent salt. It agrees approximately with the simulation result. This model also predicts that 2:2 salt is more effective than 2:1 salt. This result is inconsistent with our simulation result, as illustrated in Fig. 2. Nonetheless, the differences between these two salts are quite small.

In a solution containing charged particles, counterions are always accumulated in the vicinity of the surface of the charged particle with intrinsic charge Z. The effective charge Z^* is therefore invoked to represent the physical behavior of such a highly charged colloidal dispersion. The value of the effective charge changes with its definition. In other words, it is model dependent. Unfortunately, a thermodynamic definition of Z^*, which can also be experimentally measurable, is generally lacking. In this study, we propose an operational definition of the effective charge, which is thermodynamically well defined and can be determined by potentiometry.[7] On the basis of this definition, we are able to investigate the effect of multivalent salt addition on the effective charge of a dilute colloidal dispersion.

In order to explore the influence of adding multivalent salt, Monte Carlo simulations are performed in a spherical Wigner-Seitz cell to obtain the effective charge of the colloid. In terms of the charge ratio $\alpha = Z^*/Z$, the effects of added salt concentration, counterion valency, and particle charge are examined. The effective charge declines with increasing salt concentration and the multivalent salt is much more efficient in reducing the effective charge of the colloidal solution. Moreover, the extent of effective charge reduction is decreased with increasing intrinsic charge for a given concentration of added salt. Those results are qualitative con-sistent with experimental observations by electrophoresis. We have investigated dilute systems containing particle size of about 4 nm (a=10) with surface charge density of about 0.1 C/m2(Z~100). They correspond to the micellar or protein solution. Nonetheless, our results can be generally applied to typical colloidal dispersions with much larger a and Z.

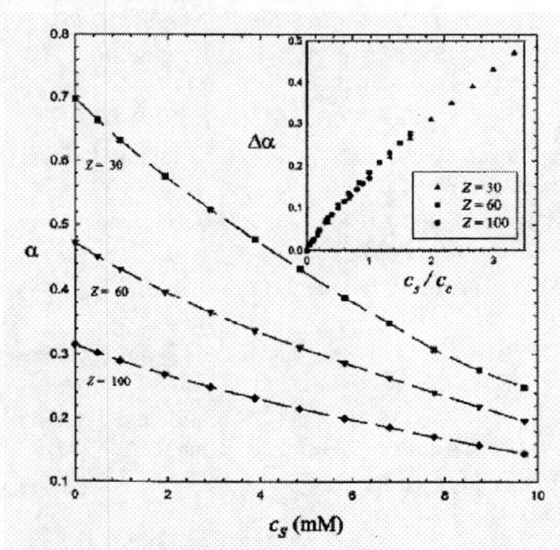

FIGURE 1. The variation of the charge ratio _ with the monovalent salt concentration c_s for colloid with different values of intrinsic charge Z. The dotted curves are drawn to guide the eyes. The data points are replotted in the inset for $\Delta\alpha=\alpha_0-\alpha$ vs c_s/c_c.

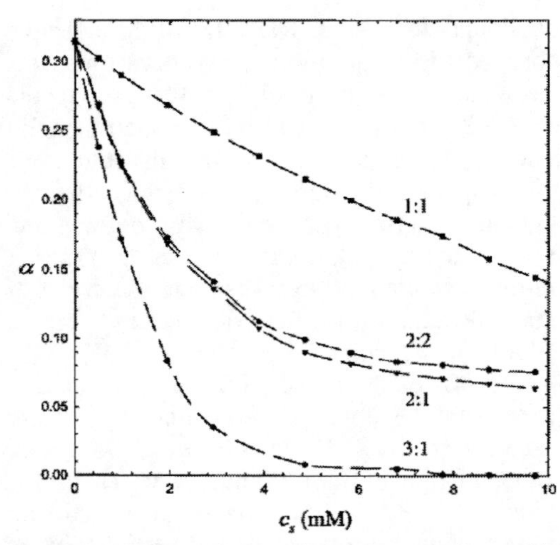

FIGURE 2. The variation of the charge ratio α with the salt concentration c_s for a=10, R=40, and Z=100. The simple added salts include 2: 1, 2: 2, and 3:1 multivalent counterions. The dotted curves are drawn to guide the eyes.

ACKNOWLEDGMENTS

This research is supported by the National Council of Science of Taiwan under Grant No. NSC-95-2221-E-008-146-MY3. Computing time provided by the National Center for High-Performance Computing of Taiwan is gratefully acknowledged.

REFERENCES

1. S. Alexander, P. M. Chaikin, P. Grant, G. J. Morales, and P. Pincus, *J. Chem. Phys.* **80**, (1984), 5776-5781.
2. E. Trizac, L. Bocquet, and M. Aubouy, *Phys. Rev. Lett.* **89**, (2002), 248301-248304.
3. R. D. Groot, *J. Chem. Phys.* **95**, (1991), 9191-9203.
4. M. J. Stevens, M. L. Falk, and M. O. Robbins, *J. Chem. Phys.* **104**, (1996), 5209-5219.
5. J. M. Roberts, J. J. O'Dea, and J. G. Osteryoung, *Anal. Chem.* **70**, 3667 (1998).
6. V. K. Aswal and P. S. Goyal, *Phys. Rev. E* **67**, (2003), 051401-051408.
7. C. C. Hsiao, T.-Y. Wang, and H.-K. Tsao, *J. Chem. Phys.* **122**, (2005), 144702-1-144702-10
8. R. D. Void and M. J. Void, "Colloid and Interface Chemistry" (Addison - Wesley, Reading, MA, 1983).
9. Y.-J. Sheng and H.-K. Tsao, *Phys. Rev. Lett.* **87**, (2001), 185501-185504.

PART III

POLYMER DYNAMICS

Addressing Unsolved Mysteries of Polymer Viscoelasticity

Ronald G. Larson

Department of Chemical Engineering, University of Michigan, Ann Arbor, MI 48109

Abstract. By using coarse-grained bead-spring and entanglement tube models, much progress has been made over the past 50 years in understanding and modeling the dynamics and rheology of polymers, both in dilute solution state and in entangled solutions and melts. However, several major issues have remained unresolved, and these are now being addressed using microscopic simulations resolved at the level of the monomer. In the dilute solution state, the dynamics can be described by a coarse-grained bead-spring model, with each spring representing around 100 backbone bonds, even at frequencies high enough that one expects to see modes of relaxation associated with local motions of smaller numbers of bonds. The apparent absence of these local modes has remained a mystery, but microscopic simulations now indicate that these modes are slowed down by torsional barriers to the extent that they are coincident with much longer ranged spring-like modes. Other mysteries of dilute solution rheology include extension-thinning behavior observed at very high extension rates, an apparent lack of complete stretching of polymers in fast extensional flows as measured by light scattering experiments, and the unusual molecular weight dependence of polymer scission in fast flows. In entangled solutions, it is still not entirely clear how, or even if, the rheology can be mapped onto that of a "dynamically equivalent" melt, and, if so, what the scaling laws are for choosing the appropriate renormalized monomer size and renormalized time and modulus scales. It is also not yet clear to what extent "dynamic dilution" can be used to simplify and organize constraint release effects in the relaxation of monodisperse and polydisperse linear and long-chain branched polymers. For multiply-branched polymers, the motion of the branch point is critical in determining the rate of relaxation of the molecule, and theories for this motion have not been adequately tested. As with dilute solutions, simulations resolved at the level of the monomer are now helping to settle these issues. For example, molecular dynamics simulations of branched polymers show that ideas of hierarchical relaxation, introduced by McLeish and coworkers, appear to be valid. Similar simulations indicate that the effective "tube diameter" increases gradually starting at times as short as the "equilibration time" at which the polymer first "feels" the presence of the tube, and that this slow increase in effective tube diameter can help explain some anomalies in the relaxation of asymmetric star branched polymers.

Keywords: Local dynamics, Polymers, Rheology, Tube Model
PACS: 83.80.Rs, 83.80.Sg

INTRODUCTION

The viscoelastic properties of both polymer melts and dilute solutions are reasonably well understood, due to theoretical advances of the last 50 years. For dilute solutions, the Rouse/Zimm model provides a rather complete, and accurate, description of the dynamics over the low to mid frequency range. For melts, the "tube model" of de Gennes [1] and Doi and Edwards [2,3] has been provided an increasingly quantitative description of the viscoelasticity of linear, and some long-chain-branched architectures. Despite this progress, however, some major gaps remain to be closed. Here, we review some of these "gaps" and attempt to provide tentative explanations for a couple of them.

DILUTE SOLUTIONS

Dilute solutions are characterized by polymer mass concentrations c well below the so-called "overlap" concentration $c^* \equiv M/[(4/3)\pi N_A R_g^{\,3}]$, where M is the polymer molecular weight which for polymers of high molecular weight (in the millions of Daltons), N_A is Avogadro's number, and R_g is the polymer radius of gyration [3,4]. Achieving diluteness usually requires that the polymer concentration be less than around 0.1% by mass. While dilute solutions avoid chain-chain interactions (including entanglements), the effects of hydrodynamic interactions and excluded volume are not screened in dilute solutions, and so obtaining a thorough understanding of their dynamics has not been trivial. The following are a few of the

CP982, *Complex Systems, 5th International Workshop on Complex Systems*
edited by M. Tokuyama, I. Oppenheim, and H. Nishiyama

remaining "problem areas" in which understanding is incomplete.

High-Frequency Linear Viscoelasticity – the "Missing Modes"

The Rouse/Zimm theory for the linear viscoelastic properties of dilute polymers is now fifty years old, and one can still credibly argue that it is the most successful theory of polymer rheological properties ever devised. The theory relies on a coarse-grained bead-spring model, in which entropic elasticity is represented by springs and viscous drag is lumped into a discrete number of beads; see Fig. 1. The effects of entropic elasticity, Brownian motion, hydrodynamic interactions, and excluded volume can all be represented through forces applied to the beads and bead mobilities [3]. The model is expected to be accurate only over long time or low frequency ranges for which the dynamics of individual, or small groups, of atoms and bonds are subsumed into "collective" effects of motion that are captured by the coarse-grained beads and springs.

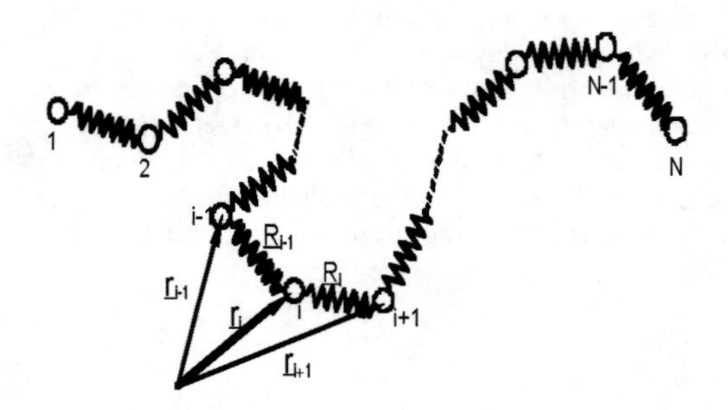

FIGURE 1. The bead-spring model.

In fact, the Rouse/Zimm model has been amazingly successful. The lines in Fig. 2 show the predictions of the Zimm theory compared to the measured linear moduli of a dilute solution of polystyrene of molecular weight 860,000 in two different theta solvents. The agreement is essentially perfect over the entire range of frequency explored.

However, in one respect, the theory is almost "too good." One expects that at frequencies high enough to probe the motions of small groups of bonds that deviations from the Rouse/Zimm theory ought to become apparent, since such motions involve details of bond length and bond angle restrictions that are not captured by the entropic spring. However, surprisingly, the bead-spring model actually describes even the viscoelasticity of dilute polymer solutions, even at frequencies high enough that the theory ought to break down, if one chooses the number of springs "correctly." For polystyrene, the "correct" number of springs is given by the molecular weight (in Daltons) divided by 5000, while for polyisoprene, it is the molecular weight divided by 2400 [6-8]. In other words, a polystyrene molecule of molecular weight 5000 acts as though it is a single spring! For both polystyrene and polyisoprene, nearly 100 backbone atoms behave collectively as a single spring, with essentially no sign of the dynamics of subsets of atoms smaller than this.

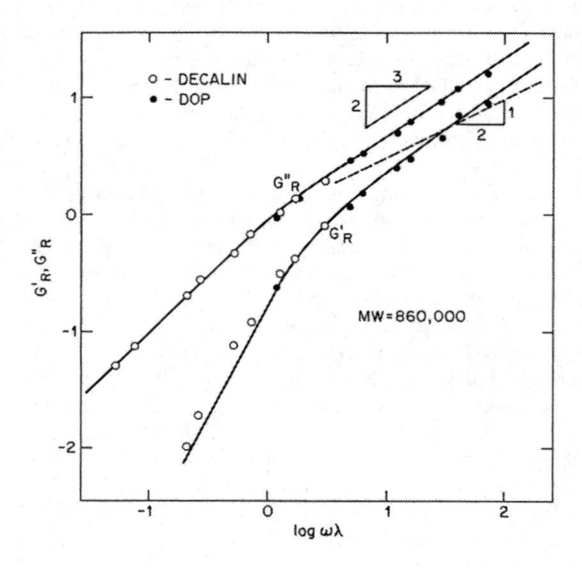

FIGURE 2. Linear viscoelastic data for dilute polystyrene in two theta solvents compared to the predictions (lines) of the Zimm theory. For dimensionless frequencies greater than 10, G' and G'' scale as $\omega^{2/3}$, in agreement with the Zimm theory rather than the Rouse theory, which predicts a proportionality of $\omega^{1/2}$ in this frequency regime. Here $G'_R \equiv G'/\nu k_B T$ is the reduced storage modulus and $G''_R \equiv (G''-\eta_s\omega)/\nu k_B T$ is the reduced loss modulus, both extrapolated to zero concentration, where $\nu = cN_A/M$ is the number of molecules per unit volume, with N_A = Avogadro's number. In addition, η_s is the solvent viscosity and $\lambda \equiv [\eta]_0\eta_s M/N_A k_B T$ is a relaxation time (Figure from Ferry [4]; data from Johnson et al. [5], used with permission from the Society of Polymer Science of Japan).

For example, Fig. 3 shows that for dilute solution of polyisoprene (closed symbols), the viscoelastic phase angle approaches -90° degrees at high frequency, which corresponds to a completely elastic response; i.e., no dissipation, and that this response is quantitatively captured by a bead-spring model that uses only one spring for every 140 backbone atoms! For the same polymer in the melt state (see Fig. 3), there is a broad spectrum of relaxation times at high frequency, leading to a leveling off of the phase angle at around -60°, which is captured qualitatively in a chain with one spring for every 18 backbone atoms. (For a precise definition of the phase angle, see Peterson et al. [8]). Thus, relaxation modes associated with motions of small groups of atoms are quite evident in the melt state, but not in dilute solutions. Why should this be the case?

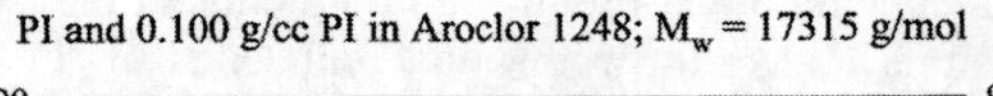

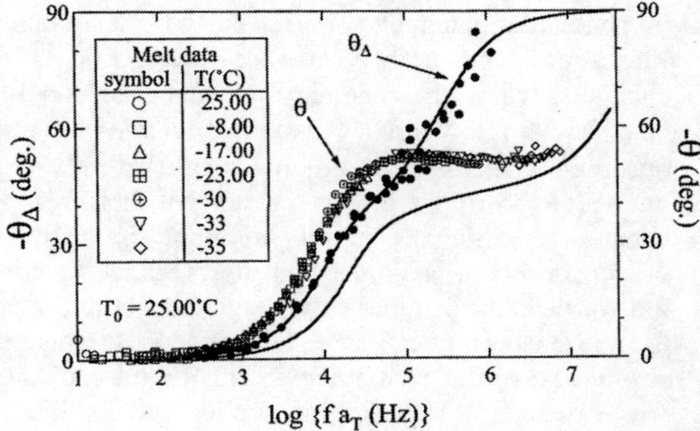

FIGURE 3. Linear viscoelastic phase angle for polyisoprene as a function of reduced frequency fa_T in the dilute solution state (closed symbols) and in the melt state (open symbols). The solid lines are predictions of the bead-spring model, using 7 springs for the dilute solution and 55 for the melt. Data at different temperatures have been shifted using shift factor a_T. From Peterson et al. [8].

As a possible explanation, we recently suggested that torsional barriers to bond rotation might confer a large 'dynamic stiffness' to polymers that slows down modes requiring fast bond motion, so that their relaxation overlaps with the torsional zone for short chains [9]. This would occur in dilute solutions but not in the melt, because in the melt the diffusive motions of the chain on length scales larger than that of bond rotation are slowed down by the high friction of the medium to such an extent that the rotational barriers can be surmounted in the time required for the diffusive motion. Computer simulations support this idea [10,11].

As an extreme example, consider a polymer whose torsional barriers are so high that it cannot change its configuration at all over the time scale of the measurement. This polymer would be effectively a stiff, though bent, wire. This stiff polymer would relax its overall orientation on a timescale set by its rotational diffusion constant, which would be comparable to that of a sphere having similar radius as the "bent-wire" polymer. Thus, if the time scale for bond rotations is slowed by internal torsional barriers to the extent that they are no faster than the time scale for overall rotational diffusion of the coil, then the molecule will seem to have only a single relaxation time, since the internal dynamical processes for the coil will be too slow to measure before the coil as a whole has relaxed through overall rotational diffusion.

Recent simulations in our group support this idea, and show that as the torsional barrier is increased, the polymer relaxation spectrum of a polymer with up to 100 bonds resembles more and more that of a single spring [11].

Extensional Viscosity at High Strain Rates

Theory and Brownian dynamics simulations [12] predict that the uniaxial extensional viscosity of a dilute polymer solution should be a monotonically increasing function of the extensional velocity gradient. (A uniaxial extensional flow can be generated by extending a filament of liquid so that its length grows exponentially in time.) At high extension rates, the extensional viscosity is predicted to approach a plateau value that can be very much higher (by three decades or so) than the value at low extension rates. While the large increase in extensional viscosity with increasing extension rate was observed experimentally long ago, Sridhar and coworkers [13] observed more recently that, surprisingly, at very high extension rates, the extensional viscosity actually *decreases* with increasing extension rate; see Fig. 4. No explanation for this phenomenon has yet been offered. A straightforward interpretation would seem to require that at high enough extension rate, the polymer chains are somehow prevented from fully extending. James and Sridhar [14] have suggested that chain extension might be block by "self entanglements," while others have suggested that tight folds or kinks might impede unraveling of the chain.

Polymer Extension Measured by Light Scattering

Another unusual phenomenon that suggests that in some cases polymer chains are somehow impeded from extending fully in extensional flows is the very low deformation of polymer chains measured by light scattering in strong extensional flows. Every attempt to measure chain deformation by light scattering has shown that the chain is never stretched more than a factor of two or three beyond its equilibrium coil dimension [15-18], even in situations where high chain segment alignment can be confirmed by simultaneous birefringence measurements [15]. Both direct visualization of DNA molecules in extensional flow

[19, 20] and in simulations [21] show, on the other hand, that the polymers have no difficulty becoming fully extended in strong extensional flow, and that the quantity measured by light scattering ought to show this as well [22]. This anomalously low apparent stretch in light scattering experiments could perhaps be partly explained if the chain is folded into a "kinked state" [21,23], in which chain segments are oriented, but not many are highly stretched, because the strain required to stretch most chains fully is not attained everywhere in the finite region around the stagnation point sampled by the laser beam used to scatter light. However, anomalously low deformation is also seen in uniform steady-state shearing flows in which the molecules can be strained indefinitely [17,18], and so this explanation does not seem to be completely correct. Both the results of Sridhar and coworkers indicating a decrease in extensional viscosity at high extension rates, and the anomalously small apparent stretch shown in light scattering experiments seem to indicate that in some situations, polymer coils for some reason fail to stretch completely even in fast flows. It is worth exploring the possibility that the restrictions to torsional rotations might somehow hinder polymer extension in very fast flows, or in very low viscosity solvents.

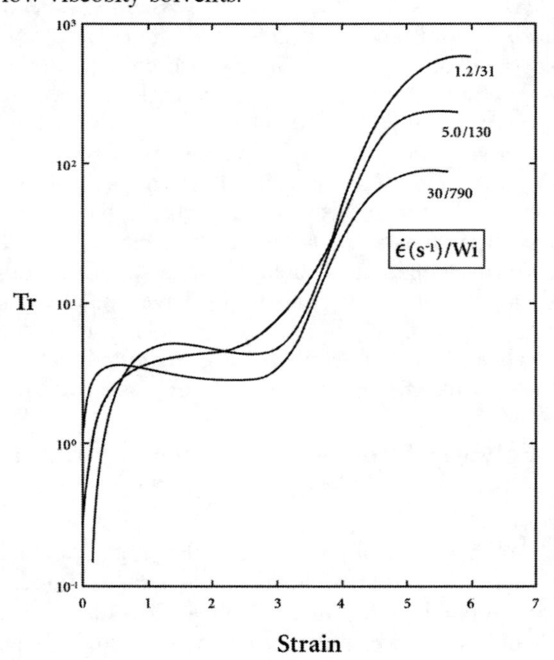

FIGURE 4. Extensional viscosity data for start-up of steady extension at high Weissenberg number for a dilute Boger solution of 69 ppm 20 million molecular-weight polystyrene in a "theta-like" solvent. The strain rates/Weissenberg numbers are given in the figure (from Gupta et al. [13], reprinted with permission from Physics of Fluids).

Polymer Scission

It has been known for many years that very long polymers are vulnerable to chain scission under strong flows, especially extensional flows [24-30]. In the work of Keller and Odell [28], long polymer chains were fractured in flow through a "cross-slot" device, in which two straight channels cross each other at right angles, with injection through both ends of one straight channel and outflow through both ends of the other, perpendicular, channel. The resulting flow contains a stagnation point at the center of intersection of the channels, where the incoming streams meet head-to-head, and the flow turns 90°, exiting out one of the two outflow channels. Scission measured by gel permeation chromatography has shown that breakage occurs predominantly at the midpoint of the chain. For years, it was assumed that the breakage occurred mostly near the laminar-flow stagnation point, where the extension rate (and birefringence) was the strongest, and the accumulated strain on the molecules was expected to be highest. Keller and Odell [28] determined that the critical extensional velocity gradient ($\dot{\varepsilon}_f$) at the onset of chain scission scales with the molecular weight of the polymer chain (M_w) as $\dot{\varepsilon}_f \propto M_w^{-2}$ in a cross-slot device. However, Nguyen and Kausch [29] observed a different scaling law, $\dot{\varepsilon}_f \propto M_w^{-1}$, for a polymer solution passing through a contraction. This difference has led to suggestions that the mechanism of chain extension and scission differs between the two geometries (cross-slot vs. contraction). However, in recent work by Vanapalli et al. [31], it was noted that virtually all experiments on polymer scission, including those in the cross-slot and contraction devices, were conducted at Reynolds numbers too high for the flow to be laminar. In fact, Vanapalli, et al. showed the experimental data, taken from all available literature, actually correlates well with the Reynolds number, with the critical chain contour length above which breakage occurs roughly following a power law $L_c \propto \mathrm{Re}^{-3/2}$. Not only was this correlation predicted from known turbulent velocity statistics, but the critical chain tension at fracture deduced from the above correlation accorded well with known polymer bond strengths. These findings indicate that the difference in scaling laws, $\dot{\varepsilon}_f \propto M_w^{-2}$ and $\dot{\varepsilon}_f \propto M_w^{-1}$ are the result of the differences in geometry that lead to different dependences of Reynolds number on apparent extension rate. Scission induced by truly laminar would require using much more viscous solvents than those used in all previous experiments. A recent simulation study [32], indicates that if laminar flow were to be achieved, then in the cross-slot device, contrary to conventional wisdom, more scission would actually occur at the wall, near the corner of the cross-slot, than at the stagnation point. The scission at the corner is predicted to arise from the combination of high, persistent, shear rates at the wall, which stretch the polymer, and extensional flow near the corner, which, while not very persistent, delivers the "coup de grace" to molecules that have been stretched (but not broken) by shear alone. These predictions have not yet been confirmed, but the predictions could be checked indirectly through comparison of predicted scission rates with those from experiments using polymers dissolved in solvents viscous enough to produce scission in laminar flow.

Non-dilute Behavior in "Dilute" Solutions

Another "mystery" related to dilute solution rheology lies in the dependence of viscoelastic properties, such as viscosity and relaxation time, on solution concentration. If it is truly dilute, each polymer molecule will contribute additively to the viscoelastic stresses and independently of the presence of other polymer molecules in the solution. This means that the viscosity should be a linear function of concentration and the polymer relaxation time should be independent of concentration, since each polymer should relax without being affected by the other polymers in the solution. One expects that this dilute behavior should hold for all concentrations that are well below the "overlap" concentration c* defined above. However, in 1994 Kalashnikov [33] found that the apparent relaxation time of "dilute" solutions of polyethylene oxide (PEO), inferred from both the zero-shear viscosity and the shear rate at the onset of shear thinning, depends on concentration to roughly the half power, even at concentrations as low as 10 ppm, two decades below c*. Similar behavior was recently found by McKinley and coworkers [34] for "dilute" polystyrene solutions in styrene oligomer, using a "capillary break-up extensional rheometer" (or "CABER"). These concentrations are well below any reasonable condition for coil overlap, or polymer entanglements, to be significant. While the possibility of a "gel-like" network structure at very dilute concentrations cannot be overlooked, the existence of such similar phenomena in very different polymer solutions (PEO in water vs. polystyrene in styrene oligomer) suggests the existence of some unsuspected polymer-polymer interactions even under very dilute conditions.

ENTANGLED POLYMERS

Leaving the region of at least nominally "dilute" behavior, we enter the much richer, more complex, field of entangled polymer dynamics. Over the last three decades, the great complexity of this field has been addressed primarily through the concept of the entanglement "tube" – a region of space surrounding a given polymer chain within which its motion is relatively free of entanglements with neighboring chains, but beyond which the chain is not able to penetrate due to those entanglements [3]. The shape of the tube is that of a one-dimensional random walk, but with step size much larger (typically by a factor of 5 or more) than that of the polymer itself, and with each "tube segment" of this walk being much thicker (by a factor of 20 or more) than the chain itself. The chain moves readily within this tube, but must "reptate" – i.e., randomly slide back and forth along the tube axis - to escape the tube and completely relax its configuration [1-3]. There are additional relaxation processes within the tube besides reptation that affect the rheological properties. These additional processes include "primitive path fluctuations," which are breathing motions of the chain within the tube, which allow the ends of the tube to be vacated without center of mass motion of the chain. These fluctuations enhance the rate of relaxation of linear polymers, and dominate the relaxation of star polymers [35,36], since star-branched polymers cannot reptate. Over the years, the various processes of relaxation within the tube, including reptation and primitive path fluctuations, have been modeled with increasing accuracy using the "tube" ansatz, allowing quite refined predictions to be made of linear and even nonlinear viscoelastic properties of linear and long-chain-branched polymers [37-39].

Despite the numerous successes, there remain a number of bedeviling problems with the "tube" model. One problem is in linking clearly the "tube" with the underlying structure of the melt, which can now be modeled at the monomer or even atomistic level by direct molecular dynamic simulations [40,41]. Such a link is needed to establish from a microscopic physical picture the effective "diameter" of the tube, the effective length of the step size of tube, as well as the distribution of tube lengths. The latter is needed to determine the statistical fluctuations in the length of tube occupied by a given chain, and thus determine the contribution to relaxation of primitive path fluctuations, alluded to above. In addition, there are problems in determining how the tube migrates, dilates, or is deformed, in response to diffusion or flow-induced motion of the chains that create the constraints making up the tube. While ideas have been put forward on all these topics [37,42], controversy and uncertainty remain. The following sections describe briefly just a few of the important areas that need further work, and, indeed, where rapid progress will hopefully soon be made.

Are Melts and Entangled Solutions Fundamentally Different?

Theories for entangled polymers, such as the tube theory, do not change fundamentally when a diluting solvent is considered. Of course, addition of a solvent enlarges the tube diameter, since polymer molecules are then, on average, farther apart from each other than they are in the melt. Also, the solvent can "lubricate" the motion of the molecules, speeding up their relaxation. However, both of these effects - the diluting effect and the "lubricating" effect - are readily accounted for in the tube model by simple adjustments of two parameters, namely the tube diameter, to account for the increased inter-molecular spacing, and the friction constant, to account for the "lubrication" effect. In fact, within the tube model, the linear viscoelastic properties of polymers in an entangled solution can theoretically be "mapped" onto those in a melt by simply rescaling the modulus to account for the increased tube diameter and the time to account for the change in friction [43, 3]. The "mapping" only works if, in addition to the above rescaling of the axes, the molecular weight of the polymer in the "equivalent" melt is smaller than that in the solution so that it occupies the same number of tube segments in the melt as the longer chain does in the solution. (This difference in molecular weight between solution and melt is required because the tube segments are longer, and contain more monomers, in the solution than in the melt.) Thus, given the validity of the tube model, for every entangled solution, there should be an "equivalent" melt, whose G' and G'' curves on a log-log scale are identical to that in the solution, except for shifts along the two logarithmic axes. (There should, however, be some difference between solution data and data for an "equivalent" melt at very high frequency where the motion of individual monomers along the chain contribute to the viscoelasticity.) Apart from this caveat, one can ask if this theoretical expectation of "equivalence" is actually met.

Surprisingly, this fundamental question is only now being carefully addressed, although there have long been anecdotal observations that the decrease in G'' with frequency above the terminal region for nearly monodisperse chains is steeper for melts than for "equivalent" entangled solutions. Recent careful comparisons of nearly monodisperse polystyrene solutions and nearly equivalent melts should provide strong tests of whether entangled solutions really are equivalent to melts [44]. If so, the standard tube model, which makes no fundamental distinction between entangled solutions and melts, should work equally well for both. If not, some important modifications of the tube model will be required for it to apply to both solutions and melts.

"Dynamic Dilution" and the Dilution Exponent

One of the most powerful tools in the toolbox of the "tube" model is the idea of "dynamic dilution." This concept, invented by Marrucci [45], was first successfully employed to help explain the relaxation of star polymers by Ball and McLeish [46]. The idea is that as the chains defining the tube enclosing a given polymer move, that polymer is able to explore farther from the axis of the tube than would be the case if the surrounding chains were not undergoing long-range diffusion themselves. In effect, the tube "dilates" in time, owing to the loss of effectiveness of constraints imposed by surrounding chains due to their motion. This "dynamic dilution" is especially important when polymer diffusion is occurring over a wide range of time scales, since on time scales where one chain has not yet moved much, many of the surrounding chains may have moved a great deal. Hence, the tube surrounding the "sluggish" chain (or sluggish portion of a chain) "sees" at late times a "fat" tube defined by only those entanglements that persist over the long time scales required for the sluggish test chain to encounter them. Thus, when there is a wide range of relaxation times characterizing a given polymer mixture, the slow chains relax on long time scales in effectively "wider" tubes than at short time scales and the "tube diameter" characterizing the melt is effectively time dependent. The process of tube "dilation" is especially significant for polymer branches that are constrained at one end by a branch point, and so relax by primitive path fluctuations, pulling their free end into the tube. This process rather rapidly relaxes the tip of the branch, but only very slowly relaxes the "inner core" of the branch nearest the branch point. Hence, dynamic dilution or "tube dilation" will widen the tube continually as the branch relaxation proceeds.

By postulating a time-dependent mean-field tube diameter, the "dynamic dilution" ansatz provides a potentially powerful means of simplifying the extremely complex relaxation processes occurring in commercial mixtures of long and short chains, some of which might be linear molecules, and some branched.

Typical models that employ the dynamic dilution concept simply divide the melt into relaxed and unrelaxed components, where the fraction of melt that is unrelaxed, $\Phi(t)$ is a function of time after some small deformation is imposed. While each molecule in a complex mixture of polymers may relax at a rate controlled by its length and branching architecture, in the dynamic dilution ansatz all molecules are assumed to experience a common tube diameter a(t) at any given time t after the deformation. Thus, the (in principle) very complex influence of the mixture of chains on the relaxation of each chain in that mixture is reduced to keeping track of single parameter a(t) that characterizes the density of entanglements that prevails at any time t after the deformation. The time-dependent tube diameter a then increases as $a(t) \propto [\Phi(t)]^{-\alpha/2}$, where α is the "dilution exponent." Since the rate of polymer relaxation accelerates as the tube enlarges, this leads to even faster dilution. Thus, the process of dynamic dilution is in a sense autocatalytic, and highly sensitive to the value of the dilution exponent α. Theoretical considerations have yielded the predicted values α=1 (47) and α=4/3 [43]. Fits to experimental data for the concentration dependence of the "plateau modulus" sometimes give the former [48] and sometimes the latter [49].

Because of its autocatalytic effect, a small difference in the value of α has a large effect on the rate of relaxation of entangled polymers, and getting the value of α right is important for quantitative predictions. Unfortunately, while neither theory nor experiment adjudicate convincingly between the proposed values of 1 and 4/3, the best theories of polymer relaxation seem sometimes to work best using α=1 and sometimes using α=4/3. Fig. 5, for example, shows the measured storage and loss moduli for a blend of a linear and a star-branched 1,4-polybutadiene, compared to the predictions of a tube model, using both α=1 and α=4/3 [52]. For this case, the value α=4/3 (solid lines in Fig. 5) gives the better prediction of the data. However, using a similar theory for blends of one linear polymer in another, the choice α=4/3 works poorly, while α=1 gives good agreement with data [53]. It is not clear at this point which value of α is correct, and whether the theory has a flaw that can be repaired, or if the idea of dynamic dilution is simply not a sufficiently accurate or reliable concept to be applied generally to polymers of arbitrary architecture. Clearly, more work will be needed to clear up this issue.

Branch Point Motion

Describing the rheology of polymer molecules with more than one branch point requires accounting for the relaxation of "backbone" segments lying between two branch points, which cannot relax by either reptation or fluctuations of the free end. Modern ideas, arising from McLeish and coworkers [37], model backbone relaxation by a kind of "hopping" motion of the branch points, wherein each "hop" occurs whenever an arm executes a complete retraction, and is then temporarily effectively unentangled, at least near the branch point, and thus the branch point is free to move roughly a tube diameter (within a factor of order unity constant "p") along the tube containing the backbone. These ideas, combined with "dynamic dilution" have proven successful in predicting the rheology of branched polymers with two or more branch points, with regular, or even irregular, branching structures [37,54,55], similar to those of commercial melts.

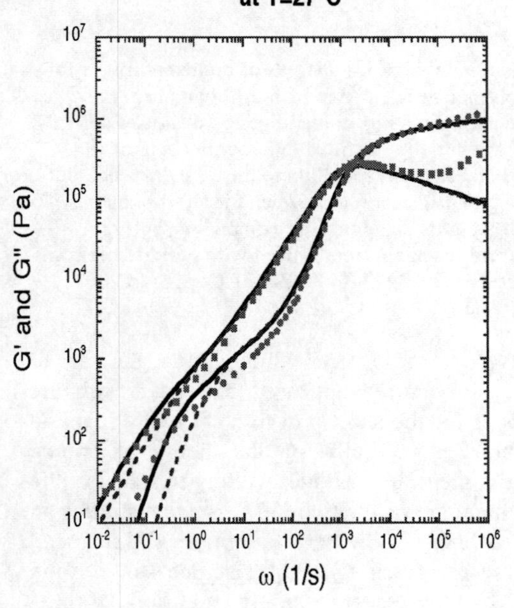

FIGURE 5. Comparison of predictions of the theory of Milner *et al.* (50) using α=1 and α=4/3 with measurements of G' (circles) and G" (squares) for a polybutadiene star-linear blend containing linear molecules of molecular weight 23600 and 4-arm stars of molecular weight $M=4M_a$=1367000 at a star volume fraction of 0.025 at T=27 ^{0}C. The symbols are the experimental data from Roovers (51); the solid lines are model predictions using α=4/3, and the dashed lines are model predictions using α=1 (modified from Park and Larson [52])

However, given the highly simplified model of branch point motion assumed above, and the need to introduce an adjustable prefactor "p" into the theory, determined by fits to data, it would be highly desirable to obtain direct microscopic evidence of the validity of this model of branch point motion. Fortunately, in very recent molecular dynamics simulations of asymmetric star polymers, having two long arms and one short arm, the postulated "hopping" motion was observed directly [56]; see Fig. 6.

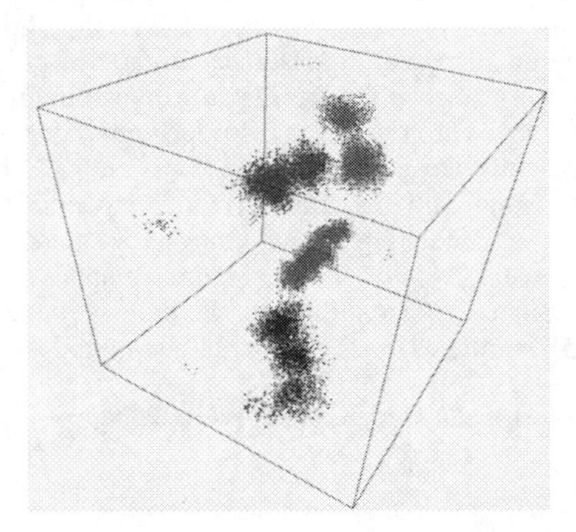

FIGURE 6. Scatter plot of locations of branch point for four different polymers, each shown by a different cluster of points, selected from a box containing 200 chains. Each branch point is initially confined to a roughly spherical volume of diameter corresponding to the tube diameter, but then "hops" to a different region when the short branch retracts, releasing its entanglement constraints, before becoming confined again (reproduced with permission from *Macromolecules* **40**, 3443-3449 (2007). Copyright 2007 Am. Chem. Soc.).

Moreover, in these recent simulations, the rate at which "hops" occurred appeared correlated with the time required for the short arm to completely relax its configuration, as predicted by the theory of branch point motion mentioned above. At long timescales, the backbone made up of the two long arms constitutes an effectively "linear" polymer containing a "fat" bead whose drag represents the resistance to motion presented by the short arm. In addition, it was observed in the simulations that the tube is effectively "thinner" at short timescales than at longer ones, which is consistent with experimental rheological data on such molecules [57].

Tube Confining Potential

Finally, a long-standing challenge in the theory of entangled polymer dynamics has been to determine the effective diameter and length of the tube, not from empirical fits of the model to rheological data, but directly from microscopic simulations. Progress in towards this goal is recently being made, stimulated by a method developed by Kremer, Grest, and coworkers (41) for calculating the effective length of the tube, i.e., its "primitive path." In their method, an equilibrated melt simulated on the computer is subjected to a "cooling" procedure in which the ends of all molecules are held fixed, and their lengths simultaneously all shortened, with the constraint that they not pass through each other. This procedure collapses all chains down to minimum lengths subject to preservation of their entanglement constraints. Using this procedure to determine the primitive path of each chain, Zhou and Larson [58] then sampled the distribution of distances that monomers of a given chain drifted from the primitive path of that chain. This yielded an effective probability density of monomers, an example of which is plotted in Fig. 7. From such results, the effective diameter of the tube can be determined as well as the "tube confining potential" created by the entanglements with other chains. While the methods used to create the primitive paths and tube "confining potentials" are not yet rigorous, they do point the way towards what in the future might well be definitive methods for grounding the tube model, and the parameters describing the tube, directly in microscopic physics.

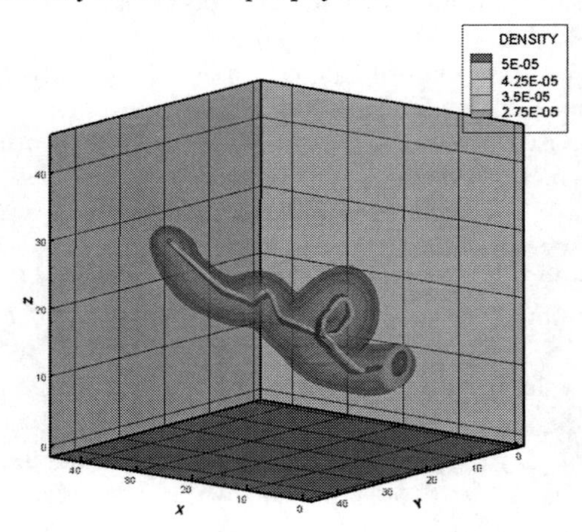

FIGURE 7. Direct visualization of a confining tube, using contours of equal probability from molecular dynamics simulations [58]. Lines are the primitive paths determined from "cooling" procedure. (reproduced with permission from *Macromolecules* **39**, 6737-6743 (2006). Copyright 2006 Am. Chem. Soc.).

These results indicate that molecular dynamics simulations offer the potential to generate deep insight into the validity of various assumptions underlying the tube model, and ways to improve upon this model.

SUMMARY

The dynamics of polymers in the dilute and entangled regimes have been modeled in coarse-grain fasion, with a great deal of success, by, respectively, the Rouse/Zimm bead spring chain model, and the de Gennes/Doi-Edwards tube model. Despite successes, a number of "unsolved mysteries" remain.

Among these mysteries is the question of why in dilute polymer solutions, but not in melts, there seems to be an absence in the linear viscoelastic spectrum of modes corresponding to short portions of the chain, a few backbone atoms in length. Other mysteries are the unexpected decrease in extensional viscosity at high extension rates, and the apparent lack of polymer stretching indicated by light scattering experiments in extensional and shearing flows. Some of these "mysteries" seem to be resolvable by advanced simulations with high-speed computers that are able to realistically model long polymer chains at the monomer level, and indicate that bond torsional potentials are likely responsible, at least for the "missing" high frequency modes, if not for some of the other unsolved mysteries.

In melts and entangled solutions, there remain uncertainties regarding whether and how such entangled solutions might be regarded as "equivalent" to a melt having rescaled chain length, modulus, and time scale. There is in addition questions as to whether the "dynamic dilution" concept, which envisions as mean-field time-dependent tube diameter, is really broadly applicable to linear and branched polymers, both monodisperse and polydisperse, and, if so, what the correct "dilution exponent" might be. Finally, there are questions about how branch point motion occurs in multiply branched polymers. As in dilute solutions, these issues are beginning to be resolved through the use of powerful computers. For instance, recent simulations indicate that the process of disentanglement by arm fluctuation, allowing branch-point hopping, as envisioned by McLeish and coworkers, seems to be valid, although the tube diameter in which the hopping occurs may depend significantly on the branch length.

ACKNOWLEDGMENTS

Parts of this work were supported by the National Science Foundation, under grant DMR-0305437, and by the National Computational Science Alliance under PHY040025N, and by the American Chemical Society under the Petroleum Research Fund. Any opinions, findings and conclusions or recommendations expressed in this material are those of the authors and do not necessarily reflect the views of the National Science Foundation (NSF).

REFERENCES

1. P. G. de Gennes, *J. Chem. Phys.* **55**, .572-579 (1971).
2. M. Doi, and S. F. Edwards *J. Chem. Soc. Faraday Trans. II* **74**, 1789-1832 (1978).
3. M. Doi, and S. F. Edwards, *The Theory of Polymer Dynamics*, New York: Oxford, 1986.
4. J. D. Ferry, *Viscoelastic Properties of Polymers,* New York: Wiley, (1980).
5. R.M. Johnson, J.L. Schrag, and J.D. Ferry, *Polym. J.* **1**, 742-749 (1970).
6. D. W. Hair, and E. J. Amis, Macromolecules, **22**, 4528-4536 (1989).
7. S. Amelar, C. E. Eastman, R. L. Morris, M. A. Smeltzly, T. P. Lodge, and E. D. von Meerwall, *Macromolecules* **24**, 3505-3516 (1991).
8. S. C. Peterson, I. Echeverría, S. F. Hahn, D. A. Strand, and J. L. Schrag, *J. Polym. Sci. B: Polym. Phys. Ed.*, **39**, 2860-2873 (2001).
9. R. G. Larson, *Macromolecules*, **37**, 5110-5114 (2004).
10. M. Fixman, *Journal of chemical physics*, **69**, 1538-1545 (1978).
11. S. Jain and R. G. Larson, *Macromolecules*, to be submitted.
12. C. C. Hsieh, and R. G. Larson, *J. Rheol.*, **48**, 995-1021 (2004).
13. R. K. Gupta, D. A. Nguyen, and T. Sridhar, *Phys. Fluids* **12**, 1296-1318 (2000).
14. D. F. James, and T. Sridhar, *J. Rheol.* **39**, 713-724 (1995).
15. M. J. Menasveta, and D.A. Hoagland, *Macromolecules* **24**, 3427-3433 (1991).
16. R. C. Armstrong, S. K. Gupta, and O. Basaran, *Polym. Eng. Sci.* **20**, 466-472 (1980).
17. E. C. Lee, M.J. Solomon, and S.J. Muller, *Macromolecules* **30**, 7313-7321 (1997).
18. F. R. Cottrell, E.W. Merrill, and K.A. Smith, *J. Polym. Sci., Polym. Phys. Ed.* **7**, 1415-1434 (1969).
19. T. T. Perkins, D. E. Smith, and S. Chu, *Science* **276**, 2016-2021 (1997).
20. D. E. Smith, and S. Chu, *Science* **281**, 1335-1340 (1998).
21. R. G. Larson, H. Hu, D. E. Smith, and S. Chu, *J. Rheol.* **43**, 276-304 (1999).
22. L. Li, and R. G. Larson, *Macromolecules* **33**, 1411-1415 (2000).

23. E. J. Hinch, *J. Non-Newt. Fluid Mech.* **54**, 209-230 (1994).
24. J. F. S. Yu, J. L. Zakin, and G. K. Patterson, *J. Applied Polym. Sci.* **23**, 2493-2512 (1979).
25. M. J. Miles and A. Keller, *Polymer* **21**, 1295-1298 (1980).
26. E. W. Merrill, and A. F. Horn, *Polymer Commun.* **25**, 144-146 (1984).
28. A. Keller, and J. A. Odell, *Colloid and Polymer Science* **263**, 181-201 (1985).
29. T. Q. Nguyen, and H. H. Kausch, *Polymer*, **33**, 2611-2621 (1992).
30. M. T. Islam, S. A. Vanapalli, and M. J. Solomon, *Macromolecules* **37**, 1023-1030 (2004).
31. S. A. Vanapalli S. L. Ceccio and M. J. Solomon *Proc. Nat. Acad. Sci. USA,* **103**, 16660-16665 (2006).
32. C. C. Hsieh, S. J. Park, and R. G. Larson *Macromolecules,* **38**, 1456-1468 (2005).
33. V. N. Kalashnikov, *J. Rheol.*, **38**, 1385-1403 (1994).
34. C. Clasen, M. Verani, J. P. Plog, G. H. McKinley, W.-M. Kulicke, Proc. XIVth Int. Congr. Rheol. Seoul, Korea, Aug. 2-27, 2004, Korean Soc. of Rheol.
35. M. Doi, and N. Y. Kuzuu *J. Polym. Sci. C – Polym. Letts.* **18**, 775-780 (1980).
36. S. Shanbhag, and R. G. Larson Phys. Rev. Letts. **94**, 076001 (2005).
37. T. C. B. McLeish *Adv. Phys.* **51**, 1379-1527 (2002).
38. J. M. Dealy and R. G. Larson *Structure and Rheology of Molten Polymers,* Munich: Carl Hanser Verlag, 2006.
39. R. G. Larson, Q. Zhou, S. Shanbhag, and S. J. Park, *AIChE J.* **53**, 542-548 (2006)
40. K. Kremer, G. S. Grest I. Carmesin Phys. Rev. Letts. **61**, 566-569 (1988).
41. R. Everaers, S. K. Sukumaran, G. S. Grest, C. Svaneborg, A. Sivasubramanian, and K. Kremer *Science* **303**, 823-826 (2004).
42. R. G. Larson, Q. Zhou, S. Shanbhag, and S. J. Park, *AIChE J.* **53**, 542-548 (2006).
43. R. H. Colby, and M. Rubinstein, *Macromolecules* **23**, 2753-2757 (1990).
44. Y. Heo, and R.G. Larson, *Macromolecules,* to be submitted (2007).
45. G. Marrucci, *J. Polym. Sci. B – Polym. Phys.* **23**, 159-177 (1985).
46. R. C. Ball, and T. C. B. McLeish *Macromolecules* **22**, 1911-1913 (1989).
47. F. Brochard, and P. G. de Gennes, *Macromolecules* **10**, 1157-1161 (1977).
48. H. Tao, C. Huang, and T. P. Lodge, *Macromolecules* **32**, 1212-1217 (1999).
49. V. R. Raju, E. V. Menezes, G. Marin, and W. W. Graessley, *Macromolecules* **14**, 1668-1676 (1981).
50. S. T. Milner, T. C. B. McLeish, R. N. Young, A. Hakiki, and J. M. Johnson, *Macromolecules* **31**, 9345-9353. (1998).
51. J. Roovers, *Macromolecules,* **20**, 148-152 (1987).
52. S. J. Park, and R. G. Larson, *J. Rheol.* **47,** 199-211 (2003).
53. S. J. Park, and R. G. Larson, *J. Rheol.* **50**, 21-39 (2006).
54. R. G. Larson, *Macromolecules* **34**, 4556-4571 (2001).
55. C. Das, N. J. Inkson, D. J. Read, M. A. Kelmanson, T. C. B. McLeish, *J. Rheol.*, **50**, 207-235 (2006).
56. Q. Zhou, and R. G. Larson, *Macromolecules* **40**, 3443-3449 (2007).
57. A. L. Frischknecht, S. T. Milner, A. Pryke, R. N. Young, R. Hawkins, and T. C. B. McLeish, *Macromolecules* **35**, 4801-4820 (2002).
58. Q. Zhou, and R. G. Larson, *Macromolecules* **39**, 6737-6743 (2006).

Polymer Dynamics from Synthetic to Biological Macromolecules

D. Richter*, K. Niedzwiedz*, M. Monkenbusch*, A. Wischnewski*, R. Biehl*, B. Hoffmann**, and R. Merkel**

**Institut für Festkörperforschung; **Institut für Bio- und Nanosysteme, Forschungszentrum Jülich, D-52425 Jülich, Germany*

Abstract. High resolution neutron scattering together with a meticulous choice of the contrast conditions allows to access the large scale dynamics of soft materials including biological molecules in space and time. In this contribution we present two examples. One from the world of synthetic polymers, the other from biomolecules. First, we will address the peculiar dynamics of miscible polymer blends with very different component glass transition temperatures. Polymethylmetacrylate (PMMA), polyethyleneoxide (PEO) are perfectly miscible but exhibit a difference in the glass transition temperature by 200K. We present quasielastic neutron scattering investigations on the dynamics of the fast component in the range from angströms to nanometers over a time frame of five orders of magnitude. All data may be consistently described in terms of a Rouse model with random friction, reflecting the random environment imposed by the nearly frozen PMMA matrix on the fast mobile PEO. In the second part we touch on some new developments relating to large scale internal dynamics of proteins by neutron spin echo. We will report results of some pioneering studies which show the feasibility of such experiments on large scale protein motion which will most likely initiate further studies in the future.

Keywords: Aggregate dynamics, Alcohol Dehydrogenase, Dynamic miscibility, Neutron spin echo, Polyethyleneoxide, Polymethylmetacrylate, Random environment, Random friction, Rouse model, Solution structure
PACS: 61.05.fg, 61.25.he, 61.25.hk, l87.15.hp, 87.15.bk

INTRODUCTION

The diffusional motion of long chain flexible polymers constitute fascinating physics and at the same time represent one of the great challenges of modern material sciences which try to connect materials properties with the underlying molecular and atomic constituents of materials. Neutron spin echo (NSE) spectroscopy, as the highest resolution neutron technique, provides the necessary time resolution (in the 100ns range), in order to access molecular motions on mesoscopic time scales. At these nanoscales the motions in molecules and molecular aggregates takes place that underlie their macroscopic behaviour. Similarly the large scale motions of biomolecules occur on similar space-time-frames [1].

In this work we will address two recent examples for the investigation of large scale molecular dynamics - one from the synthetic and one from the biological world. We start with synthetic polymers and address the dynamics of a polymer chain (polyethyleneoxide, PEO) in a frozen matrix of polymethylmetacrylate (PMMA) [2-5]. The scientific interest emerges from the extreme difference of the glass transition temperatures in both materials which result in a completely different dynamics of both molecules at a given temperature. Both polymers are perfectly miscible, i.e. the chains are neighbours on a molecular scale. We will show that locally the motions of PEO appear to be nearly completely decoupled from the matrix constraints of the surrounding PMMA. At larger scales the dynamics is greatly slowed down and we are confronted with the dynamics of a chain undergoing random friction under spatial constraints [6].

While the dynamics of synthetic polymers is largely responsible for their macroscopic properties, large scale protein motions are critical for the coordination of precise biological function. Such dynamics are invoked in genome regulatory proteins, motor proteins, signalling proteins and structural proteins. Structural studies have documented the conformational flexibility in proteins, accompanying their activity. Results from macroscopic studies such as biochemical kinetics [7] and single molecule detections have also shown the importance of conformational dynamics and Brownian thermal fluctuations within the proteins. However, the time

CP982, *Complex Systems, 5th International Workshop on Complex Systems*
edited by M. Tokuyama, I. Oppenheim, and H. Nishiyama

dependent dynamical processes that facilitate protein domain rearrangements remain poorly understood and are experimentally little investigated [8]. We will display some first NSE experiments on the dynamics of alcohol dehydrogenase, a protein which plays a key role in the fermentation process by yeast [9]. We show that by NSE spectroscopy some first insight into the rotational dynamics as well as some internal motion of the alcohol dehydrogenase tetramer may be achieved.

THE DYNAMICS OF PEO IN THE MISCIBLE PEO/PMMA BLEND

The investigation of the dynamic miscibility in polymer blends, i.e. the question of how friction arises in a chemically heterogeneous environment is a very active area of research combining the concepts of polymer physics with that of the glass transition. The question is, to what extent, though perfectly miscible, the different components retain their individuality concerning their dynamic response. At temperatures well above the component glass transition temperatures T_g, the concept of segment self concentration provides a rather successful description of the component dynamics [10]. Due to the chain connectivity this self concentration is always enhanced locally and determines the component glass transition behaviour. However, in systems with greatly different component T_g's a decoupling of the dynamics of both components has been reported. In the system PEO/PMMA ($\Delta T_g = 200$K) NMR studies have reported 12 orders of magnitude difference in local relaxation times [2]. In such a situation the low-T_g component moves in the random environment created by the frozen high-T_g component [6] – a qualitatively different scenario compared to that where both components move on similar time scales.

In the following we will present high resolution quasielastic neutron scattering results on the space time evolution of the fast component dynamics (PEO) at local and intermediate scales. Thereby we will address the self and chain dynamics focusing on the incoherent as well as the scattering from a single chain [11].

THE ROUSE MODEL

On intermediate length scales the dynamics of a polymer chain in a melt to a good approximation can be described in terms of the Rouse model [11]. This model treats the dynamics of a Gaussian chain in a heat bath which is experiencing entropic forces that originate from the conformational chain entropy and viscous friction between the segments and their surroundings. At scales $QR_e > 1$ where R_e is the chain end to end distance and at time scales sufficiently shorter than the longest internal relaxation time of a chain, the Rouse time τ_R, the intermediate self dynamic structure factor has the Gaussian form

$$S_{self}(Q,t) = \exp\left[-\frac{Q^2}{6}\langle r^2(t)\rangle\right] \quad (1)$$

where $\langle r^2(t)\rangle = 2\left(W\ell^4 t/\pi\right)^{1/2}$ is the mean square displacement (MSD) and $W = 3k_BT/\left(\zeta\ell^2\right)$ is the elementary Rouse rate. It is given by the ratio of the entropic force $3k_BT/\ell^2$ with ℓ the segment length and the friction coefficient ζ. Also the single chain dynamic structure factor $S_{chain}(Q,t)$ can be written as the function of the same scaling variable combining spatial and temporal scales

$$S(u) = \int_0^\infty \exp\left(-s-\left(\frac{u}{3\pi}\right)\int_0^\infty\left[\cos\left(\frac{6xs}{u}\right)\right]\right)/x^2 \left(1-\exp\left(-x^2\right)dx\right)\right)ds \quad (2)$$

The scaling property is a consequence of the Rouse model that does not involve intrinsic length scales in the range of its validity. In the following we will use the Rouse model with some average friction coefficient in order to rationalize the neutron data from different time and length scales.

THE SELF CORRELATION FUNCTION AT INTERMEDIATE TIME SCALES:

In order to study the PEO self correlation we exploit the large difference between the scattering length of the proton and the deuteron in preparing samples with a protonated PEO component in a deuterated PMMA matrix at different PEO concentrations. Under such circumstances the scattering signal has two contributions: an incoherent part originating from the h-PEO and the coherent scattering from d-PMMA as well as that from the contrast between h-PEO and d-PMMA. Since during incoherent scattering two thirds of the neutrons undergo a spin flip, the coherent and incoherent contributions to the scattering may be separated by neutron polarization analysis. Figure 1 presents the results of such measurements for a $PMMA_{0.65}$ $PEO_{0.35}$ blend. As may be seen in the higher Q range the incoherent scattering dominates while at low Q below

$Q \cong 0.3$ the coherent single chain structure factor arising from the PEO-PMMA contrast prevails.

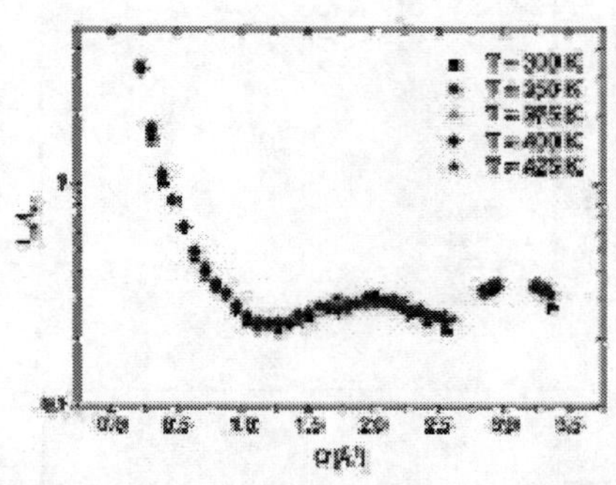

FIGURE 1. Ratio of coherent and incoherent intensities from a 35% PEO/PMMA sample at different temperatures as determined by neutron polarization analysis performed at DNS in Jülich.

Figure 2 presents some representative data taken at the backscattering instrument IN16 at the ILL for a number of small momentum transfers at $T = 400\text{K}$. The spectra consists of an elastic contribution relating to the coherent scattering from PMMA – within the time window of IN16 the PMMA dynamics is frozen – and quasielastic wings originating from the PEO motion. The data are described by a superposition of two components:

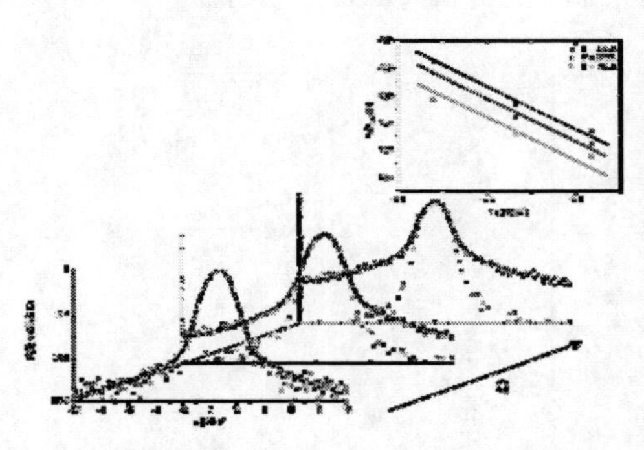

FIGURE 2. Backscattering spectra obtained at $T = 375\text{K}$ from the 35% PEO/PMMA sample for the Q values (0.19; 0.29; 0.43 A^{-1}). The solid lines are the fit results (see text), the dashed lines the instrumental resolution function.

An elastic fraction representing the contribution from the rigid PMMA matrix and a quasielastic contribution describing the PEO motion which was modelled by a stretched exponential function. The intensity ratio of both components was taken from the polarization analysis results (see Figure 1). The resulting fits are presented by solid lines in Figure 2. The insert displays the obtained Kohlrausch-William-Watts (KWW) times τ_{KWW} for the mobile PEO component. In the low Q regime the observed self motion displays a Q^{-4} dependence for the relaxation time.

This result is in accordance with what is predicted by the Rouse model. Table 1 presents the corresponding Rouse relaxation rates $W\ell^4$ for the different samples at different temperatures and compares them with those obtained for pure PEO. With increasing amount of PMMA in the blend, the characteristic rate decreases i.e. the dynamics of PEO is slowed down. The same effect occurs also with decreasing temperature. The observed temperature dependence thereby is weak and does not depend on the blend composition.

TABLE 1. Rouse rate as deduced from incoherent data and distribution width σ for all samples, obtained from the analysis of the dynamic structure factor.

Sample	$W\ell^4$(nm^4/ns)	σ
Pure PEO ($T = 400\text{K}$)	1.29	...
25% PEO/PMMA ($T = 400\text{K}$)	0.1	2.1
35% PEO/PMMA ($T = 400\text{K}$)	0.22	1.6
35% PEO/PMMA ($T = 375\text{K}$)	0.12	1.8
35% PEO/PMMA ($T = 350\text{K}$)	0.05	1.9
50% PEO/PMMA ($T = 350\text{K}$)	0.13	1.6

From Eq.[1] we know that in Gaussian approximation the self correlation function bears direct information on the time dependent mean square displacement. In order to access this quantity in a more direct way, the backscattering data were Fourier transformed into time space. In the time regime the resolution correction amounting to a deconvolution in ω-space becomes a simple division such that the changes of the intermediate scattering function with time and scattering vector may be seen much more directly. Nevertheless one has to keep in mind that after performing the Fourier transformation, the data are still containing the elastic PMMA component which needs to be corrected for.

Figure 3 presents the such obtained intermediate scattering functions for a PEO from the d-$PMMA_{0.65}$ h-$PEO_{0.35}$ sample at $T = 375\text{K}$ at different Q values.

Following Eq.[1] the intermediate scattering function may be directly transformed into the time dependent mean square displacement. The corresponding results are displayed in Figure 3b. The solid line presents the expected square root of time law. Other than expected the data taken at different Q values do not collapse on a single curve but exhibit deviations which increase with increasing Q value. Obviously we need to deal with deviations from the Gaussian assumption underlying Eq.[1]. A first order correction is provided by the so called non-Gaussianity parameter $A(t)$ [12].

$$S_{self}(Q,t) = \exp\left\{ \begin{array}{l} -\dfrac{\langle \Delta r^2(t) \rangle}{6} Q^2 + \\ \dfrac{A(t)\left[\langle \Delta r^2(t) \rangle\right]^2}{72} Q^4 \ldots \end{array} \right\} \quad (3)$$

$A(t)$ is defined in terms of the second and fourth moments of the particle displacement

$$A(t) = \frac{3\langle \Delta r^4(t) \rangle}{5\langle \Delta r^2(t) \rangle^2} - 1 \quad (4)$$

such that for Gaussian behaviour $A(t) = 0$. A non-zero non-Gaussianity parameter may relate to an anharmonic potential, local anisotropy of the environment or a heterogeneous environment.

The data in Figure 3b may now be used in order to experimentally determine the time dependent non-Gaussianity parameter. For this purpose Eq.[3] may be rewritten in the following form

$$-\frac{6}{Q^2} \ell n\, S_{self}(Q,t) = \langle \Delta r^2(t) \rangle - \frac{1}{12} A(t) \langle \Delta r^2(t) \rangle^2 Q^2 \quad (5)$$

plotting the MSD for constant t vs. Q^2 allows an extraction of both, the true mean square displacement as well as $A(t)$. Figure 4 presents the such obtained non-Gaussianity parameters for the 35% PEO sample at different temperatures. The insert displays representative plots according to Eq.[5] which allow an extraction of $A(t)$ as well as of the "true" mean square displacement. During the first 0.5ns with increasing time $A(t)$ decreases very strongly and then reaches a nearly constant level. Comparing the results for different temperatures we realize that the temperature dependence of $A(t)$ is only weak. However, as one would expect with increasing t the non-Gaussianity parameter becomes smaller. The same effect is also observed if the amount of PEO is increased in the blend. Eq.[5] may also be used in order to correct the apparent mean square displacements for non-Gaussianity.

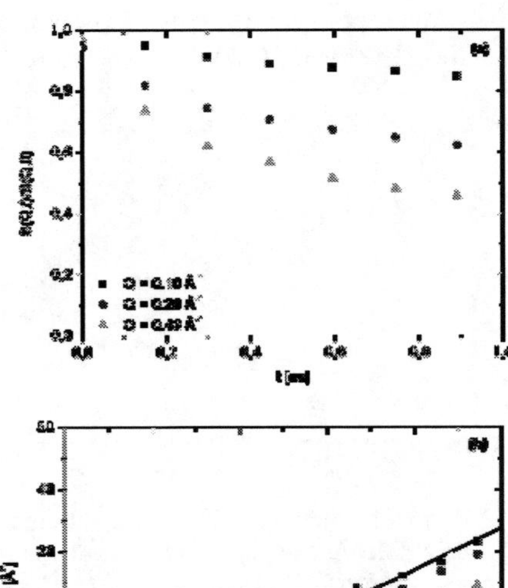

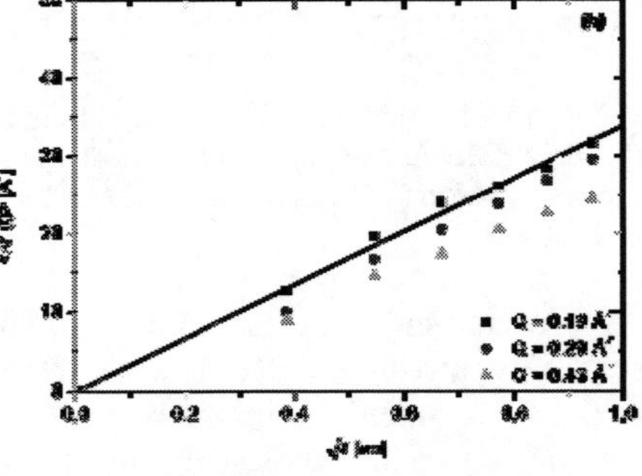

FIGURE 3. (a) Intermediate scattering function for the PEO component in the d-PMMA$_{0.65}$ h-PEO$_{0.35}$ sample at 375K from a Fourier transformation of the IN16 backscattering data. (b) Mean square displacements obtained from the data displayed in (a) in applying Eq.[1]. The solid line shows the expected $\sqrt{t}$ -law.

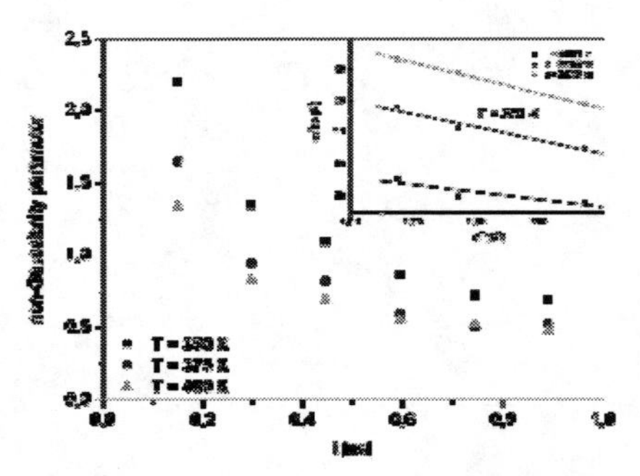

FIGURE 4. Time dependent non-Gaussianity parameter $A(t)$ for the d-PMMA$_{0.65}$ h-PEO$_{0.35}$ sample at different temperatures. The insert shows representative plots according to Eq.[5] which were used in order to extract $A(t)$.

As an example Figure 5 displays the "true" mean square displacements corresponding to the data in Figure 3. As may be seen the correction procedure is very effective resulting in consistent MSD values for all times and Q values. Furthermore the data measured at the smallest Q are nearly identical to the corrected results. There remains an assessment of the time dependence of $A(t)$. The strong decrease of $A(t)$ with increasing time indicates that in this time frame we observe a transition from a more heterogeneous to a more homogenous dynamics like in the decaging process characteristic for a glassy behaviour [13].

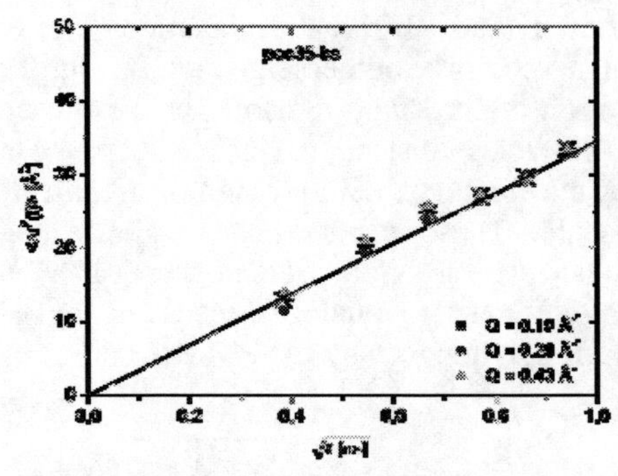

FIGURE 5. "True" MSD for the d-PMMA$_{0.65}$ h-PEO$_{0.35}$ sample at 375K obtained from the correction procedure for different Q-values. The stars represent the "true" MSD's while the other symbols give the corrected Q-dependent results. The solid line shows the expected $\sqrt{t}$ dependence.

In her thesis K. Niedzwiedz [11] has shown that qualitatively the observed behaviour may already be produced by a particle jumping in a double well potential with Gaussian distributions of particle probabilities at each site.

At the smallest Q value the non-Gaussianity corrections are nearly negligible. Thus, so far we may state that the intermediate scale dynamics of PEO in a matrix of PMMA at temperatures where PMMA is nearly frozen – at 400K for a 35% PEO system the PMMA local relaxation time is about 10ns compared to 6ps for PEO – is in good agreement with Rouse dynamics characterized by an average relaxation rate [5].

LOCAL PEO DYNAMICS: THE PICOSECOND REGIME

Figure 6 compares Fourier transformed spectra from pure PEO and d-PMMA$_{0.65}$ h-PEO$_{0.35}$ obtained from the time-of-flight (TOF) instrument FOCUS at 400K. For the blend material also Fourier transformed backscattering data are included [6].

The data were taken at $Q = 19\text{nm}^{-1}$. At this large Q neutron scattering is sensitive to the more local dynamics in a range shorter than the interchain distances (the structure factor maximum in PMMA is observed at $Q = 8\text{nm}^{-1}$ in PEO around 15nm^{-1}). The data from pure PEO may be described well by a stretched exponential with a stretching exponent $\beta = 0.5$ covering the time range from subpicoseconds to about 10ps. Compared to the spectrum of pure PEO the blend data are strongly stretched demonstrating a broad distribution of local relaxation processes in the random PMMA environment. Both spectra merge around $T = 0.4$ps indicating the transition to microscopic dynamics which is not altered by blending.

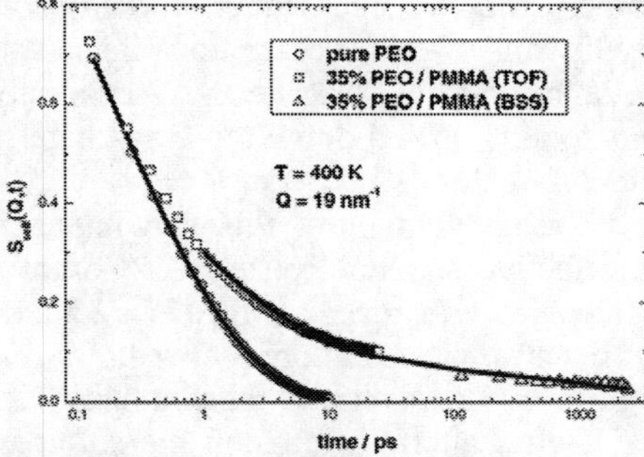

FIGURE 6. Fourier transformed TOF data from pure PEO and 35% PEO in a PMMA matrix (d-PMMA$_{0.65}$ h-PEO$_{0.35}$) at $T = 400$K. For the blend system Fourier transformed backscattering data are also included. The solid line through the PEO data describes a fit with a stretched exponential ($\beta = 0.5$). Line through the blend data, see text.

From the remaining decay we realize that a shift of the characteristic time of 1ps in pure PEO to about 6ps for this composition is found – in excellent agreement with NMR data and also the ratio of the corresponding Rouse rates in table 1.

LONGER TIME DYNAMICS: THE SINGLE CHAIN DYNAMIC STRUCTURE FACTOR

In order to increase the time sensitivity of the neutron scattering experiments neutron spin echo (NSE) needs to be employed extending the accessible time range by two orders of magnitude compared to backscattering. In order to achieve the necessary intensity the experiment has to switch from incoherent to coherent scattering measuring thereby the single chain dynamic structure factor. This may be achieved in replacing the purely protonated PEO in the blend by a mixture of h-PEO/d-PEO retaining the overall PMMA/PEO ratios. For all samples 12.5% hydrogenated PEO was used. Deuterated PEO nearly matches the scattering length density of pure PMMA such that the contrast arises from the protonated PEO against the rest. We note that under such circumstances also an elastic component arises which originates from cross terms between the protonated PEO and the deuterated PMMA. This elastic contribution to the scattering signal was calculated by dynamic RPA for an incompressible ternary system.

Figure 7 displays measurements of the single chain dynamic structure factor for the sample containing 35% PEO at 400K [6]. The shaded region in Figure 7 indicates the elastic contribution determined by dynamic RPA calculations. This contribution was considered explicitly in all further evaluations. The dotted lines in Figure 7 are a result of a Rouse description based on the average Rouse relaxation rates obtained from the backscattering data shown in table 1. Obviously this description fails strongly predicting a by far too fast decay. Apparently the motion is strongly slowed down towards longer times or larger length scales.

We may roughly quantify this slowing down in fitting an effective smaller Rouse rate. Compared to the 1ns scale such a fit reveals a further retardation of the PEO dynamics by factors between 4 and 20 depending on PEO content and temperature. Recently forced Rayleigh scattering diffusion measurements of dilute short PEO chains in a PMMA matrix displayed an even much stronger enhancement of the associated effective friction coefficient [14]: to $\zeta = 10^{-5} N/m$ compared to the here observed value of $\zeta = 3.2\ 10^{-9} N/m$ measured at a scale comparable to the size of the small chains in the diffusion experiment. The global dynamics of chain diffusion on a μm scale studied in a Rayleigh scattering experiments is apparently not related to the chain modes on a 80ns time and about 1.5nm spatial scale addressed with neutrons.

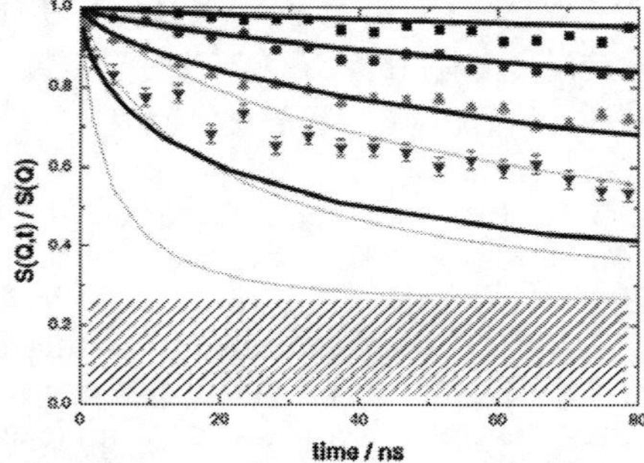

FIGURE 7. Single chain dynamic structure factor of the 35% PEO/PMMA system at $T = 400K$. Q values are 1, 1.5, 2, and 3nm^{-1} from top to bottom. The dashed region represents the elastic contribution of 0.265 as calculated by dynamic RPA [6]. The dotted lines illustrate the Rouse theory with a Rouse relaxation rate obtained from the backscattering data. The solid lines show the results of the Rouse model with random friction.

One could conclude that the neutron observations at longer times may relate to confinement, where locally fast motions could take place but globally the dynamics would be strongly reduced. In this picture the hardly moving PMMA would create "pockets of mobility" where PEO may move. We may ask whether the NSE data bear evidence for confinement effects exhibiting a characteristic confinement length. Figure 8 displays the presentation of the data in Figure 7 as a function of the Rouse scaling variable combining length and time scales. If there would be a characteristic length then the Rouse scaling i.e. the merging of all Q values to one master curve would break down and different relaxation curves for different Q vectors would emerge as it its e.g. the case for reptation or the chain dynamic in the presence of random obstacles. Instead, all data superimpose to a common master curve with no sign of a characteristic length scale. These results bear evidence that the constraints on the PEO motion in the PMMA matrix are not dominantly random obstacles but rather random friction imposed by the PMMA.

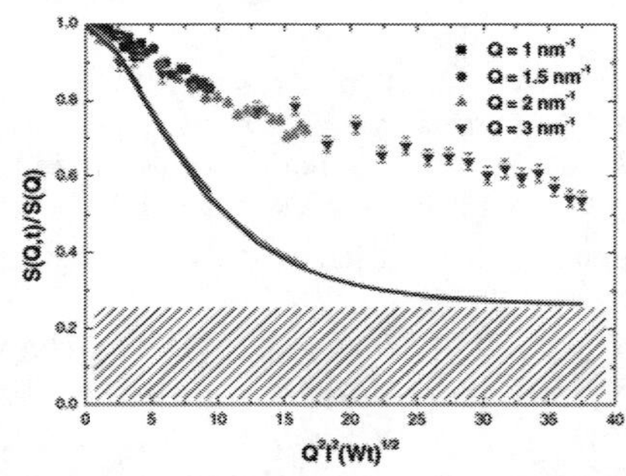

FIGURE 8. The same data as in Figure 7 in the Rouse scaling representation. The dashed region is again the elastic contribution, the lines represent the Rouse theory with a friction factor as obtained from the backscattering experiment.

RANDOM ROUSE MODEL

Summarizing what we have learned so far, we have to find a model which is able to describe

- the apparent strong discrepancies between self and single chain dynamics measured in different space-time-frames. While in the 1nm/1ns range the Rouse model gave a good description, the single chain dynamics at longer times was found to be strongly retarded;
- that there is no characteristic length scale for a PEO motion;
- that at local scales a very broad distribution of relaxation times is found.

In the following we show that a model of a Gaussian chain where the beads undergo random friction with a distribution of friction coefficients close to the distribution of relaxation times observed at high Q is able to resolve most of the apparently contradictory experimental findings. In order to calculate the dynamic structure factor for such a chain we solve a Langevin equation for a polymer chain with N beads and segmental friction coefficients distributed randomly over the Rouse chain. For the distribution function we used a lognormal distribution.

What will be the effect of such a random distribution of mobility? Imagine a chain, where each segment fluctuates with its own speed. At short times where the fluctuations are small (MSD < 1nm^2) segments only feel the presence of their next neighbours. If we have only a small number of sites with high friction then most of the segments at short times are not affected and move according to their friction environment. Thus, if we look at short times the measurement observes many of those regions and the randomness of the friction factors lead to the observation of an average value of the randomly distributed friction factors. At longer times the situation changes. A fast segment will be retarded attempting to reach larger distances by a slow segment in the neighbourhood undergoing high friction. In this way the sites of higher frictions will gain more and more importance. Even if there is only one segment in the whole chain which is slow, the full chain motion will be retarded.

To calculate the structure factor of a Rouse chain with random friction we solve the Rouse equation with the above specified random distribution of mobilities. This leads to a set of eigenvalues and eigenvectors which may be obtained by the usual normal mode Ansatz. The subsequent calculation of the dynamic structure factor is known from standard Rouse theory. The main ingredient of the calculation is a set of beat mobilities following a lognormal distribution centred around the average mobility obtained from the backscattering data. The only adjustable parameter is the width σ. The eigenvalues are then obtained by standard numerical routines.

Structural averaging is realized in creating several chains and computing the average of the structure factors. The solid lines in Figure 7 show the best results which could be achieved for the 35% PEO/PMMA sample at $T = 400\text{K}$ using $\sigma = 1.6$. We note that for the three lowest Q values the calculations are in a very good agreement with the experimental data. For the highest Q value the data are reproduced only for short times. The origin of this significant deviation at the highest Q and longer times is still unclear, though it could be related to additional confinement effects imposed by the matrix. The thus obtained widths parameters obtained for different compositions and temperatures are given in table 1. We note that the widths of the distribution function increases with increasing PMMA concentration and within one composition σ increases with decreasing temperatures. We further note that this approach also reproduces the short time Rouse behaviour as observed on the backscattering instrument in a very consistent way.

Having in mind that at high Q we observe directly the distribution of relaxation times (Figure 6), we now compare the results from the random Rouse model with the time-of-flight (TOF) spectra displayed in this figure. Again we shift the average relaxation rate from the backscattering evaluation relative to a pure PEO and take also into account a small PMMA contribution which we obtain directly from high Q backscattering data. Then we convolute the thus shifted PEO spectra with the lognormal distribution function using the result of $\sigma = 1.6$ from the random Rouse model.

$$\frac{S(Q,t)}{S(Q,0)} = \frac{1}{\sqrt{2\pi\sigma^2}} \int d(\log\sigma) \exp\left(-\left(\left(\log(\tau) - \log(\tau_{av})\right)^2 / 2\sigma^2\right)\right) \times \exp\left(-\left(\frac{t}{\tau}\right)^\beta\right) \quad (6)$$

where τ_{av} is the average relaxation rate. We obtain the solid line in Figure 6 in perfect agreement with the observed spectrum. Having in mind that σ has been obtained from the NSE data and the shift of the average relaxation rate from the backscattering measurements, this agreement combines data over five orders of magnitude in time. The large difference between the measured diffusion motion and the internal chain dynamics most likely also relates to the broad distributions of mobilities. Diffusion is dominated by the lowest existing mobility and rare very low mobilities will determine the diffusion coefficient.

LARGE SCALE DYNAMICS IN BIOPOLYMERS

Now, with the study of polymer dynamics reaching some maturity, the next challenge will be to unravel the large scale motion of biopolymers and to find out to what extend these dynamics play a role in their function. While at present the conformational dynamics on local scales have been successfully approached by e.g. time dependent crystallography [7], large scale dynamics such as protein domain motions remain basically untouched experimentally, because of the lack of techniques to study these large scale correlated motions. A first successful NSE approach on the dynamics of thermus acquaticus polymerase (Taq) has been reported recently [8]. Here we concentrate on alcohol dehydrogenase (ADH) which other than Taq polymerase in solution forms a tetrameric aggregate [9]. The alcohol dehydrogenases are enzymes that are present in many organisms, allowing the interconversion between alcohols and ketones. In humans it catalyzes the oxidation of ethanol and is always present in a form of a dimer allowing thereby the consumption of alcoholic beverages. In yeast it is at the basis of the fermentation process and converts acetaldehyde into ethanol. In the process the cofactor NAD is needed assisting the oxidation reaction at the Zinc catalytic site.

Figure 9 displays a schematic structure of a dimer based on crystallographic data. The figure displays the dimer with the two monomeric units clearly visible. Each monomer is build from two domains with a small opening in between where the cofactor NAD is placed. One of the questions to answer is to what extend the cofactor may modify the domain dynamics.

FIGURE 9. Dimer of Alcohol dehydrogenase. The molecule presented by spherical caps is the NAD cofactor used in the chemical reaction.

In our studies we used ADH from yeast which forms a tetramer structure. The crystallographic data suggest a crossed arrangement of the two dimers. In order to verify whether in solution a similar tetrameric aggregate is present neutron small angle scattering experiments were performed as a function of concentration. Figure 10 presents the SANS data together with a comparison with a number of models for solution structures. Let us commence with the insert: here for different concentrations the low Q data are presented and there the increasing influence of the structure factor on the SANS data is visible. In the main figure the low concentration data are presented over the full Q range and are compared with different structural models. The solid line reflects the crystal structure which appears to be in perfect agreement with the crystal data. Testing the sensitivity of the approach we compare with a situation, where we bring the two dimers in a planar configuration into close contact. The results are presented by the dashed line which in particular at high momentum transfers leads to an underestimation of the scattering data. Further more using the programme DAMMIN [15] which builds SANS scattering curves from an arrangement of small mesoscopic spheres also allows a perfect fit of the experimental data (result superimposes with the crystal results) and confirms the crystal based perpendicular arrangement of the two planar structures. Thus, we conclude that as in the crystal the ADH tetramer in solution is present in a crossed dimer configuration.

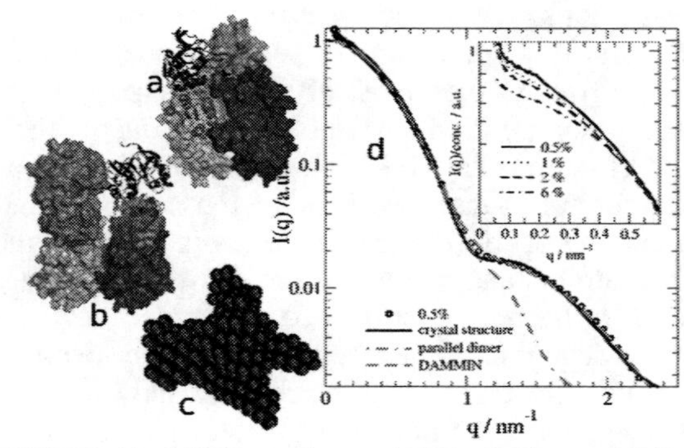

FIGURE 10. SANS results on ADH solutions of different concentration(d). Insert: concentration dependent results. The different curves in the main figure display modelling results for various dimer arrangements a) solid line: crystal structure; b) thin solid line: DAMMIN result c) dashed dotted line: planar dimers at close contact.

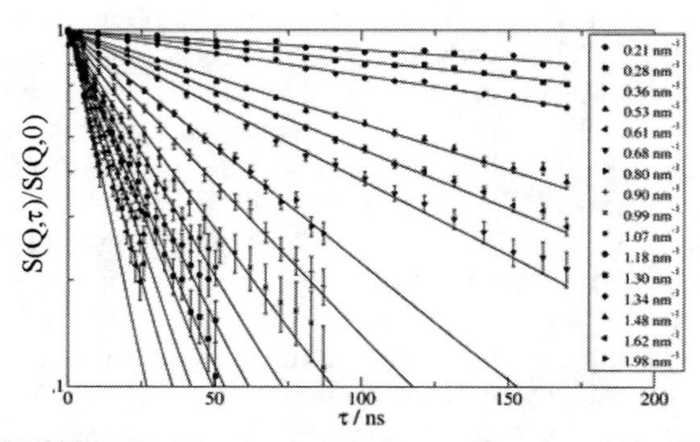

FIGURE 11. Neutron spin echo results on a 1% ADH resolution at 5°C without cofactor for various momentum transfers.

The overall molecular diffusion was measured by dynamic light scattering and was found to be independent of concentration. The translational diffusion coefficient amounts to $D_{DLS} = (2.35 \pm 0.2)10^{-2}$ nm^2/ns at 5°C corresponding to a hydrodynamic radius of $R_H \cong 4.5$ nm.

Neutron spin echo data were measured at different concentrations with and without the cofactor NAD. Figure 11 displays NSE results in a logarithmic fashion for a large number of different momentum transfers Q. In each case single exponential fits are included showing that the measured structure factor with great accuracy may be described in terms of a single exponential decay.

Comparing the covered Q range with the SANS data, one realizes that both the range of the structure factor, where intermolecular interactions are important as well as the regime of internal structure are covered.

Considering that all data may be described in terms of a single exponential decay, we approximate the spectra in terms of a first cumulant expansion as

$$\ell n \frac{S(Q,t)}{S(Q,0)} = -\Gamma(Q)t + \frac{1}{2}K_2 t^2 \; ... \tag{7a}$$

where the decay rate of the dynamic structure factor is

$$\Gamma(Q) = -\lim_{t\to 0} \frac{\partial}{\partial t} \ell n\left[S(Q,t)\right] \tag{7b}$$

This gives an effective diffusion coefficient

$$D_{eff}(Q) = \frac{\Gamma(Q)}{Q^2} \tag{7c}$$

Figure 12a presents the thus obtained effective diffusion coefficients as the function of Q for the different concentrations with and without NAD. The line at low Q indicates the level of the light scattering result.

The experimental data show a strong Q modulation exhibiting a maximum around $Q = 1\text{nm}^{-1}$. There is significant concentration dependence, though at low Q the data are in agreement with the concentration independent light scattering results. We also see that beyond the statistical error in the low Q flank of the 5% data the relaxation without NAD is somewhat faster than that including the cofactor. Thus, in the dynamics of ADH on the scale of the molecule itself we observe significant contributions beyond translational diffusion.

First, we concentrate on the low Q data which are affected by the interactions between the molecules. There the effective diffusion coefficient relates to the diffusion coefficient D_0 at infinite dilution by

$$D_{eff}(Q) = D_0 \frac{H(Q)}{S(Q)} \tag{8}$$

where $H(Q)$ is the hydrodynamic factor and $S(Q)$ is the structure factor. With the measured structure factor the data may immediately be corrected for $S(Q)$. This correction removes the low Q increase of the 5% data but leaves the results at higher Q untouched. The hydrodynamic factor cannot be measured directly. A first approximation in terms of a Percus Yevic model shows that (i) at a 1% level the correction factor $H(Q)/S(Q)$ leaves the experimental data practically untouched and (ii) at 5% the correction is somewhat weaker than the experimentally observed effect. Beyond $Q = 0.7\text{nm}^{-1}$ the ratio of $H(Q)/S(Q)$ remains constant. Thus, the observed higher Q structure is entirely determined by intra aggregate effects.

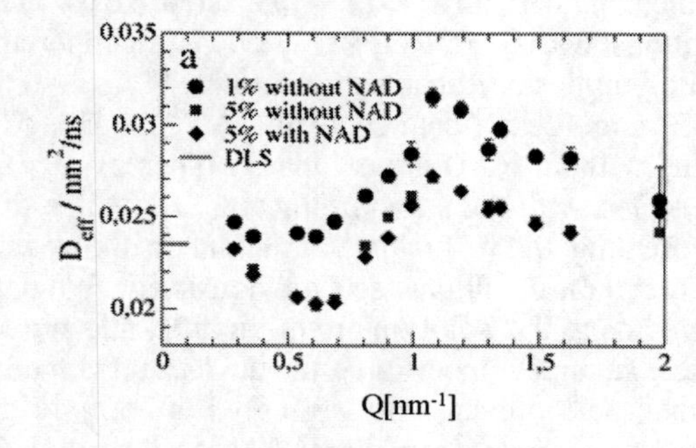

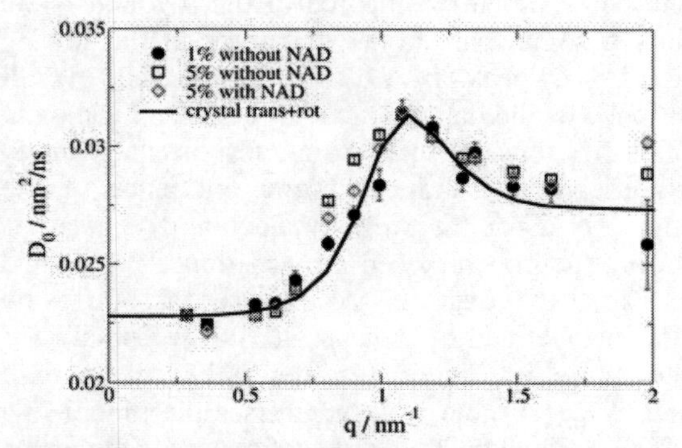

FIGURE 12. (a) Effective diffusion coefficient $D_{eff}(Q)$ for 3 different solutions of ADH, (b) comparison of $D_{eff}(Q)$ corrected for $S(Q)$ with different models (see text).

The prime reason for a Q dependent structure in $D_{eff}(Q)$ are rotational motions of the molecule. In a first cumulant approximation the effective diffusion coefficient of a rigid body undergoing translational and rotational diffusion has the following form [16]

$$D_{eff}(Q) = \frac{k_B T}{Q^2} \times \frac{\sum_{jK} \left\langle b_j e^{iQ\underline{r}_j} \begin{pmatrix} Q \\ Q\times \underline{r}_j \end{pmatrix} \overleftrightarrow{H} \begin{pmatrix} Q \\ Q\times \underline{r}_K \end{pmatrix} b_K e^{-iQ\underline{r}_K} \right\rangle}{\sum_{jK} \left\langle b_j e^{iQ\underline{r}_j} \; b_K e^{-iQ\underline{r}_K} \right\rangle} \tag{9}$$

Here $\underline{r}_i$ and $\underline{r}_K$ are the atomic coordinates, b_i and b_K the corresponding neutron scattering lengths and $\overleftrightarrow{H}$ the mobility tensor. The sums run over all atoms of the molecule or molecular aggregate and the pointed brackets indicate an ensemble average. The denominator resembles the aggregate form factor. The mobility matrix $\overleftrightarrow{H}$ is a 6x6 tensor involving translational and rotational parts including a translational rotational coupling. The further evaluation of Eq.[9] was performed using the program HYDROPRO created by Garcia de la Torre and coworkers [17]. The rotational averaging was performed numerically with small step sizes. The calculations with HYDROPRO need as an input the crystallographic coordinates of all atoms.

In Figure 12b all data sets for $D_{eff}(Q)$ at 1%, 5%, with and without NAD are compared. Thereby $D_{eff}(Q)$ was rescaled with the structure factor. We realize that after rescaling the diffusion coefficient in the region above $Q = 0.6\text{nm}^{-1}$ all data sets are consistent. We note that the data at 5% solution are of significantly higher statistical accuracy underlining the deviation from pure rotational motion even more strongly.Furthermore the sample with cofactor compared to that without NAD exhibits a somewhat slower dynamics in the low Q flank of the peak in $D_{eff}(Q)$. Given the overall consistency of the data points and the size of the error bars this difference in the dynamic response appears to be significant and indicates some influence of the binding of NAD on the overall molecular dynamics

Using the crystal based crossed dimer model and normalizing the data to the translational diffusion coefficient obtained by light scattering we obtain the dashed curve in Figure 12b describing the overall shape of $D_{eff}(Q)$ quite well but misses the proper low and high Q-flanks. These deviations indicate some additional dynamics which is not depicted by the rigid body rotation. At present we evaluate these differences in terms of a normal mode analysis indicating that they are due to fluctuations of the outer more flexible part of the tetramer [9]. In such a picture the differences in the low Q flank of the data from the aggregate with and without the cofactor seem to indicate a significant difference of the configurational stiffness of the protein depending on the cofactor presence.

CONCLUSION AND OUTLOOK

We have presented some representative results from neutron spin echo spectroscopy on the dynamics of macromolecules. In the case of synthetic polymers we have displayed some recent results on the unusual dynamics within a blend of miscible polymers exhibiting very different component glass transition temperatures. The study revealed detailed information on the space time evolution of the fast component dynamics (PEO) in a nearly frozen PMMA matrix at intermediate scales. Both the self as well as the chain dynamics were addressed. Using incoherent neutron scattering, the mean square displacement of the PEO segments were followed in a nanosecond time window with displacements up to in the order of 1nm. A Q dependent measurement of the self correlation functions allowed direct access to the time dependent non-Gaussianity parameter in this time regime - a first experimental exploit so far!

While the local dynamics was found to be fast, the chain dynamics which could be followed to about two orders of magnitude longer times appeared to be significantly slower. An appropriate scaling of the data did not show any characteristic length of confinement separating small and larger scale dynamics. Implementing a distribution of friction coefficients into a Rouse model, we were able to quantitatively and consistently describe the local, the intermediate as well as the more global dynamics.

Compared to the investigations on the dynamics of the synthetic polymers the study of the large scale relaxation dynamics of biopolymers is still in its early stages. We have presented some first experimental data on the component fluctuations of a tetrameric aggregate formed by alcohol dehydrogenase. It became possible to directly measure the Q dependent effective diffusion coefficient which bears information on the detailed rotational diffusion dynamics. This quantity is very sensitive to the actual solution structure and underpins the results from SANS solution structure determination. Furthermore, some additional dynamic effects appear to be related to motions of the outer more flexible parts of each dimer. Finally, the binding of the cofactor NAD reduces the dynamic response in the low Q -flank of $D_{eff,}$ an effect possibly related to changes in the configurational stiffness induced by the cofactor.

Aside of the earlier published dynamic data on Taq polymerase [8], where significant domain fluctuations have been observed this is the second experiment of its kind. Future experiments will need to resolve the internal dynamics of further proteins and protein complexes trying to resolve the different relaxation modes separately. Such experiments need to be accompanied by computer simulation, in order to enhance the level of interpretation. Furthermore, the experiments need to address proteins where domain motion is functionally important. We hope that in the future such NSE studies will make an important contribution to a better understanding of protein function based not only on the structure but also on the dynamics.

REFERENCES

1. For a recent review, see: D. Richter, M. Monkenbusch, A. Arbe and J. Colmenero, *Neutron Spin Echo in Polymer Science, Advances*

in Polymer Science, Springer Publisher Berlin, Heidelberg, 2005, Vol. 174.
2. Y. He, T.R. Lutz and M.D. Ediger, *J. Chem. Phys.* **119**, 9956-9965 (2003).
3 C. Lorthioir, A. Alegría and J. Colmenero, *Phys. Rev. E* **68**, 031805-031805-9 (2003).
4. T.R. Lutz, Y. He, M.D. Ediger, H. Cao, G. Lin and A.A. Jones, *Macromolecules* **36**, 1724-1730 (2003).
5. A.-C. Genix, A. Arbe, F. Alvarez, J. Colmenero, L. Willner and D. Richter, *Phys. Rev. E* **72**, 031808-031808-20 (2005).
6. K. Niedzwiedz, A. Wischnewski, M. Monkenbusch, D. Richter, A.-C. Genix, A. Arbe, J. Colmenero, M. Strauch and E. Straube, *Phys. Rev. Lett.* **98**, 168301-168301-4 (2007).
7. F. Schotte, M. Lim, T.A. Jackson, A.V. Smirnov, J. Soman, J.S. Olson, G.N. Jr Philips, M. Wulff and P.A. Aninfrud, *Science* **300**, 1944-1947 (2003).
8. Z. Bu, R. Biehl, M. Monkenbusch, D. Richter and J.E. Callaway, *Proc. Natl. Acad. Sci. U.S.A.* **102**, 17646-17651 (2005).
9. R. Biehl, M. Monkenbusch, D. Richter, B. Hoffmann and R. Merkel, *to be published.*
10. T.P. Lodge and T.C.B. McLeish, *Macromolecules* **33**, 5278-5284 (2000).
11. K. Niedzwiedz, *PhD Thesis: Polymer dynamics in miscible polymer blends*, University of Münster, Germany, 2007.
12. R. Zorn, Phys. Rev. **B55**, 6249-6259 (1997)
13. A. Arbe, J. Colmenero, F. Alvarez, M. Monkenbusch, D. Richter, B. Farago and B. Frick, Phys. Rev. Lett. **89**, 245701 (2002)
14. J.C. Haley and T.P. Lodge, *J. Chem. Phys.* **122**, 234914-234914-10 (2005).
15. D.I. Svergun, *Biophys. J.* **76**, 2879-2886 (1999).
16. N. Brown, *Dynamic Light Scattering, Monographs on the Physics and Chemistry of Materials*, Oxford Science Publications, Oxford, Vol. 49, 1993.
17. J. Garcia de la Torre, M. L. Huertas and B. Carrasco, *Biophys. J.* **78**, 719-730 (2000).

Thermal and Dielectric Behavior of Liquid-Crystalline Polybutadiene-Diols with Mesogenic Groups in Side Chains

Michal Ilavský [a,b], Jan Nedbal[a], Lenka Poláková[b], and Zdeňka Sedlákova[b]

[a]*Faculty of Mathematics and Physics, Charles University, 180 00 Prague 8, Czech Republic*
[b]*Institute of Macromolecular Chemistry, Academy of Sciences of the Czech Republic, 162 06 Prague 6, Czech Republic*

Abstract. Ordered polybutadiene diols (LCPBDs) with the comb-like architecture were prepared by radical reaction of a thiol with azobenzene mesogenic group with double bonds of telechelic HO-terminated polybutadiene (PBD); several polymers with various initial molar ratios of thiol to double bonds of PBD, R_0, in the range from 0 to 1 were prepared. DSC, polarizing microscopy, WAXS and dielectric relaxation spectroscopy (DRS) were employed to investigate their thermal and dielectric behavior in relation to morphology. DRS of the LCPBDs have revealed both collective and individual dynamic motions of molecules. Secondary β- and segmental α-relaxation were observed in unmodified PBD. In the LCPBDs, two secondary γ - and β - and two high-temperature α- and δ-relaxations were observed and assigned to specific molecular motions; all relaxations were analyzed and discussed in terms of time scale (Arrhenius diagram), magnitude (relaxation strength) and shape of the response.

Keywords: Dielectric spectroscopy , Liquid-crystalline polybutadiene-diols, Mesophase transitions
PACS: 71.20.Rv

INTRODUCTION

Introduction of rigid mesogenic groups into a polymer (backbone or side chains) usually leads to liquid-crystalline polymers (LCP), which show an intermediate state of aggregation between the crystalline and amorphous structures [1]. In liquid-crystalline side-chain polymers (LCSCPs) the mesogens are decoupled from the main chain by a spacer, which usually consists of aliphatic segments [2]. From polymers, processable films could be developed for photonic, ferroelectric and antiferroelectric applications and second harmonic generation in non-linear optics (NLO) [2,3]. Such systems, especially with azobenzene moieties in side chains, form a class of photochromic materials in which usually birefringence and optical dichroism, based on cis/trans photoisomerisation of azochromophores, can be initiated by polarized light since the transition dipole moment of azobenzene groups is oriented along the molecular axis of elongated trans-isomer. Stability of dichroism and birefringence depend on molecular motions in polymer. Dielectric relaxation spectroscopy in broad frequency and temperature regions is one of the most effective tools for characterization of molecular motions in polymer systems [4] as various relaxation mechanisms can be expressed in terms of relaxation times and their temperature dependences, magnitude and shape of dielectric response.

In this paper physical properties of a new comb-like LC polybutadiene-diols (LCPBDs) with azobenzene moieties in the side chain mesogens are described; LCPBDs were made by radical addition of thiol group onto the double bonds of PBD with various initial molar ratios of thiol to double bonds. Thermal structural transitions in LCPBDs are investigated by DSC, WAXS and polarizing microscopy; the broad-band dielectric spectroscopy is used for investigation of molecular dynamics in these LCPBDs.

EXPERIMENTAL

Preparation and Characterization of LCPBDs

PBD: Telechelic OH-terminated polybutadiene diol - Krasol LBH 3000 (PBD) (Kaučuk Kralupy, M_n~2400, 60 mol-% of 1,2 and 40 mol-% of 1,4 monomer units, number-average OH functionality f_n=2) of the structure:

CP982, *Complex Systems, 5th International Workshop on Complex Systems*
edited by M. Tokuyama, I. Oppenheim, and H. Nishiyama

was purified by addition of silica gel Silpearl (Glass Works Kavalier).

LC Thiol: We have synthesised of LC thiol - 5(4[(4-(octyloxy)phenyl)azo]phenoxy)pentanethiol (TH1) of the structure:

More details about TH1 synthesis can be found in [5].

LCPBDs: Liquid-crystalline polybutadiene-diols with the comb-like architecture were prepared by reaction of the SH groups of thiol TH1 with azobenzene grouping with the double bonds of PBD. Several polymers with various initial molar ratios of thiol to double bonds of PBD, R_0, in the range from 0 to 1 were made; the initiator - 2,2'-azobis(2-methyl propionitrile) (ABIN) was used at radical reaction.

Experimental degrees of modification, D_m ([bound thiols]/[double bonds], mol/mol), after the addition and purification, were determined from elemental analysis using the weight fractions of sulfur bonded in LCPBDs. The ^1H and ^{13}C NMR spectroscopy (300.1 MHz and 75 MHz, respectively, 60 °C) were also used for $D_{m(NMR)}$ determination; for calculation the integrated intensity of the signal of OCH_2 of LC-thiol protons at 4.0 ppm (or the signal of aromatic protons) was used. In this case also the detailed structure of LCPBDs (amount of 1,2 and 1,4 butadiene units, hydrogenated-like PBD units and OH end groups [6]) could be evaluated. The number- (M_n) and weight-average (M_w) molecular weights were determined by GPC (modular LC system with refractive index detection, column 30×8 SDV 10000) calibrated with PS standards. THF was used as solvent and measurements were carried out at ambient temperature. From M_n, values the degrees of modification, $D_{m(GPC)}$ were determined.

Dielectric, DSC, WAXS and Polarizing Microscopy Measurements

Dielectric measurements were carried out with a Schlumberger 1260 frequency-response analyzer with a Chelsea dielectric interface, in combination with the Quatro cryosystem of Novocontrol. The dielectric permittivity $\varepsilon^*(f) = \varepsilon'(f)-i\varepsilon''(f)$ (f is frequency, ε' is the real (storage) and ε'' is the imaginary (loss) component) was measured in the frequency range 10^{-1}-10^6 Hz at temperatures between -150 and 130 °C upon heating (after cooling the sample to -150 °C from the isotropic state at 140 °C).

The frequency dependence of the complex dielectric function $\varepsilon^*(f)$ originates from fluctuations of molecular dipoles and/or dipoles induced by the charge separation at boundary layers inside the material (Maxwell-Wagner-Sillars polarization, MWS) or between the material and electrodes (electrode polarization), and from the propagation of mobile charge carriers [11]. The frequency dependence of the dipole contribution to the complex dielectric function, ε_d^*, is usually described by the non-symmetrical Havriliak–Negami empirical equation [7]

$$\varepsilon_d^* = \varepsilon_\infty + \Delta\varepsilon / [1 + i(f/f_r)^a]^b \quad (1)$$

with five, generally temperature-dependent parameters: high-frequency (unrelaxed) value of the real part of permittivity ε_∞, the relaxation strength $\Delta\varepsilon = \varepsilon_0 - \varepsilon_\infty$ (ε_0 being the relaxed low-frequency value of permittivity), frequency f_r corresponding to the most probable relaxation time τ_r ($2\pi f_r \tau_r = 1$) and two shape parameters a and b. The parameter f_r is related to the peak frequency f_m at which the loss component ε'' attains its maximum In the case of symmetrical Cole-Cole distribution (b = 1), the frequencies f_r and f_m are identical [16]. A computer program based on the Marquardt procedure [8] was developed for determination of all parameters from the frequency dependence of ε'' at various temperatures.

The temperature dependence of frequency f_m could be described either by the Arrhenius equation

$$f_m = f_\infty \exp(E_a / kT) \quad (2)$$

where E_a is the activation energy and f_∞ is the pre-exponential frequency factor, or by the Vogel-Fulcher-Tammann equation (VFT) [4,9]

$$\log f_m = \log f_\infty - \frac{B}{T - T_0} \quad (3)$$

where B is the apparent activation energy, f_∞ the pre-exponential frequency factor, and T_0 the Vogel temperature ($T_0 \sim T_g$ - 50 °C, where T_g is the glass transition temperature).

The frequency dependence of the conductivity contribution to the complex dielectric permittivity, ε_c^*, can be described by the following equation [9]

$$\varepsilon_c^* = (i \tfrac{f}{f_0})^\eta \quad (4)$$

where f_0 is an adjustable parameter and the exponent η would have the value -1 in the ideal case of pure time-independent dc conductivity.

The frequency dependence of the complex conductivity $\sigma^*(f)$ is given by the equation [9]

$$\sigma^*(f) = i2\pi f \varepsilon_0 \varepsilon^*(f) \quad (5)$$

Thermal properties were measured using a Perkin-Elmer differential scanning calorimeter DSC-7e. Reproducible data were collected on cooling and subsequent heating at a rate of 10K/min (Fig. 2).

Wide-angle X-ray diffractograms were taken on a HZG4A diffractometer (Freiberger Präzisions mechanik, Germany) using Ni-filtered CuKα radiation. For high-temperature measurements a heating chamber was attached. The texture of crystalline and LC phases was determined also by a polarizing optical microscope (Nicon Eclipse 80i, crossed polarizer's) equipped with a heating stage. The heating/cooling rate was 3 K/min.

RESULTS AND DISCUSSION

Modification of LCPBDs

Comparison of experimental degrees of modification, D_m, obtained from GPC, $D_{m(GPC)}$, sulfur content, $D_{m(S)}$, and ^{1}H NMR analysis, $D_{m(NMR)}$ in dependence on initial ratio R_0 is shown in Fig. 1. As expected, modification degrees D_m, determined by all three methods, increase with increasing initial ratio R_0. While for the lowest initial thiol ratios R_0 ($R_0 \leq 0.4$), $R_0 \sim D_m$, for higher R_0, $D_m < R_0$ were found. Lower D_m values suggest that at the higher R_0 values steric hindrance probably prevents the addition reaction and thiol is subsequently removed from polymer during purification. The lowest $D_{m(GPC)}$ values suggest that some of these dimmers remain in the modified PBD even after purification.

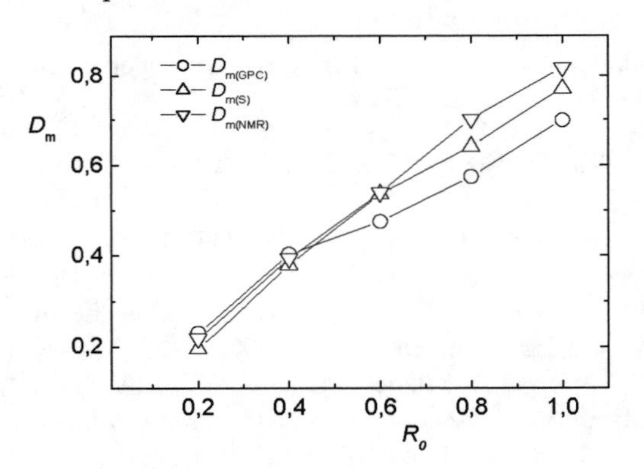

FIGURE 1. Determined degrees of modification D_m (from S content - $D_{m(S)}$, from GPC - $D_{m(GPC)}$ and from ^{1}H NMR - $D_{m(NMR)}$) in dependence on on initial ratio of thiol/double bonds - R_0

Thermal Behavior of Unmodified PBD, Thiol and LCPBDs

An example of measured DSC traces of neat thiol and LCPBD with $R_0 = 1.0$ is shown in Fig. 2. From Fig. 2 it follows that on cooling of the thiol first mesophase formation takes place at ~ 116 °C (ΔH_m ~ - 4.6 J/g, SmA phase in which the centers of mass of the rods are in one dimensional periodic order parallel to long axes of the rods); next transition is at 100 °C (ΔH_m ~ - 2.6 J/g, SmC phase with tilted molecules to long axes); two smectic phases has been usually found in the case that azobenzene grouping is present in LC

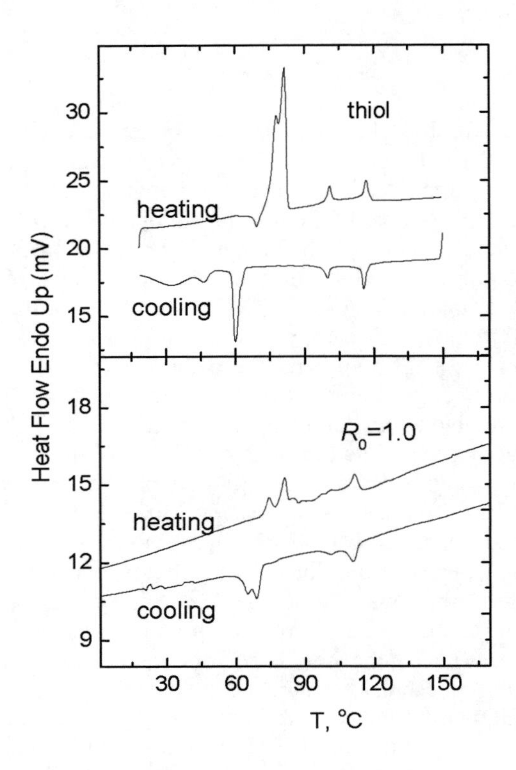

FIGURE 2. Example of measured DSC traces for neat thiol and LCPBDs with R_0 = 1.0. Cooling and heating rate was 10 K/min

organic molecules with flexible spacer [10]. WAXS measurements showed that at 95 and 77 °C diffractograms of thiol consist of one strong, narrow reflection the position of which corresponds to 3.04 nm and a broad amorphous halo. Theoretical atomistic simulations of thiol using software BIOSYM led to the length of ~ 3.2 nm, which agrees well with the found one. Due to this we assume that a smectic structures exist in thiol; this periodicity stays in crystalline phase at lower temperatures. The crystallization of TH1 on cooling takes place at ~ 60 °C (ΔH_m ~ -17.7 J/g). On subsequent heating, small recrystallization (Fig. 2) followed by melting of the crystalline phase at temperature ~ 81 °C (ΔH_m ~ 61 J/g) was observed. At temperatures higher than 85 °C, a SmC structure appears; this structure melts at 101 °C (ΔH_m ~ 2.7 J/g) to SmA mesophase which finally melts to an isotropic state at 117 °C (ΔH_m ~ 3.2 J/g).

From DSC traces of LCPBD with R_0 = 1.0 shown in Fig. 2 it follows that modification leads to simpler thermal behavior in comparison with neat thiol. On cooling only one mesophase seems to be formed at ~ 111 °C (ΔH_m ~ -8.2 J/g); crystallization starts at ~ 75 °C (ΔH_m ~ -12.4 J/g) and final structure formation takes place at ~ 65 °C (ΔH_m ~ -5.9 J/g). On heating crystalline structure melts at ~ 75 °C (ΔH_m ~ 3.7 J/g) and ~ 81 °C (ΔH_m ~ 11.4 J/g) to smectic mesophase which melts at ~ 111 °C (ΔH_m ~ 7 J/g). As expected, lower ΔH_m values were found for modified LCPBDs in comparison with those of neat thiol. The glass transition T_g could be observed on DSC thermograms only for neat PBD and LCPBD with R_0 =0.2; the higher modifications exhibit only crystal/ mesophase/isotropic state transitions (it is in agreement with dielectric results shown in Fig. xxx). As T_g for LCPBD with R_0 = 0.2 has increased for about 30 °C we believe that for LCPBDs with R_0> 0.2 the values of T_g have increased into temperature region of ordered phases formation. The amount of ordered phase in the polymers increases with increasing modification.

Dielectric Behavior of Unmodified PBD, Thiol and LCPBDs

All glass-forming polymers exhibit primary relaxation – α; this relaxation is assigned to motion of polymer main chain segments and is related to the glass transition. The temperature dependence of its relaxation time is described by the Vogel-Fulcher-Tammann (VFT) eq. (3), which describes the slowing down of the relaxation approaching to the glass transition temperature. Besides the α-relaxation, usually additional (secondary) relaxations (β- or γ-relaxations) of faster time scales take place at lower temperatures. The β- and γ-processes generally occur in the glassy state, and have different molecular nature compared with the α-relaxation [9]. These - secondary relaxations usually exhibit an Arrhenius temperature dependence of relaxation times (eq. (2)), in contrast to the much stronger VFT temperature dependence found for primary α-relaxation.

The overall dynamics in side chain liquid crystalline polymers is characterized by four dielectric relaxation processes, γ, β, α and δ, in the order of increasing temperature or decreasing frequency. The γ process is related to rotational fluctuations of the terminal groups of the side chain, the β relaxation corresponds to liberation fluctuations of mesogens around the long molecular axis and the α-relaxation is assigned to fluctuations of segments of the polymer main chain as they are observed in glass-forming systems. The δ-relaxation is assigned to liberation fluctuations of mesogens around the short molecular axis; presumably it is a rather multistep process with motional averaging rather than a 180 flip-flop jump of the mesogens [9].

Neat PBD and Thiol

A simple dielectric behavior with segmental α- and secondary β-relaxation was found for unmodified PBD in accord with previously published results [11].

Summarized results of HN parameters for TH1 in the temperature region from 0 to 110 °C are shown in Fig. 3 where the maxima of absorption peaks (log f_m), their magnitude ($\Delta\varepsilon.T$) and shape parameters a vs. reciprocal temperature are shown. The f_m vs. $1/T$ dependences of both high temperature (from 110 to 60 °C) absorptions are practically the same and do not

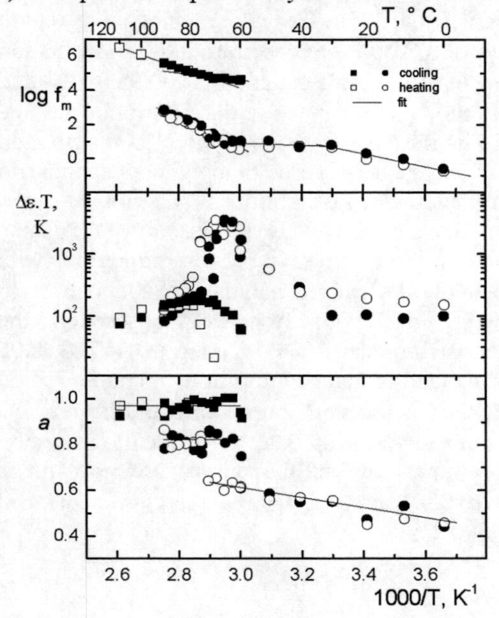

FIGURE 3. Temperature dependence of relaxation parameters: f_m, $\Delta\varepsilon$.T and shape parameter a permittivity for selected temperatures

depend on cooling or heating regime; both are described by the Arrhenius equation (eq. 2). For the first (high frequency) absorption we have found the activation energy E_a=0.13 eV (in the temperature range from 70 to 110 ° C).. The low frequency absorption is characterized by activation energy of 1.62 eV (60 °C– 90 °C); two absorptions correlate with two mesomorphic states of TH1 found in DSC measurements. Below 60 °C the crystallization starts and activation energies change from E_a=0.11 eV (30 – 60 ° C) to 0.71 eV (0– 30 ° C). Different energies ΔU are related to the changes in motion in crystalline structure of TH1 which were also observed in DSC, WAXS and polarizing microscopy measurements.

The magnitude $\Delta\varepsilon.T$ of high frequency absorption is roughly constant in temperature region from 110 to 80 °C; on further cooling the decrease in $\Delta\varepsilon.T$ is observed and absorption disappears at 60 °C. The magnitude of low frequency absorption is strongly dependent on temperature. While in temperature region from 90 to 75 °C it is comparable with high frequency one, in the temperature region from 75 to 60 °C, $\Delta\varepsilon.T$ suddenly increases more than one order of magnitude. As E_a substantially decreases in this temperature region we suppose that motion involves extended ordered regions in not well developed crystalline state. At temperatures lower than 60 ° C the molecular motion is hindered and the magnitude $\Delta\varepsilon.T$ decreased to previous values.

From Fig. 3 it can be seen that the high frequency absorption is narrow in the whole temperature region from 60 to 110°C with temperature independent a and b parameters of HN equation (a=0.95±0.02 and b=0.98±0.05); this means that the absorption can be described by the Cole-Cole rather than HN distribution function. In such a case motion corresponds to Debye's law of almost uniformly distributed polar units in SmA state of TH1. The low frequency absorption in SmC phase is also symmetrical with experimentally determined parameter b=0.998±0.1. In comparison with high frequency absorption the distribution is broader. Found parameter a=0.83±0.05 does not depend on temperature in temperature region 90–65 °C and is the same on cooling and heating. At temperatures lower than 60 °C the shape of absorption is still symmetrical but the curves are now much broader (a=0 .61 at 60 °C) and temperature dependent (a=0.44 at 0°C).

LCPBDs

Complex temperature and frequency dependences of permittivity were observed in the LCPBDs. In Fig. 4 an example of measured frequency dependences of the loss component of permittivity ε″ at various temperatures is shown for polymer with R_0=0.2. Two secondary β- and γ-relaxations were observed and characterized for all compositions; they were assigned to local motion of the octyloxy end groups of the side chains and to the motion of the azobenzene moieties of the side chains around their long molecular axis. At higher temperatures, α-relaxation was observed only for the composition with the lowest R_0 (0.2), shifted to higher temperatures/lower frequencies with respect to unmodified PBD. The δ-relaxation, assigned to liberation fluctuations of mesogens around their short molecular axis, was observed at temperatures higher than those of the α-relaxation in the LCPBDs with R_0 = 0.4, 0.6 and 0.8.

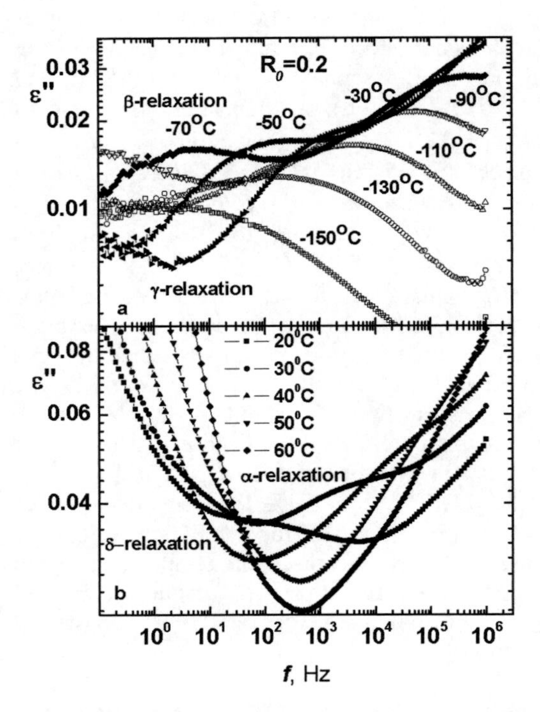

FIGURE 4. Dielectric loss component ε″ as a function of frequency for the LCPBD with R_0=0.2 at low (γ and β relaxation (a)) and high (α and δ relaxations (b)) temperatures

All relaxations were analyzed in terms of time scale (Arrhenius diagram), magnitude of the relaxation

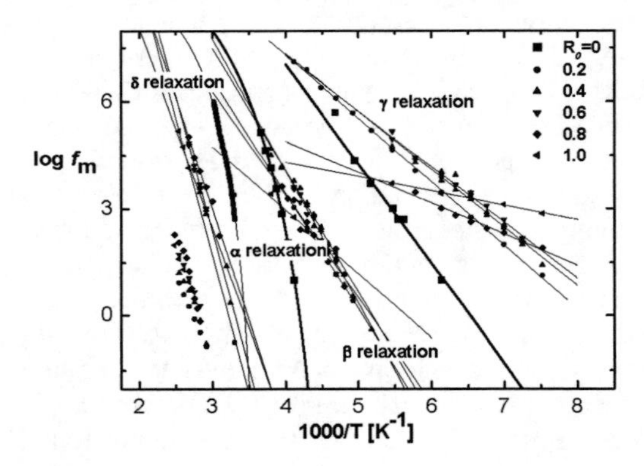

FIGURE 5. Logarithm of the peak frequencies f_m as a function of inverse temperature. The lines are fittings of the dielectric data to the VFT or Arrhenius equation (depending on the relaxation process)

strength and shape of the response. Calculated parameters E_a and f_m of eq. 2 are shown in Table 1.

TABLE 1. Parameters of Arrhenius equation for the dielectric relaxations of LCPBDs

[SH]/[BD]	γ relaxation		β relaxation		δ relaxation	
R_0	f_∞	E_a	f_∞	E_a	f_∞	E_a
mol/mol	Hz	eV	Hz	eV	Hz	eV
0.2	$3x10^{14}$	0.35	$2x10^{16}$	0.64	-	-
0.4	$6x10^{13}$	0.32	$5x10^{17}$	0.68	$1x10^{19}$	1.1
0.6	$3x10^{13}$	0.31	$1x10^{19}$	0.74	$1x10^{25}$	1.5
0.8	$2x10^{8}$	0.17	$1x10^{15}$	0.57	$1x10^{22}$	1.2
1.0	$1x10^{6}$	0.08	$1x10^{10}$	0.35	$1x10^{25}$	1.5

Low values of E_a and f_∞ of the γ-relaxation indicate a locally activated process. The E_a values decrease with increasing modification due to a decrease in the vinyl group content. The strength of the γ-relaxation was found to increase slightly with increasing temperature, with the HN shape parameter *a* between 0.17 and 0.20. The mean value of the *b* parameter varies between 0.25 and 0.30 indicating a very broad relaxation.

Similar high values of E_a and f_∞ as in Table 1have been reported for the β-relaxation in literature for side-chain liquid-crystalline polymers with the polar azobenzene moieties in side chains [12]. It was argued that this relaxation is associated not only with local motion of the mesogen along its long axis in the potential formed by the surrounding molecules. Instead, it was suggested that neighboring molecules are involved as well in the liberation dynamics of the β-relaxation. The values of E_a and f_∞ were found to increase with increasing R_0, except for the materials with $R_0 = 0.8$ and 1.0.

The α-relaxation was observed only for the neat PBD and composition with the lowest R_0=0.2; this finding corresponds to DSC data. The f_m vs. $1/T$ dependences follows the VTF-eq. 3, indicating a change in the molecular dynamics of the systems (Fig. 5). Fitting of the data the to VTF equation gives for the parameters of the equation, f_∞, B and T_0, the values 10^{11} Hz, 1409 K and 127 K (for PBD) and 10^{11} Hz, 808 K and 258 K (R_0=0.2), respectively.

The fitting procedure for the δ-relaxation was possible for the materials with R_0=0.4, 0.6, 0.8 and 1.0 only in limited temperature region (Fig. 5). Assuming the Arrhenius behavior, extraordinarily high values were calculated for the E_a and f_∞ (Table 3), indicating a cooperative process. The cooperative character of this relaxation has been discussed in the literature [13]. In the SmA phase this processes is interpreted as being due to the side group flipping around the polymer backbone as the mesogen is hopping from one smectic layer to another [14]. This interpretation agrees well with the fact that in isotropic phase no such relaxation is observed, while the absence of this process in the material with R_0 = 0.2 may reflect the lower order within this sample as a result of the low thiol content.

LCPBD with R_0=0.4 was used also for optical investigation at room temperature. Photochemically induced trans-cis-trans isomerization of azobenzene groups suggests that LCPBDs can be used for reversible optical data storage in photonics.

ACKNOWLEDGMENTS

Financial support of the Grant Agency of the Academy of Sciences (grant No. IAA4112401) and of the Ministry of Education, Youth and Sports of the Czech Republic (grant MSM 0021620835) is gratefully acknowledged.

REFERENCES

1. V. P..Shibaev and L. Lam, Eds, *Liquid Crystalline and Mesomorphic Polymers*, Berlin, Springer-Verlag, 1994.
2. W. Meier and H. Finkelmann, *Makromol Chem Rapid Commun* **11,** 1253-1259 (1990).
3. P. Bladon and M. Warner, *Macromolecules* **26,** 1078-1085 (1993).
4. N. G. McCrum, B. E. Read and G. Williams, Eds, *Anelastic and Dielectric Effects in Polymeric Solids* Berlin, Springer-Verlag, 1991.
5. D. Rais, Y. Zakrevskyy, J. Stumpe, S. Nešpurek and Z. Sedláková, *Opt Mater*, in press.
6. J. Podešva, J. Spěváček and J. Dybal, *J Appl Polym Sci* **74**, 3214-3219 (1999).
7. S. Havriliak and S. Negami, *Polymer* **8,** 161-168 (1967).
8. D. W. Marquardt, *J Soc Indian Appl Math* **11**,:431-439 (1963).
9. F. Kremer and A. Schoenhals, Eds, *Broadband Dielectric Spectroscopy,* Berlin, Springer-Verlag, 2003.
10. A.S. Govind and N. V. Madhusudana, *Eur Phys J* **E9**, 107-121 (2002).
11. R. Zorn, F. I. Mopsik, G. B. McKenna, L. Willner and D. Richter, *J Chem Phys*;**107**, 645-658 (1997).
12. A. Schoenhals and H. E. Carius, *Int J Polym Mater* **45**, 239-251 (2000).
13. J. F. Mano, *J. Macromol. Sci Part B* **42**, 1169-1182 (2003).
14. M. Mierzwa, G. Floudas and A. Wewerka, *Phy. Rev. E* **64**, 031703 (2001).

Complex Structure and Dynamics of Diblock Copolymers in a Mixture of Partially Miscible Solvents

Petr Štěpánek[1], Zdeněk Tuzar[1], Petr Kadlec[1], Jaroslav Kříž[1], Frédéric Nallet[2], Laurence Noirez[3], and Yeshayahu Talmon[4]

[1]*Institute of Macromolecular Chemistry, Heyrovský Sq. 2, 1620 6 Prague 6, Czech Republic*
[2]*Centre de recherche Paul-Pascal, CNRS, 115 Avenue du Docteur-Schweitzer, 33600 Pessac, France*
[3]*Laboratoire Léon-Brillouin, CEA-Saclay, 91191 Gif-sur-Yvette CEDEX, France*
[4]*Department of Chemical Engineering, Technion—Israel Institute of Technology, Haifa 32000, Israel*

Abstract. We have investigated the structure of self-organized microemulsions made of two partially miscible solvents emulsified by a diblock copolymer. We have found that above the phase separation temperature T_c the solution is almost homogeneous while at temperatures below T_c the solution is microphase separated into a periodic lattice generally with body centered morphology. The lattice points are occupied by micelles that contain the minority solvent. Transmission electron microscopy shows that the micelles are first formed at random positions in the sample in the vicinity of T_c and on further cooling they self-organize into the three-dimensional network. The dynamical properties of this system were investigated by dynamic light scattering and pulsed-field gradient NMR. Seven processes could be identified that correspond to (in order of increasing relaxation time) thermal diffusion, self-diffusion of solvent molecules, cooperative diffusion, self-diffusion of polymer chains, self-diffusion of micelles, cluster diffusion and diffusion of grains.

Keywords: Dynamic light scattering, NMR, Partially miscible solvents, Pulsed-field gradient, Self-organized microemulsions, Small-angle neutron scattering, Transmission electron microscopy
PACS: 61.12.Ex, 61.25 Hq, 61.41+e, 68.37 Lp

INTRODUCTION

Colloidal behavior of diblock copolymers in selective solvents is in many respects similar to that of soaps and surfactants in water [1, 2]. Diblock copolymers can form micelles, which are able to, e.g., solubilize otherwise insoluble substances or stabilize particles of colloidal dimensions. In this contribution we consider the case when the insoluble substance is another liquid.

When a diblock copolymer A-B is dissolved in a partially miscible mixture of solvents a and b, where a is a solvent selective for block A and b is a solvent selective for block B (and a precipitant for the block A), multilayered anisotropic nanostructures are formed consisting of periodically arranged domains of the solvents a and b, stabilised by the diblock copolymer forming a double brush on the a/b interface [3]. The model system discussed in this paper consists of a diblock copolymer polystyrene-*b*-poly(ethylene-*co*-propylene) dissolved in a mixture of cyclohexane and dimethylformamide. The thermodynamic state of such a system depends on many variables, in particular on temperature, polymer concentration, solvent composition, molecular weight and composition of the diblock copolymer, interfacial tension between the two immiscible solvents, and on six thermodynamic interaction parameters representing all possible combinations of interaction between the two monomer types and the two solvents making up the system. The structural and thermodynamic properties of such systems have been treated theoretically by Dan and Tirrell [4]. Because of the complexity involved in such a system, experimental studies of such solutions are rather scarce.

Capillary wave properties have been studied [5] by light scattering, and interfacial viscoelastic moduli determined, for a film of polybutadiene-b–poly(ethylene oxide) at the (macroscopic) interface between cyclohexane and water. Neutron reflectivity was used [6] to study the properties of the same diblock copolymer at the hexadecane/water interface and it was shown that the copolymer segregates at the interface with the blocks forming a concentrated

CP982, *Complex Systems, 5th International Workshop on Complex Systems*
edited by M. Tokuyama, I. Oppenheim, and H. Nishiyama

region adjacent to a dilute region on either side of the nominal dividing surface.

We have previously described [3] the self-organization properties of a model system consisting of a mixture of two liquids immiscible at room temperature, cyclohexane (solvent **a**) and dimethylformamide (solvent **b**) and several diblock copolymers satisfying the conditions mentioned above, in particular polystyrene-*b*-polyisoprene, polystyrene-*b*-polybutadiene and polystyrene-*b*-poly(ethylene-*co*-propylene). We have shown, using small-angle neutron scattering (SANS), that there is a range of temperatures and concentrations, somewhat above as well as below the coexistence curve of the two solvents where the samples, though remaining macroscopically homogeneous, are microphase-segregated, presumably between **a**-rich and **b**-rich domains separated by interfaces covered with the diblock copolymer. For the various copolymers studied, we found cubic or hexagonal structures with local order only at high temperatures, but long-range order below the coexistence curve. The characteristic structural dimensions were in the range 60—110 nm and depend both on polymer concentration and relative amount of solvents. The macroscopic viscosity of the solutions in the low-temperature ordered phase is about 4 orders of magnitude larger than that of the liquid high-temperature disordered phase. We have also shown that the chains at the interface between the **a**-rich and **b**-rich domains are slightly extended, as described by the value $a = 0.65$ of the Flory exponent that exceeds the value of a for free chains with excluded volume, $a = 0.60$.

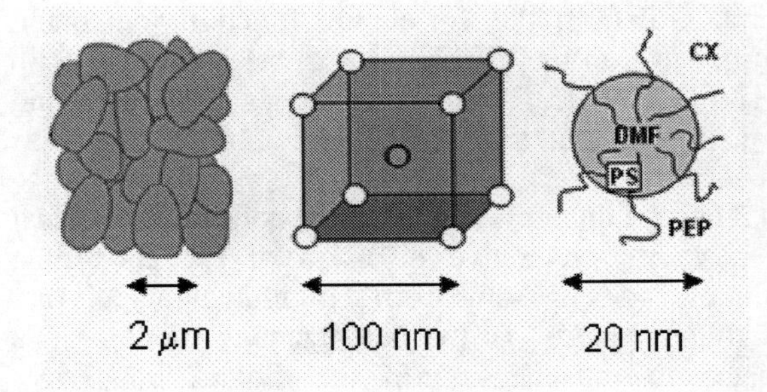

FIGURE 1. Structure of microphase-separated solution: randomly oriented grains of typical size 2 μm (left), typical morphology of each grain – body centered cubic, bcc (center), and arrangement of solvents and polymers in each of the micelles (right). Typical lattice size is 100 nm, typical size of a micelle is 20 nm.

The two-dimensional neutron scattering diffractograms were in the majority of cases isotropic which means that the structural long-range order does not extend to macroscopic dimensions. To assess the extent of the long-range order we have performed ultra small angle neutrons scattering experiments. We have shown [7] that the ordered solution consists of grains of typical size approximately 2 μm and can be schematically represented as in Figure 1.

EXPERIMENTAL

The polymers used are diblock copolymers polystyrene-*b*-poly(ethylene-*co*-propylene), commercial products of the Shell company (Shellvis®) labeled as SV-50 with molecular weight M_w = 96 000 and styrene weight fraction $f_S = 0.43$, and SV-40 with M_w = 200 000 and f_S = 0.25. The solvents, dimethylformamide (DMF) and cyclohexane (CX) of p.a. grade were purchased from Aldrich and used as received. Figure 2 show the experimental coexistence curve for the mixture of these two solvents.

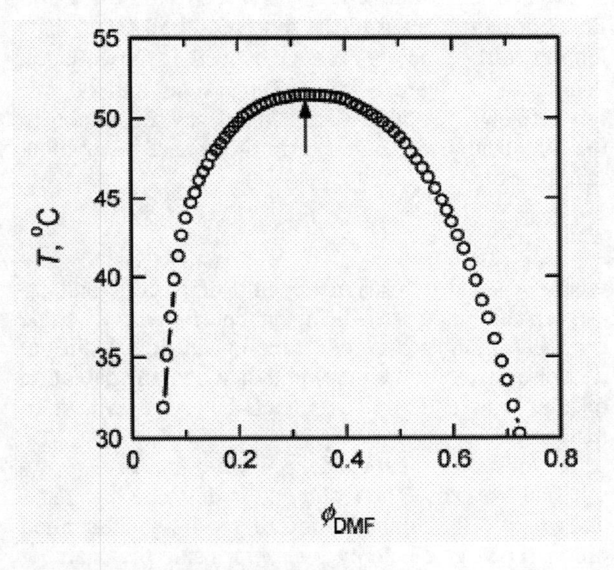

FIGURE 2. Temperature-composition diagram of the mixture of solvents *h*-cyclohexane (CX) and *h*-dimethylformamide (DMF), constructed with experimental data from ref 3. ϕ_{DMF} is the volume fraction of DMF in the mixture. The arrow indicates the critical composition at T_c (0.327); the critical temperature is T_c (51.4 °C).

The dynamic light scattering (DLS) instrument consists of an ALV CGE photogoniometer equipped with a Uniphase 22mW HeNe laser, an ALV6010 correlator and a pair of avalanche photodiodes operated in a pseudo-crosscorrelation mode using an optical fiber splitter. The instrument is described in more detail elsewhere [8]. All measurements were made at angle 90°.

The measured intensity correlation function $g^2(t)$ was analyzed using the program REPES performing the inverse Laplace transformation (ILT) according to

$$g^2(t) = 1 + \beta\left[\int A(\tau)\exp(-t/\tau)d\tau\right]^2 \qquad (1)$$

and yielding the distribution of relaxation times $A(\tau)$ which in the numerical calculations is represented by the set of amplitudes A_i corresponding to the relaxation times τ_i. In graphs with logarithmic scale on the relaxation time axis, the distributions are shown in the equal area representation [9] $\tau A(\tau)$ vs. $log(\tau)$. The relaxation time τ is related to the diffusion coefficient D by the relation $D = (1/\tau q^2)$ where q is the scattering vector.

Self-diffusion PFG NMR experiments were measured at 16-44 °C with an upgraded Bruker Avance DPX300 spectrometer using a z-gradient inverse-detection diffusion probe connected to a BGU2 gradient unit. A Tanner pulsed-gradient stimulated echo sequence was used with the field-gradient incremented in the range 0 – 650 G/cm. The lengths of the gradient pulses δ as well as the diffusion delay time Δ were held constant, 1 ms and 60 ms, respectively. The signal intensity decay was fitted to the exponential dependence on the square of gradient magnitude g

$$I(g)/I(0) = \exp[-\gamma^2 D_{\mathrm{S,NMR}}\delta^2(\Delta - \delta/3)g^2] \qquad (2)$$

where γ is the gyromagnetic ratio of protons and $D_{s,\mathrm{NMR}}$ is the fitted self-diffusion coefficient. In some cases the monoexponential function in Eq.(5) did not describe correctly the measured data and an adaptation of Eq.(2) to a double-exponential process was used instead.

Cryo-TEM measurements were performed on vitrified samples. The technique is described in detail elsewhere [10]. A small drop of the liquid is applied onto a TEM grid held by tweezers inside the chamber of the controlled environment vitrification system (CEVS). The drop is blotted by filter paper supported on a metal strip, leaving behind thin liquid films, ideally 300-nm thick, spanning the holes in the grid. The temperature in the chamber and the composition of the atmosphere in it are controlled to allow quenching of the liquid from a desired temperature and to minimize volatile loss from the sample as it is prepared. The grid is then plunged into a liquid nitrogen (in the case on an organic solvent, as here) and is stored under liquid nitrogen. The vitrified specimen is transferred into a TEM cooling holder with minimal warming and exposure to air and is observed at about -180 °C. We used an FEI T12 G^2 TEM and a Gatan 626 cryo-system. Images were recorded by a high-resolution cooled-CCD camera (Gatan US1000).

RESULTS AND DISCUSSION

The dynamical properties of semidilute homopolymer solutions in a good solvent are simple and usually described in terms of a single process, the cooperative diffusion. For a semidilute solution of a diblock copolymer in a good solvent the situation becomes more complex and a typical situation can be described by Figure 3.

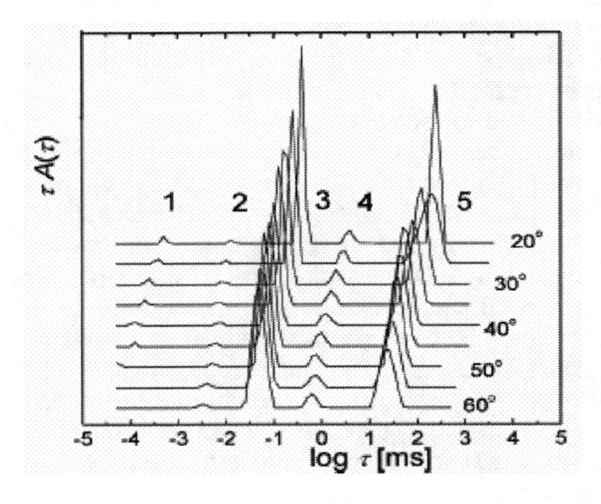

FIGURE 3. Distribution of decay times for a 5 wt % solution of SV-50 in THF at 25 °C and for the indicated scattering angles. (From Ref. 8).

With careful analysis, five dynamic processes can be identified in the correlation function, numbered from 1 to 5 in the distributions of relaxation times in Figure 3. The decay rate of all modes has been found to be proportional to q^2; thus they can be considered as diffusive.

The dominant mode 3 corresponds to classical process of cooperative diffusion of the semidilute solution and is the same as in semidilute homopolymer solutions. Mode 4 is the heterogeneity mode that is related to self-diffusion of the diblock copolymer chain as theoretically described [11]. Its relaxation rate is proportional to the self-diffusion coefficient of the copolymer chains

The relative amplitude of mode 5 increases at small scattering angles; therefore it corresponds to diffusion of objects with size comparable to or larger than q^{-1}. Applying the Stokes-Einstein relation to the relaxation rate of mode 5, we obtain an estimate for the hydrodynamic size of this object R_h = 120 nm. This rather slow diffusive mode has been observed a number of times in various polymer systems and is usually referred to as cluster of which typical sizes systematically appear to be in the range 100-200 nm [12, 13].

For solutions of diblock copolymers its existence has been plausibly explained by Lodge [14] who shows that even in a solvent that is good for both blocks the interaction between the blocks leads to a tendency of the system to create clusters.

The two fast modes 1 and 2 have been observed in all cases of dilute and semidilute solutions of a homopolymer and a diblock copolymer in various solvents. The fastest mode 1 has a diffusion coefficient of the order of 10^{-8} m^2/s. This mode has been observed also in neat solvent at very small scattering angles between 1 and 10°. It corresponds to the thermal diffusion process of the solvent that was studied previously mainly by heterodyne photon correlation spectroscopy [15] or forced Rayleigh scattering [16].

The slower of these fast modes (mode 2) is also diffusive with a diffusion coefficient of the order of 10^{-9} m^2/s. We have performed PFG NMR measurements on several samples studied also by dynamic light scattering in such a way that the PFG NMR response represents the dynamics of the solvent molecules. Very good agreement between the solvent self-diffusion coefficients obtained by PFG NMR and DLS is observed both in the solutions and in the neat solvent. Therefore this mode is thought to correspond to the self-diffusion of the solvent. It is only visible in the scattered light when there is polymer present in the system. Although this mode is not predicted by the theory of light scattering from polymer solutions, several additional experiments support the interpretation as solvent self-diffusion [8].

We now consider the dynamics of semidilute block copolymer solutions in a mixture of two partially miscible solvents, as described above. The dynamic properties of such a system are even more complex and it is of great help to use a three-dimensional representation [17] of the distributions of decay times to study the temperature dependence of various processes. Figure 4 shows a typical example of such a representation for a solution of a PS-PEP copolymer in a mixed solvent CX/DMF.

The system has a characteristic temperature T_c of approximately 27 °C. Above this temperature the system behaves as a polymer in good solvent where the three dynamic processes 3, 4, 5 mentioned above are visible. Modes 1 and 2 have a too small amplitude for a scattering angle of 90° to be seen in this representation. Below T_c, the solution becomes self-organized as was explained above and a new mode appears in Figure 4 as a result of a split of the heterogeneity mode 4 into modes 4a and 4b. We have performed PFG NMR experiments on the same solution. Above T_c, the PFG-NMR data can be described by a single self-diffusion process, representing the center of mass motion of block copolymer chains in the semi-dilute solution. To allow

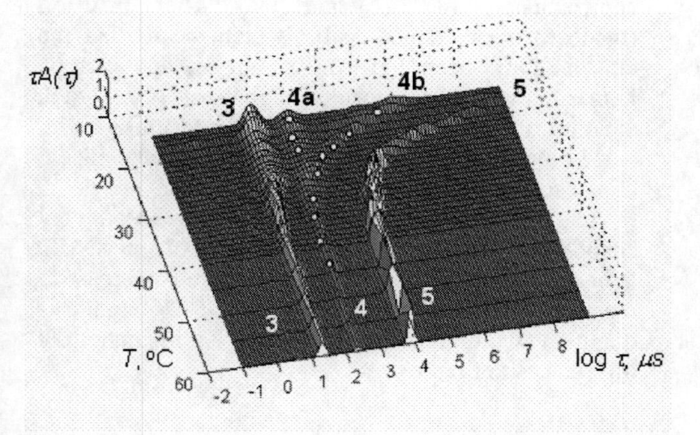

FIGURE 4. Three-dimensional representation of temperature dependence of the distributions of relaxation times $\tau A(\tau)$ vs. log τ, for a 4 wt% solution of SV-40 in a mixed solvent 96% CX and 4% DMF, measured at a scattering angle of 90°. Open circles represent PFG NMR data.

comparison with light scattering, the values of the polymer self-diffusion coefficient obtained from Equation (2) were converted into relaxation times τ and are shown as open dots in Figure 4. Below T_c, two processes are needed to describe the PFG NMR results, and they coincide very closely with the relaxation times determined by DLS, both above and below T_c.

Modes 4, 4a, and 4b thus correspond to a self-diffusion process. The temperature dependence of the faster self-diffusion process (mode 4a) follows that of mode 3 (the cooperative diffusion) both above and below T_c and is governed only by the changes in local viscosity as a function of temperature. Mode 4a therefore corresponds to polymer self-diffusion in the self-organized solution below T_c. By comparison with PFG NMR results, mode 4b is also a self-diffusion process. Its dynamics, however, closely follows that of mode 5, the cluster diffusion that remarkably slows down by four orders of magnitude over a temperature interval of about 10°C below T_c. Mode 4b thus corresponds to diffusion of larger objects, the dynamics of which is governed by the increase of macroscopic viscosity due to microscopic self-organization of the solution below T_c. The appearance of ordering in the vicinity of T_c is not a phase transition but a gradual process, as we have shown by SANS experiments [3]. Mode 4b represents diffusion of micelle-like particles of the block copolymer that coexist with a loose copolymer network and increase in number as the temperature decreases, to finally build a fully self-organized solution, as schematically shown in Figure 1. This happens for the solution referred to in Figure 4 at temperatures below approximately 18 °C, where the cluster diffusion (mode 5) is no longer visible which means that the clusters have been localized in space, and can no longer move. SANS results show that under such conditions the long-range order is already fully built, since below 18 °C the 2nd and 3rd order peaks appear on the scattering curve [3].

The solution referred to in Figure 4 had a very small content (4%) of the minority solvent DMF. When the content of DMF in the mixture is increased,, e.g. to ϕ_{DMF} = 20%, see Figure 5, the 3-D diagram is qualitatively similar but we notice some important differences to Figure 4.

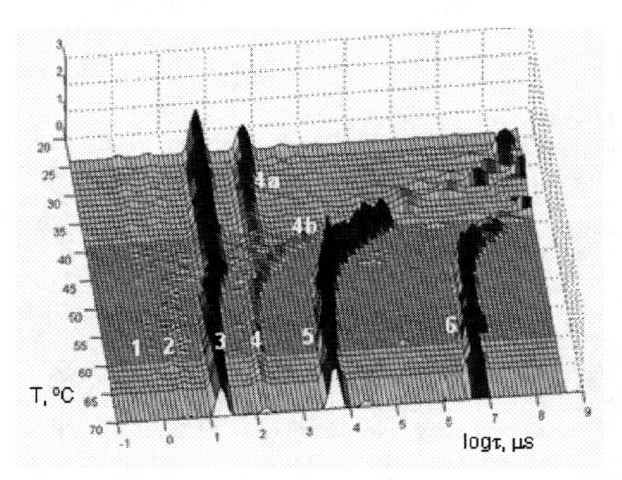

FIGURE 5. Dependence of the distributions of relaxation times $\tau A(\tau)$ vs. log τ, for a 4 wt% solution of SV-40 in a mixed solvent 80% CX and 20% DMF

The characteristic temperature T_c below which long-range order starts to appear is now higher, approximately 45 °C. This is in agreement with the shape of the coexistence curve shown in Figure 2 where for the solvent composition φ_{DMF} = 0.2 the phase separation temperature is about 48 °C. Also an additional very slow dynamic process is present (mode 6) with relaxation time of the order of 10 s. We assign it to the structural precursors of the of the grains that are to be the large-scale building blocks at lower temperatures as schematically represented in Figure 2 (left). There are a very small number of them and although the solution should be homogeneous at temperatures above T_c, their presence can be explained by selective sorption of the solvents to the polymeric blocks. Selective sorption of the solvents in these systems has been discussed in ref. [3].

Finally, we observe again in Figure 5 the splitting of the polymeric self-diffusion mode 4 into modes 4a and 4b below the characteristic temperature T_c where the micelles have formed. These micelles of typical diameter 20 nm contain in their core part the minority solvent DMF.

Figure 6 shows a cryo-TEM image of a 4% solution of SV-40 and solvent composition CX/4%DMF. The solution was kept at a temperature of 40 °C, i.e., 17°C above T_c (according to Figure 2) and then vitrified. The TEM image therefore corresponds to the structure of the solution at 40 °C.

The spherical objects in Figure 6 have an average diameter of 11 nm and are thought to represent the micelles that contain a part of the minority solvent when the microphase separation near T_c begins. The size of the micelles seen by TEM is smaller than that obtained from SANS. This can be explained by a different contrast that the system provides for the two techniques. In SANS, both solvents were deuterated and the polymer protonated, so that the scattering contrast was given only by the block copolymer, even if the total polymer concentration was small. In TEM, the electron density of the solvent inside the micelles is different from that of the outer solvent so that it is predominantly the difference between the two solvents

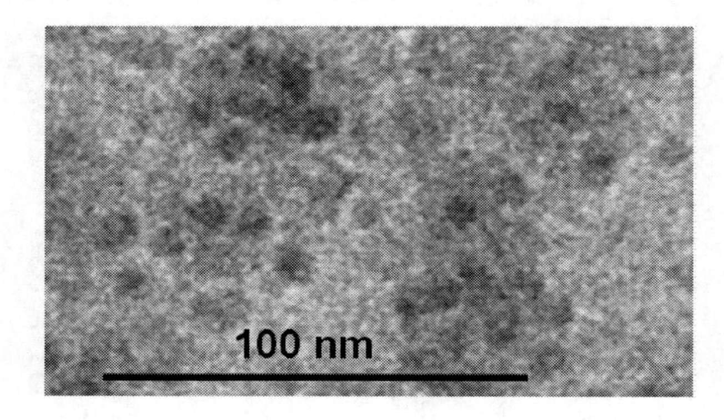

FIGURE 6. TEM image of the vitrified solution of SV-40, concentration 4% in the mixed solvent CX/4%DMF. Specimen was vitrified from 40 °C.

that is seen in Figure 6. The small concentration of the polymer does not contribute significantly to the image.

CONCLUSIONS

Using several experimental techniques, the complex behavior of self-organized solutions of diblock copolymers in partially miscible liquids was elucidated. At temperatures below the coexistence curve of the two solvents microphase separation of the system occurs generating regions rich in solvent **a** and regions rich in solvent **b**. The interphase between these regions is covered with the diblock copolymer chains. In the majority of studied cases the morphology of the microphase separated system consists of a BCC lattice made by periodic arrangement of micelles that contain the minority solvent. As demonstrated by TEM, disordered micelles are formed during the cooling process that progressively form the spatial lattice until a long-range order structure is built. The long-range order extends only to the boundaries of grains of micrometric size. The macroscopic volume of the polymeric solution is formed by the randomly oriented grains.

The dynamic properties of such self-organized solutions are very complex and the observable relaxation times cover 8 orders of magnitude. Beginning with the smallest, these relaxation times describe the thermal diffusion in the system, self-diffusion of solvent molecules, cooperative diffusion, self-diffusion of diblock copolymer chains, self-diffusion of polymeric micelles, diffusion of polymeric clusters and diffusion of grains.

ACKNOWLEDGMENTS

We acknowledge support of this work by the Grant Agency of the Academy of Sciences of the Czech Republic (grant 04050403). Cryo-TEM was performed at the Hannah and George Krumholz Laboratory for Advanced microscopy, part of the Technion project on Complex Fluids, Microstructure and Macromolecules.

REFERENCES

1. K.A.Cogan and A.P.Gast, *Macromolecules* **23**, 745-753 (1990).
2. A.Halperin, M.Tirrell, and T.P.Lodge, *Adv. Polym. Sci.* **100,** 31-71 (1992).
3. P.Stepanek, Z.Tuzar, F.Nallet, and L.Noirez, *Macromolecules* **38**, 3426-3431 (2005).
4. N.Dan and M.Tirrell, *Macromolecules*, **26**, 637-642 (1993).
5. A.J.Milling, L. R.Hutchings, and R.W.Richards, *Langmuir* **17**, 5305-5313 (2001).
6. J.Bowers, A.Zarbakhsh, J.R.P.Webster, L.R.Hutchings, and R.W.Richards, *Langmuir* **17**, 140-145 (2001).
7. V.Ryukhtin, P.Stepanek, Z.Tuzar, K.Pranzas, and D. Bellmann, *Physica* B, *Condens. Matter* **385–386** (Part 1), 762–765 (2006).
8. P.Stepanek, Z.Tuzar, P.Kadlec, and J.Kriz, *Macromolecules* **40,** 2165-2171 (2007).
9. P.Stepanek, "Data Analysis in Dynamic Light Scattering," in *Dynamic Light Scattering: The Method and Some Applications*, edited by W.Brown, Oxford: Oxford Science Publications, 1993, pp.177-241.
10. D. Danino,R. Gupta, J. Satyavolu,Y. Talmon, *Journal of Colloid and Interface Science* **249,** 180–186 (2002)
11. T.Jian, S.H. Anastasiadis, A.N. Semenov, G. Fytas, K.Adachi, and T.Kotaka, *Macromolecules* **27**, 4761-4773 (1994).
12. P. Stepanek and T.P. Lodge, *Macromolecules* **29**, 1244-1251 (1996).
13. T.Kanaya, A.Patkowski, E.W.Fischer J. Seils, H.Glaser, and K.Kaji, *Macromolecules* **28**, 7831-7836 (1995).
14. Z.Liu, K.Kobayashi, and T.P Lodge, *J.Polym.Sci., Part B: Polym.Phys.* **36**, 1831-1837 (1998).
15. S.Will and A.Leipertz, *Int. J. Thermophys.* **22**, 317-338 (2001).
16. C. Allain and P.Lallemand,. *J. Phys. (Paris)* **40**, 693-700 (1979).
17. P.Stepanek, Z.Tuzar, P.Kadlec, and J.Kriz, *Int.J.Polym.Anal.Charact.* **12**, 3-12 (2007).

Cooperativity and Materials' Dynamics – New Insights and Quantitative Predictions

José Joaquim C. Cruz Pinto

CICECO/Department of Chemistry, University of Aveiro, 3810-193 Aveiro, PORTUGAL

Abstract. This research is leading to a detailed model of materials dynamics that accounts for its cooperative nature. The main results so far are the (1) prediction of individual process and average frequencies within materials' structure; (2) prediction of their relative weight (materials frequency, or characteristic time, spectrum), including a quantitative measure of the changes in the effective cooperativity with temperature and the time scale of the experiment, as measured by the sizes of active clusters of relevant elements of the structure; (3) prediction of time-, temperature- and excitation-dependent responses in *e.g.* mechanical creep that naturally turn out consistent with KWW dynamics and time-temperature superposition at not too low temperatures - VTF or WLF at low to moderate temperatures (within a range narrower than 100 K), to Arrhenius at high temperatures - as experimentally observed; (4) prediction of dynamic (temperature scanning rate- or excitation frequency-dependent) responses to thermal and *e.g.* mechanical excitations and (5) a proposition for the microscopic clarification of the origin of the crossover region. The temperature-dependent average characteristic time (or frequency) of the structure gradually turns out super-Arrhenius at low temperature. The theory may be extrapolated to any fast or slow time scales, with no additional computational burden, and explicitly includes the equilibrium (infinite time scale) thermodynamic behavior.

Keywords: Cooperativity, Crossover, Glass transition, Glassy dynamics, Super-Arrhenius behavior
PACS: 61.20.Lc, 61.43.Er, 61.43.Fs, 64.70.Pf, 83.60.Bc, 83.60.Df

INTRODUCTION

Materials utilization and design strongly depend on the understanding and prediction of the behavior under a wide range of temperatures, forced excitations and observation/utilization time scales. Their properties and dynamics show nearly universal features also dependent on temperature, time scale, and excitation type and intensity. Materials dynamics and their glass transitions remain the most challenging fields of research in physics and materials science. Comprehensive and microscopic answers to (1) the surprising and mysterious similarity of the behavior of different glass formers and (2) the very existence of a characteristic crossover region are still lacking [1]. More recently, Ngai highlighted the reasons why the glass transition problem remains unsolved [2], clearly linking them to the fact that conventional theories and models of glass transition bypass the solution of the many-body relaxation problem. On the other hand, one must be realistic, and some coarse-graining will be unavoidable in solving such intractable problem.

Materials of any kind, and polymers in particular, show very complex dynamic behavior under physical (thermal, mechanical, electrical, etc) excitations. Looking closely, we recognize the presence of a very large variety of possible and specifiable elementary processes (responses), spanning an extremely wide range of frequencies and possible cluster sizes within the structure, at any given temperature. Further, that range significantly widens and shifts to lower and lower frequencies and (seemingly) larger and larger cluster sizes as the temperature is lowered and the experimental time scale expanded.

Our driving insight and method is based on the assumption that *each and all* such *elementary processes are activated*, with specifiable activation energies and Arrhenius-like temperature dependence, but may randomly associate in *clusters* of varying dimension, n, in addition to some association driven by topological constraints/entanglements. In polymers, each elementary motion (of the smallest possible cluster, corresponding to n = 1, which we will call *segment*) may involve a small chain crankshaft made of 4 or 6 main chain atoms each. Individual chain-end and branch or pendant group contributions will also lead to other structural effects. In atomic or molecular materials, a functional segment may be a pair of atoms or molecules, or an atom/molecule-hole pair, capable of interchanging their positions, and so the proposed

CP982, *Complex Systems, 5th International Workshop on Complex Systems*
edited by M. Tokuyama, I. Oppenheim, and H. Nishiyama

theory may be completely general, as indeed required to address the universal features of the behavior.

THE DYNAMICS OF COOPERATIVE MOLECULAR MOTIONS

The basic development within this *cooperative segmental theory of materials dynamics* (CSTMD) has been the derivation of the basic characteristic *equilibrium frequencies* (in the absence of stress or any forced excitation) of each type of cluster of size n as

$$\nu_n^*(T) = \nu_1^{\#} \cdot \left[\left(\frac{z_{\#,r}}{z} \right) \cdot \frac{\exp\left(-\dfrac{E_{01} - h\nu_1^{\#}/2}{k_B T} \right)}{\sinh\left(\dfrac{h\nu_1^{\#}}{2k_B T} \right)} \right]^n , \quad (1)$$

where h and k_B are Planck's and Boltzmann's constants, respectively, and the other symbols are explained below. For $n = 1$, $\nu_1^*(T)$ is the frequency of the *primitive relaxation* as defined by Ngai in [2] and earlier references, from which global many-body relaxation originates and develops.

The only assumptions to obtain the above equation were that (1) each elementary (segmental) process is activated, E_{01} being the corresponding activation energy at 0 K, and $\nu_1^{\#}$ the structure-specific frequency at which each activated segment completes its relaxation, (2) local (but not necessarily overall) thermal equilibrium holds before and up to activation, exactly according to the canonical distribution and, for strict statistical mechanical convenience, (3) volume is considered constant. The latter assumption will be seen to yield surprising consequences. As for $z_{\#,r}/z$, it stands for the ratio of the residual (after factoring out the relevant vibrational degree of freedom of the activated state) to full segment partition functions; it may vary somewhat depending on the detailed structure, but will be relatively insensitive to changes in temperature.

The Crossover

Equation 1 is found to yield a definition for the critically important *crossover temperature*, T_c, as the temperature at which the expression within brackets exactly equals 1 and all clusters move exactly in tune, which may come as a surprise, but may be understood as a logical condition leading to enhanced clustering. The crossover has been rightly defined and named as a transition between different dynamic behaviors, but the present theory will in fact make clear that, exactly at T_c, as defined above, a significant change in the structure's dynamics indeed sets-in, as experimentally observed. T_c may always be explicitly calculated (cf. Equation 1) for a given structure from its characteristic E_{01}, $\nu_1^{\#}$ and $(z_{\#,r}/z)$ values, the three structure-dependent parameters so far. The only physical high temperature and frequency limits will be those determined by the structure's stability and durability.

Figure 1 shows the $\nu_n^*(T)$ with n = 1, 2, 5, 10, 50 and ∞, for a typical, moderately flexible, polymer structure with E_{01} = 40 kJ mol^{-1} and $\nu_1^{\#} = 3.16 \cdot 10^6$ Hz, and they all intersect at $T = T_c$ and $\nu = \nu_c = \nu_1^{\#}$. The same Figure illustrates that, if we roughly locate the glass transition relaxation peak half-way between the ergodicity-making ($\nu_1 = \nu_{exp}$) and the ergodicity-breaking ($\nu_\infty = \nu_{exp}$) temperatures, as defined by Angell [3], a typical super-Arrhenius relaxation map may immediately be expected (cf. $*$ and $\bullet$ symbols). Its accurate calculation will be presented further below.

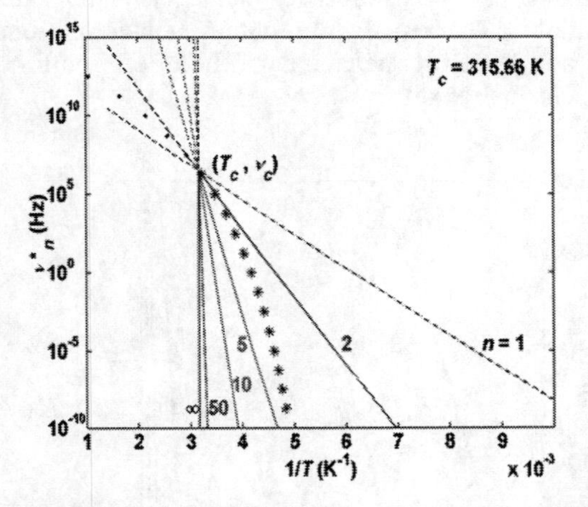

FIGURE 1. (Color Online) Cluster-specific equilibrium frequencies and approximate super-Arrhenius map.

Response to Thermal Scans

When a material is subject to a pure thermal scan at a rate $r = dT/dt$ (rather than to any other kind of forced excitation), one may calculate the scanning rate at which a given mode, corresponding to a given cluster size, n, will be activated (in heating) or de-activated (in cooling) at any given temperature, T, from $d\tau_n^*/dt = \mp 1$ with $\tau_n^* = 1/(2\pi\nu_n^*)$, the minus sign applying to heating scans. The calculation may be shown to yield $|r_n| = 2\pi\nu_n^*(T) \cdot k_B T^2/E_{a,n}$, with $E_{a,n} \sim n\,(E_{01} - h\nu_1^{\#}/2 + k_B T)$, which follow the dotted lines above and to the left of the ν_n^* ones in Figure 2, in which the temperature scale was made linear and the frequency scale expanded, for better readability. We may visualize the dynamic paths followed by the structure at varying rates, as illustrated for 0.01, 0.1 and 1 K·s^{-1}, which cover the most commonly used range of thermal scanning rates. We stress that $T_1(|r|)$ is the ergodicity-making temperature, T_{EM}, and $T_\infty(|r|)$ is the ergodicity-breaking temperature, T_{EB} [3]; this is also consistent with mode coupling theory (MCT) interpretations [4, 5]. The calculations were accurately carried out for $1 \leq n \leq 20$, 50 and 100, and Figure 2 shows all individual $(T_n,\nu_n{}^*)$ values, or frequency paths, and the detail of the calculation for 0.01 K·s^{-1} (cf. horizontal and vertical dotted lines). The emergence of such detailed, microscopic dynamical interpretation is a relevant result. It should however be pointed out that these calculated frequency paths refer to equilibrium, *i.e.* they implicitly assume that the material is heated/cooled up/down at the specified rate to the activation/de-activation of each individual mode, *n*, and then allowed to fully equilibrate, before proceeding to the following mode activations/de-activations. The strategy to obtain non-equilibrium paths and hysteresis behavior will be briefly outlined in the closing remarks.

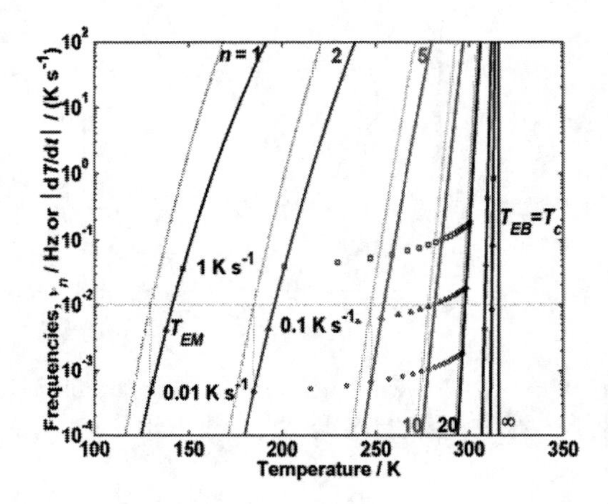

FIGURE 2. (Color Online) Equivalent equilibrium frequency paths at various temperature scanning rates.

Response to Forced Excitations (Creep)

For a number of years before we developed CSTMD, we have been measuring and theoretically modeling amorphous and semi-crystalline polymer non-linear creep behavior [6, 7]. The strategy was based on an analogy with a detailed analysis of the stress- and temperature-dependent dynamics of gauche/trans conformational transitions in the main chain of a polymer. This was generalized to the whole expected range of motions within the material's microstructure. This approach has recently shown consistence with the present CSTMD theory, after appropriate modification of the basic cluster frequencies, $\nu_n{}^*(T)$, to their stress-corrected values, $\nu_n^{"}(\sigma_0,T)$ [8], where σ_0 is the tensile stress, giving

$$\nu_n^{"}(\sigma_0,T)=\frac{c_0\nu_n^*(T)}{2\pi\sigma_0}\sinh\left(\frac{n\upsilon_1^{\#}\sigma_0}{k_BT}\right), \qquad (2)$$

which includes the limiting, low stress, linear viscoelastic case, $\nu_n^{"}(T)=n\nu_n^*(T)/(2\pi k_0T)$, with $k_0 = k_B/(c_0\upsilon_1^{\#})$, an additional parameter, c_0 being a constant (with units of pressure) and $\upsilon_1^{\#}$ the activation volume of each primitive segment. The basis of this result lies in the fact that each primitive relaxation is well represented by a non-linear, stress- and temperature-dependent solid or modified Voigt-Kelvin unit [8]. These frequencies (for the linear case) are plotted in Figure 3, showing that, in the crossover region, somewhat below T_c, the cluster frequencies tend to coalesce down to a very limited range of values. Curiously enough, at very high stress, this region is further reduced, degenerating in the limit to a single point, below T_c. In our opinion, this is identical to the so-called and experimentally well known phenomenon of compensation [9, 10].

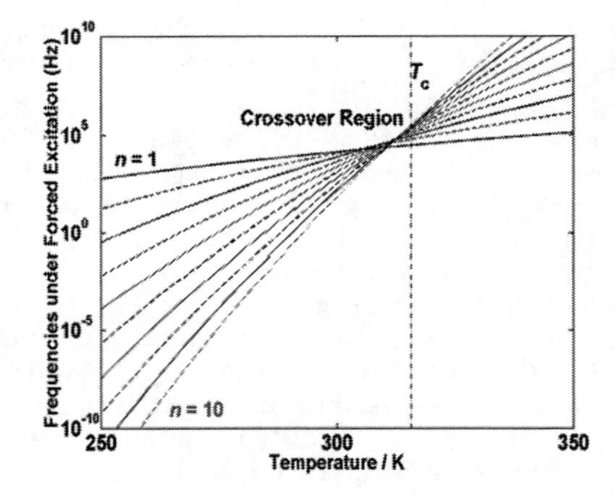

FIGURE 3. (Color Online) Cluster frequencies under a linear forced excitation.

The actual creep response function of each specific cluster will have the usual form $1-\exp[-t/\tau_n(T)]$, with $\tau_n(T)=1/[2\pi\nu_n^{"}(T)]$, and we now have to specify the relative weight or contribution of the various clusters.

Cluster Size and Characteristic Time Distributions

This has proved the most subtle and difficult part of the theoretical work. It started with conventional and conceptually simple combinatorial arguments, because clustering is assumed to be a simple random association of segments or primitive relaxors, as defined in the introduction.

Summarizing the main line of thought, the cluster weights in a mass of material containing 1 mole of individual segments should be proportional to the total number of distinguishable combinations of n *individual segments*, ${}^{N_A}C_n$, multiplied once by n (proportionately to a uniformly distributed cluster internal energy content), and by n again, because each cluster type may be expected to contribute to the material's physical response also proportionately to n. These weights, G_n, however, are not final yet, as one further expects that the effective clusters' contributions should also be proportional to the time during which the clusters remain active, *i.e.* capable of positively contributing to the response, itself proportional to its specific characteristic time, $\tau_n(T)$. After normalization, this yields explicit new temperature-dependent weights, $G'_n\ (T)$, as formulated in Reference 8 and plotted here in Figure 4.

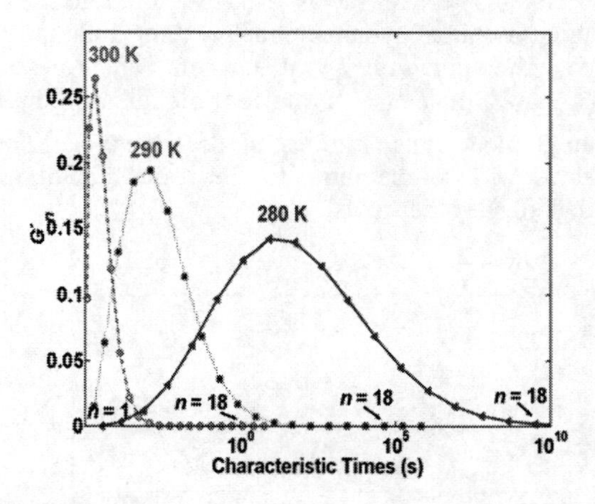

FIGURE 4. (Color Online) Cluster size and characteristic time distributions under a linear forced excitation.

This takes into account that the same segments that may have already taken part in the response of the smallest and fastest clusters will retain the ability to participate, at longer times, in larger and slower cluster motions. Bearing in mind the form of the cluster's response functions, the fastest and most probable processes thus turn out the least/most significant to the dynamic behavior for long/short time scales.

MODELING CREEP

Linear Creep, Storage and Loss Compliances

Modeling each cluster as a standard linear solid, the various compliance functions may be formulated by simply adding the clusters' response functions multiplied by their G'_n, for all n. The storage and loss compliances are as plotted in Figures 5 and 6, respectively. The theory is even able to predict the exact location of the absorption maxima of each cluster size (cf. small digits on the peaks of Figure 6).

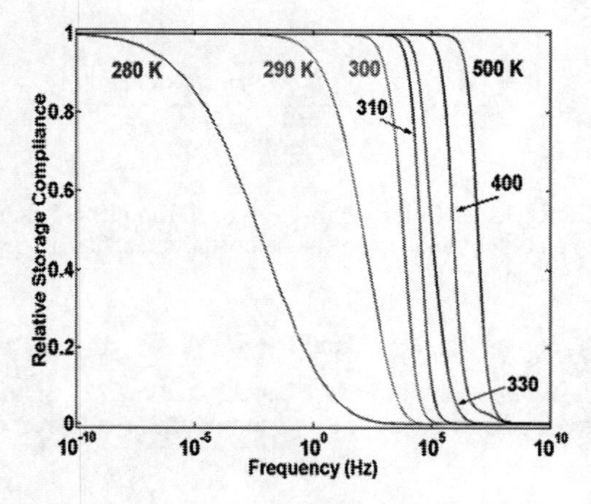

FIGURE 5. (Color Online) Linear relative storage compliances.

The variability of the structural parameters, especially E_{01}, will tend to yield a smoother spectrum, with much reduced or no ripples in the loss peaks, but their assignment to specific clusters will remain valid.

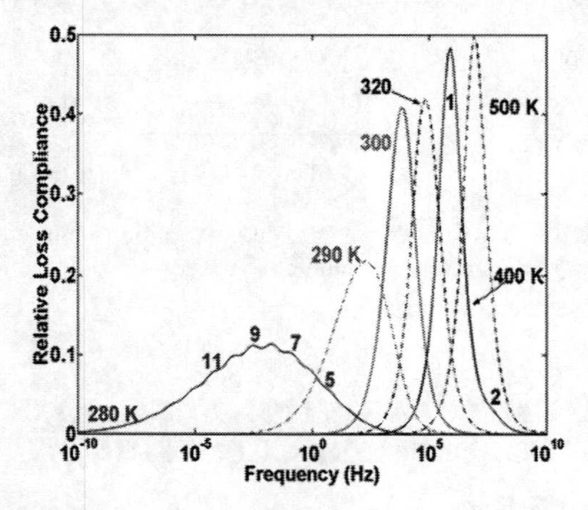

FIGURE 6. (Color Online) Linear relative loss compliances.

Super-Arrhenius (VTF/WLF) Dynamics

The creep response follows VTF/WLF superpositions (Figure 7), thus becoming super-Arrhenius below T_c and gradually Arrhenius above T_c (Figure 8),

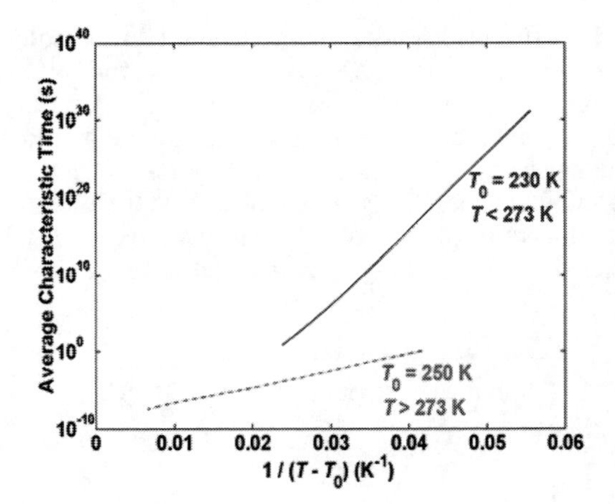

FIGURE 7. (Color Online) Vogel-Tammann-Fulcher (VTF) plots for the average characteristic times, under linear forced excitation.

with $\nu_{avg} = 1/(2\pi\tau_{avg})$ and $\tau_{avg}(T) = \sum_{n=1}^{\infty} G_n \tau_n(T)$. Average and maximum cluster sizes (*i.e.* spatial heterogeneity) also increase as temperature is lowered.

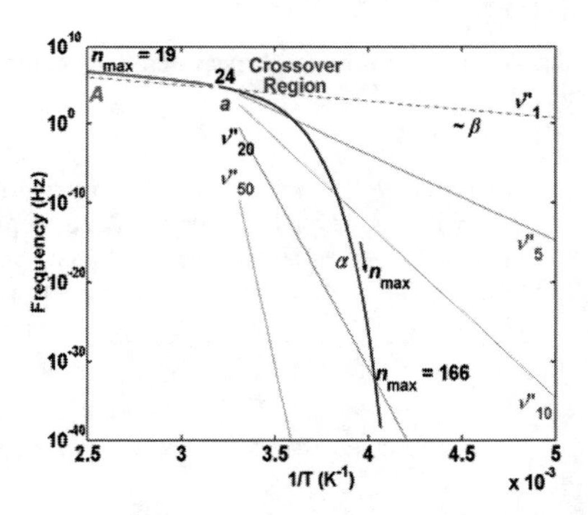

FIGURE 8. (Color Online) Cluster-specific frequencies under linear forced excitation, ν''_n, and super-Arrhenius average relaxation map.

SEGMENT PARTITION FUNCTIONS' COLLAPSE BY CLUSTERING

As CSTMD openly assumes that activation and activated states are real and indeed central to the understanding of many physical processes – as they unquestionably are to the quantitative understanding of a host of useful chemical reactions – the total number of dynamically effective n-segment activated clusters per mole of single segments may actually be calculated as

$$N_n^{\#} = {}^{N_A}C_n\, 2^n \exp\left(-\frac{\Delta f_n^{\#}}{k_B T}\right) = {}^{N_A}C_n \frac{\nu_n^*(T)}{\nu_1^{\#}}, \quad (3)$$

yielding for the n-segment clusters' contributions to the equilibrium molar free energy, relative to the corresponding number of fully relaxed segments,

$$\overline{F}^*_{coop,n} = N_n^{\#} \Delta f^{\#}_{n,total} = -{}^{N_A}C_n k_B T \frac{\nu_n^*(T)}{\nu_1^{\#}} \ln\left[\frac{\nu_n^*(T)}{\nu_1^{\#}}\right], \quad (4)$$

with $\Delta f^{\#}_{n,total} = \Delta f_n^{\#} - nk_B T \ln 2 = -k_B T \ln\left[\nu_n^*(T)/\nu_1^{\#}\right]$, which accounts for the increased entropy brought about by the 2^n (for a polymer or, generally, $z_p{}^n$) different activation paths for each type of cluster. It therefore results for the effective cooperative partition function of each n-segment cluster, at equilibrium,

$$z^*_{coop,n}(T) = \left[\frac{\nu_n^*(T)}{\nu_1^{\#}}\right]^{\frac{\nu_n^*(T)}{\nu_1^{\#}}}, \quad (5)$$

a curious result that yields a value equal to unity (meaning a unique dynamic configuration) only at the crossover temperature, T_c (all clusters synchronized, with $\nu_n^* = \nu_1^{\#}$), and at 0 K (no motions of any kind). Figure 9 plots the variations of the above partition functions with temperature, for the same conditions and data of the other plots.

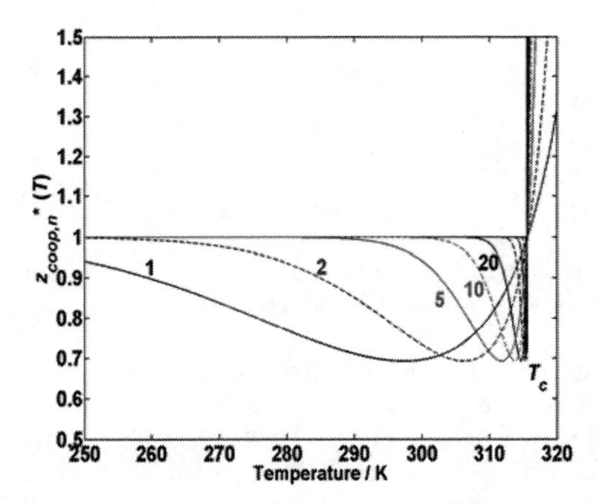

FIGURE 9. (Color Online) Equilibrium cluster-specific cooperative partition functions for the indicated n values, as functions of temperature.

The assumed segment association in clusters of varying but specific sizes must at any instant in time occur at the expense of, and in competition with, their association in clusters of others sizes, as well as of the independent, uncorrelated, transition/relaxation of their individual constitutive segments. Clustering must therefore lead to, first of all, loss of independence by the single segments that participate in any given cluster and, at the same time, also to independence of behavior relative to any other clusters or single segments at the same instant in time and at any different location within the structure, or otherwise any assumed cluster would not in fact exist and respond, and would instead need to be redefined as a still larger cluster. And in fact, the above formulation satisfies these physical requirements, as may be seen from

$$z^*_{coop,n} = \left[\frac{\nu^*_n(T)}{\nu^\#_1}\right]^{\frac{\nu^*_n(T)}{\nu^\#_1}} \neq \left(z^*_{coop,1}\right)^n = \left[\frac{\nu^*_n(T)}{\nu^\#_1}\right]^{\frac{\nu^*_1(T)}{\nu^\#_1}}, \quad (6)$$

because $\nu^*_n(T)$ widely differ from $\nu^*_1(T)$ for all $n > 1$ (cf. Figure 9). This result is, of course, critically important to the soundness of the theory and to its further development.

CLOSING REMARKS

The main specific conclusions from CSTMD were listed in the Abstract, but some closing remarks should be made. Although extensive experimental and theoretical work will still be necessary to achieve a complete evaluation of this theory, encouragingly sound predictions have already been obtained.

With reference to CSTMD's performance relative to the mode coupling theory (MCT), in full agreement with the experimental behavior, CSTMD does not predict near complete dynamic arrest at T_c, as a wide range of increasingly cooperative motions at the molecular scale persist and gradually become dominant as the temperature decreases and time scale expands. While MCT calls attention to and has the greatest impact on temperatures above T_g, CSTMD seems at its best below T_c and even T_g, no matter how low the temperature might be, in addition to its also promising performance above T_c.

Another relevant comparison is with the much older free volume theories, which seem to neglect the dominant role of temperature relative to free volume, and describe the glass transition (and the behavior of condensed matter in general) mostly as a result of congestion due to a lack of free volume. By contrast, CSTMD highlights and quantifies in detail *the role of temperature* on materials' dynamics *at constant volume*! VTF/WLF and super-Arrhenius behavior (cf. Figures 7 and 8) show no undisputable association to free volume changes, contrary to many claims. Volume should, in our opinion, be viewed and duly taken into account as a secondary (not primary) variable when it changes, as at constant pressure conditions.

The fact that the equilibrium clusters' cooperative partition functions could be obtained (cf. Equation 6), combined with the more than reasonable assumption that all cluster-specific Helmholtz free energy contributions will evolve along non-equilibrium paths following a simple and known first-order relaxation kinetics, might justify (and is indeed impelling us) to attempt the development of a general equilibrium and *non-equilibrium theory of glasses* (NETG).

Finally, CSTMD's calculations are extremely fast for any range of temperatures and time scales, including those experimentally and computationally inaccessible - *ca.* 1s in an ordinary Pentium IV processor to compute cluster frequencies, characteristic time spectra and all forms of response (transient, storage, and loss functions). This will allow future calculation of the few physical parameters from experimental response data - a total of only four, plus the instantaneous and infinite time compliances.

ACKNOWLEDGMENTS

The financial support by CICECO, FCT (through Project POCTI/CTM/46270/2002), FEDER and POCI 2010 is gratefully acknowledged.

REFERENCES

1. E. Donth, *The Glass Transition. Relaxation Dynamics in Liquids and Disordered Materials*, Springer-Verlag, Berlin, 2001, pp. VI, 377-380.
2. K. L. Ngai, *J. Non-Cryst. Solids*, **353**, 709-718 (2007).
3. C. A. Angell, *Curr. Opin. Solid State Mat. Sci.*, **1**, 578-585 (1996).
4. W. Götze and L. Sjögren, *Transport Theor. Stat. Phys.*, **24**, 801-853 (1995).
5. W. Götze and L. Sjögren, *Rep. Progr. Phys.*, **55**, 241-376 (1992).
6. J. R. S. André, *Fluência de Polímeros – Fenomenologia e Modelação Dinâmica Molecular*, Ph.D. Thesis, University of Aveiro, Portugal (2004).
7. J. R. S. André and J. J. C. Cruz Pinto, *e-Polymers* **79**, pp. 1-18(2004), URL http://www.e-polymers.org/journal/.
8. J. R. S. André and J. J. C. Cruz Pinto, *Macromol. Symp.*, **247** (1), 21-27 (2007).
9. A. Dufresne, C. Lavergne and C. Lacabanne, *Solid State Commun.*, **88,** 753-756 (1993).
10. C. Lacabanne, A. Lamure, G. Teyssèdre, A. Bernes and M. Mourgues, *J. Non-Cryst. Solids*, **172-4**, 884-890 (1994).

Water Flow through a Stimuli-Responsive Hydrogel under Mechanical Constraint

Go Kondo, Tatsuya Oda, and Atsushi Suzuki

Faculty of Environment and Information Sciences & Department of Materials Science, Yokohama National University, 79-7 Tokiwadai, Hodogaya-ku, Yokohama 240-8501, Japan

Abstract. Friction between the polymer network and the solvent water was measured under the conditions that the thermoresponsive hydrogel was mechanically constrained in a glass microcapillary. The water-flow through the hydrogel could be continuously controlled by more than 1×10^2 times only by adjusting the temperature in the vicinity of the transition temperature. The principles to control the solvent flow and the switching velocity by the temperature jump were discussed on the basis of the material parameters and the experimental conditions.

Keywords: Friction coefficients, Hydrogels, Mechanical constraint, Polyacrylamide, Water flow
PACS: 83.80.Kn, 64.70.Nd, 64.60.Ak

INTRODUCTION

Hydrogels are three-dimensional polymer network containing a large amount of water (Fig. 1), and they exhibit many unique properties in connection with the phase transition (Fig. 2) [1]. Many researchers in various disciplines have extensively studied the fundamental science and technology of hydrogels. In many industrial fields, hydrogels have been used for practical purposes as retainers of water and solutes. For example, disposable diapers contain gel absorbents, and gels are used as drug carriers in anti-inflammatory analgesic cataplasms. Unlike a sponge and a dustcloth, hydrogels do not allow water and solutes to escape easily because the friction between the polymer network and water is extremely large.

There are several reports related with deswelling by an externally applied pressure on hydrogels [2-6]. Among them, in 1991 Tokita and Tanaka investigated the relationship between the frictional property and static fluctuations of polyacrylamide (AAm) gel, and reported that the water flow is described by a simple model of the flux of water flow in a capillary based on the Hagen-Poiseuille equation [5,6]. They also confirmed that the friction between a polymer network and water in poly(*N*-isopropylacrylamide) (NIPA) gel decreased many orders of magnitude as the critical temperature was approached. They discussed the principle of this behavior on the basis of the critical phenomena in gels; the polymer network density fluctuates dynamically near the critical point, and water can pass through the dilute region of the network.

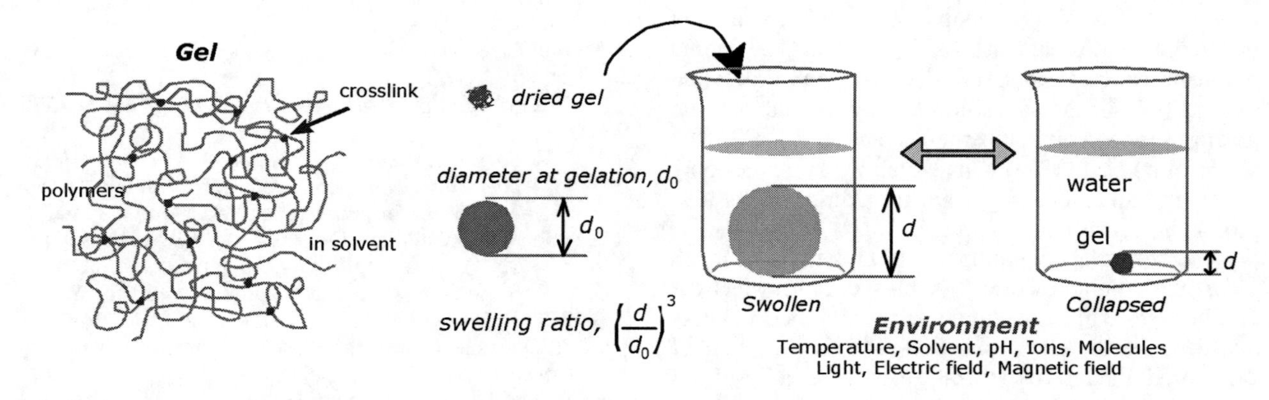

FIGURE 1. Network structure of gels.

FIGURE 2. Swelling behavior of gels and volume phase transition.

CP982, *Complex Systems, 5th International Workshop on Complex Systems*
edited by M. Tokuyama, I. Oppenheim, and H. Nishiyama

Recently, we developed a simple technique to obtain the friction coefficient between a polymer network and water in a hydrogel (Fig. 3) [7]. Using this technique, in which the gel was mechanically constrained in a glass capillary at gelation, it was much easier to control the experimental conditions than it was in the above pioneering works. This simple technique was used to study the friction between a polymer network of AAm gel and water. The effects of gel size (length and cross-section area) and the pressure applied to the solvent, Δp, on the friction coefficient, f, in AAm gel were extensively examined and showed good agreement with the above capillary model. The results suggested that the gel could be assumed to be a bundle of microcapillaries with an inner diameter, ξ (N microcapillaries per unit area), which can be applicable to the water flow through the hydrogel in general;

$$\Delta p = \frac{32\eta}{\xi^2} u\alpha l_0 \equiv f u l_0, \quad f = \frac{32\alpha\eta}{\xi^2} \propto \frac{\eta}{\xi^2}, \qquad (1)$$

where η is the dynamic viscosity of the solvent water, u is the flow velocity and the length αl_0 indicates the average length of the total water path in the gel (α>1).

More recently, we measured the friction between the polymer network and water for NIPA gels with different lengths, l_0, at gelation [8]; the temperature dependence of u was measured in the vicinity of the transition point. With increasing the temperature, the friction slightly decreased at the transition point and increased rapidly in the collapsed phase, which was well scaled by l_0. The macroscopic deformations caused not only by the pressure applied to the solvent but also by the shrinking force due to the temperature increment did not affect the model and the water flow, which can be determined only by the length of the gel (l_0) at gelation and the applied pressure (Δp).

In this paper, the effects of network concentration and inhomogeneity introduced at gelation on the temperature dependence of the friction of NIPA gels are presented. The inhomogeneity of polymer networks was introduced at gelation by changing the cross-linking density. The effects of network inhomogeneity on f were examined by measuring the absolute f and its change in the vicinity of the transition temperature. In addition, we performed the kinetic experiments on NIPA gels with different degrees of inhomogeneity; the flow velocity was measured as a function of time in the respective isothermal process after a step-like temperature change. The principles to control the solvent flow and the relaxation times in response to a temperature change will be discussed on the basis of the material parameters (intrinsic) and the experimental conditions (extrinsic).

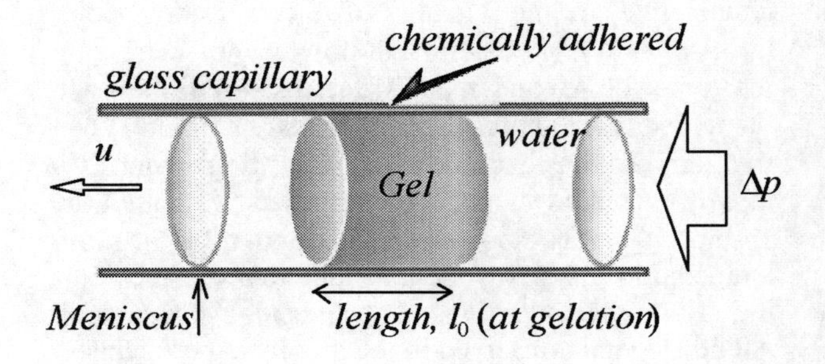

FIGURE 3. Water flow through a gel under mechanical constraint onto the inside wall of the glass capillary.

EXPERIMENTAL PROCEDURE

The pregel solution of gel was a mixture of NIPA monomers (main constituent, Kohjin), *N, N'*-methylenebisacrylamide (BIS, cross-linker, Wako), 2,2'-Azobis [2-methyl- *N*-(2-hydroxyethyl) propionamide] (initiator, VA-086, Wako) and *N,N,N',N'*-tetramethylethylenediamine (accelerator, Wako) dissolved in pure water (deionized, distilled water). The total monomer concentration of NIPA and BIS was fixed to 700mM and the amount of VA-086 was 0.03wt% of the solvent water. The gels are designated here as x/y gels, where x and y are the ratio of the monomer concentrations of NIPA and BIS, respectively ($x+y$=100). The pregel solution was injected into a glass capillary with an inner diameter of 1.35mm using a microsyringe. In order to chemically clamp the gel onto the inner surface of the capillary, bind silane (Pharmacia) was used to rinse the inner surface before the pregel solution was drawn into the microcapillary. The gelation was initiated in the microcapillary containing the pregel solution by irradiating with UV light [7-9] for 30min at 25°C (below the cloud point of NIPA polymers).

The glass capillary with a gel adhered to it was encapsulated in a transparent square glass cell. One end of the capillary was connected to a reservoir of pure water. The temperature of water was regulated using a water bath with an accuracy of ± 0.05°C, and it was circulated within the square glass cell to control the temperature of the gel as well as the flow water. The pressure could be adjusted by changing the height of the water from the water-out flow position to the top surface of the reservoir in the range between 0 and 1.35m. The top surface area was much larger than the cross-sectional area of the capillary; therefore, the decrease in the height of the water caused by the water flowing through the gel could be neglected. The position of the meniscus was measured by an optical microscope that was connected to a calibrated charge coupled device (CCD) apparatus. The pressure loss in the water passing through the capillary and the tubing before and after the gel was estimated as less than 0.25Pa (corresponded to 2.55×10^{-7}m of the height of the water) using the water viscosity, the minimum inner diameter, the maximum flow velocity, and the maximum pass length. This value was negligibly small. On the other hand, the pressure resisting the movement of the meniscus due to the surface tension between the water and air could be approximately estimated as typically the order of 10^{2}Pa. This value could be neglected in the case of the typical height of the water, but not in the case of the lower pressures. In the latter case, the pressure loss in the gel (Δp) was calculated by subtracting the pressure loss due to the movement of the meniscus.

RESULTS AND DISCUSSIONS

Swelling Ratio vs. Temperature

The temperature dependence of the equilibrium linear swelling ratio, d/d_0 of the gels (see Fig. 2) with different BIS concentrations is shown in Fig. 4 where the gels were without mechanical constraint and the temperature was gradually increased.

The diameter changed continuously and the swelling behavior could well be characterized not only by the flexion point in the swelling curve but also by the absolute diameters in the swollen and collapsed

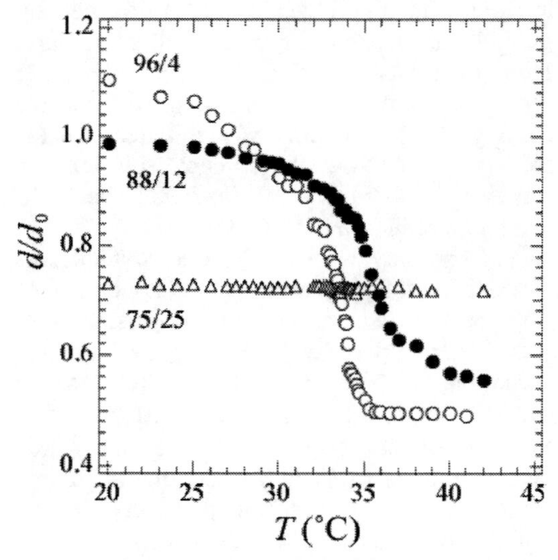

FIGURE 4. Swelling curve of 96/4, 88/12 and 75/25 gels (the respective molar ratio of NIPA:BIS=98:2, 90:10 and 75:25) gels without mechanical constraint where d_0 denotes the gel diameter at gelation.

phases. These characteristic parameters depended on the BIS concentration. In the case of the gel with lowest BIS concentration (96/4), the gel shrunk continuously but rapidly at around 33.5°C due to the change in the balance between hydrophilic and hydrophobic interactions of NIPA polymers. With increasing BIS concentration, the change during the phase change from swollen to collapsed states decreased, and the flection point slightly shifted to the higher temperature, which resulted from the decrement of negative osmotic pressure due to the inhomogeneous network. Finally the macroscopic size did not strongly depend on the temperature in the case of the gel with highest BIS concentration (75/25). Although the NIPA polymer has the lower critical solution temperature (LCST) with the cloud point of around 31°C, the microscopic changes in the network density during heating could not strongly induce the macroscopic volume change and the swelling ratio was almost constant. This evidence is related with the permanent network inhomogeneity introduced at gelation; since BIS is hydrophobic, the inhomogeneity due to the aggregation of BIS in the pregel solution could be permanently quenched into the gel network during gelation [9]. The inhomogeneity of NIPA polymer network introduced by the aggregation of BIS can transfer the large volume change to the small one when the BIS concentration increased.

Friction Coefficient vs. Temperature

Regarding the temperature dependence of f, it has been well established that f depends on the temperature because of the change in dynamic viscosity, η, and it is reasonable to plot f/η against temperature. Using the results for u, f/η can be calculated based on Eq. (1), assuming $\alpha = 1$. The temperature dependence of the normalized friction coefficient, f/η of the gels with lower and higher BIS concentrations is shown in Fig. 5. In accordance with the previous report, f/η of the gel with lower BIS concentration (94/6) depended on the temperature in the vicinity of the transition point (around 34°C); f/η decreased slowly with increasing the temperature, and rapidly decreased a little when the temperature approached the transition point. When the temperature exceeded the transition point, f/η rapidly increased more than two orders. Although the absolute f/η was close to the reported value at room temperature (far below the transition temperature) [5,6], the critical behavior [5] at the transition point could not be evidently observed in the present experiments.

This difference resulted from several technical factors, especially different clamping methods. In the present system the mechanical constraint is large, since the ratio of the diameter of cross-section to the length of gel is relatively large (>1). It was proved that the present simple apparatus gives similar results for friction properties in the swollen state to those obtained by the conventional technique with a complicated setup [5,6], but not in the vicinity of the transition temperature.

The absolute friction coefficients both in the swollen and collapsed states decreased with increasing the BIS concentration. When the BIS concentration exceeded a threshold, f/η decreased continuously with increasing the temperature even after the phase transition. In the case of the gel with the higher BIS concentration (75/25) (Fig. 5), the absolute value of f/η decreased one order smaller than that of 94/6 gel, and the flexion point of f/η was around 32°C, which is apparently smaller than the characteristic temperature (35°C) of 94/6 gel where f/η shows a minimum. Considering that the cloud point of NIPA polymer is around 31°C, the decrement of flexion point should be related with the conformation changes of the linear polymers instead of the network polymers.

From these results, the velocity of the water-flow through the hydrogel in the vicinity of the transition temperature can be continuously increased or decreased by more than one digit by simply adjusting the temperature. In addition, the absolute f/η covers more than a few orders by selecting the BIS concentration and adjusting the temperature. By the use of 94/6 and 75/25 gels shown in Fig.5, for example, f/η can be changed from around 3×10^{11} to 2×10^{15}cm^{-2}. According to our previous reports [7,8], it is reasonable to consider the network of gels as a bundle of microcapillaries of inner diameter ξ, although the real network structure is much more complicated and totally different from the capillary model. Assuming α=1 in Eq. 1, ξ is approximately from 10 to 100nm on the basis of the model. The calculated ξ gives a measure of the average pore size, but it should be a minimum value since α>1 in a real system.

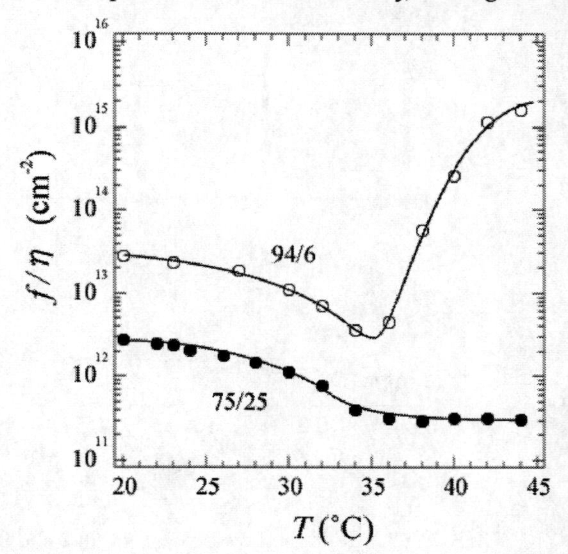

FIGURE 5. Temperature dependence of f/η of 94/6 (molar ratio of NIPA:BIS=98:2) and 75/25 (molar ratio of NIPA:BIS=75:25) gels.

Inhomogeniety due to Phase Transition

There appeared two interesting features in the friction properties of the present system. One is that the increment of the BIS concentration resulted in the suppress of the macroscopic phase transition. The other is that the direction (increase or decrease) of the change in f/η during the phase transition depended on the BIS concentration. These observations should be related with the hierarchical network structure of the present sample [10].

It has been reported [10,11] that the network inhomogeneities in NIPA gel introduced at gelation can transfer discontinuous volume phase transition to continuous volume change by increasing the gelation temperature. This is because the phase transition of gels could be related with the phase-separation of the NIPA polymer. The effect of the network inhomogeneity on the swelling ratio is evident in the collapsed phase, which means that the local shrinking network of the gels prepared above 31°C could not induce the macroscopic collapsed state. Therefore, the formation of network structure through a polymerization process at gelation should be affected by the gelation temperature; it should be remarkably different in the respective case that the gelation temperature is below or above the cloud point. In the present experiment, this permanent inhomogeneity was introduced by changing the BIS concentration (Fig. 6). In spite of the different methods to introduce the inhomogeneity, the swelling behavior in Fig. 4 is strongly affected by the permanent inhomogeneity due to the different BIS concentrations. The structural inhomogeneity can produce defects [12] of polymer networks such as the aggregation of network elements, the distribution of lengths between cross-linkers, and so forth. All these defects will affect the flow properties.

In addition to the permanent (static) inhomogeneity, the dynamic inhomogeneity due to the fluctuation of network density [13] should be taken into account in the vicinity of the transition point; the polymer network density fluctuates dynamically near the critical point, and water can pass through the dilute region of the polymer network. The small dip observed at the transition point (Fig. 5) is attributed to this dynamic inhomogeneity.

In the present experimental setup, an additional static inhomogeneity is introduced in the collapsed state; the gel was mechanically constrained onto the inner surface of the capillary, therefore the large inhomogeneity should be introduced in the network along the diameter direction of the cross-section when the gel entered in the collapsed state (Fig. 6). According to the capillary model, the inhomogeneity should enhance the water flow, since f is inversely proportional to ξ^2 (Eq. 1). From the present results, however, the water flow was strongly restrained in the case of the gel with lower BIS concentration, which indicates that the microscopic change in the local density fluctuations (dense/dilute) along the diameter direction of the cross-section due to the constraint by the side surface was found not to affect the effective ξ along the flow direction.

Relaxation Time to Reach Equilibrium

In order to examine the relaxation time to reach equilibrium, we performed kinetic experiments on the gels with lower and higher BIS concentrations; the flow velocity was measured as a function of time in the respective isothermal process after a step-like temperature change. Typical examples of the time evolution of the flow velocity are shown in Fig. 7 when the step-like temperature change was conducted between 20 and 40°C.

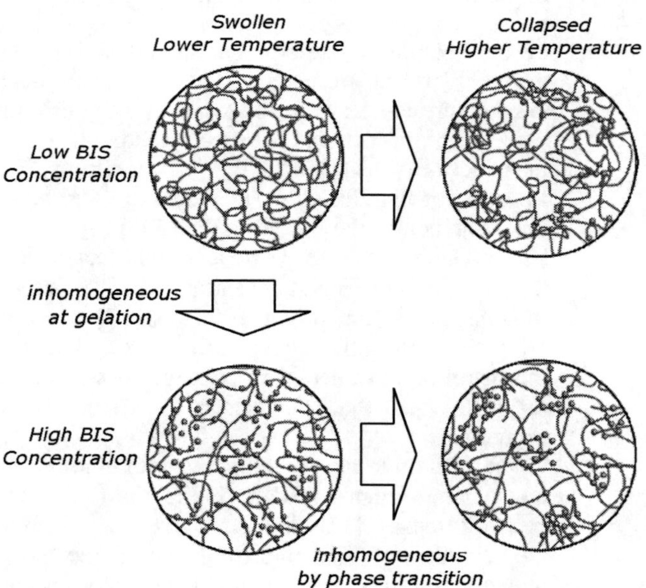

FIGURE 6. Cross-section of water flow through a gel under mechanical constraint onto the inside wall of the glass capillary. Switching from the homogeneous (left) to inhomogeneous (right) networks of the homogeneous (upper) to inhomogeneous (lower) networks at gelation.

The flow velocity instantly changed and reached the steady state within several minutes in the case of the gels with lower BIS concentration. As shown in this figure, the absolute velocity of the 98/2 gel reversibly changed by repeated temperature change. The relative shorter relaxation time in the present experiment may be attributed to the intrinsic properties of water flow through hydrogels. In the present system, the gel was largely deformed not only by the pressure applied to the solvent but also by the shrinking force caused by the temperature increment. As mentioned before, the macroscopic deformation did not affect the friction between the three-dimensional polymer network and water [8]; although the degree of macroscopic deformation should affect the water path length, the water flow can be predicted by the capillary model and α/ξ^2 should be kept constant to give a constant f for the different degrees of deformation [7,8]. This static property could be possibly related with the relative shorter relaxation; the slow collective diffusion could not affect the macroscopic water flow in nature. This small relaxation time was caused by the network diffusion at the inflow and outflow of the solvent water induced by the temperature change since the network at each boundary between the gel and water usually swell or collapse at the phase-transition point in minutes (or within an hour) [14,15].

In the case of a gel with higher BIS concentration, on the other hand, the flow velocity changed in a short period when the temperature was jumped from 40 to 20°C, that is, collapsed to swollen phase. On this

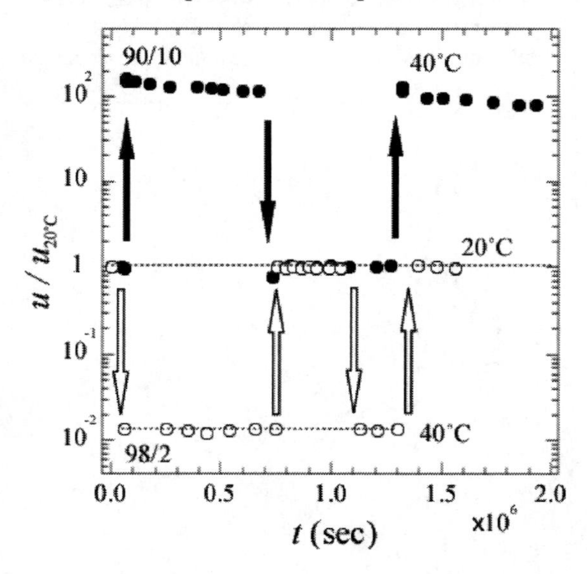

FIGURE 7. Change in the flow velocity of 98/2 and 90/10 gels by the rapid temperature change between 20 and 40°C. $u_{20°C}$ indicates the flow velocity at 20°C; $u_{20°C}$ =6.79x10^{-7} for 98/2 gel and 6.02x10^{-7} (cm/sec) for 90/10 gel.

cooling process, the absolute velocity reversibly changed by repeated temperature change. When the temperature was jumped from 20 to 40°C, however, the flow velocity gradually decreased after the flow velocity changed in a short period. The time to reach the final steady state seems more than 5×10^5sec. This period was much larger than the relaxation time for the collective diffusion of polymer networks [14-17], i.e., it is an order of 10^4sec for a gel with a few millimeter size [15,16]. It has been well established that the relaxation time for the swelling and shrinking of gels under mechanical constraints [18] is much larger than the collective diffusion of the polymer networks [14]. However, these estimations are valid to the homogeneous gel and there is no systematic study on the gel with different inhomogeneities. Although we do not have a decisive picture to explain the origin of the extremely large relaxation time, it could be related with the cooperative change of the water path length by the temperature-induced deformation of polymer networks. Considering the fact that the absolute flow velocity in the second heating is apparently smaller than that in the first heating, the relaxation should be accompanied by the irreversible deformation of polymer networks. If this assumption is correct, the irreversible deformation will finally stop after the repeated switchings of on-off transition and the gel becomes mechanically a steady state. This is an important subject for future investigations.

CONCLUSIONS

In this study, the friction between the polymer network and water of NIPA gel was measured under the mechanical constraint; the effects of network inhomogeneity on f and its temperature dependence were extensively examined. The inhomogeneity of polymer networks was introduced at gelation by changing the cross-linking density. The velocity of the water-flow through the hydrogel in the vicinity of the transition temperature can be continuously increased or decreased by more than one digit by simply adjusting the temperature. The absolute f/η covers more than a few orders by selecting the BIS concentration and adjusting the temperature. Using the capillary model based on the Hagen-Poiseuille equation, the average pore size of the gels could be changed two orders in magnitude by introducing the inhomogeneity.

In addition, the kinetic experiments were performed by conducting an isothermal measurement after a step-like temperature increase or decrease beyond the transition temperature. The flow velocity was measured as a function of time in the respective isothermal process after the temperature jump. The flow velocity instantly changed and reached the steady state within several minutes, and the absolute flow velocity reversibly changed by repeated temperature change. The principles to control the solvent flow and the relaxation time in response to a temperature change were discussed on the basis of the material parameters (intrinsic) and the experimental conditions (extrinsic).

It is noteworthy that the present switching of on-off transition with the shorte relaxation time is completely reversible by repeated temperature change. In that sense, the present findings can be applied to the design of a device (soft microvalve) for controlling water flow through a hydrogel as a soft, simple, and micro actuator.

ACKNOWLEDGMENTS

This work was supported in part by a MEXT Grant-in-Aid for Scientific Research on Priority Areas, No. 438, "Next-Generation Actuators Leading Breakthroughs" of Japan in 2005-2006.

REFERENCES

1. T. Tanaka, *Phys. Rev. Lett.*, **40**, 820-823 (1978).
2. A. M. Hecht and E. Geissler, *J. Chem. Phys.*, **73**, 4077-4080 (1980).
3. A. M. Hecht and E. Geissler, *Polymer*, **21**, 1358-1359 (1980).
4. E. Geissler and A. M. Hecht, *J. Chem. Phys.*, **77**, 1548-1553 (1982).
5. M. Tokita and T. Tanaka, *Science*, **253**, 1121-1123 (1991).
6. M. Tokita and T. Tanaka, *J. Chem. Phys.*, **95**, 4613-4619 (1991).
7. M. Yoshikawa, R. Ishii, J. Matsui, A. Suzuki and M. Tokita, *Jpn. J. Appl. Phys.*, **44**, Part 1, 8196-8200 (2005).
8. A. Suzuki and M. Yoshikawa, *J. Chem. Phys.*, **125**, 174901-174906 (2006).
9. Y. Doi and M. Tokita, *Langmuir*, **21**, 5285-5289 (2005).
10. A. Suzuki, M. Yamazaki, Y. Kobiki and H. Suzuki, *Macromolecules*, **30**, 2350-2354 (1997).
11. A. Suzuki, T. Ejima, Y. Kobiki and H. Suzuki , *Langmuir*, **13**, 7039-7044 (1997).
12. Y. Li and T. Tanaka, *Annu. Rev. Mater. Sci.*, **22**, 243-277 (1992).
13. T. Tanaka, S. Ishiwata and C. Ishimoto, *Phys. Rev. Lett.*, **38**, 771-774 (1977).
14. T. Tanaka and D. J. Fillmore, *J. Chem. Phys.*, **70**, 1214-1218 (1979).
15. E. S. Matsuo and T. Tanaka, *J. Chem. Phys.*, **89**, 1695-1703 (1988)
16. A. Suzuki, S. Yoshikawa, and G. Bai, *J. Chem. Phys.*, **111**, 360-367 (1999).
17. A. Suzuki, *Adv. Polym. Sci.*, **110**, 199-240 (1993).
18. A. Suzuki and T. Hara, *J. Chem. Phys.*, **114**, 5012-5015 (2001).

Anomalous Diffusion in Polymer Solution as Probed by Fluorescence Correlation Spectroscopy and Its Universal Importance in Biological Systems

Kiminori Ushida

Eco- soft materials research unit. Riken (The Institute of Physical and Chemical Research), 2-1 Hirosawa, Wako, Saitama, 351-0198 Japan

Abstract. Experimental evidence of anomalous diffusion occurring in an inhomogeneous media (hyaluronan aquous solution) was obtained by use of fluorescence correlation spectroscopy (FCS) combined with other techniques (PFG-NMR and Photochemical reactions). The diffusion coefficient was obtained as a function of diffusion time or diffusion distance. Since this polymer solution can be regarded as a model system of extracellular matrices (ECMs), intercellular communication, which takes part in ECM, is greatly influenced by this anomalous diffusion mode. Therefore universal importance of anomalous diffusion in biological activity is identified in this series of independent experiments to measure diffusion coefficients.

Keywords: Anomalous diffusion, Extracellular matrix, Fluorescence correlation spectroscopy, Inhomogeneous medium, Intercellular communication, Mean-square displacement, Pulsed field gradient NMR
PACS: 61.41.+e, 66.10.Cb, 82.35.Lr, 82.35.Pq

INTRODUCTION

Inhomogeneous media or materials are found everywhere in the earth, especially in the world of life. Many biological systems, including individual cells, intracellular space, extracellular matrices (ECMs), various organs, and others are totally regarded as an inhomogeneous medium in the physical point of view.

In order to sustain individual lives or their community, material transports through this inhomogeneous space are important in showing the evidence of activity of each organ. Respiration (exchanging oxygen and carbon dioxide), providing nutrition, and sweeping exhausts are most popular material transports found in biological systems.

Particularly, various signaling molecules are secreted at one spot, transport to other place, and then, are accepted at another point to trigger some other biological activities. This kind of sequential processes (here we call this process in material traffics secretion-transport-acceptance (STA) sequence) are found in both inside and outside of cells. The case of intercellular communication outside cells is visually indicated in Fig.1. "Transport" step is normally driven by the thermal librations of media. i.e. diffusion of which typical mode is Brownian motion which is random and stochastic.

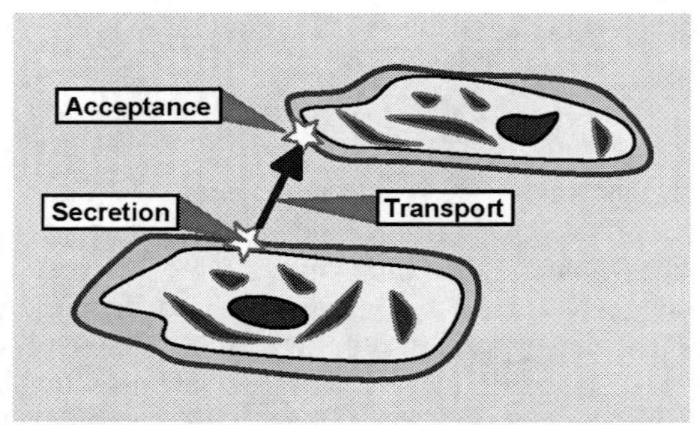

FIGURE 1. (Color online) A typical mode of secretion-transport-acceptance (STA) process in the case of intercellular communication by signaling molecules.

Signaling molecules, such as growth factors, RNAs, vectors, etc have some difficulty to be used in communication free from errors. They must be synthesized and secreted swiftly without a large loss of vital energy. They must reach at appropriate spot within a sufficiently short moment. They must have keen selectivity to induce appropriate response at other points away from the original point. After the communication is over or turns to be useless, rest of the materials, which are not involved in the reaction, must disappear or lose their activity quickly. However, the most serious problem is that they transports by

CP982, *Complex Systems, 5th International Workshop on Complex Systems*
edited by M. Tokuyama, I. Oppenheim, and H. Nishiyama

diffusion. This random and stochastic process never guarantees the completion of STA sequence in 100 %.

If we use a very primitive model of chemical reaction of two molecules (second order reaction) without any geometrical factors, the reaction probability P is expressed as

$$P \propto 4\pi DRC_S N_A \quad (1)$$

where D, R, C_S, and N_A are relative diffusion coefficient, reaction distance, the concentration of secreted materials and the number of accepting points, respectively. For communication in biological system of which reaction time is longer than nanoseconds R can be neglected as

$$P \propto 4\pi DC_S N_A . \quad (2)$$

Therefore if the biological system requires high reaction probability, three parameters, D, C_S, and N_A must be enhanced. However, this operation is not easy. If C_s is increased, the load on the system is increased because the biological system must synthesize a large number of reactants. Since only small number of initiators is needed to trigger the objective action, then the rest of secreted materials turn to be wastes after triggered. In the closed biological system, other activity to clean up the rest as gavages is additionally required. Therefore C_S must be involved in an appropriate range. Similarly, N_A must not be too large to perform selective STA process without errors.

Then, how about D? The magnitude of D depends on the size of molecule and the viscosity of the media through which the molecules transport. Actually, small ions and molecules like Na^+, K^+, Ca^{2+} CO, NO, O_2, and CO_2 are known to react dramatically fast. However, normal signaling molecules such as proteins, nucleotides, and glycol-molecules are relatively large and tend to transport very slowly.

As the results, we pay attention to the diffusion in inhomogeneous media [1-10] such as, in practical, cytoplasm (cell sap), membranes, and ECMs [11]. These media provide environment for material transports as random motions driven by the thermal fluctuation energy. In general, the diffusion mode in inhomogeneous media is regarded as "anomalous diffusion" where the mean-square displacement (MSD) is not in proportion to the time progression [1-6]. In other words, the diffusion coefficient is not constant and should be regarded as a function of time (diffusion period) or space (diffusion distance). Providing the dependence on time and space to the effect of eq. (2) by anomalous diffusion, the reactivity of STA communication mode can be regulated suitable to sustain the life.

In this report, we focus on ECMs [11] as the media for diffusion existing in the intercellular space to realize material transports in inhomogeneous (anomalous) diffusing mode. Several examples of experimental results on hyaluronan (HA) [12] model system [13-20] are reviewed and summarized. ECM contributes to the enhancement of effectiveness of intercellular communication by signaling molecules using smallest C_S and N_A and accompanied by the control in space. For example, the effect of anomalous diffusion in ECM may regulate the dynamics in cell-cell or cell-ECM adhesion.

EXPERIMENTAL AND DATA ANALYSES

Since details of experimental procedures have been presented in our literatures, here we only describe the outline briefly [13-20].

HA sample labeled molecular weight 300000, of which real values in polydispersity was described elsewhere, was obtained from Denki Kagaku Co., Ltd. HA was dissolved in a buffer solution (pH = 7, ionic strength = 0.1 M). The preparations of the aqueous HA solutions were described in our previous papers [13-20].

Two solutes, a dye molecule (Alexa 488: Alexa) and a globular protein (Cytochrome c: cytc) were used as a diffusing probe. For FCS measurement cytc was labeled with Alexa 488[15-18].

The experimental setups for FCS and Sampling-Volume-Controlled (SVC) FCS were also presented in previous literatures [17-19]. In the latter method, the size of the confocal volume (CV) is variable from 200 to 700 nm in radius.

In both methods, the fluctuation of fluorescence intensity was analyzed by the normalized autocorrelation function $G(\tau)$ [21].

$$G(\tau) = \langle I(t)I(t+\tau)\rangle / \langle I\rangle^2 = 1 + \frac{\langle \delta I(t)\delta I(t+\tau)\rangle}{\langle I\rangle^2} \quad (3)$$

Here, $I(t)$ is the detected fluorescence intensity, and $\delta I(t)$ denotes the fluctuations of $I(t)$ around its mean, $\langle I\rangle$. The brackets indicate time average. If fluorescence fluctuations arise only from the translational diffusion of the fluorescent molecules and the photochemical contribution of the dye molecules, the correlation function takes the form

$$G(\tau) = 1 + \frac{1}{N}\left\{\left[1+\frac{\tau}{\tau_D}\right]^{-1}\left[1+\frac{\tau}{q^2\tau_D}\right]^{-0.5} + f\exp(-\frac{\tau}{\tau_T})\right\} \quad (4)$$

where q is a structure parameter and is defined as $q = z / w$ (z and w are the axial and radial radii of the volume element, respectively), fraction f is the contribution of triplet-singlet conversion with a time constant τ_T, and N is the number density. By analyzing the observed correlation function using Eq. (4), the translational diffusion time, τ_D, of the molecules in the volume element was obtained. The observed diffusion coefficient, D_{obs}, was determined using,

$$D_{obs} = w^2 / 4\tau_D, \quad (5)$$

which is derived from the definition of τ_D.

The measurement of D_{obs} was also performed on photochemical bimolecular reaction (PCBR) with a second-order rate constant k_b as

$$A^* + Q \rightarrow (\text{products}) \quad (6)$$

where A* is an excited molecule, of which lifetime is τ_{ex}, and Q is a quencher. The decay rate of A* was plotted against the concentration of Q to determine k_b. Obtained k_b was also expressed as

$$k_b = 4\pi(D_A + D_Q)(R_A + R_B). \quad (7)$$

After appropriate correction, D_A and D_Q were obtained independently. In this research cytc was used as a quencher for excited $[Ru(bpy)_3]^{3+}$. Details were described in literatures.

Pulsed field gradient (PFG) NMR method was also used to determine D_{obs} of cytc. In PFG-NMR experiments, the diffusional attenuation of the echo amplitude is generally given [22] by

$$\frac{I(G)}{I(0)} = \exp[-\gamma^2 G^2 \delta^2 D(\Delta - \frac{\delta}{3})] \quad (8)$$

where $I(G)/I(0)$, γ, G, δ, and Δ are the ratio of echo amplitudes in the presence and the absence of applied field gradient (FG), the gyromagnetic ratio, the magnitude of FG, the duration of each FG pulse, and the interval between the two FG pulses (corresponding to the diffusion time), respectively. A plot of $\ln(I(G)/I(0))$ against $\gamma^2 G^2 \delta^2 (\Delta - \delta/3)$ gave a straight line with a slope of $-D$.

In this study using three different type of measurement, observable value is diffusion coefficient D_{obs}, not mean-square displacement (MSD) $<x^2>$. In normal diffusion, MSD increases in proportion to the time progression and D is constant as

$$<x^2> = 6Dt. \quad (9)$$

On the other hand in anomalous diffusion, MSD does not increase in proportion to t and observed D (D_{obs}) is a function of time as

$$<x^2> = 6D_{obs}(t)t. \quad (10)$$

If we define the diffusion distance L as

$$L = \sqrt{<x^2>} = \sqrt{6D_{obs}(t)t}, \quad (11)$$

one-to-one correspondence between t and L leads us to another form of $D_{obs}(L)$ as a function of L [15-20,23].

Throughout this study, each single result of D_{obs} observation is treated as only a single point value depending on each experimental condition, i.e. sampling time (diffusion time t) and sampling space (diffusion distance L). The sampling time for FCS, PCBR, and PFG-NMR are τ_D in (4) and (5), the lifetime of excited molecule (τ_{ex}), and Δ in (8). L of each measurement can be obtained from eq. (11). In SVC-FCS where w can be varied from 200 nm to 700 nm, L is directly related to w as

$$L = \sqrt{6D_{obs}\tau_D} = \sqrt{3/2}w = 1.225w. \quad (12)$$

Using (3)-(12), each single point data of D_{obs} can be plotted against both t and L to make "Time dependence of diffusion coefficient" (TDDC) and "Distance dependence of diffusion coefficient" (DDDC) plots.

TABLE 1. Summary of Observed Diffusion Coefficient Without Perturbation of HA Mesh by Several Independent Methods (PCBR, FCS, SVC-FCS, and PFG-NMR) and Corresponding Ranges of Diffusion Time and Diffusion Distance. (From ref.15,16, and 18)

Diffusing Probe Molecules	Estimated Molecular Diameter (nm)	Method	Non-Perturbed Diffusion Coefficient (Without HA) D_0 ($\times 10^{10}$ m^{2}s^{-1})	Range of Diffusion time T	Range of Diffusion Distance L
Cytochrome c	3.4	PCBR	1.3	300-400 ns	15 - 18 nm
		FCS	1.4	67 μs	237 nm
		SVC-FCS	1.4	90 – 400 μs	270 – 857 nm
		PFG-NMR	1.4	10-100 ms	3 – 9 μm
Alexa 488	1.4	FCS	2.5	42 μs	246 nm
		SVC-FCS	2.5	50 - 280 μs	270 – 857 nm

RESULTS AND DISCUSSION

In Table 1, we summarized the experimental conditions, especially for diffusion time and diffusion distance, of four different methods. cytc and Alexa were used as diffusing probes dissolved in water or HA solution. These four methods cover a wide area of t and L in total.

We used HA matrix (aqueous solution) as a model medium for ECM. Typical ECMs are composed of two categories of compounds, proteins and glyco materials. They form a hybrid mesh system as shown in Fig.2.

Collagens are typical component belonging to protein, which are stiff and HA is a typical glycopolymer which is soft and movable. The effect of HA seems rather important in molecular diffusion inside ECM. Since HA can solely form a soft mesh structure in aqueous solution, we used a simple system of HA aqueous solution to investigate the anomalous diffusion occurring inside ECM.

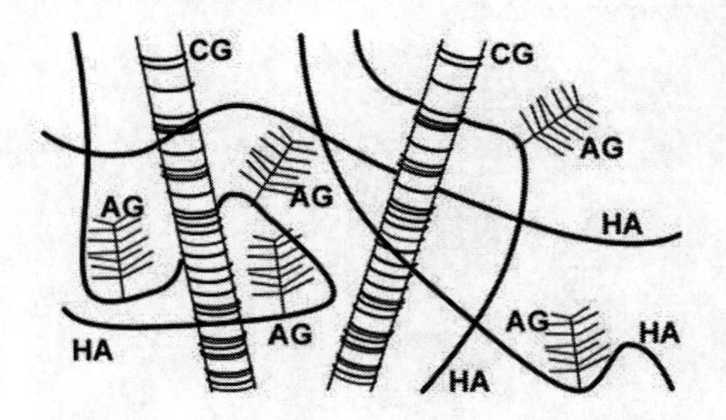

FIGURE 2. Typical composition of ECM : Stiff chains of collagens (CG), soft chains of hyaluronan (HA), and smaller molecular group composed of protein and Gglycosaminoglucans (Agrican: AG) form a hybrid mesh structure.

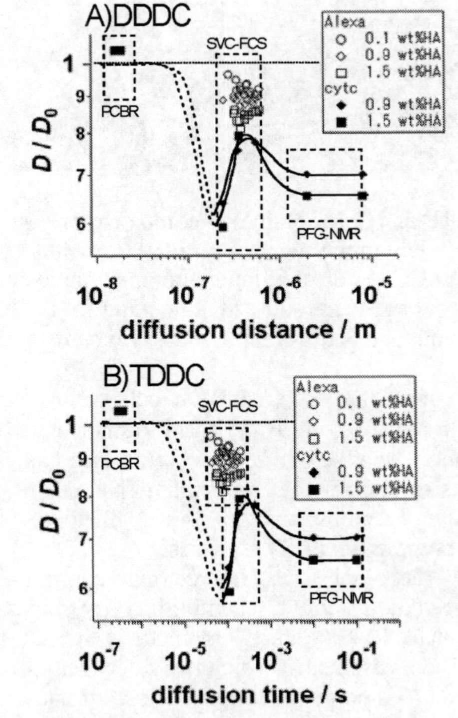

FIGURE 3. A) Distance dependence and B) time dependence of diffusion coefficient of Alexa 488 and cytc in HA aqueous solution (0.1, 0.9, 1.5 wt%) obtained by PCBR, SVC-FCS, PFG-NMR methods. The values are normalized with those obtained without HA (D_0). The figures are reproduced from ref. 19.

We measured D_{obs} of cytc and Alexa in aqueous solution of HA with various concentration of HA. The results are expressed in both DDDC and TDDC plots as shown in Fig. 3.

Since the data points for Alexa are packed in a very small area, we made another figure showing only FCS and SVC-FCS results for Alexa with linear distance (L) and time scales in Fig.4.

The values of D_{obs} dramatically changed in every case that is regarded as anomalous diffusion phenomena.

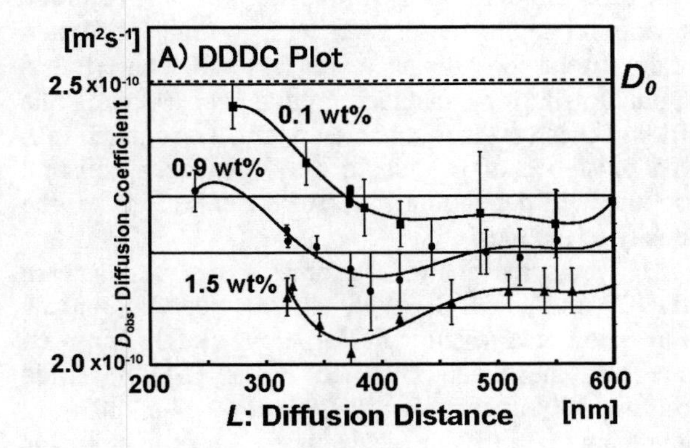

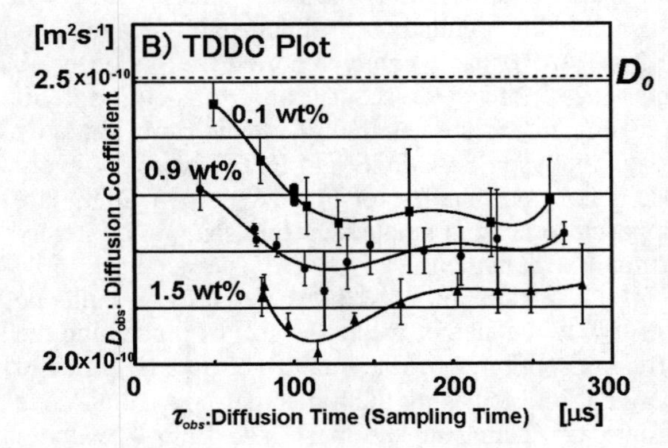

FIGURE 4. A) Distance dependence and B) time dependence of diffusion coefficient of Alexa 488 in HA aqueous solution (0.1, 0.9, 1.5 wt%) obtained by SVC-FCS. The figures are reproduced from ref.18.

In long distance area ($L \gg 10^{-7}$ m = 100 nm) D_{obs} decreased in the existence of HA, the concentration dependence of which follows the famous Ogston's law [24] as

$$D/D_0 = \exp(-\alpha[\mathrm{HA}]^{0.5}) \qquad (13)$$

where [HA] is the concentration of HA. This equation is applicable to the diffusion in 3-dimesional mesh space such as polymer gel solutions. This indicates that the diffusion of probes are interfered with by mesh structure formed from HA polymer chain.

The D_{obs} of Alexa decreased only 15-20 % of D_0 while that of cytc decreased up to 40 % on addition of

HA. The molecular sizes of Alexa and cytc are 1.4 and 3.4 nm, respectively. The estimated mesh sizes from eq (13) were 33, 15, 7 nm for 0.1, 0.9, 1.5 wt% of HA, respectively.

On the other hand in the short distance area ($L \ll 10^{-7}$ m = 100 nm), D_{obs} seemed approximately equal to D_0 independent of HA addition. In this region, the average diffusion distance is too short and most of the solutes have no opportunity to interact with the mesh.

In these two extreme areas, the value of D_{obs} becomes constant where each MSD increase linearly with time progression. Locally, the transport of probe is approximately described with normal diffusion model. In the short distance area ($L \ll 100$ nm), this is trivial. In the long distance area ($L \gg 100$ nm), the order of mesh sizes is sufficiently small compared to L. And existence of HA chain only provide a constant friction field depending on HA concentration, i.e. the density of the mesh.

Between these two extremes (two plateaus in DDDC and TDDC plot), there exist a narrow conversion area was found. For Alexa clearly appeared in Fig.4, the value obtained by SVC-FCS varied continuously on changing the size of CV, i.e. diffusion time and diffusion distance. This area is conventionally called as "Transient Anomalous Diffusion Area" as recently reported by Saxton et al. This area for cytc seems unfortunately out of measurement as shown in Fig.3 which are probably located between SVC-FCS and PCBR areas. In every case, the line shapes of TDDC and DDDC are supposed to show a single step line shape with a very narrow transient area.

The position of this transient area should be sensitive to the size of mesh, the size of probe, and the diffusion coefficient of the probe. The area is shifted to shorter L on increasing the molecular size from Alexa to cytc, or decreasing the mesh size from 0.9wt% to 1.5 wt%. The width of the transient area can be also related to the distribution of mesh size. A sharp line shape in the transient area indicates a uniform mesh structure.

If we introduce dynamic motions to the HA mesh frame, the line shape should exhibit more complicated behavior because that kind of motions are stochastic, diffusive, and having characteristic length and time constant which are comparable with that of diffusion of probes.

In this series of study, we found several properties of anomalous diffusion occurring in the HA aqueous solution as the model system of ECM. 1) The mesh or mesh-like structure of polymer solution induces the anomalous diffusion of solute molecules. 2) The DDDC or TDDC plot shows a step-like function with a sharp transient area the horizontal position of which in DDDC is one order larger than the mesh size.

The experimental condition in this study, where the HA mesh size is 5-50 nm, diffusing molecular size is 1-5 nm is similar to the condition of natural ECM and intercellular communication within it. Therefore we concluded that appearance of transient anomalous diffusion area at 10-100 nm area is universally correct for various animal systems.

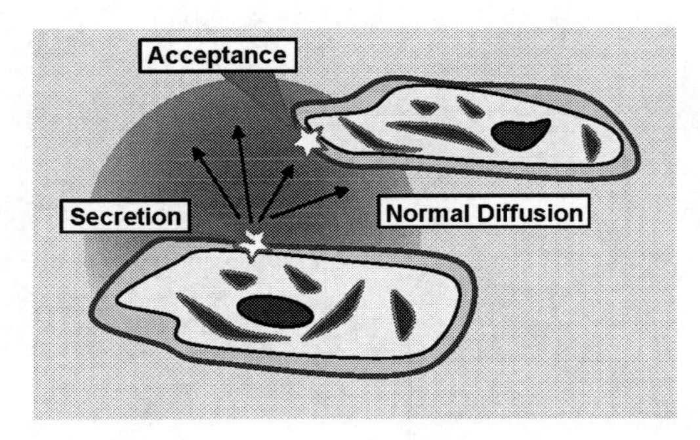

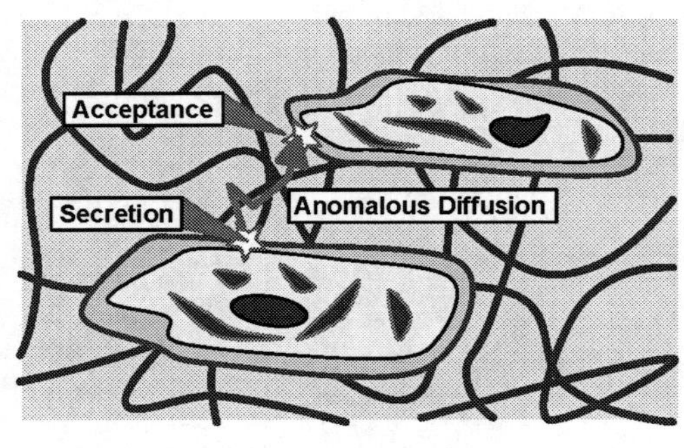

FIGURE 5. (Color online) A comparison figure of intercellular communication without (top) and with (bottom) ECM surrounding communicating cells. The STA sequence is the same as shown in Fig. 1. The diffusion mode is different in these two cases.

The materials secreted from cell surface are not likely to disperse out to the space further than 10-100 nm while the reaction activity is as large as that in aqueous solution within the smaller space than 10-100 nm scale. Precious material is not likely to be lost before completing its STA process.

On the other hand, intercellular communication must be more effective when both cell surfaces come closer than 10-100 nm. This scale is similar to the intercellular distance at which two cells recognize each other and start communication in their adhesion.

Our simple conclusion here based on our present experiments is that ECM plays an important role in intercellular communication using the transient anomalous diffusion area in controlling chemical reactions depending on the scale in space. In Fig. 5, this situation is depicted for the cases with and without ECM. When ECM exists, diffusion mode is anomalous as shown in the lower figure. The secreted molecules are involved in ECM space and tend to reach adjacent spots quickly. However, they have no tendency disperse into outer space further than 10-100 nm.

ACKNOWLEDGMENTS

This manuscript summarized the results already published in several previous papers in collaboration with following people to whom I express my thanks: Dr. Akiko Masuda, Dr. Takayuki Okamoto, Dr. Hiroyuki Koshino, Dr. Koichi Yamashita, Dr. Thomas Kluge (RIKEN), Prof. Mamoru Tamura, Prof. Masataka Kinjo, Dr. Goro Nishimura (Hokkaido University). This research is partly supported by Grants-In-Aid for Scientific Research (Kakenhi) No. 17034067 in the Priority Area "Molecular Nano Dynamics" and No.17300166 from the Ministry of Education, Culture, Sports, Science and Technology (MEXT) of Japan.

REFERENCES

1. J. Klafter and I. M. Sokolov, *Phys. World*, **2005-8**, 29-32 (2005).
2. R. Metzler and J. Klafter, *Biophys. J.* **85**, 2776-2779 (2003).
3. M. J. Saxton, *Biophys. J.* **66**, 394-401 (1994).
4. M. J. Saxton, *Biophys. J.* **89**, 3678-3679 (2005).
5. M. J. Saxton, *Biophys. J.* **92**, 1178-1191 (2007).
6. P. A. Netz and T. Dorfmüller, *J. Chem. Phys.* **107**, 9221-9233 (1997).
7. A. Aharony and D. Stauffer, *Phys. Rev. Lett.* **52**, 2368-2730 (1984).
8. D. Stauffer and A. Aharony, *Introduction to Percolation Theory*. London: Taylor & Francis, 1994.
9. T. Odagaki and Y. Hiwatari, *Phys. Rev. A* **41**, 929-937 (1990).
10. T. Odagaki, J. Matsui and Y. Hiwatari, *Phys. Rev. E* **49**, 3150-3158 (1994).
11. B. Alberts, A. Johnson, J. Lewis, M. Raff, K. Roberts, and P. Walter, *Molecular Biology of the Cell*, 4th ed. New York: Garland Science, 2001, pp 1092-1112.
12. L. Lapčík, Jr., L. Lapčík, S. De Smedt, J. Demeester, and P. Chabreek. *Chem. Rev.* **98**, 2663-2684 (1998).
13. T. Kluge, A. Masuda, K. Yamashita, and K. Ushida. *Photochem. Photobiol.* **68**, 771-775 (1998).
14. T. Kluge, A. Masuda, K. Yamashita and K. Ushida *Macromolecules*, **33**, 375-381 (2000).
15. A. Masuda, K. Ushida, H. Koshino, K. Yamashita and T. Kluge, *J. Am. Chem. Soc.* **123**, 11468-11471 (2001).
16. A. Masuda, K. Ushida, G. Nishimura, M. Kinjo, M. Tamura, H. Koshino, K. Yamashita, T. Kluge *J. Chem. Phys.* **121**, 10787-10793 (2004).
17. A. Masuda, K. Ushida, and T. Okamoto, *Biophys. J.* **88**, 3584-3591 (2005).
18. A. Masuda, K. Ushida, and T. Okamoto, *Phys. Rev. E*, **72** 060101-01-060101-04 (2005).
19. A. Masuda, K. Ushida, and T. Okamoto, *J. Photochem. Photobiol.A* **183** 304-308 (2006).
20. K. Ushida and A. Masuda Chapt. 11, pp 175-188 In *Nano Biophotonics Science and Technology Handai Nanaophotonics Vol.3*, edited by H. Masuhara, S. Kawata and F. Tokunaga. Amsterdam: Elsevier, 2007.
21. Fluorescence Correlation Spectroscopy: Theory and Applications, edited by R. Rigler and E. S. Elson Berlin: Springer, 2001.
22. E. O. Stejskal, J. E. Tanner, *J. Chem. Phys.*, **42**, 288-292 (1965).
23. K. Seki, A. Masuda, K. Ushida and M. Tachiya, *J. Phys. Chem. A* **109**, 2421-2427 (2005).
24. A. G. Ogston, B. N. Preston, J. D. Wells and J. M. Snowden, *Proc. R. Soc. London, A* **333**, 297-316 (1973).

Thermal, Mechanical and Dielectric Behavior of Liquid-Crystalline Polybutadiene-Diols with Cyanobiphenyl Groups in Side Chains

Jan Nedbal[a], Alexander Jigounov[a], Zdeňka Sedláková[b], and Michal Ilavský[a,b]

[a]*Faculty of Mathematics and Physics, Charles University, 180 00 Prague 8, Czech Republic*
[b]*Institute of Macromolecular Chemistry, Academy of Sciences of the Czech Republic, 162 06 Prague 6, Czech Republic*

Abstract. Liquid-crystalline polybutadiene-diols (LCPBDs) with the comb-like architecture were synthetized by radical reaction of a LC thiol containing cyanobiphenyl mesogenic group and flexible spacer with the double bonds of telechelic HO-terminated polybutadiene (PBD). LCPBDs with various initial molar ratios of SH groups to double bonds of PBD, R_0, in the range from 0.15 to 1, were prepared. The physical properties were investigated by differential scanning calorimetry and dynamic mechanical and dielectric spectroscopy. Measurements of the LCPBDs have revealed two low-temperature (secondary) γ- and β- and two high-temperature α- and δ-relaxations. These dispersions were assigned to specific molecular motions; all relaxations were analyzed and discussed in terms of activation energy, magnitude (relaxation strength) and shape of the response.

Keywords: Dynamic mechanical and dielectric spectroscopy, Liquid-crystalline polybutadiene-diols, Mesophase transitions
PACS: 71.20.Rv

INTRODUCTION

Liquid-crystalline polymers usually exhibit improved mechanical and other physical properties [1]. Interesting ordered systems are formed if mesogenic groups are combined with flexible spacers, such as $(CH_2)_n$, because both LC and isotropic states are present in such polymers. Side-chain LC polymers with flexible backbone and mesogenic groups in side chains have been an active area of research for long time [2], [3]; most of the work was based on flexible main-chain polyacrylates [4] and polysiloxanes [5]. Direct linkage of the mesogenic group to the backbone, or just through a spacer of one or two atoms usually gives only an isotropic thermal behavior above the glass transition T_g because thermally induced main-chain and side-chain motions are coupled. Finkelmann et al. [3] have suggested that in order to get LC properties it is necessary to decouple the main chain motion from the mesogenic group by a long flexible spacer.

Another type of the flexible backbone occurs in polydienes, such as polybutadiene (PB). In this case reactive double bond is in each 1,4 or 1,2 monomer unit; these double bonds can be used for further chemical modifications. Grafting can proceed through addition of thiols [6] onto double bonds of PB. In such a way the side chains of various LC structures (containing flexible spacers and mesogens of various types) can be grafted onto the polybutadiene backbone and LCPBs can be synthesized. As OH-terminated telechelic polybutadiene-diols (PBDs) with various lengths are commercially available, their grafted LC analogs can be prepared. These LCPBDs then can be further used for preparation of linear or crosslinked LC polyurethanes by the reaction of grafted LCPBDs with diisocyanates and triols.

Dynamic mechanical and dielectric spectroscopy were often used for investigation of LC polymers [7], [8]. It was found that the dependences of mechanical and dielectric functions on frequency and temperature are sensitive to the ordered state due to more or less aligned parts of macromolecules and behavior in the LC-state reflects the coupled response of ordered mesogenic groups to an applied force.

In this paper a new cyanobiphenyl-type of LC thiol and of comb-like LC polybutadiene-diols (LCPBDs) are described. LCPBDs were prepared with various initial molar ratios of thiols to double bonds of PBD and their thermal structure changes were investigated

CP982, *Complex Systems, 5th International Workshop on Complex Systems*
edited by M. Tokuyama, I. Oppenheim, and H. Nishiyama

by DSC, dynamic mechanical and dielectric spectroscopy.

EXPERIMENTAL

LCPBDs Synthesis and Characterization

Liquid-crystalline polybutadiene-diols (LCPBDs) with the comb-like architecture were prepared by reaction of the SH groups of new synthesized thiol containing cyanobiphenyl mesogen

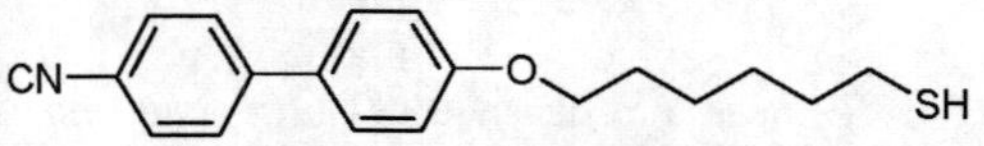

CN–(C6H4)–(C6H4)–O–(CH2)6–SH

with the double bonds of PBD. The OH-terminated polybutadiene-diol

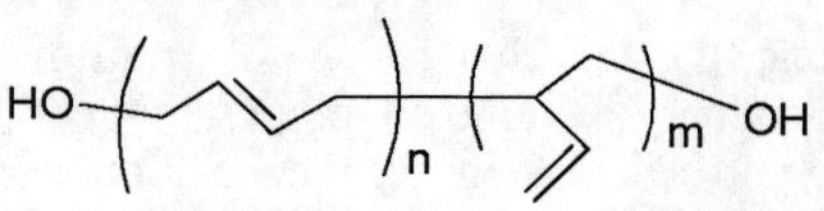

HO–(CH2–CH=CH–CH2)n–(CH2–CH(CH=CH2))m–OH

(Krasol LBH 3000, $M_n \sim 2500$ and $f_n = 2$) and the initiator -2,2'-azobis(2-methylpropionitrile) (ABIN) were used at radical reaction. LCPBDs with various initial molar ratios of SH groups to double bonds of PBD, R_0, in the range from 0.15 to 1, were prepared (Table 1).

DSC and Dynamic Mechanical and Dielectric Measurements

Thermal properties were determined by a Perkin-Elmer differential scanning calorimeter DSC-7e. Reproducible data were collected on cooling and subsequent heating at a rate of 10K/min (Table 1).
Dynamic mechanical measurements were performed with a Bohlin C-VOR apparatus with the parallel-plate geometry. Small-strain oscillatory shear measurements were performed in the frequency range f (= $\omega/2\pi$, ω is angular frequency) from 0.05 to 50 Hz at various constant temperatures; frequency dependences of the storage G' and loss G'' moduli, as well as loss tangent $tg\delta = G''/G'$, were determined. The sample was placed between the plates of the rheometer and heated to isotropic melt. Then the sample was cooled down and measurements were carried out on heating from -50 to 100°C. Using frequency-temperature superposition the superimposed curves of reduced moduli $G'_p = G'.b_T$, $G''_p = G''.b_T$ and of loss tangent $tg\delta_p = tg\delta$ vs. reduced frequency $f.a_T$ (where a_T is horizontal and b_T is vertical shift factor [9]) were obtained. The horizontal shift factor a_T was obtained mainly from superposition of the loss tangent (for tgδ no vertical shift is necessary).

TABLE 1. Degree of modification and DSC results of LCPBDs

R_0	R_e	Run	T_g °C	ΔC_p J/g.K	T_m °C	ΔH_m J/g
0	0	H	-47.3	0.49		
		C	-44.2	0.42		
0.15	0.13	H	-13.2	0.47		
		C	-15.2	0.40		
0.30	0.27	H	9.6	0.38	26.9	0.67
		C	12.4	0.41		
0.45	0.32	H	14.6	0.36	48.9	3.19
		C	11.4	0.24	42.3	-3.01
0.60	0.38	H	20.1	0.39	62.7	3.67
		C	18.3	0.24	59.2	-3.74
0.75	0.45	H	21.3	0.34	68.4	3.76
		C	19.7	0.24	65.7	-3.74
0.90	0.51	H	19.7	0.34	73.2	4.13
		C	18.6	0.23	71.8	-3.91
1.00	0.52	H	20.0	0.32	74.4	6.51
		C	19.0	0.20	72.9	-4.45

Dielectric measurements were performed with a Novocontrol α-Analyzer in the frequency range from 0.01Hz to10MHz at various constant temperatures from –100°C to +130°C. The real ε' and imaginary ε'' part of complex permittivity ε^* was obtained.

The frequency dependence of the dipole (relaxation) contribution to the complex dielectric function, ε_d^*, was described by the non-symmetrical Havriliak–Negami empirical equation [10]

$$\varepsilon_d^* = \varepsilon_\infty + \Delta\varepsilon / [1 + i(f/f_r)^a]^b \quad (1)$$

with five, generally temperature-dependent parameters: high-frequency (unrelaxed) value of the real part of permittivity ε_∞, the relaxation strength $\Delta\varepsilon = \varepsilon_0 - \varepsilon_\infty$ (ε_0 being the relaxed low-frequency value of permittivity), frequency f_r corresponding to the most probable relaxation time τ_r ($2\pi f_r \tau_r = 1$) and two shape parameters a and b. The parameter f_r is related to the peak frequency f_m at which the loss component ε'' attains its maximum. In the case of symmetrical Cole-Cole distribution ($b = 1$) and the f_r and f_m are identical [11]. A computer program based on the Marquardt procedure [12] was developed for determination of all parameters from the frequency dependence of ε' and ε'' at various temperatures.

The temperature dependence of frequency f_m could be described either by the Arrhenius equation

$$f_m = f_\infty \exp(E_a / kT) \quad (2)$$

where E_a is the activation energy and f_∞ is the pre-exponential frequency factor or by the Vogel-Fulcher-Tammann equation (VFT) [8]

$$\log f_{\mathrm{m}} = \log f_{\infty} - \frac{B}{T - T_0} \quad (3)$$

where B is the apparent activation energy, f_{∞} the pre-exponential frequency factor, and T_0 the Vogel temperature ($T_0 \sim T_g$ - 50 °C, where T_g is the glass transition temperature).

The frequency dependence of the conductivity contribution to the complex dielectric permittivity, ε_c^*, can be described by the equation [8]

$$\varepsilon_c^* = (i\frac{f}{f_0})^{\eta} \quad (4)$$

where f_0 is an adjustable parameter and the exponent η would have the value -1 in the ideal case of pure time-independent DC conductivity.

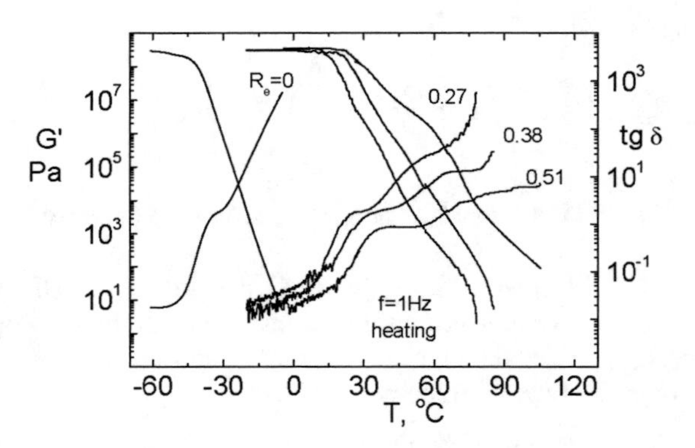

FIGURE 1. Temperature dependences of storage modulus G' and loss tangent tgδ measured at f = 1Hz on heating at 2 K/min. R_e values correspond to Table 1.

RESULTS AND DISCUSSION

The experimental degrees of modification, R_e, after the reaction and purification, were determined from elemental analysis (from the amount of sulfur bounded in LCPBDs) and from ^{1}H NMR spectroscopy. The R_e values have changed from 0.13 to 0.52 (Table 1); while for the lowest initial thiol contents ($R_0 \leq 0.3$), $R_0 \sim R_e$, for higher R_0, $R_e < R_0$ was found. Substantially lower R_e values suggest that at the highest initial degrees of modifications the steric hindrances prevent the radical reaction to go to the full conversion of SH groups; unreacted thiol is subsequently removed from the system by purification.

From DSC measurements it was found (Table 1) that with increasing R_e the glass transition temperatures T_g, and also the LC/isotropic state transition temperatures T_m of LCPBDs increase. With increasing amount of bound thiols also the change in enthalpy ΔH_m at T_m slightly increases.

Dynamic Mechanical Behavior

The temperature dependences of the storage modulus, G', and loss tangent, tgδ (measured at frequency f = 1Hz), of neat PBD and selected LCPBDs are shown in Fig. 1. The pronounced decrease in the G' (more than 7 orders of magnitude) with temperature can be clearly seen. The glass transition is seen also on temperature dependences of the tgδ as a maximum located at the lowest temperatures. While at the highest temperatures the rubbery plato and flow behavior is observed (increase in tgδ with T), at the lowest temperatures the glassy state is reached with the highest values of storage modulus G', typical of the glassy state (G'>10^8 Pa). With increasing modification of PBD (increasing R_e), the changes in the position and shape of mechanical functions on temperature can be seen (Fig. 1); with increasing R_e the position of both mechanical functions shifts to higher temperatures. Only for neat PBD, expected liquid-like behavior at the highest temperatures is observed (strong increase in tgδ and decrease in G' with temperature); for modified LCPBDs additional, not well developed maximum in the tgδ on T dependence is seen at higher temperatures. The decrease in tgδ_m is associated with increasing amount of polymer chain units involved in the ordered state; at the same time, strong physical interactions between mesogens in the side chains, which act as physical crosslinks, cause an additional maximum in the tgδ dependence at high temperatures.

The detailed dependences of superimposed curves of storage G'_p and loss G''_p moduli as well as loss tangent tgδ on the reduced frequency $f.a_T$ are shown in Fig. 2. As expected, with increasing modification the superimposed curves are shifted to lower frequencies; at the same time a broadening of the main transition region takes place. Most pronounced changes in the shape of superimposed curves are seen in the rubbery and flow regions (at the lowest frequencies). In this region the expected slopes of mechanical functions ($G'_p \sim (f.a_T)^2$, $G''_p \sim (f.a_T)^1$ and tg $\delta \sim (f.a_T)^{-1}$) vs. $f.a_T$) for Newtonian liquids [9] are found only for neat PBD (Fig. 2). For modified samples, due to strong side chains mesogens interactions, Newtonian behavior is not reached and in this frequency region a contribution of an additional (slow) relaxation process occurs. The magnitude of this process decreases with increasing modification; we believe that this process is associated with the movement of ordered chain clusters. With increasing R_e, cluster sizes increase and the motion is

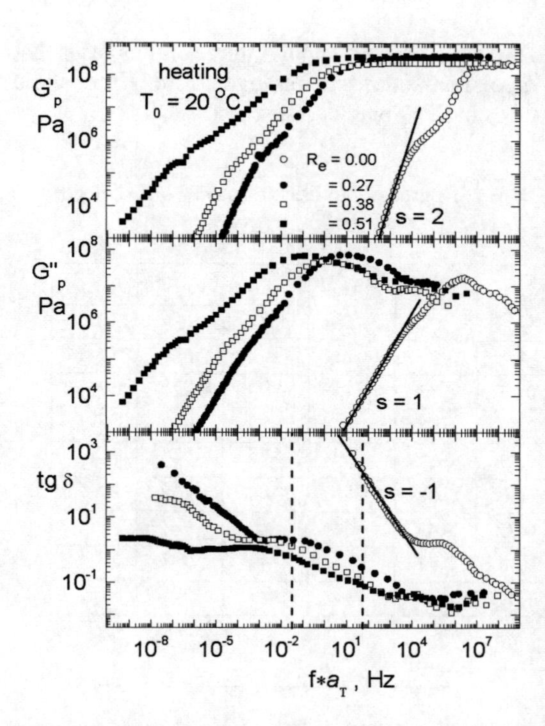

FIGURE 2. Frequency dependences of the superimposed storage G'_p, loss G''_p moduli and loss tangent tgδ, at reference temperature T_0 = 20°C. The indicated slope values, *s*, are expected for Newtonian liquid mechanical behavior. Dashed lines show used frequency interval.

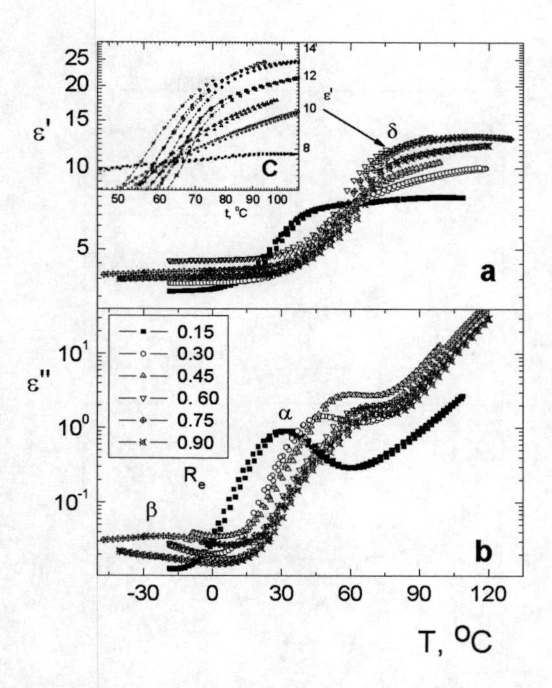

FIGURE 3. Temperature dependences of the real, ε′, and imaginary, ε″, parts of complex permittivity measured at frequency 1 Hz (cooling and heating rate was 2 K/min).

more extensively hindered. It is interesting to note that the sample with the highest modification (R_e=0.51) behaves roughly as a system with critical gel (CG) structure in the gel point for which independence of tgδ of frequency at the lowest frequencies is characteristic. This would suggest that in this sample, due to strong mesogen interactions, a first infinite physical structure was formed and this structure is stable in the measured frequency and temperature region. This is in agreement with our preceding finding [7] on ordered PU systems with mesogenic groups in the main chain where we proved that CG structure was formed by a contribution of a strong physical interactions as well as chemical junctions.

Dielectric Behavior

The temperature dependences of the real (ε′) and imaginary (ε″) parts of complex permitivity measured at frequency 1 Hz in cooling and subsequent heating regime in temperature range from –40°C up to 130°C for various degrees of mesogene modifications are shown in Fig. 3.

The frequency dependences of both parts of complex permittivity for two selected temperatures (-60°C and +70°C) are shown in Fig. 4. With regard to low values and low changes of the ε′ at temperature -60°C its frequency dependence and decomposition is shown in detail in the inset of Fig. 4. From experimental data we could separate four relaxation regions labelled γ, β, α, δ (with increasing temperature).

In the low temperature region two absorptions denoted as β and γ which are associated with the local motion of 1,2 and cis units of polybutadiene backbone (γ) and to the movement of mesogen along its long axis (β) were found. In the middle temperature region the main transition (α relaxation) was measured for all LCPBDs. At the highest temperatures, partly hidden under DC conductivity in the ε" dependence, the δ absorption, which is associated with the movement of mesogens along the short axis[8] was detected.

Temperature dependence of absorption peak of the main transition region follows the VTF equation (3) for all LCPBDs (Fig. 5); from the VTF parameters the T_S temperatures were calculated (T_S=T_g+50K). The T_S temperatures increase with increasing R_0 (Fig. 5, Table2) in accord with the increase of T_g temperature measured by DSC (Table 1). The evaluation of f_m vs. $1/T$ dependences according to the free volume theory [9]

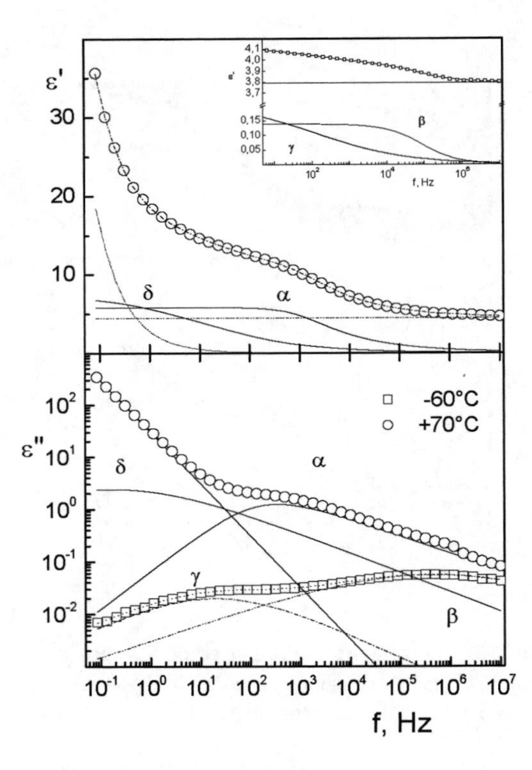

FIGURE 4. Frequency dependences of the the real, ε′, and imaginary, ε″, parts of complex permittivity, measured at temperature T = -60°C and +70°C for sample modification R_e=0.45. Symbols-experimental points, lines-corresponding decomposition to individual relaxation regions and DC conductivity.

(using T_g values from DSC) gives free volume f_g and thermal expansion coefficient α_f values shown in Table 2. Both values decrease with increasing degree of modification, R_e. Similar decrease of these parameters was found from mechanical measurements.

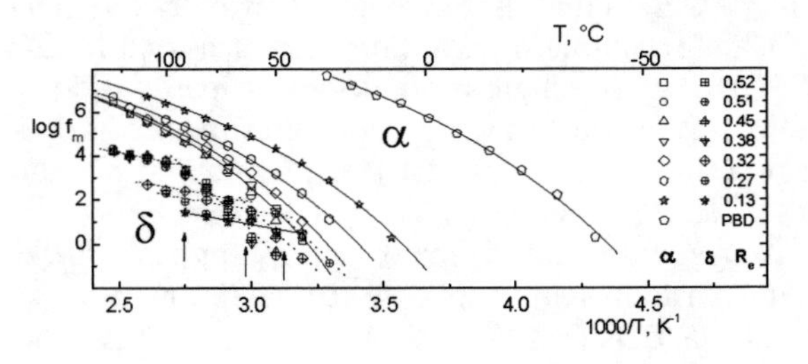

FIGURE 5. Temperature dependence of peak maxima for α and δ transitions.

The magnitude of dielectric absorption in the main transition region $\Delta\varepsilon.T$ depends on the R_e (Fig. 6). For low degrees of modification (R_e<0.3) the magnitude increases and after that reaches roughly constant value ($\Delta\varepsilon.T$~2800K). This effect of saturation suggests that dipol-dipol interactions of mesogenic units and steric hindrances in PBD play decisive role.

TABLE 2. Degree of modification, DSC and dielectric results of T_s, α_g and f_g.

R_0	R_e	T_g °C	T_s °C	$\alpha_g.10^4$ K^{-1}	f_g
0	0	-47.3	-10.5	.040	9.2
0.15	0.13	-13.2	31.5	.035	6.7
0.30	0.27	9.6	43.5	.030	6.1
0.45	0.32	14.6	54.5	.028	9.0
0.60	0.38	20.1	62	.029	5.5
0.75	0.45	21.3	63	.027	5.0
0.90	0.51	19.7	69	.022	5.3
1.00	0.52	20.0	66	.022	6.0

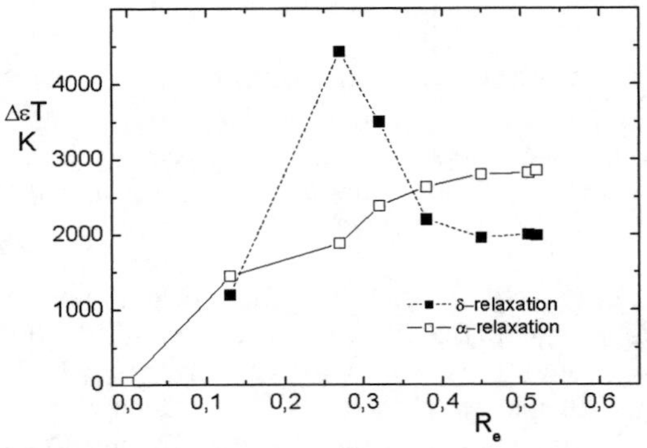

FIGURE 6. Dependence of magnitude ΔεT on degree of modification R_e for α and δ relaxations.

The shape of dielectric absorption in the main transition region is for all modified samples asymmetric and corresponds to the Havriliak-Negami distribution function with parameters a~0.7 and b~0.5. While the shape of the low-frequency part of LCPBDs curves practically does not change in comparison with PBD (a_{PBD}~0.69), the high frequency shape differs (b_{PBD} ~0.26). This means that the high frequency part of dielectric spectrum of modified samples is much narrower in comparison with neat PBD.

The δ-absorption, corresponding to the motion of mesogen along its short axis, is shifted to higher temperatures with respect to the main transition region (Fig. 5). Since the size of mesogen is rather short its movement is closely linked up to the segmental movement of PBD chain. With regard to this, the separation of α and δ peaks was slightly difficult. Nevertheless we could determine the change of activation energy at transition from LC to isotropic state as it was observed for other SCLC polymers [8].

As follows from Fig. 5 the temperature dependence of δ-absorption peaks can be described by Arrhenius equation (2) with two different activation energies. For sample with R_e=0.13 the value of activation energy was E_a~0.45eV in the whole measured temperature region. For all other modifications the change of activation energy from E_a~0.45eV (isotropic state) to E_a~2.2eV (LC state) was found. This is in agreement with DSC and mechanical measurements where the LC state was detected for LCPBDs except sample with R_e=0.13. Temperatures at which there are changes of E_a are in Fig. 5 marked by arrows; their shift to lower temperatures with decreasing R_e corresponds well to the shift of T_m measured by DSC (Table 1).

The magnitude $\Delta\varepsilon.T$ of δ-absorption is comparable to that of main transition, but it depends on temperature; with decreasing temperature the $\Delta\varepsilon.T$ decreases. The dependence of $\Delta\varepsilon.T$ on R_e determined for LC transition is shown in Fig. 6. For first two R_e values the $\Delta\varepsilon.T$ increases and after that it decreases and reaches equilibrium value $\Delta\varepsilon.T$~2000K. Also this finding is in good agreement with main role of dipol-dipol interactions of mesogenic units and of steric hindrances in PBD. The shape of δ-absorption is symmetric and rather broad in accord with the Cole-Cole distribution (a~0.5 and b=1).

The low temperature β-absorption peak was observed only for modified samples. The temperature position of absorption does not depend on R_e value (Fig. 6) and follows Arrhenius equation (2) with activation energy E_a~0.55eV. For R_e<0.35 the magnitude $\Delta\varepsilon.T$ is roughly constant ($\Delta\varepsilon.T$~200K); for higher R_e decreases to $\Delta\varepsilon.T$~70K (for R_e=0.52). The value of Cole-Cole a parameter increases with increasing temperature ($\Delta a/\Delta(1000/T)$~-0.22K) and for -30°C a~0.4 was detected. Since this absorption was not observed in neat PBD we believe that it is due to the motion of mesogens along their long axis.

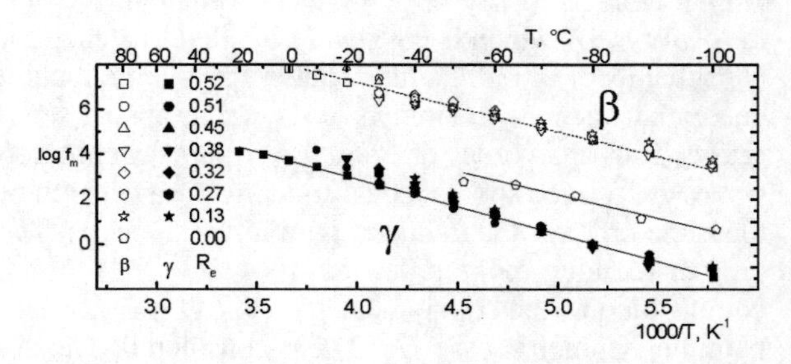

FIGURE 7. Temperature dependence of peak frequency maxima for β and γ transitions.

The last measured γ absorption was observed for all samples (including neat PBD). As in previous case its temperature position does not depend on R_e (except neat PBD whose position on temperature scale is shifted to higher frequencies) (Fig. 7) and follows the Arrhenius equation with activation energy E_a~0.4eV. The shape of Cole-Cole parameter is the same as in previous case and exhibits weak dependence on temperature; for -30°C a~0.3 was found for all samples. The magnitude practically does not depend on R_e and is $\Delta\varepsilon.T$~30K. The γ absorption measured for neat PBD has the same values of E_a, $\Delta\varepsilon.T$ and a. From this fact we conclude that this absorption originates in local movement of 1,2 and 1,4 groups of PBD [14].

ACKNOWLEDGMENTS

Financial support of the Grant Agency of the Academy of Sciences (grant No. IAA4112401) and of the Ministry of Education, Youth and Sports of the Czech Republic (grant MSM 0021620835) is gratefully acknowledged.

REFERENCES

1. P. G. De Genes, *The Physics of Liquid Crystals*, Oxford University Press, Oxford, 1974
2. H. Finkelmann, M. Happ, M. Portugall, H. Ringsdorf, *Makromol Chem* **179**, 2541-2549 (1978).
3. S. M .Aharoni and S. F. Edwards, *Adv Polym Sci* **118**, 1-89 (1994).
4. I. Nishiyama, J. W. Goodby, *J Mater Chem* **3**, 169-82 (1993).
5. C. Cesarino, L. Komitov, G. Galli, E. Chiellini, *Mol Cryst Liq Cryst* **372**, 217-27 (2001).
6. J. G. de la Campa, Q-T. Pham, *Makromol Chem* **182**, 1415-1421 (1981).
7. Y.A. Demchenko, M. Studenovský, Z. Sedláková, A. Sikora, J. Baldrian, M. Ilavský, *Europ Polym J* **39**, 1521-1532 (2003).
8. F. Kremer and A. Schoenhals, Eds, *Broadband Dielectric Spectroscopy,* Berlin, Springer-Verlag, 2003.
9. J. D. Ferry, *Viscoelastic Properties of Polymers*, 3rd Ed. New York: Wiley, 1980.
10. S. Havriliak and S. Negami, *Polymer* **8,** 161-168 (1967).
11. R. H. Cole and K. S. Cole, *J Chem Phys* **9**, 341-359 (1941)
12. D. W. Marquardt, *J Soc Indian Appl Math* **11**, 431-439 (1963).
13. H. Valentová, Z. Sedláková, J. Nedbal, M. Ilavský, *Europ Polym J* **37**, 1511-1520 (2001)
14. S. Cerveny, R. Bergman, GA. Schwartz, P. Jacobsson, *Macromolecules* **35,** 4337-4342 (2002)

Fluctuation Effects of the Microphase Separation of Diblock-Copolymers in the Presence of an Electric Field

I. Gunkel*, S. Stepanow*, S. Trimper* and T. Thurn-Albrecht†

*Martin-Luther-University, Von-Seckendorff-Platz, Halle
†Martin-Luther-University, Hoher Weg, Halle

Abstract. The Fredrickson-Helfand theory of the microphase separation of symmetric diblock-copolymer melts is generalized by including a time-independent homogeneous electric field. Within the self-consistent Hartree approximation the coupling between the electric field and the composition fluctuations is studied. The field is able to suppress the composition fluctuations and consequently the first order transition is weakened. The structure factor of the disordered phase becomes anisotropic and the critical temperature of the order-disorder transition is shifted towards its mean-field value. Due to the electric field the modulation of the order parameter along the field direction is strongly suppressed. This in accordance with the parallel orientation of the lamellae in the ordered state. The discussion is extended to membranes under an external electric fields. Likewise the fluctuations become anisotropic.

Keywords: Composition fluctuations, Diblock-copolymers, Electrical fields
PACS: 61.25.Hq, 61.41.+e,64.70.Nd

INTRODUCTION

There is a large effort in finding ways to control the alignment and the phase behavior of self-assembled structures. Shear flow, the confinement between two solid surfaces and the application of external electric fields are effective tools to macroscopically align ordered structures in diblock copolymers [1]-[12]. Here we are interested in the effect of electric fields, combined with its interplay with structural fluctuations, on the behavior of dielectric block copolymer melts. The analysis is of relevance for applications using self-assembled block copolymer structures for patterning and templating of nanostructures [13]. The driving force for electric field induced alignment is the orientation-dependent polarization in a material composed of domains with anisotropic shape. In case these inhomogeneities appear only at the interfaces of cylinders or lamellae, the polarization of the samples induces surface charges at the interfaces which depend on the relative orientation of the interfaces with respect to the field. As a consequence the system is lowering its free energy for a parallel orientation of interface and electric field. Generally, the polarization charge may be present in the whole system, when the composition of the block-copolymer varies gradually over the sample. To describe this situation one has to consider also the influence of the electric field on the composition fluctuations, which is not discussed in previous work [3]. Therefore the present paper is focused on the interplay between composition fluctuations and electric field within the correlation function of the order parameter fluctuations, see Eq. (3). To that aim the Brazovskii self-consistent Hartree approach [14] is modified. Here we apply the well established Fredrickson-Helfand formulation for the order parameter fluctuations [15], compare also [16], [17]. Intuitively it seems obvious that fluctuations become anisotropic in an electric field, and moreover fluctuations of modes with wave vectors parallel to the electric field are suppressed. The effects of an electric field on composition fluctuations are directly accessible in scattering experiments, and were studied for polymer solutions in [1] and for asymmetric diblock copolymers in [8].

MODELING

Collective Description

An AB diblock copolymer consists of a chain of N_A subunits of type A which is covalently bonded to a chain of N_B subunits of type B. A net repulsive $A-B$ interaction energy between the monomers leads to microphase separation. Thus at an order-disorder transition concentration waves are formed spontaneously, having a wavelength of the same order as the radius of gyration of the coils. The type of the long range order that forms depends on the composition of the copolymers $f = N_A/N_A + N_B$, where here the symmetric case $f = 1/2$ is considered. The deviation of the density of A-polymers from its mean value is an appropriate order parameter. Following [15] let us introduce instead of $\delta\Phi(\mathbf{r})$ the dimensionless order parameter $\psi(\mathbf{r}) = \delta\Phi_A(\mathbf{r})/\rho_m$. The effective Hamiltonian will be written in terms $\psi(\mathbf{r})$, where we follow the line

CP982, *Complex Systems, 5th International Workshop on Complex Systems*
edited by M. Tokuyama, I. Oppenheim, and H. Nishiyama

given in [15], [16] [18]. The starting expression for further discussions reads

$$H(\phi) = \frac{1}{2}\int_q \phi(-\mathbf{q})\left(\tilde{\tau}+(q-q^*)^2\right)\phi(\mathbf{q}) + \frac{\tilde{\lambda}}{4!}\int_{q_1}\int_{q_2}\int_{q_3}\phi(\mathbf{q}_1)\phi(\mathbf{q}_2)\phi(\mathbf{q}_3)\phi(\mathbf{q}_4), \quad (1)$$

where $\tilde{\lambda} = \gamma_4(\mathbf{q}^*,\mathbf{q}^*,\mathbf{q}^*,-\mathbf{q}^*-\mathbf{q}^*-\mathbf{q}^*)/c^4$ Here the vertex γ_4 is approximated by its value at $\mathbf{q}^*$. The quantity $\tilde{\tau}$ plays the role of the reduced temperature in the Landau theory of phase transitions. The scattering function is obtained from the last equation as

$$S_c^{-1}(q) = \tilde{\tau} + (q-q^*)^2\,. \quad (2)$$

Denoting as $\Gamma(\bar{\phi}$ the generating functional of the one-particle irreducible Greens's function, the second derivative of the Gibbs potential with respect to the order parameter yields the inverse correlation function [20]. The correlation functions of composition fluctuations in the ordered phase is defined by

$$S(\mathbf{x}_1,\mathbf{x}_2,\bar{\phi}) = \langle\Delta\phi(\mathbf{x}_1)\Delta\phi(\mathbf{x}_2)\rangle\,,$$

where the abbreviation $\Delta\phi(\mathbf{x}) = \phi(\mathbf{x}) - \bar{\phi}(\mathbf{x})$ is introduced. The correlation function can be obtained from the Gibbs potential Γ

$$\frac{\delta^2\Gamma(\bar{\phi})}{\delta\bar{\phi}(\mathbf{x}_1)\delta\bar{\phi}(\mathbf{x}_2)} = S^{-1}(\mathbf{x}_1,\mathbf{x}_2,\bar{\phi}). \quad (3)$$

which represents the relation between the thermodynamic quantities and the correlation function of the composition $\phi(\mathbf{x})$.
The order parameter for a symmetric diblock copolymer melt can be approximated in the vicinity of the critical temperature of the microphase separation by

$$\bar{\phi}(\mathbf{x}) = 2A\cos(q^*\mathbf{n}\mathbf{x})\,, \quad (4)$$

where $\mathbf{n}$ is an unit vector in the direction of the wave vector perpendicular to the lamellae and A is an amplitude. The Brazovskii self-consistent Hartree approach, which takes into account the fluctuation effects on the microphase separation, is based on the following expression of the derivative of the Gibbs potential with respect to the amplitude A of the order parameter

$$\frac{1}{A}\frac{\partial\Gamma(A)/V}{\partial A} = 2\left(\tilde{\tau} + \frac{\tilde{\lambda}}{2}\int_q S_0(\mathbf{q},A) + \frac{\tilde{\lambda}}{2}A^2\right). \quad (5)$$

The second term in Eq. (5) includes the propagator

$$S_0(\mathbf{q},A) = 1/\left(\tilde{\tau} + (q-q^*)^2 + \tilde{\lambda}A^2\right) \quad (6)$$

and represents the first-order correction to the thermodynamic potential owing to the self-energy. The first two terms in the brackets on the right-hand side of Eq. (5) are summarized to an effective reduced temperature denoted by $\tilde{\tau}_r$. The equation for $\tilde{\tau}_r$ becomes self-consistent by replacing $\tilde{\tau}$ in Eq. (6) for $S_0(\mathbf{q},A)$ by $\tilde{\tau}_r$. Then we get

$$\tilde{\tau}_r = \tilde{\tau} + \frac{\tilde{\lambda}}{2}\int_q S(\mathbf{q},A), \quad (7)$$

where $S^{-1}(\mathbf{q},A) = \tilde{\tau}_r + (q-q^*)^2 + \tilde{\lambda}A^2$ is the inverse of the effective propagator.

Contribution of the Electric Field to the Effective Hamiltonian

In this subsection we discuss the coupling of the diblock copolymer melt to an external time-independent electric field. The system we consider is a linear dielectric and is free of charges. The field satisfies the Maxwell equation

$$\operatorname{div}(\varepsilon(\mathbf{r})\mathbf{E}(\mathbf{r})) = 0,$$

where the inhomogeneities of the dielectric constant $\varepsilon(\mathbf{r})$ are caused by the inhomogeneities of the order parameter. According to [2] we adopt the expansion of the dielectric constant in powers of the order parameter up to quadratic terms

$$\varepsilon(\mathbf{r}) = \varepsilon_D(\mathbf{r}) + \beta\bar{\phi}(\mathbf{r}) + \frac{1}{2}\frac{\partial^2\varepsilon}{\partial\bar{\phi}^2}\bar{\phi}(\mathbf{r})^2. \quad (8)$$

In case of zero order parameter $\bar{\phi}(\mathbf{r}) = 0$ the dielectric constant is assumed to be homogeneous, i.e. $\varepsilon_D(\mathbf{r}) = \varepsilon_D$. The above Maxwell equation can be rewritten as integral equation as follows

$$\mathbf{E}(\mathbf{r}) = \mathbf{E}_0 + \frac{1}{4\pi}\nabla\int d^3r_1 G_0(\mathbf{r}-\mathbf{r}_1)\left(\mathbf{E}(\mathbf{r}_1)\nabla\right)\ln\varepsilon(\mathbf{r}_1), \quad (9)$$

where $G_0(\mathbf{r}) = 1/r$ is the Green's function of the Poisson equation. The integral equation (9) is convenient to derive iterative solutions for the electric field, and to take the dependencies of $\varepsilon(\mathbf{r})$ on the order parameter. The 2nd term in Eq. (9) takes into account the polarization due to the inhomogeneities of the order parameter. The substitution $\mathbf{E}(\mathbf{r}_1) = \mathbf{E}_0$ on the right-hand side of Eq. (9) gives the first-order correction to the external electric field as $\mathbf{E}(\mathbf{r}) = \mathbf{E}_0 + \mathbf{E}_1(\mathbf{r}) + \ldots$ with

$$\mathbf{E}_1 = \frac{1}{4\pi}\frac{\beta}{\varepsilon_D}\nabla\int d^3r_1 G_0(\mathbf{r}-\mathbf{r}_1)\left(\mathbf{E}_0\nabla\right)\bar{\phi}(\mathbf{r}_1) \quad (10)$$

The higher-order terms $\mathbf{E}_i(\mathbf{r})$ (i = 2, 3,...) in the last equation are linear in the external field, too.

In taking into account the electric energy in thermodynamic potentials one should distinguish between the thermodynamic potentials with respect to the charges or the potential [19]. These thermodynamic potentials are connected with each other by a Legendre transformation. Here, in calculating the effects of fluctuations we interpret in fact the Landau free energy as an Hamiltonian, which weights the fluctuations by the Boltzmann factor $\exp(-H)$. Therefore, the contribution of the electric field to the effective Hamiltonian corresponds to the energy of the electric field, and is given in Gaussian units by

$$k_BT\Gamma_{el} = \frac{1}{8\pi}\int d^3r\varepsilon(\mathbf{r})\mathbf{E}^2(\mathbf{r}),$$

where $\varepsilon(\mathbf{r})$ and $\mathbf{E}(\mathbf{r})$ are given by Eqs. (8,10). In the following we consider only the polarization part of Γ_{el}. The quadratic part of the latter in powers of the order parameter is given by

$$\begin{aligned} k_BT\Gamma_{el} &= \frac{1}{8\pi}\int d^3r\frac{1}{2}\frac{\partial^2\varepsilon}{\partial\bar{\phi}^2}\bar{\phi}(\mathbf{r})^2\mathbf{E}_0^2 \\ &+ \frac{1}{8\pi}\int d^3r\varepsilon_D\frac{1}{4\pi}\nabla^m \\ &\times \int d^3r_1G_0(\mathbf{r}-\mathbf{r}_1)E_0^n\frac{\beta}{\varepsilon_D}\nabla^n\bar{\phi}(\mathbf{r}_1) \\ &\times\frac{1}{4\pi}\nabla^m\int d^3r_2G_0(\mathbf{r}-\mathbf{r}_2)E_0^k\frac{\beta}{\varepsilon_D}\nabla^k\bar{\phi}(\mathbf{r}_2) \\ &\equiv \frac{1}{2}\int d^3r_1\int d^3r_2\bar{\phi}(\mathbf{r}_1)\tilde{\gamma}_2^{el}(\mathbf{r}_1,\mathbf{r}_2)\bar{\phi}(\mathbf{r}_2) \end{aligned} \quad (11)$$

Notice the sum convention over the indices m, n, $k$in the above two equations and in Eq. (13). Expressing Γ_{el} by the Fourier components of the order parameter yields

$$\Gamma_{el} = \frac{1}{2}\int_q\bar{\phi}(-\mathbf{q})\tilde{\gamma}_2^{el}(\mathbf{q})\bar{\phi}(\mathbf{q}), \quad (12)$$

whereas $\tilde{\gamma}_2^{el}(\mathbf{q})$ is given by

$$\tilde{\gamma}_2^{el}(\mathbf{q}) = \frac{1}{4\pi\rho_mk_BT}\left(\frac{1}{2}\frac{\partial^2\varepsilon}{\partial\bar{\phi}^2}\mathbf{E}_0^2+\frac{\beta^2}{\varepsilon_D}\frac{q^nq^k}{\mathbf{q}^2}E_0^nE_0^k\right). \quad (13)$$

The electric contribution to the correlation function can be obtained using Eq. (3). Note that the factor ρ_m^{-1} in Eq. (13) is due to the length redefinition $\mathbf{r} = \rho_m^{-1/3}\mathbf{x}$. Eqs. (11)-(13) are used in the subsequent section to analyze the influence of the electric field on the composition fluctuations of the order parameter. Assuming that the electric field is directed along the z-axes, and denoting the angle between the field and the wave vector $\mathbf{q}$ by θ, we obtain the quantity $\tilde{\gamma}_2^{el}(\mathbf{q})$ in Eq. (13) as

$$\tilde{\gamma}_2^{el}(\mathbf{q}) = \tilde{\alpha}\cos^2\theta+\tilde{\alpha}_2, \quad (14)$$

where the notations

$$\tilde{\alpha} = \frac{1}{4\pi\rho_mk_BT}\frac{\beta^2}{\varepsilon_D}\mathbf{E}_0^2, \quad \tilde{\alpha}_2 = \frac{1}{4\pi\rho_mk_BT}\frac{1}{2}\frac{\partial^2\varepsilon}{\partial\bar{\phi}^2}\mathbf{E}_0^2 \quad (15)$$

are used.

Hartree Treatment of Fluctuations in the Presence of the Electric Field

Whereas the Brazovskii-Hartree approach in the absence of an electric field is summarized by Eqs. (5) and (7), these set of equations is changed in case of an field. It results

$$\begin{aligned} &\frac{1}{A}\frac{\partial}{\partial A}\left(g-\tilde{\gamma}_2^{el}(\mathbf{q}^*)A^2\right) = \tilde{\tau}+ \\ &2\tilde{\lambda}\int_q\frac{1}{\tilde{\tau}+\tilde{\gamma}_2^{el}(\mathbf{q})+(q-q^*)^2+\tilde{\lambda}A^2}+\tilde{\lambda}A^2 \end{aligned} \quad (16)$$

where $g = \Gamma(A)/V$ is the thermodynamic potential per volume. The term $\tilde{\gamma}_2^{el}(\mathbf{q}^*)$ in Eq. (16) is the contribution to g associated with the lamellae in the electric field below the transition. Its orientation is defined by the angle between the wave vector $\mathbf{q}^*$ and the field strength E_0, $\mathbf{q}^*E_0 = q^*E_0\cos\theta^*$. The orientation of the lamellae in equilibrium can be obtained by minimization of the thermodynamic potential with respect to the angle θ^* resulting in $\theta^* = \pi/2$. As a consequence, the modulations of the order parameter perpendicular to the electric field possess the lowest electric energy. The term $\tilde{\gamma}_2^{el}(\mathbf{q}^*)A^2$ in Eq. (16) disappears in the equilibrium state and will therefore not considered further. The fluctuations in the presence of the electric field become anisotropic due to $\tilde{\gamma}_2^{el}(\mathbf{q})$ in Eq. (16). As above the first two terms on the right-hand side in Eq. (16) define an effective $\tilde{\tau}$, which is denoted by $\tilde{\tau}_r$. Replacing $\tilde{\tau}$ under the integral in Eq. (16) by $\tilde{\tau}_r$ we obtain a self-consistent equation for $\tilde{\tau}_r$ of the form

$$\tilde{\tau}_r = \tilde{\tau}+\frac{\tilde{\lambda}}{2}\int_q\frac{1}{\tilde{\tau}_r+\tilde{\alpha}\cos^2\theta+(q-q^*)^2+\tilde{\lambda}A^2}. \quad (17)$$

This relation generalizes Eq. (7) for non-zero field $\mathbf{E}\neq 0$. The integration can be carried out leading to

$$\begin{aligned} &\frac{\tilde{\lambda}}{2}\int_q\frac{1}{\tilde{\tau}_r+\tilde{\lambda}A^2+\tilde{\alpha}\cos^2\theta+(q-q^*)^2} \\ &= \frac{\tilde{\lambda}q_*^2}{4\pi\sqrt{\tilde{\alpha}}}\operatorname{arcsinh}\sqrt{\frac{\tilde{\alpha}}{\tilde{\tau}_r+\tilde{\lambda}A^2}} \\ &= \frac{d\tilde{\lambda}}{N\sqrt{\tilde{\alpha}}}\operatorname{arcsinh}\sqrt{\frac{\tilde{\alpha}}{\tilde{\tau}_r+\tilde{\lambda}A^2}}, \end{aligned} \quad (18)$$

where the notation $d = 3y^*/2\pi$ with $y^* = q^{*2}N/6 = 3.7852$ is used. Notice that the leading contribution to the integral in Eq. (18) comes from the peak of the structure factor at q^*. Eq. (17) shows that the fluctuations in the presence of the electric field are suppressed due to the angular dependence of the integrand. Consequently an electric field weakens the first-order phase transition. The contributions for large wave vectors, which become finite after an introduction of an appropriate cutoff at large q, are expected to renormalize the local parameters such as the χ parameter etc. These fluctuations have been considered recently in [22]. The self-consistent equation for $\tilde{\tau}_r$ in the presence of the field reads

$$\tilde{\tau}_r = \tilde{\tau} + \frac{\tilde{\lambda} d}{N\sqrt{\tilde{\alpha}}} \operatorname{arcsinh}\sqrt{\frac{\tilde{\alpha}}{\tilde{\tau}_r + \tilde{\lambda} A^2}}. \quad (19)$$

Remark the in the limiting case $\alpha \to 0$ the last equation is consistent with the Brazovskii-Fredrickson-Helfand model. In our case the thermodynamic potential is immediately obtained from Eq. (16) by integration

$$\begin{aligned} g &= \int_0^A 2t(A)dA - \frac{\lambda}{4}A^4 = \frac{1}{2\lambda}(t^2 - t_0^2) \\ &+ \frac{d}{\sqrt{N}}\left(\sqrt{t+\alpha} - \sqrt{t_0+\alpha}\right)\frac{\lambda}{4}A^4. \end{aligned} \quad (20)$$

The inverse susceptibility within the disordered phase $t_0 \equiv t(A=0)$ satisfies the equation

$$t_0 = \tau + \frac{d\lambda}{\sqrt{N\alpha}} \operatorname{arcsinh}\sqrt{\frac{\alpha}{t_0}}. \quad (21)$$

Eqs. (20, 21) generalize the Brazovskii-Fredrickson-Helfand treatment of the composition fluctuations in symmetric diblock copolymer melts in the presence of an external time-independent electric field. In the subsequent section we will discuss the results.

RESULTS

The position of the phase transition is determined by the conditions

$$g = 0, \quad \frac{\partial g}{\partial A} = 0,, \quad (22)$$

leading to

$$\frac{1}{2}(t^2 + t_0^2) = \frac{d\lambda}{\sqrt{N}}\left(\sqrt{t+\alpha} - \sqrt{t_0+\alpha}\right). \quad (23)$$

The perturbative solution of Eqs. (21) and (23) yields for the transition temperature in first order of α

$$\tau_t = -2.0308\,(d\lambda)^{2/3} N^{-1/3} + 0.48930\,\alpha. \quad (24)$$

In the same manner we get

$$\begin{aligned} (\chi N)_t &= (\chi N)_s + 1.0154\,c^2\,(d\lambda)^{2/3} N^{-1/3} \\ &- 0.48930\,\alpha\,\frac{c^2}{2}. \end{aligned} \quad (25)$$

An analogous calculation (for details see [18]) offers the amplitude of the order parameter at the transition point:

$$\begin{aligned} A_t &= 1.4554\,\left(d^2/\lambda\right)^{1/6} N^{-1/6} \\ &- 0.42701\,\left(d^2\lambda^5\right)^{-1/6} N^{1/6}\,\alpha, \end{aligned} \quad (26)$$

where α is defined by $\alpha = N\beta^2/(4\pi\rho_m k_B T \varepsilon_D)\mathbf{E}_0^2$. The corrections associated with the electric field in Eqs. (24-26) are controlled by the dimensionless expansion parameter $\alpha N^{1/3}/(d\lambda)^{2/3}$. Inserting the values of the coefficients for $f = 1/2$ gives

$$(\chi N)_t = 10.495 + 41.018 N^{-1/3} - 0.29705\,\alpha, \quad (27)$$

and

$$A_t = 0.81469 N^{-1/6} - 0.0071843 N^{1/6}\,\alpha. \quad (28)$$

Note that for $N \to \infty$ the strength of the electric field E_0 should tend to zero in order that the dimensionless expansion parameter $\alpha N^{1/3}/(d\lambda)^{2/3}$ remains small. Eqs. (27-28) are our main results, which had been not considered in the previous studies [2]-[3]. The additional terms in Eqs. (27 - 28) in powers of α describe the effects of the electric field on the fluctuations. The coupling between fluctuations and the electric field is originated from the term $\gamma_2^{el}(\mathbf{q})$ under the integral in Eq. (16). From Eq. (27) one observes a shift of the interaction parameter $(\chi N)_t$ to lower values due to the electric field and correspondingly a shift of the transition temperature to higher values towards its mean-field value. In so far the electric field favors a demixing with respect to the free field case. According to Eq. (28) the electric field lowers the value of the order parameter at the transition point. The electric field weakens the fluctuations, and consequently the first order phase transition.

Let us now discuss the limits of the Brazovskii-Hartree approach in the presence of the electric field. The general conditions on the validity of the Brazovskii-Hartree approach, discussed already in [15] and [23], hold also in the presence of the electric field. Essentially, while taking into account the fluctuations, the peak of the structure factor at the transition should remain sufficiently sharp, i.e. the transition should be of (weak) first-order. This requires large values of N [15]. The approximation made in the presence of an electric field consists in adopting the expansion of the dielectric constant in powers of the order parameter until second order. Such an expansion is however only justified in the vicinity of the

transition. The smallness of the linear term in powers of α in Eqs. (27-28) imposes a condition on E_0, namely $E_0^2 \ll \lambda^{2/3} \rho_m k_B T \varepsilon_D / \beta^2 N^{4/3}$. The applicability of the linear dielectric theory yields likewise a limitation of the strength of the electric field.
The dependence of the propagator on α in our field theoretical treatment in terms of an effective Hamiltonian enables us to make a general conclusion on the fluctuations within the Brazovskii-Hartree approach and beyond. Our studies indicate that fluctuations are suppressed for large α which can be traced back to the term $\tilde{\gamma}_2^{el}(\mathbf{q})$ in Eq. (13). According to this fact one expects that the order-disorder transition tends to a phase transition of second order for stronger fields which is in contrast to the common behavior. The numerical solution of Eqs. (22) yields that the mean-field behavior is recovered only in the limit $\alpha \to \infty$, which is obviously beyond the applicability of linear electrodynamics. The angular dependence of γ_4, which has not been taken into account in the present work, gives rise to corrections which are beyond the linear order in α.
The main prediction of the present work namely that the electric field weakens fluctuations agrees qualitatively with the behavior of diblock copolymer melts in shear flow studied in [24], where the shear also suppresses the fluctuations. Due to the completely different couplings to the order parameter the calculation schemes are different in both cases.
Now we estimate the shift of the critical temperature for the diblock copolymer Poly(styrene-block-methylmethacrylat) (molecular weight 31000 g/mol due to the electric field. The transition temperature without the field is 182°C. Following [3] we use the following values of the parameters:

$$\begin{aligned} \varepsilon_{PS} &= 2.5, \varepsilon_{PMMA} = 5.9, \quad \beta = \varepsilon_{PMMA} - \varepsilon_{PS} \\ \varepsilon_D &= (\varepsilon_{PS} + \varepsilon_{PMMA})/2, \quad \chi = 0.012 + 17.1/T. \end{aligned}$$

The estimation of the number of statistical segments N using the relation $(\chi N)_t = 10.495 + 41.018 N^{-1/3}$ for $T_t = 182°C$ yields $N \approx 331$. The relation $b = \rho_m^{-1/3}$ gives the value for the statistical segment length $b \approx 5.2 Å$. For the field strength $E_0 = 40 V/\mu m$ the shift is obtained using Eq. (27) as

$$\Delta T_t = 2.5\,\mathrm{K}.$$

The numerical value of the dimensionless parameter $\alpha N^{1/3}/(d\lambda)^{2/3}$ is found to be 0.059. Note that the shift is very sensitive to the value of N. It would be very desirable for further theoretical studies to determine the shift of the transition experimentally.
The scattering function in the disordered phase is obtained by taking into account the composition fluctuations in the presence of the electric field as

$$S_{dis}(\mathbf{q}) \simeq \frac{1}{t_0 + \alpha \cos^2\theta + (q - q^*)^2}. \tag{29}$$

The fluctuation part of the scattering function in the ordered state is obtained as

$$S_{ord}(\mathbf{q}) \simeq \frac{1}{t + \alpha \cos^2\theta + (q - q^*)^2}. \tag{30}$$

At the transition point the expansion of t_0 and t in powers of the field strength is derived from Eq. (21) as

$$t_{0,t} = 0.20079\,(\lambda\, d)^{2/3} N^{-1/3} - 0.20787\,\alpha + \ldots \tag{31}$$

and

$$t_t = 1.0591\,(\lambda\, d)^{2/3} N^{-1/3} - 0.62147\,\alpha + \ldots \tag{32}$$

The difference between $t_{0,t}$ and t_t, which is due to the finite value of the order parameter at the transition, results in the jump of the peak at the transition point. Owing to the term $\alpha \cos^2\theta$ the structure factor becomes anisotropic in the presence of the electric field. The dependence of the structure factor on the electric field is twofold, namely via $t_{0,t}$ (t_t) and the angular-dependent term $\alpha \cos^2\theta$. The suppression of $t_{0,t}$ (t_t) in an electric field according to Eqs. (31, 32) results in an increase of the peak. Thus, for wave vectors perpendicular to the field direction, where the angular part disappears, the peak is more pronounced than that for $\mathbf{E}_0 = 0$. In the opposite case for wave vectors parallel to $\mathbf{E}_0$ the anisotropic term ($= \alpha \cos^2\theta$) dominates and the peak is less pronounced as for $\mathbf{E}_0 = 0$. Composition fluctuations can be associated with fluctuation modulations of the order parameter. According to Eqs. (29-30) fluctuate modulations of the order parameter with wave vectors parallel to the field are mostly suppressed. The latter correlates with the behavior in the ordered state where the lamellae with the wave vector perpendicular to the field direction possess the lowest energy.

MEMBRANES IN ELECTRIC FIELDS

Let us shortly discuss that also membranes are able to limit the fluctuations when they subjected to an external field. In that case the energy of the membrane is determined by the bending rigidity of the membrane and the external field, where in general the electric field should be coupled to the local rigidity of the membrane. Denoting $h(x,y)$ the height of a almost planar membrane the energy can be written in lowest order as

$$H = \frac{\kappa}{2} \int (\nabla^h)^2 d^2x \tag{33}$$

Obviously in case the surface consists of charged material an electric field should influence the local curvature of the membrane. The Hamiltonian of the simplest version reads

$$H = \frac{1}{2} \int d^2x \left(\kappa (\nabla^2 h)^2 + \lambda (\nabla h \cdot \mathbf{E})^2\right) \tag{34}$$

From here we can find the mean square displacement

$$< h^2 >= \frac{1}{(2\pi)^2} \int d^2q \frac{k_BT}{\kappa q^4 + \lambda (Eq)^2 \cos^2 \Theta} \quad (35)$$

The integration has to be performed in the interval $q_i \leq q \leq q_m$ and yields

$$< h^2 >= \frac{k_BT}{2\pi\lambda E^2} \left[\sqrt{1 + \frac{\lambda E^2}{\kappa q_i^2}} - \sqrt{1 + \frac{\lambda E^2}{\kappa q_m^2}} \right] \quad (36)$$

There is no divergence at the upper cut-off $q_m \to \infty$. The effect of the electric field consists of a limitation of the fluctuations. In case of a weak field

$$\frac{\lambda E^2}{\kappa_i} << 1$$

leads to

$$< h^2 > \approx \frac{k_BT}{4\pi\kappa q_i^2} \left[1 - \frac{\lambda E^2}{4\kappa q_i^2} \right] . \quad (37)$$

For zero field the mean square displacement diverges at the lower cut-off $q_i \sim L^{-1} \to 0$, where L is the linear size of the system. The electric field weakens the fluctuation of the width. In the opposite case of a strong field

$$\frac{\lambda E^2}{\kappa_i} >> 1$$

the situation is changed and it results

$$< h^2 > \approx \frac{k_BT}{2\pi\sqrt{\lambda\kappa} E q_i} . \quad (38)$$

With increasing field the width of the membrane becomes smaller. The field leads to a limitation of the fluctuations.

CONCLUSIONS

We have generalized the Fredrickson-Helfand theory of microphase separation in symmetric diblock copolymer melts by taking into account the effects of the electric field on the composition fluctuations. We have shown that an electric field suppresses the fluctuations and therefore weakens the first-order phase transition. However, the mean-field behavior is recovered in the limit $\alpha \to \infty$, which is therefore outside the applicability of the linear electrodynamics. The collective structure factor in the disordered phase becomes anisotropic in the presence of the electric field. Fluctuation modulations of the order parameter along the field direction are mostly suppressed. Thus, the anisotropy of fluctuation modulations in the disordered state correlates with the parallel orientation of the lamellae in the ordered state. The paper is finished by considering a comparable problem, namely the influence of an electric field on a membrane with charged constituents. The field is directly coupled to the curvature of the almost flat membrane. Due to the electric field the roughness is limited.

REFERENCES

1. D. Wirtz, K. Berend, and G. G. Fuller, *Macromolecules* **25**, 7234-7246 (1992).
2. K. Amundson, E. Helfand, X. Quan, and S. D. Smith, *Macromolecules* **26**, 2698-2703 (1993).
3. K. Amundson, E. Helfand, X. Quan, S. D. Hudson, and S. D. Smith, *Macromolecules* **27**, 6559-6570 (1994).
4. A. Onuki and J. Fukuda, *Macromolecules* **28**, 8788-8795 (1995).
5. Y. Tsori and D. Andelman, *Macromolecules* **35**, 5161-5170 (2002).
6. T. Thurn-Albrecht, J. DeRouchey, T. P. Russell, *Macromolecules* **33**, 3250-3253 (2000).
7. Y. Tsori, D. Andelman, C.-Y. Lin, M. Schick, *Macromolecules* **39**, 289-293 (2006).
8. T. Thurn-Albrecht, J. DeRouchey, T. P. Russel, and R. Kolb, *Macromolecules* **35**, 8106-8110 (2002).
9. Y. Tsori and D. Andelman, *J. Polym. Scie. Part B - Polym. Phys.* **44**, 2725-2732 (2006).
10. A. Böker, A. Knoll, H. Elbs, V. Abetz, A. H. E. Müller, G. Krausch, *Macromolecules* **35**, 1319-1325 (2002).
11. A. Böker, K. Schmidt, A. Knoll, H. Zettl, H. Hansel, V. Urban, V. Abetz, G. Krausch, *Polymer* **47**, 849-859 (2006).
12. M. W. Matsen, *J. Chem. Phys.* **124**, 074906-1-074906-9 (2006).
13. C. Park, J. Yoon, E. L. Thomas, *Polymer*, **44**, 6725-6732 (2003).
14. S. A. Brazovskii, *Sov. Phys. JETP* **41**, 85-95 (1975).
15. G. H. Fredrickson and E. Helfand, *J. Chem. Phys.* **87**, 697-705 (1987).
16. L. Leibler, *Macromolecules* **13**, 1602-1617 (1980).
17. Y. Tsori, F. Tournilhac, and L. Leibler, *Nature* **430**, 544-547 (2004).
18. I. Gunkel, S. Stepanow, T. Thurn-Albrecht, and S. Trimper, *Macromolecules* **40**, 2186-2191 (2007).
19. L. D. Landau and E. M. Lifshitz, *Electrodynamics of Continuous Media*, Pergamon Press, Oxford, 1984. §10.
20. J. J. Zinn-Justin, *Quantum Field Theory and Critical Phenomena*, Oxford University Press, Oxford, 2002.
21. G. H. Fredrickson and K. Binder, *J. Chem. Phys.* **91**, 7265-7275 (1989).
22. A. Kudlay and S. Stepanow, *J. Chem. Phys.* **118**, 4272-4276 (2003).
23. J. Swift and P. C. Hohenberg, *Phys. Rev. A* **15**, 319-328 (1977).
24. M. E. Cates and S. T. Milner, *Phys. Rev. Lett.* **62**, 1856-1859 (1989).

Numerical and Experimental Studies on Alignment Dynamics of Lamellae of Block Copolymer under an Electric Field

T. Taniguchi*,†, R. Uchino*, M. Sugimoto*, and K. Koyama*

*Dept. of Polym. Sci. and Eng., Yamagata University., 4-3-16, Jonan, Yonezawa, Yamagata
†CREST, Japan Science and Technology Agency

Abstract. Dynamics of a structural change in block copolymer system by applying an electric field near a temperature of order disorder transition ($T_{\rm ODT}$) is investigated using a time resolved Small Angle X-ray Scattering technique (SAXS). The sample used in this study is nearly symmetric poly-isoprene-block-poly(isobutyl-methacrylate) copolymer (P(I-b-iBuMA)) having an order disorder phase transition temperature $T_{\rm ODT} = 92.5°$C. Grains in which lamellar are oriented along the direction of applied electric field of 2.0 kV/mm AC are observed by Atomic Force Microscopy of P(I-b-iBuMA) in 90°C. Results of continuous SAXS measurements of P(I-b-iBuMA) under an applying electric field in 90°C indicate that alignments of lamellar domains come from rotations of grains. In order to confirm whether a grain that consists of a lamella can rotate under an electric field or not, computer simulations are performed. The result indicates that alignment of lamellae along the direction of electric field can be taken place not only by rotation but also by nucleation in which lamella has already been aligned along the direction of electric field.

Keywords: Block Copolymer, Electric Field, Nucleation
PACS: 61.43.Bn, 64.60.Q-, 64.75.Va

INTRODUCTION

Recently, increasing attentions has been paid to control of meso-phase structures of polymeric systems in order to apply them industrially. It has been investigated that an orientation of a meso-phase such a lamellar or cylinder structure of block copolymer melt can be controlled by applying an electric field [1, 2, 3, 4, 5]. For instance, an orientation of cylinder structure can be aligned along the direction of applied electric field. Previous works mainly focus on a control of orientation by applying an electric field to an ordered state. In such a control of orientation of ordered structure by using electric field, rather larger strength of electric field (order of 10kV/mm) is needed to realize it. It is also reported [6] that such a control of orientation of ordered structure can be realized by applying an electric field of a few kV/mm just after quenching a block copolymer melt to the state slightly below the order-disorder transition temperature. It is considered that such a difference of necessary strength of electric field to control an orientation of an ordered structure comes from the difference in kinetic pathway to reach a final oriented state. However studies from a point of view of kinetic pathway have not been done in depth, and therefore it is not clear yet how kinetic pathway is realized to reach an oriented state by applying an electric field after quenching just below the order disorder transition temperature. So the aim of this paper is to reveal the kinetic pathway from a disordered state to an oriented state by applying an electric field for a block copolymer melt quenched to a temperature slightly below the order-disorder transition temperature.

EXPERIMENTAL

Sample we used is (P(I-b-iBuMA)) copolymer: poly-isoprene-block-poly(isobutyl-methacrylate) copolymer. The sample is mainly composed of a nearly symmetric PI/PiBuMA copolymer with a number averaged molecular weight Mn =15,800 and a block ratio PI/PiBuMA= 52/48 (wt%), but contains 9 wt% of a higher molecular weight component of PI/PiBuMA with Mn=92,800, PI/PiBuMA=10/90 (wt%). >From result of SAXS and AFM, order-disorder transition occurs at $T = 92.5 \pm 2.5°$C, and ordered state is found to be a lamellar structure. We observed an equilibrium state under applying an electric field after quenching just below the order disorder transition temperature. The sample is shaped to be a film with a thickness 500 μm. We applied an electric field (2.0kV/mmAC) along the direction of thickness. The followings are procedures of our experiments. Firstly, (1) the sample is heated to 140°C and (2) is equilibrated for 1 hour to be realized a disordered state. Then, (3) it is quenched to 90°C that is slightly below the order disorder transition temperature, and (4) the electric field is applied to the sample for 12 hours with keeping $T = 90°$C.

CP982, *Complex Systems, 5th International Workshop on Complex Systems*
edited by M. Tokuyama, I. Oppenheim, and H. Nishiyama

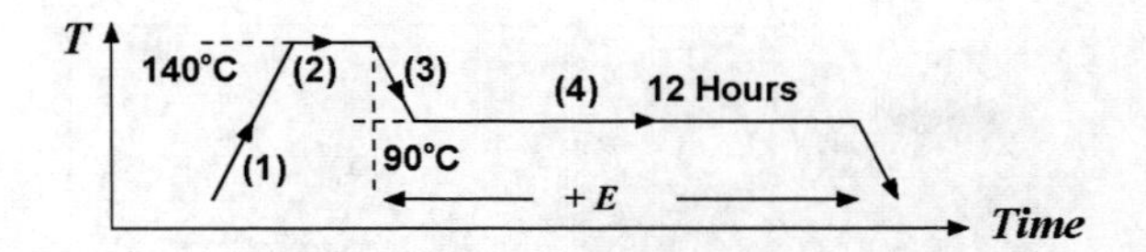

FIGURE 1. Thermal and Electric History of the sample

Finally the sample is immersed into an ice water to freeze the meso structure. The above processes (1)-(4) are performed in a Nitrogen atmosphere. The sample obtained by the above mentioned processes is analyzed by Small Angle X-ray Scattering (SAXS) (Rigaku, MicroMax007).

Time Resolved SAXS Analysis

We also investigate orientation dynamics of lamella under an electric field by performing a sequential SAXS analysis during the step (4). For this purpose, we prepare a sample cell that we can apply an electric field and control the temperature.

By time resolved SAXS analyses we investigate orientation dynamics of lamella under the electric field 2kV/mm. Figure 2 shows circularly averaged scattering profiles at times $t = 114$, 1183, 2965, 10095, 20076, 40113 sec during the process (4) in Fig.1. The maximum peak gradually grows and is sharpened with time and the peak position reaches to $q = 0.4$/nm at around $t = 2 \times 10^4$ sec. Figure 2 shows time dependences of scattering intensities of first order peak for circularly averaged (+), horizontal (filled triangle) and vertical direction (filled circle). In Fig. 3, the scattering intensity at the first peak along a radial direction in $\boldsymbol{q}$-space is plotted as a function of an azimuthal angle ϕ, where ϕ is defined by the angle from the axis perpendicular to the direction of electric field.

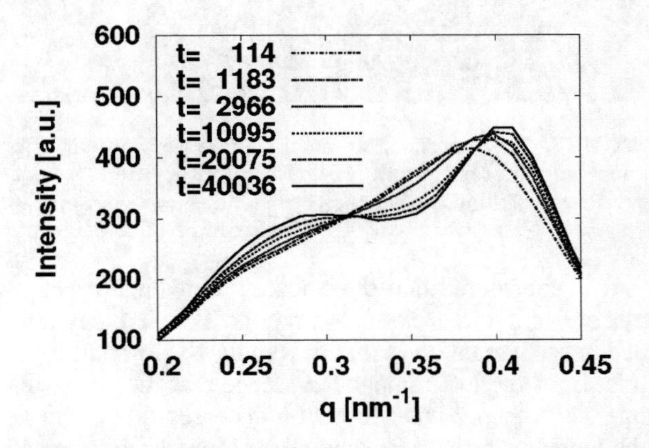

FIGURE 2. Circularly average scattering profiles at several times during the process (4).

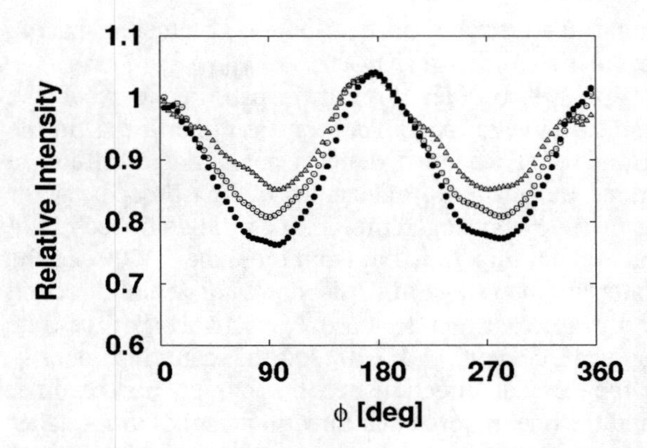

FIGURE 3. Scattering intensity at the first peak along a radial direction with an azimuthal angle ϕ in the reciprocal space, at the time $t = 114$ sec (△), $t = 2965$ sec (○) and $t = 40113$ sec (●)

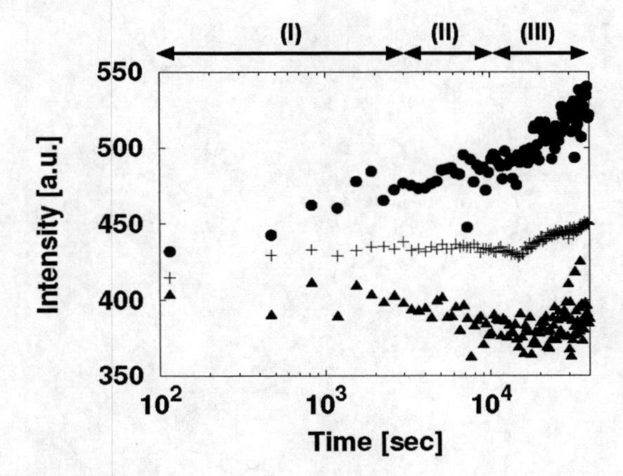

FIGURE 4. Time dependence of scattering intensities of first order peak for the circularly averaged one (+), horizontal (filled triangle) and vertical direction (filled circle).

The higher intensity at $\phi = 0$ and $\phi = 180$ than the other positions indicates that lamellae are oriented along the direction of electric field. In addition, the degree of the orientation becomes higher with time. The scattering intensities of first order peak for the circularly averaged one, horizontal and vertical direction are shown as function of time in Fig. 4. >From the time evolution of the circularly averaged scattering intensity in Figs. 2 and 4, it is found that the first order peak increases up to around 3,000 sec (Region I), and equilibrate up to 10,000 sec (Region II), then the scattering intensity restarts to increase after around 10,000 sec (Region III). By comparing Fig. 2 with the Fig. 4, it is found that in the beginning of the Region III, the broad peak around $q = 0.3$/nm is sharpened in the same way to the behavior of the first order peak around $q = 0.4$/nm. It is considered that in the Region I nucleation and growth take place, in the Region II structures are equilibrated, and in the Region III the component of block copolymers with a higher molecular weight (Mn = 82,800, 9wt%) goes out from lamellar domains, and therefore the distribution of lamellar spacing becomes sharper and sharper.

Next we consider the kinetic pathway to reach the oriented lamellar state. As seen from the Fig. 4, in the beginning of Region I there is not so large difference among the scattering intensities of first order peak for the circularly averaged profile, the profiles in horizontal and vertical direction, but from around $t = 3000$ sec the scattering intensities in vertical and horizontal direction start to increase and decrease, respectively. For the time region after about $t = 1 \times 10^4$ sec, the scattering intensity in the vertical direction becomes larger, on the other hand the one in horizontal direction reaches to a smaller constant value.

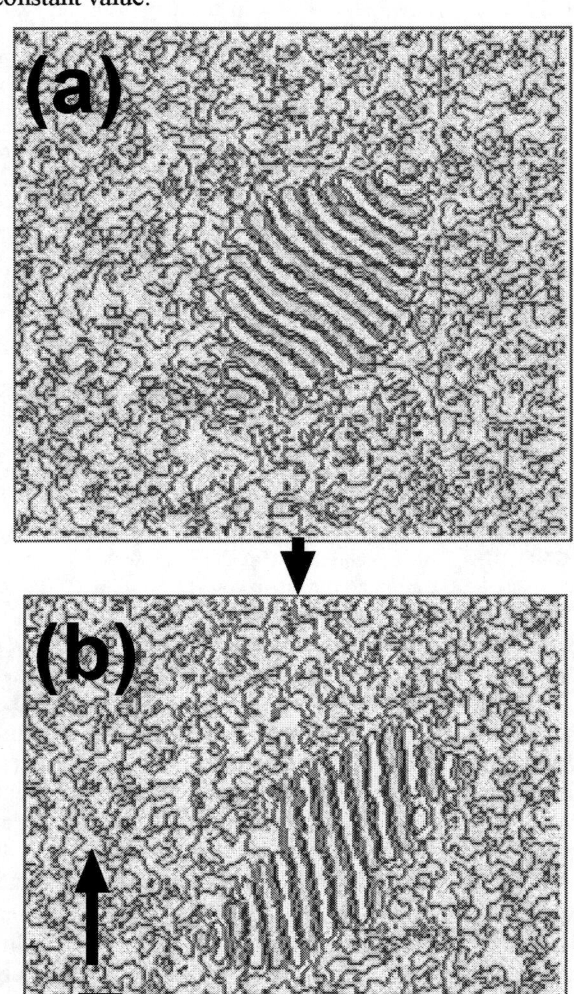

FIGURE 5. (a) A lamellar grain obtained by a numerical simulation under no electric field. (b) The grain in a certain time period after applying electric field to the state (a).

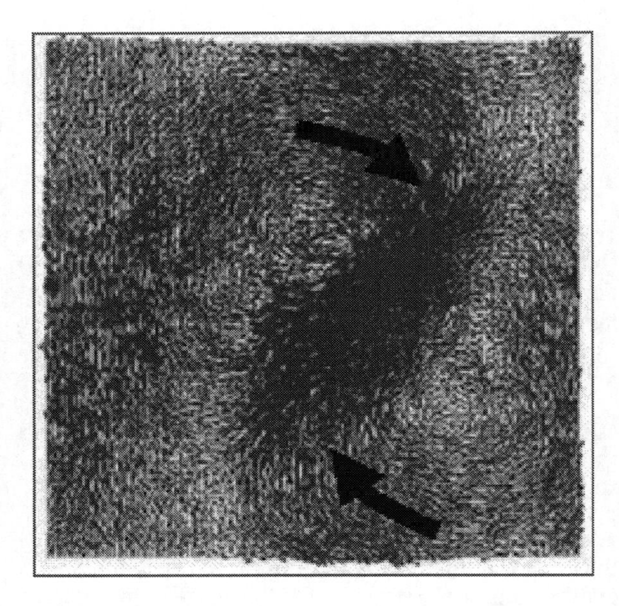

FIGURE 6. Flow field in the state Fig.5(b)

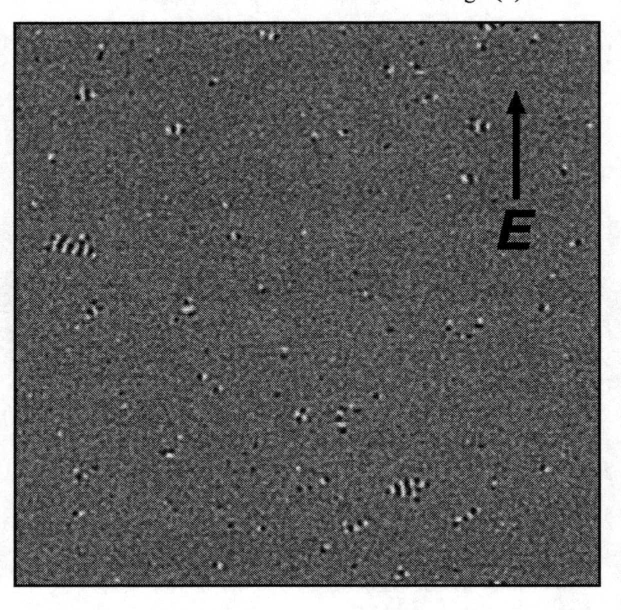

FIGURE 7. Snapshots of block copolymer system after quenching the system to a slightly below T_{ODT} from a disorder state. Simultaneously, an electric field is applied. Some of nucleated grains are initially oriented to the direction of electric field.

It is considered that the orientation of lamellar takes place mainly in Region II and the lamellar structure becomes sharpen and clear in Region III. In addition to this, from the experimental evidence that the scattering intensity of horizontal direction deceases in Region II, the lamellar in the nucleated domain randomly oriented initially, and as the grain size increases the grain starts to rotate.

NUMERICAL SIMULATION

We performed numerical simulations of a block copolymer system under an electric field in order to confirm whether the direction of lamella in each grain can be aligned by a rotation or not, because our experimental data support that the alignment of lamella along the electric field comes from rotations of lamellar grains. The dynamics of meso phase formed by block copolymer can be described by following coupled equations [7]:

$$\frac{\partial}{\partial t}\psi(\boldsymbol{r},t) = -\nabla\cdot(\boldsymbol{v}\psi) + L\nabla^2\mu(\boldsymbol{r}) + \xi(\boldsymbol{r},t) \quad (1)$$

$$\langle\xi(\boldsymbol{r},t)\xi(\boldsymbol{r}',t')\rangle = -2k_BT\nabla^2\delta(\boldsymbol{r}-\boldsymbol{r}')\delta(t-t') \quad (2)$$

$$\nabla\cdot(\varepsilon(\psi)\boldsymbol{E}) = 0, \qquad \varepsilon(\psi) = \bar{\varepsilon} + \Delta\varepsilon\psi \quad (3)$$

$$\eta\nabla^2\boldsymbol{v} - \nabla p - \psi\nabla\mu(\boldsymbol{r}) = 0, \qquad \nabla\cdot\boldsymbol{v} = 0 \quad (4)$$

where ψ is the order parameter of this system and is given by the local volume fraction difference between the constituent components of block copolymer, $\boldsymbol{v}$ the velocity field, p the pressure, η the viscosity, L the Onsager kinetic coefficient, ξ the thermal noise. In the equation for electric field, ε is a local dielectric constant that is given by a function of ψ, $\bar{\varepsilon}$ and $\Delta\varepsilon$ being constants. The local chemical potential $\mu(\boldsymbol{r})$ is given by the functional derivative of free energy of this system F with respect to ψ. The free energy functional is given by

$$F = \frac{k_BT}{a^3}\int[\frac{e}{2}\{(\nabla^2+q_o^2)\psi\}^2 + f(\psi) - \frac{\varepsilon}{2}E^2]d\boldsymbol{r}. \quad (5)$$

This free energy functional for block copolymer is called Brazovskii free energy functional. Using these equations, we performed numerical calculation. Firstly, we examined whether a grain can be rotated by the effect of electric field, or not. We prepared an initial state where a grain of lamellar orients like in Fig. 5(a) without electric field. Then we applied an electric field. As seen from Fig. 5(b), the grain is aligned along the electric field. We plot the velocity field obtained by this calculation in Fig. 6. This figure 6 indicates that a flow field is induced to yield a rotational motion of the grain. In case that an electric field is applied to the block copolymer system just after quenching from a disorder state to a slightly below the order disorder phase transition temperature, we found the direction of lamellae in some of nucleated domain have already been aligned along the direction of electric field, as shown in Fig.7. Our simulations imply that the aligning of lamellar direction in grain is induced not only by rotation of grain, but also by initially oriented lamellar nuclei.

CONCLUSION

The direction of lamellar structure of nearly symmetric block copolymer (P(I-b-iBuMA)) melt settled at a temperature just below the order disorder transition temperature could be oriented by applying an electric field of 2kV/mm AC for about 12 hours. >From the results of the time resolve SAXS analysis, we found that the scattering intensities along horizontal and vertical direction are almost same in the nucleation and growth region I, but in the region II, the scattering intensity along the horizontal and vertical direction decreases and increases, respectively. In the later stage, Region III, the scattering intensity along the horizontal direction becomes a smaller constant, on the other hand, the scattering intensity along the vertical direction continues to increase. >From the observed experimental evidences, it is considered that randomly oriented lamellar grains are nucleated and grows up to a certain size, and then rotate to orient along the direction of electric field. On the other hand, our performed numerical simulations imply that alignments of lamella in grains are induced not only by rotation of grains, but also by initially oriented lamellar nuclei.

ACKNOWLEDGMENTS

This work was supported by KAKENHI (Grant-in-Aid for Scientific Research) Grant No.19340117 from the Ministry of Education, Culture, Sports, Science and Technology of Japan.

REFERENCES

1. K. Amundson et al., *Macromolecules* **26**, 2698-2703 (1993).
2. K. Amundson et al., *Macromolecules* **27**, 6559-6570 (1994).
3. T. L. Morkved et al., *SCIENCE* **273**, 931-933 (1996).
4. T. Thurn-Albrecht et al., *Macromolecules* **33**, 3250-3253 (2000).
5. G. G. Pereira and D. R. M. Williams, *Macromolecules* **32**, 8115-8120 (1999).
6. K. Amundson et al., *Macromolecules* **24**, 6546-6548 (1991).
7. T. Hashimoto et al., *Phys. Rev. E* **54**, 5832-5835 (1996).

The Thermal Conductivity of Amorphous Polymers Calculated by Non-Equilibrium Molecular Dynamics Simulation

T. Terao*, E. Lussetti†, and F. Müller-Plathe†

*Department of Mathematical and Design Engineering, Gifu University, Gifu 501-1193, Japan
†Eduard-Zintl-Institut für Anorganische und Physikalische Chemie, Technische Universität Darmstadt, Petersenstrasse 20, 64287 Darmstadt, Germany

Abstract. We develop two novel non-equilibrium simulation methods which are suitable for calculation of thermal conductivity with good accuracy. These methods are based on simple algorithms, and it will be very easy to extend their range of application. In particular, there are no restrictions (from e.g. the force-field) to apply them to a variety of systems. Here, they are applied to the calculation of the thermal conductivity of amorphous polyamide-6,6. We treat two models of the polymer with different degrees of freedoms constrained. The results suggest that the methods are quite efficient, and that thermal conductivity strongly depends on the number of degrees of freedom in the model.

Keywords: Molecular dynamics, Polymers, Thermal conductivity
PACS: 02.70.Ns, 61.20.Ja, 82.70.-y

INTRODUCTION

In recent decades, numerical methods for non-equilibrium molecular dynamics (NEMD) have been proposed for studying transport properties of systems, such as the thermal conductivity and the shear viscosity [1-7]. The reverse non-equilibrium molecular dynamics (RNEMD) method in particular has the advantage that total energy and total linear momentum are conserved, and it has been successfully applied to various systems such as Lennard-Jones fluids, molecular liquids and their mixtures, as well as polymer systems [2,4,5,8]. In the RNEMD method, a heat flux through the system is artificially generated by suitably exchanging particle velocities in different regions. The periodic system is divided equally into slabs along one direction, with one of these slabs defined as a "hot slab" and another as a "cold slab". At intervals of several hundred timesteps, the center-of-mass Cartesian velocity vectors of the "coldest" particle in the hot slab and the "hottest" particle in the cold one of equal mass are swapped (for details of the method, see Ref. [2]). However, it is difficult to apply the RNEMD method to force fields with all bond distances constrained, since velocity-exchange procedures between different atoms often violate the constraint conditions of the SHAKE method. These problems have been discussed in detail elsewhere [4,5].

To overcome this difficulty, novel numerical methods especially suited for fully constrained models are proposed here for the calculation of thermal conductivity, termed the *Dual-Thermostat Method* and *Heat-Injection Method*. Using these numerical algorithms, we can obtain the thermal conductivity of fully-constrained polymer systems with high accuracy. Because no assumption is made concerning the details of force-fields, these new NEMD methods are widely applicable. In this paper, we demonstrate the accuracy and efficiency of these algorithms by examining the thermal conductivity of amorphous polyamide 6,6 (PA66) systems [9].

METHODS

The thermal conductivity κ is defined by the linear-response relation

$$\mathbf{j} = -\kappa \nabla T \quad (1)$$

where ∇T is a temperature gradient and $\mathbf{j}$ is a heat flux vector. In the following, we describe the methods for calculating the thermal conductivity κ. The first one is the Dual-Thermostat Method, with a setup as shown in Figure 1(a). In this method, slab H and slab C are coupled with Berendsen thermostats locally, and each slab is set to remain at a constant temperature [10]. In the Berendsen thermostat, the velocity of each atom j is updated at each time step such as

$$\mathbf{v}_j(t+\Delta t) = \eta \left[\mathbf{v}_j(t) + \Delta t \frac{\mathbf{F}_j(t)}{m_j} \right] \quad , \quad (2)$$

where m_j is the mass of the j-th atom, $\mathbf{v}_j$ is its velocity, $\mathbf{F}_j$ the force acting on it, andgη is a velocity scaling

CP982, *Complex Systems, 5th International Workshop on Complex Systems*
edited by M. Tokuyama, I. Oppenheim, and H. Nishiyama

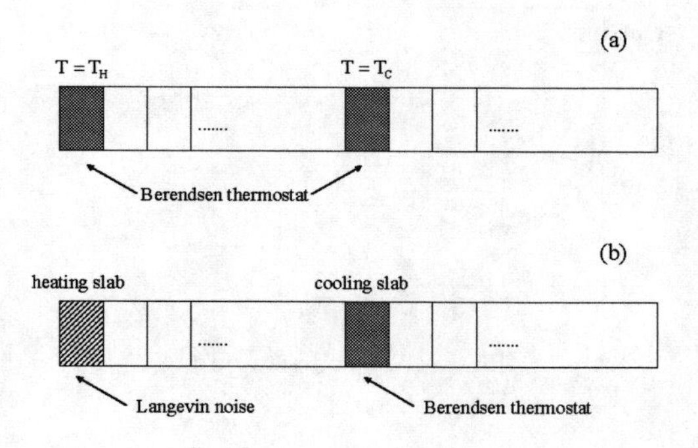

FIGURE 1. Schematic view of the numerical methods (a) the Dual-Thermostat Method (b) the Heat-Injection Method.

factor given by

$$\eta = \left[1 + \frac{\Delta t}{\tau}\left(\frac{T_0}{T} - 1\right)\right]^{1/2}. \quad (3)$$

Here T_0, T, τ, and Δt are the target temperature, the actual temperature of the system, the coupling time, and the time step, respectively. The temperatures of the thermostatted slabs are set to be $T_0=T_H$ for slab H and $T_0=T_C$ for slab C ($T_H > T_C$); the system reaches a steady state after sufficient time, and a linear temperature profile is obtained in the intervening slabs.

In some of the previous computer simulations of the thermal conductivity, the magnitude of heat flux was numerically obtained by the fluctuation-dissipation theorem [11]. In these methods, the statistical error of the calculated thermal conductivity can become quite large. We employ a different numerical method here: in slab H, the Berendsen thermostat creates energy in the system (on average), while the thermostat in slab C removes energy from the system. We did not perform any temperature control in the intervening unthermostatted slabs, so the change in total energy per time step $\langle\Delta E_{total}\rangle$ becomes $\langle\Delta E_{total}\rangle = \langle\Delta E_H\rangle + \langle\Delta E_C\rangle + \langle\Delta E_{err}\rangle$, where $\langle\Delta E_{err}\rangle$ is the change due to numerical errors per MD step, and $\langle\cdots\rangle$ denotes time-averaging. After the system reaches a stationary state($\langle\Delta E_{total}\rangle = 0$), $\langle\Delta E_H\rangle + \langle\Delta E_C\rangle$ must be zero if the integration error can be neglected. In practice, a nonzero value of $\langle\Delta E_H\rangle + \langle\Delta E_C\rangle$ results from the integration error in the simulation, and it shows the accuracy of the calculated thermal conductivity.

In this method, both the energy creation rate and the energy removal rate can be calculated. In each time step, we evaluate the energy changes ΔE_H and ΔE_C due to the Berendsen thermostat coupled with slabs H and C, which are given by

$$\Delta E_i \equiv \left(\eta_i^2 - 1\right) E_i^K \qquad (i = H, C) \quad (4)$$

where η_i, ΔE_i, and E_i^K are the velocity scaling factor (Eq. (3)) in slab i, the energy change due to velocity scaling by the Berendsen thermostat, and the kinetic energy in the slab i given by

$$E_i^K \equiv \sum_{j \in i} \frac{1}{2} m_j \mathbf{v}_j^2 \quad (5)$$

respectively. Here, the summation is performed over atoms j contained in slab i. As a result, the thermal conductivity κ is obtained from the above calculations as

$$\kappa = \frac{1}{2S} \frac{\langle |\Delta E_i / \Delta t| \rangle}{\langle |dT(z)/dz| \rangle}, \quad (6)$$

where $\langle\cdots\rangle$ is the time-averaging, S is the cross-sectional area of the simulation box perpendicular to flow direction z, and the factor 2 arises from the periodicity.

We also propose an alternative method called the Heat-Injection method. The setup of this method is shown in Figure 1(b). The main difference between the Heat-Injection method and Dual-Thermostat method is the treatment of slab H (hot slab). Using a well-equilibrated sample, slab H is heated by adding a random Langevin-noise term to the equation of motion for each atom j in the slab,

$$m_j \frac{d^2}{dt^2} \mathbf{r}_j(t) = \mathbf{F}_j(t) + \mathbf{W}_j(t), \quad (7)$$

where $\mathbf{W}_j(t) \equiv \left(W_j^x(t), W_j^y(t), W_j^z(t)\right)$ is a white noise for heat generation. Here $W_j^\alpha(t)$ $(\alpha = x,y,z)$ are Gaussian random numbers with zero mean and variance σ. In addition, slab C is set to remain at a constant temperature by coupling a Berendsen thermostat locally. In this geometry, the heat generated by the Langevin noise (slab H) flows to the heat sink (slab C). After a sufficiently long period of time, the system reaches a steady state with a finite temperature gradient $dT(z)/dz$. Next, we calculate the rate of increase in total energy $\Delta E_{tot}(t)/\Delta t$ with the Berendsen thermostat on slab C turned *off*. In principle, this is equal to the amount of thermal energy added to the system per unit of time, caused by the Langevin noise in slab H.

From these two independent calculations, thermal conductivity κ is obtained as

$$\kappa = \frac{1}{2S} \frac{\langle \Delta E_{tot}(t)/\Delta t \rangle}{\langle |dT(z)/dz| \rangle}. \quad (8)$$

Eqs. (6) and (8) have the same form, although the physical origin of the numerator in Eqs. (6) and (8) differs; in Eq. (8), $\langle \Delta E_{tot}(t)/\Delta t \rangle$ indicates the strength of the externally injected heat, which is obtained from independent simulation. It should also be noted that in both methods, thermostatting is limited to the hot and the cold slab. In the intervening slabs, which constitute the majority of the system and where heat conduction takes place, and where the calculation of the temperature gradient is performed, the system follows the pure Newtonian dynamics. It would therefore appear that, firstly, also other thermostats are permissible with both algorithms and that, secondly, the use of the sometimes criticized Berendsen thermostat is not critical here.

NUMERICAL RESULTS

To check the accuracy of these two numerical methods described above, the thermal conductivity of monodisperse Lennard-Jones system is calculated with rescaled temperature $T^* = 0.7$ and rescaled density $\rho^* = 0.85$. The number of particle is taken to be 2,592. The calculated results show that the rescaled thermal conductivity κ^* (in dimensionless unit) becomes $\kappa^* = 7.1 \pm 0.4$ by the Dual-Thermostat method, and $\kappa^* = 6.6 \pm 0.7$ by the Heat-Injection method. These results are in good agreement with those in previous study using RNEMD (6.4 – 6.9, depending on the conditions)[2].

In the following, we calculate the thermal conductivity of amorphous PA66 systems by the Dual-Thermostat method, and demonstrate its efficiency. We study an all-atom model of PA66 with bond constraints (referred to as the AA model), which is described in Ref. [9]. The intramolecular force-field contains harmonic bond angle bending and periodic cosine-type torsional potentials. The nonbonded potential includes Lennard-Jones terms, with Lorentz-Berthelot mixing rules, and electrostatic interactions for the models with partial atomic charges. The latter were treated using the reaction-field method, with a dielectric constant of 5. In addition, we also consider a fully bond-constrained united-atom model (referred to as UA model) derived from the AA model, in which all of the CH_2 and CH_3 groups are treated as single atoms and all of the intramolecular bonds are rigid. The Lennard-Jones parameters ε and σ of UA model are given by $\varepsilon = 0.48$ kJ/mol and $\sigma = 0.393$ nm for CH_2 groups, and $\varepsilon = 0.73$ kJ/mol and $\sigma = 0.393$ nm for CH_3 groups, respectively. The density of a sample is set to 1.11 g/cm^{3}for the AA model, and 1.07 g/cm^{3}for the UA model. All molecular dynamics simulations were carried out with the YASP package [6]. The equation of motion was solved by a leap-frog algorithm, and bond constraints, if present, were solved by the SHAKE method [10,12]. Non-bonded interactions were evaluated from a Verlet neighbor list, which was updated every 30 time steps using a link-cell method. In the calculation of the local temperature in each slab, the number of degrees of freedom N_i^{free} in slab i is estimated to be $N_i^{free} = N_i\left(N^{free}/N\right)$, where N, N_i, and N^{free} denote the total number of atoms, the number of atoms in slab$i$, and the total number of degrees of freedom, respectively. I.e. the number of degrees of freedom per atom has been assumed to be constant in the whole simulation cell. The simulation cells were elongated in the z-direction ($L_x = L_y = L_z/2$), and periodic boundary conditions were imposed in all directions.

The system size of the UA model was taken to be L_x, $L_y = 5.49$ nm and $L_z = 10.97$ nm, and the total num-

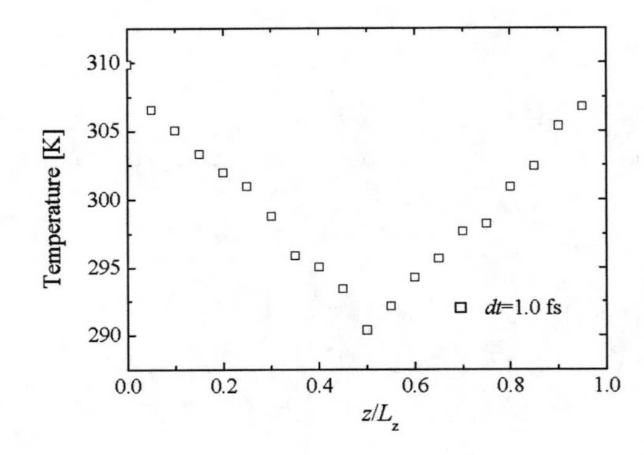

FIGURE 2. Temperature profile of the UA model obtained by the Dual-Thermostat Method.

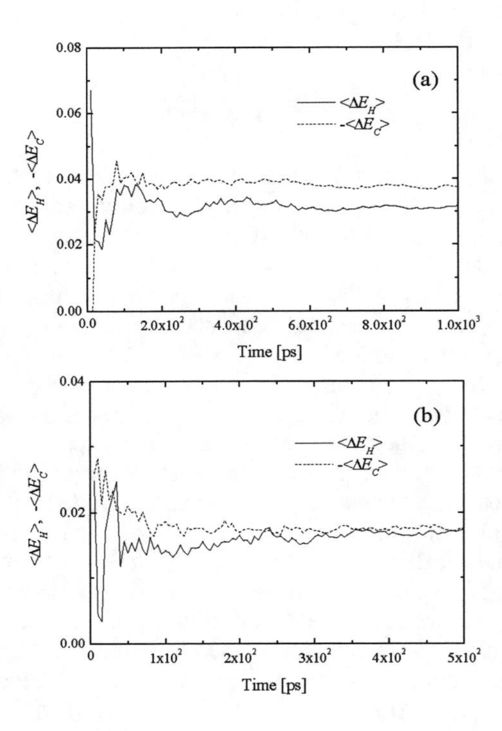

FIGURE 3. Dual-thermostat method: Cumulative average of the energy change per time in the thermostatted slabs $\langle \Delta E_i \rangle$ $(i = H, C)$ with the UA model . (a) dt = 1.0 fs (b) dt = 0.5 fs

ber of (united) atoms was N=17,328. We calculated the energy changes within the 2 thermostatted slabs ΔE_H and ΔE_C at each MD step. To confirm the numerical accuracy of the Dual-Thermostat Method, calculations were performed with 10^6 MD steps of two different step lengths, Δt=0.5 fs and 1.0 fs, after the system reaches steady state. The temperatures in the thermostatted slabs were set at T_H=310 K and T_C=290 K, and the coupling time of the local Berendsen thermostat in both slabs was 0.1ps to keep the measured average temperature within 0.3 K of the target temperature. The resulting temperature profile is shown in Figure 2. The energy changes per step due to the Berendsen thermostat in each slab became $\langle \Delta E_H \rangle = 3.18 \times 10^{-2}$ kJ/mol and $\langle \Delta E_C \rangle = -3.73 \times 10^{-2}$ kJ/mol with Δt =1.0 ps, and $\langle \Delta E_H \rangle = 1.75 \times 10^{-2}$ kJ/mol and $\langle \Delta E_C \rangle = -1.76 \times 10^{-2}$ kJ/mol with Δt =0.5 ps. Here, the difference between the energy creation rate $\langle \Delta E_H \rangle$ and the energy removal rate $-\langle \Delta E_C \rangle$ shows the magnitude of integration errors due to the finite time step Δt. This difference becomes smaller with the time step Δt. The convergence of the Dual-Thermostat method, with time steps Δt=1.0 fs and 0.5 fs, is shown in Figure 3. The solid line and dashed line in Fig. 3 show $\langle \Delta E_H \rangle$ and $-\langle \Delta E_C \rangle$, respectively, where the cumulative average of the energy change in the thermostatted slab $\langle \Delta E_i \rangle$ $(i = H, C)$ is defined as

$$\langle \Delta E_i \rangle \equiv \frac{1}{t} \int_0^t \Delta E_i (t')\, dt' \quad , \tag{9}$$

where $\Delta E_i (t')$ is the energy change at time t'. The averaged energy change $\langle \Delta E_i \rangle$ due to the Berendsen thermostat coupled to the slabs was found to converge after $t \geq 4 \times 10^2$ ps. In addition, the difference between $\langle \Delta E_H \rangle$ and $-\langle \Delta E_C \rangle$ also remained constant after this period of time, see Fig. 3(a). As a result, the thermal conductivity of the UA model was $\kappa = 0.27 \pm 0.03$ W/mK for Δt=1.0 fs, and κ= 0.27 ± 0.01 W/mK for Δt=0.5 fs.

We also examined the thermal conductivity of the AA model by the Dual-Thermostat method. The system sizes were taken to be L_x, L_y = 5.46 nm and L_z = 10.92 nm, and the total number of atoms became N=36,720. Calculations are performed with 10^6 MD steps of length Δt=0.5 fs, after the system reached steady the state. With slab temperatures set at T_H=310 K and T_C=290 K, the thermal conductivity in the AA model became $g\kappa$ = 0.38 ± 0.01 W/mK.

In addition, the thermal conductivity of the fully bond-constrained UA model of amorphous PA66 was calculated by the Heat-Injection Method (Figure 1(b)), in order to compare the two methods for the same polymer model. Firstly, we calculated the rate of increase in total energy $\Delta E_{tot}(t)/\Delta t$ for different magnitudes of the Langevin noise term (variance σ = 30.0, 15.0 and 10.0 in Eq. (7), denoted as cases I, II, and III) with the Berendsen thermostat in slab C turned *off*. The resulting linear increases of the total energy are shown in Figure 4, from which the value of $\langle \Delta E_{tot}(t)/\Delta t \rangle$ is estimated to be 0.14 kJ/mol/fs, 0.042 kJ/mol/fs, and 0.024 kJ/mol/fs in cases I, II, and III, respectively. Secondly, the three calculations were repeated with the the same settings of the Langevin noise term and with the Berendsen thermostat in the cooling slab turned *on*. From the resulting temperature profiles the corresponding temperature gradients $\langle dT(z)/dz \rangle$ (Eq. (8)) were obtained. The number of timesteps was 10^6 MD and their length Δt=1.0 fs. The resulting temperature profile in case III is shown in Figure 5. The heat conductivity of the UA model was 0.26 ± 0.02 W/mK, 0.25 ± 0.03 W/mK, and 0.28 ± 0.03 W/mK for cases I, II, and III, respectively. The thermal conductivities obtained by the two methods are, thus, consistent with each other. Moreover, the results obtained with the Heat-Injection Method at the three different heating rates agree within their statistical error.

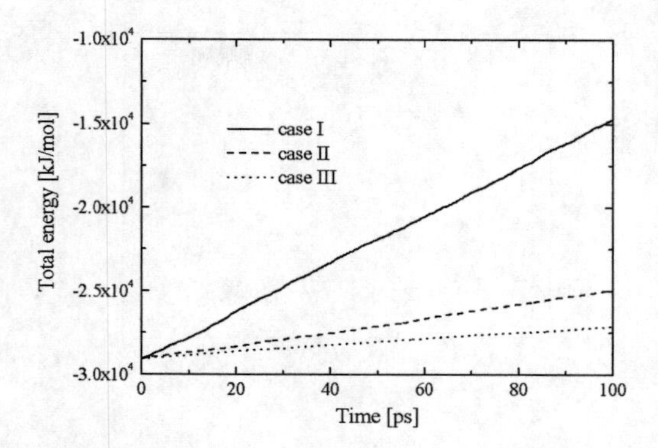

FIGURE 4. Heat-Injection Method: Increase of total energy of the UA model $E_{tot}(t)$due to the Langevin noise term in the heating slab, with no thermostat being present. The curves represent three levels of noise heating.

The findings indicate that the thermal conductivity calculated by classical MD simulation strongly depends on the number of degrees of freedom of the system, with the thermal conductivity found to grow larger with this number. Our results are also consistent with a simulation study of completely flexible models of amorphous PA66 systems [8]: The thermal conductivity of our AA model, but with flexible, harmonic bonds, was found to be $\kappa \approx 0.45$W/mK, where the number of degrees of freedom is larger than that in the AA model in our study. Similarly, the thermal conductivity of the UA model with flexible bonds became $\kappa \approx 0.32$ W/mK, where the number of degrees of freedom lies between that of AA model and UA model studied in this paper. The experimental result is approximately 0.25 W/mK [13]. The disagreement

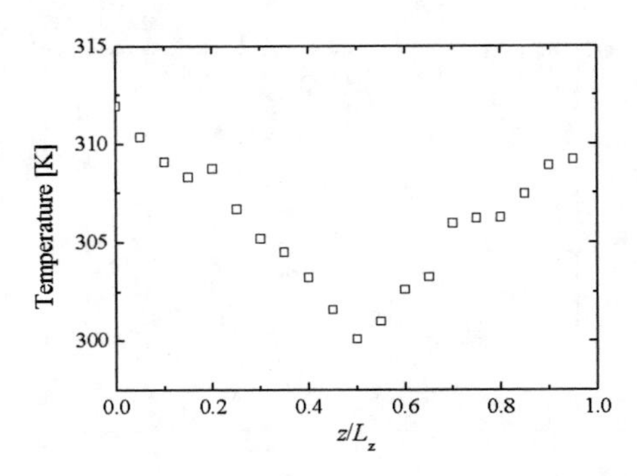

FIGURE 5. Temperature profile of the UA model obtained by the Heat-Injection Method.

is explained by the fact that in reality high-frequency vibrations are quantum-mechanical oscillators in their ground state and are not available for heat conduction. Removing these degrees of freedom form the classical MD simulation by the united-atom description (no vibrations involving hydrogens) and bond constraints (no bond vibration) appears to be a good choice.

CONCLUSIONS

In summary, we have developed two novel non-equilibrium simulation methods termed the Dual-Thermostat method and the Heat-Injection method, which are both suitable for the calculation of thermal conductivities with good accuracy. They are based on simple algorithms, and it will be very easy to extend their range of application. In particular, there are no restrictions (e.g. force-field) to application of them to a variety of types of systems, including models with bond constraints. [One should note at this point that the action of the two thermostats used here does not yield atom velocities, which comply strictly with constrained equations of motion. Still, both method work in practice. The reason is the same as for the successful use of thermostats with constraints in general: the velocity rescalings are very small, and so are the errors introduced into the constraints. Hence, they are easily compensated by iterative constraints solvers like SHAKE].

As a test, the methods have been applied to the calculation of the thermal conductivity of amorphous PA66. We have treated two different fully bond-constrained models with different degrees of freedoms. The thermal conductivities obtained for the AA model and the UA model were $\kappa = 0.38 \pm 0.01$ W/mK and $\kappa = 0.27 \pm 0.01$ W/mK, respectively. The results suggest that both numerical methods are robust and efficient, and that thermal conductivity strongly depends on the number of degrees of freedom in the model. Especially, the Dual-Thermostat method is preferable because it is much easier to set the average temperature of the system by selecting the temperatures of the two thermostatted slabs (T_H and T_C) symmetric around the desired average temperature.

Both non-equilibrium MD methods are widely applicable not only to PA66 but to various other materials as well. In addition, they are based on very simple algorithms, making extension of them easy, as noted above. For example, it will be of great interest to incorporate other widely used types of thermostats [14-16], instead of the Berendsen thermostat, with the novel simulation methods described in this paper.

ACKNOWLEDGMENTS

One of the authors (T. T.) thanks the Research Center for Computational Science, Okazaki, Japan, for the use of their facilities. This work was supported by the "Schwerpunktprogramm 1155: Molecular Simulation in Chemical Engineering" of the Deutsche Forschungsgemeinschaft.

REFERENCES

1. W. G. Hoover, *Molecular dynamics* (Lecture notes in physics 258) (Berlin, Springer-Verlag, 1986).
2. F. Müller-Plathe, *J. Chem. Phys.* **106**, 6082–6085 (1997).
3. P. Bordat and F. Müller-Plathe, *J. Chem. Phys.* **116**, 3362–3369 (2002).
4. F. Müller-Plathe and P. Bordat, in *Novel Methods in Soft Matter Simulations*, edited by M. Karttunen, I. Vattulainen, and A. Lukkarinen, Lecture Notes in Physics Vol. 640 (Springer: Heidelberg, Germany, 2004).
5. M. Zhang, E. Lussetti, L. E. S. de Souza, and F. Müller-Plathe, *J. Phys. Chem. B* **109**, 15060–15067 (2005).
6. F. Müller-Plathe, *Comp. Phys. Comm.* **78**, 77–94 (1993).
7. T. Terao and F. Müller-Plathe, *J. Chem. Phys.* **122**, 081103(1)–081103(3) (2005).
8. E. Lussetti, T. Terao, and F. Müller-Plathe, *J. Phys. Chem. B*, **111**, 11516–11523 (2007).
9. S. Goudeau, M. Charlot, C. Vergelati, and F. Müller-Plathe, *Macromolecules* **37**, 8072–8081 (2004).
10. M. P. Allen and D. J. Tildesley, *Computer Simulation of Liquids* (Oxford University Press, Oxford, 1987).
11. D. J. Evans and G. P. Morris, *Statistical Mechanics of Nonequilibrium Liquids* (Academic Press, San Diego, 1990).
12. F. Müller-Plathe and D. Brown, *Comp. Phys. Comm.* **64**, 7–14 (1991).
13. G. de Carvalho, E. Frollini, and W. N. dos Santos, *J. Appl. Polym. Sci.* **62**, 2281–2285 (1996).
14. For example, C. P. Lowe, *Europhys. Lett.* **47**, 145–151 (1999).
15. S. Nosé, *J. Chem. Phys.* **81**, 511–519 (1984).
16. W. G. Hoover, *Phys. Rev. A* **31**, 1695–1697 (1985).

Modelling Complex Systems in Full Detail: A New Approach

G. J. A. Sevink and J. G. E. M. Fraaije

Soft Matter Chemistry, Faculty of Mathematics and Natural Sciences, PO Box 9502, 2300 RA Leiden, The Netherlands

Abstract. We introduce a new hybrid method as a merger of two well-established mesoscopic methods, dynamic density functional theory (DDFT) and Brownian Dynamics (BD), by defining a coupling between particles and fields. The method is particularly suited for the efficient computational modelling of complex systems with a large variety of constituents. We discuss the potential of this method for typical systems and show the results of simulations for a nanoparticle/polymer mixture.

Keywords: Block Copolymers, Coarse-Grained, Field-Theoretic Model, Hybrid Method, Particle Model, Phase Transitions
PACS: 64.70.qj,64.60.De

INTRODUCTION

Computational studies aimed at understanding the behaviour of complex systems represent the stepping stones to promising prospects of structural and functional manipulation and/or design in nanotechnology. In general, complexity is attributed to systems that consist of a large number of elements and cannot be adequately described by studying their constituents. Naturally, the initial stage in such studies involves making choices, each of which can be viewed in different ways. Since most systems cannot be modelled from first principles alone, initially the scientist has to decide on the level of detail that is required to model the particular system (in the remainder, 'system' will used to represent both structure and structure evolution, since we aim at 'in silico' reproduction of both static and dynamic properties of a complex structure). If one considers the system as a whole, and not merely a combination of non-interacting and much smaller sub-units that can be treated seperately, this important choice is often dictated. In particular, some sort of coarse graining is required when one is interested in material properties that depend on structure formation on the mesoscopic level (1-1000 nm). However, the dynamics of most complex systems (for instance, the human cell) involves a large variety of time and length scales, and one therefore has to choose on beforehand which elementary processes could be essential for the larger-scale phenomena that one is interested in. Consequently, one has to keep in mind how the averaging procedure, that is applied to reduce the degrees of freedom in the system, affects the detail and validity of the description. Then, since even this coarse-grained description may give rise to high computational burdens, the modeller has to select only the necessary ingredients and build a 'minimal' model, in a reductionist approach. This obvious trade-off between level of detail and computational restrictions has had a serious impact on computational modelling. It has restricted computational approaches to either full atomistic models for detailed evaluation of microscopic phenomena on small scales (phenomena that are supposed to have no effect on the entire system) or rather specific mesoscopic models for phenomena on the scale of the entire system. The elementary length scales used in the averaging procedure in the latter mesoscopic methods often limit the applicability to systems of relatively homogeneous composition.

In reality, most complex systems in 'soft' nanotechnology are (self) assembled from a large number of interacting compounds, ranging from solvents, surfactants, (block) copolymers, colloids to active ingredients like drugs, in a broad range of concentrations. A detailed model for these multi-component systems has to take into account the elementary length and time scales associated with each of the compounds, which may deviate by orders of magnitude. Recently, there have been several attempts to narrow down the gap between detail/specificity and computational limits. New developments in computer power and algorithms have pushed the boundaries of full atomistic (or weakly coarse-grained) methods to larger (length and time) scales, but they are still much below the ones required for the computation of a complete (mesoscopic) system. Stepwise procedures, like the one initiated by the group of Masao Doi[1], are based on switching between different levels of description by so-called 'mappings'[2], and, in principle, allow for the determination of coarse-grained parameters directly from lower-level (ab initio or atomistic) calculations. Although this zooming approach is efficient and potentially very powerful, it does not address the problem of the heterogeneity of constituents, since it still relies on existing (mesoscopic) methods.

A number of computational methods have been very successful in describing structure formation in 'pure' block

CP982, *Complex Systems, 5th International Workshop on Complex Systems*
edited by M. Tokuyama, I. Oppenheim, and H. Nishiyama

copolymer (BCP) or colloidal systems on the mesoscale. These models describe the system in terms of coarse-grained variables, based on either field [3] or particle (see for example, Ref. [4]) descriptions. A particular field method is the dynamic density functional theory (DDFT), that describes the collective diffusional dynamics of an ensemble of molecules (represented by Gaussian chains) interacting via a mean-field, based on the assumption that the time scale is such that the chains are always locally equilibrated. This model is well suited to describe mesoscopic structure evolution in melts or concentrated solutions of flexible BCP in the weak to intermediate segregation regime. A recent comparative study has shown that the experimental dynamics can be well captured by the diffusional model in DDFT[5]. A typical particle model like dissipative particle dynamics (DPD) or Brownian Dynamics (BD) is also based on a coarse-grained chain representation for the individual molecules. Here, chain elements (or beads) interact via effective soft-core potentials, and the dynamics is based on Newton's equation of motion. Although the approaches are somewhat complimentary, field methods possess the benefit of computational efficiency, while particle methods are more flexible and allow for detail. Here, we propose a 3D hybrid method, based on these well-founded mesoscopic methods (DDFT and BD) and capable of describing pattern formation dynamics in systems containing continuous fields *and* discrete particles (or particle chains) in a single volume V. The new method allows for a natural separation in abundant (fields) and sparse (particles) constituents, and even larger colloids can be included as tight-bonded particle clusters. Below, we introduce the method and briefly illustrate it by considering application examples.

METHOD

We shortly review the two mesoscopic methods that from the basis of the new method. In DDFT[6], BCP molecules of linear or branched multiblock architecture are represented as necklaces of beads (Gaussian chains) of length N. Each bead in this chain represents a number of monomers, and differences in chemical nature give rise to different bead types, labelled by I. For instance, we have previously represented a SBS triblock copolymer by a $A_3B_{12}A_3$ chain, where each A bead represents several styrene monomers and each B bead several butadiene monomers[7]. Different BCP molecules can be included in a system, and solvent is often represented by a single bead (a number of solvent molecules). Since the systems are dense, each chain is surrounded by many other chains, and the interaction can be sufficiently captured by a mean-field model, with a strength given by the well-known Flory-Huggins parameter χ. One should realize that in the mean-field model the above mentioned parameters are the only relevant parameters. Any system that can be represented by the same parameters will behave in exactly the same way. For a solution of a single BCP in a volume V, the free energy model is given by[6]:

$$F^f[\rho_I] = -kT\ln\frac{\Phi^n}{n!} - \sum_I \int_V d\mathbf{r}U_I(\mathbf{r})\rho_I(\mathbf{r}) + F_{MF}[\rho_I]\,. \tag{1}$$

Here Φ is the intramolecular partition function for Gaussian chains, U_I is the external potential conjugate to the particle density ρ_I, n is the number of BCP molecules and $F_{MF}[\rho_I]$ is the mean-field contribution due to the non-ideal interactions, given by [6]

$$\begin{aligned} F_{MF}[\rho_I] &= \frac{1}{2}\sum_{I,J}\int_V\int_V d\mathbf{r}d\mathbf{r}'\varepsilon_{IJ}(|\mathbf{r}-\mathbf{r}'|)\rho_I(\mathbf{r})\rho_J(\mathbf{r}') \\ &+ \frac{\kappa v^2}{2}\int_V d\mathbf{r}\left(\sum_I \rho_I(\mathbf{r}) - \rho_I^0\right)^2 . \end{aligned} \tag{2}$$

The first term in equation (2) describes the cohesive mean-field interaction in the spirit of Flory-Huggins theory. By choosing the bead-bead interaction potential as

$$\varepsilon_{IJ}(|\mathbf{r}-\mathbf{r}'|) = \varepsilon_{IJ}^0\left(\frac{3}{2\pi a^2}\right)^{\frac{3}{2}} \exp\left[-\frac{3}{2a^2}(\mathbf{r}-\mathbf{r}')^2\right] . \tag{3}$$

the interaction parameters ε_{IJ}^0 (in kJ/mol) are directly related to the Flory-Huggins χ[8]. The second term in equation (2) is added to keep the system (nearly) incompressible (the bead volume v is introduced to make the concentration field dimensionless). The parameter κ determines the compressibility of the system, and ρ_I^0 is the concentration of the I block averaged over the simulation volume. The introduction of incompressibility through a penalty function is chosen in favor of exact compressibility, which would lead to slow convergence of the iterative solver due to the stiffness of the set of dynamic equations. The value of κ should be chosen large enough; for $\kappa \to \infty$ the system is incompressible. The minimization of the free energy is by dynamic iteration, and several algorithms have been used in the past. The considered equations here describe the collective diffusion of concentration fields using the same kinetic coefficients as collective Rouse dynamics, also known as density dynamics (see Ref. [8]). Details on parametrization and numerical implementation can be found in Ref.[6]. We note that any stationary solution of the dynamic equations represents a solution to self-consistent field theory (SCF).
Compared to DDFT, mesoscopic particle methods like Dissipated Particle Dynamics (DPD) and Brownian Dynamics (BD) are better suited to track the behavior of individual molecules, and, since it is not a mean-field

method, can also be applied for relative low concentrations. They describe the dynamics of a collection of soft-core particle beads that may be connected by springs, each representing a number of small molecules or monomers, based on effective soft-core interaction potentials. Using the argument that hydrodynamic modes become irrelevant for the dynamics of dense and strongly structured systems[9], the Newton's equation of motion can be rewritten into the Langevin equations for particle positions

$$\frac{\partial \mathbf{r}_k}{\partial t} = \frac{1}{\xi}\mathbf{F}_k^C + \mathbf{r}_k^R \tag{4}$$

where F_k^C is the conservative force felt by bead k and $\mathbf{r}_k^R$ a random displacement with the proper statistical weights. The constant friction coefficient ξ is related to the diffusion constant D by the Einstein relation. Using $\mathbf{r}_{kl} = |\mathbf{r}_k - \mathbf{r}_l|$ and $\hat{\mathbf{r}}_{kl} = \mathbf{r}_k - \mathbf{r}_l/\mathbf{r}_{kl}$, the conservative force $\mathbf{F}_k^C = \mathbf{F}_k^r + \mathbf{F}_k^s$ experienced by bead k includes contributions from repulsive interactions with surrounding beads

$$\mathbf{F}_k^r = \sum_l a_{kl}\left(1 - \frac{\mathbf{r}_{kl}}{d}\right)\hat{\mathbf{r}}_{kl}\,, \tag{5}$$

and due to the harmonic string connecting k to connecting beads $\in B_k$ in the same molecule

$$\mathbf{F}_k^s = -\sum_{l \in B_k} b_{kl}\mathbf{r}_{kl}\hat{\mathbf{r}}_{kl}\,. \tag{6}$$

In these expressions, d is the cut-off of the repulsive force (which is often taken equal to the bead diameter) and b_{kl} the spring constant. The repulsion strength a_{kl} is an important parameter, as it sets the degree of bead-bead penetration and therefore captures the chemical nature of the system. The value of a_{kk} ('like' beads) and the relation of $\Delta a = a_{kk} - a_{kl}$ ('unlike' beads) to the Flory-Huggins χ depends on temperature and concentration, and the determination for specific constituents is the subject of many studies (for instance, see Ref. [4]).
In the new hybrid method, the free energy F is defined as

$$F[\rho_I, \mathbf{r}_k] = F^f[\rho_I] + E^B(\mathbf{r}_k) + F^{Cp}[\rho_I, \mathbf{r}_k]\,. \tag{7}$$

The first term in expression (7) is the free energy of the field-based method of equation (1) and the second term the potential energy of the collection of particles, which is irrelevant for the field-dynamics, since it does not contribute to the thermodynamic driving force $\nabla\mu_I$. The influence of the particles on the evolution of the field is included in a new *coupling* term

$$F^{Cp}[\rho_I, \mathbf{r}_k] = \sum_{I,k} c_{Ik} \int_V d\mathbf{r}\rho_I(\mathbf{r})K(|\mathbf{r} - \mathbf{r}_k|) \tag{8}$$

chosen in terms of binary interaction, with a strength determined by the coupling parameters c_{Ik}. Consequently, the additional contribution to the chemical potential in the field model is

$$\mu_I^{Cp} = \sum_k c_{Ik}K(|\mathbf{r} - \mathbf{r}_k|) \tag{9}$$

The smoothing kernel K, which is taken Gaussian

$$K(\mathbf{r}) = \frac{1}{\pi\sigma^2}e^{-\frac{\mathbf{r}^2}{\sigma^2}} \tag{10}$$

is used to map the particles located at discrete positions $\mathbf{r}_k$ to a continuous field centred around this position, with a decay determined by the value of spreading factor σ. The particles experience the fields via the new interaction potential, leading to a contribution to the conservative force experienced by bead k of

$$\mathbf{F}_k^{Cp} = -\sum_I c_{Ik} \int_V d\mathbf{r}K(|\mathbf{r} - \mathbf{r}_k|)\nabla\rho_I(\mathbf{r})\,. \tag{11}$$

The interaction between fields and particles is completely determined by the parameters c_{Ik} and σ. A reasonable value for the spreading factor σ, that determines the range of the interaction between particles and fields, is the particle bead radius. Considering this value, we see from equation (11) that the elementary coupling force on particle bead k is given as an average (over the kernel K) of minus the gradient of the density field over the bead volume. The parameters c_{Ik}, the prefactor of this elementary force, may be negative or positive. Setting up a theoretical framework for estimating their values in varying conditions remains future work, and we rely on generic values here. Physically, a positive coupling parameter gives rise to a coupling force that drives the particle beads away from high density field values, and a chemical potential that does not favor high local concentrations of particle beads. For negative values, the opposite effect can be anticipated. Special attention should be given to the Helfand parameter κ, introduced in the excluded volume term of (2), for instance if one is interested in replacing particle molecules by a field description.

APPLICATIONS AND RESULTS

We illustrate the benefit of the new hybrid method by listing possible applications, and a condensed study of phase separation in a nanoparticle/BCP mixture.

Surfactant/BCP Melts And Solutions

Computational studies of mixtures can be carried out with 'pure' particle or field methods, but the hybrid method offers several benefits. In many structures, the ratio between the dimension of the structure (for instance vesicles) and its elementary components (surfactants or BCP) is large. Moreover, the influence of the boundary conditions for small simulation volumes can be substantial, and may even hinder phenomena from taking place[10]. Field methods such as DDFT are suited for considering large simulation volumes and long equilibration times, up to a factor of 10^3 the elementary bead volume v and 10^6 the molecular equilibration time. On the other hand, DDFT is not well suited for considering constituents in low concentration and molecular conformations, such as the detailed structure of a surfactant layer. Particle methods have been applied for low concentrations and obtaining these details, but are less efficient in considering large volumes. A typical hybrid approach would be to include the low concentration component (surfactant) as chains of particle beads and the solvent as a continuous field. One could even consider different elementary length scales in both of the coarse-grained models, by taking different bead volumes in the particle and field methods. An alternative application of the hybrid method is to retrieve information on the local chain conformations that can be important for mechanical material properties or structural transitions. One can sample specific chain evolution or conformations such as bridging by including a number of 'tracer' chains into a field description of a BCP melt. As mentioned, particularly in mixing field and particle descriptions for the same BCP, one should strive for consistency in terms of physical properties. For equal volume fractions, the overall compressibility can be made equal by demanding that the pressures of each 'pure' phase are equal. Using the expression for p in Ref. [11] for a field description of a monomeric solvent, and matching it to the pressure calculated from a DPD simulation for the same system (assuming that the field diffuses faster than the particle beads), one can compute that a dimensionless $\kappa^* = 45$ [6] must be chosen for correct physical behavior.

BCP Gels

Modelling gels in a 'pure' DDFT approach may be difficult, since the gel network in principle consists of a single, highly branched chain. The hybrid method allows for a natural separation in sparse and dense components, and one would be able to model diffusion of single particle chains through a particle chain gel network, that is swollen by a solvent, represented by a continuous field. Moreover, one could include other constituents (surfactant and/or additional solvent) in either a field or particle description. In principle, switching between field and particle descriptions through mappers, originally developed for astrophysics, is possible, allowing for zooming.

Nanoparticle Polymer Composites

Mixtures of (B)CP and nanoparticles, nanoparticle polymer composites, are in the focus of experimental, theoretical and simulational nanotechnology research for their advantageous electrical, optical and mechanical properties[12]. We refer to a recent review[13] for an overview of this field. A number of methods have been developed and/or applied to study these systems computationally. Models based on a phenomenological free energy (a generalized Cahn-Hilliard-Cook (CHC)[14] model and fluid particle dynamics[15]) and Lattice Boltzmann (LB)[16] have been applied to study systems containing, apart from particles, simple liquids or BCP blends. Phase diagrams of nanoparticle/BCP mixtures in various conditions and parameter ranges were calculated with a mixed SCF/DFT formulation[17]. We note that the current report is too short for a detailed discussion and a true comparison to the results in literature is not made. Our aim here is to illustrate the flexibility and potential of the hybrid model for a relevant system.
For simplicity, we have chosen an $A_3B_{12}A_3$ triblock copolymer system that was used in DDFT calculation before, with $\chi_{AB} = 1.9$. In earlier publications, we have shown that this coarse-grained representation is a good model for experimental polystyrene-*block*-polybutadiene-*block*-polystyrene (SBS) triblock copolymer swollen in chloroform vapor ($M_{w,PS} = 14k$, $M_{w,PB} = 73k$ and $M_{w,PS} = 15k$)[7]. In the bulk, both the experimental BCP self-assembles into PS-rich cylinders in a PB-rich matrix. By varying the χ_{AB} between $[1.7, 2.05]$ in 32^3 simulation volumes V in DDFT, the following microstructures were found after equilibration: disorder ($\chi_{AB} < 1.8$), spheres ($1.8 \leq \chi_{AB} < 1.85$), cylinders ($1.85 \leq \chi_{AB} < 1.95$) and a gyroid phase ($\chi_{AB} \geq 1.95$). We note that these values may deviate from the ones found by self-consistent field theory (SCF), since, in contrast to SCF, the kinetic pathway does play a role in DDFT (just like in experiments) and can lead to trapping of metastable structures[18]. The K nanoparticles are modelled as single particle beads with repulsion strength $a^*_{kk} = da_{kk}/kT = 25$. In experiments, nanoparticles are often covered by a functional layer that allows for small inter-penetration[13]. We assume that the diffusion of the particles is faster than the diffusion of the field, $D\xi = 0.3$. The coupling parameter between the A concentration field and the particle beads is set to slightly repulsive $c_{Ak} = 5$. The coupling parameter between the B concentration field and the particle beads, c_{Bk}, should be higher than c_{Ak} if we want to load the minority field compo-

nent A (the cylinders) with particles. The influence of the remaining coupling parameter is studied by varying c_{Bk} from 10 (system α) to 15 (system β). We first use $K = 2$ in a two dimensional system. In the early stages (not shown here), we observe that the two particles act as moving (α) or rather stationary (β) nucleation centers, like in constrained BCP systems[19]. Later stages (5000 dimensionless steps, Figure 1) show that adding particles may lead to permanent domain swelling (strongly preferential, system β), and in general mediate defect annihilation (both system, at different rates) towards perfect structures. As a comparison, in the absence of nanoparticles the structure is much less developed after the same number of simulation time steps. Next we briefly touch upon the experimentally observed phase transition that is induced by loading the system with many selective nanoparticles[20]. We consider the $A_3B_{12}A_3$ system for $\chi = 1.75$ and the same coupling parameters as in system β. The system is quenched several times with an increasing amount of particle beads K (added at random positions at $t = 0$, into an initially homogeneous BCP mixture). In the absence of particles ($K = 0$) an unstructured melt is found (see Figure 2). For an increasing amount of particles, we obtain the following sequence of phases: spheres ($K = 16$ and $K = 128$), rod-like micelles ($K = 256$), cylinders ($K = 512$), networks ($K = 1024$) and networks with small lamellar patches ($K = 4096$). We conclude that adding only a small number of particle beads ($K = 16$; 0.05 volume %) sufficiently perturbs the system, and induces phase separation. Increasing the amount of particles has the effect of swelling the minority component. These findings complement existing results obtained with static SCF/DFT methods[21], by providing a realistic dynamic picture of the detailed phase separation process. Since the fundamental mechanisms in these systems are still poorly understood, this can be very valuable in strategies aimed at design. Finally, the same method offers full flexibility to add additional constituents, like solvent or surfactants, and can also be used for nanoparticles with a heterogeneous size distribution. In the latter case, large colloids of arbitrary shape can be build as tight bonded clusters (with proper dynamic behavior) of small particle beads.

DISCUSSION

The purpose of this short contribution is to introduce and embed the new simulation method. Validation is of course important, but the dynamics in our method complicates a direct comparison between the results above and obtained by static approaches [21]. Hence, a different optimization scheme or a detailed comparitive simulational/experimental study is required, which is clearly outside the current scope. We argue, however, that the fundamental issue for validation is whether the coupling (8) between the two validated methods is appropriate. We therefore refer to a conceptually similar hybrid approach (a BD particle method combined with a time-dependent Ginzburg Landau field method), developed in the nineties by Kawakatsu and Kawasaki[22], that derived a coupling expression strikingly similar to the one we postulated in (8). Although the method seems to have been abandoned by this group after 1993, their method was applied to binary mixtures containing surfactants, and showed a very good match to experimental data in terms of scaling behavior for structure coarsening [23].

CONCLUSION AND OUTLOOK

We have introduced a new hybrid method as a merger of two well-established mesoscopic methods, DDFT and BD. The method is particularly suited for the efficient computational modelling of complex systems with a large variety of constituents. These systems are abundant in 'soft' nanotechnology. Instead of zooming between different levels of description, this method allows for switching between field and particle descriptions at the same mesoscopic level. We have introduced a new coupling term between the particle and field descriptions, containing so-called coupling parameters. In the considered simulation study, these values were chosen based on physical considerations; setting up a theoretical framework is left for the future. We have also listed a number of systems were this new method can be of added value. Although the simulation study is obviously not complete, it illustrates the potential of the method and the detailed kinetic phenomena that it is able to describe. As an outlook, we believe that a similar strategy can also be successful in merging methods acting at different scales, such as mesoscopic and microscopic methods like molecular dynamics (MD).

ACKNOWLEDGMENTS

The authors thank dr Jan van Male for his contributions to the development of the new hybrid method and valuable discussions. Part of the simulation results were obtained by Heiko Schoberth, Bayreuth University, during a month internship in Leiden (2007). All simulations were performed with the Culgi software package (courtesy of Culgi BV). This work was partially supported by the European Commission through a NMP STREP project (MULTIMATDESIGN).

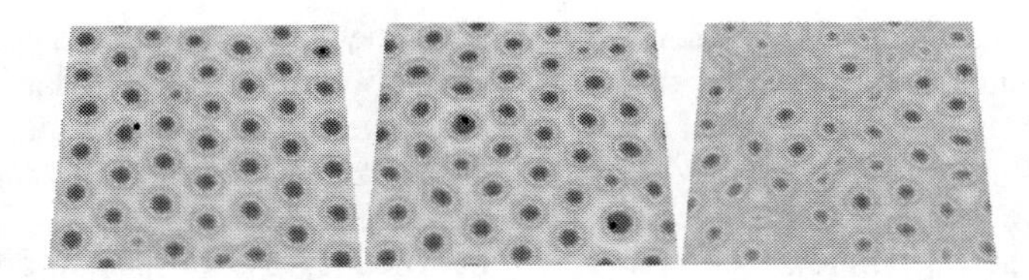

FIGURE 1. (Color online) From left to right: same time stage for system α, system β and BCP system without nanoparticles. The density/concentration of the A component and the two particles are shown.

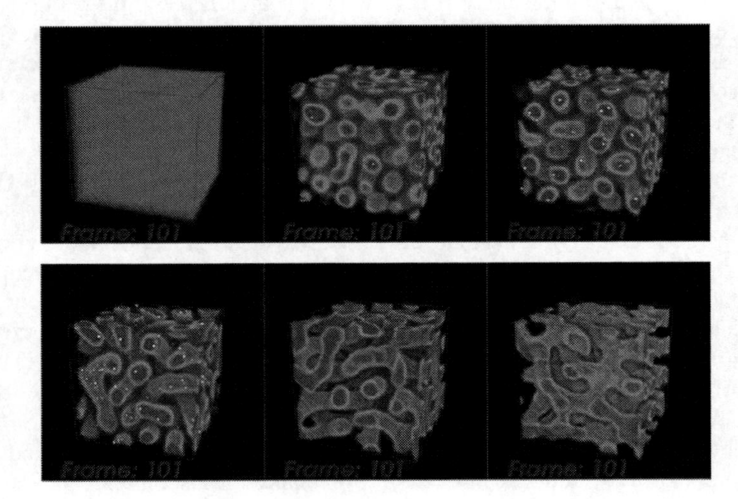

FIGURE 2. (Color online) From left to right, top to bottom: stable 3D microstructures from the simulations after 10000 time steps with varying number of particles K: 0, 16, 256, 512, 1024 and 4096. Isodensity surfaces show the structure of the A component; particles are omitted for $K = 1024$ and 4096 for better visibility.

REFERENCES

1. M. Doi, *Macromol. Symp.* **195**, 101–107 (2003).
2. J. Baschnagel, K. Binder, P. Doruker, A. A. Gusev, O. Hahn, K. Kremer, W. L. Mattice, F. Muller-Plathe, M. Murat, W. Paul, S. Santos, U. W. Suter, and V. Tries, *Adv. Pol. Sci.* **152**, 41–156 (2000).
3. G. H. Fredrickson, V. Ganesan, and F. Drolet, *Macromolecules* **35**, 16–39 (2002).
4. R. D. Groot, and P. B. Warren, *J. Chem. Phys.* **107**, 4423–4435 (1997).
5. A. Knoll, K. S. Lyakhova, A. Horvat, G. Krausch, G. J. A. Sevink, A. V. Zvelindovsky, and R. Magerle, *Nat. Mat.* **3**, 886–890 (2004).
6. B. A. C. van Vlimmeren, N. M. Maurits, A. V. Zvelindovsky, G. J. A. Sevink, and J. G. E. M. Fraaije, *Macromolecules* **32**, 646–656 (1999).
7. A. Knoll, A. Horvat, K. S. Lyakhova, G. Krausch, G. J. A. Sevink, A. V. Zvelindovsky, and R. Magerle, *Phys. Rev. Lett.* **89**, 035501-1–035501-4 (2002).
8. J. G. E. M. Fraaije, and G. J. A. Sevink, *Macromolecules* **36**, 7891–7893 (2003).
9. I. M. de Schepper, and E. G. D. Cohen, *J. Stat. Phys.* **27**, 223–281 (1982).
10. G. J. A. Sevink, and A. V. Zvelindovsky, *Macromolecules* **38**, 7502–7513 (2005).
11. N. M. Maurits, A. V. Zvelindovsky, and J. G. E. M. Fraaije, *J. Chem. Phys.* **108**, 2638–2650 (1998).
12. Y. Lin, A. Boker, J. B. He, K. Sill, H. Q. Xiang, C. Abetz, X. F. Li, J. Wang, T. Emrick, S. Long, Q. Wang, A. Balazs, and T. P. Russell, *Nature* **434**, 55–59 (2005).
13. A. C. Balazs, T. Emrick, and T. P. Russell, *Science* **314**, 1107–1110 (2006).
14. B. P. Lee, J. F. Douglas, and S. C. Glotzer, *Phys. Rev. E* **60**, 5812–5822 (1999).
15. T. Araki, and H. Tanaka, *Phys. Rev. E* **73**, 061506-1–061506-7 (2006).
16. K. Stratford, R. Adhikari, I. Pagonabarraga, J. C. Desplat, and M. E. Cates, *Science* **309**, 2198–2201 (2005).
17. R. B. Thompson, V. V. Ginzburg, M. W. Matsen, and A. C. Balazs, *Science* **292**, 2469–2472 (2001).
18. A. Horvat, K. S. Lyakhova, G. J. A. Sevink, A. V. Zvelindovsky, and R. Magerle, *J. Chem. Phys.* **120**, 1117–1126 (2004).
19. G. J. A. Sevink, A. V. Zvelindovsky, B. A. C. van Vlimmeren, N. M. Maurits, and J. G. E. M. Fraaije, *J. Chem. Phys.* **110**, 2250–2256 (1999).
20. S. W. Yeh, K. H. Wei, Y. S. Sun, U. S. Jeng, and K. S. Liang, *Macromolecules* **38**, 6559–6565 (2005).
21. J. Y. Lee, R. B. Thompson, D. Jasnow, and A. C. Balazs, *Macromolecules* **35**, 4855–4858 (2002).
22. T. Kawakatsu, and K. Kawasaki, *Physica A* **167**, 690–735 (1990).
23. T. Kawakatsu, K. Kawasaki, M. Furusaka, H. Okabayashi, and T. Kanaya, *J. Chem. Phys.* **99**, 8200–8217 (1993).

Monte Carlo Simulation for Ternary System of Water/Oil/ABA Triblock Copolymers

Natsuko Nakagawa and Kaoru Ohno

Department of Physics, Graduate School of Engineering, Yokohama National University, 79-5 Tokiwadai, Hodogayaku, Yokohama, Japan

Abstract. We performed Monte Carlo simulation of water/oil/ABA triblock amphiphiles with hydrophilic A and hydrophobic B, on a 51×51×51 simple cubic lattice with periodic boundary conditions in *x*- and *y*-directions. Bond fluctuation model(BFM) was used to describe amphiphile, where excluded volume effect and flexible movement of polymer bonds are ascertained. The ratio of water to oil was changed from 10/85 to 50/45 with 5% fixed dilute amphiphiles. For small water/oil ratio, many cubic water clusters surrounded by loop-shaped amphiphiles dispersed in oil and they were bridged between each other by bridge-shaped amphiphiles. On the other hand, phase separation between water and oil plus amphiphiles clearly occurred for large water/oil ratio, where water formed one grid-shaped cluster. At some intermediate water/oil ratio, water cells aggregated into two water clusters with shape of rectangular rod, spread in *x*- and *y*-directions and separated in *z*-direction by mainly loop-shaped amphiphiles, where water clusters and oil plus amphiphiles formed layers perpendicular to *z*-direction.

Keywords: ABA triblock copolymer, Bond fluctuation model, Monte Carlo simulation, Phase separation, Water/oil/amphiphiles ternary system
PACS: 64.60.Cn, 64.70.Nd, 82.70.Uv, 61.20.Ja, 83.10.Tv

INTRODUCTION

In a ternary mixture of water/oil/ABA triblock amphiphiles having hydrophobic chain B and hydrophilic ends A, water and oil can coexist due to the attractive interactions via amphiphiles. Many experimental and theoretical studies have been performed over the past decade for water/oil/PEO-PPO-PEO (PEO = poly(ethylene oxide), PPO = poly(propylene oxide)) system, while the difference of polarity between PPO and PEO is not so large as that of the normal surfactant consisting of head and tail groups. Phase diagrams based on SAXS spectrum and mean field calculations by Alexandridis et al.[1][2] have shown existence of various structural polymorphism such as normal and reverse micelles, micellar cubic and hexagonal phases, normal lamellar and so on, particularly, in oil-lean system or in dense amphiphile systems.

Moreover, even for dilute amphiphile systems, it has been reported that PEO-PI(poly(isoprene))-PEO forms water-in-oil micelles together with AOT (sodium bis(2-ethyl-1-hexyl) sulfosuccinate) surfactant and that the network structures interconnecting micelles via PI chain induce the gelation[3][4].

On the other hand, simulation approaches, particularly retaining molecular structures of amphiphiles, have been used to investigate in water/oil/block copolymers. Examples are the lattice MC(Monte Carlo) simulation by Larson[5] et al., and the DPD(dissipative particle dynamics) simulation used also in a commercial product[6]. In the former MC simulation, the resulting phase diagram of water/oil/AB block copolymers was simpler than, but bore many similarities to phase diagrams typical in real systems suggested by Davis et al.[7], where microstructures evolved from spherical micelles to cylindrical micelles and to lamellar as amphiphile concentrations increased along water-lean or oil-lean line. MC scheme is valid to catch the outlines of equilibrium phase structures in ternary system with no a priori limitation, and lattice model is useful to represent mesoscopic volume scale structures requiring quite long simulation steps to fully equilibrate.

We now perform lattice MC simulation in idealized water/oil/ABA triblock copolymers at low copolymer concentrations. Our aim is to systematically elucidate the dependence of phase patterns on the water concentration in such systems from the view point as basic as possible.

CP982, *Complex Systems, 5th International Workshop on Complex Systems*
edited by M. Tokuyama, I. Oppenheim, and H. Nishiyama

Furthermore, the shear simulation of such system is expected to present the information about viscosity. Then, we test the MC simulation under the shear flow by means of a kind of shear potential. For this purpose we use free boundary condition in z-direction. Using this confined system, we can expect that some ordering structures appear more easily than bulk system.

This paper is organized as follows. In Sec. II, we describe details of the model, methodology and conditions of present simulation. Results for equilibrium state are presented in Sec. III. Test calculations under shear flow are reported in Sec. IV. Sec. V containing our summary and discussion concludes this paper.

MODEL AND METHODOLOGY

Our simulation is performed on a $51\times51\times51$ simple cubic lattice, with periodic boundary condition in x- and y-directions and free boundary condition in z-direction. All lattice cells are fully filled with water, oil and copolymer cells. Each copolymer consists of a set of mean 10 cells, which are connected by bonds with variable lengths according to the bond fluctuation model(BFM)[8]. Although 108 discrete bond vectors and 5 discrete bond lengths are allowed in original model, we now restrict to 60 bond vectors (±2, 0, 0), (±2, ±1, 0), (±2, ±1, ±1), (±3, 0, 0), corresponding to 4 bond lengths of 2, $\sqrt{5}$, $\sqrt{6}$, 3. Such conditions never permit any intersections of bonds provided that eight vertexes of each cell are not sheared by the other copolymer cells. Both end cells of a copolymer are hydrophilic A-blocks, and the rest cells and whole bonds are hydrophobic B-blocks. Any water cells cannot sit in between two adjacent copolymer cells, because these parts, namely bonds, are hydrophobic.

Concerning the configuration energy, the attractive interaction strength of hydrogen bonds between water molecules or between water and copolymer ends is larger than the attractive interaction strength of van der Waals bonds among copolymers and oil or between water and oil. We just pick up the interaction energy E_{pw} between A-block and nearest-neighbor water cells and the interaction energy E_{ww} between two nearest-neighbor water cells in present simulation. Now, we put E_{pw} and E_{ww} as 3.2 $k_{\mathrm{B}}T$ and 1.6 $k_{\mathrm{B}}T$, respectively.

To generate thermal equilibrium configurations from arbitrary configurations, we use MC method according to Metropolis algorithm[9]. This algorithm sets the cell exchange probability at $\min\{1, \exp(-\Delta U/k_{\mathrm{B}}T)\}$, where ΔU is a change of total configuration energy in exchanging. Starting configuration is now a random mixture of water, oil and copolymers.

Rearrangement, namely exchanging two cells, is preformed as follows. Firstly, one picks unit site at random and secondly picks another site at random within less than 3 cells apart from the first site. If more than 3 cells, a copolymer bond would be certainly severed in exchanging. Therefore '3' is the maximum value keeping all distances of possible movement of three components the same. If the first site is oil cell and the second is water cell or vice versa, the exchange is performed according to Metropolis algorithm. If the first is copolymer cell, the exchange occurs according to Metropolis algorithm provided that the second is water or oil cell[10].

We change the volume ratio of water to oil under fixed 5% dilute copolymers, corresponding to 630 copolymers. The cell number per each copolymer takes a Gaussian distribution of mean 10 and variance 4. The statistical mean contour length of copolymer is about 21.3. The ratio of water/oil is assumed to be 10/85, 20/75, 30/65, 40/55 and 50/45. The change in pattern due to the change in the ratio of water/oil is investigated with emphasis on size distribution and the positions of water aggregations, and the configurations of copolymers bonding with aggregations.

The change in the ratio of water to oil must also induce the change in viscosity. Then, we prepare the simulation under shear flow. In Metropolis algorithm, one has to use some potential in order to deal with flow dynamics, because rearrangements depend only on the configuration 'energies'. We now test the Kramers potential[11], a kind of the velocity potential proposed by Kramers. Kramers potential depends on the only positions of components, regardless of whether the arrangement is homogeneous or not. Although MC simulation using Kramers potential has been succeeded in simple polymer solutions by Ohno et al.[12], and Xu et al.[13], it has not yet been reported in ternary systems.

The shear is now added in y-direction and its magnitude is proportional to z-height, then velocity v is $(0, -\gamma z, 0)$ in position (x, y, z), with shear rate γ. Force F exerted on copolymer or water cell against the velocity of solvent(oil), v, is $-\xi v$ with friction coefficient ξ, according to the Stokes low. Then the potential energy U_k causing this shear must be $-Fy$, that is, ξvy. Moreover, such v is realized as the sum of irrotational $1/2(0, -\gamma z, -\gamma y)$ and rotational $1/2(0, -\gamma z, \gamma y)$. If the shear can be approximated by the only irrotational part, that is, $v \sim 1/2(0, -\gamma z, -\gamma y)$, the potential U_k turns out to be $1/2\xi\gamma zy$. We now assume $1/2\xi\gamma$ is always 0.01, corresponding to moderate shear. Then the change in the Kramers potential ΔU_k in rearrangement depends only to the change of the y-component Δy and the mean z-component before and after exchanging. During the rearrangement under the

Kramers shear flow, the second site is selected only among the neighbors of the first site, while, in the process to equilibrate, the exchange between cells is allowed within maximum 3 cells apart. This restriction is needed to perform the correct dynamical process. For each volume condition, moderate shear flow is added to investigate the ease of movement of water and copolymers in the oil sea.

In all MC processes, we used the pseudorandom numbers from 0 to 1 generated by Tausworthe method[14], which confirms good uniformity. The all calculations were performed by using the supercomputers HITACHI SR11000 at the Information Technology Center of the University of Tokyo and at the Information Initiative Center of Hokkaido University.

RESULT; EQUILIBRIUM STATE

The ternary equilibrium phase patterns about 10/85/5, 30/65/5, and 50/45/5 of water/oil/amphiphiles are depicted in FIGURE 1, where black and grey indicate copolymer and water cells, respectively, while oil cells are white(transparent) in left three figures to avoid complications. In right three figures, only amphiphiles are picked up from each left figure. All snapshots are taken after 300,000 MC steps.

The patterns dramatically change with the ratio of water/oil. Water aggregates into many cubic clusters, two clusters with shape of rectangular rod, and one grid-shaped cluster, as the water concentration increases. And now we classify each amphiphile by the types of bonding with clusters; Bridge connecting its both ends with two different clusters, Loop connecting its both ends with the same cluster, One-side connecting only one end with a cluster, and Dangling not connecting both ends with any cluster. In every case of water concentrations, Bridges and Loops outnumber One-sides and Danglings. As the water increases, Bridges decrease and Loops increase. The number of Bridges, Loops and so on, and the number of water clusters and cluster sizes are tabulated in TABLE 1 and TABLE 2, respectively.

When water/oil is 10/85, water gathers to form many cubic clusters dispersed homogeneously in the system. The number of Bridges is 285, Loops 277, One-sides 64 and Danglings 4, among 630 amphiphiles. Particularly, Bridges mediating between relatively large-sized clusters are 175. One-sides and Danglings are very few.

In 20/75 of water/oil, the water clusters grow up and the phase separation between water clusters and oil plus amphiphiles occurs. In 30/65 of water/oil, water forms two clusters with shape of rectangular rod perpendicular to each other, parallel to x-y plane and separated in z-direction. Water clusters and oil plus amphiphiles regularly align in z-direction. The number of Bridges is 111, Loops 491. Particularly, in the space between two large clusters, Loop is 159 while Bridge is 56. Two water clusters are separated mainly Loop bundles.

Above 40/55 of water/oil, all water cells aggregate into one large cluster in also z-direction. In 50/45 water forms one grid-shaped cluster parallel to x-y plane. Both ends of almost all of amphiphiles are attached with this cluster except 6 One-sides. Particularly, 301 amphiphiles exist in the hollow of the cluster as Loops bonding both ends with this cluster.

TABLE 1. Classification of amphiphiles by the types of bonding with clusters. ( ) shows the number of Bridges or Loops between large clusters. [] shows the number of Loops in the hollow of grid-shaped cluster. Total number of amphiphiles is 630.

w/o/ polymer	Bridge	Loop	One-side	Dangling
10/85/5	285(175)	277	64	4
30/65/5	111(56)	491(159)	27	1
50/45/5	13	611[301]	6	0

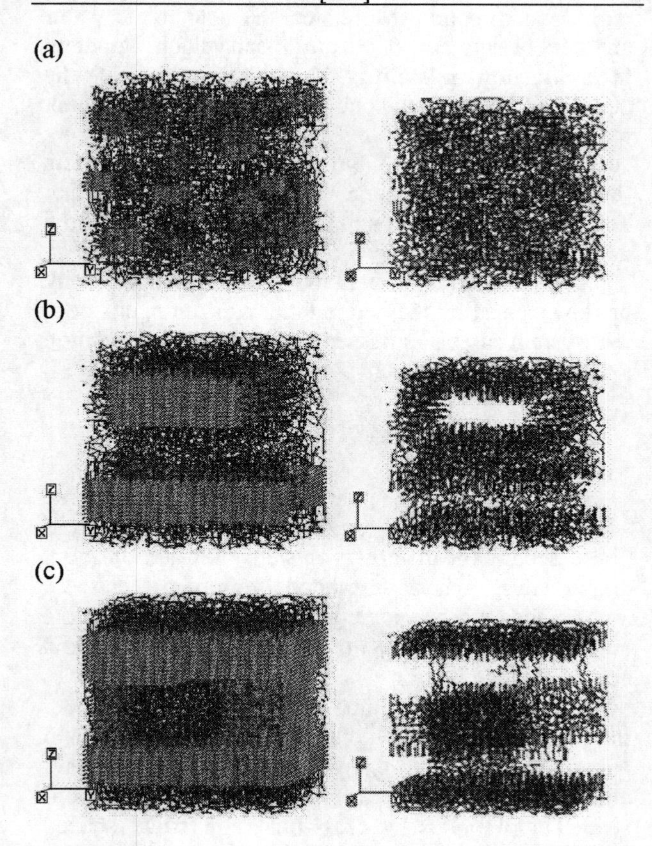

FIGURE 1. Typical equilibrium snapshots of (a)10/85/5, (b)30/65/5 and (c)50/45/5 of water/oil/amphiphiles. In left three figures, black, grey and white(transparent) mean amphiphiles, water and oil, respectively. In right three figures, only amphiphiles are picked up from each left figure.

TABLE 2. Number of clusters for different cluster sizes and different ratios of water/oil (w/o). Total number of water in each case is shown in {water = }. (1 =) is the number of unit water cube. Size of large cluster over 1001 is written in {size= }.

w/o \ size	1~10	11~100	101~1000	over
10/85{water=13265}	987 (1=826)	2	17	1{size=3557}
30/65{water=39795}	570 (1=490)	0	0	2{size=17184,21912}
50/45{water=66325}	230 (1=207)	0	0	1{size=66062}

RESULT; UNDER SHEAR FLOW

We attempt to deal with the shear by using Kramers potential. The shear flow is added in y-direction, with velocity gradient in z-direction, over 21,000 steps after 20,000 steps from random configurations. After 20,000 steps, the each phase pattern is similar to some extent to the pattern in equilibrium states(in the former section). Then we consider the time-averaged velocity of water and copolymers in y-direction over last 500 steps for w/o $\leq$ 30/65 and over last 100 steps for w/o $\geq$ 40/55. The velocity of water and copolymers decreases as the water concentration increases. In 50/45 of water/oil, particularly, copolymers and water are difficult to move in y-direction. Moreover, the velocity of water and copolymers deviates from linear velocity gradient in z-direction reflecting the heterogeneity of the density of water plus copolymers, which is we think realistic. Copolymers are excluded from not only copolymers themselves but also large water clusters (high density). And large water clusters are rather difficult to move than small water aggregations (low density).

This test simulation shows that the heterogeneity of flow caused by the heterogeneity of arrangement in real system can be expressed in also ternary simulation system by using Kramers potential.

SUMMARY AND DISCUSSION

In this paper, we applied the lattice MC scheme according to Metropolis algorithm to the simulation of phase separation in ternary system of water/oil/ABA amphiphiles, where excluded volume effects of amphiphiles are considered. Particularly, we used the lattice with periodicity in x- and y- directions and confined in z-direction.

We changed the volume ratio of water to oil under the constant dilute amphiphiles. In small water/oil ratio, water aggregated into many cubic clusters. Each water cluster dispersed homogeneously surrounded by Loop-shaped amphiphiles and cross-linked by Bridge-shaped amphiphiles. But the phase separation between water clusters and oil plus amphiphiles clearly occurred for large water/oil ratio. Water formed one grid-shaped cluster. At some intermediate water/oil ratio, water aggregated into two clusters with shape of rectangular rod parallel to x-y plane and separated in z-direction.

As the water increased, the number of Bridge-shaped amphiphiles decreased and the number of Loop-shaped amphiphiles increased. Particularly, two water clusters in intermediate water/oil ratio were separated by mainly Loop-bundles of amphiphiles.

It is an interesting subject to investigate the interrelation between the water/oil ratio, phase pattern and viscosity. For example, in dilute water, the networks between water clusters by the cross-linking of amphiphiles must govern viscosity[2]. The result of the simulation by using Kramers potential suggested that it is an effective tool in expressing moderate shear flow in ternary system. Substantial shear simulation will be reported in our forthcoming paper.

REFERENCES

1. P. Alexandridis, U. Olsson, and B. Lindman, *Macromolecules* **28**, 7700-7710 (1995).
2. M. Svensson, P. Alexandridis, and P. Linse, *Macromolecules* **32**,5435-5443 (1999).
3. C. Quellet, H. Eicke, G. Xu and Y. Hauger, *Macromolecules* **23**, 3347-3352 (1990).
4. A. Holmberg, P. Hansson, L. Piculell and P. Linse, *J. Phys. Chem. B* **103**, 10807-10815 (1999).
5. R. Larson, *J.Chem.Phys.* **91**, 2479-2488 (1989).
6. 'MS-DPD' produced by Accelrys Software Inc. ; URL `http://www.accelrys.com/reference/cases/studies/oilwater.html`; for concrete algorithm, D. Nicolaides, *Molecular Simulation* **26**, 51-76 (2001), R. Groot and P. Warren, *J.Chem.Phys.* **107**, 4423-4435 (1997).
7. H. Davis, J. Bodet, L. Scriven and W. Miller, "Microemulsions and their Precursors," in *Physics of amphiphilic layers:proceedings,* edited by J. Meunier, D. Langevin and N. Bociara, *Springer proceedings in physics* **21** (1987).
8. I. Carmesin and K. Kremer, *Macromolecules* **21**, 2819-2823 (1988).
9. N. Metropolis, A. Rosenbluth, M. Rosenbluth, A. Teller and E. Teller, *J.Chem.Phys.* **21**, 1087-1092 (1953).
10. N. Nakagawa, S. Maeda, S. Ishii and K. Ohno, *Mater. Trans.* **48**, 653-657 (2007).
11. H. Kramers, *Macromolecules* **14**, 415-424 (1946).
12. G. Xu, J. Ding and Y. Yang, *J.Chem.Phys.***107**, 4070-4084 (1997).
13. K. Ohno, M. Schulz, K. Binder and H. Frisch, *J.Chem.Phys.* **101**, 4452-4460 (1994)
14. R. Tausworthe, *Math.Coput* **19**, 201-209 (1965).

Effect of Viscosity of an Epoxy near or over Its Gel Point on Foaming Structures

Osamu Takiguchi, Daisaku Ishikawa,
Masataka Sugimoto, Takashi Taniguchi, and Kiyohito Koyama

Department of Polymer Science and Engineering, Yamagata University, Yonezawa 992-8510, Japan

Abstract. Effect of viscosity of an epoxy near or over its gel point on foaming structures was investigated. Rheological properties of the sample with a weight ratio 100/1 of epoxy/curing agent were measured by a simple shear experiment with sinusoidally varying shear. The magnitude of complex viscosity η^* as a function of time increases rapidly after a definite time period from the time when the temperature is set to 90°C and then increases slowly after η^* reached about 1×10^6Pa·s. The sample is found to gelate in the range of time $t = 15.0 \sim 15.6$min by the oscillatory measurement. The samples with a weight ratio 100/1/0.5 of epoxy/curing agent/blowing agent were foamed by putting them for 5min in a furnace of 230°C after various precuring conditions at 90°C. There are roughly two sizes of bubbles, large bubbles (>100μm) and small ones (≅30μm), when the sample is precured for relatively short time before foaming. On the other hand, foams precured for a long time before foaming, large sizes of bubbles disappear and the average diameter of bubbles becomes small, while the porosity is low.

Keywords: Chemical blowing agent, Epoxy; Foam, Gel point, Rheology
PACS: 82.70.Gg, 83.80.Jx, 83.80.Sg

INTRODUCTION

Epoxy has outstanding characteristics in chemical resistance, heat resistance, electrical properties and so on. Therefore, epoxy has widely been used in various applications. If foamings of epoxy are realized, which makes the field of application extend largely. However, applications of epoxy to foaming are quite difficult since the viscoelastic property of epoxy system highly and rapidly increases near and over its gelation point during the solidification process. So, in order to realize foamings in an epoxy, it is needed to elucidate the time dependence of the viscoelastic property and to control it. The aim of this study is to investigate the effect of viscosity of an epoxy near or over its gelation point on foaming structures. In order to clarify the role of viscosity in a foaming of epoxy, we study the effect of precured time at a lower temperature than a decomposition temperature of chemical blowing agent on structures of foams.

EXPERIMENTAL

Materials

The epoxy used in this study is EPIKOTE®834 made by Japan Epoxy Resin Co., Ltd.. We used as a curing agent Curezol®2E4MZ produced by SHIKOKU CHEMICALS CORPORATION, and as the blowing agent VINYFOR-AC#3C-K2 made by EIWA CHEMICAL IND. CO., LTD..

Rheological Properties

Rheological properties of the sample with a weight ratio 100/1 of epoxy/curing agent were measured for time and frequency dependence in a nitrogen atmosphere. The time dependence of magnitude of complex viscosity η^* was measured at 90°C, angular frequency $\omega = 1.0$rad/s and strain $\gamma = 1\%$. The storage modulus $G'(\omega)$ and loss modulus $G''(\omega)$ were measured at 90°C, in the range of $\omega = 1 \sim 100$rad/s, at the times $t = 14.6$, 15.0, and 15.6min when η^* reached 1×10^2, 1×10^3, 1×10^4Pa·s in Fig. 1, respectively.

CP982, *Complex Systems, 5th International Workshop on Complex Systems*
edited by M. Tokuyama, I. Oppenheim, and H. Nishiyama

Foaming

The samples with a weight ratio 100/1/0.5 of epoxy/curing agent/blowing agent were precured for 16, 21, and 27min at 90°C. Then, foamings of the samples were performed by setting them for 5min in a furnace of 230°C. The foamed samples were observed by SEM (scanning electron microscope).

RESULT AND DISCUSSION

Gel Point of Epoxy

Figure 1 shows curves of η^* for the sample as a function of time at 90°C. The curves of η^* rapidly increased after a definite time period from the time ($t = 0$) when the temperature is set to 90°C and then slowly increased after reaching about 1×10^6Pa·s. The definite time period up to the onset of the rapid increase of viscosity is found to be around $t = 15.0$min. In Figure 2, the storage modulus $G'(\omega)$ and loss modulus $G''(\omega)$ are plotted as a function of ω with precuring times 14.6, 15.0, and 15.6min at 90°C. Then, the $\tan\delta$ is described as a function of ω with precuring time 14.6, 15.0, and 15.6min at 90°C in Figure 3. Winter et al. suggested that $G'(\omega)$ and $G''(\omega)$ at a gelation point become as follows [1]:

$$G'(\omega) = G'_c\omega^n,\ G''(\omega) = G''_c\omega^n \ \left(0 < \omega < \infty\right) \quad (1)$$

$$\tan\delta = \tan\left(n\pi/2\right) \ \left(0 < n < 1\right) \quad (2)$$

where n is an exponent, G'_c and G''_c are constants. In our experimental situations, the state of system changes with time during the solidification process from a liquid state to a solid state. In addition to this, since the change rate of state in time near the gelation point is quite high, it is too difficult to measure $G'(\omega)$ and $G''(\omega)$ of our epoxy samples at the exact gelation point. In order to estimate the gelation point (the gelation time) of the system, we focus on the change of function shape of G' and G'' in $\log(G)$- $\log(\omega)$ plot in the similar manner to Matêjka's way [2], since eq.(1) and (2) state that, at the gelation point, the graphs of G' and G'' in $\log(G)$- $\log(\omega)$ plot becomes straight lines with the same slope. The curves of $G'(\omega)$ at 14.6 and 15.0min represent to be convex downward and then the curve of $G'(\omega)$ at 15.6min represents to be convex upward in Fig. 2. The change in $G'(\omega)$ from a convex downward function to a convex upward one denotes the change of the sample from viscoelastic liquids to viscoelastic solids. Furthermore, the curves $\tan\delta$ at 14.6 and 15.0min represent to be convex upward and then the curve of $\tan\delta$ at 15.6min represents to be convex downward in Fig.3. Therefore, we also determined the gelation point to be in the range of time $t = 15.0 \sim 15.6$min.

Foaming Structures

Figure 4 shows SEM micrographs of fracture cross-sections of the foamed samples obtained by setting them for 5min in a furnace of 230°C after a precuring for (a) 16, (b) 21, and (c) 27min at 90°C. Strings running from the bubbles are cracks that occur at making a cut of specimen using an edged tool. From Fig. 4, there are roughly two sizes of bubbles in the case (a), i.e., large bubbles (>100μm) and small ones (≅30μm). On the other hand, in (b) and (c), large sizes of bubbles disappear and the average diameter of bubbles becomes small, while the porosity is low.

From the result of rheological properties, samples after precuring at all conditions have already gelated, i.e., the samples change from viscoelastic liquids to viscoelastic solids. In addition, η^* becomes higher as the precuring time is longer. Therefore, we can understand that bubble size and porosity become smaller and lower as precuring time is longer. But, we cannot explain the reason why the size distribution of bubbles is roughly bimodal in the case (a). In future, we will investigate the origin of this bimodal distribution in bubble size for the case (a).

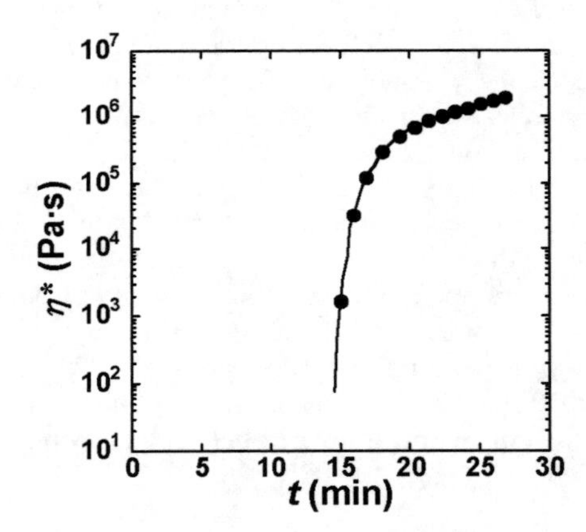

FIGURE 1. η^* for the sample as a function of time at 90°C.

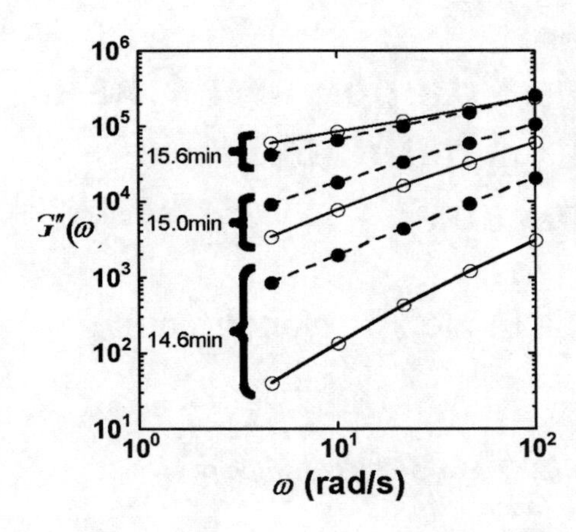

FIGURE 2. (—○—) $G'(\omega)$; (-●-) $G''(\omega)$ as a function of ω with precuring times 14.6, 15, and 15.6 min at 90°C.

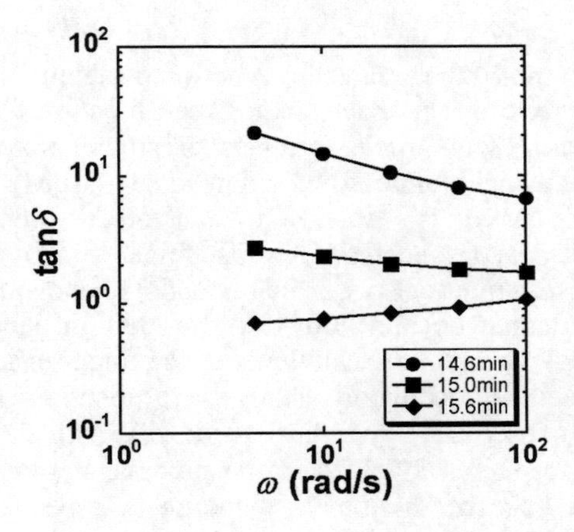

FIGURE 3. $\tan\delta$ as a function of ω with precuring times 14.6, 15.0, and 15.6 min at 90°C.

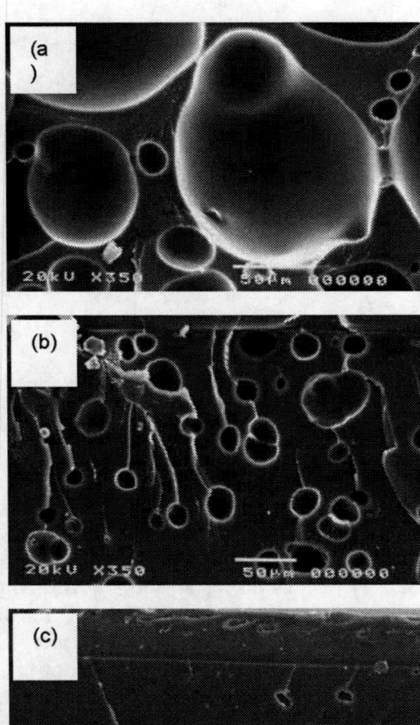

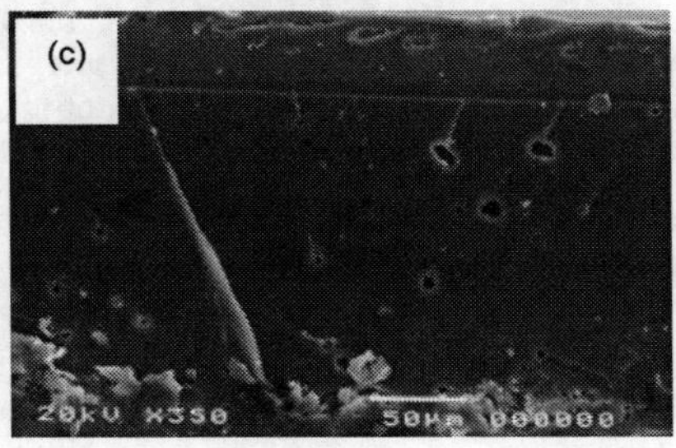

FIGURE 4. SEM micrographs of fracture cross-sections of the foamed samples obtained by setting them for 5min in a furnace of 230°C after precuring for (a) 16min, (b) 21min, and (c) 27min at 90°C.

REFERENCES

1. H. H. Winter, and F. Chambon, *J. Rheol.*, **30**, 367-382 (1986).
2. L. Matêjka, *Polym. Bull.*, **26**, 109-116 (1991).

Viscoelasticity and Dynamics of Single Biopolymer Chain Measured with Magnetically Modulated Atomic Force Microscopy

M. Kageshima[1,2], Y. Nishihara[1], Y. Hirata[3], T. Inoue[3], Y. Naitoh[1], and Y. Sugawara[1]

[1]*Department of Applied Physics, Osaka University, Suita, Osaka, 565-0871, Japan*
[2]*PRESTO, JST, Tokyo, 102-0075, Japan*
[3]*AIST, Tsukuba, Ibaraki, 305-8568, Japan*

Abstract. Viscoelactic response of a titin single molecule chain during the course of forced unraveling was studied using atomic force microscopy. Effect of transition to/from an unfolding intermediate onto the measured elasticity data was analyzed. The result hinted that emergence of the transition onto the elasticity depends on the experimental condition such as modulation force amplitude. Anomalous slow unraveling of one domain was observed and its viscoelastic response was discussed from the viewpoint of internal friction of polymer.

Keywords: Atomic force microscopy, Biopolymer, Folding, Intermediate, Nonequilibrium, Viscoelasticity
PACS: 82.37.Gk, 87.15.He, 87.15.La

INTRODUCTION

Diversity and richness of the macroscopic properties of a polymer system is highly dependent on this complexity. Recent rapid development of so-called single-molecule analysis methods have opened up a new approach to the polymer systems. A common feature among these methods is to capture the both end of chain and measure its spatial or mechanical responses. A question arises whether the complexity is retained even if the polymer system is scaled down to a single chain. In the typical time-scale of the above experimental methods the single polymer chain is still complex enough; substantial part of its quasi-equilibrium mechanical response is well-described in terms of entropic elasticity without referring to its internal microscopic variables. However, rapid transition from one equilibrium state to another is a non-equilibrium process in which microscopic variables play an essential role. Thus a single-molecule analysis is a significant approach to dynamics or transitional phenomena.

The single-molecule analysis is more essential in studying biopolymers than the rest of polymer systems, for interaction of an isolated biopolymer with different molecules or groups is quite relevant in a biological system. In addition, conformations of a protein molecule and the transition between them are especially relevant in understanding their function.

In the present article, analysis of single protein molecule using atomic force microscopy (AFM) is described. AFM is superior to the other single-molecule analysis methods in its force sensitivity typically reaching to 1 pN or below and its bandwidth often extending beyond 100 kHz. In addition, since AFM employs an elastic cantilever as the force sensor, it is suited to oscillatory analysis to discriminate between conservative (or elastic) and dissipative (or viscous) responses involved in the process. It should be noted that the dissipative response is especially relevant in describing the non-equilibrium processes.

Analysis of single biopolymer chain using AFM, in which response in DC force is measured, has been carried out by many research groups[1]. Using a spatial modulation of the cantilever sensor, the method was extended to discriminate between elastic[2-6] and dissipative[3-6] responses, hence the viscoelastic analysis of a single biopolymer chain. Origin of viscous property of polymer system, especially a single chain in dilute solution, has been controversial for decades[7]. Single-molecule analysis may provide these discussions with relevant experimental results.

CP982, *Complex Systems, 5th International Workshop on Complex Systems*
edited by M. Tokuyama, I. Oppenheim, and H. Nishiyama

EXPERIMENTAL

AFM with Magnetic Modulation

In the present analysis, the cantilever of AFM was excited magnetically via an AC magnetic force acting between a magnetic particle attached onto its backside and a solenoid placed in the vicinity of the sample holder. Oscillatory component of the excitation frequency in the deflection signal of the cantilever was detected via a lock-in amplifier and the viscoelastic profile during the course of the stretching process was recorded simultaneously with the DC force component. Schematic picture of the AFM detection is shown in Fig. 1. Detail of the analysis is described elsewhere[4].

As the simplest picture of the system, the molecule was expressed as a parallel combination of elastic and viscous mechanical elements, i.e., a Voigt-Kelvin model. The cantilever was modeled as a harmonic oscillator with its intrinsic spring constant and effective mass, while the viscous drag force exerted on the cantilever by the liquid is expressed as another viscous element. Data analysis is carried out based on the steady-state solution of the cantilever's equation of motion.

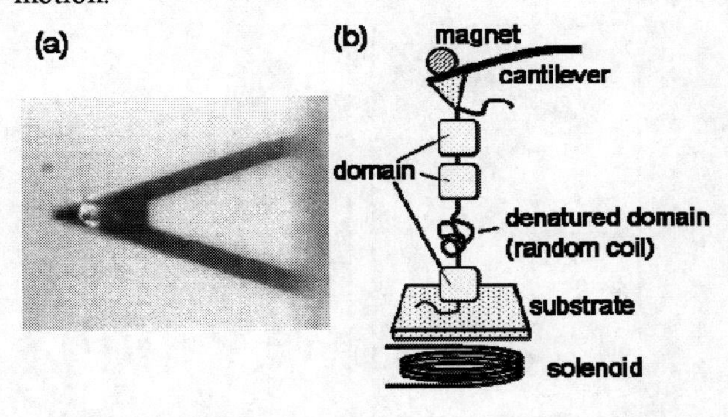

FIGURE 1. (a) AFM cantilever with spring constant of 0.03 N/m and an attached magnetic particle with a diameter of about 20 µm. (b) Schematic picture of magnetic modulation and forced denaturing of titin Ig domains.

Sample

Titin, also known as connectin, a giant protein found in striated muscle sarcomere was chosen as the sample molecule. Since titin is believed to play a crucial role in elastic properties of muscle, its response to external force has drawn attention. It has a typical modular structure in which fibronectin type III (Fn3) and immunoglobulin (Ig) domains are serially chained. Since these domains are structurally similar to each other, when an tensile force is applied on the both ends of the molecular chain, each domain is stochastically unraveled to a denatured state as shown in Fig. 1. The force response plotted with respect to the elongation length exhibits a saw-tooth like pattern with an interval of 200-300 Å, each peak in which corresponding to forced denaturing of each domains[8]. It was found that the forced unfolding of each domain can occur instantaneously while the typical time required for refolding is as long as about 1 s[8]. This large difference in unfolding/refolding is attributed to an extreme asymmetry in energy landscape between folded and unfolded states described in a simple two-state model[9]. On the other hand, also the possibility of more complex energy landscape; i.e., existence of an intermediate state emerging prior to complete unraveling of the domain has been reported[10-13]. In this intermediate state, only two hydrogen bonds are broken and the entire domain is extended by 5 Å[10]. Thus this molecule is an interesting system to investigate the transition dynamics of biopolymers.

RESULTS AND DISCUSSION

Unfolding Intermediate and Elasticity

Figure 2 shows profiles of DC force F_{DC} and elasticity k_m of the molecular chain extracted from a whole profile to show only the initial domain unraveling with a modulation force of 30 pN_{p-p} applied to the cantilever-molecule system. The DC force profile has an undulated feature at elongation around 700-800 Å as a result of multiple transition of serially-connected domains to the intermediate. However, the elasticity data integrated with the elongation length does not reconstruct the above feature and exhibits a simple downward-convex, entropic character. This implies that the transition to the intermediate is nonreversible, or in other words, nonequilibrium at the present modulation frequency of 600 Hz. This interpretation seems consistent with the report refolding time of the order of 10 ms[10].

However, data measured with a modulation force of 110 pN_{p-p} showed peaks in the elasticity data synchronized with the transition to the intermediate as shown in Fig.3. This result is consistent with recent publications by other research groups with modulation frequencies much higher than the present one[12,13].

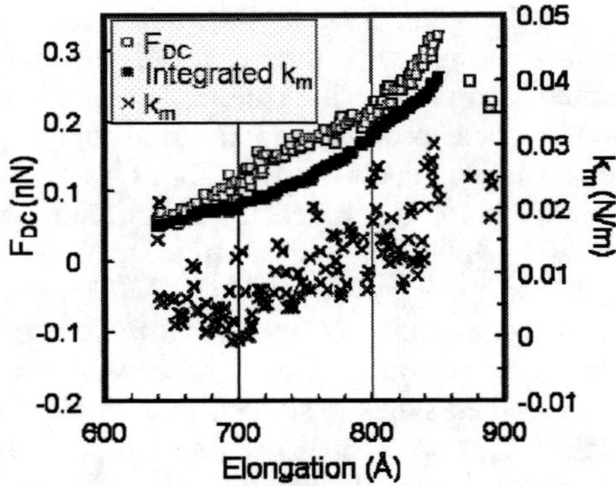

FIGURE 2. Profiles of DC force (FDC), molecular elasticity (km) and its integral with elongation during stretching of one denatured domain of a single titin molecule. Transitions to the intermediate of other domains are superimposed onto the DC force profile around 700-800 Å.

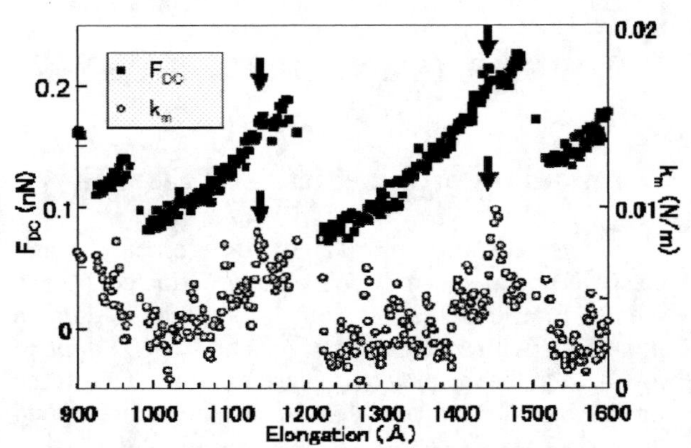

FIGURE 3. Profiles of DC force (F_{DC}) and molecular elasticity (k_m) showing unraveling of two domains. Transitions to the intermediate is indicated with arrows.

Considering the effect of DC bias force, however, it's a mystery why unfolding/refolding can take place repeatedly with an oscillation much faster than the reported refolding time. Here the two state model presented by Rief et al.[9] is extended to include the intermediate as shown in Fig. 4, where Fig. 4(b) corresponds to the case under the present consideration. Assuming typical threshold DC force for the transition F=100 pN and the distance required to refold from the intermediate d=2.5 Å, the refolding rate must be scaled by $\exp(-Fd/k_BT)$=0.002, where k_B is the Boltzman constant and T the temperature[9]. Thus, refolding from intermediate must be almost prohibited at the present frequency or in the conditions in the above two articles[12,13]. This suggests that the dynamics of single biopolymer chain might be dominated by complex effect of frequency or intensity of perturbation, local distribution of strain etc.

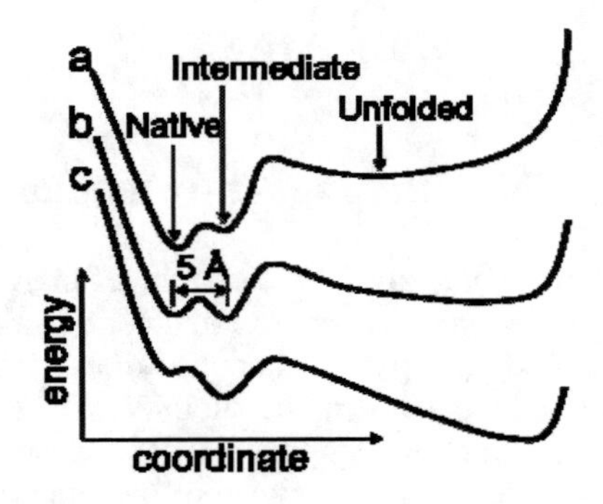

FIGURE 4. Schematic diagram of energy landscape for titin Ig domain deformed by tensile load. (a) Without load, (b) with medium load, and (c) with high load. Extension due to transition from native to intermediate was reported to be about 5 Å in Ref. 9.

Anomalous Slow Unraveling and Its Viscoelastic Response

Usually unraveling of Ig domain under a tensile load takes place instantaneously. However, in the same data profile as shown in Fig.1, only one domain was observed to unravel quite slowly. Figure 5 shows an enlarged part of this profile.

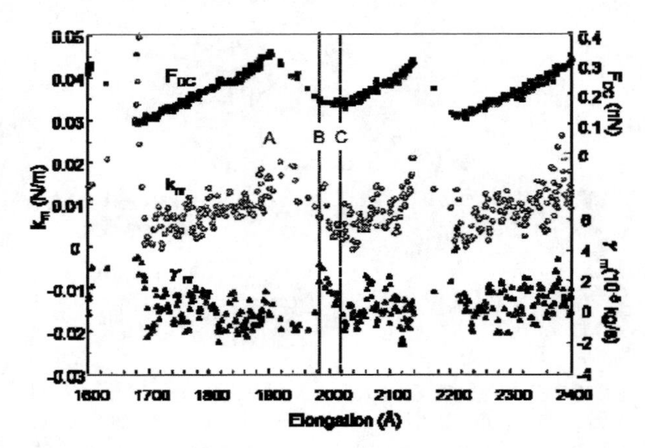

FIGURE 5. Viscoelasticity and DC force profile of titin single molecule showing anomalous (1900-2020 Å) and normal (2140-2200 Å) unraveling events. The interval between each sampling point is 5 ms.

A closer look reveals that the unraveling consists of two parts having different nature. The first one occurring from A to B in Fig. 5, and the second from B to C. In the first process taking about 50 ms the drag coefficient shows no significant value, while it rose at point B and decays throughout the second process for about 60 ms. The elastic property does not exhibit significant difference between the two processes and is decaying throughout. Although it is unclear what was happening during these processes, the first one should be regarded as simply elastic while the second dissipative. Since both the elasticity and viscosity decays in the second process, the characteristic relaxation time for this process is almost constant, indicating a consistent viscoelastic mechanism throughout the process, and is roughly of the order of 10^{-4} s. Assuming a viscous coefficient of water η_s=1.0x10^{-2} P, the length of a single amino acid residue a=3.4 Å and a segment number N=70 derived by dividing the spacing between peaks at 1900 and 2140 Å in Fig. 5 by a, the relaxation time due to the solvent friction[7] $\tau \cong \eta_s a^3 N^{9/5} / k_B T$ is about 2x10^{-6} s. This indicates additional friction mechanism like barrier friction or so-called Cerf friction is dominant. One of the proposed mechanisms to explain Cerf friction attributed to friction between the monomers happening to interact with each other as the chain intersects itself. This seems to be consistent with the present observation, since such intersection must decrease as the chain is stretched. For a more detailed discussion a more quantitative measurement and its comparison with theoretical calculation seem to be indispensable.

CONCLUSION

In contrast to conventional macroscopic measurement, single-molecule approach can simplify the phenomena to a certain extent as shown in the present study. On the other hand, a common problem among those single-molecule measurement method is that their time resolution is not always high compared to the characteristic time-scales involved in polymer dynamics such as relaxation time or folding/unfolding time. Extension of measurement bandwidth and acquisition of frequency-resolved information would be the next stage for further progress of the approach.

Another common problem among the single-molecule measurements is that only a molecule with its both ends tethered can be treated. Progress in theory taking such effects as entropic cost by tethering[14] into account is eagerly awaited.

Although a single polymer chain should be still regarded as a complex system, its nonequilibrium properties obviously emerge in its responses to external field as shown in the present article. Further improvement in the experimental technique as well as progress in theoretical approach may provide access to an unexplored realm in polymer physics.

ACKNOWLEDGMENTS

The authors express their gratitude to Prof. Sumiko Kimura for the titin sample and discussion.

REFERENCES

1. See, for example, T. Hugel and M. Seitz; *Macromol. Rapid Commun.* **22**, 989-1016 (2001); T. R. Strick, M.-N. Dessinges, G. Charvin, N. H. Dekker, J.-F. Allemand, D. Bensimon and V. Croquette; *Rep. Prog. Phys.* **66**, 1-45 (2003); W. Zhang and X. Zhang; *Prog. Polym. Sci* **28**, 1271-1295 (2003).
2. M. A. Lantz, S. P. Jarvis, H. Tokumoto, T. Martynski, T. Kusumi, C. Nakamura and J. Miyake, *Chem. Phys. Lett.* **315**, 61-68 (1999).
3. A. D. L. Humphris, J. Tamayo and M. J. Miles, *Langmuir* **16**, 7891-7894 (2000).
4. M. Kageshima, S. Takeda, A. Ptak, C. Nakamura, S. P. Jarvis, H. Tokumoto and J. Miyake, *Jpn. J. Appl. Phys.* **43**, L1510-L1513 (2004).
5. M. Kawakami, K. Byrne, B. Khatri, T. C. B. Mcleish, S. E. Radford and D. A. Smith, *Langmuir* **20**, 9299-9303 (2004).
6. H. Janovjak, D. J. Müller and A. D. L. Humphris, *Biophys. J.* **88**, 1423-1431(2005).
7. P.-G. de Gennes, "*Scaling Concepts in Polymer Physics*", Cornell University Press, Ithaca, 1979.
8. M. Rief, M. Gautel, F. Oesterhelt, J. M. Fernandez and H. E. Gaub, *Science* **276,** 1109-1112 (1997).
9. M. Rief, J. M. Fernandez and H. E. Gaub, *Phys. Rev. Lett.* **81**, 4764-4767 (1998).
10. P. E. Marszalek, H. Lu, H. Li, M. Carrion-Vazquez, A. F. Oberhauser, K. Schulten and J. M. Fernandez, *Nature* **402**, 100-103 (1999).
11. J. G. Forbes and K. Wang, *J. Vac. Sci. Technol. A*, **22**, 1439-1443 (2004).
12. M. J. Higgins, J. E. Sader and S. P. Jarvis, *Biophys. J.* **90**, 640-647 (2006).
13. M. Kawakami, K. Byrne, D. J. Brockwell, S. E. Radford and D. A. Smith, *Biophys. J.* **91**, L16-L18 (2006).
14. M. Carrion-Vazquez, A. F. Oberhauser, S. B. Fowler, P. E. Marszalek, S. E. Broedel, J. Clarke and J. M. Fernandez. *Proc. Natl. Acad. Sci. USA* **96**, 3694-3699 (1999).

Nonequilibrium Dynamics of a Manipulated Polymer: Stretching and Relaxation

Takahiro Sakaue

Department of Physics, Graduate School of Science, Kyoto University, Kyoto 606-8502, JAPAN

Abstract. Owing to their inherent softness, long flexible polymers may exhibit nonequilibrium responses upon manipulations. We attempt to analyze such processes based on the "interface" description associated with the tension propagation. Two illustrative examples, i.e., the dynamics of stretching and its reverse (shrinkage) process, are presented along with the basic formulation.

Keywords: Nonequilibrium dynamics, Polymer conformation, Single molecule experiment
PACS: 82.35.Lr, 36.20.Ey, 83.50.-v, 87.15.H-

INTRODUCTION

Thanks to the advance in experimental techniques in small scales, it is now possible to manipulate and/or control the behaviors of polymer chains in the single molecule level. For instance, one can deform and/or transport the chain by applying a force to a bead attached to the one end of the chain using optical tweezers or atomic force microscopes [1, 2, 3]. One can also make use of a geometrical confinement, i.e., the behaviors of a single chain confined in, for instance, narrow channels are controlled by applying a certain level of pressure gradient (or electric field for charged polymers) [4]. These techniques are obviously important for the practical use, but would also be useful to investigate the nonlinear-nonequilibrium dynamics of polymers in a single chain level. Indeed, a long polymer may be readily deformed and/or brought away from equilibrium upon the manipulation with a comparative ease owing to its inherent softness.

For long polymers, the induced change may not occur all at once usually. Rather, the evolution would be accompanied by a progressive spatial conformational modulation. The resultant nonequilibrium dynamics seems to be yet poorly understood in spite of the potential importance for various polymeric systems. In such problems, it is useful to consider the motion of "*interface*" explicitly. Below, we first provide a basic formulation, then, apply it to two simple examples.

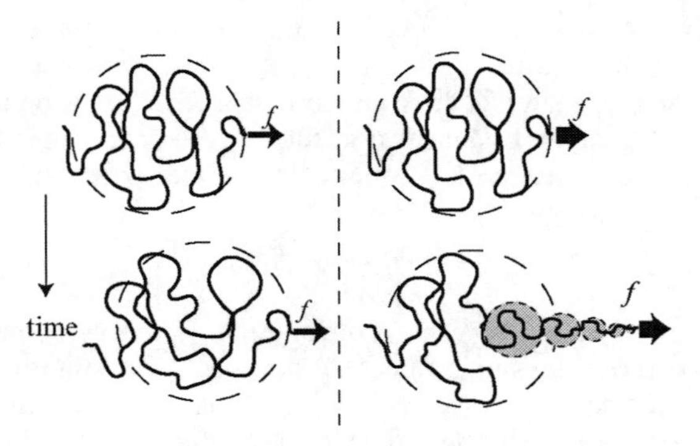

FIGURE 1. Responses of a flexible polymer to the pulling force applied to the one end. In the left, a weak force drags the chain without distorting its equilibrium conformation. The right is a snapshot of the chain at a certain moment during the transient period, where the force is strong enough to entail the chain elongation. The chain portion under the influence of the tensile force is designated by gray zone.

BASIC FORMULATION

Let us first set our system of interest in a dynamically stable state. Then, suppose that external forces are suddenly switched on, which makes the current state become unstable. The system is now out of equilibrium and begins to evolve towards a new stable state, during which the system often exhibits a characteristic spatial heterogeneity and organization. The dynamics of the transient period can be, therefore, described by the motion of interface, which is known to be very successful in various problems, perhaps most notably, in the phase transition dynamics.

We shall apply a similar idea to investigate the transient dynamics of a single manipulated polymer [5]. In this context, we can, for instance, imagine the situation, in which a relaxed coil in its equilibrium state is suddenly pulled by its one end. If the pulling force is weak enough, nothing interesting is expected: the chain will just follow the pulled direction keeping its equilibrium conformation (Fig. 1 left). Taking into the soft elasticity of long flexible polymers, however, we may expect more exotic responses in most of practical cases. Indeed, under

CP982, *Complex Systems, 5th International Workshop on Complex Systems*
edited by M. Tokuyama, I. Oppenheim, and H. Nishiyama

the action of the rather moderate force (typically on the order of $\sim$ pico Newton), the chain as a whole cannot follow the pulling force at once. In the immediate aftermath of the switching on, only an anterior part of the chain is under the influence of the pulling tension, thus deformed, but the rear part does not feel a force yet (Fig. 1 right). One can indeed define the time dependent "*interface*" or "*boundary*" between a region under the substantial influence of tension and an equilibrium state region still at rest. As time goes on, this boundary propagates along the chain backbone, as a result of which the chain progressively deforms towards a steady state conformation in its entirety.

How to mathematically formulate this kind of nonequilibrium processes? In principle, one needs the followings as basic equations; (i) *energy balance* per time, i.e., the power is balanced with the free energy change and the energy dissipation rate and (ii) *mass conservation*, i.e., how monomers are distributed between regions forward and backward of the interface at each moment. In addition, an equation of (iii) *local force balance* is required to describe the conformational properties during the transient.

TRANSIENT DYNAMICS: TWO EXAMPLES

In this proceeding, we shall illustrate our method using two examples, which were already studied based on more intuitive considerations. One is the dynamics of a polymer stretching induced by the strong pulling force [6], and the other is the reverse process, i.e., the shrinkage (or relaxation) dynamics of an initially stretched polymer upon switching off the pulling force [7]. In the first example, we shall find some merits in ours, i.e., not only the duration time (which was discussed in ref. [6] by numerical simulations and a scaling estimate), but also the time evolution of the conformation can be predicted. On the other hand, the conformational evolution during the relaxation was already discussed in ref. [7], and our approach is essential equivalent to it. Nonetheless, the second example may still deserve the presentation as an illustrative example for the application.

Stretching Dynamics

Let us consider the stretching dynamics of an initially relaxed polymer by applying a pulling force f to its one end. We are interested in (a) the duration of stretching period τ to reach a new steady state and (b) the physical characteristic of this transient dynamics, i.e., how the pulling tension is transmitted along the chain and the

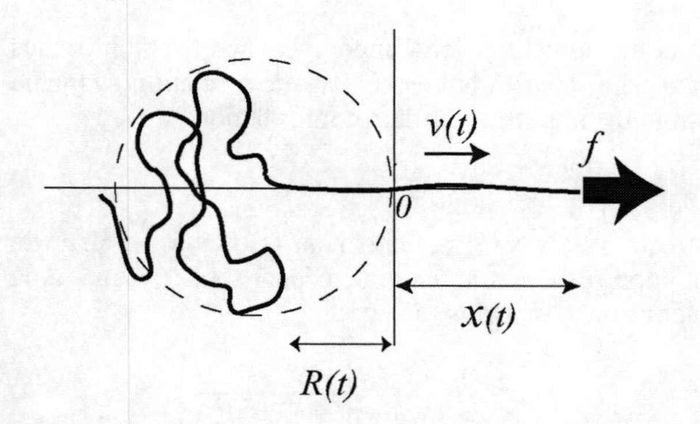

FIGURE 2. A transient period of the chain stretching by pulling its one end. A strong pulling force makes the chain portion under tension completely straightened. A dotted circle designates the shape and the position of the relaxed chain at $t = 0$.

chain conformation evolves. We shall only consider the strong pulling limit, so that a new steady state conformation is a completely stretched line. This limit is realized with $fb/(k_BT) > N_0$ (k_BT is the thermal energy and a polymer is modelled as a succession of N_0 monomers of size b). Since the conformation of the region under tension is a trivial straight line, the equation for the local force balance is unnecessary, which, in turn, makes the mass conservation equation simplified. As a result, the analysis becomes very simple, which, therefore, provides a nice example for an intuitive understanding.

At $t = 0$, we start to apply a force f to the one end of the chain, which is positioned at the origin. Figure 2 shows a schematic of a polymer at time t during the transient period of stretching. Here the polymer is divided into two region: one completely stretched and pulled with a velocity $v(t)$ by the force and the other still at rest. These are separated by a front with a radius $R(t)$ which is centered at origin. The number of monomers $N(t)$ in the tense region increases with time. These quantities are connected with the pulling force through the energy balance relation per time. As the change in the free energy due to the conformational change is negligible [1], the energy balance relation reduces to the total force balance between the pulling and the dragging force;

$$f \simeq \eta_s b N(t) v(t), \tag{1}$$

where η_s is the solvent viscosity. The mass conservation indicates the pulling velocity

$$v(t) = \frac{d}{dt}\left[bN(t) - R(t)\right], \tag{2}$$

[1] The rate of the free energy change is $\Delta\dot{F} \simeq (k_BT/b)v$, while the dissipation rate is $T\dot{S} \simeq fv >> \Delta\dot{F}$.

and the number of tense monomers and the front radius are connected through the statistical relation available from the initial equilibrium conformation:

$$R(t) = bN(t)^{\nu}, \tag{3}$$

where ν is a swelling exponent ($\simeq 0.6$ for a chain in a good solvent in three dimensional space). Combining above three equation, one gets

$$f \simeq \eta_s b^2 N(t) \frac{dN(t)}{dt}\left[1 - \frac{1}{N(t)^{1-\nu}}\right]. \tag{4}$$

The contribution from the second term becomes negligible except for the very initial momonet, therefore, one finds the solution for the tension propagation

$$N(t) \simeq \left[\frac{f}{\eta_s b^2} t\right]^{1/2}, \tag{5}$$

which, then, gives the evolution of the front;

$$R(t) \simeq b\left[\frac{f}{\eta_s b^2} t\right]^{\nu/2}. \tag{6}$$

The part of the chain contributing the dissipation grows in time, which results in the square root growth of the "stem" length (eq. (5)). The duration of the stretching period is obtained by setting $N(\tau) = N_0$;

$$\tau \simeq \frac{\eta_s b^2}{f} N_0^2. \tag{7}$$

Since the induced flow effect becomes nearly negligible (within logarithmic factors) in this strong pulling limit (see eq. (1)), the duration time, eq. (7) is the same as the result obtained by numerical simulations and scaling argument for the chain without hydrodynamic interactions [6].

It is also possible to study the dynamics with weaker (moderate) forcing [5]. In such cases, the deforming part of the chain is not completely stretched any more, and the evolution of the conformation should be determined by the local force balance equation. This makes the role of the hydrodynamics becomes evident.

Relaxation Dynamics

Now, we discuss a reverse process, i.e., how the initially stretched polymer relaxes into equilibrium state upon the release of the pulling force. Here we consider the polymer whose one end is fixed in space (origin) and the other end is pulled by a strong force. For simplicity, we assume that the initial conformation is a completely straight line. At $t = 0$, this pulling force is switched off,

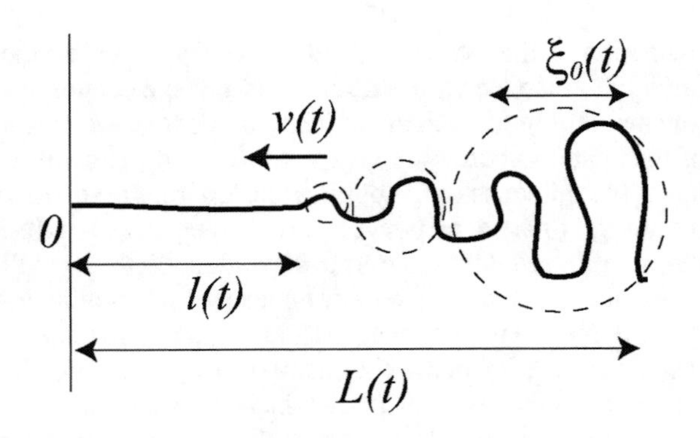

FIGURE 3. A conformation of the chain during the transient period of the shrinkage from a straight line. One end is fixed in space (origin) and the other end is released at $t = 0$.

then, the chain starts to relax from the end. Just in case of a stretching dynamics, we may divide the relaxing polymer into two regions; one assuming the initial state (complete stretching) yet, and the other starting to relax. The latter part would assume a locally steady state at each moment, therefore, can be viewed as a sequence of blobs with growing sizes (Fig. 3).

Then, the mass conservation reads

$$\frac{l(t)}{b} + \int_{l(t)}^{L(t)} \frac{g(x)}{\xi(x)} dx = N_0, \tag{8}$$

where $L(t)$ is the total length, $l(t)$ is the length of the yet completely stretched part. The blob size $\xi(x) = bg(x)^{\nu}$ at the location x and the retraction velocity $v(t) = -dl(t)/dt$ are related through the local force balance equation;

$$\frac{k_B T}{\xi(x)} \simeq \eta_s [L(t) - x] v(t) \qquad \text{[at } x > l(t)]. \tag{9}$$

At the free end, the size of largest blob ξ_0 is determined by

$$\frac{k_B T}{\xi_0} \simeq \eta_s \xi_0 v(t) \Leftrightarrow \xi_0 \simeq \left(\frac{k_B T}{\eta_s v(t)}\right)^{1/2}, \tag{10}$$

and at the boundary, the size of the blob is $\xi(l) = b$, which indicates

$$\frac{k_B T}{b} \simeq \eta_s [L(t) - l(t)] v(t) \Leftrightarrow L(t) - l(t) \simeq \left(\frac{k_B T}{\eta_s b v(t)}\right). \tag{11}$$

Equation (11), indeed, corresponds to the energetic balance relation in this case, i.e., the rate of free energy change balances with the dissipation rate[2]. Combining eq. (8) and (9) supplemented with "boundary conditions" eq. (10) and (11), one obtains the integral equation

$$\frac{k_B T}{\eta_s b v(t)} \simeq \int_0^t v(t) dt, \quad (12)$$

the solution of which is

$$v(t) \simeq \left(\frac{\eta_s b}{k_B T} t \right)^{-1/2}. \quad (13)$$

Upon an integration, this leads to the shrinkage equation of the "stem" length:

$$l(t) = l(0) - c_1 \left(\frac{k_B T}{\eta_s b} t \right)^{1/2}, \quad (14)$$

and by inserting into eq. (11), then, one finds the time evolution of the length of the relaxing (sequence of blobs) part:

$$L(t) - l(t) = c_2 \left(\frac{k_B T}{\eta_s b} t \right)^{1/2} \quad (15)$$

where $l(0) = bN_0$ is the initial length and c_1, c_2 are numerical coefficients of order unity ($c_1 > c_2$). The total length, therefore, decreases according to

$$L(t) = l(0) - (c_1 - c_2) \left(\frac{k_B T}{\eta_s b} t \right)^{1/2} \quad (16)$$

It is interesting to note that the retraction dynamics of the "stem" and the growth of the relaxing part (therefore the shrinkage of the total length) obeys the same scaling low, which also coincides with the growth of the "stem" in the stretching dynamics by a very strong force (eq. (5)). The relaxation time τ is obtained by setting $l(\tau) = 0$, thus,

$$\tau \simeq \tau_0 N^2. \quad (17)$$

All these results are consistent with the earlier study [7] (see also ref. [8] for the experimental verification). The density profile evolves according to

$$\rho(x,t) \simeq \frac{g(x,t)}{\xi(x,t)^3} \simeq \frac{1}{b^3} \left(\frac{\xi}{b} \right)^{(1-3\nu)/\nu} \quad (18)$$

where $\xi(x,t)$ is calculated from eq. (13) and (16):

$$\xi(x,t) \simeq b \left(\frac{t}{\tau_0} \right)^{1/2} \left[\frac{l(0) - x}{b} - (c_1 - c_2) \left(\frac{t}{\tau_0} \right)^{1/2} \right]^{-1} \quad (19)$$

[2] The free energy ΔF (relative to the relaxed state) of the polymer during the transient may be written as $\Delta F(l) = F_s(l) + F_r(l)$, where $F_s(l)/(k_B T) \simeq l/b$ (for the straight part) and $F_r(l)/(k_B T) \simeq \int_l^L dx/\xi(x)$ (for the relaxing part). On the other hand, the dissipation rate is given by $T\dot{S} \simeq \eta_s (L-l) v^2$. Then, $\Delta \dot{F} = -T\dot{S}$ leads to eq. (11).

SUMMARY

We have discussed the transient dynamics of a single polymer, which is associated with the propagation of the tensile force along the chain backbone based on the "*interface*" description. Two simple cases, i.e., the stretching and the relaxation, have been demonstrated as applications. We expect that the present framework would be useful for many other problems. The dynamics of polymer translocation (transport of the polymer through nanosize pore) is one of important examples, which has been studied in detail recently based on the present formulation [5].

Our argument mostly relies on the universality concept, in particular the dynamical scaling developed for long polymer chains. Therefore, the analysis reported here as well as in ref. [5] predicts, if any, some generic dynamical aspect of the problem, which would be model-independent. However, there should be certainly phenomena which require more inputs to analyze. In particular, once the biological factor comes in, one should most probably consider the structural details and resultant specific effects carefully. One nice example is the problem of DNA translocation through a protein pore [9]. There would be a barrel of interesting questions in this direction, which would open up a new interdisciplinary field among physics, chemistry, biology and various industrial technologies.

ACKNOWLEDGMENTS

This research was supported by JSPS Research Fellowships for Young Scientists (No. 01263).

REFERENCES

1. T.T. Perkins, D.E. Smith and S. Chu, *Science* **264**, 819–822 (1994).
2. Y. Murayama Y. Sakamaki and M. Sano, *Phys. Rev. Lett.* **90**, 018102-1–018102-4 (2003).
3. M. Rief, et al., *Science* **276**, 1109–1112 (1997).
4. C.H. Reccius, et al., *Phys. Rev. Lett.* **95**, 268101-1–268101-4 (2005).
5. T. Sakaue, *Phys. Rev. E* **76**, 021803-1–021803-7 (2007).
6. Y. Kantor and M. Kardar, *Phys. Rev. E* **69**, 021806-1–021806-12 (2004).
7. F. Brochard-Wyart, *Europhys. Lett.* **30**, 387–392 (1995).
8. S. Manneville, et al., *Europhys. Lett.* **36**, 413–418 (1996).
9. D.K. Lubensky, D.R. Nelson, *Biophysical J.* **77**, 1824–1838 (1999).

Morphologies of Multicompartment Micelles formed by Triblock Copolymers

Shih-Hao Chou[1], Heng-Kwong Tsao[2], and Yu-Jane Sheng[1]

[1]*Department of Chemical Engineering, National Taiwan University, Taipei, Taiwan 106, R.O.C.*

[2]*Department of Chemical and Materials Engineering, Institute of Materials Science and Engineering, National Central University, Jhongli, Taiwan 320, R.O.C.*

Abstract: Multicompartment micelles are desirable for advanced applications such as drug delivery. Recently, core-shell-corona (*CSC*) and segmented-worm (*SW*) micelles formed by *ABC* triblock terpolymers with three mutually immiscible blocks are observed in experiments. We have performed dissipative particle dynamics simulations to study the effects of molecular architecture, block length and solution concentration on the morphologies of *ABC* triblock terpolymers. The formation of *CSC* and *SW* micelles for linear and miktoarm star *ABC* terpolymers is confirmed in this work. In addition, we predict that different multicompartment micellar morphologies (e.g. incomplete skin-layered micelles and segmented worms) can be formed by linear copolymer with different arrangements of the three blocks.

Keywords: ABC terpolymer, Dissipative particle dynamics, Micelles, Molecule simulation
PACS: 31.15.at

INTRODUCTION

The self-assembly of block copolymers in a selective solvent is of fundamental interest[1] and offers tremendous promise for advanced applications, such as nanomaterial synthesis, potential delivery vehicles for pharmaceuticals, and gene therapy agent. Despite the particular morphology (e.g. spheres, cylinders, and vesicles), those aggregates generally consist of two compartments: the micellar core and the micellar corona. However, the formation of multicompartment micelles is desirable[2].

It is appealing to study whether rather simple multiblock copolymer architectures can be designed to create structures within structures[3]~[6]. Since aggregates formed by amphiphilic *AB* diblock or *ABA* triblock copolymers provide only two compartments, *ABC* triblock polymer terpolymers with three mutually immiscible blocks can serve as a minimal model system to form several compartments in one micelle. For example, *A* is hydrophobic block based on fluorocarbons, *B* is a hydrophobic block based on hydrocarbons, and *C* is a hydrophilic block. The structured micelles are possibly formed as assemblies in which separate *A* and *B* core domains exist within a solvated *C* corona[7]. The formation of "three-layer" micelles in water by linear *ABC* triblocks have been observed. These micelles, referred to as core-shell-corona (*CSC*) micelles, include spheres and disks[4][8].The subdivision of the micellar interior contains two concentric nanodomains. On the other hand, depending on the relative block lengths, miktoarm star *ABC* terpolymer can form discrete, multidomain cores or extended, segmented worm (*SW*) structures[3][5].The lateral structure consists of approximately hexagonally packed fluorocarbon channels immersed in a continuous, two-dimensional hydrocarbon bilayer.

Morphological studies of multicompartment micelles are of great importance to complement experimental observations and advanced biomedical applications. In this work we perform *DPD* to explore the morphologies of multicompartment micelles formed by *ABC* terpolymers in aqueous solutions. The effects of molecular architecture, block length, and polymer concentration on the morphologies are examined.

MODEL AND SIMULATION METHODS

he DPD method[9] is a particle-based, mesoscale simulation technique. This method combines some of the detailed description of the *MD* but allows the

CP982, *Complex Systems, 5th International Workshop on Complex Systems*
edited by M. Tokuyama, I. Oppenheim, and H. Nishiyama

simulation of hydrodynamic behavior in much larger, complex systems, up to the microsecond range. In the system, there are four different species of *DPD* particles, including solvent (*S*), highly hydrophobic particle (*A*), hydrophobic particle (*B*) and hydrophilic particle (*C*). The force is composed of three different pairwise-additive forces: conservative ($\mathbf{F}^C$), dissipative ($\mathbf{F}^D$) and random forces ($\mathbf{F}^R$),

$$\mathbf{f}_i = \sum_{j \neq i} \left(\mathbf{F}_{ij}^C + \mathbf{F}_{ij}^D + \mathbf{F}_{ij}^R \right) \quad (1)$$

These forces conserve net momentum and all acts along the line joining the two particles. The conservative force $\mathbf{F}^C$ for non-bonded beads is a soft repulsive force,

$$\mathbf{F}_{ij}^C = \begin{cases} a_{ij}(1 - r_{ij})\hat{\mathbf{r}}_{ij} & r_{ij} < 1 \\ 0 & r_{ij} > 1 \end{cases}, \quad (2)$$

where a_{ij} is a maximum repulsion between particles *i* and *j* and r_{ij} is the magnitude of the bead-bead vector. $\hat{\mathbf{r}}_{ij}$ is the unit vector joining beads *i* and *j*. The dissipative or drag force has the form,

$$\mathbf{F}_{ij}^D = -\gamma w^D(r_{ij})(\hat{\mathbf{r}}_{ij} \cdot \mathbf{v}_{ij})\hat{\mathbf{r}}_{ij} \quad (3)$$

where $\mathbf{v}_{ij}=\mathbf{v}_i-\mathbf{v}_j$ and w^D is a *r*-dependent weight function. The form is chosen to conserve the total momentum of each pair of particles and therefore the total momentum of the system is conserved. The dissipative force acts to reduce the relative momentum between particles *i* and *j*, while random force is to impel energy into the system. The random force also acts between all pairs of particles as

$$\mathbf{F}_{ij}^R = \sigma w^R(r_{ij})\theta_{ij}\hat{\mathbf{r}}_{ij}, \quad (4)$$

where w^R is also a *r*-dependent weight function and θ_{ij} is a randomly fluctuating variable with Gaussian statistics:

$$\langle \theta_{ij}(t) \rangle = 0, \quad \langle \theta_{ij}(t)\theta_{kl}(t') \rangle = (\delta_{ik}\delta_{jl} + \delta_{il}\delta_{jk})\delta(t - t'). \quad (5)$$

For polymer systems, we also have to consider the interaction forces between bonded beads,

$$\mathbf{F}_{i,spring}^C = -\sum_j C(r_{ij} - r_{eq})\hat{\mathbf{r}}_{ij}, \quad (6)$$

where the sum runs over all particles to which particle *i* is connected. In this work, we have chosen $C = 4$ and $r_{eq} = 0.0$. This choice is only a coarse-graining selection to show the constraint imposed upon connected beads of a polymeric chain. These forces conserve net momentum and all acts along the line joining the two particles. Details of the DPD method are given by the former papers[9]-[11].

On the basis of the algorithm described, the dynamics of 24000 *DPD* particles, starting from random distribution, was simulated in a cubic box (20^3 or 30^3) under periodic boundary conditions. The equation of motion are integrated with a modified velocity Verlet algorithm[11] with $\lambda = 0.65$ and Δt=0.05. In this work, each simulation takes at least 200,000 steps and the first 80,000 steps are for equilibration. Note that *DPD* simulation utilizes soft-repulsive potentials and the systems studied are allowed to evolve much faster than the "brute-force" molecular dynamics. Therefore a typical *DPD* simulation needs only about 50,000 to 100,000 steps to equilibrate. After that the resulting morphologies are analyzed. The simulation is performed by using a serial Fortran code on a PC with CPU of Pentium d-930. It takes about 72 hours of CPU time for a given simulation run with box size equal to 30.

RESULTS AND DISCUSSION

Molecular simulation is an alternative approach to gain more direct and microscopic level information than experiments. Mesoscale simulations such as dissipative particle dynamics can treat a wide range of length and time scales by many orders of magnitude compared to atomistic simulations. In this paper, *DPD* is performed to explore the morphologies of structured micelles formed by *ABC* triblock terpolymer with three mutually immiscible blocks, including miktoarm star and linear architectures.

It was experimentally reported that miktoarm star terpolymers *OEF*, as schematically illustrated in Fig. 1(a), can form "hamburger" micelles and "segmented-worm" micelles by varying the length of hydrophilic block (*O*). For the miktoarm star with a fixed length of *O* block, i.e. *OEF* (*20-6-7*), both discrete hamburger micelles (Fig. 1(*b*)) and segmented-worm micelles (Fig. 1(*c*)) can be observed depending on the polymer concentration. The discrete micelle is formed at very low concentration and consists of a *F* core, surrounded by a few *E* nanodomains. In general, two *E* nanodomains are found on top and bottom of the *F* core to form *EFE* hamburger micelles (Fig. 1(*b*)). The *O* blocks emanate from the *E-F* interface and curl around to protect the hydrophobic core. As polymeric concentration increases, elongated, wormlike structures emerge (Fig. 1(*c*)). The worms are layered with alternating section of *F* and *E* blocks along the long axis. The O coronas are shared by E and F layers to shield them from the highly unfavorable exposure to water. The segmented-worm micelle is always formed for star (x-6-7) as long as the polymer concentration is high enough. The onset concentration for the formation of segmented-worm micelles is increased with the hydrophilic O block length. When the hydrophilic O block is large (x=20), the O block screen the hydrophobic core effectively and the hamburger micelles can survive at higher concentration. On the other hand, for stars with shorter O blocks (x=10), segmented worms form at a much lower concentration. This is because by forming into segmented worms can the hydrophobic blocks (E and F) be able to share their O coronas and avoid direct contact with the solvents. When the concentration is even higher, the segmented micelles may join together to form segmented network, as illustrated in Fig. 1(d).

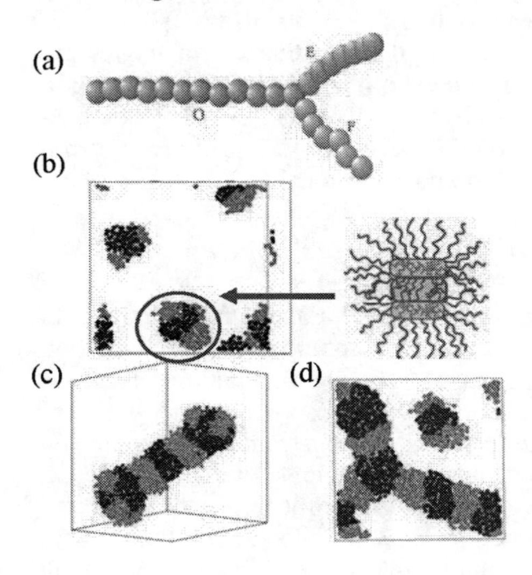

FIGURE 1. (a) representation of a miktoarm star *OEF*. (b) morphologies of hamburger micelles formed by *OEF* (20-6-7) stars. (c) segmented-worm (d) network formation by segmented worms.

In addition to miktoarm star, various morphologies of multicompartment micelles can be disclosed by linear triblock copolymers with different arrangement of the three blocks, *OEF*, *OFE*, and *EOF*. The schematic representations of linear *OEF*, *OFE*, and *EOF* are shown in Figs. 2(a), 3(a) and 4(a), respectively. For linear *OEF* (x-6-7) triblocks, the core-shell-corona structure can be evidently identified with the F core, E shell, and O corona, as shown in Fig. 2(b) and 2(c). In our simulations, we directly observe this CSC morphology, which is consistent with the experimental findings. Nonetheless, the spread of the E blocks on the surface of the F core is not uniform. This result is mainly due to the mutual incompatibility between E and F. The competition between energetic and entropic effects leads to a partial spread of the E blocks around F cores. This outcome is in fact energetically favorable as long as a_{OE} is comparable with a_{OF} and the O blocks can serve as a shielding layer preventing the E and F blocks from interacting with water. In general, the aggregation number of the CSC micelle declines as the length of O block (x-6-7) is increased.

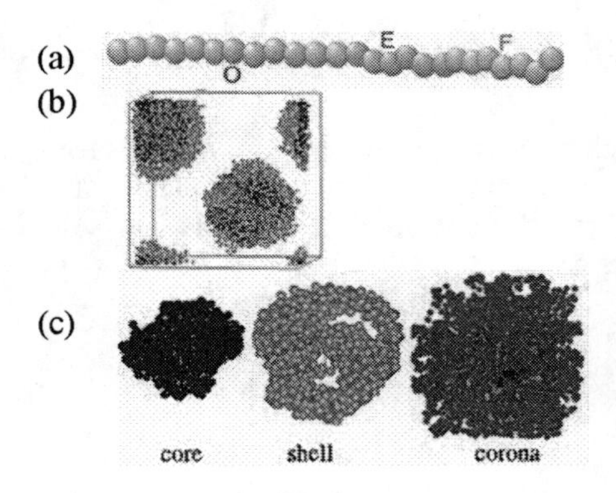

FIGURE 2. (a) representation of a linear triblock copolymer *OEF*. (b) Simulation results of micelles formed by linear *OEF* (20-6-7). (c) Morphology of a single core/shell/corona micelle formed by linear *OEF* (20-6-7).

It is interesting to find that rather different morphologies can be observed by changing the arrangement of the O, E, and F blocks even for the same composition in a triblock copolymer. For linear *OFE* (x-7-6) with shorter O blocks, the E blocks form a core that is surrounded by an incomplete skin layer of the F block as illustrated in Fig. 3(b). Since the O blocks protect mainly the F skin, part of the E core is exposed to water. However, when the O block is long enough to curl around to shield the E domain, the core consists of two separate but adjacent domains (E and F), as depicted in Fig. 3(c). When the polymer concentration is high enough, the hamburger micelle (F-E-F) can be formed by merging two F skin layer micelles (Fig. 3(d)). Its formation can diminish the exposure of the E core of the skin layer micelle to water. Nonetheless, the segmented-word micelle is never observed. The current experimental approaches might be difficult to identify those micellar morphologies undoubtedly.

Finally, a possible architecture can be obtained by placing the solvophilic O block in the middle of the linear triblock copolymer, that is the linear *EOF*. For linear *EOF* (6-x-7) with longer O blocks, a micelle with two neighboring E and F layers is shielded by the O loops, as shown in Fig. 4(b). However, when the O

block is too short, only the combination of several two-layer micelles can the solvophobic *E* and *F* blocks receive enough protection by the *O* blocks. Consequently, segmented-worm micelles are formed as shown in Fig. 4(*c*). Note that there is a difference between the segmented worm structures formed from the miktoarm star and linear *EOF* triblock copolymers. The solvophobic *O* blocks for the segmented-worm micelles of the miktoarm star have free ends and these ends spread out freely into the solvents.

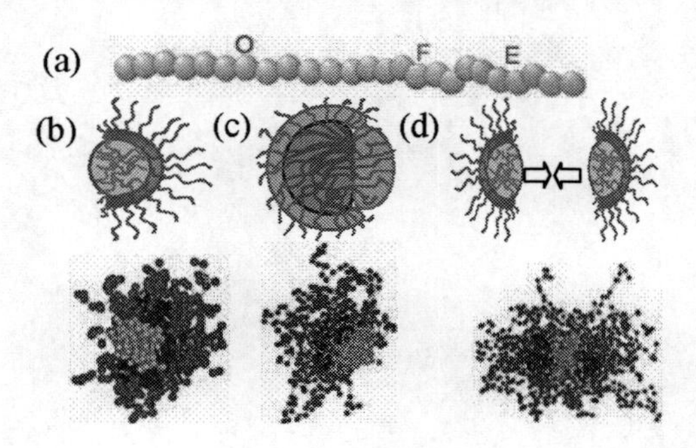

FIGURE 3. (a) representation of a linear triblock copolymer *OFE* . morphology of (b) the incomplete skin-layered micelles formed by linear *OFE* (10-7-6) , (c) the bi-core micelles formed by linear *OFE* (20-7-6) and (d) the hamburger micelles formed by linear *OFE* (10-7-6) at high concentration.

However, the ends of the solvophobic *O* blocks for the segmented-worm micelles formed by linear *EOF*s are connected to *E* and *F* blocks. Therefore the corona is not as expansive. A summary of the morphologies observed for linear triblock copolymers with different arrangement of the three blocks is presented in Table 1.

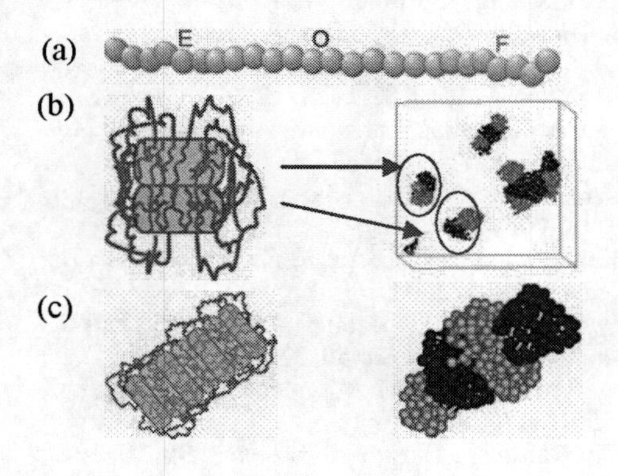

FIGURE 4. (a) representation of a linear triblock copolymer *OFE* . morphology of (b) the incomplete skin-layered micelles formed by linear *OFE* (10-7-6) , (c) the bi-core micelles formed by linear *OFE* (20-7-6) and (d) the hamburger micelles formed by linear *OFE* (10-7-6) at high concentration.

TABLE 1. Summary of the morphologies of micelles observed from the dilute solutions of linear triblock copolymers with various architectures

architecture		morphology
OEF		Core/shell/corona
OFE	short O	incomplete skin-layered micelle
	long O	bi-core micelle
EOF	short O	segmented-worm micelle
	long O	bi-layer micelle

In summary, copolymers with three chemically distinct blocks are able to form various multicompartment micelles. We have performed dissipative particle dynamics simulations to investigate the effects of molecular architecture, block length, and solution concentration on the morphologies of *ABC* triblock copolymers. The complex structures within the multidomain micelles are examined. In this work, we have observed the segmented-worm and hamburger micelles for miktoarm star and the core-shell-corona micelle for linear copolymer, which were reported based on experimental observation[3][4][8]. In addition, we predict the existence of other different morphologies, which can be formed by linear copolymers with different arrangements of the three blocks. Our simulation results indicate that mesoscale dissipative particle dynamics simulation is a powerful tool to explore the morphologies associated with self-assembled aggregates of copolymers and thus can potentially be used to design novel multicompartment micelles.

ACKNOWLEDGMENTS

This research is supported by National Council of Science of Taiwan. Computing time provided by the National Center for High-Performance Computing of Taiwan is gratefully acknowledged.

REFERENCES

1. I. W. Hamley, *The Physics of Blocks Copolymers*, (Oxford University Press: Oxford, 1998).
2. Laschewsky, *Curr. Opin. Colloids Interface Sci.* **8**, 274-281 (2003).
3. Z. Li, E. Kesselman, Y. Talmon, M. Hillmyer, and T. P. Lodge, *Science* **306**, 98-101 (2004).
4. Z. Zhou, Z. Li, Y. Ren, M. A. Hillmyer, and T. P. Lodge, *J. Am. Chem. Soc.* **125**, 10182-10183 (2003).
5. T. P. Lodge, A. Rasdal, Z. Li, and M. A. Hillmyer, *J. Am. Chem. Soc.* **127**, 17608-17609 (2005).
6. Z. Li, M. A. Hillmyer, and T. P. Lodge, *Macromolecules* **39**, 765-771 (2006).
7. M. A. Hillmyer, and T. P. Lodge, *J. Polym. Sci. Part A: Polym. Chem.* **40**, 1-8 (2002).
8. J.-F. Gohy, N. Willet, S. Varshney, J.-X. Zhang, and R. Jérome, *Angew. Chem. Int. Ed.* **40**, 3214-3216 (2001).
9. P. J. Hoogerbrugge, and J. M. V. A. Koelman, *Europhys. Lett.* **19**, 155-160 (1992).
10. Español P.; Warren, P. B. *Europhys. Lett.* **1995**, 30, 191-196.
11. R. D. Groot, and P. B. Warren, *J. Chem. Phys.* **107**, 4423-4435 (1997).
12. Materials Studio is a software developed by Accelrys Software Inc.
13. J. Xia and C. Zhong, *Macromol. Rapid Commun.* **27**, 1110-1114 (2006).

Structure Analysis of Jungle-Gym-Type Gels by Brownian Dynamics Simulation

Noriyoshi Ohta*, Kohki Ono*, Masako Takasu‡*, and Hidemitsu Furukawa†

*Department of Computational Science, Kanazawa University, Kakuma, Kanazawa, 920-1192, Japan
†Graduate School of Science, Hokkaido University, Sapporo 060-0810, Japan

Abstract. We investigated the structure and the formation process of two kinds of gels by Brownian dynamics simulation. The effect of flexibility of main chain oligomer was studied. From our results, hard gel with rigid main chain forms more homogeneous network structure than soft gel with flexible main chain. In soft gel, many small loops are formed, and clusters tend to shrink. This heterogeneous network structure may be caused by microgels. In the low density case, soft gel shows more heterogeneity than the high density case.

Keywords: Brownian dynamics, Gel, Loop, Simulation, Structure
PACS: 83.10.Mj, 82.70.Gg, 89.75.Fb, 81.07Nb, 36.40.Sx

INTRODUCTION

Gels form three-dimensional network structure which has many useful properties applied to various products. The network structure is important for the strength and the diffusion in gels. However, micro-scale structure of gels is still not fully understood. The experimental studies of characterization of the structure of gels have been performed[1-5]. Furukawa et al. investigated the distribution of network size by scanning microscopic light-scattering (SMILS)[5].

The properties and the structure of gels are also dependent on the progress of the reaction. In particular, the structure formation after the gelation has attracted much attention[6,7]. It is interesting to study the structure formation before and after gelation.

Vinyl-polymer gel is synthesized by radical polymerization and tends to have heterogeneity caused by forming microgels[5]. Recently, studies of new gel[8,9] by controlling structure have been reported. He et al.[8] synthesized polyimide gels by end-crosslinking main chain oligomers and trifunctional crosslinkers. Those polyimide gels have high mechanical strength with homogeneous network structure and are called jungle-gym-type gel.

Gels and related materials have been studied by simulation (see, for example, [10-14]), and it is hoped that the simulation gives insight to problems unsolved by experiments. In this paper, we study the formation process and the structure of polyimide gels by simulation. In our model, we use two types of angle potentials to express different flexibilities of oligomers.

MODEL AND METHOD

We use coarse grained model as shown in Fig. 1. In experiments[8], rigid chain was obtained by combining PMDA (pyromellitic dianhydride) and PDA (*p*-phenylenediamine), and flexible chain was obtained by PMDA and ODA (4,4'-oxydianiline), which has -o- bond. The main chain oligomer is designated by a linear unit (D) shown in Fig. 1(a). The triangular crosslinker TAPB (1,3,5-tris (4-aminophenyl) benzene) is designated by a triangular unit (T) shown in Fig. 1(b). One edge of the triangular unit is about 12.5[Å]. In simulation, we expressed the difference of flexibility with angle potentials as explained below.

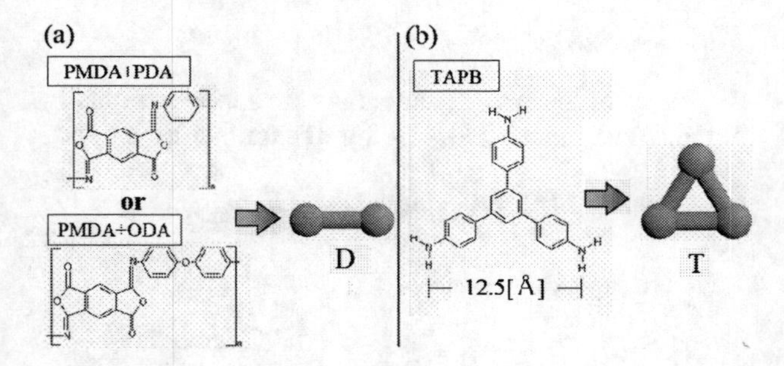

FIGURE 1. Coarse grained model used in our simulation. (a) Liner unit D (dimer) for main chain oligomer. (b) Triangular unit T for trifunctional crosslinker.

We perform Brownian dynamics simulation[12,14] using the following Langevin equation:

‡corresponding author: takasu@cphys.s.kanazawa-u.ac.jp

CP982, *Complex Systems, 5th International Workshop on Complex Systems*
edited by M. Tokuyama, I. Oppenheim, and H. Nishiyama

$$m_i \frac{d^2\mathbf{r}_i}{dt^2} = -\frac{\partial U}{\partial \mathbf{r}_i} - \zeta \frac{d\mathbf{r}_i}{dt} + \mathbf{g}_i(t) \quad (1)$$

The second term of the right hand side is the effect of the viscosity, and the third term is the random force caused by solvent molecules.

$$\langle \mathbf{g}_i(t) \rangle = 0, \langle \mathbf{g}_i(t)\mathbf{g}_j(t') \rangle = 6k_BT\zeta\delta_{ij}\delta(t-t') \quad (2)$$

The equations of motion are integrated by velocity Verlet method. We use repulsive potential between the same type of units (T and T, D and D).

$$U_{rep}(r) = \varepsilon \left(\frac{\sigma}{r}\right)^{12} \quad (3)$$

The interaction between the different types of units (D and T) is Lennard-Jones type potential.

$$U_{LJ}(r) = \varepsilon \left\{ \left(\frac{\sigma}{r}\right)^{12} - 2\left(\frac{\sigma}{r}\right)^{6} \right\} \quad (4)$$

When two particles of different types of units are within a certain distance r_{link}, a bond is formed. The interaction of the bonding particles is given by the spring potential.

$$U_{bond}(r) = \frac{1}{2} k_{bond}(r - d_0)^2 \quad (5)$$

When a triangular unit T and a linear unit D come close, we apply angle potentials, which favor the directions shown in Fig. 2(a). If we have a triangle ABC, and if unit D is approaching A, a vector $\mathbf{b}_\alpha$ is an arrow from the middle point of BC to A. Vector $\mathbf{b}_\alpha$ is on the triangular unit as shown is Fig. 2(b). Vector $\mathbf{b}_\beta$ is on the linear unit. Vector $\mathbf{b}_\gamma$ is from one particle of trianglar unit to a particle on the linear unit. The potentials in Eq. 6a and Eq. 6b favor the configuration in Fig. 2(a).

$$U_{angle}(\mathbf{b}_\alpha, \mathbf{b}_\beta) = k_{angle}(|\mathbf{b}_\alpha||\mathbf{b}_\beta| - \mathbf{b}_\alpha \cdot \mathbf{b}_\beta)^2 \quad (6a)$$

$$U_{angle}(\mathbf{b}_\alpha, \mathbf{b}_\gamma) = k_{angle}(|\mathbf{b}_\alpha||\mathbf{b}_\gamma| - \mathbf{b}_\alpha \cdot \mathbf{b}_\gamma)^2 \quad (6b)$$

The forces of these angle potentials are applied gradually by the cutoff function $h(r)$ before the units form a bond.

$$h(r) = 1 - \tanh(1.5r - 7)/2 \quad (7)$$

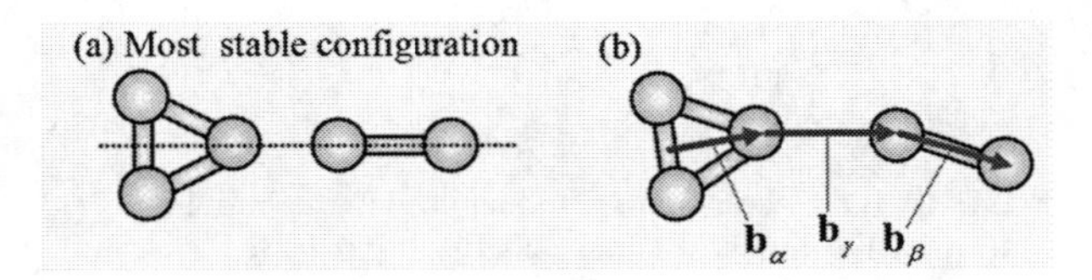

FIGURE 2. Vectors used in angle potentials.

From the mass and size of TAPB, we have $m_0 = 9.72 \times 10^{-26}$ [kg] and $l_0 = 6.25 \times 10^{-10}$ [m]. We use room temperature $T_0 = 298$ [K]. The other basic quantities are $\varepsilon_0 = 1.64 \times 10^{-21}$ [J], $t_0 = 4.8$ [ps] and $\zeta_0 = 1.9 \times 10^{-14}$ [kg/s].

In terms of the above basic quantities, we use the mass of the particle $m/m_0 = 2.0$, the diameter of the particle $\sigma/l_0 = 0.89$, bond length $d_0/l_0 = 2.0$, crosslink length $r_{link}/l_0 = 2.5$ and simulation box length $L/l_0 = 50$, 55, 60. Time step is $\Delta t/t_0 = 0.04$. Thus, one time step in our simulation corresponds to 0.19ps. We used viscosity constant $\zeta/\zeta_0 = 20.0$, energy constant $\varepsilon/\varepsilon_0 = 2.5$, spring constant $k_{bond}/(\frac{\varepsilon}{l_0^2}) = 24.0$ and angle potential constant $k_{angle}/(\frac{\varepsilon}{l_0^2}) = 1.5$ for rigid model and 0 for flexible model. These values were chosen to have the gelation within computational time. Compared with experimental values, ζ is much smaller.

In Fig. 3, snapshots of our simulations are shown. Black particles and lines belong to percolated cluster, and gray particles and lines belong to other clusters.

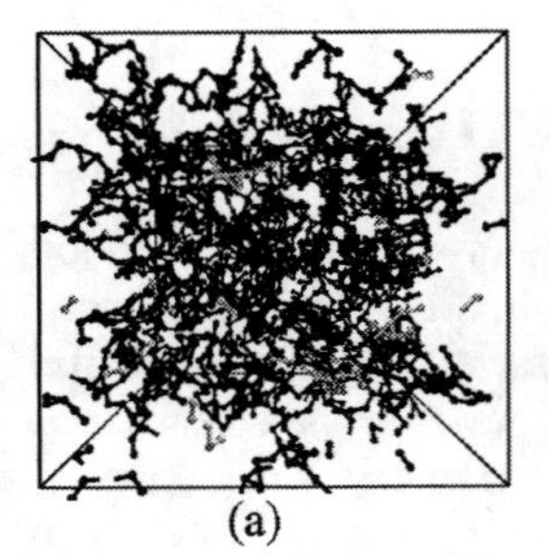

(a)

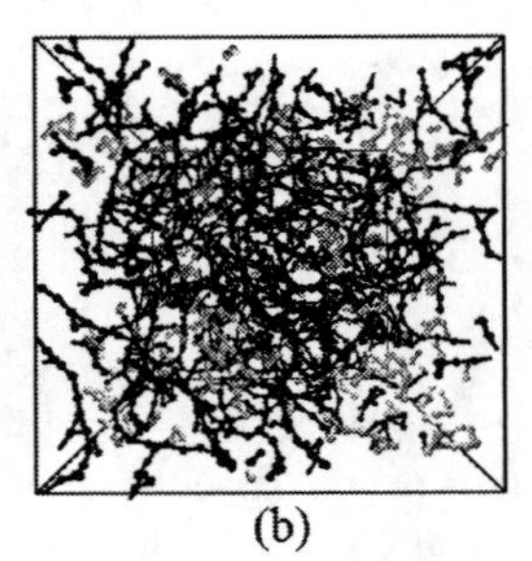

(b)

FIGURE 3. Snapshots of simulations (a) for soft gel after 50,000 steps and (b) for hard gel after 100,000 steps.

RESULTS

We analysed the structure and the gelation process from our simulation data. In the following subsections, we show our results on the distribution of crosslinking points, loop structure and clusters. We applied periodic boundary condition to our simulation box of length L. The number of triangular units N_T is 400, and the number of linear units N_D is 600. We averaged over ten samples for $L = 50$ for most of our calculations. In the last subsection, three samples are averaged for $L = 55$ and 60.

We define gelation as follows. We find the largest cluster in the system. We analyze the large loops, and if the loops percolate the system in x, y and z directions, the gelation has occurred.

Crosslinking Points

The radial distributions of crosslinking points for soft gel and hard gel are shown in Fig. 4. The crosslinking points are assumed as the middle point of a bond, shown by black points in Fig. 4.

The hard gel shows peaks at $r/l_0 = 3.8$-3.9 and 7.1-7.5. The first peak corresponds to d_1 or d_2 in Fig. 5, and the second peak corresponds to d_3. For hard gel, both the first and second peaks are sharper than soft gel, showing stronger correlation between crosslinking points. From the different peak positions, soft gel has more compact structure than hard gel. Compared with soft gel, hard gel forms more homogeneous network structure with ordered crosslinking points.

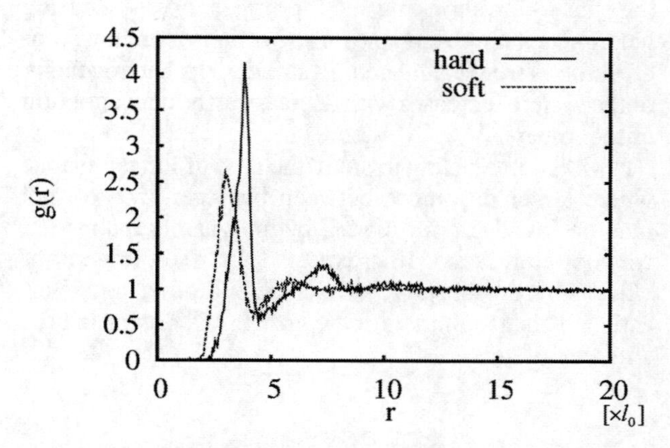

FIGURE 4. Radial distribution function of crosslinking points for soft gel (dashed line) and hard gel (solid line), averaged over 10 samples.

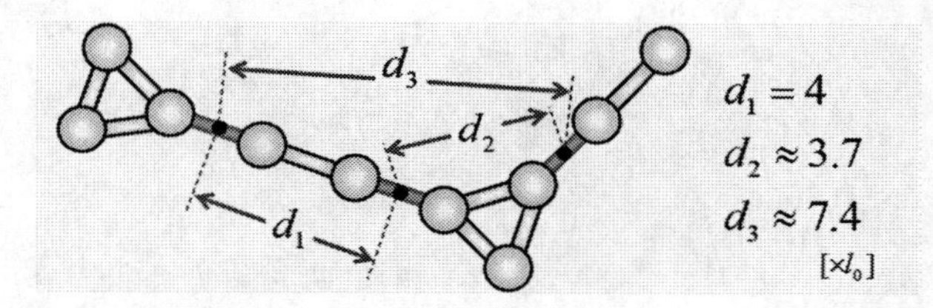

FIGURE 5. The distance between crosslinking points at the stable configuration with angle potentials.

Loop Structure

For characterization of network structure, we analyzed the closed loop structure. Snapshots of loop structure for soft gel and hard gel are shown in Fig. 6. Loops in hard gel tend to expand, while the loops in soft gel tend to shrink. In Fig. 7, the number of closed loops in each system is shown. In soft gel, we find more loops than in hard gel. In particular, the small loops are more abundant for soft gel than for hard gel. Most of the small loops are formed in early stage of the reaction in soft gel.

In soft gel, microgels are formed, shown by the increase of small loops. In hard gel, clusters extend spatially, and the small loops are formed more slowly than the soft gel. In both types of gels, formation of the network structure continues after gelation with increasing the number of loops.

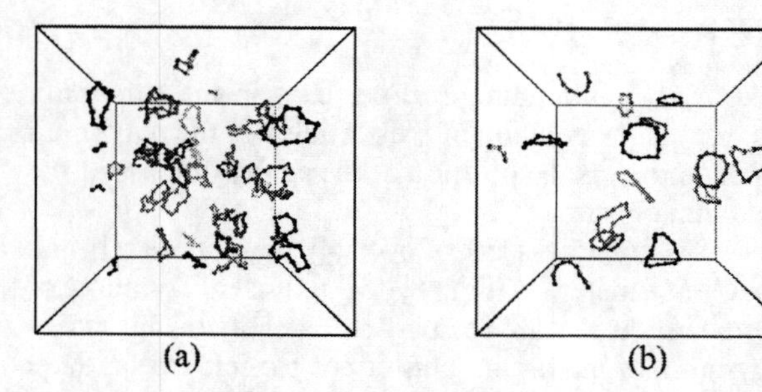

FIGURE 6. Snapshots of loops for soft gel (a) and hard gel (b).

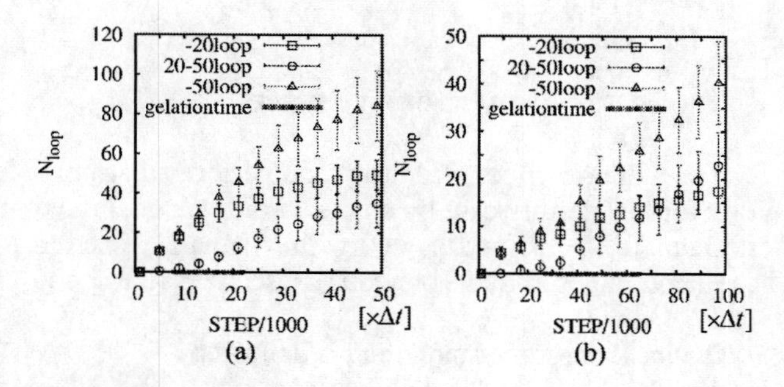

FIGURE 7. The number of loops for soft gel (a) and hard gel (b), averaged over 10 samples. □ is for the small loops with $\ell_{size} \leq 20$, ○ is for the large loops with $20 < \ell_{size} \leq 50$, △ is for all the loops with $\ell_{size} \leq 50$. ℓ_{size} is the number of particles constructing the loop. (One step corresponds to 0.19ps)

Clusters

To study the formation process of gels, we analyzed the growth of maximum clusters.

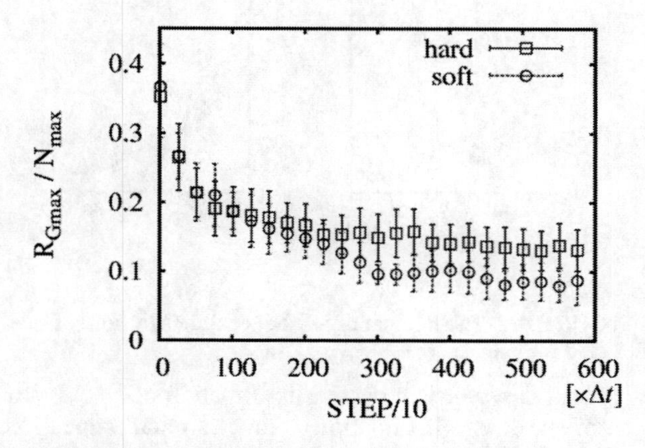

FIGURE 8. The radius of gyration of maximum cluster per particle. □ is for hard gel, ○ is for soft gel.

The radius of gyration of the maximum cluster is

defined by

$$R_{G_{max}} = \sqrt{\frac{1}{N_{max}} \sum |\mathbf{r}_i - \mathbf{r}_G|^2} \quad (8)$$

where N_{max} is the number of particles in the maximum cluster, $\mathbf{r}_i$ is the position of the particle i in the maximum cluster, and $\mathbf{r}_G$ is the position of the center of mass of the maximum cluster.

The summation is taken over all the particles in the maximum cluster. In Fig. 8, the radius of gyration of the maximum cluster per particle is shown. The radius of gyration of maximum cluster per particle for hard gel is larger than for soft gel. In hard gel, growing cluster extends spatially, and in soft gel growing cluster shrinks and forms microgel.

Low Density Case

To study the effect of density, we performed simulation for low density case by changing the size of simulation box to $L = 55$ and $L = 60$ with N_T and N_D fixed to the same values as in the previous subsections ($N_T = 400$ and $N_D = 600$).

The local density of monomer is defined by

$$\rho(r_{sphere}) = N(r_{sphere}) / \frac{4}{3}\pi r_{sphere}{}^3 \quad (9)$$

where $N(r_{sphere})$ is the number of particles in sphere of radius r_{sphere}. We place the spheres of radius r_{sphere} randomly in space, and calculate the number density $\rho(r_{sphere})$ in Eq. 9. Then we obtain the standard deviation $\Delta\rho = \sqrt{(\rho - \langle\rho\rangle)^2}$. $\lambda = \Delta\rho/\rho_{all}$ is plotted for various values of density in Fig. 9. ρ_{all} is density of the whole system.

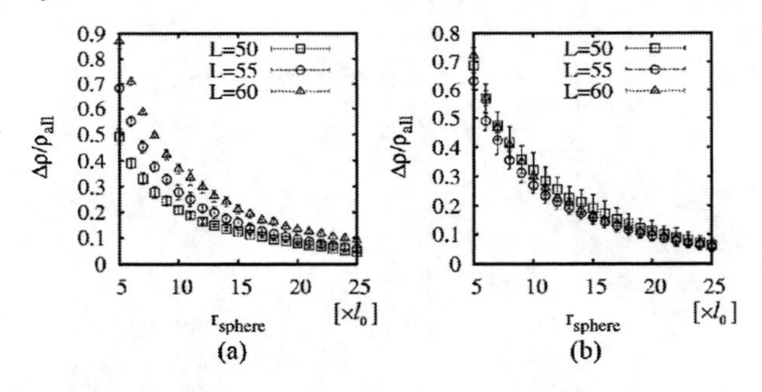

FIGURE 9. The fluctuation of the local density for soft gel (a) and for hard gel (b).

In all cases, λ is a decreasing function of r_{sphere} . In hard gel, density fluctuation λ gives similar values for all values of L. In soft gel, with decreasing the monomer concentration, that is, with larger L, the local fluctuation increases, showing greater heterogeneity due to microgels.

SUMMARY AND DISCUSSION

We investigated models of jungle-gym-type polyimide gels by Brownian dynamics simulation. The difference of flexibility of main chain oligomers was studied by two kinds of angle potentials. We analyzed the structure of two models of hard gel and soft gel. In these simulation results, hard gel has more homogeneous network structure than soft gel, and this supports experimental results[15]. This suggests that, in the formation of soft gel, the clusters are shrinking and forming microgels, causing heterogeneity. Moreover, cases of various monomer concentrations were calculated. In soft gel, the heterogeneity of the system increases with decreasing the concentration of monomers.

It will be interesting to study the case of longer chains, where larger difference between hard gel and soft gel may be found. In our model, hydrodynamic interaction was not considered. It is expected that the diffusion of clusters become faster if we include hydrodynamic interaction. This may influence the growth of clusters and the structure of gels.

REFERENCES

1. M. Shibayama, *Macromol. Chem. Phys.*, **19**, 1-30 (1998).
2. M. Shibayama, T. Norisuye, S. Nomura, *Macromolecules*, **29**, 8746-8750 (1996).
3. M. Shibayama, Y. Shirotani, Y.Shiwa, *J. Chem. Phys.*, **112**, 442-449 (2000).
4. H. Furukawa, S. Hirotsu, *J. Phys. Soc.*, **71**, 2873-2880 (2002).
5. H. Furukawa, K. Horie, R. Nozaki, M. Okada, *Phys. Rev. E*, **68**, 031406, 1-14 (2003).
6. M. Ishida, F. Tanaka, *Macromolecules*, **30**, 3900-3909 (1997).
7. I. E. Pacios, I. F. Pierola, *Macromolecules*, **39**, 4120-4127 (2006).
8. J. He, S. Machida, H. Kishi, K. Horie, H. Furukawa, R. Yokota, *J. Polym. Sci.*, **40**, 2501-2512 (2002).
9. N. Hosono, H. Furukawa, Y. Masubuchi, T.Watanabe, K. Horie, *Colloids and Surfaces B:Biointerfaces*, **56**, 285-289 (2007).
10. Y. Liu, R. B. Pandey, *Phys. Rev. E*, **54**, 6609-6617 (1996).
11. M. Nosaka, M. Takasu, K. Katoh, *J. Chem. Phys.*, **115**, 11333-11338 (2001).
12. M. Takasu, J. Tomita, *AIP Conf. Proc.*, **708**, 263-264 (2004).
13. E. Del Gado, W. Kob, *Europhys. Lett.*, **72**, 1032-1038 (2005).
14. M. Takasu, K. Ono, N. Ohta, H. Furukawa, *Bussei Kenkyu*, **87**, 84-85 (2006).
15. N. Tan, *Light Scattering Study of Network Structure in Polyimide Gels*, Master thesis, Tokyo University of Agriculture and Technology (2004) in Japanese.

Brownian Dynamics Simulation of Diffusion in Slide-Ring Gel

Masako Takasu and Takahiro Mimura

Department of Computational Science, Faculty of Science, Kanazawa University, Kakuma, Kanazawa 920-1192, Japan
taksau@cphys.s.kanazawa-u.ac.jp

Abstract. We investigate the diffusion in the slide-ring gel by Brownian dynamics simulation. A coarse-grained model is constructed and the effect of solvent is expressed by the random forces acting on both the gel and the diffusing particle. The diffusion is analyzed with various concentration of cross-linkers. It is found that the diffusion constant is a decreasing function of the cross-linker concentration with a plateau at the intermediate concentration.

Keywords: Brownian dynamics, Diffusion, Simulation, Slide-ring gel
PACS: 82.70Gg, 82.75Fq, 83.10Rs, 83.10.Mj

INTRODUCTION

Slide-ring gel is a new type of gel with mobile cross-linking points, developed by Ito and coworkers [1-3]. Compared with chemical gels [4], slide-ring gel can swell, deform or extend more easily [1]. It will be interesting to investigate the diffusion properties in the slide-ring gel.

MODEL AND CALCULATION METHOD

In Fig. 1, the coarse-grained model used in our simulation is shown. The poly(ethylene glycol) used in experiments [1-3] is modeled by a polymer chain shown in Fig. 1(a). A pair of two α -cyclodextrin (CD) molecules is expressed by two spheres, as shown in Fig. 1(b). In Fig. 2, we show two types of angle potentials used for CDs. In Fig. 3, the movements of CDs along the polymer chain are shown..

With this model, we performed Brownian dynamics simulation [5-6] described by the following equation.

$$m_i \frac{d^2\mathbf{r}_i}{dt^2} = -\frac{\partial U}{\partial \mathbf{r}_i} - \zeta \frac{d\mathbf{r}_i}{dt} + \mathbf{g}_i(t) \qquad (1)$$

The first term on the right hand side includes the potentials between monomers, described below. The second term is the viscosity term. The third term is the random force caused by the solvent molecules. The random forces are dependent on the temperature as follows.

$$< \mathbf{g}_i(t)\mathbf{g}_j(t') >= 6k_B T \varsigma \delta_{ij} \delta(t-t') \qquad (2)$$

The potentials used in our simulation are the following.
We have repulsive interaction to consider the excluded volume of the monomers.

$$U_{rep}(r_{ij}) = \varepsilon \left(\frac{\sigma}{r_{ij}} \right)^{12} \qquad (3)$$

The adjacent monomers are connected by spring potentail.

$$U_{bond}(r_{ij}) = \frac{1}{2} k_{bond} (r_{ij} - r_0)^2 \qquad (4)$$

where r_{ij} is the distance between connected monomers i and j.
We have angle potential between three adjacent monomers, and also between CD and monomer, as shown in Fig. 2.

$$U_{bend}(\theta_{ijk}) = \frac{1}{2} k_{bend} (\theta_{ijk} - \theta_0)^2 \qquad (5)$$

CP982, *Complex Systems, 5th International Workshop on Complex Systems*
edited by M. Tokuyama, I. Oppenheim, and H. Nishiyama

where θ_{ij} is the angle formed by three monomers, i, j and k.

In our simulation, if the length is estimated by CD size, the mass is estimated by CD mass, and temperature is taken as room temperature, the viscosity coefficient in our simulation is much smaller than the experimental value of the water. This is intended to make the simulation faster.

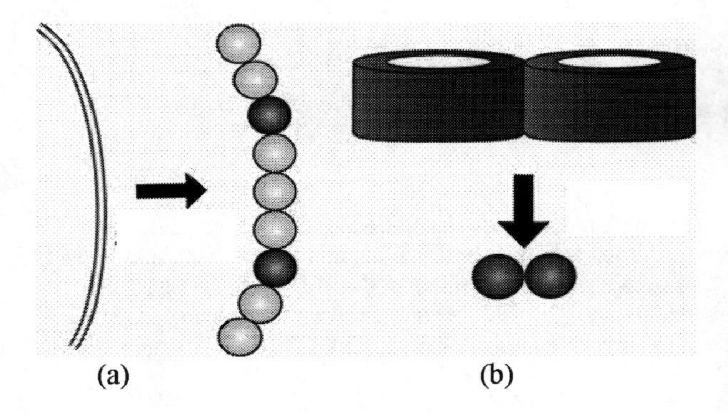

(a) (b)

FIGURE 1 .The coarse grained model in our simulation. The polymer is expressed as a string of spheres in (a). The shaded spheres are where the polymer has flexibility. The model of cross-linker is shown in (b).

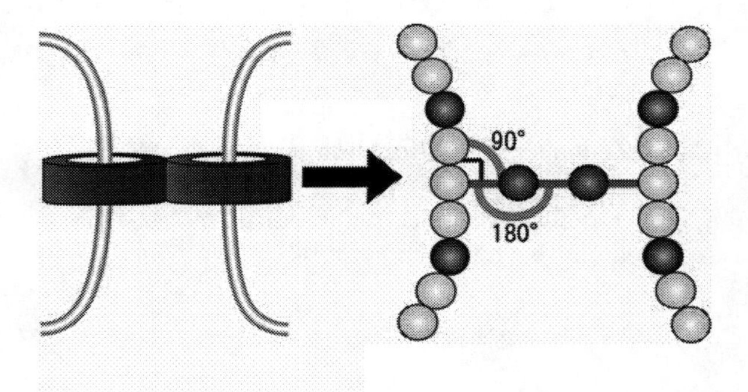

FIGURE 2. The angle potentials for crosslinkers.

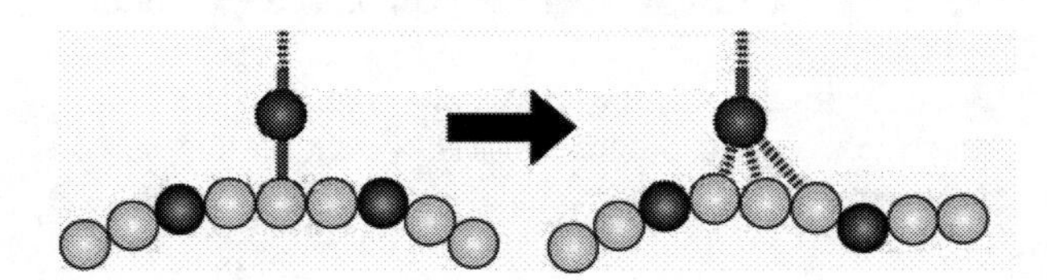

FIGURE 3. The movement of slide-ring.

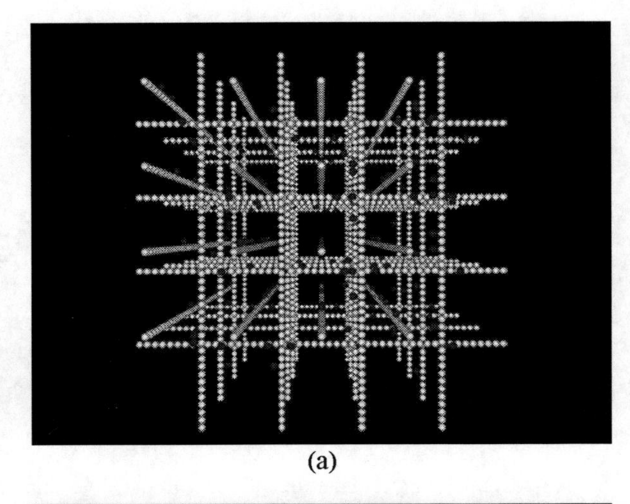

(a)

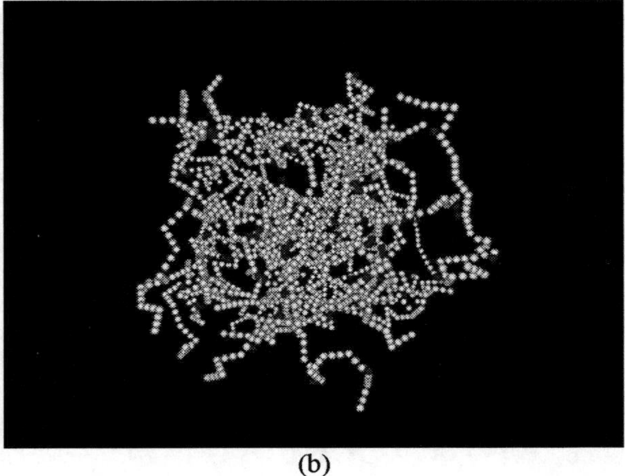

(b)

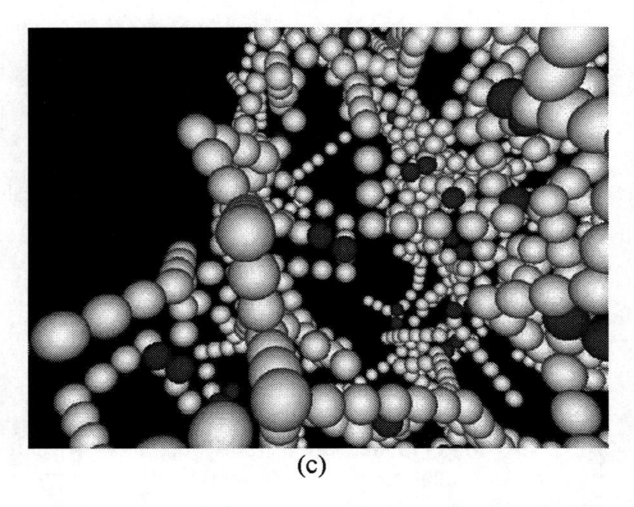

(c)

FIGURE 4. The initial configuration of our system (a), a snapshot of the slide-ring gel (b) and an enlarged version (c). The shaded spheres are CDs.

RESULTS AND DISCUSSIONS

In Fig. 4 (a), the initial configuration of the system is shown. As preparation, we perform the molecular dynamics simulation of polymers without moving the CDs. Then we let the CDs combine. After sufficient CD pairs are formed, we start the production run and let the inserted particle diffuse and we take statistics.

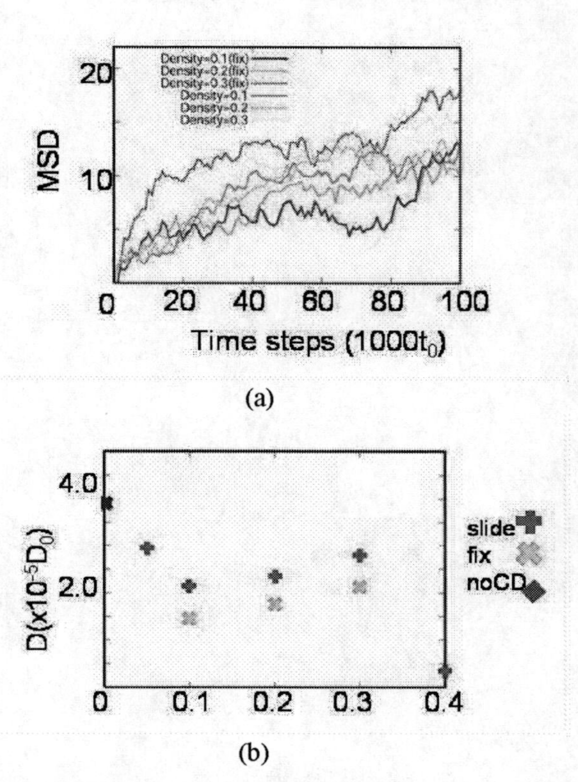

FIGURE 5. The mean-square displacement of the diffusing particle (a) and the diffusion constant (b) for various densities of CDs.

In Fig. 5, the mean-squared displacement of the diffusing particle and the diffusion constant is shown. The diffusion in slide-ring gel is faster than the case of fixed CDs. As the density of CDs is increased, the diffusion decreases. A plateau is observed at an intermediate density. The slight increase of diffusion constant at density 0.2 and 0.3 is within the range of errorbars.

The fact that the larger density of CD gives slower diffusion is schematically explained in Fig. 6. If we compare the distance between the nearest pairs of CDs for density 0.1 and 0.4 (Fig. 7), we find lower value and much smaller fluctuation for density 0.4.

The plateau in Fig. 5 (b) may be useful, when one wants to achieve certain average diffusion constant with some dispersity of the density of CDs in materials.

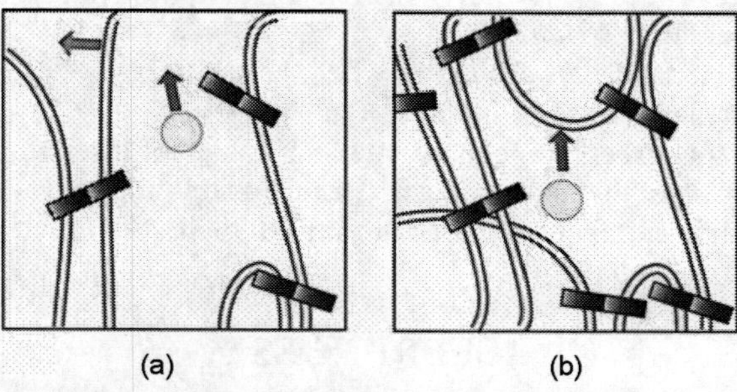

FIGURE 6. The diffusion with small density of CD (a) and large density of CD (b).

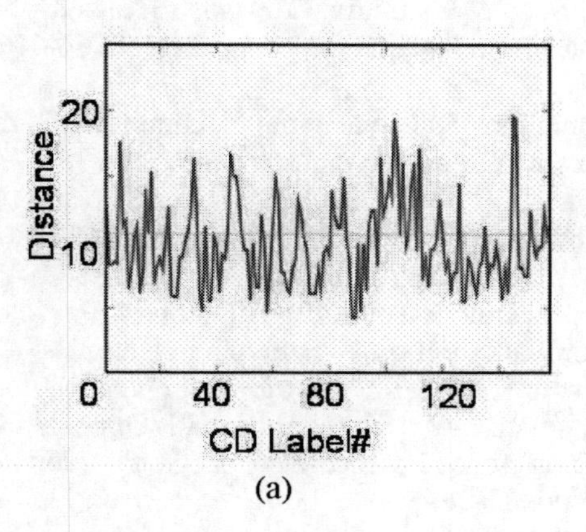

(a)

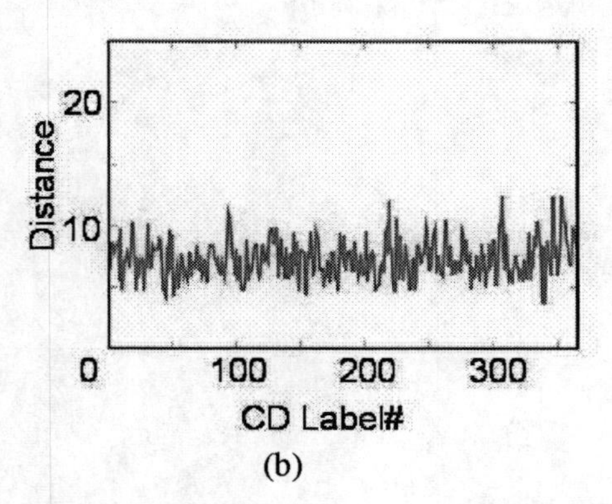

(b)

FIGURE 7. The fluctuation of the distance between the nearest pairs of CDs for CD density 0.1 (a) and 0.4 (b).

FUTURE PROBLEMS

In our simulation, the length and the width of the polymer were fixed. It may be interesting to study the effect of size of the polymer. The temperature effect should be also studied.

The change of properties when the slide-ring gel is stretched is also of interest.

In this paper, we focused on the diffusion of inserted particle in the slide-ring gel. The diffusion properties of slide-ring gel itself are currently studied.

REFERENCES

1. T. Karino, Y. Okumura, K. Ito and M. Shibayama, *Macromol.* **37**, 6177-6182 (2004).
2. T. Karino, Y. Okumura. C. Zhao, T. Kataoka, K. Ito and M. Shibayama, *Macromol.* **38**, 6161-6167 (2005).
3. Y. Shinohara, K. Kayashima, Y. Okumura, C. Zhao, K. Ito and Y. Amemiya, *Macromol.* **39**, 7386-7391 (2006).
4. M. Nosaka, M. Takasu and K. Katoh, *J. Chem. Phys.* **115**, 11333-11338 (2001).
5. M. Takasu and J. Tomita, "Diffusion of Particle in Hyaluronan Solution, a Brownian Dynamics Simulation" in *Slow Dynamics in Complex Systems*, 2003, edited by M. Tokuyama and I. Oppenheim, AIP Conference Proceedings 708, Sendai, Japan, 2004, pp263-264.
6. M. Takasu, K. Ono, N. Ohta, H. Furukawa, *Bussei Kenkyu,* **87**, 84-85 (2006).

Studies of Drug Delivery and Drug Release of Dendrimer by Dissipative Particle Dynamics

Chun-Min Lin[1], Yi-Fan Wu[1], Heng-Kwong Tsao[2], and Yu-Jane Sheng[1]

[1]*Department of Chemical Engineering, National Taiwan University, Taipei, Taiwan 106, R.O.C.*
[2]*Department of Chemical and Materials Engineering, National Central University, Jhongli, Taiwan 320, R.O.C.*

Abstract. Dendrimers, like unimolecular micelles, may encapsulate guest biomolecules (drug) and therefore are attractive candidates as carriers in drug delivery applications. Hydrophobic drugs can be complexed within the hydrophobic dendrimer interior to make them water-soluble. The equilibrium partition of hydrophobic solutes into a dendrimer with hydrophobic interior from aqueous solutions is studied by dissipative particle dynamics. The drug is mainly distributed in the vicinity of the interface between hydrophobic interior and hydrophilic exterior within a dendrimer. The partition coefficient, which is defined as the concentration ratio of the drug distributed within dendrimer to aqueous phases, depends on the interaction between drug and hydrophilic dendrimer exterior. Increasing the repulsion between them reduces the solubilization ability associated with the dendrimer.

Keywords: Dendrimers, Dissipative particle dynamics, Drug carriers, Partition
PACS: 31.15.-at

INTRODUCTION

Dendrimers that resemble a Cayley tree are regularly branched polymers with a tree-like cascade topology. Their unique structural characteristics comprise highly branched and well-defined structure, globular shape, and controlled surface functionalities. Their unique molecular architectures make them attractive materials for the development of carriers of drug delivery systems and partitioning agents for hydrophobic toxins [1-3]. In a variety of biological and biomedical applications, the bioactive guest molecules may be encapsulated into the interior of the dendrimers or chemically attached/physically adsorbed onto the dendrimer surface. The latter involves modifiable surface group functionality which provides the interactions between guest molecules and dendrimers by covalent or electrostatic interactions.

The open nature of the dendritic architecture has led to the possibility of encapsulating guest molecules within the branches of a dendrimer. This offers the potential of dendrimers to incorporate labile or poorly soluble drugs. The nature of drug encapsulation within a dendrimer may be simple physical entrapment, or can involve non-bonding interactions (i.e. hydrophobic, electrostatic and hydrogen bond interactions) with specific structures within the dendrimer. A dendrimer containing an apolar core and polar shell has been referred to as "unimolecular micelles" [4]. In other words, the dendrimers act as micelles, capable of solubilizing various hydrophobic compounds in aqueous solution. However, unlike conventional micelles self-assembled by surfactant molecules, the dendritic structure is independent of dendrimer concentration.

Dendrimer-mediated solubility enhancement depends on a few factors such as dendrimer concentration, generation size, polymeric architecture, pH, core, and temperature [4]. Dendrimers with hydrophobic core and hydrophilic periphery may exhibit micelle-like behavior and have container properties in solution. In order to compare the hydrophobic encapsulation between dendrimer and micelle, we consider conjugated dendrimer and amphiphilic dendrimer in this study. PEGylated dendrimer is an example of the former, which is formed by attaching hydrophilic chains to a third generation dendrimer core with hydrophobic building blocks. The latter can be constructed by growing hydrophilic outer generations upon hydrophobic inner generations (hydrophobic generation 1-3 and hydrophilic generation 4-G). The distribution of hydrophobic solutes within these dendrimers is investigated by dissipative particle dynamics (DPD). The influence of the interaction between hydrophobic solute and hydrophilic shell on the partition coefficient is examined.

CP982, *Complex Systems, 5th International Workshop on Complex Systems*
edited by M. Tokuyama, I. Oppenheim, and H. Nishiyama

DISSIPATIVE PARTICLE DYNAMICS

The DPD method, introduced by Hoogerbrugge and Koelman in 1992 [5], is a particle-based, mesoscale simulation technique. This method combines some of the detailed description of the molecular dynamics (MD) but allows the simulation of hydrodynamic behavior in much larger, complex systems, up to the microsecond range. Like MD, the DPD particles obey Newton's equation of motion. The interparticle force F_{ij} exerted on particle i by particle j is made of a conservative term (F_{ij}^C), a dissipative term, and a random term. Typically, the conservative force is represented by a soft-repulsive interaction, $F^C(r_{ij})=a_{ij}\max\{1-r_{ij}/r_c, 0\}e_{ij}$, where r_{ij} is the distance between particles, e_{ij} the unit vector in the direction of the separation, and a_{ij} represents the maximum repulsion between particle i and j. A polymer is modeled by connecting ordinary DPD particles with springs. In our simulation, the volume density of the monomer associated with a dendrimer and the solute are fixed at 0.018 and 0.01, respectively. The dynamics of 24000~81000 DPD particles, starting from random distribution, was simulated in a cubic box (20^3~30^3) under periodic boundary conditions. The equation of motion are integrated with a modified velocity Verlet algorithm with Δt=0.05. In this work, each simulation takes at least 160,000 steps and the first 80,000 steps are for equilibration. Note that DPD simulation utilizes soft-repulsive potentials and the systems studied are allowed to evolve much faster than the "brute-force" molecular dynamics. Therefore a typical DPD simulation needs only about 50,000 to 100,000 steps to equilibrate.

We consider an aqueous solution (W) of hydrophobic solutes (S) and a dendrimer with three inner hydrophobic generations (B) and hydrophilic outer generation (A). The interaction parameters are given as table 1.

RESULTS AND DISCUSSION

Chemical Potential

At thermodynamic equilibrium, the chemical potential is a constant everywhere. As a result, one can evaluate the chemical potential associated with hydrophobic solutes to examine whether the system reaches true equilibrium or not. In the simulation, the chemical potential is calculated by $\mu = \mu_i + \mu_r$, where the ideal chemical potential is given by $\mu_i = \mu_0 + k_BT \ln c(\mathrm{r})d^3$ and $\mu_r(\mathrm{r})$ denotes the residual chemical potential. For convenience, the reference chemical potential is set to be zero, μ_0=0. The ideal (solute concentration) and residual chemical potentials vary with the radial position from the center of mass of the dendrimer r. The residual chemical potential is obtained by the Widom's particle insertion method, which is the reversible work required to add a solute to the system, $\mu_i = -k_BT\ln\langle\exp(-U_i/k_BT)\rangle$. Figure 1 depicts the typical distributions of solute concentration cd^3, its residual chemical potential, and total chemical potential for a conjugated dendrimer with a_{AS}=48. The conjugated, hydrophilic polymer consists of 15 beads. Owing to hydrophobic attraction between solute and dendrimer core, the ideal chemical potential corresponding to the solute concentration (ln c(r)) is higher within the core region and decays to a lower value in the bulk solution. On the contrary, the residual chemical potential is increased from the core to the bulk. Nonetheless, the total chemical potential is essentially constant for all radial position r. Note that there exists a maximum solute concentration near $r \approx$ 2.5.

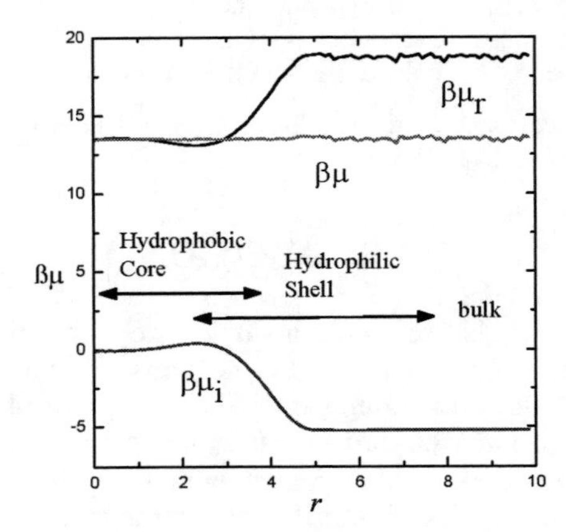

FIGURE 1. Distribution of chemical potential.

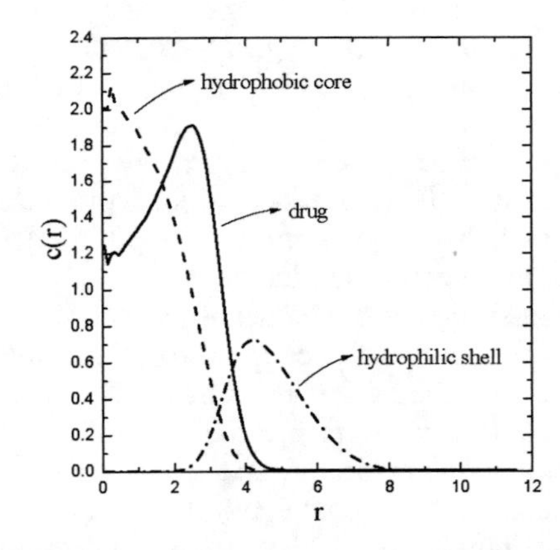

FIGURE 2. Distribution of solute & monomer.

TABLE 1. The interaction parameters

	A(hydrophilic)	B(hydrophobic)	W(solvent)	S(drug)
A(hydrophilic)	25	40	26	A_{AS} (28-48)
B(hydrophobic)	40	25	60	26
W(solvent)	26	60	25	50
S(drug)	A_{AS} (28-48)	26	50	25

Concentration Distribution of Hydrophobic Solute

By examining the radial distribution of solute and dendrimer monomer simultaneously, one is able to know where the hydrophobic solute prefers to stay. Figure 2 demonstrates that although the hydrophobic interaction prefers the solute reside inside the core, most of the solutes are distributed in the vicinity of the interface between hydrophobic interior and hydrophilic exterior within a dendrimer. It can be attributed to the balance between hydrophobic attraction and translational entropy. Inside the hydrophobic core, the monomer (B) density is high but the interaction energy is low. On the other hand, within the hydrophilic shell, the monomer (A) density is low but the interaction energy is high. The interfacial region provides the lowest free energy and thus the highest solute concentration.

Partition Coefficient

Since a substantial amount of hydrophobic solutes resides in the hydrophilic shell, the interaction between solute and hydrophilic dendrimer exterior may affect the solubilization ability of the dendrimer significantly. The encapsulating capacity can be expressed in terms of the partition coefficient, which is defined as $P = c_{in} / c_{bulk}$, where c_{in} denotes the mean solute concentration with a dendrimer. Figure 3 shows the variation of the partition coefficient with the interaction parameter a_{AS} for both conjugated dendrimer g3-S15 (hydrophilic chain length = 15) and amphiphilic dendrimer g5-S5 (hydrophilic spacer = 5). The partition coefficient declines with increasing the solute-hydrophilic monomer repulsion. Consequently, the enhancement of solubilization ability of a dendrimer requires improving the compatibility between drug and hydrophilic monomer.

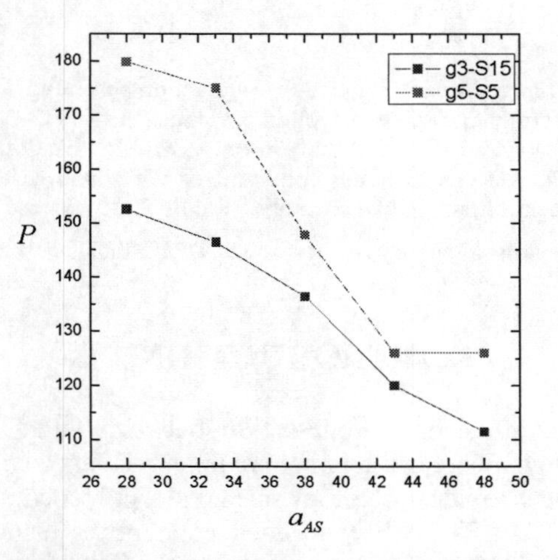

FIGURE 3. Partition coefficient as a function of the solute-shell interaction.

References

1. L. A. Fernandez, M. Gonzalez, H. Cerecetto, M. Santo, and J. J. Silber, Supramolecular Chemistry **18**, 633-643(2006).
2. M. King, X. Duan, and H. Sheardown, Biotechnol. Bioeng. **86**, 512-519 (2004).
3. A. D'Emanuele and D. Attwood, Advanced Drug Delivery Reviews **57**, 2147-2162 (2005).
4. U. Gupta, H. B. Agashe, A. Asthana, and N. K. Jain, Biomacromolecules **7**, 649-658 (2006).
5. P. J. Hoogerbrugge, and J. M. V. A. Koelman, Europhys. Lett. **19**, 155-160 (1992).

Coarse Grained Simulation of Lipid Membrane and Triblock Copolymers

Masaomi Hatakeyama* and Roland Faller†

*School of Knowledge Science, Japan Advanced Institute of Science and Technology, 1-1, Asahidai, Nomi, Ishikawa, 923-1211, Japan
†Department of Chemical Engineering and Material Science, University of California at Davis, Davis, California, 95616, USA

Abstract. We investigated the interaction between DPPC (Dipalmitoyl phosphatidylcholine) bilayer and polyethylene oxide-polypropylene oxide-polyethylene oxide (PEO-PPO-PEO) triblock copolymers using coarse grained simulation. We simulated two systems of DPPC bilayer and PEO–PPO–PEO triblock copolymer containing different mole fractions, and simulated DPPC vesicle with the copolymers. We found different adsorption mechanisms of triblock copolymers depending on concentration. And we also observed docking process between a lipid vesicle and a micelle of the copolymers.

Keywords: Coarse grained simulation, DPPC, Lipid membrane, PEO-PPO-PEO, Triblock copolymer
PACS: 82.35.Gh

INTRODUCTION

We investigated the interaction between DPPC (Dipalmitoyl phosphatidylcholine) lipid bilayers and polyethylene oxide–polypropylene oxide–polyethylene oxide (PEO–PPO–PEO) triblock copolymers using coarse grained molecular dynamics simulations. PEO–PPO–PEO triblock copolymers are non–ionic surface active agents commonly known as Pluronics® (trade mark by BASF Corp.) which are widely used for industrial applications, such as detergents, stabilizers, emulsifiers, and drug delivery systems [1]. There is great interest in the interactions between PEO–PPO–PEO and liposomes from the point of view of drug delivery because this copolymer has an effect of stabilizing liposomes [2]. It is necessary to observe the interaction between lipid membrane and the copolymers at the molecular level. It is, however, not yet feasible to simulate directly biological membrane and PEO–PPO–PEO triblock copolymers in atomistic details. Therefore, we use a coarse grained model for lipid bilayers and the copolymers to study their interactions in an approximate manner.

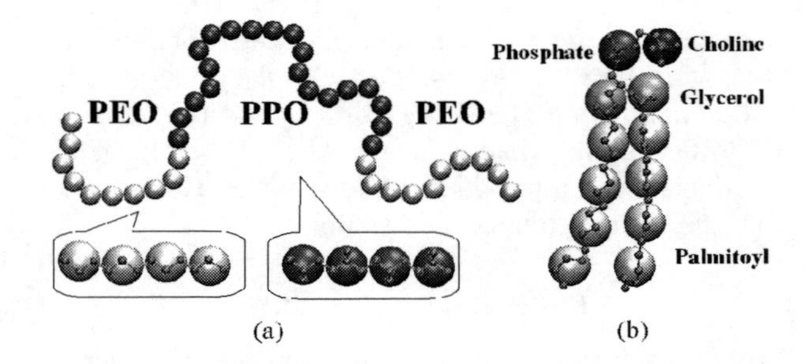

FIGURE 1. The images of coarse grained model of PEO–PPO–PEO and DPPC. (a) PEO–PPO–PEO, (b) DPPC.

MODEL AND SYSTEMS

DPPC is one of the most abundant phospholipids in naturally occuring biomembranes. We use it therefore as our model lipid. We use the coarse grained lipid model proposed by Marrink for DPPC [4], and our PEO–PPO–PEO triblock copolymer is based on Marrink's lipid model as well. PEO monomers are represented as hydrophilic superatoms whereas PPO monomers are hydrophobic superatoms [3]. Both molecules and their representations are illustrated in Figure 1. PEO and PPO monomers are represented by single interaction sites such that one chain of PEO_{10}–PPO_{20}–PEO_{10}, which is the copolymer we are using in our studies, is represented by 40 superatoms in total. Moreover, every water superatom represents 4 atomistic water molecules. In this coarse grained model four types of interaction sites are considered: polar (P), nonpolar (N), apolar (C), and charged (Q). For the nonpolar particles 4 subtypes ($0, d, a$, and da) are further distinguished. Subtype 0 has no hydrogen bonding capability, subtype d and a are hydrogen bond donors or acceptors, respectively, and subtype da is both, donor and acceptor. We use type Na particles for ethylene oxide, and type C particles for propylene oxide monomers. Water molecules are represented by P particles. The details of the triblock copolymer model can be found in [3].

The non–bonded interactions between particles i and

CP982, *Complex Systems, 5th International Workshop on Complex Systems*
edited by M. Tokuyama, I. Oppenheim, and H. Nishiyama

j are described by a standard Lennard–Jones (LJ) potential. The interactions are directly transferred from ref. [4]. For all interaction types the same range $\sigma_{ij} = 0.47$ nm is used. Bonded interactions between chemically connected particles are described by a harmonic potential $V_{\text{bond}}(R)$ with an equilibrium distance equal to the monomer size $R_{\text{bond}} = \sigma = 0.47$ nm. The force constant of the harmonic potential is $K_{\text{bond}} = 1250$ kJ mol^{-1} nm^{-2}. To represent chain–stiffness, a harmonic angle potential is used where the equilibrium bond angle is 180 ° with a force constant of $K_{\text{angle}} = 25$ kJ mol^{-1}rad^{-2}. Directly bonded atoms do not interact by non–bonded interactions.

We simulated two systems of DPPC bilayer and PEO–PPO–PEO triblock copolymer containing different mole fractions, 1) 9.489×10^{-4} [mol fraction], 2) 1.896×10^{-3} [mol fraction]. A DPPC vesicle with triblock copolymers was simulated at concentration 8.527×10^{-4} [mol fraction]. The complete characteristics of all systems are summarized in Table 1. The temperature of all systems is held constant at 350 K by Berendsen thermostat[5] with a correlation time $\tau_T = 2.0$ ps. The simulations are executed for 1.0 μs in the isothermal and isobaric NPT ensemble where pressure is kept constant by Berendsen's method with a coupling strength of 1.0 bar $^{-1}$.

We use Gromacs version 3.2.1. for the bilayer simulations [6], and Gromacs version 3.3.1 for the vesicle simulation. The simulations of a DPPC bilayer with the triblock copolymers were carried out using Red Hat Linux 8.0 on Pentium IV (3.2 GHz), and for the DPPC vesicle and the copolymers on SGI Altix3700 as well.

RESULTS AND DISCUSSION

We focus here on the conformation of the lipid bilayers and the concentration effect of PEO–PPO–PEO on the lipid bilayer. Figures 2 show snapshots of DPPC bilayers with copolymers. If the concentration of the copolymer is low, a block of PEO–PPO–PEO copolymers is embedded into the bilayer (Figure. 2(a)) and after the adsorption the block of copolymers is moving laterally on the bilayer. On the other hand, for the higher concentration the copolymers form a layered structure and cover the lipid surface (Figure. 2(b)) in which the hydrophilic layer is on the water-side and the hydrophobic layer is inside.

We calculate the density profile along the Y direction (bilayer normal) in order to quantify the morphologies shown in Figures 2. Figure. 3(a) corresponds to Figure. 2(a), and Figure. 3(b) to Figure. 2(b). In both cases, some lipids flip–flop and the head and tail groups are reversed in the layer to which triblock copolymers adsorb. This can be more clearly observed in the case of concentration 1.896×10^{-3} [mol fraction].

We also studied the interaction of copolymers with vesicles. We use the lipid vesicle from Marrink's website[1] as initial configuration and put the copolymers and water at random positions and orientations into the simulation box around the vesicle. There are 877 lipid molecules and 152 copolymer molecules in the box. The snapshots of the DPPC vesicle and the triblock copolymer at various time steps are shown in Figure 4. Within 100 ns, some polymers are embedded into the vesicle, and then the rest of copolymers assemble themselves around this embedded nucleus on the vesicle. Finally, the forming PEO–PPO–PEO micelle and the DPPC vesicle are docking with each other. This actual docking process takes place within a very short period between 215 ns and 218 ns.

CONCLUSION

We studied the interaction between DPPC membrane and PEO–PPO–PEO triblock copolymers. In particular, we focused on the effect of the triblock copolymer on the lipid membrane. We found different adsorption mechanisms of triblock copolymers depending on concentration. We also could observe the interaction between pluronics and a vesicle. Here we find the embedding of a anchoring point into the vesicle and formation of an independent micelle. The micelle eventually docks at the vesicle at this anchor. We will study the dynamics of these processes quantitatively and study the concentration effect over a wider range of concentrations in the near future.

REFERENCES

1. N. Rapoport, *Colloids and Surfaces B: Biointerfaces*, **16**, 93–111 (1999).
2. N. Bergstrand and K. Edwards, *Journal of Colloid and Interface Science*, **276**, 400–407 (2004).
3. M. Hatakeyama and R. Faller, *Physical Chemistry Chemical Physics*, **9**, 4662–4672 (2007).
4. S. J. Marrink, A. H. de Vries, and A. E. Mark, *Journal of Physical Chemistry B*, **108**, 750–760 (2004).
5. H. J. C. Berendsen, J. P. M. Postma, W. F. van Gunsteren, A. DiNola, and J. R. Haak, *Journal of Chemical Physics*, **81**, 3684–3690 (1984).
6. E. Lindahl, B. Hess, and D. van der Spoel, *Journal of Molecular Modeling*, **7**, 306–317 (2001).

---

[1] http://md.chem.rug.nl/ marrink/coarsegrain.html

TABLE 1. The systems

System No.	Lipid Molecules	Polymer chains	Water Molecules	Concentration [mol fraction]
1	128	10	2600	9.489×10^{-4}
2	128	20	2600	1.896×10^{-3}
3	877	152	44308	8.527×10^{-4}

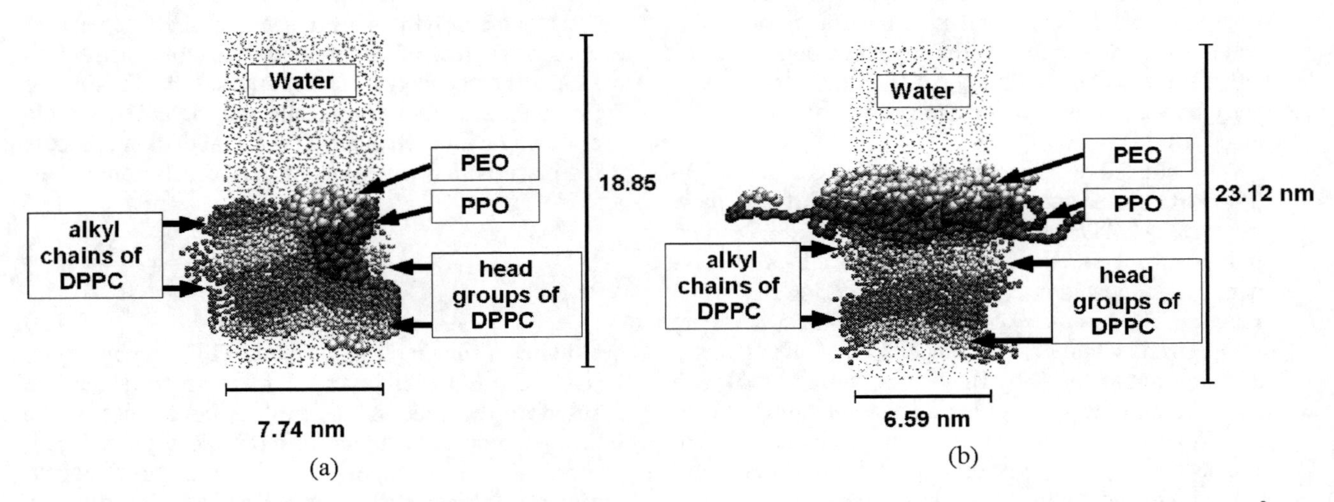

FIGURE 2. Snapshots of DPPC and Pluronics. Concentration of Pluronic, (a): 9.489×10^{-4} [mol fraction], (b): 1.896×10^{-3} [mol fraction]. PPO is black and PEO white. Periodic boundary conditions apply.

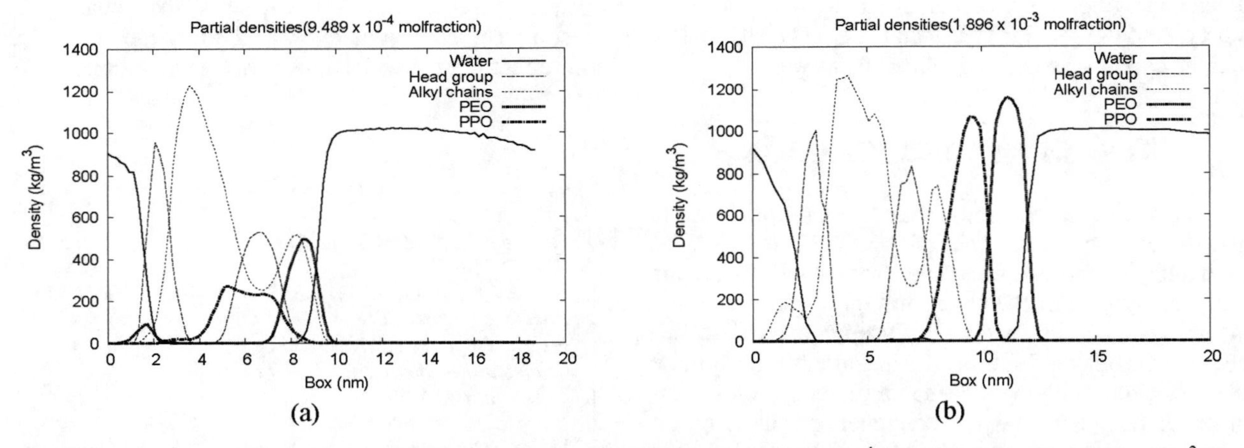

FIGURE 3. Density profiles along Y–axis. Concentration of Pluronic, (a): 9.489×10^{-4} [mol fraction], (b): 1.896×10^{-3} [mol fraction].

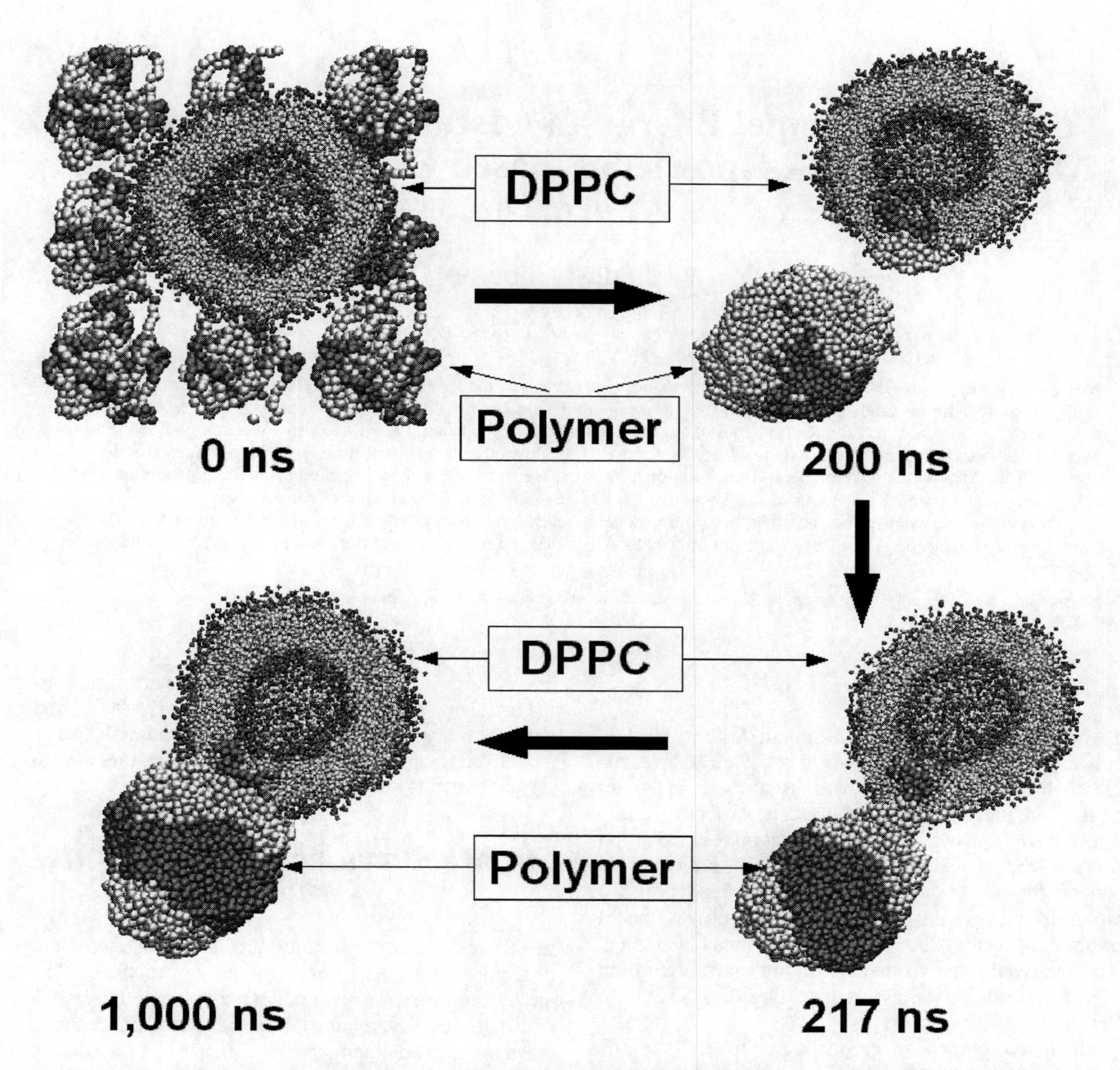

FIGURE 4. Snapshots of DPPC vesicle and triblock copolymers in each time step. Concentration of triblock copolymer: 8.527×10^{-4} [mol fraction]. Water molecules are not shown for clarity.

The Double-Funnel Energy Landscape of an Off-Lattice Model Protein: A Knowledge-Based Evolution Algorithm Approach

Chung-I Chou

Department of Physics, Chinese Culture University, Taipei 111, Taiwan

Abstract. We apply an off-lattice minimal energy model of proteins and the Knowledge-based Evolution Algorithm (KEA)to study the double-funnel landscape of protein folding problem. The testing off-lattice minimal energy model is composed of a polypeptide chain, and has the Ramachandran angles as its degrees of freedom. The force field of this model is based on hydrogen bonds and the anisotropic hydrophobicity forces. The Knowledge-based Evolution Algorithm tries to simulate the process of the knowledge development. The evolutionary knowledge database will help to direct searching processes and to reach the global minimum faster.Our results show that the KEA can improve the searching efficiency.

In this report, we will illustrate an off-lattice minimal energy model and how to apply KEA to this model to study its double-funnel energy landscape. The phase transition of α-helix and β-hairpin structures and the evolution of guiding function are reported.

Keywords: Double-funnel energy landscape, Knowledge-based evolution algorithm, Optimization algorithm
PACS: 36.20.Ey, 36.20, 31.15.Ar

INTRODUCTION

The fundamental problem of protein folding[1][2] is how a protein sequence could fold into their unique stable conformation. But in some unusual cases, a protein could fold into more than one stable structure in different environments. The famous example is the prion protein. One interesting hypothesis is these proteins possess double-funnel energy landscapes. In this report, we would like to use an off-lattice minimal energy model of proteins to study the idea of double-funnel landscape. And we would apply our new optimization algorithm, "The Knowledge-based Evolution Algorithm (KEA)", to this problem.

Firstly, we propose to introduce our model. The off-lattice minimal energy model is composed of a polypeptide chain, and has the Ramachandran angles as its degrees of freedom. The force field of this model is based on hydrogen bonds and the anisotropic hydrophobicity forces.

Secondly, we try to apply KEA algorithm to double-funnel energy landscapes. The KEA method tries to simulate the process of the knowledge development. The whole searching procedure is composed of several generations of local search. At each generation of sampling, we retrieve certain useful knowledge and construct a knowledge database. This knowledge database is then used to guide the search in the next generation in addition to the consideration of minimum energy. In this case of off-lattice protein model, we use the simulated annealing (SA) method for local searching and choose the distribution of Ramachandran angles as the guiding function.

Lastly, we report a phase transition of α-helix and β-hairpin structures of this off-lattice minimal energy model, and show the evolution of guiding function in several generations of searches.

AN OFF-LATTICE MINIMAL ENERGY MODEL

In this section, we try to introduce an off-lattice model that can be applied to both α and β structures. This model is constrained by a polypeptide chain. Each peptide unit with seven atoms, and the side chain reduced one atom. All bond lengths, bond angels and the peptide torsion angels (except two Ramachandran torsion angels, ϕ, ψ) are fixed. The only degrees of freedom are the two Ramachandran torsion angels. The polypeptide chain is illustrated in Figure 1.

The force filed is similar to Irback's model[3]. Firstly, the backbone hydrogen bonds potential is considered. The potential of hydrogen bonds will lead the model to fold to a helix structure. Secondly, We divide the amino acids into two different classes, hydrophobic and polar. The hydrophobic side chains support an anisotropic hydrophobicity force.

Our energy function,

$$E = E_{HC} + E_{HB} + \lambda E_{RR}, \tag{1}$$

is composed of three terms. The first term E_{HC} is a hard core (self-avoidance) potential. The second term E_{HB} is

CP982, *Complex Systems, 5th International Workshop on Complex Systems*
edited by M. Tokuyama, I. Oppenheim, and H. Nishiyama

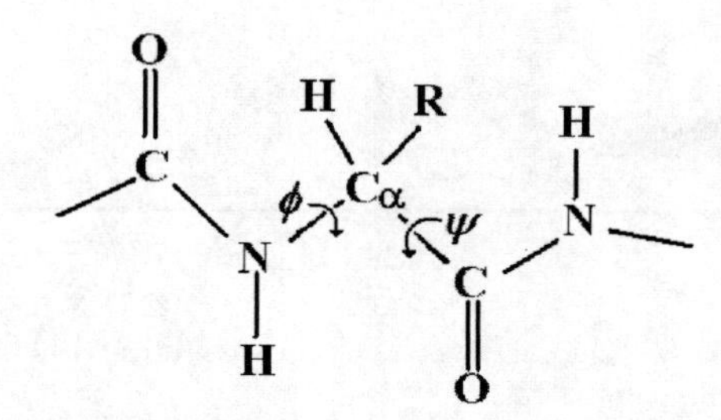

FIGURE 1. The polypeptide chain.

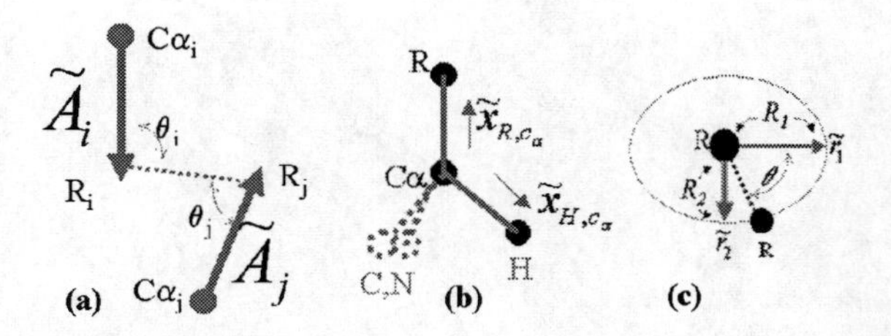

FIGURE 2. The definitions of (a) the side chain direction, (b) the unit vectors $\tilde{x}_{R,C_\alpha}$, $\tilde{x}_{H,C_\alpha}$. (c) The angle of the R-R vector and the β-sheet favored direction.

the backbone hydrogen bond potential. This potential may drive the backbone Oxygen atom to connect to the nearest backbone Hydrogen. The last term E_{RR} is the anisotropic residue-residue interaction potential (or the hydrophobicity force field potential). The parameter λ is the ratio of the term E_{HB} and the term E_{RR}.

The main change in the force field compared with that described previously is the anisotropic residue-residue term E_{RR}. This term of potential is

$$E_{RR} = \sum_{i,j} \varepsilon_1 \varepsilon_2 \varepsilon_3 E_{HP}. \quad (2)$$

The potential is anisotropy, it has some favored directions that controlled by four parameters. The first parameter is

$$\varepsilon_1 = \begin{cases} (\tilde{A}_i \cdot \tilde{A}_j)^2 \ , if \, \tilde{A}_i \cdot \tilde{A}_j \geq 0 \\ 0.5(\tilde{A}_i \cdot \tilde{A}_j)^2 \ , else \end{cases} . \quad (3)$$

Figure 2(a) shows the definition of the side chain direction , and both are unit vectors.

The second parameter is

$$\varepsilon_2 = (\sin^4\theta_i + \sin^4\theta_j)/2 \ , \quad (4)$$

θ_i, θ_j are defined in Figure 2(a).

The third parameter is

$$\varepsilon_3 = (\cos^2(2\theta) + 1)/2 \ . \quad (5)$$

Now we try to describe how to compute the variable in eq. 5. Firstly we should define two β-sheet favored directions, $\tilde{r}_1 = \tilde{x}_{R,C\alpha} \times \tilde{x}_{H,C\alpha}$ and $\tilde{r}_2 = \tilde{r}_1 \times \tilde{x}_{R,C\alpha}$ (Figure 2(b)). And we compute the angle of the $R-R$ vector and the β-sheet favored direction(Figure 2(c)).

At last, we consider the hydrophobic type energy. There are two favored equilibrium lengths in this energy term. This changing is for simulating the real β-sheet structure. The energy is

$$E_{HP} = \varepsilon_{i,j} \left(\left(\frac{R_0}{R_{i,j}} \right)^{12} - 2 \left(\frac{R_0}{R_{i,j}} \right)^{6} \right) , \quad (6)$$

where $R_{i,j}$ is the distance between two residues, $R_0 = R_1 + \sin^2\theta \cdot (R_2 - R_1)$ and $R_2 = 5.3\dot{A}$, $R_1 = 6.6\dot{A}$. In our simplified model, we only consider the hydrophobic interaction. So we choose a set of index like $\varepsilon_{HH} = 1, \varepsilon_{PP} = \varepsilon_{HP} = 0$. It means only two hydrophobic residues provide energy. This residue-residue interaction potential (E_{RR}) leads our model to like to contract to a β-sheet structure. These two effects (E_{RR} ,E_{HB}) give our possibility to generate both α-helix and β-sheet structures by controlling the ratio of two potentials.

THE KNOWLEDGE-BASED EVOLUTION ALGORITHM

Several years ago, we developed a newly optimization algorithm, the guided simulated annealing method [4][5][6], and applied it to several topics. Such as the X-ray crystallography, the atomic cluster problem and the protein folding problem. This method is based on the simulated annealing method. But we inject a new quality function into the searching process. This improvement influences both local and the global searching directions in traditional simulated annealing. Recently we generalized the Guided Simulated Annealing method into the "Knowledge-based Evolution Algorithm".

This KEA method tries to simulate the process of the knowledge development. The whole searching procedure is composed of several generations of local search. At each generation of sampling, we retrieve certain useful knowledge and construct a knowledge database (called guiding functions). These functions are then used to guide the search in the next generation in addition to the consideration of minimum energy. In this kind of process, the guiding functions (or the knowledge) will evolve from one generation to another. The evolutional guiding function will help to direct searching processes and to reach the global minimum faster.

Figure 3 shows a cartoon of KEA method. In the first generation(Figure 3(a)), we got a lot of local minima (white dots). And we make a distribution of these local minima as a guiding function(Figure 3(b)). In the next generation, we sample the configuration space by guid-

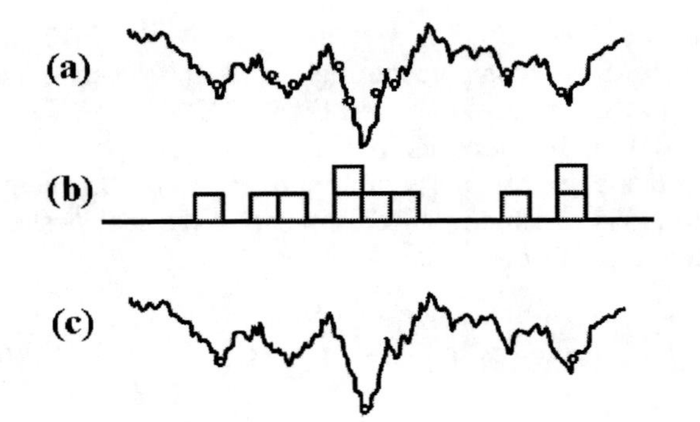

FIGURE 3. A cartoon of KEA method.(a) Results of first generation, (b) Guiding function(distribution of local minima), (c) Results of second generation.

ing function. High distribution area means more samplings. And we may find some better results. In this cartoon, the global minimum is found in second generation(Figure 3(c)).

The central issue of implementing KEA is how to retrieve the useful information or knowledge to construct the guiding function from earlier generations. In general, there are three ways to construct guiding functions: selection, generalization and classification. The simplest guiding function is just the distributions of variables. For this, we only need to select the parent generation's data and make a histogram. The high level guiding function would be the relation of variables, the sub structures, the symmetry or any useful knowledge. In other words, the high level guiding functions may be some kinds of the "order parameter" of the system. In our experience, a higher level guiding function may speed up the searching process effectively. Table 1 shows a comparison with other algorithms for several HP lattice testing proteins. Using KEA, we can find all of the best results in the records.

Following is the computing schedule of KEA.

(1). Find a lot of local minima by using certain local searching methods,

(2). Select some minima with lower energy.

(3). Retrieve useful knowledge and construct guiding functions.

(4). Guiding functions are used to guide the local searches of next generation.

(5). Back to step (2) till the global minimum is reached or convergence.

In this case of off-lattice protein model, we use the simulated annealing (SA) method for local searching and choose the distribution of Ramachandran angles as the guiding function. By using KEA with this model, we can fold several different secondary structures of proteins in simulations, including the α-helix, β-sheet and the α/β structures. Depending on the controlling parameter, the

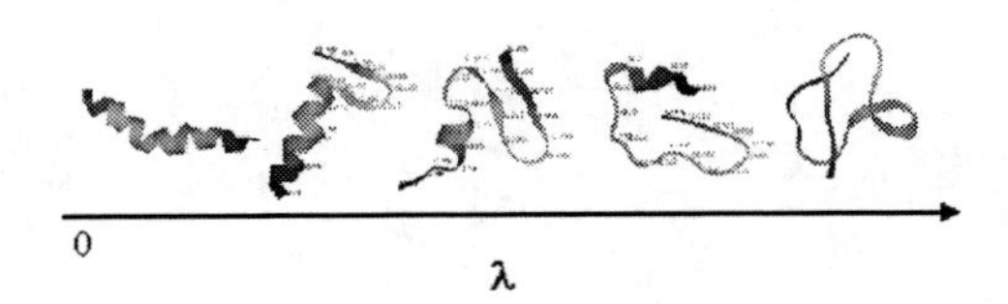

FIGURE 4. The scheme of protein structures with different controlling parameters.

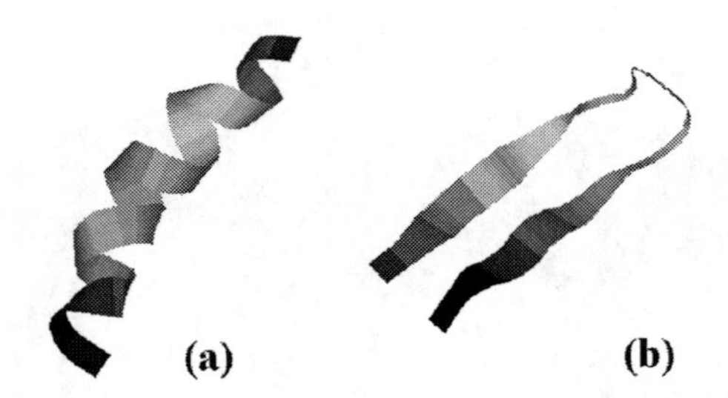

FIGURE 5. (a) A α-helix structure ($\lambda < \lambda_c$), (b) A β-hairpin structure ($\lambda > \lambda_c$).

protein will have different structures (Figure 4).

RESULTS

The testing case is a 18-mers off-lattice protein It is a short sequence but it is long enough to fold helix and hairpin structures. The sequence of amino acides is "H P H H H P H P P P P H P H H H P H".

By using KEA, we have several results of different ground states in different control papameters. After we studied the relation between the controlling parameter and the shapes of protein structure, we find there are some kinds of phase transition of α-helix and β-hairpin structures. If λ less than the critical point λ_c, the ground state structure is a α-helix(Figure 5(a)),else it will be a β-hairpin(Figure 5(b)). Figure 8 shows the evolution of guiding function. In Figure 6(a,b,c), λ is less than λ_c. Figure 6(a) shows the Ramachandran plot which created by survived children local minima structures of the first generation. Figure 6(b,c) are the Ramachandran plots of second and third generations. We can see the guiding functions concentrate to the α-helix structure. In Figure 6(g,h,i), λ is large than λ_c, and with 1st, 2nd and 4th generations' Ramachandran plots. The guiding functions concentrate to the β-hairpin structure. In Figure 6(d,e,f), λ is close to λ_c, and with 1st, 3rd and 7th generations' plots. In the beginning generations, the guiding functions concentrate to the α-helix and β-hairpin structures both. But the guiding functions still concentrate to one global minimum finally. In this case, we can see a competition of two main funnels in the energy landscape.

TABLE 1. The values of the lowest energy reported by several authors for ten sequences of HP protein model.

	N=20	**24**	**25**	**36**	**48**	**50**	**60**	**64**	**100a**	**100b**
Genetic Algorithm(Unger, 1993)[7]	**-9**	**-9**	**-8**	**-14**	-22	**-21**	-34	-37	—	—
Contact interactions method(Toma, 1996)[8]	-9	-9	-8	-14	**-23**	-21	-35	-40	—	—
CCG model(Beutler, 1996)[9]	-9	-9	-8	-14	-23	-21	-35	**-42**	—	—
DMC method(Ramakrishn, 1996)[10]	—	—	—	—	—	—	—	—	-44	-46
PERM(Frauenkron, 1998)[11]	—	—	—	—	—	—	**-36**	—	-47	-49
MSO Ensemble method(Chikenji, 1999)[12]	—	—	—	—	—	—	—	-42	-47	**-50**
EMC method(Liang, 2001)[13]	-9	-9	-8	-14	-23	-21	-35	-39	—	—
Improved PERM(Hsu, 2003)[14]	-9	-9	-8	-14	-23	-21	-36	-42	**-48**	-50
KEA(Chou, 2003)[5]	**-9**	**-9**	**-8**	**-14**	**-23**	**-21**	**-36**	**-42**	**-48**	**-50**

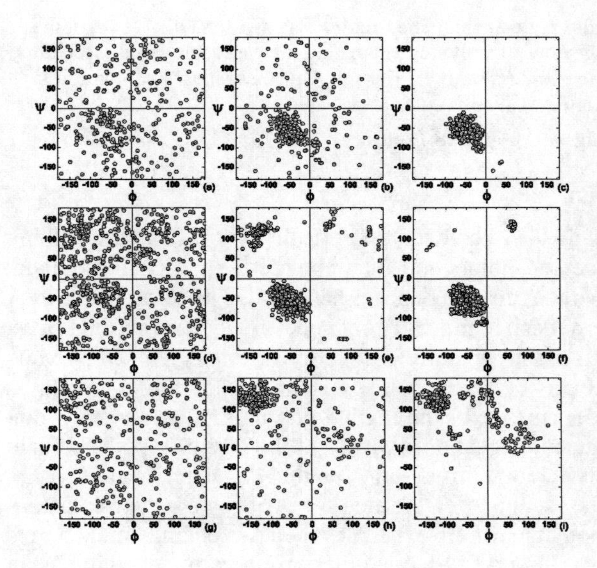

FIGURE 6. The evolution of guiding functions.

SUMMARY

In this report, we introduce an off-lattice minimal energy model of proteins. And we apply the Knowledge-based Evolution Algorithm (KEA) to study double-funnel landscape of protein folding.

The force field of the testing off-lattice minimal energy model is based on hydrogen bonds and the anisotropic hydrophobicity forces. The potential of the backbone hydrogen bonds will lead the model to fold to a helix structure, and the potential of the anisotropic hydrophobicity will lead the model to fold to a β-sheet structure. By controlling the ratio of two potentials, we can get many different shapes of protein.

The Knowledge-based Evolution Algorithm can be applied to this off-lattice protein model. The ground structure will be found in several searching generations of KEA. The evolutionary guiding function will help to direct searching processes and to reach the global minimum faster. The guiding functions of generations exhibit some kinds of funnel structure. Applying KEA, this model could show several types of funnels. If control parameter λ is less than λ_c , the bottom of the main funnel is α-helix. If control parameter λ is large than λ_c, the guiding functions of generations will concentrate to a β-hairpin structure. In case of the control parameter λ is close to transition point λ_c, the evolution of guiding function shows that a double-funnel energy landscape exists.

ACKNOWLEDGMENTS

This work was supported in part by the National Science Council, Taiwan, R.O.C. (grant no. 95-2745-M-034-003-URD)

REFERENCES

1. D. J. Wales, and H. A. Scheraga,*Science*, **285**, 1368-1372 (1999).
2. C. B. Anfinsen,*Science*, **181**, 223-230 (1973).
3. G. Favrin, A. Irback and S. Wallin,*Proteins: Struct. Funct. Genet.*, **47**, 99-105 (2002).
4. C. I. Chou and T. K. Lee,*Acta Cryst. A*, **58**, 42-46 (2002).
5. C. I. Chou, R. S. Han, S. P. Li and T. K. Lee,*Phys. Rev. E*, **67**, 066704-1-066704-6 (2003).
6. C. I. Chou, R. S. Han, T. K. Lee and S. P. Li,*LNCS*, **2690**, 447-451 (2003).
7. R. Unger and J. Moult, *J. Mol. Biol.*, **231**, 75-81 (1993).
8. L. Toma and S. Toma, *Protein Sci.*, **5**, 147-153 (1996).
9. T. C. Beutler and K. A. Dill, *Protein Sci.*,**5**, 2037-2043 (1996).
10. R. Ramakrishnan, B. Ramachandran and J. F. Pekny, *J. Chem. Phys.*, **106**, 2418-2425 (1997).
11. H. Frauenkron , U. Bastolla, E. Gerstner, P. Grassberger and W. Nadler, *Phys. Rev. Lett.*, **80**, 3149-3152 (1998).
12. G. Chikenji , M. Kikuchi and Y. Iba, *Phys. Rev. Lett.*, **83**, 1886-1889 (1999).
13. F. Liang and W. H. Wong, *Chem. Phys.*, **115**, 3374-3380 (2001).
14. H.P Hsu, V. Mehra, W. Nadler, and P. Grassberger, *J. Chem. Phys.*, **118**, 444-451 (2003).

Viscoelasticity of Entangled Semiflexible Polymers via Primitive Path Analysis

N. Uchida*, Gary S. Grest†, and Ralf Everaers**

*Department of Physics, Tohoku University, Sendai 980-8578, Japan
†Sandia National Laboratories, Albuquerque, NM 87185, USA
**Laboratoire de Physique, ENS Lyon, 46, allée d'Italie, 69364 Lyon cedex 07, France

Abstract. Viscoelasticity of entangled semiflexible polymers in the plateau frequency regime is investigated. We combine computer simulations and scaling arguments to develop a unified view of polymer entanglement based on primitive path analysis. The plateau modulus is obtained as a function of the dimensionless segment density. Our theoretical and numerical results agree well with the experimental data over a wide range of stiffness/density.

Keywords: Brownian dynamics, Molecular dynamics, Rheology, Scaling theory, Semiflexible polymers
PACS: 83.10.Kn, 83.10.Mj

INTRODUCTION

Modern theories of the rheology of concentrated polymer solutions and melts are based on the idea that entanglements confine individual chains to a one-dimensional, tube-like regions [1]. While this picture has been successfully applied to loosely entangled flexible polymers, less attention has been paid to tightly entangled semiflexible polymers, which are biologically important as the building blocks of the cell and extra-cellular matrix. How strongly polymers are entangled to each other depends on both their stiffness and contour length density[2]. Stiffer polymers have larger spatial extension and hence get entangled at smaller density. The stiffness is parametrized by the Kuhn length $\ell_K = \lim_{\ell\to\infty}\langle R^2\rangle/\ell$, where R^2 is the square end-to-end distance and ℓ is the contour length of the chain. The number of Kuhn segments between consecutive entanglements depends on the dimensionless parameter $\rho_K\ell_K^3$, where ρ_K is the number density of the Kuhn segments. If $\rho_K\ell_K^3 < 1$, the chain behaves like a random coil between entanglements ("loosely entangled") while for $\rho_K\ell_K^3 \gg 1$, the chain is almost straight and exhibits small bending fluctuations between entanglements ("tightly entangled"). Here we utilize the primitive path analysis (PPA) [3, 4] as a way to renormalize a loosely to tightly entangled system, which enable us to predict the plateau modulus of entangled semiflexible polymers over a wide range of stiffness.

THEORY

Primitive paths are defined as the shortest paths that connect the end-points of individual chains and that are topologically equivalent to the original chain configurations. This idea has been numerically implemented by fixing the end-points and then contracting the chain contours without crossing each other [3, 4]. The result of the PPA is a mesh of mutually entangled, piecewise straight lines. To them we apply the theory of entangled semiflexible chains by Semenov [5], regarding the primitive paths as renormalized semiflexible polymers. Our ansatz is that the conformation statistics of the primitive paths contain sufficient information to predict the plateau modulus. On the scaling level, the area swept out by transverse fluctuation of a primitive path between two entangelment points is transversed by one other primitive path serving as an obstacle. This gives the estimate of the mesh size ξ,

$$\xi^2 = c_\xi L_e (L_e^3/L_K)^{1/2} \tag{1}$$

where L_e is the average distance between two entanglements, L_K is the Kuhn length of the primitive path, and c_ξ is a constant. The contour length ℓ_e of the original chain between entanglements is related to L_e via the usual worm-like chain statistics; we use the simple formula $L_e^2 \approx \ell_e^2/(1+\ell_e/\ell_K)$ interpolating the flexible and stiff limits. In the meantime, the contraction ratio of the chain length is given by $\ell_e/L_e = \ell/L = L_K/\ell_K$, where ℓ and L are the lengths of the original chains and the primitive paths and we note that the end-to-end distance is unchanged by the PPA. Also, the mesh size is trivially related to the chain length density as

$$\xi^{-2} = \rho_{chain} L = \rho_K(\ell_K/\ell)L \tag{2}$$

Finally, the entropic plateau modulus G_N^0 is estimated by assigning k_BT per entanglement. Since each entanglement is, schematically, contained in a cylindrical region with the height L_e and diameter ξ, we have

$$G_N^0 = c_G k_B T/(L_e\xi^2) \tag{3}$$

CP982, *Complex Systems, 5th International Workshop on Complex Systems*
edited by M. Tokuyama, I. Oppenheim, and H. Nishiyama

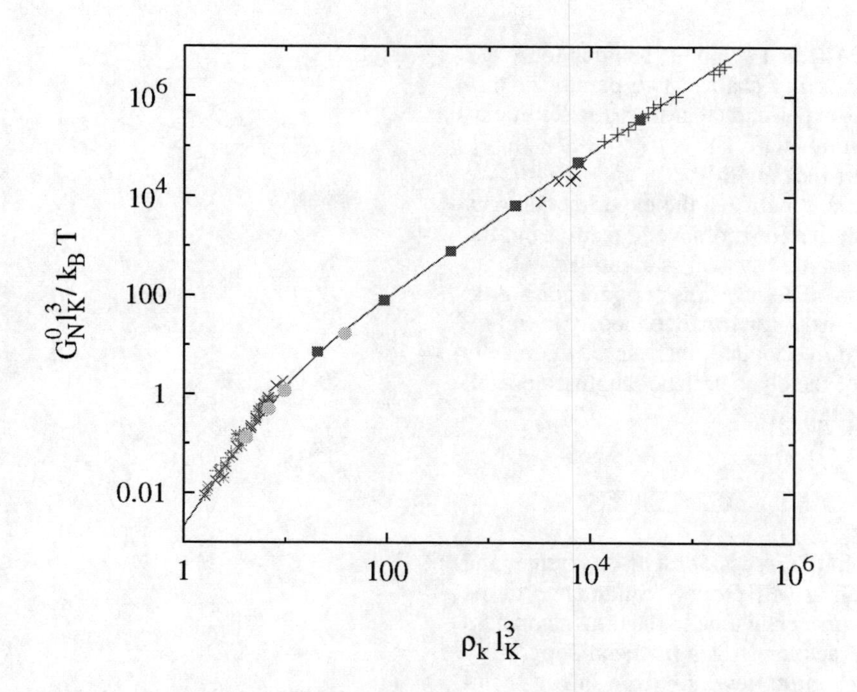

FIGURE 1. (Color online) Dimensionless plateau modulus $G_N^0 \ell_K^3 / k_B T$ as a function of the dimensionless segments density. Results of numerical PPA for WLCs (filled squares) and bead-spring chains (filled circles) and two parameter theory (solid line) are compared with experimental data for f-actin (+) [6], fd-phages (×) [7], and melts of polydienes (+), polyolefines (∗), and polyacrylates (×) [8].

where the prefactor c_G remains unknown. Combining these, we obtain the plateau modulus in terms of the dimensionless segment density $\rho_K \ell_K^3$, as

$$G_N^0 \ell_K^3 / k_B T = c_G c_\xi^{2/5} (\rho_K \ell_K^3)^{7/5} (L/\ell)^{8/5} \tag{4}$$

where the chain contraction ratio is given by

$$(\ell/L)^2 \approx 1 + (c_\xi \rho_K \ell_K^3)^{2/5} + (c_\xi \rho_K \ell_K^3)^2. \tag{5}$$

This gives a two-parameter formula for the plateau modulus, while we can also determine the chain contraction from numerical simulation instead of Eq.(5).

NUMERICAL SIMULATION

We numerically implemented the PPA using two types of model polymer liquids: tightly entangled solutions of zero-diameter worm-like-chains with $10 < \rho_K \ell_K^3 < 10^5$ [8] and dense melts of flexible bead-spring chains with $1 < \rho_K \ell_K^3 < 40$ [3, 4]. For the former, bending of each chain was penalized by the Hamiltonian $\mathscr{H} = -\kappa \sum_i \boldsymbol{u}_i \cdot \boldsymbol{u}_{i+1}$ where $\boldsymbol{u}_i$ is the unit vector along the i-th segment. Depending on the bending rigidity κ, we varied the number of segments per chain N between 40 and 320 and the number of chains between 500 and 10^4, ensuring multiple entanglements per chain. Equilibrium configurations of zero-diameter chains are obtained by randomly placing the initial segment and then generating the other segments one by one with a simple sampling. For the PPA, we replaced the rod-like segment by a linear spring with zero natural length and used a force-biased Monte-Carlo technique to reject chain crossings. Bending rigidity is killed and temperature is gradually reduced to zero to obtain the primitive paths. For the loosely entangled polymers, we used bead-spring chains with the FENE spring and the truncated Lennard-Jones 6-12 potential for the excluded volume interaction. The bead density is set to $\rho\sigma^3 = 0.85$ where σ is the Lennard-Jones bead diameter. PPA is done by setting the temperature to zero [3, 4].

RESULTS AND DISCUSSION

>From the numerical simulation we obtained the chain contraction ratio directly and combined it with Eq.(4) to estimate the plateau modulus, which was best fitted to experimental data with $c_G = 0.6$. We also got a two-parameter fitting with the purely theoretical formulae (4,5) with $c_\xi = 0.06$. Plotted in Figure 1 are the results

of the numerical PPA (filled symbols, squares for WLC and circles for bead-spring chains), two-parameter theory (solid line), and experimental results for semiflexible biopolymer solutions with $10^3 < \rho_K \ell_K^3 < 10^6$ [6, 7]. and flexible polymers melts with $1 < \rho_K \ell_K^3 < 10$ [8]. We find excellent agreement between the experimental data and our results. Note that the crossover density from the loosely to tightly entangled system is identified with no adjustable parameters. These results support our ansatz that the PPA is a means to renormalize loosely to tightly entangled system. More detailed analysis of the results including the plots of the characteristic lengthscales will be given elsewhere [9].

ACKNOWLEDGMENTS

We acknowledge helpful discussions with K. Kremer and C. Svaneborg as well as help from S. Sukumaran in the conversion of the exprimental data to the dimensionaless form in Fig. 1. NU acknowledges financial support by Grand-in-Aid for Scientific Research from Japan's Ministry of Education, Culture, Sports, Science and Technology, and the hospitality of the Max-Planck-Institute for Polymer Research in Mainz where we started to develop the simulation code. Sandia is a multiprogram laboratory operated by Sandia Corporation, a Lockheed Martin Company, for the United States department of Energy's National Nuclear Security Administration under contract de-AC04-94AL85000.

REFERENCES

1. M. Doi and S.F. Edwards, *The Theory of Polymer Dynamics* (Clarendon, Oxford, 1986).
2. W.W. Graessley and S.F. Edwards, Polymer **22**,1329-1334 (1981).
3. R. Everaers, S.K. Sukumaran, G.S. Grest, C. Svaneborg, A. Sivasubramanian, and K. Kremer, Science **303**, 823-826 (2004).
4. S.K. Sukumaran, G.S. Grest, K. Kremer, and R. Everaers, J. Poly. Sci. B: Polymer Physics **43**, 917-933 (2005).
5. A.N. Semenov, J. Chem. Soc. Faraday Trans. **82**, 317-329 (1986).
6. B. Hinner, M. Tempel, E. Sackmann, K. Kroy, and E. Frey, Phys. Rev. Lett. **81**, 2614-2617 (1998).
7. F.G. Schmidt, B. Hinner, E. Sackmann, and J.X. Tang, Phys. Rev. E **62**, 5509-5517 (2000).
8. L.J. Fetters, D.J. Lohse, D. Richter, T.A. Witten, and A. Zirkel, Macromolecules **27**, 4639-4647 (1994).
9. N. Uchida, G.S. Grest, and R. Everaers, preprint.

PART IV

NANO-MEGA SCALE FLOW DYNAMICS IN COMPLEX SYSTEMS

Electrorheological Fluid Dynamics

Jianwei Zhang[1], Chun Liu[2], and Ping Sheng[1]

[1]Department of Physics, Hong Kong University of Science and Technology, Clear Water Bay, Kowloon, Hong Kong, China
[2]Department of Mathematics, Pennsylvania State University, PA, USA 16802

Abstract. We present the formulation of a two-phase, electrical-hydrodynamic model for the description of electrorheological fluid dynamics. By considering the energetics of (induced) dipole-dipole interaction between the solid particles in terms of a field variable $n(\vec{x})$, we employ the Onsager principle to derive the relevant coupled hydrodynamic equations, together with a continuity equation for $n(\vec{x})$. Numerical solution of the relevant equations yields predictions that display very realistic behaviors as seen experimentally.

Keywords: Electrorheological fluid, Dynamics, Colloid, Hydrodynamics
PACS: 47.65.Gx, 83.80.Gv, 47.50.-d, 83.10.Bb, 47.57.J-

INTRODUCTION

Electrorhoelogical (ER) fluids [1-10] constitute a class of colloids whose rheological characteristics can be controllably varied through the application of an external electric field. They have broad applications potential in active dampers, valves, etc [2,3]. However, many of the ER fluid applications involve flows with high shear-rates. While the static characteristics of the ER fluids have been studied successfully with the effective dielectric constant formulation [1,6], the dynamic behavior of ER fluids still represents a challenging area. Since ER fluid is a colloid, it is inherently a two-component material consisting of solid particles and carrier fluid (usually an insulating dielectric oil). A direct simulation involving a number of discrete, electrically interacting particles would be time consuming and computationally limited by the particle number [11-17], hence difficult to apply to realistic systems for the simulation of their dynamic behavior. Here we present a phenomenological two-fluid model in which the electrical part of the interaction energy is treated from first principles, on the basis of (induced) dipole-dipole interaction. The dynamics, on the other hand, is treated by homogenizing the solid component, leading to two coupled Navier-Stokes equations plus a continuity equation for the solid component.

In what follows, we first present the basic model and the relevant continuum hydrodynamic equations. The part on the particle-particle interaction will be examined in some detail. Some numerical results will be presented.

MODEL

Consider identically-sized solid microspheres of radius a (= 5 μm in our calculations), dielectric constant ε_s (= 10.0 in our calculations), and mass m (= $1.2\text{x}10^{-9}$ g in our calculations) suspended in oil with dielectric constant ε_f (= 2.0 in our calculations), viscosity η_f (= 10 cP in our calculations), and density ρ_f (= 0.96 g/cc in our calculations). Due to the difference between ε_s and ε_f, in the presence of external field the solid particles will be polarized with an induced dipole moment

$$\vec{p} = \frac{\varepsilon_s - \varepsilon_f}{\varepsilon_s + 2\varepsilon_f} a^3 \vec{E}_l \,. \tag{1}$$

Here E_l denotes the local electric field, which is the sum of the externally applied electrical field $\vec{E}_{ext}$, plus the field from all the other induced dipoles, both at the position of the microsphere. The accurate knowledge of the latter requires a description of the induced dipole distribution in space, which represents the global self-consistent solution of the problem. To facilitate the construction of the model, we first assume that the point dipole $\vec{p}$ is situated at the center of the microsphere. To prevent microspheres from

CP982, *Complex Systems, 5th International Workshop on Complex Systems*
edited by M. Tokuyama, I. Oppenheim, and H. Nishiyama

overlapping in space, we introduce a repulsive interaction potential between any two spheres i and j, situated at $\vec{x}$ and $\vec{y}$, respectively, as

$$\varepsilon_0 \left(\frac{a}{|\vec{x}-\vec{y}|} \right)^{12}, \qquad (2)$$

where ε_o is a suitably chosen energy constant. Second, we treat the solid particles collectively by regarding their density $n(\vec{x}) = f_s(\vec{x})(4\pi a^3/3)^{-1}$ as a field variable, where $f_s(\vec{x})$ denotes the dimensionless, local volume fraction of solid microspheres. This component of our model is denoted the "s" component. It is obviously not a solid, but rather a homogenized colloidal phase. One can write down the total energy for this component, including the interaction between the particles and between the particles and the external field, as a functional of $n(\vec{x})$:

$$F[n(\vec{x})] = \frac{1}{2}\int G_{ij}(\vec{x},\vec{y}) p_i(\vec{x}) n(\vec{x}) p_j(\vec{y}) n(\vec{y}) d\vec{x} d\vec{y} \\ -\int \vec{E}_{ext}(\vec{x}) \cdot \vec{p}(\vec{x}) n(\vec{x}) d\vec{x} + \frac{\varepsilon_0}{2}\int \left(\frac{a}{|\vec{x}-\vec{y}|} \right)^{12} n(\vec{x}) n(\vec{y}) d\vec{x} d\vec{y}, \qquad (3)$$

where

$$G_{ij}(\vec{x},\vec{y}) = \frac{\vec{\vec{I}}_{i,j}}{|\vec{x}-\vec{y}|^3} - \frac{3(\vec{x}-\vec{y})_i(\vec{x}-\vec{y})_j}{|\vec{x}-\vec{y}|^5} = \vec{\nabla}_i \vec{\nabla}_j \frac{1}{|\vec{x}-\vec{y}|} \qquad (4)$$

is the dipole interaction operator, and the Einstein summation convention is followed in Eq. (3), where the repeated indices imply summation. A variation of F [18-22] with respect to n leads to $\delta F = \int \mu(n) \delta n d\vec{x}$, where

$$\mu[n(\vec{x})] = -\vec{E}_{ext}(\vec{x}) \cdot \vec{p}(\vec{x}) + \int G_{ij}(\vec{x},\vec{y}) p_i(\vec{x}) p_j(\vec{y}) n(\vec{y}) d\vec{y} \\ +\varepsilon_0 \int \left(\frac{a}{|\vec{x}-\vec{y}|} \right)^{12} n(\vec{y}) d\vec{y} \qquad (5)$$

is the chemical potential for the "s" component. It should be noted that the first two terms on the right-hand side of Eq. (5) may be interpreted as $-\vec{E}_l \cdot \vec{p}$, where $[\vec{E}_l(\vec{x})]_i = [\vec{E}_{ext}(\vec{x})]_i - \int G_{ij}(\vec{x},\vec{y}) p_j(\vec{y}) n(\vec{y}) d\vec{y}$. Since n is a locally conserved variable, there is a continuity equation for n, given by

$$\dot{n} + \nabla \cdot \vec{J} = \frac{\partial n}{\partial t} + V_s \cdot \nabla n + \nabla \cdot \vec{J} = 0, \qquad (6)$$

where V_s is the "s" phase velocity, and $\vec{J}$ is a convective-diffusive current density.

Besides the "s" component, the model consists of another "f", or fluid, component, together with a coupling term that characterizes the dissipative coupling between the two components.

Here we first give the complete coupled equations of motion for the two fluids model. Their derivation will be given in the following section. Besides the continuity equation (6), the coupled equations of motion for the "s" phase and the "f" phase are given by

$$\rho_s\left(\frac{\partial \vec{V}_s}{\partial t} + \vec{V}_s \cdot \nabla \vec{V}_s\right) = -\nabla p_s + \nabla \cdot \tau^s_{visc} + \nabla \cdot \tau_s + K(\vec{V}_f - \vec{V}_s), \qquad (7)$$

$$\rho_f\left(\frac{\partial \vec{V}_f}{\partial t} + \vec{V}_f \cdot \nabla \vec{V}_f\right) = -\nabla p_f + \nabla \cdot \tau^f_{visc} + K(\vec{V}_s - \vec{V}_f), \qquad (8)$$

with the supplementary incompressibility conditions $\nabla \cdot \vec{V}_{s,f} = 0$. Here $\rho_s = mn(\vec{x}) + (1-f_s)\rho_f$ is the local mass density of the "s" phase, , p_s and p_f are the pressures in the two phases, $\nabla \cdot \tau_s$ is the force density arising from the energy functional (3), and $\tau^s_{visc} = \eta_s(\nabla \vec{V}_s + \nabla^T \vec{V}_s)/2$, $\tau^f_{visc} = \eta_f(\nabla \vec{V}_f + \nabla^T \vec{V}_f)/2$ are the viscous stresses of the two components [23,24]. While η_f is just the fluid viscosity, for η_s we use the concentration-dependent colloidal viscosity, to be given later. In Eqs. (7) and (8) K is a constant which characterizes the relative drag force density between the "s" and "f" components, in the linear approximation. Hence if we consider only the Stokes drag of the "s" phase by the fluid, then $K = 9f_s\eta_f/2a^2$ [24].

In Eqs. (7) and (8), the two crucial expressions, $\vec{J}$ and $\nabla \cdot \tau_S$, are to be specified. This can be done by using the Onsager principle, together with the forms of Eqs. (7) and (8), as shown below.

DERIVATION THROUGH THE ONSAGER PRINCIPLE

The Onsager principle of minimum energy dissipation [25-27] is one of the most basic elements in the statistical mechanics of dissipative systems. It underlies almost all the linear response phenomena. Simply stated, the Onsager principle consists of a variational functional, the minimization of which would (1) guarantee the balance of the dissipative and mechanical forces, and (2) insure the predictions of the resulting equations of motion to represent the statistically most probable course of motion. For the "*s*" component the Onsager variational functional is given by

$$A(\vec{J},\vec{V}_S) = \dot{F} + \Phi\,, \tag{9}$$

where

$$\begin{aligned}\dot{F} &= \int \mu \frac{\partial n}{\partial t} d\vec{x} = \int \mu\left(\dot{n} - \vec{V}_S \cdot \nabla n\right) d\vec{x} \\ &= -\int \mu\left(\nabla \cdot \vec{J} + \vec{V}_S \cdot \nabla n\right) d\vec{x} = \int\left(\nabla \mu \cdot \vec{J} + n \nabla \mu \cdot \vec{V}_S\right) d\vec{x}\end{aligned} \tag{10}$$

and Φ is a quadratic function of rates, given as ½ the energy dissipation rate,

$$\Phi = \int\left(\frac{1}{4}\eta_S[\partial_i(\vec{V}_S)_j + \partial_j(\vec{V}_S)_i]^2 + \frac{\gamma}{2n}J^2 + \frac{1}{2}K(\vec{V}_f - \vec{V}_s)^2\right) d\vec{x}, \tag{11}$$

together with the constraint of $\nabla \cdot \vec{V}_S = 0$, which can be implemented by using a Lagrange multiplier λ. In Eq. (10), we have used the integration by parts as well as the incompressibility condition to reach the final desired form. In Eq. (11), γ is a frictional coefficient related to the convective-diffusive current's dissipation. The form of the convective-diffusive dissipation can be simply obtained by realizing that $\vec{J} = n\vec{V}_d$, where $\vec{V}_d$ denotes the drift velocity. The dissipative force acting on a single microsphere is $\gamma \vec{V}_d$. Hence the force density is given by $n\gamma \vec{V}_d$, and the energy dissipation rate per unit volume is $n\gamma V_d^2 = \gamma J^2 / n$. Taking into account the factor of ½ leads directly to the expression shown in Eq. (11). The other two terms of Φ are simply the well-known viscous dissipation and the dissipation caused by the friction between the two components. Minimization of the variational functional with respects to the rates $(\vec{J},\vec{V}_S)$ leads to the desired expression for $\vec{J}$, and the Stokes equation for the "*s*" component (i.e., where the inertial effects are negligible, or the left-hand side of Eq. (7) equals zero). That is,

$$\vec{J} = -\frac{n}{\gamma}\nabla \mu, \tag{12}$$

and

$$0 = -\nabla p_s + \nabla \cdot \tau_{visc}^s + n\nabla \mu + K(\vec{V}_f - \vec{V}_s), \tag{13}$$

where $\lambda = -2p_S$. Comparison of the right-hand sides of Eqs. (13) and (7) leads to the conclusion that $\nabla \cdot \tau_S = n\nabla \mu$. When the inertial effects are not negligible, momentum balance requires the left-hand side of Eq. (13) be replaced by $\rho_S \dot{\vec{V}}_S$, which is precisely Eq. (7).

For the frictional coefficient γ, we propose the Stokes drag form $\gamma = 6\pi\eta_s a$, where it is noted that the viscosity used is that of the effective colloidal viscosity of the "*s*" component, owing to the interaction between the different microspheres that would determine the drift velocity of a microsphere inside the "*s*" component. This effective viscosity has been a topic of extensive study both theoretically and experimentally. When solid particle density is lower than $f_s \le 0.55$, Pade approximants [28] can be used to represent viscosity variation with f_s. In the lowest order, the viscosity can be written as $\eta_s / \eta_f = 1 + \frac{5}{2} f_s + O(f_s^2)$. For f_s near the random close pack fraction $f_s^{max} = 0.698$, experimental results [29] shows an exponential divergence: $\eta_s / \eta_f \propto \exp\left[0.6 / \left(f_s^{max} - f_s\right)\right]$. In order to cover both the lower and higher ends of the solid density, we have matched the Pade approximation at lower volume fraction, $f_s \le 0.45$, and exponential divergence at higher volume fractions $f_s \ge 0.45$. Figure 1 shows the matched relation.

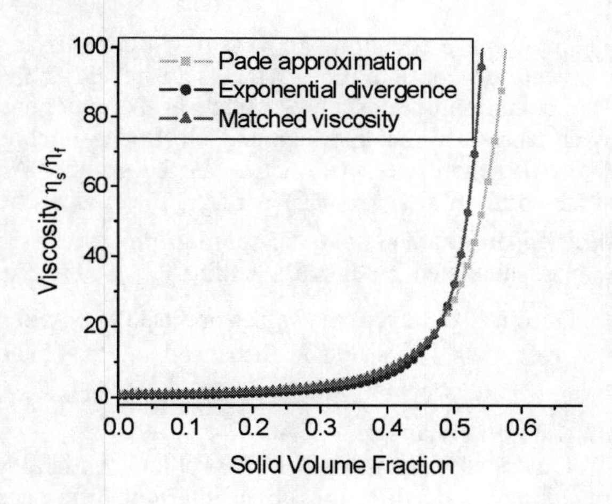

FIGURE 1. (Color online) The "*s*" component viscosity variation with the solid particles volume fraction. Green shows the low density Pade approximation; blue shows the high density variation; red shows the matched variation through the whole range of solid densities.

For equation (8), which is much simpler than Eq. (7), almost identical application of the Onsager principle would lead to the desired result.

NUMERICAL IMPLEMENTATION

Numerical solution of the above scheme consists of two main elements that underlie the dynamics of ER fluids: coupled hydrodynamics of the two components, together with the electrical interactions. The geometry used is that of a channel formed by two plates, parallel to the *xy* plane, separated by a distance Z_0 (=650 μm in our calculations). The channel is filled with ER fluid. Periodic boundary conditions are imposed on the calculational sample boundaries along the *x* and *y* directions. Non-slip boundary condition is used at the fluid-solid interfaces. The upper plate is assumed to be either moving at a constant speed along the *x* direction, or moved with some incremental distance along the *x* direction after the electric field is applied.

The electrical element of the problem enters through the local electric field $[\vec{E}_l(\vec{x})]_i = [\vec{E}_{ext}(\vec{x})]_i - \int G_{ij}(\vec{x},\vec{y})p_j(\vec{y})n(\vec{y})d\vec{y}$, and Eq. (1). Here $\vec{E}_{ext} = -\nabla\phi$, ϕ being the solution of the Laplace equation $\nabla\bar{\varepsilon}\nabla\phi = 0$, with the local effective dielectric constant $\bar{\varepsilon}$ obtained from the Maxwell-Garnett equation

$$\frac{\bar{\varepsilon}(\vec{x})-\varepsilon_f}{\bar{\varepsilon}(\vec{x})+2\varepsilon_f} = f_s(\vec{x})\frac{\varepsilon_s-\varepsilon_f}{\varepsilon_s+2\varepsilon_f}. \tag{14}$$

The Laplace equation can be solved by specifying the electrode configuration, which can be either the usual condition of constant potentials at the upper and lower plates, or the interdigitated electrodes on one plate, with the upper plate acting as the ground. An initial configuration of $n(\vec{x})$ (or $f_S(\vec{x})$) need to be specified, in order to start the solution process. Then $\vec{p}(\vec{x})$ is calculated by initially letting $\vec{E}_l = \vec{E}_{ext}$ in Eq. (1). Once it is obtained, the values are used to obtain a new value for $\vec{E}_l$, which is then used in Eq. (1) to obtain a new $\vec{p}(\vec{x})$, etc, until consistency is achieved. A few iterations suffice.

Numerically, we solve the 2D problem (variations only along *x* and *z* directions) by using finite difference [30-32] with spectral differentiation along the *x* direction, and explicit in time. Starting from a random initial configuration of $n(\vec{x})$, we first apply the external potential to the problem, and with local field (and thus $\vec{p}(\vec{x})$ through Eq. (1)) obtained as described above, $n(\vec{x})$ is updated through Eqs. (6) and (12). The updated $n(\vec{x})$ is used to calculated $\bar{\varepsilon}(\vec{x})$ through Eq. (14), and the process is iterated til consistency. Thus starting from a random configuration, it is easy to see the formation of chain-like columns in the "*s*" component when the external field is applied (see below). This is the intuitively desired consequence of an external field, as required by energetics.

The moving boundary condition of the upper plate (or the incremental displacement) is then applied, and the coupled hydrodyanmic equations (7) and (8), together with continuity equation (6), are solved with the incompressibility conditions. The boundary conditions for both $\vec{V}_S$ and $\vec{V}_f$ are the non-slip conditions for the tangential components at the upper and lower solid boundaries, and zero normal components. For *n*, the boundary condition is that the normal component of the convective-diffusive current density $\vec{J}$ be zero at the solid boundaries. By time-stepping forward the solution, at each time step iterating the electrical solution to insure that consistency is achieved in $n(\vec{x})$, we obtain the time evolution of the ER dynamics. Below we show some of our results.

RESULTS

In Fig. 2a, we show that by applying an electric field along the *z* direction, with a magnitude of 1kV/mm to an initial configuration of $n_S(\vec{x})$ with a uniform f_s = 0.3 can lead to the formation of columns. Here the upper plate is stationary. When the upper plate is moved relative to the lower plate with a uniform speed, the upper halves of the "solid" columns will move with the upper boundary, while the lower halves are fixed by the lower boundary. Hence the columns will initially be stretched and eventually be brokern. When that happens, the measured shear stress at the upper plate experiences a drop. However, at a later time the the upper halves of the columns will reconnect to the lower halves (since periodic boundary condition is applied along the *x* direction), and when that happens the measured shear stress at the upper boundary experiences a jump.

By averaging the measured shear stress on the upper plate over one period of time, the averaged shear stress variation with shear rate (which can be changed by changing the velocity of the upper plate) is shown in Fig. 3. Four different curves are shown, corresponding to four different applied electric fields. It is seen that up to 800/s, the shear stress increases; but beyond that the shear stress decreases. This is the commonly observed shear-thinning effect in ER fluids, and can be simply explained heuristically on the geometric basis [33]. That is, since it is the electric field that holds the solid particles together to form the columns, the "adhesive force" for the column formation is necessarily along the field direction *z*.

Initially, when the shear rate is small, the shear stress should increase with the shear rate, since it takes a larger force to break the column within a short time. However, as the shear rate increases, the steady-state tilting of the columns as seen in Fig. 2b becomes more pronounced. Hence the adhesive force decreases as cosine of the tilting angle. It follows that there is a peak. This is precisely the behavior seen in Fig. 3.

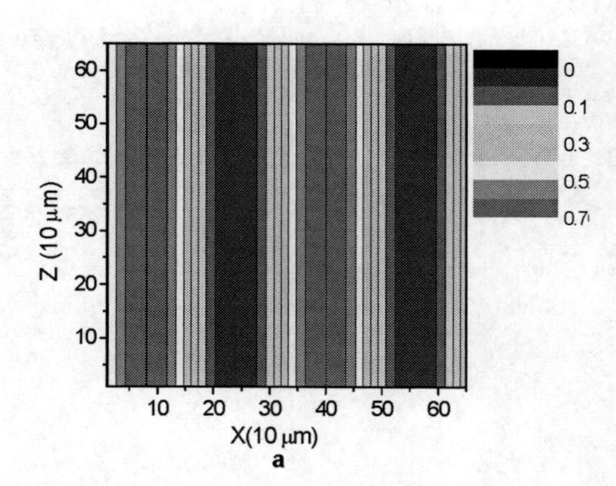

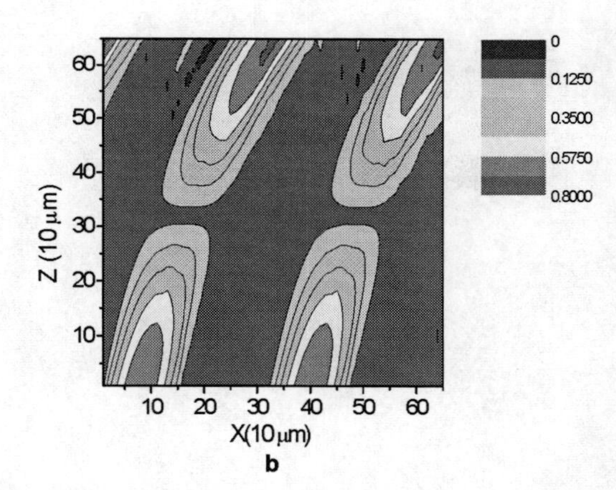

FIGURE 2. (Color online) (a) Pattern of $n_S(\vec{x})$ when an electric field (1 kV/mm) is applied between the upper and lower plates, and the upper plate is not moving. Here red indicates high density, and blue indicates low density. (b) A snapshot of the $n_S(\vec{x})$ pattern when the upper plate is moving at a constant speed. It is seen that the columns are tilted along the shearing direction and broken in the middle. However, the tilted columns can re-attach at a later time. Breaking and re-attachment implies fluctuations in the measured yield stress, also seen.

It is noteworthy that if one extrapolates the stress curves to zero shear rate in Fig. 3, the intersection is at a non-zero shear stress. This is denoted static yield stress, as measured dynamically. The static yield stress varies quadratically as a function of the applied field. It is part of the Bingham fluid behavior. However, for the Bingham fluid there is no shear-thinning, which is purely an effect that results from the interaction of the fluid dynamics with the applied electric field, and allows a simple, heuristic geometric explanation as described above.

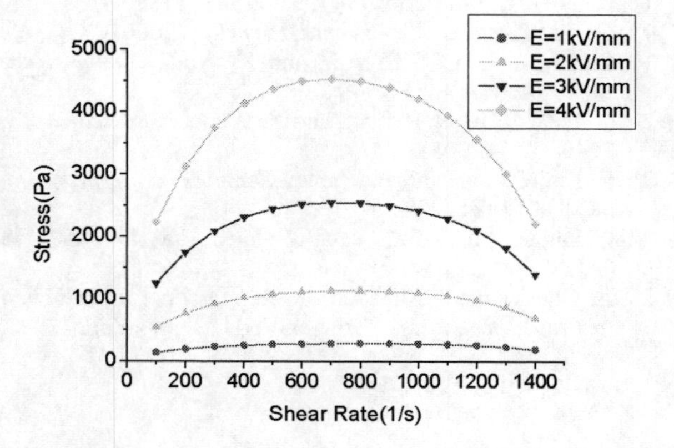

FIGURE 3. (Color online) Time-averaged stress plotted as a function of shear rate at different electrical fields. It is seen that at higher electric field, one can extrapolate to a non-zero stress at zero shear rate, just as implied by the Bingham fluid behavior.

In summary, we present a two-component, electrical-hydrodynamic continuum model for ER fluid dynamics based on the energetics of the induced dipole-dipole interaction and the Onsager principle. Numerical implementation of the scheme has demonstrated realistic behaviors.

REFERENCES

1. H. R. Ma, W. J. Wen, W. Y. Tam, and P. Sheng, *Adv. in Phys.* **52**, 343-383 (2003).
2. M. Whittle and W. Bullough, *Nature* **358**, 373-373 (1992).
3. T. Halsey, *Science* **258**, 761-766 (1992).
4. H. Block and J. P. Kelly, *J. Phys. D: Appl. Phys.* **21**, 1661-1677 (1988).
5. R. Tao and J. M. Sun, *Phys. Rev. Lett.* **67**, 398-401 (1991).
6. H. R. Ma, W. J. Wen, W. Y. Tam, and P. Sheng, *Phys. Rev. Lett.* **77**, 2499-2502 (1996).
7. W. Y. Tam, G. H. Yi, W. Wen, H. R. Ma, M. M. T. Loy, and P. Sheng, *Phys. Rev. Lett.* **78**, 2987-2990 (1997).
8. W. J. Wen, X. Huang, S. Yang, K. Lu, and P. Sheng, *Nature Materials* **2**, 727-730 (2003).
9. W. Wen, X. Huang, and P. Sheng, *Appl. Phys. Lett.* **85**, 299-301 (2004).
10. X. Huang, W. Wen, S. Yang, and P. Sheng, *Sol. Stat. Comm.* **139**, 581-588 (2006).
11. R. T. Bonnecaze and J. F. Brady, *J. Rheol.* **36**, 73-115 (1992).

12. R. T. Bonnecaze and J. F. Brady, *J. Chem. Phys.* **96**, 2183-2202 (1992).
13. D. J. Klingenberg, F. V. Swol, and C. F. Zukoski, *J. Chem. Phys.* **91**, 7888-7895 (1989).
14. D. J. Klingenberg, F. V. Swol, and C. F. Zukoski, *J. Chem. Phys.* **94**, 6160-6169 (1991).
15. D. J. Klingenberg, *J. Rheol.* **37,** 199-214 (1993).
16. D. J. Klingenberg, C. F. Zukoski, and J. C. Hill, *J. Appl. Phys.* **73**, 4644-4648 (1993).
17. K. C. Hass, *Phys. Rev. E* **47**, 3362-3373 (1993).
18. Q. Du, C. Liu, and R. Ryham, *Preprint*, (2006).
19. Q. Du, C. Liu, R. Ryham, and X. Wang, *Comm. Pure Appl. Analysis* **4**, 537-548 (2005).
20. Q. Du, C. Liu, R. Ryham, and X. Wang, *Nonlinearity* **18**, 1249-1267 (2005).
21. F. Lin, C. Liu, and P. Zhang, *Comm. Pure Appl. Math.* **58**, 1437-1471 (2005).
22. Q. Du, C. Liu, and X. Wang, *J. Comp. Phy.* **198**, 450-468 (2005).
23. A. Chorin and J. Marsden, *A Mathematical Introduction to Fluid Mechanics*, Springer-Verlag, New York, 1993, pp. 1-169.
24. G. K. Batchelor, *An Introduction to Fluid Dynamics*, Cambridge University Press, Cambridge, 2000, pp. 1-635.
25. L. Onsager, *Phys. Rev.* **37**, 405-426 (1931); *Phys. Rev.* **38**, 2265-2279 (1931).
26. L. Onsager and S. Machlup, *Phys. Rev.* **91**, 1505-1512 (1953).
27. L. D. Landau and E. M. Lifshitz, *Statistical Physics, 2nd Ed.*, Addison-Wesley Publishing Co., London, 1969, pp. 1-484.
28. R. Verberg, I. de Schepper, and E. Cohen, *Phys. Rev. E* **55**, 3143-3158 (1997).
29. Z. Cheng, J. Zhu, P. Chaikin, S. Phan and W. Russell, *Phys. Rev. E* **65**, 041405-041412 (2002).
30. T. Qian, X. Wang, and P. Sheng, *Phys. Rev. E* **68**, 016306-016320 (2003).
31. J. W. Cahn and J. E. Hilliard, *J. Chem. Phys.* **28**, 258-267 (1958).
32. W. Ren and X. P. Wang, *J. Comput. Phys.* **159**, 246-273 (2000).
33. L. Liu, X. Huang, C. Shen, Z. Liu, J. Shi, W. Wen, and P. Sheng, *Appl. Phys. Lett.* **87**, 104106-104108 (2005).

Flow Dynamics in Arc Welding

John J. Lowke* and Manabu Tanaka¶

*CSIRO Materials Science and Engineering, PO Box 218, Lindfield, Sydney NSW 2070 Australia
¶Joining and Welding Research Institute, Osaka University, 11-1 Mihogaoka, Ibaraki, Japan.

Abstract. The state of the art for numerical computations has now advanced so that the capability is within sight of calculating weld shapes for any arc current, welding gas, welding material or configuration. Inherent in these calculations is "flow dynamics" applied to plasma flow in the arc and liquid metal flow in the weld pool. Examples of predictions which are consistent with experiment, are discussed for (1) conventional tungsten inert gas welding (2) the effect of a fraction of a percent of sulfur in steel, which can increase weld depth by more than a factor of two through changes in the surface tension (3) the effect of a flux, which can produce increased weld depth due to arc constriction (4) use of aluminium instead of steel, when the much larger thermal conductivity of aluminium greatly reduces the weld depth and (5) addition of a few percent of hydrogen to argon, which markedly increases weld depth.

Keywords: Welding, Flow, Weld-pool , Arcs, Convection, Plasmas
PACS: 52.77Fv

INTRODUCTION

Manufacturing, (e.g. for cars or hot water systems), and construction industries, (e.g. for buildings, bridges or pipelines), is crucially dependent on the technology of electric arc welding. Central to the science of arc welding is the understanding of flow dynamics. The flows of arc welding are in two regions. Firstly, the electric arc itself is a flowing plasma. Secondly, the arc produces a pool of molten metal for the production of the weld and flow in the weld pool is crucial in determining the shape, strength and thickness of the weld. Advances in numerical modelling techniques have enabled increased understanding of arc welding and also made possible detailed predictions of weld thickness for various welding parameters such as arc current, and also for different weld materials.

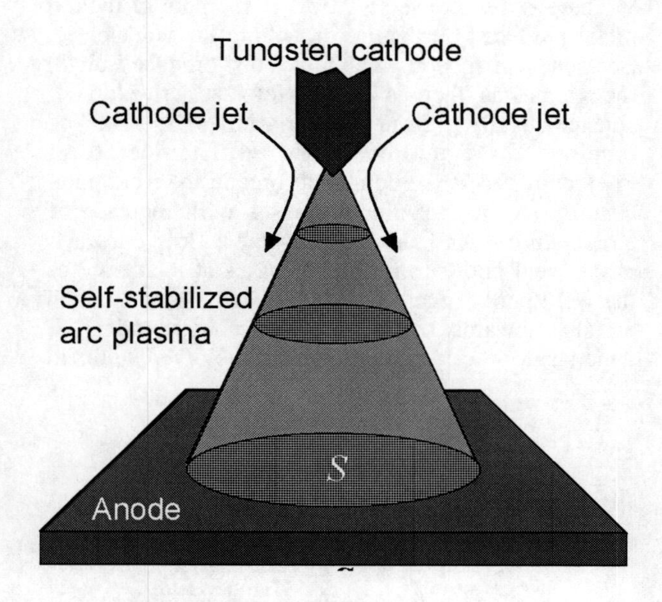

FIGURE 1. Welding electrode configuration.

FLOW DYNAMICS

Flow Dynamics in the Arc Plasma

The fluid dynamics of arc plasmas is the same as conventional fluid dynamics for gases or liquids, where one solves the usual coupled equations for the conservation of mass, energy and momentum under the influence of forces due to pressure, gravity and buoyancy. But there is an additional force to account for arc behaviour, and it happens that this is usually the dominant force. This force is the magnetic pinch force due to the magnetic field produced by the arc [1].

In any arc welding system there is generally one electrode which is a pointed cylindrical rod, of about a mm in diameter, for example a tungsten rod in Tungsten Inert Gas (TIG) welding, as illustrated in Fig. 1, or a metal wire in Metal Inert Gas (MIG) welding. The arc produces a series of molten metal drops from

CP982, *Complex Systems, 5th International Workshop on Complex Systems*
edited by M. Tokuyama, I. Oppenheim, and H. Nishiyama

the wire in MIG welding, assisting the weld. The other electrode in the welding system is the "work-piece", consisting of the two pieces of metal that are to be welded together. Surrounding the electric current flowing through the arc plasma that bridges the electrode and the work-piece is a concentric magnetic field. This magnetic field has a pinching effect on the plasma, tending to increase the pressure inside the plasma. This increase in pressure is larger if the current density is larger, as occurs at the tungsten or wire electrode, where the arc diameter is smaller than at the "work-piece" electrode. The plasma pressure at the electrode is then higher than the plasma pressure at the work-piece, inducing plasma flow from the narrow electrode to the work-piece. The increase in pressure is usually only about 1% of atmospheric pressure, but this pressure induces plasma flows of the order of several hundred metres per second. It is this flow which defines the arc diameter and its temperature.

Flow Dynamics in the Weld Pool

There is also convective flow in the pool of molten metal produced by the arc in melting the work-piece. Besides the magnetic pinch forces that produce flow in the arc plasma, there is also the force at the weld pool surface resulting from the variation of the surface tension of the molten liquid with temperature. Frequently this is the dominant force in the weld pool. Usually surface tension decreases with increase of temperature, thus inducing convective flow outwards at the weld pool surface and upwards at the centre of the weld pool. Together with buoyancy forces, which are also upwards, the effect of convection produces a wide shallow weld pool. However, dissolved sulfur in molten stainless steel can change the dependence of the surface tension on temperature so that it increases instead of decreases with temperature. Then the maximum surface tension force is at the weld pool centre, inducing radially inward flow, as illustrated in Fig. 2. The convective flow at the centre of the weld pool is then downwards, leading to a deeper weld.

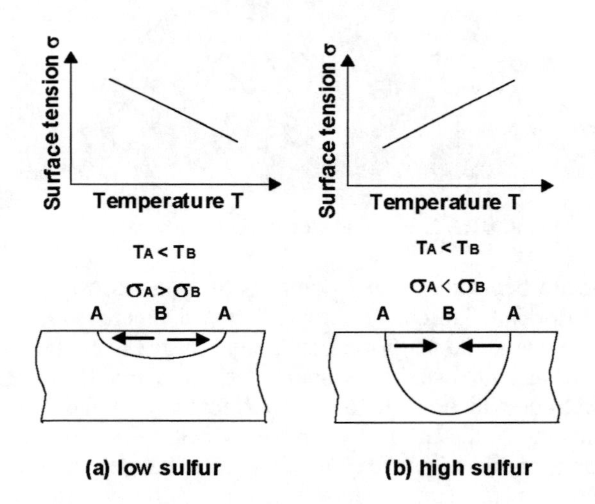

FIGURE 2. Effect of surface tension on the weld-pool.

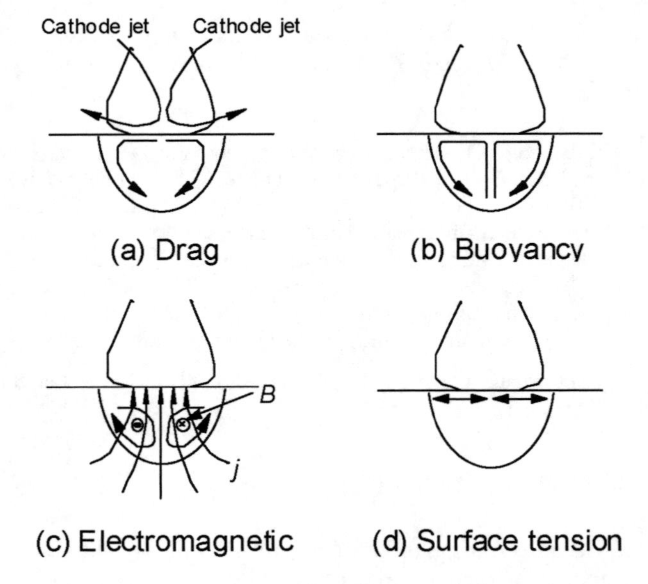

FIGURE 3. Different forces producing flow in the weld pool.

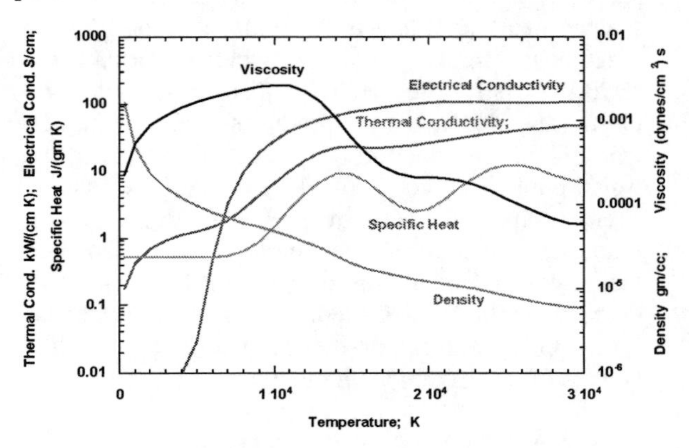

FIGURE 4. (Color online) Calculated transport coefficients for argon [1].

Another important force on the weld pool is the drag force of the plasma on the surface of the weld pool. To account for this force it is necessary to account for the flow dynamics of the arc and the weld pool in a unified system. Fig. 3 illustrates the four forces: the drag force, buoyancy, electromagnetic or magnetic pinch forces and the surface tension force in the weld pool. For depression of the weld-pool surface, there are further surface tension forces normal to the surface.

Modelling of the Flow System

It is possible to model the electrodes, plasma flow and weld pool flow in a unified system. Interactions between the arc flow and weld pool flow are treated in a consistent way, for example in accounting for the surface drag force of the arc plasma on flow in the molten metal. The current density distribution of the arc over the surface of the weld pool is coupled directly into the weld pool calculation.

The numerical procedure is that of Patankar [2], in which the two-dimensional transport equations are solved "row by row" and "column by column" through the inversion of tri-diagonal matrices. Transport coefficients, namely thermal conductivity, electrical conductivity, viscosity, density and specific heat as a function of temperature are required for the arc plasma, the molten weld pool and also the solid electrodes. For the arc plasma, it is now possible to calculate these quantities as a function of temperature for any given pressure, for almost any gas or mixture of gases [3,4]. Such transport coefficients are shown in Fig. 4 for argon at atmospheric pressure. Usually radiation effects are small, so that radiation emission effects are not crucial. Effects of metal vapour on plasma transport coefficients can also be calculated, and advances are being made for effects of metal vapour transport in welding arcs [5].

Fig. 5 shows a comparison of theoretical profiles for the temperature and electron density for a current of 150 A compared with experimental results obtained by measurements of laser scattering [6]. The example is for a water-cooled anode of copper because then there is greater arc stability than for a steel anode as is generally used in welding. It is seen that there is good agreement between the experimental results and the theoretical predictions. For these calculations the continuity equation for the electron density was also solved to account for electron diffusion across the non-equilibrium plasma sheath at the electrodes [1].

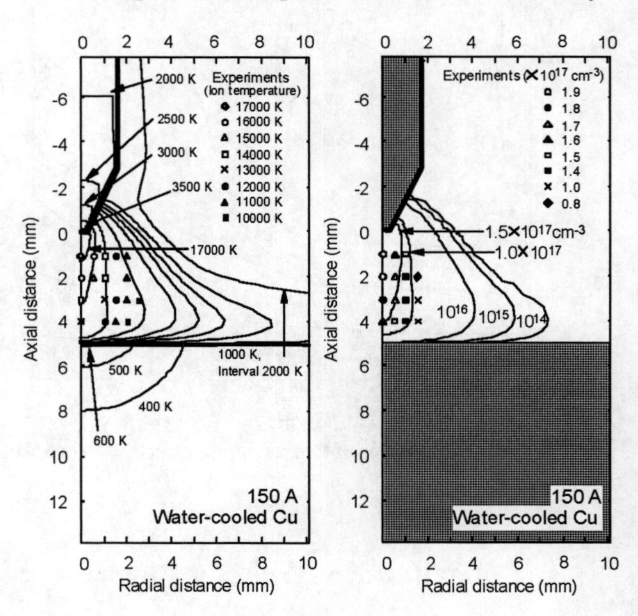

FIGURE 5. Temperatures and electron densities; theory-curves, experiment-points .

TUNGSTEN INERT GAS WELDING

In Fig. 6 we show a similar calculation to Fig. 5, but for a stainless steel electrode as used in TIG welding [1]. The region of the anode with temperatures above the melting point of steel of 1750 K is the predicted weld shape. The predicted flow in the weld pool is radially outward at the weld pool surface, driven by surface tension forces and the viscous drag force of the plasma.

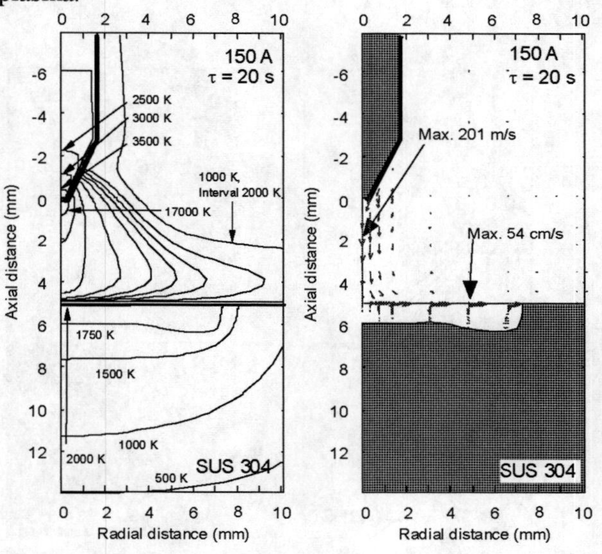

FIGURE 6. Similar calculation to Fig. 5 but for a stainless steel electrode showing the predicted weld shape [1]. Also shown are plasma and weld-pool velocities.

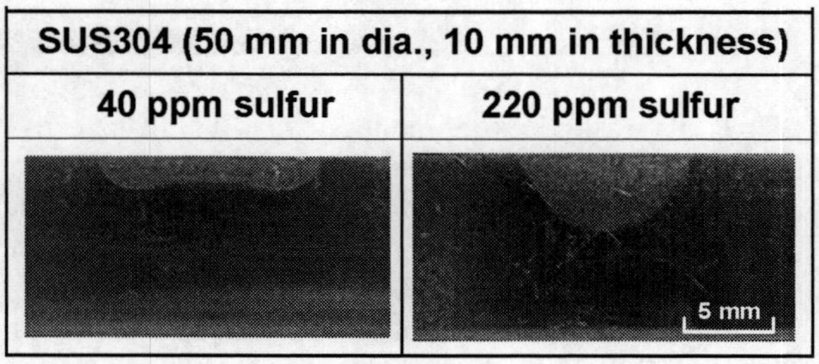

SUS304 (50 mm in dia., 10 mm in thickness)	
40 ppm sulfur	220 ppm sulfur
	5 mm

FIGURE 7. Weld shapes for low and high sulfur steel [1].

Surface Tension Effects, e.g. Due to Sulfur in Steel

The predicted weld shape of Fig. 6 is in fair agreement with experimental weld shapes as shown for low sulfur steels of 40 ppm in the left-hand half of Fig. 7. However for steels with a high concentration of sulfur, e.g. 220 ppm, the weld profiles are quite different, as shown in the right-hand half of Fig. 7.

In Fig. 8 we show predicted temperature and weld profiles as in Fig. 6, but calculated with the surface tension coefficient increasing instead of decreasing with temperature for temperatures near the melting point. Such a temperature dependence of the surface tension coefficient tends to produce inward flow at the surface of the weld pool as shown in Fig. 2. Such a temperature dependence of surface tension coefficients for steel with impurities of sulfur or oxygen is obtained from both experimental [7] measurements and theoretical [8] calculations.

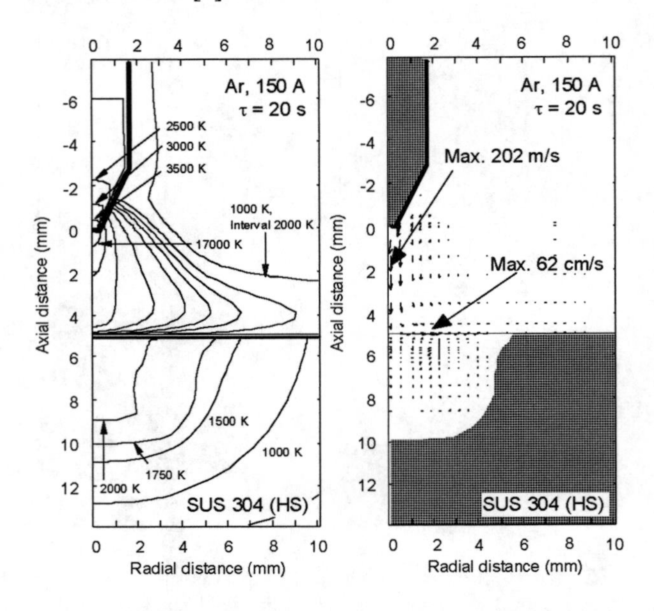

FIGURE 8. Predicted weld shape with high sulfur steel [1].

Flux Effects, e.g. Due to Arc Constriction

It is also found experimentally that a thin layer of flux consisting of the oxide of a metal on the surface of the work-piece can cause a significant increase in weld depth. Such an increase due to flux is shown in the left hand half of Fig. 9, compared with the right-hand figure for no flux, each figure corresponding to a current of 200 A. Such an increase in weld depth could well be due, at least in part, to changes in the temperature dependence of the surface tension due to dissolved flux in the molten weld metal, as described for Figs. 2 and 8.

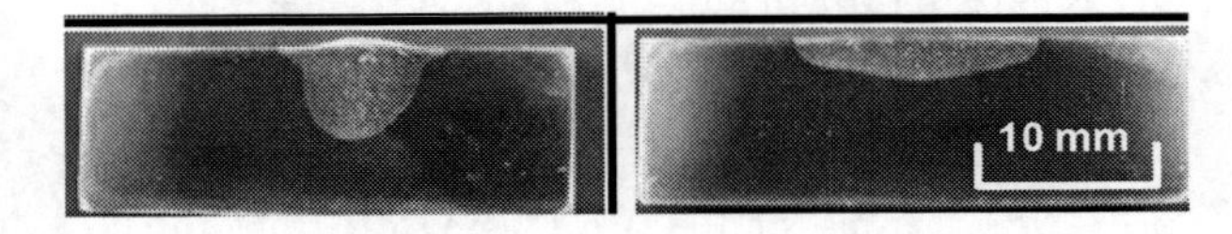

FIGURE 9. (Color online) Influence of flux on the surface of the work piece, left hand figure, which increases weld depth.

But there is a further possible effect due to the flux. The effect of a flux on the surface of the weld pool will produce an insulating layer at the edge of the weld pool thus impeding current flow in the outer region of the arc and increasing the current density and magnetic pressure at the centre of the weld pool. Such an increased current density will result in increased magnetic pressure at the arc centre at the top of the weld pool and produce a downward force on the convective flow in the weld pool.

The insulating effect of such a flux is simulated in the calculations [9] of Fig, 10, where the electrical current into the weld pool is blocked for radii greater than 2 mm. There is significantly increased magnetic pressure at the centre of the weld pool that produces downward convective flow and a weld that is narrower and deeper than for upward flow, sometimes by more than a factor of two.

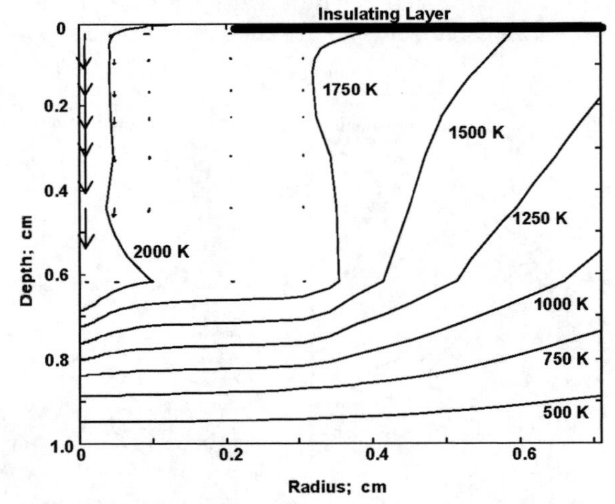

FIGURE 10. Simulation of the insulating effect of a flux at the work-piece surface [9].

Magnitudes of Weld Pool Forces

Methods of numerical analysis can enable an assessment to be made of the relative importance of the various forces on convective flow in the weld pool. In Fig. 11 we show the calculations of weld pool shape of Fig. 6, but including, in turn, only one of each of the four forces of Fig. 4, setting all other forces in the computer code to zero [1]. It is seen that the buoyancy force is relatively unimportant. Viscous drag and buoyancy forces produce radially outward flow; magnetic pinch, also known as Lorentz forces, produce radially inward flow; and surface tension, also known as Marangoni forces, produces flow whose direction depends on whether the surface tension coefficient increases or decreases with increase in temperature.

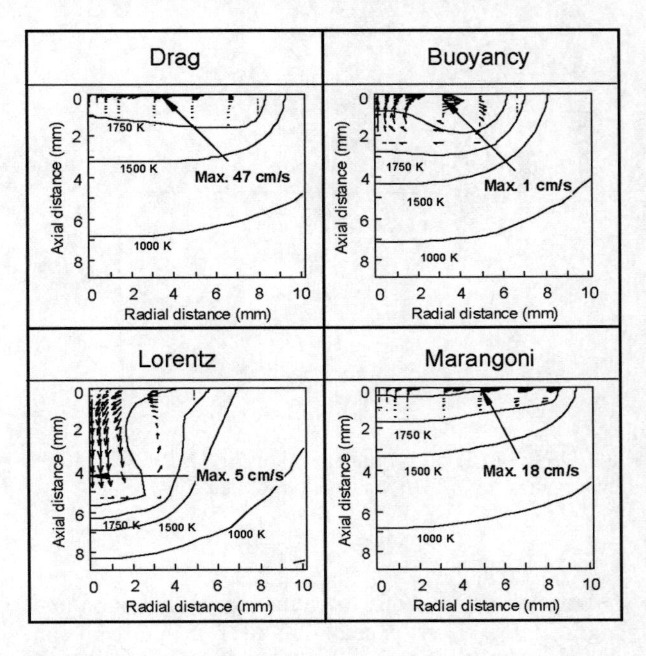

FIGURE 11. Relative importance of the various forces on convective flow in the weld pool.

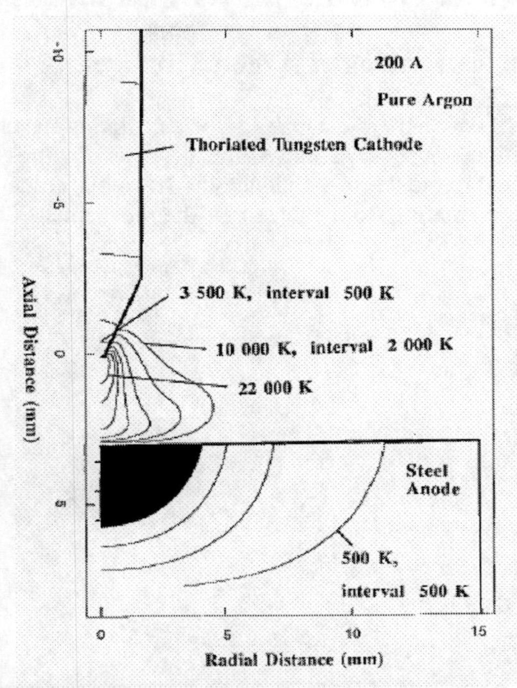

FIGURE 12(a). Comparison of calculated molten metal, shown in black, for steel and aluminium for a 200 A arc; anode stainless steel [10].

EFFECT OF DIFFERENT WORK PIECE METAL

Aluminium Instead of Steel

Weld depths are critically dependent on the properties of the metal of the work piece, in particular the thermal conductivity of the work-piece. The thermal conductivity of aluminium is about five times larger than that of stainless steel so that heat is conducted away much more readily, greatly reducing the size of molten metal and thus the weld depth, even though the melting point of aluminium, 933 K, is much lower than that of steel, 1750 K. This effect is illustrated in the calculation of Fig. 12 for a 200 A arc, which does not include convective flow in the weld pool [10]. The predicted weld-pool volume is almost zero for the calculation of Fig. 12(b) for aluminium.

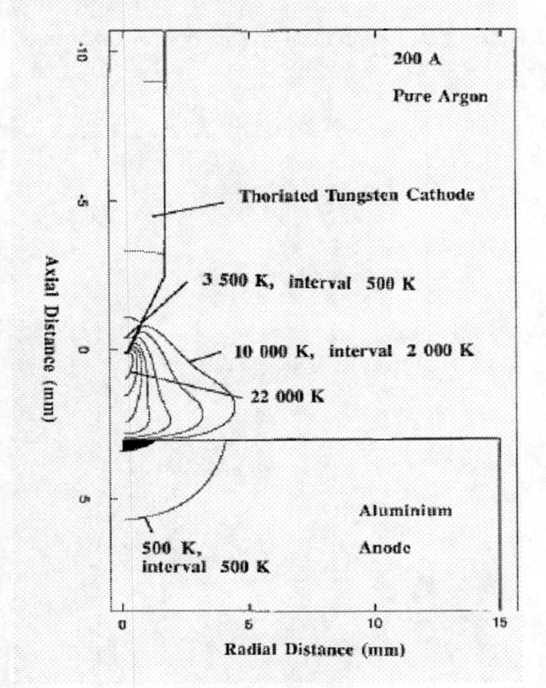

FIGURE 12(b). Comparison of calculated molten metal, shown in black, for steel and aluminium for a 200 A arc; anode aluminium [10].

EFFECT OF DIFFERENT PLASMA GAS

Hydrogen-Argon Mixtures

We now have the capability of calculating material functions for most gases and for most gas mixtures. In particular, the thermal conductivity has a big influence on conduction from the arc to the work piece. We

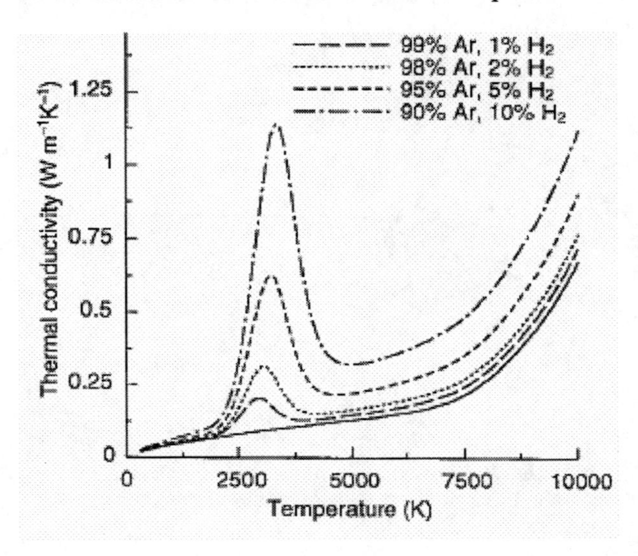

FIGURE 13. Calculated thermal conductivity of mixtures of hydrogen and argon [10].

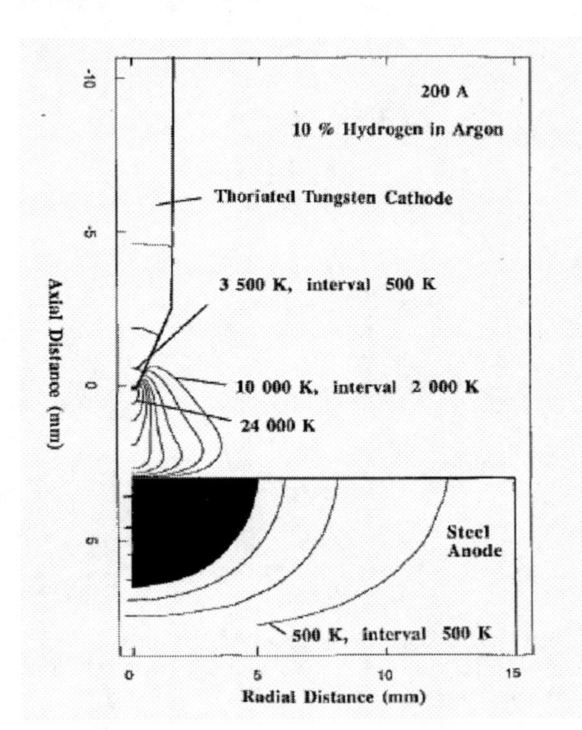

FIGURE 14(a) Calculations as for Fig. 12, but with 10 % hydrogen in argon; anode stainless steel [10].

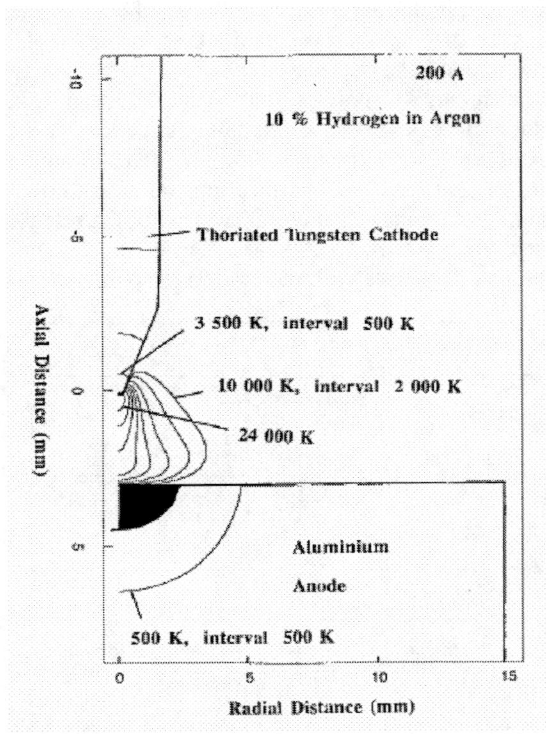

FIGURE 14 (b). Calculations as for Fig.12, but with 10 % hydrogen in argon [10]; anode aluminium [10].

show such calculations for mixtures of hydrogen and argon in Fig. 13. It is seen that 10% of hydrogen can increase the thermal conductivity of argon at 3000 K by a factor of ten. There is then a significant increase in the calculated volume of the molten weld pool for both stainless steel and aluminium, as shown in Figs. 14(a) and 14(b).

Carbon Dioxide

It is now possible to make predictions of weld properties for arc welding in different gases and in different gas mixtures which have markedly different properties than those using argon. Fig. 15 compares the calculated thermal conductivity of argon, carbon dioxide and helium. It is seen that the thermal conductivity of carbon dioxide and helium is about a factor of 10 larger than argon at low temperatures corresponding to positions near the work-piece. Calculated heat flux densities at the work-piece of a carbon dioxide arc found to be significantly higher than those in pure argon, as shown in Fig. 16. Thus new welding systems involving carbon dioxide are being proposed, partly as a result of new expertise in flow dynamics [11].

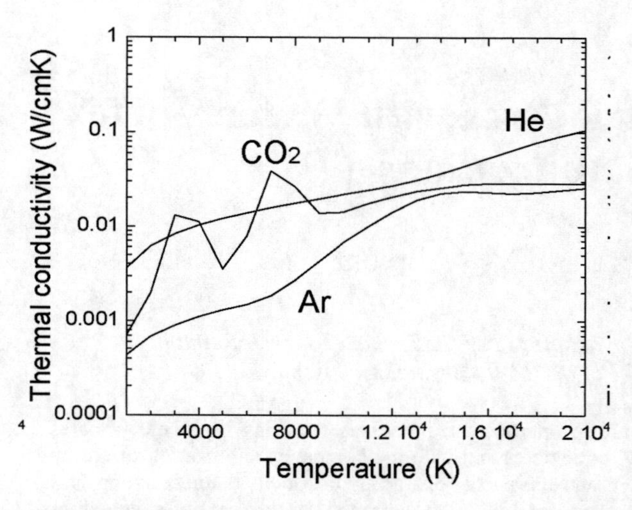

FIGURE 15. Calculated thermal conductivities of carbon dioxide and helium compared with argon [11].

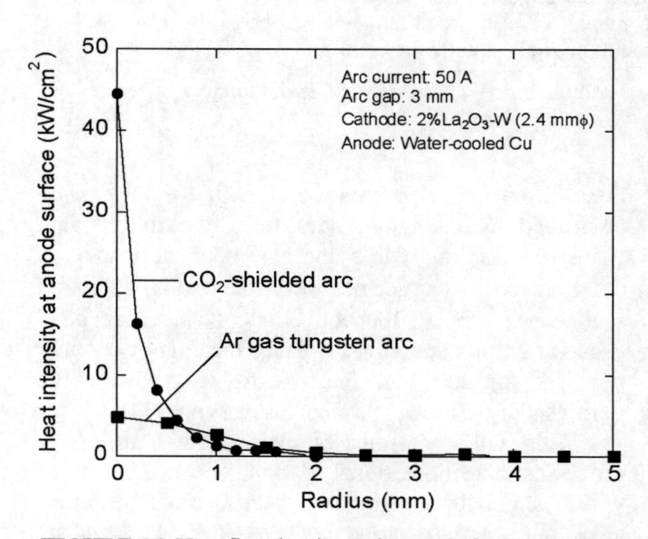

FIGURE 16. Heat flux density at the anode for 50 A arc [1].

SUMMARY

Electric arc welding is a process in which 'flow dynamics' are crucial, both for the plasma flow in the arc and for the flow in the molten weld pool. Increased expertise, particularly in the modelling of such flows in a unified system involving both the arc and the weld pool, has resulted in significantly improved understanding of the basic processes of arc welding. Furthermore we now have a greatly improved predictive capability to account for different plasma gases, arc currents and work piece materials.

ACKNOWLEDGEMENTS

The authors are grateful to J. Haidar and A.B. Murphy for comments on the manuscript.

REFERENCES

1. M. Tanaka and J. J. Lowke, *J. Phys. D: Appl. Phys.* **40**, R1-R23 (2007).
2. S. V. Patankar, *Numerical Heat Transfer and Fluid Flow,* McGraw Hill, New York, 1980, pp. 79-137.
3. M. I. Boulos, P. Fauchais and E. Pfender, *Thermal Plasmas,* Plenum, New York, 1994, pp. 213-263.
4. A. B. Murphy and C.,J. Arundell, *Plasma Chem. Plasma Proc.* **34**, 451-490 (1994).
5. M. Tanaka, K. Yamamoto, S. Tashiro, N. Nakata, M. Ushio, K. Yamazaki, E. Yamamoto, K. Suzuki, A. B. Murphy and J. J. Lowke, *IIW Doc.* 212-1107-07 (2007).
6. M. Tanaka and M. Ushio, *J. Phys. D: Appl. Phys.* **32**, 1153-1162 (1999).
7. B. J. Keene, *Int. Materials Reviews* **33**, 1-36 (1988).
8. P. Sahoo, T. DebRoy and M. McNallan, *Metall. Trans.* **198**, 483 (1988).
9. J. J. Lowke, M. Tanaka and M. Ushio, *J. Phys. D: Appl. Phys.* **38**, 3438-3445 (2005).
10. J. J. Lowke, R. Morrow, J. Haidar and A.B. Murphy, *IEEE Trans. Plasma Science* **25**, 925-930 (1997).
11. M. Tanaka, S. Tashiro, M. Ushio, T. Mita, A. B. Murphy and J. J. Lowke, *Vacuum* **80**, 1195-1198 (2006).

Numerical Modeling of Electric Arcs with Water Vortex and Hybrid Stabilizations

J. Jeništa[1], M. Bartlová[2], and V. Aubrecht[2]

[1]*Institute of Plasma Physics AS CR, v.v.i., Za Slovankou 3, 182 00 Praha 82, Czech Republic*
[2]*Brno University of Technology, Technická 8, 616 00 Brno, Czech Republic*

Abstract. In this paper we deal with numerical investigation of properties and processes occurring in the electric arcs with tangential stabilization of electric arc by water vortex (Gerdien arc) and with the combined stabilization of arc by axial gas flow and water vortex. The net emission coefficient and the partial characteristics method for radiation loss from these arcs are employed. Results carried out for the water arc for 150 - 600 A proved that typical outlet velocities are 0.7 - 8 km s^{-1}, temperatures 14 000 K - 26 000 K, the voltage drop 110 - 200 V, the pressure drop 0.02 - 0.4 atm. and the Mach numbers range from 0.1 to 0.8. The partial characteristics model gives a lower value of radiation loss from the arc than the net emission model, implying higher outlet velocities and temperatures, closer to experimental values. The hybrid arc exhibits higher outlet velocities under the practically unchanged plasma enthalpy compared to Gerdien arc. The contribution of O_2, H_2 and OH molecular bands to the amount of reabsorbed radiation has been also discussed. Comparison between present calculation and available experiments carried out at the Institute shows a good agreement.

Keywords: Electric arc, Water vortex stabilization, Hybrid stabilization, Partial characteristics, Radiation loss
PACS: 52.65.Kj, 52.75.Hn

INTRODUCTION

Plasma generators with arc discharge stabilization by water vortex exhibit special performance characteristics; such as high outlet plasma velocities (~ 7 000 m/s), temperatures (~ 30 000 K), plasma enthalpy and, namely, high powder throughput, compared to commonly used gas-stabilized (Ar, He) torches [1]. In a water-stabilized arc, the stabilizing wall is formed by the inner surface of water vortex which is created by tangential water injection under high pressure (~ 10 atm.) into the arc chamber [Fig. 1].

The so-called *hybrid stabilized electric arc* utilizes combination of gas and vortex stabilization [Fig. 2]. In the hybrid H_2O-Ar plasma torch the arc chamber is divided into the short cathode part, where the arc is stabilized by tangential argon flow, and the longer part is water-vortex stabilized. This arrangement not only provides additional stabilization of the cathode region and protection of the cathode tip, but also offers the possibility of controlling plasma jet characteristics in wider range than that of pure gas or liquid stabilized arcs [2]. The arc is attached to the external water-cooled rotating disc anode a few mm downstream of the torch orifice. Experiments made on this type of torch [2] showed that plasma mass flow rate, velocity and momentum flux in the jet can be controlled by changing mass flow rate in the gas-stabilized section, while thermal characteristics are determined by processes in the water-stabilized section. At present, the both kind of arcs have been used for plasma spraying [3] using metallic or ceramic powders injected into the plasma jet and for the pyrolysis of waste and production of syngas [4].

This work presents numerical simulations of processes and operational parameters in the water vortex and hybrid stabilized electric arcs. Section I gives information about the model assumptions, boundary conditions and the applied numerical scheme. Section II, divided into three subsections, reveals the most important findings. In Section III we summarize the obtained results.

PHYSICAL MODEL AND NUMERICAL APPROACH

The following *assumptions for the water stabilized arc* are applied: 1) the numerical model is two-dimensional with the discharge axis as the axis of symmetry; 2) plasma flow is laminar and compressible in the state of local thermodynamic equilibrium; 3) only self-generated magnetic field by the arc itself is considered; 4) the net emission coefficient and the

CP982, *Complex Systems, 5th International Workshop on Complex Systems*
edited by M. Tokuyama, I. Oppenheim, and H. Nishiyama

partial characteristics methods for radiation loss from the arc are employed; 5) cathode phenomena and space charge near the cathode are neglected. For the *hybrid stabilized electric arc* we assume, in addition, that 6) argon and water create a uniform mixture in the arc chamber.

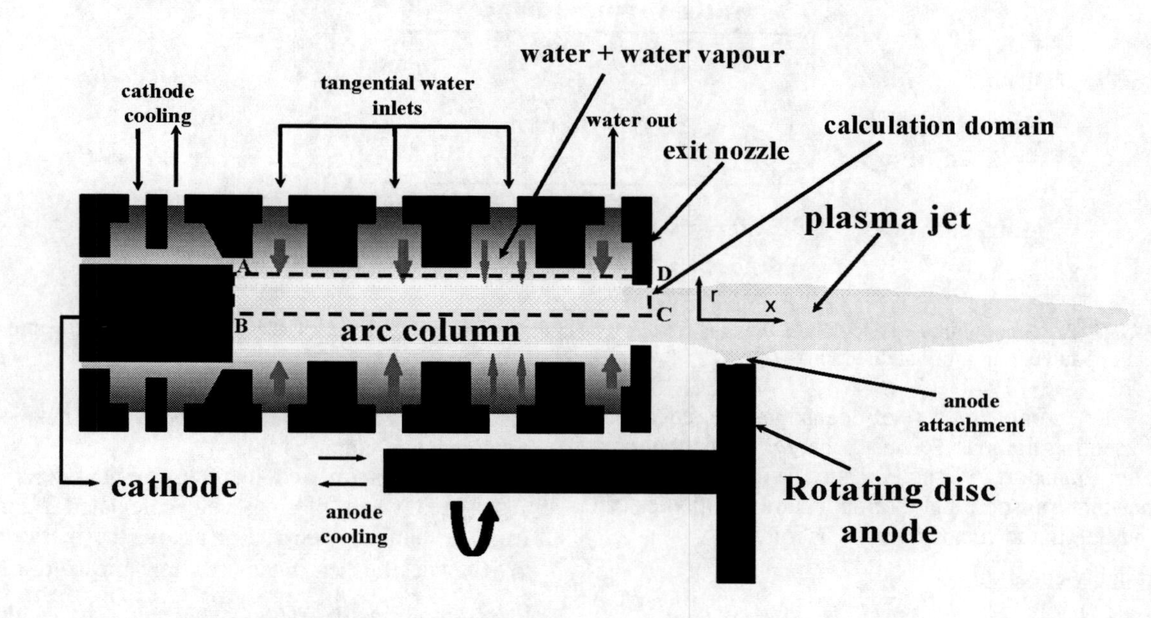

FIGURE 1. (Color online) Principle drawing of the experimental arc chamber of the water vortex stabilized arc. Water is injected tangentially into the sections and creates the vortex in the chamber. The arc burns between the cathode, made of a small piece of zirconium pressed into a copper rod, and the water-cooled anode rotating disc. The calculation domain is shown by a dashed line.

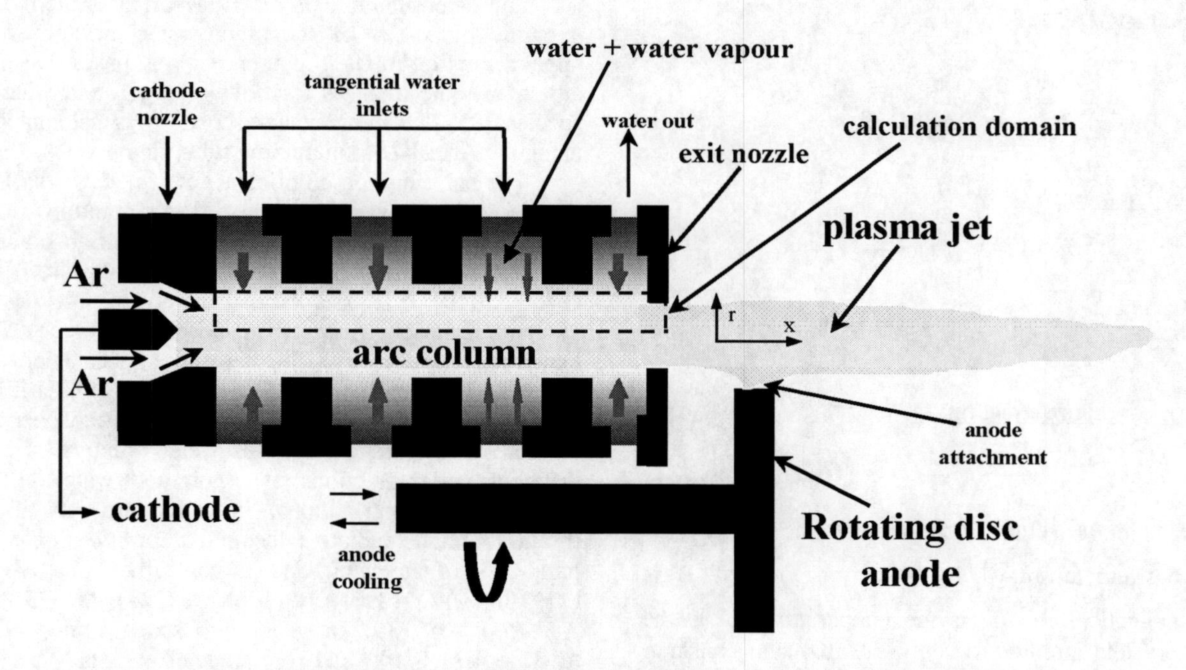

FIGURE 2. (Color online) Principle of the hybrid plasma torch with combined gas (Ar) and vortex (water) stabilizations.

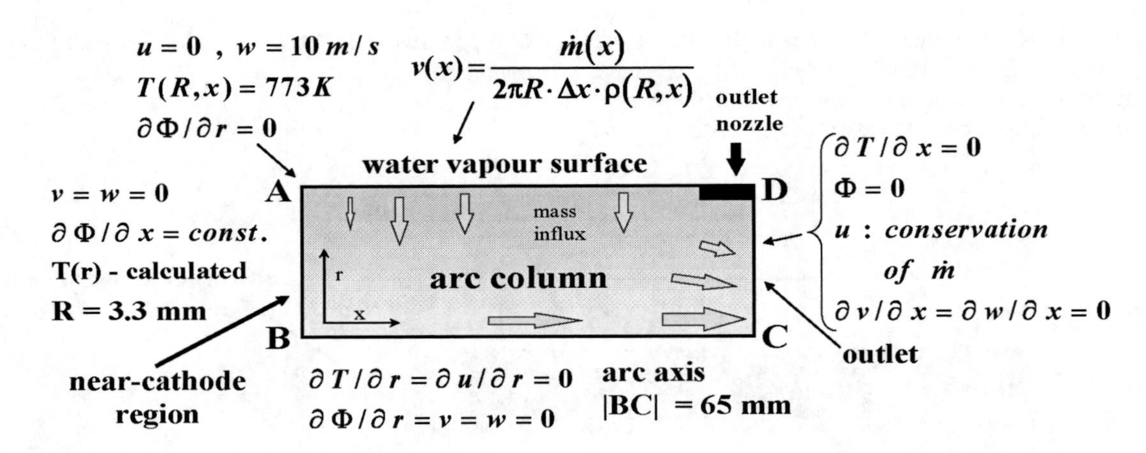

FIGURE 3. (Color online) Calculation domain for the water discharge. Dimensions of the outlet nozzle are 5 mm (axial direction) and 0.3 mm (radial direction).

The complete set of conservation equations representing the mass, electric charge, momentum and energy transport of such plasma with temperature-dependent transport and thermodynamic properties can be written in the vector notation as follows:

continuity equation:

$$\frac{\partial}{\partial t}\rho + \nabla \cdot (\rho \vec{u}) = 0 \quad , \tag{1}$$

momentum equations:

$$\frac{\partial}{\partial t}(\rho \vec{u}) + \nabla \cdot (\rho \vec{u} \vec{u}) = -\nabla p + \nabla \cdot \vec{\tau} + \vec{j} \times \vec{B} \quad ,$$

$$\tau_{ij} = \eta \left(\frac{\partial u_i}{\partial x_j} + \frac{\partial u_j}{\partial x_i} - \frac{2}{3}\delta_{ij}\frac{\partial u_l}{\partial x_l} \right) \quad , \tag{2}$$

energy equation:

$$\frac{\partial}{\partial t}(\rho c_p T) + \nabla \cdot (\rho \vec{u} c_p T) - \frac{\partial p}{\partial t} = -\nabla \cdot (\lambda \nabla T) + \vec{j}\vec{E} + \vec{u}\nabla p + \frac{5}{2}\frac{k}{e}(\vec{j} \cdot \nabla T) + \Phi_{diss} - \dot{R} \quad , \tag{3}$$

charge continuity equation:

$$\nabla \cdot (\sigma \nabla \Phi) = 0 \quad . \tag{4}$$

Here $\vec{u}$ is the velocity vector, p is the pressure, $\vec{\tau}$ is the stress tensor, $\vec{j}$ is the current density, $\vec{B}$ is the magnetic field, T is the temperature, $\vec{E}$ is the electric field strength, k is the Boltzmann constant, e is the elementary charge of electron, Φ_{diss} is the viscous dissipation term and $\dot{R}$ means the divergence of radiation flux.

The transport and thermodynamic properties of argon and water plasma were calculated rigorously from the kinetic theory. For argon, the mass density ρ, the specific heat under constant pressure c_p and the sonic velocity were taken from [5], the thermal conductivity λ, the electrical conductivity σ and the dynamical viscosity η from [6]. For water plasma, the transport and thermodynamic properties are based on the results published in [7].

For determination of the transport and thermodynamic properties of the argon-water mixture we applied mixing rules for non-reacting gases based either on mole or mass fractions of argon and water species [8]. The dynamical viscosity was calculated using the Armaly-Sutton mixing rule [9].

The net emission coefficient for water plasma is composed of three components: the continuum, the resonance lines which are partially self-absorbed within the plasma and the other lines which are not absorbed [10]. It is assumed that the optical thickness of the arc is 3.3 mm, corresponding to the real experimental geometry. The net emission coefficient for argon plasma was taken from [11]. Partial characteristics for plasmas containing atmospheric pressure water and the argon-water mixture with different molar fractions of argon and water [12] include these contributions: 1) continuum radiation (bremsstrahlung, radiative recombination). 2) Several hundreds of oxygen and argon lines, namely, O (93 lines), O^+ (296), O^{2+} (190), Ar (224), Ar^+ (178), Ar^{2+} (53), Ar^{3+} (35). Broadening of spectral lines due to Doppler, Stark and pressure effects has been considered.

Boundary conditions for the problem are represented in Fig. 3. The rectangular calculation region has been chosen with the dimensions of 3.3 mm for the radius and 58.32 mm (65 mm) for the axial coordinate of the hybrid (water) discharge. This domain represents the discharge region of the plasma torch with water vortex type of stabilization. The task was solved numerically by the compressible modification of the control volume method using the iteration procedure SIMPLER [13], elaborated by the author. A non-equidistant rectangular grid with 60 control volumes in the axial direction and 40 in the radial direction was used. Calculation was carried out for currents 150-600 A. Mass flow rate for water-stabilized section of the discharge for each current between 300 and 600 A was taken either from experiment [1] or from our previously published work [14], where it was determined iteratively as a minimum difference between numerical and experimental outlet values of the temperature, velocity and electric potential drop.

RESULTS OF CALCULATION

Water-Stabilized Electric Arc

Calculations for the currents 300, 400, 500 and 600 A have been carried out with evaporation mass flow rates taken from experiments. The net emission coefficient radiation model has been considered here. The typical outlet velocities for this range of currents are 2.5 - 7 km s^{-1}, temperatures 19 000 K - 27 000 K, the voltage and pressure drops are about 200 V and 0.06 - 0.25 atm. respectively. The magnitudes of all calculated characteristics increase with current as can be expected for increasing powers to the arc and evaporation mass flow rates. Despite the axial velocities are very high, the Mach numbers at the point C (Fig. 1) range from 0.3 to 0.7 because of the temperature-dependent speed of sound so the plasma flow is mildly compressible. A satisfactory correspondence between the numerical and available experimental data is obtained. The higher differences are found in temperature, namely for higher currents. The lower numerical values are caused by neglecting reabsorption of radiation in the discharge. The input power $I \cdot \Phi$ increased more than twice for the 600 A arc with respect to the 300 A value. The power loss from the arc is the sum of the radial conduction power and radiation power leaving the discharge, which are considered as the two principal processes responsible for power loss. They change slowly with current and represent around 50 % of the input power.

The temperature contours for all currents reveal that there exists an approximately thermal fully-developed region at the distance between 2 cm and 5 cm from the cathode, i.e. temperature is the function of only the radial coordinate. An indirect implication of this fact is that the current-voltage characteristics of this discharge calculated using the one-dimensional Elenbaas-Heller equation with the radiation source term approximates the real current-voltage characteristics within a few per cent.

The dimensionless fluid-dynamic and heat transfer numbers M, Re, Pr, Pe as a function of current have been also calculated to characterize processes within the arc discharge For higher current, plasma becomes more compressible and the role of thermal diffusion as well as the convection increases. The viscous stresses are more significant at lower currents.

From the set of magnetogasdynamic numbers (the magnetic Reynolds number, the magnetic pressure number, the Hartmann number and the magnetic parameter, i.e. the square root of the ratio: magnetic force/inertial force) it was found that magnetic field is partially influenced by the arc motion (the magnetic Reynolds number is ~ 0.1); the magnetic pressure represents ~ 0.05 - 0.1 of the dynamic pressure; magnetic forces and viscous forces are comparable (the Hartmann number ~ 1) and inertial forces are of the order of 10^3 with respect to magnetic forces.

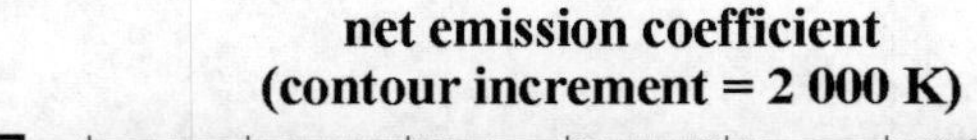

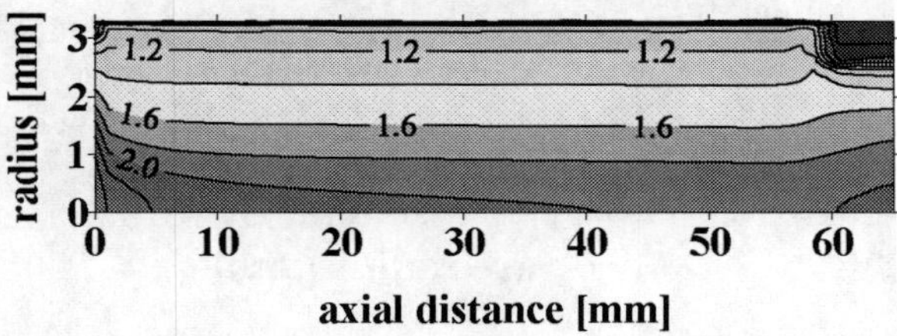

partial characteristics, reabsorption = 14.0 %
(contour increment = 2 000 K)

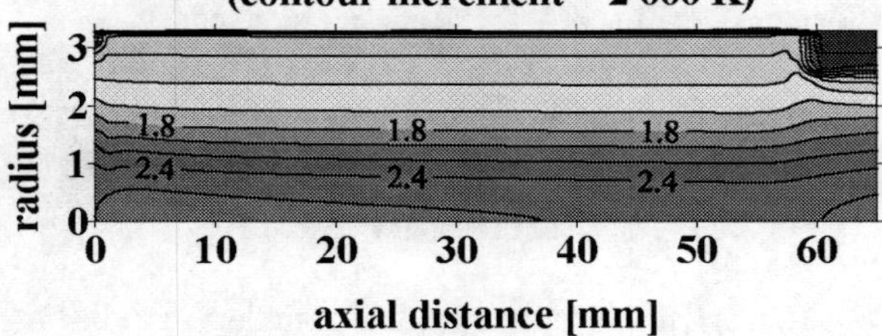

FIGURE 4. (Color online) Temperature contours, 600 A, water mass flow rate 0.363 g s^{-1}. Comparison between the net emission coefficients and partial characteristics models.

Figure 4 shows temperature contours for 600 A and water mass flow rate of 0.363 g s^{-1} for the net emission and partial characteristics models. The axial position 0 mm (65 mm) corresponds to the cathode (outlet). Orientation of the domain is the same as in Fig. 1. It is obvious that temperature in the axial discharge region for the partial characteristics is higher and the maximum difference is about 5 000 K. The total amount of reabsorbed radiation is 14 %. Temperatures obtained with the partial characteristics method are closer to experimental values.

Comparison of radial profiles of temperature at the nozzle exit for the present calculation [15] and experiment [1] shows satisfactory agreement (Fig. 5). The radial position 0 is the axis, 3 mm is the nozzle edge (the CD line in Fig. 1). Radiation loss from the water plasma is calculated in this case by the net emission.

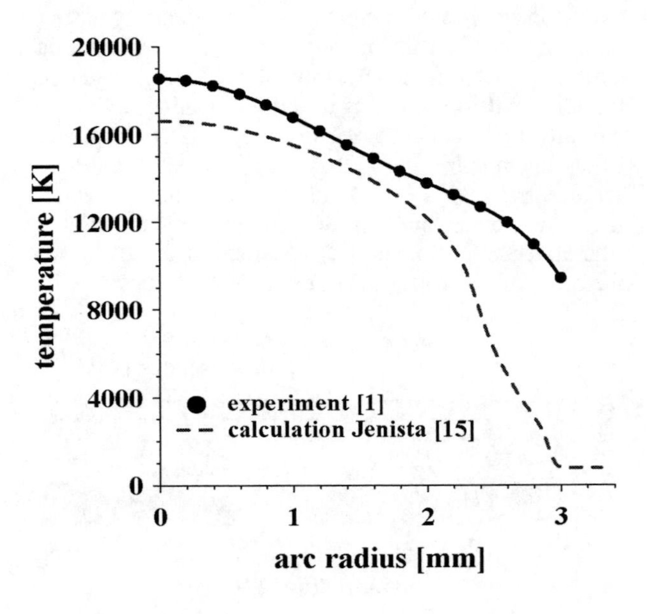

FIGURE 5. Radial distribution of temperature at the nozzle exit for 300 A. A dashed line is calculation [15], a dotted solid line is experiment [1]. The net emission model.

Comparison of Performance Between the Hybrid and Water Vortex Stabilized Electric Arcs

The calculation domain and boundary conditions for the hybrid stabilized arc remain nearly the same as in Fig. 3. The first change concerns geometry – the length of the BC line is now 58.32 mm, the length of the nozzle is 6 mm. The second change is along the AB line which now represents the outlet orifice for the argon (see Fig. 2). Here we specify the inlet argon velocity profile approximately from the one dimensional fully developed flow. The amplitude of this profile has been determined to satisfy experimental values of argon mass flow rate. All necessary transport and thermodynamic properties, as well as partial characteristics, have been recalculated for argon-water mixtures with specified argon molar fraction. Calculations have been carried out for the currents 150, 200, 250, 300, 350, 400, 500 and 600 A. Argon mass flow rate was varied in agreement with experiments in the interval from 7.5 slm to 27.5 slm (standard liter per minute), namely 7.5, 12.5, 17.5, 22.5 and 27.5 slm.

The main results can be summarized as follows:

1) the hybrid-stabilized arc provides higher outlet velocities than the water-stabilized one; the differences range from 400 m s^{-1} to 3 000 m s^{-1}. Comparison with available experimental data shows relative difference below 10 %.

2) Outlet axial temperatures for both kind of discharges are nearly the same due to relatively low enthalpy of argon compared to water plasma.

3) The amount of reabsorbed radiation within the discharge divided to the total divergence of radiation flux is between 35-26 % for the hybrid arc and 25-16 % for the water arc.

4) Mach numbers for currents higher or equal to 500 A for the hybrid arc are higher than 1 in the axial free-stream region near the outlet, i.e. there is a supersonic region in the discharge.

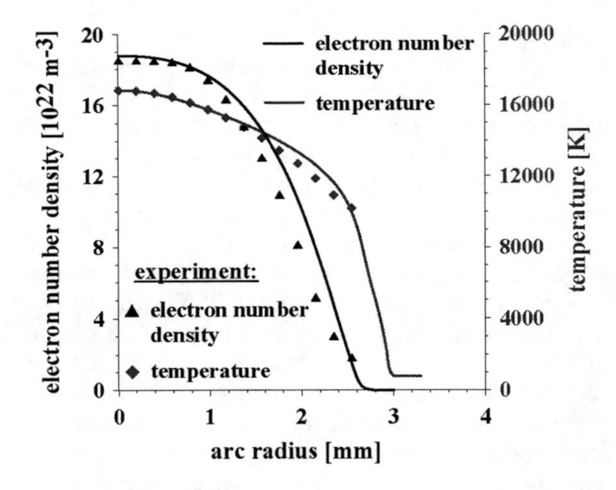

FIGURE 6. Profiles of temperature and electron number density at the nozzle exit for 300 A, argon mass flow rate is 22.5 slm; the partial characteristics method.

Finally, comparison of radial profiles of electron number density and plasma temperature for 300 A and argon mass flow rate of 22.5 slm at the torch exit reveals good agreement between calculation and experiment (Fig. 6) - a few per cent of relative difference occurs in the outer regions, but the two sets of numerical and experimental data nearly overlap in the axial region. Electron number densities for pure water (argon) plasma were taken from pre-calculated equilibrium compositions for atmospheric pressure. Electron number density for a given mixture argon/water was calculated by mole, i.e.

$$N_e(T,p)|_{mixture} = X \cdot N_e(T,p)|_{Ar} + (1-X) \cdot N_e(T,p)|_{H_2O} \quad ,$$

where X denotes argon mole fraction. This approach provides a good estimate of electron density for the case shown here. In experiment [16], the temperature was calculated from the ratios of various argon atomic to ionic line emission coefficients by using Saha equation and the measured electron number density. Concentrations of atomic species at the torch exit were determined from emission coefficients of various argon and oxygen atomic lines and H_β line assuming Boltzmann distribution of atomic level population.

Latest Improvements in the Radiation Model

Recently, the partial characteristics for water and argon-water mixtures have been more elaborated. The following contributions have been added to the original spectra referred in Sec. I:

1) molecular bands of O_2 (Schuman-Runge system), H_2 (Lyman and Verner systems) and OH (transition $A^2\Sigma^+ \rightarrow X^2\Pi_i$).

2) 3 506 new lines of argon atom and ions, so the total number of lines included is now 3 996, namely, Ar (739 lines), Ar^+ (2 781), Ar^{2+} (403), Ar^{3+} (73).

3) Photoionization cross sections for the ground states of Ar atom and its ions are calculated using analytic fits [17], based on the Opacity Project theoretical cross sections [18] interpolated and smoothed over resonances. Dissociation of molecules and radiation from hydrogen lines have been neglected.

We will call this radiation model the *"new" partial characteristics,* while the original model the *"old" partial characteristics*. Calculations carried out for the currents 150-600 A proved that reabsorption of radiation is changed more considerably in the hybrid arc. With the "old" partial characteristics, reabsorption in the hybrid arc ranges between 34.3-19.6 %, while with the "new" partial characteristics it presents 16.3-10.6 %, i.e. the difference is between – 18% and – 7.5 %. The result is shown in Fig. 7, where the difference $\delta = \Psi_{new} - \Psi_{old}$; Ψ_{new} and Ψ_{old} are the amount of reabsorption obtained with the "new" and "old" partial characteristics respectively. The reason for decrease of reabsorption in the hybrid arc consists most probably in a different method applied for calculation of photoionization cross sections for the ground states of Ar atom and its ions [17, 18]. For water arc a nearly-constant positive difference of 3.5 % is obvious for all currents which can be attributed to the influence of molecular bands of H_2, O_2 and OH involved in the radiation model.

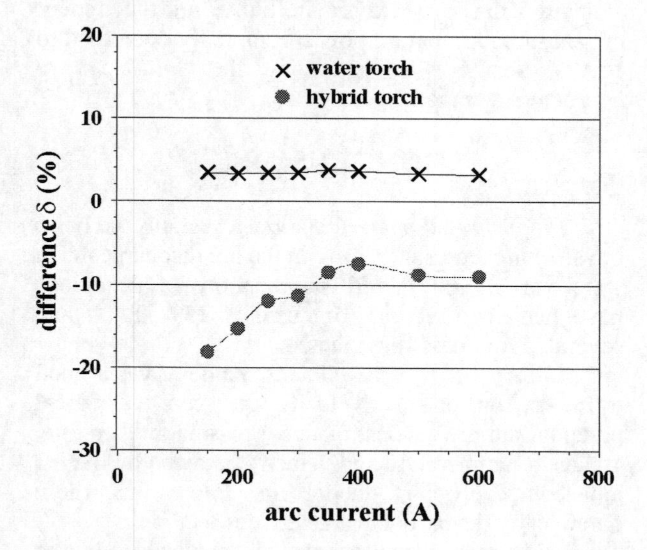

FIGURE 7. Difference δ in the amount of reabsorbed radiation within the arc obtained using the "new" and "old" partial characteristics for water and hybrid plasmas.

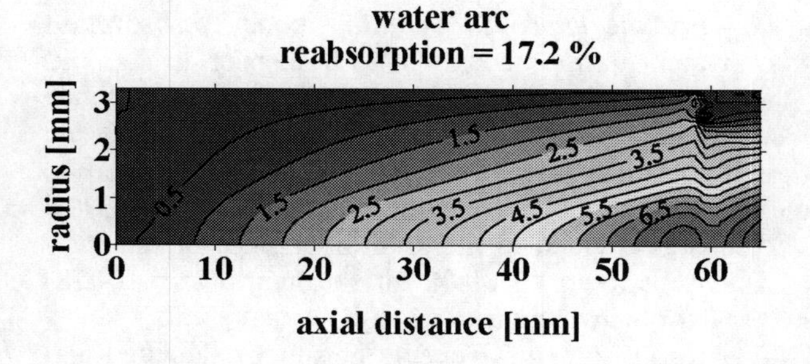

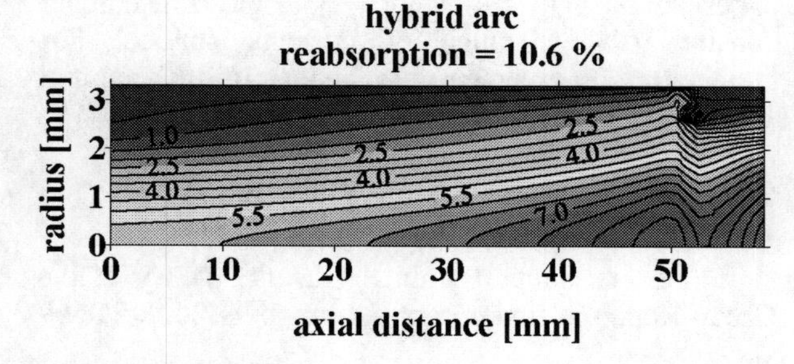

FIGURE 8. (Color online) Contours of the axial velocity for the water and hybrid arcs for 600 A; partial characteristics

method. Mass flow rates: Ar = 0.312 g s^{-1}, water vapor = 0.363 g s^{-1}. 1 000 m s^{-1} corresponds to 1 in the plot.

The calculations also proved that radial distance from the arc axis where the amount of emitted radiation is compensated by reabsorption (the region of zero divergence of radiation flux) range between 2.2 – 2.8 mm and increases with current.

Finally, Fig. 8 presents contours of the axial velocity for the water and hybrid arcs for 600 A. Velocities in the hybrid arc are higher but the amount of reabsorbed radiation now substantially decreased to 10.6%, nearly twice regarding the *"old" partial characteristics* model.

CONCLUSIONS

The proposed numerical model explains the basic physical processes occurring in the arc discharges with water-vortex and hybrid stabilizations. Calculations have been carried out for currents 150-600 A and several argon mass flow rates.

The net emission coefficients model gives a good estimate for plasma velocity, pressure, electrical potential but a worse estimate for plasma temperature. Values of temperature and velocity obtained by the net emission coefficient model are lower that those obtained by the partial characteristics method.

Addition of argon to water plasma leads to higher outlet velocities compared to water vortex stabilized arc. The outlet becomes supersonic in the hybrid torch for currents higher than or equal to 500 A. Outlet plasma temperatures remain practically unchanged.

For the improved radiation model, reabsorption in water plasma increases with respect to the old model due to contribution in the molecular bands about 3.5% nearly independently of arc current. In the hybrid arc, reabsorption is changed more considerably. The amount of reabsorbed radiation within the discharge divided to the total divergence of radiation flux is between 16-11 % for the hybrid arc and 28-17 % for the water arc.

Good agreement occurs between calculated and experimental values of velocity and electron number density for the range of currents studied. For temperature, agreement is better in the partial characteristics method.

ACKNOWLEDGMENTS

Financial support from the Grant Agency of the Czech Republic under contract numbers 202/05/0669 and 202/06/0898 is greatfully acknowledged. The author would like also to thank to the Academy of Sciences of the Czech Republic, v.v.i., and the METACENTRUM project for granting their computational resources.

REFERENCES

1. M. Hrabovský, M. Konrád, M. Kopecký and V. Sember, *IEEE Trans. Plasma Sci.* **25**, 833-839 (1997).
2. V. Březina, M. Hrabovský, M. Konrád, V. Kopecký and V. Sember, "New Plasma Spraying Torch with Combined Gas-Liquid Stabilization of Arc" in *15th Int. Symp. on Plasma Chemistry (ISPC 15)*, Orleans, France, July 9-13, 2001, pp. 1021-1026.
3. P. Chráska, B. Kolman, J. Dubský, P. Ctibor and K. Neufuss, *Acta Technica CSAV* **46**, 323-336 (2001).
4. G. Van Oost, M. Hrabovský, V. Kopecký, M. Konrád, M. Hlína, T. Kavka, A. Chumak, E. Beeckman and J. Verstraeten, *Vacuum* **80**, 1132-1137 (2006).
5. K. S. Drellishak, report *AEDC TDR-64-22*, 10, 1964.
6. R. S. Devoto, *Phys. Fluids* **16**, 616-623 (1973).
7. P. Křenek and M. Hrabovský, "Properties of Thermal Plasma Generated by the Torch with Water Stabilized Arc" in *11th Int. Symp. on Plasma Chemistry (ISPC-11)*, Loughborough, UK, 1993, pp. 315-319.
8. J. M. Bauchire, Ph.D. Thesis. Centre de Physique des Plasmas et de leurs Applications de Toulouse, Université Paul Sabatier, Toulouse, France 1997.
9. G. E. Palmer and M. J. Wright, *J. Thermophysics and Heat Transfer* **17**, 232-239 (2003).
10. A. Gleizes, J. J. Gonzalez and H. Riad, "Net Emission Coeficient for Thermal Plasmas in H_2, O_2, C, H_2O, CF_4 and CH_4", in *12th Int. Symp. on Plasma Chemistry (ISPC 12)*, Minneapolis, USA, August 21-25, 1995, pp. 1731-1736.
11. V. Aubrecht, *Czechoslovak Journal of Physics* **52** (2002), Suppl. D (CD-ROM).
12. V. Aubrecht and M. Bartlová, *Czechoslovak Journal of Physics* **54**, 759-765 (2004).
13. S. V. Patankar, *Numerical Heat Transfer and Fluid Flow*, McGraw-Hill, NY, 1980.
14. J. Jeništa, *J. Phys. D: Appl. Phys.* **36**, 2995-3006 (2003).
15. J. Jeništa, *J. Phys. D: Appl. Phys.* **32,** 2763-2776 (1999).
16. V. Sember, T. Kavka, V. Kopecký and M. Hrabovský, "Comparison of Spectroscopic and Enthalpy Probe Measurements in H_2O-Ar Thermal Plasma Jet" in *16th Int. Symp. on Plasma Chemistry (ISPC 16)*, Taormina, Italy, June 22-27, 2003 (CD-ROM).
17. D. A. Verner, G. J. Ferland and K. T. Korista, *Astrophysical Journal* **465**, 487-498 (1996).
18. W. Cunto, C. Mendoza, F. Ochsenbein and C. J. Zeippen, *Astronomy and Astrophysics* **275**, L5-L8 (1993).

Spreading and Solidification of Hollow Molten Droplet under Its Impact onto Substrate: Computer Simulation and Experiment

Oleg P. Solonenko, Andrey V. Smirnov, and Igor' P. Gulyaev

Khristianovich Institute of Theoretical and Applied Mechanics, Siberian Branch of Russian Academy of Sciences, 4/1 Institutskaya str, Novosibirsk, 630090, RUSSIA

Abstract. The peculiarities of plasma treatment of agglomerated (spray dry) YSZ powder and subsequent plasma spraying of produced HOSP powder (consisting of the hollow spherical particles) are analyzed. Formation of splats of hollow YSZ droplets deserves a special attention within thermal spraying. In this case, immediately prior to particle – substrate collision, we have a droplet that consists of a liquid shell enclosing a gas cavity heated to a temperature close to the melt temperature. The paper presented includes the results of computer simulation and model experiments carried out under full control of the key physical parameters: temperature, velocity, external diameter of droplet, thickness of its shell, and temperature of polished substrate. It was shown that formation of splats of hollow droplets proceeds in a manner more stable compared to the case of "dense" molten particles obtained from fused and crushed compacts, and this provides superior coating-substrate interface.

Keywords: Agglomerate, Spray dry powder, Plasma treatment, Hollow sphere, HOSP powder, Splat, Coating formation
PACS: 52.30-q, 52.77.Fv, 81.15.Rs

INTRODUCTION

HOSP powders of metals, metal alloys and oxide ceramics are of great interest for powder metallurgy, material science and thermal spraying. HOSP powders can be produced by different methods [1-3], etc.

Application of such powders allows creating a new generation of the heat-resistant and thermal-insulation materials with adjustable thermal conductivity. For instance, the advanced materials can be produced of the spherical hollow metal oxide powder particles (ZrO_2, Al_2O_3 etc.) with equal or different particles' size. Absolutely spherical hollow particles of 10-150 μm in size can be produced by thermal plasma processing of spray dry (agglomerated) powders [2]. These materials can be applied for operation in temperature range from 300°C till 1500°C and higher, and will have a binary porosity: (i) porosity appeared due to laying the particles under their joining, and (ii) an own porosity of the particles defined by the inner size of gas cavities. Because of binary porosity, the material will have a low thermal conductivity. Regulation of the porosity will be possible by changing the method of hollow particles packing as well as by applying the HOSP powders with polydisperse particles. Besides, it will be possible to control the thermal conductivity by force of filling of inner cavity of particles using the gases with different thermal conductivity, for instance, argon, helium etc. In this case a heat will be transferred through the material of hollow particles as well as through the gas filling their cavities. The filling of hollow particles by specific gas composition can be realized under plasma treatment of a spray dry powder in controlled atmosphere. As a result of the powder particles' structure and gas presence, the thermal conductivity of such kind of material increases with the rise of temperature as against the other heat-resistant and thermal-insulation materials.

Great interest has been recently expressed in studying the characteristics of thermal barrier coatings (TBC) sprayed from YSZ powders prepared as HOSP powders in comparison with coatings sprayed from agglomerated and sintered powders [4-6], etc.

Nonetheless, the typical morphology of commercially available HOSP powders, illustrated by Fig. 1, fails to provide, in our opinion, sufficiently comprehensive reliable data for such a comparison since commercially available HOSP powders often contain considerable fraction of non-hollow particles, a noticeable fraction of spray dry particles in such powders escapes complete processing, the particles shell thickness varies in a wide range, etc.

In this connection, of great fundamental and practical importance are model studies aimed on investigation of the processes underlying at HOSP powders producing, involving plasma processing of spray dry powders, and evaluation of the potentialities of obtained powders for plasma spraying.

In plasma spraying, the main disadvantage of YSZ powders made up by dense particles is that, even with known particle surface temperature, one cannot be certain that the particles were molten completely in the plasma jet because the temperature gradients inside the

CP982, *Complex Systems, 5th International Workshop on Complex Systems*
edited by M. Tokuyama, I. Oppenheim, and H. Nishiyama

particles are high. The reason for the latter is low thermal conductivity of the particle material and different dwell times of particles in the plasma jet ("effect of trajectory"). The latter makes the aggregate state of the particles spatially non-uniform over the spraying spot, which in turn results in non-optimal characteristics of sprayed coatings.

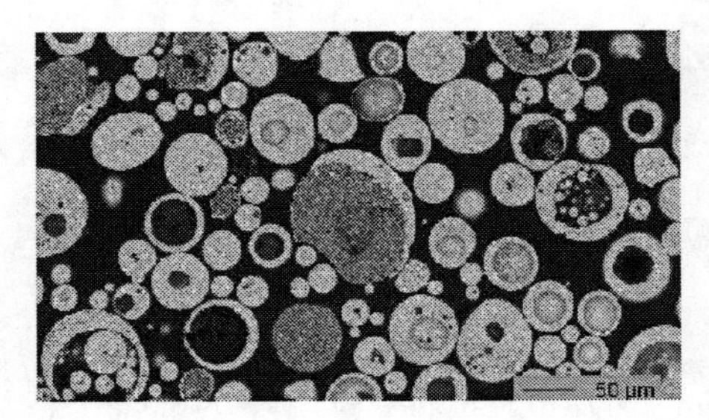

FIGURE 1. Typical morphology of commercially available HOSP powders.

On the contrary, in the spraying of hollow particles-balloons, it can be guaranteed that the 5-10 µm thick shell material in such particles undergoes complete melting, with the temperature being quite uniform across the particle. This provides strong arguments in favor of such powders. The surface temperature of hollow particles is close to their mean-mass temperature, enabling more correct interpretation of experimental data regarding the splat and coating formation process.

The present work is a continuation of our studies commenced in [2, 7], aimed at gaining a deeper insight into the plasma spheroidization of agglomerated particles and splat formation of hollow YSZ droplets under complete control of pre-impact key physical parameters (KPPs), including the impact particle velocity u_{p0}, temperature T_{p0}, diameter D_{p0} and shell thickness Δ_{p0}, as well as the substrate temperature T_{b0} and its surface state. The model physical experiments were carried out on a setup described elsewhere [2].

PECULIARITIES OF AGGLOMERATED POWDERS PLASMA TREATMENT

As mentioned above, different methods are possible to produce the HOSP powder using thermal plasma.

A highly efficient method for obtaining the hollow particles is processing of spray-dried and sintered powder in plasma.

The morphology of so-processed particles is largely defined by process conditions, and also by the thermal and gas-dynamic characteristics of the used plasma flow.

Figure 2 shows photographs of the initial (*a*) and plasma-processed (*b* and *c*) spray dry YSZ powders. Figure 2,*b* refers to the case in which YSZ particles were injected radially into the plasma torch exit region to undergo spheroidization in the nitrogen plasma jet. Figure 2,*c* refers to the case of spheroidization performed in a two-jet plasma torch (Fig. 3). In the latter case, axial injection of the powder was applied, made from the distance L_i=25 mm from the mixing zone of the nitrogen current-conducting jets. In both cases, the thermal power of the plasma flows was about 45 kW, and the powder feed rate was about 20 kg/hour.

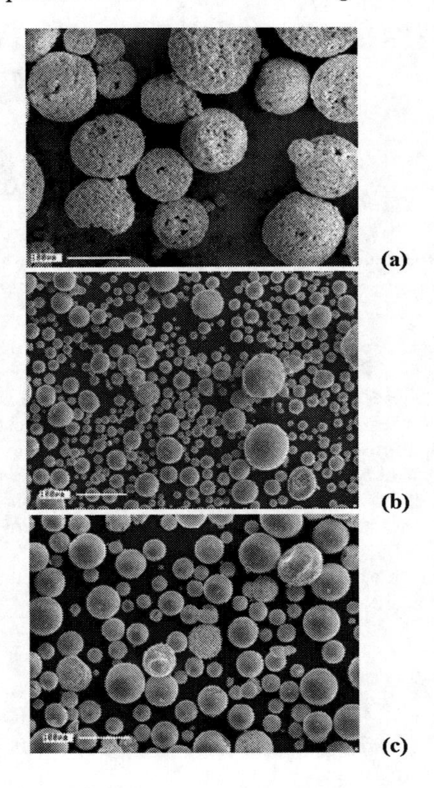

FIGURE 2. SEM images of YSZ powders: (a) – as-received powder; (b) – plasma jet treated powder; (c) – powder processed using the two-jet plasma torch.

In the *first case* the mean diameter of processed particles is seen to be much smaller that in the second one.

In this connection, consider first why the above processes yield powders so differing in particle sizes. First, the particles forming the initial spray dry powder experience in plasma the following mechanical stresses (Fig. 4): 1) the stress due to strain induced by non-uniform thermal expansion of the spray dry material in non-uniformly heated particles, 2) the baric stress due to increased gas and organic-binder-vapor pressure in microvoids present in any spray dry material; the amount of the residual gas seized in a hollow sphere is conditioned by the gas losses due to the radial filtration flow inside spray dry particle that disappears when a continuous outer melt film forms around the particle.

Thus, the spray dry particles of the initial YSZ powder experienced a considerable thermal and baric shocks during their radial injection into the plasma jet (case 1), since the mean-mass temperature of the nitrogen plasma at the nozzle exit plane of the used plasma torch approximated 8000 K. Hence, the initial particles could be disintegrated in several smaller particles undergoing subsequent spheroidization. In the *second case* the particles of the initial powder are subject to rather soft radiative-convective heating for a time of several milliseconds prior to their entering the mixing zone of the current-conducting jets. As a consequence, the most probable pre-treatment scenario for these particles involves sintering of their constituent micro-particles. Under such heating conditions, the spray dry particles display less pronounced tendency toward disintegration.

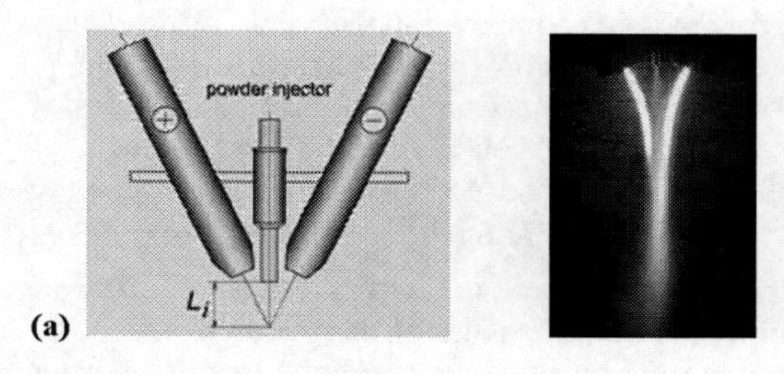

FIGURE 3. Diagram of the two-jet plasma torch with axial injection of powder particles into the mixing zone of the cathode and anode jets - (a), plasma torch in operation – (b).

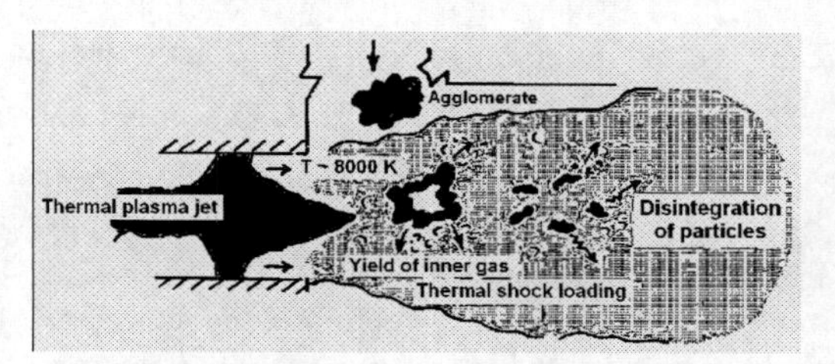

FIGURE 4. Disintegrating spray dry particles in plasma jet.

Two other factors deserve mention which may cause disintegration of already formed hollow melt droplets, thus resulting in decreased particle size of obtained powder. These factors, provoking fractionation of droplets into smaller particles, which subsequently undergo spheroidization and in-flight solidification, are:

- dynamic force acting upon hollow droplets from the side of the carrier gas flow due to the appreciable velocity difference in the "particle – plasma" system;
- gas heating in already formed liquid hollow particles leads to increase of the outer droplet diameter, decrease of the inner gas pressure, thinning and possible rupture of droplet shell.

As far as our knowledge goes, these points were never addressed with reference to hollow droplets; they therefore call for further study.

SPLATS AND COATING FORMATION OF HOSP YSZ POWDER

Micro-metallurgy of splats and plasma coatings formation of HOSP powders, in comparison with powders consisting of dense particles [8], has a number of specific features.

In this connection, it should be noted that, unlike the case of spreading of dense droplets impinging onto substrate, the high-velocity impact of hollow droplets involves adiabatic compression of the gas contained in the droplets since the heat losses to environment and into substrate can be neglected with satisfactory accuracy within characteristic droplet deformation time.

		Impact of the hollow melted particle with substrate at specific key physical parameters.
	Main scenarios	Commentary
1		Deformation, gas compression and consequent rebound at rather small diameter and moderate impact velocity of droplet without its solidification and adhesion to the substrate.
2		Deformation, gas compression and consequent rebound at rather small diameter and moderate impact velocity of droplet, its simultaneous partial solidification on substrate causing some material loss without disintegration of hollow droplet.
3		This scenario is similar to previous one, but due to the higher rate of solidification and larger material loss the hollow droplet disintegrates.
4	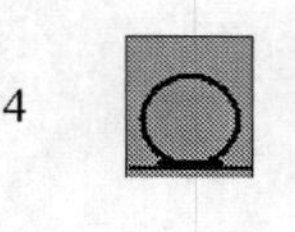	Deformation and gas compression with simultaneous solidification in the vicinity of stagnation point without consequent rebound; as a result of possible oscillations of liquid shell the final shape of splat is a spherical calotte.
5	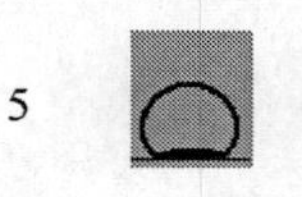	Scenario is similar to previous one, but due to higher rate of solidification at the initial stage of deformation and higher surface tension of melt the droplets' material is accumulated in contact zone.
6	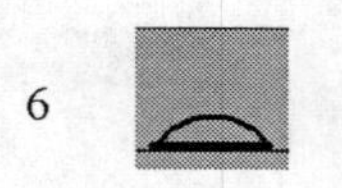	Scenario is similar to previous one, but it is realized at higher rate of solidification and higher droplet velocity.
7	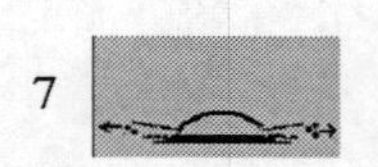	Explosive rupture of droplet liquid shell at the stage of deformation and gas compression causing peripheral jets of gas outflow, realized at increased impact velocity.
8	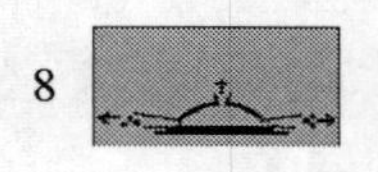	Similar scenario, but accompanied by simultaneous vertical and peripheral jets of gas outflow.
9	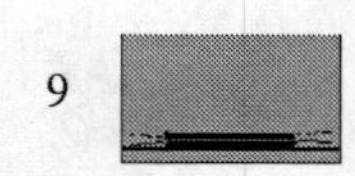	High-velocity impact of hollow droplet causing the intense peripheral gas jets, gas cavity collapse and consequent viscous radial spreading of cylindrical layer with its simultaneous solidification.

FIGURE 5. Schematic presentation of basic scenarios of hollow droplet – substrate interaction.

The basic scenarios, possible at melt hollow droplet impingement with substrate, are presented above (see Fig. 5). Collision of hollow droplets with the substrate under conditions of plasma spraying mainly occurs as follows.

At the *first stage* of the impingement process the lower hemispherical shell comes into contact with the substrate surface and progressively deforms. At the same time, the upper hemispherical shell proceeds the motion with initial velocity.

At the *second stage*, as a rule, at rather high impact

velocity there occurs rupture at the periphery of the upper hemisphere due to the sharp rise of pressure, and this hemisphere then collides with the lower liquid film that spreads and solidifies on the substrate surface. The final splat morphology depends on the particular values of the KPPs.

Let's estimate characteristic time of thermal relaxation of gas (air) cavity $t_g \approx (D_p - 2\Delta_p)^2 / 4a_g$, where Δ_p is a thickness of shell of a hollow particle of diameter D_p, a_g is thermal diffusivity of the gas. For the droplet of D_p=50 μm and Δ_p=5-10 μm in which air cavity is contained with temperature being equal to melting point of YSZ (i.e. $T_p \approx 3000$ K), this time is about $3.5 \cdot 10^{-7}$ sec. Namely, this time is close enough to the typical time of deformation of the impinging droplet with velocity u_p=100 m/sec, $t_d = D_p/u_p \sim 5 \cdot 10^{-7}$ sec.

It is of special importance also to compare characteristic time t_{sol} of solidification of thin melt layer being in the contact with the substrate during impact and that $t_d = D_p/u_p$ of droplet deformation. In accordance with [8], the time of solidification is found from equation $\Delta_p = c_\varsigma \sqrt{a_{pm}^{(l)} t}$, therefore $t_{sol} = \Delta_p^2 / c_\varsigma^2 a_{pm}^{(l)}$, where c_ς is the constant of solidification, and $a_{pm}^{(l)}$ is thermal diffusivity of particle material in liquid state,

$$c_\varsigma = P[\sqrt{1 + 4Q/P^2} - 1]/2, \quad (1)$$

$$P = \frac{\pi \lambda_{p,p}^{(s,l)} \mathrm{Ku}_p^{(l)} + 2\mathrm{K}_\varepsilon^{(b,p)} (\vartheta_{p0} - 1)}{\sqrt{\pi}\, \mathrm{K}_\varepsilon^{(b,p)} \mathrm{Ku}_p^{(l)}}, \quad (2)$$

$$Q = \frac{2\lambda_{p,p}^{(s,l)} (1 - \vartheta_{b0})}{\mathrm{Ku}_p^{(l)}} \cdot \left[1 - \frac{\vartheta_{p0} - 1}{(1 - \vartheta_{b0}) \mathrm{K}_\varepsilon^{(b,p)}}\right]. \quad (3)$$

Here and below, $\vartheta_{p0} = T_{p0}/T_{pm}$, $\vartheta_{b0} = T_{b0}/T_{pm}$. Definition of other parameters in the relations (2) and (3) is possible to find in [8].

It can be shown that t_{sol} is at least two-three values of t_d in the considered case. For instance, the thickness of the shell of YSZ hollow particles obtained after plasma spheroidization was found 5-10 μm. The constant of solidification within the melt temperature range from the melting point up to the boiling one is satisfied the following inequality $1 < c_\varsigma < 2.5$. Consequently, $9.1 \cdot 10^{-6} < t_{sol} < 2.3 \cdot 10^{-4}$ sec, which means that the time of solidification of the lower part of droplets' shell can be varies in a very wide range being subject to specified values of the KPPs. Meanwhile, for 150 μm – particle with velocities within 50 – 200 m/sec prior to impact, $t_d < 3 \cdot 10^{-6}$ sec. One can conclude that there is 2 stage process of splat formation from YSZ hollow droplets. As mentioned above, at the first stage, the lower part of hollow droplet shell flows onto substrate surface. Here this part of the droplet simultaneously solidifies as the front of solidification moves inside the liquid layer. The gas contained in the droplet is adiabatic compressed since environment heat loss is negligible. At the second stage, if the deforming droplet still has a stable form, the upper part of its liquid shell impinges with either solidified or partially liquid film of the solidifying lower part depending on KPPs.

In order to study the process of deformation of hollow particles of YSZ the numerical model of the motion dynamics of particle shell at impingement with substrate was developed. The matter of numerical calculation presents computation the forces of inside gas pressure, surface tension for the elementary section of the shell and following determination of coordinates of given section and its velocity in the next moment of time. The most essential parameters of the particle are u_{p0}, R_p, R_b - the velocity of the particle, its outer radius and radius of the inside gas cavity ("bubble"). It was determined that on the stage of deformation of the gas cavity (up to the break of the shell) the main parameter governing the compression is dimensionless complex

$$\mathrm{We} \cdot \left(1 - \frac{\Delta}{2 - \Delta}\right), \quad (4)$$

where Weber number $\mathrm{We} = 2\rho_p R_p u_{p0}^2 / \sigma_p$ includes ρ_p, σ_p – density and surface tension coefficient at melting point of ZrO_2, $\Delta = (R_p - R_b)/R_p$ – dimensionless thickness of the shell. In the paper [2] it was shown, that HOSP YSZ powders of 45-90 μm in size is characterized by $\Delta = 0.2$. Therefore the complex (4) could be approximated by $\mathrm{We} \cdot (1 - \Delta/2)$. If there is no break of the shell during deformation of the particle, the top of the particle moves down towards the substrate, stops and starts to move backwards. Figure 6 shows dependence of the minimal height of the liquid shells top (related to the initial radius of the gas bubble R_b) and the time needed to reach it (related to the characteristic deformation time R_b/u_{p0}) on characteristic complex (4).

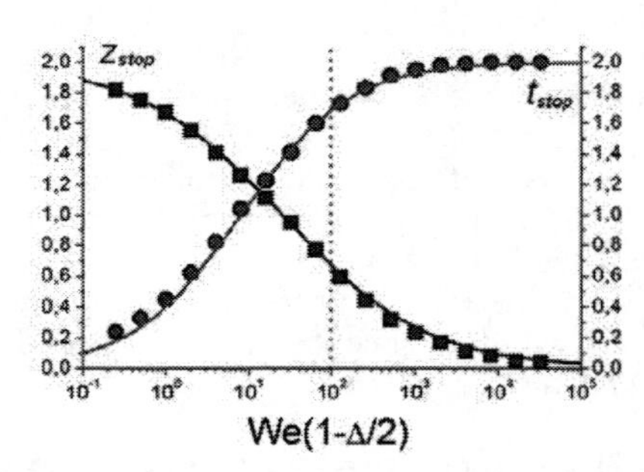

FIGURE 6. Dependence of the parameters of particle top stop on complex $\mathrm{We} \cdot (1 - \Delta/2)$.

In Fig. 6 there are presented also the graphs of approximating curves (5) and (6)

$$z_{stop} = \frac{2}{1 + 0.2\sqrt{\mathrm{We}(1 - \Delta/2)}}, \ (\sigma = 0.02); \quad (5)$$

$$t_{stop} = \frac{2}{1 + 4/\sqrt[3]{[\mathrm{We}(1 - \Delta/2)]^2}}, \ (\sigma = 0.03), \quad (6)$$

where σ is a standard deviation. Vertical dotted line in Fig. 6 separates zone of the parameters, realized under plasma spraying: $\mathrm{We} \cdot (1 - \Delta/2) > 10^2$. For example,

YSZ particle of diameter 50 μm, shell width 5 μm and velocity of 15 m/s has $We\cdot(1-\Delta/2)=130$. Usually in plasma spraying this number belongs to the range 10^3-10^5. When $We\cdot(1-\Delta/2)>10^4$ it is reasonable to assume, that upper part of hollow droplet during its deformation on substrate has the shape of segment of the sphere of initial radius at every moment up to the stop or up to rapture, which is more probable. It is explained by small value of surface tension and gas pressure compared to the value of dynamic pressure, so they are not able to sufficiently change the shape of the upper part of liquid shell at characteristic time of deformation. To the left of dotted line lies area of more elastic collision, occurring at lower velocity and/or lower particle size. Study of this criterial area is of a great interest in connection of possibility obtaining coating formed of "unexploded" hollow particles, which opens up the new potentials of plasma spraying.

It should be noted, that after the break of upper part of a liquid shell the inertial spreading of flat film of the melt on the substrate occurs. At that, the final shape and size of splat will depend on action of viscous and surface tension forces (i.e. Reynolds and Weber numbers) as well as the rate of cooling and solidification of melt.

Regarding the thermophysical phenomena under hollow droplet – substrate interaction, by analogy with dense droplets, the following four basic scenarios of splats formation are possible, depending on contact temperature T_{c0} [8]:

(1) – spreading of the droplet over the solid surface of the base and its simultaneous solidification;

(2) – spreading of the droplet over the surface, its simultaneous solidification, and partial sub-melting of the base at the base/particle contact spot;

(3) – spreading of the liquid droplet over the surface with simultaneous sub-melting of the base, followed by their cooling and solidification;

(4) – spreading of the liquid particle over the solid surface, followed by cooling and solidification of the flattened drop with possible simultaneous recoiling.

Here $\vartheta_{c0}=T_{c0}/T_{pm}$ is the dimensionless contact temperature without taking into consideration the latent heat of melting, $T_{c0}=(T_p+K_\varepsilon^{(b,p)}T_b)/(1+K_\varepsilon^{(b,p)})$, T_{pm} is melting point of particles' material. In consideration of particular scenario, the temperature ϑ_{c0} should be corrected for specific phase transitions [8].

For each of the above-listed scenario, two characteristic regimes of heat transfer between the melt and the solid wall (either the particle solidification front surface $\zeta(r,t)$ or the substrate z=0) are possible depending on the ratio between the thicknesses of the dynamic and thermal boundary layers in the melt flow, whose value, in a first-order approximation, can be estimated as $\delta_v/\delta_a\approx\sqrt{\Pr}$, $\Pr=\nu_{pm}/a_{pm}$ is the Prandtl number, ν_{pm}, a_{pm} is the kinematic viscosity and thermal diffusivity of melt at melting temperature T_{pm}.

To perform the model studies of the splat formation process involving hollow YSZ droplets, we had to thoroughly prepare narrow-fraction HOSP powders. Similar to [2], the starting model HOSP powder was a custom-prepared spray dry YSZ powder which, when processed in plasma, provided for a 100-% yield of hollow particles (see Table 1 and Fig. 7). Note that only spray dry YSZ powder was processed in plasma (Fig. 8).

The effective density of powder particles, evaluated by the procedure [2], was used to determine the average values of the gas cavity diameter and the relative shell thickness of the HOSP particles in each fraction of the powder (Table 1)

$$(D_b/D_p)_{av}=\sqrt[3]{1-\rho_{eff}/\rho_{real}}\,, \qquad (7)$$

$$(\Delta_p/D_p)_{av}=[1-(D_b/D_p)_{av}]/2\,. \qquad (8)$$

Here, $D_{p,av}$, D_b, and Δ_p are the mean particle diameter measured with the help of an SALD–2101 instrument (SHIMADZU, Japan), the gas-cavity diameter, and the shell thickness, respectively; ρ_{eff} and ρ_{real} are the effective and truth densities of ZrO_2 particles. It is seen that, for all fractions of the HOSP powder, the relative shell thickness was about 10% of the outer particle diameter.

Table 1. Mean values of the measured gas-cavity diameter and the particle shell thickness in the obtained narrow fractions of HOSP powder.

D_p, μm	45-50	50-56	63-71	81-90
$D_{p,av}$, μm	49	55	69	85
ρ_{eff}, g/cm^3	3.04	2.89	2.68	2.67
$(D_b/D_p)_{av}$	0.77	0.79	0.80	0.81
$(\Delta/D_p)_{av}$	0.11	0.11	0.10	0.10

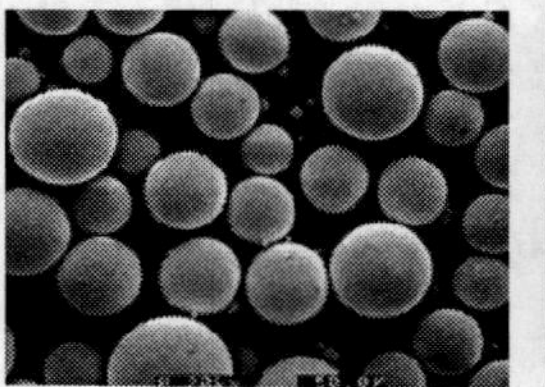
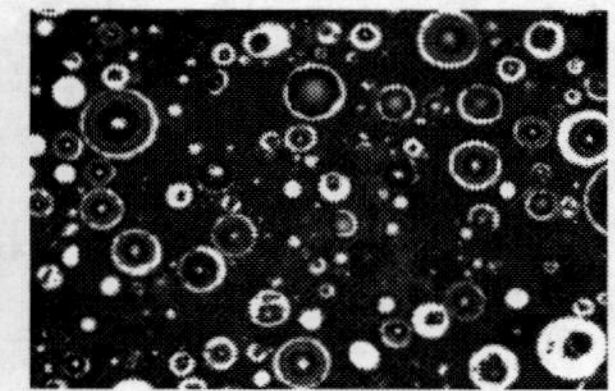

FIGURE 7. General appearance (a) and cross-sectional view (b) of hollow YSZ particles of specially prepared HOSP powder.

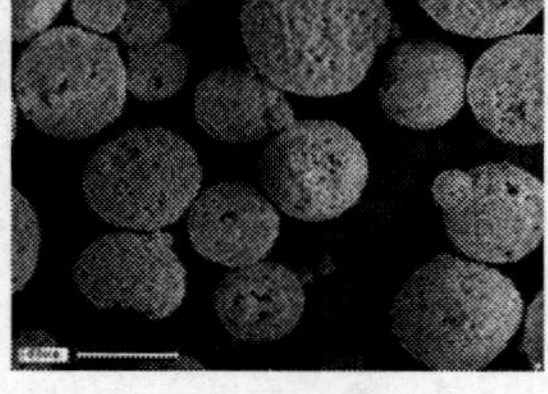

FIGURE 8. Typical view of agglomerated (spray dry) and sintered particles.

To study the impingement of single melted particle onto a substrate under full control of KPPs, we used an experimental setup that included the following components: a 50 kW nominal power air plasma torch with interelectrode insertion [9]; a TWIN-2 powder feeder; a unit for separating out single plasma jet heated particles that consisted of a water-cooled diaphragm, a rotating disk, and a solenoid-driven shutter; a thermally controlled pedestal for mounting the substrate and its controllable heating to a desired temperature; a diagnosing complex for measuring the impact velocity, surface temperature and size of a single particle impinging onto the substrate. A schematic diagram of the setup is shown in Fig. 9.

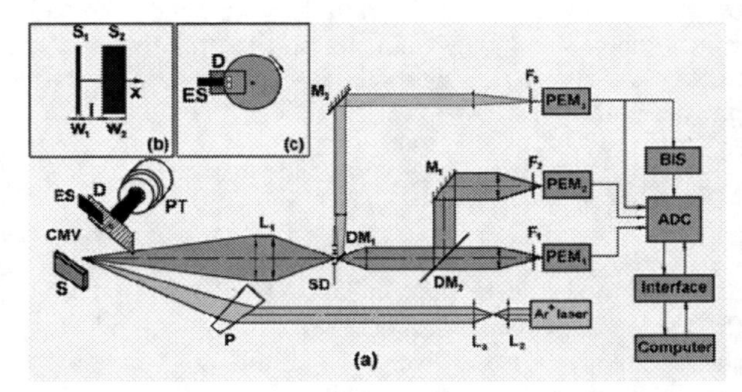

FIGURE 9. Schematic of the experimental setup. (a) – diagnosing complex, (b) – multi-slit diaphragm, (c) – rotating disk with a solenoid-driven shutter. PT – plasma torch, D – water-cooled diaphragm, ES – solenoid-driven shutter, S – substrate; L_1, L_2, L_3 – lenses; P – turn prism; SD – diaphragm; DM_1, DM_2 – dichroic mirrors; M_1, M_2 – mirrors; F_1, F_2, F_3 – light filters; PEM_1, PEM_2, PEM_3 – photoelectronic multipliers; BIS – discriminator; ADC – analog-to-digital converter.

It should be noted here that, in the plasma spraying of HOSP powders with fixed initial solid particle diameters D_{p0} and shell thickness Δ_{p0}, the melting and subsequent overheating of particles may result in increase of the liquid-sphere diameter and in decrease of the shell thickness. The outer diameter D_p and the thickness Δ_p of the liquid shell answering the equilibrium state of the hollow sphere under particular given conditions can be found from the system of equations

$$P_{g\infty} + \frac{2\sigma_p}{R_p}\cdot\left(1+\frac{1}{1-\Delta_p/R_p}\right) - \frac{3m_{g0}RT_g}{4\pi R_p^3 M_g}\left(1-\frac{\Delta_p}{R_p}\right)^{-3} = 0, \tag{9}$$

$$\frac{\Delta_p}{R_p} = 1-\sqrt[3]{1-\frac{3m_{p0}}{4\pi\rho_p R_p^3}}, \tag{10}$$

where, m_{p0}, m_g, T_g, R_g, and M_g are, respectively, mass of follow droplet material, mass of the residual gas inside sphere, the current gas temperature inside the particle, the universal gas constant, and the molecular mass of the gas.

The features of splat morphology are determined by presence of above described processes and strongly depend on specified KPPs prior to impact. In the course of the experiments the hollow particles of above cited sizes impinging at $50 \leq u_{p0} \leq 200$ m/sec with either polished stainless steel or quartz substrates were studied. The particles' temperature here varied from the zirconia melting point (T_{pm}=2960 K) up to 3850 K. The substrates were heated at most $T_{b0} \leq 715$ K.

Without going into unnecessary details one should point out that there are basic scenarios of irregular splat formation as result of competition of droplet deformation, cooling and solidification processes. They are as follows (see Fig. 10, corresponding values of the KPPs are given in Table 2):

(*a*) – formation of "ragged" periphery of the splat because of blowing out of the flat gas jet dragging a part of melt (Fig. 10, # 1-3); in our opinion, such scenario could be dealt with the following factors: 1) the impact of the droplet not being normal one; 2) irregularity of substrate surface, and 3) varying thickness of the liquid shell at the periphery of the expanding droplet owing to rise of its internal pressure; all these factors promote the droplet spreading to become unstable.

(*b*) – splat is formed according to previous scenario, but it should be notified that upper part of the melt film, interacting with lower one, forms quasi second splat layer with its own blowing out of the melt jet (Fig. 10, #5-7); this implies that instability of droplet deformation has occurred at the early stage when upper part of liquid shell being far enough from lower one.

(*c*) – high-speed impingement of overheated droplet with further axially symmetrical flowing of numerous gas radial jets at the splat periphery with melt drag-out ('fingering') (Fig. 10, #8 and 9); this happens at relatively low level of substrate heating; whereas substrate temperature being as much as 600 K and higher allows one to suppress the process of fingering essentially even for under heated particle.

(*d*) – breaking of stable spreading of the lower part of the melt film on low heat-conductivity substrate with simultaneous vertical flowing of gas jet from the top of the droplet being deformed and then solidified (Fig. 10, #4); this is the case when temperature of hollow droplet which comes into collision with rather cold substrate is close to zirconia melting point.

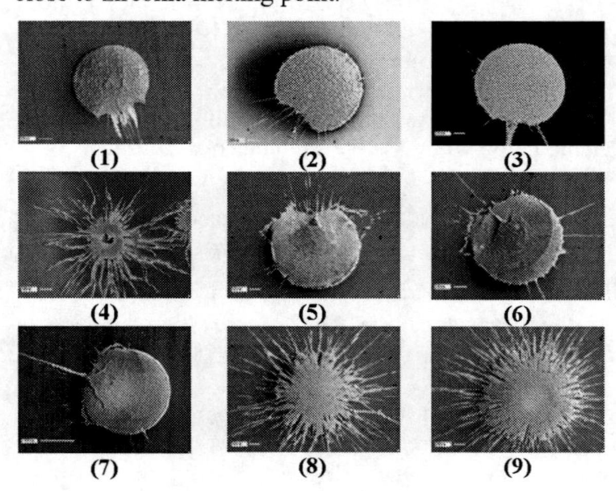

FIGURE 10. Irregular YSZ splats formed as a consequence of jet gas emission at the periphery of flattened hollow droplets.

It is quite clear that analysis conducted is not comprehensive and does not include all factors favoring unstable YSZ splat formation in plasma spraying. The splat morphology is determined by presence of above described processes and in many respects depends on specified KPPs prior to impact.

Table 2. The values of KPPs characterizing the irregular YSZ splats' formation on the stainless steel substrates presented in Fig. 4 (splat #4 has been deposited on quartz substrate).

#	T_{b0}, K	u_{p0}, m/sec	T_{p0}, K	D_p, μm	$D_{s, max}$, μm
		45-50 μm			
1	523	178	2920	74.4	168.6
2	622	170	2920	68.8	227.3
		56-63 μm			
3	582	194	3300	75.8	238.5
4	473	154	2950	74.4	-
		80-90 μm			
5	490	186	3410	75.8	225.6
6	611	178	3180	73.0	251.7
7	686	170	3610	78.6	259.6
8	513	154	3450	77.2	-
9	546	194	3640	74.4	-

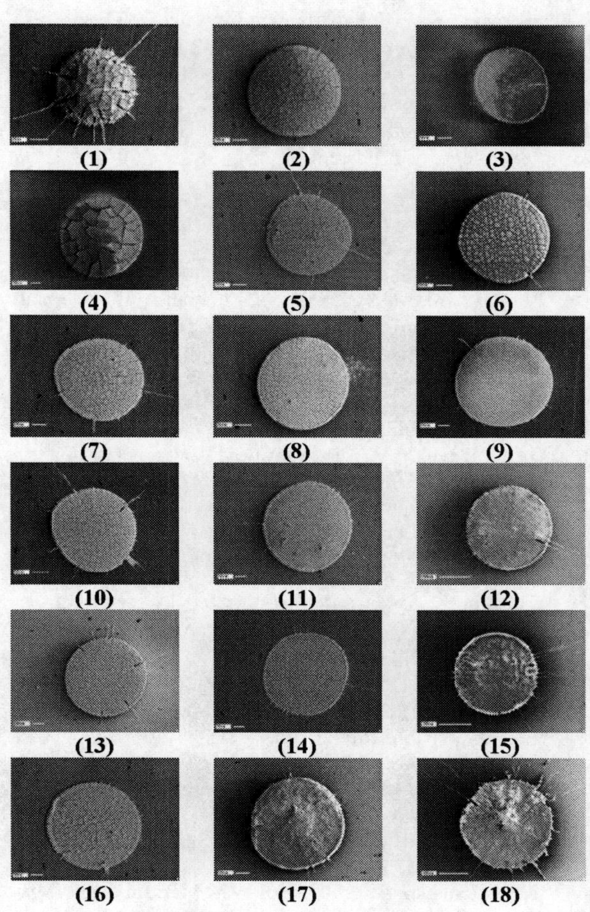

FIGURE 11. YSZ splats with regular structure.

It is of special interest to investigate conditions providing the stable formation of splats for hollow YSZ powder under plasma spraying. In Fig. 11 the photos of the regular splats deposited on the polished substrates at KPPs values noted in Table 3 are presented. The data obtained enable one to conclude that a dominant number of regular splats is set at both higher droplet velocities and elevated temperatures of substrates (splats #5, 6, 8-10 and 12-18). The samples #1 and 2 demonstrate the fact that elevated substrate temperature is valuable factor in order to form regular splats at other factors being equal. It is mainly concerning the splat microstructure.

Table 3. The values of KPPs and splats' diameter characterizing the regular YSZ splats presented in Figure 11. Splats #3 and 4 were deposited onto quartz substrate, all other onto stainless steel substrates.

#	T_{b0}, K	u_{p0}, m/sec	T_{p0}, K	D_{p0}, μm	D_s, μm
		45-50 μm			
1	511	130	2940	68.8	136.5
2	692	130	2810	80.0	191.1
3	499	194	3450	54.8	163.9
4	569	178	2920	60.4	159.7
5	702	178	3130	71.6	194.6
6	716	218	3190	68.8	179.2
		56-63 μm			
7	572	162	3000	70.2	189.7
8	712	186	3190	71.6	171.2
9	596	194	3630	78.6	255.4
10	709	202	3160	77.4	159.4
		80-90 μm			
11	500	146	3060	75.8	265.7
12	596	146	3200	74.4	270.5
13	620	178	3460	75.8	219.0
14	712	178	3510	75.8	243.3
15	679	138	3110	77.2	264.5
16	686	138	3090	77.2	232.2
17	608	186	3430	37.2	224.0
18	608	162	3630	81.4	320.1

In [3], substantial difference was revealed in the formation pattern of YSZ splats from hollow and equivalent dense liquid spheres of identical masses impinging onto substrates at identical impact velocities. In particular, the spreading degrees were shown to differ by factor of two.

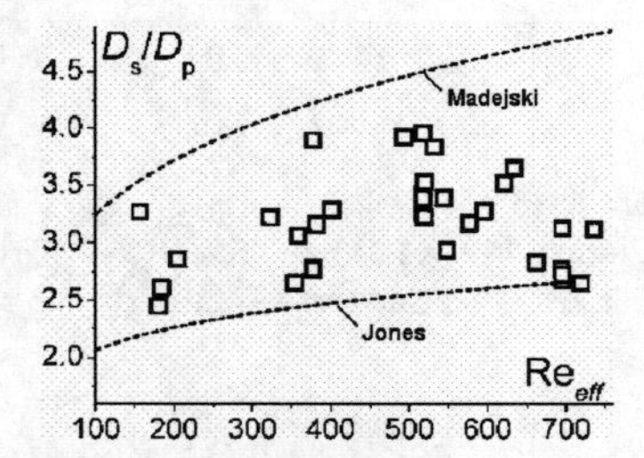

FIGURE 12. Experimentally determined deformation factor of hollow droplets vs Reynolds number (squares), and similar dependences for dense droplets calculated by Madejski [10] and Jones formulas [11].

The diameter of the equivalent dense droplet and the effective Reynolds number based on this diameter were calculated by the formulas

$$D_{p,eq} = \sqrt[3]{D_{p,av}^3 - D_b^3}\,, \tag{11}$$

$$D_b = (D_b / D_p)_{av} \cdot D_{p,av}\,,\ \mathrm{Re} = D_{p,eq} u_{p0} / \nu_{pm}\,, \tag{12}$$

Such comparison is illustrated by Fig. 12. The upper

dashed curve corresponds to Madejski formula $D_s / D_{p,eq} = 1.29\mathrm{Re}_{eff}^{0.2}$ [10], while the lower curve corresponds to Jones $D_s / D_{p,eq} = 1.16\mathrm{Re}_{eff}^{0.125}$ [11].

Figure 13 illustrates our attempt to construct a splat morphology map in the coordinates ($u_p - T_p$). Indicated under each SEM image are the fraction-mean particle size and the dimensionless splat diameter.

The presented splat morphology map illustrates the fact that if there is some initial overheating of the particle over the melting point there is no significant relation between the spreading degree of the particle and the particle temperature. At the same time, for identically sized particles the final splat size is predominantly defined by the impact particle velocity. The formation of radial spikes is defined by particular impingement conditions, including such factors as the local substrate roughness, the distribution of crystallization sites, and the particle shell non-uniformity. It should be noted here that all experiments in the present study were performed using heated substrates since the particles did not stick to the cold substrate. Apparently, the overheating of the particles after which the spreading behavior of droplets and the final splat shape no longer depend on the initial temperature is defined by substrate temperature. At some particle temperature close to the melting points crystallization may proceed simultaneously with particle deformation, the splats in such cases displaying an irregular morphology and structure.

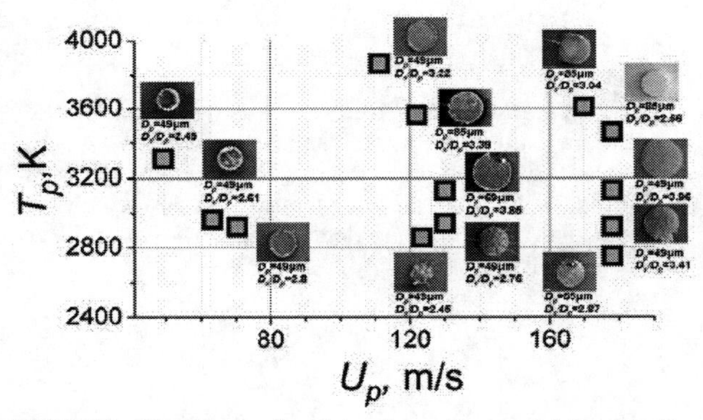

FIGURE 13. Map of splats formed on stainless-steel substrates (T_{b0}=600-700 K) from YSZ HOSP particles with different impact velocities and temperatures (squares).

CONCLUSIONS

From the performed analysis of the process of splat formation from droplets of YSZ HOSP powders the following conclusions can be drawn concerning the advantages of such powders over fused and crushed powders:

1) in spraying of HOSP powders with the help of low-velocity plasma jet and, in particular, with the help of RF – plasma, it is possible to obtain the coatings with controllable porosity whose magnitude can be varied in a broad range (up to 30% and greater);

2) at atmospheric plasma spraying, acceleration of hollow particles to velocities far exceeding the velocity of dense particles of the same mass is possible. With high Reynolds numbers of the plasma jet, this seems to be capable of ensuring a record-breaking low coating porosity (near 1-3%);

3) in contrast to dense particles of spray dry powders, which guaranteed spheroidization can only be achieved with plasma torches producing extended high-enthalpy plasma jets with low Reynolds numbers, the requirements to plasma torches for spraying coatings from hollow oxide powders are much less stringent.

Since the splat formation process involving droplets of HOSP particles proceeds, following their high-velocity impact deformation on the base/substrate, in an explosive manner, this type of spraying can be termed the bubble–explosive thermal spraying (BETS) or bubble–explosive plasma spraying (BEPS).

ACKNOWLEDGMENTS

The present work was supported in part by Presidium of SB RAS under Program No. 8 for the years 2006-2007 (Project No. 6), and also by the Siberian Branch of the Russian Academy of Sciences and the Ukrainian National Academy of Sciences under Complex Integration Project No. 2.9.

REFERENCES

1. M.F. Zhukov and O.P. Solonenko, *Plasma Jets in the Powder Materials Treatment*, Institute of Thermal Physics USSR Academy of Sciences, Novosibirsk, 1990, 516 p. (in Russian).
2. O.P. Solonenko, A.A. Mikhalchenko, E.V. Kartaev, *Proc. of ITSC'05*, 2-4 May 2005, Bazel, Switzerland (Electronic publication).
3. O.P. Solonenko, V.A. Poluboyarov, A.N. Cherepanov et. al, *Proc. of ISPC18*, 26-31 August 2007, Kyoto, Japan (Electronic publication).
4. Y. Tan, J.P. Longtin, S. Sampath, *Proc. of ITSC'06*, 15-18 May 2006, Seattle, Washington, USA (Electronic publication).
5. F. Ladru, H. Reymann, M. Mensing et al, *Proc. of ITSC'06*, 15-18 May 2006, Seattle, Washington, USA (Electronic publication).
6. N. Markocsan, P. Nylen, J. Wigren, X.-H. Li, *Proc. of ITSC'06*, 15-18 May 2006, Seattle, Washington, USA (Electronic publication).
7. O.P. Solonenko, I.P. Gulyaev, A.V. Smirnov, E.V. Kartaev, *Proc. of ISPC18*, 26-31 August 2007, Kyoto, Japan (Electronic publication).
8. O.P. Solonenko, V.V. Kudinov, A.V. Smirnov et al, *JSME Int. J. Ser. B* **48**, 3, 366-380 (2005).
9. O.P. Solonenko, A.P. Alkhimov, V.V. Marusin et al., *High Energy Processes of Materials Treatment*, Novosibirsk, Nauka, Siberian Publishing House of RAS, 2000, pp. 63-163 (in Russian).
10. J. Madejski, *Int.. J. Heat Mass Transfer* **19**, 1009 (1976).
11. H. Jones, *J. Phys. D: Appl. Phys.* **4**, 1657 (1971).

Mathematical Modelling of Advanced Thermal Plasma Reactors and Application to Nanoparticle Production

Pierre Proulx and Mbark El Morsli

Department of Chemical Engineering, Universite de Sherbrooke, Canada.

Abstract. When using plasma torches or reactors, the high thermal energy available for melting and evaporation of refractory materials is a significant advantage over conventional flame or furnace heating because of the higher temperatures in part but also because of the flexibility to use oxidizing, neutral or reducing environment at will. It is therefore possible to use plasmas produce nonoxide materials, carbides, nitrides as well as pure metal nanoparticles. The rapid chemistry and active species characteristics of plasmas offers chemical routes that differ largely from more conventional chemical processing. It also offers the possibility of creating materials that are far from thermodynamic equilibrium through the use of rapid quenching. With the high temperatures obtained in thermal plasmas, combined to the possibility to control the reactivity of the plasma gas, processing of complex nanopowders can be obtained within a single processing unit. In the present study, it is found by mathematical modelling that a double induction coil could be used to improve the material feed in the inductively coupled plasma, thus improving the treatment of the material. Furthermore it is also found the the mathematical model indicates that the supersonic plasma expansions could offer significant advantages over subsonic plasma processing.

Keywords: Induction torch, Classical single phase and double phase induction, Electromagnetic, Supersonic flow, Plasma, Nanoparticles
PACS: 50., 51., 51.30.+i, 51.35.+a

INTRODUCTION

In a nanoparticle thermal plasma reactor, when a particle is created in a section of the plasma and gets recirculated, its trajectory in the different sections of the reactor is reflected through changes in morphology (aggregation, agglomeration, sintering, phase transformation, etc,) so that the nanoparticle reactor overall production keeps track of the trajectories of the nanoparticles. In the innovative induction system [1] based on the original ideas of [2], the feedstock powder moves on the contrary way compared with the classical single phase induction system, meaning that it moves on a centripetal way from the peripheral walls to the plasma central area where it melts easily, which implies an improvement of the treatment quality. The importance of the particle trajectories, stream lines and temperature histories in the nanoparticle plasma reactor is therefore of great importance in the design of such reactors [3], [4]. The industrial interest in nanoparticle production using thermal plasmas has lead, in the last few years, to a number of scientific papers addressing the possibility of creating various types of materials using thermal plasma technology [5]. It has brought even more studies directed more specifically the design of better plasma reactors, with the size, morphology, chemical and structural aspects of the powders as a function of the operating conditions of the reactors. Flow rates of powders, reactants, quench configuration, plasma power, reactor pressures and other design parameters have been studied using on-line diagnostic of the plasma reactor, as well as on-line monitoring of the powders. In particular, Inductively Coupled Plasmas (ICP) offer unique advantages for the synthesis of many types of nanopowders, for example advanced ceramic powders, due to their high temperatures and energy densities and electrode free power generation. In addition, a high temperature gradient exists between the hot plasma flame and the surrounding gas phase, a situation which may be enhanced through quench gas injection. The rapid evolution of the computational fluid dynamics (CFD) and the so-called multi-physics modelling methods, including the powerful methods describing the evolution of size and morphology of the nanopowders, can be applied to the thermal plasma reactors. These modelling tools are becoming available not only to the specialists but to a larger and larger body of process engineers, so that design studies of plasma nanopowder using CFD are beginning to be much more used in industry.

The present work presents the state of current scientific progress in the modelling of nanoparticle production through induction thermal plasma technology, including more traditional subsonic inductively coupled plasma torches and also supersonic plasma torches. Indeed, supersonic plasma expansions offer significant advantages in many applications such as projection by plasma and the deposition by thermal evaporation of the materials. The supersonic plasma jet is already a useful tool used in the spraying technology to produce high-quality coatings, and the interest for its use in nanopowder production is growing steadily. Modelling results are also presented, dealing with an alternative method to manipulate the plasma recirculation eddies that could be used to con-

CP982, *Complex Systems, 5th International Workshop on Complex Systems*
edited by M. Tokuyama, I. Oppenheim, and H. Nishiyama

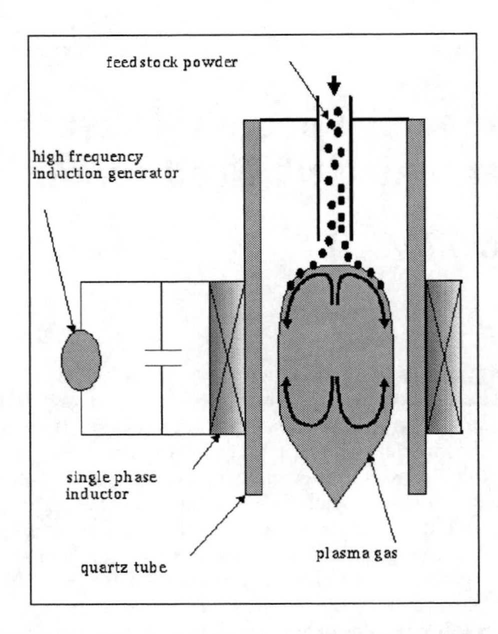

FIGURE 1. Classical induction configuration (Ref. [1] and [2]).

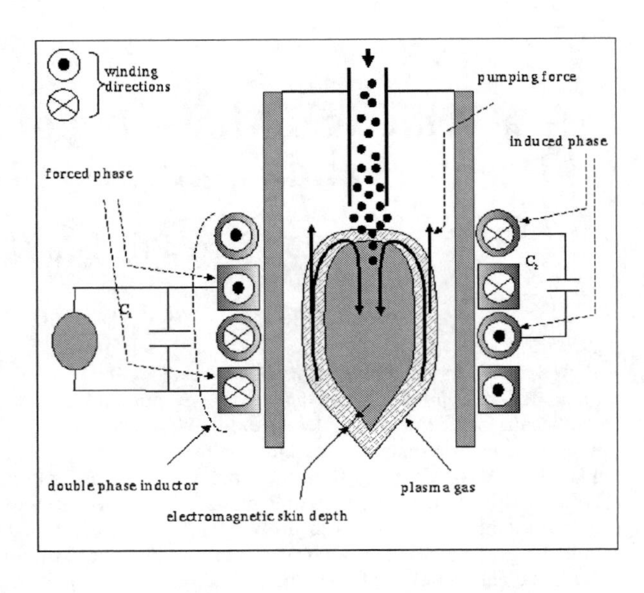

FIGURE 2. Double phase induction configuration (Ref. [1] and [2]).

trol nanoparticles sizes and morphology, an alternative that is inspired largely by the work of [2] on inductive mixing of liquid metals and the recent work of [3] on the configuration effect of advanced inductively torches.

In inductive plasma torch installations, the direction of the recirculation eddy in the upper section of the coil is centrifugal (Fig. 1), causing the feedstock powder to move from the center of the plasma to the peripheral wall which is generally cold compared with the central plasma area. The solution to this problem is to increase the flow rate of the carrier gas in order to penetrate this backflow, or to introduce the powder injection probe into the plasma flame itself. This in turn decreases the residence time of the particles in the hot plasma zone, so that the engineer has to choose an optimum between the desired central particulate trajectories and sufficient residence time. In the double induction system that is proposed in the following, a vertical ascending electromagnetic pumping force is created in the peripheral plasma electromagnetic skin depth. This pumping force (Fig. 2) can invert the stirring direction on the upper plasma zone so that the feedstock powder moves on the contrary way compared with the classical single phase induction system [2]. This modification of the flow fields could mean a much easier and better treatment of the material in the plasma.

BASIC ELECTROMAGNETIC EQUATIONS AND SIMPLIFYING ASSUMPTIONS

The description of the new double phase inductor configuration is well detailed by [1]. The double phase inductor is made of two oscillating circuits and then several resonance frequencies are possible, which makes the determination of the working conditions for the generator more complicated. In order to determine theses resonance frequencies, the electric equivalent diagram of the inductor given by Fig. 3 is considered as presented by [2].

In this simplified diagram, it is admitted that if the plasma charge is present, the two primary and secondary resistances and inductances and resistances R_1, R_2, L_1 and L_2 include the coupling on this plasma charge. Figure 3 shows that the whole system is made of two parallel impedances: Z_1 which is the impedance of the C_1 capacitor, and Z_2 which is the impedance of the set of the two inductors whose second one (the induced phase) is connected to the C_2 capacitor. Z_1 and Z_2 are easily obtained after writing the Kirchhoff's equations which give also the secondary to primary currents ratio called also the modulus ratio ($K = I_2/I_1$):

$$Z_1 = \frac{1}{C_1 \omega} \tag{1}$$

$$Z_2 = Z_{2r} + iZ_{2i} \tag{2}$$

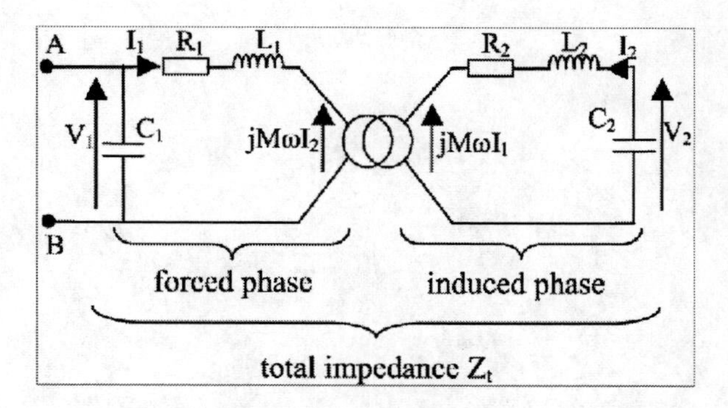

FIGURE 3. Electrical equivalent (Ref. [2]).

with

$$Z_{2r} = R_1 + \frac{M^2\omega^2}{R_2^2 + \left(L_2\omega - \frac{1}{C_2\omega}\right)^2} R_2 \tag{3}$$

and

$$Z_{2i} = L_1\omega - \frac{M^2\omega^2\left(L_2\omega - \frac{1}{C_2\omega}\right)}{R_2^2 + \left(L_2\omega - \frac{1}{C_2\omega}\right)^2} \tag{4}$$

$$\frac{\vec{I}_2}{\vec{I}_1} = \frac{-iM\omega}{R_2 + i\left(L_2\omega - \frac{1}{C_2\omega}\right)} \tag{5}$$

In the above equations M is the mutual inductance:

$$|M| = \frac{K}{\omega}\sqrt{R_2^2 + \left(L_2\omega - \frac{1}{C_2\omega}\right)^2} \tag{6}$$

The whole system is equivalent to a total impedance Z_t which is "seen" by the generator. As this circuit is made of two coupled oscillating circuits, there are several possible resonance frequencies. In order to determine the working conditions for the generator, it is convenient to express the evolution of Z_t with respect to the pulsation ω. An analytical calculation is made with some simplifications like assuming that the primary elements and the secondary elements are similar ($R = R_1 = R_2$, $L = L_1 = L_2$, $C = C_1 = C_2$) and neglecting the resistive term (like R) compared with the reactive terms (like $L\omega$, $1/C\omega$ or $M\omega$). The expression of Z_t are as follow:

$$\frac{Z_1 * Z_2}{Z_1 + Z_2} = Z_{tr} + Z_{ti} \tag{7}$$

Z_t is a pure imaginary number ($Z_{tr} = 0$). The resonance frequencies are given by equating Z_t (or Z_{ti}) to zero. The method of calculation of these frequencies are described by [2]. The values of the possible resonance pulsations are listed in Table 1.

TABLE 1. Values of the possible resonance pulsations.

ω_1	ω_2	ω_3	ω_4
$\frac{1}{\sqrt{(L+M)C}}$	$\frac{1}{\sqrt{LC}}$	$\frac{1}{\sqrt{LC\left(1-\left(\frac{M}{L}\right)^2\right)}}$	$\frac{1}{\sqrt{(L-M)C}}$

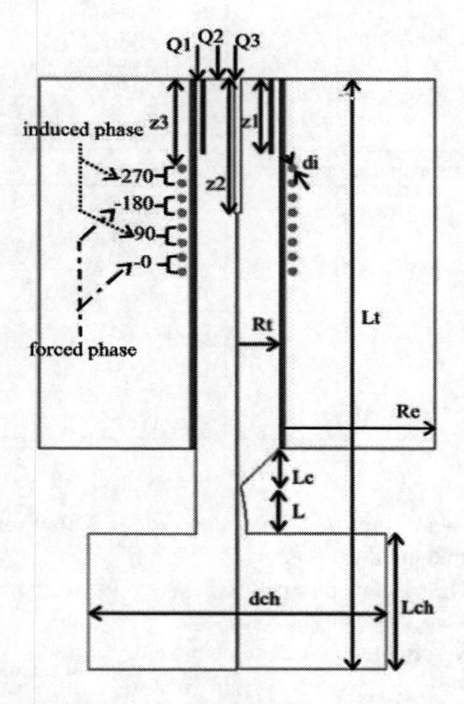

FIGURE 4. (Color online) Torch geometry.

NUMERICAL MODEL

A commercial Tekna plasma torch, PL-50, is employed for the numerical simulations. The torch geometry are shown in Fig. 4 and its dimensions (in meter) are listed in Table 2. The metal tube inserted centrally into the plasma for injecting the particles has a inner diameter of 4 mm and an outer diameter of 9 mm. Eight-turn coil are assumed for the torch rather than the normally used 5-turn coil. The phase angles of the coil current are shown in the Fig. 4. For performing the simulation the following assumptions have been taken:

Hypothesis

The basic assumptions used in the present work are as follow :

1. The plasma flow is axisymetric, steady and turbulent of pure argon.

TABLE 2. Characteristics of the torch, principal dimensions and operating conditions.

Torch dimension	
$d_0 = 0.0982(m)$	$dc = 0.03(m)$
$dch = 0.07(m)$	$di = 0.006(m)$
$L = 0.0326(m)$	$Lch = 0.08(m)$
$Lc = 0.012(m)$	$Lt = 0.3746(m)$
$Re = 0.085(m)$	$R_{tor} = 0.0175(m)$
$Z_1 = 0.05(m)$	$Z_2 = 0.09(m)$
$P_{Ohmic} = 16(kW)$	$f = 300(kH_z)$
$Z_3 = 0.057(m)$	
Operating conditions	
Supersonic torch	$Q_1 = 10^{-4}\frac{Kg}{s}$ $Q_2 = 1.74 * 10^{-3}\frac{Kg}{s}$ $Q_3 = 1.74 * 10^{-3}\frac{Kg}{s}$ Pressure = 1800 Pa
Subsonic torch	$Q_1 = 1.89 * 10^{-4}\frac{Kg}{s}$ $Q_2 = 8 * 10^{-4}\frac{Kg}{s}$ $Q_3 = 2.7 * 10^{-3}\frac{Kg}{s}$ Pressure = 101325 Pa

2. Two-dimensional velocity, temperature and EM field conditions apply.
3. The viscous dissipation is taken into account in the energy equation.
4. Negligible displacement current.
5. Plasma is optically thin, under Local Thermodynamic Equilibrium (LTE) condition.
6. Coil is transparent for electromagnetic field by assuming the zero coil electrical conductivity.
7. The current density is uniformly distributed on the cross-section of rings of coil.

Governing Eequations

Using the hypothesis mentioned above, the argon supersonic ICP mathematical model can be developed, giving rise to the governing equations described in the following :

Mass Conservation

$$\frac{\partial(\rho u)}{\partial z} + \frac{1}{r}\frac{\partial(r\rho v)}{\partial r} = 0 \quad (8)$$

Momentum

$$\frac{\partial(\rho uu)}{\partial z} + \frac{1}{r}\frac{\partial(r\rho vu)}{\partial r} = \frac{\partial}{\partial z}\left(\mu_{eff}\frac{\partial u}{\partial z}\right) + \frac{1}{r}\frac{\partial}{\partial r}\left(r\mu_{eff}\frac{\partial u}{\partial r}\right) - \frac{\partial P}{\partial z} + \frac{\partial}{\partial z}\left(\mu_{eff}\frac{\partial u}{\partial z}\right) + \frac{1}{r}\frac{\partial}{\partial r}\left(r\mu_{eff}\frac{\partial v}{\partial z}\right) + F_z \quad (9)$$

$$\frac{\partial(\rho uv)}{\partial z} + \frac{1}{r}\frac{\partial(r\rho vv)}{\partial r} = \frac{\partial}{\partial z}\left(\mu_{eff}\frac{\partial v}{\partial z}\right) + \frac{1}{r}\frac{\partial}{\partial r}\left(r\mu_{eff}\frac{\partial v}{\partial r}\right) - \frac{\partial P}{\partial r} + \frac{\partial}{\partial z}\left(\mu_{eff}\frac{\partial u}{\partial r}\right) + \frac{1}{r}\frac{\partial}{\partial r}\left(r\mu_{eff}\frac{\partial v}{\partial r}\right) - 2\mu_{eff}\frac{v}{r^2} + F_r \quad (10)$$

where u and v are, respectively, the axial and radial velocity components. F_z and F_r are the axial and the radial components of the Lorentz force. $\mu_{eff} = \mu_l + \mu_t$ is the effective viscosity, which is the combination of molecular and turbulent viscosities (μ_l and μ_t) and ρ is the density.

Energy

$$\nabla\cdot(\vec{v}(\rho E + p)) = \nabla\cdot\left(\lambda_{\text{eff}}\nabla T + (\overline{\overline{\tau}}_{\text{eff}}\cdot\vec{v})\right) + P_J - P_{rad} \quad (11)$$

where $\lambda_{\text{eff}} = \lambda + \lambda_t$ is the effective conductivity. λ_t is the turbulent thermal conductivity. C_p is the specific heat at constant pressure. The first two terms on the right-hand side of 11 represent energy transfer due to conduction and viscous dissipation, respectively. P_J and P_{rad} are the volumetric Joule heating and the volumetric radiation heat loss, respectively.

In equation (11), the total energy E is related to the static enthalpy by the following relation :

$$E = h - \frac{p}{\rho} + \frac{|V|^2}{2} \quad (12)$$

Vector Potential

Two equations are required to describe the flow of electric charges. The first is the current continuity equation and is given by:

$$\nabla\cdot j = 0 \quad (13)$$

and the second is one of Maxwell's equations:

$$\nabla \times B = \mu_0 j \tag{14}$$

where j is the current density, B is the magnetic field induced by the current and μ_0 is the permeability of free space.

In order to simplify the solution of the above electromagnetic equations, the vector potential formulation is used to obtain electromagnetic field.

$$\nabla^2 A = -\mu_0 \left(J_{coil} + J_{ind}\right) \tag{15}$$

where A is the vector potential, J_{coil} is the current density induced in the coil by the voltage applied to the ends of the coil and J_{ind} is the current density developed in the plasma and the coil by the induced electric field. If J_{coil} is given, A can be calculated from equation (15), since J_{ind} can be expressed in terms of A. Assuming that all the fields A, E and H are sinusoidal with a frequency f in the r.f. plasma, the time variation of A and J can be omitted and the equation (15) remains the same but A and J will not change with time.

Assuming that the coil is made of parallel rings (an hypothesis necessary to keep the symmetry, the vector potential A and the electric field E has only an azimuthal component $A = (0, A_\theta, 0)$ and $E = (0, E_\theta, 0)$.

A_θ is a phasor quantity and has a real and an imaginary part :

$$A_\theta = A_{\theta R} + iA_{\theta I} \tag{16}$$

and the equation 15 becomes:

$$\nabla^2 A_\theta - \frac{A_\theta}{r^2} = \mu_0 \left(J_{coil} + J_{ind}\right) \tag{17}$$

where

$$\nabla^2 = \frac{1}{r}\frac{\partial}{\partial r}\left(r\frac{\partial}{\partial r}\right) + \frac{\partial^2}{\partial z^2} \tag{18}$$

In different zones the vector potential equations take different forms:
-In plasma:

$$\nabla^2 A_{\theta R} - \frac{A_{\theta R}}{r^2} + \omega\mu_0\sigma A_{\theta I} = 0 \tag{19}$$

$$\nabla^2 A_{\theta I} - \frac{A_{\theta I}}{r^2} - \omega\mu_0\sigma A_{\theta R} = 0 \tag{20}$$

-In coil:

$$\nabla^2 A_{\theta R} - \frac{A_{\theta R}}{r^2} = \mu_0 J_{coil} cos(\phi) \tag{21}$$

$$\nabla^2 A_{\theta I} - \frac{A_{\theta I}}{r^2} = \mu_0 J_{coil} sin(\phi) \tag{22}$$

-In the quartz tubes and outer region:

$$\nabla^2 A_{\theta R} - \frac{A_{\theta R}}{r^2} = 0 \tag{23}$$

$$\nabla^2 A_{\theta I} - \frac{A_{\theta I}}{r^2} = 0 \tag{24}$$

The electric field and the magnetic field intensity can then be determined using:

$$E_\theta = -i\omega A_\theta \tag{25}$$

$$\mu_0 H_z = \frac{1}{r}\frac{\partial}{\partial r}\left(rA_\theta\right) \tag{26}$$

$$\mu_0 H_r = -\frac{\partial}{\partial z}\left(A_\theta\right) \tag{27}$$

These equations are coupled to the plasma flow and energy equations (9–11) through the radial and axial components of the Lorentz force and through the Joule heating rate :

$$F_r = \frac{1}{2}\mu_0\sigma Re\left[E_\theta H_z^*\right] \tag{28}$$

$$F_z = -\frac{1}{2}\mu_0\sigma Re\left[E_\theta H_r^*\right] \tag{29}$$

$$P = \frac{1}{2}\sigma Re\left[E_\theta E_\theta^*\right] \tag{30}$$

where the superscript * denotes the complex conjugate.

RESULTS AND DISCUSSION

In this section, the computational results from double phase induction for both supersonic and subsonic torch are presented and compared the classical induction one. The simulation is carried out with the two different models: the single phase induction model called also the classical model corresponds to the case where the current densities in all rings of coil are in phase whereas the double phase induction model corresponds to the case where the induced phase and forced phase are out of phase. The double phase induction results are obtained by imposing the coil current phases in different rings of the coil.

The vector potential contours show the main differences between the coupling to the plasma and the effect on the plasma stream lines and temperature fields. Figures 5 and 6 show the classical results that have been published by many authors since the work of Xue et al [6] and [1], however the double induction coil presents very different patterns. Figures 7 and 8 show the temperature and stream lines in the single and double induction

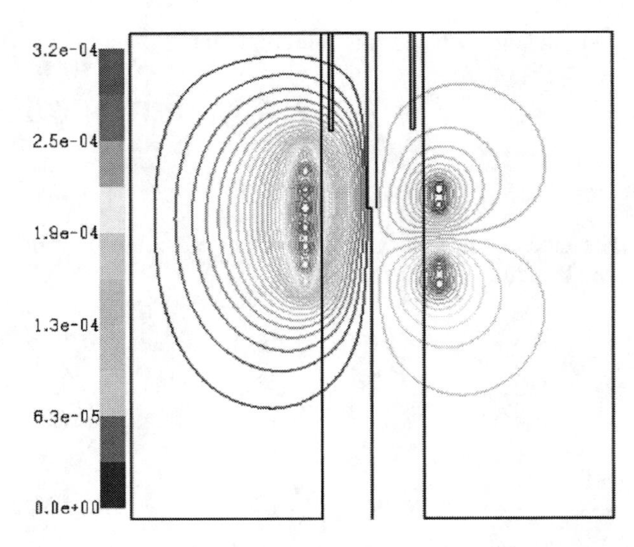

FIGURE 5. (Color online) Contours of real part of vector potential obtained by classical induction model (left side) and double phase induction model (right side).

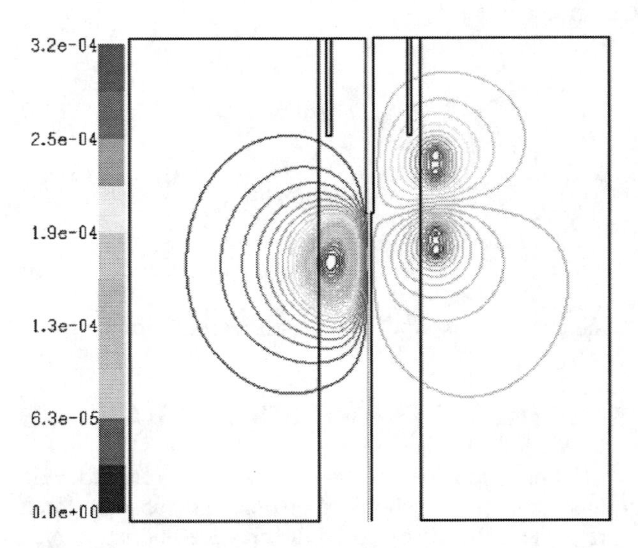

FIGURE 6. (Color online) Contours of imaginary part of vector potential obtained by classical induction model (left side) and double phase induction model (right side).

subsonic system. Although the authors did not study in depth the performance of the simulated plasma torches as a function of the operating parameters, these results show an interesting flow pattern modification that could be advantageous for nanoparticle production. As a matter of fact, the flow pattern is directed more toward the central part of the plasma high temperature, compared to the single phase inductor. The same pattern is observed in the supersonic case presented in Figs. 9 and 10. It is not surprising to detect difference in the temperature field because of the observed difference in the Joule heating

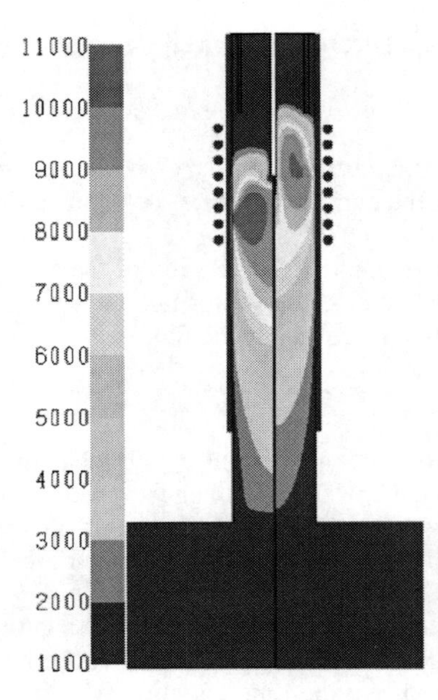

FIGURE 7. (Color online) Comparison of the the temperature obtained by single phase classical model (left side) and double phase induction model (right side) in argon subsonic plasma.

as can be seen from Fig. 11. As shown in Fig. 11, the Joule heating obtained from the double phase induction model show a double peak which cause the presence of the double peak in the temperature field. Also, because of the observed differences in the electromagnetic fields predicted by the classical model and the double phase induction model, the radial Lorentz force vary between the two models as shown in Fig. 12. The obtained radial force is negative, which means that the direction of its points to the plasma center. this radial Lorentz force pushes the sheath gas to the central plasma region and it causes the plasma circulation upstream. It can clearly be seen that the flow patterns in the double inductor coil are more favorable to a central trajectory, where higher temperatures are encountered. Even more importantly, there is much less tendency to project material toward the walls of the induction torch as is well known to happen in poorly tuned induction plasma treatment systems. The presence of the central injection probe is used to prevent the plasma treated materials to be entrained in the strong backflow present in the induction coil, a problem that could be avoided using double inductors.

CONCLUSIONS

In this work, a mathematical model for an innovative induction system (double phase induction) for both subsonic and supersonic torch is presented and is compared

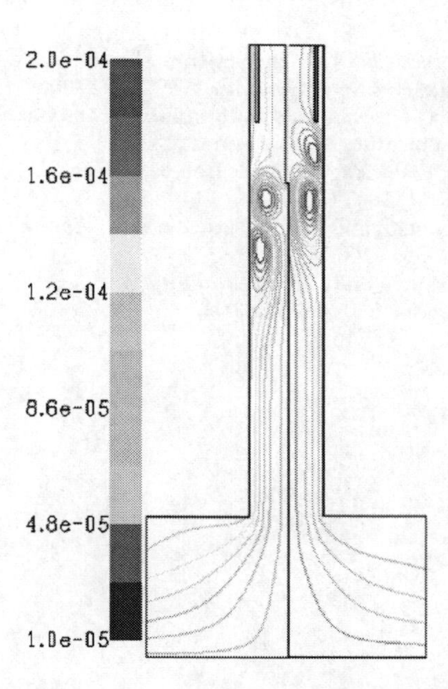

FIGURE 8. (Color online) The stream function obtained by single phase classical model (left side) and double phase induction model (right side) in argon subsonic plasma.

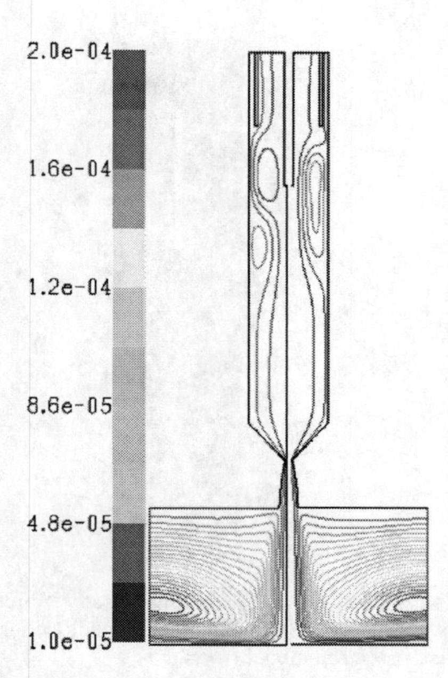

FIGURE 10. (Color online) The stream function obtained by single phase classical model (left side) and double phase induction model (right side) in argon supersonic plasma.

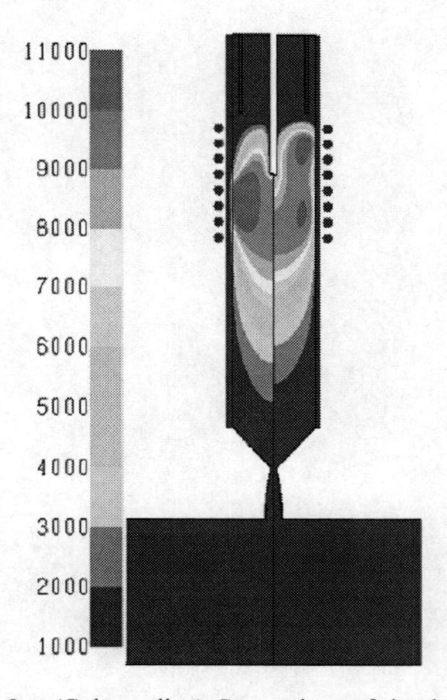

FIGURE 9. (Color online) Comparison of the the temperature obtained by single phase classical model (left side) and double phase induction model (right side) in argon supersonic plasma.

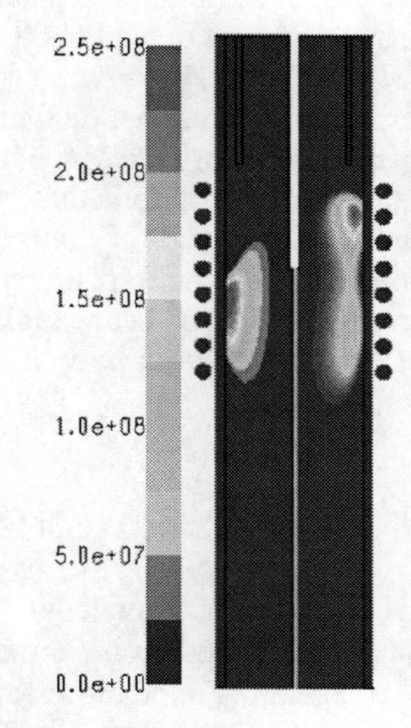

FIGURE 11. (Color online) Isocontour of power dissipation obtained by the single phase induction model (left side) and double phase induction model (right side).

to the classical single phase induction one. The obtained result show that the flow patterns in the double inductor coil are more favorable to a central trajectory, where higher temperatures are encountered. Even more importantly, there is much less tendency to project material toward the walls of the induction torch as is well known to happen in induction plasma treatment. The presence of the central injection probe is used to prevent the plasma treated materials to be entrained in the strong backflow present in the induction coil, a problem that could be avoided using double inductors. It is of interest to note that to the authors' knowledge, up to now, no industrial application of this double induction plasma has been de-

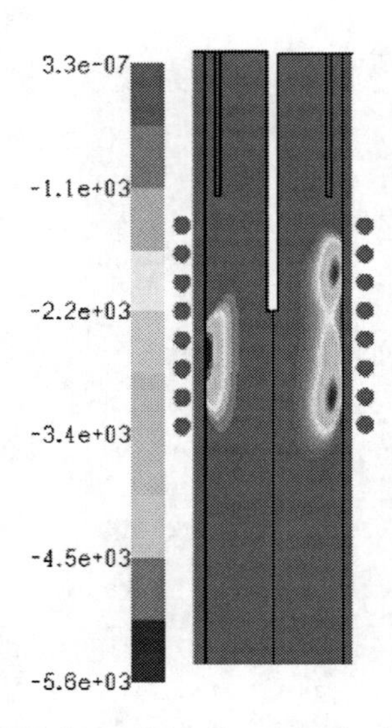

FIGURE 12. (Color online) Contour of radial force lorentz obtained by the single phase induction model (left side) and double phase induction model (right side).

velopped. Furthermore in this work, we did not study in depth the performance of the simulated plasma torches as a function of the operating parameters such as power, frequency, mass flow rate, etc. The presented results show an interesting flow pattern modification that could be advantageous for nanoparticle production, but that remains to be examined much more in details through systematic parametric modelling study and, of course, experimentally. Further modelling work will include more realistic gas mixtures (an aspect that has been studied by the authors [7] and [8]), and better description of the induction coupling between the coils.

ACKNOWLEDGMENTS

The authors wish to acknowledge the National Sciences and Engineering Research Council of Canada (NSERC), the Canadian Foundation for Innovation (CFI) and Tekna Plasma Systems for the financial support.

REFERENCES

1. S. Xue, R. Ernst, C. Trassy, and P. Proulx, *4th Inter. Conf. on Electromagnetic Proc. of Mat, (EMP 2003), Lyon (France)* (2003).
2. R. Ernst, D. Perrier, P. Brun, and J. Lacombe, *Int J for Computation and Maths. in Electrical and Electronic Eng.* **24**, 334–343 (2005).
3. M. Shigeta, and H. Nishiyama, *Transactions of ASME, Journal of Heat Transfer* **127**, 1222–1230 (2005).
4. B. Goortani, N. Y. Mendoza Gonzalez, S. Xue, and P. Proulx, *Powder Technology* **175**, 22–32 (2007).
5. P. Proulx, *Thermal Plasma Synthesis of Nanoparticles*, Von Karman Institute Lecture Series, 2007.
6. S. Xue, P. Proulx, and M. I. Boulos, *J. Phys. D.: Applied Physics* **34**, 1897 (2001).
7. M. El Morsli, and P. Proulx, *J. Phys. D.: Applied Physics* **40**, 380–394 (2007).
8. M. El Morsli, and P. Proulx, *Journal of Physics D: Applied Physics* **40**, 4810–4828 (2007).

Fluid Complex Plasmas – Studies at the Particle Level

A. V. Ivlev, G. E. Morfill, V. Nosenko, R. Pompl,
M. Rubin-Zuzic, and H. M. Thomas

Max Planck Institute for Extraterrestrial Physics, 85741 Garching, Germany

Abstract. Complex plasmas are ideal laboratory systems to investigate kinetics of strongly coupled many-particle ensembles. In contrast to colloidal suspensions, the particle dynamics in complex plasmas is virtually undamped. This makes complex plasmas particularly suited to study kinetics of fluids, by observing fully resolved motion of individual particles. In this paper we focus on three major experimental highlights characterizing kinetics of fluid plasmas – laminar shear flows, onset and development of hydrodynamic instabilities, and heterogeneous nucleation in supercooled fluids. Analysis of elementary processes observed in these experiments provides important insights into fundamental generic processes governing fluid behavior, demonstrating significant interdisciplinary potential of the complex plasma research.

Keywords: Complex plasmas, Fluids, Kinetics
PACS: 52.27.Lw, 47.57.-s

INTRODUCTION

Laboratory complex (dusty) plasmas are low-pressure (rf or dc) gas-discharge plasmas containing microparticles (dust grains) which are charged due to absorption of surrounding electrons and ions [1, 2]. Microparticles interact electrostatically with each other, and the strength of the interaction, Γ (ratio of the average interaction energy to the temperature) can be tuned by changing discharge parameters and/or dust density. In laboratory conditions, the magnitude of Γ can vary in extremely wide range: The interaction is proportional to the square of the grain charge, which, in turn, is proportional to the grain size and is usually quite large ($\sim 3 \times 10^3\ e$ for 1 μm particles). This unique feature allows us to observe transitions between fluid and crystalline phases occurring in complex plasmas.

Although complex plasmas are intrinsically multi-species systems, the rate of momentum exchange through mutual interactions between the microparticles can significantly exceed that of other interactions (e.g., neutral gas drag) [1]. That is one of the important properties which distinguishes complex plasmas from colloidal suspensions. Therefore, complex plasmas (viz., charged microparticles) can act as an essentially single-species system where the fully resolved dynamics of each individual particle can be observed. This gives us a unique opportunity to investigate various phenomena occurring in strongly coupled many-particle systems at the kinetic (individual particle) level and go beyond the limits of continuous media, down to the smallest length scale available – the interparticle distance.

FLUID BEHAVIOR AT THE KINETIC LEVEL

How relevant are liquid plasmas for the study of conventional liquids? The implication is clear – if they are relevant, we have opened up a completely new approach to "nanohydrodynamics", the kinetic approach, which will then have a major impact in a field of great future potential and relevance. As was pointed out in the introduction, complex plasmas can be considered as an essentially single-species system (e.g., fluids) observable at the particle level. Moreover, comparison in terms of similarity parameters – Reynolds and Mach numbers – suggests that liquid complex plasmas can be like conventional liquids (e.g., water) [1, 3, 4]. This suggests that complex plasmas can indeed serve as a powerful new tool for investigating fluid flows on (effectively) nanoscales, including the all-important mesoscopic transition from collective hydrodynamic behavior to the dynamics of individual particles, as well as nonlinear processes on scales that have not been accessible for studies so far.

Examples of Fluid Behavior

In this paper we focus on three major experimental highlights characterizing kinetics of fluid complex plasmas – laminar shear flows, onset and development of hydrodynamic instabilities, and heterogeneous nucleation in supercooled fluids. (We have to omit the discussion of many other important issues, such as microstructure and microdynamics of fluids, onset of cooperative phenomena, etc. – there is simply not enough space for that!) By considering these examples, the main emphasis is made

CP982, *Complex Systems, 5th International Workshop on Complex Systems*
edited by M. Tokuyama, I. Oppenheim, and H. Nishiyama

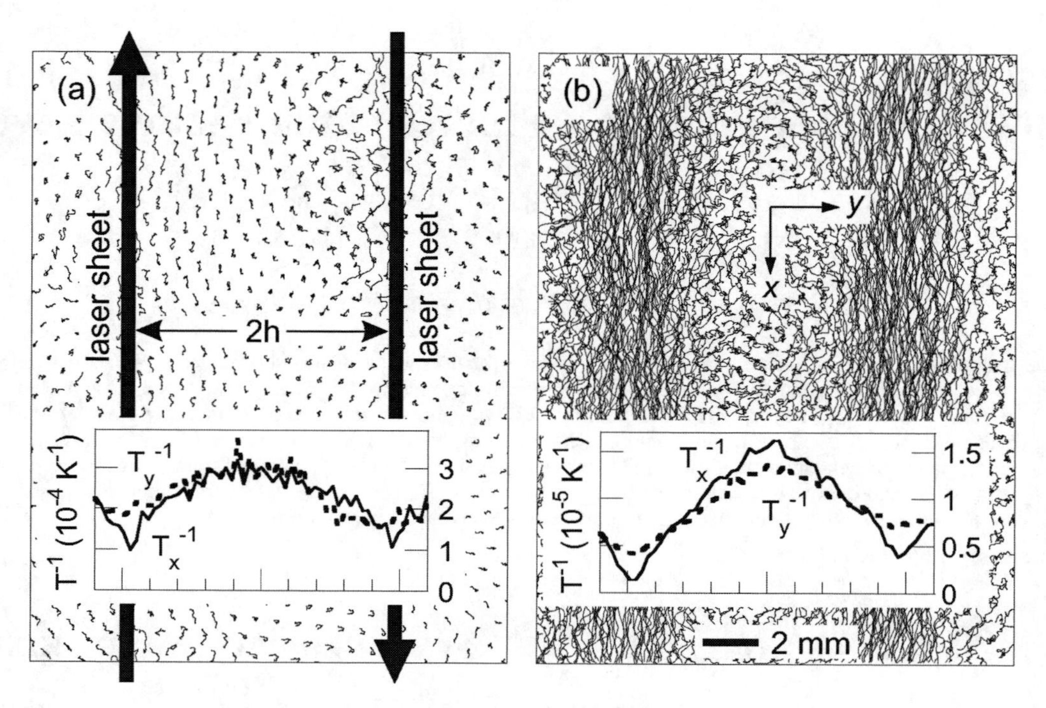

FIGURE 1. Planar Couette flow in a two-dimensional complex plasma [3]. Initially crystalline complex plasma is sheared by two counter-propagating laser sheets. At the onset of plastic flow (a), the particles hop between equilibrium lattice sites. In a fully developed shear flow (b), the particle motion is highly irregular on a small scale compared to the interparticle spacing, but on a larger scale, it is like a laminar flow in a fluid. By fitting the particle velocity profiles in the shear flow to a Navier-Stokes model, the shear viscosity of the complex plasma can be calculated.

on unique experimental properties of complex plasmas as the model system for kinetic investigations of fluids.

Laminar Shear Flows

Shear flows appear as an almost inevitable ingredient of more complicated flows. Even in the simplest case of laminar shear flows, when one tries to imagine how they look like at the kinetic level, many fundamental questions immediately arise: What is the kinetic structure of the flow? How is momentum transferred perpendicular to the flow? What is the kinetics of non-Newtonian and viscoelastic fluids? What happens (kinetically) at sheared fluid boundaries? etc. To address these issues, a series of experiments has been performed in 2D and 3D fluid complex plasmas.

In the 2D case [3], particles formed an (almost perfectly) hexagonal crystalline monolayer. To induce a shear flow, we applied two counter-propagating laser sheets, as shown in Fig. 1. By increasing the laser power, we were able to increase the level of shear stress, and observed that the particle suspension passed through four stages: elastic deformation, defect generation while in a solid state, onset of plastic flow, and fully developed shear flow. We present data for the latter two stages. At the onset of plastic flow, Fig. 1a, the particles hopped between equilibrium lattice sites. Domain walls developed, and they moved continuously. The crystalline order of the lattice in the shearing region deteriorated, broadening the peaks in the static structure factor (not shown here). At still higher levels of shear stress, the lattice fully melted everywhere, and a shear flow developed, Fig. 1b. The particle motion was highly irregular on a small scale compared to the interparticle spacing, but on a larger scale, it was like a laminar flow in a fluid. In this case, the liquid-like order of the particle suspension can be clearly indicated by the diffusiveness of the structure factor. Particles were confined so that after flowing out of the field of view on one side, they circulated around the suspension's perimeter and reentered on the opposite side. Within the field of view, more than 95 % of the time-averaged flow velocity was directed in the x-direction, with less than 5 % of the flow velocity diverted in the y-direction. It is worth noting that for all values of the laser power used in the experiment the local velocity distribution of particles was (with very good accuracy) a Maxwellian one, although at highest shear rates the mismatch between the longitudinal and transverse temperatures was as high as ~ 30 %.

Numerical simulations [5, 6] predict that the shear viscosity of complex plasmas depends on the concentration

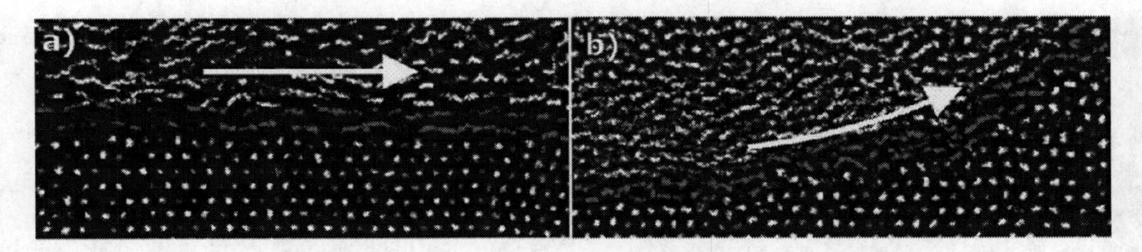

FIGURE 2. (Color online) Two examples of highly resolved complex plasma flows [10]. The figures show (a) a shear flow over a flat-surface plasma crystal and (b) a flow over a curved-surface plasma crystal. Note the small angle perturbations in the particle trajectories in (a), and the considerably larger scattering in the curved flow in (b). The flow velocity is $\simeq 1$ mm/s.

of microparticles, which is one of the essential features of complex fluids. Experiments with 2D shear flows discussed above as well as recent simulations [7] pointed out at shear thinning suggesting that complex plasmas behave like non-Newtonian fluids. In order to address the issue of non-Newtonian behavior, we performed dedicated experiments with 3D complex plasmas [8]. The shear flow was induced by means of either inhomogeneous flow of background gas or by laser beams. The combination of these two methods allowed us to measure the shear viscosity in the entire range of shear rates – up to the hydrodynamic limit where the discreteness enters and a fluid cannot be formally considered as a continuous medium.

Based on the obtained results [8], one can identify three distinct regimes for a qualitative dependence of the viscosity ν and the shear stress $\sigma = \nu\gamma$ on the velocity shear rate γ: At sufficiently low γ the viscosity remains constant and stress grows linearly with γ, which corresponds to Newtonian fluids (regime I). Above a certain threshold, the shear-thinning is observed (regime II), which can be quite significant – the viscosity can decrease by an order of magnitude. At even higher γ the crossover to the shear-thickening occurs (regime III). A remarkable rheological feature is that the viscosity decrease in the second regime can be so rapid that the $\sigma(\gamma)$ dependence may have an anomalous N-shaped profile. In this case the part of the curve with $d\sigma/d\gamma < 0$ becomes unstable and the flow is accompanied by a discontinuity in γ. This causes the formation of shear "bands" – the phenomenon often observed in complex fluids. Another important result is that at "extreme" shear rates (up to $\gamma \sim U/\Delta$, where U is the magnitude of the flow velocity and Δ is the interparticle distance), the formal hydrodynamic description with the Navier-Stokes equation still provides fairly good agreement with the experiment [8].

Hydrodynamic Instabilities

Kinetic investigations of instabilities in fluids and the transition to turbulence are of particular interest (e.g., Ref. [9]). Individual particle observations can provide crucial new insights – e.g., whether the basic hydrodynamical instabilities (Kelvin-Helmholtz, Rayleigh-Taylor, Tollmien-Schlichting, etc.) survive on interparticle distance scales and whether the transition to turbulence can be seen at the particle (kinetic) level, is there a kinetic (particle) trigger for hydrodynamic instabilities, etc.

Let us consider examples of highly resolved shear flows observed in complex plasmas [10] and shown in Fig. 2. In different regions of the microparticle cloud we observed different flow topologies, with the (average) flow lines being either straight (a) or curved (b), the lower part was at rest. The observations suggest that the width and the structure of the transition (mixing) layer strongly depends on the geometry. For the planar flow the interface is remarkably smooth, with the flow along a particular monolayer. The trajectories of individual flowing particles experience only weak deflections and the overall flow appears to be stable and laminar. In contrast, the curved flow interface has a curious "rough" structure, the flow is not laminar, a "mixing layer" is formed – and this is where the momentum transfer takes place. It is also apparent that the mixing layer becomes unstable at the individual particle level. The microscopic behavior seen in Fig. 2b may be interpreted as the centrifugally driven Rayleigh-Taylor instability. Analyzing a sequence of such images, one can quantify "elementary" (discrete) perturbations in two ways – the fraction of interpenetrating (say, $\geq \Delta$) particles, and the fraction of particles undergoing "large angle" (say, $\geq 30^\circ$) collisions in the surface layer. For instance, for the straight flow the quantities are (almost) 0 % and $\simeq 5$ %; for the curved flow, $\simeq 5$ % and $\simeq 25$ %. The latter can be understood kinetically in terms of the higher collision frequency with smaller impact parameter due to particle inertia at a curved surface. This has also been confirmed by numerical simulations conducted for similar geometry and flow conditions as in the experiments. The topology of the mixing layer found in the simulations corresponds closely to the measurements, which supports the kinetic interpretation.

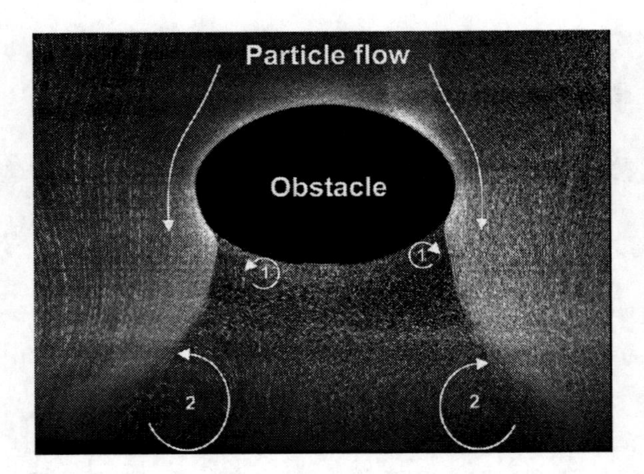

FIGURE 3. Topology of the particle flow around the void [4]. The flow leads to a compressed laminar layer, which becomes detached at the outer perimeter of the wake. The steady vortex flow patterns in the wake are illustrated. The boundary between the laminar flow and wake becomes unstable; a mixing layer is formed, which grows in width with distance downstream. The system is approximately symmetric around the vertical axis, the vortices are tori and the wake has the shape of a flaring funnel (exposure time 1 s).

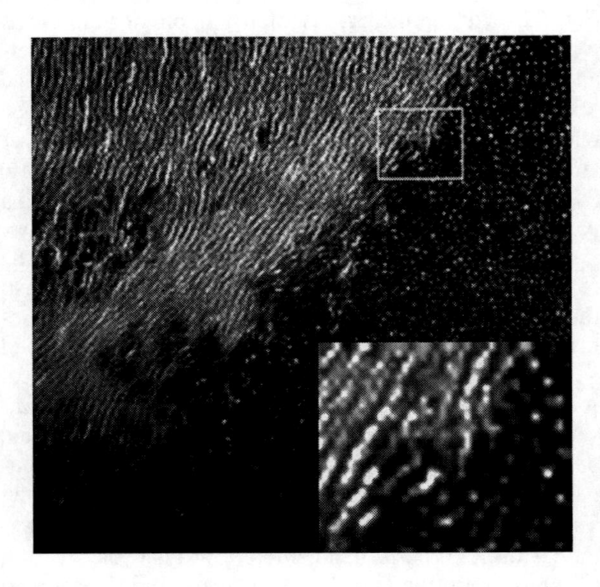

FIGURE 4. An example of the mixing layer – an enlargement of the left side of the flow regime shown in Fig. 3 (exposure time 0.05 s). The points (lines) represent traces of slow (fast) moving microparticles.

Another example of the hydrodynamic behavior of liquid complex plasmas [4] is shown in Fig. 3. Particles were flowing around an "obstacle" – a void of size $\sim 100\,\Delta$ (equivalent to ~ 100 "molecular" sizes). One can see a stable laminar shear flow around the obstacle, the development of a "wake" exhibiting stable vortex flows, and a mixing layer between the flow and wake. The mixing layer is observed to be quite unstable at the kinetic level, with instabilities becoming rapidly nonlinear. The width of the mixing layer grows monotonically with distance from the border where the laminar flow becomes detached from the obstacle. The growth length scale is of the order of a few Δ, i.e., much smaller than the hydrodynamic scales $n(dn/dx)^{-1}$ or $u(du/dx)^{-1}$, which would be expected macroscopically in fluids and refer to the Rayleigh-Taylor or Kelvin-Helmholtz instability, respectively. This rapid onset of surface instabilities followed by mixing and momentum exchange at scales $\sim \Delta$, i.e., the smallest interaction length scale available, is not consistent, therefore, with conventional macroscopic fluid instability theories. While this could not be rightfully expected at the kinetic level, it clearly points to new physics and, possibly, a hierarchy of processes that is necessary to describe interacting fluid flows: first, binary collision processes provide particle and momentum exchange on kinetic scales (a few Δ), then collective effects (due to the correlations defining fluid flows) take over and "propel" this "seed" instability to macroscopic scales. The latter create cascades of clumps, similar to the vortex merging that occurs in, e.g., turbulent jets (see, e.g., [11]).

These examples suggest an "elementary" microscopic picture of the hydrodynamic instabilities: It is not unreasonable to conclude that many instabilities have a kinetic analog or trigger and that the most effective trigger mechanism is provided by binary large-angle scattering in localized structures and/or inhomogeneities of scales comparable to the "effective" particle interaction size. However, the mathematical techniques required to quantify the kinetic behavior and to transfer this to macroscopic scales still need to be developed.

Heterogeneous Nucleation

While the atomic structure of crystals can be relatively easily measured, the detailed dynamics of nucleation (including the evolution of self-organization, structure formation, and the associated kinetic and thermodynamic development) remains one of the most important topics of solid state physics [12, 13]. Very little is known, for instance, about the self-organization principles governing heterogeneous nucleation, the resultant surface structure and its temporal evolution, the microscopic (kinetic) structure of interfaces, etc.

Complex plasmas are perfectly suited for such studies [1]. We illustrate the nucleation process in 3D complex plasmas with the recently reported results obtained on the ground [14]. First, the particles were brought into a disordered liquid-like phase (by a short pulse of increased discharge power). Afterwards, the system starts recrystallizing. Usually this results in homogeneous nucleation (with nucleation sites appearing spontaneously at different locations in a liquid), but sometimes this occurs heterogeneously, in the form of a front. Figure 5 shows the nucleation front with color-coded particle traces, which also gives an impression of the particle temperature. Average displacement of individual particles is also characterized by the "displacement area" [14] shown in Fig. 6. The front is fairly narrow (about 3–4 interparticle distances) and has a well developed fractal structure. The relative temperature drop across the front is about a factor of a few. This indicates that the observed nucleation is strongly nonequilibrium. One can also see the interface between different crystalline domains, which has a narrow width (2–3 lattice planes) and a substantially higher temperature than the crystal domains themselves – direct evidence for interfacial melting.

A very important feature of the crystallization process is that the relaxation of the particle energy proceeds much more slowly than in weakly coupled states (when the energy of each particle decays independently, due to weak neutral gas friction) [15]. The transition from one metastable crystalline state to another, lower energy level can take minutes. MD simulations show that the decay of kinetic energy occurs over much longer time than the conventional dissipation scale [15]. Presumably, this is because most of the energy is stored in the mutual electrostatic coupling, and each local transition between neighboring energy levels releases only a small fraction for dissipation. This finding is of great significance for self-organization processes, since it clearly marks the difference between cooperative and non-cooperative systems. In some cases, even when the system reaches overall lattice equilibrium, it can remain noisy – caged particles oscillate with rather low frequencies. This can be due to the possible existence of a few shallow metastable states of the same levels (separated by potential hills of the order of the particle thermal energy). Then the whole system can continuously jump from one state to another, yet keeping the same type of lattice.

Let us discuss peculiarities of the fluid-crystalline transition in detail. Surprisingly, analysis of the nucleation front (see Fig. 6) reveals the existence of small "temperature islands" – transient "cold" crystallites (inside the fluid regime) and "hot" droplets (inside the crystalline regime) whose lifetimes depend on their size, as shown in Fig. 7. Typical crystallites contain 30–150 particles, and droplets contain 50–250 particles.

The evolution of the crystallites can be understood in terms of the thermodynamics [14]. Random fluctuations may produce local temperature decreases, which, in turn, induce the formation of seed crystallites. However, if the size of a given crystallite is smaller than the critical (nucleation) threshold, the resulting decrease of the bulk free energy is compensated by an increase in the surface energy, that is, the surface tension exceeds the internal pressure. Then the crystallite will shrink, and eventually

FIGURE 5. (Color online) The nucleation front under gravity conditions [14]. Each image is a superposition of ten consecutive video frames (side view, about 0.7 s) showing the front propagating upwards, particle positions are color-coded from green to red, i.e., cooler particles appear redder, hotter are multicolored.

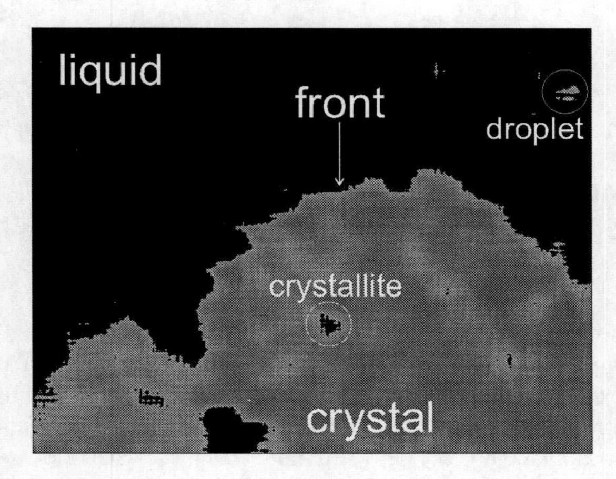

FIGURE 6. (Color online) Color-coded image of the nucleation front [14]. The so-called "displacement area" of individual particles in four consecutive video frames is shown: Red implies high crystalline order, black denotes the fluid phase, and yellow indicates transitional regions. Along with the crystallization front propagating upwards, droplets and crystallites are seen that may grow and then dissolve again.

disappear. Naturally, larger crystallites produced by a stronger local perturbation will survive longer. Figure 7a clearly demonstrates this trend. Note that many large crystallites (presumably above the nucleation threshold) may have artificially (non-thermodynamically determined) short lifetimes, because they are absorbed by the propagating front.

The mechanism responsible for the formation of the droplets must be quite different, because thermodynamically, both the bulk and the surface contributions cause the free energy to increase, as in the case of bubbles in

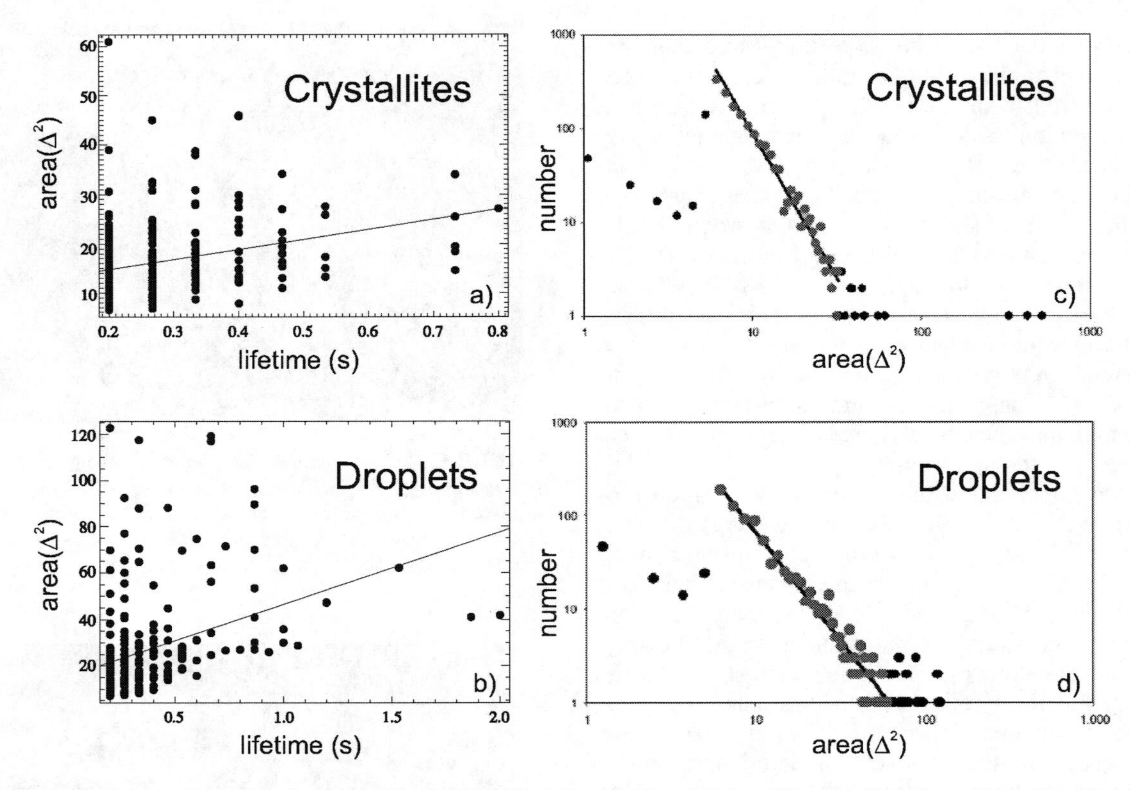

FIGURE 7. (Color online) Characteristics of crystallites and droplets [14, 16]. Area of crystallites (a) and droplets (b) measured in units of a single particle cell (squared interparticle distance) versus their lifetimes, and histograms showing number of crystallites (c) and droplets (d) versus their areas.

boiling water. It was proposed [14] that after the initial solidification, a gradual relaxation to a ground state (for example, to bcc or fcc lattice domains) occurs downstream from the nucleation front. This is naturally accompanied by a release of latent heat, so that the observed "hot" droplets could be a local manifestation of this relaxation. The larger the droplet, the longer it takes to dissipate the released heat, and the longer its lifetime, as shown in Fig. 7b. (That this relaxation process may be operating is also seen at the domain boundary in Figs. 5 and 6, when interfacial melting of only a few lattice planes occurs – possibly also supported by energy released in reorganizing the lattices into an energetically favorable state.)

Another remarkable peculiarity of the crystallites and droplets is a scale-free power-law size distribution down to the size of a single particle cell, as shown in Fig. 7c,d. Such behavior is not in line with the classic (macroscopic) theory of nucleation kinetics [17] and has to be studied further.

Presumably, both mechanisms – for the crystallite and droplet formation – may operate in any medium undergoing non-equilibrium nucleation. This suggests that our findings may represent generic kinetic properties of solidification in supercooled liquids. The discovery of the droplets (or local excited regions) could be important in the context of annealing processes and technology. Generally, the nucleation and the subsequent growth processes in complex plasmas look very similar to conventional crystallization in supercooled media (e.g., in semiconductors, see Refs [18]). Therefore, space- and time-resolved investigations of elementary processes accompanying the nucleation and growth of plasma crystals can be very useful for understanding some basic microscopic processes in liquid-solid phase transitions.

CONCLUSIONS

The physics of complex plasmas has already made significant progress in the investigations of some fundamental processes – ranging from nanofluidics [19] and formation of clusters [20] to hydrodynamic instabilities [4], sublimation and crystallization [14] at the kinetic level. Along with remarkable achievements in studying various aspects of the plasma physics itself, these investigations help us to make substantial advances in other fields. In this paper we showed just a few examples that allow us to consider complex plasmas as the model system for such kinetic investigations. We regard this interdisciplinary research as the major direction for our future work.

ACKNOWLEDGMENTS

This work was partly supported by DLR/BMWi Grant No. 50WP0203.

REFERENCES

1. V. E. Fortov *et al.*, *Phys. Rep.* **421**, 1-103 (2005).
2. P. K. Shukla and A. A. Mamun, *Introduction to Dusty Plasma Physics*, IOP Publ., Bristol, 2001.
3. V. Nosenko and J. Goree, *Phys. Rev. Lett.* **93**, 155004/1-4 (2004).
4. G.E. Morfill *et al.*, *Phys. Rev. Lett.* **92**, 175004/1-4 (2004).
5. T. Saigo and S. Hamaguchi, *Phys. Plasmas* **9**, 1210-1216 (2002).
6. G. Salin and J.-M. Caillol, *Phys. Rev. Lett.* **88**, 065002/1-4 (2002).
7. Z. Donko *et al.*, *Phys. Rev. Lett.* **96**, 145003/1-4 (2006).
8. A. V. Ivlev *et al.*, *Phys. Rev. Lett.* **98**, 145003/1-4 (2007).
9. M. Lesieur, *Turbulence in Fluids*, Kluwer, Dordrecht, 1991.
10. G. E. Morfill *et al.*, *Phys. Scripta* **T107**, 59-64 (2004).
11. F. F. Grinstein, F. Hussain, and E. S. Oran, *Europ. J. Mech. B* **9**, 499-525 (1990).
12. A. C. Levi and M. Kotrla, *J. Phys. Condens. Matter* **9**, 299-344 (1997).
13. T. Palberg, *J. Phys. Condens. Matter* **11**, R323-R360 (1999).
14. M. Rubin-Zuzic *et al.*, *Nature Physics* **2**, 181-185 (2006).
15. A.P. Nefedov *et al.*, *New J. Phys.* **5**, 33/1-10 (2003).
16. A. V. Ivlev *et al.*, "Complex Plasmas – New Discoveries in Strong Coupling Physics", *Appl. Phys. B* (2007) (to be published).
17. E. M. Lifshitz and L. P. Pitaevskii, *Physical Kinetics*, Pergamon, Oxford, 1981.
18. M. Kästner and B. Voigtländer, *Phys. Rev. Lett.* **82**, 2745-2748 (1999).
19. L. W. Teng, P. S. Tu, and I. Lin, *Phys. Rev. Lett.* **90**, 245004/1-4 (2003).
20. A. Melzer, M. Klindworth, and A. Piel, *Phys. Rev. Lett.* **87**, 115002/1-4 (2001).

Radio-Frequency, Atmospheric-Pressure Glow Discharges: Producing Methods, Characteristics and Applications in Bio-Medical Fields

He-Ping Li[*], Guo Li[*], Wen-Ting Sun[*], Sen Wang[*], Cheng-Yu Bao[*],
Liyan Wang[**], Ziliang Huang[**], Nan Ding[**], Hongxin Zhao[**], and Xin-Hui Xing[**]

[*]*Department of Engineering Physics, Tsinghua University, Beijing 100084, P. R. China*
[**]*Department of Chemical Engineering, Tsinghua University, Beijing 100084, P. R. China*

Abstract. Radio-frequency (RF), atmospheric-pressure glow discharge (APGD) plasmas with bare metallic electrodes have shown their promising prospects in different fields. In this paper, based on the induced gas discharge approach, the discharge characteristics of RF, APGD plasmas using helium/oxygen mixture as the plasma working-gas are presented. The bio-medical effects of the helium RF APGD plasma jet acting on the *gfp* DNA and *E. coli* are also reported. Studies concerning the lethal and sub-lethal effects of the RF APGDs on the molecular and cell levels, which are related with the characteristics of the plasmas and their operation conditions are necessary in the future work based on a closer cooperation between the researchers in the field of the plasma science & technology and of the bio-medical science.

Keywords: Atmospheric pressure, Radio-frequency, Glow discharge, Bare metal electrode, Oxygen, Biomedical effects
PACS: 52.80.-s, 52.80.Pi

INTRODUCTION

In last decades, low-pressure plasmas have been widely used in materials processing fields, such as plasma-aided chemical vapor deposition, etching, etc., due to their features of low gas temperature, high electron energy, and large amounts of chemically active species which result in high chemical reaction rates in actual plasma materials processing. However, due to the usage of the vacuum system, on one hand, the capital costs for construction and maintenance of the complicated vacuum system are very high; and on the other hand, the geometrical sizes of the treated materials are also limited by the volume of the vacuum chamber, and also accompanied by a very complex robotic assembly used to shuttle materials in and out of the vacuum chamber. In recent years, different kinds of atmospheric-pressure non-equilibrium discharge (APNED) plasma sources have been developed with their outstanding advantages compared to the low-pressure plasmas, such as the lower capital costs, no limitations on the sizes of the treated materials, convenient operation processes to shuttle materials, etc., due to the removal of the vacuum system. The APNED plasma sources reported in previously published papers include the radio-frequency (RF) atmospheric-pressure plasma jet (APPJ) [1, 2], dielectric barrier discharge (DBD) [3, 4], the one atmosphere uniform glow discharge plasma (OAUGDP) [5, 6], the resistive barrier discharge (RBD) [7, 8], the RF DBD plasmas using flexible electrodes of fabric structure [9], hollow-cathode discharges [10-12], and floating electrode discharges [13], etc. In the preceding studies, most of the researchers used power supplies with frequencies ranging from kHz to MHz, but recently, power supplies with frequencies of GHz or periods on the order of nanosecond have been employed to generate APNED plasmas [14].

Among different kinds of APNED plasma sources, the atmospheric-pressure glow discharge (APGD) plasmas using bare-metallic electrodes and driven by RF power supply (which is called RF APGD hereafter in this paper) have contracted much attention of researchers all over the world for their unique features [1, 2] except for those compared with low-pressure discharge plasmas. For example, compared with the atmospheric-pressure DBD plasmas, the breakdown voltage of RF APGD plasmas can be reduced significantly, and more homogeneous discharge can be obtained resulting from eliminations of the dielectric barriers covered on the electrodes or placed between

CP982, *Complex Systems, 5th International Workshop on Complex Systems*
edited by M. Tokuyama, I. Oppenheim, and H. Nishiyama

electrodes [1]. With the foregoing unparallel capabilities, RF APGD plasma sources have been or will be used in much wider fields, including the industries (e.g. plasma-aid etching [15-17], deposition [18, 19], etc), military fields (e.g. decontamination of chemical and biological warfare agents [20], etc), biomedical and public health fields (e.g., plasma-based disinfection and sterilization, in-door atmosphere purification, food safety [21], etc), and even being extended to national securities.

Although studies concerning the discharge mechanisms and characteristics of RF APGD plasmas, as well as their novel applications in the bio-medical fields, have been extensively conducted, there still exist many problems need to be solved for promoting actual applications of RF APGDs in different fields. From the plasma physics side, the understandings to the discharge mechanisms, including the discharge processes, discharge modes, key factors influencing the discharge characteristics, etc, are very limited. For example, studies on the discharge modes presented by different authors are very different. In Refs. [22, 23], it was reported that the RF APGD plasmas could work in a α mode or γ mode; while in Refs. [24, 25], it was indicated that the RF APGD plasmas could work in more different modes, e.g. the normal glow mode, the transition phase between the normal and abnormal mode, the abnormal glow mode, and the recovery mode; and in Ref. [26], arcing mode could occur with a larger RF power input, instead of the γ mode discharge. Another challenging problem in actual applications of RF APGDs for materials processing is the very high capital costs resulting from using helium or argon as the primary plasma working gas. In previous papers, most of the researchers employed helium or argon as the primary plasma-forming gas, to which a small fraction (0.5%-3%) of reactive gases (e.g. O_2, N_2, CF_4 or water vapor, etc) is added in order to generate chemically active species [2, 22, 23, 26-28]. Up to now, it is very difficult to directly initiate the RF APGDs using cheaper gases, such as oxygen, nitrogen or air, with bare metallic electrodes due to the very high breakdown voltages and the intense avalanche of electrons under so large electric field at atmospheric pressure [3]. From the bio-medical science side, although theories [29-32] and applications of RF APGD plasmas in the bio-medical fields, such as sterilization of reusable heat-sensitive medical instruments contaminated with micro-organisms without thermal damages [33], blood coagulation and living tissue sterilization [13], etc, are being developed, the plasma-microbe interaction mechanisms are ambiguous, and the action efficiency is a very complex function of the operation parameters of the RF APGDs, the nature of the microorganisms and their existing environment, etc. Furthermore, most of the previous studies focus on the factors, such as heat, ultraviolet (UV) radiation, charged particles and reactive species [29, 30, 34], which influence the lethal effects of the APNED plasmas. In our opinion, not only the lethal effects, but also the sub-lethal effects [33], such as noticeable changes in the enzyme activities, metabolic behaviors without lethal damages on the cells, especially on the DNA levels, are important for the actual applications of the APNED plasmas in the bio-medical fields.

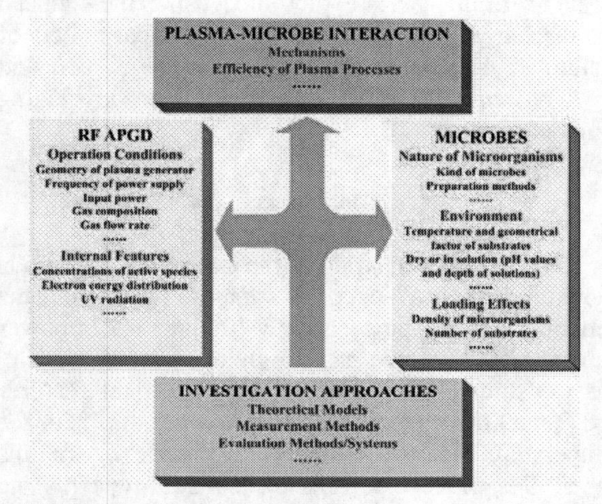

FIGURE 1. (Color online) Schematic diagram of the cooperative studies on the biological effects of RF APGDs.

Based on the preceding discussions, for reducing the capital costs, as well as for extending the potential applications, of the RF APGD plasma sources, cooperative studies by the researchers from both the plasma science & technology research community and from the bio-medical research community are necessary [34]. As illustrated in Fig. 1, to better understand the influences of different parameters on the efficiency of the plasma processes acting on microorganisms, the cooperative work includes plasma physics and actions of plasmas on microbes. On the plasma physics side, studies concerning the design of the RF APGD plasma generators, the control of the discharge modes, the internal parameters of plasmas and the optimization of the operation conditions are undergoing, while on the bio-medical side, interactions between plasmas and microbes are being analyzed on both the molecular and cell levels. In addition, the development of theoretical models, measurement and/or evaluation methods for the biological effects and the efficiency of plasmas on the treated microbes are also indispensable.

In this laboratory, based on the experimental studies on the discharge characteristics of helium RF APGDs [35], the discharge features of RF APGD plasmas using argon/ethanol mixture as the plasma-forming gas were reported [36], and the gas induced discharge approach was proposed for obtaining the uniform glow discharges with air [37] or argon/nitrogen mixture as the plasma working-gas [38] at atmospheric pressure using water-cooled bare metallic electrodes, and a preliminary experimental result concerning the bio-molecular effects of helium RF APGD plasmas was also reported recently [39].

In this paper, based on the descriptions of the experimental setup, the electrical features of the RF APGD plasmas with helium/oxygen mixture as the plasma-forming gas are presented using the induced gas discharge approach, then, the biological effects of helium RF APGD plasmas acting on the *E. coli* and *gfp* DNA are discussed, and finally, the concluding remarks are given.

EXPERIMENTAL SETUP

The schematic diagram of the experimental setup is shown in Fig. 2 (a), which is composed of the plasma generator, the RF power supply (the maximum power output is 500 W), the gas supply and control system, the measurement systems for the electrical features and optical features, as well as a thermal couple for the temperature measurement of the plasma jet. In this paper, the planar-type and the co-axial-type plasma generators, as shown in Figs. 2 (b) and (c), are employed for the studies of the electrical features and the biological effects of the RF APGDs. The planar-type plasma generator, shown in Fig. 2 (b), is composed of two 5×8 cm^2, planar, water-cooled, bare-copper electrodes, which are connected to the RF (13.56 MHz) power supply (which is the so-called RF electrode) and the ground (which is the so-called grounded electrode), respectively. Teflon spacers are used to seal the plasma generator on both sides and adjust the distance between the electrodes. The co-axial-type plasma generator consists of two 95 mm long, coaxial, water-cooled copper electrodes. The inner diameter of the outer electrode, which is employed as the grounded electrode, is 19.2 mm, while the central electrode is RF (13.56 MHz) powered with a diameter of 16.0 mm, as shown in Fig. 2 (c). The plasma forming-gas (99.99% or better for helium and argon, 99.999% or better for nitrogen and oxygen from gas cylinders, or air from an oil-free compressor made in China, or the gas mixtures of them) is admitted into the discharge region of the plasma generator, ionized by the applied RF electric field, and flows out of the plasma generator forming a non-thermal plasma jet. In the biological experiments, the co-axial-type plasma generator is employed with helium as the plasma-working gas, and a transparent, plastic cap is placed close to the nozzle exit of the plasma torch in order to prevent from the back-diffusion effect of air [35], as shown in Fig. 3.

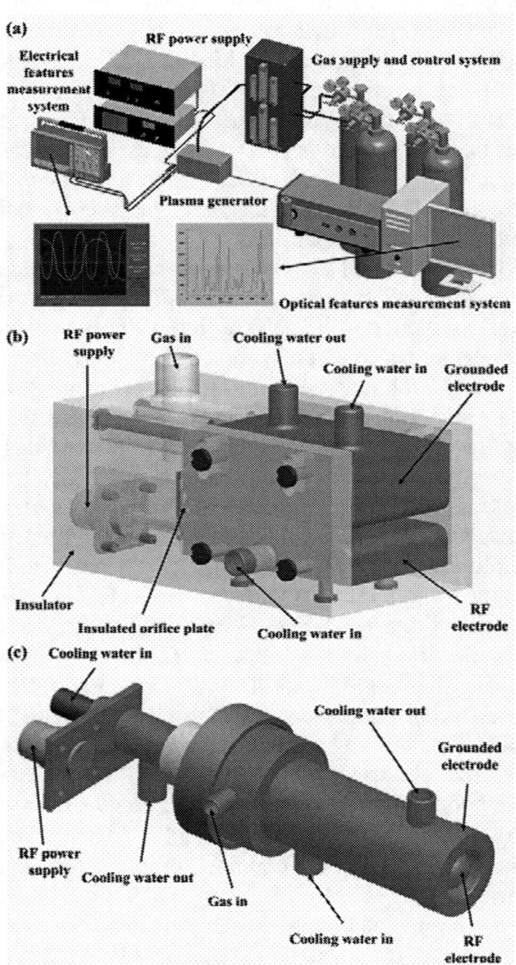

FIGURE 2. (Color online) Schematic diagrams of the experimental setup (a) and the planar-type (b) and co-axial-type (c) plasma generators.

In this study, the root-mean-square (rms) values of the discharge current and voltage, as well as the current-voltage phase difference (θ), are measured using a current probe (Tektronix TCP202) and a high voltage probe (Tektronix P5100), and are recorded on a digital oscilloscope (Tektronix DPO4034). The RF power input can be expressed as $P_{in}=V_{rms}\cdot I_{rms}\cdot \cos\theta$, where V_{rms} and I_{rms} represent the rms values of the discharge voltage and current, respectively. The discharge images are taken by a digital camera (Fujifilm S9600) and an Andor iStar 734 intensified CCD (iCCD) camera, respectively. Due to the very short time acquisition of the discharges with the Andor iStar 734 iCCD (down to 5 ns exposure time), the presence of streamers in the discharge region can be detected. The emission spectrum of the discharges are acquired by a monochromator (WDG30, Beijing Optical Instrument Factory, China) working with a photomultiplier tube (CR 131-01, Beijing Hamamatsu Photon Techniques Inc., China) with wavelengths ranging from 200 to 800 nm.

In the present paper, for the convenience of adjusting the gap spacing between electrodes and of taking pictures with the iCCD, we use the planar-type plasma generator, as shown in Fig. 2 (b), to investigate the electrical characteristics of RF APGD plasmas working with different plasma-working gases. For the study of the biological effects of RF APGD plasmas acting on the *E. coli* and the *gfp* DNA, the co-axial-type plasma torch, shown in Fig. 2 (c), is employed to generate a non-thermal plasma jet issued from the torch nozzle and impinging upon the bio-materials as shown in Fig. 3.

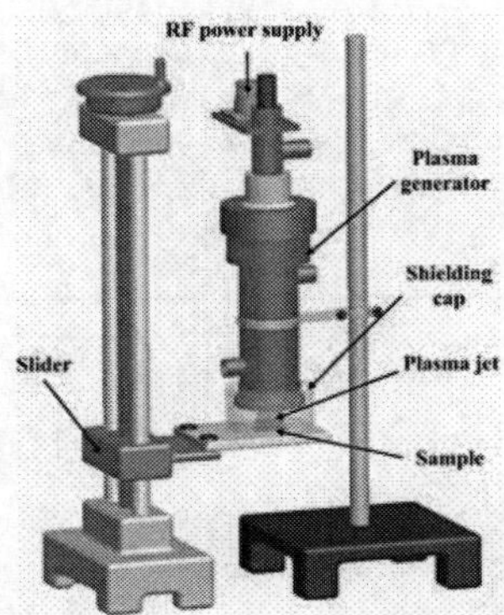

FIGURE 3. (Color online) Schematic diagram of the experimental setup used for studies of the biological effects of RF APGDs.

RESULTS AND DISCUSSIONS

Electrical Features of RF APGDs

Brief Description of the Induced Gas Discharge Approach

As indicated in the Introduction Section, due to the very high sparking field of the ordinary gases, e.g. ~30 kV/cm for air [40] at atmospheric pressure, as well as the tendency of transferring into a filamentary arc after breakdown resulting from the intense avalanche of electrons under so large electric field, it is very difficult to obtain the RF APGDs of the ordinary gases (e.g. nitrogen, oxygen, or air) and using the plasma generators with bare metallic electrodes at one atmosphere. Thus, one of the key points to obtain glow discharges at atmospheric pressure is to control the avalanche amplification so as to avoid its rapid growth. In previous studies concerning the atmospheric-pressure DBDs, on one hand, the dielectric barrier layers were employed to inhibit the occurrence of a filamentary arc with relatively high breakdown voltages; while on the other hand, the discharge with many thin filaments or micro-discharges could transfer to a glow discharge provided that there are enough seed electrons to turn on the discharge under a low electric field [3, 4]. Based on the preceding discussions, the induced gas discharge approach was proposed for obtaining the RF APGD plasmas with air or nitrogen as the plasma working-gas in Refs. [37, 38, 41]. If we define the gas which can be ignited directly to form the RF APGD plasmas as the plasma-inducing gas, e.g. helium or argon, while on the other hand, the gas which cannot be used to generate the RF APGD plasmas directly at the present time, e.g. air, nitrogen, oxygen, etc, as the plasma-forming gas, the key point of the induced gas discharge approach is that a γ or α-γ co-existing mode discharge, instead of a pure α mode discharge, with the plasma-inducing gas must be obtained first, then, by introducing the plasma-forming gas, and simultaneously decreasing the flow rate of the plasma-inducing gas, a stable glow discharge plasma operating in the γ mode can be obtained when no plasma-inducing gas is added into the plasma-forming gas any more [37, 38, 41].

Discharge Characteristics of RF APGDs with Helium/Oxygen mixtures

In this section, the planar-type plasma generator working in atmosphere is employed to study the discharge characteristics of the helium/oxygen RF APGDs. As reported in previous papers, the percentage of oxygen, as an admixed gas, in the helium/oxygen mixtures is usually very low (less than several percentage) for sustaining a stable RF APGD [1, 2]. In this paper, the oxygen concentration is defined as the mixing ratio $\chi=Q_{O2}/(Q_{O2}+Q_{He})$, where Q_{O2} and Q_{He} represent the flow rates of oxygen and helium, and χ=0 and 1 stand for pure helium and pure oxygen, respectively. The experimental measurements in this study show that for a lower concentration of oxygen (e.g. $\chi\leq3\%$), a helium/oxygen mixture can be ignited directly and an uniform RF APGD plasma can be obtained (shown in Fig. 4); and the oxygen concentration can influence not only the breakdown voltages/currents of the gas mixtures, but also the discharge modes and mode transitions after breakdown.

Figure 5 shows that with the increase of the oxygen concentration, the breakdown voltages (V_b) and

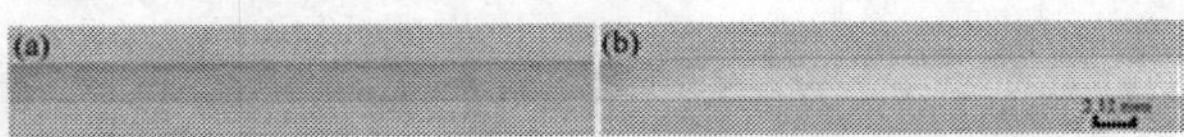

FIGURE 4. (Color online) Discharge images of the RF APGDs operating in the α mode. (a) pure helium, Q_{He}=10.0 slpm, gap spacing d=2.12 mm, exposure time T_{ex}=10 ms; (b) helium/oxygen mixture, Q_{He}=10.0 slpm, d=2.12 mm, χ=0. 50%, T_{ex}=20 ms.

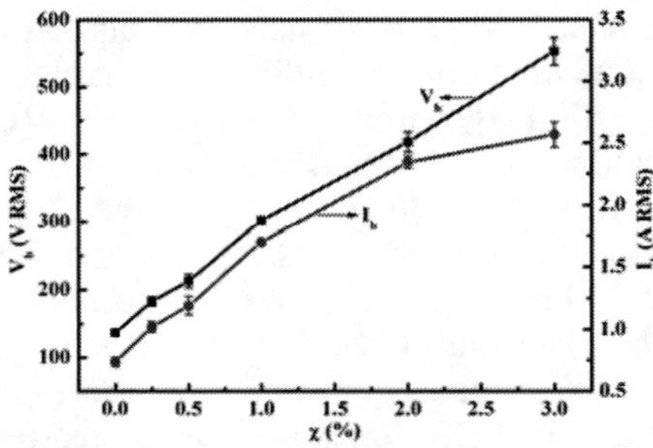

FIGURE 5. Variations of the breakdown voltages and currents of the RF APGDs operating in the α mode with the increase of the oxygen concentrations.

TABLE 1. Discharge modes after breakdown.

χ(%)	Discharge Mode after Breakdown	α-to-γ Mode Transition
0	α	Yes
0.25	α	Yes
0.5	α	No
1.0	α	No
2.0	α	No
3.0	γ	/

currents (I_b) increase monotonously. The measurements are repeated three times, and the maximum standard deviations for the voltages and currents are 20.0 V and 0.1 A, respectively.

The discharge modes after breakdown for the helium/oxygen mixtures are listed in Table 1. From Table 1 and the experimental results presented in Ref. [42], it can be concluded that: (1) for the glow discharges with pure helium or helium/oxygen mixture (χ=0.25%), a α-to-γ mode transition can occur with increasing the RF power input to a critical value; (2) when the oxygen concentration is higher than 0.5%, due to the limitation of the maximum power output of the RF power supply used in this study, an uniform α mode discharge is obtained and sustained with increasing the RF power input, and no α-to-γ mode transitions occur, which is qualitatively consistent with the results presented in Ref. [26]; (3) keeping the gap spacing between electrodes constant (d=1.24 mm in this study), with the increase of the oxygen concentration, only the γ mode discharge can occur after breakdown companied by a large voltage drop (e.g. ~265 V for χ=3% in this study); (4) with increasing the oxygen concentrations, the input RF powers also increase after breakdown, and the operation windows for the α mode discharges are enlarged since more energies are consumed to dissociate/ionize the oxygen molecules [42].

For the cases with higher oxygen concentrations, the uniform RF APGDs of helium/oxygen mixtures cannot be initiated directly. But using the induced gas discharge approach, the glow discharges of helium/oxygen mixtures operating in the γ mode can be obtained with any mixing ratios. The discharge images for the cases of helium/oxygen mixture (χ=50%) and pure oxygen (χ=1) taken by a digital camera, as well as the corresponding image for the pure oxygen discharge taken by an iCCD, are shown in Fig. 6. It can be seen clearly from Fig. 6 that the RF APGDs are laterally uniform without any streamers.

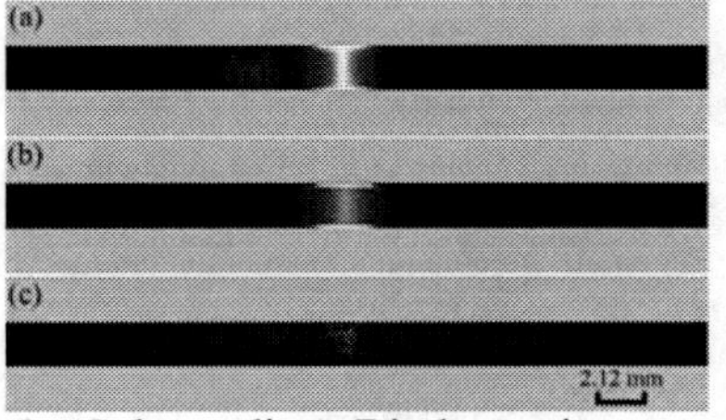

FIGURE 6. (Color online) Discharge images of the RF APGDs operating in the γ mode. d=2.12 mm, (a) helium/oxygen mixture, Q_{He}= Q_{O2}=1.0 slpm, T_{ex}=3 ms; (b) pure oxygen, Q_{O2}=1.0 slpm, T_{ex}=2.5 ms; (c) pure oxygen, Q_{O2}=1.0 slpm, T_{ex}=10 ns.

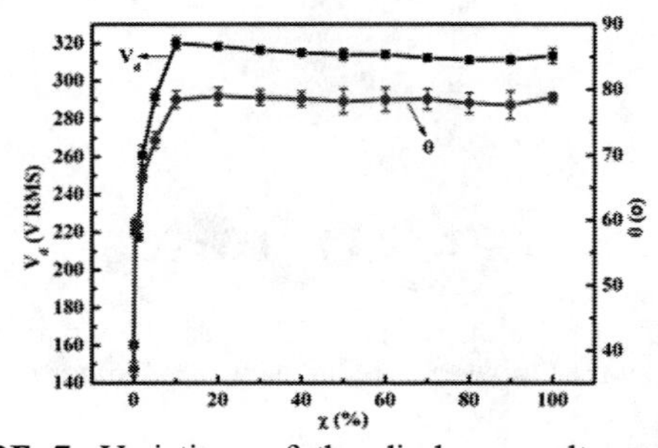

FIGURE 7. Variations of the discharge voltage and the current-voltage phase difference with the oxygen concentrations for the RF APGDs operating in the γ mode. Q=2.0 slpm, d=2.12 mm, P_{in}=270 W.

The variations of the discharge voltage (V_d) and the current-voltage phase difference (θ) with the mixing ratios are shown in Fig. 7 for the cases with the total gas flow rate Q=Q_{O2}+Q_{He}=2.0 slpm, gap spacing d=2.12 mm, and RF power input P_{in}=270 W being unchanged. It can be seen from Fig. 7 that the gas discharge voltage and the current-voltage phase difference increase sharply with increasing the oxygen concentration for the cases with lower values of χ (e.g. χ<10% in this study), and then, V_d and θ reach stable values with further increase of the mixing ratios (χ).

The influences of the RF power input on the current-voltage phase difference (θ) for different helium/oxygen mixtures with other parameters (e.g. the total gas flow rate and the gap spacing between electrodes) being unchanged are illustrated in Fig. 8. Figure 8 shows that: (1) for the γ mode discharges with pure gas (helium or oxygen) or gas mixtures, the current-voltage phase difference decreases with increasing the RF power input due to the increase of the ionization degrees of the plasma-working gases; (2) the values of θ for the γ mode discharges with helium/oxygen mixtures (including pure oxygen) are higher than those for the case of pure helium discharges, while the corresponding variation ranges of θ for the cases of helium/oxygen mixtures are smaller than those for the pure helium discharges.

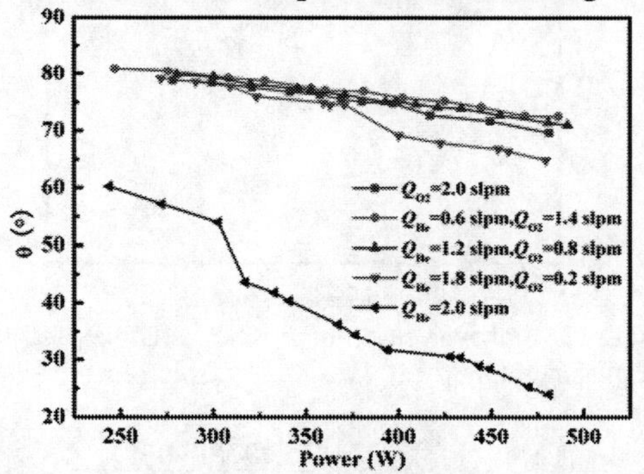

FIGURE 8. (Color online) Variations of the current-voltage phase difference with the power input for the RF APGDs operating in the γ mode. Q=2.0 slpm, d=2.12 mm.

Biological Effects of RF APGD Plasmas

As discussed in the Introduction Section, the studies on the plasma-microbe interactions on both the molecular and cell levels are indispensable for promoting the actual applications of the RF APGD plasma sources in the bio-medical fields. In this section, the co-axial-type plasma torch (shown in Figs. 2(c) and 3) is employed to generate the non-thermal plasma jet at atmosphere, and the biological effects of the helium RF APGDs on the *E. coli* and the plasmid DNA are studied experimentally.

Effects of RF APGDs on DNA

In this section, a loopfull recombinant plasmid (pP-GFP) is employed, which is constructed by linkage of pMD-18T with *gfp*, and then, is transformed into *E. coli* DH5α.The physical map of the pP-GFP is shown in Fig. 9. *E. coli* DH5α, as host of the pP-GFP, which is cultivated in Luira-Bertani (LB) medium (10.0 g of tryptone, 5.0 g of yeast extract, and 5.0 g of NaCl in 1.0 liter of distilled water) or on LB agar plates containing 100.0 μg·ml^{-1} of ampicillin at 37 °C. After cultured over night, the cells are harvested by centrifugation at 1000×g for 5 minutes. The plasmid DNA is extracted from the cells using a Plasmid Purification Mini Kit (Tiangen, China).

An 8.0-mm-in-diameter stainless steal disk spread with 3 μl of pP-GFP plasmid (conc: 100 ng·μl^{-1}) is placed on a horizontal platform. The distance between the torch nozzle exit and the sample (D) can be adjusted by sliding the platform up/down along the vertical support (shown in Fig. 3). After the DNA was treated by the helium plasma jet, it is analyzed on 1.0% agarose gel electrophoresis in 1x TAE buffer at 80 V for 40 minutes. The experimental process is illustrated in Fig. 10.

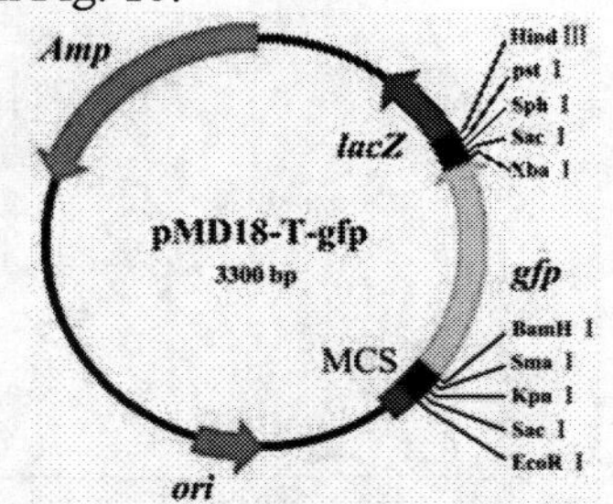

FIGURE 9. (Color online) Physical map of pP-GFP plasmid.

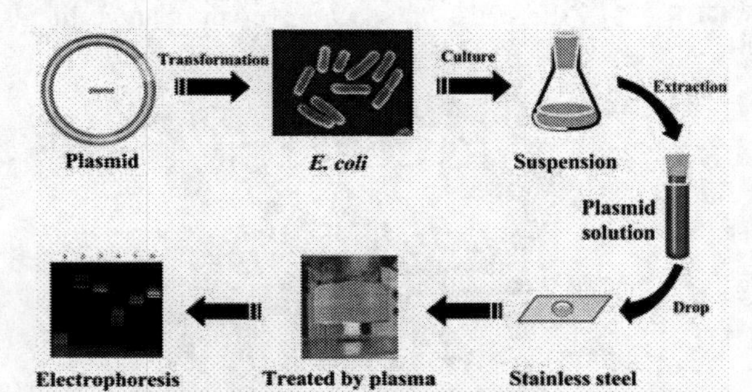

FIGURE 10. (Color online) Illustration of the experimental process for studying the biological effects of RF APGDs on DNA.

From the bio-medical side, micro-organisms may be inactivated by one of the four factors, i. e. heat, UV radiation, charged particles, and reactive neutral species, or by a synergistic combination of them [30]. In this section, the pure helium plasma jet produced by the co-axial-type plasma torch is used to treat the plasmid DNA. The operation parameters are as follows: P_{in}=180 W, Q_{He}=15.0 slpm, D=2.0 mm. The temperature of the helium plasma jet measured using a thermal couple at 2.0 mm downstream of the plasma torch is below 50 °C under such conditions [39]. UV radiation and the active species in the plasma jet under the same operation conditions were detected using the optical emission spectroscopy. As shown in Fig. 11, the UV radiation at 220–280 nm range (with doses of several mW·s/cm^2 are known to have the optimum effect on the bio-materials [30]) is very weak, while the emission intensity of helium at 706.5 nm is the strongest line than other helium spectra, and the oxygen atom lines at 777.4 and 615.8 nm are also obvious. In addition, the concentration of charged species in the jet effluent is relatively low [1]. The argarose gel electrophoresis images of the pP-GFP after treated by plasmas, as well as by UV radiation and heat, are shown in Fig. 12, which shows that it is the active species in the plasma jet region (e.g. helium and oxygen atoms, OH radicals), instead of the UV radiation, heat, and/or charged particles, that breaks the double-chain of the pP-GFP in a short time.

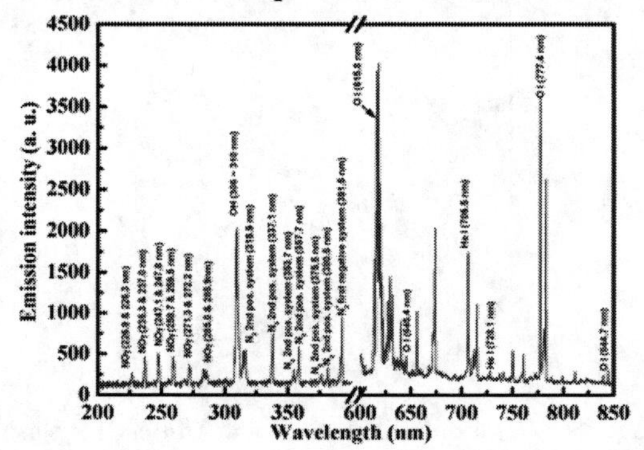

FIGURE 11. The emission spectrum of the pure helium discharge. P_{in}=180 W, Q_{He}=15.0 slpm.

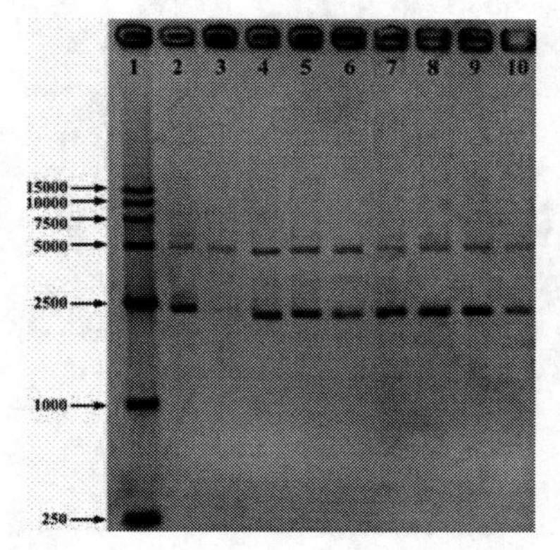

FIGURE 12. Electrophoresis analysis of the pP-GFP (Lane 1: DNA ladder 15000; Lane 2: control; Lane 3: treated by he helium RF APGD plasma for 3 min; Lane 4-6: treated by UV radiation for 3, 5 and 10 min, respectively; Lane 7-10: treated by hot gas for 10 min in the temperature range 40~100 °C with the interval of 20 °C).

Effects of RF APGDs on the E. Coli

In this sub-section, *E. coli* K12 w3110 is used to investigated the lethiferous ability of the RF APGD plasma jet. After cultured over night, *E. coli* cells are diluted to an appropriate concentration. The 20 μl bacteria suspension are spread on the stainless steel disk and dry by clean air in a laminar flow cabinet for about 20 minutes; and then, the stainless steal disk with cells is placed on platform located 3.0 mm downstream of the torch nozzle exit and treated in the condition of P_{in}=120 W and Q_{He}=15.0 slpm of the helium RF APGD plasma jet. After treated for assay time, the steal disk with cells is put into a tube with 2 ml physiological saline and quivers the tube in order to wash the cell into the suspension. The number of colony is calculated after spread cell suspension on LB agar medium and cultured for about 10 hours. The survival curve of *E. coli* exposed to the helium plasma jet is shown in Fig. 13, which shows a very high killing efficiency.

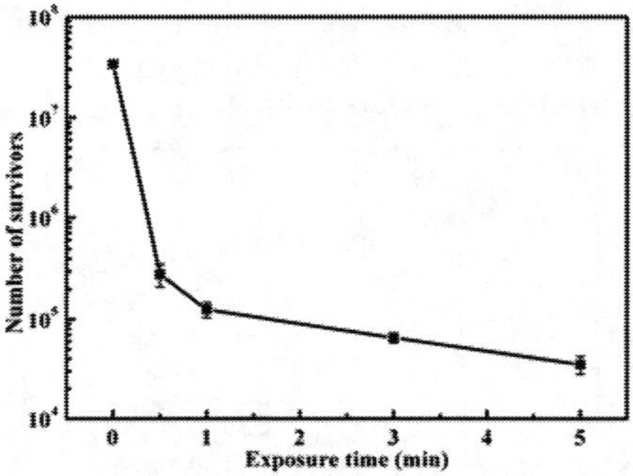

FIGURE 13. Survival curve of *E. coli* cells exposed to the helium plasma jet. P_{in}=120 W, Q_{He}=15.0 slpm.

CONCLUDING REMARKS

In the present paper, the recent research work in this lab concerning the discharge characteristics of RF APGDs and their bio-medical effects acting on the bio-materials (e.g. the *gfp* DNA and *E. coli*) on the molecular and cell levels are reported. The main conclusions obtained in this study are as follows: (1) the induced gas discharge approach can be used to generate uniform RF APGD plasmas using different ordinary gases, including oxygen, nitrogen, air, etc; (2) the discharge characteristics of the plasmas, including the breakdown/discharge voltage, the voltage-current curve, the current-voltage phase different, etc, could be influenced by the plasma-working gas, the RF power input, and so on; (3) for the case of helium glow discharges with low gas temperatures (below 50 °C), it is the active species in the plasma jet region, instead of the UV radiation, heat and/or charged species, that breaks the double-chain of the pP-GFP or kill the *E. coli* cells in a short time. Thus, it is necessary to establish the relationship of the plasma operation parameters – discharge characteristics – and biological effects by a closer cooperation between the researchers in the field of the plasma science & technology and of the bio-medical science in the future work.

ACKNOWLEDGEMENT

This work was supported by the National Key Technology R&D Program (2006EP805), the projects sponsored by Beijing Municipal Sciences & Technology Commission (Z07000200540705), and SRF for ROCS, SEM. The iCCD images presented in this paper were taken using the iCCD camera of the Plasma Lab, the Institute of Mechanics, Chinese

Academy of Sciences.

REFERENCES

1. A. Schütze, J. Y. Jeong, S. E. Babayan, J. Park, G. S. Selwyn and R. F. Hicks, *IEEE Trans. Plasma Sci.* **26**, 1685-1694 (1998).
2. J. Park, I. Henins, H. W. Herrmann, G. S. Selwyn, J. Y. Jeong, R. F. Hicks, D. Shim and C. S. Chang, *Appl. Phys. Lett.* **76**, 288-290 (2000).
3. F. Massines, A. Rabehi, P. Decomps, R. B. Gadri, P. Segur and C. Mayoux, *J. Appl. Phys.* **83**, 2950-2957 (1998).
4. N. Gherardi, G. Gouda, E. Gat, A. Ricard and F. Massines, *Plasma Sources Sci. Technol.* **9**, 340-346 (2000).
5. J. R. Roth, D. M. Sherman, R. B. Gadri, F. Karakaya, Z. Y. Chen, T. C. Montie, K. Kelly-Wintenberg and P. P.-Y. Tsai, *IEEE Trans. Plasma Sci.* **28**, 56-63 (2000).
6. J. R. Roth, J. Rahel, X. Dai and D. M. Sherman, *J. Phys. D: Appl. Phys.* **38**, 555-567 (2005).
7. M. Laroussi, I. Alexeff, J. P. Richardson, F. F. Dyer, *IEEE Trans. Plasma Sci.* **30**, 158-159 (2002).
8. M. Thiyagarajan, I. Alexeff, S. Parameswaran and S. Beebe, *IEEE Trans. Plasma Sci.* **33**, 322-323 (2005).
9. O. Sakai, T. Shirafuji and K. Tachibana, "Atmospheric-Pressure Discharge Ignited on Flexible Electrodes of Fabric Structure" in *Proc. 18th Int. Symp. Plasma Chem.*, edited by K. Tachibana, O. Takai, K. Ono and T.Shirafuji, Kyoto, Japan, 2007, p. 208.
10. O. Sakai, Y. Kishimoto and K. Tachibana, *J. Phys. D: Appl. Phys.* **38**, 431-441 (2005).
11. H. Baránková and L. Bardos, "Hollow Cathode and Hybrid Atmospheric Plasma Sources" in *Proc. 18th Int. Symp. Plasma Chem.*, edited by K. Tachibana, O. Takai, K. Ono and T.Shirafuji, Kyoto, Japan, 2007, p. 2.
12. H. Baránková and L. Bardos, *Surf. Coat. Technol.*, **174-175**, 63-67 (2003).
13. G. Fridman, A. Shereshevsky, M. M. Jost, A. D. Brooks, A. Fridman, A. Gutsol, V. Vasilets and G. Friedman, *Plasma Chem. Plasma Process.* **27**, 163-176 (2007).
14. X. Duten, M. Redolfi, N. Aggadi, A. Michau and K. Hassouni, "Experimental Studies and Modeling of an Atmospheric Pressure Plasma working in the Nanosecond Regime Coupled with a Catalytic Bed" in *Proc. 18th Int. Symp. Plasma Chem.*, edited by K. Tachibana, O. Takai, K. Ono and T.Shirafuji, Kyoto, Japan, 2007, p. 149.
15. J. Y. Jeong, S. E. Babayan, V. J. Tu, J. Park, I. Hennis, R. F. Hicks and G. S. Selwyn, *Plasma Sources Sci. Technol.* **7**, 282-285 (1998).
16. J. Y. Jeong, S. E. Babayan, A. Schütze, V. J. Tu, J. Park, I. Hennis, G. S. Selwyn and R. F. Hicks, *J. Vac. Sci. Technol. A* **17**, 2581-2585 (1999).
17. V. J. Tu, J. Y. Jeong, A. Schütze, S. E. Babayan, G. Ding, G. S. Selwyn and R. F. Hicks, *J. Vac. Sci. Technol. A* **18**, 2799-2805 (2000).
18. S. E. Babayan, J. Y. Jeong, A. Schütze, V. J. Tu, M. Moravej, G. S. Selwyn and R. F. Hicks, *Plasma Sources Sci. Technol.* **10**, 573-578 (2001).
19. G. R. Nowling, S. E. Babayan, V. Jankovic and R. F. Hicks, *Plasma Sources Sci. Technol.* **11**, 97-103 (2002).
20. H. W. Herrmann, I. Henins, J. Park and G. S. Selwyn, *Phys. Plasmas* **6**, 2284-2289 (1999).
21. M. Vleugels, G. Shama, X. T. Deng, E. Greenacre, T. Brocklehurst and M. G. Kong, *IEEE Trans. Plasma Sci.* **33**, 824-828 (2005).
22. J. Laimer, S. Haslinger, W. Meissl, J. Hell and H. Störi, *Vacuum* **79**, 209-214 (2005).
23. X. Yang, M. Moravej, G. R. Nowling, S. E. Babayan, J. Panelon, J. P. Chang and R. F. Hicks, *Plasma Sources Sci. Technol.* **14**, 314-320 (2005).
24. J. J. Shi, X. T. Deng, R. Hall, J. D. Punnett and M. G. Kong, *J. Appl. Phys.* **94**, 6303-6310 (2003).
25. S. Y. Moon, J. K. Rhee, D. B. Kim and W. Choe, *Phys. Plasmas* **13**, 033502 (2006).
26. J. Park, I. Henins, H. W. Herrmann, G. S. Selwyn and R. F. Hicks, *J. Appl. Phys.* **89**, 20-28 (2001).
27. J. Park, I. Henins, H. W. Herrmann, and G. S. Selwyn, *J. Appl. Phys.* **89**, 15-19 (2001).
28. S. Wang, V. Schulz-von der Gathen and H. F. Döbele, *Appl. Phys. Lett.* **83**, 3272-3274 (2003).
29. M. Laroussi, J. P. Richardson and F. C. Dobbs, *Appl. Phys. Lett.* **81**, 772-774 (2002).
30. M. Laroussi, F. Leipold, *International Journal of Mass Spectrometry* **233**, 81-86 (2004).
31. M. K. Boudam, M. Moisan, B. Saoudi, C. Popovici, N. Gherardi and F. Massines, *J. Phys. D: Appl. Phys.* **39**, 3494-3507 (2006)
32. D. Purevdorj, N. Igura, M. Shimoda, O. Ariyada, I. Hayakawa, *Acta Biotechnol.* **21**, 333-342 (2001).
33. M. Laroussi, *Plasma Process. Polym.* **2**, 391-400 (2005).
34. S. Lerouge, M. R. Wertheimer and L'H. Yahia, *Plasmas and Polymers* **6**, 175-188 (2001).
35. W.-T. Sun, T.-R. Liang, H.-B. Wang, H.-P. Li and C.-Y. Bao, *Plasma Sources Sci. Technol.* **16**, 290–296 (2007).
36. W.-T. Sun, G. Li, H.-P. Li, C.-Y. Bao, H.-B. Wang, S. Zeng, X. Gao and H.-Y. Luo, *J. Appl. Phys.* **101**, 123302 (2007).
37. H.-B. Wang, W.-T. Sun, H.-P. Li, C.-Y. Bao and X.-Z. Zhang, *Appl. Phys. Lett.* **89**, 16502 (2006).
38. H.-B. Wang, W.-T. Sun, H.-P. Li, C.-Y. Bao, X. Gao and H.-Y. Luo, *Appl. Phys. Lett.* **89**, 16504 (2006).
39. W.-T. Sun, G. Li, H.-P. Li, C.-Y. Bao, L.-Y. Wang, Z.-X. Gou, H.-X. Zhao and X.-H. Xing, "Biomolecular Effects Using Radio-Frequency, Atmospheric-Pressure, Bare-Metallic-Electrode Glow Discharge Plasmas" in *Proc. 18th Int. Symp. Plasma Chem.*, edited by K. Tachibana, O. Takai, K. Ono and T.Shirafuji, Kyoto, Japan, 2007, p. 309.
40. T. C. Montie, K. Kelly-Wintenberg and J. R. Roth, *IEEE Trans. Plasma Sci.* **28**, 41-50 (2000).
41. H.-P. Li, W.-T. Sun, H.-B. Wang, G. Li, C.-Y. Bao, *Plasma Chem. Plasma Process.* 2007 (in press).
42. W.-T. Sun, G. Li, S. Wang, H.-P. Li and C.-Y. Bao, "Influences of Oxygen on the Features of Radio-Frequency, Atmospheric-Pressure, Bare-Metallic-Electrode Glow Discharges" in *Proc. 18th Int. Symp. Plasma Chem.*, edited by K. Tachibana, O. Takai, K. Ono and T.Shirafuji, Kyoto, Japan, 2007, p. 288.

Evaluation of Sealing Characteristics in MR Fluid Channel Flow Relating to Magneto-Rheological Properties

Hideya Nishiyama, Hidemasa Takana, Kotoe Mizuki,
Tomoki Nakajima, and Kazunari Katagiri

Institute of Fluid Science, Tohoku University, 2-1-1 Katahira, Aoba-ku, Sendai 980-8577, JAPAN

Abstract. Experimental analyses are conducted on the rheological properties of a commercial MR fluid in a rotating shear mode and on sealing characteristics of MR fluid with correlating the MR fluid structure in pressure mode. The time evolution of axial pressure distribution in pressure mode under a magnetic field and pressure loss at steady flow are also discussed in detail to provide the fundamental data for the novel application of MR seal to a medical or safety devices.

Keywords: Magneto-rheological fluid, Seal, Pressure driven flow, Visualization
PACS: 83.80.Gv, 47.50.-d, 83.85.Cg, 47.85.L-, 47.80.Jk

INTRODUCTION

Magnetorheological fluid (MR fluid) is one of the functional fluids which can respond to magnetic field [1]. Compared with a magnetic fluid, MR fluid has a controllable yield stress and shows a significant viscous force under applied magnetic field [2,3]. Moreover, MR fluid exhibits a semi-rigid behavior even under a small magnetic field [2,3]. These functions of MR fluid come from the robust chain-like cluster formation by magnetic dipolar interaction under even small magnetic field strength [4].

The researches are now actively conducted on the synthesis of a novel MR fluid and the evaluation of its rheological properties, and also on the industrial applications of MR fluid to damper, shock absorber, break, clutch, actuator, and seismic isolation rubber [2-8]. Furthermore, MR seal is now expanding its application to medical field or safety devices, such as blood flow sealing [9] or steam shut-off valve [10], in these days. However, there are few researches on the analysis of rheological properties by considering the complex flow structure of MR fluid and furthermore the researches from the integrated viewpoint through the performance evaluation of MR fluid system are hardly seen [11].

Therefore, in this study, the experimental analyses are conducted on the rheological properties of commercial MR fluid in a rotating shear mode and on sealing performance of MR fluid with correlating the MR fluid structure in pressure mode. The main objective of this study is to provide a fundamental data for the application of MR seal to the medical or safety devices.

TABLE 1. Typical properties of MRF-132DG and MRF-132AD.

MR fluid	MRF-132DG	MRF-132AD
Appearance	Dark Gray Liquid	Dark Gray Liquid
Particle Material	Iron	Iron
Base Oil	Hydrocarbon	Hydrocarbon
Viscosity (Pa s) at 40 °C	0.092 ± 0.015	0.09 ± 0.02
Density (g/cm^3)	2.98 – 3.18	3.09
Solids Content by Weight (%)	80.98	81.64
Flash Point (°C)	> 150	> 150
Operating Temperature (°C)	-40 to +130	-40 to + 130

EXPERIMENTAL ANALYSIS

Test Sample

In this study, MRF-132DG and MRF-132AD manufactured by LORD Corporation are utilized as test samples. The physical properties of MRF-132DG and MRF-132AD officially released by LORD Corp. are listed in the Table 1.

CP982, *Complex Systems, 5th International Workshop on Complex Systems*
edited by M. Tokuyama, I. Oppenheim, and H. Nishiyama

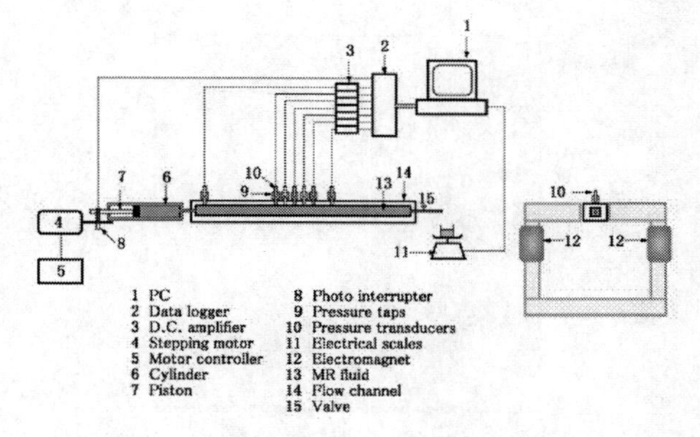

FIGURE 1. Schematic illustration of experimental setup.

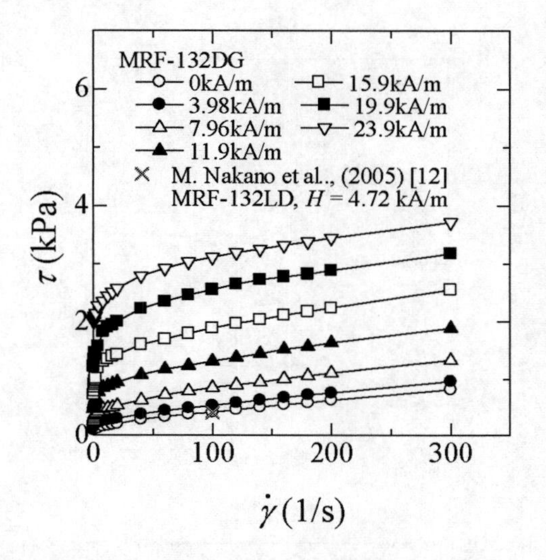

FIGURE 2. Shear stress as a function of shear rate at various magnetic field strengths.

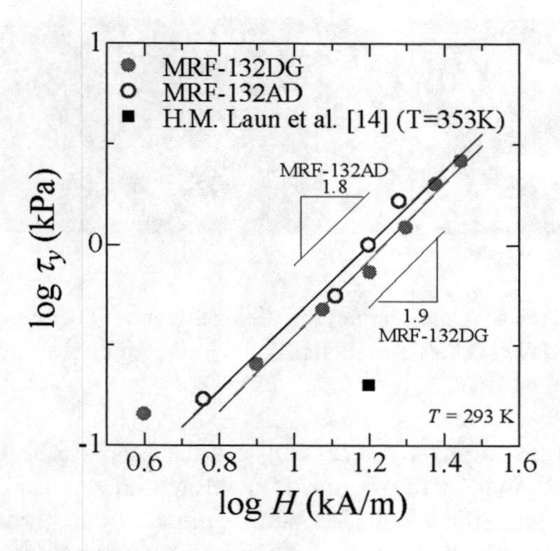

FIGURE 3. Yield stress versus magnetic field strength.

Measurement of Rheological Properties

The rheological properties of MR fluid, such as yield stress, apparent viscosity, etc. are measured by a cone-plate rheometer in a rotating shear mode. A solenoidal coil is installed in the rheometer for the measurement under a magnetic field. A magnetic field is applied vertically to the sample at almost uniform magnetic field strength up to 23.9 kA/m.

Experimental Apparatus

Figure 1 shows the schematic illustration of MR fluid channel flow system. The horizontal MR seal system has been originally manufactured. In this system, MR fluid is driven by piston controlled by a stepping motor. The MR channel is made of acrylics with rectangular cross section of 10 mm x 10 mm. The flow shutdown characteristics under applied magnetic field and also time evolutions of pressure distributions are clarified experimentally. Magnetic field is applied perpendicular to the flow as shown in Fig. 1. The maximum applied magnetic field strength in this experiment is 95.5 kA/m. With MR fluid filled in the channel, the magnetic field strength in the channel reduces by 13.8% in case of applied magnetic field strength of 63.6 kA/m.

Experimental Procedures

The experimental procedures can be summarized as follows.

(1) MR fluid is filled inside the piston cylinder and flow channel. After forming MR fluid seal by applying magnetic field, piston is driven at the constant head velocity of 0.5 mm/s controlled by stepping motor.
(2) A magnetic field is applied by electrical magnet. Pressures at inlet of channel, in the vicinity of applied magnetic field are measured by transducer piezoresistive pressure sensors.
(3) The breakdown of MR fluid seal is filmed by digital video camera from the top.
(4) Exit of flow channel is opened to atmospheric pressure. The mass of exhaust MR fluid is measured by electrical scale for the estimation of flow rate.

RESULTS AND DISCUSSION

Rheological Characteristics of MR Fluid

Figure 2 shows shear stress versus shear rate at various magnetic field strengths. The shear stress

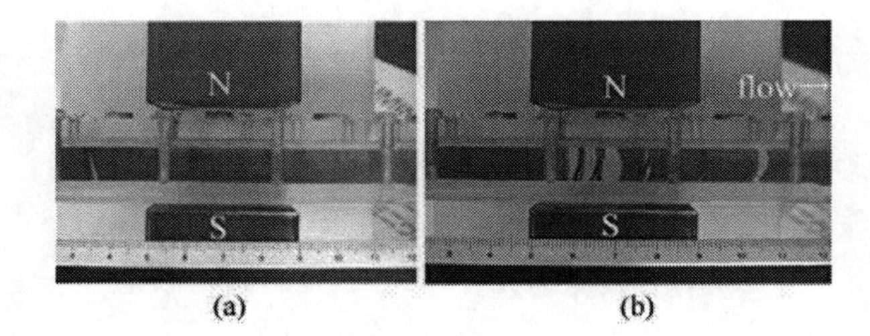

FIGURE 4. (Color online) MR fluid structure (a) before seal breakdown, (b) after seal breakdown in pressure driven flow (H = 63.8 kA/m, v_{piston} = 0.5 mm/s).

obtained at shear rate of 100 1/s for MRF-132LD is also shown in the figure as a referential value [12]. The yield stress increases with magnetic field strength and plastic viscosities are almost the same for any magnetic field strength. The rounded shapes of the curves at low shear rate can be inferred to come from wall slippage of cluster due to nonmagnetic cone and bottom plate installed in rheometer [13].

Figure 3 shows yield stress as a function of magnetic field strength for MRF-132DG and MRF-132AD manufactured by Lord Corp. The yield stress is defined as the shear stress at the shear rate of 0.2 1/s in this study. The yield stresses of both fluids show a power law dependence on magnetic field. As shown in the figure, MRF-132DG and MRF-132AD have almost the same power law index and the obtained power law indexes are 1.9 for MRF-132DG and 1.8 for MRF-132AD, respectively. Compared to MRF-132AD, MRF-132DG shows a little smaller yield stress especially under higher magnetic field strength. Thus the fluidity of MRF-132DG under magnetic field is better than that of MRF-132AD. Compared with other MR fluid containing 55 weight % of carbonyl iron as a magnetic particle [14], both MRF-132DG and MRF-132AD show larger yield stress mainly due to higher concentration of magnetic particles.

Sealing Performance of MR Fluid

Visualization of MR Fluid Structure in Pressure Driven Flow

Figures 4 (a) and (b) show pictures of MR fluid structure near magnet in pressure driven flow. Figure 4(a) shows the picture just before the seal breakdown and (b) shows the picture after the seal breakdown and MR fluid flows out at constant flow rate. These pictures are taken at piston head velocity of 0.5 mm/s under magnetic field strength of H = 63.8 kA/m. In these pictures, MR fluid is driven toward the right. For the clear visualization of cluster structure in pressure

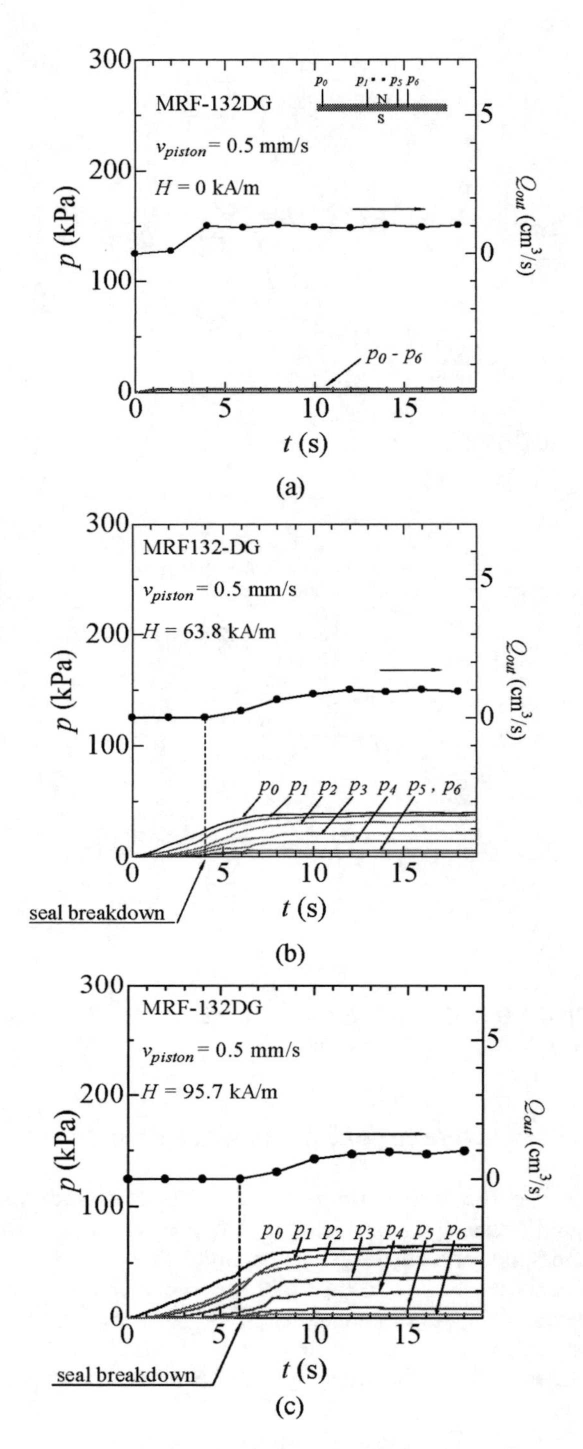

FIGURE 5. Time evolution of pressures and flow rate at (a) H = 0 kA/m, (b) H = 63.8 kA/m and (c) H = 95.7 kA/m.

mode, MRF-132DG diluted with silicon oil (dynamic viscosity of 1.0 x 10^{-7} m^2/s) at volume ratio of 1:1 is utilized here. By applying magnetic field without

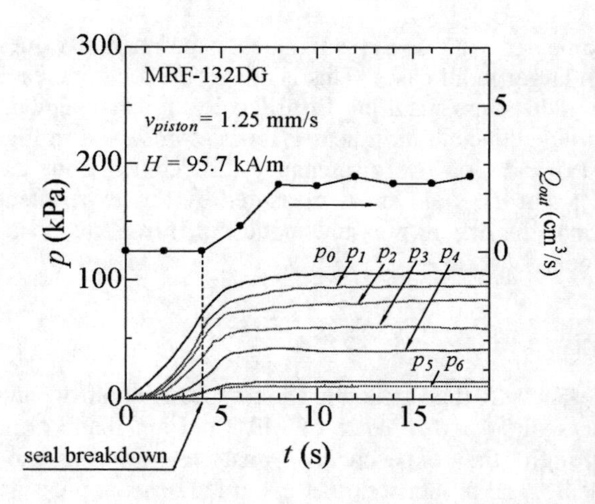

FIGURE 6. Time evolution of pressures and flow rate at H = 95.7 kA/m under piston head velocity of 1.25 mm/s.

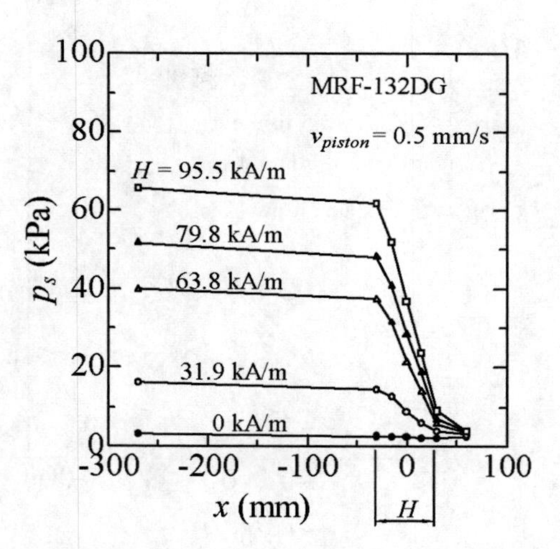

FIGURE 8. Axial pressure distribution for various magnetic field strengths.

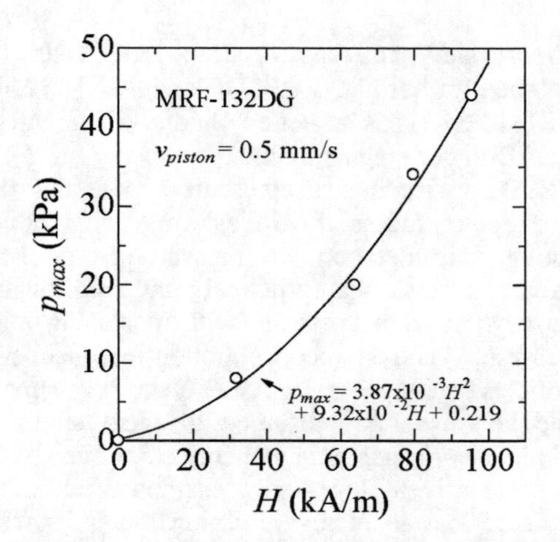

FIGURE 7. MR seal pressure as a function of magnetic field strength.

moving the piston, region with high magnetic particle density is formed in the range of 90 mm outward from the center of magnet. When moving the piston, MR clusters are gradually gathered toward the magnet from upstream with increasing inlet pressure and local concentration of magnetic particles. Then the concentration of magnetic particles becomes locally higher near the magnet. Once the upstream pressure reaches up to seal breakdown pressure, aggregated clusters near magnet suddenly starts to flow without breaking up. Under the steady state at constant flow rate, clusters are repeatedly formed and are flown out. Under the steady state in pressure driven flow, the cluster forms a bow-like configuration as observed in Fig. 4 (b) due to the phase separation especially under the dilute condition.

Time Evolution of Channel Pressure and Flow Rate in Pressure Driven Flow

Figures 5 (a), (b) and (c) show the time evolutions of channel pressures and exhaust flow rate at piston head velocity of 0.5 mm/s for H = 0 kA/m, 63.8 kA/m and 95.7 kA/m, respectively. Piston starts at 0 s. Each experiment is carried out after 3 minutes in resting state for the cluster formation. Breakdown of MR seal occurs when pressure difference between p_5 and p_6 appears as indicated by change in flow rate. The pressure difference between the most upstream pressure p_0 and the most downstream pressure p_6 is defined as seal pressure p_{max}. The pressures increase drastically with applied field. The seal breakdown is observed at 4 s in case of H = 63.8 kA/m. Seal pressure and the time before seal breakdown increase with applied magnetic field strength.

Figure 6 shows the time evolutions of pressures and exhaust flow rate at piston velocity of 1.25 mm/s for H = 95.7 kA/m. Compared with Fig. 5 (c), the piston head velocity is 2.5 times faster under the same magnetic field strength. With increasing piston head velocity, inlet pressure becomes higher at seal breakdown and the larger difference is observed between p_0 and p_6. Furthermore, the time before seal breakdown is shortened. The increase in seal pressure is due to larger inertia of MR fluid and also due to local agglomeration of magnetic particles near the magnet with increasing pressure as clarified from the visualization shown in Fig. 4 (a).

Figure 7 shows the correlation between applied magnetic field strength and seal pressure. Seal pressure increases quadratically corresponding to yield stress as

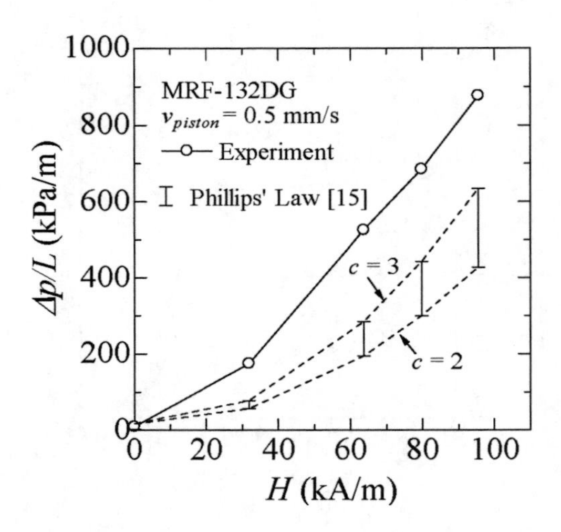

FIGURE 9. Pressure loss in magnetic region as a function of magnetic field strength.

shown in Fig. 3. The seal pressure of 45 kPa can be obtained for magnetic field strength of 95.7 kA/m.

Figure 8 shows axial pressure distributions for various magnetic field strengths at steady state after seal breakdown. Significant pressure drop is observed in the applied magnetic region. The pressure decreases quite linearly in the magnetic region. The gradient of pressure drop in the magnetic field becomes larger with magnetic field strength.

Pressure Loss in MR Fluid Flow

Figure 9 shows the relation between applied magnetic field and pressure loss in a magnetic region. The pressure loss in magnetic region increases with magnetic field strength related to the dependence of yield stress on applied magnetic field. The analytical solution of pressure loss in pressure driven MR fluid flow is given by Phillips and is described as the following relation [15]:

$$\Delta p = \Delta p_\eta + \Delta p_\tau(H) = \frac{12\eta QL}{h^3 w} + \frac{c\tau_y(H)L}{h} \quad (1)$$

where L, h and w are channel length, height and width, respectively. Q and η appeared in the above equation are volume flow rate and viscosity under no magnetic field. The parameter c in eq. (1) varies from 2 to 3 depending on the velocity profile. The first term in eq. (1) shows pressure loss due to viscosity and second term is due to yield stress. Under magnetic field, pressure loss resulting from yield stress is much larger than that from viscosity. In the case of H = 95.7 kA/m, pressure loss due to yield stress is 40 times larger. Both experiment and analytical values increase with magnetic field strength and qualitatively agree well. Compared with analytical values, experimental values are larger in all cases. This is due to the local increase in yield stress resulting from locally higher magnetic particle concentration near magnet as observed in Fig. 4 (a) and (b). The discrepancy also comes from the fact that the yield stress measured by this experiment is not the one in pressure mode but in rotating shear mode.

CONCLUSIONS

The MR fluid structure in pressure driven flow and the sealing performance of MR fluid are made clear through the experimental analyses of magneto-rheological properties, time evolution of axial pressure distribution in pressure driven flow under a magnetic field and pressure loss at steady state with utilizing MRF-132DG and MRF-132AD as test samples. The obtained results can be summarized as follows.

(1) Almost the same power low index can be obtained for both MRF-132DG and MRF-132AD. MRF-132DG has a better fluidity than MRF-132AD under magnetic field.
(2) MR fluid structure in pressure mode and the breakdown process of MR seal are clearly shown through visualization. In pressure driven MR flow, MR clusters are gradually gathered toward the magnet from upstream with increasing inlet pressure. Then the concentration of magnetic particles becomes locally higher near the magnet.
(3) The pressure loss attributed to yield stress is dominantly larger than that due to viscosity in pressure driven flow under magnetic field. The pressure loss in magnetic region increases with magnetic field strength related to the dependence of yield stress on applied magnetic field strength.

ACKNOWLEDGMENTS

The present study was partly supported by a Grant-in-aid for Exploratory Research from the Japan Society for Promotion of Science and also by a 21st Century COE program Grant of the International COE of Flow Dynamics from the Ministry of Education, Culture, Sports, Science and Technology.

REFERENCES

1. R. G. Larson, *The Structure and Rheology of Complex Fluids*, London: Oxford University Press, 1999, pp. 376-378.
2. M. R. Jolly, J. W. Bender and J. D. Carlson, *J. Intell. Mater. Syst. Struct.* **10**, 5-13 (1999).
3. J. D. Carlson, D. M. Catanzarite and K. A. St Clair, *Int. J. Mod. Phys. B* **10**, 2857-2865 (1996).

4. J. Liu, *JSME Int J., Ser. B* **45**, 55-60 (2002).
5. J. D. Carlson and M. R. Jolly, *Mechatronics* **10**, 555-559 (2000).
6. L. Zipser, L. Richter and U. Lange, *Sens. Actuators A* **95**, 318-325 (2001).
7. M. D. Symans and M. C. Contantinou, *Engineering Structures* **21**, 469-487 (1999).
8. F. Gordaninejad and S. P. Kelso, *J. Intell. Mater. Syst. Struct.* **11**, 395-406 (2000).
9. R. Sheng, G. A. Flores and J. Liu, *J. Magn. Magn. Mater.* **194**, 167-175 (1999).
10. S. Shimizu, T. Saito and H. Ikeda, "Development of a Fail-safe Safety Valve and Safety Control System for Boilers", *Specific Research Reports of the National Institute of Industrial Safety*, NIIS-PR-2002, 2003, pp. 29-39 (in Japanese) .
11. H. Nishiyama, K. Katagiri, K. Hamada and K. Kikuchi, *Int. J. Mod Phys B* **19**, 1437-1442 (2005).
12. M. Nakano, A. Satou, Y. Sugamata and H. Nishiyama, *JSME Int J., Ser. B* **48**, 494-500 (2005).
13. J. Popplewell, R. E. Rosensweig, J. K. Siller, *J. Magn. Magn. Mater.* **149**, 53-56 (1995).
14. H. M. Laun, C. Kormann and N. Willenbacher, *Rheol. Acta* **35**, 417-432 (1996).
15. R. W. Phillips, "Engineering Application of Fluids with a Variable Yield Stress", Ph.D. Thesis, University of California, Berkley, 1969.

Numerical Simulation of Structure Formation of Magnetic Particles and Nonmagnetic Particles in MAGIC Fluids under Steady Magnetic Field

Yasushi Ido and Takafumi Inagaki

Department of Engineering Physics, Electronics and Mechanics, Nagoya Institute of Technology, Gokiso-cho, Showa-ku, Nagoya 466-8555, Japan

Abstract. Distribution and microstructure formation of both magnetic particles and nonmagnetic particles in MAGIC (MAGnetic Intelligent Compound) fluids are investigated numerically. We show the particles distribution in MAGIC fluid under applied steady uniform magnetic field. In order to arrange nonmagnetic particles in the magnetic field direction, the diameter of nonmagnetic particles should be the same as the diameter of magnetic particles. Distribution uniformity of particles under applied magnetic field is almost the same compared with the uniformity of the initial state.

Keywords: Magnetic intelligent fluid, Magnetorheological fluid, Cluster, Microstructure, Numerical simulation
PACS: 47.65.Cb, 47.57.J-, 47.57eb, 47.57.E-

INTRODUCTION

It is well known that magnetic particles in magnetorheological fluids form clusters under applied steady uniform magnetic field [1-2]. There are several applications using this phenomenon and MAGIC (MAGnetic Intelligent Compound) polishing pad is one of examples. MAGIC is a solidified magnetorheological fluid containing both micron size magnetic particles and nonmagnetic abrasive particles under applied magnetic field [3-4]. Cluster formation is used for arranging nonmagnetic abrasive particles in the field direction. Distribution of nonmagnetic abrasive particles can be controlled by changing physical conditions of microstructure formation process. Distribution of nonmagnetic abrasive particles has an influence on the finishing state of polishing. Thus, it is important to know how these particles are arranged in MAGIC fluid under magnetic field. And distribution and microstructure formation of suspended particles in magnetorheological fluids are one of the interesting subjects in statistical physics as a kind of phase transition phenomenon. The authors have reported distribution of particles in MAGIC fluids under magnetic field [5-6]. In the previous papers, we have shown the effects of both magnetic field intensity and fractions of particles on microstructure formation.

In this paper, microstructure formation of particles in MAGIC fluids under applied magnetic field is investigated by using numerical simulation of the simple particle method [1,6]. In particular, we take the ratio of diameter of nonmagnetic particles to diameter of magnetic particles, as one of the changing physical conditions. Influence of this ratio on distribution of particles is studied. Effects of the ratio of volume fraction of magnetic particles to volume fraction of all particles on microstructure formation are also examined. Distribution of particles and microstructure formation are analyzed statistically.

NOMENCLATURE

$\boldsymbol{B}$: magnetic flux density
CU_L : averaged Christiansen's uniformity coefficient
C_v : contact coefficient
D_i^r : diffusion coefficient for rotational motion
D_i^t : diffusion coefficient for translational motion
d_i : diameter of the i th particle
$d_{i,L}$: mean distance between the target particle and the neighbor particles on the Lth plane
d_{mag} : diameter of magnetic particles
d_{non} : diameter of nonmagnetic particles
$\boldsymbol{F}_i$: total force acting on the particle

CP982, *Complex Systems, 5th International Workshop on Complex Systems*
edited by M. Tokuyama, I. Oppenheim, and H. Nishiyama

$\boldsymbol{F}_{ij}^{M}$: magnetic dipole interaction force

$\boldsymbol{F}_{ij}^{rep}$: repulsive force

$\boldsymbol{H}$: magnetic intensity vector

I_i : inertia moment of the particle

k_B : Boltzmann constant

m_i : mass of the particle

N_L : number of particles existing on the Lth plane

$\boldsymbol{r}_i$: position vector of the particle

S : distance between the surface of two particles

T : absolute temperature

$\boldsymbol{T}_i$: total torque acting on the particle

$\boldsymbol{T}_{ij}^{M}$: magnetic torque due to the dipole-dipole interaction

$\boldsymbol{T}_i^{field}$: magnetic torque due to the magnetic field

ΔCU : standard deviation of Christiansen's uniformity coefficient

β_i : ratio of diameter of the particle to the diameter of magnetic particles

η : viscosity of fluid

κ : Debye-Hückel parameter

μ : magnetic permeability of magnetic particles

μ_0 : magnetic permeability in vacuum

μ_S : saturated magnetization of magnetic particles

$\boldsymbol{\mu}_i$: magnetic moment of a magnetic particle

ρ : density of magnetic particles

ϕ_{all} : volume fraction of all particles

ϕ_{mag} : volume fraction of magnetic particles

$\boldsymbol{\Omega}_i$: angular velocity of particle

SIMULATION METHOD

Basic Equations

The motion of ith particle having mass m_i and inertia moment I_i at time t and position $\boldsymbol{r}_i$ is described by the following equations:

$$m_i \frac{d^2\boldsymbol{r}_i}{dt^2} = \boldsymbol{F}_i - \frac{k_B T}{D_i^t}\frac{d\boldsymbol{r}_i}{dt}, \qquad (1)$$

$$I_i \frac{d\boldsymbol{\Omega}_i}{dt} = \boldsymbol{T}_i - \frac{k_B T}{D_i^r}\boldsymbol{\Omega}_i, \qquad (2)$$

where k_B is the Boltzmann constant, T is the absolute temperature, $\boldsymbol{\Omega}_i$ is the angular velocity, D_i^t and D_i^r are the diffusion coefficients for translational and rotation motion of sphere particles [7], $\boldsymbol{F}_i$ is the total force acting on the particle and it includes the dipole-dipole interactions and repulsive forces based on the DLVO theory. $\boldsymbol{T}_i$ is the total torque acting on the particle and it is the sum of the torque due to the dipole-dipole interactions and the torque due to the interaction between applied magnetic field and the magnetic moment of the particle. In our simulations, Brownian force and Brownian torque are ignored. The total force $\boldsymbol{F}_i$ is given by

$$\boldsymbol{F}_i = \sum_{j(\neq i)} \boldsymbol{F}_{ij}^{M} + \sum_{j(\neq i)} \boldsymbol{F}_{ij}^{rep} + \sum_{k} \boldsymbol{F}_{ik}^{rep}, \qquad (3)$$

where subscript j indicates the magnetic particles and subscript k indicates the nonmagnetic particles. The magnetic dipole interaction $\boldsymbol{F}_{ij}^{M}$ and the repulsive force $\boldsymbol{F}_{ij}^{rep}$ are expressed, respectively, by

$$\boldsymbol{F}_{ij}^{M} = \frac{3}{4\pi\mu_0 r_{ij}^4}\left\{(\boldsymbol{\mu}_i \bullet \boldsymbol{\mu}_j)\frac{\boldsymbol{r}_{ij}}{r_{ij}} - 5(\boldsymbol{\mu}_i \bullet \boldsymbol{r}_{ij})(\boldsymbol{\mu}_j \bullet \boldsymbol{r}_{ij})\frac{\boldsymbol{r}_{ij}}{r_{ij}^3} + \left[(\boldsymbol{\mu}_i \bullet \boldsymbol{r}_{ij})\boldsymbol{\mu}_j + (\boldsymbol{\mu}_j \bullet \boldsymbol{r}_{ij})\boldsymbol{\mu}_i\right]\frac{1}{r_{ij}}\right., \qquad (4)$$

$$\boldsymbol{F}_{ij}^{rep} = \frac{3\mu_S^2}{4\pi\mu_0 d_{\text{mag}}^4}\exp(-\kappa S)\frac{\boldsymbol{r}_{ij}}{r_{ij}}, \qquad (5)$$

where μ_0 is the magnetic permeability in vacuum, μ_S is the saturated magnetization of the magnetic particles, S is the distance between the surfaces of two particles, d_{mag} is the diameter of magnetic particles, κ is the coefficient corresponding to the Debye-Hückel parameter ($\kappa = 40$ in our simulations), $\boldsymbol{r}_{ij} = \boldsymbol{r}_i - \boldsymbol{r}_j$, $\boldsymbol{\mu}_i$ is the magnetic moment given by

$$\boldsymbol{\mu}_i = \frac{\pi d^3}{2} \frac{\mu - \mu_0}{\mu + 2\mu_0} \boldsymbol{B}, \tag{6}$$

in a weak magnetic flux density $\boldsymbol{B}$, where μ is the magnetic permeability of magnetic particles. The total torque $\boldsymbol{T}_i$ is given by the following equation.

$$\boldsymbol{T}_i = \sum_{j(\neq i)} \boldsymbol{T}_{ij}^M + \boldsymbol{T}_i^{field}, \tag{7}$$

where $\boldsymbol{T}_{ij}^M$ is the dipole-dipole torque and $\boldsymbol{T}_i^{field}$ is the magnetic torque due to the applied magnetic field. They are given, respectively, by

$$\boldsymbol{T}_{ij}^M = \frac{1}{4\pi\mu_o r_{ij}^3}\left[\boldsymbol{\mu}_i \times \boldsymbol{\mu}_j - (\boldsymbol{\mu}_j \bullet \boldsymbol{r}_{ij})\boldsymbol{\mu}_i \times \boldsymbol{r}_{ij}\frac{3}{r_{ij}^2}\right], \tag{8}$$

$$\boldsymbol{T}_i^{field} = \boldsymbol{\mu}_i \times \boldsymbol{H}, \tag{9}$$

where $\boldsymbol{H}$ is the intensity vector of applied magnetic field.

We introduce scale unit of some physical quantities for dimensionless form: the unit of length as d_{mag}, force as $F_0 = 3\mu_S^2 / 4\pi\mu_o d_{\text{mag}}{}^4$, time as $t_0 = \rho d_{\text{mag}}{}^2 / 18\eta$, where η is the viscosity of the fluid, and angular velocity as $1/t_0$. The basic equations (1) and (2) can be rewritten as the following dimensionless forms:

$$\frac{d\boldsymbol{r}_i^*}{dt} = \frac{A\boldsymbol{F}_i^*}{\beta_i}, \tag{10}$$

$$\boldsymbol{\Omega}_i^* = 3A\boldsymbol{T}_i^* / \beta_i^3, \tag{11}$$

where the parameter β_i is the ratio of diameter of the particle to the diameter of magnetic particles and it is defined by $\beta_i = d_i / d_{\text{mag}}$. The parameter A is defined by

$$A = \frac{\rho d_{\text{mag}}{}^2}{288\mu_o \eta^2}\left(\frac{\mu - \mu_0}{\mu + 2\mu_0} B\right)^2, \tag{12}$$

where ρ is the density of the magnetic particles.

Analytical Model and Assumptions

A three-dimensional system to be simulated is based on a dipole model for magnetic particles and a particle method [4-5]. We use a hard sphere picture without full hydrodynamics effects through the medium such as many-body hydrodynamics interactions, lubrication force, and buoyancy. Figure 1 shows the analytical model of our simulations. Sphere particles are arranged randomly at the initial state. An external steady uniform magnetic field is applied to the z-direction. Periodic boundary condition is imposed in all directions.

In our simulations, we have two parameters. One is the ratio of the volume fraction of magnetic particles to the volume fraction of all particles $\phi_{\text{mag}} / \phi_{\text{all}}$, and the other is the ratio of the diameter of nonmagnetic particles to that of magnetic particles $d_{\text{non}} / d_{\text{mag}}$. Simulations are conducted on eight different ratios of volume fraction of magnetic particles to that of all particles: 0.125, 0.25, 0.375, 0.5, 0.625, 0.75, 0.875 and 1.0 as $\phi_{\text{mag}} / \phi_{\text{all}}$, and six different ratios of the diameter of nonmagnetic particles to the diameter of magnetic particles, $d_{\text{non}} / d_{\text{mag}} = 0.5$, 0.8, 1.0, 1.2, 1.5 and 2.0. The volume ratio of all particles is constant of 0.32. Number of particles is, for example, 600 when all particles are magnetic particles. Numerical region is $10d_{\text{mag}} \times 10d_{\text{mag}} \times 10d_{\text{mag}}$ in cases of $d_{\text{non}} / d_{\text{mag}}$ =1.0, 1.2, 1.5 and 2.0, while numerical region is $8d_{\text{mag}} \times 8d_{\text{mag}} \times 8d_{\text{mag}}$ in case of $d_{\text{non}} / d_{\text{mag}}$ =0.5 and 0.8. Applied magnetic field is constant (approximately 70 mT) in our simulations, so the parameter A is constant of 10^4.

The simple Euler method with a time step $10^{-9}t_0$ is used to integrate the above mentioned equations (10) and (11).

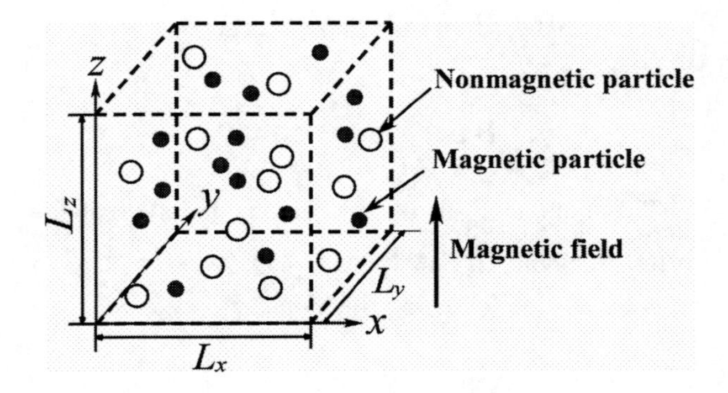

FIGURE 1. Analytical model for numerical simulations.

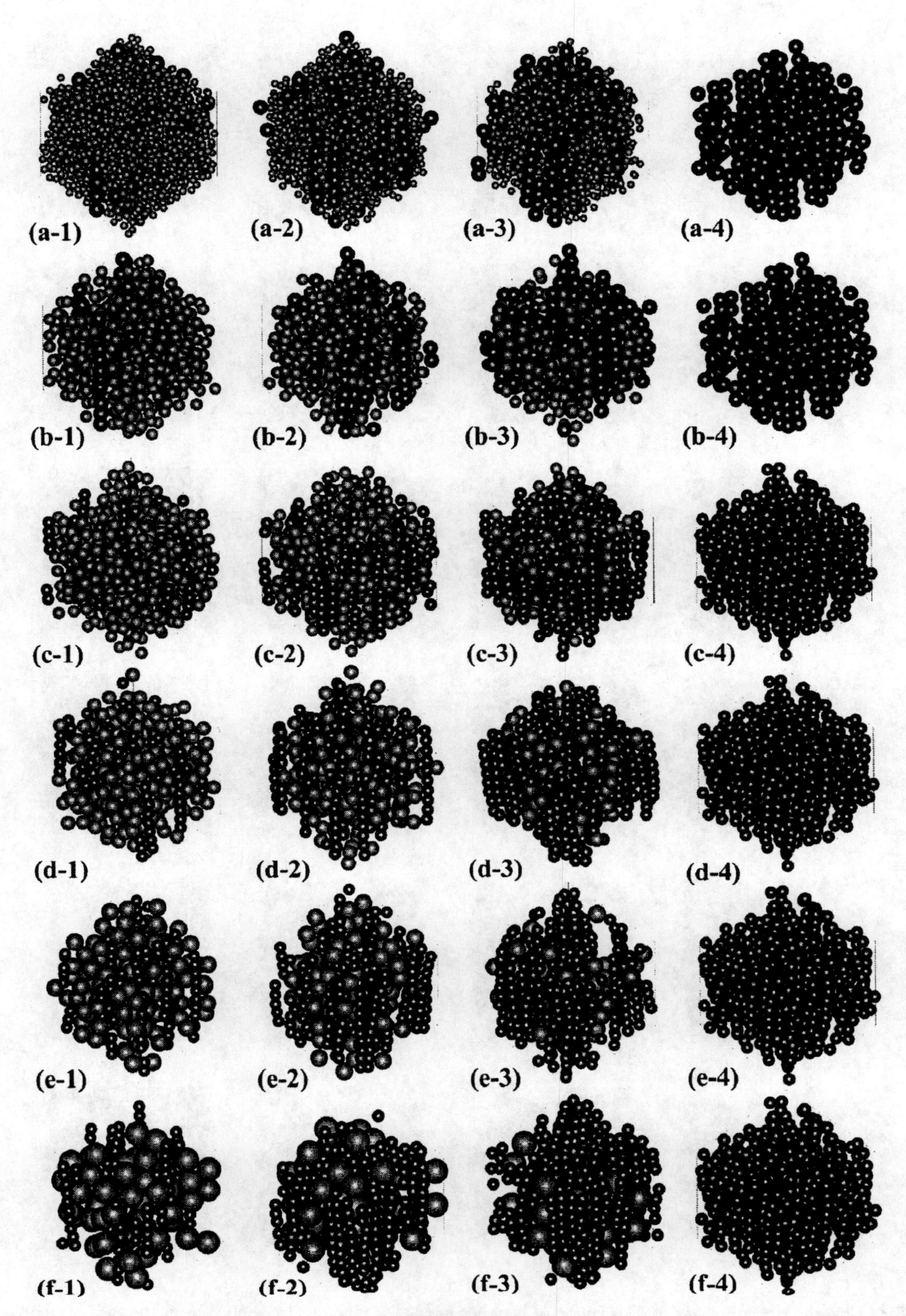

FIGURE 2. Snapshots of bird's eye views of distribution of particles after 10^5 steps in MAGIC fluids under applied magnetic field. The ratio of the diameters of particles is (a) 0.5, (b) 0.8, (c) 1.0, (d) 1.2, (e) 1.5 and (f) 2.0, respectively. The ratio of volume fraction of magnetic particles to that of all particles is (1) 0.25, (2) 0.5, (3) 0.75 and (4) 1.0, respectively.

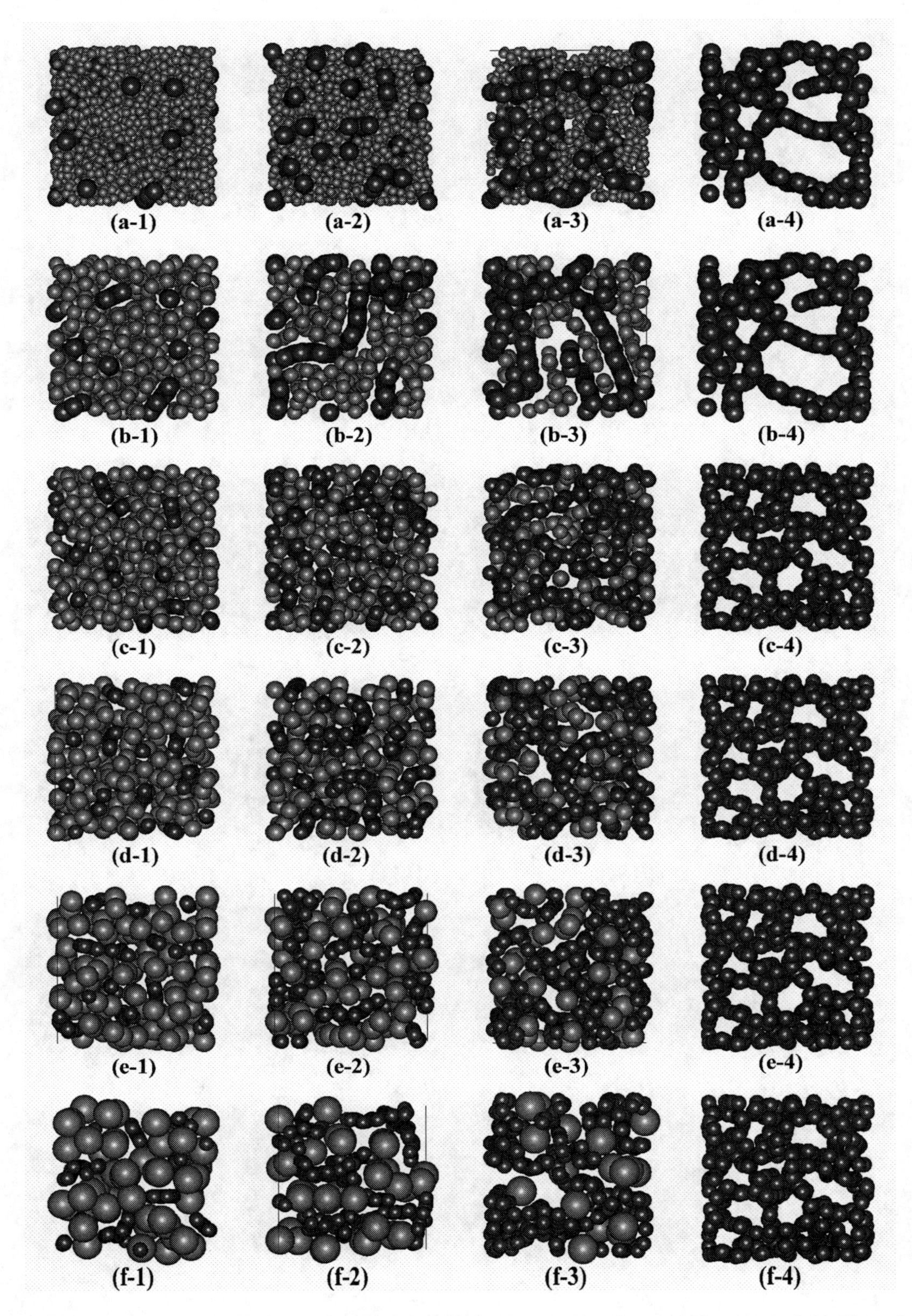

FIGURE 3. Snapshots of top views of distribution of particles in MAGIC fluid under applied magnetic field. The ratio of the diameters of particles is (a) 0.5, (b) 0.8, (c) 1.0, (d) 1.2, (e) 1.5, and (f) 2.0, respectively. The ratio of volume fraction of nonmagnetic particles to that of magnetic particles is (1) 0.25, (2) 0.5, (3) 0.75, and (4) 1.0. The dark color balls are magnetic particles and the light color balls are nonmagnetic particles.

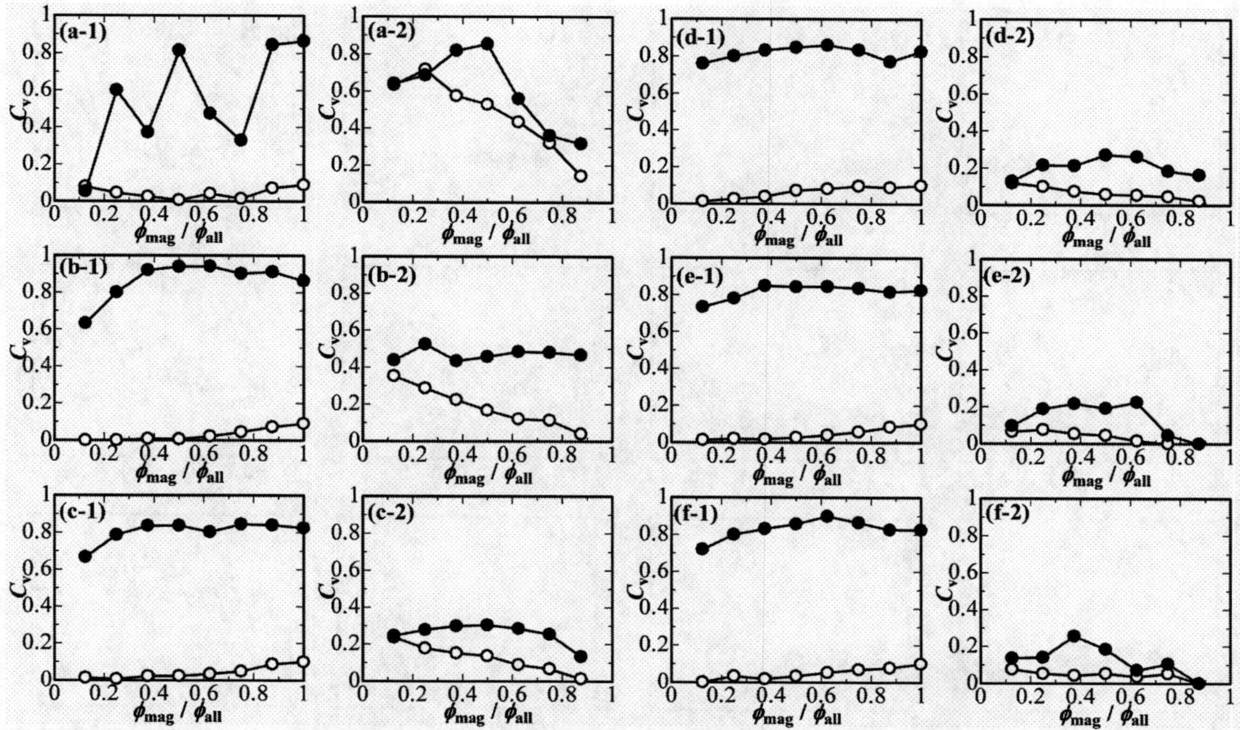

FIGURE 4. Contact coefficients of (1) magnetic particles and (2) nonmagnetic particles. The ratio of the diameter of nonmagnetic particles to that of magnetic particles is (a) 0.5, (b) 0.8, (c) 1.0, (d) 1.2, (e) 1.5 and (f) 2.0, respectively.

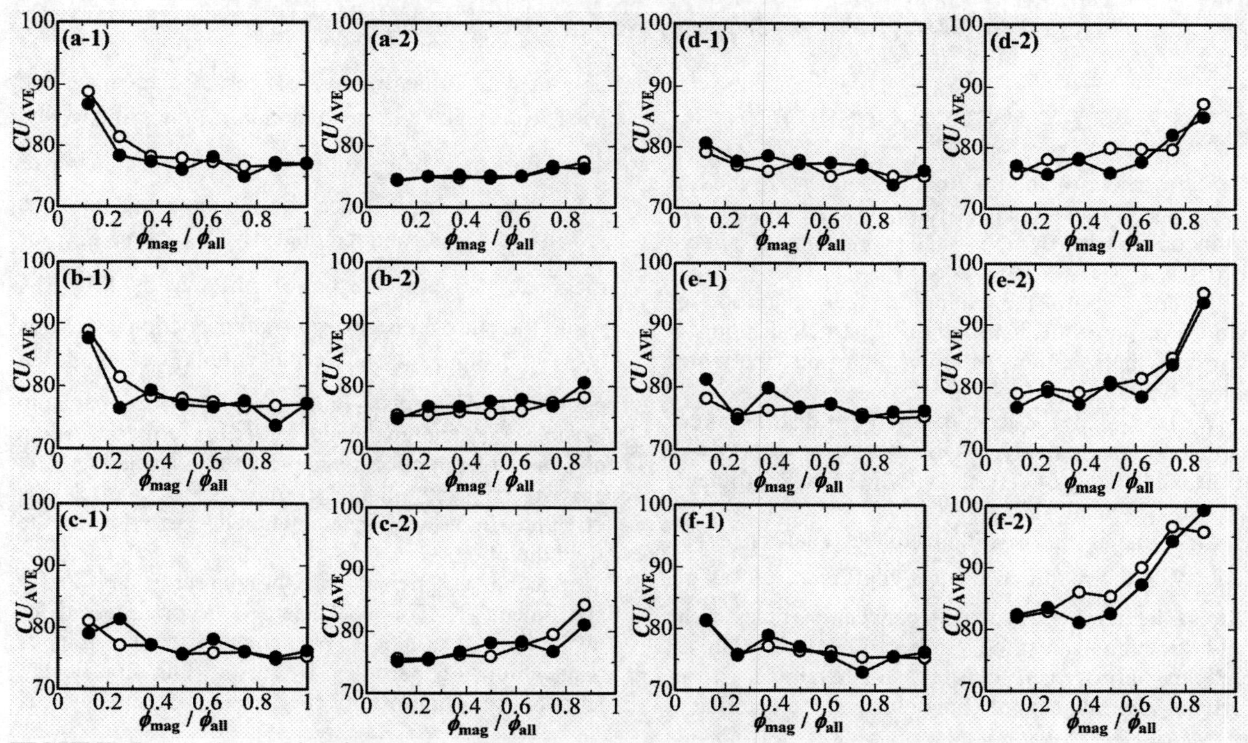

FIGURE 5. Average Christiansen's uniformity coefficients for (1) magnetic particles and (2) nonmagnetic particles. The ratio of the diameter of nonmagnetic particles to the diameter of magnetic particles is (a) 0.5, (b) 0.8, (c) 1.0, (d) 1.2, (e) 1.5 and (f) 2.0, respectively.

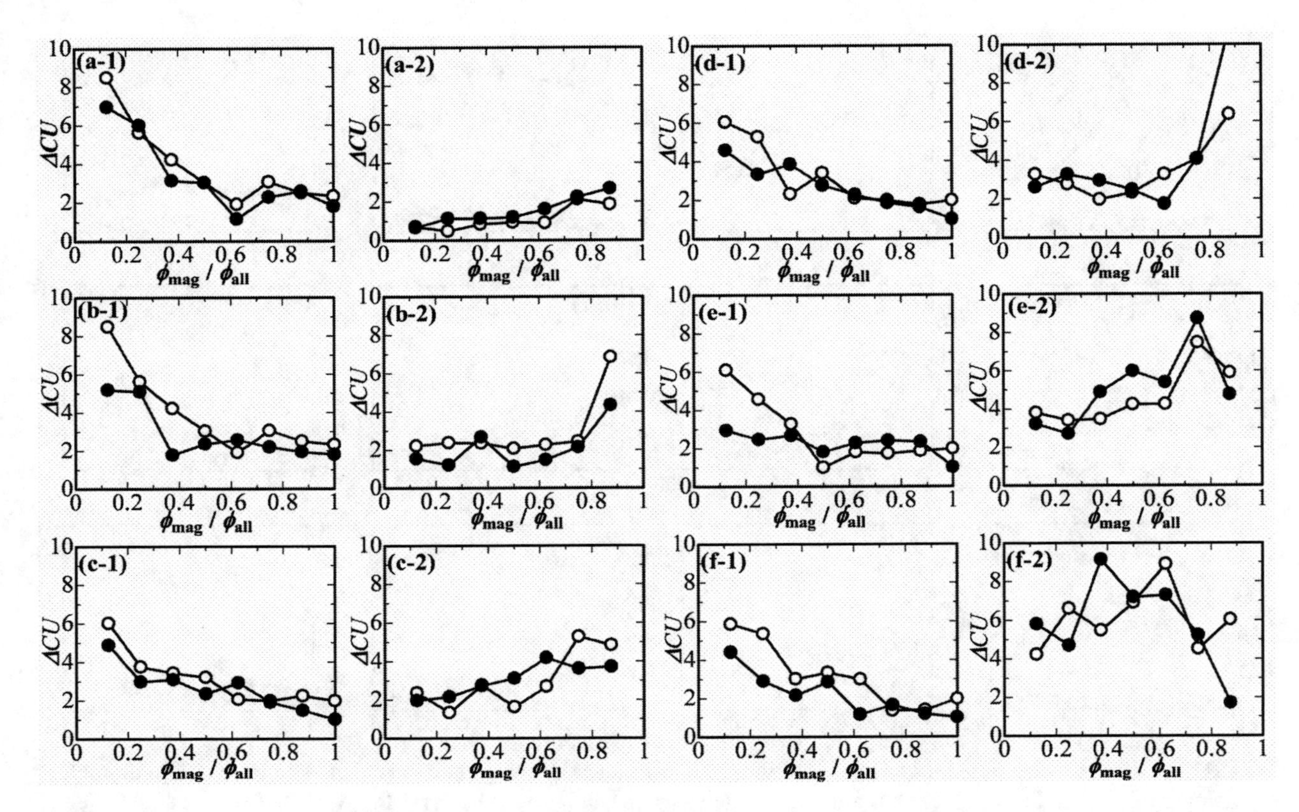

FIGURE 6. Standard deviation of Christiansen's uniformity coefficients for (1) magnetic particles and (2) nonmagnetic particles. The ratio of the diameter of nonmagnetic particles to the diameter of magnetic particles is (a) 0.5, (b) 0.8, (c) 1.0, (d) 1.2, (e) 1.5 and (f) 2.0, respectively.

RESULT AND DISCUSSION

The particle method simulations are performed to investigate microstructure formation of interacting magnetic sphere particles and nonmagnetic sphere particles.

Figure 2 shows the examples of the snapshots of bird's eye views of distribution of particles. Figure 3 illustrates the snapshots of top views of distribution of particles.

We have analyzed from the views of distribution of particles qualitatively. However, we need a quantitative analysis. Thus, we define two coefficients for statistical analysis.

First one is the modified contact coefficient [1]. We define the modified contact coefficient C_{v} as the ratio of the number of contacts between particles in the field direction against the number of contacts when all particles form chain clusters, for example, if all particles are members of chain clusters, the coefficient C_{v} equals to 1. When the neighbor particle is located inside the cut-off distance r_{c} from the target particle and its direction is in the limited direction, we count it as a contact. When the cut-off distance for magnetic particles is $1.5d_{\mathrm{mag}}$, we obtain almost the same results as the case of $1.1d_{\mathrm{mag}}$ of the cut-off distance, however, when the cut-off distance for nonmagnetic particles is $1.1d_{\mathrm{mag}}$, we cannot count any contacts. So, for magnetic particles, the cut-off distance is $1.1d_{\mathrm{mag}}$, while the cut-off distance for nonmagnetic particles is $1.5d_{\mathrm{mag}}$. Figure 4 shows the variation of modified contact coefficients. From Fig.4, both magnetic particles and nonmagnetic particles form chain-like clusters under applied magnetic field. Especially, using almost the same diameter of magnetic particles is effective in arranging nonmagnetic particles in the field direction.

Second coefficient is Christiansen's uniformity coefficient [8]. This coefficient is originally used for evaluation of sprinklers. We can analyze uniformity of distribution of particles by using this coefficient. Christiansen's uniformity coefficient is defined by

$$CU_L = \left(1 - \frac{D_L}{M_L}\right) \times 100 \text{ [\%]}, \qquad (13)$$

where

$$D_L = \sum_{i=1}^{N_L} \left(d_{i,L} - M_L\right) / N_L \,, \tag{14}$$

$$M_L = \sum_{i=1}^{N_L} \frac{d_{i,L}}{N_L} \,, \tag{15}$$

and N_L is the number of particles existing on the Lth plane and $d_{i,L}$ is the mean distance between the particle and the neighbor particles on the Lth plane. For example, if the particles are distributed completely uniform, then CU_L equals to 100%. CU_L is calculated on the plane perpendicular to the field sliced every $0.5d_{\text{mag}}$. Figure 5 demonstrates the average Christiansen's uniformity coefficients for magnetic particles and for nonmagnetic particles. From these figures, in almost all cases we calculated, Christiansen's uniformity coefficients do not change after applied magnetic field compared with the coefficients in the initial state. Thus, the uniformity of particles distribution is maintained even after particles distribution is changed by applying magnetic field.

The standard deviation of Christiansen's uniformity coefficient is calculated by the following equation.

$$\Delta CU = \sqrt{\sum_{i=1}^{N_L} \left(CU_{\text{AVE}} - CU_L\right)^2} \,. \tag{16}$$

If the standard deviation is small, almost the same surfaces appear at any time when we use it as a polishing pad. Figure 6 illustrates the standard deviation of Christiansen's uniformity coefficients. As can be seen in Fig. 6, standard deviation of CU is not affected by applying magnetic field in MAGIC fluids with sphere suspended particles even when the diameter of nonmagnetic particles and that of magnetic particles is different.

From results mentioned above, physical conditions using nonmagnetic particles with diameter of $0.8d_{\text{mag}}$ or d_{mag} and taking the ratio of volume fraction $\phi_{\text{mag}} / \phi_{\text{all}}$ = 0.5 or 0.625 are slightly effective to increase uniformity and to arrange nonmagnetic particles in the field direction, compared with other conditions.

CONCLUSIONS

We have simulated various ordering processes and microstructure formation in MAGIC fluids by performing the particle method. In order to arrange nonmagnetic abrasive particles in the field direction, using magnetic particles with almost the same diameter is suitable. The uniformity of particles distribution is kept even after particles distribution is changed by applying magnetic field.

ACKNOWLEDGMENTS

This work was partly supported by the Hosokawa Powder Technology Foundation and a Grant-in-Aid for Scientific Research of Japan Society for the Promotion of Science.

REFERENCES

1. M. Mohebi, N. Jamasbi and J. Liu, *Phys. Rev. Letters* **54**, 5407-5413 (1996).
2. G. L. Gulley and R. Tao, *Int. J. Modern. Phys.* **15**, 851-858 (2001).
3. N. Umehara, I. Shibata and K. Edamura, *J. Intelli. Mat. Syst. Struct.* **10**, 620-623 (2003).
4. N. Umehara, *J. Magn. Magn. Mat.* **252**, 341-343 (2003).
5. Y. Zhu, N. Umehara, Y. Ido and A. Sato, *J. Magn. Magn. Mat.* **302**, 96-104 (2006).
6. Y. Ido, T. Inagaki and N. Umehara, *11th Int. Conf. Magnetic Fluids,* 7P11, 2007.
7. J. G. Torre and V. A. Bloomfield, *Quart. Rev. Biophys.* **14**, 81-139 (1981).
8. J. E. Christiansen, *Agricultural Experimental Station Bulletin*, University of California, Berkeley, CA (1942), p.670.

Interfacial Phenomena of Magnetic Fluid with Permanent Magnet in a Longitudinally Excited Container

Seiichi Sudo, Hirofumi Wakuda, and Tetsuya Yano

Department of Machine Intelligence and Systems Engineering, Akita Prefectural University, Yurihonjyo 015-0055, Japan

Abstract. This paper describes the magnetic fluid sloshing in a longitudinally excited container. Liquid responses of magnetic fluid with a permanent magnet in a circular cylindrical container subject to vertical vibration are investigated. Experiments are performed on a vibration- testing system which provided longitudinal excitation. A cylindrical container made with the acrylic plastic is used in the experiment. A permanent magnet is in the state of floating in a magnetic fluid. The disk-shaped and ring-shaped magnets are examined. The different interfacial phenomena from the usual longitudinal liquid sloshing are observed. It is found that the wave motion frequency of magnetic fluid with a disk-shaped magnet in the container subject to vertical vibration is exactly same that of the excitation. In the case of ring-shaped magnet, the first symmetrical mode of one-half subharmonic response is dominant at lower excitation frequencies. The magnetic fluid disintegration of the free surface was also observed by a high-speed video camera system.

Keywords: Interfacial phenomena, Magnetic fluid, Magnetic fluid sloshing, Surface wave, Longitudinal vibration
PACS: 80

INTRODUCTION

Wave motions of liquid in moving containers is a fascinating subject that has attracted attention of geophysicists and seismologists, engineers, mathematicians, and other scientific workers for a period of many years [1]. The interfacial phenomena of liquids in moving containers are called liquid sloshing. In particular, extensive investigations on the excitation of Faraday waves have been conducted and reported [2]. On the other hand, magnetic fluids have been synthesized in the early to mid-1960s [3]. Magnetic fluid behaves remarkably like a true liquid having concomitant fluid and magnetic properties [4]. Therefore, some investigations on magnetic fluid sloshing have been also conducted and reported [5-8]. In our previous papers, liquid responses of magnetic fluids under magnetic and vibrating fields were investigated [6-8]. In the previous research, a permanent magnet and the cylindrical container were mounted on the vibrating table of the electrodynamic shaker [7-8]. The research data on the surface responses of a magnetic fluid which included the ferrite permanent magnet are insufficient, and there are many points which must be clarified.

In this paper, the dynamic behavior of magnetic fluid including a permanent magnet in the cylindrical container excited longitudinally was investigated. The free surface of magnetic fluid in the cylindrical container subject to the vertical vibration showed different responses according to different shape magnets.

EXPERIMENTAL APPARATUS

Experiments were performed on a vibration-testing system (IMV Model VE-3201) as shown in Fig. 1. The electro dynamic shaker was operated at a given frequency, displacement and acceleration within the range of maximum exciting force 4.1kN, maximum exciting acceleration 784m/s^2, maximum exciting amplitude 25mm, and maximum exciting frequency 3000Hz. A cylindrical container used in the experiment was made of acrylic plastic. The dimensions of the container were 200mm long, 50mm inner diameter, and 5mm thick wall with 10 mm thick flat bottom and an open top. The permanent magnet was put in the container, and the container was partially filled with the magnetic fluid.

Two kinds of ferrite permanent magnets as shown in Fig. 2 were examined, that is, disk-shaped and ring-shaped permanent magnets were used. The dimensions

CP982, *Complex Systems, 5th International Workshop on Complex Systems*
edited by M. Tokuyama, I. Oppenheim, and H. Nishiyama

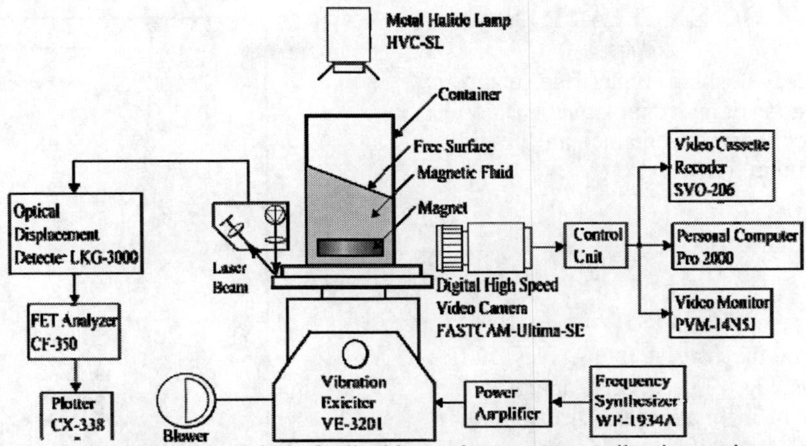

FIGURE 1. The experimental apparatus consists of a liquid container system, a vibration-testing system, and a measuring device system. The liquid container is partially filled with the magnetic fluid.

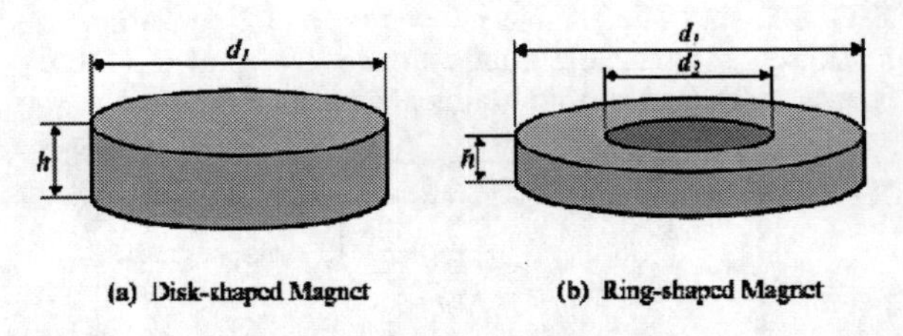

FIGURE 2. Two kinds of ferrite magnets were used in the experiment. The permanent magnet floats in a magnetic fluid.

TABLE 1. Dimensions of two ferrite magnets used in the experiment.

	Disk-shaped Magnet	Ring-shaped Magnet
h (mm)	10	6
d_1 (mm)	38	45
d_2 (mm)		22
B (mT)	120	110

of permanent magnets are shown in Table 1. In Table 1, B is the magnetic flux density at the surface of magnet. Fig. 3 shows the outline of magnetic flux density distributions of two permanent magnets. Three kinds of liquid depths were examined. Sample magnetic fluid was water-based ferricolloid W-40. The cylindrical container was mounted on the vibrating table of the electrodynamic shaker. The vibration frequency was set as desired, and the vibration table connecting to the exciter was vibrated vertically at an arbitrary level of input vibration acceleration.

The dynamic behavior of magnetic fluid surface was analyzed by the high-speed video camera system. In the experiment, the dynamic behavior of magnetic fluid was recorded at 750 frames per second, and recording frames were analyzed with the personal computer.

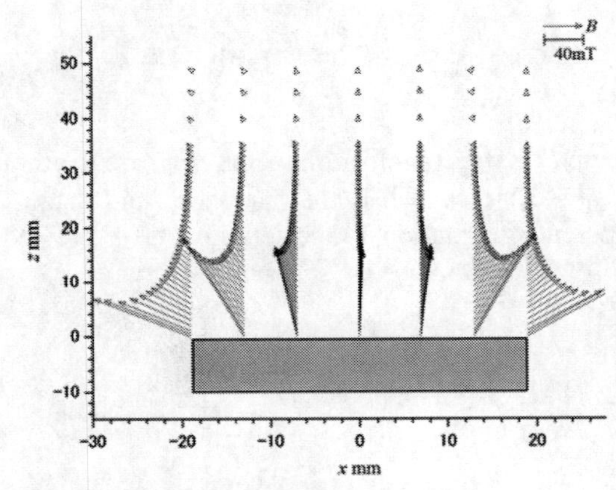

(a) Disk-shaped magnet

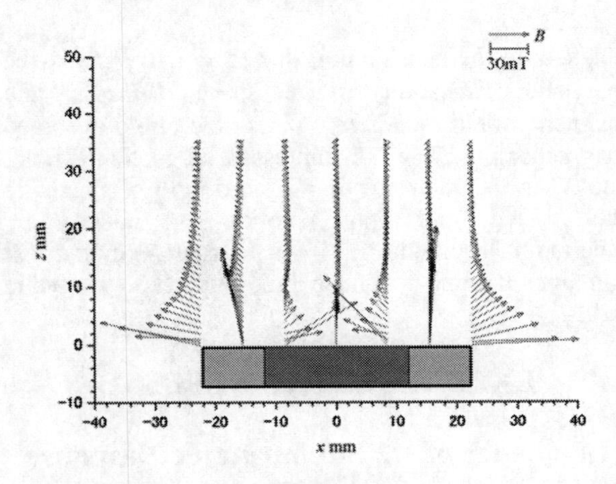

(b) Ring-shaped magnet

FIGURE 3. Distribution of the magnetic flux density on two ferrite permanent magnets. Magnetic flux density decreases rapidly with the distance from the magnet surface. The rectangles in figures show the magnet.

PARAMETRIC EXCITATION

In magnetic fluid sloshing under the magnetic fields, the small motions of an ideal magnetic fluid in a vertically vibrated cylindrical container are described by Mathieu's equation in the form [8]

$$\frac{d^2T_{mn}}{dt^2}+\left(\omega_{mn}{}^2-\omega^2{}_{mn}\frac{\omega^2 z_0}{\bar{g}}\cos\omega t\right)T_{mn}=0 \quad (1)$$

where ω_{mn} is one of the natural frequencies of free magnetic fluid sloshing, T_{mn} is the amplitude of corresponding mode of magnetic fluid surface motion, ω is the forcing frequency, g is the apparent gravitational acceleration, and z_0 is the amplitude of the container motion. The natural frequencies of magnetic fluid sloshing ω_{mn} are described as follows

$$\omega_{mn}{}^2=\frac{2\xi_{mn}\bar{g}}{D}\tanh\frac{2\xi_{mn}L}{D} \quad (2)$$

where D is the container diameter, ξ_{mn} are the roots of $J_m'(\xi_{mn})=0$, and L is the magnetic fluid depth. The apparent gravitational acceleration is expressed as follows

$$\bar{g}=g-\frac{\mu_0}{\rho}\overline{M}\frac{dH}{dz} \quad (3)$$

$$\overline{M}=\frac{1}{H}\int_0^H MdH \quad (4)$$

where g is the acceleration due to gravity, μ_0 is the magnetic permeability of free space, dH/dz is the magnetic field gradient, $\overline{M}$ is the field-averaged magnetization, H is the impressed magnetizing field, and M is the intensity of magnetization of magnetic fluid. The fluid wave motion is induced by longitudinal vibration of a circular cylindrical container holding magnetic fluid subject to magnetic fields.

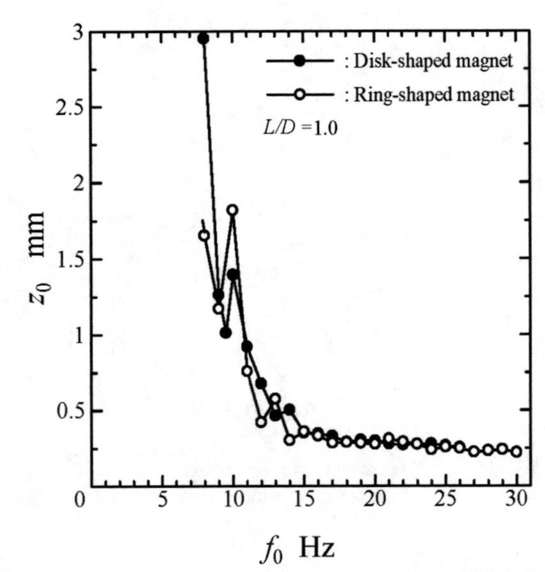

FIGURE 4. The incipience points of 1/2-subharmonic responses of magnetic fluid surface are plotted. These results are predicted from Mathieu's equation (1).

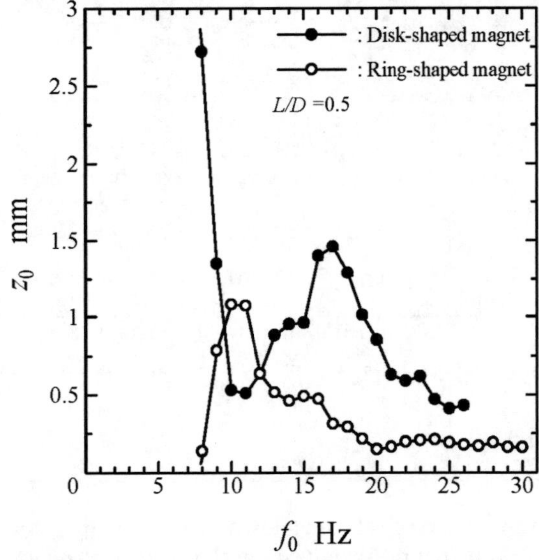

FIGURE 5. The incipience amplitudes of wave motion of magnetic fluid with the permanent magnet are plotted. Two permanent magnets differ in the threshold amplitude of wave motion.

EXPERIMENTAL RESULTS

Incipience of 1/2 Subharmonic Response

The liquid wave motion induced by longitudinal vibration of a circular cylindrical container holding magnetic fluid and a ferrite permanent magnet is investigated in this section. Starting with an initially quiescent surface, as the excitation amplitude z_0 is increased from zero for a given frequency, a point is reached at which the interface suddenly begins to exhibit 1/2-subharmonic response.

A few tests were conducted in the following manner: (1) The magnetic fluid and the permanent magnet to be tested was put into the test tank with sufficient depth (L/D=1.0). (2) The magnetic fluid-tank system was excited longitudinally at a given frequency with the amplitude being carefully varied until the minimum amplitude necessary to give surface wave motion.

The threshold of wave motion was obtained in the above manner for water-based magnetic fluid (ferricolloid W-40). The range of excitation frequency was 8 to 30 Hz for most tests, although some tests were conducted somewhat above or below this range. The results of two tests are shown in Fig. 4. This figure corresponds to the general stability boundary for principal parametric resonance of magnetic fluid surface in the longitudinally excited rigid cylindrical container under the magnetic fields. It can be seen from Fig. 4 that there is almost no difference in the shape difference between two magnets. These results correspond to the unstable regions for several 1/2-subharmonic modes of the Mathieu diagram. At higher excitation amplitudes, the unstable regions for various sloshing modes overlap considerably. However, the first antisymmetrical and first symmetrical modes can be fairly well isolated, at least for small values of z_0. They are convenient modes to study experimentally.

Characteristic Responses in Magnetic Fluid Sloshing

As stated above, when the cylindrical container holding magnetic fluid with sufficient depth and a permanent magnet was vibrated, the free surface of magnetic fluid exhibited 1/2-subharmonic response at the certain excitation amplitude. At lower liquid depth, however, the unique sloshing phenomena were observed in the cylindrical container excited longitudinally.

The threshold amplitudes of wave motion of magnetic fluid with the permanent magnet in the longitudinally excited cylindrical container are shown in Fig. 5. The influence of the magnet shape in magnetic fluid on generation of wave motion is clear. The effects of liquid depth are also clear by the comparison between Fig.4 and Fig. 5.

In the Case of Disk-Shaped Magnet

The dynamic behavior of magnetic fluid included the disk-shaped ferrite magnet in the circular cylindrical container excited longitudinally was examined with the high-speed video camera system. The free surface deflection time history of magnetic fluid is presented in Fig. 6. The experimental conditions were a liquid depth of L/D=0.5, excitation frequency of f_0=10Hz, and excitation amplitude of z_0=0.5mm. In Fig. 6, Z_L is the free surface deflection measured at the center axis of the container. It can be seen from Fig.6 that the response on the surface of

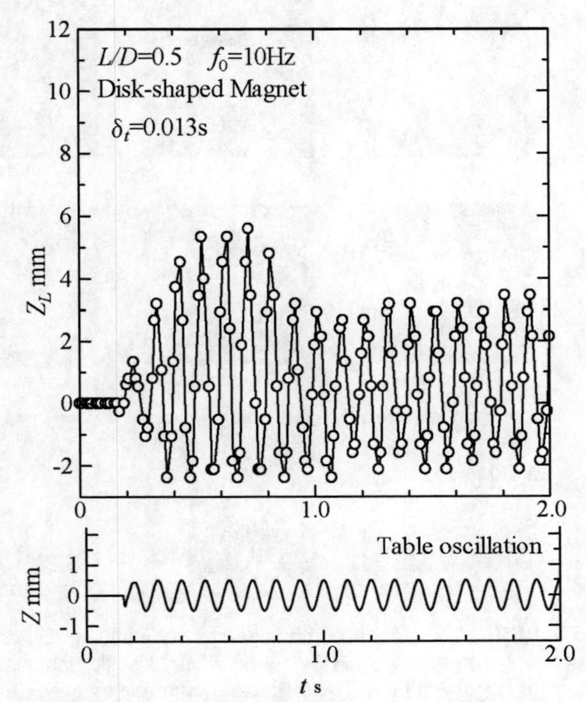

FIGURE 6. Time history of the amplitude of the interface of magnetic fluid is plotted with the high-speed video camera system. The lower figure shows table oscillation.

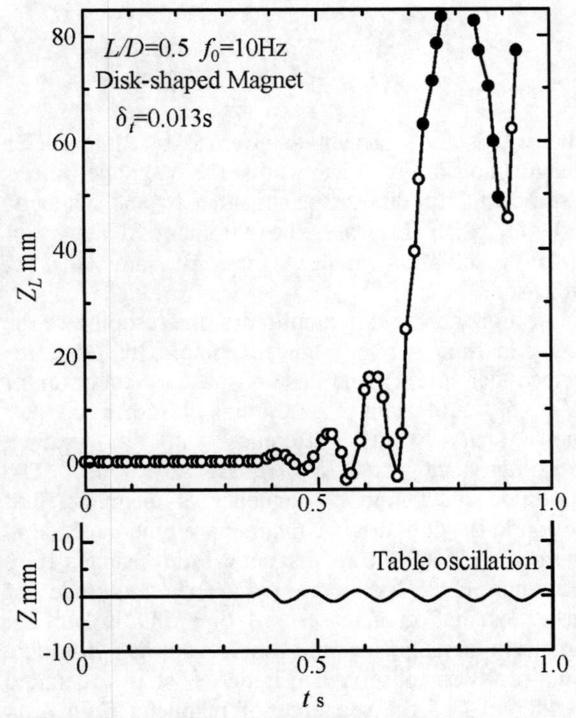

FIGURE 7. Time evolution of free surface motion of magnetic fluid is plotted. The amplitude and dominant frequency of surface oscillation change with time.

FIGURE 8. A sequence of photographs showing the generation of magnetic fluid jet and the formation of magnetic fluid droplets. It can be seen that the vertical height of the magnetic fluid column formed by the growth of mode(0,1) wave rapidly becomes large with time. The disintegration of the magnetic fluid column is caused by axisymmetric waves on the liquid column.

magnetic fluid is not 1/2- subharmonic but harmonic. In this case, the sum of magnetic body force and gravity is larger than oscillating force;

$$1-\frac{\mu_0}{\rho g}\overline{M}\frac{dH}{dz} > K\frac{\omega_0^2 z_0}{g} \tag{5}$$

where K is the constant determined by fluids. This phenomenon in Fig. 6 shows that the magnetic fluid is adsorbed to the disk-shaped magnet by the magnetic body force. In this case, the dominant frequency of surface oscillation mode is exactly same that of excitation.

At higher excitation amplitudes, the response of the magnetic fluid surface showed complexity. The free surface deflection time history at the condition of L/D=0.5, f_0=10Hz, and z_0=1.0mm is presented in Fig. 7. There are obvious differences in the excited phenomena at z_0=0.5mm and z_0=1.0mm. The amplitude and dominant frequency of magnetic fluid motion in the cylindrical container are time-dependent. The surface response of magnetic fluid changes from harmonic to 1/2-subharmonic. The amplitude of surface oscillation increases with time. In this case, the formation of magnetic fluid droplets in the cylindrical container was observed. Figure 8 shows detailed photographs of the sequence of magnetic fluid drop formation from the thin liquid column. The growth rate of the top part of the fluid column is very large compared to horizontal width of the column, that is, cylindrical jet column is formed on the magnetic fluid surface. Some swells appear immediately at the top of the magnetic fluid jet. Subsequently, some necks appear just below the swells, and the swells grow rapidly. Finally, the liquid column disintegrated into some liquid droplets at the neck points. In this experimental condition, the following inequality is realized;

$$1-\frac{\mu_0}{\rho g}\overline{M}\frac{dH}{dz} < K\frac{\omega_0^2 z_0}{g} \tag{6}$$

that is, oscillating force is larger than the sum of magnetic body force and gravity.

In the Case of Ring-Shaped Magnet

In the case of ring-shaped magnet, the threshold of wave motion at f_0=8Hz was lower as compared with the case of disk-shaped magnet as shown in Fig. 5. The wave motion response in the cylindrical container with a ring-shaped magnet showed the first symmetrical mode(0,1) at f_0=8Hz. In the case of ring-shaped magnet, the first symmetrical mode of magnetic fluid in the cylindrical container is dominant at lower excitation frequencies. This reason is based on the magnet shape. Vibration of the combination of a

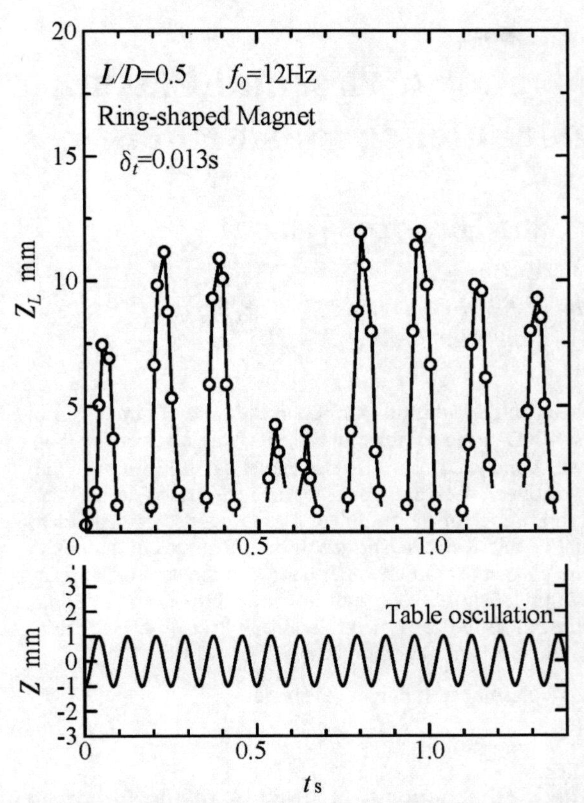

FIGURE 9. Relation between wave motion of magnetic fluid with ring-shaped magnet in cylindrical container and table oscillation. The surface motion of magnetic fluid shows one-half subharmonic response.

cylindrical container and a ring-shaped magnet causes an axisymmetric flow of magnetic fluid easily. The dynamic behavior of magnetic fluid included the ring-shaped ferrite magnet in the circular cylindrical container excited longitudinally is presented in Fig. 9. It can be seen from Fig. 9 that the surface motion of magnetic fluid is one-half subharmonic response. The liquid surface of magnetic fluid is oscillating in the half of excitation frequency.

As mentioned above, the surface response of magnetic fluid is dependent not only on the excitation frequency and amplitude but the shape of ferrite permanent magnet.

CONCLUSIONS

The interfacial phenomena of magnetic fluid with permanent magnet in a cylindrical container subject to vertical vibration were studied. The results obtained are summarized as follows.

(1) At lower liquid depth, the wave motion frequency of magnetic fluid with the disk-shaped magnet in cylindrical container subject to vertical vibration is exactly same that of the excitation. However, increase of excitation amplitude causes the 1/2-subharmonic responses of magnetic fluid surface.

(2) In the case of ring-shaped magnet, the first symmetrical mode of 1/2-subharmonic response is dominant at lower excitation frequencies.

REFERENCES

1. H. N. Abramson, *NASA SP-106,* 1-467 (1966).
2. J. Miles and D. Henderson, *Annu. Rev. Fluid Mech.* **22**, 143-165 (1990).
3. R. E. Rosensweig, *Chem. Eng. Comm.* **67,** 1-18(1988).
4. R. E. Rosensweig, *Ferrohydrodynamics*, Cambridge University Press, Cambridge, 1985, pp.1-344.
5. R. E. Zelazo and J.R.Melchen, *J. Fluid Mech.* **39**, 1-24 (1969).
6. S. Sudo, M.Ohaba, K. Katagiri, and H. Hashimoto, *J. Magn. Magn. Mater.* **122**, 248-253 (1993).
7. S. Sudo, M.Ohaba, K. Katagiri, and H. Hashimoto, *Magnetohydrodynamics* **33**, 480-485 (1997).
8. S. Sudo, H. Nishiyama, K. Katagiri, and J. Tani, *J. Intell. Mater. Syst. Struct.* **10**, 498-504 (1999).

TiO_2 Film Deposition by Atmospheric Thermal Plasma CVD Using Laminar and Turbulence Plasma Jets

Yasutaka Ando*, Shogo Tobe*, and Hirokazu Tahara**

*Ashikaga Institute of Technology, 268-1 Omae, Ashikaga, Tochigi 326-8558, Japan
**Osaka Institute of Technology, 5-16-1 Omiya, Asahi-Ku, Osaka 535-8585, Japan

Abstract. In this study, to provide continuous plasma atmosphere on the substrate surface in the case of atmospheric thermal plasma CVD, TiO_2 film deposition by thermal plasma CVD using laminar plasma jet was carried out. For comparison, the film deposition using turbulence plasma jet was conducted as well. Consequently, transition of the plasma jet from laminar to turbulent occurred on the condition of over 3.5 l/min in Ar working gas flow rate and the plasma jet became turbulent on the condition of over 10 l/min. In the case of the turbulent plasma jet use, anatase rich titanium oxide film could be obtained though plasma jet could not contact with the surface of the substrate continuously even on the condition that feedstock material was injected into the plasma jet. On the other hand, , in the case of laminar gas flow rate, the plasma jet could contact with the substrate continuously without melt down of the substrate during film deposition. Besides, titanium oxide film could be obtained even in the case of the laminar plasma jet use. From these results, this technique was thought to have high potential for atmospheric thermal plasma CVD.

Keywords: TiO_2, Thermal plasma CVD, Photo catalysis, Turbulent plasma jet, Laminar plasma jet
PACS: 52.77.Fv

INTRODUCTION

Since titanium oxide (TiO_2) has many attractive properties, this material has been successfully utilized for numerous industrial fields. Especially, its photo-catalytic property was thought to be useful for photovoltaic device of the dye sensitized solar cell (DSC) as well as the coatings for anti-bacillus and cleaning polluted water and so on. Although TiO_2 used as a photo-catalyst was only anatase type with large band gap (3.2 eV) so far, it was proven that rutile type which has been used for the colorant also had sufficient photo-catalytic property to clean water. Furthermore, photo-catalytic property could appear under an ultraviolet irradiated circumstances until now, the property could appear under a low intensity visible light circumstances by doping of Pt, V, and Fe [1]-[2] to anatase and rutile and industrialization of brookite. Although sputtering [3], MOCVD [4] and a sol-gel process [5] have been practically used as film deposition methods of TiO_2, sputtering and MOCVD have some disadvantages such as necessity of vacuum equipment, low deposition rate, and sol-gel process also has some problems, for example difficulty of thick film deposition due to break down by internal stress occurred during deposition and so on. Although thermal spraying has been widely used as high rate thick film deposition process in the atmosphere environment, difficulty of film component and structure control and high starting material powder cost should be solved in this process. Therefore,

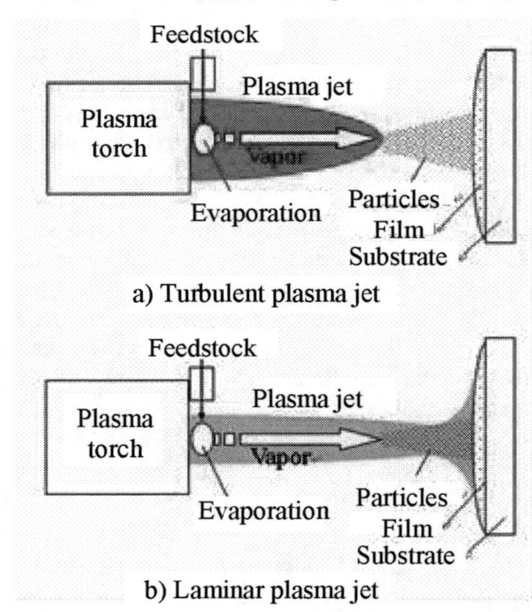

FIGURE 1. (Color online) Illustration of film deposition by thermal plasma CVD.

CP982, *Complex Systems, 5th International Workshop on Complex Systems*
edited by M. Tokuyama, I. Oppenheim, and H. Nishiyama

development of a simpler high rate film deposition method can control film's structure and component is desired. However, since turbulent plasma jet has been generally used in the case of atmospheric TPCVD, plasma jet tail is short and high temperature. Therefore, it is difficult to make the plasma jet contact with the surface of the substrate continuously in order to provide a steady plasma environment on the surface of the substrate in the case of turbulent plasma jet use. On the other hand, in the case of the laminar plasma jet, since plasma jet tail is long and its end is relatively low temperature, plasma jet can contact with the surface of the substrate continuously for a long time. So, it is thought to be easy to provide a continuous plasma environment on the surface of the substrate (Fig. 1).

In this study, in order to develop an atmospheric TPCVD process with plasma chemical reaction on the surface of the substrate, titanium oxide film deposition by atmospheric thermal plasma chemical vapor deposition (TPCVD) using laminar Ar plasma jet as well as turbulent Ar plasma jet was carried out.

EXPERIMENTAL PROCEDURE

The TPCVD equipment consisted of plasma torch, DC power supplying system, micro tube pump (feedstock material supplying system) and working gas supplying system as shown in Fig. 2. The plasma torch has water cooled electrodes. The anode is made from copper, has alkoxide feeding port at the head and has the constrictor which is 6 mm in diameter. A cylindrical cathode made from 2 % thoriated tungsten has a diameter of 3 mm. Ar was used as the working gas. Mass flow rate of the gas was varied 1-20 l/min.

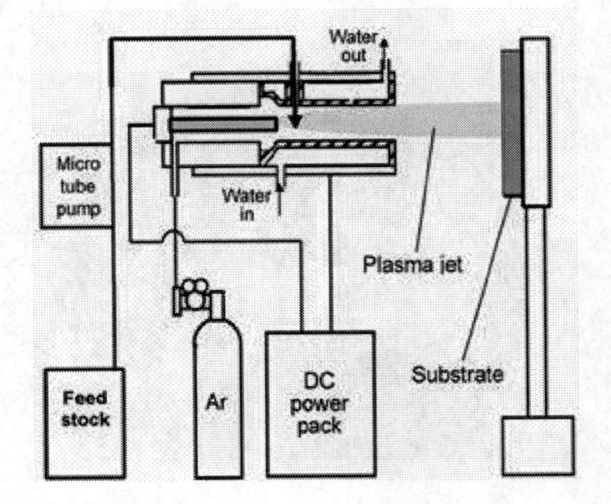

FIGURE 2. Schematic diagram of Thermal plasma CVD equipment..

In this TiO_2 film deposition, ethanol diluted titanium tetra butoxide ($Ti(OC_4H_9)_4$) was used as starting material. Substrate was 15mm x 15mm x 1mmt 430 stainless steel plate with grit blasted surface. The substrate was horizontally set on the substrate holder and the central area of the sample was placed as the axial center of the plasma jets irradiate this sample. The distance between the nozzle outlet of the plasma torch and the surface of the substrate was fixed at 100mm. The deposition time was 7 min in the case of the turbulent plasma jet use and/or 15 min in the case of the laminar plasma jet use. The discharge power was 20-25 V, 125 A. After TiO_2 film deposition, investigations of the microstructures of the films by X-ray diffraction (CuKa, 40 kV, 100 mA). In addition, in order to confirm photo-catalytic property of the film, dye sensitized solar cells using these deposited TiO_2 films were fabricated and their photovoltaic testing was conducted. Table 1. shows the film deposition condition.

TABLE 1. Experimental condition.

Working gas (Flow rate)	Ar (1-20SLM)
Discharge Current	125A, 25V
Deposition distance	30-100mm
Deposition time	7min (Turbulent) 15min(Laminar)
Feedstock material	C_2H_5OH diluted $Ti(OC_4H_9)_4$
Feedstock feed rate	0-100 ml /h
Substrate	430 stainless steel

RESULTS AND DISCUSSION

Relationship Between Appearance of Plasma Jet and Working Gas Flow Rate

Figure 3 shows the appearance of the plasma jet on each condition of Ar working gas flow rate (Q_{Ar}). In the case of Q_{Ar} < 3.5 l/min., the length of the plasma jet increased from 70 mm (at Q_{Ar}=1 l/min.) to 100 mm (at Q_{Ar} =3.5 l/min.) with increasing Q_{Ar}. However, in the case of Q_{Ar} >3.5 l/min., since plasma jet varied from laminar flow to turbulent flow gradually, the length of the plasma jet decreased to 15 mm (at Q_{Ar} =10 l/min.) with increasing Q_{Ar}. After variation of plasma jet to turbulent flow (in the case of Q_{Ar} >10 l/min.), the length of the plasma jet increased to 20 mm (at Q_{Ar} =20 l/min.) with increasing Q_{Ar} again since the pressure in arc chamber of the plasma torch. As for the discharge voltage, the voltage was almost 20 V on the condition of Q_{Ar} < 3.5 l/min and almost 25 V on the condition of Q_{Ar} >10l/min, though the voltage increased slightly with increasing gas flow rate in each case. However, the voltage increased linearly with increasing Q_{Ar} on the condition of 3.5 l/min< Q_{Ar} <10 l/min. This phenomenon occurred because anode point of the arc column was moved to downstream by pressure of the working gas with increasing Q_{Ar}.

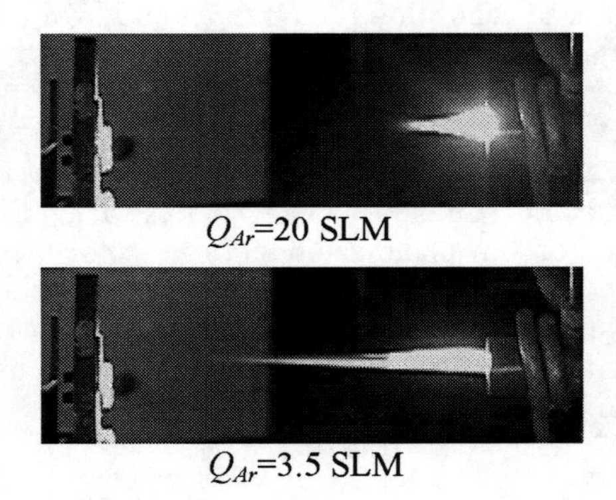

FIGURE 3. (Color online) Appearances of the Ar plasma jets. (Q_{Ar}: Ar gas flow rate).

TiO_2 Deposition Using Turbulent Plasma Jet

Figures 4-5 show substrate temperature and XRD pattern of the sample on each condition of deposition distance (d_s). Although degree of crystallinity of the film was low in the case of d_s=150, 200 mm, anatase rich titanium oxide film could be obtained. In our previous study [6], it was confirmed that the anatase rich film had enough photo catalytic property to dissociate methylene blue under an ultraviolet irradiation environment. As for the film deposition mechanism, though nucleation was thought to be occurred in the plasma jet generally in the case of TPCVD [7], from the result of the relationship between crystal structure of the film and d_s, it was thought that the particle precipitated in the plasma jet was not crystallized but amorphous and crystallization of the particle occurred on the substrate by the heat from the substrate and/ or plasma jet in this study. Accordingly the substrate temperature, the temperature decreased with increasing d_s and the temperature was raised approximately 100 K by feedstock material injection into plasma jet. Since evaporated feedstock material played a roll of working gas in the case that the feedstock material was injected, total working gas flow rate was raised. Hence, the plasma jet was elongated and the plasma tail reached the surface of the substrate shown in Fig. 6. However, since feedstock injection was intermittent and fluctuation of the plasma jet occurred frequently, it was difficult to make the plasma jet contact with the surface of the substrate continuously and provide continuous plasma environment on the surface of the substrate in the case of turbulent plasma jet use.

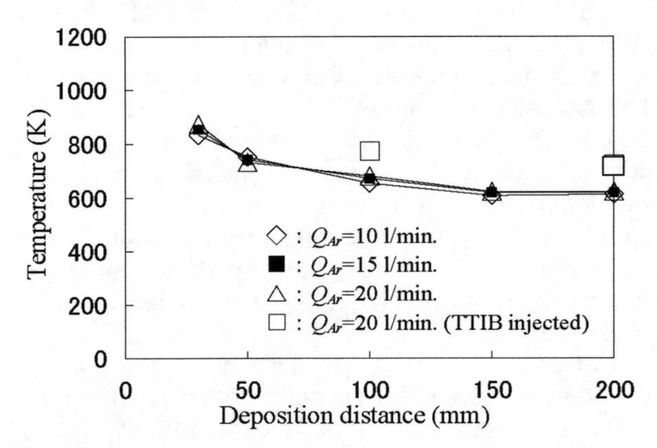

FIGURE 4. Substrate temperature of the sample on each condition of deposition distance.

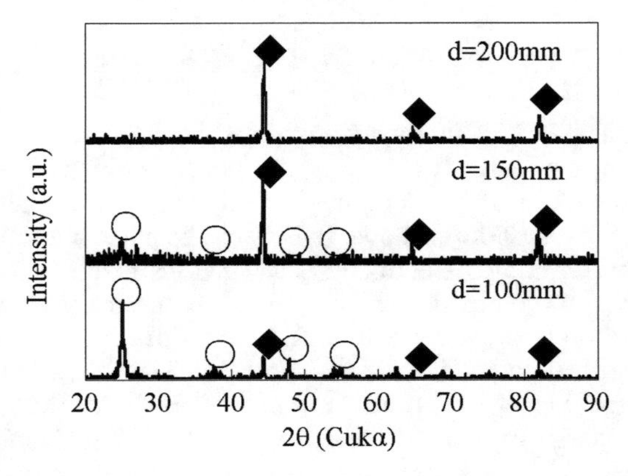

FIGURE 5. XRD pattern of the TiO_2 film on each condition of deposition distance. (d: Deposition distance, ○: Anatase, ◆: Fe (Substrate)).

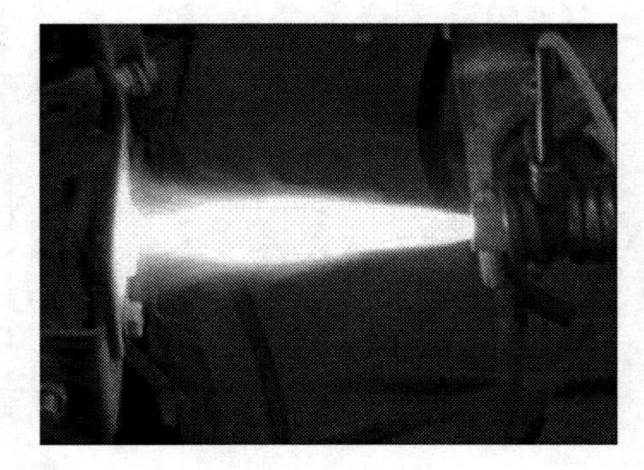

FIGURE 6. (Color online) Appearances of the TTIB introduced Ar plasma jet on the condition of 20 l/min in Ar gas flow rate.

Figure 7 shows the fracture cross-section of the sample on each condition of TTIB concentration in feedstock material. In the case of dense TTIB feedstock material (20 vol% TTIB), 120 μm thick anatase rich film could be obtained even on the condition of 7 min in deposition time though only 60 μm thick film could be obtained in the case of dulute TTIB feedstock material (5 vol% TTIB). However, relation between deposition rate and concentration of TTIB in feedstock material was parabolic though consumption quantity of TTIB during film deposition linearly increased with increasing concentration of TTIB. The reason, which is thought as follows, should be investigated as a future work. As for the crystal structure of the films, as shown in the Fig. 8. anatase rich films could be deposited on any conditions of TTIB concentration in the feedstock materials.

* Sintering of the film occurred during film deposition.
* Infiltration of the titanium oxide particle into the film occurred during film deposition.

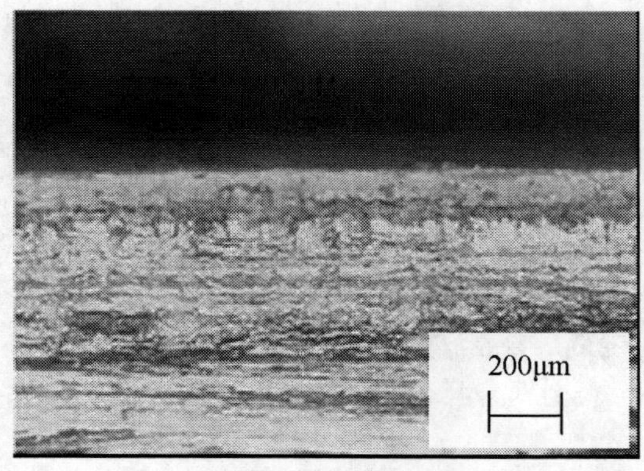

a) 20 vol% TTIB

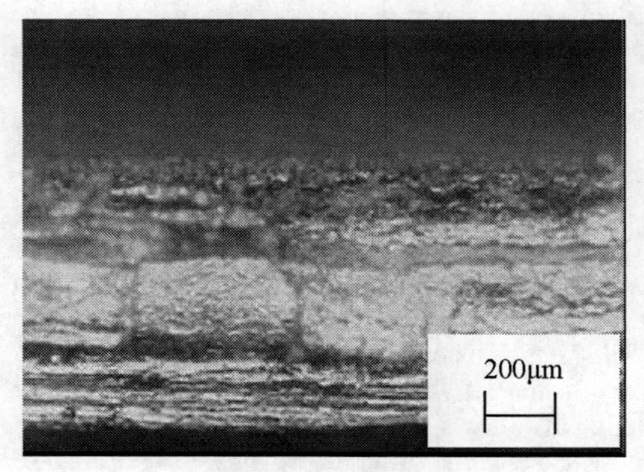

b) 5 vol% TTIB

FIGURE 7. (Color online) Optical micrographs of the fracture cross-sections of the TiO_2 films.

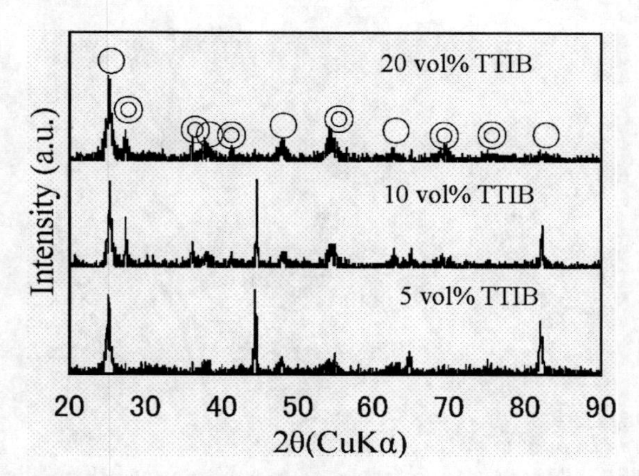

FIGURE 8. XRD patterns of the TiO_2 films deposited on the condition of 20 l/min in Ar working gas flow rate. (d: Deposition distance, ○ : Anatase, ◎ : Rutile, ▲ : Fe (Substrate)).

TiO_2 Deposition Using Laminar Plasma Jet

Figure 9 shows the appearance of the plasma jet without feedstock material injection. In this case, plasma jet could contact with the surface of the substrate continuously for 10 min without melt down of the substrate. Hence, from this result, it was proved that using laminar plasma jet was effective to provide plasma atmosphere on the surface of the substrate in atmospheric TPCVD.

Figure 10 shows the appearance of the plasma jet on each condition of the feedstock material (10 vol% TTIB) feed rate (F_s). Laminar flow was maintained in the case F_s >40 ml/h though non-evaporated feedstock appeared in the plasma jet and fluctuation of the plasma jet occurred frequently in the case of F_s <40 ml/h.

Figure 11 shows substrate temperature on each condition of deposition distance. Since the length of the plasma jet increased by injection of feedstock material into the plasma jet also in the case of laminar plasma jet use, the substrate temperature was raised by feedstock material injection. However, though 100 K temperature increase occurred in the case of the turbulent plasma jet use, temperature increase was only 20-80 K (except the condition of 35 mm in deposition distance) in the case of the laminar plasma jet use.

Figures 12-13 show the appearance and XRD pattern of the sample on the condition of 70 mm in deposition distance. Although the degree of crystallinity of the film was very low in comparison with that in the case of the turbulent plasma jet use, titanium oxide film with the same appearance as that

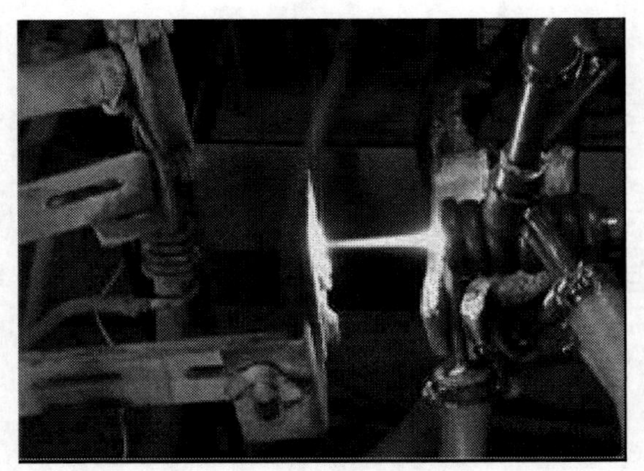

FIGURE 9. Appearances of the Ar plasma jet on the condition of 3.5 l/min in Ar gas flow rate.

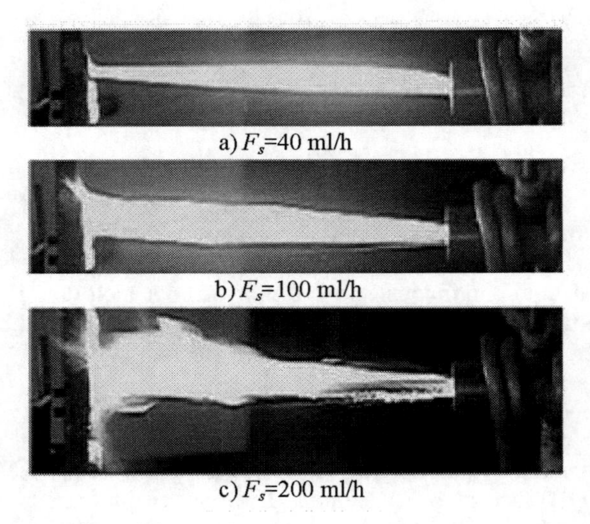

a) F_s=40 ml/h

b) F_s=100 ml/h

c) F_s=200 ml/h

FIGURE 10. Appearances of TTIB introduced Ar plasma jets on the condition of 3.5 l/min in Ar working gas flow rate. (F_s: Feedstock feed rate).

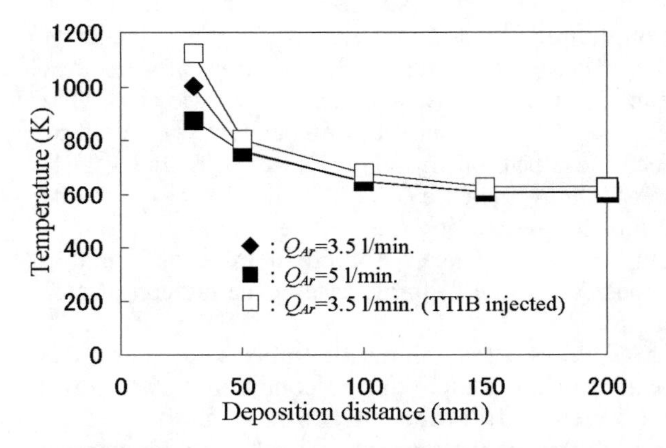

FIGURE 11. Substrate temperature of the sample on each condition of deposition distance.

in the case of the turbulent plasma jet use could be deposited even in this case. From these results, this technique using laminar plasma jet was found to have high potential for high rate film deposition process.

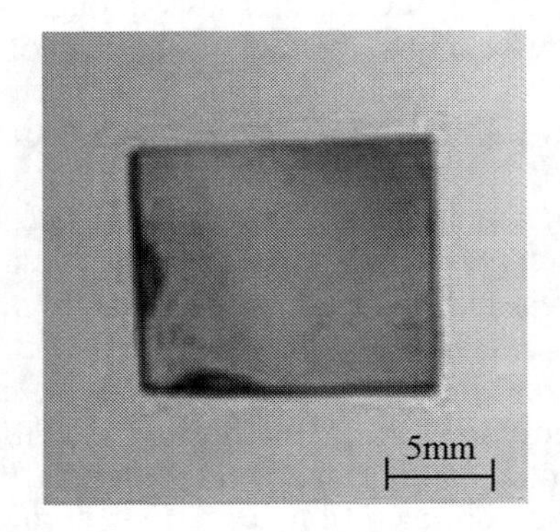

FIGURE 12. Appearances of the TTIB introduced Ar plasma jet on the condition of 3.5 l/min in Ar working gas flow rate. (Deposition distance: 50mm).

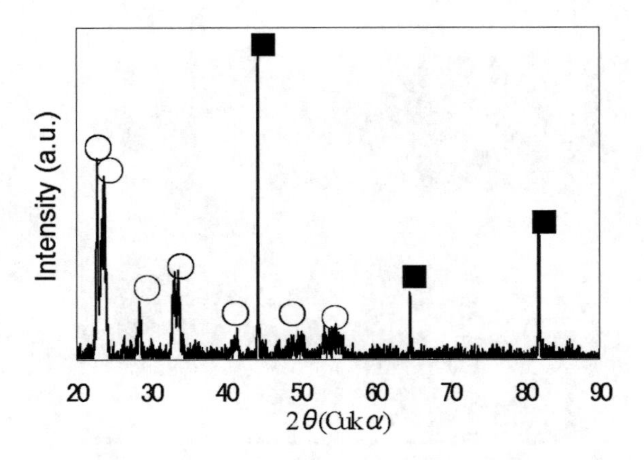

FIGURE 13. XRD pattern of the TiO_2 film on the condition of 3.5 l/min in Ar working gas flow rate. (Deposition distance: 50mm, ○: Titanium oxide, ■ : Fe substrate).

CONCLUSION

In order to provide continuous plasma atmosphere on the substrate surface in atmospheric thermal plasma CVD, laminar thermal plasma jet was generated and TiO_2 film deposition by thermal plasma CVD using laminar plasma jet. In addition, to Consequently, we concluded as follows;

1) In the case of the turbulent plasma jet use, anatase rich titanium oxide film could be obtained though plasma jet could not contact with the surface of the substrate continuously even on the condition that feedstock material was injected into the plasma jet.

2) In the case of laminar gas flow rate, the plasma jet could contact with the substrate continuously without melt down of the substrate during film deposition. Besides, titanium oxide film could be obtained even in the case of the laminar plasma jet use.

REFERENCES

1. M. Akanuma, H. Tanaka, N. Katayama, M. Kakimoto, Y. Tamura, K. Tominaga, *Technical Reports of Hokkaido Industrial Research Institute,* **299**, 31-38 (2000).
2. M. Anpo, *Catal. Surv. Japan*, **1**, 169-179 (1997).
3. H. Nanto, T. Minami, S. Shoji, S. Tanaka, *J. Appl. Phys* .**55**, 1029-1034 (1984).
4. M. I. B. Bernardi, E. J. H. Lee, P. N. Lisboa-Filho, E. R. Leite, E. Longo, A. G. Souga, *Ceramica* **48**, 192-198 (2002).
5. N. Negishi, K. Takeuchi, T. Ibusuki, *J. Mater. Sci.* **33,** 5789-5794 (1998).
6. Y. Ando, S. Tobe, H. Tahara, *IEEE Trans.on Plasma Sci.* **34**, 1229-1234 (2006).
7. E. Bouyer, M. Muller, R. H. Henne and G. Schiller, *Proceedings of 1st International Thermal Spray Conference*, 2000, pp. 919-928.

Non-Equilibrium Plasma MHD Electrical Power Generation at Tokyo Tech

T. Murakami, Y. Okuno, and H. Yamasaki

Tokyo Institute of Technology
4259-G3-38, Nagatsuta, Midori-ku, Yokohama 226-8502, Japan

Abstract. This paper reviews the recent activities on radio-frequency (rf) electromagnetic-field-assisted magnetohydrodynamic (MHD) power generation experiments at the Tokyo Institute of Technology. An inductively coupled rf field (13.56 MHz) is continuously supplied to the disk-shaped Hall-type MHD generator. The first part of this paper describes a method of obtaining increased power output from a pure Argon plasma MHD power generator by incorporating an rf power source to preionize and heat the plasma. The rf heating enhances ionization of the Argon and raises the temperature of the free electron population above the nominally low 4500 K temperatures obtained without rf heating. This in turn enhances the plasma conductivity making MHD power generation feasible. We demonstrate an enhanced power output when rf heating is on approximately 5 times larger than the input power of the rf generator. The second part of this paper is a demonstration of a physical phenomenon of the rf-stabilization of the ionization instability, that had been conjectured for some time, but had not been seen experimentally. The rf heating suppresses the ionization instability in the plasma behavior and homogenizes the nonuniformity of the plasma structures. The power-generating performance is significantly improved with the aid of the rf power under wide seeding conditions. The increment of the enthalpy extraction ratio of around 2 % is significantly greater than the fraction of the net rf power, that is, 0.16 %, to the thermal input.

Keywords: Magnetohydrodynamics, Radio-frequency electromagnetic-field. Preionization, Dynamic stabilization
PACS: R52.75.Fk, 52.35-g, and 52.50Qt

INTRODUCTION

A closed-cycle magnetohydrodynamic (CCMHD) electrical power generator using a nonequilibrium plasma has been drawing attention as a novel energy conversion device [1]. Recently, the CCMHD energy conversion has been revived and has been the focus of great attention with respect to perceived space needs, such as multi-megawatt space power installations [2,3]. The CCMHD electrical power generator has several advantages, that is, virtually no upper limits to the temperature it can tolerate, rapid response, and the ability to achieve high thermal efficiency and power density [4], which lead to a reduction of the system mass. Since a high-temperature gas of approximately 2000 K is used, the CCMHD generator offers better energy conversion efficiency than conventional devices. Different from turbine devices, the CCMHD generator directly converts the enthalpy of a high temperature gas to electrical energy without any high-temperature rotating parts. Because of its small inertial force, the CCMHD generator can control the power output in milliseconds. Despite such great advantages, the MHD generator is still subject to unstable nonequilibrium plasma behavior.

The ability to preionize plasma, suppress nonequilibrium ionization instability and to control unstable plasma behavior would represent a very important advance in the development of a high-performance CCMHD electrical power generator. We have proposed a scheme of preionization and dynamic plasma stabilization by electromagnetic fields. We employ a homogeneous application of an inductively coupled radio-frequency (rf) electromagnetic field in space and time to control the unstable plasma in a disk-shaped Hall MHD generator. The rf inductive coupling technique is considered to be one of the best ways to generate and control an azimuthally uniform plasma in the disk generator. The physical features of the laboratory-scale nonequilibrium seeded inductively coupled plasma have been described in a previous paper [5]. Moreover, feasibility studies on preionization by the inductively coupled technique have been conducted through experiments and multidimensional numerical simulations [6-9]. The calculation in Ref. 7 has shown that the rise in the inlet

CP982, *Complex Systems, 5th International Workshop on Complex Systems*
edited by M. Tokuyama, I. Oppenheim, and H. Nishiyama

electron temperature upon rf preionization significantly contributed to improving the generator performance. The calculation in Ref. 8 has shown that additional rf heating stabilized the plasma suffering from ionization instability, and improved the generator performance.

This paper reviews the recent progress in the rf power-assisted MHD power generation in a pure-argon plasma [10] and a cesium-seeded helium plasma [11-13]. In the first part, we describe the experiment of a combined scheme of the external rf-power-assisted preionization and seed-free pure-argon-gas working MHD energy conversion. In the second part, a scheme of dynamic plasma stabilization by rf-power combined with plasma stabilization by the fully ionized seed concept is described for cesium-seeded helium gas working MHD generator.

EXPERIMENTAL SET-UP

Disk MHD Generator

The disk-shaped radial-flow MHD generator is driven by a shock tube [14,15]. The shock-tube facility consists of a shock tube, a disk MHD generator, an rf-power supply network, a magnet, and a gas supply and an exhaust system. The seed fraction is defined as the number density ratio of seed (cesium) atoms to inert gas (helium) atoms. The seed fraction and inlet total (stagnation) temperature are measured by a line reversal method at the tube end. A piezoresistive pressure transducer at the same tube end monitors the inlet total (stagnation) pressure. The shock tube supplies a high-enthalpy gas flow, Ar case: a total pressure of 0.10 ± 0.01 MPa, a total temperature of 3200-4800 K, and a thermal input to the generator is 1.53-1.88 MW, and He-Cs case: a total pressure of 0.10 ± 0.005 MPa, a total temperature of 2100 ± 50 K, and a thermal input to the generator is 3.2 ± 0.2 MW. A Helmholtz magnet supplies a pulsed field (6 ms duration). Detailed descriptions of the shock-tube operation are addressed in Refs. 4, 14, and 15. The rf power supply system consists of the rf-power source and an impedance-matching network [6].

Figure 1 shows back-wall and cross-sectional views of the present disk generator consisting of an upstream supersonic nozzle (from first to second anodes, which are electrically shorted) and a downstream MHD power-generating channel (from the second anode to the cathode, which are loaded by a resistance). The first and second electrodes are embedded only on the front wall. The magnetic flux density is 3.0 T at $\leq$ 100 mm and 0.3 T at the cathode. The gas flows perpendicular to the magnetic field in the disk. Spectroscopic measurements are performed at the radii of 100 mm (upstream window) and 145 mm (downstream window) using four photomultiplier tubes with band-pass filters (half-width of 3 nm). Two pairs of photomultipliers detect cesium-ion recombination radiation, namely, $6^2P_{1/2}$ (continuum radiation up to 494.4 nm). The electron temperature is estimated from the cesium ion recombination continuum intensity at the wavelengths of 406.0 and 490.0 nm. A high-speed camera (6000 frames per second with an exposure time of 0.8 μ s) photographs plasma structures from the back wall [16].

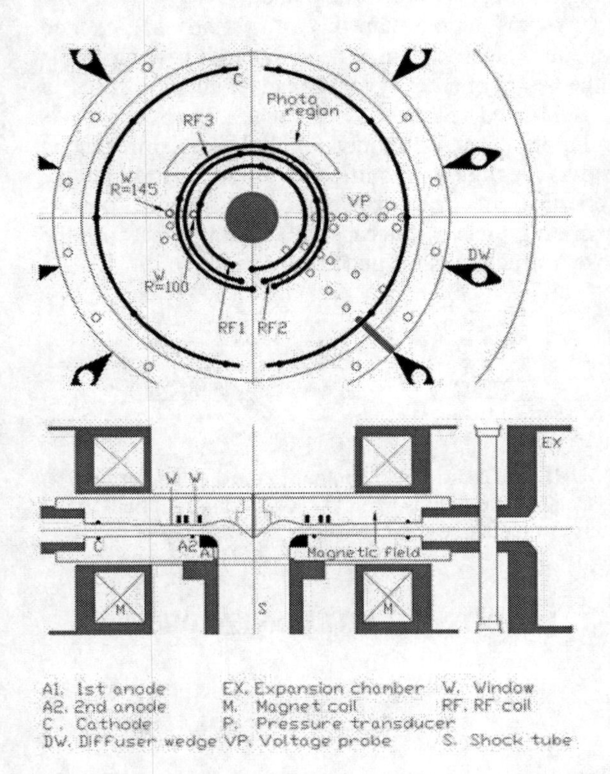

FIGURE 1. Back-wall and cross-sectional views of the disk MHD generator.

Inductive Coupling Rf Field

As shown in Fig. 1, three rf inductive coupling coils, RF1, RF2, and RF3 (6-mm-wide copper rings), are embedded in the back wall. Radii of the coils are r=92 mm (RF1), r=116 mm (RF2), and r=129 mm (RF3). The rf electromagnetic field (the excitation frequency is 13.56 MHz and the net power supplied to the plasma is approximately 5 kW) is applied through the outermost coil (RF3). Electromagnetic matching between the rf field and the plasma is adjusted using the impedance-matching network. The net rf power supplied to the plasma is estimated from the forward and reflected powers monitored.

Figure 2 shows a calculated contour map of the electric field strength induced by the rf current (5 A) in the outermost inductive coupling coil (RF3), where the two-dimensional vector potential model [5] was used. The region simulated is from r=85.0 to r=315.0 mm in the radial direction, and from the bottom to upper wall in the height direction. The detailed numerical procedure is described in Ref. 5. The rf field is distributed from the upstream supersonic nozzle to the downstream MHD channel in the radial direction. The strength of the rf electric field at the radius of 129 mm (RF3 position) and at the center of the channel height is approximately 180 V/m, which is weaker by one order of magnitude than that of a steady self-excited tangential electric field. The skin depth is comparable to the height of the generator channel under the present rf field and plasma conditions; therefore, it is negligible. The distribution of the rf electromagnetic field is negligibly modified by the variation in the properties of the nonequilibrium MHD power-generating plasma, because the skin effect on the plasma structure is insignificant in our case [5,7,8].

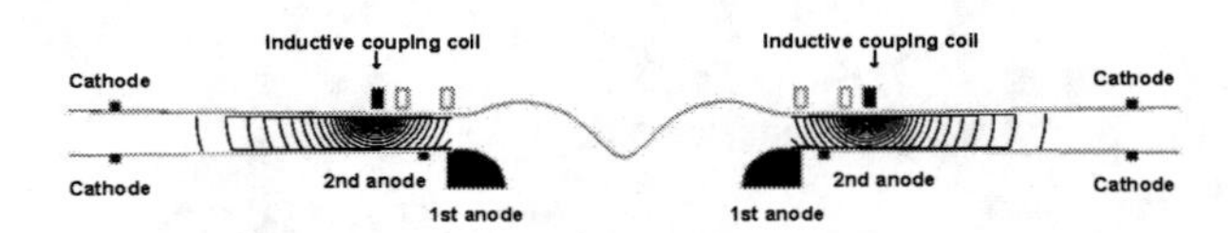

FIGURE 2. Electric-field contour in the disk generator. The electromagnetic field is induced by applying rf current to the outermost inductive coupling coil.

RF-POWER PREIONIZATION

Theoretical Concept

The concept of seed-free pure-inert-gas working MHD generation is based on the fact that a three-body recombination coefficient has a low value at electron temperatures above 5000 K due to the specific structure of the electron levels of inert gases. As a result, at high electron temperatures, the electron number density is "frozen" in gas flow. The plasma slowly recombines with the relaxation time much greater than the propulsion time in a generator. As we know, the frozen inert-gas-plasma MHD generator is feasible only in the case that the initial degree of ionization at the generator entrance is high enough to attain electrical conductivity.

Figure 3 shows the result of a local steady-state calculation regarding nonequilibrium MHD plasma at the generator inlet [14,15,17]; ionization degree of argon atoms (solid curve), characteristic time of ion recombination (dashed curve), and resident time of plasma in the generator (dotted curve), as a function of electron temperature. Calculation conditions: total argon gas pressure of 0.1 MPa, total argon gas temperature of 4000 K, static argon gas temperature of 2000 K, and Mach number of 1.7 (radial gas velocity of 1120 m/s). As indicated by the theoretical prediction shown in Fig. 3, the present scheme is feasible as long as the following conditions are realized: electron temperature upstream of the generator is increased to more than 6000 K by rf heating, the resulting degree of argon ionization is at the 0.01 % level, and the characteristic e-folding time of argon ion recombination is greater than the plasma residence time.

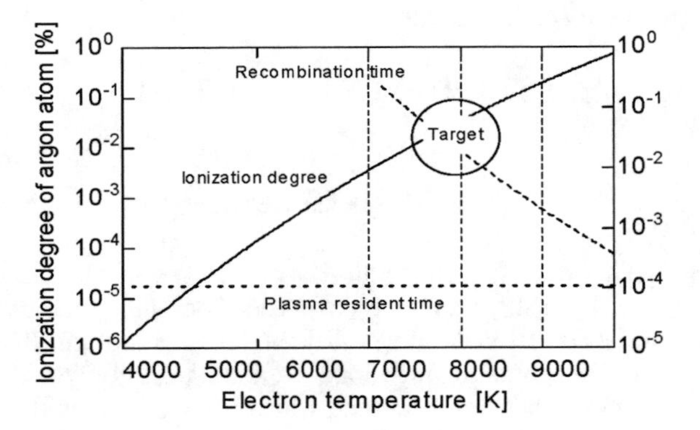

FIGURE 3. Ionization degree of Ar atoms, and characteristic time of ion recombination and plasma residence as a function of electron temperature.

Plasma Ignition and Energy Conversion

Figure 4 shows time variation of experimental results obtained without rf power (left-hand side) and with rf power (right-hand side). (a) Inflow total pressure and magnetic flux density, (b) net rf power, (c) Hall current, and (d) intensity of argon line spectrum upstream and (e) downstream. The inflow total temperature is 4300 K. Figure 5 shows power output as a function of inflow total temperature. The power output is averaged over an interval of 0.5 ms in the rf-assisted phase (t=0.7-1.3 ms).

The left-hand side of Fig. 4 shows that nominal radiation intensity and Hall current are monitored; the pure argon gas is negligibly excited at an inflow total temperature of 4300 K. Even when the total temperature is increased to 4800 K, little power is extracted (Fig. 5). Thermal equilibrium ionization induced within the present total temperature range is insufficient to attain the electrical conductivity required for power generation.

The right-hand side of Fig. 4 clearly shows the positive effect of the continuous rf power (from 0.5 to 1.5 ms) on plasma initiation. The turning on of the rf power ignites the plasma both upstream and downstream. The continuous plasma generation assisted by rf power generates the Hall current via the MHD interaction. The resulting power output of approximately 25 kW at the total temperature of 4300 K is larger by a factor of five than the input rf power and corresponds to an enthalpy extraction ratio ([power output] / [thermal input]) of 1.35 % (Fig. 5). The measured electron temperatures are 7500±400 K and 7600±200 K at the total temperatures of 3800 K and 4300 K, respectively, which are significantly higher than the static argon gas temperature of 2000 K.

The power-generation plasma initiated by rf power (rf-assisted phase, t=0.5-1.5 ms) continuously maintains itself by the self-excited Joule heating mechanism (self-sustained phase, t=1.5-3.2 ms) until the natural decay limit of the present shock tube experiment (Fig. 4). This transient phenomenon from the rf-assisted state to self-sustained state can be explained by an inherent positive feedback of the Hall current from the downstream cathode to the upstream anode via the external resistance; this suggests the possibility that MHD energy conversion will continue after the initial rf-preionization phase.

We can determine the difference between the transient behavior (Fig. 4) of the rf-assisted plasma and the self-sustained plasma. The temporal behavior of the upstream radiation in the rf-assisted phase (Fig. 4(d)) indicates that the slowly recombining plasma is relatively stable; although the radiation is perturbed in a narrow range with a characteristic frequency of 3 kHz, the intensity is continuously high (fluctuations defined as [standard deviation / average] are 0.34 in the upstream and 0.47 in the downstream). The electron system does not interact thermodynamically with the heavy particle system, so the electron number density established upstream could be continuously sustained in the frozen state.

By contrast, as indicated by the rather large fluctuation of the radiation intensity in the self-sustained duration (oscillating frequency is 2 kHz; fluctuations are 0.56 and 0.66 in the upstream and downstream, respectively) in Fig. 4, the plasma without rf power is rather unstable. Although the initial rf-assisted state could be accompanied by the self-sustained state, it is preferable, for stable plasma generation and stable energy conversion, to employ continuous rf power assistance throughout the entire operation.

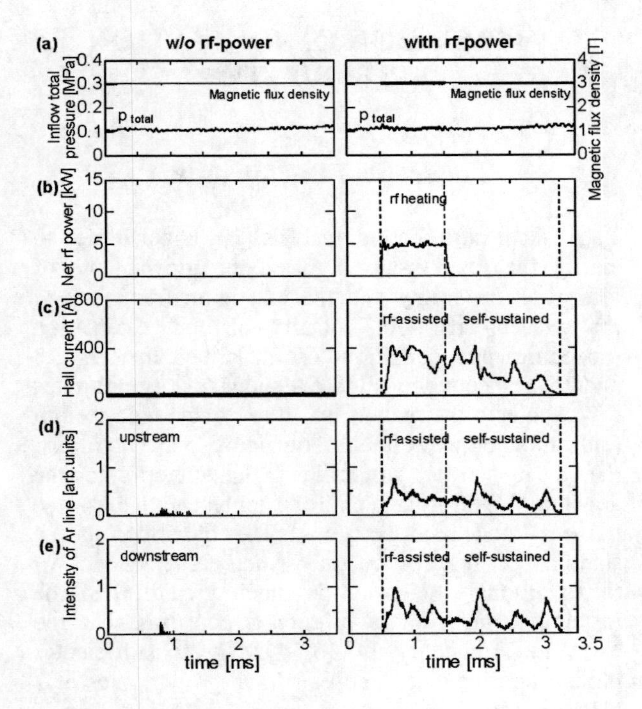

FIGURE 4. Time variation of experimental results obtained without rf power (left-hand side) and with rf power (right-hand side). (a) Inflow total pressure and magnetic flux density, (b) net rf power, (c) Hall current, and (d) intensity of argon line spectrum upstream and (e) downstream.

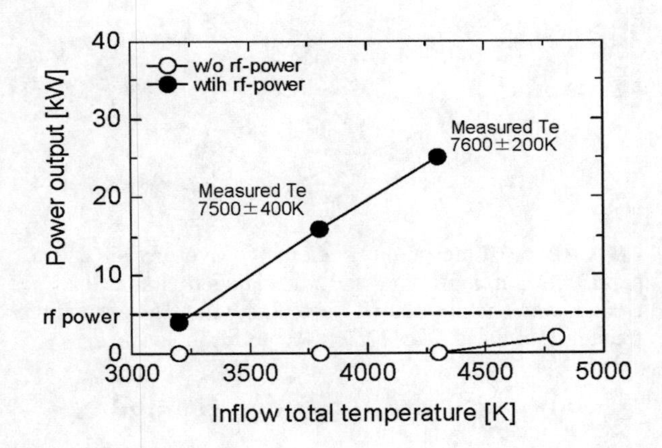

FIGURE 5. Power output as a function of inflow total temperature.

SUPPRESSION OF IONIZATION INSTABILITY

Theoretical Background

A linear perturbation analysis [18] is performed to clarify the relationship between the growth rate of ionization instability and the excitation frequency of the present rf field (details of the calculation procedures and conditions are described in Refs. 18 and 15, respectively). Figure 6 shows the growth rates of an electrothermal wave due to partial cesium ionization and weak helium ionization, which is one of the most serious instabilities that deteriorate the generator performance, for representative wave numbers and seed fractions. The dynamic plasma stabilization is effective only when the frequency of the additional oscillation is much greater than the instability growth rate. Figure 6 confirms that the excitation frequency of 13.56 MHz is sufficiently higher than the electrothermal wave growth rates of 1 kHz to 1 MHz.

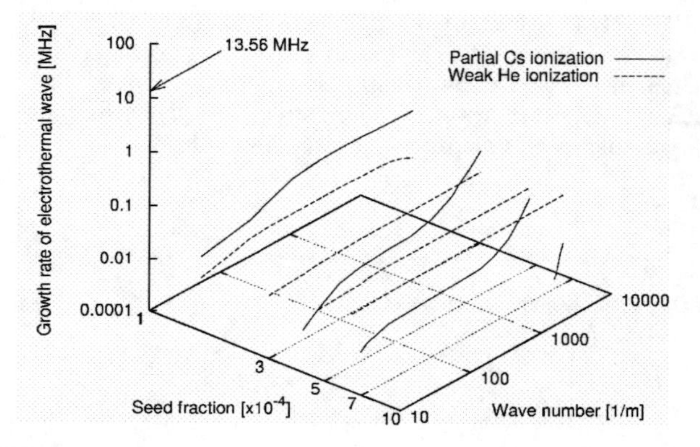

FIGURE 6. Rate of electrothermal wave growth due to partial cesium ionization and weak helium ionization for typical wave numbers and seed fractions. The excitation frequency of the rf field, 13.56 MHz, is also indicated.

Suppression of Ionization Instability

Figure 7 shows (a) a photograph of the plasma, and time variations of (b) the electron temperature and (c) the function of electron and cesium ion number densities $(n_e.n_{Cs+})^{1/2}$ upstream, which were obtained without the rf power under a seed fraction of 13×10^{-4}. In the case that free electrons are provided only from singly ionized cesium ions, $n_e.n_{Cs+})^{1/2}$ is proportional to the electron number density. Figure 8 shows the schematic illustration of the plasma (Fig. 7) with the disturbance vector diagram of wavefront orientations of ionization perturbation. The angle ϕ between Hall electric field $\boldsymbol{E}_{Hall}$ and conduction current density $\boldsymbol{j}$ is estimated from the Hall parameter β of 1 to 2, that is $\boldsymbol{j}.\boldsymbol{E}_{Hall} = jE_{Hall}. \phi_{1,2}$, where $\phi_1 = \tan^{-1}\beta = \tan^{-1}1$ and $\phi_2 = \tan^{-1}\beta = \tan^{-1}2$. The determination of angle θ between the Hall electric field $\boldsymbol{E}_{Hall}$ and wave vector $\boldsymbol{k}$ is based on the plasma structure.

It is seen from Fig. 7(a) that the plasma is inhomogeneous and is associated with distorted layer-like luminosities. As predicted on the basis of a nonlinear theory [18-20], the nonequilibrium seeded plasma was conditionally unstable when the angle $\theta + \phi$ was closed to $\phi/2$ under the situation that the initial disturbance was rather large (Fig. 8). The concentration and distribution of the luminosity is influenced by the temperature and density profile in the plasma. We can observe, in Figs. 7(b) and 7(c), that the density changes in close correlation with the electron temperature. In the fluctuating unstable plasma, the low "background" electron temperature of 3000-3500 K and low "background" density results in the weak brightness. In these dark regions, the seeded cesium atoms are partially ionized. In contrast, high temperature and density (the peak electron temperatures reach 6000 K and 8500 K) lead to the high-luminosity layers. The present ionization instability is originally caused by partial seed ionization due to the low electron temperature. We can identify the electrothermal wave from six wavelike peaks in the temporal temperature and density behavior (Figs. 7(b) and 7(c)). The measured characteristic frequency of electrothermal wave propagation is 20-25 kHz, which is related to the time-periodic interval required for the high-luminosity wavelengths to pass through the optical window. Since the distance between the neighboring luminous layers is approximately 20 mm, the propagation velocity is estimated to be approximately 400-500 m/s, which is low compared with the gas velocity of approximately 3000 m/s at a local Mach number of 1.

Figure 9 shows the noteworthy effects of rf-power assistance on the improvement of the plasma structure and the plasma behavior. Due to the additional rf heating, the plasma is homogenized and the ionization instability is suppressed. The plasma structure becomes symmetric in the azimuthal direction and is accompanied by stationary luminosities (radially expanding trails). The temporal temperature and density behaviors are relatively stable; the electron temperature is only slightly pretreated in a narrow range from 3900 to 5200 K. The decline of the electron temperature due to rf heating seems to be inconsistent with the concept of the superimposition of additional heating on self-excited Joule heating. However, this is the core of the present study. The slight but prominent enhancement of the background electron temperature from 3000 K to 3900 K prevents the growth of electrothermal waves.

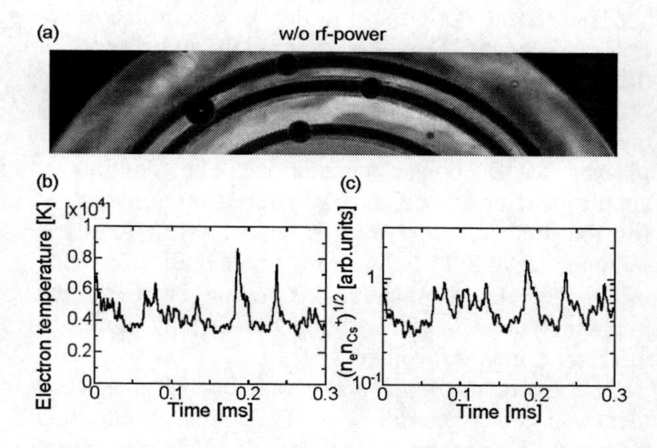

FIGURE 7. (Color online) The rf power is not applied. (a) A photograph of the plasma, and time variations of (b) the electron temperature and (c) the function of electron and cesium ion number densities.

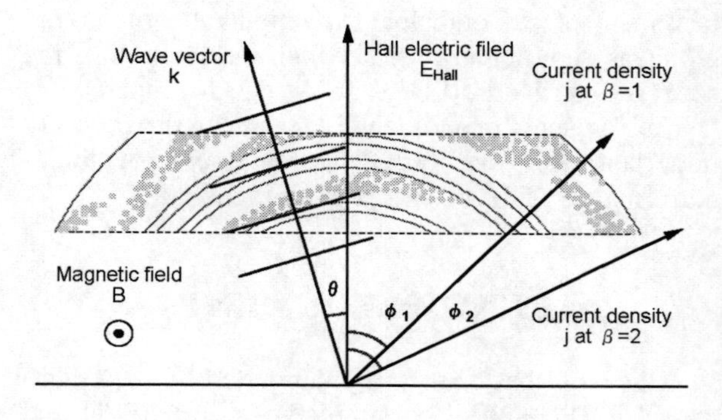

FIGURE 8. (Color online) Schematic illustration of the unstable plasma with disturbance vector diagram of wavefront orientations of ionization perturbation.

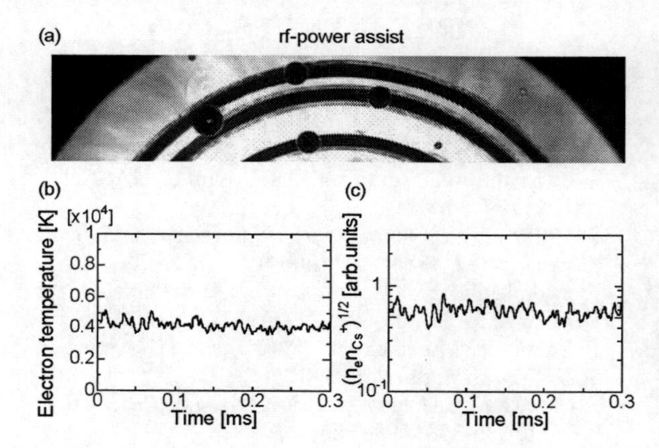

FIGURE 9. (Color online) The rf power is applied. (a) A photograph of the plasma, and time variations of (b) the electron temperature and (c) the function of the electron and cesium ion number densities.

Improvement of Generator Performance

Figure 10 shows the influence of the rf-power assistance on the plasma behavior and the power-generating performance: (a) the electron temperature, (b) the fluctuation of the electron temperature, its characteristic frequencies (c) upstream and (d) downstream, and (e) the enthalpy extraction ratio as functions of seed fraction. In Fig. 10 (a), the plots indicate the average electron temperature and vertical bars represent the range of its fluctuation. The electron temperature could not be measured at the lowest seed fraction because of insufficient radiation intensity. The plasma condition changes with increasing electron temperature, namely, a partial cesium ionization state at less than approximately 4000 K, a full cesium ionization state in the range from 4000 to 9000 - 10000 K, and a weak helium ionization state at more than 10000 K. The fully ionized seed (cesium) is described as a state in which seed atoms are almost fully ionized without the ionization of the inert (helium) gas atoms, whereby the nonequilibrium plasma becomes stable against infinitesimal perturbations. The dashed lines indicate the theoretical threshold of ionization instability. The fluctuating behavior of the electron temperature is quantified by fast Fourier transform and shown as the normalized standard deviation in Fig. 10(b). The fluctuating electron temperature is also quantified by its characteristic frequency in Figs. 10(c) and 10(d).

The characteristic frequency of the upstream electron temperature of the rf-assisted plasma is 30 kHz under the seed fractions of 3 and 5×10^{-4} (Fig. 10(c)). As indicated by this relatively low frequency, which is the same level as that in the no-rf case, rf heating might not completely suppress the upstream large-scale wavelike perturbation under an insufficient seed fraction. However, the slight but prominent enhancement of the low-level upstream electron temperature from 3100 to 3300 K prevents the generation of electrothermal waves. Thus, the resulting fluctuation level declines from 0.22 (no-rf case) to 0.12 (rf-assisted case) (Fig. 10(b)). Furthermore, the fluctuation in the rf-assisted plasma decreases from 0.12 to 0.03 with increasing seed fraction both upstream and downstream (Figs. 10(c) and 10(d)). As the seed fraction increases to moderate and high levels ($6\text{-}10\times10^{-4}$), the plasma structure becomes finer (frequency increases from 30 to 50-60 kHz) and the plasma becomes stable (fluctuation decreases from 0.12 to 0.07). Even under very high seeding conditions of $13\text{-}27\times10^{-4}$, the rf power suppresses ionization instability both upstream and downstream. The upstream plasma, which greatly influences the entire plasma behavior, is improved with the aid of rf power. Subsequently, downstream in the MHD channel, the electron temperature increases and its fluctuation level becomes low.

The nonuniformities of the plasma properties in the direction perpendicular to the magnetic field and the flow cause the reduction of "effective" electrical conductivity or the "effective" Hall parameter [1]. The deterioration of the effective electrical conductivity or effective Hall parameter due to the plasma nonuniformity would seriously affect the power-generating performance. Thus, the present plasma stabilization and homogenization effects directly contribute to the energy conversion efficiency. As shown in Fig. 10(e), the enthalpy extraction ratio is improved from 0.4 to 1.0 % at the minimum seed fraction of 1.7×10^{-4}; from 3.2 to 5.8 % at the insufficient seed fraction of 5×10^{-4}; from 6.7 to 8.8 % at the optimum seed fraction of 10×10^{-4}; and from 5.7 to 7.9 % at the excessive seed fraction of 13×10^{-4}. The increase of the enthalpy extraction ratio, i.e., 2.1 % at the optimum seed fraction, is significantly greater than the fraction of additional rf-power input, that is, [rf-power input] / [thermal input]=0.16 %.

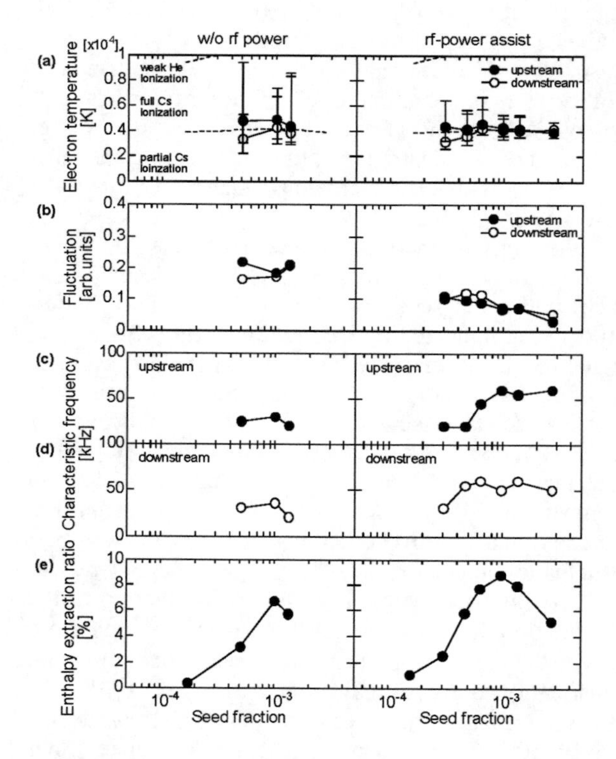

FIGURE 10. The (a) electron temperatures upstream and downstream (plots: average value; vertical bars: fluctuation range), (b) fluctuation of the electron temperature (standard deviation normalized by average) upstream and downstream, characteristic frequencies of the fluctuating electron temperature (c) upstream and (d) downstream, and (e) enthalpy extraction ratio as functions of seed fraction. The plots on the left- and right-hand sides are for without and with rf power, respectively.

SUMMARY

The recent activities on the rf electromagnetic-field-assisted MHD power generation experiments at the Tokyo Institute of Technology have been reviewed.

In summary, we successfully demonstrated rf-electromagnetic-field-assisted seed-free pure-argon-plasma MHD power generation. The continuous supply of rf power enhanced the electron temperature (on the order of 7500-7600 K) under a low total gas temperature (4300 K). Due to the nonequilibrium state established by rf heating, the rf-assisted plasma behavior was relatively stable, which contributed to efficient energy conversion.

We overcame ionization instability of the MHD electrical power-generating He-Cs plasma by coupling with an rf electromagnetic field. The rf heating suppressed the ionization instability in the plasma behavior and homogenized the nonuniformity of the plasma structure. As a result, the power-generating performance was significantly improved. The increment of the enthalpy extraction ratio of around 2 % was significantly greater than the fraction of the net rf power, that is, 0.16 %, to the thermal input.

The present experimental results will be very important for the future development of high-performance MHD electrical power generation.

ACKNOWLEDGMENTS

This work has been partly supported by the Grant-in-Aid for Scientific Research (B) No. 16360134 and (B) No.19360127 of the Japan Society for the Promotion of Science (JSPS).

REFERENCES

1. R. J. Rosa, *Magnetohydrodynamic Energy Conversion,* McGraw-Hill, New York, 1968.
2. L. Bitteker, *J. Appl. Phys.* **38**, 4018-4023 (1998).
3. R. J. Litchford, L. Bitteker, and J. Jones, NASA/TP-2001-211274 (2001).
4. T. Murakami, Y. Okuno, and H. Yamasaki, *IEEE Trans. Plasma Sci.* **32**, 1886-1892 (2004).
5. T. Murakami, Y. Okuno, and S. Kabashima, *IEEE Trans. Plasma Sci.* **25**, 769-775 (1997).
6. T. Fujino, T. Murakami, Y. Okuno, and H. Yamasaki, *IEEE Trans. Plasma Sci.* **31**, 166-173 (2003).
7. T. Murakami, Y. Okuno, and S. Kabashima, *IEEJ Trans. PE.* **118**, 333-338 (1998) (in Japanese).
8. T. Murakami, Y. Okuno, and S. Kabashima, *IEEE Trans. Plasma Sci.* **27**, 604-612 (1999).
9. G. Lou, T. Murakami, T. Fujino, and Y. Okuno, *IEEE Trans. Plasma Sci.* **33**, 997-1004 (2005).

10. T. Murakami, Y. Okuno, and H. Yamasaki, *Appl. Phys. Lett.* **86**, 171502 (2005).
11. T. Murakami, Y. Okuno, and H. Yamasaki, *Appl. Phys. Lett.* **86**, 191502 (2005).
12. T. Murakami, Y. Okuno, and H. Yamasaki, *Phys. Plasma* **12**, 113503 (2005).
13. T. Murakami, Y. Okuno, and H. Yamasaki, *J. Appl. Phys.* **98**, 113306 (2005).
14. T. Murakami and H. Yamasaki, *IEEE Trans. Plasma Sci.* **32**, 1752-1759 (2004).
15. T. Murakami, Y. Okuno, and H. Yamasaki, *J. Appl. Phys.* **96**, 5441-5449 (2004).
16. T. Murakami and H. Yamasaki, *IEEE Trans. Plasma Sci.* **33**, 520-521 (2005).
17. H. Kobayashi, Y. Satou, and Y. Okuno, *IEEE Trans. Plasma Sci.* **30**, 2152-2159 (2002).
18. M. L. Hougen and J. E. McCune, *AIAA J.* **9**, 1947-1956 (1981).
19. A. Yu. Sokolov, S. Kabashima, and V. M. Zubstov, *J. Plasma Phys.* **54**, 105-118 (1995).
20. K. Yasui, V. M. Zubtov, K. Yoshikawa, and S. Kabashima, *Jpn. J. Appl. Phys.* **34**, 683-689 (1995).

Can Microscale Wall Roughness Trigger Unsteady/Chaotic Flows?

Fathollah Varnik and Dierk Raabe

Max-Planck Institut für Eisenforschung, Max-Planck Straße 1, 40237 Düsseldorf, Germany

Abstract. Results of lattice Boltzmann simulations on the flow through narrow channels are presented. The focus of the work is the effect wall roughness on microscale flows at moderate Reynolds numbers, i.e. at Reynolds numbers significantly larger than unity ($100 \leq \mathrm{Re} \leq 2000$), but not large enough in order to give rise to fully developed turbulence. It was shown in a previous work [1] that the presence of wall roughness may significantly alter the flow behavior, triggering a transition from a laminar to a time dependent flow. Further studies underlined the significance of the roughness wave length on the observed phenomenon [2]. In the present work, we study this issue further providing new support for the occurrence of flow instability as well as on the chaotic nature of the flow. In particular, via a study of probability density of the velocity field as well as spatial correlations between the velocity fluctuations along the flow direction, we demonstrate that wall roughness may give rise to a variety of situations such as simple oscillatory modes as well as fully chaotic flows. The latter point is further evidenced by visualizing the flow field via passive tracers.

Keywords: Flow instability, Wall roughness effects, Lattice Boltzmann
PACS: 47.11.-j, 47.11.Qr, 47.20.-k, 47.85.Np

INTRODUCTION

In macroscopic channels, the wall roughness is known to be relevant for high Reynolds number turbulent flows. However, in such situations, the effect of the roughness is not to trigger the instability (the flow remains turbulent if the walls are made smooth). Rather, it may significantly alter characteristic features of turbulence such as the distribution of the Reynolds stress and other turbulent quantities across the channel [3, 4].

In microchannels, on the other hand, the flow is usually assumed to be laminar and roughness effects are often studied in the context of wetting phenomena (super hydro-phobic/philic behavior) [5].

Even though the assumption of a Stokes flow is true for many applications dealing with the flow at small scales, there are examples where the characteristic Reynolds number may reach values as high as a few thousands even at microscales. In metal forming (a focus of research within our department in Max Planck Institut für Eisenforschung), for instance, the flow of lubricant during the so-called stamping process provides such an example: A metallic stamp hits the work piece leading to a significant plastic deformation of the latter.

During this process, the lubricant fluid covering the work piece and the stamp is pushed out of the contact zone with a high velocity of the order of a few hundred meters per second [2]. Note that this occurs as soon as the work piece and the stamp come into contact. The vertical dimension of the so formed 'channel' along which the lubricant flow occurs is thus of the order of long wave length corrugations of the work piece and the stamp. Our measurements via white light confocal microscopy give values of a few tens of microns for these corrugations and thus for the channel height. Putting numbers, $U = 200$m/s for the characteristic fluid velocity, $H = 50\mu$m for the channel height and $\nu = 10^{-5}$m^2/s for dynamic viscosity of oil, one obtains a typical Reynolds number of $\mathrm{Re} = 1000$ (see Ref. [2] for a more detailed discussion of this issue).

A Reynolds number of the order of 1000 can be classified as "moderate", since it is low in the sense that the flow is stable with respect to infinitesimally small perturbations but high in the sense that finite amplitude perturbations (originating e.g. from the wall roughness) may grow to some extent. This brings about the possibility of a roughness-induced transition toward an unsteady/chaotic flow. Indeed, results of our lattice Boltzmann simulations indicate that such a transition from laminar toward unsteady flow may be observed via an increase of the wall roughness at constant Reynolds number [2].

The present work builds upon these studies. While the previous studies focused on the very possibility for the occurrence of non-trivial flow behavior triggered by wall roughness, we provide here an analysis of the normalized probability density for the velocity field. Moreover, we also investigate spatial correlations between velocity fluctuations and discuss the effects of the (average) distance between roughness elements on these correlations.

In order to focus on new aspects, we directly start with simulation results. As to our computer simulation approach, namely the lattice Boltzmann method, the reader is referred to excellent monographs [6, 7, 8] and com-

CP982, *Complex Systems, 5th International Workshop on Complex Systems*
edited by M. Tokuyama, I. Oppenheim, and H. Nishiyama

prehensive review articles [9, 10, 11] on this issue. A brief introduction to the specific D2Q9-lattice Boltzmann model can also be found in one of our recent articles [1].

RESULTS

We begin with a brief survey of the basic observation. For this purpose, we study the flow through a channel made by a flat wall on the top and a wall with an idealized type of roughness, a zig-zag one on the bottom. A schematic view of the simulation box is shown in the upper panel of Fig. 1. A zig-zag surface may easily be characterized by its wave length, λ, and its height h. Therefore, we refer to a given zig-zag surface by indicating its height and the half wave length in the form $h:\lambda/2$ [$h/(\lambda/2)$ being the magnitude of the surface's slope]. Flow velocity versus time is recorded at 9 points placed equidistantly along the line $x = 40$ (array of circles in the upper panel of Fig. 1). It must be emphasized here that, the phenomenon we are going to discuss is not restricted to this specific choice of the surface roughness but does also occur for other choices of (periodic) distribution of triangular obstacles as well as for a random distribution of obstacle elements [1].

The lower panel of Fig. 1 illustrates how a variation of the roughness height alone may give rise to a time dependent flow. For this purpose, the fluid velocity in the middle of the flow region (midway between the roughness tip and the top wall) is shown versus time for the flow over a 20:20 zig-zag wall and the flow over a 40:20 zig-zag ("rougher") surface. In the both cases, the simulation is started with a quiescent fluid ($\boldsymbol{U}(t=0) = \boldsymbol{0}$) and an external body force of $g = 8\nu U_0/(L_y - h)^2$ is switched on at time $t = 0$. This choice ensures that, in the steady state, the maximum velocity in the channel reaches a value close to U_0, provided that the flow is laminar (the subtraction of the roughness height, h, from L_y takes into account the reduction of the effective channel width due to the wall roughness).

In the case of the channel with a 20:20 zig-zag surface, the fluid velocity in the direction parallel to the wall, U_x, increases continuously and smoothly until it reaches the steady state after a time of order $t_{\rm d} = (L_y - h)^2/(8\nu)$ as estimated from a study of the Stokes flow in a planar channel of effective width $L_y - h$. The vertical component of the fluid velocity, U_y, on the other hand, remains zero at all times.

A qualitative change in the flow behavior is, however, observed in the lower panel of Fig. 1 as the roughness height is increased from 20 to 40. Now, both U_x and U_y exhibit strong fluctuations indicative of instability. The flow instability also manifests itself in a drop (by more than a factor of two) of the mean mid-channel velocity, signaling a higher friction force and thus a higher energy dissipation rate.

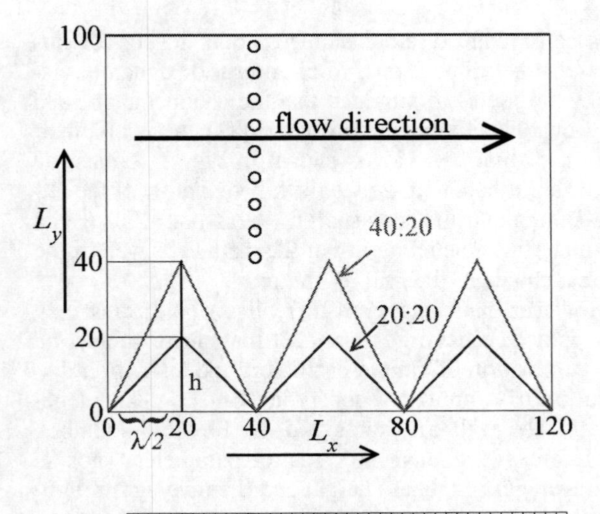

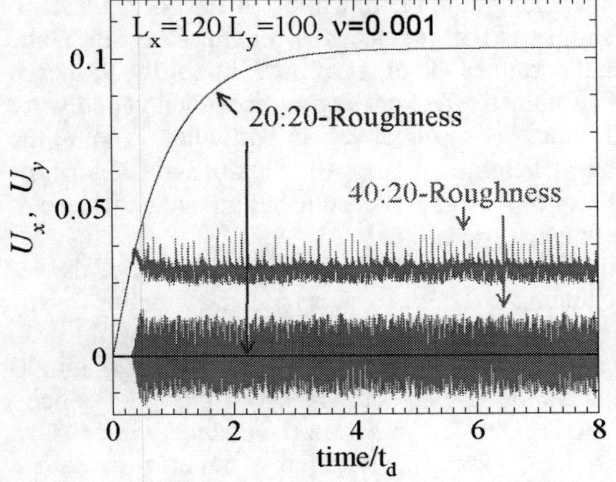

FIGURE 1. (Color online) Top: A sketch of the simulation box. In the notation $a:b$, the first argument indicates the height of the roughness tip and the second the half width of the baseline (a/b=slope). Flow velocity versus time is recorded at 9 points placed equidistantly along the line $x = 40$ (array of circles). Note that the region below the zig-zag line is filled with solid particles (not shown for clarity). Bottom: Time evolution of U_x and U_y, measured at a point in the middle of the flow region. From Ref. [1].

The reliability of our results is checked by computing various quantities for which exact behavior can be derived from the Navier-Stokes equation. An example is the sum of the viscous stress $\nu\rho\partial < U_x > /\partial y$ and the so-called Reynolds stress $-\rho < \delta U_x \delta U_y >$ which obeys a straight line with a slope, ρg, (g=imposed external force per unit mass): $\nu\rho < \frac{\partial U_x}{\partial y} > -\rho < u_x u_y > = \tau_{\rm w} + \rho g (y - y_{\rm w})$ ($y_{\rm w}$=the position of the wall, $\tau_{\rm w}$=viscous stress at the wall). Indeed, this equation is very well satisfied by our simulation results (see Fig. 6 in Ref. [1]).

Furthermore, the obtained solutions obey the transformation rules of the Navier-Stokes equation: When expressed in reduced (dimensionless) units, results for various channel dimensions, forcing term or dynamic viscosity are identical provided that the channel shape and the Reynolds number are unchanged [1]. An example of such a scaling behavior is shown in Fig. 2. It must be emphasized here that, due to the kinetic nature of the lattice Boltzmann method, such a test is not at all trivial but underlines the reliability of the method as well as the correct choice of the grid resolution.

However, it is important to realize that an unsteady flow is not necessarily chaotic. It may, for example, be a superposition of simple oscillatory modes with a distribution of amplitudes depending on the channel geometry, the wall roughness and the Reynolds number. Therefore, we studied the effect of parameters such as the (average) roughness height and the (average) roughness wave length on the transition. A results of these investigations is that it is not the roughness height alone which determines the onset of flow instability. Rather, it is the combined effect of the roughness height and wave length which is essential [2]. In particular, keeping the roughness height constant, it is possible to trigger flow instability or increase its chaotic nature by an increase of the roughness wave length [2].

This observation is corroborated by a study of the velocity fluctuations. For this purpose, we monitor for the case of a planar channel of size $L_x = 240$ and $L_y = 50$ (LB units) with 12:20 zig-zag walls both the longitudinal and the transverse components of the fluid velocity for various points (x,y) across the channel ($x = 40$ and $y = 14, 18, \cdots, 26, 34$). Sampling is done in the course of simulation at regular time intervals. The Reynolds number based on the average mid-channel flow velocity $\bar{U} \approx 0.058$, the effective channel width $L_y - 2h = 50 - 2 \times 12 = 26$ (h denotes the roughness height) and the kinematic viscosity $\nu = 0.001$ is $\mathrm{Re} = (\mathrm{L_y} - 2\mathrm{h})\bar{\mathrm{U}}/\nu \approx 1500$.

Independent simulations show that the flow at this Reynolds number is laminar for the case of smooth walls, but develops a time dependent character as the triangular obstacles are introduced. However, it is seen from the upper panel of Fig. 3 that the probability density for the transverse velocity (identical to velocity fluctuations, since the average transverse velocity vanishes) do not show any sign of a chaotic motion. Rather, the distribution exhibits a wide plateau and two pronounced peaks, suggesting strong correlations between velocities at different times.

Interestingly, a qualitatively different behavior is observed, if the average distance between the obstacles is increased (by e.g. removing a certain number of triangles). The lower panel of Fig. 3 shows the probability distribution for the transverse component of the fluid velocity for this type of situation. Despite the relatively large statistical noise, the qualitative difference to the data shown in the upper panel is rather obvious. From these data, one could expect a more chaotic flow behavior in the channel with a less number of obstacles. Less obstacles, may, therefore, perturb the flow more efficiently.

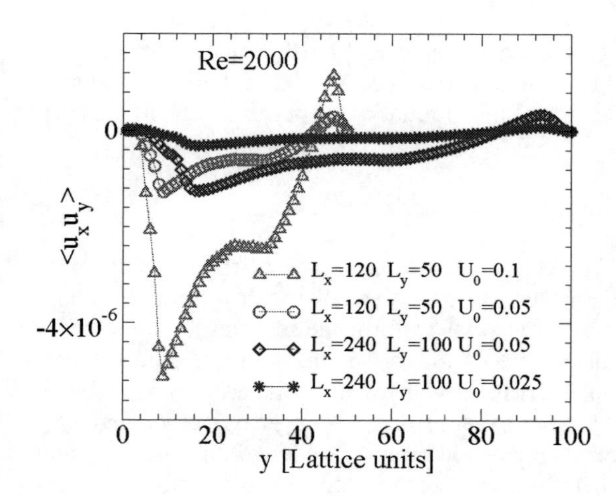

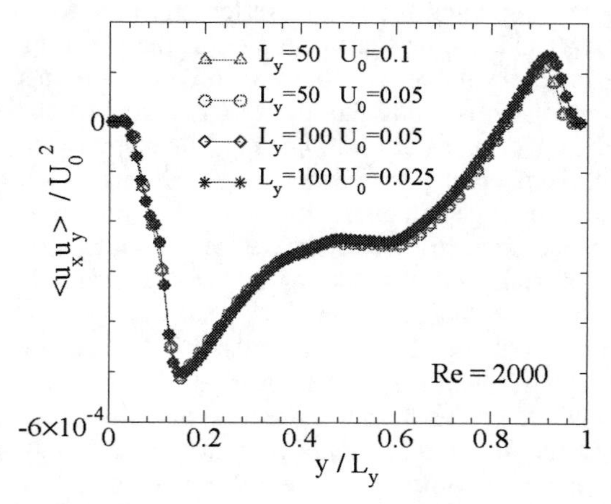

FIGURE 2. (Color online) Top: The Reynolds stress for four different combinations of the channel size and the desired mid-channel velocity, U_0, as indicated. These parameters are chosen in a way to keep both the channel geometry and the expected Reynolds number $\mathrm{Re} = L_y\bar{U}_0/\nu$ ($\nu = 0.001$) approximately constant. $\bar{U}_0$ is the time-averaged mid-channel velocity. It is in general smaller than U_0 (which enters the determination of driving force; see the text), $\bar{U}_0 = U_0$ holding in the case of laminar flow. The wall roughness stdied here is a 5:20 zig-zag wall for a system size of $L_x = 120$ and $L_y = 50$ and a 10:40 one for $L_x = 240$ and $L_y = 100$. All quantities are expressed in lattice Boltzmann units. Bottom: The rescaled version of the same quantities as in the upper panel. From Ref. [1].

We repeat the same type of analysis in Fig. 4 but for the distribution of the velocity along the longituidinal (x) direction. Again, we compare the effect of obstacle removal (increase of the roughness wave length) on the characteristic features of the flow. Also here, an increase of the random character of the velocity distribution upon obstacle removal is clearly visible.

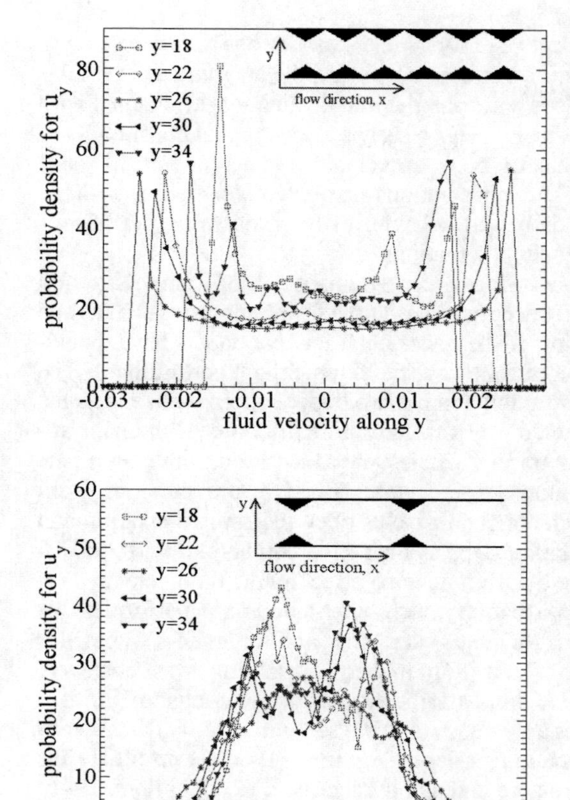

FIGURE 3. (Color online) Top: Probability density, $P(u_y)$, for velocity fluctuations along transverse direction, u_y. The velocity is sampled at a constant rate at various positions (x,y) across the channel with $x = 40$ and transverse coordinates, y, as indicated ($y = 26$ roughly corresponds to the midway between the walls). The inset shows a schematic view of the corresponding channel: $L_x = 240$, $L_y = 50$ confined via two 12:20 zigzag walls. The Reynolds number based on the average mid-channel flow velocity $\bar{U} \approx 0.058$, the effective channel width $L_y - 2h = 50 - 2 \times 12 = 26$ (h denotes the roughness height) and the kinematic viscosity $\nu = 0.001$ is $\mathrm{Re} = (\mathrm{L_y} - 2\mathrm{h})\bar{\mathrm{U}}/\nu \approx 1500$. Bottom: The same quantity as in the upper panel but now after having increased the obstacle distance by a factor of three (via removal of 2/3 of all obstacles). This leads to a stronger chaoticity and the corresponding increase in the flow resistence. As a result, the average mid-channel fluid velocity drops to $\bar{U} \approx 0.035$. More interestingly and in contrast to the upper panel, the distribution of velocity fluctuations is no longer a simple bimodal.

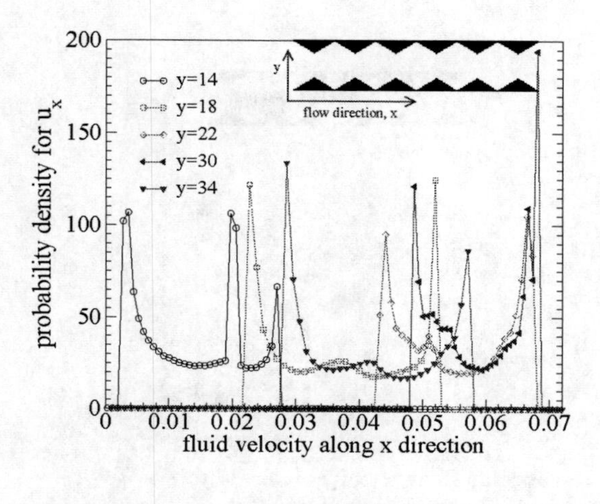

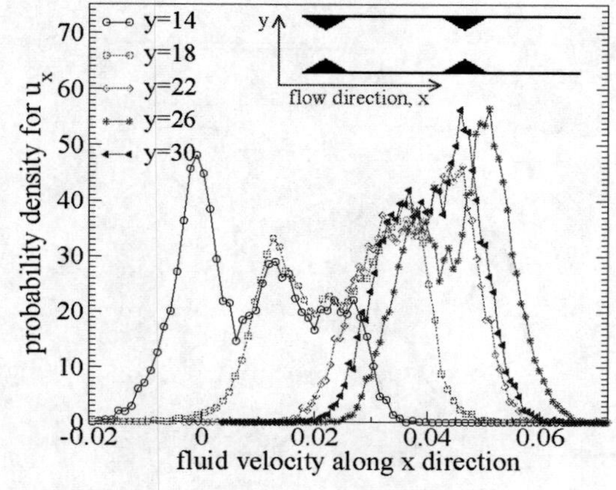

FIGURE 4. (Color online) Same as in Fig. 3 but for the component of the fluid velocity along the streaming direction, U_x. Obviously, distribution of U_x shows a similar behavior as the probability distribution of U_y upon a variation of the obstacle distance. A more chaotic flow is thus expected in the case of the lower panel.

For cases studied here, an increase of statistical accuracy via a longer simulation time is quite time consuming (these simulations were running typically for 2 days on an Itanium 1.3GHz CPU). A far better statistical accuracy is, however, obtained when studying the velocity averaged over the longituidinal coordinate, i.e. $\bar{U}_x(y) = \sum_x U_x(x,y)/L_x$. This quantity is, of course, not identical to the, more appropriate, local velocity, but may still contain interesting information about the chaotic nature of the flow.

Figure 5 depicts the probability distribution for $\bar{U}_x(y)$ for various transverse coordinates. Interestingly, $\bar{U}_x(y)$ seems to show a well defined value with negligible fluctuations in the case of the upper panel, where obstacles cover the whole surface of the walls. In contrast to this, $\bar{U}_x(y)$ follows an approximately gaussian distribution whose width is roughly 15% of the mid-channel velocity. This underlines further the qualitative change in the flow behavior triggered via an increase of the distance between roughness elements.

Further evidence for the effect of roughness wave length is found via an investigation of spatial correlations between velocity fluctuations. For this purpose, we compute the normalized correlation function [12]

$$R_{xx}(x) = \langle u_x(x_0)u_x(x+x_0)\rangle \,/\, \langle u_x(x_0)^2\rangle \qquad (1)$$

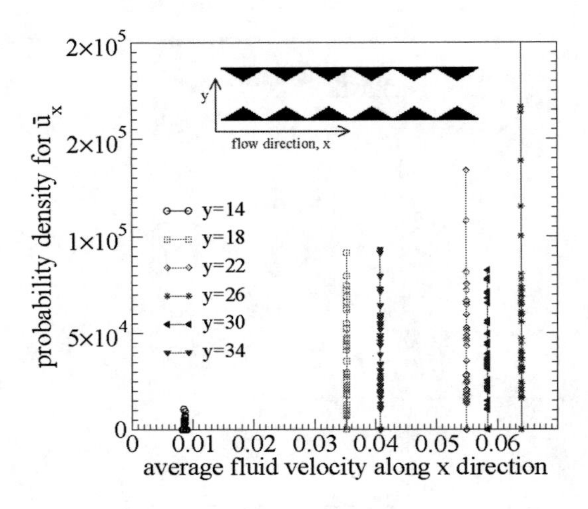

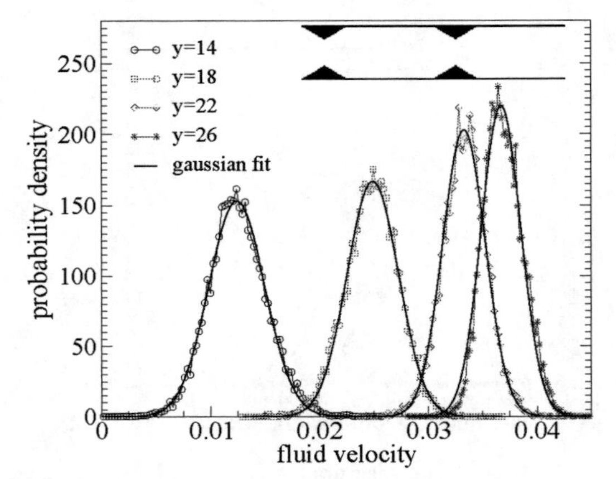

FIGURE 5. (Color online) Probability distribution for the x-component of the fluid velocity at a transverse position y but averaged along the flow direction, $\bar{U}_x(y) = \sum_x U_x(x,y)/L_x$. The thick solid lines in the lower panel give the corresponding gaussian fits.

as well as

$$R_{yy}(x) = \langle u_y(x_0)u_y(x+x_0)\rangle / \langle u_y(x_0)^2\rangle . \quad (2)$$

Here, $u_x = \delta U_x = U_x - \langle U_x\rangle$ is the fluctuation of the x-component of the fluid velocity. A similar definition holds for u_y.

Note that both R_{xx} and R_{yy} also depend on the transverse coordinate y (not shown in order to keep the notation as simple as possible). However, it follows from the symmetry of the channel, that $R_{xx}(x,y_1) = R_{xx}(x,y_2)$ for transverse coordinates y_1 and y_2 corresponding to the same distance to the middle of the channel, $y_{\text{center}} = (50+1)/2 = 25.5$. As will be shown below (Figs. 6 and 7), this property is nicely born out in our simulations.

Results on spatial correlation between velocity fluctuations are depicted in Figs. 6 and 7. As seen from these figures, in the case of a dense array of obstacles, R_{xx} roughly oscillates between the two limiting values 1 and -1, indicative of long range correlations between velocity fluctuations. In contrast to this, the magnitude of correlations does not reach the initial value for the channel with less obstacles. This clearly shows a partial but significant decorrelation between velocity fluctuations in the latter case. These observations are qualitatively in line with the above discussed behavior of the probability density for the velocity fluctuations.

A way to visualize the chaoticity of a flow is to add passive tracer particles to the flow and to survey the time evolution of the tracer field. Passive tracers have the advantage of following the flow without perturbing it. The time evolution of a passive tracer pattern thus reflects the genuine flow dynamics. Within the lattice Boltzmann approach, tracer particles are added by introducing a new population density as described in our previous work [13]. The tracer variables obey the same relaxation and propagation steps as their fluid counterpart with the difference that they have no effect on the fluid velocity.

Using this approach, simulations are performed for various Reynolds numbers and roughness types for which a transition towards instability was observed. First, we simulate the flow in a given channel until it reaches a steady state in the sense that its statistical properties no longer depend on time. Then we reset the time clock to zero and add a tracer field to the bottom half of the channel. The simulation then goes on while monitoring the dynamic of the tracer field.

A typical result of these simulations is shown in Fig. 8, where the tracer field (dark color) at various times is shown in two channels differing in the wall geometry only, all the other parameters of the simulation being nearly identical. As discussed above, the flow is unsteady for the both channel geometries shown. The degree of chaoticity, on the other hand, strongly varies with roughness wave length. While the tracer field remains roughly demixed in the left channel, it is fully mixed in the right one. Since the tracer diffusivity is identical in the case of the both channels, it follows that the observed mixing is purely a result of chaotic nature of the flow.

DISCUSSION AND CONCLUSIONS

In this report, we present results of lattice Boltzmann simulations on the possibility for the occurrence of non-trivial flow behavior, triggered by the wall roughness/channel geometry. We first briefly recall our central observation on the very possibility of a roughness-induced flow instability [1] then switch to an analysis of various quantities, relevant for a characterization of the observed phenomenon.

This includes a computation of the probability density for the velocity field along the flow and the transverse directions (Figs. 3-5) as well as spatial correlation func-

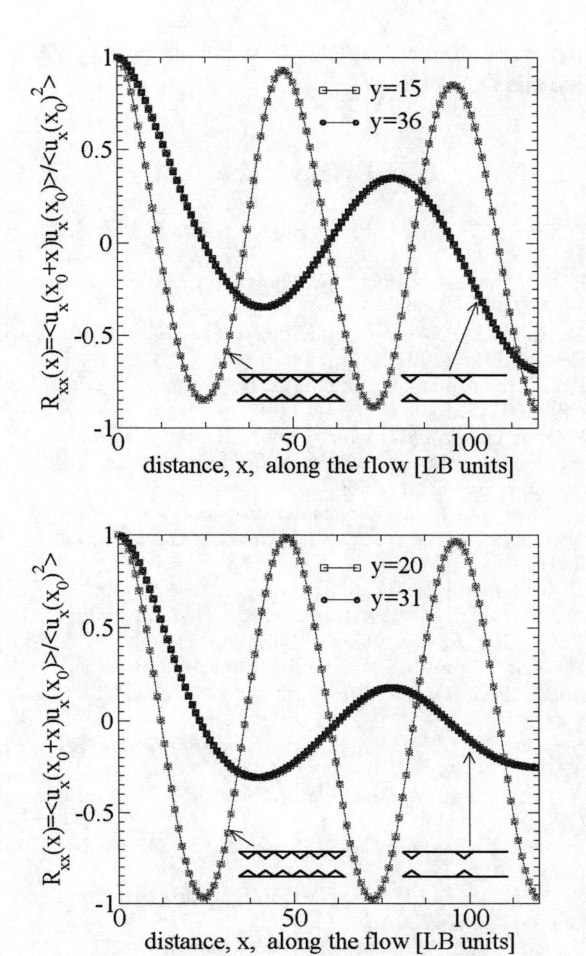

FIGURE 6. (Color online) Top: Normalized correlation function, $R_{xx}(x)$, between the *longituidinal* velocity fluctuations at two points separated by a distance x (see Eq. 1). In each panel, results on R_{xx} are shown for two different channels as indicated. For each channel type, the correlation function is computed at two transverse coordinates $y = 15$ and $y = 36$ having the same distance to the channel center ($y_{\mathrm{center}} = (50+1)/2 = 25.5 \Longrightarrow |y - y_{\mathrm{center}}| = 10.5$). Bottom: The same as in the upper panel for a smaller distance from the channel center $|y - y_{\mathrm{center}}| = 5.5$.

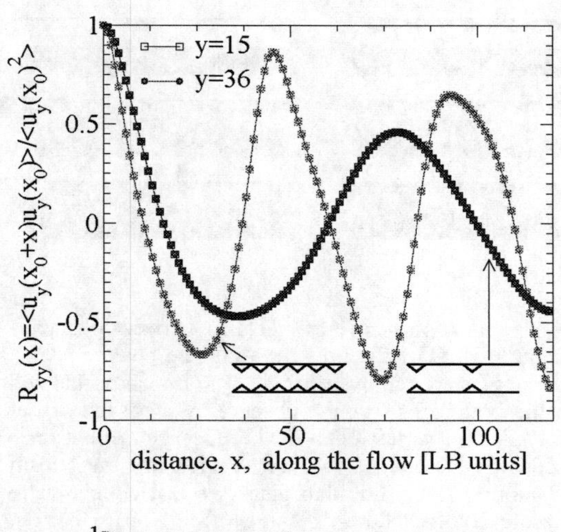

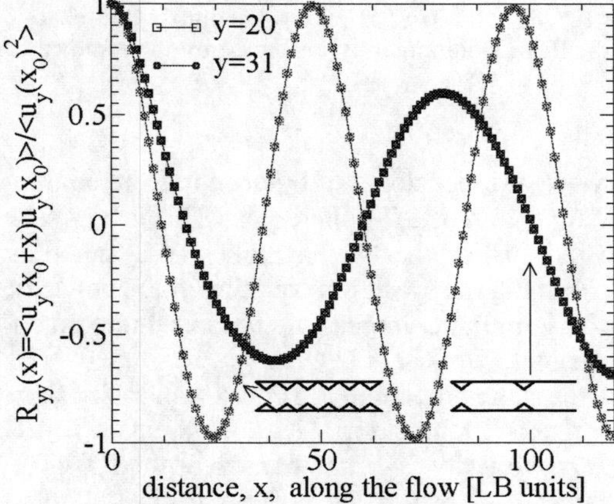

FIGURE 7. (Color online) The same type of comparison as shown in Fig. 6, but for the normalized correlation function between the *transverse* velocity fluctuations, $R_{yy}(x)$ (see Eq. 2).

tions between velocity fluctuations (Figs. 6 and 7).

The overall picture emerging from this analysis is that, depending on the roughness parameters, wall roughness may give rise to quite different types of time dependent behavior ranging from a combination of simple oscillatory modes to fully chaotic flows, the latter best visualized via the use of a tracer field (Fig. 8).

The situation considered here may be relevant in all cases where the dimensions of wall roughness is a finite fraction of the channel height and where the Reynolds number is in intermediate range, of the order of a few hundreds up to a few tousands.

In fact, our studies show that the 'critical' Reynolds number for the onset of non-trivial flow behavior may considerably be reduced via an increase of the roughness height as well as via an optimum choice of the distance between roughness elements (roughness wave length).

Following this route in very recent simulations, we were able to observe chaotic flow at a Reynolds number as low as 100 (estimated using effective channel width and the average mid-channel flow velocity) by increasing the height of obstacles to 30% of the channel width in a symmetric channel while at the same time increasing the distance between the roughness elements, in a way similar to what discussed above. Results of these simulations will be presented elsewhere.

An important issue regarding the above presented results relates to the dimensionality of our simulations. Indeed, in order to concentrate on the principal question addressed above, we had to restrict our attention to the case of two dimensional flows. We are aware of

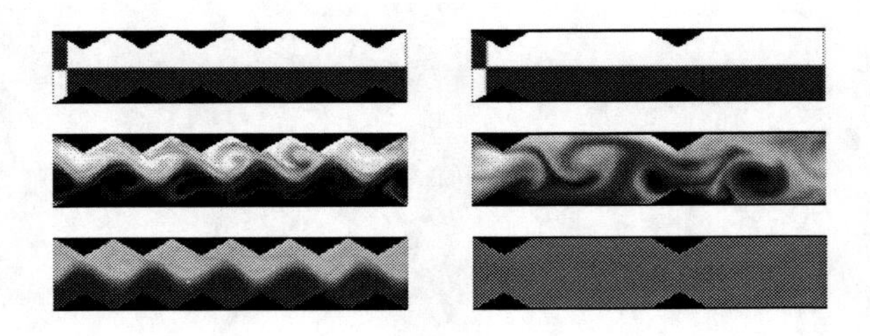

FIGURE 8. Effect of the average distance between obstacles on the chaotic nature of the flow. In a zig-zag channel (left), 2/3 of all triangles are removed (right). The flow is visualized by filling the bottom half of the channel by a passive tracer (dark color). A thin vertical slab of the tracer at the entrance of the channel is also added in order to visualize the stream wise fluid motion. This thin slab disperses fast compared to the time scale resolved here. The time from top to bottom $t = 1$, 5×10^3, 3×10^4 (in LB units). Obviously, the chaotic nature of the flow is significantly enhanced by the removal of obstacles.

the fact that two-dimensional turbulence may be qualitatively different from its 3D-counterpart. This is best seen by taking the curl of the Navier-Stokes (NS) equation and thus obtaining the evolution equation of the vorticity, $\boldsymbol{\omega} = \boldsymbol{\nabla} \times U$. For an incompressible fluid in three dimension, this leads to $\partial \boldsymbol{\omega}/\partial t + \boldsymbol{U} \cdot \boldsymbol{\nabla}\boldsymbol{\omega} = \nu\nabla^2\boldsymbol{\omega} + \boldsymbol{\omega} \cdot \boldsymbol{\nabla}\boldsymbol{U}$. Repeating the same procedure in 2D, one can easily show that no vortex stretching term $(\boldsymbol{\omega} \cdot \boldsymbol{\nabla}\boldsymbol{U})$ exists in 2D [12, 14]. The vorticity is hence conserved in 2D inviscid flow ($\nu = 0$). Close to the walls, however, the first term on the right-hand-side of the above equation becomes dominant even at low viscosities thus giving rise to a complex scenario for the time evolution of the vorticity.

Despite the limitation of our studies to two dimensional case, we expect similar effects also in three dimension. In fact, there are many more path ways for the advection of fluctuations in a higher dimensional space. Furthermore, the vortex stretching term present in 3D leads to a more complex spatial and temporal variation of the vorticity presumably enhancing the chaotic (unstable) character of the flow. The perturbations generated at the wall roughness are, therefore, expected to grow faster in the case of 3D flows as the transition regime is approached. This suggests that, in 3D, the phenomenon discussed here may be observable even at lower Reynolds numbers than in 2D. This issue is a subject of our present studies.

ACKNOWLEDGMENTS

This work was supported by the Deutsche Forschungsgemeinschaft (DFG) within the priority programme Nano-& microfluidics (SPP1164) under the grants VA 205/3-2 and Do 794/1-1.

REFERENCES

1. F. Varnik, D. Dorner, and D. Raabe, *J. Fluid Mech.* **573**, 191–209 (2007).
2. F. Varnik and D. Raabe, *Modelling Simul. Mater. Sci. Eng.* **14**, 857–873 (2006).
3. M. Raupach, R. Antonia, and S. Rajagopalan, *Appl. Mech. Rev.* **44**, 1–25 (1991).
4. P.-A. Krogstad and R. Antonia, *Experiments in Fluids* **27**, 450–460 (1999).
5. P. G. de Gennes, F. Brochard, and D. Quere, *Capillarity and Wetting Phenomena: Drops, Bubbles, Pearls, Waves*, Springer, New York, 2004.
6. S. Succi, *The Lattice Boltzmann Equation for Fluid Dynamics and Beyond*, Oxford University Press, Oxford, 2001.
7. D. Rothman and S. Zaleski, *Lattice-Gas Cellular Automata (Simple Models of Complex Hydrodynamics)*, Cambridge University Press, Cambridge, 1997.
8. D. Wolf-Gladrow, *Lattice-Gas Cellular Automata and Lattice Boltzmann Models*, Springer, Berlin Heidelberg, 2000.
9. S. Chen and G. D. Doolen, *Ann. Rev. Fluid Mech.* **30**, 329–364 (1998).
10. A. Ladd and R. Verberg, *J. Stat. Phys.* **104**, 1191–1251 (2001).
11. D. Raabe, *Modelling Simul. Mater. Sci. Eng.* **12**, R13–R46 (2004).
12. S. Pope, *Turbulent Flows*, Cambridge University Press, Cambridge, UK, 2000.
13. F. Varnik and D. Raabe, *Molecular Simulation* **33**, 583–587 (2007).
14. J. Mathieu and J. Scott, *An Introduction to Turbulent Flow*, Cambridge University Press, Cambridge, UK, 2000.

Drop Impact on a Solid Surface Comprising Micro Groove Structure

R. Kannan and D. Sivakumar

Department of Aerospace Engineering, Indian Institute of Science, Bangalore 560012, India

Abstract. Spreading and receding processes of water drops impacting on a stainless steel surface comprising rectangular shaped parallel grooves are studied experimentally. The study was confined to the impact of drops in inertia dominated flow regime with Weber number in the range 15 – 257. Measurements of spreading drop diameter and drop height were obtained during the impact process as function of time. Experimental measurements of spreading drop diameter and drop height obtained for the grooved surface were compared with those obtained for a smooth surface to elucidate the influence of surface grooves on the impact process. The grooves definitely influence both spreading and receding processes of impacting liquid drops. A more striking observation from this study is that the receding process of impacting liquid drops is dramatically changed by the groove structure for all droplet Weber number.

Keywords: Droplet impact, Droplet spreading, Wetting
PACS: 47.55.nd, 61.30.Hn, 68.08.Bc

INTRODUCTION

The impact of liquid drops on a dry solid surface is encountered in several applications. Typical example is plasma spray coating on metal surfaces to provide thermal barrier. During the impact process, the liquid drop spreads rapidly in the radial direction, reaches a maximum spreading diameter and then recedes. The forces driving the impact process are droplet inertia force, surface wettability force, and viscous force. The comprehensive reviews of drop impact phenomenon by Rein [1] and Yarin [2] provide salient details of the impact process.

A target surface influences the flow behavior of a spreading liquid drop via surface energy caused by the molecular structure seen in the top layer of the surface and surface roughness. The surface energy controls wettability characteristics of the surface and this topic of surface wettability has been studied well in the literature [3,4]. Early studies on the role of surface roughness on the drop impact process were confined to description of the relationship between the surface roughness and the droplet splashing [5-8]. These studies gave a convincing demonstration that roughening a target surface by increasing the mean surface roughness, R_a promotes the splashing of impacting drops. A similar conclusion was obtained by Range and Feuillebios [9] by carrying out a systematic experimental study of drop impacts on rough surfaces with a wide range of surface structures. The random crest and trough patterns of a rough surface may be introducing perturbations to the lamella front formed immediately after the start of impact, and these perturbations grow with time and lead to an enhanced droplet splashing [10]. The enhancement of droplet splashing with increasing Ra has been noticed in several other studies [11,12], however, this behavior is restricted to target surfaces with small values of nondimensional surface roughness, $R_a^* = R_a/D_o$, where D_o is the drop diameter prior to the impact. A larger R_a^* tends to reduce the number of perturbations developing from the lamella front and hence suppresses the intensity of splashing [9,13].

Random distribution of asperities and grooves on a conventional rough surface poses difficulties in understanding the precise role of surface roughness on the drop impact process. In recent years, model rough surfaces have been used in the place of conventional rough surfaces to extract physical details on droplet spreading on the conventional rough surfaces [14-17]. A majority of these works were confined to the effect of surface roughness on surface wettability characteristics by conducting experiments with almost zero Weber number. The experiments conducted by Hitchcock et al. [14] with different combinations of experimental surfaces and liquids reveal that roughening of a surface modifies the wettability behavior of the surface. A strong correlation between

CP982, *Complex Systems, 5th International Workshop on Complex Systems*
edited by M. Tokuyama, I. Oppenheim, and H. Nishiyama

the surface geometric features of the rough surface and the surface wetting characteristics exists as per some recently reported studies [16,18]. For the impact of high Weber number (*We*) drops, Sivakumar et al. [17] showed that the surface texture pattern determines the spreading pattern of impacting drops. The present study deals with the impact of liquid drops on a model rough surface comprising unidirectional parallel grooves of rectangular cross section. The groove depth of the model rough surface is of a few microns and is comparable to the height of surface asperities seen with conventional rough surfaces. The experiments of drop impact were confined to high *We* as seen with several practical applications. The outcomes of this study are helpful to understand the spreading of droplet liquid on conventional rough surfaces.

EXPERIMENTAL DETAILS

A schematic of the experimental apparatus used in this study is shown in Fig. 1. Photographic techniques were employed to capture the physical events occurring during drop impact process. Image sequences illustrating spreading and receding processes of a particular drop impact case were

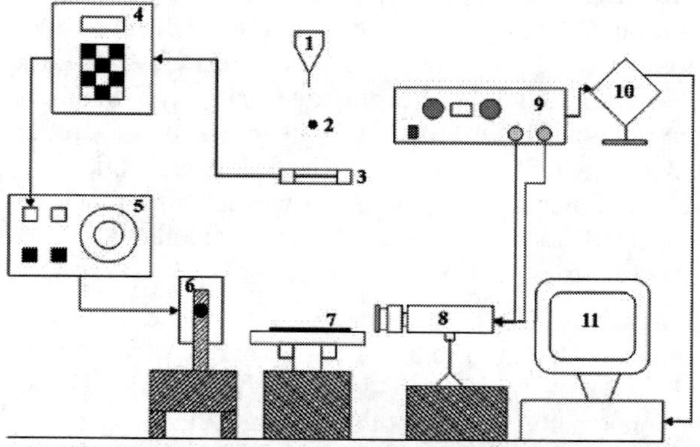

FIGURE 1. A Schematic of the experimental apparatus used in the study.

constructed from the images captured from several identical drop impact experiments with different time delays from the start of impact. This procedure was followed in several earlier studies [11,13,19,20]. Liquid drops were generated by using a flat-tipped hypodermic needle of internal and external diameters 0.25 mm and 0.37 mm respectively. Distilled water with density, $\rho = 996$ kg/m^3, surface tension, $\sigma = 0.073$ N/m, and viscosity, $\mu = 0.00089$ kg/ms was used as the droplet liquid. Drop diameter prior to impact, D_o was measured from its image captured near the target surface. Velocity of an impacting drop, U_o was estimated from the distance between the needle tip and the target surface, *H*, and different values of U_o were simulated by varying *H*. The details of an image capturing system, a lighting arrangement, a time delay setup, and an optical signal interrupter used in the experimental apparatus are highlighted in Fig. 1. In a typical experimental run, the optical signal interrupter identifies a liquid drop released from the hypodermic needle by means of the light intensity obscured by the moving liquid drop. A sudden drop in the light intensity makes the optical signal interrupter to send an output electric signal, which in turn triggers the light source for a single flash after a specific time interval set by the user. The flash duration of light source is 10 – 15 μs (as per the user manual), which is sufficient enough to freeze the moving drop, and the camera captures a single image during the light flash.

Solid target surfaces, a smooth surface and a grooved surface, made of stainless steel were used in

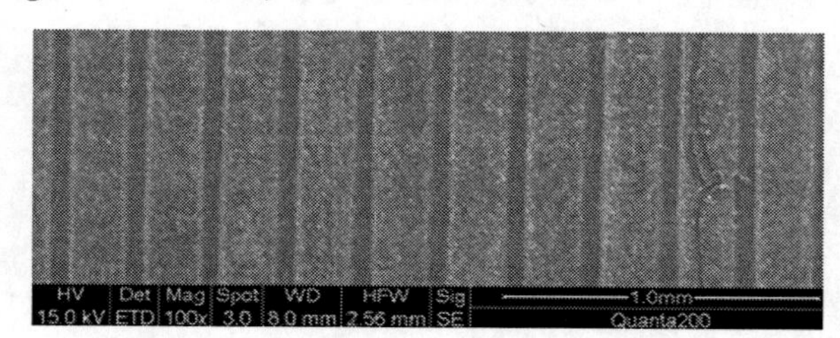

FIGURE 2. SEM image of the grooved surface. Lighter region in the image corresponds to the grooves.

the study. The smooth surface was prepared using a diamond paste polishing machine and the value of R_a was measured from surface profile plot as 0.013 μm. The grooved surface was prepared using photolithography techniques. The grooves were unidirectional and were formed on finely polished steel surfaces of size 20 mm x 20 mm via a chemical etching process. Major geometrical dimensions of the grooved surface, groove depth, *d*, groove width, *b* and width of solid pillar separating any two successive grooves, *w*, were kept at 7.5 μm, 136 μm, and 66 μm respectively. Figure 2 shows a scanning electron microscopy (SEM) image of the grooved surface. The above geometrical parameters of the grooved surface were measured from both the SEM image and surface profile plot.

An impacting drop was characterized in terms of *We* which was calculated using the expression

$$\mathrm{We} = \frac{\rho U_o^2 D_o}{\sigma}. \tag{1}$$

The simulated values of *We* were in the range 15 - 257. The images of spreading drops were captured with two different camera views, an inclined camera view to capture the image of spreading drop from the top and a normal camera view to capture the image of spreading drop from the side. The images captured with the inclined camera view were used to construct image sequences to illustrate qualitative details of the drop impact process and those with the normal camera view were used to extract quantitative measurements of spreading drop diameter, $2R$ and drop height at the center, Z. The images were recorded with 768 × 576 pixel resolution to a computer. The measurements of $2R$ and Z were nondimensionalized with D_o and the time lapse, t measured from the instant at which the impacting drop touches the top of the target surface was nondimensionalized with time scale D_0/U_0 as $\tau = tU_0/D_0$. Since the grooves formed on the target surface are unidirectional, the images of spreading drop were obtained to illustrate both the spreading on the surface along the groove direction and the spreading on the surface perpendicular to the groove direction.

RESULTS AND DISCUSSION

Spreading of Impacting Drops on Grooved Surface

Figure 3 shows image sequences of droplet spreading process on the grooved surface for different We. The corresponding image sequences of droplet spreading process on the smooth surface are included in Fig. 3 for the sake of comparison purposes. It is observed that the droplets spreading on the grooved surface develop a larger sized rim in the periphery of lamella. This behavior is well seen in Fig. 3 for the impact of high We drops (We = 84). In other words a larger liquid volume is collected in the lamella front for the droplets spreading on the grooved surface. During high We drop impact process, the impact develops frontal undulations in the lamella front. The undulations grow with time and develop liquid fingers from the lamella front. It is observed in the present study that the grooves of the target surface augment the growth of undulations developed in the lamella front as seen in the image sequence given in Fig. 3 for We = 162. For conventional rough surfaces, Bussmann et al. [10] indicated that surface asperities enhance magnitude of perturbation levels of lamella front during early stages of the impact process and hence augment the formation of liquid fingers. The behavior of frontal undulations observed for the grooved surface in the current study may be similar to the one seen with conventional rough surfaces. Note that the order of magnitude for the values of simulated by Bussmann et al. [10] was around 10^{-4}, which is comparable to the value of (10^{-3}) for the grooved surface used in the present study.

It was reported earlier that grooves of a textured surface develop thin liquid jets from the lamella front and the jets move ahead of the lamella front along the groove direction [17]. Similar liquid jets are observed

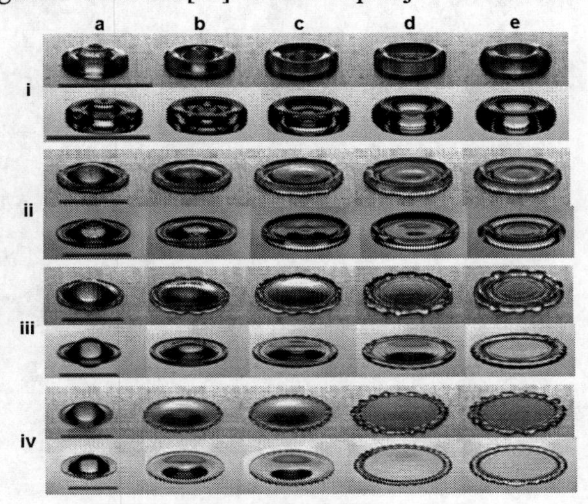

	Surfaces	a	b	c	d	e
i	Grooved	1.46	1.92	2.38	2.92	3.69
	Smooth	1.43	1.90	2.38	2.94	3.73
ii	Grooved	1.40	1.81	2.74	3.49	4.42
	Smooth	1.41	1.83	2.81	3.56	4.50
iii	Grooved	1.15	1.67	2.20	3.23	4.64
	Smooth	1.17	1.71	2.16	3.27	4.71
iv	Grooved	1.03	2.06	2.70	4.89	6.26
	Smooth	1.01	2.02	2.65	4.75	6.43

FIGURE 3. Spreading process of impacting liquid drops on the grooved surface for different We. (i) We = 16, (ii) We = 48, (iii) We = 84, and (iv) We = 162. The second image sequence given for each We corresponds to spreading process on the smooth surface. Values of τ for the images are given in the grid. Groove direction is from top to bottom. The line shown in the images corresponds to a scale length of 5 mm.

in the present study of drop impact on the grooved surface. This is illustrated by the image sequences given in Fig. 4 for two different We values. The jets are seen during early stages of the impact process at which droplet inertia is driving the spreading process. As the droplet spreading continues with time, these liquid jets gradually lose their momentum and eventually disappear. It is observed that the jet

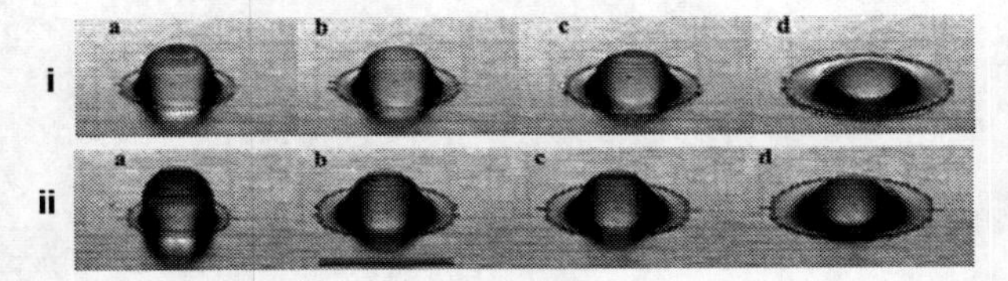

FIGURE 4. Images of spreading drops showing thin liquid jets ahead of lamella front. (i) We = 162, and (b) We = 250. Values of τ for the images are ranging from 0.4 to 1.2. Groove direction is from left to right. The line shown in the images corresponds to a scale length of 5 mm.

splashes which occur on conventional rough surfaces at very high *We* without corona formation [19].

Receding of Impacting Drops on Grooved Surface

Figure 5 shows image sequences illustrating the receding process of impacting liquid drops on the grooved surface for different *We*. Surface wettability forces drive the lamella flow during the receding process because of the absence of inertia and viscous forces. A striking observation from the image sequences given in Fig. 5 is that spreading lamella on the grooved surface is not receding fully, which is in contrast with that seen on the smooth surface, particularly for the impact of high *We* drops. A larger sized lamella is seen on the grooved surface even at high values of τ. Image sequences illustrating the receding process of liquid drops along the groove direction show a similar result.

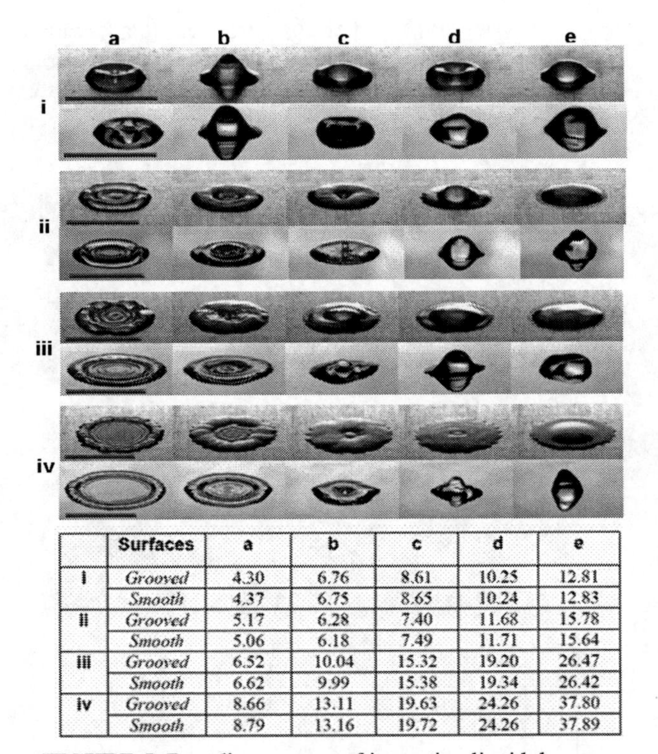

	Surfaces	a	b	c	d	e
i	*Grooved*	4.30	6.76	8.61	10.25	12.81
	Smooth	4.37	6.75	8.65	10.24	12.83
ii	*Grooved*	5.17	6.28	7.40	11.68	15.78
	Smooth	5.06	6.18	7.49	11.71	15.64
iii	*Grooved*	6.52	10.04	15.32	19.20	26.47
	Smooth	6.62	9.99	15.38	19.34	26.42
iv	*Grooved*	8.66	13.11	19.63	24.26	37.80
	Smooth	8.79	13.16	19.72	24.26	37.89

FIGURE 5. Receding process of impacting liquid drops on the grooved surface for different *We*. (i) *We* = 16, (ii) *We* = 48, (iii) *We* = 84, and (iv) *We* = 162. The second image sequence given for each *We* corresponds to spreading process on the smooth surface. Values of τ for the images are given in the grid. Groove direction is from top to bottom. The line shown in the images corresponds to a scale length of 5 mm.

formation process depends on the flow conditions of impacting drops. At low *We*, the lack of sufficient momentum to push the droplet liquid through the grooves inhibits the formation of these liquid jets. The formation and the development of these liquid jets in the lamella front may be the precursor for prompt

Temporal Variation of *2R* and *Z* Obtained during Spreading and Receding Processes of Impacting Drops on Grooved Surface

Figure 6 shows the variation of *2R* with τ obtained during the impact process of liquid drops with different *We* on the grooved surface. The measurements of *2R* obtained along the groove direction and perpendicular to the groove direction are shown in the plots of Fig. 6. For the impact of low *We* drops on the grooved surface, the measured values of *2R* remain almost constant after reaching $(2R)_{max}$, which indicates a complete absence of the receding of droplet liquid. As *We* increases, some amount of receding of droplet liquid are seen on the grooved surface, however, with lower retraction rate (seen from the slope of *2R* versus τ observed beyond $(2R)_{max}$) in comparison with that seen on the smooth surface. It is

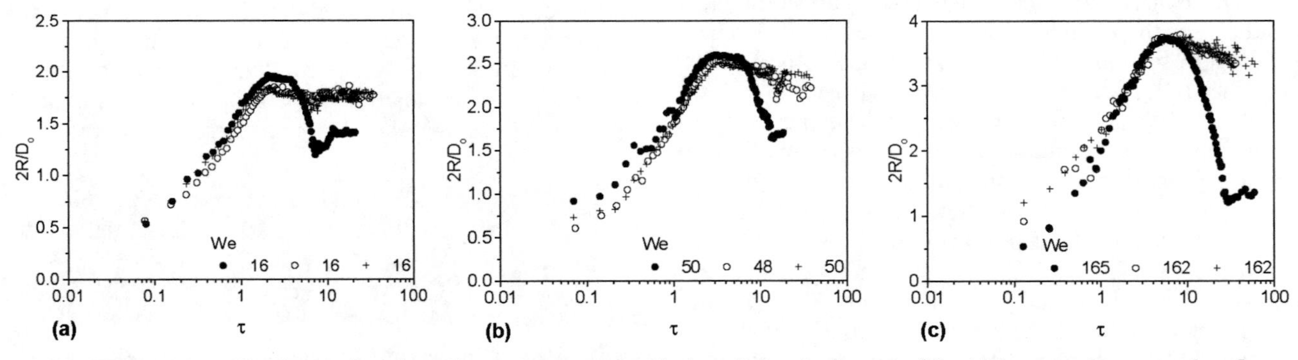

FIGURE 6. Normalized spreading drop diameter measured with time during the impact of liquid drops on the grooved surface for different *We*. • Smooth surface, ○ Grooved surface (along the groove direction), and + Grooved surface (perpendicular to the groove direction).

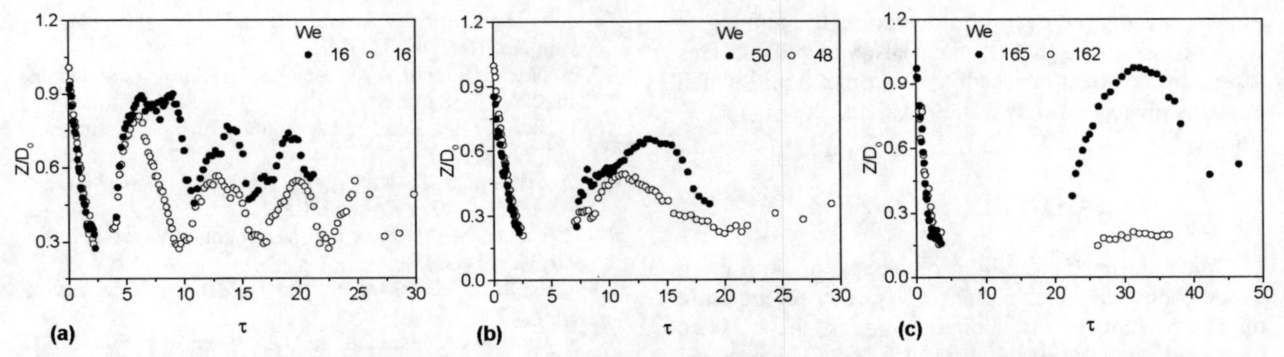

FIGURE 7. Normalized droplet height measured at the droplet center with time during the impact of liquid drops on the grooved surface for different We. ● Smooth surface, and ○ Grooved surface.

observed from Fig. 6 that the groove direction of the target surface does not influence the variation of 2R for all *We* conditions studied here. In general, it is quite clear from Fig. 6 that the grooved surface shows a dramatic decrease in the degree of recoiling for impacting liquid drops.

It is common that an impacting liquid drop exhibits post-spreading droplet oscillations on target surface without showing any significant variation in *2R*. During this process, droplet liquid in the central region of lamella undergoes vertical and radial motions caused by momentum differences between the central and periphery liquid volumes of the lamella. This feature is dominantly observed for low *We* impact conditions. It was characterized in this study by recording the measurements of droplet height, *Z* at the droplet center. Figure 7 shows the instantaneous variation of nondimensionalized droplet height obtained during the impact of liquid drops on the grooved surface for different *We*. At low *We*, post-spreading droplet oscillations seen on the grooved surface are similar to those observed on the smooth surface with the rebounding droplet height on the grooved surface is smaller than that measured on the smooth surface. As *We* increases, the oscillations are gradually suppressed on the grooved surface. At high *We*, in contrast to a strong droplet rebounding seen on the smooth surface, the liquid volume spreading on the grooved surface is not showing any sign of rebounding at the droplet center.

Discussion

The experimental results of drop impact process on stainless steel target surfaces summarized in the previous sections clearly indicate that the groove structure influences the impact process. The grooves modify the outcomes of drop impact process at *We* conditions where surface wettability forces play a dominant role, for instance the spreading and receding processes of low *We* drops and the receding process of high *We* drops. An interesting observation from this experimental study is that the grooved surface dramatically reduces the receding and rebounding of droplet liquid. Wettability behavior of rough surfaces, both conventional rough surfaces and textured/grooved surfaces, is described in the literature by the contact angle models proposed by Wenzel [21] and Cassie and Baxter [22]. The Wenzel's model [21] assumes that the droplet liquid wets and fills surface grooves completely. The model by Cassie and Baxter [22] considers an interface composed of both solid and ambient gas under the liquid drop. Additional experiments were carried out to understand the wettability behavior of target surfaces used in the study. A liquid drop of volume 8 μL was placed on the target surface with almost zero impact velocity using a micro pipette during these experiments and the droplet shapes were captured to extract equilibrium contact angle, θ_e made by the liquid drop on the surface. The image analysis revealed that the grooves were filled with the droplet liquid. This shows that drops on the grooved surface acquire Wenzel's free energy configuration [21]. For the grooved surface, the measurements of θ_e were obtained for droplet liquid along the groove direction ($\theta_{e,//}$) and perpendicular to the groove direction ($\theta_{e,\perp}$)from the images captured in their respective viewing planes. The estimated values were: $(\theta_e)_{smooth}$ = 80 deg, $\theta_{e,//}$ = 68.3 deg, and $\theta_{e,\perp}$ = 76.8 deg. The values of θ_e obtained on the grooved surface are in qualitative agreement with the previous works reported in the literature [16,23,24]. On a target surface comprising parallel grooves, droplet liquid pins on the edge of solid pillars [16,23,24]. Since the value of θ_e for droplet liquid measured on the grooved surface along the groove direction ($\theta_{e,//}$) is much lower than θ_e measured on the grooved surface perpendicular to the groove direction ($\theta_{e,\perp}$), the droplet liquid may be pinning on the pillar edges of the present grooved surface. The spreading droplet on the grooved surface may be lacking the energy to overcome the pinning of liquid contact line at the edges of solid pillars. This results in the arrest of receding of droplet liquid which explains the results shown in Figs. 6 and 7.

CONCLUSIONS

Spreading and receding processes of impacting water drops on a stainless steel surface comprising micron sized unidirectional grooves have been presented with We in the range 15- 257. The grooved surface was prepared using photolithographic techniques. The temporal plots of spreading droplet diameter, droplet height at the center and dynamic contact angle were obtained from the images captured as function of time during the impact process. The results obtained for the grooved surface were compared with those obtained for a smooth stainless steel surface to highlight the influence of grooves on the outcomes of drop impact process. The grooves influence both the spreading and receding processes of impacting drops, however, the effect is more pronounced during the receding process. The grooves completely arrest the receding of droplet liquid on the grooved surface for the impact of low *We* drops. Some degree of droplet receding occurs for high *We* drops, however, with lesser intensity than those seen on the smooth surface. In addition the grooves strongly suppress the rebounding of droplet liquid at the drop center which may be attributed to the pinning of liquid contact line on the edges of solid pillars of the grooved surface.

REFERENCES

1. M. Rein M, *Fluid Dyn. Res.* **12**, 61-93 (1993).
2. A. L. Yarin, *Ann. Rev. Fluid Mech.* **38**, 159-192 (2006).
3. E. B. Dussan, *Ann. Rev. Fluid Mech.* **11**, 371-400 (1979).
4. P. G. de Gennes, *Rev. Mod. Phys.* **57**, 827-842 (1985).
5. O. G. Engel, *J. Res. Natl. Bur. Stand.* **54**, 281-298 (1955).
6. Z. Levin and P. V. Hobbs, *Philos. Trans. R. Soc. London Ser. A* **269**, 555-585 (1971).
7. C. D. Stow and M. G. Hadfield, *Proc. R. Soc. London Ser. A* **373**, 419-441 (1981).
8. C. Mundo, M. Sommerfeld, and C. Tropea, *Int. J. Multiphase Flow* **21**, 151-173 (1995).
9. K. Range and F. Feuillebois, *J. Colloid Interface Sci.* **203**, 16-30 (1998).
10. M. Bussmann, S. Chandra, and J. Mostaghimi, *Phys. Fluids* **12**, 3121-3132 (2000).
11. S. Šikalo, M. Marengo, C. Tropea, and G. N. Ganic, *Exp. Thermal Fluid Sci.* **25**, 503-510 (2002).
12. R. L. Vander Wal, G. M. Berger, and S. D. Mozes, *Exp. Fluids* **40**, 23-32 (2006).
13. S. Shakeri and S. Chandra, *Int. J. Heat Mass Transf.* **45**, 4561-4575 (2002).
14. S. J. Hitchcock, N. T. Carroll, and M. G. Nicholas, *J. Mat. Sci.* **16**, 714-732 (1981).
15. H. Nakae, R. Inui, Y. Hirata, and H. Saito, *Acta Mater.* **46**, 2313-2318 (1998).
16. J. Bico, C. Tordeux, and D. Quéré, *Europhys. Lett.* **55**, 214-220 (2001).
17. D. Sivakumar, K. Katagiri, T. Sato, and H. Nishiyama, *Phys. Fluids* **17**, 100608 (2005).
18. H. Nakae, M. Yoshida, and M. Yokota, *J. Mat. Sci.* **40**, 2287-2293 (2005).
19. R. Rioboo, M. Marengo, and C. Tropea, *Atomization Sprays* **11**, 155-166 (2001).
20. R. Rioboo, M. Marengo, and C. Tropea, *Exp. Fluids* **33**, 112-124 (2002).
21. R. N. Wenzel, *Ind. Eng. Chem.* **28**, 988-995 (1936).
22. A. B. D. Cassie and S. Baxter, *Trans. Faraday Soc.* **40**, 546-549 (1944).
23. J. F. Oliver, C. Huh, and S. G. Mason, *J. Adhesion* **8**, 223-234 (1976).
24. Y. Chen, B. He, J. Lee, and N. A. Patankar, *J. Colloid Interface Sci.* **281**, 458-464 (2005).

Surface Tension Effect to Die-Swell Extrusion of Viscoelastic Fluid

S. Bunditsaovapak*, T. Fagon*, and S. Thenissara**

*Department of Mathematics and Computer Science, Faculty of Science
King Mongkut's Institute of Technology Ladkrabang, Thailand
** Department of Mathematics, Faculty of Science and Liberal Arts,
Rajamagala university of Technology Isan Nakornratchasima, Thailand

Abstract. The die-swell problem for Viscoelastic fluid by using the finite element methods (FEM) under the semi-implicit Taylor-Galerkin pressure correction principle with consistent streamline upwinding is presented. The assumptions of incompressible fluid, creeping flow, no gravitational effect, temperature independence ,and no slip effect are used. The two-dimensional governing equations, in which the equations are nonlinear partial differential equations, are consists of the conservation of mass and momentum, and Oldroy–B model with $We < 1$ .The deformation of fluid at free surface will be affect by surface tension force, thus both dynamic and kinematic boundary conditions are considered. These conditions make the simulation result looked reasonable. Evolution of die-swell flow: the variations of velocities, pressure, stresses, shear rate and extension rate. The simulation program has been created to generate grid and compute the solutions, which utilize remeshing and interpolating techniques as well as the gradient recovery for increased accuracy and stability of solutions. The results exhibit the same trend as the experimental solutions.

Keywords: Finite element method, Die-swell extrusion, Oldroy–B model
PACS: 47.10.ad, 47.11.Fg, 47.50Gj, 47.85md

INTRODUCTION

Research in die-swell extrusion for viscoelastic fluid (non fixed viscosity) is the study of fluid deformation behaviour and its properties while the fluid flows and swells at the outside in time. Because processes in the factory such as sheet extrusion, cable production, pipe process and wire coating process need to use knowledge in rheology; the research is very popular and many people have attempted to describe and make a mathematical model. Since the first attempt in 1967 by Fenner and Williams [1], they have been attempted to simulate wire coating flow. In 1978, Caswell and Tanner [2] studied the characteristic of die and the change of stress within by using finite element method (FEM). In 1988, Mitsoulis et al. [3] has studied the high-speed coating of low-density polyethylene using FEM with Streamline-Upwind/Petrov-Galerkin (SUPG) technique. In recent year, there were many researcher, such as Binding et al. [4], Gunter et al. [5], Mutlu et al. [6], Ngamaramvaranggul and Webster [7-8], studied about the topic using numerical methods.

This paper studies the surface tension effect to the die-swell deformation in pressure tooling of HDPE by using Oldroy–B model and showing the change of velocities, pressure, stresses, shear rate and extension rate of die-swell extrusion. The six nodes triangular mesh is used in FEM with the semi-implicit Taylor-Galerkin pressure correction scheme in two dimensionless cylindrical coordinate systems. The assumptions of incompressible flow, no gravitational effect, no inertia force, no slip effect and isothermal case are supposed to adopt. Tension forces at the free surface have an effect for setting both dynamic and kinematic boundary conditions. Some techniques of remeshing, interpolation, and reprojection also with the SUPG technique are considered in order to enhance stability and accuracy over Galerkin method. In the current study, the comparison of all results has referred extensively to the fundamental work in Ngamaramvaranggul and Webster [9].

GOVERNING EQUATIONS

For two-dimensional isothermal incompressible viscoelastic fluid, the generalized momentum and continuity equations can be expressed as:

CP982, *Complex Systems, 5th International Workshop on Complex Systems*
edited by M. Tokuyama, I. Oppenheim, and H. Nishiyama

$$\rho\frac{D\vec{u}}{Dt}=\nabla\cdot\tilde{\sigma}+\rho\vec{f} \tag{1}$$

$$\nabla\cdot\vec{u}=0 \tag{2}$$

where ρ, $\vec{u}$, t, ∇, $\tilde{\sigma}$ and $\vec{f}$ are the fluid density, the velocity vector, time, the spatial differential operator, Cauchy stress tensor and body force vector respectively. The Cauchy stress tensor is given in the form

$$\tilde{\sigma}=-p\delta+\tilde{T} \tag{3}$$

where p is an isotropic pressure and the Kronecker delta tensor is $\delta_{ij}=\begin{cases}1, & i=j\\ 0, & i\neq j\end{cases}$

For viscoelastic fluid, the extra stress tensor ($\tilde{T}$) is obtained by

$$\tilde{T}=2\mu_2\tilde{D}+\tilde{\tau}_v \tag{4}$$

$$\tilde{D}=\frac{\nabla\vec{u}+\nabla\vec{u}^t}{2} \tag{5}$$

where $\tilde{D}$ is the rate of deformation tensor and $\tilde{\tau}_v$ is gained from the constitutive equation.

Oldroy-B model is one of the constitutive equations that gives the result of the stress tensor of viscoelastic fluid ($\tilde{\tau}_v$), namely,

$$\tilde{T}_v+\lambda_1\overset{\nabla}{\tilde{T}}_v=2\mu(D+\lambda_2\overset{\nabla}{D})$$

$$\overset{\nabla}{\tilde{T}}=\frac{\partial\tilde{T}}{\partial t}+\vec{u}.\nabla\tilde{T}-\tilde{T}.\nabla\vec{u}-(\tilde{T}.\nabla\vec{u})^t \tag{6}$$

$$\mu=\mu_1+\mu_2$$

where μ is the fluid viscosity, that may be split to additive contributions from a polymeric solute μ_1 and Newtonian solvent μ_2, λ_1 is the relaxation time and λ_2 is the retardation time as $\lambda_2=\lambda_1(\mu_2/\mu)$.

For inelastic isothermal non-Newtonian flow, the extra stress tensor is defined as a function of the rate of deformation tensor (Rivlin and Eriksen [10], Reiner [11]) that is:

$$T_{ij}=2\mu\left(\dot{\gamma},\dot{\varepsilon}\right)D_{ij} \tag{7}$$

where the shear rate ($\dot{\gamma}$) for simple shear flow is given by:

$$\dot{\gamma}=2\sqrt{\mathrm{II}_d} \tag{8}$$

and elongation rate ($\dot{\varepsilon}$) for elongational flow as

$$\dot{\varepsilon}=3\frac{\mathrm{III}_d}{\mathrm{II}_d} \tag{9}$$

where II_d and III_d are the second and third invariants of the rate of deformation tensor (D_{ij}) respectively. In cylindrical coordinate system II_d and III_d are obtained by

$$\mathrm{II}_d=\frac{1}{2}\mathrm{tr}\left(D^2\right)$$
$$=\frac{1}{2}\left\{\left(\frac{\partial u_r}{\partial r}\right)^2+\left(\frac{\partial u_z}{\partial z}\right)^2+\left(\frac{u_r}{r}\right)^2+\frac{1}{2}\left(\frac{\partial u_r}{\partial z}+\frac{\partial u_z}{\partial r}\right)^2\right\} \tag{10}$$

and

$$\mathrm{III}_d=\det\left(D\right)=\frac{u_r}{r}\left\{\frac{\partial u_r}{\partial r}\frac{\partial u_z}{\partial z}-\frac{1}{4}\left(\frac{\partial u_r}{\partial z}+\frac{\partial u_z}{\partial r}\right)^2\right\} \tag{11}$$

For convenience of comparison and representation, the dimensionless form is considered. The non-dimensional variables $r^*,z^*,u^*,p^*,T^*,t^*,\mu^*,\frac{D}{Dt^*},\lambda^*,\mu_i^*,\chi^*$ are the rate of dimensional variables per characteristic factor as follows:

$$r^*=\frac{r}{L},\ z^*=\frac{z}{L},\ u^*=\frac{u}{V},\ p^*=\frac{L}{\mu_0 V}p,\ T^*=\frac{L}{\mu_0 V}T,$$
$$t^*=\frac{V}{L}t,\quad \nabla^*=L\nabla,\quad \frac{D}{Dt^*}=\frac{L}{V}\frac{D}{Dt},\quad \lambda^*=\frac{V}{L}\lambda,$$
$$\mu_i^*=\frac{1}{\mu_0}\mu_i;\ i=1,2,\ \dot{\gamma}^*=\frac{L}{V}\dot{\gamma}$$

where L is a characteristic length, V is a characteristic velocity, μ_0 is reference viscosity with index $i=1,2$.

NUMERICAL METHOD

The computation converts non-linear partial differential equations into the algebraic system for the convenience of computation by converting the velocity and stress expressions in terms of a time–dependent differential expression into the forward finite different form using Taylor's series expansion in time (with half time step method). For the pressure, the semi-implicit pressure correction method was used. After that, equations of pressure and velocities are separated into two by Galerkin weak formulation which includes integration by part so that the algebraic equations appear as equations (12)-(17) and this method is called Taylor-Galerkin method. The way to discretise time and spatial derivatives was introduced by Ngamaramvaranggul and Webster [8].

Stage 1a

$$\frac{2\,\mathrm{Re}}{\Delta t}\left(\tilde{u}^{n+\frac{1}{2}}-\tilde{u}^{n}\right)=\left[\nabla.\left(\tilde{\tau}+2\mu_N\tilde{D}\right)-\mathrm{Re}\,\tilde{u}.\nabla\tilde{u}-\nabla p\right]^{n}$$
$$+\mu_N\nabla.\left(\tilde{D}^{n+\frac{1}{2}}-\tilde{D}^{n}\right) \tag{12}$$

$$\frac{2We}{\Delta t}\left(\tilde{\tau}^{n+\frac{1}{2}}-\tilde{\tau}^{n}\right)=\left[2\mu_V\tilde{D}\right]^{n}$$
$$-We\left\{\tilde{u}.\nabla\tilde{\tau}-\tilde{\tau}.\nabla\tilde{u}-\left(\tilde{\tau}.\nabla\tilde{u}\right)^{t}+\xi\left[\left(\tilde{D}.\tilde{\tau}\right)+\left(\tilde{D}.\tilde{\tau}\right)^{t}\right]\right\}^{n} \tag{13}$$

Stage 1b

$$\frac{\mathrm{Re}}{\Delta t}\left(\tilde{u}^{*}-\tilde{u}^{n}\right)=\left[\nabla.\tilde{\tau}-\mathrm{Re}\,\tilde{u}.\nabla\tilde{u}\right]^{n+\frac{1}{2}}$$
$$+\left[\nabla.\left(2\mu_N\tilde{D}\right)-\nabla p\right]^{n}+\nabla.\left(\tilde{D}^{*}-\tilde{D}^{n}\right) \tag{14}$$

$$\frac{We}{\Delta t}\left(\tilde{\tau}^{n+1}-\tilde{\tau}^{n}\right)=\left[2\mu_V\tilde{D}\right]^{n+\frac{1}{2}}$$
$$-We\left\{\tilde{u}.\nabla\tilde{\tau}-\tilde{\tau}.\nabla\tilde{u}-\left(\tilde{\tau}.\nabla\tilde{u}\right)^{t}+\xi\left[\left(\tilde{D}.\tilde{\tau}\right)+\left(\tilde{D}.\tilde{\tau}\right)^{t}\right]\right\}^{n+\frac{1}{2}} \tag{15}$$

Stage 2 $(\theta=1/2)$ (Crank-Nicolson [12])

$$\theta\nabla^{2}q^{n+1}=\frac{\mathrm{Re}}{\Delta t}\nabla.\tilde{u}^{*};\quad q^{n+1}=p^{n+1}-p^{n} \tag{16}$$

Stage 3

$$\frac{\mathrm{Re}}{\Delta t}\left(\tilde{u}^{n+1}-\tilde{u}^{*}\right)=-\theta\nabla q^{n+1} \tag{17}$$

where Re and We are Reynolds number and Weissenberg number respectively. For this research, Oldroy-B model is considered for predict behavior of the stresses, thus the suitable Weissenberg number is in $0 < We \le 1$ [7,23].

The solutions of all stages have been solved by Jacobi iterative method, with Penalty approach for handling boundary condition. The computation of integral was approximated by a triangular 4-points Gaussian quadrature approach (Reddy [13]). The Gradient recovery is a strategy to improve the stability of solution, and has been stated by Hawken et al. [14], Levine [15-16], Boroomand and Zienkiewicz [17], Zienkiewicz and Zhu [18] and Matallah [19], who applied it to adjust the smooth convergence. At each time step, the fixed-connectivity remeshing technique is employed by adjusting values in domain with interpolation technique while the constant values of the outside part are kept. The free surface computation was considered by condition of surface tension and solve by using Taylor-Galerkin. Equations (18) from the method of time dependent prediction.

$$\frac{1}{\Delta t}\frac{\partial h}{\partial t}=u_r-u_z\left(\frac{\partial h}{\partial z}\right) \tag{18}$$

For each computational step, the conservation of flow rate is verified from the summation of frustum of a cone compared with the initial volume.

The convergent criteria is considered by

$$\left\|E(x)\right\|_{\infty}=\frac{\left\|\left(x^{n}-x^{n-1}\right)\right\|_{\infty}}{\left\|x^{n}\right\|_{\infty}}\le \mathrm{TOL} \tag{19}$$

Here $\mathrm{TOL}=10^{-6}$ for velocities and pressure and stresses, $\mathrm{TOL}=10^{-8}$ for Jacobi computation and $\mathrm{TOL}=10^{-4}$ for displacement of free surface.

In a 2-D and a 3-D problem, the parameters of SUPG are proved by Shakib [20]. After that Baaijens et all [21] and Carew [22] used SUPG in viscoelastic problem which the streamline upwinding test functions are obtained from an addition of a scalar multiplicative element dependent factor of the advection operator to the standard Galerkin test functions as follow:

$$\phi_i^{Petrov}=\phi_i+\alpha^{h}u.\nabla\phi_i \tag{20}$$

where ϕ_i is a standard Galerkin shape function, α^h is the arbitrary elementwise parameter which depends upon the size of the mesh and direction of velocity defined as:

$$\alpha^{h}=\begin{cases}0 & ;\ g=0\\ \left.\begin{array}{ll}\frac{\Delta t}{2}+\Delta t^{2}\sqrt{g} & \text{implicit}\\ \Delta t^{2}\sqrt{g} & \text{explicit}\end{array}\right\} & ;\ g<1\\ \left.\begin{array}{ll}\frac{\Delta t}{2}+\frac{1}{\sqrt{g}} & \text{implicit}\\ \frac{1}{\sqrt{g}} & \text{explicit}\end{array}\right\} & ;\ g\ge 1\end{cases}$$

where $g=\frac{\partial L_i}{\partial x_j}\frac{\partial L_i}{\partial x_k}v_j v_k$

where L_i is the local coordinates for the triangular element.

v_j, v_k are the component of velocity, i = 1, 2, 3 and j, k = 1, 2.

PROBLEM SPECIFICATION

Initial Condition

Initially, the rectangular domain for computational is divided into 1944 triangular-elements mesh as shown in Fig. 1.

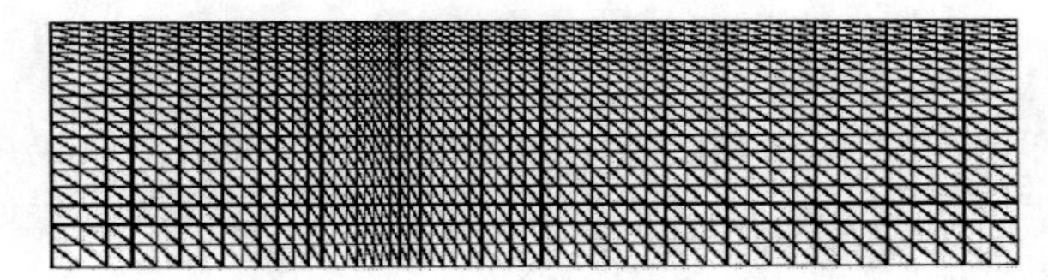

FIGURE 1. Domain of the problem.

Boundary Condition

The boundary condition of the die-swell model consists of 2 regions, inside and outside the die. Inside the die is sticked at the boundary wall where the input flow is of Poiseuille's type and plug flow at outlet of the die. The swelling area lies outside the die where its pressure is zero. Dirichlet boundary condition and Neumann boundary condition will be used as Fig. 2.

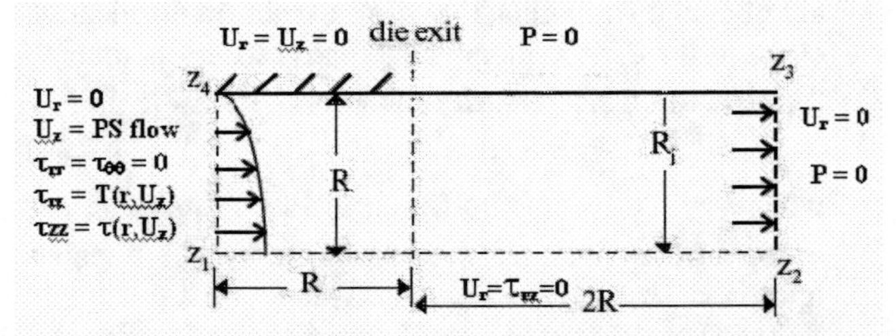

FIGURE 2. Boundary condition of the die-swell [23].

In additional, the free surface will be affected by surface tension. The surface tension itself, improved by Anastasiadis et al. [24], defined by the following dynamic and kinematic boundary conditions:

$$\frac{d\phi}{dS} = \frac{2}{B} + Z - \frac{\sin\phi}{x}$$

$$\frac{dX}{dS} = \cos\phi \qquad (21)$$

$$\frac{dZ}{dS} = \sin\phi$$

$$X(0) = Z(0) = \phi(0) = 0$$

Material Parameters

Parameters of material for computing the viscoelastic case are shown as in Table 1.

TABLE 1. Material parameters for a poly-eterene fluid.

ρ (density)	0.0001
η_N (Newtonian viscosity)	0.01
η_V (Viscoelastic viscosity)	0.99
λ_1(relaxation time)	1
L(length)	1
R_0(initial radius)	1

RESULTS

As it will be shown in this section, the result trend is in agreement with our hypothesis. The followed figures will illustrate computational results where Weissenberg Number (We) = 0.25 and 0.75.

As shown in Fig. 3, the radial velocity (Ur) is very small except at the region around the die exit . The velocity of the fluid in the r-direction is increasing, thus outside of die is swell. Furthermore, Ur is decreasing when We increases.

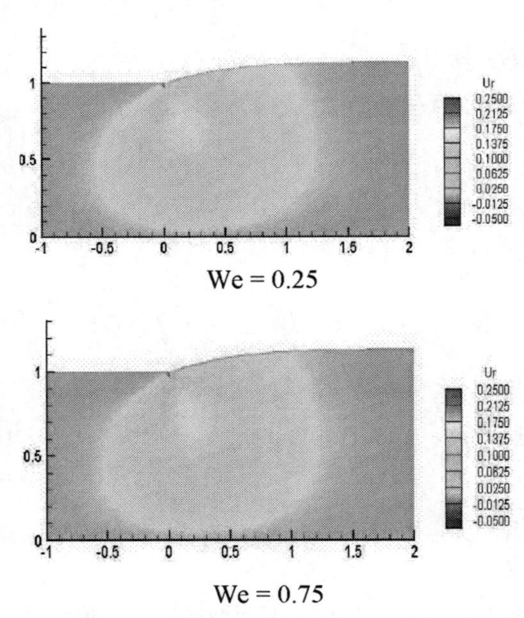

FIGURE 3. (Color online) The radial velocity of We=0.25 and 0.75.

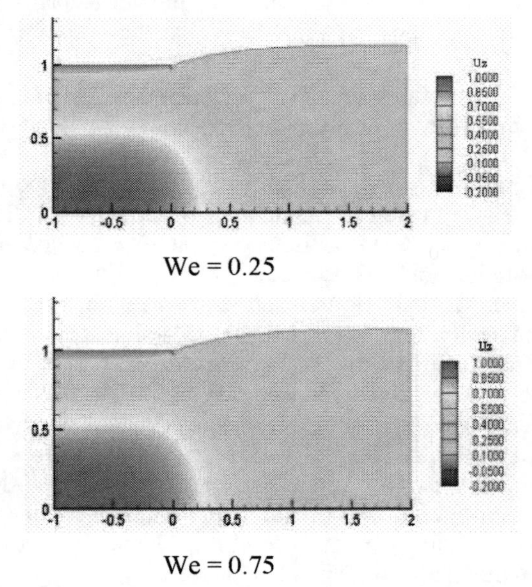

FIGURE 4. (Color online) The axial velocity of We=0.25 and 0.75.

The axial velocity (U_z) contours in Fig. 4. shown that at the region around the die wall is small. It is then increase roughly to 1.0 at the upstream die center.

The value of pressure (P) will be changed according to the distance from inlet except for when the singularity reaches a peak value of 5.68 where We=0.25. Then, the singularity pressure is decreased as We increased (Fig. 5).

The shear stress (τ_{rz}) established in Fig. 6. can be observed that its localized maximum lies within the neighborhood of the die exit.

The axial extra stress (τ_{zz}) is generally small everywhere except at the die-exit where the change is noticeable (shown in Fig. 7).

Stress $\tau_{\theta\theta}$ contour demonstrates in Fig. 8 is very small and reach a maximum at 0.49 at the center of die.

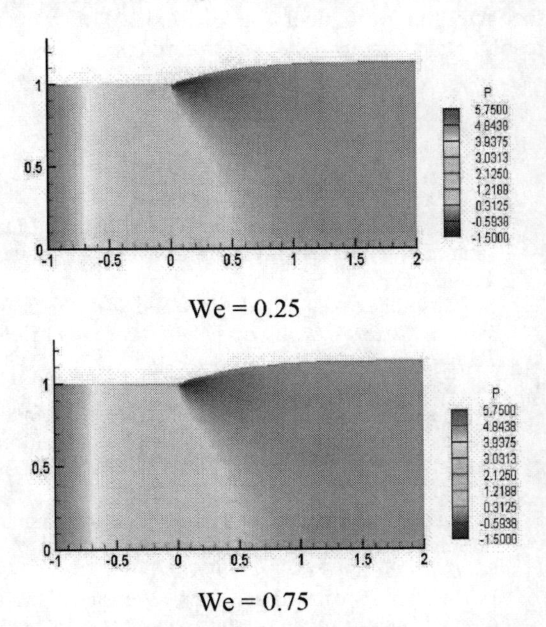

FIGURE 5. (Color online) The pressure of We=0.25 and 0.75.

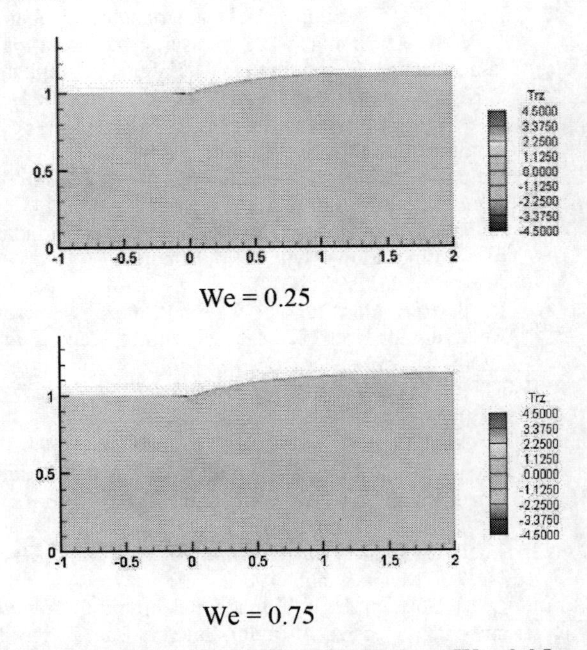

FIGURE 6. (Color online) The shear stress We=0.25 and 0.75.

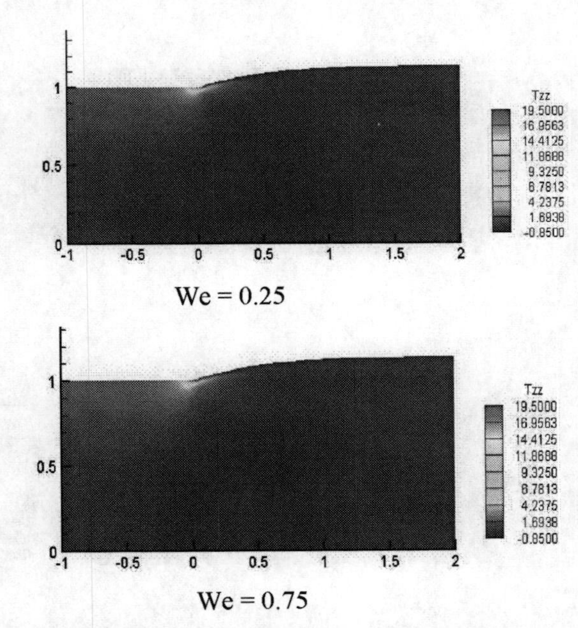

FIGURE 7. (Color online) The axial extra stress of We=0.25 and 0.75.

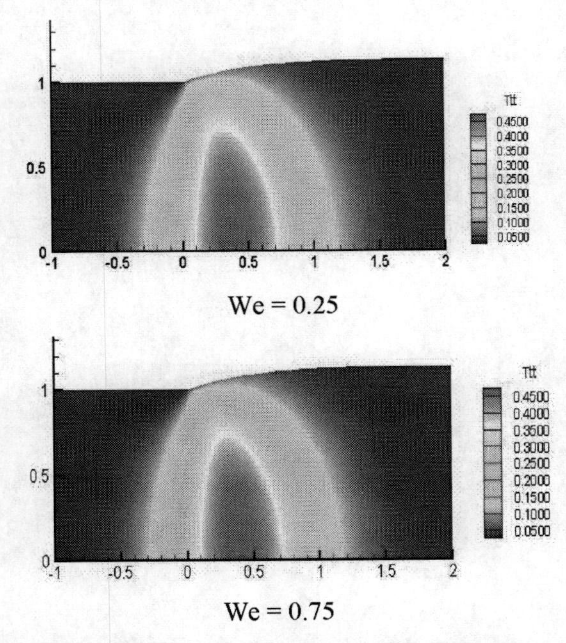

FIGURE 8. (Color online) Stress $\tau_{\theta\theta}$ of We=0.25 and 0.75.

The shear rate ($\dot{\gamma}$) is shown in Fig. 9. The numerical solution suggests that $\dot{\gamma}$ reaches its maximum value at the die exit resulted from the change of direction flow and swelling of fluid.

The extension rate ($\dot{\varepsilon}$) is highest at the free surface region or swelling area because of the change of direction flow.

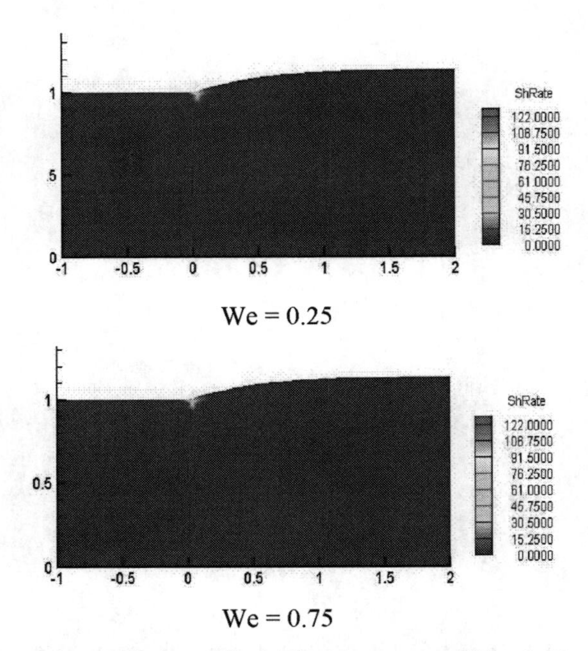

FIGURE 9. (Color online) The shear rate of We=0.25 and 0.75.

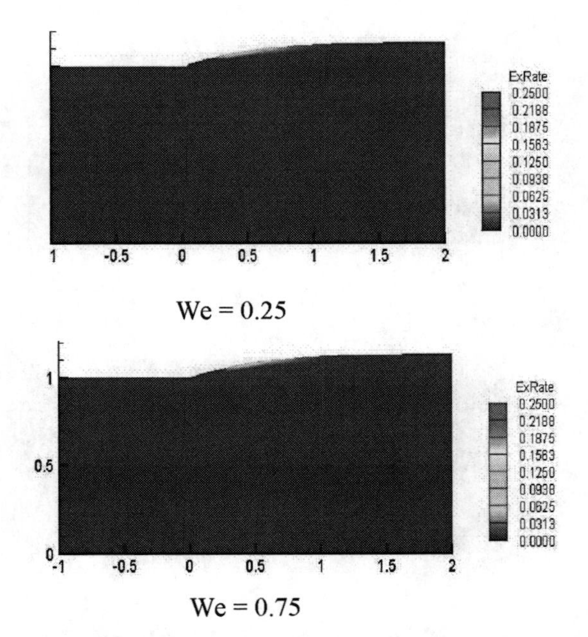

FIGURE 10. (Color online) The extension rate of We=0.25 and 0.75.

CONCLUSION

Surface tension and the variation of Weissenberg Number give a lot of effects to computation of velocities, pressure, stress, shear rate and extension rate. However, we still have a smaller effect to swelling ratio than the surface tension because range of Weissenberg Number is narrow. On the other hand, the surface tension is considered in order to decrease the swelling ratio near the die exit. The numerical results yield similar result as experiment.

REFERENCES

1. R. T. Fenner and J. G. Williams, *Trans. J. Plastics Inst.*, 701-706 (1967).
2. B. Caswell and R. I. Tanner, Wirecoating Die Design Using Finite Element Methods, *Polymer Technology* **18**, 416-421 (1978).
3. E. Mitsoulis et al., Fluid Flow and Heat Transfer in Wire Coating: A Review, *Advances in Polymer Technology* **6**, 467-487 (1986).
4. D. M. Binding, A. R. Blythe, S. Gunter, A. A. Mosquera, P. Townsend and M. F. Webster, Modelling Polymer Melt Flows in Wire-coating Processes Related Fields, *J. Non-Newtonian Fluid Mech.* **64**, 191-206 (1996).
5. S. Gunter, P. Townsend and M. F. Webster, Simulation of Some Model Viscoelastic Extensional Flows, *J. Num. Meth. Fluids* **23**, 691-710 (1996).
6. I. Mutlu, P. Townsend and M. F. Webster, Simulation of Cable-coating Viscoelastic Flows with Coupled and Decoupled Schemes, *J. Non-Newtonian Fluid Mech.* **74**, 1-23 (1998).
7. A. Baloch, H. Matallah, V. Ngamaramvaranggul and M. F. Webster, Simulation of Pressure- and Tube-tooling Wire-Coating Flows through Distributed Computation, *J. Num. Meth. Heat Fluid Flow* **12**, 458-493 (2002).
8. V. Ngamaramvaranggul and M. F. Webster., Computation of Free Surface Flows with a Taylor-Galerkin/Pressure-Correction Algorithm, *J. Num. Meth. Fluids* **33**, 993-1026 (2000).
9. V. Ngamaramvaranggul and M. F. Webster, Simulation of Coating Flows with Slip Effects, *J. Num. Meth. Fluids* **33**, 961-992 (2000).
10. R. S. Rivlin and J. L. Eriksen, Stress Deformation Relations for Isotropic Material, *J. Rat. Mech. Anal.* **4**, 323-425 (1955).
11. M. Reiner, *Deformation Strain and Flow*, Wiley, NY, 1960.
12. J. Crank and P. Nicolson, A Practical Method for Numerical Evaluation of Solution of Patial Differential equations of the Heat-conduction Type, *Proc. Camb. Phil. Soc.*, 1947, pp. 50-67.
13. J. N. Reddy, *An Introduction to the Finite Element Method*, McGraw-Hill, 1984.
14. D. M. Hawken, H. R. Tamaddon-Jahromi, P. Townsend and M. F. Webster, A Taylor-Galerkin Based Algorithm for Viscous Incompressible Flow, *J. Num. Meth. Fluids* **10**, 327-351 (1990).

15. N. Levine, Superconvergent Recovery of the Gradient from Finite Element Approximation on Triangles, *Technical report Num. Anal. Rep. 6/83,* University of Reading, U.K., 1983.
16. N. Levine, *Superconvergent Estimation of the Gradient from Linear Finite Element Approximation of Triangular Elements* (Ph.D. Thesis, University of Reading, U.K., 1985).
17. B. Boroomand and O. C. Zienkiewicz, An Improve REP Recovery and the Effectively Robustness Test, *J. Num. Meth. Eng.* **40**, 3247-3277 (1997).
18. O. C. Zienkiewicz and J. Z. Zhu, Superconvergent and Superconvergent Patch Recovery, *Finite Elements in Analysis and Design* **19**, 11-23 (1995).
19. H. Matallah, *Numerical Simulation of Viscoelastic Flows* (Ph.D. Thesis, University of Wales Swansea, U.K., 1998).
20. F. Shakib, (Ph.D. Dissertation, Stanford University, 1987).
21. F. P. T. Baaijens, S. H. A. Selen, H. P. W. Baaijens, G. W. N. Peters and H. E. H. Meijer, Viscoelastic Flow Past a Confined Cylinder of a LDPE Melt, *J. Non-Newtonian Fluid Mech.* **68**, 173-203 (1997).
22. E. O. A. Carew, P. Townsend, and M. F. Webster, A Taylor-Petrov-Galerkin Algorithm for Viscoelastic Flows, *J. Non-Newtonian Fluid Mech.* **75**, 139-166 (1998).
23. V. Ngamavamvaranggul and M. F. Webster, Viscoelastic Simulations of Stick-Slip and Die-Swell Flows, *J. Num. Meth. Fluids* **36**, 539-595 (2001).
24. S. H. Anastasiadis and S. G. Hatzikiriakkos, The work of Adhesion of Polymer/wall Interfaces and its Association with the Onset of Wall Slip, *J. Rheol.* **42**, 795-812 (1998).

Linear Analysis for a Finite Liquid Jet

Chien-Chi Chao

Department of Electrical Engineering, Chien-Kuo Technology University, No.1 Chieh-Sou N. Rd., Changhua City 500, Taiwan

Abstract. A 3-D boundary element method with linear triangular element has been developed for the simulation of a finite jet subjected to the surface tension force. The codes include 3D Laplace solver, grid generation of a single liquid jet, and free surface module for the calculation of surface normal vector, surface curvature, and tangential velocity. The comparison of computational results and the predicted values from the dispersion equation, which serves as the analytical solution for the growth rate, for the temporal instability analysis on a liquid jet shows a very good agreement. This is shown the proposed model is capable for the complex 3-D liquid jet simulation.

Keywords: 3-D BEM, Liquid jet, Temporal instability analysis, Surface tension force
PACS: 47.20.Dr

INTRODUCTION

The bulk of investigations of deforming liquid jet have been two-dimensional analyses. In most practical applications, the breakup of the liquid jet or the secondary atomization of droplets are three-dimensional flows. Studies conducted by Chahine et al.[1-4] have concentrated on the problem of the dynamics of bubbles and their interaction with other bodies. They applied the boundary element method in their investigations and extended the method to three-dimensional bubble dynamics problems. Oguz and Zeng[5] developed a potential-flow, boundary-integral technique to simulate the formation of a bubble from a hypodermic needle.

The boundary element methods (BEM) is now recognized as a powerful tool for solving problems in mechanics which are posed in domains of complex shape. In particular, because only the boundary of the domain needs to be discretized, the preprocessing effort required is less than that for domain methods such as the finite difference and finite element techniques. Recent developments of numerical models based on the use of Boundary Element Methods (BEMs)[6,7] permit us to analyze the acoustically-driven interactions. The use of the BEM approach provides high resolution of the interface (under very large distortions), as well as the capability to predict unsteady behavior. Chao[8] presents simulations of transverse acoustic wave interactions with an initially-cylindrical column of fluid. Transverse simulations include coupling of gas and liquid velocity and pressure fields. In this paper, a 3-D BEM with the linear elements was developed for the liquid jet. In this approach, the velocity potential and the normal derivative are varied linearly over the element and are expressed as a function of the values at the three vertices which delimit a particular element. Thus, the movement of a element can be precisely determined by its three vertices. Another advantage of using linear element assumption is the integrals which result from the solution of Laplace's equation can be calculated analytically. This provides an improvement of the accuracy and efficacy of the 3-D model.

The processes to solve the problem of linear deformation of an infinite jet by using 3-D BEM model begins with the description of the governing equations and the associated boundary conditions. Then, followed by a description of an analytic temporal instability analysis by Yang[9]. The resulting dispersion equation has been simplified to fit our assumptions and serves as an analytic solution to compare to the numerical results from 3-D BEM model. Then, a grid generation process for the liquid column is presented next. The liquid column serves as a portion of an infinite liquid jet. The initial shape of jet can be arbitrary specified by manipulated the longitudinal (streamwise) wave number and transverse (circumferential) wave number. Then, a free surface module is developed specifically for the jet geometry. The process of dealing with the corner problem is described in this section as well. Finally, the discussion of the results from both computational and analytical will be presented.

CP982, *Complex Systems, 5th International Workshop on Complex Systems*
edited by M. Tokuyama, I. Oppenheim, and H. Nishiyama

GOVERNING EQUATIONS

A single column will be considered by the 3-D BEM model. The driving force for this case is the presence of surface tension. The assumptions are made for this simulation are: liquid is inviscid and incompressible fluid, gas phase has been ignored, and no body forces are present. Based on the above assumptions, the governing equations are given by Laplace equation:

$$\nabla'^2 \phi' = 0 \tag{1}$$

Primes denote dimensional quantities and ϕ' is dimensional liquid velocity potential. Two boundary conditions are given by:

Kinematic condition:

$$\frac{Dx'}{Dt'} = \frac{\partial \phi'}{\partial x'} \tag{2}$$

$$\frac{Dy'}{Dt'} = \frac{\partial \phi'}{\partial y'} \tag{3}$$

$$\frac{Dz'}{Dt'} = \frac{\partial \phi'}{\partial z'} \tag{4}$$

and the Dynamic condition:

$$\frac{\partial \phi'}{\partial t'} + \frac{P_l'}{\rho_l'} + \frac{1}{2}(\nabla' \phi')^2 = 0 \tag{5}$$

where P_l' is liquid pressure which can be represented by the surface tensions in the form:

$$P_l' = \sigma' \kappa' \tag{6}$$

Nonlinearity is appeared in the terms of boundary conditions. The suitable characteristic dimensions have been chosen to nondimensionlize the above equations are: Undisturbed radius, a' of column radius, liquid density, ρ_l' , and surface tension, σ' . The dimensionless forms of the governing equations and boundary equations are given by:

$$\nabla^2 \phi = 0 \tag{7}$$

$$\frac{\partial \phi}{\partial x} + \kappa + \frac{1}{2}(\nabla \phi)^2 = 0 \tag{8}$$

$$\frac{Dx}{Dt} = \frac{\partial \phi}{\partial x} \tag{9}$$

$$\frac{Dy}{Dt} = \frac{\partial \phi}{\partial y} \tag{10}$$

$$\frac{Dz}{Dt} = \frac{\partial \phi}{\partial z} \tag{11}$$

In order to integrate the above equations, the Bernoulli equation (Equation (8)) must be expressed in the Lagrangian forms:

$$\frac{D\phi}{Dx} = \frac{1}{2}(\nabla \phi)^2 - \kappa \tag{12}$$

Since $D/Dt = \partial/\partial t + \nabla\phi \cdot \nabla$ assuming nodes are tracked along the local velocity vector, Equation (7), Equation (12), and Equations (9)-(11) constitutes time-dependent relations to describe the behavior of liquid jet movement.

A model based on Boundary Element Method is developed to solve the set of aforementioned equations. The model begins with an integral representation of Laplace equation[10]:

$$\alpha' \phi(\vec{r}_i) + \int_{\Gamma} (\phi \frac{\partial G}{\partial n} - qG) d\Gamma = 0 \tag{13}$$

where α' is the contribution from the singularity which occurs when the integration passes over the base point $(\vec{r}_i)$ where the prime is added to distinguish the variable α ,which represents the grouping of several variables included to be used later. Here, Γ is the boundary of domain which is a surface in the 3-D case, q is the velocity normal to the surface which is given by $\partial\phi / \partial n$, and, G denotes the three-dimensional Green's function of Laplace equation in an unbounded domain and given by:

$$G = \frac{1}{4} \frac{1}{|\vec{r} - \vec{r}_i|} \tag{14}$$

with $\vec{r}$ represents the field point that lies on the triangular element of interest. The distance between the base point ($\vec{r}_i$) and the field points ($\vec{r}$) is given by the denominator of the Green's function which is $|\vec{r} - \vec{r}_i|$.

We discretize the boundary domain into elements which can approximate the boundary domain. After the discretization, we get:

$$\alpha' \phi(\vec{r}_i) + \sum_{j=1}^{e} \int_{\Gamma} (\phi \frac{\partial G}{\partial n} - qG) d\Gamma = 0 \tag{15}$$

where e represents the number of elements on the surface.

We assume the velocity potential (ϕ) and normal velocity (q) are distributed linearly within the triangular element. The numbering order of the vertices on a triangular element warrants careful attention since this order has to provide the normal direction pointing outward from the local surface. The discretized integral Equation (15) becomes:

$$\alpha\phi_i + \sum_{j=1}^{e} \phi_{j1} D_{1,ij} + (\phi_{j2} - \phi_{j1}) D_{2,ij} + (\phi_{j3} - \phi_{j1}) D_{3,ij}$$

$$= -\sum_{j=1}^{e} q_{j1} S_{1,ij} + (q_{j2} - q_{j1}) S_{2,ij} + (q_{j3} - q_{j1}) S_{3,ij} \tag{16}$$

where:

$$\alpha = -4\pi\alpha' \tag{17}$$

$$S_{1,ij} = \int_0^1 \int_0^{1-\eta_2} \frac{1}{|\vec{r} - \vec{r}_i|} |J| d\eta_1 d\eta_2 \tag{18}$$

$$S_{2,ij} = \int_0^1 \int_0^{1-\eta_2} \frac{\eta_1}{|\vec{r} - \vec{r}_i|} |J| d\eta_1 d\eta_2 \quad (19)$$

$$S_{3,ij} = \int_0^1 \int_0^{1-\eta_2} \frac{\eta_2}{|\vec{r} - \vec{r}_i|} |J| d\eta_1 d\eta_2 \quad (20)$$

$$D_{1,ij} = n_i \int_0^1 \int_0^{1-\eta_2} \frac{1}{|\vec{r} - \vec{r}_i|^3} |J| d\eta_1 d\eta_2 \quad (21)$$

$$D_{2,ij} = n_i \int_0^1 \int_0^{1-\eta_2} \frac{\eta_1}{|\vec{r} - \vec{r}_i|^3} |J| d\eta_1 d\eta_2 \quad (22)$$

and,

$$D_{3,ij} = n_i \int_0^1 \int_0^{1-\eta_2} \frac{\eta_2}{|\vec{r} - \vec{r}_i|^3} |J| d\eta_1 d\eta_2 \quad (23)$$

with $n_i = (\vec{r} - \vec{r}_i) \cdot \hat{n}$ and $\hat{n}$ is the normal vector of a given element. Here, $D_{1,ij}$, $D_{2,ij}$, $D_{3,ji}$, $S_{1,ij}$, $S_{2,ij}$, and $S_{3,ij}$ are the kernel functions corresponding to doublet (D) and source (S) terms in Equation (16). Note that these quantities are only a function of the current geometry.

In order to solve Equation (16), the kernels have to be evaluated either numerically or analytically. Fortunately, the closed form solution of each kernel can be determined for this linear element approximation. Once the values of each of the kernels have been calculated, the results are substituted into Equation (16). The summation processes in Equation (16) will require information about the surrounding elements and nodes around a given node which is denoted as base point. Three types of connectivity will provide the sufficient surrounding data for a base point. The first type of connectivity needed is which three nodes constitute a given element. The second type of connectivity will provide information about what elements surround a base point. The last connectivity needed to develop is the contiguous nodes next to a base point.

With the aid of this connection, the Equation (16) can be reduced to a simple linear algebra equation:

$$[D]\{\phi\} = [S]\{q\} \quad (24)$$

Elements of the D matrix will be composed of several kernels Equations (21)-(23) which provide information regarding a particular vertex location. Similarly, the S matrix contains grouping of kernels from Equations (18)-(20).

Either ϕ or q must be known on each vertex around the boundary in order to form a well-posed boundary value problem for solving Laplace's equation. Thus, after applying boundary conditions, the matrices can be pivoted such that all known boundary values (and associated D_{ij}, S_{ij} values) lie on the right-hand-side of the equation. Using this approach, we end up with a standard linear algebra problem:

$$[A]\{x\} = \{b\} \quad (25)$$

where the *nnode x nnode* matrix *[A]* contains portions of *[D]* and *[S]* where *nnode* is the total number of nodes, x is the vector of all unknown ϕ and q values, and b is a vector obtained by multiplying out all known boundary values by the corresponding D and S elements.

Here, *[A]* is a fully-populated matrix as compared to the sparse matrix system resulting from the use of finite difference methods. By solving Equation (25), which is solved by an LU-decomposition method taken from Numerical Recipes in Fortran [11], the unknown values of ϕ or q at each node can be determined.

DISPERSION EQUATION

The mathematical model used for the temporal instability analysis is based on linear analysis conducted by Yang[9]. The resulting dispersion equation has been simplified to fit our assumptions of inviscid flow with negligible gas-phase effects. Surface tension is a dominant force for the instability of the low speed jet which is also called the capillary jet. Rayleigh[12] showed that only unstable mechanism to cause the breakup of a low speed jet is the growth of axisymmetric waves which are also referred to as dilation waves. For the high speed jet, the transverse modes are introduced by viscous and turbulent effects at the nozzle exit. Asymmetric waves are also called sinuous waves. In order to understand the behavior of the sinuous wave, the theory has to include the transverse mode oscillation.

Yang's theory begins with a two stream flow, liquid jet and gas stream, which are assumed to be incompressible and inviscid fluid. Liquid jet with density ρ_1 and issued at a uniform speed U_1, while a coaxial gas with density ρ_2 flowed with a uniform speed U_2. The interface between liquid and gas phase has been perturbed by:

$$\eta = \eta_1 = \eta_2 = \eta_0 e^{i(kz+m\theta)+\alpha t} \quad (26)$$

where η_0 is perturbation magnitude, k and m are the wave number in streamwise or longitudinal and azimuthal or transversal direction, respectively. Here, α is a complex variable, $\alpha = \alpha_r + i\alpha_i$. The real part of α, α_r, is the growth rate. The positive valve of α_r indicates the wave growth, while the negative value indicates the wave is stable and will eventually damp out. The imaginary part of α, α_i, is connected to the wave propagation speed. In order to fit in our previous assumptions, the resulting characteristic equation, i.e., dispersion equation is given by:

$$\left(\alpha_r^*\right)_m^2 = ka\left[1 - m^2 - (ka)^2\right]\frac{I'_m(ka)}{kI_m(ka)} \quad (27)$$

And, a is the radius of nozzle. I_m is the modified Bessel function of the first kind and order m. The prime on I_m denotes the first derivative with respective to their parameters. For an axisymmetric wave oscillation, the azimuthal wave number m is simply equal to zero. Thus, the asymmetric instability is introduced by providing nonzero value of m. Figure 1 shows the plot of growth rate, $\left(\alpha_r^*\right)_m^2$ vs. ka at different values of m. It indicates that the only unstable wave for a quiescent, incompressible, and inviscid liquid jet is axisymmetric wave. All of non-axisymmetric modes ($m>0$) are stable due to the negative values of growth rate over all of dimensionless streamwise wave number ka.

Thus, the dispersion equation is used for the temporal instability analysis later in this paper is given by:

$$\left(\alpha_r^*\right)_0^2 = ka\left[1 - (ka)^2\right]\frac{I'_0(ka)}{kI_0(ka)} \quad (28)$$

which is identical to the result of Rayleigh's analysis.

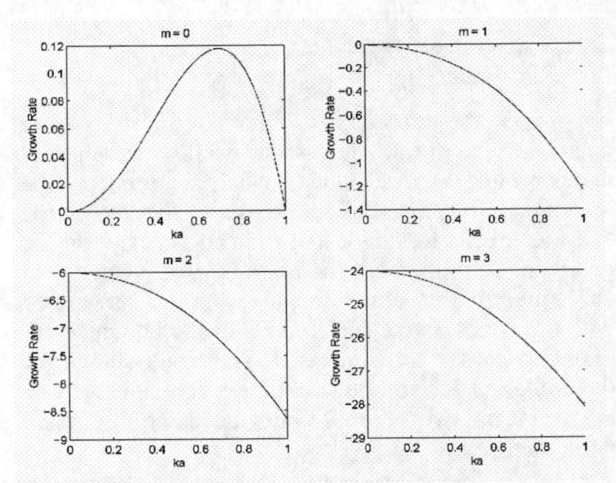

FIGURE 1. Growth rate vs dimensionless streamwise wave number at different azimuthal wave number.

GRID GENERATION

The first step in the numerical analysis is to set up the grid points for a liquid jet. A cylinder with a small perturbation on the surface is used to represent the initial shape of the liquid jet. Since the liquid jet is assumed to be of infinite length, the computational domain is chosen to be one wavelength (λ) of the perturbation.

A grid generation scheme has been applied on a portion of a liquid jet which is described by a cylindrical column. The number of intervals in the r, θ , and z directions (*int_r*, *int_z*, and *int_the*, respectively) must be input in order to generate the grid. The initial shape of the liquid column is given by:

$$r = 1 + \delta\cos(kz + m\theta) \quad (29)$$

where δ is the perturbation magnitude. For an undisturbed column, δ is given by zero, and an axisymmetric column is given by $m=0$. Figure 2 highlights the geometrical perspective of there three grid setups.

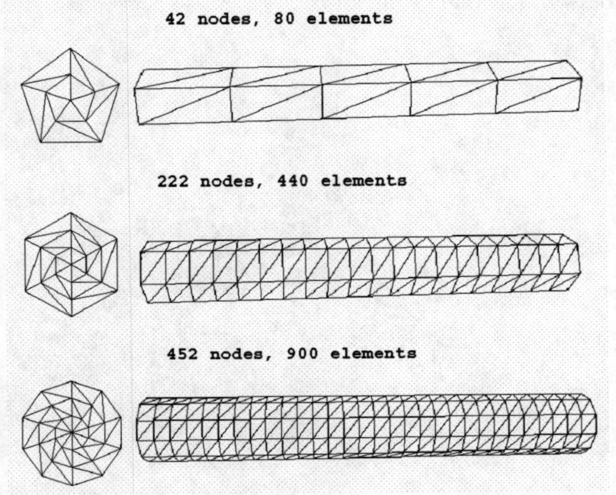

FIGURE 2. Three different grid setups for a liquid jet.

FREE SURFACE MODULES

The least square method is applied for the calculations of the surface normal vector, surface curvature, and tangential velocity. Here, the fit function is chosen to be

$$F(x,y,z) = a_1x^2 + a_2y^2 + a_3x + a_4y + a_5z - 1 \quad (30)$$

And, the fit function for the velocity potential fitting in the liquid jet case is chosen to be:

$$\phi(x,y,z) = G(x,y,z) = b_1x^2 + b_2y^2 + b_3x + b_4y + b_5z - 1 \quad (31)$$

The accuracy of the fitting function can be determined by observing the results of surface curvature for an undisturbed jet. For an undisturbed jet which is a smooth cylindrical column, the surface curvature is equal to one at every node on the surface side which does not include the nodes on the two end faces.

The corner nodes warrant a careful treatment when dealing with the geometry of liquid column. The corner nodes are referred to the nodes that lie on the intersection potion between the surface side and two end faces. Those nodes have multiple value of q which is $q = \partial\phi/\partial n$. Since these two end faces are the cross sections of an infinite jet which serve as symmetry planes. By the definition, the values of q are set to be zero for the nodes on these symmetry planes. Thus, the elements lying on one of the two end faces will have no contribution to the multiple value problem of q since q is zero on these elements. However, for the elements on the surface side, an assumption is made to deal with the multiple value of q. Since the multiple value problem is arisen due to the different normal vectors associated with those surrounding elements, these differences in the normal vectors can be assumed to be small when a finer grid is used. Therefore, multiple value of q for those elements becomes a single value of q.

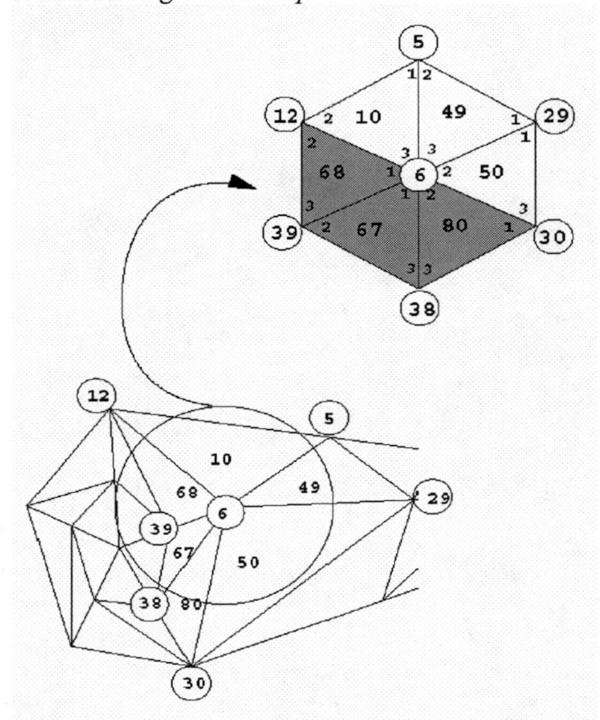

FIGURE 3. Schematic description of surrounding grids and elements around a corner node.

An example is given to further explain how to deal with the multiple value of q on the corner nodes. For grid setup number 1, the corner node and their surrounding elements are shown on the Fig. 3. The zoom-in picture in Fig. 3 shows the node numbers and element numbers surrounding node 6. The dark portion of this zoom-in picture indicates the end face while the white portion of this picture refers to the surface side of column. The summation process in Equation (16) gives the following equation around node 6:

$$\begin{aligned} \cdots \;=\; & \cdots + [q^6_{10.3} S_{3,6.10} + q^6_{49.3} S_{3,6.49} + q^6_{50.2} S_{3,6.50} \\ & + q^6_{67.1} S_{1,6.67} - q^6_{67.1} S_{2,6.67} - q^6_{67.1} S_{3,6.67} \\ & + q^6_{68.1} S_{1,6.68} - q^6_{68.1} S_{2,6.68} - q^6_{68.1} S_{3,6.68} \\ & + q^6_{80.2} S_{2,6.80}] \end{aligned} \tag{32}$$

The notation of $q^i_{j.k}$ is that i refers to node number, j indicates the element number and k is the node order in the j element. For example, $q^6_{68.1}$ indicates the q value at node 6 which lies on the first node of element 68. And, $S_{l,m.n}$ is the value of kernel function S_l which the integration is performed on the base point m over element n. Thus, $S_{3,6.10}$ is the value of kernel function S_3 which is based on the base point 6 and integrating over element 10. According to our discussion and assumption, we have:

By assumption:

$$q^6_{10.3} = q^6_{50.2} = q^6_{49.3} = q^6 \tag{33}$$

By definition:

$$q^6_{67.1} = q^6_{68.1} = q^6_{80.1} = 0 \tag{34}$$

Thus, Equation (46) becomes:

$$\cdots \;=\; \cdots + q^6 [S_{3,6.10} + S_{3,6.49} + S_{3,6.50}] + \cdots \tag{35}$$

A special attention is warranted on the corner nodes for the fitting process as well. The adjacent nodes around a corner node, which are used for the least square method, can not be used if they lie on the end face of the jet due to the dramatic change on the geometrical properties such as normal vector and curvature. In this case, the image points associated with the nodes on the surface side of the jet which are around a corner node are used. The result shown for the curvature for an undisturbed jet, the value of one are given on every corner nodes by using the image point method.

RESULTS OF LINEAR ANALYSIS

The mathematical model developed by Yang[9] is used for the temporal instability analysis on a liquid jet which is assumed as incompressible, inviscid, and quiescent fluid. The dispersion Equation (28) serves as the analytical solution for the growth rate. The analytical solution is used to compare to the results from 3D BEM model. The calculation of growth rate for the numerical result is based on the derivation by Mansour[13] which he utilized the prediction of Rayleigh's analysis. The extension of derivation by Mansour for 3D jet is described as follow.

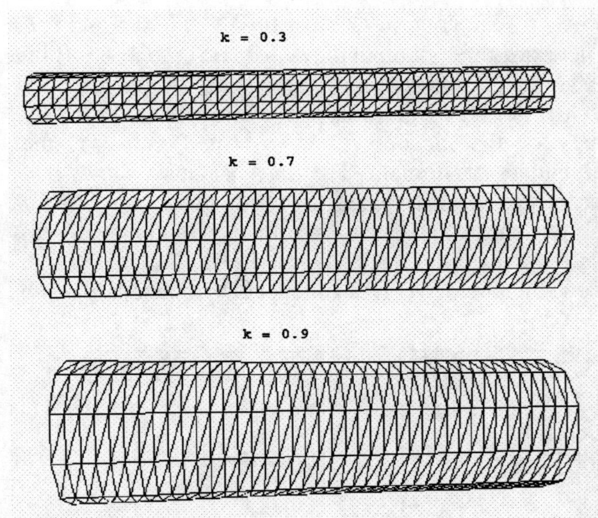

FIGURE 4. *int_z*=10, *int_the*=10 and *int_r*=3 for *k*=0.3, 0.7, and 0.9. Perturbation magnitude is 0.01 for *m*=0.

The development of jet is assumed to be in the form of:

$$r = 1 + \delta \cos(kz + m\theta)\cosh(\alpha_r t) \quad (36)$$

where α_r is the growth rate which is the real part of α which is defined previously. And, the velocity in radius direction (V_r) is calculated by:

$$V_r = \frac{dr}{dt} = \delta \cos(kz + m\theta)\sinh(\alpha_r t) \quad (37)$$

Thus the growth rate α_r is given by:

$$\alpha_r^2 = \frac{V_r^2}{(r-1)^2 - \delta^2 \cos^2(kz + m\theta)} \quad (38)$$

Since the only unstable wave in the incompressible, inviscid, and quiescent fluid is the axisymmetric wave, m is set to be zero. Along the longitudinal direction, the nodes lying on the trough of wave are chosen to be the location which the temporal analysis is conducted (thus $kz = \pi$). Therefore, the preceding equation is simplified to the equation which was used in the Mansour's analysis which is given by:

$$\alpha_r^2 = \frac{V_r^2}{(r-1)^2 - \delta^2} \quad (39)$$

In this case, only the axisymmetric wave develops, the nodes on the surface side of jet will move only in the radial direction and V_r is given by q which is calculated by 3D BEM model. Thus, the temporal instability analysis for axisymmetric oscillation can be conducted based on Equation (28) for analytical solution and Equation (39) for numerical solution.

The temporal instability analysis is performed by picking several longitudinal wave numbers (k) and letting transverse wave number (m) to be zero for axisymmetric case. The total wavelength is set to be one. A single grid setup is not appropriate to run this analysis due to the aspect ratio of grid element varying for different values of k. Figure 4 shows the grid setup with *int_z* = 40, *int_the* = 10, and *int_r* = 3 for three different *k* values: *k* = 0.3, *k* = 0.7and *k* = 0.9. The aspect ration $\Delta\theta / \Delta z$ will increases when Δz decreases due to increasing *k* values. Note that $\lambda = 2\pi / k$ and $\Delta z = \lambda / \mathrm{int\_}z$. The total length of jet is chosen to be one which results in a large aspect ratio for higher *k* value. The shape of jet will looks bigger for those higher *k* values due to the higher aspect ratio. A grid setup with nonuniform aspect ratio for element will effect the accuracy of results. In order to get better results, the optimized grid needs to be used for each *k* value.

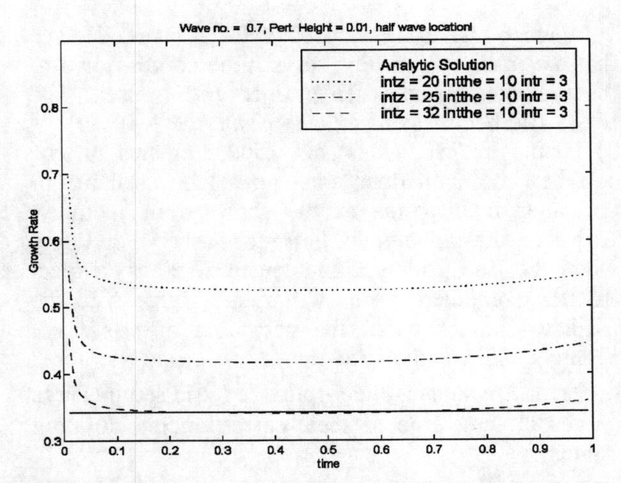

FIGURE 5. Track of growth rate with time for *k*=0.7, perturbation magnitude = 0.01.

Figure 5 shows that the growth rate calculated by 3D BEM model reaches the growth rate predicted by Equation (28) when grid setup approaches the optimized grid for *k* = 0.7. The growth rate converges quickly to a constant. The same behavior will show for different *k* values. The convergent constant will be the desired value of growth rate.

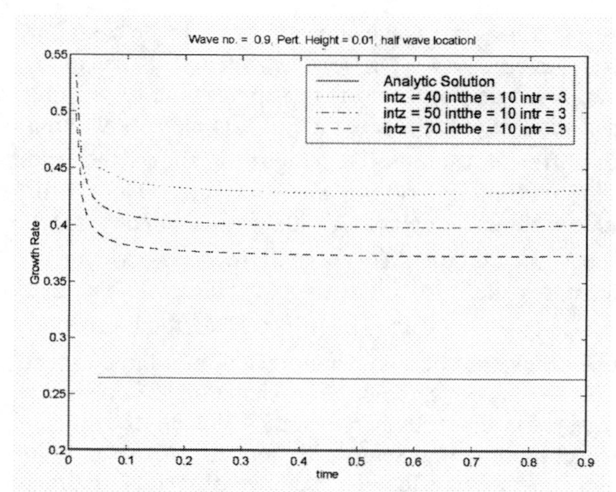

FIGURE 6. Track of growth rate with time for k=0.9, perturbation magnitude = 0.01.

Figure 6 shows the same plot for k = 0.9. We see that the track of growth rate vs. time comes down to the predicted value when a finer grid is used. The finest grid been tested (*int_z* = 70,*int_the* = 10,*int_r* = 3) results in 752 nodes and 1500 elements on the surface. The even finer grid can not be used in our computer facility due to the shortage of memory. However, the convergence behavior is shown on Fig. 6 when the fine grid is implemented. We may expect that the computed results will finally agree with the predicted values when the very fine grid is used. Figure 7 shows the growth rate vs. time for three different k values at their optimized grid setup. These curves all converge to their corresponding constant value.

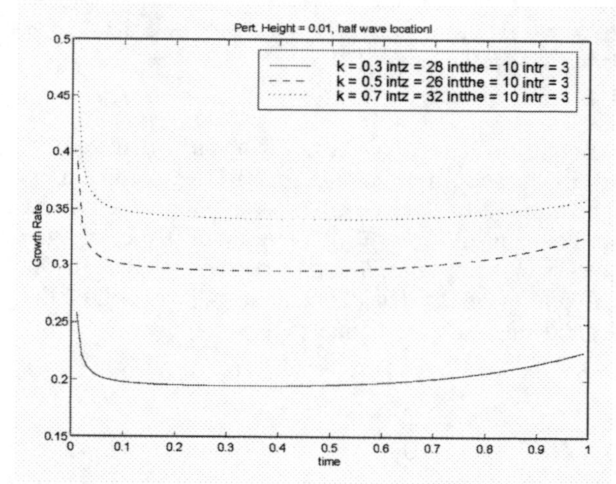

FIGURE 7. Track of growth rate with time for three different k values at their optimized grid setup.

Figure 8 shows the comparison of computational results and predicted values for linear analysis. It shows a very good agreement for k = 0.3, 0.5 and 0.7. There are three computations for the case k = 0.9 by using three different grid setups as in the case of shown in Fig. 6. The reason for error for k = 0.9 may due to the desire of even fine grid for the computational domain. The linear analysis shows that the predicted growth rate for a liquid column without the presence of gas phase are matched very well with the computational results from 3D BEM model. Thus, the validation of 3D BEM model is completed.

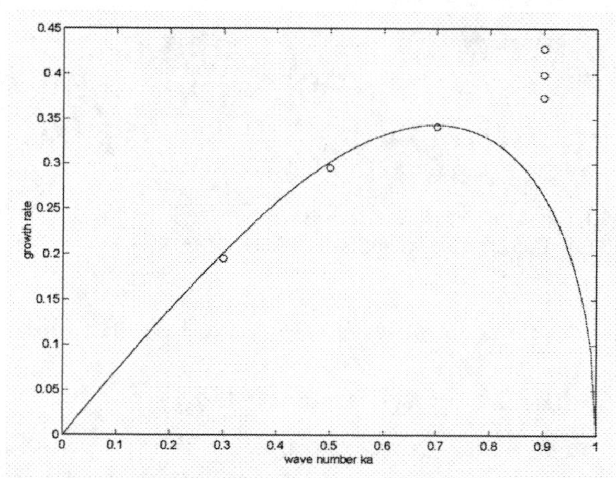

FIGURE 8. Comparison of analytic and numerical solution for linear temporal analysis.

CONCLUSIONS

The topic of this research is the development of a 3-D BEM model. The need of 3-D model is obvious for the most practical engineering application where the atomization of liquid jets into droplets is three-dimensional. A 3-D BEM model was developed using linear triangular elements to solve the Laplace equation. The 3-D model has been validated for analytic temporal instability analysis on an infinite jet. A good agreement between computational results and predicted linear analysis results.

REFERENCES

1. G. L. Chahine, *22nd American Towing Tank Conference*, 1989.
2. G. L. Chahine and T. O. Purdue, A.I.P Conference Preceedings, edited by T. G. Wang, 1989, pp. 169-187.
3. G. L. Chahline, K. M. Kalumuck and R. Duraiswami, *Bubble Noise and Cavitation and Multiphase Flow Forum* **109**, 49-54 (1991).
4. G. L. Chahline, R. Duraiswami and A. N. Lakshminarsimha, *Cavitation and Multiphase Flow Forum* **109**, 49-54 (1991).
5. H. Oguz and J. Zeng, *Engineering Analysis with Boundary Elements* **19**, 319-330 (1997).

6. C. A. Spangler, J. H. Hilbing and S. D. Heister, *Atomization and Sprays* **5**, 621-638 (1995).
7. C. A. Spangler, J. H. Hibing and S. D. Heister, *Phys. Fluids* **7**, 964-971 (1995).
8. C. C. Chao, S. D. Heister, *Engineering Analysis with Boundary Element* **28**, 1045-1053 (2004).
9. H. Q. Yang, *Phys. Fluids* **4**, 681-689 (1992).
10. J. A. Liggett and P. L. F. Liu, *The Boundary Integral Equation Method for Porous Media Flow*, George Allen and Unwin, London, 1983.
11. W. H. Press, S. A. Teukolsky, W. T. Vetterling and B. P. Flannery, *Numerical Recipes in FORTRAN*, The Art of Scientific Computing, Cambridge University Press, New York, 2nd ed., 1992.
12. W. S. Rayleigh, *Proc. London Math. Soc.* **10**, 4 (1878).
13. N. N. Mansour, *Phys. Fluids A* **2**, 1141-1144 (1990).

Bubble Rising Velocity in Sodium Chloride Aqueous Solution under Horizontal DC High Magnetic Field

Kazuhiko Iwai[1] and Ippei Furuhashi[2]

[1]*Department of Materials, Physics and Energy Engineering*
Graduate School of Engineering, Nagoya University
Nagoya, Aichi 464-8603, Japan
[2] *Graduate Student, Department of Materials, Physics and Energy Engineering*
Graduate School of Engineering, Nagoya University
Nagoya, Aichi 464-8603, Japan

Abstract. In a continuous casting of steel, argon bubbles are injected from a nozzle to prevent nozzle clogging. However, this sometimes causes a problem of the entrapment of inclusions in a solidifying metal front. On the other hand, an electromagnetic brake has been utilized to control molten metal flow in the continuous casting process. Therefore, the understanding of bubble behavior in molten steel under the electromagnetic brake in which inertial force, Lorentz force and buoyancy force play an important role is essential for the optimization of the continuous casting process of steel. A water model experiment is one of the typical methods for direct observation of bubble behavior while it is impossible to use the water model experiment for this purpose because the Lorentz force is not induced by the bubble motion in the water. The Lorentz force is excited when a molten metal with low melting temperature is used instead of the water, however, the direct observation of the bubble motion is impossible because of opaque nature of metals. In order to overcome this problem and to get useful information for the bubble behavior under the electromagnetic brake, the bubble behavior has been simulated by use of a strong electrolyte under a high magnetic field. The principle of the simulation is based on that the ratios among those forces in the simulation system are the nearly same as the ratios in a practical operation. New knowledge about the effect of Lorentz force on the bubble behavior is discussed in this manuscript.

Keywords: Continuous casting process, Simulation, Bubble, Electromagnetic brake
PACS: 47.55.dd

INTRODUCTION

Control of second phases in molten steel such as gas bubbles and inclusions is important for production of high quality steel. Water model experiments and numerical simulations have been used for investigating the behavior of the second phases in steelmaking processes [1,2]. In continuous casting process, electromagnetic brake (EMBr) has been used to control the molten steel flow in the mold though the second phase distribution affects the metal flow through the change of Lorentz force in the metal. Hence, it is necessary to clarify the behavior of gas bubbles and inclusions in the molten steel under a magnetic field produced by the electromagnetic brake.

Water model experiment is the typical experimental method for investigating the behavior of the gas bubbles and the inclusions in molten steel and then it has been used until now. In the water model experiments, however, the effect of the Lorentz force on liquid motion is negligible because water is an electrically insulating material. Therefore, to reveal the behavior of electrically non-conductive second phase moving in a liquid metal under the magnetic field, some investigations have been done by using computational fluid dynamics (CFD) [3] and theoretical analyses [4-7]. To clarify the second phase behavior experimentally, the second phase behavior exposed to the Lorentz force should be measured. If a liquid metal is used as a medium to excite the Lorentz force, a non-optical method is required for measuring the second phase behavior because of opaque nature of the metal. Dresden group [8] developed a new method in which an argon gas bubble behavior in a liquid metal was measured using Ultrasonic Doppler Velocimetry. However, a measuring method with a

CP982, *Complex Systems, 5th International Workshop on Complex Systems*
edited by M. Tokuyama, I. Oppenheim, and H. Nishiyama

high resolution is desired because a rising gas bubble in a liquid metal must easily deform its shape and its rising behavior highly depends on the bubble shape.

To increase the effect of the Lorentz force, we propose use of a transparent strong electrolyte solution. Motion of a second phase in the solution under the imposition of a strong static magnetic field produces the Lorentz force, which affects the behavior of the second phase.

The ratio of the Lorentz force to inertia acting on the second phase in the model experiment proposed here can be adjusted to that in the continuous casting of steel by controlling intensity of the magnetic field and/or size of the second phase in the proposed experimental system. That is, the similarity law can be satisfied. The present study gives new knowledge about the effect of the Lorentz force on the bubble behavior in an electrically conductive liquid under the imposition of a static magnetic field.

SIMILARITY LAW

For the estimation of a gas bubble behavior in the continuous casting process, water model experiments have been used under the same Froude number, which is the square root of the ratio of inertial force to buoyancy force acting on the bubble defined as

$$F_r = \frac{V}{\sqrt{gD}}$$

where V is characteristic velocity, g is acceleration of gravity and D is diameter of the gas bubble, respectively.

TABLE 1. Physical properties of molten steel, service water and NaCl aqueous solution.

	Electrical conductivity	Density
NaCl aqueous solution	23S/m	1230 kg/m^3
Molten steel	$7.1*10^5$S/m	7010 kg/m^3
service water	10^{-1}~10^{-3}S/m	1000 kg/m^3
distilled water	10^{-3}~10^{-4}S/m	1000 kg/m^3

The density and electrical conductivity of saturated sodium chloride aqueous solution, molten steel and service water are shown in Table1. The bubble rising velocities in the water and in the molten steel are the same magnitude under the same Froude number condition because density of water is the same order with that of molten steel. On the other hand, the Lorentz force acting on the bubble in the water model experiment is smaller than that in the molten steel because of the difference in the electrical conductivity between them. For intensification of the reaction of the Lorentz force acting on the bubble in the model experiment, increase in electrical conductivity of the model liquid and/or increase in the magnetic field are required as mentioned above. One of the solutions of this obstacle is the use of a transparent strong electrolyte such as sodium chloride aqueous solution. The order of the bubble size and its rising velocity in the strong electrolyte can be fit to those in the molten steel under the similarity law of the Froude number. Since the Lorentz force per unit volume acting on an electrically conductive liquid submerged in a magnetic field is estimated as $\sigma v B^2$ where σ is the electrical conductivity of liquid, v is the velocity of the gas bubble, B is the magnetic field intensity, the order of σB^2 in the model experiment should be adjusted to that in the continuous casting process to satisfy the similarity law relating the reaction of the Lorentz force and the other forces such as the buoyancy force and the inertial force acting on the bubble. This can be achieved when the magnitude of the magnetic field in the model experiment is 10^2 times larger than that in the continuous casting process because the ratio of the electrical conductivities is about 10^4. Then the required magnetic field is roughly 10T in the model experiment if the magnetic field in the continuous casting process is 0.1T. The model experimental system can be built up using a super-conducting magnet which can supply a 10T magnetic field.

In this study, a new model experimental system in which the similarity law relating the Lorentz force to other forces acting on a gas bubble can be satisfied by using a strong electrolyte submerged in a high magnetic field is proposed and the argon gas bubble behavior with different size is observed to investigate the effect of the magnetic field on the bubble behavior.

EXPERIMENTAL SETUP

A schematic view of the experimental apparatus is shown in Fig. 1. A rectangular acrylic vessel whose width, depth and height were 64mm, 54mm, 890mm, respectively was filled with saturated sodium chloride aqueous solution. And it was set in the bore of a super-conducting magnet generating a horizontal static magnetic flux density of 7T. An argon gas bubble of 3mm diameter was injected from the bottom of the vessel through a copper nozzle. The bubble motion was recorded by a high speed camera to measure the passing time of 30mm distance in both the cases with and without the magnetic field. The effect of the magnetization force on the bubble motion is negligible in this experiment because the ratio of the magnetization force to the Lorentz force acting on the bubble is estimated at 0.05 under the 7T magnetic field.

Buoyancy force, inertial force and reaction of the Lorentz force acting on the bubble in this experiment and in the continuous casting of steel are calculated using the physical properties shown in Table 1, and the similarity among these forces are estimated.

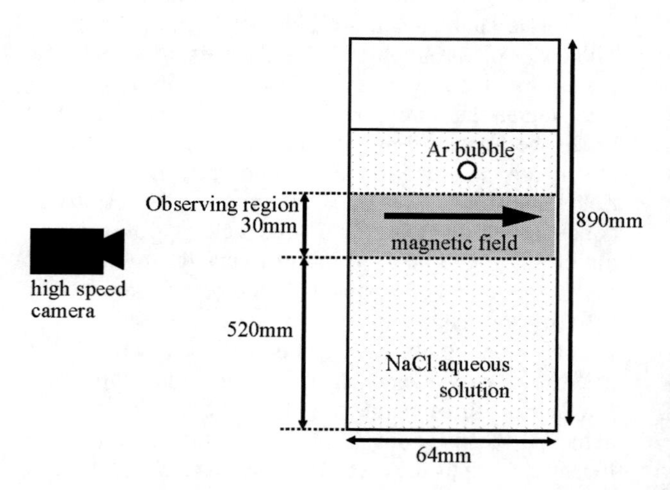

FIGURE 1. Experimental setup.

The 3mm bubble diameter in this experiment corresponds to the 2mm diameter bubble in the continuous casting process from the viewpoint of the similarity of the Froude number, and it corresponds to the 0.3mm diameter bubble from the viewpoint of the similarity of the ratio of the Lorentz force to the buoyancy force, respectively.

EXPERIMENTAL RESULTS

Rising bubble shapes with and without the horizontal magnetic field are shown in Fig. 2. Resolution of the pictures is not so good because the high speed camera should be set far away from the super conducting magnet to prevent the attraction between the superconducting magnet and iron parts in the high speed camera and to prevent the effect of the magnetic field to the electrical circuit in the high speed camera. Thus, the schematic view of the pictures is also shown in Fig. 2. The bubble slightly deforms from spherical shape and elongates to the horizontal direction in the both cases. The elongation of the bubble in the case without the magnetic field is intensive in comparison with the bubble elongation without the magnetic field.

Table 2 shows average passing times of 30mm distance with and without the imposition of the magnetic field and their 95% confidence interval. Without the imposition of the 7T magnetic field on the sodium chloride aqueous solution, the averaged passing velocity of the bubbles is 236mm/s, while it decreases to 229mm/s by imposing the 7T static magnetic field in the horizontal direction. The 95% confidence interval of bubble passing time in the case without the magnetic field is between 236mm/s and 238mm/s, while it is between 226mm/s and 232mm/s in the case with the 7T static horizontal magnetic field. The 95% confidence interval with the magnetic field is smaller than the maximum of the 95% confidence interval without the magnetic field. Thus, it is concluded that the averaged passing velocity becomes slower by imposing a magnetic field. This might be the result that the liquid motion surrounding the bubble is suppressed by the Lorentz force.

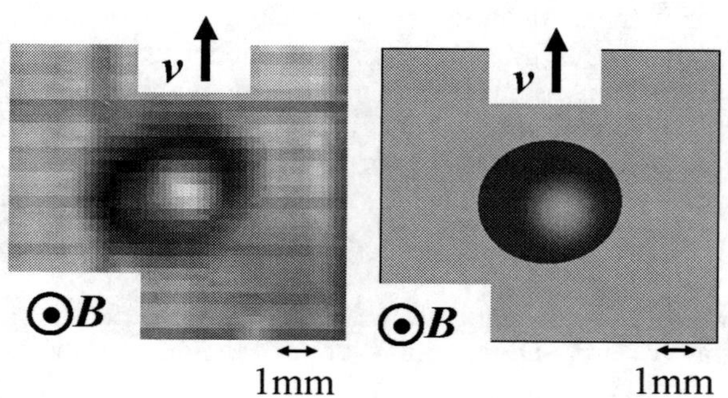

(a) with 7T magnetic field

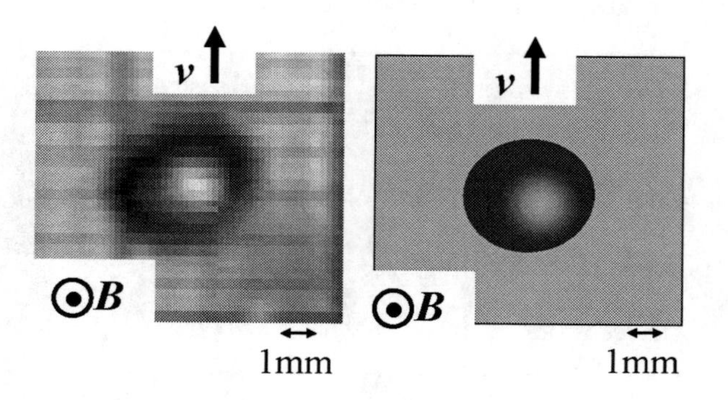

(b) without magnetic field

FIGURE 2. Ar gas bubble shape with and without magnetic field.

TABLE 2. Average bubble passing velocity and its 95% confidence interval with and without a magnetic field.

	B=0T	B=7T
Average bubble passing velocity	236mm/s	229mm/s
95% confidence interval of bubble passing velocity	236-238mm/s	226-232mm/s

CONCLUSION

To clarify the gas bubble behavior in a liquid metal submerged in a magnetic field, we proposed the model experimental system in which a high magnetic field is imposed on a transparent strong electrolyte such as saturated sodium chloride aqueous solution. By using this model experimental system, the argon gas bubble rising velocity in the steel was simulated. The rising velocity of an argon gas bubble in a saturated sodium chloride solution is suppressed by imposing a horizontal high magnetic field.

ACKNOWLEDGMENTS

This work was partially supported by JSPS Asian Core Program "Construction of the World Center on Electromagnetic Processing of Materials".

REFERENCES

1. K. Takatani, *ISIJ Int.* **43**, 915-922 (2003).
2. M. Iguchi and N. Kasai, *Metallurgical and Materials Transactions B* **31B**, 453-460 (2000).
3. T. Kato, T. Tagawa and H. Ozoe, *Proc. 17th Symp. on Chem. Eng.*, 2004, p.113.
4. W. Chester, *J. Fluid. Mech.* **3**, 304-308 (1957).
5. W. Chester, *J. Fluid. Mech.* **10**, 459-465 (1961).
6. W. Chester and D. W. Moore, *J. Fluid. Mech.* **10**, 466-472 (1961).
7. K. Ueno, and H. Yasuda, *Magnetohydrodynamics*, **39**, 547-556 (2003).
8. C. Zhang, S. Eckert, G. Gerbeth, *Int. J. Multiphase Flow*, **31**, 824-842 (2005).

Study on the CO_2 Solid-Gas Two Phase Flow with Particle Sublimation and Its Basic Applications

Xin-Rong Zhang[a,b,*], Hiroshi Yamaguchi[a], and Minoru Masuda[a]

[a]*Department of Mechanical Engineering, Doshisha University, Kyo-Tanabeshi, Kyoto 610-0321, Japan*
[b]*Department of Energy and Resources Engineering, College of Engineering, Peking University, Beijing, China*

Abstract. A basic study was carried out on the CO_2 solid-gas two phase flow with particle sublimation. The CO_2 two phase flow is achieved by liquid CO_2 expansion process throughout the CO_2 triple point. A cryogenic refrigeration below -56.6℃ is possible from the CO_2 solid particle sublimation in the two phase fluid flow. An experiment was conducted in order to investigate some basic points related to the liquid CO_2 expanding into the solid-gas fluid flow, which reveals the CO_2 particle size and so on. Based on the primary experiment, a new refrigeration method is introduced by using the CO_2 solid-gas two phase flow with particle sublimation. A CO_2 heat pump, which can achieve a cryogenic refrigeration below -56.6℃ is designed, constructed and tested. In the paper, details of the CO_2 heat pump system are presented and the obtained results show that a continuous operation is possible with CO_2 solid-gas flow in the closed loop of the heat pump system. Furthermore, the performance of the new CO_2 heat pump system is also presented in this paper, which utilize the flow dynamic of liquid CO_2 expanding into the solid-gas fluid.

Keywords: CO_2, Solid-gas, Two phase fluid, Flow dynamics, CO_2 heat pump, Heat transfer
PACS: 44.35.+c, 47.55.-t

INTRODUCTION

Fluorocarbon and HFCs were widely used as a refrigerant of refrigerators and heat pump. However, from the viewpoint of protecting the ozone layer and preventing global warming, now there is strong demand for technology based on ecologically safe 'natural' working fluids, i.e. fluids like water, ammonia and carbon dioxide. Among these natural refrigerants, carbon dioxide is a byproduct from the burning of fossil fuels to generate electricity. CO_2 emissions from these combustion processes are main reasons of environment deterioration such as global warming in the world. Based on such a background, the interests in carbon dioxide as a refrigerant increased considerably during the past decade [1-4].

CO_2 is a non-flammable natural fluid with no Ozone Depletion Potential (ODP) and a negligible Global Warming Potential (GWP). CO_2 is responsible for over 60% of the greenhouse effect. It is always a good way of relieving the greenhouse effect by recycling CO_2 and using it as refrigerant, which can also be considered as a kind of CO_2 capture and storage. In addition, the CO_2 thermodynamic and transport properties seem to be favorable in terms of heat transfer and pressure drop, compared to other typical refrigerant, where the critical pressure and temperature of CO_2 are 7.38 MPa (73.8 bar) and 31.1℃ respectively [5-7]. Because of the advantages above, CO_2 fluid has received much attention in recent years in some new energy systems [5,7-8,9-13], especially in CO_2 trans-critical compression refrigeration thermodynamics cycle of air conditioners, and heat pumps.

Among the refrigeration process using CO_2, the refrigeration temperature range is about -30.0~0.0℃ achieved by CO_2 evaporation process. As far as the authors are aware, the refrigeration below -30.0℃ using CO_2 as working fluid is rarely seen in the existing studies. Here, a kind of refrigeration using CO_2 is introduced and this can achieve a cryogenic temperature below CO_2 triple point temperature -56.6 ℃. Figure 1 show a schematic of the CO_2 refrigeration principle, in which the refrigeration is achieved by liquid CO_2 expanding into solid-gas two phase fluid. The process of 1-2 represents the liquid CO_2 expansion into the two phase flow, the dry ice region, shown in Fig. 1, which goes down through the CO_2 triple point in CO_2 P-h diagram. By the expansion process, the CO_2 solid-gas two phase fluid is obtained, which is below -56.6℃. The refrigeration method introduced

CP982, *Complex Systems, 5th International Workshop on Complex Systems*
edited by M. Tokuyama, I. Oppenheim, and H. Nishiyama

here is that CO_2 solid particles from the expansion process sublimate and absorb heat when flowing through a pipe. The process is shown in the 2-3 process in CO_2 P-h diagram in Fig. 1. The process shown in Fig. 1 presents a possibility of CO_2 refrigeration, which can achieve a temperature environment below -56.6℃. As the authors aware, there are no existing studies, which cover the proposed refrigeration by liquid CO_2 expanding into solid-gas two phase fluid flow. As the first step of the study on the CO_2 refrigeration, in the present paper, an experiment study is carried out in order to investigate the feasibility and characteristics of liquid CO_2 expanding into solid-gas two phase fluid in a horizontal tube by expansion valve. Furthermore, as an application example, a new CO_2 heat pump system is designed and built, which uses the CO_2 solid-gas two phase fluid in an evaporator.

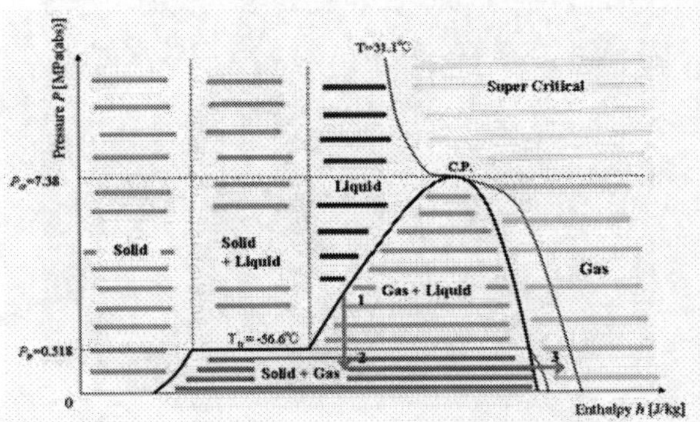

FIGURE 1. (Color online) P-h diagram for carbon dioxide to show a schematic of the new refrigeration method.

EXPERIMENTAL STUDY

The experimental set-up is made in order to investigate the feasibility of achieving CO_2 liquid expanding into solid-gas fluid in a horizontal circular tube by expansion valve. Figure 2 shows a flow diagram of the experimental set-up. The experimental set-up is mainly comprised of CO_2 container, pressure control valve, expansion valve, visualization section and heating section (CO_2 particle sublimation section) and orifice flow meter. The visualization section is a Pyrex circular pipe for visualizing the solid-gas two phase fluid flow obtained from the CO_2 liquid expansion. The heating section is a SUS316 circular pipe for heating the CO_2 solid-gas fluid, which makes CO_2 particle sublimate and CO_2 gas exits the heating section. As the first step of observing the feasibility, an open loop is taken in the experiment.

The carbon dioxide at the outlet of the CO_2 container is controlled in the condition of gas-liquid two-phase flow by pressure control valve. The gas-liquid two-phase flow is separated in Gas-liquid separator and only liquid phase fluid is introduced to the expansion valve. Separated gas phase is recycled to the container. Obviously, the type of expansion valve has an obvious influence on the liquid CO_2 expansion. As the first step of the present study field, a needle valve is used as an expansion valve with a maximum aperture diameter of 12.7 mm. The discussion on types of expansion valves is not included here, which will be another topic in the future. By the expansion valve, the CO_2 fluid expands, and the solid CO_2 particles are produced by Joule-Thomson effect. Thus in the visualization section, the CO_2 fluid is in solid-gas two phase flow. In this study, aperture opening ratio of the expansion valve can be controlled based on the needle valve, and therefore the flow rate is controlled. The heating section is heated under the constant heat flux condition by the sheath heater and solid phase CO_2 sublimes in the heating tube. CO_2 flow is controlled in the gas phase condition at orifice flow meter and it is then discharged outside the system. Orifice diameter is 4mm. The pressure is measured in the each test section. The temperature is measured only at the inlet of the visualization section and the outlet of the heating section to know the condition of CO_2 fluid. The absolute pressures are measured in the following points by pressure transmitters: expansion valve inlet P_1, visualization section inlet P_2, orifice inlet P_3 and outlet P_4. Total flow rate is measured based on P_3 and P_4. There is ±1500 Pa measurement error for pressure gauge. In addition, a data acquisition system is used, which can achieve real-time data measurement and acquisition with a sampling time period of 1.0 second.

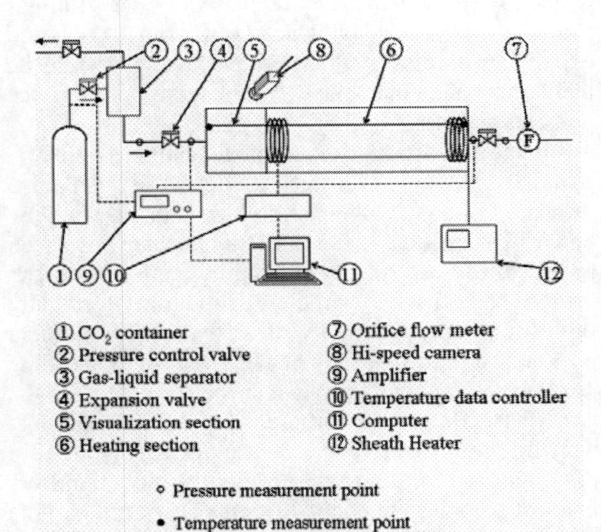

FIGURE 2. Schematic of experiment set up to investigate the basic characteristics of liquid CO_2 expanding into solid-gas two phase fluid flow.

Visualization observation is achieved by the high-speed video camera.

Although the loop is open one, both the visualization and heating section are set long enough to observe the solid-gas flows without CO_2 sublimation and with CO_2 sublimation, respectively. Tube diameter of the visualization section and heating section is 0.04 m. Pipe length is 0.59 m for the visualization section, 1.34 m for the heating section. Pipe wall thickness is 2.5 mm. Because of low temperature characteristics in this study, vacuum thermal insulation structure of double cylinder is installed for the visualization and heating section to reduce heat transfer between the piping and the ambient.

Total flow rate in the test section is measured by orifice flow meter. Visualization measurement is carried out in the case that the total flow is fixed.

RESULTS AND DISCUSSION

In the experimental investigation, focus is put on the feasibility of liquid CO_2 expanding into solid-gas fluid flow by the expansion valve. The experiment test is carried out at the inlet pressure 1.0 MPa and the inlet temperature -45℃ of the expansion valve. The opening of the expansion valve can be adjusted to control CO_2 flow rate. The pictures of the solid-gas two phase fluids in the visualization test are shown in Fig. 3. Because the interest is only focused on the feasibility of expanding liquid CO_2 into solid-gas two phase fluid flows by the expansion valve, the results of the expansion process are given only for the two flow rates in the paper: 2.50 m/s and 0.78 m/s. From Fig. 3, it is seen that by the expansion valve, the solid-gas fluid flow is successfully achieved from CO_2 liquid expanding process by the needle valve. It should be mentioned here that black colors represents dry ice particles and white one CO_2 gas phase.

In the visualization experiment, particle velocity, particle density and particle behaviors are also measured using the high-speed video camera. The flow behavior is observed, and particle diameter and speed are obtained by averaging 100 representative particles.

It is seen that the particle distribution is generally uniform at a flow velocity of 2.50 m/s. The particle size is also approximately same. Because of the flow complexity, laminar flow cannot be imagined for the two phase flow. The turbulent behavior in the test section may contribute to the uniform particle distribution. From the visualization results, diameters of most particles are about 1.0 mm. The mean particle size is measured to be 1.023 mm. When the flow velocity is about 0.78 m/s, as shown in Fig. 3, it is seen that there is a sedimentation phenomena and also large particles is seen. From the photographs taken at the flow velocity of 0.78 m/s, the sedimentations are seen by the collision of the CO_2 solid particles. In the case that total flow velocity is fast, the phenomenon that the particle deposited in the test section is not observed. When total flow velocity is small, as seen in Fig. 3, there is the sedimentation by the collision of the CO_2 particles. Based on the result, particle production in the test section was achieved by needle valve in this experiment. It may be concluded that the particle becomes easier to adhere on the wall surface, when flow velocity decreases.

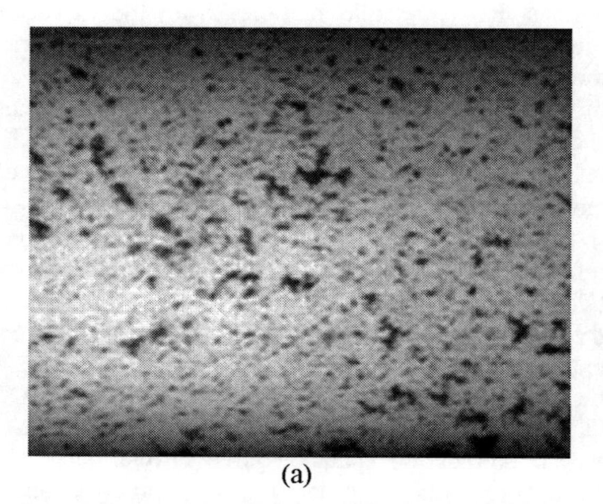

(a)

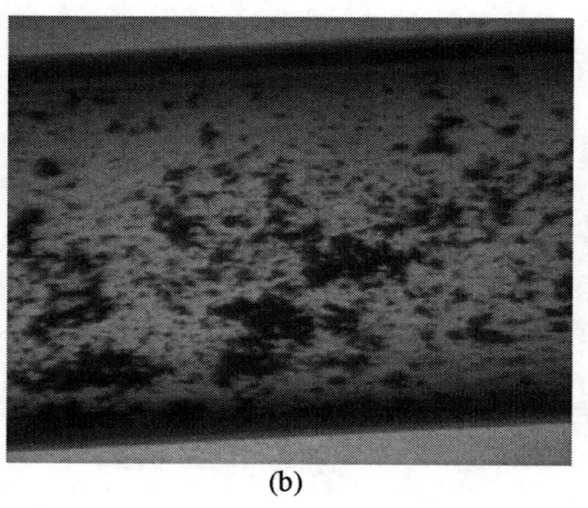

(b)

FIGURE 3. Pictures of CO_2 solid-gas two phase fluid flow achieved from liquid CO_2 expansion throughout CO2 triple point. (a) CO_2 solid-gas flow at a CO_2 flow velocity measured at 2.50 m/s; (b) CO_2 solid-gas flow at a CO_2 flow velocity measured at 0.78 m/s.

Here, it should be mentioned that CO_2 particle size, density and behavior etc. strongly depend on the type of the expansion valve used, CO_2 pressure and temperature, and the opening size of the expansion valve. In the present study, only focuses on the feasibility of liquid CO_2 expansion by a needle valve. The above influences on the liquid CO_2 expansion are not intended to be included in the present paper. The influences need to have a much more detailed and closer investigation in the future.

Application of The CO_2 Solid-Gas Fluid to Build Up A New CO_2 Heat Pump

Here, an application example is given in CO_2 heat pump system in order to build up a new system, which utilize CO_2 solid-gas fluid flow and can achieve a low-temperature refrigeration below -56.6℃. A schematic diagram of a cascaded CO_2 heat pump system has been illustrated in Fig. 4. Low condensing temperature is needed for having dry ice out in expanding process. So this system is designed from two heat pump cycles, which is here called low temperature side (LTS) and high temperature side (HTS), respectively. Brine cycle connects the evaporator of HTS with the gas cooler of LTS. The cycle of HTS is designated for cooling the brine and the cycle of LTS is used for low temperature (below -56.6℃) refrigeration with CO_2 gas-liquid two phase flow. Operating the heat pump of HTS can cool the brine, and the brine can cool CO_2 in the gas cooler of LTS enough to obtain dry ice through the expanding valve.

In HTS, CO_2 gas compressed by the compressor is condensed into liquid CO_2 through two heat exchangers, with hot water (80℃) and cool water (30℃). Then liquid CO_2 is expanded to low pressure and low temperature by the expanding valve. The cold CO_2 has a heat exchange with the brine and changed to the CO_2 gas.

In LTS, CO_2 gas compressed by the compressor is condensed to liquid CO_2 by heat exchanging with the hot water, cool water and the brine (-20℃), respectively. Then through the expanding process, dry ice-gas two phase fluid is realized in the test section. The shape in the test section is the horizontal circular straight pipe whose length, diameter and inner diameter are 5000 mm, 50 mm and 45 mm respectively. The heater is twisted around the pipe as the thermal source of the LTS. The opening of expanding valve can be adjusted between 0.00 mm and 30.00 mm. The amount of heating power at the test section can be controlled by a volt slider. CO_2 temperatures and pressures are measured in this study.

Here behavior characteristic of the cascade system is described. Figure 5 shows the variations of measured CO_2 pressures of HTS and LTS with time. Figure 6 shows the variations of measured CO_2 temperatures of HTS and LTS with time. The condition is that heating power input is 1000 W and opening of expansion valve is 15 mm.

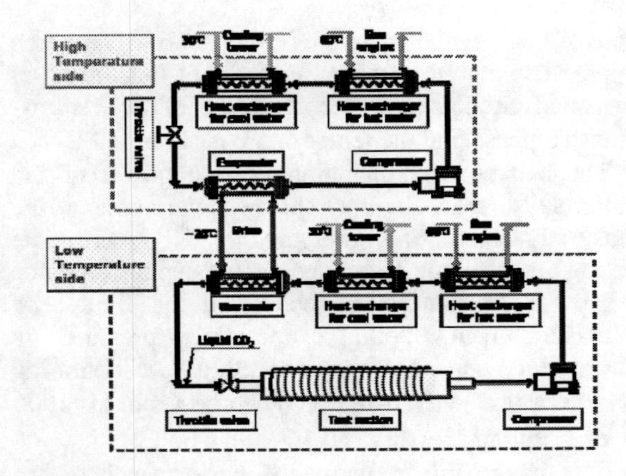

FIGURE 4. (Color online) Schematic of a new CO_2 heat pump system utilizing CO_2 solid-gas two phase fluid.

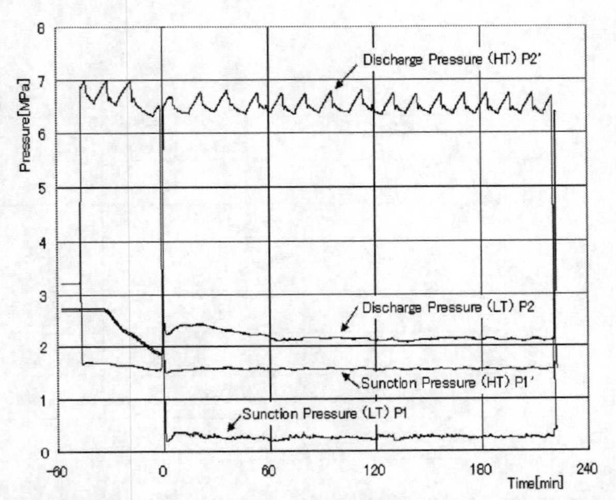

FIGURE 5. (Color online) Variations of measured CO_2 pressures of HTS and LTS with time.

First, the brine is cooled to -20℃ by working of HTS and after that LTS starts working. 180 minutes later from starting working HTS and LTS, pressure and temperature become steady. After that, average pressure and temperature during 26 minutes are adopted. In the cycle of HTS, the discharge pressure of compressor is 6.60 MPa, discharge temperature is 136℃, condensation temperature is 25.1℃, the inlet pressure of compressor is 1.65 MPa and inlet temperature is -15℃. It should be mentioned here that the oscillation of the discharge is due to the automatic valve opening and closing in cooling tower side and therefore, temperature change of cooling water in the heat exchanger for cool water. In the cycle of LTS, the discharge pressure of compressor is 2.20 MPa, discharge temperature is 136℃, condensation temperature is -17℃, the inlet pressure of compressor is 0.36 MPa and inlet temperature is -30℃. In the cycle of LTS, evaporating pressure is 0.36 MPa, which is under triple point pressure. Because of this, inside of horizontal circular pipe, the condition of CO_2 is gas-solid two phase and the temperature is about -62℃.

The behavior by changing heating power of the heater is also seen. Figure 7 show the variations of the measured evaporating pressure of LTS with the heating power input. Here, opening of expansion valve is kept at 15 mm. It is seen from the figure, by decreasing heat input, evaporating pressure is decreased. When heat input is 2000W, evaporating pressure is 0.47 MPa (-58℃). When heat input is 1000 W, evaporating pressure is 0.36 MPa (-62℃).

It is interesting to know the performance of the built heat pump system. The performance, COP, is estimated based on the measured data as follows,

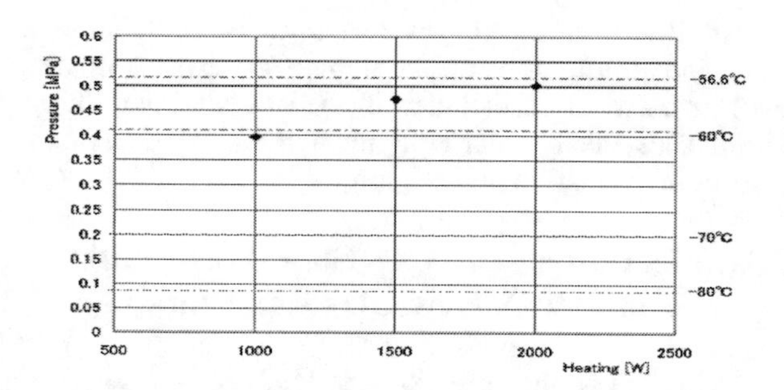

FIGURE 7. (Color online) Variations of pressures with heating power.

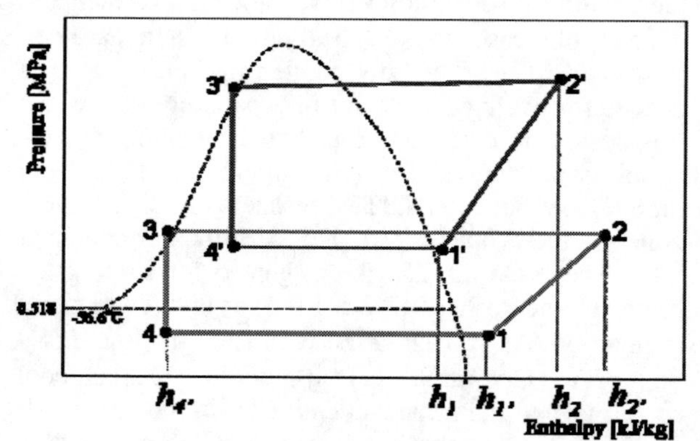

FIGURE 8. (Color online) P-h diagram for the CO_2 heat pump system.

$$\mathrm{COP}_{\mathrm{system}} = \frac{Q_{cool}}{W_h + W_l} = \frac{h_1 - h_4}{(h_{2'} - h_{1'}) + (h_2 - h_1)} \quad (1)$$

$$\mathrm{COP}_{\mathrm{LT}} = \frac{Q_{cool}}{W_l} = \frac{h_1 - h_4}{h_2 - h_1} \quad (2)$$

Where W_h, W_l are compressor power consumption of HTS and LTS, respectively. Q_{cool} is refrigeration capacity of LTS. The expansion loss is assumed to be ignored. The thermo-physical property of PROPATH 12.1 is used to calculate the enthalpy values of CO_2. Based on the measured data and calculations, the system performance is obtained at COP_{system}=1.14 and COP_{LT}=2.00. The obtained values are not high enough to attract industrial focus, mainly because the present study and the built CO_2 heat pump system are focused on the feasibility of the refrigeration method using CO_2 solid-gas two phase fluid. In the future, the more studies may be carried out in order to optimize the system and increase the system efficiency. But the above values may give us an image of the future application feasibility and potentials.

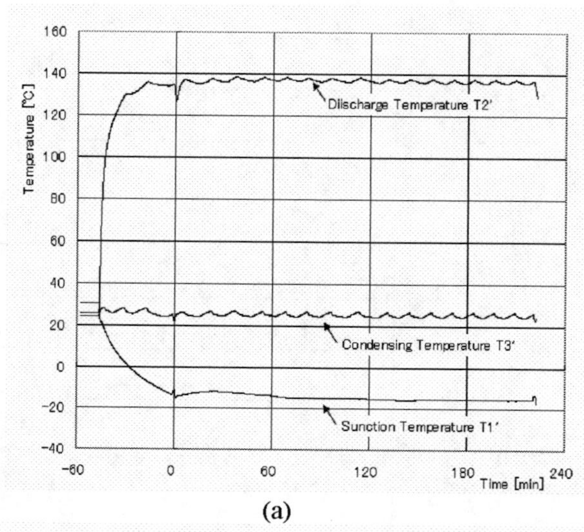

(a)

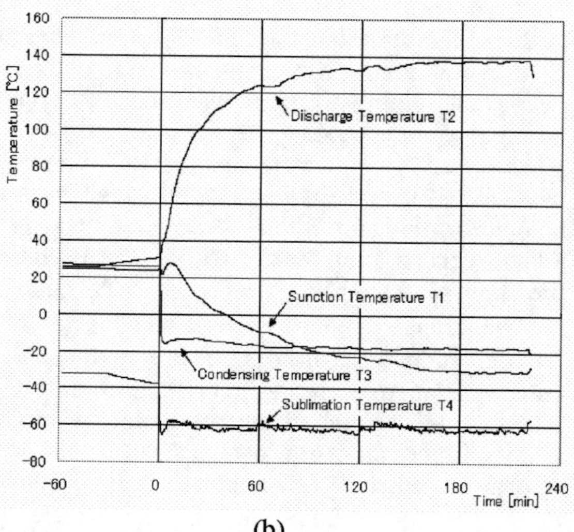

(b)

FIGURE 6. Variations of measured temperatures with time. (a) HTS; (b)LTS.

CONCLUSION

In the present study, a new CO_2 heat pump system is introduced and investigated experimentally, which can utilize the CO_2 solid-gas fluid flow and achieve a refrigeration below -56.6℃. Regarding the new system, basic characteristics of CO_2 liquid expanding into solid-gas two phase flow are studied and feasibility of the two phase fluid flow utilized in CO_2 heat pump is verified. In the future, the more detailed investigations are needed not only in studying the process of CO_2 liquid expanding into solid-gas fluid, but also in enhancing CO_2 heat pump system performance.

ACKNOWLEDGMENTS

This study was supported by the Academic Frontier Research Project on "Next Generation Zero-Emission Energy Conversion System" of Ministry of Education, Culture, Sports, Science and Technology, Japan.

REFERENCES

1. G. Lorentzen, Trans-critical Vapour Compression Cycle Device, *International Patent Publication* WO 90/07683, 1990.
2. G. Lorentzen and J. Pettersen, New Possibilities for Non-CFC Refrigeration, *IIR International Symposium on Refrigeration, Energy and Environment*, Trondheim, Norway, 1992, pp. 147-163.
3. K. Hashimoto and M. Saikawa, Trend Report of Development of Residential CO_2 Heat Pump Hot Water Heater in Japan, *JSME 9th Power Energy Tech. Symp*, 2004, pp. 425-430.
4. M. Saikawa, Development of Home CO_2 Heat Pump Hot Water Supplying Apparatus, *Science of Machine* **56**, 446-451 (2004).
5. M. H. Kim, J. Pettersen and C. W. Bullard, Fundamental Process and System Design Issues in CO_2 Vapor Compression Systems, *Progress in Energy and Combustion Science* **30**, 119-174 (2004).
6. S. M. Liao and T. S. Zhao, An Experimental Investigation of Convection Heat Transfer to Supercritical Carbon Dioxide in Miniature Tubes, *Int. J. Heat Mass Transfer* **45**, 5025-5034 (2002).
7. X. R. Zhang and H. Yamaguchi, Forced Convection Heat Transfer of Supercritical Carbon Dioxide in a Horizontal Circular Tube, *J. Supercritical Fluids* **41**, 412-420 (2007).
8. A. Hafner, Compact Heat Exchangers for Mobile CO_2 Systems, *IIR-5th Gustav Lorentzen Conference on Natural Working Fluids*, Guangzhou, China, September 17-20, 2002, pp. 177-184.
9. X. R. Zhang, H. Yamaguchi, K. Fujima, M. Enomoto and N. Sawada, A Feasibility Study of CO_2-based Rankine Cycle Powered by Solar Energy, *JSME Int. J., Series. B (Fluids and Thermal Engineering)* **48**, 540-547 (2005).
10. H. Yamaguchi, X. R. Zhang, K. Fujima, M. Enomoto and N. Sawada, A Solar Energy Powered Rankine Cycle Using Supercritical Carbon Dioxide, *Appl. Thermal Eng.* **26**, 2345-2354 (2006).
11. P. Nekså, H. Rekstad, G. R. Zakeri and P. A. Schiefloe, CO_2-heat Pump Water Heater: Characteristics, System Design and Experimental Results, *Int. J. Refrigeration* **21**, 172-179 (1998).
12. P. Nekså, CO_2 Heat Pump Systems, *Int. J. Refrigeration* **25**, 421-427 (2002).
13. J. Stene, Residential CO_2 Heat Pump System for Combined Space Heating and Hot Water Heating, *Int. J. Refrigeration* **28**, 1259-1265 (2005).

PART V

CROSS DISCIPLINARY PHYSICS

Shape Deformation of Vesicle Coupled with Phase Separation

M. Yanagisawa, M. Imai, and T. Taniguchi*

Department of Physics, Ochanomizu University
Otsuka, Bunkyo, Tokyo 112-8610, Japan
**Department of Polymer Science and Engineering, Yamagata University*
Yonezawa, Yamagata 992-8510, Japan

Abstract. In the presence of the osmotic pressure difference between inside and outside of a vesicle, spherical vesicles deform to various shapes, such as prolate, discocyte, starfish and so on, depending on an excess area and an area difference. When we couple the shape deformation with the phase separation, the heterogeneity in membranes brings new deformation pathways, shape convergence and directional budding. We discuss the shape deformation pathways in terms of the membrane elasticity model for multi-component vesicles.

Keywords: Budding, Osmotic pressure, Phase separation, Shape deformation, Vesicle
PACS: 87.14.Cc, 87.15.Zg, 87.16.dt

INTRODUCTION

One of the most fascinating properties of lipid membranes is that the membranes easily deform their shapes according to the internal and external circumstances, which brings unique functionalities of cell membranes. The basic shape deformation patterns of model vesicles can be integrated into a systematic diagram shown in Fig. 1. In this context, there are extensive studies on the deformation of lipid membranes to understand what governs the vesicle shape [1]. About one decade ago, Seifert and Lipowsky revealed the shape deformations based on membrane elasticity energy model using two key parameters, an excess area and an area difference. The excess area is the area-to-volume ratio and measured by a dimensionless parameter, ξ, defined by $\xi \equiv R_s / R_v - 1$, where R_s and R_v are the radii of spheres with the same area and volume, respectively. The area difference is given by $\Delta A = A_{out} - A_{in}$ where A_{out} and A_{in} are areas of outer and inner leaflets, respectively and the intrinsic area difference is given by $\Delta A_0 = (N_{out} - N_{in}) a_0$ where a_0 is the cross area of a lipid and N_{out} and N_{in} are numbers of lipids in outer and inner leaflets, respectively. They succeeded to map the various morphologies of the vesicle onto a simple phase diagram by optimizing the free energy F_{ADE} for a given area A:

$$F_{ADE} = \kappa\left[\frac{1}{2}\oint dA(2H)^2 + \frac{\alpha\pi}{2Ad^2}(\Delta A - \Delta A_0)^2\right], \quad (1)$$

where κ is the bending rigidity of membrane, H is the mean curvature, d is the distance between the two mono-layers and α is the numerical constant balancing the bending energy and the area difference energy (ADE). The complicated morphologies such as rackets, boomerangs and starfish are also mapped on the phase diagram based on the ADE model [2,3].

Another interesting feature is that the biomembranes are a multi-component system composed of proteins, polysaccharides, phospholipids, cholesterols and so on. These constituents form heterogeneities on the membranes, which plays important roles in the biological functionalities, so-called lipid raft model. The origin of the raft formation is believed to be a phase separation of lipid components. For example, it has been shown that the model vesicles composed of saturated phospholipids, unsaturated phospholipids and cholesterols exhibit lateral phase separation between L_o (liquid-ordered) phase and L_d (liquid-disordered) phase below a miscibility temperature T_{mix} [4]. In this case the domain boundary energy governs the total vesicle energy and leads to the domain coarsening and the vesicle deformation (budding).

In this study we deal with the shape deformation kinetics of the multi-component vesicle under constant osmotic pressure difference. We investigate the shape

CP982, *Complex Systems, 5th International Workshop on Complex Systems*
edited by M. Tokuyama, I. Oppenheim, and H. Nishiyama

transitions systematically and interpret them based on the ADE model combined with the phase separation.

EXPERIMENTS AND ANALYSIS

Commercial Reagents

1,2-dipalmitoyl-sn-glycero-3-phosphocholine (DPPC) (> 99 % purity) and 1, 2-dioleoyl-sn-glycero-3-phosphocholine (DOPC) (> 99 % purity) were obtained in a powder form from Avanti Polar Lipid, Inc. (Alabaster, AL). Cholesterol (Chol) (> 99 % purity) and D-sorbitol was purchased from Sigma-Aldrich Co. (St. Louis, Mo). All lipids were used without further purification. Mother solutions of lipids were stored in chloroform at -20°C until use. Texas Red 1,2-dihexanoyl-sn-glycero-3-phosphatidylethanolamine (TR-DHPE) and Perylen were obtained from Molecular Probes (Eugene, OR). TR-DHPE is localized in a liquid disordered (DOPC-rich) phase and shows a red color, whereas Perylene partitions preferentially into a liquid ordered (DPPC-and-Chol-rich) phase and shows a blue color.

Formation of Giant Vesicles

In this work, we fixed the ratio between the components comprising the vesicles at DPPC/DOPC/Chol=4/4/2(mole ratio). Giant vesicles (GVs) were prepared by the gentle hydration method. At first we dissolved the prescribed amounts of lipids, DPPC, DOPC and Chol in 10 μl of chloroform (10mM). In order to dye the domains and matrix, we added TR-DHPE and Perylene at the ratio of 0.8/10 and 1/500 (dye/lipid), respectively. The solvent was evaporated in a stream of nitrogen gas and the obtained lipid film was then kept under vacuum for one night to remove the remaining solvent completely. The dried lipid film was pre-warmed at 60 °C, and then we added 30 μl of pure water of 60 °C. After the 1 minute prehydration, the sample was hydrated with 970 μl of pure water of 60°C. During the hydration process, the lipid films spontaneously form GVs with diameters of 10-150 μm.

Observation of Giant Vesicle Deformation Using Fluorescence Microscope

The vesicle suspension was put on a glass plate with a silicon rubber spacer having the thickness of 0.5mm, and then sealed with a cover glass immediately. This sample cell was set on the temperature control stage (Carl Zeiss) with an accuracy of ±0.2 °C. To avoid the domain formation before the observation, we paid a special attention to keep the sample temperature above the melting temperature of DPPC (T_m: 41°C). First we kept the temperature at 60 °C in the homogeneous one phase region and then added the 1 mM of salt (D-sorbitol) to the vesicle suspension. We monitored the vesicle deformation process induced by the addition of the salt and at prescribed shape we dropped the temperature from 60 °C to 24 °C in the coexistence phase region. The shape deformation coupled with the phase separation was followed by an inverted conformal microscope (Carl Zeiss, LSM 5) with either the laser scanning mode or the Hg lamp mode with a CCD camera (Carl Zeiss, Axio Cam). To avoid the photo-oxidation, we minimized the exposure to light.

RESULTS AND DISCUSSIONS

Shape Deformation of Vesicle Induced by Osmotic Pressure Difference

According to the phase diagram proposed by Seifert and Lipowsky, we can control the shape of vesicle by changing the excess area and the area difference. The intrinsic area difference is determined by the spontaneous vesiculation process and keeps constant during the deformation process due to the extremely slow lipid flip-flop rate. On the other hand, the excess area can be changed by applying an osmotic pressure difference between the inside and outside of the vesicle. Under the constant osmotic pressure difference, ξ increases with elapse of time. Figure 1 shows a parade of the morphology transitions induced by the addition of sorbitol (1 mM) outside of GVs. Such a series of shape deformations induced by the osmotic pressure difference were first demonstrated by Hotani et al [5]. We revisited the parade of the shape deformation and rearranged the tree of the shape deformation based on the Seifert free energy landscape as shown in Fig. 1. By addition of the salt, GVs start to show fluctuations around the spherical shape using the obtained excess area [6]. When the excess area is beyond a threshold value, the spherical shape transforms to the prolate/discocyte probably depending on the intrinsic area difference [7]. By further increase of the excess area, the prolate/discocyte shapes bifurcate to stomatocyte/starfish and tube/pear, respectively, which is consistent with the ADE model. It should be noted that we observed no jump pathway to another branch, indicating the constant intrinsic area difference ΔA_0 during the deformation.

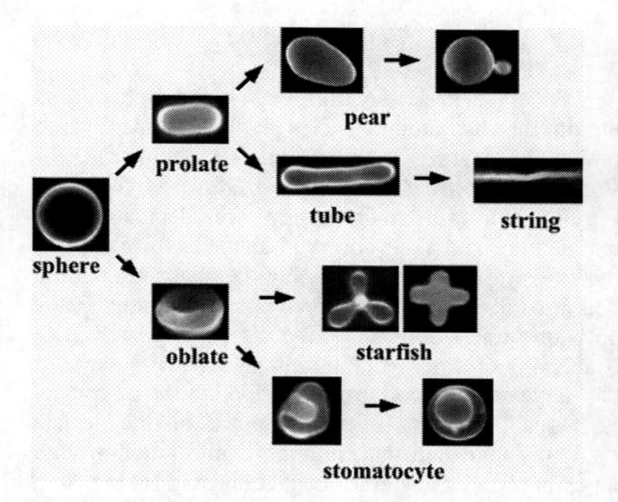

FIGURE 1. Shape deformation pathways of homogeneous vesicle under constant osmotic pressure difference.

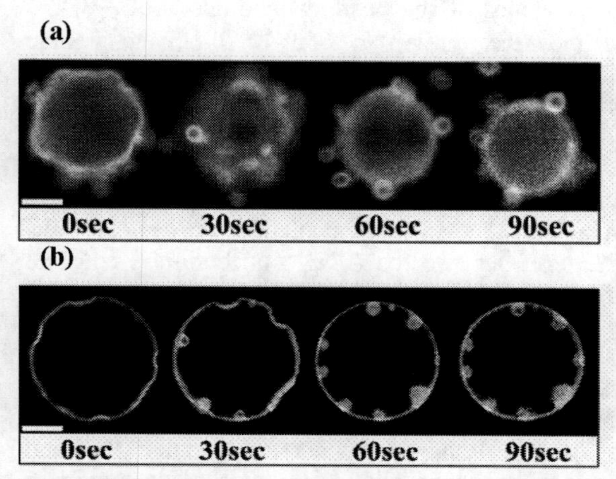

FIGURE 2. Time evolution of the shape deformation induced by phase separation in the sphere stage, (a) phase separation just after addition of salt and (b) phase separation just before sphere to prolate/discocyte transition.

Shape Deformation of Multi-component Vesicle

It is well known that the mixture of phospholipids having different melting temperatures shows a phase separation below T_{mix} [4]. The domains on the vesicle generated by the phase separation become unstable at a certain size where the line energy of the domain boundary surpasses the bending energy and then undergo the budding [8]. Thus phase separation of the multi-component vesicle brings another kind of the shape deformation. In this study we deal with the deformation pathways of the multi-component vesicle under the constant osmotic pressure difference. Here we classify the shape deformation into two categories, the phase separation in spherical vesicle (sphere stage) and polygonal (prolate, discocyte, stomatocyte, starfish, tube and pear) vesicle (polygon stage).

Sphere Stage

Figure 2 shows time evolution of the shape deformation of GVs induced by the phase separation in the sphere stage, where we set the origin of time at the onset time of the budding. First we prepared spherical vesicle with ξ~0 in the homogeneous one-phase region and then added the salt (sorbitol). When we decreased the temperature to 24 °C (< T_{mix}) just after the addition of salt, small domains began to appear and showed coarsening due to diffusion and coalescence. At a certain domain size, domains budded toward outside of the vesicle using the excess area to decrease the line energy (Fig. 2(a)). With elapse of time the budding domains grew to small spherical capsules on the mother matrix vesicle and sometimes pinched off from the matrix (complete budding). After the complete budding the mother homogeneous vesicle started to deform following the standard deformation pathway shown in Fig. 1.

On the other hand, when we decreased the temperature just before the sphere to the prolate/discocyte transition, the domains budded toward inside of the vesicle (Fig. 2(b)). In this case, the domains formed small capsules inside the mother vesicle and the homogeneous mother vesicle containing small vesicles deformed following the standard pathway. Taking into account that the excess area of the vesicle is proportional to the waiting time, the budding direction may be determined by the excess area. Then we calculated the total energy of a vesicle having budding domains as a function of ξ.

To describe the shape of the phase separated vesicle we adopt the following multi-domain ADE model

$$F_{mADE} = \sum_i \tfrac{1}{2}\int dA_d^i \kappa_d (2H_d)^2 + \tfrac{1}{2}\int dA_m \kappa_m (2H_m)^2 + \sigma \sum_i \oint ds^i + \frac{\alpha\pi}{2Ad^2}(\Delta A - \Delta A_0)^2 \qquad (2)$$

where subscripts d and m denote domain and matrix, respectively, A_d^i is the area of the ith domain, σ is the line tension, and s^i is the length of the ith domain boundary.

The first term and the second term represent the bending energy of n domains and the matrix, respectively. Here we assume that all domains have cap-like shape of the same size and the matrix is represented as a perforated spherical film. The third term denotes the line energy of the domain boundary. The fourth term represents the ADE. Comparing typical values of the bending modulus κ of ~10^{-19} J and the line tension σ of ~10^{-11} N, the vesicle deformation is governed by the boundary energy of domains, which causes the domain coarsening and the budding. Under the conservation laws for total area and total domain area, we calculated the total energies for outside budding and inside budding vesicles as a function of ξ. In this calculation we neglected the ADE term because the ADE contribution is constant during the shape deformation. In Fig. 3 we compare the energies of outside budding and inside budding vesicles with n=20 domains (n: number of domains). First the total energies of both vesicles monotonically decrease with increase of ξ due to the decrease of the line energy. It is worthwhile to note that the total energy of a vesicle with n domains has a minimum when the domains bud completely and the minimum value is given by $8\pi\kappa_m + 8n\pi\kappa_d$. The excess area giving the complete budding of domains ξ^* is obtained by simple geometrical calculation and ξ^* for inside budding is always larger than that for outside budding, *i.e.* $\xi^*_{out} < \xi^*_{in}$. This means that in the region of $\xi < \xi^*_{out}$, the total energy for the outside budding vesicle is always smaller than that for the inside budding vesicle. After the complete budding the total energy of the vesicle increases with increase of ξ, because the mother vesicle starts to deform to prolate/discocyte shape. In Fig. 3 the lines after the complete budding are the energies for the prolate vesicles. The total energy crossover point is in $\xi^*_{out} < \xi < \xi^*_{in}$ region. Thus for the vesicle with small excess area, the domains tend to deform toward outside, whereas for the vesicle with large excess area, they start to deform inside of the vesicle. This theoretical prediction agrees with our experimental observation.

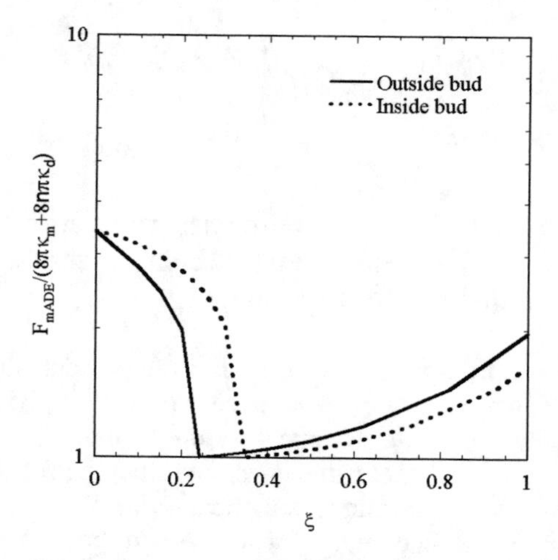

FIGURE 3. Total energies of outside and inside budding vesicles as a function of ξ. The number of domains n is 20.

Polygon Stage

When we keep the sample in the homogeneous region, the fluctuating vesicles show the transitions to prolate/discocyte and then tube/pear or stomatocyte/starfish with elapse of time as shown in Fig. 1. At each morphology we decreased the temperature and then observed shape deformation of the polygonal vesicle induced by the phase separation. A unique feature of the shape deformation is that each polygon vesicle deforms to the discocyte vesicle with coarsening of domains. Figure 4 shows an example of the shape deformation from the tripod to the discocyte. First small domains begin to appear on the vesicle randomly and then they migrate to the junction with repeating collisions and coalescences. Taking into account that the liquid-ordered domain has large bending modulus, the domains tend to move toward the flat region, *i.e.* the junction. Simultaneously, the tripod arms shrink and finally we obtain a discocyte vesicle with two large domains at both sides of the discocyte. Similar shape deformations to the discocyte triggered by the phase separation are observed for other polygon vesicles. Unfortunately at present we can not make clear the origin of the shape convergence toward the discocyte, although we propose the following two candidates to explain the observed shape deformation. One is that the multi-component vesicle shape is determined by optimization of the energy model expressed by eq. (2), which is essentially different from that for the homogeneous vesicle expressed by eq. (1). The dominant term for the multi-component vesicle is the line energy term, which may make the discocyte vesicle the most stable because the liquid ordered domains have larger bending modulus than the matrix. Another candidate is the area difference. According to the phase diagram proposed by Ziherl, the polygon vesicles transform to the discocyte vesicle by decreasing the area difference. Thus the phase separation may decrease the area difference between the outer and the inner monolayers.

After the shape deformation to the discocyte, the two large domains at both sides start to bud as shown in Fig. 5. Interestingly the budding shows two characteristic pathways. One is that the two domains bud toward outside of the vesicle and the other is that one domain buds toward inside whereas another

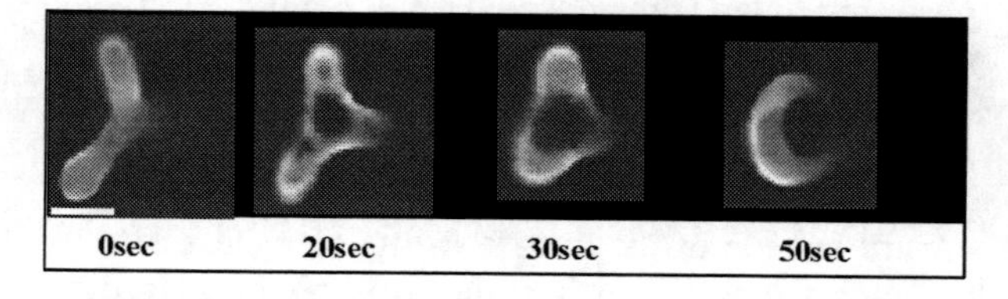

FIGURE 4. Shape deformation of tripod vesicle toward discocyte induced by phase separation.

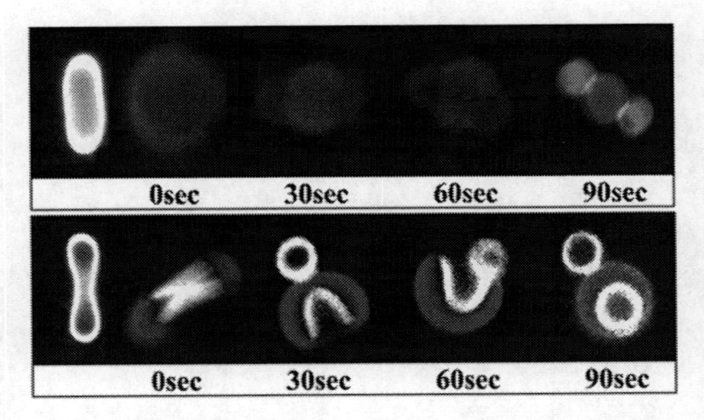

FIGURE 5. Two types of budding of discocyte vesicle with two domains.

domain buds toward outside. From the geometrical constraint, both domains can not bud toward inside at this total area fraction of domains, $\phi_d \approx 0.6$. This behavior is also interpreted by the multi-domain ADE model combined with the excess area constraint. Taking into that the two complete budding vesicles (out/out and in/out) have the same energy of $8\pi\kappa_m + 16\pi\kappa_d$, the vesicle having small excess area prefers out/out budding whereas the vesicle having large excess area prefers in/out budding. The theoretical calculation shows the crossover excess area $\xi_c^{\mathrm{the}} \approx 0.3$. On the other hand, the experimentally obtained cross over point is $\xi_c^{\mathrm{exp}} \approx 0.3$, which agrees well with the theoretical prediction. After the complete budding, the mother vesicle started to deform following the homogeneous deformation pathways shown in Fig. 1.

Conclusion

By coupling the shape deformation induced by the osmotic pressure difference with the phase separation, the multi-component vesicle show unique shape transitions. Using the excess area introduced by the osmotic pressure difference, the vesicle transforms its shape to optimize the total elastic energy consisting of bending energy and the area difference energy under the geometrical constraints. The shape deformation is modified by the phase separation. In the domain coarsening stage, the bifurcations to polygonal shape vesicles converge to the discocyte branch. On the other hand, in the budding stage, the budding direction is determined by the excess area. If the vesicle has small excess area the domains bud toward outside, whereas domains with large excess area show the inside buding. The multi-component ADE model with the geometrical constraints describes the observed shape transition.

REFERENCES

1. U. Seifert, Adv. Phys. **46**, 13-137 (1997).
2. W. Wintz, H.-G. Döbereiner and U. Seifert, Europhys. Lett. **33**, 403-408 (1996).
3. P. Ziherl and S. Svetina, Europhys. Lett. 70, 690-696 (2005).
4. S. L. Veatch and S.L. Keller, Biochim. Biophys. Acta **1746**, 172-185 (2005).
5. H. Hotani, J. Mol. Biol. **178**, 113-120 (1984).
6. S.T. Milner and S.A. Safran, Phys. Rev. A **36**, 4371-4379 (1987).
7. T. Taniguchi, K. Kawasaki, D. Andelman and T. Kawakatsu, J. Phys. II France **4**, 133-1362 (1994).
8. R. Lipowsky and R. Dimova, J. Phys. Condens. Matter **15**, S31-S45 (2003).

Anomalous Diffusion in Microrheology: A Comparative Study

I. Santamaría-Holek

Facultad de Ciencias, Universidad Nacional Autónoma de México. Circuito exterior de Ciudad Universitaria. 04510, D. F., México.

Abstract. We present a comparative study on two theoretical descriptions of microrheological experiments. Using a generalized Langevin equation (GLE), we analyze the origin of the power-law behavior of the main properties of a viscoelastic medium. Then, we discuss the equivalence of the GLE with a generalized Fokker-Planck equation (GFPE), and how more general GFPE's can be derived from a thermo-kinetic formalism. These complementary theories lead to a justification for the physical nature of the Hurst exponent of fractional kinetics. Theory is compared with experiments.

Keywords: Anomalous diffusion, Brownian motion, Microrheology
PACS: 05.70.Ln; 05.10.Gg; 87.17.Aa

INTRODUCTION

In the last years, microrheology has become one of the most important experimental techniques in soft condensed matter [1, 2, 3, 4]. It is a powerful technique to determine the viscoelastic properties of complex fluids at time and length scales complementary to those of classical rheological methods [2, 3]. In these complex fluids, ranging from polymer solutions and colloidal suspensions to the intracellular medium, the presence of elastic forces and molecular motors make anomalous the dynamics of the Brownian particles [4, 5, 6]. Hence, anomalous diffusion becomes a central question that must be carefully analyzed due to the presence of confinement and finite-size effects related with particle dimensions. The dynamics of these particles can be described through both, non-Markovian Langevin and Fokker-Planck equations [7, 8, 9, 10, 11]. Here, we use and compare two different formalisms of anomalous diffusion that in the linear force case are equivalent [4].

The article is organized as follows. In Sec. **II** we analyze the generalized Langevin equation and its equivalence with a non-Markovian Fokker-Planck equation. In Sec. **III** we introduce the thermokinetic description, show its equivalence with the GLE formalism in the linear force case, and how it can be generalized to the case of non-linear forces. Sec. **IV** is devoted to the conclusions.

THE LANGEVIN APPROACH TO MICRORHEOLOGY

Consider a test Brownian particle of mass m and radius a with position $\mathbf{x}(t)$ and velocity $\mathbf{u}(t) = d\mathbf{x}(t)/dt$. The particle performs its Brownian motion through a viscoelastic medium which can be made up by a solution of polymers or a suspension of particles at sufficiently high concentration. In these conditions, the motion of the particle could be restricted to a small volume and, consequently, its dynamics may manifest confinement and finite-size effects modifying the anomalous behavior of the mean square displacement (MSD). This last quantity is very important because the viscoelastic properties of the medium can be inferred from it through the Stokes-Einstein relation [4].

In a first approximation, the heat bath can be assimilated as an effective medium that interacts with the particle by means of elastic forces. In first approximation, these forces may be represented through a harmonic force $\mathbf{F}_h = -\omega_m^2\mathbf{x}$, in which ω_m is a characteristic frequency [4]. Then the dynamics of the particle can be described by means of the generalized Langevin equation

$$\frac{d\mathbf{u}(t)}{dt} = -\omega_m^2\mathbf{x}(t) - \int_0^\infty \beta(t-\tau)\mathbf{u}(\tau)d\tau + F(t), \quad (1)$$

where $F(t)$ is a random force per unit mass and $\beta(t)$ is a memory function called the friction kernel [7]. In the Markovian case it takes the form: $\beta(t) = \beta_0\delta(t-\tau)$ with $\beta_0 = 6\pi\eta a/m$ the Stokes friction coefficient per mass unit, η is the viscosity of the solvent and $\delta(t-\tau)$ the Dirac delta function. Hence, in this case Eq. (1) recovers its usual phenomenological form [9].

To describe the diffusion of the particle at sufficiently long times, $t \gg \beta_0^{-1}$, two approximations can be followed.

a) The Overdamped Case. In this case it is assumed that the acceleration term of Eq. (1) can be neglected. As

CP982, *Complex Systems, 5th International Workshop on Complex Systems*
edited by M. Tokuyama, I. Oppenheim, and H. Nishiyama

a consequence the GLE takes the approximate form

$$\omega_m^2 \mathbf{x}(t) = -\int_0^\infty \beta(t-\tau)\mathbf{u}(\tau)d\tau + F(t). \qquad (2)$$

It is convenient to stress that in Eq. (2), the memory term involves the velocity of the particle.

b) Adiabatic Elimination. In this second approximation, one must first solve Eq. (1) in order to obtain

$$\mathbf{u}(t) = -\omega_m^2 \int_0^\infty \chi(t-\tau)\mathbf{x}(\tau)d\tau + F^*(t), \qquad (3)$$

where we have used Laplace transforms, assumed the initial velocity of the particle as equal to zero for simplicity, and defined the memory function $\chi(t)$ as the inverse Laplace transform $\chi(t) = \mathscr{L}^{-1}\left\{[s+\tilde{\beta}(s)]^{-1}\right\}$. This definition implies that $\chi(t)$ is the relaxation function of the velocity of the particle [7]. Moreover, we have defined the time-scaled random force $F^*(t)$ by

$$F^*(t) = \int_0^\infty \chi(t-\tau)F(t)d\tau. \qquad (4)$$

Using now the identity $\mathbf{u}(t) = d\mathbf{x}(t)/dt = \dot{\mathbf{x}}$, one then obtains the following generalized Langevin equation for the position vector of the particle

$$\dot{\mathbf{x}}(t) = -\omega_m^2 \int_0^\infty \chi(t-\tau)\mathbf{x}(\tau)d\tau + F^*(t). \qquad (5)$$

In the diffusion regime, $t \gg \beta_0^{-1}$, Eqs. (2) and (5) can be considered as the GLE's for the position of the particle and then use them to describe anomalous diffusion of a Brownian particle in a viscoelastic heat bath. They constitute two different models because Eq. (5) differs from (2) in the fact that it can be obtained through an adiabatic elimination of variables, whereas the approximated Eq. (2) has been obtained by neglecting the inertial term. Physically, the difference lies in the fact that the relaxation function of the particle determining the time dependence of the MSD, is different.

Quantitatively, these equations are not simply related one to each other, since the memory kernel and the random force are not equivalent due to the scaling introduced through $\chi(t)$. The fluctuation-dissipation theorem (FDT) associated with Eq. (2) is

$$\langle F(t)F(0)\rangle = \beta(t), \qquad (6)$$

where the bracket $\langle\rangle$ means the average over noise realizations. The corresponding FDT for the random force in Eq. (5) is given by the expression $\langle F^*(t)F^*(t')\rangle = \int_0^t d\tau \int_0^{t'} d\tau' \chi(t-\tau)\chi(t'-\tau')\beta(\tau-\tau')$, [9]. However, in the case when the friction kernel is proportional to a Dirac delta function, Eq. (6) defines a thermal (white) noise. After performing the adiabatic elimination of variables, one obtains the relation $\langle F^*(t)F^*(0)\rangle \sim exp(-\beta_0 t)$ for the time-scaled random force. This implies that the position $\mathbf{x}(t)$ of the particles satisfies a non-Markovian equation with a exponentially decaying time correlation of the random force, [9]. Only in the Markovian case Eqs. (2) and (5) are equal.

At certain time scales, anomalous diffusion is characterized by a power-law behavior of the MSD of the particle as a function of time. Within a Langevin description, this behavior is a consequence of the statistical properties of the random force, that is, of the heat bath as follows from the relations already obtained. In particular, by using Eqs. (2) and (6), it can be shown that anomalous diffusion is a consequence of a fractionary Gaussian noise (FGN) that satisfies the FDT [10]

$$\langle F(t)F(0)\rangle = (k_BT/m)\beta_0^2 2H(2H-1)|\tilde{t}|^{2H-2}, \qquad (7)$$

where the exponent must satisfy the relation: $1/2 \leq H \leq 1$, the factor β_0^2 gives the correct dimensions and we have introduced the dimensionless time $\tilde{t} = t_0^{-1}t$, with t_0 a characteristic time. In particular, t_0 can be taken as the characteristic diffusion time $\tau_D = a^2 D_0^{-1}$, with $D_0 = k_BT/(m\beta_0)$. Hence, using Eqs. (2) and (7) one may calculate the time dependence of the MSD by using the method of Laplace transforms. Here, it is convenient to write (2) in the dimensionless form

$$\tilde{\mathbf{x}}(\tilde{t}) = -\alpha \int_0^\infty \tilde{\beta}(\tilde{t}-\tilde{\tau})\tilde{\mathbf{u}}(\tilde{\tau})d\tilde{\tau} + \tilde{F}(\tilde{t}). \qquad (8)$$

where $\tilde{\beta}(\tilde{t}) = |\tilde{t}|^{2H-2}$ and $\tilde{F} = a^{-1}\omega_m^{-2}F$ are dimensionless quantities in accordance with Eqs. (7) and (6), and $\alpha = 2H(2H-1)\beta_0^2\omega_m^{-2}$. In writing (8) we have used $\tilde{t}$ and the dimensionless position vector $\tilde{\mathbf{x}} = a^{-1}\mathbf{x}$.

Now, by Laplace transforming Eq. (8), one may obtain the expression for the transformed velocity $\hat{\mathbf{u}}(s)$, which in turn can be substituted into the definition $s\hat{\mathbf{x}}(s) = \hat{\mathbf{u}}(s)$ with $\mathbf{x}(0) = 0$ for simplicity. After solving for $\hat{\mathbf{x}}(s)$ one obtains

$$\hat{\mathbf{x}}(s) = \frac{\alpha^{-1}\Gamma^{-1}(2H-1)s^{2H-2}}{1+\alpha^{-1}\Gamma^{-1}(2H-1)s^{2H-2}}\hat{F}(s). \qquad (9)$$

where $\Gamma(2H-2)s^{1-2H}$ and $\hat{F}(s)$ are the Laplace transforms of $\tilde{\beta}$ and $\tilde{F}$, respectively. The inverse Laplace transform of Eq. (9) is given in terms of the convolution of the noise $\tilde{F}$ and the Mittag-Leffler polynomial $z_*^{1-2H}E_{2-2H,2-2H}\left(-z_*^{2-2H}\right)$, with the scaled variable $z_* = [\alpha\Gamma(2H-1)]^{-1/(2-2H)}\tilde{z}$, [12].

The expression for the MSD can now be obtained by calculating $\langle x^2\rangle(t)$ in terms of the inverse Laplace transform of Eq. (9), using the FDT (7) and performing the corresponding integrals. The general expression involving Mittag-Leffler polynomials is complicated, however at times t satisfying $\beta_0^{-1} \ll t \sim t_0$, one obtains the following short-time power law behavior

$$\langle x^2 \rangle(t) \sim \left(k_B T \beta_0^2 / m\omega_m^4\right) (t/t_0)^{2-2H}, \tag{10}$$

where some constants have been dropped for the sake of simplicity.

In Figure 1, we compare the MSD given by Eq. (10) (dash-dotted lines) with experiments (symbols) of $0.965\mu m$ diameter latex microspheres imbedded in F-actin solutions of increasing concentrations, [3]. As expected, the agreement between theory and experiments is good for short times $\beta_0^{-1} \ll t \sim t_0$ with $t_0 \sim \tau_D$ playing the role of a crossover time. If we take $\beta_0 \sim 10^6 s^{-1}$ then $\tau_D \sim 10^{-2} s$ in typical situations.

The MSD can be used to obtain the microrheological properties of the medium by means of the generalized Stokes-Einstein relation [2]. From Eq. (1) in the zero inertia approximation, one may show that the complex shear modulus of the medium $G''(\omega)$ satisfies the relation

$$G''(\omega) \simeq \left(k_B T / m\omega \langle \hat{x}^2 \rangle(\omega)\right), \tag{11}$$

where ω is the frequency. Now, by taking the Fourier transform of Eq. (10) and substituting the result into (11), one obtains the scaling relation $G''(\omega) \sim (t_0\omega)^{2-2H}$.

From this analysis it follows that the power-law behavior of the MSD of the particles and the related complex shear modulus of the effective medium, are a consequence of the fractionary Gaussian noise of the GLE (1). Notice that Eq. (5) can also be used to describe anomalous diffusion, however this will be explained after Eq. (15), below.

The Generalized Fokker-Planck Equation. An equivalent approach to study anomalous diffusion in a viscoelastic bath can be performed by means of a generalized Fokker-Planck equation .

By using the Laplace transform technique, in Ref. [7] it has been shown that the GLE (1) is equivalent to the following non-Markovian Fokker-Planck equation containing time dependent coefficients

$$\frac{\partial}{\partial t} f + \frac{\partial f}{\partial \mathbf{x}} \cdot (\mathbf{u} f) - \tilde{\omega}^2(t)\mathbf{x} \cdot \frac{\partial f}{\partial \mathbf{u}} = \psi(t) \frac{\partial}{\partial \mathbf{u}} \cdot \frac{\partial f}{\partial \mathbf{x}} + \frac{\partial}{\partial \mathbf{u}} \cdot \tilde{\beta}(t) \left[\mathbf{u} f + \frac{k_B T}{m} \frac{\partial f}{\partial \mathbf{u}}\right]. \tag{12}$$

Here, $f(\mathbf{x}, \mathbf{x}_0; \mathbf{u}, \mathbf{u}_0, t)$ is the probability distribution function depending on the instantaneous position $\mathbf{r}$ and velocity $\mathbf{u}$ of the Brownian particle. $\mathbf{r}_0$ and $\mathbf{u}_0$ are the corresponding initial conditions. In this equation, memory effects are incorporated through the time dependent coefficients $\tilde{\beta}(t)$ and $\tilde{\omega}(t)$ and $\psi(t) = (k_B T / m\omega_m^2)\left[\tilde{\omega}^2(t) - \omega_m^2\right]$. They are given in terms of combinations of the relaxation functions $\chi_u(t) = m(3k_B T)^{-1} \langle \mathbf{x}(t) \cdot \mathbf{u}_0 \rangle$ and

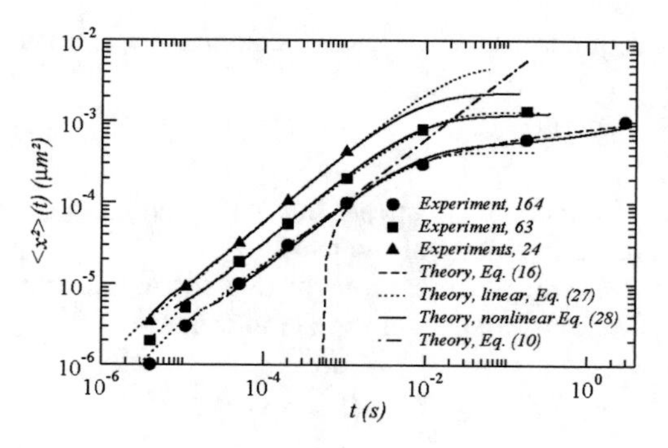

FIGURE 1. MSD vrs time. Comparison between theory (lines) and experiments with actin filament networks (symbols) (taken from Ref. [3]). The dashed line corresponds to the power law behavior valid for long times, Eq. (16) with $t_0 = 2 \cdot 10^{-3} s$, $\omega_m \sim 2.6 \cdot 10^2 s^{-1}$, and $b \sim 1 \cdot 10^{-2}$. The dotted lines correspond to a stretched exponential behavior Eq. (27), with $\omega_T^2/\omega_m^2 \sim 5.61 \cdot 10^{-4}$, $2\omega_m^2 B_2 t_0/\alpha\beta_0 \sim 35$ and $2\beta_0/\alpha t_0 \omega_m^2 = 3/4$. The solid lines correspond to the nonlinear model Eq. (28). The values of the parameters are indicated in the text. The dash-dotted line corresponds to Eq. (10) with $H = 5/8$. The numbers 24, 63 and 124 correspond to increasing concentrations of actin filaments, [3].

$\chi_x(t) = \langle \mathbf{x}_0^2 \rangle^{-1} \langle \mathbf{x}(t) \cdot \mathbf{x}_0 \rangle$, and their explicit form is given in Ref. [7]. It is worth to stress here that an equation similar to (12) has also been obtained by using projector operator techniques in [8].

The result expressed in Eq. (12) is important, because it implies that generalized Fokker-Planck equations incorporating memory effects through memory functions constitute models not simply related with Eq. (1), [4, 1]. Passing from one model to the other implies an approximation that imposes conditions on the rate of change of the fields involved in the description [4, 9, 11].

At times $t \gg \beta_0^{-1}$, from Eq. (12) one may derive a generalized Smoluchowski equation for the mass density $\rho(\mathbf{x}, t) = m \int f d\mathbf{u}$. To derive it, one may follow the adiabatic elimination of variables procedure by calculating the evolution equations for the first three moments of $f(\mathbf{x}, \mathbf{u}, t)$. Then, after imposing the condition of large times $t \gg \beta_0^{-1}$, a reduced description is obtained in terms of an evolution equation for ρ.

Averaging Eq. (12) over $\mathbf{u}$ assuming that the currents vanish at the boundaries, we obtain the continuity equation

$$\partial \rho / \partial t = -\nabla \cdot (\rho \mathbf{v}), \tag{13}$$

where $\mathbf{v}(\mathbf{x}, t) = m\rho^{-1} \int \mathbf{u} f d\mathbf{u}$ is the average velocity field.

At long times one obtains the following constitutive relation for the diffusion current [13]

$$\rho \mathbf{v} \simeq -\tilde{\beta}^{-1} \tilde{\omega}^2 \mathbf{x} \rho - \tilde{D} \nabla \rho, \tag{14}$$

where we have defined the effective diffusion coefficient $\tilde{D}(t) = (k_B T / m\omega_m^2) \tilde{\beta}^{-1} \tilde{\omega}^2$ and used the definition of

$\psi(t)$. Substitution of Eq. (14) into (13) leads to

$$\partial\rho/\partial t = \tilde{D}(t)\left[\nabla^2\rho - (m\omega_m^2/k_BT)\nabla\cdot(\rho\mathbf{x})\right], \quad (15)$$

which constitutes the generalized Smoluchowski equation (GSE) for ρ. An equation similar to (15) can also be obtained from Eq. (1) by using Laplace transforms. In Ref. [14], the authors found that $(m/k_BT)\tilde{D}(t)$ is proportional to $-d\ln|R(t)|/dt$, with $R(t) = \chi(t)/\chi(t_0)$ and $\chi(t) = \langle\mathbf{x}\cdot\mathbf{x}_0\rangle(t)$ the relaxation function of the position of the particle. If one considers the case of FGN, the relaxation function $R(t)$ is given in terms of a Mittag-Leffler polynomial leading to a complicated form for the effective diffusion coefficient $\tilde{D}(t)$, [10]. By evaluating numerically the obtained relation, it can be shown that at short times the function follows a power law behavior with an exponent equal to that calculated by simply taking the time derivative of Eq. (10). At larger times, a crossover arises in which the relaxation function can be assumed to decay as the power law $R(\tilde{t}) \sim \tilde{t}^{-b}$. Using the definition $\langle x^2\rangle = \int x^2\rho\, d\mathbf{x}$, Eq. (15) and assuming $\tilde{\beta}^{-1}\tilde{\omega}^2 = \beta_0^{-1}\omega_m^2\tilde{\gamma}(\tilde{t})$, with $\tilde{\gamma}$ dimensionless, one may calculate the following expression for the MSD

$$\langle x^2\rangle \sim 3(k_BT/m\omega_m^2)\left[1-(t/t_0)^{-2b}\right]. \quad (16)$$

At times $t \gg t_0$, Eq. (16) can be used to fit the experimental results, as shown by the dashed line in Figure **1**. See the discussion after Eq. (10).

This analysis shows that, in the linear force case the GLE with FGN offers a good explanation of the subdiffusion observed in microrheological experiments. A different approach in order to have deeper understanding of the physical mechanisms responsible for subdiffusion in a viscoelastic fluid will be presented in the following section.

THERMOKINETIC-DESCRIPTION OF ANOMALOUS DIFFUSION

The Fokker-Planck type Eqs. (12) and (15) for the probability distribution function, admit a different interpretation in terms of kinetic equations. This reinterpretation enriches and simplifies the description of anomalous diffusion, because allows one to formulate phenomenological models for the coefficients $\tilde{\beta}$ and $\tilde{\omega}$ which not do depend on the *a priori* election of the statistical properties of the random force of Eq. (1).

Fokker-Planck type kinetic equations for different physical phenomena can be obtained by using two general principles. The first one is the conservation of the probability, expressed by the equation

$$\partial f(\underline{\Gamma},t)/\partial t = -\nabla_\Gamma\cdot\left[f(\underline{\Gamma},t)\mathbf{V}_\gamma\right], \quad (17)$$

where $\underline{\Gamma}$ represents the set of variables necessary to describe the state of the system at a mesoscopic level. Here, $f(\underline{\Gamma},t)\mathbf{V}_\gamma$ represents a diffusion current in $\underline{\Gamma}$-space and ∇_Γ the corresponding gradient operator. The second principle is the so-called generalized Gibbs entropy postulate [6, 15]

$$\rho\Delta s(t) = -k_B\int f(\underline{\Gamma},t)\ln\left|f(\underline{\Gamma},t)/f_{leq}(\underline{\Gamma})\right| d\underline{\Gamma}, \quad (18)$$

where Δs is the difference of the specific entropy with respect to a local equilibrium reference state, characterized through the probability distribution $f_{leq}(\underline{\Gamma})$. The assumption of local equilibrium allows one to calculate $f_{leq}(\underline{\Gamma})$ by using the techniques of equilibrium statistical mechanics.

To obtain the evolution equation for $f(\underline{\Gamma},t)$, we may follow the rules of mesoscopic nonequilibrium thermodynamics (MNET) [6, 16]. Schematically, by taking the time derivative of Eq. (18) and inserting (17), after integrating by parts assuming the usual boundary conditions, one obtains the entropy production $\sigma(t)$

$$\sigma(t) = -\frac{m}{T}\int f(\underline{\Gamma},t)\mathbf{V}_\gamma\cdot\nabla_\Gamma\left[\frac{k_BT}{m}\ln\left|f/f_{leq}\right|\right] d\underline{\Gamma}. \quad (19)$$

In a similar way as in linear irreversible thermodynamics [15], we can assume linear relationships between the forces $(k_BT/m)\ln\left|f/f_{leq}(\underline{\Gamma})\right|$ and the conjugated currents $f\mathbf{V}_\gamma$. Using them, one finally arrives at the multivariate Fokker-Planck equation

$$\partial f/\partial t = \nabla_\Gamma\cdot\underline{\underline{\tilde{\xi}}}(t)\left[(k_BT/m)\nabla_\Gamma f - (f\mathbf{X})\right], \quad (20)$$

where the generalized force $\mathbf{X}(\underline{\Gamma})$ is defined in the usual form: $\mathbf{X} = -k_BT\nabla_\Gamma\ln f_{leq}$, [9]. Here, memory effects are incorporated through the time dependent Onsager coefficients introduced by the linear coupling between forces and fluxes, and contained in $\underline{\underline{\tilde{\xi}}}(t)$, [4, 13]. The tensorial character of $\underline{\underline{\tilde{\xi}}}(t)$ accounts for the anisotropy of the medium.

If we consider a Brownian particle under the action of a linear force and whose mesoscopic state is determined by its instantaneous velocity $\mathbf{u}$ and position $\mathbf{x}$, Eq. (20) takes the form (12). In this case, the local equilibrium distribution is $f_{leq} = f_0 exp\left[-(k_BT/2m)\left(\mathbf{u}^2+2\phi\right)\right]$, where $\phi(\mathbf{x})$ is the potential associated with the harmonic force, f_0 is a normalization factor and we have assumed that in Γ-space entropic forces are coupled to the currents independently from those arising from an energy potential, reflecting its different physical origin [13].

At larger times, we may assume that the state of the particle is only determined by its position vector, then the local equilibrium distribution is $\rho_{leq} = \rho_0 exp\left[-(k_BT/m)\phi(\mathbf{x})\right]$, where the potential $\phi(\mathbf{x})$ is arbitrary. By following the procedure indicated above, the following GSE can be derived

$$\partial\rho/\partial t = D(t)\left\{\nabla^2\rho - (m/k_BT)\nabla\cdot[\rho\mathbf{X}(\mathbf{x})]\right\}, \quad (21)$$

where we have assumed an isotropic medium when writing the scalar effective diffusion coefficient $D(t)$ and defined $\mathbf{X} = -\nabla\phi(\mathbf{x})$. Hence, the thermokinetic formalism allows one to derive generalized Fokker-Planck equations of the form (21) which contain nonlinear forces. No assumptions on the statistical nature of the random force have been done until now. This is an important difference with respect to the GLE description, because the method presented in the second section gives exact Fokker-Planck equations only in the case of linear forces.

During its motion through the viscoelastic medium, the Brownian particle induces perturbations on the velocity field of the host fluid surrounding it. These perturbations propagate and are reflected by the local boundaries which can be made up by a polymer network or another suspended particles. As a consequence, they modify the velocity field of the fluid around the particle in a later time, introducing memory effects. In this form, hydrodynamic interactions change the distribution of stresses used to calculate the force over the surface of the particle. In first approximation, this force is characterized by the effective mobility coefficient $\gamma(t) = (m/k_BT)D(t)$, in the form [4]

$$\gamma(t) = \beta_0^{-1}\alpha\tilde{\gamma}(t). \quad (22)$$

Here, $\tilde{\gamma}(t)$ is a dimensionless function of time and $\alpha = 1 + B_1 a/y$ is the mentioned correction [4, 17, 18]. The coefficient B_1 depends on the nature of the boundary (solid wall, membrane, polymer network or a cage formed by surrounding particles) and y represents a characteristic length of the medium.

By using the dimensionless variables previously introduced and defining the time scaling $\tau(t) = \int\tilde{\gamma}(t)dt$, from (21) one obtains

$$\partial\rho/\partial\tau = \alpha\left(\omega_T^2 t_0/\beta_0\right)\tilde{\nabla}^2\rho - \alpha\left(F_0 t_0/\beta_0 a\right)\tilde{\nabla}\cdot\left(\rho\tilde{\mathbf{X}}\right), \quad (23)$$

where $\omega_T^2 \equiv k_BT/ma^2$ and F_0 is the magnitude of the force per unit mass $\mathbf{X}$ due, for instance, to the polymer network.

The time dependence of τ can be obtained by using the evolution equation for the time correlation function $\chi(\tilde{t}) = \langle\tilde{\mathbf{x}}\cdot\tilde{\mathbf{x}}_0\rangle(\tilde{t})$ (see, Eq. 15), [4]

$$d\chi(\tilde{t})/d\tilde{t} = -\alpha\left(F_0 t_0/\beta_0 a\right)\tilde{\gamma}(\tilde{t})\langle\tilde{\mathbf{X}}\cdot\tilde{\mathbf{x}}_0\rangle(\tilde{t}). \quad (24)$$

Eq. (24) involves a complete hierarchy for the moments of the distribution ρ. However, in the linear force case it takes a closed form leading to the relation

$$\gamma(\tilde{t}) = d\ln\left|R^{-\beta_0/\alpha t_0\omega_m^2}\right|/d\tilde{t}, \quad (25)$$

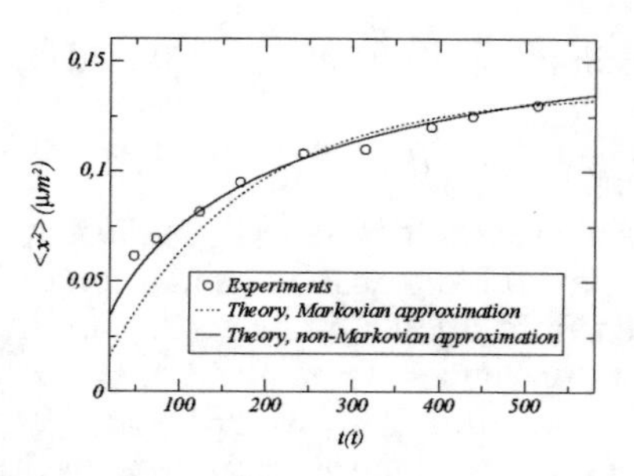

FIGURE 2. MSD vrs time. Comparison between theory (lines) and experiments of chromatin diffusion in living cells. The dotted line corresponds to a linear model in the Markovian approximation. The data were taken from Ref. [5] (circles) and the figure from [20].

where we have used $\mathbf{X} = -\omega_m^2\mathbf{x}$ with $\omega_m^2 = a^{-1}F_0$ and $R(\tilde{t}) = \chi(\tilde{t})/\chi(0)$. Using the definition and Eq. (25) one gets $\tau(\tilde{t}) = \ln\left|R^{-\beta_0/\alpha t_0\omega_m^2}\right|$. These relations imply: $D = \tilde{D}$, see the discussion after Eq. (15).

At short times, we can infer the time dependence of $\tau(\tilde{t})$. Expanding the logarithm in its argument around the unity: $\ln|R^{-\beta_0/\alpha t_0\omega_m^2}| \simeq R^{-\beta_0/\alpha t_0\omega_m^2} - 1 + O(R^2)$ and taking into account that $R(\tilde{t})$ must be an even function of time, a first order expansion leads to: $R^{-1}(\tilde{t}) \simeq 1 + B_2^*\tilde{t}^2 + O(\tilde{t}^4)$. Thus, we obtain the approximate expression

$$\tau(\tilde{t}) \sim B_2\tilde{t}^{2\beta_0/\alpha t_0\omega_m^2}, \quad (26)$$

where the parameter $B_2 \propto B_2^*$ in general depends on the characteristic length of the medium [4]. Important to notice is that the exponent depends on β_0, t_0, ω_m and α. Eq. (26) is consistent with a stretched exponential relaxation of the correlations [19]. Hence, in the linear case, Eq. (23) gives the MSD

$$\langle\tilde{x}^2\rangle(\tau) = 3(\omega_T^2/\omega_m^2)\left[1 - e^{-2\alpha\omega_m^2 t_0\tau(t)/\beta_0}\right], \quad (27)$$

which at short times yields $\langle\tilde{x}^2\rangle \simeq 6B_2\frac{\alpha\omega_T^2}{\beta_0 t_0^{-1}}\tilde{t}^{2\beta_0/\alpha t_0\omega_m^2}$. Comparison with Eq. (10) leads to the identifications $t_0 \sim \beta_0^3\omega_m^{-4}$ and $H = 1 - \beta_0/\alpha t_0\omega_m^2$. This expression for H constitutes the central result of this work. It shows that the so-called Hurst exponent is a consequence of hydrodynamic and elastic interactions between the particle and the viscoelastic medium.

This model explains the behavior of the MSD in the case of constrained diffusion [4]. Fig. **2** shows a comparison between theory (solid line) and experiments of constrained diffusion of chromatin in living cells [5]. The dashed line corresponds to a similar model in the Markovian case. As expected, the short time behavior of the MSD is subdiffusive. These effects are also present in Fig. **1** (dotted lines). At short times the behavior of MSD

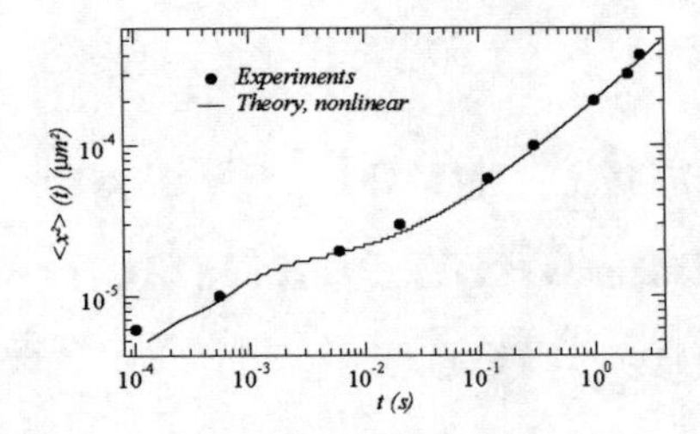

FIGURE 3. MSD vrs time. Comparison between experiments (circles) and theory (solid line) for the case of actin filament networks. The data were taken from Ref. [3]. The theoretical model corresponds to Eq. (28) with $\langle x^2\rangle_0 \sim 4.4\cdot 10^{-6}\mu m^2$, $D_0 \sim 9.3\cdot 10^{-5}\mu m^2 s^{-1}$, $\beta_0 \sim 5.5\cdot 10^9 s^{-1}$, $F_0 \sim 3.8\cdot 10^{-4}\mu m s^{-2}$, $t_0 \sim 0.1 s$, $\lambda \sim 1.5\cdot 10^{-3}\mu m$, $B_2 \sim 60$, $\alpha \sim 1$.

follows a power law on the scaled time $\tilde{t}$ and saturates at long times, where confinement becomes significant. A more detailed model is reported in [4].

A nonlinear model can also be discussed in terms of Eq. (21). If we assume that the force exerted on the particle by the medium is of the form $\cos\left[\lambda^{-1}\langle x\rangle(\tau)\right]$, with λ the average distance between the particle and the local boundaries, it can be shown that [4]

$$\langle x^2\rangle(t) \simeq \langle x^2\rangle_0 + 2D_0\tau(t) + 4\lambda^2 \tanh\left[\frac{F_0 D_0}{2\lambda k_B T}\tau(t)\right]^2, \tag{28}$$

where $\langle x^2\rangle_0$ is some initial value. The behavior of the MSD (28) as a function of time (solid line) is shown in Figs. **1** and **3**, and compared with experiments (symbols). The plateau is a signature of the existence of cage effects, which in our model are related with the maximum value of the elastic force **X**. The agreement between theory and experiments is good.

For Fig. **1** in the case with label 164, the values of the parameters are: $\langle x^2\rangle_0 \sim 5\cdot 10^{-7}\mu m^2$, $D_0 \sim 1.3\cdot 10^{-4}\mu m^2 s^{-1}$, $\beta_0 \sim 5.5\cdot 10^9 s^{-1}$, $F_0 \sim 1\cdot 10^{-3}\mu m s^{-2}$, $t_0 \sim 0.1 s$, $\lambda \sim 1.1\cdot 10^{-2}\mu m$, $B_2 \sim 7.2$, $\alpha \sim 1$ and we have defined $\omega_m^2 \sim F_0\lambda^{-1}$. Finally, it is interesting to notice that the effective mobility $\gamma(t)$ may be interpreted phenomenologically as the inverse Laplace transform: $\mathcal{L}^{-1}\left[\omega/G''(\omega)\right]$. Taking into account the relation between the MSD and the compliance J: $\langle x^2\rangle(t) \sim J(t)$, [3], one obtains $\gamma(t) = \mathcal{L}^{-1}\left[\omega^2 J(\omega)\right]$.

CONCLUSIONS

By following two different approaches, in this article we have described the anomalous diffusion of a Brownian particle in microrheological experiments. In the linear force case, we have shown that the generalized Langevin description with fractionary Gaussian noise explains well the behavior of the microrheological properties of the viscoelastic medium. We have also shown the equivalence of this description with generalized Fokker-Planck equations having time dependent coefficients. The comparison with a thermokinetic formalism gives a plausible justification for the physical nature of the exponent $H = 1 - \beta_0/\alpha t_0 \omega_m^2$ used in the GLE description. The thermokinetic formalism constitutes a powerful generalization of the theory allowing the formulation of non-Markovian models even in the case of non-linear forces. The agreement between theory and experiment is good.

Acknowledgments. I am indebted to Prof. J. Miguel Rubí for illuminating discussions concerning this work. Thanks go also to Profs. R. F. Rodríguez and A. Gadomski and to Drs. A. Pérez-Madrid, M. Mayorga and L. Romero-Salazar. This work was supported by UNAM-DGAPA under the grant IN-108006.

REFERENCES

1. S. Trimper, K. Zabrocki, M. Schulz, Phys. Rev. E **70**, 056133-1-056133-7 (2004).
2. T. G. Mason, Rheol. Acta **39**, 371-378 (2000).
3. J. Xu, V. Viasnoff, D. Wirtz Rheol. Acta **37**, 387-398 (1998).
4. I. Santamaría-Holek, J. M. Rubí, J. Chem. Phys. **125**, 064907-1-064907-4 (2006); I. Santamaría-Holek, J. M. Rubí, A. Gadomski, J. Phys. Chem. B **111**, 2293-2298 (2007).
5. W. F. Marshall, *et al.*, Current Biology **7**, 930-939 (1997).
6. D. Reguera, J. M. G. Vilar, J. M. Rubí, J. Phys. Chem. B **109**, 21502-21515 (2005).
7. S. A. Adelman, J. Chem. Phys. **64**, 124-130 (1976).
8. M. Tokuyama and H. Mori, Prog. Theor. Phys. **55**, 411-429 (1976).
9. R. Zwanzig, *Nonequilibrium Statistical Mechanics*, Oxford University Press, New York, 2001.
10. S. C. Kou, X. S. Xie, Phys. Rev. Lett. **93**, 180603-1-180603-4 (2004).
11. R. Balescu *Statistical Dynamics*, Imperial College Press, Singapore, 1997.
12. A. M. Mathai, R. K. Saxena, H. J. Haubold, Astrophys. Space Sci. **305**, 283-288 (2006).
13. I. Santamaría-Holek and J. M. Rubí, Physica A **326**, 384-389 (2003).
14. S. Okuyama and D. W. Oxtoby, J.Chem. Phys. **84**, 5824-5829 (1986).
15. S. R de Groot and P. Mazur, *Non-Equilibrium Thermodynamics*, Dover, New York, 1984.
16. A. Gadomski, Physica A **373**, 43-57 (2007).
17. J. Happel, H. Brenner, *Low Reynolds number hydrodynamics*, Kluwer, Dodrecht, 1991.
18. C. W. J. Beenakker, W. van Saarloos, and P. Mazur, Physica A **127**, 451-472 (1984).
19. A. Pérez-Madrid, J. Chem. Phys. **122**, 214914-1-214914-6 (2005).
20. I. Santamaría-Holek, J. M. Rubí, "Anomalous Diffusion in Intracellular Transport" in *Physics of Complex Systems and Life Sciences*, edited by A. Wagemakers and M. A. F. Sanjuan, Research Signpost, Kerala, 2008.

Markov-Chain Monte Carlo Simulation of Inverse-Halftoning for Error Diffusion based on Statistical Mechanics of the Q-Ising Model

Yohei Saika

Wakayama National College of Technology, 77 Noshima, Nada, Gobo, Wakayama 644-0023, Japan

Abstract. On the basis of statistical mechanics of the Q-Ising model we formulate the problem of inverse-halftoning for the halftone image which is obtained by the error diffusion method using the Floyd-Steinburg and two weight kernels. Then using the Markov-Chain Monte Carlo simulation both for a set of the snapshots of the Q-Ising model and a gray-level standard image, we estimate the performance of our method based on the mean square error and the edge structures observed both in the halftone image and reconstructed images, such as the edge length and the gradient of the gray-level. We clarify that our method reconstructs the gray-level image from the halftone image by suppressing the gradient of the gray-level on the edges embedded in the halftone image and by removing a part of the edges if we appropriately set parameters of our model.

Keywords: Digital halftoning, Inverse-halftoning, Probabilistic information processing, Statistical mechanics

INTRODUCTION

For many years, a lot of researchers have investigated information processing, such as image analysis, spatial data and the Markov-random fields [1-5]. In recent years on the basis of the analogy between probabilistic information processing and statistical mechanics, a lot of statistical-mechanical methods have been applied to image restoration [6] and error-correcting codes [7]. For instance, Pryce and Bruce [8] have constructed the threshold posterior marginal (TPM) estimate based on statistical mechanics of the classical spin system. Then, Sourlas [9,10] has pointed out the analogy between error correction of the Sourlas' codes and statistical mechanics of spin glasses. Based on these studies Nishimori and Wong [11] have constructed the unified framework of image restoration and error-correcting codes based on statistical mechanics of the Ising model. Recently the approaches have been are applied to various problems, such as mobile communication [12].

On the other hand, in the field of the print technology, a lot of techniques in information processing have been playing important roles to print a multi-level image with high quality. In particular, the technique which is referred as digital halftoning is essential to convert a multi-level image into a bi-level dot pattern which is visually similar to the original multi-level image [13]. Various techniques of digital halftoning have been established, such as the threshold mask method [14], the dither method [15], the blue noise mask method [16] and the error diffusion method [17,18]. The technique which is referred as inverse-halftoning is also important to reconstruct the multi-level image from the halftone image [19]. For this purpose, the conventional filters [20] established for image restoration have been practically applied to inverse-halftoning. In recent years, the MAP estimate [21] based on the Bayes inference has been applied to the problem of inverse-halftoning for the threshold mask method and the error diffusion method. Recently the maximizer of the posterior marginal (MPM) estimate based on statistical mechanics of the Q-Ising model has applied to the problem of inverse-halftoning for the threshold mask method [21-23].

In this article, on the basis of statistical mechanics of the Q-Ising model with ferromagnetic interactions under the random fields, we formulate the problem of inverse-halftoning for the error diffusion method using the Floyd-Steinburg and two weight kernels. We note that our method is regarded as the maximizer of the posterior marginal (MPM) estimate in the field of information sciences.

CP982, *Complex Systems, 5th International Workshop on Complex Systems*
edited by M. Tokuyama, I. Oppenheim, and H. Nishiyama

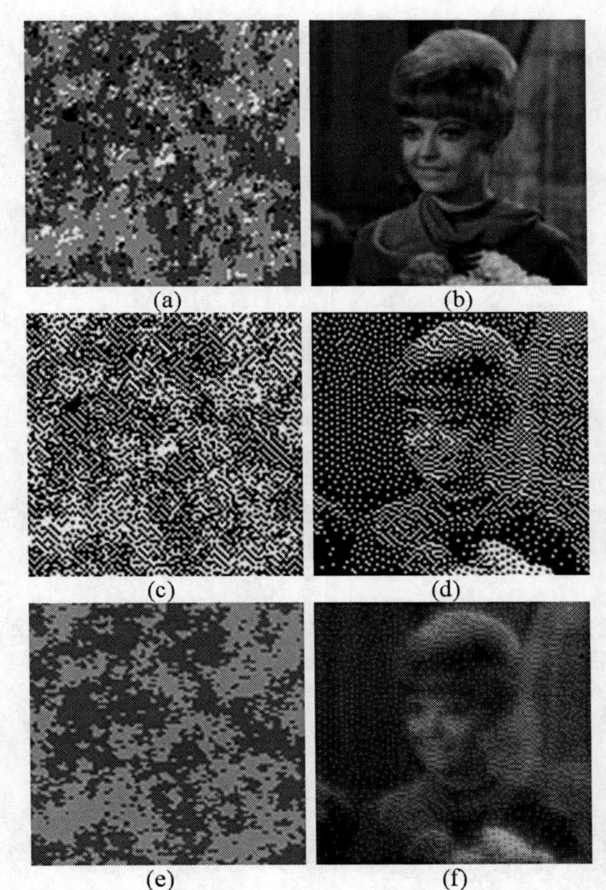

FIGURE 1. (a) a snapshot of the Q=4 Ising model with 100×100 pixels, (b) the 256-level standard image "girl" with 100×100 pixels, (c) the halftone image obtained from the gray-level image (a) by the error diffusion method using the Floyd-Steinburg kernel, (d) the halftone image obtained from the 256-level standard image "girl" which is shown in (b), (d) a sample of the 4-level image obtained from the halftone image shown in (c) by the MPM estimate when hs=1, Ts=1, h=1, T=0.1, J=0.875, (f) a sample of the 256-level image reconstructed from the halftone image shown in (d) by the MPM estimate when h=1, T=0.1, J=1.4.

In this study we use the posterior probability which is composed both of the model prior expressed by the Boltzmann factor of the Q-Ising model and of the likelihood expressed by the Boltzmann factor of the random fields enhancing the halftone image. Then we estimate the performance of the MPM estimate using the Monte Carlo simulation both for a set of the snapshots of the Q-Ising model and a 256-level standard image. The estimate based on the mean square error clarifies that our method works effectively for inverse-halftoning for the error diffusion method using the Floyd-Steinburg and two weight kernels if we set the parameters appropriately. Next we estimate the performance of our method based on the edge

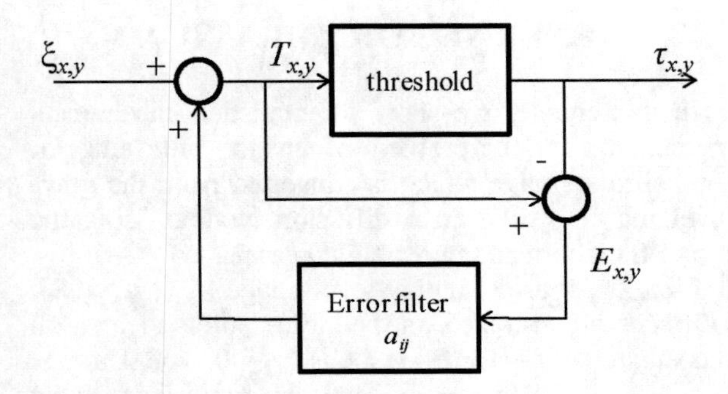

FIGURE 2. The block diagram of the error diffusion method.

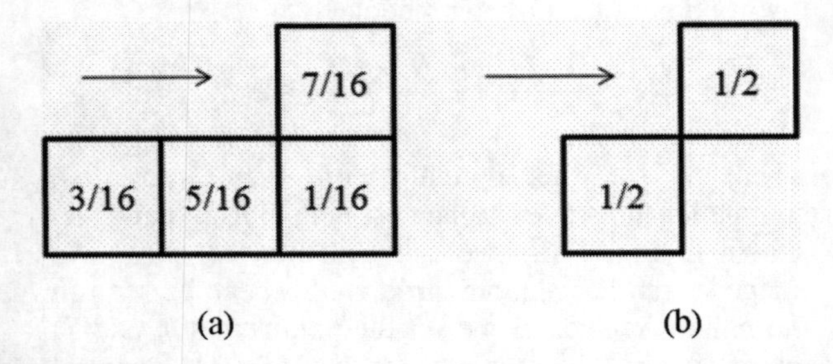

FIGURE 3. The error diffusion kernels, (a) the Floyd-Steinburg kernel, (b) the two weight kernel.

structures which are observed both in the halftone and reconstructed images, such as the edge length and the gray-level difference between neighboring pixels. The simulations clarify that the MPM estimate reconstructs the gray-level image by suppressing the gradient of the gray-level on the edges embedded in the halftone image and by removing a part of the edges of the halftone image. Further we also estimate the dynamical properties of the MPM estimate using the Monte Carlo simulation. If the parameters of the model are set appropriately, the mean square error smoothly converges to the optimal value irrespective of the choice of the initial condition. On the other hand, if the parameters are not appropriately set, the convergent value depends on the choice of the initial condition.

This article is organized as follows. In chapter 2, we show the statistical-mechanical formulation for the problem of inverse-halftoning for the error diffusion method using the Floyd-Steinburg and two weight kernels. Then using the Monte Carlo simulation both for the set of the snapshots of the Q-Ising model and the gray-level standard image, we estimate the performance of the MPM estimate based on the mean square error and the edge structures observed both in the halftone and reconstructed images. The chapter 4 is devoted to summary and discussion.

GENERAL FORMULATION

In this chapter we show the statistical-mechanical formulation for the problem of inverse-halftoning for the halftone image which is converted from the gray-level image by the error diffusion method using the Floyd-Steinberg and two weight kernels.

First, we consider a gray-level image $\{\xi_{x,y}\}$ in which all pixels are arranged on the lattice points located on the square lattice. Here we set as $\xi_{x,y}= 0,\ldots,Q-1$ and $x, y = 1,\ldots, L$. In this study, we treat two kinds of the original images. One is the set of the gray-level images generated by a true prior expressed by the Boltzmann distribution of the Q-Ising model:

$$\Pr(\{\xi_{x,y}\}) = \frac{1}{Z_s}\exp\left[-\frac{h_s}{T_s}\sum_{n.n.}(\xi_{x,y}-\xi_{x',y'})^2\right] \quad (1)$$

where Z_s is the normalization factor and the summation runs over all nearest neighboring pairs. As shown in Fig. 1 (a), we can generate the gray-level image which has smooth structures, as can be seen in the natural images, if we set the parameters h_s and T_s appropriately. Then the other is the 256-level standard image "girl" shown in Fig. 1(b).

Next, when we print the gray-level image, we convert the gray-level image $\{\xi_{x,y}\}$ into a halftone image $\{\tau_{x,y}\}$ which is visually similar to the original gray-level image due to the error diffusion method using the Floyd-Steinburg and two weight kernels. Here $\tau_{x,y}=0, 255$ and $x, y = 1, \ldots, L$. The samples of the halftone images are shown in Fig. 1 (c) and (d). We carry out the halftone conversion using the block diagram of the error diffusion algorithm given in Fig. 2 and the Floyd-Steinburg and two weight kernels given in Fig. 3. As shown in these figures, the error diffusion algorithm is performed through the image in a raster scan. Then at each pixel, a binary decision is made based on the gray-level of the input pixel and filtered error from the previous threshold samples. At the (x,y)-th pixel the gray-level $u_{x,y}$ is rewritten into the modified gray-level $u'_{x,y}$ as

$$u'_{x,y} = u_{x,y} - \sum_{(k,l)\in S} h_{k,l}e_{x-k,y-l}. \quad (2)$$

Here $\{h_{k,l}\}$ is the Floyd-Steinburg and two weight kernels shown in Figs. 3 (a) and (b) and then S is the region which supports the site (x,y) by the kernels. Then $e_{x,y}$ is the error of the halftone image $\tau_{x,y}$ to the gray-level one $u_{x,y}$ at the site (x, y) as

$$e_{x,y} = \tau_{x,y} - u'_{x,y}. \quad (3)$$

Here the pixel value $\tau_{x,y}$ of the halftone image is obtained using the threshold procedure as

$$\tau_{x,y} = \begin{cases} Q-1 & (z'_{x,y} \geq (Q-1)/2) \\ 0 & (\text{otherwize}) \end{cases}. \quad (4)$$

Next, using the MPM estimate based on statistical mechanics of the Q-Ising model, we reconstruct a gray-level image for the halftone image converted by the error diffusion method using the Floyd-Steinburg and two weight kernels. Here we use the model system given by a set of Q-Ising spins $\{z_{x,y}\}$ ($z_{x,y}= 0,\ldots, Q-1$, $x, y = 1,\ldots, L$) located on the square lattice. The procedure of inverse-halftoning is carried out so as to maximize the posterior marginal probability as

$$\hat{z}_{x,y} = \arg\max_{z_{x,y}} \sum_{\{z\}\neq z_{x,y}} P(\{z\}\,|\,\{J\}), \quad (5)$$

where the posterior probability is estimated based on the Bayes formula:

$$P(\{z\}\,|\,\{J\}) = P(\{z\})P(\{J\}\,|\,\{z\}) \quad (6)$$

using the model of the true prior and the likelihood. In this study, we assume the model of the true prior which is expressed by the Boltzmann factor of the Q-Ising model as

$$P(\{z\}) = \frac{1}{Z_m}\exp\left[-\frac{J}{T_m}\sum_{n.n.}(z_{x,y}-z_{x',y'})^2\right]. \quad (7)$$

This model prior is considered to enhance smooth structures which can be seen in natural images. Then, we assume the likelihood as

$$P(\{z\}\,|\,\{\tau\}) \propto \exp\left[-\frac{h}{T_m}\sum_{x,y}(z_{x,y}-\hat{\tau}_{x,y})^2\right]. \quad (8)$$

This likelihood shows the property to enhance the gray-level image:

$$\hat{\tau}_{x,y} = \sum_{i,j=-1}^{1} a_{i,j}\tau_{x+i,y+j} \quad (9)$$

where $\{a_{i,j}\}$ is the kernel of the conventional filter. In this study we set as

$$a_{ij} = \delta_{i,0}\,\delta_{j,0}. \quad (10)$$

We note that our method corresponds to the MAP estimate in the limit of $T\to 0$.

Next, in order to estimate the performance of our method for a standard image, we use the mean square error as

$$\sigma = \frac{1}{NQ^2}\sum_{x,y=1}^{L}(\hat{z}_{x,y}-\xi_{x,y})^2 \quad (11)$$

where $\xi_{x,y}$ and $\hat{z}_{x,y}$ are the pixel values of the original and reconstructed images at the site (x,y). When we estimate the performance for the set of the gray-level images which are generated by the true prior $P(\{\xi\})$, we evaluate the mean square error averaged over the true prior as

$$\sigma = \sum_{\{\xi\}} P(\{\xi\})\frac{1}{NQ^2}\sum_{x,y=1}^{L}(\hat{z}_{x,y}-\xi_{x,y})^2. \quad (12)$$

This value becomes zero if each pixel value of all reconstructed images is completely same with that of the corresponding original images. Then in order to estimate whether the edge structures embedded in the halftone image survives in the gray-level image reconstructed by the MPM estimate, we evaluate the edge length observed both in the halftone and reconstructed images as:

$$L_{\text{edge}} = \sum_{\{\xi\}} P(\{\xi\}) \left[\sum_{\text{n.n.}} \left(1 - \delta\tau_{x,y}, \tau_{x',y'}\right)\left(1 - \delta z_{x,y}, z_{x',y'}\right) \right] \tag{13}$$

averaged all over the snapshots of the Q-Ising model. Also we estimate the gray-level difference between neighboring pixels observed both in the halftone and reconstructed images as:

$$\left|\nabla z_{x,y}\right| = \sum_{\{\xi\}} P(\{\xi\}) \left[\sum_{\text{n.n.}} \left(1 - \delta\tau_{x,y}, \tau_{x',y'}\right) | z_{x,y} - z_{x',y'} | \right]. \tag{14}$$

PERFORMANCE

In order to estimate the performance of the MPM estimate for inverse-halftoning for the error diffusion method using the Floyd-Stein

burg and two weight kernels, we carry out the Monte Carlo simulation both for the set of gray-level images generated by the Q-Ising model and the 256-level standard image "girl".

The Monte Carlo simulation is carried out under the following conditions. As shown in Fig. 1 (a), we consider 10 snapshots of the Q=4 Ising model which is generated by the true prior when h_s=1 and T_s=1. Then we convert the gray-level images into the halftone images as shown in Fig. 1 (d) by the error diffusion method using the Floyd-Steinburg and two weight kernels. Next when we carry out the Monte Carlo simulation for inverse-halftoning, we use the Q-Ising model $\{z_{x,y}\}$ under the periodic boundary condition, that is $z_{L-1,y}=z_{0,y}$ and $z_{x,L-1}=z_{x,0}$ where $x,y=0,\ldots,L-1$. When we estimate the performance of our method, we take averages on the mean square error and the edge structures, such as the edge length and the gray-level difference between neighboring pixels, over 10 snapshots of the Q-Ising model.

From now on we investigate the static properties of the Q=4 model for the case of the error diffusion using the Floyd-Steinburg kernel. When J=0, the MPM estimate reconstructs the gray-level image which is approximately same with the halftone image because the likelihood has the characteristics to enhance the halftone image itself. As shown in Fig. 4, the mean square error is given by σ=0.10596±0.00045. Then as shown in Fig. 5, the length of the edges observed both in the halftone and reconstructed images is given by L_{edge}=12049.6 which is longer than that observed in

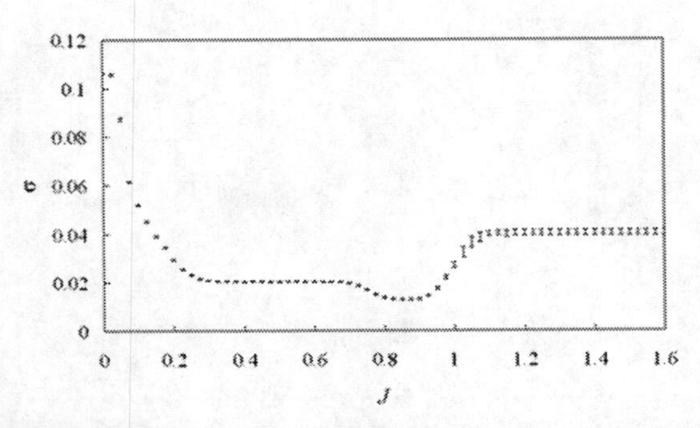

FIGURE 4. The mean square error as a function of the parameter J obtained by the MPM estimate for the halftone images converted from the set of the snapshots of the Q=4 Ising model by the error diffusion method using the Floyd-Steinburg kernel when h_s=1, T_s=1, h=1, T=0.1.

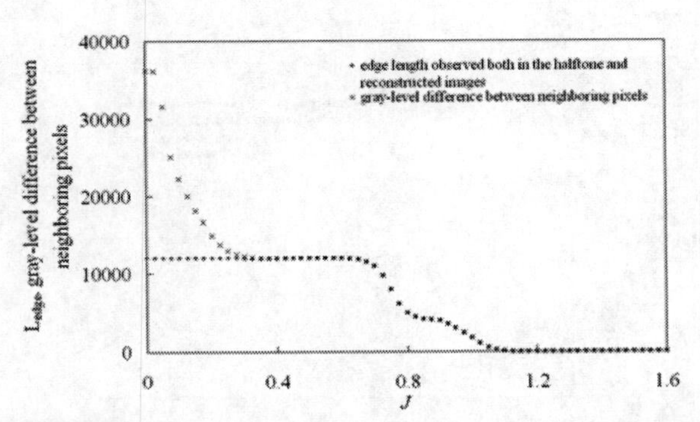

FIGURE 5. The length of the edges and the gray-level differences between neighboring pixels as a function of the parameter J obtained by the MPM estimate for the halftone images converted from the set of the snapshots of the Q=4 Ising model by the error diffusion method using the Floyd-Steinburg kernel when h_s=1, T_s=1, h=1, T=0.1.

the original image because the halftone image is expressed by the dot pattern which is visually similar to the original image. Therefore the absolute value of the gray-level gradient of the reconstructed image is $|\delta z_{x,y}|$=36149.3 which is larger than that of the original image $|\delta z_{x,y}|$=6383.6. On the other hand, in the limit of $J\to\infty$, the MPM estimate reconstructs the flat pattern because the model prior which strongly suppresses the edges and enhances flat pattern. Here the mean square error is given by σ =0.0404 ± 0.00011. Then $L_{\text{edge}}=|\delta z_{x,y}|=0$ respectively. Next we estimate the performance of the MPM estimate when we move the parameter J from 0 to infinity. Figure 4 shows how the mean square error depends on the parameter J. In this figure we reveal two points as follows. The first point is that the optimal performance is achieved by the MPM estimate if the parameters are appropriately set.

That is, as shown in Fig. 4, the mean square error takes its minimum σ=0.01286±0.00039 if we set as h=1, J=0.875 and T=0.1. This means that the gradient

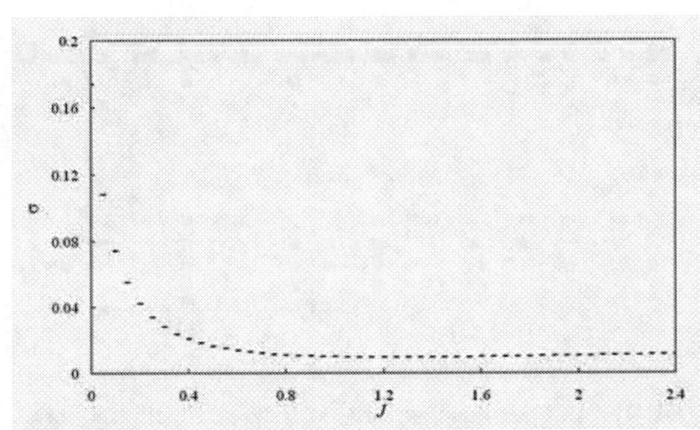

FIGURE 6. The mean square error as a function of the parameter J obtained by the MPM estimate for the halftone images converted from the 256-level standard image "girl" by the error diffusion method using the Floyd-Steinburg kernel when h_s=1, T_s=1, h=1, T=0.1.

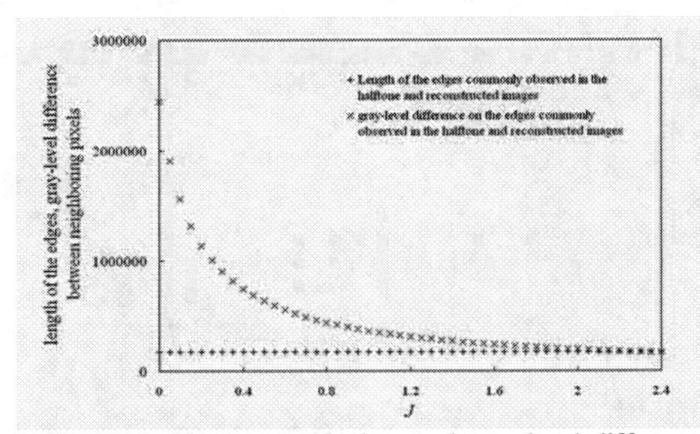

FIGURE 7. The length of edges and gray-level difference between neighboring pixels mean square error as a function of the parameter J obtained by the MPM estimate for the halftone image converted from the 256-level standard image "girl" by the MPM estimate when h_s=1, T_s=1, h=1, T=0.1.

of the gray-level enhanced by the likelihood is appropriately suppressed by the model prior which enhances smooth structures. Then the second point is that the rapid drop of the mean square error is observed if we increase J from 0.025 to 0.3. The origin of this behavior will be clarified in the following part of this chapter.

Next in order to clarify the behavior of the dot pattern of the halftone image appearing in the reconstructed image, we pay attention to several variables related to the edge structures which are observed both in the halftone and reconstructed images, such as the edge length and the absolute value of the gradient of the gray-level. Figure 5 shows how the edge length and the gray-level difference between neighboring pixels commonly observed both in the halftone and reconstructed images depends on the parameter J. This indicates that the reconstructed image keeps the edges embedded in the halftone image in $0<J<0.6$ although the gradient of the gray-level on the edges is rapidly suppressed with the increase in J from 0.025 to 0.3. On the other hand, the figure also indicates that the edge length observed both in the halftone and reconstructed images becomes short with the increase in J from 0.6 and it becomes zero at J=1.175. The optimal point determined based on the mean square error is located at J=0.875, as we have shown in Fig. 4 and $L_{\mathrm{edge}}=4096.3\pm92.1$ (= $|\delta z_{x,y}|$) at the point. Further by comparison of the patterns of the original and reconstructed images, we find that 2/3 of the edges of the halftene image is not observed in the original image but is artificially embedded in the halftone image to generate the dot pattern which is visually similar to the original image. These results leads to the statement that the optimal performance of the MPM estimate is achieved so that the gradient of the gray-level appearing in the halftone image is suppressed and the edges of the halftone image are appropriately removed by the model prior which enhances smooth structures.

Next we investigate the dynamical properties of the Q=4 model for the error diffusion method using the Floyd-Steinburg kernel. The Monte Carlo simulation derives the results that the mean square error smoothly converges irrespective of the choice of the initial condition if $J<J_{\mathrm{opt}}$ and that the convergent value depends on the initial condition if $J>J_{\mathrm{opt}}$.

In the case of the two weight kernel, the Monte Carlo simulation clarifies that the optimal performance $\sigma=0.0131\pm0.00032$ is achieved when h=1 and J=0.875. This means that the MPM estimate achieves same optimal performance both for the cases of the Floyd-Steinburg and two weight kernels within the statistical uncertainty.

Having clarified the characteristics of the Q=4 model, we now move to the performance estimation of the Q=256 model using the Monte Carlo simulation for the 256-level standard image "girl". As shown in Figs. 6 and 7, we obtain the results that the optimal performance is achieved by tuning the parameters appropriately and that our method reconstructs images by suppressing the edge structures which is embedded in the halftone image.

SUMMARY AND DISCUSSION

In the previous chapters, we formulate the problem of inverse-halftoning for the error diffusion method using the Floyd-Steinburg and two weight kernels on the basis of statistical mechanics of the Q-Ising model. Then using the Monte Carlo simulation for the set of the snapshots of the Q=4 Ising model and the 256-level standard image "girl", we numerically estimate the performance of the MPM estimate based on the mean square error and the edge structures which are observed both of the halftone and reconstructed images, such as the edge length and the gradient of the gray-level. The Monte Carlo simulation for the Q=4 model clarifies that the optimal performance of the MPM estimate is achieved by suppressing the gradient of the gray-level and by removing parts of the edges embedded in the halftone image if we appropriately set the parameters. Similar behavior is also observed by the Monte Carlo simulation for the Q=256 model for the standard image "girl". We have also clarified that the mean square error smoothly converges irrespective of the choice of the initial condition if J<Jopt and that the convergent value depends on the initial condition of the Monte Carlo simulation if $J>J_{\text{opt}}$.

As a future problem, in order to reconstruct the gray-level image with high quality, we are going to construct the statistical-mechanical formulation with line process for the problem of inverse-halftoning for the error diffusion method.

ACKNOWLEDGMENTS

The author was financially supported by Grant-in-Aid Scientific Research on Priority Areas "Deepening and Expansion of Statistical Mechanical Informatics (DEX-SMI)" of the Ministry of Education, Culture, Sports, Science and Technology (MEXT) No.18079001.

REFERENCES

1. J. Besag, *Journal of the Royal Statistical Society* B, **48**, 259-302 (1986).
2. G. Winkler, Image analysis, *Random fields and Dynamic Monte Carlo Methods, A Mathematical Introduction*, Berlin, Springer-Verlag, 1995.
3. N. A. Cressie, *Statistics for Spatial Data*, New York, Wiley, 1993.
4. *Spatial Statistics and Imaging, Lecture Notes-monograph Series, Institute of Mathematical Statistics*, ed. A. Possio, Vol. 20, California, Hayward, 1991.
5. Y. Ogata and M. Tanemura, *Ann. Inst. Stat. Math.* **B 33**, 131 (1981).
6. H. Nishimori, *Statistical Physics of Spin Glasses and Information Processing: An Introduction*, Oxford University Press, Oxford, 2001.
7. K. Tanaka, *Journal of Physics* A: Mathematics and General, **35**, R38-R150 (2002).
8. J. M. Pryce and A. D. Bruce, *Journal of Physics* A **28**, 511-532 (1995).
9. N. Sourlas, *Nature*, **339**, pp. 693-695 (1989).
10. N. Sourlas, *Europhys. Letters*, **25**, 159-164 (1994).
11. H. Nishimori and K. Y. M. Wong, *Statistical mechanics of image restoration and error-correcting codes*, Physical Review E, **60**, 132-144 (1999).
12. T. Tanaka, *Statistical mechanics of CDMA multiuser demodulation*, Europhysics Letters, **54**, 540-546 (2001).
13. R. Ulichney, *Digital Halftoning*, Massachusetts, The MIT Press, 1987.
14. B. E. Bayer, "An optimum method for two-level rendition of continuous-tone pictures," ICC CONF. RECORD **26**, 1973, pp. 11-15.
15. R. Ulichney, "Dithering with blue noise," the proceeding of IEEE, **76**, Vol. 1, 1988, pp. 56-79.
16. T. Mitsa and K. J. Parker, Digital halftoning technique using a blue noise mask, Journal of the Optical Society of America, A/9, **11**, 1920-1929 (1992).
17. M. Yao and K. J. Parker, "The blue noise mask and its comparison with error diffusion," SID 94 Digest, **37**, 3, 805-807 (1994).
18. M. Yao and K. J. Parker, *Journal of Electric Imaging*, **3**, 1, 92-97 (1994).
19. C. M. Miceli and K. J. Parker, *Journal of Electric Imaging*, **1**, 143-151 (1992).
20. P. W. Wong, *IEEE Trans. Image Processing*, **4**, 486-498 (1995).
21. Y. Saika and J. Inoue, "Probabilistic inference to the problem of inverse-halftoning based on statistical mechanics of spin systems," *in the proceeding of SICE-ICCAS 2006*, 2006, pp. 4563-4568.
22. Y. Saika and J. Inoue, "Probabilistic inference to the problem of invere-halftoning based on the Q-Ising model," *in the proceeding of the First IEEE Symposium on Foundations of Computational Intelligence* (FOCI'07), 2007, pp. 429-433.
23. J. Inoue, Y. Saika and M. Okada, "Statistical-Mechanical Analysis of Inverse Digital Halftoning", the proceeding of the 7th International Conference on Intelligent the System Design and Applications (ISDA'07), 2007, pp. 617-622.

Aging, Rejuvenation and Memory Effects in Systems far from Equilibrium

Philipp Maass*, Falk Scheffler†, and Hajime Yoshino**

*Institut für Physik, Technische Universität Ilmenau, 98684 Ilmenau, Germany
†Fachbereich Physik, Universität Konstanz, 78457 Konstanz, Germany
**Faculty of Science, Osaka University, Toyonaka, 560-0043 Osaka, Japan

Abstract. The slow non-equilibrium dynamics in various systems show common physical aging features and sometimes surprising rejuvenation and memory effects. We present two theoretical approaches to understand these phenomena on the basis of coarse grained models. In the first approach a system in configuration space is considered that changes its state by thermally activated jumps in a random energy landscape. In the second approach an effective dynamics for the Edwards-Anderson spin-glass model on a hierarchical lattice is studied by means of real-space renormalization concepts.

Keywords: Aging, Non-equilibrium processes, Rejuvenation, Renormalization, Slow dynamics, Spin glasses, Stochastic processes
PACS: 02.50.-r,05.10.Cc,05.20.-y,75.10.Nr,75.40.Gb

INTRODUCTION

If a physical system is brought into a non-equilibrium state by a change of external control parameter and subsequently reorganizes itself to restore thermal equilibrium, its relaxation dynamics generally slows down with progressing time. This aging phenomenon is of particular interest when the equilibration time becomes very large, as, for example, in glasses, where below the glass transition temperature equilibrium never is reached within time windows accessible by experiment. Prominent examples are shear-stress relaxations in polymeric glasses [1, 2] and relaxations of the magnetization (or slow decays of ac susceptibilities) in spin glasses [3, 4]. Aging effects are observed also in many other systems, as, for example, colloidal gels [5, 6] and granular materials [7].

In addition to aging, non-equilibrium systems can show so-called rejuvenation and memory effects. These effects are less ubiquitous and mainly observed in spin glasses [8, 9]. Rejuvenation is attributed to the fact that even a small temperature shift can cause the system to appear as if it had a much shorter thermalization period. On the other hand, by reversing the temperature shift at some later time, the system can show memories to the thermalization period before the rejuvenation.

In this work we will discuss these non-equilibrium phenomena based on two different coarse-grained approaches. In the first approach [10] the aging dynamics of a system is described by thermally activated jumps between valleys of a complex potential surface (or between metastable basins of a free energy surface), which is represented by a random energy landscape. Models of this type have become a key tool to describe dynamical phenomena in a large variety of non-equilibrium systems. In the second approach [11] an effective dynamics of the Edward-Anderson (EA) spin glass model on a hierarchical lattice is considered to study aging, rejuvenation and memory effects.

AGING IN A RANDOM ENERGY LANDSCAPE

For simplicity we consider an energy landscape with valleys on a hypercubic lattice in d dimensions. To each valley i is assigned a random energy E_i drawn from the exponential density

$$\psi(E) = T_0^{-1}\exp(E/T_0), \quad -\infty < E \leq 0 \tag{1}$$

Given this density of states, the canonical distribution $\psi(E)\exp(-E/T)$ is normalizable only for $T > T_0$, while for $T < T_0$ an equilibrium state can never be reached (in the thermodynamic limit). The system is considered to evolve by thermally activated transitions between neighboring valleys i and j with hopping rates $w_{i,j} = w(E_i, E_j)$ obeying detailed balance.

To study aging features, we consider a quench from $T = \infty$ to $T < T_0$ at time 0, let the system evolve for a waiting time t_w and investigate the subsequent behavior during an observation time t. A 'generic correlation function' $C(t+t_w, t_w)$ can be defined by the probability for the system to be in the same valley at time $t+t_w$ as it was at time t_w. This correlation exhibits normal aging behavior $C(t+t_w, t_w) \sim \mathscr{C}(t/t_w)$ for both t and t_w becoming large (for details, including forms of the scaling function in $d = 1$, see [12]).

A richer behavior is obtained when considering the probability $\Pi(t+t_w, t_w)$ that the system undergoes no

CP982, *Complex Systems, 5th International Workshop on Complex Systems*
edited by M. Tokuyama, I. Oppenheim, and H. Nishiyama

transition between t_w and $t+t_w$. By using a partial equilibrium concept, it was shown [10] that this probability, for certain forms of jump rates $w(E_i,E_j)$, shows subaging behavior, $\Pi(t+t_w,t_w) \sim F(t/t_w^{\nu_1})$ with $0<\nu_1<1$, and exhibits a second 'short-time' scaling regime associated with a generalized scaling law $[1-\Pi(t+t_w,t_w)] \sim t_w^{-\nu_1}F(t/t_w^{\nu_2})$ with $0<\nu_2<\nu_1<1$. These features are absent in the mean-field version of the model [13], where the site energies E_i are drawn anew after each transition.

Here we will show that, dependent on the form of the jump rates $w(E,E')$ (E energy of initial site, E' energy of target site involved in a transition), it is possible to obtain *(i)* a scaling behavior $\Pi(t+t_w,t_w) \sim F(t/\tau(t_w))$ with an arbitrary dependence of the correlation time $\tau(t_w)$ on the waiting time t_w, and *(ii)* multiple scaling regimes in the $t-t_w$ plane, as they were conjectured on the basis of mean-field spin glass models [14].

To determine $\tau(t_w)$ we use simple scaling arguments. After time t_w, the system has visited $S=S(t_w)$ distinct valleys and the occupation probability of these valleys is dominated by the one with lowest energy $E_{\min}$ (in the 'glassy phase' $T<T_0$ the extreme values are dominant). If we confine ourself to dimensions $d>2$, we have (see the analysis in [10]).

$$S(t_w) \sim t_w^{T/T_0}\,. \tag{2}$$

The typical value of $E_{\min}$ can be estimated from $S(t_w)\int_{-\infty}^{E_{\min}} \psi(E) = S(t_w)\exp(E_{\min}/T_0) \simeq 1$, yielding

$$E_{\min}(t_w) \sim -T_0 \ln S(t_w) \sim -T \log t_w\,. \tag{3}$$

Being at a valley with energy $E_{\min}$ at time t_w, the system typically encounters a situation where the neighboring valleys have energies close to zero. Accordingly, the characteristic time $\tau(t_w)$ to leave the valley reached after t_w should be given by the inverse hopping rate for initial energy $E=E_{\min}$ and target energy $E'=0$, i.e.

$$\tau(t_w) \sim w(E_{\min},0)^{-1} \sim w^{-1}(-T\log t_w,0) \tag{4}$$

Hence, to obtain a scaling with $\tau(t_w)$, we can choose jump rates satisfying detailed balance and eq. (4):

$$w(E,E') = \frac{\exp(-E'/T)}{\tau(\exp[-(E+E')/T])}\,. \tag{5}$$

For example, to obtain $\tau(t_w) = \log t_w$ we may use $w(E,E') = \exp(-E'/T)/|E+E'|$. Indeed, results from Monte-Carlo simulations and numerical calculations using the partial equilibrium concept [10] confirm this ultraslow logarithmic aging dynamics as demonstrated in Fig. 1.

To show the possible occurrence of multiple scaling regimes, we consider the rates used in [10],

$$w(E,E') = \exp\left(-\frac{[\alpha E' - (1-\alpha)E]}{T}\right), \tag{6}$$

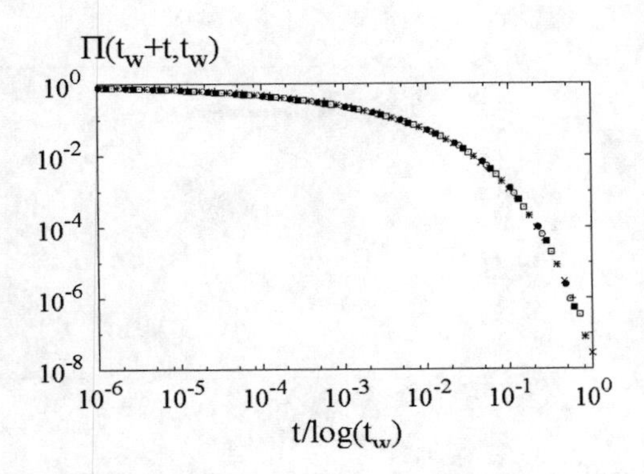

FIGURE 1. Numerical results for the probability $\Pi(t_w+t.t_w)$ as a function of $t/\log t_w$ for various waiting times ranging from 10^4 to 10^{16}. The results were obtained for a simple cubic lattice ($d=3$) with hopping rates $w(E,E') = \exp(-E'/T)/|E+E'|$.

where the parameter α, $0\le\alpha<1$, weights the initial and target valley. For $\alpha=0$, eq. (6) defines a trap model, where the jump rates depend on the initial energy only, while the case $\alpha=1/2$ defines a "force model", where the jump rates are determined by the energy difference between the initial and target site.

In a real energy landscape one cannot expect the transition rates to be of unique form. Let us investigate the behavior, if to each link between valleys a parameter α_1 is assigned with probability $1-p$ and a parameter $\alpha_2<\alpha_1$ with probability p. Using (4), the characteristic correlation time for the rates (6) scales as $\tau(t_w) \sim t_w^{1-\alpha}$, i.e. the escape form the valley reached after t_w becomes the faster the larger α is. Hence, two situations can occur after time t_w: Either at least one of the transitions to the neighboring states is characterized by the larger value α_1, which gives $\tau \sim t_w^{1-\alpha_1}$, or all transitions to the neighboring states have the smaller α_2 and $\tau \sim t_w^{1-\alpha_2}$. On the hypercubic lattice these cases occur with probabilities $1-p^{2d}$ and p^{2d}, respectively, yielding

$$\Pi(t_w+t,t_w) \sim (1-p^{2d})F_1\left(\frac{t}{t_w^{1-\alpha_1}}\right) + p^{2d}F_2\left(\frac{t}{t_w^{1-\alpha_2}}\right)\,. \tag{7}$$

This means that, when we take $t=\Lambda t_w^{1-\alpha_1}$ and consider the limit $t_w\to\infty$ with Λ fixed, we would observe the first scaling function, i.e. $\Pi(t_w+\Lambda t_w^{1-\alpha_1},t_w) \sim (1-p^{2d})F_1(\Lambda)+p^{2d}$. By contrast, taking the limit $t_w\to\infty$ with $t=\Lambda t_w^{1-\alpha_2}$ and Λ fixed, yields the second scaling function $\Pi(t_w+\Lambda t_w^{1-\alpha_2},t_w) \sim p^{2d}F_2(\Lambda)$. As shown in Fig. 2, numerical calculations confirm this prediction. It is straightforward to extend the procedure outlined here to more α values. In principle one may generate an infinite number of scaling regimes in this way.

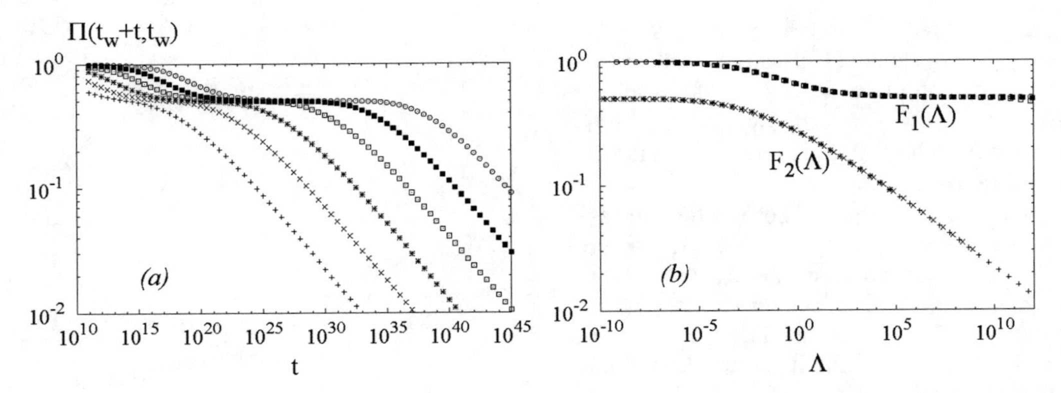

FIGURE 2. Numerical results for the probability $\Pi(t_w+t,t_w)$ in $d=3$ for the jump rates (6) and a bimodal distribution of the parameter α ($\alpha_1=0.5$, $\alpha_2=0$, $p=0.89$, see text). The waiting times range from 10^4 to 10^{16}. In *(a)* the un-scaled data are shown and in *(b)* the scaled data with the scaling functions $F_{1,2}(\Lambda)$ ($\Lambda=t/t_w^{1-\alpha_1}$ for F_1 and $\Lambda=t/t_w^{1-\alpha_2}$ for F_2).

EFFECTIVE SPIN DYNAMICS ON THE HIERARCHICAL LATTICE

The hierarchical lattice for the EA model in three dimensions is constructed iteratively as depicted in Fig. 3 and reflects the Migdal-Kadanoff approximation for a real lattice. The spins $\{S^{(0)}\}$ with only two neighbors introduced in the last iteration step are associated with a length scale $L_0=1$, while the spins $\{S^{(n)}\}$ at increasingly higher levels n have 2×4^n nearest neighbors and are associated with a length scale $L_n=2^n$. Random exchange interactions are assigned to each bond in this lattice. They are drawn from a Gaussian distribution with zero mean and variance J^2, giving rise to a spin glass transition at $T_c\simeq0.88J$.

To study the slow non-equilibrium dynamics below T_c, a coarse-grained approach is used [11], by which the systems evolves in successive epochs $n=1,2,\ldots$. In the nth epoch, the spins $\{S^{(n+1,n+2,\ldots)}\}$ are considered to be frozen and the spins $\{S^{(n,n-1,\ldots,0)}\}$ are successively equilibrated. First the spins $S_i^{(n)}$ are thermalized (aligned with Boltzmann weights) in their local fields $h_i^{(n)}=\sum_j J_{ij}^{(n)}S_j^{(n+1,n+2,\ldots)}$, where $J_{ij}^{(n)}$ are the effective couplings given by the exact real space renormalization group (RSRG) transformation $J_{ij}^{(n)} = T\sum_{k=1}^{4}\tanh^{-1}[\tanh(J_{ik}^{(n-1)}/T)\tanh(J_{jk}^{(n-1)}/T)]$. In the second step, the spins $S_i^{(n-1)}$ are thermalized in their local fields $h_i^{(n-1)}$, which depend on the spins $S_i^{(n)}$ updated in the first step (and the effective couplings $\{J^{(n-1)}\}$). By repeating this procedure, the spins $\{S^{(n)}\},\ldots,\{S^{(0)}\}$ are updated one after the other in the nth epoch. In the spirit of the droplet model [15], a period $t_n\propto\exp(E_n/T)$ is assigned to the nth epoch, where $E_n\sim JL_n^{\psi}$, $\psi>0$, is the energy barrier for excitations (flips of domains or 'droplets') of linear size L_n.

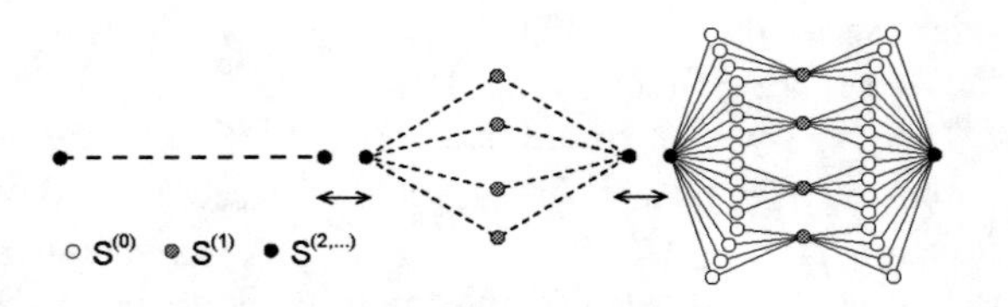

FIGURE 3. Sketch of the hierarchical lattice. The iterative construction corresponds to going from the left to the right in the figure. Each bond in one iterative step is replaced by four pairs of bonds with a new Ising spin in between.

Aging, Rejuvenation and Memory Effects

We consider a quench from $T=\infty$ to $T<T_c$ at time zero and study the spin autocorrelation function

$$C(t_m,t_n)=\sum_{\alpha}w_\alpha\sum_{i_\alpha}\left[\left\langle S_{i\alpha}^{(\alpha)}(t_n+t_m)S_{i\alpha}^{(\alpha)}(t_n)\right\rangle\right]_{\text{av}}, \quad (8)$$

where $\langle\ldots\rangle$ denotes a thermal average and an average over random initial spin orientations for a fixed realization of the disorder, and $[\ldots]_{\text{av}}$ denotes the disorder average over the random bonds. The $w_\alpha>0$, $\sum_\alpha N_\alpha w_\alpha=1$ (N_α: number of spins at level α), are weighting factors, which are chosen proportional to the connectivity $\propto4^\alpha$ of spins $\{S^{(\alpha)}\}$ (a per-site weighting with $w_\alpha=const.$ yields analogous results).

In the isothermal protocol (same temperature till time t_n+t_m) the correlation function $C(t_m,t_n)$ shows first a slow decay for times $t_m\leq t_n$, since spins up to levels m are revisited (again thermalized), which during the waiting time t_n already have been equilibrated with respect to the local field generated by the spins at higher levels $n'>n$. In this regime $C(t_m,t_n)$ is close to the quasi-equilibrium correlation function $C_{\text{eq}}(t_m)=$

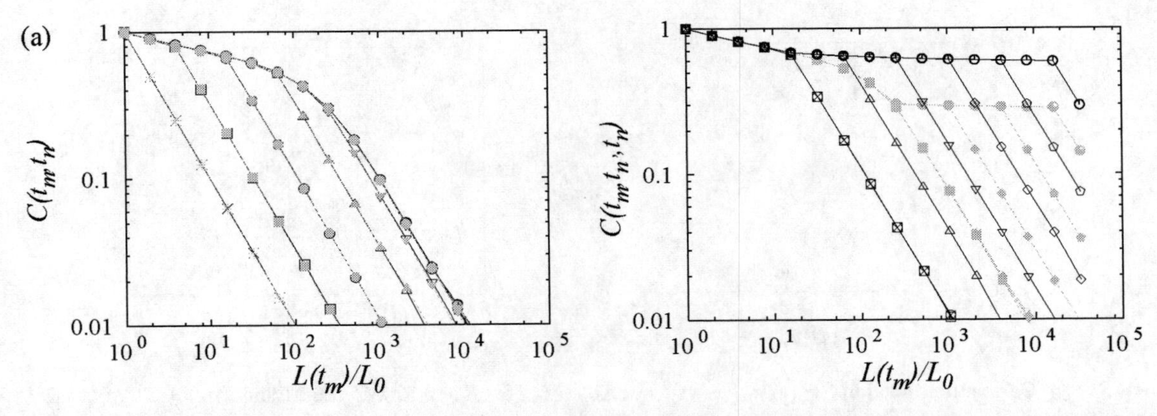

FIGURE 4. (Color online) *(a)* Correlation function $C(t_m,t_n)$ as a function of $L(t_m)$ in twin temperature shift protocols for $L(t_n)=2^0,2^2,\ldots,2^{14}$ (from left to right). The open symbols (black) refer to the case, where during the waiting time the temperature is hold at $T_1=0.3T_c$ and then switched to $T_2=0.7T_c$, while the filled symbols (green) refer to the reversed protocol ($T_1=0.7T_c$ during waiting time t_n, $T_2=0.3T_c$ during observation time t_m). *(b)* Correlation function $C(t_m,t_{n'},t_n)$ as a function of $L(t_m)$ after temperature cycling $T_1\to T_2\to T_1$ with $T_1=0.3T_c$ and $T_2=0.7T_c$. The duration of the intermediate stage $t_{n'}$ is fixed to $L(t_{n'})=2^4$ (open symbols, black) and $L(t_{n'}=2^8$ (filled symbols, green). The duration t_n of the first stage is varied as $L(t_n)=2^0,2^2,\ldots,2^{14}$ (from left to right). Both figures *(a)* and *(b)* were redrawn from [11].

$\lim_{n\to\infty}C(t_m,t_n)$ for infinite waiting time t_n, which for increasing t_m gradually approaches the EA order parameter q_{EA}. It can be shown [11] that in the low temperature limit $T\to 0$, $C_{\mathrm{eq}}(t_m)\sim q_{\mathrm{EA}}+c(T/J)L(T_m)^{-\theta}$, where $\theta\simeq 0.26$ is the stiffness exponent that characterizes the increase of the typical effective couplings (standard deviation of the distribution of couplings at level n) with the length scale in the spin glass phase, $J^{(n)}\sim L_n^{\theta}$. The constant c is related to the distributions $\rho_\alpha(\Delta_\alpha)$ of energy gaps $\Delta_\alpha=2|h_i^\alpha|$ associated with the effective local fields h_i^α at level α. These distributions for different α were found [11] to exhibit an analogous scaling form as the distribution of renormalized bonds [15] with respect to the typical energy scale $\tilde{\Delta}_\alpha=JL_\alpha^\theta$ at level α, i.e. $\rho_\alpha(\Delta_\alpha)=\tilde{\Delta}_\alpha^{-1}\tilde{\rho}(\Delta_\alpha/\tilde{\Delta}_\alpha)$. Calculations give $c=\tilde{\rho}(0)/[2(2^\theta-1)]$, where $\tilde{\rho}(0)>0$ is the finite value of the scaled distribution for zero energy gaps.

For times $t_m>t_n$ one enters the aging regime, where spins at levels $n<\gamma\leq m$ are newly thermalized and flip with probabilities $1/2$. The spins $S^{(\gamma)}$ are connected to bonds $J^{(\gamma)}$, which in turn are associated with unique clusters (or "droplets") of lower level spins that are traced out in the RSRG transformation to give $J^{(\gamma)}$. It can be shown [11] that a flip of $S^{(\gamma)}$ triggers flips of an $O(1)$ fraction of these spins in the associated clusters. As a consequence, the correlation functions decays as $C(t_m,t_n)=C(t_n,t_n)/L(t_m)$ in the aging regime $t_m>t_n$.

If the temperature is shifted from a value T_1 after the waiting time t_n to a temperature T_2, this aging behavior seemingly stops and the system rejuvenates when the length scale $L(t_n)$ explored during the waiting time t_n becomes large. This can be seen in Fig. 4a: For $L(t_n)>2^8$ a further slowing down of the decay of correlations with increasing t_n is hardly visible. More precisely speaking, the aging does not stop but it becomes progressively irrelevant as the limiting curve $C_\infty(t_m)=\lim_{n\to\infty}C(t_m,t_n)$ now decays to zero for $t_m\to\infty$ (i.e. not to a plateau value q_{EA} as in the isothermal protocol).

This behavior of $C_\infty(t_m)$ is caused by the change ("chaos") of the effective couplings induced by the temperature shift (predominantly through the T dependence of the RSRG transformations and often referred to as the "temperature chaos effect"). When the induced changes $\Delta J^{(\alpha)}$ become comparable to the characteristic energy gaps $\tilde{\Delta}_\alpha\sim JL_\alpha^\theta$ at some level α, the spins $\{S^{(\alpha)}\}$ have a high probability to flip and thereby trigger many further flips of lower level spins in the associated clusters as discussed above. Using droplet scaling theory, one can estimate $\Delta J^{(\alpha)}\sim\delta L_\alpha^{(d-1)/2}$, where $\delta=|T_1-T_2|/T_c$ is the perturbation and $L_\alpha^{(d-1)}$ the number of random links in the "domain wall" of the associated clusters. Accordingly, when comparing the two energy scales $\delta L_\alpha^{(d-1)/2}$ and JL_α^θ, one finds the characteristic length scale $L^\star(\delta)$ associated with the perturbation δ, the so-called overlap length, to scale as $L^\star(\delta)\sim\delta^{-1/\zeta}$, where $\zeta=(d-1)/2-\theta\simeq 0.74$ is the chaos exponent [15]. It is possible to scale the $C_\infty(t_m)$ for different δ by taking $L(t_m)/L^\star(\delta)$ as scaling variable and to calculate the behavior of the scaling function in the weakly perturbed regime $L(t_m)/L^\star(\delta)\ll 1$ and strongly perturbed regime $L(t_m)/L^\star(\delta)\gg 1$ [11]. These issues, however, will not further discussed here.

Let us point out that the behavior of $C(t_m,t_n)$ seen in Fig. 4a remains unchanged when T_1 and T_2 are inter-

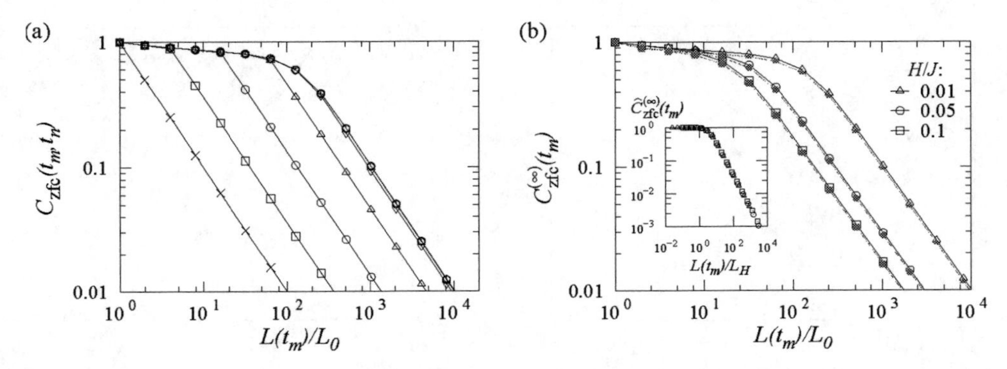

FIGURE 5. *(a)* Correlation function $C_{\text{zfc}}(t_m,t_n)$ as a function of $L(t_m)$ for $T=0.3T_c$, $H=0.01$ and $L(t_n)=2^0,2^2,\dots,2^{14}$ (from left to right); *(b)* Limiting curve $C^{(\infty)}_{\text{zfc}}(t_m)$ as a function of $L(t_m)$ for three different values of H (open symbols are for per-bond weighting, $w_\alpha \propto 4^\alpha$, and in addition the filled symbols mark data for per-site weighting, $w_\alpha = const.$) The inset demonstrates the scaling with respect to $L(t_m)/L_H$.

changed (for $t_m \geq t_n$ this is indeed exactly true). This symmetry in temperature-twin protocols has also been observed experimentally [16].

Next we study the spin correlations under temperature cycling, where the system evolves at a temperature T_1 for a time t_n and subsequently at another temperature T_2 for a time $t_{n'}$. Then the temperature is put back to T_1 and we measure the correlation function $C(t_m,t_{n'},t_n)$, which is the overlap between the spin configuration at $(t_n+t_{n'})$ and $(t_n+t_{n'}+t_m)$. Figure 4b shows the simulated data. The initial decay shows rejuvenation: The correlation function behaves in the same way as after a temperature shift $T_2 \to T_1$. If $L_{T_1}(t_n) > L_{T_2}(t_{n'})$, this regime terminates after the recovery time $t_{m^\star}$ determined by $L_{T_1}(t_{m^\star}) = L_{T_2}(t_{n'})$ [17]. There, a plateau region appears since the system's evolution recovers length scales already thermalized at temperature T_1 during the first stage. Eventually, for $t_m > t_n$, the system also memorizes the limits of this first thermalization and $C(t_m,t_{n'},t_n)$ decays as in the isothermal case.

It is interesting to note that analogous rejuvenation and memory effects are observed if the bare coupling J_{ij} are shifted randomly rather than the temperature. Indeed, in recent experiments on spin glasses [18] changes of the spin-spin interactions could be induced by photoillumination. The results of these experiments support the chaos picture. Also in recent Monte-Carlo simulations of the EA model on a simple cubic lattice the chaos picture could be clearly identified by using perturbations of the couplings [19]. The advantage of perturbing the couplings rather than the temperature is that one needs not to deal with the large changes of time scales caused by temperature shifts. In the effective dynamics defined for the hierarchical lattice this problem is circumvented by considering in fact length scales (or associated levels) through the updating of spins in epochs.

Magnetic Field Effects

In the presence of an external magnetic field H, the RSRG transformations get modified and include the changes of the field when tracing out the spin degrees of freedom of the next lower level. For two spins $S_i^{(n)}$, $S_j^{(n)}$ connected by 4 pairs of bonds J_{ik}, J_{kj}, $k=1,\dots,4$ with in between 4 spins $S_k^{(n-1)}$ in effective fields $H_k^{(n-1)}$ (cf. Fig. 3), they read [20]

$$J_{ij}^{(n)} = \frac{1}{4}\sum_{k=1}^{4} \log\left[\frac{A_{ikj}^{++}A_{ikj}^{+-}}{A_{ikj}^{-+}A_{ikj}^{--}}\right], \tag{9}$$

$$H_i^{(n)} = H_i^{(n-1)} + \frac{1}{4}\sum_{k=1}^{4} \log\left[\frac{A_{ikj}^{++}A_{ikj}^{-+}}{A_{ikj}^{+-}A_{ikj}^{--}}\right], \tag{10}$$

where $A_{ikj}^{++} = \cosh(J_{ik}^{(n-1)} + J_{kj}^{(n-1)} + H_k^{(n-1)})$, $A_{ikj}^{+-} = \cosh(J_{ik}^{(n-1)} + J_{kj}^{(n-1)} - H_k^{(n-1)})$, etc. The field $H_j^{(n)}$ is obtained by interchanging i and j in (10). Using these RSRG transformations, the systems evolves as in the field-free case by thermalization of spins in epochs.

Figure 5a shows the correlation function $C_{\text{zfc}}(t_m,t_n)$ in the zero-field cooling protocol, where after a quench from $T=\infty$ to $T<T_c$ the system first evolves for the time t_n without the field, and thereafter for the time t_m in a field $H>0$. The overall behavior $C_{\text{zfc}}(t_m,t_n)$ is similar to the one in Fig. 4a found for the temperature shift protocol. The limiting curve $C^{(\infty)}_{\text{zfc}}(t_m) = \lim_{n\to\infty} C_{\text{zfc}}(t_m,t_n)$ decreases to zero and the aging becomes irrelevant for $L(t_m)$ exceeding a characteristic length scale L_H. To derive the scaling of L_H with H we can compare the typical field energy $HL_\alpha^{d/2}$ at level α with the typical energy gap JL_α^θ in the field-free case. This yields $L_H \sim H^{-1/\zeta_H}$ with $\zeta_H = d/2 - \theta \simeq 1.24$. As shown in Fig. 5b, using $L(t_m)/L_H$ as scaling variable allows one to collapse the

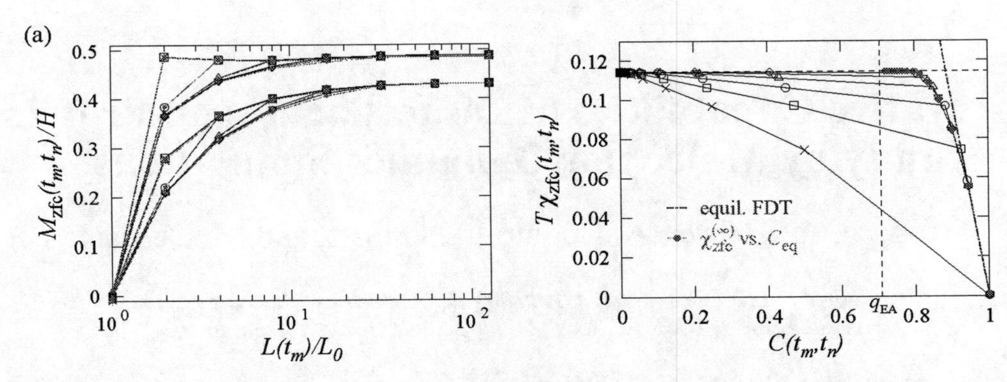

FIGURE 6. *(a)* Susceptibility $\chi_{zfc}(t_m,t_n) = M_{zfc}(t_m,t_n)/H$ as a function of $L(t_m)$ for $T = 0.3T_c$, $H = 0.1$ (open symbols), $H = 0.05$ (filled symbols) and $L(t_n) = 2^0, 2^1, \ldots, 2^5$ (from left to right). The lowest curve of each set correspond to the limit $t_n \to \infty$. The upper set of curves refers to per-site weighting with $w_\alpha = const.$, while the lower one refers to per-bond weighting with $w_\alpha = 4^\alpha$. *(b)* Parametric plot of $C(t_m,t_n)$ versus $T\chi_{zfc}(t_m,t_n)$ with t_m as parameter for the same temperature, magnetic fields and waiting times as in *(a)* ($L(t_n)$ increasing from below; only data for per-bond weighting are shown). Also shown are the limit $t_n \to \infty$ (•), the EA order parameter for the simulated temperature, and the line $y = 1 - x$ pertaining to the equilibrium FDT.

$C^{(\infty)}_{zfc}(t_m)$ for different H onto a common scaling curve.

Finally, we shortly discuss the behavior of the magnetization $M_{zfc}(t_m,t_n) = \sum_\alpha w_\alpha \sum_{i_\alpha} [\langle S^{(\alpha)}_{i_\alpha}(t_m)\rangle]_{av}$ in the zero-field cooling protocol and the violation of the fluctuation-dissipation theorem (FDT)

$$\chi_{zfc}(t_m,t_n) = \frac{M_{zfc}(t_m,t_n)}{H} = \frac{1}{T}\left[1 - C(t_m,t_n)\right]. \quad (11)$$

$M_{zfc}(t_m,t_n)/H$ as a function of $L(t_m)$ is shown in Fig. 6a (lower set of curves refers to per-bond weighting): It grows more slowly with increasing waiting time t_n and the collapse of data for the two different fields proofs that the linear response regime is tested. The parametric plot of $\chi_{zfc}(t_m,t_n)$ vs. $C(t_m,t_n)$ with t_m as a parameter shown in Fig. 6b appears as a mixture of what is expected for different types of systems: The plateau part is similar to the behavior in simple coarsening models, while the bent part is reminiscent of mean-field spin glass models or the EA model on ordinary lattices.

ACKNOWLEDGMENTS

We thank J. P. Bouchaud and B. Rinn for many stimulating discussions.

REFERENCES

1. L. C. E. Struick, *Physical Aging in Amorphous Polymers and Other Materials*, Elsevier, Houston, 1978.
2. K. S. Sinnathamby, H. Oukris and N. E. Israeloff, *Phys. Rev. Lett.*, **95**, 067205-1-067205-4 (2005).
3. E. Vincent, J. Hammann, M. Ocio, J.-P. Bouchaud and L. Cugliandolo, in *Complex Behaviour of Glassy Systems*, edited by M. Rubi, Lecture Notes in Physics, Springer Verlag, Berlin, 1997, Vol. 492, pp. 184-219, and refs. therein; P. Nordblad and P. Svendlish, in: *Spin Glasses and Random Fields*, edited by P. Young, World Scientific, Singapore, 1998, pp. 1-27.
4. P. E. Jönsson, R. Mathieu, P. Nordblad, H. Yoshino, H. A. Katori and A. Ito, *Phys. Rev. B*, **70**, 174402-1-174402-27 (2004).
5. L. Cipelletti, S. Mansley, R. C. Ball and D. A. Weitz, *Phys. Rev. Lett.*, **84**, 2275-2278 (2000).
6. S. Manley *et al.*, *Phys. Rev. Lett.*, **95**, 048302-1-048302-4 (2005).
7. A. Barrat and V. Loreto, *Europhys. Lett.*, **53**, 297-303 (2001).
8. K. Jonason, E. Vincent, J. Hamann, J.-P. Bouchaud and P. Nordblad, *Phys. Rev. Lett.*, **81**, 3243-3246 (1998).
9. T. Jonsson, K. Jonason, P. Jönsson and P. Nordblad, *Phys. Rev. B*, **59**, 8770-8777 (1999).
10. B. Rinn, P. Maass and J.-P. Bouchaud, *Phys. Rev. Lett.*, **84**, 5403-5406 (2000); *ibid.*, *Phys. Rev. B*, **64**, 104417-1-104417-15 (2001).
11. F. Scheffler, H. Yoshino and P. Maass, *Phys. Rev. B*, **68**, 060404-1-060404-14 (2003).
12. E. M. Bertin and J.-P. Bouchaud, *Phys. Rev. E*, **67**, 026128-1-026128-20 (2003).
13. C. Monthus and J.-P. Bouchaud, *J. Phys. A*, **29**, 3847-3869 (1996).
14. L. Cugliandolo and J. Kurchan, *J. Phys. A*, **27**, 5749-5772 (1994).
15. D. S. Fisher and D. A. Huse, *Phys. Rev. B*, **38**, 386-411 (1988); *ibid.* 373-383 (1988).
16. P. E. Jönsson, H. Yoshino and P. Nordblad, *Phys. Rev. Lett.*, **89**, 097201-1-097201-4 (2002); *ibid.* **90**, 059702-1-059702-4 (2003).
17. H. Yoshino, A. Lemaitre and J.-P. Bouchaud, *Eur. Phys. J. B*, **20**, 367-395 (2001).
18. R. Arai, K. Katsuyoshi and T. Sato, *Phys. Rev. B*, **75**, 144424-1-144424-8 (2007).
19. H. G. Katzgraber and F. Krzącała, *Phys. Rev. Lett.*, **98**, 017201-1-017201-4 (2007).
20. F. Scheffler, *Spin Dynamics in Disordered Systems*, dissertation, Universität Konstanz, Germany, 2006.

Biopreservative Capabilities of Disaccharides on Proteins : A Study by Molecular Dynamics Simulations

F. Affouard, A. Lerbret, A. Hédoux, Y. Guinet, and M. Descamps

Laboratoire de Dynamique et Structure des Matériaux Moléculaires
Université Lille 1, UMR CNRS 8024, 59655 Villeneuve d'Ascq, France

Abstract. A comparative investigation of lysozyme in trehalose, sucrose and maltose aqueous solutions has been performed using Molecular Dynamics simulations. The vibrational properties in the low frequency spectral range [0 - 350] cm^{-1} were mainly analyzed. This study confirms that the hydrogen bond (HB) network of water is highly dependent on the presence of sugars and contributes to the stabilization of lysozyme. The favored interaction of trehalose with water is confirmed below a threshold weight sugar concentration of about 50 %. Above this concentration and unlikely to the sugar/water binary mixtures, trehalose becomes less efficient to distort the tetra-bonded HB network of water than maltose.

Keywords: Biopreservation, Lysozyme, Molecular dynamics, Trehalose
PACS: 61.20.Ja, 64.70.Pf

INTRODUCTION

Biopreservation is an important problem for pharmaceutical industries to use peptides or proteins as therapeutic agents. Indeed, biomolecules are only stable in very limited conditions. During heating, cooling, dehydration or a pH change of the solvent [1, 2], they may exhibit an irreversible transformation to a denaturated state in which they are no longer active. Sugars are often used as additives in pharmaceutical formulations to protect biomaterials during freeze-drying processes and to improve long term storage [3]. However, these procedures are highly empirical and the microscopic preservation mechanisms are not yet fully understood. In recent years, trehalose, [α-D-glucopyranosyl-α-D-glucopyranoside], a non-reducing disaccharide, has received considerable attention for its exceptional biopreservative efficiency under extreme conditions. Several hypotheses have been suggested to explain the superior effectiveness of trehalose and more generally to decipher the bioprotection mechanisms. However, none of these hypotheses is fully satisfactory (see [4] and references within).

Green and Angell [5] have suggested that the high glass transition temperature of trehalose $T_g \simeq 120°$ C compared to other protectants could explain its exceptional ability to protect. Indeed, at high concentrations, trehalose may form an amorphous solid matrix in which biomolecules would be embedded. Therefore, biomolecules would be caged by the glassy solvent and their flexibility highly reduced. However this hypothesis does not explain why a 0.95:0.05 trehalose/glycerol mixture, with a lower T_g than pure trehalose, is more effective [6, 7].

Furthermore, it is well known that the activity of biomolecules highly depends on their 3-dimensional structure. Their native folded structure is stabilized by the hydrogen bonds (HBs) they form with water molecules. Crowe *et al.* have proposed that sugars would replace water molecules during dehydration to prevent the denaturing stress caused by the removal of the first water hydration shell of biomolecules, thus preserving their native structure [4, 8]. According to this scheme, trehalose would form a larger number of HBs with biomolecules owing its larger hydration number [9]. Although many experiments [10] and simulations [11] have shown that trehalose can be directly hydrogen bonded to proteins, trehalose cannot enter into internal or confined regions of proteins [12]. In addition, Timasheff *et al.* [13] have shown that many osmoprotectants are excluded from the first hydration shell of proteins at moderate concentrations. This latter behavior has also been extended to higher sugar concentrations by Belton and Gil [14] and recently by Cottone *et al.* who demonstrated that sugars are preferentially excluded from the protein surface [15–17].

The destructuring effect proposed by Magazù *et al.* [18, 19] is based on the fact that the presence of sugars leads to a perturbation on the tetrahedral HB network of water. The main reason is the possibility of HBs formation between sugars and water molecules. Therefore, sugars could inhibit ice formation in water at low temperatures. Trehalose would be the most efficient in destructuring the tetrahedral HB network of water by promoting a more extended hydration than the other disaccharides. This hypothesis seems relevant to explain the cryoprotection but not the lyoprotection of biomolecules.

Raman scattering and modulated differential scanning calorimetry investigations of the sugar-induced ther-

CP982, *Complex Systems, 5th International Workshop on Complex Systems*
edited by M. Tokuyama, I. Oppenheim, and H. Nishiyama

mostabilization of hen egg-white lysozyme have been recently performed by the authors [20, 21]. This study revealed that sugars distort the HB network of water and strongly contribute to the stabilization of the native tertiary structure of lysozyme. They also reduce the first stage of denaturation, *i.e.* the transformation of the tertiary structure into a highly flexible state with intact secondary structure. Trehalose was shown to exhibit superior capabilities to distort the tetra- bonded HB network of water. In [21], it was found that this effect is responsible for the stability of the tertiary structure.

In order to better understand the physical properties of sugars on the HB network of water and their influence on the protein stability we have performed a Molecular Dynamics (MD) investigation of hen egg-white lysozyme in presence of three homologous disaccharides : trehalose, sucrose [β-D-fructofuranosyl-α-D-glucopyranoside] and maltose [4-O-(α-D-glucopyranosyl)-β-D-glucopyranoside]. We have especially focused our study on the influence of disaccharides on the protein vibrational properties and on the strengthening of water inter-molecular HBs. It should be noted that the HB capabilities of these three sugars are directly comparable since they possess the same chemical formula $C_{12}H_{22}O_{11}$ and the same number of hydroxyl groups (8).

COMPUTATIONAL DETAILS

Numerous simulation investigations have been performed on the structure and the dynamics of binary sugar/water solutions but fewer have been focused on ternary biomolecule/sugar/water systems (see [9] and references within). In the present study, Molecular Dynamics simulations of lysozyme in trehalose, sucrose and maltose aqueous solutions have been performed using the CHARMM program [22], version 29b1. The all-atom CHARMM22 force field [23] has been used to model the protein. The CSFF carbohydrate force field [24] has been considered for disaccharides and water molecules were represented by the SPC/E model [25]. The production simulations were performed in the isochoric-isothermal (N,V,T) ensemble. The length of all covalent bonds involving an hydrogen atom as well as the geometry of water molecules were kept fixed using the SHAKE algorithm [26], with a relative tolerance of 10^{-5}. A 2-fs timestep has been used to integrate the equations of motion with the verlet leapfrog algorithm [27].

During the different stages of the simulations, the temperatures have been maintained constant with weak coupling to a heat bath (Berendsen thermostat [28]) with a relaxation time of 0.2 ps. A cutoff radius of 10 Å has been used to account for van der Waals interactions, which were switched to zero between 8 and 10 Å . A Lennard-Jones potential has been employed to represent van der Waals interactions and Lorentz-Berthelot mixing-rules have been used for cross-interaction terms. Electrostatic interactions have been handled by the particle mesh Ewald (PME) [29] method with $\kappa = 0.32$ Å^{-1} and the fast-Fourier grid densities set to $\sim$ 1/Å (48 and 64 grid points in the X/Y and Z directions, respectively).

The starting structure of lysozyme was obtained from the X-ray crystal structure solved at 1.33 Å (193L entry of the Brookhaven Protein Data Bank) [30]. Most probable charge states at pH 7 were chosen for ionizable residues. The total charge of lysozyme (+8 *e*) was then neutralized by uniformly rescaling the charge of protein atoms, similarly to ref. [31]. The disaccharide initial conformations have been deduced from neutron and X-ray studies of trehalose [32], maltose [33] and sucrose [34]. The sugar concentrations on a protein-free basis are 37, 50 and 60 wt %. These concentrations have been purposefully chosen based on our previous study of sugar/water solutions [9, 35]. Indeed, we showed that the relative effect of sugars on water may be distinguished above a threshold concentration of about 40 wt %. Therefore, their relative influence on lysozyme at ambient temperature may be characterized above this concentration. Lysozyme and its 142 crystallographic hydration water molecules were first placed in an orthorhombic box with cell parameters a = b = 46.7 Å and c = 62.2 Å. Then, disaccharide molecules were located and oriented randomly around lysozyme, with minimum sugar-protein and sugar-sugar distance criteria, which ensure an isotropic distribution of sugars around lysozyme. Finally, water molecules non-overlapping with either lysozyme or sugars were randomly added in the simulation box. Initial configurations were minimized in three steps, keeping first lysozyme and sugars fixed, then keeping only lysozyme fixed and finally keeping free all molecules. This minimized configuration was heated to 473 K in the canonic ensemble during 1 to 3 ns, while maintaining fixed the conformation of lysozyme to prevent conformational changes. This aimed at equilibrating solvent configurations, particularly the position and orientation of sugars. Then, the resulting configurations were thermalized at 300K and simulated in the isobaric-isothermal (N,P,T) ensemble. The stabilized volume of the simulation box during this simulation was considered to compute the averaged density of the system and used to perform the subsequent simulations in the (N,V,T) ensemble. A steepest-descent minimization procedure of 1000 iterations was then performed, while applying a decreasing harmonic potential on atoms of lysozyme. After the minimization procedure, the temperature was raised from 0 to 300 K, with a 5-K temperature increase every 250 steps. Then, an equilibration at 300 K was performed during about 80 ps. Finally, simulations of 10, 12 and 17

TABLE 1. System compositions (where N_L, N_S and N_W denote the number of lysozyme, sugar and water molecules, respectively), densities, and equilibration/simulation times for the different sugar concentrations ϕ (on a protein-free basis). Data corresponding to $\phi = 0$ wt % result from only one simulation of the lysozyme/pure water solution. T, M and S denote trehalose, maltose and sucrose respectively.

ϕ (wt %)	$N_L/N_S/N_W$	density ($g.cm^{-3}$)			Eq./Sim. time (ns)
		T	M	S	T, M, S
0	1/0/3800	1.04	1.04	1.04	2/8
37	1/85/2800	1.16	1.16	1.15	2/8
50	1/125/2400	1.20	1.21	1.20	2/10
60	1/165/2100	1.24	1.25	1.24	4/13

ns were performed for the systems at concentrations of 37, 50 and 60 wt %, respectively, and configurations were saved every 0.25 ps. A control simulation of lysozyme in pure water was done in an analogous way as the one described above. In this simulation, the orthorhombic box was directly filled with water molecules. Moreover, this system was not heated at 473 K, since water molecules are much more mobile than sugars. The first two and four ns were not considered to compute the structural and dynamical properties presented in this paper for the 0-50 and 60 wt % systems, respectively. Table 1 summarizes some simulation data for the different systems considered in the present study.

RESULTS AND DISCUSSION

The effects of the studied sugars at a concentration of 40 wt % on the thermal denaturation of lysozyme have recently been investigated by Raman scattering and modulated differential scanning calorimetry experiments [20, 21]. Trehalose has been found the most effective sugar for stabilizing the tertiary structure of lysozyme and its enhanced efficiency has been primarily attributed to its greater ability to distort and strengthen the HB network of water.

In order to get useful complementary results on the influence of sugars on the vibrational properties of lysozyme and water, we have computed the low-frequency ($\omega < 350$ cm^{-1}) vibrational densities of states (VDOS) of lysozyme and water. They can be calculated from the Fourier transform of the autocorrelation function of atomic velocities :

$$c_{vv}(t) = \frac{1}{M}\sum_{i=1}^{N} m_i < \mathbf{v}_i(0).\mathbf{v}_i(t) > \quad (1)$$

where $\mathbf{v}_i$ is the velocity vector of atom i and m_i its mass, and $M = \sum_i m_i$ is either the mass of lysozyme or the total mass of all water molecules. The VDOS is then written as :

$$g_{vv}(\omega) = \int_0^{\infty} cos(\omega t)c_{vv}(t)dt \quad (2)$$

The figure 1 shows the VDOS of lysozyme in the [0-350 cm^{-1}] range for the different studied systems. It is composed of two main bands centered on about 60 cm^{-1} and 230 cm^{-1}. The general shape is in qualitative agreement with the one obtained experimentally by Colaianni and Nielsen [36], although the position of the peaks are located at lower frequencies in our study. The first peak has often been attributed to the *cage* effect, and is observed in many liquids [37, 38]. It is sometimes referred to as the *boson peak*. Its width represents the heterogeneity of the local environments experienced by individual atoms. Another very broad band centered at about 230 cm^{-1} appears. This band could arise from librational motions of atoms of lysozyme surface residues exposed to the solvent. The presence of sugars seems to induce two significant changes on the VDOS of lysozyme. The first one is the sharpening of the boson peak, which amplifies with sugar concentration. This effect is well in line with the reduction of protein atomic fluctuations observed in [39]. Sugars would therefore be likely to reduce absolute differences of cage effects among the different lysozyme atoms. In pure water, the amplitude of atomic motions is limited by steric constraints imposed by the other residues of the polypeptidic chain. Motions of the surface residues are thus more free, and may be favored by HBs with water molecules. Sugars would strongly reduce the amplitude of motions of residues with which they form HBs, thus imposing an additional constraint on them in comparison to the pure water solution case. The second main consequence of the presence of sugars is that the second band is much more defined. The width of this band is lower and its amplitude raises and becomes comparable to that of the first band. Furthermore, its position shifts from $\sim$ 220 cm^{-1} up to $\sim$ 250 cm^{-1} when the sugar concentration raises from 37 to 60 wt %. This band is located in about the same frequency range as the second main low-frequency band of pure water, which is generally considered to be representative of the formation of HBs. This observed effect on the second band could therefore result from the strengthening of water-lysozyme HBs, as well as the formation of sugar-lysozyme HBs.

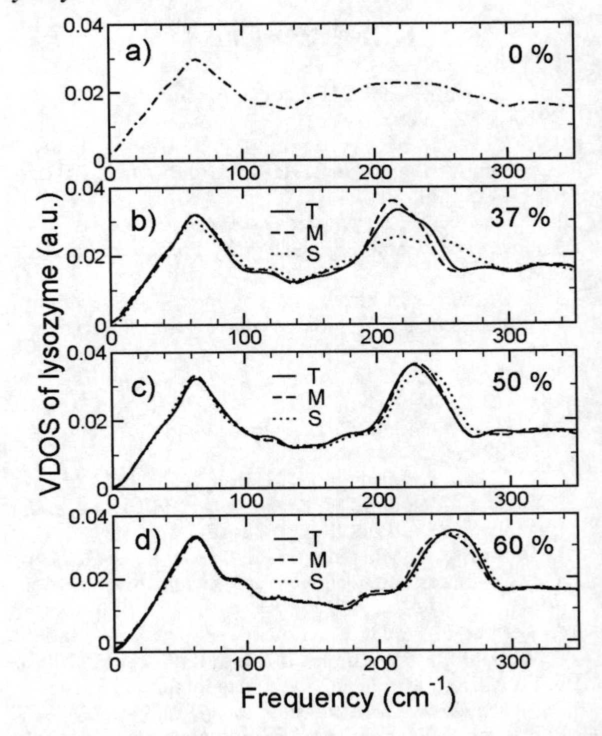

FIGURE 1. Vibrational density of states (VDOS) of lysozyme for the pure water/lysozyme solution (a) and for the different ternary solutions (b-d). Curves have been smoothed with the Savitzky-Golay algorithm [40] to make easier the comparison of results.

The VDOS of water has been calculated to know if the effects observed on lysozyme could actually arise from a strengthening of the HB network of water. It is displayed in figure 2. The VDOS shape is in agreement with the results of previous experimental [41] and numerical studies [38, 42] performed on pure water. The broadening of the first band induced by the addition of sugars may reflect the increased heterogeneity of the local environments sampled by water molecules. In addition, the amplitude of the second band decreases when the sugar concentration increases, in line with the destructuring effect of sugars on the HB network of water [18, 43] and with the results of Raman scattering experiments [21]. The concomitant increase of the frequency position of the two bands of water suggests an increased constraint imposed by sugars on water molecules. Their lower mobility may lead to tighter cages and stronger HBs. A quantitative comparison between the different solutions has been performed by arbitrarily fitting the [0-350] cm^{-1} range with a log-normal and a Gaussian functions, as shown in figure 2a.

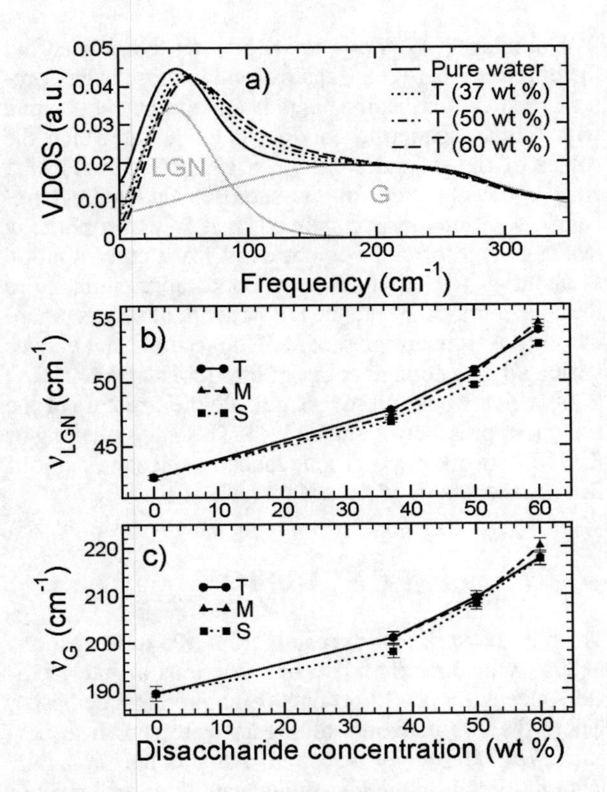

FIGURE 2. (a) Vibrational density of states (VDOS) of water for the pure water/lysozyme solution and for the different trehalose solutions in the [0-350] cm^{-1} frequency range. Curves have been smoothed with the Savitzky-Golay algorithm [40] to simplify the comparison of results. The low-frequency VDOS has been fitted with a log-normal (LGN) and a Gaussian (G) curves. The dependences on the sugar concentration of the frequency positions ν_{LGN} and ν_G of these two functions are displayed in (b) and (c).

The positions of the two bands ν_{LGN} and ν_G are similar for the different sugar solutions at all concentrations (see figures 2b,c) but some tendencies clearly emerge. The position of the first band ν_{LGN} is always found at a lower frequency in the sucrose solutions than in the trehalose and maltose ones. Therefore, the addition of sucrose leads to a weaker distortion of the HB network of water than that of trehalose or maltose. Except for the 60 wt % concentration, the highest frequencies of the water ν_{LGN} band are found in presence of trehalose. It confirms the favored interaction of this sugar with water and its capability to perturb the HB water network more efficiently. It should be mentioned that this result is consistent with the data obtained from the Raman scattering measurements of lysozyme in disaccharide/water solutions at the 40 wt % concentration [44]. However, an inversion is detected at the 60 wt % concentration, so that the position of the first band ν_{LGN} is at a higher frequency in the maltose solution than in the trehalose one.

A similar trend is observed for the second band but the difference between trehalose and maltose is less pronounced than for the first band. A fair agreement is found with Raman scattering data at 40 wt % for which the values of the second band ν_G = 193, 194.5 and 198.5 cm^{-1} were obtained for the sucrose, maltose and trehalose solutions respectively [44]. It is worth noticing that a change of slope is observed for a concentration of about 40 wt %, which corresponds approximately to the concentration where the HB network of sugars percolates [9]. A frequency increase of the second band may be attributed to a strengthening of intermolecular O-H···O HBs, which would make less flexible the structure of the protein in presence of sugars [21]. This strengthening of the HB network of water may indeed reduce the motions that are precursors of the unfolding process.

CONCLUSION

The present work reports results from MD investigations of lysozyme/disaccharide/water solutions at intermediate concentrations (37-60 wt % on a protein-free basis), where the HB networks of sugars were shown to percolate [9]. This study reveals that sugars have a significant influence on the low-frequency vibrational density of states (VDOS) of lysozyme. They induce a sharpening of the first band ($\sim$ 60 cm^{-1}), related to the cage effect, and the emergence of the second band ($\sim$ 230 cm^{-1}), which may reflect a strengthening of solvent-protein interactions associated to an increase of the solvent viscosity. The analysis of the VDOS of water indeed clearly shows a strengthening of its HB network. These results are in agreement with those recently reported on the denaturation of lysozyme in presence of sugars at 40 wt % by Raman scattering and modulated differential scanning calorimetry investigations [21]. They suggest that sugars may hinder the unfolding of lysozyme by strengthening the protein-solvent HB network, thus making lysozyme less flexible and less sensitive to a temperature increase. Below the 50 wt % sugar concentration, trehalose is found the most efficient to distort the tetra- bonded HB network of water as already reported in the literature for binary sugar/water mixtures. Above 50 wt %, a change between maltose and trehalose is detected and trehalose becomes less efficient than maltose. Additional experimental investigations are clearly needed in order to fully validate the MD results.

ACKNOWLEDGMENTS

The authors wish to acknowledge the use of the facilities of the IDRIS (Orsay, France) and the CRI (Villeneuve d'Ascq, France) where calculations were carried out. This work was supported by the INTERREG III (FEDER) program (Nord-Pas de Calais/Kent).

REFERENCES

1. J. F. Carpenter, and J. H. Crowe, *Biochemistry* **28**, 3916–3922 (1989).
2. J. V. Ricker, N. M. Tsvetkova, W. F. Wolkers, C. Leidy, F. Tablin, M. Longo, and J. H. Crowe, *Biophys. J.* **84**, 3045–3051 (2003).
3. F. Franks, editor, *Biophysics and Biochemistry at low Temperatures*, Cambridge University Press, Cambridge, 1985.
4. J. H. Crowe, L. M. Crowe, A. E. Oliver, N. Tsvetkova, W. Wolkers, and F. Tablin, *Cryobiology* **43**, 89–105 (2001).
5. J. L. Green, and C. A. Angell, *J. Phys. Chem.* **93**, 2880–2882 (1989).
6. M. T. Cicerone, and C. L. Soles, *Biophys. J.* **86**, 3836–3845 (2004).
7. G. Caliskan, D. Mechtani, J. H. Roh, A. Kisliuk, A. P. Sokolov, S. Azzam, M. T. Cicerone, S. Lin-Gibson, and I. Peral, *J. Chem. Phys.* **121**, 1978–1983 (2004).
8. L. M. Crowe, R. Mouradian, J. H. Crowe, S. A. Jackson, and C. Womersley, *Biochim. Biophys. Acta* **769**, 141–150 (1984).
9. A. Lerbret, P. Bordat, F. Affouard, M. Descamps, and F. Migliardo, *J. Phys. Chem. B.* **109**, 11046–11057 (2005).
10. S. D. Allison, B. Chang, T. W. Randolph, and J. F. Carpenter, *Arch. Biochem. Biophys.* **365**, 289–298 (1999).
11. R. D. Lins, C. S. Pereira, and P. H. Hünenberger, *Proteins: Structure, Function, and Bioinformatics* **55**, 177–186 (2004).
12. G. M. Sastry, and N. Agmon, *Biochemistry* **36**, 7097–7108 (1997).
13. S. N. Timasheff, *Biochemistry* **41**, 13473–13482 (2002).
14. P. S. Belton, and A. M. Gil, *Biopolymers* **34**, 957–961 (1994).
15. G. Cottone, G. Ciccotti, and L. Cordone, *J. Chem. Phys.* **117**, 9862–9866 (2002).
16. G. Cottone, S. Giuffrida, G. Ciccotti, and L. Cordone, *Proteins: Structure, Function, and Bioinformatics* **59**, 291–302 (2005).
17. G. Cottone, *J. Phys. Chem. B* **111**, 3563–3569 (2007).
18. C. Branca, S. Magazù, G. Maisano, and P. Migliardo, *J. Chem. Phys.* **111**, 281–287 (1999).
19. C. Branca, S. Magazù, G. Maisano, P. Migliardo, and E. Tettamanti, *J. Mol. Struct.* **480-481**, 133–140 (1999).
20. R. Ionov, A. Hédoux, Y. Guinet, P. Bordat, A. Lerbret, F. Affouard, D. Prévost, and M. Descamps, *J. Non. Cryst. Solids* **352**, 4430–4436 (2006).
21. A. Hédoux, J.-F. Willart, R. Ionov, F. Affouard, Y. Guinet, L. Paccou, A. Lerbret, and M. Descamps, *J. Phys. Chem. B* **110**, 22886–22893 (2006).
22. B. R. Brooks, R. E. Bruccoleri, B. D. Olason, D. J. States, S. Swaminathan, and M. Karplus, *J. Comp. Chem.* **4**, 187–217 (1983).
23. A. D. Mackerell, D. Bashford, R. L. Bellott, R. L. Dunbrack, J. D. Evanseck, M. J. Field, S. Fischer, J. Gao, H. Guo, S. Ha, D. Joseph-McCarthy, L. Kuchnir, K. Kuczera, F. T. K. Lau, C. Mattos, S. Michnick,

T. Ngo, D. T. Nguyen, B. Prodhom, W. E. Reiher, B. Roux, M. Schlenkrich, J. C. Smith, R. Stote, J. Straub, M. Watanabe, J. Wiorkiewicz-Kuczera, D. Yin, and M. Karplus, *J. Phys. Chem. B* **102**, 3586–3616 (1998).

24. M. Kuttel, J. W. Brady, and K. J. Naidoo, *J. Comput. Chem.* **23**, 1236–1243 (2002).
25. H. J. C. Berendsen, J. R. Grigera, and T. P. Straatsma, *J. Phys. Chem.* **91**, 6269–6271 (1987).
26. J. P. Ryckaert, G. Ciccotti, and H. J. C. Berendsen, *J. Comput. Phys.* **23**, 327–341 (1977).
27. R. W. Hockney, *Meth. Comp. Phys.* **9**, 136–211 (1970).
28. H. J. C. Berendsen, J. P. M. Postma, W. F. van Gunsteren, A. DiNola, and J. R. Haak, *J. Chem. Phys.* **81**, 3684–3690 (1984).
29. U. Essmann, L. Perera, M. L. Berkowitz, T. Darden, H. Lee, and L. G. Pedersen, *J. Chem. Phys.* **103**, 8577–8593 (1995).
30. M. C. Vaney, S. Maignan, M. Riès-Kautt, and A. Ducruix, *Acta Cryst. D* **52**, 505–517 (1996).
31. F. Sterpone, M. Ceccarelli, and M. Marchi, *J. Mol. Biol.* **311**, 409–419 (2001).
32. T. Taga, M. Senma, and K. Osaki, *Acta Cryst.* **B28**, 3258–3263 (1972).
33. M. E. Gress, and G. A. Jeffrey, *Acta Cryst.* **B33**, 2490–2495 (1977).
34. G. M. Brown, and H. A. Levy, *Acta Cryst.* **B29**, 790–797 (1973).
35. P. Bordat, A. Lerbret, J.-P. Demaret, F. Affouard, and M. Descamps, *Europhys. Lett.* **65**, 41–47 (2004).
36. S. E. M. Colaianni, and O. F. Nielsen, *J. Mol. Struct.* **347**, 267–284 (1995).
37. A. Idrissi, S. Longelin, and F. Sokolic, *J. Phys. Chem. B* **105**, 6004–6009 (2001).
38. J. A. Padro, and J. Marti, *J. Chem. Phys.* **118**, 452–453 (2003).
39. A. Lerbret, P. Bordat, F. Affouard, A. Hédoux, Y. Guinet, and M. Descamps, *to appear in J. Phys. Chem. B* (2007).
40. A. Savitzky, and M. J. E. Golay, *Anal. Chem.* **36**, 1627–1639 (1964).
41. G. E. Walrafen, M. R. Fisher, M. S. Hokmabadi, and W.-H. Yang, *J. Chem. Phys.* **85**, 6970–6982 (1986).
42. I. Ohmine, and S. Saito, *Acc. Chem. Res.* **32**, 741–749 (1999).
43. A. Lerbret, P. Bordat, F. Affouard, Y. Guinet, A. Hédoux, L. Paccou, D. Prévost, and M. Descamps, *Carbohydr. Res.* **340**, 881–887 (2005).
44. A. Hédoux, F. Affouard, M. Descamps, Y. Guinet, and L. Paccou, *J. Phys.: Cond. Matter* **19**, 205142–205149 (2007).

Ice Film Morphologies and the Structure Zone Model

Julyan H. E. Cartwright*, Bruno Escribano*, Oreste Piro†, C. Ignacio Sainz-Diaz*, Pedro A. Sánchez†, and Tomás Sintes†

*Laboratorio de Estudios Cristalográficos, CSIC, E-18100 Armilla, Granada, Spain
†Dept de Física e Instituto de Física Interdisciplinar y Sistemas Complejos IFISC (CSIC-UIB), Universitat de les Illes Balears, E-07122 Palma de Mallorca, Spain

Abstract. Ice, the solid phase of water, is ubiquitous. A knowledge of ice helps us to comprehend water, a simple molecule, but one with much complex behaviour. Our aim is to understand the morphologies and physics of thin icy films. To treat this complex system we have developed new experimental capabilities with an environmental scanning electron microscope (ESEM) capable of working with ice films, at the same time as new simulation approaches to understanding the physics of ice morphology. A comprehension of the physics of thin-film morphologies has applicability beyond ice to thin films of metals, ceramics, and other materials.

Keywords: Amorphous ice, Phase transitions, Structure zone model
PACS: 68.37.Hk, 68.55.Jk, 81.15.Kk, 96.25.hf

INTRODUCTION

Water exists all around us, as vapour, liquid, and solid. Despite its ubiquity and importance, there remains much to be discovered about this simple molecule. The anomalous nature of water is manifest to anyone who has pondered why, unlike the great majority of substances, the solid floats on the liquid. Although there are myriad forms of the solid phase of water — ice — found on Earth, from snowflakes to icebergs, with very different bulk properties, at the molecular scale they are all composed of the same polymorph, hexagonal ice — ice Ih — which is the stable version at the temperatures and pressures in our biosphere. However, in the laboratory, beyond the range of natural terrestrial temperatures and pressures, ice crystallizes as many other polymorphs [1]. It also appears to have more than one amorphous state. At least two apparently quite distinct amorphous ices have been described: low density amorphous (LDA) and high density amorphous (HDA). These amorphous forms are glasses, and have been investigated intensively, being phases that help us to understand the structure on the molecular scale of liquid water [2, 3, 4].

There is considerable debate surrounding amorphous ice obtained in the laboratory. LDA and HDA have different densities and other physical properties, and the transition between them is reminiscent of a first-order phase transition. This has led to the suggestion that there may be two associated liquid phases, low density liquid (LDL) and high density liquid (HDL), together with a second critical point in the phase diagram of water. LDA and HDA, as structurally arrested versions of these two liquid phases, would then constitute genuinely distinct amorphous solid phases, or polyamorphs. This viewpoint is challenged by others who note that amorphous ices produced using different methods have somewhat different physical properties such as density and thermal conductivity. Do these differences simply reflect differences in bulk morphologies, while the underlying phase is either LDA or HDA, just as both snowflakes and icebergs are composed of ice Ih, or are LDA and HDA not really distinct polyamorphs, but just the two extremes of a continuum of amorphous states?

On the other hand, the study of the thin-film deposition of materials such as ceramics and metals is an extremely active field. For some time it has been known that, independently of the material being deposited, similar deposition conditions lead to similar characteristic morphologies. A structure zone model has been been developed, which predicts the bulk morphology depending on the film surface temperature and the gas pressure [5, 6]. The main experimental aim of this work has been to determine whether the morphologies characteristic of thin-film deposition of other materials are seen in ice deposited by ballistic deposition. Alongside our experimental work, we have investigated with computer simulations the dynamical basis of the structure zone model of thin-film deposition. With this we wish to gain insights into the physical basis of the different morphologies observed in thin-film deposition experiments and eventually to answer the question: How does the structure zone model arise from the basic physics of the deposition process? The applicability of these results is much more general than to ice alone, as understanding the structure-zone model applies to metallic, ceramic, and other solid films.

Water has a complex phase diagram, due to hydrogen bonding and proton disorder effects. It exhibits an ex-

CP982, *Complex Systems, 5th International Workshop on Complex Systems*
edited by M. Tokuyama, I. Oppenheim, and H. Nishiyama

tensive range of crystalline solid phases, or polymorphs, most of which are stable only under high pressure. The physical properties of the polymorph, such as density, conductivity, vapour pressure and sublimation rate, are dictated by its crystalline structure. The phases can be distinguished from each other by the arrangement of water molecules in the crystal lattice, and by the degree of proton disorder within the network. In all crystalline ices, the water molecules have four-fold coordination, donating two hydrogen bonds, and accepting two others, even if the bonds are distorted. Thirteen ice polymorphs are known at present: ices I to XIII (the phases are numbered with Roman numerals in the order they were discovered). Most phases are thermodynamically stable under some range of pressure–temperature conditions, with some phases also exhibiting metastable zones, and a few having no regions of absolute stability at all (Ices Ic, IV, IX, and XII are only ever metastable).

If we disregard now the high-pressure polymorphs, and consider only those phases stable at normal atmospheric pressure and below, we find that hexagonal ice, Ih, that encountered on Earth, is the thermodynamically stable phase down to around 72 K. Below this temperature, ice XI is the thermodynamically stable phase, but in pure ice Ih the molecular relaxation rate is too slow for the transformation process to be observed, and ice Ih continues to be metastable. The phase change can however be mediated by doping the ice with OH^- ions, which are relatively mobile in the ice, even at 72 K. Ice Ic should be mentioned at this point. It is always metastable to ice Ih; there is the same relationship between ices Ic and Ih as between cubic and hexagonal close packing structures in metals. In vapour deposition experiments, ice Ih is formed above $\sim$ 150 K, while ice Ic appears between $\sim$ 130–150 K. Below $\sim$ 130 K, the deposit produced is of amorphous ice.

Ice is one of the materials that has been hypothesized to possess more than one amorphous phase, a phenomenon termed amorphous polymorphism, or polyamorphism [7]. LDA and HDA are seen as glassy versions of two proposed liquid phases, low density liquid (LDL) and high density liquid (HDL), which, in turn, are associated with a putative liquid–liquid phase transition in water [8]. A similar amorphous–amorphous transition has been reported in silicon [9], and associated with a liquid–liquid phase transition in that system [10]. While the coexistence of two liquid phases is well known for liquid crystals and multicomponent systems (e.g., protein solutions), it has only recently been described for pure isotropic substances. Apart from water, other liquids show signs of such a phase transition [11]; there is strong experimental evidence for the phenomenon in liquid phosphorus [12].

The high density amorphous ice HDA was first observed on compressing ice Ih at low temperature and high pressure (77 K, 1 GPa) [13]. Upon heating this material to around 120 K at atmospheric pressure (0.1 MPa), it transforms to a low density amorphous ice termed LDA [14]. At 135 K and 0.6 GPa, the transition can be reversed [15]. Large quantities of liquid water cannot be cooled into the glassy solid state of an amorphous ice, as crystallization always occurs first. Supercooled liquid droplets, however, can form amorphous solid particles when cooled rapidly ($\sim 10^6$ K s^{-1}) in a cold air or cryogenic liquid flow: this phase is referred to as hyperquenched glassy water (HGW) [16, 17]. This, after annealing at around 130 K, relaxes to a form that some consider to be LDA [2], while others see as essentially different [18]. On the other hand, rapid cooling of emulsified liquid water under pressure ($\sim 10^3$–10^4 K s^{-1}; 0.5 GPa) produces HDA [19].

As well as production from the liquid or solid phases, amorphous ice is also obtained from condensation from the gas phase. In fact, amorphous ice was first found in vapour deposition experiments. In 1935 Burton & Oliver [20] announced that below about 163 K, the condensate in their experiments was no longer crystalline, but rather amorphous ice (their proposed transition temperature is rather high, and most investigators take 130 K as more typical). This ice was later termed amorphous solid water (ASW). Much later, it was shown that ASW differs when deposited at lower and higher temperatures. At low temperatures (below $\sim$ 30 K), a higher density amorphous ice is formed, while at higher temperatures ($\sim$ 30–130 K) a low density amorphous ice is deposited [21, 22]. These ices are similar to the HDA and LDA produced in solid and liquid state transformations. As before, while some see essential differences, others propose that any variations arise from differing annealing times and microporosities [22].

Amorphous ice is also of interest to astronomers and astrochemists [23], as at the low temperatures of interstellar space, water adsorbing onto dust grains solidifies in an amorphous state [22]. This ice is built up by ballistic deposition; molecules stick directly where they land, without surface diffusion, owing to the very low temperature (3–90 K) of the interstellar medium. It has been estimated that most of the ice in the universe is to be found on these interstellar dust grains, which implies that the majority of the ice in existence is in this amorphous form.

For many years thin films of many different materials, both crystalline and amorphous, have been deposited from the vapour phase onto substrates. The field is driven by a huge number of technological applications, but also has much scientific interest. There are many deposition techniques: sputtering, chemical vapour depostion, molecular beam epitaxy, etc, based on three fundamental processes for vapour generation and subsequent deposition — temperature, bombardment, and chemistry.

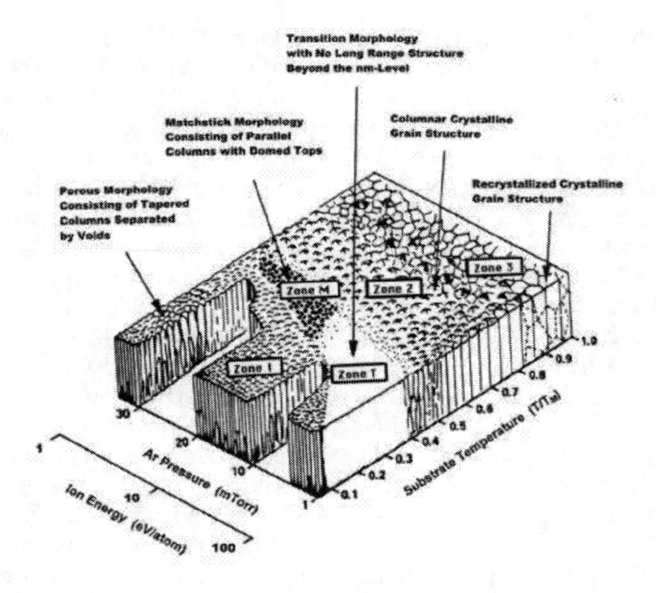

FIGURE 1. A recent version of the structure zone model of film deposition [24].

One of the key differences between thin films and bulk materials is in their morphologies. The study of thin-film morphology has a long history. From the 1960s on efforts have been made to construct a classification of the morphology of a film depending on its deposition conditions. This has culminated in a structure-zone model showing the morphology obtained for different film surface temperatures and gas pressures. This model has been progressively revised and updated, and there are now recognized to be at least five distinct zones with differing morphologies: see Figure 1. From lower to higher substrate temperatures and ion energies: Zone 1 has a porous morphology consisting of tapered columns separated by voids, and from above looks like the surface of a cauliflower; Zone M has a matchstick morphology consisting of parallel columns with domed tops; Zone T is a transition morphology with no long-range structure beyond the nanometre level; Zone 2 has a columnar cystalline grain structure; and finally Zone 3 has a recrystallized crystalline grain structure [5, 6]. There is also recent mention of a spongelike morphology between zones T and 2 [25].

The two axes of the structure-zone model are the ratio of the substrate temperature to the melting point of the deposited material, and the specific energy of the arriving growth units (molecules, ions, atoms, etc). This latter variable is inversely related to the gas pressure used in sputtering deposition. The structure-zone model describes well the trends seen in experiments with many different materials. So far, however, there has not been a comprehensive explanation of the patterns seen in the model in terms of simple dynamical effects.

EXPERIMENTS

In our experiments we used a FEI Quanta 200 environmental scanning electron microscope (ESEM) equipped with a liquid helium cold stage to grow ice films *in situ* at low pressures and at temperatures of 6–220 K. We began by evacuating the chamber in the high-vacuum mode of the microscope and lowering the substrate to the working temperature. The microscope was set up so that the helium cold finger, together with a thermostat, was directly beneath the substrate on which we grew the ice film. We used several substrates; we did not detect differences in film morphology between them once the film coverage was complete over the surface. To grow an ice film we switched the microscope to low-vacuum mode, in which we could inject water vapor into the chamber for a given length of time and at a determined target pressure. We either used demineralized water alone, or else bubbled helium through the water prior to injection to provide an inert auxiliary gas in the chamber and to reduce the partial pressure of water vapour. The ice film was then deposited on top of the substrate, which was the coldest point within the microscope. We switched again to high-vacuum mode to image the results.

The region of lowest substrate temperature in the structure zone model is occupied by zone 1. Zone 1, or cauliflower morphology, consists of competing void-separated tapered columns whose diameters expand with the film depth according to a power-law [24]. The surface resembles a cauliflower, showing self-similarity over a range of scales. We produced this morphology in the example in Fig. 2 using water vapor accompanied by helium as an auxiliary gas in the chamber. By depositing

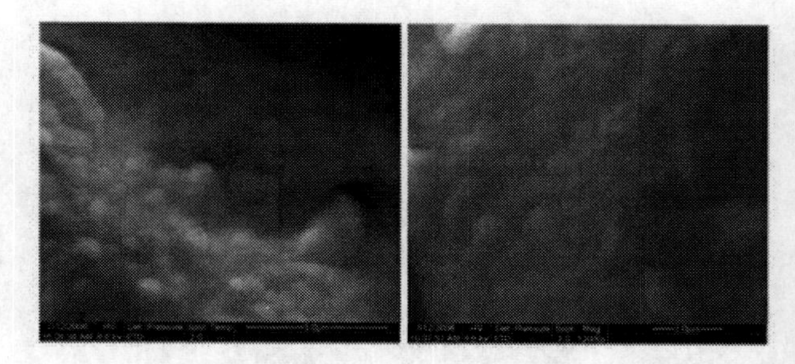

FIGURE 2. Zone 1, cauliflower morphology, film produced with water bubbled through with helium injected with a target pressure of 10 Pa for 6 s deposited on a titanium substrate at 6 K.

the ice film at the same substrate temperature as in Fig. 2, but this time at a higher water vapor pressure for a shorter time without an auxiliary gas, we obtained zone T, or transition morphology, in which there is no long-range structure above the nanoscale. In qualitative terms, the morphology of zone 1 is clearly driven by a competitive process of growth of clusters at all scales, leading to a fractal geometry, while greater gas-induced mobility of admolecules in zone T than in zone 1 smoothes out the surface. As they occur at the lowest temperatures, both zone 1 and zone T morphologies can be presumed to be composed of high-density amorphous ice. We noted when heating our samples that at close to 30 K the water vapor pressure in the chamber increased sharply, which we attribute to sublimation during the transition from high- to low-density amorphous ice. We did not however see any mesoscale structural change associated with this transformation at the molecular scale.

At substrate temperatures above those of the transition morphology of zone T, a spongelike morphology has been described for metallic films [25]. We were able to reproduce this morphology with an ice film, by injecting for a longer time than for zone T. This morphology is characterized by a three-dimensional open network of material like a sponge. On the other hand at substrate temperatures above those corresponding to cauliflower morphology (zone 1), and for relatively high gas-induced admolecule mobility there appears the final zone in the low-temperature region of the structure zone model: zone M, or matchstick morphology. In the example of this morphology we reproduce in Fig. 3 we see very large columns tens of micrometers in diameter resulting from ice deposition over several minutes using helium as an auxiliary gas. The columns display the domed tips characteristic of matchstick morphology, and also show interesting substructures both on the tip and along their length, presumably from their growth process; their shape is biomimetic, like an icy worm. The transition from zone 1 to zone M we can view qualitatively in terms of competitive growth leading to a fractal morphology giving way to noncompetitive growth producing a columnar morphology. These large matchsticks may have been the morphology seen by Laufer et al. [26], who described an ice film grown at somewhere between 20–100 K as looking like "a shaggy woolen carpet". With regard to the temperatures involved for zone M, it should typically be composed of low-density amorphous ice.

FIGURE 3. Zone M, matchstick morphology, film produced with water bubbled through with helium injected with a target pressure of 133 Pa for 6 minutes deposited on a platinum substrate at 6 K.

FIGURE 4. Film with a dendritic morphology resembling a palm tree produced with water injected with a target pressure of 10 Pa for 15 minutes deposited on a carbon substrate at 6 K.

The propensity of hexagonal crystalline ice to form complex dendritic structures is familiar from examples such as snowflakes. This knowledge should forewarn us that the structures we should encounter in zones 2 and 3 of the structure zone model, at high substrate temperatures corresponding to the deposition of crystalline ice, would reflect this complexity. Nevertheless we were taken aback by the beauty of the structures we grew in these conditions, seen in Fig. 4. The hexagonal crystalline nature of this ice is reflected in the angles of growth of the branches, shown in Fig. 4(c). This fascinating morphology of branched whiskers is intermediate between whiskers and dendritic growth. One interesting open question regarding these higher temperature zones is how the amorphous to crystalline transition at the molecular scale relates to the transition from zones 1 and T to zones 2 and 3 on the mesoscale.

SIMULATIONS

In our model atoms are described as hard disks of radius a located by the positions of their centers. Disks are deposited onto a one-dimensional substrate with periodic lateral boundary conditions via ballistic deposition at a given mean rate R. Ballistic trajectories are taken normal to the substrate, which is formed of a monolayer of fixed adatoms in a close-packed arrangement. Bonds between adatoms are limited to nearest neighbors. On making a first contact, either with the adatoms on the film surface or directly with the substrate, incoming particles are accommodated instantaneously to the closest available site on the microstructure. For this purpose, we developed a generalization of the schema presented by Savaloni and Shahraki [28]. In their model, fixed lattice positions are replaced by a set of relative positions, *active positions*, defined around each film particle, representing the possible locations of bonded neighbors, so lattice sites are delocalized with respect to the substrate and instead are localized around adatoms for first nearest neighbor positions. Savaloni and Shahraki proposed this approach to deal with a complex microstructure formed by a superposition of two elementary bidimensional cells corresponding to the simple square and hexagonal lattices. However, it can be used for arbitrarily simple or complex microstructures. We present results for four different microstructures corresponding to the simple square (*sq*) and the simple hexagonal (*hex*) lattices; the hexagonal–square microstructure (*hex–sq*) comprising the superposition of the previous ones; and the square–square microstructure (*sq–sq*) obtained by the superposition at 45° of two simple square microstructures with horizontal and oblique orientations respectively. The available sites for accommodating the incoming particles from their first contact position with the film surface are the unoccupied and non-shadowed active positions associated with the surrounding surface adatoms. After a new surface adatom is accommodated, bonds with its nearest neighbors are established and the corresponding new unoccupied active positions are created around it. Simultaneously with the deposition of new particles, surface adatoms are allowed to diffuse thermally to any of the close active positions that are neither occupied nor shadowed. Adatoms are considered as belonging to the surface when they are not "buried" by the presence of neighbors in the top active positions. In our simulations, diffusive hops to active positions are restricted to a neighborhood of size $2a$ around the original adatom position, except for the square microstructure. In this case, the low number of available active positions makes difficult the compacting of mesoscopic structures without allowing diffusive hops to second-nearest neighbors. Hence we allow "over-a-step" hops for this microstructure. Simulations were performed with the deposition of 15 000 particles onto a flat substrate 160 adatoms wide disposed in a close packed arrangement.

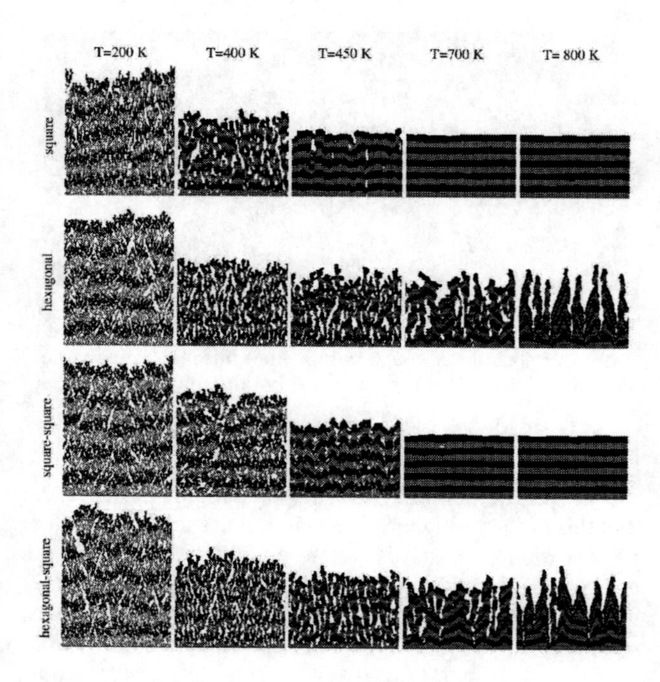

FIGURE 5. Film structures obtained at different temperatures and a fixed deposition rate of 1 monolayer/s for each given microstructure; the stripes represent 10 monolayers.

By varying the main deposition parameters: the substrate temperature and the deposition rate, one can observe the development of distinct characteristic mesoscopic film morphologies. In general, an essential equivalence has been observed between deposition rates and substrate temperatures as the parameter. In Fig. 5 we show different film structures obtained at distinct substrate temperatures for a fixed deposition rate of one monolayer per second. In all cases we found a minimum substrate temperature below which are obtained low-density dendritic structures with a self-affine surface typical of ballistic deposition with negligible surface diffusion. As the substrate temperature increases, or equivalently at lower deposition rates, surface diffusion becomes significant and dendritic morphologies give way to more compact fibrous structures with a preferential growth axis parallel to the deposition trajectories. With greater surface diffusion, the fibrous structures compact to densely packed grains that grow vertically as competing columns whose thickness increases with the diffusion. Finally, when column thickness equals the system size, the resultant films are formed by a single compact structure without interstitial voids. This behavior, which is qualitatively independent of the underlying microstructure, corresponds to the typical morphologies of zones 1 and 2 of the structure zone model [24]. Since no significant parameters for grain boundary evolution other than the underlying microstructure and the competition between deposition rates and surface diffusion are present in our model, it is not suitable for characterizing other structure zones in which effects such as activated diffusion by ion bombardment of the surface, bulk diffusion, or the presence of impurities are determinant [24].

CONCLUSIONS

It is clear that structure zone morphologies do appear in ice films. Can this knowledge contribute to understanding the physics of the structure zone model? We are moving towards a physical understanding of the structure zone model as a consequence of the competition between the spatially disordered deposition of particles on the growing film surface and the ordering effect of activated particle mobility processes. An intriguing aspect of this work is the finding that ice on its own can form biomimetic structures under extreme conditions. Knowledge of this phenomenon is important for astrobiologists searching for life in similar extreme conditions in space, and is a reminder that biomimetic forms are not in themselves evidence of life.

ACKNOWLEDGMENTS

We thank Daniel Araujo and Russ Messier for useful discussions, and acknowledge the financial support of the CSIC project Hielocris. The simulations were carried out by PAS as part of his PhD studies under the supervision of TS and OP; the experimental work by BE as part of his PhD studies under the supervision of JHEC and CISD.

REFERENCES

1. V. F. Petrenko, and R. W. Whitworth, *Physics of Ice*, Oxford University Press, 1999.
2. O. Mishima, and H. E. Stanley, *Nature* **396**, 329–335 (1998).
3. P. G. Debenedetti, and H. E. Stanley, *Phys. Today* **56**, 40–46 (2003).
4. P. G. Debenedetti, *J. Phys. Condens. Matter* **15**, 1669–1726 (2003).
5. R. Messier, V. C. Venugopal, and P. D. Sunal, *J. Vac. Sci. Technol. A* **18**, 1538–1545 (2000).
6. R. Messier, and S. Trolier-McKinstry, "Processsing of Ceramic Thin Films," in *Encylopedia of Materials: Science and Technology*, edited by K. H. I. Buschow, R. W. Cahir, M. C. Flemings, B. Ilschner, E. J. Kramer, and S. Mahajan, Elsevier, 2001.
7. P. H. Poole, T. Grande, F. Sciortino, H. E. Stanley, and C. A. Angell, *Comput. Mater. Sci.* **4**, 373–382 (1995).
8. H. E. Stanley, S. V. Buldyrev, M. Canpolat, M. Meyer, O. Mishima, M. R. Sadr-Lahijany, A. Scala, and F. W. Starr, *Physica A* **257**, 213–232 (1998).
9. S. K. Deb, M. Wilding, M. Somayazulu, and P. F. McMillan, *Nature* **412**, 514–517 (2001).
10. I. Salka-Volvod, P. H. Poole, and F. Sciortino, *Nature* **414**, 528–530 (2001).
11. P. H. Poole, T. Grande, C. A. Angell, and P. F. McMillan, *Science* **275**, 322–323 (1997).
12. Y. Katayama, T. Mizutuni, W. Utsumi, O. Shimomura, M. Yamakata, and K. Funakoshi, *Nature* **403**, 170–173 (2000).
13. O. Mishima, L. D. Calvert, and E. Whalley, *Nature* **310**, 393–395 (1984).
14. O. Mishima, L. D. Calvert, and E. Whalley, *Nature* **314**, 76–78 (1985).
15. O. Mishima, *J. Chem. Phys.* **100**, 5910–5912 (1994).
16. P. Brüggeller, and E. Mayer, *Nature* **288**, 569–571 (1980).
17. E. Mayer, and P. Brüggeller, *Nature* **298**, 715–718 (1982).
18. G. P. Johari, A. Hallbrucker, and E. Mayer, *Science.* **273**, 90–92 (1996).
19. O. Mishima, and Y. Suzuki, *J. Chem. Phys.* **115**, 4199–4202 (2001).
20. E. F. Burton, and W. F. Oliver, *Proc. Roy. Soc. Lond. A* **153**, 166–172 (1935).
21. P. Jenniskens, and D. F. Blake, *Science* **265**, 753–756 (1994).
22. P. Jenniskens, D. F. Blake, M. A. Wilson, and A. Pohorille, *Astrophys. J.* **455**, 389–401 (1995).
23. P. Ehrenfreund, H. J. Fraser, J. Blum, J. H. E. Cartwright, J. M. García-Ruiz, E. Hadamcik, A. C. Levasseur-Regourd, S. Price, F. Prodi, and A. Sarkissian, *Planet. Space Sci.* **51**, 473–494 (2003).
24. A. Lakhtakia, and R. Messier, *Sculptured Thin Films: Nanoengineered Morphology and Optics*, SPIE, 2005.
25. A. F. Jankowski, and J. P. Hayes, *J. Vac. Sci. Technol. A* **21**, 422–425 (2003).
26. D. Laufer, E. Kochavi, and A. Bar-Nun, *Phys. Rev. B* **36**, 9219–9227 (1987).
27. K. G. Libbrecht, *Rep. Prog. Phys.* **68**, 855–895 (2005).
28. H. Savaloni, and M. G. Shahraki, *Nanotechnol.* **15**, 311–319 (2004).

A Study of α-Relaxations in Trehalose Super Cooled Liquids

Jeong-Ah Seo, Hyun-Joung Kwon, Hyung Kook Kim, and Yoon-Hwae Hwang

Department of Nanomaterials & BK21 Nano Fusion Technology Division, Pusan National University, Miryang 627-706, Korea

Abstract. We measured the α-relaxations in trehalose super cooled liquids by using photon correlation spectroscopy (PCS). The α-relaxations of trehalose super cooled liquids showed a crossover from stretched to compressed-exponential relaxations as the temperature increased [1]. From the Raman scattering measurements, we found that the unusual compressed-exponential relaxation in trehalose super cooled liquids may be caused by the change of glycosidic linkage structure in trehalose molecule.

Keywords: Disaccharide, Photon correlation spectroscopy (PCS), Raman scattering sepetroscopy, Trehalose super cooled liquid
PACS: 61.43.Fs, 64.70.Pf, 65.60.+a, 78.30.-j, 81.70.Pg

INTRODUCTION

The systems of sugar and sugar containing materials are a matter of common interest to many researchers because sugar is a main constituent of the biological system. Moreover, glass-forming sugars have a great significance in nature. In many biological materials, the vitrification phenomena were observed when they were allowed to the stressed environments like dried and/or cooled state. Nowadays, these vitrification phenomena were widely accepted that as the bio-protection effects by the sugars [2, 3, 4, 5, 6].

Among the sugars, we are interested in trehalose because the bio-protection ability of trehalose was known to be the most effective [7, 8, 9, 10, 11, 12, 13, 14, 15, 16, 17, 18, 19]. Alpha, alpha - trehalose (α-D-glucopyranosyl α-D-glucopyranoside) is a well-known, non-reducing disaccharide that is commonly found in yeast, fungi, bacteria, mushrooms, and desert plants. Trehalose consists of two glucose molecules and has several unique characteristics. The glass transition temperature of trehalose is higher than that of other disaccharides [7]. Therefore, the viscosity of trehalose is higher than that of other sugars at a given temperature. Several researchers have pointed out that the highest glass-transition temperature of trehalose may affect to the preservation of biological molecules through the control of water mobility [8]. The flexibility of α,α-(1→1)-glycosidic linkage in trehalose could be an important clue in explaining its biological functions [9, 10, 11]. Also, trehalose has a hydration characteristic which displays extremely interesting features, including the ability to protect and reversibly reconstitute proteins and bio-membranes from dehydration and freeze-drying. There are many studies about the protection ability of trehalose during drying and ensuing storage [12, 13, 14]. These studies reported several possible origins of their protection ability, which include water replacement processes [15, 16, 17], vitrification [18, 19], and dynamic reducers [11].

In this study, we are interested in the relaxation process of trehalose super cooled liquids. The study of the relaxation process of trehalose super cooled liquids is an interesting topic to understand the dynamics of molecules in a glass state. In addition, we believed that it will be helpful to understand the bio-protection mechanism of trehalose super cooled liquides.

EXPERIMENTS

Trehalose dihydrate was sponsored by Cargill Corporation and was dried before the experiments by using moisture analyzer(Sartorius MA100, Germany) [20, 21, 22] at 130 °C for 5 hours. The trehalose glass was prepared by

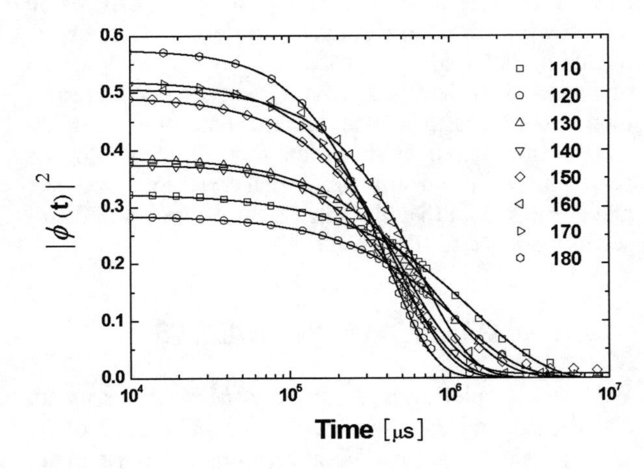

FIGURE 1. The $|\phi(t)|^2$ of depolarized (VH) components of trehalose super cooled liquids for different temperatures from 110°C to 180°C.

CP982, *Complex Systems, 5th International Workshop on Complex Systems*
edited by M. Tokuyama, I. Oppenheim, and H. Nishiyama

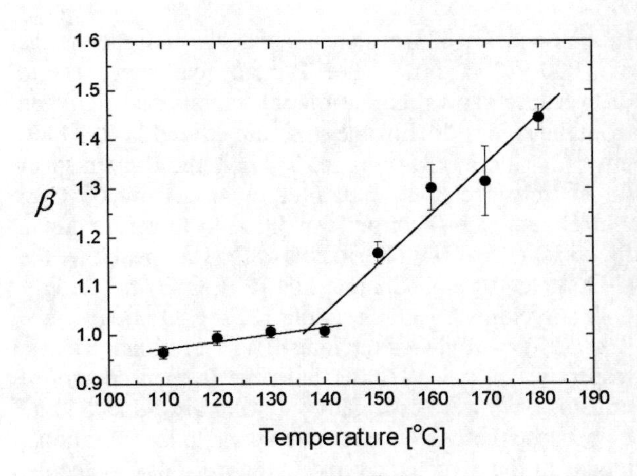

FIGURE 2. The $|\phi(t)|^2$ of depolarized (VH) components of trehalose super cooled liquids at different temperatures ranging from 110°C to 180°C.

using microwave oven because the microwave oven is an effective method to heat the trehalose quickly and uniformly without caramelization process during the heating [23]. The glass phase of trehalose was confirmed by X-ray diffraction and differential scanning calorimeter (DSC; MAC science, DSC3100, Japan) measurements. The range of glass transition temperature of trehalose glass was 105 $\sim$ 110°C for different heating rates 2 $\sim$ 8°C/min.

In a photon correlation spectroscopy(PCS) experiment, we used the Brookhaven BI-9000AT digital correlator (Brookhaven Instruments Corp., U.S.A.) for measuring the correlation function, *G(t)*, of trehalose super cooled liquid. The Brookhaven correlator can cover ten decades of time($10^{-1}\mu s \sim 10^{9}\mu s$) and the time range used in this study was $10^3\mu s \sim 10^8\mu s$. The incident beam was a vertically polarized 514.5 nm green light Ar-ion laser (I90-C, Coherent, USA) at 200 mW. For detecting the scattered light, we used a single mode fiber optic. The depolarized (VH) components of the scattered light were selected by a Glan-Thompson Analyzer with a leakage factor of less than 1%. A scattering geometry was 90°. We can get an instrumental coherent factor of A$\sim 0.95 \pm 0.01$ with an aqueous suspension of polystylene spheres.

From the measured correlation function by the digital correlator, we can calculate the intensity-intensity correlation function $g_2(t)$. The $g_2(t)$ is related to the intermediate-scattering function $\phi_q(t)$ by $G(t)/G(\infty)$ = $g_2(t) = 1 + A|\phi(t)|^2$ where A is an instrumental factor. The subscript q which means the scattering wave number was omitted because q was fixed at 2.4×10^5 cm^{-1} ($\theta = 90^o$) in this study.

We used Raman scattering spectroscopy to find the molecular structure change in trehalose. The incident beam was a vertically polarized 514.5 nm green light Ar-ion laser (I90-C, Coherent, USA) at 100 mW and we used back-scattering geometry. The scattered light was measured by using a monochromator (Acton Research, Spectra Pro-750, USA) and a charge-coupled device (Andor MCD, USA) at wavenumbers ranging from 1000 to 1200 cm^{-1}. The exposure time was 1 second and the spectrum was accumulated 1000 times. The temperature range used in this study was from room temperature up to the melting temperature of trehalose. The melting temperatures of trehalose was 213°C. Trehalose was ground to fine powder and pressed down into a glass vessel for Raman scattering measurement.

RESULTS & DISCUSSION

We measured α-relaxations in trehalose super cooled liquids by using photon correlation spectroscopy(PCS) at temperatures ranging from 110 ^{o}C to 180 ^{o}C which covered the super-cooled liquid state. Figure 1 shows the $|\phi(t)|^2$ of VH components in trehalose super cooled liquid. The symbols are experimental results and the solid lines are fit the data to the square of stretched-exponential function $|\phi(t)|^2 = f_c^2 exp(-2(t/\tau)^\beta)$ with three fitting parameters, non-ergodicity parameter f_c, relaxation time τ, and stretched exponent β.

Figure 2 shows the stretched exponent β of trehalose super cooled liquids at temperatures ranging from 110°C to 180°C. The stretched exponent β increased from 0.96 to 1.44 with increasing temperature and the slope of the β was changed around 140°C.

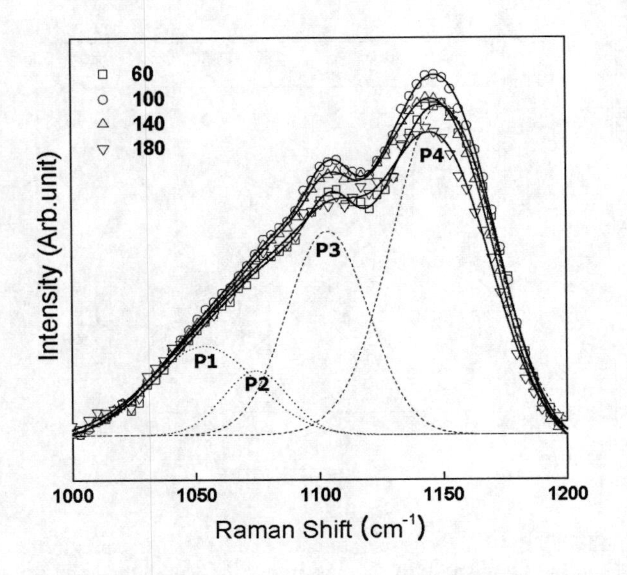

FIGURE 3. The Raman spectra of trehalose at different temperatures ranging from 60°C to 180°C.

Moreover, the stretched exponent β is bigger than 1 above 140°C. This result indicates that the relaxation process of trehalose super cooled liquids changed from stretched- to compressed-exponential relaxations around 140°C. While such compressed-exponential relaxation with $\beta > 1$ may seems unusual, similar behavior has been observed on a number of soft solids including colloidal gels[24], micellar polycrystals[25], and clays[26]. Recently, Bouchaud and Pitard[27] suggested a specific model about the compressed-exponential relaxation which associates with the local rearrangements or micro-collapses of particles. In this study, we believe that the compressed-exponential behavior in trehalose super cooled liquids could be caused from an intra molecular structure change of trehalose molecule based on the X-ray diffraction experiment results which will be described in another publications.

We expected that the structure change in trehalose molecule may originated from the change of glycosidic linkage structure between glucose molecules. To find the structure change of glycosidic linkage, we measured the

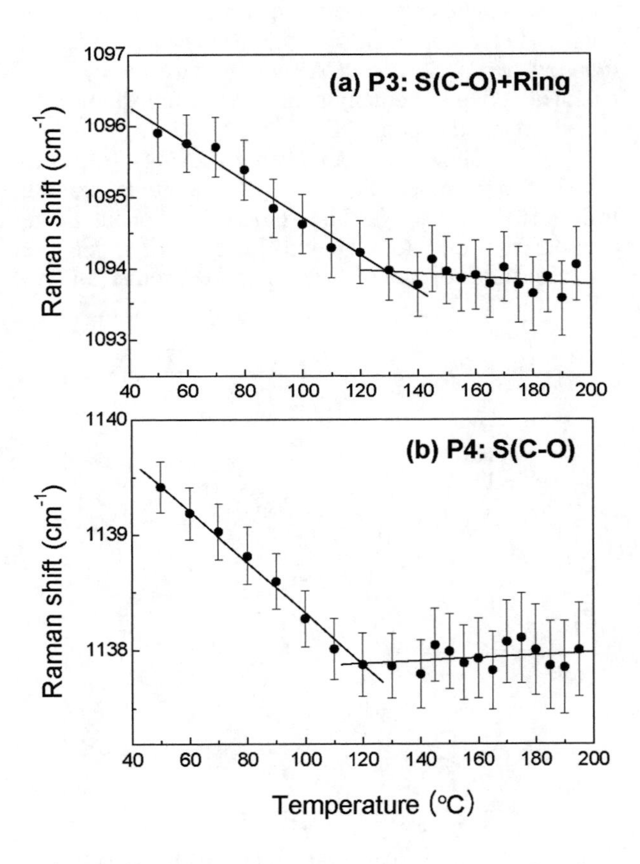

FIGURE 4. The temperature dependent Raman shifts in trehalose at temperatures ranging from 50°C to 200°C. The origins of the four peaks are (a) C-O stretching + Ring (P3), and (b) C-O stretching (P4) vibrational modes.

Raman spectra of trehalose in the wavenumber range of 1000$\sim$1200 cm^{-1}. The Raman scattering spectra of trehalose shows the out-of-ring vibrations originated from the glycosidic linkage structure around 1140$\sim$1150 cm^{-1}[28, 29, 30, 31]. Figure 3 shows the Raman spectra of trehalose at temperatures ranging from 60^{o}C to 180^{o}C and each spectrum was fitted to four Lorentzian functions (P1$\sim$ P4). The origins of the four peaks are the C-O stretching + C-C stretching (P1), C-O stretching + C-C stretching + COH bending (P2), C-O stretching + Ring (P3), and C-O stretching (P4) vibrational modes, respectively [28, 29, 30, 31]. In the Raman spectra of trehalose, we were especially interested in peaks P3 and P4 because these two peaks correspond to the vibrational modes of the glycosidic bond in the trehalose molecule. Figures 4(a) and 4(b) show the two temperature dependent Raman modes of trehalose at temperatures ranging from 50°C to 200°C. In Figures 4(a) and 4(b), the Raman shifts of peaks T3 and T4 decreased with increasing temperature up to a temperature of around 120°C. At temperatures above 120°C, peaks T3 and T4 were almost temperature independent. This clearly indicates that the vibrational modes of the glycosidic bonds in trehalose molecules changed at 120°C. From this result, we concluded that the unusual compressed-exponential relaxation in trehalose super cooled liquids was caused by the change of the out-of-ring vibrations(glycosidic linkage structure) in trehalose molecule.

CONCLUSION

We measured the α-relaxations in trehalose super cooled liquids by using photon correlation spectroscopy(PCS) at temperatures ranging from 110 ^{o}C to 180 ^{o}C. The α-relaxations of trehalose super cooled liquids showed a crossover from stretched- to compressed-exponential relaxations as the temperature increased. From the Raman scattering measurements in the range of 1000$\sim$1200 cm^{-1}, we found that the slope of temperature dependent Raman shift of glycosidic linkage in trehalose changed around 120 ^{o}C. From this result, we concluded that the unusual compressed-exponential relaxation in trehalose super cooled liquid maybe caused by the change of glycosidic linkage structure in trehalose molecule.

ACKNOWLEDGMENTS

We thank H. Z. Cummins for suggesting sugars for a glass transition study. This study was financially supported by Pusan National University in program Post-Doc. 2007. This work was also supported by the Korea Research Foundation Grant KRF-2006-521-C00060.

REFERENCES

1. J.-A. Seo, H.-J. Kwon, H. M. Lee, H. K. Kim, and Y.-H. Hwang, *Rep. Inst. Fluid Science*, **19**, 41–44 (2007).
2. J. F. Carpenter, L. M. Crowe, and J. H. Crowe, *Biochim. Biophys. Acta*, **923**, 109–115 (1987).
3. L. M. Crowe, J. H. Crowe, J. F. Carpenter, and C. A. Wistrom, *Biochem. J.*, **242**, 1–10 (1987).
4. S. Rossi, M. P. Buera, S. Moreno, and J. Chirife, *J. Biotechnol. Prog.*, **13**, 609–616 (1997).
5. C. Schebor, L. Burin, M. P. Buera, and J. Chirife, *J. Biotechnol. Prog.*, **13**, 857–863 (1997).
6. A. Patist, and H. Zoerb, *Col. & Surf. B: Biointerfaces*, **40**, 107–113 (2005).
7. J. L. Green, and C. A. Angell, *J. Phys. Chem.*, **93**, 2880–2882 (1989).
8. L. M. Crowe, D. S. Reid, J. H. Crowe, *Biophys. J.*, **71**, 2087–2093 (1996).
9. M. K. Dowd, P. J. Reilly, A. D. French, *J. Comp. Chem.*, **12**, 102–114 (1992).
10. M. M. uttel, K. J. Naidoo, *Carbohyr. Res.*, **340**, 875–879 (2005).
11. Y. Choi, K. w. Cho, k. Jeong, s. Jung, *Carbohyr. Res.*, **341**, 1020–1028 (2006).
12. J. H. Crowe, L. M. Crowe, D. Champman, *Science*, **223**, 701–703 (1984).
13. K. Koster, M. S. Webb, G. Bryant, D. V. Lynch, *Biochim. Biophys. Acta*, **1193**, 143–150 (1994).
14. M. F. Mazzobre, M. P. Buera, J. Chirife, *Biotechnol. Prog.*, **13**, 195–199 (1997).
15. P. Bordat, A. Lerbret, J.-P. Demaret, F. Affouard, M. Descamps, *Europhys. Lett.*, **65(1)**, 41–47 (2004).
16. R. Giangiacomo, *Food Chemistry*, **96**, 371–379 (2006).
17. A. Lerbret, P. Bordat, F. Affouard, Y. Guinet, A. Hedoux, L. Paccou, D. Prevost, M. Descamps, *Carbohyr. Res.*, **340**, 881–887 (2005).
18. J. H. Crowe, J. F. Carpenter, L. M. Crowe, *Annu. Rev. Physiol.*, **60**, 73–103 (1998).
19. J. H. Crowe, S. B. Leslie, L. M. Crowe, *Cryobiology*, **31(4)**, 355–366 (1994).
20. F. Sussich, S. Bortoluzzi, and A. Cesaro, *Thermochem. Acta*, **391**, 137–150 (2002).
21. F. Sussich, F. Princivalle, and A. Cesaro, *Carbohydr. Res.*, **322**, 113–119 (1999).
22. H. Nagase, T. Endo, H. Ueda, and M. Nakagaki, *Carbohydr. Res.*, **337**, 167–173 (2002).
23. J.-A. Seo, J. Oh, D. J. Kim, H. K. Kim, and Y.-H. Hwang, *J. Non-Crystal. Solids*, **333**, 111–114 (2004).
24. L. Cipelletti, S. Manley, R. C. Ball, and D. A. Weitz, *Physical Review Letter*, **84**, 2275–2278 (2000).
25. L. Cipelletti, L. Ramos, S. Manley, E. Pitard, D. A. Weitz, E. E. Pashkovski, and M. Johansson, *Faraday Discuss*, **123**, 237–251 (2003).
26. R. Bandyopadhyay, D. Liang, H. Yardimci, D. A. Sessoms, M. A. Borthwick, S. G. J. Mochrie, J. L. Harden, and R. L. Leheny, *Physical Review Letter*, **93**, 228–302 (2004).
27. J.-P. Bouchaud and E. R. Pitard, *Euorphisics Letter*, **6**, 231–236 (2001).
28. M. Dauchez, P. Derreumaux, P. Lagant, G. Vergoten, *Spectrochim. Acta Part A*, **50A(1)**, 87–104 (1994).
29. O. R. Fennema, “Carbohydrates,” in *Food Chemistry*, Marcel Dekker, New York, 1985, pp. 69–137.
30. A. M. Gil, P. S. Belton, V. Felix, *Spectrochim. Acta Part A*, **52**, 1649–1659 (1996).
31. M. Kacurakova, M. Mathlouthi, *Carbohyr. Res.*, **284**, 145–157 (1996).

Hydration Water Dynamics in Solutions of Hydrophilic Polymers, Biopolymers and Other Glass Forming Materials by Dielectric Spectroscopy

Silvina Cerveny[1], Angel Alegría[1-2], and Juan Colmenero[1-3]

[1] *Centro de Física de Materiales, Centro Mixto CSIC- Universidad del País Vasco (UPV/EHU)*
[2] *Departamento de Física de Materiales, UPV/EHU, Facultad de Química, Apartado 1072, 20018, San Sebastián, Spain*
[3] *Donostia International Physics Center, Paseo Manuel de Lardizabal 4, 20018, San Sebastián, Spain*

Abstract. Broadband dielectric spectroscopy (10^{-2} Hz – 10^{6} Hz) and differential scanning calorimetry measurements have been performed to study the molecular dynamics of several water solutions in the water concentration region from 30wt% to 50wt % and in the temperature range from 140 to 250K. The analysis of the dielectric data in these water solutions revealed an universality in the temperature dependence of the relaxation times at temperatures lower and higher than the calorimetric glass transition (T_g). At temperatures lower than T_g, an universal Arrhenius behaviour with an average activation energy of (0.54 ± 0.10) eV is obtained, whereas at temperatures higher than T_g an universal VFT (Vogel-Fulcher-Tamman) behaviour is found.

Keywords: Confinement, Supercooled water, Water dynamics, Water solutions
PACS: 61.41.+e, 77.22.Gm

INTRODUCTION

Water is necessarily present at the surface of all proteins and other biomolecules and hydration plays a decisive role in the function, structure and stability of these systems [1]. Water molecules in direct contact with the protein surface are part of the so-called "first hydration shell". In living systems, water-related phenomena occur in restricted environments in the cells and also at active sites of proteins and membranes [2-4]. Consequently, hydration water is an essential component of any biological systems. Since water and bio-polymers dynamics seem to be strongly coupled [5], the study of water dynamics in diverse solutions provides valuable information for understanding the structure and the dynamics of many biological processes.

The water dynamic in solutions of polymers, small organic molecules and bio-polymers in the low temperature range (150–250K) have been investigated by different experimental techniques such as neutron scattering, dielectric spectroscopy, NMR etc [6]. In particular, by dielectric spectroscopy, at least two relaxation processes in the liquid-poor side (water concentrations lower than 50 wt%) are usually observed [7-11]. The relaxation times of the slower process (process I) usually follows a VFT equation [12]. On the other hand, the relaxation times for the faster process (process II) show an Arrhenius behavior and it was associated to the reorientation of the water molecules in the solutions [6,7,11,13]. In addition, at high water concentration (higher than 30 wt%) the dielectric intensity of process II increases more rapidly than at low water content and in some cases its intensity [9] is higher than that of process I. In this case the dielectric response of process II is dominated by water-water interactions [14].

In this work we will investigate on the role of the hydration water (process II) in several solutions of polymers, small organic glass formers liquids, sugars and biopolymers (proteins and DNA). We will show that, despite the fact that water imbibed in these systems is surrounded by different environments and different molecular interactions between water and solute takes place, hydration water dynamic presents universal features.

CP982, *Complex Systems, 5th International Workshop on Complex Systems*
edited by M. Tokuyama, I. Oppenheim, and H. Nishiyama

EXPERIMENT

To study the water dynamics we have selected different systems such as: water soluble polymers (Poly (vinyl methyl ether) (PVME) M_n = 21.900g/mol, poly (vinyl pyrrolidone) M_n = 160.000gr/mol (PVP)), low molecular weight organic glass-formers (5EG (pentaethylene glycol)), sugars (fructose, lyxose and ribose) and biopolymers (myoglobin from heart horses and deoxyribonucleic acid from herring testes (DNA)). All the pure materials were purchased from Sigma-Aldrich. We have considered the water concentration range between 30 to 55 wt%. This is the proper range to follow the water dynamics by dielectric spectroscopy and to avoid water crystallization effects as well. For PVME and PVP, the final hydration value was obtained by evaporating water from the 50 wt% sample under normal room conditions. Aqueous solutions of 5EG were prepared varying the water concentration from 30 to 50 wt% whereas sugars solutions were prepared at a water concentration of 30 wt%. Recipients with the different solutions were sealed and put in an ultrasonic cleaner for 30 min to ensure a good microdispersion and homogeneity at molecular level. For Myoglobin and DNA, the desirable concentration was obtained by mixing the dry material with the solvent for 48Hs at 4°C. In mixtures of polymers, sugars and low molecular weight organic glass-formers there is no evidence of crystallization on cooling. In the case of DNA and myoglobin only about 5% of the water in solution crystallizes at 235K on cooling. However, all the mixtures show on heating a clear calorimetric glass transition indicating that there is a glass phase at low temperatures. We have to note that on heating, all the samples have a cold crystallization at about 250K. For this reason we have restricted our study in temperatures between 140 and 250K. Therefore, we were able to study the water dynamics of these systems in both the supercooled liquid and glassy state. To do that, a broadband dielectric spectrometer, Novocontrol Alpha analyzer, was used to measure the complex dielectric function, $\varepsilon^*(\omega) = \varepsilon'(\omega) - i\,\varepsilon''(\omega)$, $\omega = 2\pi f$, in the frequency (f) range from $f = 10^{-2}$ Hz to $f = 10^{6}$ Hz. The samples were placed between parallel gold-plated electrodes with a diameter of 30 mm and were typically 0.1 mm thick. After cooling at a rate of 10 K/min, isothermal frequency scans recording $\varepsilon^*(\omega)$ were performed every 5 degree over the temperature range 140-290K. The sample temperature was controlled with stability better than ± 0.10K. In addition, PVP was measured also in a higher frequency range (10^6-10^9 Hz) by using an Agilent RF impedance analyzer 4192B. In this case parallel gold plated electrodes with a diameter of 10 mm were used.

A DSC Q1000 TA Instrument was used in standard mode. Standard DSC measurements were performed using cooling and heating rates of 10K/min and the glass transition values were determined as the onset of the heat flow curve. Hermetic aluminum pans were used for all the materials. The sample weights were about 20 mg.

RESULTS

Since the water dynamic in different solutions is rather similar, in this section we will show the results related to fructose water mixtures as representative of all the materials studied in this work.

Figure 1 shows the imaginary part of the dielectric response at T = 215K for fructose-water mixture at a water concentration (w_c) of 30 wt%. Clearly, the dielectric spectra shows two main relaxation processes (the slower Process I and the faster Process II). Exceptionally, the polymeric solutions only presents one of these relaxations (Process II) due to the fact that both dc conductivity and electrode polarization mask the relaxation process of the chain motion (process I).

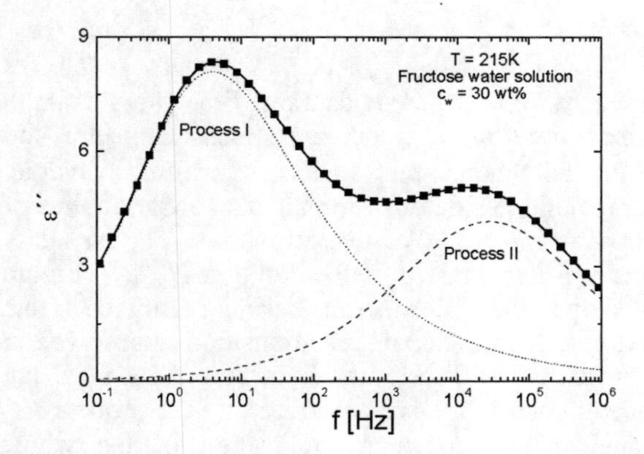

FIGURE 1. Frequency dependence of the imaginary part of the complex dielectric permittivity of fructose-water solution. The solid line is a least square fit using a superposition of a power law for conductivity, the imaginary part of a HN function for process I and the imaginary part of a CC function for process II (see text).

Since most literature dielectric data are collected in the frequency domain, semi-empirical frequency dependent functions are often used to describe the data. The most common one is the Havriliak-Negami (HN) expression [15],

$$\varepsilon^*(\omega)=\varepsilon_\infty+\frac{\varepsilon_s-\varepsilon_\infty}{\left[1+(i\omega\tau_{HN})^\alpha\right]^\gamma} \quad (1)$$

where ε_∞ and ε_s are the unrelaxed and relaxed values of the dielectric constant, τ_{HN} is a relaxation time and ω is the angular frequency. In eq. (1) α and γ are shape parameters ($0 < \alpha$, $\alpha\,\gamma \le 1$). Setting $\gamma = 1$ a symmetrical function is obtained (Cole-Cole function (CC)) which is widely used to describe secondary relaxations in glassy materials. An example of the fitting procedure is also represented in figure 1. A HN function plus a CC function were used to fit the dielectric loss. A power law term was added to account for small effects at low frequencies. The broadening of process II results in the range of 2-3.5 decades for all the materials analyzed.

Dielectric relaxation times were calculated using the maximum of the loss curve and, in this way, they are independent of any fitting frame. Representative relaxation times for fructose are shown in figure 2b. The time scale of the slower relaxation process (process I, open symbols) over the whole temperature range, can be well described by the Vogel-Fulcher-Tamman equation (VFT) [12],

$$\tau_\alpha=\tau_o\exp(DT_o/(T-T_o)) \quad (2)$$

where τ_o is the relaxation time in the high temperature limit, D is a dimensionless parameter, and T_o is the temperature where τ_α would diverge. Extrapolation using this formula to a relaxation time of ~100s gives a dielectric estimation of the glass transition temperature of the solution ($T_{g,100s}$). Figure 2a shows the calorimetric behavior for the same mixture where a global glass transition is observed at T_g = 202K. As $T_{g,100s}$ and T_g measured by DSC are close to each other, we conclude that the process I is related to the α-primary relaxation of the whole mixture.

In figure 2b we also show the temperature dependence of the relaxation times of process II, $\tau_{max} = 1/(2\pi f_{max})$. The logarithm of the relaxation times as a function of the reciprocal temperature, $log(\tau_{max})$, shows a lineal behavior at low temperatures. As a consequence, in this temperature range, the relaxation times were fitted by an Arrhenius temperature dependence: $\tau(T)=\tau_o\, exp(E_a/kT)$, where for a simple activated process τ_o corresponds to a molecular vibration time and E_a is the activation energy. The so obtained activation energy is (0.55 ± 0.03) eV. From the comparison between Fig 2a and Fig 2b, it is immediately clear that the crossover from non-Arrhenius to Arrhenius is found at around the calorimetric T_g. Thus, this crossover can be identified with the frozen-in of the mixture at the glass transition temperature [14,16].

Summarizing, process I is related to the α-primary relaxation of the whole mixture whereas process II is related to water molecules in the mixtures and its relaxation times show a crossover at T_g. This behavior is common to all systems analyzed in this work as well as to other water mixtures investigated in the literature such as (sucrose [17], Poly(ethylene glycol) (PEG600, M_n =600 g/mol) [17], tri- penta- and hexa- (ethylene glycol)[7] and water in purple membrane [18]).

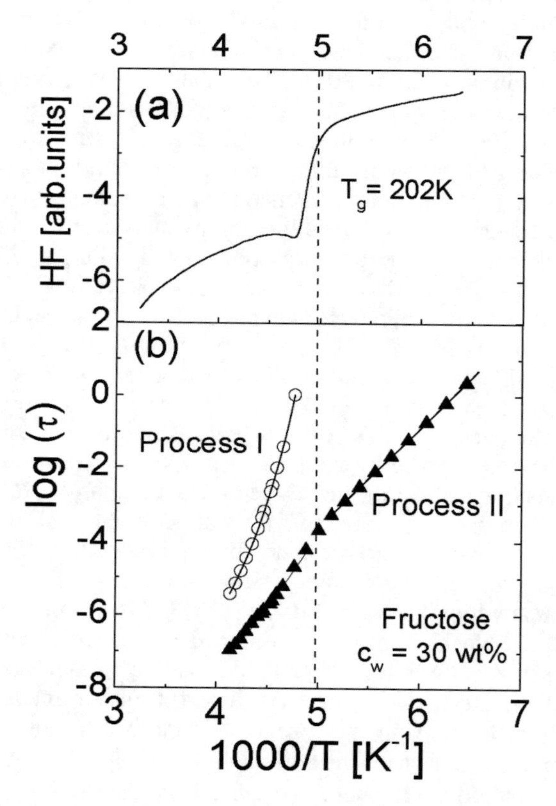

FIGURE 2. (a) Heat flow of fructose-water solution during heating at a rate of 10K/min. (b) Temperature dependence of the relaxation times obtained from dielectric spectroscopy on fructose-water solution. Process I follows a VFT behavior whereas the slower one Process II exhibits a crossover from non-Arrhenius to Arrhenius when the system reach T_g. The crossover temperature coincides with the calorimetric T_g.

DISCUSSION

In this section, we will focus on the slower process II observed in all the water mixtures. All the solutes used in this work are hydrophilic since it is possible to avoid crystallization on cooling in the temperature range between 130K and 250K. In addition, note that we will considered our own results which means that all of them were treated in the same way.

Figure 3 shows the temperature dependence of the relaxation times for the different water-solutions investigated at w_c = 30 wt%. Thus, we corroborate that the experimental findings showed before for fructose water solutions are not unique since is the most common behavior for the water dynamics in mixtures of polymers, glasses and bio-polymers. As the crossover is produced at the glass transition of the mixtures, the crossover temperature is different for the each sample. Since the crossover temperature coincides with the calorimetric glass transition temperature, this varies from 165K to 220K. Consequently, the crossover temperature is not a characteristic of water dynamics since depends on both the characteristics of the solute and the water concentration. Contrarily, when the low temperature behavior of water dynamics in the different mixtures was compared it was found that the activation energy describing it was very similar in all the cases. Figure 4 summarizes the value of the activation energy for water in different systems (polymers, glasses and bio-molecules) at temperature lower than the corresponding T_g. An almost constant value – within the experimental uncertainty- can be deduced of E_a = (0.54 ±0.06) eV which can be considered as an universal value characteristic of the water dynamics in restricted geometries. Note that the activation energy of water in a well defined confined system [19] (molecular sieves) is about 0.51 eV which also agree with the general trend.

To find an universal value of E_a implies that a master curve for the temperature dependence of the relaxation times of water dynamics, below the crossover temperature, could be obtained by properly shifting the relaxation times of all systems in the Y-axis. The master curve obtained is shown in Figure 5 and summarizes the universal behavior of water dynamics (process II) in hydrophilic systems of very different nature. This result suggests that water dynamics in this regime is in fact controlled by the confinement character (restricted geometry) which seems to be the only common feature of those systems.

On the other hand, by direct observation of Figure 3 we can envisage the possibility that a master curve could be constructed at temperatures higher than T_g. This supposition in confirmed as shown in Figure 6 where a master plot was constructed by shifting the relaxation times in the temperature region higher than T_g. We have also note that the master curve could also be extended at temperatures close to room temperature since data of water in molecular sieves [19] and water in PVP and PVME at higher frequencies were also included in this graph.

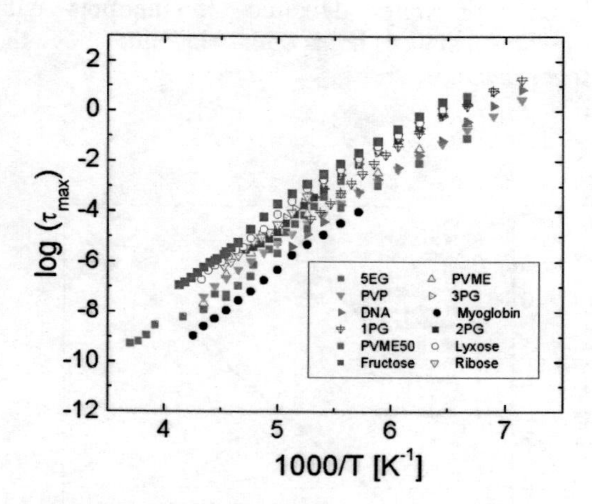

FIGURE 3. Temperature dependence of the relaxation times for different water solution (see legend) at a water concentration of c_w = 30 wt%.

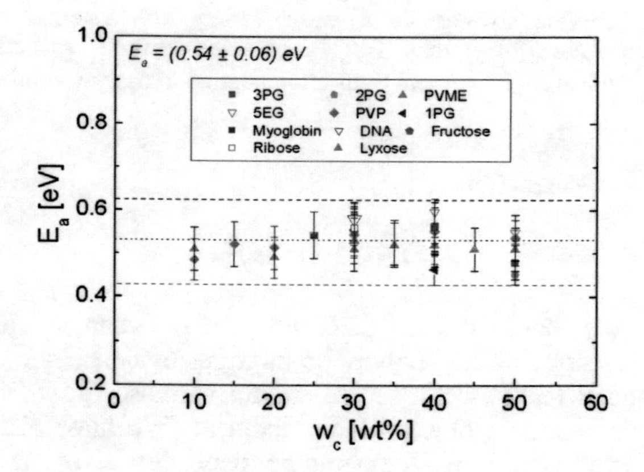

FIGURE 4. Activation energy (E_a) for hydration water at temperatures lower than $T_{g,DSC}$ in all the studied systems (see legend).

The master plot can be used to estimate the parameters of a VFT equation [12], $\tau \propto exp\ (B/(T-T_o))$ where $B = DT_o$ and D is related with the fragility, m, index introduced by Angell [20] to classify the temperature behavior of the supercooled liquids. The so calculated value of T_o was (96 ± 5) K. This T_o can be fixed and then used to calculate the VFT parameters for each investigated mixture. An average value of B =

(2393 ± 362) K can be deduced. From B and T_o a value of the fragility parameter m = (40 ± 5) is obtained. This value suggests that water in hydrophilic mixtures above $T_{g,DSC}$ seems to behave as a relatively strong glass former in this temperature regime. This finding evidences that the universality of the temperature dependence of water dynamics in mixtures with hydrophilic substances holds both below and above the crossover range.

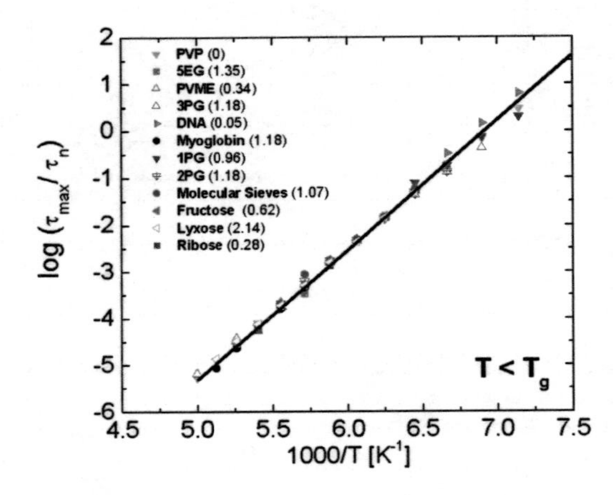

FIGURE 5. Master curve of the dielectric relaxation time for confined water in a wide variety of systems at temperature lower than T_g. The number between brackets (τ_n) indicates the vertical translation applied to each system.

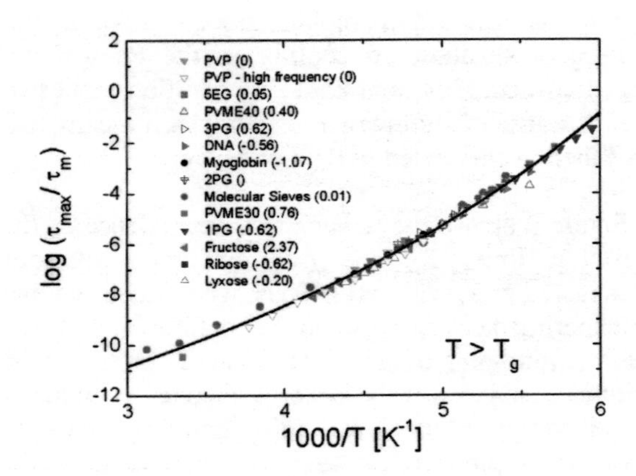

FIGURE 6. Master curve of the dielectric relaxation time for hydration confined water in a wide variety of systems at temperature higher than T_g. The solid line represents the VFT equation (see text). The number between brackets (τ_m) indicates the vertical translation applied to each system.

CONCLUSION

We have shown that the water dynamics in polymeric and non-polymeric mixtures evidence some general features. These are similar to those found in water under geometrical confinement. We have also showed that the temperature dependence of the relaxation times of water dynamics, display a crossover for non-Arrhenius to Arrhenius behavior at about the global glass transition of the system. Nevertheless, our main finding is that the temperature dependence of the relaxation times characterizing the water dynamics in mixtures of hydrophilic substances is universal both below and above the crossover.

ACKNOWLEDGMENTS

The authors acknowledge the support of the University of Basque Country and the Basque Government: project UPV/EHU, 9/UPV 00206.215-13568/2001 and the Spanish Ministry of Education project MAT2004-01017.

REFERENCES

1. A V Finkelstein and O B Ptitsyn, Protein Physics, 2002th edition (Academic Press, London, 2002), Vol. 1, Chap. 6, p.57-60.
2. D. Vitkup, D. Ringe, G. A. Petsko, and M. Karplus, *Nat. Struc. Biol.*, **7**, 34-38, (2000).
3. M. C. Bellissent-Funel, *J. Phys.: Condens. Matter,* 13, 9165-9177, (2001).
4. M. Weik, *Eur. Phys. J E*, **12**, 153-158 (2003).
5. P.W. Fenimore et al, *Proc. Natl. Acad. Sci. U.S.A*, **99**, 16047-16051, (2002).
6. K. Bhattacharyya and B. Bagchi, *J. Phys. Chem. A*, **104**, 10603-10613, (2000).
7. S. Sudo, S. Tsubotani, M. Shimomura, N. Shinyashiki, and S. Yagihara, *J. Chem. Phys.*, **121**, 7332-7340, (2004).
8. S. Cerveny, G.A. Schwartz, R. Bergman, and J. Swenson, *Phys. Rev. Lett.* , **93**, 245702-1-245702-4 (2004).
9. S. Cerveny, G.A. Schwartz, A. Alegría, R. Bergman, and J. Swenson, *J. Chem. Phys.*, **124**, 194501-1-194501-9 (2006).
10. M.Y. Sun, S. Pejanovic, and J. Mijovic, *Macromolecules,* **38**, 9854-9864, (2005).
11. S. Cerveny, J. Colmenero, A. Alegría, *Macromolecules,* **38**, 7056-7063 (2005)
12. H Vogel, Phys. Z. **22**, 645-645 (1921) ; G. S. Fulcher, *J. Am. Chem. Soc.* **8**, 339-355 (1925)

13. N. Shinyashiki, S. Sudo, S. Yagihara, A. Spanoudaki, A. Kyritsis, P. Pissis, *J. Phys.: Condens. Matter,* **19**, 205113-1-205113-12 (2007)
14. S. Cerveny, A. Alegría, J. Colmenero *J. Non-Cryst. Solids,* **353**, 4523–4527, (2007).
15. S. Havriliak, S. Negami. *Polymer*, **8,** 161-163 (1967).
16. S. Cerveny, A. Alegría, J. Colmenero, *Phys. Rev. Lett.*, **97**, 189802-1, (2006)
17. S M Tyagy and SSN Murthy, *Carb. Res.*, **341**, 650-662 (2006).
18. P. Berntsen et al, *Biophys. J*, **89**, 3120-3128 (2005).
19. H. Jansson and J. Swenson, *Eur. Phys. J. E*, **12**, S51-S54 (2003).
20. C.A. Angell, *J. Non-Cryst. Solids*, **131-133**, 13-31 (1991).

Structure of $Te_{1-x}Cl_x$ Liquids

D. Le Coq [a], B. Beuneu [b], and E. Bychkov [a]

[a] *Laboratoire de PhysicoChimie de l'Atmosphère – CNRS/UMR 8101 Université du Littoral-Côte d'Opale – 189A Av. M. Schumann, 59140 Dunkerque, France*
[b] *Laboratoire Léon Brillouin – CEA Saclay – Bât. 563, 91191 Gif-sur-Yvette Cedex, France*

Abstract. The structure of $Te_{1-x}Cl_x$ liquids having composition outside the glassy domain has been investigated using neutron diffraction. Te-rich (x = 0.0, 0.1, 0.2, 0.3) and Cl-rich (x = 0.7, 0.8) liquids have been measured just above their respective melting points, 240 °C ≤ T ≤ 490 °C. The neutron structure factor $S_N(Q)$ exhibits a pronounced First Sharp Diffraction Peak (FSDP) for the Cl-rich compositions suggesting a significant intermediate range order in these molecular liquids. In contrast, the FSDP appears to be weak for the Te-rich liquid alloys presumably having a chain network structure. As expected, two different nearest-neighbour distances have been found in the total correlation function $T_N(r)$. The first peak at $r_1 \approx 2.4$ Å, corresponding to Cl-Te contacts, increases with x. The second at $r_2 \approx 2.8$ Å is related to the Te-Te first neighbours and decreases with x. A detailed analysis of the $T_N(r)$ and difference $\Delta T_N(r)$ is given in the contribution and allows between three possible structural models to be chosen.

Keywords: Chalcohalide liquids, Liquid structure, Neutron diffraction
PACS: 61.20.-p; 61.12.Ex

INTRODUCTION

The structure of non-crystalline phases in the tellurium – chlorine system has been previously studied for vitreous compositions by using ^{129m}Te emission and ^{125}Te absorption Mössbauer spectroscopy [1]. Wells et al. reported that the specific glass composition g-Te_3Cl_2 consists of chain fragments similar to those found in the crystalline compound c-Te_3Cl_2. A typical glassy chain exhibits 3 types of Te sites: site A, two-fold coordinated to two Te nearest neighbors (NNs); site B, four-fold coordinated to two Te and two Cl NNs; and site A', intrinsic to a glass, identified as a chain terminating Te-site having one Te and one Cl NNs. Our recent X-ray and neutron diffraction experiments for $Te_{1-x}Cl_x$ glasses over a wide composition range, $0.35 \le x \le 0.65$, are in good agreement with the chain model [2]. We have found that the Te chain structure is preserved in the glasses since the Te-Te coordination number, N_{Te-Te}, is close to 2 at $r_2 \approx 2.8$ Å (typical NN distance in liquid and amorphous Te). All chlorine species were found to be non-bridging and bound with Te, $N_{Cl-Te} \approx 1$ at $r_1 \approx 2.4$ Å. An absence of any Cl-Cl first-neighbour correlations at $r < 2.4$ Å was observed. Nevertheless, details of the glass structure still need to be clarified. The questions arise (i) whether the existence of sites B and A' can be proved by direct structural methods, and (ii) what is the final stoichiometry in the Cl-rich region.

The tellurium – chlorine system has the only congruently melting compound, $TeCl_4$. Solid $TeCl_4$ crystallises in a cubane-like structure and consists of Te_4Cl_{16} tetramers [3-5]. Each Te atom in the tetramer is attached to three terminal Cl atoms with an average distance of 2.32 Å and three bridging Cl with a much longer distance of 2.92 Å. There are no direct Te-Te NN contacts in the tetramer structure. The average intramolecular Te-Te second neighbour distance is ≈ 4.1 Å. In contrast, electron diffraction [6] and gas-phase Raman spectroscopy [7] indicate that gaseous $TeCl_4$ is a monomer having trigonal bipyramidal structure with an equatorial lone pair of electrons, hence the Te is coordinated to four Cl NNs.

Recently, using neutron and high-energy X-ray diffraction, we have found that liquid TeX_4 (X = Cl or Br) can be considered as a molecular liquid consisting of Te_2X_8 dimers [8]. Two TeX_4 monomers in the dimer are connected by a short Te-Te bond (≈ 2.75 Å). Thus, the liquid structure appears to be intermediate between the monomeric gas and tetrameric crystalline solid. The last finding also suggests that the Te_2Cl_8 dimer could be a final member of the $Te_{1-x}Cl_x$ chain-like family.

In this contribution, we present neutron diffraction experiments for $Te_{1-x}Cl_x$ liquids having composition

CP982, *Complex Systems, 5th International Workshop on Complex Systems*
edited by M. Tokuyama, I. Oppenheim, and H. Nishiyama

outside the glass-forming domain. These liquids belong to two regions: Te-rich region 1 (x = 0.0, 0.1, 0.2, 0.3) and Cl-rich region 2 (x = 0.7, 0.8). The obtained results will be discussed in order to get a better understanding of the liquid and glass structure in this binary system.

EXPERIMENTAL

The polycrystalline samples were synthesized from commercial $TeCl_4$ powder (Aldrich, 99%) and elemental Te (Fluka, 99.999 %) in evacuated silica tubes (ID 8 mm, OD 10 mm). Since $TeCl_4$ is moisture-sensitive, all manipulations with the starting materials were performed in a glove box (Argon environment). The maximum temperature of the synthesis was 300°C higher than the liquidus temperature for a given composition to ensure melt homogenization.

Neutron scattering experiments have been carried out on the hot source of the Orphée reactor at the Laboratoire Léon Brillouin (Saclay, France) using the 7C2 spectrometer with a 0.707 Å incident wavelength [9]. Diffraction pattern were recorded with a 640-cell multidetector which provides an accurate determination of the spectra over the whole momentum transfer range of 0.3 $Å^{-1} < Q < 16$ $Å^{-1}$. Spectra were collected at a temperature just above the respective melting temperature of the samples (240 °C $\leq T \leq$ 490 °C) by using a vanadium furnace consisting in a vertical cylinder (30 mm diameter, 0.01 mm thickness, and 300 mm height). The neutron structure factors $S_N(Q)$ were obtained from the scattered intensities. Conventional data reduction was applied for absorption, incoherent and multiple scattering. Inelastic effects were treated by a Placzek type correction. The total pair correlation functions $g(r)$ were obtained by a Fourier transformation of $S_N(Q)$ as follows:

$$g(r) = 1 + \frac{1}{2\pi^2 \rho r} \int (S_N(Q) - 1) Q \sin(Qr)\, dQ \tag{1}$$

where ρ is the average number density of the samples. The total correlation functions $T_N(r)$,

$$T_N(r) = 4\pi r \rho g(r), \tag{2}$$

were used for obtaining the correlation distances r and coordination numbers N_{ij} in the system using a least-square fitting of the experimental data to Gaussian functions.

RESULTS

Figure 1 shows the neutron structure factors $S_N(Q)$ for two specific compositions of $Te_{1-x}Cl_x$ liquids in the 0.5-16 $Å^{-1}$ range, characteristic of the Te-rich (x = 0.1, Region 1) and Cl-rich (x = 0.7, Region 2) domains. Distinct oscillations are observed in the $S_N(Q)$ for liquid L-$Te_{0.9}Cl_{0.1}$ even at $Q \geq 12$ $Å^{-1}$. The oscillations are shifted to higher Q and significantly damped for L-$Te_{0.3}Cl_{0.7}$. A decrease in the average interatomic distances and a change from a network-like (Region 1) to a molecular-like (Region 2) liquid seem to be responsible for the observed features. The x = 0.7 composition exhibits also a well-pronounced First Sharp Diffraction Peak (FSDP) at $Q_1 \approx 1.15$ $Å^{-1}$, whereas the FSDP is broad and weak for the Te-rich counterpart.

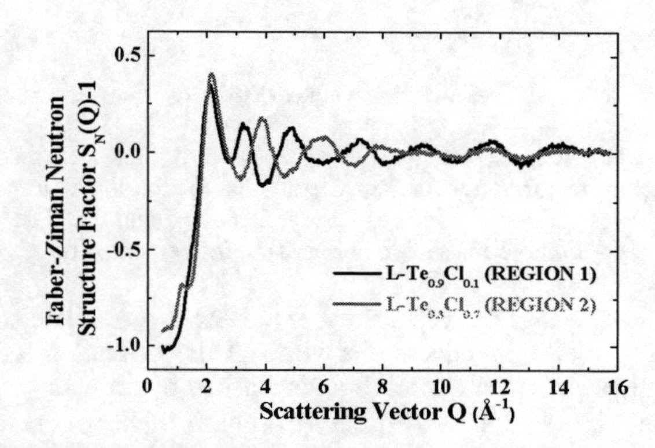

FIGURE 1. Total Faber-Ziman neutron structure factor $S_N(Q)$ for two specific compositions, $Te_{0.9}Cl_{0.1}$ and $Te_{0.3}Cl_{0.7}$. The first one corresponds to the Te-rich domain (Region 1) and the second one corresponds to the Cl-rich domain (Region 2).

Figure 2 represents the low Q-part of $S_N(Q)$ for all investigated compositions, i.e., x = 0.0, 0.1, 0.2, 0.3, 0.7, and 0.8. Systematic changes with x and a notable increase of the FSDP amplitude between Region 1 and 2 liquids are clearly seen.

The respective total correlation functions $T_N(r)$ are shown in Figure 3. The $T_N(r)$ for liquid Te at 490 °C is very similar to that published earlier [10-12]. As expected, two different NN distances are observed for $Te_{1-x}Cl_x$ liquids. The first peak at 2.35 Å $\leq r_1 \leq$ 2.45 Å, increasing with x, is attributed to Cl-Te first neighbours. This distance is consistent with the sum of Te and Cl covalent radii equal to 1.35 Å and 0.99 Å, respectively. Similar Cl-Te bond length was also found in crystalline [3-5] and liquid $TeCl_4$ [8] as well as in the $Te_{1-x}Cl_x$ glasses [2]. All the Cl atoms are terminal, $N_{Cl\text{-}Te} = 0.9 \pm 0.1$, as those in the glasses and L-$TeCl_4$.

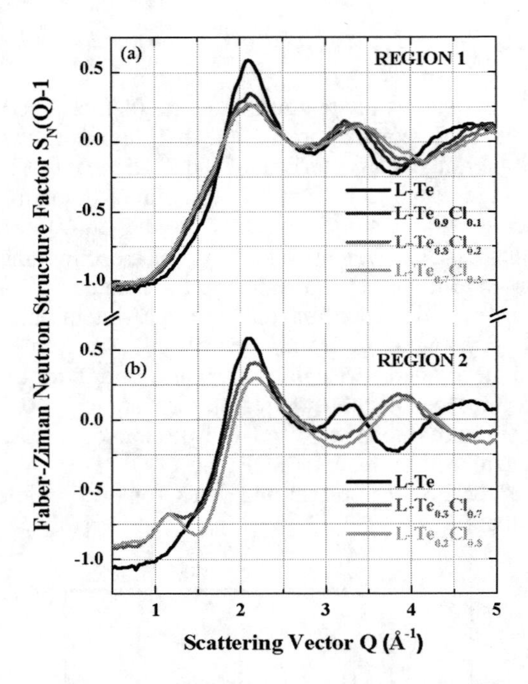

FIGURE 2. Low-Q part of the Faber-Ziman neutron structure factor $S_N(Q)$ for $Te_{1-x}Cl_x$ liquids; (a) Region 1, x = 0.0, 0.1, 0.2, and 0.3, and (b) Region 2, x = 0.7 and 0.8. The $S_N(Q)$ for liquid Te is shown in Fig. 2 (b) for comparison.

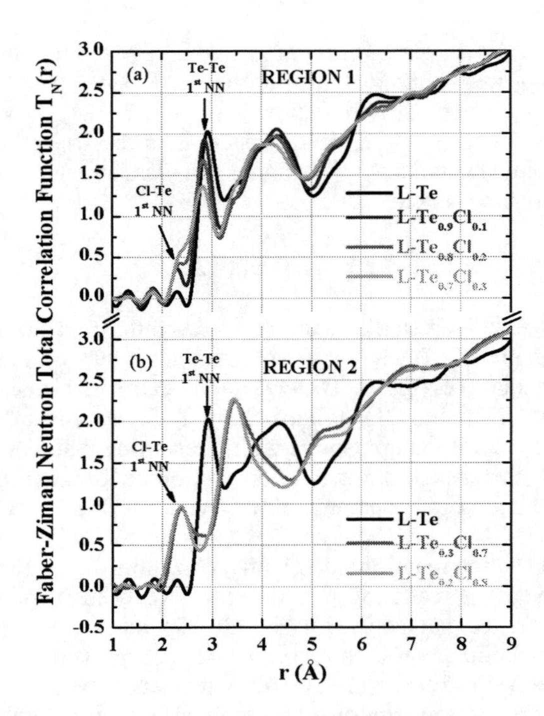

FIGURE 3. Faber-Ziman neutron total correlation function $T_N(r)$ for $Te_{1-x}Cl_x$ liquids; (a) Region 1, x = 0.0, 0.1, 0.2, and 0.3, and (b) Region 2, x = 0.7 and 0.8. The $T_N(r)$ for liquid Te is shown in Fig. 3 (b) for comparison.

The second peak at 2.75 Å $\leq r_2 \leq$ 2.92 Å is related to Te-Te NNs and decreases with x. This correlation is hardly seen in the Cl-rich liquids masked by an intense peak at $r \approx$ 3.4 Å (Fig. 3 (b)). Nevertheless, one observes a distinct difference between the x = 0.7 and 0.8 liquids in the above r-range. The Te-Te coordination number, $N_{Te\text{-}Te}$, is slightly above 2 for the Region 1 liquids caused by overlapping with interchain Te-Te correlations (similar to that in liquid Te [10,11]) and, possibly, with intra/interchain Cl-Te and/or Cl-Cl contacts. The $N_{Te\text{-}Te} \leq 2$ for the Cl-rich liquids is consistent with a molecular-type liquid structure built-up by Te_2Cl_8 dimers (x = 0.8, $N_{Te\text{-}Te}$ = 1.0) or/and short chains Te_kCl_m (1.33 (k = 3) $\leq N_{Te\text{-}Te} \leq$ 1.71 (k = 7)).

The Te-rich (Fig. 3 (a)) and Cl-rich (Fig. 3 (b)) compositions are also clearly different at higher r, 3 Å $\leq r \leq$ 5 Å. A broad asymmetric feature, centred at ≈ 4.4 Å and overlapping with the Te-Te NN peak, is characteristic of liquid Te. This feature corresponds to Te-Te second neighbour intrachain contacts as well as to multiple interchain correlations [12]. Its variations with x are observable however the feature at 4.4 Å remains dominant in the above r-range of $T_N(r)$, consistent with a chain network structure for the Region 1 liquids.

The peak at ≈ 3.4 Å is intrinsic to the Cl-rich liquids. Short Cl-Cl second neighbour intramolecular contacts and also intermolecular correlations of the Te_2Cl_8 dimer are responsible for this peak. A chain-like contribution to the $T_N(r)$ is hardly visible for the x = 0.7 liquid, confirming an essentially molecular-like structure for the Region 2 liquids.

DISCUSSION

Three possible structural models for the $Te_{1-x}Cl_x$ liquids are shown in Figure 4. Hypothesis A suggests a mixture of $TeCl_4$ and Te liquids, phase-separated or homogeneous on mesoscopic scale. A modified Te chain having terminal $TeCl_4$–like units is assumed to be a principal building block in model B. Finally, hypothesis C is similar to the Wells et al. model [1] but instead of a single terminal Cl species the $-TeCl_4$ terminal units are proposed.

A two-component liquid A seems to be incompatible with compositional trends of the Cl-Te and Te-Te NN distances (Figure 5). The Cl-Te bond length increases with x from 2.35 Å to 2.45 Å for the Region 1 liquids whereas the Te-Te one contracts simultaneously by 0.07 Å. In addition, the two distances are different from those in Region 2. In model A, one should expect either invariant NN distances or those changing similarly.

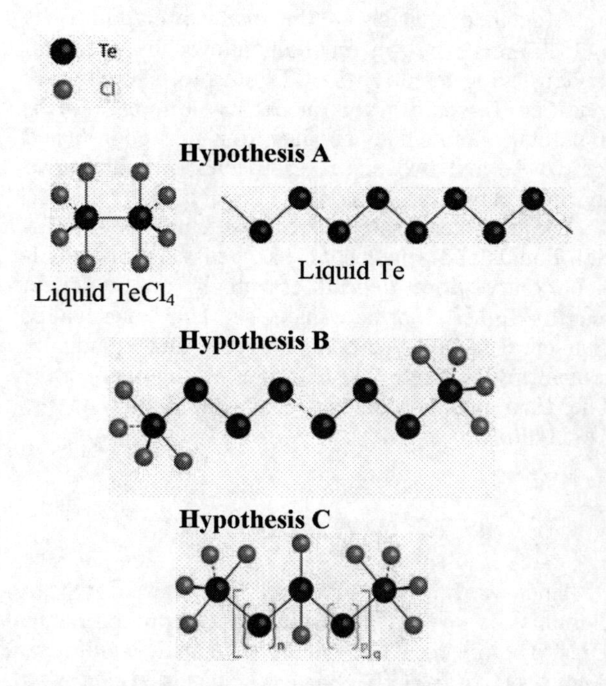

FIGURE 4. Schematic representation of three structural hypotheses for the $Te_{1-x}Cl_x$ liquids: (A) Mixture of $TeCl_4$ and Te liquids, phase-separated or homogeneous on mesoscopic scale; (B) Te-chain structure with terminal $TeCl_4$–like units; (C) Te-chain with terminal $-TeCl_4$ units and intermediate Te species, four-fold coordinated to two Te and two Cl NNs; n, p, and q are integer numbers.

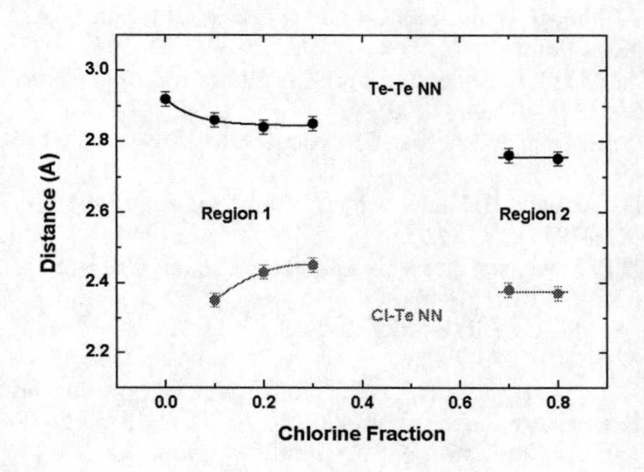

FIGURE 5. Compositional dependence of the Cl-Te and Te-Te NN distances as a function of the chlorine content *x*.

Short and intermediate range ordering at $r < 5$ Å for a modified Te chain B can be reproduced by scaled, expanded or contracted, $T_N(r)$ functions for $TeCl_4$ and Te. As an example, Figure 6 (a) shows experimental $T_N(r)$ for liquid $Te_{0.7}Cl_{0.3}$ plotted together with those functions.

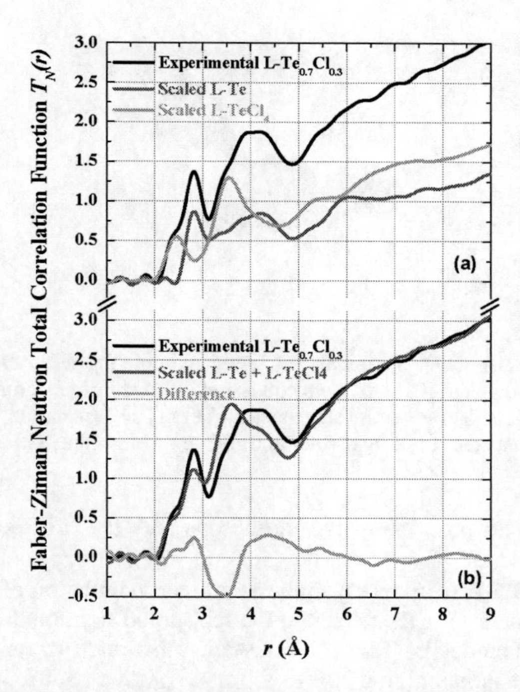

FIGURE 6. Experimental total correlation function $T_N(r)$ for liquid $Te_{0.7}Cl_{0.3}$ plotted together with (a) scaled expanded (L-$TeCl_4$) and contracted (L-Te) $T_N(r)$ functions for these components; (b) directly superposes experimental $T_N(r)$ and the sum of scaled functions, $\Delta_{TeCl4}T_N(r)$ + $\Delta_{Te}T_N(r)$. Difference between the two functions is also shown in (b).

All the main features, peaks at ≈ 2.4 Å, ≈ 2.8 Å, ≈ 4.4 Å, and even a shoulder at ≈ 3.4 Å, are well reproduced. However, the sum of scaled functions exhibits important differences compared to experimental $T_N(r)$ (Fig. 6 (b)). We should note a much lower amplitude of the correlations at ≈ 3.5 Å for the experimental function as well as more intense correlations at 4-5 Å.

These findings indicate that two tellurium sites in the chain B (terminal $-TeCl_4$ and chain-like Te) are not sufficient to reproduce the experimental data. For better understanding the observed difference, let us focus on second neighbour correlations in the terminal $-TeCl_4$ unit, in fact, half the Te_2Cl_8 dimer. There are four short Cl1-Cl2 (3.4 Å), one intermediate Cl2-Cl2 (4.1 Å), and one long Cl1-Cl1 (4.8 Å) contacts in addition to two short Cl1-Te2 (3.6 Å) and two long Cl2-Te2 (4.5 Å) correlations (Fig. 7). It is also necessary to note that the Cl-Cl correlations have a larger weighting factor since the chlorine coherent neutron scattering length ($\overline{b}_{Cl}$ = 9.58 fm) is significantly higher than the tellurium one ($\bar{b}_{Te}$ = 5.80 fm).

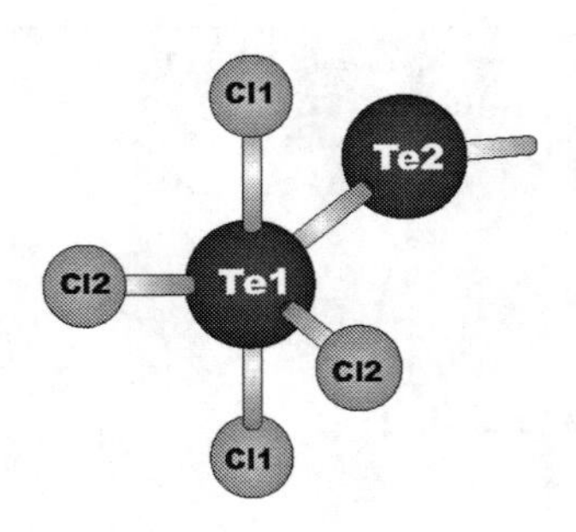

FIGURE 7. Terminal–$TeCl_4$ unit completing a Te chain. Two apical (Cl1) and two equatorial (Cl2) chlorine atoms are attached to a central Te species (Te1). The last one is also connected to a two-fold coordinated chain-like Te atom (Te2).

Strongly negative amplitude of the difference function, $T_N(r)[\exp] - \Delta_{TeCl4}T_N(r) - \Delta_{Te}T_N(r)$, peaked at ≈ 3.5 Å (Fig. 6 (b)) suggests a lower population of the equatorial Cl2 species in the real liquid than predicted by model B. These species are substituted by apical Cl1 atoms giving rise to a broad positive contribution between 3.8 Å and 5.2 Å. In other words, the third Te site, four-fold coordinated to two Te and two apical Cl species is needed to explain the experimental results. Thus, hypothesis C seems to be the most appropriate model.

Within the framework of this model, difference between the Te-rich and Cl-rich liquids appears to be quantitative, i.e., depending on chain length. Higher chlorine content shortens the chains or decreases n, p, and q (see Fig. 4 (c) for details). In the limit of $x \rightarrow 0.8$, q becomes 0.

In order to improve details of the proposed liquid structure in the $Te_{1-x}Cl_x$ system, we are planning complementary high-energy X-ray diffraction experiments. In the case of Te/Cl compositions, the situation is extremely favourable since the neutron and X-ray coherent scattering factors are significantly different, $\bar{b}_{Cl} / \bar{b}_{Te}$ = 1.65 but Z_{Cl} / Z_{Te} = 0.327. The contrast between neutron and X-ray weighted structure factors and r-space functions will improve data analysis by eliminating one of the three partials (Cl-Te, Te-Te, and Cl-Cl) at once, also giving access to interchain/intermolecular correlations.

CONCLUSIONS

The structure of $Te_{1-x}Cl_x$ liquids having composition outside the glassy domain has been investigated using neutron diffraction. Te-rich (x = 0.0, 0.1, 0.2, 0.3) and Cl-rich (x = 0.7, 0.8) liquids have been measured just above their respective melting points at 240 °C ≤ T ≤ 490 °C.

A detailed analysis of the total correlation $T_N(r)$ and difference $\Delta T_N(r)$ functions allows the structural model having three types of Te sites to be chosen. A modified Te-chain in this model has terminal $-TeCl_4$ units and intermediate Te sites, four-fold coordinated to two Te and two apical Cl species, in addition to 'normal' chain-like Te atoms.

The observed chain network structure for the Te-rich liquids and a molecular-like behavior for the Cl-rich compositions depend essentially on the chain length. Higher chlorine content shortens long chains, characteristic of Te-rich liquids, and gradually transforms them into the short ones or oligomeric units. The final member of this chain-like family is the Te_2Cl_8 dimer (x = 0.8).

ACKNOWLEDGMENTS

This work was supported by the European Commission within the Interreg III program (the CTMM project). The authors thank Jean-Pierre Ambroise (LLB, CEA Saclay), Francis Hindle and Marc Fourmentin (LPCA, Dunkerque) for their assistance.

REFERENCES

1. J. Wells, W.J. Bresser, P. Boolchand, and J. Lucas, *J. Non-Cryst. Solids*, **195,** 170-175 (1996).
2. D. Le Coq, E. Bychkov, C.J. Benmore, and A.C. Hannon, *13th Intern. Symp. on Non-Oxide Glasses*, Florida USA, Nov 2004.
3. A. Alemi, E. Soleimani, and Z.A. Starikova, *Acta Chim. Slov.*, **47** , 89-98 (2000).
4. G. A. Ozin and A. Van der Voet, *J. Mol. Struct.*, **10**, 397-403 (1971).
5. B. Buss and B. Krebs, *Angew. Chem. Int. Ed. Engl.*, **9**, 463 (1970).
6. D.P. Stevenson and V. Schomaker, *J. Amer. Chem. Soc.*, **62**,1267-1270 (1940).
7. I.R. Beattie, J.R. Horder, and P.J. Jones, *J. Chem. Soc. A*, 329-330 (1970).
8. D. Le Coq, A. Bytchkov, V. Honkimäki, B. Beuneu, and E. Bychkov, *J. Non-Cryst. Solids* (2007), under press.
9. J.-P. Ambroise and R. Bellissent, *Rev. Phys. Appl.*, **19**, 731-734 (1984).
10. S. Takeda, S. Tamaki, Y. Waseda, *J. Phys. Soc. Japan* **53**, 3830-3836 (1984).
11. M. Misawa, *J. Phys. Condens. Matter,* **4**, 9491-9500 (1992).
12. C. Bichara, J.-Y. Raty, J.-P. Gaspard, *J. Non-Cryst. Solids*, **205-207**, 361-364 (1996).

Dynamics of Nano-Meter-Sized Domains on a Vesicle

Y. Sakuma, N. Urakami*, Y. Ogata*, M. Nagao†¶, S. Komura§, and M. Imai

Department of Physics, Ochanomizu University, Bunkyo, Tokyo112-8610, Japan
**Department of Physics and Information Sciences, Yamaguchi University, Yoshida, Yamaguchi 753-8512, Japan*
†Cyclotron Facility, Indiana University, Bloomington, IN 47408-1398, USA
¶Center for Neutron Research, National Institute of Standards and Technology, Gaithersburg, MD 20899-6102, USA
§Department of Chemistry, Tokyo Metropolitan University, Minami Osawa, Tokyo 192-0397, Japan

Abstract. We have investigated the dynamics of nano-meter-sized domains on a vesicle composed of saturated phospholipids, unsaturated phospholipids and cholesterols by means of a neutron spin echo technique. The diffusion coeffisient of the domains obtained from the intermediate scattering function is about two orders of magnitude larger than that calculated by Saffman and Delbrück law. From molecular dynamics simulations we found another type of the domain dynamics where the domains are agitated by thermal fluctuation and repeats coalescence and rupture. The relaxation rate of the new mode is about two orders of magnitude larger than that of the diffusion mode. The coalescence and rupture mode of domains may be responsible for the observed domain dynamics.

Keywords: Domain dynamics, Lipid raft, Nano-meter-sized domain, Neutron spin echo, Small unilamellar vesicle
PACS: 87.14.Cc, 87.14.dt, 87.16.dr

INTRODUCTION

In cell membranes, the constituents form characteristic lateral heterogeneities arising from the immiscibility of the lipid components such as sphingomyelin, unsaturated phospholipids, and cholesterol (Chol). The heterogeneitis coupled with the specific protein distribution are responsible for the important biological functionalities such as endocytosis, adhesion, signaling, protein transport and apoptosis and called lipid rafts [1,2]. Due to the wealth of biological significance of the lipid raft model, the structure of lipid rafts in cell membranes has been investigated using many techniques, which have shown that the size of lipid raft is nano-meter length scale. On the other hand, the dynamical nature of the raft is still controversial [3-5]. One of the important points of the raft is whether the raft is stable entity in the cell membranes or dynamically fluctuating platform, which is closely related to the mechanism of the biological functionalities.

In order to understand the statical and dynamical nature of the lipid raft, the model biomembrane systems consisting of phospholipids and cholesterol have also been investigated [6-8]. For example, it has been shown that the model vesicles composed of saturated phospholipids, unsaturated phospholipids and cholesterols exhibit lateral phase separation between L_o (liquid-ordered) phase and L_d (liquid-disordered) phase below a miscibility temperature T_m, which is observed as micro-meter-sized liquid domain structure by a fluorescence microscopy. The diffusion of micro-meter-sized domains on a giant vesicle was visited by Keller and our groups. The former group showed that the diffusion of domains follows the Saffman and Delbrück law [9]

$$D(r) = \frac{k_B T}{4\pi\eta\delta}[\ln(\frac{\eta\delta}{\eta_w r}) - \gamma] \qquad (1)$$

where D is the diffusion coefficient, η and η_w are the viscosities of the membrane and water, respectively, δ is the membrane thickness, γ=0.5772 and r is the radius of domain [10], whereas we claimed that the hydrodynamic flow in the membrane affects the diffusion of domains [11]. The behavior of the micro-meter-sized liquid domains on the model biomembranes is well understood by the direct observations, while that of the nano-meter-length scale

CP982, *Complex Systems, 5th International Workshop on Complex Systems*
edited by M. Tokuyama, I. Oppenheim, and H. Nishiyama

domains is poorly understood due to the optical resolution limit of the microscopy.

In this study we prepare small unilamellar vesicles (SUVs) with nano-meter length scale for spatial confinement, which makes it possible to focus our attention on the nano-meter-sized domains. In order to reveal the dynamical structure we adopt the small angle neutron scattering (SANS) and the neutron spin echo (NSE) techniques. A unique feature of the neutron scattering is that a hydrogen atom and a deuterium atom have different scattering lengths. We prepare SUVs using deuterated saturated lipid, protonated unsaturated lipid and protonated cholesterol and obtain the information on the deutrated domains in the SUV. In addition we perform molecular dynamics simulations with a coarse grained model to reproduce the phase separation on a SUV. By comparing the SANS, NSE and the simulation results we elucidate the statical and dynamical nature of nano-meter-sized domains in the SUV.

EXPERIMENTS

Commercial Reagents

1,2-dioleoyl-sn-glycero-3-phosphocholine (DOPC) (> 99 % purity), 1,2-dipalmitoyl-sn-glycero-3-phosphocholine (*h*-DPPC) (> 99 % purity) and 1,2-dipalmitoyl-d62-sn-glycero-3-phosphocholine (*d*-DPPC) (> 99 % purity) were obtained in a powder form from Avanti Polar Lipid, Inc. (Birmingham, AL) [12]. Cholesterol (Chol) (> 99 % purity) was purchased from Sigma-Aldrich Co. (St. Louis, MO) [12]. In this study we fixed the vesicle composition of *d*-DPPC/DOPC/cholesterol = 4/4/2 (mole ratio) having T_m = 29°C, because it shows a nucleation-growth type phase separation in observed temperature region.

Sample Preparation

We prepared SUVs by the following procedure. First we dissolved the lipids (DPPC, DOPC and Chol) in chloroform (5mM) and then the solutions were mixed at the prescribed ratio of DOPC, DPPC and Chol in a vial. The chloroform was evaporated in a stream of nitrogen gas and then the obtained lipid film was kept under vacuum for one night to remove the remaining organic solvent completely. The dried lipid film was dispersed in 1 ml of pure water at 60 °C, which results in the formation of micro-meter-sized vesicles (giant vesicles). The suspension of vesicles with milky appearance was sonicated using a ultrasonic homogenizer with 20kHz frequency (Yamato, Powersonic Model 50 [12]) for 30 min. After the sonication we obtained SUVs with radius of ≈100Å and the suspension became transparent.

Small Angle Neutron Scattering Measurements

In this study we performed SANS experiments under two experimental conditions, one is a film contrast condition where we can obtain whole shape of the SUVs and the other is a contrast matching condition where we can estimate the lateral heterogeneity in the SUVs. In the film contrast condition, we used *h*-DPPC/DOPC/Chol ternary vesicles in D_2O solvent, whereas in the contrast matching experiment, we substituted *d*-DPPC for *h*-DPPC to extract the domain structure on the vesicle. At this matching condition (the H_2O volume fraction of the matching solvent; ϕ^*_{H2O}=0.548) we observed no scattering from SUVs in the homogeneous one-phase region. Decreasing temperature from the one-phase region to two-phase region, the lateral phase separation takes place at the miscibility temperature, where *d*-DPPC rich L_o phase coexists with DOPC rich L_d phase. Then the scattering length densities (SLDs) of the domain ρ_d and the matrix ρ_m do not match with SLD of the solvent ρ_{solv}. These SLD differences bring a characteristic scattering function.

The SANS measurements were performed using the SANS-U instrument of the Institute for Solid State Physics, The University of Tokyo at JRR-3M of the Japan Atomic Energy Agency [13,14].

Neutron Spin Echo Experiments

In order to reveal the dynamics of nano-meter-sized domains we performed NSE experiments under two experimental conditions, one is the film contrast condition where we can obtain the dynamics of whole shape of the SUVs and the other is the contrast matching condition where we can estimate the dynamics of nano-meter-sized heterogeneity in the SUVs.

The NSE experiments were performed using NG5-NSE spectrometer at the NIST Center for Neutron Research [15,16]. Incident neutron wavelength of 6 Å was selected by a mechanical neutron velocity selector with a wavelength resolution of about 20 %. The measured momentum transfer, q, covered the range (0.047 to 0.09) $Å^{-1}$ and the time range was (0.05 to 15) ns. The DAVE software package was used for elements of the data reduction and analysis [17].

RESULTS AND DISCUSSIONS

Domain Formation on SUVs

First, we checked the shape and the size of the SUVs. We obtained the SANS profiles from SUVs at the film contrast condition at 333.2 K (homogeneous phase) which is well described by the model scattering function for the spherical shell structure. From the fitting we extracted mean outer radius of vesicles R_{out} = (112±11) Å, the bilayer thickness δ = (42±6) Å and the size polydispersity ($p^2 = \langle R_{in}^2 \rangle / \langle R_{in} \rangle^2 - 1$) p= (0.32±0.02), and these values were independent of the temperature.

Next, we investigated the nano-meter-sized domain formation on the SUVs with $R \approx 110$Å using the contrast matching technique of SANS experiments. The SANS profiles of the SUV are shown in Fig. 1. At 333.2K where the SUVs are in one phase homogeneous region, we observed no significant scattering from the vesicles, indicating that the DOPC, Chol and DPPC are mixed molecular level and each SUV has almost uniform composition. When we decreased temperature to the phase separation region (285.2 K), excess scattering intensity due to the domain formation was observed as shown in Fig. 1. The scattering intensity profile has a maximum in the vicinity of q = 0.03 Å^{-1}, which is a characteristic scattering pattern from domains on a vesicle at the matching condition and the peak position depends on the vesicle size or number of domains.

Dynamics of Nano-meter-sized Domains

In order to make clear the dynamical nature of the nano-meter-sized domains, we measured the intermediate scattering function of the ternary SUVs with *d*-DPPC/DOPC/Chol = 4/4/2 at the matching condition. The intermediate scattering function is given by

$$S(q,t) = \langle \sum_{i,j} e^{i\mathbf{q}\cdot[\mathbf{R}_i(t)-\mathbf{R}_j(0)]} \rangle \quad (2)$$

where $\mathbf{R}_i(t)$ is the coordinates of individual lipid molecules at time t. In Fig. 2 we show the obtained intermediate scattering function for the ternary SUVs at the contrast matching condition at 285.2 K. By assuming that the intermediate scattering function consists of two components, the whole vesicle diffusion mode and the nano-meter-sized domain relaxation mode, the intermediate scattering functions were fitted by a double exponential function

$$\frac{S(q,t)}{S(q,0)} = A(q)\exp(-D_0 q^2 t) + (1 - A(q))\exp(-\Gamma t) \quad (3)$$

where D_0 is the diffusion coefficient of the whole SUV, Γ is the relaxation rate of the domain relaxation mode and A is the numerical constant. The value of D_0 used here is obtained from the intermediate scattering function of the SUV at the film contrast condition.

Under the assumption that the relaxation profile of spherical vesicle is expressed by a sum of the whole vesicle diffusion mode and the vesicle deformation mode, we fitted the relaxation profile measured at the film contrast condition by a double exponential function

$$\frac{S(q,t)}{S(q,0)} = B(q)\exp(-D_0 q^2 t) + (1 - B(q))\exp(-\Gamma_d t) \quad (4)$$

where Γ_{d} is the relaxation rate of the shape deformation mode [18]. The relaxation rate for the vesicle diffusion mode show linear relationship against q^2 and the slope gives the diffusion coefficient D_0=(1.4 ±0.01)$\times 10^{-11}$ m^2/s. The diffusion coefficient can be estimated from the Stokes-Einstein equation

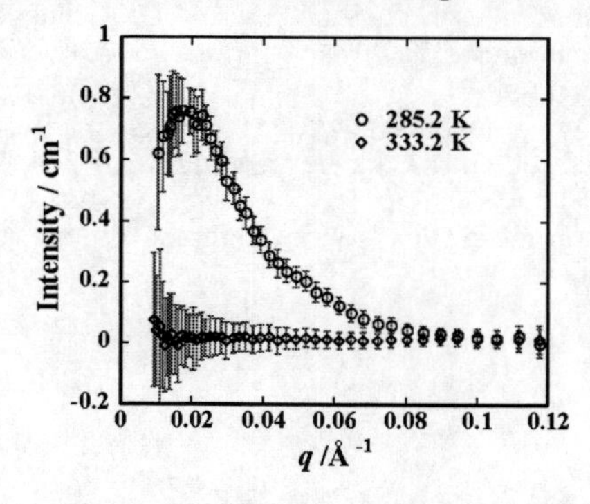

FIGURE 1. SANS profiles from SUVs at the contrast matching condition at 333.2 and 285.2 K [19]

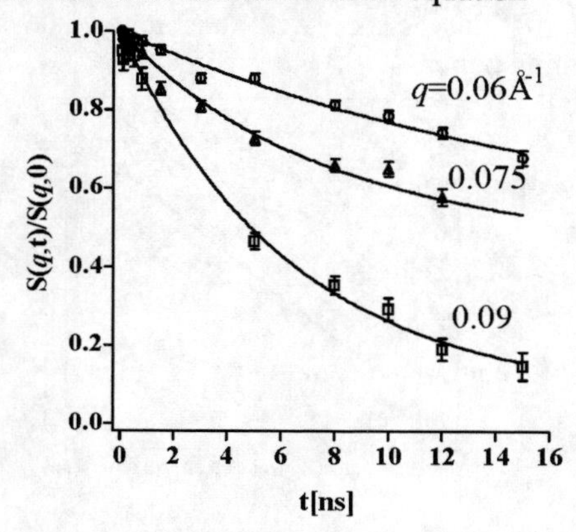

FIGURE 2. Intermediate scattering function for the ternary SUVs at the contrast matching condition at 285.2 K [19]

$$D_0 = \frac{k_B T}{6\pi\eta_w R_{out}} \qquad (5)$$

Using the values of T=285.2 K, η_w=1.3 × 10 Pa·s and R_{out}=112Å, the calculated D_0 is 1.6×10^{-11} m²/s, which agrees well with the value estimated by the NSE measurement. Then the eq. (3) contains only two adjustable parameters, A and Γ, and the fitting results are shown in Fig. 2 by solid lines. The relaxation rate of the nano-meter-sized domain mode obtained from the fitting are plotted against q^2 in Fig. 3, which gives good linear relationship although we have only three points. The q dependence of Γ indicates that the vesicle deformation mode does not affect the domain relaxation mode significantly[18]. When we assume that the domain relaxation mode corresponds to the domain diffusion, the slope gives the diffusion coefficient of the domains of $(3.7 \pm 0.2) \times 10^{-10}$ m²/s. The Saffman and Delbrück law (eq. (1)) gives the diffusion coefficient of the domains of 2.7×10^{-12} m²/s, which is smaller than the experimental value by about two orders of magnitude. Thus the domain diffusion model may be inadequate to explain the observed domain mode.

In order to interpret the domain dynamics on a SUV, we performed the molecular dynamics simulation. For a shallow quenched vesicle from the homogeneous state, the mono-domain is agitated by thermal fluctuations and repeats coalescence and rupture, whereas for the deep quenched vesicle, the mono-domain shows the Brownian motion on the SUV. From the time resolved simulation data in the equilibrium state we calculated the intermediate scattering function for two states. The obtained relaxation rate for the domain diffusion mode is about two orders of magnitude smaller than that of the coalescence and rupture mode. Thus the coalescence and rupture mode may correspond to the observed relaxation mode of the nano-meter-sized domains. Details of this issue will be reported in the forthcoming paper.

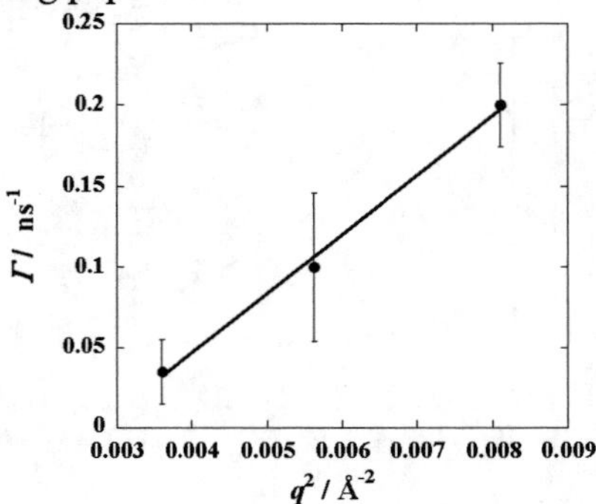

FIGURE 3. The relaxation rate of the nano-meter-sized domain mode against q^2 [19]

ACKNOWLEDGMENTS

This work utilized facilities supported in part by the National Science Foundation under Agreement No. DMR-0454672. This work was performed using SANS-U of the Institute for Solid State Physics, the University of Tokyo (Proposal No. 7606).

REFERENCES

1. K. Simons and E. Ikonen, *Nature*. **387**, 569-572 (1997).
2. M. Edidin, *Annu. Rev. Biophys. Biomol. Struct.* **32**, 257-283 (2003).
3. A. Kusumi, I. K. Honda, and K. Suzuki, *Traffic*. **5** 213-230 (2004).
4. A. Pralle, P. Keller, E. L. Florin, K. Simons. and J. K. H. Hörber, *J. Cell. Biol.* **148**, 997-1007 (2000).
5. K.Simons and W L. C. Vaz, *Annu. Rev. Biophys. Biomol. Struct.* **33**, 269-295 (2004).
6. G. W. Feigenson, *Annu. Rev. Biophys. Biomol. Struct.* **36**, 63-77 (2007).
7. S. L. Veatch and S. L. Keller, *Phys. Rev. Lett.* **89**, 268101 (2002).
8. S. L. Veatch and S. L. Keller, *Biophys. J.* **85**, 3074-3083 (2003).
9. P. G. Saffman and M. Delbrück, *Proc. Nat. Acad. Sci. USA* **72**, 3111-3113 (1975).
10. P. Cicuta, S.L. Keller and S.L. Veatch, *J. Phys. Chem. B* **111**, 3328-3331 (2007).
11. M. Yanagisawa, M. Imai, T. Masui, S. Komura and T. Ohta, *Biophysical J.* **92**, 115-125 (2007).
12. Certain commercial equipment, instruments, materials, or suppliers are identified in this paper to foster understanding. Such identification does not imply recommendation or endorsement by the National Institute of Standards and Technology, nor does it imply that the materials or equipment identified are necessarily the best available for the purpose.
13. Y. Ito, M. Imai and S. Takahashi, *Physica B* **213&214**, 889-891 (1995).
14. S. Okabe, M. Nagao, T. Karino, S. Watanabe, M. Shibayama, □*J. Appl. Cryst.* **38**, 1035-1037 (2005).
15. N. Rosov, S. Rathgeber, and M. Monkenbusch, *ACS Symp. Series* **739**, 103-116 (2000).
16. M. Monkenbusch, R. Schätzler, and D. Richter, *Nuclear Instruments and Methods in Physics Research Section A* **399**, 301 (1997).
17. http://www.ncnr.nist.gov/dave
18. T. Hellweg and D. Langevin, *Physica. A* **264**, 370-387 (1999).
19. Error bars in this paper represent ± one standard deviation.

Chain Structures of Microparticles Induced by Focusing a Laser Beam near the Liquid-Air Interface of a Droplet

Hiroto Adachi and Kenji Miyakawa

Department of Applied Physics, Fukuoka University, Fukuoka, 814-0180, Japan

Abstract. We investigate optically induced dynamics of micron-sized particles by focusing a laser beam near the liquid-air interface of a suspension droplet. We find three distinct regimes in behaviors of particles depending on the contact angle of the droplet: convection, linear flow and formation of a closed-packed array. These behaviors are governed by the most dominant of effects arising from a focused laser beam, such as local heating and radiation pressure. We find that micron-sized polystyrene beads can be assembled into a chain structure by taking advantage of linear flows. This novel method is applicable to the formation of a chain of carbon nanotubes, non-transparent objects which are difficult to trap.

Keywords: Carbon nanotube, Chain structure, Micro-scale flow, Optical tweezers, Suspension
PACS: 47.61.-k, 81.16.Dn, 82.70.Dd, 87.80.Cc

INTRODUCTION

When a moderately powerful laser is focused to a diffraction-limited spot, a steep gradient in light is produced in the focal region. Then transparent microparticles such as polystyrene beads experience two form of radiation pressure. One is the gradient force that tends to attract them to the high intensity region and the other is the scattering force that tends to push objects down along the beam, in the direction of propagation of the light. Stable trapping takes place when the effect of the gradient force is large enough to overcome the effect of the scattering force [1, 2]. On the basis of such laser trapping, assemblies of micron-size objects, such as a crystallization of polystyrene beads and patterning of the folded DNA on the substrate, have been attempted in the last decade [3, 4]. In contrast, it is difficult to trap and assemble non-transparent objects, such as metallic particles and carbon nanotubes (CNT), because the effect of the scattering force becomes dominant to the effect of the gradient force [5]. Recently, it has been reported that metallic micron-size particles self-assemble into a chain in poorly conducting liquid subjecting to a constant electric field [6].

In this study, we report the creation of a chain structure of microparticles by making use of an optically induced flow. Here we make flows of microparticles by focusing the laser beam near the liquid-air interface of a suspension droplet, *i.e.*, a liquid droplet containing microparticles. The resulting linear flow is found to serve to fabricate a chain of polystyrene beads of 1 μm. This novel method enables CNTs to assemble into a long chain.

EXPERIMENT

We prepared two kinds of suspensions of microparticles. One was obtained by dispersing polystyrene beads of 1 μm in diameter into distilled water. The other consisted of single-walled CNT dispersed in water by sonication with 0.5% sodium dodecyl sulfate (SDS) for 30 minutes [7]. These solutions were placed on the glass cover-slip on the inverted microscope to form suspension droplets with the contact angle of θ_c. Here θ_c was controlled by treating the surface of the glass cover-slip. In order to focus laser beam near the liquid-air interface of the droplet, we used optical tweezers system; the Nd:YVO_4 laser (Spectra-Physics, BL106C) operating at λ=1064 nm was introduced into the inverted microscope and focused by the oil immersion objective lens (100×, NA=1.3). We specify the focus position of the beam near the liquid-air interface of the droplet by (X_0, Z_0), where X_0 is the distance of the optical axis from the contact line and Z_0 is the distance between the focus position and the cover-slip along the optical axis. The microscope image was monitored by a charge coupled device (CCD) camera connected to the video recorder. In order to measure the velocity of

CP982, *Complex Systems, 5th International Workshop on Complex Systems*
edited by M. Tokuyama, I. Oppenheim, and H. Nishiyama

the particle up to the order of 1 mm/s, we captured images at rates up to 1000 FPS using high-speed CCD camera.

RESULTS AND DISCUSSION

The focus position of the laser beam was fixed at the position $(X_0, Z_0) = (16.5\ \mu m, 12.0\ \mu m)$ by adjusting the sample stage and the height of the objective lens. When the contact angle of the droplet was varied, the resulting motions of polystyrene beads were classified into three distinct regimes. At $\theta_c=55°$, we see the occurrence of the convection of particles, as shown in Fig. 1(a). Particles flow along streamlines, and form a butterfly-like pattern, as depicted by arrows in the figures. The cross-sectional view of this flow is illustrated in Fig. 1(d). The position of the focus is below the air-liquid interface of the droplet. This convection arose with increasing the laser power P beyond 150 mW. The convective flow is probably driven by two effects. One is thermocapillarity as characterized by the Marangoni number M. The other is buoyancy as characterized by the Rayleigh number R. The ratio M/R characterizes which effect is dominant; $M/R>1$ provides a necessary condition for thermocapillarity to dominate over buoyancy. Assuming that in our case d corresponds to the distance between the air-liquid interface and the point of the beam focus along the optical axis [8, 9], the value of M/R is estimated to be about 10^5. Therefore, we can say that the observed flow is governed by the Marangoni effect. The butterfly-like pattern is the streamlines characteristic of the surface tension driven convection [10].

When the contact angle was decreased to $\theta_c=37°$, the motion of particles changed into a linear flow, as shown in Fig. 1(b). There was no threshold power for the appearance of such flows. Here the thickness of the droplet along optical axis is thinned down to ca. 12.4 μm. This means that the focus position is almost at the air-liquid interface of the droplet. Then the critical temperature difference ΔT_c required to give rise to Marangoni convection in such a droplet is extremely large, on the assumption that the critical Marangoni number M_c is 80 [11]. Hence motions of microparticles cannot be due to thermal convective flow. Instead, it is instructive to consider that the linear flow results from optical forces. The cross-sectional view of this flow is illustrated in Fig. 1(e). A part of the focused beam is totally reflected and results in a thin light flux toward the center of the droplet, because the one-half angular aperture of the objective with NA=1.3 is large enough that the phenomenon of total internal reflection occurs. The scattering force of such light flux will push out particles attracted to the focal point and induce a linear flow of particles toward the center of the droplet.

When the contact angle was further decreased to $\theta_c=21°$, flow motions of beads disappeared and beads were trapped in a two-dimensional (2D) close-packed array on the air-liquid interface (Fig. 1(c)). The focus position is above the air-liquid interface of the droplet. The cross-sectional view of this 2D trapping is illustrated in Fig. 1(f). Beads were pushed against the air-liquid interface and trapped, and arranged in a close-packed array by the gradient force and the scattering force. Trapping in 2D close-packed arrays was also achieved at the water-coverslip interface, by taking advantage of Fresnel diffraction of the focused Gaussian beam [12, 13].

The linear flow can be controlled by adjusting the ΔZ, the focus position relative to the free surface along the optical axis, as shown in Fig. 2, where X_0 and the laser power are fixed at 16.5 μm and 200 mW, respectively. When ΔZ=8 μm, that is, the focus

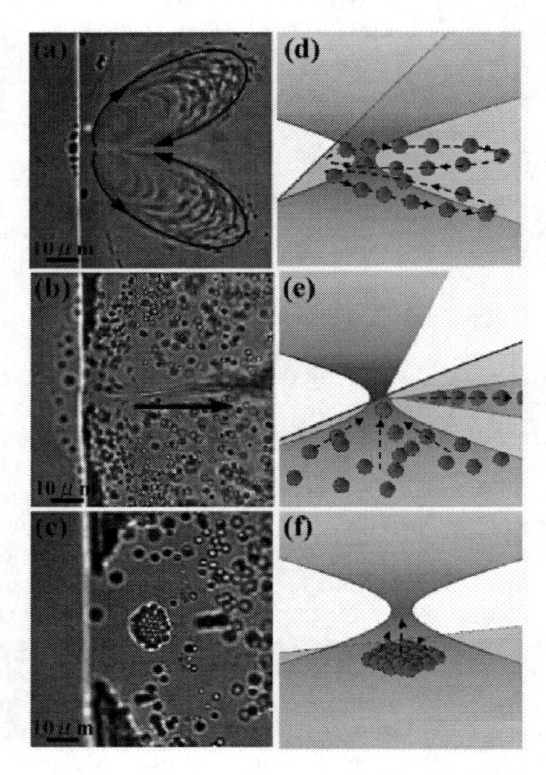

FIGURE 1. Flow patterns of 1-μm-diameter polystyrene beads in the droplet at the laser power of 200 mW, with a change of the contact angle θ_c; (a) $\theta_c = 55°$, (b) $\theta_c = 37°$, (c) $\theta_c = 22°$. The arrow in each picture outlines streamlines of particles. The white line in each picture is the contact line of the droplet. Schematic side views of the sample cell and the trap laser light path corresponding to (a), (b), and (c) are illustrated in (d), (e), and (f), respectively; (d) convection at $\theta_c = 55°$, (e) a linear flow at $\theta_c = 37°$, and (f) a close-packed array at $\theta_c = 22°$.

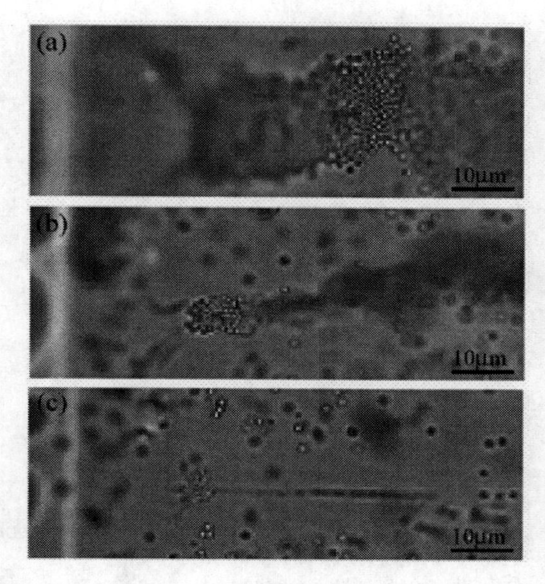

FIGURE 2. Flow patterns in the droplet with the contact angle of 37° with a change in ΔZ, the distance between the air-liquid interface and the focus position along the optical axis: (a) $\Delta Z = 8$ μm, (b) $\Delta Z = 4$ μm, and (c) $\Delta Z = 0$ μm. The white line at left side in each picture is the contact line of the droplet.

position is far above the free surface of the droplet, the flow pattern of particles is broad, as shown in Fig. 2(a). Here the apparently close-packed structure of particles moves toward the center of the droplet. That is, this pattern is quite unlike the stationary close-packed structure in Fig. 1(c). With decreasing ΔZ, the width of the flow monotonously decreases and finally comes to be of the order of the particle diameter, i.e., a linear flow, as shown in Figs. 2(b) and (c). The radius of the Gaussian beam W at the distance of ΔZ from the focus

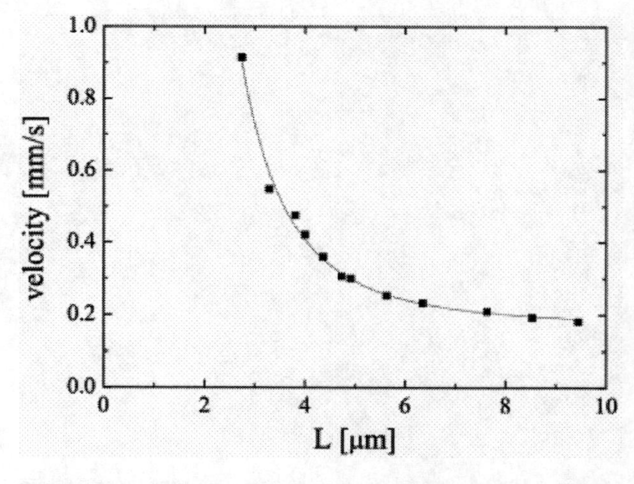

FIGURE 3. Particle velocity as a function of the distance from the focus position in the linear flow at the laser power of 180 mW. The solid line is drawn to guide the eye.

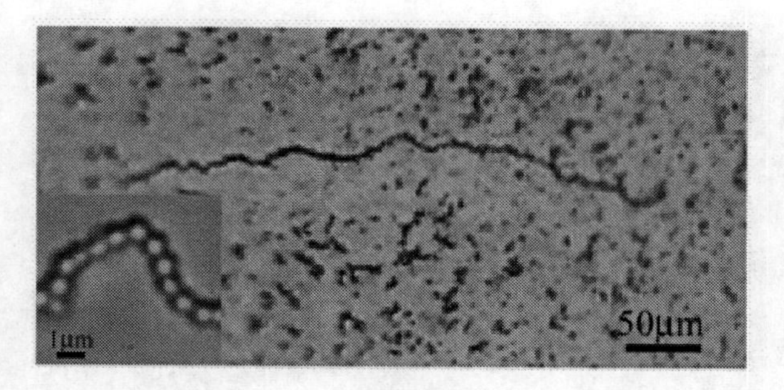

FIGURE 4. Image of a chain of polystyrene particles assembled by utilizing a linear flow at the laser power of 180 mW. The inset shows a close-up of bonding between particles.

position is expressed as $W = W_0(1+(\Delta Z/Z_R)^2)^{1/2}$ with the waist radius W_0 and the Rayleigh length Z_R. The increase in ΔZ leads to the increase in W. This results in the increase of the size of the total internal reflected beam on the free surface. Naturally, the flow width is determined by the size of the beam driving the flow. Thus the change in the flow width with ΔZ can be accounted for by the change in size of the beam totally reflected at the air-liquid interface. It is thus concluded that the linear flow of microparticles is driven by radiation pressure.

The particle velocity in a linear flow will be determined through a balance between the optical force arising from total internal reflection and the force arising from the viscous drag. The optical force proportional to the laser power will decrease with the horizontal distance L from the focal point. On the other hand, the viscous drag is proportional to the velocity. As a result, the velocity is expected to rapidly decrease with L. The result is in accordance with the expectation, as shown in Fig. 3.

As shown in Fig. 2(c), it is possible to induce a linear flow by adjusting ΔZ. We tried to fabricate a chain of polystyrene beads of 1 μm in the suspension droplet, utilizing a linear flow. An aqueous solution of 0.3 μl of 30 mM $CaCl_2$ was added to a suspension droplet in order to facilitate an adhesion of particles at the focal point. Figure 4 shows a 3D chain structure consisting of several hundred individual beads. We see that single beads are in a face to face linkage (see the inset).

This method was applied to the formation of a chain of CNT. CNT strongly absorbs light of 1064 nm, so that it is difficult to trap it by a focused Gaussian beam. Experiments were carried out just after adding 0.2 μl of 30 mM $CaCl_2$ to the suspension droplet. When the laser power was kept at 40 mW, CNT particles were attracted to the focal point by the gradient force and assembled into the chain structure. Figure 5 shows the CNT chain to grow with the elapse

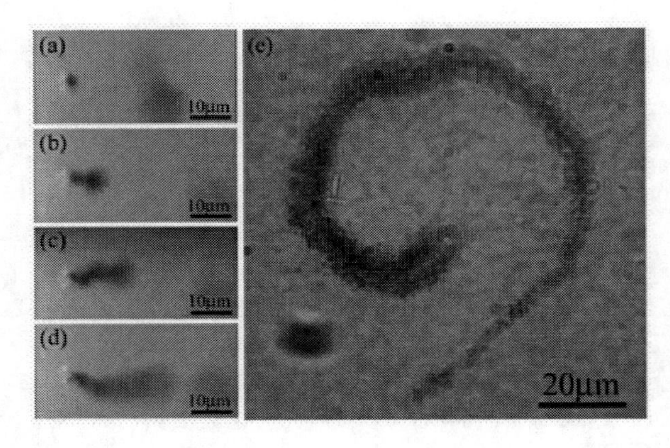

FIGURE 5. Images of carbon nanotube chain at (a) 1 s, (b) 4 s, (c) 7 s, (d) 10 s after starting irradiation of the trapping laser. The image (e) is the generated chain structure.

of time, where the growth rate of the chain is 2.7 μm/s. The final length of the chain structure attained to about 220 μm. Here the minimum width of the chain was about 5 μm, which can be controlled by adjusting the height of the focus position. To the best of our knowledge, this is the first report of optically fabricated structures of CNT. It should be emphasized that fabricated structures are 3D linear chains suspended in the bulk solution, which is different from 2D laser-trapped structures adhered to substrates [14, 15].

CONCLUSION

We have investigated behaviors of micron-sized particles induced by focusing a Gaussian laser beam near a liquid-air interface of a suspension droplet. We have demonstrated three distinct regimes in dynamical behaviors, such as convection, linear flow and a closed-packed array, by controlling contact angles at the fixed position of the beam focus. We have shown that micron-sized particles can be assembled into a chain structure by taking advantage of a linear flow. This novel method is applicable to not only transparent but also non-transparent particles which are difficult to optically trap.

ACKNOWLEDGMENTS

This work was supported in part by a Grant-in-Aid for Scientific Research from the Ministry of Education, Culture, Sports, Science and Technology in Japan (Grant Nos. 15540377 and 18540385), and was also supported in part by the Central Research Institute of Fukuoka University.

REFERENCES

1. A. Ashkin, *Phys. Rev. Letters* **24**, 156–159 (1970).
2. A. Ashkin, *Science* **210**, 1081–1088 (1980).
3. S. Duhr and D. Braun, *Appl. Phys. Letters* **86**, 131921-1–131921-3 (2005).
4. M. Ichikawa, Y. Matsuzawa, Y. Koyama, and K. Yoshikawa, *Langmuir* **19**, 5444–5447 (2003).
5. H. Rubinsztein-Dunlop, T. A. Nieminen, M. E. J. Friese, and N. R. Heckenberg, *Adv. Quant. Chem.* **30**, 469–492 (1998).
6. M.V. Sapozhnikov, I. S. Aranson, W. -K. Kwok, and Y.V. Tolmachev, *Phys. Rev. Letters* **93**, 084502-1–084502-5 (2004).
7. M. J. O'Connell, S. M. Bachilo, C. B. Huffman, V. C. Moore, M. S. Strano, E. H. Haroz, K. L. Rialon, P. J. Boul, W. H.Noon, C. Kittrell, J. Ma, R. H. Hauge, R. B. Weisman, R. E. Smalley, *Science* **297**, 593–596 (2002).
8. Erwin J. G. Peterman, Frederick Gittes, and Christoph F. Schmidt, *Biophys. J.* **84**, 1308–1316 (2003).
9. Michael F. Schatz, Stephen J. Vanhook, W. D. McCormick, J. B. Swift, and Harry L. Swinney, *Phys. Fluids* **11**, 2577–2582 (1999).
10. Y. Kamotani and S. Ostrach, A. Pline, *Phys. Fluids* **6**, 3601–3609 (1994).
11. J. R. A. Pearson, *J. Fluid Mech* **4**, 489-500 (1958).
12. K. Miyakawa, H. Adachi and Y. Inoue, *Appl. Phys. Letters* **84**, 5440–5442 (2004).
13. H. Adachi, K. Miyakawa, *Phys. Rev. A* **75**, 063409-1–0603409-5 (2007).
14. J. Won, T. Inaba, H. Masuhara, H. Fujiwara, K. Sasaki, S. Miyawaki, and S. Sato, *Appl. Phys. Letters* **75**, 1506-1508 (1999).
15. J. Plewa, E. Tanner, D. M. Mueth, and D. G. Grier, *Opt. Express* **12**, 1978-1981 (2004).

Study of Molecular Dynamics in Liquid-Crystalline Phases of CBOOA by High-Resolution ^{13}C NMR

Shoko Hagiwara, Yoko Iwama, and Hiroki Fujimori

Graduate School of Integrated Basic Sciences, Nihon University, Sakurajosui, Setagaya-ku, Tokyo 156-8550, Japan

Abstract. High-resolution ^{13}C NMR experiments were performed in smectic Ad (SAd), nematic (N), and isotropic- liquid (I) phases of a liquid crystal, 4-octyloxy-*N*-(4-cyanobenzylidene)aniline (CBOOA). The temperature dependency of the spin-lattice relaxation times varied smoothly in the liquid-crystalline phases for almost all of the carbon atoms. However, for the carbon atom in the cyano-end group, a sharp change was detected at the SAd–N transition temperature. It is suggested that intermolecular interaction, which is not observed in the N phase, occurs in the SAd phase around the cyano-end group.

Keywords: CBOOA, Liquid crystal, NMR, Phase transition
PACS: 64.70.mj, 76.60.-k

INTRODUCTION

A liquid-crystalline phase is a medium that exhibits both the mobility of a liquid and the anisotropy of a crystal. In the liquid phase (isotropic liquid), complex physical properties are observed due to position and orientation disorders. In the liquid-crystalline phase, molecular orientations are ordered, but position disorders are present; therefore, physical properties of the liquid-crystalline phase should not be as complex as in the liquid phase. However, the details of the molecular dynamics are still difficult to understand.

As the temperature is lowered at atmospheric pressure, 4-Octyloxy-*N*-(4-cyanobenzylidene)aniline (CBOOA, Fig. 1) exhibits the following thermotropic phase sequence: isotropic liquid–nematic–smectic Ad–crystal (I–N–SAd–Cry). McMillan [1] showed that the SAd–N phase transition is of the second order, but the associated latent heat was observed by calorimetry [2, 3]. As the temperature is lowered under high pressure, CBOOA exhibits the following phase sequence: I–N–SAd–N–Cry [4, 5]. The second N phase is called a reentrant nematic (RN) phase. It also appears at atmospheric pressure when CBOOA is mixed at a particular rate with 4-cyanobenzylidene-*N*-(4-hexyloxy)aniline (HBAB) [6]. To understand molecular dynamics and interactions in the liquid-crystalline phases of CBOOA, high-resolution ^{13}C NMR experiments were performed, which is a powerful method in the analysis of molecular orientation orders and molecular dynamics in liquid crystals [7]. In this research, we report the complete spectral assignment of CBOOA and discuss orientational ordering and molecular dynamics in the SAd and N phases of CBOOA at atmospheric pressure.

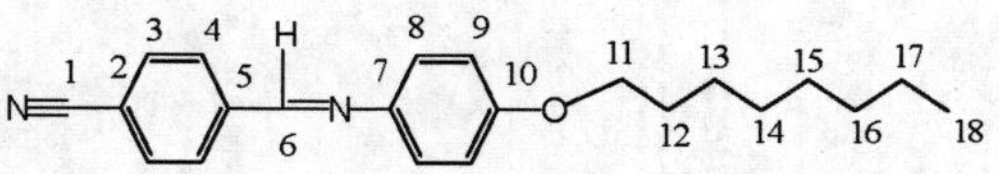

FIGURE 1. Molecular structure of CBOOA and numbering of carbon atoms.

EXPERIMENT

CBOOA synthesized by Miyajima et al. [8] was used. The SAd–N and N–I phase transition temperatures were 356.2 and 382.1 K, respectively.

The sample was evacuated and sealed in a 5-mm-dia Pyrex tube under helium gas at a pressure of 3 kPa. The sample was then fixed tightly in the probe head. Field-aligned spectra of CBOOA were obtained by high-resolution ^{13}C NMR using a JEOL EX-270 spectrometer operating at 67.94 MHz in the temperature range of 340 to 390 K. Continuous-wave proton decoupling was applied. The spin-lattice relaxation times (T_1) of each CBOOA carbon atom were measured by applying the π–τ–$\pi/2$ pulse sequence.

RESULTS AND DISCUSSION

Figure 2 shows the high-resolution ^{1}H-decoupled ^{13}C NMR spectra of CBOOA in a field-aligned condition.

CP982, *Complex Systems, 5th International Workshop on Complex Systems*
edited by M. Tokuyama, I. Oppenheim, and H. Nishiyama

The spectrum of the I phase was very similar to that obtained by ^{13}C NMR of CBOOA in deuterated chloroform at room temperature. Therefore, the NMR lines in the I phase were easily assigned using the ^{1}H, ^{13}C, and ^{13}C-^{1}H COSY NMR spectra of CBOOA in deuterated chloroform [9]. On the other hand, the assignment of field-aligned spectra in liquid-crystalline phases is generally difficult. In liquid-crystalline phases, because molecules align along the magnetic field, alignment-induced shifts (AIS) occur in the ^{13}C NMR spectra. Because the I–N phase transition is of the first order, AIS suddenly appear at the phase transition, and the lines of each carbon cannot be tracked. Therefore, it is not possible to assign field-aligned spectra in the liquid-crystalline phase even if the assignment is complete in the I phase.

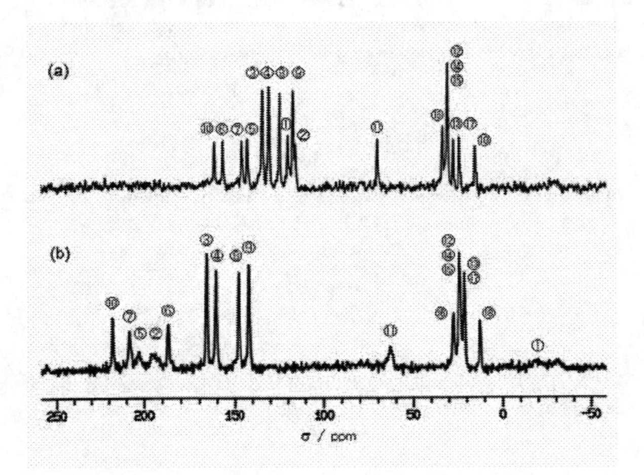

FIGURE 2. ^{1}H-decoupled ^{13}C NMR spectra of CBOOA in field-aligned condition in I phase (a), and in SAd phase (b). Numbers indicate results of line assignment.

Nakai *et al.* demonstrated the method of assigning ^{13}C NMR lines in liquid-crystalline phases using the temperature dependency of AIS [7]. This method was applied to the ^{13}C NMR spectrum of CBOOA [9]. The numbers in Fig. 2b show the results of the NMR line assignments in the liquid-crystalline phases. Figure 3 shows the temperature dependencies of the chemical shifts for CBOOA. Almost continuous chemical shifts were observed in the carbons of the alkyl-chain part during the I–N phase transition. This indicates that the intermolecular interaction in the alkyl chain is weak, and the chain rotates freely in the liquid-crystalline phase (similar to the I phase). The AIS values in the core part were positive, excluding the carbon of No. 1, C(1), in the cyano-end group. This indicates that in the liquid-crystalline phases, the molecules align along the magnetic field and the paramagnetic susceptibility of the carbons increases. This is a characteristic property of liquid crystals that have a spread of π electrons in the direction of a molecular short axis. The AIS of the C(1) was negative. This is appropriate considering the values of the shift tensor of the carbon in the cyano group of cyanobenzene ($\sigma_{xx} = 231$, $\sigma_{yy} = 213$, and $\sigma_{zz} = -88$ [10]). The temperature dependencies of the AIS of all the carbons varied smoothly in the liquid-crystalline phases.

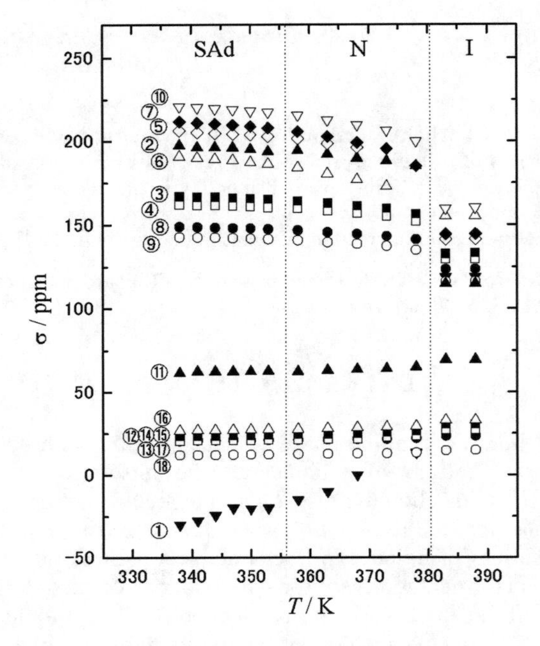

FIGURE 3. Temperature dependencies of chemical shift in ^{1}H-decoupled ^{13}C NMR spectrum of CBOOA in field-aligned condition. Numbers indicate results of line assignment. Dotted lines represent phase transition temperatures.

Figure 4 shows the temperature dependencies of T_1 for typical carbons—the C(1) in the cyano-end group, C(3) in the core part, and C(16) in the alkyl-chain part. For the carbons in the core and in the alkyl chain, the temperature dependencies of T_1 are almost continuous in the liquid-crystalline phases. T_1 is affected by the fluctuation of the local magnetic field produced by molecular motion, and the intra- and intermolecular interaction of atoms. This indicates that the molecular dynamics and interactions involve no remarkable changes between the N and the SAd phases. The activation energies obtained by the temperature dependency of T_1 in the liquid-crystalline phases were found to be 29 and 9 kJmol^{-1} for C(3) and C(16), respectively. However, the temperature dependency of T_1 for C(1) in the cyano-end group, changes at the SAd–N transition temperature. Cladis *et al.* showed that the molecular overlapping at the core was formed in the SAd phase [4, 5]. Therefore, it is suggested that at the N–SAd phase transition, a new intermolecular interaction occurs between the cyano-end group and the oxygen atom that is between the core and the alkyl-chain. If the N–SAd transition is of the first order, the T_1 of C(1) may show discontinuity at the transition temperature. However, the precision of the T_1 values obtained was not adequate to extract the details of the phase transition.

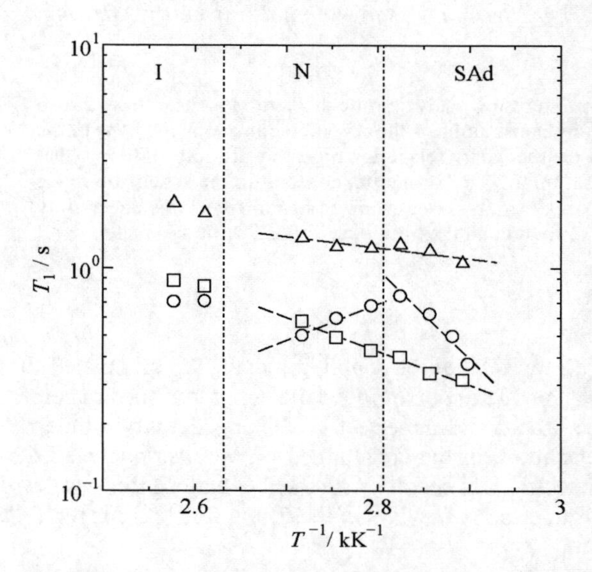

FIGURE 4. Temperature dependencies of spin-lattice relaxation time (T_1). Circles, squares, and triangles represent T_1 of C(1), C(3), and C(16), respectively. Dotted lines represent phase transition temperatures.

ACKNOWLEDGMENTS

We thank Dr. Seiichi Miyajima for providing the sample, CBOOA, and for valuable discussions. We also thank Prof. Tetsuo Asaji for valuable discussions. This work was partly supported by a grant from the Institute of National Sciences, College of Humanities and Sciences, Nihon University.

REFERENCES

1. W. L. McMillan, *Phys. Rev. A*, **7**, 1419-22 (1973).
2. K. Ema, T. Yamaguchi, H. Yao, S. Miyajima, and T. Chiba (private communication).
3. K. Tsuji, R. Kamei, M. Wakabayashi, and S. Seki (private communication).
4. P. E. Cladis, R. K. Bogardus, and D. Aadsen, *Phys. Rev. A*, **18,** 2292-2306 (1978).
5. D. Guillon, P. E. Cladis, D. Aadsen, and W. B. Daniels, *Phys. Rev. A*, **21,** 658-665 (1980).
6. P. E. Cladis, *Phys. Rev. Lett.*, **35,** 48-51 (1975).
7. T. Nakai, H. Fujimori, D. Kuwahara, and S. Miyajima, *J. Phys. Chem. B*, **103,** 417-425 (1999).
8. S. Miyajima, T. Enomoto, T. Kusanagi, and T. Chiba, *Bull. Chem. Soc. Jpn.*, **64,** 1679-1681 (1991).
9. H. Fujimori, T. Jinbo, and T. Asaji, *Proceedings of the Institute of National Sciences, Nihon University*, **39,** 405-409 (2004).
10. B. M. Fung, *P J. Am. Chem. Soc.*, **105,** 5713-5714 (1983).

A Study of the Secondary Relaxation in Galactose-Water Mixtures

Hyun-Joung Kwon, Jeong-Ah Seo, Hyung Kook Kim, and Yoon-Hwae Hwang*

Department of Nanomaterials & BK21 Nano Fusion Technology Division, Pusan National University, Miryang 627-706, Korea

Abstract. We studied the secondary relaxation in galactose-water mixtures by analyzing the dielectric loss spectra with two different fitting methods. The first method was the free fit without any constraint and the second method was the fit with the coupling relation in coupling model (CM) [1, 2, 3]. The behavior of the secondary relaxation process were very similar to that of the secondary relaxation process with changing the rotational-traslational (RT) coupling constant in the schematic mode coupling theory (MCT) [4, 5]. The secondary relaxation times (τ_{JG}) obtained by the constrained fit contain a large uncertainty and were consistent with τ_{sec} within the experimental errors. We also found that the fitting quality of free fit was better.

Keywords: Sugar, Glass, Dielectric, Secondary relaxation
PACS: 61.43.Fs,61.20.Lc,77.22.Gm

INTRODUCTION

The relaxation spectra of glass forming materials usually exhibit at least two relaxation processes, the α- and fast β relaxations. The α-relaxation is related to a long time-scale and corresponds to the overall structural rearrangement of a glass-forming materials. The fast β relaxation, usually called β relaxation in the mode coupling theory [4, 5], is related to a short time-scale and corresponds to local dynamics. The existence of an extra relaxation [6, 7, 8] at frequencies between α- and the fast β-relaxations was recognized about 35 years ago. In this study, we refer to this extra relaxation as the secondary relaxation. Although considerable progress has been made in recent years, the question of the microscopic origin of the secondary relaxation and the structural peculiarities of the glasses responsible for this relaxation process are still the subject of controversy.

In this study, we analyzed the secondary relaxation in the dielectric loss spectra of galactose-water mixtures with two methods. The first method is the fit the dielectric loss spectrum without any constraint. The second method is the fit data together with the coupling relation in the coupling model [2, 3, 6]. In coupling model, the secondary relaxation is called genuine Johari-Goldstein relaxation if it obeys the coupling relation and shows effects of aging or pressure. The fitting function of the free fit consists of the power law for the DC part, the Havriliak-Negami [9] function for the α-relaxation and the Cole-Cole [10] function for the secondary relaxation. A more detailed data analysis processes can be found in our previous study [11]. The fitting function of the constrained fit consists of the CC function for the α-relaxation and the HN function for secondary relaxation with the relation of coupling model. We eliminated the DC conductivity part in the dielectric loss spectra before the fit to avoid the effect of DC conductivity on the α-relaxation. In the constrained fit, we also analyzed the dielectric loss spectra with the DC part. In this case, we include the power law in the fitting function for the DC part.

EXPERIMENTS

Galactose was purchased from the Sigma-Aldrich Chemical Co. and used without further purification. The weight percent of water in galactose-water mixtures was measured by using a moisture analyzer (Sartorius MA100, Germany) with 0.1 wt. % precision. The glass phase of galactose-water mixture was made by using a microwave method [12]. The frequency dependent dielectric loss spectra at different temperatures were carried out by using a cylindrical shape stainless steel cell and separated with a Teflon plate. The geometric capacitance C_0 was 0.943 pF. Two different measurement schemes were employed depending on the frequency range under consideration. For the frequency range of 0.1 Hz to 10 kHz, the in-phase and out-of-phase currents through the capacitor were measured using a Dual Phase Lock-in Amplifier (Stanford SR 830, U.S.A.). For the frequency range of 100 Hz to 10 MHz, an Impedance Analyzer (HP 4194A, Hewlett-Packard U.S.A.) with a standard four-terminal pair configuration was used to measure the capacitance and conductance of the sample. The data obtained from the two methods were in excellent agreement where they overlapped. The temperature of the sample was controlled by using a closed-cycle He refrigerator and a tem-

CP982, *Complex Systems, 5th International Workshop on Complex Systems*
edited by M. Tokuyama, I. Oppenheim, and H. Nishiyama

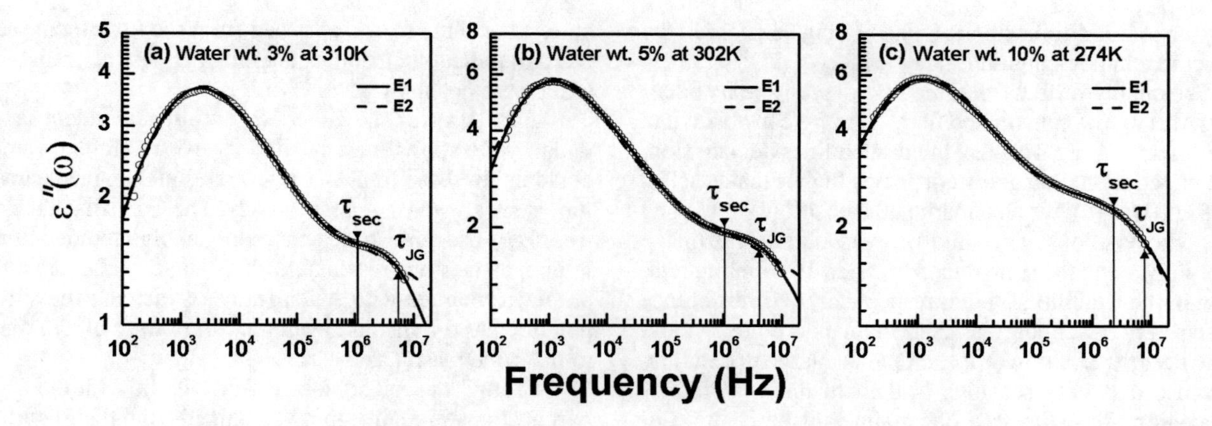

FIGURE 1. Dielectric loss spectra of galactose-water mixtures with different water contents and temperatures. The water content and the temperature of each figure are (a) 3 wt.% and 310K, (b) 5 wt.% and 302K, and (c) 10 wt.% and 274K.

perature controller (Lakeshore 330, U.S.A.) with 0.1 K precision.

RESULTS & DISCUSSION

Figures 1(a)-(c) show the dielectric loss spectra of galactose-water mixtures with different water contents and temperatures. The water content and the temperature of each figure are (a) 3 wt.% and 310K, (b) 5 wt.% and 302K, and (c) 10 wt.% and 274K. In order to compare two different fitting methods, we arbitrary selected dielectric loss spectra which show a clear separation between the α-relaxation and the secondary relaxation process for three different water contents.

The solid lines represent the fitted results by the free fit;τ_{sec} and dashed lines represent the fitted results by the constrained fit;τ_{JG} which includes the relation of coupling model. The DC part in the dielectric loss spectra was removed before the both fits. From this data analysis, we obtained the secondary relaxation times from the free fit, τ_{sec} and τ_{JG} from the constrained fit, at different temperatures and with different water contents. As we can more easily observed in Fig.2, the deviation of the constrained fit at high frequency region. However the difference of two relaxation times was less than 1 decades for all three cases with different water contents at different temperatures.

In our previous study [11], we analyzed the temperature and water content dependent secondary relaxation times in the galactose-water mixture by the free fit. In that study, we found that as the water content or temperature decreased, the relaxation time difference between α- and secondary relaxations increased and the relative relaxation strength between the α- and the secondary relaxation decreased. The observed behavior of the secondary relaxation was similar to that of decreas-

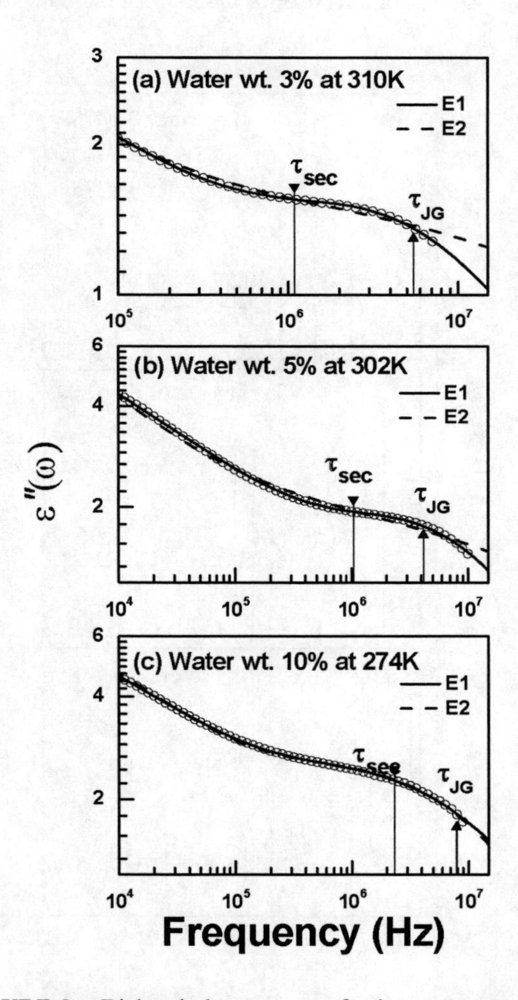

FIGURE 2. Dielectric loss spectra of galactose-water mixtures with different water contents and temperatures at high frequency ranges. (Enlarged version of Figure 1.)

ing the rotation-translation (RT) coupling constant in the schematic mode-coupling thoey [5].

The quality of fit by the free fit;τ_{sec} seems to be better than that by the constrained fit;τ_{JG}. As we can observe in the insets of Figs. 1(a)-(c), the dashed lines deviate from the experimental data at frequencies higher that 1 MHz. In Figs. 1(a)-(c), we also marked the positions of τ_{sec} and τ_{JG}. The positions of τ_{sec} and τ_{JG} were obtained from the free fit;τ_{sec} and the constrained fit;τ_{JG} with coupling relation in the Coupling Model, respectively. The difference between two secondary relaxation times is quite evident. However the position of τ_{JG} contains a large uncertainty because it is very sensitive to the broadness of the α-relaxation. To estimate a uncertainty in the position of the secondary relaxations, we analyzed the dielectric loss spectra of galactose-water mixture with the DC part. We found that τ_{sec} obtained from the fit with the DC part is almost the same to τ_{sec} obtained from the fit without the DC part indicating that the position of τ_{sec} is very reliable (data were not shown).

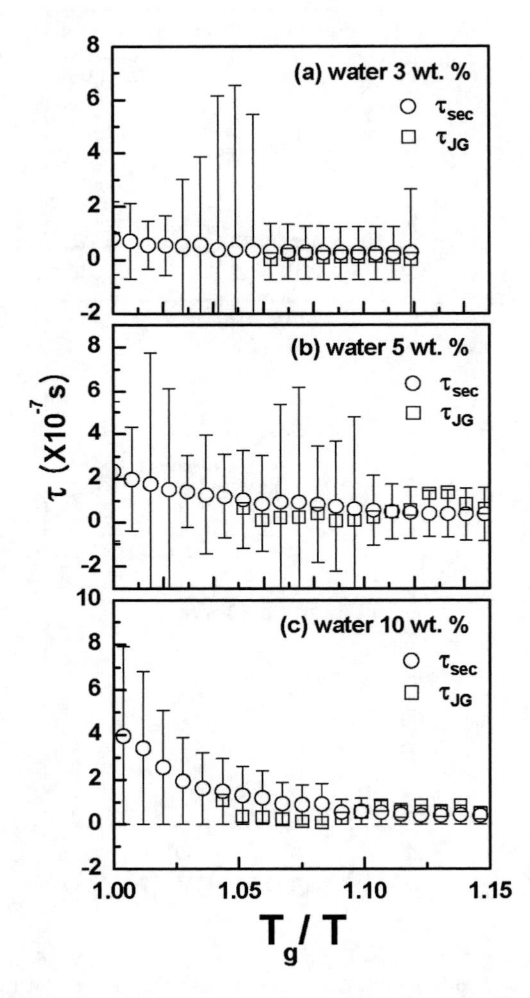

FIGURE 3. Temperature dependent secondary relaxation times (τ_{sec} & τ_{JG}) obtained from different fitting methods with the water contents of (a) 3 wt.%, (b) 5 wt.% , and (c) 10 wt.%

Figures 3(a)-(c) show the temperature dependent secondary relaxation times, τ_{sec} and τ_{JG} with different water contents. In those figures, open circles and open squares represent τ_{sec} and τ_{JG}, respectively. The experimental error of τ_{sec} has quite large uncertainty at high temperature region. It is quite reasonable that the secondary relaxation was merged with α-relaxation as increase the temperature. And it also allowed the uncertainty of τ_{JG} due to the same reason as the τ_{sec}.

We found that τ_{sec} obtained from the fit with the DC part is almost the same to τ_{sec} obtained from the fit without the DC part indicating that the position of τ_{sec} is very reliable. However, the positions of τ_{JG} can be different depends on whether the DC part in the dielectric loss spectra was included in the fitting process or not. We can observe in Figs. 3 (a)-(c) that two relaxation times are consistent with each other which is quite different from the result in Fig. 1 indicating that the position of τ_{JG} naturally contained a large uncertainty. Therefore, there is a possibility that the secondary relaxation observed in galactose-water mixture might be the genuine JG relaxation in the Coupling Model. According to the Coupling Model, the genuine JG relaxation has to show an aging effect because the origin of the JG relaxation based on inter molecular interactions. The aging experiment of the galactose-water mixture is currently in progress.

CONCLUSIONS

We analyzed the secondary relaxation of galactose-water mixtures in the supercooled and glassy states at frequencies from 10 mHz to 10MHz with two different data analysis methods. We found that the properties of the secondary relaxation obtained from the free fit; τ_{sec} are very similar to those with different rotational-translational coupling constant in the schematic MCT. The secondary relaxation times obtained from both the free fit;τ_{sec} and the constrained fit;τ_{JG} with the coupling relation in Coupling Model are consistent with each other within the experimental error bars. However, there is a possibility that the observed secondary process may not be directly related to the genuine JG relaxation process of a pure galactose system which exist at higher frequencies [13]. Therefore, it seems to be possible that the secondary relaxation in galactose-water mixtures may be related to the genuine JG relaxation process.

ACKNOWLEDGMENTS

We would like to thank H.Z. Cummins for many helpful discussion and for suggesting us to study the sugar glasses. This work was supported by Korea Research Foundation Grant (KRF-2006-521-C00060) and partially supported by BK21 project.

REFERENCES

1. K.L. Ngai, *Physica A*, **261**, 36-50 (1998).
2. S. Capaccioli, D. Prevosto, M. Lucchesi, P.A. Rolla, R. Casalini, and K.L. Ngai, *J. Non-Crystalline Solids*, **351**, 2643-2651 (2005).
3. S. Capaccioli, K. Kessairi, D. Prevosto, M. Lucchesi, and K.L. Ngai, *J. Non-Crystalline Solids*, **352**, 4643-4648 (2006).
4. W. Göze, in Liquids, Freezing and Glass transition, edited by J.P. Hansen, D. Levesque, and J. Zinn-Justin, North-Holland, Amsterdam, 1992, ; W. Götze and L. Sjogren, *Rep. Prog. Phys.*, **55**, 241-376 (1992). W. Götze and M. Sperl, *Phys. Rev. Lett.*, **92**, 105701- (2004).
5. H. Z. Cummins, *J. Phys. : Condens. Matter*, **17**, 1457-1470 (2005)
6. U. Schneider, R. Brand, P. Lunkenheimer, and A. Loidl, *Physical Review Letters*, **84**, 5560-5563 (2000).
7. Jiyoung Oh, Jeong-Ah Seo, Hyung Kook Kim, and Yoon-Hwae Hwang, *J. Non-Crystalline Solids*, **352**, 4679-4684 (2006).
8. Herman Z. Cummins, Hepeng Zhang, Jiyoung Oh, Jeong-Ah Seo, Hyung Kook Kim, Yoon-Hwae Hwang, Y.S. Yang, Yun Sik Yu, and Yongwoo Inn, *J. Non-Crystalline Solids*, **352**, 4464-4474 (2006).
9. S. Havriliak and Jr. S. Negami, *Polymer*, **8**, 161-210 (1967).
10. K. S. Cole and R. H. Cole, *J. Chem. Phys.*, **9**, 341-351 (1941).
11. Hyun-Joung Kwon, Jeong-Ah Seo, Hyung Kook Kim, and Yoon-Hwae Hwang, *Rep. Inst. Fluid Science*, **19**, 35-39 (2007).
12. Jeong-Ah Seo, Jiyoung Oh, Dong Jin Kim, Hyung Kook Kim, and Yoon-Hwae Hwang, *J. Non-Crystalline Solids*, **333**, 111-114 (2004).
13. K. Kaminski, E. Kaminska, M. Paluch, J. Ziolo, and K. L. Ngai, *J. Phys. Chem. B*, **110**, 25045-25049 (2006).

Fluctuations in Human's Walking (IV)

T. Obata, T. Mashiyama, T. Kogure, S. Itakura, T. Obata, T. Sato, K. Takahashi*, H. Oshima†, and H. Hara**

*Department of Electronic Media Technology, Gunma National College of Technology, Maebashi 371-0845, Japan
†Department of Physics, Toho University School of Medicine, Ota-ku, Tokyo 143-8540, Japan
**10-21-406 Tsutsumidori Amamiya, Aoba-ku, Sendai 981-0914, Japan

Abstract. A field experiment of ring-wandering is executed on a wide playground. Blindfolded and stoppled subjects are observed to do ring-wandering rather than random-walking. This experiment simulates the phenomenon of ring-wandering that climbers encounter in snowy mountains. 15 samples of walking for 13 subjects are reported. Their walking periods are about 40 minutes or 2 hours. The walking data are acquired every second, using a GPS receiver. The discrete velocity $v(t)$ and discrete angular velocity $\omega(t)$ of the data are analyzed, using Hurst's R/S analysis and Fourier spectrum analysis. The Hurst exponents of $v(t)$ show long-range correlations. The Hurst exponents of $\omega(t)$ show anti-correlations in short-ranges and correlations in long-ranges. These characteristics of the Hurst exponents in the present data in addition to previous data in this study series describe the ring-wandering phenomena very well. Significant differences are not seen between 40-minutes walking and 2-hours walking.

Keywords: GPS, Hurst exponent, Power spectrum, Random walk, Ring wandering
PACS: 01.80.+b, 05.40.Fb, 05.45.Tp

INTRODUCTION

Our research group has been studying fluctuating walks of healthy persons whose eyes are blindfolded and ears are stoppled [1][2][3]. Persons control their direction of walking by getting external information through physical sensors such as eyes, ears, nose, and so on. Usually main sensors for controlling the direction of walking are firstly eyes and secondly ears. Other sensors are auxiliary in general. Our experiment condition for eyes and ears remarkably weakens the direction sense of subjects. As a result it is almost difficult for most subjects to walk straight. Such fluctuating walks are expected to elicit internal features of subjects explicitly, as the main external information controlling walking motion is screened.

Analogous situations naturally happen in whiteouts such as in snowy mountains veiled in dense fog or mist. Climbers then fall into difficulties in getting external information for controlling their direction of walking, and they wander along ring-like orbits and return near their starting points. The phenomenon, called ring-wandering, sometimes brought climbers into unfortunate accidents. It is very important to reveal stochastic characteristics of ring-wandering in averting such mountain accidents. Physiologists had an interest in the phenomenon of ring-wandering. Exact experimental simulations of ring-wandering have been thought to be impossible. So most researchers investigated properties of ring-wandering through experiments on walking bias, that is, a deviation of a blindfolded walker from a straight line when the subject were asked to walk along the line. Especially Nigorikawa examined walking bias from a straight line of 100 meter [4]. An important conclusion of the study is that there were no significant relationships between the walking bias direction and the various dominant physical factors affecting the subjects.

Our research group of statistical physicists also started a similar study on human's walking without noticing previous works by physiologists [1]. After that we succeeded in realizing a field-experiment simulation of ring-wandering on a playground and also elicited some features of ring-wandering, studying Hurst statistics about the time series of Cartesian position coordinates, velocity and angular velocity [2][3]. In [2] some features of Hurst exponents and power spectra of Cartesian position coordinates were shown by studying 9 walking samples for 2 subjects on a wide playground. In the following article [3] the statistical features of velocities and angular velocities as well as position coordinates were revealed by investigating 13 walking samples of 10 subjects. The most important result is that the velocities of walking have long-range correlations but that the angular velocities have anticorrelations in short ranges and correlations in long ranges. These characteristics of walking describe the ring-wandering phenomenon very well, as has been discussed in [3].

In the present article, we want firmly to establish the previous results, investigating more walking samples for more subjects. In addition it is studied whether the results depend on walking periods.

CP982, *Complex Systems, 5th International Workshop on Complex Systems*
edited by M. Tokuyama, I. Oppenheim, and H. Nishiyama

TABLE 1. Hurst exponents H of velocity time series $v(t)$ and angular velocity time series $\omega(t)$ for 15 walking samples of 13 subjects. The last letter L's in subject nos. 9L and 13L stand for long time walking of about 2 hours. Every RS graph for the time series fits in a sequential line graph very well, which consists of two lines. H_1 is the slope of a regression line in an early region $0 \leq t \leq T_m$ and H_2 is the slope of a regression line in a late region $T_m \leq t \leq T$. The transition time T_m between two regions depends on samples. The T_m is not evaluated in any exact method, but it is here judged only by looking at the data.

subject no.	period T[sec]	H of $v(t)$		H of $\omega(t)$	
		H_1	H_2	H_1	H_2
1	2223	0.71	1.01	0.33	0.65
2	2289	0.65	0.89	0.40	0.69
3	2264	0.68	1.13	0.38	1.04
4	2318	0.68	1.31	0.31	0.80
5	2023	0.62	0.75	0.40	1.06
6	2317	0.68	0.99	0.40	1.13
7	2261	0.67	1.25	0.40	1.16
8	2371	0.71	1.19	0.39	1.01
9	2670	0.71	1.08	0.34	0.65
10	2375	0.63	1.14	0.34	0.82
11	2381	0.76	1.20	0.42	0.85
12	2417	0.67	1.74	0.34	0.47
13	2183	0.69	1.14	0.42	1.12
mean		0.68	1.14	0.37	0.88
s. d. (σ)		0.03	0.23	0.04	0.21
9L	7365	0.71	1.06	0.38	0.87
10L	7020	0.71	1.31	0.44	0.85

METHODS AND RESULTS

The methods of experiment and data processing are the same as before (see [2][3]). The field experiment is executed on a wide playground; 120 meters in W-E direction, 285 meters in N-S direction. Subjects are blindfolded and stoppled. We here report 15 samples of walking for 13 subjects executed from October 2005 to January 2007. Each subject is a 20 or 21 years old healthy male. 13 samples have the walking periods of about 40 minutes and the other two samples have very long walking periods of about 2 hours. The walking data of such subjects are acquired every second, using a GPS receiver. The discrete velocity $v(t)$ and discrete angular velocity $\omega(t)$ of the subjects are calculated from position data $(x(t), y(t))$ recorded every second. The $v(t)$ and $\omega(t)$ are defined as the walking distance per second and the deflection angle of walking direction per second, respectively. The time series $v(t)$ and $\omega(t)$ are analyzed, using Hurst's R/S analysis [5] and Fourier spectrum analysis. The results of the Hurst exponents are summarized in Table 1. Walking trajectories, time series $v(t)$ and $\omega(t)$, their power spectra and R/S statistics are given for two typical samples in Figs. 1 and 2.

DISCUSSIONS AND CONCLUSIONS

Correlations or anticorrelations in a time series can be judged by whether its Hurst exponent is greater or less than 0.5 [5]. The Hurst exponents for velocity $v(t)$ are greater than 0.5, which means correlations in $v(t)$. Meanwhile the Hurst exponents for angular velocity $\omega(t)$ are less than 0.5 in early times and greater than 0.5 in late times. This means that there are short-term anticorrelations and long-term correlations in $\omega(t)$. Let's take a close look at the exponents. As for 13 samples of 40-minutes walking, the early time Hurst exponents H_1 have the small standard deviations of σ=0.03 for $v(t)$ and σ=0.04 for $\omega(t)$, which are one digit smaller than those in late times. So we can say that the early-time Hurst statistics of walking are almost independent of individuals, while the late-time Hurst statistics largely depend on each personality. These result are nicely consistent with a previous data analysis of 13 samples for 10 subjects executed from March 2004 to January 2005 [3].

As for very-long-term walking of 2 hours, their H_1 and H_2 are approximately within one-σ deviation from the means of 40-minutes walking. That is no significant difference between 40-minutes walks and 2-hours walks.

Correlations or anticorrelations in a time series can be identified by whether the slope of its power spectrum is negative or positive, too. The power spectra of $v(t)$ have negative slopes in the frequency span, while the power spectra of $\omega(t)$ have negative slopes at low frequencies and positive slopes at high frequencies, as seen in Figs. 1 and 2. Hence the power spectra of $v(t)$ and $\omega(t)$ are consistent with the characteristics of their Hurst exponents.

In conclusion, these characteristics of the Hurst exponents in the present data in addition to previous data describe the ring-wandering phenomena very well, as discussed in [3]. There are no significant differences between 2-hours walking and 40-minutes walking. A more detailed analysis will be published elsewhere.

REFERENCES

1. T. Obata, T. Ohyama, J. Shimada, H. Oshima, H. Hara, and S. Fujita, "FLUCTUATIONS IN HUMAN'S WALKING," in *Statistical Physics*, edited by M. Tokuyama and H. E. Stanely, AIP Conference Proceedings 519, American Institute of Physics, New York, 1999, pp. 720–722.
2. T. Obata, T. Shimizu, H. Osaki, H. Oshima, and H. Hara, *J. Korean Phys. Soc.* **46**, 713-718 (2005).
3. T. Obata, T. Shimizu, T. Mashiyama, D. Takei, H. Oshima, and H. Hara, "Fluctuations in Human's Walking (III)," in *Statistical and Condensed Matter Physics: Over the Horizon*, edited by S. Fujita, T. Obata and A. Suzuki, Nova Science Publishers, New York, 2007, pp. 135-160.
4. T. Nigorikawa, *Ann. Physiol. Anthrop.* **7**, 99-106 (1988).
5. J. Feder, *FRACTAL*, Plenum Press, New York, 1988, chapters 8, 9.

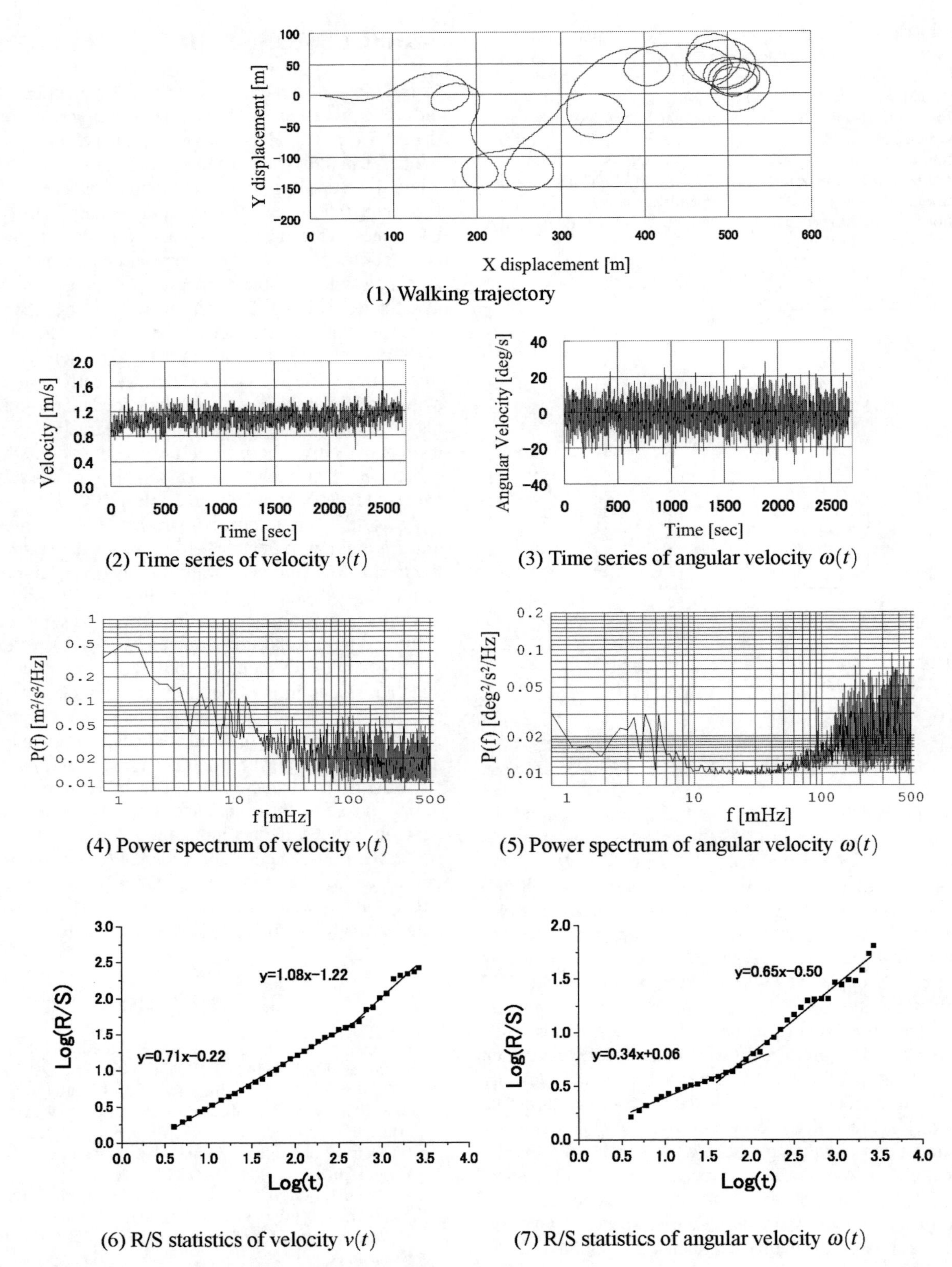

FIGURE 1. Sample no. 9: Walking trajectory, time series, power spectra, R/S statistics. The (X, Y) coordinate system in (1) is set up such that the sum of $Y(t_i)$'s becomes zero.

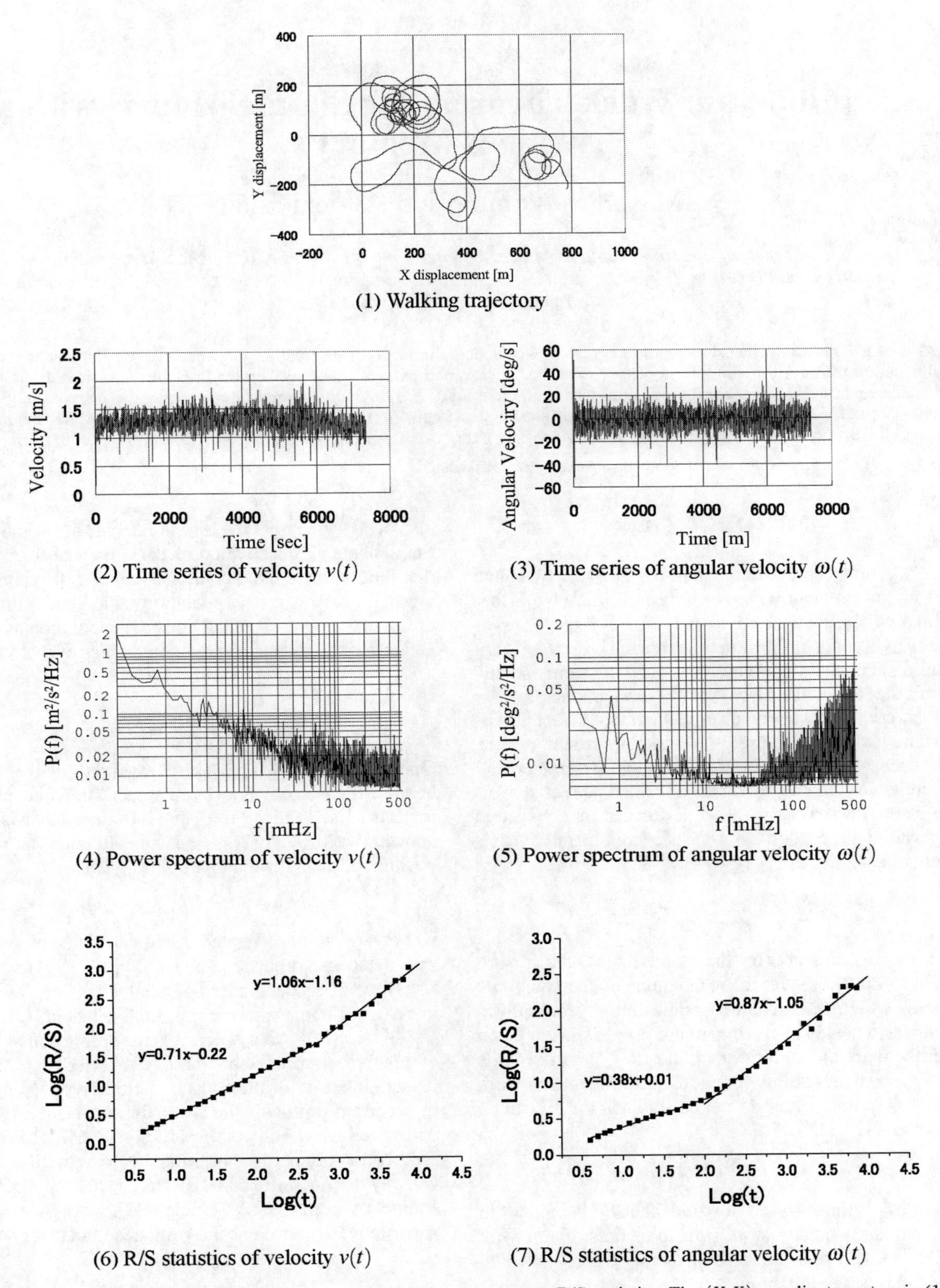

FIGURE 2. Sample no. 9L: Walking trajectory, time series, power spectra, R/S statistics. The (X,Y) coordinate system in (1) is set up such that the sum of $Y(t_i)$'s becomes zero.

Nonequilibrium Work Theorem for Simple Models with and without Memory

Mitsuhiro Akimoto and Kei-ichi Endo

Department of Electronics and Computer Science, Tokyo University of Science, Yamaguchi, Liquid Crystal Institute, Tokyo University of Science, Yamaguchi, 1-1-1 Daigaku-Dori, San-Yo-Onoda, Yamaguchi 756-0884, Japan

Abstract. This study investigates the validity of the nonequilibrium equality called the Jarzynski equality or the work fluctuation theorem for a simple stochastic model of a Brownian particle in a trap harmonic potential. A viscoelastic fluid is adopted as well as a purely viscous fluid for the ambient fluid. It is found that there is a case where the Jarzynski equality does not hold. The key factor for the validity of the Jarzynski equality is the time-scale of the dynamics, which should be determined when the system is modeled in the stochastic Langevin dynamics.

Keywords: Fluctuation phenomena, Jarzynski equality, Viscoelastic fluid

INTRODUCTION

In 1997 Jarzynski showed a non-trivial equality, called the Jarzynski equality (JE) or nonequilibrium work theorem, which is expected to hold for a large class of systems in nonequilibrium situations[1]. Let us consider a small thermodynamic system. The term "small" means that molecular discreteness causes fluctuations for any externally manipulated quantities such as thermodynamic work, which are usually non-fluctuating in the macroscopic limit. The JE describes that the free energy difference ΔF can be obtained from an exponential average of the fluctuating work W performed on the system by switching the external parameter from its initial to its final value:

$$e^{-\beta \Delta F} = \left\langle e^{-\beta W} \right\rangle = \int_{-\infty}^{\infty} p(W) e^{-\beta W} dW. \tag{1}$$

The average, denoted by the angular brackets, is supposed to be taken over all realizations of the switching process which starts from a system initially in equilibrium with a heat bath at temperature $\beta = 1/k_{\mathrm{B}}T$, or taken over the work probability function $p(W)$. When $p(W)$ is the Gaussian probability function, the cumulants of work higher than third order do not exist. Then the JE can be rewritten as,

$$\Delta F = \langle W \rangle - \frac{\beta}{2} \sigma_W^2, \tag{2}$$

where σ_W^2 is the variance of work. The JE (1) reduces to the familiar inequality of the principle of maximum work $\Delta F \leq \langle W \rangle$ by use of Jensen's inequality which generally holds for convex functions.

The JE (1) has been proven for several types of dynamics ranging from Newtonian and thermostatted dynamics to Langevin and Monte Carlo dynamics. It has also been shown that the essential factor of its validity is the dynamical preservation of the canonical form of the distribution for a fixed control parameter. It is unsolved problem, however, what violation will occur if this condition may not be satisfied. To see the violation of the JE and to pursuit its reason are main themes of this study.

MODELS

One typical model for which the validity of the JE has been verified exactly is an overdamped Brownian particle in a harmonic (optical) trap, first introduced by Mazonka and Jarzynski (MJ) [2]. The Langevin equation for the position of the Brownian particle is given by

$$\dot{x} = -\gamma^{-1} \partial_x U(x,t) + f(t), \tag{3}$$

where $f(t)$ is uncorrelated, Gaussian white noise with zero mean and variance $\langle f(t) f(t') \rangle = (2/(\beta\gamma))\delta(t-t')$. The time-dependent trap potential $U(x,t)$ is assumed to be a harmonic one with an uniform field, $U(x,t) = (k/2)(x-u(t))^2 + \alpha x$, where k is the spring constant and $u(t)$ is the controllable bottom of the harmonic potential, α is an intensity of the constant field. By sweeping $u(t)$ in a certain protocol, the work by an outside agent is performed on a Brownian particle. In MJ's case such protocol is a linear one in time, $u(t) = vt$ with constant velocity v. According to the framework of stochastic energetics pioneered by Sekimoto [3], work as a random variable is introduced as the work accumulated one,

$$\dot{w}(x,t) = \partial_t U(x,t). \tag{4}$$

The total work done during a whole process is expressed by $W = \int_0^{t_f} \dot{w}(x,t) dt = \int du (\partial U / \partial u)$. The work accumulated variable w is a random variable because of the

CP982, *Complex Systems, 5th International Workshop on Complex Systems*
edited by M. Tokuyama, I. Oppenheim, and H. Nishiyama

randomness of the position variable x and then the work can fluctuate. Hence the simultaneous stochastic differential equations of (3) and (4) are to be solved. Based on this model, Mazonka and Jarzynski showed exactly several nonequilibrium relations including the JE.

We treat here a slightly generalized model which is a direct extension of Mazonka-Jarzynski's one (3). We examine the validity of the JE for that model. The model can be described by a one-dimensional generalized Langevin equation for the position x of a Brownian particle of unit mass,

$$\ddot{x} = -\partial_x U(x,t) - \int_{-\infty}^{t} \gamma(t-t')\dot{x}(t')dt' + \xi(t), \quad (5)$$

where $\gamma(t)$ is the memory function which reflects the viscous and/or viscoelastic nature of ambient fluids. $\xi(t)$ is the Gaussian random force which stems from the same fluids. These two quantities are related to each other through the fluctuation-dissipation theorem $\langle \xi(t)\xi(t')\rangle = (1/\beta)\gamma(t-t')$.

The difference of the present model (5) from the original MJ's one (3) is threefold. First, we use the periodic protocol $u(t) = A\sin\omega t$, instead of the linear one as treated by MJ. Here A is the amplitude of ac force, and ω is its angular frequency. The second is concerned on the ambient fluid, which is always assumed to be purely viscous in MJ's case. For a Brownian particle embedded in a Newtonian fluid, the memory function takes the delta function, $\gamma_0(t) = 2\gamma\delta(t)$, that leads to the linear friction in velocity seen as time derivative of the position in Eq.(3). But the surrounding fluids are not necessarily purely Newtonian. Thus it is interesting to consider also a viscoelastic fluid as the surrounding. In this study, we consider a single-relaxation time Maxwell fluid as one of the simplest models for viscoelastic fluids, in addition to the purely Newtonian fluid. It has been shown theoretically and experimentally that the memory function of exponential type is appropriate for a single-relaxation time Maxwell fluid [4]. Then, the memory function reads

$$\gamma(t) = \gamma_0(t) + \gamma_{\mathrm{M}}(t) = 2\gamma\delta(t) + (\gamma_p/\tau)\exp(-|t|/\tau), \quad (6)$$

where τ is a relaxation time and $\gamma_p = \int_0^\infty \gamma_{\mathrm{M}} dt$ is the long-time friction constant. For this choice of the memory function, the random force turns out to be composed of two parts $\xi(t) = \xi_0(t) + \xi_{\mathrm{M}}(t)$ from the fluctuation-dissipation theorem. $\xi_0(t)$ comes from the Newtonian part, and $\xi_{\mathrm{M}}(t)$ from the Maxwellian part which is no longer a delta-correlated white, but colored. Throughout the paper, we postulate that these two random forces are statistically independent of each other. The third and final point, that may be the most important, is on the time coarse-graining. MJ considered only the overdamped case, that means that the system is modelled in a certain time-scale where the relaxation of velocity is accomplished so rapidly that the acceleration term of the Langevin equation can be dropped. The resulting equation of motion becomes first order in time for the position of the Brownian particle. By contrast, we study the underdamped cases as well. The last two conditions are associated with the time-scale of the dynamics. Therefore it is fascinating to verify whether or not these factors affect the validity of the JE.

In the present study, we can see from the equation for the model, especially for the work accumulated variable w, Eq.(4), that the solution w should be a linear combination of the position x owing to the harmonicity of the potential. Hence w must be Gaussian distributed, so that it is sufficient to follow the dynamics of the mean $\langle w\rangle$ and the variance σ_w^2 as already mentioned in Eq.(2). In the following sections, we examine the respective models one by one.

BROWNIAN OSCILLATOR IN A VISCOUS FLUID

Underdamped Case

First we consider the case where the ambient fluid is purely Newtonian. The memory function should be the delta function $\gamma(t) = 2\gamma_0\delta(t)$ and then the Langevin equation (5) reduces to

$$\dot{x} = v, \; m\dot{v} = -\gamma_0 v - \partial_x U(x,t) + \xi(t). \quad (7)$$

Here, the random force is the delta-correlated white noise with zero mean from the fluctuation-dissipation theorem: $\langle \xi(t)\xi(t')\rangle = (2\gamma_0/\beta)\delta(t-t')$. Adding the equation for the work accumulated variable, the equation which we have to solve is a set of stochastic differential equations:

$$\left\{ \begin{array}{l} \dot{w} = \partial_t U(x,t) \\ \dot{x} = v \\ \dot{v} = -\partial_x U(x,t) - \gamma_0 v + \sqrt{\frac{2\gamma_0}{\beta}}\tilde{\xi}(t), \end{array} \right. \quad (8)$$

where $\tilde{\xi}(t)$ is the normalized random force with zero mean and its variance is $\left\langle \tilde{\xi}(t)\tilde{\xi}(t')\right\rangle = \delta(t-t')$. The corresponding Fokker-Planck equation (FPE) which gives a conditional probability of all variables can be obtained by the ordinary method [5] as

$$\partial_t p = -\sum_i \partial_i [A_i p] + \frac{1}{2}\sum_{i,j}\partial_i\partial_j \left\{[\underline{BB}^T]_{ij} p\right\}, \quad (9)$$

with the drift vector $\overrightarrow{A}(w,x,v,t)$

$$\overrightarrow{A} = \begin{pmatrix} A_w \\ A_x \\ A_v \end{pmatrix} = \begin{pmatrix} -Ak\omega\cos(\omega t)(x - A\sin\omega t) \\ v \\ -k(x - A\sin\omega t) - \alpha - \gamma_0 v + \frac{1}{\gamma} v \end{pmatrix}, \quad (10)$$

and the diffusion matrix $\underline{B}$,

$$\underline{B} = \begin{pmatrix} 0 & 0 & 0 \\ 0 & 0 & 0 \\ 0 & 0 & \sqrt{\frac{2\gamma_0}{\beta}} \end{pmatrix}. \tag{11}$$

>From this FPE (9), we can derive two sets of evolution equations for moments of variables w, x, v. One is for the first moment,

$$\begin{cases} \frac{d}{dt}\langle w\rangle = -c_1(t)\langle x\rangle - c_2(t) \\ \frac{d}{dt}\langle x\rangle = -c_3\langle v\rangle \\ \frac{d}{dt}\langle v\rangle = -c_4\langle x\rangle - c_5(t) - c_6\langle v\rangle, \end{cases} \tag{12}$$

and the other is for the second moment,

$$\begin{cases} \frac{d}{dt}\sigma_w^2 = -2c_1(t)C_{wx} \\ \frac{d}{dt}\sigma_x^2 = -2c_3C_{xv} \\ \frac{d}{dt}\sigma_v^2 = -2c_4C_{xv} - 2c_6\sigma_v^2 + 2c_7 \\ \frac{d}{dt}C_{wx} = -c_1(t)\sigma_x^2 - c_3c_{wv} \\ \frac{d}{dt}C_{wv} = -c_1(t)C_{xv} - c_4C_{wx} - c_6C_{wv} \\ \frac{d}{dt}C_{xv} = -c_3\sigma_v^2 - c_4\sigma_x^2 - c_6C_{xv}, \end{cases} \tag{13}$$

where the coefficients are given by

$$\begin{aligned} &c_1(t) = kA\omega\cos\omega t,\ c_2(t) = -(1/2)kA^2\omega\sin 2\omega t, \\ &c_3 = -1,\ c_4 = k, \\ &c_5(t) = \alpha - kA\sin\omega t,\ c_6 = \gamma_0,\ c_7 = \gamma/\beta. \end{aligned} \tag{14}$$

The solutions of these equations fully determine the time-dependent Gaussian probability distributions $p(w,x,v,t)$ which is a general solution of the FPE (9). Here, we are not going to show an explicit solution of the FPE. Our subject is to verify the validity of the JE. Here we define the equation

$$I(t) = \Delta F(t) - \langle w\rangle + \frac{\beta}{2}\sigma_w^2. \tag{15}$$

>From Eq.(2), the function $I(t)$ should be always equal to zero if the JE holds at every time. In the model present here, the free energy difference $\Delta F(t)$ at arbitrary time is given by

$$\Delta F(t) = F(t) - F(0) = \alpha A\sin\omega t. \tag{16}$$

We numerically solve Eq.(13) and the solutions of $\langle x\rangle$, σ_x^2, and $I(t)$ are shown in Figure 1. The mean $\langle x\rangle$ and variance σ_x^2 of the position variable rapidly reach stationary states and the function $I(t)$ is zero at all times, which shows the validity of the JE in this case.

Overdamped Case

In the overdamped limit $\gamma_0 \to \infty$, we can neglect the acceleration term in Eq.(7) . Then the Langevin equation

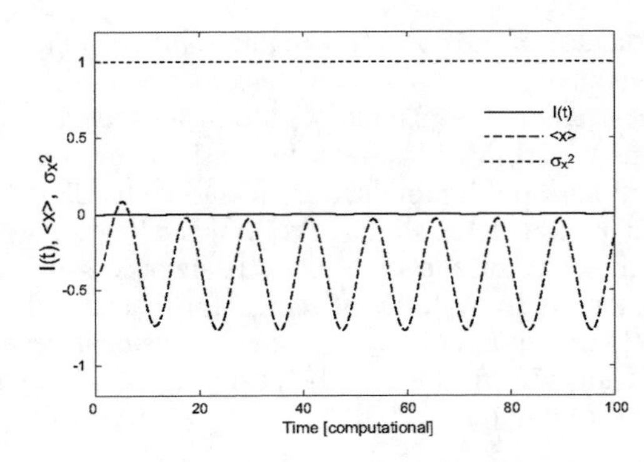

FIGURE 1. Solutions of the mean $\langle x\rangle$ and variance σ_x^2 of the position variable. In addition, the function $I(t)$ are included in the same plane. Parameter settings: $\alpha = 0.4, \beta = 1.0, \gamma_0 = 5.0, k = 1.0, A = 1.0, \omega = \pi/6$.

becomes a first-order stochastic differential equation for the position variable x only:

$$\gamma_0\dot{x} = -\partial_x U(x,t) + \xi(t), \tag{17}$$

that is almost equivalent to the model treated by MJ. According to the same procedure as the previous subsection, we can show that the JE holds at arbitrary time. Since there are many other works which treat this case, we do not mention the details and not go further.

BROWNIAN OSCILLATOR IN A VISCOELASTIC FLUID

Underdamped Case

In this section, we treat the case where the ambient fluid of a Brownian oscillator has viscoelastic nature in the meaning that we use the memory function of Eq.(6) for the generalized Langevin equation (5), instead of the delta function. If a simple exponential dependence is chosen for the memory function, as is the case considered here, the non-Markovian equation can be replaced by two Markovian equations involving an auxiliary variable y. Then the simultaneous stochastic differential equation including one for the work accumulated variable is rewritten by

$$\begin{cases} \dot{w} = \partial_t U(x,t) \\ \dot{x} = v \\ \dot{v} = -\partial_x U(x,t) + y - \gamma_0 v + \sqrt{\frac{2\gamma_0}{\beta}}\tilde{\xi}_1(t) \\ \tau\dot{y} = -y - \frac{\gamma_p}{\tau}v + \sqrt{\frac{2\gamma_p}{\beta}}\tilde{\xi}_2(t), \end{cases} \tag{18}$$

where $\tilde{\xi}_i(t), (i = 1,2)$ are the Gaussian white noises with the variances $\left\langle\tilde{\xi}_i(t)\tilde{\xi}_j(t')\right\rangle = \delta_{ij}\delta(t-t')$. In fact, this case is already investigated in detail by Mai and Dhar [6]. They show analytically several nonequilibrium equalities including the JE for almost the same system as Eq.(18) for an arbitrary sweeping protocol. Since their analysis includes the case considered here, it can be concluded that the JE holds at arbitrary time for Eq.(18). Of course it is possible to show the validity of the JE just following the procedure used here, i.e., by deriving the evolution equations for moments from Eq. (18) and solving them.

Overdamped Case

Next, we treat the overdamped case. Taking the overdamped limit for Eq.(5) , we obtain a set of differential equations,

$$\begin{cases} \dot{w} = \partial_t U(x,t) \\ \gamma_0 \dot{x} = -\partial_x U(x,t) - \frac{\gamma_p}{\tau} x + z + \sqrt{\frac{2\gamma_0}{\beta}}\,\tilde{\xi}_1(t) \\ \tau \dot{z} = -z + \frac{\gamma_p}{\tau} x + \sqrt{\frac{2\gamma_p}{\beta}}\,\tilde{\xi}_2(t), \end{cases} \tag{19}$$

where z is an auxiliary variable which is slightly different from one in the underdamped case. The relaxation-time τ and the angular frequency of the protocol ω are the most important parameters for this system of a Brownian particle embedded in a viscoelastic fluid. Following the procedure used throughout this paper, we derive the evolution equations for the moments from the Langevin equations (19), and solve them. The results of the function $I(t)$ are shown in Figure 2 and 3. Figure 2 shows the behavior of $I(t)$ for a variety of ω, and Figure 3 shows $I(t)$ for a variety of τ with a fixed angular frequency. We can see that the relaxation of the system is slowly accomplished compared to the underdamped case considered previously, since there are two characteristic relaxation-times associated with the two types of fluids. At first period, $I(t)$ is not zero, and hence the JE does not hold. We might expect that the validity of the JE would be recovered for the long-time limit, since the mean position $\langle x \rangle$ and variance σ_x^2 reach the stationary state at long time as shown in the insets of those figures. It is true that the function $I(t)$ attains the time-independent stationary state for a long time, but its value generally differs from zero. As we can see from Fig.3, the JE may be valid when the relaxation-time τ of the Maxwell fluid goes to zero. It is reasonable because the Maxwell fluid reduces to a purely Newtonian in the limit $\tau \to 0$. On the other hand, the JE loses its validity even for the case where the angular frequency becomes quite low. Thus we can conclude that the JE does not hold for the system which shows the intrinsic slow-relaxational behavior, which is endowed by introducing the overdamped limit for this case. The violation of the JE may be attributed to the fact that the system (19) does not satisfy the fully detail balance condition which is inevitable for the validity of the JE [7], whereas the other three cases which are already analyzed fulfill the condition. To postulate the overdamped limit corresponds to the choice of the time-scale of the dynamics of the system which we make when we model the system by a stochastic Langevin equation. In order to clarify fully the reason why the JE violates, further investigation is needed.

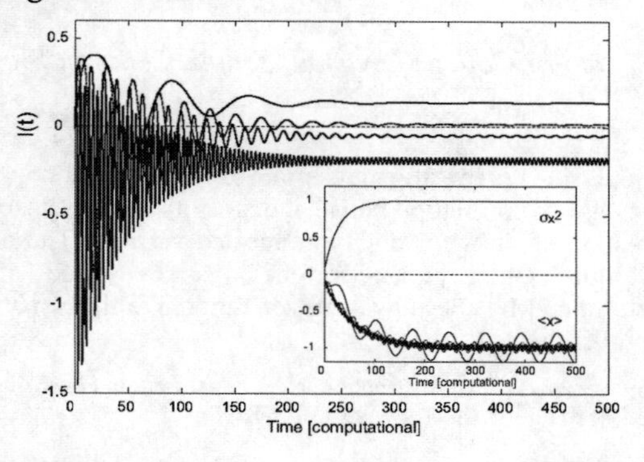

FIGURE 2. The function $I(t)$ for different values of angular frequency ω. From bottom to top, $\omega = \pi/2, \pi/6, \pi/12, \pi/36$. Other parameter settings: $\alpha = 1.0, \beta = 1.0, k = 1.0, A = 1.0, \gamma_0 = 5.0, \gamma_p = 50.0$.

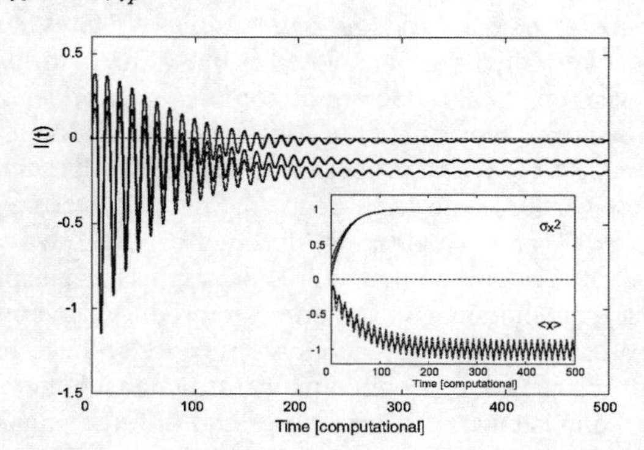

FIGURE 3. The function $I(t)$ for different values of the relaxation time τ and for fixed angler frequency $\pi/6$. From top to bottom, $\tau = 1, 5, 10$. Other parameter settings are the same ones as Figure 2.

REFERENCES

1. C. Jarzynski, *Phys. Rev. Lett.* **78**, 2690–2693 (1997).
2. O. Mazonka and C. Jarzynski, e-print cond-mat/9912121.
3. K. Sekimoto, *J. Phys. Soc. Japan* **66**, 1234–1237 (1997).
4. J van der Gucht, *et.al.*, *Phys. Rev. E* **67**, 051106-1-051106-10 (2003).
5. C. W. Gardiner, *Handbook of Stochastic Methods: for Physics, Chemistry and the Natural Sciences 3rd. ed.* Springer-Verlag, Berlin, 2004.
6. T. Mai and A. Dhar, *Phys. Rev. E* **75**, 061101-1-061101-7 (2007).
7. R. D. Astumian, *J. Chem. Phys.* **126**, 111102-1-111102-4 (2007).

Explanation of Correlation between Adjacent Vertices in Network Formed by Traces of Random Walkers

Nobutoshi Ikeda

Junior College Division, Tohoku Seikatsu Bunka College, 1-18-2, Niji-no-Oka, Izumi-ku, Sendai, Miyagi 981-8585, Japan

Abstract. We investigated certain kinds of vertex correlation in networks formed by traces of random walkers. In this paper, the sign of correlation between degree and local clustering on adjacent vertices is interpreted based on the capacity for new creation of links for highly connected vertices. These interpretations are consistent with numerical simulation of network evolution. Other types of correlation can be related with each other. It is interesting that the local rules of network evolution determined by the movement of random walkers naturally provide large clustering coefficients and various finite vertex correlations.

Keywords: Degree distribution, Network structure, Phase diagram, Random walker, Shortest path, Vertex correlation
PACS: 02.50.-r, 05.40.Fb, 89.75.Fb, 87.23.Kg

INTRODUCTION

The importance of network theory is increasingly being recognized recently for studying complex systems, since the network theory is considered as one of tools for them [1]. Mathematical modeling of complex networks based on network theory [2] is useful for understanding the emergence of typical properties of real networks such as "a few vertices with large degree", "large clustering coefficients", and "small mean distance between two vertices". It is important that simple principles such as preferential attachment can sometimes reproduce such properties easily. However, each system can also have individual properties especially in local structures. For example, the existence of communities and different signs of vertex correlations have been observed according to different kinds of real networks [3, 4]. The study of correlations between vertices is an important tool for describing the local structure of these systems [5].

In our recent works, we proposed a network evolution model in which the rise and fall of connections determined by random transports and a physical configuration of network are considered [6]. The principle based on the movement of random walkers is expected to provide local structures intrinsic to the network. In this model, random walkers representing transports abstractly move in a network initialized as a square lattice and leave edges behind their traces. The details of creation rules on edges are that a vertex where a walker currently stays and vertices where the walker stays one and two time steps before are newly linked or increase in strength of edges by one, while all links except the initially existed links tend to be extinct with probability p_d per unit time. The fixed lattice on time describes a simple geographical relation between elements in the system. The interesting points of this model are that various network structures can be found by time observation or by adjusting values of parameters [6].

In the present paper, we report on a time observation of various correlations between adjacent vertices. The calculation is carried out on clustering coefficient, degree-degree (d-d) correlation, cluster-cluster (c-c) correlation, and degree-cluster (d-c) correlation between adjacent vertices. Correlation between degree and local clustering on the same vertices is also calculated. The behavior of d-c correlation is strongly dependent on time, number of walkers w, extinction probability of edges per time p_d, and dimension of the initial lattice. The aim of this paper is to interpret these results by the local evolution of the network, which is determined by the movement of random walkers. In the next section, such interpretation is explained. The subsequent sections explain the results of numerical calculation from the above viewpoint.

INTERPRETATION OF THE SIGN OF CORRELATION COEFFICIENTS

Let us introduce the next ratio defined on each vertex i, (Number of triangles including vertex i)/(Number of all combination of vertices joined to vertex i), which we call in this paper "*local clustering*". Clustering coefficient is defined as an average of the above values over all vertices. In our model, the local clustering on a vertex can be related to the capacity for new creation of links for the adjacent vertex as follows.

Let us consider a movement of a random walker. Suppose that a random walker moves in the order vertex 1,

CP982, *Complex Systems, 5th International Workshop on Complex Systems*
edited by M. Tokuyama, I. Oppenheim, and H. Nishiyama

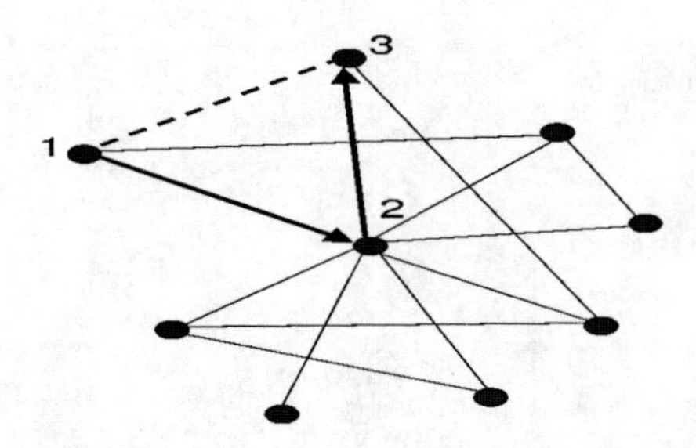

FIGURE 1. Relation between local clustering and capacity of new creation of links. If a random walker moves in the order vertex 1 and 2, then the walker will go to arbitrary vertices joined to vertex 2. (Vertex 3 is accidentally selected in this figure.) Whether an edge incident to vertex 1 is created or not is dependent on the local clustering for vertex 2.

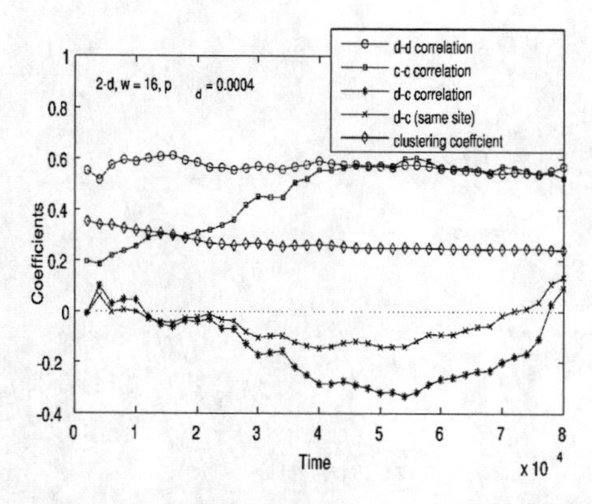

FIGURE 2. Time observation of various correlations and clustering coefficient for a case where 16 random walkers start from the same vertex in 2-d squared lattice under the condition of $p_d = 0.0004$. This calculation only covers vertices with edges created by movements of random walkers.

vertex 2, and vertex 3 (see Figure 1), then whether vertex 1 and 3 are newly linked is determined by whether vertex 1 and 3 have already been linked or not. The probability of the existence of a link between 1 and 3 is equal to the local clustering for vertex 2. Taking this property of clustering strength into consideration, a vertex whose adjacent vertices are low clustered is thought to have the advantage of gaining new links. In order to understand this, suppose that a random walker moves in the order of vertex 1 and 2, since the walker will go to arbitrary vertices joined to vertex 2, the probability of the new creation of an edge incident to vertex 1 in the next time step is given by the formula

$$1 - L_{(1 \to s(2))}/(L_{s(2)} - 1), \tag{1}$$

where $L_{(1 \to s(2))}$ means the number of edges between vertex 1 and all the vertices joined to vertex 2, and $L_{s(2)}$ means the degree of vertex 2. If the degree of vertex 2 is sufficiently large, the second term of probability (1) can roughly be estimated as the local clustering for vertex 2.

It should be noted that this discussion does not make sense if random walkers do not come to these vertices at all, because all edges are extinct regardless of the network topology if no walker comes at all. This notation is necessary to understand the results in the following sections.

RESULTS FOR NETWORKS ON TWO-DIMENSIONAL LATTICE

In order to study multiplier effects of a few random walkers, let us consider a special case in which certain walkers depart from the same vertex in a two-dimensional (2-d) squared lattice.

Figure 2 shows a time observation of various correlations and clustering coefficient for 16 random walkers.

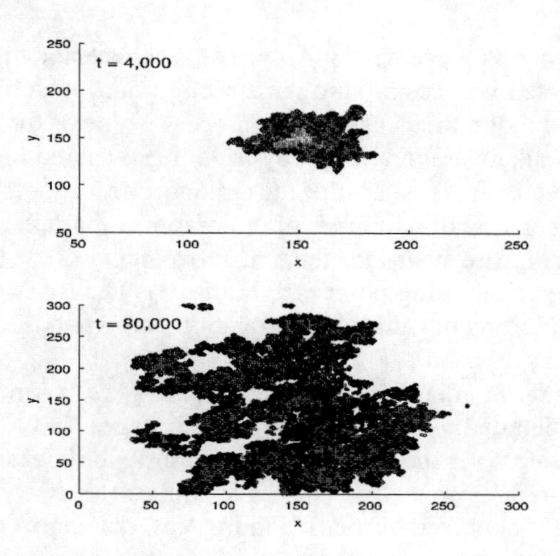

FIGURE 3. Spatial distribution of vertices with edges created by movements of random walkers corresponding to the situation Figure 2. The starting point of random walkers is (x, y)=(150,150). Above is $t = 4,000$. Below is $t = 80,000$.

Large clustering coefficient is one of the characteristics of our model, based on the fact that triangles composed of edges can easily be produced by the movement of random walkers. It can be seen from the figure that the d-c correlation coefficient can take negative values only at certain time intervals. Negative d-c correlation means that highly clustered vertices tend to have adjacent vertices with low vertex degree.

According to our previous work[7], network evolution undergoes three stages. The first stage is the situation in which all walkers wander around the starting point, and the resulting connected network (Figure 3) exhibits a degree distribution with clear power law. The sec-

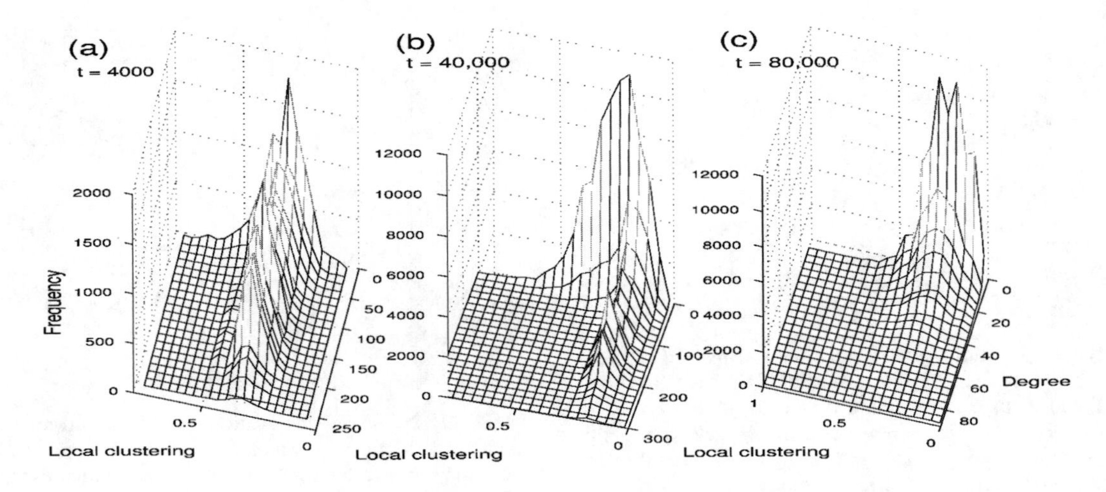

FIGURE 4. Scatter diagrams for degree and local clustering on adjacent vertices corresponding to the situation Figure 2. (a)$t = 4,000$. (b)$t = 40,000$. (c) $t = 80,000$.

ond stage is a collapse process of the cluster of highly connected vertices around the starting point, which was formed in the first stage. In the process, some of the random walkers are moving away from the clustered highly connected vertices. In the third stage, the formed network consists of a number of emigrating subgraphs corresponding to near-independent movements of walkers leaving the starting point (see Figure 3). The time interval exhibiting negative d-c correlation corresponds to the second stage.

Scattering diagrams presented in Figure 4 provide more detailed information about these stages. In the first stage, most vertices are concentrated around the starting point of random walkers and have adjacent vertices with strongly clustered vertices (Figure 4(a)) owing to richness of edges around the area. This situation results in the observation of positive correlations. As time passes, vertices with highly clustered adjacent vertices cannot survive especially for highly connected vertices (Figure 4(b)). In other words, vertices whose adjacent vertices have small local clustering oppose the extinction of edges as expected by the explanation in the second section. This situation results in negative d-c correlation. However, even local structures exhibiting negative d-c correlation never maitain highly connected vertices without frequent visitation of random walkers, because edges are never created without the existence of random walkers. Therefore, vertices with large degree themselves disappear after random walkers have distanced far from the starting point (Figure 4(c)).

Negative d-c correlation produced by above mechanism can be observed only when a number of random walkers start from the same vertex, because only one walker never exhibits the second stage. In the case of one walker, the first stage proceeds to the third stage directly although the results are omitted here.

RESULTS FOR NETWORKS ON ONE-DIMENSIONAL LATTICE

In this section, networks evolving in a one-dimensional (1-d) squared lattice are considered. According to our previous works[7], the network evolution on 1-d lattice is different from that on 2-d lattice in that random walkers starting from the same vertex are confined to a connected subgraph with finite size formed near the starting point (see Figure 5). As a result, random walkers can provide the new creation of edges to all vertices in the connected network without distinction. Owing to this structure, the situation corresponding to the second stage observed in 2-d case does not occur.

The time observation of the correlation and clustering coefficients for 1-d case is shown in Figure 6. The figure shows an rapid up-and-down motion of d-c correlation around null value.

Figure 7 shows two scattering diagrams, each of which, respectively, leads to negative and positive d-c correlation coefficients. This figure shows that vertices with large degree only have adjacent vertices with relatively small local clustering regardless of the sign of d-c clustering coefficient. The condition of vertices with small degree determines the sign of the d-c correlation coefficient. There is another fact to be noted, which was also revealed in previous works. Vertices with a few % largest degree have a life time of about a few hundred time steps when $p_d = 0.03$ [7]. In other words, the rise and fall of vertex degree occur inside the connected

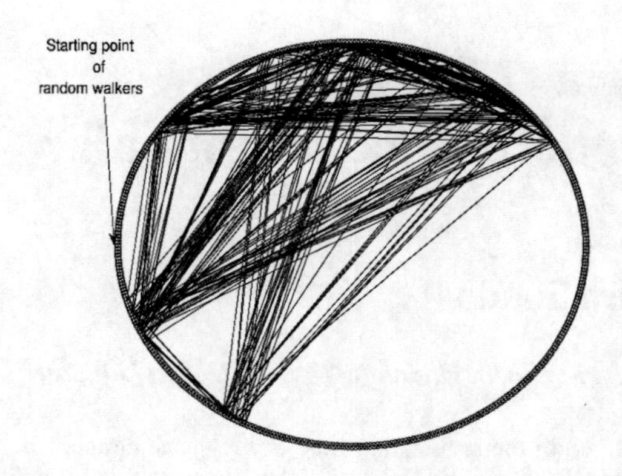

FIGURE 5. Example of formed network on 1-d lattice. All walkers are confined to a connected subgraph with small mean vertex-vertex distance. (The situation is $w = 8$, $p_d = 0.03$ and $t = 200,000$.)

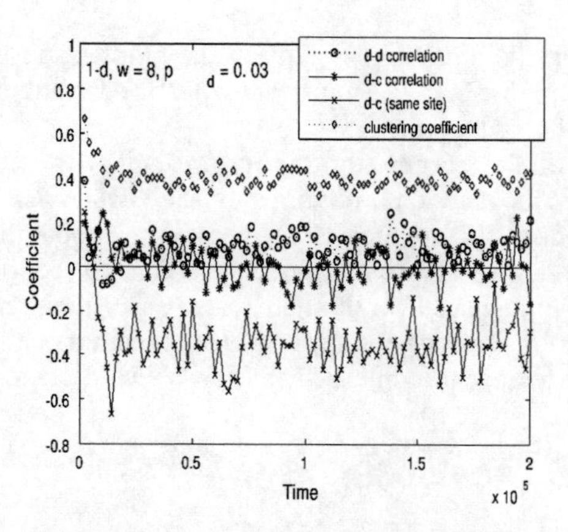

FIGURE 6. Clustering coefficient and correlation coefficients when 8 random walkers start from the same vertex in 1-d squared lattice under the condition of $p_d = 0.03$.

network. Although these results are not described here, when p_d is small enough to increase the life time of highly connected vertices, the time dependence of the d-c correlation coefficient will change moderately compared to the results shown in Figure 6. Temporal state exhibiting negative d-c correlation indicated in Figure 6 implies the gathering of random walkers into the small degree area and future increases of edges in the area.

It is interesting that the d-d correlation behaves like symmetrical motion to the motion of d-c correlation. This behavior means that vertices with small degree tend to be joined each other in the temporal state exhibiting negative d-c correlation and vice versa in the temporal state exhibiting positive d-c correlation. This relation between various correlations can be understood by the behavior of correlation between degree and local clustering

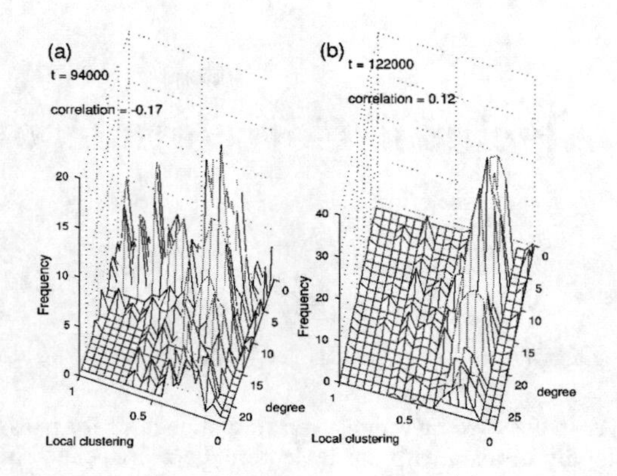

FIGURE 7. Scattering diagrams for 1-d case corresponding to (a) negative and (b) positive d-c correlation coefficient.

on the same vertex, which always takes negative values as shown in Figure 6.

CONCLUSION

Vertices with low clustered adjacent vertices tend to maintain edges opposing to continuous extinction of edges with probability p_d per time. This fact is supported by theoretical consideration based on the movement of random walkers and by numerical calculations for correlation coefficients and scattering diagrams. It is interesting that the local rule of network evolution determined by the movements of random walkers provides large clustering coefficients and various vertex correlations. However, this interpretation is applicable only when random walkers visit the area with adequate frequency, and situations where negative d-c correlation can be observed are very limited. Finally, we briefy mentioned that various correlations (d-d, c-c, d-c) can be related with each other via the correlation between degree and local clustering on the same vertices.

REFERENCES

1. L.A.N. Amaral and M. Ottino, *Eur. Phys. J. B* **38**, 147–162 (2004).
2. S.N. Dorogovtsev and J.F.F. Mendes, *Adv. Phys.* **51**, 1079-1145 (2002).
3. M.E.J. Newman and J. Park, *Phys. Rev. E* **68**, 036122-1-036122-9 (2003).
4. S. Zhou and R. J. Mondragón, *IEEE Communication Lett.* **8**, 180-182 (2004).
5. M.E.J. Newman, *Phys. Rev. Lett.* **89**, 208701-1-208702-5 (2002).
6. N. Ikeda, *Physica A* **379**, 701-713 (2007).
7. N. Ikeda, submitted.

Algebraic Correlations of Hard Spheres in a Rectangular Box

Akira Akaishi and Akira Shudo

Department of Physics, Tokyo Metropolitan University, 1-1 Minami-Osawa, Hachioji, Tokyo 192-0397, Japan

Abstract. We investigate the slow dynamics for hard-disk systems confined in the rectangular box. Phase space contains a family of marginally unstable periodic orbits, called bouncing ball orbits (BBOs), in which all particles move only in either the vertical or the horizontal direction. We derive the exponent $\gamma = 4$ for the asymptotic power law decay in the recurrence time distribution by estimating residence time around BBOs in the two hard-disk system. On the basis of a dimensionality argument, we also give the exponent as $\gamma = N + 2$ for generic N hard-disk systems. Numerical calculations are in good agreement with these theoretical predictions.

Keywords: Algebraic decay of correlations, Bouncing ball orbits, Hamiltonian chaos, Hard-disk systems
PACS: 05.45.-a

INTRODUCTION

Slow dynamics is a ubiquitously observed phenomenon and is regarded as one of the fundamental problems in the statistical physics. Slow motions for glass forming or supercooled liquids which induce divergence of viscosity without the thermodynamic transition are typical examples and extensively studied for the past several decades. In particular, in the system with many degrees of freedom, it is believed that the potential energy landscape picture is the most plausible scenario to capture the essence of the slow dynamics [1]. Complicated structures of the potential energy surface on multidimensional *configuration space* lead to the long time correlations even though the hopping between minima of the landscape is stochastic.

On the other hand, in the theory of Hamiltonian dynamics, there is another argument which also explains the slow dynamics [2]: the existence or remnants of regular structures in *phase space* prevents exponentially fast mixing processes, and thus causes the algebraically slow relaxation. The relation or consistency between two scenarios has not been seriously discussed so far. The present report is aimed at clarifying its connection especially by examining how the invariant structures with marginal stability in phase space affect the slow dynamics in the hard-disk system, which is a typical toy model studied especially in the former field.

Phase space of generic Hamiltonian systems is composed of regular and irregular regions. On the regular region, the orbits are stable and are quasi-periodic while the irregular components show chaotic behaviors typically leading to the exponentially fast mixing. Since the chaotic trajectories stay for a long time in the neighborhood of invariant tori, and exhibit *sticky motions* around them [3]; the boundaries between the regular and irregular components play significant roles for the algebraic decay of correlations.

One of the commonly studied measures to see the algebraic decay of correlations is the *recurrence time distribution* $P(T)$, which is defined as the probability to return to a given region in phase space with a recurrence time T. In generic Hamiltonian systems in which regular and chaotic motions coexist in phase space, it is believed that the recurrence time distribution obeys a power law [4]

$$P(T) \sim T^{-\gamma}. \tag{1}$$

However, complicated hierarchical nature of phase space in generic systems makes it hard to specify the exponent γ up to a quantitative level. It is still under discussions and has not been clearly concluded [4, 5]. For analytical and numerical convenience, the cumulative recurrence time distribution is introduced as

$$Q(T) = \sum_{t=T}^{\infty} P(t). \tag{2}$$

Note that summation is replaced by integration in case of continuous dynamics. The asymptotic form is a power law $Q(T) \sim T^{-\gamma'}$ with the exponent $\gamma' = \gamma - 1$.

Algebraic decay of correlations is also observed in Hamiltonian systems with rather simpler phase space, that is, without hierarchical phase space. Some of the examples are realized in the billiard system, which is a frictionless motion of a point particle in a given hard boundary. The stadium billiard, that is, billiard with a boundary which is composed of two half circles and two parallel straight segments inserted between the half circles is a well-studied system, and exhibits the algebraic decay of

CP982, *Complex Systems, 5th International Workshop on Complex Systems*
edited by M. Tokuyama, I. Oppenheim, and H. Nishiyama

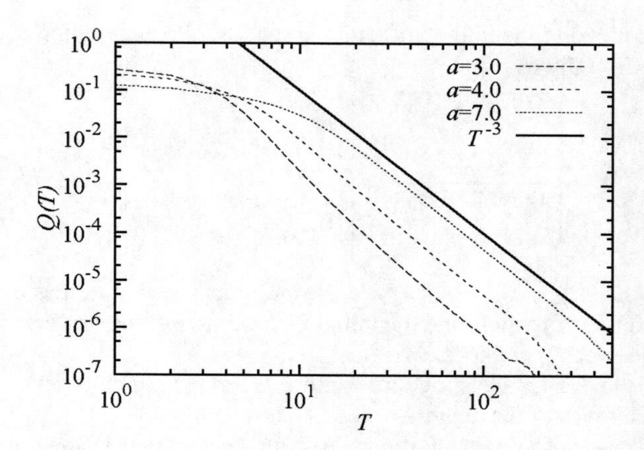

FIGURE 1. The cumulative recurrence time distribution for the two-disk system with $b = 1.9$

correlations although it belongs to the K-system [6]. Periodic orbits which bounce between the parallel segments show marginal stability and form a one-parameter family in phase space. These are called *bouncing ball orbits (BBOs)*, and such orbits also slow down the mixing rate as hierarchical invariant tori do so in the generic system. There are several analytical investigations to evaluate the exponent γ of the recurrence time distribution [7, 8].

In the present note, for a possible link between two different pictures mentioned above, we shall particularly focus on the effect of BBOs in the hard-disk system. In contrast to usual billiard systems, which are mostly studied as issues of dynamical systems, the particles inside the container have finite radii, but it also has BBOs if the container has the parallel straight lines. We here derive the exponent γ for the two hard-disk system by analyzing the residence time around the BBOs, and then extend the argument to general N hard-disk systems. Our final claim is summarized in (14), which provides the relation between γ and N. Our result implies that the effect of BBOs becomes weaker and weaker as the number of the particles increases.

ALGEBRAIC DECAYS IN THE TWO HARD-DISK SYSTEM

Let first us consider the system in which two hard disks with the same diameter move in the rectangular shaped hard wall. Denoting the length of the horizontal and vertical sides of the rectangular by a and b respectively, we assume that either a or b is longer than a twice of the diameter length. Otherwise, there is no enough space that BBOs exist. If BBOs do not exist, no algebraic behavior is observed since collisions between the particles merely produce the exponential instability due to the convexity of hard disks. To simplify our argument, we limit ourselves to the cases satisfying $a > 2, b < 2$, which ensures that only the BBOs bouncing in the vertical direction can appear.

In Fig. 1, we show the cumulative recurrence time distribution obtained by numerical simulations with several a and a fixed b. Here, the recurrence time is defined as the time interval between the collisions of two particles. In the long time regime, we can see that the distribution decays algebraically with $\gamma' = 3$ except for small deviations, which is due to numerical errors. It should be noted that the exponent obtained here differs from that for the one particle billiard system [8].

In order to make a local analysis around the BBOs, we first introduce a suitable phase space coordinate. Let $\{\mathbf{x}^{(i)}, \mathbf{v}^{(i)}\}$ be the Euclidian coordinate and velocity of i-th particle respectively. Phase space is 8-dimensional but the conservation of the total energy $E = (|\mathbf{v}^{(1)}|^2 + |\mathbf{v}^{(2)}|^2)/2$ reduces it to 7-dimensional space. Note that the phase space reduction applies only to the momentum space. We here assume that the invariant measure on the energy surface is expressed as a product form:

$$d\mu = d\mu_x d\mu_v, \tag{3}$$

where μ_x and μ_v represent the invariant measures for the configuration and momentum space, respectively.

The velocities only change either at the moment when a particle bounces at the boundary or when two particles collide with each other. In analogy with the Birkhoff coordinate in the ordinary two-dimensional billiard system, a new coordinate for velocities, denoted by $(\sin\theta_{\rm coll}, \theta_{\rm free}, u)$, is introduced in the following way. When one particle bounces at the boundary, we take $\theta_{\rm coll}$ as the angle between the velocity of the bouncing particle and the normal vector at the bouncing point, and $\theta_{\rm free}$ and u respectively as the angle and the absolute value of the velocity of the other particle which moves freely. At the moment when two particles collide, denoting respectively by $\theta^{\pm}, u^{\pm}$ the angles and the absolute values for the center-of-mass and relative velocity, that is,

$$\mathbf{v}^{+} = \frac{1}{\sqrt{2}}(\mathbf{v}^{(1)} + \mathbf{v}^{(2)}), \quad \mathbf{v}^{-} = \frac{1}{\sqrt{2}}(\mathbf{v}^{(2)} - \mathbf{v}^{(1)}), \tag{4}$$

we take $\theta_{\rm coll} = \theta^{-}$ and $\theta_{\rm free} = \theta^{+}$, and $u = u^{+}$. [1]

On this coordinate system, we again put an ansatz that the invariant measure for the velocity is decomposed into a product form,

$$d\mu_v(\sin\theta_{\rm coll}, \theta_{\rm free}, u) = d(\sin\theta_{\rm coll})d\theta_{\rm free}d(f(u)), \tag{5}$$

[1] θ^{-} must be taken as the incident angle of the collision in the relative coordinate.

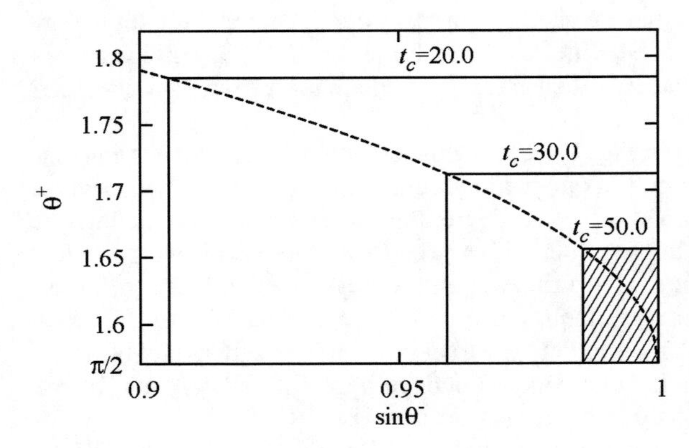

FIGURE 2. Contour lines (solid lines) for t_c on $(\sin\theta^-, \theta^+)$ plane. The closer the contour line approaches to the point $(1, \pi/2)$, the larger t_c becomes. The dashed curve indicates a trace of the corner of rectangular-shaped contour lines and is given by $Lu^+ \sin\varepsilon^+ = Xu^- \sqrt{2\varepsilon^- - \varepsilon^{-2}}$ where the two cases in (8) coincide. The parameters are chosen as $L = 6, X = 3$, and $u^+ = u^- = 1$.

where $f(u)$ is some function of u. Under the present setting, there is no rigorous proof for ergodicity of the hard-disk systems, which would be necessary to justify the assumptions (3) and (5). Their validity, however, has been checked numerically [9]. This ansatz allows us to analyze the residence time distribution around BBOs in the space $(\sin\theta_{\rm coll}, \theta_{\rm free})$, the time evolution of which is essentially given as a discrete dynamics.

As mentioned, we define the recurrence time distribution as the distribution of time intervals between successive collisions. Since we may expect that its asymptotically long-time behavior is determined by the orbits staying in the vicinity of BBOs, we estimate the recurrence time of the trajectories around BBOs.

Let the initial condition of the particles be inside the recurrence region, that is, two particles touch with each other. Note that the point satisfying $(\sin\theta^-, \theta^+) = (1, \pi/2)$ corresponds to BBOs. Thus, the initial condition for the trajectories starting close to BBOs can be expressed as $(1-\varepsilon^-, \pi/2+\varepsilon^+, u_0^+)$ using small constants ε^- and ε^+ and the initial condition u_0^+. The relation to the velocities in the cartesian coordinates is given as

$$v_x^{(1)}(0) = \frac{1}{\sqrt{2}}\left(-u_0^+ \sin\varepsilon^+ - u_0^- \sqrt{2\varepsilon^- - \varepsilon^{-2}}\right), \quad (6)$$

$$v_x^{(2)}(0) = \frac{1}{\sqrt{2}}\left(-u_0^+ \sin\varepsilon^+ + u_0^- \sqrt{2\varepsilon^- - \varepsilon^{-2}}\right), \quad (7)$$

where $(u_0^{+2} + u_0^{-2})/2 = E$. For sufficiently small ε^- and ε^+, the particles move fast in the vertical direction and slow in the horizontal direction. Therefore, t_c in which the particles collide again is approximated by the collision time in the horizontal direction. We then obtain

$$t_c = \begin{cases} \frac{2X}{|v_x^{(1)}(0)| - |v_x^{(2)}(0)|} = \frac{\sqrt{2}X}{u_0^+ \sin\varepsilon^+} \\ \qquad \text{if } |v_x^{(1)}(0)| > \frac{L+X}{L-X}|v_x^{(2)}(0)|, \\ \frac{2L}{|v_x^{(1)}(0)| + |v_x^{(2)}(0)|} = \frac{\sqrt{2}L}{u_0^- \sqrt{2\varepsilon^- - \varepsilon^{-2}}} \\ \qquad \text{if } |v_x^{(1)}(0)| \le \frac{L+X}{L-X}|v_x^{(2)}(0)|, \end{cases} \quad (8)$$

where $X = x_c - 2r, L = a - 2r$, and r and x_c denote the radius of particle and the initial x-coordinate of the center between two particles.

In Fig. 2, we show the contour lines of t_c on a upper half part of the neighborhood of $(\sin\theta^-, \theta^+) = (1, \pi/2)$. The set of initial conditions inside the rectangular-shaped region whose boundary is given by $t_c =$ const represents the orbits that take larger time than t_c to return back to the recurrence region. Therefore, using the ansatz (5), it is enough to estimate the area of the rectangular to obtain the probability of the recurrence with time larger than t_c. The area of the rectangular is indeed evaluated as a function of t as,

$$\begin{aligned} S(t) &= \varepsilon^+(t) \times \varepsilon^-(t) \\ &= \sin^{-1}\left(\frac{\sqrt{2}X}{u_0^+}\cdot\frac{1}{t}\right) \times \left(1 - \sqrt{1 - \left(\frac{\sqrt{2}L}{u_0^-}\cdot\frac{1}{t}\right)^2}\right) \\ &\sim \frac{\sqrt{2}XL^2}{u_0^+ u_0^{-2}}\cdot\frac{1}{t^3} + \mathcal{O}\left(\frac{1}{t^4}\right). \end{aligned} \quad (9)$$

In the long time regime, *i.e.* for small $\varepsilon^+, \varepsilon^-$, the cumulative function asymptotically coincide with $S(t)$. As a result,

$$Q(T) \sim S(T) \sim T^{-3}, \quad (10)$$

and we conclude $\gamma' = 3$ for the two hard-disk system.

IN N HARD-DISK SYSTEM

We next examine the effect of BBOs in the general N hard-disk system. Since the coordinate system introduced in two-disk system becomes complicated in general cases, we only argue the dimensionality of sticky regions around BBOs and deduce the exponent from it. We assume again that the length of the horizontal sides is longer than the total length of the particle diameter. Let $\mathbf{v}^{(i)}$ denote the velocity of the i-th particle. BBOs for the N-disk system is defined by the condition $v_x^{(1)} = \cdots = v_x^{(N)} = 0$. This condition leads that BBOs form a $3N$-dimensional subset in $4N$-dimensional full phase space. Due to the marginal stability, the orbit whose distance from BBOs is $\varepsilon \ll 1$ is a linear flow keeping ε constant

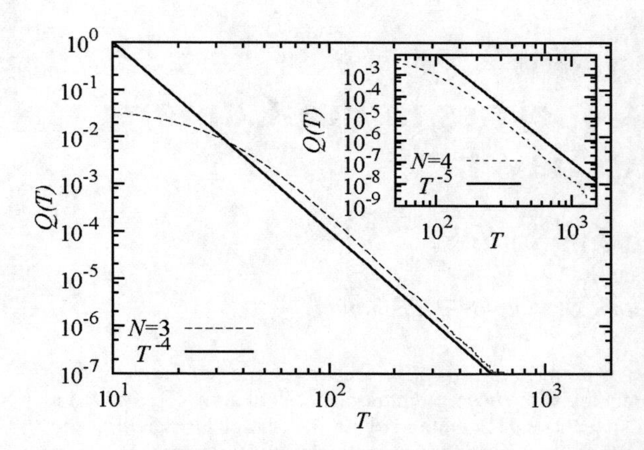

FIGURE 3. The cumulative recurrence time distribution for the three-disk system. Inset: that for the four-disk system. The parameters of the rectangular box are chosen as $a = 13.0, b = 1.8$ for the three-disk system and $a = 20.0, b = 1.5$ for the four-disk system.

along BBOs and a displacement by unit time step is proportional to ε. Therefore, the time scale in which the orbits spend in the neighborhood of BBOs can be estimated as

$$T \sim \frac{1}{\varepsilon}. \tag{11}$$

Taking into account that the number of the linearly independent directions perpendicular to the linear flows is N, we have the probability that the trajectories move along the neighborhood of BBOs with distance ε as

$$p(\varepsilon) \sim \varepsilon^N. \tag{12}$$

Then, we can derive the residence time distribution as

$$P_{\rm res}(T) = p(\varepsilon)\left|\frac{d\varepsilon}{dT}\right| \sim p(\varepsilon)\varepsilon^2 \sim T^{-(N+2)}. \tag{13}$$

If we follow the argument made in [8], in which it is claimed that the recurrence time distribution in the long time regime is determined by the residence time distribution for the neighborhood of BBOs in case where there exist no stable invariant structures in phase space except for BBOs. Therefore, we can conclude that the exponent for the recurrence time is derived as

$$\gamma = N+2, \tag{14}$$

and $\gamma' = N+1$ for the cumulative function.

We numerically confirm the validity of these arguments. In Fig. 3, we show the cumulative recurrence time distributions for the three and four hard-disk system. The recurrence time distribution is defined similarly as the two-disk system, that is, $P(T)dT = \{$ the probability of collisions between the two neighboring particles within $[T, T+dT]$ $\}$. Note that, in the present setting, the particles cannot exchange their ordering in the horizontal direction since the length of vertical sides of the rectangular box is not large enough to allow the exchange of the particles. The algebraic parts of the distributions show good agreements with the theoretical predictions which were obtained based on the previous argument.

SUMMARY

We have studied the nature of the decay of correlations in the hard-disk system in the rectangular box. The recurrence time distribution, which is often used because of numerical reliability of its convergence, has particularly been examined. Due to the presence of BBOs, the recurrence time distributions have power law decaying parts. We have made, by introducing a proper coordinate around BBOs, an analytical argument to provide its exponent in the two hard-disk system. We further predicted the exponents for general N-disk systems, which have been derived from a dimensionality argument of the sticky orbits around BBOs, and checked its validity by numerical experiments.

In the theory of Hamiltonian dynamics, the mechanism to yield the slow relaxation is essentially explained by the existence of stable (or marginally unstable) invariant structures in phase space. On the other hand, as far as the authors' knowledge, one has not so closely examined their effects in the system modeling glass forming and supercooled liquids, although the hard-disk system in the rectangular box has been studied so extensively. The present analysis suggests that the effect of marginally unstable periodic orbits is attenuated as a function of the system size.

REFERENCES

1. D. J. Wales, *Energy Landscapes*, Cambridge University, Cambridge, 2003.
2. A. L. Lichtenberg and M. A. Lieberman, *Regular and Chaotic Dynamics*, Springer-Verlag, New York, 1983.
3. C. F. F. Karney, *Physica D* **8**, 360–380 (1983).
4. B. V. Chirikov and D. L. Shepelyansky, *Phys. Rev. Lett.* **82**, 528–531 (1999).
5. M. Weiss, L. Hufnagel, and R. Ketzmerick, *Phys. Rev. E* **67**, 046209-1-046209-6 (2002).
6. R. Markarian, *Ergod. Th. & Dynam. Sys.* **24**, 177–197 (2004).
7. D. N. Armstead, B. R. Hunt, and E. Ott, *Physica D* **193**, 96–127 (2004).
8. E. G. Altmann, A. E. Motter, and H. Kantz, *Phys. Rev. E* **73**, 026207-1-026207-10 (2006).
9. A. Akaishi and A. Shudo, to be submitted.

Anomaly of Transmission Properties in Pre-Cantor Dielectric Multilayers

Keimei Kaino and Jun Sonoda

Sendai National College of Technology, Sendai 989-2128, Japan

Abstract. Using the transmission-line theory, we investigate wave propagation in a pre-Cantor multilayer. Transmission spectra of the low stages of pre-Cantor media show good agreement with those of numerical calculation of Maxwell's equations using the FDTD method. Numerical results obtained using the FDTD method show that the electric field at the midpoint of the nth stage pre-Cantor medium has sharp resonance and broad attenuation at transmission bands that are newly generated in attenuation bands of the $(n-1)$th stage. Using an expression of transmittance of the high stage of pre-Cantor multilayer, we show that the transmittance t becomes a two-valued function of $t = 0/1$ and the collection of points for $t = 1$ is a power set of positive integers whose cardinal number is $3^{\aleph_0}$.

Keywords: Cantor multilayer, FDTD method, Fractal, Transmission-line, Resonance
PACS: 42.25.Bs, 78.20.Bh

INTRODUCTION

In random media and deterministic media, the idea of light localization has been explored intensively. In the third stage of a Menger sponge fabricated from epoxy resin, attenuation of both reflection and transmission intensity was observed at 12.8 GHz [1]. The intersection of the sponge with medians or diagonals of the initial cube are triadic Cantor sets. The fractal Cantor set, which has the topological dimension $D_T = 0$ and the Hausdorff dimension $D = 0.631$ is thin and spare to the point of being invisible [2]. We thicken it into what might be called a Cantor bar.

In the Cantor bar, the initiator is the interval $[0, L]$, whereas the next three stages are shown in Fig. 1 where $d_n = L/3^n$ is the thickness of the layer for the nth stage pre-Cantor, and ε_i is the dielectric constant and μ_i the magnetic permeability for $i = 1$ (light region) and $i = 2$ (dark region). The Cantor bar has an infinitely larger number of infinitely thin layers of infinitely high density. Many studies have addressed electromagnetic waves in pre-Cantor bars [3, 4, 5]. We specifically examine the anomaly of transmission properties in the Cantor multilayer.

First, we obtain a formula for transmittance of a pre-Cantor multilayer using the transmission-line theory. Then we investigate resonance and attenuation in transmission spectra calculated using the Finite-Difference Time-Domain (FDTD) method of Maxwell's equations [6]. Finally, after deriving an expression of transmittance of a high stage of the pre-Cantor multilayer, we consider anomalous features of the transmittance of the Cantor multilayer.

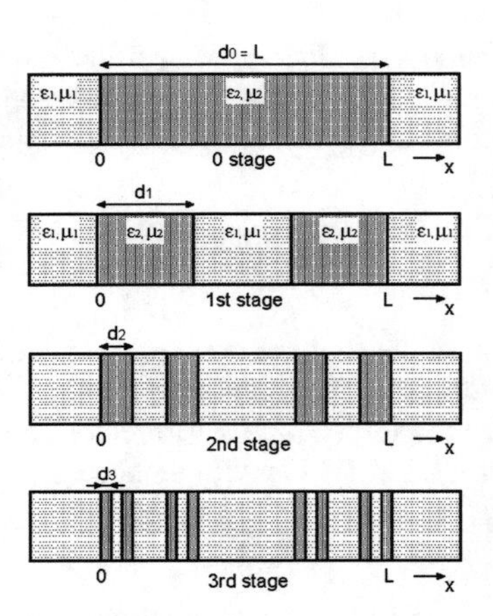

FIGURE 1. Pre-Cantor Multilayers.

TRANSMISSION-LINE THEORY

Electromagnetic energy can be transmitted along the pre-Cantor bar of a dielectric multilayer. We regard the pre-Cantor bar as a transmission line, as shown in Fig. 2. For the pre-Cantor multilayer of 0 stage, we can describe the frequency behavior of the transmission line given its inductance per unit length, L_i, and its capacity per unit length, C_i, where $i = 1$ for $x < 0$, $i = 2$ for $0 < x < L$ and $i = 3$ for $x > L$ with $L_3 = L_1$ and $C_3 = C_1$. The basic equations for the voltage $V(x)$ and the current $I(x)$ of a transmission line are given as

$$\frac{dV(x)}{dx} = -j\omega L_i I(x), \quad \frac{dI(x)}{dx} = -j\omega C_i V(x). \tag{1}$$

For $C_i = \varepsilon_i$ and $L_i = \mu_i$, those equations are the same as the Maxwell equations. The voltage and current propa-

CP982, *Complex Systems, 5th International Workshop on Complex Systems*
edited by M. Tokuyama, I. Oppenheim, and H. Nishiyama

gate along the line as a wave,

$$V(x) = A_i e^{-\gamma_i x} + B_i e^{\gamma_i x} \quad (2)$$

$$I(x) = (A_i e^{-\gamma_i x} - B_i e^{\gamma_i x})/Z_{0i} \quad (3)$$

where the propagation constant and the characteristic impedance are given as

$$\gamma_i = j\beta_i = j\omega\sqrt{L_i C_i}, \quad Z_{0i} = \sqrt{L_i/C_i}, \quad (4)$$

where $i = 1$ in the white layers and $i = 2$ in the gray layers in Fig. 1. For the discussion in this article, we assume that $\mu_1 = \mu_2$ and denote the ratio Z_{01}/Z_{02} by e^{α}. Consequently, we have the relation

$$e^{\alpha} = Z_{01}/Z_{02} = \sqrt{\varepsilon_2/\varepsilon_1} = \beta_2/\beta_1. \quad (5)$$

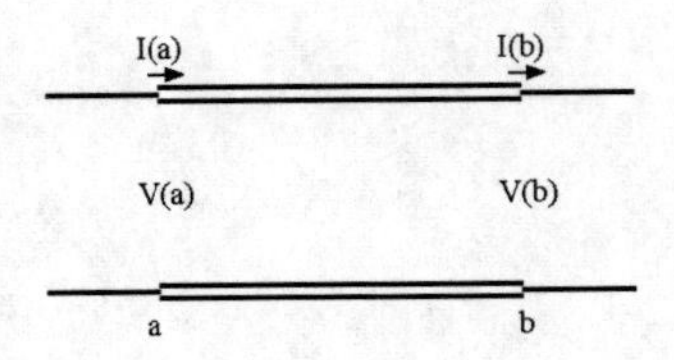

FIGURE 2. Transmission line.

For the transmission line in Fig. 2, the voltage and current are continuous at the boundaries of layers: $x = a$ and $x = b$. Amplitudes A_1 and B_1 for $x < a$ are linear combinations of A_3 and B_3 for $x > b$ and in the matrix from the relation is

$$\begin{pmatrix} A_1 \\ B_1 \end{pmatrix} = T(a,b) \begin{pmatrix} A_3 \\ B_3 \end{pmatrix}. \quad (6)$$

The transmission matrix T is given as

$$T(a,b) = \begin{pmatrix} t_{11}\, e^{-\gamma_i(b-a)} & t_{12}\, e^{\gamma_i(a+b)} \\ t_{21}\, e^{-\gamma_i(a+b)} & t_{22}\, e^{\gamma_i(b-a)} \end{pmatrix}, \quad (7)$$

where the matrix elements t_{ij} are $t_{11} = c + j\cosh\alpha s = t_{22}{}^*$ and $t_{12} = j\sinh\alpha s = t_{21}{}^*$ with $c = \cos\beta_2(b-a)$ and $s = \sin\beta_2(b-a)$. The determinant of matrix $T(a,b)$ is unity.

The pre-Cantor multilayer is generated using the recurrence relation [5]

$$T^{(k)}(0,L) = T^{(k-1)}(0,L/3)\, T^{(k-1)}(2L/3,L). \quad (8)$$

The initial value $T^{(0)}(a,b) = T(a,b)$. Consequently, the transmission matrix of the nth stage is

$$T^{(n)}(0,L) = \begin{pmatrix} \xi_n(d_n)e^{-\gamma_1 L} & \eta_n(d_n)e^{\gamma_1 L} \\ \eta_n^*(d_n)e^{-\gamma_1 L} & \xi_n^*(d_n)e^{\gamma_1 L} \end{pmatrix}, \quad (9)$$

where $d_n = L/3^n$ is the thickness of the layers of the nth stage pre-Cantor. Elements $\xi_n(d_n)$ and $\eta_n(d_n)$ are obtained using the recurrence relation

$$\xi_{k+1} = \xi_k^2\, e^{\gamma_1 3^k d_n} + |\eta_k|^2\, e^{-\gamma_1 3^k d_n}, \quad (10)$$

$$\eta_{k+1} = \eta_k \left(\xi_k\, e^{\gamma_1 3^k d_n} + \xi_k^*\, e^{-\gamma_1 3^k d_n} \right), \quad (11)$$

and the initial values

$$\xi_0 = \cos\beta_2 d_n + j\cosh\alpha \sin\beta_2 d_n, \quad (12)$$

$$\eta_0 = j\sinh\alpha \sin\beta_2 d_n. \quad (13)$$

Note that $\det T^{(n)}(0,L) = |\xi_n|^2 - |\eta_n|^2 = 1$. The transmittance $t^{(n)}$ and the reflectance $r^{(n)}$ are given as

$$t^{(n)} = \frac{1}{|\xi_n(d_n)|^2}, \quad r^{(n)} = 1 - t^{(n)}. \quad (14)$$

The transmittance of the pre-Cantor multilayer of the zeroth stage is shown as

$$t^{(0)}(\beta_1 d_0) = 1/(1 + \sinh^2\alpha\, s^2) \quad (15)$$

where $s = \sin\beta_2 d_0$. The period of $t^{(0)}$ is $\beta_2 d_0 = \pi$. The transmittance of the first stage is

$$t^{(1)}(\beta_1 d_1) = 1/(1 + 4\sinh^2\alpha\, |\xi|^2 s^2 C_0^2), \quad (16)$$

where $|\xi|^2 = 1 + \sinh^2\alpha\, s^2$, $s = \sin\beta_2 d_1$ and $C_0 = \cos(\beta_1 d_1 + \theta)$ with $\tan\theta = \coth\alpha \tan\beta_1 d_1$. The transmittance $t^{(1)}$ is an almost periodic function with two periods $\beta_1 d_1 = \pi$ and $\beta_1 d_1 = e^{-\alpha}\pi$. Figure 3 shows that, in the special case of $\beta_2/\beta_1 = 2$, $t^{(1)}$ has the period of $\beta_1 d_1 = \pi$, where the solutions of $t^{(1)} = 1$ consist of $\beta_1 d_1 = 0, \pi/2$ for $s = 0$ and $\beta_1 d_1 = 0.156\pi$, $\pi/2$ and 0.844π for $C_0 = 0$.

The curve of $t^{(n)}$ has the period of $\beta_1 d_n = \pi$ when β_2/β_1 is an integer. It is useful to regard the transmittance of the nth stage as a function of $\beta_1 d_n$. We can prove that a set of solutions of $t^{(n)} = 1$ includes a set of solutions of $t^{(n-1)} = 1$.

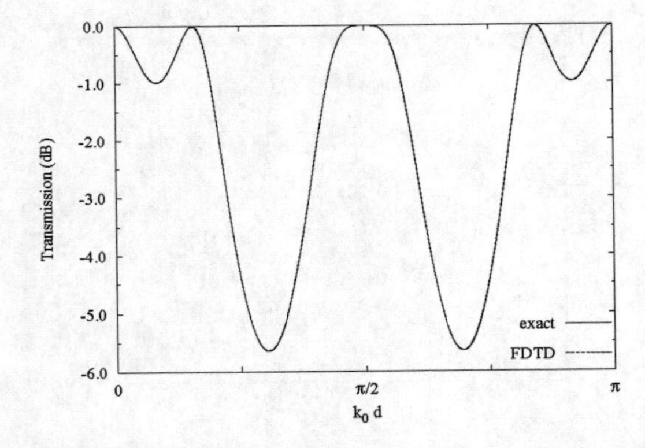

FIGURE 3. Transmittance of the first stage pre-Cantor with $\beta_2/\beta_1 = 2$ and $k_0 d = \beta_1 d_1$.

FDTD METHOD AND NUMERICAL RESULTS

Maxwell's equations are replaced by a set of finite differential equations. We use the Finite-Difference Time-Domain (FDTD) method to obtain solutions of one-dimensional multilayers numerically [6]. The problem is a scattering of electromagnetic pulses by a pre-Cantor multilayer. First, we obtain the solutions of time-dependent response using the FDTD method. Then we apply Fourier analysis to those solutions and obtain the transmission coefficients. We set the source at a point in $x < 0$ and denote by $E_i(t)$ a response of electric field to the incident wave. From the response of transmitted electric field $E_t^{(n)}(t)$ at $x > L$ and the response of reflected electric field $E_r^{(n)}(t)$ at $x < 0$ in the nth stage pre-Cantor media, we can obtain the transmission coefficient $\tau^{(n)}$ and the reflection coefficient $\Gamma^{(n)}$ given as

$$\Gamma^{(n)} = \frac{\mathrm{FT}[E_r^{(n)}(t)]}{\mathrm{FT}[E_i^{(n)}(t)]}, \quad \tau^{(n)} = \frac{\mathrm{FT}[E_t^{(n)}(t)]}{\mathrm{FT}[E_i^{(n)}(t)]}, \tag{17}$$

where $\mathrm{FT}[E(t)]$ means the Fourier transform of $E(t)$. Finally, the transmittance $t^{(n)}$ is given as

$$t^{(n)} = |\tau^{(n)}|^2. \tag{18}$$

In Figs. 3–5, the transmittances are shown in $10\log|t^{(n)}|$. In each stage of the pre-Cantor multilayer, the transmittance of numerical calculation of the FDTD method shows good agreement with the exact one of the transmission-line theory. Those curves are similar. The higher the stage of pre-Cantor media becomes, the larger the number of sharper transmitted bands that appear in attenuation bands. It seems that 3^n transmission bands appear as we move from the $(n-1)$th stage to the nth stage.

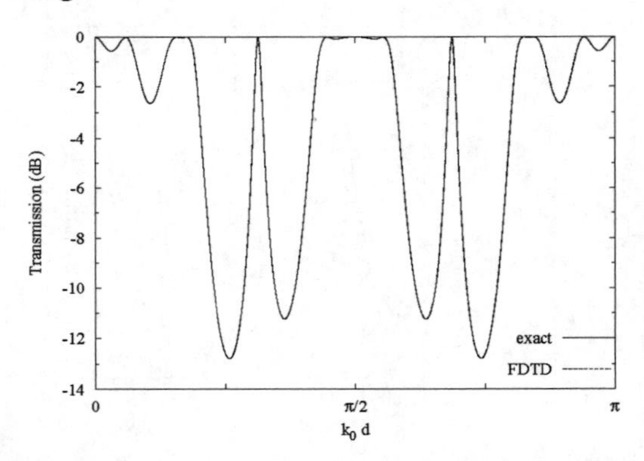

FIGURE 4. Comparison of transmittances of the second stage with $\beta_2/\beta_1 = 2$ and $k_0 d = \beta_1 d_2$.

Figure 6 presents the minimum values of $|t^{(n)}|$ for the stage number of pre-Cantor n in three values of relative dielectric constants $\varepsilon_r = \varepsilon_2/\varepsilon_1$. Minimum values of transmittance, as measured in decibels, are halved because the stage number increases by one.

The most interesting problem in fractal media is a localization of light waves [7] and electromagnetic waves [1]. We next consider electric fields in a pre-Cantor multilayer. Figures 7 and 8 show amplitudes of electric fields

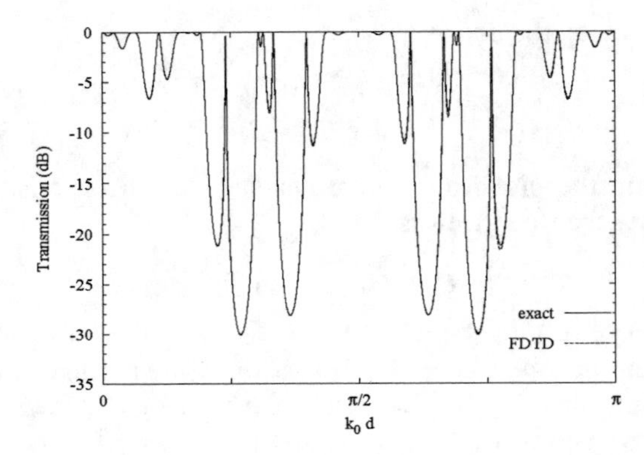

FIGURE 5. Comparison of transmittances of the third stage with $\beta_2/\beta_1 = 2$ and $k_0 d = \beta_1 d_3$.

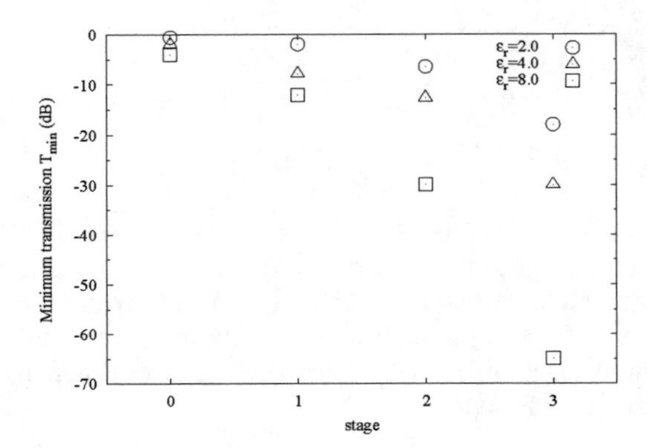

FIGURE 6. Minima of transmittances versus stage number for ε_r =2, 4 and 8.

at $x = L/2$ inside the multilayer and at $x = L$ outside the multilayer. Both sharp resonance and broad attenuation inside the multilayer appear at transmission bands that are newly generated in the attenuation bands with increment of the stage number by one. Because the stage number increases, the maximum value of the strongest resonance increases rapidly.

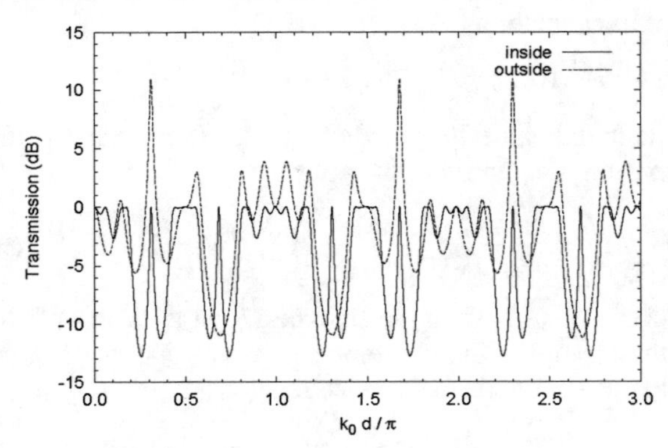

FIGURE 7. Amplitudes of electric field of the second stage.

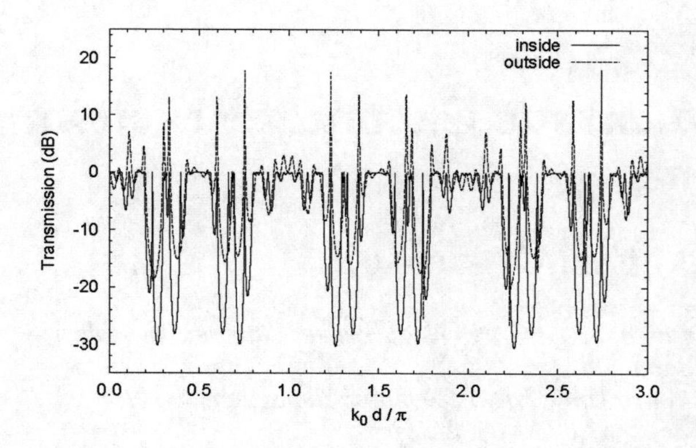

FIGURE 8. Amplitudes of the electric field of the third stage.

ANOMALY OF TRANSMITTANCE OF CANTOR MEDIA

We consider the asymptotic formula of the transmittance of the high stage of pre-Cantor media. Using $\xi_k = |\xi_k| e^{j\theta_k}$, we can rewrite the recurrence relation (9) and (10) as

$$\xi_{k+1} = e^{j\theta_k}\left\{(1+2|\eta_k|^2)C_k + jS_k\right\}, \quad (19)$$

$$\eta_{k+1} = \eta_k |\xi_k| C_k, \quad (20)$$

where $C_k = \cos(3^k \beta_1 d_n + \theta_k)$ and $S_k = \sin(3^k \beta_1 d_n + \theta_k)$. Then, transmittance $t^{(n)}$ is given as

$$t^{(n)} = 1/\left(1 + 4^n \sinh^2 \alpha\, s^2 \prod_{k=0}^{n-1} |\xi_K|^2 C_k{}^2\right). \quad (21)$$

Using an inequality

$$4^n \sinh^2 \alpha \prod_{k=0}^{n-1} |\xi_K|^2 \le \sinh^2(2^n \alpha),$$

we obtain an asymptotic formula

$$t^{(n)} \simeq 1/\left(1 + \sinh^2(2^n \alpha)\, s^2 \prod_{k=0}^{n-1} C_k{}^2\right). \quad (22)$$

Using this formula, we have $\log|t^{(n)}| \ge -2^{n+1}\alpha$, which explains that the minimum of $t^{(n)}$ goes to zero rapidly, as shown in Fig. 6.

The number of roots of $C_k = 0$ is about 3^n in the region $0 \le \beta_1 d_n < \pi$. Therefore, the number of roots of $t^{(n)} = 1$ is $O(3^n)$. Consequently, the set of roots is a power set. In the limit of n as infinity, this set of roots becomes an uncountable set with the cardinal number $3^{\aleph_0}$. Because the stage number n goes to infinity, $t^{(n)}$ becomes a two-valued function of 0/1. The transmittance of the Cantor multilayer differs greatly from that of a monolayer of equal thickness [5, 3].

The other interesting property in the pre-Cantor multilayer is the strong resonance and attenuation shown in Figs. 7 and 8. Our theoretical analysis is now in progress. Detailed calculations will be published in another paper.

CONCLUSIONS

We studied the anomaly of transmission properties in the pre-Cantor dielectric multilayer using the transmission-line theory and the FDTD method of Maxwell's equations. The obtained results demonstrate the following.

1. Using the transmission-line theory, we derived exact formulas of the transmittance of low nth stage of the pre-Cantor multilayer.
2. In low stages of pre-Cantor media, transmittances that have been calculated with the FDTD method show good agreement with those of the transmission-line theory.
3. Using the FDTD method, we calculated the amplitude of the electric field inside a pre-Cantor multilayer and showed that both strong resonance and broad attenuation appear in the attenuation bands with the increment of the stage number.
4. From the asymptotic expression of transmittance of the pre-Cantor multilayer of the high stage, we predicted that the transmittance of the Cantor multilayer is a two-valued function of 0/1 and differs greatly from that of a monolayer of the same thickness.

ACKNOWLEDGMENTS

The authors wish to thank Prof. H. Hara and Dr. N. Ikeda for helpful discussions.

REFERENCES

1. M. W. Takeda, S. Kirihara, Y. Miyamoto, K. Sakoda, and K. Honda, *Phys. Rev. Lett.* **92**, 093902–1–093902–4 (2004).
2. B. B. Mandelbrot, *The Fractal Geometry of Nature*, Freeman, New York, 1977, pp. 74–83.
3. U. Sangawa, *IEICE Trans. Electron.* **E88-C**, pp. 1981–1991 (2005).
4. N. Hatano, *J. Phys. Soc. Jpn.* **74**, 3093–3111 (2005).
5. K. Honda, and Y. Otobe, *J. Phys. A* **39**, L315–L322 (2006).
6. K. S. Yee, *IEEE Trans. Antennas Propagat.* **AP-14**, pp. 302–307 (1966).
7. W. Gellermann, M. Kohmoto, B. Sutherland, and P. C. Taylor, *Phys. Rev. Lett.* **72**, 633–636 (1994).

The Fluctuation Theorem in Stochastic Chemical Reaction Systems

Naohiro Akuzawa* and Mitsuhiro Akimoto†

*Department of Physics, Faculty of Science, Tokyo University of Science, 1-3 Kagurazaka, Shinjuku-ku, Tokyo 162-8601, Japan
†Department of Electronics and Computer Science, Tokyo University of Science, Yamaguchi, 1-1-1 Daigaku-Dori, San-Yo-Onoda, Yamaguchi 756-0884, Japan

Abstract. In this study, we consider a simple discrete model of isomerization-type reaction, which can be described in Markovian kinetics. Then we investigate the validity of FT in that model by using the Gillespie method which gives an ensemble of stochastic paths of successive reactions explicitly. We obtain a result that the validity of the FT is determined by time-scale of the dynamics.

INTRODUCTION

In small systems like living cells, one of the important elements to understand their functions is chemical reactions. When the number of the molecules in a system of interest as well as the system size is small, molecular discreteness becomes evident, so that the chemical reactions are under the strong influence of fluctuations. Namely, the probabilistic description is needed to model accurately the chemical reactions in such small systems. Many methods has been proposed among which the chemical Langevin equation, the chemical master equation (or its Fokker-Planck approximated equation), etc.[1] are well-known. In this study, we are interested in the characterization of fluctuations in such environments. It is a fundamental problem to describe energetics of fluctuations under non-equilibrium condition on the basis of the probabilistic models mentioned above.

In the mid 90's, a non-equilibrium equality for systems where fluctuations play a significant role was discovered[2]. This equality is called the fluctuation theorem(FT),

$$\frac{P(\Delta S)}{P(-\Delta S)} = e^{\Delta S}, \tag{1}$$

and there is a growing interest in their importance and implication. Here, ΔS is the entropy production of the system. It should be noted that the term of "entropy production" means in this study that the total amount of entropy generated from an initial state to a final state. The FT embodies recent developments toward a unified treatment of arbitrarily large fluctuations in small systems. The FTs can be brought together in one expression[3]:

$$\langle A[\chi]\rangle = \langle A^{\dagger}[\chi]e^{-\Delta S}\rangle. \tag{2}$$

Here, $\langle A[\chi]\rangle$ indicates the average of an arbitrary stochastic-path function $A[\chi]$ and χ is a stochastic trajectory generated by a certain stochastic process. The average of a path function is taken over an ensemble of paths. $\langle A^{\dagger}[\chi]e^{-\Delta S}\rangle$ is $A^{\dagger}$, the time reversal of a A, averaged over all the reverse processes and weighted by the exponential of ΔS. The dagger † is a time-reversal operator. Not only the FTs, but also several relations including the Kawasaki nonlinear response relation and the Jarzynski equality can be derived from Eq.(2). The entropy production can be regarded as a quantity which expresses the extent of non-equilibrium. Since the relation $\langle A[\chi]\rangle = \langle A^{\dagger}[\chi]\rangle$ realizes in equilibrium, Eq.(2) can be regarded as an extended form of the detailed balance condition. Then, we expect that these relations are fundamental equalities of systems like living cells. Our aim here is to investigate the validity of Eq.(2) by solving a simple stochastic model of isomerization reaction explicitly.

MODEL

Consider a vessel with its volume Ω, in which N chemical species $X_j (j = 1,\cdots,N)$ exist. The number of molecules X_j is denoted by n_j. Among these chemical species, the following chemical reactions take place[4, 5, 6]:

$$\sum_{j=1}^{N} s_j^{(\rho)} X_j \underset{k_-^{(\rho)}}{\overset{k_+^{(\rho)}}{\rightleftarrows}} \sum_{j=1}^{N} r_j^{(\rho)} X_j, \tag{3}$$

with $\rho = 1,\cdots,M$ labeling the reaction events. k is a bare reaction constant and $+$ denotes a forward reaction, $-$ denotes a reverse reaction. The stoichiometric coefficients $s_j^{(\rho)}$ and $r_j^{(\rho)}$ are non-negative integer. Let $\nu_j^{(\rho)}$ be

CP982, *Complex Systems, 5th International Workshop on Complex Systems*
edited by M. Tokuyama, I. Oppenheim, and H. Nishiyama

$v_j^{(\rho)} = -v_j^{-(\rho)} = r_j^{(\rho)} - s_j^{(\rho)}$ and (ρ) be $(-\rho) = -(\rho)$. The state of the system corresponds to a vector of the number of chemical species $\mathbf{n} = (n_1, \cdots, n_N)$. The probability distribution $P(\mathbf{n}, t)$ that the system is in a state $\mathbf{n}$ at time t follows the master equation,

$$\frac{\partial P(\mathbf{n},t)}{\partial t} = \sum_{\rho=-M}^{M} \big[w^{(\rho)}(\mathbf{n}+\mathbf{v}^{(\rho)})P(\mathbf{n}+\mathbf{v}^{(\rho)},t) - w^{(\rho)}(\mathbf{n})P(\mathbf{n},t) \big]. \tag{4}$$

We assume a dilute solution of reacting species in the system and therefore the transition probabilities per unit time for the reaction events take the forms[4]:

$$\begin{aligned} w^{+(\rho)}(\mathbf{n}) \equiv w_+^{(\rho)}(\mathbf{n}) &= \Omega k_+^{(\rho)} \Pi_j \frac{n_j!}{(n_j - s_j^{(\rho)})! \Omega^{s_j^{(\rho)}}}, \\ w^{-(\rho)}(\mathbf{n}) \equiv w_-^{(\rho)}(\mathbf{n}) &= \Omega k_-^{(\rho)} \Pi_j \frac{n_j!}{(n_j - r_j^{(\rho)})! \Omega^{r_j^{(\rho)}}}. \end{aligned} \tag{5}$$

The bare rates $k_\pm^{(\rho)}$ are the transition probabilities per unit time per unit volume. We define the bare rates as $k_+^{-(\rho)} \equiv k_-^{(\rho)}$ and $k_-^{-(\rho)} \equiv k_+^{(\rho)}$. Under equilibrium condition, distribution function should obey the grand canonical distribution, and the detailed balanced condition $P_{eq}(\mathbf{n}) w_+^{(\rho)}(\mathbf{n}) = P_{eq}(\mathbf{n}+\mathbf{v}^{(\rho)}) w_-^{(\rho)}(\mathbf{n}+\mathbf{v}^{(\rho)})$ should hold. Thus, the relation of energy changes under reactions is given by

$$\sum_j \varepsilon_j v_j^{(\rho)} = -\frac{1}{\beta} \ln \frac{k_+^{(\rho)}}{k_-^{(\rho)}}, \tag{6}$$

where $\beta \equiv 1/T$ is the inverse temperature. We suppose that a molecule of each compound has the energy ε_j. When the reaction ρ takes place, the state jumps from $\mathbf{n}$ to $\mathbf{n}+\mathbf{v}^{(\rho)}$ and the energy transfer arises between the system and bath.

STOCHASTIC TRAJECTORY FRAMWORK

The Gillespie Method

We do not intend to obtain the explicit solution of the chemical master equation at arbitrary time but consider the stochastic trajectories. Here, we consider the situation where the state of system changes during the times $[0,t]$. t is the final time of the dynamics under consideration, which we call the time-scale parameter since it determines the mean numbers of chemical reactions during $[0,t]$. The Gillespie method as a numerical simulation generates realizations of a stochastic trajectory χ directly[7]. This method allows us to obtain a equivalent result of the master equation Eq.(4) with Eq.(5)[8]. Let us consider a stochastic path, $\chi \equiv \mathbf{N}(0) \to \mathbf{N}(\tau_1) \to \cdots \to \mathbf{N}(t)$ corresponding to a forward successive reactions, each of which takes place randomly at $\tau_l (l = 1, \cdots, L)$ with different l. Here, $\mathbf{N}(\tau_l)$ is a stochastic vector at time τ_l and when the state is decided, it becomes $\mathbf{n}(\tau_l)$.

Then, let $P[\chi]$ be the probability of the trajectory in stochastic reaction systems

$$P[\chi] = P_{eq}(\mathbf{N}(0)) \Pi_{l=1}^{L} w^{(\rho_l)}(\mathbf{N}(\tau_l)) \times \exp\left[-\int_0^t \sum_{\rho=-M}^{M} w^{(\rho)}(\mathbf{N}(\tau)) d\tau \right], \tag{7}$$

given that the system starts at equilibrium state $\mathbf{N}(0)$. Here it is noted that Eq.(7) itself should be regarded as a stochastic variable, though it might look strange since it is quite contrary to mathematical usage of probability distributions. However, a probability distribution as a stochastic variable is required for the stochastic trajectory framework invented by Seifert[9], which will be introduced in the following sections. The stochastic trajectory framework is effective to see the validity of the FT within the framework of the Gillespie method.

Entropy Production

In Ref.[9], Seifert proposed a theoretical framework within which thermodynamic quantities can be well-defined even for a stochastic path χ. In this subsection, we implement a concise explanation of this formalism. First, we define the total amount of entropy generated from an initial state to a final state. We define Q as the heat released into the heat bath. We distinguish an entropy production of the system from the entropy production of the bath. The internal energy change of the system in a reaction event is given by $\Delta E^{(\rho)} = \sum_j \varepsilon_j \Delta n_j^{(\rho)} = \sum_j \varepsilon_j v_j^{(\rho)}$, where $\Delta n_j^{(\rho)}$ is the increment of the j-th species in a single reaction event ρ. We identify the first law of thermodynamics with $\Delta E^{(\rho)} = -Q^{(\rho)}$. Thus, the dissipative heat is given by $Q^{(\rho)} = \frac{1}{\beta} \ln(k_+^{(\rho)}/k_-^{(\rho)})$. For a finite time interval $[0,t]$, the dissipated heat as a stochastic-path function can be written as

$$Q[\chi] = \frac{1}{\beta} \int_0^t ds \sum_{l=1}^{L} \ln \frac{k_+^{(\rho_l)}}{k_-^{(\rho_l)}} \delta(s - \tau_l), \tag{8}$$

with ρ_l expressing the reaction events at times τ_l. When forward reaction or reverse reaction occurs, ρ_l becomes positive or negative, respectively. Then, "external" entropy production ΔS_e is

$$\Delta S_e[\chi] \equiv \frac{Q[\chi]}{T} = \int_0^t ds \sum_{l=1}^{L} \ln \frac{k_+^{(\rho_l)}}{k_-^{(\rho_l)}} \delta(s - \tau_l). \tag{9}$$

Next, we need to define the entropy production in the system. According to the framework of Seifert [9], the entropy of the system for a state $\mathbf{N}(t)$ at time t reached through a stochastic path χ reads

$$S_i[\chi]_t = -\ln P(\mathbf{N}(t),t) + \ln g(\mathbf{N}(t)), \qquad (10)$$

where $\ln g(\mathbf{N}(t))$ stems from the degeneracy[6]. The entropy production for a single stochastic trajectory is $\Delta S_i[\chi] = S_i[\chi]_t - S_i[\chi]_0$. This $P(\mathbf{N}(t),t)$ is a stochastic variable while its functional form is the same as the general solution of the master equation Eq.(4). The total amount of entropy generated from an initial state to a final state given by

$$\Delta S[\chi] = \Delta S_e[\chi] + \Delta S_i[\chi]. \qquad (11)$$

The total entropy production Eq.(11) is also the stochastic-path function, so that it can be negative for the time reversed path.

Detailed Fluctuation Theorem

From Eq.(7) we can derive

$$\frac{P[\chi]}{P[\chi^\dagger]} = e^{\Delta S[\chi]}, \qquad (12)$$

by computing the ratio of probability of trajectories χ to one of time-reversed trajectories $\chi^\dagger$. Here, $\Delta S[\chi]$ can be identified with Eq. (11). Since we regard $P[\chi]$ as a stochastic variable in this study, we can consider Eq.(12) corresponds to Eq.(2) $\langle A[\chi]\rangle = \langle A[\chi^\dagger]e^{-\Delta S}\rangle$; If we put $P[\chi] \equiv \delta(\Delta S - \Delta S[\chi])$ and take the average of both sides of Eq.(12) over all realizations, we can obtain the FT, Eq.(1),

$$\frac{P(\Delta S)}{P(-\Delta S)} = e^{\Delta S}, \qquad (13)$$

where $P(\pm\Delta S)$ are the probability distributions of positive entropy production ΔS the negative entropy production $-\Delta S$ respectively. This relation is the detailed fluctuation theorem. It is called from including the meaning of the FT and statement of detailed balance condition[10].

Stationary State Ansatz

Here, we restrict ourselves to consider the isomerization-type reaction between X and Y only:

$$X \underset{k_-}{\overset{k_+}{\rightleftarrows}} Y. \qquad (14)$$

In the stochastic trajectory framework, the entropy production are defined as (9) and (10). To obtain an explicit expression of Eq.(10), the solution $P(\mathbf{N}(t),t)$ of the master equation is needed. This solution cannot be obtained in general, however. Instead, we assume as a first approximation the stationary solution for $P(\mathbf{N}(t),t)$ of Eq.(10)[11]. The entropy production in our model is then given by

$$\begin{aligned}\Delta S = & \int_0^t ds \sum_{l=1}^{L} \ln \frac{k_+^{(\rho_l)}}{k_-^{(\rho_l)}} \delta(s-\tau_l) \\ & + (N_Y(0) - N_Y(t)) \ln \frac{k_+}{k_-},\end{aligned} \qquad (15)$$

where $N_Y(t)$ is the number of Y molecules at time t that is a stochastic variable, the first term of r.h.s is the external entropy production and the second term is the internal entropy production. This expression of the entropy production is also a stochastic variable, and we verify the FT by generating N_Y numerically with the aid of the Gillespie method.

RESULTS AND DISCUSSION

As is shown in Figure 1 and 2, we obtain a result that the validity of the FT depends on the time-scale of the dynamics, which is determined by parameters, the time-scale parameter t and the rate constant k. Here the time-scale parameter t represents the time when we follow the dynamics of the system and also controls the number of reactions within the duration $[0,t]$. If the time-scale parameter is large, the number of reactions is also large. The rate constant is related to the relaxation-time of the system to a stationary state. If the rate constant is large, the system quickly relaxes to the final stationary state. From Table 1, it is necessary to choose appropriate time-scale parameters and rate constants for the validity of the FT. There is the case where the FT does not hold. It has been expected that the FT has formal validity for arbitrary time-scale. The violation of the FT shown in the present study could be attributed to the assumption that the final state is a stationary state which was made in the previous subsection. We assume the stationary state solution of the master equation for the stochastic probability $P(\mathbf{N},t)$ to obtain the explicit expression of the system entropy production, Eq.(10). This assumption may cause the violation of the FT in the transient state. If we use instead the full solution of the master equation, we would obtain a different expression for the system entropy. It is quite difficult, however, to obtain the full solution of the master equation even for a simple case considered here. Rather, we can conclude that the present results show that the possibility of violation of the FT is far limited even if we use the strong assumption that the stationary state is always reached.

Additionally, we compare the situations in which the numbers of reaction molecules are different, while the

TABLE 1. The validity of the FT

	Time scale large	Time scale small
rate constant small	×	○
rate constant large	○	○

time-scale parameter and the rate constants are appropriately chosen, as is shown in Figure 3. The inset shows the probability distribution of the entropy production. As the number of particles increases gradatim, negative entropy production can be neglected. From these results, we expect that the FT can be valid even for small systems. In other words the validity of the FT does not depend on the number of molecules. The extension of the present results for more general case such as temperature-dependent rate constants will be reported elsewhere.

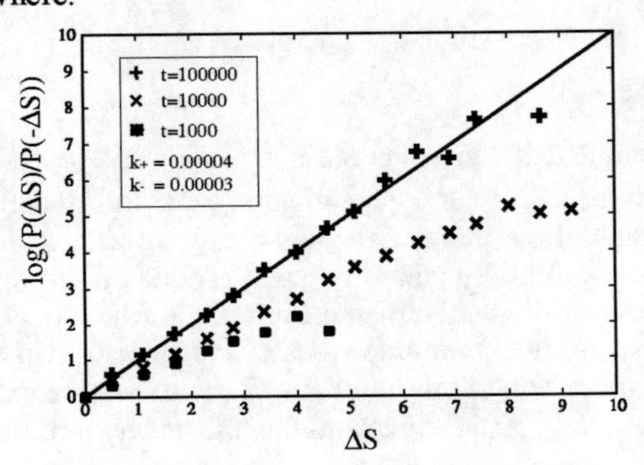

FIGURE 1. Logarithm of the ratio of probability with entropy production $-\Delta S$ to one with entropy production ΔS versus ΔS. The validity of the FT is dependent on the time scale. Here, k is the rate constant and t is the time scale parameter. A total number of molecules is 100.

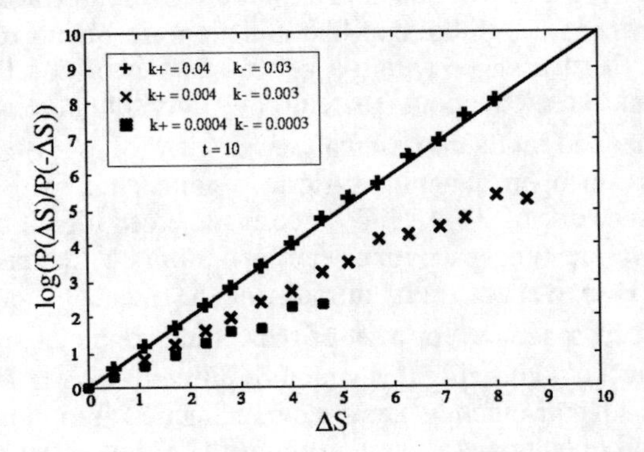

FIGURE 2. Logarithm of the ratio of probability with entropy production $-\Delta S$ to one with entropy production ΔS versus ΔS. The validity of the FT is dependent on the rate constant. Here, k is the rate constant and t is the time scale parameter. A total number of molecules is 100.

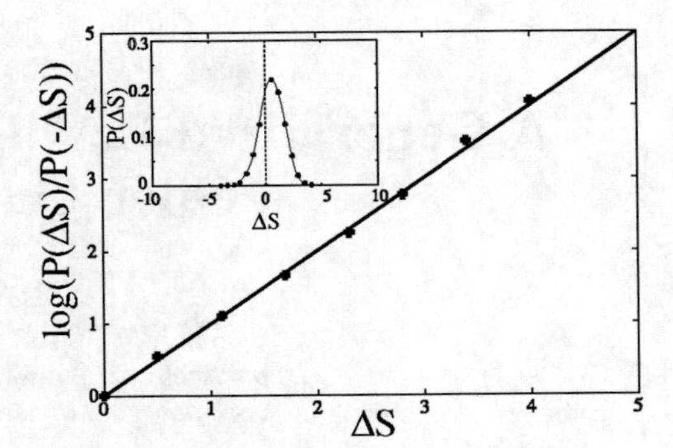

FIGURE 3. The total number of particle is 14. The rate constant and time scale parameter is $k_+ = 0.004$, $k_- = 0.003$ and 500. The inset shows probability distribution of entropy production.

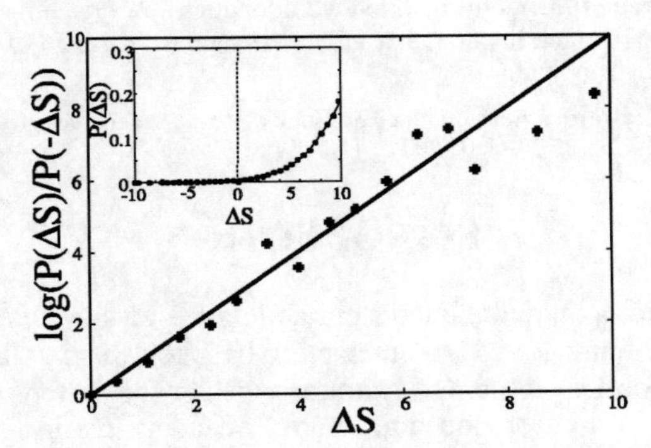

FIGURE 4. The total number of particle is 600. The rate constant and time scale parameter is $k_+ = 0.004$, $k_- = 0.003$ and 500. The inset shows probability distribution of entropy production.

REFERENCES

1. D. T. Gillespie, *J. Chem. Phys.* **113**, 297-306 (2000).
2. D. J. Evans, E. G. D. Cohen and G. P. Morriss, *Phys. Rev. Lett.* **71**, 2401-2404 (1993).
3. G. E. Crooks, *Phys. Rev. E.* **61**, 2361-2366 (2000).
4. N. G. van Kampen, *Stochastic processes in physics and chemistry*,Elsevier Science Publishers, Amsterdam,1992,pp.180-193.
5. T. Shibata, cond-mat/0012404, 1-4(2000).
6. T. Schmiedl and U. Seifert, *J. Chem. Phys.* **126**, 044101-1-044101-12 (2007).
7. D. T. Gillespie, *J. Comp. Phys.* **22**, 403-434 (1976).
8. L. Vereecken, G. Huyberechts, and J. Peeters, *J. Chem. Phys.* **106** 6564-6573 (1997).
9. U. Seifert, *Phys. Rev. Lett.* **95**, 040602-1-040602-4 (2005).
10. C. Jarzynski, *J. Stat. Phys.* **98**, 77-102(2000).
11. C. W. Gardiner, *Handbook of Stochastic Methods: for Physics, Chemistry and the Natural Sciences 3rd. ed.*, Springer-Verlag, Berlin, 2004,pp.235-241.

A Generalized Entropy-Hamiltonian Relation in Open Quantum Systems

Hiroki Majima

Department of General Education, Salesian Polytechnic,
4-6-8, Oyamagaoka, Machida-city, Tokyo, 194-0215 JAPAN

Abstract. The state of a system changes as a consequence of its internal dynamics and of the interaction with the environment that leads to a system-environment correlation and cannot be represented in terms of unitary Hamiltonian dynamics of the system. In order to investigate the role of interaction between the system and its environment, we derive a general formula which relates the Hamiltonian of the system to entropy. This formula enables us to discriminate a type of interactions which causes the change in the state of the system in a spontaneous process. In other words, this formula enables us to know the type of interactions which would induce a new equilibrium state of the system in the process. It is found that the certain type of interactions does not affect the state (dynamics) of the system even if there are interactions between the system and its environment.

Keywords: Entropy, Open quantum systems, Second law, Thermo field dynamics
PACS: 05.30.-d, 05.70.Ln, 11.10.Wx

INTRODUCTION

In an open quantum system comprised of a system, its environment and the interaction between them, it is an important problem to examine whether the system is in equilibrium or nonequilibrium. We adopt thermo field dynamics (TFD) [1] as a means for treating open quantum systems [2] and derive a formula which connects an entropy of a system and a Hamiltonian of an open quantum system [3]. In this paper, we develop a theory that connects the entropy of the system to the Hamiltonian for an open quantum system, where the interaction between the system and its environment plays an essential role in determining whether the system is in equilibrium or non-equilibrium. In the formulation of the problem, we shall utilize the theoretical method developed in TFD, where the theoretical structure of TFD (namely, doubling the degrees of freedom by introducing the tilde Hilbert space $\widetilde{\mathscr{H}}$) plays the essential role for the formulation.

We shall show how the entropy of the system is connected to the microscopic description of the interaction between the system and its environment in the representation space in TFD. Once the entropy of the system is expressed in $\mathscr{H}$ and is connected to the system-environment interaction expressed in the representation space $\mathscr{H} \otimes \widetilde{\mathscr{H}}$, we can examine the behavior of the entropy from the form of the interaction Hamiltonian $H_{\text{sys-environ}}$. In other words, we can study the behavior of the entropy of the system from the interaction between the system represented on the space $\mathscr{H}$ and its environment represented on the space $\widetilde{\mathscr{H}}$ within the framework of TFD. By studying the interactions between the system and its environment mapped on the representation space $\mathscr{H} \otimes \widetilde{\mathscr{H}}$, we can gain some information on the type of the interactions that changes the system into a non-equilibrium state in a spontaneous or irreversible processes. We shall obtain the criteria for the types of interactions that determine whether the system is in equilibrium or non-equilibrium according to the second law of thermodynamics by examining the behavior of the entropy.

ENTROPY-HAMILTONIAN RELATION

In this section, the equation of motion for entropy and the generalized Schrödinger-like equation are obtained for the thermal states within a general framework of TFD. From these equations, the entropy-Hamiltonian relation is derived on the representation space $\mathscr{H} \otimes \widetilde{\mathscr{H}}$. We discuss an open quantum system in connection with the framework of TFD. TFD is constructed by doubling a representation space for a set of operators $\{A\}$ defined in the Hilbert space $\mathscr{H}$ by introducing the tilde conjugation Hilbert space $\widetilde{\mathscr{H}}$ for a set of those tilde conjugate operators $\{\widetilde{A}\}$. Utilizing this structure in the theory of TFD, we can conveniently express the interaction Hamiltonian of an open quantum system in terms of those operators on $\mathscr{H}$ and on $\widetilde{\mathscr{H}}$ in the representation space $\mathscr{H} \otimes \widetilde{\mathscr{H}}$ as will be shown shortly. Those operators $\{A\}$ on $\mathscr{H}$ and $\{\widetilde{A}\}$ on $\widetilde{\mathscr{H}}$ are assumed to be independent each other: $[A, \widetilde{B}] = 0$, where [,] denotes usual commutation (anticommutation) rules for bosons (fermions). Each opera-

CP982, *Complex Systems, 5th International Workshop on Complex Systems*
edited by M. Tokuyama, I. Oppenheim, and H. Nishiyama

tor in $\{\widetilde{A}\}$ satisfies the tilde conjugation rules. Thermal state $|\Psi(t)\rangle$ defined in TFD is expressed by

$$|\Psi(t)\rangle = \rho^{1/2}(t)|I\rangle, \tag{1}$$

where $\rho(t)$ is a density operator (matrix) describing a system and a state $|I\rangle$ is defined by $|I\rangle := \sum_n |n,\widetilde{n}\rangle$. By using the thermal state $|\Psi(t)\rangle$, we can obtain the mean value (finite temperature expectation value) of the microscopic description of physical quantity (denoted by A) at time t:

$$\langle A\rangle_t = \mathrm{Tr}\{\rho(t)A\} = \langle\Psi(t)|A|\Psi(t)\rangle. \tag{2}$$

Here and in the following, time dependence of the expectation values $\langle\cdots\rangle_t$ is indicated by a subscript t. Entropy $\langle S(t)\rangle_t$ may be expressed by the density operator $\rho(t)$:

$$\begin{aligned}\langle S(t)\rangle_t &= -\mathrm{Tr}\{\rho(t)\ln\rho(t)\} \\ &= \mathrm{Tr}\{\rho(t)(-\ln\rho(t))\} = \langle -\ln\rho(t)\rangle_t. \end{aligned} \tag{3}$$

It should be noted that we have set the Boltzmann constant k_B equal to 1. Eq. (3) defines entropy for nonequilibrium processes.

In what follows, it will be convenient to consider a dynamical variable

$$S(t) = -\ln\rho(t). \tag{4}$$

as an *entropy operator* since its mean value $\langle S(t)\rangle_t$ coincides with the nonequilibrium thermodynamic (von Neumann) entropy $\langle S(t)\rangle_t$ given by Eq. (3). Noticing that the density operator $\rho(t)$ can be written in terms of the entropy operator $S(t)$ as

$$\rho(t) = e^{-S(t)}, \tag{5}$$

the thermal state $|\Psi(t)\rangle$ in Eq. (1) can be expressed in terms of the entropy operator as

$$|\Psi(t)\rangle = e^{-S(t)/2}|I\rangle. \tag{6}$$

The time derivative of Eq. (6) is given by

$$\begin{aligned}\frac{\mathrm{d}}{\mathrm{d}t}|\Psi(t)\rangle &= \frac{\mathrm{d}}{\mathrm{d}t}\left\{e^{-S(t)/2}|I\rangle\right\} \\ &= \left[-\frac{1}{2}\int_0^1 e^{-\lambda S(t)}\frac{\mathrm{d}S(t)}{\mathrm{d}t}e^{\lambda S(t)}\mathrm{d}\lambda\right]|\Psi(t)\rangle. \end{aligned} \tag{7}$$

Accordingly, if we assume $[S(t),\mathrm{d}S(t)/\mathrm{d}t] = 0$, we obtain the relation:

$$\frac{\mathrm{d}}{\mathrm{d}t}|\Psi(t)\rangle = -\frac{1}{2}\frac{\mathrm{d}S(t)}{\mathrm{d}t}|\Psi(t)\rangle. \tag{8}$$

According to quantum mechanics, the state vector $|\psi(t)\rangle$ evolves with time according to the Schrödinger equation,

$$i\frac{\mathrm{d}}{\mathrm{d}t}|\psi(t)\rangle = H(t)|\psi(t)\rangle, \tag{9}$$

where $H(t)$ is the Hamiltonian of the system and Planck's constant $\hbar$ has been set equal to 1. If the dynamics of the system can be formulated in terms of a possibly time-dependent Hamiltonian $H(t)$, the system will be said to be *closed*, while we reserve the term *isolated* to mean that the Hamiltonian of the system is time independent.

Evolution of the system at finite temperature can be described by a density operator (that will be denoted by $\rho(t)$). It is straightforward to derive an equation of motion for the density operator starting from the Schrödinger equation (9) and it is given by the quantum Liouville-von Neumann equation:

$$i\frac{\mathrm{d}}{\mathrm{d}t}\rho(t) = [H(t),\rho(t)], \tag{10}$$

where Planck's constant $\hbar$ has been set equal to 1 and $H(t)$ is given by the same Hamiltonian as in Eq. (9).

Now we consider the thermal state $|\Psi(t)\rangle$ defined in TFD. The thermal state evolves with time according to the Schrödinger-like equation [4]:

$$\frac{\mathrm{d}}{\mathrm{d}t}|\Psi(t)\rangle = -i\widehat{H}(t)|\Psi(t)\rangle, \tag{11}$$

where $\widehat{H}(t)$ $(:= H(t) - \widetilde{H}(t))$ is defined by the Hamiltonian of the system $H(t)$ on $\mathscr{H}$ and the tilde conjugation Hamiltonian $\widetilde{H}(t)$ (*i.e.*, the copy of the system $H(t)$) on $\widetilde{\mathscr{H}}$. It should be noted that according to the theory of TFD the representation space of $\widehat{H}$ in Eq. (11) is on $\mathscr{H}\otimes\widetilde{\mathscr{H}}$. In this section, we shall derive a general formula which relates entropy $\langle S(t)\rangle_t$ and the Hamiltonian $H(t)$ for an open quantum system. The general relation between the entropy operator $S(t)$ (defined by Eq. (4)) and the time dependent total Hamiltonian $\widehat{H}(t)\,(= H(t) - \widetilde{H}(t))$ is easily obtained from Eqs. (8) and (11) as

$$\frac{\mathrm{d}S(t)}{\mathrm{d}t} = 2i\widehat{H}(t) = 2i[H(t) - \widetilde{H}(t)]. \tag{12}$$

It should be noted that the time-dependent total Hamiltonian $\widehat{H}(t)$ on $\mathscr{H}\otimes\widetilde{\mathscr{H}}$ depends on the particular physical situation under study.

In order to treat an open system within the framework of TFD, it is worth reconsidering the physical foundation of TFD. Let us consider a system described by the Hamiltonian $H(t) = H_{\mathrm{sys}}(t) + H_{\mathrm{environ}}(t)$, where we assume $[H_{\mathrm{sys}}, H_{\mathrm{environ}}] = 0$ (*i.e.*, the system and its environment are uncorrelated). When this example system is in a thermal equilibrium state, the system is equilibrium with its environment at same temperature. In such a case, it is by no means to discriminate the state of the system thermodynamically from that of its environment, meaning that the environment in equilibrium is of deindividuation from the system in equilibrium. Therefore, we do not care about the specific form of the environment Hamiltonian [1] as far as those systems (the system of interest and its environment) are in the same thermal equilibrium state. Accordingly. the environment having the same thermodynamic property of the system can be regarded as the copy of the system of interest. Indeed, this allows us to assign the system of interest to the non-tilde space $\mathscr{H}$ and its environment to the tilde space $\widetilde{\mathscr{H}}$ in the representation space $\mathscr{H} \otimes \widetilde{\mathscr{H}}$ introduced in TFD. Regarding the environment as the copy of the system, we can make the transformation of the system under consideration onto the non-tilde Hilbert space $\mathscr{H}$ (*i.e.*, $H_{\rm sys} \mapsto H$) and the environment onto the tilde Hilbert space $\widetilde{\mathscr{H}}$ (*i.e.*, $H_{\rm environ} \mapsto \widetilde{H}$). It is important to notice that we do not know the effect of interactions between the system and its environment in equilibrium. It can be easily proved that $S(t)$ in Eq. (12) describes the entropy operator of the system in this case since $[H_{\rm sys}, H_{\rm environ}] = 0$. The above mapping (assumption) is valid when there is *no* interaction between the system and its environment. since the entropy of the system is then given by

$$\left\langle \frac{\mathrm{d}S(t)}{\mathrm{d}t} \right\rangle_t = 2i(\langle H(t)\rangle_t - \langle \widetilde{H}(t)\rangle_t) = 0, \qquad (13)$$

meaning that the system is in equilibrium with its environment and the entropy of the system does not change with time.

Next we consider how to generalize the previous case to an open quantum system described by the Hamiltonian $H_{\text{open sys}}(t) = H_{\rm sys}(t) + H_{\rm environ}(t) + H_{\rm I}(t)$ and make it clear for the applicability of the transformation ($H_{\rm sys} \mapsto H$ and $H_{\rm environ} \mapsto \widetilde{H}$ in the representation space $\mathscr{H} \otimes \widetilde{\mathscr{H}}$) described above to the open quantum system within the framework of TFD. Open quantum system consists of a system ($H_{\rm sys}$), its environment ($H_{\rm environ}$) and the interaction between them ($H_{\rm I}$). Let us now consider the transformation of such an open quantum system into the representation space in TFD. Mapping the open quantum system to the representation space in TFD can be done in the following way: It is natural to transform the system of interest to the non-tilde (original) space $\mathscr{H}$ and the environment to the tilde conjugate space $\widetilde{\mathscr{H}}$ where the environment can be regarded as the copy of the system. That is, the mapping $H_{\rm sys} \mapsto H$ in $\mathscr{H}$ and $H_{\rm environ} \mapsto \widetilde{H}$ in $\widetilde{\mathscr{H}}$ is a natural mapping in equilibrium when there exists no interaction between them. With this prescription, a natural transformation of an open quantum system within the theoretical structure of TFD may be done by the following way: the interaction Hamiltonian $H_{\rm I}$ is mapped to the representation space $\mathscr{H} \otimes \widetilde{\mathscr{H}}$ by the mapping $H_{\rm I} \mapsto H_{\text{tilde}-\text{nontilde}}$. By using the correspondence rules (mapping) for $H_{\rm sys}$ and $H_{\rm environ}$, *i.e.*, $H_{\rm sys} \mapsto H$ and $H_{\rm environ} \mapsto \widetilde{H}$, we can establish a mapping for an open quantum system with the help of the theoretical structure in TFD as will be shown below. Consider the case where there is some interaction between the system and its environment. For an open quantum system, the interaction between the system and its environment possibly plays a special role in the dynamics (or the spontaneous, irreversible thermodynamic processes) of the system. Obviously if there is some interaction between them, the system may be in non-equilibrium. Whether the system is equilibrium or non-equilibrium due to the interaction can be judged from the study of entropy $\langle S(t)\rangle_t$. Indeed, when the system evolves with time due to the presence of the interaction, how can we modify the theory developed above for an open quantum system? In the following, we assume the case where the system is very close to the equilibrium state of its environment (in other words, the interaction between the system and its environment is very weak). In such a case we could employ the mapping described above for the system and its environment and we generalize Eq. (12) for the case where there exists interactions between the system and its environments. In this case, the total Hamiltonian $\widehat{H}$ in Eq. (12) may be expressed by [2]

$$\widehat{H} \;\rightarrow\; \widehat{H}(t) = \widehat{H}_0 + \theta(t)\widehat{H}_{\rm I}(t), \qquad (14)$$

where $\widehat{H}_0$ denotes the same total Hamiltonian expressed by $\widehat{H}$ in Eq. (12) and $\theta(t)$ is the Heaviside step function. Here we assumed that the interaction (denoted by $\widehat{H}_{\rm I}$) is turned on at a time $t = 0$. From Eq. (11), thermal state $|\Psi(t)\rangle$ obeys the equation of motion:

$$\frac{\mathrm{d}}{\mathrm{d}t}|\Psi(t)\rangle = -i\widehat{H}(t)|\Psi(t)\rangle. \qquad (15)$$

Taking tilde conjugate of this equation, we obtain the equation of motion for the same thermal state $|\Psi(t)\rangle$:

$$\begin{aligned}\frac{\mathrm{d}}{\mathrm{d}t}|\Psi(t)\rangle &= i(\widehat{H}(t))^{\sim}|\Psi(t)\rangle \\ &= (-i\widehat{H}(t))^{\sim}|\Psi(t)\rangle. \qquad (16)\end{aligned}$$

Since the thermal state $|\Psi(t)\rangle$ defined in TFD has to obey the same equation of motion in $\mathscr{H}$ as in $\widetilde{\mathscr{H}}$, the generator $\widehat{H}$ for the time evolution of the system must

[1] The Hamiltonian of the environment $H_{\rm environ}$ may be expressed by that of simple harmonic oscillators or ideal gas or more complex composite particle.

[2] If the interaction is turned on adiabatically, the total Hamiltonian $\widehat{H}$ may be given by

$$\widehat{H}(t) = \widehat{H}_0 + e^{-\varepsilon|t|}\widehat{H}_{\rm I}(t).$$

satisfy the following condition:

$$(i\widehat{H})^{\sim} = i\widehat{H}. \tag{17}$$

In TFD, this condition is called the "tildian" relation. It should be noted that the generator $\widehat{H}$ describing the time evolution of the total system must be tildian and not Hermitian. Therefore the interaction Hamiltonian $\widehat{H}_I$ does not necessarily have the structure of $\widehat{H}_I = H_I - \widetilde{H}_I$ defined in TFD but rather have the structure $V^{\sim} = V$ if we write the interaction $\widehat{H}_I$ as a tildian by introducing iV for $\widehat{H}_I$. In physical applications one often encounters the situation that the system under consideration is driven by external forces. If in such a case the dynamics of the system can still be formulated in terms of a possibly time-dependent Hamiltonian $H(t)$ the system will be said to closed, while we reserve the term isolated to mean that the Hamiltonian of the system is time independent. For a time-dependent Hamiltonian $H(t)$ for an open quantum system, the time-evolution operator $U(t)$ may be represented as a time-ordered exponential,

$$U(t) = \mathrm{T}_{\leftarrow} \exp\left[-i\int_0^t d\tau H(\tau)\right], \tag{18}$$

where $T_{\leftarrow}$ denotes the chronological time-ordering operator which orders products of time-dependent operators such that their time-arguments increase from right to left as indicated by the arrow. The state of the system at time t is given by the density operator:

$$\rho(t) = U(t)\rho(0)U^{\dagger}(t). \tag{19}$$

Therefore, the entropy operator defined by $S = -\ln\rho$ also evolves with time as

$$S(t) = U(t)S(0)U^{\dagger}(t). \tag{20}$$

By using this time-dependent entropy operator $S(t)$, we can obtain the same expression as in Eq. (8) for the case where there exists some interactions between the system and its environments. Therefore, the entropy-Hamiltonian relation

$$\frac{\mathrm{d}S(t)}{\mathrm{d}t} = 2i\widehat{H}(t) \tag{21}$$

holds for a general case where the Hamiltonian includes the interaction Hamiltonian.

Taking the expectation value of Eq. (21) and noticing that $\langle\widehat{H}_0(t)\rangle_t = \langle H_0(t)\rangle - \langle\widetilde{H}_0(t)\rangle = 0$, we obtain the entropy-Hamiltonian relation for open quantum systems:

$$\left\langle\frac{\mathrm{d}S(t)}{\mathrm{d}t}\right\rangle_t = 2i\langle\widehat{H}(t)\rangle_t = 2i\langle\widehat{H}_I(t)\rangle_t, \qquad (t>0) \tag{22}$$

where $\langle A\rangle_t$ expresses the expectation value $\langle\Psi(t)|A|\Psi(t)\rangle$ at time t. It should be noted that this interaction Hamiltonian $\widehat{H}_I(t)$ can be represented by the transformation $\widehat{H}_I \mapsto H_{\text{tilde-nontilde}}$ in the representation space $\mathscr{H}\otimes\widetilde{\mathscr{H}}$ of TFD as expressed by tildian comprised of non-tilde and tilde operators.

Hereafter we assume $\left\langle\frac{\mathrm{d}S(t)}{\mathrm{d}t}\right\rangle_t \equiv \frac{\mathrm{d}\langle S(t)\rangle_t}{\mathrm{d}t}$. Eq. (22) expresses an entropy production rate for open quantum systems in the spontaneous or irreversible process and relates the entropy for the system to a possible interaction described by the microscopic interaction Hamiltonian $\widehat{H}_I$ expressed by a tildian interaction operator iV. The general formula (21) will be used to investigate the effect of interaction directly on the system in the spontaneous or irreversible process for open quantum systems by giving the explicit form of interactions expressed by non-tilde and tilde operators.

SUMMARY AND CONCLUDING REMARKS

We studied the role of interactions between the system and its environment in open quantum systems. We derived the entropy-Hamiltonian relation which connects Hamiltonian of the total system and the entropy of the system in a open quantum system. We obtained the criterion for the type of interactions between the system and its environment, that discriminates whether the system is equilibrium or non-equilibrium according to the second law of thermodynamics by the entropy-Hamiltonian relation (21). This result also shows that thermal interaction of a system and its environment (*i.e.*, heat flow) could be expressed in terms of a microscopic quantum interaction.

ACKNOWLEDGMENTS

We would like to thank Prof. A. Suzuki for helpful discussion.

REFERENCES

1. H. Umezawa, H. Matsumoto and M. Tachiki, *Thermo Field Dynamics and Condensed States*, Elsevier Science, North-Holland, 1982.
2. H. Majima and A. Suzuki, *Nuovo Cimento*, **121**, 57-64 (2006).
3. H. Majima and A. Suzuki, *J. Kor. Phys. Soc.*, **46**, 678-683 (2005).
4. M. Suzuki, *J. Phys. Soc. Jpn.* **54**, 4483-4485 (1985).

Interaction of Adjacent Amino Acids

Sheh-Yi Sheu[1] and Dah-Yen Yang[2]

[1] *Faculty of Life Sciences, National Yang-Ming University, Taipei 112, Taiwan*
[2] *Institute of Atomic and Molecular Science, Academia Sinica, Taipei 106, Taiwan*

Abstract. Ramachandran plots display the dihedral angles of a single protein residue. We here propose a crossed torsion angle plot called SSY-plot between two neighboring amino acids and demonstrate that a special coherence motion can exist between some very special amino acid pairs leading to spontaneous unusual structures. We also suggest that the existence of two domains corresponds to a bifurcation between two different protein structures and that the special pair is the key to producing these two structures. These are two different structures and are produced spontaneously without an external agent.

Keywords: Alpha helix, Beta sheet, Ramachandran plot
PACS: 87.64.Aa

INTRODUCTION

The structure motion of proteins is a field of great interest, leading to protein folding, but also to aberrant structure behavior leading to dozens of grave human diseases such as prion etc. In such case we typically supposed that the errant behavior is due to misfolding of the protein without the action external agents [1] – although propagation and may perhaps be infected by other aberrant proteins.

One of the common methods for labeling the structure behavior of individual amino acids in a protein chain has been the use of the Ramanchandran plot (R-plot)[2] which identifies the characteristic dihedral angles for each site in the protein. These angles are typically derived from the crystal structures. It should be emphasized however that this R-plots are not constants for the protein but rather evolve in time.

In the present work, we want to investigate the time development of structural features of pairs of residues of the protein. The ideal is that such structural interactions, if they occur, should happen in the early time history of the folding process. The final crystal structure may not show such an interaction readily. If for example we choose a short peptide 6-mer with sequence, FSRSDE, for a studied structure and focus on a central site like Arg_3, we notice that the dihedral angles of the R-plot cover a wider area than one would surmise from the equilibrium structures (see Figure. 2).

These R-plots relate only to a single amino acid and only indirectly provide information about the next nearest neighbors. In case we have interaction between neighbors we might expect that angles of neighboring residues might be correlated. To test for such cases we propose to plot the dihedral angles of contacting pairs. We refer to this as SSY-plot. In particular for a given pair, two dihedral angles ψ_i and ϕ_{i+1} will point toward each other called an endo form and two dihedral angles ϕ_i and ψ_{i+1} called an exo form (Figure. 1).

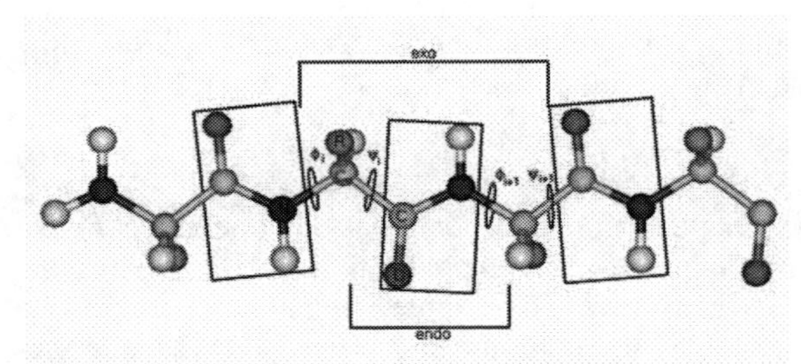

FIGURE 1. Endo and exo structures of polypeptide chain. Endo-plot is defined by (ψ_i, ϕ_{i+1}) torsion angles. The other pair of (ϕ_i, ψ_{i+1}) is denoted as exo-plot.

CP982, *Complex Systems, 5th International Workshop on Complex Systems*
edited by M. Tokuyama, I. Oppenheim, and H. Nishiyama

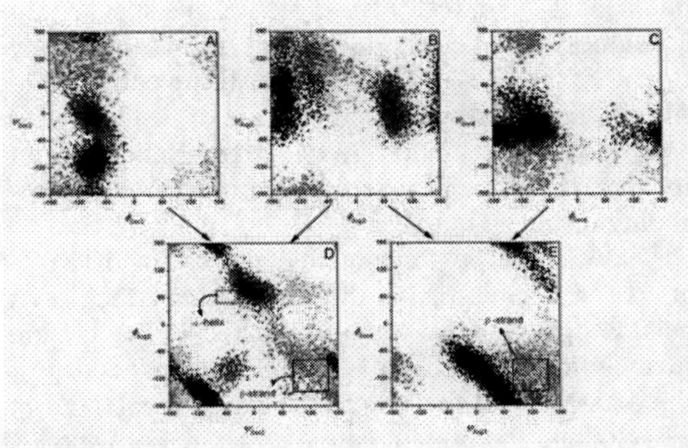

FIGURE 2. The Ramachandran plot and SSY-plot for 6-mer within 1 ns at 1000 K. The dots in red, green, blue, and black denote the simulation times 0-25 ps, 25.02-50 ps, 50.02-550 ps, and 550.02-1060 ps, respectively. (A-C): show R-plots of Ser_2, Arg_3 and Ser_4 in 6-mer, separately. (D-E): show SSY-plots of Ser_2-Arg_3 and Arg_3-Ser_4 pairs in 6-mer. {(A) R-plot at Ser_2. At the early stage, there is only one red domain, then the red domain splits into two green domains and finally there are three black domains. Each domain in R-plot is an entropy trap. To overcome the entropy barrier, the peptide requires energy. Since there is a large side chain hindrance between Ser_2-Arg_3, a communication between these two amino acids to allow the coherent changing their torsion angles assists the changing domains inside R-plot. (B) R-plot at Arg_3. Corresponding to (A), the red domain in (B) also split in the same manner as previous figure. This already shows a coherence or correlation motion of two nearby amino acids. (C) R-plot at Ser_4. Basically there is only one domain due to the periodic behavior of the phase space of R-plot. Clearly there is almost none coherence motion. This Ser_4 R-plot is very different from Ser_2 R-plot. Certainly this is caused by the protein sequence. (D) Ser_2 and Arg_3 SSY-plot or endo-plot of 6-mer. This figure just simply plots the correlation relationship in figure (B). The α-helix domain and the β-strand domain in SSY-plots are marked by squares. In (E), the SSY plot of Arg_3-Ser_4. There is only one domain in this figure and therefore no correlation relationship exists. This clearly shows that the pair correlation merely exists for special pairs.}

METHODS

Model and Simulation Procedures

In this work, the sequence of 6-mer peptide FSRSDE was extracted from a designed protein BBA1 (code 1HCW) with 23 residues. During the simulation the initial structure of the 6-mer peptide chain was generated in an extended form. We performed the molecular dynamics simulations and energy minimization using the program CHARMM [3] and the CHARMM 22 force field [4]. In this simulation, the protonation states of ionizable residues were set for a pH of 7. In addition, both N and C termini were considered ionized. Such peptide was first executed by minimization of the 1000 steps for the steepest descent algorithm. In the following step, the periodic boundary condition was used and the minimized structure was orient in the central box of ~30 x 30 x 30 $Å^3$. There were no counter-ions added to the system. Typically all of the temperature of the system was set at 1000 K. One of the constrain we used is SHAKE algorithm [5] which was applied to fix the bond lengths involving the hydrogen atoms. All of our MD simulations were adopted with time step 2 fs and a force-switched non-bond cutoff of 13.0 Å was used for the trajectory production. To save the simulation time, we update the non-bond pair list every 50 steps. In the final stage of our simulation, we saved the dynamic trajectories every 10 fs for early 50 ps and 100 fs for late 50 ps and used to perform the R plot and SSY plot. Since our observation of the coherence motion in R-plot corresponding to a long time scale protein folding process, our MD simulation was up to 100 ns.

Analyses of MD Trajectories

As is well known, the secondary structure of the residue can be described by a Ramachandran-plot. Here the R-plot was used to measure ϕ_i and ψ_i dihedral angle (in degrees) for the backbone atoms C-N-C_α-C and N-C_α-C-N, respectively, at residue *i*. The method we choose will be related to the analysis of the time development of the dihedral angles of the protein as often reflected in the R-plot. Such R-plots reflect different structural regions for the protein and clearly distinguishes between regions of the β sheet and region of the α helix.

SSY-Plot

Here we want to introduce another scheme for structural analysis of the protein which emphasizes the possible interaction between neighboring amino acids—the pair-wise effects, if any. If we consider the dihedral angles of two adjacent proteins, we find that two angles face each other in an endo form and two angles diverge in an exo form (see Fig. 1). We will now generate plots based on the endo dihedral angles of the pair. This would show if a specific angle of the first amino acid directly correlates with the angle of the second amino acid, or if these are free to choose any arbitrary angle. In contrast to the R-plot, we will use these SSY-plots to show possible correlated motions between neighboring amino acids. If there are sides chain interactions these might show up in such a SSY-plot.

The SSY-plot was used to measure the cross correlation of ψ_i and ϕ_{i+1} dihedral angle between neighboring residues i and $i+1$. It can be understood as a propagation of interaction between two nearby amino acids. This kind of correlation occurred for special amino acid pair due to the van der Waal interaction and hydrogen bonding interaction of their side chain groups.

In our work, it is interesting to show that the coherent structure between the Ser_2-Arg_3 pair have complementary dihedral angles – suggesting a local symmetry of the dots inside the SSY-plot. The time evolutions of the torsion angles in the plots were plotted in different colored dots. Furthermore, it should be noted that in the SSY-plot the top left structure is in the region of α-helices whereas the bottom right structure is in the region of β-strands. Hence we have shown that these are coherent motions of the Ser_2-Arg_3 pair that lead to two distinct structures. Our observation of the coherence motion in a short peptide 6-mer containing the Ser_2-Arg_3 pair determines the bifurcation of protein structure, and hence, decides the final folded state on protein folding.

DISCUSSION

The proposal has been made here that if certain short sequences are contained in a protein, then it has a strong tendency to a misfolding structure. As an initial example we will take the short sequence FSRSDE cut from BBA. We do these studies at elevated temperatures to enable the results to show up with a reasonable amount of computer time. One run was performed at room temperature for many months, to show the equivalence of the results. In Fig. 2, we display the SSY-plot between Ser_2 and Arg_3 as a function of time. We now see regions of interaction and hence a correlation in the resulting SSY-plot. Interestingly the movement between adjacent residues for these special pairs is not completely independent, but the interactions do occur as shown in the SSY-plot. If instead we now focus on the Ser_2-Arg_3 pair in this sequence, the SSY-plot shows something quite different. In Fig 1, at early times it is again completely benign---but after some 20 ps, we find the surprising result that the SSY-plot bifurcates into two active regions demonstrating that the structure oscillates between two states, where the second structure now has substantial importance. Hence for this simulation we have found the interesting result that a single structure at a single pair of adjacent amino acid residues, after some induction time bifurcates into two structures spontaneously. Furthermore, these structures occur at complimentary angles---hence local mirror symmetry. Hence two equivalent structures spontaneously manifest themselves for one special pair of residues in one peptide sequence without any external influences. This is most revealing and confirms the hypothesis that one chain of residues with special pairs can, under given circumstances, after some time, produce two definitely different structures without outside action.

Now with the fundamental definition of SSY-plot, we can investigate the pair wise effect of a special sequence pair. First let us recall the torsion angles motion inside R-plot. In our previous work [6], we understood that there exists an entropy potential in the R-plot and the protein native structure corresponds to certain domains. The special pair is Ser_2-Arg_3-Ser_4 of 6-mer. Their R-plots are shown in Figs. 2(A-C). In Fig. 2A, the R-plot at Ser_2 shows that at the early stage, there is only one red domain, then the red domain splits into two green domains and finally there are three black domains. One of the interesting features is that the domains for Ser_2 are vertical to the ϕ-axis. But the domains in Arg_3 are horizontal to the ϕ-axis. Each domain in R-plot is an entropy trap. To overcome the entropy barrier, the peptide requires energy. Since there is a large side chain hindrance between Ser_2-Arg_3, a communication between these two amino acids to allow the coherent changing their torsion angles assists the changing domains inside R-plot. Fiigure 2B shows that the R-plot at Arg_3, similar to Fig. 2A, the red domain in Fig. 2B also split in the same manner as Fig. 2A. This already shows a coherence or correlation motion of two nearby amino acids. It is clear that this is due to the coherence motion between Ser_2-Arg_3 pair. In other words, once Ser_2 domain in its R-plot is switched from the upper part to the lower part, the accompanied left domain in the R-plot of Arg_3 is also shifted to its right hand part. The R-plot at Ser_4 shows that basically there is only one domain due to the periodic behavior of the phase space of R-plot. Clearly here there is almost no coherence motion. This Ser_4 R-plot is very different from Ser_2 R-plot. Certainly this is caused by the protein sequence. In Fig. 2D, the SSY-plot of the Ser_2-Arg_3 pair of 6-mer just simply plots the correlation relationship in Figs. 2A and 2B. The α-helix domain and the β-strand domain in SSY-plots are marked by squares. In Fig. 2E, the SSY plot of Arg_3-Ser_4 shows that there is only one domain in this figure and therefore no correlation relationship exists. This clearly shows that the pair correlation only exists for special pairs. Once we plot their SSY-plot in Fig. 2D, we see two domains diagonal in the endo form. This kind of coherence motion is unusual, because this phenomenon does not show up in the next pair Arg_3-Ser_4 (see Fig. 2E) or other pairs. Recall that R-plot is a phase diagram, after shifting the angles in R-plot, in the SSY-plot of Arg_3-Ser_4 pair, we only observed one domain. This confirmed the requirement of special pair for observing two domains in SSY-plot.

Next the standard α-helices and β-strands domains inside SSY-plot are indicated in Figs. 2D and 2E. According to the dot points in Fig. 2, we plot the peptide three dimensional and ribbon structures corresponding to the dot inside SSY-plot in Figure. 3.

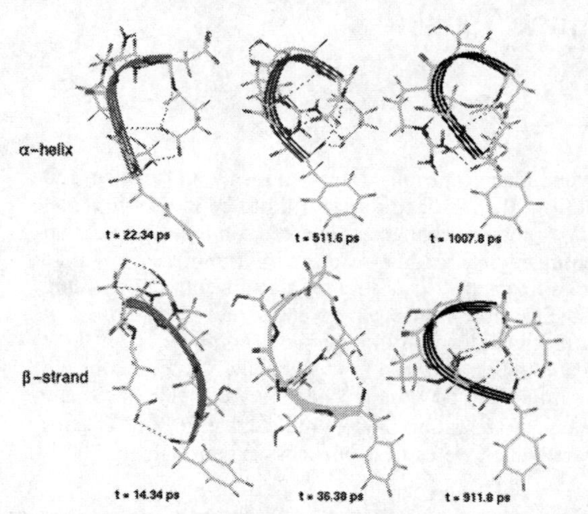

FIGURE 3. The α-helix and β-strands structures of 6-mer corresponding to the dot of SSY-plots shown in Fig 3(D). The ribbon colors just consistent with their dot colors, i.e. evolution time in our MD simulation.

CONCLUSION

We here suggest a new type of correlation plot for neighboring amino acids in a protein which uniquely demonstrate pair-wise interaction between sites as they evolve in time. Interestingly these interactions are not present at the start nor are they seen in final equilibrium structure. They clearly show the evolution of two distinct structures of the time evolution of the protein. This also uniquely demonstrates the pair-wise interaction of Ser_2 with Arg_3, as no other pair in this chain reveals a bifurcated SSY-plot with more than one structure.

In short these pair correlation plots form an interesting new approach to the discovery of pairwise interactions in proteins and the spontaneous bifurcation of structures in the evolution of protein folding.

ACKNOWLEDGMENTS

This work was supported by the Volkswagen Foundation and the Taiwan/Germany program at the NSC/Deutscher Akademischer Austauschdienst. SYS and DYY gratefully thank the Grant Nos NSC-94-2113-M-010-001 and NSC-94-2113-M-001-043, respectively. The authors would like to thank the National Center for High-Performance Computing of Taiwan. This work was supported by a gratious availability of computer time at the Pacific Northwest Laboratory of the Department of Energy.

REFERENCES

1. S. B. Prusiner, Human prion diseases and neurodegeneration. *Curr Top Microbiol Immunol* 207, 1-17 (1996); Molecular biology and genetics of prion diseases. *Cold Spring Harb Symp Quant Biol* 61, 473-493 (1996); Molecular biology and pathogenesis of prion diseases. *Trends Biochem Sci* 21, 482-487 (1996); Some speculations about prions, amyloid, and Alzheimer's disease. *N Engl J Med* 310, 661-663 (1984).
2. G. N. Ramachandran and V. Sasisekharan, *Adv. Prot. Chem.* 28, 283-437 (1968).
3. B. R. Brooks, R. E. Bruccoleri, B. D. Olafson, D. J. States, S. Swaminarhan, and M. Karplus, *J. Comput. Chem.* 4, 187 -217 (1983).
4. A. D. Jr. MacKerell, D. Bashford, M. Bellott, R. L. Jr. Dunbrack, J. D. Evanseck, M. J. Field, S. Fischer, J. Gao, H. Guo, S. Ha, D. Joseph-McCarthy, L. Kuchnir, K. Kuczera, F. T. K. Lau, C. Mattos, S. Michnick, T. Ngo, D. T. Nguyen, B. Prodhom, W. E. Reiher, III B. Roux, M. Schlenkrich, J. C. Smith, R. Stote, J. Straub, M. Watanabe, J. Wiorkiewicz-Kuczera, D. Yin and M. Karplus, *Phys. Chem. B* 102, 3586-3616 (1998).
5. J. P. Ryckaert, G. Ciccotti, and H. J. C. Berendsen, *J. Comput. Phys.* 23, 327-341 (1977).
6. E. W. Schlag, Sheh-Yi Sheu, Dah-Yen Yang, H. L. Selzle, and S. H. Lin, *J. Phys. Chem. B* 104, 7790-7794 (2000).

Violation of the Incompressibility of Liquid by Simple Shear Flow

Akira Furukawa and Hajime Tanaka

Institute of Industrial Science, University of Tokyo, 4-6-1 Komaba, Tokyo 153-8505, Japan

Abstract. Fluid dynamics, which describes the flow of gas and liquid, has contributed tremendously to science and technology. If we can safely ignore the density change associated with flow, then we can regard fluid to be incompressible. For simple shear flow, for example, it has been established that there is no pressure change associated with flow and thus no violation of the incompressibility. This is because the flow does not accompany any volume deformation (no pressure change due to viscous stress) and inertia effects can be neglected (no inertial pressure drop). According to this conventional wisdom, any flow-induced instability such as cavitation are unexpected for simple shear flow. However, if we take into account the fact that the viscosity is a function of the density, this scenario is drastically changed. Contrary to the above common belief, here we demonstrate that the incompressibility condition can be violated by a coupling between flow and density fluctuations via the density dependence of viscosity η even for simple shear flow and a liquid can become mechanically unstable above the critical shear rate, $\dot{\gamma}_c = (\partial\eta/\partial p)_T^{-1}$, where p is the pressure and T is the temperature. Our model predicts that for very viscous liquids this shear-induced instability should occur at a moderate shear rate, which we can easily access experimentally.

INTRODUCTION

We often experience that bubbles vigorously spout out while pouring or shaking a bottle of champagne or bear. When on a boat, we also see that enormous gas bubbles are formed around a rotating screw. These phenomena of flow-induced cavitation have been qualitatively explained by the Bernoulli's theorem as follows: When a pressure drop due to high-speed flow makes the pressure in a liquid lower than its vapour pressure, cavitation occurs. This scenario of inertial pressure drop and the resulting cavitation works for flow at rather high Reynolds number Re.

Even for very low Re flow, flow-induced instability of liquids, which appears as either cavitation as the above examples or shear banding (or shear localization), is quite widely observed in various materials: Very viscous liquids such as glassy materials often exhibit shear banding under flow (flow-induced instability of material) [1]. Liquid lubricants cavitate under high pressures and high shear stresses. Bair *et al.* [2, 3, 4] revealed by systematic experimental studies combining rheological measurements and direct microscopy observations that the shear-induced cavitation occurs accompanying marked shear-thinning behaviour. In low molecular weight unentangled polymer melts, cavity formation under simple shear were also observed [5]. It was shown that liquids just before cavitation behave as (linear) Newtonian fluids. This indicates that this phenomenon is not due to constitutive (viscoelastic) instability. Thus, a new scenario of liquid instability under simple shear is necessary to account for these phenomena, which are of significant fundamental and practical importance.

Previously, a cavitation criterion for sheared fluids was proposed by several researchers independently [3, 6]. According to their criterion, cavitation occurs if the maximum principal stress created by shear deformation, which is estimated as $\sim (\eta\dot{\gamma} - p)$, becomes positive. Their analyses are based on the standard Navier-Stokes equation of incompressible fluids in a steady state. In this framework, however, simple shear flow does not accompany any volume change (density fluctuations); in other words, a shear mode never couples with a longitudinal one. Thus, although their explanation is intuitively appealing, we cannot expect instability of liquid by simple shear. Here, we focus on a fundamental question of how simple shear deformation can be coupled with density fluctuations. We will propose a new mechanism of flow-induced instability of liquid matter.

A MECHANISM OF SHEAR-INDUCED INSTABILITY

The dynamics of compressible fluids can be described by the following hydrodynamic equations [7]. The density

CP982, *Complex Systems, 5th International Workshop on Complex Systems*
edited by M. Tokuyama, I. Oppenheim, and H. Nishiyama

$\rho(\mathbf{r},t)$ obeys the continuity equation:

$$\frac{\partial}{\partial t}\rho = -\nabla\cdot(\rho\mathbf{v}), \tag{1}$$

where v is the velocity field. Then the velocity field obeys the Navier-Stokes equation:

$$\rho\left(\frac{\partial}{\partial t}+\mathbf{v}\cdot\nabla\right)\mathbf{v} = -\nabla\cdot\overleftrightarrow{\pi}+\nabla\cdot\overleftrightarrow{\sigma}. \tag{2}$$

In Eq. (2), $\overleftrightarrow{\pi}(\mathbf{r},t)$ is the pressure tensor, whose explicit form is given in Methods. We note that if we ignore the nonlinear off-diagonal terms, it reduces to the familiar relation: $\Pi_{ij} = p\delta_{ij}$. $\overleftrightarrow{\sigma}(\mathbf{r},t)$ is the viscous stress tensor expressed as

$$\sigma_{ij} = \eta(\rho)\left(\nabla_i v_j+\nabla_j v_i-\frac{2}{d}\delta_{ij}\partial_m v_m\right) + \delta_{ij}\zeta(\rho)\partial_m v_m, \tag{3}$$

where d is the spatial dimension and $\eta(\rho)$ and $\zeta(\rho)$ are the shear and bulk viscosities, respectively, which in general depend on the density ρ. We emphasize that the constitutive relation in Eq. (3), which is commonly used for simple fluids, does not involve any viscoelastic (or structural) relaxation. We will show below that it is this density dependence of η that leads to a coupling between density fluctuations and shear flow.

Now we consider a mechanism of shear-induced instability of liquid on the basis of a linear analysis of the above set of standard equations. For shear flow, we set the x axis along the direction of the mean flow and the y axis along the direction of the mean velocity gradient. Then, the mean velocity profile is given by $\langle\mathbf{v}(\mathbf{r},t)\rangle = \dot{\gamma}y\hat{\mathbf{x}}$, where $\hat{\mathbf{x}}$ is the unit vector along the x axis. For a small density fluctuation $\delta\rho$ from the average ρ_0, the pressure tensor becomes $\pi_{ij} = (p_0+K_T^{-1}\delta\rho/\rho_0)\delta_{ij}$, where p_0 is the average pressure and $K_T = (\partial\rho/\partial p)_T/\rho$ is the isothermal compressibility. Then, from Eq. (2), the linearized equation of motion for the volume dilation rate $Z = \nabla\cdot\mathbf{v}$ is derived in the Fourier representation as

$$\rho_0\left(\frac{\partial Z_{\mathbf{k}}}{\partial t}-\dot{\gamma}k_x\frac{\partial Z_{\mathbf{k}}}{\partial k_y}+2i\dot{\gamma}k_x v_{y\mathbf{k}}\right) = -\left(\zeta_0+\frac{2d-2}{d}\eta_0\right)k^2 Z_{\mathbf{k}}+\frac{1}{\rho_0 K_T}k^2\rho_{\mathbf{k}} -2\left(\frac{\partial\eta}{\partial\rho}\right)\dot{\gamma}k_x k_y\rho_{\mathbf{k}}, \tag{4}$$

where $\eta_0 = \eta(\rho_0)$ and $\zeta_0 = \zeta(\rho_0)$. Here, $Z_{\mathbf{k}}(t) = \int d\mathbf{r}\, Z(\mathbf{r},t)e^{-i\mathbf{k}\cdot\mathbf{r}}$ and $\rho_{\mathbf{k}}(t) = \int d\mathbf{r}\, \rho(\mathbf{r},t)e^{-i\mathbf{k}\cdot\mathbf{r}}$. In Eq. (4), there are several terms depending on the shear rate $\dot{\gamma}$. The terms on the left hand side only represent the advection by the mean flow, whereas the last term on the right hand side causes a coupling between density fluctuations and the mean shear flow. By arranging the last two terms on the right hand side in Eq. (4), we here introduce the following effective compressibility K_T^{eff}:

$$\frac{1}{K_T^{\text{eff}}} = \frac{1}{K_T}\left[1-2\rho_0 K_T\left(\frac{\partial\eta}{\partial\rho}\right)\dot{\gamma}\hat{k}_x\hat{k}_y\right], \tag{5}$$

where $\hat{k}_x$ and $\hat{k}_y$ are the x and y components of a unit wave vector $\hat{k}$. >From this, we can see that the system softens in the direction of $\hat{k}_x = \hat{k}_y$ (45° from the mean flow direction) under shear flow. Above the critical shear rate $\dot{\gamma}_c$, which is given by

$$\dot{\gamma}_c = \left[\rho_0 K_T\left(\frac{\partial\eta}{\partial\rho}\right)\right]^{-1} = \left(\frac{\partial\eta}{\partial p}\right)_T^{-1}, \tag{6}$$

the effective isothermal compressibility becomes "negative"; i.e., the homogeneous state becomes unstable (shear-induced instability). Our theory indicates that this instability is induced by a coupling of shear flow to the longitudinal modes, namely, the density fluctuations, which has been overlooked so far.

We emphasize that our argument relies solely on the well-established classical hydrodynamic theory. In very viscous liquids, which exhibit severe pressure- (or density-) dependencies of viscosities, this shear-induced instability should occur at a moderate shear rate, which we can easily access experimentally.

In Table 1 we compare our prediction with the experimental data on the shear-induced cavitation for a liquid lubricant, phenyl ether oligomer (molecular weight $M_w = 450$), by Bair *et al* [2, 3, 4]. $\dot{\gamma}_c^{ex}$ is the critical shear rate, above which the linear Newtonian behaviour breaks down. The values of both $\dot{\gamma}_c^{ex}$ and $\dot{\gamma}_c^{th} = (\partial\eta/\partial p)_T^{-1}$ are estimated from the data in Refs. [2, 3, 4]. Our prediction is well consistent with the experimental results. We note that in all conditions the structural relaxation time τ_α estimated from the data using the relation $\tau_\alpha \sim \eta/G$ (G: shear elastic modulus) is sufficiently shorter than $1/\dot{\gamma}_c^{ex}$. We stress that this occurrence of the instability for $\dot{\gamma}\tau_\alpha < 1$ cannot be explained by the conventional knowledge of rheology such as viscoelastic instability.

NUMERICAL SIMULATION

In the following we check the validity of our prediction by 2-dimensional numerical simulation of shear-induced instability of a liquid. In order to retain the generality, we use a simple phenomenological fluid model, where the gas-liquid phase transition is expressed by employing the van der Waals theory. (See Ref.[8] for details of our simulation method.)

TABLE 1. Comparison between the critical shear rate estimated from experimental rheological data measured by a Couette flow device, $\dot{\gamma}_c^{ex}$, and our theoretical prediction, $\dot{\gamma}_c^{th} = (\partial\eta/\partial p)_T^{-1}$, for a lubricant, phenyl ether oligomer ($M_w = 450$) [2, 3, 4]. We estimated $\dot{\gamma}_c^{th}$ directly from the experimental data of the the pressure dependent viscosity reported in Refs. [4]. The agreement is remarkable despite of the large range of $\dot{\gamma}$.

T(°C)	p(GPa)	$\dot{\gamma}_c^{ex}(s^{-1})$	$\dot{\gamma}_c^{th}(s^{-1})$
5	0.0207	$\sim 2\times 10^3$	$\sim 5\times 10^3$
5	0.0310	$\sim 6\times 10^2$	$\sim 2\times 10^3$
5	0.0414	$\sim 2\times 10^2$	$\sim 5\times 10^2$
20	0.0970	$\sim 2\times 10^3$	$\sim 4\times 10^3$
20	0.110	$\sim 5\times 10^2$	$\sim 10^3$
40	0.260	~ 5	~ 5
60	0.390	~ 3	~ 1

In Figure 1, we show the time evolution of the normalized density. Note that in the absence of shear flow one-phase state is thermodynamically stable. For $\dot{\gamma} > \dot{\gamma}_c$, we find that a liquid indeed becomes mechanically unstable even under simple shear flow: The liquid is *in a thermodynamically stable state*, but mechanically destabilized by simple shear flow. After some transient of the initial process of instability, the large-amplitude density fluctuations are steadily maintained.

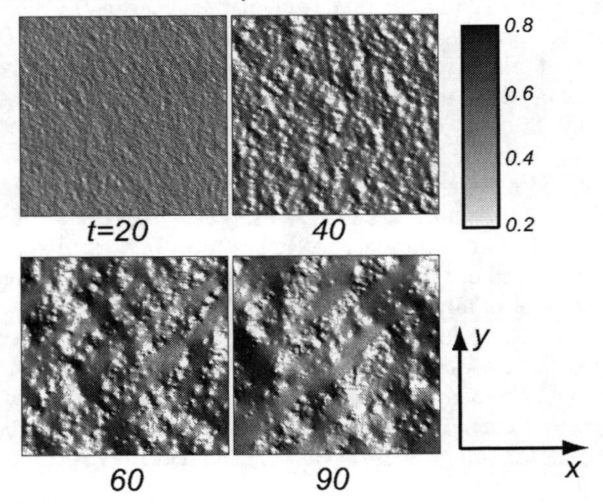

FIGURE 1. We plot the time evolution of the normalized density. The bright and dark regions indicate low and high density regions, respectively. Initially the density fluctuations develop along the direction of $\hat{k}_x = \hat{k}_y$. Then its amplitude grows. In the late stage, advection and nonlinear effects play a crucial role.

Figure 2 shows the average variance of the normalized density $\sqrt{\langle\delta\psi^2\rangle}$, the averaged shear stress $\langle\Sigma_{xy}\rangle$, the averaged normal stress difference $\langle N_1\rangle$, and the average viscosity $\langle\eta\rangle(=\langle\Sigma_{xy}\rangle/\dot{\gamma})$ as functions of the shear rate $\dot{\gamma}$, where both the time and spatial averages are taken for the dynamical steady states. For low enough shear rates, $\dot{\gamma} \lesssim \dot{\gamma}_c$, the system exhibits the simple Newtonian behaviour, that is $\Sigma_{xy} \propto \dot{\gamma}$. For $\dot{\gamma} \gtrsim \dot{\gamma}_c$, on the other hand, the shear-thinning behaviour appears due to the mechanical instability. What is intriguing is that even a simple liquid without any internal degrees of freedom exhibits such highly-nonlinear behaviour including the appearance of the first normal stress difference above $\dot{\gamma}_c$. This is because, due to the density dependence of viscosity, the velocity gradient is concentrated on the lower density regions having a lower viscosity, while the stress is supported mainly by the higher density regions; asymmetric stress division. The critical shear rate of the onset of the nonlinear behaviour is consistent with the theoretically estimated value of $\dot{\gamma}_c$. The slight discrepancy (a factor of 1.4) may be due to advection effect. Thus, we conclude that a thermodynamically stable liquid indeed becomes mechanically unstable even for simple shear flow around $\dot{\gamma} = \dot{\gamma}_c$. We note that, for a sufficiently high shear rate, the inhomogeneous state is maintained even in a thermodynamically stable one-phase region, by nonlinear interactions among the fluctuations even without thermal fluctuations.

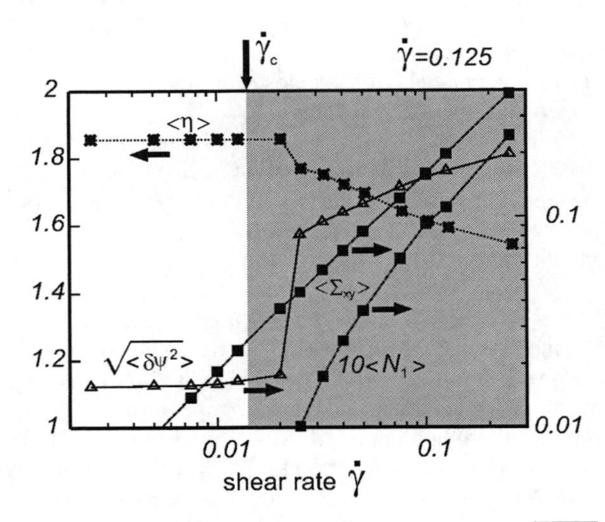

FIGURE 2. $\dot{\gamma}$-dependence of steady state values of $\sqrt{\langle\delta\psi^2\rangle}$, $\langle\Sigma_{xy}\rangle$, $\langle\eta\rangle$, and $\langle N_1\rangle$. The unstable region predicted by our linear analysis ($\dot{\gamma} \geq \dot{\gamma}_c$) is shaded. For $\dot{\gamma} \geq \dot{\gamma}_c$, strong nonlinear behaviour is clearly seen even for a simple liquid which does not have any internal degrees of freedom. We note that these results well reproduces the basic rheological features experimentally observed for a liquid lubricant, phenyl ether oligomer ($M_w = 450$). [9].

SUMMARY AND CONCLUDING REMARKS

We have demonstrated that the incompressibility condition can be violated by a coupling between flow and density fluctuations via the density dependence of viscosity η even for simple shear flow and a liquid can become mechanically unstable above the critical shear rate, $\dot{\gamma}_c = (\partial\eta/\partial p)_T^{-1}$.

Our mechanism not only offers a new condition for the incompressibility ($\dot{\gamma} \ll \dot{\gamma}_c$), but also may shed new light on poorly understood phenomena associated with the mechanical instability of liquid at a low Reynolds number: instability of lubricants under shear, shear-induced cavitation and bubble growth, shear banding of very viscous liquids including metallic glasses and mantle.

REFERENCES

1. F. Spaepen: Acta Metal. **25**, 407-415 (1977).
2. S. Bair and W.O. Winer: J. Tribol. **112**, 246-253 (1990).
3. S. Bair and W.O. Winer: J. Tribol. **114**, 1-13 (1992).
4. S. Bair, F. Qureshi, and W.O. Winer: J. Tribol. **115**, 507-514 (1993).
5. L.A. Archer, D. Ternet, and R.G. Larson: Rheol. Acta. **36**, 579-584 (1997).
6. D.D. Joseph: J. Fluid. Mech. **366**, 367-378 (1998).
7. L. D. Landau and E.M. Lifshitz: *Fluid Mechanics* (Pergamon, New York, 1959).
8. A. Furukawa and H. Tanaka: Nature. **443**, 434-438 (2006).
9. S. Bair: Rheol. Acta. **35**, 13-23 (1996).

Robustness of Attractor States in Complex Networks

Shu-ichi Kinoshita*, Kazumoto Iguchi†, and Hiroaki S. Yamada**

*Graduate School of Science and Technology, Niigata University, Niigata 950-2181, Japan
†KazumotoIguchi Research Laboratory, 70-3 Shinhari, Hari, Anan, Tokushima 774-0003, Japan
**Yamada Physics Research Laboratory, Nishi-ku Aoyama 5-7-14, Niigata 950-2002, Japan

Abstract. We study the intrinsic properties of attractors in the Boolean dynamics in complex network with scale-free topology, comparing with those of the so-called random Kauffman networks. We have numerically investigated the frozen and relevant nodes for each attractor, and the robustness of the attractors to the perturbation that flips the state of a single node of attractors in the relatively small network ($N = 30 \sim 200$). It is shown that the rate of frozen nodes in the complex networks with scale-free topology is larger than that in the random Kauffman model. Furthermore, we have found that in the complex scale-free networks with fluctuations of in-degree number the attractors are more sensitive to the state flip of a highly connected node than to the state flip of a less connected node.

Keywords: Attractor, Boolean dynamics, Frozen nodes, Intrinsic property, Robustness, Scale-free network
PACS: 05.10.-a, 05.45.-a, 64.60.-i, 87.18.Sn

INTRODUCTION

Dynamics of Boolean networks is often used as a model for genetic networks inside cells, in which the genetic states are represented in terms of the language of attractors [1, 2]. A Boolean network consists of N nodes, each of which receives k_i inputs such that the degree is k_i. In the so-called *Kauffman model* – a random Boolean network (RBN) model, each node receives a certain fixed number of inputs such that the degree is $k_i \equiv K$.

In real systems each node has a different number of inputs, however. As a more realistic modeling of biological systems we expect that the fluctuation of the number of input-degree is treated as a random variable with a probability distribution function such as the inverse power-law distribution or the exponential distribution or the Poisson distribution. In fact, the in-degree distribution appears to be exponential in *E.coil* and to be power-law in yeast [3, 4, 5, 6, 7, 8, 9, 10, 11, 12].

Recently, the relationship between the Boolean dynamics and the network topology has been investigated by many authors from the view point of stability and evolvability of the network systems [10, 11, 12, 13, 14, 15, 16, 17]. They considered stability of the network dynamics when one node is noisily perturbed, and showed that the stability depends on the connectivity of nodes in a relatively small network $N = 19$ with scale-free topology.

In the present report, the number of in-degree k_i at the i–th node is determined by the preferential attachment rule that makes the system a complex network with scale-free topology (SFRBN)[9]. We do not deal with large networks(i.e. $N = 30 \sim 200$), focusing on the intrinsic properties of the Boolean dynamics on the complex networks. The details of the method for generating networks have been given in our previous paper [16]. As is numerically seen in the relatively small networks, the degree distribution functions $P(k)$ are not different from each other among networks with different topology. However, we would like to mention that the network size in our study is larger than the one used in the related previous studies by other researchers [10, 11, 12, 13, 14, 15, 16, 17], and that our study provides more details of the robustness of attractors and the frozen nodes in the attractors.

MODEL

The initial values for the nodes are chosen randomly and are synchronously updated in the time steps, according to the connectivity $\{k_i\}$ and the Boolean functions $\{f_i\}$ assigned for each node in the network as,

$$\sigma_i(t+1) = f_i(\sigma_{i_1}(t), \sigma_{i_2}(t), \cdots, \sigma_{i_{k_i}}(t)), \qquad (1)$$

where $i = 1, ..., N$ and $\sigma_i \in \{0, 1\}$ is the binary state. All trajectories starting at any initial state run into a certain number of attractors(i.e. points or cycles). We study the directed RBNs, the directed SFRBNs, and the directed SFRBNs throughout this paper.

Figure 1 shows the time-dependence of the state which constitutes an typical attractor in the SFRBN with $\langle k \rangle = 2$.

Fig. 2(a) shows the histogram of the length ℓ_c of the attractors in the RBN with $K = 2$ and that in the SFRBN where the average degree of nodes $\langle k \rangle = 2$. Fig. 2(b) shows the median value $\bar{m}$ of the distribution of state cycle lengths with respect to N, the total number of nodes.

CP982, *Complex Systems, 5th International Workshop on Complex Systems*
edited by M. Tokuyama, I. Oppenheim, and H. Nishiyama

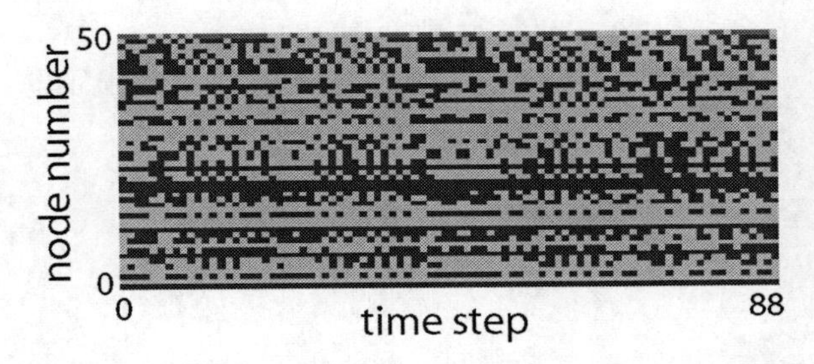

FIGURE 1. (Color online) Space-time diagram for an typical attractor with $\ell_c = 88$. in the SFRBN with $\langle k \rangle = 2$. The vertical axis denotes the node number that is in the high connectivity order. The dense or gray color corresponds to the binary values of the state. The static nodes of the attractor are frozen nodes. The network size is $N = 50$.

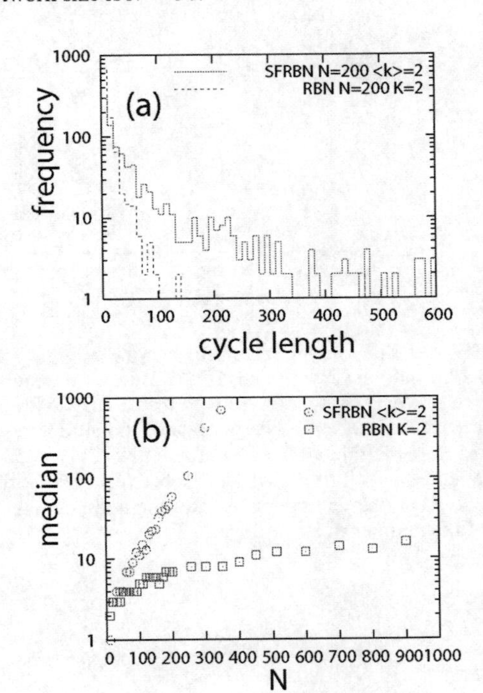

FIGURE 2. (Color online) (a) Histogram of the length ℓ_c of state cycles is shown for the RBNs and the SFRBNs, respectively, where the network size is $N = 200$. Each histogram is generated by 10^3 different sets of the Boolean functions and five different network structures. The maximum iteration number of the Boolean dynamics is 10^5 until the convergence to the cycle is realized. (b) Semi-log plots of the median value $\bar{m}$ of 1000 samples of the lengths of the state cycles with respect to the total number N of nodes for the directed RBNs and SFRBNs.

Apparently, the distribution of the attractor lengths in the SFRBN is much wider than that in the RBN, and the attractor length has longer period than that in the RBN. This is directly related to the diversity of attractors in the SFRBNs, which is of great importance for the stability of living cells. We investigated the function form $\bar{m}(N)$ in more detail in the previous paper [16], and found that the function form $\bar{m}(N)$ asymptotically changes from the algebraic type $\bar{m}(N) \propto N^{\alpha}$ to the exponential one as the average degree $\langle k \rangle$ goes to $\langle k \rangle = 2$.

Here we have a question: *How does the characteristics of the attractors depend on the network topology?* In the present paper we study the difference between attractors in the RBN and the SFRBN without a bias, focusing on the frozen nodes and the robustness of attractors against the external perturbations.

FROZEN NODES OF ATTRACTORS

In this section, we investigate the so-called frozen nodes of attractors whose values remain constant through a given trajectory of the attractors [2]. Frozen nodes arise through canalizing the Boolean functions and the homogeneity bias.

We count the number of frozen nodes N_f for each attractor and plot the histograms for some cases in Fig.3. The remarkably different peak structure exists between the cases in the SFRBN of $\langle k \rangle = 2$ and the RBN of $K = 2$. The distributions in the SFRBN have a peak around $N_f \sim N/2$, while the distributions in the RBN are broad with a peak at $N_f = N$. Note that $N_f = N$ corresponds to the point attractors that all nodes are frozen.

The fact that N_f is relatively smaller in the SFRBN would make the attractor period larger than that in the RBN, as seen in Fig.2. However, the role of the frozen nodes is not so clear in the network because all frozen nodes also are connected to the network and might influence the attractors. In the next section, we investigate the significance of each node in the respective attractor.

ROBUSTNESS OF ATTRACTORS TO STATE FLIP

One of the important properties in the scale-free topology is the existence of the highly connected hub node as seen in the yeast synthetic network and so on. In this section, we investigate the robustness of attractors to an external perturbation caused by an inversion of the binary state of a single node. We consider an attractor of period ℓ_c and flip the state of the single node at time $t \in [1, \ell_c]$ as a perturbation. The perturbation to the trajectory of the attractor may leap from the trajectory of the original attractor to another one, i.e. the attractor shift. The high homeostatic stability implies low reachability among different attractors.

We investigate the probability R_s that the attractor remains in the original attractor under the inversion of the single node state [1, 18], which is the rate returning

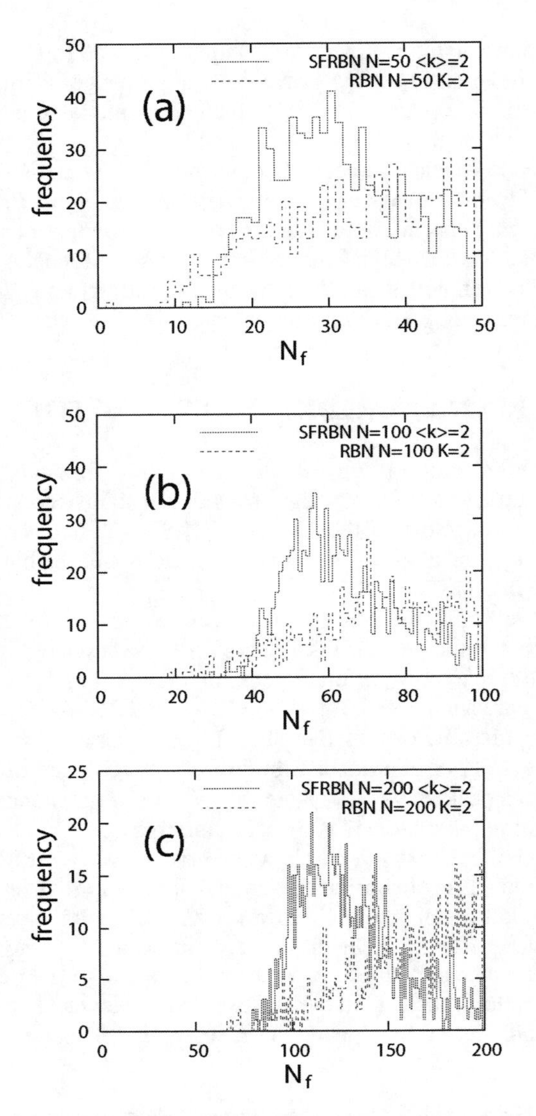

FIGURE 3. (a) Histograms of the number of frozen nodes N_f for 1000 attractors in the RBNs with $K = 2$ and the SFRBNs with $\langle k \rangle = 2$. The network size is (a) $N = 50$, (b) $N = 100$ and (c) $N = 200$. The scale out data at $N_f = N$ are not shown in the figures.

to the original attractor under the inversion. Here we call the rate the *robustness* of the attractor.

Figure 4 shows the robustness R_s of attractors with $\ell_c = 55$ in the SFRBN and $\ell_c = 48$ in the RBN, respectively. It follows that in the SFRBN the number of "active nodes" ($R_s < 1$) is much more larger than that in the RBN. On the other hand, in the RBN the perturbation to the active nodes influences effectively the shift of attractor ($R_s < 0.6$) although the number of active nodes is not so many [18]. As a result, in the SFRBN the perturbation to the highly connected hubs may give rise to the attractor shift, comparing with the one to the less connected

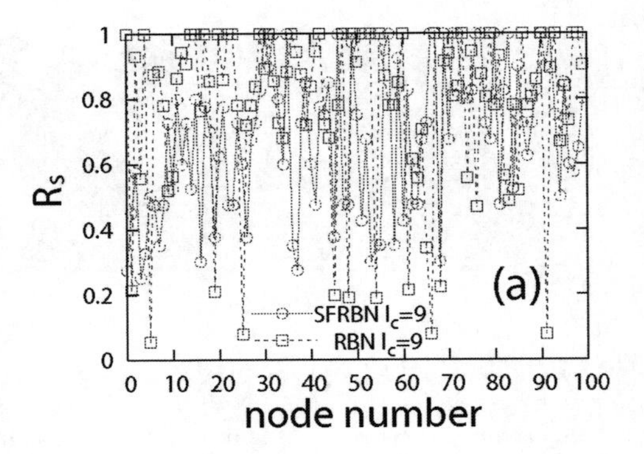

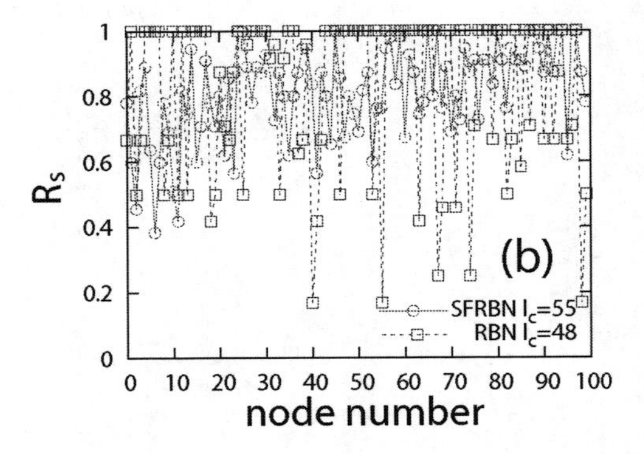

FIGURE 4. The rate R_s of returning to the original attractor as a function of numbered nodes in the order of the number of in-degree k_i. We used $N = 100$ in both the SFRBN with $\langle k \rangle = 2$ and the RBN with $K = 2$. The periods of the attractors are (a) $\ell_c = 9$ in the SFRBN and $\ell_c = 9$ in the RBN, (b) $\ell_c = 55$ in the SFRBN and $\ell_c = 48$ in the RBN, respectively. The horizontal axis denoted as "Node number" shows the node number in the order of the number of input-degree.

nodes.

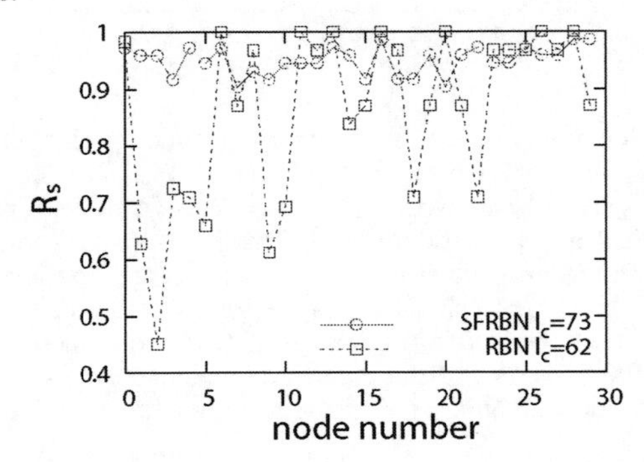

FIGURE 5. Robustness R_s for some attractors of the SFRBN with $\langle k \rangle = 4$ and the RBN with $K = 4$ in the network size of $N = 30$. The periods of attractors are $\ell_c = 73$ in the SFRBN and $\ell_c = 62$ in the RBN.

The robustness of some attractors in the SFRBN with $\langle k \rangle = 4$ and in the RBN with $K = 4$ is given in Fig.5. Although the whole structure is almost similar to that for the cases in Fig. 4, it is found that the effect of inversion of the single site state on the attractor shift is relatively small compared to the cases of $\langle k \rangle = 2$ and $K = 2$. There is a tendency that the attractors become more robust to the perturbation as the average number of input-degree $\langle k \rangle$ increases.

SUMMARY AND DISCUSSION

In summary, we have studied the Boolean dynamics of the Kauffman model with the directed SFRBN, comparing with the ones with the directed RBN for the relatively small network size. In this study we investigated some intrinsic properties of attractors between the RBNs and the SFRBNs, focusing on the frozen nodes and the robustness to a perturbation. The obtained results are as follows. (i) The number of frozen nodes in the SFRBN is smaller than that in the RBN and the property reflects on the much more widely distributed attractor lengths. (ii) The perturbation to the highly connected hubs may give rise to the attractor shift in comparison to the less connected nodes. (iii) The attractors becomes more robust to the perturbation as the average number of input degree $\langle k \rangle$ increases.

Although in this report we did not show the details of the attractor shifts by the perturbation, we will present the details of the numerical results for the diagram of transition among the attractors and the robustness to perturbation in our forthcoming paper [18].

Robustness against genetic mutations and environmental perturbations is one of the universal features of biological systems. And the robustness is important for understanding evolutionary processes and homeostasis of gene regulatory networks [18, 19, 20, 21, 22]. However, in this report we investigated only robustness to the single site inversion. For the purpose of the study, the other robustness of the SFRBN might become significant; for instance, robustness of attractors to the change of Boolean functions and to the breakdown of the network structures. We expect that such a study on the robustness of attractors provide some insights into important biological phenomena such as cellular homeostasis and apotosis.

Actually Aldana *et al* investigated the small SFRBN ($N \sim 15 - 20$) and found that the robustness of the ordered phase to the network damage in the SFRBN is lower than that in the RBN [19]. The result implies that there exists a possibility of evolution though mutation even in the ordered phase, despite of the Kauffman's conjecture that life evolves in "edge of chaos".

Moreover, recently, an interesting network model, the so called *feedback network* has been proposed by White *et al* [23]. The feedback networks can be a good model for describing the autocatalytic chemical reactions and the kinship, and so on, because the node selection, the search distance and the search path of networks are controlled by the attachment, the distance decay and the cycle formation parameters. It is interesting to investigate the features of the Boolean dynamics in the feedback networks from the point of view of frozen nodes and robustness [18].

REFERENCES

1. S. A. Kauffman, J. Theor. Biol. **22**, 437-467 (1969).
2. S. A. Kauffman, *Origins of Order : Self-Organization and Selection in Evolution* , Oxford University Press, Oxford, 1993,pp.441-522.
3. A.-L. Barabási, R. Albert and H. Jeong, Physica A **272**, 173-187 (1999).
4. T.I. Lee *et al.*, Science **298**, 799-804 (2002).
5. P. Sen *et al.*, Phys. Rev. E **67**, 036106-1-036106-5(2003).
6. S. A. Kauffman *et al.*, Proc. Nat. Acad. Sci. USA **100**, 14796-14799 (2003).
7. M. Skarja, B. Remic, and I. Jerman, Chaos **14**, 205-216(2004).
8. R. Albert, Journal of Cell Science **118**, 4947-4957 (2006).
9. R. Albert and A.-L.Barabási, Phys. Rev. Lett. **84**, 5660-5663(2000).
10. J. J. Fox and C. C. Hill, Chaos **11**, 809-815 (2001).
11. C. Oosawa and A. Savageau, Physica **D 170**, 143-161 (2002).
12. R. Serra, M. Villani and L. Agostini, WIRN VIETRI 2003, Lecture Notes in Computer Science **2859**, 43-49 (2003).
13. M. Aldana, Physica **D 185**, 45-66 (2003).
14. M. Aldana and P. Cluzel, PNAS **100**, 8710-8714 (2003).
15. M. Aldana *et al.*, J. Theor. Biol. **245**, 433-448 (2007).
16. K. Iguchi *et al.*, J. Theor. Biol. **247**, 138-151 (2007). cond-mat/0510430.
17. C. Handrey *et al.*, Physica **A 373**, 770-776 (2007).
18. S. Kinoshita, K. Iguchi, and H.S. Yamada, in preparation (2007).
19. M. Aldana *et al.*, in "*Perspectives and Problems in Nonlinear Science*", edited by E. Kaplan, J. E. Marsden, and K. R. Screenivasan, Springer-Verlag, NY, 2003, pp.23-90.
20. Y. Bar-Yam and I.R. Epstein, PNAS **101**, 4341-4345(2004).
21. J.M. Monte *et al.*, Report of research work for CSSS05, July 2005
22. S.A. Kauffman, *Complexity and Genetic Networks*, Existence Project News, 2003.
23. . R. White *et al.*, Phys. Rev. E **73**, 016119-1-016119-8 (2006).

Vortex Nucleation Effects on Vortex Dynamics in Corbino Disk at Zero Field

Y. Enomoto, M.Ohta, and Y.Yamada

Department of Environmental Technology,
Nagoya Institute of Technology, Nagoya 466-8555, Japan

Abstract. We study the radial current driven vortex dynamics in the Corbino disk sample at zero field, by using a logarithmically interacting point vortex model involving effect of temperature, random pinning centers, and disk wall confinement force. We also take into account both the current induced vortex pair nucleation and the vortex pair annihilation processes in the model. Simulation results demonstrate that the vortex motion induced voltage exhibits almost periodic pulse behavior in time, observed experimentally, for a certain range of the model parameters. Such an anomalous behavior is thought to originate from large fluctuations of the vortex number due to the collective dynamics of this vortex system.

Keywords: Corbino Disk, Vortex Pair Nucleation, Vortex Pair Annihilation
PACS: 74.60Ge, 74.60Mj

INTRODUCTION

The Corbino disk of type-II superconductors is a circular thin disk sample [1-3]. When a transport current is injected at the Corbino disk center and removed at its perimeter, the current flows radially in the disk toward the sample boundary that decays as $1/r$ with r the distance from the disk center. Magnetic vortices in the Corbino disk under a magnetic field applied along the disk axis experience a Lorentz force along the azimuthal direction that decreases as $1/r$. Thus, the Corbino disk geometry introduces the shear stress to a vortex lattice due to the spatially inhomogeneous Lorentz force, leading to peculiar transport properties compared with those in usual stripe-shaped samples [1-3].

Recently, Okuma and co-workers [4-6] have measured the current-voltage characteristics of amorphous superconducting films at zero field in the Corbino disk geometry. They have shown that large voltage pulses appear almost periodically when the radial current is larger than a critical value, and also have found that both the pulse period t_p and pulse width t_w are much larger than the time scale t_f for free vortices to rotate half in the circle. They have speculated that the observed voltage pulses may originate from collective motion of many vortices which leads to large number fluctuations of moving vortices.

After then, Hayashi and Ebisawa [7] have qualitatively supported the above speculation analytically and numerically by using a simple model involving the vortex nucleation and annihilation effects [8]. They have pointed out an important role of the pair annihilation process, but they have not observed the voltage pulse phenomenon.

Stimulated by these works [4-7], we here simulate the radial current driven vortex dynamics of amorphous superconducting film at zero field in the Corbino disk by using a logarithmically interacting point vortex model. The present model includes effects of Lorentz force, disk wall confinement force, thermal noise, and random pinning force, as well as the vortex pair nucleation and annihilation effects. The effect of random pinning force is introduced in the model as random potential [9] to realize the nature of amorphous superconductors. Current induced vortex-antivortex pair nucleation process is treated stochastically with a Metropolis algorithm, which has been developed by our group [10,11]. From large scale simulations by changing values of current, temperature, and pinning strength, we find that experimentally observed voltage pulses do appear in a certain range of these parameters.

THE MODEL

We consider a Corbino disk sample with radius D and thickness w under an injected radial current at zero field. Under the thin thickness assumption $w \ll D$, we consider a two-dimensional vortex system. For simplicity we neglect the Magnus force. Hereafter we assign a vortex to a magnetic vortex having positive circulation along the disk axis and an antivortex to one having negative circulation, and also we often use the word 'vortex' as a generic term including both vortex and antivortex if confusion does not occur.

The equation of motion for the i–th vortex can be described by the overdamped equation as [9-12]

$$\eta \frac{d}{dt}\mathbf{r}_i = \mathbf{F}_i^V + \mathbf{F}_i^L + \mathbf{F}_i^C + \mathbf{F}_i^T(t) + \mathbf{F}_i^P \qquad (1)$$

CP982, *Complex Systems, 5th International Workshop on Complex Systems*
edited by M. Tokuyama, I. Oppenheim, and H. Nishiyama

where $\mathbf{r}_i$ denotes a two-dimensional position vector of the i–th vortex and η is a viscous coefficient. The inter-vortex force, $\mathbf{F}_i^V$, is given by

$$\mathbf{F}_i^V = f_V \sum_{k(\neq i)}^{N(t)} \frac{\lambda e_i e_k}{|\mathbf{r}_i - \mathbf{r}_k|} \frac{\mathbf{r}_i - \mathbf{r}_k}{|\mathbf{r}_i - \mathbf{r}_k|} \tag{2}$$

with $f_V \equiv \phi_0^2/(8\pi^2\lambda^3)$, the number of vortices $N(t)$ at time t having $N(0) = 0$ due to the zero field condition, the magnetic penetration depth λ, and $e_i = 1(-1)$ for a vortex (an antivortex). The Lorentz force due to the applied radial current, $\mathbf{F}_i^L$, is given by

$$\mathbf{F}_i^L = f_L e_i \frac{I}{|\mathbf{r}_i|} \hat{\theta} \tag{3}$$

with $f_L \equiv \phi_0/(2\pi c w)$, the current magnitude I, the light velocity c, and the unit vector $\hat{\theta}$ along the azimuthal direction. The disk wall confinement force due to the sample edge barrier, $\mathbf{F}_i^C$, is assumed to be given by

$$\mathbf{F}_i^C = -f_C \exp(-(D - |\mathbf{r}_i|)/d)\hat{r} \tag{4}$$

with $f_C \equiv f_V \lambda/d$, the effective barrier range d, and the unit vector $\hat{r}$ along the radial direction. The α–component ($\alpha = x$ or y) of the thermal fluctuation force, $F_i^{T,\alpha}(t)$, is taken to be a Gaussian white noise characterized by zero mean ($< F_i^{T,\alpha}(t) >= 0$) and the correlation $< F_i^{T,\alpha}(t) F_k^{T,\beta}(t') >= 2(\eta k_B T/\lambda)\delta(t-t')\delta_{\alpha\beta}\delta_{ik}$ with temperature T. For simplicity we assume that the temperature effect is involved into the model only through the thermal noise and other physical quantities have no temperature dependence.

To realize the nature of amorphous superconductors, the random pinning effect is introduced into the model as follows [9]. A simulation system of the Corbino disk is divided by a $0.5\lambda \times 0.5\lambda$ square piece, and then random numbers having uniform distribution in the interval $[-1,0]$ are set on these lattice points. The α–component ($\alpha = x$ or y) of the random pinning force, $F_i^{P,\alpha}$, acting on the i–vortex of the position $\mathbf{r}_i = (x_i, y_i)$ is assumed to be given by [9]

$$F_i^{P,x} = -p f_V \left(V([x_i]+1,[y_i]) - V([x_i]-1,[y_i])\right) \tag{5}$$

$$F_i^{P,y} = -p f_V \left(V([x_i],[y_i]+1) - V([x_i],[y_i]-1)\right) \tag{6}$$

where p is a dimensionless pinning strength, $V(m,n)$ is the random number on a lattice point (m,n) defined above, $[x]$ denotes the nearest integer of x.

Effects of the current induced vortex–antivortex pair nucleation and also the pair annihilation process can be incorporated into the above model as follows [10,11]. In the presence of the radial transport current, the lowest energy barrier to create a vortex–antivortex pair is given by [13]

$$U(\mathbf{r}_c) = \lambda f_V \left[\ln\left(\frac{|\mathbf{r}_c|}{\lambda}\right) - 1\right] \tag{7}$$

where $\mathbf{r}_c \equiv \lambda f_V R/(f_L I)\hat{\theta}$ is a critical relative position vector orienting from the antivortex to the vortex. Only vortex pairs nucleated at a relative distance larger than $|\mathbf{r}_c|$ will be free to move apart and thereby contribute to the transport. Thus, the rate of production of the vortex pair, able to contribute to the resistivity, Γ, will be estimated as $\Gamma \sim \exp[-U(\mathbf{r}_c)/k_B T]$ [13]. From these results, the vortex pair nucleation process can be treated stochastically as follows [10,11,14]. We first pick up a position A at random and its neighboring position B with the relative position vector $\mathbf{r}_c$. The radial coordinates of both positions A and B are assumed to be the same value R where R is an uniformly distributed random number in the interval $[0,D]$. Only if both positions are unoccupied by vortices, the pair is nucleated with probability $\exp[-U(\mathbf{r}_c)/k_B T]$ obeying the Metropolis rule. One trial for the pair nucleation discussed above is carried out after every characteristic time interval t_0. On the other hand, vortices and antivortices with their distance less than $\lambda/2$ are interpreted to annihilate. Then, such pairs are removed out from the system considered.

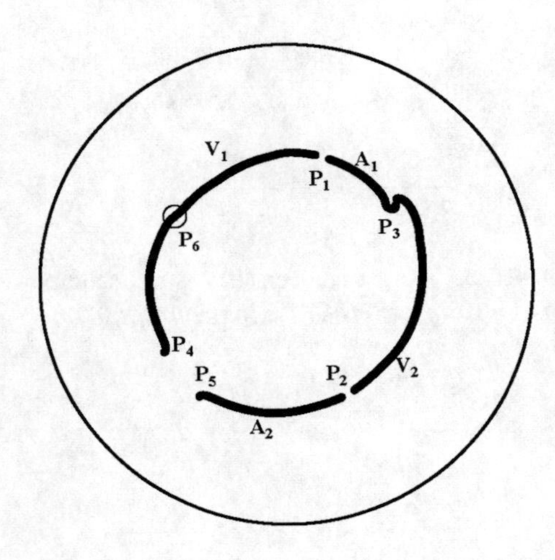

FIGURE 1. Trajectories of two vortex pairs nucleated at different times and different points for $I/I_0 = 1$, $T/T_0 = 0.5$, and $p = 0.01$. For abbreviations, see the text.

Throughout this work, all lengths are measured in units of λ, time in units of $t_0 \equiv \eta\lambda/f_V$, current magnitude in units of $I_0 \equiv \lambda f_V/f_L$, and temperature in units of

$T_0 \equiv \lambda f_V / k_B$. For $\lambda = 200$nm whose value is reasonable for YBCO samples, we can estimate as $t_0 \simeq 4 \times 10^{-4}$sec, $I_0 \simeq 1$A, and $T_0 \simeq 10^5$K.

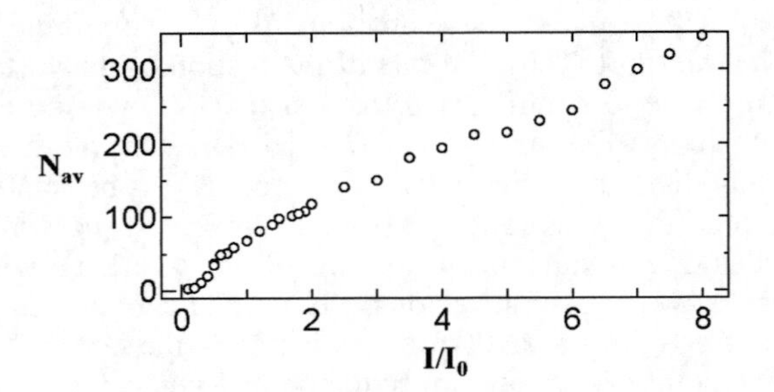

FIGURE 2. Mean number of vortices N_{av} as a function of current I/I_0 for $T/T_0 = 0.5$ and $p = 0.01$.

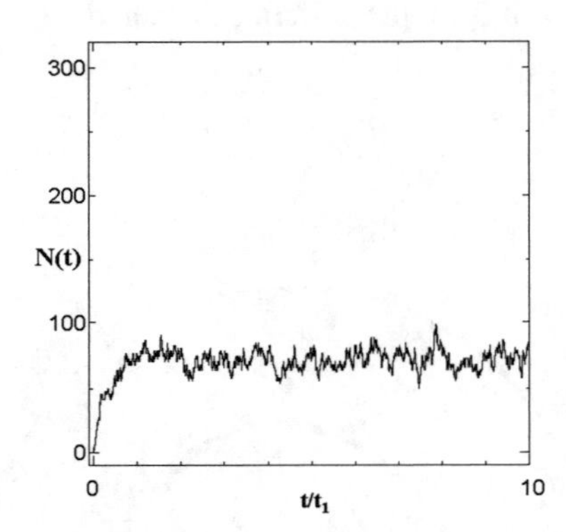

FIGURE 3. Vortex number $N(t)$ as a function of time t/t_1 with $t_1 = 10^5 t_0$ for $(I/I_0, T/T_0, p) = (1, 0.5, 0.01)$.

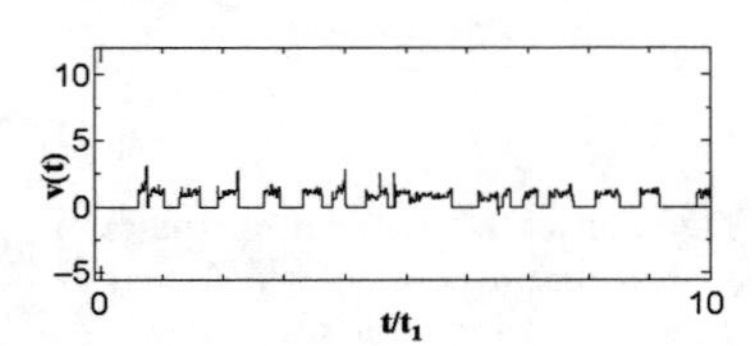

FIGURE 4. Averaged vortex velocity $v(t)$ as a function of time t/t_1 with $t_1 = 10^5 t_0$ for $(I/I_0, T/T_0, p) = (1, 0.5, 0.01)$.

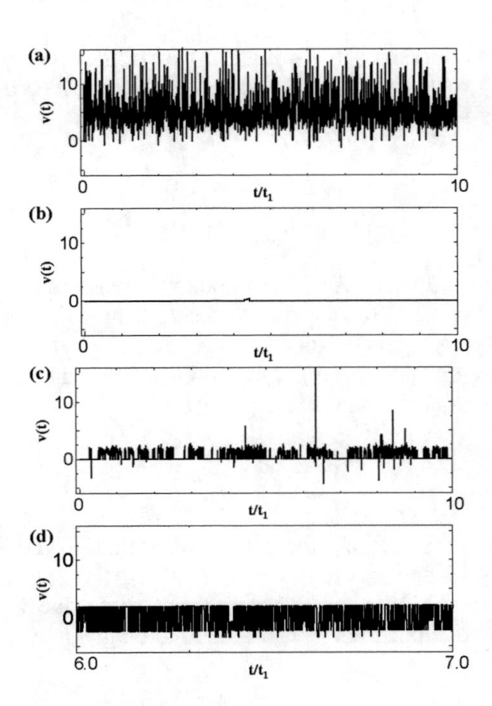

FIGURE 5. Averaged vortex velocity $v(t)$ as a function of time t/t_1 with $t_1 = 10^5 t_0$ for (a) $(I/I_0, T/T_0, p) = (8, 0.5, 0.01)$, (b) $(0.1, 0.5, 0.01)$, (c) $(1, 5, 0.01)$, and (d) $(1, 0.5, 0.5)$.

SIMULATION RESULTS

We carry out molecular dynamics simulations of eq.(1) in the Corbino disk sample, taking into account the vortex pair nucleation and annihilation processes. To do so, we use a simple Euler method with a time step $0.01 t_0$. In actual simulations, we consider a system with $D/\lambda = 50$, $w/\lambda = 0.1$ and $d/\lambda = 0.1$.

First of all, we explain the qualitative behavior of nucleation and annihilation processes of vortex pairs. In Fig.1 we show the trajectories of two vortex pairs nucleated at different times and different positions P_1 and P_2 for the case of current $I/I_0 = 1$, temperature $T/T_0 = 0.5$, and pinning strength $p = 0.01$. Any other vortices are not shown in this figure. From this figure it is found that (1) an antivortex A_1 nucleated at P_1 and a vortex V_2 nucleated at P_2 annihilate each other at position P_3, (2) a vortex V_1 (an antivortex A_2) nucleated at P_1 (P_2) annihilates at P_4 (P_5) with an invisible another antivortex (vortex), (3) a vortex V_1 is pinned and depinned at P_6 marked by a open circle, and (4) these trajectories are disturbed from circular ones by themselves, other invisible vortices, and pinning centers. In Fig.2 the mean number of vortices N_{av} is plotted as a function of the current I/I_0 for $T/T_0 = 0.5$ and $p = 0.01$, where each data point represents a time-average of the vortex number over 10 simulation runs. In this figure we can see a sudden increase of the vortex number near $I/I_0 = 0.5$. It is because that for $I/I_0 > 0.5$ at $T/T_0 = 0.5$ the probability of the vortex pair nucleation, $\exp[-U(\mathbf{r}_c)/(k_B T)]$, is estimated as the order of one, and thus the vortex nucleation takes place frequently for this case.

Next, we discuss the time evolution of the vortex motion induced voltage by changing values of three parameters, (I/I_0, T/T_0, p). To do so, we calculate the averaged vortex velocity as a function of time, because the averaged vortex velocity is proportional to the induced voltage. Hereafter, we consider only the azimuthal component $v(t)$ of the averaged vortex velocity, defined by $v(t) = \frac{1}{n(t)} \sum_i^{n(t)} \frac{d\mathbf{r}_i}{dt} \cdot \hat{\theta}$, which is proportional to the induced voltage along the radial direction. Here, $n(t)$ denotes the vortex number crossing the line $\theta = 0$ during the time interval $[t, t+t_0]$. We should note that the radial component of the averaged vortex velocity is almost zero or small for the present simulations. In Figs.3 and 4 the vortex number $N(t)$ and the averaged vortex velocity $v(t)$ are plotted as a function of time t, respectively, for the case of $I/I_0 = 1$, $T/T_0 = 0.5$, and $p = 0.01$. In Fig.3 we can see that the vortex number fluctuation in time is rather large (~ 27 percent of its average value) [7]. In Fig.4 we can see the voltage pulse phenomenon, and also find that the pulses occur almost periodically with the pulse period $t_p \simeq 3 \times 10^4 t_0$ and pulse width $t_w \simeq 3 \times 10^4 t_0$ on the average. For this case the time scale t_f for free vortex to rotate half in the circle having radius $D/2$ is estimated as $t_f \simeq 0.2 \times 10^4 t_0$. Thus, two time scales t_p and t_w are larger than t_f. This is similar to the experimental result [4-6]. However, the ratios t_p/t_f and t_w/t_f for this case are about 10, but the experimental ones are about 10^5. To resolve this discrepancy, further analyses are needed.

Finally, we study the parameter region for the periodic voltage pulse behavior. To do so, we have simulated the vortex dynamics for some values of three parameters. From these simulations, it has been found that the vortex motion induced voltage exhibits the periodic pulse-like behavior in time only for $0.5 < I/I_0 < 5$, $0.1 < T/T_0 < 3$, and $p < 0.1$. Indeed, as are shown in Figs.5(a)-(d), no periodic voltage pulses can be observed for the cases of (a) large current, (b) small current, (c) high temperature, and (d) strong pinning strength, respectively. Moreover, we have confirmed that for these cases of Figs.5(a)-(d) the fluctuations of the vortex number in time are not so large (less than 12 percent of each average value). Thus, large fluctuation of the vortex number may be the key to understand the observed voltage pulse behavior [7]. Studying the detailed mechanism to enhance or reduce the vortex number fluctuation is a future work.

CONCLUSIONS

We have numerically studied the radial current driven vortex dynamics in two-dimensional Corbino disk samples at zero field, based on a logarithmically interacting point vortex model with effects of the current induced vortex pair nucleation and the vortex annihilation processes. From molecular dynamics simulations of the model by changing values of current, temperature, and pinning strength, it has been found that the periodic voltage pulse behavior occurs for a certain range of these parameters. We have expected that such an anomalous behavior appears when the vortex number fluctuation is rather large due to the collective dynamics of the vortex system.

The present simulations have qualitatively reproduced some properties of experimental data [4-6]. At present, to understand experimental results completely, further simulations are under the way.

REFERENCES

1. D.Lopez, W.K.Kwok, H.Safar, R.J.Olsson, A.M.Petrean, L.Paulius, G.W.Crabtree, *Phys. Rev. Lett.* **82**, 1277-1280 (1999).
2. D.Benetatos, M.C.Marchetti, *Phys. Rev. B* **65**, 134517-134522 (2002).
3. M.C.Miguel, S.Zapperi, *Nat. Mater.* **2**, 477-482 (2003).
4. M.Kamada, Y.Arakawa, S.Okuma, *Phyica C* **392-396**, 424-427 (2003).
5. M.Kamada, Y.Watanabe, S.Okuma, *Phyica C* **412-414**, 535-538 (2004).
6. S.Okuma, M.Kameda, *J.Phys.Soc.Jpn.* **73**, 2807-2811 (2004).
7. M.Hayashi, H.Ebisawa, *J.Phys.Chem.Solids* **66**, 1380-1385 (2005).
8. J.M.Kosterlitz, D.J.Thouless, *J.Phys.C* **6**, 1181-1190 (1973).
9. Y.Enomoto, K.Katsumi, S.Maekawa, *Physica C* **215**, 51-57(1993).
10. Y.Enomoto, S.Maekawa, *Physica C* **247**, 156-160 (1995).
11. Y.Enomoto, S.Maekawa, *Physica C* **274**, 351-359 (1997).
12. Y.Enomoto, M.Ohta, *Physica C* **445-448**, 228-232 (2006).
13. A.M.Kadin, K.Epstein, A.M.Goldman, *Phys.Rev.B* **27**, 6691-6699 (1983).
14. Y.Saito, H.M.Drumbhaar,*Phys.Rev.B* **23**, 308-311 (1981).

Formation of Boundary Structures in $L1_0$ Type Ordering

R. Oguma[*], S. Matsumura[†], and T. Eguchi[*]

[*]Department of Applied Physics, Fukuoka University, Fukuoka 814-0180, Japan
[†]Department of Applied Quantum Physics and Nuclear Engineering, Kyushu University, Fukuoka 819-0395, Japan

Abstract. TDGL formulation has been applied for $L1_0$ type ordering in a binary alloy, taking account of the crystal symmetry. Kinetic equations for time-evolution of the order parameters and the concentration are derived from the Ginzburg-Landau type potential consisting of the mean-field free energy density and the interfacial energy terms. Three-dimensional simulation of the kinetic equations was performed to simulate time-evolution of off-phase domain structures in real alloy systems. The microstructures obtained are compared with the experimental results of TEM observation.

Keywords: Continuous medium, Order parameter, TDGL formulation, Tweed pattern, Twin boundary
PACS: 64.60.Cn

INTRODUCTION

The $L1_0$ type ordered structure is sometimes formed in *fcc*-based AB alloys such as CuAu. In the $L1_0$ structure, A atoms occupy corner-sites and one kind of face-centered sites preferentially, while B atoms are on the other two kinds of face-centered sites. The structure is characterized by alternating stacking of two different (100) planes. Therefore three orientational variants can be formed depending on which of <100> directions corresponds to *c*-axis and two translational variants are possible for each of them; six distinct crystallographic variants exist in this type of order. The $L1_0$ type ordering process from an *fcc* disordered state ($A1$) involves cubic-tetragonal transition. This structural change causes lattice mismatch between the parent cubic and the product tetragonal phases as well as between different orientation variants of the product phase. It has been observed in the cubic-tetragonal transition of some alloys that tweed structure at first appears and then it continuously transforms into twinning structure with alternate arrays of plates of two different orientational variants [1-2].

The present authors have developed a time-dependent Ginzburg-Landau (TDGL) formulation on a continuous medium for ordering process of $L1_0$ type in binary alloys. The authors apply the formulation to simulate time-evolution of boundary structures with the characteristic anisotropy observed in images by transmission electron microscopy (TEM).

FORMULATION

Atomic Occupation Probabilities

The fundamental *fcc* lattice is divided into four simple cubic sublattices, two of which are preferentially occupied by A atoms in the $L1_0$ structure (see Fig. 1).

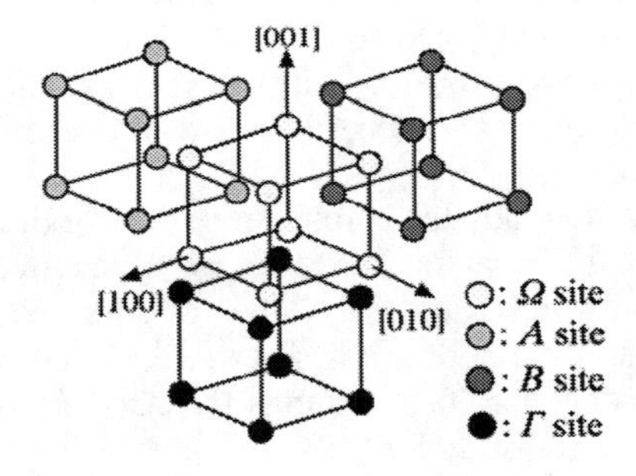

FIGURE 1. The crystal lattice of $L1_0$ structure is decomposed into four simple cubic sublattices Ω, A, B and Γ to treat six types of variants equivalently.

If four sites are grouped into two of equivalent sites, two translational variants with the same direction of *c*-axis are defined;

$A \equiv \Omega \neq B \equiv \Gamma$: c // [100], $B \equiv \Omega \neq \Gamma \equiv A$: c // [010], $\Gamma \equiv \Omega \neq A \equiv B$: c // [001].

CP982, *Complex Systems, 5th International Workshop on Complex Systems*
edited by M. Tokuyama, I. Oppenheim, and H. Nishiyama

In disordered $A1$ state, on the other hand, four sites are equivalent to each other, such as $\Omega \equiv A \equiv B \equiv \Gamma$. Then atomic occupation probabilities for each site of $A_{2(1-\varepsilon)}B_{2(1+\varepsilon)}$ alloy are defined with three order parameters ξ, η and ζ, and a composition parameter ε, as shown in Table 1. The values of (ξ,η,ζ) for the variants are as follows;

$(\pm S,0,0)$: c // [100], $(0,\pm S,0)$: c // [010]
and $(0,0,\pm S)$: c // [001], (1)

where S is the degree of order. If the state of order of atomic arrangement is represented by a point in the three-dimensional Euclidean space spanned by the three order parameters, the six equivalent variants are defined by the six tips of a regular octahedron centered on the origin for the disordered state. The axes of ξ, η and ζ can be parallel to [100], [010] and [001], respectively and pass through midpoints of line segments between two of the corner sites of the sublattices (See Fig. 2). Since the six midpoints are also related to the tips of the regular octahedron, vector $\vec{S}(\xi,\eta,\zeta)$ elongates from the origin in the direction to one of the six points as ordering proceeds.

TABLE 1. Atomic occupation probabilities for each site.

Site	A atoms	B atoms
Ω	$\frac{1}{2}(1-\varepsilon-\xi-\eta-\zeta)$	$\frac{1}{2}(1+\varepsilon+\xi+\eta+\zeta)$
A	$\frac{1}{2}(1-\varepsilon-\xi+\eta+\zeta)$	$\frac{1}{2}(1+\varepsilon+\xi-\eta-\zeta)$
B	$\frac{1}{2}(1-\varepsilon+\xi-\eta+\zeta)$	$\frac{1}{2}(1+\varepsilon-\xi+\eta-\zeta)$
Γ	$\frac{1}{2}(1-\varepsilon+\xi+\eta-\zeta)$	$\frac{1}{2}(1+\varepsilon-\xi-\eta+\zeta)$

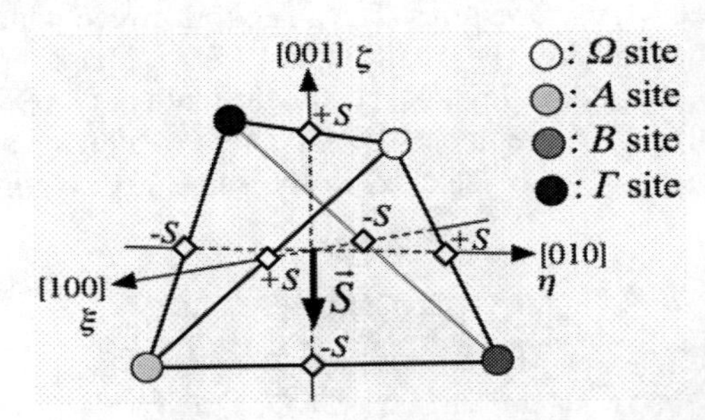

Figure 2. Vector $\vec{S}(\xi,\eta,\zeta)$ in the Euclidian space spanned with the order parameters. Midpoints of line segments between two of the corner sites of the sublattices correspond to six tips of a regular octahedron, indicated by open rhombuses.

These order parameters are measurable, since the structure factors $F(hkl)$ for reciprocal lattice points are given as a function of them. For example,

$F(100)=2(f_A-f_B)\xi$, $F(010)=2(f_A-f_B)\eta$
and $F(110)=2(f_A-f_B)\zeta$, (2)

where f_A and f_B are atomic scattering factors for A and B atoms. Therefore dark-field TEM images taken with 100, 010 and 110 Bragg reflections show spatial variations of ξ^2, η^2 and ζ^2, respectively.

TDGL Formulation

In TDGL formulation thermodynamic potential Φ is given in a form of

$$\Phi\{\xi,\eta,\zeta,\varepsilon,T\} = \int \left\{F(\xi,\eta,\zeta,\varepsilon,T)+G\left(\vec{\nabla}\xi,\vec{\nabla}\eta,\vec{\nabla}\zeta,\vec{\nabla}\varepsilon,T\right)\right\}d\vec{r}. \quad (3)$$

Here the first term F is a free energy density and the second term means the interface energy density. We introduce the free energy density in such a Landau expansion as,

$$F(\xi,\eta,\zeta,\varepsilon,T)=A_0(T)f(\xi,\eta,\zeta,\varepsilon,T) \quad (4)$$

$$f(\xi,\eta,\zeta,\varepsilon,T)=\frac{1}{2}b(\varepsilon-\varepsilon_0)^2+\frac{1}{2}S_e^{\,3}\sigma\left(\xi^2+\eta^2+\zeta^2\right)$$
$$-S_e(S_e+\sigma)\left(\xi^4+\eta^4+\zeta^4\right)+\frac{1}{6}\left(\xi^2+\eta^2+\zeta^2\right)^3, \quad (5)$$

where A_0 and b are positive parameters depending on temperature T. The equilibrium order parameter $S_e(\varepsilon,T)$ and the parameter $\sigma(\varepsilon,T)$ satisfy the condition $S_e(\varepsilon,T)=3\sigma(\varepsilon,T)$ at the transition temperature T_0. The parameter S_e increases and σ decreases with decreasing temperature for $T< T_0$. When $S_e(\varepsilon,T)^3\sigma(\varepsilon,T)<0$ and $S_e(\varepsilon,T)(S_e(\varepsilon,T)+\sigma(\varepsilon,T))>0$, the function f has its maximal value at the origin and the minimal value at the points corresponding to the six variants, $\vec{S}(\xi,\eta,\zeta)=(\pm S_e,0,0)$, $(0,\pm S_e,0)$, $(0,0,\pm S_e)$.

The interface energy density is given in a gradient square approximation [3]. Cubic symmetry should be satisfied in this energy density, because interfaces formed between two of six variants are equivalent to each other;

$$G\left(\vec{\nabla}\xi,\vec{\nabla}\eta,\vec{\nabla}\zeta,\vec{\nabla}\varepsilon,T\right)=\frac{1}{2}\mu Q^2+\frac{1}{2}\nu\left(P_x^{\,2}+P_y^{\,2}+P_z^{\,2}\right)$$
$$+\frac{1}{2}\gamma\left(V_1^2+V_2^2+V_1V_2\right)+\frac{1}{2}\omega\left(W_1^{\,2}+W_2^{\,2}+W_3^{\,2}\right)+\frac{1}{2}\chi\left(\vec{\nabla}\varepsilon\right)^2 \quad (6)$$

where

$$Q\equiv\frac{\partial\xi}{\partial x}+\frac{\partial\eta}{\partial y}+\frac{\partial\zeta}{\partial z}, \quad (7)$$

$$P_x\equiv\frac{\partial\zeta}{\partial y}-\frac{\partial\eta}{\partial z}, P_y\equiv\frac{\partial\xi}{\partial z}-\frac{\partial\zeta}{\partial x}, P_z\equiv\frac{\partial\eta}{\partial x}-\frac{\partial\xi}{\partial y}, \quad (8)$$

$$V_1\equiv\frac{2}{3}\left(2\frac{\partial\xi}{\partial x}-\frac{\partial\eta}{\partial y}-\frac{\partial\zeta}{\partial z}\right), V_2\equiv\frac{2}{3}\left(-\frac{\partial\xi}{\partial x}+2\frac{\partial\eta}{\partial y}-\frac{\partial\zeta}{\partial z}\right), \quad (9)$$

and

$$W_1 \equiv \frac{\partial \zeta}{\partial y} + \frac{\partial \eta}{\partial z}, W_2 \equiv \frac{\partial \xi}{\partial z} + \frac{\partial \zeta}{\partial x}, W_3 \equiv \frac{\partial \eta}{\partial x} + \frac{\partial \xi}{\partial y}. \quad (10)$$

Here $\mu(T)$, $\nu(T)$, $\gamma(T)$, $\omega(T)$ and $\chi(T)$ are positive, and the directions of ξ, η and ζ agree with those of x, y and z, respectively. Tetragonal symmetry of the interfaces is satisfied when the vector $\vec{S}$ (ξ,η,ζ) lengthens towards one of the tips of the regular octahedron as illustrated in Fig. 2. The same type of the interface energy density was applied to the previous work of present authors for $L1_2$ type ordering [4].

TDGL kinetic equations for ξ, η, ζ, and ε are obtained from the thermodynamic potential Φ. Thus

$$\frac{\partial \xi}{\partial t} = -L\frac{\delta \Phi}{\delta \xi} = -L\left\{S_e^{\,3}\sigma\xi - S_e(S_e+\sigma)\xi^3 + \xi\left(\xi^2+\eta^2+\zeta^2\right)^2\right\} + Ll\frac{\partial^2 \xi}{\partial x^2} + Lm\left(\frac{\partial^2 \xi}{\partial y^2} + \frac{\partial^2 \xi}{\partial z^2}\right) + Ln\left(\frac{\partial^2 \eta}{\partial x \partial y} + \frac{\partial^2 \zeta}{\partial z \partial x}\right), \quad (11)$$

$$\frac{\partial \eta}{\partial t} = -L\frac{\delta \Phi}{\delta \eta} = -L\left\{S_e^{\,3}\sigma\eta - S_e(S_e+\sigma)\eta^3 + \eta\left(\xi^2+\eta^2+\zeta^2\right)^2\right\} + Ll\frac{\partial^2 \eta}{\partial y^2} + Lm\left(\frac{\partial^2 \eta}{\partial z^2} + \frac{\partial^2 \eta}{\partial x^2}\right) + Ln\left(\frac{\partial^2 \zeta}{\partial y \partial z} + \frac{\partial^2 \xi}{\partial x \partial y}\right), \quad (12)$$

$$\frac{\partial \zeta}{\partial t} = -L\frac{\delta \Phi}{\delta \zeta} = -L\left\{S_e^{\,3}\sigma\zeta - S_e(S_e+\sigma)\zeta^3 + \zeta\left(\xi^2+\eta^2+\zeta^2\right)^2\right\} + Ll\frac{\partial^2 \zeta}{\partial z^2} + Lm\left(\frac{\partial^2 \zeta}{\partial x^2} + \frac{\partial^2 \zeta}{\partial y^2}\right) + Ln\left(\frac{\partial^2 \xi}{\partial z \partial x} + \frac{\partial^2 \eta}{\partial y \partial z}\right) \quad (13)$$

and

$$\frac{\partial \varepsilon}{\partial t} = M\nabla^2\frac{\delta \Phi}{\delta \varepsilon} = M\left[\nabla^2\left\{a\varepsilon + \frac{1}{2}\frac{\partial S_e^{\,3}\sigma}{\partial \varepsilon}\left(\xi^2+\eta^2+\zeta^2\right) - \frac{\partial S_e(S_e+\sigma)}{\partial \varepsilon}\left(\xi^4+\eta^4+\zeta^4\right)\right\} - \chi\nabla^4\varepsilon\right]. \quad (14)$$

Here $L(T)$ and $M(T)$ are reaction constants and

$$l = \mu + \frac{4}{3}\gamma,\ m = \nu + \omega \text{ and } n = \mu - \nu - \frac{2}{3}\gamma + \omega. \quad (15)$$

We evaluate some types of the energy density of according to eqs. (6)-(10). Evaluated energy densities for interfaces parallel to {110} planes are

$$G_1 \sim \mu + \frac{1}{3}\gamma + \omega \text{ and } G_2 \sim \nu + \gamma, \quad (16)$$

when the displacement vectors of atomic arrangements are normal and parallel to the interface, respectively. The coefficient n in eqs. (11)-(13) is thus found to be the difference in values between G_1 and G_2 according to eq. (15). The coefficients l and m, on the other hand, correspond to evaluated energy values of interfaces parallel to {100} planes. We assumed that the ratio of G_1 to G_2, l and m is small in the simulation so that the interfaces parallel to {110} are preferentially formed.

SIMULATION AND RESULTS

The simulation of $A1$-$L1_0$ transition at stoichiometric composition ε=0 was performed on a three-dimensional cubic grid of 120x120x120 mesh with periodic boundary conditions. The initial $A1$ state is prepared by distributing random numbers around zero level to the parameters of order and composition in the grid cells. Time-evolution of the parameters is obtained by numerically solving difference type of the TDGL equations at integer multiples of the time step. Here the value of the dimensionless time step was chosen to be 0.002.

A result of the simulation at 2.2 k steps is represented in Fig. 3(a). Here the values of ξ^2, η^2 and ζ^2 in ten layers of computational grids were integrated along x-direction, and were plotted in two-dimensionally projected views. Three types of representations show similar tweed patterns with traces almost parallel to <110> directions, indicating coexistence of variants whose c-axis lies on one of the three <100> directions. The analogous tweed pattern is observed in the dark-field TEM image of a $Cu_{50}Au_{40}Pd_{10}$ alloy taken by Matsumura et al. (See Fig. 3(b)) [2]. Here most of Au and Pd atoms can be considered to be stacked in the same {100} plane in

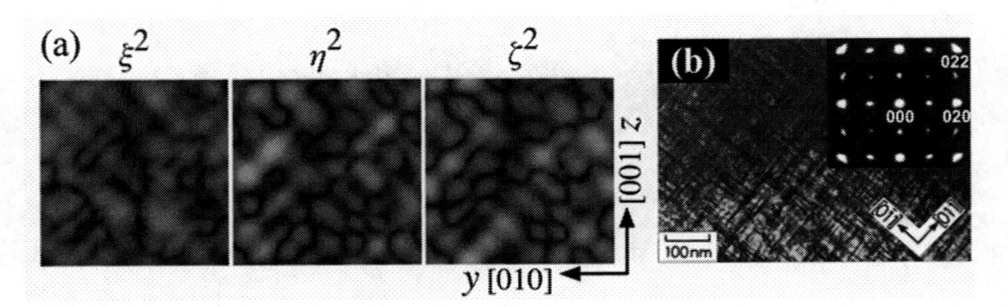

FIGURE 3. (a) Representations of ξ^2, η^2 and ζ^2 obtained by simulation at 2.2 k steps. The gray level varies from dark to bright with an increase in their values. (b) 010 dark-field TEM image and the diffraction pattern of CuAuPd alloy [2]. The TEM image corresponds to η^2 representation according to eq. (2).

the alloy system [5-6]. In the corresponding diffraction pattern, superlattice spots arise 010, 001 and 011 positions in addition to fundamental lattice reflections of *fcc*. Therefore the authors demonstrated that three *c*-axes along the <100> directions are preferred equivalently.

In the later stage of the simulation, variants with *c*-axis alien to [010] and [001] are found predominantly, and twin boundaries along one of the <011> directions are formed between the variants (See Fig. 4(a)). The comparable TEM image by Matsumura et al. is presented in Fig. 4(b) [2]. You can see alternate arrays of dark and bright bands oriented to one of the <011> directions. The diffraction pattern in an upper area of Fig. 4(b) lacks 011 superlattice reflections, and involves splitting of fundamental spots except for 022 systematic reflections. The authors showed clearly that the dark and bright bands correspond to twined $L1_0$ plates of the [010] and [001] variants with {022} habit planes.

Time-evolution of boundary structures is shown in Fig. 5. One may notice that the {011} twin boundaries are continuously formed from the tweed structure with mixture of variants whose *c*-axis lies on one of the <100> directions in the preceding stage.

CONCLUSIONS

In the present study, we have formulated coupled TDGL equations for $L1_0$ type ordering, taking account of the crystal symmetry. Time evolutions of the derived TDGL equations were obtained by three-dimensional simulations, and explain well characteristic features in boundary structures of $L1_0$, which have been observed in dark-field TEM images.

REFERENCES

1. B. Zhang, M. Lelovic and W. A. Soffa, Scripta Metal. **25**, 1577-1582 (1991).
2. S. Matsumura, T. Furuse, Y. Sasano and K. Oki, *Solid-Solid Phase Transformations*, edited by W. C. Johnson et al., TMS, Warrendale, 1994, pp. 485-490.
3. J. W. Cahn and E. J. Hilliard, J. Chem. Phys. **28**, 258-267 (1959).
4. R. Oguma, T. Eguchi, S. Matsumura and S. K. Son, Acta Mater. **54**, 1533-1539 (2006).
5. S. Matsumura, T. Morimura and K. Oki, Mater. Trans. JIM **32**, 905-910 (1991).
6. S. Matsumura, T. Furuse, and K. Oki, Mater. Trans. JIM **39**, 159-168 (1998).

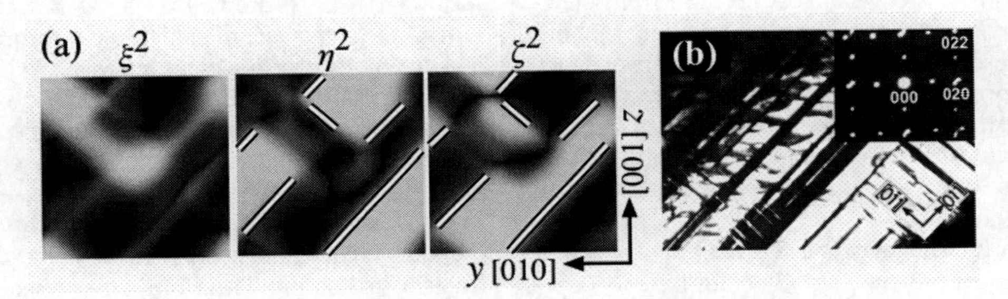

FIGURE 4. (a) Representations of ξ^2, η^2 and ζ^2 obtained by simulation at 20 k steps. The gray level varies from dark to bright with an increase in their values. Comparing the three representations and considering eq. (1), you can notice that the [010] and [001] variants are preferentially formed. Double lines indicate twin boundaries between the variants. (b) 010 dark-field TEM image and the diffraction pattern of $Cu_{50}Au_{40}Pd_{10}$ alloy [2]. The TEM image corresponds to η^2 representation according to eq. (2).

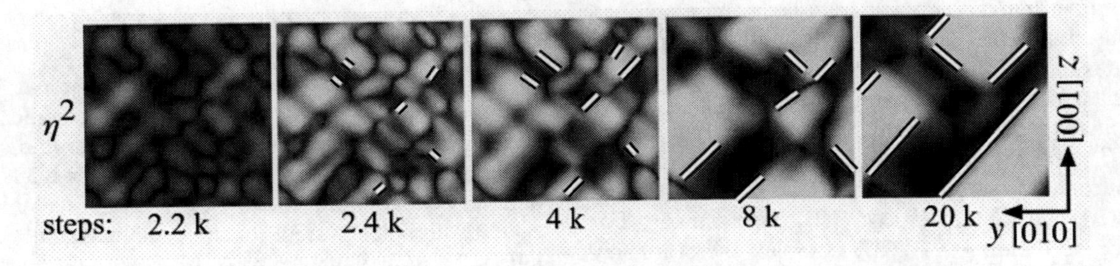

FIGURE 5. Time-evolution of boundary structures is shown by η^2 representation. The gray level varies from dark to bright with an increase in its value. Double lines indicate twin boundaries between variants with c-axis alien to [010] and [001] directions.

Free Energy Calculation of Docking Structure of Azurin(I)-Cytochrome c_{551}(III) Complex Systems by Using the Energetic Representation

Keigo Nishikawa*, Tetsunori Yamamoto*, Ayumu Sugiyama*, Acep Purqon*, Taku Mizukami†, Hideto Shimahara†, Hidemi Nagao*,†, and Kiyoshi Nishikawa*

*Division of Mathematical and Physical Science, Graduate School of Natural Science and Thechnorogy, Kanazawa University, Kakuma, Kanazawa 920-1192, Japan
†School of Materials Science, Japan Advanced Institute of Science and Technology, Ishikawa 923-1211, Japan

Abstract. We investigate the docking stability of Azurin(I) - Cytochrome c_{551}(III) complex by molecular dynamic simulations. The charge distribution around the active sites for Azurin(I) and Cytochrome c_{551}(III) is estimated by quantum chemical calculation to simulate the complex system by molecular dynamic simulations. We estimete some physical properties such as the root square mean deviation, distance between iron ion in active site of Cytochrome c_{551}(III) and copper ion in active site of Azurin(I), the dynamical cross-correlation map and free energy in the energetic representation. We discuss the stability of the complex system of Azurin(I) - Cytochrome c_{551}(III) from these properties.

Keywords: Azurin, Cytochrome c_{551}, Energy representation, Free energy
PACS: 65.40.Gr, 87.15.He

INTRODUCTION

Metalloproteins have important roles in biological cell such as electron transfer, chemical reaction from viewpoint of catalysis and so on. Electron transfer (ET) between metalloproteins has much attracted interests in relation to the formation of complexes in which the partners assemble through complementary contact surface. However, because of the difficulty of crystalization of such complexs, some theoretical approaches have been presented.

An electron transfer between Azurin (Az), which is one of blue copper proteins, and Cytochrome c_{551} (Cyt-c_{551}), which is redox protein with heme ion can, be expressed as a chemical reaction of the following scheme:

$$\mathrm{Az(II)} - \mathrm{Cyt}{-}c_{551}\mathrm{(II)} \Leftrightarrow \mathrm{Az(I)} - \mathrm{Cyt}{-}c_{551}\mathrm{(III)}.$$

Theoretical and experimental works for Azurin (II)-Cytochrome c_{551}(II) complex have been presented and have discussed the docking comformation, chemical properties around the active site and so on by quantum chemical and molocular dynamics approaches [1].

>From experimental results for third structures of Azurin(I) and Cytochrome-c_{551}(III), we can find the little different structures around their active site from Azurin(II) and Cytochrome-c_{551}(II), respectively. For Azurin(I)-Cytochrome-c_{551}(III) complex there are few experimental studies on only the third structure of those proteins because of the difficulty of crystallization of proteins, and the predict the complex structure of Azurin(I) - Cytochrome-c_{551}(III) is important to understand the electron transfer between the proteins. In previous study [2], we have discussed dynamics of complex Azurin(II) - Cytochrome-c_{551}(II) and electron structure around the active site of each protein by quantum chemical and molecular dynamics approach.

In this study, we estimate the docking conformation of the complex system of Azurin (I) and Cytochrome c_{551} and discuss the docking stability of the complex in relation to physical properties such as free energy by using the energy representation [3-5], the distance between the ions in each active site, functionality of the electron transfer and so on. Therefore we estimate average distance between one copper and one heme ions, root-mean-square-deviations for each proteins and a predicted complex system, dynamical cross-correlation, and free energy by using the energy representation.

COMPUTATIONAL METHOD

In this section, we present the computational methods for Azurin(I)-Cytochrome c_{551}(III) complex system. First, we approximately calculate charge distribution around each active site by using cluster models. Second, we decide initial structure of Azurin(I)-Cytochrome c_{551}(III) complex system for molocular dynamics simulation. Third, we analyze some physical properties such as RMSD, dynamical cross correlation and so on from the simulation of the complex system. Finaly, we present

CP982, *Complex Systems, 5th International Workshop on Complex Systems*
edited by M. Tokuyama, I. Oppenheim, and H. Nishiyama

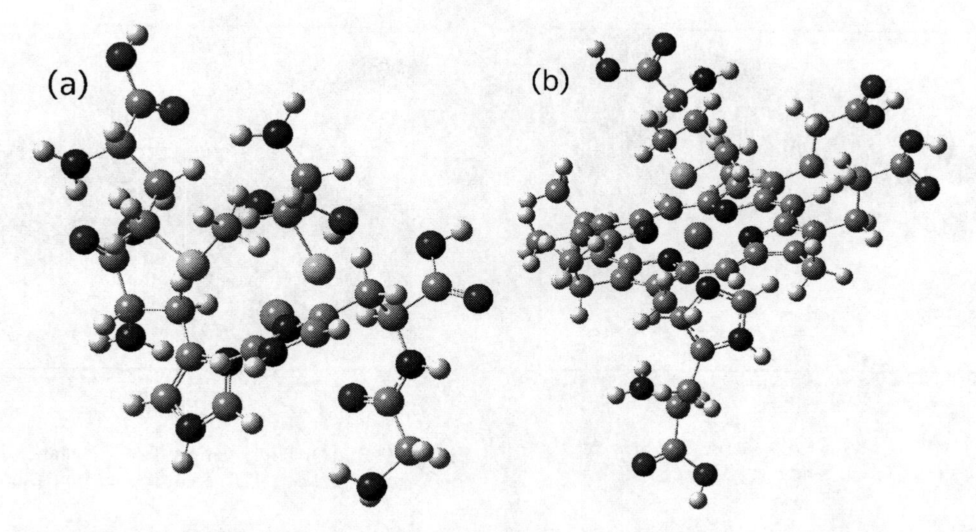

FIGURE 1. Cluster models of the active site of Azurin(II) and Cytochrome c_{551}(II). (a)Active site model of Azurin. (b)Active site model of Cytochrome c_{551} used for calculating the charge distribution.

free energy calculated by using the energetic representation and discuss docking stability by molocular dynamics simulation of Azurin(I)-Cytochrome c_{551}(III) complex system. We calculate charge distribution around active sites of Azurin(I) and Cytochrome c_{551}(III). To obtain charge distribution around the active site, we use a cluster model of each active site as shown Fig.1. The cluster model of Azurin(I) shown in Fig.1(a) consists of one copper ion and five residues which are Gly45, His46, Cys112, His117, and Met121. Figure 1(b) shows the cluster model of Cytochrome c_{551}(III) which consists of one iron ion, the porphyrin ring, and two residues of His16 and Met61. The configuration of the atoms in the active site cluster models of Azurin(I) and Cytochrome c_{551}(III) is from the Brookhaven Protein Data Bank (PDB) of 1E5Y and 351C, respectively. For the calculation of charge distributions, we use Gausssian 03 program package. In this study, the restrained electrostatic potential (RESP) charge, which is well known charges in molocular dynamics simulation, is adopted for atomic charges.

Next,as the initial structure of Azurin(I)-Cytochrome c_{551}(III) complex system we adopt the structure determined from last snapshot of molecular dynamics simulation of Azurin(II)-Cytochrome c_{551}(II)[2] with the calculated charge distribution around each active sites. To represent the water enviroment, 6204 TIP5P [6] water molecules was used in our simulation. Water box size is 79.464 × 50.437 × 54.032. SHAKE [7] algorithm is adopted for the bonds involving hydrogen, and the vibration of hydrogen is fixed. All MD simulations are carried out using the AMBER8 program package with force field 03 [8].

To confirm the docking stability, we calculate the root mean square deviation (RMSD) by calculating the following equation:

$$RMSD(t) = \sqrt{\frac{1}{N}\sum_{i}^{N}(R(i,t) - R(i,0))}. \qquad (1)$$

In addition, we discuss the correlation motion between protein atoms i and j by calculating the following equation:

$$C_{i,j} = \frac{< \Delta r_i \cdot \Delta r_j >}{\sqrt{< \Delta r_i^2 \cdot \Delta r_j^2 >}}. \qquad (2)$$

The dynamical cross-correlation map is a matrix representation of the correlation motion. Positive values are indicate motions in the same direction, and negative values mean motions in the opposite direction.

To estimate the stability of the complex system, we calculate the distance between copper ion and heme ion in the Azurin(I) and Cytochrome c_{551}(III) active sites, respectively.

In this study, we calculate free energy of the complex system according to the energy representation [3-5] by calculating the difference of chemical potentials $\Delta\mu$. $\Delta\mu$ is given by

$$\Delta\mu = -k_B T \int d\varepsilon[(\rho(\varepsilon) - \rho_0(\varepsilon)) + \frac{1}{k_B T} w(\varepsilon)\rho(\varepsilon) \\ -\{\alpha(\varepsilon)F(\varepsilon) + (1-\alpha(\varepsilon))F_0(\varepsilon)\}(\rho(\varepsilon) - \rho_0(\varepsilon))] \qquad (3)$$

where ε, k_B, and T mean energy, Boltzmann constant and temperature, respectively. $\rho(\varepsilon)$ and $\rho_0(\varepsilon)$ are energy distribution function corresponding to the solution and pure solvent, respectively, while, $w(\varepsilon)$ in solute condition and $w_0(\varepsilon)$ in non solute condition for the indirect part of the solute-solvent potential of mean force.

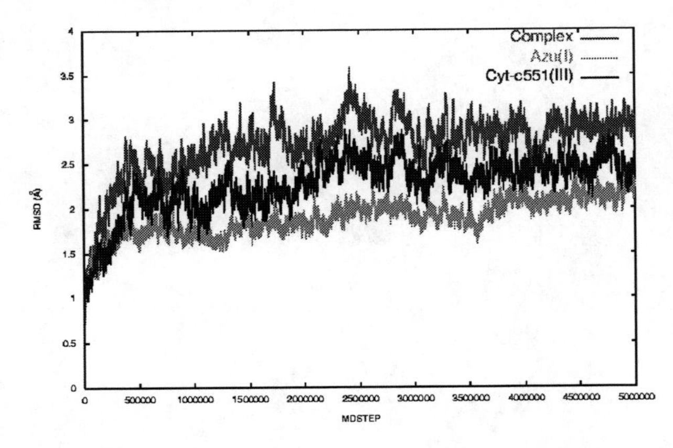

FIGURE 2. RMSDs versus MD steps. RMSDs of complexes, Azurin(I) and Cytochrome c_{551}(III)).

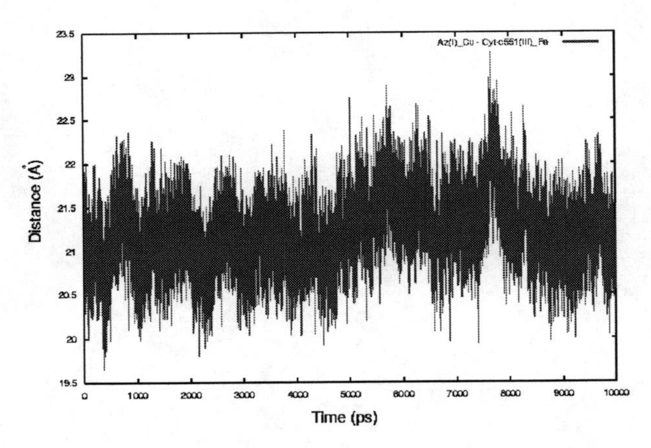

FIGURE 3. Distance versus Time. Distance Between Iron Ion(Cytochrome c_{551}(III)) And Copper Ion(Azurin(I))

$F(\varepsilon)$ and $F_0(\varepsilon)$ of the right hand side in Eq.(1) that are functions only of $w(\varepsilon)$ and $w_0(\varepsilon)$ are written as

$$F(\varepsilon)=\begin{cases}\beta w(\varepsilon)+1+\frac{\beta w(\varepsilon)}{exp(-\beta w(\varepsilon))-1} & (w(\varepsilon)\leq 0)\\ \frac{1}{2}\beta w(\varepsilon) & (w(\varepsilon)\geq 0).\end{cases} \tag{4}$$

$$F_0(\varepsilon)=\begin{cases}-\ln(1-\beta w_0(\varepsilon))+1+\frac{\ln(1-\beta w_0(\varepsilon))}{\beta w_0(\varepsilon)} & (w_0(\varepsilon)\leq 0)\\ \frac{1}{2}\beta w_0(\varepsilon) & (w_0(\varepsilon)\geq 0).\end{cases} \tag{5}$$

the weight factor $\alpha(\varepsilon)$ is written by

$$\alpha(\varepsilon)=\begin{cases}1 & (\rho(\varepsilon)\leq\rho_0(\varepsilon))\\ 1-\left(\frac{\rho(\varepsilon)-\rho_0(\varepsilon)}{\rho(\varepsilon)+\rho_0(\varepsilon)}\right)^2 & (\rho(\varepsilon)\leq\rho_0(\varepsilon)).\end{cases} \tag{6}$$

RESULT AND DISCUSSION

We carry out molocular dynamics simulation of Azurin(I)-Cytochrome c_{551}(III). RMSD of Azurin(I)-Cytochrome c_{551}(III) complex system is shown in Fig.2 with RMSDs of Az(I) and Cyt-c_{551}(III). >From Fig.2 we can find a stable complex system after 2,500,000 MD steps (after 5.0ns). The distance between iron and copper ion is shown in Fig.3. The average distance is 21.211 ± 0.385 Åfrom 0.5 ns to 10.0 ns. For analysis of dynamical cross correlation and free energy, we use simulation data from 9.95 ns to 10.0 ns when the structure of complex system is stable.

We show the dynamical cross correlation maps in Fig.4. Figure 4(a) shows the dynamical cross correlation map between Azurin(II) versus Cytochrome c_{551}(II). In the case of Azurin(II)-Cytochrome-c_{551}(II), we can find the positive correlation regions which indicate residue numbers 40-45 in Az(II) and 30-40 in Cty-c_{551}(II) (Docking part A) and 64-72 in Az(II) and 53-58 in Cyt-c_{551}(II) (Docking part B) as shown Fig.5. The results of Az(II)-Cyt-c_{551}(II) suggests that these two docking sites collectively and strongly move or vibrate and that the dynamics of each protein is influenced by the protein docking.

Figure 4(b) shows the dynamical cross-correlation function for Azurin(I) versus Cytochrome c_{551}(III). In Figure 4(b), we can also find positive correlation around both docking parts A and B, and another region of the positive correlation appears around residue number 50-60 in Az(I) and 1-18 and 30-83 in Cyt-c_{551}(III) (Docking part C) as shown in Fig.5. Docking part C in Az(I) corresponds to the alpha helix. These results suggest that docking structure of Azurin(I)-Cytochrome c_{551}(III) is more stable than that of Azurin(II)-Cytochrome c_{551}(II) for the docking stability of Azurin-Cytochrome c_{551} complex system.

$\Delta\mu$(Az(II)-Cyt c_{551}(II)) becomes -162.9 kcal/mol and $\Delta\mu$(Az(I)-Cyt c_{551}(III)) is -722.3 kcal/mol. Therefore the difference of free energy between $\Delta\mu$(Az(I)-Cyt c_{551}(III)) and $\Delta\mu$(Az(II)-Cyt c_{551}(II)) is 559.4 kcal/mol.

CONCLUSION

We have investigated the docking stability of Azurin(I) - Cytochrome c_{551}(III) complex by molecular dynamic simulations. The charge distribution around the active sites for Azurin(I) and Cytochrome c_{551}(III) has been estimated by quantum chemical calculation to simulate the complex system by molecular dynamic simulations. We have estimated some physical properties such as the root square mean deviation, distance between iron ion in active site of Cytochrome c_{551}(III) and copper ion in active site of Azurin(I), the dynamical cross-correlation map and free energy in the energetic representation.

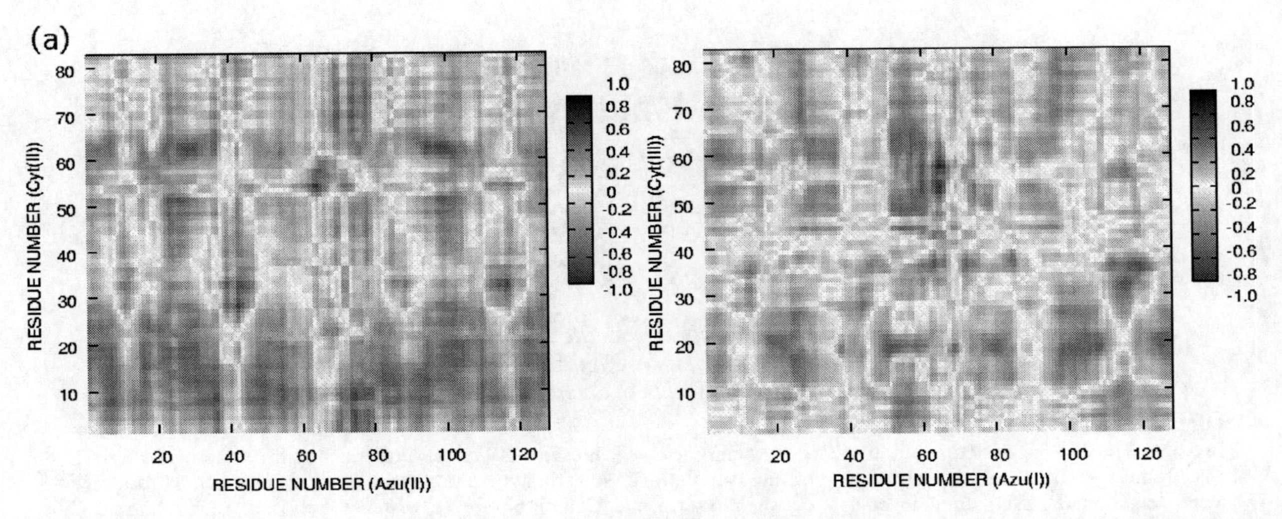

FIGURE 4. Dynamical cross-correlation maps. (a) Azurin(II) versus Cytochrome c_{551}(II). (b) Azurin(I) versus Cytochromec_{551}(III)

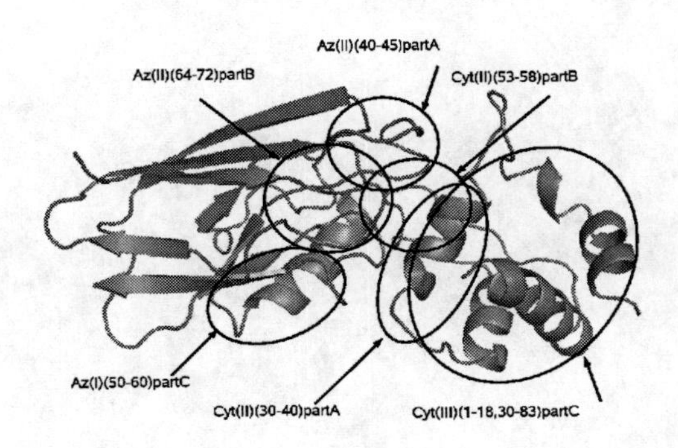

FIGURE 5. Docking Part of dynamical cross correlation maps of Azurin - Cytochrome c_{551} complex.

We have determined the docking conformation of Azurin(I) - Cytochrome c_{551}(III) and have found one docking site of the alpha helix in Azurin(I) becomes important for the docking their proteins. The dynamics of the part of the alpha helix strongly and collectively vibrates with Cytchrome c_{551}(III). We have discussed the stability of the complex system of Azurin(I) - Cytochrome c_{551}(III) in relation to these properties.

$\Delta\mu$(Az(II)-Cyt c_{551}(II)) becomes -162.9 kcal/mol and $\Delta\mu$(Az(I)-Cyt c_{551}(III)) is -722.3 kcal/mol. Therefore the difference of free energy between $\Delta\mu$(Az(I)-Cyt c_{551}(III)) and $\Delta\mu$(Az(II)-Cyt c_{551}(II)) is 559.4 kcal/mol. $\Delta\mu$(Az(I)-Cyt c_{551}(III)) lower than $\Delta\mu$(Az(II)-Cyt c_{551}(II)) and differential is 559.4 kcal/mol.

ACKNOWLEDGEMENTS

H.N is grateful for financial support from the Ministry of Education, Science and Culture of Japan (grant 19029014).

REFERENCES

1. F. Cutruzzol, *et al.*, *J. Inorganic Biochem.*, **88**, 353-361 (2002).
2. A. Sugiyama, *et al.*, *Int J Quantum Chem*, **106**, 3071-3078 (2006).
3. N. Matubayasi, and M. Nakahara, *J. Chem. Physics*, **113**, 6070-6081 (2000).
4. N. Matubayasi, and M. Nakahara, *J. Chem. Physics*, **117**, 3605-3616 (2002).
5. N. Matubayasi, and M. Nakahara, *J. Chem. Physics*, **119**, 9689-9702 (2003).
6. M. Mahoney, and W. Jorgensen, *J. Chem. Phys.*, **112**, 8910-8922 (2000).
7. J. P. Ryckaert, G. Ciccotti, and H. Berendsen, *J. Comput. Phys.*, **23**, 327-341 (1977).
8. D. A. Pearlman, *et al.*, *Comp. Phys. Commun*, **91**, 1-41 (1995).

Theoretical Study of Free Energy in Docking Stability of Azurin(II)-Cytochrome c_{551}(II) Complex System

Tetsunori Yamamoto*, Keigo Nishikawa*, Ayumu Sugiyama*, Acep Purqon*, Taku Mizukami†, Hideto Shimahara†, Hidemi Nagao*,†, and Kiyoshi Nishikawa*

*Division of Mathematical and Physical Science, Graduate School of Natural Science and Technology, Kanazawa University, Kakuma, Kanazawa 920-1192, Japan
†School of Materials Science, Japan Advanced Institute of Science and Technology, Ishikawa 923-1211, Japan

Abstract. The docking structure of the Azurin-Cytochrome c_{551} is presented. We investigate a complex system of Azurin(II)-Cytochrome c_{551}(II) by using molecular dynamics simulation. We estimate some physical properties, such as root-mean-square deviation (RMSD), binding energy between Azurin and Cytochrome c_{551}, distance between Azurin(II) and Cytochrome c_{551}(II) through center of mass and each active site. We also discuss docking stability in relation to the configuration by free energy between Azurin(II)-Cytochrome c_{551}(II) and Azurin(I)-Cytochrome c_{551}(III).

Keywords: Azurin, Cytochrome, Electron transfer, Free energy
PACS: 65.40.Gr, 87.15.He

INTRODUCTION

Electron transfer (ET) between proteins has much attracted interests to understand biological system such as oxidized and reduced chemical reaction from a viewpoint of catalysis and so on. Many proteins have metal ions around their active site, and the metal ions and their enviroment are important for electron transfer. These functions are the follows : (1) allosteric interactions between the active sites in different subunits, (2) surface recognition sites for interactions with donor and acceptor proteins, (3) forming a substrate binding pocket near the metal active site.

The investigations of complex structure of metalloproteins have attracted interests in relation to the ET function. In previous work [1-3], we have investigated docking structure of Azurin(II)-Cytochrome c_{551}(II) complex systems and have also analyzed the dynamics of the complex system by quantum chemical and molecular dynamics approaches. The ET between Azurin(Azu) and Cytochrome c_{551}(Cyt c_{551}) can be expressed as

$$\mathrm{Azu(II)} + \mathrm{Cyt}\ c_{551}\mathrm{(II)} \Leftrightarrow \mathrm{Azu(I)} + \mathrm{Cyt}\ c_{551}\mathrm{(III)}. \quad (1)$$

This ET reaction has been studied for wild-type and some mutant protein[4,12].

Azurin from *Pseudomonas aeruginosa* is one of the blue copper proteins that consists of 128 residues and has copper ion in the active site. Cytochrome c_{551} from *Pseudomonas aeruginosa* is a monomeric redox protein that consists of 82 residues and has a ferric iron in the active site. We have estimated electrical charge distribution and some parameters around the active sites from quantum chemical calculation.

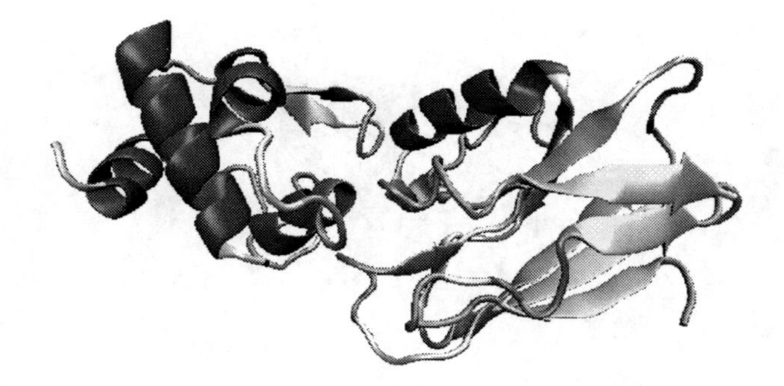

FIGURE 1. Higher-order structure of complex system of Azurin(II) - Cytochrome c_{551}(II). Left side is Cytochrome c_{551}(II) and right side is Azurin(II).

We investigate a complex system (Figure 1) of Azurin (II) (PDBID: 4AZU) [5] and Cytochrome c_{551} (PDBID: 451C) [6] by using molecular dynamics simulation, and discuss docking stability in relation to the configuration by free energy. We estimate the difference of free energy [7] between complex systems Azurin(II) - Cytochrome c_{551}(II) and Azurin(I) - Cytochrome c_{551}(III).

In this study, to discuss docking stability of Azurin-Cytochrome c_{551}, we consider free energy difference ΔG. Free energy is one of the most important general physical properties that describes the tendency to associate and react. We also calculate root-mean-square deviation (RMSD), binding free energy between Azurin and Cytochrome c_{551} by mm_pbsa method[11], distance between Azurin(II) and Cytochrome c_{551}(II) through center of mass and each active site. We pay attention to the docking stability of the complex system of Azurin(II)

CP982, *Complex Systems, 5th International Workshop on Complex Systems*
edited by M. Tokuyama, I. Oppenheim, and H. Nishiyama

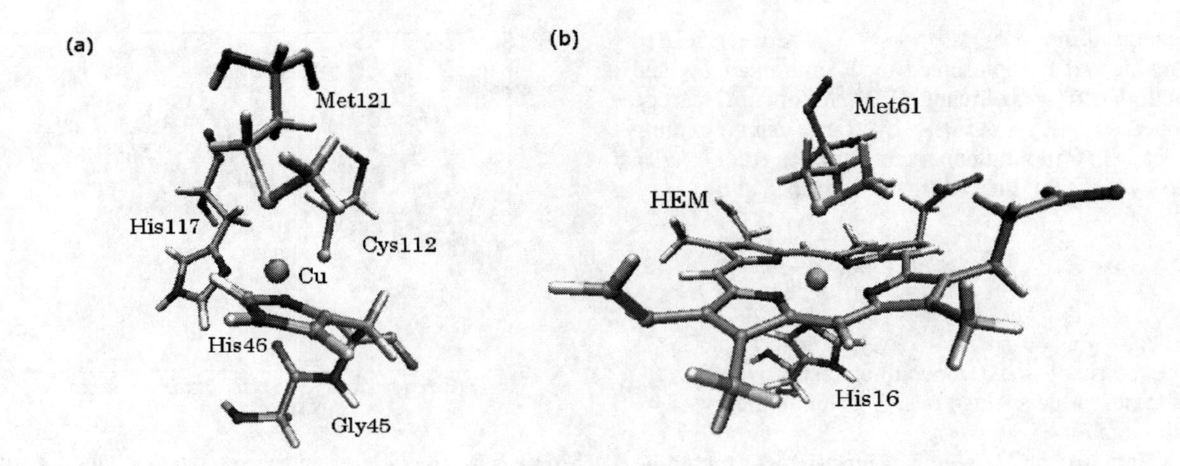

FIGURE 2. Cluster models of the active site of Azurin(II) and Cytochrome c_{551}(II). (a)Active site model of Azurin. (b)Active site model of Cytochrome c_{551}.

and Cytochrome c_{551}(II) in relation to free energy.

COMPUTATIONAL METHOD

We consider cluster models of 4AZU and 451C, which are registered in the Brookhaven Protein Data Bank (PDB). These cluster models consist of the copper atom or iron atom, amino acid residues, and porphyrin ring. The cluster model of Azurin is constituted of the copper atom and five residues (Gly45, His46, Cys112, His117, and Met121). Cytochrome c_{551} is also constituted of porphyrin ring and two residues (His16 and Met61). We show the cluster models of Azurin and Cytochrome c_{551} for models (a) and (b) in Figure 2, respectively.

To estimate the charge distribution and the force field parameters of Azurin(II) and Cytochrome c_{551}(II) for each cluster model, we adopted the same method of our previous study, using the Gaussian program package. From the geometry-optimized structures of models (a) and (b) in Figure 2, the atomic charges for MD simulations are calculated. The restrained electrostatic potential (RESP)[13,14] charge, generated by the electrostatic potential and produced by Kollman, is adopted for atomic charges. Furthermore, we estimated the charge distribution of Azurin(I) and Cytochrome c_{551}(III) by the similar method [3] for Azurin(II)-Cytochrome c_{551}(II), using cluster models of Azurin(I) (PDBID:1E5Y) and Cytochrome c_{551}(III) (PDBID:351C). Table 1 shows atomic charge distribution of Azurin(II)-Cytochrome c_{551}(II) and Azurin(I)-Cytochrome c_{551}(III).

For MD simulations of Azurin-Cytochrome c_{551} complex, we adopted the RESP charge of each active site in Table 1. MD simulations were carried out under a constant volume condition with a periodic boundary at room temperature (300K), using the AMBER8 program package with force field 03[11]. To represent the water enviroment, a 10 Å-thick TIP5P [9] water molecule layer was applied in each simulation. SHAKE [10] algorithm was adopted for the bonds involving hydrogen, and the vibration of hydrogen was fixed. Total run time is 2.5 ns (1MD steps=0.002 ps); the cutoff radii for nonbonded interactions are 8 Å.

TABLE 1. Atomic charge distribution of Azurin and Cytochrome c_{551} from RESP method.

	Azu(II)-Cyt(II)	Azu(I)-Cyt(III)
Azurin		
Cu	0.591	0.243
Gly45	0.213	0.115
His46	0.403	0.124
Cys112	-0.281	-0.663
His117	0.403	0.732
Met121	0.311	-0.130
Cytochrome c_{551}		
Fe	1.004	0.860
His16	0.418	0.717
Met61	0.311	0.570
Porphyrin	0.314	0.850

We calculated free energy differences ΔG. The statistical mechanical definition of free energy is in terms of the partition function, a sum of the Bolzmann weights of all the energy levels of the systems. However, we had to calculate an integration over all $3N$ degrees of freedom, where N means number of atoms in the system.

In this study, we adopted the method of evaluation of

free energy differences $\Delta G = G_B - G_A$ between related systems A and B represented by Hamiltonian H_A and H_B, introduced by Kollman[11]. Hamiltonian H can be described as $H(\lambda) = \lambda H_B + (1-\lambda)H_A$ using coupling parameter λ. Coupling parameter λ varies from 0 ($H = H_A$) to 1 ($H = H_B$). Free energy is described as

$$\Delta G = G_B - G_A = -\sum_{\lambda=0}^{1} RT \ln\left\langle \exp\left(\frac{-\Delta H'}{RT}\right)\right\rangle_\lambda, \quad (2)$$

where $\Delta H' = H(\lambda + d\lambda) - H(\lambda)$, R and T in Eq.(2) are gas constant and temperature, respectively. $\langle\ldots\rangle_\lambda$ refers to ensemble average over a system represented by Hamiltonian $H(\lambda)$.

The Hamiltonian H_A means Azurin(II) - Cytochrome c_{551}(II) state when $\lambda = 0$ and the Hamiltonian H_B means Azurin(I) - Cytochrome c_{551}(III) state when $\lambda = 1$.

RESULT AND DISCUSSION

We carried out MD simulations of the Azurin-Cytochrome c_{551} complex system coupled from Azurin(II) - Cytochrome c_{551}(II) to Azurin(I) - Cytochrome c_{551}(III) with coupling parameter λ. Hamiltonian H of Eq.(2) containes potential energy term U, which consisted of bond, angle, dihedral potentials, van der Walls interaction and electrostatic interaction:

$$V_{\text{elec}} = \sum_{i<j}^{\text{atoms}} \frac{q_i q_j}{\varepsilon R_{ij}}, \quad (3)$$

where q_i and ε are charges and dielectric constant, R_{ij} means distance between site i and j. We change the charge q, because the ET between Azurin and Cytochrome c_{551} results in changing charge q in Eq.(3). The charge parametars are determed from $q(\lambda) = q_0 + \lambda(q_1 - q_0)$ [8], where q_0 and q_1 mean charge parameters of Azurin(II) - Cytochrome c_{551}(II) and Azurin(I) - Cytochrome c_{551}(III) in Table 1, respectively. In this study, we choose $d\lambda$ as 0.25 , coupling parameter λ gose from 0 to 1. We carried out five MD simulations with each coupling parameter λ.

RMSD of Azurin-Cytochrome c_{551} complex system of each λ are shown in Figure 3. From the RMSD of all atoms in the complexes, a stable doucking state is achieved after 600,000 MD steps (1.2ns).

Figure 4 shows (1) l_{mass}, the distance between the center of mass of Azurin and Cytochrome c_{551} and (2) l_{metal}, the distance between the copper atom in Azurin and the iron atom of HEM in Cytochrome c_{551}. l_{mass} and l_{metal} are 29.16 ± 0.16 Å and 22.93 ± 0.27 Å when $\lambda = 0$, 28.99 ± 0.15 Å and 21.27 ± 0.27 Å when $\lambda = 1$ during 50 ps (from 2.45 to 2.50 ns). As λ goes from 0 to 1, distances between Azurin and Cytochrome c_{551} become short.

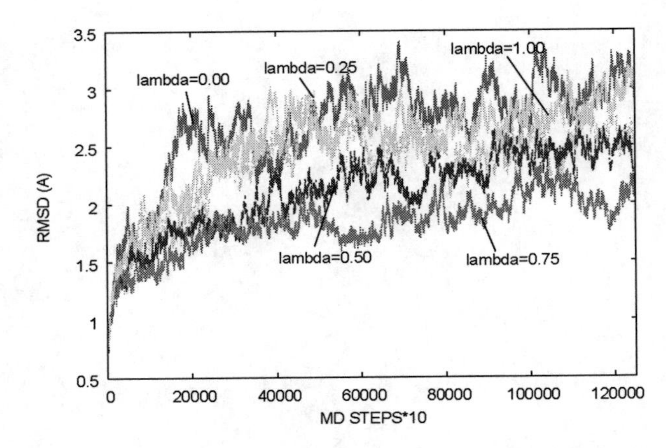

FIGURE 3. RMSDs of complexes versus MD steps with each coupling parameter λ. RMSD was calculated from all atoms of docking model.

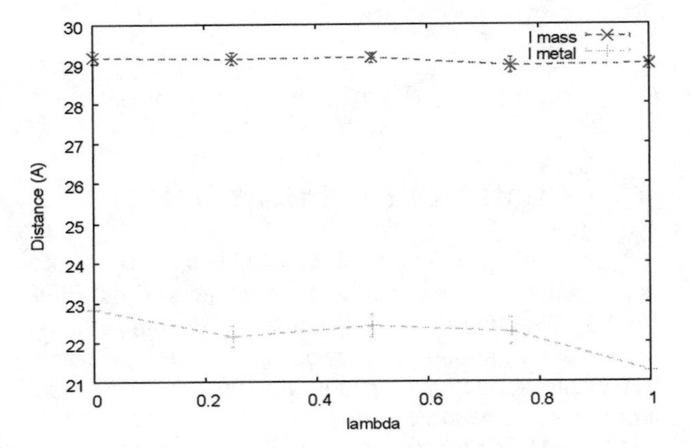

FIGURE 4. Distance between the center of mass of (l_{mass}) Azurin and Cytochrome c_{551} and between the copper atom in Azurin and the iron atom of HEM in Cytochrome c_{551} (l_{metal}).

We computed free energy changes ΔG from Eq.(2) during 100 ps (from 2.40 to 2.50 ns) which were presented in Table 2 and in Figure 5. Simulations were run for λ_i=0.00, 0.25, 0.50, 0.75 and 1.00. In order to check the self-consistency, the simulations were run in both directions, i.e. $\lambda_i \rightarrow \lambda_{i+1}$ and $\lambda_{i+1} \rightarrow \lambda_i$ except at the two end points. This is known as "double-ended" sampling[15]. As λ goes from 0 to 1, we found that ΔG decreased. ΔG between Azurin(II) - Cytochrome c_{551}(II) and Azurin(I) - Cytochrome c_{551}(III) is -46.32 kcal/mol in $\lambda = 0 \rightarrow 1$ process, and 44.49 kcal/mol in $\lambda = 1 \rightarrow 0$ process. >From these results, we concluded Azurin(I) - Cytochrome c_{551}(III) complex system is more stable than Azurin(II) - Cytochrome c_{551}(II) complex system.

We also caluculated the binding free energy during

TABLE 2. Free energy change for interconversion of Azurin(II) - Cytochrome c_{551}(II) and Azurin(I) - Cytochrome c_{551}(III). Coupling parameters λ_i and λ_j mean intermediate state between Azurin(II) - Cytochrome c_{551}(II) and Azurin(I) - Cytochrome c_{551}(III).

λ_i	λ_j	ΔG (kcal/mol) $i \rightarrow j$	$j \rightarrow i$
0.00	0.25	-18.95 ± 2.14	18.95 ± 1.97
0.25	0.50	-4.70 ± 1.97	4.70 ± 1.82
0.50	0.75	-9.33 ± 1.52	9.33 ± 1.55
0.75	1.00	-13.34 ± 1.37	11.52 ± 1.57

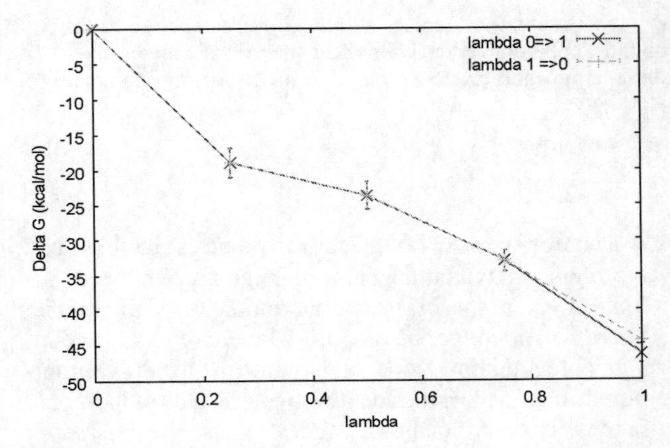

FIGURE 5. Computed free energy changes for the interconversion of Azurin(II) - Cytochrome c_{551}(II) (λ=0) and Azurin(I) - Cytochrome c_{551}(III) (λ=1).The solid and dashed lines correspond to forward ($\lambda = 0 \rightarrow 1$) and reverse $\lambda = 1 \rightarrow 0$ processes, respectively.

200 ps (from 2.30 to 2.50 ns) between Azurin and Cytochrome c_{551} by the mm_pbsa method. >From the MD simulation, binding free energy is -80.73 ± 1.21 kcal/mol when $\lambda = 0$ and is -63.16 ± 0.36 kcal/mol when $\lambda = 1$. Therefore, the binding of Azurin(II) - Cytochrome c_{551}(II) complex system is stronger than that of Azurin(I) - Cytochrome c_{551}(III) complex system.

The calculation of ΔG (Figure 5) includes all nonbond interactions, however that of the binding free energy does not include water-protein interaction. Thus, we suggest the contribution of water behavior made ΔG decrese in $\lambda = 0 \rightarrow 1$ process, although the protein-protein binding free energy is increasing.

CONCLUDING REMARKS

We caluculated the charge distributions of the cluster models of Azurin(I) and Cytochrome c_{551}(III) from RESP method. The free enegy differences ΔG were calculated from five MD simulations which have a variety of the coupling parameters λ in Eq.(2) to produce "mixed" charge distributions. To discuss the docking stability between Azurin and Cytochrome c_{551}, we estimate physical properties, such as RMSD, distance between Azurin and Cytochrome c_{551}, free energy and binding free energy. >From these results, Azurin(I) - Cytochrome c_{551}(III) complex system is more stable than Azurin(II) - Cytochrome c_{551}(II) complex system.

ACKNOWLEDGMENTS

H. N. is grateful for financial support from the Ministry of Education, Science and Culture of Japan (grant 19029014).

REFERENCES

1. T. Shuku, *et al.*, *Polyhedron*, **24**, 2665-2670 (2005).
2. A. Sugiyama, *et al.*, *Int. J. Quantum. Chem.*,**105**, 588-595 (2005).
3. A. Sugiyama, *et al.*, *Int. J. Quantum. Chem.*, **106**, 3071-3078 (2006).
4. F. Cutruzzola, *et al.*, *Biochem. J.*, **322**, 35-42 (1997).
5. H. Nar, *et al.*, *J. Mol. Biol.*, **221**, 765-772 (1991).
6. Y. Matsuura, T. Takano, and R. E. Dickerson, *J. Mol. Biol.*, **156**, 389-409 (1982).
7. P. Kollman, *Chem. Rev.*, **93**, 2395-2417 (1993).
8. W. Jorgensen, and C. Ravimohan, *J. Chem. Phys.*, **83**, 3050-3054 (1985).
9. M. Mahoney, and W. Jorgensen, *J. Chem. Phys.*, **112**, 8910-8922 (2000).
10. J. P. Ryckaert, G. Ciccotti, and H. Berendsen, *J. Comput. Phys.*, **23**, 327-341 (1977).
11. D. A. Pearlman, *et al.*, *Comp. Phys. Commun.*, **91**, 1-41 (1995).
12. F. Cutruzzola, *et al.*, *J. Inorganic. Biochem.*, **88**, 353-361 (2002).
13. C. I. Bayly, *et al.*, *J. Phys. Chem.*, **97**, 10269-10280 (1993).
14. P. Cieplak, *et al.*, *J. Comput. Chem.*, **16**, 1357-1377 (1995).
15. C. H. Bennett, *J. Comp. Phys.*, **22**, 245-268 (1976).

Effect of Breather Existence on Reconstructive Transformations in Mica Muscovite

J. F. R. Archilla*, J. Cuevas*, and F. R. Romero†

*Grupo de Física No Lineal. Departamento de Física Aplicada I. ETSI Informática.
Universidad de Sevilla. Avda. Reina Mercedes, s/n. 41012-Sevilla, Spain
†Grupo de Física No Lineal. Facultad de Física.
Universidad de Sevilla. Avda. Reina Mercedes, s/n. 41012-Sevilla, Spain

Abstract. Reconstructive transformations of layered silicates as mica muscovite take place at much lower temperatures than expected. A possible explanation is the existence of breathers within the potassium layer. Numerical analysis of a model shows the existence of many different types of breathers with different energies and existence ranges which spectrum coincides approximately with a statistical theory for them.

Keywords: Discrete breathers, Intrinsic localized modes, Reconstructive transformations
PACS: 63.20.Ry, 63.20.Pw, 64.70.Kb, 82.20.Db

INTRODUCTION

Some silicates experience reconstructive transformations, which implies the breaking of the bond between silicon and oxygen, a particularly strong one. Therefore, high activation energies and a very slow reaction speed are expected. In the laboratory, temperatures above 1000°C are necessary. In nature, many years are required. However, recent experiments in some layered silicates such as mica muscovite have been performed at temperatures 600°C below the lowest experimental results previously reported [1, 2]. The authors of these articles have performed experiments with layered silicates and, in particular, with mica muscovite, in an aqueous solution with lutetium nitrate during 3 days at 300°C. After that time about 36% of muscovite has been transformed into lutetium disilicate [3]. The lack of explanation from an approach based on conventional Chemical Kinetics suggested the exploration of new hypotheses. Reactions of this type will be referred hereafter as *Low Temperature Reconstructive Transformations* (LTRT).

>From Transition State Theory, a transition state with higher energy than the reactants has to be formed for the reaction to take place. The activation energy E_a is the height of the energy barrier that has to be overcome. The rate of reaction is, therefore, proportional to the number of linear vibration modes or phonons with energy above E_a. This number is proportional to the Boltzmann factor $\exp(-E_a/RT)$, i.e., the fraction of phonons with energy above E_a, bringing about Arrhenius' law:

$$k = A\exp(-E_a/RT), \quad (1)$$

where k is the reaction rate constant and the pre-exponential factor A is known as the frequency factor.

There is, however a different type of excitations than the phonons, which appears for large amplitudes of vibration where the intrinsic anharmonicity of the atomic bonds can no longer be ignored. They have received considerable attention from the Nonlinear Physics community during the last decade and are known as anharmonic modes, intrinsic localized modes or *discrete breathers*. The mathematical proof of their existence and the methods to obtain them with machine precision in mathematical models has been firmly stablished in Ref. [4]. In this seminal article the authors also suggest that breathers could produce an apparent violation of Arrhenius' law. Breathers are localized, that is, they involve only a few particles or atoms. The conditions for their existence are the anharmonicity of the potentials and that their frequency has to be outside the phonon band. They are called *soft* if their frequency is below the phonon band, which is only possible in systems with an optical phonon band, of *hard* if their frequency lies above the phonon band. Although the subject is still under discussion it seems very unlikely that they can be observed by spectroscopic means due to their localized nature, the small number of them and the basic principles of Physics [5].

We have tried to explore the *breather hypothesis*, that is, that in mica muscovite, there exist breathers, that they have enough energy to overcome the activation energy, and that there are enough of them to influence the reaction speed [3].

THE MODEL

We have considered vibration in the cation layer, where the potassium ions form a rough hexagonal lattice (see

CP982, *Complex Systems, 5th International Workshop on Complex Systems*
edited by M. Tokuyama, I. Oppenheim, and H. Nishiyama

FIGURE 1. Interlayer sheet of the mica muscovite. The circles represent the potassium ions.

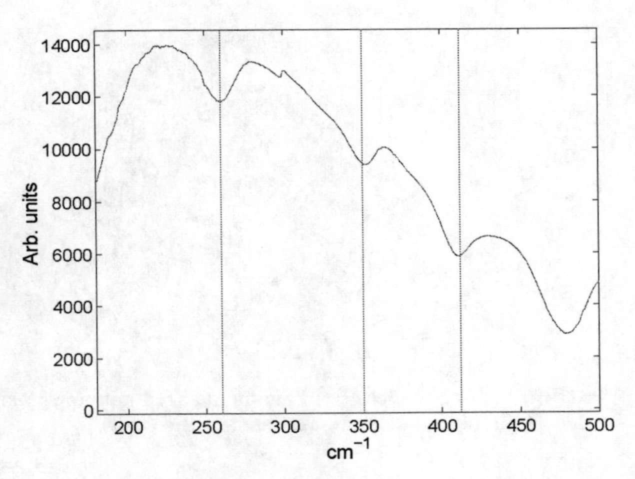

FIGURE 2. Muscovite far infrared spectrum

Fig. 1). So far, we have only taken into account off-plane vibrations. The model is a classical one with Hamiltonian:

$$H = \sum_{\vec{n}} \left(\frac{1}{2} m\dot{u}_{\vec{n}}^2 + V(u_{\vec{n}}) + \frac{1}{2}\kappa \sum_{\vec{n}' \in NN} (u_{\vec{n}} - u_{\vec{n}'})^2 \right), \quad (2)$$

where $m = 39.1$ amu is the mass of a potassium cation, κ is the elastic constant of the cation-cation bond, $V(u_{\vec{n}})$ is an on–site potential, and the second sum is extended to the nearest-neighbours.The value of the elastic constant κ is taken as 10 ± 1 N/m after Ref. [6].

For the on-site potential, the linear frequency is known after Ref. [7], where a band at 143 cm^{-1} is assigned to the K^+ vibration perpendicular to the K^+-plane in infrared spectra from 30 to 230 cm^{-1}. To obtain more characteristics of the nonlinear potential V, we have performed far infrared spectra above 200 cm^{-1} in CNRS-LADIR [1]. We observe bands at 260, 350 and 420 cm^{-1} as shown in Fig. 2, which we tentatively assign to higher order transitions of the same vibration.

Using standard numerical methods to solve the Schrödinger equation for the K^+ vibrations, with a potential composed of the linear combination with three gaussians and a polynomial of degree six, with adjustable parameters, we have been able to find a suitable potential that fits these bands and their intensities. It is given by:

$$V(u) = D(1 - \exp(-b^2u^2)) + \gamma u^6, \quad (3)$$

with $D = 453.11$ cm^{-1}, $b^2 = 36.0023$ Å^{-2} and $\gamma = 49884$ cm^{-1}Å^{-6}. To determine the potential, we have also taken into account the limitation to the K^+ displacement due to the muscovite structure.

BREATHERS IN MUSCOVITE

With this model we can obtain breathers with machine precision, using numerical methods based on the anticon-

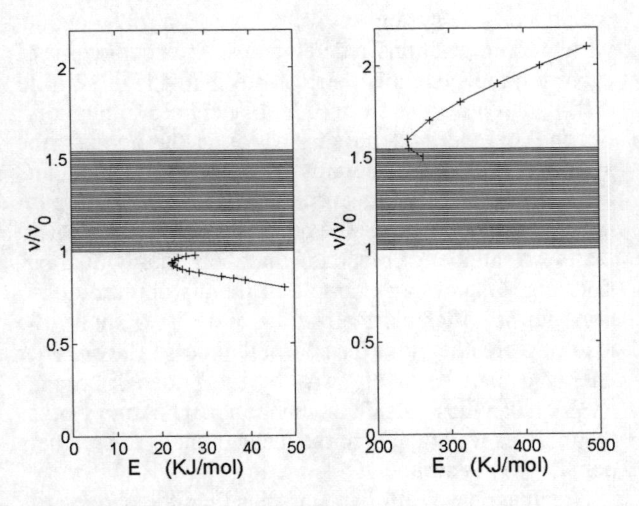

FIGURE 3. Relative frequency versus energy for soft (left) and hard (right) breathers in the muscovite model. Note the different energy scales. The phonon band is also shown. $\nu_0 = 167.5$ cm^{-1} $\simeq 5 \cdot 10^{12}$ Hz.

[1] Laboratoire de Dynamique, Interactions et Réactivité at CNRS-Thiais, Paris

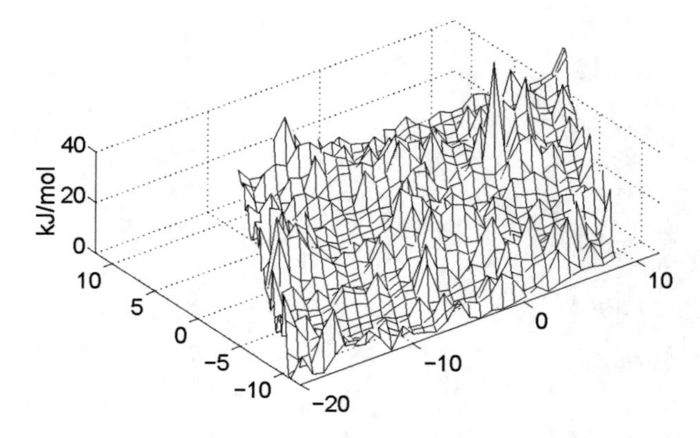

FIGURE 4. Energy density in the thermalized muscovite's model. The units in the x and y axes are in lattice units.

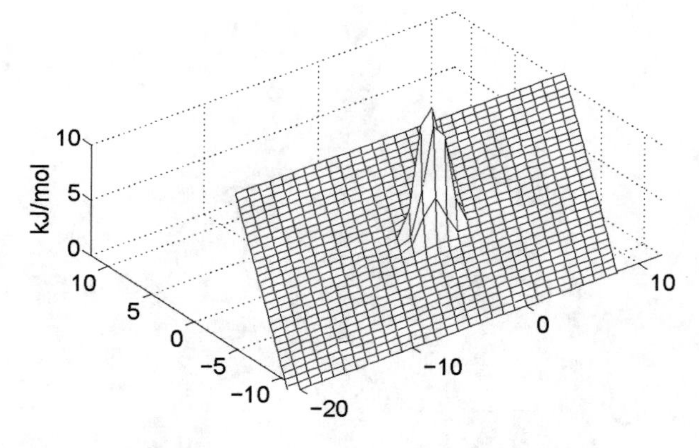

FIGURE 5. Energy density in muscovite's model after cooling. An asymmetric multibreather can be observed. The units in the x and y axes are in lattice units.

tinuous limit. That is, starting with a coupling parameter $\kappa = 0$ and a single excited atom, we obtain the exact solution using the Newton method in the frequency space. By path continuation, we obtain breathers at the physical value of κ.

Due to the characteristic of the potential V, there are both types of breathers, soft and hard ones. They are symmetric and their energies and frequencies can be obtained and are shown in Fig. 3. The expected activation energy for the reconstructive transformation is about 100-200 KJ/mol or higher [8]. Therefore we can see that hard breathers in this model may have enough energy to overcome the energy barrier.

BREATHER STATISTICS

In order to obtain the breather statistics numerically, first we deliver to a system of 50×50 atoms a given energy with random positions and velocities. After some time of evolution the system is thermalized, but it is difficult to distinguish breathers from the subjacent sea of phonons. Second, by adding some dissipation at the borders, the system is cooled, the phonons disappear but the breathers are left in place. We repeat this procedure several hundreds of times, calculate the breather energies and obtain the mean number of breathers and their energy distribution. Fig. 4 shows an example of the distribution of energy density after thermalization, and Fig. 5 shows the resulting breather after the cooling process. The breather energy cannot be taken away by the phonons because its frequency lies outside the phonon band. Among other magnitudes we obtain that the mean number of breathers per site $\langle n_b \rangle$ is around 10^{-3}.

Breathers have different statistics than the phonons as they tend to populate higher energies [9]. In this reference, a theory for breather statistics has been developed. It is based on the following assumptions:

- 2D breathers have a minimum energy Δ. This is an established fact proven in Ref. [10]
- Breathers are created through an activation process, i.e., the creation rate is proportional to $\exp(-E/RT)$
- Large breathers have longer lives than smaller ones as it has been observed in numerical experiments. A destruction rate of the form $1/(E-\Delta)^z$ is proposed, where z is a parameter to be determined.
- At thermal equilibrium, the number of breathers created and destroyed are equal for each energy. That is $\exp(-E/RT) = CP_b(E)/(E-\Delta)^z$, with C a constant independent of the energy and $P_b(E)\,\mathrm{d}E$ is the probability of existence (or the mean fraction) of breathers with energy between E and $E+\mathrm{d}E$.

With this hypothesis the following magnitudes are obtained:

The mean energy per atom is given by:

$$\langle E \rangle = \Delta + (z+1)\mathrm{k_B}T, \tag{4}$$

with $\mathrm{k_B}$ being the Boltzmann constant.

The probability density $P_b(E)$ is:

$$P_b(E) = \frac{\beta^{z+1}}{\Gamma(z+1)}(E-\Delta)^z \exp[-\beta(E-\Delta)], \tag{5}$$

with $\beta = 1/\mathrm{k_B}T$.

The cumulative probability $C_b(E)$, i.e., the probability that a breather has energy higher than E, is given by:

$$C_b(E) = \frac{\Gamma(z+1, \beta(E-\Delta))}{\Gamma(z+1)}, \tag{6}$$

where $\Gamma(z+1,x) = \int_x^\infty y^z \exp(-y)\mathrm{d}y$ is the first incomplete Gamma function.

However, the distribution of breathers obtained numerically cannot be fitted with this theory with the only adjustable parameter being z. This should be obvious by detailed observations of Fig. 5, where it can be seen that the breather is not a single symmetric breather, but a multibreather, i.e., several atoms have similar energies. In other simulations we obtain different multibreathers or single breathers with different symmetries or lack of them. Therefore, there is a *zoo* of breathers, each one of them with a different minimum energy Δ, parameter z and probability of appearance. Moreover, breathers may also have a maximum energy, due to different types of bifurcations. It appears in Fig. 3-left (for the hard breathers on the right hand side, the path has not been continued further, so the maximum energy is not known). Also, if they are or become unstable they will not be observed in numerical simulations or in nature.

Therefore, we have modified the theory, with a maximum energy E_M, the previous magnitudes become:

The probability density:

$$P_b(E) = \frac{\beta^{z+1}(E-\Delta)^z \exp[-\beta(E-\Delta)]}{\gamma(z+1, \beta(E_M - \Delta))}, \quad (7)$$

where $\gamma(z+1,x) = \int_0^x y^z \exp(-y)\,dy$ is the second incomplete gamma function.

The cumulative probability:

$$C_b(E) = 1 - \frac{\gamma(z+1, \beta(E-\Delta))}{\gamma(z+1, \beta(E_M - \Delta))}. \quad (8)$$

With these modifications we can fit the probability density using six different types of breathers, each one with a different probability of existence P_i. This is, however an approximation, as it is likely that there are many more involved. However, even with several hundreds of simulations it is not possible to obtain a good definition of the numerical $P_b(E)$ and only an approximation for large energies above 100 KJ/mol because the probability of occurrence of each breather with a given energy is very low.

EFFECT ON THE REACTION SPEED

The number of breathers is much smaller than the number of phonons, about one to a thousand, but only the excitations with energy above the activation energy, which is estimated to be $\simeq$100-200 KJ/mol will influence the reaction rate. The increase of the reaction rate with breathers will be roughly equal to the ratio between the number of breathers and the number of phonons above the activation energy, i.e., $\langle n_b \rangle C_b(E)/C_{ph}(E)$. We estimate it at about $10^4 - 10^5$, in other words, as the three days experimental time leads to about 30% of the transformation performed, the time without breathers to obtain the same result would be $10^4 - 10^5$ times larger and thus, completely unobservable. Furthermore, breather localization will increase the probability of delivering the energy to break a bond, which will increase the reaction rate.

We certainly cannot consider the breather hypothesis as proven, as the statistic theory is only a heuristic one and it is not based on first principles. We also need a theory to obtain the different probabilities of existence of different breathers, and obtain them numerically. We do not know how the energy is delivered and how a quantum treatment would modify our conclusions. At present we are working on these problems. But the fact remains that there is presently no other explanation and that breathers are localized and tend to populate states with higher energy than phonons. Recently, it has been observed that localized vibrations, produced by alpha radiation, travel extremely long distances along the lattice directions, bringing about the sputtering of an atom at the crystal surface [11]. This further reinforces our hypothesis. We can conclude that breathers are good candidates to explain LTRT.

ACKNOWLEDGMENTS

The authors acknowledge sponsorship by the Ministerio de Educación y Ciencia, Spain, project FIS2004-01183. They also acknowledge Prof. R. Livi, from Florence University, and Profs. J.M. Trillo and M.D. Alba from CSIC, Sevilla, for useful discussions.

REFERENCES

1. M. D. Alba, A. I. Becerro, M. A. Castro, and A. C. Perdigón, *Amer. Mineral.* **86**, 115–123 (2001).
2. A. I. Becerro, M. Naranjo, M. D. Alba, and J. M. Trillo, *J. Mater. Chem.* **13**, 1835–1842 (2003).
3. J. Archilla, J. Cuevas, M. Alba, M. Naranjo, and J. Trillo, *J. Phys. Chem. B* **110**, 24112–24120 (2006).
4. R. S. MacKay, and S. Aubry, *Nonlinearity* **7**, 1623–1643 (1994).
5. F. Fillaux, "Vibrational Spectroscopy and Quantum Localization," in *Energy Localization and Transfer*, World Scientific, Singapore, 2004, chap. 2, pp. 73–148.
6. D. R. Lide, editor, *Handbook of Chemistry and Physics*, CRC Press, Boca Raton, Florida, USA, 2007, 88th edn.
7. M. Diaz, V. C. Farmer, and R. Prost, *Clays Clay Miner.* **48**, 433–438 (2000).
8. A. Putnis, *An Introduction to Mineral Sciences*, Cambridge U. Press, Cambridge, UK, 1992.
9. F. Piazza, S. Lepri, and R. Livi, *Chaos* **13**, 637–645 (2003).
10. S. Flach, K. Kladko, and R. S. MacKay, *Phys. Rev. Lett.* **78**, 1207–1210 (1997).
11. F. M. Russell, and J. C. Eilbeck, *Eur. Phys. Lett.* **78**, 10004–1–1004–5 (2007).

A Simple Efficient Molecular Dynamics Scheme for Evaluating Electrostatic Interaction of Particle Systems

Ikuo Fukuda*, Yasushige Yonezawa†, and Haruki Nakamura†

*RIKEN (The Institute of Physical and Chemical Research) 2-1 Hirosawa, Wako, Saitama 351-0198, Japan

†Institute for Protein Research, Osaka University, 3-2 Yamadaoka, Suita, Osaka 565-0871, Japan

Abstract. We present results from a recently developed new molecular dynamics method to calculate electrostatic interaction of point particle systems. This method is very simple, which is based on the charge neutralized pairwise summation developed by Wolf et al., while solves problems presented in previous similar approaches and consistently gives the force, potential, and the total energy of the system.

Keywords: Charge neutralized summation, Coulombic energy, Cutoff method, Molecular dynamics, Wolf method
PACS: 02.70.-c, 31.15.Qg, 71.15.Nc

INTRODUCTION

For a computational study of materials, an appropriate treatment of coulombic interaction is very important. The handling of this interaction is, however, tough because it is long ranged and takes both positive and negative signatures. Various methods have thereby been developed [1, 2, 3], but it is still difficult to achieve high accuracy, low computational cost, an ease in the implementation, and free from artifacts [4, 5].

Wolf et al. [6, 7] proposed an effective way to calculate the coulombic interaction in compatible with the cutoff method. The point is that to ensure the charge neutrality in the cutoff sphere is critical to reach the true total energy. To calculate the coulombic energy, instead of a bare potential $V_0(r) \equiv 1/r$ and the total energy represented by the pairwise sum $E = \frac{1}{2}\sum_{i\neq j} q_i q_j V_0(r_{ij})$, Wolf et al. considered a damped form potential

$$V(r) \equiv \mathrm{erfc}(\alpha r)/r \tag{1}$$

and proposed the following simple summation approximate to E:

$$\frac{1}{2}\sum_{\substack{i\neq j\\ r_{ij}<r_c}} q_i q_j [V(r_{ij}) - V(r_c)] - \left[\frac{V(r_c)}{2} + \frac{\alpha}{\sqrt{\pi}}\right]\sum_i q_i^2, \tag{2}$$

where q_i is a charge of particle i and r_c is a cutoff length. The first term of Eq. (2) is defined by a pairwise sum of $q_i q_j$-times the potential

$$\begin{cases} \bar{V}_{\mathrm{Wolf}}(r) \equiv V(r) - V(r_c) & \text{for} \quad r < r_c \\ 0 & \text{for} \quad r \geq r_c \end{cases}, \tag{3}$$

and accompanies an additional interaction $-q_j V(r_c)$, producing the mirror image of q_j on q_i's cutoff sphere with the opposite-sign charge [7]. The second term proportional to $V(r_c)$ in Eq. (2) is a constant and represents a contribution from the mirror image of q_i itself. These terms effectively form the charge neutrality.

Direct application of Eq. (2) to molecular dynamics (MD) scheme, however, contains a critical problem, that Eq. (3) cannot be differentiated at $r = r_c$. Such a problem is emphasized as the parameter α reduces, whereas the choice of small α is in accord with one of the Wolf's strategies, i.e., to neglect the contribution corresponding to the Fourier reciprocal part of the Ewald sum (except constant) [2]. This problem was reduced by Wolf et al. themselves, but a new problem arose. After that, several methods have been proposed e.g., by Zahn, Schilling, and Kast (ZSK) [8] and Fennell and Gezelter (FG) [9]. However, they still involve several problems, including an issue of the total energy correction accompanied by the potential deformation and suitability of the choice of the force function.

In this work, we propose a novel MD protocol to overcome these problems. We analyze these theoretical aspects and discuss the difference between the current and other methods. We exhibit numerical

CP982, *Complex Systems, 5th International Workshop on Complex Systems*
edited by M. Tokuyama, I. Oppenheim, and H. Nishiyama

simulation results on the application of the current method to a molten ion system.

MOLECULAR DYNAMICS SCHEME

New Protocol

Our assertion is that we can nicely reformulate the Wolf summation method by using a suitably constructed force switching technique. On the basis of the derivative of the Wolf effective potential,

$$F(r) \equiv -D\bar{V}_{\mathrm{Wolf}}(r) \equiv \frac{\mathrm{erfc}(\alpha r)}{r^2} + \frac{2\alpha}{\sqrt{\pi}} \frac{\exp(-\alpha^2 r^2)}{r}, \tag{4}$$

our method defines a force

$$\tilde{F}(r) \equiv \begin{cases} F(r) & \text{for} \quad 0 < r < r_1 \\ F^*(r) & \text{for} \quad r_1 \le r \le r_{\mathrm{c}} \\ 0 & \text{for} \quad r_{\mathrm{c}} < r < \infty \end{cases}, \tag{5}$$

where r_1 is chosen slightly smaller than r_{c}. F^* is a suitable function, which smoothly (class C^1; a function is of class C^k if it has a continuous k-th derivative) and monotonically connects F at r_1 and 0 at r_{c}, realized by e.g., a polynomial function. By its integral, we define a pair potential $\tilde{V}$, which eventually has a simple form:

$$\tilde{V}(r) \equiv \int_r^{\infty} \tilde{F}(t)dt \tag{6}$$

$$= \begin{cases} V(r) + c & \text{for} \quad 0 < r < r_1 \\ V^*(r) & \text{for} \quad r_1 \le r \le r_{\mathrm{c}} \\ 0 & \text{for} \quad r_{\mathrm{c}} < r < \infty \end{cases}, \tag{7}$$

where $V^*(r) \equiv \int_r^{r_{\mathrm{c}}} F^*(t)dt$ and $c \equiv V^*(r_1) - \mathrm{erfc}(\alpha r_1)/r_1$. Obviously $-D\tilde{V} = \tilde{F}$.

The deviation between $\tilde{V}$ and $\bar{V}_{\mathrm{Wolf}}$ is relevant to the correction of the total energy. We can show that the total Coulomb energy using $\tilde{V}$ has a consistent formula with respect to the charge neutrality. To obtain the energy for N charged particle system, we basically obey the Wolf's scheme: an approximation holds by a neutralized sum of the damped potential $\mathrm{erfc}(\alpha r)/r$ and the self term, as

$$E = \frac{1}{2} \sum_{i \in \mathcal{N}} \sum_{j \in \mathcal{N}_i} \frac{q_i q_j}{r_{ij}} \tag{8}$$

$$\approx \frac{1}{2} \sum_{i \in \mathcal{N}} \sum_{j \in \mathcal{L}_i} \frac{q_i q_j}{r_{ij}} \mathrm{erfc}(\alpha r_{ij}) - \frac{\alpha}{\sqrt{\pi}} \sum_{i \in \mathcal{N}} q_i^2, \tag{9}$$

where $\mathcal{N} \equiv \{1, ..., N\}$ and $\mathcal{N}_i \equiv \mathcal{N} - \{i\}$. The summation with respect to j in the first term of Eq. (9) is a well-defined neutralized summation and can be shown to be

$$\sum_{j \in \mathcal{L}_i} \frac{q_j}{r_{ij}} \mathrm{erfc}(\alpha r_{ij}) \simeq \sum_{\substack{j \in \mathcal{N}_i \\ r_{ij} < r_{\mathrm{c}}}} q_j \tilde{V}(r_{ij}) - \left[\frac{\mathrm{erfc}(\alpha r_1)}{r_1} - V^*(r_1) \right] q_i. \tag{10}$$

Thus the total energy is

$$E \approx \frac{1}{2} \sum_{i \in \mathcal{N}} \sum_{\substack{j \in \mathcal{N}_i \\ r_{ij} < r_{\mathrm{c}}}} q_i q_j \tilde{V}(r_{ij}) - \left[\frac{1}{2} \frac{\mathrm{erfc}(\alpha r_1)}{r_1} - \frac{1}{2} V^*(r_1) + \frac{\alpha}{\sqrt{\pi}} \right] \sum_{i \in \mathcal{N}} q_i^2. \tag{11}$$

Namely, it can obtained by performing MD simulation using the pair potential $q_i q_j \tilde{V}$ and by adding the constant term proportional to $\sum_{i \in \mathcal{N}} q_i^2$.

Comparison with Other Methods

The previous approaches [7, 9] use the force

$$F_{\mathrm{FG}}(r) \equiv \begin{cases} F_{\mathrm{Wolf}}(r) & \text{for} \quad r < r_{\mathrm{c}} \\ 0 & \text{for} \quad r \ge r_{\mathrm{c}} \end{cases}, \tag{12}$$

where

$$F_{\mathrm{Wolf}}(r) \equiv F(r) - F(r_{\mathrm{c}}) = -[D\bar{V}_{\mathrm{Wolf}}(r) - D\bar{V}_{\mathrm{Wolf}}(r_{\mathrm{c}})], \tag{13}$$

to handle the differentiability problem. As if it reflects the mirror image charge effect on the cutoff sphere, unfortunately, it lacks a theoretical foundation. On the basis of this force, Fennell and Gezelter [9] construct the corresponding pair potential:

$$\bar{V}_{\mathrm{FG}}(r) \equiv \begin{cases} \bar{V}_{\mathrm{Wolf}}(r) - D\bar{V}_{\mathrm{Wolf}}(r_{\mathrm{c}})(r - r_{\mathrm{c}}), & r < r_{\mathrm{c}} \\ 0, & r \ge r_{\mathrm{c}} \end{cases}. \tag{14}$$

In contrast, the force in our method is uniquely derived by differentiating the right hand side of Eq. (11) with respect to each particle coordinate, yielding the pairwise force (5). It should be noted that Eq. (11) is obtained through a neutralized summation principle, so the force function is based on this, not on the intuitive definition. The difference about the force magnitude is an intrinsically distinguishable point from all the other methods. However, the difference generated by an actual simulation will, sometimes strongly, depend on the parameters r_c and α, along with r_1.

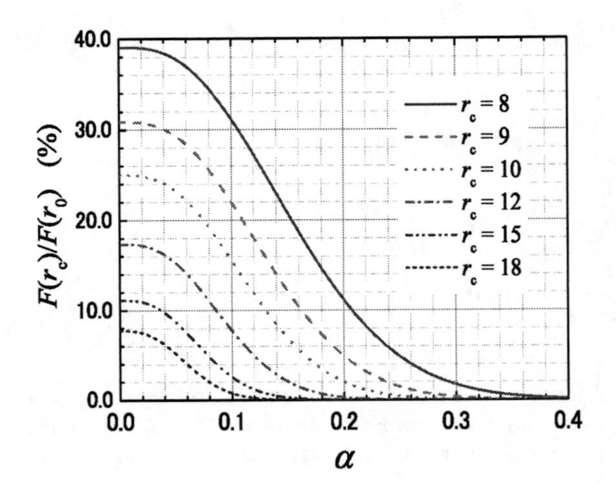

FIGURE 1. (Color online) $F(r_c)/F(r_0)$ is shown, where $F(r_c)$ is equal to the major difference between force $\tilde{F}$ and force F_{FG}. $r_0 = 5$ is set as a characteristic length, and r_c is varied from 8 to 18. Length unit can be thought as Å. Note that the ratio is an increasing function of r_0.

The difference with respect to various values of r_c and α is shown in figure 1. While the difference disappears as r_c and α are large, the difference is clearly visible for small values of r_c and α, e.g., when $r_c \lesssim 10$ and $\alpha \lesssim 0.2$. In those cases, the differences in the dynamics between our method and the others will be clear.

Furthermore, we can estimate the neutralized summation in the FG approach. In a similar consideration affirmed, we get

$$\begin{aligned}\sum_{j\in\mathcal{L}_i} q_j V(r_{ij}) &\simeq \sum_{\substack{j\in\mathcal{N}_i\\ r_{ij}<r_c}} q_j[V(r_{ij}) - V(r_c)] - q_i V(r_c) \\ &= \sum_{\substack{j\in\mathcal{N}_i\\ r_{ij}<r_c}} q_j \bar{V}_{\mathrm{FG}}(r_{ij}) + \sum_{\substack{j\in\mathcal{N}_i\\ r_{ij}<r_c}} q_j \delta V_{\mathrm{FG}}(r_{ij}) - q_i V(r_c),\end{aligned} \tag{15}$$

where

$$\delta V_{\mathrm{FG}}(r) \equiv \bar{V}_{\mathrm{Wolf}}(r) - \bar{V}_{\mathrm{FG}}(r) = D\bar{V}_{\mathrm{Wolf}}(r_c)(r - r_c) \tag{16}$$

for $r < r_c$. In the original FG approach [9], δV_{FG} is not considered (in fact, its accurate estimation in MD is nontrivial), at least in an explicit manner. If we incorporate this term to their approach, as is immediately seen, we can recover the force function $F(r) = -DV(r)$ for $r < r_c$. In this region, as stated above, the difference between the natural force $F = -D\bar{V}_{\mathrm{Wolf}}$ (equals to the force in the principal region in our approach) and force F_{Wolf} (employed in the previous approaches) cannot be neglected according to the choice of the parameter values. Thus, the difference between the potential function corresponding to the former force [viz., corrected potential: $\bar{V}_{\mathrm{FG}} + \delta V_{\mathrm{FG}} = \bar{V}_{\mathrm{Wolf}}$] and the potential function corresponding to the latter force [viz., FG original potential: $\bar{V}_{\mathrm{FG}}$] cannot be possibly neglected in such a case, resulting in some difference of the energetics.

Compared with the previous similar approaches in MD scheme, the current method thus provides (i) the consistency between the force function and the potential function (different point from the approach of Wolf et al.), (ii) an exempt from the nontrivial wide-range deformation in the effective potential (different point from the approaches of ZSK and FG), (iii) the force that is theoretically justified and natural such that in an almost all the region in the cutoff sphere it is just the minus of the derivative of the Wolf original effective potential (different point from the approaches of Wolf et al., ZSK, and FG), (iv) a sufficient smoothness (C^2) of the potential (different from the incompatibility problem in the Wolf et al., being not C^0 in the ZSK, and being C^1 in the FG), which ensures good properties of the ordinary differential equation for MD scheme and could keep stability in the numerical time integration, and (v) a suitable correction of the total energy due to the deviation from the Wolf effective potential (different point from the approaches of ZSK and FG).

Numerical Results

We performed numerical simulations of the current method using a molten NaCl system composed of 2304 ions obeying the three-dimensional periodic boundary condition. We prepared the charged particle configurations x by performing NTV simulation with 1200 K, employing the particle mesh Ewald (PME) method [10]. Sampling these configurations, we calculated the average of the difference between the energy obtained from the current method, $E(x)$, and that from the PME method, $E_{\mathrm{PME}}(x)$. Here, we confirmed in a number of preliminary runs that r_1 dependence was moderate, and so practically set $r_1 = r_c - 1$ Å.

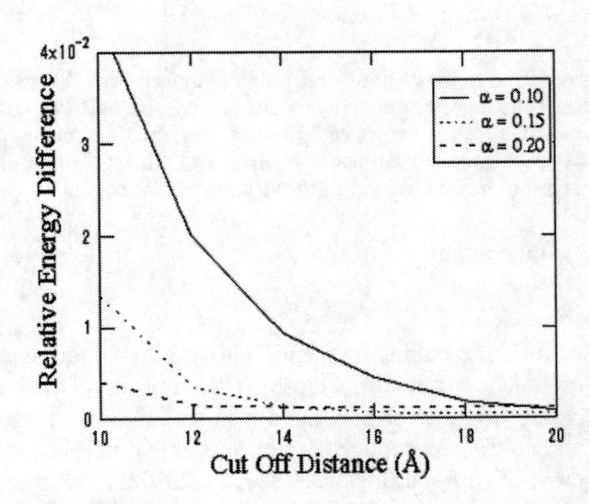

FIGURE 2. Averaged relative difference of the total energy, obtained from the molten NaCl system, between the current and the PME methods.

Figure 2 shows the relative difference $\langle |E(x) - E_{\mathrm{PME}}(x)|/|E_{\mathrm{PME}}(x)|\rangle_x$ and exhibits its dependence on the parameters α and r_c. They show that correct absolute total energies on a number of configurations can be attained using small α and reasonable r_c within the relative difference less than 1 percent. For example, we get 3×10^{-3} for $\alpha = 0.15$ and $r_c = 12$ Å. We can see the controllable parameter dependences. Clearly, the error becomes smaller for larger r_c. Note that as α tends smaller, the error slowly converges but falls in lower value. In addition, we verified the consistency of MD simulation with a good energy conservation.

SUMMARY

For evaluating coulombic interaction we have presented a very simple MD scheme for which complications and queries in the previous approaches are completely solved. The current method presents an $O(N)$ scheme for large N, and the simplicity can adapt to various high performance computational architectures. We expect that the current technique will be efficiently used in many applications, especially for large systems such as biological macromolecules.

ACKNOWLEDGMENTS

This work was supported by the grants from the New Energy and Industrial Technology Development Organization, the Ministry of Economy, Trade, and Industry, and the Ministry of Education, Culture, Sports, Science and the Technology, of Japan.

REFERENCES

1. C. L. Brooks III, B. M. Pettitt, and M. Karplus, *J. Chem. Phys.*, **83**, 5897-5908 (1985).
2. P. P. Ewald, *Ann. Phys. (Leipzig)*, **64**, 253-287 (1921).
3. L. Greengard, and V. Rokhlin, *J. Comput. Phys.*, **73**, 325-348 (1987).
4. D. J. Tobias, *Curr. Opin. Struct. Biol.*, **11**, 253-261 (2001).
5. W. Weber, P. H. Hünenberger, and J. A. McCammon, *J. Phys. Chem.*, B **104**, 3668-3675 (2000).
6. D. Wolf, *Phys. Rev. Lett.*, **68**, 3315-3318 (1992).
7. D. Wolf, P. Keblinski, S. R. Phillpot, and J. Eggebrecht, *J. Chem. Phys.*, **110**, 8254-8282 (1999).
8. D. Zahn, B. Schilling, and S. M. Kast, *J. Phys. Chem.*, B **106**, 10725-10732 (2002).
9. C. J. Fennell, and J. D. Gezelter, *J. Chem. Phys.*, **124**, 234104 (2006).
10. U. Essmann, L. Perera, M. L. Berkowitz, T. Darden, H. Lee, and L. G. Pedersen, *J. Chem. Phys.*, **103**, 8577-8593 (1995).

Effectiveness of Slow Relaxation Dynamics in Finite-Time Optimization by Simulated Annealing

M. Hasegawa

Graduate School of Systems and Information Engineering, University of Tsukuba, Tsukuba 305-8573, Japan

Abstract. The origin of the specific temperature beneficial to finite-time optimization by simulated annealing is discussed on the analogy of the dynamics of complex physical systems. Rate-cycling experiments are introduced and performed on practical time scales on the random Euclidean traveling salesman problems. In the present systems, the effective relaxation dynamics and the resulting good optimization performance are not only dependent on but also sensitive to the search around an intermediate temperature. This influential temperature is understood to be determined from the temperature dependence of the Deborah number used to identify glass transition.

Keywords: Optimization, Relaxation, Simulated annealing, Traveling salesman problem
PACS: 02.60.Pn, 61.20.Lc, 05.90.+m

INTRODUCTION

Motivated by computer experiments on physical systems at low temperatures, simulated annealing (SA) was proposed as a heuristic solution method for optimization problems [1, 2, 3, 4, 5]. By replacing the microscopic state and its energy with the feasible solution and its cost, respectively, a Monte Carlo simulation using the Metropolis algorithm (MA) [4, 5] is performed with a slow decrease in effective temperature. On the analogy of the formation of crystalline solid, a careful search around the freezing (or freezing-like) point determined from a specific heat peak has sometimes been recommended in the articles [1, 3, 6]. The presence of such specific temperature has been remarked also in the real optimization process [7, 8]; however, this temperature was distinctly lower than the above recommendation. Moreover, it has been shown that SA without cooling (i.e. MA) can outperform standard SA with cooling for some moderately sized problem instances [9]. These findings motivate us to reconsider the design of the finite-time SA algorithm.

In connection with this problem, in our previous study [10], the hidden search characteristics of the MA was investigated for the random Euclidean traveling salesman problem (TSP) [6] using a mapping-onto-minima method [11, 12]. The results show that, after the system is quenched to the target temperature, unidirectional transitions from high-laying basin to low-laying one appear at low temperatures as has been observed in the Hamiltonian dynamics of glassy solid [13]. With a decrease in the target temperature, the transition becomes substantially slow and the cost of the reachable basin goes down. Consequently, on practical time scales, the optimum target temperature is determined from the trade-off between the quickness of the unidirectional transition and the fitness of the reachable basin. This optimum temperature in MA is sufficiently close to the optimum final temperature which affects the optimization performance of adaptive SA introduced by the author [8]. In this method, most of the search time is spent near the final temperature. Hence, it is suggested that slow relaxation dynamics observed as unidirectional transition play a *positive* role in finite-time optimization by SA.

Originally, the MA was devised as a computational method for constructing the canonical ensemble at a prescribed temperature with a Markov process [14]; therefore, the probability for finding better solutions to the minimization problem increases with the decrease in temperature. If the system is designed so that relaxation proceeds quickly even at low temperatures, SA, and also MA, works out satisfactorily for finding globally optimal solutions. In this conventional view, it is implied that slow relaxation dynamics due to effectively broken ergodicity [15], resulting in the slow convergence to the equilibrium distribution, play a *negative* role in the optimization process.

The verification of the paradoxical statement "slow relaxation dynamics play a *positive* role in finite-time optimization by SA" will enable us to interpret the origin of the specific temperature beneficial to finite-time optimization by SA because, on the analogy of glassy solid, the occurrence of slow relaxation dynamics (unidirectional interbasin transition) can be characterized by temperatures such as glass transition temperature. This is the reason why we here investigate the role of slow relaxation dynamics in finite-time optimization by SA. For the present purpose, a simple but illuminating experiment is designed by using a new framework of the cooling schedule. In this framework, the cooling rate is changed cyclically so that the search in a specified tem-

CP982, *Complex Systems, 5th International Workshop on Complex Systems*
edited by M. Tokuyama, I. Oppenheim, and H. Nishiyama

perature range is controlled on a fixed observation time scale. The experiments with this framework of the cooling schedule are called rate-cycling experiments for short here. We consider also here the solution for the random Euclidean TSPs. The present goal is not a proposal of a better optimization algorithm but a better understanding of the origin of the effectiveness in the existent method.

EXPERIMENTAL METHOD

The goal in the Euclidean TSP is to find the shortest tour that passes through each of the given cities once and returns to the starting city; the intercity distance is computed under the Euclidean metric. The locations of the N cities are sampled uniformly in a unit square.

Let x be a tour (a feasible solution) and $f(x)$ be the cost function defined by the tour length. Furthermore, let $\mathcal{N}(x)$ be the neighborhood function; here the neighborhood is defined as a set of tours constructed by any change of two intercity paths (2-opt neighborhood). We use the notation x_n $(n = 0, 1, 2, \cdots)$ to represent the (actual) search history. A framework of the SA algorithm is described as follows:

1. Fix the cooling schedule $T(t)$.
2. Set $n := 0$ and generate randomly an initial solution x_0.
3. Select a trial solution x_n' randomly from the neighborhood $\mathcal{N}(x_n)$ and let $\Delta := f(x_n') - f(x_n)$. If $\Delta \leq 0$, set $x_{n+1} := x_n'$. If $\Delta > 0$, set $x_{n+1} := x_n'$ with probability $\exp(-\Delta/T(n))$ and $x_{n+1} := x_n$ with the rest probability.
4. Increase n by one. If the stopping criterion is satisfied, then terminated.
5. Go to step 3.

In all the present experiments, the algorithm is terminated when n reaches a prescribed total number of search steps.

The cooling schedule used in rate-cycling experiments is designed in the following way. Let Θ denote the logarithmic temperature $\log_{10} T$. First, we choose the values of the initial logarithmic temperature Θ_s, the final logarithmic temperature Θ_e, and the total length t_e of search time (or the total number n_e of search steps). Next, we divide the cooling process into three stages so that the search in a specified temperature range is performed as its second stage; this temperature range is specified by the target logarithmic temperature Θ_c as its middle point and its width $\Delta\Theta$ on the logarithmic temperature scale. The cooling rate is kept constant in each of the three stages, and the rate in the first stage is equalized to that in the third and is λ times of that in the second on the logarithmic temperature scale. The cooling schedule is thus described by the six parameters.

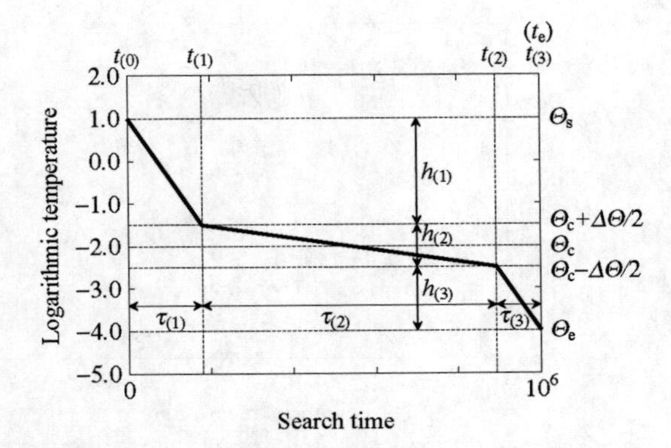

FIGURE 1. Example of the cooling schedule $\Theta(t)$ ($\Theta_s = 1.0$, $\Theta_e = -4.0$, $\Theta_c = -2.0$, $\Delta\Theta = 1.0$, $\lambda = 10$, and $t_e = 10^6$). See text for symbols.

Let $[t_{(i-1)}, t_{(i)}]$ be the time range of the i-th stage. The cooling schedule for the i-th stage is described as follows:

$$\Theta(t) = \Theta(t_{(i-1)}) - h_{(i)} \cdot \frac{t - t_{(i-1)}}{\tau_{(i)}} \quad (t_{(i-1)} \leq t \leq t_{(i)}), \tag{1}$$

where $h_{(i)}$ is the width of the logarithmic temperature range of the i-th stage and $\tau_{(i)}$ is the length of the search time spent there; consequently, $t_{(i)}$ is the cumulative value of $\tau_{(j)}$'s up to $j = i$ except for $t_{(0)} = 0$. An example of the cooling schedule is shown in Fig. 1. Note that $h_{(i)}$'s are determined from Θ_s, Θ_e, Θ_c, and $\Delta\Theta$ as shown in the figure and $\tau_{(i)}$'s are determined from $h_{(i)}$'s, λ, and t_e as follows: $\tau_{(1)} = (h_{(1)}/H)t_e$, $\tau_{(2)} = (\lambda h_{(2)}/H)t_e$, and $\tau_{(3)} = (h_{(3)}/H)t_e$, where $H = h_{(1)} + \lambda h_{(2)} + h_{(3)}$. The search around the target temperature is carefully performed when $\lambda > 1$ and carelessly done when $\lambda < 1$. This cooling scheme reduces to the geometric schedule [4] when $\lambda = 1$ irrespective of the values of Θ_c and $\Delta\Theta$.

Besides the optimization performance the hidden search dynamics $f(y_n)$ are investigated using a mapping-onto-minima method [11, 12]; here y_n represents the locally optimal solution whose basin includes the solution x_n. The mapping process can easily be realized by an independent run of a simple local search, where the move to the best solution in the neighborhood is repeated.

We examined a single instance of $N = 10^2$ and that of $N = 18^2$, and preliminarily measured the specific heat [1] of these systems using our stepwise adaptive cooling method [8]. We found a single specific heat peak in each of the systems: the peak is located at $\hat{\Theta} \approx -1.2$ for $N = 10^2$ and $\hat{\Theta} \approx -1.4$ for $N = 18^2$. On the basis of these results, the parameter values for rate-cycling experiments were selected as follows. (i) The value of Θ_s was fixed at

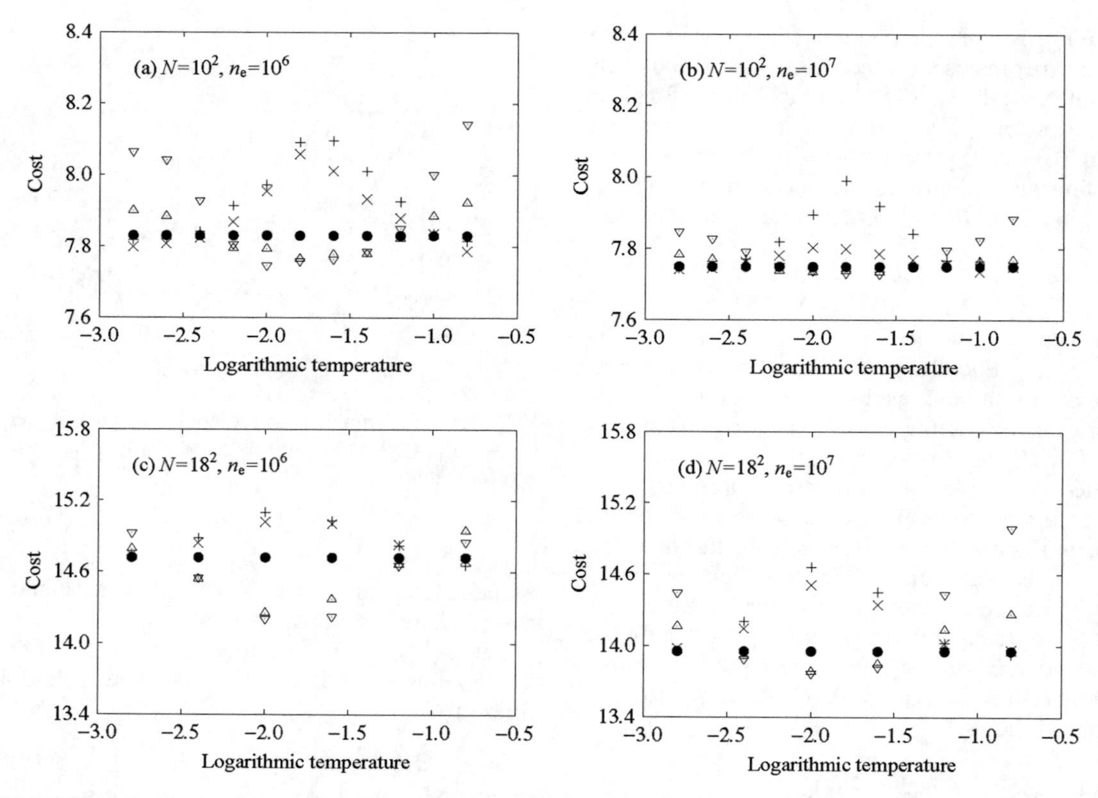

FIGURE 2. Optimization performance ($\Delta\Theta = 1.0$). The average final cost $f(y_{n_e})$ of the hidden search dynamics is plotted as a function of the target temperature Θ_c. $+$: $\lambda = 0.01$, $\times$: $\lambda = 0.1$, $\bullet$: $\lambda = 1$, $\triangle$: $\lambda = 10$, $\triangledown$: $\lambda = 100$.

1.0 and that of Θ_e at -4.0; the former is a high enough temperature where the average acceptance ratio is more than 98% and the latter is a low enough temperature where the ratio is less than 0.001%. (ii) All combinations of the following values were selected for Θ_c, λ, and n_e: $\Theta_c = -0.8 - 0.2m$ ($m = 0, 1, 2, \cdots, 10$ for $N = 10^2$ and $m = 0, 2, 4, \cdots, 10$ for $N = 18^2$); $\lambda = 0.01$, 0.1, 1, 10, and 100; and $n_e = 10^6$ and 10^7. (iii) The value of $\Delta\Theta$ was fixed at 1.0 and changed in the range from 0.2 to 1.8 for comparison purpose.

RESULTS AND DISCUSSION

In this section, the average result of 50 independent runs will be shown in each of the figures.

The results for the optimization performance are shown in Fig.2 for four combinations of two N's and two n_e's: the average final cost $f(y_{n_e})$ of the hidden search dynamics is plotted for $\Delta\Theta = 1.0$. The results show that the average performance is maximized when the search is carefully performed around an intermediate temperature. Furthermore, the performance is sensitive to the search around this temperature. This influential temperature is distinctly below $\hat{\Theta}$ and is close to the optimum temperature found in our previous experiments [10] on the hidden search dynamics. These features appeared commonly in every $\Delta\Theta$; however, the narrower $\Delta\Theta$ seemed not to improve the performance.

The change in the hidden search dynamics with a decrease in the target temperature is shown in Fig.3 for $N = 10^2$, $\Delta\Theta = 1.0$, $\lambda = 100$, and $n_e = 10^6$. Under these conditions, most of the search time is spent in the second stage ($\tau_{(2)} > 0.96t_e$). When $\Theta_c \geq -1.2$, the unidirectional dynamics appear late in the second stage; they begin when the temperature falls below $\hat{\Theta}$ and continue until the end of the observation. In contrast, when $\Theta_c \leq -2.4$, the system is almost frozen in the middle of the second stage. Consequently, the average optimization performance is maximized at an intermediate target temperature around which the unidirectional dynamics appearing below $\hat{\Theta}$ work successfully. As seen in Fig.2(a), this effective dynamics improve the performance of the geometric cooling without any additional effort in computation, only with a simple deformation of the cooling schedule. These features were observed also in the other three cases corresponding to Figs.2(b)-(d).

The present and the previous [10] observations provide useful materials for explaining the origin of the influential temperature of the present systems. These

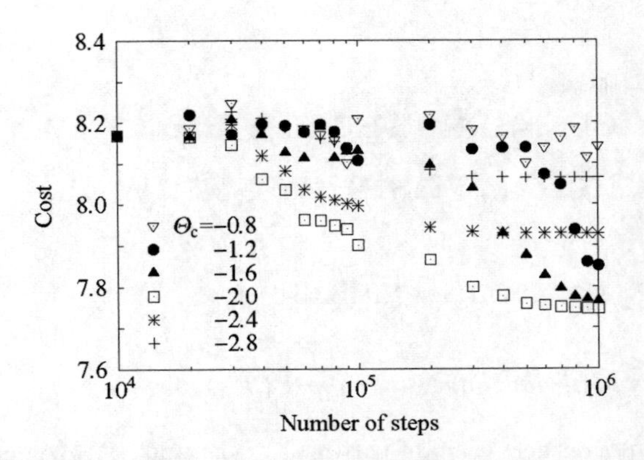

FIGURE 3. Hidden search dynamics $f(y_n)$ for various target temperatures ($N = 10^2, \Delta\Theta = 1.0, \lambda = 100, n_e = 10^6$).

results on the hidden search dynamics remind us that the present systems go into the "landscape-influenced" regime [16] near $\hat{\Theta}$. If accepting this analogy, we can consider that each system goes into the supercooled liquid state below $\hat{\Theta}$ and that the unidirectional dynamics is related with slow dynamics characteristic of glassy solid. Hence, the present specific temperature is understood to be determined not from the temperature dependence of the specific heat but from that of the Deborah number [17]. This number is defined by the ratio of relaxation time to observation time and can be used to identify glass transition [18]. The present view reflects also the previous observations that the optimum temperature decreases with the extension of the search time [8, 9, 10].

With regard to the random Euclidean TSP with the 2-opt neighborhood, it has been suggested that the landscape of the cost function has a rugged "big valley" structure [19]. If this picture is accepted, as has been done so far, we can sketch the following scenario as a possible cause of the present observations. When the temperature falls below $\hat{\Theta}$, the system goes into the landscape-influenced regime mentioned above and a slow downhill search occurs on the ramified valley slope. At lower temperatures, the system can descend deep into the big valley; however, it is apt to get stuck in holes on the valley slope. Hence, the average optimization performance is maximized when this slow downhill search works fully on the observation time scale.

CONCLUSIONS

The origin of the specific temperature in finite-time optimization by SA was discussed on the analogy of the dynamics of complex physical systems. Rate-cycling experiments were systematically performed on practical time scales on the random instances of the Euclidean TSP. The results showed that the effective relaxation dynamics and the resulting good optimization performance are dependent on and also sensitive to the search around an intermediate temperature. For the present systems, this influential temperature is understood to be determined from the temperature dependence of the Deborah number used to identify glass transition.

REFERENCES

1. S. Kirkpatrick, C. D. Gelatt, Jr., and M. P. Vecchi, *Science*, **220**, 671-680 (1983).
2. V. Černý, *J. Optim. Theor. Appl.*, **45**, 41-51 (1985).
3. M. Mezard, G. Parisi, and M. Virasoro (Eds.), *Spin Glass Theory and Beyond*, World Scientific, Singapore, 1987, Chapter VIII.
4. E. H. L. Aarts, J. H. M. Korst, and P. J. M. van Laarhoven, "Simulated Annealing," in *Local Search in Combinatorial Optimization*, edited by E. H. L. Aarts and J. K. Lenstra, John Wiley & Sons, Chichester, 1997, pp. 91-120.
5. E. H. L. Aarts and J. H. M. Korst, "Selected Topics in Simulated Annealing," in *Essays and Surveys in Metaheuristics*, edited by C. C. Ribeiro and P. Hansen, Kluwer Academic Publishers, Boston, 2002, pp. 1-37.
6. D. S. Johnson and L. A. McGeoch, "The traveling salesman problem: a case study," in *Local Search in Combinatorial Optimization*, edited by E. H. L. Aarts and J. K. Lenstra, John Wiley & Sons, Chichester, 1997, pp. 215–310.
7. K. H. Hoffmann, D. Würtz, C. de Groot and M. Hanf, "Concepts in Optimizing Simulated Annealing Schedules: An Adaptive Approach for Parallel and Vector Machines," *Parallel and Distributed Optimization*, edited by M. Grauer and D. B. Pressmar, Springer-Verlag, Heidelberg, 1991, pp. 155-175.
8. M. Hasegawa, "A Concept of Effectively Global Search in Optimization by Local Search Heuristics," in *Slow Dynamics in Complex Systems*, edited by M. Tokuyama and I. Oppenheim, AIP Conference Proceedings 708, American Institute of Physics, Melville, New York, 2004, pp. 747–748.
9. M. Fielding, *SIAM J. Optim.*, **11**, 289-307 (2000).
10. M. Hasegawa, *J. Phys. Soc. Jpn.*, **74**, 2872-2873 (2005).
11. F. H. Stillinger and T. A. Weber, *Science*, **225**, 983-989 (1984).
12. F. H. Stillinger and T. A. Weber, *Phys. Rev. A*, **28**, 2408-2416 (1983).
13. K. Shinjo, *Phys. Rev. B*, **40**, 9167-9175 (1989).
14. N. Metropolis, A. W. Rosenbluth, M. N. Rosenbluth, A. H. Teller, and E. Teller, *J. Chem. Phys.*, **21**, 1087-1092 (1953).
15. R. G. Palmer, "Broken Ergodicity," in *Lectures in the Sciences of Complexity, SFI Studies in the Sciences of Complexity*, edited by D. L. Stein, Addison-Wesley, Reading, 1989, pp. 275-300.
16. S. Sastry, P. G. Debenedetti, and F. H. Stillinger, *Nature*, **393**, 554-557 (1998).
17. M. Reiner, *Phys. Today*, **17**, 62 (1964).
18. J. M. Stevels, *J. Non-Cryst. Solids*, **6**, 307-321 (1971).
19. K. D. Boese, A. B. Kahng, and S. Muddu, *Oper. Res. Lett.*, **16**, 101-113 (1994).

Structure and Dynamics of Water Surrounding the Poly (Methacrylic Acid):A Molecular Dynamics Study

Ching-I Huang, Wei-Zen Cheng, and Yong-Ting Chung

Institute of Polymer Science and Engineering, National Taiwan University, Taipei 106, Taiwan

Abstract. All-atom molecular dynamics simulations are used to study a single chain of poly(methacrylic acid) (PMAA) in aqueous solutions at various degrees of charge density. We observe that local arrangements of water molecules, surrounding the functional groups of COO^- and COOH in the chain, behave differently and correlated well to the resulting chain conformation behavior. Furthermore, water molecules often act as a bridging agent between two neighboring COO^- groups. These bridged water molecules are observed to stabilize the rod-like chain conformation that the highly charged chain reveals, as they significantly limit torsional and bending degrees of the backbone monomers. In addition, they display different dynamic properties from the bulk water. Both the resulting oxygen and hydrogen spectra are greatly shifted due to the presence of strong H-bonded interactions.

Keywords: ENCAD, Molecular Dynamics, PMAA, Polyelectrolyte
PACS: 02.70.Ns, 36.20.Ey, 47.11.Mn

INTRODUCTION

Polyelectrolytes are of great importance due to their extensive presence throughout biological systems and use in a wide range of practical applications, such as sensors, detergents, drug delivery, and gene therapy. In general, polyelectrolytes, when in a solution, dissociate into polyions and counterions. The associated electrostatic interactions that involve the polyions, counterions, and the solvent, make the chain conformation behavior of polyelectrolytes quite distinct from that of neutral polymers. Polyelectrolyte conformation has been, and continues to be, an important research area, not only for its relevance to physical properties but also for a deeper understanding of biological phenomena and applications.

Experiments have shown that polyelectrolyte conformation in dilute solutions depends on the charge density that exists along the chain, salt concentration, ionic strength, and the solution pH value [1-3]. For example, highly charged polyelectrolyte chains in low ionic strength monovalent salt solutions exhibit an extended conformation due to the net repulsion between the charged monomers. When the added salts are multivalent, a series of transitions from extended → collapsed → re-expanded conformations often occurs upon increasing the salt concentrations. Though current theoretical studies have captured most of the phenomena associated with the conformational variation in polyelectrolytes [2,4-7], solvent effects that involve direct electrostatic interactions between solvent and polyelectrolyte, the local arrangement of the solvent molecules around the polyelectrolyte chain, and the formation of the hydrogen bonds between the solvent molecules and monomers, have not yet been addressed.

We thus employ all-atom molecular dynamics (MD) simulations to study the local structure and dynamics of water in the vicinity of a single polyelectrolyte chain. We consider a single molecule of syndiotactic poly(methacrylic acid) (PMAA) in an aqueous solution. We show that due to the strong attractive interactions between water and charged monomers, the water molecules form highly bonded structures surrounding the chain via the formation of hydrogen bonds. These bridged water molecules significantly affect the resulting chain conformation behavior as well as their dynamic properties.

MODEL AND SIMULATION METHODS

We employ an Energy Calculations and Dynamics (ENCAD) simulation program [8], which is an all-atom model, to calculate the atomic interaction parameters between the PMAA and the solvent water. The total potential energy function U is given as follows,

CP982, *Complex Systems, 5th International Workshop on Complex Systems*
edited by M. Tokuyama, I. Oppenheim, and H. Nishiyama

$$U = U_{bond} + U_{bend} + U_{torsion} + U_{vdw} + U_{els} \quad (1)$$

where U_{bond}, U_{bend}, and $U_{torsion}$ describe the bonded interactions contributed from bond stretching, bond angle bending, and torsion angle twisting, respectively, with the remaining two terms, U_{vdw} and U_{els} representing van der Waals interactions and electrostatic Coulomb potential energy, respectively. The detailed terms were shown in our previous work [9].

All the parameters of the ENCAD are derived from ab-initio quantum mechanics, spectroscopy, and crystallography, and can be obtained from refs. [8] and [10]. In addition, the solvent water is treated explicitly via a flexible three-centered (F3C) model, in which the associated water potential still adopts the same type of the force fields as in the ENCAD program [10]. This F3C water model has proved suitable for describing the structural and dynamic properties of liquid water.

The system contains a single chain of syndiotactic PMAA with a degree of polymerization N equal to 48 in the presence of approximately 1700 water molecules. In our simulations, we used Materials Studio (MS) molecular modeling software to construct the initial atomistic structure of the PMAA, and varied the number of charged monomers N_C, which were equally distributed along the chain. In particular, the charge density f (= N_C /N) for the PMAA chain is equal to 1, 0.5, 0.33, 0.25, and 0, respectively. We assumed that no counterions were present in the study. The molecular dynamics simulations were performed by integrating the positions and velocities of all atoms according to the velocity-Verlet algorithm [11,12]. The initial temperature was set at 300 K and the integration time step was chosen as 1 fs. The PMAA chain was first put in a vacuum box with a side length of 40 Å and a periodic boundary condition, then the box was gradually filled with water molecules to a density of 0.791g/cm^3. Prior to simulation, the steepest descent minimization method was adopted to relax and equilibrate the initial structure. To clarify, the PMAA chain was fixed and the water solvent was relaxed to populate the relevant hydration sites on the PMAA for 30 ps in a canonical NVT ensemble. This minimized structure was then compressed to a density of 0.987 g/cm^3 at a rate of 0.1Å/fs, followed by a series of annealing processes. The annealing temperature was first raised from 300 to 600 K at a rate of 5 K/1ps and kept at 600 K for 100 ps. It was then quenched to 300 K at the same rate and kept at 300 K for 20 ps. This annealing cycle was repeated 4 to 6 times to assure that the system had been equilibrated. Finally, it was followed by a long relaxation period of 100 ps at 300 K. After these processes were performed so that the system energy has reached the equilibrium value, we analyzed the radius of gyration (R_g) of the PMAA chain and the radial distribution function (RDF) every 100 fs as well as the self-velocity autocorrelation function (VACF) of water molecules every 5 fs, which were averaged out for the data collection interval of 10 ps.

RESULTS AND DISCUSSION

Fig. 1 displays the variation in the radius of gyration (R_g) with f, where we also designate the maximum and minimum values of R_g with error bars. In order to clearly manifest the significant effects contributed from water, we repeat the simulation on the same system; however, the water is treated implicitly as a comparison. To clarify, the solvent is taken into account as a continuum dielectric with a dielectric constant. It clearly shows that with the same charge density f, the value of R_g for the PMAA chain when the water is treated via a F3C model is greater than that when the water is treated implicitly. This indicates that the existence of real water causes a greater degree of stretching in the charged polymer chain. As the charge density f increases, due to the fact that the repulsive degree of the electrostatic interactions between the COO^- groups becomes more significant, R_g shows an increasing behavior. Later we will show that the organization of water molecules surrounding the PMAA chain also plays an important role in the PMAA chain conformation behavior.

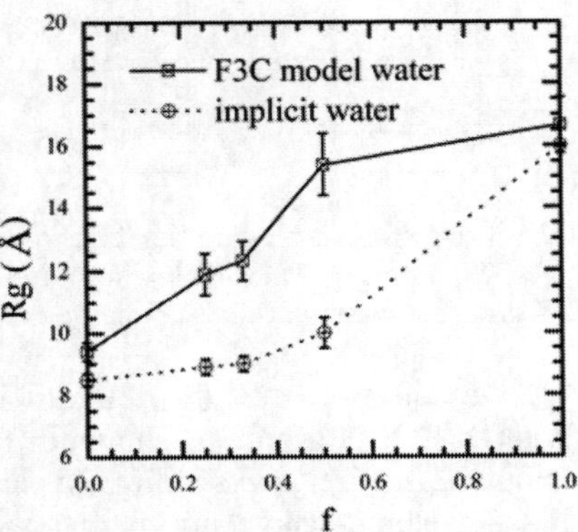

FIGURE 1. Plot of the radius of gyration (R_g) of PMAA versus charge density f when the solvent water is treated explicitly via a F3C model and implicitly via a continuum dielectric, respectively.

In aqueous solutions, the hydration behavior of molecules is a complex subject, but worth further exploration. In order to investigate the distribution of water molecules surrounding the PMAA chain, we analyze the radial distribution functions of the oxygen (O) and hydrogen (H) atoms of water with respect to the O atom in the COO^- group, the O atom of carbonyl (-C=O) in the COOH group, and the O atom of hydroxyl (-OH) in the COOH group of the PMAA at various values of charge density f, in Figs. 2 and 3, respectively, This radial distribution function $g_{A-B}(r)$ indicates the local probability density of finding B atoms at a distance r from A atoms averaged over the equilibrium density. In Fig. 2, where the distribution of water (H and O atoms) is analyzed from the central O atom in the COO^- group, we observe two prominent peaks for the hydrogen RDF profiles and one for the oxygen RDF profiles regardless of the charge density f values. The observed high and sharp first peaks as well as the first minimum values close to 0 for both $g_{O(COO^-)-H}$ and $g_{O(COO^-)-O}$ manifest the fact that these COO^- groups are strongly hydrophilic in nature and therefore attract a large amount of water molecules to form shell-like layers surrounding them. In addition, the first peak of the hydrogen RDF profiles occurs at 1.4 Å, which is less than the normal hydrogen bonding length of 1.8 Å [13], indicating that the interaction between the O atom of the COO^- group and the H atom of the water molecule is stronger than the strength of hydrogen bonds in bulk water. This is expected as polar water molecules and negatively charged oxygen atoms have a stronger interaction than in bulk water.

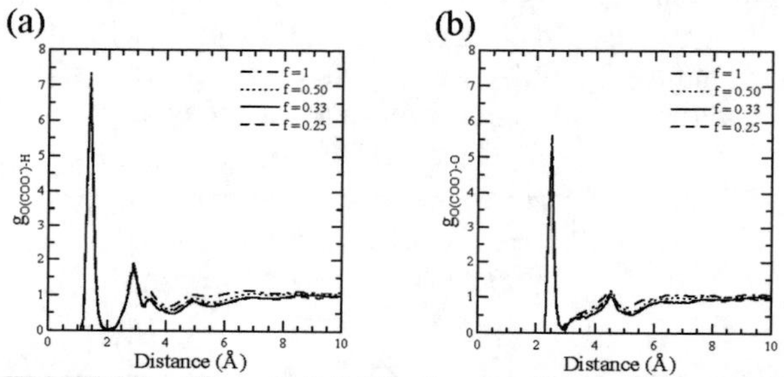

FIGURE 2. The radial distribution functions of oxygen (O) and hydrogen (H) atoms of water with respect to the O atom in the COO^- groups at various values of charge density f.

Next, we discuss the distribution of water surrounding the COOH group, as shown in Fig. 3. we find the height of the first peaks in Fig. 3 is less than 1, i.e., the local water density surrounding the COOH groups is even smaller than that of bulk water. To clarify, only a small amount of hydrogen bonds form between the water molecules and the COOH groups. Indeed, for the noncharged PMAA case (f = 0), we observe that when the distance from the COOH group is smaller than R_g ($\fallingdotseq$10Å), all the water distribution profiles are far less than 1.0. This indicates that the COOH groups appear to be less hydrophilic in nature and therefore fewer water molecules could remain inside the coiled PMAA chain.

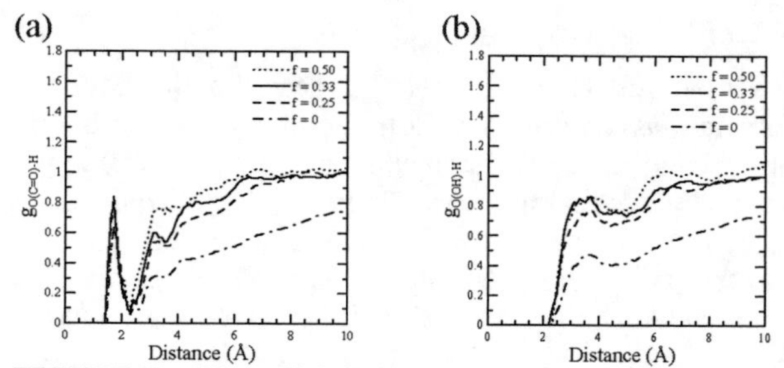

FIGURE 3. The radial distribution functions of hydrogen (H) atoms of water with respect to the O atom of -C=O and -OH in the COOH groups at various values of charge density f.

Fig. 4(a) presents the vibration spectra of the water oxygen types, O0, O1, and O2, which are obtained by applying the Fourier transformation to the VACF profiles. O0, O1, and O2 belong to the water molecules in the bulk state, with a single hydrogen bond formed with the COO^- group, and with two hydrogen bonds formed with the COO^- groups, respectively. The O0 spectrum shows a major peak centered around 58cm^{-1} and a broad shoulder peak at around 200-300cm^{-1}, which has been observed in other MD studies [14,15]. When water molecules have stronger interactions with charged COO^- groups, through the formation of the hydrogen bonds (O2 > O1 > O0), we observe that both spectrum peaks shift to higher wavenumbers. Worthy of note, is that the second peak of the O2 spectrum (i.e., the bridged water molecules between two neighboring COO^- groups) moves significantly toward 400cm^{-1}. Moreover, the second peak becomes more significant while the first peak shows an opposite trend. Similar to the low frequency Raman spectra of liquid water reported by Walrafen et al.[16], these results imply that the second peak is primarily associated with the H-bonded O-O intermolecular stretching vibration, whereas the first peak is attributed to the non-H-bonded molecules. In Fig. 4(b) we present the hydrogen spectrum, which is typically related to the libration (400-1200cm^{-1}), intramolecular bending (1200-2200cm^{-1}) and intramolecular stretching (2200-4000cm^{-1}) motions. It is apparent that the presence of the charged COO^- groups has a significant influence on the resulting hydrogen spectrum. The hydrogen atoms are divided into four types of H0, H1y, H1n, and H2. Note, for the water molecules with only one hydrogen bond formed with the COO^- group, the two types of hydrogen-bonded and non hydrogen-bonded H atoms are denoted as H1y and H1n, respectively. We find that for the H1y and H2 atoms, which are strongly connected to the oxygen atoms of the COO^- groups, both the libration and bending peaks shift significantly from a lower frequency for H0 towards a higher frequency, whereas the stretching peak shifts oppositely. For the H1n atoms, which are not directly adsorbed into the oxygen atoms of the COO^- groups, the shifting degrees of the main peak positions are not as significant as those obtained from the strongly bonded H1y and H2 atoms.

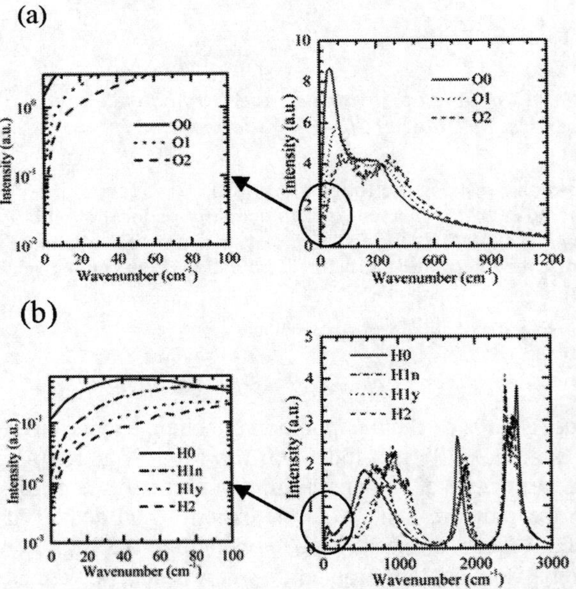

FIGURE 4. Vibration spectra of (a) oxygen and (b) hydrogen atoms of water in different interaction states with the COO^- groups.

CONCLUSIONS

We employ all-atom molecular dynamics simulations to study a single molecule of PMAA at various charge densities f in aqueous solutions. We find that in addition to the repulsive electrostatic interactions between the charged COO^- groups, the strong attractive interaction between the charged COO^- group and water via the formation of hydrogen bonds holds great influence on both the PMAA chain conformation and the dynamic properties of water. Since bridged water molecules significantly limit torsional and bending degrees of the backbone monomers, it is reasonable to conclude that the rod-like chain conformation, exhibited by the strongly charged polymer chain, is significantly enhanced via the bridged water. Furthermore, this influence slows down water diffusion and enables the two characteristic peaks of the oxygen spectrum to shift to higher frequencies. When water molecules bridge between two neighboring COO^- groups, we observe a significant increase to the second peak of the corresponding oxygen spectrum ($\approx 100 cm^{-1}$). This is not surprising since the second peak is mainly attributed to the H-bonded O-O intermolecular stretching vibration. In addition, the strong H-bonded interaction also causes a respective increase of 386 cm^{-1} and 111cm^{-1} to the libration and bending peaks of the hydrogen spectrum and a decrease of 151cm^{-1} to the stretching peak.

ACKNOWLEDGMENTS

This work was supported by the National Science Council of the Republic of China under Grant Number NSC-94-2212-E-110-005, NSC-095-SAF-I-564-623-TMS, and NSC 95-2221-E-002-155.

REFERENCES

1. P. Ravi, C. Wang, K. C. Tam, and L. H. Gan, *Macromolecules*, **36**, 173–179 (2003).
2. V. Sfica, and C. Tsitsilianis, *Macromolecules*, **36**, 4983–4988 (2003).
3. C. Wang, P. Ravi, K. C. Tam, and L. H. Gan, *J. Phys. Chem. B*, **108**, 1621–1627 (2004).
4. J. Wittner, A. Johner, and J. F. Joanny, *J. Phys.II*, **5**, 635–642 (1995).
5. T. T. Nguyen, I. Rouzina, and B. I. Shklovskii, *J. Chem. Phys.*, **112**, 2562–2568 (2000).
6. F. J. Solis, *J. Chem. Phys.*, **117**, pp. 9009–9015 (2002).
7. P. Y. Hsiao and E. Luijten, *Phys. Rev. Lett.*, **97**, 148301-1-148301-4 (2006).
8. M. Levitt, M. Hirshberg, R. Sharon, and V. Daggett, *Comput. Phys. Commun.*, **91**, 215–231 (1995).
9. S. P. Ju, W. J. Lee, C. I. Huang, W. Z. Cheng, and Y. T. Chung, *J. Chem. Phys.*, **126**, 224901-1-224901-10 (2007).
10. M. Levitt, M. Hirshberg, R. Sharon, K. E. Laiding, and V. Daggett, *J. Phys. Chem. B*, **101**, 5051–5061 (1997).
11. J. M. Haile, *Molecular Dynamics Simulation*, Wiley–Interscience, New York, 1992.
12. D. C. Rapaport, *The Art of Molecular Dynamics Simulations,* Cambridge University Press, Cambridge, 1997.
13. M. Praprotnik and D. Janežič, *J. Chem. Phys.*, **122**, 174103-1-174103-10 (2005).
14. J. Martí, J. A. Padro, and E. Guárdia, *J. Chem. Phys.*, **105**, 639–649 (1996).
15. J. A. Padro, J. Marti, and E. Guardia, *J. Phys. Condens. Matter*, **6**, 2283–2290 (1994).
16. G. E. Walrafen, M. S. Hokmabadi, W. H. Yang, and Y. C. Chu, *J. Phys. Chem.*, **93**, 2909–2917 (1989).

Structural Heterogeneity and Non-Exponential Relaxation in Supercooled Liquid Silicon

Tetsuya Morishita

Research Institute for Computational Sciences (RICS), National Institute of Advanced Industrial Science and Technology (AIST), Central 2, 1-1-1 Umezono, Tsukuba, Ibaraki 305-8568, Japan

Abstract. Isothermal-isobaric first-principles molecular-dynamics simulations were performed to investigate structural heterogeneity in supercooled liquid silicon (l-Si). Using the order parameter that measures the tetrahedrality of local atomic configurations, we demonstrate considerable structural fluctuations in l-Si at 1000 K. We also find that structural relaxation is described by the stretched-exponential form, which can be attributed to the intermittent structural fluctuations between high- and low-tetrahedral configurations.

Keywords: First-Principles, Heterogeneity, Molecular-Dynamics, Silicon, Structural relaxation, Supercooled liquid
PACS: 61.25.-f, 61.20.Ja, 71.15.Pd, 64.60.My

INTRODUCTION

Molecular-dynamics (MD) simulations are a powerful tool to obtain microscopic information of bulk liquids. Many of the recent MD studies have focused on structural and dynamical heterogeneity in supercooled liquids, which is believed to account for the slow dynamics in the supercooled state [1]. A variety of liquid models, e.g., Lennard-Jones (LJ) mixtures and SPC/E water, have been employed to clarify the mechanism of the slow dynamics [2, 3]. However, the validity of some empirical liquid models for the supercooled state has been called into question [4].

Here, we report on a first-principles molecular-dynamics (FPMD) study of structural heterogeneity in supercooled liquid silicon (l-Si). We find that highly tetrahedral configurations are intermittently formed in the deeply supercooled state (1000 K). The formation of the highly tetrahedral configurations is responsible for the anomalous structural relaxation characterized by the stretched-exponential function. The temporal structural fluctuation is found to give rise to the $1/f$ dependence in the corresponding power spectral density.

COMPUTATIONAL METHOD

FPMD simulations under constant temperature and pressure [5, 6] were performed for supercooled l-Si. The supercell contained 64 Si atoms and periodic boundary conditions were imposed. The electronic state calculation was performed within the local density approximation of density functional theory. The electronic wave functions were expanded in a plane wave basis with an energy cutoff of 21.5 Ry at the Γ-point in the Brillouin zone. Details of the computational techniques are given in Ref. 7. A 200 ps production run for l-Si at 1000 K was performed after an equilibration stage of $\sim$ 8 ps under the pressure that yields the experimental density ρ of 2.59 g/cm^3 at the melting temperature (1687 K). For comparison, another production run was also performed to calculate the properties of moderately supercooled l-Si (1600 K) under the same pressure.

RESULTS AND DISCUSSION

In order to examine structural fluctuations in supercooled l-Si, we calculated the order parameter q_t that measures the degree of tetrahedrality in the arrangement of an atom and its four nearest neighbors: q_t is defined for each atom as $q_t = 1 - \frac{3}{8}\sum_{i=1}^{3}\sum_{j=i+1}^{4}\left(\cos\theta_{ij} + \frac{1}{3}\right)^2$, where θ_{ij} is the angle between the vectors that join a central atom with its ith and jth nearest neighbors ($j \leq 4$) [7, 8, 9]. Note that q_t is 1 in a perfect tetrahedron. Figure 1 shows the time evolution of q_t, averaged over all atoms, for 1000 K and 1600 K. Although its time average is $\sim$ 0.68 at 1000 K, the fluctuation of q_t is remarkable. Interestingly, q_t intermittently remains very high ($\sim$ 0.75) during a few ps. Such a high-q_t state is, for instance, found at time step $\sim 2.7 \times 10^5$. This dynamical behavior of q_t indicates significant structural heterogeneity at 1000 K. In contrast, at 1600 K, q_t fluctuates around $\sim$ 0.55 and its fluctuation is considerably suppressed. This indicates that l-Si at 1600 K is much more homogeneous than at 1000 K.

The spatial correlations of high-q_t atoms and of low-q_t atoms are examined by the pair correlation function $g(r)$. In Fig. 2, $g(r)$ between the high-q_t atoms ($q_t \geq$ 0.8), $g_h(r)$, is shown together with $g(r)$ between the

CP982, *Complex Systems, 5th International Workshop on Complex Systems*
edited by M. Tokuyama, I. Oppenheim, and H. Nishiyama

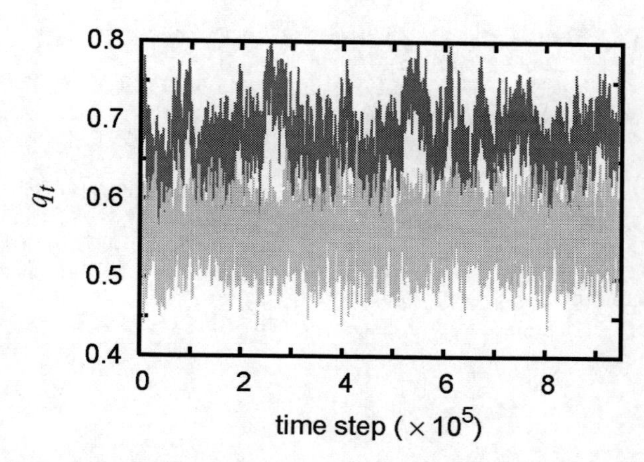

FIGURE 1. (Color online) Time evolution of q_t averaged over all atoms for 1000 K (upper lines) and 1600 K (lower lines). A time step of 0.121 fs was used to integrate equations of motion.

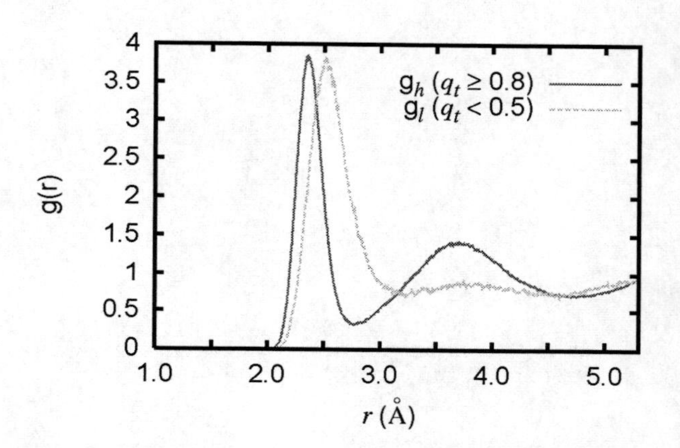

FIGURE 2. (Color online) Pair correlation function between high-q_t atoms, $g_h(r)$, and that between low-q_t atoms, $g_l(r)$, for l-Si at 1000 K.

low-q_t atoms ($q_t < 0.5$), $g_l(r)$. We see a striking difference between $g_h(r)$ and $g_l(r)$. The sharp first peak in $g_h(r)$ reflects a highly tetrahedral configuration, while the broader peak in $g_l(r)$ indicates weak atomic bonding compared with covalent (tetrahedral) bonding. The second peak in $g_h(r)$ is much more prominent than that in $g_l(r)$, indicating that the high-q_t atoms tend to form the tetrahedral order extending beyond the first neighbor shell. The low-q_t atoms, in contrast, exhibit almost no correlation beyond ~ 3 Å. The coexistence of these different types of structures is a notable feature of the deeply supercooled state.

It should be remarked here that our visual inspection confirms that the high- and low-q_t atoms definitely coexist during the MD run. That is, the heterogeneity indicated in Figs. 1 and 2 does not result from the fluctuation of the whole system. In fact, we calculated the atomic

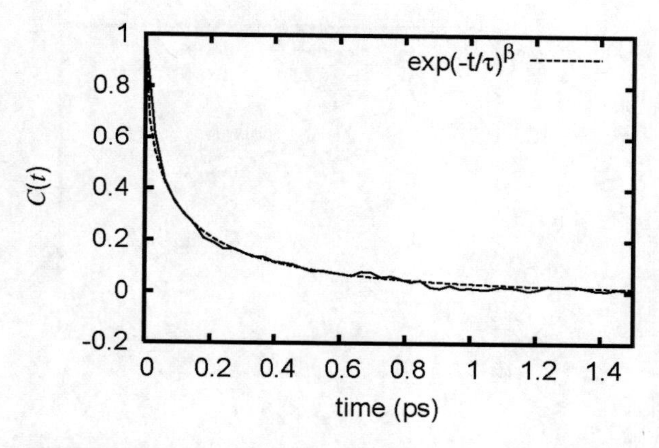

FIGURE 3. Time correlation function $C(t)$ for l-Si at 1000 K.

displacement during the time interval in which the non-Gaussian parameter takes a maximum, and found that there exists strong correlation between the structural heterogeneity and single-atom dynamics.

Structural relaxation in l-Si at 1000 K was examined by the auto correlation function $C(t)$ defined by

$$C(t) = \frac{\langle \delta q_t(t) \delta q_t(0) \rangle}{\langle \delta q_t(0) \delta q_t(0) \rangle}, \tag{1}$$

where $\delta q_t(t) = q_t(t) - \langle q_t(t) \rangle$, and $\langle ... \rangle$ denotes an average over atoms and time. We found that $C(t)$ can be well fitted by the stretched-exponential form, $e^{-(t/\tau)^\beta}$ (Fig. 3): the relaxation timescale τ is ~ 0.085 ps and β is ~ 0.5. This relaxation timescale is relatively short probably due to the fact that $C(t)$ reflects many body correlation, not two body correlation only; τ is not necessarily coincident with α relaxation timescale. Detailed analyses suggest that the non-exponential relaxation arises from the inherent atomic-scale dynamics, i.e., intermittent structural fluctuations between the high- and low-q_t states.

The power spectral density $S(f)$ of the q_t fluctuation (Fig. 1) is shown in Fig. 4. It is known that the $1/f^\alpha$ dependence with $0.5 < \alpha < 1.8$ indicates multiple timescale behavior, while that with $\alpha = 2$ indicates single timescale behavior. We see that the spectrum for 1000 K yields the $1/f^\alpha$ dependence with $\alpha \sim 1.3$ in a range 0.4 – 100 cm^{-1}. In contrast, the $1/f$ range is much narrower for 1600 K. The spectrum is mostly a white-noise below 10 cm^{-1}, indicating the lack of long-time correlation observed at 1000 K. It has been shown that water also exhibits the $1/f$ type dependence, which is attributed to the global rearrangement of the tetrahedral network on a timescale of several tens of ps [10, 11]. The present result implies that global structural rearrangements take place on a timescale of 100 ps or less also in deeply supercooled l-Si.

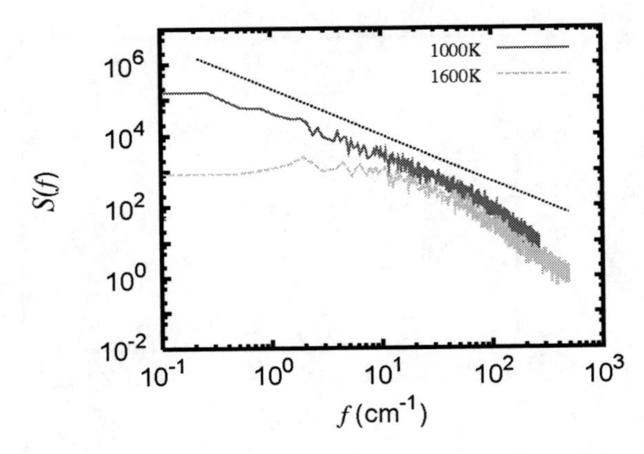

FIGURE 4. (Color online) Power spectra of the q_t fluctuation (Fig. 1) for 1000 K and 1600 K. Results for 1600 K are shifted downward by $|\log 0.4|$ for clarity. Dotted lines indicate $1/f^{1.3}$.

SUMMARY

We have demonstrated the structural heterogeneity in deeply supercooled *l*-Si. The formation and collapse of highly tetrahedral configurations are found to be responsible for this heterogeneity. We have found that the structural relaxation can be described by the stretched-exponential relaxation, which is closely related with the intermittent fluctuation between high- and low-q_t states. It should be pointed out that even *l*-Si at 1000 K would exhibit rather homogeneous character on the timescale of laboratory experiments [1] due to the lack of long-lived heterogeneity (beyond $\sim$ ns or μs). The experimental confirmation of this heterogeneity would thus be a challenging task.

ACKNOWLEDGMENTS

This work is supported in part by a Grant-in-Aid for Scientific Research from the Ministry of Education, Culture, Sports, Science and Technology, and Molecular-based New Computational Science program, NINS. The computations were carried out at the Research Center for Computational Science, National Institute of Natural Sciences.

REFERENCES

1. M. D. Ediger, *Annu. Rev. Phys. Chem.* **51**, 99-128 (2000).
2. C. Donati, S. C. Glotzer, P. H. Poole, W. Kob, and S. J. Plimpton, *Phys. Rev. E* **60**, 3107-3119 (1999).
3. E. La Nave and F. Sciortino, *J. Phys. Chem. B* **108**, 19663-19669 (2004).
4. P. Beaucage and N. Mousseau, *J. Phys.: Condens. Matter* **17**, 2269-2279 (2005).
5. P. Focher *et al.*, *Europhys. Lett.* **26**, 345-351 (1994).
6. T. Morishita, *Mol. Simul.* **33**, 5-12 (2007).
7. T. Morishita, *Phys. Rev. Lett.* **97**, 165502-1-165502-4 (2006).
8. J. R. Errington and P. G. Debenedetti, *Nature* **409**, 318-321 (2001).
9. T. Morishita, *Phys. Rev. E* **72**, 021201-1-021201-4 (2005).
10. E. Shiratani and M. Sasai, *J. Chem. Phys.* **104**, 7671-7680 (1996); *ibid.* **108**, 3265-3276 (1998).
11. I. Ohmine, *J. Phys. Chem.* **99**, 6767-6776 (1995).

Selectivity Principle of the Ligand Escape Process from a Two-Gate Tunnel in Myoglobin: Molecular Dynamics Simulation

Sheh-Yi Sheu

Department of Life Sciences and Institute of Genome Sciences
National Yang-Ming University, Taipei 112, Taiwan

Abstract. We proposed a selectivity principle for the ligand escape process from two fluctuating bottlenecks in a cavity with a multi-gate inside a myoglobin pocket. Our analytical theory proposed a fluctuating bottleneck model for a Brownian particle passing through two gates on a cavity surface of an enzyme protein and has determined the escape rate in terms of the time-dependent gate function and the competition effect. It illustrated that with two (or more than two) gates on a cavity surface the gate modulation, which is controlled by protein fluctuation, dominates the ligand escape pathway. We have performed a molecular dynamics simulation to investigate the selectivity principle of the ligand escape process from two-gate tunnel in myoglobin. The simulation results confirm our theoretical conjecture. It indicates that the escape process is actually entropy driven and the ligand escape pathway is chosen via the gate modulation. This suggests an interesting intrinsic property, which is that the oxy-myoglobin tertiary structure is favorable to the departure of the ligand from one direction rather than through a biased random walk.

Keywords: Molecular dynamics simulation, Ligand escape, Entropy driven, Selectivity principle
PACS: 87.64.Aa.

INTRODUCTION

Gating behavior is one of the most important and fundamental processes in nature which includes ligand escape, enzyme-substrate reaction, ion channel, ion pumps, drug delivery from a vesicle to a quantum dot, spintronics, and more.[1] Several proteins such as fluctuating bottleneck enzyme protein and heme proteins contain many modulated gates. As we know the active site of an enzyme molecule is usually buried inside the protein with a cavity gated by a protein side chain. Ligand motion inside these proteins is regulated by gate modulation. The escape tunnel usually consists of a sequence of connected cavities. This leads us to consider that there should be a multigate control of the tunneling process of ligand not only in heme proteins but also in other biological channels. On the other hand, most biological channels provide an efficient and energetically favorable pathway for the passage of ions through a cell membrane and this is due to their gating behavior. Hence, ion permeation is selected by the chemical affinity of the protein.

A cavity or channel with multi-gate, like a junction, can control the direction of the particle's motion. This occurs quite often in a molecular motor problem.[2] For some natural membrane proteins, such as ion channels and ion pumps, the gate modulation is regulated by protein dynamics. An analytical theory of entropy selectivity and uni-directional motion was developed in our previous paper.[3] In that work, the diffusion motion inside the cavity is entropy driven. The surface gate cuts the entropy potential surface and generates a fluctuating barrier height. The controlling factors are gate size and gate modulation rate. In this work, we show the selectivity principle as applied to the myoglobin tunnel using molecular dynamics simulation (MD). According to the gate modulation, the ligand escape rate of each individual gate is obtained. Hence, the entropy selectivity principle is based upon the competition effect between gates.

MODEL

We constructed a two-gate model, where the Brownian particle moves inside the spherical blob and flows unidirectionally by entropy-driven force. The Brownian particle is thermally distributed inside a two-gate spherical cavity as Fig.1. On the surface of this cavity, there are two punctures that act as gates whose undulations are regulated by protein dynamics. The motion of the Brownian particle (ligand) follows a 3D-Smoluchowski equation with undulated radiation boundary condition

$$\frac{\partial}{\partial t} n(\vec{r},t) = D_0 \nabla^2 n(\vec{r},t) + \int d\Omega\, \delta(\vec{r}-\vec{R})\, \sigma(\Omega,t) \quad (1)$$

CP982, *Complex Systems, 5th International Workshop on Complex Systems*
edited by M. Tokuyama, I. Oppenheim, and H. Nishiyama

where Ω is the solid angle originated from the cavity center and R is the cavity radius; the σ term denotes the surface reaction, and can be solved by a self-consistent method. We assume that the Brownian particle diffuses inside the cavity with a diffusion constant D_0 and is bounced back from the cavity surface until it reaches the gates. On the right hand side of Eq. (1), the second term describes the surface reaction with a surface reaction kernel σ, which can be solved by employing the method of Sheu and Yang. [3,5]

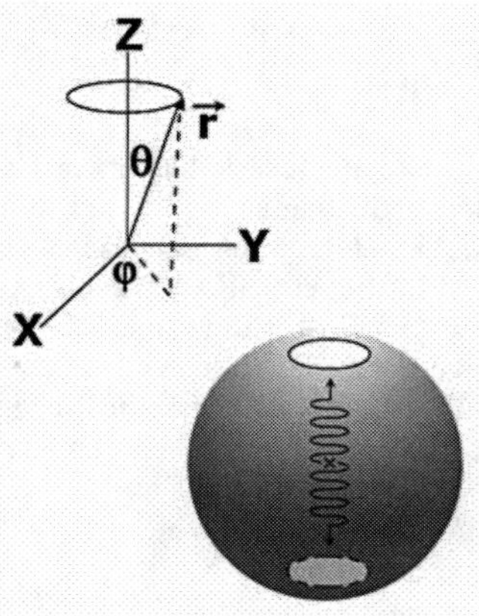

FIGURE 1. A two-gate spherical cavity model. r is the radial coordinate, θ is the polar angle and φ is the azimuth angle. The modulation of gates is regulated by the polar angles $\theta_1(t)$ and $\theta_2(t)$ in the ranges of $0 \le \theta_i(t) \le \theta_{i,\max}$, $(i = 1, 2)$.

COMPUTATIONAL METHOD

All the MD simulations and the analysis were performed with the program CHARMM and the CHARMM22 force field. The inital structure of myoglobin was taken from the PDB 1mbo structure. The molecule was first subjected to 500 steps of steepest descents energy minimization. For the detailed equilibration procedures of the system, we have described these in Sec. III in our previous paper.[4,5]

Sequentially, at the equilibrated system of the molecular dynamics simulations the O_2 ligand is still bounded to the Fe^{+2} in the myoglobin. In order to simulate the ligand escape process and the experimental system, we neglected to consider the detailed process by which the ligand is released from Fe^{+2}, and just broke the Fe^{+2}- O_2 bond by directly increasing the bond length of $Fe^{+2} - O_2$ by about 1Å. Therefore, the O_2 ligand could move freely inside the protein. Meanwhile, to enhance the configurations of the distribution of the O_2 ligand in a specific cavity, the locations of O_2 ligands were initially randomly generated inside the specific cavity based on the thermal distribution. Actually, we discarded those configurations in which the O_2 ligand was out of the specific cavity. The trajectories were saved every 100 fs and each simulation duration average was at least 100 ps or over. Eventually, almost 300 dynamic trajectories runs were collected and used to perform the complete analysis.

RESULTS AND DISCUSSION

The MD simulation showed that there were several ligand escape tunnels inside myoglobin. It showed the trajectories of the ligand moving inside the cavity I, II, and III and the ligand escape pathway through a gate area. To find out whether this issue could play a critical role in the communication between protein and ligand, we investigate the selectivity of the escape pathway in this work (see Fig. 2). For more details of the procedures, please see our papers. [4,5] In the present simulation, there were total 279 sets of dynamic trajectories and three pathways that correspond to the ligands escape from the cavity II; 1) 204 trajectories through gate I to cavity I, 2) 69 trajectories through gate II to cavity III, and 3) 6 trajectories through other pathways.

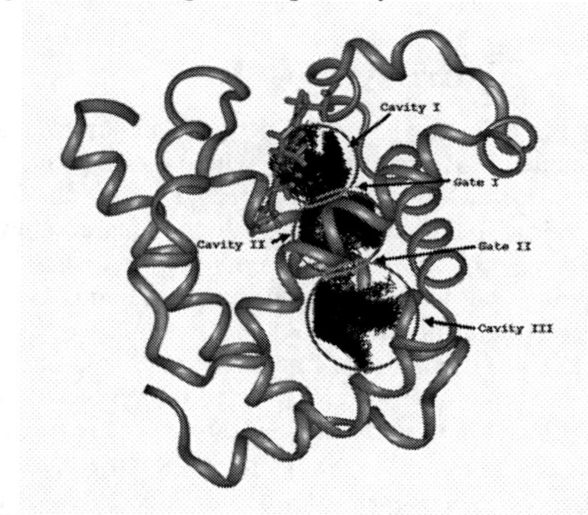

FIGURE 2. The tertiary structure of myoglobin and the trajectories of the O_2 ligand. The cavities and gates area are shown as circle.

Based on the pathway analysis, there are several different O_2 escape tunnels. In Fig. 3, we focused on the trajectories of the ligand moving inside the heme pocket and its escape from the cavity I to cavity II through gate I, and portrayed the gate area and the gate opening and closure processes.

It is important to know how gate modulation dominates the ligand escape pathway and escape rate. In fact, the gate modulation or radius of gyration of gate, $R_g(t)$, involves the protein dynamic information. Its modulation pattern and initial state determine the escape rate. Since the gate initial state is open, within the diffusion characteristic time scale, the gate closes and opens several times within the major escape process, i.e. 5 ps (see Fig. 4a).

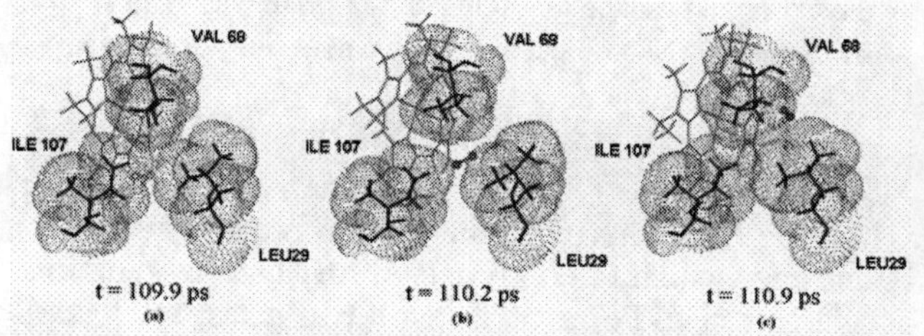

FIGURE 3. The gate area residues 29, 68, and 107 and the gate opening and closure processes. The ligand O_2 is represented by ball and stick. From time 109.9 ps to 110.9 ps, we observed that the O_2 escaped out of the gate area. At t=109.9 ps, the O_2 was inside the cavity I. At t=110.2 ps, the ligand was passing through the gate (gate opening). At t=110.9 ps, the ligand was in cavity II (gate closed).

Now we turn to consider the effect of gate modulation frequency and gate size on the ligand escape rate. Here, gate I and gate II have different static gate sizes and modulation frequencies (see Table I). Our theoretical prediction indicates that the ligand escape pathway is mainly through gate I, which has a larger gate size rather than through gate II with a smaller gate size because the small gate size has a higher $\widetilde{F}_0^{-1}$ value, which corresponds to a fast gate modulation frequency. The estimated theoretical total survival time is 3.1 ps. Here the characteristic time scale, τ_D $(= R^2 / D_0)$ that we adopted was 2.759 ps, where R is the averaged radius of cavity II at 2.39Å and D_0 is the ligand mean diffusion constant in cavity II at 2.07Å^2/ps.

Our MD simulation results obtained so far indicate that the survival time calculated by a time integration of the first passage time distribution for ligand to escape through gate I is 4.32 ps and through gate II is 7.623 ps; therefore, the escape rates of k_1 and k_2 are 0.2315 ps^{-1} and 0.1312 ps^{-1}, respectively (see Fig. 5). Accordingly, the mean first passage time of ligand escape through gates I and II can be obtained by a time integration of the first passage time distribution multiplied by time and are 60.06 and 54.05 ps, independently. Based on the definition of the total escape rate k, the MD simulation results show that the two-gate total escape rate k is 0.271 ps^{-1}. In other words, for the two-gate cavity, the simulated survival time is ca. 3.69 ps, which is in reasonably good agreement with the theoretical value of about 3.1 ps. This confirms our argument for the existence of a competition effect between two gates on the same cavity surface. The other factor supporting our argument is the negative value for Δ. This is because according to our simulation, our two-gate model supports the selectivity of the two gates of the ligand escape pathways as being strongly dependent on the gate size and gate modulation frequency.

The MD simulation ligand survival time in a one-gate cavity was calculated at about 4.37 ps, which is longer than that of two-gate cavity at 3.69 ps.[4,5] Actually, cavity size also plays a key role to influence the ligand escape time (or survival time). Here, the radius of gyration of the one-gate cavity I is about 2.47 Å and that of the two-gate cavity II is about 2.39 Å. In this case the cavity size effect will contribute only a little to the escape rate and therefore we can neglect the cavity size effect. The major dominant factor affecting the enhancement of the ligand escape rate is that the two-gate cavity provides two opportunities for the ligand to escape.

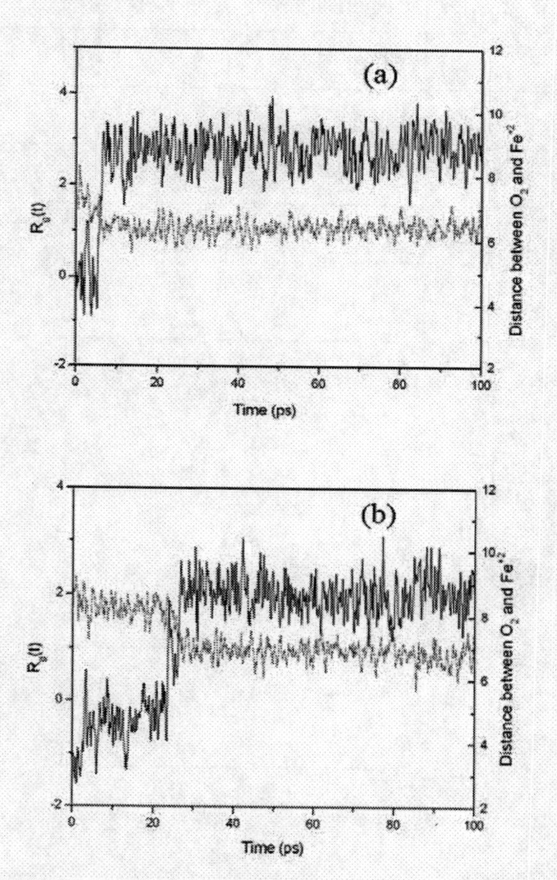

FIGURE 4. (a) The modulation of $R_g(t)$ for a short time escape process (dashed line and the scale on the left side). The relative distance between ligand and Fe^{+2} verses time, indicating the ligand position within cavity I or cavity II, is plotted with a solid line (the scale on the right side). Within 5 ps, the ligand – Fe^{+2} distance is about 4 Å and the ligand stays inside cavity I. After 5 ps, the ligand - Fe^{+2} distance jumps to ca. 9 Å. Ligand escapes to cavity II. Here the gate closing time is consistent with ligand escape time. (b) The modulation of $R_g(t)$ for a longer time escape process. After

the ligand escaped, the gate area tends to close for a longer time.

TABLE I. Comparison of theoretical and numerical escape times

	ONE-GATE CAVITY	TWO-GATE CAVITY
D_0 ($Å^2$/ ps)	1.9	2.07
$\tilde{F}_0^{-1}$	4.234	3.33
τ_D (ps)	3.21	2.759
τ_0 (ps) (theoretical)	4.34	3.1
τ_0 (ps) (simulation)	4.37	3.69
$\tau_{0,1}$ (ps) (theoretical), k_1^{-1}		4.73
$\tau_{0,1}$ (ps) (simulation)		4.32
$\tau_{0,2}$ (ps) (theoretical), k_2^{-1}		8.92
$\tau_{0,2}$ (ps) (simulation)		7.623
Δ		-0.0102

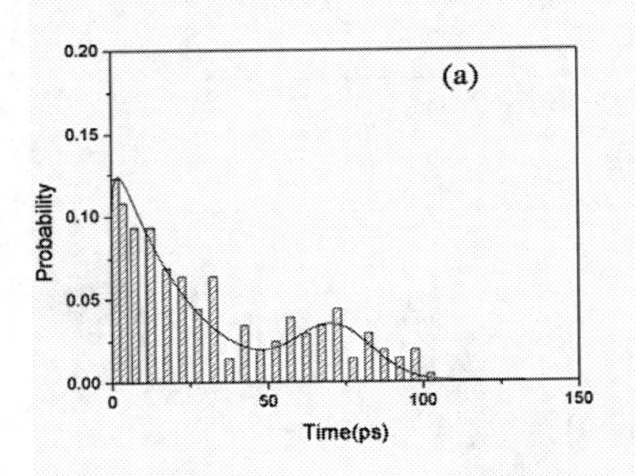

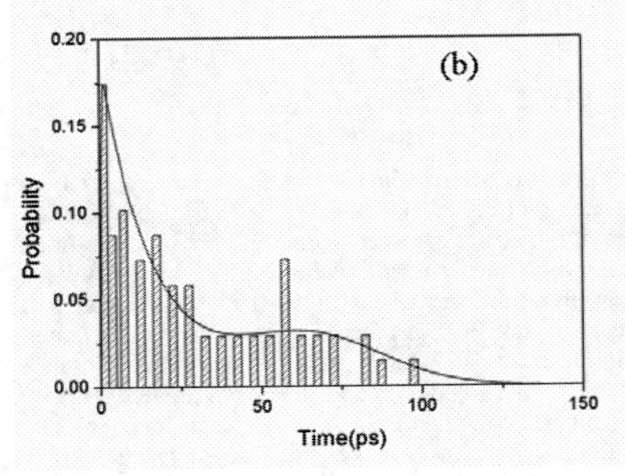

FIGURE 5. (a) First passage time distribution versus time for the O_2 ligand escape through gate I from cavity II to cavity I. A total 204 MD trajectories were carried out for analysis. (b) First passage time distribution versus time for the O_2 ligand escape through gate II from cavity II to cavity III. A total 69 MD trajectories were carried out for analysis.

CONCLUSION

The basic driving force for the escape process is entropy, which is controlled by the gate modulation and gate size. For a cavity with two gates, the ligand escape pathway is selected by these two gates due to the gate modulation and exhibits a strong competition effect with a negative Δ value when these two gates are neither static nor equal in gate size and gate modulation frequency. The escape rate for each gate can be approximated by an equal sharing of the competition effect. Once each gate has different escape rate, these two gates, due to the gate regulation, select the ligand escape pathway. Our simulation results coincide well with the previous analytical theory.

For an ion channel warped by protein bundles, the gate is regulated by the protein fluctuation. With external non-equilibrium fluctuation forces, such an electrical field there exits a net flow inside the ion channel.

ACKNOWLEDGMENTS

This work was supported by grants from the National Science Council of Taiwan under Grant Nos. NSC-93-2113-M-010-001 and NSC-94-2113-M-010-001 and the National Center for High-Performance Computing of Taiwan.

REFERENCES

1. Auerbach, *Proc. Natl. Acad. Sci. U.S.A.* **102**, 1408-1412 (2005); D. A. Doyle, J. Morais Cabral, R. A. Pfuetzner, A. Kuo, J. M. Gulbis, S. L. Cohen, B. T. Chait and R. MacKinnon, *Science* **280**, 69-77 (1998); Y. Jiang, A. Lee, J. Chen, M. Cadene, B. T. Chait and R. MacKinnon, *Nature* **417**, 523-526 (2002).
2. P. Reimann, *Phys. Rep.* **361,** 57-265 (2002); S. Matthia and F. Müller, *Nature*, **424**, 53-57 (2003); Y. Tu and G. Grinstein, *Phys. Rev. Lett.*, **94**, 208101-208105 (2005).
3. Sheh-Yi Sheu and Dah-Yen Yang, *J. Chem. Phys.*, **114**, 3325-3329 (2001).
4. Sheh-Yi Sheu, *J. Chem. Phys.*, **122,** 104905-104911 (2005).
5. Sheh-Yi Sheu, *J. Chem. Phys,.* **124**, 154711-154719 (2006).

Participants

Hiroto Adachi	Miyakawa lab, Department of Applied Physics, Fukuoka University, 8-19-1 Nanakuma, Jonan-ku, Fukuoka, 814-0180, Japan
Frederic Affouard	Laboratoire de Dynamique et Structure des Materiaux Moleculaires - UMR CNRS 8024 BAT P5 - UFR DE PHYSIQUE - USTL - CITE SCIENTIFIQUE - VILLENEUVE D, 59655, France
Akira Akaishi	Department of Physics, Tokyo Metropolitan University, 1-1 Minami-osawa, Hachioji-shi, Tokyo, 192-0397, Japan
Mitsuhiro Akimoto	Department of Electronics and Computer Science, Tokyo University of Science Yamaguchi, 1-1-1 Daigaku-Dori, San-yo-Onoda, Yamaguchi, 756-0884, Japan
Naohiro Akuzawa	Tokyo University of Science, Room 0810 Bldg. 1 1-3 Kagurazaka, Shinjyuku-ku, Tokyo, 162-8601, Japan
Benjamin John Anderson	Department of Chemical & Biomolecular Engineering, University of Illinois at Urbana-Champaign, 114 ROGER ADAMS LAB, 600 SOUTH MATHEWS AVE, 61802, USA
Yasutaka Ando	Department of Mechanical Engineering, Faculty of Engineering, Ashikaga Institute of Technology, 268-1 Omae, Ashikaga, Tochigi 326-8558, Japan
Juan FR Archilla	Grupo de Fisica No Lineal, Departamento de Fisica Aplicada I, University of Sevilla, ETSI Informatica-G0-44, Avda Reina Mercedes s/n, 41012, Spain
Robert Botet	Laboratoire de Physique des Solides - bat. 510, Universite Paris-Sud, Orsay, 91405, France
Sirikul Bunditsaovapak	Department of Mathematics and Computer Science, Faculty of Science, King Mongkut's Institute of Technology, Ladkrabang, Chalongkrung Road, Ladkrabang, Bangkok, 10520, Thailand
Simone Capaccioli	INFM-CNR SOFT, Roma & Dip. Fisica Università di Pisa, Pisa, Largo Pontecorvo 3, Pisa, 56127, Italy

Julyan Cartwright	Laboratorio de Estudios Cristalograficos, CSIC, Edificio BIC Granada, Avenida de la Innovacion, 1, P.T. Ciencias de la Salud, E-18100 Armilla, Granada, E-18100, Spain
Silvina Cerveny	Consejo Nacional de Investigaciones Cientificas y tecnicas ,CSIC, Paseo Manuel de Lardizabal 4, San Sebastian, 20018, Spain
Feng Ming Chang	Department of Chemical and Materials Engineering, National Central University, Jhongli, Taiwan 320, Republic of China, Taiwan
Chien-Chi Chao	Department of Electrical Engineering, Chien-Kuo Technology University, No.1 Chieh-Sou N. Rd., Changhua City 500, Taiwan
Sow-Hsin Chen	Department of Nuclear Science and Engineering, Massachusetts Institute of Technology, 24-209, MIT, Cambridge, MA 2139, USA
Wei Chen	Institute of Physics, Academia Sinica, Taipei 11529, Taiwan
Song-Ho Chong	Institute for Molecular Science, Myodaiji 38, Okazaki, 444-8585, Japan
Chuen-Shii Chou	Department of Mechanical Engineering, National Pingtung University of Science & Technology, 1, Hseuh Fu Road, Nei Pu Hsiang, Pingtung, 912, Taiwan
Chung-I Chou	Department of Physics, Chinese Culture University, 55, Hwa-Kang Road, Yang-Ming-Shan, Taipei, 111, Taiwan
Shih-Hao Chou	National Taiwan University, Department of Chemical Engineering, Room 401a, No.1, Sec. 4, Roosevelt Rd., Da-an District, Taipei City 106, Taiwan
Tage Christensen	DNRF centre "Glass and Time", IMFUFA, Department of Sciences, Roskilde University, Universitetsvej 1, Postbox 260, DK-4000, Roskilde, DK-4000, Denmark
Natalia Correia	REQUIMTE/CQFB Departamento de Quimica, Faculdade de Ciencias e Tecnologia Universidade Nova de Lisboa, Caparica, 2829-516, Portugal
Jose Joaquim Cruz Pinto	Department of Chemistry, University of Aveiro / CICECO Campus of Santiago, 3810-193, Portgal

Cristiano De Michele	Department of Physics, University of Rome "La Sapienza", Piazzale Aldo Moro N.2, I-00185, Italy
Marc Descamps	UFR Physique, University Lille1, Villeneuve d, 59655, France
Silvia Cecilia Di Fonzo	Sincrotrone Trieste, s.s. 14 km 163.5, 34 012 Trieste, Italy
Jeppe C Dyre	DNRF centre for Viscous Liquid Dynamics "Glass and Time", Roskilde University, Postbox 260, DK-4000, Denmark
Toru Ekimoto	Department of Physics, Faculty of Science, Kyushu University, Hakozaki, Higashi-ku, Fukuoka, 812-8581, Japan
Yoshihisa Enomoto	Department of Mechanical Engineering, Nagoya Institute of Technology, Nagoya, 466-8555, Japan
Eigo Ezaki	Tokyo University of Science, Room 0810, Bldg. 1, 1-3 Kagurazaka, Shinjuku-ku, Tokyo, 162-8601, Japan
Roland Faller	Department of Chemical Engineering and Materials, University of California Davis, 1 Shields Ave. Davis, CA, 95616, USA
Giuseppe Foffi	Ecole Polytechnique Federale de Lausanne (EPFL), EPFL SB IRRMA-GE, PPH 331 (Bâtiment PPH (CRPP)), Station 13, CH-1015 Lausanne, Switzerland
John J Fontanella	Physics Department, United States Naval Academy, Annapolis, MD 21402-5026, USA
Hiroki Fujimori	Nihon University, 3-25-40 Sakurajosui, Setagaya-ku, Tokyo, 156-8550, Japan
Ikuo Fukuda	RIKEN, 2-1 Hirosawa, Wako, Saitama, 351-0198, Japan
Takaaki Furubayashi	Tokuyama Lab., Institution of Fluid Science, Tohoku University, 2-1-1 Katahira, Aoba-ku, Sendai, 980-8577, Japan
Akira Furukawa	Institute of Industrial Science, The University of Tokyo, Komaba 4-6-1, Meguro-ku, Tokyo, 153-8505, Japan
Gary S. Grest	Sandia National Laboratories, Albuquerque, NM 87185, USA
Hua Guo	Van der Waals-Zeeman Ins, Emmastraat 9. Amsterdam, 1071JA, The Netherlands

Junko Habasaki	Tokyo Institute of Technology, 4259 Nagatsuta-cho, Yokohama, 226-8502, Japan
Shoko Hagiwara	Graduate School of Integrated Basic Sciences, Nihon University, 3-25-40 Sakurajosui, Setagaya-ku, Tokyo, 156-8550, Japan
Ken Harris	School of Physical, Environmental and Mathematical Sciences, University of New South Wales, Australian Defence Force Academy, Canberra, ACT, 2600, Australia
Peter Harrowell	School of Chemistry, Univeristy of Sydney, New South Wales, 2006, Australia
Manabu Hasegawa	Graduate School of Systems and Information Engineering, University of Tsukuba, 1-1-1 Tennoudai, Tsukuba, Ibaraki, 305-8573, Japan
Eiji Hashimoto	Kojima Lab., Institute of Materials Science, University of Tsukuba, 1-1-1 Tennodai, Tsukuba, Ibaraki, 305-8573, Japan
Masaomi Hatakeyama	Japan Advanced Institute of Science and Technology, 2-413 1-8, Asahidai, Nomi, Ishikawa, 923-1211, Japan
Yosio Hiki	Department of Physics, Tokyo Institute of Technology , 39-3-303 Motoyoyogi, Shibuya-ku, Tokyo, 151-0062, Japan
Shinya Hosokawa	Hiroshima Institute of Technology, Faculty of Engineering, 2-1-1 Miyake, Saeki-ku, Hiroshima, 731-5193, Japan
Ching-I Huang	Institute of Polymer Science and Engineering, National Taiwan University, No. 1, Roosevelt Road, Sec. 4, 10617, Taiwan
Yoon-Hwae Hwang	Department of Nanomaterials, Pusan National University, San 30, Jangjeon-Dong, Keumjeong-Ku, 609-735, South Korea
Yasushi Ido	Quality Innovation Techno-Center, Nagoya Institute of Technology, Gokiso-cho, Showa-ku, Nagoya 466-8555, Japan
Nobutoshi Ikeda	Junior College division, Tohoku Seikatsu Bunka College, 1-18-2, Niji-no-Oka, Izumi-ku, Sendai, Miyagi, 981-8585, Japan
Michal Ilavsky	Faculty of Mathematics and Physics, Charles University, V Holesovickach 2, Prague 8, 180 00, Czech Republic

Masayuki Imai	Department of Physics, Ochanomizu University, 2-1-1 Otsuka, Bunkyo, Tokyo, 112-8610, Japan
Ryo Imayama	Statistical Mechanics Laboratory, Kyoto Institute of Technology, Matsugasaki, Sakyo-ku, Kyoto, 606-8585, Japan
Akihisa Inoue	Institute for Material Research, Tohoku University, 2-1-1 Katahira, Aoba-Ku, Sendai, 980-8577, Japan
Kikujiro Ishii	Department of Chemistry, Gakushuin University, 1-5-1 Mejiro, Toshimaku, Tokyo, 171-8588, Japan
Alexey Ivlev	Max Planck Institute for Extraterrestrial Physics, Giessenbachstrasse, 85741 Garching, Germany
Kazuhiko Iwai	Department of Materials, Physics and Energy Engineering, Graduate School of Engineering, Nagoya University, Furo-cho Chikusa-ku, Nagoya, Aichi, 464-8603, Japan
Jiři Jeništa	Institute of Plasma Physics, ASCR, v. v. i., Za Slovankou 3, 182 00 Praha 82, Czech Republic
Masami Kageshima	Department of Applied Physics, Osaka University, 2-1, Yamada-oka, Suita, Osaka, 565-0871, Japan
Keimei Kaino	Sendai National College of Technology, 4-16-1 Ayashichuo, Aobaku, Sendai, 989-3128, Japan
Mitsuhiro Kanakubo	National Institute of Advanced Industrial Science and Technology (AIST), 4-2-1 Nigatake, Miyagino-ku, Sendai, 983-8551, Japan
Toshihiro Kawakatsu	Department of Physics, Tohoku University, Aza-Aoba, Aramaki, Aoba-ku, Sendai, 980-8578, Japan
Junichi Kawamura	Institute of Multidisciplinary Research for Advanced Materials, Tohoku University, 2-1-1 Katahira, Aoba-Ku, Sendai, 980-8577, Japan
Takeshi Kawasaki	Institute of Industrial Science, University of Tokyo, 4-6-1 Komaba, Meguro-ku, Tokyo, 153-8505, Japan
Bongsoo Kim	Institute for Molecular Science, Myodaiji Cho, Okazaki, 444-8585, Japan

Kang Kim	Institute for Molecular Science, 38 Nishigo-Naka Myodaiji Okazaki, 444-8585, Japan
Yuto Kimura	Tokuyama Lab., Institute of Fluid Science, Tohoku Univeristy, 2-1-1 Katahira, Aoba-ku Sendai, 980-8577, Japan
Shu-ichi Kinoshita	Gracuate School of Science and Technology, Niigata University, A-509, Kawabatatyo, Tyuo-ku, Niigata-shi, Niigata, 951-8053, Japan
Seiji Kojima	Institute of Materials Science, University of Tsukuba, 1-1-1 Tennoudai, Tsukuba, Ibaraki, 305-8573, Japan
Sotiria Kripotou	National Technical University of Athens, Department of Applied Mathematics and Physics, Heroon Polytechneiou 9, 157 80, Greece
Hyun-Joung Kwon	Dept. Nano Fusion Technology, 2418 NanoScience building, Miryang Camp. Pusan Nat. Univ., Samnangjin-eup Mirying-si, 627-706, South Korea
Ronald Larson	Departments of Chemical and Mechanical Engineering, University of Michigan, 2300 Hayward Ave., 48103, USA
David Le Coq	LPCA - UMR CNRS 8101 - Universite du Littoral Cote d, 189A Avenue Maurice Schumann - DUNKERQUE, 59140, France
He-Ping Li	Department of Engineering Physics, Tsinghua University, Beijing 100084, P. R. China
Chun-Min Lin	Department of Chemical Engineering, National Taiwan University, room 401a, No. 1, sec. 4, Roosevelt Rd, Da-an District, Taipei, 106, Taiwan
Hartmut Löwen	Institut für Theoretische Physik II: Weiche Materie, Heinrich-Heine-Universität Düsseldorf, Universitätsstraße 1, D-40225 Düsseldorf, Germany
John J. Lowke	Materials Science and Engineering, CSIRO, PO Box 218, Lindfield, Sydney NSW 2070, Australia
Philipp Maass	Institute of Physics, Technical University Ilmenau, Germany PF 100565, 98684 Ilmenau, 98684, Germany

Hiroki Majima	Department of General Education, Salesian Polytechnic, 4-6-8, Oyamagaoka, Machida-city, Tokyo, 194-0215, Japan
Yu Matsuda	Kojima Lab., Institute of Materials Sciences, University of Tsukuba, 1-1-1 Tennodai, 305-8573, Japan
Masakazu Matsumoto	Nagoya University Recearch Center for Materials Science, Furo, Chikusa, Nagoya, 464-8602, Japan
Shigeki Matsunaga	Nagaoka National College of Technology, Nishikatakai 888, Nagaoka, 940-8532, Japan
Katsuyoshi Matsushita	National Institute for Materials Science, 1-2-1, Sengen, Tsukuba, Ibaraki, 305-0047, Japan
Masamichi Miyama	Sasa Lab., Graduate School of Arts and Sciences Department of Basic Science, The University of Tokyo, 3-8-1 Komaba, Meguro-ku, Tokyo, 153-8902, Japan
Kunimasa Miyazaki	The Research Institute of Kochi University of Technology, Miyanokuchi 185, Tosa-Yamada, Kochi, 782-8502, Japan
Tomoko Mizuguchi	Department of Physics, Kyushu University, 6-10-1 Hakozaki Higashi-ku Fukuoka, Fukuoka, 812-8581, Japan
Tetsuya Morishita	National Institute of Advanced Industrial Science and Technology (AIST), Central 2, 1-1-1 Umezono,Tsukuba, Ibaraki, 305-8568, Japan
Kei Morohoshi	Toyota Motor Corporation, 1200, Mishuku, Susono, Shizuoka, 410-1193, Japan
Tomoyuki Murakami	Department of Energy Sciences, Interdisciplinary Graduate School of Science and Engineering, Tokyo Institute of Technology, 4259-G3-38, Nagatsuta-cho, Midori-ku, Yokohama 226-8502, Japan
Tadashi Muranaka	General Education, Aichi Institute of Technology, Yachigusa Yagusa-cho, Toyota-city, Aichi Pref., 470-0392, Japan
Takahiro Murashima	Department of Polymer Science and Engineering, Yamagata University, 4-3-16 Jonan, Yonezawa, Yamagata, 992-8510, Japan

Ken-ichiro Murata	Institute of Industrial Science, University of Toyko, Komaba 4-6-1, Meguro-ku, Tokyo, 153-8505, Japan
Atsushi Nagoe	Tokyo Institute of Technology, 2-12-1 Oookayama, Meguroku, Tokyo, 152-8550, Japan
Natsuko Nakagawa	Department of Physics, Graduate School of Engineering, Yokohama National University, 79-5 Tokiwadai, Hodogaya-ku, Yokohama, 240-8501, Japan
Masahiro Nakanishi	Depertment of Physics, Faculty of Science, Hokkaido University, Kita 10, Nishi 8, Kita-Ku, Sapporo, 060-0810, Japan
Hideyuki Nakayama	Department of Chemistry, Gakushuin University, 1-5-1 Mejiro, Toshimaku, Tokyo, 171-8588, Japan
Takayuki Narumi	Tokuyama Lab., Institute of Fluid Science, Tohoku University, 2-1-1 Katahira, Aoba-ku, Sendai, Miyagi 980-8577, Japan
Jan Nedbal	Faculty of Mathematics and Physics, Charles University, V Holesovickach 2, Prague 8, 180 00, Czech Republic
Kia Ling Ngai	Naval Research Laboratory, 4555 Overlook Ave., SE Washington, DC, 20375-5320, USA
Keigo Nishikawa	Graduate School of Natural Science and Technology, Kanazawa University, Kakuma, Kanazawa-shi, Ishikawa-ken, 920-1192, Japan
Yoshihito Nishioka	Graduate School of Integrated Basic Sciences, Nihon University, 3-25-40 Sakurajosui, Setagaya-ku, Tokyo, 156-8550, Japan
Tsunehiro Obata	Gunma National College of Technology, 580 Toriba, Maebashi, 371-0845, Japan
Ryuichiro Oguma	Department of Applied Physics, Fukuoka University, Nanakuma 8-19-1, Jonanku, Fukuoka, 814-0180, Japan
Masaharu Oguni	Department of Chemistry, Tokyo Institute of Technology, 2-12-1 O-okayama, Meguro-ku, Tokyo, 152-8551, Japan
Hiroki Ohta	Department of Pure and Applied Sciences, University of Tokyo, 153-8902, Japan

Noriyoshi Ohta	Graduate School of Natural Science and Technology, Kanazawa University, Kakuma, Kanazawa-shi, Ishikawa-ken, 920-1192, Japan
Irwin Oppenheim	Department of Chemistry, Massachusetts Institute of Technology, 77 Massachusetts Ave., Room 6-223, 2139, USA
Hiroshi Oshima	School of Medicine, Toho University, 5-21-16 Ota-ku Tokyo, 143-8540, Japan
Marian Paluch	Institute of Physics, Silesian University, ul. Universytecka 4, 40-007 Katowice, 40-007, Poland
Ulf Rörbæk Pedersen	DNRF centre "Glass and Time", Roskilde University, Postbox 260, DK-4000, Denmark
Dvora Perahia	Chemsitry Department, Clemson University, Clemson SC 29634-0973, USA
Pierre Proulx	Department of Chemical Engineering, Universite de Sherbrooke, 2500, Boul Université, Sherbrooke, Québec, J1J 2Z8, Canada
Acep Purqon	Graduate School of Natural Science and Technology, Kanazawa University, Kanazawa, Ishikawa, 920-1192, Japan
Salima Rafaí	Van der Waals - Zeeman Instituut, Valckenierstraat 65, 1018XE Amsterdam, The Netherlands
Carlos Rascon	Departamento de Matematicas, Universidad Carlos III de Madrid, Av de la Universidad 30, 28911 Leganes, Spain
Dieter Richter	IFF Forschungszentrum Juelich, Juelich Germany, 52425, Germany
Christopher Patrick Royall	Institute of Industrial Science, The University of Tokyo, Komaba 4-6-1, Meguro-ku, Tokyo, 153-8505, Japan
Yohei Saika	Wakayama National College of Technology, 77 Noshima, Nada, Gobo, Wakayama, 644-0023, Japan
Takahiro Sakaue	Department of Physics, Kyoto University, Kitashirakawa-Oiwake-cho, Sakyo-ku, Kyoto, 606-8502, Japan
Yuka Sakuma	Department of Physics, Ochanomizu University, 2-1-1 Otsuka, Bunkyo, Tokyo, 112-8610, Japan

Ivan Santamaria-Holek	Facultad de Ciencias, Universidad Nacional Autonoma de Mexico, Circuito exterior de Ciudad Universitaria. DF, 4510, Mexico
Keita Sasanuma	Kojima Lab, University of Tsukuba, Institute of Materials Science, Tennoudai 1-1-1,Tsukuba, Ibaraki, 305-8573, Japan
Katsuhiko Satoh	Dept. of Chemistry, Osaka Sangyo University, 3-1-1, Nakagaito, Daito City, Osaka, 574-8530, Japan
T. B. Schrøder	IMFUFA, Department of Sciences, Roskilde University, PO Box 260, DK-4000 Roskilde, 4000, Denmark
Jeong-Ah Seo	Department of Nano Fusion Technology, Pusan National University Pusan National University, #50 Cheonghak-ri, Samnangjin-eup, Miryang-si, Gyeongnam, 627-706, South Korea
Yuichi Seshimo	Kojima lab, Institute of Materials Science, University of Tsukuba, Tennoudai1-1-1, Tsukuba, Ibaraki, 305-8573, Japan
Geert Jan Agur Sevink	Soft Matter Chemistry, Faculty of Mathematics and Natural Science, Leiden University, PO Box 9502 2300 RA Leiden, The Netherlands
Ping Sheng	Department of Physics, Hong Kong University of Science and Technology, Clear Water Bay, Kowloon, Hong Kong, China
Sheh-Yi Sheu	Department of Life Sciences, National Yang-Ming University, 155, sec.2, Li-Long St. Shin-Pai, Pai-Tao 112, Taipei 112, Taiwan
Hiroshi Shintani	Institute of Industrial Science, University of Tokyo, Komaba 4-6-1, Meguro-ku, Tokyo, 153-8505, Japan
Naoki Shinyashiki	Department of Physics, Tokai University, 1117 Kitakaname, Hiratsuka, Kanagawa, 259-1292, Japan
Yasuhiro Shiwa	Kyoto Institute of Technology, Matsugasaki, Sakyo-ku, Kyoto, 606-8585, Japan
D. Sivakumar	Department of Aerospace Engineering, Indian Institute of Science, Bangalore 560012, India
Oleg P. Solonenko	Institute of Theoretical and Applied Mechanics, Siberian Branch of Russian Academy of Sciences, 4/1 Institutskaya str, Novosibirsk, 630090, Russia

Anna Spanoudaki	Department of Apllied Mathematics and Physics, National Technical University of Athens, Heroon Polytechneiou 9, Zografou, Athens, 15780, Greece
H. Eugene Stanley	Department of Physics, Boston University, 590 Commonwealth Ave, Boston, MA, 02215, USA
Petr Stepanek	Institute of Macromolecular Chemistry, Heyrovsky Sq. 2, Prague 6, 16206, Czech Republic
Anatoliy Strybulevych	Deptartment of Physics and Astronomy, University of Manitoba, 301 Allen Bldg, Winnipeg, R3T 2N2, Canada
Seiichi Sudo	Department of Machine Intelligence and Systems Engineering, Akita Prefectural University, 84-4, Ebinokuchi, Yurihonjo 015-0055, Japan
Akira Suzuki	Center for Solid State Physics & Department of Physics, Faculty of Science, Tokyo University of Science, 1-3 Kagurazaka, Shinjuku-ku, Tokyo, 162-8601, Japan
Atsushi Suzuki	Faculty of Environment and Information Sciences, Yokohama National University, 79-7 Tokiwadai, Hodogaya-ku, Yokohama, 240-850, Japan
Stephen F Swallen	Department of Chemistry, University of Wisconsin, Madison 1101 University Ave., Madison, WI, 53706, USA
Grzegorz Szamel	Department of Chemistry, Colorado State University, Fort Collins, Colorado, 80523, USA
Tadashi Toyoda	Department of Physics, Tokai University, 1117 Kitakaname, Hiratsuka, Kanagawa, 259-1292, Japan
Masako Takasu	Dept. Comutational Science, Kanazawa University, Kakuma Kanazawa, 920-1192, Japan
Osamu Takiguchi	Department of Polymer Science and Engineering, Yamagata University, 4-3-16 Jonan Yonezawa, 992-8510, Japan
Hajime Tanaka	Institute of Industrial Science, University of Tokyo, 4-6-1 Komaba, Meguro-ku, Tokyo, 153-8505, Japan
Hidemasa Takana	Institute of Fluid Science, Tohoku University, 2-1-1 Katahira, Aoba-ku, Sendai, 980-8577, Japan

Takashi Taniguchi
Department of Polymer Science and Engineering, Yamagata University, 4-3-16, Jonan, Yonezawa, Yamagata, 992-8510, Japan

Gilles Tarjus
Laboratoire de Physique Théorique de la Matière Condensée, CNRS-UMR 7600, Université Pierre et Marie Curie, Boîte 121, 4 Place Jussieu, 75252 – Paris Cedex 05, France

Piero Tartaglia
Dipartimento di Fisica, Università di Roma La Sapienza, P.le Aldo Moro 5, I-00185, Italy

Yayoi Terada
Insititute of Fluid Science, Tohoku University, 2-1-1 Katahira, Sendai, 980-8577, Japan

Takamichi Terao
Department of Mathematical and Design Engineering, Gifu university, Gifu, 501-1193, Japan

Michio Tokuyama
WPI Advanced Institute for Materials Research and Institute of Fluid Science, Tohoku University, 2-1-1 Katahira, Aoba-ku, Sendai, Miyagi, 980-8577, Japan

John Torkelson
Dept. of Chemical and Biological Engineering, Northwestern University, TECH E136, 2145 Sheridan Rd., Evanston, 60208, USA

Steffen Trimper
Institute of Physics, University Halle Martin-Luther-University, D-06099 Halle, Germany

Kun Hsin Tsai
Department of Electronphysics, Institute of Physics, National Chiao-Tung University, ShinChu, Taiwan, 300, Taiwan

Remco Tuinier
IFF Forschungszentrum Juelich, Leo Brandt strasse 1, Juelich, 52425, Germany

Nariya Uchida
Department of Physics, Tohoku University, Aramaki Aoba, Aoba-ku, Sendai, 980-0813

Takahiro Ueno
Department of Physics and Mathematics, Aoyama Gakuin University, Fuchinobe 5-10-1, Sagamihara, Kanagawa, 229-8558, Japan

Kiminori Ushida
Eco- Soft Materials Research Unit, Riken, 2-1 Hirosawa, Wako, Saitama, 351-0198, Japan

Fathollah Varnik
Max-Planck-Institut für Eisenforschung. Max-Planck-Straße 1, D-40237 Düsseldorf, Germany

Tzu-Yu Wang	Department of Chemical and Materials Engineering, National Central University, Jhongli, Taiwan 320, Republic of China, Taiwan
Keisuke Watanabe	Department of Chemistry, Graduate School of Science and Engineering, Tokyo Institute of Technology, O-okayama 2-12-1, Meguro-ku, Tokyo, 152-8550, Japan
David Weitz	Department of Physics, Harvard University, 29 Oxford St., Cambridge, MA 2138, USA
Jean-Francois Willart	CNRS - University of Lille, Laboratoire de Dynamique et Structure des Materiaux Moleculaires - UFR de Physique - bât P5, 59655, France
Stephen Rodney Williams	Research School of Chemistry, The Australian National University, Canberra, ACT 200, Australia
Mary Wintersgill	Physics Department, U. S. Naval Academy, 572C Holloway Road, Annapolis, MD 21402-5026, USA
Long Wu	Graduate School of Science and Technology, Kobe University, 1-1 Rokko-dai, Nada-ku, Kobe, 675-8501, Japan
Yuko Yamada	Department of Environmental Technology, Nagoya Institute of Technology, Nagoya, 466-8555, Japan
Tetsunori Yamamoto	Graduate School of Natural Science and Technology, Kanazawa University, Kakuma, Kanazawa-shi, Ishikawa-ken, 920-1192, Japan
Dah Yen Yang	Institute of Atomic and Molecular Science Academia Sinica, Taipei, P.O.Box 23-166, Taipei Taiwan, ROC, 106, Taiwan
Naohiro Yasuda	Department of Chemistry, Gakushuin University, 1-5-1 Mejiro, Toshimaku, Tokyo, 171-8588, Japan
Yoshiki Yomogida	Department of Physics, Faculty of Science, Hokkaido University, Kita-ku, Kita-10 Nishi-8, Sapporo, 060-0810, Japan
Xin-Rong Zhang	Department of Mechanical Engineering, Doshisha University, 1-3, Tatara-Miyakodani, Kyo-Tanabeshi, Kyoto, 610-0321, Japan
Reiner Zorn	IFF Forschungszentrum Juelich, PO Box 1913, 52425, Germany

AUTHOR INDEX

A

B

C

D

E

F

G

H

I

J

K

L

M

N

O

P

R

S

T

U

V

W

X

Y

Z